HANDBOOK OF ENVIRONMENTAL ENGINEERING CALCULATIONS

ABOUT THE EDITORS

DR. C. C. LEE is the Fuel Cell Research Program Manager at the National Risk Management Research Laboratory of the U.S. Environmental Protection Agency in Cincinnati, Ohio.* He has more than 30 years of experience in conducting various engineering and research projects which often involve multimedia environmental issues ranging from air and water pollution control to solid waste disposal. He has been recognized as a worldwide expert in the thermal treatment of medical and hazardous wastes, witness his leading discussions on medical waste disposal technologies at a meeting conducted by the Congressional Office of Technology Assessment. He initiated and served as the Chairman of the First and Second International Congresses on Toxic Combustion Byproducts (ICTCB) in 1989 and 1991, respectively. The ICTCB has been holding its meetings every two years since its creation in 1989. Also, at the initiation of the U.S. State Department, he served as head of the U.S. delegation to the Conference on National Focal Points for Low- and Non-Waste Technology (it was sponsored by the United Nations and held in Geneva, Switzerland on August 28–30, 1978). He has been invited to lecture on various issues regarding solid waste disposal in numerous national and international conferences, and he has authored 18 books and has published more than 175 papers and reports in various environmental areas. Until recently, he was an adjunct professor with the University of Cincinnati in Ohio and was an assistant professor at the North Carolina State University before joining EPA in 1974. He received a B.S. from the National Taiwan University in 1964, and a MS and PhD from the North Carolina State University in 1968 and 1972, respectively.

SHUN DAR LIN is Emeritus faculty, State Water Survey Division of University of Illinois. He received his PhD in Sanitary Engineering from Syracuse University. Dr. Lin holds an MS in Sanitary Engineering from University of Cincinnati and a BS in Civil Engineering from National Taiwan University. Dr. Lin has taught and conducted research since 1960 at the Institute of Public Health of National Taiwan University and is a registered professional engineer in Illinois. He has published nearly 100 articles and reports on water and wastewater engineering. In 1986, Dr. Lin received the Water Quality Division Best Paper Award for "*Giardia lamblia* and Water Supply" from the American Water Works Association. Dr. Lin has over 40 years of experience in teaching, research, field and laboratory work, and practical engineering experience. Dr. Lin is a life member of the American Society of Civil Engineers, the American Water Works Association, and the Water Environment Federation.

Note: This book was written and edited by Dr. C. C. Lee in his private capacity. No official support or endorsement by the U.S. Environmental Protection Agency is intended nor should be inferred.

HANDBOOK OF ENVIRONMENTAL ENGINEERING CALCULATIONS

C. C. Lee Editor in Chief

Shun Dar Lin Associate Editor

Second Edition

McGraw Hill

New York Chicago San Francisco Lisbon London Madrid
Mexico City Milan New Delhi San Juan Seoul
Singapore Sydney Toronto

The McGraw-Hill Companies

Library of Congress Cataloging-in-Publication Data

Handbook of environmental engineering calculations / C. C. Lee, editor in chief;
Shun Dar Lin, associate editor.—2nd ed.
p. cm.
Includes bibliographical references and index.
ISBN-13: 978-0-07-147583-9 (alk. paper)
ISBN-10: 0-07-147583-4 (alk. paper)
1. Environmental engineering—Handbooks, manuals, etc. I. Lee, C. C.
II. Lin, Shun Dar.

TD145.H27 2007
628—dc22

2006053099

McGraw-Hill books are available at special quantity discounts to use as premiums and sales promotions, or for use in corporate training programs. For more information, please write to the Director of Special Sales, McGraw-Hill Professional, Two Penn Plaza, New York, NY 10121-2298. Or contact your local bookstore.

1 2 3 4 5 6 7 8 9 0 DOC/DOC 0 1 2 1 0 9 8 7

ISBN-13: 978-0-07-147583-9
ISBN-10: 0-07-147583-4

This book is printed on acid-free paper.

Sponsoring Editor
Larry S. Hager
Editorial Supervisor
David E. Fogarty
Project Manager
Gita Raman
Copy Editor
Ragini Pandey
Proofreader
Anju Panthari
Indexer
Steve Ingle
Production Supervisor
Richard C. Ruzycka
Composition
International Typesetting and Composition
Art Director, Cover
Handel Low

CONTENTS

Part 3 Air Pollution Control Calculations

CONTRIBUTORS

J. C. S. Chang *U.S. Environmental Protection Agency, Research Triangle Park, North Carolina* (CHAP. 3.1)

Floyd Hasselriis *Consulting Engineer, Forest Hills, New York* (CHAPS. 2.2, 2.3, 2.4)

Thomas C. Ho *Department of Chemical Engineering, Lamar University, Beaumont, Texas* (CHAP. 2.2)

G. L. Huffman *U.S. Environmental Protection Agency, Cincinnati, Ohio* (CHAPS. 2.1, 2.2, 2.6, 3.1, 3.2, 3.3, 5.1)

Carl F. Isonhart *Mixer Systems, Inc., Pewaukee, Wisconsin* (CHAP. 2.5)

C. C. Lee *U.S. Environmental Protection Agency, Cincinnati, Ohio* (EDITOR IN CHIEF; CHAPS. 2.1, 2.2, 2.6, 3.1, 3.2, 3.3, 5.1)

Shun Dar Lin *Illinois State Water Survey, Peoria, Illinois* (ASSOCIATE EDITOR; CHAPS. 1.1 TO 1.7)

Christopher J. Nagel *Quantum Catalytics, Fall River, Massachusetts* (CHAP. 2.5)

Jesse L. Thé *Lakes Environmental Software and University of Waterloo, Canada* (CHAP. 4.1)

David A. Weeks. *Risk Management & Engineering Ltd., Garland, Texas* (CHAP. 4.1)

PREFACE

Because environmental problems are uncovered almost daily, this second edition has been created to meet the challenge of solving them. Environmental engineering encompasses many areas. It ranges from solid waste disposal, wastewater treatment, air pollution control, to life analysis. Although there are many publications relative to the descriptions of concepts and methodologies in the environmental control area, the actual calculations relative to the field seldom appear in these publications. In addition to the scarcity of environmental calculations thousands of environmental regulations from federal, state, and local regulators impact environmental engineering design every day. Just keeping abreast of such regulations is an enormous task for engineers. The main objective of this book is, therefore, to provide step-by-step, practical calculational procedures on various environmental subjects. More importantly, this book integrates the regulatory requirements into environmental designs so the result can make these designs more acceptable to regulators. The major subjects covered in the second edition include:

1. Calculations of water quality assessment and control
2. Solid waste treatment calculations
3. Air pollution control calculations
4. Air toxic risk assessment
5. Fuel cell technologies

A majority of the calculational examples provided in this book were developed by the authors themselves and the materials were excerpted from previous USEPA publications. Since its creation in 1970, the USEPA has published many environmental regulations and engineering reports. Many very interesting calculational examples were scattered throughout these publications. The huge volume of EPA regulations and reports makes a search for example calculations extremely difficult and time consuming. To help resolve this difficulty, many of these examples were collected and edited in a format for readers to easily understand. The citing of references for each example calculation is provided herein. This is to expeditiously assist users in locating additional information, if needed.

This book is intended to be a reference tool for those who are involved in the protection of air, water, and land resources. It is believed that the book will make many environmental jobs much easier. Lastly, the editor wishes to express his deep appreciation to the contributing authors who have spent so many of their days and nights to make this book possible.

C. C. LEE

HANDBOOK OF ENVIRONMENTAL ENGINEERING CALCULATIONS

P • A • R • T • 1

CALCULATIONS OF WATER QUALITY ASSESSMENT AND CONTROL

Part 1 of this book is written for use by the following readers: students taking coursework relating to public water supply, waste-water engineering or stream sanitation, practicing environmental (sanitary) engineers; regulatory officers responsible for the review and approval of engineering project proposals; operators, engineers, and managers of water and/or wastewater treatment plants; and any other professionals, such as chemists and biologists, who have gained some knowledge of water/wastewater issues. This work will benefit all operators and managers of public water supply and of wastewater treatment plants, environmental design engineers, military environmental engineers, undergraduate and graduate students, regulatory officers, local public works engineers, lake managers, and environmentalists.

The chapters in Part 1 present the basic principles and concepts relating to water/wastewater engineering and provide illustrative examples of the subject. To the extent possible, examples rely on practical field data. Each of the calculations provided herein are solved step-by-step in a streamlined manner that is intended to facilitate understanding. Calculations (step-by-step solutions) range from calculations commonly used by operators to more complicated calculations required for research or design.

Advances and improvements in many fields are driven by competition or the need for increased profits. It may be fair to say, however, that advances and improvements in environmental engineering are driven instead by regulation. The US Environmental Protection Agency (EPA) sets up maximum contaminant levels, which research and project designs must reach as a goal. The step-by-step solution examples provided in this book are informed by the integration of rules and regulations on every aspect of waters and wastewaters. The author has performed an extensive survey of literature on surface and groundwaters encountered in environmental engineering and compiled them in the following chapters. Rules and regulations are described as simply as possible, and practical examples are given.

The following chapters include calculations for basic science, surface waters ground water, drinking water treatment, and wastewater engineering. Chapter 1.1 covers conversion factors between the two measurement systems, the United States (US) customary system and the System International (SI), basic mathematics for water and wastewater plant operators, fundamental chemistry and physics, and basic statistics for environmental engineers.

Chapter 1.2 comprises calculations for river and stream waters. Stream sanitation had been studied for nearly 100 years. By the mid-twentieth century, theoretical and empirical models for assessing waste assimilating capacity of streams were well developed. Dissolved oxygen and biochemical oxygen demand in streams and rivers have been comprehensively illustrated in this chapter. Apportionment of stream users and pragmatic approaches for stream dissolved oxygen models are also covered. From the 1950s through the 1980s, researchers focused extensively on wastewater treatment. In 1970s, rotating biological contactors also became a hot subject. Design criteria and examples for all of these are included. Some treatment and management technologies are no longer suitable in the United States. However, they are still of some use in developing countries.

Chapter 1.3 is a compilation of adopted methods and documented research. In the early 1980s, the USEPA published Guidelines for Diagnostic and Feasibility Study of Public Owned Lakes (Clean Lakes Program, or CLP). This was intended to be used as a guideline for lake management. CLP and its calculation (evaluation) methods are present in this chapter. Hydrological, nutrient, and sediment budgets are presented for reservoir and lake waters. Techniques for classification of lake water quality, assessment of the lake trophic state index, and of lake use support are presented.

Calculations for groundwater are given in Chapter 1.4. They include groundwater hydrology, flow in aquifers, pumping and its influence zone, setback zone, and soil remediation. Well setback zone is regulated by the state EPA. Determinations of setback zones are also included in the book. Well function for confined aquifers is presented in Appendix B.

Hydraulics for environmental engineering is included in Chapter 1.5. This chapter covers fluid (water) properties and definitions; hydrostatics; fundamental concepts of water flow in pipes, weirs, orifices, and in open channel; and of flow measurements. Pipe networks for water supply distribution systems and hydraulics for water and wastewater treatment plants are included.

Chapters 1.6 and 1.7 cover each unit process for drinking water and wastewater treatments, respectively. The USEPA developed design criteria and guidelines for almost all unit processes. These two chapters depict the integration of regulations (or standards) into water and wastewater design procedure. Water fluoridation and the CT values are incorporated in Chapter 1.6. Biosolids are discussed in detail in Chapter 1.7. These two chapters are the heart Part 1, providing the theoretical considerations of unit processes, traditional (or empirical) design concepts, and integrated regulatory requirements.

Most calculations provided herein use U.S. Customary units. Readers who use the International System (SI) may apply the conversion factors listed in Chapter 1.1. Answers are also generally given in SI for most of problems solved using U.S. units.

The current edition corrects certain computational, typographical, and grammatical errors found in the previous edition. Drinking water quality standards, wastewater effluent standards, and several new examples have also been added. *The author also wishes to acknowledge Meiling Lin, Heather Lin, Robert Greenlee, Luke Lin, Kevin Lin, Jau-hwan Tzeng, and Lucy Lin for their assistance.* *Any reader suggestions and comments will be greatly appreciated.*

Shun Dar Lin

CHAPTER 1.1
BASIC SCIENCE AND FUNDAMENTALS

Shun Dar Lin

1 CONVERSION FACTORS

The units most commonly used by water and wastewater professionals in the United States are based on the complicated U.S. Customary System of Units. However, laboratory work is usually based on the metric system due to the convenient relationship between milliliters (mL), cubic centimeters (cm^3), and grams (g). The International System of Units (SI) is used in all other countries. Factors for converting U.S. units to the SI are given below (Table 1.1) to four significant figures.

EXAMPLE 1: Find degrees in Celsius of water at 68°F.

Solution:

$$°C = (°F - 32) \times \frac{5}{9} = (68 - 32) \times \frac{5}{9} = 20$$

TABLE 1.1 Factors for Conversions

U.S. Customary units	Multiply by	SI or U.S. Customary units
Length		
inches (in)	2.540	centimeters (cm)
	0.0254	meters (m)
feet (ft)	0.3048	m
	12	in
yard (yd)	0.9144	m
	3	ft
miles	1.609	kilometers (km)
	1760	yd
	5280	ft
Area		
square inch (sq in, in^2)	6.452	square centimeters (cm^2)
square feet (sq ft, ft^2)	0.0929	m^2
	144	in^2
acre (a)	4047	square meters (m^2)
	0.4047	hectare (ha)
	43,560	ft^2
	0.001562	square miles
square miles (mi^2)	2.590	km^2
	640	acres
Volume		
cubic feet (ft^3)	28.32	liters (L)
	0.02832	m^3
	7.48	US gallons (gal)
	6.23	Imperial gallons
	1728	cubic inches (in^3)
cubic yard (yd^3)	0.7646	m^3
gallon (gal)	3.785	L
	0.003785	m^3
	4	quarts (qt)
	8	pints (pt)
	128	fluid ounces (fl oz)
	0.1337	ft^3
million gallons (Mgal)	3785	m^3
quart (qt)	32	fl oz
	946	milliliters (mL)
	0.946	L
acre · feet (ac · ft)	1.233×10^{-3}	cubic hectometers (hm^3)
	1233	m^3
Weight		
pound (lb, #)	453.6	grams (gm or g)
	0.4536	kilograms (kg)
	7000	grains (gr)
	16	ounces (oz)
grain	0.0648	g
ton (short)	2000	lb
	0.9072	tonnes (metric tons)
ton (long)	2240	lb
gallons of water (US)	8.34	lb
Imperial gallon	10	lb

TABLE 1.1 Factors for Conversions (*contd.*)

U.S. Customary units	Multiply by	SI or U.S. Customary units
Unit weight		
ft^3 of water	62.4	lb
	7.48	gallon
pound per cubic foot (lb/ft^3)	157.09	newton per cubic meter (N/m^3)
	16.02	kg force per square meter (kgf/m^2)
	0.016	grams per cubic centimeter (g/cm^3)
Concentration		
parts per million (ppm)	1	mg/L
	8.34	lb/Mgal
grain per gallon (gr/gal)	17.4	mg/L
	142.9	lb/Mgal
Time		
day	24	hours (h)
	1440	minutes (min)
	86,400	seconds (s)
hour	60	min
minute	60	s
Slope		
feet per mile	0.1894	meter per kilometer
Velocity		
feet per second (ft/sec)	720	inches per minute
	0.3048	meter per second (m/s)
	30.48	cm/s
	0.6818	miles per hour (mph)
inches per minute	0.043	cm/s
miles per hour (mi/h)	0.4470	m/s
	26.82	m/min
	1.609	km/h
knot	0.5144	m/s
	1.852	km/h
Flowrate		
cubic feet per second (ft^3/s, cfs)	0.646	million gallons daily (MGD)
	448.8	gallons per minutes (gpm)
	28.32	liter per second (L/s)
	0.02832	m^3/s
million gallons daily (MGD)	3.785	m^3/d (CMD)
	0.04381	m^3/s
	157.7	m^3/h
	694	gallons per minute
	1.547	cubic feet per second (ft^3/s)
gallons per minute (gpm)	3.785	liters per minute (L/min)
	0.06308	liters per second (L/s)
	0.0000631	m^3/s
	0.227	m^3/h
	8.021	cubic feet per hour (ft^3/h)
	0.002228	cubic feet per second (cfs, ft^3/s)
gallons per day	3.785	liters (or kilograms) per day
MGD per acre · ft	0.4302	gpm per cubic yard
acre · feet per day	0.01427	m^3/s
Application (loading) rate		
pounds per square foot (lb/ft^2)	4.8827	kilograms per square meter (kg/m^2)

TABLE 1.1 Factors for Conversions (*contd.*)

U.S. Customary units	Multiply by	SI or U.S. Customary units
pounds per 1000 square foot per day (lb/1000 $ft^2 \cdot d$)	0.00488	kilograms per square meter per day ($kg/m^2 \cdot d$)
pounds per cubic foot (lb/ft^3)	16.017	kilograms per cubic meter (kg/m^3)
pounds per 1000 cubic foot per day (lb/1000 $ft^3 \cdot d$)	0.016	kilograms per cubic meter per day ($kg/m^3 \cdot d$)
pounds per foot per hour ($lb/ft \cdot h$)	1.4882	kilograms per meter per hour ($kg/m \cdot h$)
pounds per horse power per hour ($lb/hp \cdot h$)	0.608	kilograms per kilowatts per hour ($kg/kW \cdot h$)
pounds per acre per day ($lb/acre \cdot d$)	1.121	kilograms per hectare per day ($kg/ha \cdot d$)
gallons per acre (gal/acre)	0.00935	m^3/ha
million gallons per acre (Mgal/acre)	0.93526	m^3/m^2
million gallons per acre · ft (Mgal/acre · ft)	0.43	gpm/yd^3
gallons per square foot per day ($gal/ft^2 \cdot d$)	0.04074	cubic meter per square meter per day ($m^3/m^2 \cdot d$)
	0.04356	Mgal/acre · d
gallons per minute per square foot (gpm/ft^2)	58.674	$m^3/m^2 \cdot d$
square root of gpm per square foot $(gal/min)^{0.5}/ft^2$	2.7	$(L/s)^{0.5}/m^2$
gallons per day per foot ($gal/d \cdot ft$)	0.01242	$m^3/d \cdot m$
square foot per cubic foot (ft^2/ft^3)	3.28	m^2/m^3
cubic foot per gallon (ft^3/gal)	7.48	m^3/m^3
cubic foot per pound (ft^3/lb)	0.06243	m^3/kg
	62.43	L/kg
cubic foot per 1000 cubic foot per minute ($ft^3/1000\ ft^3 \cdot min$)	1	$L/m^3 \cdot min$
Force		
pounds	0.4536	kilograms force (kgf)
	453.6	grams(g)
	4.448	newtons (N)
Pressure		
pounds per square inch (lb/in^2, psi)	2.309	feet head of water
	2.036	inches head of mercury
	51.71	mmHg
	6895	newtons per square meter (N/m^2) = pascal (Pa)
	0.0703	kgf/cm^2
	703.1	kgf/m^2
	0.0690	bars
pounds per square foot (lb/ft^2)	4.882	kgf/m^2
	47.88	N/m^2 (Pa)
pounds per cubic inch	0.01602	gmf/cm^3
	16.017	gmf/L
tons per square inch	1.5479	kg/mm^2
millibars (mb)	100	N/m^2
inches of mercury	345.34	kg/m^2
	0.0345	kg/cm^2
	0.0334	bar
	0.491	psi (lb/in^2)

TABLE 1.1 Factors for Conversions (*contd.*)

U.S. Customary units	Multiply by	SI or U.S. Customary units
inches of water	248.84	pascals (Pa)
atmosphere	101,325	Pa
	1013	millibars (1 mb = 100 Pa)
	14.696	psi (lb/in^2)
	29.92	inches of mercury
	33.90	feet of water
pascal (SI)	1.0	N/m^2
	1.0×10^{-5}	bar
	1.0200×10^{-5}	kg/m^2
	9.8692×10^{-6}	atmospheres (atm)
	1.40504×10^{-4}	psi (lb/in^2)
	4.0148×10^{-3}	in, head of water
	7.5001×10^{-4}	cm head of mercury
Mass and density		
slug	14.594	kg
	32.174	lb (mass)
pound	0.4536	kg
slug per $foot^3$	515.4	kg/m^3
density (γ) of water	62.4	lb/ft^3 at 50°F
	980.2	N/m^3 at 10°C
specific wt (ρ) of water	1.94	$slugs/ft^3$
	1000	kg/m^3
	1	kg/L
	1	gram per milliliter (g/mL)
Viscosity		
pound-second per $foot^3$ or slug per foot second	47.88	newton second per square meter (Ns/m^2)
square feet per second (ft^2/s)	0.0929	m^2/s
Work		
British thermal units (Btu)	1.0551	kilo joules (kj)
	778	ft lb
	0.293	watt-h
	1	heat required to change 1 lb of water by 1°F
hp-h	2545	Btu
	0.746	kW-h
kW-h	3413	Btu
	1.34	hp-h
Power		
horsepower (hp)	550	ft lb per sec
	746	watt
	2545	Btu per h
kilowatts (kW)	3413	Btu per h
Btu per hour	0.293	watt
	12.96	ft lb per min
	0.00039	hp
Temperature		
degree Fahrenheit (°F)	(°F − 32) × (5/9)	degree Celsius (°C)
(°C)	(°C) × (9/5) + 32	(°F)
	°C + 273.15	Kelvin (K)

EXAMPLE 2: At a temperature of 4°C, water is at its greatest density. What is the degree of Fahrenheit?

Solution:

$$
\begin{aligned}
°F &= (°C) \times \frac{9}{5} + 32 \\
&= 4 \times \frac{9}{5} + 32 \\
&= 7.2 + 32 \\
&= 39.2
\end{aligned}
$$

2 PREFIXES FOR SI UNITS

The prefixes commonly used in the SI system are based on the power 10. For example, a kilogram means 1000 grams, and a centimeter means one-hundredth of 1 meter. The most used prefixes are listed in Table 1.2, together with their abbreviations, meanings, and examples.

3 MATHEMATICS

Most calculations required by the water and wastewater plants operators and managers are depended on ordinary addition, subtraction, multiplication, and division. Calculations are by hand, by calculator, or by a computer. Engineers should master the formation of problems: daily operations require calculations of simple ratio, percentage, significant figures, transformation of units, flow rate, area and volume computations, density and specific gravity, chemical solution, and mixing of solutions.

Miscellaneous Constants and Identities

$$
\begin{aligned}
\pi &= 3.14 \text{ (pi)} \\
e &= 2.7183 \text{ (Napierian)} \\
x^o &= 1 (x > 0) \\
0^x &= 0 \ (x > 0)
\end{aligned}
$$

TABLE 1.2 Prefixes for SI Units

Prefix	Abbreviation	Multiplication factor	Example
tera	T	$1\,000\,000\,000\,000 = 10^{12}$	
giga	G	$1\,000\,000\,000 = 10^9$	
mega	M	$1\,000\,000 = 10^6$	
myria	my	$10\,000 = 10^4$	
kilo	k	$1\,000 = 10^3$	km, kg
hecto	h	$100 = 10^2$	
deka	da	$10 = 10^1$	
		$1 = 10^0$	meter (m), gram (g)
deci	d	$0.1 = 10^{-1}$	
centi	c	$0.01 = 10^{-2}$	cm
milli-	m	$0.001 = 10^{-3}$	mm, mg
micro	μ	$0.000001 = 10^{-6}$	μm, μg
nano	n	$0.000\,000\,001 = 10^{-9}$	
pico	p	$0.000\,000\,000\,001 = 10^{-12}$	
femto	f	$0.000\,000\,000\,000\,001 = 10^{-15}$	
atto	a	$0.000\,000\,000\,000\,000\,001 = 10^{-18}$	

3.1 Logarithms

Every positive number x can be expressed as a base b raised to a power y:

$$x = b^y$$

y is called the logarithm of x to the base b and is symbolized as

$$y = \log_b x$$

The logarithm of a number is the power to which the base must be raised to equal that number. Logarithms and exponential functions have numerous uses in engineering. Logarithms simplify arithmetic and algebraic computations according to the following rules:

$$\log_b(xy) = \log_b x + \log_b y$$

$$\log_b \frac{x}{y} = \log_b x - \log_b y$$

$$\log_b x^y = y\log_b x$$

The two most common bases are 10 and e. The exponential e is an irrational number resulting from an infinite series and is approximately equal to 2.71828.

Definitions:

$$\text{log base } 10 = \log_{10} = \log$$

$$\text{log base } e = \log_e = \ln$$

$$\text{log base } q = \log_q$$

where q is any positive number except 0 or 1.

Antilog$_q$, or $\log_q^{-1}$, is the number whose log value being stated for $\log_q x = m$.

EXAMPLE 1:

$$\log x = a \rightarrow x = 10^a$$

$$\text{since } 1000 = 10^3, \quad \text{i.e. } \log 1000 = 3$$

$$0.001 = 10^{-3}, \quad \text{i.e. } \log 0.001 = -3$$

$$\ln y = b \rightarrow y = e^b$$

$$\log_q z = c \rightarrow z = q^c$$

Given m, find

$$x = \log_q^{-1} m$$

EXAMPLE 2: Find log 3.46 and log 346.

Solution: The answer can be found in $\log_{10}$ tables or using a calculator or computer. We can obtain

$$\log 3.46 = 0.539, \quad \text{i.e. } 3.46 = 10^{0.539}$$

$$\log 346 = 2.539, \quad \text{i.e. } 346 = 10^{2.539}$$

If a positive number N is expressed as a power of e, we can write it as $N = e^p$. In this case, p is the logarithm of N to the base e (2.7183) or the natural logarithm of N and can be written as $p = \ln N$ or $p = \ln_e N$. More examples are as follows:

$$\ln 2 = 0.6931$$

$$\ln 10 = 2.3026$$

$$\ln y = 2.3026 \log y$$
$$\log x = 0.4343 \ln x$$
$$\log ab = \log a + \log b$$
$$\log \frac{a}{b} = \log a - \log b$$
$$\log a^n = n \log a$$
$$\log \sqrt[n]{a} = \frac{1}{n} \log a$$

3.2 Basic Math

This section involves the basics of plus, minus, multiplication, and division, which may be useful for basic-level certification examinations of water or wastewater operators.

Addition

EXAMPLE 1: If a trickling filter has a recirculation ratio of 0.25 to 1, how many times the flow of raw wastewater passes through the filter?

Solution:

$$1 + 0.25 = 1.25 \times \text{the flow of raw wastewater}$$

EXAMPLE 2: The field crew drives the company's car 18.8 miles on Monday, 45 miles on Tuesday, 0 miles on Wednesday, 22.2 miles on Thursday, and 36.5 miles on Friday. What is the total mileage driven for the week and what is the average per day?

Solution:

$$\begin{aligned} \text{Total} &= (18.8 + 45 + 0 + 22.2 + 36.5) \text{ miles} \\ &= 122.5 \text{ miles} \\ \text{Average} &= \text{total/5 day} = 122.5 \text{ miles/5 day} \\ &= 24.5 \text{ miles/day} \end{aligned}$$

EXAMPLE 3: If an activated sludge unit has a recirculation of 0.4 to 1, how many times the flow of raw wastewater passes over the unit?

Solution:

$$\text{Answer} = 1 + 0.4 = 1.4 \text{ (times)}$$

Subtraction

EXAMPLE 1: A flow meter reads 00023532 gal on June 1 and 06040872 gal on July 1. What is the average daily flow?

Solution:

$$(6040872 - 23532) \text{ gal/30 days} = 200{,}578 \text{ gal/day (gpd)}$$

EXAMPLE 2: If 5.2 mg/L of chlorine (Cl_2) is added to a water sample, after 30 min of contact time the residual chlorine concentration is 0.9 mg/L. What is the chlorine demand of the water?

Solution:

$$\begin{aligned} Cl_2 \text{ demand} &= 5.2 \text{ mg/L} - 0.9 \text{ mg/L} \\ &= 4.3 \text{ mg/L} \end{aligned}$$

Multiplication

EXAMPLE 1: The weight of a sand bag averages 46.5 lb. What is the total weight of 1200 bags of sand to fill a hole in the bank?

Solution:

$$\begin{aligned} \text{Weight} &= 46.5 \text{ lb/bag} \times 1200 \text{ bags} \\ &= 55{,}800 \text{ lb} \end{aligned}$$

EXAMPLE 2: If a water treatment plant uses 124 chlorine cylinders (cy) during a 30-day period, the average supply from each cylinder is 135 lb. What is the total weight of chlorine used in 30 days?

Solution:

$$\begin{aligned} \text{Total weight} &= 135 \text{ lb/cy} \times 124 \text{ cy} \\ &= 16{,}740 \text{ lb} \end{aligned}$$

EXAMPLE 3: How many pounds in 5 gal of water? A gallon (gal) of water weighs 8.34 lb.

Solution:

$$\begin{aligned} \text{Weight} &= 8.34 \text{ lb/gal} \times 5 \text{ gal} \\ &= 41.7 \text{ lb} \end{aligned}$$

EXAMPLE 4: Since 1 cubic foot (ft^3) of a liquid is equal to 7.48 gal, how many gallons are equivalent to 25 ft^3 of the liquid?

Solution:

$$25 \text{ ft}^3 \times 7.48 \frac{\text{gal}}{\text{ft}^3} = 187 \text{ gal}$$

EXAMPLE 5: The minimum velocity (v) at average flow in a sanitary sewer is 2.0 ft/s. What is it in terms of m/s? (1 ft = 0.3 m)

Solution:

$$\begin{aligned} v &= 2.0 \text{ ft/s} \times 0.3 \text{ m/ft} \\ &= 0.6 \text{ m/s} \\ \text{or} &= 60 \text{ cm/s} \end{aligned}$$

EXAMPLE 6: Calculate the molecular weight of alum, $Al_2(SO_4)_3.14H_2O$. The approximate atomic weights are aluminum = 27, sulfur = 32, oxygen = 16, and hydrogen = 1.

Solution:

$$\begin{aligned} \text{Atomic weight} &= 2(27) + 3(32 + 4 \times 16) + 14(1 \times 2 + 16) \\ &= 54 + 3(96) + 14(18) \\ &= 594 \end{aligned}$$

EXAMPLE 7: A chemical storage room is to be painted. The room size is 16 ft wide, 20 ft long, and 10 ft high. The walls and the ceiling will use the same paint. The floor paint will be a different paint. Assume 1 gal of paint covers an area of 400 ft^2. How many gallons of wall and floor paints are required for the painting job?

Solution:

Step 1: Determine floor paint needed:

$$\text{Floor area} = 16 \text{ ft} \times 20 \text{ ft} = 320 \text{ ft}^2 < 400 \text{ ft}^2$$

One gallon of floor paint is enough.

Step 2: For wall and ceiling:

$$\begin{aligned}\text{Wall area} &= 10 \text{ ft} \times (16 + 16 + 20 + 20) \text{ ft, including doors}\\ &= 720 \text{ ft}^2\\ \text{Ceiling area} &= 320 \text{ ft}^2\text{, same as the floor}\\ \text{Total area} &= 720 \text{ ft}^2 + 320 \text{ ft}^2\\ &= 1040 \text{ ft}^2\\ \text{Gallon cans} &= 1040 \text{ ft}^2/(400 \text{ ft}^2/\text{gal})\\ &= 2.6 \text{ gal}\end{aligned}$$

Therefore, purchase three 1-gal cans.

EXAMPLE 8: If the water rates are

$1.846 per 100 cubic foot (cf) for the first 3000 cf
$1.100 per 100 cf from 3001 cf to 60,000 cf
$0.755 per 100 cf from 60,001 cf to 1,300,000 cf, and
$0.681 per 100 cf for 1,300,001 cf and over

In addition to the above monthly charge, there is a customer charge depending on the meter size provided and public fire protection service charge. Their rates are as follows:

Size of meter, inches:	5/8	3/4	1	1.5	2	4	8
Customer charge, $/M:	9.50	12.5	20.5	45.5	70.5	225.5	710.50
Fire protection, $/M:	1.96	2.94	4.90	9.80	9.80	9.80	9.80

What is the monthly charge (excluding taxes, if any) for

(a) 575 cf with 5/8-in meter
(b) 86,000 cf with 2-in meter
(c) 1,500,000 cf with 8-in meter

Solution:

Step 1: Solve for (a)

(1) Water used $= \$1.846 \times \dfrac{575}{100} = \$10.61$
(2) Customer charge $= \$9.50$
(3) Fire protection $= \$1.96$

Total monthly charge $= \$22.07$

Step 2: Solve for (b)

(1) Water charge:
First 3000 cf $= \$(1.846 \times 3000)/100 = \$55.38$
Next 57,000 cf $= \$1.100(600 - 30) = \$627.00$
Next 26,000 cf $= \$0.755(860 - 600) = \$196.30$
Total water charge $= \$878.68$ (sum of above)
(2) Customer charge (2-in meter) $= \$70.50$
(3) Fire protection $= \$9.80$

Total monthly charge $= \$958.98$

Step 3: Solve for (c)

(1) Water charge:

First 3000 cf		= $55.38
Next 57,000 cf		= $627.00
Next 1,240,000 cf	= $0.755(13000 − 600)	= $9,362.00
Next 200,000 cf	= $0.681 × 2000	= $1,362.00
Total for 1,500,000 cf		= $11,406.38
(2) Customer charge (8-in meter)		= $710.50
(3) Fire protection		= $9.80
Total charge for $1,500,000 cf		= $12,125.68

Division

EXAMPLE 1: The storage tank of your plant has a 1.20-Mgal capacity and the demand is 0.40 million gallons per day (MGD). If the pump was broken, how many more days could your plant supply the water?

Solution:

$$1.2 \text{ Mgal}/0.4 \text{ MGD} = 3.0 \text{ days}$$

EXAMPLE 2: The intermittent sand filter needs to be drained and cleaned. If the treated wastewater volume in the filter is 18,000 gal, what is the time to empty the filter if the withdrawal rate is 500 gal/min?

Solution:

$$\text{Time } t = \frac{18{,}000 \text{ gal}}{500 \text{ gal/min}}$$
$$= 36 \text{ min}$$

EXAMPLE 3: If a membrane filter count is 69 fecal coliform (FC) colonies and 5.00 mL of a river water sample is filtered, what is the reported FC density?

Solution:

The bacterial density (D) is commonly reported as number of bacteria per 100 mL of water:

$$D = (69 \text{ FC}/5 \text{ mL}) = 13.8 \text{ FC/mL}$$
$$= 13.8 \times 100 \text{ FC}/100 \text{ mL}$$
$$= 1380 \text{ FC}/100 \text{ mL}$$

Report: $D = 1400$ FC/100 mL

Note: Use two significant figures for reporting bacterial density.

3.3 Threshold Odor Measurement

Taste and odor in drinking water supplies are the most common customer complaint. Test procedures are listed in *Standard methods* (APHA, AWWA, and WEF, 1995). The 'threshold odor number' (TON) is the greatest dilution of sample with odor-free water yielding a definitely perceptible odor. The TON is the dilution ratio at which taste or odor is just detectable. The sample is diluted with odor-free water to a total volume of 200 mL in each flask. The TON can be computed as

$$\text{TON} = \frac{A + B}{A}$$

where A = sample size, mL, and B = volume of odor-free water, mL. If the total volume is 200 mL, then

$$\text{TON} = 200/A$$

EXAMPLE: The first detectable odor is observed when a 25 mL sample is diluted to 200 mL with odor-free water. What is the TON of the water sample?

Solution:

$$\text{TON} = \frac{200}{A} = \frac{200\text{ mL}}{25\text{ mL}}$$
$$= 8$$

3.4 Simple Ratio

A ratio is one number divided by another number. It is a pure number, such as 2, 5, or 4.6. A pure number multiplied by a physical unit (mg/L, or miles, etc.) is called a concrete number, such as 2.5 mg/L or 20 ft. The division of two concrete numbers may be either a pure number (if the same physical unit) or a rate (with different physical units). The ratios are shown in the following examples.

$$\frac{3}{4}\text{ is the ratio of 3 to 4, a pure number (or 0.75)}$$

$$\frac{12\text{ m}}{3\text{ m}} = 4\text{; the ratio is a pure number}$$

$$\frac{120\text{ mg}}{5\text{ L}} = 24\text{ mg/L} = 24\text{ times}\left(\frac{1\text{ mg}}{1\text{ L}}\right)\text{, a rate}$$

$$\frac{1\text{ mg}}{1\text{ L}} = 1\text{ mg of a constituent in 1L of liquid, a physical unit}$$

$$\frac{24\text{ oz}}{4\text{ gal}} = 6\text{ oz/gal, a rate}$$

$$\frac{4\text{ ft}}{16\text{ s}} = 0.25\text{ ft/s, a rate}$$

$$\frac{3.2\text{ lb/d}}{8.0\text{ lb/d}} = 0.4\text{, a pure number}$$

EXAMPLE: If 1.5 L of activated sludge with volatile suspended solids (VSS) of 1880 mg/L is mixed with 7.5 L of raw domestic wastewater with BOD_5 of 252 mg/L, what is the F/M (food/microbes) ratio?

Solution:

$$\frac{\text{F}}{\text{M}} = \frac{\text{amount of BOD}_5}{\text{amount of VSS}}$$
$$= \frac{252\text{ mg/L} \times 7.5\text{ L}}{1880\text{ mg/L} \times 1.5\text{ L}} = \frac{1890}{2820}$$
$$= \frac{0.67}{1}$$
$$\text{or } = 0.67$$

3.5 Percentage

A percentage is a ratio with the denominator of 100; while the numerator is called the percent (%). It is also called parts per hundred. A proportion is the equality of two ratios. This equality with one denominator of 100 is used in calculating percentage (ratio times 100%).

EXAMPLE 1: What percentage is 4/25?

Solution:

$$\frac{4}{25} = \frac{16}{100} \qquad \text{therefore } 4 = 16\% \text{ of } 25$$

EXAMPLE 2: What percentage is 18/40?

Solution 1:

$$\frac{18}{40} = 0.45$$

$$\text{then} \qquad 100\% \times 0.45 = 45\%$$

Solution 2:

$$\frac{18}{40} = \frac{x}{100}$$

Cross-multiplying: $40x = 18 \times 100$

$$x(\%) = \frac{1800}{40} = 45$$

EXAMPLE 3: What percentage is 121/94?

Solution: $\frac{121}{94} = \frac{y}{100}$

$$y(\%) = \frac{121 \times 100}{94} = 128.7$$

One liter of water weighs 1 kg (1000 g = 1,000,000 mg). Hence milligrams per liter is parts per million (ppm); i.e. pounds per million pounds, grams per million grams, and milligrams per million milligrams are parts per million.

EXAMPLE 4: What is the percentage of 1 ppm?

Solution: Since 1 part per million (ppm) =1 mg/L

$$\begin{aligned} 1 \text{ mg/L} &= \frac{1 \text{ mg}}{1 \text{ L} \times 1{,}000{,}000 \text{ mg/L}} \times 100\% \\ &= \frac{1}{10{,}000}\% \\ &= 0.0001\% \end{aligned}$$

EXAMPLE 5: How many mg/L is a 1.2% solution?

Solution:

$$\begin{aligned} 1.2\% &= \frac{1.2}{100}, \text{ since the weight of 1 L water is } 10^6 \text{ mg} \\ &= \frac{1.2}{100} \times 1{,}000{,}000 \text{ mg/L} \\ &= 12{,}000 \text{ mg/L} \end{aligned}$$

EXAMPLE 6: Compute pounds per million gallons (Mgal) for 1 ppm (1 mg/L) of water.

Solution: Since 1 gal of water = 8.34 lb

$$
\begin{aligned}
1 \text{ ppm} &= \frac{1 \text{gal}}{10^6 \text{ gal}} \\
&= \frac{1 \text{ gal} \times 8.34 \text{ lb/gal}}{\text{Mgal}} \\
&= 8.34 \text{ lb/Mgal}
\end{aligned}
$$

EXAMPLE 7: For a laboratory-scale test, 2.4 lb of activated carbon (AC) and 32.0 lb of sand are mixed. Compute the percentage of activated carbon in the mixture.

Solution: Total weight = (2.4 + 32.0) lb = 34.4 lb. Let percent of AC = x, therefore

$$
\begin{aligned}
\frac{x}{100\%} &= \frac{2.4 \text{ lb}}{34.4 \text{ lb}} \\
x &= 100\% \times 2.4/34.4 \\
&= 7.0\%
\end{aligned}
$$

EXAMPLE 8: How many kilograms of AC are required to make 38 kg of mixture that is to be 28% AC?

Solution: Let y be the weight of AC

$$
\begin{aligned}
\frac{y \text{ kg}}{38 \text{ kg}} &= 28\% = \frac{28}{100} = 0.28 \\
y &= 38 \text{ kg} \times 0.28 = 10.64 \text{ kg}
\end{aligned}
$$

EXAMPLE 9: How many pounds of AC are needed with 38 lb of sand so that the mixture is 28% AC?

Solution:

$$
\begin{aligned}
\frac{y}{38 + y} &= 0.28 \\
y &= 0.28(38 + y) \\
y &= 10.64 + 0.28y \\
(1 - 0.28)y &= 10.64 \\
y &= \frac{10.64}{0.72} = 14.78 \text{ (lb)}
\end{aligned}
$$

EXAMPLE 10: A softening unit removes total hardness from 258 to 62 mg/L as $CaCO_3$. What is the efficiency of the process (percent removed)?

Solution:

$$
\begin{aligned}
x &= \frac{\text{amount removed}}{\text{original concentration}} \times 100\% \\
&= \frac{258 - 62}{258} \times 100\% \\
&= 76\% \text{ removal}
\end{aligned}
$$

EXAMPLE 11: A water treatment plant treats 996,000 gal/d (gpd). On average it uses 28,000 gpd for backwashing the filters. What is the net production of the plant and what is the percentage of backwash water?

Solution:

Step 1:

$$\begin{aligned}\text{Net production of plant} &= 996{,}000 \text{ gpd} - 28{,}000 \text{ gpd} \\ &= 968{,}000 \text{ gpd}\end{aligned}$$

Step 2:

$$\begin{aligned}\text{Percent of backwash water} &= \frac{28{,}000 \text{ gpd}}{996{,}000 \text{ gpd}} \times 100\% \\ &= 2.9\%\end{aligned}$$

EXAMPLE 12: A pipe is laid at a rise of 120 mm in 20 m. What is the grade?

Solution:

$$\begin{aligned}\text{Grade} &= \frac{120 \text{ mm}}{20 \text{ m}} \times 100(\%) \\ &= \frac{120 \text{ mm}}{20 \times 1000 \text{ mm}} \times 100\% \\ &= 0.6\%\end{aligned}$$

EXAMPLE 13: In the reverse osmosis process, water recovery is the percent of product water flow Q_p (in gpd or m^3/d) divided by feedwater flow Q_f. If $Q_f = 100{,}000$ gpd and $Q_p = 77{,}000$ gpd, what is the water recovery?

Solution:

$$\begin{aligned}\text{Recovery} &= \frac{Q_p}{Q_f} \times 100\% \\ &= \frac{77{,}000}{100{,}000} \times 100\% \\ &= 77\%\end{aligned}$$

EXAMPLE 14: The laboratory data show that the total suspended solids (TSS) for the influent and effluent of a sedimentation basin are 188 and 130 (mg/L), respectively. What is the efficiency of TSS removal?

Solution:

$$\begin{aligned}\text{Percent removal} &= \frac{\text{TSS removal}}{\text{TSS influent}} \times 100\% \\ &= \frac{\text{TSS}_{\text{inf}} - \text{TSS}_{\text{eff}}}{\text{TSS}_{\text{inf}}} \times 100\% \\ &= \frac{188 - 130}{188} \times 100\% \\ &= 30.9\%\end{aligned}$$

EXAMPLE 15: A motor is rated as 30 horsepower (hp). However, the output horsepower of the motor is only 24.6 hp. What is the efficiency of the motor?

Solution:

$$\begin{aligned}\text{Efficiency} &= \frac{\text{hp output}}{\text{hp input}} \times 100\% \\ &= \frac{24.6\text{ hp}}{30\text{ hp}} \times 100\% \\ &= 82\%\end{aligned}$$

EXAMPLE 16: If 1000 m^3 of residual (sludge) containing 0.8% (P_1) of solids have to be thickened to 4% (P_2) what is the final volume?

Solution:
Since the amount of solids is the same before and after thickening,

$$\begin{aligned}V_1P_1 &= V_2P_2 \\ 1000\text{ m}^3 \times 0.8\% &= V_2 \times 4\% \\ V_2 &= \frac{1000 \times 0.8}{4}\text{ m}^3 \\ &= 200\text{ m}^3\end{aligned}$$

EXAMPLE 17: Sludge analysis shows that weights of water and dry solids are 88.46 and 1.22 g, respectively. What is the moisture content of the sludge?

Solution:

$$\begin{aligned}\text{Moisture content (\%)} &= \frac{\text{weight of water}}{\text{weight of wet sludge}} \times 100 \\ &= \frac{88.46}{88.46 + 1.22} \times 100 \\ &= 98.6\end{aligned}$$

EXAMPLE 18: The Surface Water Treatment Rule requires that water treatment must achieve at least 3-log removal or inactivation of *Giardia lamblia* cysts and 4-log removal of viruses. What are the percents removal?

Solution:

Let C_i = initial concentration of cysts or virus plaques

C_f = final concentration of cysts or virus plaques after treatment

(a) For *G. lamblia* cysts removal/inactivation:

$$\log C_i - \log C_f = 3$$

$$\log \frac{C_i}{C_f} = 3$$

$$\text{taking anti-log } \frac{C_i}{C_f} = \log^{-1} 3 = 1000$$

$$\frac{C_f}{C_i} = \frac{1}{1000}$$

$$\begin{aligned}\text{Removal} &= \frac{C_i - C_f}{C_i} = \frac{1000 - 1}{1000} = \frac{999}{1000} \\ &= 0.999 \times 100\% \\ &= 99.9\%\end{aligned}$$

(b) For viruses removal:

$$\log C_\mathrm{i} - \log C_\mathrm{f} = 4$$

$$\log \frac{C_\mathrm{i}}{C_\mathrm{f}} = 4$$

$$\frac{C_\mathrm{i}}{C_\mathrm{f}} = 10{,}000$$

$$\frac{C_\mathrm{f}}{C_\mathrm{i}} = \frac{1}{10{,}000}$$

$$\text{Removal} = \frac{C_\mathrm{i} - C_\mathrm{f}}{C_\mathrm{i}} = \frac{10{,}000 - 1}{10{,}000}$$

$$= 0.9999 \times 100\%$$

$$= 99.99\%$$

3.6 Significant Figures

In calculation, numbers may be either absolutes or measurements: absolutes indicate exact values, while measurements are readings of meters, balances, gages, scales, and manometers. Each measurement has a sensitivity limit below which a change of measurement is not registered; thus, every measurement is not exact or incomplete.

All digits in a reported value are expected to be known definitely except for the last digit, which may be doubtful. Each correct digit of the approximation, except a zero which serves only to fix the decimal point, is called a significant figure. For example 321, 83.8, 7.00, 0.925, and 0.0375 each contain three significant figures.

During measurements, a rounded-off value is usually recorded. Rounding off is carried out either by the operator or using an instrument with minimum error. As a general rule, the magnitude of the error is less than or equal to one-half unit of the place of the nth digit in the rounded number. For example 8.05349 = 8.0535, 8.053, 8.05, 8.1, and 8; but 8.0535 = 8.054 and 8.05 = 8.0.

A general rule is to assume that a reading is correct within one-half unit in the last place recorded. For example, a reading of 38.8 means 38.8 $\pm$ 0.05 or any number between 38.75 and 38.85: rounding off, by dropping digits which are not significant. If the digit 0, 1, 2, 3, or 4 is dropped, do not change the number to the left. If the number 6, 7, 8, or 9 is dropped, increase by one to the preceding digit. If the digit 5 is dropped, round off preceding digit to the nearest even number.

EXAMPLE 1: Round off to one decimal

$$38.73 \rightarrow 38.7$$

$$38.77 \rightarrow 38.8$$

$$38.75 \rightarrow 38.8$$

$$38.45 \rightarrow 38.4$$

$$38.35 \rightarrow 38.4$$

The digit 0 may be recorded as a measured value of zero or may serve merely as a space to locate the decimal point. For example, in 420, 32.08, and 6.00, all the zeros are significant. The rounded-off values may be 9.00, 9.0, or 9. The zeros are also significant. For 38,500, the zeros may or may not be significant: if written as 38.5×10^3, the two zeros are not significant.

When a calculated value (x) with a standard deviation (s) is known, it is recommended to report the value as $x \pm$ s.

If numbers are added or subtracted, the number which has the fewer decimal places, not necessarily the fewer significant figures, makes the limit on the number of places for the sum or difference.

EXAMPLE 2: Add the following numbers

Solution:

$$
\begin{array}{r}
3.1472 \\
32.05 \\
1234 \\
8.9426 \\
0.0032 \\
+\quad 9.00 \\
\hline
1287.1430
\end{array}
$$

The sum should be rounded off to 1287, no decimals.

EXAMPLE 3: Subtraction 78.3 − 3.14 − 0.388

Solution:

$$
\begin{array}{r}
78.3 \\
3.14 \\
-\ 0.388 \\
\hline
74.772
\end{array}
\quad \rightarrow \quad
\begin{array}{r}
78.3 \\
3.1 \\
-\ 0.4 \\
\hline
74.8
\end{array}
$$

The answer is 74.8

In case several numbers are multiplied or divided, the number of significant figures of the calculated value should use the fewest significant figures presented in the factors; i.e. the weakest link in the chain.

EXAMPLE 4: For the residual chlorine titration with the iodometric method, use a 250-mL water sample, 2.3 mL of 0.01N $Na_2S_2O_3$ added to the end point. While 0.1 mL of 0.01N $Na_2S_2O_3$ is titrated for the blank. What is the residual chlorine in the water sample?

Solution: The result of residual chlorine analysis is computed by

$$
\text{MgCl as } Cl_2/L = \frac{(2.3 - 0.1) \times 0.01 \times 35.450}{250}
$$

$$
= 3.1196
$$

The answer is 3.1 mg/L of residual chlorine: only take two significant figures.

After considering significant figures or rounding off, the maximum relative error can be estimated by how far it can be trusted.

EXAMPLE 5: The flow chart of the water plant reads 3.8 million gallons per day (MGD). What is the maximum relative error for the reading?

Solution:

$$
3.8 \text{ MGD} = 3.8 \pm 0.05 \text{ MGD}
$$

$$
\text{The maximum relative error} = \frac{0.05}{3.8}
$$

$$
= 0.01316
$$

$$
= 0.013 \times 100\%
$$

$$
= 1.3\%
$$

EXAMPLE 6: When we multiply 3.84 × 0.36, the arithmetic product is 1.3824. Evaluate the maximum relative errors.

Solution:

Step 1: The product 1.3824 is rounded off to 1.4

Step 2: Calculate errors.

The product 1.4 means 1.4 ± 0.05

has

$$\frac{0.05}{1.4} \times 100\% = 3.6\% \text{ error}$$

While 3.84 has

$$\frac{0.005}{3.84} \times 100\% = 1.3\% \text{ error}$$

and 0.36 has

$$\frac{0.005}{0.36} \times 100\% = 1.4\% \text{ error}$$

Total 2.7% error

3.7 Transformation of Units

Since all measurements are products of pure numbers and physical units, through transformation by multiplication and/or division, the computed results may be a ratio (pure number) or a concrete number with a physical unit. Physical units, like numbers, are also included in calculation.

EXAMPLE 1:

Length $2 \text{ ft} \times 32 = 64 \text{ ft}$

Area $12 \text{ ft} \times 14 \text{ ft} = 168 \text{ ft}^2$

Volume $6 \text{ ft} \times 8 \text{ ft} \times 3.5 \text{ ft} = 168 \text{ ft}^3$

Ratio $\frac{25 \text{ ft}}{5 \text{ ft}} = \frac{25}{5} = 5$

$$\frac{88 \text{ ft}^3}{4 \text{ ft}^3} = \frac{22 \text{ ft} \times \text{ft} \times \text{ft}}{1 \text{ ft} \times \text{ft} \times \text{ft}} = \frac{22}{1}$$

$$\frac{99 \text{ ft}^3/\text{h}}{3 \text{ ft}^2} = 99\,\frac{\text{ft}^3}{\text{h}} \times \frac{1}{3 \text{ ft}^2} = 33 \text{ ft/h}$$

Division $\frac{72 \text{ ft}^2}{3} = 24 \text{ ft}^2$

$\frac{36 \text{ ft/s}}{36 \text{ ft/s}} = 1$ (any quantity divided by its equivalent equals one.)

Multiplication $28 \text{ ft}^3 \times 5 = 140 \text{ ft}^3$

$$68\,\frac{\text{ft}^3}{\text{s}} \times 6 \text{ s} = 408 \text{ ft}^3$$

EXAMPLE 2: An operator recorded that the water treatment plant treated 2.8 Mgal of water during his shift. What is the plant's total volume treated per day, assuming the same treatment rate?

Solution:

$$\begin{aligned} &\frac{2.8 \text{ Mgal}}{8 \text{ h}} \times \frac{24 \text{ h}}{\text{day}} \\ &= \frac{2.8 \times 24}{8} \text{ MGD} \\ &= 8.4 \text{ MGD} \end{aligned}$$

EXAMPLE 3: 1 MGD equals how many
(a) cubic feet per second (cfs)?
(b) gallons per minute (gpm)?
(c) cubic meters per day (or metric tons per day)?
(d) cubic meters per second (m^3/s)

Solution:

(a)

$$\begin{aligned} 1 \text{ MGD} &= \frac{10^6 \text{ gal}}{1 \text{ day}} \\ &= \frac{10^6 \text{ gal} \times 0.1337 \text{ ft}^3/\text{gal}}{1 \text{ day} \times 86{,}400 \text{ s/day}} \\ &= \frac{133{,}700 \text{ ft}^3}{86{,}400 \text{ s}} \\ &= 1.547 \text{ cfs} \end{aligned}$$

(b)

$$\begin{aligned} 1 \text{ MGD} &= \frac{10^6 \text{ gal}}{1 \text{ day} \times 24 \text{ h/day} \times 60 \text{ min/h}} \\ &= \frac{1{,}000{,}000 \text{ gal}}{24 \times 60 \text{ min}} \\ &= 694.4 \text{ gpm} \end{aligned}$$

(c)

$$\begin{aligned} 1 \text{ MGD} &= \frac{10^6 \text{ gal}}{1 \text{ day}} \\ &= \frac{10^6 \text{ gal} \times 3.785 \text{ L/gal} \times 1 \text{ kg/L}}{1 \text{ day}} \\ &= \frac{3.785 \times 10^6 \text{ kg}}{\text{day}} \\ &= \frac{3.785 \times 10^6 \text{ kg} \times 10^{-3} \text{ metric ton/kg}}{\text{day}} \\ &= 3785 \text{ metric tons per day (m}^3\text{/day)} \end{aligned}$$

(d)

$$\begin{aligned} 1 \text{ MGD} &= 1 \text{ Mgal/day} \\ &= \frac{1 \text{ Mgal} \times 3785 \text{ m}^3/\text{Mgal}}{1 \text{ day} \times 86400 \text{ s/day}} \\ &= 0.0438 \text{ m}^3/\text{s} \end{aligned}$$

EXAMPLE 4: A blower has a capacity of 200 cfs. Compute the capacity in
(a) cubic meters per second?
(b) cubic meters per minute?

Solution:

(a)
$$200\,\frac{\text{ft}^3}{\text{s}} = \frac{200\ \text{ft}^3 \times (0.3048)^3\ \text{m}^3/\text{ft}^3}{\text{s}}$$
$$= 200(0.02832)\ \text{m}^3/\text{s}$$
$$= 5.66\ \text{m}^3/\text{s}$$

(b)
$$200\,\frac{\text{ft}^3}{\text{s}} = 5.66\ \text{m}^3/\text{s} \times 60\ \text{s/min}$$
$$= 340\ \text{m}^3/\text{min}$$

EXAMPLE 5: How many cubic inches are in 1 cubic foot?

Solution:

$$1\ \text{ft}^3 = 1\ \text{ft} \times 1\ \text{ft} \times 1\ \text{ft}$$
$$= \left(1\ \text{ft} \times \frac{12\ \text{in}}{1\ \text{ft}}\right)^3$$
$$= 12^3\ \text{in}^3$$
$$= 1728\ \text{in}^3$$

EXAMPLE 6: A wastewater treatment plant receives an average flow of 626 gpm. What is the daily total flow to plant?

Solution:

$$\text{Flow} = 626\ \text{gpm (gallons per minute)}$$
$$= 626\,\frac{\text{gal}}{\text{min}} \times 60\,\frac{\text{min}}{\text{h}} \times 24\,\frac{\text{h}}{\text{day}}$$
$$= 901{,}440\,\frac{\text{gal}}{\text{day}}$$
$$= 0.90 \times 10^6\ \text{gallons per day (gpd)}$$
$$= 0.90\ \text{million gallons per day (MGD)}$$

EXAMPLE 7: What is the weight (in pounds, in tons, and in kilograms) of 200 gal of water?

Solution:

(a)
$$\text{Weight (lb)} = 200\ \text{gal} \times 8.34\,\frac{\text{lb}}{\text{gal}}$$
$$= 1668\ \text{lb}$$

(b)
$$\text{Weight (tons)} = 1668\ \text{lb} \times \frac{\text{tons}}{2000\ \text{lb}}$$
$$= 0.834\ \text{tons}$$

(c)
$$\text{Weight (kg)} = 1668\ \text{lb} \times 0.454\,\frac{\text{kg}}{\text{lb}}$$
$$= 757\ \text{kg}$$

EXAMPLE 8: How many gallons per minute/cubic yard (gpm/yd^3) for 1 MGD/acre · ft (million gallons per minute/acre · foot)?

Solution:

$$\begin{aligned} 1 \text{ MGD} &= 1{,}000{,}000 \text{ gal/day} \times (1 \text{ day}/1440 \text{ min}) \\ &= 694.4 \text{ gpm} \\ 1 \text{ acre} \cdot \text{ft} &= 43{,}560 \text{ ft}^2 \times 1 \text{ ft} \\ &= 43{,}560 \text{ ft}^3 \times 1 \text{ yd}^3/27 \text{ ft}^3 \\ &= 1613.3 \text{ yd}^3 \end{aligned}$$

therefore

$$\begin{aligned} 1 \text{ MGD/acre} \cdot \text{ft} &= 694.4 \text{ gpm}/1613 \text{ yd}^3 \\ &= 0.430 \text{ gpm/yd}^3 \end{aligned}$$

EXAMPLE 9: How many grains per gallon for 1 mg/L?

Solution:

$$\begin{aligned} 1 \text{ mg/L} &= 1 \frac{\text{mg}}{\text{L}} \times 0.0154 \frac{\text{grain}}{\text{mg}} \times 3.785 \frac{\text{L}}{\text{gal}} \\ &= 0.0583 \text{ grain/gal} \\ \text{or } 1 \text{ grain/gal} &= \frac{1 \text{ mg/L}}{0.0583} = 17.1 \text{ mg/L} \end{aligned}$$

EXAMPLE 10: How many lb/Mgal for 1 grain/gal?

Solution: Since 1 lb = 7000 grains

$$\begin{aligned} 1 \text{ grain/gal} &= 1/7000 \text{ lb/gal} \\ &= 10^6/7000 \text{ lb/Mgal} \\ &= 142.9 \text{ lb/Mgal} \end{aligned}$$

EXAMPLE 11: How many lb/Mgal for 1 mg/L?

Solution:

$$\begin{aligned} 1 \text{ mg/L} &= 1 \text{ gal/Mgal} \\ &= 1 \text{ gal} \times 8.34 \text{ lb/gal}/1 \text{ Mgal} \\ &= 8.34 \text{ lb/Mgal} \end{aligned}$$

EXAMPLE 12: How many gallons is 50 lb of water?

Solution: Since 1 gal of water weighs 8.34 lb

$$\begin{aligned} V &= 50 \text{ lb} \times \frac{1 \text{ gal}}{8.34 \text{ lb}} \\ &= 6.0 \text{ gal} \end{aligned}$$

EXAMPLE 13: A 5-gal empty tank weighs 2.3 lb. What is the total weight of the tank filled with 5 gal of water?

Solution:

$$\begin{aligned} \text{Weight of water} &= 5 \text{ gal} \times 8.34 \text{ lb/gal} \\ &= 41.7 \text{ lb} \end{aligned}$$

$$\begin{aligned}\text{Total weight} &= 41.7\text{ lb} + 2.3\text{ lb}\\ &= 44.0\text{ lb}\end{aligned}$$

EXAMPLE 14: The discharge rate is 1.2 ft³/s. How many gallons of water is discharged in 5 min?

Solution:

$$\begin{aligned}\text{Total discharge} &= 1.2\text{ ft}^3/\text{s} \times 60\text{ s/min} \times 5\text{ min}\\ &= 360\text{ ft}^3\\ &= 360\text{ ft}^3 \times 7.48\text{ gal/ft}^3\\ &= 2693\text{ gal}\end{aligned}$$

EXAMPLE 15: In a small activated sludge plant, the maximum BOD_5 loading is in the range 25–35 lb BOD/1000 ft³/d. What is the BOD/m³ · d range in grams?

Solution:

Step 1: Convert 25 lb BOD/1000 (ft³ · d) to g BOD/(m³ · d)

$$\begin{aligned}\frac{25\text{ lb BOD}}{1000\ (\text{ft}^3\cdot\text{d})} &= \frac{25\text{ lb BOD} \times 453.6\text{ g/lb}}{1000\ (\text{ft}^3\cdot\text{d}) \times 0.02832\text{ m}^3/\text{ft}^3}\\ &= \frac{25}{1000} \times 16{,}017\text{ g BOD/(m}^3\cdot\text{d)}\\ &= 400\text{ g BOD/(m}^3\cdot\text{d)}\end{aligned}$$

Step 2: Convert 35 lb BOD/1000 ft³/d

$$\begin{aligned}\frac{35\text{ lb BOD}}{1000\ (\text{ft}^3\cdot\text{d})} &= \frac{35}{1000} \times 16{,}017\text{ g BOD/(m}^3\cdot\text{d)}\\ &= 560\text{ g BOD/(m}^3\cdot\text{d)}\end{aligned}$$

Answer: 400–560 g BOD/(m³ · d)

EXAMPLE 16: In an activated sludge process, 2 ft³ of compressed air is needed for treating each gallon of domestic wastewater. What is this in terms of liter of air per liter of wastewater (L/L)?

Solution:

$$\begin{aligned}2\text{ ft}^3/\text{gal} &= \frac{2\text{ ft}^3 \times 28.32\text{ L/ft}^3}{1\text{ gal} \times 3.785\text{ L/gal}}\\ &= 2 \times 7.48\text{ L air/L wastewater}\\ &= 14.96\text{ L/L}\\ &\cong 15\text{ L/L}\end{aligned}$$

Note: 1 ft³/gal = 7.48 L/L.

EXAMPLE 17: The depth of sludge usually applied to the sludge drying bed each time is 8–12 in. What is the depth in centimeters?

Solution:

$$\begin{aligned}8\text{ in} &= 8 \times 2.54\text{ cm (as 2.54 cm = 1 in)}\\ &= 20.32\text{ cm}\\ 12\text{ in} &= 12 \times 2.54 = 30.48\text{ cm}\end{aligned}$$

For simplicity, the beds are 20–30 cm in depth.

3.8 Geometrical Formulas

1. Oblique triangle

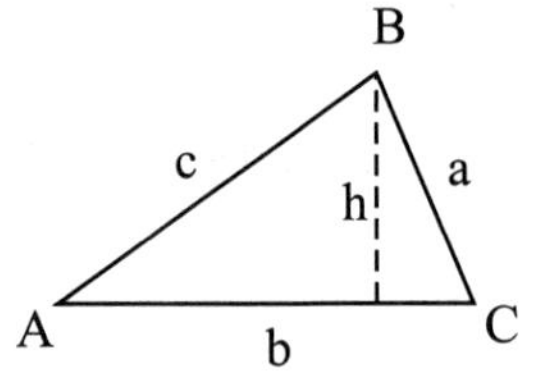

$$\text{Area} = \tfrac{1}{2}b \times h$$
$$= \tfrac{1}{2}b \times c \sin A$$
$$= \sqrt{[l(l - a)(l - b)(l - c)]}$$

where b = base
h = altitude
$1 = (a + b + c)/2$

2. Right-angled triangle

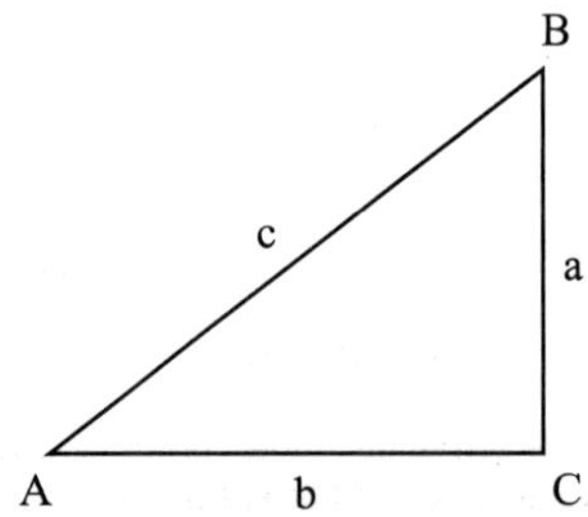

$$\text{Area} = \tfrac{1}{2}a \times b$$

where $\angle C = 90°$
$c^2 = a^2 + b^2$

3. Rectangle

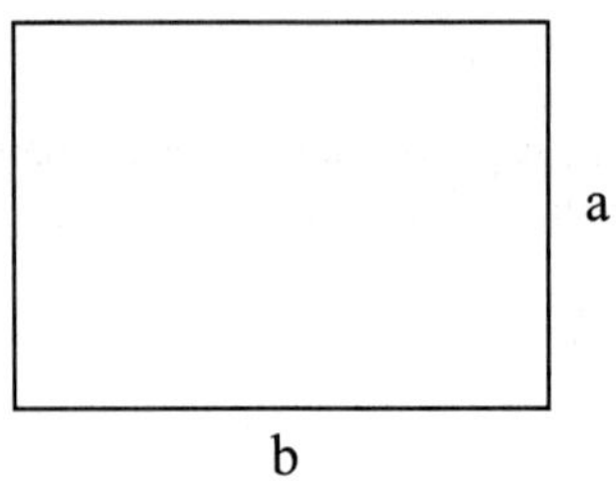

$$\text{Area} = a \times b$$

4. Trapezoid

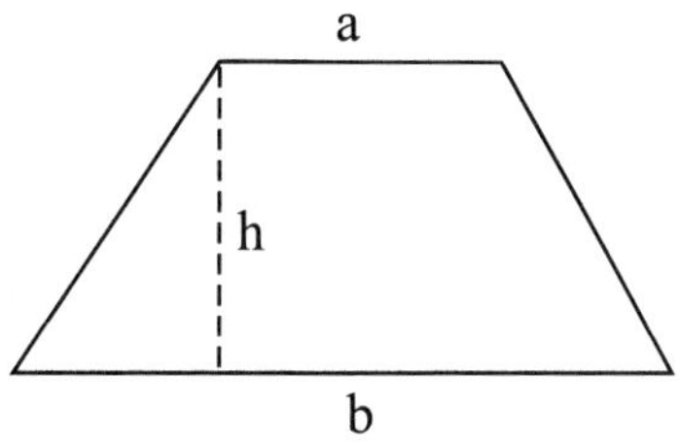

$$\text{Area} = \tfrac{1}{2}(a + b)h$$

5. Circle and sphere

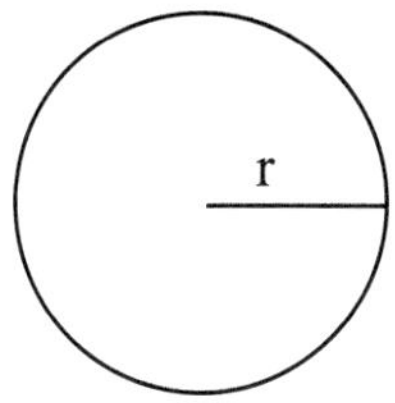

$$\text{Area of circle} = \pi r^2$$
$$\text{Area of sphere} = 4\pi r^2$$
$$\text{Volume of sphere} = \tfrac{4}{3}\pi r^3$$
$$\text{Circumference} = 2\pi r$$
where $r =$ radius

6. Cylinder

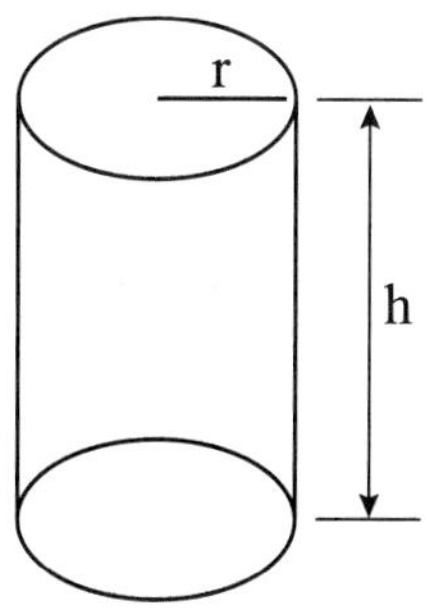

$$\text{Lateral area} = 2\pi rh$$
$$\text{Volume} = \pi r^2 h$$

7. Cone

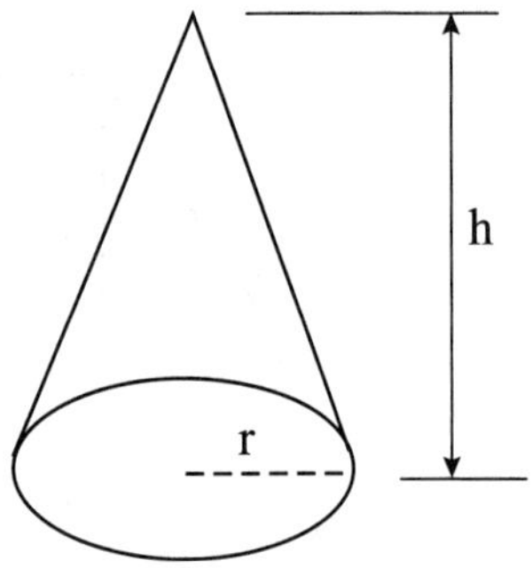

$$\text{Volume of cone} = \tfrac{1}{3}\pi r^2 h$$

8. Frustrum

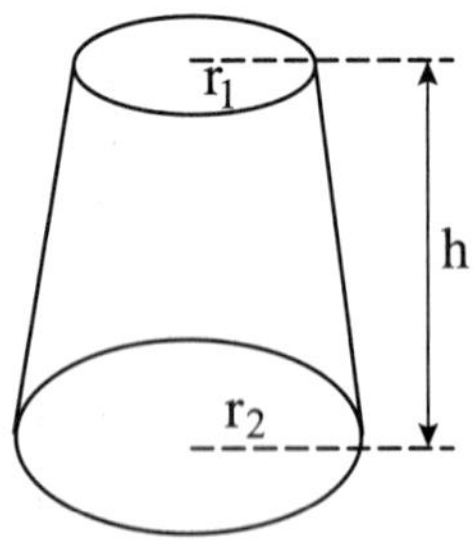

$$\text{Volume of frustrum} = \tfrac{1}{3}(A_1 + A_2 + \sqrt{A_1 \times A_2})h$$

$$\text{where } A_1 = \pi r_1^2$$
$$A_2 = \pi r_2^2$$

EXAMPLE 1: The dimensions of a primary settling basin are 22 ft wide, 33 ft deep, and 66 ft long. Compute the volume of the basin.

Solution:

$$\begin{aligned}\text{Volume} &= 22\text{ ft} \times 33\text{ ft} \times 66\text{ ft}\\ &= 47{,}916\text{ ft}^3\end{aligned}$$

EXAMPLE 2: A circular sedimentation tank has a radius of 18 ft. Its wastewater depth is 20 ft. What is the volume of the tank?

Solution:

$$\begin{aligned}\text{Volume} &= \pi r^2 h\\ &= 3.14 \times (18\text{ ft})^2 \times 20\text{ ft}\\ &= 20{,}347\text{ ft}^3\end{aligned}$$

EXAMPLE 3: The cone-shape lime applicator has a diameter of 18 ft and a height of 7 ft. What is the capacity of the tank?

Solution:

$$r = \tfrac{1}{2}d = 1/2 \times 18 \text{ ft} = 9 \text{ ft}$$

$$\begin{aligned}\text{Volume of cone} &= \tfrac{1}{3}\pi r^2 h \\ &= 1/3 \times 3.14 \times (9 \text{ ft})^2 \times 7 \text{ ft} \\ &= 593.5 \text{ ft}^3\end{aligned}$$

EXAMPLE 4: A sedimentation tank has a diameter of 66 ft. Its effluent weir is located along the rim of the tank. What is the length of the weir?

Solution:

$$\begin{aligned}L &= 2\pi r \quad \text{or } \pi d \\ &= 3.14 \times 66 \text{ ft} \\ &= 207 \text{ ft}\end{aligned}$$

EXAMPLE 5: Copper sulfate is commonly used for controlling algae in lakes and reservoirs. If an area of 2200 ft (670 m) long and 790 ft (240 m) wide is to be treated with an application rate of 1.5 lb per surface acre (1.83 kg/ha), how much copper sulfate is needed?

Solution:

$$\begin{aligned}\text{Surface area } A &= 2200 \text{ ft} \times 790 \text{ ft} \\ &= 1{,}738{,}000 \text{ ft}^2 \\ &= 1{,}738{,}000 \text{ ft}^2 \times \frac{1 \text{ acre}}{43{,}560 \text{ ft}^2} \\ &= 39.9 \text{ acre} \\ &= 16.1 \text{ ha} \\ \text{Weight of CuSO}_4 \text{ needed} &= 1.5 \text{ lb/acre} \times 39.9 \text{ acre} \\ &= 59.85 \text{ lb} \\ &= 27.1 \text{ kg}\end{aligned}$$

say, 60 lb are needed.

EXAMPLE 6: A cylinder tank has a diameter of 20 ft. Liquid alum is filled to a depth of 7.8 ft. What is the volume (gal) of alum?

Solution: Step 1. Compute the volume (v) in cubic feet:

$$\begin{aligned}v &= \text{area} \times \text{height} = \pi r^2 h \\ &= 3.14 \times (20/2)^2 \times 7.8 \\ &= 3.14 \times 100 \times 7.8 \\ &= 2449.2 \ (\text{ft}^3)\end{aligned}$$

Step 2. Convert to gallons:

$$\begin{aligned}v &= 2449.2 \text{ ft}^3 \times \frac{7.48 \text{ gal}}{1 \text{ ft}^3} \\ &= 18{,}320 \text{ gal}\end{aligned}$$

EXAMPLE 7: A standpipe has an inside diameter of 96 in and is 18 ft high. How many gallons does it contain if the water is 15 ft deep?

Solution:

$$\text{Radius} \quad r = \frac{96}{2}\text{ in} \times \frac{1\text{ ft}}{12\text{ in}} = 4\text{ ft}$$

$$\begin{aligned} v &= \pi r^2 h = 3.14 \times 4\text{ ft} \times 4\text{ ft} \times 15\text{ ft} \\ &= 753.6\text{ ft}^3 \\ &= 753.6\text{ ft}^3 \times 7.48\text{ gal/ft}^3 \\ &= 5637\text{ gal} \end{aligned}$$

EXAMPLE 8: The diameter (d) of a circular sedimentation tank is 88 ft. What is the circumference (c) and area of the tank?

Solution:

Step 1.

$$\begin{aligned} c &= \pi d = 3.14 \times 88\text{ ft} \\ &= 376.3\text{ ft} \end{aligned}$$

Step 2.

$$\text{Radius } r = d \div 2 = 88\text{ ft}/2 = 44\text{ ft}$$

$$\begin{aligned} \text{Area} &= \pi r^2 = 3.14(44)^2 \\ &= 6079\text{ ft}^2 \end{aligned}$$

EXAMPLE 9: The circumference of a storage tank is 113 ft. What is radius (r) of the tank?

Solution:

$$\begin{aligned} c &= \pi d = 2\pi r \\ 113 &= 2 \times 3.14 \times r \\ r &= 113/6.28 \\ &= 18\text{ (ft)} \end{aligned}$$

EXAMPLE 10: Your treatment plant occupies a rectangular area with a length of 488 ft (l) and a width of 95 ft (w). What is the length of fence needed for the lot (including gates)? What is the area of the lot (in square feet and acres)?

Solution:

Step 1.

$$\begin{aligned} \text{Fence length} &= 2(l + w) = 2(488 + 195) = 2 \times 683 \\ &= 1366\text{ (ft)} \end{aligned}$$

Step 2.

$$\begin{aligned} \text{Area} &= l \times w = 488 \times 195 = 95{,}160\text{ (ft}^2\text{)} \\ &= 95{,}160\text{ ft}^2 \times \frac{1\text{ acre}}{43{,}560\text{ ft}^2} \\ &= 2.18\text{ acres} \end{aligned}$$

EXAMPLE 11: A rectangle has a perimeter of 1234 ft and a length of 390 ft. What is its width?

Solution:

$$\begin{aligned} \text{Perimeter} &= 2(l + w) \\ 1234 &= 2(390 + w) \\ 617 &= 390 + w \\ w &= 617 - 390 = 227\text{(ft)} \end{aligned}$$

EXAMPLE 12: A rectangular sedimentation basin is 20 m wide, 40 m long, and 4 m water depth. What is its volume in m^3, ft^3, and in gallons?

Solution:

$$\begin{aligned}\text{Volume} &= 20\text{ m} \times 40\text{ m} \times 4\text{ m} \\ &= 3200\text{ m}^3\end{aligned}$$

or

$$\begin{aligned}&= 3200\text{ m}^3 \times \frac{1\text{ ft}^3}{0.02832\text{ m}^3} \\ &= 112{,}990\text{ ft}^3\end{aligned}$$

or

$$\begin{aligned}&= 112{,}990\text{ ft}^3 \times 7.48\text{ gal/ft}^3 \\ &= 845{,}200\text{ gal}\end{aligned}$$

EXAMPLE 13: A trench is 2 ft wide, 5 ft deep, and $\frac{1}{2}$ mile long. How many cubic yards of soil are excavated?

Solution:

$$\begin{aligned}0.5\text{ mile} &= 0.5 \times 43{,}560\text{ ft} = 21{,}780\text{ ft} \\ \text{Volume} &= \text{width} \times \text{depth} \times \text{length} \\ &= 2\text{ ft} \times 5\text{ ft} \times 21{,}780\text{ ft} \\ &= 217{,}800\text{ ft}^3 \\ &= 217{,}800\text{ ft}^3 \times \frac{1\text{ yd}^3}{27\text{ ft}^3} \\ &= 24{,}200\text{ yd}^3 \\ &= 18{,}500\text{ m}^3\end{aligned}$$

4 BASIC CHEMISTRY AND PHYSICS

4.1 Density and Specific Gravity

Density (ρ) is a weight for a unit volume at a particular temperature. It generally varies with temperature. The weight may be expressed in terms of pounds, ounces, grams, kilograms, etc. The volume may be liters, milliliters, gallons, or cubic feet, etc. The densities for fresh water under different temperature conditions are shown in Table 1.3. For salt water, density is 64 lb/ft^3 at 0°C. The densities of plain and reinforced concretes are 144 and 150 lb/ft^3, respectively.

Specific gravity (sp gr) is defined as the ratio of the density of a substance to the density of water. The temperature of both the substance and water be stated. The specific gravity of fresh water at 0°C is 1.00, while that of salt water is 1.02. The specific gravities of vegetable oil, fuel oil, and lubricant oil are 0.91–0.94, 1.0, and 0.9, respectively.

EXAMPLE 1: The specific gravity of an oil is 0.98 at 62°F. What is the weight of 1 gal of oil?

Solution:

$$\begin{aligned}\text{Weight} &= \text{specific gravity} \times \text{weight of water} \\ &= 0.98 \times 8.34\text{ lb/gal} \\ &= 8.17\text{ lb}\end{aligned}$$

TABLE 1.3 Temperature and Density of Fresh Water

Temperature		Density (unit weight)		
°C	°F	lb/ft^3	lb/gal	N/m^3
0	32	62.417		9805
4	39.2	62.424		9806
16.7	62	62.355	8.34	9795
21	69.8	52.303		9787
100	212	59.7		9378

EXAMPLE 2: An empty 250-mL graduated cylinder weighs 260.8 g. The cylinder filled with 250 mL of water weighs 510.8 g. The cylinder filled with a given solution (250 mL) weighs 530.4 g. Calculate the specific gravity of the solution.

Solution:

$$\text{Weight of water} = 510.8 - 260.8 = 250.0 \text{ g}$$

$$\text{Weight of solution} = 530.4 - 260.8 = 269.6 \text{ g}$$

$$\text{Specific gravity} = \frac{\text{weight of solution}}{\text{weight of water}} = \frac{269.6 \text{ g}}{250.0 \text{ g}} = 1.078$$

The density of water varies slightly with temperature. Simon (1976) reported the relationship of water temperature and density (Table 1.4).

EXAMPLE 3: A liquid has a specific gravity of 1.13. How many pounds is 52 gal of this liquid?

Solution:

$$\text{Weight} = 52 \text{ gal} \times 8.34 \text{ lb/gal} \times 1.13 = 490 \text{ lb}$$

TABLE 1.4 Effect of Water Temperature on Density, Viscosity, Surface Tension, and Vapor Pressure

Temperature, °C	Density, g/mL	Viscosity, poise	Surface tension, dyne/cm	Vapor pressure, atmosphere
0	0.99987	0.01792	75.6	0.006
4	1.00000	0.01568		
5		0.01519		
10	0.99973	0.01308	74.22	0.012
15	0.99913	0.01140		0.0168
20	0.99823	0.01005	72.75	0.0231
25	0.99707	0.00894		0.0313
30	0.99567	0.00801	71.18	0.0419
40		0.00656	69.56	0.0728
50		0.00549	67.91	0.1217
60		0.00469	66.18	0.1965
70		0.00406	64.4	0.4675
80		0.00357	62.6	
90		0.00317		
100	0.95838	0.00284	58.9	1.000

EXAMPLE 4: If a solid in water has a specific gravity of 1.25, how many percent is it heavier than water?

Solution:

$$\text{Percent heavier} = \frac{\text{sp gr of solid} - \text{sp gr of water}}{\text{sp gr of water}} \times 100$$

$$= \frac{1.25 - 1.0}{1.0} \times 100$$

$$= 25$$

4.2 Chemical Solutions

Solids, liquids, and gases may dissolve in water to form solutions. The amount of solute present may vary below certain limits, so-called solubility. The strength of a solution can be expressed in two ways: (1) weight (lb) of active solute per 100 pounds (i.e. %) and (2) weight of active solute per unit volume (gallons or liters) of water. Either expression can be computed to the other if the density or specific gravity is known. If the solution is dilute (less than 1%), the specific gravity can be assumed to be 1.0; i.e. 1 L of solution is equal to 1 kg and 1 gal of solution equals 8.34 lb.

In water chemistry, molarity is defined as the number of gram-molecular weights or moles of substance present in a liter of the solution. If solutions have equal molarity, it means that they have an equal number of molecules of dissolved substance per unit volume. The weight of substance in the solution can be determined as follows:

The molarity (M) of a solution can be expressed as:

$$\text{M (mol/L)} = \frac{\text{moles of solute (mole)}}{\text{1.0 L of solution}}$$

For uniform purpose, the normality (N) is used for preparation of laboratory solutions. The normality can be written as:

$$\text{N (eq/L or meq/L)} = \frac{\text{equivalent of solute (eq or meq)}}{\text{1.0 L of solution}}$$

where

$$\text{Equivalent mass (g/eq or mg/meq)} = \frac{\text{molecular (or atom) mass (g or mg)}}{v \text{ (eq or meq)}}$$

Here v is the equivalence or the number of replacement hydrogen atoms; for oxidation–reduction reactions, v = change in valence. It should be noted that 1.0 equivalent (eq) is 1000 milliequivalents (meq) and that 1.0 eq/m^3 = 1.0 meq/L.

In environmental engineering practices, parts per million (ppm) is a term frequently used. It may be expressed by

$$\text{ppm} = \frac{\text{mass of solute (g)}}{10^6 \text{ g of solution}}$$

or

$$\text{ppm} = \frac{\text{concentration, mg/L}}{\text{specific gravity of liquid}}$$

The relationship of molarity and weight of solute and concentrations can be expressed as follows:

$$\text{Weight in grams} = \text{molarity} \times \text{volume of solution in liters} \times \text{gram-molecular weight}$$

where

$$\text{unit of molarity} = \text{mole per liter (mol/L)}$$
$$\text{unit of the gram-molecular weight} = \text{grams per mole (g/mol)}$$

It may be converted to milligrams per liter (mg/L). The relationship between mg/L and molarity is

$$\begin{aligned}\text{mg/L} &= \text{molarity (mol/L)} \times \text{L} \times \text{gram-molecular weight (g/mol} \times \text{mg/g)}\\ &= \text{molarity} \times \text{gram-molecular weight} \times 10^3\end{aligned}$$

It also may be converted to concentration in parts per million.

$$\text{Since mg/L} = (\text{ppm}) \times \text{density of solution}$$
$$(\text{ppm}) \times \text{density of solution} = \text{molarity} \times \text{gram-molecular weight} \times 10^3$$

$$\text{ppm} = \frac{\text{molarity} \times \text{gram-molecular weight} \times 10^3}{\text{density of solution}}$$

EXAMPLE 1: How much of reagent grade calcium oxide (CaO) and aluminum sulfate [$Al_2(SO_4)_3 \cdot 14H_2O$] is needed to prepare 0.10 mole solutions?

Solution:

(a) The molecular weight of CaO = 40.1 + 16.0 = 56.1

0.10 mole = 0.10 × 56.1 = 5.61 (g/L)

(b) The molecular weight of $Al_2(SO_4)_3 \cdot 14H_2O$

= 27 × 2 + 3(32 + 16 × 4) + 14(1 × 2 + 16)

= 594

0.10 mole = 0.10 × 594 = 59.4 (g/L)

Note: The two solutions have equal numbers of solute molecules.

EXAMPLE 2: Laboratory-use sulfuric acid has 96.2% strength or purity. Its specific gravity is 1.84. Compute (a) the total weight in grams per liter, (b) the weight of acid in grams per liter, (c) pounds of sulfuric acid per cubic foot, (d) pounds of acid per gallon, (e) the molecular weight of acid (H_2SO_4), and (f) the volume (mL) of concentrated acid required to prepare 1.0 L of 0.1 N solution.

Solution:

(a)

$$\begin{aligned}\text{Total weight} &= 1000 \text{ g/L} \times \text{sp gr}\\ &= 1000 \times 1.84 \text{ g/L}\\ &= 1840 \text{ g/L}\end{aligned}$$

(b)

$$\begin{aligned}\text{Weight of acid} &= \text{Total weight} \times \frac{\%}{100}\\ &= 1840 \times \frac{96.2}{100}\\ &= 1770 \text{ g/L}\end{aligned}$$

(c) Since 1 ft^3 of water weighs 62.4 lb,

$$\begin{aligned}\text{Weight of acid} &= 62.4 \text{ lb/ft}^3 \times \text{sp gr} \times \frac{\%}{100}\\ &= 62.4 \times 1.84 \times 0.962 \text{ lb/ft}^3\\ &= 110 \text{ lb/ft}^3\end{aligned}$$

(d) Since 1 gallon of water weighs 8.34 lb

$$\begin{aligned}\text{Weight of acid} &= 8.34 \text{ lb/gal} \times \text{sp gr} \times \frac{\%}{100} \\ &= 8.34 \times 1.84 \times 0.962 \text{ lb/gal} \\ &= 14.8 \text{ lb/gal}\end{aligned}$$

(e) The standard atomic weights of elements are available from *Standard methods* (APHA, AWWA, and WEF, 1995).

$$\begin{aligned}\text{The molecular weight of } H_2SO_4 &= (1 \times 2) + 32 + (16 \times 4) \\ &= 98\end{aligned}$$

(f) For H_2SO_4, valence = 2

$$\begin{aligned}\text{Equivalent mass} &= \frac{98}{2} = 49 \text{ g/eq} \\ 0.1 \text{ N (eq/L)} &= \frac{0.1 \text{ eq}}{1.0 \text{ L of solution}} \\ &= \frac{0.1 \text{ eq} \times 49 \text{ g/eq}}{1.0 \text{ L}} \\ &= \frac{4.9 \text{ g}}{1 \text{ L}} \\ \text{Divided by sp gr} &= \frac{4.9 \text{ g}}{1 \text{ L} \times 1.84 \text{ g/mL}} \\ &= 2.66 \text{ mL/L}\end{aligned}$$

With correction for impurity, the volume needed is

$$\text{Volume} = \frac{2.66 \text{ mL/L}}{0.962} = 2.77 \text{ mL/L}$$

EXAMPLE 3: After adding 0.72 lb of chemical to 2 gal of water, what is the percent strength of the solution?

Solution:

$$\begin{aligned}\text{Total weight of the solution} &= 2 \text{ gal} \times 8.34 \text{ lb/gal} + 0.72 \text{ lb} \\ &= 17.40 \text{ lb}\end{aligned}$$

$$\begin{aligned}\text{Percent of solution} &= \frac{0.72}{17.40} \times 100\% \\ &= 4.14\%\end{aligned}$$

EXAMPLE 4: Sodium hydroxide (NaOH) is used as the base in acid–base titrations. It has a molecular weight of 40 (23 for Na + 16 for O + 1 for H). How many grams of NaOH are needed to prepare 500 mL of a 0.25 N (normal) solution?

Solution:

$$\begin{aligned}1 \text{ N NaOH} &= 40 \text{ g/L} \\ 0.25 \text{ N NaOH} &= 0.25 \times 40 \text{ g/L} \\ &= 10 \text{ g/1000 mL}\end{aligned}$$

$$500 \text{ mL } 0.25 \text{ N} = \frac{10/2 \text{ g}}{1000/2 \text{ mL}}$$
$$= 5 \text{ g}/500 \text{ mL}$$

Answer: 5 g NaOH will be required to prepare 500 mL 0.25 N solution.

EXAMPLE 5: Caustic soda, NaOH, is an important chemical for pH adjustment and acid titration. It is often manufactured by the reaction of slaked lime, $Ca(OH)_2$, and soda ash, Na_2CO_3.
(a) What weight in kilograms of NaOH will be generated if 26.5 kg of soda ash is used? (b) How many kilograms of lime, CaO, is needed for the reaction? The atomic weights are Na = 23, C = 12, O = 16, Ca = 40.1, and H = 1.

Solution for (a):

Step 1. Balance the formula and compute molecular weights (MWs):

	Na_2CO_3	+	$Ca(OH)_2$	$\rightarrow$ 2NaOH	+	$CaCO_3$
MW	$(23 \times 2) + 12 + (16 \times 3)$		$40.1 + 2(16 + 1)$	$2(23 + 16 + 1)$		
	$= 106$		$= 74.1$	$= 80$		

Step 2. Solve NaOH generated, x, by proportion:

$$\frac{106}{26.5} = \frac{80}{x}$$

$$x = \frac{80 \times 26.5}{106} = 20 \text{ kg}$$

Solution for (b):

Step 3. Compute the weight of CaO, y:

	$CaO + H_2O \rightarrow$	$Ca(OH)_2$
MW	$40.1 + 16$	
	$= 56.1$	74.1

Also, by proportion, weight of CaO needed is y

$$\frac{y}{x} = \frac{56.1}{80}$$

therefore

$$y = \frac{56.1x}{80} = \frac{56.1 \times 20}{80}$$
$$= 14.03 \text{ (kg)}$$

EXAMPLE 6: If calcium (Ca) is 56.4 mg/L and Mg is 8.8 mg/L, what is the total hardness of the water in mg/L as $CaCO_3$? The atomic weights are Ca = 40.05, Mg = 24.3, C = 12.01, and O = 16.

Solution:

Step 1. Compute molecular weight (MW) of $CaCO_3$:

$$\text{MW} = 40.05 + 12.01 + (16 \times 3) = 100.06$$

Step 2. Compute the factor (f_1) of Ca equivalent to $CaCo_3$:

$$f_1 = \frac{\text{MW of } CaCO_3}{\text{MW of Ca}} = \frac{100.06}{40.05}$$
$$= 2.498$$

Step 3. Compute the factor (f_2) of Mg equivalent to $CaCO_3$:

$$f_2 = \frac{\text{MW of } CaCO_3}{\text{MW of Mg}} = \frac{100.06}{24.3}$$
$$= 4.118$$

Step 4. Compute total alkalinity (T.alk.)

$$\text{T.alk.} = \text{Ca (mg/L)} \times f_1 + \text{Mg (mg/L)} \times f_2$$
$$= 56.4 \times 2.498 + 8.8 \times 4.118$$
$$= 177 \text{ (mg/L as } CaCO_3)$$

Note: According to *Standard methods* (APHA, AWWA, and WEF, 1995):

$$\text{T.alk. (mg/L as } CaCO_3) = 2.497 \times \text{Ca (mg/L)} + 4.118 \times \text{Mg (mg/L)}$$

4.3 pH

The acid–base property of water is a very important factor in water and wastewater. Water ionizes to a slight degree to produce both hydrogen ion (H^+) and hydroxyl ion (OH^-) as below:

$$H_2O \leftrightarrow H^+ + OH^-$$

Water may be considered both as an acid and a base. The concentration of H^+(aq) in a solution is often expressed as pH. The pH is defined as the negative log to the base 10 of the hydrogen ion concentration (molecular weight in g/L). It can be written as

$$\text{pH} = -\log [H^+] = \log(1/[H^+])$$

It is noted that a change in $[H^+]$ by a factor of 10 results in 1 unit change in pH as shown in Table 1.5.

TABLE 1.5 Hydrogen Ion Concentration and pH Value

$[H^+]$, g/L		pH
1.0		0
0.1		1
0.01	increasing	2
0.001	acidity	3
0.0001		4
0.00001	↑	5
0.000001		6
0.0000001 (10^{-7})	neutral	7
10^{-8}	↓	8
10^{-9}		9
10^{-10}	increasing	10
10^{-11}	basicity	11
10^{-12}		12
10^{-13}		13
10^{-14}		14

When $[H^+] = [OH^-] = 1 \times 10^{-7}$, the pH of the solution is neutral at 7.0

$$\begin{aligned} pH &= -\log [H^+] = -\log (1 \times 10^{-7}) \\ &= 0 - (-7) \\ &= 7 \end{aligned}$$

Because pH is simply another means of expressing $[H^+]$, acidic and basic solutions have the relationships:

$$pH < 7 \text{ in acidic solutions}$$
$$pH = 7 \text{ in neutral solutions}$$
$$pH > 7 \text{ in basic solutions}$$

The number of $[H^+]$ ions multiplied by the number of $[OH^-]$ ions gives the same value (constant), i.e.

$$K = [H^+]\,[OH^-] = 1.0 \times 10^{-14}$$

Therefore, if the number of $[H^+]$ ions is increased tenfold, then the number of $[OH^+]$ ions will be automatically reduced to one-tenth.

EXAMPLE 1: Calculate the pH values for two solutions: (a) in which $[OH^-]$ is 0.001 mole; (b) in a solution $[OH^-]$ is 2.5×10^{-10} mole.

Solution:

Step 1. Solve for question (a)

$$[H^+]\,[OH^-] = 1.0 \times 10^{-14}$$

$$\begin{aligned} [H^+] &= \frac{1.0 \times 10^{-14}}{[OH^-]} = \frac{1.0 \times 10^{-14}}{0.001} \\ &= 1.0 \times 10^{-11} \text{ mole} \end{aligned}$$

then

$$\begin{aligned} pH &= -\log [H^+] = -\log (1.0 \times 10^{-11}) = -(-11.00) \\ &= 11.00 \qquad \text{basic} \end{aligned}$$

Step 2. Solve for question (b)

$$\begin{aligned} [H^+] &= \frac{1.0 \times 10^{-14}}{[OH^-]} = \frac{1.0 \times 10^{-14}}{[2.5 \times 10^{-10}]} \\ &= 4.0 \times 10^{-5} \text{ mol} \\ pH &= -\log(4.0 \times 10^{-5}) \\ &= -(\log 4 + \log 10^{-5}) \\ &= -(0.602 - 5.00) \\ &= 4.40 \qquad \text{acidic} \end{aligned}$$

EXAMPLE 2: Alum coagulated water has a pH of 4.56. Calculate $[H^+]$.

Solution:

$$pH = -\log [H^+] = 4.56$$

thus

$$\log [H^+] = -4.56$$

Find antilog

$$[H^+] = \log^{-1}(-4.56)$$
$$= 2.75 \times 10^{-5} \text{ mole}$$

Or

$$[H^+] = \log^{-1}(-5 + 0.44)$$
$$= 1 \times 10^{-5} \times 2.75$$
$$= 2.75 \times 10^{-5} \text{ mole}$$

EXAMPLE 3: How many moles of water is contained in 1 L of water?

Solution: 1 L of water is 1000 g
1 mole of water is 18 g

Thus

$$\frac{1000 \text{ g/L}}{18 \text{ g/mol}} = 55.55 \text{ mol/L}$$

EXAMPLE 4: Demonstration of ionization of water—pH

Solution:

Step 1.

$$H_2O \leftrightarrow H^+ + OH^-$$

In 10,000,000 (or 10^7) liters of water, 1 gram-ion each of hydronium (H^+) and hydroxyl (OH^-) will be formed; i.e. $1/(1 \times 10^7) = 10^{-7}$ mol H^+ and 10^{-7} mol OH^- in 1 liter of water.

Step 2. Compute equilibrium constant K

In 1 liter of water

$$[H_2O] = \frac{1000}{18} = 55.56 \text{ mol/L of water} = W$$

$$K = \frac{[H^+]\,[OH^-]}{[H_2O]} = \frac{(10^{-7})(10^{-7})}{55.56} = 1.80 \times 10^{-16}$$

or

$$K \times W = 1.80 \times 10^{-16} \times 55.56 = 1 \times 10^{-14}$$

Step 3. Compute pH

In pure water, $[H^+] = [OH^-] = 1 \times 10^{-7}$

By definition,

$$\text{pH} = \log \frac{1}{[H^+]} = \log \frac{1}{10^{-7}} = \log 10^7 = 7.0 \text{ (pure water)}$$

EXAMPLE 5: What are the hydrogen ion concentration (in gram-ions per liter) and pH value of a 0.2 mol acetic acid (CH_3COOH) solution at 25°C? The ionization constant of acetic acid is 1.8×10^{-5} at 25°C.

Solution:

Step 1. Determine H^+

$$CH_3COOH \leftrightarrow H^+ + CH_2COOH^-$$

$$K = \frac{[H^+]\,[CH_2COOH^-]}{[CH_3COOH]}$$

Let $[H^+] = y$ mole

then $[CH_2COOH] = y$ mole

and $[CH_3COOH] = (0.2 - y)$ mole

$\cong$ 0.2 mole, since y is very small compared with the concentration of non-ionized acid.

Simply

$$K = 1.8 \times 10^{-5} = \frac{(y)\cdot(y)}{0.2}$$

$$y^2 = 0.2 \times 1.8 \times 10^{-5} = 3.6 \times 10^{-6}$$

$$y = 1.9 \times 10^{-3}$$

$$[H^+] = 1.9 \times 10^{-3} \text{ gram-ion/L}$$

Step 2. Determine pH

$$\text{pH} = \log \frac{1}{[H^+]} = \log \frac{1}{1.9 \times 10^{-3}} = \log 526 = 2.72$$

4.4 Mixing Solutions

Solutions may be prepared by diluting a strong stock solution with distilled water or deionized (DI) water or by diluting a strong solution with a weak solution. The following relationships can be expressed for the mixture of solutions:

$$Q = Q_1 + Q_2$$

$$Q \times C = Q_1 \times C_1 + Q_2 \times C_2$$

where Q = quantity (weight or volume) of mixture
Q_1 = quantity (weight or volume) of solution 1
Q_2 = quantity (weight or volume) of solution 2
C = concentration of mixture
C_1 = concentration of solution 1
C_2 = concentration of solution 2

EXAMPLE 1: In order to prepare a solution of 3.0 mg/L from 0.10% stock solution, calculate how many milliliters of stock solution are needed to prepare 500 mL of solution from deionized water.

Solution:

$$\text{Initial concentration } C_i = 0.1\% = \frac{1 \times 10^3}{10^6}$$

$$\text{Final concentration } C_f = 3.0 \text{ mg/L} = \frac{3}{10^6}$$

Let V_i and V_f be the initial and final volumes, respectively.

$$V_f = 500 \text{ mL}$$

Since $V_i \times C_i = V_f \times C_f$

$$V_i \times \frac{1 \times 10^3}{10^6} = 500 \times \frac{3}{10^6}$$

$$V_i = \frac{1500}{1000} = 1.50 \text{ mL}$$

EXAMPLE 2: 50 gallons of a 4.0% solution is made by mixing of 6.8% and 2.8% solutions. (a) In what ratio should the 6.8% and 2.8% solutions be mixed? (b) How much of each is needed?

Solution:

Step 1:

$$Q = Q_1 + Q_2$$
$$50 = Q_1 + Q_2$$
$$Q_1 = 50 - Q_2$$
$$\frac{Q_1}{Q_2} = \frac{50 - Q_2}{Q_2}$$

Step 2:

$$Q \times C = Q_1 \times C_1 + Q_2 \times C_2$$
$$50 \times 4 = Q_1 \times 6.8 + Q_2 \times 2.8$$
$$200 = (50 - Q_2)6.8 + 2.8Q_2$$
$$200 = 340 - 6.8Q_2 + 2.8Q_2$$
$$4Q_2 = 140$$
$$Q_2 = 35 \text{ gal}$$

Step 3:

(a)
$$\frac{Q_1}{Q_2} = \frac{50 - Q_2}{Q_2} = \frac{50 - 35}{35} = \frac{15}{35} = \frac{3}{7}$$

Step 4:

(b)
$$Q_2 = 35 \text{ gal}$$
$$Q_1 = 50 - Q_2 = 50 - 35 = 15 \text{ (gal)}$$

EXAMPLE 3: In the laboratory, 500 mL of hydrochloric acid (HCl) is being prepared by reaction of sulfuric acid (H_2SO_4) and sodium chloride. The density of HCl will be 1.20 with 40% purity by weight. The atomic weights are H = 1, O = 16, Na = 23, S = 32, and Cl = 35.5. What are the weights of H_2SO_4 and NaCl needed?

Solution:

Step 1. Compute HCl generated

$$\text{HCl} = 0.5 \text{ L} \times 1.20 \times 1000 \text{ g/L} \times 0.40$$
$$= 240 \text{ g}$$

Step 2. Write the chemical equation and calculate the molecular weight (MW)

	H_2SO_4	+	NaCl	→	$2HCl + Na_2SO_4$
MW	(1 × 2) + 32 + (16 × 4) = 98		23 + 35.5 = 58.5		2(1 + 35.5) = 73
Required	*X*		*Y*		240

Step 3. Determine H_2SO_4 needed

$$X = 240 \text{ g} \times \frac{98}{73} = 322.2 \text{ g}$$

Step 4. Determine NaCl needed

$$Y = 240 \text{ g} \times \frac{58.5}{73} = 192.3 \text{ g}$$

Note: 0.5 mole of H_2SO_4 and 1 mole of NaCl will produce 1 mole of HCl.

4.5 Chemical Reactions and Dosages

Equations of chemical reactions commonly encountered in water and wastewater treatment plants are

$$Cl_2 + H_2O \leftrightarrow HCl + HOCl$$
$$NH_3 + HOCl \leftrightarrow NH_2Cl + H_2O$$
$$NH_2Cl + HOCl \leftrightarrow NHCl_2 + H_2O$$
$$NHCl_2 + HOCl \leftrightarrow NCl_3 + H_2O$$
$$Ca(OCl)_2 + Na_2CO_3 \leftrightarrow 2NaOCl + CaCO_3$$
$$Al_2(SO_4)_3 + 3CaCO_3 + 3H_2O \leftrightarrow Al_2(OH)_6 + 3CaSO_4 + 3CO_2$$
$$CO_2 + H_2O \leftrightarrow H_2CO_3$$
$$H_2CO_3 + CaCO_3 \leftrightarrow Ca(HCO_3)_2$$
$$Ca(HCO_3)_2 + Na_2CO_3 \leftrightarrow CaCO_3 + 2NaHCO_3$$
$$CaCO_3 \leftrightarrow Ca^{2+} + CO_3^{2-}$$
$$CaCO_3 + H_2SO_4 \leftrightarrow CaSO_4 + H_2CO_3$$
$$Ca(HCO_3)_2 + H_2SO_4 \leftrightarrow CaSO_4 + 2H_2CO_3$$
$$2S^{2-} + 2O_2 \rightarrow SO_4^{2-} + S^\circ \downarrow$$
$$H_2S + Cl_2 \rightarrow 2HCl + S^\circ \downarrow$$
$$H_2S + 4Cl_2 + 4H_2O \rightarrow H_2SO_4 + 8HCl$$
$$SO_2 + H_2O \rightarrow H_2SO_3$$
$$HOCl + H_2SO_3 \rightarrow H_2SO_4 + HCl$$
$$NH_2Cl + H_2SO_3 + H_2O \rightarrow NH_4HSO_4 + HCl$$
$$Na_2SO_3 + Cl_2 + H_2O \rightarrow Na_2SO_3 + 2HCl$$

EXAMPLE 1: Chlorine dosage at a treatment plant averages 114.7 lb/d. Its average flow is 12.5 MGD. What is the chlorine dosage in mg/L?

Solution:

$$\begin{aligned}\text{Dosage} &= \frac{114.7\ \text{lb/d}}{12.5\ \text{MGD}} = \frac{114.7\ \text{lb}}{12.5 \times 10^6\ \text{gal}} \\ &= \frac{114.7\ \text{lb}}{12.5 \times 10^6\ \text{gal} \times 8.34\ \text{lb/gal}} \\ &= \frac{1.1}{10^6} \\ &= 1.1\ \text{ppm} \\ &= 1.1\ \text{mg/L}\end{aligned}$$

EXAMPLE 2: Twenty-seven pounds (27 lb) of chlorine gas is used for treating 750,000 gallons of water. The chlorine demand of the water is measured to be 2.6 mg/L. What is the residual chlorine concentration in the treated water?

Solution:

$$\begin{aligned}\text{Total dosage} &= 27\ \text{lb/0.75 Mgal} = 36\ \text{lb/Mgal} \\ &= 36\ \text{lb/Mgal} \times \frac{1\ \text{mg/L}}{8.34\ \text{lb/Mgal}} \\ &= 4.3\ \text{mg/L}\end{aligned}$$

$$\text{Residual } Cl_2 = 4.3 \text{ mg/L} - 2.6 \text{ mg/L}$$
$$= 1.7 \text{ mg/L}$$

EXAMPLE 3: What is the daily amount of chlorine needed to treat an 8 MGD of water to satisfy 2.8 mg/L chlorine demand and provide 0.5 mg/L residual chlorine.

Solution:

$$\text{Total } Cl_2 \text{ needed} = 2.8 + 0.5 = 3.3 \text{ mg/L}$$
$$\text{Daily weight} = 8 \times 10^6 \text{ gal/d} \times 8.34 \text{ lb/gal} \times 3.3 \text{ mg/L} \times 1 \text{ L}/10^6 \text{ mg}$$
$$= 220 \text{ lb/d}$$

EXAMPLE 4: At a 11-MGD water plant, a hydrofluosilicic acid (H_2SiF_6) with 23% by wt solution is fed by a pump at the rate of 0.03 gpm. The specific gravity of the H_2SiF_6 solution is 1.191. What is the fluoride (F) dosage?

Solution:

$$\text{Pump rate} = 0.03 \text{ gal/min} \times 1440 \text{ min/day} = 43.2 \text{ gal/day}$$
$$\text{F appl. rate} = 43.2 \text{ gal/day} \times 0.23 = 9.94 \text{ gal/day}$$
$$\text{Wt of F} = 9.94 \text{ gal/day} \times 8.34 \text{ lb/gal} \times 1.191$$
$$\text{Wt of water} = 11 \times 10^6 \text{ gal/day} \times 8.34 \text{ lb/gal} \times 1.0$$
$$\text{Dosage} = \text{wt of F/wt of water}$$
$$= \frac{9.94 \times 8.34 \times 1.191}{11 \times 10^6 \times 8.34 \times 1.0}$$
$$= \frac{1.08}{10^6}$$
$$= 1.08 \text{ mg/L}$$

EXAMPLE 5: The average specific gravity of 25% hydrofluosilicic acid (H_2SiF_6) is 1.208 (Reeves, 1986). How much of 25% acid is required to dose 1 (mg/L) of fluoride (F) to 1 Mgal of water?

Solution:

Step 1. Determine the density ρ of the acid

$$\rho = 8.345 \text{ lb/gal} \times \text{sp gr} = 8.345 \text{ lb/gal} \times 1.208$$
$$= 10.08 \text{ lb/gal}$$

Step 2. Determine F required

Since 1 mg/L = 1 gal/Mgal

This means that 1 gal of full strength F is needed; however, the acid is only 25% strength. Therefore, the amount needed q is

$$q = \frac{1 \text{ gal}}{0.25} = 4 \text{ gal}$$

or

$$= 4 \text{ gal} \times 10.08 \text{ lb/gal}$$
$$= 40.32 \text{ lb}$$

EXAMPLE 6: 12 mg/L of liquid alum with 60% strength is continuously fed to a raw water flow that averages 8.8 MGD. How much liquid alum will be used in a month (assume 30 days)?

Solution:

$$\text{Since } 1 \text{ mg/L} = 1 \text{ gal/Mgal}$$

$$\begin{aligned}\text{Required/day} &= \frac{12}{0.60}\,\frac{\text{gal}}{\text{Mgal}} \times 8.8\,\frac{\text{Mgal}}{\text{day}} \\ &= 176 \text{ gal/day} \\ \text{Required/month} &= 176\,\frac{\text{gal}}{\text{day}} \times 30\,\frac{\text{day}}{\text{month}} \\ &= 5{,}280 \text{ gal/month}\end{aligned}$$

4.6 Pumpage and Flow Rate

Operators of water and wastewater treatment plants encounter pumpage and flow rate during daily operations. In this chapter, some fundamental knowledge of pumpings and flow rates are illustrated for operators. Hydraulics for environmental engineering will be presented in Chapter 1.5.

EXAMPLE 1: What is the pressure gage reading at the base of 1 ft (0.3048 m) water column?

Solution: Since 1 cubic foot of water is 62.4 lb

$$\begin{aligned}\frac{62.4 \text{ lb}}{1 \text{ cu ft}} &= \frac{62.4 \text{ lb}}{1 \text{ ft} \times 1 \text{ ft} \times 1 \text{ ft}} \\ &= \frac{62.4 \text{ lb}}{12 \text{ in} \times 12 \text{ in} \times 1 \text{ ft}} \\ &= \frac{0.433 \text{ lb}}{\text{sq in} \times 1 \text{ ft}} \\ &= 0.433 \text{ psi/ft} \qquad \text{pounds per square inch (lb/in}^2\text{) per foot}\end{aligned}$$

Note: At the base of 1 ft depth of water is 0.433 psig (g means gage).

EXAMPLE 2: How many pounds of pressure (psi) will be produced by 50 ft of head?

Solution:

$$\begin{aligned}\text{Pressure} &= 0.433 \text{ psi/ft} \times 50 \text{ ft} \\ &= 21.65 \text{ psi}\end{aligned}$$

EXAMPLE 3: A fire hydrant needs a nozzle pressure of 100 psi. What is the head of water to provide this pressure?

Solution:

$$100 \text{ psi} = H \text{ ft} \times 0.433 \text{ psi/ft}$$

$$\begin{aligned}H &= \frac{100 \text{ psi}}{0.433 \text{ psi/ft}} \\ &= 231 \text{ ft}\end{aligned}$$

EXAMPLE 4: A high-service pump is pumping the finish water to a storage tank of 120 ft elevation difference. The pressure gauge reading at the discharge line of the pump is 87.6 psi. Determine the head loss due to friction.

Solution:

$$\begin{aligned}\text{Gauge reading} &= \text{elevation pressure} + \text{friction loss}\\ \text{Friction head loss} &= 87.6 \text{ psi} - 0.433 \text{ psi/ft} \times 120 \text{ ft}\\ &= 87.6 \text{ psi} - 52.0 \text{ psi}\\ &= 35.6 \text{ psi}\end{aligned}$$

EXAMPLE 5: What is the height (h) of a water column that produces one pound per square inch (1 psi) pressure at the bottom?

Solution: Since the pressure at the base of a 1-ft water column is 0.433 psi,

$$\frac{1 \text{ psi}}{h} = \frac{0.433 \text{ psi}}{1 \text{ ft}}$$

$$h = \frac{1}{0.433} \quad \text{or} \quad \frac{144}{62.4}$$

$$= 2.308 \text{ (ft)}$$

EXAMPLE 6: When a water tank is filled with water to a height of 46.2 ft and the tank is 22 ft in diameter, what is the pressure (psi) at the bottom of the tank?

Solution:

$$\begin{aligned}\text{Pressure} &= \frac{\text{height of water column}}{2.308 \text{ ft/psi}}\\ &= \frac{46.2 \text{ ft}}{2.308 \text{ ft/psi}}\\ &= 20.0 \text{ psi}\end{aligned}$$

Note: Pressure is affected by the height of the water column only, not the diameter.

EXAMPLE 7: A positive displacement pump delivers 48 gpm at 25 strokes per minute. What does the pump deliver at 30 strokes per minute which is within the accurate regulated operating speed?

Solution: By proportioning

$$\frac{x \text{ gpm}}{48 \text{ gpm}} = \frac{30 \text{ strokes}}{25 \text{ strokes}}$$

$$x = \frac{30 \times 48}{25}$$

$$= 57.6 \text{ (gpm)}$$

EXAMPLE 8: The static groundwater level, before pumping a well, is 18 ft below the ground surface. During pumping, the dynamic water level is 52 ft below the ground surface. What is the drawdown?

Solution:

$$\begin{aligned}\text{The drawdown} &= \text{dynamic level} - \text{static level}\\ &= 52 \text{ ft} - 18 \text{ ft}\\ &= 34 \text{ ft}\end{aligned}$$

EXAMPLE 9: A pump discharges at a rate of 125 gpm (Q). How long will it take (t) to fill up a 5000-gal (V) tank?

Solution:

$$t = \frac{V}{Q} = \frac{5000 \text{ gal}}{125 \text{ gal/min}}$$
$$= 40 \text{ min}$$

EXAMPLE 10: A rectangular basin is 50 ft wide, 120 ft long, and 18 ft deep. When 187,000 gal of water are pumped into the basin, what is the height of water raised?

Solution:

Step 1. Compute volume V in ft^3

$$V = 187{,}000 \text{ gal} \times \frac{1 \text{ ft}^3}{7.48 \text{ gal}}$$
$$= 25{,}000 \text{ ft}^3$$

Step 2. Compute surface area A in ft^2

$$A = 50 \text{ ft} \times 120 \text{ ft} = 6000 \text{ ft}^2$$

Step 3. Determine water raised h in ft (or inches)

$$h = V/A = 25{,}000 \text{ ft}^3/6000 \text{ ft}^2$$
$$= 4.17 \text{ ft}$$
$$= 4.17 \text{ ft} \times 12 \text{ in/ft}$$
$$= 50 \text{ in}$$

EXAMPLE 11: A wastewater plant has eight activated sludge units. It is designed so that each pipe from the primary settling basin to the secondary units has a 1.25 MGD rate capacity. The activated sludge units are capable of operating to a flow of 10% in excess of rate capacity. However, the velocity of sewage in the pipe is limited to the regulatory requirement of 2 fps. What standard pipe size from the basin should be selected? (1 MGD = 1.547 cfs)

Solution:

Step 1. Determine the maximum allowable flow Q in cfs

$$Q = 1.25 \text{ MGD} \times 1.547 \text{ cfs/MGD} \times 8 \times (1 + 0.1)$$
$$= 17.02 \text{ cfs}$$

Step 2. Compute required area A

$$A = Q/V = 17.02 \text{ cfs}/2 \text{ fps}$$
$$= 8.51 \text{ ft}^2$$
$$= 8.51 \text{ ft}^2 \times 144 \text{ in}^2/\text{ft}^2$$
$$= 1225 \text{ in}^2$$

Step 3. Select standard pipe size

$$A = \pi r^2 \text{ or } r^2 = A/\pi$$
$$r^2 = 1225/3.14 = 390$$
$$r = 19.7 \text{ in or diameter } d = 39.4 \text{ in}$$

However, the standard pipe sizes are 24, 30, 36, 42, and 48 inches. Thus, a 42-inch pipe will be selected.

EXAMPLE 12: A 5-hp motor runs 24 hours per day for aeration purpose. It has 88% efficiency. The electricity cost is 0.018 \$/kW-h. What is the monthly (M) power cost?

Solution:

$$
\begin{aligned}
\text{Work} &= \frac{5\text{ hp}}{0.88} \times 0.746\,\frac{\text{kW}}{\text{hp}} \\
&= 4.24\text{ kW} \\
\text{M use} &= 4.24\text{ kW} \times 24\text{ h/day} \times 30\text{ days} \\
&= 3052.8\text{ kW}\cdot\text{h} \\
\text{M cost} &= 0.018\text{ \$/kW}\cdot\text{h} \times 3052.8\text{ kW}\cdot\text{h} \\
&= \$54.95
\end{aligned}
$$

EXAMPLE 13: A pump delivers a flow of 600 gallons per minute against a head of 30.5 ft (10 m). What is the horsepower of the pump delivered?

Solution:

Step 1. Convert gpm into lb/s

$$
\begin{aligned}
Q &= 600\,\frac{\text{gal}}{\text{min}} \times 8.34\,\frac{\text{lb}}{\text{gal}} \times \frac{1\text{ min}}{60\text{ s}} \\
&= 83.4\text{ lb/s}
\end{aligned}
$$

Step 2. Compute work (W) done by the pump

$$
\begin{aligned}
W = Q \times h &= 83.34\text{ lb/s} \times 30.5\text{ ft} \\
&= 2543.7\text{ ft}\cdot\text{lb/s}
\end{aligned}
$$

Step 3. Convert to horsepower hp

$$
\begin{aligned}
W &= 2543.7\text{ ft}\cdot\text{lb/s} \times \frac{1\text{ hp}}{550\text{ ft}\cdot\text{lb/s}} \\
&= 4.62\text{ hp}
\end{aligned}
$$

EXAMPLE 14: A sodium hypochlorite (NaOCl) solution containing 20% chlorine is used for temporary disinfection when the gas chlorinator is being repaired. Five mg/L of chlorine are required for 2.4 million gallons pumped in 12 hours. How many gallons per hour of hypochlorite solution (assume its specific gravity is 1.0) should be used?

Solution:

$$
\begin{aligned}
5\text{ mg/L} &= 5\text{ gal of Cl}_2\text{/Mgal of water} \\
\text{Pumpage} &= 5\,\frac{\text{gal}}{\text{Mgal}} \times \frac{2.4\text{ Mgal}}{12\text{ h}} \\
&= 1\text{ gal/h of Cl}_2 \\
&= \frac{1}{0.2}\text{ gal/h of NaOCl} \\
&= 5\text{ gal/h of NaOCl}
\end{aligned}
$$

EXAMPLE 15: A filter has a filter bed area of 22 ft × 22 ft. It is filled to the water level of 15 inches above the backwash-water trough with a water surface area of 22 ft × 24 ft. After closing the influent valve, the water dropped 10 in in 4 min. What is the filter rate in gallons per minute per square foot?

Solution:

Step 1. Compute volume filtered per minute, Q

$$Q = 22 \text{ ft} \times 24 \text{ ft} \times (10/12) \text{ ft}/4 \text{ min}$$
$$= 110 \text{ ft}^3/\text{min}$$
$$= 110 \text{ ft}^3/\text{min} \times 7.48 \text{ gal/ft}^3$$
$$= 822.8 \text{ gpm}$$

Step 2. Compute filter area A

$$A = 22 \text{ ft} \times 22 \text{ ft}$$
$$= 484 \text{ ft}^2$$

Step 3. Determine filter rate R

$$R = Q/A = 822.8 \text{ gpm}/484 \text{ ft}^2$$
$$= 1.7 \text{ gpm/ft}^2$$

EXAMPLE 16: A rectangular sedimentation basin has a bottom of 60 ft × 100 ft and wastewater height of 18 ft. The flow rate is 9.7 MGD. What is the detention time in the basin?

Solution:

Step 1. Compute volume (V) of the basin:

$$V = Ah = 60 \text{ ft} \times 100 \text{ ft} \times 18 \text{ ft}$$
$$= 108{,}000 \text{ ft}^3$$
$$= 108{,}000 \text{ ft}^3 \times 7.48 \text{ gal/ft}^3$$
$$= 808{,}000 \text{ gal}$$

Step 2. Compute detention time t

$$t = V/Q$$
$$= 0.808 \times 10^6 \text{ gal}/9.7 \times 10^6 \text{ gal/day}$$
$$= 0.0833 \text{ days}$$
$$= 0.0833 \text{ days} \times 24 \text{ h/day}$$
$$= 2.0 \text{ h}$$

EXAMPLE 17: The dimensions of a filter are 25 ft × 20 ft. What is the backwash rise rate in inches per minute, when the backwash rate is 8230 gpm (gallons per minute)?

Solution:

Step 1. Calculate the area of the filter A

$$A = 25 \text{ ft} \times 20 \text{ ft} = 500 \text{ ft}^2$$

Step 2. Convert gpm to cubic feet per minute

$$\text{Wash rate} = 8230 \text{ gal/min} = 8230 \text{ gal/min} \times \frac{1 \text{ ft}^3}{7.48 \text{ gal}}$$
$$= 1100 \text{ ft}^3/\text{min}$$

Step 3. Determine the rise rate

$$\text{Rise rate} = \frac{\text{wash rate}}{A} = \frac{1100 \text{ ft}^3/\text{min}}{500 \text{ ft}^2}$$
$$= 2.2 \text{ ft/min}$$
$$= 2.2 \frac{\text{ft}}{\text{min}} \times 12 \frac{\text{in}}{\text{ft}}$$
$$= 26.4 \text{ in/min}$$

EXAMPLE 18: A filter has a surface area of 18 ft (6 m) × 27 ft (9 m). The filtration rate (Q) is 800 gpm (0.0505 m^3/s). What is the surface loading rate?

Solution:

Step 1. Determine surface area A

$$A = 18 \text{ ft} \times 27 \text{ ft} = 486 \text{ ft}^2 = 45.1 \text{ m}^2$$

Step 2. Calculate loading rate (LR)

$$\text{LR} = \frac{Q}{A} = \frac{800 \text{ gpm}}{486 \text{ ft}^2}$$

$$= 1.646 \text{ gpm/ft}^2$$

or

$$= 0.000935 \text{ m}^3/(\text{s} \cdot \text{m}^2)$$

EXAMPLE 19: After a new main of 4500 ft of 24-in diameter pipeline is installed, it has to be chlorinated, then flushed at a rate of 96 gpm. How long will it take to flush the pipeline completely?

Solution:

Step 1. Determine the volume (V) of the water in the pipeline.

$$\text{Diameter} = 24 \text{ in} = 24 \text{ in/12 in/ft} = 2 \text{ ft}$$

$$\text{radius, } r = 1 \text{ ft}$$

$$\text{Volume, } V = \pi r^2 l$$

$$= 3.14(1)^2 \times 4500 \text{ ft}^3$$

$$= 14{,}130 \text{ ft}^3 \times 7.48 \text{ gal/ft}^3$$

$$= 105{,}692 \text{ gal}$$

Step 2. Compute time of flushing (t)

$$t = V/Q = 105{,}692 \text{ gal/96 gal/min}$$

$$= 1101 \text{ min}$$

$$= 1101 \text{ min/60 min/h}$$

$$= 18.35 \text{ h}$$

or

$$= 18 \text{ h } 21 \text{ min}$$

EXAMPLE 20: It is generally accepted that the average daily rate of water demand (consumption) is 100–200 gallons per capita per day (gpcpd). Assume that the maximum daily rate is 1.5 times the average daily rate, and the maximum hourly rate is approximately 2.0 times the maximum daily rate, or 300% of the average daily rate. If a city has a population of 140,000 and its water consumption is 130 gpcpd, what are the average and maximum daily rates and maximum hourly rate of water consumption?

Solution:

Step 1. Compute average daily rate Q_a

$$Q_a = 130 \text{ gpcpd} \times 140{,}000$$

$$= 18.2 \times 10^6 \text{ gallons per day, gpd}$$

or

$$= 18.2 \text{ million gallons per day, MGD}$$

Step 2. Compute maximum daily rate Q_m

$$Q_m = 1.5\, Q_a = 1.5 \times 18.2 \text{ MGD}$$

$$= 27.3 \text{ MGD}$$

Step 3. Compute maximum hourly rate H_m

$$H_m = 2 \times Q_m = 2 \times 2.73 \text{ MGD}$$
$$= 54.6 \text{ MGD}$$

5 STATISTICS

5.1 Measure of Central Value

Central value refers to the location of the center of the distribution. There are several measures of central tendency, such as the median, the mid, and the geometric mean; most are used by environmental professionals.

5.2 The Arithmetic Mean

The arithmetic mean or the mean (or average) of a set of N numbers $X_1, X_2, X_3, \ldots X_n$ is denoted by $\overline{X}$. The mean is the sum of the values of observations divided by the number of observations and is given by

$$\overline{X} = \frac{X_1 + X_2 + X_3 + \cdots + X_n}{N} = \frac{\sum_{i=1}^{n} X_i}{N}$$

EXAMPLE: The mean of the values 9, 7, 21, 16, 11, 9, and 18 is

$$\overline{X} = \frac{9 + 7 + 21 + 16 + 11 + 9 + 18}{7} = \frac{91}{7} = 13$$

5.3 The Medium

The medium (Md) of a set of numbers is the middle value or the mean of the two middle values when they have been arranged in order of magnitude.

EXAMPLE 1: The set of numbers 7, 9, 9, 11, 16, 18, and 21 (previous example) has a median 11.

EXAMPLE 2: The medium of 33, 41, 43, 44, 48, 51, 53, and 54 is

$$\frac{44 + 48}{2} = 46$$

5.4 The Mode

The mode M_o of a set of numbers is the value which occurs most frequently and is hence the most common value. It is the value corresponding to the maximum of the frequency curve of best fit. The mode may not exist, and even if it does exist it may not be unique. An important example of the use of the mode is in the determination of the most probable number of coliform bacteria in a water sample.

EXAMPLE 1: The mode of numbers 9, 11, 11, 17, 18, 18, 18, and 21 is 18.

EXAMPLE 2: The set of numbers 31, 38, 41, 44, 46, 50, and 51 has no mode.

EXAMPLE 3: The set of numbers 3, 3, 8, 8, 8, 9, 11, 14, 14, 14, and 17 has two modes. The numbers 8 and 14 are called bimodal.

5.5 Moving Average

Moving averages are averages (simple or weighted) of a convenient number of successive terms. Given a set of numbers $Y_1, Y_2, Y_3, \ldots Y_n$, the moving average of order k is calculated by the sequence of arithmetic means, i.e.

$$\frac{Y_1 + Y_2 + \cdots + Y_k}{k}, \quad \frac{Y_2 + Y_3 + \cdots Y_k + Y_{k+1}}{k}, \quad \frac{Y_3 + Y_4 + \cdots + Y_{k+2}}{k}, \ldots$$

The sums in the numerators above are called moving totals of order k.

EXAMPLE 1: Given a set of numbers 4, 6, 9, 5, 7, 8, 2, 3, 5, 7, and 9, find the moving averages of order 4 by the sequence.

Solution: Moving averages are

$$\frac{4+6+9+5}{4}, \quad \frac{6+9+5+7}{4}, \quad \frac{9+5+7+8}{4}, \quad \frac{5+7+8+2}{4}$$

$$\frac{7+8+2+3}{4}, \quad \frac{8+2+3+5}{4}, \quad \frac{2+3+5+7}{4}, \quad \frac{3+5+7+9}{4}$$

or 6.00, 6.75, 7.25, 5.50, 5.00, 4.50, 4.25, 6.00

EXAMPLE 2: If the weights 2, 1, 3, and 1 are used for the number set in Example 1, a weighted moving average of order 4 is calculated by the sequence

Solution:

$$\frac{2(4) + 1(6) + 3(9) + 1(5)}{2+1+3+1}, \quad \frac{2(6) + 1(9) + 3(5) + 1(7)}{7}, \quad \frac{2(9) + 1(5) + 3(7) + 1(8)}{7},$$

$$\frac{2(5) + 1(7) + 3(8) + 1(2)}{7}, \quad \frac{2(7) + 1(8) + 3(2) + 1(3)}{7}, \quad \frac{2(8) + 1(2) + 3(3) + 1(5)}{7},$$

$$\frac{2(2) + 1(3) + 3(5) + 1(7)}{7}, \quad \frac{2(3) + 1(5) + 3(7) + 1(9)}{7}$$

or 6.57, 6.14, 7.43, 6.14, 4.43, 4.57, 4.14, 5.86

Moving averages tend to reduce the amount of variable in a set of data: they eliminate unwanted fluctuations and smooth time series. Using moving averages of appropriate orders, the trend movement can be found, which eliminates seasonal, cyclical, and irregular patterns. One disadvantage of the moving average method is that data at the beginning and the end portion of a series are lost. Another disadvantage is that the method may generate cycles or other movements and is strongly affected by extreme values.

In the water and wastewater industries and water resources, data extending over a long period are usually available. Using the moving average method with a proper order may provide valuable trends and results for management purposes. If data are given daily, monthly, or yearly, a moving average of order k is a k-day moving average, a k-month moving average, or a k-year moving average, respectively.

5.6 The Geometric Mean

The geometric mean M_g of a set of numbers is the Nth root of the product of the number:

$$M_g = \sqrt[N]{(X_1 \cdot X_2 \cdot X_3 \cdots X_n)}$$

or common logarithm transformation to (Fair *et al.*, 1963)

$$\log M_g = \frac{1}{N}\left(\sum_{i=1}^{i=N} \log X_i\right)$$

EXAMPLE 1: Find the M_g of the numbers, 5, 17, 58, 88, 150, and 220.

Solution:

$$M_g = \sqrt[6]{(5 \times 17 \times 58 \times 88 \times 150 \times 220)}$$

$$= \sqrt[6]{14{,}316{,}720{,}000}$$

$$= 49.28$$

This can be calculated with a computer or calculator; however, it also can be done using common logarithms:

$$\log M_g = \frac{1}{6} \log 14{,}316{,}720{,}000$$

$$= \frac{1}{6}(10.15584)$$

$$= 1.69264$$

then

$$M_g = 49.28$$

Another solution:

$$\log M_g = \frac{1}{6}(\log 5 + \log 17 + \log 58 + \log 88 + \log 150 + \log 220)$$

$$= \frac{1}{6}(0.69897 + 1.23045 + 1.76343 + 1.94448 + 2.17609 + 2.34242)$$

$$= \frac{1}{6}(10.15584)$$

$$= 1.69264$$

$$M_g = 49.28$$

EXAMPLE 2: The arithmetic mean of the same set of numbers is

$$\bar{X} = \frac{1}{6}(5 + 17 + 58 + 88 + 150 + 220)$$

$$= \frac{1}{6}(538)$$

$$= 89.67$$

As shown in Examples 1 and 2, the arithmetic mean of a set of numbers is greater than the geometric mean. However, the two means can be equal to each other for the same cases. Analysis of geometrically normal distribution can also be plotted as the straight-line summation of a frequency distribution on logarithmic probability paper to determine the geometric mean and geometric standard deviation.

5.7 The Variance

Because an average in itself does not give a clear picture of a distribution of a set of numbers, many different distributions may all have the same arithmetic mean. A measure of dispersion, spread, or variability is the variance. The variance is defined as the sum of squares of deviations of the observation results from their mean ($\overline{X}$) divided by one less than the total number of observations (N). The variance denoted as s^2 is given by

$$s^2 = \frac{\sum_{i=1}^{N}(X_i - \overline{X})^2}{N - 1}$$

For simpler calculation, an equivalent formula for s^2 is used:

$$s^2 = \frac{\sum_{i=1}^{N} X_i^2 - \frac{\left(\sum_{i=1}^{N} X_i\right)^2}{N}}{N - 1}$$

Many books (Spiegel, 1961; REA, 1996; Clarke and Cooke, 1992) use N instead of $N - 1$ in the denominator for s^2 and later adjust this by multiplying by $N/(N - 1)$. For large values of N (generally $N > 30$), there is practically no difference between the two definitions.

5.8 The Standard Deviation

The standard deviation of a set of numbers $X_1, X_2, X_3, \ldots X_N$ is denoted by s and is defined as the positive square root of the variance. Thus the standard deviation is given by

$$s = \sqrt{\left[\frac{\sum_{i=1}^{N}(X_i - \overline{X})^2}{N - 1}\right]}$$

For normal distribution, 68.27% of observations (samples) are in the range between $\overline{X} - s$ and $\overline{X} + s$, i.e. one standard deviation on either side of the mean. Two and three standard deviations on either side of the mean include 95.45% and 99.73% of the observations, respectively.

The standard deviation for the data of observations defined with $(N - 1)$ in the denominator is better than defined with N to estimate the standard deviation of a population.

EXAMPLE: Find the standard deviation of monthly chemical usages (5, 4, 6, 8, 10, 12, 13, 11, 9, 7, 5, and 6 tons) in a year.

Solution:

$$\text{Mean} = \overline{X} = \frac{\Sigma X}{N}$$

$$= \frac{5 + 4 + 6 + 8 + 10 + 12 + 13 + 11 + 9 + 7 + 5 + 6}{12}$$

$$= 8.0 \text{ (tons)}$$

$$\Sigma(X - \overline{X})^2 = (5 - 8)^2 + (4 - 8)^2 + (6 - 8)^2 + (8 - 8)^2 + (10 - 8)^2 + (12 - 8)^2$$

$$+ (13 - 8)^2 + (11 - 8)^2 + (9 - 8)^2 + (7 - 8)^2 + (5 - 8)^2 + (6 - 8)^2$$

$$= 98$$

$$s = \sqrt{\left[\frac{\Sigma(X - \overline{X})^2}{N - 1}\right]}$$

$$= \sqrt{\frac{98}{12 - 1}}$$

$$= 2.985 \text{ (tons)}$$

5.9 The Geometric Standard Deviation

The geometric standard deviation of a set of numbers $X_1, X_2, X_3, \ldots X_N$ is denoted by σ_g and is defined by

$$\log \sigma_g = \sqrt{\left[\frac{\sum_{i=1}^{N}(\log X_i - \log M_g)^2}{N-1}\right]}$$

$$= \sqrt{\left(\frac{1}{N-1}\right)} \times \sqrt{\left(\sum_{i=1}^{N}\log^2 X_i - N\log^2 M_g\right)}$$

The value of σ_g can be determined by a graphic method. On logarithmic probability paper, the straight line of best fit passes through the intersection of the geometric mean with the 50% percentile point and through the intersection of $M_g \times \sigma_g$ with the 84.1 percentile frequency, and M_g/σ_g with the 15.9 percentile, respectively.

EXAMPLE: Find the geometric mean and standard deviation for the example in Sec. 5.8.

Solution:

$$\log M_g = \frac{\Sigma \log X}{N}$$

$$= (\log 5 + \log 4 + \log 6 + \log 8 + \log 10 + \log 12 + \log 13 + \log 11 + \log 9 + \log 7 + \log 5 + \log 6)/12$$

$$= 0.87444$$

$$M_g = 7.489 \text{ (tons)}$$

$$\Sigma \log^2 X = 9.481$$

$$\log \sigma_g = \sqrt{\left(\frac{\Sigma\log^2 X - N\log M_g}{N -}\right)}$$

$$= \sqrt{\left(\frac{9.481 - 12(0.87444)^2}{12 - 1}\right)}$$

$$= \sqrt{0.02775}$$

$$= 0.16659$$

$$\sigma_g = 1.468 \text{ (tons)}$$

5.10 The Student's *t* Test

There are some useful statistical tools to compare means. The Student's *t* test can be used for statistical comparison of two means to determine whether or not they are the same. Two sets of data may be such as one from a new experiment and one for control; two results of split samples from two laboratories; or a new analytical and an existing analytical method.

To test difference of two means, the *t* value is calculated by

$$t = \frac{\bar{X}_1 - \bar{X}_2}{\sigma\sqrt{1/N_1 + 1/N_2}}$$

where N_1, N_2 = the numbers of sample sizes in the first and second sets

$\bar{X}_1, \bar{X}_2$ = the means, respectively

s_1, s_2 = the standard deviations, respectively

σ = universal standard deviation

$$= \sqrt{\left(\frac{N_1 s_1^2 + N_2 s_2^2}{N_1 + N_2 - 2}\right)}$$

$\nu = N_1 + N_2 - 2$ degrees of freedom

EXAMPLE: In a water treatment plant, 12-day observations are made on two parallel flocculation–sedimentation units using a new type of coagulant and an old coagulant. The results of turbidity removal (%) are given in Table 1.6. Find any improvement of treatment with new chemical.

$$\bar{X}_1 = \frac{\Sigma X_1}{N} = \frac{1088}{12} = 90.7$$

$$\bar{X}_2 = \frac{\Sigma X_2}{N} = \frac{1051}{12} = 87.6$$

The s_1^2 and s_2^2 are calculated by

$$s_1^2 = \frac{\Sigma X_1^2 - (\Sigma X_1)^2/N}{N - 1} = \frac{98{,}980 - (1088)^2/12}{12 - 1} = 30.42$$

$$s_2^2 = \frac{\Sigma X_2^2 - (\Sigma X_2)^2/N}{N - 1} = \frac{92{,}367 - (1051)^2/12}{12 - 1} = 28.81$$

The universal standard deviation is

$$\sigma = \sqrt{\left[\frac{(N_1 - 1)s_1^2 + (N_2 - 1)s_2^2}{N_1 + N_2 - 2}\right]} = \sqrt{\left[\frac{(12 - 1)30.42 + (12 - 1)28.81}{12 + 12 - 2}\right]} = 5.442$$

The Student's t value for the data is calculated by

$$t = \frac{\bar{X}_1 - \bar{X}_2}{\sigma\sqrt{(1/N_1 + 1/N_2)}} = \frac{90.7 - 87.6}{5.442\sqrt{(1/12 + 1/12)}} = 1.395$$

TABLE 1.6 Results of Turbidity Removal (%)

	New chemical used		Old chemical used	
Day	X_1	X_1^2	X_2	X_2^2
1	95	9025	90	8100
2	90	8100	92	8464
3	88	7744	80	6400
4	96	9216	90	8100
5	80	6400	81	6561
6	93	8649	90	8100
7	99	9801	95	9025
8	90	8100	85	7225
9	86	7396	84	7056
10	97	9409	94	8836
11	88	7744	90	8100
12	86	7396	80	6400
Σ	1088	98980	1051	92367

The critical value of t for 22 ($N_1 + N_2 - 2 = 12 + 12 - 2$) degrees of freedom is 1.717 at a 95% confidence level, $t_{0.05}(22)$. The critical value can be found from the t distribution table of any statistical textbook.

The calculated t value of 1.395 is less than the critical value of 1.717. Therefore, it can be concluded that the results from the two types of coagulants used are not significantly different at the 95% confidence level.

5.11 Multiple Range Tests

Testing significance difference for a group of ranked means can employ multiple range tests. The multiple range tests may include the lsd (least significant difference) or the multiple t test, Student—Newman—Keuls test, Tukey's test based on allowances, and Duncan's multiple range test (Duncan, 1955; Federer, 1955). An example of Duncan's multiple test is given below.

EXAMPLE: A lake area has implemented the best management practices (soil conservation) in the watershed in 1984 and 1985. Water quality monitoring programs were carried out in 1981 (pre-implementation); in 1994–95 during implementation; in 1986 (after implementation); and in 1992–94 (long-term post-implementation). The observed total phosphorus (TP) concentrations and yearly mean values for a sampling station are listed in Table 1.7 (Lin and Raman, 1997).

Question: Is there any water quality improvement (decrease concentration) after implementation?

Solution: This can be solved by using Duncan's multiple range test to determine any significant differences among the means of each year. It is necessary to determine the standard error of the difference between two treatment means (S_d). The S_d is calculated as the square root of two times the error mean square divided by the number of replication (Carmer and Walker, 1985) as follows:

$$S_d = \sqrt{\left(\frac{2s^2}{r}\right)}$$

where s^2 is the estimated error variance and r is the number of replications.

If the ith and jth groups (treatments) have different number of observations (this example) or unequal variances, the standard error of the difference between the ith and jth treatment means (each pair) is calculated by

$$S_{d(ij)} = \sqrt{\left(\frac{s^2}{r_i} + \frac{s^2}{r_j}\right)}$$

TABLE 1.7 Total Phosphorus Concentrations at Station 1

	1981	1984	1985	1986	1992	1993	1994
May	0.74	0.088	0.125	0.051	0.065	0.135	0.048
	0.28	0.074	0.17	0.131			0.027
Jun	0.33	0.268	0.194	0.128	0.319	0.874	0.131
	0.6	0.801	0.151	0.124			
Jul	0.77	0.857	0.453	0.115	0.626	0.167	
	1.2	0.91	0.389	0.214			
Aug	1.11	0.873	0.473	0.289	0.971	0.119	0.656
	0.96	1.671	0.555	0.33			
Sep	1.44	0.733	0.557	0.204	1.066	0.162	
	1.43	0.143		0.757			
Oct	0.26	0.209	0.312	0.265	0.096	0.083	0.146
Mean $\bar{X}$	0.829	0.602	0.338	0.237	0.524	0.257	0.202
Total T	9.12	6.627	3.379	2.608	3.143	1.54	1.008
n	11	11	10	11	6	6	5
T^2/n	7.5613	3.9925	1.1418	0.6183	1.6464	0.3953	0.2032

TABLE 1.8 Analysis of Variance Table for One Variable

	Sum of squares	Degree of freedom (df)	Mean square	F ratio
Category means	$\Sigma\frac{T_{i+}}{n_i} - \frac{T^2_{++}}{N}$	$k - 1$	s^2_m	$F = \frac{S^2_m}{S^2_p}$
Within group	$\Sigma\Sigma X^2_{ij} - \Sigma\frac{T^2_{i+}}{n_i}$	$N - k$	s^2_p	
Total	$\Sigma\Sigma X^2_{ij} - \frac{T^2_{++}}{N}$	$N - 1$		

where k = number of categories or treatments
n = number of measurements (observations, or replication r) in each category
i = the ith category
j = the jth measurement in each category
$T_{1+} = X_{11} + X_{12} + X_{13} + \cdots + X_{1n}$, etc.
$T_{++} = T_{1+} + T_{2+} + \cdots + T_{k+}$ = the grand total
$N = n_1 + n_2 + n_3 + \cdots + n_k$

Step 1. We shall make two-way analyses of variance by using the formulas in Table 1.8 (Dixon and Massey, 1957) to estimate the population σ^2 (the variance of the universe) in two ways, then compare these two estimates.

The *mean square for (or between) categories*, s^2_m, is the *sum of square for means (or between categories)* divided by the degree of freedom for s^2_m. The computing formula for s^2_m if there are k categories is

$$s^2_m = \frac{\sum_{i=1}^{k}\frac{T^2_{i+}}{n_i} - \frac{T^2_{++}}{N}}{k - 1}$$

The pooled variance s^2_p is sometimes called the within-groups mean square, or within-groups variance. The s^2_p is the within-groups sum of squares divided by the degrees of freedom (k categories) in s^2_p and is calculated by

$$s^2_p = \frac{\Sigma\Sigma X^2_{ij} - \Sigma(T^2_{i+}/n_i)}{\Sigma n_i - k}$$

These two values, s^2_m and s^2_p, may be tested for significant difference by using the F ratio ($F = s^2_m/s^2_p$).

Step 2. We shall estimate σ^2. The computations are simple to perform with a calculator or a computer. The hypothesis of no difference in categories (i.e. equal means) will be tested at 5% level of significance (or 95% of confidence).

Computation: Using data from Table 1.7

$$\sum\frac{T^2_{i+}}{n_i} = (7.5613 + 3.9925 + 1.1418 + 0.6183 + 1.6464 + 0.3953 + 0.2032)$$
$$= 15.5588$$
$$T_{++} = 9.12 + 6.627 + 3.379 + 2.608 + 3.143 + 1.14 + 1.008$$
$$= 27.425 \text{ grand total}$$
$$N = 11 + 11 + 10 + 11 + 6 + 6 + 5 = 60$$
$$\Sigma\Sigma X^2_{ij} = (0.74)^2 + (0.28)^2 + (0.33)^2 + \cdots + (0.146)^2$$
$$= 22.2167$$

Sum of squares category means

$$= \sum \frac{T_{i+}^2}{n_i} - \frac{T_{++}^2}{N}$$

$$= (15.5588) - (27.425)^2/60$$

$$= 3.0233$$

Degree of freedom (df) for category $= k - 1 = 7 - 1 = 6$

$$s_{\mathrm{m}}^2 = 3.0233/6 = 0.5039$$

Sum of squares within group

$$= \Sigma\Sigma X_{ij}^2 - \Sigma \frac{T_{it}^2}{n_i}$$

$$= 22.2167 - 15.5588$$

$$= 6.6579$$

df for within $= N - k = 60 - 7 = 53$

$$s_{\mathrm{p}}^2 = 6.6579/53 = 0.1256$$

$$\text{Calculated } F = \frac{s_{\mathrm{m}}^2}{s_{\mathrm{p}}^2} = \frac{0.5039}{0.1256} = 4.012$$

$$\text{Critical } F_{.95}(6.53) = 2.28$$

The results of computations are summarized in Table 1.9. The calculated F ratio is greater than the critical F ratio. We will reject the hypothesis of no difference in means. In other words, the groups are from populations having unequal means. The procedure for Steps 1 and 2 is called two-way analysis of variance.

Step 3. We determine the difference in the means among the group. First, the means are rearranged with the order of magnitude. Each possible difference in means is determined. Table 1.10 presents the ranked means and all possible differences among the seven means.

Step 4. We determine s_{p}^2/r_i for each year. These are 0.0114, 0.0114, 0.0209, 0.0126, 0.0209, 0.0114, and 0.0251 for 1981, 1984, 1992, 1985, 1993, 1986, and 1994, respectively. Since $s^2 = s_{\mathrm{p}}^2$, we then determine the standard error of the difference between each two means by

$$S_{\mathrm{d}(ij)} = \sqrt{\left[\frac{1}{2}\left(\frac{s_{\mathrm{p}}^2}{r_i} + \frac{s_{\mathrm{p}}^2}{r_j}\right)\right]}$$

The results are presented in Table 1.11, which is similar to Table 1.10.

Step 5. For the present example, choose the significance level of 5%. From Duncan's tables or by interpolation from the table, significant ranges (SR) for 5% significant level with df = 53 are 3.25, 3.21, 3.15, 3.09, 2.99, and 2.88, respectively, for $n = 7, 6, 5, 4, 3$, and 2. The critical ranges are computed by SR times $S_{\mathrm{d}(ij)}$ as shown in Table 1.12. Values of $S_{\mathrm{d}(ij)}$ are from the corresponding matrix in Table 1.11.

Step 6. Comparing respective values in Tables 1.10 and 1.12, we may conclude the difference is significant with 95% confidence if the difference of mean is greater than the critical value of the multiple range test. It is concluded that the mean value in 1981 was significantly higher than that in 1994, 1986, 1993, and 1995. Also, the mean in 1984 was greater than that in 1986. There is no signficant difference in mean among 1985, 1986, and 1992–94.

Note: This example not only illustrates Duncan's multiple range test but also analysis of variance.

TABLE 1.9 Results of Analysis of Variance

	Sum of squares	df	Mean square	*F* ratio
Category means	3.0233	6	0.5039	$F = \frac{0.5039}{0.1256} = 4.012$
Within-group	6.6579	53	0.1256	$F_{0.95}(6.53) = 2.28$
Total	9.6811	59		

TABLE 1.10 Differences of Total Phosphorus Means

Year	Year Mean	1981 0.829	1984 0.602	1992 0.524	1985 0.338	1993 0.257	1996 0.237
1994	0.202	0.627	0.400	0.322	0.136	0.055	0.035
1986	0.237	0.592	0.365	0.287	0.101	0.020	
1993	0.257	0.572	0.345	0.267	0.081		
1985	0.338	0.491	0.264	0.186			
1992	0.524	0.305	0.078				
1984	0.602	0.227					

TABLE 1.11 Standard Error of Means

Year	Year s_p^2/r_i	1981 0.0114	1984 0.0114	1992 0.0209	1985 0.0126	1993 0.0209	1986 0.0114
1994	0.0251	0.1351	0.1351	0.1517	0.1372	0.1517	0.1351
1986	0.0114	0.1068	0.1068	0.1272	0.1096	0.1272	
1993	0.0209	0.1272	0.1272	0.1446	0.1293		
1985	0.0126	0.1096	0.1096	0.1295			
1992	0.0209	0.1272	0.1272				
1984	0.0114	0.1068					

TABLE 1.12 Critical Ranges for Multiple Range Test

Year	SR	1981	1984	1992	1985	1993	1986
1994	3.25	0.439	0.439	0.479	0.424	0.454	0.380
1986	3.21	0.343	0.343	0.393	0.328	0.362	
1993	3.15	0.401	0.401	0.433	0.368		
1985	3.09	0.339	0.339	0.368			
1992	2.99	0.381	0.381				
1984	2.88	0.343					

5.12 Regression Analysis

Data collected from studies of water and wastewater engineering may be evaluated from the relationships between two or more parameters measured: there are many mathematics functions to fit the data. Those most commonly encountered by environmental engineers for curve fitting are linear, semi-log, and log-log relations between parameters (or variables) *X* and *Y*. Curve fitting can be estimated by the graphical (plot) method and can be calculated by least-square regression analysis.

Linear equation (the straight line). A straight line is described by the equation

$$Y = a + bX$$

where a and b are constants and are computed from normal equations.

Given any two points (X_1, Y_1) and (X_2, Y_2) on a straight line, the constants a and b can be computed. The following relation exists:

$$\frac{Y - Y_1}{X - X_1} = \frac{Y_2 - Y_1}{X_2 - X_1}$$

then

$$Y - Y_1 = \left(\frac{Y_2 - Y_1}{X_2 - X_1}\right)(X - X_1)$$

$$Y - Y_1 = m(X - X_1)$$

where m is called the slope of the line. The constant a, which is the value of Y when $X = 0$, is called the Y intercept.

The least-square regression. Given the set of points (X_1, Y_1), (X_2, Y_2), (X_3, Y_3) . . . (X_n, Y_n) is described by the equation

$$Y = a + bX$$

The normal equations for the least-square line are

$$\Sigma Y = aN + b\Sigma X$$

$$\Sigma XY = a\Sigma X + b\Sigma X^2$$

The constants a and b can be determined by solving simultaneously the above equations

$$a = \frac{n(\Sigma Y)(\Sigma X^2) - (\Sigma X)(\Sigma XY)}{n\Sigma X^2 - (\Sigma X)^2}$$

$$b = \frac{n\Sigma XY - (\Sigma X)(\Sigma Y)}{n\Sigma X^2 - (\Sigma X)^2}$$

The coefficient of correlation is given by

$$r = \sqrt{\left[\frac{\Sigma(Y_{\text{est.}} - \overline{Y})^2}{\Sigma(Y - \overline{Y})^2}\right]}$$

where $Y_{\text{est.}}$ is the estimated Y value from the regression line and $\overline{Y}$ is the mean of Y values. The r lies between zero and one. It is more frequently used as r^2, because it is always non-negative (US Department of Interior, 1968).

EXAMPLE: Determine the regression line of Y (new chemical used) on X (old chemical) and coefficient of regression for the data of the example given in Sec. 5.10 (Student's t test section)

Solution:

$$\text{Let } Y = a + bX$$

Step 1. Compute the following values (Table 1.13)

TABLE 1.13 Determination of Constants

Y	X	X^2	XY	$Y_{est.}$
95	90	8100	8550	92.7
90	92	8464	8280	94.3
88	80	6400	7040	84.3
96	90	8100	8640	91.7
80	81	6561	6480	85.1
93	90	8100	6370	92.7
99	95	9025	9405	96.9
90	85	9225	7650	88.5
86	84	7056	7224	87.6
97	94	8836	9118	96.0
88	90	8100	7920	92.7
86	80	6400	6880	84.3
Σ 1088	1051	92367	95557	

$\overline{Y} = 1088/12 = 90.7$

Step 2. Compute a and b

$$a = \frac{(\Sigma Y)(\Sigma X^2) - \Sigma(X)\Sigma(XY)}{N\Sigma X^2 - (\Sigma X)^2}$$

$$= \frac{(1088)(92{,}367) - (1051)(95{,}557)}{(12)(92{,}367) - (1051)^2}$$

$$= 17.06$$

$$b = \frac{N\Sigma XY - (\Sigma X)(\Sigma Y)}{N\Sigma X^2 - (\Sigma X)^2}$$

$$= \frac{12(95{,}557) - (1051)(1088)}{(12)(92{,}367) - (1051)^2}$$

$$= 0.84$$

The regression equation is

$$Y = 17.06 + 0.84X$$

Step 3. Calculate $Y_{est.}$ by the above equation and shown in Table 1.13.

Step 4. Compute coefficient of correlation with Table 1.14. Then, coefficient of correlation is given by

$$r = \sqrt{\frac{\Sigma(Y_{est.} - \overline{Y})^2}{\Sigma(Y - \overline{Y})^2}} = \sqrt{\frac{223.22}{334.68}} = 0.817$$

and

$$r^2 = 0.667$$

Nonlinear equations. For two parameters without a linear relationship, three mathematical models are usually used in an effort to formulate their relationship. They include

TABLE 1.14 Determination of Coefficient of Correlation

$Y_{est.}$	$(Y_{est.} - \bar{Y})^2$	Y	$(Y - \bar{Y})^2$
92.7	4	95	18.49
94.3	12.96	90	8.49
84.3	40.96	88	7.29
92.7	4	96	28.09
85.1	31.36	80	114.49
92.7	4	93	5.29
96.9	38.44	99	68.89
88.5	4.84	90	0.49
87.6	9.61	86	22.09
96.0	28.09	97	39.69
92.7	4	88	7.29
84.3	40.96	86	22.09
Σ	223.22		334.68

quadratic $Y = a + bX + cX^2$

logarithmic $Y = aX^b$

and geometric $Y = a \log X + b$

where Y is a dependent variable, X is an independent variable, and a, b, c are constants determined by regression analysis.

EXAMPLE: A river water is used as a supply source for a city. The jar test results show the algal removal with various alum dosage as below:

Alum dosage, mg/L	Algal reduction, %
0	9–51
10	51–83
20	69–100
25	73–97
30	87–100
40	93–100

Determine the relationship of algal removal (X) versus alum dosage (Y).

Solution: (see Lin *et al.*, 1971)

Applying the three models stated above for 25 test runs using regression analysis, it was found that model 1 (quadratic equation) expressed best the relationship of algal (and turbidity) removal with coagulant dosage. The relationships are

for algal removal

$$Y = 36.54 + 3.2325X + 0.04256X^2$$

for turbidity removal

$$Y = 37.08 + 2.9567X - 0.04355X^2$$

Semi-log regression. Nonlinear relationships can sometimes be reduced to linear relationships by appropriate transformation of variables. If the data of two variables show a semi-log relationship, the formula is given by

$$Y = ae^{bx}$$

Taking logarithm transformation, it can be rewritten as

$$\log Y = \log a + bx \log e$$

$$\log Y = \log a + (0.4343b)X$$

This is similar to a linear equation. The constants a and b and the coefficient of correlation can be determined with the steps mentioned earlier (linear regression) after transformation of Y values to log Y values. On the other hand, a and b can also be roughly estimated by plotting data points (X_1, Y_1), $(X_2, Y_2) \ldots (X_n, Y_n)$ on semi-log paper and drawing a straight line.

Log-log regression. If the data of two variables are plotted on log-log paper and show a straight-line relationship, the data can be expressed as

$$Y = aX^b$$

or

$$\log Y = \log a + b(\log X)$$

Similarly, constants a and b as well as the coefficient of correlation r can be computed using the same manner of linear regression.

The least-square parabola. A set of points (X_1, Y_1), (X_2, Y_2), $(X_3, Y_3) \ldots (X_n, Y_n)$ may show a parabolic relationship. The equation can be written

$$Y = a + bX + cX^2$$

where the constants a, b, and c are calculated by solving normal equations simultaneously for the least-square parabola as follows:

$$\Sigma Y = aN + b\Sigma X + c\Sigma X^2$$

$$\Sigma XY = a\Sigma X + b\Sigma X^2 + c\Sigma X^3$$

$$\Sigma X^2 Y = a\Sigma X^2 + b\Sigma X^3 + c\Sigma X^4$$

These three equations are obtained by multiplying equation $Y = a + bX + cX^2$ by 1, X, and X^2, respectively, and summing both sides of the result.

This technique can be extended to obtain normal equations for least-square cubic curves, least-square quartic curves, and so on. The relationship of some chemical parameters in water, population increase with time periods, and wastewater production against time may fit least-square parabola.

EXAMPLE: The following table presents data on temperature and dissolved solids in lake water samples. Find the equation of a least-square parabola fitting the data.

Temperature, °C	6	9	12	15	18	21	24	27	30
DS, mg/L	38	48	56	63	69	76	83	88	92

Solution:

Step 1. Let the variables X and Y denote the water temperature and the dissolved solids concentration, respectively. The equation of a least-square parabola fitting the data can be written as

$$Y = a + bX + cX^2$$

where constants a, b, and c can be determined from the normal equations.

TABLE 1.15 Calculation of Data

	Y	X	X^2	X^3	X^4	XY	X^2Y
	38	6	36	216	1296	228	1368
	48	9	81	729	6561	432	3888
	56	12	144	1728	20736	672	8064
	63	15	225	3375	50625	945	14175
	69	18	324	5832	104976	1242	22356
	76	21	441	9261	194481	1596	33516
	83	24	576	13824	331776	1992	47808
	88	27	729	19683	531441	2376	64152
	92	30	900	27000	810000	2760	82800
Σ	613	162	3456	81648	2051892	12243	278127

Step 2. Perform the calculations shown in Table 1.15 to solve power of 2 equations.

$$\Sigma Y = aN + b\Sigma X + c\Sigma X^2$$
$$a(9) + b(162) + c(3456) = 613$$
$$\Sigma XY = a\Sigma X + b\Sigma X^2 + c\Sigma X^3$$
$$a(162) + b(3456) + c(81{,}648) = 12{,}243$$
$$\Sigma X^2Y = a\Sigma X^2 + b\Sigma X^3 + c\Sigma X^4$$
$$a(3456) + b(81{,}648) + c(2{,}051{,}892) = 278{,}127$$

The equations are simplified as

$$a + 18b + 384c = 68.11$$
$$a + 21.33b + 504c = 75.57$$
$$a + 23.63b + 593.7c = 80.48$$

Eliminate a

$$3.33b + 120c = 7.46$$
$$56.3b + 209.7c = 12.37$$

or

$$b + 36c = 2.24$$
$$b + 37.2c = 2.20$$

Eliminate b

$$1.2c = -0.04$$
$$c = -0.033$$

Determine b

$$b + 36(-0.033) = 2.24$$
$$b = 3.44$$

Determine a

$$a + 18(3.44) + 384(-0.033) = 68.11$$
$$a = 18.86$$

Step 4. The equation for the data set is

$$Y = 18.86 + 3.44X - 0.033X^2$$

where the origin $X = 0$ is at temperature of 0°C.

Another method used to solve the least-square parabola will be presented in an example of population projection in the section on water treatment (Chapter 1.6).

Multiple linear regression analysis. In environmental engineering, a water quality parameter may be influenced by several factors. To evaluate the observed results, one can use multiple linear regression analysis. This analysis can be easily performed with a readily available computer program. The standard multiple linear relationship can be expressed as

$$Z = C_0 + C_1U_1 + C_2U_2 + \cdots + C_nU_n$$

where Z = a parameter, such as algal density
$C_0 \ldots C_n$ = regression constants
$U_1 \ldots U_n$ = independent water quality

For example, Lin *et al.* (1971) evaluated the algal density after the coagulation process and the factors that were possibly influenced, such as initial turbidity, total solids, nonfilterable residue, pH of untreated water, water temperature, time of floc formation, and initial algal concentration. A step-wise multiple linear regression analysis was used to evaluate data for all observed 25 runs. The result gave

$$Z = 1132 - 104.85U_1 + 0.23U_2 + 1.80U_3 + 33.25U_4$$

where Z = algal concentration in the treated water, cells/mL
U_1 = alum dosage, mg/L
U_2 = initial algal concentration, cells/mL
U_3 = nonfilterable residue, mg/L
U_4 = water temperasture, °C

The coefficient of correlation is 0.857 and the standard error of estimate is 896 cells/mL.

5.13 Calculation of Data Quality Indicators

Quality assurance. Quality assurance (QA) is a set of operating principles that should be followed during sample collection and analysis. It includes quality control and quality assessment. Any data collection activities performed for the US Environmental Protection Agency (EPA) are required to have a QA project plan to produce data with a high level of confidence. *Standard methods* (APHA, AWWA, and WEF, 1995), Simes (1991), and EPA (1991) have recommended guidelines for QA programs. The QA plan specifies the correction factors to be applied, as well as the procedures to be followed, to validate the result.

Quality control. Quality control (QC) data production may be either or both internal and external. A good-quality control program should include at least certification of operator competence, recovery of known additions, analysis of externally supplied standards, analyses of reagent blank, calibration with standards, analyses of duplicates, and control charts.

The main indicators of data quality are bias and precision. Bias is consistent deviation of the measured value from the true value caused by systematic errors which are due to the analytical method and to the laboratory's use of the method. Precision is a measurement of closeness, in which the results of multiple analyses of a given sample agree with each other; it is usually expressed as the standard deviation.

The requirements of data quality indicators of EPA projects should at least include calculations for precision, accuracy, completeness, and method detection limit. In addition, equations must be given for other project-specific calculations, such as mass balance, emission rates, confidence ranges, etc.

Precision. Precision includes the random error in sampling and the error in sample preparation and analysis. It is specified by the standard deviation of the results. Precision of the laboratory analysis should be assessed by comparing the analytical results for a sample and its duplicate. If calculated from duplicate measurements, the relative percent difference (RPD) is the normal measurement of precision (USEPA, 1996, 1997):

$$\text{RPD} = \frac{(C_1 - C_2) \times 100\%}{(C_1 + C_2)/2}$$

where RPD = relative percent difference
C_1 = larger of the observed values
C_2 = smaller of the observed values

EXAMPLE 1: Find the RPD of the total alkalinity measurements (152 and 148 mg/L as $CaCO_3$) of a duplicate water sample.

Solution:

$$\text{RPD} = \frac{(C_1 - C_2)(100\%)}{(C_1 + C_2)/2}$$

$$= \frac{(152 - 148)(100\%)}{(152 + 148)/2}$$

$$= 2.67\%$$

If calculated from three or more replicates, relative standard deviation (RSD) is used rather than RPD. The RSD is given by

$$\text{RSD} = (s/Y) \times 100\%$$

where RSD = relative standard deviation
s = standard deviation
Y = mean of replicate analysis

Standard deviation, defined previously, is

$$s = \sqrt{\left[\sum_{n=1}^{n}\left[\frac{(Y_i - \bar{Y})^2}{n - 1}\right]\right]}$$

where s = standard deviation
Y_i = measured value of ith replicate
$\bar{Y}$ = mean of replicate measurements
n = number of replicates

EXAMPLE 2: Find the RSD for 6 replicates (56, 49, 54, 48, 47, and 51 mg/L) of a sample.

Solution:

Step 1. Determine the mean

$$\bar{Y} = \Sigma Y_i/n = 305/6 = 51$$

Step 2. Determine s

$$s = \sqrt{\left[\frac{\Sigma(Y_i - \overline{Y})^2}{n - 1}\right]} = \sqrt{(5^2 + 2^2 + 3^2 + 3^2 + 4^2 + 0)/5}$$
$$= 3.55$$

Step 3. Calculate RSD

$$\text{RSD} = (s/Y) \times 100\% = (3.55/51) \times 100\%$$
$$= 7.0\%$$

EXAMPLE 3: A known concentration (1.10 mg/L of nitrate-nitrogen) is added to each of 25 water samples. For spiked samples, the precision calculation can be computed (Table 1.16).

Solution:

Step 1. Calculate the concentration of recovery which is spiked (1) minus sample concentraction (2): shown in column (3).

Step 2. The differences between calculated recovery (3) and known addition (4) are shown in column (5).

TABLE 1.16

(1) Concentration for spiked sample	(2) Concentration for sample only	(3) Calculated recovery (1) − (2)	(4) Known concentration added	(5) Deviation from expected (3) − (4)
3.58	2.33	1.25	1.1	0.15
3.5	2.29	1.21	1.1	0.11
2.7	1.58	1.12	1.1	0.02
2.55	1.49	1.06	1.1	−0.04
4.81	3.56	1.25	1.1	0.15
2.52	1.44	1.08	1.1	−0.02
4.8	3.57	1.23	1.1	0.13
5.4	4.46	0.94	1.1	−0.16
2.53	1.6	0.93	1.1	−0.17
2.77	1.51	1.26	1.1	0.16
4.56	3.45	1.11	1.1	0.01
5.05	3.99	1.06	1.1	−0.04
2.45	1.3	1.15	1.1	0.05
3.6	2.54	1.06	1.1	−0.04
4.78	3.48	1.3	1.1	0.2
3.9	2.79	1.11	1.1	0.01
4.66	4.48	0.18	1.1	−0.92
3.8	2.79	1.01	1.1	−0.09
3.25	2.11	1.14	1.1	0.04
5.55	4.55	1	1.1	−0.1
6.17	5.01	1.16	1.1	0.06
5.66	4.39	1.27	1.1	0.17
5.44	4.39	1.05	1.1	−0.05
4.28	3.27	1.01	1.1	−0.09
5.78	4.66	1.12	1.1	0.02
4	2.79	1.21	1.1	0.11
3.07	1.99	1.08	1.1	−0.02

Σ (deviation)^2 = 1.1245

Step 3. Calculate s by

$$s = \sqrt{\frac{(\Sigma \text{ deviation}^2)}{n - 1}} = \sqrt{\frac{1.1245}{25 - 1}} = 0.216$$

Accuracy. Accuracy is the combination of bias and precision of an analytical procedure and reflects the closeness of a measured value to a true value. In order to assure the accuracy of the analytical procedures, matrix spikes are analyzed. The spike sample is made by adding a known amount of analyte to a water sample or to deionized water. The percent recovery (%R) for the spiked sample is calculated by the following formula:

$$\%R = \frac{S - U}{C} \times 100\%$$

where $\%R$ = percent recovery
S = measured concentration in spiked aliquot
U = measured concentration in unspiked aliquot
C = actual concentration of known addition

EXAMPLE: In the previous (precision) example, the percent recovery for the first sample is

$$\%R = \frac{3.58 - 2.33}{1.10} \times 100\% = 13.6\%$$

When a standard reference material (SRM) is used, the percent can be computed by

$$\%R = \frac{C_{\rm m}}{C_{\rm srm}} \times 100\%$$

where $\%R$ = percent recovery
$C_{\rm m}$ = measured concentration of SRM
$C_{\rm srm}$ = actual concentration of SRM

Completeness. Completeness for all measures is defined as follows:

$$\%C = \frac{V}{T} \times 100\%$$

where $\%C$ = percent completeness
V = number of measurements judged valid
T = total number of measurements

Method detection limit. The method detection limit (MDL) for all measurements is defined as follows:

$$\text{MDL} = t_{(n-1,1-\alpha=0.99)} \times s$$

where MDL = method detection limit
s = standard deviation of replicate analyses
$t_{(n-1,1-\alpha=0.99)}$ = Student's t value for a one-side 99% confidence level and a standard deviation estimate with $n - 1$ degree of freedom. They are as follows:

df	2	3	4	5	6	7	8	10	20
$t_{(n-1,1-\alpha=0.99)}$	6.965	4.541	3.747	3.365	3.143	2.998	2.896	2.764	2.528

TABLE 1.17 Determination of s

Concentration measured	Standard concentration	Deviation	$(\text{Deviation})^2$
0.022	0.020	0.002	0.000004
0.019	0.020	−0.001	0.000001
0.025	0.020	0.005	0.000025
0.026	0.020	0.006	0.000036
0.016	0.020	−0.004	0.000016
0.017	0.020	−0.003	0.000009
0.020	0.020	0	0.000000
0.022	0.020	0.002	0.000004
Σ			0.000095

To determine the MDL, it is required that a standard solution must be measured for at least seven replicates. The concentration of standard solution should be as close as the MDL; generally 2 to 5 times of the MDL.

EXAMPLE: A 0.02 mg/L of nitrate nitrogen was measured for eight replicates. The measured results are 0.022, 0.019, 0.025, 0.026, 0.016, 0.017, 0.020, and 0.022 mg/L. Find the MDL of nitrate nitrogen for your laboratory.

Solution:

Step 1. Determine s (Table 1.17)

From Table 1.17

$$\Sigma\,(\text{Deviation})^2 = 0.00095$$

$$s = \sqrt{\frac{\Sigma(\text{Deviation})^2}{n-1}} = \sqrt{\frac{0.000095}{8-1}} = 0.00368$$

Step 2. Compute MDL

$$\begin{aligned}\text{MDL} &= t_{(n-1,1-\sigma=0.99)} \times s \\ &= 2.998 \times 0.00368 \\ &= 0.011\end{aligned}$$

It will be reported that the MDL of nitrate nitrogen is 0.011 mg/L for the laboratory.

REFERENCES

American Public Health Association (APHA), American Water Works Association (AWWA), and Water Environment Federation (WEF). 1995. *Standard methods for the examination of water and wastewater.* Washington, DC: American Public Health Association.

Carmer, S. G. and Walker, W. M. 1985. Pair wise multiple comparisons of treatment means in agronomic research. *J. Agronomic Educ.* **14**(1): 19–26.

Clark, G. M. and Cooke, D. 1992. *A basic course in statistics*, 3rd edn. London: Edward Arnold.

Dixon, J. and Masey, Jr., J. 1957. *Introduction to statistical analyses*, 2nd edn. New York: McGraw-Hill.

Duncan, B. 1955. Multiple range and multiple F-tests. *Biometrics*, **11**(1): 1–42.

Fair, G. M., Geyer, J. C. and Morris, J. C. 1963. *Water supply and waste-water disposal*. New York: John Wiley.

Federer, W. T. 1955. *Experimental design, theory, and application*. New York: W. W. Norton.

Friedman, D., Pisani, R., Purves, R. and Adhikor, A. 1991. *Statistics*, 2nd edn. New York: W. W. Norton.

Lin, S. D. and Raman, R. K. 1997. *Phase III, post-restoration monitoring of Lake Le-Aqua-Na*. Contract Report 610, Champaign: Illinois State Water Survey.

Lin, S. D., Evans, R. L. and Beuscher, D. B. 1971. *Algal removal by alum coagulation*. Report of Investigation 68. Urbana: Illinois State Water Survey.

New York State Department of Health. 1961. *Manual of instruction for water treatment plant operators*. Albany, New York: State of New York.

Reeves, T. G. 1986. *Water fluoridation—A manual for engineers and technicians*. Atlanta, Georgia: US Department of Health and Human Service.

Research and Education Association. 1996. *The statistics problem solvers*. Piscataway, New Jersey: REA.

Simon, A. L. 1976. *Practical hydraulics*. New York: John Wiley.

Simes, G. F. 1991. *Preparation aids for the development of category III quality assurance project plans*. EPA/600/8-91/005. Washington DC: USEPA.

Spiegel, M. R. 1961. *Schaum's outline of theory and problems of statistics*. New York: McGraw-Hill.

US Department of the Interior. 1968. *Data evaluation and analysis—Training manual*. Washington, DC: US Department of the Interior.

US Environmental Protection Agency. 1991. *Preparation aids for the development of category I quality assurance project plans*. EPA/600/8–91/1003. Washington, DC: USEPA.

US Environmental Protection Agency. 1996. *EPA guidance for quality assurance project plans*. EPA QA/G–5. External workshop draft, US Environmental Protection Agency, Washington, DC: USEPA.

US Environmental Protection Agency. 1997. *EPA request for quality assurance project plans for environmental data operations*. EPA QA/R–5. Washington, DC: USEPA.

CHAPTER 1.2
STREAMS AND RIVERS

Shun Dar Lin

1 GENERAL

This chapter presents calculations on stream sanitation. The main portion covers the evaluation of water assimilative capacities of rivers or streams. The procedures include classical conceptual approaches and pragmatic approaches: the conceptual approaches use simulation models, whereas Butts and his coworkers of the Illinois State Water Survey use a pragmatic approach. Observed dissolved oxygen (DO) and biochemical oxygen deman (BOD) levels are measured at several sampling points along a stream reach. Both approaches are useful for developing or approving the design of wastewater treatment facilities that discharge into a stream.

In addition, biological factors such as algae, indicator bacteria, diversity index, macroinvertebrate biotic index, and fish habitat analysis are also presented.

2 POINT SOURCE DILUTION

Point source pollutants are commonly regulated by a deterministic model for an assumed design condition having a specific probability of occurrence. A simplistic dilution and/or balance equation can be written as

$$C_d = \frac{Q_u C_u + Q_e C_e}{Q_u + Q_e} \tag{2.1}$$

where C_d = completely mixed constituent concentration downstream of the effluent, mg/L
Q_u = stream flow upstream of the effluent, cubic feet per second, cfs or m³/s
C_u = constituent concentration of upstream flow, mg/L
Q_e = flow of the effluent, cfs or m³/s
C_e = constituent concentration of the effluent, mg/L

Under the worst case, a 7-day, 10-year low flow is generally used for stream flow condition, for design purposes.

EXAMPLE: A power plant pumps 27 cfs from a stream, with a flow of 186 cfs. The discharge of the plant's ash-pond is 26 cfs. The boron concentrations for upstream water and the effluent are 0.051 and 8.9 mg/L, respectively. Compute the boron concentration in the stream after completely mixing.

Solution: By Eq. (2.1)

$$\begin{aligned} C_d &= \frac{Q_u C_u + Q_e C_e}{Q_u + Q_e} \\ &= \frac{(186 - 27)(0.051) + 26 \times 8.9}{(186 - 27) + 26} \\ &= 1.29 \text{ (mg/L)} \end{aligned}$$

3 DISCHARGE MEASUREMENT

Discharge (flow rate) measurement is very important to provide the basic data required for river or stream water quality. The total discharge for a stream can be estimated by float method with wind and other surface effects, by die study, or by actual subsection flow measurement, depending on cost, time, man-power, local conditions and so on. The discharge in a stream cross-section can be measured from a sub-section by the following formula:

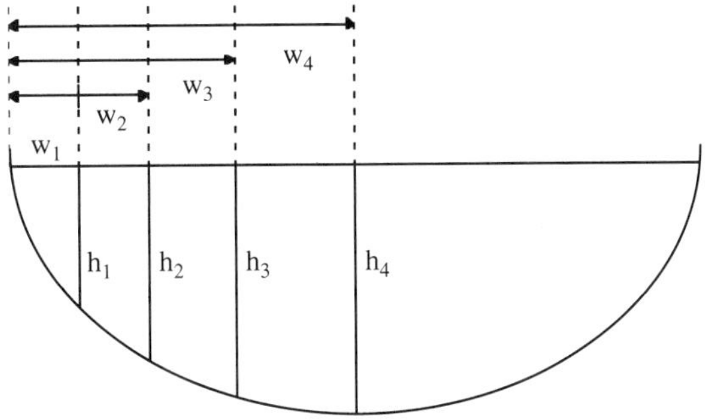

$$Q = \text{Sum (mean depth} \times \text{width} \times \text{mean velocity)}$$

$$Q = \sum_{n=1}^{n} \frac{1}{2}(h_n + h_{n-1})(w_n - w_{n-1}) \times \frac{1}{2}(v_n + v_{n-1}) \tag{2.2}$$

If equal width w

$$Q = \sum_{n=1}^{n} \frac{w}{4}(h_n + h_{n-1})(v_n + v_{n-1}) \tag{2.2a}$$

where Q = discharge, cfs
w_n = nth distance from initial point 0, ft
h_n = nth water depth, ft
v_n = nth velocity, ft/s

Velocity v is measured by a velocity meter, of which there are several types.

The flow rate is 1559 cfs.

EXAMPLE: Data obtained from the velocity measurement are listed in the first three columns of Table 2.1. Determine the flow rate at this cross-section.

Solution: Summarized field data and complete computations are shown in Table 2.1.

4 TIME OF TRAVEL

The time of travel can be determined by dye study or by computation. The river time of travel and stream geometry characteristics can be computed using a volume displacement model. The time of travel is determined at any specific reach as the channel volume of the reach divided by the flow as follows:

$$t = \frac{V}{Q} \times \frac{1}{86{,}400} \tag{2.3}$$

where t = time of travel at a stream reach, days
V = stream reach volume, ft^3 or m^3
Q = average stream flow in the reach, ft^3/sec (cfs) or m^3/s
86,400 = a factor, s/d

EXAMPLE: The cross-section areas at river miles 62.5, 63.0, 63.5, 64.0, 64.5, and 64.8 are, respectively, 271, 265, 263, 259, 258, and 260 ft^2 at a surface water elevation. The average flow is 34.8 cfs. Find the time of travel for a reach between river miles 62.5 and 64.8.

TABLE 2.1 Velocity and Discharge Measurements

1 Distance from 0, ft	2 Depth, ft	3 Velocity, ft/s	4 Width, ft	5 Mean depth, ft	6 Mean velocity, ft/s	7 = 4 × 5 × 6 Discharge, cfs
0	0	0				
2	1.1	0.52	2	0.55	0.26	0.3
4	1.9	0.84	2	1.50	0.68	2.0
7	2.7	1.46	3	2.30	1.15	7.9
10	3.6	2.64	3	3.15	2.05	19.4
14	4.5	4.28	4	4.05	3.46	56.1
18	5.5	6.16	4	5.00	5.22	104.4
23	6.6	8.30	5	6.05	7.23	349.9
29	6.9	8.88	6	6.75	8.59	302.3
35	6.5	8.15	6	6.70	7.52	302.3
40	6.2	7.08	5	6.35	6.62	210.2
44	5.5	5.96	4	5.85	6.52	152.2
48	4.3	4.20	4	4.90	5.08	99.6
50	3.2	2.22	2	3.75	3.21	24.1
52	2.2	1.54	2	2.70	1.88	10.2
54	1.2	0.75	2	1.45	1.15	3.3
55	0	0	1	0.35	0.38	0.1
						1559.0*

*The discharge is 1559 cfs.

Solution:

Step 1. Find average area in the reach

$$\text{Average area} = \frac{1}{6}(271 + 265 + 263 + 259 + 258 + 260)\text{ ft}^2$$
$$= 262.7\text{ ft}^2$$

Step 2. Find volume

$$\text{Distance of the reach} = (64.8 - 62.5)\text{ miles}$$
$$= 2.3\text{ miles} \times 5280\,\frac{\text{ft}}{\text{mile}}$$
$$= 12{,}144\text{ ft}$$
$$V = 262.7\text{ ft}^2 \times 12{,}144\text{ ft}$$
$$= 3{,}190{,}000\text{ ft}^3$$

Step 3. Find t

$$t = \frac{V}{Q} \times \frac{1}{86{,}400}$$
$$= \frac{3{,}190{,}000\text{ ft}^3}{34.8\text{ ft}^3/\text{s} \times 86{,}400\text{ s/d}}$$
$$= 1.06\text{ days}$$

5 DISSOLVED OXYGEN AND WATER TEMPERATURE

Dissolved oxygen (DO) and water temperature are most commonly in situ monitored parameters for surface waters (rivers, streams, lakes, reservoirs, wetlands, ocean, etc.). Dissolved oxygen concentration in milligrams per liter (mg/L) is a measurement of the amount of oxygen dissolved in water. It can be determined with a DO meter or by a chemical titration method.

The DO in water has an important impact on aquatic animals and plants. Most aquatic animals, such as fish, require oxygen in the water to survive. The two major sources of oxygen in water are from diffusion from the atmosphere across the water surface and the photosynthetic oxygen production from aquatic plants such as algae and macrophytes. Important factors that affect DO in water (Fig. 2.1) may include water temperature, aquatic plant photosynthetic activity, wind and wave mixing, organic contents of the water, and sediment oxygen demand.

Excessive growth of algae (bloom) or other aquatic plants may provide very high concentration of DO, so called supersaturation. On the other hand, oxygen deficiencies can occur when plant respiration depletes oxygen beyond the atmospheric diffusion rate. This can occur especially during the winter ice cover period and when intense decomposition of organic matter in the lake bottom sediment occurs during the summer. These oxygen deficiencies will result in fish being killed.

5.1 Dissolved Oxygen Saturation

DO saturation (DO_{sat}) values for various water temperatures can be computed using the American Society of Civil Engineers' formula (American Society of Civil Engineering Committee on Sanitary Engineering Research, 1960):

$$DO_{sat} = 14.652 - 0.41022T + 0.0079910T^2 - 0.000077774T^3 \tag{2.4}$$

where DO_{sat} = dissolved oxygen saturation concentration, mg/L
T = water temperature, °C

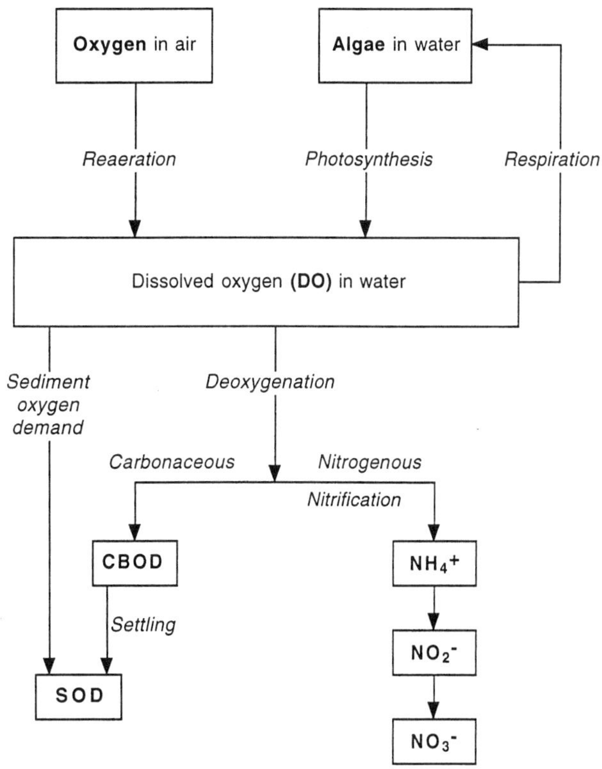

FIGURE 2.1 Factors affecting dissolved oxygen concentration in water.

This formula represents saturation values for distilled water ($\beta = 1.0$) at sea level pressure. Water impurities can increase the saturation level ($\beta > 1.0$) or decrease the saturation level ($\beta < 1.0$), depending on the surfactant characteristics of the contaminant. For most cases, β is assumed to be unity. The DO_{sat} values calculated from the above formula are listed in Table 2.2 (example: DO_{sat} = 8.79 mg/L, when $T = 21.3$°C) for water temperatures ranging from zero to 30°C (American Society of Engineering Committee on Sanitary Engineering Research, 1960).

EXAMPLE 1: Calculate DO saturation concentration for a water temperature at 0, 10, 20 and 30°C, assuming $\beta = 1.0$.

Solution:
(a) at $T = 0$°C

$$\begin{aligned} DO_{sat} &= 14.652 - 0 + 0 - 0 \\ &= 14.652 \text{ mg/L} \end{aligned}$$

TABLE 2.2 Dissolved Oxygen Saturation Values in mg/L

Temp., °C	0.0	0.1	0.2	0.3	0.4	0.5	0.6	0.7	0.8	0.9
0	14.65	14.61	14.57	14.53	14.49	14.45	14.41	14.37	14.33	14.29
1	14.25	14.21	14.17	14.13	14.09	14.05	14.02	13.98	13.94	13.90
2	13.86	13.82	13.79	13.75	13.71	13.68	13.64	13.60	13.56	13.53
3	13.49	13.46	13.42	13.38	13.35	13.31	13.28	13.24	13.20	13.17
4	13.13	13.10	13.06	13.03	13.00	12.96	12.93	12.89	12.86	12.82
5	12.79	12.76	12.72	12.69	12.66	12.62	12.59	12.56	12.53	12.49
6	12.46	12.43	12.40	12.36	12.33	12.30	12.27	12.24	12.21	12.18
7	12.14	12.11	12.08	12.05	12.02	11.99	11.96	11.93	11.90	11.87
8	11.84	11.81	11.78	11.75	11.72	11.70	11.67	11.64	11.61	11.58
9	11.55	11.52	11.49	11.47	11.44	11.41	11.38	11.35	11.33	11.30
10	11.27	11.24	11.22	11.19	11.16	11.14	11.11	11.08	11.06	11.03
11	11.00	10.98	10.95	10.93	10.90	10.87	10.85	10.82	10.80	10.77
12	10.75	10.72	10.70	10.67	10.65	10.62	10.60	10.57	10.55	10.52
13	10.50	10.48	10.45	10.43	10.40	10.38	10.36	10.33	10.31	10.28
14	10.26	10.24	10.22	10.19	10.17	10.15	10.12	10.10	10.08	10.06
15	10.03	10.01	9.99	9.97	9.95	9.92	9.90	9.88	9.86	9.84
16	9.82	9.79	9.77	9.75	9.73	9.71	9.69	9.67	9.65	9.63
17	9.61	9.58	9.56	9.54	9.52	9.50	9.48	9.46	9.44	9.42
18	9.40	9.38	9.36	9.34	9.32	9.30	9.29	9.27	9.25	9.23
19	9.21	9.19	9.17	9.15	9.13	9.12	9.10	9.08	9.06	9.04
20	9.02	9.00	8.98	8.97	8.95	8.93	8.91	8.90	8.88	8.86
21	8.84	8.82	8.81	8.79	8.77	8.75	8.74	8.72	8.70	8.68
22	8.67	8.65	8.63	8.62	8.60	8.58	8.56	8.55	8.53	8.52
23	8.50	8.48	8.46	8.45	8.43	8.42	8.40	8.38	8.37	8.35
24	8.33	8.32	8.30	8.29	8.27	8.25	8.24	8.22	8.21	8.19
25	8.18	8.16	8.14	8.13	8.11	8.10	8.08	8.07	8.05	8.04
26	8.02	8.01	7.99	7.98	7.96	7.95	7.93	7.92	7.90	7.89
27	7.87	7.86	7.84	7.83	7.81	7.80	7.78	7.77	7.75	7.74
28	7.72	7.71	7.69	7.68	7.66	7.65	7.64	7.62	7.61	7.59
29	7.58	7.56	7.55	7.54	7.52	7.51	7.49	7.48	7.47	7.45
30	7.44	7.42	7.41	7.40	7.38	7.37	7.35	7.34	7.32	7.31

Source: American Society of Civil Engineering Committee on Sanitary Engineering Research, 1960.

(b) at $T = 10°C$

$$\begin{aligned} DO_{sat} &= 14.652 - 0.41022 \times 10 + 0.0079910 \times 10^2 - 0.000077774 \times 10^3 \\ &= 11.27 \text{ (mg/L)} \end{aligned}$$

(c) at $T = 20°C$

$$\begin{aligned} DO_{sat} &= 14.652 - 0.41022 \times 20 + 0.0079910 \times 20^2 - 0.000077774 \times 20^3 \\ &= 9.02 \text{ (mg/L)} \end{aligned}$$

(d) at $T = 30°C$

$$\begin{aligned} DO_{sat} &= 14.652 - 0.41022 \times 30 + 0.0079910 \times 30^2 - 0.000077774 \times 30^3 \\ &= 7.44 \text{ (mg/L)} \end{aligned}$$

The DO saturation concentrations generated by the formula must be corrected for differences in air pressure caused by air temperature changes and for elevation above the mean sea level (MSL). The correction factor can be calculated as follows:

$$f = \frac{2116.8 - (0.08 - 0.000115A)E}{2116.8} \tag{2.5}$$

where f = correction factor for above MSL
A = air temperature, °C
E = elevation of the site, feet above MSL

EXAMPLE 2: Find the correction factor of DO_{sat} value for water at 620 ft above the MSL and air temperature of 25°C? What is DO_{sat} at a water temperature of 20°C?

Solution:

Step 1. Using Eq. (2.5)

$$\begin{aligned} f &= \frac{2116.8 - (0.08 - 0.000115A)E}{2116.8} \\ &= \frac{2116.8 - (0.08 - 0.000115 \times 25)620}{2116.8} \\ &= \frac{2116.8 - 47.8}{2116.8} \\ &= 0.977 \end{aligned}$$

Step 2. Compute DO_{sat}
From Example 1, at $T = 20°C$

$$DO_{sat} = 9.02 \text{ mg/L}$$

With an elevation correction factor of 0.977

$$DO_{sat} = 9.02 \text{ mg/L} \times 0.977 = 8.81 \text{ mg/L}$$

5.2 Dissolved Oxygen Availability

Most regulatory agencies have standards for minimum DO concentrations in surface waters to support indigenous fish species in surface waters. In Illinois, for example, the Illinois Pollution Control Board stipulates that dissolved oxygen shall not be less than 6.0 mg/L during at least 16 hours of any 24-hour period, nor less than 5.0 mg/L at any time (IEPA, 1990).

The availability of dissolved oxygen in a flowing stream is highly variable due to several factors. Daily and seasonal variations in DO levels have been reported. The diurnal variations in DO are primarily induced by algal productivity. Seasonal variations are attributable to changes in temperature that affect DO saturation values. The ability of a stream to absorb or reabsorb oxygen from the atmosphere is affected by flow factors such as water depth and turbulence, and it is expressed in terms of the reaeration coefficient. Factors that may represent significant sources of oxygen use or oxygen depletion are biochemical oxygen demand (BOD) and sediment oxygen demand (SOD). BOD, including carbonaceous BOD (CBOD) and nitrogenous BOD (NBOD), may be the product of both naturally occurring oxygen use in the decomposition of organic material and oxygen depletion in the stabilization of effluents discharged from wastewater treatment plants (WTPs). The significance of any of these factors depends upon the specific stream conditions. One or all of these factors may be considered in the evaluation of oxygen use and availability.

6 BIOCHEMICAL OXYGEN DEMAND ANALYSIS

Laboratory analysis for organic matter in water and wastewater includes testing for biochemical oxygen demand, chemical oxygen demand (COD), total organic carbon (TOC), and total oxygen demand (TOD). The BOD test is a biochemical test involving the use of microorganisms. The COD test is a chemical test. The TOC and TOD tests are instrumental tests.

The BOD determination is an empirical test that is widely used for measuring waste (loading to and from wastewater treatment plants), evaluating the organic removal efficiency of treatment processes, and assessing stream assimilative capacity. The BOD test measures (1) the molecular oxygen consumed during a specific incubation period for the biochemical degradation of organic matter (carbonaceous BOD); (2) oxygen used to oxidize inorganic material such as sulfide and ferrous iron; and (3) reduced forms of nitrogen (nitrogenous BOD) with an inhibitor (trichloromethylpyridine). If an inhibiting chemical is not used, the oxygen demand measured is the sum of carbonaceous and nitrogenous demands, so-called total BOD or ultimate BOD.

The extent of oxidation of nitrogenous compounds during the 5-day incubation period depends upon the type and concentration of microorganisms that carry out biooxidation. The nitrifying bacteria usually are not present in raw or settleable primary sewage. These nitrifying organisms are present in sufficient numbers in biological (secondary) effluent. A secondary effluent can be used as 'seeding' material for an NBOD test of other samples. Inhibition of nitrification is required for a CBOD test for secondary effluent samples, for samples seeded with secondary effluent, and for samples of polluted waters.

The result of the 5-day BOD test is recorded as carbonaceous biochemical oxygen demand, $CBOD_5$, when inhibiting nitrogenous oxygen demand. When nitrification is not inhibited, the result is reported as BOD_5 (with 5-day incubation at 20°C).

The BOD test procedures can be found in *Standard methods for the examination of water and wastewater* (APHA, AWWA, and WEF, 1995). When the dilution water is seeded, oxygen uptake (consumed) is assumed to be the same as the uptake in the seeded blank. The difference between the sample BOD and the blank BOD, corrected for the amount of seed used in the sample, is the true BOD. Formulas for calculation of BOD are as follows (APHA, AWWA, and WEF, 1995):

When dilution water is not seeded:

$$\text{BOD, mg/L} = \frac{D_1 - D_2}{P} \tag{2.6}$$

When dilution water is seeded:

$$\text{BOD, mg/L} = \frac{(D_i - D_e) - (B_i - B_e)f}{P} \tag{2.7}$$

where D_1, D_i = DO of diluted sample immediately after preparation, mg/L
D_2, D_e = DO of diluted sample after incubation at 20°C, mg/L
P = decimal volumetric fraction of sample used; mL of sample/300 mL (for Eq. (2.6))
B_i = DO of seed control before incubation, mg/L

B_e = DO of seed control after incubation, mg/L
f = ratio of seed in diluted sample to seed in seed control
P = percent seed in diluted sample/percent seed in seed control (for Eq. (2.7))

If seed material is added directly to the sample and to control bottles:

$$f = \text{volume of seed in diluted sample/volume of seed in seed control}$$

EXAMPLE 1: For a BOD test, 75 mL of a river water sample is used in the 300 mL of BOD bottles without seeding with three duplications. The initial DO in three BOD bottles read 8.86, 8.88, and 8.83 mg/L, respectively. The DO levels after 5 days at 20°C incubation are 5.49, 5.65, and 5.53 mg/L, respectively. Find the 5-day BOD (BOD_5) for the river water.

Solution:

Step 1. Determine average DO uptake

$$\begin{aligned} x &= \frac{\Sigma(D_1 - D_2)}{3} \\ &= \frac{[(8.86 - 5.49) + (8.88 - 5.65) + (8.83 - 5.53)]}{3} \\ &= 3.30 \text{ (mg/L)} \end{aligned}$$

Step 2. Determine P

$$P = \frac{75}{300} = 0.25$$

Step 3. Compute BOD_5

$$BOD_5 = \frac{x}{P} = \frac{3.30 \text{ mg/L}}{0.25} = 13.2 \text{ mg/L}$$

EXAMPLE 2: The wastewater is diluted by a factor of 1/20 using seeded control water. DO levels in the sample and control bottles are measured at 1-day intervals. The results are shown in Table 2.3. One milliliter of seed material is added directly to diluted and to control bottles. Find daily BOD values.

Solution:

Step 1. Compute f and P

$$f = \frac{1 \text{ mL}}{1 \text{ mL}} = 1.0$$

$$P = \frac{1}{20} = 0.05$$

TABLE 2.3 Change in Dissolved Oxygen and Biochemical Oxygen Demand with Time

	Dissolved oxygen, mg/L		
Time, days	Diluted sample	Seeded control	BOD, mg/L
0	7.98	8.25	—
1	5.05	8.18	57.2
2	4.13	8.12	74.4
3	3.42	8.07	87.6
4	2.95	8.03	96.2
5	2.60	7.99	102.4
6	2.32	7.96	107.4
7	2.11	7.93	111.0

Step 2. Find BODs using Eq. (2.7)
Day 1:

$$\mathrm{BOD}_1 = \frac{(D_\mathrm{i} - D_\mathrm{e}) - (B_\mathrm{i} - B_\mathrm{e})f}{P}$$

$$= \frac{(7.98 - 5.05) - (8.25 - 8.18)1}{0.05}$$

$$= 57.2 \text{ (mg/L)}$$

Similarly
Day 2:

$$\mathrm{BOD}_2 = \frac{(7.98 - 4.13) - (8.25 - 8.12)1}{0.05}$$

$$= 74.4 \text{ (mg/L)}$$

For other days, BOD can be determined in the same manner. The results of BODs are also presented in the above table. It can be seen that BOD_5 for this wastewater is 102.4 mg/L.

7 STREETER–PHELPS OXYGEN SAG FORMULA

The method most widely used for assessing the oxygen resources in streams and rivers subjected to effluent discharges is the Streeter–Phelps oxygen sag formula that was developed for the use on the Ohio River in 1914. The well-known formula is defined as follows (Streeter and Phelps, 1925):

$$D_t = \frac{K_1 L_\mathrm{a}}{K_2 - K_1}[e^{-K_1 t} - e^{-K_2 t}] + D_\mathrm{a} e^{-K_2 t} \tag{2.8a}$$

or

$$D_t = \frac{k_1 L_\mathrm{a}}{k_2 - k_1}[10^{-k_1 t} - 10^{-k_2 t}] + D_\mathrm{a} 10^{-k_2 t} \tag{2.8b}$$

where D_t = DO saturation deficit downstream, mg/L or lb ($\mathrm{DO}_{\mathrm{sat}} - \mathrm{DO}_\mathrm{a}$) at time t
t = time of travel from upstream to downstream, days
D_a = initial DO saturation deficit of upstream water, mg/L or lb
L_a = ultimate upstream biochemical oxygen demand (BOD), mg/L
e = base of natural logarithm, 2.7183
K_1 = deoxygenation coefficient to the base e, per day
K_2 = reoxygenation coefficient to the base e, per day
k_1 = deoxygenation coefficient to the base 10, per day
k_2 = reoxygenation coefficient to the base 10, per day

In the early days, K_1 or K_2 and k_1 or k_2 were used classically for values based on e and 10, respectively. Unfortunately, in recent years, many authors have mixed the usage of K and k. Readers should be aware of this. The logarithmic relationships between k and K are $K_1 = 2.3026\,k_1$ and $K_2 = 2.3026\,k_2$; or $k_1 = 0.4343\,K_1$ and $k_2 = 0.4343\,K_2$.

The Streeter–Phelps oxygen sag equation is based on two assumptions: (1) at any instant the deoxygenation rate is directly proportional to the amount of oxidizable organic material present; and (2) the reoxygenation rate is directly proportional to the dissolved oxygen deficit. Mathematical expressions for assumptions (1) and (2) are

$$\frac{\mathrm{d}D}{\mathrm{d}t} = K_1(L_\mathrm{a} - L_t) \tag{2.9}$$

and

$$\frac{dD}{dt} = -K_2 D \tag{2.10}$$

where $\frac{dD}{dt}$ = the net rate of change in the DO deficit, or the absolute change of DO deficit (D) over an increment of time dt due to stream waste assimilative capacity affected by deoxygenation coefficient K_1 and due to an atmospheric exchange of oxygen at the air/water interface affected by the reaeration coefficient K_2

L_a = ultimate upstream BOD, mg/L

L_t = ultimate downstream BOD at any time t, mg/L

Combining the two above differential equations and integrating between the limits D_a, the initial upstream sampling point, and t any time of flow below the initial point, yields the basic equation devised by Streeter and Phelps. This stimulated intensive research on BOD, reaction rates, and stream sanitation.

The formula is a classic in sanitary engineering stream work. Its detailed analyses can be found in almost all general environmental engineering texts. Many modifications and adaptations of the basic equation have been devised and have been reported in the literature. Many researches have been carried out on BOD, K_1, and K_2 factors. Illustrations for oxygen sag formulas will be presented in the latter sections.

8 BOD MODELS AND K_1 COMPUTATION

Under aerobic conditions, organic matter, and some inorganics can be used by bacteria to create new cells, energy, carbon dioxide, and residue. The oxygen used to oxidize total organic material and all forms of nitrogen for 60 to 90 days is called the ultimate BOD (UBOD). It is common that measurements of oxygen consumed in a 5-day test period called 5-day BOD or BOD_5 is practiced. The BOD progressive curve is shown in Fig. 2.2.

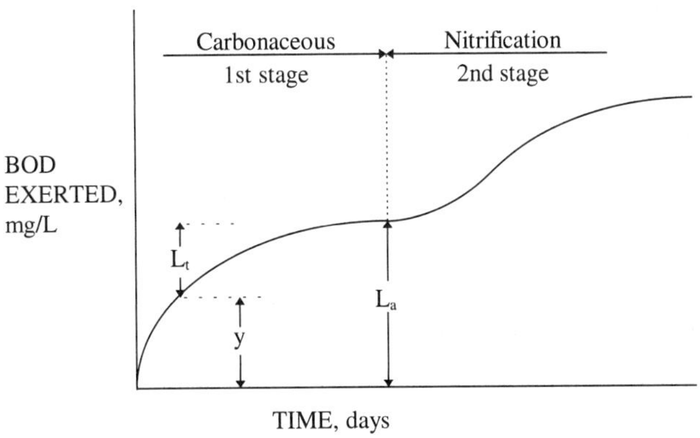

FIGURE 2.2 BOD progressive curve.

8.1 First-order Reaction

Phelps law states that the rate of biochemical oxidation of organic matter is proportional to the remaining concentration of unoxidized substance. Phelps law can be expressed in differential form as follows (monomolecular or unimolecular chemical reaction):

$$-\frac{dL_t}{dt} = K_1 L_t$$

$$\frac{dL_t}{L_t} = -K_1 dt \tag{2.11}$$

by integration

$$\int_{L_a}^{L_t} \frac{dL_t}{L_t} = -K_1 \int_0^t dt_1$$

$$\ln \frac{L_t}{L_a} = -K_1 t \quad \text{or} \quad \log \frac{L_t}{L_a} = -0.434 K_1 t = -k_1 t$$

$$\frac{L_t}{L_a} = e^{-K_1 t} \quad \text{or} \quad \frac{L_t}{L_a} = 10^{-k_1 t}$$

$$L_t = L_a \cdot e^{-K_1 t} \quad \text{or} \quad L_t = L_a \cdot 10^{-k_1 t} \tag{2.12}$$

where L_t = BOD remaining after time t days, mg/L
L_a = first-stage BOD, mg/L, per day
K_1 = deoxygenation rate, based on e, $K_1 = 2.303\, k_1$, per day
k_1 = deoxygenation rate, based on 10, $k_1 = 0.4343\, K_1$ ($k_1 = 0.1$ at 20°C)
e = base of natural logarithm, 2.7183

Oxygen demand exerted up to time t, y, is a first-order reaction:

$$y = L_a - L_t \text{ (see Fig. 2.2)}$$

$$y = L_a(1 - e^{-K_1 t}) \tag{2.13a}$$

or based on $\log_{10}$

$$y = L_a(1 - 10^{-k_1 t}) \tag{2.13b}$$

When a delay occurs in oxygen uptake at the onset of a BOD test, a lag-time factor, t_0 should be included and Eqs. (2.13a) and (2.13b) become

$$y = L_a[1 - e^{-K_1(t-t_0)}] \tag{2.14a}$$

or

$$y = L_a[1 - 10^{-k_1(t-t_0)}] \tag{2.14b}$$

For the Upper Illinois Waterway study, many of the total and nitrogenous BOD curves have an S-shaped configuration. The BOD in waters from pools often consists primarily of high-profile second-stage or nitrogenous BOD, and the onset of the exertion of this NBOD is often delayed 1 or 2 days. The delayed NBOD and the total BOD (TBOD) curves, dominated by the NBOD fraction, often exhibit an S-shaped configuration. The general mathematical formula used to simulate the S-shaped curve is (Butts *et al.*, 1975)

$$y = L_a[1 - e^{-K_1(t-t_0)^m}] \tag{2.15}$$

where m is a power factor, and the other terms are as previously defined.

Statistical results show that a power factor of 2.0 in Eq. (2.15) best represents the S-shaped BOD curve generated in the Lockport and Brandon Road areas of the waterway. Substituting $m = 2$ in Eq. (2.15) yields

$$y = L_a[1 - e^{-K_1(t-t_0)^2}] \tag{2.15a}$$

EXAMPLE 1: Given $K_1 = 0.25$ per day, $BOD_5 = 6.85$ mg/L, for a river water sample. Find L_a when $t_0 = 0$ days and $t_0 = 2$ days.

Solution:

Step 1.

$$\begin{aligned} \text{When } t_0 &= 0 \\ y &= L_a(1 - e^{-K_1 t}) \\ 6.85 &= L_a(1 - e^{-0.25\times 5}) \\ L_a &= \frac{6.85}{1 - 0.286} \\ &= 9.60 \text{ (mg/L)} \end{aligned}$$

Another solution using k_1: Since

$$\begin{aligned} k_1 &= 0.4343\,K_1 = 0.4343 \times 0.25 \text{ per day} = 0.109 \text{ per day} \\ y &= L_a(1 - 10^{-k_1 t}) \\ 6.85 &= L_a(1 - 10^{-0.109\times 5}) \\ L_a &= \frac{6.85}{1 - 0.286} \\ &= 9.60 \text{ (mg/L)} \end{aligned}$$

Step 2. When $t_0 = 2$ days, using Butts *et al.*'s equation [Eq. (2.15a)]:

$$y = L_a[1 - e^{-K_1(t-t_0)^2}]$$

or

$$\begin{aligned} L_a &= \frac{y}{1 - e^{-K_1(t-t_0)^2}} \\ &= \frac{6.85}{1 - e^{-0.25(5-2)^2}} \\ &= \frac{6.85}{1 - e^{-2.25}} \\ &= \frac{6.85}{1 - 0.105} \\ &= 7.65 \text{ (mg/L)} \end{aligned}$$

EXAMPLE 2: Compute the portion of BOD remaining to the ultimate BOD ($1 - e^{-K_1 t}$ or $1 - 10^{-k_1 t}$) for $k_1 = 0.10$ (or $K_1 = 0.23$).

Solution: By Eq. (2.13b)

$$y = L_a(1 - 10^{-k_1 t})$$

$$\frac{y}{L_a} = 1 - 10^{-k_1 t} \quad \text{or} \quad 1 - e^{-K_1 t}$$

TABLE 2.4 Relationship between t and the Ultimate BOD ($1 - 10^{-k_1 t}$ or $1 - e^{-K_1 t}$)

t	$1 - 10^{-k_1 t}$ or $1 - e^{-K_1 t}$	t	$1 - 10^{-k_1 t}$ or $1 - e^{-K_1 t}$
0.25	0.056	4.5	0.646
0.50	0.109	5.0	0.684
0.75	0.159	6.0	0.749
1.00	0.206	7.0	0.800
1.25	0.250	8.0	0.842
1.50	0.291	9.0	0.874
1.75	0.332	10.0	0.900
2.0	0.369	12.0	0.937
2.5	0.438	16.0	0.975
3.0	0.500	20.0	0.990
3.5	0.553	30.0	0.999
4.0	0.602	∞	1.0

when $t = 0.25$ days

$$\frac{y}{L_a} = 1 - 10^{-0.10 \times 0.25} = 1 - 0.944 = 0.056$$

Similar calculations can be performed as above. The relationship between t and $(1 - 10^{-k_1 t})$ is listed in Table 2.4.

8.2 Determination of Deoxygenation Rate and Ultimate BOD

Biological decomposition of organic matter is a complex phenomenon. Laboratory BOD results do not necessarily fit actual stream conditions. BOD reaction rate is influenced by immediate demand, stream or river dynamic environment, nitrification, sludge deposit, and types and concentrations of microbes in the water. Therefore laboratory BOD analyses and stream surveys are generally conducted for raw and treated wastewaters and river water to determine BOD reaction rate.

Many investigators have worked on developing and refining methods and formulas for use in evaluating the deoxygenation (K_1) and reaeration (K_2) constants and the ultimate BOD (L_a). There are several methods proposed to determine K_1 values. Unfortunately, K_1 values determined by different methods given by the same set of data have considerable variations. Reed–Theriault least-squares method published in 1927 (US Public Health Service, 1927) give the most consistent results, but it is time consuming and tedious. Computation using a digital computer was developed by Gannon and Downs (1964).

In 1936, a simplified procedure, the so-called log-difference method of estimating the constants of the first-stage BOD curve, was presented by Fair (1936). The method is also mathematically sound, but is also difficult to solve.

Thomas (1937) followed Fair *et al.* (1941a, 1941b) and developed the "slope" method, which, for many years, was the most used procedure for calculating the constants of the BOD curve. Later, Thomas (1950) presented a graphic method for BOD curve constants. In the same year, Moore *et al.* (1950) developed the "moment method" that was simple, reliable, and accurate to analyze BOD data; this soon became the most used technique for computing the BOD constants.

Researchers found that K_1 varied considerably for different sources of wastewaters and questioned the accepted postulate that the 5-day BOD is proportional to the strength of the sewage. Orford and Ingram (1953) discussed the monomolecular equation as being inaccurate and unscientific in its relation to BOD. They proposed that the BOD curve could be expressed as a logarithmic function.

Tsivoglou (1958) proposed a 'daily difference' method of BOD data solved by a semi-graphical solution. A "rapid ratio" method can be solved using curves developed by Sheehy (1960), O'Connor (1966) modified least-squares method using BOD_5.

This book describes Thomas's slope method, method of moments, logarithmic function, and rapid methods calculating K_1 (or k_1) and L_a.

Slope method. The slope method (Thomas, 1937) gives the BOD constants via the least-squares treatment of the basic form of the first-order reaction equation or

$$\frac{dy}{dt} = K_1(L_a - y) = K_1 L_a - K_1 y \tag{2.16}$$

where dy = increase in BOD per unit time at time t
K_1 = deoxygenation constant, per day
L_a = first stage ultimate BOD, mg/L
y = BOD exerted in time t, mg/L

This differential equation (Eq. (2.16)) is linear between dy/dt and y. Let $y' = dy/dt$ to be the rate of change BOD and n be the number of BOD measurements minus one. Two normal equations for finding K_1 and L_a are

$$na + b\Sigma y - \Sigma y' = 0 \tag{2.17}$$

and

$$a\Sigma y + b\Sigma y^2 - \Sigma yy' = 0 \tag{2.18}$$

Solving Eqs. (2.17) and (2.18) yields values of a and b, from which K_1 and L_a can be determined directly by following relations:

$$K_1 = -b \tag{2.19}$$

and

$$\mathrm{L_a} = -a/b \tag{2.20}$$

The calculations include first determinations of y', $y'\ y$, y^2 for each value of y. The summation of these gives the quantities of $\Sigma y'$, $\Sigma y'y$, and Σy^2 which are used for the two normal equations. The values of the slopes are calculated from the given data of y and t as follows:

$$\frac{dy_i}{dt} = y_i' = \frac{(y_i - y_{i-1})\left(\frac{t_{i+1} - t_i}{t_i - t_{i-1}}\right) + (y_{i+1} - y_i)\left(\frac{t_i - t_{i-1}}{t_{i+1} - t_i}\right)}{t_{i+1} - t_{i-1}} \tag{2.21}$$

For the special case, when equal time increments $t_{i+1} - t_i = t_3 - t_2 = t_2 - t_1 - \Delta t$, y' becomes

$$\frac{dy_i}{dt} = \frac{y_{i+1} - y_{i-1}}{2\Delta t} \quad \text{or} \quad \frac{y_{i+1} - y_{i-1}}{t_{i+1} - t_{i-1}} \tag{2.21a}$$

A minimum of six observations ($n > 6$)of y and t are usually required to give consistent results.

EXAMPLE 1: Equal time increments, BOD data at temperature of 20°C, t and y are shown in Table 2.5. Find K_1 and L_a.

Solution:

Step 1. Calculate y', $y'y$, and y^2

Step 2. Determine a and b.

Writing normal equations (Eqs. (2.17 and (2.18)), $n = 9$

$$na + b\Sigma y - \Sigma y' = 0$$

$$9a + 865.8b - 89.6 = 0$$

$$a + 96.2b - 9.96 = 0 \tag{1}$$

TABLE 2.5 Calculations for y', $y'y$, and y^2 Values

t, day (1)	y (2)	y' (3)	$y'y$ (4)	y^2 (5)
0	0			
1	56.2	37.2*	2090.64	3158.44
2	74.4	15.7	1168.08	5535.36
3	87.6	10.9	954.84	7673.76
4	96.2	7.4	711.88	9254.44
5	102.4	5.6	573.44	10485.76
6	107.4	4.3	461.82	11534.76
7	111.0	3.3	366.30	12321.00
8	114.0	2.8	319.20	12996.00
9	116.6	2.4	279.84	13595.56
10	118.8			
Σ	865.8†	89.6	6926.04	86555.08

$*y'_1 = \frac{y_3 - y_1}{t_3 - t_1} = \frac{74.4 - 0}{2 - 0} = 37.2$

†Sum of first nine observations.

and

$$a\Sigma y + b\Sigma y^2 - \Sigma yy' = 0$$

$$865.8\text{a} + 86{,}555\text{b} - 6926 = 0$$

$$\text{a} + 99.97\text{b} - 8.0 = 0 \qquad (2)$$

Eq. (2) – Eq. (1)

$$3.77b + 1.96 = 0$$

$$b = -0.52$$

From Eq. (2)

$$a + 99.97(-0.52) - 8.0 = 0$$

$$a = 59.97$$

Step 3. Calculate K_1 and L_a with Eqs. (2.19) and (2.20)

$$K_1 = -b = -(-0.52)$$

$$= 0.52 \text{ (per day)}$$

$$L_a = -a/b$$

$$= -59.97/(-0.52)$$

$$= 115.3 \text{ (mg/L)}$$

EXAMPLE 2: Unequal time increments, observed BOD data, t and y are given in Table 2.6. Find K_1 and L_a.

Solution:

Step 1. Calculate Δt, Δy, y', yy', and y^2; then complete Table 2.6.

From Eq. (2.21) (see Table 2.6)

$$*y'_i = \frac{(y_i - y_{i-1})\left(\frac{t_{i+1} - t_i}{t_i - t_{i-1}}\right) + (y_{i+1} - y_i)\left(\frac{t_i - t_{i-1}}{t_{i+1} - t_i}\right)}{t_{i+1} - t_{i-1}}$$

$$= \frac{(\Delta y_{i-1})\left(\frac{\Delta t_{i+1}}{\Delta t_{i-1}}\right) + (\Delta y_{i+1})\left(\frac{\Delta t_{i-1}}{\Delta t_{i+1}}\right)}{(\Delta t_{i-1}) + (\Delta t_{i+1})}$$

$$y'_1 = \frac{(28.8)\left(\frac{0.6}{0.4}\right) + (27.4)\left(\frac{0.4}{0.6}\right)}{0.4 + 0.6} = 61.47$$

TABLE 2.6 Various t and y Values

t	Δt	y	Δy	y'	yy'	y^2
0		0				
	0.4		28.8			
0.4		28.8		61.47*	1770.24	829.44
	0.6		27.4			
1		56.2		30.90	1736.75	3158.44
	0.5		9.3			
1.5		65.5		19.48	1276.00	4290.25
	0.7		14.5			
2.2		80.0		15.48	1238.48	6400.00
	0.8		7.6			
3		87.6		9.10	797.16	7673.76
	1		8.6			
4		96.2		7.40	711.88	9254.44
	1		6.2			
5		102.4		5.57	570.03	10485.76
	2		8.6			
7		111.0		3.55	394.05	12321.00
	2		5.6			
9		116.6		2.43	282.95	13595.56
	3		5.6			
12		122.2				
Sum		744.3		155.37	8777.53	68008.65

*See text for calculation of this value.

Step 2. Compute a and b; while $n = 9$

$$na + b\sum y - \sum y' = 0$$
$$9a + 744.3b - 155.37 = 0$$
$$a + 82.7b - 17.26 = 0 \quad (1)$$

and

$$a\sum y + b\sum y^2 - \sum yy' = 0$$
$$744.3a + 68{,}008.65b - 8777.53 = 0$$
$$a + 91.37b - 11.79 = 0 \quad (2)$$

Eq. (2) − Eq. (1)

$$8.67b + 5.47 = 0$$
$$b = -0.63$$

with Eq. (2)

$$a + 91.37(-0.63) - 11.79 = 0$$
$$a = 69.35$$

Step 3. Determine K_1 and L_a

$$K_1 = -b = 0.63 \text{ (per day)}$$
$$L_a = -a/b$$
$$= -69.35/-0.63$$
$$= 110.1 \text{ (mg/L)}$$

Moment method. This method requires that BOD measurements must be a series of regularly spaced time intervals. Calculations are needed for the sum of the BOD values, Σy, accumulated to the end of a series of time intervals and the sum of the product of time and observed BOD values, Σty, accumulated to the end of the time series.

The rate constant K_1 and the ultimate BOD L_a can then be easily read from a prepared graph by entering values of $\Sigma y/\Sigma ty$ on the appropriate scale. Treatments of BOD data with and without lag phase will be different. The authors (Moore *et al.*, 1950) presented three graphs for 3-, 5-, and 7-day sequences (Figs. 2.3, 2.4, and 2.5) with daily intervals for BOD value without lag phase. There is another chart presented for a 5-day sequence with lag phase (Fig. 2.6).

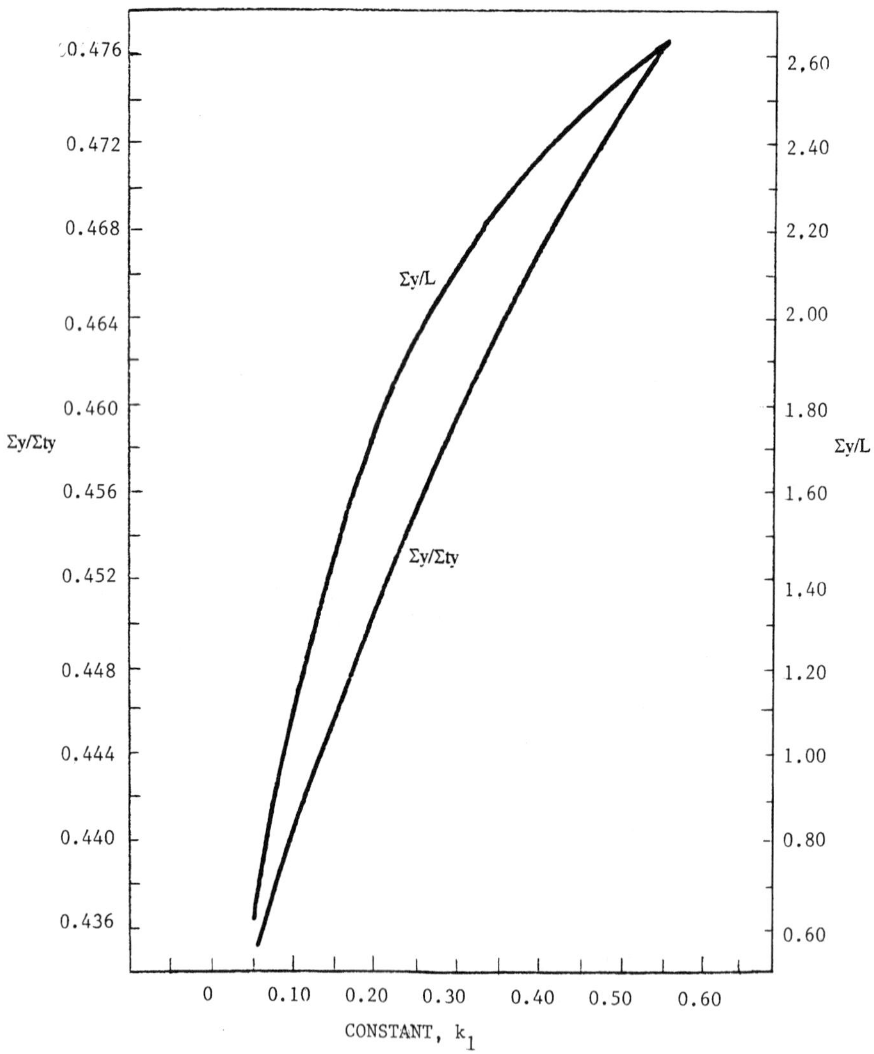

FIGURE 2.3 $\Sigma y/L$ and $\Sigma y/\Sigma ty$ for various values of k_1 in a 3-day sequence.

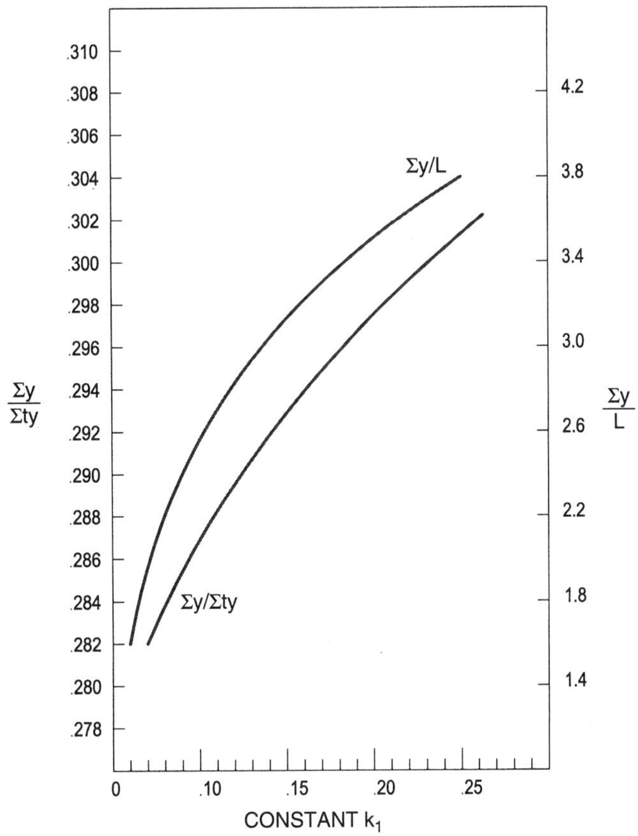

FIGURE 2.4 $\Sigma y/L$ and $\Sigma y/\Sigma ty$ for various values of k_1 in a 5-day sequence.

EXAMPLE 1: Use the BOD (without lag phase) on Example 1 of Thomas' slope method, find K_1 and L_a.

Solution:

Step 1. Calculate Σy and Σty (see Table 2.7).

Step 2. Compute $\Sigma y/\Sigma ty$

$$\Sigma y/\Sigma ty = 635.2/2786 = 0.228$$

Step 3. Find K_1 and L_a

On the 7-day time sequence graph (Fig. 2.5), enter the value 0.228 on the $\Sigma y/\Sigma ty$ scale, extend a horizontal line to the curve labeled $\Sigma y/\Sigma ty$, and from this point follow a vertical line to the k_1 scale. A value of $k_1 = 0.264$.

$$K_1 = 2.3026 \times k_1 = 2.3026 \times 0.264$$

$$= 0.608 \text{ (per day)}$$

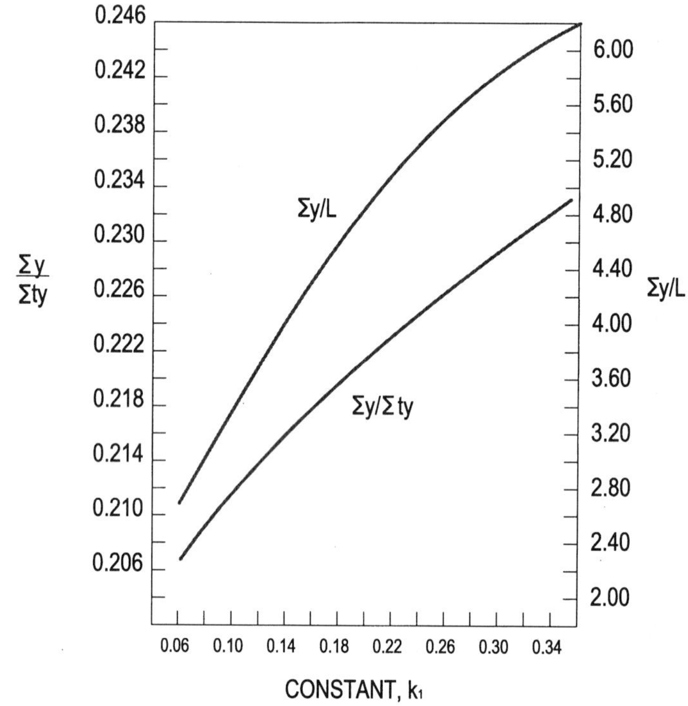

FIGURE 2.5 $\Sigma y/L$ and $\Sigma y/\Sigma ty$ for various values of k_1 in a 7-day sequence.

Extend the same vertical line to the curve labeled $\Sigma y/L_a$, obtaining a value of 5.81. Since

$$L_a = \Sigma y/5.81 = 635.2/5.81$$
$$= 109.3 \text{ (mg/L)}$$

The technique of the moment method for analyzing a set of BOD data containing a lag phase is as follows:

1. Compute Σt, Σy, Σty
2. Compute t^2 and t^2y and take the sum of the values of each quantity as Σt^2 and $\Sigma t^2 y$
3. Compute the derived quantity:

$$\frac{\Sigma ty/\Sigma t - \Sigma y/n}{\Sigma(t^2 y)/\Sigma t^2 - \Sigma y/n}$$

   in which n is the number of observations
4. Enter the above quantity on appropriate curve to find k_1
5. Project to other curves for values and solve equations for C and L_a. The BOD equation with log phase is expressed

$$y = L[1 - 10^{-k_1(t-t_0)}] \tag{2.14b}$$

$$y = L_a[1 - C10^{-k_1 t}] \tag{2.14c}$$

in which t_0 is the lag period and $C = 10^{k_1 t_0}$

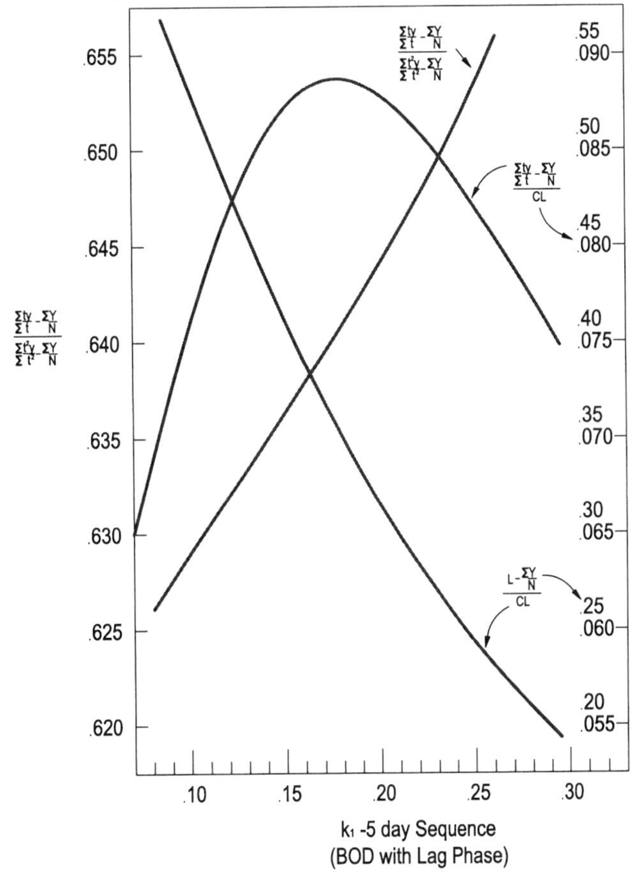

FIGURE 2.6 Curves for BOD computation with lag phase, based on a 5-day time sequence.

EXAMPLE 2: With a lag phase BOD data, BOD values are shown in Table 2.8. Find K_1 and L_a and complete an equation of the curve of best fit for the BOD data.

Solution:

Step 1. Compute Σt, Σy, Σty, Σt^2, and $\Sigma t^2 y$ (Table 2.8)

TABLE 2.7 Calculation of Σy and Σty

t	y	ty
1	56.2	56.2
2	74.4	148.8
3	87.6	262.8
4	96.2	384.8
5	102.4	512.0
6	107.4	644.4
7	111.0	777.0
Sum	635.2	2786.0

TABLE 2.8 Computations of Σt, Σy, Σty, Σt^2, and $\Sigma t^2 y$

t	y	ty	t^2	t^2y
1	12	12	1	12
2	74	148	4	296
3	101	303	9	909
4	126	504	16	2016
5	146	730	25	3650
15	459	1697	55	6883

Step 2. Compute some quantities

$$\Sigma ty/\Sigma t = 1697/15 = 113.13$$
$$\Sigma y/n = 459/5 = 91.8$$
$$\Sigma t^2 y/\Sigma t^2 = 6883/55 = 125.14$$

and

$$\frac{\Sigma ty/\Sigma t - \Sigma y/n}{\Sigma t^2 y/\Sigma t^2 - \Sigma y/n} = \frac{113.13 - 91.8}{125.14 - 91.8} = 0.640$$

Step 3. Enter Fig. 2.6 on the vertical axis labeled

$$\frac{\Sigma ty/\Sigma t - \Sigma y/n}{\Sigma t^2 y/\Sigma t^2 - \Sigma y/n}$$

with the value 0.640, and proceed horizontally to the diagonal straight line. Extend a vertical line to the axis labeled k_1 and read $k_1 = 0.173$ per day ($K_1 = 0.173$ per day $\times$ 2.3026 = 0.398 per day). Extend the vertical line to curve labeled

$$\frac{\Sigma ty/\Sigma t - \Sigma y/n}{CL}$$

and proceed horizontally from the intersection to the scale at the far right; read

$$\frac{\Sigma ty/\Sigma t - \Sigma y/n}{CL} = 0.0878$$

Step 4. Find CL

$$CL = \frac{113.13 - 91.8}{0.0878} = 242.94$$

Step 5. Find L_a, $L_a = L$ for this case. Continue to the same vertical line to the curve labeled

$$\frac{L - \Sigma y/n}{CL}$$

and read horizontally on the inside right-hand scale

$$\frac{L - \Sigma y/n}{CL} = 0.353$$

Then

$$L = L_a = 0.353(242.94) + 91.8$$
$$= 177.6 \text{ (mg/L)}$$

Also

$$C = \frac{CL}{L} = \frac{242.94}{177.6} = 1.368$$

and lag period

$$t_0 = \frac{1}{k_1}\log_{10} C$$

$$= \frac{1}{0.173}\log_{10} 1.368$$

$$= 0.787 \text{ (days)}$$

Step 6. Write complete equation of the best fit for the data

$$y = 177.6\left[1 - 10^{-0.173(t-0.787)}\right]$$

or

$$y = 177.6\left[1 - e^{-0.398(t-0.787)}\right]$$

Logarithmic formula. Orford and Ingram (1953) reported that there is a relationship between the observed BOD from domestic sewage and the logarithm of the time of observation. If the BOD data are plotted against the logarithm of time, the resultant curve is approximately a straight line. The general equation is expressed as

$$y_t = m \log t + b \tag{2.22}$$

where m is the slope of the line and b is a constant (the intercept).

The general equation can be transformed by dividing each side by the 5-day BOD (or BOD_5) intercept of the line to give

$$\frac{y_t}{s} = \frac{m}{s}\log t + \frac{b}{s}$$

$$\frac{y_t}{s} = M \log t + B \tag{2.22a}$$

or

$$y_t = s(M \log t + B) \tag{2.22b}$$

where s = BOD_5 intercept of the line
M = m/s, BOD rate parameter
B = b/s, BOD rate parameter

For domestic sewage oxidation at 20°C, the straight line through the observed plotted points, when extrapolated to the log t axis, intercepts the log t axis at 0.333 days. The general equation is

$$y_t = s(0.85 \log t + 0.41) \tag{2.23a}$$

where 0.85 and 0.41 are the BOD rate parameters for domestic sewage.

For any observed BOD curve with different oxidation time, the above equation may generalize as

$$y_t = S(0.85 \log at + 0.41) \tag{2.23b}$$

where S = BOD intercept (y-axis) of the line at $5/a$ days
= 5-day BOD at the standardized domestic sewage oxidation rate, when $a = 1$
= strength factor
a = log t at x-axis intercept of normal domestic sewage BOD curve divided by the x-axis intercept of
$= \frac{0.333}{x\text{-axis intercept}}$
= 1 for a standardized domestic sewage BOD_5 curve

The oxidation rate

$$\frac{dy}{dt} = \frac{0.85S}{t} = K_1(L_a - y_t) \tag{2.24a}$$

or

$$\frac{0.85S}{2.303t} = k_1(L_a - y_t) \tag{2.24b}$$

For the logarithmic formula method you need to determine the two constants S and a. Observed BOD data are plotted on semilogarithmic graph paper. Time in days is plotted on a logarithmic scale on the x-axis and percent of 5-day BOD on a regular scale on the y-axis. The straight line of best fix is drawn through the plotted points. The time value of the x-axis intercept is then read, x_1 (for a standardized domestic sewage sample, this is 0.333). The a value can be calculated by $a = 0.333/x_1$.

EXAMPLE: At 20°C, $S = 95$ mg/L and $x_1 = 0.222$ days. What is the BOD equation and K_1 for the sample?

Solution:

Step 1. Compute a

$$\begin{aligned} a &= 0.333/x_1 = 0.333/0.222 \\ &= 1.5 \end{aligned}$$

Step 2. Determine the simplified Eq. (2.23b)

$$\begin{aligned} y_t &= S(0.85 \log at + 0.41) \\ &= 95(0.85 \log 1.5t + 0.41) \end{aligned}$$

Step 3. Compute the oxidation rate at $t = 5$ days, using Eq. (2.24a)

$$\begin{aligned} \frac{dy}{dt} &= \frac{0.85S}{t} = \frac{0.85 \times 95 \text{ mg/L}}{5 \text{ days}} \\ &= 16.15 \text{ mg/(L}\cdot\text{d)} \end{aligned}$$

Step 4. Determine L_a when $t = 20$ days, using Eq. (2.24b) from Step 2

$$\begin{aligned} L_a = y_{20} &= 95(0.85 \log (1.5 \times 20) + 0.41) \\ &= 158 \text{ (mg/L)} \end{aligned}$$

Step 5. Compute K_1

$$\frac{dy}{dt} = K_1(L_a - y_5)$$

$$\begin{aligned} K_1 &= \frac{16.15 \text{ mg/(L}\cdot\text{d)}}{(158 - 95) \text{ mg/L}} \\ &= 0.256 \text{ per day} \end{aligned}$$

Rapid methods. Sheehy (1960) developed two rapid methods for solving first-order BOD equations. The first method, using the 'BOD slide rule,' is applicable to the ratios of observations on whole day intervals, from 1 to 8 days, to the 5-day BOD. The second method, the graphical method, can be used with whole or fractional day BOD values.

The BOD slide rule consists of scales A and B for multiplication and division. Scales C and D are the values of $1 - 10^{-k_1 t}$ and $10^{-k_1 t}$ plotted in relation to values of $k_1 t$ on the B scale. From given BOD values, the k_1 value is easily determined from the BOD slide rule. Unfortunately, the BOD slide rule is not on the market.

The graphical method is based on the same principles as the BOD slide rule method. Two figures are used to determine k_1 values directly. One figure contains k_1 values based on the ratios of BOD at time t(BOD_t) to the 5-day BOD for t less than 5 days. The other figure contains k_1 values based on the ratio of BOD_t to BOD_5 at times t greater than 5 days. The value of k_1 is determined easily from any one of these figures based on the ratio of BOD_t to BOD_5.

8.3 Temperature Effect on K_1

A general expression of the temperature effect on the deoxygenation coefficient (rate) is

$$\frac{K_{1a}}{K_{1b}} = \theta^{(T_a - T_b)} \tag{2.25}$$

where K_{1a} = reaction rate at temperature T_a, per day
K_{1b} = reaction rate at temperature T_b, per day
θ = temperature coefficient

On the basis of experimental results over the usual range of river temperature, θ is accepted as 1.047. Therefore the BOD reaction rate at any T in Celsius (working temperature), deviated from 20°C, is

$$K_{1(T)} = K_{1(20°\text{C})} \times 1.047^{(T-20)} \tag{2.26}$$

or

$$k_{1(T)} = k_{1(20°\text{C})} \times 1.047^{(T-20)} \tag{2.26a}$$

Thus

$$L_{\text{a}(T)} = L_{\text{a}(20°\text{C})}[1 + 0.02(T - 20)] \tag{2.27}$$

or

$$L_{\text{a}(T)} = L_{\text{a}(20°\text{C})}(0.6 + 0.02T) \tag{2.28}$$

EXAMPLE 1: A river water sample has $k_1 = 0.10$ (one base 10) and $L_\text{a} = 280$ mg/L at 20°C. Find k_1 and L_a at temperatures 14°C and 29°C.

Solution:

Step 1. Using Eq. (2.26a)

$$k_{1(T)} = k_{1(20°\text{C})} \times 1.047^{(T-20)}$$

at 14°C

$$\begin{aligned} k_{1(14°\text{C})} &= k_{1(20°\text{C})} \times 1.047^{(14-20)} \\ &= 0.10 \times 1.047^{-6} \\ &= 0.076 \text{ (per day)} \end{aligned}$$

at 29°C

$$\begin{aligned} k_{1(29°\text{C})} &= k_{1(20°\text{C})} \times 1.047^{(29-20)} \\ &= 0.10 \times 1.047^{9} \\ &= 0.15 \text{ (per day)} \end{aligned}$$

Step 2. Find $L_{a(T)}$, using Eq. (2.28)

$$L_{a(T)} = L_{a(20°C)}(0.6 + 0.02T)$$

at 14°C

$$\begin{aligned} L_{a(14°C)} &= L_{a(20°C)}(0.6 + 0.02T) \\ &= 280(0.6 + 0.02 \times 14) \\ &= 246 \text{ (mg/L)} \end{aligned}$$

at 29°C

$$\begin{aligned} L_{a(29°C)} &= 280(0.6 + 0.02 \times 29) \\ &= 330 \text{ (mg/L)} \end{aligned}$$

The ultimate BOD and K_1 values found in the laboratory at 20°C have to be adjusted to river temperatures using the above formulas. Three types of BOD can be determined: i.e. total, carbonaceous, and nitrogenous. TBOD (uninhibited) and CBOD (inhibited with trichloromethylpyridine for nitrification) are measured directly, while NBOD can be computed by subtracting CBOD values from TBOD values for given time elements.

EXAMPLE 2: Tables 2.9 and 2.10 show typical long-term (20-day) BOD data for the Upper Illinois Waterway downstream of Lockport (Butts and Shackleford, 1992). The graphical plots of these BOD progressive curves are presented in Figs. 2.7–2.9. Explain what are their unique characteristics.

TABLE 2.9 Biochemical Oxygen Demand at 20°C in the Upper Illinois Waterway

Station sample:	Lockport 18			Station sample:	Lockport 36		
Date:	01/16/90			Date:	09/26/90		
pH:	7.03			pH:	6.98		
Temp:	16.05°C			Temp:	19.10°C		
Time, days	TBOD, mg/L	CBOD, mg/L	NBOD, mg/L	Time, days	TBOD, mg/L	CBOD, mg/L	NBOD, mg/L
0.89	1.09	0.98	0.11	0.72	0.19	0.18	1.01
1.88	1.99	1.73	0.25	1.69	0.66	0.18	0.48
2.87	2.74	2.06	0.68	2.65	0.99	0.92	0.07
3.87	3.39	2.25	1.14	3.65	1.38	1.25	0.13
4.87	4.60	3.08	1.52	4.72	1.81	1.25	0.56
5.62	5.48	3.45	2.04	5.80	2.74	1.62	1.12
6.58	7.41	4.17	3.24	6.75	3.51	2.09	1.43
7.62	9.87	4.83	5.04	7.79	4.17	2.53	1.64
8.57	12.62	5.28	7.34	8.67	4.65	2.92	1.74
9.62	16.45	6.15	10.30	9.73	5.59	2.92	2.67
10.94	21.55	6.62	14.94	10.69	6.22	3.08	3.14
11.73	25.21	6.96	18.25	11.74	6.95	3.08	3.88
12.58	29.66	7.26	22.40	12.66	7.28	3.12	4.16
13.62	34.19	7.90	26.29	13.69	7.79	3.56	4.23
14.67	35.68	8.14	27.55	14.66	7.79	3.56	4.23
15.67	36.43	8.38	28.05	15.65	8.36	3.56	4.80
16.58	37.08	8.75	28.33	16.64	8.56	3.57	4.99
17.81	37.45	8.92	28.53	17.76	8.80	3.82	4.99
18.80	37.64	9.16	28.47	17.67	8.80	3.82	4.98
19.90	38.19	9.39	28.79	19.66	8.92	3.87	5.05
20.60	38.56	9.55	29.01				

TABLE 2.10 Biochemical Oxygen Demand at 20°C in the Kankakee River

Station sample:	Kankakee 11
Date:	08/13/90
pH:	8.39
Temp:	23.70°C

Time, days	TBOD, mg/L	CBOD, mg/L	NBOD, mg/L
0.78	0.81	0.44	0.37
2.79	1.76	1.13	0.63
3.52	2.20	1.26	0.93
6.03	2.42	1.62	0.79
6.77	2.83	1.94	0.89
7.77	3.20	2.22	0.98
8.76	3.54	2.49	1.06
9.49	3.83	2.81	1.02
10.76	4.19	3.09	1.10
13.01	4.85	3.76	1.10
13.43	5.08	3.94	1.14
14.75	5.24	4.02	1.22
15.48	5.57	4.42	1.15
17.73	6.64	5.30	1.17
19.96	7.03	5.85	1.19

Solution:

These three sets of data have several unique characteristics and considerations. Figure 2.7 demonstrates that S-shaped NBOD and TBOD curves exist at Lockport station (Illinois) that fit the mathematical formula represented by Eq. (2.8). These curves usually occur at Lockport during cold weather periods, but not always.

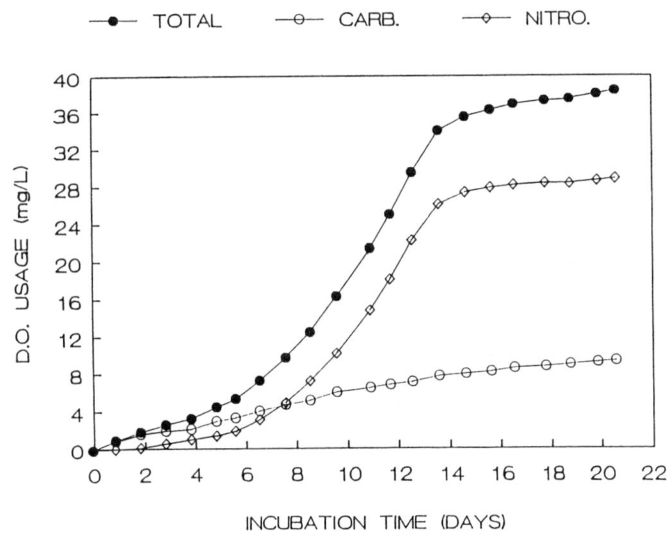

FIGURE 2.7 BOD progressive curves for Lockport 18, January 16, 1990 (*Butts and Shackleford, 1992*).

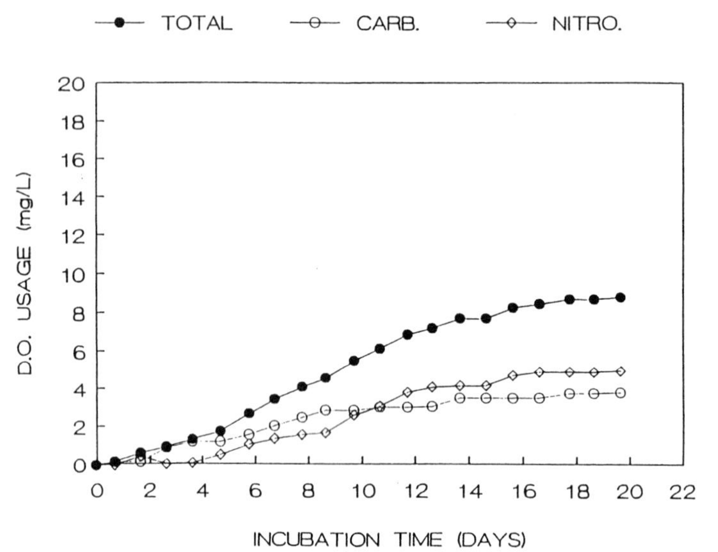

FIGURE 2.8 BOD progressive curves for Lockport 36, September 26, 1990 (*Butts and Shackleford, 1992*).

Figure 2.8 illustrates Lockport warm weather BOD progression curves. In comparisons with Figs. 2.7 and 2.8, there are extreme differences between cold and warm weather BOD curves. The January $NBOD_{20}$ represents 75.3% of the $TBOD_{20}$, whereas the September $NBOD_{20}$ contains only 56.6% of the $TBOD_{20}$. Furthermore, the September $TBOD_{20}$ is only 23.3% as great as the January $TBOD_{20}$ (see Table 2.9).

Figure 2.9 shows most tributary (Kankakee River) BOD characteristics. The warm weather $TBOD_{20}$ levels for the tributary often approach those observed at Lockport station (7.03 mg/L versus 8.92 mg/L), but the fraction of $NBOD_{20}$ is much less (1.19 mg/L versus 5.05 mg/L) (Tables 2.9 and 2.10). Nevertheless, the $TBOD_{20}$ loads coming from the tributaries are usually much less than those originating from Lockport since the tributary flows are normally much lower.

8.4 Second-order Reaction

In many cases, researchers stated that a better fit of BOD data can be obtained by using a second order chemical reaction equation, i.e. the equation of a rotated rectangular hyperbola. A second-order chemical reaction is characterized by a rate of reaction dependent upon the concentration of two reactants. It is defined as (Young and Clark, 1965):

$$-\frac{dC}{dt} = KC^2 \tag{2.29}$$

When applying this second-order reaction to BOD data, K is a constant; C becomes the initial substance concentration; L_a, minus the BOD; y, at any time, t; or

$$-\frac{d(L_a - y)}{dt} = K(L_a - y)^2 \tag{2.30}$$

rearranging

$$\frac{d(L_a - y)}{(L_a - y)^2} = -K dt$$

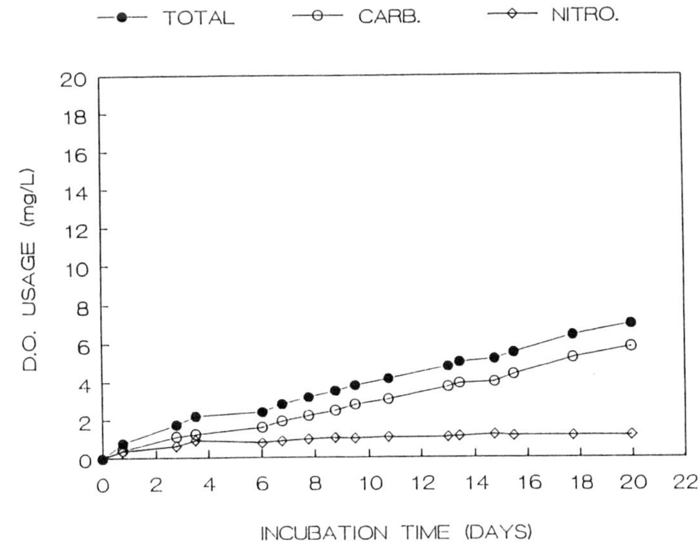

FIGURE 2.9 BOD progressive curves for Kankakee 10, August 13, 1990 (*Butts and Shackleford, 1992*).

Integrating yields

$$\int_{y=0}^{y=y} \frac{\mathrm{d}(L_a - y)}{(L_a - y)^2} = \int_{t=0}^{t=t} -K\mathrm{d}t$$

or

$$\frac{1}{L_a} - \frac{1}{L_a - y} = -Kt \tag{2.31}$$

Multiplying each side of the equation by L_a and rearranging,

$$1 - \frac{L_a}{L_a - y} = -KL_a t$$

$$\frac{L_a - y - L_a}{L_a - y} = -KL_a t$$

$$\frac{-y}{L_a - y} = -KL_a t$$

$$y = KL_a t(L_a - y) \tag{2.32}$$

$$y = KL_a^2 t - KL_a ty$$

$$(1 + KL_a t)y = KL_a^2 t$$

$$y = \frac{KL_a^2}{1 + KL_a t} t$$

$$y = \frac{t}{\frac{1}{KL_a^2} + \frac{1}{L_a} t} \tag{2.33}$$

or

$$y = \frac{t}{a + bt} \tag{2.34}$$

where a is $1/KL_a^2$ and b represents $1/L_a$. The above equation is in the form of a second-order reaction equation for defining BOD data. The equation can be linearized in the form

$$\frac{t}{y} = a + bt \tag{2.35}$$

in which a and b can be solved by a least-squares analysis. The simultaneous equations for the least-squares treatment are as follows:

$$\sum\left(a + bt - \frac{t}{y}\right) = 0$$

$$\sum\left(a + bt - \frac{t}{y}\right)t = 0$$

$$at + b\sum t - \frac{\sum t}{y} = 0$$

$$a\sum t + b\sum t^2 - \frac{\sum t^2}{y} = 0$$

$$a + b\left(\frac{\sum t}{t}\right) - \left(\frac{\sum t/y}{t}\right) = 0$$

$$a + b\left(\frac{\sum t^2}{\sum t}\right) - \left(\frac{\sum t^2/y}{\sum t}\right) = 0$$

To solve a and b using five data points, $t = 5$, $\Sigma t = 15$, and $\Sigma t^2 = 55$; then

$$a + b\left(\frac{15}{5}\right) - \left(\frac{\sum t/y}{5}\right) = 0$$

$$a + b\left(\frac{55}{15}\right) - \left(\frac{\sum t^2/y}{15}\right) = 0$$

Solving for b

$$\left(\frac{55}{15} - \frac{15}{5}\right)b - \left[\left(\frac{\sum t^2/y}{15}\right) - \left(\frac{\sum t/y}{5}\right)\right] = 0$$

$$\left(\frac{55 - 45}{15}\right)b - \frac{1}{15}\left(\frac{\sum t^2}{y} - \frac{3\sum t}{y}\right) = 0$$

$$10\text{b} = \frac{\sum t^2}{y} - \frac{3\sum t}{y}$$

or

$$b = 0.10\left(\frac{\sum t^2}{y} - \frac{3\sum t}{y}\right) \tag{2.36}$$

Solving for a, by substituting b in the equation

$$\begin{aligned} a &= \frac{1}{5}\left(\frac{\sum t}{y} - 15b\right) \\ &= \frac{1}{5}\left[\frac{\sum t}{y} - 15 \times 0.1\left(\frac{\sum t^2}{y} - \frac{3\sum t}{y}\right)\right] \\ &= \frac{1}{5}\left(\frac{5.5\sum t}{y} - \frac{1.5\sum t^2}{y}\right) \\ &= 1.1\left(\frac{\sum t}{y}\right) - 0.3\left(\frac{\sum t^2}{y}\right) \end{aligned} \tag{2.37}$$

The velocity of reaction for the BOD curve is y/t. The initial reaction is $1/a$ (in mg/(L · d)) and is the maximum velocity of the BOD reaction denoted as v_m.

The authors (Young and Clark, 1965) claimed that a second-order equation has the same precision as a first-order equation at both 20°C and 35°C.

Dougal and Baumann (1967) modified the second-order equation for BOD predication as

$$y = \frac{t}{a + bt} = \frac{t/bt}{\frac{a}{bt} + 1} = \frac{1/b}{\frac{a}{bt} + 1} = \frac{L_a}{\frac{1}{K'\,t} + 1} \tag{2.38}$$

where

$$L_a = 1/b \tag{2.39}$$

$$K' = b/a \tag{2.40}$$

This formula has a simple rate constant and an ultimate value of BOD. It can also be transformed into a linear function for regression analysis, permitting the coefficients a and b to be determined easily.

EXAMPLE: The laboratory BOD data for wastewater were 135, 198, 216, 235, and 248 mg/L at days 1, 2, 3, 4, and 5, respectively. Find K, L_a, and v_m for this wastewater.

Solution:

Step 1. Construct a table for basic calculations (Table 2.11).

Step 2. Solve a and b using Eqs. (2.36) and (2.37)

$$\begin{aligned} b &= 0.10\left(\frac{\sum t^2}{y} - \frac{3\sum t}{y}\right) \\ &= 0.10(0.238167 - 3 \times 0.068579) \\ &= 0.003243 \\ a &= 1.1\left(\frac{\sum t}{y}\right) - 0.3\left(\frac{\sum t^2}{y}\right) \\ &= 1.1 \times 0.068579 - 0.3 \times 0.238167 \\ &= 0.003987 \end{aligned}$$

Step 3. Calculate L_a, K, v_m, and K' using Eqs. (2.38)–(2.40).

$$\begin{aligned} L_a &= 1/b \\ &= 1/0.003243 \\ &= 308 \text{ (mg/L)} \\ K &= 1/aL_a^2 \\ &= 1/0.003987(308)^2 \\ &= 0.00264 \text{ (per mg/(L} \cdot \text{d))} \\ v_m &= 1/a \\ &= 1/0.003987 \text{ mg/(L} \cdot \text{d)} \\ &= 250.8 \text{ mg/(L} \cdot \text{d)} \\ &= 10.5 \text{ mg/(L} \cdot \text{h)or} \end{aligned}$$

or

$$\begin{aligned} K' &= b/a \\ &= 0.003243/0.003987 \\ &= 0.813 \text{ (per day)} \end{aligned}$$

TABLE 2.11 Data for Basic Calculations for Step 1

t	y	t/y	t^2/y
1	135	0.007407	0.007407
2	198	0.010101	0.020202
3	216	0.013889	0.041667
4	235	0.017021	0.068085
5	248	0.020161	0.100806
Sum		0.068579	0.238167

9 DETERMINATION OF REAERATION RATE CONSTANT K_2

9.1 Basic Conservation

For a stream deficient in DO but without BOD load, the classical formula is

$$\frac{dD}{dt} = -K_2 D \tag{2.41}$$

where dD/dt is the absolute change in DO deficit D over an increment of time dt due to an atmospheric exchange of oxygen at air/water interface; K_2 is the reaeration coefficient, per day; and D is DO deficit, mg/L.

Integrating the above equation from t_1 to t_2 gives

$$\int_{D_1}^{D_2} \frac{dD}{D} = -K_2 \int_{t_1}^{t_2} dt \qquad \text{or} \qquad \int_{D_a}^{D_t} \frac{dD}{D} = -K_2 \int_0^t dt$$

$$\ln \frac{D_2}{D_1} = -K_2(t_2 - t_1) \qquad \text{or} \qquad \ln \frac{D_t}{D_a} = -K_2 t$$

$$\ln \frac{D_2}{D_1} = -K_2\, \Delta t \qquad \text{or} \qquad D_t = D_a\, e^{-K_2 t}$$

$$K_2 = -\frac{\ln \frac{D_2}{D_1}}{\Delta t} \tag{2.42}$$

The K_2 values are needed to correct for river temperature according to the equation

$$K_{2(@T)} = K_{2(@20)}(1.02)^{T-20} \tag{2.43}$$

where $K_{2(@T)} = K_2$ value at any temperature T°C and $K_{2(@20)} = K_2$ value at 20°C.

EXAMPLE 1: The DO deficits at upstream and downstream stations are 3.55 and 2.77 mg/L, respectively. The time of travel in this stream reach is 0.67 days. The mean water temperature is 26.5°C. What is the K_2 value for 20°C?

Solution:

Step 1. Determine K_2 at 26.5°C

$$\begin{aligned} K_{2(@26.5)} &= -\frac{\ln \frac{D_2}{D_1}}{\Delta t} \\ &= -\left(\ln \frac{2.77}{3.55}\right) \Big/ 0.67 \\ &= -(-0.248)/0.67 \\ &= 0.37 \text{ (per day)} \end{aligned}$$

Step 2. Calculate $K_{2(@20)}$

$$K_{2(@T)} = K_{2(@20)}(1.02)^{T-20}$$

therefore

$$\begin{aligned} K_{2(@20)} &= 0.37/(1.02)^{26.5-20} = 0.37/1.137 \\ &= 0.33 \text{ (per day)} \end{aligned}$$

9.2 From BOD and Oxygen Sag Constants

In most stream survey studies involving oxygen sag equations (discussed later), the value of the reoxygenation constant (K_2) is of utmost importance. Under different conditions, several methods for determining K_2 are listed below:

I. K_2 may be computed from the oxygen sag equation, if all other parameters are known; however, data must be adequate to support the conclusions. A tria-and-error procedure is generally used.

II. The amount of reaeration (r_m) in a reach (station A to station B) is equal to the BOD exerted ($L_{aA} - L_{aB}$) plus oxygen deficiency from station A to station B, ($D_A - D_B$). The relationship can be expressed as

$$r_m = (L_{aA} - L_{aB}) + (D_A - D_B) \tag{2.44}$$

$$K_2 = r_m/D_m \tag{2.45}$$

where r_m is the amount of reaeration and D_m is the mean (average) deficiency.

EXAMPLE 2: Given L_a for the upper and lower sampling stations are 24.6 and 15.8 mg/L, respectively; the DO concentration at these two stations are 5.35 and 5.83 mg/L, respectively; and the water temperature is 20°C; find the K_2 value for the river reach.

Solution:

Step 1. Find D_A, D_B, and D_m at 20°C.

$$\begin{aligned} DO_{sat} \text{ at } 20°C &= 9.02 \text{ mg/L (Table 2.2)} \\ D_A &= 9.02 - 5.35 = 3.67 \text{ (mg/L)} \\ D_B &= 9.02 - 5.83 = 3.19 \text{ (mg/L)} \\ D_m &= (3.67 + 3.19)/2 = 3.43 \text{ (mg/L)} \end{aligned}$$

Step 2. Calculate r_m and K_2

$$\begin{aligned} r_m &= (L_{aA} - L_{aB}) + (D_A - D_B) \\ &= (24.6 - 15.8) + (3.67 - 3.19) \\ &= 9.28 \text{ (mg/L)} \\ K_2 &= r_m/D_m = 9.28/3.43 \\ &= 2.71 \text{ (per day)} \end{aligned}$$

III. For the case, where DO = 0 mg/L for a short period of time and without anaerobic decomposition (O'Connor, 1958),

$$K_2 D_{max} = K_2 C_s \tag{2.46}$$

where D_{max} is the maximum deficit, mg/L, and C_s is the saturation DO concentration, mg/L. The maximum deficiency is equal to the DO saturation concentration and the oxygen transferred is oxygen utilized by organic matter.

Organic matter utilized $= L_{aA} = L_{aB}$, during the time of travel t; therefore

$$\text{rate of exertion} = (L_{aA} - L_{aB})/t = K_2 D_{max}$$

and

$$(L_{aA} - L_{aB})/t = K_2 C_s$$

then

$$K_2 = (L_{aA} - L_{aB})/C_s t \tag{2.47}$$

EXAMPLE 3: Given that the water temperature is 20°C; L_{aA} and L_{aB} are 18.3 and 13.7 mg/L, respectively; and the time of travel from station A to station B is 0.123 days. Compute K_2.

Solution: At 20°C,

$$\begin{aligned} C_s &= 9.02 \text{ mg/L} \\ K_2 &= (L_{aA} - L_{aB})/C_s t \\ &= (18.3 - 13.7)/(9.02 \times 0.123) \\ &= 4.15 \text{ (per day)} \end{aligned}$$

IV. At the critical point in the river (O'Connor, 1958)

$$\frac{dD}{dt} = 0 \tag{2.48}$$

and

$$K_d L_c = K_2 D_c \tag{2.49}$$

$$K_2 = K_d L_c / D_c \tag{2.50}$$

where K_d is the deoxygenation rate in stream conditions, L_c is the first-stage ultimate BOD at critical point, and D_c is the DO deficit at the critical point. These will be discussed in a later section.

V. Under steady-state conditions, at a sampling point (O'Connor, 1958):

$$\frac{dD}{dt} = 0 \tag{2.48}$$

or at this point

$$K_2 D = K_d L \tag{2.51}$$

The value of D can be obtained from a field measurement. The value of $K_d L$ can be obtained by measuring oxygen uptake from a sample taken from a given point on a river. Thus, K_2 can be computed from the above equation.

9.3 Empirical Formulas

The factors in the Streeter–Phelps equation has stimulated much research on the reaeration rate coefficient K_2. Considerable controversy exists as to the proper method of formulas to use. Several empirical and semi-empirical formulas have been developed to estimate K_2, almost all of which relate stream velocity and water depth to K_2, as first proposed by Streeter (1926). Three equations below are widely known and employed in stream studies:

$$K_2 = \frac{13.0V^{0.5}}{H^{1.5}} \quad \text{O'Connor and Dobbins (1958)} \tag{2.52}$$

$$K_2 = \frac{11.57V^{0.969}}{H^{1.673}} \quad \text{Churchill } et\ al. \text{ (1962)} \tag{2.53}$$

$$K_2 = \frac{7.63V}{H^{1.33}} \quad \text{Langbein and Durum (1967)} \tag{2.54}$$

where K_2 = reaeration rate coefficient, per day
V = average velocity, ft/s (fps)
H = average water depth, ft

On the basis of K_2-related physical aspects of a stream, O'Connor (1958) and Eckenfelder and O'Connor (1959) proposed K_2 formulas as follows:
for an isotropic turbulence (deep stream)

$$K_2 = \frac{(D_L V)^{0.5}}{2.3\, H^{1.5}} \tag{2.55}$$

for a nonisotropic stream

$$K_2 = \frac{480 D_L^{0.5} S^{0.25}}{H^{1.25}} \tag{2.56}$$

where D_L = diffusivity of oxygen in water, ft²/d
H = average depth, ft
S = slope
V = velocity of flow, fps = $B\sqrt{(HS)}$ (nonisotropic if $B < 17$; isotropic if $B > 17$)

The O'Connor and Dobbins (1958) equation, (Eq. (2.52)), is based on a theory more general than the formula developed by several of the other investigators. The formula proposed by Churchill (Eq. (2.53)) appears to be more restrictive in use. The workers from the US Geological Survey (Langbein and Durum, 1967) have analyzed and summarized the work of several investigators and concluded that the velocities and cross-section information were the most applicable formulation of the reaeration factor.

EXAMPLE 4: A stream has an average depth of 9.8 ft (2.99 m) and velocity of flow of 0.61 ft/s (0.20 m/s). What are the K_2 values determined by the first three empirical formulas in this section (Eqs. (2.52)–(2.54))? What is K_2 at 25°C of water temperature (use Eq. (2.54) only).

Solution:

Step 1. Determine K_2 at 20°C
(a) by the O'Conner and Dobbins formula (Eq. (2.52))

$$K_2 = \frac{13.0V^{0.5}}{H^{1.5}} = \frac{13.0 \times (0.61)^{0.5}}{(9.8)^{1.5}}$$
$$= 0.33 \text{ (per day)}$$

(b) by the Churchill *et al.* formula (Eq. (2.53))

$$K_2 = \frac{11.57 \times (0.61)^{0.969}}{(9.8)^{1.673}} = 0.16 \text{ (per day)}$$

(c) by the Langbein and Durum formula (Eq. (2.54))

$$K_2 = \frac{7.63 \times 0.61}{(9.8)^{1.33}} = 0.22 \text{ (per day)}$$

Step 2. For a temperature of 25°C

$$\begin{aligned} K_{2(@25)} &= K_{2(@20)}(1.02)^{T-20} \\ &= 0.22(1.02)^{25-20} \\ &= 0.24 \text{ (per day)} \end{aligned}$$

9.4 Stationary Field Monitoring Procedure

Larson *et al.* (1994) employed physical, biochemical, and biological factors to analyze the K_2 value. Estimations of physical reaeration and the associated effects of algal photosynthetic oxygen production (primary productivity) are based on the schematic formulations as follows:

$$\text{Physical aeration} = \text{ambient DO} - \text{light chamber DO} + \text{SOD}$$

Algal productivity:

$$\text{Gross} = \text{light chamber DO} - \text{dark chamber DO}$$

$$\text{Net} = \text{light chamber DO at end} - \text{light chamber DO at beginning}$$

The amount of reaeration (REA) is computed with observed data from DataSonde as

$$\text{REA} = C_2 - C_1 - \text{POP} - \text{PAP} + \text{TBOD} + \text{SOD} \tag{2.57}$$

where REA = reaeration (= $D_2 - D_1 = dD$), mg/L
C_2 and C_1 = observed DO concentrations in mg/L at t_2 and t_1, respectively
POP = net periphytonic (attached algae) oxygen production for the time period ($t_2 - t_1 = \Delta t$), mg/L
PAP = net planktonic algae (suspended algae) oxygen production for the time period $t_2 - t_1$, mg/L
TBOD = total biochemical oxygen demand (usage) for the time period $t_2 - t_1$, mg/L
SOD = net sediment oxygen demand for the time period $t_2 - t_1$, mg/L

The clear periphytonic chamber is not used if periphytonic productivity/respiration is deemed insignificant, i.e. POD = 0. The combined effect of TBOD − PAP is represented by the gross output of the light chamber, and the SOD is equal to the gross SOD chamber output less than dark chamber output.

The DO deficit D in the reach is calculated as

$$D = \frac{S_1 + S_2}{2} - \frac{C_1 + C_2}{2} \tag{2.58}$$

where S_1 and S_2 are the DO saturation concentration in mg/L at t_1 and t_2, respectively, for the average water temperature (T) in the reach. The DO saturation formula is given in the preview section (Eq. (2.4)).

Since the Hydrolab's DataSondes logged data at hourly intervals, the DO changes attributable to physical aeration (or deaeration) are available for small time frames, which permits the following modification of the basic natural reaeration equation to be used to calculate K_2 values (Broeren *et al.*, 1991)

$$\sum_{i=1}^{24}(R_{i+1} - R_i) = -K_2\sum_{i=1}^{24}\left(\frac{S_i + S_{i+1}}{2} - \frac{C_i + C_{i+1}}{2}\right) \tag{2.59}$$

where

R_i = an ith DO concentration from the physical aeration DO-used "mass diagram" curve
R_{i+1} = a DO concentration 1 hour later than R_i on the physical reaeration DO-used curve
S_i = an ith DO saturation concentration

S_{i+1} = a DO saturation concentration 1 hour later than S_i
C_i = an ith-observed DO concentration
C_{i+1} = an observed DO concentration 1 hour later than C_i

Note: All units are in mg/L.

Thus, the reaeration rate can be computed by

$$K_2 = \frac{\text{REA}}{D(t_2 - t_1)} \tag{2.60}$$

Both the algal productivity/respiration (P/R) rate and SOD are biologically associated factors which are normally expressed in terms of grams of oxygen per square meter per day (g/(m^2 · day)). Conversion for these areal rates to mg/L of DO usage for use in computing physical aeration (REA) is accomplished using the following formula (Butts *et al*., 1975):

$$U = \frac{3.28Gt}{H} \tag{2.61}$$

where U = DO usage in the river reach, mg/L
G = SOD or P/R rate, g/(m^2 · day)
t = time of travel through the reach, days
H = average depth, ft

EXAMPLE 5: Given that C_1 = 6.85 mg/L, C_2 = 7.33 mg/L, POP = 0, at T = 20°C, gross DO output in light chamber = 6.88 mg/L, dark chamber DO output = 5.55 mg/L, gross SOD chamber output = 6.15 mg/L, and water temperature at beginning and end of field monitoring = 24.4 and 24.6°C, respectively, $t_2 - t_1$ = 2.50 days, calculate K_2 at 20°C.

Solution:

Step 1. Determine REA

$$\text{TBOD} - \text{PAP} = \text{gross} = 6.88 \text{ mg/L} - 5.55 \text{ mg/L} = 1.33 \text{ mg/L}$$
$$\text{SOD} = 6.15 \text{ mg/L} - 5.55 \text{ mg/L} = 0.60 \text{ mg/L}$$

From Eq. (2.57)

$$\begin{aligned}\text{REA} &= C_2 - C_1 - \text{POP} + (-\text{PAP} + \text{TBOD}) + \text{SOD} \\ &= 7.33 = 6.85 - 0 + 1.33 + 0.60 \\ &= 2.41 \text{ (mg/L)}\end{aligned}$$

Step 2. Calculate S_1, S_2, and D

S_1 at T = 24.4°C

$$\begin{aligned}S_1 &= 14.652 - 0.41022(24.4) + 0.007991(24.4)^2 - 0.00007777(24.4)^3 \\ &= 8.27 \text{ mg/L or from Table 2.2}\end{aligned}$$

S_2 at T = 24.6°C, use same formula

$$S_2 = 8.24 \text{ (mg/L)}$$

Note: The elevation correction factor is ignored, for simplicity.

From Eq. (2.29)

$$\begin{aligned}D &= \frac{S_1 + S_2}{2} - \frac{C_1 + C_2}{2} \\ &= \frac{1}{2}(8.27 + 8.24 - 6.85 - 7.33) \\ &= 1.17 \text{ (mg/L)}\end{aligned}$$

Step 3. Compute K_2 with Eq. (2.60)

$$K_2 = \frac{\text{REA}}{D(t_2 - t_1)}$$

$$= \frac{2.41 \text{ mg/L}}{1.17 \text{ mg/L} \times 2.50 \text{ days}}$$

$$= 0.82 \text{ per day}$$

10 SEDIMENT OXYGEN DEMAND

Sediment oxygen demand (SOD) is a measure of the oxygen demand characteristics of the bottom sediment which affects the dissolved oxygen resources of the overlying water. To measure SOD, a bottom sample is especially designed to entrap and seal a known quantity of water at the river bottom. Changes in DO concentrations (approximately 2 h, or until the DO usage curve is clearly defined) in the entrapped water are recorded by a DO probe fastened in the sampler (Butts, 1974). The test temperature should be recorded and factored for temperature correction at a specific temperature.

SOD curves can be plotted showing the accumulated DO used (y-axis) versus elapsed time (x-axis). SOD curves resemble first-order carbonaceous BOD curves to a great extent; however, first-order kinetics are not applicable to the Upper Illinois Waterway's data. For the most part, SOD is caused by bacteria reaching an "unlimited" food supply. Consequently, the oxidation rates are linear in nature.

The SOD rates, as taken from SOD curves, are in linear units of milligrams per liter per minute (mg/(L · min)) and can be converted into grams per square meter per day (g/(m^2 · d)) for practical application. A bottle sampler SOD rate can be formulated as (Butts, 1974; Butts and Shackleford, 1992):

$$\text{SOD} = \frac{1440\, SV}{10^3 A} \tag{2.62}$$

where SOD = sediment oxygen demand, g/(m^2 · d)
S = slope of stabilized portion of the curve, mg/(L · min)
V = volume of sampler, L
A = bottom area of sampler, m^2

EXAMPLE 1: An SOD sampler is a half-cylinder in shape (half-section of steel pipe). Its diameter and length are $1\frac{1}{6}$ ft (14 in) and 2.0 ft, respectively.
(1) Determine the relationship of SOD and slope of linear portion of usage curve.
(2) What is SOD in g/(m^2 · d) for S = 0.022 mg/(L · min)?

Solution:

Step 1. Find area of the SOD sampler

$$A = 1\tfrac{1}{6} \text{ (ft)} \times 2.0 \text{(ft)} \times 0.0929 \text{ (m}^2\text{/ft}^2)$$

$$= 0.217 \text{ m}^2$$

Step 2. Determine the volume of the sampler

$$V = \tfrac{1}{2}\pi r^2 l$$

$$V = \tfrac{1}{2} \times 3.14 \times (7/12 \text{ ft})^2 \times (2.0 \text{ ft}) \times (28.32 \text{ L/ft}^3)$$

$$= 30.27 \text{ L}$$

Step 3. Compute SOD with Eq. (2.62)

$$\text{SOD} = \frac{1440\,SV}{10^3\,A}$$

$$= \frac{1400 \times S \times 30.27}{10^3 \times 0.217}$$

$$= 195\,S \qquad \text{ignoring the volume of two union connections}$$

Note: If two union connections are considered, the total volume of water contained within the sampler is 31.11 L. Thus, the Illinois State Water Survey's SOD formula is

$$\text{SOD} = 206.6\ \text{S}$$

Step 4. Compute SOD

$$\text{SOD} = 206.6 \times 0.022$$

$$= 4.545\ [\text{g}/(\text{m}^2 \cdot \text{d})]$$

10.1 Relationship of Sediment Characteristics and SOD

Based on extensive data collections by the Illinois State Water Survey, Butts (1974) proposed the relationship of SOD and the percentage of dried solids with volatile solids.

The prediction equation is

$$\text{SOD} = 6.5(\text{DS})^{-0.46}\,(\text{VS})^{0.38} \tag{2.63}$$

where SOD = sediment oxygen demand g/($m^2 \cdot d$)
DS = percent dried solids of the decanted sample by weight
VS = percent volatile solids by weight

EXAMPLE 2: Given DS = 68% and VS = 8%, predict the SOD value of this sediment.

Solution:

$$\text{SOD} = 6.5(\text{DS})^{-0.46}(\text{VS})^{0.38}$$

$$= \frac{6.5 \times (8)^{0.38}}{(68)^{0.46}} = \frac{6.5 \times 2.20}{6.965}$$

$$= 2.05\ [\text{g}/(\text{m}^2 \cdot \text{d})]$$

10.2 SOD Versus DO

An SOD–DO relationship has been developed from the data given by McDonnell and Hall (1969) for 25-cm deep sludge. The formula is

$$\text{SOD}/\text{SOD}_1 = \text{DO}^{0.28} \tag{2.64}$$

where SOD = SOD at any DO level, g/($m^2 \cdot d$)
SOD_1 = SOD at a DO concentration of 1.0 mg/L, g/($m^2 \cdot d$)
DO = dissolved oxygen concentration, mg/L

This model can be used to estimate the SOD rate at various DO concentration in areas having very high benthic invertebrate populations.

EXAMPLE 3: Given $SOD_1 = 2.4\ g/(m^2 \cdot d)$, find SOD at DO concentrations of 5.0 and 8.0 mg/L.

Solution:

Step 1. For DO = 5.0 mg/L, with Eq. (2.64)

$$\begin{aligned} SOD_5 &= SOD_1 \times DO^{0.28} \\ &= 2.4 \times 5^{0.28} \\ &= 2.4 \times 1.57 \\ &= 3.77\ (mg/L) \end{aligned}$$

Step 2. For DO = 8.0 mg/L

$$\begin{aligned} SOD_8 &= 2.4 \times 8^{0.28} \\ &= 2.4 \times 1.79 \\ &= 4.30\ (mg/L) \end{aligned}$$

Undisturbed samples of river sediments are collected by the test laboratory to measure the oxygen uptake of the bottom sediment: the amount of oxygen consumed over the test period is calculated as a zero-order reaction (USEPA, 1997):

$$\frac{dC}{dt} = -\frac{SOD}{H} \tag{2.65}$$

where dC/dt = rate change of oxygen concentration, g $O_2/(m^2 \cdot d)$
SOD = sediment oxygen demand, g $O_2/(m^2 \cdot d)$
H = average river depth, m

11 ORGANIC SLUDGE DEPOSITS

Solids in wastewaters may be in the forms of settleable, flocculation or coagulation of colloids, and suspended. When these solids are being carried in a river or stream, there is always a possibility that the velocity of flow will drop to some value at which sedimentation will occur. The limiting velocity at which deposition will occur is probably about 0.5–0.6 ft/s (15–18 cm/s).

Deposition of organic solids results in a temporary reduction in the BOD load of the stream water. Almost as soon as organic matter is deposited, the deposits will start undergoing biological decomposition, which results in some reduction in the DO concentration of the water adjacent to the sediment material. As the deposited organic matter increases in volume, the rate of decomposition also increases. Ultimately, equilibrium will be established. Velz (1958) has shown that at equilibrium the rate of decomposition (exertion of a BOD) equals the rate of deposition.

In order to properly evaluate the effects of wastewater on the oxygen resources of a stream, it may be necessary to account for the contribution made by solids being deposited or by sediments being scoured and carried into suspension.

From field observations, Velz (1958) concluded that enriched sediments will deposit and accumulate at a stream velocity of 0.6 ft/s (18 cm/s) or less, and resuspension of sediments and scouring will occur at a flow velocity of 1.0–1.5 ft/s (30–45 cm/s). Velz (1958) has reported that the effect of sludge deposits can be expressed mathematically. The accumulation of sludge deposition (L_d) is

$$L_d = \frac{P_d}{2.303k}(1 - 10^{-kt}) \tag{2.66}$$

where L_d = accumulation of BOD in area of deposition, lb/d
P_d = BOD added, lb/d
k = rate of oxidation of deposit on base 10
t = time of accumulation, days

EXAMPLE: Given (A)$k = 0.03$; $t = 2$ days
(B)$k = 0.03$; $t = 30$ days
(C)$k = 0.03$; $t = 50$ days

Find the relationship of L_d and P_d

Solution: For A with Eq. (2.66)

$$\begin{aligned} L_d &= \frac{P_d}{2.303k}(1 - 10^{-kt}) \\ &= \frac{P_d}{2.303 \times 0.03}(1 - 10^{-0.03\times2}) \\ &= \frac{P_d}{0.0691}(0.13) \\ &= 1.88P_d \end{aligned}$$

This means that in 2 days, 188% of daily deposit (P_d) will have accumulated in the deposit area, and 13% of P_d will have been oxidized. The rate of utilization is about 6.9% per day of accumulated sludge BOD.

Solution for B:

$$\begin{aligned} L_d &= \frac{P_d}{0.0691}(1 - 10^{-0.03\times30}) \\ &= \frac{P_d}{0.0691}(0.874) \\ &= 12.65P_d \end{aligned}$$

In 30 days, 1265% of daily deposit will have accumulated and the rate of utilization is about 87.4% of daily deposit.

Solution for C:

$$\begin{aligned} L_d &= \frac{P_d}{0.0691}(1 - 10^{-0.03\times50}) \\ &= \frac{P_d}{0.0691}(0.97) \\ &= 14.03P_d \end{aligned}$$

In 50 days, 1403% of daily deposit will have accumulated and the rate of utilization is about 97% of daily deposit (98% in 55 days, not shown). This suggests that equilibrium is almost reached in approximately 50 days, if there is no disturbance from increased flow.

12 PHOTOSYNTHESIS AND RESPIRATION

Processes of photosynthesis and respiration by aquatic plants such as phytoplankton (algae), periphyton, and rooted aquatic plants (macrophytes) could significantly affect the DO concentrations in the water column. Plant photosynthesis consumes nutrients and carbon dioxide under the light, and produces oxygen. During dark conditions, and during respiration, oxygen is used. The daily average oxygen production and reduction due to photosynthesis and respiration can be expressed in the QUALZE model as (USEPA, 1997):

$$\frac{dC}{dt} = P - R \tag{2.67}$$

$$\frac{dC}{dt} = (\alpha_3\mu - \alpha_4\rho)\text{Ag} \tag{2.68}$$

where

dC/dt = rate of change of oxygen concentration, mg O_2/(L · d)
P = average gross photosynthesis production, mg O_2/(L · d)
R = average respiration, mg O_2/(L · d)
α_3 = stoichiometric ratio of oxygen produced per unit of algae photosynthesis, mg/mg
α_4 = stoichiometric ratio of oxygen uptake per unit of algae respired, mg/mg
μ = algal growth rate coefficient, per day
ρ = algal respiration rate coefficient, per day
Ag = algal mass concentration, mg/L

13 NATURAL SELF-PURIFICATION IN STREAMS

When decomposable organic waste is discharged into a stream, a series of physical, chemical, and biological reactions are initiated and thereafter the stream ultimately will be relieved of its pollutive burden. This process is so-called natural self-purification. A stream undergoing self-purification will exhibit continuously changing water quality characteristics throughout the reach of the stream.

Dissolved oxygen concentrations in water are perhaps the most important factor in determining the overall effect of decomposable organic matters in a stream. It is also necessary to maintain mandated DO levels in a stream. Therefore the type and the degree of wastewater treatment necessary depend primarily on the condition and best usage of the receiving stream.

Recently the US Environmental Protection Agency has revised the water quality model QUALZE for total maximum daily loads to rivers and streams (US EPA, 1997). Interested readers may refer to an excellent example of total maximum daily load analysis in Appendix B of the US EPA technical guidance manual.

13.1 Oxygen Sag Curve

The dissolved oxygen balance in a stream which is receiving wastewater effluents can be formulated from a combination of the rate of oxygen utilization through BOD and oxygen transfer from the atmosphere into water. Many factors involved in this process are discussed in the previous sections. The oxygen sag curve (DO balance) is as a result of DO added minus DO removed. The oxygen balance curve or oxygen profile can be mathematically expressed (Streeter and Phelps, 1925) as previously discussed:

$$D_t = \frac{k_1 L_a}{k_2 - k_1}(10^{-k_1 t} - 10^{-k_2 t}) + D_a \times 10^{-k_2 t} \tag{2.8b}$$

where k_1 and k_2 are, respectively, deoxygenation and reoxygenation rates to the base 10 which are popularly used. Since k_1 is determined under laboratory conditions, the rate of oxygen removed in a stream by oxidation may be different from that under laboratory conditions. Thus, a term k_d is often substituted. Likewise, the rate of BOD removal in a stream may not equal the deoxygenation rate in a laboratory bottle and the oxidation rate in a stream, so the term k_r is used to reflect this situation. Applying these modified terms, the oxygen sag equation becomes

$$D_t = \frac{k_d L_a}{k_2 - k_r}(10^{-k_r t} - 10^{-k_2 t}) + D_a \times 10^{-k_2 t} \tag{2.69}$$

If deposition occurs, k_r will be greater than k_1; and if scour of sediment organic matter occurs, k_r will be less than k_1.

Computation of organic waste-load capacity of streams or rivers may be carried out by using Eq. (2.69) (the following example), the Thomas method, the Churchill–Buckingham method, and other methods.

EXAMPLE: Station 1 receives a secondary effluent from a city. BOD loading at station 2 is negligible. The results of two river samplings at stations 3 and 4 (temperature, flow rates, BOD_5, and DO) are adopted from Nemerow (1963) and are shown below. Construct an oxygen sag curve between stations 3 and 4.

Station	Temp °C	Flow Q	BOD_5		DO, mg/L		DO deficit	
		cfs	mg/L	lb/d	Measure	Saturated	mg/L	lb/d
(1)	(2)	(3)	(4)	(5)	(6)	(7)	(8)	(9)
3	**20.0**	**60.1**	**36.00**	11662	**4.1**	9.02	4.92	1594
	17.0	**54.0**	**10.35**	3012	**5.2**	9.61	4.41	1284
			Mean	*7337*				*1439*
4	20.0	**51.7**	**21.2**	5908	**2.6**	9.02	6.42	1789
	16.5	**66.6**	**5.83**	2093	**2.8**	9.71	6.91	2481
			Mean	*4000*				*2135*

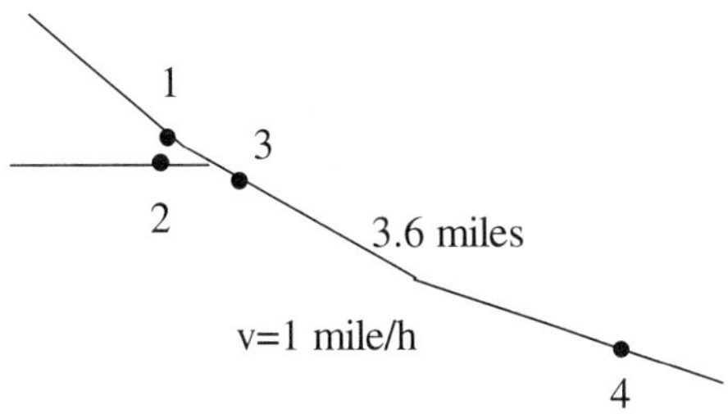

Solution:

Step 1. Calculation for columns 5, 7, 8, and 9

Col. 5: lb/d = 5.39 × col. 3 × col. 4

Col. 7: taken from Table 2.2

Col. 8: mg/L = col. 7 − col. 6

Col. 9: lb/d = 5.39 × col. 3 × col. 8

Note: 5.39 lb/d = 1 cfs × 62.4 lb/ft^3 × 86,400 s/d × (1 mg/10^6 mg)

Step 2. Determine ultimate BOD at stations 3 (L_{a3}) and 4 (L_{a4}). Under normal deoxygenation rates, a factor of 1.46 is used as a multiplier to convert BOD_5 to the ultimate first-stage BOD.

$$L_{a3} = 7337 \text{ lb/d} \times 1.46 = 10{,}712 \text{ lb/d}$$

$$L_{a4} = 4000 \text{ lb/d} \times 1.46 = 5840 \text{ lb/d}$$

Step 3. Compute deoxygenation rate k_d

$$\Delta t = \frac{3.6 \text{ mi}}{1.0 \text{ mi/h} \times 24 \text{ h/d}} = 0.15 \text{ day}$$

$$k_d = \frac{1}{\Delta t} \log \frac{L_{a3}}{L_{a4}} = \frac{1}{0.15 \text{ day}} \log \frac{10{,}712}{5840}$$

$$= 1.76 \text{ (per day)}$$

Step 4. Compute average BOD load $\bar{L}$, average DO deficit $\bar{D}$, and the difference in DO deficit ΔD in the reach (stations 3 and 4) from above

$$\begin{aligned}\bar{L} &= \tfrac{1}{2}(L_{a3} + L_{a4}) = \tfrac{1}{2}(10{,}712 + 5840) \text{ lb/d} \\ &= 8276 \text{ lb/d} \\ \bar{D} &= \tfrac{1}{2}(1439 + 2135) \text{ lb/d} \\ &= 1787 \text{ lb/d} \\ \Delta D &= (1789 + 2481) - (1594 + 1284) \\ &= 1392 \text{ (lb/d)}\end{aligned}$$

Step 5. Compute reaeration rate k_2

$$\begin{aligned}k_2 &= k_d \frac{\bar{L}}{\bar{D}} - \frac{\Delta D}{2.303 \Delta t \bar{D}} \\ &= 1.76 \times \frac{8276}{1787} - \frac{1392}{2.303 \times 0.15 \times 1787} \\ &= 5.90 \text{ (per day)} \\ f &= k_2/k_d = 5.9/1.76 = 3.35\end{aligned}$$

Step 6. Plot the DO sag curve

Assuming the reaction rates $k_d(k_d = k_1)$ and k_2 remain constant in the reach (stations 3–4), the initial condition at station 3:

$$\begin{aligned}L_a &= 10{,}712 \text{ lb/d} \\ D_a &= 1439 \text{ lb/d}\end{aligned}$$

Using Eq. (2.69)

$$D_t = \frac{k_d L_a}{k_2 - k_r}(10^{-k_r t} - 10^{-k_2 t}) + D_a \times 10^{-k_2 t}$$

When $t = 0.015$

$$\begin{aligned}D_{0.015} &= \frac{1.76 \times 10{,}712}{5.9 - 1.76}(10^{-1.76 \times 0.015} - 10^{-5.9 \times 0.015}) + 1439 \times 10^{-5.9 \times 0.015} \\ &= 1745 \text{ (lb/d)}\end{aligned}$$

D values at various t can be computed in the same manner. We get D values in lb/d as below. It was found that the critical point is around t = 0.09 days. The profile of DO deficit is depicted as below (Fig. 2.10):

$$\begin{aligned}&D_0 = 1439 && D_{0.09} = 2245 \\ &D_{0.015} = 1745 && D_{0.105} = 2227 \\ &D_{0.03} = 1960 && D_{0.12} = 2190 \\ &D_{0.045} = 2104 && D_{0.135} = 2137 \\ &D_{0.06} = 2192 && D_{0.15} = 2073 \\ &D_{0.075} = 2235 &&\end{aligned}$$

Thomas method. Thomas (1948) developed a useful simplification of the Streeter–Phelps equation for evaluating stream waste—assimilative capacity. His method presumes the computation of the

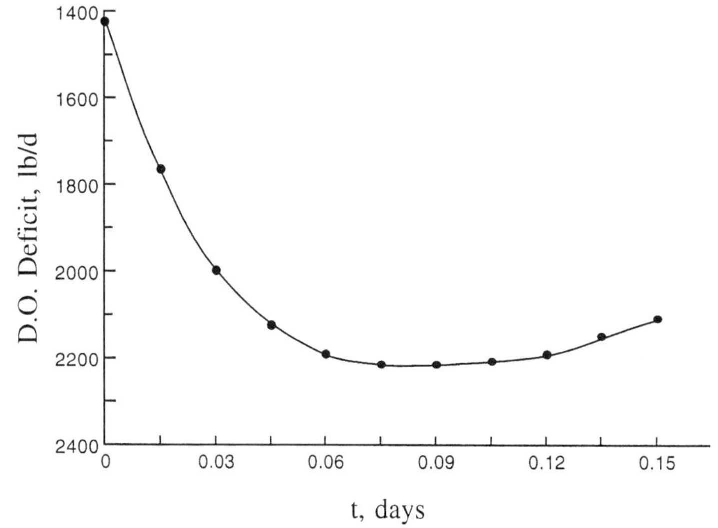

FIGURE 2.10 Profile of DO deficit.

stream deoxygenation coefficient (k_1) and reoxygenation coefficient (k_2) as defined in previous sections (also example). He developed a nomograph to calculate the DO deficit at any time, t (DO profile), downstream from a source of pollution load. The nomograph (not shown, the interested reader should refer to his article) is plotted as D/L_a versus $k_2\,t$ for various ratios of k_2/k_1. In most practical applications, this can be solved only by a tedious trial-and-error procedure. Before the nomograph is used, k_1, k_2, D_a, and L_a must be computed (can be done as in the previous example). By means of a straight edge, a straight line (isopleth) is drawn, connecting the value of D_a/L_a at the left of the point representing the reaeration constant × time-of-travel ($k_2\,t$) on the appropriate k_2/k_1 curve (a vertical line at k_2t and intercept on the k_2/k_1 curve). The value D_a/L_a is read at the intersection with the isopleth. Then, the value of the deficit at time t (D_t) is obtained by multiplying L_a with the interception value.

Nemerow (1963) claimed that he had applied the Thomas nomograph in many practical cases and found it very convenient, accurate, and timesaving. He presented a practical problem-solving example for an industrial discharge to a stream. He illustrated three different DO profiles along a 38-mile river: i.e. (1) under current loading and flow conditions; (2) imposition of an industrial load at the upstream station under current condition; and (3) imposition of industrial load under 5-year low-stream conditions.

Churchill–Buckingham method. Churchill and Buckingham (1956) found that the DO values in streams depend on only three variables: temperature, BOD, and stream flow. They used multiple linear correlation with three normal equations based on the principle of least squares:

$$\sum x_1y = ax_1^2 + bx_1x_2 + cx_1x_3 \tag{2.70}$$

$$\sum x_2y = ax_1x_2 + bx_2^2 + cx_2x_3 \tag{2.71}$$

$$\sum x_3y = ax_1x_3 + bx_2x_3 + cx_3^2 \tag{2.72}$$

where y = DO dropped in a reach, mg/L
x_1 = BOD at sag, mg/L
x_2 = temperature, °C
x_3 = flow, cfs (or MGD)
a, b, and c = constants

TABLE 2.12 DO Sag between Stations 2 and 4

Observed DO, mg/L		DO drop, mg/L	BOD @ sag, mg/L	Temperature, °C	Flow, 1000 cfs
Station 2	Station 4	y	x_1	x_2	x_3
7.8	5.7	2.1	6.8	11.0	13.21
8.2	5.9	2.3	12.0	16.5	11.88
6.1	4.8	1.9	14.8	21.2	9.21
6.0	3.2	2.8	20.2	23.4	6.67
5.5	2.3	3.2	18.9	28.3	5.76
6.2	2.9	3.3	14.3	25.6	8.71
Sum		15.6	87.0	126.0	55.44
Mean		2.6	14.5	21.0	9.24

By solving these three equations, the three constants a, b, and c can be obtained. Then, the DO drop is expressed as

$$y = ax_1 + bx_2 + cx_3 + d \tag{2.73}$$

This method eliminates the often questionable and always cumbersome procedure for determining k_1, k_2, k_d, and k_r. Nemerow (1963) reported that this method provides a good correlation if each stream sample is collected during maximum or minimum conditions of one of the three variables. Only six samples are required in a study to produce practical and dependable results.

EXAMPLE: Data obtained from six different days' stream surveys during medium- and low-flow periods showed that the DO sag occurred between stations 2 and 4. The results and computation are as shown in Table 2.12. Develop a multiple regression model for DO drop and BOD loading.

Solution:

Step 1. Compute element for least-squares method (see Table 2.13).

Step 2. Apply to three normal equations by using the values of the sum or average of each element.

$$229.84a + 326.7b + 122.81c = 39.04$$
$$326.7a + 474.42b + 179.77c = 56.98$$
$$122.81a + 179.77b + 92.33c = 23.07$$

Solving these three equations by dividing by the coefficient of a:

$$a + 1.421b + 0.534c = 0.1699$$
$$a + 1.452b + 0.550c = 0.1744$$
$$a + 1.464b + 0.752c = 0.1879$$

TABLE 2.13 Elements for Least-Squares Method

	y^2	yx_1	yx_2	yx_3	x_1^2	x_1x_2	x_1x_3	x_2^2	x_2x_3	x_3^2
	4.41	14.28	23.10	27.74	46.24	74.8	89.83	121.00	145.31	174.50
	5.29	27.60	37.95	27.32	144.00	198.0	142.56	272.25	176.02	141.13
	3.61	28.12	40.28	17.50	219.04	313.76	136.31	449.44	195.25	84.82
	7.84	56.56	65.52	18.68	408.04	472.68	134.74	547.56	196.08	44.49
	10.24	60.48	90.56	18.43	357.21	534.87	108.86	800.89	163.01	33.18
	10.89	47.19	84.48	28.74	204.49	366.08	124.55	655.36	222.98	75.86
Sum	42.28	234.23	341.89	138.42	1379.02	1960.19	736.85	2846.50	1078.64	553.99
Mean	7.05	39.04	56.98	23.07	229.84	326.70	122.81	474.42	179.77	92.33

Subtracting one equation from the other:

$$0.043b + 0.218c = 0.0180$$
$$0.012b + 0.202c = 0.0135$$

Dividing each equation by the coefficient of b:

$$b + 5.07c = 0.4186$$
$$b + 16.83c = 1.125$$

Subtracting one equation from the other:

$$11.76c = 0.7064$$
$$c = 0.0601$$

Substituting c in the above equation:

$$b + 5.07 \times 0.0601 = 0.4186$$
$$b = 0.1139$$

Similarly

$$a = 0.1699 - 1.421 \times 0.1139 - 0.534 \times 0.0601$$
$$= -0.0241$$

Check on to the three normal equations with values of a, b, and c:

$$d = Y - (ax_1 + bx_2 + cx_3)$$
$$= 39.04 - (-229.84 \times 0.0241 + 326.7 \times 0.1139 + 122.81 \times 0.0601)$$
$$= 0$$

The preceding computations yield the following DO drop:

$$Y = -0.0241x_1 + 0.1139x_2 + 0.0601x_3$$

Step 3. Compute the DO drop using the above model:

The predicted and observed DO drop is shown in Table 2.14.

It can be seen from Table 2.14 that the predicted versus the observed DO drop values are reasonably close.

Step 4. Compute the allowable BOD loading at the source of the pollution.

The BOD equation can be derived from the same least-squares method by correlating the upstream BOD load (as x_1) with the water temperature (as x_2), flow rate (as x_3) and resulting BOD (as z) at the

TABLE 2.14 Predicted and Observed DO Drop

	Dissolved oxygen drop, mg/L	
Sampling date	Calculated	Observed
1	1.93	2.1
2	2.35	2.3
3	2.61	1.9
4	2.62	2.8
5	3.07	3.2
6	3.04	3.3

sag point in the stream. Similar to steps 1 and 2, the DO drop is replaced by BOD at station 4 (not shown). It will generate an equation for the allowable BOD load:

$$z = ax_1 + bx_2 + cx_3 + d$$

Then, applying the percent of wastewater treatment reduction, the BOD in the discharge effluent is calculated as x_1. Selecting the design temperature and low-flow data, one can then predict the BOD load (lb/d or mg/L) from the above equation. Also, applying z mg/L in the DO drop equation gives the predicted value for DO drop at the sag.

13.2 Determination of k_r

The value k_r can be determined for a given reach of stream by determining the BOD of water samples from the upper (A) and lower (B) end of the section under consideration:

$$\log \frac{L_{aB}}{L_{aA}} = -k_r t \tag{2.74}$$

$$k_r = -\frac{1}{t} \log \frac{L_{aB}}{L_{aA}} = \frac{1}{t} \log \frac{L_{aA}}{L_{aB}} \tag{2.75a}$$

or

$$k_r = \frac{1}{t} (\log L_{aA} - \log L_{aB}) \tag{2.75b}$$

While the formula calls for first-stage BOD values, any consistent BOD values will give satisfactory results for k_r.

EXAMPLE 1: The first-stage BOD values for river stations 3 and 4 are 34.6 and 24.8 mg/L, respectively. The time of travel between the two stations is 0.99 days. Find k_r for the reach.

Solution:

$$\begin{aligned} k_r &= \frac{1}{t} (\log L_{aA} - \log L_{aB}) \\ &= \frac{1}{0.99} (\log 34.6 - \log 24.8) \\ &= 0.15 \text{ (per day for log base 10)} \\ K_r &= 2.3026\, k_r = 0.345 \text{ per day for log base } e \end{aligned}$$

EXAMPLE 2: Given

$$\begin{aligned} L_{aA} &= 32.8 \text{ mg/L}, \qquad L_{aB} = 24.6 \text{ mg/L} \\ D_A &= 3.14 \text{ mg/L}, \qquad D_B = 2.58 \text{ mg/L} \\ t &= 1.25 \text{ day} \end{aligned}$$

Find k_2 and k_r

Solution:

Step 1. Determine k_2 using Eq. (2.44) and Eq. (2.45)

Reaeration

$$\begin{aligned} r_m &= (L_{aA} - L_{aB}) + (D_A - D_B) \\ &= (32.8 - 24.6) + (3.14 - 2.58) \\ &= 8.76 \text{ (mg/L)} \end{aligned}$$

Mean deficit

$$D_m = (3.14 + 2.58)/2 = 2.86$$

$$K_2 = \frac{r_m}{D_m} = \frac{8.76}{2.86} = 3.06 \text{ (per day)} \quad \text{or} \quad k_2 = \frac{3.06}{2.303} = 1.33 \text{ (per day)}$$

Step 2. Compute k_r

$$\begin{aligned} k_r &= \frac{1}{t}(\log L_{aA} - \log L_{aB}) \\ &= \frac{1}{1.25}(\log 32.8 - \log 24.6) \\ &= 0.10 \text{ (per day)} \end{aligned}$$

13.3 Critical Point on Oxygen Sag Curve

In many cases, only the lowest point of the oxygen sag curve is of interest to engineers. The equation can be modified to give the critical value for DO deficiency (D_c) and the critical time (t_c) downstream at the critical point. At the critical point of the curve, the rate of deoxygenation equals the rate of reoxygenation:

$$\frac{dD}{dt} = K_1 L_t - K_2 D_c = 0 \tag{2.76}$$

thus

$$K_1 L_t = K_2 D_c$$

$$D_c = \frac{K_1}{K_2} L_t \tag{2.77a}$$

or

$$D_c = \frac{K_d}{K_2} L_t \tag{2.77b}$$

where L_t = BOD remaining.

Use K_d if deoxygenation rate at critical point is different from K_1. Since

$$L_t = L_a \times e^{-K_r t_c}$$

then

$$D_c = \frac{K_d}{K_2}(L_a \times e^{-K_r t_c}) \tag{2.78a}$$

or

$$D_c = \frac{k_d}{k_2}(L_a \times 10^{-k_r t_c}) \tag{2.78b}$$

Let

$$f = \frac{k_2}{k_d} = \frac{K_2}{K_d} \tag{2.79}$$

or

$$D_c = \frac{1}{f}(L_a \times e^{-K_r t_c}) \tag{2.80}$$

and from Thomas (1948)

$$t_c = \frac{1}{k_2 - k_r} \log \frac{k_2}{k_r}\left[1 - \frac{D_a(k_2 - k_r)}{k_d L_a}\right] \tag{2.81}$$

then

$$t_c = \frac{1}{k_r(f - 1)} \log \left\{f\left[1 - (f - 1)\frac{D_a}{L_a}\right]\right\} \tag{2.82}$$

Substitute t_c in D_c formula (Thomas, 1948):

$$L_a = D_c\left(\frac{k_2}{k_r}\right)\left[1 + \frac{k_r}{k_2 - k_r}\left(1 - \frac{D_a}{D_c}\right)^{0.418}\right] \tag{2.83}$$

or

$$\log L_a = \log D_c + \left[1 + \frac{k_r}{k_2 - k_r}\left(1 - \frac{D_a}{D_c}\right)^{0.418}\right] \log\left(\frac{k_2}{k_r}\right) \tag{2.84}$$

Thomas (1948) provided this formula which allows us to approximate L_a: the maximum BOD load that may be discharged into a stream without causing the DO concentration downstream to fall below a regulatory standard (violation).

EXAMPLE 1: The following conditions are observed at station A. The water temperature is 24.3°C, with $k_1 = 0.16$, $k_2 = 0.24$, and $k_r = 0.19$. The stream flow is 880 cfs (ft^3/s). DO and L_a for river water is 6.55 and 5.86 mg/L, respectively. The state requirement for minimum DO is 5.0 mg/L. How much additional BOD (Q = 110, cfs, DO = 2.22 mg/L) can be discharged into the stream and still maintain 5.0 mg/L DO at the flow stated?

Solution:

Step 1. Calculate input data with total flow $Q = 880 + 110 = 990$ cfs; at $T = 24.3$°C

Saturated DO from Table 1.2

$$DO_S = 8.29 \text{ mg/L}$$

After mixing

$$DO_a = \frac{6.55 \times 880 + 2.22 \times 110}{990} = 6.07 \text{ (mg/L)}$$

Deficit at station A

$$D_a = DO_s - DO_a = 8.29 \text{ mg/L} - 6.07 \text{ mg/L} = 2.22 \text{ mg/L}$$

Deficit at critical point

$$D_c = DO_s - DO_{min} = 8.29 \text{ mg/L} - 5.00 \text{ mg/L} = 3.29 \text{ mg/L}$$

Rates

$$k_d = k_r = 0.19 \text{ (per day)}$$

$$f = \frac{k_2}{k_1} = \frac{0.24}{0.16} = 1.5$$

$$f - 1 = 0.5$$

Step 2. Assume various values of L_a and calculate resulting D_c.

$$t_c = \frac{1}{k_r(f-1)} \log \left\{ f \left[1 - (f-1)\frac{D_a}{L_a} \right] \right\}$$
$$= \frac{1}{0.19(0.5)} \log \left\{ 1.5 \left[1 - (0.5)\frac{2.22}{L_a} \right] \right\}$$
$$= 10.53 \log \left(1.5 - \frac{1.665}{L_a} \right)$$

Let $L_a = 10$ mg/L

$$t_c = 10.53 \log(1.5 - 0.1665) = 1.32 \text{ (days)}$$
$$D_c = \frac{k_d}{k_2}(L_a \times 10^{-k_r t_c})$$
$$= \frac{0.19}{0.24}(10 \times 10^{-0.19\times 1.32})$$
$$= 0.792(10 \times 0.561)$$
$$= 4.45 \text{ (mg/L)}$$

Similarly, we can develop a table

L_a, mg/L	T_c, days	D_c, mg/L
10.00	1.32	4.45
9.00	1.25	4.13
8.00	1.17	3.80
7.00	1.06	3.48
6.40	**0.90**	**3.29**
6.00	0.92	3.17

Therefore, maximum $L_{a,\,max} = 6.40$ mg/L

Step 3. Determine effluent BOD load (Y_e) that can be added.

$$\text{BOD that can be added} = \text{maximum load} - \text{existing load}$$
$$110Y_e = 6.40 \times 990 - 5.86 \times 880$$
$$Y_e = 10.72 \text{ mg/L}$$

This means that the first-stage BOD of the effluent should be less than 10.72 mg/L.

EXAMPLE 2: Given: At the upper station A of the stream reach, under standard conditions with temperature of 20°C, $BOD_5 = 3800$ lb/d, $k_1 = 0.14$, $k_2 = 0.25$, $k_r = 0.24$, and $k_d = k_r$. The stream temperature is 25.8°C with a velocity 0.22 miles/h. The flow in the reach (A → B) is 435 cfs (including effluent) with a distance of 4.8 miles. $DO_A = 6.78$ mg/L, $DO_{min} = 6.00$ mg/L. Find how much additional BOD can be added at station A and still maintain a satisfactory DO level at station B?

Step 1. Calculate L_a at $T = 25.8$°C.

$$y = L_a(1 - 10^{-k_1 t})$$

When $t = 5$ days, $T = 20$°C (using loading unit of lb/d):

$$3800 = L_a(1 - 10^{-0.14\times 5})$$
$$L_a = 3800/0.80 = 4750 \text{ (lb/d)} = L_{a(20)}$$

Convert to $T = 25.8°C$:

$$
\begin{aligned}
L_{a(T)} &= L_{a(20)}[1 + 0.02(T - 20)] \\
L_{a(25.8)} &= 4750[1 + 0.02(25.8 - 20)] \\
&= 5300 \text{ (lb/d)}
\end{aligned}
$$

Step 2. Change all constants to 25.8°C basis.
At 25.8°C,

$$
\begin{aligned}
k_{1(T)} &= k_{1(20)} \times 1.047^{T-20} \\
k_{1(25.8)} &= 0.14 \times 1.047^{(25.8-20)} \\
&= 0.18 \text{ (per day)} \\
k_{2(T)} &= k_{2(20)} \times (1.02)^{T-20} \\
k_{2(25.8)} &= 0.25 \times (1.02)^{25.8-20} \\
&= 0.28 \text{ (per day)} \\
k_{r(25.8)} &= 0.24 \times 1.047^{(25.8-20)} \\
&= 0.31 \text{ (per day)} \\
k_{d(25.8)} &= 0.31 \text{ (per day)}
\end{aligned}
$$

Step 3. Calculate allowable deficit at station B.
At 25.8°C,

$$
\begin{aligned}
DO_{sat} &= 8.05 \text{ mg/L} \\
DO_B &= 6.00 \text{ mg/L} \\
D_B &= 8.05 \text{ mg/L} - 6.00 \text{ mg/L} = 2.05 \text{ mg/L} \\
D_A &= 8.05 \text{ mg/L} - 6.78 \text{ mg/L} = 1.27 \text{ mg/L}
\end{aligned}
$$

Step 4. Determine the time of travel (t) in the reach.

$$
t = \frac{\text{distance}}{V} = \frac{4.8 \text{ miles}}{0.22 \text{ mph} \times 24 \text{ h/d}} = 0.90 \text{ days}
$$

Step 5. Compute allowable L_a at 25.8°C using Eq. (2.69).

$$
D_t = \frac{k_d L_a}{k_2 - k_r}(10^{-k_r t} - 10^{-k_2 t}) + D_a \times 10^{-k_2 t}
$$

Here $D_t = D_B$ and $D_a = D_A$

$$
\begin{aligned}
2.05 &= \frac{0.31 L_a}{0.28 - 0.31}(10^{-0.31 \times 0.9} - 10^{-0.28 \times 0.9}) + 1.27 \times 10^{-0.28 \times 0.9} \\
2.05 &= (-10.33)L_a(0.526 - 0.560) + 0.71 \\
L_a &= 1.34/0.35 = 3.83 \text{ (mg/L)}
\end{aligned}
$$

Allowable load:

$$
\begin{aligned}
\text{lb/day} &= 5.39 \times L_a \cdot Q \\
&= 5.39(3.83 \text{ mg/L})(435 \text{ cfs}) \\
&= 8980 \text{ (lb/d)}
\end{aligned}
$$

Step 6. Find load that can be added at station A.

$$\begin{aligned}\text{Added} &= \text{allowable} - \text{existing}\\ &= 8980 - 5300\\ &= 3680\ (\text{lb/d})\end{aligned}$$

Step 7. Convert answer back to 5-day 20°C BOD.

$$\begin{aligned}3680 &= L_{a(20)}[1 + 0.02(25.8 - 20)]\\ L_{a(20)} &= 3297\ (\text{lb/d})\end{aligned}$$

EXAMPLE 3:

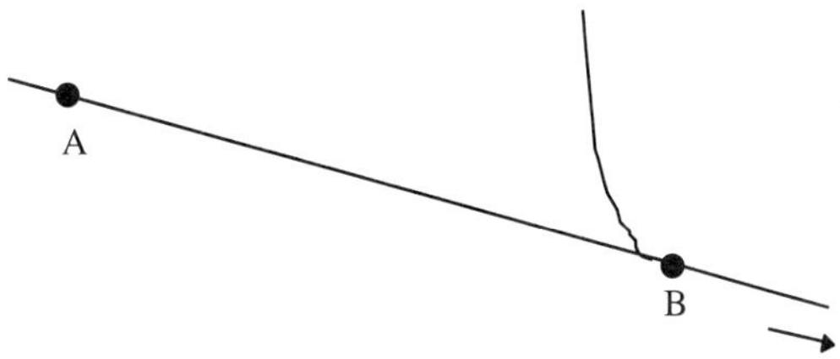

Given: The following data are obtained from upstream station A.

$k_1 = 0.14@20°\text{C}$ BOD$_5$ load at station A = 3240 lb/d

$k_2 = 0.31@20°\text{C}$ Stream water temperature = 26°C

$k_r = 0.24@20°\text{C}$ Flow θ = 188 cfs

$k_d = k_r$ Velocity V = 0.15 mph

$\text{DO}_A = 5.82$ mg/L A − B distance = 1.26 miles

Allowable $\text{DO}_{min} = 5.0$ mg/L

Find: Determine the BOD load that can be discharged at downstream station B.

Solution:

Step 1. Compute DO deficits.
At T = 26°C,

$$\begin{aligned}\text{DO}_s &= 8.02\text{ mg/L and } D_a \text{ at station A}\\ D_a &= \text{DO}_s - \text{DO}_A = 8.02\text{ mg/L} - 5.82\text{ mg/L} = 2.20\text{ mg/L}\end{aligned}$$

Critical deficit at station B.

$$D_c = 8.02 - \text{DO}_{min} = 8.02\text{ mg/L} - 5.0\text{ mg/L} = 3.02\text{ mg/L}$$

Step 2. Convert all constants to 26°C basis.
At 26°C,

$$\begin{aligned}k_{1(26)} &= k_{1(20)} \times 1.047^{(26-20)}\\ &= 0.14 \times 1.047^6\\ &= 0.18\ (\text{per day})\end{aligned}$$

$$
\begin{aligned}
k_{2(26)} &= k_{2(20)} \times (1.02)^{26-20} \\
&= 0.31 \times (1.02)^{6} \\
&= 0.34 \text{ (per day)} \\
k_{r(26)} &= 0.24 \times 1.047^{6} \\
&= 0.32 \text{ (per day)} \\
\frac{k_{2(26)}}{k_{r(26)}} &= \frac{0.34}{0.32} = 1.0625
\end{aligned}
$$

Step 3. Calculate allowable BOD loading at station B at 26°C.

From Eq. (2.84):

$$
\log L_{aB} = \log D_c + \left[1 + \frac{k_r}{k_2 - k_r}\left(1 - \frac{D_a}{D_c}\right)^{0.418}\right] \log\left(\frac{k_2}{k_r}\right)
$$

$$
\begin{aligned}
\log L_{aB} &= \log 3.02 + \left[1 + \frac{0.32}{0.34 - 0.32}\left(1 - \frac{2.2}{3.02}\right)^{0.418}\right] \log 1.0625 \\
&= 0.480 + [1 + 16 \times (0.272)^{0.418}]0.0263 \\
&= 0.480 + 10.28 \times 0.0263 \\
&= 0.750 \\
L_{aB} &= 5.52 \text{ (mg/L)}
\end{aligned}
$$

Allowable BOD loading at station B:

$$
\begin{aligned}
\text{lb/d} &= 5.39 \times L_{aB} \text{ (mg/L)} \cdot Q \text{ (cfs)} \\
&= 5.39 \times 5.52 \times 188 \\
&= 5594
\end{aligned}
$$

Step 4. Compute ultimate BOD at station A at 26°C.

$$
\begin{aligned}
L_{aA} &= \text{BOD}_5 \text{ loading}/(1 - 10^{-k_1 t}) \\
&= 3240/(1 - 10^{-0.14 \times 5}) \\
&= 4050 \text{ (lb/d at 20°C)} \\
L_{aA(26)} &= L_{aA}[1 + 0.02 \times (26 - 20)] \\
&= 4050(1 + 0.12) \\
&= 4536 \text{ (lb/d)}
\end{aligned}
$$

Step 5. Calculate BOD at 26°C from station A oxidized in 1.26 miles.

Time of travel,

$$
\begin{aligned}
t &= \frac{1.26 \text{ miles}}{0.15 \text{ mph} \times 24 \text{ h/d}} = 0.35 \text{ days} \\
y_B &= L_{aA}(1 - 10^{-k_r t}) \\
&= 4536(1 - 10^{-0.32 \times 0.35}) \\
&= 4536 \times 0.227 \\
&= 1030 \text{ (lb/d)}
\end{aligned}
$$

Step 6. Calculate BOD remaining from station A at station B.

$$
L_t = 4536 \text{ lb/d} - 1030 \text{ lb/d} = 3506 \text{ lb/d}
$$

Step 7. Additional BOD load that can be added in stream at station B (=x)

$$\begin{aligned} x_{(26)} &= \text{allowable BOD} - L_t \\ &= 5594 \text{ lb/d} - 3506 \text{ lb/d} \\ &= 2088 \text{ lb/d} \end{aligned}$$

This is first-stage BOD at 26°C.

At 20°C,

$$\begin{aligned} x_{(26)} &= x_{(20)}[1 + 0.02 \times (26 - 20)] \\ x_{(20)} &= 2088/1.12 = 1864 \text{ (lb/d)} \end{aligned}$$

For BOD_5 at 20°C (y_5)

$$\begin{aligned} y_5 &= x_{20}(1 - 10^{-0.14\times5}) \\ &= 1864(1 - 10^{-0.7}) \\ &= 1492 \text{ (lb/d)} \end{aligned}$$

EXAMPLE 4: Given: Stations A and B are selected in the main stream, just below the confluences of tributaries. The flows are shown in the sketch below. Assume there is no significant increased flow between stations A and B, and the distance is 11.2 miles (18.0 km). The following information is derived from the laboratory results and stream field survey:

k_1 = 0.18 per day @20°C
k_2 = 0.36 per day @20°C
k_r = 0.22 per day @20°C
$k_d = k_r$
Velocity, V = 0.487 mph

stream water temperature = 25°C
DO above station A = 5.95 mg/L
L_a just above A = 8870 lb/d @20°C
L_a added at A = 678 lb/d @20°C
DO deficit in the tributary above A = 256 lb/d

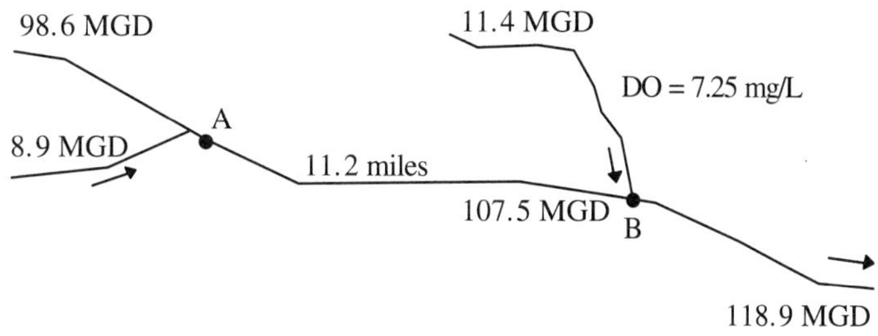

Find: Expected DO concentration just below station B.

Solution:

Step 1. Calculate total DO deficit just below station A at 25°C.

$$\begin{aligned} \text{DO deficit above station A } D_A &= 8.18 \text{ mg/L} - 5.95 \text{ mg/L} = 2.23 \text{ mg/L} \\ \text{lb/d of DO deficit above A} &= D_A \text{ (mg/L)} \times Q \text{ (MGD)} \times 8.34 \text{ (lb/gal)} \\ &= 2.23 \times 98.6 \times 8.34 \\ &= 1834 \\ \text{DO deficit in tributary above A} &= 256 \text{ lb/d} \\ \text{Total DO deficit below A} &= 1834 \text{ lb/d} + 256 \text{ lb/d} = 2090 \text{ lb/d} \end{aligned}$$

Step 2. Compute ultimate BOD loading at A at 25°C.

$$\begin{aligned}\text{At station A at 20°C, } L_{aA} &= 8870 \text{ lb/d} + 678 \text{ lb/d} = 9548 \text{ lb/d}\\ \text{At station A at 25°C, } L_{aA} &= 9548[1 + 0.02(25 - 20)]\\ &= 10{,}503 \text{ (lb/d)}\end{aligned}$$

Step 3. Convert rate constants for 25°C.

$$\begin{aligned}k_{1(25)} &= k_{1(20)} \times 1.047^{(25-20)}\\ &= 0.18 \times 1.258\\ &= 0.23 \text{ (per day)}\\ k_{2(25)} &= k_{2(20)} \times 1.024^{25-20}\\ &= 0.36 \times 1.126\\ &= 0.40 \text{ (per day)}\\ k_{r(25)} &= k_{r(20)} \times 1.047^{25-20}\\ &= 0.22 \times 1.258\\ &= 0.28 \text{ (per day)} = k_{d(25)}\end{aligned}$$

Step 4. Calculate DO deficit at station B from station A at 25°C.
Time of travel $t = 11.2/(0.487 \times 24) = 0.958$ day

$$\begin{aligned}D_B &= \frac{k_d L_a}{k_2 - k_r}(10^{-k_r t} - 10^{-k_2 t}) + D_a \times 10^{-k_2 t}\\ &= \frac{0.28 \times 10{,}503}{0.40 - 0.28}(10^{-0.28\times0.958} - 10^{-0.4\times0.958}) + 2090 \times 10^{-0.4\times0.958}\\ &= 24{,}507(0.5392 - 0.4138) + 2090 \times 0.4138\\ &= 3073 - 865\\ &= 2208 \text{ (lb/d)}\end{aligned}$$

Convert the DO deficit into concentration.

$$\begin{aligned}D_B &= \text{amount, lb/d/(8.34 lb/gal} \times \text{flow, MGD)}\\ &= 2208/(8.34 \times 107.5)\\ &= 2.46 \text{ (mg/L)}\end{aligned}$$

This is DO deficiency at station B from BOD loading at station A.

Step 5. Calculate tributary loading above station B.

$$\begin{aligned}\text{DO deficit} &= 8.18 \text{ mg/L} - 7.25 \text{ mg/L} = 0.93 \text{ mg/L}\\ \text{Amount of deficit} &= 0.93 \times 8.34 \times 11.4 = 88 \text{ (lb/d)}\end{aligned}$$

Step 6. Compute total DO deficit just below station B.
Total deficit

$$D_B = 2208 \text{ lb/d} + 88 \text{ lb/d} = 2296 \text{ (lb/d)}$$

or

$$\begin{aligned}&= 2296/(8.34 \times 118.9)\\ &= 2.23 \text{ (mg/L)}\end{aligned}$$

Step 7. Determine DO concentration just below station B.

$$\text{DO} = 8.18 \text{ mg/L} - 2.23 \text{ mg/L} = 5.95 \text{ mg/L}$$

EXAMPLE 5: A treated wastewater effluent from a community of 108,000 persons is to be discharged into a river which is not receiving any other significant wastewater discharge. Normally, domestic wastewater flow averages 80 gal (300 L) per capita per day. The 7-day, 10-year low flow of the river is 78.64 cfs. The highest temperature of the river water during the critical flow period is 26.0°C. The wastewater treatment plant is designed to produce an average carbonaceous 5-day BOD of 7.8 mg/L; an ammonia–nitrogen concentration of 2.3 mg/L, and DO for 2.0 mg/L. Average DO concentration in the river upstream of the discharge is 6.80 mg/L. After the mixing of the effluent with the river water, the carbonaceous deoxygenation rate coefficient (K_{rC} or K_C) is estimated at 0.25 per day (base e) at 20°C and the nitrogenous deoxygenation coefficient (K_{rN} or K_N) is 0.66 per day at 20°C. The lag time (t_0) is approximately 1.0 day. The river cross-section is fairly constant with mean width of 30 ft (10 m) and mean depth of 4.5 ft (1.5 m). Compute DO deficits against time t.

Solution:

Step 1. Determine total flow downstream Q and V.

$$\begin{aligned}
\text{Effluent flow } Q_e &= 80 \times 108{,}000 \text{ gpd} \\
&= 8.64 \times 10^6 \text{ gpd} \times 1.54/10^6 \text{ cfs/gpd} \\
&= 13.31 \text{ cfs} \\
\text{Upstream flow } Q_u &= 78.64 \text{ cfs} \\
\text{Downstream flow } Q_d &= Q_e + Q_u = 13.31 \text{ cfs} + 78.64 \text{ cfs} \\
&= 91.95 \text{ cfs} \\
\text{Velocity } V = Q_d/A &= 91.95 \text{ ft}^3\text{/s}/(30 \times 4.5)\text{ft}^2 \\
&= 0.681 \text{ ft/s}
\end{aligned}$$

Step 2. Determine reaeration rate constant K_2.

The value of K_2 can be determined by several methods as mentioned previously. From the available data, the method of O'Connor and Dobbins (1958), Eq. (2.52), is used at 20°C:

$$\begin{aligned}
K_2 &= 13.0\text{V}^{1/2}\text{H}^{-3/2} \\
&= 13.0(0.681)^{1/2}(4.5)^{-3/2} \\
&= 1.12 \text{ (per day)}
\end{aligned}$$

Step 3. Correct temperature factors for coefficients.

$$\begin{aligned}
K_{2(26)} &= 1.12 \text{ per day} \times 1.024^{26-20} = 1.29 \text{ per day} \\
K_{C(26)} &= 0.25 \text{ per day} \times 1.047^{6} = 0.33 \text{ per day}
\end{aligned}$$

Zanoni (1967) proposed correction factors for nitrogenous K_N in wastewater effluent at different temperatures as follows:

$$K_{N(T)} = K_{N(20)} \times 1.097^{T-20} \text{ for } 10 - 22°\text{C} \tag{2.85}$$

and

$$K_{N(T)} = K_{N(20)} \times 0.877^{T-22} \text{ for } 22 - 30°\text{C} \tag{2.86}$$

For 26°C,

$$\begin{aligned}
K_{N(26)} &= 0.66 \text{ per day} \times 0.877^{26-22} \\
&= 0.39 \text{ per day}
\end{aligned}$$

Step 4. Compute ultimate carbonaceous BOD (L_{aC}).

At 20°C,

$$
\begin{aligned}
K_{1C} &= 0.25 \text{ per day} \\
BOD_5 &= 7.8 \text{ mg/L} \\
L_{aC} &= BOD_5/(1 - e^{-K_{1C}\times 5}) \\
&= 7.8 \text{ mg/L}/(1 - e^{-0.25\times 5}) \\
&= 10.93 \text{ mg/L}
\end{aligned}
$$

At 26°C, using Eq. (2.28)

$$
\begin{aligned}
L_{aC(26)} &= 10.93 \text{ mg/L}(0.6 + 0.02 \times 26) \\
&= 12.24 \text{ mg/L}
\end{aligned}
$$

Step 5. Compute ultimate nitrogenous oxygen demand (L_{aN}). Reduced nitrogen species (NH_4^+, NO_3, and NO_2^-) can be oxidized aerobically by nitrifying bacteria which can utilize carbon compounds but always require nitrogen as an energy source. The two-step nitrification can be expressed as

$$NH_4^+ + \tfrac{3}{2}O_2 \xrightarrow{\text{Nitrosomonas}} NO_2^- + 2H^+ + H_2O \tag{2.87}$$

and

$$NO_2^- + \tfrac{1}{2}O_2 \xrightarrow{\text{Nitrobacter}} NO_3^- \tag{2.88}$$

$$NH_4^+ + 2O_2 \longrightarrow NO_3^- + 2H^+ + H_2O \tag{2.89}$$

Overall 2 moles of O_2 are required for each mole of ammonia; i.e. the N:O_2 ratio is 14:64 (or 1:4.57). Typical domestic wastewater contains 15–50 mg/L of total nitrogen. For this example, effluent NH_3 = 2.3 mg/L as N; therefore, the ultimate nitrogenous BOD is

$$L_{aN} = 2.3 \text{ mg/L} \times 4.57 = 10.51 \text{ mg/L}$$

Step 6. Calculate DO deficit immediately downstream of the wastewater load.

$$
\begin{aligned}
DO &= \frac{13.31 \times 2 + 78.64 \times 6.8}{13.31 + 78.64} \\
&= 6.10 \text{ (mg/L)}
\end{aligned}
$$

From Table 2.2, the DO saturation value at 26°C is 8.02 mg/L.

The initial DO deficit D_a is

$$D_a = 8.02 \text{ mg/L} - 6.10 \text{ mg/L} = 1.92 \text{ mg/L}$$

Step 7. Compute D_t and DO values at various times t.

Using

$$D_t = \frac{K_C L_{aC}}{K_2 - K_C}(e^{-K_C t} - e^{-K_2 t}) + D_a e^{-K_2 t} + \frac{K_N L_{aN}}{K_2 - K_N}[e^{-K_N(t-t_0)} - e^{-K_2(t-t_0)}] \tag{2.90}$$

and at 26°C

$$
\begin{aligned}
K_2 &= 1.29 \text{ per day} \\
K_C &= 0.33 \text{ per day} \\
K_N &= 0.39 \text{ per day} \\
L_{aC} &= 12.24 \text{ mg/L} \\
L_{aN} &= 10.51 \text{ mg/L} \\
D_a &= 1.92 \text{ mg/L} \\
t_0 &= 1.0 \text{ day}
\end{aligned}
$$

For $t \leq 1.0$ day:

When $t = 0.1$ day

$$
\begin{aligned}
D_{0.1} &= \frac{0.33 \times 12.24}{1.29 - 0.33}(e^{-0.33\times 0.1} - e^{-1.29\times 0.1}) + 1.92e^{-1.29\times 0.1} \\
&= 4.2075(0.9675 - 0.8790) + 1.6876 \\
&= 2.060 \text{ (mg/L)} \\
D_{0.2} &= 2.17 \text{ mg/L and so on}
\end{aligned}
$$

For $t > 1.0$ day

$$
\begin{aligned}
D_{1.1} &= 4.2075(e^{-0.33\times 1.1} - e^{-1.29\times 1.1}) + 1.92e^{-1.29\times 1.1} \\
&\quad + \frac{0.39 \times 10.51}{1.29 - 0.39}\left[e^{-0.39(1.1-1)} - e^{-1.29(1.1-1)}\right] \\
&= 4.2075(0.6956 - 0.2420) + 0.4646 \\
&\quad + 4.5543(0.9618 - 0.8790) \\
&= 2.75 \text{ (mg/L)} \\
D_{1.2} &= 3.04 \text{ mg/L and so on}
\end{aligned}
$$

Step 8. Produce a table for DO sag.

Table 2.15 gives the results of the above calculations for DO concentrations in the stream at various locations.

$$
\begin{aligned}
V &= 0.681 \text{ ft/s} = 0.681 \text{ ft/s} \times 3600 \text{ s/h} \times \frac{1 \text{ mile}}{5280 \text{ ft}} \\
&= 0.463 \text{ miles/h} \\
\text{When } t &= 0.1 \text{ days} \\
\text{Distance} &= 0.463 \text{ miles/h} \times 24 \text{ h/d} \times 0.1 \text{ day} \\
&= 1.11 \text{ miles}
\end{aligned}
$$

TABLE 2.15 DO Concentrations at Various Locations

t, days	Distance below outfall, miles	DO deficit, mg/L	Expected DO* mg/L
0.0	0.00	1.92	6.10
0.1	1.11	2.06	5.96
0.2	2.22	2.17	5.85
0.3	3.33	2.26	5.74
0.4	4.44	2.32	5.70
0.5	5.55	2.37	5.64
0.6	6.67	2.40	5.62
0.7	7.78	2.41	5.61
0.8	**7.89**	**2.42**	**5.60**
0.9	10.00	2.41	5.61
1.0	11.11	2.40	5.62
1.1	12.22	2.75	5.27
1.2	13.33	3.04	4.98
1.3	14.44	3.27	4.75
1.4	15.55	3.45	4.57
1.5	16.67	3.59	4.43
1.6	17.78	3.69	4.33
1.8	18.89	3.81	4.21
2.0	22.22	3.83	4.19
2.1	**23.33**	**3.81**	**4.21**
2.2	24.44	3.79	4.23

*Expected DO = Saturated DO (8.02 mg/L) − DO deficit.

Step 9. Explanation

From Table 2.15, it can be seen that minimum DO of 5.60 mg/L occurred at t = 0.8 days due to carbonaceous demand. After this location, the stream starts recovery. However, after t = 1.0 days, the stream DO is decreasing due to nitrogenous demand. At t = 2.0, it is the critical location (22.22 miles below the outfall) with a critical DO deficit of 3.83 mg/L and the DO level in the water is 4.19 mg/L.

13.4 Simplified Oxygen Sag Computations

The simplified oxygen sag computation was suggested by Le Bosquet and Tsivoglou (1950). As shown earlier at the critical point

$$\frac{dD}{dt} = 0 \tag{2.48}$$

and

$$K_d L_c = K_2 D_c \tag{2.49}$$

Let the lb/day of first-stage BOD be C

$$C\text{ (lb/day)} = \frac{L_a\text{ (mg/L) } Q\text{ (cfs)} \times 62.38\text{ (lb/cf)} \times 86{,}400\text{ (s/d)}}{1{,}000{,}000}$$

$$= 5.39 L_a Q$$

or

$$L_a = \frac{C}{5.39Q} = \frac{C_1}{Q} \tag{2.91}$$

where C_1 = constant

If $D_a = 0$, then D_c is a function of L_a, or

$$D = L_a C_2 = \frac{C_1}{Q} \times C_2 = \frac{C_3}{Q} \tag{2.92}$$

where C_2, C_3 = constants

$$DQ = C_3 \tag{2.93}$$

Since $K_d L_c = K_2 D_c$

$$D_c = \frac{K_d}{K_2} L_c = L_a C_2 = \frac{C_3}{Q} \tag{2.94}$$

Hence the minimum allowable DO concentration at the critical point (DO_C) is the DO saturation value (S, or DO_{sat}) minus D_c. It can be written as

$$DO_C = S - D_c \tag{2.95}$$

$$= S - C_3\left(\frac{1}{Q}\right) \tag{2.95a}$$

From the above equation, there is a linear relationship between DO_C and $1/Q$. Therefore, a plot of DO_C versus $1/Q$ will give a straight line, where S is the y-intercept and C_3 is the slope of the line. Observed values from field work can be analyzed by the method of least squares to determine the degree of correlation.

14 SOD OF DO USAGE

The SOD portion of DO usage is calculated using the following formula:

$$DO_{sod} = \frac{3.28Gt}{H} \tag{2.96}$$

where DO_{sod} = oxygen used per reach, mg/L
G = the SOD rate, g/(m^2 · d)
t = the retention time for the reach
H = the average water depth in the reach, ft

All biological rates, including BOD and SOD, have to be corrected for temperature using the basic Arrhenius formula:

$$R_T = R_A(\theta^{T-A}) \tag{2.97}$$

where R_T = biological oxygen usage rate at a temperature, T°C
R_A = biological oxygen usage rate at ambient or standard temperature (20°C as usual), A°C
θ = proportionality constant, 1.047 for river or stream

EXAMPLE: A stream reach has an average depth of 6.6 ft. Its SOD rate was determined as 3.86 g/(m^2 · day) at 24°C. The detention time for the reach is 0.645 days. What is the SOD portion of DO usage at the standard temperature of 20°C?

Solution:

Step 1. Determine SOD rate G at 20°C.

$$G_{24} = G_{20}(1.047^{24-20})$$

$$G_{20} = G_{24}/(1.047)^4 = 3.86/1.2017 = 3.21 \text{ (g/(m}^2\cdot\text{d))}$$

Step 2. Calculate DO_{sod}:

$$\begin{aligned} DO_{sod} &= \frac{3.28Gt}{H} \\ &= \frac{3.28 \times 3.21 \times 0.645}{6.6} \\ &= 1.03 \text{ (mg/L)} \end{aligned}$$

15 APPORTIONMENT OF STREAM USERS

In the United States, effluent standards for wastewater treatment plants are generally set for BOD, total suspended solids (TSS), and ammonia-nitrogen (NH_3-N) concentrations. Each plant often should meet standards of 10–10–12 (mg/L for BOD, TSS, and NH_3-N, respectively), right or wrong. It is not like classically considering the maximum use of stream-assimilative capacity. However, in some parts of the world, maximum usage of stream natural self-purification capacity may be a valuable tool for cost saving of wastewater treatment. Thus, some concepts of apportionment of self-purification capacity of a stream among different users are presented below.

The permissible load equation can be used to compute the BOD loading at a critical point L_c, as previously stated (Eq. (2.84); Thomas, 1948):

$$\log L_c = \log D_c + \left[1 + \frac{k_r}{k_2 - k_r}\left(1 - \frac{D_a}{D_c}\right)^{0.418}\right]\log\left(\frac{k_2}{k_r}\right) \tag{2.98}$$

where L_c = BOD load at the critical point
D_c = DO deficit at the critical point
D_a = DO deficit at upstream pollution point
k_r = river BOD removal rate to the base 10, per day
k_2 = reoxygenation rate to the base 10, per day

L_c also may be computed from limited BOD concentration and flow. Also, if city A and city B both have the adequate degree of treatment to meet the regulatory requirements, then

$$L_c = p\alpha L_A + \beta L_B \tag{2.99}$$

where p = part or fraction of L_A remaining at river point B
α = fraction of L_A discharged into the stream
β = fraction of L_B discharged into the stream
L_A = BOD load at point A, lb/d or mg/L
L_B = BOD load at point B, lb/d or mg/L

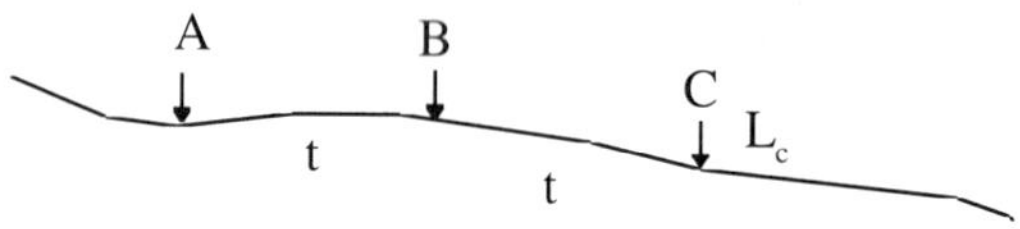

If $L_A = L_B = L$

$$L_c = p\alpha L + \beta L \tag{2.99a}$$

A numerical value of L_c is given by Eq. (2.98) and a necessary relationship between α, β, p, L_A, and L_B is set up by Eq. (2.99). The value p can be computed from

$$\log p = -k_r t$$

Since

$$\log \frac{Lt}{L_a} = -k_r t = \log \text{fraction remaining}$$

Therefore

$$\log p = -k_r t \tag{2.100}$$

or

$$p = 10^{-k_r t} \tag{2.100a}$$

The apportionment factors α and β can be determined by the following methods.

15.1 Method 1

Assume: City A is further upstream than city B, with time-of-travel t given, by nature, a larger degree of stream purification capacity. The time from city B to the lake inlet is also t. Since

$$y = L_a(1 - 10^{-k_r t})$$

$$\frac{y}{L_a} = 1 - 10^{-k_r t} = 1 - p \tag{2.101}$$

The proportion removed by the stream of BOD added by city A is

$$\frac{y}{L_A} = 1 - 10^{-k_r t \times 2} = 1 - 10^{-2k_r t} \tag{2.102}$$

$$= 1 - p^2 \tag{2.102a}$$

The proportion removed by the stream of BOD added by city B is

$$\frac{y}{L_B} = 1 - 10^{-k_r t} = 1 - p \tag{2.102b}$$

Then the percent amount removed BOD from cities A and B, given as P_A and P_B, respectively, are

$$P_A = \frac{1 - p^2}{(1 - p^2) + (1 - p)} = \frac{(1 - p)(1 + p)}{(1 - p)(1 + p) + (1 - p)}$$

$$= \frac{(1 - p)(1 + p)}{(1 - p)(1 + p + 1)}$$

$$P_A = \frac{1 + p}{2 + p} \tag{2.103}$$

$$P_B = \frac{1 - p}{(1 - p^2) + (1 - p)} = \frac{(1 - p)}{(1 - p)(1 + p + 1)}$$

$$P_B = \frac{1}{2 + p} \tag{2.104}$$

Calculate the degree of BOD removal proportion:

$$\frac{\alpha}{\beta} = \frac{P_A}{P_B} = \frac{(1 + p)/(2 + p)}{1/(2 + p)} = \frac{1 + p}{1}$$

$$\alpha = (1 + p)\beta \tag{2.105}$$

Solving Eqs. (2.99a) and (2.105) simultaneously for α and β:

$$L_c = p\alpha L + \beta L$$

$$\frac{L_c}{L} = p\alpha + \beta$$

$$\frac{L_c}{L} = p(1 + p)\beta + \beta = (p + p^2 + 1)\beta$$

$$\beta = \frac{L_c}{L}\left(\frac{1}{1 + p + p^2}\right) \tag{2.106}$$

$$\alpha = (1 + p)\beta$$

$$= \frac{L_c}{L} \cdot \frac{1 + p}{1 + p + p^2} \tag{2.107}$$

The required wastewater treatment plant efficiencies for plants A and B are $1-\alpha$ and $1-\beta$, respectively. The ratio of cost for plant A to plant B is $(1-\alpha)$ to $(1-\beta)$.

EXAMPLE 1: Assume all conditions are as in Method 1. The following data are available: $k_r = 0.16$ per day; $t = 0.25$ days; flow at point C, $Q_c = 58$ MGD; $L = 6910$ lb/d; and regulation required $L_c \leq 10$ mg/L. Determine the apportionment of plants A and B, required plant BOD removal efficiencies, and their cost ratio.

Solution:

Step 1. Compute allowable L_c.

$$L_c = 10 \text{ mg/L} \times 8.34 \text{ lb/d/MGD} \cdot \text{mg/L} \times 58 \text{ MGD}$$

$$= 4837 \text{ lb/d}$$

Step 2. Compute p

$$p = 10^{-k_r t} = 10^{-0.16\times 0.25}$$
$$= 0.912$$

Step 3. Calculate α and β by Eqs. (2.106) and (2.107):

$$\alpha = \frac{L_c}{L}\left(\frac{1 + p}{1 + p + p^2}\right) = \frac{4837}{6910}\left(\frac{1 + 0.912}{1 + 0.912 + 0.912^2}\right)$$
$$= 0.70 \times 0.697$$
$$= 0.49$$
$$\beta = \frac{L_c}{L}\left(\frac{1}{1 + p + p^2}\right) = 0.70 \times \frac{1}{2.744}$$
$$= 0.255$$

Step 4. Determine percent BOD removal required

For plant A:

$$1 - \alpha = 1 - 0.49 = 0.51 \qquad \text{i.e. 51\% removal needed}$$

For plant B:

$$1 - \beta = 1 - 0.255 = 0.745 \qquad \text{i.e. 74.5\% removal needed}$$

Step 5. Calculate treatment cost ratio

$$\frac{\text{Cost, plant A}}{\text{Cost, plant B}} = \frac{1 - \alpha}{1 - \beta} = \frac{0.51}{0.745} = \frac{0.68}{1} = \frac{1}{1.46}$$

15.2 Method 2

Determine a permissible BOD load for each city as if it were the only city using the river or stream. In other words, calculate maximum BOD load L_A for plant A so that dissolved oxygen concentration is maintained above 5 mg/L (most state requirements), assuming that plant B does not exist. Similarly, also calculate maximum BOD load L_B for plant B under the same condition. It should be noted that L_B may be larger than L_A and this cannot be the case in Method 1.

The BOD removed by the stream from that added

at point A: referring to Eq. (2.102a)

$$(1 - p^2)L_A$$

at point B: referring to Eq. (2.102b)

$$(1 - p)L_B$$

The percentage removed is

at point A,

$$P_A = \frac{(1 - p^2)L_A}{(1 - p^2)L_A + (1 - p)L_B} \tag{2.108}$$

at point B,

$$P_B = \frac{(1 - p)L_B}{(1 - p^2)L_A + (1 - p)L_B} \tag{2.109}$$

The degree of BOD removal is

$$\frac{\alpha}{\beta} = \frac{P_A}{P_B} \tag{2.105}$$

Solve equations (2.99), (2.105), (2.108), and (2.109) for α and β. Also determine $(1-\alpha)$ and $(1-\beta)$ for the required wastewater treatment efficiencies and cost ratios as $(1-\alpha)/(1-\beta)$.

EXAMPLE 2: All conditions and questions are the same as for Example 1, except use Method 2.

Solution:

Step 1. Compute allowable L_A and L_B.

For point B, using Eq. (1.100)

$$\log p = \log \frac{L_t}{L_B} = -k_r t = -0.16 \times 0.25 = -0.04$$

$$p = \frac{L_t}{L_B} = 10^{-0.04} = 0.912$$

$$\begin{aligned} L_B &= L_t/0.912 \; (L_t = L_c \text{ in this case}) \\ &= 4837/0.912 \\ &= 5304 \text{ (lb/d)} \end{aligned}$$

For point A,

$$\frac{L_c}{L_A} = 10^{-0.16\times0.25\times2} = 10^{-0.08} = 0.8318$$

$$\begin{aligned} L_A &= 4837 \text{ lb/d}/0.8318 \\ &= 5815 \text{ lb/d} \end{aligned}$$

Step 2. Calculate P_A and P_B.

$$\begin{aligned} P_A &= \frac{(1 - p^2)L_A}{(1 - p^2)L_A + (1 - p)L_B} \\ &= \frac{(1 - 0.912^2)5815}{(1 - 0.912^2)5815 + (1 - 0.912)5304} \\ &= \frac{978.4}{978.4 + 466.8} \\ &= 0.677 \\ P_B &= 1 - 0.677 = 0.323 \end{aligned}$$

Step 3. Determine α and β.

$$\begin{aligned} \alpha/\beta &= P_A/P_B = 0.677/0.323 = 2.1/1 \\ \alpha &= 2.1\beta \\ L_c &= p\alpha\, L_A + \beta\, L_B \\ 4837 &= 0.912 \times (2.1\beta) \times 5815 + \beta \times 5304 \\ 4837 &= 11{,}137\beta + 5304\beta \\ \beta &= 4837/16{,}441 = 0.29 \\ \alpha &= 2.1 \times 0.29 = 0.61 \end{aligned}$$

Step 4. Determine degree of BOD removal

For plant A

$$1 - \alpha = 1 - 0.61 = 0.39 \qquad \text{i.e. 39\% removal efficiency}$$

For plant B

$$1 - \beta = 1 - 0.29 = 0.71 \qquad \text{i.e. 71\% removal efficiency}$$

Step 5. Determine cost ratio

$$\frac{\text{Plant A}}{\text{Plant B}} = \frac{1 - \alpha}{1 - \beta} = \frac{0.39}{0.71} = \frac{1}{1.82} \text{ or } \frac{0.55}{1}$$

15.3 Method 3

The principle of this method is to load the river to the utmost so as to minimize the total amount of wastewater treatment. The proportion of P_A and P_B can be computed by either Method 1 or Method 2. The cost of treatment can be divided in proportion to population.

There are two cases of initial BOD loading:

Case 1. $L_A > L$ (L is the BOD load of plant A influent)

- Let plant A discharge L without a treatment.
- Have plant B treat an amount equal to $L - L_c$.
- City A is assessed for $P_A(L - L_c)$.
- City B is assessed for $P_B(L - L_c)$.

Case 2. $L_A < L$

- Let plant A discharge L_A and treat $L - L_A$.
- Have plant B treat an amount equal to $(L - L_c) - pL_A$.
- Then the total treatment will remove $(L - L_A) + (L - L_c - pL_A) = 2L - (1 + p)L_A - L_c$.
- City A is assessed for $P_A[2L - (1 + p)L_A - L_c]$.
- City B is assessed for $P_B[2L - (1 + p)L_A - L_c]$.

EXAMPLE 3: Using data listed in Examples 1 and 2, determine the cost ratio assessed to plants A and B by Method 3.

Solution:

Step 1. Select case

Since $L_A = 5815$ lb/d $< L = 6910$ lb/d

Case 2 will be applied

Step 2. Determine amounts needed to be treated.

For plant A:

$$6910 - 5815 = 1095 \text{ (lb/d treated)}$$
$$5815 = \text{lb/d discharged without treatment}$$

For plant B:

$$L - L_c - pL_A = 6910 - 4837 - 0.912 \times 1095$$
$$= 1074 \text{ (lb/d)}$$

Total treatment = 1095 + 1074 = 2169 (lb/d)

Step 3. Compute cost assessments

For Example 2,

$$P_A = 0.677$$

$$P_B = 0.323$$

For plant A

$$\text{Cost} = 0.677 \times \text{treatment cost of 2169 lb/d}$$

For plant B

$$\text{Cost} = 0.323 \times \text{treatment cost of 2169 lb/d}$$

16 VELZ REAERATION CURVE (A PRAGMATIC APPROACH)

Previous sections discuss the conceptual approach of the relationships between BOD loading and DO in the stream. Another approach for determining the waste-assimilative capacity of a stream is a method first developed and used by Black and Phelps (1911) and later refined and used by Velz (1939). For this method, deoxygenation and reoxygenation computations are made separately, then added algebraically to obtain the net oxygen balance. This procedure can be expressed mathematically as

$$\text{DO}_{\text{net}} = \text{DO}_{\text{a}} + \text{DO}_{\text{rea}} - \text{DO}_{\text{used}} \tag{2.110}$$

where DO_{net} = dissolved oxygen at the end of a reach
DO_{a} = initial dissolved oxygen at beginning of a reach
DO_{rea} = dissolved oxygen absorbed from the atmosphere
DO_{used} = dissolved oxygen consumed biologically

All units are in lb/d or kg/d.

16.1 Dissolved Oxygen Used

The dissolved oxygen used (BOD) can be computed with only field-observed dissolved oxygen concentrations and the Velz reoxygenation curve. From the basic equation, the dissolved oxygen used in pounds per day can be expressed as:

$$\text{DO}_{\text{used}} = (\text{DO}_{\text{a}} - \text{DO}_{\text{net}}) + \text{DO}_{\text{rea}} \text{ (in lb/d)}$$

The terms DO_{a} and DO_{net} are the observed field dissolved oxygen values at the beginning and end of a subreach, respectively, and DO_{rea} is the reoxygenation in the subreach computed by the Velz curve. If the dissolved oxygen usage in the river approximates a first-order biological reaction, a plot of the summations of the DO_{used} versus time-of-travel can be fitted to the equation:

$$\text{DO}_{\text{used}} = L_a[1 - \exp(-K_1 t)] \tag{2.111}$$

and L_a and K_1 are determined experimentally in the same manner as for the Streeter–Phelps method. The fitting of the computed values to the equations must be done by trial and error, and a digital computer is utilized to facilitate this operation.

16.2 Reaeration

The reaeration term (DO_{rea}) in the equation is the most difficult to determine. In their original work, Black and Phelps (1911) experimentally derived a reaeration equation based on principles of gas transfer of diffusion across a thin water layer in acquiescent or semiquiescent system. A modified but equivalent form of the original Black and Phelps equation is given by Gannon (1963):

$$R = 100\left(\frac{1 - B_0}{100}\right)(81.06)\left(e^{-K} + \frac{e^{-9K}}{9} + \frac{e^{-25K}}{25} + \cdots\right) \tag{2.112}$$

where R = percent of saturation of DO absorbed per mix
B_0 = initial DO in percent of saturation
K = $\pi^2\, am/4L^2$ in which m is the mix or exposure time in hours, L is the average depth in centimeters, and a is the diffusion coefficient used by Velz

The diffusion coefficient a was determined by Velz (1947) to vary with temperature according to the expression:

$$a_T = a_{20}(1.1^{T-20}) \tag{2.113}$$

where a_T = diffusion coefficient at T°C
a_{20} = diffusion coefficient at 20°C

when the depth is in feet,

$$a_{20} = 0.00153$$

when the depth is in centimeters,

$$a_{20} = 1.42$$

Although the theory upon which this reaeration equation was developed is for quiescent conditions, it is still applicable to moving, and even turbulent, streams. This can be explained in either of two ways (Phelps, 1944):

> (1) Turbulence actually decreases the effective depth through which diffusions operate. Thus, in a turbulent stream, mixing brings layers of saturated water from the surface into intimate contact with other less oxygenated layers from below. The actual extent of such mixing is difficult to envision, but it clearly depends upon the frequency with which surface layers are thus transported. In this change-of-depth concept therefore there is a time element involved.
> (2) The other and more practical concept is a pure time effect. It is assumed that the actual existing conditions of turbulence can be replaced by an equivalent condition composed of successive periods of perfect quiescence between which there is instantaneous and complete mixing. The length of the quiescent period then becomes the time of exposure in the diffusion formula, and the total aeration per hour is the sum of the successive increments or, for practical purposes, the aeration period multiplied by the periods per hour.

Because the reaeration equation involves a series expansion in its solution, it is not readily solved by desk calculations. To facilitate the calculations, Velz (1939, 1947) published a slide-rule curve solution to the equation. This slide-rule curve is reprinted here as Fig. 2.11 (for its use, consult Velz, 1947). Velz's curve has been verified as accurate by two independent computer checks, one by Gannon (1963) and the other by Butts (1963).

Although the Velz curve is ingenious, it is somewhat cumbersome to use. Therefore, a nomograph (Butts and Schnepper, 1967) was developed for a quick, accurate desk solution of R at zero initial DO (B_0 in the Black and Phelps equation). The nomograph, presented in Fig. 2.12, was constructed on the premise that the exposure or mix time is characterized by the equation

$$M = (13.94)\log_e(H) - 7.45 \tag{2.114}$$

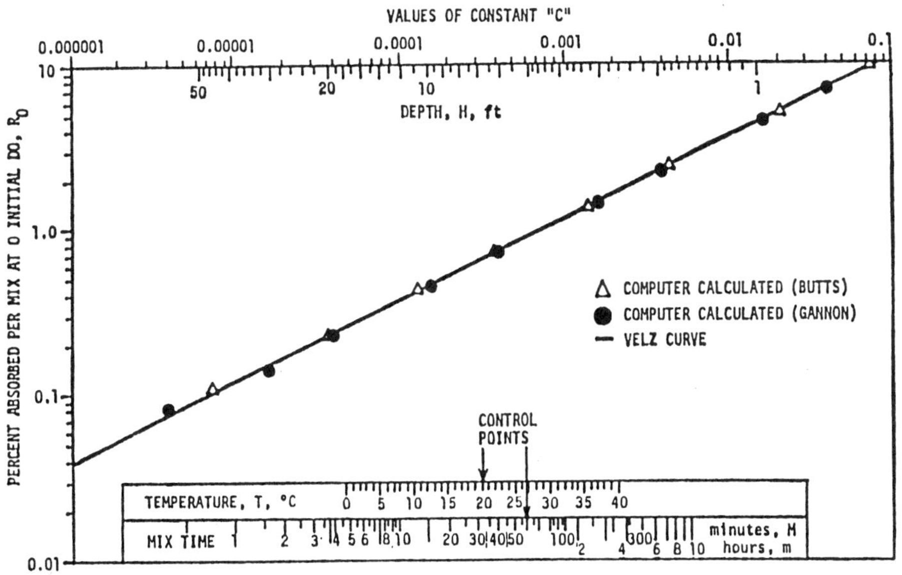

FIGURE 2.11 Standard reoxygeneration curve (*Butts et al., 1973*).

in which M is the mix time in minutes and H is the water depth in feet. This relationship was derived experimentally and reported by Gannon (1963) as valid for streams having average depths greater than 3 ft. For an initial DO greater than zero, the nomograph value is multiplied by $(1 - B_0)$ expressed as a fraction because the relationship between B_0 and the equation is linear.

The use of the nomograph and the subsequent calculations required to compute the oxygen absorption are best explained by the example that follows:

EXAMPLE: Given that $H = 12.3$ ft; initial DO of stream = 48% of saturation; $T = 16.2$°C; DO load at saturation at 16.2°C = 8800 lb; and time-of-travel in the stream reach = 0.12 days. Find the mix time (M) in minutes; the percent DO absorbed per mix at time zero initial DO (R_0); and the total amount of oxygen absorbed in the reach (DO_{rea}).

Solution:

Step 1. Determine M and R_0.

From Fig. 2.12, connect 12.3 ft on the depth-scale with 16.2°C on the temperature-scale. It can be read that:

$$M = 27.5 \text{ min on the mix time-scale, and}$$

$$R_0 = 0.19\% \text{ on the } R_0\text{-scale}$$

Step 2. Compute DO_{rea}

Since R_0 is the percent of the saturated DO absorbed per mix when the initial DO is at 100% deficit (zero DO), the oxygen absorbed per mix at 100% deficit will be the product of R_0 and the DO load at saturation. For this example, lb/mix per 100% deficit is

$$8800 \text{ lb} \times \frac{0.19}{100} = 16.72 \text{ (lb/mix per 100\% deficit)}$$

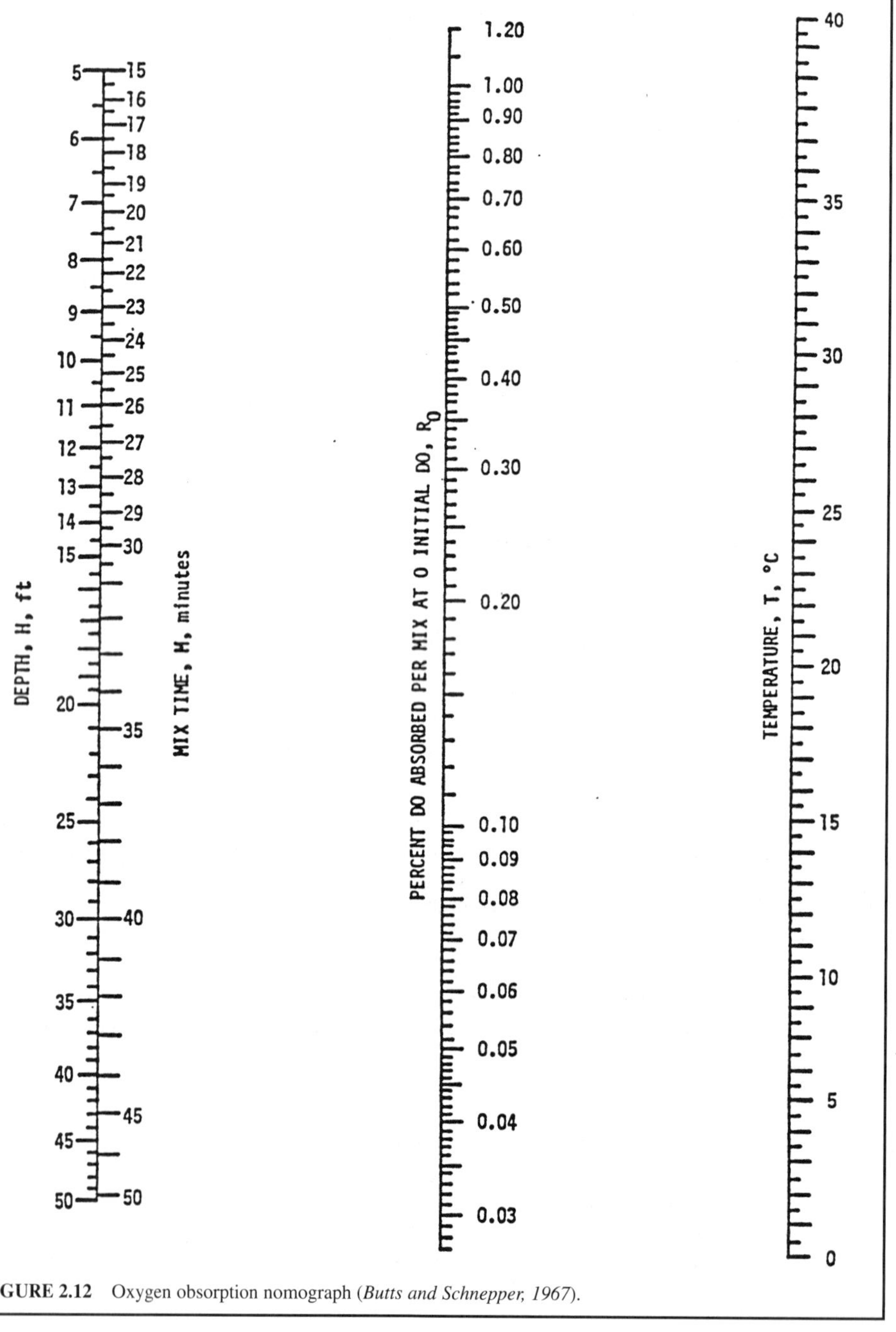

FIGURE 2.12 Oxygen obsorption nomograph (*Butts and Schnepper, 1967*).

However, the actual DO absorbed (at 48% saturation) per mix is only

$$16.72 \times [1 - (48/100)] = 8.7 \text{ (lb/mix per 100\% deficit)}$$

Because the amount absorbed is directly proportional to the deficit, the number of mixes per reach must be determined:

The time-of-travel

$$\begin{aligned} t &= 0.12 \text{ days} \\ &= 0.12 \text{ days} \times 1440 \text{ min/d} \\ &= 173 \text{ min} \end{aligned}$$

The mix time

$$M = 27.5 \text{ min}$$

Consequently, the number of mixes in the reach = 173/27.5 = 6.28.

Reaeration in the reach is therefore equal to the product of the DO absorbed per mix × the number of mixes per reach. DO_{rea} is computed according to the expression:

$$\begin{aligned} DO_{rea} &= \left(1 - \frac{\%\text{ DO saturation}}{100}\right)\left(\frac{R_0}{100}\right)\left(\frac{t}{M}\right)(\text{DO saturation load}) \qquad (2.115) \\ &= \left(1 - \frac{48}{100}\right)\left(\frac{0.19}{100}\right)\left(\frac{173}{27.5}\right)(8800) \\ &= 8.7 \times 6.28 \\ &= 54.6 \text{ (lb)} \end{aligned}$$

17 STREAM DO MODEL (A PRAGMATIC APPROACH)

Butts and his coworkers (Butts *et al.*, 1970, 1973, 1974, 1975, 1981; Butts, 1974; Butts and Shackleford, 1992) of the Illinois State Water Survey expanded Velz's oxygen balance equation. For the pragmatic approach to evaluate the BOD/DO relationship in a flowing stream, it is formulated as follows:

$$DO_n = DO_a - DO_u + DO_r + DO_x \qquad (2.116)$$

where DO_n = net dissolved oxygen at the end of a stream reach
DO_a = initial dissolved oxygen at the beginning of a reach
DO_u = dissolved oxygen consumed biologically within a reach
DO_r = dissolved oxygen derived from natural reaeration within a reach
DO_x = dissolved oxygen derived from channel dams, tributaries, etc.

Details of computation procedures for each component of the above equation have been outlined in several reports mentioned above. Basically the above equation uses the same basic concepts as employed in the application of the Streeter–Phelps expression and expands the Velz oxygen balance equation by adding DO_x. The influences of dams and tributaries are discussed below.

17.1 Influence of a Dam

Low head channel dams and large navigation dams are very important factors when assessing the oxygen balance in a stream. There are certain disadvantages from a water quality standpoint associated with such structures but their reaeration potential is significant during overflow, particularly if a low DO concentration exists in the upstream pooled waters. This source of DO replenishment must be considered.

Procedures for estimating reaeration at channel dams and weirs have been developed by researchers in England (Gameson, 1957; Gameson *et al.*, 1958; Barrett *et al.*, 1960; Grindrod, 1962). The methods are easily applied and give satisfactory results. Little has been done in developing similar procedures for large navigation and power dams. Preul and Holler (1969) investigated larger structures.

Often in small sluggish streams more oxygen may be absorbed by water overflowing a channel dam than in a long reach between dams. However, if the same reach were free flowing, this might not be the case; i.e. if the dams were absent, the reaeration in the same stretch of river could conceivably be greater than that provided by overflow at a dam. If water is saturated with oxygen, no uptake occurs at the dam overflow. If it is supersaturated, oxygen will be lost during dam overflow. Water, at a given percent deficit, will gain oxygen.

The basic channel dam reaeration formula takes the general form

$$r = 1 + 0.11qb(1 + 0.046T)h \tag{2.117}$$

where r = dissolved oxygen deficit ratio at temperature T
q = water quality correction factor
b = weir correction factor
T = water temperature, °C
h = height through which the water falls, ft

The deficit dissolved oxygen ratio is defined by the expression

$$r = (C_s - C_A)/(C_s - C_B) = D_A/D_B \tag{2.118}$$

where C_A = dissolved oxygen concentration upstream of the dam, mg/L
C_B = dissolved oxygen concentration downstream of the dam, mg/L
C_s = dissolved oxygen saturation concentration, mg/L
D_A = dissolved oxygen deficit upstream of the dam, mg/L
D_B = dissolved oxygen deficit downstream of the dam, mg/L

Although Eqs. (2.117) and (2.118) are rather simplistic and do not include all potential parameters which could affect the reaeration of water overflowing a channel dam, they have been found to be quite reliable in predicting the change in oxygen content of water passing over a dam or weir (Barrett *et al.*, 1960). The degree of accuracy in using the equations is dependent upon the estimate of factors q and b.

For assigning values for q, three generalized classifications of water have been developed from field observations. They are $q = 1.25$ for clean or slightly polluted water; $q = 1.0$ for moderately polluted water; and $q = 0.8$ for grossly polluted water. A slightly polluted water is one in which no noticeable deterioration of water quality exists from sewage discharges; a moderately polluted stream is one which receives a significant quantity of sewage effluent; and a grossly polluted stream is one in which noxious conditions exist.

For estimating the value of b, the geometrical shape of the dam is taken into consideration. This factor is a function of the ratio of weir coefficients W of various geometrical designs to that of a free weir where

$$W = (r - 1)/h \tag{2.119}$$

Weir coefficients have been established for a number of spillway types, and Gameson (1957) in his original work has suggested assigning b values as follows:

Spillway type	b
Free	1.0
Step	1.3
Slope (ogee)	0.58
Sloping channel	0.17

For special situations, engineering judgment is required. For example, a number of channel dams in Illinois are fitted with flashboards during the summer. In effect, this creates a free fall in combination with some other configuration. A value of b, say 0.75, could be used for a flashboard installation on top of an ogee spillway (Butts *et al.*, 1973). This combination would certainly justify a value less than 1.0, that for a free weir, because the energy dissipation of the water flowing over the flashboards onto the curved ogee surface would not be as great as that for a flat surface such as usually exists below free and step weirs.

Aeration at a spillway takes place in three phases: (1) during the fall; (2) at the apron from splashing; and (3) from the diffusion of oxygen due to entrained air bubbles. Gameson (1957) has found that little aeration occurs in the fall. Most DO uptake occurs at the apron; consequently, the greater the energy dissipation at the apron the greater the aeration. Therefore, if an ogee spillway is designed with an energy dissipator, such as a hydraulic jump, the value of b should increase accordingly.

EXAMPLE: Given $C_A = 3.85$ mg/L, $T = 24.8°C$, $h = 6.6$ ft (2.0 m), $q = 0.9$, and $b = 0.30$ (because of hydraulic jump use 0.30).

Find DO concentration below the dam C_B.

Solution:

Step 1. Using dam reaeration equation to find r

$$\begin{aligned} r &= 1 + 0.11qb(1 + 0.046T)h \\ &= 1 + 0.11 \times 0.9 \times 0.30(1 + 0.046 \times 24.8) \times 6.6 \\ &= 1.42 \end{aligned}$$

Step 2. Determine DO saturation value. From Table 2.2, or by calculation, for C_s at 24.8°C this is

$$C_s = 8.21 \text{ mg/L}$$

Step 3. Compute C_B. Since

$$r = (C_s - C_A)/(C_s - C_B)$$

therefore

$$\begin{aligned} C_B &= C_s - (C_s - C_A)/r \\ &= 8.21 - (8.21 - 3.85)/1.42 \\ &= 8.21 - 3.07 \\ &= 5.14 \text{ (mg/L)} \end{aligned}$$

17.2 Influence of Tributaries

Tributary sources of DO are often an important contribution in deriving a DO balance in stream waters. These sources may be tributary streams or outfalls of wastewater treatment plants. A tributary contribution of DO_x can be computed based on mass balance basis of DO, ammonia, and BOD values from tributaries. The downstream effect of any DO input is determined by mass balance computations: in terms of pounds per day the tributary load can simply add to the mainstream load occurring above the confluence. The following two examples are used to demonstrate the influence of tributary sources of DO. Example 1 involves a tributary stream. Example 2 involves the design of an outfall structure to achieve a minimum DO at the point of discharge.

EXAMPLE 1: Given:

Tributary flow	$Q_1 = 123$ cfs
Tributary DO	$= 6.7$ mg/L
Mainstream flow	$Q_2 = 448$ cfs
Mainstream DO	$= 5.2$ mg/L

Find: DO concentration and DO load at the confluence.

Solution:

Step 1. Determine DO concentration.

Assuming there is a complete mix immediately below the confluence, the DO concentration downstream on the confluence can be determined by mass balance.

$$\begin{aligned} DO &= \frac{Q_1 \times DO_1 + Q_2 \times DO_2}{Q_1 + Q_2} \\ &= (123 \times 6.7 + 448 \times 5.2)/(123 + 448) \\ &= 5.5\ (\text{mg/L}) \end{aligned}$$

Step 2. Compute DO loadings. Since

$$\begin{aligned} \text{DO load} &= (\text{DO, mg/L}) \times (\text{Q, cfs}) \\ &= \text{DO} \times \text{Q}\left(\frac{\text{mg}}{\text{L}} \times \text{cfs} \times 28.32\,\frac{\text{L/s}}{\text{cfs}}\right) \\ &= 28.32 \times \text{DO} \times \text{Q}\left(\frac{\text{mg}}{\text{s}} \times \frac{1\ \text{lb}}{454{,}000\ \text{mg}} \times \frac{86{,}400\ \text{s}}{\text{day}}\right) \\ &= 5.39 \times \text{DO} \times \text{Q lb/d} \\ \text{Tributary load} &= 5.39 \times 6.7 \times 123\ \text{lb/d or} = 5.39\ \text{lb/(d} \cdot \text{mg/L} \cdot \text{cfs)} \times \text{DO mg/L} \times \text{Q cfs} \\ &= 4442\ \text{lb/d} \end{aligned}$$

Note: The factor 5.39 is the same as that in Eq. (2.91)

Similarly,

$$\begin{aligned} \text{Mainstream load} &= 5.39 \times 5.2 \times 448 \\ &= 12{,}557\ \text{(lb/d)} \\ \text{Total DO load} &= 4442\ \text{lb/d} + 12{,}557\ \text{lb/d} \\ &= 16{,}999\ \text{lb/d} \end{aligned}$$

EXAMPLE 2: The difference in elevation between the outfall crest and the 7-day, 10-year low flow is 4.0 ft (1.22 m). Utilizing this head difference, determine if a free-falling two-step weir at the outfall will insure a minimum DO of 5.0 mg/L at the point of discharge. The flow of wastewater effluent and the stream are, respectively, 6.19 and 14.47 cfs (175 and 410 L/s, or 4 MGD and 10 MGD). DO for the effluent and stream are 1.85 and 6.40 mg/L, respectively. The temperature of the effluent is 18.9°C. The stream is less than moderately polluted. Determine the stream DO at the point of effluent discharge and the oxygen balance.

Solution:

Step 1. The water quality correction factor is selected as $q = 1.1$

$$\begin{aligned} b &= 1.3 \text{ for free fall step-weir} \\ C_s &= 9.23 \text{ mg/L, when } T = 18.9°\text{C from Table 2.2} \\ C_A &= 1.85 \text{ mg/L as effluent DO} \\ h &= 4.0 \text{ ft} \end{aligned}$$

Step 2. Determine r with the channel dam reaeration model

$$\begin{aligned} r &= 1 + 0.11qb(1 + 0.046T)h \\ &= 1 + 0.11 \times 1.1 \times 1.3(1 + 0.046 \times 18.9) \times 4 \\ &= 2.18 \end{aligned}$$

Step 3. Compute C_B

$$r = (C_s - C_A)/(C_s - C_B)$$
$$C_B = C_s - (C_s - C_A)/r$$
$$= 9.23 - (9.23 - 1.85)/2.18$$
$$= 5.85 \text{ (mg/L)}$$

Note: In this case, C_B is the effluent DO as it reaches the stream.

Step 4. Compute the resultant DO in the stream assuming there is a complete mixing with the stream water, the resultant DO at the point of discharge is

$$\text{DO} = \frac{5.85 \times 6.19 + 6.40 \times 14.47}{6.19 + 14.47}$$
$$= 6.23 \text{ (mg/L)}$$

Step 5. Calculate oxygen mass balance

$$\text{DO load from effluent} = 5.39 \times \text{DO} \times Q$$
$$= 5.39 \times 5.85 \times 6.19$$
$$= 195 \text{ (lb/d)}$$

DO load in stream above the point of effluent discharge

$$= 5.39 \times \text{DO} \times Q$$
$$= 5.39 \times 6.40 \times 14.47$$
$$= 499 \text{ (lb/d)}$$

The total amount of oxygen below the point of discharge is 694 lb/d.

17.3 DO Used

The term DO_{used} (DO_u) in Eq. (2.111) represents the oxygen consumed biologically within a stream reach. This term can be determined by three methods (Butts *et al.*, 1973): (1) observed DO concentrations in conjunction with reaeration estimate; (2) bottle BOD and deoxygenation rate determinations of river water samples; and (3) long-term bottle BOD progression evaluation of a wastewater effluent. The method used will probably be dictated by the existing data or the resources available for collecting usable data.

The term DO_u includes DO usage due to carbonaceous and nitrogenous BOD and to SOD, as stated previously. The ratio of DO contribution by algal photosynthesis to DO consumption by algal respiration is assumed to be unity, although it can handle values greater or less than one when derived on a diurnal basis. For a series of stream reaches, each incremental DO_u value is added and the accumulated sums, with the corresponding time-of-travel in the stream, are fitted to the first-order exponential expression

$$y = L_a(1 - e^{-K_d t}) \tag{2.120}$$

where y = oxygen demand exerted (DO_u)
L_a = ultimate oxygen demand, including carbonaceous and nitrogenous
K_d = in-stream deoxygenation coefficient to the base e, per day
t = time-of-travel, days

Note: The coefficient K_d is comparable to the composite of the terms K_C and K_N previously defined in the discussion regarding the conceptual approach to waste-assimilative analysis.

The use of field DO values in estimating the waste-degradation characteristics of a stream has certain advantages: the need for laboratory BOD tests is eliminated which saves time and cost; the reliability of the results should be better since the measurement of dissolved oxygen is far more precise and accurate than the BOD test; also, stream measured DO concentrations take into account the in situ oxygen demand in the stream, which includes both dissolved and benthic demand, whereas laboratory BOD results generally reflect only the oxygen demand exerted by dissolved matter.

Estimations of K_d *and* L_a. The data obtained from stream survey, a summation of DO_u values and corresponding time-of-travel t at an observed water temperature, have to be computed along the reach of a stream. Nowadays, for determinations of K_d and L_a one can simplify the calculation by using a computer or a programmable calculator. However, the following example illustrates the use of the Thomas slope method for determining K_d and L_a.

17.4 Procedures of Pragmatic Approach

The steps of the pragmatic approach involved in estimating the waste-assimilative capacity of a stream based upon a pragmatic approach can be summarized as follows (Butts *et al.*, 1973):

1. Develop a full understanding of the stream length, its channel geometry, water stage and flow patterns, and the general hydrologic features of the watershed.
2. Determine the 7-day, 10-year flow of the stream and select a design water temperature.
3. Define the location of all dams and their physical features; define also the location of all tributary flows and relevant data regarding them.
4. Divide the stream into reaches consistent with significant changes in cross-sections and determine the volumes and average depth in each reach.
5. At the beginning and end of each reach, during low-flow conditions and summer temperatures, undertake a series of field determinations for at least water temperature and dissolved oxygen concentrations and, if desired, collect water samples for BOD determinations.
6. Compute the time-of-travel within each reach at stream flows observed during the time-of-sampling as well as that during 7-day, 10-year low flow.
7. From the observed DO values, flow, and time-of-travel compute $DO_a - DO_n$, as demonstrated in Table 2.17 (see later).
8. Select DO saturation values from Table 2.2 for observed stream temperature conditions and compute the natural reaeration for each reach using Figs. 2.12 and 2.13 in conjunction with appropriate equations for finding the mix time M and the percent absorption at 100% deficit R_0. Keep in mind the need to make adjustments in accordance with weir and mass balance formulas where dams and tributaries are encountered.
9. Calculate, by summation, the DO_u for each reach as demonstrated in Table 2.18 (see later).
10. From an array of the DO_u versus t data, determine L_a and K_d, preferably by the methods of Reed–Theriault, steepest descent, or least squares. For a graphical solution, the Thomas slope method is satisfactory. Adjust the values for L_a and K_d for the selected design water temperature by the use of Eqs. (2.26) and (2.28).
11. Apply the removal efficiency anticipated to the computed ultimate oxygen demand L_a and develop the required expression $DO_u = L_a(1 - e^{-Kt})$.
12. From the values developed in step 6 for time-of-travel and depth at 7-day, 10-year low flow, use the observed DO just upstream of the discharge point (D_a at the beginning of the reach) as a starting-point, and follow the computation.
13. Note whether or not the removal efficiency selected will permit predicted DO concentrations in the stream compatible with water quality standards.

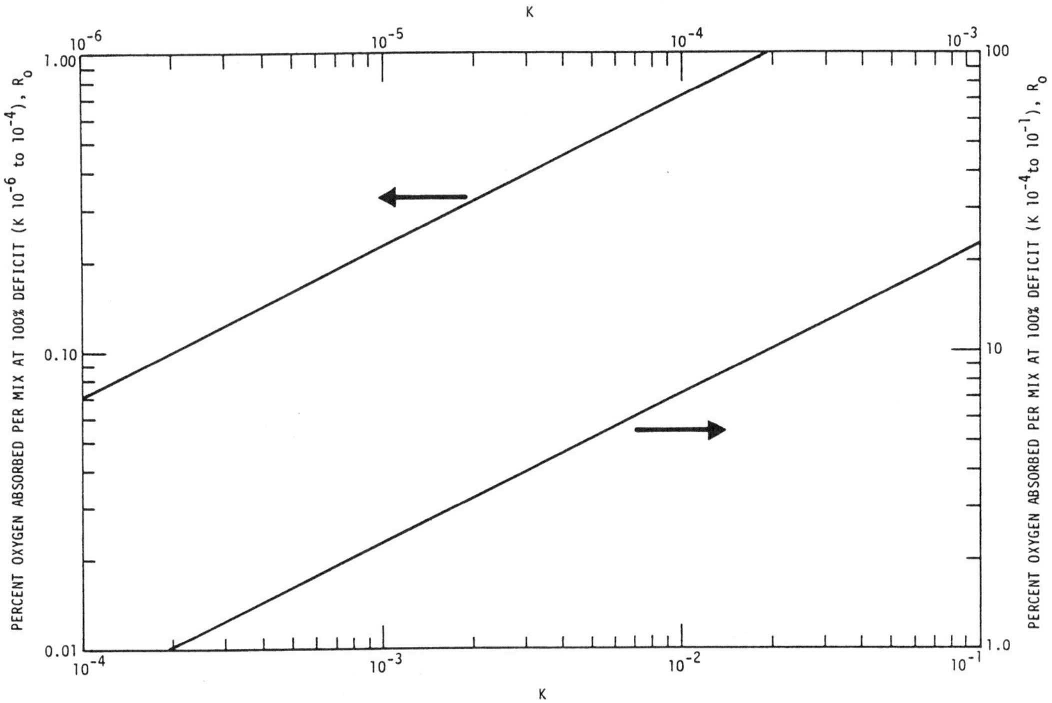

FIGURE 2.13 Modified Velz curve (*Butts et al., 1973*).

EXAMPLE: A long reach of a moderately sized river is investigated. There is an overloaded secondary wastewater treatment plant discharging its effluent immediately above sampling station 1 (upstream). The plant, correctly, has BOD removal efficiency of 65%. A total of 12 sampling stations are established in reaches of the study river. Downstream stations 3 and 4 are above and below an ogee channel dam, respectively. Stations 10 and 11 are immediately above and below the confluence of the main stream and a tributary which has relatively clean water. The 7-day, 10-year low flow of the main receiving stream and the tributary are 660 and 96 cfs, respectively. The design water temperature is selected as 27°C.

A stream field survey for flow, water temperature, DO, average depth, and time-of-travel is conducted during summer low-flow conditions. During the field survey, the receiving stream in the vicinity of the effluent discharge was 822 cfs and the tributary flow was 111 cfs. The DO values are 5.96 and 6.2 mg/L, respectively. Data obtained from field measurements and subsequent computations for DO values are presented in Tables 2.16, 2.17, and 2.18. The DO levels in the stream are not less than 5.0 mg/L, as required.

Question: Determine the DO concentrations along the reach of the stream at design temperature and design flow conditions, if the efficiency of sewage treatment plant is updated to at least 90% BOD removal.

Answer: The solution is approached in two sessions: (1) The deoxygenation rate coefficients K_d and L_a should be determined under the existing conditions; this involves steps 1 to 5. (2) A predictive profile of DO concentrations will be calculated under 7-day 10-year low flow and 27°C design conditions (steps 6 and 7).

TABLE 2.16 Field Data and DO Computations

Station	Flow, cfs at station	Flow, cfs mean	DO_a observed† mg/L	DO_a observed† lb/d*	$DO_a - DO_n$, lb/day	DO averaged mg/L	DO averaged lb/d
1	822		5.96	26406			
2	830	826	3.44	15390	11016	4.70	20925
3	836	833	2.40	10814	4575	2.92	13110
Dam							
4	836		3.96	17844			
5	844	840	3.90	17741	103	3.93	17793
6	855	849	3.70	17051	690	3.80	17389
7	871	863	3.60	16900	151	3.65	16978
8	880	875	3.90	18498	−1598	3.75	17686
9	880	880	4.04	19163	−665	3.97	18831
10	888	884	4.44	21251	−2288	4.27	20202
Tributary							
11	999		5.18	27892	−6641	4.81	24474
12	1010	1005	5.28	28743	−851	5.23	28330

*DO (lb/d) = 5.39 × flow (cfs) × *DO (mg/L)*.
†DO_a at the end of a reach is the DO_n of that reach.

Solution:

Step 1. Compute DO values.

In Table 2.16, data are obtained from field work and subsequent DO computations for other purposes. The term $DO_a - DO_n$ (lb/d) will be used in the equation for Table 2.18. A plot of DO measured and time-of-travel t can be made (not shown). It suggests that a profound DO sag exists in the stream below the major pollution source of effluent discharge.

Step 2. Compute the natural reaeration, DO_r.

In Table 2.17, values of saturated DO (DO_s) at the specified water temperature are obtained from Table 2.2. The values of M and R_0 are taken from Figs. 2.12 and 2.13 on the basis of the physical dimension of the stream listed in Table 2.16 or calculated from Eq. (2.114) ($M = 13.94 \ln H - 745$).

Step 3. Calculate DO used, DO_u.

The DO_U is calculated from equation $DO_u = (DO_a - DO_n) + DO_r + DO_x$. Without consideration of DO_x, the values of DO_u for each reach of the stream are given in Table 2.18. It should be noted that DO_x does not apply yet.

Step 4. Determine $K_{d(23)}$ and $L_{d(23)}$ for $T = 23°C$ (during the field survey).

Using the Thomas (1950) slope method, the factor $(\Sigma t/\Sigma DO_u)^{1/3}$ is plotted on arithmetic graph paper against Σt as shown in Fig. 2.14. A line of best fit is then drawn, often neglecting the first two points. The ordinate intercept b is read as 0.0338; and the slope of the line S is computed as 0.00176 [i.e. (0.0426 − 0.0338)/5]. Then determine K_d and L_a at 23°C.

$$K_{d(23)} = 6S/b = 6 \times 0.00176/0.0338 = 0.31 \text{ (per day)}$$

$$L_{a(23)} = 1/(K_d \cdot b^3) = 1/(0.31 \times 0.0338^3)$$

$$= 83{,}500 \text{ (lb/d)}$$

The curve is expressed as

$$y = 83{,}500(1 - e^{-0.31t})$$

TABLE 2.17 Computations of Natural Reaeration

Station	Temp., °C	DO_s mg/L	Mean D_s, mg/L	Mean $D_a^{\dagger}$, mg/L	Depth h, ft	M, min	t', min	$1 - \frac{DO_a}{DO_s}$	$\frac{R_o}{100}$	$\frac{t'}{M}$	$5.39Q \times DO_s$ lb/d	DO_r,* lb/d
1	22.8	8.53										
2	22.9	8.52	8.53	4.70	6.6	18.8	846.7	0.449	0.0042	45.0	37977	3225
3	23.0	8.50	8.51	2.92	6.8	19.4	1201.0	0.657	0.0041	61.9	38209	6370
4	23.0	8.50										
5	23.0	8.50	8.50	3.93	2.9	4.4	396.0	0.538	0.0046	90.0	38485	8566
6	23.1	8.48	8.49	3.80	3.8	11.2	792.0	0.552	0.0052	70.7	38851	7892
7	23.1	8.48	8.48	3.65	5.2	15.5	1591.2	0.57	0.0049	102.7	39445	11301
8	23.2	8.46	8.47	3.75	6.1	17.6	1630.0	0.557	0.0045	92.6	39946	9277
9	23.2	8.46	8.46	3.97	6.3	18.2	1118.9	0.531	0.0044	61.5	40127	5761
10	23.3	8.45	8.46	4.24	6.5	18.6	911.5	0.499	0.0043	49.0	40310	4237
11	23.1	8.48	8.47	4.81				0.432				
12	23.1	8.48	8.48	5.23	7.2	19.9	656.6	0.383	0.0040	33.0	45936	2324

*$DO_r = \left(1 - \frac{DO_a}{DO_s}\right)\left(\frac{R_0}{100}\right)\left(\frac{t'}{M}\right)\left(5.39Q \times DO_s\right)$ and is determined as the same manner of example in Section 16.2.

†Mean D_a is from Table 2.16.

TABLE 2.18 Calculation of DO Used

Station	t, days	Σt, days	$DO_a - DO_n$, lb/d	DO_r, lb/d	DO_u,* lb/d	ΣDO_u, lb/d	$(\Sigma t/\Sigma DO_u)^{1/3}$
1							
2	0.588	0.588	11016	3225	14241	14241	0.03456
3	0.834	1.422	4575	6370	10945	25186	0.03836
4						25186	
5	0.275	1.697	103	8566	8669	33855	0.03687
6	0.550	2.247	690	7892	8582	42437	0.03755
7	1.105	3.352	151	11301	11452	53889	0.03962
8	1.132	4.484	−1598	9277	7679	61568	0.04176
9	0.777	5.261	−665	5761	5096	66664	0.04289
10	0.633	5.894	−2888	4237	1349	68013	0.04425
11						68013	
12	0.456	6.350	−851	2324	1473	69486	0.04504

*$DO_u = DO_a - DO_n + DO_r$

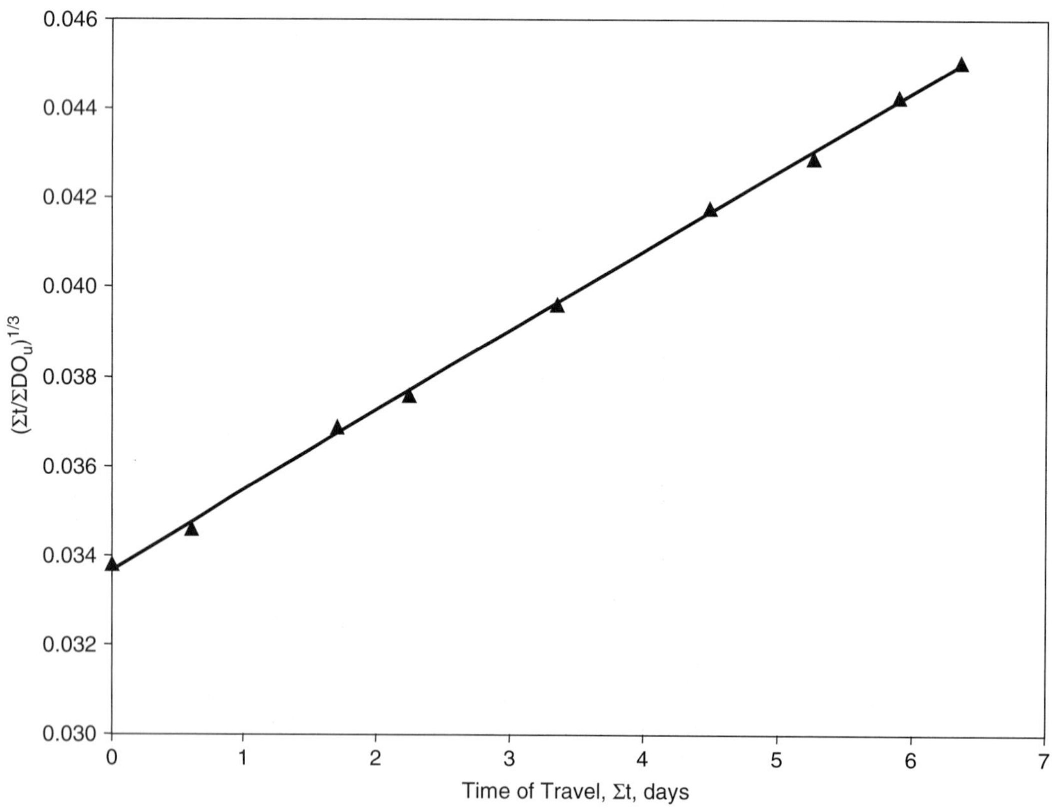

FIGURE 2.14 Plot of $(\Sigma t/\Sigma\ DO_u)^{1/3}$ versus time of travel.

Step 5. Compute $K_{d(27)}$ and $L_{a(27)}$ for design temperature 27°C. Using Eq. (2.26)

$$K_{d(27)} = K_{d(20)} \times 1.047^{27-20}$$
$$K_{d(23)} = K_{d(20)} \times 1.047^{23-20}$$
$$K_{d(27)} = K_{d(23)} \times 1.047^7/1.047^3$$
$$= 0.31 \times 1.3792/1.1477$$
$$= 0.37 \text{ (per day)}$$

Similarly, using Eq. (2.28)

$$L_{a(T)} = L_{a(20)} \times (0.6 + 0.02T)$$
$$L_{a(27)} = L_{a(23)} \times (0.6 + 0.02 \times 27)/(0.6 + 0.02 \times 23)$$
$$= 83{,}500 \times 1.14/1.06$$
$$= 89{,}800 \text{ (lb/d)}$$

Next step goes to design phase.

Step 6. Calculate the design BOD loading after plant improvement.

Since the current overloaded wastewater treatment plant only removes 65% of the incoming BOD load, the raw wastewater BOD load is

$$89{,}800 \text{ lb/d}/(1 - 0.65) = 256{,}600 \text{ lb/d}$$

The expanded (updated) activated sludge process may be expected to remove 90% to 95% of the BOD load from the raw wastewater. For the safe side, 90% removal is selected for the design purpose. Therefore the design load L_a at 27°C to the stream will be 25,660 lb/d (256,600 × 0.1).

Step 7. Gather new input data for DO profile.

In order to develop the predictive profile for DO concentration in the receiving stream after the expanded secondary wastewater treatment plant is functioning, the following design factors are obtained from field survey data and the results computed:

Stream flow	= 660 cfs	At the dam	$q = 1.1$
Tributary flow	= 96 cfs		$b = 0.58$
Tributary DO	= 6.2 mg/L		$h = 6.1$ ft
DO at station 1	= 5.90 mg/L		$K_{d(27)} = 0.37$ per day
Water temperature	= 27°C		$L_{a(27)} = 25{,}660$ lb/d
H and t are available from cross-section data			

Step 8. Perform DO profile computations.

Computations are essentially the same as previous steps with some minor modifications (Table 2.19).

For columns 9, 10, 13, 19, 20, and 21, calculate iteratively from station by station from upstream downward. The other columns can be calculated independently. At station 4 (below the dam), the dam aeration ratio is given as Eq. (1.117) and Eq. (1.118)

$$r = 1 + 0.11qb(1 + 0.046\,T)h$$
$$= 1 + 0.11 \times 1.1 \times 0.58(1 + 0.046 \times 27) \times 6.1$$
$$= 1.960$$
$$r = (C_s - C_A)/(C_s - C_B)$$
$$1.96 = (7.87 - 5.09)/(7.87 - C_B)$$
$$C_B = -2.78/1.96 + 7.87 = 6.45 \text{ (mg/L)}$$

This DO value of 6.41 mg/L is inserted into column 21 at station 4. The corresponding DO_n (in lb/d) is computed by multiplying 5.39Q to obtain 23,148 lb/d, which is inserted into column 20 at station 4.

At station 11, immediately below the tributary, DO contribution is 3208 lb/d (i.e. 5.39 × 6.2 × 96). This value plus 28,068 lb/d (31,276 lb/d) is inserted into column 20. Then computations continue with iterative procedures for station 12.

It should be noted that the parameters H, M, t', t'/M, and $R/100$ can be computed earlier (i.e. in earlier columns). Therefore, the other DO computations will be in the latter part of Table 2.19. It seems easier for observation. Also, these steps of computation can be easily programmed using a computer.

The results of computations for solving equation, $DO_n = DO_a - DO_u + DO_r + DO_x$ are given in Table 2.19, and the DO concentrations in column 21 of Table 2.19 are the predictive DO values for 7-day, 10-year low flow with secondary wastewater treatment. All DO values are above the regulatory standard of 5.0 mg/L, owing to expanded wastewater treatment and dam aeration.

TABLE 2.19 DO Profile Computation at 27°C and a 7-Day, 10-Year Low Flow

(1)	(2) Flow, cfs	(3)	(4)	(5)	(6)	(7) Total	(8) Subreach
Station	at station	mean	t, days	Σt days	$1 - e^{-K_d \Sigma t}$	DO_u, lb/d	DO_u, lb/d
1	660					0	
2	666	663	0.750	0.750	0.242	6218	6218
3	670	668	1.076	1.826	0.491	12603	6385
Dam							
4	670						
5	674	672	0.345	2.171	0.551	14168	1565
6	682	678	0.626	2.797	0.645	16544	2376
7	696	689	1.313	4.110	0.781	20052	3508
8	702	699	1.370	5.48	0.868	22282	2230
9	704	703	0.938	6.418	0.907	23272	990
10	708	706	0.845	7.263	0.932	23913	941
Tributary	96						
11	804						
12	810	806	0.608	7.871	0.946	24265	351

TABLE 2.19 (*contd.*)

	(9)	(10) Mean	(11) DO_s	(12)	(13)	(14) Mean	(15)
Station	$DO_a - DO_u$, lb/d	DO_a, lb/d	mg/L	lb/d	$1 - \frac{DO_a}{DO_s}$	depth H, ft	M, min
1				21084			
2	14866	17975	7.87	28124	0.361	5.9	17.4
3	11977	15170	7.87	28336	0.465	6.1	17.5
Dam							
4							
5	21584	22366	7.87	28506	0.215	2.5	4.1
6	23895	25083	7.87	28760	0.128	3.4	9.6
7	22908	24662	7.87	29227	0.165	4.6	13.8
8	24556	25671	7.87	29651	0.134	5.5	16.2
9	26367	26862	7.87	29821	0.099	5.7	16.6
10	27093	27404	7.87	29948	0.085	6.0	17.4
Tributary							
11							
12	30924	31100	7.87	34190	0.090	6.5	18.6

TABLE 2.19 *(contd.)*

Station	(16) t' = 1440t, min	(17) t'/M	(18) $R_0/100$	(19) DO_r, lb/d	(20) DO_n (or DO_a) lb/d	(21)	Remark
1					21084	5.90	
2	1080	62.1	0.00555	3696	18362	5.14	
3	1549	88.5	0.00545	6353	18330	5.09	Above dam
Dam							
4					23148	6.41	Below dam
5	497	121.2	0.00630	4687	26271	7.25	
6	901	93.9	0.00730	2521	26416	7.23	
7	1891	137.0	0.00620	3878	26786	7.21	
8	1973	121.8	0.00578	2802	27357	7.26	
9	1351	81.4	0.00568	1367	27734	7.32	
10	1217	69.9	0.00550	975	28067	7.38	Above trib.
Tributary							
11					31276		Below trib.
12	876	47.1	0.00521	788	31682	7.29	

Notes: Col 7 = L_a × Col 6 = 25,660 × Col 6
Col 8_n = Col 7_n − Col 7_{n-1}
Col 9_n = Col 20_{n-1} − Col 8_n
Col 10_n = (Col 20_{n-1} + Col 9_n)/2
Col 11 = DO_S at T of 27°C
Col 12 = 5.39 × Col 3 × Col 11
Col 15 = Calculated or from monogram Fig. 2.12
Col 18 = Calculated or from monogram Fig. 2.12
Col 19 = Col 12 × Col 13 × Col 17 × Col18
Col 20 = Initial value is 663 × 5.90 × 5.39 and thereafter Col 9 + Col 19
Col 21 = Col 20/Col 3/5.39
trib. = Tributary

18 BIOLOGICAL FACTORS

18.1 Algae

Algae are most commonly used to assess the extent that primary productivity (algal activity) affects the DO resources of surface waters. Many factors affect the distribution, density, and species composition of algae in natural waters. These include the physical characteristics of the water, length of storage, temperature, chemical composition, in situ reproduction and elimination, floods, nutrients, human activities, trace elements, and seasonal cycles.

Of the many methods suggested for defining the structure of a biological community, the most widely used procedure has been the diversity index. A biological diversity index provides a means of evaluating the richness of species within a biological community using a mathematical computation. Although different formulas have been used, the one determined by the information theory formula, the Shannon–Weiner diversity index, is widely used for algal communities or communities of other organisms. The formula is (Shannon and Weaver, 1949):

$$D = -\sum_{i=1}^{m} P_i \log_2 P_i \tag{2.121}$$

where D = diversity index
$P_i = n_i/N$
n_i = the number (density) of the ith genera or species
N = total number (density) of all organisms of the sample

$$i = 1, 2, \cdots, \mathrm{m}$$
$$m = \text{the number of genera or species}$$
$$\log_2 P_i = 1.44 \ \ln P_i$$

or

$$D = -1.44 \sum_{i-1}^{m} (n_i/N) \ln(n_i/N) \tag{2.122}$$

The index D has a minimum value, when a community consisting solely of one ($m = 1$) species has no diversity or richness and takes on a value of unity. As the number of species increases and as long as each species is relatively equal in number, the diversity index increases numerically. It reaches a maximum value when $m = N$.

EXAMPLE: Seven species of algae are present in a stream water sample. The numbers of seven species per mL are 32, 688, 138, 98, 1320, 424, and 248, respectively. Compute the diversity index of this algal community.

Solution: Using formula (Eq. (2.122))

$$D = -1.44 \sum_{i-1}^{m} (n_i/N) \ln(n_i/N)$$

and construct a table of computations (Table 2.20). Answer: $D = 2.15$

18.2 Indicator Bacteria

Pathogenic bacteria, pathogenic protozoan cysts, and viruses have been isolated from wastewaters and natural waters. The sources of these pathogens are the feces of humans and of wild and domestic animals. Identification and enumeration of these disease-causing organisms in water and wastewater are not recommended because no single technique is currently available to isolate and identify all the pathogens. In fact, concentrations of these pathogens are generally low in water and wastewater. In addition, the methods for identification and enumeration of pathogens are labor-intensive and expensive.

Instead of direct isolation and enumeration of pathogens, total coliform (TC) has long been used as an indicator of pathogen contamination of a water that poses a public health risk. Fecal coliform (FC), which is more fecal-specific, has been adopted as a standard indicator of contamination in natural waters in Illinois, Indiana, and many other states. Both TC and FC are used in standards for drinking-water and natural waters. Fecal streptococcus (FS) is used as a pollution indicator in Europe. FC/FS ratios have been employed for identifying pollution sources in the United States. Fecal streptococci are present in the intestines of warm-blooded animals and of insects, and they are present in the environment (water, soil, and vegetation) for long periods of time. *Escherichia coli* bacteria have also been used as an indicator.

TABLE 2.20 Values for Eq. (2.122)

i	n_i	n_i/N	$-1.44 \ln (n_i/N)$	$-1.44(n_i/N) \ln (n_i/N)$
1	32	0.0109	6.513	0.071
2	688	0.2334	2.095	0.489
3	138	0.0468	4.409	0.206
4	98	0.0332	4.902	0.163
5	1320	0.4478	1.157	0.518
6	424	0.1438	2.792	0.402
7	248	0.0841	3.565	0.300
Σ	2948(N)	1.0000		2.149

Calculation of bacterial density. The determination of indicator bacteria, total coliform, fecal coliform, and fecal streptococcus or enterococcus is very important for assessing the quality of natural waters, drinking-waters, and wastewaters. The procedures for enumeration of these organisms are presented elsewhere (APHA *et al.*, 1995). Examinations of indicator bacterial density in water and in wastewater are generally performed by using a series of four-decimal dilutions per sample, with 3 to 10, usually five, tubes for each dilution. Various special broths are used for the presumptive, confirmation, and complete tests for each of the bacteria TC, FC, and FS, at specified incubation temperatures and periods.

MPN method. Coliform density is estimated in terms of the most probable number (MPN). The multiple-tube fermentation procedure is often called an MPN procedure. The MPN values for a variety of inoculation series are listed in Table 2.21. These values are based on a series of five tubes for three dilutions. MPN values are commonly determined using Table 2.21 and are expressed as MPN/100 mL. Only two significant figures are used for the reporting purpose. Other methods are based on five 20 mL or ten 10 mL of the water sample used.

Table 2.21 presents MPN index for combinations of positive and negative results when five 10-mL, five 1-mL, and five 0.1-mL volumes of sample are inoculated. When the series of decimal dilution is

TABLE 2.21 MPN Index for Various Combinations of Positive Results when Five Tubes are Used per Dilution (10 mL, 1.0 mL, 0.1 mL)

Combination of positives	MPN index 100 mL	Combination of positives	MPN index 100 mL
		4-2-0	22
0-0-0	<2	4-2-1	26
0-0-1	2	4-3-0	27
0-1-0	2	4-3-1	33
0-2-0	4	4-4-0	34
		5-0-0	23
1-0-0	2	5-0-1	30
1-0-1	4	5-0-2	40
1-1-0	4	5-1-0	30
1-1-1	6	5-1-1	50
1-2-0	6	5-1-2	60
2-0-0	4	5-2-0	50
2-0-1	7	5-2-1	70
2-1-0	7	5-2-2	90
2-1-1	9	5-3-0	80
2-2-0	9	5-3-1	110
2-3-0	12	5-3-2	140
3-0-0	8	5-3-3	170
3-0-1	11	5-4-0	130
3-1-0	11	5-4-1	170
3-1-1	14	5-4-2	220
3-2-0	14	5-4-3	280
3-2-1	17	5-4-4	350
		5-5-0	240
4-0-0	13	5-5-1	300
4-0-1	17	5-5-2	500
4-1-0	17	5-5-3	900
4-1-1	21	5-5-4	1600
4-1-2	26	5-5-5	≥1600

Source: APHA, AWWA, and WEF (1995).

different from that in the table, select the MPN value from Table 2.21 for the combination of positive tubes and compute the MPN index using the following formula:

$$\frac{\text{MPN}}{100\text{ mL}} = \text{MPN value from table} \times \frac{10}{\text{largest volume tested in dilution series used}} \tag{2.123}$$

For the MPN index for combinations not appearing in Table 2.21 or for other combinations of dilutions or numbers of tubes used, the MPN can be estimated by the Thomas equation:

$$\frac{\text{MPN}}{100\text{ mL}} = \frac{\text{No. of positive tubes} \times 100}{\sqrt{(\text{mL sample in negative tubes})(\text{mL sample in all tubes})}} \tag{2.124}$$

Although the MPN tables and calculations are described for use in the coliform test, they are equally applicable for determining the MPN of any other organisms for which suitable test media are available.

EXAMPLE 1: Estimate the MPN index for the following six samples.

Solution:

	Positive/five tubes, mL used					
Sample	1	0.1	0.01	0.001	Combination of positive	MPN index/ 100 mL
A-raw	5/5	**5/5**	**3/5**	**1/5**	5-3-1	11000
B × 10^{-3}	5/5	**5/5**	**3/5**	**1/5**	5-3-1	11000000
C	**5/5**	**3/5**	**2/5**	0/5	5-3-2	1400
D	**5/5**	**3/5**	**1/5**	1/5	5-3-2	1400
E	**4/5**	**3/5**	**1/5**	0/5	4-3-1	330
F	**0/5**	**1/5**	**0/5**	0/5	0-1-0	20

Step 1. For sample A
For 5-3-1; from Table 2.21, gives MPN = 110. Using Eq. (2.123)

$$\text{For sample: MPN} = 110 \times \frac{10}{0.1} = 11{,}000$$

Step 2. For sample B
Same as sample A; however, it is a polluted water and a 10^{-3} dilution is used. Thus,

$$\text{MPN} = 11{,}000 \times 10^3 = 11{,}000{,}000$$

Step 3. For sample C
From Table 2.21, MPN = 140 for 5-3-2 combination

$$\text{For sample: MPN} = 140 \times \frac{10}{1} = 1400$$

Step 4. For sample D
Since one is positive at .001/mL used, add 1 positive into 0.01 dilution. Therefore, the combination of positive is 5-3-2.
Thus, MPN = 1400

Step 5. For sample E with a 4-3-1 combination

$$\text{MPN} = 33 \times \frac{10}{1} = 330$$

Step 6. For sample F, 0-1-0 positive

$$\text{MPN} = 2 \times \frac{10}{1} = 20$$

Note: Unit for MPN is the number per 100 mL sample.

EXAMPLE 2: Calculate the MPN value by the Thomas equation with data given in Example 1.

Solution:

Step 1. For sample A

Number of positive tubes for the last two dilutions	$= 3 + 1 = 4$
mL sample in negative tubes	$= 2 \times 0.01 + 4 \times 0.001 = 0.024$
mL sample in all tubes	$= 5 \times 0.01 + 5 \times 0.001 = 0.055$

With the Thomas equation:

$$\frac{\text{MPN}}{100\text{ mL}} = \frac{\text{No. of positive tubes} \times 100}{\sqrt{(\text{mL sample in negative tubes})(\text{mL sample in all tubes})}}$$
$$= \frac{4 \times 100}{\sqrt{0.024 \times 0.055}}$$
$$= 11{,}000 \text{ (for 10, 1, and 0.1 mL)}$$

Step 2. For sample B

As step 1, in addition multiply by 10^3

$$\frac{\text{MPN}}{100\text{ mL}} = 11{,}000 \times 10^3 = 11{,}000{,}000$$

Step 3. For sample C

$$\frac{\text{MPN}}{100\text{ mL}} = \frac{(3 + 2) \times 100}{\sqrt{0.23 \times 0.55}}$$
$$= 1406$$
$$= 1400$$

Note: Only use two significant figures for bacterial count.

Step 4. For sample D

The answer is exactly the same as for sample C

$$\frac{\text{MPN}}{100\text{ mL}} = 1400$$

Step 5. For sample E

$$\frac{\text{MPN}}{100\text{ mL}} = \frac{(4 + 3 + 1) \times 100}{\sqrt{1.24 \times 5.55}}$$
$$= 305$$
$$= 310$$

Step 6. For sample F

$$\frac{\text{MPN}}{100\text{ mL}} = \frac{100}{\sqrt{(5.45 \times 5.55)}}$$
$$= 18$$

EXAMPLE 3: The results of positive tubes of 4 tubes with five dilutions (100, 10, 1, 0.1, and 0.01 mL) are 4/4, 3/4, 1/4, 1/4, and 0/4. Estimate the MPN value of the sample.

Solution:

Step 1. Adjust the combination of positive tubes
It will be 4-3-2 at 100, 10, and 1 mL dilutions.

Step 2. Find MPN by the Thomas formula

$$\frac{\text{MPN}}{100\text{ mL}} = \frac{(3 + 2) \times 100}{\sqrt{(10 + 2 \times 1) \times 4(10 + 1)}}$$
$$= 22$$

Membrane filter method. The membrane filtration (MF) method is the most widely used method. Calculations of MF results for these indicators or other organisms are presented here (Illinois EPA, 1987):

Step 1.

Select the membrane filter with the number of colonies in the acceptable range. The acceptable range for TC is 20–80 TC colonies and no more than 200 colonies of all types per membrane. Sample quantities producing MF counts of 20–60 colonies of FC or FS are desired.

Step 2. Compute count per 100 mL, according to the general formula:

$$\frac{\text{Colonies}}{100\text{ mL}} = \frac{\text{No. colonies counted} \times 100}{\text{Volume of sample, mL}}$$

EXAMPLE 1: Counts with the acceptable limits:
Assume that filtration of volumes 75, 25, 10, 3, and 1 mL produced FC colony counts of 210, 89, 35, 11, and 5, respectively. What is the FC density for the sample?

Solution:
The MF with 35 FC colonies is the best MF for counting.

Thus, the FC density is

$$\frac{\text{FC}}{100\text{ mL}} = \frac{35 \times 100}{10} = 350$$

which will be recorded as 350 FC/100 mL.

It should be noted that an analyst would not count the colonies on all five filters. After inspection of five membrane filters, the analyst would select the MF(s) with 20–60 FC colonies and then limit the actual counting to such a membrane. If there are acceptable counts on replicate filters, count colonies on each MF independently and determine final reporting densities. Then compute the arithmetic mean of these densities to obtain the final recorded density.

EXAMPLE 2: For duplicate samples:
A duplicate filtration of 2.0 mL of water sample gives plate counts of 48 and 53 TC colonies. Compute the recorded value.

Solution:

Step 1. Determine FS densities for two membranes independently, with the general formula:

$$\frac{\text{TC}}{100\text{ mL}} = \frac{48 \times 100}{2} = 2400$$

and

$$\frac{\text{TC}}{100\text{ mL}} = \frac{53 \times 100}{2} = 2650$$

Step 2. Taking the average

$$\frac{2400 + 2650}{2} = 2525 \text{ [reporting 2500 A]}$$

The FS density for the water sample is reported as 2500 A/100 mL. Code A represents two or more filter counts within acceptable range and same dilution.

Note: Use two significant figures for all reported bacteria densities.

If more than one dilution is used, calculate the acceptable range results to final reporting densities separately, then average for final recorded value.

EXAMPLE 3: More than one dilution:
Volumes of 1.00, 0.30, 0.10, 0.03, and 0.01 mL generate total coliform colony counts of TNTC (too numerous to count), 210, 78, 33, and 6. What is the total coliform density in the water?

Solution:

Step 1. In this case, only two volumes, 0.10 and 0.03 mL, produce TC colonies in the acceptable limit. Calculate each MF density, separately.

$$\frac{\text{TC}}{100\text{ mL}} = \frac{78 \times 100}{0.10} = 78{,}000$$

and

$$\frac{\text{TC}}{100\text{ mL}} = \frac{33 \times 100}{0.03} = 110{,}000$$

Step 2. Compute the arithmetic mean of these two densities to obtain the final recorded TC density:

$$\frac{\text{TC}}{100\text{ mL}} = \frac{78{,}000 + 110{,}000}{2} = 94{,}000$$

which would be recorded as 94,000 TC per 100 mL C. Code C stands for the calculated value from two or more filter colony counts within acceptable range but different dilution.

If all MF colony counts are below the acceptable range, all counts should be added together. Also, all mL volumes should be added together.

EXAMPLE 4: All results below desirable range:
Volumes of 25, 10, 4, and 1 mL produced FC colony counts of 18, 10, 3, and 0, respectively. Calculate the FC density for the water sample.

Solution:

$$\frac{\text{FC}}{100\text{ mL}} = \frac{(18 + 10 + 3 + 0) \times 100}{25 + 10 + 4 + 1}$$

$$= 77.5$$

which would be recorded as 77 FC per 100 mL B, where code B represents results based on colony counts outside the acceptable range.

If colony counts from all membrane are zero, there is no actual calculation possible, even as an estimated result. If there had been one colony on the membrane representing the largest filtration volume, the density is estimated from this filter only.

If all MF counts are above the acceptable limit, calculate the density as shown in Example 4.

EXAMPLE 5: Filtration volumes 40, 15, 6, and 2 mL produced FC colony counts of 1, 0, 0, and 0, respectively. Estimate the FC density for the sample.

Solution:

$$\frac{\text{FC}}{100\text{ mL}} = \frac{1 \times 100}{40} = 2.5$$

which would be recorded as 2.5 FC per 100 mL K1, where K1 means less than 2.5/100 mL using one filter to estimate the value.

EXAMPLE 6: Membrane filtration volumes of 0.5, 1.0, 3.0, and 10 mL produced FC counts of 62, 106, 245, and TNTC, respectively. Compute the FC density.

Solution:

$$\frac{\text{FC}}{100\text{ mL}} = \frac{(62 + 106 + 245) \times 100}{0.5 + 1 + 3} = 9180 \text{ (reporting 9200 B)}$$

which would be reported as 9200 FC per 100 mL B.

If bacteria colonies on the membrane are too numerous to count, use a count of 200 colonies for the membrane filter with the smallest filtration volume. The number at which a filter is considered TNTC, is when the colony count exceeds 200 or when the colonies are too indistinct for accurate counting.

EXAMPLE 7: Given that filtration volumes of 2, 8, and 20 mL produce TC colony counts of 244, TNTC, and TNTC, respectively, what is the TC density in the sample?

Solution:

Using the smallest filtration volume, 2 mL in this case, and TC for 200 per filter, gives a better representative value. The estimated TC density is

$$\frac{\text{TC}}{100\text{ mL}} = \frac{200 \times 100}{2} = 10{,}000 \text{ L}$$

Code L, for bacterial data sheet description has to be changed to *greater than*; i.e. all colony counts greater than acceptable range, calculated as when smallest filtration volume had 200 counts.

If some MF colony counts are below and above the acceptable range and there is a TNTC, use the useable counts and disregard any TNTCs and its filtration volume.

EXAMPLE 8: Assume that filtration volumes 0.5, 2, 5, and 15 mL of a sample produced FC counts of 5, 28, 67, and TNTC. What is the FC density?

Solution:

The counts of 5 and TNTC with their volume are not included for calculation.

$$\frac{\text{FC}}{100\text{ mL}} = \frac{(28 + 67) \times 100}{2 + 5} = 1357$$
$$= [1400 \text{ B}]$$

Bacterial standards. Most regulatory agencies have stipulated standards based on the bacterial density for waters. For example, in Illinois, the Illinois Department of Public Health (IDPH) has promulgated the indicator bacteria standards for recreational-use waters as follows:

- A beach will be posted 'Warning—Swim At Your Own Risk' when bacterial counts exceed 1000 TC per 100 mL or 100 FC per 100 mL.
- A beach will be closed when bacterial densities exceed 5000 TC per 100 mL or 500 FC per 100 mL in water samples collected on two consecutive days.

The Illinois Pollution Control Board (IEPA, 1999) has adopted rules regarding FC limits for general-use water quality standards applicable to lakes and streams. The rules of Section 302.209 are

> a. During the months May through October, based on a minimum of five samples taken over not more than a 30-day period, fecal coliforms (STORET number 31616) shall not exceed a geometric mean of 200 per 100 mL, nor shall more than 10 percent of the samples during any 30-day period exceed 400 per 100 mL in protected waters. Protected waters are defined as waters that, due to natural characteristics, aesthetic value, or environmental significance, are deserving of protection from pathogenic organisms. Protected waters must meet one or both of the following conditions:

TABLE 2.22 Fecal Coliform Densities

	FC density/100 mL	
Date, 1997	Observed	Moving geometric mean
5 May	42	
12 May	20	
19 May	300	
26 May	50	
2 June	19	47
9 June	20	41
16 June	250	68
23 June	3000	110
30 June	160	140
7 July	240	220
14 July	20	220
21 July	130	200
28 July	180	110

(1) They presently support or have the physical characteristics to support primary contact.

(2) They flow through or adjacent to parks or residential areas.

EXAMPLE: FC densities were determined weekly for the intake water from a river. The results are listed in Table 2.22. Calculate the 30-day moving geometric mean of FC densities and complete the table.

Solution:

Step 1. Check the minimum sampling requirement: Five samples were collected in less than a 30-day period.

Step 2. Calculate geometric means (M_g) from each consecutive five samples.

$$M_g = \sqrt[5]{42 \times 20 \times 300 \times 50 \times 19} = 47$$

Bacterial die-off in streams. There have been many studies on the die-off rate of bacteria in streams. Most of the work suggests that Chick's law, one of the first mathematics formulations to describe die-off curves, remains quite applicable for estimating the survival of pathogens and nonpathogens of special interest in stream sanitation investigations. Chick's law is

$$N = N_0 10^{-kt} \tag{2.125}$$

or

$$\log N/N_0 = -kt \tag{2.126}$$

where N = bacterial density at time t, days
N_0 = bacterial density at time 0
k = die-off or death rate, per day

Before Chick's law is applied, the river data are generally transformed to bacterial population equivalents (BPE) in the manner proposed by Kittrell and Furfari (1963). The following expressions were used for TC and FC data:

$$\text{BPE} = Q\,(\text{cfs}) \times \text{TC/100 mL} \times 6.1 \times 10^{-5} \tag{2.127}$$

$$\text{BPE} = Q\,(\text{cfs}) \times \text{FC/100 mL} \times 6.1 \times 0.964 \times 10^{-5} \tag{2.128}$$

Here the FC:TC ratio is assumed to be 0.964, as reported by Geldreich (1967).

Average streamflows and geometric means for each stream sampling station are used for BPE calculation. The relationship of BPE versus time-of-travel are plotted on semi-log paper for both TC and FC. Reasonably good straight lines of fit can be developed. The decay or death rates for TC and FC can be estimated from the charts.

18.3 Macroinvertebrate Biotic Index

The macroinvertebrate biotic index (MBI) was developed to provide a rapid stream-quality assessment. The MBI is calculated at each stream station as a tool to assess the degree and extent of wastewater discharge impacts. The MBI is an average of tolerance rating weighted by macroinvertebrate abundance, and is calculated from the formula (IEPA, 1987):

$$\text{MBI} = \sum_{i=1}^{n} (n_i t_i)/N \tag{2.129}$$

where n_i = number of individuals in each taxon i
t_i = tolerance rating assigned to that taxon i
N = total number of individuals in the sediment sample

Most macroinvertebrate taxa known to occur in Illinois have been assigned a pollution tolerance rating, ranging from 0 to 11 based on literature and field studies. A 0 is assigned to taxa found only in unaltered streams of high water quality, and an 11 is assigned to taxa known to occur in severely polluted or disturbed streams. Intermediate ratings are assigned to taxa that occur in streams with intermediate degrees of pollution or disturbance. Appendix A presents a list of these tolerance ratings for each taxon.

EXAMPLE: The numbers of macroinvertebrates in a river sediment sample are 72, 29, 14, 14, 144, 445, 100, and 29 organisms/m^2 for *Corbicula, Perlesta placida, Stenonema, Caenis, Cheumatopsyche, Chironomidae, Stenelmis*, and *Tubificidae*, respectively. The tolerance values for these organisms are, respectively, 4, 4, 4, 6, 6, 6, 7, and 10 (Appendix A). Compute MBI for this sample.

Solution:

$$n = 8$$

$$N = 72 + 29 + 14 + 14 + 144 + 445 + 100 + 29 = 847$$

$$\begin{aligned}
\text{MBI} &= \sum_{n=1}^{8} (n_i t_i)/N \\
&= \frac{1}{847}(72 \times 4 + 29 \times 4 + 14 \times 4 + 14 \times 6 \\
&\quad + 144 \times 6 + 445 \times 6 + 100 \times 7 + 29 \times 10) \\
&= 5.983 \\
&\cong 6.0
\end{aligned}$$

REFERENCES

American Public Health Association (APHA), American Water Works Association (AWWA), and Water Environment Federation (WEF). 1995. *Standard methods for the examination of water and wastewater,* 19th edn. Washington, DC: American Public Health Association.

American Society of Civil Engineering Committee on Sanitary Engineering Research. 1960. Solubility of atmospheric oxygen in water. *J. Sanitary Eng Div.* **86**(7): 41–53.

Barrett, M. J., Gameson, A. L. H. and Ogden, C. G. 1960. Aeration studies at four weir systems. *Water and Water Eng.* 64: 407.

Black, W. M. and Phelps, E. B. 1911. *Location of sewer outlets and discharge of sewage in New York Harbor*. New York City Board of Estimate, March 23.

Broeren, S. M., Butts, T. A. and Singh, K. P. 1991. *Incorporation of dissolved oxygen in aquatic habitat assessment for the upper Sangamon River*. Contract Report 513. Champaign: Illinois State Water Survey.

Butts, T. A. 1963. *Dissolved oxygen profile of Iowa River*. Masters Thesis. Iowa City: University of Iowa.

Butts, T. A. 1974. *Measurement of sediment oxygen demand characteristics of the Upper Illinois Waterway*. Report of Investigation 76 Urbana: Illinois State Water Survey.

Butts, T. A. and Schnepper, D. H. 1967. Oxygen absorption in streams. *Water and Sewage Works*, **114**(10): 385–386.

Butts, T. A. and Shackleford, D. B. 1992. *Reduction in peak flows and improvements in water quality in the Illinois Waterway downstream of Lockport due to implementation of Phase I and II of TARP*, Vol. 2: Water Quality. Contract Report 526. Urbana: Illinois State Water Survey.

Butts, T. A., Evans, R. L. and Stall, J. B. 1974. *A waste allocation study of selected streams in Illinois*. Contract Report prepared for Illinois EPA. Urbana: Illinois State Water Survey.

Butts, T. A., Evans, R. L. and Lin, S. D. 1975. *Water quality features of the Upper Illinois Waterway*. Report of Investigation 79. Urbana: Illinois State Water Survey.

Butts, T. A., Kothandaraman, V. and Evans, R. L. 1973. *Practical considerations for assessing waste assimilative capacity of Illinois streams*. Circular 110. Urbana: Illinois State Water Survey.

Butts, T. A., Roseboom, D., Hill, T., Lin, S. D., Beuscher, D., Twait, R. and Evans, R. L. 1981. *Water quality assessment and waste assimilative analysis of the LaGrange Poole, Illinois River*. Contract Report 260. Urbana: Illinois State Water Survey.

Butts, T. A., Schnepper, D. H. and Evans, R. L. 1970. *Dissolved oxygen resources and waste assimilative capacity of a LaGrange pool, Illinois River*. Report of Investigation 64. Urbana: Illinois State Water Survey.

Churchill, M. A. and Buckingham, R. A. 1956. Statistical method for anlaysis of stream purification capacity. *Sewage and Industrial Wastes* **28**(4): 517–537.

Churchill, M. A., Elmore, R. L. and Buckingham, R. A. 1962. The prediction of stream reaeration rates. *J. Sanitary Eng. Div.* **88**(7): 1–46.

Dougal, M. D. and Baumann, E. R. 1967. Mathematical models for expressing the biochemical oxygen demand in water quality studies. *Proc. 3rd Ann. American Water Resources Conference*, pp. 242–253.

Eckenfelder, W. and O'Conner, D. J. 1959. Stream analysis biooxidation of organic wastes—theory and design. Civil Engineering Department, Manhattan College, New York.

Fair, G. M. 1936. The log-difference method of estimating the constants of the first stage biochemical oxygen-demand curve. *Sewage Works J.* **8**(3): 430.

Fair, G. M., Moore, W. E. and Thomas, Jr., H. A. 1941a. The natural purification of river muds and pollutional sediments. *Sewage Works J.* **13**(2): 270–307.

Fair, G. M., Moore, W. E. and Thomas, Jr., H. A. 1941b. The natural purification of river muds and pollutional sediments. *Sewage Works J.* **13**(4): 756–778.

Gameson, A. L. H. 1957. Weirs and the aeration of rivers. *J. Inst. Water Eng.* **11**(6): 477.

Gameson, A. L. H., Vandyke, K. G. and Ogden, C. G. 1958. The effect of temperature on aeration at weirs. *Water and Water Eng.* **62**: 489.

Gannon, J. J. 1963. *River BOD abnormalities*. Final Report USPHS Grants RG-6905, WP-187(C1) and WP-187(C2). Ann Arbor, Michigan: University of Michigan School of Public Health.

Gannon, J. J. and Downs, T. D. 1964. Professional Paper Department of Environmental Health. Ann Arbor, Michigan: University of Michigan.

Geldreich, E. E. 1967. Fecal coliform concepts in stream pollution. *Water and Sewage Works* **114**(11): R98–R110.

Grindrod, J. 1962. British research on aeration at weirs. *Water and Sewage Works* **109**(10): 395.

Illinois Environmental Protection Agency. 1987. *Quality Assurance and Field Method Manual*. Springfield, Illinois: IEPA.

Illinois Environmental Protection Agency. 1990. *Title 35: Environmental Protection, Subtitle C: Water pollution, Chapter I: Pollution Control Board.* Springfield, Illinois: IEPA.

Kittrel, F. W. and Furfari, S. A. 1963. Observation of coliform bacteria in streams. *J. Water Pollut. Control Fed.* **35**: 1361.

Langbein, W. B. and Durum, W. H. 1967. Aeration capacity of streams. Circular 542. Washington, DC: US Geological Survey, 67.

Larson, R. S., Butts, T. A. and Singh, K. P. 1994. *Water quality and habitat suitability assessment: Sangamon River between Decatur and Petersburg.* Contract Report 571, Champaign: Illinois State Water Survey.

Le Bosquet, M. and Tsivoglou, E. C. 1950. Simplified dissolved oxygen computations. *Sewage and Industrial Wastes* **22**: 1054–1061.

McDonnell, A. J. and Hall, S. D. 1969. Effect of environmental factors on benthal oxygen uptake. *J. Water Pollut. Control Fed. Research Supplement, Part 2* **41**: R353–R363.

Moore, E. W., Thomas, H. A. and Snow, W. B. 1950. Simplified method for analysis of BOD data. *Sewage and Industrial Wastes* **22**(10): 1343.

Nemerow, N. L. 1963. *Theories and practices of industrial waste treatment.* Reading, Massachusetts: Addison-Wesley.

O'Connor, D. J. 1958. The measurement and calculation of stream reaeration ratio. In: *Oxygen relation in streams.* Washington, DC: US Department of Health, Education, and Welfare, pp. 35–42.

O'Connor, D. J. 1966. *Stream and estuarine analysis.* Summer Institute in Water Pollution Control, Training Manual. New York: Manhattan College.

O'Connor, D. J. and Dobbins, W. E. 1958. The mechanics of reaeration in natural streams. *Trans. Am. Soc. Civil Engineers* **123**: 641–666.

Orford, H. E. and Ingram, W. T. 1953. Deoxygenation of sewage. *Sewage and Industrial Waste* **25**(4): 419–434.

Perry, R. H. 1959. *Engineering manual*, 2nd edn. New York: McGraw-Hill.

Phelps, E. B. 1944. *Stream sanitation.* New York: John Wiley.

Preul, H. C. and Holler, A. G. 1969. *Reaeration through low days in the Ohio river.* Proceedings of the 24th Purdue University Industrial Waste Conference. Lafayette, Indiana: Purdue University.

Shannon, C. E. and Weaver, W. 1949. *The mathematical theory of communication.* Urbana: University of Illinois Press, 125 pp.

Sheehy, J. P. 1960. Rapid methods for solving monomolecular equations. *J. Water Pollut. Control Fed.* **32**(6): 646–652.

Streeter, H. W. 1926. The rate of atmospheric reaeration of sewage polluted streams. *Trans. Amer. Soc. Civil Eng.* **89**: 1351–1364.

Streeter, H. W. and Phelps, E. B. 1925. *A study of the pollution and natural purification of the Ohio River.* Cincinatti: US Public Health Service, Bulletin No. 146.

Thomas, H. A., Jr. 1937. The 'slope' method of evaluating the constants of the first-stage biochemical oxygen demand curve. *Sewage Works J.* **9**(3): 425.

Thomas, H. A. 1948. Pollution load capacity of streams. *Water and Sewage Works* **95**(11): 409.

Thomas, H. A., Jr. 1950. Graphical determination of BOD curve constants. *Water and Sewage Works* **97**(3): 123–124.

Tsivoglou, E. C. 1958. *Oxygen relationships in streams.* Technical Bulletin W-58-2.

US Environmental Protection Agency. 1997. *Technical guidance manual for developing total maximum daily loads, book 2: streams and rivers. Part 1: Biological oxygen demand/dissolved oxygen and nutrients/eutrophication.* EPA 823-B-97-002. Washington, DC: US EPA.

US Public Health Service. 1927. *The oxygen demand of polluted waters.* Public Health Bulletin No. 172. Washington, DC: US Public Health Service.

Velz, C. J. 1958. Significance of organic sludge deposits. In: Taft, R. A. (ed.), *Oxygen relationships in streams.* Sanitary Engineering Center Technical Report W58-2. Cincinnati, Ohio: US Department of Health, Education, and Welfare.

Velz, C. J. 1939. Deoxygenation and reoxygenation. *Trans. Amer. Soc. Civil Eng.* **104**: 560–572.

Wetzel, R. G. 1975. *Limnology.* Philadelphia, Pennsylvania: Saunders.

Young, J. C. and Clark, J. W. 1965. Second order equation for BOD. *J. Sanitary Eng. Div.*, ASCE **91**(SA): 43–57.

Zanoni, A. E. 1967. *Effluent deoxygenation at different temperatures.* Civil Engineering Department Report 100-5A. Milwaukee, Wisconsin: Marquette University.

CHAPTER 1.3
LAKES AND RESERVOIRS

Shun Dar Lin

This chapter includes mainly lake morphometry, evaporation, and the Clean Lakes Program (CLP). Since most lake management programs in the United States are based on the CLP, the CLP is discussed in detail. Regulatory requirements and standardization of research and application are provided with a focus on the Phase 1, diagnostic/feasibility study.

1 LAKES AND IMPOUNDMENT IMPAIRMENTS

Lakes are extremely complex systems whose conditions are a function of physical, chemical, and biological (the presence and predominance of the various plants and organisms found in the lake) factors. Lakes inherently function as traps or sinks for pollutants from tributaries watershed (drainage basins) or from precipitation and atmospheric deposits.

Like streams, lakes are most often impaired by agricultural activities (main sources in the United States), hydrologic/habitat modification (stream channelization), and point pollution sources. These activities contribute to nutrient and sediment loads, suspended solids, and organic matter, and subsequently cause overgrowth of aquatic plants. The resulting decline in water quality limits recreation, impairs other beneficial uses, and shortens the expected life span of a lake.

Common lake problems are eutrophication, siltation, shoreline erosion, algal bloom, bad taste and/or odor, excessive growth of aquatic vegetation, toxic chemicals, and bacterial contamination. Eutrophication, or aging, the process by which a lake becomes enriched with nutrients, is caused primarily by point and nonpoint pollution sources from human activities. Some man-made lakes and impoundments may be eutrophic from their birth. These problems impact esthetic and practical uses of the lake. For example, the growth of planktonic algae in water-supply impoundments may cause tase and odor problems, shortened filter runs, increased chlorine demand, increased turbidity, and, for some facilities, increased trihalomethane precursors. The effects ultimately lead to increased water treatment costs and, in some instances, even to abandonment of the lake as a public water-supply source.

Lakes and reservoirs are sensitive to pollution inputs because they flush out their contents relatively slowly. Even under natural conditions, lakes and reservoirs undergo eutrophication, an aging process caused by the inputs of organic matters and siltation.

2 LAKE MORPHOMETRY

Lake morphometric data can be calculated from either a recent hydrographic map or a pre-impoundment topographic map of the basin. In general, pre-impoundment maps may be obtainable from the design engineering firm or local health or environmental government agencies. If the map is too old and there is evidence of significant siltation, sections of the lake, such as the upper end and coves with inflowing streams, will need to be remapped.

In some cases it is necessary to create a new lake map. The procedures include the following:

1. The outline of the shoreline is drawn either from aerial photographs or from United States Geological Survey (USGS) 7.5-minute topographic maps.
2. The water depth for transacts between known points on the shoreline of the lake is measured by a graphing sonar.
3. The sonar strip chart of the transacts is interpreted and drawn on to an enlarged copy of the lake outline.
4. Contours may be drawn by hand or with a computer on the map.
5. All maps are then either digitized or scanned and entered into a geographic information system (GIS).
6. Coverages are then converted into sea level elevations by locating the map on a 7.5-minute USGS topographic map and digitizing reference points both on the map and on the quadrangle for known sea level elevations.
7. Depths are assigned to the contour. The GIS gives the length of each contour and areas between adjacent contour lines.

From this data, surface area, maximum depth of the lake, and shoreline length can be computed. Lake volume (V), shoreline development index (SDI) or shoreline configuration ratio, and mean depth ($\overline{D}$) can be calculated by the formula in Wetzel (1975):

1. Volume

$$V = \sum_{i=0}^{n} \frac{h}{3}(A_i + A_{i+1} + \sqrt{A_i \times A_{i+1}}) \qquad (3.1)$$

where V = volume, ft^3, acre·ft, or m^3
h = depth of the stratum, ft or m
i = number of depth stratum
A_i = area at depth i, ft^2, acre, or m^2

2. Shoreline development index

$$\mathrm{SDI} = \frac{L}{2\sqrt{\pi \times A_0}} \tag{3.2}$$

where L = length of shoreline, miles or m
A_0 = surface area of lake, acre, ft^2, or m^2

3. Mean depth

$$\overline{D} = \frac{V}{A_0} \tag{3.3}$$

where $\overline{D}$ = mean depth, ft or m
V = volume of lake, ft^3, acre·ft or m^3
A_0 = surface area, ft^2, acre or m^2

Other morphometric information also can be calculated by the following formulas:

4. Bottom slope

$$S = \frac{\overline{D}}{D_{\mathrm{m}}} \tag{3.4}$$

where S = bottom slope
$\overline{D}$ = mean depth, ft or m
D_{m} = maximum depth, ft or m

5. Volume development ratio, V_{d} (Cole, 1979)

$$V_{\mathrm{d}} = 3 \times \frac{\overline{D}}{D_{\mathrm{m}}} \tag{3.5}$$

6. Water retention time

$$\mathrm{RT} = \frac{\text{storage capacity, acre} \cdot \text{ft or m}^3}{\text{annual runoff, acre} \cdot \text{ft/year or m}^3\text{/year}} \tag{3.6}$$

where RT = retention time, year

7. Ratio of drainage area to lake capacity R

$$R = \frac{\text{drainage area, acre or m}^2}{\text{storage capacity, acre} \cdot \text{ft or m}^3} \tag{3.7}$$

EXAMPLE: A reservoir has a shoreline length of 9.80 miles. Its surface area is 568 acres. Its maximum depth is 10.0 ft. The areas for each foot depth are 480, 422, 334, 276, 205, 143, 111, 79, 30, and 1 acres. Annual rainfall is 38.6 inches. The watershed drainage is 11,620 acres. Calculate morphometric data with the formulas described above.

Solution: Compute the following parameters:

1. The volume of the lake

$$V = \sum_{i=0}^{10} \frac{h}{3}(A_i + A_{i+1} + \sqrt{A_i \times A_{i+1}})$$

$$= \frac{1}{3}[(568 + 480 + \sqrt{568 \times 480}) + (480 + 422 + \sqrt{480 \times 422})$$

$$+ (422 + 324 + \sqrt{422 \times 324}) + \cdots + (30 + 1 + \sqrt{30 \times 1})]$$

$$= \frac{1}{3}[7062]$$

$$= 2354 \text{ (acre·ft)}$$

$$= 2{,}902{,}000 \text{ m}^3$$

2. Shoreline development index or shoreline configuration ratio

$$A_0 = 568 \text{ acres} = 568 \text{ acres} \times \frac{1 \text{ sq miles}}{640 \text{ acres}} = 0.8875 \text{ sq miles}$$

$$\text{SDI} = \frac{L}{2\sqrt{\pi \times A_0}}$$

$$= \frac{9.80 \text{ miles}}{2\sqrt{3.14 \times 0.8875 \text{ sq miles}}}$$

$$= \frac{9.80}{3.34}$$

$$= 2.93$$

3. Mean depth

$$\overline{D} = \frac{V}{A_0}$$

$$= \frac{2354 \text{ acre·ft}}{568 \text{ acres}}$$

$$= 4.14 \text{ ft}$$

4. Bottom slope

$$S = \frac{\overline{D}}{D_m} = \frac{4.13 \text{ ft}}{10.0 \text{ ft}} = 0.41$$

5. Volume development ratio

$$V_d = 3 \times \frac{\overline{D}}{D_m} = 3 \times 0.41 = 1.23 \text{ (or 1.23:1)}$$

6. Water retention time

$$\text{Storage capacity } V = 2354 \text{ acre}\cdot\text{ft}$$

$$\begin{aligned}\text{Annual runoff} &= 38.6 \text{ inches/year} \times 11{,}620 \text{ acres} \\ &= 38.6 \text{ inches/year} \times \frac{1 \text{ ft}}{12 \text{ inches}} \times 11{,}620 \text{ acres} \\ &= 37{,}378 \text{ acre}\cdot\text{ft/year}\end{aligned}$$

$$\begin{aligned}\text{RT} &= \frac{\text{storage capacity}}{\text{annual runoff}} \\ &= \frac{2354 \text{ acre}\cdot\text{ft}}{37{,}378 \text{ acre}\cdot\text{ft/year}} \\ &= 0.063 \text{ years}\end{aligned}$$

7. Ratio of drainage area to lake capacity

$$\begin{aligned}R &= \frac{\text{drainage area}}{\text{storage capacity}} \\ &= \frac{11{,}620 \text{ acres}}{2354 \text{ acre}\cdot\text{ft}} \\ &= \frac{4.94 \text{ acres}}{1 \text{ acre}\cdot\text{ft}} \quad \text{or} \quad 4.94 \text{ acre/acre}\cdot\text{ft}\end{aligned}$$

3 WATER QUALITY MODELS

Lakes and reservoirs are usually multipurpose, serving municipal and industrial water supplies, recreation, hydroelectric power, flood control, irrigation, drainage, and/or agriculture, due to the importance of protecting these natural resources. Water quality involves the physical, chemical, and biological integrity of water resources. Water quality standards promulgated by the regulatory agencies define the water quality goals for protection of water resources in watershed management.

Modeling the water quality in lakes and reservoirs is very different from that in rivers (Chapter 1.2) or estuaries. A variety of models are based on some of physical, chemical, and biological parameters and/or combinations. Physical models deal with temperature, dissolved oxygen (DO), energy budget diffusion, mixing, vertical and horizontal aspects, seasonal cycles, and meteorological data setting. Chemical models involve mass balance and toxic substances. Biological models deal with nutrients and the food chain, biological growth, perdition, oxygen balance, and tropical conditions, etc.

A detailed modeling of lake water quality is given in a book by Chapra and Reckhow (1983). Clark *et al.* (1977), Tchobanoglous and Schroeder (1985), and James (1993) also present modeling water quality in lakes and reservoirs. However, most of those models are not used for lake management practices.

For construction of a new lake or reservoir, it is required that the owner submit the prediction of water quality in the future new lake to the US Army Corps of Engineers. For example, Borah *et al.* (1997) used the US Army Corps of Engineers' HEC-5Q model (US Army Corps of Engineers, 1986, 1989) for a central Illinois lake and two proposed new lakes. The model was calibrated and verified on the existing water supply lake used monitored data in the draft years of 1986 and 1988. Eight water quality constituents were simulated. Calibration, verification, and predictions were made only for water temperature, DO, nitrate nitrogen, and phosphate phosphorus. The calibrated and verified model was used to predict water surface elevations and constituent concentrations in the three lakes, individually and in combination, with meteorological and estimated flow data for a 20-month period from May 1953 through December 1954, the most severe drought of record. The HEC-5Q model provided a useful tool in the water quality evaluation of the three lakes. The HEC-5Q model was developed by the Hydraulic Engineering Center in Davis, California, for simulation of flood control and conservation systems with water quality analysis. The details of these models are beyond the scope of this book.

The major portion of this chapter will cover evaporation and lake management programs in the Clean Lakes Program. Because these subjects are general, they are not usually included in environmental engineering textbooks.

4 EVAPORATION

Evaporation converts water in its liquid or solid state into water vapor which mixes with the atmosphere. The rate of evaporation is controlled by the availability of energy at the evaporating surface, and the ease with which water vapor can diffuse into the atmosphere.

Shuttleworth (1993) presented a detailed description with modeling of evaporation and transpiration from soil surfaces and crops. Modelings include fundamental and empirical equations. Measurements of evaporation and methods for estimating evaporation are also given.

Evaporation of water from a lake surface uses energy provided by the sun to heat the water. The rate of evaporation is controlled primarily by water temperature, air temperature, and the level of moisture saturation in the air. Knowledge of evaporative processes is important in understanding how water losses through evaporation from a lake or reservoir are determined. Evaporation increases the storage requirement and decreases the yield of lakes and reservoirs.

Determination of evaporation from a lake surface can be modeled by the water budget method, the mass transfer method, and the energy budget method (Robert and Stall, 1967). Many empirical equations have been proposed elsewhere (Fair *et al.*, 1966, Linsley and Franzinni, 1964).

4.1 Water Budget Method

The water budget method for lake evaporation depends on an accurate measurement of the inflow and outflow of the lake. The simple mathematical calculation shows that the change in storage equals the input minus output. It is expressed as

$$\Delta S = P + R + GI - GO - E - T - O \tag{3.8}$$

where ΔS = change in lake storage, mm (or in)
P = precipitation, mm
R = surface runoff or inflow, mm
GI = groundwater inflow, mm
GO = groundwater outflow, mm
E = evaporation, mm
T = transpiration, mm
O = surface water release, mm

For the case of a lake with little vegetation and negligible groundwater inflow and outflow, lake evaporation can be estimated by

$$E = P + R - O \pm \Delta S \tag{3.9}$$

The water budget method has been used successfully to estimate lake evaporation in some areas.

4.2 Energy Budget Method

The principal elements in energy budget of the lake evaporation are shown in Fig. 3.1. The law of conservation of energy suggests that the total energy reaching the lake must be equal to the total energy leaving the lake plus the increase in internal energy of the lake. The energy budget in Fig. 3.1 can be expressed as (US Geological Survey, 1954)

$$Q_e = Q_s - Q_r - Q_b - Q_h - Q_\theta \pm Q_v \tag{3.10}$$

where Q_e = energy available for evaporation
Q_s = solar radiation energy
Q_r = reflected solar radiation
Q_b = net long-wave radiation
Q_h = energy transferred from the lake to the atmosphere
Q_θ = increase in energy stored in the lake
Q_v = energy transferred into or from the lake bed

The energy budget shown in Fig. 3.1 can be applied to evaporation from a class A pan (illustrated in Fig. 3.2). The pan energy budget (Fig. 3.2) contains the same elements as the lake budget (Fig. 3.1) except that heat is lost through the side and the bottom of the pan. Equation (3.10) is also applied to pan energy budget calculation.

A pioneering and comprehensive research project centered on techniques for measuring evaporation was conducted at Lake Hefner, Oklahoma, in 1950–51 with the cooperation of five federal agencies (US Geological Survey, 1954). The energy budget data from a Weather Bureau class A evaporation pan was used to evaluate the study. Growing out of that project was the US Weather Bureau nomograph, a four-gradient diagram (Fig. 3.3) for general use for lake evaporation (Kohler *et al.*, 1959). Subsequently, the formula for a mathematical solution of the nomograph was adapted

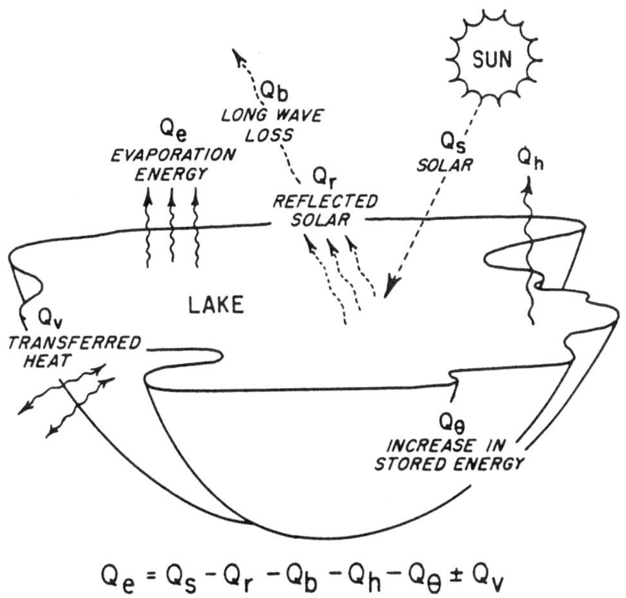

FIGURE 3.1 Elements of energy budget method for determining lake evaporation (*Robert and Stall, 1967*).

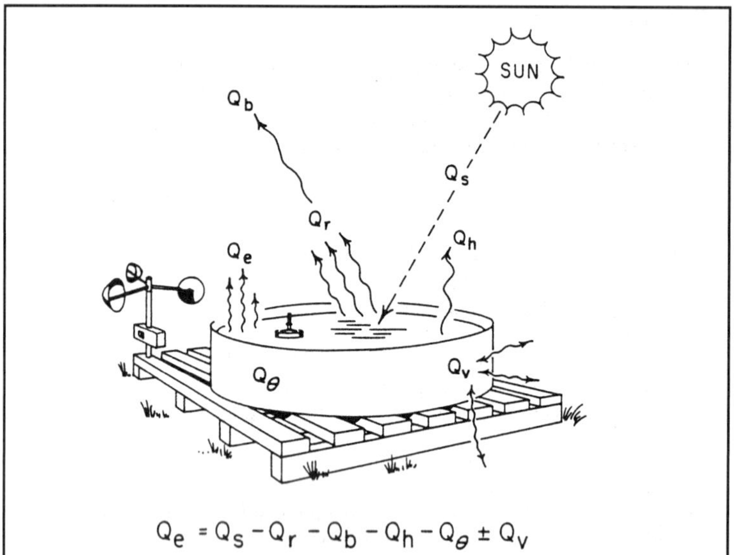

FIGURE 3.2 Energy budget method applied to class A pan (*Robert and Stall, 1967*).

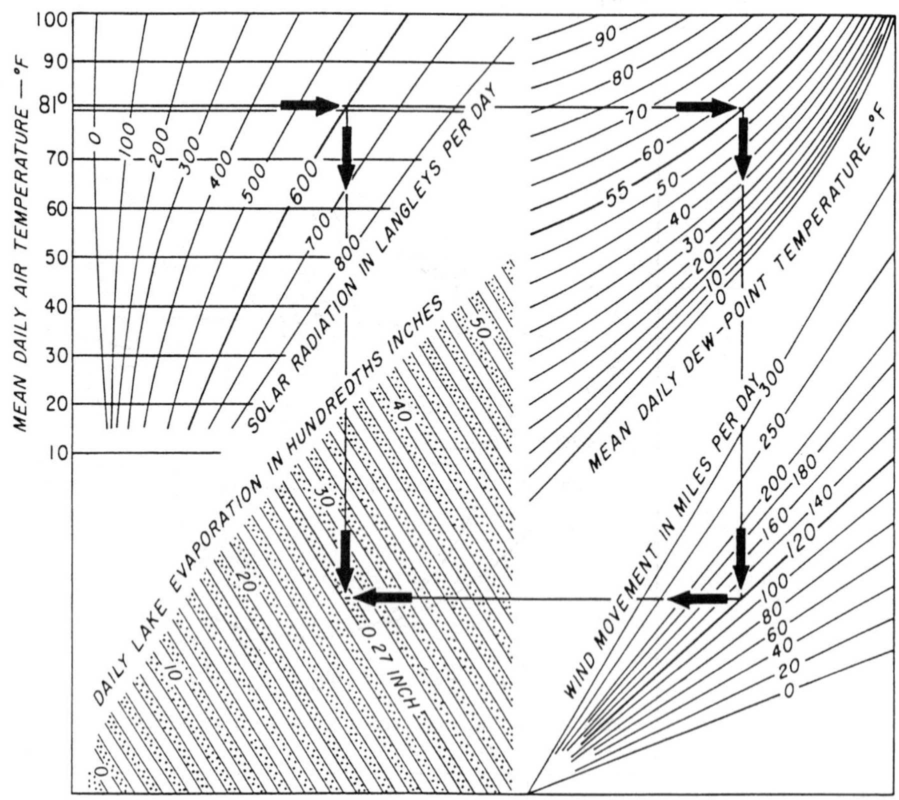

FIGURE 3.3 Lake evaporation nomograph (*Kohler et al., 1959*).

from computer use (Lamoreux, 1962). These procedures allow lake and pan evaporation to be computed from four items of climate data, i.e. air temperature, dew point temperature, wind movement, and solar radiation.

Although the nomograph (Fig. 3.3) was developed for the use with daily values, it is also usable with monthly average values and gives reasonably reliable results with annual average values. The four-quadra diagram may be used to determine a lak evaporation rate is illustrated in Figure 3.3. For example, at the upper left quadrant, a horizontal line passed from the mean daily air temperature 81°F to the first intersection at the solar radiation 600-langleys curve, and project toward the right to intersection of the 55°F mean daily Dew-point curve. At the stop, a vertical line runs downward to the line of 120-mile per day wind movement, from where a horizontal line projects toward to the left to interset a vertical line from the first intersection. The average daily lake evaporation, 0.27 in, is read at this point.

Since long-term climate data are mostly available, these techniques permit extended and refined evaporation determination. The following two equations for evaporation can be computed by computer (Lamoreux, 1962) as exemplified by the Weather Bureau computational procedures for the United States.

For pan evaporation:

$$E_p = \{\exp[(T_a - 212)(0.1024 - 0.01066 \ln R)] - 0.0001 + 0.025(e_s - e_a)^{0.88}(0.37 + 0.0041\, U_p)\} \times \{0.025 + (T_a + 398.36)^{-2}\, 4.7988 \times 10^{10} \exp[-7482.6/(T_a + 398.36)]\}^{-1} \tag{3.11}$$

For lake evaporation, the expression is

$$E_L = \{\exp[(T_a - 212)(0.1024 - 0.01066 \ln R)] - 0.0001 + 0.0105(e_s - e_a)^{0.88} \times (0.37 + 0.0041 U_p)\} \times \{0.015 + (T_a + 398.36)^{-2}\, 6.8554 \times 10^{10} \times \exp[-7482.6/(T_a + 398.36)]\}^{-1} \tag{3.12}$$

where E_p = pan evaporation, in
E_L = lake evaporation, in
T_a = air temperature, °F
e_a = vapor pressure, in of mercury at temperature T_a
e_s = vapor pressure, in of mercury at temperature T_d
T_d = dew point temperature, °F
R = solar radiation, langleys per day
U_p = wind movement, miles per day

Pan evaporation is used widely to estimate lake evaporation in the United States. The lake evaporation E_L is usually computed for yearly time periods. Its relationship with the pan evaporation E_p can be expressed as

$$E_L = p_c E_p \tag{3.13}$$

where p_c is the pan coefficient. The pan coefficient on an annual basis has been reported as 0.65 to 0.82 by Kohler *et al.* (1955).

Robert and Stall (1967) used Eqs. (3.9) and (3.10) for 17 stations in and near Illinois over the May–October season for a 16-year period to develop the general magnitudes and variability of lake evaporation in Illinois.

Evaporation loss is a major factor in the net yield of a lake or reservoir. The net evaporation loss from a lake is the difference between a maximum expected gross lake evaporation and the minimum expected precipitation on the lake surface for various recurrence intervals and for critical periods having various durations. This approach assumes maximum evaporation and minimum precipitation would occur simultaneously. The following example illustrates the calculation of net draft rate for a drought recurrence interval of 40 years.

EXAMPLE: Calculate the net yield or net draft for 40-year recurrence interval for a reservoir (333 acres, 135 ha). Given: drainage area = 15.7 miles^2, unit RC = 2.03 in/miles^2, draft = 1.83 MGD, E = 31 in/acre.

Solution:

Step 1. Compute reservoir capacity (RC) in million gallons (Mgal)

$$\begin{aligned} \text{RC} &= 2.03 \text{ in} \times 15.7 \text{ miles}^2 \\ &= 2.03 \text{ in} \times (1 \text{ ft}/12 \text{ in}) \times 15.7 \text{ miles}^2 \times (640 \text{ acres/miles}^2) \\ &= 1700 \text{ acre}\cdot\text{ft} \\ &= 1700 \text{ acre}\cdot\text{ft} \times 43{,}560 \text{ ft}^2 \text{ acre} \times 7.48 \text{ gal/ft}^3 \\ &= 554 \text{ Mgal} = 2{,}097{,}000 \text{ m}^3 \end{aligned}$$

Note: 1 Mgal = 3785 m^3

Step 2. Compute the total gross draft

$$\begin{aligned} 16 \text{ months} &= 365 \text{ days}/12 \times 16 = 486 \text{ days} \\ \text{Draft} &= 1.83 \text{ MGD} \times 486 \text{ days} \\ &= 888 \text{ Mgal} \end{aligned}$$

Step 3. Compute the inflow to the reservoir (total gross draft minus RC)

$$\begin{aligned} \text{Inflow} &= \text{draft} - \text{RC} = 888 \text{ Mgal} - 554 \text{ Mgal} \\ &= 334 \text{ Mgal} \end{aligned}$$

Step 4. Compute the evaporation loss E
Effective evaporation surface area of the lake is 333 acres × 0.65 = 216 acres

$$\begin{aligned} E &= 31 \text{ in} \times (1 \text{ ft}/12 \text{ in}) \times 216 \text{ acres} \\ &= 558 \text{ acre}\cdot\text{ft} \\ &= 182 \text{ Mgal} \end{aligned}$$

Step 5. Compute the net usable reservoir capacity (the total reservoir minus the evaporation loss)

$$\begin{aligned} \text{RC} - E &= 554 \text{ Mgal} - 182 \text{ Mgal} \\ &= 372 \text{ Mgal} \end{aligned}$$

Step 6. Compute the total net draft which the reservoir can furnish (the net usable reservoir capacity plus the inflow)

$$\begin{aligned} \text{Net draft} &= 372 \text{ Mgal} + 334 \text{ Mgal} \\ &= 706 \text{ Mgal} \end{aligned}$$

Step 7. Compute the net draft rate, or the net yield, which the reservoir can furnish (the total net draft divided by total days in the critical period)

$$\begin{aligned} \text{Net draft rate} &= 706 \text{ Mgal}/486 \text{ days} \\ &= 1.45 \text{ MGD} \\ &= 5490 \text{ m}^3/\text{d} \end{aligned}$$

Note: 1 MGD = 3785 m^3/d

5 THE CLEAN LAKES PROGRAM

In the 1960s, many municipal wastewater treatment plants were constructed or upgraded (point source pollution control) in the United States. However, the nation's water quality in lakes and reservoirs has not been improved due to nonpoint sources of pollution. In 1972, the US Congress amended Section 314 of the Federal Water Pollution Control Act (Public Law 92-500) to control the nation's pollution sources and to restore its freshwater lakes. In 1977, the Clean Lakes Program (CLP) was established pursuant to Section 314 of the Clean Water Act. Under the CLP, publicly owned lakes can apply and receive financial assistance from the US Environmental Protection Agency (USEPA) to conduct diagnostic studies of the lakes and to develop feasible pollution control measures and water-quality enhancement techniques for lakes. This is the so-called Phase 1 diagnostic/feasibility (D/F) study.

The *objectives* of the CLP are to (USEPA, 1980):

1. classify publicly owned freshwater lakes according to trophic conditions;
2. conduct diagnostic studies of specific publicly owned lakes, and develop feasible pollution control and restoration programs for them; and
3. implement lake restoration and pollution control projects.

5.1 Types of Funds

The CLP operates through four types of financial assistance by cooperative agreements. They are

State lake classification survey
Phase 1—diagnostic/feasibility study
Phase 2—implementation
Phase 3—post-restoration monitoring

Funding is typically provided yearly for the state lake classification survey, but on a long-term basis for Phases 1, 2, and 3. Phase 1 funds are used to investigate the existing or potential causes of decline in the water quality of a publicly owned lake; to evaluate possible solutions to existing or anticipated pollution problems; and to develop and recommend the most 'feasible' courses of action to restore or preserve the quality of the lake. Activities typically associated with sample collection, sample analyses, purchase of needed equipment, information gathering, and report development are eligible for reimbursement. During Phase 1, total project costs cannot exceed $100,000. Fifty percent of this funding will come from the CLP, while the rest must come from non-federal or local sources.

5.2 Eligibility for Financial Assistance

Only states are eligible for CLP financial assitance, and cooperative agreements will be awarded to state agencies. A state may in turn make funds available to a sub-state agency or agencies for all or any portion of a specific project. The D/F study is generally carried out by a contracted organization, such as research institutes or consulting engineers.

For nearly two decades, 46 percent of the US lake acres were assessed for their water qualities and impairments (USEPA, 1994). Since the early 1990s, federal funds for CLP were terminated due to the budget cuts. Some states, such as Illinois, continued the CLP with state funds.

5.3 State Lake Classification Survey

In general, most states use state lake classification survey funding to operate three types of lake survey activities. These include the volunteer lake monitoring program (VLMP), the ambient lake monitoring program (ALMP), and lake water quality assessment (LWQA) grant. All three programs are partially supported by a Section 314 Federal CPL LWQA grant. State funds matched equally with federal grant funds are used to improve the quantity and quality of lake information reported in the annual 305(b) report to the US Congress.

Volunteer lake monitoring program. The VLMP is a statewide cooperative program which volunteers to monitor lake conditions twice a month from May through October and transmit the collected data to the state agency (EPA or similar agency) for analysis and report preparation. In Illinois, the VLMP was initiated by Illinois Environmental Protection Agency (IEPA) in 1981 (IEPA, 1984). The volunteers can be personnel employed by the lake owner (a water treatment plant, Department of Conservation or other state agency, city, lake association, industry, etc.), or they can be local citizens. (The lakes monitored are not limited to publicly owned lakes.) Volunteers must have a boat and an anchor in order to perform the sampling. Volunteers receive a report prepared by the state EPA, which evaluates their sampling results.

Volunteers measure Secchi disc transparencies and total depths at three sites in the lake. For reservoirs, site 1 is generally located at the deepest spot (near the dam); site 2 is at midlake or in a major arm; and site 3 is in the headwater, a major arm, or the tributary confluence. The data are recorded on standard forms. In addition, the volunteers also complete a field observation form each time the lake is sampled and record the number that best describes lake conditions during sampling for each site on the lake. Observations include color of the water, the amount of sediment suspended in the water, visible suspended algae, submerged or floating aquatic weeds, weeds near the shore, miscellaneous substances, odors, cloudiness, precipitation, waves, air temperature at the lake, lake water levels, recreational usage, and lake management done since the last sampling (IEPA, 1983).

Ambient lake monitoring program. The ALMP is also a statewide regular water quality monitoring program. Water quality samples are collected and analyzed annually by IEPA personnel on selected lakes throughout the state. The major objectives of the ALMP are to (IEPA, 1992):

- characterize and define trends in the condition of significant lakes in the state;
- diagnose lake problems, determine causes/sources of problems, and provide a basis for identifying alternative solutions;
- evaluate progress and success of pollution control/restoration programs;
- judge effectiveness of applied protection/management measures and determine applicability and transferability to other lakes;
- revise and update the lake classification system;
- meet the requirements of Section 314 CLP regulation and/or grant agreements.

The ALMP provides a much more comprehensive set of chemical analyses on the collected water samples than does the VLMP.

In Illinois, the ALMP was initiated in 1977. Water samples are analyzed for temperature, dissolved oxygen, Secchi disc transparency, alkalinity, conductivity, turbidity, total and volatile suspended solids, dissolved and total phosphorus, and nitrogen (ammonia, nitrite/nitrate, and total kjeldahl). In addition, 9 metals and 11 organics analyses are performed on water and sediment samples.

Lake water quality assessment. The LWQA grant is focused on non-routinely monitored lakes to gather basic data on in-lake water quality and sediment quality. This information increases the efficiency, effectiveness, and quality of the state EPA's lake data management and 305(b) report.

LWQA fieldwork involves two major tasks: namely, collection of lake assessment information and in-lake water and sediment sampling.

Lake assessment information collected includes lake identification and location, morphometric data, public access, designated uses and impairments, lake and shoreline usages, watershed drainage area usage, water quality and problems, status of fisheries, causes and sources of impairment (if any), past lake protection and management techniques, and lake maps.

In general, only one sampling trip per lake is made in order to evaluate a maximum number of lakes with the limited resources. The water and sediment samples are collected in summer by contracted agencies and analyzed by the state EPA laboratory. Chemical analyses for water and sediment samples are the same as in the ALMP.

5.4 Phase 1: Diagnostic/Feasibility Study

A diagnostic/feasibility study is a two-part study to determine a lake's current condition and to develop plans for its restoration and management. Protocol for the D/F study of a lake (Lin, 1994) is a digest from the *CLP Guidance Manual* (USEPA, 1980).

For a Phase 1 D/F study, the following activities should be carried out and completed within three years (USEPA, 1980):

1. development of a detailed work plan;
2. study of the natural characteristics of the lake and watershed;
3. study of social, economic, and recreational characteristics of the lake and watershed;
4. lake monitoring;
5. watershed monitoring;
6. data analysis;
7. development and evaluation of restoration alternatives;
8. selection and further development of watershed management plans;
9. projection of benefits;
10. environmental evaluation;
11. public participation;
12. public hearings (when appropriate);
13. report production.

Activities 1 through 6 are part of the diagnostic study; the other activities pertain to the feasibility study.

Diagnostic study. The diagnostic study is devoted to data gathering and analysis. It involves collecting sufficient limnological, morphological, demographic, and socioeconomic information about the lake and its watershed.

The study of lake and watershed natural characteristics, and most of the social, economic, and recreational information of the lake region can often be accomplished by obtaining and analyzing secondary data; i.e. data already available from other sources.

Baseline limnological data include a review of historical data and 1 year of current limnological data. Lake monitoring is expensive. Monthly and bimonthly samples are required for assessing physical and chemical characteristics. At least three sites are chosen for each lake, as mentioned in the discussion of the ALMP (Sec. 5.3). In addition, biological parameters, such as chlorophylls, phytoplankton, zooplankton, aquatic macrophytes, indicator bacteria, benthic macroinvertebrate, and fish surveys, must be assessed at different specified frequencies. At least one surficial and/or core sediment sample and water sample at each site must be collected for heavy metals and organic compounds analyses. Water quality analyses of tributaries and ground water may also be included, depending on each lake's situation.

Watershed monitoring is done by land-use stream monitoring manually or automatically to determine the nutrient, sediment, and hydraulic budget for a lake.

Primary and secondary data collected during the diagnostic study are analyzed to provide the basic information for the feasibility portion of the Phase 1 study. Data analysis involves:

1. inventory of point source pollutant discharges;
2. watershed land use and nonpoint nutrient/solids loading;
3. analyses of lake data—water and sediment quality;
4. analyses of stream and ground-water data;
5. calculating the hydrologic budget of the watershed and lake;
6. calculating the nutrient budget of the lake;
7. assessing biological resources and ecological relationships—lake fauna, terrestrial vegetation, and animal life;
8. determining the loading reductions necessary to achieve water quality objectives.

The following analyses of lake data are typically required: (1) identification of the limiting nutrient based on the ratio of total nitrogen to total phosphorus; (2) determination of the trophic state index based on total phosphorus, chlorophyll *a*, Secchi depth, and primary productivity; (3) calculation of the fecal coliform to fecal streptococcus ratios to identify causes of pollution in the watershed; and (4) evaluation of the sediment quality for the purpose of dredging operations.

A description of the biological resources and ecological relationships is included in the diagnostic study based on information gathered from secondary sources. This description generally covers lake flora (terrestrial vegetation, such as forest, prairie, and marsh) and fauna (mammals, reptiles and amphibians, and birds).

Feasibility study. The feasibility study involves developing alternative management programs based on the results of the diagnostic study. First, existing lake quality problems and their causes should be identified and analyzed. Problems include turbid water, eroding shorelines, sedimentation and shallow water depths, low water levels, low dissolved oxygen levels, excessive nutrients, algal bloom, unbalanced aquatic vegetative growth (excessive growth of macrophytes), non-native exotic species, degraded or unbalanced fishery and aquatic communities, bad taste and odor, acidity, toxic chemicals, agricultural runoff, bacterial contamination, poor lake esthetics, user conflicts, and negative human impacts.

Once the lake problems have been defined, a preliminary list of corrective alternatives needs to be established and discussed. Commonly adopted pollution control and lake restoration techniques can be found in the literature (USEPA, 1980, 1990). The corrective measures include in-lake restoration and/or best management practices in the watershed. Each alternative must be evaluated in terms of cost; reliability; technical feasibility; energy consumption; and social, economic, and environmental impacts. For each feasible alternative, a detailed description of measures to be taken, a quantitative analysis of pollution control effectiveness, and expected lake water quality improvement must be provided.

There are usually many good methods of achieving specific objectives or benefits. The Phase 1 report should document the various alternatives by describing their relative strengths and weaknesses and showing how the one chosen is superior. Final selection should be discussed in a public meeting.

Once an alternative has been selected, work can proceed on developing other program details required as part of Phase 1. These include developing the Phase 2 monitoring program, schedule and budget, and source of non-federal funds; defining the relationship of Phase 2 plans to other programs; developing an operation and maintenance plan; and obtaining required permits. In other words, the feasibility study portion of the Phase 1 report actually constitutes the proposal for Phase 2 study. The report also must include the projection of the project benefits, environmental evaluations, and public participation.

Final report. The final report will have to follow strictly the format and protocol stipulated in the *Clean Lakes Program Guidance Manual*. Requirements for diagnostic feasibility studies and environmental evaluations are at a minimum, but not limited, as follows:

1. Lake identification and location
2. Geologic and soils description of drainage basins
   a. Geologic description
   b. Ground-water hydrology
   c. Topography
   d. Soils
3. Description of public access
4. Description of size and economic structure of potential user population
5. Summary of historical lake uses
6. Population segments adversely affected by lake degradation
7. Comparison of lake uses to uses of other lakes in the region
8. Inventory of point-source pollution discharges
9. Land uses and nonpoint pollutant loadings
10. Baseline and current limnological data
    a. Summary analysis and discussion of historical baseline limnological data
    b. Presentation, analysis, and discussion of one year of current limnological data
    c. Tropic condition of lake
    d. Limiting algal nutrient
    e. Hydraulic budget for lake
    f. Nutrient sediment budget
11. Biological resources and ecological relationships
12. Pollution control and restoration procedures
13. Benefits expected from restoration
14. Phase II monitoring program
15. Schedule and budget
16. Source of matching funds
17. Relationship to other pollution control programs
18. Public participation summary
19. Operation and maintenance plan
20. Copies of permits or pending applications
21. Environmental evaluation
    a. Displacement of people
    b. Defacement of residential area
    c. Changes in land use patterns
    d. Impacts on prime agricultural land
    e. Impacts on parkland, other public land and scenic resources
    f. Impacts on historic, architectural, archaeological, or cultural resources
    g. Long range increases in energy demand
    h. Changes in ambient air quality or noise levels
    i. Adverse effects of chemical treatments
    j. Compliance with executive order 11988 on flood plain management
    k. Dredging and other channel bed or shoreline modifications
    l. Adverse effects on wetlands and related resources
    m. Feasible alternatives to proposed project
    n. Other necessary mitigative measures requirements

5.5 Phase 2: Implementation

A Phase 1 D/F study identifies the major problems in a lake and recommends a management plan. Under the CLP, if funds are available, the plan can then be implemented and intensive monitoring of the lake and tributaries conducted either by the state EPA or a contracted agency.

The physical, chemical, and biological parameters monitored and the frequencies of monitoring are similar to that for Phase 1 D/F study (or more specific, depending on the objectives). One year of water quality monitoring is required for post-implementation study. Non-federal cost-share (50%) funds are usually provided by state and local landowners. A comprehensive monitoring program should be conducted to obtain post-implementation data for comparison with pre-implementation (Phase 1) data.

The implementation program may include keeping point- and nonpoint-source pollutants from entering a lake, implementing in-lake restoration measures to improve lake water quality, a monitoring program, environmental impact and cost evaluation, and public participation.

5.6 Phase 3: Post-Restoration Monitoring

The CLP has recently begun setting up a system for managing and evaluating lake project data, designed to derive benefit from past projects. Quantitative scientific data on long-term project effectiveness will become increasingly available through post-restoration monitoring studies being conducted under Phase 3 CLP grants instituted in 1989.

Several years after Phase 2 implementation, a lake may be selected for Phase 3 post-restoration monitoring with matching funds. For Phase 3 monitoring, the water quality parameters monitored and the frequency of sampling in the lake and tributaries are similar to that for Phases 1 and 2. However, the period of monitoring for Phase 3 is generally 3 years with a total project period of 4 to 5 years. The purpose is to build upon an extensive database of information gathered under Phases 1 and 2 investigations.

Data gathered from Phases 1 are compared with Phases 2 and 3 databases to determine the long-term effectiveness of watershed protection measures and in-lake management techniques implemented during and since Phase 2 completion.

5.7 Watershed Protection Approach

The watershed protection approach (WPA) is an integrated strategy for more effective restoration and protection of aquatic ecosystems and human health; i.e. drinking water supplies and fish consumption (USEPA, 1993). The WPA focuses on hydrologically defined drainage basins ('watersheds') rather than on areas arbitrarily defined by political boundaries. Local decisions on the scale of geographic units consider many factors including the ecological structure of the basin, the hydrologic factors of groundwaters, the economic lake uses, the type and scope of pollution problems, and the level of resources available for protection and restoration projects (USEPA, 1991, 1993).

The WPA has three major principles (USEPA, 1991, 1993):

1. problem identification—identify the primary risk to human health and ecosystem within the watershed;
2. stakeholder involvement—involve all parties most likely to be concerned or most able to take action for solution;
3. integrated actions—take corrective actions in a manner that provides solutions and evaluates results.

The WPA is not a new program, and any watershed planning is not mandated by federal law. It is a flexible framework to achieve maximum efficiency and effect by collaborative activities. Everyone—individual citizens, the public and private sectors—can benefit from a WPA.

The USEPA's goal for the WPA is to maintain and improve the health and integrity of aquatic ecosystems using comprehensive approaches that focus resources on the major problems facing these systems within the watershed (USEPA, 1993).

For more than two decades, the CLP has emphasized using the watershed protection approach. A long-standing program policy gives greater consideration to applicants who propose restoration and protection techniques that control pollutants at the source through watershed-wide management rather than dealing with symptoms in the lake.

5.8 In-lake Monitoring

In Phase 1, 2, and 3 studies, in-lake water quality parameters that need to be monitored generally include water temperature, dissolved oxygen, turbidity, Secchi transparency, solids (total suspended and volatile), conductivity, pH, alkalinity, nitrogen (ammonia, nitrite/nitrate, and total kjeldahl), phosphorus (total and dissolved), total and fecal coliforms, fecal streptococci, algae, macrophytes, and macroinvertebrates. Except for the last two parameters, samples are collected bimonthly (April–October) and monthly (other months) for at least 1 year. At a minimum, three stations (at the deepest, the middle, and upper locations of the lake) are generally monitored (Lin *et al.*, 1996, 1998).

Lake water quality standards and criteria. The comprehensive information and data collected during Phase 1, 2, or 3 study are evaluated with state water quality standards and state lake assessment criteria. Most states in the United States have similar state standards and assessment criteria. For example, Illinois generally use the water quality standards applied to lake water listed in Table 3.1; and Illinois EPA's lake assessment criteria are presented in Table 3.2.

The data obtained from Phase 1, 2, and 3 of the study can be plotted from temporal variations for each water quality parameter, as shown in the example given in Fig. 3.4. Similarly, the historical data also can be plotted in the same manner for comparison purposes.

Indicator bacteria (total and fecal coliforms, and fecal streptococcus) densities in lakes and reservoirs should be evaluated in the same manner as that given for rivers and streams in Chapter 1.2. The density and moving geometric mean of FC are usually stipulated by the state's bacterial quality standards. The ratio of FC/FS is used to determine the source of pollution.

Temperature and dissolved oxygen. Water temperature and dissolved oxygen in lake waters are usually measured at 1 ft (30 cm) or 2 ft (60 cm) intervals. The obtained data is used to calculate values of the percent DO saturation using Eq. (2.4) or from Table 2.1. The plotted data is shown in Fig. 3.5: observed vertical DO and temperature profiles on selected dates at a station. Data from long-term observations can also be depicted in Fig. 3.6 which shows the DO isopleths and isothermal plots for a near-bottom (2 ft above the bottom station).

TABLE 3.1 Illinois General Use Water Quality Standards

1.	Dissolved oxygen:	> 5 mg/L at any time > 6 mg/L at 16 h/24 h
2.	Temperature: < 30°C	
3.	Total dissolved solids:	100 mg/L (conductivity: 1700 $\mu\Omega^{-1}$/cm)
4.	Chloride:	< 500 mg/L for protection of aquatic life
5.	pH:	6.5–9.0, except for natural causes (alkalinity in Illinois lakes 20–200 mg/L as $CaCO_3$)
6.	NH_3-N:	< 1.5 mg/L at 20°C (68°F) and pH of 8.0 < 15 mg/L under no conditions
7.	NO_3-N:	< 10 mg/L in all waters < 1.0 mg/L in public water supply
8.	Total P:	< 0.05 mg/L in any lake > 8.1 ha (20 acres)

Source: IEPA, 1987.

TABLE 3.2 Illinois EPA Lake Assessment Criteria

Parameter	Minimal	Slight	Moderate	High	Indication
Secchi depth	> 79	49–79	18–48	< 18	lake use impairment
TSS, mg/L	< 5	5–15	15–25	> 25	lake use impairment
Turbidity, NTU	< 3	3–7	7–15	> 15	suspended sediment
COD, mg/L	< 10	10–20	20–30	> 30	organic enrichment

Source: IEPA, 1987.

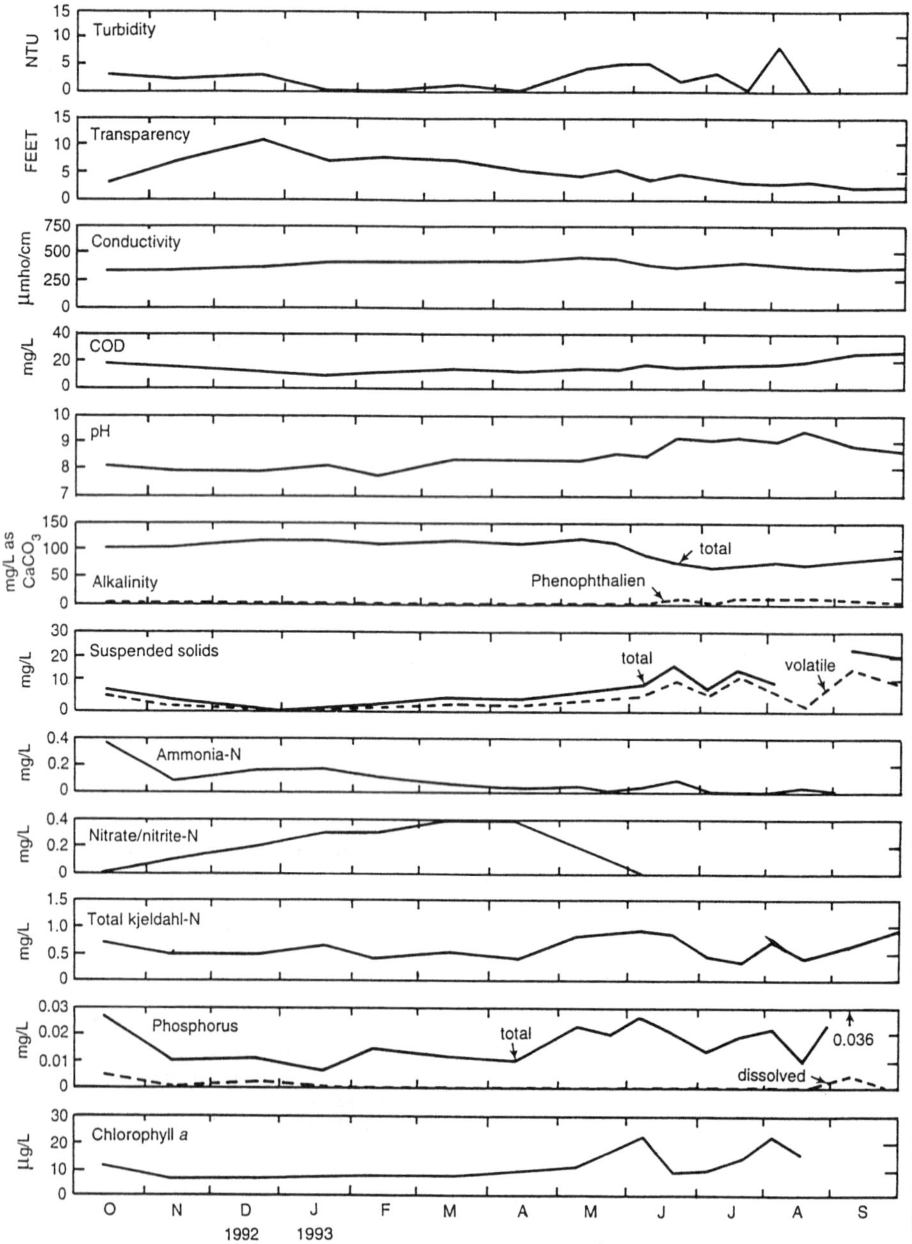

FIGURE 3.4 Temporal variations of surface water characteristics at RHA-2 (*Lin et al., 1996*).

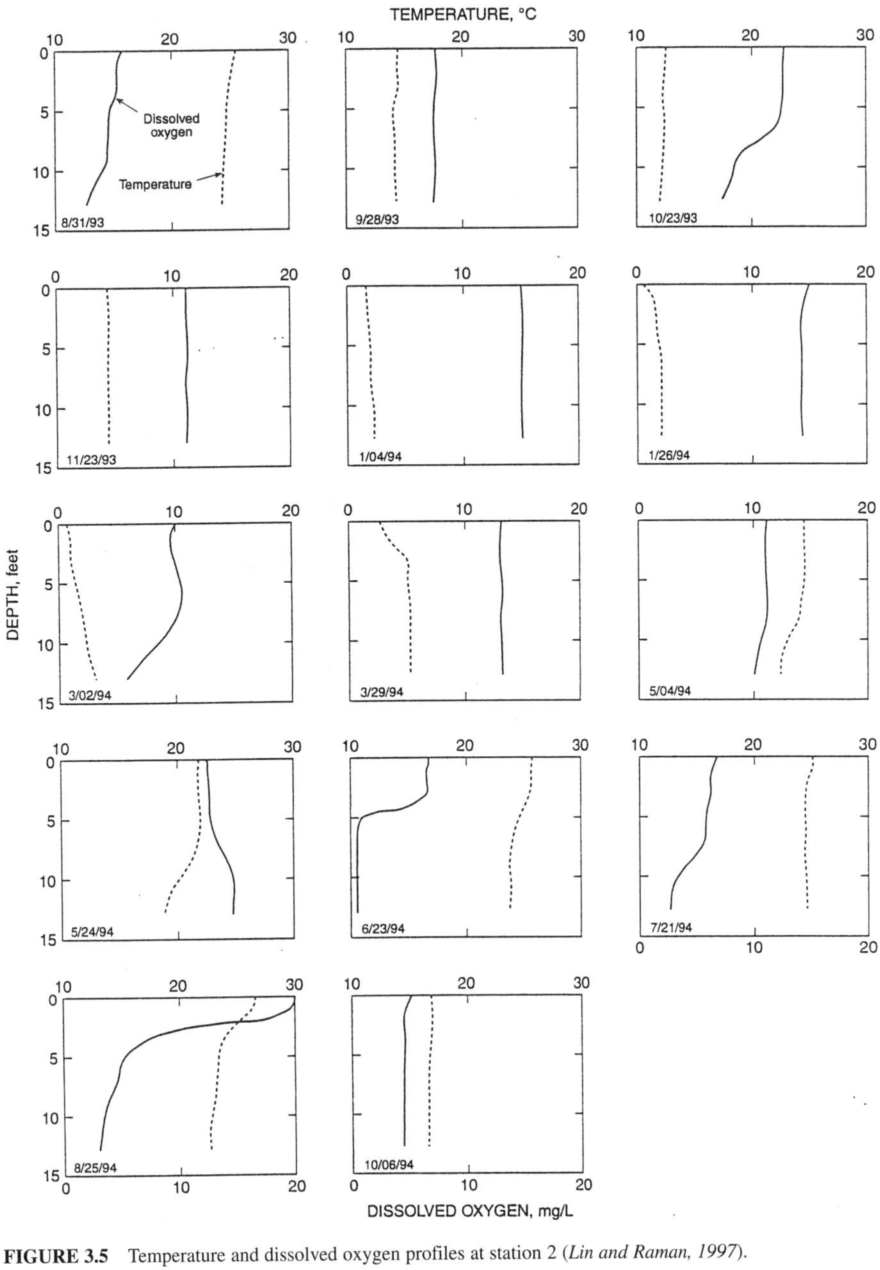

FIGURE 3.5 Temperature and dissolved oxygen profiles at station 2 (*Lin and Raman, 1997*).

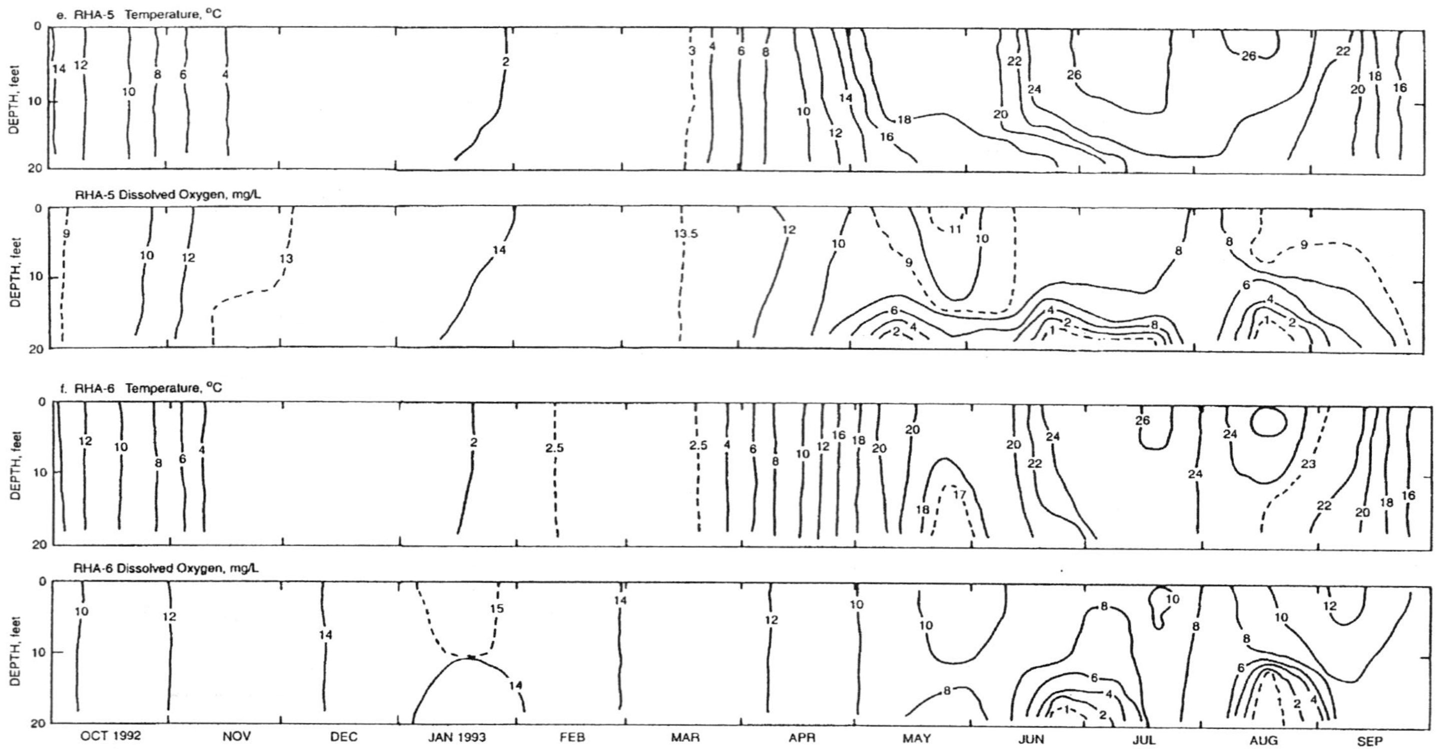

FIGURE 3.6 Isothermal and iso-dissolved oxygen plots for the deep stations (*Lin et al., 1996*).

Algae. Most algae are microscopic, free-floating plants. Some are filamentous and attached ones. Algae are generally classified into four major types: blue-greens, greens, diatoms, and flagellates. Through their photosynthesis processes, algae use energy from sunlight and carbon dioxide from bicarbonate sources, converting it to organic matter and oxygen (Fig. 3.7). The removal of carbon dioxide from the water results in an increase in pH and a decrease in alkalinity. Algae are important sources of dissolved oxygen in water. Phytoplanktonic algae form at the base of the aquatic food web and provide the primary source of food for fish and other aquatic insects and animals.

Excessive growth (blooms) of algae may cause problems such as bad taste and odor, increased color and turbidity, decreased filter run at a water treatment plant, unsightly surface scums and esthetic problems, and even oxygen depletion after die-off. Blue-green algae tend to cause the worst problems. To prevent such proliferation (upset the ecological balance), lake and watershed managers often try to reduce the amounts of nutrients entering lake waters, which act like fertilizers in promoting algal growth. Copper sulfate is frequently used for algal control, and sometimes it is applied on a routine basis during summer months, whether or not it is actually needed. One should try to identify the causes of the particular problems then take corrective measures selectively rather than on a shot-in-the-dark basis.

Copper sulfate ($CuSO_4 \cdot 5H_2O$) was applied in an Illinois impoundment at an average rate of 22 pounds per acre (lb/a); in some lakes, as much as 80 lb/a was used (Illinois State Water Survey, 1989). Frequently, much more copper sulfate is applied than necessary. Researchers have shown that 5.4 pounds of copper sulfate per acre of lake surface (6.0 kg/ha) is sufficient to control problem-causing blue-green algae in waters with high alkalinity (> 40 mg/L as $CaCO_3$). The amount is equivalent to the rate of 1 mg/L of copper sulfate for the top 2 ft (60 cm) of the lake surface (Illinois State Water Survey, 1989). The literature suggests that a concentration of 0.05–0.10 mg/L as Cu^{2+} is effective in controlling blue-green algae in pure cultures under laboratory conditions.

$$\underset{\text{Carbon dioxide}}{CO_2} + \underset{\text{Water}}{H_2O} \xrightarrow[\text{Enzymes}]{\text{Sunlight}} \underset{\text{Organic matter}}{(CH_2O)_n} + \underset{\text{Oxygen}}{O_2}$$

a) The photosynthetic process

$$CO_2 + H_2O \rightleftharpoons H_2CO_3 \rightleftharpoons HCO_3^- + H^+$$

$$HCO_3^- \rightleftharpoons CO_3^= + H^+$$

$$CO_3^= + H_2O \rightleftharpoons HCO_3^- + OH^-$$

b) Carbonate equilibria

Removed by photosynthesis ($\Uparrow$ from CO_2)

$$2HCO_3^- \rightleftharpoons CO_2 + CO_3^= + H_2O$$

($\Downarrow$ from $CO_3^=$)

$$Ca^{++} + CO_3^= \rightarrow CaCO_3 \downarrow \text{ (Marl deposits)}$$

c) Chemical processes that occur during photosynthesis

FIGURE 3.7 Photosynthesis and its chemical processes (*Illinois State Water Survey, 1989*).

EXAMPLE 1: What is the equivalent concentration of Cu^{2+} of 1 mg/L of copper sulfate in water?

Solution:

$$\text{MW of } CuSO_4 \cdot 5H_2O = 63.5 + 32 + 16 \times 4 + 5(2 + 16)$$

$$= 249.5$$

$$\text{MW of } Cu^{2+} = 63.5$$

$$\frac{Cu^{2+}, \text{mg/L}}{CuSO_4 \cdot 5H_2O, \text{mg/L}} = \frac{63.5}{249.5} = 0.255$$

$$Cu^{2+} = 0.255 \times 1 \text{ mg/L}$$

$$= 0.255 \text{ mg/L}$$

EXAMPLE 2: Since 0.05–0.10 mg/L of Cu^{2+} is needed to control blue-green algae, what is the theoretical concentration expressed as $CuSO_4 \cdot 5H_2O$?

Solution:

$$\text{Copper sulfate} = 0.05 \text{ mg/L} \times \frac{249.5}{63.5}$$

$$\cong 0.20 \text{ mg/L}$$

Answer: This is equivalent to 0.20–0.4 mg/L as $CuSO_4 \cdot 5H_2O$.

Note: for field application, however, a concentration of 1.0 mg/L as $CuSO_4 \cdot 5H_2O$ is generally suggested.

EXAMPLE 3: For a lake with 50 acres (20.2 ha), 270 lb (122 kg) of copper sulfate is applied. Compute the application rate in mg/L on the basis of the top 2 ft of lake surface.

Solution:

Step 1. Compute the volume (V) of the top 2 ft

$$V = 2 \text{ ft} \times 50 \text{ acre}$$

$$= 2 \text{ ft} \times 0.3048 \text{ m/ft} \times 50 \text{ acre} \times 4047 \text{ m}^2/\text{acre}$$

$$= 123.350 \text{ m}^3$$

$$= 123.3 \times 10^6 \text{ L}$$

Step 2. Convert weight (W) in pounds to milligrams

$$W = 270 \text{ lb} \times 453{,}600 \text{ mg/lb}$$

$$= 122.5 \times 10^6 \text{ mg}$$

Step 3. Compute copper sulfate application rate

$$\text{Rate} = \frac{W}{V} = \frac{122.5 \times 10^6 \text{ mg}}{123.3 \times 10^6 \text{ L}}$$

$$= 0.994 \text{ mg/L}$$

$$\cong 1.0 \text{ mg/L}$$

Secchi disc transparency. The Secchi disc is named after its Italian inventor Pietro Angelo Secchi, and is a black and white round plate used to measure water clarity. Values of Secchi disc transparency are used to classify a lake's trophic state.

Secchi disc visibility is a measure of a lake's water transparency, which suggests the depth of light penetration into a body of water (its ability to allow sunlight to penetrate). Even though Secchi disc transparency is not an actual quantitative indication of light transmission, it provides an index for comparing similar bodies of water or the same body of water at different times. Since changes in water color and turbidity in deep lakes are generally caused by aquatic flora and fauna, transparency is related to these entities. The euphotic zone or region of a lake where enough sunlight penetrates to allow photosynthetic production of oxygen by algae and aquatic plants is taken as two to three times the Secchi disc depth (USEPA, 1980).

Suspended algae, microscopic aquatic animals, suspended matter (silt, clay, and organic matter), and water color are factors that interfere with light penetration into the water column and reduce Secchi disc transparency. Combined with other field observations, Secchi disc readings may furnish information on (1) suitable habitat for fish and other aquatic life; (2) the lake's water quality and esthetics; (3) the state of the lake's nutrient enrichment; and (4) problems with and potential solutions for the lake's water quality and recreational use impairment.

Phosphorus. The term total phosphorus (TP) represents all forms of phosphorus in water, both particulate and dissolved forms, and includes three chemical types: reactive, acid-hydrolyzed, and organic. Dissolved phosphorus (DP) is the soluble form of TP (filterable through a 0.45 μm filter).

Phosphorus as phosphate may occur in surface water or groundwater as a result of leaching from minerals or ores, natural processes of degradation, or agricultural drainage. Phosphorus is an essential nutrient for plant and animal growth and, like nitrogen, it passes through cycles of decomposition and photosynthesis.

Because phosphorus is essential to the plant growth process, it has become the focus of attention in the entire eutrophication issue. With phosphorus being singled out as probably the most limiting nutrient and the one most easily controlled by removal techniques, various facets of phosphorus chemistry and biology have been extensively studied in the natural environment. Any condition which approaches or exceeds the limits of tolerance is said to be a limiting condition or a limiting factor.

In any ecosystem, the two aspects of interest for phosphorus dynamics are phosphorus concentration and phophorus flux (concentration $\times$ flow rate) as functions of time and distance. The concentration alone indicates the possible limitation that this nutrient can place on vegetative growth in the water. Phosphorus flux is a measure of the phosphorus transport rate at any point in flowing water.

Unlike nitrate-nitrogen, phosphorus applied to the land as a fertilizer is held tightly to the soil. Most of the phosphorus carried into streams and lakes from runoff over cropland will be in the particulate form adsorbed to soil particles. On the other hand, the major portion of phosphate-phosphorus emitted from municipal sewer systems is in a dissolved form. This is also true of phosphorus generated from anaerobic degradation of organic matter in the lake bottom. Consequently, the form of phosphorus, namely particulate or dissolved, is indicative of its source to a certain extent. Other sources of dissolved phosphorus in the lake water may include the decomposition of aquatic plants and animals. Dissolved phosphorus is readily available for algae and macrophyte growth. However, the DP concentration can vary widely over short periods of time as plants take up and release this nutrient. Therefore, TP in lake water is the more commonly used indicator of a lake's nutrient status.

From his experience with Wisconsin lakes, Sawyer (1952) concluded that aquatic blooms are likely to develop in lakes during summer months when concentrations of inorganic nitrogen and inorganic phosphorus exceed 0.3 and 0.01 mg/L, respectively. These critical levels for nitrogen and phosphorus concentrations have been accepted and widely quoted in scientific literature.

To prevent biological nuisance, the IEPA (1990) stipulates, 'Phosphorus as P shall not exceed a concentration of 0.05 mg/L in any reservoir or lake with a surface area of 8.1 hectares (20 acres) or more or in any stream at the point where it enters any reservoir or lake.'

Chlorophyll. All green plants contain chlorophyll *a*, which constitutes approximately 1 to 2 percent of the dry weight of planktonic algae (APHA *et al.*, 1992). Other pigments that occur in phytoplankton include chlorophyll *b* and *c*, xanthophylls, phycobilius, and carotenes. The important chlorophyll degration products in water are the chlorophyllides, pheophorbides, and pheophytines. The concentration of photosynthetic pigments is used extensively to estimate phytoplanktonic biomass. The presence or absence of the various photosynthetic pigments is used, among other features, to identify the major algal groups present in the water body.

Chlorophyll *a* is a primary photosynthetic pigment in all oxygen-evolving photosynthetic organisms. Extraction and quantification of chlorophyll *a* can be used to estimate biomass or the standing crop of planktonic algae present in a body of water. Other algae pigments, particularly chlorophyll *b* and *c*, can give information on the type of algae present. Blue-green algae (Cyanophyta) contain only chlorophyll *a*, while both the green algae (Chlorophyta) and the euglenoids (Euglenophyta) contain chlorophyll *a* and *c*. Chlorophyll *a* and *c* are also present in the diatoms, yellow-green and yellow-brown algae (Chrysophyta), as well as dinoflagellates (Pyrrhophyta). These accessory pigments can be used to identify the types of algae present in a lake. Pheophytin *a* results from the breakdown of chlorophyll *a*, and a large amount indicates a stressed algal population or a recent algal die-off. Because direct microscopic examination of water samples is used to identify and enumerate the type and concentrations of algae present in the water samples, the indirect method (chlorophyll analyses) of making such assessments may not be employed.

Nutrient in lake water will impact the aquatic community in the surface water during summer. Dillon and Rigler (1974) used data from North American lakes to drive the relationship between spring phosphorus and chlorophyll *a* concentration in summer as follows:

$$\text{CHL} = 0.0731\,(\text{TP})^{1.449} \tag{3.14}$$

where CHL = summer average chlorophyll *a* concentration at lake surface water, mg/m^3
TP = spring average total phosophorus concentration at lake surface water, mg/m^3

EXAMPLE: Estimate the average chlorophyll *a* concentration in a North American lake during the summer if the average spring total phosphorus is 0.108 mg/L.

Solution:

$$\begin{aligned}\text{TP} &= 0.108 \text{ mg/L} \\ &= 0.108 \text{ mg/L} \times 1000 \text{ L/m}^3 \\ &= 108 \text{ mg/m}^3\end{aligned}$$

Estimate summer CHL using Eq. (3.14)

$$\begin{aligned}\text{CHL} &= 0.0731\,(\text{TP})^{1.449} \\ &= 0.0731\,(108)^{1.449} \text{ mg/m}^3 \\ &= 64.6 \text{ mg/m}^3\end{aligned}$$

5.9 Trophic State Index

Eutrophication is a normal process that affects every body of water from its time of formation (Walker, 1981a, 1981b). As a lake ages, the degree of enrichment from nutrient materials increases. In general, the lake traps a portion of the nutrients originating in the surrounding drainage basin. Precipitation, dry fallout, and groundwater inflow are the other contributing sources.

A wide variety of indices of lake trophic conditions have been proposed in the literature. These indices have been based on Secchi disc transparency; nutrient concentrations; hypolimnetic oxygen depletion; and biological parameters, including chlorophyll *a*, species abundance, and diversity. In its *Clean Lake Program Guidance Manual*, the USEPA (1980) suggests the use of four parameters as trophic indicators: Secchi disc transparency, chlorophyll *a*, surface water total phosphorus, and total organic carbon.

In addition, the lake trophic state index (TSI) developed by Carlson (1977) on the basis of Secchi disc transparency, chlorophyll *a*, and surface water total phosphorus can be used to calculate a lake's trophic state. The TSI can be calculated from Secchi disc transparency (SD) in meters (m), chlorophyll *a* (CHL) in micrograms per liter (μg/L), and total phosphorus (TP) in μg/L as follows:

$$\text{on the basis of SD, TSI} = 60 - 14.4 \ln(\text{SD}) \tag{3.15}$$

$$\text{on the basis of CHL, TSI} = 9.81 \ln(\text{CHL}) + 30.6 \tag{3.16}$$

$$\text{on the basis of TP, TSI} = 14.42 \ln(\text{TP}) + 4.15 \tag{3.17}$$

The index is based on the amount of algal biomass in surface water, using a scale of 0 to 100. Each increment of 10 in the TSI represents a theoretical doubling of biomass in the lake. The advantages and disadvantages of using the TSI were discussed by Hudson *et al.* (1992). The accuracy of Carlson's index is often diminished by water coloration or suspended solids other than algae. Applying TSI classification to lakes that are dominated by rooted aquatic plants may indicate less eutrophication than actually exists.

Lakes are generally classified by limnologists into one of three trophic states: oligotrophic, mesotrophic, or eutrophic (Table 3.4). Oligotrophic lakes are known for their clean and cold waters and lack of aquatic weeds or algae, due to low nutrient levels. There are few oligotrophic lakes in the Midwest. At the other extreme, eutrophic lakes are high in nutrient levels and are likely to be very productive in terms of weed growth and algal blooms. Eutrophic lakes can support large fish populations, but the fish tend to be rougher species that can better tolerate depleted levels of DO. Mesotrophic lakes are in an intermediate stage between oligotrophic and eutrophic. The great majority of Midwestern lakes are eutrophic. A hypereutrophic lake is one that has undergone extreme eutrophication to the point of having developed undesirable esthetic qualities (e.g. odors, algal mats, and fish kills) and water-use limitations (e.g. extremely dense growths of vegetation). The natural aging process causes all lakes to progress to the eutrophic condition over time, but this eutrophication process can be accelerated by certain land uses in the contributing watershed, streams, and from unprotected areas (e.g. agricultural activities, application of lawn fertilizers, and erosion from construction sites, streams, and from unprotected areas). Given enough time, a lake will grow shallower and will eventually fill in with trapped sediments and decayed organic matter, such that it becomes a shallow marsh or emergent wetland.

EXAMPLE 1: Lake monitoring data shows that the Secchi disc transparency is 77 inches; total phosphorus, 31 μg/L; and chlorophyll *a* 3.4 μg/L. Calculate TSI values using Eqs. (3.15)−(3.17).

Solution:

Step 1. Using Eq. (3.15)

$$\begin{aligned} \text{SD} &= 77 \text{ in} \times 2.54 \text{ cm/in} \times 0.01 \text{ m/cm} = 1.956 \text{ m} \\ \text{TSI} &= 60 - 14.4 \ln(\text{SD}) = 60 - 14.4 \ln(1.956) \\ &= 50.3 \end{aligned}$$

Step 2. Using Eq. (3.16)

$$\begin{aligned} \text{TSI} &= 9.81 \ln(\text{CHL}) + 30.6 = 9.81 \ln(3.4) + 30.6 \\ &= 42.6 \end{aligned}$$

Step 3. Using Eq. (3.17)

$$\begin{aligned} \text{TSI} &= 14.42 \ln(\text{TP}) + 4.15 = 14.42 \ln(31) + 4.15 \\ &= 53.7 \end{aligned}$$

EXAMPLE 2: One-year lake monitoring data for a Secchi disc transparency, total phosphorus, and chlorophyll *a* in a southern Illinois lake (at the deepest station) are listed in Table 3.3. Determine the trophic condition of the lake.

TABLE 3.3 Trophic State Index and Trophic State of an Illinois Lake

Date	Secchi disc trans		Total phosphorus		Chlorophyll *a*	
	in	TSI	μg/L	TSI	μg/L	TSI
10/26/95	77	50.3	31	53.7	3.4	42.6
11/20/95	104	46.0	33	54.6	4.7	45.8
12/12/95	133	42.5	21	48.1	3	41.4
1/8/96	98	46.9	29	52.7		
2/14/96	120	44.0	26	51.1	2.1	37.9
3/11/96	48	57.1	22	48.7	6.2	48.5
4/15/96	43	58.7	49	60.3	10.4	53.6
5/21/96	52	56.0	37	56.2	12.4	55.3
6/6/96	51	56.3	36	55.8	7.1	49.8
6/18/96	60	53.9	30	53.2	14.6	56.9
7/1/96	60	53.9	21	48.1	7.7	50.6
7/16/96	42	59.1	38	56.6	12.4	55.3
8/6/96	72	51.3	19	46.6	6	48.2
8/20/96	48	57.1	20	47.3	1.7	35.8
9/6/96	54	55.4	32	54.1	9.8	53.0
9/23/96	69	51.9	22	48.9	11.3	54.4
10/23/96	42	59.1	42	58.0		
Mean		52.9		52.6		48.6
Trophic state		eutrophic		eutrophic		mesotrophic
Overall mean		51.4				
Overall trophic state		eutrophic				

Source: Bogner *et al.* (1997).

Solution:

Step 1. Calculate TSI values:

As example 1, the TSI values for the lake are calculated using Eqs. (3.15)−(3.17) based on SD, TP, and chlorophyll *a* concentration of each sample. The TSI values are included in Table 3.3.

Step 2. Determine trophic state:

a. Calculate average TSI based on each parameter.
b. Classify trophic state based on the average TSI and the criteria listed in Table 3.4.
c. Calculate overall mean TSI for the lake.
d. Determine overall trophic state. For this example, the lake is classified as *eutrophic*. The classifications are slightly different if based on each of the three water quality parameters.

TABLE 3.4 Quantitative Definition of a Lake Trophic State

Trophic state	Secchi disc transparency		Chlorophyll *a*	Total phosphorus, lake surface	TSI
	in	m	(μg/L)	(μg/L)	
Oligotrophic	> 157	> 4.0	< 2.6	< 12	< 40
Mesotrophic	79–157	2.0–4.0	2.6–7.2	12–24	40–50
Eutrophic	20–79	0.5–2.0	7.2–55.5	24–96	50–70
Hypereutrophic	< 20	< 0.5	> 55.5	> 96	> 70

5.10 Lake Use Support Analysis

Definition. An analysis of a lake's use support can be carried out employing a methodology developed by the IEPA (1994). The degree of use support identified for each designated use indicates the ability of the lake to (1) support a variety of high-quality recreational activities, such as boating, sport fishing, swimming, and esthetic enjoyment; (2) support healthy aquatic life and sport fish populations; and (3) provide adequate, long-term quality and quantity of water for public or industrial water supply (if applicable). Determination of a lake's use support is based upon the state's water quality standards as described in Subtitle C of Title 35 of the State of Illinois Administrative Code (IEPA, 1990). Each of four established use-designation categories (including general use, public and food processing water supply, Lake Michigan, and secondary contact and indigenous aquatic life) has a specific set of water quality standards.

For the lake uses assessment, the general use standards—primarily the 0.05 mg/L TP standard—shall be used. The TP standard has been established for the protection of aquatic life, primary-contact (e.g. swimming) and secondary-contact (e.g. boating) recreation, agriculture, and industrial uses. In addition, lake-use support is based in part on the amount of sediment, macrophytes, and algae in the lake and how these might impair designated lake uses. The following is a summary of the various classifications of use impairment:

- *Full* = full support of designated uses, with minimal impairment.
- *Full/threatened* = full support of designated uses, with indications of declining water quality or evidence of existing use impairment.
- *Partial/minor* = partial support of designated uses, with slight impairment.
- *Partial/moderate* = partial support of designated uses, with moderate impairment; only partial is used after the year of 2000.
- *Nonsupport* = no support of designated uses, with severe impairment.

Lakes that fully support designated uses may still exhibit some impairment, or have slight-to-moderate amounts of sediment, macrophytes, or algae in a portion of the lake (e.g. headwaters or shoreline); however, most of the lake acreage shows minimal impairment of the aquatic community and uses. *It is important to emphasize that if a lake is rated as not fully supporting designated uses, it does not necessarily mean that the lake cannot be used for those purposes or that a health hazard exists.* Rather, it indicates impairment in the ability of significant portions of the lake waters to support either a variety of quality recreational experiences or a balanced sport fishery. Since most lakes are multiple-use water bodies, a lake can fully support one designated use (e.g. aquatic life) but exhibit impairment of another (e.g. swimming).

Lakes that partially support designated uses have a designated use that is slightly to moderately impaired in a portion of the lake (e.g. swimming impaired by excessive aquatic macrophytes or algae, or boating impaired by sediment accumulation). So-called nonsupport lakes have a designated use that is severely impaired in a substantial portion of the lake (e.g. a large portion of the lake has so much sediment that boat ramps are virtually inaccessible, boating is nearly impossible, and fisheries are degraded). However, in other parts of the same nonsupport lake (e.g. near a dam), the identical use may be supported. *Again, nonsupport does not necessarily mean that a lake cannot support any uses, that it is a public health hazard, or that its use is prohibited.*

Lake-use support and level of attainment shall be determined for aquatic life, recreation, swimming, fish consumption, drinking water supply, secondary contact, and overall lake use, using methodologies described in the IEPA's *Illinois Water Quality Report 1994–1995* (IEPA, 1996).

The primary criterion in the aquatic life use assessment is an aquatic life use impairment index (ALI), while in the recreation use assessment the primary criterion is a recreation use impairment index (RUI). While both indices combine ratings for TSI (Carlson, 1977) and degree of use impairment from sediment and aquatic macrophytes, each index is specifically designed for the assessed use. ALI and RUI relate directly to the TP standard of 0.05 mg/L. If a lake water sample is found to have a TP concentration at or below the standard, the lake is given a 'full support' designation. The aquatic life use rating reflects the degree of attainment of the 'fishable goal' of the Clean Water Act, whereas the recreation use rating reflects the degree to which pleasure boating, canoeing, and esthetic enjoyment may be obtained at an individual lake.

The assessment of swimming use for primary-contact recreation was based on available data using two criteria: (1) Secchi disc transparency depth data and (2) Carlson's TSI. The swimming use rating reflects the degree of attainment of the 'swimmable goal' of the Clean Water Act. If a lake is rated 'nonsupport' for swimming, it does not mean that the lake cannot be used or that health hazards exist. It indicates that swimming may be less desirable than at those lakes assessed as fully or partially supporting swimming.

Finally, in addition to assessing individual aquatic life, recreation, and swimming uses, and drinking water supply, the overall use support of the lake is also assessed. The overall use support methodology aggregates the use support attained for each of the individual lake uses assessed. Values assigned to each use-support attainment category are summed and averaged, and then used to assign an overall lake-use attainment value for the lake.

Designated uses assessment. Multiple lakes designated are assessed for aquatic life, recreation, drinking water supply, swimming, fish consumption, and overall use. Specific criteria for determining attainment of these designated lake uses are described below. The degree of use support attainment is described as full, full/threatened, partial/minor impairment, partial/moderate impairment, or nonsupport.

Aquatic life. An aquatic life use impairment index (ALI) which combines ratings for trophic state index and the amount of use impairments from aquatic macrophytes and sediment is used as the primary criteria for assessing aquatic life lake use (Table 3.5). The higher the ALI number, the more impaired the lake. Specific critiera used for each level of aquatic life use support attainment are presented in Table 3.6.

Recreation. A recreation use impairment index (RUI), which combines TSI and the amount of use impairments from aquatic life and from sediment is utilized as the primary criteria for assessing recreation lake use (Table 3.7). Lake uses include pleasure boating, canoeing, skiing, sailing, esthetic enjoyment, and fishing. The higher the RUI number, the more impaired the lake. Specific criteria used for each level of attaining recreation use support are listed in Table 3.8.

Swimming. The assessment criteria for swimming use is based primarily on the Secchi disc transparency depth and on the fecal coliform (FC) density—percent that exceed the 200 FC/100 mL standard. If FC data are not available, a TSI is calculated and used to make the assessment. The degree of swimming use support attainment is presented in Table 3.9.

Drinking water supply. Drinking water supply use assessment for a lake is determined on the basis of water supply advisories or closure issued through state regulatory public water supply

TABLE 3.5 Aquatic Life Use Impairment Index (ALI)

Evaluation factor	Parameter	Weighting criteria	Points
1. Mean trophic state index (Carson, 1977)	Mean TSI value between 30 and 100	a. TSI < 60 b. $60 \leq$ TSI < 85 c. $85 \leq$ TSI < 90 d. $90 <$ TSI	a. 40 b. 50 c. 60 d. 70
2. Macrophyte impairment	Percent of lake surface area covered by weeds, or amount of weeds recorded on form	a. $15 \leq \% < 40$; or minimal (1) b. $10 \leq \% < 15$ & $< 40 \leq \% < 50$; or slight (2) c. $5 \leq \% < 10$ & $50 \leq \% < 70$; or moderate (3) d. $\% < 5$ & $70 \leq \%$; or substantial (4)	a. 0 b. 5 c. 10 d. 15
3. Sediment impairment	Concentration of non-volatile suspended solids (NVSS); or amount of sediment value reported on form	a. NVSS < 12; or minimal (1) b. $12 \leq$ NVSS < 15; or slight (2) c. $15 \leq$ NVSS < 20; or moderate (3) d. $20 \leq$ NVSS; or substantial (4)	a. 0 b. 5 c. 10 d. 15

Source: Modified from Illinois EPA (1996).

TABLE 3.6 Assessment Criteria for Aquatic Life and Overall Use in Illinois Lakes

Degree of use support	Criteria
Full	a. Total ALI points are < 75 b. Direct field observations of minimal aquatic life impairment
Full/threatened	a. Total ALI point are < 75 and evidence of a declined water quality trend exists b. Specific knowledge of existing or potential threats to aquatic life impairment
Partial/minor	a. 75 ≤ total ALI points < 85 b. Direct field observations of slight aquatic life impairment
Partial/moderate	a. 85 ≤ total ALI points < 95 b. Direct field observations of moderate aquatic life impairment
Nonsupport	a. Total ALI points ≥ 95 b. Direct field observation of substantial aquatic life impairment

Source: Modified from Illinois EPA (1996).

TABLE 3.7 Recreation Use Impairment Index (RUI)

Evaluation factor	Parameter	Weighting criteria	Points
1. Mean trophic state index (Carlson, 1977)	Mean TSI value (30–110)	a. Actual TSI value	Actual TSI value
2. Macrophyte impairment	Percent of lake surface area covered by weeds; or amount of weeds value reported on form	a. % < 5; or minimal (1) b. 5 < % < 15; or slight (2) c. 15 < % < 25; or moderate (3) d. 25 < %; or substantial (4)	a. 0 b. 5 c. 10 d. 15
3. Sediment impairment	Concentration of nonvolatile suspended solids (NVSS); or amount of sediment value reported on form	a. NVSS % < 3; or minimal (1) b. 3 ≤ NVSS < 7; or slight (2) c. 7 ≤ NVSS < 15; or moderate (3) d. 15 ≤ %; or substantial (4)	a. 0 b. 5 c. 10 d. 15

Source: Modified from Illinois EPA (1996).

TABLE 3.8 Assessment Criteria for Recreation Use in Illinois Lakes

Degree of use support	Criteria
Full	a. Total RUI points are less than 60 b. Direct field observation of minor recreation impairment
Full/threatened	a. Total RUI points are < 60, and evidence of a decline in water quality trend exists b. Specific knowledge or potential threats to recreation impairment
Partial/minor moderate	a. 60 ≤ total RUI points < 75 b. Direct field observation of slight impairment
Partial/moderate	a. 75 ≤ total RUI points < 90 b. Direct field observation of moderate recreation impairment
Nonsupport	a. Total RUI points ≥ 90 b. Direct field observation of substantial recreation impairment

Source: Illinois EPA (1996).

TABLE 3.9 Assessment Criteria for Swimming Use in Illinois Lakes

Degree of use support	Criteria
Full	a. No Secchi depths are < 24 in b. 10% ⩾ fecal coliforms (FC) samples exceed the standard c. TSI ⩽ 50
Partial/minor	a. ⩽ 50% of Secchi depths were < 24 in b. 10% < FC ⩽ 25% exceed the standard c. TSI ⩽ 65
Partial/moderate	a. 50–100% of Secchi depths were < 24 in b. 10% < FC sample ⩽ 25% exceed the standard TSI ⩽ 75
Nonsupport	a. 100% of Secchi depths were < 24 in b. 25% < FC sample exceed the standard c. TSI > 75

Source: Modified from Illinois EPA (1996).

programs. For example, in Illinois, the primary criteria used is the length of time (greater than 30 days) nitrate and/or atrazine concentration exceeding the public water supply standards of 10 mg/L and 3 μg/L, respectively. Other problems which affect the quality of finished waters, such as chemical or oil spills or severe taste and odor problems requiring immediate attention, are also included in assessing use support. Specific criteria used for assessing the drinking water supply and the degree of use support are shown in Table 3.10.

Fish consumption. The assessment of fish consumption use is based on fish tissue data and resulting sport fish advisories generated by the state fish contaminant monitoring program. The degree of fish consumption use support attainment can be found in Table 3.11.

Overall use. After assessing individual lake uses, the overall use support of a lake can be determined. The overall use support methodology aggregates the use support attained for each of the individual lake uses assessed (i.e. aquatic life, recreation, swimming, drinking water supply, and fish consumption). The aggregation is achieved by averaging individual use attainments for a lake. For instance, individual uses meeting full, full/threatened, partial/minor, partial/moderate, or nonsupport are assigned values from five (5) to one (1), respectively. The values assigned to each individual use are subsequently summed and averaged. The average value is rounded down to the next whole number, which is then applied to assign an overall lake use attainment. Full support attainment is assigned to an average value of 5; full/threatened, 4; partial/minor, 3; partial/moderate, 2; and nonsupport, 1.

TABLE 3.10 Assessment Criteria for Drinking Water Supply in Illinois Lakes

Degree of use support	Criteria
Full	No drinking water closures or advisories in effect during reporting period; no treatment necessary beyond 'reasonable levels' (copper sulfate may occasionally be applied for algae/taste and odor control).
Partial/minor	One or more drinking water supply advisory(ies) lasting 30 days or less; or problems not requiring closure or advisories but adversely affecting treatment costs and quality of treated water, such as taste and odor problems, color, excessive turbidity, high dissolved solids, pollutants requiring activated carbon filters.
Partial/moderate	One or more drinking water supply advisories lasting more than 30 days per year.
Nonsupport	One or more drinking water supply closures per year.

Source: Illinois EPA (1996).

TABLE 3.11 Assessment Criteria for Fish Consumption Use in Illinois Lakes

Degree of use support	Criteria
Full	No fish advisories or bans are in effect.
Partial/moderate	Restricted consumption fish advisory or ban in effect for general population or a subpopulation that could be at potentially greater risk (pregnant women, children). Restricted consumption is defined as limits on the number of meals or size of meals consumed per unit time for one or more fish species. In Illinois, this is equivalent to a Group II advisory.
Nonsupport	No consumption fish advisory or ban in effect for general population for one or more fish species; commercial fishing ban in effect. In Illinois this is equivalent to a Group III advisory.

Source: Illinois EPA (1996).

EXAMPLE: The mean TSI is determined by averaging 18 months' SD-TSI, TP-TSI, and CHL-TSI values at station 1-surface (deepest station) and station 2-surface (mid-lake station) of a central Illinois lake, for values of 54.3 and 58.9, respectively. All secchi disc reading were greater than 24 inches. The mean of nonvolatile suspended solids concentrations observed during 1996–97 at stations 1-surface and 2-surface are 5 and 7 mg/L, respectively. Estimated macrophyte impairment and other observed data are given in Table 3.12. Determine support of designated uses in the lake based on Illinois lake-use support assessment criteria.

TABLE 3.12 Use Support Assessment for Otter Lake, 1996–97

	Station 1		Station 2	
	value	ALI points*	value	ALI points
I. Aquatic life use				
1. Mean trophic state index	54.3	40	58.9	40
2. Macrophyte impairment	< 5%	10	< 5%	10
3. Mean NVSS	5 mg/L	0	7 mg/L	5
Total points:		50		55
Criteria points:		< 75		< 75
Use support:		*Full*		*Full*
	Value	RUI points*	Value	RUI points
II. Recreation use				
1. Mean trophic state index	54.3	54	58.9	59
2. Macrophyte impairment	<5%	0	< 5%	0
3. Mean NVSS	5 mg/L	5	7 mg/L	5
Total points:		59		64
Criteria points:		<60		65 ≤ R ≤ 75
Use support:		*Full*		*Partial*
	Value	Degree of use support	Value	Degree of use support
III. Swimming use				
1. Secchi depth < 24 in	0%	Full	0%	Full
2. Fecal coliform > 200/100 mL	0%	Full	0%	Full
3. Mean trophic state index	54.3	Full	58.9	Full
Use support:		*Full*		*Full*
IV. Drinking water supply		*Full*		*Full*
V. Overall use	5		4	
Use support:		*Full*		*Full/threatened*

*ALI, aquatic life use impairment index; RUI, recreation use impairment index.

Solution: Solve for station 1 (Steps 1−5); then station 2 (Step 6) can be solved in the same manner.

Step 1. Assess aquatic life use

1. **Find ALI points for TSI**
   Since mean TSI = 54.3, TSI points = 40 is obtained from Table 3.5, section 1.
2. **Determine macrophyte impairment (MI) of ALI points**
   The estimated macrophyte impairment is 5% of the lake surface area. From section 2 of Table 3.5, the MI point is 10.
3. **Determine mean nonvolatile suspended solids (NVSS) of ALI point**
   The NVSS = 5 mg/L. From section 3 of Table 3.5, we read NVSS point = 0.
4. **Calculate total ALI points**

$$\begin{aligned}\text{Total ALI points} &= \text{TSI points} + \text{MI points} + \text{NVSS points}\\ &= 40 + 10 + 0\\ &= 50\end{aligned}$$

5. **Determine the degree of aquatic life index use support**
   Since ALI points = 50 (i.e. < 75) of the critical value, and there are potential threats to aquatic life impairment, the degree of use support is *full* from Table 3.6.

Step 2. Assess recreation use

1. **Determine the mean TSI of RUI point**
   Since mean TSI value is = 54.3, then RUI points for TSI = 54, from section 1 of Table 3.7.
2. **Determine macrophyte impairment of RUI point**
   Since < 5% of the lake's surface area is covered by weeds, from section 2 of Table 3.7, the RUI point for MI = 0.
3. **Determine sediment impairment**
   The mean NVSS is 5 mg/L. From section 3 of Table 3.7, for sediment impairment (NVSS), RUI = 5.
4. **Compute total RUI points**

$$\begin{aligned}\text{Total RUI points} &= \text{TSI value} + \text{MI points} + \text{NVSS points}\\ &= 54 + 0 + 5\\ &= 59\end{aligned}$$

5. **Determine the degree of recreation use support**
   Total RUI points < 60, therefore the degree of recreation use is considered as *Full* from Table 3.8.

Step 3. Assess swimming use

1. **Based on Secchi disc transparency**
   Since no Secchi depth is less than 24 in, from Table 3.9, the degree of use support is classified as *Full*.
2. **Fecal coliform criteria**
   No fecal coliform density is determined.
3. **Based on trophic state index**
   Mean TSI = 54.3. From Table 3.9, it can be classified as *Full*.
4. **Determine swimming use support**
   Based on the above use analysis, it is assessed as *Full* for swimming use support.

Step 4. Assess drinking water supply use

There was no drinking water supply closure or advisories during the 1996–97 study period. Also no chemicals were applied to the lake. It is considered as *Full* use support for drinking water supply.

Step 5. Assess overall use for station 1

On the basis of overall use criteria, the assigned score values for each individual use are as follows:

Aquatic life use	5
Recreation use	5
Swimming use	5
Drinking water use	5
Total:	20

The average value is 5. Then the overall use for station 1 is attained as *Full* (Table 3.12).

Step 6. Assess for station 2. Most assessments are as for station 1 except the following:

1. Recreation use:

   From Table 3.12, we obtain RUI points $= 64$ which is in the critical range $60 \leqslant \text{RUI} < 75$; therefore, the degree of research on use is considered as partial/minor from Table 3.8.
2. The assigned scores for overall use are

Aquatic life use	5
Recreation use	3
Swimming use	5
Drinking water supply	5
Total	18

The average is 4.5, which is rounded doun to the next whole number, 4. Then, the overall use for station 2 is attained as *full/threatened.*

5.11 Lake Budgets

Calculation of lake budgets (fluxes) usually includes the hydrologic budget, nutrients (nitrogen and phosphorus) budgets, and the sediment budget. These data should be generated in Phases I, II, and III studies.

Hydrologic budget. A lake's hydrologic budget is normally quantified on the basis of the height of water spread over the entire lake surface area entering or leaving the lake in a year period. The hydrologic budget of a lake is determined by the general formula:

$$\text{Storage change} = \text{inflows} - \text{outflows} \tag{3.8}$$

or

$$\Delta S = P + I + U - E - O - R \tag{3.8a}$$

(Eqs. (3.8) and (3.8a) are essentially the same.)

In general, inflow to the lake includes direct precipitation (P), watershed surface runoff (I), subsurface groundwater inflow through lake bottom (U), and pumped input if any. Outflows include lake surface evaporation (E), discharge through surface outlet (O), outflow through lake bottom (groundwater recharge, R), and pumped outflow for water supply use, if any.

The storage term is positive if the water level increases in the period and negative if it decreases. The unit of the storage can be, simply, in (or mm) or acre-foot (or ha-cm). The hydrologic budget is used to compute nutrient budgets and the sediment budget and in selecting and designing pollution control and lake restoration alternatives.

EXAMPLE: Table 3.13 illustrates the hydrologic analysis from Vienna Correctional Center Lake, a water-supply lake (Bogner et al., 1997). Date needed for evaluating various parameters (Eq. (3.8)) to develop a hydrologic budget for the lake were collected for a 1-year period (October 1995 to September 1996). Table 3.13 presents monthly and annual results of this monitoring. Data sources and methods are explained as below:

Solution:

Step 1. Determine lake storage change

Lake storage change was determined on the basis of direct measurement of the lake level during the study period. Lake-level data were collected by automatic water-level recorder at 15-min intervals and recorded at 6-h intervals or less. On the basis of water-level record frequency, changes in storage were estimated by multiplying the periodic change in lake storage from the water-level recorder by the lake surface area (A) to determine net inflow or outflow volume.

Step 2. Compute direct precipitation

Direct precipitation was obtained from the precipitation record at the University of Illinois' Dixon Spring Experiment Station. The volume of direct precipitation input to the lake was determined by multiplying the precipitation depth by A.

TABLE 3.13 Summary of Hydrologic Analysis for Vienna Correction Center Lake, October 1995–September 1996

Date	Storage change (acre·ft)	Direct precipitation (acre·ft)	Ground-water inflow outflow(–) (acre·ft)	Surface inflow (acre·ft)	Monthly evaporation (acre·ft)	Spillway discharge (acre·ft)	Water supply withdrawal (acre·ft)
1995							
October	–56	9	7	3	14	0	61
November	–14	16	–13	40	7	0	49
December	–13	14	–2	31	4	0	52
1996							
January	40	19	18	63	4	0	55
February	–11	4	24	19	6	0	51
March	153	27	36	154	12	0	52
April	74	34	31	221	20	142	50
May	–8	33	24	262	28	244	55
June	–37	21	–3	76	30	42	59
July	–55	30	–6	11	33	0	58
August	–76	3	4	4	29	0	59
September	–36	34	–18	27	21	0	58
Annual	–39	244	102	911	208	428	658

Step 3. Compute spillway discharge Q

The general spillway rating equation was used as below:

$$Q = 3.1\ LH^{1.5} = 3.1\ (120)(H^{1.5})$$

where L = spillway length, ft
H = the height of water level exceeding the spillway

Step 4. Estimate evaporation

Evaporation was estimated using average monthly values for Carbondale in Lake Evaporation in Illinois (Robert and Stall, 1967). Monthly evaporation rates were reduced to daily rates by computing an average daily value for each month. The daily lake surface evaporation volume was determined for the study period by multiplying the daily average evaporation depth by A.

Step 5. Obtain water-supply withdraw rate

The daily water-supply withdraw rates were taken directly from the monthly reports of the water treatment plant.

Step 6. Estimate groundwater inflow and surface inflow

Groundwater inflow and surface water inflow could not be estimated from direct measurements and instead were determined on the basis of a series of sorting steps:

a. If the daily spillway discharge was zero and no rainfall occurred during the preceding 3 days, all inflow was attributed to seepage from the groundwater system.
b. If there was precipitation during the preceding 3-day period, or if there was discharge over the spillway, the groundwater input was not determined in step 6a. In this case, a moving average was used for the groundwater parameter based on the step 6a values determined for the preceding 5-day and following 10-day periods.
c. If the daily balance indicated an outflow from the lake, it was attributed to seepage into the groundwater inflow/outflow to the lake; the surface water inflow to the lake was determined to be any remaining inflow volume needed to achieve a daily balance.

Step 7. Summary

Table 3.14 summarizes the hydrologic budget for the 1-year monitoring period. The inflows and outflows listed in Table 3.14 are accurate within the limits of the analysis. During the study period, 18.9,

TABLE 3.14 Annual Summary of the Hydrologic Budget for Vienna Correction Center Lake, October 1995–September 1996

Source	Inflow volume (acre·ft)	Outflow volume (acre·ft)	Inflow(%)	Outflow(%)
Storage change	37.1		2.9	
Direct precipitation	243.8		18.9	
Surface inflow	910.2		70.4	
Groundwater inflow	100.6		7.8	
Spillway discharge		426.9		33.0
Evaporation		207.8		16.1
Water supply withdrawal		656.9		50.9
Totals	1291.7	1291.7	100.0	100.0

Note: Blank spaces—not applicable, 1 acre·ft = 1233 m^3.

70.4, 7.8, and 2.9% of the inflow volume to the lake were, respectively, direct precipitation on the lake surface, watershed surface runoff, groundwater inflow, and decrease in storage. Outflow volume was 59.0%, water-supply withdrawal, 33.0% spillway overflow, and 16.1% evaporation.

Nutrient and sediment budgets. Although nitrogen and phosphorus are not the only nutrients required for algal growth, they are generally considered to be the two main nutrients involved in lake eutrophication. Despite the controversy over the role of carbon as a limiting nutrient, the vast majority of researchers regard phosphorus as the most frequently limiting nutrient in lakes.

Several factors have complicated attempts to quantify the relationship between lake trophic status and measured concentrations of nutrients in lake waters. For example, measured inorganic nutrient concentrations do not denote nutrient availability but merely represent what is left over by the lake production process. A certain fraction of the nutrients (particularly phosphorus) becomes refractory while passing through successive biological cycles. In addition, numerous morphometric and chemical factors affect the availability of nutrients in lakes. Factors such as mean depth, basin shape, and detention time affect the amount of nutrients a lake can absorb without creating nuisance conditions. Nutrient budget calculations represent the first step in quantifying the dependence of lake water quality on the nutrient supply. It is often essential to quantify nutrients from various sources for effective management and eutrophication control.

A potential source of nitrogen and phosphorus for lakes is watershed drainage, which can include agricultural runoff, urban runoff, swamp and forest runoff, domestic and industrial waste discharges, septic tank discharges from lakeshore developments, precipitation on the lake surface, dry fallout (i.e. leaves, dust, seeds, and pollen), groundwater influxes, nitrogen fixation, sediment recycling, and aquatic bird and animal wastes. Potential sink can include outlet losses, fish catches, aquatic plant removal, denitrification, groundwater recharge, and sediment losses.

The sources of nutrients considered for a lake are tributary inputs from both gaged and ungaged streams, direct precipitation on the lake surface, and internal nutrient recycling from bottom sediments under anaerobic conditions. The discharge of nutrients from the lake through spillway is the only readily quantifiable sink.

The flow weighted-average method of computing nutrient transport by the tributary are generally used in estimating the suspended sediments, phosphorus, and nitrogen loads delivered by a tributary during normal flow conditions. Each individual measurement of nitrogen and phosphorus concentrations in a tributary sample is used with the mean flow values for the period represented by that sample to compute the nutrient transport for the given period. The total amount of any specific nutrient transported by the tributary is given by the expression (Kothandaraman and Evans, 1983; Lin and Raman, 1997):

$$T\ \text{lb} = 5.394\ \Sigma q_i c_i n_i \tag{3.18a}$$

$$T\ \text{kg} = 2.446\ \Sigma q_i c_i n_i \tag{3.18b}$$

where T = total amount of nutrient (nitrogen or phosphorus) or TSS, lb or kg
q_i = average daily flow in cfs for the period represented by the ith sample
c_i = concentration of nutrient, mg/L
n_i = number of days in the period represented by the ith sample

A similar algorithm with appropriate constant (0.0255) can be used for determining the sediment and nutrient transport during storm events. For each storm event, n_i is the interval of time represented by the ith sample and q_i is the instantaneous flow in cfs for the period represented by the ith sample. The summation is carried out for all the samples collected in a tributary during each storm event. An automatic sampler can be used to take storm water samples.

The level of nutrient or sediment input is expressed either as a concentration (mg/L) of pollutant or as mass loading per unit of land area per unit time (kg/ha-yr). There is no single correct way to express the quantity of nutrient input to a lake. To analyze nutrient inputs to a lake, the appropriate averaging time is usually 1 year, since the approximate hydraulic residence time in a lake is of this order of magnitude and the concentration value is a long-term average.

5.12 Soil loss rate

Specific soil loss rate from the watershed (for agricultural land, large construction sites, and other land uses of open land) can be estimated through the universal soil loss equation or USLE (Wischmeier and Smith, 1965):

$$A = R \times LS \times K \times C \times P \tag{3.19}$$

where A = average soil loss rate, tons/(acre · year)
R = rainfall factor
L = slope length factor, ft
S = slope steepness factor, %
LS = length-slope factor or topographic factor (not L times S)
K = soil erodibility factor
C = cropping and management factor
P = conservation practices factor

The slope steepness, slope length, cropping factors, and erodibility of each soil type cropping factor can be determined for various land uses in consultation with the local (county) USDA Soil and Water Conservation Service.

The $R \times P$ factor value is assigned as 135 for Illinois agricultural cropland and 180 for other land uses. Wischmeier and Smith (1965) and Wischmeier *et al.* (1971) are useful references. The value of P is one for some CRP (Conservation Reserve Program) implementation after 1985.

Based on the soil information compiled in the watershed or subwatersheds, the soil loss rates can be computed. The soil loss for each soil type for each subwatershed is obtained by multiplying the rate and soil acreage. The total soil loss for the watershed is the sum of soil loss in all subwatersheds expressed as tons per year. Excluding the lake surface area, the mean soil erosion rate for the watershed is estimated in terms of tons/(acre · year).

The USLE accounts for a series of factors that are the most significant influences on the erosion of soil by precipitation. The USLE is a calculation of in-field soil losses and does not account for deposition from the field to the stream or lake. Redeposition within the field or drainage system is accounted for by a sediment delivery ratio that defines the proportion of the upstream soil losses that actually pass through the stream then to the lake.

The RUSLE (Revised Universal Soil Loss Equation) is an updated version of the USLE and was adopted recently by the USDA and many states. The original USLE has been retained in RUSLE; however it has been put into a computer program to facilitate calculations, the technology for factor evaluation has been altered, and new data has been introduced to evaluate each factor under more specific conditions. There are three tables for LS value determinations.

For the following example, taking the sum of annual loss for each soil type, the total annual soil loss in the Nashville City Reservoir watershed (984.3 acres) is 2.953 tons (Table 3.16). This analysis shows that the average annual soil loss in the watershed in 3.5 tons per acre per year.

EXAMPLE: Estimate the watershed soil loss rate for the Nashville City Reservoir watershed in Washington County, Illinois (southern). Use USLE instead of RUSLE, since new factors are not yet available.
The following values are given for each of these factors for Nashville City Reservoir:

- The rainfall factor (R) is set to 200, the value applicable for southern Illinois.
- The soil classifications for the watershed and related information are presented in Table 3.15.
- The soil erodibility factor (K), the slope length (L), slope steepness (S) and the cropping factor (C) are provided in Table 3.16 by the Washington County Soil and Water Conservation District.
- The conservation factor (P) is one.

Solution: (Assistance in developing these estimates can be obtained through the county conservation district or directly from the Soil and Water Conservation District of US Department of Agriculture. Both are in the same office).

TABLE 3.15 Soil Classifications in the Nashville City Reservoir Watershed

Soil type	Soil name	Slope	Area (acres)
2	Cisne silt loam	–	38.0
3A	Hoyleton silt loam	0–2	14.9
3B	Hoyleton silt loam	2–5	11.9
5C2	Blair silt loam, eroded	5–10	36.1
5C3	Blair silt loam, severely eroded	5–10	31.0
5D3	Blair silt loam, severely eroded	10–15	9.8
8D2	Hickory silt loam	10–15	6.2
12	Wynoose silt loam	–	186.0
13A	Bluford silt loam	0–2	208.4
13B	Bluford silt loam	2–5	73.7
13B2	Bluford silt loam, eroded	2–5	17.5
14B	Ava silt loam	2–5	14.2
48	Ebbert silt loam	–	7.5
912A	Dartstadt-Hoyleton complex	0–2	61.4
912B2	Dartstadt-Hoyleton complex, eroded	2–5	164.7
991	Huey-Cisne complex	–	32.6
1334	Birds silt loam	–	10.9
3415	Orian silt loam, wet	–	17.5
Water			42.0
Total			984.3

Note: Means flat, using 0.5% for calculation.
Sources: USDA, 1998; J. Quinn, personal communication, 2000.

TABLE 3.16 Summary of Soil Losses in the Nashville City Reservoir Watershed Due to Precipitation Runoff

Soil type	Areal coverage (acres)	*K*	*L*(ft)	*S*(%)	*LS*	*C*	Annual soil loss rate (tons/acre)	Total annual loss (tons)
2	38.0	0.37	175	0.5	0.11	0.37	3.0	114
3A	14.9	0.32	175	1.1	0.24	0.20	3.1	46
3B	11.9	0.32	125	3.5	0.37	0.13	3.1	37
5C2	36.1	0.37	100	7.5	0.91	0.06	4.0	146
5C3	31.0	0.37	100	7.0	0.82	0.05	3.0	94
5D3	9.8	0.37	75	12.5	1.67	0.02	2.5	24
8D2	6.2	0.37	75	12.5	1.67	0.04	4.9	31
12	186.0	0.43	175	0.5	0.11	0.32	3.0	563
13A	208.4	0.43	175	1.1	0.24	0.15	3.1	645
13B	73.7	0.43	125	3.5	0.37	0.09	2.9	211
13B2	17.5	0.43	100	4.0	0.40	0.09	3.1	54
14B	14.2	0.43	125	3.5	0.37	0.13	4.1	59
48	7.5	0.32	175	0.5	0.11	0.71	5.0	37
912A	61.4	0.37	175	1.1	0.24	0.17	3.0	185
912B2	164.7	0.37	125	3.5	0.37	0.11	3.0	496
991	32.6	0.43	175	0.5	0.11	0.21	2.0	65
1334	10.9	0.43	175	0.5	0.11	0.57	5.4	59
3415	17.5	0.37	175	0.5	0.11	0.61	5.0	87
Water	42.0							
Total	984.3							2953
Average		0.38					3.5	

Note: Blank spaces—not applicable.
Source: Lin, 2001.

Step 1. Determine watershed boundary

The watershed boundary of the lake can be determined from a 7.5-minute US Geological Survey topographic map and digitized into a GIS coverage.

Step 2. Obtain the values for, K, L, S, and C

Obtain "Soil Survey of Washington County, Illinois" from the USDA office. The soil types, values of K and S, and soil classifications, etc. are listed in the book. Values for L and C can be obtained by consultation with USDA or the county's Soil and Water Conservation District personnel. These values for each type of soil are presented in Table 3.16.

Step 3. Measure the area coverage for each type of soil

The area coverage for each type of soil in the watershed can be directly measured from Soil Survey of Washington County, Illinois. The area coverage for each type of soil is listed in Table 3.15.

Step 4. Find LS value

The LS value can be determined from tables of topographic factor based on L, S, R, and land use; and then listed in Table 3.16 for calculations.

Step 5. Calculate annual soil loss rate for a type of soil

For soil type 2: Using Eq. (3.19),
$$\begin{aligned} A &= R \times K \times LS \times C \times P \\ &= 200 \times 0.37 \times 0.11 \times 0.37 \times 1.0 \\ &= 3.0 \text{ (tons/acre/year) (see Table 3.16)} \end{aligned}$$

Step 6. Calculate total annual soil loss for each type of soils

The total annual soil loss for each type of soil is calculated from annual soil loss rates for that soil type multiplying by the area coverage, such as

For soil type 2:
$$\begin{aligned} \text{Total soil loss} &= 3.0 \text{ tons/acre/year} \times 38 \text{ acres} \\ &= 114 \text{ tons/year} \end{aligned}$$

Step 7. Calculate the total annual soil loss (2953 tons, from Table 3.16) in the watershed

REFERENCES

American Public Health Association, American Water Works Association and Water Environment Federation. 1992. *Standard methods for the examination of water and wastewater*, 18th edn. Washington, DC: APHA.

Bogner, W. C., Lin, S. D., Hullinger, D. L. and Raman, R. K. 1997. *Diagnostic—feasibility study of Vienna Correctional Center Lake Johnson County, Illinois*. Contract report 619. Champaign: Illinois State Water Survey.

Borah, D. K., Raman, R. K., Lin, S. D., Knapp, H. V. and Soong, T. W. D. 1997. *Water quality evaluations for Lake Springfield and proposed Hunter Lake and proposed Lick Creek Reservoir*. Contract Report 621. Champaign: Illinois State Water Survey.

Carlson, R. E. 1977. A trophic state index for lakes. *Limnology and Oceanography* **22**(2): 361−369.

Chapra, S. C. and Reckhow, K. H. 1983. *Engineering approaches for lake management*. Vol. 2: *Mechanistic modeling*. Woborn, Massachusetts: Butterworth.

Clark, J. W., Viessman, Jr., W. and Hammer, M. J. 1977. *Water supply and pollution control*, 3rd edn. New York: Dun-Donnelley.

Cole, G. A. 1979. *Textbook of limnology*, 2nd edn. St. Louis, Missouri: Mosby.

Dillon, P. J. and Rigler, F. H. 1974. The phosphorus-chlorophyll relationship in lakes. *Limnology Oceanography* **19**(5): 767–773.

Fair, G. M., Geyer, J. C. and Okun, D. A. 1966. *Water and wastewater engineering*, Vol. 1: *Water supply and wastewater removal*. New York: John Wiley.

Hudson, H. L., Kirschner, R. J. and Clark, J. J. 1992. *Clean Lakes Program, Phase 1: Diagnostic/feasibility study of McCullom Lake, McHenry County, Illinois*. Chicago: Northeastern Illinois Planning Commission.

Illinois Environmental Protection Agency. 1983. *Illinois Water Quality Report 1990–1991*. IEPA/WPC/92-055. Springfield, Illinois: IEPA.

Illinois Environmental Protection Agency. 1984. *Volunteer Lake Monitoring Program—Report for 1983 Wolf Lake/Cook Co.* IEPA/WPC/84-015. Springfield, Illinois: IEPA.

Illinois Environmental Protection Agency. 1990. *Title 35: Environmental protection, Subtitle C: Water pollution*. State of Illinois, Rules and Regulations, Springfield, Illinois: IEPA.

Illinois Environmental Protection Agency. 1992. *Illinois Water Quality Report 1990–1991*. IEPA/WPC/92-055. Springfield, Illinois: IEPA.

Illinois Environmental Protection Agency. 1996. *Illinois Water Quality Report 1994–1995*. Vol. 1. Springfield, Illinois: Bureau of Water, IEPA.

Illinois State Water Survey. 1989. *Using copper sulfate to control algae in water supply impoundment*. Misc. Publ. 111. Champaign: Illinois State Water Survey.

James, A. 1993. Modeling water quality in lakes and reservoirs. In: James, A. (ed.), *An introduction to water quality modeling*, 2nd edn. Chichester: John Wiley.

Kohler, M. A., Nordenson, T. J. and Fox, W. E. 1955. *Evaporation from pan and lakes*. US Weather Bureau Research Paper 38. Washington, DC.

Kohler, M. A., Nordenson, T. J. and Baker, D. R. 1959. Evaporation maps for the United States. US Weather Bureau Technical Paper 37, Washington, DC.

Kothandaraman, V. and Evans, R. L. 1983. *Diagnostic-feasibility study of Lake Le-Aqua-Na.*, Contract Report 313. Champaign: Illinois State Water Survey.

Lamoreux, W. M. 1962. Modern evaporation formulae adapted to computer use. *Monthly Weather Review*, January.

Lin, S. D. 1994. Protocol for diagnostic/feasibility study of a lake. *Proc. Aquatech Asia '94*, November 22–24, 1994, Singapore, pp. 165–76.

Lin, S. D. and Raman, R. K. 1997. *Phase III, Post-restoration monitoring of Lake Le-Aqua-na*. Contract Report 610. Champaign: Illinois State Water Survey.

Lin, S. D. *et al.* 1996. *Diagnostic feasibility study of Wolf Lake, Cook County, Illinois, and Lake County, Indiana*. Contract Report 604. Champaign: Illinois State Water Survey.

Lin, S. D., Bogner, W. C. and Raman, R. K. 1998. *Diagnostic-feasibility study of Otter Lake, Illinois*. Contract Report (draft). Champaign: Illinois State Water Survey.

Linsley, R. K. and Franzinni, J. B. 1964. *Water resources engineering*. New York: McGraw-Hill.

Reckhow, K. H. and Chapra, S. C. 1983. *Engineering approaches for lake management. Vol. 1: Data analysis and empirical modeling*. Woburn, Massachusetts: Butterworth.

Robert, W. J. and Stall, J. B. 1967. *Lake evaporation in Illinois*. Report of Investigation 57. Urbana: Illinois State Water Survey.

Sawyer, C. N. 1952. Some aspects of phosphate in relation to lake fertilization. *Sewage and Industrial Wastes* **24**(6): 768–776.

Shuttleworth, W. J. 1993. Evaporation. In: Maidment, D. R. (ed.) *Handbook of hydrology*. New York: McGraw-Hill.

Tchobanoglous, G. and Schroeder, E. D. 1985. *Water quality*. Reading, Massachusetts: Addison-Wesley.

US Environmental Protection Agency. 1980. *Clean lakes program guidance manual*. USEPA-440/5-81-003. Washington, DC: USEPA.

US Environmental Protection Agency. 1990. *The lake and reservoir restoration guidance manual*, 2nd edn. USEPA-440/4-90-006. Washington, DC: USEPA.

US Environmental Protection Agency. 1991. *The watershed protection approach—an overview*. USEPA/503/9-92/002. Washington, DC: USEPA.

US Environmental Protection Agency. 1993. *The watershed protection approach—annual report 1992.* USEPA840-S-93-001. Washington, DC: USEPA.

US Environmental Protection Agency. 1994. *The quality of our nation's water: 1992.* USEPA841-S-94-002. Washington, DC: USEPA.

US Geological Survey. 1954. *Water loss investigations: Lake Hefner Studies.* Technical Report, Professional paper 269.

US Army Corps of Engineers. 1986. *HEC-5: simulation of flood control and conservation systems: Appendix on water quality analysis.* Davis, California: Hydrologic Engineering Center, US Army Corps of Engineers.

US Army Corps of Engineers. 1989. *HEC-5: Simulation of flood control and conservation systems: Exhibit 8 of user's manual: input description.* Davis, California: Hydrologic Engineering Center, US Army Corps of Engineers.

Walker, Jr., W. W. 1981a. *Empirical methods for predicting eutrophication in impoundments. Part 1. Phase I: data base development.* Environmental Engineers. Tech. Report E-81-9, Concord, Massachusetts.

Walker, Jr., W. W. 1981b. *Empirical methods for predicting eutrophication in impoundments. Part 2. Phase II: model testing.* Environmental Engineers Tech. Report E-81-9, Concord, Massachusetts.

Wetzel, R. G. 1975. *Limnology.* Philadelphia, Pennsylvania: Saunders.

Wischmeier, W. H. and Smith, D. D. 1965. *Predicting rainfall—erosion losses from cropland east of the Rocky Mountains.* Agriculture Handbook 282. Washington, DC: US Department of Agriculture.

Wischmeier, W. H., Johnson, C. B. and Cross, B. V. 1971. A soil-erodibility monograph for farmland and construction sites. *J. Soil and Water Conservation*, **26**(5), 189–93.

CHAPTER 1.4
GROUNDWATER

Shun Dar Lin

1 DEFINITION

1.1 Groundwater and Aquifer

Groundwater is subsurface water which occurs beneath the earth's surface. In a hydraulic water cycle, groundwater comes from surface waters (precipitation, lake, reservoir, river, sea, etc.) and percolates into the ground beneath the water table. The groundwater table is the surface of the groundwater exposed to an atmospheric pressure beneath the ground surface (the surface of the saturated zone). A water table may fluctuate in elevation.

An aquifer is an underground water-saturated stratum or formation that can yield usable amounts of water to a well. There are two different types of aquifers based on physical characteristics. If the saturated zone is sandwiched between layers of impermeable material and the groundwater is under pressure, it is called a confined aquifer (Fig. 4.1). If there is no impermeable layer immediately above the saturated zone, that is called an unconfined aquifer. In an unconfined aquifer, the top of the saturated zone is the water table defined as above.

Aquifers are replenished by water infiltrated through the earth above from the upland area. The area replenishing groundwater is called the recharge area. In reverse, if the groundwater flows to the lower land area, such as lakes, streams, or wetlands, it is called discharge.

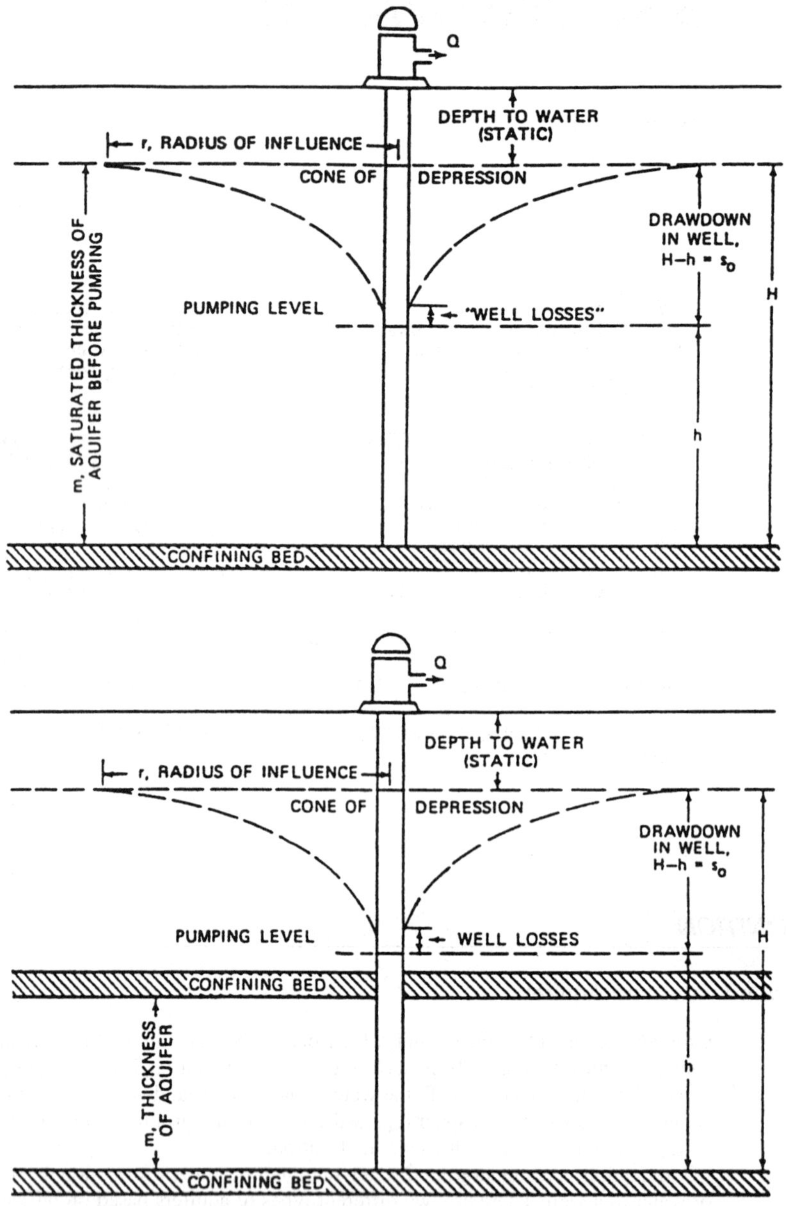

FIGURE 4.1 Drawdown, cone of depression, and radius of influence in unconfined and confined aquifers (*Illinois EPA, 1995*).

On the earth, approximately 3 percent of the total water is fresh water. Of this, groundwater comprises 95 percent, surface water 3.5 percent, and soil moisture 1.5 percent. Out of all the fresh water on earth, only 0.36 percent is readily available to use (Leopold, 1974).

Groundwater is an important source of water supply. Fifty-three percent of the population of the United States receives its water supply from groundwater or underground sources (USEPA, 1994). Groundwater is also a major source of industrial uses (cooling, water supply, etc.) and agricultural uses (irrigation and livestock). The quantity of groundwater available is an important value. The so-called safe yield of an aquifer is the practicable rate of withdrawing water from it perennially. Such a safe amount does not exist, however.

The quantity of groundwater is also affected by water engineering. For decades and centuries, through improper disposal of wastes (solid, liquid, and gaseous) to the environment and subsurface areas, many groundwaters have become contaminated. Major sources of contaminants are possibly from landfill leachate, industrial wastes, agricultural chemicals, wastewater effluents, oil and gasoline (underground tanks, animal wastes, acid-mine drainage, road salts, hazardous wastes spillage, household and land chemicals, etc.). Illegal dumpings of wastes and toxic chemicals to waterways and to lands are major environmental problems in many countries.

Efforts to protect the quantity and quality of groundwater have been made by cooperation between all government agencies, interested parties, and researchers. Most states are responsible for research, education, establishment of minimum setback zones for public and private water supply wells, and contamination survey and remediation.

1.2 Zones of Influence and Capture

The withdrawal of groundwater by a pumping well causes a lowering of the water level. Referring to Fig. 4.2, the difference between water levels during non-pumping and pumping is called drawdown. The pattern of drawdown around a single pumping well resembles a cone. The area affected by the pumping well is called the cone of depression, the radius of influence, or lateral area of influence (LAI). Within the LAI, the flow velocity continuously increases as it flows toward the well due to gradually increased slope or hydraulic gradient.

As a well pumps groundwater to the surface, the groundwater withdrawn from around the well is replaced by water stored within the aquifer. All water overlaid on the non-pumping potentiometric surface will eventually be pulled into the well. This area of water entering the area of influence of the well is called the zone of capture (ZOC), zone of contribution, or capture zone (Fig. 4.2). The ZOC generally extends upgradient from the pumping well to the edge of the aquifer or to a groundwater divide. The zone of capture is usually asymmetrical. It is important to identify the ZOC because any pollution will be drawn toward the well, subsequently contaminating the water supply.

A zone of capture is usually referred to the time of travel as a time-related capture zone. For instance, a "ten-year" time-related capture zone is the area within which the water at the edge of the zone will reach the well within ten years. The state primacy has established setback zones for wells which will be discussed later.

EXAMPLE 1: Groundwater flows into a well at 4 in/d (10 cm/d) and is 365 feet (111 m) from the well. Estimate the capture time.

Solution:

$$\text{Time} = 365 \text{ ft}/(4/12 \text{ ft/d})$$

$$= 1095 \text{ days}$$

or

$$= 3\text{-year time related capture zone}$$

EXAMPLE 2: Calculate the distance from a well to the edge of the 3-year capture zone if the groundwater flow is 2 ft/d (0.61 m/d).

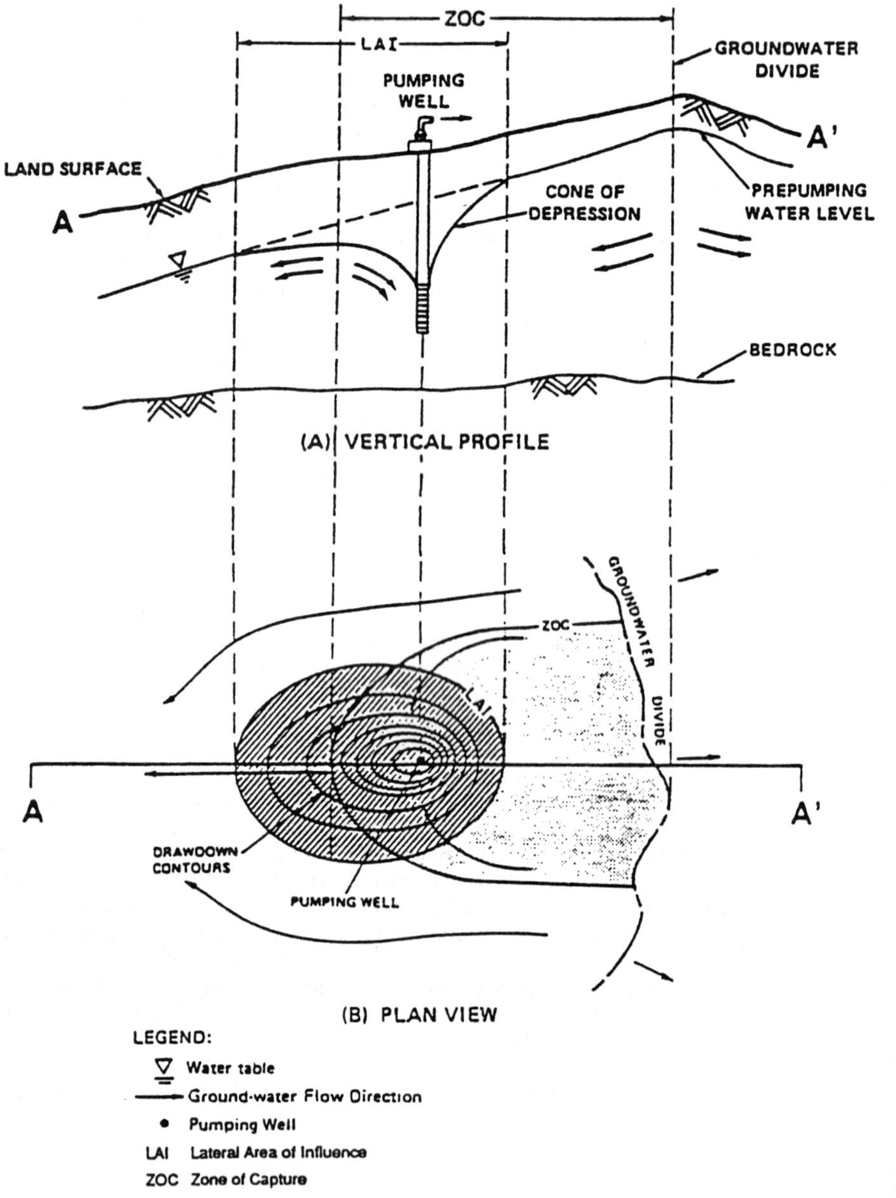

FIGURE 4.2 Relationship between the cone of depression and the zone of capture within a regional flow field (*Illinois EPA, 1995*).

Solution:

$$\begin{aligned} \text{Distance} &= 2 \text{ ft/d} \times 3 \text{ years} \times 365 \text{ d/year} \\ &= 2190 \text{ ft} \end{aligned}$$

or

$$= 668 \text{ m}$$

1.3 Wells

Wells are classified according to their uses by the US Environmental Protection Agency (EPA). A public well is defined as having a minimum of 15 service connections or serving 25 persons at least 60 days per year. Community public wells serve residents the year round. Noncommunity public wells serve nonresidential populations at places such as schools, factories, hotels, restaurants, and campgrounds. Wells that do not meet the definition of public wells are classified as either semi-private or private. Semi-private wells serve more than one single-family dwelling, but fewer than 15 connections or serving 25 persons. Private wells serve an owner-occupied single-family dwelling.

Wells are also classified into types based on construction. The most common types of wells are drilled and bored dependent on the aquifer to be tapped, and the needs and economic conditions of the users (Babbitt *et al.*, 1959; Forest and Olshansky, 1993).

2 HYDROGEOLOGIC PARAMETERS

There are three critical aquifer parameters: porosity, specific yield (or storativity for confined aquifers), and hydraulic conductivity (including anisotropy). These parameters required relatively sophisticated field and laboratory procedures for accurate measurement.

Porosity and specific yield/storativity express the aquifer storage properties. Hydraulic conductivity (permeability) and transmissivity describe the groundwater transmitting properties.

2.1 Aquifer Porosity

The porosity of soil or fissured rock is defined as the ratio of void volume to total volume (Wanielista, 1990; Bedient and Huber, 1992):

$$n = \frac{V_v}{V} = \frac{(V - V_s)}{V} = 1 - \frac{V_s}{V} \tag{4.1}$$

where n = porosity
V_v = volume of voids within the soil
V = total volume of sample (soil)
V_s = volume of the solids within the soil
= dry weight of sample/specific weight

The ratio of the voids to the solids within the soil is called the void ratio e expressed as

$$e = V_v/V_s \tag{4.2}$$

Then the relationship between void ratio and porosity is

$$e = \frac{n}{1 - n} \tag{4.3}$$

or

$$n = \frac{e}{1 + e} \tag{4.4}$$

The porosity may range from a small fraction to about 0.90. Typical values of porosity are 0.2 to 0.4 for sands and gravels depending on the grain size, size of distribution, and the degree of compaction; 0.1 to 0.2 for sandstone; and 0.01 to 0.1 for shale and limestone depending on the texture and size of the fissures (Hammer, 1986).

When groundwater withdraws from an aquifer and the water table is lowered, some water is still retained in the voids. This is called the specific retention. The quantity drained out is called the specific yield. The specific yield for alluvial sand and gravel is of the order of 90 to 95 percent.

EXAMPLE: If the porosity of sands and gravels in an aquifer is 0.38 and the specific yield is 92 percent, how much water can be drained per cubic meter of aquifer?

Solution:

$$\begin{aligned}\text{Volume} &= 0.38 \times 0.92 \times 1 \text{ m}^3 \\ &= 0.35 \text{ m}^3\end{aligned}$$

2.2 Storativity

The term storativity (S) is the quantity of water that an aquifer will release from storage or take into storage per unit of its surface area per unit change in land. In unconfined aquifers, the storativity is in practice equal to the specific yield. For confined aquifers, storability is between 0.005 and 0.00005, with leaky confined aquifers falling in the high end of this range (USEPA, 1994). The smaller storativity of confined aquifers, the larger the pressure change throughout a wide area to obtain a sufficient supply from a well. However, this is not the case for unconfined aquifers due to gravity drainage.

2.3 Transmissivity

Transmissivity describes the capacity of an aquifer to transmit a water. It is the product of hydraulic conductivity (permeability) and the aquifer's saturated thickness:

$$T = Kb \tag{4.5}$$

where T = transmissivity of an aquifer, gpd/ft or $\text{m}^3/(\text{d} \cdot \text{m})$
K = permeability, gpd/ft^2 or $\text{m}^3/(\text{d} \cdot \text{m}^2)$
b = thickness of aquifer, ft or m

A rough estimation of T is by multiplying specific capacity by 2000 (USEPA, 1994).

EXAMPLE: If the aquifer's thickness is 50 ft, estimate the permeability of the aquifer using data in the example of the specific capacity (in Sec. 2.7).

Solution:

$$\begin{aligned}T &= 2000 \times \text{specific capacity} = 2000 \times 15 \text{ gpm/ft} \\ &= 30{,}000 \text{ gpm/ft}\end{aligned}$$

Rearranging Eq. (4.5)

$$\begin{aligned}K &= T/b = (30{,}000 \text{ gpm/ft})/50 \text{ ft} \\ &= 600 \text{ gpm/ft}^2\end{aligned}$$

2.4 Flow Nets

Many groundwater systems are two or three dimensional. Darcy's law was first derived in a one dimensional equation. Using Darcy's law can establish a set of streamlines and equipotential lines to develop a two dimensional flow net. The details of this concept are discussed elsewhere in the text (Bedient *et al.*, 1994).

A flow net is constructed by flow lines that intersect the equipotential lines or contour lines at a right angle. Equipotential lines are developed based on the observed water levels in wells penetrating an isotropic aquifer. Flow lines are then drawn orthogonally to indicate the flow direction.

Referring to Fig. 4.3, the horizontal flow within a segment in a flow net can be determined by the following equation (USEPA, 1994):

$$q_a = T_a \Delta H_a W_a / L_a \tag{4.6}$$

where q_a = groundwater flow in segment A, m^3/d or ft^3/d
T_a = transmissivity in segment A, m^3/d or ft^3/d
ΔH_a = drop in groundwater level across segment A, m or ft
W_a = average width of segment A, m or ft
L_a = average length of segment A, m or ft

The flow in the next segment, B, is similarly computed as Eq. (4.7)

$$q_b = T_b \Delta H_b W_b / L_b \tag{4.7}$$

Assuming that there is no flow added between segments A and B by recharge (or that recharge is insignificant), then

$$q_b = q_a$$

or

$$T_b \Delta H_b W_b / L_b = T_a \Delta H_a W_a / L_a$$

solving T_b computation of which allows T_b from T_a

$$T_b = T_a (L_b \Delta H_a W_a / L_a \Delta H_b W_b) \tag{4.8}$$

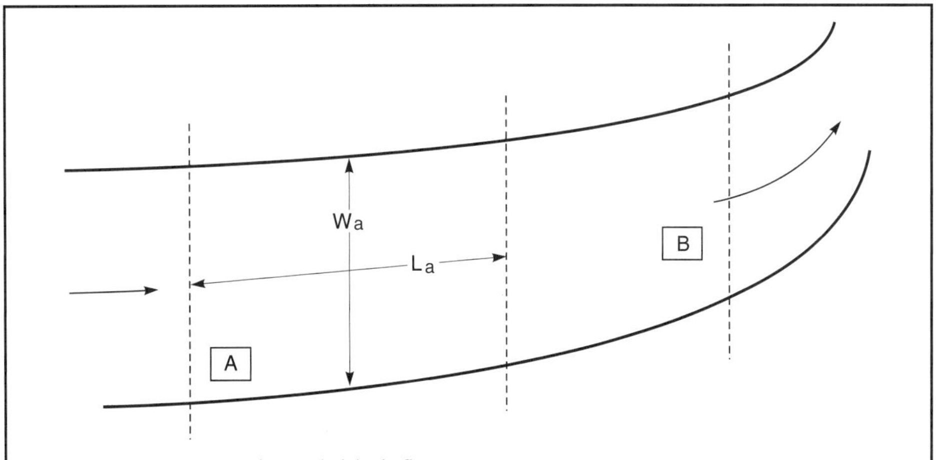

FIGURE 4.3 Distribution of transmissivity in flow nets.

Measurement or estimation of transmissivity (T) for one segment allows the computation of variations in T upgradient and downgradient. If variations in aquifer thickness are known, or can be estimated for different segments, variation in hydraulic conductivity can also be calculated as

$$K = T/b \tag{4.9}$$

where K = hydraulic conductivity, m/d or ft/d
T = transmissivity, m^2/d or ft^2/d
b = aquifer thickness, m or ft

Eq. (4.9) is essentially the same as Eq. (4.5).

2.5 Darcy's Law

The flow movement of water through the ground is entirely different from the flow in pipes and in an open channel. The flow of fluids through porous materials is governed by Darcy's law. It states that the flow velocity of fluid through a porous medium is proportional to the hydraulic gradient (referring to Fig. 4.4):

$$v = Ki \tag{4.10a}$$

or

$$v = K\frac{(h_1 + z_1) - (h_2 + z_2)}{L} \tag{4.10b}$$

where v = Darcy velocity of flow, mm/s or ft/s
K = hydraulic conductivity of medium or coefficient of permeability, mm/s or ft/s
i = hydraulic gradient, mm or ft/ft
h_1, h_2 = pressure heads at points 1 and 2, m or ft
z_1, z_2 = elevation heads at points 1 and 2, m or ft
L = distance between points (piezometers) 1 and 2, m or ft

The pore velocity v_p is equal to the Darcy velocity divided by porosity as follows:

$$v_p = v/n \tag{4.11}$$

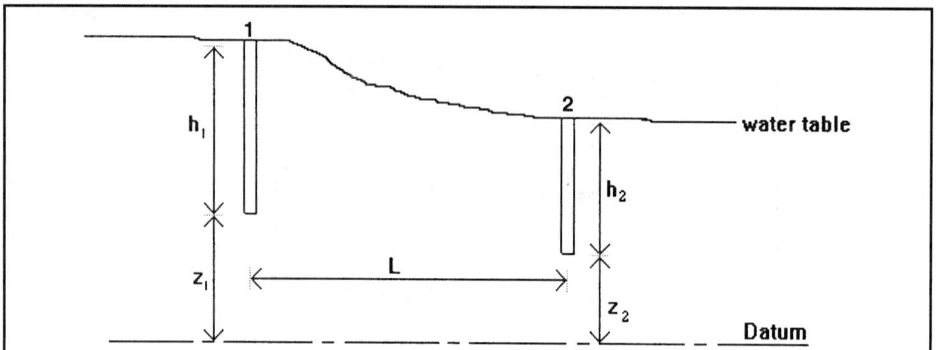

FIGURE 4.4 Groundwater flow (one dimensional).

Darcy's law is applied only in the laminar flow region. Groundwater flow may be considered as laminar when the Reynolds number is less than unity (Rose, 1949). The Reynolds number can be expressed as

$$\mathbf{R} = \frac{VD}{v} \tag{4.12}$$

where $\mathbf{R}$ = Reynolds number
V = velocity, m/s or ft/s
D = mean grain diameter, mm or in
v = kinematic viscosity, m^2/d or ft^2/d

EXAMPLE 1: Determine the Reynolds number when the groundwater temperature is 10°C (from Table 5.1*a*, $v = 1.31 \times 10^{-6}$ m^2/s); the velocity of flow is 0.6 m/d (2 ft/d); and the mean grain diameter is 2.0 mm (0.079 in).

Solution:

$$\begin{aligned} V &= 0.6 \text{ m/d} = (0.6 \text{ m/d})/(86{,}400 \text{ s/d}) \\ &= 6.94 \times 10^{-6} \text{ m/s} \\ D &= 2 \text{ mm} = 0.002 \text{ m} \\ \mathbf{R} &= \frac{VD}{\nu} = \frac{6.94 \times 10^{-6} \text{ m/s} \times 0.002 \text{ m}}{1.31 \times 10^{-6} \text{ m}^2\text{/s}} \\ &= 0.011 \end{aligned}$$

Note: it is a laminar flow ($\mathbf{R} < 1$)

EXAMPLE 2: If the mean grain diameter is 0.12 in (3.0 mm), its porosity is 40%, and the groundwater temperature is 50°F (10°C). Determine the Reynolds number for a flow 5 ft (1.52 m) from the centerline of the well with a 4400 gpm in a confined aquifer depth of 3.3 ft (1 m) thick.

Solution:

Step 1. Find flow velocity (V)

$$\begin{aligned} Q &= 4400 \text{ gpm} \times 0.002228 \text{ cfs/gpm} \\ Q &= 9.80 \text{ cfs} \\ n &= 0.4 \\ r &= 5 \text{ ft} \\ h &= 3.3 \text{ ft} \end{aligned}$$

Since

$$\begin{aligned} Q &= nAV = n(2\pi rh)\,V \\ V &= Q/n(2\pi rh) \\ &= 9.8 \text{ cfs}/(0.4 \times 2 \times 3.14 \times 5 \text{ ft} \times 3.3 \text{ ft}) \\ &= 0.236 \text{ fps} \end{aligned}$$

Step 2. Compute $\mathbf{R}$

$$\begin{aligned} v &= 1.41 \times 10^{-5} \text{ ft}^2\text{/s} \quad \text{(see Table 5.1}b\text{ at 50°F)} \\ D &= 0.12 \text{ in} = 0.12 \text{ in/(12 in/ft)} = 0.01 \text{ ft} \\ \mathbf{R} &= \frac{VD}{v} = \frac{0.236 \text{ fps} \times 0.01 \text{ ft}}{1.41 \times 10^{-5} \text{ ft}^2\text{/s}} \\ &= 167 \end{aligned}$$

EXAMPLE 3: The slope of a groundwater table is 3.6 m per 1000 m. The coefficient of permeability of coarse sand is 0.51 cm/s (0.2 in/s). Estimate the flow velocity and the discharge rate through this aquifer of coarse sand 430 m (1410 ft) wide and 22 m (72 ft) thick.

Solution:

Step 1. Determine the velocity of flow, v, using Eq. (4.10a)

$$i = 3.6 \text{ m}/1000 \text{ m} = 0.0036$$

$$\begin{aligned} v &= Ki = 0.51 \text{ cm/s}(0.0036) \\ &= 0.00184 \text{ cm/s } (86{,}400 \text{ s/d})(0.01 \text{ m/cm}) \\ &= 1.59 \text{ m/d} \\ &= 5.21 \text{ ft/d} \end{aligned}$$

Step 2. Compute discharge

$$\begin{aligned} Q &= vA = 1.59 \text{ m/d} \times 430 \text{ m} \times 22 \text{ m} \\ &= 15{,}040 \text{ m}^3\text{/d} \\ &= 15{,}040 \text{ m}^3\text{/d} \times 264.17 \text{ gal/m}^3 \\ &= 3.97 \text{ MGD (million gallons per day)} \end{aligned}$$

EXAMPLE 4: If the difference in water level between two wells 1.6 miles (2.57 km) apart is 36 ft (11 m), and the hydraulic conductivity of the media is 400 gpd/ft^2 (16,300 L/(d · m^2)). The depth of the media (aquifer) is 39 ft (12 m). Estimate the quantity of groundwater flow moving through the cross-section of the aquifer.

Solution:

$$\begin{aligned} i &= 36 \text{ ft}/1.6 \text{ miles} = 36 \text{ ft}/(1.6 \text{ miles} \times 5280 \text{ ft/miles}) \\ &= 0.00426 \\ A &= 1.6 \times 5280 \text{ ft} \times 39 \text{ ft} \\ &= 329{,}500 \text{ ft}^2 \end{aligned}$$

then

$$\begin{aligned} Q &= KiA = 400 \text{ gpd/ft}^2 \times 0.00426 \times 329{,}500 \text{ ft}^2 \\ &= 561{,}400 \text{ gpd} \\ &= 0.561 \text{ MGD} \\ &= 2123 \text{ m}^3\text{/d} \end{aligned}$$

EXAMPLE 5: If the water moves from the upper to the lower lake through the ground. The following data is given:

difference in elevation $\Delta h = 25$ m (82 ft)

length of flow path $L = 1500$ m (4920 ft)

cross-sectional area of flow $A = 120$ m^2 (1290 ft^2)

hydraulic conductivity $K = 0.15$ cm/d

porosity of media $n = 0.25$

Estimate the time of flow between the two lakes.

Solution:

Step 1. Determine the Darcy velocity v

$$v = Ki = K\Delta h/L = 0.0015 \text{ m/s } (25 \text{ m}/1500 \text{ m})$$
$$= 2.5 \times 10^{-5} \text{ m/s}$$

Step 2. Calculate pore velocity

$$v_p = v/n = (2.5 \times 10^{-5} \text{ m/s})/0.25$$
$$= 1.0 \times 10^{-4} \text{ m/s}$$

Step 3. Compute the time of travel t

$$t = L/v_p = 1500 \text{ m}/(1 \times 10^{-4} \text{ m/s})$$
$$= 1.5 \times 10^{7}\text{s} \times (1 \text{ d}/86{,}400 \text{ s})$$
$$= 173.6 \text{ days}$$

2.6 Permeability

The terms permeability (P) and hydraulic conductivity (K) are often used interchangeably. Both are measurements of water moving through the soil or an aquifer under saturated conditions. The hydraulic conductivity, defined by Nielsen (1991), is the quantity of water that will flow through a unit cross-sectional area of a porous media per unit of time under a hydraulic gradient of 1.0 (measured at right angles to the direction of flow) at a specified temperature.

Laboratory measurement of permeability. Permeability can be determined using permeameters in the laboratory. Rearranging Eq. (4.5) and $Q = vA$, for constant head permeameter, the permeability is

$$K = \frac{LQ}{H\pi R^2} \tag{4.13}$$

where K = permeability, m/d or ft/d
L = height of sample (media), m or ft
Q = flow rate at outlet, m^3/d or ft^3/d
H = head loss, m or ft
R = radius of sample column

The permeability measure from the falling-head permeameter is

$$K = \left(\frac{L}{t}\right)\left(\frac{r}{R}\right)^2 \ln\left(\frac{h_1}{h_2}\right) \tag{4.14}$$

where r = radius of standpipe, m or ft
h_1 = height of water column at beginning, m or ft
h_2 = height of the water column at the end, m or ft
t = time interval between beginning and end, d
other parameters are the same as Eq. (4.13)

Groundwater flows through permeable materials; such as sand, gravel, and sandstone; and is blocked by less permeable material, such as clay. Few materials are completely impermeable in nature. Even solid bedrock has fine cracks, so groundwater can flow through. Groundwater recharge occurs when surface water infiltrates the soil faster than it is evaporated, used by plants, or stored as soil moisture.

Field measurement of permeability. Ideal steady-state flow of groundwater is under the conditions of uniform pump withdrawal, a stable drawdown curve, laminar and horizontal uniform flow, a flow velocity proportional to the tangent of the hydraulic gradient, and a homogeneous aquifer. Assuming these ideal conditions, the well flow is a function of the coefficient of permeability, the shape of the drawdown curve, and the thickness of the aquifer. For an unconfined aquifer the well discharge can be expressed as an equilibrium equation (Steel and McGhee, 1979; Hammer and Mackichan, 1981):

$$Q = \pi K \frac{H^2 - h_w^2}{\ln (r/r_w)} \tag{4.15}$$

where Q = well discharge, L/s or gpm
π = 3.14
K = coefficient of permeability, mm/s or fps
H = saturated thickness of aquifer before pumping, m or ft (see Fig. 4.1)
h_w = depth of water in the well while pumping, m or ft
= h + well losses in Fig. 4.1
r = radius of influence, m or ft
r_w = radius of well, m or ft

Also under ideal conditions, the well discharge from a confined aquifer can be calculated as

$$Q = 2\pi Km \frac{H - h_w}{\ln (r/r_w)} \tag{4.16}$$

where m is the thickness of the aquifer, m or ft. Other parameters are the same as Eq. (4.15). Values of Q, H, and r may be assumed or measured from field well tests, with two observation wells, often establishing a steady-state condition for continuous pumping for a long period. The coefficient of permeability can be calculated by rearranging Eqs. (4.14) and (4.15). The K value of an unconfined aquifer is also computed by the equation:

$$K = \frac{Q \ln(r_2/r_1)}{\pi (h_2^2 - h_1^2)} \tag{4.17}$$

and for a confined aquifer:

$$K = \frac{Q \ln (r_2/r_1)}{2m\pi (h_2 - h_1)} \tag{4.18}$$

where h_1, h_2 = depth of water in observation wells 1 and 2, m or ft
r_1, r_2 = centerline distance from the well and observation wells 1 and 2, respectively, m or ft

EXAMPLE 1: A well is pumped to equilibrium at 4600 gpm (0.29 m^3/s) in an unconfined aquifer. The drawdown in the observation well at 100 ft (30.5 m) away from the pumped well is 10.5 ft (3.2 m) and at 500 ft (152 m) away is 2.8 ft (0.85 m). The water table is 50.5 ft (15.4 m). Determine the coefficient of permeability.

Solution:

$$\begin{aligned}
h_1 &= 50.5 - 10.5 = 40.0 \text{ (ft)} \\
h_2 &= 50.5 - 2.8 = 47.7 \text{ (ft)} \\
r_1 &= 100 \text{ ft} \\
r_2 &= 500 \text{ ft} \\
Q &= 4600 \text{ gpm} = 4600 \text{ gpm} \times 0.002228 \text{ cfs/gpm} \\
&= 10.25 \text{ cfs}
\end{aligned}$$

Using Eq. (4.17)

$$K = \frac{Q \ln (r_2/r_1)}{\pi (h_2^2 - h_1^2)}$$

$$= \frac{(10.25 \text{ cfs}) \ln(500 \text{ ft}/100 \text{ ft})}{3.14 [(47.7 \text{ ft})^2 - (40 \text{ ft})^2]}$$

$$= 0.00778 \text{ ft/s}$$

or

$$= 0.00237 \text{ m/s}$$

EXAMPLE 2: Referring to Fig. 4.1, a well with a diameter of 0.46 m (1.5 ft) in a confined aquifer which has a uniform thickness of 16.5 m (54.1 ft). The depth of the top impermeable bed to the ground surface is 45.7 m (150 ft). Field pumping tests are carried out with two observation wells to determine the coefficient of permeability of the aquifer. The distances between the test well and observation wells 1 and 2 are 10.0 and 30.2 m (32.8 and 99.0 ft), respectively. Before pumping, the initial piezometric surface in the test well and the observation wells are 10.4 m (34.1 ft) below the ground surface. After pumping at a discharge rate of 0.29 m^3/s (4600 gpm) for a few days, the water levels in the wells are stabilized with the following drawdowns: 8.6 m (28.2 ft) in the test well, 5.5 m (18.0 ft) in the observation well 1, and 3.2 m (10.5 ft) in the observation well 2. Compute (a) the coefficient of permeability of the aquifer and (b) the well discharge with the drawdown in the well 10 m (32.8 ft) above the impermeable bed if the radius of influence (r) is 246 m (807 ft) and ignoring head losses.

Solution:

Step 1. Let a datum be the top of the aquifer, then

$$H = 45.7 \text{ m} - 10.4 \text{ m} = 35.3 \text{ m}$$

$$h_w = H - 8.6 \text{ m} = 35.3 \text{ m} - 8.6 \text{ m} = 26.7 \text{ m}$$

$$h_1 = 35.3 \text{ m} - 5.5 \text{ m} = 29.8 \text{ m}$$

$$h_2 = 35.3 \text{ m} - 3.2 \text{ m} = 32.1 \text{ m}$$

$$m = 16.5 \text{ m}$$

Using Eq. (4.18)

$$K = \frac{Q \ln (r_2/r_1)}{2\pi m (h_2 - h_1)}$$

$$= \frac{0.29 \text{ m}^3\text{/s} \ln (30.2 \text{ m}/10 \text{ m})}{2 \times 3.14 \times 1.65 \text{ m} \times (32.1 \text{ m} - 29.8 \text{ m})}$$

$$= 0.00134 \text{ m/s}$$

or

$$= 0.00441 \text{ ft/s}$$

Step 2. Estimate well discharge

$$h_w = 10 \text{ m}, \qquad r_w = 0.46 \text{ m}$$

$$H = 35.3 \text{ m}, \qquad r = 246 \text{ m}$$

Using Eq. (4.16)

$$Q = 2\pi K m \frac{H - h_w}{\ln(r/r_w)}$$

$$= 2 \times 3.14 \times 0.00134 \text{ m/s} \times 16.5 \text{ m} \times (35.3 - 10) \text{ m}/\ln(264/0.46)$$

$$= 0.553 \text{ m}^3\text{/s}$$

or

$$= 8765 \text{ gpm}$$

2.7 Specific Capacity

The permeability can be roughly estimated by a simple field well test. The difference between the static water level prior to any pumping and the level to which the water drops during pumping is called drawdown (Fig. 4.2). The discharge (pumping) rate divided by the drawdown is the specific capacity. The specific capacity gives the quantity of water produced from the well per unit depth (ft or m) of drawdown. It is calculated by

$$\text{Specific capacity} = Q/wd \tag{4.19}$$

where Q = discharge rate, gpm or m^3/s
wd = well drawdown, ft or m

EXAMPLE: The static water elevation is at 572 ft (174.3 m) before pumping. After a prolonged normal well pumping rate of 120 gpm (7.6 L/s), the water level is at 564 ft (171.9 m). Calculate the specific capacity of the well.

Solution:

$$\begin{aligned}\text{Specific capacity} &= Q/wd = 120 \text{ gpm}/(572 - 564) \text{ ft} \\ &= 15 \text{ gpm/ft}\end{aligned}$$

or

$$\begin{aligned}\text{Specific capacity} &= (7.6 \text{ L/s})/(174.3 - 171.9) \text{ m} \\ &= 3.17 \text{ L/s} \cdot \text{m}\end{aligned}$$

3 STEADY FLOWS IN AQUIFERS

Referring to Fig. 4.3, if $z_1 = z_2$, for an unconfined aquifer,

$$Q = KA\, dh/dL \tag{4.20}$$

and let the unit width flow be q, then

$$\begin{aligned} q &= Kh\, dh/dL \\ qdL &= Kh\, dh \end{aligned} \tag{4.21}$$

by integration:

$$\begin{aligned} q\int_0^L dL &= K\int_{h_2}^{h_1} hdh \\ qL &= (K/2)(h_1^2 - h_2^2) \\ q &= \frac{K(h_1^2 - h_2^2)}{2L} \end{aligned} \tag{4.22}$$

This is the so-called Dupuit equation.

For a confined aquifer, it is a linear equation

$$q = \frac{KD(h_1 - h_2)}{L} \tag{4.23}$$

where q = unit width flow, m^2/d or ft^2/d
K = coeffcient of permeability, m/d or ft/d
h_1, h_2 = piezometric head at locations 1 and 2, m or ft
L = length of aquifer between piezometric measurements, m or ft
D = thickness of aquifer, m or ft

EXAMPLE: Two rivers are located 1800 m (5900 ft) apart and fully penetrate an aquifer. The water elevation of the rivers are 48.5 m (159 ft) and 45.6 m (150 ft) above the impermeable bed. The hydraulic conductivity of the aquifer is 0.57 m/d. Estimate the daily discharge per meter of width between the two rivers, neglecting recharge.

Solution:

This case can be considered as an unconfined aquifer:

$$K = 0.57 \text{ m/d}$$
$$h_1 = 48.5 \text{ m}$$
$$h_2 = 45.6 \text{ m}$$
$$L = 1800 \text{ m}$$

Using Eq. (4.22) the Dupuit equation:

$$q = \frac{K(h_1^2 - h_2^2)}{2L} = \frac{0.57 \text{ m/d}[(48.5 \text{ m})^2 - (45.6 \text{ m})^2]}{2 \times 1800 \text{ m}}$$
$$= 0.0432 \text{ m}^2/\text{d}$$

4 ANISOTROPIC AQUIFERS

Most real geologic formations tend to have more than one direction for the movement of water due to the nature of the material and its orientation. Sometimes a soil formation may have a hydraulic conductivity (permeability) in the horizontal direction, K_x, radically different from that in the vertical direction, K_z. This phenomenon ($K_x \neq K_z$), is called anisotropy. When hydraulic conductivities are the same in all directions, ($K_x = K_z$), the aquifer is called isotropic. In typical alluvial deposits, K_x is greater than K_z. For a two-layered aquifer of different hydraulic conductivities and different thicknesses, applying Darcy's law to horizontal flow can be expressed as

$$K_x = \frac{K_1 z_1 + K_2 z_2}{z_1 + z_2} \tag{4.25}$$

or, in general form

$$K_x = \frac{\sum K_i z_i}{\sum z_i} \tag{4.26}$$

where K_i = hydraulic conductivity in layer i, mm/s or fps
z_i = aquifer thickness of layer i, m or ft

For a vertical groundwater flow through two layers, let q_z be the flow per unit horizontal area in each layer. The following relationship exists:

$$dh_1 + dh_2 = \left(\frac{z_1}{K_1} + \frac{z_2}{K_2}\right)q_z \tag{4.27}$$

Since

$$(dh_1 + dh_2)K_z = (z_1 + z_2)q_z$$

then

$$dh_1 + dh_2 = \left(\frac{z_1 + z_2}{K_z}\right)q_z \tag{4.28}$$

where K_z is the hydraulic conductivity for the entire aquifer. Comparison of Eqs. (4.27) and (4.28) yields

$$\frac{z_1 + z_2}{K_z} = \frac{z_1}{K_1} + \frac{z_2}{K_2}$$
$$K_z = \frac{z_1 + z_2}{z_1/K_1 + z_2/K_2} \tag{4.29}$$

or in general form

$$K_z = \frac{\sum z_i}{\sum z_i/K_i} \tag{4.30}$$

The ratios of K_x to K_z for alluvium are usually between 2 and 10.

5 UNSTEADY (NONEQUILIBRIUM) FLOWS

Equilibrium equations described in the previous sections usually overestimate hydraulic conductivity and transmissivity. In practical situations, equilibrium usually takes a long time to reach. Theis (1935) originated the equation relations for the flow of ground water into wells and was then improved on by other investigators (Jacob, 1940, 1947; Wenzel 1942; Cooper and Jacob, 1946).

Three mathematical/graphical methods are commonly used for estimations of transmissivity and storativity for nonequilibrium flow conditions. They are the Theis method, the Cooper and Jacob (straight-line) method, and the distance-drawdown method.

5.1 Theis Method

The nonequilibrium equation proposed by Theis (1935) for the ideal aquifer is

$$d = \frac{Q}{4\pi T}\int_m^\infty \frac{e^{-u}}{u}\,du = \frac{Q}{4\pi T}W(u) \tag{4.31}$$

and

$$u = \frac{r^2S}{4Tt} \tag{4.32}$$

where d = drawdown at a point in the vicinity of a well pumped at a constant rate, ft or m
Q = discharge of the well, gpm or m^3/s
T = transmissibility, ft^2/s or m^2/s
r = distance from pumped well to the observation well, ft or m
S = coefficient of storage of aquifer
t = time of the well pumped, min
$W(u)$ = well function of u

The integral of the Theis equation is written as $W(u)$, and is the exponential integral (or well function) which can be expanded as a series:

$$W(u) = -0.5772 - \ln u + u - \frac{u^2}{2\cdot 2!} + \frac{u^3}{3\cdot 3!} - \frac{u^4}{4\cdot 4!} + \cdots \tag{4.33a}$$
$$= -0.5772 - \ln u + u - u^2/4 + u^3/18 - u^4/96 + \cdots + (-1)^{n-1}\frac{u^n}{n\cdot n!} \tag{4.33b}$$

Values of $W(u)$ for various values of u are listed in Appendix B, which is a complete table by Wenzel (1942) and modified from Illinois EPA (1990).

If the coefficient of transmissibility T and the coefficient of storage S are known, the drawdown d can be calculated for any time and at any point on the cone of depression including the pumped well. Obtaining these coefficients would be extremely laborious and is seldom completely satisfied for field conditions. The complete solution of the Theis equation requires a graphical method of two equations (Eqs. (4.31) and (4.32)) with four unknowns.

Rearranging as:

$$d = \frac{Q}{4\pi T} W(u) \tag{4.31}$$

and

$$\frac{r^2}{t} = \frac{4T}{S} u \tag{4.34}$$

Theis (1935) first suggested plotting *W(u)* on log-log paper, called a *type curve*. The values of S and T may be determined from a series of drawdown observations on a well with known times. Also, prepare another plot, of values of d against r^2/t on transparent log-log paper, with the same scale as the other figure. The two plots (Fig. 4.5) are superimposed so that a match point can be obtained in the region of which the curves nearly coincide when their coordinate axes are parallel. The coordinates of the match point are marked on both curves. Thus, values of u, $W(u)$, d, and r^2/t can be obtained. Substituting these values into Eqs. (4.31) and (4.32), values of T and S can be calculated.

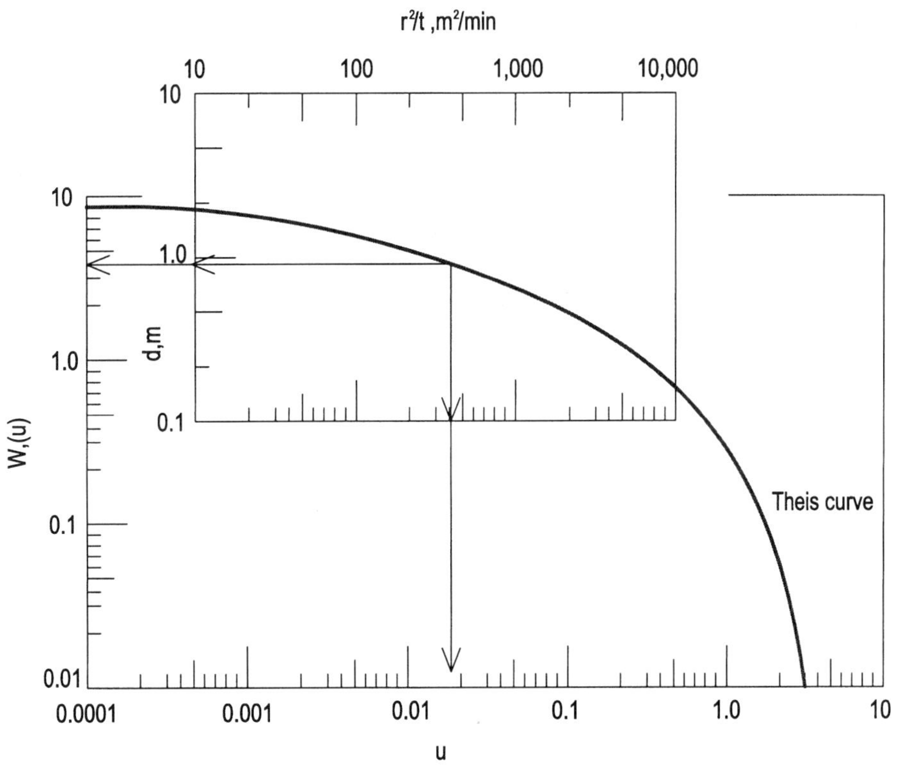

FIGURE 4.5 Observed data and Theis type curve.

EXAMPLE: An artesian well is pumped at a rate of 0.055 m^3/s for 60 h. Observations of drawdown are recorded and listed below as a function of time at an observation hole 90 m away. Estimate the transmissivity and storativity using the Theis method.

Solution:

Time t, min	Drawdown d, m	r^2/t, m^2/min
1	0.12	8100
2	0.21	4050
3	0.33	2700
4	0.39	2025
5	0.46	1620
6	0.5	1350
10	0.61	810
15	0.75	540
20	0.83	405
30	0.92	270
40	1.02	203
50	1.07	162
60	1.11	135
80	1.2	101
90	1.24	90
100	1.28	81
200	1.4	40.5
300	1.55	27
600	1.73	13.5
900	1.9	10

Step 1. Calculate r^2/t and construct a table with t and d.

Step 2. Plot the observed data r^2/t versus d on log-log (transparency) paper in Fig. 4.5.

Step 3. Plot a Theis type curve u versus $W(u)$ on log-log paper (Fig. 4.5) using the data listed in Appendix B (Illinois EPA, 1988).

Step 4. Select a match point on the superimposed plot of the observed data on the type curve.

Step 5. From the plots, the coordinates of the match points on the two curves are

$$\text{Observed data: } r^2/t = 320 \text{ m}^2/\text{min}$$
$$d = 0.90 \text{ m}$$
$$\text{Type curve: } W(u) = 3.5$$
$$u = 0.018$$

Step 6. Substitute the above values to estimate T and S.
Using Eq. (4.31)

$$d = \frac{Q}{4\pi T} W(u)$$
$$T = \frac{QW(u)}{4\pi d} = \frac{0.055 \text{ m}^3/\text{s} \times 3.5}{4\pi \times 0.90 \text{ m}}$$
$$= 0.017 \text{ m}^2/\text{s}$$

Using Eq. (4.34)

$$\frac{r^2}{t} = \frac{4Tu}{S}$$

$$S = \frac{4Tu}{r^2/t} = \frac{4 \times 0.017 \text{ m}^2/\text{s} \times 0.018}{320 \text{ m}^2/\text{min} \times (1 \text{ min}/60 \text{ s})}$$

$$= 2.30 \times 10^{-4}$$

5.2 Cooper–Jacob Method

Cooper and Jacob (1946) modified the nonequilibrium equation. It is noted that the parameter u in Eq. (4.32) becomes very small for large values of t and small values of r. The infinite series for small u, $W(u)$, can be approximated by

$$W(u) = -0.5772 - \ln u = -0.5772 - \ln \frac{r^2 S}{4Tt} \qquad (4.35)$$

Then

$$d = \frac{Q}{4\pi T} W(u) = \frac{Q}{4\pi T}\left(-0.5772 - \ln \frac{r^2 S}{4Tt}\right) \qquad (4.36)$$

Further rearrangement and conversion to decimal logarithms yields

$$d = \frac{2.303Q}{4\pi T} \log \frac{2.25Tt}{r^2 S} \qquad (4.37)$$

Therefore, the drawdown is to be a linear function of log t. A plot of d versus the logarithm of t forms a straight line with the slope $Q/4\pi T$ and an intercept at $d = 0$, when $t = t_0$, yields

$$0 = \frac{2.3Q}{4\pi T} \log \frac{2.25Tt_0}{r^2 S} \qquad (4.38)$$

Since $\log(1) = 0$ therefore

$$1 = \frac{2.25Tt_0}{r^2 S}$$

$$S = \frac{2.25Tt_0}{r^2} \qquad (4.39)$$

If the slope is measured over one log cycle of time, the slope will equal the change in drawdown Δd, and Eq. (4.37) becomes

$$\Delta d = \frac{2.303Q}{4\pi T}$$

then

$$T = \frac{2.303Q}{4\pi \Delta d} \qquad (4.40)$$

The Cooper and Jacob modified method solves for S and T when values of u are less than 0.01. The method is not applicable to periods immediately after pumping starts. Generally, 12 h or more of pumping are required.

EXAMPLE: Using the given data in the above example (by the Theis method) with the Cooper and Jacob method, estimate the transmissivity and storativity of a confined aquifer.

Solution:

Step 1. Determine t_0 and Δd.

Values of drawdown *(d)* and time *(t)* are plotted on semilog paper with the t in the logarithmic scale as shown in Fig. 4.6. A best fit straight line is drawn through the observed data. The intercept

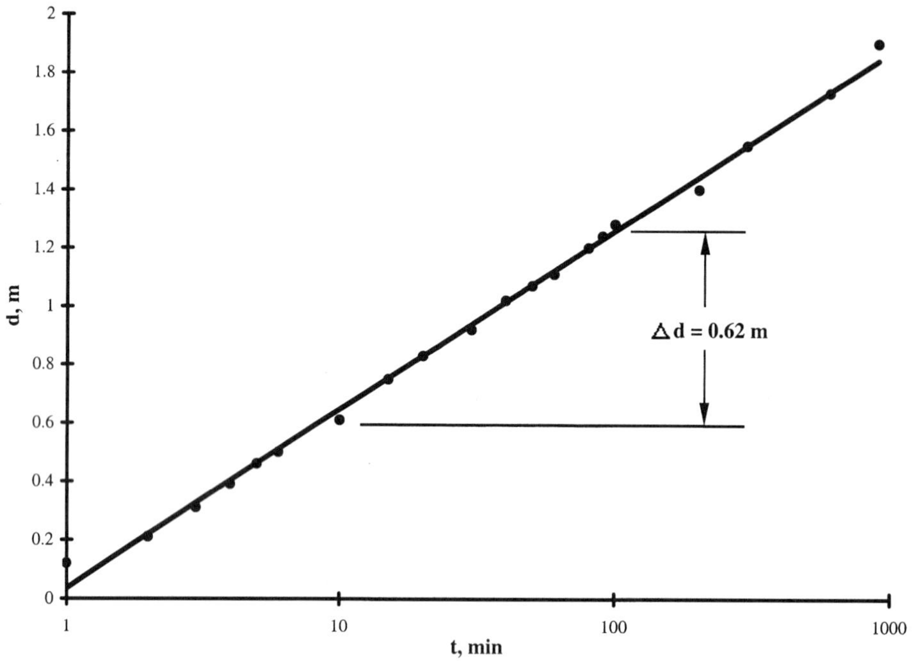

FIGURE 4.6 Relationship of drawdown and the time of pumping.

of the t axis is 0.98 min. The slope of the line Δd is measured over 1 log cycle of t from the figure. We obtain:

$$t_0 = 0.98 \text{ min}$$
$$= 58.8 \text{ s}$$

and, for a cycle ($t = 10$ to 100 min)

$$\Delta d = 0.62 \text{ m}$$

Step 2. Compute T and S

Using Eq. (4.40)

$$T = \frac{2.303Q}{4\pi\Delta d} = \frac{2.303 \times 0.055 \text{ m}^3\text{/s}}{4 \times 3.14 \times 0.62 \text{ m}}$$
$$= 0.016 \text{ m}^2\text{/s}$$

Using Eq. (4.39)

$$S = \frac{2.25Tt_0}{r^2} = \frac{2.25 \times 0.016 \text{ m}^2\text{/s} \times 58.8 \text{ s}}{(90 \text{ m})^2}$$
$$= 2.61 \times 10^{-4}$$

5.3 Distance-drawdown Method

The distance-drawdown method is a modification of the Cooper and Jacob method and is applied to obtain quick information about the aquifer characteristics while the pumping test is in progress. The method needs simultaneous observations of drawdown in three or more observation wells. The aquifer properties can be determined from pumping tests by the following equations (Watson and Burnett, 1993):

$$T = \frac{0.366Q}{\Delta(h_0 - h)} \text{ for SI units} \tag{4.41}$$

$$T = \frac{528Q}{\Delta(h_0 - h)} \text{ for U.S. customary units} \tag{4.42}$$

and

$$S = \frac{2.25Tt}{r_0^2} \text{ for SI units} \tag{4.43}$$

$$S = \frac{Tt}{4790r_0^2} \text{ U.S. customary units} \tag{4.44}$$

where T = transmissivity, m^2/d or gpd/ft
Q = normal discharge rate, m^3/d or gpm
$\Delta(h_0 - h)$ = drawdown per log cycle of distance, m or ft
S = storativity, unitless
t = time since pumping when the simultaneous readings are taken in all observation wells, d or min
r_0 = intercept of the straight-line plot with the zero-drawdown axis, m or ft

The distance-drawdown method involves the following procedures:

1. Plot the distance and drawdown data on semi-log paper; drawdown on the arithmetic scale and the distance on the logarithmic scale.
2. Read the drawdown per log cycle in the same manner for the Cooper and Jacob method: this gives the value of $\Delta(h_0 - h)$.
3. Draw a best-fit straight line.
4. Extend the line to the zero-drawdown and read the value of the intercept, r_0.

EXAMPLE: A water supply well is pumping at a constant discharge rate of 1000 m^3/d (11,000 gpm). It happens that there are five observation wells available. After pumping for 3 h, the drawdown at each observation well is recorded as below. Estimate transmissivity and storativity of the aquifer using the distance-drawdown method.

Distance		
m	ft	Drawdown, m
3	10	3.22
7.6	25	2.21
20	66	1.42
50	164	0.63
70	230	0.28

Solution:

Step 1. Plot the distance-drawdown data on semi-log paper as shown in Fig. 4.7.

Step 2. Draw a best-fit straight line over the observed data and extend the line to the X axis.

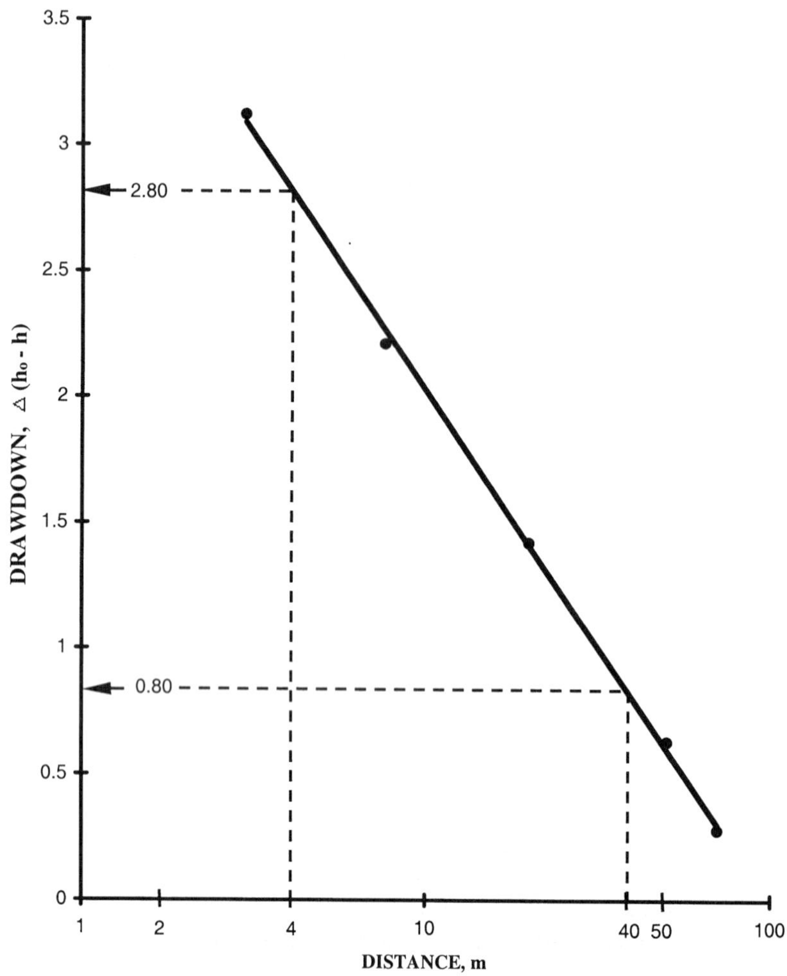

FIGURE 4.7 Plot of observed distance-drawdown data.

Step 3. Read the drawdown value for one log cycle. From Fig. 4.7, for the distance for a cycle from 4 to 40 m, the value of the drawdown, then $\Delta(h_0 - h)$ is read as 2.0 (2.8 − 0.8) m.

Step 4. Read the intercept on the X axis for r_0,

$$r_0 = 93 \text{ m.}$$

Step 5. Determine T and S by Eqs. (4.41) and (4.43).

$$T = \frac{0.366Q}{\Delta(h_0 - h)} = \frac{0.366 \times 1000 \text{ m}^3\text{/d}}{2.0 \text{ m}}$$

$$= 183 \text{ m}^2\text{/d}$$

$$\text{Time of pumping: } t = 3 \text{ h}/24 \text{ h/d}$$
$$= 0.125 \text{ days}$$

$$S = \frac{2.25Tt}{r_0^2} = \frac{2.25 \times 183 \text{ m}^2/\text{d} \times 0.125 \text{ day}}{(93 \text{ m})^2}$$
$$= 0.006$$

5.4 Slug Tests

In the preceding sections, the transmissivity T and storativity S of the aquifer, and permeability K of the soil are determined by boring one or two more observation wells. Slug tests use only a single well for the determination of those values by careful evaluation of the drawdown curve and information of screen geometry. The tests involve either raising or lowering the water level in the well and measuring the return to a static water level as a function of time.

A typical test procedure requires introducing an object to record the volume (the slug) of the well. The Hvorslev (1951) method using a piezometer in an confined aquifer is widely used in practice due to it being quick and inexpensive. The procedures of conducting and analyzing the Hvorslev test are as follows:

1. Record the instantaneously raised (or lowered) water level to the static water level as H_0.
2. Read subsequently changing water levels with time as h. Thus $h = H_0$ at $t = 0$.
3. Measure the final raised head as H at infinite time.
4. There is a relationship as

$$\frac{H - h}{H - H_0} = e^{-t/T_0} \tag{4.45}$$

where

$$T_0 = \frac{\pi r^2}{Fk} \tag{4.46}$$
$$= \text{Hvorslev-defined basic time lag}$$
$$F = \text{shape factor}$$

5. Calculate ratios $(H - h)/(H - H_0)$ as recovery.
6. Plot on semi-logarithmic paper $(H - h)/(H - H_0)$ on the logarithmic scale versus time on the arithmetic scale.
7. Find T_0 at recovery equals 0.37 (37 percent of the initial change caused by the slug).
8. For piezometer intake length divided by radius (L/R) greater than 8, Hvorslev evaluated the shape factor F and proposed an equation for hydraulic conductivity K as

$$K = \frac{r^2 \ln (L/R)}{2LT_0} \tag{4.47}$$

where K = hydraulic conductivity, m/d or ft/d
r = radius of the well casing, cm or in
R = radius of the well screen, cm or in
L = length of the well screen, cm or in
T_0 = time required for water level to reach 37 percent of the initial change, s

Other slug test methods have been developed for confined aquifers (Cooper *et al.*, 1967; Papadopoulous *et al.*, 1973; Bouwer and Rice, 1976). These methods are similar to Theis's type curves in that a curve-matching method is used to determine T and S for a given aquifer. A family of type curves H_t/H_0 versus Tr/r_c^2 were published for five values of the variable α, defined as $(r_s^2/r_c^2)\ S$, to estimate transmissivity, storativity, and hydraulic conductivity.

The Bouwer and Rice (1976) slug test method is most commonly used for estimating hydraulic conductivity in groundwater. Although the method was originally developed for unconfined aquifers, it can also be applied for confined or stratified aquifers if the top of the screen is some distance below the upper confined layer. The following formula is used to compute hydraulic conductivity:

$$K = \frac{r^2 \ln (R/r_w)}{2L} \frac{1}{t} \ln \frac{y_0}{y_t} \tag{4.48}$$

where K = hydraulic conductivity, cm/s
r = radius of casing, cm
y_0, y_t = vertical difference in water levels between inside and outside the well at time $t = 0$, and $t = t$, m
R = effective radius distance over which y is dissipated, cm
r_w = radius distance of undisturbed portion of aquifer from well centerline (usually r plus thickness of gravel).
L = length of screen, m
t = time, s

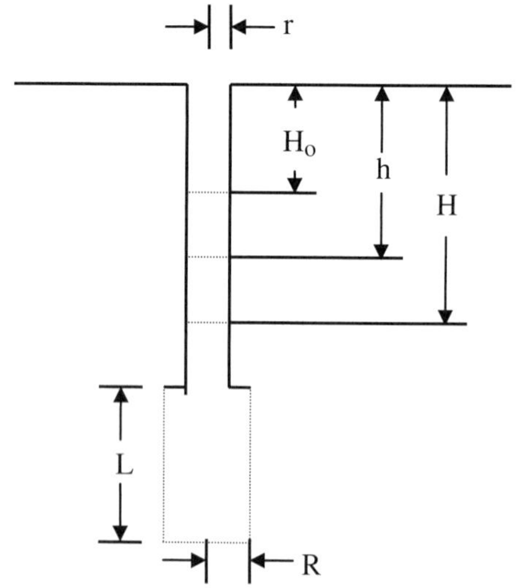

FIGURE 4.8 Hvorslev slug test.

EXAMPLE 1: The internal diameters of the well casing and well screen are 10 cm (4 in) and 15 cm (6 in), respectively. The length of the well screen is 2 m (6.6 ft). The static water level measured from the top of the casing is 2.50 m (8.2 ft). A slug test is conducted and pumped to lower the water level to 3.05 m (10 ft). The time-drawdown h in the unconfined aquifer is recorded every 3 s as shown in the following table. Determine the hydraulic conductivity of the aquifer by the Hvorslev method.

Solution:

Step 1. Calculate $(h - H_0)/(H - H_0)$
Given: $H_0 = 2.50$ m, $H = 3.05$ m
Then $H - H_0 = 3.05\text{ m} - 2.50\text{ m} = 0.55$ m

Time t, s	h, m	$h - H_0$, m	$(h - H_0)/0.55$
0	3.05	0.55	1.00
3	2.96	0.46	0.84
6	2.89	0.39	0.71
9	2.82	0.32	0.58
12	2.78	0.28	0.51
15	2.73	0.23	0.42
18	2.69	0.19	0.34
21	2.65	0.15	0.27
24	2.62	0.12	0.22
27	2.61	0.10	0.18
30	2.59	0.09	0.16

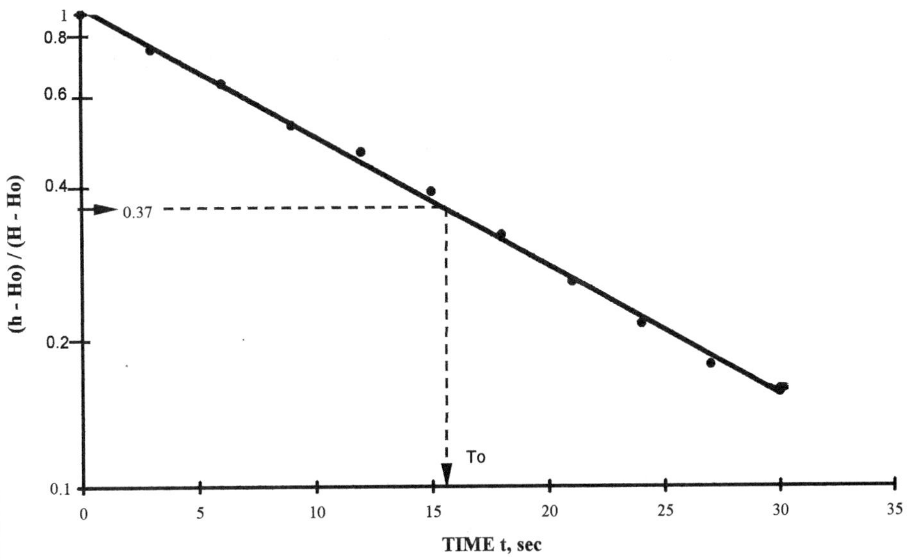

FIGURE 4.9 Plot of Hvorslev slug test results.

Step 2. Plot t versus $(h - H_0)/(H - H_0)$ on semi-log paper as shown in Fig. 4.9, and draw a best-fit straight line.

Step 3. Find T_0.
From Fig. 4.9, read 0.37 on the $(h - H_0)/(H - H_0)$ scale and note the time for the water level to reach 37 percent of the initial change T_0 caused by the slug. This is expressed as T_0. In this case $T_0 = 16.2$ s.

Step 4. Determine the L/R ratio

$$R = 15\text{ cm}/2 = 7.5\text{ cm}$$

$$L/R = 200\text{ cm}/7.5\text{ cm} = 26.7 > 8$$

Thus Eq. (4.47) can be applied.

Step 5. Find K by Eq. (4.47)

$$r = 10\text{ cm}/2 = 5\text{ cm}$$

$$K = \frac{r^2 \ln(L/R)}{2LT_0}$$

$$= (5\text{ cm})^2 \ln 26.7/(2 \times 200\text{ cm} \times 16.2\text{ s})$$

$$= 0.0127\text{ cm/s}$$

$$= 0.0127\text{ cm/s} \times (1\text{ m}/100\text{ cm}) \times 86{,}400\text{ s/d}$$

$$= 11.0\text{ m/d}$$

$$= 36\text{ ft/d}$$

EXAMPLE 2: A screened, cased well penetrates a confined aquifer with gravel pack 3.0 cm thickness around the well. The radius of casing is 5.0 cm and the screen is 1.2 m long. A slug of water is injected and water level raised by *m*. The effective radial distance over which *y* is dissipated is 12 cm. Estimate hydraulic conductivity for the aquifer. The change of water level with time is as follows:

t, s	y_t, cm	*t*, s	y_t, cm
1	30	10	4.0
2	24	13	2.0
3	17	16	1.1
4	14	20	0.6
5	12	30	0.2
6	9.6	40	0.1
8	5.5		

Solution:

Step 1. Plot values of *y* versus *t* on semi-log paper as shown in Fig. 4.10. Draw a best-fit straight line. The line from $y_0 = 36$ cm to $y_t = 0.1$ cm covers 2.5 log cycles. The time increment between the two points is 26 s.

Step 2. Determine *K* by Eq. (4.48), the Bouwer and Rice equation:

$$r = 5 \text{ cm}$$

$$R = 12 \text{ cm}$$

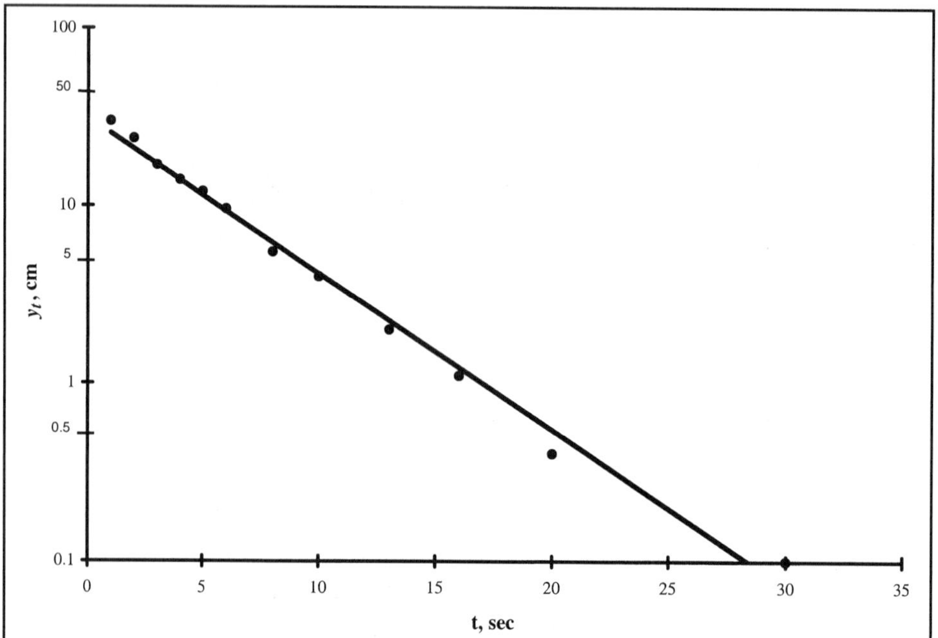

FIGURE 4.10 Plot of y_t versus *t*.

$$r_w = 5 \text{ cm} + 3 \text{ cm} = 8 \text{ cm}$$

$$K = \frac{r^2 \ln (R/r_w)}{2L} \frac{1}{t} \ln \frac{y_0}{y_t}$$

$$= \frac{(5 \text{ cm})^2 \ln(12 \text{ cm}/8 \text{ cm})}{2(120 \text{ cm})} \frac{1}{26 \text{ s}} \ln \frac{36 \text{ cm}}{0.1 \text{ cm}}$$

$$= 9.56 \times 10^{-3} \text{ cm/s}$$

6 GROUNDWATER CONTAMINATION

6.1 Sources of Contamination

There are various sources of groundwater contamination and various types of contaminants. Underground storage tanks, agricultural activity, municipal landfills, abandoned hazardous waste sites, and septic tanks are the major threats to groundwater. Other sources may be from industrial spill, injection wells, land application, illegal dumps (big problems in many countries), road salt, saltwater intrusion, oil and gas wells, mining and mine drainage, municipal wastewater effluents, surface impounded waste material stockpiles, pipelines, radioactive waste disposal, and transportation accidents.

Large quantities of organic compounds are manufactured and used by industries, agriculture, and municipalities. These man-made organic compounds are of most concern. The inorganic compounds occur in nature and may come from natural sources as well as human activities. Metals from mining, industries, agriculture, and fossil fuels also may cause groundwater contamination.

Types of contaminants are classified by chemical group. They are metals (arsenic, lead, mercury, etc.), volatile organic compounds (VOCs, gasoline, oil, paint thinner), pesticides and herbicides, and radionuclides (radium, radon). The most frequently reported groundwater contaminants are the VOCs (Voelker, 1984; Rehfeldt *et al.*, 1992).

Volatile organic compounds are made of carbon, hydrogen, and oxygen and have a vapor pressure less than one atmosphere. Compounds such as gasoline or dry cleaning fluid evaporate when left open to the air. They are easily dissolved in water. There are numerous incidences of VOCs contamination caused by leaking underground storage tanks. Groundwater contaminated by VOCs poses cancer risks to humans either by ingestion of drinking water or inhalation.

6.2 Contaminant Transport Pathways

The major contaminant transport mechanisms in groundwater are advection, diffusion, dispersion, adsorption, chemical reaction, and biodegradation. Advection is the movement of contaminant(s) with the flowing groundwater at the seepage velocity in the pore space and is expressed as Darcy's law:

$$v_x = \frac{K}{n}\frac{dh}{L} \tag{4.49}$$

where v_x = seepage velocity, m/s or pps
K = hydraulic conductivity, m/s or fps
n = porosity
dh = pressure head, m or ft
L = distance, m or ft

Eq. (4.49) is similar to Eq. (4.10b).

Diffusion is a molecular-scale mass transport process that moves solutes from an area of higher concentration to an area of lower concentration. Diffusion is expressed by Fick's law:

$$F_x = -D_d \frac{dc}{dx} \tag{4.50}$$

where F_x = mass flux, mg/m^2 s or lb/ft^2 s
D_d = diffusion coefficient, m^2/s or ft^2/s
dc/dx = concentration gradient, mg/(m^3 m) or (lb/ft^3 ft)

Diffusive transport can occur at low or zero flow velocities. In a tight soil or clay, typical values of D_d range from 1 to 2×10^9 m^2/s at 25°C (Bedient *et al.* 1994). However, typical dispersion coefficients in groundwater are several orders of magnitude greater than that in clay.

Dispersion is a mixing process caused by velocity variations in porous media. Mass transport due to dispersion can occur parallel and normal to the direction of flow with two dimensional spreading.

Sorption is the interaction of a contaminant with a solid. It can be divided into adsorption and absorption. An excess concentration of contaminants at the surfaces of solids is called adsorption. Absorption refers to the penetration of the contaminants into the solids.

Biodegradation is a biochemical process that transforms contaminants (certain organics) into simple carbon dioxide and water by microorganisms. It can occur in aerobic and anaerobic conditions. Anaerobic biodegradation may include fermentation, denitrification, iron reduction, sulfate reduction, and methane production.

Excellent and complete coverage of contaminant transport mechanisms is presented by Bedient *et al.* (1994). Theories and examples are covered for mass transport, transport in groundwater by advection, diffusion, dispersion, sorption, chemical reaction, and by biodegradation.

Some example problems of contaminant transport are also given by Tchobanoglous and Schroeder (1985). Mathematical models that analyze complex contaminant pathways in groundwater are also discussed elsewhere (Canter and Knox, 1986; Willis and Yeh, 1987; Canter *et al.*, 1988; Mackay and Riley, 1993; Smith and Wheatcraft, 1993; Watson and Burnett, 1993; James, 1993; Gupta, 1997).

The transport of contaminants in groundwater involves adsorption, advection, diffusion, dispersion, interface mass transfer, biochemical transformations, and chemical reactions. On the basis of mass balance, the general equation describing the transport of a dissolved contaminant through an isotropic aquifer under steady-state flow conditions can be mathematically expressed as (Gupta 1997)

$$\begin{aligned}\frac{\partial c}{\partial t} &+ \frac{\rho_b}{\rho}\frac{\partial s}{\partial t} + ck_\ell + Sk_s\frac{\rho_b}{\varphi} \\ &= D_x\frac{\partial^2 c}{\partial x^2} + D_y\frac{\partial^2 c}{\partial y^2} + D_z\frac{\partial^2 c}{\partial z^2} - V_x\frac{\partial c}{\partial x} - V_y\frac{\partial c}{\partial y} - V_z\frac{\partial c}{\partial z}\end{aligned} \tag{4.51}$$

where c = solute (contaminant in liquid phase) concentration, g/m^3
S = concentration in solid phase as mass of contaminant per unit mass of dry soil, g/g
t = time, d
ρ_b = bulk density of soil, kg/m^3_x
φ = effective porosity
k_ℓ, k_s = first-order decay rate in the liquid and soil phases, respectively, d^{-1}
x, y, z = Cartesian coordinates, m
D_x, D_y, D_z = directional hydrodynamic dispersion coefficients, m^2/d
V_x, V_y, V_z = directional seepage velocity components, m/d

There are two unknowns (c and S) in Eq. (4.51). Assuming a linear adsorption isotherm of the form

$$S = K_d\, c \tag{4.52}$$

where K_d = distribution coefficient due to chemical reactions and biological degradation and substituting Eq. (4.52) into Eq. (4.51), we obtain

$$R\frac{\partial c}{\partial t} + kc = D_x\frac{\partial^2 c}{\partial x^2} + D_y\frac{\partial^2 c}{\partial y^2} + D_z\frac{\partial^2 c}{\partial z^2} - V_x\frac{\partial c}{\partial x} - V_y\frac{\partial c}{\partial y} - V_z\frac{\partial c}{\partial z} \tag{4.53}$$

where $R = 1 + k_d(\rho_b\varphi)$
= retardation factor which slows the movement of solute due to adsorption

$$k = k_1 + k_s K_d(\rho_b \varphi)$$
$$= \text{overall first-order decay rate, d}^{-1}$$

The general equation under steady-state flow conditions in the x direction, we modify from Eq. (4.53).

6.3 Underground Storage Tank

There are 3 to 5 million underground storage tanks (UST) in the United States. It is estimated that 3 to 10 percent of these tanks and their associated piping systems may be leaking (USEPA 1987). The majority of UST contain petroleum products (gasoline and other petroleum products). When the UST leaks and is left unattended, and subsequently contaminates subsurface soils, surface and groundwater monitoring and corrective actions are required.

The migration or transport pathways of contaminants from the UST depend on the quantity released, the physical properties of the contaminant, and characteristics of the soil particles.

When a liquid contaminant is leaked from a UST below the ground surface, it percolates downward to the unconfined groundwater surface. If the soil characteristics and contaminant properties are known, it can be estimated whether the contaminant will reach the groundwater. For hydrocarbons in the unsaturated (vadose) zone it can be estimated by the equation (USEPA 1987)

$$D = \frac{R_v V}{A} \tag{4.54}$$

where D = maximum depth of penetration, m
R_v = a coefficient of retention capacity of the soil and the viscosity of the product (contaminant)
V = volume of infiltrating hydrocarbon, m^3
A = area of spill, m^2

The typical values for R_v are as follows (USEPA 1987):

Soil	Gasoline	Kerosene	Light fuel oil
Coarse gravel	400	200	100
Gravel to coarse sand	250	125	62
Coarse to medium sand	130	66	33
Medium to fine sand	80	40	20
Fine sand to silt	50	25	12

Retention (attenuative) capacities for hydrocarbons vary approximately from 5 L/m^3 in a coarse gravel to more than 40 L/m^3 in silts. Leaked gasoline can travel 25 ft (7.6 m) through unsaturated, permeable, alluvial, or glacial sediments in a few hours, or at most a few days, which is extremely site specific.

The major movement of nonaqueous-phase liquids that are less dense than water (such as gasoline and other petroleum products) in the capillary zone (between unsaturated and saturated zones) is lateral. The plume will increase thickness (vertical plane) and width depending on leakage rates and the site's physical conditions. The characteristic shape of the flow is the so-called "oil package." Subsequently it will plug the pores of the soil or be diluted and may be washed out into the water table.

The transport of miscible or dissolved substances in saturated zones follows the general direction of groundwater flow. The transport pathways can be applied to the models (laws) of advection and dispersion. The dispersion may include molecular diffusion, microscopic dispersion, and macroscopic dispersion.

Immiscible substances with a specific gravity of less than 1.0 (lighter than water) are usually found only in the shallow part of the saturated zone. The transport rate depends on the groundwater gradient and the viscosity of the substance.

Immiscible substances with a specific gravity of more than 1.0 (denser than water) move downward through the saturated zone. A dense immiscible substance poses a greater danger in terms of migration potential than less dense substances due to its deeper penetration into the saturated zone. When the quantity of released contaminant exceeds the retention capacity of the unsaturated and saturated zones, the denser nonaqueous-phase liquid continues its downward migration until it reaches an impermeable boundary. A liquid substance leaking from a UST enters the vapor phase in the unsaturated zone according to its specific vapor pressure. The higher the vapor pressure of the substance, the more it evaporates. The contaminant in the vapor phase moves by advection and by diffusion. Vapor moves primarily in a horizontal direction depending on the slope of the water table and the location of the impermeable bedrock. If the vapor, less dense than air, migrates in a vertical direction, it may accumulate in sewer lines, basements, and such areas.

6.4 Groundwater Treatment

Once a leaking UST is observed, corrective action should be carried out, such as tank removal, abandonment, rehabilitation, removal/excavation of soil and sediment, onsite and/or offsite treatment and disposal of contaminants, product and groundwater recovery, and groundwater treatment, etc.

Selection of groundwater treatment depends on the contaminants to be removed. Gasoline and volatile organic compounds can be removed by air stripping and stream stripping processes. Activated carbon adsorption, biological treatment, and granular media filtration can be used for removal of gasoline and other organics. Nonvolatile organics are removable by oxidation/reduction processes. Inorganic chemicals can be treated by coagulation/sedimentation, neutralization, dissolved air floatation, granular media filtration, ion exchange, resin adsorption, and reverse osmosis.

These treatment processes are discussed in detail in Chapter 1.6, Public Water Supply and in Chapter 1.7, Wastewater.

7 SETBACK ZONES

Section 1428 of the Safe Drinking Water Act (SDWA) requires each state to submit a wellhead protection program to the USEPA. For example, Illinois EPA promulgated the Illinois Groundwater Protection Act (IGPA) in 1991. The Act assigned the responsibilities of existing state agencies and established groundwater classifications. On the basis of classification, different water quality standards, monitoring, and remedial requirements can be applied. Classifications are based on the PCB levels which may be associated with hazardous wastes.

Groundwater used as a public water supply source is called potable groundwater. This requires the highest degree of protection with the most stringent standards. The groundwater quality standards for potable groundwater are generally equal to the USEPA's maximum contamination levels applicable at-the-tap pursuant to the SDWA. The rationale is that potable groundwater should be safe for drinking water supply without treatment.

The state primacy agency establishes a comprehensive program for the protection of groundwater. Through inter-agency cooperation, local groundwater protection programs can help to prevent unexpected and costly water supply systems. An Illinois community experienced a leaking gasoline underground tank, operated by the city-owned garage. It contaminated one well and threatened to contaminate the entire well field. The city has spent more than $300,000 in an attempt to replace the water supply.

Some parts of the groundwater protection programs, such as minimum and maximum setback zones for wellhead protection, are used to protect public and private drinking water supplies from potential sources of groundwater contamination. Each community well must have a setback zone to restrict land use near the well. The setback zone provides a buffer between the well and potential contamination sources and routes. It will give time for cleanup efforts of contaminated groundwater or to obtain an alternative water supply source before the existing groundwater source becomes unfit for use.

A minimum setback zone is mandatory for each public well. Siting for new potential primary or secondary pollution sources or potential routes is prohibited within the setback zone. In Illinois, the

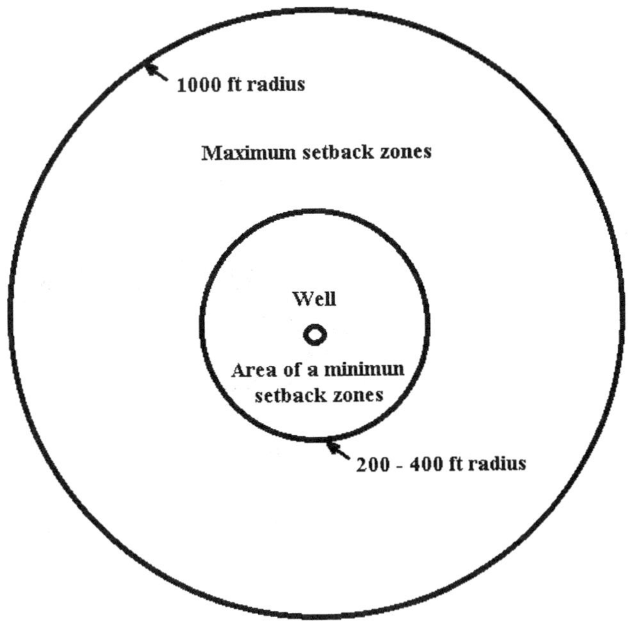

FIGURE 4.11 Maximum and minimum setback zones.

minimum setback zone is a 200 ft (61 m) radius area around the wellhead for every water supply well. For some vulnerable aquifers, the zone may be 400 ft in radius.

The maximum setback zone is a second level of protection from pollution. It prohibits the siting of new potential primacy pollution sources within the area outside the minimum setback zone up to 1000 ft (305 m) from a wellhead in Illinois (Fig. 4.11). Maximum setback zones allow the well owners, county or municipal government, and state to regulate land use beyond the minimum setback zone.

The establishment of a maximum zone is voluntary. A request to determine the technical adequacy of a maximum setback zone determination must first be submitted to the state by a municipality or county. Counties and municipalities served by community water supply wells are empowered to enact maximum setback zone ordinances. If the community water supply wells are investor or privately owned, a county or municipality served by that well can submit an application on the behalf of the owner.

7.1 Lateral Area of Influence

As described in the previous section, the lateral radius of influence (LRI) is the horizontal distance from the center of the well to the outer limit of the cone of depression (Figs. 4.1 and 4.12). It is the distance from the well to where there is no draw of groundwater (no reduction in water level). The lateral area of influence (LAI) outlines the extent of the cone of depression on the land surface as shown in the hatched area in Fig. 4.12.

The lateral area of influence in a confined aquifer is generally 4000 times greater than the LAI in an unconfined aquifer (Illinois EPA, 1990). If a pollutant is introduced within the LAI, it will reach the well faster than other water replenishing the well. The slope of the water table steepens toward the well within the LAI. Therefore, it is extremely important to identify and protect the lateral area of influence. The LAI is used to establish a maximum setback zone.

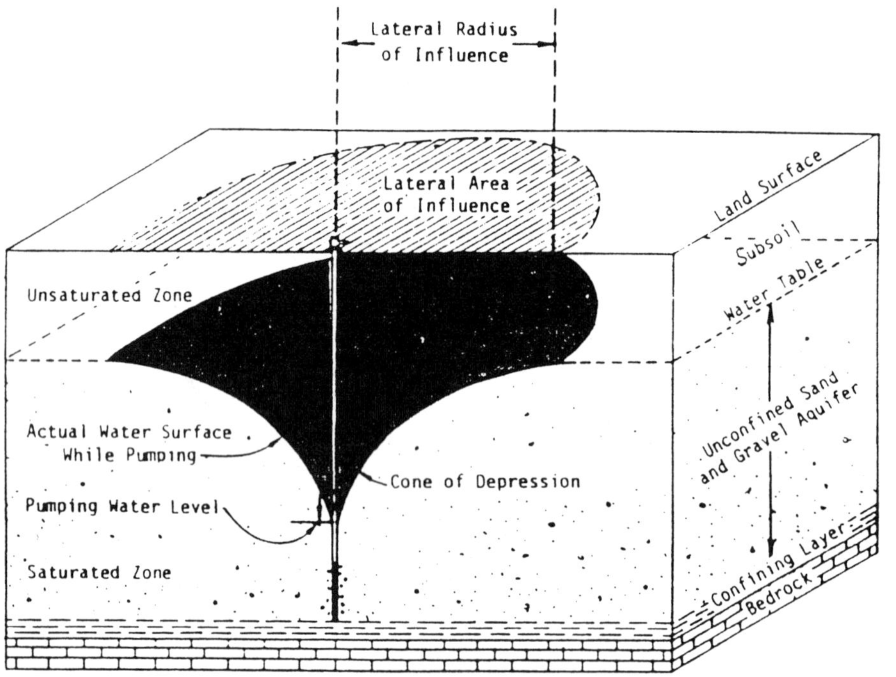

FIGURE 4.12 Lateral area of influence and the cone of depression under normal operation conditions (*Illinois EPA, 1990*).

7.2 Determination of Lateral Radius of Influence

There are three types of determination methods for the lateral radius of influence. They are the direct measurement method, the use of the Theis equation or volumetric flow equation, and the use of a curve-matching technique with the Theis equation. The third method involves the interpretation of pump test data from observation wells within the minimum setback zone using a curve-matching technique and the Theis equation to determine the transmissivity and storativity of the aquifer. Once these aquifer constants have been obtained, the Theis equation is then used to compute the lateral radius of influence of the well. It should be noted that when the observation well or piezometer measurement was made outside of the minimum setback zone, the determination could have been conducted by the direct measurement method.

Volumetric flow equations adopted by the Illinois EPA (1990) are as follows. For unconfined unconsolidated or unconfined nonfractured bedrock aquifers, the lateral radius of influence can be calculated by

$$r = \sqrt{\frac{Qt}{4524nH}} \tag{4.55}$$

where r = radius of influences, ft
Q = daily well flow under normal operational conditions, cfs
t = time that the well is pumped under normal operational conditions, min
H = open interval or length of well screen, ft
n = aquifer porosity, use following data unless more information is available:

Sand	0.21
Gravel	0.19
Sand and gravel	0.15
Sandstone	0.06
Limestone primary dolomites	0.18
Secondary dolomites	0.18

If pump test data is available for an unconfined/confined unconsolidated or nonfractured bedrock aquifer, the LRI can be calculated by

$$r = \sqrt{\frac{u\,T\,t}{2693\,S}} \tag{4.56}$$

where r = radius of influence, ft
T = aquifer transmissivity, gpd/ft
t = time that well is pumped under normal operational conditions, min
S = aquifer storativity or specific yield, dimensionless
u = a dimensionless parameter related to the well function W(u)

and

$$W(u) = \frac{T(h_0 - h)}{114.6\,Q} \tag{4.57}$$

where $W(u)$ = well function
$h_0 - h$ = drawdown in the piezometer or observation well, ft
Q = production well pumping rate under normal operational conditions, gpm

Direct measurement method. The direct measurement method involves direct measurement of drawdown in an observation well piezometer or in an another production well. Both the observation well and another production well should be located beyond the minimum setback zone of the production well and also be within the lateral area of influence.

EXAMPLE: A steel tape is used to measure the water level of the observation well located 700 ft from a drinking water well. The pump test result shows that under the normal operational conditions, 0.58 ft of drawdown compared to the original nonpumping water level is recorded. Does this make the well eligible for applying maximum setback zone?

Answer:

Yes. A municipality or county may qualify to apply for a maximum setback zone protection when the water level measurements show that the drawdown is greater than 0.01 ft and the observation well is located beyond (700 ft) the minimum setback zone (200 to 400 ft). It is also not necessary to estimate the lateral radius of influence of the drinking water well.

Theis equation methods. The Theis equation (Eqs. (4.56) and (4.57)) can be used to estimate the lateral radius of influence of a well if the aquifer parameters, such as transmissivity, storativity, and hydraulic conductivity, are available or can be determined. The Theis equation is more useful if the observation well is located within the minimum setback zone when the direct measurement method can not be applied. Equations (4.52) and (4.53) can be used if pump test data is available for an unconfined/confined, unconsolidated or nonfractured bedrock aquifer (Illinois EPA, 1990).

EXAMPLE: An aquifer test was conducted in a central Illinois county by the Illinois State Water Survey (Walton 1992). The drawdown data from one of the observations located 22 ft from the production well was analyzed to determine the average aquifer constants as follows:

$$\text{Transmissivity: } T = 340{,}000 \text{ gpd/ft of drawdown}$$

$$\text{Storativity: } S = 0.09$$

The pumping rate under normal operational conditions was 1100 gpm. The volume of the daily pumpage was 1,265,000 gal. Is this aquifer eligible for a maximum setback zone protection?

Solution:

Step 1. Determine the time of pumping, t

$$t = \frac{v}{Q} = \frac{1{,}265{,}000 \text{ gal}}{1100 \text{ gal/min}}$$

$$= 1150 \text{ min}$$

Step 2. The criteria used to represent a minimum level of drawdown $(h_0 - h)$ is assumed to be

$$h_0 - h = 0.01 \text{ ft}$$

Step 3. Calculate $W(u)$ using Eq. (4.57):

$$W(u) = \frac{T(h_0 - h)}{114.6\, Q}$$

$$= \frac{340{,}000 \text{ gpm/ft} \times 0.01 \text{ ft}}{114.6 \times 1100 \text{ gpm}}$$

$$= 0.027$$

Step 4. Find u associated with a $W(u) = 0.027$ in Appendix B:

$$u = 2.43$$

Step 5. Compute the lateral radius of influence r using Eq. (4.56):

$$r = \sqrt{\frac{u\,T\,t}{2693\, S}}$$

$$= \sqrt{\frac{2.43 \times 340{,}000 \text{ gpd/ft} \times 1150 \text{ min}}{2693 \times 0.09}}$$

$$= 1980 \text{ ft}$$

Answer:
Yes. The level of drawdown was estimated to be 0.01 ft at a distance of 1980 ft from the drinking water wellhead. Since r is greater than the minimum setback distance of 400 ft from the well head, the well is eligible for a maximum setback zone.

7.3 The Use of Volumetric Flow Equation

The volumetric flow equation is a cost-effective estimation for the radius of influence and is used for wells that pump continuously. It is an alternative procedure where aquifer constants are not available and where the Theis equation cannot be used. The volumetric flow equation (Eq. (4.55)) may be utilized for wells in unconfined unconsolidated or unconfined nonfractured bedrock aquifers.

EXAMPLE: A well in Peoria, Illinois (Illinois EPA, 1995), was constructed in a sand and gravel deposit which extends from 40 to 130 ft below the land surface. The static water level in the aquifer is 65 ft below the land surface. Because the static water level is below the top of the aquifer, this would be considered an unconfined aquifer. The length of the well screen is 40 ft. Well operation records indicate that the pump is operated continuously for 30 days every other month. The normal discharge rate is 1600 gpm (0.101 m^3/s). Is the well eligible for a maximum setback zone protection?

Solution:

Step 1. Convert discharge rate from gpm to ft^3/d:

$$\begin{aligned} Q &= 1600 \text{ gpm} \\ &= 1600 \frac{\text{gal}}{\text{min}} \times \frac{1 \text{ ft}^3}{7.48 \text{ gal}} \times \frac{1440 \text{ min}}{1 \text{ day}} \\ &= 308{,}021 \text{ ft}^3/\text{d} \\ &= 8723 \text{ m}^3/\text{d} \end{aligned}$$

Step 2. Calculate t in min:

$$\begin{aligned} t &= 30 \text{ day} \\ &= 30 \text{ day} \times 1440 \text{ min/d} \\ &= 43{,}200 \text{ min} \end{aligned}$$

Step 3. Find the aquifer porosity n:

$$n = 0.15 \text{ for sand and gravel}$$

Step 4. Calculate r using Eq. (4.55):

$$\begin{aligned} H &= 40 \text{ ft} \\ r &= \sqrt{\frac{Qt}{4524\, nH}} \\ &= \sqrt{\frac{308{,}021 \text{ ft}^3/\text{d} \times 43{,}200 \text{ min}}{4524 \times 0.15 \times 40 \text{ ft}}} \\ &= 700 \text{ ft} \\ &= 213.4 \text{ m} \end{aligned}$$

Yes, the well is eligible for maximum setback zone.

Note: The volumetric flow equation is very simple and economical.

REFERENCES

Babbitt, H. E., Doland, J. J. and Cleasby, J. L. 1959. *Water supply engineering.* 6th edn. New York: McGraw-Hill.

Bedient, P. B. and Huber, W. C. 1992. *Hydrology and floodplain analysis.* 2nd edn. Reading, Massachusetts: Addison-Wesley.

Bedient, P. B., Rifai, H. S. and Newell, C. J. 1994. *Ground water contamination transport and remediation.* Englewood Cliffs, New Jersey: Prentice Hall.

Bouwer, H. and Rice, R. C. 1976. A slug test for determining hydraulic conductivity of unconfined aquifers with completely or partially penetrating wills. *Water Resources Res*, **12**: 423–428.

Canter, L. W. and Knox, R. C. 1986. *Ground water pollution control.* Chelsea, Michigan: Lewis.

Canter, L. W., Knox, R. C. and Fairchild, D. M. 1988. *Gound water quality protection.* Chelsea, Michigan: Lewis.

Cooper, H. H., Jr., Bredehoeft, J. D. and Papadopoulos, I. S. 1967. Response of a finite-diameter well to an instantaneous charge of water. *Water Resources Res*, **3**: 263–269.

Cooper, H. H. Jr. and Jacob, C. E. 1946. A generalized graphical method for evaluating formation constants and summarizing well field history. *Trans. Amer. Geophys. Union*, **27**: 526–34.

Forrest, C. W. and Olshansky, R. 1993. Groundwater protection by local government. University of Illinois at Urbana-Champaign. Illinois.

Gupta, A. D. 1997. Groundwater and the environment. In: Biswas, A. K. (ed.) *Water resources*. New York: McGraw-Hill.

Hammer, M. J. 1986. *Water and wastewater technology*. New York: John Wiley.

Hammer, M. J. and Mackichan, K. A. 1981. *Hydrology and quality of water resources*. New York: John Wiley.

Hvorslev, M. J. 1951. Time lag and soil permeability in groundwater observations. U.S. Army Corps of Engineers Waterways Experiment Station, Bulletin 36, Vicksburg, Mississippi.

Illinois Environmental Protection Agency (IEPA). 1987. *Quality assurance and field method manual*. Springfield: IEPA.

Illinois Environmental Protection Agency (IEPA). 1988. *A primer regarding certain provisions of the Illinois groundwater protection act*. Springfield: IEPA.

Illinois Environmental Protection Agency (IEPA). 1990. *Maximum setback zone workbooks*. Springfield: IEPA.

Illinois Environmental Protection Agency (IEPA). 1994. *Illinois water quality report 1992–1993*. Springfield: IEPA.

Illinois Environmental Protection Agency (IEPA). 1995. *Guidance document for groundwater protection needs assessments*. Springfield: IEPA.

Illinois Environmental Protection Agency (IEPA). 1996. *Illinois Water Quality Report 1994–1995*. Vol. I. Springfield, Illinois: Bureau of Water, IEPA.

Jacob, C. E. 1947. Drawdown test to determine effective radius of artesian well. *Trans. Amer. Soc. Civil Engs.*, **112**(5): 1047–1070.

James, A. (ed.) 1993. *Introduction to water quality modelling*. 2nd edn. Chichester: John Wiley.

Leopold, L. B. 1974. *Water: A primer*. New York: W. H. Freeman.

Mackay, R. and Riley, M. S. 1993. Groundwater quality modeling. In: James, A. (ed.) *An introduction to water quality modeling*. 2nd edn. New York: John Wiley.

Nielsen, D. M. (ed.). 1991. *Practical handbook of groundwater monitoring*. Chelsea, Michigan: Lewis.

Papadopoulous, I. S., Bredehoeft, J. D. and Cooper, H. H. 1973. On the analysis of slug test data. *Water Resources Research*, **9**(4): 1087–1089.

Rehfeldt, K. R., Raman, R. K., Lin, S. D. and Broms, R. E. 1992. *Assessment of the proposed discharge of ground water to surface waters of the American Bottoms area of southwestern Illinois*. Contract report 539. Champaign: Illinois State Water Survey.

Rose, H. E. 1949. On the resistance coefficient-Reynolds number relationship for fluid flow through a bed of granular materials. *Proc. Inst. Mech. Engrs. (London)* **154**, 160.

Smith, L. and Wheatcraft, S. W. 1993. In: Maidment, D. R. (ed.) *Handbook of hydrology*, New York: McGraw-Hill.

Steel, E. W. and McGhee, T. J. 1979. *Water supply and sewerage*. 5th edn. New York: McGraw-Hill.

Tchobanoglous, G. and Schroeder, E. D. 1985. *Water quality, characteristics, modeling, modification*. Reading, Massachusetts: Addison-Wesley.

Theis, C. V. 1935. The relation between the lowering of the piezometric surface and the rate and duration of discharge of a well using ground storage. *Trans. Am. Geophys. Union* **16**: 519–524.

US Environmental Protection Agency (EPA). 1987. Underground storage tank corrective action technologies. EPA/625/6-87-015. Cincinnati, Ohio: USEPA.

US Environmental Protection Agency (EPA). 1994. Handbook: Ground water and wellhead protection, EPA/625/R-94/001. Washington, DC: USEPA.

Voelker, D. C. 1984. *Quality of water in the alluvial aquifer, American Bottoms, East St. Louis, Illinois*. US Geological Survey Water Resources Investigation Report 84-4180. Chicago, Illinois.

Walton, W. C. 1962. *Selected analytical methods for well and aquifer evaluation*. Illinois State Water Survey. Bulletin 49, Urbana, Illinois.

Wanielista, M. 1990. *Hydrology and water quality control*. New York: John Wiley.

Watson, I., and Burnett, A. D. 1993. *Hydrology: An environmental approach*. Cambridge, Ft. Lauderdale, Florida: Buchanan Books.

Wenzel, L. K. 1942. *Methods for determining the permeability of water-bearing materials*. US Geological Survey. Water Supply Paper 887, p. 88.

Willis. R., and Yeh, W. W-G. 1987. *Groundwater systems planning & management*. Englewood Cliffs, New Jersey: Prentice-Hall.

CHAPTER 1.5

FUNDAMENTAL AND TREATMENT PLANT HYDRAULICS

Shun Dar Lin

1 DEFINITIONS AND FLUID PROPERTIES

1.1 Weight and Mass

The weight (W) of an object, in the International System of Units (SI), is defined as the product of its mass (m, in grams, kilograms, etc.) and the gravitational acceleration ($g = 9.81$ m/s^2 on the earth's surface) by Newton's second law of motion: $F = ma$. The weight is expressed as

$$W = mg \tag{5.1a}$$

The unit of weight is kg · m/s^2 and is usually expressed as newton (N).

In the SI system, one newton is defined as the force needed to accelerate 1 kg of mass at a rate of 1 m/s^2. Therefore

$$\begin{aligned} 1\text{ N} &= 1\text{ kg}\cdot\text{m/s}^2 \\ &= 1 \times 10^3\text{ g}\cdot 10^2\text{ cm/s}^2 \\ &= 10^5\text{ g}\cdot\text{cm/s}^2 \\ &= 10^5\text{ dynes} \end{aligned}$$

In the U.S. customary system, mass is expressed in slugs. One slug is defined as the mass of an object which needs one pound of force to accelerate to one ft/s^2, i.e.

$$m = \frac{\mathrm{W\ (lb)}}{g\ (\mathrm{ft/s^2})} = \frac{w}{g}\ \text{(in slugs)} \tag{5.1b}$$

EXAMPLE: What is the value of the gravitational acceleration (g) in the U.S. customary system?

Solution:

$$\begin{aligned} g &= 9.81\ \mathrm{m/s^2} = 9.81\ \mathrm{m/s^2} \times 3.28\ \mathrm{ft/m} \\ &= 32.174\ \mathrm{ft/s^2} \end{aligned}$$

Note: This is commonly used as $g = 32.2\ \mathrm{ft/s^2}$.

1.2 Specific Weight

The specific weight (weight per unit volume) of a fluid such as water, γ, is defined by the product of the density (ρ) and the gravitational acceleration (g), i.e.

$$\begin{aligned} \gamma &= \rho g\ (\text{in kg/m}^3 \cdot \text{m/s}^2) \\ &= \rho g\ (\text{in N/m}^3) \end{aligned} \tag{5.2}$$

Water at 4°C reaches its maximum density of 1000 kg/m^3 or 1.000 g/cm^3. The ratio of the specific weight of any liquid to that of water at 4°C is called the specific gravity of that liquid.

EXAMPLE: What is the unit weight of water at 4°C in terms of N/m^3 and dyn/cm^3?

Solution:

Step 1. In N/m^3, $\rho = 1000\ \mathrm{kg/m^3}$

$$\begin{aligned} \gamma &= \rho g = 1000\ \mathrm{kg/m^3} \times 9.81\ \mathrm{m/s^2} \\ &= 9810\ \mathrm{N/m^3} \end{aligned}$$

Step 2. In dyn/cm^3, $\rho = 1.000\ \mathrm{g/cm^3}$

$$\begin{aligned} \gamma &= 1\ \mathrm{g/cm^3} \times 981\ \mathrm{cm/s^2} \\ &= 981\ (\mathrm{g \cdot cm/s^2})/\mathrm{cm^3} \\ &= 981\ \mathrm{dyn/cm^3} \end{aligned}$$

1.3 Pressure

Pressure (P) is the force (F) applied to or distributed over a surface area (A) as a measure of force per unit area:

$$P = \frac{F}{A} \tag{5.3a}$$

In the SI system, the unit of pressure can be expressed as barye, bar, N/m^2, or pascal. One barye equals one dyne per square centimeter (dyn/cm^2). The bar is one megabarye, 10^6 dynes per square centimeter (10^6 barye). One pascal equals one newton per square meter (N/m^2). In the U.S. customary system the unit of pressure is expressed as lb/in^2 (psi) or lb/ft^2 etc.

Pressure is also a measure of the height (h) of the column of mercury, water, or other liquid which it supports. The pressure at the liquid surface is atmospheric pressure (P_a). The pressure at any point in the liquid is the absolute pressure (P_{ab}) at that point. The absolute pressure is measured with respect to zero pressure. Thus P_{ab} can be written as

$$P_{ab} = \gamma h + P_a \tag{5.3b}$$

where γ = specific weight of water or other liquid.

In engineering work gage pressure is more commonly used. The gage pressure scale is designed on the basis that atmospheric pressure (P_a) is zero. Therefore the gage pressure becomes:

$$P = \gamma h \tag{5.3c}$$

and pressure head

$$h = \frac{P}{\gamma} \tag{5.3d}$$

According to Pascal's law, the pressure exerted at any point on a confined liquid is transmitted undiminished in all directions.

EXAMPLE: Under normal conditions, the atmospheric pressure at sea level is approximately 760 mm height of mercury. Convert it in terms of m of water, N/m², pascals, bars, and psi.

Solution:

Step 1. Solve for m of water

Let γ_1 and γ_2 be specific weights of water and mercury, respectively, and h_1 and h_2 be column heights of water and mercury, respectively; then approximately

$$\gamma_1 = 1.00, \quad \gamma_2 = 13.5936$$

From Eq. (5.3c):

$$P = \gamma_1 h_1 = \gamma_2 h_2$$

$$h_1 = \frac{\gamma_2}{\gamma_1} h_2 = \frac{13.5936}{1} \times 760 \text{ mm} = 10{,}331 \text{ mm}$$

$$= 10.33 \text{ m (of water)}$$

Step 2. Solve for N/m²

Since $\gamma_1 = 9810 \text{ N/m}^3$

$$P = \gamma_1 h_1 = 9810 \text{ N/m}^3 \times 10.33 \text{ m} = 101{,}337 \text{ N/m}^2 = 1.013 \times 10^5 \text{ N/m}^2$$

Step 3. Solve for pascals

Since 1 pascal = 1 N/m², from step 2:

$$P = 1.013 \times 10^5 \text{ N/m}^2$$

$$= 1.013 \times 10^5 \text{ Pa}$$

Note: 1 atmosphere approximately equals 10^5 N/m² or 10^5 pascals.

Step 4. Solve for bars
Since approximately $\gamma_1 = 981$ dyn/cm^3

$$P = \gamma_1 h_1 = 981 \text{ dyn/cm}^3 \times 1033 \text{ cm}$$
$$= 1{,}013{,}400 \text{ dyn/cm}^2$$
$$= 1.013 \times 10^6 \text{ dyn/cm}^2 \times \frac{1 \text{ bar}}{1 \times 10^6 \text{ dyn/cm}^2}$$
$$= 1.013 \text{ bars}$$

Note: 1 atmosphere pressure is approximately equivalent to 1 bar.

Step 5. Solve for lb/in^2 (psi)
From step 1,

$$h_1 = 10.33 \text{ m} = 10.33 \text{ m} \times 3.28 \text{ ft/m}$$
$$= 33.88 \text{ ft of water}$$
$$P = 33.88 \text{ ft} \times 1 \text{ (lb/in}^2\text{)/2.31 ft}$$
$$= 14.7 \text{ lb/in}^2$$

1.4 Viscosity of Water

All liquids possess a definite resistance to change of their forms and many solids show a gradual yielding to force (shear stress) tending to change their forms. Newton's law of viscosity states that, for a given rate of angular deformation of liquid, the shear stress is directly proportional to the viscosity. It can be expressed as

$$\tau = \mu \frac{du}{dy} \tag{5.4a}$$

where τ = shear stress
μ = proportionality factor, viscosity
du/dy = velocity gradient
u = angular velocity
y = depth of fluid

The angular velocity and shear stress change with y.

The viscosity can be expressed as absolute (dynamic) viscosity or kinematic viscosity. From Eq. (5.4a), it may be rewritten as

$$\mu = \frac{\tau}{du/dy} \tag{5.4b}$$

The viscosity μ is frequently referred to absolute viscosity, or mass per unit length and time. In the SI system, the absolute viscosity μ is expressed as poise or in dyne-second per square centimter. One poise is defined as the tangent shear force (τ) per unit area (dyn/cm^2) required to maintain unit difference in velocity (1 cm/s) between two parallel planes separated by 1 cm of fluid. It can be written as

$$1 \text{ poise} = 1 \text{ dyne}\cdot\text{s/cm}^2$$
$$= 1 \text{ dyne}\cdot\text{s/cm}^2 \times \frac{1 \text{ g}\cdot\text{cm/s}^2}{1 \text{ dyne}}$$
$$= 1 \text{ g/cm}\cdot\text{s}$$

or

$$\begin{aligned} 1 \text{ poise} &= 1 \text{ dyne}\cdot\text{s/cm}^2 \\ &= 1 \text{ dyne}\cdot\text{s/cm}^2 \times \frac{1 \text{ N}}{10^5 \text{dyn}} \Big/ \frac{1 \text{ m}^2}{10^4 \text{ cm}^2} \\ &= 0.1 \text{ N}\cdot\text{s/m}^2 \end{aligned}$$

also

$$1 \text{ poise} = 100 \text{ centipoise}$$

The usual measure of absolute viscosity of fluid and gas is the centipoise. The viscosity of fluids is temperature dependent. The viscosity of water at room temperature (20°C) is one centipoise. The viscosity of air at 20°C is approximately 0.018 centipoise.

The U.S. customary system unit of viscosity is 1 lb · s/ft² or 1 slug ft · s.

EXAMPLE 1: How much in U.S. customary units is 1 poise absolute viscosity?

Solution:

$$\begin{aligned} 1 \text{ poise} &= 0.1 \text{ N}\cdot\text{s/m}^2 \\ &= 0.1 \text{ N} \times \frac{1 \text{ lb}}{4.448 \text{ N}}\cdot\text{s/m}^2 \Big/ \frac{(3.28 \text{ ft})^2}{\text{m}^2} \\ &= 0.02248 \text{ lb}\cdot\text{s}/10.758 \text{ ft}^2 \\ &= 0.00209 \text{ lb}\cdot\text{s/ft}^2 \end{aligned}$$

On the other hand:

$$\begin{aligned} 1 \text{ lb}\cdot\text{s/ft}^2 &= 479 \text{ poise} \\ &= 47.9 \text{ N}\cdot\text{s/m}^2 \end{aligned}$$

The kinematic viscosity ν is the ratio of viscosity (absolute) to mass density ρ. It can be written as

$$\nu = \mu/\rho \tag{5.5}$$

The dimension of ν is length squared per unit time. In the SI system, the unit of kinematic viscosity is the stoke. One stoke is defined as 1 cm²/s. However, the standard measure is the centistoke (= 10^{-2} stoke or 10^{-2} cm²/s). Kinematic viscosity offers many applications, such as the Reynolds number $\mathbf{R} = VD/\nu$. For water, the absolute viscosity and kinematic viscosity are essentially the same, especially when the temperature is less than 10°C. The properties of water in SI units and U.S. customary units are respectively given in Tables 5.1a and 5.1b.

EXAMPLE 2: At 21°C, water has an absolute viscosity of 0.00982 poise and a specific gravity of 0.998. Compute (a) the absolute (N · s/m²) and kinematic viscosity in SI units and (b) the same quantities in U.S. customary units.

Solution:

Step 1. For (a), at 21°C (69.8°F)

$$\mu = 0.00982 \text{ poise} = 0.00982 \text{ poise} \times \frac{0.1 \text{ N}\cdot\text{s/m}^2}{1 \text{ poise}}$$

$$= 0.000982 \text{ N}\cdot\text{s/m}^2 \quad \text{or}$$

$$= 0.000982, \text{N}\cdot\text{s/m}^2 \times \frac{1 \text{ kg}\cdot\text{m/s}^2}{1 \text{ N}}$$

$$= 0.000982 \text{ kg/m}\cdot\text{s}$$

$$\nu = \frac{\mu}{\rho} = \frac{0.000982 \text{ kg/m}\cdot\text{s}}{0.998 \times 1000 \text{ kg/m}^3}$$

$$= 0.984 \times 10^{-6} \text{m}^2\text{/s}$$

TABLE 5.1a Physical Properties of Water—SI Units

Temperature T, °C	Specific gravity	Specific weight γ, N/m^3	Absolute viscosity μ, N · s/m^2	Kinematic viscosity ν, m^2/s	Surface tension σ, N/m^2	Vapor pressure P_v, N/m^2
0	0.9999	9805	0.00179	1.795×10^{-6}	0.0756	608
4	1.0000	9806	0.00157	1.568×10^{-6}	0.0750	809
10	0.9997	9804	0.00131	1.310×10^{-6}	0.0743	1226
15	0.9990	9798	0.00113	1.131×10^{-6}	0.0735	1762
21	0.9980	9787	0.00098	0.984×10^{-6}	0.0727	2504
27	0.9966	9774	0.00086	0.864×10^{-6}	0.0718	3495
38	0.9931	9739	0.00068	0.687×10^{-6}	0.0700	6512
93	0.9630	9444	0.00030	0.371×10^{-6}	0.0601	79,002

Source: Brater *et al.* (1996).

TABLE 5.1b Physical Properties of Water—U.S. customary Units

Temperature T, °F	Density ρ, slug/ft^3	Specific weight γ, lb/ft^3	Absolute viscosity $\mu \times 10^5$, lb · s/ft^2	Kinematic viscosity $\nu \times 10^5$, ft^2/s	Surface tension σ, lb/ft	Vapor pressure P_v, psia
32	1940	62.42	3.746	1.931	0.00518	0.09
40	1938	62.43	3.229	1.664	0.00514	0.12
50	1936	62.41	2.735	1.410	0.00509	0.18
60	1934	62.37	2.359	1.217	0.00504	0.26
70	1931	62.30	2.050	1.059	0.00498	0.36
80	1927	62.22	1.799	0.930	0.00492	0.51
90	1923	62.11	1.595	0.826	0.00486	0.70
100	1918	62.00	1.424	0.739	0.00480	0.95
120	1908	61.71	1.168	0.609		
140	1896	61.38	0.981	0.514		
160	1890	61.00	0.838	0.442		
180	1883	60.58	0.726	0.385		
200	1868	60.12	0.637	0.341		
212	1860	59.83	0.593	0.319		

Source: Benefield *et al.* (1984) and Metcalf & Eddy, Inc. (1972).

Step 2. For (b), from Step 1 and Example 1

$$\mu = 0.000982 \text{ N} \cdot \text{s/m}^2 \times \frac{1 \text{ slug/ft} \cdot \text{s}}{47.9 \text{ N} \cdot \text{s/m}^2}$$

$$= 2.05 \times 10^{-5} \text{ slug/ft} \cdot \text{s}$$

$$\nu = 0.984 \times 10^{-6} \text{ m}^2/\text{s} \times \left(\frac{3.28 \text{ ft}}{1 \text{ m}}\right)^2$$

$$\nu = 1.059 \times 10^{-5} \text{ ft}^2/\text{s}$$

1.5 Perfect Gas

A perfect gas is a gas that satisfies the perfect gas laws, such as the Boyle–Mariotte law and the Charles–Gay-Lussac law. It has internal energy as a function of temperature only. It also has specific heats with values independent of temperature. The normal volume of a perfect gas is 22.4136 liters/mol (commonly quoted as 22.4 L/mol).

Boyle's law states that at a constant temperature the volume of a given quantity of any gas varies inversely as the pressure applied to the gas. For a perfect gas, changing from pressure P_1 and volume V_1 to P_2 and V_2 at constant temperature, the following law exists:

$$P_1V_1 = P_2V_2 \tag{5.6}$$

According to the Boyle–Mariotte law for perfect gases, the product of pressure P and volume V is constant in an isothermal process.

$$PV = nRT \tag{5.7}$$

or

$$P = \frac{1}{V}RT = \rho RT$$

where P = pressure, pascal (P_a) or lb/ft^2
V = volume, m^3 or ft^3
n = number of moles
R = gas constant
T = absolute temperature, K = °C + 273, or °R = °F + 459.6
ρ = density, kg/m^3 or slug/ft^3

On a mole (M) basis, a pound mole (or kg mole) is the number of pounds (or kg) mass of gas equal to its molecular weight. The product MR is called the universal gas constant and depends on the units used. It can be

$$MR = 1545 \text{ ft} \cdot \text{lb/lb} \cdot {}^\circ\text{R}$$

The gas constant R is determined as

$$R = \frac{1545}{M} \text{ ft lb/lb} \cdot {}^\circ\text{R} \tag{5.8a}$$

or, in slug units

$$R = \frac{1545 \times 32.2}{M} \text{ ft} \cdot \text{lb/slug } {}^\circ\text{R} \tag{5.8b}$$

For SI units

$$R = \frac{8312}{M}n \text{ N/kg} \cdot \text{K} \tag{5.8c}$$

EXAMPLE: For carbon dioxide with a molecular weight of 44 at a pressure of 12.0 psia (pounds per square inch absolute) and at a temperature of 70°F (21°C), compute R and its density.

Solution:

Step 1. Determine R from Eq. (5.8b)

$$R = \frac{1545 \times 32.2}{M} \text{ ft} \cdot \text{lb/slug °R}$$

$$= \frac{1545 \times 32.2}{44}$$

$$= 1130 \text{ ft} \cdot \text{lb/slug} \cdot \text{°R}$$

Step 2. Determine density ρ

$$\rho = \frac{P}{RT} = \frac{12 \text{ lb/in}^2 \cdot 144 \text{ in}^2/\text{ft}^2}{(1130 \text{ ft} \cdot \text{lb/slug°R})\,(460 + 70\text{°R})}$$

$$= 0.00289 \text{ slug/ft}^3$$

Based on empirical generation at constant pressure in a gaseous system (perfect gas), when the temperature varies the volume of gas will vary approximately in the same proportion. If the volume is exactly one mole of gas at 0°C (273.15 K) and at atmosphere pressure, then, for ideal gases, these are the so-called standard temperature and pressure, STP. The volume of ideal gas at STP can be calculated from Eq. (5.7) (with $R = 0.08206$ L · atm/K · mol):

$$V = \frac{nRT}{P} = \frac{(1 \text{ mole})\,(0.08206 \text{ L} \cdot \text{atm/mol} \cdot \text{K})\,(273.15 \text{ K})}{1 \text{ atm}}$$

$$= 22.41 \text{ liters.}$$

Thus, according to Avogadro's hypothesis and the ideal-gas equation, one mole of any gas will occupy 22.41 L at STP.

2 WATER FLOW IN PIPES

2.1 Fluid Pressure

Water and wastewater professionals frequently encounter some fundamentals of hydraulics, such as pressure, static head, pump head, velocity of flow, and discharge rate. The total force acting on a certain entire space, commonly expressed as the force acting on unit area, is called intensity of pressure, or simply pressure, P. The U.S. customary system of Units generally uses the pound per square inch (psi) for unit pressure. This quantity is also rather loosely referred to simply as pounds pressure; i.e., "20 pounds pressure" means 20 psi of pressure. The International System of Units uses the kg/cm^2 (pascal) or g/cm^2.

To be technically correct, the pressure is so many pounds or kilograms more than that exerted by the atmosphere (760 mm of mercury). However, the atmospheric pressure is ignored in most cases, since it is applied to everything and acts uniformly in all directions.

2.2 Head

The term *head* is frequently used, such as in energy head, velocity head, pressure head, elevation head, friction head, pump head, and loss of head (head loss). All heads can be expressed in the dimension of length, i.e. ft × lb/lb = ft, or m × kg/kg = m, etc.

The pump head equals the ft · lb (m · kg) of energy put into each pound (kg) of water passing through the pump. This will be discussed later in the section on pumps.

Pressure drop causes loss of head and may be due to change of velocity, change of elevation, or to friction loss. Hydraulic head loss may occur at lateral entrances and is caused by hydraulic components such as valves, bends, control points, sharp crested weirs, and orifices. These types of head loss have been extensively discussed in textbooks and handbooks of hydraulics. Bend losses and head losses due to dividing and combining flows are discussed in detail by James M. Montgomery, Consulting Engineers, Inc. (1985).

Velocity head. The kinetic energy (KE) of water with mass m is its capacity to do work by reason of its velocity V and mass and is expressed as $\frac{1}{2}mV^2$.

For a pipe with mean flow velocity V (m/s or ft/s) and pipe cross-sectional area A (cm^2 or sq. in), the total mass of water flowing through the cross section in unit time is $m = \rho VA$, where ρ is the fluid density. Thus the total kinetic energy for a pipe flow is

$$\text{KE} = \tfrac{1}{2}mV^2 = \tfrac{1}{2}(\rho VA)V^2 = \tfrac{1}{2}\rho AV^3 \tag{5.9}$$

The total weight of fluid $W = m \cdot g = \rho AVg$, where g is the gravitational acceleration. It is commonly expressed in terms of energy in a unit weight of fluid. The kinetic energy in unit weight of fluid is

$$\frac{\text{KE}}{\text{W}} = \frac{\frac{1}{2}\rho AV^3}{\rho gAV} = \frac{V^2}{2g} \tag{5.10}$$

This is the so-called velocity head, i.e. the height of the fluid column.

EXAMPLE 1: Twenty-two pounds of water are moving at a velocity of 2 ft/s. What are the kinetic energy and velocity head?

Solution:

Step 1.

$$\begin{aligned}\text{KE} &= \tfrac{1}{2}mV^2 = \tfrac{1}{2} \times 22 \text{ lb} \times (2 \text{ ft/s})^2 \\ &= 44 \text{ lb} \cdot \text{ft}^2/\text{s}^2 = 44 \text{ ft} \cdot \text{lb} \cdot \text{ft/s}^2 \\ &= 44 \text{ slug} \cdot \text{ft, (since 1 slug} = 1 \text{ lb} \cdot \text{ft/s}^2)\end{aligned}$$

Step 2.

$$\begin{aligned}h_v &= \frac{V^2}{2g} = \frac{(2 \text{ ft/s})^2}{2 \times 32.2 \text{ ft/s}^2} \\ &= 0.062 \text{ ft}\end{aligned}$$

EXAMPLE 2: Ten kilograms of water are moving with a velocity of 0.61 m/s. What are the kinetic energy and velocity head?

Solution:

Step 1.

$$\begin{aligned}\text{KE} &= \tfrac{1}{2}mV^2 = \tfrac{1}{2}(10 \text{ kg})(0.61 \text{ m/s})^2 \\ &= 1.86 \text{ kg} \cdot \text{m}^2/\text{s}^2 = 1.86 \text{ m} \cdot \text{kg} \cdot \text{m/s}^2 \\ &= 1.86 \text{ N} \cdot \text{m, since 1 N} = 1 \text{ kg} \cdot \text{m/s}^2\end{aligned}$$

Step 2.

$$h_v = \frac{V^2}{2g} = \frac{(0.61\text{ m/s})^2}{2 \times 9.81\text{ m/s}^2}$$
$$= 0.019\text{ m}$$
$$= 0.019\text{ m} \times 3.28\text{ ft/m}$$
$$= 0.062\text{ ft}$$

Note: Examples 1 and 2 are essentially the same.

Pressure head. The pressure energy (PE) is a measure of work done by the pressure force on the fluid mass and is expressed as

$$\text{PE} = pAV \tag{5.11}$$

where p = pressure at a cross section
A = pipe cross-sectional area, cm^2 or in^2
V = mean velocity

The pressure head (h) is the pressure energy in unit weight of fluid. The pressure is expressed in terms of the height of the fluid column h. The pressure head is

$$h = \frac{\text{PE}}{W} = \frac{pAV}{\rho gAV} = \frac{p}{\rho g} = \frac{p}{\gamma} = \frac{\text{pressure}}{\text{sp. wt.}} \tag{5.12}$$

where γ is the weight per unit volume of fluid or its specific weight in N/m^3 or lb/ft^3. The general expression of unit pressure is

$$p = \gamma h \tag{5.13}$$

and the pressure head is $h_p = p/\gamma$ in ft or m.

EXAMPLE: At the bottom of a water storage tank, the pressure is 31.2 lb/in^2. What is the pressure head?

Solution:

Step 1. Convert the pressure to lb/ft^2

$$p = 31.2\text{ lb/in}^2 = 31.2\text{ lb/in}^2 \times 144\text{ in}^2\text{/ft}^2$$
$$= 31.2 \times 144\text{ lb/ft}^2$$

Step 2. Determine h_p; $\gamma = 62.4\text{ lb/ft}^3$ for water

$$h_p = \frac{p}{\gamma} = \frac{31.2 \times 144\text{ lb/ft}^2}{62.4\text{ lb/ft}^3}$$
$$= 72\text{ ft}$$

Elevation head. The elevation energy (EE) of a fluid mass is simply the weight multiplied by the height above a reference plane. It is the work to raise this mass W to elevation h and can be written as

$$\text{EE} = Wz \tag{5.13a}$$

The elevation head, z, is EE divided by the total weight W of fluid:

$$\text{Elevation head} = \frac{\text{EE}}{W} = \frac{Wz}{W} = z \tag{5.13b}$$

EXAMPLE: (1) Ten kilograms and (2) one pound of water are at 50 feet above the earth's surface. What are their elevation heads?

Solution:

Step 1. For 10 kg

$$h_e = \frac{Wh}{W} = \frac{10 \text{ kg} \times 50 \text{ ft}}{10 \text{ kg}}$$
$$= 50 \text{ ft}$$

Step 2. For 1 lb

$$h_e = \frac{Wh}{W} = \frac{1 \text{ lb} \times 50 \text{ ft}}{1 \text{ lb}} = 50 \text{ ft}$$

Both have 50 feet of head.

Bernoulli equation. The total energy (H) at a particular section of water in a pipe is the algebraic sum of the kinetic head, pressure head, and elevation head. For sections 1 and 2, they can be expressed as

$$H_1 = \frac{V_1^2}{2g} + \frac{p_1}{\gamma} + z_1 \tag{5.14}$$

and

$$H_2 = \frac{V_2^2}{2g} + \frac{p_2}{\gamma} + z_2 \tag{5.15}$$

If water is flowing from section 1 to section 2, friction loss (h_f) is the major loss. The energy relationship between the two sections can be expressed as the Bernoulli equation

$$\frac{V_1^2}{2g} + \frac{p_2}{\gamma} + z_1 = \frac{V_2^2}{2g} + \frac{p_2}{\gamma} + z_2 + h_f \tag{5.16}$$

where h_f is the friction head. This is also called the continuity equation.

EXAMPLE: A nozzle of 12 cm diameter is located near the bottom of a storage water tank. The water surface is 3.05 m (10 ft) above the nozzle. Determine (a) the velocity of the effluent from the nozzle and (b) the discharge.

Solution:

Step 1. (a) Determine V_2
Let point 1 be at the water surface and point 2 at the center of the nozzle. By the Bernoulli equation

$$\frac{V_1^2}{2g} + \frac{p_1}{\gamma} + z_1 = \frac{V_2^2}{2g} + \frac{p_2}{\gamma} + z_2$$

Since $p_1 = p_2 = 0$, and $V_1^2 = 0$

$$\frac{V_2^2}{2g} = z_1 - z_2 = H = 3.05 \text{ m}$$
$$V_2 = \sqrt{2gH} = \sqrt{2 \times 9.806 \times 3.05}$$
$$= 7.73 \text{ m/s}$$

Step 2. Solve for (b) flow

$$Q = A_2V_2 = \pi r^2 V_2 = 3.14\ (0.06\ \text{m})^2 \times 7.73\ \text{m/s}$$
$$= 0.087\ \text{m}^3/\text{s}$$

Friction head. Friction head (h_f) equals the loss of energy by each unit weight of water or other liquid through frction in the length of a pipe, in which the energy is converted into heat. The values of friction heads are usually obtained from the manufacturer's tables.

Darcy–Weisback equation. The Darcy–Weisback formula can be calculated from the friction head:

$$h_f = f\left(\frac{L}{D}\right)\frac{V^2}{2g} \tag{5.17}$$

where h_f = head of friction loss, cm or ft
f = friction factor, dimensionless
L = length of pipeline, cm or ft
D = diameter of pipe, cm or ft
$\frac{V^2}{2g}$ = velocity head, cm or ft
V = average velocity of flow, cm/s or ft/s

EXAMPLE 1: Rewrite the Darcy–Weisback formula for h_f in terms of flow rate Q instead of velocity V.

Solution:

Step 1. Convert V to Q.
In a circular pipe,

$$A = \pi(D/2)^2 = \pi D^2/4 = 0.785\ D^2$$

The volumetric flow rate may be expressed in terms of velocity and area (A) as

$$Q = VA$$

then

$$V = \frac{Q}{A} = 4Q/\pi D^2$$
$$V^2 = 16Q^2/\pi^2 D^4$$

Step 2. Substitute in the formula

$$h_f = f\left(\frac{L}{D}\right)\frac{V^2}{2g} = f\left(\frac{L}{D}\right)\frac{16Q^2}{2g\pi^2 D^4}$$
$$= f\left(\frac{L}{D^5}\right)\frac{8\ Q^2}{\pi^2 g} \tag{5.18}$$

The head of loss due to friction may be determined in three ways:

1. The friction factor f depends upon the velocity of flow and the diameter of the pipe. The value of f may be obtained from a table in many hydraulic textbooks and handbooks, or from the Moody chart (Fig. 5.1) of Reynolds number **R** versus f for various grades and sizes of pipe. $\mathbf{R} = DV/\nu$, where ν is the kinematic viscosity of the fluid.

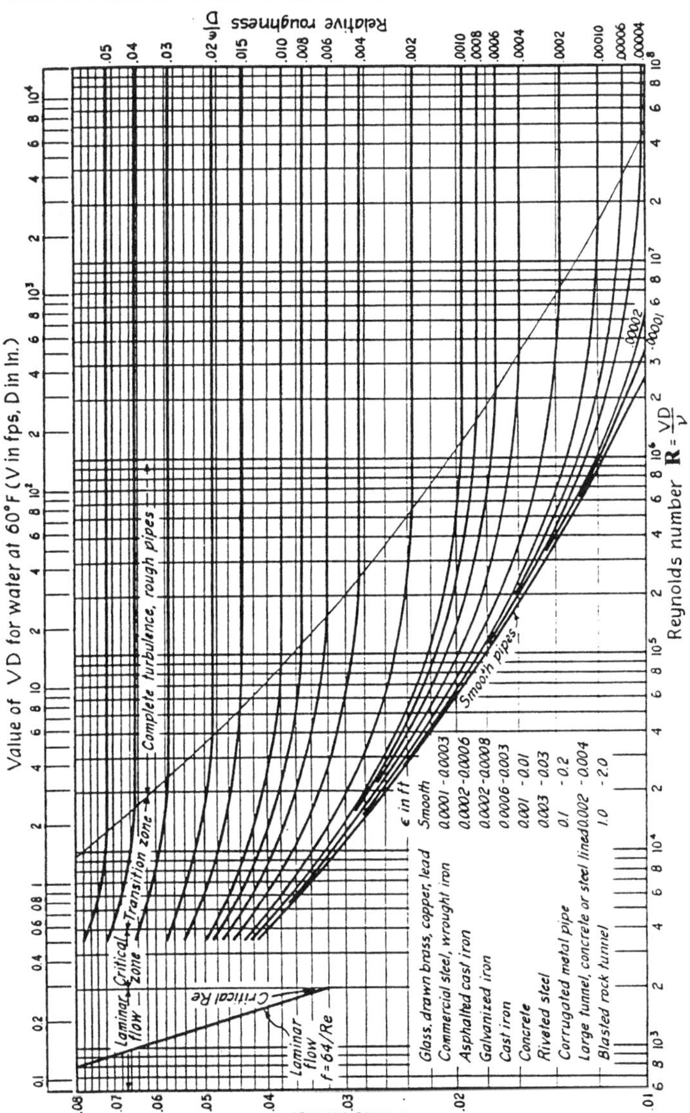

FIGURE 5.1 Moody diagram (*Metcalf and Eddy, Inc., Wastewater Engineering Collection and Pumping of Wastewater, Copyright 1990, McGraw-Hill, New York, reproduced with permission of McGraw-Hill*).

TABLE 5.2 Values of Equivalent Roughness *e* for New Commercial Pipes

Type of pipe	*e*, in
Asphalted cast iron	0.0048
Cast iron	0.0102
Concrete	0.01–0.1
Drawn tubing	0.00006
Galvanized iron	0.006
PVC	0.00084
Riveted steel	0.04–0.4
Steel and wrought iron	0.0018
Wood stave	0.007–0.04

Source: Benefield *et al.* (1984).

2. The friction loss of head h_f per 1000 ft of pipe may be determined from the Hazen–Williams formula for a pipe flow. A constant C for a particular pipe, diameter of pipe, and either the velocity or the quantity of flow should be known. A nomograph chart is most commonly used for solution by the Hazen–Williams formula.

3. Another empirical formula, the Manning equation, is also a popular formula for determining head loss due to free flow.

The friction factor f is a function of Reynolds number $\mathbf{R}$ and the relative roughness of the pipe wall e/D. The value of e (Table 5.2), the roughness of the pipe wall (equivalent roughness), is usually determined from experiment.

For laminar pipe flow ($\mathbf{R} < 2000$) f is independent of surface roughness of the pipe. The f value can be determined from

$$f = 64/\mathbf{R} \tag{5.19}$$

When $\mathbf{R} > 2000$ then the relative roughness will affect the f value. The e/D values can be found from the manufacturer or any textbook. The relationship between f, $\mathbf{R}$, and e/D are summarized in graphical expression as the Moody diagram (Fig. 5.1).

EXAMPLE 2: A pumping station has three pumps with 1 MGD, 2 MGD, and 4 MGD capacities. Each pumps water from a river at an elevation of 588 ft above sea level to a reservoir at an elevation of 636 ft, through a cast iron pipe of 24 in diameter and 2600 ft long. The Reynolds number is 1600. Calculate the total effective head supplied by each pump and any combinations of pumping.

Solution:

Step 1. Write Bernoulli's equation

Let stations 1 and 2 be river pumping site and discharge site at reservoir, respectively; h_p is effective head applied by pump in feet.

$$\frac{V_1^2}{2g} + \frac{p_1}{\gamma} + z_1 + h_p = \frac{V_2^2}{2g} + \frac{p_2}{\gamma} + z_2 + h_f$$

Since $V_1 = V_2$,

p_1 is not given (usually negative), assumed zero

p_2 = zero, to atmosphere

then

$$
\begin{aligned}
z_1 + h_p &= z_2 + h_f \\
h_p &= (z_2 - z_1) + h_f \\
&= (636 - 588) + h_f \\
&= 48 + h_f
\end{aligned}
$$

Step 2. Compute flow velocity (V) for the 1 MGD pump

$$
\begin{aligned}
A &= \pi r^2 = 3.14\ (1\ \text{ft})^2 = 3.14\ \text{ft}^2 \\
Q &= 1\ \text{MGD} \\
&= \frac{10^6\ \text{gal} \times 0.1337\ \text{ft}^3/\text{gal}}{1\ \text{day} \times 24 \times 60 \times 60\ \text{s/day}} \\
&= 1.547\ \text{ft}^3/\text{s (cfs)} \\
V &= \frac{Q}{A} = \frac{1.547\ \text{cfs}}{3.14\ \text{ft}^2} \\
&= 0.493\ \text{ft/s}
\end{aligned}
$$

Step 3. Find h_f and h_p for 1 MGD pump
From Eqs. (5.19) and (5.17),

$$
\begin{aligned}
f &= 64/\mathbf{R} = 64/1600 = 0.04 \\
h_f &= f\left(\frac{L}{D}\right)\frac{V^2}{2g} \\
&= 0.04\left(\frac{2600}{2}\right)\frac{(0.493)^2}{2 \times 32.2} \\
&= 0.20\ \text{ft} \\
h_p &= 48\ \text{ft} + 0.2\ \text{ft} = 48.2\ \text{ft}
\end{aligned}
$$

Step 4. Similar calculations for the 2-MGD pump

$$
\begin{aligned}
V &= 0.493\ \text{ft/s} \times 2 = 0.986\ \text{ft/s} \\
h_f &= 0.2\ \text{ft} \times 2^2 = 0.8\ \text{ft} \\
h_p &= 48\ \text{ft} + 0.8\ \text{ft} = 48.8\ \text{ft}
\end{aligned}
$$

Step 5. For the 4 MGD pump

$$
\begin{aligned}
V &= 0.493\ \text{ft/s} \times 4 = 1.972\ \text{ft/s} \\
h_f &= 0.2\ \text{ft} \times 4^2 = 3.2\ \text{ft} \\
h_p &= 48\ \text{ft} + 3.2\ \text{ft} = 51.2\ \text{ft}
\end{aligned}
$$

Step 6. For the 3 MGD: using the 1 and 2 MGD capacity pumps

$$
\begin{aligned}
h_f &= (0.2 + 0.8)\ \text{ft} = 1.0\ \text{ft} \\
h_p &= (48.2 + 48.8)\ \text{ft} = 97\ \text{ft}
\end{aligned}
$$

Step 7. For the 5 MGD: using the 1 and 4 MGD capacity pumps

$$h_f = (0.2 + 3.2) \text{ ft} = 3.4 \text{ ft}$$
$$h_p = (48.2 + 51.2) \text{ ft} = 99.4 \text{ ft}$$

Step 8. For the 6 MGD: using the 2 and 4 MGD capacity pumps

$$h_f = (0.8 + 3.2) \text{ ft} = 4.0 \text{ ft}$$
$$h_p = (48.8 + 51.2) \text{ ft} = 100.0 \text{ ft}$$

Step 9. For the 7 MGD: using all pumps simultaneously

$$h_f = (0.2 + 0.8 + 3.2) \text{ ft} = 4.2 \text{ ft}$$
$$h_p = (48.2 + 48.8 + 51.2) \text{ ft} = 148.2 \text{ ft}$$

Hazen–Williams equation. Due to the difficulty of using the Darcy–Weisback equation for pipe flow, engineers continue to make use of an exponential equation with empirical methods for determining friction losses in Pipe flows. Among these the empirical formula of the Hazen–Williams equation is most widely used to express flow relations in pressure conduits, while the Manning equation is used for flow relations in free-flow conduits and in pipes partially full. The Hazen–Williams equation, originally developed for the British measurement system, has the following form:

$$V = 1.318\, CR^{0.63}S^{0.54} \tag{5.20a}$$

where V = average velocity of pipe flow, ft/s
C = coefficient of roughness (see Table 5.3)
R = hydraulic radius, ft
S = slope of the energy gradient line or head loss per unit length of the pipe ($S = h_f/L$)

The hydraulic radius R is defined as the water cross-sectional area A divided by the wetted perimeter P. For a circular pipe, if D is the diameter of the pipe, then R is

$$R = \frac{A}{P} = \frac{\pi D^2/4}{\pi D} = \frac{D}{4} \tag{5.21}$$

The Hazen–Williams equation was developed for water flow in large pipes ($D \geq 2$ in, 5 cm) with a moderate range of velocity ($V \leq 10$ ft/s or 3 m/s). The coefficient of roughness C values range from 140 for very smooth (new), straight pipe to 90 or 80 for old, unlined tuberculated pipe. The value of 100 is used for the average conditioned pipe. It is not a function of the flow condition (i.e. Reynolds number **R**). The major limitation of this equation is that the viscosity and temperature are considered. A nomograph (Fig. 5.2) can be used to solve the Hazen–Williams equation.

In the SI system, the Hazen–Williams equation is written as

$$V = 0.85\, C\, R^{0.63}S^{0.54} \tag{5.20b}$$

The units are R in m, S in m/1000 m, V in m/s, and D in m, and discharge Q in m^3/s (from chart).

Manning equation. Another popular empirical equation is the Manning equation, developed by the Irish engineer Robert Manning in 1889, and employed extensively for open channel flows. It is also commonly used for pipe free-flow. The Manning equation is

$$V = \frac{1.486}{n} R^{2/3} S^{1/2} \tag{5.22a}$$

where V = average velocity of flow, ft/s
n = Manning's coefficient of roughness (see Table 5.4)
R = hydraulic radius (ft) (same as in the Hazen–Williams equation)
S = slope of the hydraulic gradient, ft/ft

TABLE 5.3 Hazen–Williams Coefficient of Roughness C for Various Types of Pipe

Pipe material	C value
Brass	130–140
Brick sewer	100
Cast iron	
tar coated	130
new, unlined	130
cement lined	130–150
uncertain	60–110
Cement–asbestos	140
Concrete	130–140
Copper	130–140
Fire hose (rubber lined)	135
Galvanized iron	120
Glass	140
Lead	130–140
Plastic	140–150
Steel	
coal-tar enamel lined	145–150
corrugated	60
new unlined	140–150
riveted	110
Tin	130
Vitrified clay	110–140
Wood stave	110–120

Sources: Perry (1967), Hwang (1981), and Benefield *et al.* (1984).

This equation can be easily solved with the nomograph shown in Fig. 5.3. Figure 5.4 is used for partially filled circular pipes with varying Manning's n and depth.

The Manning equation for SI units is

$$V = \frac{1}{n} R^{2/3} S^{1/2} \tag{5.22b}$$

where V = mean velocity, m/s
n = same as above
R = hydraulic radius, m
S = slope of the hydraulic gradient, m/m

EXAMPLE 1: Assume the sewer line grade gives a sewer velocity of 0.6 m/s with a half full sewer flow. The slope of the line is 0.0081. What is the diameter of the uncoated cast iron sewer line?

Solution:

Step 1. Using the Manning equation to solve R

$$V = \frac{1}{n} R^{2/3} S^{1/2}$$

$$R^{2/3} = Vn/S^{1/2}$$

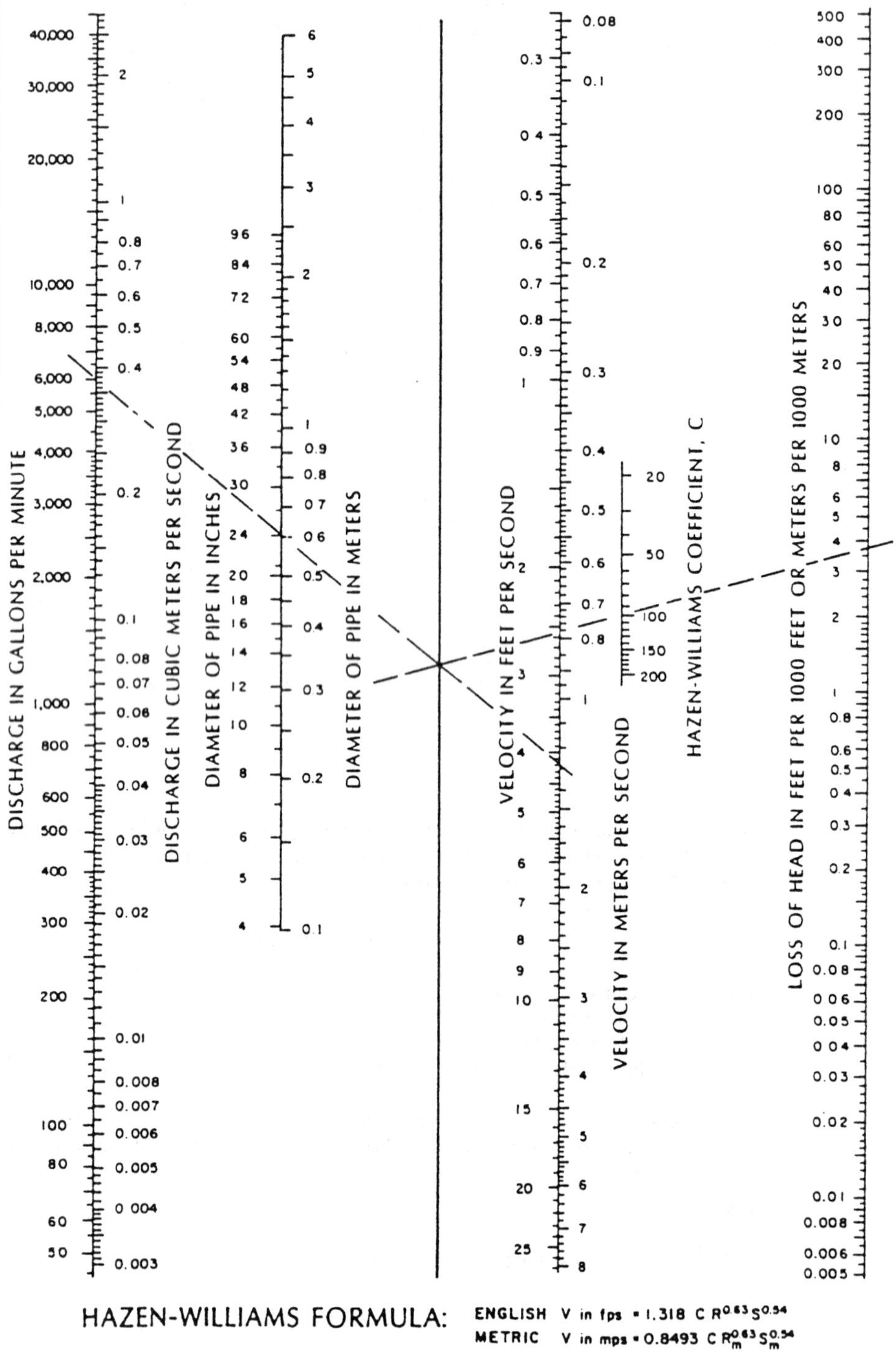

FIGURE 5.2 Nomograph for Hazen–Williams formula.

TABLE 5.4 Manning's Roughness Coefficient n for Various Types of Pipe

Type of pipe	n
Brick	
open channel	0.014–0.017
pipe with cement mortar	0.012–0.017
brass copper or glass pipe	0.009–0.013
Cast iron	
pipe uncoated	0.013
tuberculated	0.015–0.035
Cement mortar surface	0.011–0.015
Concrete	
open channel	0.013–0.022
pipe	0.010–0.015
Common clay drain tile	0.011–0.017
Fiberglass	0.013
Galvanized iron	0.012–0.017
Gravel open channel	0.014–0.033
Plastic pipe (smooth)	0.011–0.015
Rock open channel	0.035–0.045
Steel pipe	0.011
Vitrified clay	
pipes	0.011–0.015
liner plates	0.017–0.017
Wood, laminated	0.015–0.017
Wood stave	0.010–0.013
Wrought iron	0.012–0.017

Sources: Perry (1967), Hwang (1981), and ASCE & WEF (1992).

From Table 5.4, $n = 0.013$

$$R^{2/3} = V\,n/S^{1/2}$$
$$= 0.6 \times 0.013/0.0081^{1/2}$$
$$= 0.0864$$
$$R = 0.0254$$

Step 2. Calculate diameter (D) of sewer line using (Eq. (5.21))

$$D = 4\,R = 4 \times 0.0254$$
$$= 0.102\text{ (m)} = 4\text{ (in)}$$

EXAMPLE 2: Determine the energy loss over 1600 ft in a new 24-in cast iron pipe when the water temperature is 60°F and the flow rate is 6.25 cfs, using (1) the Darcy–Weisback equation, (2) the Hazen–Williams equation, and (3) the Manning equation.

Solution:

Step 1. By the Darcy–Weisback equation
From Table 5.2, $e = 0.0102$ in
(a) Determine e/D, $D = 24$ in $= 2$ ft

$$\frac{e}{D} = \frac{0.0102}{24} = 0.000425$$

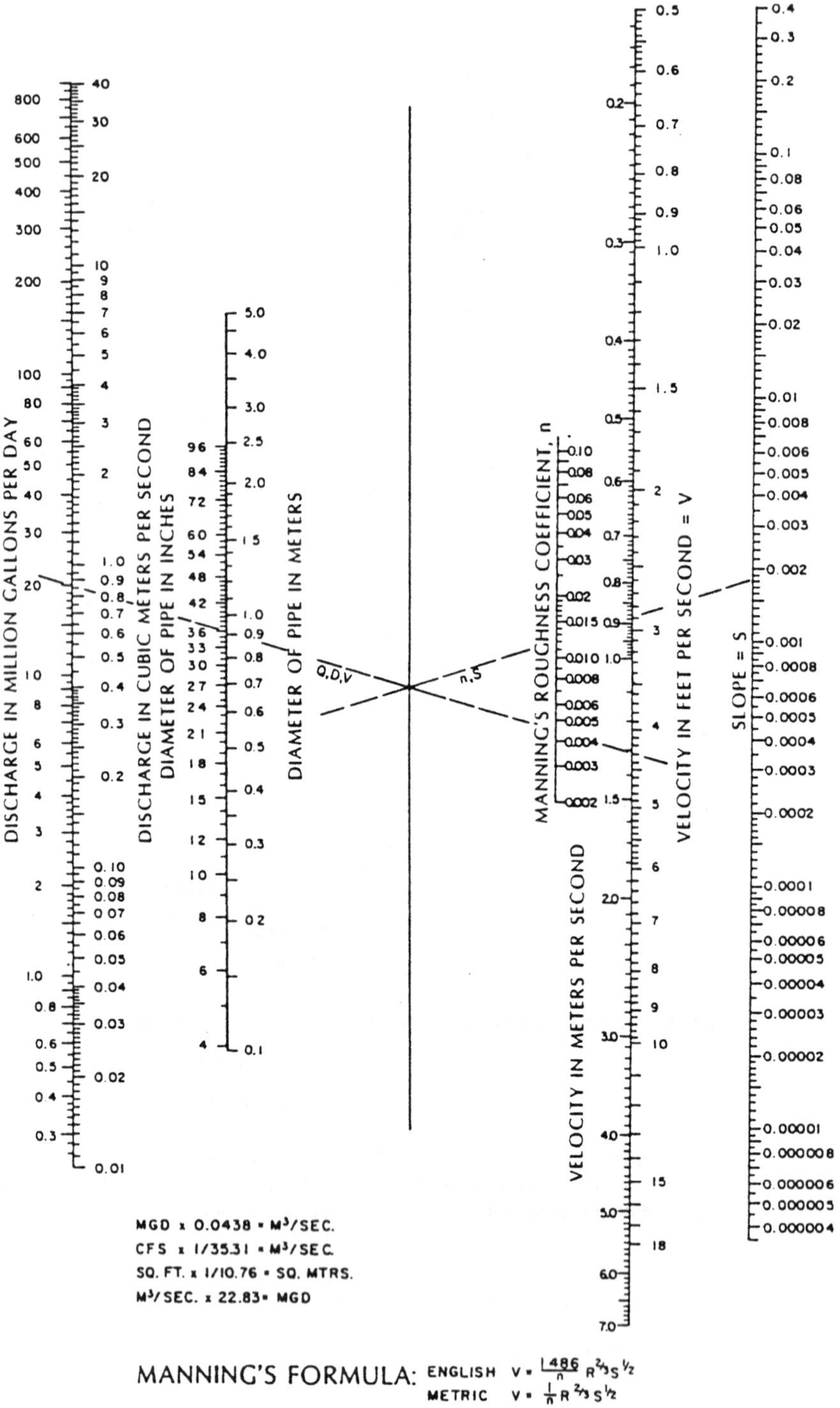

MANNING'S FORMULA: ENGLISH $V = \frac{1.486}{n} R^{2/3} S^{1/2}$
METRIC $V = \frac{1}{n} R^{2/3} S^{1/2}$

FIGURE 5.3 Nomograph for Manning formula.

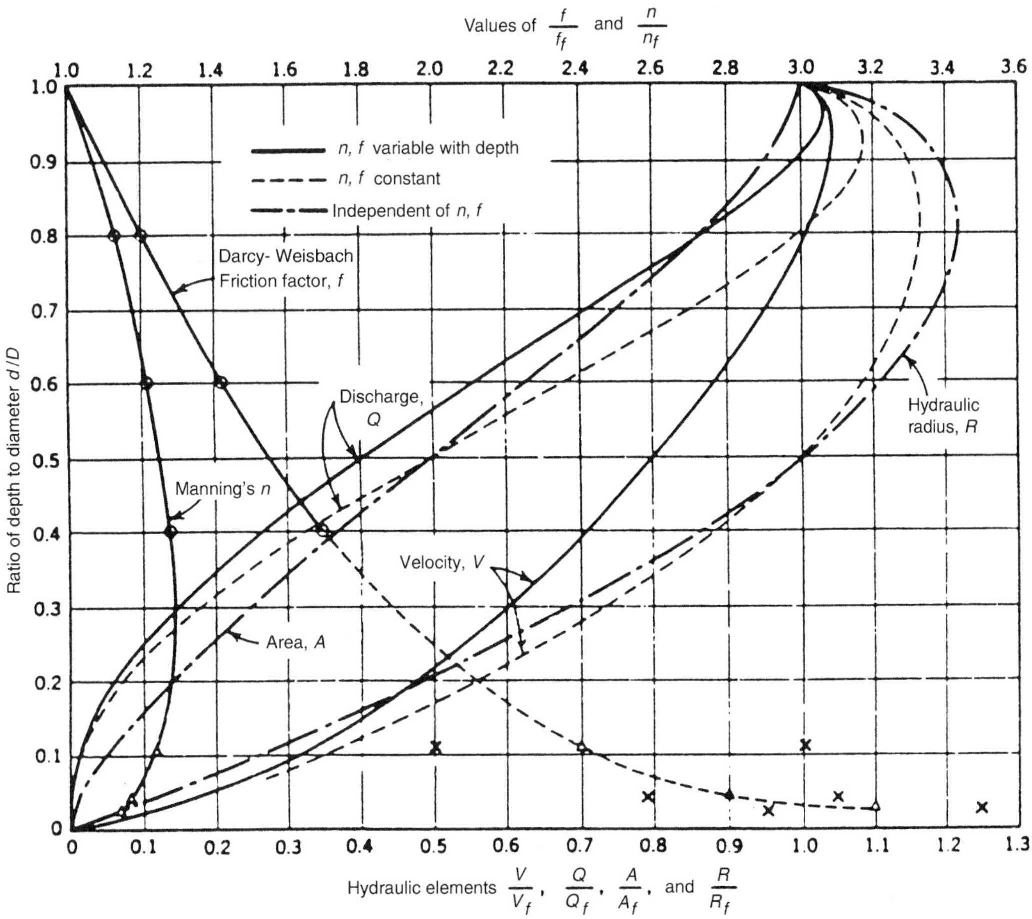

FIGURE 5.4 Hydraulic elements graph for circular sewers (*Metcalf and Eddy, Inc., Wastewater Engineering: Collection and Pumping of Wastewater, Copyright, 1990, McGraw-Hill, New York, reproduced with permission of McGraw-Hill*).

(b) Compute velocity V, $r = 1$ ft

$$V = \frac{Q}{A} = \frac{6.25 \text{ ft}^3/\text{s}}{\pi (1 \text{ ft})^2}$$
$$= 1.99 \text{ ft/s}$$

(c) Compute Reynolds number **R**
From Table 5.1b, $\nu = 1.217 \times 10^{-5}$ ft²/s for 60°F

$$\mathbf{R} = \frac{VD}{\nu} = \frac{1.99 \text{ ft/s} \times 2 \text{ ft}}{1.217 \times 10^{-5} \text{ ft}^2/\text{s}}$$
$$= 3.27 \times 10^5$$

(d) Determine f value from the Moody diagram, (Fig. 5.1)

$$f = 0.018$$

(e) Compute loss of head h_f

$$h_f = f\left(\frac{L}{D}\right)\frac{V^2}{2g} = 0.018\left(\frac{1600}{2}\right)\frac{(1.99)^2}{2 \times 32.2}$$
$$= 0.885 \text{ (ft)}$$

Step 2. By the Hazen–Williams equation

(a) Find C

From Table 5.3, for new unlined cast pipe

$$C = 130$$

(b) Solve S

$$\text{Eq. 5.21: } R = D/4 = 2/4 = 0.5$$
$$\text{Eq. 5.20a: } V = 1.318\ CR^{0.63}\ S^{0.54}$$

Rearranging,

$$S = \left[\frac{V}{1.318CR^{0.63}}\right]^{1.852}$$
$$= \left[\frac{1.99}{1.318 \times 130 \times (0.5)^{0.63}}\right]^{1.852}$$
$$= 0.000586 \text{ (ft/ft)}$$

(c) Compute h_f

$$h_f = S \times L = 0.000586 \text{ ft/ft} \times 1600 \text{ ft}$$
$$= 0.94 \text{ ft}$$

Step 3. By the Manning equation

(a) Find n

From Table 5.4, for unlined cast iron pipe

$$n = 0.013$$

(b) Determine slope of the energy line S

$$\text{From Eq. (5.21), } R = D/4 = 2/4 = 0.5$$
$$\text{From Eq. (5.22a), } V = \frac{1.486}{n} R^{2/3}\ S^{1/2}$$

Rearranging

$$S = \left(\frac{Vn}{1.486\ R^{2/3}}\right)^2 = \left[\frac{1.99 \times 0.013}{1.486 \times (0.5)^{2/3}}\right]^2$$
$$= 0.000764 \text{ (ft/ft)}$$

(c) Compute h_f

$$h_f = S \times L = 0.000764 \text{ ft/ft} \times 1600 \text{ ft}$$
$$= 1.22 \text{ ft}$$

EXAMPLE 3: The drainage area of the watershed to a storm sewer is 10,000 acres (4047 ha). Thirty-eight percent of the drainage area has a maximum runoff of 0.6 $ft^3/s \cdot$ acre (0.0069 $m^3/s \cdot$ ha), and the rest of the area has a runoff of 0.4 $ft^3/s \cdot$ acre (0.0046 $m^3/s \cdot$ ha). Determine the size of pipe needed to carry the storm flow with a grade of 0.12 percent and $n = 0.011$.

Solution:

Step 1. Calculate the total runoff Q

$$Q = 0.6 \text{ cfs/acre} \times 0.38 \times 10{,}000 \text{ acre} + 0.4 \text{ ft}^3/(\text{s} \cdot \text{acre}) \times 0.62 \times 10{,}000 \text{ acre}$$
$$= 4760 \text{ ft}^3/\text{s}$$

Step 2. Determine diameter D of the pipe by the Manning formula

$$A = \frac{\pi}{4}D^2 = 0.785\ D^2$$
$$R = D/4$$
$$Q = AV = A\frac{1.486}{n}R^{2/3}S^{1/2}$$
$$4760 = 0.785\ D^2 \times \frac{1.486}{0.011}\left(\frac{D}{4}\right)^{2/3}(0.0012)^{1/2}$$
$$3265 = D^{8/3}$$
$$D = 20.78 \text{ (ft)}$$

Minor head losses due to hydraulic devices such as sharp-crested orifices, weirs, valves, bend, construction, enlargement, discharge, branching, etc. have been extensively covered in textbooks and handbooks elsewhere. Basically they can be calculated from empirical formulas.

2.3 Pipeline Systems

There are three major types of compound piping system: pipes connected in series, in parallel, and branching. Pipes in parallel occur in a pipe network when two or more paths are available for water flowing between two points. Pipe branching occurs when water can flow to or from a junction of three or more pipes from independent outlets or sources. On some occasions, two or more sizes of pipes may connect in series between two reservoirs. Figure 5.5 illustrates all three cases.

The flow through a pipeline consisting of three or more pipes connected in various ways can be analyzed on the basis of head loss concept with two basic conditions. First, at a junction, water flows in and out should be the same. Second, all pipes meeting at the junction have the same water pressure at the junction. The general procedures are to apply the Bernoulli energy equation to determine the energy line and hydraulic gradient mainly from friction head loss equations. In summary, using Fig. 5.5:

1. For pipelines in series
Energy: $H = h_{f1} + h_{f2}$
Continuity: $Q = Q_1 = Q_2$
Method: (a) Assume h_f
(b) Calculate Q_1 and Q_2
(c) If $Q_1 = Q_2$, the solution is correct
(d) If $Q_1 \neq Q_2$, repeat step (a)

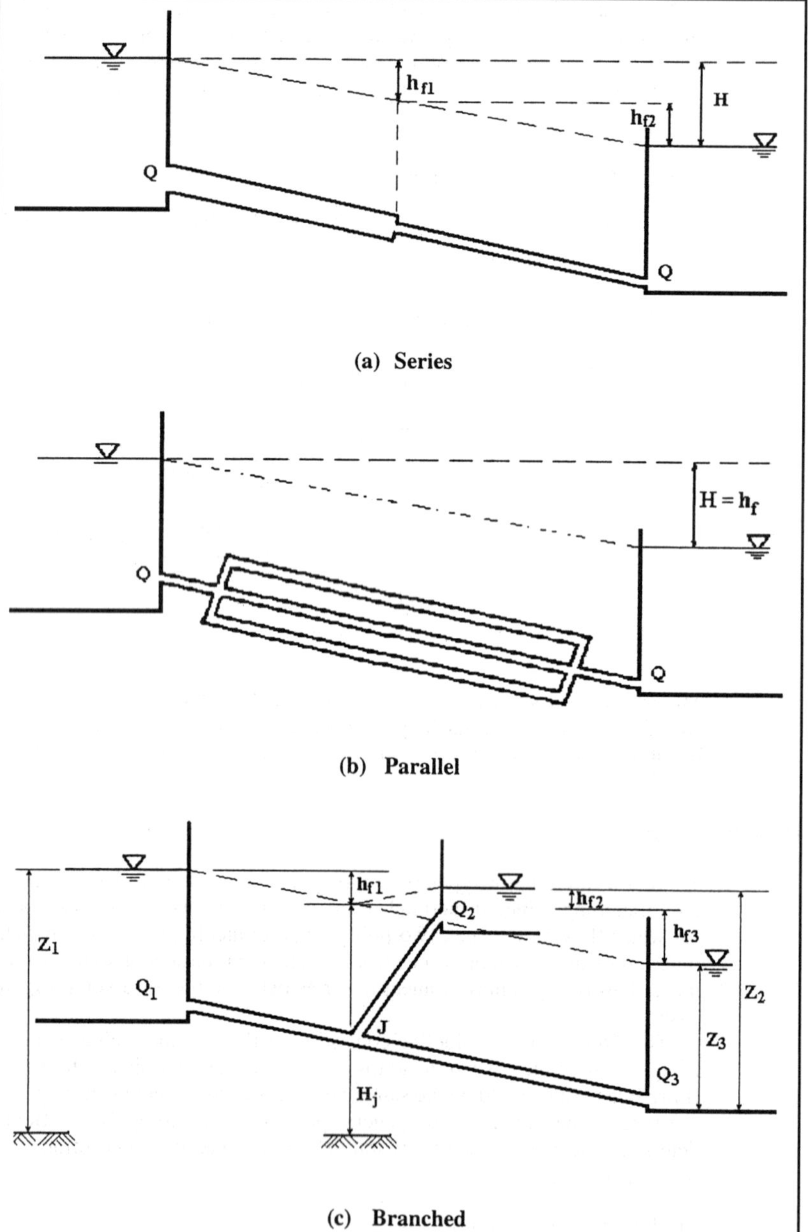

FIGURE 5.5 Pipeline system for two reservoirs.

2. For pipelines in parallel

Energy: $H = h_{f1} = h_{f2}$

Continuity: $Q = Q_1 + Q_2$

Method: Parallel problem can be solved directly for Q_1 and Q_2

3. For branched pipelines

Energy: $h_{f1} = z_1 - H_J$

$h_{f2} = z_2 - H_J$

$h_{f3} = H_J - z_3$

where H_J is the energy head at junction J.

Continuity: $Q_3 = Q_1 + Q_2$

Method: (a) Assume H_J

(b) Calculate Q_1, Q_2, and Q_3

(c) If $Q_1 + Q_2 = Q_3$, the solution is correct

(d) If $Q_1 + Q_2 \neq Q_3$, repeat step (a)

Pipelines in series

EXAMPLE: As shown in Fig. 5.5a, two concrete pipes are connected in series between two reservoirs A and B. The diameters of the upstream and downstream pipes are 0.6 m (D_1 = 2 ft) and 0.45 m (D_2 = 1.5 ft); their lengths are 300 m (1000 ft) and 150 m (500 ft), respectively. The flow rate of 15°C water from reservoir A to reservoir B is 0.4 m³/s. Also given are coefficient of entrance $K_e = 0.5$, coefficient of contraction $K_c = 0.13$, and coefficient of discharge $K_d = 1.0$. Determine the elevation of the water surface of reservoir B when the elevation of the water surface of reservoir A is 100 m.

Solution:

Step 1. Find the relative roughness e/D from Fig. 5.1 for a circular concrete pipe, assume $e = 0.003$ ft $= 0.0009$ m

$$e/D_1 = 0.0009/0.6 = 0.0015$$

$$e/D_2 = 0.0009/0.45 = 0.0020$$

Step 2. Determine velocities (V) and Reynolds number (**R**)

$$V_1 = \frac{Q}{A_1} = \frac{0.4\ \text{m}^3/\text{s}}{\frac{\pi}{4}(0.6\ \text{m})^2} = 1.41\ \text{m/s}$$

$$V_2 = \frac{Q}{A_2} = \frac{0.4}{\frac{\pi}{4}(0.45)^2} = 2.52\ \text{m/s}$$

From Table 5.1a, $\nu = 1.131 \times 10^{-6}$ m²/s for $T = 15°$C

$$\mathbf{R}_1 = \frac{V_1 D_1}{\nu} = \frac{1.41\ \text{m/s} \times 0.6\ \text{m}}{1.131 \times 10^{-6}\,\text{m}^2/\text{s}} = 7.48 \times 10^5$$

$$\mathbf{R}_2 = \frac{V_2 D_2}{\nu} = \frac{2.51 \times 0.45}{1.131 \times 10^{-6}} = 9.99 \times 10^5$$

Step 3. Find f value from Moody chart (Fig. 5.1) corresponding to **R** and e/D values

$$f_1 = 0.022$$
$$f_2 = 0.024$$

Step 4. Determine total energy loss for water flow from A to B

$$h_e = K_e\frac{V_1^2}{2g} = 0.5\frac{V_1^2}{2g}$$

$$h_{f1} = f_1\frac{L_1}{D_1}\frac{V_1^2}{2g} = 0.022 \times \frac{300}{0.6}\frac{V_1^2}{2g} = 11\frac{V_1^2}{2g}$$

$$h_c = K_c\frac{V_2^2}{2g} = 0.13\frac{V_2^2}{2g}$$

$$h_{f2} = f_2\frac{L_2}{D_2}\frac{V_2^2}{2g} = 0.024\frac{150}{0.45}\frac{V_2^2}{2g} = 8\frac{V_2^2}{2g}$$

$$h_d = K_d\frac{V_2^2}{2g} = 1 \times \frac{V_2^2}{2g}$$

Step 5. Calculate total energy head H

$$H = h_e + h_{f1} + h_c + h_{f2} + h_d$$
$$= (0.5 + 11)\frac{V_1^2}{2g} + (0.13 + 8 + 1)\frac{V_2^2}{2g}$$
$$= 11.5 \times \frac{(1.41)^2}{2 \times 9.81} + 9.13\frac{(2.52)^2}{2 \times 9.81}$$
$$= 4.12 \text{ (m)}$$

Step 6. Calculate the elevation of the surface of reservoir B

$$\text{Ele.} = 100 \text{ m} - H = 100 \text{ m} - 4.12 \text{ m}$$
$$= 95.88 \text{ (m)}$$

Pipelines in parallel

EXAMPLE: Circular pipelines 1, 2, and 3, each 1000 ft (300 m) long and of 6, 8, and 12 in (0.3 m) diameter, respectively, carry water from reservoir A and join pipeline 4 to reservoir B. Pipeline 4 is 2 ft (0.6 m) in diameter and 500 ft (150 m) long. Determine the percentages of flow passing pipelines 1, 2, and 3 using the Hazen–Williams equation. The C value for pipeline 4 is 110 and, for the other three, 100.

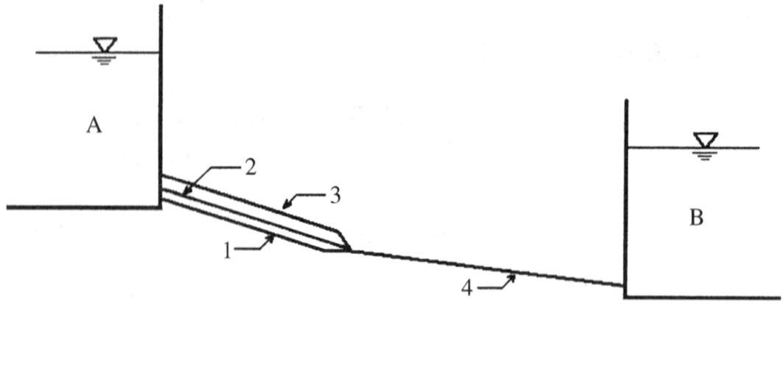

Solution:

Step 1. Calculate velocity in pipelines 1, 2, and 3.

Since no elevation and flow are given, any slope of the energy line is OK. Assuming the head loss in the three smaller pipelines is 10 ft per 1000 ft of pipeline:

$$S = 10/1000 = 0.01$$

For a circular pipe, the hydraulic radius is

$$R = \frac{D}{4}$$

then, from the Hazen–Williams equation, Eq. (5.20a)

$$\begin{aligned} V_1 &= 1.318\, C\, R^{0.63}\, S^{0.54} = 1.318 \times 100 \times \left(\frac{0.5}{4}\right)^{0.63} (0.01)^{0.54} \\ &= 10.96 \times \left(\frac{0.5}{4}\right)^{0.63} \\ &= 2.96 \text{ (ft/s)} \\ V_2 &= 10.96\left(\frac{0.667}{4}\right)^{0.63} = 3.54 \text{ (ft/s)} \\ V_3 &= 10.96(\tfrac{1}{4})^{0.63} = 4.58 \text{ (ft/s)} \end{aligned}$$

Step 2. Determine the total flow Q_4

$$\begin{aligned} Q_1 &= A_1V_1 = \frac{\pi}{4}D_1^2V_1 = 0.785(0.5)^2 \times 2.96 \\ &= 0.58 \text{ (ft}^3\text{/s)} \\ Q_2 &= A_2V_2 = 0.785(0.667)^2 \times 3.54 \\ &= 1.24 \text{ (ft}^3\text{/s)} \\ Q_3 &= A_3V_3 = 0.785(1)^2 \times 4.58 \\ &= 3.60 \text{ (ft}^3\text{/s)} \\ Q_4 &= Q_1 + Q_2 + Q_3 = 0.58 + 1.24 + 3.60 \\ &= 5.42 \text{ (ft}^3\text{/s)} \end{aligned}$$

Step 3. Determine percentage of flow in pipes 1, 2, and 3

For

$$\begin{aligned} \text{pipe 1} &= (Q_1/Q_4) \times 100\% = 0.58 \times 100/5.42 \\ &= 10.7\% \\ \text{pipe 2} &= 1.24 \times 100\%/5.42 = 22.9\% \\ \text{pipe 3} &= 3.60 \times 100\%/5.42 = 66.4\% \end{aligned}$$

2.4 Distribution Networks

Most waterline or sewer distribution networks are complexes of looping and branching pipelines. The solution methods described for the analysis of pipelines in series, parallel, and branched systems are not suitable for the more complex cases of networks. A trial and error procedure should be used. The most widely used is the loop method originally proposed by Hardy Cross in 1936. Another method, the nodal method, was proposed by Cornish in 1939 (Chadwick and Morfett, 1986).

For analysis flows in a network of pipes, the following conditions must be satisfied:

1. The algebraic sum of the pressure drops around each circuit must be zero.
2. Continuity must be satisfied at all junctions; i.e., inflow equals outflow at each junction.
3. Energy loss must be the same for all paths of water.

Using the continuity equation at a node, this gives

$$\sum_{i=1}^{m} q_i = 0 \tag{5.23}$$

where m is the number of pipes joined at the node. The sign conventionally used for flow into a joint is positive, and outflow is negative.

Applying the energy equation to a loop, we get

$$\sum_{i=1}^{n} h_{fi} = 0 \tag{5.24}$$

where n is the number of pipes in a loop. The sign for flow (q_i) and head loss (h_{fi}) is conventionally positive when clockwise.

Since friction loss is a function of flow

$$h_{fi} = \phi(q_1) \tag{5.25}$$

Equations (5.23) and (5.24) will generate a set of simultaneous nonlinear equations. An iterative solution is needed.

Method of equivalent pipes. In a complex pipe system, small loops within the system are replaced by single hydraulically equivalent pipes. This method can also be used for determination of diameter and length of a replacement pipe; i.e., one that will produce the same head loss as the old one.

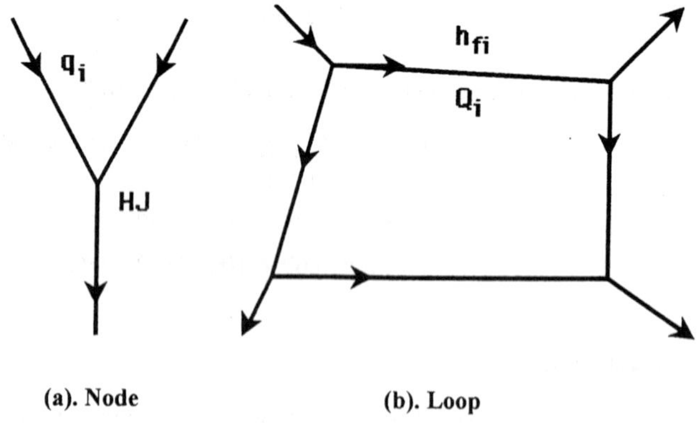

FIGURE 5.6 Two types of pipe network.

EXAMPLE 1: Assume $Q_{AB} = 0.042$ m³/s (1.5 ft³/s), $Q_{BC} = 0.028$ m³/s (1.0 ft³/s), and a Hazen–Williams coefficient $C = 100$. Determine an equivalent pipe for the network of the following figure.

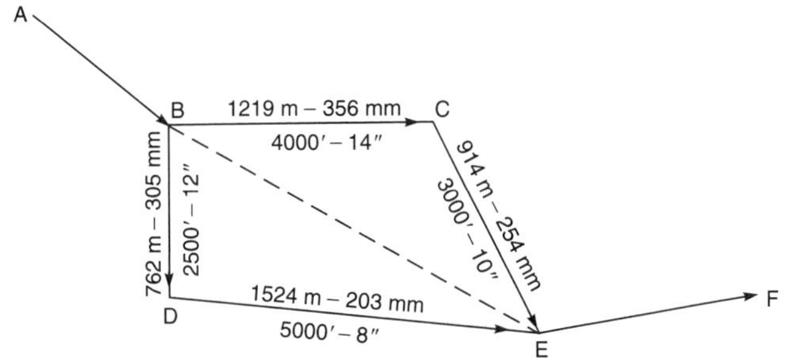

Solution:

Step 1. For line BCE, $Q = 1$ ft³/s. From the Hazen–Williams chart (Fig. 5.2) with $C = 100$, to obtain loss of head in ft per 1000 ft, then determine the total head loss for the pipe proportionally

(a) Pipe BC, 4000 ft, 14 in ϕ

$$h_{BC} = \frac{0.44}{1000} \times 4000 \text{ ft} = 1.76 \text{ ft}$$

(b) Pipe CE, 3000 ft, 10 in ϕ

$$h_{CE} = \frac{2.40}{1000} \times 3000 \text{ ft} = 7.20 \text{ ft}$$

(c) Total

$$h_{BCE} = 1.76 \text{ ft} + 7.20 \text{ ft} = 8.96 \text{ ft}$$

(d) Equivalent length of 10-in pipe: $1000 \text{ ft} \times \dfrac{8.96}{2.4} = 3733 \text{ ft}$

Step 2. For line BDE, $Q = 1.5 - 1.0 = 0.5$ ft³/s or 0.014 m³/s

(a) Pipe BD, 2500 ft, 12 in ϕ

$$h_{BD} = \frac{0.29}{1000} \times 2500 \text{ ft} = 0.725 \text{ ft}$$

(b) Pipe DE, 5000 ft, 8 in ϕ

$$h_{DE} = \frac{1.80}{1000} \times 5000 \text{ ft} = 9.0 \text{ ft}$$

(c) Total

$$h_{BDE} = h_{BD} + h_{DE} = 9.725 \text{ ft}$$

(d) Equivalent length of 8 in pipe: $1000 \text{ ft} \times \dfrac{9.725}{1.80} = 5403 \text{ ft}$

Step 3. Determine equivalent line BE from results of Steps 1 and 2, assuming $h_{BE} = 8.96$ ft

(a) Line BCE, 3733 ft, 10 in ϕ

$$S = \frac{8.96}{3733} = 2.4/1000 = 24\%$$

$$Q_{BCD} = 1.0 \text{ ft}^3\text{/s}$$

(b) Line BDE, 5403 ft, 8 in ϕ

$$S = \frac{8.96}{5403} = 1.66/1000$$

From Fig. 5.2 $$Q_{\text{BDE}} = 0.475 \text{ ft}^3/\text{s} = 0.0134 \text{ m}^3/\text{s}$$

(c) Total

$$Q_{\text{BE}} = 1 \text{ ft}^3/\text{s} + 0.475 \text{ ft}^3/\text{s} = 1.475 \text{ ft}^3/\text{s} = 0.0417 \text{ m}^3/\text{s}$$

(d) Using $Q = 1.475$ ft³/s, select 12-in pipe to find equivalent length
From chart

$$S = 1.95$$

Since assumed total head loss $H_{\text{BE}} = 8.96$

$$L = \frac{8.96}{1.95} \times 1000 \approx 4600 \text{ ft}$$

Answer: Equivalent length of 4600 ft of 12-in pipe.

EXAMPLE 2: The sewer pipeline system below shows pipe diameter sizes (ϕ), grades, pipe section numbers, and flow direction. Assume there is no surcharge and full flow in each of sections 1, 2, 3, and 4. All pipes are fiberglass with $n = 0.013$. A, B, C, and D represent inspection holes. Determine

(a) the flow rate and minimum commercial pipe size for section AB;
(b) the discharge, sewage depth, and velocity in section BC; and
(c) the slope required to maintain full flow in section CD.

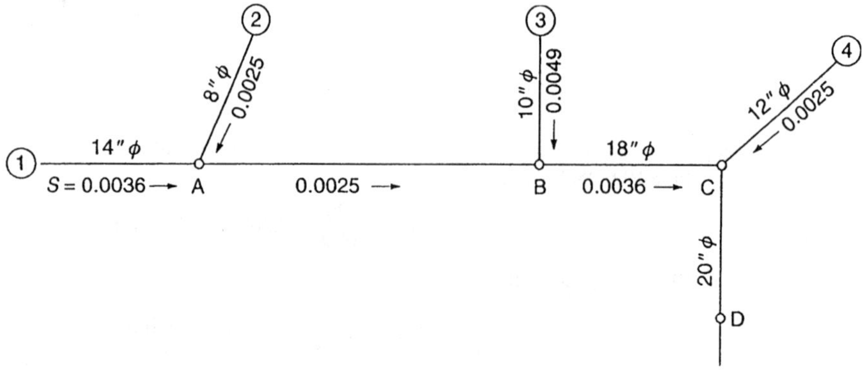

Solution:

Step 1. Compute cross-sectional areas and hydraulic radius

Section	Diameter, D (ft)	A (ft)	$R = D/4$(ft)
1	1.167	1.068	0.292
2	0.667	0.349	0.167
3	0.833	0.545	0.208
4	1.0	0.785	0.250
BC	1.5	1.766	0.375
CD	1.667	2.180	0.417

Step 2. Compute flows with Manning formula

$$Q = A\frac{1.486}{n}R^{2/3}S^{1/2}$$

$$Q_1 = 1.068 \times \frac{1.486}{0.013}(0.292)^{2/3}(0.0036)^{1/2}$$
$$= 3.22\ (\text{ft}^3/\text{s})$$

$$Q_2 = 0.349 \times 114.3\ (0.167)^{2/3}\ (0.0025)^{1/2}$$
$$= 0.60\ (\text{ft}^3/\text{s})$$

$$Q_{AB} = Q_1 + Q_2 = 3.22 + 0.60$$
$$= 3.82\ (\text{ft}^3/\text{s})$$

Step 3. Compute D_{AB}

$$Q_{AB} = 3.82 = 0.785\ D^2 \times 114.3\left(\frac{D}{4}\right)^{2/3} \times (0.0025)^{1/2}$$

$$D^{8/3} = 2.146$$
$$D = 1.33\ (\text{ft})$$
$$= 15.9\ \text{in}$$

Use 18-in commercially available pipe for answer (a).

Step 4. Determine Q_3, Q_{BC}, Q_{BC}/Q_f

$$Q_3 = 0.545 \times 114.3 \times (0.208)^{2/3}(0.0049)^{1/2}$$
$$= 1.53\ (\text{ft}^3/\text{s})$$

$$Q_{BC} = Q_{AB} + Q_3 = 3.82\ \text{ft}^3/\text{s} + 1.53\ \text{ft}^3/\text{s}$$
$$= 5.35\ \text{ft}^3/\text{s}$$

At section BC: velocity for full flow

$$V_f = 114.3 \times (0.375)^{2/3}\ (0.0036)^{1/2}$$
$$= 3.57\ (\text{ft/s})$$
$$Q_f = AV_f = 1.766 \times 3.57$$
$$= 6.30\ (\text{ft}^3/\text{s})$$

Percentage of actual flow

$$\frac{Q_{BC}}{Q_f} = \frac{5.35\ \text{ft}^3/\text{s}}{6.30\ \text{ft}^3/\text{s}} = 0.85$$

Step 5. Determine sewage depth d and V for section BC from Fig. 5.4 (page 1.285)
We obtain $d/D = 0.70$
Then

$$d = 0.70 \times 18\ \text{in} = 12.6\ \text{in}$$
$$= 1.05\ \text{ft}$$

and

$$V/V_f = 1.13$$

Then

$$V = 1.13 \times V_f = 1.13 \times 3.57 \text{ ft/s}$$
$$= 4.03 \text{ ft/s}$$

Step 6. Determine slope S of section CD

$$Q_4 = 0.785 \times 114.3 \times (0.25)^{2/3}(0.0025)^{1/2}$$
$$= 1.78 \text{ (ft}^3\text{/s)}$$
$$Q_{CD} = Q_4 + Q_{BC} = 1.78 \text{ ft}^3\text{/s} + 5.35 \text{ ft}^3\text{/s}$$
$$= 7.13 \text{ ft}^3\text{/s}$$

Section CD if flowing full:

$$7.13 = 2.18 \times 114.3 \, (0.417)^{2/3} S^{1/2}$$
$$S^{1/2} = 0.05127$$
$$S = 0.00263$$

Hardy–Cross method. The Hardy–Cross method (1936) of network analysis is a loop method which eliminates the head losses from Eqs. (5.24) and (5.25) and generates a set of discharge equations. The basics of the method are:

1. Assume a value for Q_i for each pipe to satisfy $\Sigma Q_i = 0$
2. Compute friction losses h_{fi} from Q_i; find S from Hazen–Williams equation
3. If $\Sigma h_{fi} = 0$, the solution is correct
4. If $\Sigma_{fi} \neq 0$, apply a correction factor ΔQ to all Q_i, then repeat step 1.
5. A reasonable value of ΔQ is given by Chadwick and Morfett (1986)

$$\Delta Q = \frac{\Sigma h_{fi}}{2\Sigma h_{fi}/Q_i} \tag{5.26}$$

This trial and error procedure solved by digital computer program is available in many textbooks on hydraulics (Hwang 1981, Streeter and Wylie 1975).

The nodal method. The basic concept of the nodal method consists of the elimination of discharges from Eqs. (5.23) and (5.25) to generate a set of head loss equations. This method may be used for loops or branches when the external heads are known and the heads within the networks are needed. The procedure of the nodal method is as follows:

1. Assume values of the head loss H_j at each junction.
2. Compute Q_i from H_j.
3. If $\Sigma Q_i = 0$, the solution is correct.
4. If $\Sigma Q_i \neq 0$, adjust a correction factor ΔH to H_j, then repeat step 2.
5. The head correction factor is

$$\Delta H = \frac{2\Sigma Q_i}{\Sigma Q_i/h_{fi}} \tag{5.27}$$

2.5 Sludge Flow

To estimate every loss in a pipe carrying sludge, the Hazen–Williams equation with a modified C value and a graphic method based on field experience are commonly used. The modified C values for various total solids contents are as follows (Brisbin 1957)

Total solids (%)	0	2	4	6	8.5	10
C	100	81	61	45	32	25

2.6 Dividing-Flow Manifolds and Multiport Diffusers

The related subjects of dividing-flow manifolds and multiport diffusers are discussed in detail by Benefield *et al.* (1984). They present basic theories and excellent design examples for these two subjects. In addition, there is a design example of hydraulic analysis for all unit processes of a wastewater treatment plant.

3 PUMPS

3.1 Types of Pump

The centrifugal pump and the displacement pump are most commonly used for water and wastewater works. Centrifugal pumps have a rotating impeller which imparts energy to the water. Displacement pumps are often of the reciprocating type in which a piston draws water or slurry into a closed chamber and then expels it under pressure. Reciprocating pumps are widely used to transport sludge in wastewater treatment plants. Air-lift pumps, jet pumps, and hydraulic rams are also used in special applications.

3.2 Pump Performance

The Bernoulli equation may be applied to determine the total dynamic head on the pump. The energy equation expressing the head between the suction (s) and discharge (d) nozzles of the pumps is as follows:

$$H = \frac{P_\mathrm{d}}{\gamma} + \frac{V_\mathrm{d}^2}{2g} + z_\mathrm{d} - \left(\frac{P_\mathrm{s}}{\gamma} + \frac{V_\mathrm{s}^2}{2g} + z_\mathrm{s}\right)$$

where H = total dynamic head, m or ft
P_d, P_s = gage pressure at discharge and suction, respectively, N/m^2 or lb/in^2
γ = specific weight of water, N/m^3 or lb/ft^3
V_d, V_s = velocity in discharge and suction nozzles, respectively, m/s or ft/s
g = gravitational acceleration, 9.81 m/s^2 or 32.2 ft/s^2
z_d, z_s = elevation of discharge and suction gage above the datum, m or ft

The power P_w required to pump water is a function of the flow Q and the total head H and can be written as

$$P_\mathrm{w} = jQH \tag{5.28}$$

where P_w = water power, kW (m · m^3/min) or hp (ft · gal/min)
j = constant, at 20°C
= 0.163 (for SI units)

$= 2.525 \times 10^{-4}$ (for US customary units)
Q = discharge, m³/min or gpm
H = total head, m or ft

EXAMPLE 1: Calculate the water power for a pump system to deliver 3.785 m³/min (1000 gpm) against a total system head of 30.48 m (100 ft) at a temperature of 20°C.

Solution:

$$
\begin{aligned}
P_w &= jQH \\
&= 0.163 \times 3.785 \text{ m}^3/\text{min} \times 30.48 \text{ m} \\
&= 18.8 \text{ kW}
\end{aligned}
$$

or $= 25.3$ hp

The power output of a pump is the work done to lift the water to a higher elevation per unit time. The theoretical power output or horsepower (hp) required is a function of a known discharge and total pump lift. For US customery units, it may be written as

$$P_o \text{ (in hp)} = \frac{Q\gamma H}{550} \tag{5.28a}$$

For SI units:

$$P_o \text{ (in N} \cdot \text{m/s or watt)} = Q\gamma H \tag{5.28b}$$

where Q = discharge, ft³/s (cfs) or m³/s
γ = specific weight of water or liquid, lb/ft³ or N/m³
H = total lift, ft or m
550 = conversion factor from ft · lb/s to hp

EXAMPLE 2: Determine how many watts equal one horsepower.

$$
\begin{aligned}
1 \text{ hp} &= 550 \text{ ft} \cdot \text{lb/s} \\
&= 550 \text{ ft} \times 0.305 \text{ m/ft} \cdot 1 \text{ lb} \times 4.448 \text{ N/(lb} \cdot \text{s)} \\
&= 746 \text{ N} \cdot \text{m/s (joule/s = watt)} \\
&= 746 \text{ watts (W) or } 0.746 \text{ kilowatts (kW)}
\end{aligned}
$$

Conversely, 1 kW = 1.341 hp

The actual horsepower needed is determined by dividing the theoretical horsepower by the efficiency of the pump and driving unit. The efficiencies for centrifugal pumps normally range from 50 to 85 percent. The efficiency increases with the size and capacity of the pump.

The design of a pump should consider total dynamic head which includes differences in elevations. The horsepower which should be delivered to a pump is determined by dividing Eq. (5.28b) by the pump efficiency, as follows:

$$\text{hp} = \frac{\gamma QH}{550 \times \text{efficiency}} \tag{5.28c}$$

The efficiency of a pump (e_p) is defined as the ratio of the power output ($P_o = \gamma QH/550$) to the input power of the pump ($P_i = \omega\tau$). It can be written as:

for U.S. customary units

$$e_p = \frac{P_o}{P_i} = \frac{\gamma QH}{h_p \times 550} \tag{5.28d}$$

for SI units

$$e_p = \frac{\gamma QH}{P_i} = \frac{\gamma QH}{\omega\tau} \tag{5.28e}$$

where e_p = efficiency of the pump, %
γ = specific weight of water, kN/m^3 or lb/ft^3
Q = capacity, m^3/s or ft^3/s
h_p = brake horsepower
550 = conversion factor for horsepower to ft · lb/s
ω = angular velocity of the turbo-hydraulic pump
τ = torque applied to the pump by a motor

The efficiency of the motor (e_m) is defined as the ratio of the power applied to the pump by the motor (P_i) to the power input to the motor (P_m), i.e.

$$e_m = \frac{P_i}{P_m} \tag{5.29}$$

The overall efficiency of a pump system (e) is combined as

$$e = e_p e_m = \frac{P_o}{P_i} \times \frac{P_i}{P_m} = \frac{P_o}{P_m} \tag{5.30}$$

EXAMPLE 3: A water treatment plant pumps its raw water from a reservoir next to the plant. The intake is 12 ft below the lake water surface at elevation 588 ft. The lake water is pumped to the plant influent at elevation 611 ft. Assume the suction head loss for the pump is 10 ft and the loss of head in the discharge line is 7 ft. The overall pump effiency is 72 percent. The plant serves 44,000 persons. The average water consumption is 200 gallons per capita per day (gpcpd). Compute the horsepower output of the motor.

Solution:

Step 1. Determine discharge Q

$$\begin{aligned} Q &= 44{,}000 \times 200 \text{ gpcpd} = 8.8 \text{ Mgal/day (MGD)} \\ &= 8.8 \text{ Mgal/day} \times 1.547 \text{ (ft}^3\text{/s)/(Mgal/day)} \\ &= 13.6 \text{ ft}^3\text{/s} \end{aligned}$$

Step 2. Calculate effective head H

$$H = (611 - 588) + 10 + 7 = 40 \text{ (ft)}$$

Step 3. Compute overall horsepower output
Using Eq. (5.28c),

$$\begin{aligned} \text{hp} &= \frac{Q\gamma H}{550 e_p} = \frac{13.6 \text{ ft}^3\text{/s} \times 62.4 \text{ lb/ft}^3 \times 40 \text{ ft}}{550 \text{ ft} \cdot \text{lb/hp} \times 0.72} \\ &= 85.7 \text{ hp} \end{aligned}$$

EXAMPLE 4: A water pump discharges at a rate of 0.438 m^3/s (10 Mgal/day). The diameters of suction and discharge nozzles are 35 cm (14 in) and 30 cm (12 in), respectively. The reading of the suction gage located 0.3 m (1 ft) above the pump centerline is 11 kN/m^2 (1.6 lb/in^2). The reading of the discharge gage located at the pump centerline is 117 kN/m^2 (17.0 lb/in^2). Assume the pump efficiency is 80 percent and the motor efficiency is 93 percent; water temperature is 13°C. Find (a) the power input needed by the pump and (b) the power input to the motor.

Solution:

Step 1. Write the energy equation

$$H = \left(\frac{P_d}{\gamma} + \frac{V_d^2}{2g} + z_d\right) - \left(\frac{P_s}{\gamma} + \frac{V_s^2}{2g} + z_s\right)$$

Step 2. Calculate each term in the above equation
At $T = 13°\text{C}$, $\gamma = 9800\ \text{N/m}^3$, refer to table 5.1a

$$\frac{P_d}{\gamma} = \frac{117{,}000\ \text{N/m}^2}{9800\ \text{N/m}^3} = 11.94\ \text{m}$$

$$V_d = \frac{Q_d}{A_d} = \frac{0.438\ \text{m}^3/\text{s}}{(\pi/4)(0.30\ \text{m})^2} = 6.20\ \text{m/s}$$

$$\frac{V_d^2}{2g} = \frac{(6.20\ \text{m/s})^2}{2(9.81\ \text{m/s}^2)} = 1.96\ \text{m}$$

Let $z_d = 0$ be the datum at the pump centerline

$$\frac{P_s}{\gamma} = \frac{11{,}000\ \text{N/m}^2}{9800\ \text{N/m}^3} = 1.12\ \text{m}$$

$$V_s = \frac{0.438\ \text{m}^3/\text{s}}{(\pi/4)(0.35\ \text{m})^2} = 4.55\ \text{m/s}$$

$$\frac{V_s^2}{2g} = \frac{(4.55\ \text{m/s})^2}{2(9.81\ \text{m/s}^2)} = 1.06\ \text{m}$$

$$z_s = +0.30\ \text{m}$$

Step 3. Calculate the total head H

$$H = [11.94 + 1.96 + 0 - (1.12 + 1.06 + 0.30)]\ \text{m}$$
$$= 11.42\ \text{m} \text{ or } 34.47\ \text{ft}$$

Step 4. Compute power input P_i by Eq. (5.28e) for question (a)

$$P_i = \frac{P_q}{e_p} = \frac{\gamma Q H}{e_p} = \frac{(9.8\ \text{kN})(0.438\ \text{m}^3/\text{s})(11.42\ \text{m})}{0.80}$$
$$= 61.27\ \text{kW}$$
$$= 61.27\ \text{kW} \times \frac{1.341\ \text{hp}}{1\ \text{kW}}$$
$$= 82.2\ \text{hp}$$

Step 5. Compute power input to the motor (P_m) for question (b)

$$P_m = \frac{P_i}{e_m} = \frac{61.27\ \text{kW}}{0.93}$$
$$= 65.88\ \text{kW}$$

or

$$= 88.4\ \text{hp}$$

3.3 Cost of Pumping

The cost of pumping through a pipeline is a function of head loss, flow rate, power cost, and the total efficiency of the pump system. It can be expressed as (Ductile Iron Pipe Research Association, 1997)

$$\text{CP} = 1.65\, HQ\$/E \tag{5.31}$$

where CP = pumping cost, \$/(yr · 1000 ft) (based on 24 h/day operation)
H = head loss, ft/1000 ft
Q = flow, gal/min
\$ = unit cost of electricity, \$/kWh
E = total efficiency of pump system, %

Velocity is related to flow by the following equation

$$V = \frac{Q}{2.448d^2} \tag{5.32a}$$

where V = velocity, ft/s
Q = flow, gal/min
d = actual inside diameter, in

Head loss is determined by the Hazen–Williams formula

$$H = 1000\left(\frac{V}{0.115\, C\, d^{0.63}}\right)^{1.852} \tag{5.32b}$$

where C = a coefficient, and other symbols are as above.

EXAMPLE: Water is pumped at a rate of 6300 gal/min (0.40 m³/s) through a 24-in pipeline of 10,000 ft (3048 m) length. The actual inside diameters for ductile iron pipe and PVC pipe are respectively 24.95 and 22.76 in. Assume the unit power cost is \$0.058/kWh; the total efficiency of pump system is 75%; the pump is operated 24 hours per day. Estimate: (a) the cost of pumping for each kind of pipeline, (b) the present value of the difference of pumping cost, assuming 50-year design pipe life (n), 6.6 percent annual rate (r) of return on the initial investment, and 3.5 percent annual inflation rate of power costs.

Solution:

Step 1. Compute the velocity of flow for each pipeline.

Let V_d and V_p represent velocity for ductile and PVC pipe respectively, using Eq. (5.32a)

$$V_d = \frac{Q}{2.448d^2} = \frac{6300\ \text{gal/min}}{2.448(24.95\ \text{in})^2}$$

$$= 4.13\ \text{ft/s}$$

$$V_p = \frac{6300}{2.448(22.76)^2}$$

$$= 4.96\ \text{ft/s}$$

Step 2. Compute head losses (H_d and H_p) for each pipe flow. Coefficients C for ductile and PVC pipes are 140 and 150, respectively.

$$H_d = 1000\left(\frac{V_d}{0.115Cd^{0.63}}\right)^{1.852}$$

$$= 1000\left(\frac{4.13}{0.115 \times 140 \times 24.95^{0.63}}\right)^{1.852}$$

$$= 1.89 \text{ (ft/1000 ft)}$$

$$H_p = 1000\left(\frac{4.96}{0.115 \times 150 \times 22.76^{0.63}}\right)^{1.852}$$

$$= 2.59 \text{ (ft/1000 ft)}$$

Step 3. Compute the costs of pumping, CP_d and CP_p

$$CP_d = 1.65\, H_d\, Q\, \$/E = 1.65 \times 1.89 \times 6300 \times 0.058/0.75$$

$$= 1519 \text{ (\$/1000 ft/yr)}$$

$$CP_p = 1.65 \times 2.59 \times 6300 \times 0.058/0.75$$

$$= 2082 \text{ (\$/1000 ft/yr)}$$

Step 4. Compute the difference of total cost for 10,000 ft annually (A)

$$A = (2082 - 1519)\$/(1000 \text{ ft/yr}) \times (10{,}000 \text{ ft})$$

$$= 5630 \text{ \$/yr}$$

Step 5. Compute the present worth (PW) of A adjusting for inflation using the appropriate equation below

When $g = r$

$$\text{PW} = An \tag{5.33a}$$

When $g \neq r$

$$\text{PW} = A\left[\frac{(1 + i)^n - 1}{i(1 + i)^n}\right] \tag{5.33b}$$

$$i = \frac{r - g}{1 + g} \tag{5.33c}$$

where PW = present worth of annual difference in pumping cost, \$

A = annual difference in pumping cost, \$

i = effective annual investment rate accounting for inflation, %

n = design life of pipe, yr

g = inflation (growth) rate of power cost, %

r = annual rate of return on the initial investment, %

In this example, $g \neq n$

$$i = \frac{r - g}{1 + g} = \frac{0.066 - 0.035}{1 + 0.035} = 0.030$$

$$\text{PW} = A\left[\frac{(1 + i)^n - 1}{i(1 + i)^n}\right] = 5630\left[\frac{(1 + 0.03)^{50} - 1}{0.03(1 + 0.03)^{50}}\right]$$

$$= 144{,}855\ (\$)$$

4 WATER FLOW IN OPEN CHANNELS

4.1 Che'zy Equation for Uniform Flow

In 1769, the French engineer Antoine Che'zy proposed an equation for uniform open channel flow in which the average velocity is a function of the hydraulic radius and the energy gradient. It can be written as

$$V = C\sqrt{RS} \tag{5.34}$$

where V = average velocity, ft/s
C = Che'zy discharge coefficient, $\text{ft}^{1/2}$/s
R = hydraulic radius, ft
= cross-sectional area A, (ft^2) divided by the wetted perimeter P (ft)
= A/P
S = energy gradient (slope of the bed, slope of surface water for uniform flow)
= head loss (h_f) over the length of channel (L) divided by L
= h_f/L

The value of C can be determined from

$$C = \sqrt{\frac{8g}{f}} \tag{5.35a}$$

where g = gravity constant, ft/s
f = Darcy–Weisback friction factor

EXAMPLE: A trapezoidal open channel has a bottom width of 20 ft and side slopes of inclination 1:1.5. Its friction factor $f = 0.056$. The depth of the channel is 4.0 ft. The channel slope is 0.025. Compute the flow rate of the channel.

Solution:

Step 1. Determine C value from Eq. (5.34)

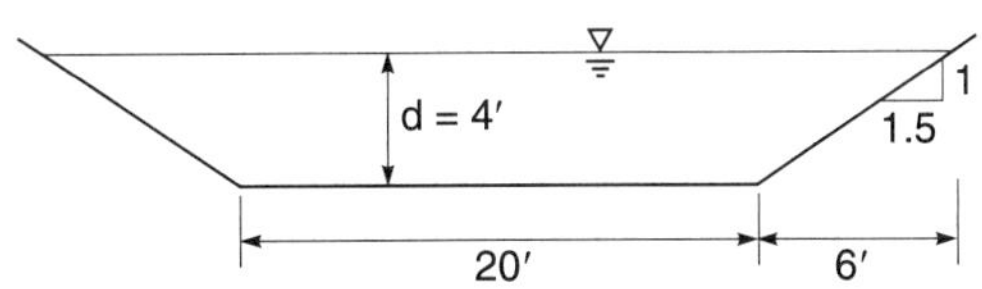

$$C = \sqrt{\frac{8g}{f}} = \sqrt{\frac{8 \times 32.2}{0.056}}$$

$$= 67.8\ (\text{ft}^{1/2}/\text{s})$$

Step 2. Compute A, P, and R
Width of water surface = 20 ft + 2 × 6 ft = 32 ft

$$A = \tfrac{1}{2}(20 + 32)\text{ ft} \times 4\text{ ft}$$
$$= 104\text{ ft}^2$$
$$P = 2\text{ ft} \times \sqrt{4^2 + 6^2} + 20\text{ ft}$$
$$= 34.42\text{ ft}$$
$$R = A/P = 104\text{ ft}^2/34.42\text{ ft}$$
$$= 3.02\text{ ft}$$

Step 3. Determine the flow rate Q

$$Q = AV = AC\sqrt{RS}$$
$$= 104\text{ ft}^2 \times 67.8\text{ ft}^{1/2}/\text{s} \times \sqrt{3.02\text{ ft} \times 0.025}$$
$$= 1937\text{ ft}^3/\text{s}$$

4.2 Manning Equation for Uniform Flow

As in the previous section, Eq. (5.22) was proposed by Robert Manning in 1889. The well-known Manning formula for uniform flow in open channel of non-pressure pipe is

$$V = \frac{1.486}{n} R^{2/3} S^{1/2} \quad \text{(for U.S. customary)} \tag{5.22a}$$

$$V = \frac{1}{n} R^{2/3} S^{1/2} \quad \text{(for SI units)} \tag{5.22b}$$

The flow rate (discharge) Q can be determined by

$$Q = AV = A\frac{1.486}{n} R^{2/3} S^{1/2} \tag{5.36}$$

All symbols are the same as in the Che'zy equation. The Manning roughness coefficient (n) is related to the Darcy–Weisback friction factor as follows:

$$n = 0.093\, f^{1/2} R^{1/6} \tag{5.37}$$

Manning also derived the relationship of n (in s/ft$^{1/3}$) to the Che'zy coefficient C by the equation

$$C = \frac{1}{n} R^{1/6} \tag{5.35b}$$

Typical values of n for various types of channel surface are shown in Table 5.4.

EXAMPLE 1: A 10-ft wide (w) rectangular source-water channel has a flow rate of 980 ft^3/s at a uniform depth (d) of 3.3 ft. Assume n = 0.016 s/ft$^{1/3}$ for concrete. (a) Compute the slope of the channel. (b) Determine the discharge if the normal depth of the water is 4.5 ft.

Solution:

Step 1. Determine A, P, and R for question (a)

$$A = wd = 10\text{ ft} \times 3.3\text{ ft} = 33\text{ ft}^2$$

$$P = 2d + w = 2 \times 3.3\text{ ft} + 10\text{ ft} = 16.6\text{ ft}$$

$$R = A/P = 33\text{ ft}^2/16.6\text{ ft} = 1.988\text{ ft}$$

Step 2. Solve S by the Manning formula

$$Q = A\frac{1.486}{n}R^{2/3}S^{1/2}$$

Rewrite

$$S = \left(\frac{Qn}{1.486AR^{2/3}}\right)^2 = \left(\frac{980 \text{ cfs} \times 0.016 \text{ s/ft}^{1/3}}{1.486 \times 33 \text{ ft}^2 \times (1.988 \text{ ft})^{2/3}}\right)^2 = 0.041$$

Answer for (a), $S = 0.041$

Step 3. For (b), determine new A, P, and R

$$A = wd = 10 \text{ ft} \times 4.5 \text{ ft} = 45 \text{ ft}^2$$

$$P = 2d + w = 2\,(4.5 \text{ ft}) + 10 \text{ ft} = 19 \text{ ft}$$

$$R = A/P = 45 \text{ ft}^2/19 \text{ ft} = 2.368 \text{ ft}$$

Step 4. Calculate Q for answer of question (b)

$$Q = A\frac{1.486}{n}R^{2/3}S^{1/2}$$

$$= 45 \times \frac{1.486}{0.016}(2.368)^{2/3}(0.041)^{1/2}$$

$$= 1500 \text{ (ft}^3\text{/s)}$$

EXAMPLE 2: A rock trapezoidal channel has bottom width 5 ft (1.5 m), water depth 3 ft (0.9 m), side slope 2:1, $n = 0.044$, 5 ft wide. The channel bottom has 0.16% grade. Two equal-size concrete pipes will carry the flow downstream. Determine the size of pipes for the same grade and velocity.

Solution:

Step 1. Determine the flow Q

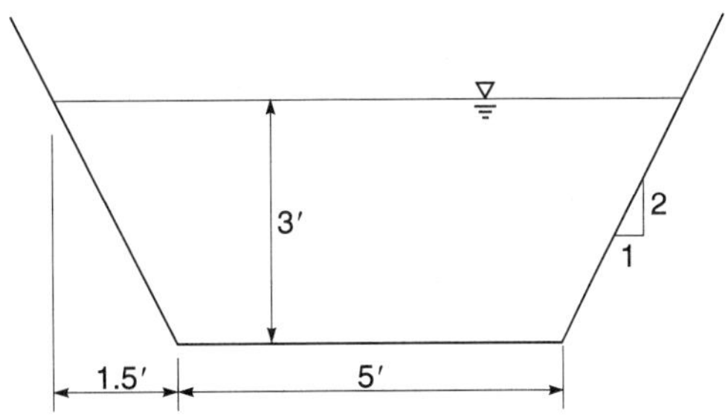

$$A = \frac{5 + 8}{2} \text{ ft} \times 3 \text{ ft}$$

$$= 19.5 \text{ ft}^2$$

$$R = 5 \text{ ft} + 2\sqrt{1.5^2 + 3^2} \text{ ft}$$

$$= 11.71 \text{ ft}$$

By the Manning formula, Eq. (5.35)

$$Q = A\frac{1.486}{n}R^{2/3}S^{1/2}$$
$$= 19.5 \times \frac{1.486}{0.044}(11.71)^{2/3}(0.0016)^{1/2}$$
$$= 135.8 \text{ (ft}^3\text{/s)}$$

Step 2. Determine diameter of a pipe D
For a circular pipe flowing full, a pipe carries one-half of the total flow

$$R = \frac{D}{4}$$
$$\frac{135.8}{2} = \frac{\pi D^2}{4} \times \frac{1.486}{0.013}\left(\frac{D}{4}\right)^{2/3}(0.0016)^{1/2}$$
$$67.9 = 1.424\, D^{8/3}$$
$$D = 4.26 \text{ (ft)}$$
$$= 51.1 \text{ in}$$

Note: Use 48-in pipe, although the answer is slightly over 48 in.

Step 3. Check velocity of flow
Cross-sectional area of pipe A_p

$$A_p = \frac{\pi(4\text{ ft})^2}{4} = 12.56 \text{ ft}^2$$
$$V = \frac{Q}{A_p} = \frac{67.9 \text{ ft}^3\text{/s}}{12.56 \text{ ft}^2}$$
$$= 5.4 \text{ ft/s}$$

This velocity is between 2 and 10 ft/s; it is thus suitable for a storm sewer.

4.3 Partially Filled Conduit

The conditions of partially filled conduit are frequently encountered in environmental engineering, particularly in the case of sewer lines. In a conduit flowing partly full, the fluid is at atmospheric pressure and the flow is the same as in an open channel. The Manning equation (Eq. (5.22)) is applied.

A schematic pipe cross section is shown in Fig. 5.7. The angle θ, flow area A, wetted perimeter P and hydraulic radius R can be determined by the following equations:

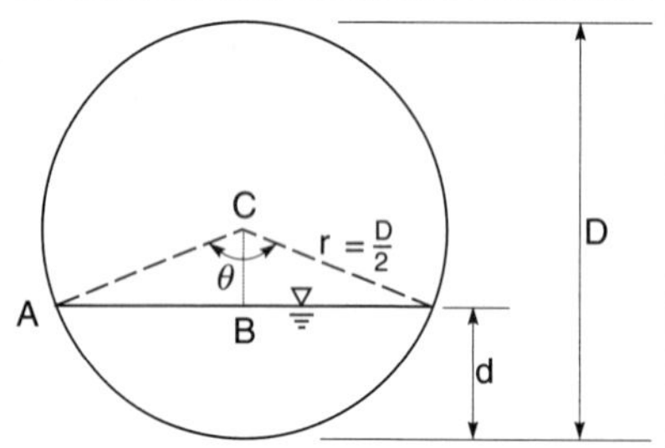

FIGURE 5.7 Flow in partially filled circular pipe.

For angle

$$\cos\frac{\theta}{2} = \frac{\overline{BC}}{\overline{AC}} = \frac{r-d}{r} = 1 - \frac{d}{r} = 1 - \frac{2d}{D} \tag{5.38a}$$

$$\theta = 2\cos^{-1}\left(1 - \frac{2d}{D}\right) \tag{5.38b}$$

Area of triangle ABC, a

$$a = \frac{1}{2}\overline{AB}\ \overline{BC} = \frac{1}{2}r\sin\frac{\theta}{2}r\cos\frac{\theta}{2} = \frac{1}{2}r^2\frac{\sin\theta}{2}$$

$$= \frac{1}{2}\frac{D^2}{4}\frac{\sin\theta}{2}$$

Flow area A

$$A = \frac{\pi D^2}{4}\frac{\theta}{360} - 2a$$

$$= \frac{\pi D^2}{4}\frac{\theta}{360} - 2\left(\frac{1}{2}\frac{D^2}{4}\frac{\sin\theta}{2}\right)$$

$$A = \frac{D^2}{4}\left(\frac{\pi\theta}{360} - \frac{\sin\theta}{2}\right) \tag{5.39}$$

For wetted perimeter P

$$P = \pi D\frac{\theta}{360} \tag{5.40}$$

For hydraulic radius R

$$R = A/P$$

Thus we can mathematically calculate the flow area A, the wetted perimeter P, and the hydraulic radius R. In practice, for a circular conduit, a chart is generally used which is available in hydraulic textbooks and handbooks (Chow, 1959; Morris and Wiggert, 1972; Zipparo and Hasen, 1993; Horvath, 1994).

EXAMPLE 1: Assume a 24-in diameter sewer concrete pipe ($n = 0.012$) is placed on a slope of 2.5 in 1000. The depth of the sewer flow is 10 in. What is the average velocity and the discharge? Will this grade produce a self-cleansing velocity for the sanitary sewer?

Solution:

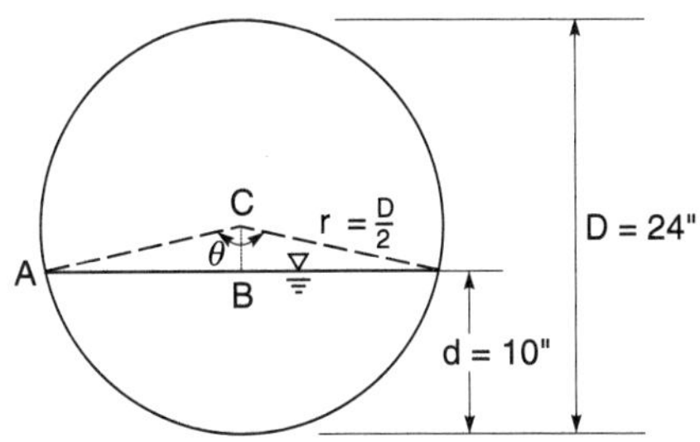

Step 1. Determine θ from Eq. (5.38), referring to Fig. 5.7 and to the figure above

$$\cos\frac{\theta}{2} = 1 - \frac{2d}{D} = 1 - \frac{2 \times 10}{24}$$
$$= 0.1667$$
$$\frac{\theta}{2} = \cos^{-1}(0.1667) = 80.4°$$
$$\theta = 160.8°$$

Step 2. Compute flow area A from Eq. (5.39)

$$D = 24 \text{ in} = 2 \text{ ft}$$
$$A = \frac{D^2}{4}\left(\pi\frac{\theta}{360} - \frac{\sin\theta}{2}\right)$$
$$= \frac{2^2}{4}\left(3.14 \times \frac{160.8}{360} - \frac{1}{2}\sin 160.8°\right)$$
$$= 1.40 - 0.16$$
$$= 1.24 \text{ (ft}^2\text{)}$$

Step 3. Compute P from Eq. (5.40)

$$P = \pi D\frac{\theta}{360} = 3.14 \times 2 \text{ ft} \times \frac{160.8}{360}$$
$$= 2.805 \text{ ft}$$

Step 4. Calculate R

$$R = A/P = 1.24 \text{ ft}^2/2.805 \text{ ft}$$
$$= 0.442 \text{ ft}$$

Step 5. Determine V from Eq. (5.22a)

$$V = \frac{1.486}{n}R^{2/3}S^{1/2}$$
$$= \frac{1.486}{0.012}(0.442)^{2/3}(2.5/1000)^{1/2}$$
$$= 123.83 \times 0.58 \times 0.05$$
$$= 3.591 \text{ (ft/s)}$$

Step 6. Determine Q

$$Q = AV = 1.24 \text{ ft}^2 \times 3.591 \text{ ft/s}$$
$$= 4.453 \text{ ft}^3\text{/s}$$

Step 7. It will produce self-cleaning since $V > 2.0$ ft/s.

EXAMPLE 2: A concrete circular sewer has a slope of 1 m in 400 m. (a) What diameter is required to carry 0.1 m^3/s (3.5 ft^3/s) when flowing six-tenths full? (b) What is the velocity of flow? (c) Is this a self-cleansing velocity?

Solution:
These questions are frequently encountered in sewer design engineering. In the partially filled conduit, the wastewater is at atmospheric pressure and the flow is the same as in an open channel, which can be determined with the Manning equation (Eq. (5.22)). Since this is inconvenient for mathematical calculation, a chart (Fig. 5.4) is commonly used for calculating area A, hydraulic radius R, and flow Q for actual values (partly filled), as opposed to full-flow values.

Step 1. Find full flow rate Q_f (subscript "f" is for full flow)
Given:

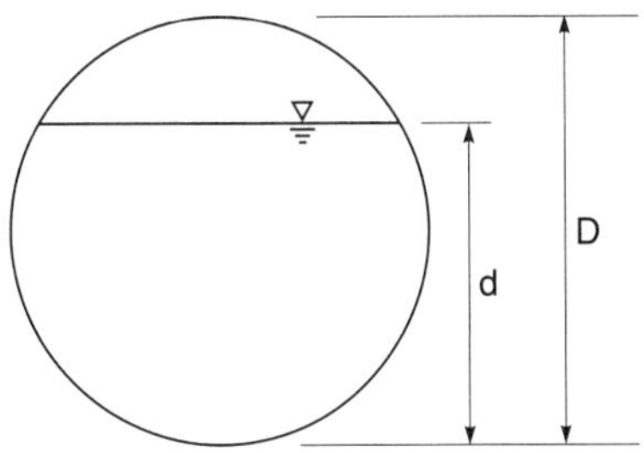

$$\frac{d}{D} = 0.60$$

From Fig. 5.4

$$\frac{Q}{Q_f} = 0.68$$

$$Q_f = \frac{Q}{0.68} = \frac{0.1 \text{ m}^3/\text{s}}{0.68}$$

$$= 0.147 \text{ m}^3/\text{s}$$

Step 2. Determine diameter of pipe D from the Manning formula
For full flow

$$A_f = \frac{\pi}{4}D^2$$

$$R_f = \frac{D}{4}$$

For concrete $n = 0.013$
Slope $S = 1/400 = 0.0025$
From Eq. (5.22b):

$$V = \frac{1}{n}R^{2/3}S^{1/2}$$

$$Q = AV = A\frac{1}{n}R^{2/3}S^{1/2}$$

$$= \frac{\pi}{4}D^2 \times \frac{1}{0.013} \times \left(\frac{D}{4}\right)^{2/3} \times (0.0025)^{1/2}$$

$$0.147 = 1.20\ D^{8/3}$$

Answer (a)

$$D = 0.455 \text{ (m)}$$
$$= 1.5 \text{ ft}$$

Step 3. Calculate V_f

$$V_f = \frac{1}{0.013}\left(\frac{0.455}{4}\right)^{2/3}(0.0025)^{1/2}$$
$$= 0.903 \text{ (m/s)}$$
$$= 2.96 \text{ ft/s}$$

From Fig. 5.4

$$\frac{V}{V_f} = 1.07$$

Answer (b)

$$V = 1.07\ V_f = 1.07 \times 0.903 \text{ m/s}$$
$$= 0.966 \text{ m/s}$$
$$= 3.17 \text{ ft/s}$$

Step 4. For answer (c)
Since

$$V = 3.27 \text{ ft/s} > 2.0 \text{ ft/s}$$

it will provide self-cleansing for the sanitary sewer.

EXAMPLE 3: A 12-in (0.3 m) sewer line is laid on a slope of 0.0036 with n value of 0.012 and flow rate of 2.0 ft^3/s (0.057 m^3/s). What are the depth of flow and velocity?

Solution:

Step 1. Calculate flow rate in full, Q_f
From Eq. (5.22a)

$$D = 1 \text{ ft}$$

$$Q_f = AV = \frac{\pi}{4}D^2\,\frac{1.486}{0.012}\left(\frac{D}{4}\right)^{2/3}(0.0036)^{1/2}$$
$$= 0.785 \times 123.8 \times 0.397 \times 0.06$$
$$= 2.32 \text{ (ft}^3\text{/s)}$$
$$V_f = 2.96 \text{ ft/s}$$

Step 2. Determine depth of flow d

$$\frac{Q}{Q_f} = \frac{2.0}{2.32} = 0.862$$

From Fig. 5.4

$$\frac{d}{D} = 0.72$$

$$d = 0.72 \times 1 \text{ ft}$$
$$= 0.72 \text{ ft}$$

Step 3. Determine velocity of flow V
From chart (Fig. 5.4)

$$\frac{V}{V_\mathrm{f}} = 1.13$$

$$V = 1.13 \times 2.96 \text{ ft/s}$$
$$= 3.34 \text{ ft/s}$$

or
$$= 1.0 \text{ m/s}$$

4.4 Self-cleansing Velocity

The settling of suspended mater in sanitary sewers is of great concern to environmental engineers. If the flow velocity and turbulent motion are sufficient, this may prevent deposition and resuspend the sediment and move it along with the flow. The velocity sufficient to prevent deposits in a sewer is called the self-cleansing velocity. The self-cleansing velocity in a pipe flowing full is (ASCE and WEF 1992):

for SI units

$$V = \frac{R^{1/6}}{n}[B(s-1)D_\mathrm{p}]^{1/2} = \left[\frac{8B}{f}g(s-1)D_\mathrm{p}\right]^{1/2} \tag{5.41a}$$

for U.S. customary units

$$V = \frac{1.486R^{1/6}}{n}[B(s-1)D_\mathrm{p}]^{1/2} = \left[\frac{8B}{f}g(s-1)D_\mathrm{p}\right]^{1/2} \tag{5.41b}$$

where V = velocity, m/s or ft/s
R = hydraulic radius, m or ft
n = Manning's coefficient of roughness
B = dimensionless constant
= 0.04 to start motion
= 0.8 for adequate self-cleansing
s = specific gravity of the particle
D_p = diameter of the particle
f = friction factor, dimensionless
g = gravitational acceleration
= 9.81 m/s^2 or 32.2 ft/s^2

Sewers flowing between 50 and 80 percent full need not be placed on steeper grades to be as self-cleansing as sewers flowing full. The reason is that velocity and discharge are functions of attractive force which depends on the friction coefficient and flow velocity (Fair *et al.* 1966). Figure 5.8 presents the hydraulic elements of circular sewers that possess equal self-cleansing effect.

Using Fig. 5.8, the slope for a given degree of self-cleansing of partly full pipes can be determined. Applying Eq. (5.41) with the Manning equation (Eq. (5.22)) or Fig. 5.3, a pipe to carry a design full discharge Q_f at a velocity V_f that moves a particle of size D_p can be selected. This same particle will be moved by a lesser flow rate between Q_f and some lower discharge Q_s.

Figure 5.8 suggests that any flow ratio Q/Q_f that causes the depth ratio d/D to be larger than 0.5 requires no increase in slope because S_s is less than S_f. For smaller flows, the invert slope must be increased to S_s to avoid a decrease in self-cleansing.

EXAMPLE: A 10 in (25 cm) sewer is to discharge 0.353 ft³/s (0.01 m³/s) at a self-cleansing velocity. When the sewer is flowing full, its velocity is 3 ft/s (0.9 m/s). Determine the depth and velocity of flow and the required sewer line slope. Assume $N = n = 0.013$.

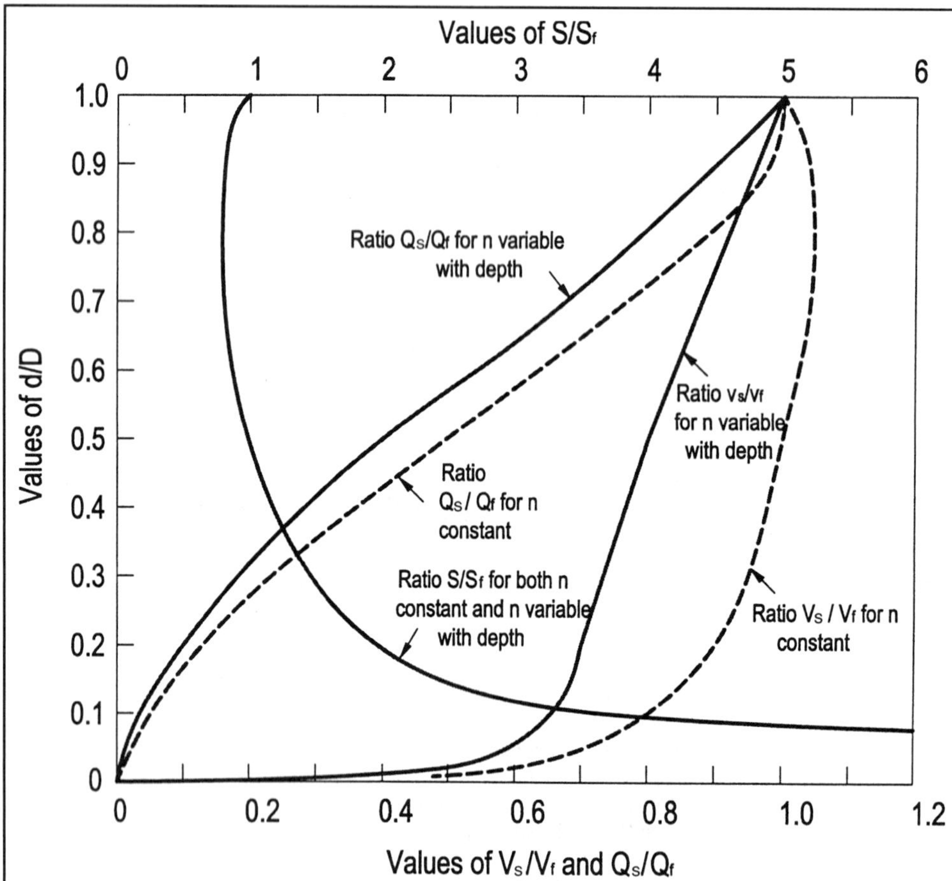

FIGURE 5.8 Hydraulic elements of circular sewers that possess equal self-cleansing properties at all depths.

Solution:

Step 1. Determine the flow Q_f and slope S_f during full flow

$$Q_f = \frac{\pi r^2}{4} V_f = 0.785(10/12 \text{ ft})^2 \times 3.0 \text{ ft/s}$$
$$= 1.64 \text{ ft}^3/\text{s}$$

Using the Manning formula

$$Q_f = \frac{1.486}{n} R^{2/3} S_f^{1/2}$$

$$1.64 = \frac{1.486}{0.013} \left(\frac{0.833}{4}\right)^{2/3} S_f^{1/2}$$

Rearranging

$$S_f^{1/2} = 0.04084$$
$$S_f = 0.00167$$
$$= 1.67\%$$

Step 2. Determine depth d, velocity V_s, and slope S for self-cleansing
From Fig. 5.8
For $N = n$ and $Q_s/Q_f = 0.353/1.64 = 0.215$, we obtain

$$d/D = 0.28$$
$$V_s/V_f = 0.95$$

and

$$S/S_f = 1.50$$

Then

$$d = 0.33D = 0.28 \times 10 \text{ in} = 2.8 \text{ in}$$
$$V_s = 0.95\ V = 0.95 \times 3 \text{ ft/s} = 2.85 \text{ ft/s}$$
$$S = 1.5\ S_f = 1.5 \times 1.67\% = 2.50\%$$

4.5 Specific Energy

For a channel with small slope (Fig. 5.9) the total energy head at any section may be generally expressed by the general Bernoulli equation as

$$E = \frac{V^2}{2g} + \frac{P}{\gamma} + z \tag{5.15}$$

where z is the elevation of the bed. For any stream line in the section, $P/\gamma + z = D$ (the water depth at the section).

When the channel bottom is chosen as the datum ($z = 0$), the total head or total energy E is called the specific energy (H_e). The specific energy at any section in an open channel is equal to the sum of the velocity head (kinetic energy) and water depth (potential energy) at the section. It is written as

$$H_e = \frac{V^2}{2g} + D \tag{5.42a}$$

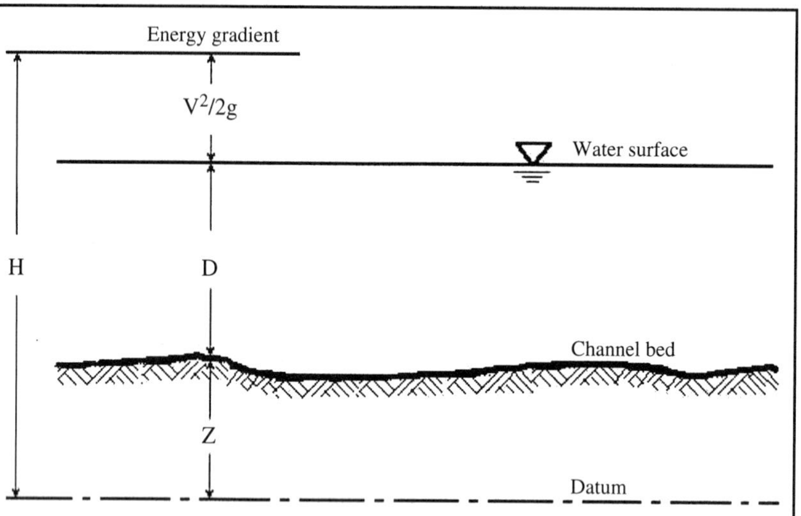

FIGURE 5.9 Energy of open channel flow.

Since $V = Q/A$ = flow/area, then

$$H_e = \frac{Q^2}{2gA^2} + D \tag{5.42b}$$

For flow rate in a rectangular channel, area A is

$$A = WD \tag{5.43}$$

where W is the width of the channel.

4.6 Critical Depth

Given the discharge (Q) and the cross-sectional area (A) at a particular section, Eq. (5.42a) may be rewritten for specific energy expressed in terms of discharge

$$H_e = \frac{Q^2}{2gA^2} + D \tag{5.42b}$$

If the discharge is held constant, specific energy varies with A and D. At the critical state the specific energy of the flow is at a minimum value. The velocity and depth associated with the critical state are called the critical velocity and critical depth, respectively. At the critical state, the first derivative of the specific energy with respect to the water depth should be zero:

$$\frac{dE}{dD} = \frac{d}{dD}\left(\frac{Q^2}{2gA^2} + D\right) = 0 \tag{5.44}$$

$$\frac{dE}{dD} = -\frac{Q^2}{gA^3}\frac{dA}{dD} + 1 = 0$$

Since near the free surface

$$dA = WdD$$

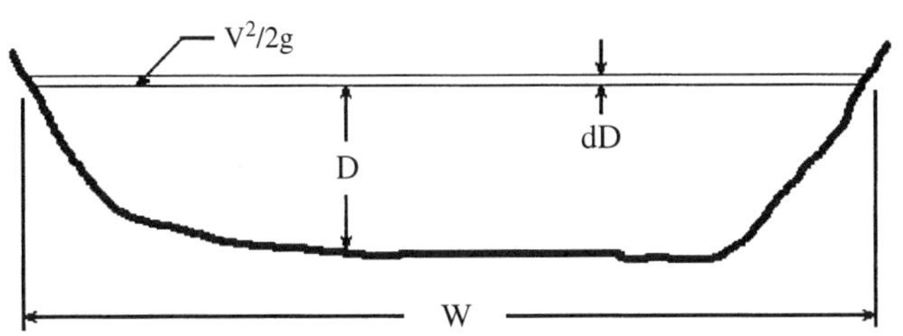

FIGURE 5.10 Channel cross-section.

then

$$W = \frac{dA}{dD}$$

where W is the top width of the channel section. For open channel flow, $D_m = A/W$ is known as the hydraulic depth (or mean depth, D_m).

Hence for nonrectangular channels

$$-\frac{Q^2W}{gA^3} + 1 = 0 \tag{5.45}$$

or

$$Q^2W = gA^3 \tag{5.45a}$$

Then the general expression for flow at the critical depth is

$$Q = \sqrt{gA^3/W} = A\sqrt{gA/W}$$

$$Q = A\sqrt{gD_m} \tag{5.46}$$

and the critical velocity V_c is

$$V_c = \sqrt{gA/W} = \sqrt{gD_m} \tag{5.47}$$

$$D_m = \frac{Q^2}{gA^2} \tag{5.48}$$

$$H_m = D_c + \frac{D_m}{2} \tag{5.49}$$

where H_m is the minimum specific energy.

For the special case of a rectangular channel, the critical depth

$$D_c = \left(\frac{Q^2}{gW^2}\right)^{1/3} = \left(\frac{q^2}{g}\right)^{1/3} \tag{5.50}$$

where q = discharge per unit width of the channel, $m^3/s/m$.

For rectangular channels, the hydraulic depth is equal to the depth of the flow. Eq. (5.45) can be simplified as

$$\frac{Q^2}{A^2}\frac{W}{Ag} = 1$$

$$\frac{V^2}{gD} = 1$$

$$\frac{V}{\sqrt{gD}} = 1 \tag{5.51}$$

The quantity $V/\sqrt{gD}$ is dimensionless. It is the so-called Froude number, F. The Froude number is a very important factor for open channel flow. It relates three states of flow as follows:

F	Velocity	Flow state
1	$V = \sqrt{gD}$	critical flow (V = speed of surface wave)
< 1	$V < \sqrt{gD}$	subcritical ($V <$ speed of surface wave)
> 1	$V > \sqrt{gD}$	supercritical ($V >$ speed of surface wave)

EXAMPLE: Determine the critical depth of a channel when the water is flowing in a trapezoidal channel with a base width of 1.5 m (5.0 ft) and side slope of 2 on 1. The flow is 4.0 m^3/s (63,400 gpm).

Solution: Let h = the critical depth of the channel

Then base = 1.5 m

$$W = 1.5\text{ m} + 2 \times (1/2\text{ h})\text{ m}$$
$$= (1.5 + \text{h})\text{ m}$$
$$A = \tfrac{1}{2} \times [1.5\text{ m} + (1.5 + h)\text{ m}] \times h\text{ m}$$
$$= (1.5\text{ h} + 0.5\text{ h}^2)\text{ m}^2$$
$$Q = 4.0\text{ m}^3/\text{s}^2$$
$$g = 9.806\text{ m/s}^2$$

Solve h using Eq. (4.45a)

$$Q^2W = gA^3$$
$$(4.0)^2 \times (1.5 + h) = 9.806 \times (1.5h + 0.5h^2)^3$$
$$(1.5h + 0.5h^2)^3/\,(1.5 + h) = 1.63$$

by trial $h = 0.818$ (m)

$= 2.68$ ft

4.7 Hydraulic Jump

When water flows in an open channel, an abrupt decrease in velocity may occur due to a sudden increase of water depth in the downstream direction. This is called hydraulic jump and may be a natural phenomenon. It occurs when the depth in an open channel increases from a depth less than the critical depth D_c to one greater than D_c. Under these conditions, the water must pass through a hydraulic jump, as shown in Fig. 5.11.

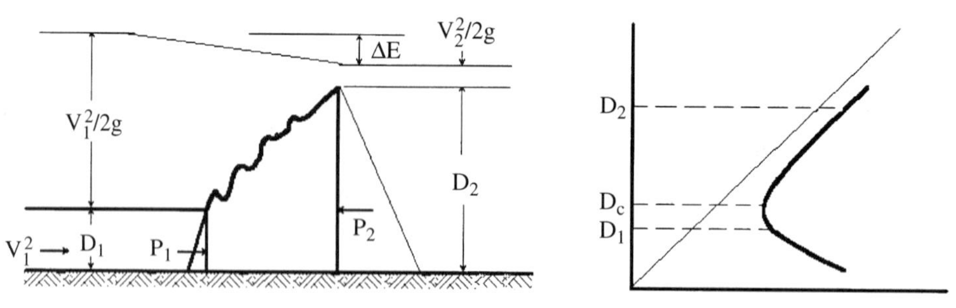

FIGURE 5.11 Hydraulic jump.

The balance between the hydraulic forces P_1 and P_2, represented by the two triangles and the momentum flux through sections 1 and 2, per unit width of the channel, can be expressed as

$$P_1 - P_2 = \rho q\,(V_2 - V_1) \tag{5.52}$$

where q is the flow rate per unit width of the channel. Substituting the following quantities in the above equation

$$P_1 = \frac{\gamma}{2}D_1^2; \quad P_2 = \frac{\gamma}{2}D_2^2$$

$$V_1 = \frac{q}{D_1}; \quad V_2 = \frac{q}{D_2}$$

we get

$$\frac{\gamma}{2}(D_1^2 - D_2^2) = \rho q\left(\frac{q}{D_2} - \frac{q}{D_1}\right)$$

$$\frac{1}{2}(D_1 + D_2)(D_1 - D_2) = \frac{\rho}{\gamma}q^2\left(\frac{D_1 - D_2}{D_1 D_2}\right)$$

since

$$\frac{\rho}{\gamma} = \frac{1}{g}$$

then

$$\frac{q^2}{g} = D_1 D_2\left(\frac{D_1 + D_2}{2}\right) \tag{5.53}$$

This equation may be rearranged into a more convenient form as follows:

$$\frac{D_2}{D_1} = \left(\frac{1}{4} + 2F_1^2\right)^{1/2} - \frac{1}{2} \tag{5.54}$$

where F_1 is the upstream Froude number and is expressed by

$$F_1 = \frac{V_1}{\sqrt{gD_1}} \tag{5.55}$$

If the hydraulic jump is to occur, F_1 must be greater than one; i.e. the upstream flow must be supercritical. The energy dissipated (ΔE) in a hydraulic jump in a rectangular channel may be calculated by the equation

$$\Delta E = E_1 - E_2 = \frac{(D_2 - D_1)^3}{4D_1 D_2} \tag{5.56}$$

EXAMPLE 1: A rectangular channel, 4 m wide and with 0.5 m water depth, is discharging 12 m^3/s flow before entering a hydraulic jump. Determine the critical depth and the downstream water depth.

Solution:

Step 1. Calculate discharge per unit width q

$$q = \frac{Q}{W} = \frac{12\ \text{m}^3/\text{s}}{4\ \text{m}} = 3\ (\text{m}^3/\text{m}\cdot\text{s or m}^2/\text{s})$$

Step 2. Calculate critical depth D_c from Eq. (5.50)

$$D_c = \left(\frac{q^2}{g}\right)^{1/3} = \left(\frac{3\ \text{m}^2/\text{s} \times 3\ \text{m}^2/\text{s}}{9.81\ \text{m/s}^2}\right)^{1/3}$$
$$= 0.972\ \text{m}$$

Step 3. Determine upstream velocity V_1

$$V_1 = \frac{q}{D_1} = \frac{3\ \text{m}^2/\text{s}}{0.5\ \text{m}} = 6.0\ \text{m/s}$$

Step 4. Determine the upstream Froude number

$$F_1 = \frac{V_1}{\sqrt{gD_1}} = \frac{6.0}{\sqrt{9.81 \times 0.5}} = 2.71$$

Step 5. Compute downstream water depth D_2 from Eq. (5.54)

$$D_2 = D_1\left[\left(\frac{1}{4} + 2F_1^2\right)^{1/2} - \frac{1}{2}\right]$$
$$= 0.5\left[\left(\frac{1}{4} + 2 \times 2.71^2\right)^{1/2} - \frac{1}{2}\right]$$
$$= 1.68\ (\text{m})$$

EXAMPLE 2: A rectangular channel 9 ft (3 m) wide carries 355 ft³/s (10.0 m³/s) of water with a water depth of 2 ft (0.6 m). Is a hydraulic jump possible? If so, what will be the depth of water after the jump and what will be the horsepower loss through the jump?

Solution:

Step 1. Compute the average velocity in the channel

$$V_1 = \frac{Q}{A_1} = \frac{355\ \text{ft}^3/\text{s}}{9\ \text{ft} \times 2\ \text{ft}}$$
$$= 19.72\ \text{ft/s}$$

Step 2. Compute F_1 from Eq. (5.55)

$$F_1 = \frac{V_1}{\sqrt{gD_1}} = \frac{19.72\ \text{ft/s}}{\sqrt{32.2\ \text{ft/s}^2 \times 2\ \text{ft}}}$$
$$= 2.46$$

Since $F_1 > 1$, the flow is supercritical and a hydraulic jump is possible.

Step 3. Compute the depth D_2 after the hydraulic jump

$$D_2 = D_1[(0.25 + 2F_1^2)^{1/2} - 0.5]$$
$$= 2[(0.25 + 2 \times 2.46^2)^{1/2} - 0.5]$$
$$= 6.03\ \text{ft}$$

Step 4. Compute velocity V_2 after jump

$$V_2 = \frac{Q}{A_2} = \frac{355 \text{ ft}^3/\text{s}}{9 \text{ ft} \times 6.03 \text{ ft}}$$
$$= 6.54 \text{ ft/s}$$

Step 5. Compute total energy loss ΔE (or Δh)

$$E_1 = D_1 + \frac{V_1^2}{2g} = 2 + \frac{19.72^2}{2 \times 32.2} = 8.04 \text{ (ft)}$$
$$E_2 = D_2 + \frac{V_2^2}{2g} = 6.03 + \frac{6.54^2}{2 \times 32.2} = 6.69 \text{ (ft)}$$
$$\Delta E = E_1 - E_2 = 8.04 \text{ ft} - 6.69 \text{ ft} = 1.35 \text{ ft}$$

Step 6. Compute horsepower (hp) loss

$$\text{Loss} = \frac{\Delta E \gamma Q}{550} = \frac{1.35 \text{ ft} \times 62.4 \text{ lb/ft}^3 \times 355 \text{ ft}^3/\text{s}}{550 \text{ ft} \cdot \text{lb/hp}}$$
$$= 54.4 \text{ hp}$$

5 FLOW MEASUREMENTS

Flow can be measured by velocity methods and direct discharge methods. The measurement flow velocity can be carried out by a current meter, Pitot tube, U-tube, dye study, or salt velocity. Discharge is the product of measured mean velocity and cross-sectional area. Direct discharge methods include volumetric gravimeter, Venturi meter, pipe orifice meter, standardized nozzle meter, weirs, orifices, gates, Parshall flumes, etc. Detail flow measurements in orifices, gates, tubes, weirs, pipes, and in open channels are discussed by Brater *et al.* (1996).

5.1 Velocity Measurement in Open Channel

The mean velocity of a stream or a channel can be measured with a current meter. A variety of current meters is commercially available. An example of discharge calculation with known mean velocity in sub-cross section is presented in Chapter 1.2.

5.2 Velocity Measurement in Pipe Flow

Pitot tube. A Pitot tube is bent to measure velocity due to the pressure difference between the two sides of the tube in a flow system. The flow velocity can be determined from

$$V = \sqrt{2g\Delta h} \tag{5.57}$$

where V = velocity, m/s or ft/s
g = gravitational acceleration, 9.81 m/s^2 or 32.2 ft/s^2
Δh = height of the fluid column in the manometer or a different height of immersible liquid such as mercury, m or ft

EXAMPLE: The height difference of the Pitot tube is 5.1 cm (2 in). The specific weight of the indicator fluid (mercury) is 13.55. What is the flow velocity of the water?

Solution 1: For SI units

Step 1. Determine the water column equivalent to Δh

$$\Delta h = 5.1 \text{ cm} \times 13.55 = 69.1 \text{ cm} = 0.691 \text{ m of water}$$

Step 2. Determine velocity

$$V = \sqrt{2g\Delta h} = \sqrt{2 \times 9.81 \text{ m/s}^2 \times 0.691 \text{ m}}$$
$$= 3.68 \text{ m/s}$$

Solution 2: For US customary units

Step 1.

$$\Delta h = 2/12 \text{ ft} \times 13.55 = 2.258 \text{ ft}$$

Step 2.

$$V = \sqrt{2g\Delta h} = \sqrt{2 \times 32.2 \text{ ft/s}^2 \times 2.258 \text{ ft}}$$
$$= 12.06 \text{ ft/s}$$
$$= 12.06 \text{ ft/s} \times 0.304 \text{ m/ft}$$
$$= 3.68 \text{ m/s}$$

5.3 Discharge Measurement of Water Flow in Pipes

Direct collection of volume (or weight) of water discharged from a pipe divided by time of collection is the simplest and most reliable method. However, in most cases it cannot be done by this method. A change of pressure head is related to a change in flow velocity caused by a sudden change of pipe cross-section geometry. Venturi meters, nozzle meters, and orifice meters use this concept.

Venturi meter. A Venturi meter is a machine-cased section of pipe with a narrow throat. The device, in a short cylindrical section, consists of an entrance cone and a diffuser cone which expands to full pipe diameter. Two piezometric openings are installed at the entrance (section 1) and at the throat (section 2). When the water passes through the throat, the velocity increases and the pressure decreases. The decrease of pressure is directly related to the flow. Using the Bernoulli equation at sections 1 and 2, neglecting friction head loss, it can be seen that

$$\frac{V_1^2}{2g} + \frac{P_1}{\gamma} + z_1 = \frac{V_2^2}{2g} + \frac{P_2}{\gamma} + z_2 \tag{5.16}$$

For continuity flow between sections 1 and 2, $Q_1 = Q_2$

$$A_1V_1 = A_2V_2$$

Solving the above two equations, we get

$$Q = \frac{A_1A_2\sqrt{2g[(h_1 - h_2) + (z_1 - z_2)]}}{\sqrt{A_1^2 - A_2^2}}$$

$$= \frac{A_1 A_2 \sqrt{2g(H + Z)}}{\sqrt{A_1^2 - A_2^2}}$$

$$= C_d A_1 \sqrt{2g(H + Z)} \tag{5.58a}$$

where Q = discharge, m³/s or ft³/s
A_1, A_2 = cross-sectional areas at pipe and throat, respectively, m² or ft²
g = gravity acceleration, 9.81 m/s² or 32.2 ft/s²
$H = h_1 - h_2$ = pressure drop in Venturi tube, m or ft
$Z = z_1 - z_2$ = difference of elevation head, m or ft
C_d = coefficient of discharge

For a Venturi meter installed in a horizontal position, $Z = 0$

$$Q = \frac{A_1 A_2 \sqrt{2gH}}{\sqrt{A_1^2 - A_2^2}} = C_d A_1 \sqrt{2gH} \tag{5.58b}$$

where

$$C_d = \frac{A_2}{\sqrt{A_1^2 - A_2^2}} = \frac{1}{\sqrt{\left(\frac{A_1}{A_2}\right)^2 - 1}} \tag{5.59}$$

EXAMPLE: A 6-cm throat Venturi meter is installed in an 18-cm diameter horizontal water pipe. The reading of the differential manometer is 18.6 cm of mercury column (sp gr = 13.55). What is the flow rate in the pipe?

Solution:

Step 1. Determine A_1/A_2

$$A_1 = \pi(18/2)^2 = 254.4 \text{ cm}^2 = 0.2544 \text{ m}^2$$

$$A_2 = \pi(6/2)^2 = 28.3 \text{ cm}^2 = 0.0283 \text{ m}^2$$

$$\frac{A_1}{A_2} = \frac{254.4}{28.3} = 8.99$$

Step 2. Calculate C_d

$$C_d = 1/\sqrt{\left(\frac{A_1}{A_2}\right)^2 - 1} = 1/\sqrt{(8.99)^2 - 1}$$

$$= 0.112$$

Step 3. Calculate H

$$H = \gamma y = 13.55 \times \frac{18.6}{100} \text{ m}$$

$$= 2.52 \text{ m}$$

Step 4. Calculate Q

$$Q = C_d A_1 \sqrt{2gH} = 0.112 \times 0.2544 \times \sqrt{2 \times 9.81 \times 2.52}$$

$$= 0.20 \text{ (m}^3\text{/s)}$$

Nozzle meter and orifice meter. The nozzle meter and the orifice meter are based on the same principles as the Venturi meter. However, nozzle and orifice meters encounter a significant loss of head due to a nozzle or orifice installed in the pipe. The coefficient (C_v) for the nozzle meter or orifice meter is added to the discharge equation for the Venturi meter, and needs to be determined by on-site calibration. The discharge for the nozzle meter and orifice meter is

$$Q = C_v C_d A_1 \sqrt{2g(H + Z)} \tag{5.60}$$

or, for horizontal installation

$$Q = C_v C_d A_1 \sqrt{2gH} \tag{5.61}$$

The nozzle geometry has been standardized by the American Society of Mechanical Engineers (ASME) and the International Standards Association. The nozzle coefficient C_v is a function of Reynolds number ($\mathbf{R} = V_2 d_2/\nu$) and ratio of diameters (d_2/d_1). A nomograph (Fig. 5.12) is available. Values of C_v range from 0.96 to 0.995 for **R** ranging from 5×10^4 to 5×10^6.

EXAMPLE: Determine the discharge of a 30-cm diameter water pipe. An ASME nozzle of 12-cm throat diameter is installed. The attached differential manometer reads 24.6 cm of mercury column. The water temperature in the pipe is 20°C.

Solution:

Step 1. Determine ratio $A_1 : A_2$

$$A_1 = \frac{\pi}{4}(30 \text{ cm})^2 = 707 \text{ cm}^2 = 0.0707 \text{ m}^2$$

$$A_2 = \frac{\pi}{4}(12 \text{ cm})^2 = 113 \text{ cm}^2 = 0.0113 \text{ m}^2$$

$$\frac{A_1}{A_2} = \frac{30^2}{12^2} = 6.25$$

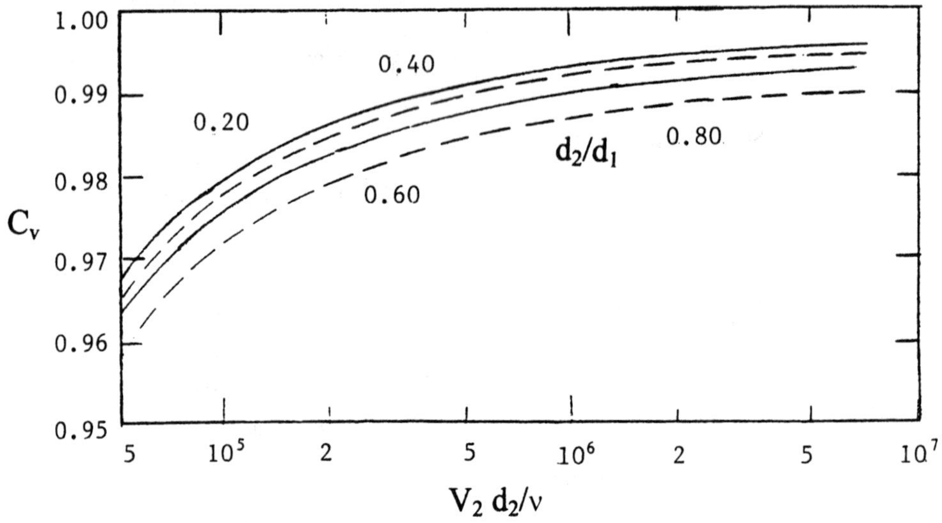

FIGURE 5.12. ASME nozzle coefficients.

Step 2. Calculate C_d

$$C_d = 1/\sqrt{(A_1/A_2)^2 - 1} = 1/\sqrt{(6.25)^2 - 1}$$
$$= 0.162$$

Step 3. Calculate H

$$\gamma = 13.55 \text{ at } 20°\text{C}$$
$$H = \gamma y = 13.55 \times 24.6 \text{ cm} = 333 \text{ cm}$$
$$= 3.33 \text{ m}$$

Step 4. Estimate discharge Q
Assuming $C_v = 0.98$

$$Q = C_v C_d A_1 \sqrt{2gH}$$
$$= 0.98 \times 0.162 \times 0.0707\sqrt{2 \times 9.81 \times 3.33}$$
$$= 0.091 \text{ (m}^3\text{/s)}$$

This value needs to be verified by checking the corresponding Reynolds number **R** of the nozzle.

Step 5. Calculate **R**

$$V_2 = Q/A_2 = 0.091 \text{ m}^3\text{/s}/0.0113 \text{ m}^2$$
$$= 8.053 \text{ m/s}$$

The kinematic viscosity at 20°C is (Table 5.1a)

$$\nu = 1.007 \times 10^{-6} \text{ m}^2\text{/s}$$

then

$$\mathbf{R} = \frac{V_2 d_2}{\nu} = \frac{8.053 \text{ m/s} \times 0.12 \text{ m}}{1.007 \times 10^{-6} \text{ m}^2\text{/s}}$$
$$= 9.6 \times 10^5$$

Using this **R** value, with $d_1/d_2 = 0.4$, the chart in Fig. 5.12 gives

$$C_v = 0.977$$

Step 6. Correct Q value

$$Q = \frac{0.977}{0.98} \times 0.091 \text{ m}^3\text{/s}$$
$$= 0.0907 \text{ m}^3\text{/s}$$

5.4 Discharge Measurements

Orifices. The discharge of an orifice is generally expressed as

$$Q = C_d A \sqrt{2gH} \tag{5.62}$$

where Q = flow rate
C_d = coefficient of discharge
A = area of orifice
g = gravity acceleration
H = water height above center of orifice
All units are as Eq. (5.58a)

EXAMPLE: The water levels upstream and downstream are 4.0 and 1.2 m, respectively. The rectangular orifice has a sharp-edged opening 1 m high and 1.2 m wide. Calculate the discharge, assuming $C_d = 0.60$.

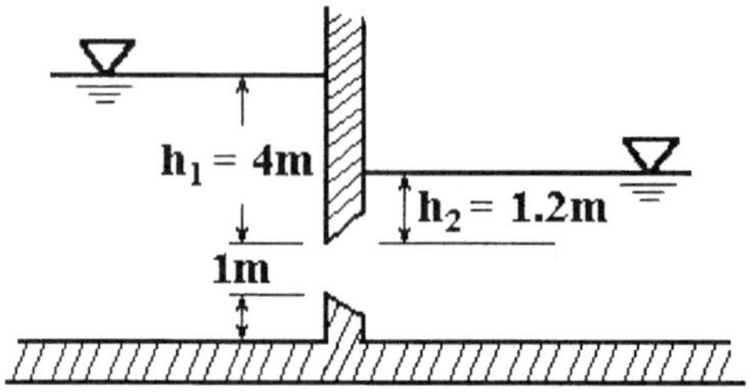

Solution:

Step 1. Determine area A

$$A = 1\text{ m} \times 1.2\text{ m} = 1.2\text{ m}^2$$

Step 2. Determine discharge Q

$$\begin{aligned} Q &= C_d A \sqrt{2g(h_1 - h_2)} \\ &= 0.60 \times 1.2\text{ m}^2 \sqrt{2 \times 9.81\text{ m/s}^2 \times (4 - 1.2)\text{ m}} \\ &= 5.34\text{ m}^3\text{/s} \end{aligned}$$

Weirs. The general discharge equation for a rectangular, horizontal weir is

$$Q = C_d L H^{3/2} \tag{5.63a}$$

where Q = flow rate, ft³/s or m³/s
L = weir length, ft or m
H = head on weir, ft or m
= the water surface above the weir crest
C_d = coefficient of discharge

In the US customary system, let y be the weir height in feet.

$$C_d = 3.22 + 0.40\frac{\text{H}}{\text{y}} \quad \text{(US customary units)} \tag{5.64a}$$

$$C_d = 1.78 + 0.24\frac{\text{H}}{\text{y}} \quad \text{(SI units)} \tag{5.64b}$$

The theoretical equation for a rectangular streamlined weir is (ASCE & WEF, 1992)

$$Q = \frac{2}{3}L\sqrt{\frac{2}{3}gH^3} \tag{5.63b}$$

where g is the gravitational acceleration rate. This formula reduces to

$$Q = 3.09\, L\, H^{3/2} \quad \text{(for US customary units system)} \tag{5.63c}$$

or

$$Q = 1.705\, L\, H^{3/2} \quad \text{(for SI units)} \tag{5.63d}$$

EXAMPLE 1: A rectangular flow control tank has an outflow rectangular weir 1.5 m (5 ft) in length. The inflow to the box is 0.283 m^3/s (10 ft^3/s). The crest of the weir is located 1.2 m above the bottom of the tank. Find the depth of water in the tank.

Solution:

Step 1. Determine head on weir H
Using Eq. 5.63d

$$\begin{aligned} Q &= 1.705\, L\, H^{3/2} \\ 0.283 &= 1.705(1.5)\, H^{3/2} \\ H^{3/2} &= 0.111 \\ H &= 0.23\ \text{(m)} \end{aligned}$$

Step 2. Calculate water depth D

$$\begin{aligned} D &= 1.2\ \text{m} + H = 1.2\ \text{m} + 0.23\ \text{m} \\ &= 1.43\ \text{m} \end{aligned}$$

EXAMPLE 2: In a rectangular channel 4 m (13 ft) high and 12 m (40 ft) wide, a sharp-edged rectangular weir 1.2 m (4 ft) high without end contraction will be installed. The flow of the channel is 0.34 m^3/s (12 ft^3/s). Determine the length of the weir to keep the head on the weir 0.15 m (0.5 ft).

Solution:

Step 1. Compute velocity of approach at channel, V

$$V = \frac{Q}{A} = \frac{0.34\ \text{m}^3/\text{s}}{12 \times 4} = 0.007\ \text{m/s (negligible)}$$

Head due to velocity of approach

$$h = \frac{V^2}{2g} \quad \text{(negligible)}$$

Step 2. Use weir formula (Eq. (5.63d))
(a) Without velocity of approach

$$\begin{aligned} Q &= 1.705\, L\, H^{3/2} \\ 0.34 &= 1.705\, L\, H^{3/2} \\ 0.34 &= 1.705\, L\, (0.15)^{1.5} \\ L &= 3.43\ \text{(m)} \\ &= 11.2\ \text{ft} \end{aligned}$$

(b) Including velocity of approach

$$\begin{aligned} Q &= 1.705\, L\, [(0.15 + h)^{1.5} - h^{1.5}] \\ L &= 3.4\ \text{m} \end{aligned}$$

For freely discharging rectangular weirs (sharp-crested weirs), the Francis equation is most commonly used to determine the flow rate. The Francis equation is

$$Q = 3.33\ LH^{3/2} \tag{5.63e}$$

where Q = flow rate, ft^3/s
L = weir length, ft
H = head on weir, ft
= the water surface above the weir crest

For constracted rectangular weirs

$$Q = 3.33\ (L - 0.1\ nH)\ H^{3/2}\ \text{(US customary units)} \tag{5.63f}$$

$$Q = 1.705\ (L - 0.1\ nH)\ H^{1/2}\ \text{(SI units)} \tag{5.63g}$$

where n = number of end constractions.

EXAMPLE 3: Conditions are the same as in Example 2, except that the weir has two end constractions. Compute the width of the weir.

Solution:

Step 1. For constracted weir, without velocity of approach

$$Q = 1.705\ (L - 0.2\ H)\ H^{3/2}$$

$$0.34 = 1.705\ (L - 0.2 \times 0.15) \times (0.15)^{1.5}$$

$$L = 3.40\ \text{(m)}$$

$$= 11.1\ \text{ft}$$

Step 2. Discharge including velocity of approach

$$Q = 1.705\ (L - 0.2H)[(H + h)^{3/2} - h^{3/2}]$$

Since velocity head is negligible

$$L = 3.4\ \text{m}$$

EXAMPLE 4: Estimate the discharge of a weir 4 ft long where the head on the weir is 3 in.

Solution:

By the Francis equation

$$Q = 3.33\ L\ H^{3/2}/\text{s}$$

$$= 3.33 \times 4 \times (3/12)^{3/2}$$

$$= 1.66\ (\text{ft}^3/\text{s})$$

For the triangular weir and V-notch weir, the flow is expressed as

$$Q = C_d\left(\tan\frac{\theta}{2}\right)H^{5/2} \tag{5.65}$$

where θ = weir angle
C_d = discharge coefficient, calibrated in place (different values have been proposed)
= 1.34 for SI units
= 2.50 for US customary units

The triangular weir is commonly used for measuring small flow rates. Several different notch angles are available. However, the 90° V-notch weir is the one most commonly used. The discharge for 90° V-notch with free flow is

$$Q = 2.5\, H^{2.5} \tag{5.66}$$

where Q = discharge, ft³/s
H = head on the weir, ft

Detailed discussion of flow equations for SI units for various types of weir is presented by Brater *et al.* (1996).

EXAMPLE 5: A rectangular control tank has an outflow rectangular weir 8 ft (2.4 m) in length. The crest of the weir is 5 ft (1.5 m) above the tank bottom. The inflow from a pipe to the tank is 10 ft³/s (0.283 m³/s). Estimate the water depth in the tank, using the Francis equation.

Solution:

$$Q = 3.33\, LH^{3/2}$$

Rewrite equation for H

$$H = \left(\frac{Q}{3.33\,L}\right)^{2/3} = \left(\frac{10}{3.33 \times 8}\right)^{2/3}$$
$$= 0.52 \text{ (ft)}$$

EXAMPLE 6: A rectangular channel 6 ft (1.8 m) wide has a sharp-crested weir across its whole width. The weir height is 4 ft (1.2 m) and the head is 1 ft (0.3 m). Determine the discharge.

Solution:

Step 1. Determine C_d using Eq. (5.64)

$$C_d = 3.22 + 0.40\frac{H}{y} = 3.22 + 0.40 \times \frac{1}{4}$$
$$= 3.32$$

Step 2. Compute discharge Q

$$Q = C_d L\, H^{3/2}$$
$$= 3.32 \times 6 \times (1)^{3/2}$$
$$= 19.92 \text{ (ft}^3\text{/s)}$$

EXAMPLE 7: A circular sedimentation basin has a diameter of 53 ft (16 m). The inflow from the center of the basin is 10 MGD (0.438 m³/s). A circular effluent weir with 90° V-notches located at 0.5 ft (0.15 m) intervals is installed 1.5 ft (0.45 m) inside the basin wall. Determine the water depth on each notch and the elevation of the bottom of the V-notch if the water surface of the basin is at 560.00 ft above mean sea level (MSL).

Solution:

Step 1. Determine the number of V-notches
Diameter of the weir

$$d = 53 \text{ ft} - 2 \times 1.5 \text{ ft} = 50 \text{ ft}$$

Weir length

$$l = \pi d = 3.14 \times 50 \text{ ft} = 157 \text{ ft}$$

No. of V-notches

$$n = 157 \text{ ft} \times \frac{2 \text{ notches}}{\text{ft}}$$
$$= 314 \text{ notches}$$

Step 2. Compute the discharge per notch

$$Q = \frac{10{,}000{,}000 \text{ gal/day}}{314 \text{ notches}} = 31{,}847 \text{ (gal/day)/notch}$$
$$= 31{,}847 \frac{\text{gal/day}}{\text{notch}} \times \frac{1 \text{ ft}^3}{7.48 \text{ gal}} \times \frac{1 \text{ day}}{86{,}400 \text{ s}}$$
$$= 0.0493(\text{ft}^3/\text{h})/\text{notch}$$

Step 3. Compute the head of each notch from Eq. (5.66)

$$Q = 2.5\, H^{2.5}$$
$$H = (Q/2.5)^{1/2.5} = (Q/2.5)^{0.4}$$
$$= (0.0493/2.5)^{0.4} = 0.21 \text{ (ft)}$$

Step 4. Determine the elevation of the bottom of the V-notch

$$\text{Elevation} = 560.00 \text{ ft} - 0.21 \text{ ft}$$
$$= 559.79 \text{ ft MSL}$$

Parshall flume. The Parshall flume was developed by R. L. Parshall in 1920 for the British measurement system. It is widely used for measuring the flow rate of an open channel. It consists of a converging section, a throat section, and a diverging section. The bottom of the throat section is inclined downward and the bottom of the diverging section is inclined upward. The dimensions of the Parshall flume are specified in the U.S. customary system, not the metric system. The geometry creates a critical depth to occur near the beginning of the throat section and also generates a back water curve that allows the depth H_a to be measured from observation well a. There is a second measuring point H_b located at the downstream end of the throat section. The ratio of H_b to H_a is defined as the submergence of the flume.

Discharge is a function of H_a. The discharge equation of a Parshall flume is determined by its throat width, which ranges from 3 in to 50 ft. The relationship between the empirical discharge and the gage reading H_a for various sizes of flume is given below:

Throat width W, ft	Discharge equation, ft³/s	Flow capacity, ft³/s
3 in (0.25 ft)	$Q = 0.992\, H_a^{1.547}$	0.03–1.9
6 in (0.5 ft)	$Q = 2.06\, H_a^{1.58}$	0.05–3.9
9 in (0.75 ft)	$Q = 3.07\, H_a^{1.53}$	0.09–8.9
1–8 ft	$Q = 4\, W\, H_a^{1.522W^{0.26}}$	up to 140
10–50 ft	$Q = (3.6875\, W + 2.5)\, H_a^{1.6}$	

Source: R. L. Parshall (1926).

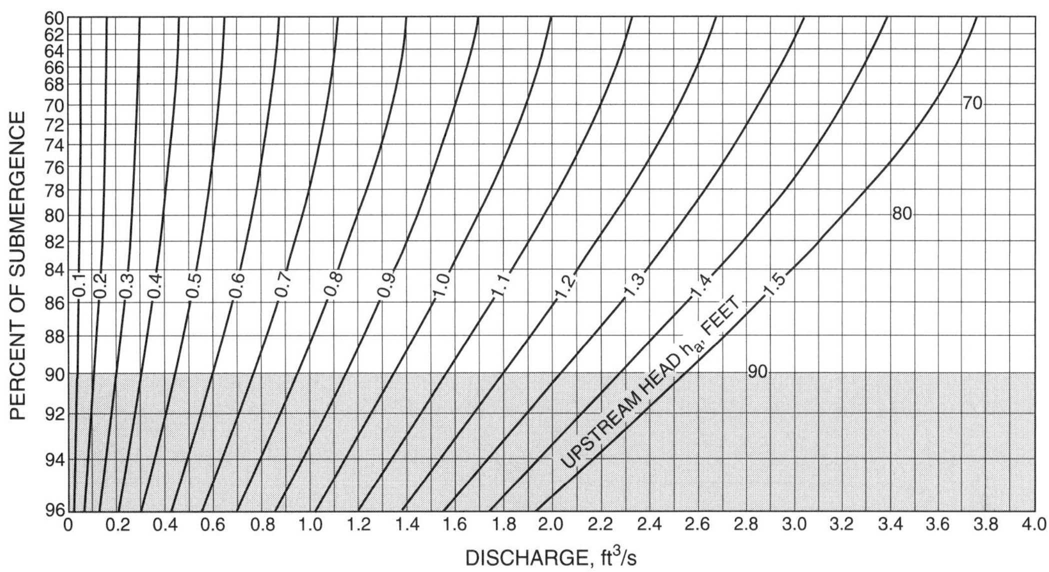

FIGURE 5.13 Diagram for determining rate of submerged flow through a 6-in Parshall flume (*U.S. Department of the Interior, 1997*).

When the ratio of H_a to H_b exceeds the following values:

0.5 for flumes of 1, 2, and 3 in width,
0.6 for flumes of 6 and 9 in width,
0.7 for flumes of 1 to 8 ft width,
0.8 for flumes of 8 to 50 ft width.

The flume is said to be submerged. When a flume is submerged, the actual discharge is less than that determined by the above equations. The diagram presented in Figs. 5.13 and 5.14 can be used to determine discharges for submerged Parshall flumes of 6 and 9 in width, respectively. Figure 5.15 shows the discharge correction for flumes with 1 to 8 ft throat width and with various percentages of submergence. The correction for the 1-ft flume can be applicable to larger flumes by multiplying by a correction factor of 1.0, 1.4, 1.8, 2.4, 3.1, 4.3, and 5.4 for flume sizes of 1, 1.5, 2, 3, 4, 6, and 8 ft width, respectively. Figure 5.16 is used for determining the correction to be subtracted from the free-flow value for a 10-ft Parshall flume. For larger sizes, the correction equals the value from the diagram multiplied by the size factor M. Figs. 5.13–5.16 are improved curves from earlier developments (Parshall 1932, 1950).

EXAMPLE 1: Calculate the discharge through a 9-in Parshall flume for the gage reading H_a of 1.1 ft when (a) the flume has free flow, and (b) the flume is operating at 72 percent of submergence.

Solution:

Step 1. Calculate Q for free flow for (a)

$$Q = 3.07\,H_a^{1.53} = 3.07\,(1.1)^{1.53}$$

$$= 3.55\ (\text{ft}^3/\text{s})$$

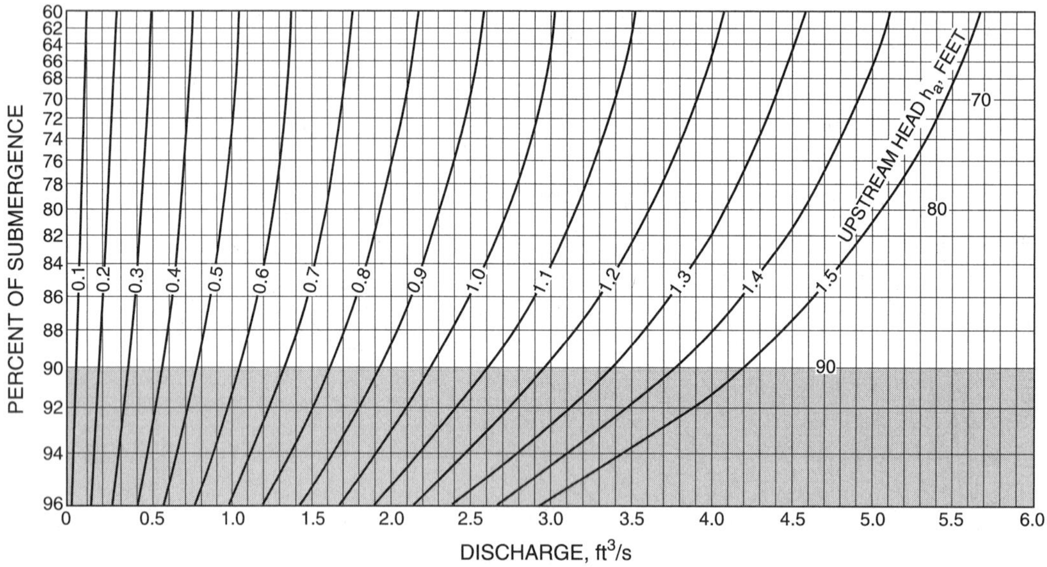

FIGURE 5.14 Diagram for determining rate of submerged flow through a 9-in Parshall flume (*U.S. Department of the Interior, 1997*).

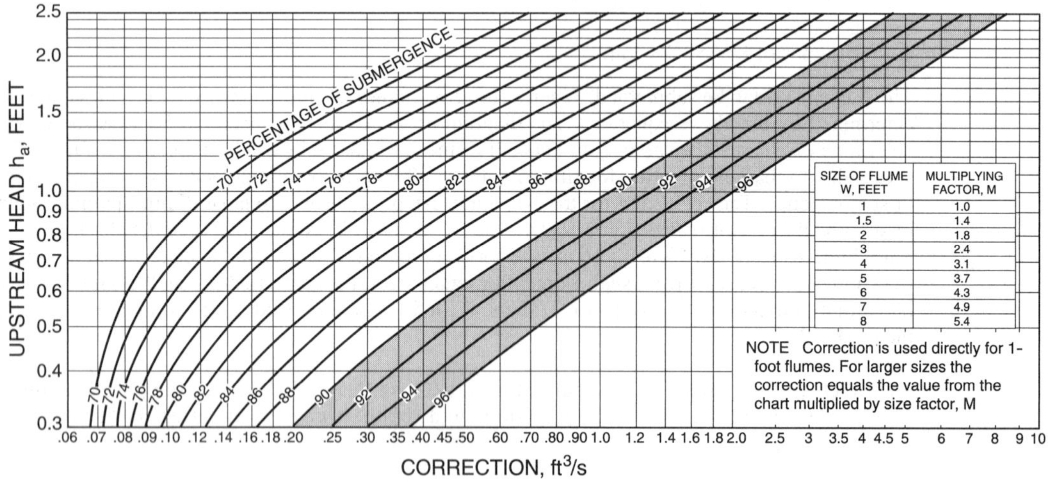

FIGURE 5.15 Diagram for determining correction to be subtracted from the free discharge to obtain rate of submerged flow through 1- to 8-ft Parshall flumes (*U.S. Department of the Interior, 1997*).

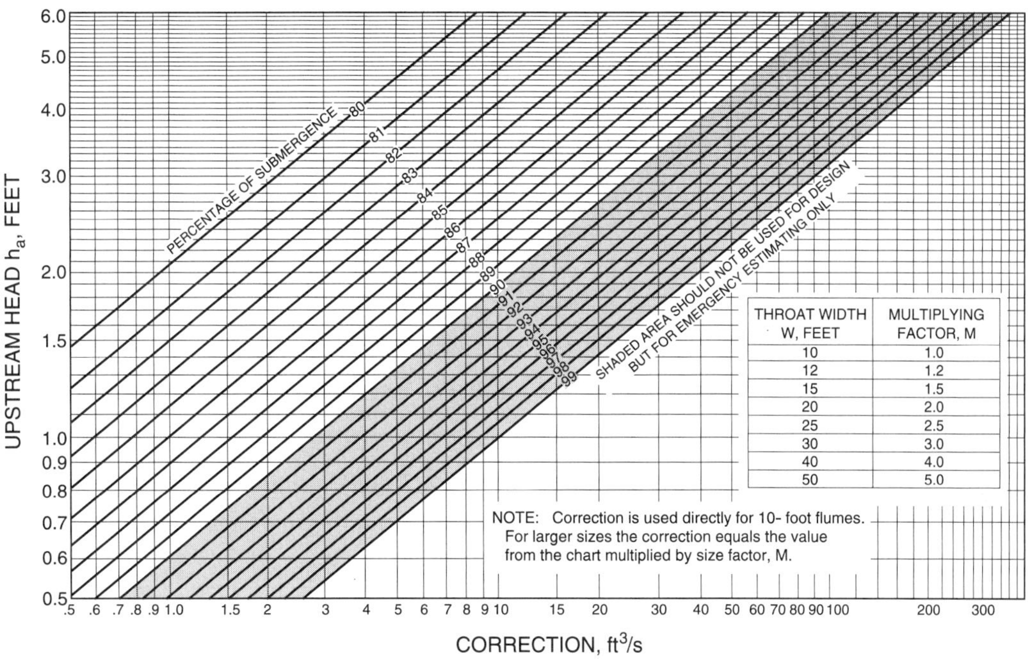

FIGURE 5.16 Diagram for determining correction to be subtracted from discharge flow to obtain rate of submerged flow through 10- to 50-ft Parshall flumes (*U.S. Department of the Interior, 1997*).

Step 2. Determine discharge with 0.72 submergence for (b)
From Fig. 5.14, enter the plot at $H_b/H_a = 0.72$ and proceed horizontally to the line representing $H_a = 1.1$. Drop downward and read a discharge of

$$Q_s = 3.33 \text{ ft}^3/\text{s}$$

EXAMPLE 2: Determine the discharge of a 2-ft Parshall flume with gage readings $H_a = 1.55$ ft and $H_b = 1.14$ ft.

Solution:

Step 1. Compute H_b/H_a ratio

$$\frac{H_b}{H_a} = \frac{1.14 \text{ ft}}{1.55 \text{ ft}} = 0.74 \text{ or } 74\%$$

This ratio exceeds 0.7, therefore the flume is submerged.

Step 2. Compute the free discharge of the 2-ft flume

$$\begin{aligned} Q &= 4WH_a^{1.522W^{0.26}} \\ &= 4(2)\,(1.55)^{1.522\times 2^{0.26}} \\ &= 8 \times 2.223 \\ &= 17.78 \text{ (ft}^3\text{/s)} \end{aligned}$$

Step 3. Determine the correction discharge for a 1-ft flume from Fig. 5.15, enter the plot at a value of $H_a = 1.55$ ft and proceed horizontally to the line representing the submergence curve of 74 percent. Then drop vertically downward and read a discharge correction for the 1-ft flume of

$$Q_{cl} = 0.40 \text{ ft}^3/\text{s}$$

Step 4. Compute discharge correction for the 2-ft flume, M = 1.8 (Fig. 5.15)

$$\begin{aligned} Q_{c2} &= 1.8 \times 0.40 \text{ ft}^3/\text{s} \\ &= 0.72 \text{ ft}^3/\text{s} \end{aligned}$$

Step 5. Compute the actual discharge

$$\begin{aligned} Q &= Q \text{ (at Step 2 − at Step 4)} \\ &= 17.78 \text{ ft}^3/\text{s} - 0.72 \text{ ft}^3/\text{s} \\ &= 17.06 \text{ ft}^3/\text{s} \end{aligned}$$

REFERENCES

American Society of Civil Engineers (ASCE) and Water Environment Federation (WEF). 1992. *Design and construction of urban stormwater management systems*. New York: ASCE & WEF.

Benefield, L. D., Judkins, J. F. Jr and Parr, A. D. 1984. Treatment plant hydraulics for environmental engineers. Englewood Cliffs, New Jersey: Prentice-Hall.

Brater, E. F., King, H. W., Lindell, J. E. and Wei, C. Y. 1996. *Handbook of hydraulics*, 7th edn. New York: McGraw-Hill.

Brisbin, S. G. 1957. Flow of concentrated raw sewage sludges in pipes. *J. Sanitary Eng. Div., ASCE*, **83**: 201.

Chadwick, A. J. and Morfett, J. C. 1986. *Hydraulics in civil engineering*. London: Allen & Unwin.

Chow, V. T. 1959. *Open-channel hydraulics*. New York: McGraw-Hill.

Cross, H. 1936. Analysis of flow in networks of conduits or conductors. University of Illinois Bulletin no. 286. Champaign-Urbana: University of Illinois.

Ductile Iron Pipe Research Association. 1997. *Ductile Iron Pipe News*, fall/winter 1997.

Fair, G. M., Geyer, J. C. and Okun, D. A. 1966. *Water and wastewater engineering, Vol. 1: Water supply and wastewater removal.* New York: John Wiley.

Fetter, C. W. 1994. *Applied hydrogeology*, 3rd edn. Upper Saddle River, New Jersey: Prentice-Hall.

Horvath, I. 1994. *Hydraulics in water and waste-water treatment technology*. Chichester: John Wiley.

Hwang, N. H. C. 1981. *Fundamentals of hydraulic engineering systems*. Englewood Cliffs, New Jersey: Prentice-Hall.

Metcalf & Eddy, Inc. 1972. *Wastewater engineering: Collection, treatment, disposal*. New York: McGraw-Hill.

Metcalf & Eddy, Inc. 1990. *Wastewater engineering: Collection and pumping of wastewater*. New York: McGraw-Hill.

James M. Montgomery, Consulting Engineers, Inc. 1985. *Water treatment principles and design*. New York: John Wiley.

Morris, H. M. and Wiggert, J. M. 1972. *Applied hydraulics in engineering*, 2nd edn. New York: John Wiley.

Parshall, R. L. 1926. The improved Venturi flume. *Trans. Amer. Soc. Civil Engineers*, **89**: 841–851.

Parshall, R. L. 1932. *Parshall flume of large size*. Colorado Agricultural Experiment Station, Bulletin 386.

Parshall, R. L. 1950. *Measuring water in irrigation channels with Parshall flumes and small weirs*. US Soil Conservation Service, Circular 843.

Perry, R. H. 1967. *Engineering manual: A practical reference of data and methods in architectural, chemical, civil, electrical, mechanical, and nuclear engineering*. New York: McGraw-Hill.

Streeter, V. L. and Wylie, E. B. 1975. *Fluid mechanics*, 6th edn. New York: McGraw-Hill.

US Department of the Interior, Bureau of Reclamation. 1997. *Water measurement manual*, 3rd edn. Denver, Colorado: US Department of the Interior.

Williams, G. S. and Hazen, A. 1933. *Hydraulic tables*, 3rd edn. New York: John Wiley.

Zipparro, V. J. and Hasen, H. 1993. *Davis handbook of applied hydraulics*. New York: McGraw-Hill.

CHAPTER 1.6
PUBLIC WATER SUPPLY

Shun Dar Lin

1 SOURCES AND QUANTITY OF WATER

The sources of a water supply may include rainwater, surface waters, and groundwater. During the water cycle, rainwater recharges the surface waters and groundwater. River, stream, natural pond, impondment, reservoir, and lake waters are major surface water sources. Some communities use groundwater, such as galleries, wells, aquifers, and springs as their water supply sources.

Water supplies may be drawn from a single source or from a number of different ones. The water from multiple sources could be mixed before distribution or separately distributed. Any new source of water has to be approved by the federal, state, and related authorities.

The quantity of water needed varies with reason, geography, size and type of community, and culture. A water supply system may provide for domestic, industrial, and commercial usage; public services; fire demand; unaccounted losses; as well as farm uses. The design flow for a water supply system is discussed in the sections of water requirement and fire demand.

The engineers should examine the availability of water sources and the quantities in the area. The prediction of the future population should be made for the design purpose. Based on long-term meteorological data, the amount of water stored in lakes and reservoirs can be estimated.

Runoff refers to the precipitation that reaches a stream or river. Theoretically, every unit volume of water passing the observation station should be measured and the sum of all these units passing in a certain period of time would be the total runoff. However, it is not available due to cost. Observations should be carried out at reasonably close intervals so that the observed flow over a long period of time is the sum of flows for the individual shorter periods of observation and not the sum of the observed rates of flow multiplied by the total period.

EXAMPLE 1: Records of observations during a month (30 days) period show that a flow rate 2.3 cfs (0.065 m^3/s) for 8 days, 3.0 cfs (0.085 m^3/s) for 10 days, 56.5 cfs (16.0 m^3/s) for 1 days, 12.5 cfs (0.354 m^3/s) for 2 days, 5.3 cfs (0.150 m^3/s) for 6 days, and 2.65 cfs (0.075 m^3/s) for 3 days. What is the mean flow rate?

Solution:

$$Q = \frac{2.3 \times 8 + 3.0 \times 10 + 56.5 \times 1 + 12.5 \times 2 + 5.3 \times 6 + 2.65 \times 3}{30}$$

$$= 5.66 \text{ ft}^3/\text{s}$$

$$= 0.160 \text{ m}^3/\text{s}$$

EXAMPLE 2: If the mean annual rainfall is 81 cm (32 in). A horizontal projected roof area is 300 m^2 (3230 ft^2). Make a rough estimate of how much water can be caught.

Solution:

Step 1. Calculate annual gross yield, V_y

$$V_y = 300 \text{ m}^2 \times 0.81 \text{ m/yr} = 243 \text{ m}^3/\text{yr}$$

$$= 0.666 \text{ m}^3/\text{d}$$

Step 2. Estimate net yield V_n

According to Fair *et al.* (1966), $V_n = \frac{2}{3} V_y$

$$V_n = \tfrac{2}{3} V_y = \tfrac{2}{3}(243 \text{ m}^3/\text{yr}) = 162 \text{ m}^3/\text{yr}$$

$$= 0.444 \text{ m}^3/\text{d}$$

Step 3. Estimate the water that can be stored and then be used, V_u

$$V_u = 0.5\ V_n = 0.5(162\ \text{m}^3/\text{yr}) = 81\ \text{m}^3/\text{yr}$$

or

$$= 2860\ \text{ft}^3/\text{yr}$$

$$= 0.5(0.444\ \text{m}^3/\text{d}) = 0.222\ \text{m}^3/\text{d}$$

or

$$= 7.83\ \text{ft}^3/\text{d}$$

EXAMPLE 3: Determine the rainfall-runoff relationship.
Using a straight line by method of average. The given values are as follows:

Year	Rainfall, in	Runoff, cfs/mile²
1988	15.9	26.5
1989	19.4	31.2
1990	23.9	38.6
1991	21.0	32.5
1992	24.8	37.4
1993	26.8	40.2
1994	25.4	38.7
1995	24.3	36.4
1996	27.3	39.6
1997	22.7	34.5

Solution:

Step 1. Write a straight line equation as $y = mx + b$ in which m is the slope of the line, and b is the intercept on the Y axis.

Step 2. Write an equation for each year in which the independent variable is X (rainfall) and the dependent variable is Y (runoff). Group the equations in two groups, and alternate the years. Then total each group.

1988	$26.5 = 15.9\ m + b$	1989	$31.2 = 19.4\ m + b$
1990	$38.6 = 23.9\ m + b$	1991	$32.5 = 21.0\ m + b$
1992	$37.4 = 24.8\ m + b$	1993	$40.2 = 26.8\ m + b$
1994	$38.7 = 25.4\ m + b$	1995	$36.4 = 24.3\ m + b$
1996	$39.6 = 27.3\ m + b$	1997	$34.5 = 22.7\ m + b$
	$180.8 = 117.3\ m + 5b$		$174.8 = 114.2\ m + 5b$

Step 3. Solve the total of each group of equations simultaneously for m and b

$$\begin{aligned} 180.8 &= 117.3m + 5b \\ -(174.8 &= 114.2m + 5b) \\ \hline 6.0 &= 3.1m \\ m &= 1.94 \end{aligned}$$

Step 4. Substituting for m and solving for b.

$$\begin{aligned} 180.8 &= 117.3(1.94) + 5b \\ 5b &= -46.76 \\ b &= -9.35 \end{aligned}$$

Step 5. The equation of straight line of best fit is

$$Y = 1.94X - 9.35$$

where X = rainfall, in
Y = runoff, cfs/mile^2

EXAMPLE 4: A watershed has a drainage area of 1000 ha (2470 acres). The annual rainfall is 927 mm (36.5 in). The expected evaporation loss is 292 mm (11.5 in) per year. The estimated loss to groundwater is 89 mm (3.5 in) annually. Estimate the amount of water that can be stored in a lake and how many people can be served, assuming 200 L/(c · d) is needed.

Solution:

Step 1. Using a mass balance

$$R \text{ (rainfall excess)} = P \text{ (precipitation)} - E \text{ (evaporation)} - G \text{ (loss to groundwater)}$$

$$R = 927 \text{ mm} - 292 \text{ mm} - 89 \text{ mm}$$

$$= 546 \text{ mm}$$

Step 2. Convert R from mm to m^3 (volume) and L

$$R = 564 \text{ mm} \times \frac{1 \text{ m}}{1000 \text{ mm}} \times 1000 \text{ ha} \times \frac{10{,}000 \text{ m}^2}{1 \text{ ha}}$$

$$= 5.46 \times 10^6 \text{ m}^3$$

$$= 5.46 \times 10^6 \text{ m}^3 \times \text{m}^3 \text{ L}/1 \text{ m}^3$$

$$= 5.46 \times 10^9 \text{ L}$$

Step 3. Compute the people that can be served

$$\text{Annual usage per capita} = 200 \text{ L/(c} \cdot \text{d)} \times 365 \text{ day}$$

$$= 7.3 \times 10^4 \text{ L/c}$$

$$\text{No. of people served} = \frac{5.46 \times 10^9 \text{ L}}{7.3 \times 10^4 \text{ L/c}}$$

$$= 74{,}800 \text{ c}$$

2 POPULATION ESTIMATES

Prior to the design of a water treatment plant, it is necessary to forecast the future population of the communities to be served. The plant should be sufficient generally for 25 to 30 years. It is difficult to estimate the population growth due to economic and social factors involved. However, a few methods have been used for forecasting population. They include the arithmetic method and uniform percentage growth rate method (Clark and Viessman, 1966; Steel and McGhee, 1979; Viessman and Hammer, 1993). The first three methods are short-term (< 10 years) forecasting.

2.1 Arithmetic Method

This method of forecasting is based upon the hypothesis that the rate of increase is constant. It may be expressed as follows:

$$\frac{dp}{dt} = k_a \tag{6.1}$$

where p = population
t = time, yr
k_a = arithmetic growth rate constant

Rearrange and integrate the above equation, p_1 and p_2 are the populations at time t_1 and t_2 respectively.

$$\int_{p_1}^{p_2} dp = \int_{t_1}^{t_2} k_a dt$$

We get

$$p_2 - p_1 = k_a(t_2 - t_1)$$

$$k_a = \frac{p_2 - p_1}{t_2 - t_1} = \frac{\Delta p}{\Delta t} \tag{6.2a}$$

or

$$p_t = p_0 + k_a t \tag{6.2b}$$

where p_t = population at future time
p_0 = present population, usually use p_2 (recently censused)

2.2 Constant Percentage Growth Rate Method

The hypothesis of constant percentage or geometric growth rate assumes that the rate increase is proportional to population. It can be written as

$$\frac{dp}{dt} = k_p p \tag{6.3a}$$

Integrating this equation yields

$$\ln p_2 - \ln p_1 = k_p (t_2 - t_1)$$

$$k_p = \frac{\ln p_2 - \ln p_1}{t_2 - t_1} \tag{6.3b}$$

The geometric estimate of population is given by

$$\ln p = \ln p_2 + k_p (t - t_2) \tag{6.3c}$$

2.3 Declining Growth Method

This is a decreasing rate of increase on the basis that the growth rate is a function of its population deficit. Mathematically it is given as

$$\frac{dp}{dt} = k_d (p_s - p) \tag{6.4a}$$

where p_s = saturation population, assume value
Integration of the above equation gives

$$\int_{p_1}^{p_2} \frac{dp}{p_s - p} = k_d \int_{t_1}^{t_2} dt$$

$$-\ln \frac{p_s - p_2}{p_s - p_1} = k_d (t_2 - t_1)$$

Rearranging

$$k_d = -\frac{1}{t_2 - t_1} \ln \frac{p_s - p_2}{p_s - p_1} \tag{6.4b}$$

The future population P is

$$P = P_0 + (P_s - P_0)(1 - e^{-k_d t}) \tag{6.4c}$$

where P_0 = population of the base year

2.4 Logistic Curve Method

The logistic curve-fitting method is used for modeling population trends with an S-shape for large population center, or nations for long-term population predictions. The logistic curve form is

$$P = \frac{p_s}{1 + e^{a+b\Delta t}} \tag{6.5a}$$

where P_s = saturation population
a, b = constants

They are:

$$P_s = \frac{2p_0 p_1 p_2 - p_1^2(p_0 + p_2)}{p_0 p_2 - p_1^2} \tag{6.5b}$$

$$a = \ln \frac{p_s - p_0}{p_0} \tag{6.6}$$

$$b = \frac{1}{n} \ln \frac{p_0(p_s - p_1)}{p_1(p_s - p_0)} \tag{6.7}$$

where n = time interval between successive censuses

Substitution of these values in Eq. (6.5a) gives the estimation of future population on P for any period Δt beyond the base year corresponding to P_0.

EXAMPLE: A mid-size city recorded populations of 113,000 and 129,000 in the April 1980 and April 1990 census, respectively. Estimate the population in January 1999 by comparing (a) arithmetic method, (b) constant percentage method, and (c) declining growth method.

Solution:

Step 1. Solve with the arithmetic method
Let t_1 and t_2 for April 1980 and April 1990, respectively

$$\Delta t = t_2 - t_1 = 10 \text{ yr}$$

Using Eq. (6.2a)

$$K_a = \frac{p_2 - p_1}{t_2 - t_1} = \frac{129{,}000 - 113{,}000}{10} = 1600$$

Predict p_t for January 1999 from t_2, using Eq. (6.2b)

$$t = 8.75 \text{ yr}$$

$$p_t = p_2 + k_a t$$

$$= 129{,}000 + 1600 \times 8.75$$

$$= 143{,}000$$

Step 2. Solve with constant percentage method, using Eq. (6.3b)

$$k_p = \frac{\ln p_2 - \ln p_1}{t_2 - t_1} = \frac{\ln 129{,}000 - \ln 113{,}000}{10}$$

$$= 0.013243$$

Then, using Eq. (6.3c)

$$\ln P = \ln P_2 + k_p (t - t_2)$$

$$= \ln 129{,}000 + 0.013243 \times 8.75$$

$$= 11.8834$$

$$p = 144{,}800$$

Step 3. Solve with declining growth method
Assuming

$$p_s = 200{,}000$$

Using Eq. (6.4b)

$$k_d = -\frac{1}{t_2 - t_1} \ln \frac{p_s - p_2}{p_s - p_1}$$

$$= -\frac{1}{10} \ln \frac{200{,}000 - 129{,}000}{200{,}000 - 113{,}000}$$

$$= 0.02032$$

From Eq. (6.4c)

$$P = P_0 + (P_s - P_0)(1 - e^{-k_d t})$$

$$= 129{,}000 + (200{,}000 - 129{,}000)(1 - e^{-0.02032 \times 8.75})$$

$$= 129{,}000 + 71{,}000 \times 0.163$$

$$= 140{,}600$$

3 WATER REQUIREMENTS

The uses of water include domestic, commercial and industrial, public services such as fire fighting and public buildings, and unaccounted pipeline system losses and leakage. The average usage in the United States for the above four categories are 220, 260, 30, and 90 liters per capita per day (L/(c · d)), respectively (Tchobanoglous and Schroeder 1985). These correspond to 58, 69, 8, and 24 gal/(c · d), respectively. Total municipal water use averages 600 L/(c · d) or 160 gal/(c · d) in the United States.

The maximum daily water use ranges from about 120 to 400 percent of the average daily use with a mean of about 180 percent. Maximum hourly use is about 150 to 12,000 percent of the annual average daily flow; and 250 to 270 percent are typically used in design.

3.1 Fire Demand

Fire demand of water is often the determining factor in the design of mains. Distribution is a short-term, small quantity but with a large flow rate. According to uniform fire code, the minimum fire flow requirement for a one- and two-family dwelling shall be 1000 gallons per minute (gpm). For the water demand for fire fighting based on downtown business districts and high-value area for communities of 200,000 people or less, the National Board of Fire Underwriters (1974) recommended the following fire flow rate and population relationship:

$$Q = 3.86\sqrt{p}\,(1 - 0.01\sqrt{p}) \quad \text{(SI units)} \tag{6.8a}$$

$$Q = 1020\sqrt{p}\,(1 - 0.01\sqrt{p}) \quad \text{(US customary units)} \tag{6.8b}$$

where Q = discharge, m^3/min or gal/min (gpm)
P = population in thousands

The required flow rate for fire fighting must be available in addition to the coincident maximum daily flow rate. The duration during the required fire flow must be available for 4 to 10 h. National Board of Fire Underwriters recommends providing for a 10-h fire in towns exceeding 2500 in population.

The Insurance Services Office Guide (International Fire Service Training Association, 1993) for determination of required fire flow recommends the formula

$$F = 18\,C\sqrt{A} \quad \text{(for US customary units)} \tag{6.9a}$$

$$F = 320\,C\sqrt{A} \quad \text{(for SI units)} \tag{6.9b}$$

where F = required fire flow, gpm or m^3/d
C = coefficient related to the type of construction

C value	Construction	Maximum flow, gpm (m^3/d)
1.5	wood frame	8000 (43,600)
1.0	ordinary	8000 (43,600)
0.9	heavy timber type building	
0.8	noncombustible	6000 (32,700)
0.6	fire-resistant	6000 (32,700)

A = total floor area, ft^2 or m^2

EXAMPLE 1: A four-story building of heavy timber type building of 715 m^2 (7700 ft^2) of ground area. Calculate the water fire requirement.

Solution:

Using Eq. (6.9b):

$$\begin{aligned} F &= 320\, C\sqrt{A} \\ &= 320 \times 0.9\sqrt{4 \times 715} \\ &= 15{,}400 \ (\text{m}^3/\text{d}) \end{aligned}$$

or

$$= 2800 \ \text{gpm}$$

EXAMPLE 2: A five-story building of ordinary construction of 7700 ft² (715 m²) of ground area communicating with a three-story building of ordinary construction of 9500 ft² (880 m²) ground area. Compute the required fire flow.

Solution:

Using Eq. (6.9a)

Total area

$$A = 5 \times 7700 + 3 \times 9500 = 67{,}000 \ \text{ft}^2$$

$$\begin{aligned} F &= 18 \times 1.0\sqrt{67{,}000} = 4660 \ (\text{gpm}) \\ &= 4750 \ (\text{gpm}) \qquad \text{rounded to nearest 250 gpm} \end{aligned}$$

or

$$= 25{,}900 \ \text{m}^3/\text{d}$$

EXAMPLE 3: Assuming a high-value residential area of 100 ha (247 acres) has a housing density of 10 houses/ha with 4 persons per household, determine the peak water demand, including fire, in this residential area.

Solution:

Step 1. Estimate population P

$$\begin{aligned} P &= (4 \ \text{capita/house})\ (10 \ \text{houses/ha})\ (100 \ \text{ha}) \\ &= 4000 \ \text{capita} \end{aligned}$$

Step 2. Estimate average daily flow Q_a

$$\begin{aligned} Q_a &= \text{residential} + \text{public service} + \text{unaccounted} \\ &= (220 + 30 + 90) \\ &= 340 \ (\text{L}/(\text{c} \cdot \text{d})) \end{aligned}$$

Step 3. Estimate maximum daily flow Q_{md} for the whole area
Using the basis of Q_{md} is 180 percent of Q_a

$$\begin{aligned} Q_{md} &= (340 \ \text{L}/(\text{c} \cdot \text{d}))(1.8)\ (4000 \ \text{c}) \\ &= 2{,}448{,}000 \ \text{L/d} \\ &\cong 2400 \ \text{m}^3/\text{d} \end{aligned}$$

Step 4. Estimate the fire demand using Eq. (5.8a)

$$\begin{aligned} Q_f &= 3.86\sqrt{p}\,(1 - 0.01\sqrt{p}) \ \text{m}^3/\text{min} \\ &= 3.86\sqrt{4}\,(1 - 0.01\sqrt{4}) \\ &= 7.57 \ \text{m}^3/\text{min} \\ &= 7.57 \ \text{m}^3/\text{min} \times 60 \ \text{min/h} \times 10 \ \text{h/d} \\ &= 4540 \ \text{m}^3/\text{d} \end{aligned}$$

Step 5. Estimate total water demand Q

$$\begin{aligned} Q &= Q_{\text{md}} + Q_{\text{f}} \\ &= 2400\ \text{m}^3/\text{d} + 4540\ \text{m}^3/\text{d} \\ &= 6940\ \text{m}^3/\text{d} \end{aligned}$$

Note: In this area, fire demand is a control factor. It is measuring to compare Q and peak daily demand.

Step 6. Check with maximum hourly demand Q_{mh}
The Q_{mh} is assumed to be 250% of average daily demand.

$$\begin{aligned} Q_{\text{mh}} &= 2400\ \text{m}^3/\text{d} \times 2.5 \\ &= 6000\ \text{m}^3/\text{d} \end{aligned}$$

Step 7. Compare Q versus Q_{mh}

$$Q = 6940\ \text{m}^3/\text{d} > Q_{\text{mh}} = 6000\ \text{m}^3/\text{d}$$

Use $Q = 6940\ \text{m}^3/\text{d}$ for the main pipe to this residential area.

EXAMPLE 4: Estimate the municipal water demands for a city of 225,000 persons.

Solution:

Step 1. Estimate the average daily demand Q_{avg}

$$\begin{aligned} Q_{\text{avg}} &= 600\ \text{L}/(\text{c} \cdot \text{d}) \times 225{,}000\ \text{c} \\ &= 135{,}000{,}000\ \text{L/d} \\ &= 1.35 \times 10^5\ \text{m}^3/\text{d} \end{aligned}$$

Step 2. Estimate the maximum daily demand Q_{md} ($f = 1.8$)

$$\begin{aligned} Q_{\text{md}} &= 1.35 \times 10^5\ \text{m}^3/\text{d} \times 1.8 \\ &= 2.43 \times 10^5\ \text{m}^3/\text{d} \end{aligned}$$

Step 3. Calculate the fire demand Q_{f}, using Eq. (5.8a)

$$\begin{aligned} Q_{\text{f}} &= 3.86\sqrt{p}\,(1 - 0.01\sqrt{p})\ \text{m}^3/\text{min} \\ &= 3.86\sqrt{225}\,(1 - 0.01\sqrt{225})\ \text{m}^3/\text{min} \\ &= 49.215\ \text{m}^3/\text{min} \end{aligned}$$

For 10 h duration of daily rate

$$Q_{\text{f}} = 49.215\ \text{m}^3/\text{min} \times 60\ \text{min/h} \times 10\ \text{h/d} = 0.30 \times 10^5\ \text{m}^3/\text{d}$$

Step 4. Sum of Q_{md} and Q_{f} (fire occurs coincident to peak flow)

$$\begin{aligned} Q_{\text{md}} + Q_{\text{f}} &= (2.43 + 0.30) \times 10^5\ \text{m}^3/\text{d} \\ &= 2.73 \times 10^5\ \text{m}^3/\text{d} \end{aligned}$$

Step 5. Estimate the maximum hourly demand Q_{mh} ($f = 2.7$)

$$\begin{aligned} Q_{\text{mh}} &= f\,Q_{\text{avg}} \\ &= 2.7 \times 1.35 \times 10^5\ \text{m}^3/\text{d} \\ &= 3.645 \times 10^5\ \text{m}^3/\text{d} = 96.3\ \text{MGD} \end{aligned}$$

Step 6. Compare steps 4 and 5

$$Q_{\text{mh}} = 3.645 \times 10^5 \text{ m}^3\text{/d} > Q_{\text{md}} + Q_{\text{f}} = 2.73 \times 10^5 \text{ m}^3\text{/d}$$

Use $Q_{\text{mh}} = 3.65 \times 10^5$ m^3/d to design the plant's storage capacity.

Fire flow tests. Fire flow tests involve discharging water at a known flow rate from one or more hydrants and examining the corresponding pressure drop from the main through another nearby hydrant. The discharge from a hydrant nozzle can be determined as follows (Hammer, 1975):

$$Q = 29.8\, C\, d^2 \sqrt{p} \qquad (6.10)$$

where Q = hydrant discharge, gpm
C = coefficient, normally 0.9
d = diameter of outlet, in
p = pitot gage reading, psi

The computed discharge at a specified residual pressure can be computed as

$$\frac{Q_{\text{p}}}{Q_{\text{f}}} = \left(\frac{\Delta H_{\text{p}}}{\Delta H_{\text{f}}}\right)^{0.54} \qquad (6.11\text{a})$$

or

$$Q_{\text{p}} = Q_{\text{f}}\left(\frac{\Delta H_{\text{p}}}{\Delta H_{\text{f}}}\right)^{0.54} \qquad (6.11\text{b})$$

where Q_{p} = computed flow rate at the specified residual pressure, gpm
Q_{f} = total discharge during fire flow test, gpm
ΔH_{p} = pressure drop from beginning to special pressure, psi
ΔH_{f} = pressure drop during fire flow test, psi

EXAMPLE: All four hydrant numbers 2, 4, 6, and 8, as shown in the following figure, have the same nozzle size of 2.5 in and discharge at the same rate. The pitot tube pressure at each hydrant during the test is 25 psi. At this discharge the residual pressure at hydrant #5 dropped from the original 100 psi to 65 psi. Compute the flow rate at a residual pressure of 30 psi based on the test.

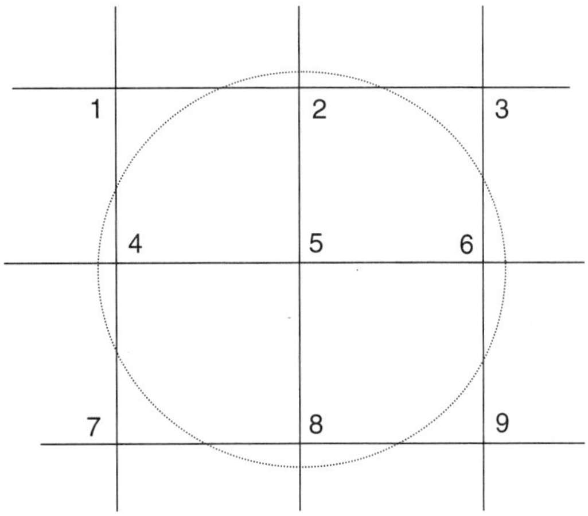

Solution:

Step 1. Determine total discharge during the test
Using Eq. (6.10):

$$Q_f = 4 \times 29.8\, C\, d^2 \sqrt{p} = 4 \times 29.8 \times 0.9 \times (2.5)^2 \times \sqrt{25}$$
$$= 3353 \text{ (gpm)}$$

Step 2. Compute Q_p at 30 psi by using Eq. (6.11b)

$$Q_p = Q_f \left(\frac{\Delta H_p}{\Delta H_f}\right)^{0.54} = 3353 \times \left(\frac{100 - 30}{100 - 65}\right)^{0.54}$$
$$= 4875 \text{ (gpm)}$$

3.2 Leakage Test

After a new main is installed, a leakage test is usually required. The testing is generally carried out for a pipe length not exceeding 300 m (1000 ft). The pipe is filled with water and pressurized (50 percent above normal operation pressure) for at least 30 min. The leakage limit recommended by the American Water Work Association is determined as:

$$L = (ND\sqrt{p})/326 \quad \text{for SI units} \tag{6.12a}$$

$$L = (ND\sqrt{p})/1850 \quad \text{for US customary units} \tag{6.12b}$$

where L = allowable leakage, L/(mm diameter · km · d) or gal/(in · mile · d)
N = number of joints in the test line
D = normal diameter of the pipe, mm or in
P = average test pressure

Leakage allowed in a new main is generally specified in the design. It ranges from 5.6 to 3.2 L/mm diameter per km per day (60 to 250 gal/(in diameter/mile/d). Nevertheless, recently, some water companies are not allowing for any leakage in a new main due to the use of better sealers.

4 REGULATION FOR WATER QUALITY

4.1 Federal Drinking Water Standards

Water Quality is a term used to describe the physical, chemical, and biological characteristics of water with respect to its suitability for a particular use. National Primary Drinking Water Regulations (NPDWRs or primary standards) are legally enforceable standards that apply to public water systems. The NPDWRs rotect drinking water quality by limiting the levels of specific contaminants that can adversely affect public health and are known or are anticipated to occur in public water systems. The NPDWRs, as shown in Table 6.1, divides these contaminants into inorganic chemicals, organic chemicals, radionuclides, and microorganisms (USEPA, 2000). All the primary standards are based on health effects to the customers; and they are mandatory standards. The USEPA (2000) also promulgated secondary drinking water regulations as shown in Table 6.2. The secondary drinking water regulations pertain to those contaminants such as taste, odor, color, and some chemicals that may adversely affect the aesthetic quality of drinking water. They are intended as guidelines for the states. The states may establish higher or lower values as appropriate to their particular circumstances, which are adequate to protect public health and welfare.

TABLE 6.1 National Primary Drinking Water Regulations

	MCLG[1] (mg/L)[2]	MCL or TT[1] (mg/L)[2]	Potential health effects from ingestion of water	Sources of contaminant in drinking water
Contaminant			**Microorganisms**	
Cryptosporidium	0	TT[3]	Gastrointestinal illness (e.g., diarrhea, vomiting, cramps)	Human and fecal animal waste
Giardia lamblia	0	TT[3]	Gastrointestinal illness (e.g., diarrhea, vomiting, cramps)	Human and animal fecal waste
Heterotrophic plate count	n/a	TT[3]	HPC has no health effects; it is an analytic method used to measure the variety of bacteria that are common in water. Lower the concentration of bacteria in drinking water, better maintained is the water system.	HPC measures a range of bacteria that are naturally present in the environment
Legionella	0	TT[3]	Legionnaire's disease, a type of pneumonia	Found naturally in water; multiplies in heating systems
Total Coliforms (including fecal coliform and *Escherichia Coli*)[5]	0	5.0%[4]	Not a health threat in itself; it is used to indicate whether other potentially harmful bacteria may be present[5]	Coliforms are naturally present in the environment, as well as feces, fecal coliforms and *E. coli* only come from human and animal fecal waste.
Turbidity	n/a	TT[3]	Turbidity is a measure of the cloudiness of water. It is used to indicate water quality and filtration effectiveness (e.g., whether disease-causing organisms are present). Higher turbidity levels are often associated with higher levels of disease-causing microorganisms such as viruses, parasites, and some bacteria. These organisms can cause symptoms such as nausea, cramps, diarrhea, and associated headaches.	Soil runoff
Viruses (enteric)	0	TT[3]	Gastrointestinal illness (e.g., diarrhea, vomiting, cramps)	Human and animal fecal waste
			Disinfection Byproducts	
Bromate	0	0.010	Increased risk of cancer	Byproduct of drinking water disinfection
Chlorite	0.8	1.0	Anemia; infants & young children: nervous system effects	Byproduct of drinking water disinfection
Haloacetic acids (HAA5)	n/a[6]	0.060	Increased risk of cancer	Byproduct of drinking water disinfection
Total Trihalomethanes (TTHMs)	none[7] n/a[6]	0.10 0.080	Liver, kidney, or central nervous system problems; increased risk of cancer	Byproduct of drinking water disinfection
	MRDL[1] (mg/L)[2]	MRDL[1] (mg/L)[2]	Potential health effects from ingestion of water	Sources of contaminant in drinking water
Contaminant			**Disinfectants**	
Chloramines (as Cl_2)	MRDLG = 4[1]	MRDL = 4.0[1]	Eye/nose irritation; stomach discomfort, anemia	Water additive used to control microbes
Chlorine (as Cl_2)	MRDLG = 4.0[1]	MRDL = 4.0[1]	Eye/nose irritation; stomach discomfort	Water additive used to control microbes
Chlorine dioxide (as ClO_2)	MRDLG = 0.8[1]	MRDL = 0.8[1]	Anemia; infants & young children: nervous system effects	Water additive used to control microbes

(*Continued*)

TABLE 6.1 National Primary Drinking Water Regulations (*Continued*)

Contaminant	MCLG[1] (mg/L)[2]	MCL or TT[1] (mg/L)[2]	Potential health effects from ingestion of water	Sources of contaminant in drinking water
			Inorganic Chemicals	
Antimony	0.006	0.006	Increase in blood cholesterol; decrease in blood sugar	Discharge from petroleum refineries; fire retardants; ceramics; electronics; solder
Arsenic	0[7]	0.010 as of 01/23/06	Skin damage or problems with circulatory systems, and may have increased risk of getting cancer	Erosion of natural deposits; runoff from orchards, runoff from glass & electronicsproduction wastes
Asbestos (fiber > 10 μm)	7 million fibers per liter	7 MFL	Increased risk of developing benign intestinal polyps	Decay of asbestos cement in water mains; erosion of natural deposits
Barium	2	2	Increase in blood pressure	Discharge of drilling wastes; discharge from metal refineries; erosion of natural deposits
Beryllium	0.004	0.004	Intestinal lesions	Discharge from metal refineries and coal-burning factories; discharge from electrical, aerospace and defense industries
Cadmium	0.005	0.005	Kidney damage	Corrosion of galvanized pipes; erosion of natural deposits; discharge from metal refineries; runoff from waste batteries and paints
Chromium (total)	0.1	0.1	Allergic dermatitis	Discharge from steel and pulp mills; erosion of natural deposits
Copper	1.3	TT[8]; Action Level = 1.3	Short term exposure: gastrointestinal distress Long term exposure: liver or kidney damage People with Wilson's Disease should consult their personal doctor if the amount of copper in their water exceeds the action level	Corrosion of household plumbing systems; erosion of natural deposits
Cyanide (as free cyanide)	0.2	0.2	Nerve damage or thyroid problems	Discharge from steel/metal factories; discharge from plastic and fertilizer factories
Fluoride	4.0	4.0	Bone disease (pain and tenderness of the bones); Children may get mottled teeth	Water additive which promotes strong teeth; erosion of natural deposits; discharge from fertilizer and aluminum factories
Lead	0	TT[8]; Action Level = 0.015	Infants and children: delays in physical or mental development; children could show slight deficits in attention span and learning abilities Adults: kidney problems; high blood pressure	Corrosion of household plumbing systems; erosion of natural deposits
Mercury (inorganic)	0.002	0.002	Kidney damage	Erosion of natural deposits; discharge from refineries and factories; runoff from landfills and croplands

TABLE 6.1 National Primary Drinking Water Regulations (*Continued*)

	MCLG[1] (mg/L)[2]	MCL or TT[1] (mg/L)[2]	Potential health effects from ingestion of water	Sources of contaminant in drinking water
Contaminant			**Inorganic Chemicals**	
Nitrate (measured as nitrogen)	10	10	Infants below the age of 6 months who drink water containing nitrate in excess of the MCL could become seriously ill and, if untreated, may die. Symptoms include shortness of breath and blue-baby syndrome.	Runoff from fertilizer use; leaching from septic tanks, sewage; erosion of natural deposits
Nitrite (measured as nitrogen)	1	1	Infants below the age of 6 months who drink water containing nitrite in excess of the MCL could become seriously ill and, if untreated, may die. Symptoms include shortness of breath and blue-baby syndrome.	Runoff from fertilizer use; leaching from septic tanks, sewage; erosion of natural deposits
Selenium	0.05	0.05	Hair or fingernail loss; numbness in fingers or toes; circulatory problems	Discharge from petroleum refineries; erosion of natural deposits; discharge from mines
Thallium	0.0005	0.002	Hair loss; changes in blood; kidney, intestine, or liver problems	Leaching from ore-processing sites; discharge from electronics, glass, and drug factories
			Organic Chemicals	
Acrylamide	0	TT[9]	Nervous system or blood problems; increased risk of cancer	Added to water during sewage/ waste water treatment
Alachlor	0	0.002	Eye, liver, kidney or spleen problems; anemia; increased risk of cancer	Runoff from herbicide used on row crops
Atrazine	0.003	0.003	Cardiovascular system or reproductive problems	Runoff from herbicide used on row crops
Benzene	0	0.005	Anemia; decrease in blood platelets; increased risk of cancer	Discharge from factories; leaching from gas storage tanks and landfills
Benzo(a)pyrene (PAHs)	0	0.0002	Reproductive difficulties; increased risk of cancer	Leaching from linings of water storage tanks and distribution lines
Carbofuran	0.04	0.04	Problems with blood, nervous system, or reproductive system	Leaching of soil fumigant used on rice and alfalfa
Carbon tetrachloride	0	0.005	Liver problems; increased risk of cancer	Discharge from chemical plants and other industrial activities
Chlordane	0	0.002	Liver or nervous system problems; increased risk of cancer	Residue of banned termiticide
Chlorobenzene	0.1	0.1	Liver or kidney problems	Discharge from chemical and agricultural chemical factories
2,4-D	0.07	0.07	Kidney, liver, or adrenal gland problems	Runoff from herbicide used on row crops
Dalapon	0.2	0.2	Minor kidney changes	Runoff from herbicide used on rights of way
1,2-Dibromo-3-chloropropane (DBCP)	0	0.0002	Reproductive difficulties; increased risk of cancer	Runoff/leaching from soil fumigant used on soybeans, cotton, pineapples, and orchards
o-Dichlorobenzene	0.6	0.6	Liver, kidney, or circulatory system problems	Discharge from industrial chemical factories

(*Continued*)

TABLE 6.1 National Primary Drinking Water Regulations (*Continued*)

Contaminant	MCLG[1] (mg/L)[2]	MCL or TT[1] (mg/L)[2]	Potential health effects from ingestion of water	Sources of contaminant in drinking water
			Organic Chemicals	
p-Dichlorobenzene	0.075	0.075	Anemia; liver, kidney, or spleen damage; changes in blood	Discharge from industrial chemical factories
1,2-Dichloroethane	0	0.005	Increased risk of cancer	Discharge from industrial chemical factories
1,1-Dichloroethylene	0.007	0.007	Liver problems	Discharge from industrial chemical factories
cis-1,2-Dichloroethylene	0.07	0.07	Liver problems	Discharge from industrial chemical factories
trans-1,2-Dichloroethylene	0.1	0.1	Liver problems	Discharge from industrial chemical factories
Dichloromethane	0	0.005	Liver problems; increased risk of cancer	Discharge from drug and chemical factories
1,2-Dichloropropane	0	0.005	Increased risk of cancer	Discharge from industrial chemical factories
Di(2-ethylhexyl) adipate	0.4	0.4	Weight loss, liver problems, or possible reproductive difficulties.	Discharge from chemical factories
Di(2-ethylhexyl) phthalat	0	0.006	Reproductive difficulties; liver problems; increased risk of cancer	Discharge from rubber and chemical factories
Dinoseb	0.007	0.007	Reproductive difficulties	Runoff from herbicide used on soybeans and vegetables
Dioxin (2,3,7,8-TCDD)	0	0.00000003	Reproductive difficulties; increased risk of cancer	Emissions from waste incineration and other combustion; discharge from chemical factories
Diquat	0.02	0.02	Cataracts	Runoff from herbicide use
Endothall	0.1	0.1	Stomach and intestinal problems	Runoff from herbicide use
Endrin	0.002	0.002	Liver problems	Residue of banned insecticide
Epichlorohydrin	0	TT[9]	Increased cancer risk; and over a long period of time, stomach problems	Discharge from industrial chemical factories; an impurity of some water treatment chemicals
Ethylbenzene	0.7	0.7	Liver or kidneys problems	Discharge from petroleum refineries
Ethylene dibromide	0	0.00005	Problems with liver, stomach, reproductive system, or kidneys; increased risk of cancer	Discharge from petroleum refineries
Glyphosate	0.7	0.7	Kidney problems; reproductive difficulties	Runoff from herbicide use
Heptachlor	0	0.0004	Liver damage; increased risk of cancer	Residue of banned termiticide
Heptachlor epoxide	0	0.0002	Liver damage; increased risk of cancer	Breakdown of heptachlor
Hexachloro-benzene	0	0.001	Liver or kidney problems; reproductive difficulties; increased risk of cancer	Discharge from metal refineries and agricultural chemical factories
Hexachlorocy-clopentadiene	0.05	0.05	Kidney or stomach problems	Discharge from chemical factories
Lindane	0.0002	0.0002	Liver or kidney problems	Runoff/leaching from insec ticide used on cattle, lumber, gardens

TABLE 6.1 National Primary Drinking Water Regulations (*Continued*)

Contaminant	MCLG[1] (mg/L)[2]	MCL or TT[1] (mg/L)[2]	Potential health effects from ingestion of water	Sources of contaminant in drinking water
			Organic Chemicals	
Methoxychlor	0.04	0.04	Reproductive difficulties	Runoff/leaching from insecticide used on fruits, vegetables, alfalfa, livestock
Oxamyl (Vydate)	0.2	0.2	Slight nervous system effects	Runoff/leaching from insecticide used on apples, potatoes, and tomatoes
Polychlorinated biphenyls (PCBs)	0	0.0005	Skin changes; thymus gland problems; immune deficiencies; reproductive or nervous system difficulties; increased risk of cancer	Runoff from landfills; discharge of waste chemicals
Pentachloro-phenol	0	0.001	Liver or kidney problems; increased cancer risk	Discharge from wood preserving factories
Picloram	0.5	0.5	Liver problems	Herbicide runoff
Simazine	0.004	0.004	Problems with blood	Herbicide runoff
Styrene	0.1	0.1	Liver, kidney, or circulatory system problems	Discharge from rubber and plastic factories; leaching from landfills
Tetrachloroe-thylene	0	0.005	Liver problems; increased risk of cancer	Discharge from factories and dry cleaners
Toluene	1	1	Nervous system, kidney, or liver problems	Discharge from petroleum factories
Toxaphene	0	0.003	Kidney, liver, or thyroid problems; increased risk of cancer	Runoff/leaching from insecticide used on cotton and cattle
2,4,5-TP (Silvex)	0.05	0.05	Liver problems	Residue of banned herbicide
1,2,4-Trichlorobenzene	0.07	0.07	Changes in adrenal glands	Discharge from textile finishing factories
1,1,1-Trichloroethane	0.20	0.2	Liver, nervous system, or circulatory problems	Discharge from metal degreasing sites and other factories
1,1,2-Trichloroethane	0.003	0.005	Liver, kidney, or immune system problems	Discharge from industrial chemical factories
Trichloroethylene	0	0.005	Liver problems; increased risk of cancer	Discharge from metal degreasing sites and other factories
Vinyl chloride	0	0.002	Increased risk of cancer	Leaching from PVC pipes; discharge from plastic factories
Xylenes (total)	10	10	Nervous system damage	Discharge from petroleum factories; discharge from chemical factories
			Radionuclides	
Alpha particles	none[7] 0	15 picocuries per liter (pCi/L)	Increased risk of cancer	Erosion of natural deposits of certain minerals that are radioactive and may emit a form of radiation known as alpha radiation
Beta particles and photon emitters	none[7] 0	4 millirems per year	Increased risk of cancer	Decay of natural and man-made deposits of certain minerals that are radioactive and may emit forms of radiation known as photons and beta radiation

(*Continued*)

TABLE 6.1 National Primary Dringking Water Regulations (*Concluded*)

Contaminant	MCLG[1] (mg/L)[2]	MCL or TT[1] (mg/L)[2]	Potential health effects from ingestion of water	Sources of contaminant in drinking water
			Radionuclides	
Radium 226 and Radium 228 (combined)	none[7] 0	5 pCi/L	Increased risk of cancer	Erosion of natural deposits
Uranium	0	30 μg/L as of 12/08/03	Increased risk of cancer, kidney toxicity	Erosion of natural deposits

Notes:

[1]Definitions:

Maximum contaminant level (MCL): The highest level of a contaminant that is allowed in drinking water. MCLs are set as close to MCLGs as feasible using the best available treatment technology and taking cost into consideration. MCLs are enforceable standards.

Maximum contaminant level goal (MCLG): The level of a contaminant in drinking water below which there is no known or expected risk to health. MCLGs allow for a margin of safety and are non-enforceable public health goals.

Maximum residual disinfectant level (MRDL): The highest level of a disinfectant allowed in drinking water. There is convincing evidence that addition of a disinfectant is necessary for control of microbial contaminants.

Maximum residual disinfectant level goal (MRDLG): The level of a drinking water disinfectant below which there is no known or expected risk to health. MRDLGs do not reflect the benefits of the use of disinfectants to control microbial contaminants.

Treatment technique: A required process intended to reduce the level of a contaminant in drinking water.

[2]Units are in milligrams per liter (mg/L) unless otherwise noted. Milligrams per liter are equivalent to parts per million.

[3]EPA's surface water treatment rules require systems using surface water or ground water under the direct influence of surface water to (1) disinfect their water and (2) filter their water or meet criteria for avoiding filtration so that the following contaminants are controlled at the following levels:

- Cryptosporidium: (as of 1/1/02 for systems serving >10,000 and 1/14/05 for systems serving < 10,000) 99 percent removal.
- *Giardia lamblia*: 99.9 percent removal/inactivation
- Viruses: 99.99% removal/inactivation
- *Legionella*: No limit, but EPA believes that if *Giardia* and viruses are removed/inactivated, *Legionella* will also be controlled.
- Turbidity: At no time can turbidity (cloudiness of water) go above 5 nephelolometric turbidity units (NTU); systems that filter must ensure that the turbidity go no higher than 1 NTU (0.5 NTU for conventional or direct filtration) in at least 95percent of the daily samples in any month. As of January 1, 2002, turbidity may never exceed 1 NTU, and must not exceed 0.3 NTU in 95 percent of daily samples in any month.
- HPC: No more than 500 bacterial colonies per milliliter.
- Long Term 1 enhanced surface water treatment (Effective Date: January 14, 2005); Surface water systems or (GWUDI) systems serving fewer than 10,000 people must comply with the applicable Long Term 1 Enhanced Surface Water Treatment Rule provisions (e.g. turbidity standards, individual filter monitoring, Cryptosporidium removal requirements, updated watershed control requirements for unfiltered systems).
- Filter Backwash Recycling; The Filter Backwash Recycling Rule requires systems that recycle to return specific recycle flows through all processes of the system's existing conventional or direct filtration system or at an alternate location approved by the state.

[4]more than 5.0 percent samples total coliform-positive in a month. (For water systems that collect fewer than 40 routine samples per month, no more than one sample can be total coliform-positive per month.) Every sample that has total coliform must be analyzed for either fecal coliforms or *E. coli* if two consecutive TC-positive samples, and one is also positive for *E. coli* fecal coliforms, system has an acute MCL violation.

[5]Fecal coliform and *E. coli* are bacteria whose presence indicates that the water may be contaminated with human or animal wastes. Disease-causing microbes (pathogens) in these wastes can cause diarrhea, cramps, nausea, headaches, or other symptoms. These pathogens may pose a special health risk for infants, young children, and people with severely compromised immune systems.

[6]Although there is no collective MCLG for this contaminant group, there are individual MCLGs for some of the individual contaminants:

- Trihalomethanes: bromodichloromethane (zero); bromoform (zero); dibromochloromethane (0.06 mg/L). Chloroform is regulated with this group but has no MCLG.
- Haloacetic acids: dichloroacetic acid (zero); trichloroacetic acid (0.3 mg/L). Monochloroacetic acid, bromoacetic acid, and dibromoacetic acid are regulated with this group but have no MCLGs.

[7]MCLGs were not established before the 1986 amendments to the Safe Drinking Water Act. Therefore, there is no MCLG for this contaminant.

[8]Lead and copper are regulated by a treatment technique that requires systems to control the corrosiveness of their water. If more than 10 percent of tap water samples exceed the action level, water systems must take additional steps. For copper, the action level is 1.3 mg/L and for lead is 0.015 mg/L.

[9]Each water system must certify, in writing, to the state (using third-party or manufacturer's certification) that when acrylamide and epichlorohydrin are used in drinking water systems, the combination (or product) of dose and monomer level does not exceed the levels specified, as follows:

- Acrylamide = 0.05 percent dosed at 1 mg/L (or equivalent)
- Epichlorohydrin = 0.01 percent dosed at 20 mg/L (or equivalent)

Source: *www.epa.gov/OGWDW/regs.html,* 2007 (modified)

TABLE 6.2 National Secondary Drinking Water Regulations

Contaminants	Concentration
Aluminum	0.05 to 0.2 mg/L
Chloride	250 mg/L
Color	15 (color units)
Copper	1.0 mg/L
Corrosivity	Non-corrosive
Fluoride	2.0 mg/L
Foaming Agents	0.5 mg/L
Iron	0.3 mg/L
Manganese	0.05 mg/L
Odor	3 threshold odor number
pH	6.5–8.5
Silver	0.10 mg/L
Sulfate	250 mg/L
Total dissolved solids	500 mg/L
Zinc	5 mg/L

Note: National Secondary Drinking Water Regulations (NSDWRs or secondary standards) are non-enforceable guidelines regulating contaminants that may cause cosmetic effects (such as skin or tooth discoloration) or aesthetic effects (such as taste, odor, or color) in drinking water. EPA recommends secondary standards to water systems but does not require systems to comply. However, states may choose to adopt them as enforceable standards.
Source: *www.epa.gov/OGWDW/regs.html,* 2007

The design engineer should be thoroughly familiar with and integrate the various regulatory requirements into a water treatment design. It is necessary to obtain a permit from the state regulatory agency. Meetings with the representatives of the necessary regulatory agencies and related parties are recommended. The state reviews the design based on meeting minimum design requirements or standards, local conditions, and good engineering practice and experience.

The most widely used standards are the Recommended Standards for Water Works promulgated by the Great Lakes–Upper Mississippi River Board of State Sanitary Engineers (GLUMRB) which are the so-called Ten State Standards. This Board includes representatives from the states of Illinois, Indiana, Iowa, Michigan, Minnesota, Missouri, New York, Ohio, Pennsylvania, and Wisconsin. Recently, the Province of Ontario (Canada) has joined the Board. Many other states have used identical or similar versions of the Ten State Standards.

Maximum contaminant level goal (MCLG) for water quality is set by the US Environmental Protection Agency (EPA). For agents in drinking water not considered to have carcinogenic potential, i.e. no effect levels for chromic and/or lifetime period of exposure including a margin of safety, are commonly referred to as reference dosages (RfDs). Reference dosages are the exposure levels estimated to be without significant risk to humans when received daily over a lifetime. They are traditionally reported as mg per (kg · d) or mg/(kg · d). For MCLG purposes, however, no effect level in drinking water is measured by mg/L which has been in terms of the drinking water equivalent level (DWEL). The DWEL is calculated as (Cotruvo and Vogt, 1990)

$$\text{DWEL} = \frac{(\text{NOAEL})(70)}{(\text{UF})(2)} \tag{6.13}$$

where NOAEL = no observed adverse effect level, mg/(kg · d)
(70) = assumed weight of an adult, kg
UF = uncertainty factor (usually 10, 100, or 1000)
(2) = assumed quantity of water consumed by an adult, L/d

When sufficient data is available on the relative contribution from other sources, i.e. from food and air, the MCLG can be determined as follows:

$$\text{MCLG} = \text{RfD} - \text{contribution from (food + air)}$$

In fact, comprehensive data on the contributions from food and air are generally lacking. Therefore, in this case, MCLG is determined as MCLG = DWEL. The drinking water contribution often used in the absence of specific data for food and air is 20 percent of RfD. This effectively provides an additional safety factor of 5.

EXAMPLE: The MCLG for nitrate-nitrogen is 10 mg/L. Assume UF = 100. What is the "no observed adverse effect level for the drinking water?"

Solution:

$$\text{DWEL} = \frac{(\text{NOAEL}) \times 70}{(\text{UF}) \times 2}$$

since

$$\text{DWEL} = \text{MCLG} = 10 \text{ mg/L}$$

$$10 \text{ mg/L} = \frac{(\text{NOAEL, mg/(kg} \cdot \text{d)}) \times (70 \text{ kg})}{100 \times 2 \text{ L/d}}$$

$$\text{NOAEL} = \frac{2000}{70}$$

$$= 28.6 \text{ (mg/(kg} \cdot \text{d))}$$

4.2 Atrazine

Atrazine is a widely used organic herbicide due to its being inexpensive compared to other chemicals. Often, excessive applications are used. When soil and climatic conditions are favorable, atrazine may be transported to the drinking water sources by runoff or by leaching into ground water.

Atrazine has been shown to affect offspring of rats and the hearts of dogs. The USEPA has set the drinking water standard for atrazine concentration to 0.003 mg/L (or 3 μg/L) to protect the public against the risk of the adverse health effects. When atrazine or any other herbicide or pesticide is detected in surfacewater or groundwater, it is not meant as a human health risk. Treated (finished) water meets the EPA limits with little or no health risk and is considered safe with respect to atrazine.

EXAMPLE: The lifetime health advisory level of 0.003 μg/L for atrazine is set at 1000 times below the amount that causes no adverse health effects in laboratory animals. The no observed adverse effect level is 0.086 mg/(kg · d). How is the 0.003 mg/L of atrazine in water derived?

Solution:

As previous Example 1,
Using Eq. (6.13)

$$\text{DWEL} = \frac{(\text{NOAEL})(70)}{(\text{UF})(2)}$$

$$= \frac{(0.086 \text{ mg/(kg} \cdot \text{d)})(70 \text{ kg})}{(1000)(2 \text{ L/d})}$$

$$= 0.003 \text{ mg/L}$$

5 *WATER TREATMENT PROCESSES*

As the raw surface water comes to the treatment plant, physical screening is the first step to remove coarse material and debris. Thereafter, following the basic treatment process of clarification, it would include coagulation, flocculation, and sedimentation prior to filtration, then disinfection (mostly by the use of chlorination). With a good quality source, the conventional treatment processes may be modified by removing the sedimentation process and to just have the coagulation and flocculation processes followed by filtration. This treatment process scheme is called direct filtration.

Groundwater is generally better quality; however, it is typically associated with high hardness, iron, and manganese content. Aeration or air stripping is required to remove volatile compounds in groundwater. Lime softening is also necessary to remove the impurities and recarbonation is used to neutralize excess lime and to lower the pH value. Ion exchange processes can also be employed for the softening of water and the removal of other impurities, if low in iron, manganese, particulates, and organics. For a groundwater source high in iron and manganese, but with acceptable hardness, the aeration process and/or chemical oxidation can be followed by filtration to remove these compounds.

The reverse osmosis process is used to remove chemical constituents and salts in water. It offers the promise of conversion of salt water to fresh water. Other membrane processes, such as microfiltration, ultrafiltration, nanofiltration, and electrodialysis, are also applied in the water industry.

6 AERATION AND AIR STRIPPING

Aeration has been used to remove trace volatile organic compounds (VOCs) in water either from surfacewater or groundwater sources. It has also been employed to transfer a substance, such as oxygen, from air or a gas phase into water in a process called gas adsorption or oxidation (to oxidize iron and/or manganese). Aeration also provides the escape of dissolved gases, such as carbon dioxide and hydrogen sulfide.

Air stripping has also been utilized effectively to remove ammonia gas from wastewater and to remove volatile tastes and other such substances in water.

The solubility of gases which do not react chemically with the solvent and the partial pressure of the gases at a given temperature can be expressed as Henry's law:

$$p = Hx \tag{6.14}$$

where x = solubility of a gas in the solution phase
H = Henry's constant
p = partial pressure of a gas over the solution

In terms of the partial pressure, Dalton's law states that the total pressure (P_t) of a mixture of gases is just the sum of the pressures that each gas (P_i) would exert if it were present alone:

$$P_t = P_1 + P_2 + P_3 + \cdots + P_j \tag{6.15}$$

Since

$$PV = nRT$$

$$P = n\left(\frac{RT}{V}\right)$$

$$P_t = \frac{RT}{V}(n_1 + n_2 + \cdots n_j)$$

$$P_1 = \frac{n_1 P_t}{n_1 + n_2 + \cdots n_j} \tag{6.16}$$

Combining Henry's law and Dalton's law, we get

$$Y_i = \frac{H_i x_i}{P_t} \tag{6.17}$$

where Y_i = mole fraction of ith gas in air
x_i = mole fraction of ith gas in fluid (water)
H_i = Henry's constant for ith gas
P_t = total pressure, atm

The greater the Henry's constant, the more easily a compound can be removed from a solution. Generally, increasing temperature would increase the partial pressure of a component in the gas phase. The Henry's constant and temperature relationship is (James M. Montgomery Consulting Engineering, 1985; American Society of Civil Engineers and American Water Works Association, 1990):

$$\log H = \frac{-\Delta H}{RT} + J \tag{6.18}$$

where H = Henry's constant
ΔH = heat absorbed in evaporation of 1 mole of gas from solution at constant temperature and pressure, kcal/kmol
R = universal gas constant,
= 1.9872 kcal/kmol · K
= 0.082057 atm · L/mol · K
T = temperature, Kelvin
J = empirical constant

Values of H, ΔH, and J of some gases are given in Table 6.3.

EXAMPLE 1: At 20°C the partial pressure (saturated) of chloroform $CHCl_3$ is 18 mm of mercury in a storage tank. Determine the equilibrium concentration of chloroform in water assuming that gas and liquid phases are ideal.

Solution:

Step 1. Determine mole fraction (x) of $CHCl_3$
From Table 6.3, at 20°C and a total pressure of 1 atm, the Henry constant for chloroform is

$$H = 170 \text{ atm}$$

TABLE 6.3 Henry's Law Constant and Temperature Correction Factors

Gas	Henry's constant at 20°C, atm	Δ h, 10^3 kcal/kmol	J
Ammonia	0.76	3.75	6.31
Benzene	240	3.68	8.68
Bromoform	35	–	–
Carbon dioxide	1.51×10^2	2.07	6.73
Carbon tetrachloride	1.29×10^3	4.05	10.06
Chlorine	585	1.74	5.75
Chlorine dioxide	54	2.93	6.76
Chloroform	170	4.00	9.10
Hydrogen sulfide	515	1.85	5.88
Methane	3.8×10^4	1.54	7.22
Nitrogen	8.6×10^4	1.12	6.85
Oxygen	4.3×10^4	1.45	7.11
Ozone	5.0×10^3	2.52	8.05
Sulfur dioxide	38	2.40	5.68
Trichloroethylene	550	3.41	8.59
Vinyl chloride	1.21×10^3	–	–

Partial pressure of $CHCl_3$, p, is

$$p = \frac{18\text{ mm}}{760\text{ mm}} \times 1\text{ atm} = 0.024\text{ atm}$$

Using Henry's law

$$p = Hx$$

$$x = p/H = 0.024\text{ atm}/170\text{ atm}$$
$$= 1.41 \times 10^{-4}$$

Step 2. Convert mole fraction to mass concentration.

In 1 L of water, there are 1000/18 mol/L = 55.6 mol/L.
Thus

$$x = \frac{n}{n + n_w}$$

where n = number of mole for $CHCl_3$
n_w = number of mole for water

Then $nx + n_w x = n$

$$n = \frac{n_w x}{1 - x} = \frac{55.6 \times 1.41 \times 10^{-4}}{1 - 1.41 \times 10^{-4}}$$
$$= 7.84 \times 10^{-3}\text{ mol/L}$$

Molecular weight of $CHCl_3$ = 12 + 1 + 3 × 35.45 = 119.4
Concentration (C) of $CHCl_3$ is

$$C = 119.4\text{ g/mol} \times 7.84 \times 10^{-3}\text{ mol/L}$$
$$= 0.94\text{ g/L}$$
$$= 940\text{ mg/L}$$

Note: In practice, the treated water is open to the atmosphere. Therefore the partial pressure of chloroform in the air is negligible. Therefore the concentration of chloroform in water is near zero or very low.

EXAMPLE 2: The finished water in an enclosed clear well has dissolved oxygen of 6.0 mg/L. Assume it is in equilibrium. Determine the concentration of oxygen in the air space of the clear well at 17°C.

Solution:

Step 1. Calculate Henry's constant at 17°C (290 K)
From Table 6.3 and Eq. (6.18)

$$\log H = \frac{-\Delta H}{RT} + J$$

$$= \frac{-1.45 \times 10^3\text{ kcal/kmol}}{(1.987\text{ kcal/kmol} \cdot \text{K})(290\text{ K})} + 7.11$$

$$= 4.59$$

$$H = 38{,}905\text{ atm} \times \frac{103.3\text{ kPa}}{1\text{ atm}}$$

$$= 4.02 \times 10^6\text{ kPa}$$

Step 2. Compute the mole fraction of oxygen in the gas phase
Oxygen (MW = 32) as gas i in water

$$C_i = 6 \text{ mg/L} \times 10^{-3} \text{ g/mg} \times 10^3 \text{ L/m}^3 \div 32 \text{ g/mol}$$
$$= 0.1875 \text{ mol/m}^3$$

At 1 atm for water, the molar density of water is

$$C_w = 55.6 \text{ kmol/m}^3$$

$$x_i = \frac{C_i}{C_w} = \frac{0.1875 \text{ mol/m}^3}{55.6 \times 10^3 \text{ mol/m}^3}$$

$$= 3.37 \times 10^{-6}$$

$$P_t = 1 \text{ atm}$$

From Eq. (6.17)

$$y_i = \frac{H_i x_i}{P_t} = \frac{38{,}905 \text{ atm} \times 3.37 \times 10^{-6}}{1 \text{ atm}}$$

$$= 0.131$$

Step 3. Calculate partial pressure P_i

$$P_i = y_i P_t = 0.131 \times 1$$
$$= 0.131 \text{ atm}$$

Step 4. Calculate oxygen concentration in gas phase $C_g = n_i/V$
Using the ideal gas formula

$$P_i V = n_i RT$$

$$\frac{n_i}{V} = \frac{P_i}{RT} = \frac{0.131 \text{ atm}}{0.08206 \text{ atm} \cdot \text{L/mol} \cdot \text{K} \times (290 \text{ K})}$$

$$= 5.50 \times 10^{-3} \text{ mol/L}$$
$$= 5.50 \times 10^{-3} \text{ mol/L} \times 32 \text{ g/mol}$$
$$= 0.176 \text{ g/L}$$
$$= 176 \text{ mg/L}$$

6.1 Gas Transfer Models

Three gas-liquid mass transfer models are generally used. They are two-film theory, penetration theory, and surface renewal theory. The first theory is discussed here. The other two can be found elsewhere (Schroeder, 1977).

The two-film theory is the oldest and simplest. The concept of the two-film model is illustrated in Fig. 6.1. Flux is a term used as the mass transfer per time through a specified area. It is a function of the driving force for diffusion. The driving force in air is the difference between the bulk concentration and the interface concentration. The flux gas through the gas film must be the same as the flux through the liquid film. The flux relationship for each phase can be expressed as:

$$F = \frac{dW}{dtA} = k_g(P_g - P_i) = k_l(C_i - C_l) \tag{6.19}$$

where F = flux
W = mass transfer
A = a given area gas-liquid transferal
t = time
k_g = local interface transfer coefficient for gas
P_g = concentration of gas in the bulk of the air phase
P_i = concentration of gas in the interface
k_l = interface transfer coefficient for liquid
C_i = local interface concentration at equilibrium
C_l = liquid-phase concentration in bulk liquid

Since P_i and C_i are not measurable, volumetric mass-transfer coefficients are used which combine the mass-transfer coefficients with the interfacial area per unit volume of system. Applying Henry's law and introducing P^* and C^* which correspond to the equilibrium concentration that would be associated with the bulk-gas partial pressure P_g and bulk-liquid concentration C_l, respectively, we can obtain

$$F = K_G\,(P_g - P^*) = K_L\,(C^* - C_l) \tag{6.20}$$

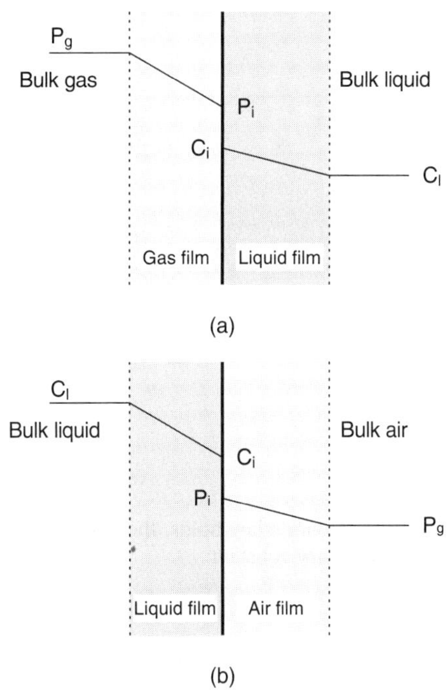

FIGURE 6.1 Schematic diagram of two-film theory for (a) gas absorption; (b) gas stripping.

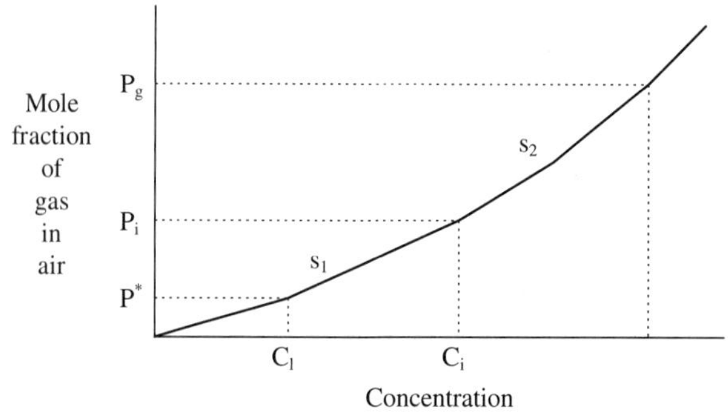

FIGURE 6.2 Concentration differences of gas in air and water phases.

The concentration differences of gas in air and water is plotted in Fig. 6.2. From this figure and Eq. (6.19), we obtain:

$$\begin{aligned} P_g - P^* &= (P_g - P_i) + (P_i - P^*) \\ &= P_g - P_i + s_1 (C_i - C_l) \\ &= \frac{F}{k_g} + \frac{s_1 F}{k_l} \end{aligned}$$

From Eq. (6.20):

$$\frac{F}{K_G} = P_g - P^*$$

Thus

$$\frac{F}{K_G} = \frac{F}{k_g} + \frac{s_1 F}{k_l}$$

$$\frac{1}{K_G} = \frac{1}{k_g} + \frac{s_1}{k_l}$$

Similarly, we can obtain

$$\frac{1}{K_L} = \frac{1}{s_2 k_g} + \frac{1}{k_l}$$

In dilute conditions, Henry's law holds, the equilibrium distribution curve would be a straight line and the slope is the Henry constant:

$$s_1 = s_2 = H$$

Finally the overall mass-transfer coefficient in the air phase can be determined with

$$\frac{1}{K_G} = \frac{1}{k_l} + \frac{H}{k_l} \tag{6.21}$$

For liquid-phase mass transfer

$$\frac{1}{K_L} = \frac{1}{k_l} + \frac{H}{Hk_g} \tag{6.22}$$

where $1/k_l$ and $1/k_g$ are referred to as the liquid-film resistance and the gas-film resistance, respectively. For oxygen transfer, $1/k_l$ is considerably larger than $1/k_g$, and the liquid film usually controls the oxygen transfer rate across the interface.

The flux equation for the liquid phase using concentration of mg/L is

$$F = \frac{dc}{dt}\frac{V}{A} = K_L\,(C_s - C_t) \tag{6.23a}$$

mass transfer rate:

$$\frac{dc}{dt} = K_L\frac{A}{V}\,(C_s - C_t) \tag{6.23b}$$

or

$$\frac{dc}{dt} = K_La\,(C^* - C_t) \tag{6.23c}$$

where K_L = overall mass-transfer coefficient, cm/h
A = interfacial area of transfer, cm^2
V = volume containing the interfacial area, cm^3
C_t = concentration in bulk liquid at time t, mg/L
C^* = equilibrium concentration with gas at time t as $P_t = HC_s$ mg/L
a = the specific interfacial area per unit system volume, cm^{-1}
K_La = overall volumetric mass-transfer coefficient in liquid, h^{-1} or $[g \cdot mol/(h \cdot cm^3 \cdot atm)]$

the volumetric rate of mass transfer M is

$$M = \frac{A}{V}N = K_La\,(C^* - C) \tag{6.24}$$

$$= K_Ga(P - P^*) \tag{6.25}$$

where N = rate at which solute gas is transferred between phases, $g \cdot mol/h$

EXAMPLE 1: An aeration tank has the volume of 200 m^3 (7060 ft^3) with 4 m (13 ft) depth. The experiment is conducted at 20°C for tap water. The oxygen transfer rate is 18.7 kg/h (41.2 lb/h). Determine the overall volumetric mass-transfer coefficient in the liquid phase.

Solution:

Step 1. Determine the oxygen solubility at 20°C and 1 atm
Since oxygen in air is 21%, according to Dalton's law of partial pressure, P is

$$P = 0.21 \times 1 \text{ atm} = 0.21 \text{ atm}$$

From Table 6.3, we find

$$H = 4.3 \times 10^4 \text{ atm}$$

The mole fraction x from Eq. (6.14) is

$$x = \frac{P}{H} = \frac{0.21}{4.3 \times 10^4} = 4.88 \times 10^{-6}$$

In 1 L of solution, the number of moles of water n_w is

$$n_w = \frac{1000}{18} = 55.6 \text{ g mol/L}$$

Let n = moles of gas in solution

$$x = \frac{n}{n + n_w}$$

since n is so small

$$n \cong n_w x$$

$$= 55.6 \text{ g} \cdot \text{mol/L} \times 4.88 \times 10^{-6}$$

$$= 2.71 \times 10^{-4} \text{ g} \cdot \text{mol/L}$$

Step 2. Determine the mean hydrostatic pressure on the rising air bubbles.
Pressure at the bottom is p_b
1 atm = 10.345 m of water head

$$p_b = 1 \text{ atm} + 4 \text{ m} \times \frac{1 \text{ atm}}{10.345 \text{ m}} = (1 + 0.386) \text{ atm}$$

$$= 1.386 \text{ atm}$$

Pressure at the surface of aeration tank is p_s = 1 atm
Mean pressure p is

$$p = \frac{p_s + p_b}{2} = \frac{1 \text{ atm} + 1.386 \text{ atm}}{2} = 1.193 \text{ atm}$$

Step 3. Compute $C^* - C$

$$C^* = 2.71 \times 10^{-4} \text{ g} \cdot \text{mol/L} \cdot \text{atm} \times 1.193 \text{ atm}$$

$$= 3.233 \times 10^{-4} \text{ g} \cdot \text{mol/L}$$

The percentage of oxygen in the air bubbles decreases, when the bubbles rise up through the solution, assuming that the air that leaves the water contains only 19 percent of oxygen.
Thus

$$C^* - C = 3.233 \times 10^{-4} \text{ g} \cdot \text{mol/L} \times \frac{19}{21}$$

$$= 2.925 \times 10^{-4} \text{ g} \cdot \text{mol/L}$$

Step 4. Compute $K_L a$

$$N = 18{,}700 \text{ g/h} \div 32/\text{mol} = 584 \text{ g} \cdot \text{mol/h}$$

Using Eq. (6.24)

$$K_L a = \frac{N}{V(C^* - C)}$$

$$= \frac{584 \text{ g} \cdot \text{mol/h}}{200 \text{ m}^3 \times 1000 \text{ L/m}^3 \times 2.925 \times 10^{-4} \text{ g} \cdot \text{mol/L}}$$

$$= 9.98 \text{ h}^{-1}$$

Equation (6.23c) is essentially the same as Fick's first law:

$$\frac{dC}{dt} = K_L a(C_s - C) \qquad (6.23d)$$

where $\frac{dC}{dt}$ = rate of change in concentration of the gas in solution, mg/L s

K_La = overall mass transfer coeffcient, s^{-1}

C_s = saturation concentration of gas in solution, mg/L

C = concentration of solute gas in solution, mg/L

The value of C_s can be calculated with Henry's law. The term $C_s - C$ is a concentration gradient. Rearrange Eq. (6.23d) and integrating the differential form between time from 0 to t, and concentration of gas from C_0 to C_t in mg/L, we obtain:

$$\int_{C_0}^{C_t} \frac{dC}{C_s - C} = K_La \int_0^t dt$$

$$\ln \frac{C_s - C_t}{C_s - C} = -K_L at \qquad (6.24a)$$

or

$$\frac{C_s - C_t}{C_s - C} = e^{-K_L at} \qquad (6.24b)$$

when gases are removed or stripped from the solution, Eq. (6.24b) becomes:

$$\frac{C_0 - C_s}{C_t - C_s} = e^{-(K_L at)} \qquad (6.24c)$$

Similarly, Eq. (6.25) can be integrated to yield

$$\ln \frac{p_t - p^*}{p_0 - p^*} = K_G aRTt \qquad (6.25a)$$

K_La values are usually determined in full-scale facilities or scaled up from pilot-scale facilities. Temperature and chemical constituents in wastewater affect oxygen transfer. The temperature effects on overall mass-transfer coefficient K_La are treated in the same manner as they were treated in the BOD rate coefficient (Section 8.3 of Chap. 1.2). It can be written as

$$K_La_{(T)} = K_La_{(20)}\theta^{T-20} \qquad (6.26)$$

where $K_La_{(T)}$ = overall mass-transfer coefficient at temperature T°C, s^{-1}

$K_La_{(20)}$ = overall mass-transfer coefficient at temperature 20°C, s^{-1}

Values of θ range from 1.015 to 1.040, with 1.024 commonly used.

mass-transfer coefficient are influenced by total dissolved solids in the liquid. Therefore, a correction factor α is applied for wastewater (Tchobanoglous and Schroeder, 1985):

$$\alpha = \frac{K_La \text{ (wastewater)}}{K_La \text{ (tap water)}} \qquad (6.27)$$

Values of α range from 0.3 to 1.2. Typical values for diffused and mechanical aeration equipment are in the range of 0.4 to 0.8 and 0.6 to 1.2.

The third correction factor (β) for oxygen solubility is due to particulate, salt, and surface-active substances in water (Doyle and Boyle, 1986):

$$\beta = \frac{C^*\text{(wastewater)}}{C^*\text{(tap water)}} \qquad (6.28)$$

Values of β range from 0.7 to 0.98, with 0.95 commonly used for wastewater.

Combining all three correction factors, we obtain (Tchobanoglous and Schroeder 1985):

$$\text{AOTR} = \text{SOTR}(\alpha)(\beta)(\theta^{T-20})\left(\frac{C_s - C_w}{C_{s20}}\right) \tag{6.29}$$

where AOTR = actual oxygen transfer rate under field operating conditions in a respiring system, $\text{kg}\cdot\text{O}_2/\text{kW}\cdot\text{h}$

SOTR = standard oxygen transfer rate under test conditions at 20°C and zero dissolved oxygen, $\text{kg}\cdot\text{O}_2/\text{kW}\cdot\text{h}$

α, β, θ = defined previously

C_s = oxygen saturation concentration for tap water at field operating conditions, g/m^3

C_w = operating oxygen concentration in wastewater, g/m^3

C_{s20} = oxygen saturation concentration for tap water at 20°C, g/m^3

EXAMPLE 2: Aeration tests are conducted with tap water and wastewater at 16°C in the same container. The results of the tests are listed below. Assume the saturation DO concentrations (C_s) for tap water and wastewater are the same. Determine the values of K_La for water and wastewater and α values at 20°C.

Assume $\theta = 1.024$

	DO concentration, mg/L		$-\ln\dfrac{C_s - C_t}{C_s - C_0}$	
Contact time, min	Tap water	Wastewater	Tap water	Wastewater
0	0.0	0.0	0	0
20	3.0	2.1	0.36	0.24
40	4.7	3.5	0.65	0.44
60	6.4	4.7	1.05	0.65
80	7.2	5.6	1.32	0.84
100	7.9	6.4	1.69	1.05
120	8.5	7.1	2.01	1.28

Solution:

Step 1. Find DO saturation concentration at 16°C
From Table 2.2

$$C_s = 9.82 \text{ mg/L}$$

Step 2. Calculate $-\ln(C_s - C_t)/(C_s - C_0)$, $C_0 = 0$
The values of $-\ln(C_s - C_t)/(C_s - C_0)$ calculated are listed with the raw test data

Step 3. Plot the calculated results in step 2 in Fig. 6.3.

Step 4. Find the K_La at the test temperature of 16°C, (T)

For tap water

$$K_La = \frac{2.01}{120 \text{ min}} \times \frac{60 \text{ min}}{1 \text{ h}}$$

$$= 1.00 \text{ h}^{-1}$$

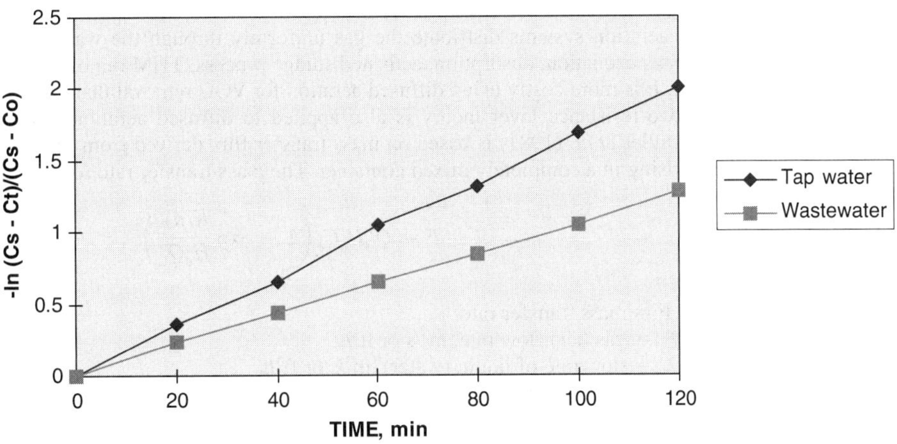

FIGURE 6.3 Functional plot of the data from aeration test.

For wastewater:

$$K_L a = \frac{1.28}{120 \text{ min}} \times \frac{60 \text{ min}}{1 \text{ h}}$$
$$= 0.64 \text{ h}^{-1}$$

Step 5. Convert the values of $K_L a$ at 20°C using $\theta = 1.024$
Using Eq. (6.26)
For tap water

$$K_L a_{(T)} = K_L a_{(20°\text{C})} \theta^{20-T}$$

or

$$K_L a_{(20)} = K_L a_{(T)} \theta^{20-T}$$
$$= 1.00 \text{ h}^{-1} (1.024)^{20-16}$$
$$= 1.10 \text{ h}^{-1}$$

For wastewater:

$$K_L a_{(20)} = 0.64 \text{ h}^{-1} (1.024)^{20-16}$$
$$= 0.70 \text{ h}^{-1}$$

Step 6. Compute the α value using Eq. (6.27)

$$\alpha = \frac{K_L a \text{ for wastewater}}{K_L a \text{ for tap water}}$$
$$= \frac{0.70 \text{ h}^{-1}}{1.10 \text{ h}^{-1}}$$
$$= 0.64$$

6.2 Diffused Aeration

Diffused aeration systems distribute the gas uniformly through the water or wastewater, in such processes as ozonation, absorption, activated sludge process, THM removal, and river or lake reaeration, etc. It is more costly using diffused aeration for VOC removal than the air stripping column.

The two-resistance layer theory is also applied to diffused aeration. The model proposed by Mattee-Müller *et al.* (1981) is based on mass transfer flux derived from the assumption of diffused bubbles rising in a completely mixed container. The mass transfer rate for diffused aeration is

$$F = Q_G H_u C_e \left(1 - \exp \frac{K_L a V}{H_u Q_L}\right) \tag{6.30}$$

where F = mass transfer rate
Q_G = gas (air) flow rate, m^3/s or ft^3/s
Q_L = flow rate of liquid (water), m^3/s or ft^3/s
H_u = unitless Henry's constant, see section Design of Packed tower
C_e = effluent (exit) gas concentration, μg/L
$K_L a$ = overall mass-transfer coefficient, per time
V = reaction volume (water), m^3 or ft^3

Assuming that the liquid volume in the reactor is completely mixed and the air rises as a steady state plug flow, the mass balance equation can be expressed as

$$\frac{C_e}{C_i} = \frac{1}{1 + H_u Q_G/Q_L[1 - \exp(-K_L aV/H_u Q_G)]} \tag{6.31a}$$

or

$$\frac{C_e}{C_i} = \frac{1}{1 + H_u Q_G/Q_L[1 - \exp(-\theta)]} \tag{6.31b}$$

where C_i = initial concentration, μg/L

$$\theta = \frac{K_L aV}{H_u Q_G}$$

If $\theta >> 1$, the transfer of a compound is with very low Henry's constant such as ammonia. Air bubbles exiting from the top of the liquid surface is saturated with ammonia in the stripping process. Ammonia removal could be further enhanced by increasing the air flow. Until $\theta < 4$, the exponent term becomes essentially zero. When the exponent term is zero, the air and water have reached an equilibrium condition and the driving force has decreased to zero at some point within the reactor vessel. The vessel is not fully used. Thus the air-to-water ratio could be increased to gain more removal.

On the other hand, if $\theta << 1$, the mass-transfer efficiency could be improved by increasing overall mass-transfer coefficient by either increasing the mixing intensity in the tank or by using a finer diffuser. In the case for oxygenation, $\theta < 0.1$, the improvements are required.

EXAMPLE : A groundwater treatment plant has a capacity of 0.0438 m^3/s (1 MGD) and is aerated with diffused air to remove trichloroethylene with 90 percent design efficiency. The detention time of the tank is 30 min. Evaluate the diffused aeration system with the following given information.

$$T = 20°C$$

$$C_i = 131 \text{ μg/L (expected } C_e = 13.1 \text{ μg/L)}$$

$$H_u = 0.412$$

$$K_L a = 44 \text{ h}^{-1}$$

Solution:

Step 1. Compute volume of reactor V

$$V = 0.0438\ \text{m}^3/\text{s} \times 60\ \text{s/min} \times 30\ \text{min}$$
$$= 79\ \text{m}^3$$

Step 2. Compute Q_G, let $Q_G = 30\ Q_L = 30\ V$

$$Q_G = 30 \times 79\ \text{m}^3 = 2370\ \text{m}^3$$

Step 3. Compute θ

$$\theta = \frac{K_L a V}{H_u Q_G} = \frac{44 \times 79}{0.412 \times 2370} = 3.56$$

Step 4. Compute effluent concentration C_e with Eq (6.31b).

$$C_e = \frac{C_i}{1 + H_u\, Q_G/Q_L[1 - \exp(-\theta)]}$$
$$= \frac{131\ \mu\text{g/L}}{1 + 0.412 \times 30[1 - \exp(-3.56)]}$$
$$= 10.0\ \mu g/L$$

$$\%\ \text{removal} = \frac{(131 - 10) \times 100}{131} = 92.4$$

This exceeds the expected 90%.

6.3 Packed Towers

Recently, the water treatment industry used packed towers for stripping highly volatile chemicals, such as hydrogen sulfide and VOCs from water and wastewater. It consists of a cylindrical shell containing a support plate for the packing material. Although many materials can be used as the packing material, plastic products with various shapes and design are most commonly used due to less weight and lower cost. Packing material can be individually dumped randomly into the cylinder tower or fixed packing. The packed tower or columns are used for mass transfer from the liquid to gas phase.

Figure 6.4 illustrates a liquid-gas contacting system with a downward water velocity L containing c_1 concentration of gas. The flows are counter current. The air velocity G passes upward through

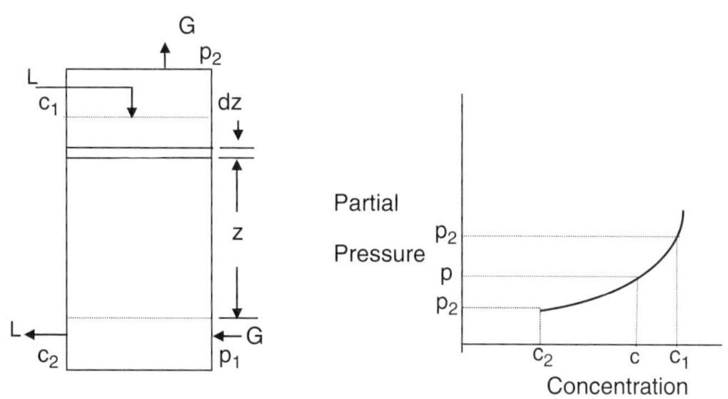

FIGURE 6.4 Schematic diagram of packed column.

the packed material containing influent p_1 and effluent p_2. There are a variety of mixing patterns, each with a different rate of mass transfer. Removal of an undesirable gas in the liquid phase needs a system height of z and a selected gas flow rate to reduce the mole fraction of dissolved gas from c_1 to c_2. If there is no chemical reaction that takes place in the packed tower, the gas lost by water should be equal to the gas gained by air. If the gas concentration is very dilute, then

$$L\Delta c = G\Delta p \tag{6.32}$$

where L = liquid velocity, m/s, $m^3/(m^2 \cdot s)$, or $mol/(m^2 \cdot s)$
G = gas velocity, same as above
Δc = change of gas concentration in water
Δp = change in gas fraction in air

From Eq. (6.23d), for stripping (c and c_s will be reversed):

$$\frac{dc}{dt} = K_L a\,(c - c_s) \tag{6.23e}$$

then

$$dt = \frac{dc}{K_L a(c - c_s)} \tag{6.33}$$

Multiplying each side by L, we obtain

$$Ldt = dz = \frac{Ldc}{K_L a(c - c_s)} \tag{6.34}$$

where dz is the differential of height.

The term $(c - c_s)$ is the driving force (DF) of the reaction and is constantly changing with the depth of the column due to the change of c with time.

Integrating the above equation yields

$$\int dz = \frac{L}{K_L a}\int_{c_1}^{c_2} \frac{dc}{d(c - c_s)}$$

$$z = \frac{L}{K_L a}\int_{c_1}^{c_2}\int_{\text{DF}} \frac{dc}{d(\text{DF})} \tag{6.35}$$

The integral of DF is the same as the log mean (DF_{lm}) of the influent (DF_i) and effluent driving forces (DF_e). Therefore

$$z = \frac{L(c_i - c_e)}{K_L a \text{DF}_{\text{lm}}} \tag{6.36}$$

and

$$\text{DF}_{\text{lm}} = \frac{\text{DF}_e - \text{DF}_i}{\ln(\text{DF}_e/\text{DF}_i)} \tag{6.37}$$

where z = height of column, m
L = liquid velocity, $m^3/m^2 \cdot h$
c_i, c_e = gas concentration in water at influent and effluent, respectively, mg/L
$K_L a$ = overall mass-transfer coefficient for liquid, h^{-1}
DF_i, DF_e = driving force at influent and effluent, respectively, mg/L
DF_{lm} = log mean of DF_i and DF_e, mg/L

EXAMPLE 1: Given

$$T = 20°C = 293 \text{ K}$$
$$L = 80 \text{ m}^3 \text{ water/(m}^2 \cdot \text{h)}$$
$$G = 2400 \text{ m}^3 \text{ air/(m}^2 \cdot \text{h)}$$
$C_i = 131$ μg/L trichloroethylene concentration in water at entrance
$C_e = 13.1$ μg/L $CCHCl_3$ concentration at exit
$$K_L a = 44 \text{ h}^{-1}$$

Determine the packed tower height to remove 90 percent of trichloroethylene by an air stripping tower.

Solution:

Step 1. Compute the molar fraction of $CCHCl_3$ in air p_2
Assuming no $CCHCl_3$ present in the air the entrance, i.e.

$$p_i = 0$$

MW of $CCHCl_3$ = 131, 1 mole = 131 g of $CCHCl_3$ per liter.

$$c_i = 131\ \mu\text{g/L} = 131\ \frac{\mu\text{g}}{\text{L}} \times \frac{1\ \text{mol/g}}{131 \times 10^6\ \mu\text{g/g}} \times \frac{10^3\ \text{L}}{1\ \text{m}^3}$$
$$= 1 \times 10^{-3}\ \text{mol/m}^3$$
$$c_e = 13.1\ \mu\text{g/L} = 0.1 \times 10^{-3}\ \text{mol/m}^3\ \text{(with 90\% removal)}$$

Applying Eq. (6.32)

$$L\Delta c = G\Delta p$$
$$80\ \text{m}^3/(\text{m}^2 \cdot \text{h})(1 - 0.1) \times 10^{-3}\ \text{mol gas/m}^3 = 2400\ \text{m}^3\ \text{air/(m}^2 \cdot \text{h)}(p_e - 0)$$
$$p_e = 3.0 \times 10^{-5}\ \text{mol gas/m}^3\ \text{air}$$

Step 2. Convert p_e in terms of mol gas/mol air
Let V = volume of air per mole of air

$$V = \frac{nRT}{P} = \frac{(1\ \text{mole})(0.08206\ \text{L atm/mol} \cdot \text{K}) \times (293\ \text{K})}{1\ \text{atm}}$$
$$= 24.0\ \text{L}$$
$$= 0.024\ \text{m}^3$$

From Step 1

$$p_e = 3.0 \times 10^{-5}\ \text{mol gas/m}^3\ \text{air}$$
$$= 3.0 \times 10^{-5}\ \frac{\text{mol gas}}{\text{m}^3\ \text{air}} \times \frac{0.024\ \text{m}^3\ \text{air}}{1\ \text{mol air}}$$
$$= 7.2 \times 10^{-7}\ \text{mol gas/mol air}$$

Step 3. Compute driving force, DF, for gas entrance and exit
At gas influent (bottom)

$$p_i = 0$$
$$c_e = 13.1\ \mu g/L = 0.0131\ mg/L$$
$$c_s = 0$$
$$DF_i = c_e - c_s = 0.0131\ mg/L$$

At gas effluent (top)

$$p_e = 7.2 \times 10^{-7}\ \text{mol gas/mol air}$$
$$c_i = 131\ \mu g/L = 0.131\ mg/L$$
$$c_s = \text{to be determined}$$

From Table 6.3, Henry's constant H at 20°C

$$H = 550\ \text{atm}$$

Convert atm to atm L/mg, H_d

$$H_d = \frac{H}{55{,}600 \times MW} = \frac{550\ \text{atm}}{55{,}600 \times 131\ mg/L}$$
$$= 7.55 \times 10^{-5}\ \text{atm} \cdot L/mg$$
$$c_s = \frac{p_e P_t}{H_d} = \frac{7.2 \times 10^{-7}\ \text{mol gas/mol air} \times 1\ \text{atm}}{7.55 \times 10^{-5}\ \text{atm} \cdot L/mg}$$
$$= 0.0095\ mg/L$$
$$DF_e = c_i - c_s = 0.131\ mg/L - 0.0095\ mg/L$$
$$= 0.1215\ mg/L$$

Step 4. Compute DF_{lm}
From Eq. (6.37)

$$DF_{lm} = \frac{DF_e - DF_i}{\ln(DF_e/DF_i)} = \frac{0.1215\ mg/L - 0.0131\ mg/L}{\ln(0.1215/0.0131)}$$
$$= 0.0487\ mg/L$$

Step 5. Compute the height of the tower z
From Eq. (6.36)

$$z = \frac{L(c_i - c_e)}{K_L a DF_{lm}} = \frac{80\ m/h \times (0.131 - 0.0131)\ mg/L}{44\ h^{-1} \times 0.0487\ mg/L}$$
$$= 4.4\ m$$

EXAMPLE 2: In a groundwater remediation study, volatile organic carbon removal through the vapor phase, the total hydrocarbons analyzer measured 1,1,1,-trichloroethane (Jones *et at.*, 2000). The extraction pump and air stripper combined air flow averaged 0.40 m^3/min (14 ft^3/min). The concentration of 1,1,1,-trichloroethane averaged 25 parts per million by volume (ppmv) of air. Determine the amount of 1,1,1,-thrichloroethane removed daily. Temperature is 20°C.

Solution:

Step 1. Calculate 1,1,1,-trichloroethane concentration using the ideal gas law

$$T = 20°\text{C} = (20 + 273)\text{K} = 293\text{ K}$$

$$n = \frac{PV}{RT} = \frac{(1\text{ atm})/1000\text{ L/m}^3}{0.082(\text{L}\cdot\text{atm/mol}\cdot\text{K}) \times 293\text{ K}}$$

$$= 41.6\text{ mol/m}^3$$

This means 41.6 mol of total gases in 1 m^3 of the air.
The MW of 1,1,1,-trichloroethane (CH_3CCl_3) = 133
One ppmv of 1,1,1,-trichloroethane = 41.6×10^{-6} mol/m^3 × 133 g/mol × 1000 mg/g
= 5.53 mg/m^3
25 ppmv of 1,1,1,-trichloroethane = 5.53 mg/m^3 × 25
= 138 mg/m^3

Step 1. Calculate the daily removal

$$\text{Daily removal} = 138\text{ mg/m}^3 \times 0.40\text{ m}^3/\text{min} \times 1440\text{ min/d}$$
$$= 79{,}500\text{ mg/d}$$
$$= 79.5\text{ g/d}$$

Design of packed tower. Process design of packed towers or columns is based on two quantities: the height of the packed column, z, to achieve the designed removal of solute is the product of the height of a transfer unit (HTU) and the number of transfer unit (NTU). It can be expressed as (Treybal, 1968):

$$z = (\text{HTU})(\text{NTU}) \tag{6.38}$$

The HTU refers the rate of mass transfer for the particular packing materials used. The NTU is a measure of the mass transfer driving force and is determined by the difference between actual and equilibrium phase concentrations. The height of a transfer unit is the constant portion of Eq. (6.35):

$$\text{HTU} = \frac{L}{K_\text{L}a} \tag{6.39}$$

The number of transfer units is the integral portion of Eq. (6.35). For diluted solutions, Henry's law holds. Substituting the integral expression for NTU with $p_1 = 0$, the NTU is:

$$\text{NTU} = \frac{R}{R-1}\ln\frac{(c_1/c_2)(R-1)+1}{R} \tag{6.40}$$

where $$R = \frac{H_\text{u}G}{L} \tag{6.41}$$

= stripping factor, unitless when H_u is unitless
c_1, c_2 = mole fraction for gas entrance and exit, respectively

Convert Henry's constant from terms of atm to unitless:

$$H_\text{u} = \left[H\frac{\text{atm(mol gas/mol air)}}{\text{mol gas/mol water}}\right]\left(\frac{1\text{ mol air}}{0.082\,T\text{ atm L of air}}\right)\left(\frac{1\text{ L of water}}{55.6\text{ mol}}\right) \tag{6.42}$$

$$= H/4.56\,T$$

When $T = 20°C = 293$ K

$$H_u = H/4.56 \times 293 = 7.49 \times 10^{-4}\ H, \text{ unitless} \tag{6.42a}$$

G = superficial molar air flow rate (k mol/s m^2)

L = superficial molar water flow rate (k mol/s m^2)

The NTU depends upon the designed gas removal efficiency, the air-water velocity ratio, and Henry's constant. Treybal (1968) plotted the integral part (NTU) of Eq. (6.35) in Fig. 6.5. By knowing the desired removal efficiency, the stripping factor, and Henry's constant, the NTU in a packed column can be determined for any given stripping factor of the air to water flow rate ratios. It can be seen from Fig. 6.5 that when the stripping factor, R, is greater than 3, little improvement for the NTU occurs.

EXAMPLE 3: Using the graph of Fig. 6.5 to solve Example 1. Given:

$T = 20°C = 293$ K

$L = 80$ m^3 water/(m^2 column cross section h)

$G = 2400$ m^3 air/(m^2 column cross section h)

$c_1 = 131$ μg/L of $CCHCl_3$

$c_2 = 13.1$ μg/L of $CCHCl_3$

$K_La = 44$ h^{-1}

Solution:

Step 1. Compute HTU with Eq. (6.39)

$$\text{HTU} = \frac{L}{K_La} = \frac{80 \text{ m/h}}{44 \text{ h}^{-1}} = 1.82 \text{ m}$$

Step 2. Compute H_u, the unitless Henry's constant at 20°C
From Table 6.1

$$H = 550 \text{ atm}$$

Using Eq. (6.42a)

$$H_u = 7.49 \times 10^{-4}\ H = 7.49 \times 10^{-4} \times 550$$
$$= 0.412$$

Step 3. Compute the stripping factor R using (Eq. 6.41)

$$R = \frac{H_u G}{L} = \frac{0.412 \times 2400 \text{ m/h}}{80 \text{ m/h}}$$
$$= 12.36$$

Step 4. Find NTU from Fig. 6.5

since $$\frac{c_2}{c_1} = \frac{13.1\ \mu\text{g/L}}{131\ \mu\text{g/L}} = 0.1$$

Using R = 12.36 and c_2/c_1 = 0.1, from the graph in Fig. 6.5, we obtain

$$\text{NTU} = 2.42$$

Step 5. Compute the height of tower z by Eq. (6.38)

$$z = (\text{HTU})(\text{NTU}) = 1.82 \text{ m} \times 2.42$$
$$= 4.4 \text{ m}$$

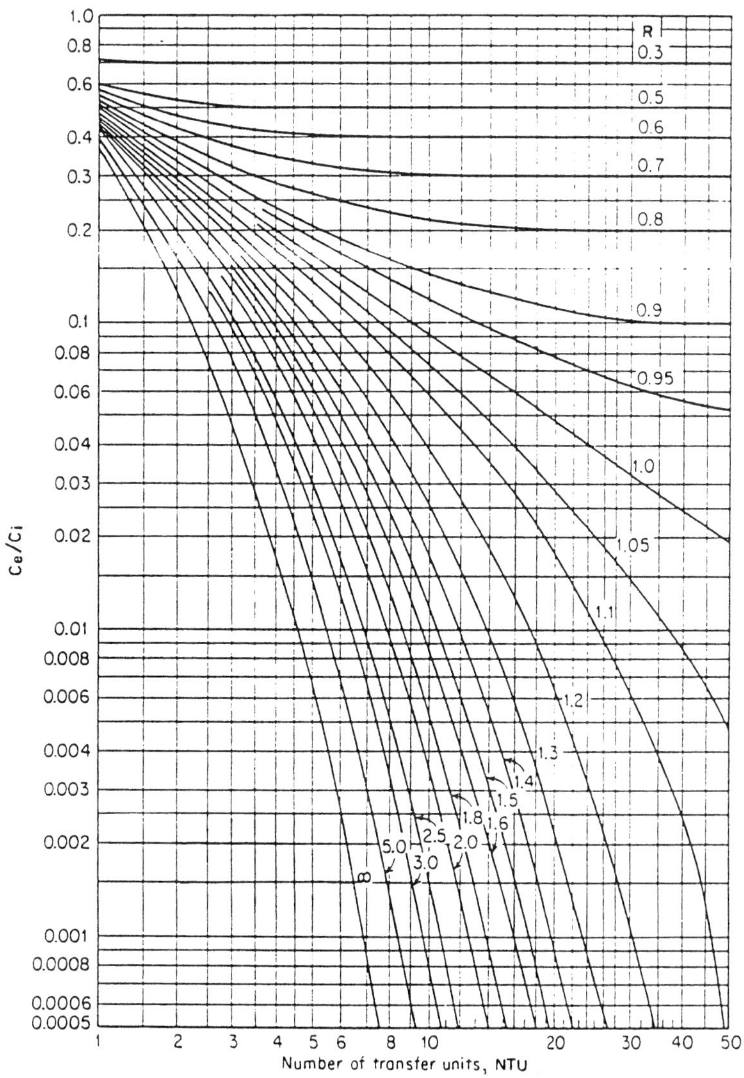

FIGURE 6.5 Number of transfer units for absorbers or strippers with constant absorption or stripping factors (*D. A. Cornwell, Air Stripping and aeration. In*: AWWA, *Water Quality and Treatment. Copyright 1990, McGraw-Hill, New York, reprinted with permission of McGraw-Hill*).

6.4 Nozzles

There are numerous commercially available spray nozzles. Nozzles and tray aerators are air-water contact devices. They are used for iron and manganese oxidation, removal of carbon dioxide and hydrogen sulfide from water, and removal of taste and odor causing materials.

Manufacturers may occasionally have mass transfer data such as K_L and a values. For an open atmosphere spray fountain, the specific interfacial area a is (Calderbrook and Moo-Young, 1961):

$$a = \frac{6}{d} \tag{6.43}$$

where d is the droplet diameter which ranges from 2 to 10,000 μm. Under open atmosphere conditions, C_s remains constant because there is an infinite air-to-water ratio and the Lewis and Whitman equation can express the mass transfer for nozzle spray (Fair *et al.*, 1968; Cornwell, 1990)

$$C_e - C_i = (C_s - C_i)\,[1 - \exp(-K_L at)] \tag{6.44}$$

where C_i, C_e = solute concentrations in bulk water and droplets, respectively, μg/L
t = time of contact between water droplets and air
= twice the rise time, t_r

$$= \frac{2V \sin \phi}{g} \tag{6.45a}$$

or

$$t_r = V \sin \phi / g \tag{6.45b}$$

where ϕ = angle of spray measured from horizontal
V = velocity of the droplet from the nozzle

$$V = C_v \sqrt{2gh} \tag{6.46}$$

where C_v = velocity coefficient, 0.40 to 0.95, obtainable from the manufacturer
g = gravitational acceleration, 9.81 m/s^2

Combine Eq. (6.45b) and Eq. (6.46) $gt_r = C_V \sqrt{2gh} \sin \phi$

The driving head h is

$$h = gt_r^2/(2C_v^2 \sin^2 \phi) \tag{6.47}$$

Neglecting wind effect, the radius of the spray circle is

$$\begin{aligned} r &= Vt_r \cos \phi \\ &= V(V \sin \phi/g) \cos \phi \\ &= V^2/g \cdot \sin \phi \cos \phi \\ &= C_v^2\, 2gh/g \cdot \sin 2\phi \text{ [from Eq. (6.46)]} \\ &= 2C_v^2\, h \sin 2\phi \end{aligned} \tag{6.48}$$

and the vertical rise of the spray is

$$h_r = \frac{1}{2} g t_r^2$$

From Eq. (6.47)

$$\begin{aligned} h_r &= \frac{1}{2} h\,(2C_v^2 \sin^2\phi) \\ &= C_v^2 h \sin^2\phi \end{aligned} \tag{6.49}$$

EXAMPLE 1: Determine the removal percentage of trichloroethylene from a nozzle under an operating pressure of 33 psi (228 kPa).
The following data is given:

$$d = 0.05\ \text{cm}$$
$$C_v = 0.50$$
$$\phi = 30°$$
$$K_L = 0.005\ \text{cm/s}$$
$$C_i = 131\ \mu\text{g/L}$$

Solution:

Step 1. Determine the volumetric interfacial area a by Eq. (6.43):

$$a = \frac{6}{d} = \frac{6}{0.05\ \text{cm}} = 120\ \text{cm}^{-1}$$

Step 2. Compute the velocity of the droplet V by Eq. (6.46)
The pressure head

$$h = 33\ \text{psi} \times \frac{70.3\ \text{cm of waterhead}}{1\ \text{psi}}$$
$$= 2320\ \text{cm}$$

Using Eq. (6.46)

$$V = C_v\sqrt{2\,gh} = 0.5\sqrt{2 \times 981 \times 2320}$$
$$= 1067\ \text{cm/s}$$

Step 3. Compute the time of contact by Eq. (6.45a)

$$t = \frac{2V\sin\phi}{g} = \frac{2 \times 1067\ \text{cm/s}\ \sin 30}{981\ \text{cm/s}^2}$$
$$= \frac{2 \times 1067 \times 0.5}{981}\ \text{s}$$
$$= 1.09\ \text{s}$$

Step 4. Compute the mass transfer $C_e - C_i$ by Eq. (6.44)

$$C_e - C_i = (C_s - C_i)[1 - \exp(-K_L at)]$$
$$C_e - 131 = (0 - 131)[1 - \exp(-0.005 \times 120 \times 1.09)]$$
$$C_e = 131 - 131 \times 0.48$$
$$= 68.1\ (\mu\text{g/L}) = 52\%\ \text{remained}$$
$$= 48\ \text{percent removal}$$

EXAMPLE 2: Calculate the driving head, radius of the spray circle, and vertical rise of spray for (a) a vertical jet and (b) a jet at 45° angle. Given the expose time for water droplet is 2.2 s and C_v is 0.90.

Solution:

Step 1. For question (a). Using Eq. (6.45b)

$$t_r = \frac{1}{2}t = \frac{1}{2} \times 2.2 \text{ s} = 1.1 \text{ s}$$
$$\phi = 90°$$
$$\sin \phi = \sin 90 = 1$$
$$\sin 2\phi = \sin 180 = 0$$

Using Eq. (6.47)

$$\begin{aligned} h &= gt_r^2/(2\, C_v^2 \sin^2 \phi) \\ &= 9.81 \text{ m/s}^2\ (1.1 \text{ s})^2/(2 \times 0.9^2 \times 1^2) \\ &= 7.33 \text{ m} \end{aligned}$$

Using Eq. (6.48)

$$\begin{aligned} r &= 2C_v^2 h \sin 2\phi \\ &= 0 \end{aligned}$$

Since $\sin 2\phi = 0$, and using Eq. (6.49)

$$\begin{aligned} h_r &= \frac{1}{2} g\, t_r^2 = \frac{1}{2} \times 9.81 \text{ m/s}^2 \times (1.1 \text{ s})^2 \\ &= 5.93 \text{ m} \end{aligned}$$

Step 2. For question (b)

$$t_r = 1.1 \text{ s}$$
$$\phi = 45°$$
$$\sin \phi = \sin 45 = 1/\sqrt{2} = 0.707$$
$$\sin 2\phi = \sin 90 = 1$$

Using Eq. (6.47)

$$\begin{aligned} h &= gt_r^2/(2C_v^2 \sin \phi) \\ &= 9.81(1.1)^2/(2 \times 0.9^2 \times 0.707^2) \\ &= 14.66 \text{ (m)} \end{aligned}$$

Using Eq. (6.48)

$$\begin{aligned} r &= 2C_v^2 h \sin 2\phi \\ &= 2(0.9)^2\ (14.66 \text{ m})\ (1) \\ &= 23.7 \text{ m} \end{aligned}$$

and using Eq. (6.49)

$$\begin{aligned} h_r &= C_v^2 h \sin \phi \\ &= (0.9)^2 \times 14.66 \text{ m} \times (0.707)^2 \\ &= 5.93 \text{ m} \end{aligned}$$

7 SOLUBILITY EQUILIBRIUM

Chemical precipitation is one of the most commonly employed methods for drinking water treatment. Coagulation with alum, ferric sulfate, or ferrous sulfate, and lime softening involve chemical precipitation. Most chemical reactions are reversible to some degree. A general chemical reaction which has reached equilibrium is commonly written by the law of mass action as

$$aA + bB \leftrightarrow cC + dD \tag{6.50a}$$

where A, B = reactants
C, D = products
a, b, c, d = stoichiometric coefficients for A, B, C, D, respectively

The *equilibrium constant* K_{eq} for the above reaction is defined as

$$K_{eq} = \frac{[C]^c[D]^d}{[A]^a[B]^b} \tag{6.50b}$$

where K_{eq} is a true constant, called the equilibrium constant, and the square brackets signify the molar concentration of the species within the brackets. For a given chemical reaction, the value of equilibrium constant will change with temperature and the ionic strength of the solution.

For an equilibrium to exist between a solid substance and its solution, the solution must be saturated and in contact with undissolved solids. For example, at pH greater than 10, solid calcium carbonate in water reaches equilibrium with the calcium and carbonate ions in solution: consider a saturated solution of $CaCO_3$ that is in contact with solid $CaCO_3$. The chemical equation for the relevant equilibrium can be expressed as

$$CaCO_3\ (s) \leftrightarrow Ca^{2+}\ (aq) + CO_3^{2-}(aq) \tag{6.50c}$$

The equilibrium constant expression for the dissolution of $CaCO_3$ can be written as

$$K_{eq} = \frac{[Ca^{2+}][CO_3^{2-}]}{[CaCO_3]} \tag{6.50d}$$

Concentration of a solid substance is treated as a constant called K_s in mass-action equilibrium, thus $[CaCO_3]$ is equal to K_s. Then

$$K_{eq}, K_s = [Ca^{2+}][CO_3^{2-}] = K_{sp} \tag{6.50e}$$

The constant K_{sp} is called the *solubility product constant*. The rules for writing the solubility product expression are the same as those for the writing of any equilibrium constant expression. The solubility product is equal to the product of the concentrations of the ions involved in the equilibrium, each raised to the power of its coefficient in the equilibrium equation. For the dissolution of a slightly soluble compound when the (brackets) concentration is denoted in moles, the equilibrium constant is called the solubility product constant. The general solubility product expression can be derived from the general dissolution reaction

$$A_xB_y\ (s) \leftrightarrow xA^{y+} + yB^{x-} \tag{6.50f}$$

and is expressed as

$$K_{sp} = [A^{y+}]^x\ [B^{x-}]^y \tag{6.50g}$$

Solubility product constants for various solutions at or near room temperature are presented in Appendix C.

If the product of the ionic molar concentration $[A^{y+}]^x[B^{x-}]^y$ is less than the K_{sp} value, the solution is unsaturated, and no precipitation will occur. In contrast, if the product of the concentrations of ions in solution is greater than K_{sp} value, precipitation will occur under supersaturated condition.

EXAMPLE 1: Calculate the concentration of OH^- in a 0.20 mole (M) solution of NH_3 at 25°C. $K_{eq} = 1.8 \times 10^{-5}$

Solution:

Step 1. Write the equilibrium expression

$$NH_3\ (aq) + H_2O\ (l) \leftrightarrow NH_4^+\ (aq) + OH^-(aq)$$

Step 2. Find the equilibrium constant

$$K_{eq} = \frac{[NH_4^+][OH^-]}{[NH_3]} = 1.8 \times 10^{-5}$$

Step 3. Tabulate the equilibrium concentrations involved in the equilibrium:
Let x = concentration (M) of OH^-

	NH_3 (aq) + H_2O (1) ↔	NH_4^+ (aq)+	OH^-(aq)
Initial:	0.2 M	0 M	0 M
Equilibrium:	$(0.20 - x)$M	x M	x M

Step 4. Inserting these quantities into step 2 to solve for x

$$K_{eq} = \frac{[NH_4^+][OH^-]}{[NH_3]}$$

$$= \frac{(x)(x)}{(0.20 - x)} = 1.8 \times 10^{-5}$$

or

$$x^2 = (0.20 - x)(1.8 \times 10^{-5})$$

Neglecting x term, then approximately

$$x^2 \cong 0.2 \times 1.8 \times 10^{-5}$$

$$x = 1.9 \times 10^{-3}\ M$$

EXAMPLE 2: Write the expression for the solubility product constant for (a) $Al(OH)_3$ and (b) $Ca_3(PO_4)_2$.

Solution:

Step 1. Write the equation for the solubility equilibrium

(a) $Al(OH)_3\ (s) \leftrightarrow Al^{3+}\ (aq) + 3OH^-(aq)$
(b) $Ca_3(PO_4)_2\ (s) \leftrightarrow 3Ca^{2+}\ (aq) + 2PO_4^{3-}\ (aq)$

Step 2. Write K_{sp} by using Eq. (6.50g) and from Appendix C

(a) $K_{sp} = [Al^{3+}]\ [OH^-]^3 = 2 \times 10^{-32}$
(b) $K_{sp} = [Ca^{2+}]^3\ [PO_4^{3-}]^2 = 2.0 \times 10^{-29}$

EXAMPLE 3: The K_{sp} for $CaCO_3$ is 8.7×10^{-9} (Appendix C). What is the solubility of $CaCO_3$ in water in grams per liter?

Solution:

Step 1. Write the equilibrium equation

$$CaCO_3\,(s) \leftrightarrow Ca^{2+}(aq) + CO_3^{2-}(aq)$$

Step 2. Set molar concentrations

For each mole of $CaCO_3$ that dissolves, 1 mole of Ca^{2+} and 1 mole of CO_3^{2-} enter the solution.

Let x be the solubility of $CaCO_3$ in moles per liter. The molar concentrations of Ca^{2+} and CO_3^{2-} are

$$[Ca^{2+}] = x \quad \text{and} \quad [CO_3^{2-}] = x$$

Step 3. Solve solubility x in moles per liter

$$K_{sp} = [Ca^{2+}][CO_3{}^{2-}] = 8.7 \times 10^{-9}$$

$$(x)(x) = 8.7 \times 10^{-9}$$

$$x = 9.3 \times 10^{-5}\ \text{mol/L}$$

1 mole of $CaCO_3$ = 100 g/L of $CaCO_3$

then

$$x = 9.3 \times 10^{-5}(100\ \text{g/L})$$

$$= 0.0093\ \text{g/L}$$

$$= 9.3\ \text{mg/L}$$

8 COAGULATION

Coagulation is a chemical process to remove turbidity and color producing material that is mostly colloidal particles (1 to 200 millimicrons, mμ) such as algae, bacteria, organic and inorganic substances, and clay particles. Most colloidal solids in water and wastewater are negatively charged. The mechanisms of chemical coagulation involve the zeta potential derived from double-layer compression, neutralization by opposite charge, interparticle bridging, and precipitation. Destabilization of colloid particles is influenced by the *Van der Waals force* of attraction and Brownian movement. Detailed discussion of the theory of coagulation can be found elsewhere (American Society of Civil Engineers and American Water Works Association 1990). Coagulation and flocculation processes were discussed in detail by Amirtharajah and O'Melia (1990).

Coagulation of water and wastewater generally add either aluminum, or iron salt, with and without polymers and coagulant aids. The process is complex and involves dissolution, hydrolysis, and polymerization; pH values play an important role in chemical coagulation depending on alkalinity. The chemical coagulation can be simplified as the following reaction equations:

Aluminum sulfate (alum):

$$Al_2(SO_4) \cdot 18H_2O + 3Ca\,(HCO_3)_2 \rightarrow 2Al(OH)_3 \downarrow + 3CaSO_4 + 6CO_2 + 18H_2O \quad (6.51a)$$

When water does not have sufficient total alkalinity to react with alum, lime or soda ash is usually also dosed to provide the required alkalinity. The coagulation equations can be written as below:

$$Al_2(SO_4)_3 \cdot 18H_2O + 3Ca(OH)_2 \rightarrow 2Al(OH)_3 \downarrow + 3CaSO_4 + 18H_2O \quad (6.51b)$$

$$Al_2(SO_4)_3 \cdot 18H_2O + 3Na_2\,CO_3 + 3H_2O \rightarrow 2Al(OH)_3 \downarrow + 3Na_2SO_4 + 3CO_2 + 18H_2O \quad (6.51c)$$

Ferric chloride:

$$2FeCl_3 + 3Ca(HCO_3)_2 \rightarrow 2Fe(OH)_3 \downarrow + 3CaCl_2 + 6CO_2 \tag{6.51d}$$

Ferric sulfate:

$$Fe(SO_4)_3 + 3Ca(HCO_3)_2 \rightarrow 2Fe(OH)_3 \downarrow + 3CaSO_4 + 6CO_2 \tag{6.51e}$$

Ferrous sulfate and lime:

$$FeSO_4 \cdot 7H_2O + Ca(OH)_2 \rightarrow Fe(OH)_2 + CaSO_4 + 7H_2O \tag{6.51f}$$

followed by, in the presence of dissolved oxygen

$$4Fe(OH)_2 + O_2 + 2H_2O \rightarrow 4Fe(OH)_3 \downarrow \tag{6.51g}$$

Chlorinated copperas:

$$3FeSO_4 \cdot 7H_2O + 1.5Cl_2 \rightarrow Fe_2(SO_4)_3 + FeCl_3 + 21H_2O \tag{6.51h}$$

followed by reacting with alkalinity as above

$$Fe_2(SO_4)_3 + 3Ca(HCO_3)_2 \rightarrow 2Fe(OH)_3 \downarrow + 3CaSO_4 + 6CO_2 \tag{6.51i}$$

and

$$2FeCl_3 + 3Ca(HCO_3)_2 \rightarrow 2Fe(OH)_3 \downarrow + 3CaCl_2 + 6CO_2 \tag{6.51j}$$

Each of the above reaction has an optimum pH range.

EXAMPLE 1: What is the amount of natural alkalinity required for coagulation of raw water with dosage of 15.0 mg/L of ferric chloride?

Solution:

Step 1. Write the reaction equation (Eq. 6.51j) and calculate MW

$$\begin{array}{ccc} 2FeCl_3 + & 3Ca(HCO_3)_2 \rightarrow 2Fe(OH)_3 + 3CaCl_2 + 6CO_2 \\ 2(55.85 + 3 \times 35.45) & 3[40.08 + 2(1 + 12 + 48)] \\ = 324.4 & = 486.2 \end{array}$$

The above equation suggests that 2 moles of ferric chloride react with 3 moles of $Ca(HCO_3)_2$.

Step 2. Determine the alkalinity needed for X

$$\frac{\text{mg/L } Ca(HCO_3)_2}{\text{mg/L } FeCl_2} = \frac{486.2}{324.4} = 1.50$$

$$\begin{aligned} X &= 1.50 \times \text{mg/L } FeCl_2 = 1.50 \times 15.0 \text{ mg/L} \\ &= 22.5 \text{ mg/L as } Ca(HCO_3)_2 \end{aligned}$$

Assume $Ca(HCO_3)_2$ represents the total alkalinity of the natural water. However, alkalinity concentration is usually expressed in terms of mg/L as $CaCO_3$. We need to convert X to a concentration of mg/L as $CaCO_3$.

$$\begin{aligned} \text{MW of } CaCO_3 &= 40.08 + 12 + 48 = 100.1 \\ \text{MW of } Ca(HCO_3)_2 &= 162.1 \end{aligned}$$

Then

$$X = 22.5 \text{ mg/L} \times \frac{100.1}{162.1} = 13.9 \text{ mg/L as } CaCO_3$$

EXAMPLE 2: A water with low alkalinity of 12 mg/L as $CaCO_3$ will be treated with the alumlime coagulation. Alum dosage is 55 mg/L. Determine the lime dosage needed to react with alum.

Solution:

Step 1. Determine the amount of alum needed to react with the natural alkalinity
From Example 1

$$\text{Alkalinity} = 12 \text{ mg/L as } CaCO_3 \frac{162.1 \text{ as } Ca(HCO_3)_2}{100.1 \text{ as } CaCO_3}$$

$$= 19.4 \text{ mg/L as } Ca(HCO_3)_2$$

$$\text{MW of } Al_2(SO_4)_3 \cdot 18H_2O = 27 \times 2 + 3(32 + 16 \times 4) + 18(2 + 16)$$

$$= 666$$

$$\text{MW of } Ca(HCO_3)_2 = 162.1$$

Equation (6.51a) suggests that 1 mole of alum reacts with 3 moles of $Ca(HCO_3)_2$. Therefore, the quantity of alum to react with natural alkalinity is Y:

$$Y = 19.4 \text{ mg/L as } Ca(HCO_3)_2 \times \frac{666 \text{ as alum}}{3 \times 162.1 \text{ as } Ca(HCO_3)_2}$$

$$= 26.6 \text{ mg/L as alum}$$

Step 2. Calculate the lime required. The amount of alum remaining to react with lime is (dosage – Y)

$$55 \text{ mg/L} - 26.6 \text{ mg/L} = 28.4 \text{ mg/L}$$

$$\text{MW of } Ca(OH)_2 = 40.1 + 2(16 + 1) = 74.1$$

$$\text{MW of CaO} = 40.1 + 16 = 56.1$$

Equation (6.51b) indicates that 1 mole of alum reacts with 3 moles of $Ca(OH)_2 \cdot Ca(OH)_2$ required:

$$28.4 \text{ mg/L alum} \frac{3 \times 74.1 \text{ as } Ca(OH)_2}{666 \text{ alum}}$$

$$= 9.48 \text{ mg/L as } Ca(OH)_2$$

Let dosage of lime required be Z

$$Z = 9.48 \text{ mg/L as } Ca(OH)_2 \times \frac{56.1 \text{ as CaO}}{74.1 \text{ as } Ca(OH)_2}$$

$$= 7.2 \text{ mg/L as CaO}$$

8.1 Jar Test

For the jar test, chemical (coagulant) is added to raw water sample for mixing in the laboratory to simulate treatment-plant mixing conditions. Jar tests may provide overall process effectiveness, particularly to mixing intensity and duration as it affects floc size and density. It can also be used for evaluating chemical feed sequence, feed intervals, and chemical dilution ratios. A basic description of the jar test procedure and calculations is presented elsewhere (APHA 1995).

It is common to use six 2 L Gator jars with various dosages of chemical (alum, lime, etc); and one jar as the control (without coagulant). Appropriate coagulant dosages are added to the 2 L samples before the rapid mixing at 100 revolution per minute (rpm) for 2 min. Then the samples and the control were flocculated at 20 rpm for 20 or more minutes and allowed to settle. Water temperature, floc size, settling characteristics (velocity, etc.), color of supernatant, pH, etc. should be recorded.

The following examples illustrate calculations involved in the jar tests; prepare stock solution, jar test solution mixture.

EXAMPLE 1: Given that liquid alum is used as a coagulant. Specific gravity of alum is 1.33. One gallon of alum weighs 11.09 lb and contains 5.34 lb of dry alum. Determine: (a) the alum concentration, (b) mL of liquid alum required to prepare a 100 mL solution of 20,000 mg/L alum concentration, (c) the dosage concentration of 1 mL of stock solution in a 2000 mL Gator jar sample.

Solution:

Step 1. Determine alum concentration in mg/mL

$$\text{Alum (mg/L)} = \frac{(5.34\text{ lb})(453{,}600\text{ mg/lb})}{(1\text{ gal})(3785\text{ mL/gal})}$$

$$= 640\text{ mg/mL}$$

Step 2. Prepare 100 mL stock solution having a 20,000 mg/L alum concentration
Let x = mg of alum required to prepare 100 mL stock solution

$$\frac{x}{100\text{ mL}} = \frac{20{,}000\text{ mg}}{1000\text{ mL}}$$

$$x = 2000\text{ mg}$$

Step 3. Calculate mL (y) of liquid alum to give 2000 mg

$$\frac{y\text{ mL}}{1\text{ mL}} = \frac{2000\text{ mg}}{640\text{ mg}}$$

$$y = 3.125\text{ mL}$$

Note: Or add 6.25 mL of liquid alum into the 200 mL stock solution, since 3.125 mL is difficult to accurately measure.

Step 4. Find 1 mL of alum concentration (z) in 2000 mL sample (jar)

$$(z\text{ mg/L})\,(2000\text{ mL}) = (20{,}000\text{ mg/L})\,(1\text{ mL})$$

$$z = \frac{20{,}000}{2000}\text{ mg/L}$$

$$= 10\text{ mg/L}$$

Note: Actual final volume is 2001 mL; using 2000 mL is still reasonable.

EXAMPLE 2: Assuming that 1 mL of potassium permanganate stock solution provides 1 mg/L dosage in a 2 L jar, what is the weight of potassium permanganate required to prepare 100 mL of stock solution? Assume 100 percent purity.

Solution:

Step 1. Find concentration (x mg/L) of stock solution needed

$$(x \text{ mg/L})\ (1 \text{ mL}) = (1 \text{ mg/L})\ (2000 \text{ mL})$$

$$x = 2000 \text{ mg/L}$$

Step 2. Calculate weight (y mg) to prepare 100 mL (as 2000 mg/L)

$$\frac{y \text{ mg}}{100 \text{ mL}} = \frac{2000 \text{ mg}}{1000 \text{ mL}}$$

$$y = 200 \text{ mg}$$

Add 200 mg of potassium permanganate into 100 mL of deionized water to give 100 mL stock solution of 2000 mg/L concentration.

EXAMPLE 3: Liquid polymer is used in a 2 L Gator jar test. It weighs 8.35 pounds per gallon. One mL of polymer stock solution provides 1 mg/L dosage. Compute how many mL of polymer are required to prepare 100 mL of stock solution.

Solution:

Step 1. Calculate (x) mg of polymer in 1 mL

$$x = \frac{(8.35 \text{ lb})(453{,}600 \text{ mg/lb})}{(1 \text{ gal})(3785 \text{ mL/gal})}$$

$$= 1000 \text{ mg/mL}$$

Step 2. As in Example 2, find the weight (y mg) to prepare 100 mL of 2000 mg/L stock solution

$$\frac{y \text{ mg}}{100 \text{ mL}} = \frac{2000 \text{ mg}}{1000 \text{ mL}}$$

$$y = 200 \text{ mg (of polymer in 100 mL distilled water)}$$

Step 3. Compute volume (z mL) of liquid polymer to provide 200 mg

$$z = \frac{200 \text{ mg}}{x} = \frac{200 \text{ mg} \times 1 \text{ mL}}{1000 \text{ mg}}$$

$$= 0.20 \text{ mL}$$

Note: Use 0.20 mL of polymer in 100 mL of distilled water. This stock solution has 2000 mg/L polymer concentration. One mL stock solution added to a 2 L jar gives 1 mg/L dosage.

8.2 Mixing

Mixing is an important operation for the coagulation process. In practice, rapid mixing provides complete and uniform dispersion of a chemical added to the water. Then follows a slow mixing for flocculation (particle aggregation). The time required for rapid mixing is usually 10 to 20 s. However, recent studies indicate the optimum time of rapid mixing is a few minutes.

Types of mixing include propeller, turbine, paddle, pneumatic, and hydraulic mixers. For a water treatment plant, mixing is used for coagulation and flocculation, and chlorine disinfection. Mixing is also used for biological treatment processes for wastewater. Rapid mixing for coagulant in raw water and activated sludge process in wastewater treatment are complete mixing. Flocculation basins after rapid mixing are designed based on an ideal plug flow using first-order kinetics. It is very

difficult to achieve an ideal plug flow. In practice, baffles are installed to reduce short-circuiting. The time of contact or detention time in the basin can be determined by:
For complete mixing:

$$t = \frac{V}{Q} = \frac{1}{K}\left(\frac{C_i - C_e}{C_e}\right) \tag{6.52a}$$

For plug flow:

$$t = \frac{V}{Q} = \frac{L}{v} = \frac{1}{K}\left(\ln\frac{C_i}{C_e}\right) \tag{6.52b}$$

where t = detention time of the basin, min
V = volume of basin, m^3 or ft^3
Q = flow rate, m^3/s or cfs
K = rate constant
C_i = influent reactant concentration, mg/L
C_e = effluent reactant concentration, mg/L
L = length of rectangular basin, m or ft
v = horizontal velocity of flow, m/s or ft/s

EXAMPLE 1: Alum dosage is 50 mg/L · K = 90 per day based on laboratory tests. Compute the detention times for complete mixing and plug flow reactor for 90 percent reduction.

Solution:

Step 1. Find C_e

$$C_e = (1 - 0.9)C_i = 0.1 \times C_i = 0.1 \times 50 \text{ mg/L}$$
$$= 5 \text{ mg/L}$$

Step 2. Calculate t for complete mixing
Using Eq. (6.52a)

$$t = \frac{1}{K}\left(\frac{C_i - C_e}{C_e}\right) = \frac{1}{90/\text{day}}\left(\frac{50 \text{ mg/L} - 5 \text{ mg/L}}{5 \text{ mg/L}}\right)$$
$$= \frac{1 \text{ day}}{90} \times \frac{1440 \text{ min}}{1 \text{ day}} \times 9$$
$$= 144 \text{ min}$$

Step 3. Calculate t for plug flow
Using Eq. (6.52b)

$$t = \frac{1}{K}\left(\ln\frac{C_i}{C_e}\right) = \frac{1440 \text{ min}}{90}\left(\ln\frac{50}{5}\right)$$
$$= 36.8 \text{ min}$$

Power requirements. Power required for turbulent mixing is traditionally based on the velocity gradient or G values proposed by Camp and Stein (1943). The mean velocity gradient G for mechanical mixing is

$$G = \left(\frac{P}{\mu V}\right)^{1/2} \tag{6.53}$$

where G = mean velocity gradient; velocity (ft/s)/distance (ft) is equal to per second
P = power dissipated, ft lb/s or N m/s (W)
μ = absolute viscosity, lb s/ft^2 or N s/m^2
V = volume of basin, ft^3 or m^3

The equation is used to calculate the mechanical power required to facilitate rapid mixing. If a chemical is injected through orifices with mixing times of approximately 1.0 s, the G value is in the range of 700 to 1000/s. In practice, G values of 3000 to 5000/s are preferable for rapid mixing (ASCE and AWWA, 1990).

Eq. (6.53) can be expressed in terms of horsepower (hp) as

$$G = \left(\frac{550\ \text{hp}}{\mu V}\right)^{1/2} \tag{6.54a}$$

For SI units the velocity gradient per second is:

$$G = \left(\frac{\text{kW} \times 10}{\mu V}\right)^{1/2} \tag{6.54b}$$

where kW = energy input, kW
V = effective volume, m^3
μ = absolute viscosity, centipoise, cp
= 1 cp at 20°C (see Table 5.1a, 1 cp = 0.001 N · s/m^2)

The equation is the standard design guideline used to calculate the mechanical power required to facilitate rapid mixing. Camp (1968) claimed that rapid mixing at G values of 500/s to 1000/s for 1 to 2 min produced essentially complete flocculation and no further benefit for prolonged rapid mixing. For rapid mixing the product of Gt should be 30,000 to 60,000 with t (time) generally 60 to 120 s.

EXAMPLE 2: A rapid mixing tank is 1 m × 1 m × 1.2 m. The power input is 746 W (1 hp). Find the G value at a temperature of 15°C.

Solution:

At 10°C, μ = 0.00113 N · s/m^2 (from Table 5.1a)

$$V = 1\ \text{m} \times 1\ \text{m} \times 1.2\ \text{m} = 1.2\ \text{m}^3$$

$$P = 746\ \text{W} = 746\ \text{N m/s}$$

Using Eq. (6.53):

$$G = (P/\mu V)^{0.5}$$

$$= \left(\frac{746\ \text{N m/s}}{0.00113\ \text{N} \cdot \text{s/m}^2 \times 1.2\ \text{m}^3}\right)^{0.5}$$

$$= 742\ \text{s}^{-1}$$

9 FLOCCULATION

After rapid mixing, the water is passed through the flocculation basin. It is intended to mix the water to permit agglomeration of turbidity settled particles (solid capture) into larger flocs which would have a mean velocity gradient ranging 20 to 70 s^{-1} for a contact time of 20 to 30 min taking place in the flocculation basin. A basin is usually designed in four compartments (ASCE and AWWA 1990).

The conduits between the rapid mixing tank and the flocculation basin should maintain G values of 100 to 150 s^{-1} before entering the basin.

For baffled basin, the G value is

$$G = \left(\frac{Q\gamma H}{\mu V}\right)^{0.5} = \left(\frac{62.4H}{\mu t}\right)^{0.5} \tag{6.55}$$

where G = mean velocity gradient, s^{-1}
Q = flow rate, ft^3/s
γ = specific weight of water, 62.4 lb/ft^3
H = head loss due to friction, ft
μ = absolute viscosity, $lb \cdot s/ft^2$
V = volume of flocculator, ft^3
t = detention time, s

For paddle flocculators, the useful power input of an impeller is directly related to the dragforce of the paddles (F). The drag force is the product of the coefficient of drag (C_d) and the impeller force (F_i). The drag force can be expressed as (Fair *et al.* 1968):

$$F = C_d F_i \tag{6.56a}$$

$$F_i = \rho A \frac{v^2}{2} \tag{6.57}$$

then

$$F = 0.5\, C_d \rho A v^2 \tag{6.56b}$$

where F = drag force, lb
C_d = dimensionless coefficient of drag
ρ = mass density, $lb\ s^2/ft^4$
A = area of the paddles, ft^2
v = velocity difference between paddles and water, fps

The velocity of paddle blades (v_p) can be determined by

$$v_p = \frac{2\pi\, r\, n}{60} \tag{6.58}$$

where v_p = velocity of paddles, fps
n = number of revolutions per minute, rpm
r = distance from shaft to center line of the paddle, ft

The useful power input is computed as the product of drag force and velocity difference as below:

$$P = Fv = 0.5\, C_d\, \rho\, A v^3 \tag{6.59}$$

EXAMPLE 1: In a baffled basin with detention time of 25 min. Estimate head loss if G is 30 per second, $\mu = 2.359 \times 10^{-5}$ $lb \cdot s/ft^2$ at $T = 60°F$ (Table 5.1b).

Solution:

Using Eq. (6.55)

$$G = \left(\frac{62.4\, H}{\mu t}\right)^{1/2}$$

$$H = \frac{G^2 \mu t}{62.4} = \frac{(30\ s^{-1})^2 (2.359 \times 10^{-5}\ lb \cdot s/ft^2) \times (25 \times 60\ s)}{62.4\ lb/ft^3}$$

$$= 0.51\ ft$$

EXAMPLE 2: A baffled flocculation basin is divided into 16 channels by 15 around-the-end baffles. The velocities at the channels and at the slots are 0.6 and 2.0 fps (0.18 and 0.6 m/s), respectively. The flow rate is 12.0 cfs (0.34 m^3/s). Find (a) the total head loss neglecting channel friction; (b) the power dissipated; (c) the mean velocity gradient at 60°F (15.6°C); the basin size is 16 × 15 × 80 ft^3; (d) the *Gt* value, if the detention (displacement) time is 20 min, and (e) loading rate in gpd/ft^3.

Solution:

Step 1. Estimate loss of head *H*

$$\text{Loss of head in a channel} = \frac{v_1^2}{2g} = \frac{0.6^2}{2 \times 32.2} = 0.00559 \text{ (ft)}$$

$$\text{Loss of head in a slot} = \frac{2^2}{64.4} = 0.0621 \text{ (ft)}$$

(a):

$$H = 16 \times 0.00559 + 15 \times 0.0621 = 1.02 \text{ (ft)}$$

Step 2. Compute power input *P*

(b):

$$P = Q\gamma H = 12 \text{ ft}^3/\text{s} \times 62.37 \text{ lb/ft}^3 \times 1.02 \text{ ft} = 763 \text{ ft lb/s}$$

Note: γ = 62.37 lb/ft^3 at 60°F (Table 5.1b)

Step 3. Compute *G*
From Table 5.1b
At 60°F, $\mu = 2.359 \times 10^{-5}$ lb · s/ft^2

$$V = 16 \text{ ft} \times 15 \text{ ft} \times 80 \text{ ft} = 19{,}200 \text{ ft}^3$$

Using Eq. (6.54a)

(c):

$$G = \left(\frac{P}{\mu V}\right)^{1/2} = \left(\frac{763 \text{ ft} \cdot \text{lb/s}}{2.359 \times 10^{-5} \text{lb} \cdot \text{s/ft}^2 \times 19{,}200 \text{ ft}^3}\right)^{1/2} = 41.0 \text{s}^{-1}$$

Step 4. Compute *Gt*

(d):

$$Gt = 41 \text{ s}^{-1} \times 20 \text{ min} \times 60 \text{ s/min} = 49.200$$

Step 5. Compute loading rate

$$Q = 12 \text{ cfs} = 12 \text{ cfs} \times 0.646 \text{ MGD/cfs} = 7.75 \text{ MGD} = 7.75 \times 10^6 \text{ gpd}$$

(e):

$$\text{Loading rate} = Q/V = 7.76 \times 10^6 \text{ gpd}/19{,}200 \text{ ft}^3 = 404 \text{ gpd/ft}^3$$

EXAMPLE 3: A flocculator is 16 ft (4.88 m) deep, 40 ft (12.2 m) wide, and 80 ft (24.4 m) long. The flow of the water plant is 13 MGD (20 cfs, 0.566 m^3/s). Rotating paddles are supported parallel to four horizontal shafts. The rotating speed is 2.0 rpm. The center line of the paddles is 5.5 ft (1.68 m) from the shaft (mid-depth of the basin). Each shaft equipped with six paddles. Each paddle blade is 10 in (25 cm) wide and 38 ft (11.6 m) long. Assume the mean velocity of the water is 28 percent of the velocity of the paddles and their drag coefficient is 1.9. Estimate:

(a) the difference in velocity between the paddles and water

(b) the useful power input

(c) the energy consumption per million gallons (Mgal)

(d) the detention time

(e) the value of G and Gt at 60°F

(f) the loading rate of the flocculator

Solution:

Step 1. Find velocity differential v
Using Eq. (6.58)

$$v_p = \frac{2\pi rn}{60} = \frac{2 \times 3.14 \times 5.5 \text{ ft} \times 2}{60 \text{ s}}$$

$$= 1.15 \text{ fps}$$

(a)

$$v = v_p (1 - 0.28) = 1.15 \text{ fps} \times 0.72$$
$$= 0.83 \text{ fps}$$

Step 2. Find P

$$A = \text{paddle area} = 4 \text{ shaft} \times 6 \text{ pads/shaft} \times 38 \text{ ft} \times 10/12 \text{ ft/pad} = 760 \text{ ft}^2$$

$$\rho = \frac{\gamma}{g} = \frac{62.4 \text{ lb/ft}^3}{32.2 \text{ ft/s}^2} = 1.938 \text{ lb} \cdot \text{s}^2/\text{ft}^4$$

Using Eq. (6.59):

(b)

$$P = 0.5C_d \rho A v^3$$
$$= 0.5 \times 1.9 \times 1.938 \text{ lb} \cdot \text{s}^2/\text{ft}^4 \times 760 \text{ ft}^2 \times (0.83 \text{ fps})^3$$
$$= 800 \text{ ft} \cdot \text{lb/s}$$

or

$$= (800/550) \text{ hp}$$
$$= 1.45 \text{ (hp)}$$

or

$$= 1.45 \times 0.746 \text{ kW}$$
$$= 1.08 \text{ kW}$$

Step 3. Determine energy consumption E

(c)

$$E = \frac{1.45 \text{ hp}}{13 \text{ Mgal/d}} \times \frac{24 \text{ h}}{\text{day}}$$
$$= 2.68 \text{ hp} \cdot \text{h/Mgal}$$

or

$$E = \frac{1.08 \text{ kW}}{13 \text{ Mgal/d}} \times \frac{24 \text{ h}}{\text{day}}$$
$$= 1.99 \text{ kW} \cdot \text{h/Mgal}$$

Step 4. Determine detention time t

$$\text{Basin volume, } V = 16 \text{ ft} \times 40 \text{ ft} \times 80 \text{ ft} = 5.12 \times 10^4 \text{ ft}^3$$
$$= 5.12 \times 10^4 \text{ ft}^3 \times 7.48 \text{ gal/ft}^3$$
$$= 3.83 \times 10^5 \text{ gal}$$

(d)
$$t = \frac{V}{Q} = \frac{3.83 \times 10^5 \text{ gal}}{13 \times 10^6 \text{ gal/d}} = 0.0295 \text{ day}$$
$$= 0.0295 \text{ day} \times 1440 \text{ min/d}$$
$$= 42.5 \text{ min}$$

Step 5. Compute G and Gt
Using Eq. (6.53):

(e)
$$G = (P/\mu V)^{0.5}$$
$$= (800/2.359 \times 10^{-5} \times 5.12 \times 10^4)^{0.5}$$
$$= 25.7 \text{ (fps/ft, or s}^{-1}\text{)}$$
$$Gt = (25.7 \text{ s}^{-1})\,(42.5 \times 60 \text{ s})$$
$$= 65{,}530$$

Step 6. Compute the loading rate

(f)
$$\text{Loading rate} = \frac{Q}{V} = \frac{13 \times 10^6 \text{ gpd}}{5.12 \times 10^4 \text{ ft}^3}$$
$$= 254 \text{ gpd/ft}^3$$

10 SEDIMENTATION

Sedimentation is one of the most basic processes of water treatment. Plain sedimentation, such as the use of a pre-sedimentation basin (grit chamber) and sedimentation tank (or basin) following coagulation—flocculation, is most commonly used in water treatment facilities. The grit chamber is generally installed upstream of a raw water pumping station to remove larger particles or objects. It is usually a rectangular horizontal-flow tank with a contracted inlet and the bottom should have at minimum a 1:100 longitudinal slope for basin draining and cleaning purposes. A trash screen (about 2-cm opening) is usually installed at the inlet of the grit chamber.

Sedimentation is a solid–liquid separation by gravitational settling. There are four types of sedimentation: discrete particle settling (type 1); flocculant settling (type 2); hindered settling (type 3); and compression settling (type 4). Sedimentation theories for the four types are discussed in Chapter 1.7 and elsewhere (Gregory and Zabel, 1990).

The terminal settling velocity of a single discrete particle is derived from the forces (gravitational force, buoyant force, and drag force) that act on the particle. The classical discrete particle settling theories have been based on spherical particles. The equation is expressed as

$$u = \left(\frac{4g(\rho_\text{p} - \rho)d}{3C_\text{d}\rho}\right)^{1/2} \tag{6.60}$$

where u = settling velocity of particles, m/s or ft/s
g = gravitational acceleration, m/s^2 or ft/s^2
ρ_p = density of particles, kg/m^3 or lb/ft^3
ρ = density of water, kg/m^3 or lb/ft^3
d = diameter of particles, m or ft
C_d = coefficient of drag

The values of drag coefficient depend on the density of water (ρ), relative velocity (u), particle diameter (d), and viscosity of water (μ), which gives the Reynolds number **R** as:

$$\mathbf{R} = \frac{\rho u d}{\mu} \tag{6.61}$$

The value of C_d decreases as the Reynolds number increases. For **R** less than 2 or 1, C_d is related to **R** by the linear expression as follows:

$$C_d = \frac{24}{\mathbf{R}} \tag{6.62}$$

At low values of **R**, substituting Eq. (6.61) and (6.62) into Eq. (6.60) gives

$$u = \frac{g(\rho_p - \rho)d^2}{18\mu} \tag{6.63}$$

This expression is known as the Stokes equation for laminar flow conditions.

In the region of higher Reynolds numbers ($2 < \mathbf{R} < 500\text{–}1000$), C_d becomes (Fair *et al.* 1968):

$$C_d = \frac{24}{\mathbf{R}} + \frac{3}{\sqrt{\mathbf{R}}} + 0.34 \tag{6.64}$$

In the region of turbulent flow ($500-1000 < \mathbf{R} < 200{,}000$), the C_d remains approximately constant at 0.44. The velocity of settling particles results in Newton's equation (ASCE and AWWA 1990):

$$u = 1.74\left[\frac{(\rho_p - \rho)gd}{\rho}\right]^{1/2} \tag{6.65}$$

When the Reynolds number is greater than 200,000, the drag force decreases substantially and C_d becomes 0.10. No settling occurs at this condition.

EXAMPLE: Estimate the terminal settling velocity in water at a temperature of 15°C of spherical silicon particles with specific gravity 2.40 and average diameter of (a) 0.05 mm and (b) 1.0 mm

Solution:

Step 1. Using the Stokes equation (Eq. (6.63)) for (a), $d = .05$ mm

From Table 5.1a, at $T = 15°\text{C}$

$$\rho = 999\ \text{kg/m}^3,\ \text{and}\ \mu = 0.00113\ \text{N}\cdot\text{s/m}^2$$

$$d = 0.05\ \text{mm} = 5 \times 10^{-5}\ \text{m}$$

(a)

$$u = \frac{g(\rho_p - \rho)d^2}{18\mu}$$

$$= \frac{9.81\ \text{m/s}^2\ (2400 - 999)\ \text{kg/m}^3\ (5 \times 10^{-5}\ \text{m})^2}{18 \times 0.00113\ \text{N}\cdot\text{s/m}^2}$$

$$= 0.00169\ \text{m/s}$$

Step 2. Check with the Reynolds number ((Eq. (6.61))

$$\mathbf{R} = \frac{\rho u d}{\mu} = \frac{999 \times 0.00169 \times 5 \times 10^{-5}}{0.00113}$$

$$= 0.075$$

(a) The Stokes' law applies, since $\mathbf{R} < 2$.

Step 3. Using the Stokes' law for (b), $d = 1$ mm $= 0.001$ m

$$u = \frac{9.81\ (2400 - 999)\ (0.001)^2}{18 \times 0.00113}$$
$$= 0.676\ \text{(m/s)}$$

Step 4. Check the Reynold number

Assume the irregularities of the particles $\phi = 0.85$

$$\mathbf{R} = \frac{\phi \rho u d}{\mu} = \frac{0.85 \times 999 \times 0.188 \times 0.001}{0.00113}$$
$$= 508$$

Since $\mathbf{R} > 2$, the Stokes' law does not apply. Use Eq. (6.60) to calculate u.

Step 5. Using Eqs. (6.64) and (6.60)

$$C_d = \frac{24}{\mathbf{R}} + \frac{3}{\sqrt{\mathbf{R}}} + 0.34 = \frac{24}{508} + \frac{3}{\sqrt{508}} + 0.34$$
$$= 0.52$$
$$u^2 = \frac{4g(\rho_p - \rho)d}{3C_d\rho}$$
$$u^2 = \frac{4 \times 9.81 \times (2400 - 999) \times 0.001}{3 \times 0.52 \times 999}$$
$$u = 0.188\ \text{(m/s)}$$

Step 6. Re-check **R**

$$\mathbf{R} = \frac{\phi \rho u d}{\mu} = \frac{0.85 \times 999 \times 0.188 \times 0.001}{0.00113}$$
$$= 141$$

Step 7. Repeat Step 5 with new **R**

$$C_d = \frac{24}{141} + \frac{3}{\sqrt{141}} + 0.34$$
$$= 0.76$$
$$u^2 = \frac{4 \times 9.81 \times 1401 \times 0.001}{3 \times 0.76 \times 999}$$
$$u = 0.155\ \text{(m/s)}$$

Step 8. Re-check **R**

$$\mathbf{R} = \frac{0.85 \times 999 \times 0.155 \times 0.001}{0.00113}$$
$$= 116$$

Step 9. Repeat Step 7

$$C_d = \frac{24}{116} + \frac{3}{\sqrt{116}} + 0.34$$
$$= 0.83$$
$$u^2 = \frac{4 \times 9.81 \times 1401 \times 0.001}{3 \times 0.83 \times 999}$$
$$u = 0.149\ \text{(m/s)}$$

(b) The estimated velocity is around 0.15 m/s based on steps 7 and 9.

10.1 Overflow Rate

For sizing the sedimentation basin, the traditional criteria used are based on the overflow rate, detention time, weir loading rate, and horizontal velocity. The theoretical detention time is computed from the volume of the basin divided by average daily flow (plug flow theory):

$$t = \frac{24V}{Q} \tag{6.66}$$

where t = detention time, h
24 = 24 h per day
V = volume of basin, m^3 or million gallon (Mgal)
Q = average daily flow, m^3/d or Mgal/d (MGD)

The overflow rate is a standard design parameter which can be determined from discrete particle settling analysis. The overflow rate or surface loading rate is calculated by dividing the average daily flow by the total area of the sedimentation basin as follows:

$$u = \frac{Q}{A} = \frac{Q}{lw} \tag{6.67}$$

where u = overflow rate, $m^3/(m^2 \cdot d)$ or gpd/ft^2
Q = average daily flow, m^3/d or gpd
A = total surface area of basin, m^2 or ft^2
l and w = length and width of basin, respectively, m or ft

For alum coagulation u is usually in the range of 40 to 60 $m^3/(m^2 \cdot d)$ (or m/d) (980 to 1470 gpd/ft^2) for turbidity and color removal. For lime softening, the overflow rate ranges 50 to 110 m/d (1230 to 2700 gpm/ft^2). The overflow rate in wastewater treatment is lower, ranging from 10 to 60 m/d (245 to 1470 gpm/ft^2). All particles having a settling velocity greater than the overflow rate will settle and be removed.

It should be noted that rapid particle density changes due to temperature, solid concentration, or salinity can induce density current which can cause severe short-circuiting in horizontal tanks (Hudson, 1972).

EXAMPLE: A water treatment plant has four clarifiers treating 4.0 MGD (0.175 m^3/s) of water. Each clarifier is 16 ft (4.88 m) wide, 80 ft (24.4 m) long, and 15 ft (4.57 m) deep. Determine: (a) the detention time, (b) overflow rate, (c) horizontal velocity, and (d) weir loading rate assuming the weir length is 2.5 times the basin width.

Solution:

Step 1. Compute detention time t for each clarifier

$$Q = \frac{4\text{ MGD}}{4} = \frac{1{,}000{,}000\text{ gal}}{\text{day}} \times \frac{1\text{ ft}^3}{7.48\text{ gal}} \times \frac{1\text{ day}}{24\text{ h}}$$

$$= 5570\text{ ft}^3/\text{h}$$

$$= 92.83\text{ ft}^3/\text{min}$$

(a)

$$t = \frac{V}{Q} = \frac{16\text{ ft} \times 80\text{ ft} \times 15\text{ ft}}{5570\text{ ft}^3/\text{h}}$$

$$= 3.447\text{ h}$$

Step 2. Compute overflow rate u, using Eq. (6.67)

(b)
$$u = \frac{Q}{lw} = \frac{1{,}000{,}000 \text{ gpd}}{16 \text{ ft} \times 80 \text{ ft}}$$
$$= 781 \text{ gpd/ft}^2 = 31.83 \text{ m}^3/\text{m}^2 \cdot \text{d}$$

Step 3. Compute horizontal velocity v

(c)
$$v = \frac{Q}{wd} = \frac{92.83 \text{ ft}^3/\text{min}}{16 \text{ ft} \times 15 \text{ ft}}$$
$$= 0.387 \text{ ft/min} = 11.8 \text{ cm/min}$$

Step 4. Compute weir loading rate u_w

(d)
$$u_w = \frac{Q}{2.5w} = \frac{1{,}000{,}000 \text{ gpd}}{2.5 \times 16 \text{ ft}}$$
$$= 25{,}000 \text{ gpd/ft} = 310 \text{ m}^2/\text{m} \cdot \text{d}$$

10.2 Inclined Settlers

Inclined (tube and plate) settlers are sedimentation units that have been used for more than two decades. A large number of smaller diameter (20 to 50 mm) tubes are nested together to act as a single unit and inclined with various angles (7°C to 60°). The typical separation distance between inclined plates for unhindered settling is 2 in (5 cm) with inclines of 3 to 6 ft (1 to 2 m) height. The solids or flocs settle by gravitational force. It is not necessary to use tubes and can take various forms or plates also. The materials are lightweight, generally PVC or ABC plastic (1 m × 3 m in size).

Tube settlers have proved as effective units. However, there is a tendency of clogging.

Inclined settling systems can be designed as cocurrent, countercurrent, and cross-flow. Comprehensive theoretical analyses of various flow geometries have been discussed by Yao (1976). The flow velocity of the settler module (v) and the surface loading rate for the inclined settler (u) (Fig. 6.6) are calculated as (James M. Montgomery Consulting Engineering, 1985):

$$v = \frac{Q}{A \sin \theta} \tag{6.68}$$

$$u = \frac{Q\,w}{A(H \cos \theta + w \cos^2 \theta)} \tag{6.69}$$

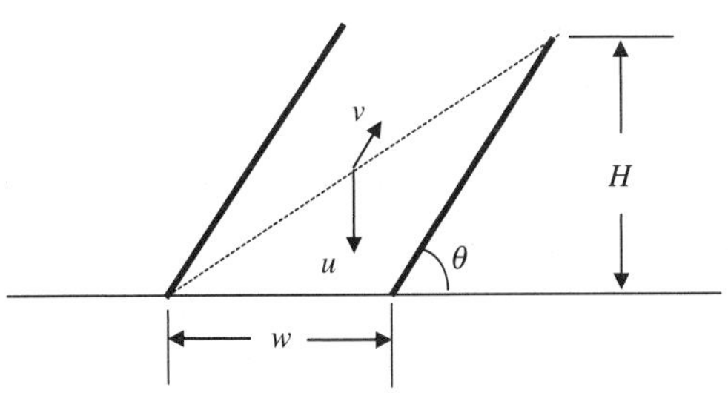

FIGURE 6.6 Schematic diagram of settling action.

where v = velocity of the water in settlers, m/s or ft/s
Q = flow rate, m^3/s or MGD
A = surface area of basin, m^2 or ft^2
θ = inclined angle of the settlers
u = settling velocity, m/s or ft/s
w = width of settler, m or ft
H = vertical height, m or ft

EXAMPLE: Two flocculators treat 1.0 m^3/s (22.8 MGD) and remove flocs larger than 0.02 mm. The settling velocity of the 0.02 mm flocs is measured in the laboratory as 0.22 mm/s (0.67 in/min) at 15°C. Tube settlers of 50.8 mm (2 in) square honeycombs are inclined at a 50° angle, and its vertical height is 1.22 m (4 ft). Determine the basin area required for the settler module and the size of each flocculator at 15°C.

Solution:

Step 1. Determine the area needed for the settler modules

$$Q = (1\ \text{m}^3/\text{s})/2 = 0.5\ \text{m}^3/\text{s} = 30\ \text{m}^3/\text{min}$$

$$w = 50.8\ \text{mm} = 0.0508\ \text{m}$$

$$H = 1.22\ \text{m}$$

$$\theta = 50^\circ$$

Using Eq. (6.69)

$$u = \frac{Qw}{A(H\cos\theta + w\cos^2\theta)}$$

$$= \frac{0.5(0.0508)}{A(1.22 \times 0.643 + 0.0508 \times 0.643^2)}$$

$$= \frac{0.315}{A}$$

Step 2. Determine A

In practice, the actual conditions in the settlers are not as good as under controlled laboratory ideal conditions.

A safety factor of 0.6 may be applied to determine the designed settling velocity. Thus

$$u = 0.6 \times 0.00022\ \text{m/s} = \frac{0.0315}{A}$$

$$A = 238.6\ \text{m}^2$$

$$(\text{use } 240\ \text{m}^2)$$

Step 3. Find surface loading rate Q/A

$$Q/A = (0.5 \times 24 \times 60 \times 60\ \text{m}^3/\text{d})/240\ \text{m}^2$$

$$= 180\ \text{m}^3/(\text{m}^2\ \text{d})$$

$$= 3.07\ \text{gpm/ft}^2\ (\textit{Note}\text{: } 1\text{m}^3/(\text{m}^2 \cdot \text{d}) = 0.017\ \text{gpm/ft}^2)$$

Step 4. Compute flow velocity in the settlers, using Eq. (6.68)

$$v = Q/A\sin\theta = 180/0.766$$

$$= 235\ (\text{m/d})$$

$$= 0.163\ \text{m/min}$$

$$= 0.0027\ \text{m/s}$$

Step 5. Determine size of the basin

Two identical settling basins are designed. Generally, the water depth of the basin is 4 m (13.1 ft). The width of the basin is chosen as 8.0 m (26.2 ft). The calculated length of the basin covered by the settler is

$$l = 240 \text{ m}^2/8 \text{ m} = 30 \text{ m} = 98 \text{ ft}$$

In practice, one-fourth of the basin length is left as a reserved volume for future expansion. The total length of the basin should be

$$30 \text{ m} \times \frac{4}{3} = 40 \text{ m (131 ft)}$$

Step 6. Check horizontal velocity

$$\begin{aligned} Q/A &= (30 \text{ m}^3/\text{min})/(4 \text{ m} \times 8 \text{ m}) \\ &= 0.938 \text{ m/min} = 3 \text{ ft/min} \end{aligned}$$

Step 7. Check Reynolds number (**R**) in the settler module

$$\text{Hydraulic radius } R = \frac{0.0508^2}{4 \times 0.0508} = 0.0127 \text{ m}$$

$$\begin{aligned} \mathbf{R} = \frac{vR}{\mu} &= \frac{(0.0027 \text{ m/s})(0.0127 \text{ m})}{0.000001131 \text{ m}^2/\text{s}} \\ &= 30 < 2000, \text{ thus it is in a lamella flow} \end{aligned}$$

11 FILTRATION

The conventional filtration process is probably the most important single unit operation of all the water treatment processes. It is an operation process to separate suspended matter from water by flowing it through porous filter medium or media. The filter media may be silica sand, anthracite coal, diatomaceous earth, garnet, ilmenite, or finely woven fabric.

In early times, a slow sand filter was used. It is still proved to be efficient. It is very effective for removing flocs of bacteria, algae, other microorganisms, and protozoa, such as *Giardia*, and *Cryptosporidium*. Rapid filtration has been very popular for several decades. Filtration usually follows the processes coagulation—flocculation—sedimentation. However, for some water treatment, direct filtration is used due to the high quality of raw water processes. Dual-media filters (sand and anthracite, activated carbon, or granite) give more benefits than single-media filters and became more popular; even triple-media filters have been used. In Russia, up-flow filters are used. All filters need to clean out the medium by backwash after a certain period (most are based on head loss) of filtration.

The filters are also classified by allowing loading rate. Loading rate is the flow rate of water applied to the unit area of the filter. It is the same value as the flow velocity approaching the filter surface and can be determined by

$$v = Q/A \tag{6.67a}$$

where v = loading rate, $\text{m}^3/(\text{m}^2 \cdot \text{d})$ or gpm/ft^2
Q = flow rate, m^3/d or ft^3/d or gpm
A = surface area of filter, m^2 or ft^2

On the basis of loading rate, the filters are classified as slow sand filters, rapid sand filters, and high-rate sand filters. With each type of filter medium or media, there are typical design criteria for the range of loading rate, effective size, uniform coefficient, minimum depth requirements, and backwash rate. The typical loading rate for rapid sand filters is 120 $m^3/(m^2 \cdot d)$, (83 $L/(m^2 \cdot min)$ or 2 gpm/ft^2). For high-rate filters, the loading rate may be four to five times this rate.

EXAMPLE: A city is to install rapid sand filters downstream of the clarifiers. The design loading rate is selected to be 160 $m^3/(m^2 \cdot d)$, (2.7 gpm/ft^2). The design capacity of the water works is 0.35 m^3/s (8 MGD). The maximum surface per filter is limited to 50 m^2. Design the number and size of filters and calculate the normal filtration rate.

Solution:

Step 1. Determine the total surface area required

$$A = \frac{Q}{v} = \frac{0.35 \text{ m}^3\text{/s } (86{,}400 \text{ s/d})}{160 \text{ m}^3\text{/m}^2 \cdot \text{d}}$$

$$= 189 \text{ m}^2$$

Step 2. Determine the number (n) of filters

$$n = \frac{189 \text{ m}^2}{50 \text{ m}} = 3.78$$

Select four filters.
The surface area (a) for each filter is

$$a = 189 \text{ m}^2/4 = 47.25 \text{ m}^2$$

We can use 7 m × 7 m or 6 m × 8 m, or 5.9 m × 8 m (exact)

Step 3. If a 7 m × 7 m filter is installed, the normal filtration rate is

$$v = \frac{Q}{A} = \frac{0.35 \text{ m}^3\text{/s} \times 86{,}400 \text{ s/d}}{4 \times 7 \text{ m} \times 7 \text{ m}}$$

$$= 154.3 \text{ m}^3/(\text{m}^2 \cdot \text{d})$$

11.1 Filter Medium Size

Before a filter medium is selected, a grain size distribution analysis should be performed. The sieve size and percentage passing by weight relationships are plotted on logarithmic-probability paper. A straight line can be drawn. Determine the geometric mean size (μ_g) and geometric standard deviation size (σ_g). The most common parameters used in the United States to characterize the filter medium are effective size (ES) and uniformity coefficient (UC) of medium size distribution. The ES is that the grain size for which 10 percent of the grain (d_{10}) are smaller by weight. The UC is the ratio of the 60-percentile (d_{60}) to the 10-percentile. They can be written as (Fair *et al.*, 1968; Cleasby, 1990):

$$\text{ES} = d_{10} = \mu_g/\sigma_g^{1.282} \tag{6.70}$$

$$\text{UC} = d_{60}/d_{10} = \sigma_g^{1.535} \tag{6.71}$$

The 90-percentile, d_{90}, is the size for which 90 percent of the grains are smaller by weight. It is interrelated to d_{10} as (Cleasby 1990)

$$d_{90} = d_{10}(10^{1.67 \log \text{UC}}) \tag{6.72}$$

The d_{90} size is used for computing the required filter backwash rate for a filter medium.

EXAMPLE: A sieve analysis curve of a typical filter sand gives $d_{10} = 0.54$ mm and $d_{60} = 0.74$ mm. What are its uniformity coefficient and d_{90}?

Solution:

Step 1. $\text{UC} = d_{60}/d_{10} = 0.74 \text{ mm}/0.54 \text{ mm}$

$= 1.37$

Step 2. Find d_{90} using Eq. (6.72)

$$\begin{aligned} d_{90} &= d_{10}(10^{1.67 \log \text{UC}}) \\ &= 0.54 \text{ mm}(10^{1.67 \log 1.37}) \\ &= 0.54 \text{ mm}(10^{0.228}) \\ &= 0.91 \text{ mm} \end{aligned}$$

11.2 Mixed Media

Mixed media are popular for filtration units. For the improvement process performance, activated carbon or anthracite is added on the top of the sand bed. The approximate specific gravity (s) of ilmenite sand, silica sand, anthracite, and water are 4.2, 2.6, 1.5, and 1.0, respectively. For equal settling velocities, the particle sizes for media of different specific gravity can be computed by

$$\frac{d_1}{d_2} = \left(\frac{s_2 - s}{s_1 - s}\right)^{2/3} \tag{6.73}$$

where d_1, d_2 = diameter of particles 1 and 2, respectively
s_1, s_2, s = specific gravity of particles 1, 2, and water, respectively

EXAMPLE: Estimate the particle sizes of ilmenite (specific gravity = 4.2) and anthracite (specific gravity = 1.5) which have same settling velocity of silica sand 0.60 mm in diameter (specific gravity = 2.6)

Solution:

Step 1. Find the diameter of anthracite by Eq. (6.73)

$$\begin{aligned} d &= (0.6 \text{ mm})\left(\frac{2.6 - 1}{1.5 - 1}\right)^{2/3} \\ &= 1.30 \text{ mm} \end{aligned}$$

Step 2. Determine diameter of ilmenite sand

$$\begin{aligned} d &= (0.6 \text{ mm})\left(\frac{2.6 - 1}{4.2 - 1}\right)^{2/3} \\ &= 0.38 \text{ mm} \end{aligned}$$

11.3 Hydraulics of Filter

Head loss for fixed bed flow. The conventional fixed-bed filters use a granular medium of 0.5 to 1.0 mm size with a loading rate or filtration velocity of 4.9 to 12.2 m/h (2 to 5 gpm/ft^2). When the clean water flows through a clean granular (sand) filter, the loss of head (pressure drop) can be estimated by the Kozeny equation (Fair *et al.*, 1968):

$$\frac{h}{L} = \frac{k\mu(1 - \varepsilon)^2}{g\rho\varepsilon^3} - \left(\frac{A}{V}\right)v \tag{6.74}$$

where h = head loss in filter depth L, m or ft
k = dimensionless Kozeny constant, 5 for sieve openings, 6 for size of separation
g = acceleration of gravity, 9.81 m/s or 32.2 ft/s
μ = absolute viscosity of water, $N \cdot s/m^2$ or $lb \cdot s/ft^2$
ρ = density of water, kg/m^3 or lb/ft^3
ε = porosity, dimensionless
A/V = grain surface area per unit volume of grain
= specific surface S (or shape factor = 6.0 to 7.7)
= $6/d$ for spheres
= $6/\psi\, d_{eq}$ for irregular grains
ψ = grain sphericity or shape factor
d_{eq} = grain diameter of spheres of equal volume
v = filtration (superficial) velocity, m/s or fps

The Kozeny (or Carmen–Kozeny) equation is derived from the fundamental Darcy—Waeisback equation for head loss in circular pipes (Eq. (5.17)). The Rose equation is also used to determine the head loss resulting from the water passing through the filter medium.

The Rose equation for estimating the head loss through filter medium was developed experimentally by Rose in 1949 (Rose, 1951). It is applicable to rapid sand filters with a uniform near spherical or spherical medium. The Rose equation is

$$h = \frac{1.067\, C_d\, L\, v^2}{\varphi\, g\, d\, \varepsilon^4} \tag{6.75}$$

where h = head loss, m or ft
φ = shape factor (Ottawa sand 0.95, round sand 0.82, angular sand 0.73, pulverized coal 0.73)
C_d = coefficient of drag (Eq. (6.64))

Other variables are defined previously in Eq. (6.74)

Applying the medium diameter to the area to volume ratio for homogeneous mixed beds the equation is

$$h = \frac{1.067\, C_d\, Lv^2}{\varphi\, g\, \varepsilon^4}\sum\frac{x}{d_g} \tag{6.76}$$

For a stratified filter bed, the equation is

$$h = \frac{1.067\, Lv^2}{\varphi\, g\, \varepsilon^4}\sum\frac{C_d x}{d_g} \tag{6.77}$$

where x = percent of particles within adjacent sizes
d_g = geometric mean diameter of adjacent sizes

EXAMPLE 1: A dual medium filter is composed of 0.3 m (1 ft) anthracite (mean size of 2.0 mm) that is placed over a 0.6 m (2 ft) layer of sand (mean size 0.7 mm) with a filtration rate of 9.78 m/h

(4.0 gpm/ft²). Assume the grain sphericity is $\psi = 0.75$ and a porosity for both is 0.40. Estimate the head loss of the filter at 15°C.

Solution:

Step 1. Determine head loss through anthracite layer
Using the Kozeny equation (Eq. (6.74))

$$\frac{h}{L} = \frac{k\mu\,(1-\varepsilon)^2}{g\,\rho\,\varepsilon^3}\left(\frac{A}{V}\right)^2 \upsilon$$

where
$k = 6$
$g = 9.81\ \text{m/s}^2$
$\mu/\rho = \nu = 1.131 \times 10^{-6}\ \text{m}^2 \cdot \text{s}$ (Table 5.1a) at 15°C
$\varepsilon = 0.40$
$A/V = 6/0.75d = 8/d = 8/0.002$
$\upsilon = 9.78\ \text{m/h} = 0.00272\ \text{m/s}$
$L = 0.3\ \text{m}$

then

$$h = 6 \times \frac{1.131 \times 10^{-6}}{9.81} \times \frac{(1-0.4)^2}{0.4^3} \times \left(\frac{8}{0.002}\right)^2 (0.00272)(0.3)$$
$$= 0.0508\ \text{(m)}$$

Step 2. Compute the head loss passing through the sand
Most input data are the same as in Step 1, except

$$k = 5$$
$$d = 0.0007\ \text{m}$$
$$L = 0.6\ \text{m}$$

then

$$h = 5 \times \frac{1.131 \times 10^{-6}}{9.81} \times \frac{0.6^2}{0.4^3}\left(\frac{8}{d}\right)^2 (0.00272)(0.6)$$
$$= 0.3387 \times 10^{-6}/d^2$$
$$= 0.3387 \times 10^{-6}/(0.0007)^2$$
$$= 0.6918\ \text{(m)}$$

Step 3. Compute total head loss

$$h = 0.0508\ \text{m} + 0.6918\ \text{m}$$
$$= 0.743\ \text{m}$$

EXAMPLE 2: Using the same data given in Example 1, except the average size of sand is not given. From sieve analysis, d_{10} (10 percentile of size diameter) = 0.53 mm, d_{30} = 0.67 mm, d_{50} = 0.73 mm, d_{70} = 0.80 mm, and d_{90} = 0.86 mm, estimate the head of a 0.6 m sand filter at 15°C.

Solution:

Step 1. Calculate head loss for each size of sand in the same manner as Step 2 of Example 1

$$h_{10} = 0.3387 \times 10^{-6}/d^2 = 0.3387 \times 10^{-6}/(0.00053)^2$$
$$= 1.206 \text{ (m)}$$
$$h_{30} = 0.3387 \times 10^{-6}/(0.00067)^2$$
$$= 0.755 \text{ (m)}$$

Similarly

$$h_{50} = 0.635 \text{ m}$$
$$h_{70} = 0.529 \text{ m}$$
$$h_{90} = 0.458 \text{ m}$$

Step 2. Taking the average of the head losses given above

$$h = (1.206 + 0.755 + 0.635 + 0.529 + 0.458)/5$$
$$= 0.717 \text{ (m)}$$

Note: This way of estimation gives slightly higher values than Example 1.

Head loss for a fluidized bed. When a filter is subject to back washing, the upward flow of water travels through the granular bed at a sufficient velocity to suspend the filter medium in the water. This is fluidization. During normal filter operation, the uniform particles of sand occupy the depth L. During backwashing the bed expands to a depth of L_e. When the critical velocity (v_c) is reached, the pressure drops ($\Delta p = h\rho g = \gamma c$) and is equal to the buoyant force of the grain. It can be expressed as

$$h\gamma = (\gamma_s - \gamma)(1 - \varepsilon_e)L_e$$

or

$$h = L_e(1 - \varepsilon_e)(\gamma_s - \gamma)/\gamma \qquad (6.78)$$

The porosity of the expanded bed can be determined by using (Fair *et al.*, 1963)

$$\varepsilon_e = (u/v_s)^{0.22} \qquad (6.79)$$

where u = upflow (face) velocity of the water
v_s = terminal settling velocity of the particles

A uniform bed of particles will expand when

$$u = v_s \varepsilon_e^{4.5} \qquad (6.80)$$

The depth relationship of the unexpanded and expanded bed is

$$L_e = L\left(\frac{1 - \varepsilon}{1 - \varepsilon_e}\right) \qquad (6.81)$$

The minimum fluidizing velocity (u_{mf}) is the superficial fluid velocity needed to start fluidization. The minimum fluidization velocity is important in determining the required minimum backwashing flow rate. Wen and Yu (1966) proposed the U_{mf} equation excluding shape factor and porosity of fluidization:

$$U_{mf} = \frac{\mu}{\rho\, d_{eq}}(1135.69 + 0.0408\, G_n)^{0.5} - \frac{33.7\, \mu}{\rho\, d_{eq}} \qquad (6.82)$$

where

$$G_n = \text{Galileo number}$$
$$= d_{eq}^3 \rho\, (\rho_s - \rho)\, g/\mu^2 \qquad (6.83)$$

Other variables used are expressed in Eq. (6.74).

In practice, the grain diameter of spheres of equal volume d_{eq} is not available. Thus the d_{90} sieve size is used instead of d_{eg}. A safety factor of 1.3 is used to ensure adequate movement of the grains (Cleasby and Fan, 1981).

EXAMPLE: Estimate the minimum fluidization velocity and backwash rate for the sand filter at 15°C. The d_{90} size of sand is 0.88 mm. The density of sand is 2.65 g/cm^3.

Solution:

Step 1. Compute the Galileo number
From Table 5.1a, at 15°C

$$\rho = 0.999 \text{ g/cm}^3$$
$$\mu = 0.00113 \text{ N} \cdot \text{s/m}^2 = 0.00113 \text{ kg/m s} = 0.0113 \text{ g/cm s}$$

(See Example 2 in Section 1.4 in Chapter 1.5, page 1.247)

$$\mu/\rho = 0.0113 \text{ cm}^2\text{/s}$$
$$g = 981 \text{ cm/s}^2$$
$$d = 0.088 \text{ cm}$$
$$\rho_s = 2.65 \text{ g/cm}^3$$

Using Eq. (6.83)

$$G_n = (0.088)^3\, (0.999)\, (2.65 - 0.999)\, (981)/(0.0113)^2$$
$$= 8635$$

Step 2. Compute U_{mf} by Eq. (6.82)

$$U_{mf} = \frac{0.0113}{0.999 \times 0.088}(1135.69 + 0.0408 \times 8635)^{0.5} - \frac{33.7 \times 0.0113}{0.999 \times 0.088}$$
$$= 0.627 \text{ (cm/s)}$$

Step 3. Compute backwash rate
Apply a safety factor of 1.3 to U_{mf} as backwash rate

$$\text{Backwash rate} = 1.3 \times 0.627 \text{ cm/s} = 0.815 \text{ cm/s}$$
$$= 0.00815 \text{ m/s}$$
$$= 0.00815 \text{ m/s} \times 86{,}400 \text{ s/d}$$
$$= 704.16 \text{ m/d or } (\text{m}^3/\text{m}^2 \cdot \text{d})$$
$$= 704 \text{ m/d} \times (0.01705 \text{ gpm/ft}^2)/1 \text{ (m/d)}$$
$$= 12.0 \text{ gpm/ft}^2$$

11.4 Washwater Troughs

In United States, in practice washwater troughs are installed at even spaced intervals (5 to 7 ft apart) above the gravity filters. The washwater troughs are employed to collect spent washwater. The total rate of discharge in a rectangular trough with free flow can be calculated by (Fair *et al.* 1968, ASCE and AWWA 1990)

$$Q = C\,w\,h^{1.5} \tag{6.84a}$$

where Q = flow rate, cfs
C = constant (2.49)
w = trough width, ft
h = maximum water depth in trough, ft

For rectangular horizontal troughs of such a short length friction losses are negligible. Thus the theoretical value of the constant C is 2.49. The value of C may be as low as 1.72 (ASCE and AWWA, 1990).

In European practice, backwash water is generally discharged to the side and no troughs are installed above the filter. For SI units, the relationship is

$$Q = 0.808\,w\,h^{1.5} \tag{6.84b}$$

where Q = flow rate, $m^3/(m^2 \cdot s)$
w = trough width, m
h = water depth in trough, m

EXAMPLE 1: Troughs are 20 ft (6.1 m) long, 18 in (0.46 m) wide, and 8 ft (2.44 m) to the center with a horizontal flat bottom. The backwash rate is 24 in/min (0.61 m/min). Estimate (1) the water depth of the troughs with free flow into the gullet, and (2) the distance between the top of the troughs and the 30 in sand bed. Assuming 40 percent expansion and 6 in of freeboard in the troughs and 8 in of thickness.

Solution:

Step 1. Estimate the maximum water depth (h) in trough

$$\begin{aligned} v &= 24 \text{ (in/min)} = 2 \text{ ft/60 s} = 1/30 \text{ fps} \\ A &= 20 \text{ ft} \times 8 \text{ ft} = 160 \text{ ft}^2 \\ Q &= VA = 160/30 \text{ cfs} \\ &= 5.33 \text{ cfs} \end{aligned}$$

Using Eq. (6.84a)

$$\begin{aligned} Q &= 2.49\,w\,h^{1.5},\ w = 1.5 \text{ ft} \\ h &= (Q/2.49\,w)^{2/3} \\ &= [5.33/(2.49 \times 1.5)]^{2/3} \\ &= 1.27 \text{ (ft)} \end{aligned}$$

Say h = 16 in = 1.33 ft

Step 2. Determine the distance (y) between the sand bed surface and the top troughs

$$\begin{aligned} \text{Sand bed} &= 30 \text{ in} = 2.5 \text{ ft} \\ \text{Height of expansion} &= 2.5 \text{ ft} \times 0.40 \\ \text{Freeboard} &= 6 \text{ in} = 0.5 \text{ ft} \\ \text{Thickness} &= 8 \text{ in} = 0.67 \text{ ft (the bottom of the trough)} \\ y &= 2.5 \text{ ft} \times 0.4 + 1.33 \text{ ft} + 0.5 \text{ ft} + 0.67 \text{ ft} \\ &= 3.5 \text{ ft} \end{aligned}$$

EXAMPLE 2: A filter unit has surface area of 16 ft (4.88 m) wide and 30 ft (9.144 m) long. After filtering 2.88 Mgal (10,900 m^3) for 50 h, the filter is backwashed at a rate of 16 gpm/ft^2 (0.65 m/min) for 15 min. Find: (a) the average filtration rate, (b) the quantity of washwater, (c) percent of washwater to treated water, and the flow rate to each of the four troughs.

Solution:

Step 1. Determine flow rate Q

$$Q = \frac{2{,}880{,}000 \text{ gal}}{50 \text{ h} \times 60 \text{ min/h} \times 16 \text{ ft} \times 30 \text{ ft}}$$

$$= 2.0 \text{ gpm/ft}^2$$

$$= \frac{2 \text{ gal} \times 0.0037854 \text{ m}^3 \times 1440 \text{ min} \times 10.764 \text{ ft}^2}{\text{min} \times \text{ ft}^2 \times 1 \text{ gal} \times 1 \text{ day} \times 1 \text{ m}^2}$$

$$= 117.3 \text{ m/d}$$

say $= 120$ m/d

Step 2. Determine quantity of wash water q

$$q = (16 \text{ gal/min} \cdot \text{ft}^2) \times 15 \text{ min} \times 16 \text{ ft} \times 30 \text{ ft}$$

$$= 115{,}200 \text{ gal}$$

Step 3. Determine percentage (%) of wash water to treated water

$$\% = 115{,}200 \text{ gal} \times 100\%/2{,}880{,}000 \text{ gal}$$

$$= 4.0\%$$

Step 4. Compute flow rate (r) in each trough

$$r = 115{,}200 \text{ gal}/(15 \text{ min} \times 4 \text{ troughs})$$

$$= 1920 \text{ gpm}$$

11.5 Filter efficiency

The filter efficiency is defined as the effective filter rate divided by the operation filtration rate as follows (AWWA and ASCE, 1998):

$$E = \frac{R_e}{R_o} = \frac{UFRV - UBWV}{UFRV} \tag{6.84c}$$

where E = filter efficiency, %
R_e = effective filtration rate, gpm/ft^2 or m^3/m^2/h (m/h)
R_0 = operating filtration rate, gpm/ft^2 or m^3/m^2/h (m/h)
$UFRV$ = unit filter run volume, gal/ft^2 or L/m^2
$UBWV$ = unit backwash volume, gal/ft^2 or L/m^2

EXAMPLE: A rapid sand filter operating at 3.7 gpm/ft^2 (9.0 m/h) for 46 hours. After this filter run, 295 gal/ft^2 (12,000 L/m^2) of backwash water is used. Find the filter efficiency.

Solution:

Step 1. Calculate operating filtration rate, R_o

$$R_o = 3.7 \text{ gal/min/ft}^2 \times 60 \text{ min/h} \times 46 \text{ h}$$

$$= 10{,}212 \text{ gal/ft}^2$$

Step 2. Calculate effective filtration rate, R_e

$$R_e = (10{,}212 - 295)\ \text{gal/ft}^2$$
$$= 9917\ \text{gal/ft}^2$$

Step 3. Calculate filter efficiency, E, using Eq. (5.84c)

$$E = 9917/10{,}212 = 0.97$$
$$= 97\%$$

12 WATER SOFTENING

Hardness in water is mainly caused by the ions of calcium and magnesium. It may also be caused by the presence of metallic cations of iron, sodium, manganese, and strontium. These cations are present with anions such as HCO_3^-, SO_4^{2-}, Cl^-, NO_3^-, and SiO_4^{2-}. The carbonates and bicarbonates of calcium, magnesium, and sodium cause carbonate hardness or temporary hardness since it can be removed and settled by boiling of water. Noncarbonate hardness is caused by the chloride and sulfate slots of divalent cations. The total hardness is the sum of carbonate and noncarbonate hardness.

The classification of hardness in water supply is shown in Table 6.4. Although hard water has no health effects, using hard water would increase the amount of soap needed and would produce scale on bath fixtures, cooking utensils, also cause scale and corrosion in hot-water heaters, boilers, and pipelines. Moderate hard water with 60 to 120 mg/L as $CaCO_3$ is generally publicly acceptable (Clark *et al.*, 1977).

The hardness in water can be removed by precipitation with lime, $Ca(OH)_2$ and soda ash, Na_2CO_3 and by an ion exchange process. Ion exchange is discussed in another section. The reactions of lime-soda ash precipitation for hardness removal in water and recarbonation are shown in the following equations:

Removal of free carbon dioxide with lime

$$\begin{bmatrix} CO_2 \\ H_2CO_3 \end{bmatrix} + Ca(OH)_2 \rightarrow CaCO_3 \downarrow + H_2O \tag{6.85}$$

Removal of carbonate hardness with lime

$$Ca^{2+} + 2HCO_3^- + Ca(OH)_2 \rightarrow 2CaCO_3 \downarrow + 2H_2O \tag{6.86}$$

$$Mg^{2+} + 2HCO_3^- + 2Ca(OH)_2 \rightarrow 2CaCO_3 \downarrow + Mg(OH)_2 \downarrow + 2H_2O \tag{6.87}$$

TABLE 6.4 Classification of Hard Water

Hardness classification	mg/L as $CaCO_3$ U.S.	mg/L as $CaCO_3$ International
Soft	0–60	0–50
Moderate soft		51–100
Slightly hard		101–150
Moderate hard	61–120	151–200
Hard	121–180	201–300
Very hard	>180	>300

Removal of noncarbonate hardness with soda ash and lime

$$Ca^{2+} + \begin{bmatrix} 2Cl^- \\ SO_4^{2-} \end{bmatrix} + Na_2CO_3 \rightarrow CaCO_3\downarrow + 2Na^+ + \begin{bmatrix} 2Cl^- \\ SO_4^{2-} \end{bmatrix} \tag{6.88}$$

$$Mg^{2+} + \begin{bmatrix} 2Cl^- \\ SO_4^{2-} \end{bmatrix} + Ca(OH)_2 \rightarrow Mg(OH)_2\downarrow + Ca^{2+} + \begin{bmatrix} 2Cl^- \\ SO_4^{2-} \end{bmatrix} \tag{6.89}$$

Recarbonation for pH control (pH $\cong$ 8.5)

$$CO_3^{2-} + CO_2 + H_2O \rightarrow 2HCO_3^- \tag{6.90}$$

Recarbonation for removal of excess lime and pH control (pH $\cong$ 9.5)

$$Ca(OH)_2 + CO_2 \rightarrow CaCO_3\downarrow + H_2O \tag{6.91}$$

$$Mg(OH)_2 + CO_2 \rightarrow MgCO_3 + H_2O \tag{6.92}$$

Based on the above equations, the stoichiometric requirement for lime and soda ash expressed in equivalents per unit volume are as follows (Tchobanoglous and Schroeder 1985):

$$\text{Lime required (eq/m}^3) = CO_2 + HCO_3^- + Mg^{2+} + \text{excess} \tag{6.93}$$

$$\text{Soda ash required (eq/m}^3) = Ca^{2+} + Mg^{2+} - \text{alkalinity} \tag{6.94}$$

Approximately 1 eq/m^3 of lime in excess of the stoichiometric requirement must be added to bring the pH to above 11 to ensure $Mg(OH)_2$ complete precipitation. After the removal of precipitates, recarbonation is needed to bring pH down to a range of 9.2 to 9.7.

The treatment processes for water softening may be different depending on the degree of hardness and the types and amount of chemical added. They may be single stage lime, excess lime, single stage lime-soda ash, and excess lime-soda ash processes.

Coldwell—Lawrence diagrams are based on equilibrium principles for solving water softening. The use of diagrams is an alternative to the stoichiometric method. It solves simultaneous equilibria equations and estimates the chemical dosages of lime-soda ash softening. The interested reader is referred to the American Water Works Association (1978) publication: Corrosion Control by Deposition of $CaCO_3$ Films, and to Benefield and Morgan (1990).

EXAMPLE 1: Water has the following composition: calcium = 82 mg/L, magnesium = 33 mg/L, sodium = 14 mg/L, bicarbonate = 280 mg/L, sulfate = 82 mg/L, and chloride = 36 mg/L. Determine carbonate hardness, noncarbonate hardness, and total hardness, all in terms of mg/L of $CaCO_3$.

Solution:

Step 1. Convert all concentration to mg/L of $CaCO_3$ using the following formula:

The species concentration in milliequivalents per liter (meq/L) is computed by the following equation

$$\text{meq/L} = \frac{\text{mg/L}}{\text{equivalent weight}} = \frac{100}{2} = 50 \text{ for } CaCO_3$$

The species concentration expressed as mg/L of $CaCO_3$ is calculated from

$$\text{mg/L as } CaCO_3 = \text{meq/L of species} \times 50$$

$$= \frac{(\text{mg/L of species}) \times 50}{\text{equivalent weight of species}}$$

Step 2. Construct a table for ions in mg/L as $CaCO_3$

Ion species	Molecular weight	Equivalent weight	Concentration mg/L	meq/L	mg/L as $CaCO_3$
Ca^{2+}	40	20.0	82	4.0	200
Mg^{2+}	24.3	12.2	33	2.7	135
Na^+	23	23.0	14	0.6	30
				Total: 7.3	365
HCO_3^-	61	61	280	4.6	230
Cl^-	35.5	35.5	36	1.0	50
SO_4^{2-}	96.1	48	82	1.7	85
				Total: 7.3	365

Step 3. Construct an equivalent bar diagram for the cationic and anionic species of the water. The diagram shows the relative proportions of the chemical species important to the water softening process. Cations are placed above anions on the graph. The calcium equivalent should be placed first on the cationic scale and be followed by magnesium and other divalent species and then by the monovalent species sodium equivalent. The bicarbonate equivalent should be placed first on the anionic scale and immediately be followed by the chloride equivalent and then by the sulfate equivalent.

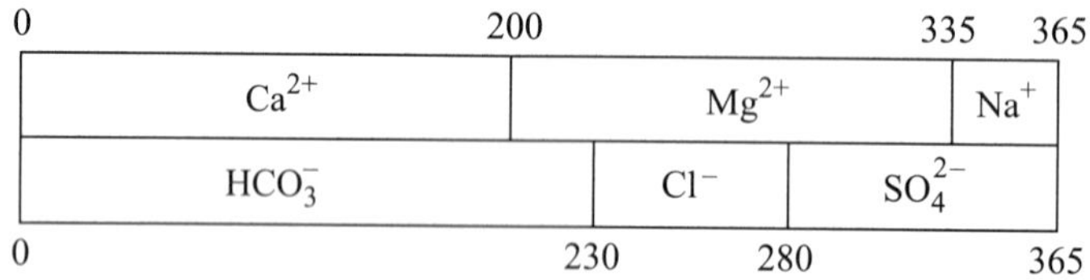

Step 4. Compute the hardness distribution

$$\text{Total hardness} = 200 + 135 = 335 \text{ (mg/L as CaCO}_3\text{)}$$

$$\text{Alkalinity (bicarbonate)} = 230 \text{ mg/L as CaCO}_3$$

$$\text{Carbonate hardness} = \text{alkalinity}$$

$$= 230 \text{ mg/L as CaCO}_3$$

$$\text{Noncarbonate hardness} = 365 - 230 = 165 \text{ (mg/L as CaCO}_3\text{)}$$

$$\text{or} = Na^+ + Cl^- + SO_4^{2-}$$

EXAMPLE 2: (single-stage lime softening): Raw water has the following composition: alkalinity = 248 mg/L as $CaCO_3$, pH = 7.0, $\alpha_1 = 0.77$ (at T = 10°C), calcium = 88 mg/L, magnesium = 4 mg/L. Determine the necessary amount of lime to soften the water, if the final hardness desired is 40 mg/L as $CaCO_3$. Also estimate the hardness of the treated water.

Solution:

Step 1. Calculate bicarbonate concentration

$$\text{Alkalinity} = [HCO_3^-] + [CO_3^{2-}] + [OH^-] + [H^-]$$

At pH = 7.0, assuming all alkalinity is in the carbonate form

$$HCO_3^- = 248\,\frac{\text{mg}}{\text{L}} \times \frac{1\text{ g}}{1000\text{ mg}} \times \frac{1\text{ mole}}{61\text{ g}} \times \frac{61\text{ eq wt of HCO}_3}{50\text{ eq wt of alkalinity}}$$

$$= 4.96 \times 10^{-3}\text{ mol/L}$$

Step 2. Compute total carbonate species concentration C_T (Snoeyink and Jenkins, 1980)

$$C_T = [HCO_3^-]/\alpha_1$$
$$= 4.96 \times 10^{-3}/0.77$$
$$= 6.44 \times 10^{-3}\ \text{mol/L}$$

Step 3. Estimate the carbonate acid concentration

$$C_T = [H_2CO_3] + [HCO_3^-] + [CO_3^{2-}]$$

Rearranging, and $[CO_3^{2-}] = 0$

$$[H_2CO_3] = (6.44 - 4.96) \times 10^{-3}\ \text{mol/L}$$
$$= 1.48 \times 10^{-3}\ \text{mol/L}$$
$$= 1.48 \times 10^{-3} \times 1000 \times 62\ \text{mg/mol} \times \text{mol/L}$$
$$= 92\ \text{mg/L}$$

or

$$= 148\ \text{mg/L as } CaCO_3\ (92 \times 50/32)$$

Step 4. Construct a bar diagram for the raw water converted concentration of Ca and Mg as $CaCO_3$

$$Ca = 88 \times 100/40 = 220\ \text{mg/L as } CaCO_3$$
$$Mg = 4 \times 100/24.3 = 16\ \text{mg/L as } CaCO_3$$

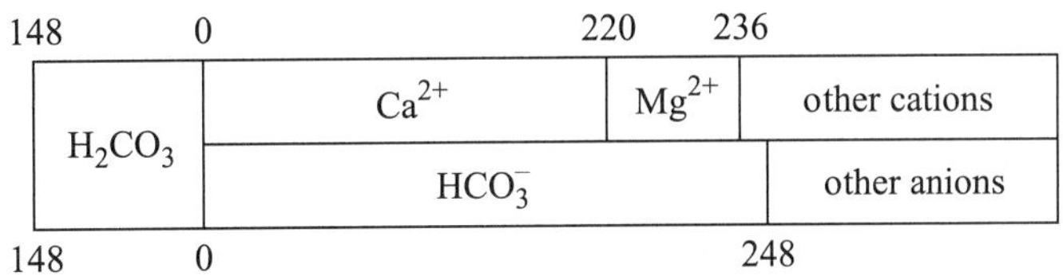

Step 5. Find hardness distribution

$$\text{Calcium carbonate hardness} = 220\ \text{mg/L}$$
$$\text{Magnesium carbonate hardness} = 16\ \text{mg/L}$$
$$\text{Total hardness} = 220 + 16 = 236\ (\text{mg/L})$$

Step 6. Estimate the lime dose needed

Since the final hardness desired is 40 mg/L as $CaCO_3$, magnesium hardness removal would not be required. The amount of lime needed (x) would be equal to carbonic acid concentration plus calcium carbonate hardness:

$$x = 148 + 220$$
$$= 368\ \text{mg/L as } CaCO_3$$

or

$$= 368\ \text{mg/L as } CaCO_3 \times \frac{74 \text{ as } Ca(OH)_2}{100 \text{ as } CaCO_3}$$
$$= 272\ \text{mg/L as } Ca(OH)_2$$

or

$$= 368 \times \frac{56}{100}\ \text{mg/L as CaO}$$
$$= 206\ \text{mg/L as CaO}$$

The purity of lime is 70 percent. The total amount of lime needed is

$$206/0.70 = 294 \text{ mg/L as CaO}$$

Step 7. Estimate the hardness of treated water

The magnesium hardness level of 16 mg/L as $CaCO_3$ remains in the softened water theoretically. The limit of calcium achievable is 30 to 50 mg/L of $CaCO_3$ and would remain in the water unless using 5 to 10 percent in excess of lime. Hardness levels less than 50 mg/L as $CaCO_3$ are seldom achieved in plant operation.

EXAMPLE 3: (excess lime softening): Similar to Example 2, except for concentrations of calcium and magnesium: pH = 7.0, $\alpha_1 = 0.77$ (at $T = 10°C$). Alkalinity, calcium, and magnesium concentrations are 248, 158, and 56 mg/L as $CaCO_3$, respectively. Determine the amount of lime needed to soften the water.

Solution:

Step 1. Estimate H_2CO_3 as Example 2

$$H_2CO_3 = 148 \text{ mg/L as } CaCO_3$$

Step 2. Construct a bar diagram for the raw water

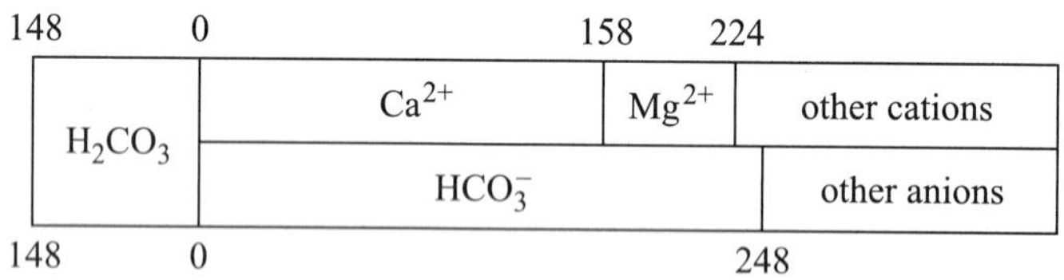

Step 3. Find hardness distribution

$$\text{Calcium carbonate hardness} = 158 \text{ mg/L}$$
$$\text{Magnesium carbonate hardness} = 66 \text{ mg/L}$$
$$\text{Total carbonate hardness} = 158 + 66 = 224 \text{ (mg/L)}$$

Comment: Sufficient lime (in excess) must be dosed to convert all bicarbonate alkalinity to carbonate alkalinity and to precipitate magnesium as magnesium hydroxide (needs 1 eq/L excess lime).

Step 4. Estimate lime required (x) is the sum for carbonic acid, total alkalinity, magnesium hardness, and excess lime (say 60 mg/L as $CaCO_3$) to raise pH = 11.0.

$$x = (148 + 248 + 66 + 60) \text{ mg/L as } CaCO_3$$
$$= 522 \text{ mg/L as } CaCO_3$$
$$= 522 \times 74/100 \text{ mg/L as } Ca(OH)_2$$
$$= 386 \text{ mg/L as } Ca(OH)_2$$

or

$$= 522 \times 0.56 \text{ mg/L as CaO}$$
$$= 292 \text{ mg/L as CaO}$$

EXAMPLE 4: (single stage lime-soda ash softening): A raw water has the following analysis: pH = 7.01, $T = 10°C$, alkalinity = 248 mg/L, Ca = 288 mg/L, Mg = 12 mg/L, all as $CaCO_3$. Determine amounts of lime and soda ash required to soften the water.

Solution:

Step 1. As in previous example, construct a bar diagram

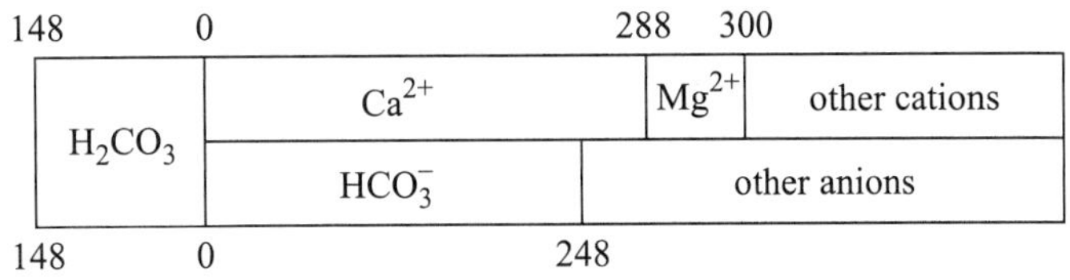

Step 2. Find the hardness distribution

$$\text{Total hardness} = 288 + 12 = 300 \text{ (mg/L as CaCO}_3\text{)}$$
$$\text{Calcium carbonate hardness, CCH} = 248 \text{ mg/L CaCO}_3$$
$$\text{Magnesium carbonate hardness} = 0$$
$$\text{Calcium noncarbonate hardness, CNH} = 288 - 248 = 40 \text{ (mg/L as CaCO}_3\text{)}$$
$$\text{Magnesium noncarbonate hardness, MNH} = 12 \text{ mg/L}$$

Step 3. Determine lime and soda ash requirements for the straight lime-soda ash process. Refer to Eqs. (6.85) and (6.86)

$$\text{Lime dosage} = H_2CO_3 + \text{CCH} = (148 + 248) \text{ mg/L as CaCO}_3$$
$$= 396 \text{ mg/L as CaCO}_3$$

or

$$= 396 \times 0.74 \text{ mg/L as Ca(OH)}_2$$
$$= 293 \text{ mg/L as Ca(OH)}_2$$

Refer to Eq. (6.88)

$$\text{and soda ash dosage} = \text{CNH}$$
$$= 40 \text{ mg/L as CaCO}_3$$
$$= (40 \times 106/100) \text{ mg/L as Na}_2\text{CO}_3$$
$$= 42.4 \text{ mg/L as Na}_2\text{CO}_3$$

EXAMPLE 5: (excess lime-soda ash process): Raw water has the following analysis: pH = 7.0, T = 10°C, alkalinity, calcium, and magnesium are 248, 288, and 77 mg/L as $CaCO_3$, respectively. Estimate the lime-and soda-ash dosage required to soften the water.

Solution:

Step 1. Construct a bar diagram for the raw water. Some characteristics are the same as the previous example.

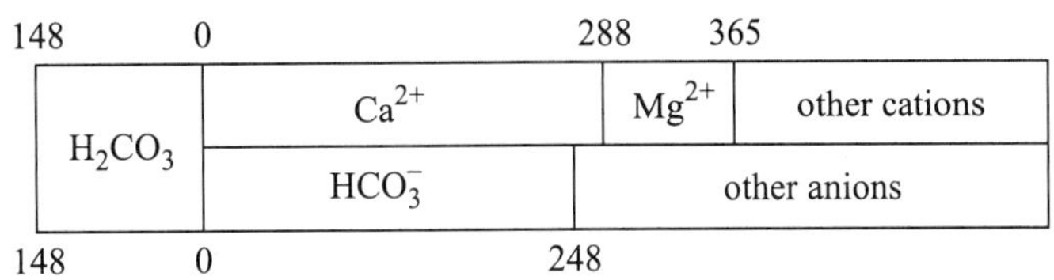

Step 2. Define hardness distribution

Total hardness = 288 + 77 = 365 (mg/L as $CaCO_3$)

Calcium carbonate hardness, CCH = 248 mg/L as $CaCO_3$

Magnesium carbonate hardness, MCH = 0

Calcium noncarbonate hardness, CNH = 288 mg/L − 248 mg/L = 40 mg/L

Magnesium noncarbonate hardness, MNH = 77 mg/L

Step 3. Estimate the lime- and soda-ash requirements by the excess lime-soda ash process.

$$\begin{aligned}\text{Lime dosage} &= \text{carb. acid} + \text{CCH} + 2(\text{MCH}) + \text{MNH} + \text{excess lime}\\ &= (148 + 248 + 2(0) + 77 + 60)\ \text{mg/L as CaCo}_3\\ &= 533\ \text{mg/L as CaCO}_3\end{aligned}$$

or

$$\begin{aligned}&= 533 \times 74/100\ [\text{mg/L as Ca(OH)}_2]\\ &= 394\ \text{mg/L as Ca(OH)}_2\end{aligned}$$

and soda ash dosage for excess lime-soda ash (Y) is

$$\begin{aligned}Y &= \text{CNH} + \text{MNH}\\ &= (40 + 77)\ \text{mg/L as CaCo}_3\\ &= 117\ \text{mg/L as CaCO}_3\end{aligned}$$

or

$$= 87\ \text{mg/L as Ca(OH)}_2$$

13 ION EXCHANGE

Ion exchange is a reversible process. Ions of a given species are displaced from an insoluble solid substance (exchange medium) by ions of another species dissolved in water. In practice, water is passed through the exchange medium until the exchange capacity is exhausted and then it is regenerated. The process can be used to remove color, hardness (calcium and magnesium), iron and manganese, nitrate and other inorganics, heavy metals, and organics.

Exchange media which exchange cations are called cationic or acid exchangers, while materials which exchange anions are called anionic or base exchangers.

Common cation exchangers used in water softening are zeolite, greensand, and polystyrene resins. However, most ion exchange media are currently in use as synthetic materials.

Synthetic ion exchange resins include four general types used in water treatment. They are strong- and weak-acid cation exchangers and strong- and weak-base anion exchangers. Examples of exchange reactions as shown below (Schroeder, 1977):

Strong acidic

$$2\text{R—SO}_3\text{H} + \text{Ca}^{2+} \leftrightarrow (\text{R—SO}_3)_2\text{Ca} + 2\text{H}^+ \quad (6.95)$$

$$2\text{R—SO}_3\text{Na} + \text{Ca}^{2+} \leftrightarrow (\text{R—SO}_3)_2\text{Ca} + 2\text{Na}^+ \quad (6.96)$$

Weak acidic

$$2\text{R—COOH} + \text{Ca}^{2+} \leftrightarrow (\text{R—COO})_2\text{Ca} + 2\text{H}^+ \quad (6.97)$$

$$2\text{R—COONa} + \text{Ca}^{2+} \leftrightarrow (\text{R—COO})_2\text{Ca} + 2\text{Na}^+ \quad (6.98)$$

Strong basic

$$2R\text{—}X_3NOH + SO_4^{2-} \leftrightarrow (R\text{—}X_3N)_2SO_4 + 2OH^- \tag{6.99}$$

$$2R\text{—}X_3NCl + SO_4^{2-} \leftrightarrow (R\text{—}X_3N)_2SO_4 + 2Cl^- \tag{6.100}$$

Weak basic

$$2R\text{—}NH_3OH + SO_4^{2-} \leftrightarrow (R\text{—}NH_3)_2SO_4 + 2OH^- \tag{6.101}$$

$$2R\text{—}NH_3Cl + SO_4^{2-} \leftrightarrow (R\text{—}NH_3)_2SO_4 + 2Cl^- \tag{6.102}$$

where in each reaction, R is a hydrocarbon polymer and X is a specific group, such as CH_2.

The exchange reaction for natural zeolites (Z) can be written as

$$Na_2Z + \left\{\begin{matrix} Ca^{2+} \\ Mg^{2+} \\ Fe^{2+} \end{matrix} \quad \leftrightarrow \quad \begin{matrix} Ca^{2+} \\ Mg^{2+} \\ Fe^{2+} \end{matrix}\right\} Z + 2Na^+ \tag{6.103}$$

In the cation-exchange water softening process, the hardness-causing elements of calcium and magnesium are removed and replaced with sodium by a strong-acid cation resin. Ion-exchange reactions for softening may be expressed as

$$Na_2R + \left.\begin{matrix} Ca \\ Mg \end{matrix}\right\} \begin{matrix} (HCO_3)_2 \\ SO_4 \\ Cl_2 \end{matrix} \rightarrow \left.\begin{matrix} Ca \\ Mg \end{matrix}\right\} R + \left\{\begin{matrix} 2\ NaHCO_3 \\ Na_2SO_4 \\ 2\ Na\ Cl \end{matrix}\right. \tag{6.104}$$

where R represents the exchange resin. They indicate that when a hard water containing calcium and magnesium is passed through an ion exchanger, these metals are taken up by the resin, which simultaneously gives up sodium in exchange. Sodium is dissolved in water. The normal rate is 6 to 8 gpm/ft^2 (350 to 470 m/d) of medium.

After a period of operation, the exchanging capacity would be exhausted. The unit is stopped from operation and regenerated by backwashing with sodium chloride solution and rinsed. The void volume for backwash is usually 35 to 45 percent of the total bed volume.

The exchange capacity of typical resins are in the range of 2 to 10 eq/kg. Zeolite cation exchangers have the exchange capacity of 0.05 to 0.1 eq/kg (Tchobanoglous and Schroeder 1985). During regeneration, the reaction can be expressed as:

$$\left.\begin{matrix} Ca \\ Mg \end{matrix}\right\} R + 2\ NaCl \rightarrow Na_2R + \left.\begin{matrix} Ca \\ Mg \end{matrix}\right\} Cl_2 \tag{6.105}$$

The spherical diameter in commercially available ion exchange resins is of the order of 0.04 to 1.0 mm. The most common size ranges used in large treatment plant are 20 to 50 mesh (0.85 to 0.3 mm) and 50 to 100 mesh (0.3 to 0.15 mm) (James M. Montgomery Consulting Engineering 1985). Details on the particle size and size range, effective size, and uniform coefficient are generally provided by the manufacturers.

The affinity of exchanges is related to charge and size. The higher the valence, the greater affinity and the smaller the effective size, the greater the affinity. For a given sense of similar ions, there is a general order of affinity for the exchanger. For synthetic resin exchangers, relative affinities of common ions increase as shown in Table 6.5.

The design of ion exchange units is based upon ion exchange equilibria. The generalized reaction equation for the exchange of ions A and on a cation exchange resin can be expressed as

$$nR^-A^+ + B^{n+} \leftrightarrow Rn^-\ B^{n+}nA^+ \tag{6.106}$$

where R^- = an anionic group attached to exchange resin

A^+, B^{n+} = ions in solution

TABLE 6.5 Selectivity Scale for Cations on Eight Percent Cross-Linked Strong-Acid Resin and for Anions on Strong-Base Resins

Cation	Selectivity	Anion	Selectivity
Li^+	1.0	HPO_4^{2-}	0.01
H^+	1.3	CO_3^{2-}	0.03
Na^+	2.0	OH^-(type I)	0.06
UO^{2+}	2.5	F^-	0.1
NH_4^+	2.6	SO_4^{2-}	0.15
K^+	2.9	CH_3COO^-	0.2
Rb^+	3.2	HCO_3^-	0.4
Cs^+	3.3	OH^-(type II)	0.65
Mg^{2+}	3.3	BrO_3^-	1.0
Zn^{2+}	3.5	Cl^-	1.0
Co^{2+}	3.7	CN^-	1.3
Cu^{2+}	3.8	NO_2^-	1.3
Cd^{2+}	3.9	HSO_4^-	1.6
Ni^{2+}	3.9	Br^-	3
Be^{2+}	4.0	NO_3^-	4
Mn^{2+}	4.1	I^-	8
Pb^{2+}	5.0	SO_4^{2-}	9.1
Ca^{2+}	5.2	SeO_4^{2-}	17
Sr^{2+}	6.5	CrO_4^{2-}	100
Ag^+	8.5		
Pb^{2+}	9.9		
Ba^{2+}	11.5		
Ra^{2+}	13.0		

Sources: James M. Montgomery Consulting Engineering (1985), Clifford (1990).

The equilibrium expression for this reaction is

$$K_{A\to B} = \frac{[R_n^- B^{n+}][A^+]^n}{[R^- A^+]^n[B^{n+}]} = \frac{q_B C_A}{q_A C_B} \tag{6.107}$$

where

$K_{A\to B} = K_A^B$ = selectivity coefficient, a function of ionic strength and is not a true constant.

$[R^- A^+]$, $[R^- B^{n+}]$ = mole fraction of A^+ and B^+ exchange resin, respectively, overbars represent the resin phase, or expressed as $[\overline{A}]$, and $[\overline{B}]$

$[A^+]$, $[B^{n+}]$ = concentration of A^+ and B^+ in solution, respectively, mol/L

q_A, q_B = concentration of A and B on resin site, respectively, eq/L

C_A, C_B = concentration of A and B in solution, respectively, mg/L

The selectivity constant depends upon the valence, nature, and concentration of the ion in solution. It is generally determined in laboratory for specific conditions measured.

For monovalent/monovalent ion exchange process such as

$$H^+R^- + Na^+ \leftrightarrow Na^+R^- + H^+$$

The equilibrium expression is (by Eq. (6.107))

$$K_{HR\to NaR} = \frac{[Na^+R^-][H^+]}{[H^+R^-][Na^+]} = \frac{q_{Na} C_H}{q_H C_{Na}} \tag{6.108}$$

and for divalent/divalent ion exchange processes such as

$$2\,(Na^+R^-) + Ca^{2+} \leftrightarrow Ca^{2+}R^{2-} + 2Na^+$$

then using Eq. (6.107)

$$K_{\mathrm{NaR \to CaR}} = \frac{[\mathrm{Ca^{2+}R}][\mathrm{Na^+}]^2}{[\mathrm{Na^+R}]^2[\mathrm{Ca^{2+}}]} = \frac{q_{\mathrm{Ca}} C_{\mathrm{Na}}^2}{q_{\mathrm{Na}}^2 C_{\mathrm{Ca}}} \tag{6.109}$$

Anderson (1975) rearranged Eq. (6.107) using a monovalent/monovalent exchange reaction with concentration units to equivalent fraction as follows:

1. In the solution phase: Let C = total ionic concentration of the solution, eq/L. The equivalent ionic fraction of ions A^+ and B^+ in solution will be

$$X_{\mathrm{A}}^{+} = [\mathrm{A^+}]/C \quad \text{or} \quad [\mathrm{A^+}] = CX_{\mathrm{A}}^{+} \tag{6.110}$$

$$X_{\mathrm{B}}^{+} = [\mathrm{B^+}]/C \quad \text{or} \quad [\mathrm{B^+}] = CX_{\mathrm{B}}^{+} \tag{6.111}$$

then

$$[\mathrm{A^+}] + [\mathrm{B^+}] = 1$$

or

$$[\mathrm{A^+}] = 1 - [\mathrm{B^+}] \tag{6.112}$$

2. In the resin phase: let $\overline{C}$ = total exchange capacity of the resin per unit volume, eq/L. Then we get

$$\overline{X}_{\mathrm{A^+}} = [\mathrm{R^-A^+}]/\overline{C} \quad \text{or} \quad [\mathrm{R^-A^+}] = \overline{CX}_{\mathrm{A}}^{+} \tag{6.113}$$

$$= \text{equivalent fraction of the A}^+ \text{ ion in the resin}$$

and

$$\overline{X}_{\mathrm{B^+}} = [\mathrm{R^-B^+}]/\overline{C} \quad \text{or} \quad [\mathrm{R^-B^+}] = \overline{CX}_{\mathrm{B}}^{+} \tag{6.114}$$

also

$$\overline{X}_{\mathrm{A^+}} + \overline{X}_{\mathrm{B^+}} = 1$$

or

$$\overline{X}_{\mathrm{A^+}} = 1 - \overline{X}_{\mathrm{B^+}} \tag{6.115}$$

Substitute Eqs. (6.110), (6.111), (6.113), and (6.114) into Eq. (6.107) which yields:

$$K_{\mathrm{A}}^{\mathrm{B}} = \frac{[\overline{CX}_{\mathrm{B^+}}][\overline{CX}_{\mathrm{A^+}}]}{[\overline{CX}_{\mathrm{A^+}}][CX_{\mathrm{B^+}}]}$$

or

$$K_{\mathrm{A}}^{\mathrm{B}} = \frac{[\overline{CX}_{\mathrm{B^+}}][CX_{\mathrm{A^+}}]}{[\overline{CX}_{\mathrm{A^+}}][CX_{\mathrm{B^+}}]}$$

$$\frac{\overline{X}_{\mathrm{B^+}}}{\overline{X}_{\mathrm{A^+}}} = K_{\mathrm{A}}^{\mathrm{B}} \frac{X_{\mathrm{B^+}}}{X_{\mathrm{A^+}}} \tag{6.116}$$

Substituting for Eqs. (6.112) and (6.115) in Eq. (6.116) gives

$$\frac{\overline{X}_{\mathrm{B^+}}}{1 - \overline{X}_{\mathrm{B^+}}} = K_{\mathrm{A}}^{\mathrm{B}} \frac{X_{\mathrm{B^+}}}{1 - X_{\mathrm{B^+}}} \tag{6.117}$$

If the valence is n, Eq. (6.117) will become

$$\frac{\overline{X}_{\mathrm{B^{n+}}}}{(1 - \overline{X}_{\mathrm{B^{n+}}})^n} = K_{\mathrm{A}}^{\mathrm{B}} \left(\frac{\overline{C}}{C}\right)^{n-1} \frac{X_{\mathrm{B^{n+}}}}{(1 - X_{\mathrm{B^{n+}}})^n} \tag{6.118}$$

EXAMPLE 1: Determine the meq/L of Ca^{2+} if Ca^{2+} concentration is 88 mg/L in water.

Solution:

Step 1. Determine equivalent weight (EW)

$$\begin{aligned} EW &= \text{molecular weight/electrical charge} \\ &= 40/2 \\ &= 20 \text{ grams per equivalent weight (or mg/meq)} \end{aligned}$$

Step 2. Compute meq/L

$$\begin{aligned} \text{meq/L} &= (\text{mg/L})/\text{EW} \\ &= (88 \text{ mg/L})/(20 \text{ mg/meq}) \\ &= 4.4 \end{aligned}$$

EXAMPLE 2: A strong-base anion exchange resin is used to remove nitrate ions from well water which contain high chloride concentration. Normally bicarbonate and sulfate is present in water (assume they are negligible). The total resin capacity is 1.5 eq/L. Find the maximum volume of water that can be treated per liter of resin. The water has the following composition in meq/L:

$$\begin{aligned} Ca^{2+} &= 1.4 & Cl^- &= 3.0 \\ Mg^{2+} &= 0.8 & SO_4^{2-} &= 0.0 \\ Na^+ &= 2.6 & NO_3^- &= 1.8 \\ \text{Total cations} &= 4.8 & \text{Total anions} &= 4.8 \end{aligned}$$

Solution:

Step 1. Determine the equivalent fraction of nitrate in solution

$$X_{NO_3^-} = 1.8/4.8 = 0.38$$

Step 2. Determine selective coefficient for sodium over chloride from Table 6.5.

$$K_{Cl}^{NO_3} = 4/1 = 4$$

Step 3. Compute the theoretical resin available for nitrate ion by Eq. (6.117)

$$\frac{\overline{X}_{NO_3^-}}{1 - \overline{X}_{NO_3^-}} = 4 \times \frac{0.38}{1 - 0.38} = 2.45$$

$$\overline{X}_{NO_3^-} = 0.71$$

It means that 71 percent of resin sites will be used.

Step 4. Compute the maximum useful capacity Y

$$Y = 1.5 \text{ eq/L} \times 0.71 = 1.065 \text{ eq/L} = 1065 \text{ meq/L}$$

Step 5. Compute the volume of water (V) that can be treated per cycle

$$\begin{aligned} V &= \frac{1065 \text{ meq/L of resin}}{1.8 \text{ meq/L of water}} \\ &= 592 \text{ L of water/L of resin} \\ &= 592 \frac{\text{L}}{\text{L}} \times \frac{1 \text{ gal}}{3.785 \text{ L}} \times \frac{28.32 \text{ L}}{1 \text{ ft}^3} \\ &= 4429 \text{ gal of water/ft}^3 \text{ of resin} \end{aligned}$$

EXAMPLE 3: A strong-acid cation exchanger is employed to remove calcium hardness from water. Its wet-volume capacity is 2.0 eq/L in the sodium form. If calcium concentrations in the influent and effluent are 44 mg/L (2.2 meq/L) and 0.44 mg/L, respectively, find the equivalent weight (meq/L) of the component in the water if given the following:

Cations		Anions	
Ca^{2+}	2.2	HCO_3^-	2.9
Mg^{2+}	1.0	Cl^-	3.1
Na^+	3.0	SO_4^{2-}	0.2
Total cations	6.2	Total anions	6.2

Solution:

Step 1. Determine the equivalent fraction of Ca^{2+}

$$X_{Ca^{2+}} = 2.2/6.2 = 0.35$$

Step 2. Find the selectivity coefficient K for calcium over sodium from Table 6.5

$$K_{Na}^{Ca} = 5.2/2.0 = 2.6$$

Step 3. Compute the theoretical resin composition with respect to the calcium ion

$$\overline{C}/C = 6.2/6.2 = 1.0$$

Using Eq. (6.118), $n = 2$

$$\frac{\overline{X}_{Ca^{2+}}}{(1 - \overline{X}_{Ca^{2+}})^2} = 2.6 \frac{0.35}{(1 - 0.35)^2} = 2.15$$

$$\overline{X}_{Ca^{2+}} = 0.51$$

This means that a maximum of 51 percent of the resin sites can be used with calcium ions from the given water. At this point, the water and resin are at equilibrium with each other.

Step 4. Compute the limiting useful capacity of the resin (Y)

$$\begin{aligned} Y &= 2.0 \text{ eq/L} \times 0.51 \\ &= 1.02 \text{ eq/L (or 1020 meq/L)} \end{aligned}$$

Step 5. Compute the maximum volume (V) of water that can be treated per cycle:

$$\begin{aligned} V &= \frac{1020 \text{ meq/L of resin (for Ca)}}{2.2 \text{ meq/L of calcium in water}} \\ &= 464 \text{ L of water/L of resin} \\ &= 3472 \text{ gal of water/ft}^3 \text{ of resin} \end{aligned}$$

13.1 Leakage

Leakage is defined as the appearance of a low concentration of the undesirable ions in the column effluent during the beginning of the exhaustion. It comes generally from the residual ions in the bottom resins due to incomplete regeneration. Leakage will occur for softening process. A water softening column is usually not fully regenerated due to inefficient use of salt to completely regenerate the resin on the sodium form. The leakage depends upon the ionic composition of the bed bottom and the composition of the influent water.

A detailed example of step-by-step design method for a fix bed ion exchange column for water softening is given by Benefield *et al.* (1982).

EXAMPLE: The bottom of the water softener is 77 percent in the calcium form after regeneration. The strong-acid cation resin has a total capacity of 2.0 eq/L and selective coefficient for calcium over sodium is 2.6. Determine the initial calcium leakage the water composition has as follows (in meq/L):

$$
\begin{array}{rr}
Ca^{2+} = 0.4 & Cl^- = 0.4 \\
Mg^{2+} = 0.2 & SO_4^{2-} = 0.4 \\
Na^+ = 1.2 & HCO^-{}_3 = 1.0 \\
\text{Total cations} = 1.8 & \text{Total anions} = 1.8
\end{array}
$$

Solution:

Step 1. Determine the equivalent fraction of calcium ion from the given:

$$\overline{X}_{Ca^{2+}} = 0.77$$
$$\overline{C} = 2.0 \text{ eq/L}$$
$$C = 2 \text{ meq/L} = 0.002 \text{ eq/L}$$
$$K_{Na}^{Ca} = 2.6$$

Using Eq. (6.118)

$$\frac{0.77}{(1 - 0.77)^2} = 2.6\left(\frac{2}{0.002}\right)\left[\frac{X_{Ca}}{(1 - X_{Ca})^2}\right]$$

$$\frac{X_{Ca}}{(1 - X_{Ca})^2} = 0.0056$$

Step 2. Calculate the initial calcium leakage y

$$
\begin{aligned}
y &= 0.0056\, c = 0.0056 \times 1.8 \text{ meq/L} \\
&= 0.010 \text{ meq/L} \\
&= 0.010 \frac{\text{meq}}{\text{L}} \times \frac{20 \text{ mg as Ca}}{\text{meq}} \\
&= 0.20 \text{ mg/L as Ca} \\
&= 0.20 \text{ mg/L as Ca} \frac{100 \text{ as } CaCO_3}{40 \text{ as Ca}} \\
&= 0.50 \text{ mg/L as } CaCO_3
\end{aligned}
$$

13.2 Nitrate Removal

Nitrate, NO_3^-, is a nitrogen—oxygen ion that occurs frequently in nature as the result of interaction between nitrogen in the atmosphere and living things on earth. It is a portion of the nitrogen cycle. When plant and animal proteins are broken down, ammonia and nitrogen gas are released. Ammonia is subsequently oxidized to nitrite (NO_2^-) and nitrate by bacterial action (Fig. 2.1).

High concentration of nitrate in drinking water may cause methemoglobinemia, especially for infants. Nitrates interfere with the body's ability to take oxygen from the air and distribute it to body cells. The United States EPA's standard for nitrate in drinking water is 10 mg/L. In Illinois, the water purveyor must furnish special bottle water (low nitrate) to infants if the nitrate level of the finished water exceeds the standard.

The sources of nitrates to water are due to human activities. Agricultural activities, such as fertilizer application and animal feedlots, are major contributors of nitrate and ammonia to be washed off (runoff) and percolate to soil with precipitation. The nitrates that are polluted can subsequently flow into surface waters (streams and lakes) and into groundwater. Municipal sewage effluents improperly treat domestic wastewaters, and draining of septic tanks may cause nitrate contamination in the raw water source of a water works. Identification of pollution sources and protective alternatives should be taken.

If the nitrate concentration in the water supply is excessive, there are two approaches for solving the problem, i.e. treatment for nitrate removal and nontreatment alternatives. Nontreatment alternatives may include (1) use new water source, (2) blend with low nitrate waters, (3) connect to other supplier(s), and (4) organize in a regional system.

Nitrate is not removed by the conventional water treatment processes, such as coagulation—flocculation, sedimentation, filtration, activated carbon adsorption, and disinfection (chlorination and ozonation). Some treatment technologies, such as ion exchange, reverse osmosis, electrodialysis, microbial denitrification, and chemical reduction, can remove nitrates from drinking water. An ion exchange process design is given as the following examples.

Two types (fixed bed and continuous) of ion exchangers are used. The fixed bed exchange is used mostly for home and industry uses and is controlled by a flow totalizer with an automatic regeneration cycle at approximately 75 to 80 percent of the theoretical bed capacity. Continuous ion exchangers are employed by larger installations, such as water works, that provide continuous product water and require minimum bed volume. A portion of the resin bed is withdrawn and regenerated outside of the main exchange vessel.

In the design of an ion exchange system for nitrate removal, raw water quality analyses and pilot testing are generally required. The type of resin and resin capacity, bed dimensions, and regenerant requirement must be determined. Basic data, such as the design flow rate through the exchanger, influent water quality, total anions, and suggested operating conditions for the resin selected (from the manufacturer) must be known. Analysis of water quality may include nitrate, sulfate, chloride, bicarbonate, calcium carbonate, iron, total suspended solids, and total organic carbon. For Duolite A-104, for example, the suggested design parameters are listed below (Diamond Shamrock Co., 1978):

Parameters	Recommended values
Minimum bed depth	30 in
Backwash flow rate	2–3 gpm/ft^2
Regenerant dosage	15–18 lb sodium chloride (NaCl) per ft^3 resin
Regenerant concentration	10–12% NaCl by weight
Regenerant temperature	Up to 120°F or 49°C
Regenerant flow rate	0.5 gpm/ft^3
Rinse flow rate	2 gpm/ft^3
Rinse volume	50–70 gal/ft^3
Service flow rate	Up to 5 gpm/ft^3
Operating temperature	Salt form, up to 180°F or 85°C
pH limitation	None

In design processes, resin capacity, bed dimensions, and regenerant requirements must be computed. Resin capacity determines the quantity of resin needed in the ion exchanger and is computed from a pilot study. Data is provided by the manufacturer. For example, the operating capacity of Duloite A-104 resin for nitrate removal is quite dependent upon the nitrate, sulfate, and total anion concentrations to calculate the corrected resin capacity. The raw or uncorrected resin capacity is determined by using the manufacturer's graph (Fig. 6.7). This capacity must be adjusted downward to reflect the presence of sulfate in the water (Fig. 6.8), because sulfate anions will be exchanged before nitrate (USEPA 1983).

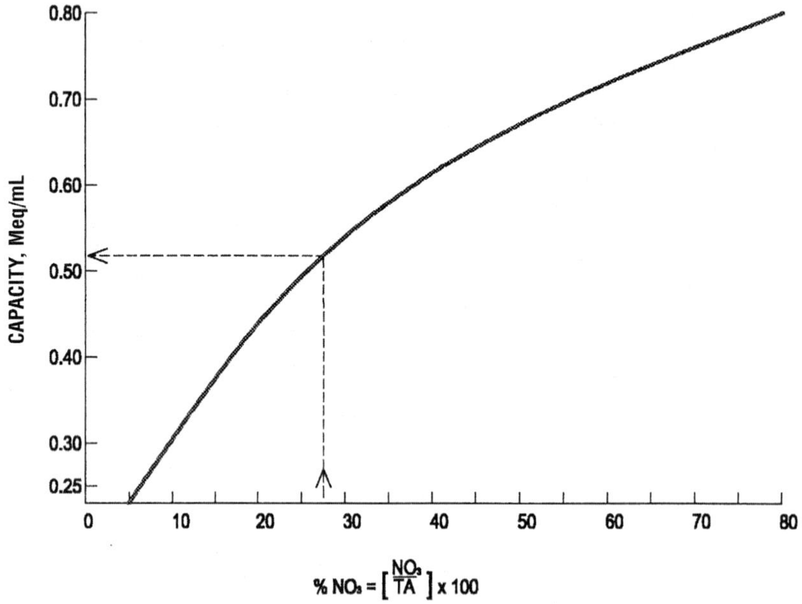

FIGURE 6.7 Relationship of NO_3/TA and unadjusted resin capacity for A-104 resin (*USEPA, 1983*).

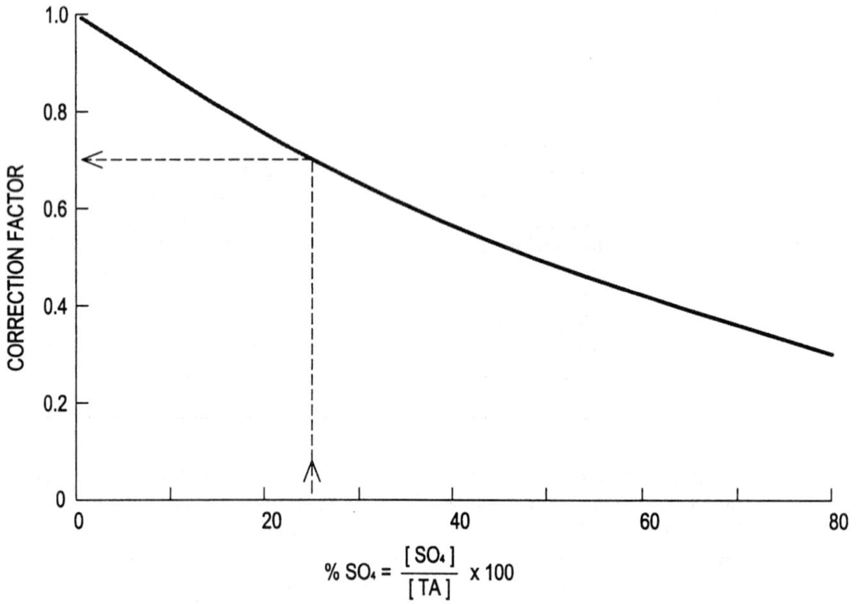

FIGURE 6.8 Correction factor for A-104 resin (*USEPA, 1983*).

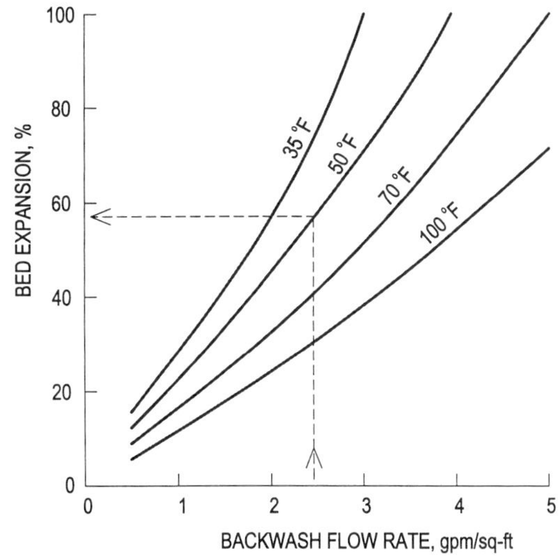

FIGURE 6.9 Bed expansion curve for A-104 resin (modified from *USEPA, 1983*).

Once the adjusted resin capacity is determined for the water to be treated, the required bed volume of the ion exchange resin can be calculated. The bed volume is the amount of nitrate that must be removed in each cycle divided by the adjusted resin capacity. Using this bed volume and minimum depth requirement, the surface area of resin can be calculated. A standard size containment vessel can be selected with the closest surface area. Using this standard size, the depth of the resin can be recalculated.The final height of the vessel should be added to a bed expansion factor during backwashing with a design temperature selected (Fig. 6.9).

Once the bed volume and dimensions are computed, the regeneration system can be designed. The regeneration system must determine: (1) the salt required per generation cycle, (2) the volume of brine produced per cycle, (3) the total volume of the brine storage tank, and (4) the time needed for the regeneration process. The design of an ion exchanger for nitrate removal is illustrated in the following three examples.

EXAMPLE 1: A small public water system has the maximum daily flow rate of 100,000 gallons per day (gpd), maximum weekly flow of 500,000 gallons per week, and maximum nitrate concentration of 16 mg/L. The plant treats 100,000 gallons of water and operates only 7 h per day and 5 days per week (assume sufficient storage capacity for weekend demand). Finished water after ion exchange process with 0.6 mg/L of nitrate-nitrogen (NO_3-N) will be blended with untreated water at 16 mg/L of NO_3-N to produce a finished water of 8 mg/L or less of NO_3-N. The NO_3-N standard is 10 mg/L. Determine the flow rate of ion exchanger in gpm and the blending rate.

Solution:

Step 1. Compute the quantity of water that passes the ion exchanger, q_1

Let untreated flow $= q_2$ and

C, C_1, and C_2 = NO_3-N concentrations in finished, treated (ion exchanger), and untreated waters, respectively.

Then

$$C = 8 \text{ mg/L}$$
$$C_1 = 0.6 \text{ mg/L}$$
$$C_2 = 16 \text{ mg/L}$$

by mass balance

$$C_1q_1 + C_2q_2 = 100{,}000\ C$$
$$0.6q_1 + 16q_2 = 100{,}000 \times 8 \qquad \text{(a)}$$

since

$$q_2 = 100{,}000 - q_1 \qquad \text{(b)}$$

Substituting (b) into (a)

$$0.6q_1 + 16\,(100{,}000 - q_1) = 800{,}000$$
$$15.4q_1 = 800{,}000$$
$$q_1 = 51{,}950 \text{ (gpd)}$$

Step 2. Compute flow rate in gpm to the exchanger (only operates 7 hours per day)

$$q_1 = 51{,}950\,\frac{\text{gal}}{\text{d}} \times \frac{24\text{ h}}{7\text{ h}} \times \frac{1\text{ day}}{1440\text{ min}}$$
$$= 124 \text{ gpm}$$

Step 3. Compute blending ratio, br

$$\text{br} = \frac{51{,}950\text{ gpd}}{100{,}000\text{ gpd}}$$
$$\cong 0.52$$

EXAMPLE 2: The given conditions are the same as in Example 1. The anion concentration of the influent of the ion exchanger are: NO_3-N = 16 mg/L, SO_4 = 48 mg/L, Cl = 28 mg/L, and HCO_3 = 72 mg/L. Determine the total quantity of nitrate to be removed daily by the ion exchanger and the ratios of anions.

Solution:

Step 1. Compute total anions and ratios

Anions	Concentration, mg/L	Milliequivalent weight, mg/meq	Concentration, meq/L	Percentage
NO_3-N	16	14.0	1.14	27.7
SO_4	48	48.0	1.00	24.4
Cl	28	35.5	0.79	19.2
HCO_3	72	61.0	1.18	28.7
Total			4.11	

In the last column above for NO_3-N

$$\% = 1.14 \times 100/4.11 = 27.7$$

Step 2. Determine nitrate to be removed daily.

The total amount of nitrates to be removed depends on the concentrations of influent and effluent, C_i and C_e and can be determined by

$$\text{nitrates removed (meq/d)} = [(C_i - C_e)\ \text{meq/L}] \times q_1(\text{L/d})$$

Since

$$C_i = 1.14\ \text{meq/L}$$
$$C_e = 0.6\ \text{mg/L} \div 14\ \text{mg/meq}$$
$$= 0.043\ \text{meq/L}$$

$$\text{Nitrates removed} = (1.14 - 0.043)\ \text{meq/L} \times 51{,}950\ \text{gpd} \times 3.785\ \text{L/gal}$$
$$= 215{,}700\ \text{meq/d}$$

If the ion exchanger is going to operate with only one cycle per day, the quantity of nitrate to be removed per cycle is 215,700 meq.

EXAMPLE 3: Design a 100,000 gpd ion exchanger for nitrate removal using the data given in Examples 1 and 2. Find resin capacity, resin bed dimension, and regenerent requirements (salt used, brine production, the total volume of the brine storage tank, and regeneration cycle operating time) for the ion exchanger.

Solution:

Step 1. Determine the uncorrected (unadjusted) volume of resin needed

(a) In Example 2, the nitrate to total anion ratio is 27.7%. Using 27.7% in Fig. 6.7, we obtain the uncorrected resin capacity of 0.52 meq/(NO_3 mL) resin.

(b) In Example 2, the sulfate to total anion ratio is 24.4%. Using 25% in Fig. 6.8, we obtain the correction factor of resin capacity due to sulfate is 0.70.

(c) The adjusted resin capacity is down-warded to

$$0.52\ \text{meq/mL} \times 0.70 = 0.364\ \text{meq/mL}$$

(d) Convert meq/mL to meq/ft^3

$$0.364\ \text{meq/mL} \times 3785\ \text{mL/gal} \times 7.48\ \text{gal/ft}^3$$
$$= 10{,}305\ \text{meq/ft}^3$$

Step 2. Determine the bed volume, BV

In Example 2, nitrate to be removed in each cycle is 215,700 meq

$$\text{BV} = 215{,}700\ \text{meq} \div 10{,}305\ \text{meq/ft}^3$$
$$= 20.93\ \text{ft}^3$$

Step 3. Check the service flow rate, SFR

In Example 1, the ion exchange unit flow rate = 124 gpm

$$\text{SFR} = 124\ \text{gpm/BV} = 124\ \text{gpm}/20.93\ \text{ft}^3$$
$$= 5.9\ \text{gpm/ft}^3$$

This value exceeds the manufacturer's maximum SFR of 5 gpm/ft^3. In order to reduce the SFR, we can modify it either by (a) increasing the exchanger operating time per cycle, or (b) increasing the bed volume.

(a) Calculate adjusted service time, AST

$$\text{AST} = \frac{7\text{ h}}{\text{cycle}} \times \frac{5.9\text{ gpm/ft}^3}{5.0\text{ gpm/ft}^3}$$
$$= 8.3\text{ h/cycle}$$

Calculate the adjusted flow rate for the new cycle

In Example 1, q_1 = 51,950 gpd or 51,950 gal/cycle

$$\text{then, flow rate} = q_1/\text{AST}$$
$$= 51{,}950\text{ gal/cycle}/(8.3\text{ h/cycle} \times 60\text{ min/h})$$
$$= 104\text{ gpm}$$

Alternatively

(b) Increase the bed volume and retain the initial unit flow rate

$$\text{Adjusted BV} = 20.93\text{ ft}^3 \times (5.9\text{ gpm/ft}^3/5.0\text{ gpm/ft}^3)$$
$$= 24.7\text{ ft}^3$$

say $= 25\text{ ft}^3$

In comparison between alternatives (a) and (b), it is more desirable to increase BV using 7 h/d operation. Thus BV = 25 ft^3 is chosen.

Step 4. Determine bed dimensions

(a) Select the vessel size

Based on the suggested design specification, the minimum bed depth (z) is 30 in or 2.5 ft. The area, A, would be

$$A = \text{BV}/z = 25\text{ ft}^3/2.5\text{ ft}$$
$$= 10.0\text{ ft}^2$$

For a circular vessel, find the diameter

$$A = \pi D^2/4$$

$$D = \sqrt{4A/\pi} = \sqrt{4 \times 10.0\text{ ft}^2/3.14}$$
$$= 3.57\text{ ft}$$

The closest premanufactured size is 3.5 ft in diameter with a corresponding cross-sectional area of 9.62 ft^2. The resin height, h, would then be re-calculated as

$$h = 25\text{ ft}^3/9.62\text{ ft}^2$$
$$= 2.6\text{ ft}$$

(b) Adjusting for expansion during backwash

Since allowable backwash flow rate is 2 to 3 gpm/ft^2. If the backwash rate is selected as 2.5 gpm/ft^2 with the backwash water temperature of 50°F, the percent bed expansion would be 58 percent as read from the curve in Fig. 6.9. Thus, the design height (H) of the vessel should be at least

$$H = 2.6\text{ ft} \times (1 + 0.58)$$
$$= 4.11\text{ ft}$$

Note: We use 3.5 ft diameter and 4.11 ft (at least) height vessel.

Step 5. Determine regenerant requirements

(a) Choice salt (NaCl) required per regeneration cycle since regenerant dosage is 15 to 18 pounds of salt per cubic foot of resin. Assume 16 lb/ft^3 is selected. The amount of salt required is

$$\begin{aligned}\text{Salt} &= 16\ \text{lb/ft}^3 \times 25\ \text{ft}^3\\ &= 400\ \text{lb}\end{aligned}$$

We need 400 pounds of salt per cycle.

(b) Calculate volume of brine used per cycle
On the basis of the specification, the regenerant concentration is 10 to 12 percent NaCl by weight. Assume it is 10 percent, then

$$\begin{aligned}\text{Salt concentration, \%} &= \frac{\text{wt. of salt} \times 100}{\text{total wt. of brine}}\\ \text{total wt. of brine} &= 400\ \text{lb} \times 100/10\\ &= 4000\ \text{lb}\end{aligned}$$

(c) Calculate total volume of brine

$$\begin{aligned}\text{Weight of water} &= (4000 - 400)\ \text{lb/cycle}\\ &= 3600\ \text{lb}\\ \text{Volume of water} &= \frac{\text{wt. of water}}{\text{density of water}} = \frac{3600\ \text{lb}}{62.4\ \text{lb/ft}^3}\\ &= 57.69\ \text{ft}^3\end{aligned}$$

The specific weight of salt is 2.165

$$\begin{aligned}\text{Volume of salt} &= 400\ \text{lb}/(62.4\ \text{lb/ft}^3 \times 2.165)\\ &= 2.96\ \text{ft}^3\\ \text{Total volume of the brine} &= 57.69\ \text{ft}^3 + 2.96\ \text{ft}^3\\ &= 60.65\ \text{ft}^3\\ \text{Total volume in gallons} &= 60.65\ \text{ft}^3 \times 7.48\ \text{gal/ft}^3\\ &= 454\ \text{gal}\end{aligned}$$

Note: The total volume of brine generated is 454 gal per cycle. The total volume of the brine storage tank should contain sufficient capacity for 3 to 4 generations. Assume 3 cycles of brine produced should be stored, then the total brine tank volume (V) would be

$$\begin{aligned}V &= 454\ \text{gal/cycle} \times 3\ \text{cycle}\\ &= 1362\ \text{gal}\end{aligned}$$

(d) Calculate time required for regeneration cycle
The regeneration flow rate (Q) is 0.5 gpm/ft^3 resin specified by the manufacturer. Then

$$\begin{aligned}Q &= 0.5\ \text{gpm/ft}^3 \times 25\ \text{ft}^3\\ &= 12.5\ \text{gpm}\end{aligned}$$

The regeneration time (t) is the volume of the brine per cycle divided by the flow rate Q, thus

$$\begin{aligned}t &= 454\ \text{gal}/12.5\ \text{gal/min}\\ &= 36.3\ \text{min}\end{aligned}$$

14 IRON AND MANGANESE REMOVAL

Iron (Fe) and manganese (Mn) are abundant elements in the earth's crust. They are mostly in the oxidized state (ferric, Fe^{3+}, and Mn^{4+}) and are insoluble in natural waters. However, under reducing conditions (i.e., where dissolved oxygen is lacking and carbon dioxide content is high), appreciable amounts of iron and manganese may occur in surface water, groundwater, and in water from the anaerobic hypolimnion of stratified lakes and reservoirs. The reduced forms are soluble divalent ferrous (Fe^{2+}) and manganous (Mn^{2+}) ions that are chemically bound with organic matter. Iron and manganese get into natural water from dissolution of rocks and soil, from acid mine drainage, and from corrosion of metals. Typical iron concentrations in ground water are 1.0 to 10 mg/L, and typical concentrations in oxygenated surface waters are 0.05 to 0.2 mg/L. Manganese exists less frequently than iron and in smaller amounts. Typical manganese values in natural water range from 0.1 to 1.0 mg/L (James M. Montgomery Consulting Engineers, 1985). Voelker (1984) reported that iron and manganese levels in groundwaters in the American Bottoms area of southwestern Illinois ranged from < 0.01 to 82.0 mg/L and < 0.01 to 4.70 mg/L, respectively, with mean concentrations of 8.4 mg/L and 0.56 mg/L, respectively.

Generally, iron and manganese in water are not a health risk. However, in public water supplies they may discolor water and cause tastes and odors. Elevated iron and manganese levels also can cause stains in plumbing, laundry, and cooking utensils. Iron and manganese may also cause problems in water distribution systems because metal depositions may result in pipe encrustation and may promote the growth of iron bacteria which may in turn cause tastes and odors. Iron and manganese may also cause difficulties in household ion exchange units by clogging and coating the exchange medium.

To eliminate the problems caused by iron and manganese, the U.S. Environmental Protection Agency (1987) has established secondary drinking water standards for iron at 0.3 mg/L and for manganese at 0.05 mg/L. The Illinois Pollution Control Board (IPCB, 1990) has set effluent standards of 2.0 mg/L for total iron and 1.0 mg/L for total manganese.

It is considered that iron and manganese are essential elements for plants and animals. The United Nations Food and Agriculture Organization recommended maximum levels for iron and manganese in irrigation water are 5.0 and 0.2 mg/L, respectively.

The techniques for removing iron and manganese from water are based on the oxidation of relatively soluble Fe(II) and Mn(II) to the insoluble Fe(III) and Mn(III, IV) and the oxidation of any organic-complex compounds. This is followed by filtration to remove the Fe(III) and processes for iron and manganese removal are discussed elsewhere (Rehfeldt *et al.*, 1992).

Four major techniques for iron and manganese removal from water are: (1) oxidation–precipitation–filtration, (2) manganese zeolite process, (3) lime softening–settling–filtration, and (4) ion exchange. Aeration–filtration, chlorination–filtration, and manganese zeolite process are commonly used for public water supply.

14.1 Oxidation

Oxidation of soluble (reduced) iron and manganese can be achieved by aeration, chlorine, chlorine dioxide, ozone, potassium permanganate ($KMnO_4$) and hydrogen peroxide (H_2O_2). The chemical reactions can be expressed as follows:

For aeration (Jobin and Ghosh, 1972; Dean, 1979)

$$Fe^{2+} + \tfrac{1}{4}O_2 + 2OH^- + \tfrac{1}{2}H_2O \rightarrow Fe(OH)_3 \qquad (6.119)$$

$$Mn^{2+} + O_2 \rightarrow MnO_2 + 2e \qquad (6.120)$$

In theory, 0.14 mg/L of oxygen is needed to oxidize 1 mg/L of iron and 0.29 mg/L of oxygen for each mg/L of manganese.

For chlorination (White, 1972):

$$2Fe\ (HCO_3)_2 + Cl_2 + Ca(HCO_3)_2 \rightarrow 2Fe(OH)_3 + CaCl_2 + 6CO_2 \tag{6.121}$$

$$MnSO_4 + Cl_2 + 4NaOH \rightarrow MnO_2 + 2NaCl + Na_2SO_4 + 2H_2O \tag{6.122}$$

The stoichiometric amounts of chlorine needed to oxidize each mg/L of iron and manganese are 0.62 and 1.3 mg/L, respectively.

For potassium permanganate oxidation (Ficek 1980):

$$3Fe^{2+} + MnO_4 + 4H^+ \rightarrow MnO_2 + 2Fe^{3+} + 2H_2O \tag{6.123}$$

$$3Mn^{2+} + 2MnO_4 + 2H_2O \rightarrow 5MnO_2 + 4H^+ \tag{6.124}$$

In theory, 0.94 mg/L of $KMnO_4$ is required for each mg/L of soluble iron and 1.92 mg/L for one mg/L of soluble manganese. In practice, a 1 to 4 percent solution of $KMnO_4$ is fed in at the low-lift pump station or at the rapid-mix point. It should be totally consumed prior to filtration. However, the required $KMnO_4$ dosage is generally less than the theoretical values (Humphrey and Eikleberry 1962, Wong 1984). Secondary oxidation reactions occur as

$$Fe^{2+} + MnO \rightarrow Fe^{3+} + Mn_2O_3 \tag{6.125}$$

$$Mn^{2+} + MnO_2 \cdot 2H_2O \rightarrow Mn_2O_3 \cdot X(H_2O) \tag{6.126}$$

and

$$2Mn^{2+} + MnO_2 \rightarrow Mn_3O_4 \cdot X(H_2O) \tag{6.127a}$$

or

$$MnO \cdot Mn_2O_3 \cdot X(H_2O) \tag{6.127b}$$

For hydrogen peroxide oxidation (Kreuz, 1962):

1. Direct oxidation

$$3Fe^{2+} + H_2O_2 + 2H^+ \rightarrow 2Fe^{3+} + 2H_2O \tag{6.128}$$

2. Decomposition to oxygen

$$H_2O_2 \rightarrow \tfrac{1}{2}O_2 + H_2O \tag{6.129}$$

Followed by oxidation

$$3Fe^{2+} + \tfrac{1}{2}O_2 + H_2O \rightarrow 2Fe^{3+} + 2OH^- \tag{6.130}$$

In either case, 2 moles of ferrous iron are oxidized per mole of hydrogen peroxide, or 0.61 mg/L of 50 percent H_2O_2 oxidizes 1 mg/L of ferrous iron.

For oxidation of manganese by hydrogen peroxide, the following reaction applies (H. M. Castrantas, FMC Corporation, Princeton, NJ, personal communication).

$$3Mn^{2+} + H_2O_2 + 2H^+ + 2e \rightarrow 2Mn^{3+} + e + 2H_2O \tag{6.131}$$

To oxidize 1 mg/L of soluble manganese, 1.24 mg/L of 50 percent H_2O_2 is required.

Studies on ground water by Rehfeldt *et al.* (1992) reported that oxidant dosage required to meet the Illinois recommended total iron standard follow the order of Na, O, Cl as $Cl_2 > KMnO_4 > H_2O_2$. The amount of residues generated by oxidants at the critical dosages follow the order of $KMnO_4 >$ $NaOCl > H_2O_2$, with $KMnO_4$ producing the largest, strongest, and most dense flocs.

For ozonation: It is one of the strongest oxidants used in the water industry for disinfection purposes. The oxidation potential of common oxidants relative to chlorine is as follows (Peroxidation Systems, 1990):

Fluoride	2.32
Hydroxyl radical	2.06
Ozone	1.52
Hydrogen peroxide	1.31
Potassium permanganate	1.24
Chlorine	1.00

Ozone can be very effective for iron and manganese removal. Because of its relatively high capital costs and operation and maintenance costs, the ozonation process is rarely employed for the primary purpose of oxidizing iron and manganese. Since ozone is effective in oxidizing trace toxic organic matter in water, pre-ozonation instead of pre-chlorination is becoming popular. In addition, many water utilities are using ozone for disinfection purposes. Ozonation can be used for two purposes: disinfection and metal removal.

Manganese zeolite process. In the manganese zeolite process, iron and manganese are oxidized to the insoluble form and filtered out, all in one unit, by a combination of sorption and oxidation. The filter medium can be manganese greensand, which is a purple-black granular material processed from glauconitic greens, and/or a synthetic formulated product. Both of these compositions are sodium compounds treated with a manganous solution to exchange manganese for sodium and then oxidized by $KMnO_4$ to produce an active manganese dioxide. The greensand grains in the filter become coated with the oxidation products. The oxidized form of greensand then adsorbs soluble iron and manganese, which are subsequently oxidized with $KMnO_4$. One advantage is that the greensand will adsorb the excess $KMnO_4$ and any discoloration of the water.

Regenerative-Batch Process. The regenerative-batch process uses manganese treated greensand as both the oxidant source and the filter medium. The manganese zeolite is made from $KMnO_4$-treated greensand zeolite. The chemical reactions can be expressed as follows (Humphrey and Eikleberry, 1962; Wilmarth, 1968):

Exchange:

$$NaZ + Mn^{2+} \rightarrow MnZ + 2Na^{+} \tag{6.132}$$

Generation:

$$MnZ + KMnO_4 \rightarrow Z \cdot MnO_2 + K^{+} \tag{6.133}$$

Degeneration:

$$Z \cdot MnO_2 + \begin{matrix} Fe^{2+} \\ Mn^{2+} \end{matrix} \rightarrow Z \cdot MnO_3 + \begin{matrix} Fe^{3+} \\ Mn^{3+} \\ Mn^{4+} \end{matrix} \tag{6.134}$$

Regeneration:

$$Z \cdot MnO_3 + KMnO_4 \rightarrow Z \cdot MnO_2 \tag{6.135}$$

where NaZ is greensand zeolite and $Z \cdot MnO_2$ is manganese zeolite. As the water passes through the mineral bed, the soluble iron and manganese are oxidized (degeneration). Regeneration is required after the manganese zeolite is exhausted.

One of the serious problems with the regenerative-batch process is the possibility of soluble manganese leakage. In addition, excess amounts of $KMnO_4$ are wasted, and the process is not economical for water high in metal content. Manganese zeolite has an exchange capacity of 0.09 lb of iron or manganese per cubic foot of material, and the flow rate to the exchanger is usually 3.0 gpm/ft^3 Regeneration needs approximately 0.18 lb $KMnO_4$ per cubic foot of zeolite.

Continuous Process. For the continuous process, 1 to 4 percent $KMnO_4$ solution is continuously fed ahead of a filter containing anthracite be (6 to 9 in. thick), manganese-treated greensand (24 to 30 in), and gravel. The system takes full advantage of the higher oxidation potential of $KMnO_4$ as compared to manganese dioxide. In addition, the greensand can act as a buffer. The $KMnO_4$ oxidizes iron, manganese, and hydrogen sulfide to the insoluble state before the water reaches the manganese zeolite bed.

Greensand grain has a smaller effective size than silica sand used in filters and can result in comparatively higher head loss. Therefore, a layer of anthracite is placed above the greensand to prolong filter runs by filtering out the precipitate. The upper layer of anthracite operates basically as a filter medium. When iron and manganese deposits build up, the system is backwashed like an ordinary sand filter. The manganese zeolite not only serves as a filter medium but also as a buffer to oxidize any residual soluble iron and manganese and to remove any excess unreacted $KMnO_4$. Thus a $KMnO_4$ demand test should be performed. The continuous system is recommended for waters where iron predominates, and the intermittent regeneration system is recommended for groundwater where manganese predominates (Inversand Co., 1987).

EXAMPLE 1: Theoretically, how many mg/L each of ferrous iron and soluble manganese can be oxidized by 1 mg/L of potassium permanganate?

Solution:

Step 1. For Fe^{2+}, using Eq. (6.123)

$$3Fe^{2+} + 4H^+ + KMnO_4 \rightarrow MnO_2 + 2Fe^{3+} + K^+ + 2H_2O$$

$$\begin{array}{ll} \text{MW } 3 \times 55.85 & 39.1 + 54.94 + 4 \times 16 \\ = 167.55 & = 158.14 \\ X \text{ mg/L} & 1 \text{ mg/L} \end{array}$$

By proportion

$$\frac{X}{1} = \frac{167.55}{158.14}$$

$$X = 1.06 \text{ mg/L}$$

Step 2. For Mn^{2+}, using Eq. (6.124)

$$3Mn^{2+} + 2H_2O + 2KMnO_4 \rightarrow 5MnO_2 + 4H^+ + 2K^+$$

$$\begin{array}{ll} M_1W_1 \; 3 \times 54.94 & 2 \times 158.14 \\ = 164.82 & = 316.28 \\ y & 1 \text{ mg/L} \end{array}$$

then

$$y = \frac{164.82}{316.28} = 0.52 \text{ mg/L}$$

EXAMPLE 2: A groundwater contains 3.6 mg/L of soluble iron and 0.78 mg/L of manganese. Find the dosage of potassium permanganate required to oxidize the soluble iron and manganese.

Solution:

From Example 1 the theoretical potassium permanganate dosage are 1.0 mg/L per 1.06 mg/L of ferrous iron and 1.0 mg/L per 0.52 mg/L of manganese. Thus

$$KMnO_4 \text{ dosage} = \frac{1.0 \times 3.6 \text{ mg/L}}{1.06} + \frac{1.0 \times 0.78 \text{ mg/L}}{0.52}$$

$$= 4.9 \text{ mg/L}$$

15 ACTIVATED CARBON ADSORPTION

15.1 Adsorption Isotherm Equations

The Freundlich isotherm equation is an empirical equation which gives an accurate description of adsorption of organic adsorption in water. The equation under constant temperature equilibrium is (USEPA, 1976; Brown and LeMay, 1981; Lide, 1996)

$$q_e = KC_e^{1/n} \tag{6.136}$$

or

$$\log q_e = \log K + (1/n)\log C_e \tag{6.136a}$$

where q_e = quantity of absorbate per unit of absorbent, mg/g
C_e = equilibrium concentration of adsorbate in solution, mg/L
K = Freundlich absorption coefficient, (mg/g) (L/mg)$^{1/n}$
n = empirical coefficient

The constant K is related to the capacity of the absorbent for the absorbate. $1/n$ is a function of the strength of adsorption. The molecule that accumulates, or adsorbs, at the surface is called an *adsorbate*; and the solids on which adsorption occurs is called *adsorbent*. Snoeyink (1990) compiled the values of K and $1/n$ for various organic compounds from the literature which is listed in Appendix D.

From the adsorption isotherm, it can be seen that, for fixed values of K and C_e, the smaller the value of $1/n$, the stronger the adsorption capacity. When $1/n$ becomes very small, the adsorption tends to be independent to C_e. For fixed values of C_e and $1/n$, the larger the K value, the greater the adsorption capacity q_e.

Another adsorption isotherm developed by Langmuir assumed that the adsorption surface is saturated when a monolayer has been absorbed. The Langmuir adsorption model is

$$q_e = \frac{a\,b\,C_e}{1 + b\,C_e} \tag{6.137}$$

or

$$\frac{C_e}{q_e} = \frac{1}{ab} + \frac{C_e}{a} \tag{6.138}$$

where a = empirical coefficient
b = saturation coefficient, m^3/g

Other terms are the same as defined in the Freundlich model. The coefficient a and b can be obtained by plotting C_e/q_e versus C_e on arithmetic paper from the results of a batch adsorption test with Eq. (6.138).

Adsorption isotherm can be used to roughly estimate the granular activated carbon (GAC) loading rate and its GAC bed life. The bed life Z can be computed as

$$Z = \frac{(q_e)_0 \times \rho}{(C_0 - C_1)} \tag{6.139}$$

where Z = bed life, L H_2O/L GAC
$(q_e)_0$ = mass absorbed when $C_e = C_0$, mg/g of GAC
ρ = apparent density of GAC, g/L
C_0 = influent concentration, mg/L
C_1 = average effluent concentration for entire column run, mg/L

C_1 would be zero for a strongly absorbed compound that has a sharp breakthrough curve, and is the concentration of the nonadsorbable compound presented.

The rate at which GAC is used or the carbon usage rate (CUR) is calculated as follows:

$$\text{CUR (g/L)} = \frac{C_0 - C_1}{(q_e)_0} \tag{6.140}$$

Contact Time. The contact time of GAC and water may be the most important parameter for the design of a GAC absorber. It is commonly used as the empty-bed contact time (EBCT) which is related to the flow rate, depth of the bed, and area of the GAC column. The EBCT is calculated by taking the volume (V) occupied by the GAC divided by the flow rate (Q):

$$\text{EBCT} = \frac{V}{Q} \tag{6.141}$$

or

$$\text{EBCT} = \frac{H}{Q/A} = \frac{\text{The depth of GAC bed}}{\text{Loading rate}} \tag{6.142}$$

where EBCT = empty-bed contact time, s
V = HA = GAC volume, ft^3 or m^3
Q = flow rate, cfs or m^3/s
Q/A = surface loading rate, cfs/ft^2 or m^3/(m^2 s)

The critical depth of a bed (H_{cr}) is the depth which creates the immediate appearance of an effluent concentration equal to the breakthrough concentration, C_b. The C_b for a GAC column is designated as the maximum acceptable effluent concentration or the minimum datable concentration. The GAC should be replaced or regenerated when the effluent quality reaches C_b. Under these conditions, the minimum EBCT (EBCT_{min}) will be:

$$\text{EBCT}_{min} = \frac{H_{cr}}{Q/A} \tag{6.143}$$

The actual contact time is the product of the EBCT and the interparticle porosity (about 0.4 to 0.5).

Powdered activated carbon (PAC) has been used for taste and odor removal. It is reported that PAC is not as effective as GAC for organic compounds removal. The PAC minimum dose requirement is

$$D = \frac{C_i - C_e}{q_e} \tag{6.144}$$

where D = dosage, g/L
C_i = influent concentration, mg/L
C_e = effluent concentration, mg/L
q_e = absorbent capacity = $KC_e^{1/n}$, mg/g

EXAMPLE 1: In a dual media filtration unit, the surface loading rate is 3 gpm/ft^2. The regulatory requirement of EBCT_{min} is 5.5 min. What is the critical depth of GAC?

Solution:

$$Q = 3.74 \text{ gpm} = 3.74 \frac{\text{gal}}{\text{min}} \times \frac{1 \text{ ft}^3}{7.48 \text{ gal}}$$

$$= 0.5 \text{ ft}^3/\text{min}$$

$$Q/A = 0.5 \text{ ft}^3/\text{min}/1 \text{ ft}^2 = 0.5 \text{ ft/min}$$

From Eq. (6.143), the minimum EBCT is

$$\text{EBCT}_{\text{min}} = \frac{H_{\text{cr}}}{Q/A}$$

and rearranging

$$\begin{aligned} H_{\text{cr}} &= \text{EBCT}_{\text{min}}\,(Q/A) \\ &= 5.5 \text{ min} \times 0.5 \text{ ft/min} \\ &= 2.75 \text{ ft} \end{aligned}$$

EXAMPLE 2: A granular activated carbon absorber is designed to reduce 12 μg/L of chlorobenzene to 2 μg/L. The following conditions are given: $K = 100$ (mg/g) (L/mg)$^{1/n}$, $1/n = 0.35$ (Appendix D), and $\rho_{\text{GAC}} = 480$ g/L. Determine the GAC bed life and carbon usage rate.

Solution:

Step 1. Compute equilibrium adsorption capacity, $(q_e)_0$

$$\begin{aligned} (q_e)_0 &= KC^{1/n} \\ &= 100 \text{ (mg/g) (L/mg)}^{1/n} \times (0.002 \text{ mg/L})^{0.35} \\ &= 11.3 \text{ mg/g} \end{aligned}$$

Step 2. Compute bed life, Z

$$\text{Assume: } C_T = 0 \text{ mg/L}$$

$$\begin{aligned} Z &= \frac{(q_e)_0 \times \rho}{(C_0 - C_1)} \\ &= \frac{11.3 \text{ mg/g} \times 480 \text{ g/L of GAS}}{(0.012 - 0.002) \text{ mg/L of water}} \\ &= 542{,}400 \text{ L of water/L of GAC} \end{aligned}$$

Step 3. Compute carbon usage rate, CUR.

$$\begin{aligned} \text{CUR} &= \frac{(C_0 - C_1)}{(q_e)_0} \\ &= \frac{(0.012 - 0.002) \text{ mg/L}}{11.3 \text{ mg/g}} \\ &= 0.00088 \text{ g GAC/L water} \end{aligned}$$

EXAMPLE 3: The maximum contaminant level (MCL) for trichloroethylene is 0.005 mg/L. $K = 28$ (mg/g) (L/mg)$^{1/n}$, and $1/n = 0.62$. Compute the PAC minimum dosage needed to reduce trichloroethylene concentration from 33 μg/L to MCL.

Solution:

Step 1. Compute q_e assuming PAC will equilibrate with 0.005 mg/L (C_e) trichloroethylene concentration.

$$\begin{aligned} q_e &= KC_e^{1/n} \\ &= 28 \text{ (mg/g) (L/mg)}^{1/n} \times (0.005 \text{ mg/L})^{0.62} \\ &= 1.05 \text{ mg/g} \end{aligned}$$

Step 2. Compute required minimum dose, D

$$D = \frac{(C_0 - C_e)}{q_e}$$

$$= \frac{0.033 \text{ mg/L} - 0.005 \text{ mg/L}}{1.05 \text{ mg/g}}$$

$$= 0.027 \text{ g/L}$$

or

$$= 27 \text{ mg/L}$$

16 MEMBRANE PROCESSES

The use of semipermeable membranes having pore size as small as 3 Å (1 angstrom = 10^{-8} cm) can remove dissolved impurities in water or wastewater. Desalting of sea water is an example. The process allowing water to pass through membranes is called "osmosis" or hyperfiltration. On the other hand, ions and solutes that pass through a membrane is a process called "dialysis" which is used in the medical field. The driving forces can be by pressure, chemical concentration, temperature, and by electrical charge. Various types and pore size of membranes incorporate the various driving forces. The membrane processes may include reverse osmosis (RO), electrodialysis (ED) or electrodialysis reversal (EDR), ultrafiltration (UF), and nanofiltration (NF).

In the United States, RO and EDR have been used for water utilities ranging from 0.5 to 12 MGD (1.89 to 45.4 metric tons/d); most are located on the east and west coasts (Ionic, 1993). The electrodialysis process is when ions from a less concentrated solution pass through a membrane to a more concentrated solution under an electric current. Impurities are electrically removed from the water. Electrodialysis reversal is an ED process. A direct current transfers ions through the membranes by reversing polarity two to four times per hour. The system provides constant, automatic self-cleaning that enables improvements in treatment efficiency and less downtime for periodic cleaning.

Ultrafiltration is to remove colloids and high molecular-weight material by pressure. The UF process retains nonionic material and generally passes most ionic matter depending on the molecular weight cutoff of the membrane.

Nanofiltration is a process that removes all extreme fine matter and was an emerging technology in the 1980s. "Nano" means one thousand millionth (10^{-9}), thus one nanometer is 1 nm = 10^{-9} m. Water treated by nanofiltration can be used by special industries, such as electronics. Nowadays, the NF process has been used at full scale for treating filter backwash waters in New York. In Europe, hundreds of NF installations have been used for drinking water treatment.

16.1 Reverse Osmosis

In the reverse osmosis process, the feedwater is forced by hydrostatic pressure through membranes while impurities remain behind. The purified water (permeable or product water) emerges at near atmospheric pressure. The waste (as brine) at its original pressure is collected in a concentrated mineral stream. The pressure difference is called the osmotic pressure which is a function of the solute concentration, characteristics, and temperature. The operating pressure ranges from 800 to 1000 psi (54.4 to 68 atm) for desalinating seawater and from 300 to 400 psi (20.4 to 27.2 atm) for brackish water desaltation (James M. Montgomery Consulting Engineering, 1985).

The osmotic pressure theoretically varies in the same manner as the pressure of an ideal gas (James M. Montgomery Consulting Engineering, 1985; Conlon, 1990):

$$\pi = \frac{nRT}{v} \tag{6.145a}$$

or (Applegate 1984):

$$\pi = 1.12\, T\, \Sigma m \tag{6.145b}$$

where π = osmotic pressure, psi
n = number of moles of solute
R = universal gas constant
T = absolute temperature, °C + 273 = K
v = molar volume of water
Σm = sum of molarities of all ionic and nonionic constituent in solution

At equilibrium, the pressure difference between the two sides of the RO membranes equals the osmotic pressure difference. In low solute concentration, the osmotic pressure (π) of a solution is given by the following equation (USEPA 1996):

$$\pi = C_s RT \tag{6.145c}$$

where π = osmotic pressure, psi
C_s = concentration of solutes in solution, mol/cm³ or mol/ft³
R = ideal gas constant, ft lb/mol K
T = absolute temperature, K = °C + 273

When dilute and concentrated solutions are separated by a membrane, the liquid tends to flow through the membrane from the dilute to the concentrated side until equilibrium is reached on both sides of the membrane. The liquid and salt passage (flux of water) through the membrane is a function of the pressure gradient (Allied Signal 1970):

$$F_w = W(\Delta P - \Delta\pi) \tag{6.146}$$

and for salt and solute flow through a membrane:

$$F_s = S(C_1 - C_2) \tag{6.147}$$

where F_w, F_s = liquid and salt fluxes across the membrane, respectively, g/(cm² · s) or lb/(ft² · s); cm³/(cm² · s) or gal/(ft² · d)
W, S = flux rate coefficients, empirical, depend on membrane characteristics, solute type, salt type, and temperature, s/cm or s/ft
ΔP = drop in total water pressure across membrane, atm or psi
$\Delta\pi$ = osmotic pressure gradient, atm or psi
$\Delta P - \Delta\pi$ = net driving force for liquid pass through membrane
C_1, C_2 = salt concentration on both sides of membrane, g/cm³ or lb/ft³
$C_1 - C_2$ = concentration gradient, g/cm³ or lb/ft³

The liquid or solvent flow depends upon the pressure gradient. The salt or solute flow depends upon the concentration gradient. Also, both are influenced by the membrane types and characteristics.

The terms water flux and salt flux are the quantities of water and salt that can pass through the membrane per unit area per unit time. The percent of feedwater recovered is given in terms of water recovery:

$$\text{Recovery} = \frac{Q_p}{Q_f} \times 100\% \tag{6.148}$$

where Q_p = product water flow, m³/d or gpm
Q_f = feedwater flow, m³/d or gpm

Water recovery for most RO plants are designed to be 75 to 80 percent for brackish water and 20 to 25 percent for seawater (James M. Montgomery Consulting Engineering 1985).

The percent of salt rejection is:

$$\text{Rejection} = 100 - \text{salt passage}$$
$$= \left(1 - \frac{\text{product concentration}}{\text{feedwater concentration}}\right) \times 100\% \quad (6.149)$$

Salt rejection is commonly measured by the total dissolved solids (TDS). Measurement of conductivity is sometime used in lieu of TDS. In the RO system, salt rejection can be achieved at 90 percent or more in normal brackish waters. Solute rejection varies with feedwater concentration, membrane types, water recovery rate, chemical valance of the ions in the solute, and other factors.

EXAMPLE 1: A city's water demand is 26.5 metric ton/d (7 MGD). What is the source water (feedwater) flow required for a brackish water RO, if the plant recovery rate is 78 percent?

Solution:

$$\text{Recovery} = \frac{Q_p}{Q_f} \times 100$$

rearranging:

$$Q_f = (26.5 \text{ ton/d}) \times 100/78$$
$$= 34.0 \text{ ton/d}$$

or

$$= 9 \text{ MGD}$$

EXAMPLE 2: For a brackish water RO treatment plant, the feedwater applied is 53.0 ton/d (14 MGD) to the membrane and the product water yields 42.4 ton/day (11.2 MGD). What is the percentage of brine rate?

Solution:

$$\text{Amount of brine produced} = \text{feed rate} - \text{product rate}$$
$$= Q_f - Q_p$$
$$= (53.0 - 42.4) \text{ ton/d}$$
$$= 10.6 \text{ ton/d}$$

$$\% \text{ of brine rate} = \frac{10.6 \text{ ton/d}}{53.0 \text{ ton/d}} \times 100\%$$
$$= 20\%$$

EXAMPLE 3: At a brackish water RO treatment plant, the total dissolved solids concentrations for the pretreated feedwater and the product water are 2860 and 89 mg/L, respectively. Determine the percentages of salt rejection and salt passage.

Solution:

Step 1. Compute salt rejection

$$\text{Salt rejection} = \left(1 - \frac{\text{product concentration}}{\text{feedwater concentration}}\right) \times 100\%$$

$$= \left(1 - \frac{89}{2680}\right) \times 100\%$$

$$= 96.7\%$$

Step 2. Compute salt passage

$$\begin{aligned}\text{Salt passage} &= 100 - \text{salt rejection}\\ &= (100 - 96.7)\%\\ &= 3.3\%\end{aligned}$$

EXAMPLE 4: A pretreated feedwater to a brackish water RO process contains 2600 mg/L of TDS. The flow is 0.25 m^3/s (5.7 MGD). The designed TDS concentration of the product water is no more than 450 mg/L. The net pressure is 40 atm. The membrane manufacturer provides that the membrane has a water flux rate coefficient of 1.8×10^{-6} s/m and a solute mass transfer rate of 1.2×10^{-6} m/s. Determine the membrane area required.

Solution:

Step 1. Compute flux of water

$$\begin{aligned}\text{Given: } \Delta P - \Delta\pi &= 40 \text{ atm} = 40 \text{ atm} \times 101.325 \text{ kPa/atm}\\ &= 4053 \text{ kPa}\\ &= 4053 \text{ kg/m} \cdot \text{s}^2\\ W &= 1.8 \times 10^{-6} \text{ s/m}\end{aligned}$$

Using Eq. (6.146)

$$\begin{aligned}F_w &= W(\Delta P - \Delta\pi) = 1.8 \times 10^{-6} \text{ s/m } (4053 \text{ kg/m} \cdot \text{s}^2)\\ &= 7.3 \times 10^{-3} \text{ kg/m}^2 \cdot \text{s}\end{aligned}$$

Step 2. Estimate membrane area

$$Q = F_w A$$

rearranging:

$$\begin{aligned}A &= Q/F_w = (0.25 \text{ m}^3\text{/s} \times 1000 \text{ kg/m}^3)/(7.3 \times 10^{-3} \text{ kg/m}^2 \cdot \text{s})\\ &= 34{,}250 \text{ m}^2\end{aligned}$$

Step 3. Determine permeate TDS (C_p) with the above area

$$\begin{aligned}C_p &= 2600 \text{ mg/L} = 2600 \text{ g/m}^3 = 2.6 \text{ kg/m}^3\\ F_s &= S(C_f - C_p)\\ QC_p &= F_s A = S(C_f - C_p)A\\ C_p &= \frac{SC_fA}{Q + SA} = \frac{1.2 \times 10^{-6}\text{m/s} \times 2.6 \text{ kg/m}^3 \times 34{,}250 \text{ m}^2}{0.25 \text{ m}^3\text{/s} + 1.2 \times 10^{-6} \text{ m/s} \times 34{,}250 \text{ m}^2}\\ &= 0.367 \text{ kg/m}^3\\ &= 367 \text{ g/m}^3\\ &= 367 \text{ mg/L}\end{aligned}$$

The TDS concentration of the product water is lower than the desired limit. For economic purposes, missing some feedwater into the product water can reduce membrane area.

Step 4. Estimate blended flows

Using $C_p = 0.367$ kg/m^3 as an estimated TDS concentration of the product water, the TDS in the blended water is 0.45 kg/m^3.

Mass balance equation would be

$$(Q_f + Q_p)(0.45 \text{ kg/m}^3) = Q_f(2.6 \text{ kg/m}^3) + Q_p(0.367 \text{ kg/m}^3)$$

$$2.15Q_f - 0.083Q_p = 0 \tag{a}$$

and

$$Q_f + Q_p = 0.25 \text{ m}^3\text{/s}$$

$$Q_p = 0.25 - Q_f \tag{b}$$

Substituting (b) into (a)

$$2.15Q_f - 0.083\,(0.25 - Q_f) = 0$$

$$2.233\,Q_f = 0.02075$$

$$Q_f = 0.009 \text{ (m}^3\text{/s)}$$

then

$$Q_p = 0.25 - 0.009 = 0.241 \text{ (m}^3\text{/s)}$$

The water supply will use a blend of 0.24 m^3/s of RO product water and 0.01 m^3/s of feedwater (4%).

Step 5. Compute the required membrane area

$$A = \frac{Q_p}{f_w}$$

$$= \frac{0.24\text{m}^3\text{/s} \times 1000 \text{ kg/m}^3}{7.3 \times 10^{-3} \text{ kg/m}^2 \cdot \text{s}}$$

$$= 32{,}880 \text{ m}^2$$

16.2 Silt Density Index

The silt density index (SDI) test involves water filtering in the direct-flow mode with no cross-flow at a constant pressure of 207 kPa (30 psig). It commonly uses disc membranes rated at 0.45 μm for 15 min test with a 500 mL water sample. The SDI is calculated as (Chellam *et al.* 1997):

$$\text{SDI} = \left(1 - \frac{t_i}{t_f}\right)\frac{100}{15} \tag{6.150}$$

where SDI = silt density index
t_i = time initially needed to filter 500 mL of sample
t_f = time needed to filter 500 mL at the end of the 15 min test period

The SDI has been widely used as a rough estimation of the potential for colloidal fouling in nanofiltration. Generally, the feedwaters for nanofiltration (pore size 0.2 μm) should have SDI of 5 or less.

EXAMPLE: The time initially required to filter 500 mL of a dual-media filter effluent is 14.5 s. The time required to filter 500 mL of the same water sample at the end of the 15 min test period is 48 s. Calculate the SDI.

Solution:

$$\text{SDI} = \left(1 - \frac{t_i}{t_f}\right)\frac{100}{15} = \left(1 - \frac{14.5 \text{ s}}{48 \text{ s}}\right) \times \frac{100}{15}$$

$$= 4.65$$

17 RESIDUAL FROM WATER PLANT

A water treatment plant not only produces drinking water, but is also a solids generator. The residual (solids or wastes) comes principally from clarifier basins and filter backwashes. These residuals contain solids which are derived from suspended and dissolved solids in the raw water with the addition of chemicals and chemical reactions.

Depending on the treatment process employed, waste from water treatment plants can be classified as alum, iron, or polymer sludge from coagulation and sedimentation; lime sludge or brine waste from water softening; backwash wastewater and spent granular activated carbon from filtration; and wastes from iron and manganese removal process, ion exchange process, diatomaceous earth filters, microstrainers, and membranes.

The residual characteristics and management of water plant sludge and environmental impacts are discussed in detail elsewhere (Lin and Green, 1987).

Prior to the 1970s, residuals from a water treatment plant were disposed of in a convenient place, mostly in surface water. Under the 1972 Water Pollution Control Act, the discharge of a water plant residuals requires a National Pollution Discharge Elimination System (NPDES) permit. Residuals directly discharged into surface water is prohibited by law in most states. Proper residual (solids) handling, disposal, and/or recovery is necessary for each water work nowadays. Engineers must make their best estimate of the quantity and quality of residuals generated from the treatment units.

EXAMPLE 1: A conventional water treatment plant treats an average flow of 0.22 m^3/s (5.0 MGD). The total suspended solids (TSS) concentration in raw (river) water averages 88 mg/L. The TSS removal through sedimentation and filtration processes is 97 percent. Alum is used for coagulation/sedimentation purpose. The average dosage of alum is 26 mg/L. Assume the aluminum ion is completely converted to aluminum hydroxide. (a) Compute the average production of alum sludge. (MW: Al = 27, S = 32, O = 16, and H = 1). (b) Estimate the sludge volume generated daily, assuming the specific gravity of 4% sludge is 1.02.

Solution:

Step 1. Determine quantity (q_1) settled for TSS

$$\text{TSS} = 88 \text{ mg/L} = 88 \text{ g/m}^3$$

$$\begin{aligned} q_1 &= \text{flow} \times \text{TSS} \times \%\text{ removal} \\ &= 0.22 \text{ m}^3/\text{s} \times 88 \text{ g/m}^3 \times 0.97 \\ &= 18.8 \text{ g/s} \end{aligned}$$

Step 2. Determine $Al(OH)_3$ generated, q_2

The reaction formula for alum and natural alkalinity in water: (Eq. (6.51a))

$$\begin{array}{ll} \text{Al}_2(\text{SO}_4)_3 \cdot 18\text{H}_2\text{O} + 3\text{Ca(HCO}_3)_2 \rightarrow & 2\text{Al(OH)}_3 + 3\text{CaSO}_4 + 6\text{CO}_2 + 18\text{H}_2\text{O} \\ 27 \times 2 + 3(32 + 16 \times 4) + 18(18) & 2(27 + 17 \times 3) \\ = 666 & = 156 \end{array}$$

The above reaction formula suggests that each mole of dissolved Al^{3+} dosed will generate 2 moles of $Al(OH)_3$

$$\begin{aligned} q_2 &= \text{alum dosage} \times \text{flow} \times \left[\frac{156 \text{ g Al(OH)}_3}{666 \text{ g alum}}\right] \\ &= 26 \text{ g/m}^3 \times 0.22 \text{ m}^3/\text{s} \times \frac{156}{666} \\ &= 1.34 \text{ g/s} \end{aligned}$$

Step 3. Compute total residual generated q

$$q = q_1 + q_2 = 18.8 \text{ g/s} + 1.34 \text{ g/s}$$
$$= 20.14 \text{ g/s}$$

or

$$= 20.14 \frac{\text{g}}{\text{s}} \times \frac{1 \text{ kg}}{1000 \text{ g}} \times \frac{86{,}400 \text{ s}}{1 \text{ day}}$$
$$= 1740 \text{ kg/d}$$

or

$$= 1740 \ \frac{\text{kg}}{\text{day}} \times \frac{2.194 \text{ lb}}{1 \text{ kg}}$$
$$= 3818 \text{ lb/day}$$

Step 4. Estimate the sludge volume generated daily for (b)

$$\text{Volume} = 1740 \text{ kg/d} \div (1000 \text{ kg/m}^3 \times 1.02 \times 0.04)$$
$$= 42.6 \text{ m}^3\text{/d}$$

or

$$\text{Volume} = 3818 \text{ lb/d} \div (62.4 \text{ lb/ft}^3 \times 1.02 \times 0.04)$$
$$= 1500 \text{ ft}^3\text{/d}$$

EXAMPLE 2: A river raw water is coagulated with a dosage of 24 mg/L of ferrous sulfate and an equivalent dosage of 24 mg/L of ferrous sulfate and an equivalent dosage of lime. Determine the sludge generated per m^3 of water treated.

Solution:

From Eqs. (6.51f) and (6.51g), 1 mole of ferrous sulfate reacts with 1 mole of hydrated lime and this produces 1 mole of $Fe(OH)_3$ precipitate.

$$\text{MW of FeSO}_4 \cdot 7\text{H}_2\text{O} = 55.85 + 32 + 4 \times 16 + 7 \times 18$$
$$= 277.85$$
$$\text{MW of Fe(OH)}_3 = 55.85 + 3 \times 17$$
$$= 106.85$$

by proportion

$$\frac{\text{sludge}}{\text{F.S. dosage}} = \frac{106.85}{277.85} = 0.3846$$
$$\text{Sludge} = \text{F.S. dosage} \times 0.3846$$
$$= 24 \text{ mg/L} \times 0.3846$$
$$= 9.23 \text{ mg/L}$$
$$= 9.23 \frac{\text{mg}}{\text{L}} \times \frac{1 \text{ g}}{1000 \text{ mg}} \times \frac{1000 \text{ L}}{1 \text{ m}^3}$$
$$= 9.23 \text{ g/m}^3$$

or

$$= 9.23 \frac{\text{g}}{\text{m}^3} \times \frac{0.0022 \text{ lb}}{1 \text{ g}} \times \frac{1 \text{ m}^3}{2.642 \times 10^{-4} \text{ Mgal}}$$
$$= 77.0 \text{ lb/Mgal}$$

Note: Mgal = million gallons

EXAMPLE 3: A 2-MGD (0.088 m^3/s) water system generates 220 lb/Mgal (26.4 mg/L) solids in dry weight basis. The residual solids concentration by weight is 0.48 percent. Estimate the rate of residual solids production and pumping rate if the settled residual is withdrawn every 5 minutes per hour.

Solution:

Step 1. Calculated residual production

$$\frac{(220 \text{ lb/Mgal})(2 \text{ MGD})}{(8.34 \text{ lb/gal})(0.0048)} = 10{,}990 \text{ gpd} = 7.63 \text{ gpm}$$

Step 2. Calculate pump capacity required

$$7.63 \text{ gpm} \times 60/5 = 92 \text{ gpm}$$

17.1 Residual Production and Density

The quantity of residuals (called sludge) generated from clarifiers are related to coagulants used, raw water quality, and process design. The residual production for whole water plant as the weight per unit volume treated (g/m^3 or lb/Mgal) and its solids concentration should be determined. These values with the average flow rate will be used for the estimation of daily residual production. Then the residual production rate can be employed for the design of residual (sludge) treatment units, commonly intermittent sand drying bed.

The density of residuals (pounds per gallon) can be estimated using the following equation:

$$w = \frac{8.34}{c_s/\rho_s + c/\rho} \quad \text{(US customary units)} \tag{6.151a}$$

$$w = \frac{8.31}{c_s/\rho_s + c/\rho} \quad \text{(SI units)} \tag{6.151b}$$

where w = density of residuals, lb/gal or kg/L
c_s = weight fractional percent of solids
c = weight fractional percent of water
ρ_s = specific gravity of solid
ρ = specific gravity of water (approximately 1.0)

EXAMPLE 1: Calculate the sludge density (lb/gal) for a 6 percent sludge composed of solids with a specific gravity of 2.48.

Solution:

Step 1. Find specific gravity of 6% residual (sludge) ρ_r
Weight fraction of sludge, solids, and water are 1.0, 0.06, and 0.94, respectively.
Using mass balance, we obtain

$$\frac{1}{\rho_r} = \frac{0.06}{\rho_s} + \frac{0.94}{\rho} = \frac{0.06}{2.48} + \frac{0.94}{1.0}$$

$$\rho_r = 1.037$$

Step 2. Determine the density of the residue

$$\begin{aligned} \text{Density} &= \rho_r \times 8.34 \text{ lb/gal} = 1.037 \times 8.34 \text{ lb/gal} \\ &= 8.65 \text{ lb/gal} \\ \text{or} \quad &= 1037 \text{ g/L} \end{aligned}$$

Note: This solution is a kind of proof of Eq. (6.151).

EXAMPLE 2: A water plant treats 4.0 MGD (0.175 m^3/s) and generates 240 lb of residuals per million gallons (Mgal) treated (28.8 g/m^3). The solid concentration of the residuals is 0.5%. Estimate: (a) the daily production of residuals, (b) the pump capacity, if the pump operates for 12 min each hour, and (c) the density of the residuals, assuming the specific density of the residual is 2.0 with 4 percent solids.

Solution:

Step 1. Determine residual production rate q

$$q = \frac{(4\text{ MGD})\,(240\text{ lb/Mgal})}{(0.005)\,(8.34\text{ lb/gal})}$$

(a)

$$= 23{,}020\text{ gpd}$$

$$= 16.0\text{ gpm}$$

Step 2. Find pump capacity q_p

$$q_\text{p} = 16\text{ gpm} \times \frac{60\text{ min}}{12\text{ min}}$$

(b)

$$= 80\text{ gpm}$$

Step 3. Estimate residual density w, using Eq. (6.151a)

$$w = \frac{8.34}{0.04/2.0 + 0.96/1.0}$$

(c)

$$= 8.51\text{ lb/gal}$$

EXAMPLE 3: In a 1 MGD (0.0438 m^3/s) water works, 1500 lb of dry solids are generated per million gallons of water treated. The sludge is concentrated to 3 percent and then applied to an intermittent sand drying bed at a 18 in depth, with 24 beds used per year. Determine the surface area needed for the sand drying bed.

Solution:

Step 1. Calculate yearly 3% sludge volume, υ

$$\upsilon = 1500\text{ lb/Mgal} \times 1\text{ Mgal/d}/0.03$$

$$= 50{,}000\text{ lb/d}$$

$$= 50{,}000\text{ lb/d} \times 365\text{ d}/(62.4\text{ lb/ft}^3)$$

$$= 292{,}500\text{ ft}^3$$

$$= 8284\text{ m}^3$$

Step 2. Calculate area A, (D = depth, n = number of applications/year)

$$\upsilon = A\,D\,n$$

or

$$A = \upsilon/D\,n = 292{,}500\text{ ft}^3/(1.5\text{ ft} \times 24)$$

$$= 8125\text{ ft}^2$$

$$= 755\text{ m}^2$$

18 DISINFECTION

Disinfection is a process to destroy disease-causing organisms, or pathogens. Disinfection of water can be done by boiling the water, ultraviolet radiation, and chemical inactivation of the pathogen. In the water treatment processes, pathogens and other organisms can be partly physically eliminated through coagulation, flocculation, sedimentation, and filtration, in addition to the natural die-off. After filtration, to ensure pathogen-free water, the chemical addition of chlorine (so called chlorination), rightly or wrongly, is most widely used for disinfection of drinking water. For the public health standpoint, chlorination of drinking water provides more benefits than its shortcoming due disinfection by-products (DBPs). The use of ozone and ultraviolet for disinfection of water and wastewater is increasing in the United States.

Chlorination serves not only for disinfection, but as an oxidant for other substances (iron, manganese, cyanide, etc.) and for taste and odor control in water and wastewater. Other chemical disinfectants include chlorine dioxide, ozone, bromine, and iodine. The last two chemicals are generally used for personal application, not for the public water supply.

18.1 Chemistry of Chlorination

Free available chlorine. Effective chlorine disinfection depends upon its chemical form in water. The influencing factors are temperature, pH, and organic content in the water. When chlorine gas is dissolved in water, it rapidly hydrolyzes to hydrochloric acid (HCl) and hypochlorous acid (HOCl)

$$Cl_2 + H_2O \leftrightarrow H^+ + Cl^- + HOCl \tag{6.152}$$

The equilibrium constant is

$$\begin{aligned} K_H &= \frac{[H^+][Cl^-][HOCl]}{[Cl_{2(aq)}]} \qquad (6.153) \\ &= 4.48 \times 10^{-4} \text{ at } 25°C \text{ (White 1972)} \end{aligned}$$

The dissolution of gaseous chlorine, $Cl_{2(g)}$, to form dissolved molecular chlorine, $Cl_{2(aq)}$ follows Henry's law and can be expressed as (Downs and Adams, 1973)

$$Cl_{2(g)} = \frac{Cl_{2(aq)}}{H\ (\text{mol/L} \cdot \text{atm})} = \frac{[Cl_{2(aq)}]}{P_{Cl_2}} \tag{6.154}$$

where $[Cl_{2(aq)}]$ = molar concentration of Cl_2

P_{Cl_2} = partial pressure of chlorine in atmosphere

The distribution of free chlorine between HOCl and OCl^- is presented in Fig. 6.10. The disinfection capabilities of HOCl is generally higher than that of OCl^- (Water, 1978).

$$\begin{aligned} H &= \text{Henry's law constant, mol/L} \cdot \text{atm} \\ &= 4.805 \times 10^{-6} \exp\left(\frac{2818.48}{T}\right) \qquad (6.155) \end{aligned}$$

Hypochlorous acid is a weak acid and subject to further dissociation to hypochlorite ions (OCl^-) and hydrogen ions:

$$HOCl \leftrightarrow OCl^- + H^+ \tag{6.156}$$

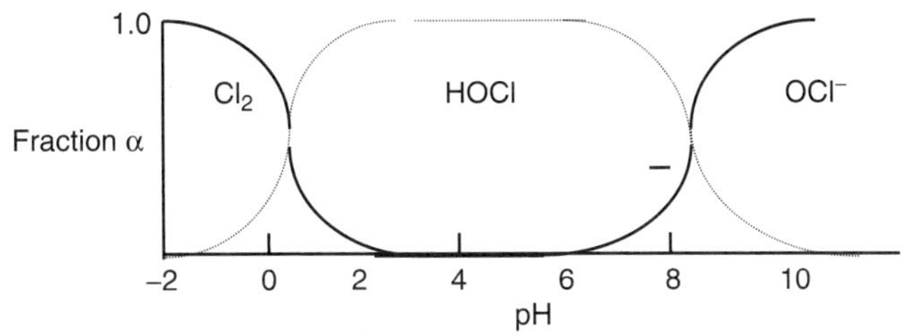

FIGURE 6.10 Distribution of chlorine species (*Snoeyink and Jenkins, 1980*).

and its acid dissociation constant K_a is

$$K_a = \frac{[OCl^-][H^+]}{[HOCl]} \tag{6.157}$$
$$= 3.7 \times 10^{-8} \text{ at } 25°C$$
$$= 2.61 \times 10^{-8} \text{ at } 20°C$$

The values of K_a for hypochlorous acid is a function of temperature in kelvins as follows (Morris 1966):

$$\text{In } K_a = 23.184 - 0.058T - 6908/T \tag{6.158}$$

EXAMPLE 1: The dissolved chlorine in the gas chlorinator is 3900 mg/L at pH 4.0. Determine the equilibrium vapor pressure of chlorine solution at 25°C (use $K_H = 4.5 \times 10^{-4}$).

Solution:

Step 1. Since pH = 4.0 < 5, from Fig. 6.10, the dissociation of HOCl to OCl^- is not occurring. The available chlorine is $[Cl_2]$ + [HOCl]:

$$[Cl_2] + [HOCl] = 3900 \text{ mg/L} = \frac{3.9 \text{ g/L}}{70.9 \text{ g/mol}}$$
$$= 0.055 \text{ mol/L}$$
$$[HOCl] = 0.055 - [Cl_2]$$

Step 2. Compute $[H^+]$

$$\text{pH} = 4.0 = \log \frac{1}{[H^+]}$$
$$[H^+] = 10^{-4}$$

Step 3. Apply K_H at 25°C

$$K_H = 4.5 \times 10^{-4} = \frac{[H^+][Cl^-][HOCl]}{[Cl_2]}$$

Substitute $[H^+]$ from Step 2

$$4.5 \times 10^{-8} = \frac{[Cl^-][HOCl]}{[Cl_2]}$$

From Eq. (6.152), it can be expected that, for each mole of HOCl produced, one mole of Cl^- will also be produced. Assuming there is no chloride in the feedwater, then

$$[Cl^-] = [HOCl] = 0.055 - [Cl_2]$$

Step 4. Solve the above two equations

$$4.5 \times 10^{-8} = \frac{(0.055 - [Cl_2])^2}{[Cl_2]}$$

$$0.000212\,[Cl_2]^{1/2} = 0.055 - [Cl_2]$$

$$[Cl_2] + 21{,}400\,[Cl_2]^{1/2} - 0.055 = 0$$

$$[\,Cl_2]^{1/2} = \frac{-0.000212 \pm \sqrt{0.000212^2 + 0.22}}{2}$$

$$= 0.2345$$

$$[Cl_2] = 0.055 \text{ (mol/L)}$$

Step 5. Compute H using Eq. (6.155)

$$H = 4.805 \times 10^{-6} \exp\left(\frac{2818.48}{25 + 273}\right)$$

$$= 0.062 \text{ mol/(L} \cdot \text{atm)}$$

Note: [HOCl] is negible

Step 6. Compute the partial pressure of chlorine gas:

$$P_{Cl_2} = [Cl_2]/H = (0.055 \text{ mol/L})/(0.062 \text{ mol/(L} \cdot \text{atm))}$$

$$= 0.89 \text{ atm}$$

or

$$= 0.89 \text{ atm} \times 101{,}325 \text{ Pa/atm}$$

$$= 90 \text{ Pa}$$

EXAMPLE 2: A water treatment plant uses 110 lb/d (49.0 kg/d) of chlorine to treat 10.0 MGD (37,850 m^3/d or 0.438 m^3/s) of water. The residual chlorine after 30 minutes contact time is 0.55 mg/L. Determine the chlorine dosage and chlorine demand of the water.

Solution:

Step 1. Calculate chlorine demand

Since 1 mg/L = 8.34 lb/mil gal

$$\text{Chlorine dosage} = \frac{110 \text{ lb/d}}{10.0 \text{ Mgal/d}} \times \frac{1 \text{ mg/L}}{8.34 \text{ lb/Mgal}}$$

$$= 1.32 \text{ mg/L}$$

or solve by SI units:

Since 1 mg/L = 1 g/m^3

$$\text{Chlorine dosage} = (49.9 \text{ kg/d}) \times (1000 \text{ g/kg})\ (37{,}850 \text{ m}^3\text{/d})$$

$$= 1.32 \text{ g/m}^3$$

$$= 1.32 \text{ mg/L}$$

Step 2. Calculate chlorine demand

$$\begin{aligned}\text{Chlorine demand} &= \text{chlorine dosage} - \text{chlorine residual}\\ &= 1.32\text{ mg/L} - 0.55\text{ mg/L}\\ &= 0.77\text{ mg/L}\end{aligned}$$

EXAMPLE 3: Calcium hypochlorite (commercial high-test calcium hypochlorite, HTH) containing 70 percent available chlorine is used for disinfection of a new main. (a) Calculate the weight (kilograms) of dry hypochlorite powder needed to prepare a 2.0 percent hypochlorite solution in a 190 L (50 gal) container. (b) The volume of the new main is 60,600 L (16,000 gal). How much of the 2 percent hypochlorite solution will be used with a feed rate of 55 mg/L of chlorine?

Solution:

Step 1. Compute kg of dry calcium hypochlorite powder needed

$$\begin{aligned}\text{Weight} &= \frac{190\text{ L} \times 1.0\text{ kg/L} \times 0.02}{0.70}\\ &= 5.42\text{ kg}\end{aligned}$$

Step 2. Compute volume of 2 percent hypochlorite solution needed

2% solution = 20,000 mg/L = 20 g/L
Let X = 2 percent hypochlorite solution needed
Then (20,000 mg/L) (X L) = (55 mg/L) (60,600 L)
X = 167 L

Combined available chlorine. Chlorine reacts with certain dissolved constituents in water, such as ammonia and amino nitrogen compounds to produce chloramines. These are referred to as combined chlorine. In the presence of ammonium ions, free chlorine reacts in a stepwise manner to form three species: monochloramine, NH_2Cl; dichloramine, $NHCl_2$; and trichloramine or nitrogen trichloride, NCl_3.

$$NH_4^+ + HOCl \leftrightarrow NH_2Cl + H_2O + H^+ \tag{6.159}$$

or

$$NH_{3(aq)} + HOCl \leftrightarrow NH_2Cl + H_2O \tag{6.160}$$

$$NH_2Cl + HOCl \leftrightarrow NHCl_2 + H_2O \tag{6.161}$$

$$NHCl_2 + HOCl \leftrightarrow NCl_3 + H_2O \tag{6.162}$$

At high pH, the reaction for forming dichloramine from monochloramine is not favored. At low pH, other reactions occur as follows:

$$NH_2Cl + H^+ \leftrightarrow NH_3Cl^+ \tag{6.163}$$

$$NH_3Cl^+ + NH_2Cl \leftrightarrow NHCl_2 + NH_4^+ \tag{6.164}$$

Free residual chlorine is the sum of [HOCl] and [OCl^-]. In practice, free residual chlorine in water is 0.5 to 1.0 mg/L. The term "total available chlorine" is the sum of "free chlorine" and "combined chlorine". Chloramines are also disinfectants, but they are slower to act than that of hypochlorite.

Each chloramine molecule reacts with two chlorine atoms. Therefore, each mole of mono-, di-, and trichloramine contains 71, 142, and 213 g of available chlorine, respectively. The molecular weight of these three chloramines are respectively 51.5, 85.9, and 120.4. Therefore, the chloramines contain 1.38, 1.65, and 1.85 g of available chlorine per gram of chloramine, respectively.

The pH of the water is the most important factor on the formation of chloramine species. In general, monochloramine is formed at pH above 7. The optimum pH for producing monochloramine is approximately 8.4.

Breakpoint chlorination. When the molar ratio of chlorine to ammonia is greater than 1.0, there is a reduction of chlorine and oxidation of ammonia. A substantially complete oxidation–reduction process occurs in ideal conditions by a 2:1 ratio and results in the disappearance of all ammonium ions with excess free chlorine residual. This is called the breakpoint phenomenon. As shown in Fig. 6.11, chlorine reacts with easily oxidized constituents, such as iron, hydrogen sulfide, and some organic matter. It then continues to oxidize ammonia to form chloramines and chloroorganic compound below a Cl_2:NH_4^+ ratio of 5.0 (which is around the peak). The destruction of chloramines and chloroorganic compounds are between the ratio of 5.0 and 7.6. The ratio at 7.6 is the breakpoint. All chloramines and other compounds are virtually oxidized. Further addition of chlorine becomes free available chlorine, HOCl, and OCl^-. At this region, it is called breakpoint chlorination.

The breakpoint chlorination can be used as a means of ammonia nitrogen removal from waters and wastewaters. The reaction is:

$$2NH_3 + 3HOCl \leftrightarrow N_2\uparrow + 3H^+ + 3Cl^- + 3H_2O \tag{6.165}$$

or

$$2NH_3 + 3Cl_2 \leftrightarrow N_2\uparrow + 6HCl \tag{6.166}$$

$$NH_3 + 4Cl_2 + 3H_2O \leftrightarrow NO_3^- + 8Cl^- + 9H^+ \tag{6.167}$$

In practice, the breakpoint does not occur at a Cl_2: NH_4^+ mass ratio of around 10:1, and mass dose ratio of 15:1 or 20:1 is applied (White, 1972; Hass, 1990).

McKee *et al.* (1960) developed the relationship of available chlorine in the form of dichloramine to available chlorine in the form of monochloramine as follows:

$$A = \frac{BM}{1 - [1 - BM(2 - M)]^{1/2}} - 1 \tag{6.168}$$

where A = ratio of available chlorine in di- to monochloramine
M = molar ratio of chlorine (as Cl_2) added to ammonia-N present
$B = 1 - 4K_{eq}[H^+]$ (6.169)

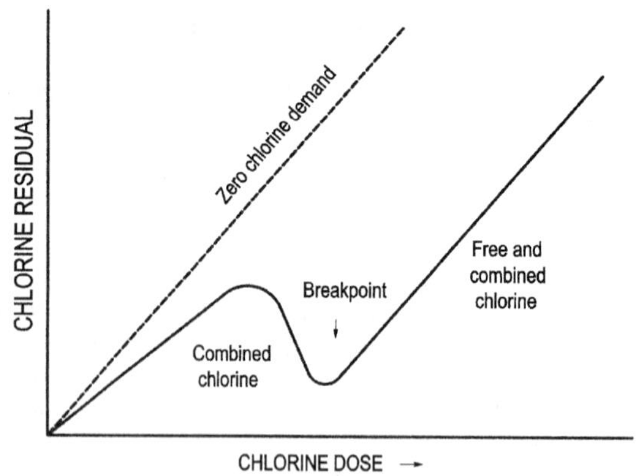

FIGURE 6.11 Theoretical drawing of breakpoint chlorination curve.

where the equilibrium constant K_{eq} is from

$$H^+ + 2NH_2Cl \leftrightarrow NH_4^+ + NHCl_2 \tag{6.170}$$

$$K_{eq} = \frac{[NH_4^+][NHCl_2]}{[H^+][NH_2Cl]^2} \tag{6.171}$$

$$= 6.7 \times 10^5 \text{ L/mol at } 25°\text{C}$$

The relationship (Eq. (6.168)) is pH dependent. When pH decreases and the Cl:N dose ratio increases, the relative amount of dichloramine also increases.

EXAMPLE: The treated water has pH of 7.4, a temperature of 25°C, and a free chlorine residual of 1.2 mg/L.

Chloramine is planned to be used in the distribution system. How much ammonia is required to keep the ratio of dichloramine to monochloramine of 0.15, assuming all residuals are not dissipated yet.

Solution:

Step 1. Determine factor B by Eq. (6.171)

$$K_{eq} = 6.7 \times 10^5 \text{ L/mol at } 25°\text{C}$$

$$\text{pH} = 7.4, [H^+] = 10^{-7.4}$$

From Eq. (6.169)

$$B = 1 - 4K_{eq}[H^+] = 1 - 4(6.7 \times 10^5)(10^{-7.4})$$

$$= 0.893$$

Step 2. Compute molar ratio M

$$A = 0.15$$

Using Eq. (6.168)

$$A = \frac{BM}{1 - [1 - BM(2 - M)]^{1/2}} - 1$$

$$0.15 = \frac{0.893\,M}{1 - [1 - 0.893\,M(2 - M)]^{1/2}} - 1$$

$$1 - [1 - 0.893\,M(2 - M)]^{1/2} = 0.777\,M$$

$$(1 - 0.777\,M)^2 = 1 - 0.893\,M(2 - M)$$

$$0.29\,M^2 - 0.232\,M = 0$$

$$M = 0.80$$

Step 3. Determine the amount of ammonia nitrogen (N) to be added

$$\text{MW of Cl}_2 = 70.9 \text{ g}$$

$$1.2 \text{ mg/L of Cl}_2 = \frac{1.2 \text{ mg/L}}{70.9 \text{ g/mol}} \times 0.001 \text{ g/mg}$$

$$= 1.7 \times 10^{-5} \text{ mol/L}$$

$$M = \frac{1.7 \times 10^{-5} \text{ mol/L}}{\text{NH}_3}$$

$$\text{NH}_3 = 1.7 \times 10^{-5} \text{ (mol/L)}/0.80$$

$$= 2.12 \times 10^{-5} \text{ mol/L}$$

$$= 2.12 \times 10^{-5} \text{ mol/L} \times 14 \text{ g of N/mol}$$

$$= 0.30 \text{ mg/L as N}$$

Chlorine dioxide. Chlorine dioxide (ClO_2) is an effective disinfectant (1.4 times that of the oxidation power of chlorine). It is an alternative disinfectant of potable water, since it does not produce significant amounts of trihalomethanes (THMs) which chlorine does. Chlorine dioxide has not been widely used for water and wastewater disinfection in the United States. In western Europe, the use of chlorine dioxide is increasing.

Chlorine dioxide is a neutral compound of chlorine in the +IV oxidation state, and each mole of chlorine dioxide yields five redox equivalents (electrons) on being reduced to chlorine ions. The chemistry of chlorine dioxide in water is relatively complex. Under acid conditions, it is reduced to chloride:

$$ClO_2 + 5e + 4H^+ \leftrightarrow Cl^- + 2H_2O \qquad (6.172)$$

Under reactively neutral pH found in most natural water it is reduced to chlorite:

$$ClO_2 + e^- \leftrightarrow ClO_2^- \qquad (6.173)$$

In alkaline solution, chlorine dioxide is disproportioned into chlorite, ClO_2^-, and chlorate, ClO_3:

$$2ClO_2 + 2OH^- \leftrightarrow ClO_2^- + ClO_3^- + H_2O \qquad (6.174)$$

Chlorine dioxide is generally generated at the water treatment plant prior to application. It is explosive at elevated temperatures and on exposure of light or in the presence of organic substances. Chlorine dioxide may be generated by either the oxygeneration of lower-valence compounds (such as chlorites or chlorous acid) or the reduction of a more oxidized compound of chlorine. Chlorine dioxide may be generated by oxidation of sodium chlorite with chlorine gas by the following reaction.

$$2NaClO_2 + Cl_2 \leftrightarrow 2NaCl + 2ClO_2 \qquad (6.175)$$

The above equation suggests that 2 moles of sodium chlorite react with 1 mole of chlorine to produce 2 moles of chlorine dioxide.

Chlorine dioxide is only 60 to 70 percent pure (James M. Montgomery Consulting Engineering, 1985). Recent technology has improved it to in excess of 95 percent. It is a patented system that reacts gaseous chlorine with a concentrated sodium chlorite solution under a vacuum. The chlorine dioxide produced is then removed from the reaction chamber by a gas ejector, which is very similar to the chlorine gas vacuum feed system. The chlorine dioxide solution concentration is 200 to 1000 mg/L and contains less than 5 percent excess chlorine (ASCE and AWWA, 1990).

EXAMPLE: A water treatment plant has a chlorination capacity of 500 kg/d (1100 lb/d) and is considering the switch to using chlorine dioxide for disinfection. The existing chlorinator would be employed to generate chlorine dioxide. Determine the theoretical amount of chlorine dioxide that can be generated and the daily requirement of sodium chlorite.

Solution:

Step 1. Find ClO_2 generated by Eq. (6.175)

$$
\begin{array}{lccc}
 & 2NaClO_2 + & Cl_2 \leftrightarrow 2NaCl + & 2ClO_2 \\
\text{MW} & 180.9 & 70.9 & 134.9 \\
 & y & 500\ \text{kg/d} & x
\end{array}
$$

$$
\begin{aligned}
x &= \frac{500\ \text{kg/d}}{70.9\ \text{g}} \times 134.9\ \text{g} \\
&= 951\ \text{kg/d} \\
&= 2100\ \text{lb/d}
\end{aligned}
$$

Step 2. Determine $NaClO_2$ needed

$$
\begin{aligned}
y &= \frac{500\ \text{kg/d}}{70.9\ \text{g}} \times 180.9\ \text{g} \\
&= 1276\ \text{kg/d} \\
&= 2810\ \text{lb/d}
\end{aligned}
$$

18.2 Ozonation

Ozone (O_3) is a blue (or colorless) unstable gas at the temperature and pressure encountered in water and wastewater treatment processes. It has a pungent odor. Ozone is one of the most powerful oxidizing agents. It has been used for the purposes of disinfection of water and wastewater and chemical oxidation and preparation for biological processing, such as nitrification.

Ozonation of water for disinfection has been used in France, Germany, and Canada. Interest in ozonation in water treatment is increasing in the United States due to HTMs problems of chlorination. Ozone is an allotrope of oxygen and is generated by an electrical discharge to split the stable oxygen–oxygen covalent bond as follows:

$$3O_2 + \text{energy} \leftrightarrow 2O_3 \tag{6.176}$$

The solubility of ozone in water follows Henry's law. The solubility constant H is (ASCE and AWWA 1990):

$$H = (1.29 \times 10^6/T) - 3721 \tag{6.177}$$

where T is temperature in Kelvin.

The maximum gaseous ozone concentration that can be generated is approximately 50 mg/L (or 50 g/m^3). In practice, the maximum solution of ozone in water is only 40 mg/L (USEPA, 1986). Ozone production in ozone generators from air is from 1 to 2 percent by weight, and from pure oxygen ranges from 2 to 10 percent by weight. One percent by weight in air is 12.9 g/m^3 ($\cong$ 6000 ppm) and one percent by weight in oxygen is 14.3 g/m^3 ($\cong$ 6700 ppm). Residual ozone in water is generally 0.1 to 2 g/m^3 (mg/L) of water (Rice, 1986). The ozone residual present in water decays rapidly.

18.3 Disinfection Kinetics

Basic models describing the rate of microorganism destruction or natural die-off is Chick's law, a first-order chemical reaction:

$$\ln(N/N_0) = -kt \tag{6.178a}$$

or

$$-dN/dt = kt \tag{6.178b}$$

where N = microbial density present at time t
N_0 = microbial density present at time 0 (initial)
k = rate constant
t = time
$-dN/dt$ = rate of decrease in microbial density

Watson refined Chick's equation and developed an empirical equation including changes in the concentration of disinfectant as the following relationship:

$$k = k'C^n \tag{6.179}$$

where k' = corrected die-off rate, presumably independent of C and N
C = disinfection concentration
n = coefficient of dilution

The combination of Eqs. (6.178a) and (6.179) yields

$$\ln(N/N_0) = -k'\,C^n\,t \tag{6.180a}$$

or

$$-dN/dt = k'\,t\,C^n \tag{6.180b}$$

The disinfection kinetics is influenced by temperature, and the Arrhenius model is used for temperature correction as (USEPA, 1986):

$$k'_T = k'_{20}\theta^{(T-20)} \tag{6.181}$$

where k'_T = rate constant at temperature T, °C
k'_{20} = rate constant at 20°C
θ = empirical constant

The Chick–Watson relationship can plot the disinfection data of the survival rate (N/N_0) on a logarithmic scale against time, and a straight line can be drawn. However, in practice, most cases do not yield a straight line relationship. Hom (1972) refined the Chick–Watson model and developed a flexible, but highly empirical kinetic formation as follows:

$$-dN/dt = k'Nt^mC^n \tag{6.182}$$

where m = empirical constant
= 2 for *E. coli*

Other variables are as previous stated

For changing concentrations of the disinfectant, the disinfection efficiency is of the following form (Fair *et al.* 1968):

$$C^n\,t_p = \text{constant} \tag{6.183}$$

where t_p is the time required to produce a constant percent of kill or die-off. This approach has evolved into a *CT* (concentration × contact time) regulation to ensure a certain percentage kills of *Crypotosporidium, Giardia,* and viruses. The percentage kills are expressed in terms of log removals.

18.4 *CT* Values

On June 29, 1989, the USEPA (1989b) published the final Surface Water Treatment Regulations (SWTR). Under these new rules, water treatment plants which use surface water as a source or groundwater under the influence of surface water, and without filtration, must calculate the *CT* values each day.

Regulations. The USEPA reaffirms its commitment to the current Safe Drinking Water Act regulations, including those related to disinfection and pathogenic organism control. The public water supply utilities must continue to comply with the current rules while new rules for microorganisms, and disinfectants/disinfection byproducts rules are being developed.

The Surface Water Treatment Rule applies to all surface water supplies and to groundwater supplies that are not protected from contamination or by surface water. The public water system is required to comply with a new operating parameter, the so-called *CT* value. The *CT* value refers to the value of disinfectant content or disinfectant residual, *C* (mg/L), multiplied by the contact time, *T* (min). *CT* is an indicator of the effectiveness of the disinfection process, depending on pH and temperature to remove or inactivate the *Giardia lamblia* (a protozoan) and viruses which could pass through the water treatment unit processes.

Requirements for treatment (using the best available technology) such as disinfection and filtration are established in place of maximum contaminant level (MCL) for *Giardia*, *Legionella*, heterotrophic plate count (HPC), and turbidity. Treatment must achieve at least 99.9 percent (3-log) removal or inactivation of *G. lamblia* cysts and 99.99 percent (4-log) removal or inactivation of viruses. Under the SWTR, the concept of *CT* values is established for estimating inactivation efficiency of disinfection practices in plants. Filtered water turbidity must at no time exceed 5 NTU (nephelometric turbidity units), and 95 percent of turbidity measurements must meet 0.5 NTU for conventional and direct filtration and 1.0 NTU for slow sand filtration and diatomaceous earth filtration. Turbidity must be measured every 4 h by grab sampling or continuous monitoring. *Cryptosporidium* was not regulated under SWTR of 1989, because sufficient data about the organism was lacking at the time.

The residual disinfectant in finished water entering the distribution system should not be less than 0.2 mg/L for at least 4 h. The residual disinfectant at any distribution point should be detectable in 95 percent of the samples collected in a month for any two consecutive months. A water plant may use HPC in lieu of residual disinfectant. If HPC is less than 500 per 100 mL, the site is considered to have a detectable residual for compliance purposes.

There are some criteria for avoiding filtration. The water plant must calculate the *CT* value each day. The water plant using filtration may or may not need to calculate *CT* values depending on the state primary requirements. The primary agency credits the slow sand filter, which produce water with turbidity of 0.6 to 0.8 NTU, with a 2-log *Giardia* and virus removal. Disinfection must achieve an additional 1-log *Giardia* and 2-log virus removal/inactivation to meet overall treatment objective.

Determination of CT values. Calculation of *CT* values is a complicated procedure depending on the configuration of the whole treatment system, type and number of point applications, and residual concentration of disinfectant (Pontius, 1993).

A water utility may determine the inactivation efficiency based on one point of disinfectant residual measurement prior to the first customer, or on a profile of the residual concentration between the point of disinfectant application and the first customer. The first customer is the point at which finished water is first consumed. To determine compliance with the inactivation (*Giardia* and viruses) requirements, a water system must calculate the *CT* value(s) for its disinfection conditions during peak hourly flow once each day that it is delivering water to its first customers.

To calculate *CT* value, *T* is the time (in minutes) that the water moves, during the peak hourly flow, between the point of disinfectant application and the point at which *C*, residual disinfectant concentration, is measured prior to the first customer. The residual disinfectant concentration, *C*, pH, and temperature should be measured each day during peak hourly flow at the effluents of treatment units (each segment or section) and at the first customer's tap. The segments may include the inflow piping system, rapid mixer, flocculators, clarifiers, filters, clearwells, and distribution systems.

The contact time, T, is between the application point and the point of residual measured or effective detention time (correction factor from hydraulic retention time, HRT). The contact time in pipelines can be assumed equivalent to the hydraulic retention time and is calculated by dividing the internal volume of the pipeline by the peak hourly flow rate through the pipeline. Due to short circuiting, the contact time in mixing tanks, flocculators, settling basins, clearwells, and other tankages should be determined from tracer studies or an equivalent demonstration. The time determined from tracer studies to be used for calculating CT is T_{10}. In the calculation of CT values, the contact time, designated by T_{10}, is the time needed for 10 percent of the water to pass through the basin or reservoir. In other words, T_{10} is the time (in minutes) that 90 percent of the water (and microorganisms in the water) will be exposed to disinfectant in the disinfection contact chamber.

The contact time T_{10} is used to calculate the CT value for each section. The CT_{cal} is calculated for each point of residual measurement. Then the inactivation ratio, $CT_{cal}/CT_{99.9}$ or CT_{cal}/CT_{99} should be calculated for each section. $CT_{99.9}$ and CT_{99} are the CT values to achieve 99.9 and 99 percent inactivation, respectively. Total inactivation ratio is the sum of the inactivation ratios for each section (sum $CT_{cal}/CT_{99.9}$). If the total inactivation ratio is equal to or greater than 1.0, the system provides more than 99.9 percent inactivation of *Giardia* cysts and meets the disinfection performance requirement. In fact, many water plants meet the CT requirement for 3-log *Giardia* inactivation because the plant configurations allow for adequate contact time.

Calculation of CT values is a complicated procedure and depends on the configuration of the entire treatment system, type and number of point applications, and residual concentration of disinfectant (USEPA, 1999). The water temperature, pH, type and concentration of disinfectant, and effective retention time (ERT) determine the CT values. Flow patterns and tank configurations of flocculators, clarifiers, and clearwells (location of influent and effluent pipe, and the baffling condition) affect the ERT. The ERT is different from the HRT in most instances.

The theoretical contact time, T_{theo}, can be determined as (assuming completely mixed tank):

$$T_{theo} = V/Q \tag{6.184}$$

where T_{theo} = time (= HRT), minutes (min)
V = volume of tank or pipe, gallons or m^3
Q = flow rate, gallons per minute, gpm or m^3/s

The flow rate can be determined by a theoretical calculation or by using tracer studies. An empirical curve can be developed for T_{10} against Q. For the distribution pipelines, all water passing through the pipe is assumed to have detention time equal to the theoretical (Eq. (6.184)) or mean residence time at a particular flow rate, i.e. the contact time is 100 percent of the time the water remains in the pipe.

In mixing basins, storage reservoirs, and other treatment plant process units, the water utility is required to determine the contact time for the calculation of the CT value through tracer studies or other methods approved by the primacy agency, such as the state EPA.

The calculation of contact time T for treatment units is not a straightforward process of dividing the volume of tank or basin by the flow through rate (Eq. 6.184). Only a partial credit is given to take into account short circuiting that occurs in tanks without baffles. The instructions on calculating the contact time (out of the scope of this handbook) are given in Appendix C of the Guidance Manual (USEPA, 1989a).

Baffling is used to maximize utilization of basin volume, increase the plug flow zone in the basin, and to minimize short circuiting. Some form of baffling is installed at the inlet and outlet of the basin to evenly distribute flow across the basin. Additional baffling may be installed within the interior of the basin. Baffling conditions generally are classified as poor, average, or superior and are based on results of tracer studies for use in determining T_{10} from the theoretical detention time of a tank. The T_{10}/T fractions associated with various degrees of baffling conditions are shown in Table 6.6. In practice, theoretical T_{10}/T values of 1.0 for plug flows and 0.1 for mixed flow are seldom achieved due to the effect of dead space. Superior baffling conditions consist of at least a baffled inlet and outlet, and possibly some intra-basin baffling to redistribute the flow throughout the basin's cross section. Average baffling conditions include intrabasin baffling and either a baffled inlet or outlet. Poor baffling conditions refer to basins without intrabasin baffling and unbaffled inlets and outlets.

TABLE 6.6 Classification of Baffling and T_{10}/T Values

Condition of baffling	Description of baffling	T_{10}/T
Unbaffled (mixed flow)	No baffle, agitated basin, very low length-to-width ratio, high inlet and outlet flow velocities	0.1
Poor	Single or multiple unbaffled inlets and outlet, no intrabasin baffles	0.3
Average	Only baffled inlet or outlet, with some intrabasin baffles	0.5
Superior	Perforated inlet baffle, serpentine or perforated intra basin baffles, outlet weir, or perforated launders	0.7
Perfect (plug flow)	Very high length-to-width ratio (pipeline flow), perforated inlet, outlet, and intra basin baffles.	1.0

Source: Table C-5 of the USEPA Guidance Manual (USEPA, 1989a).

The procedure to determine the inactivation capability of a water plant is summarized as follows (USEPA, 1989a, 1990, 1999; IEPA, 1992):

1. Determine the peak hourly flow rate (Q in gpm) for each day from monitoring records.
2. Determine hydraulic detention time: HRT $= T = V/Q$.
3. Calculate the contact time (T_{10}) for each disinfection segment based on baffling factors or tracer studies.
4. Find correction factor, T_{10}/T, look at the actual baffling conditions to obtain T_{10} from Table 6.6.
5. Compute effective retention time ERT = HRT $\times$ (T_{10}/T).
6. Measure the disinfectant residual: C in mg/L (with pH, and temperature in °C) at any number of points within the treatment train during peak hourly flow (in gpm).
7. Calculate CT value (CT_{cal}) for each point of residual measurement (using ERT for T) based on actual system data.
8. Find $CT_{99.9}$ or $CT_{99.99}$ value from Tables 6.7 or 6.9 (from Appendix E of the Handbook, USEPA, 1989a) based on water temperature, pH, residual disinfectant concentration, and $\log_{10}$ removal = 3 or 4.
9. Compute the inactivation ratio, $CT_{\text{cal}}/CT_{99.9}$ and $CT_{\text{cal}}/CT_{99.99}$ for *Giardia* and viruses, respectively.
10. Calculate the estimated log inactivation by multiplying the ratio of Step 9 by 3 for *Giardia* and by 4 for viruses, because $CT_{99.9}$ and $CT_{99.99}$ are equivalent to 3-log and 4-log inactivations, respectively.
11. Sum the segment log inactivation of *Giardia* and viruses (such as rapid mixing tanks, flocculators, clarifiers, filters, clearwell, and pipelines) to determine the plant total log inactivations due to disinfection.
12. Determine whether the inactivations achieved are adequate. If the sum of the inactivation ratios is greater than or equal to one, the required 3-log inactivation of *Giardia* cysts and 4-log viruses inactivation have been achieved.
13. The total percent of inactivation can be determined as:

$$y = 100 - 100/10^x \tag{5.185}$$

where y = % inactivation
x = log inactivation

Tables 6.7 and 6.8 present the CT values for achieving 99.9 percent and 99 percent inactivation of *G. lamblia*, respectively. Table 6.9 present CT values for achieving 2-, 3-, and 4-log inactivation

TABLE 6.7 *CT* Values [(mg/L) min] for Achieving 99.9 Percent (3 log) Inactivation of *Giardia lamblia*

Disinfectant, mg/L	pH	Temperature, °C 0.5 or < 1	5	10	15	20	25
Free chlorine ≤ 0.4	6	137	97	73	49	36	24
	7	195	139	104	70	52	35
	8	277	198	149	99	74	50
	9	390	279	209	140	105	70
1.0	6	148	105	79	53	39	26
	7	210	149	112	75	56	37
	8	306	216	162	108	81	56
	9	437	312	236	156	117	78
1.6	6	157	109	83	56	42	28
	7	226	155	119	79	59	40
	8	321	227	170	116	87	58
	9	466	329	236	169	126	82
2.0	6	165	116	87	58	44	29
	7	236	165	126	83	62	41
	8	346	263	182	122	91	61
	9	500	353	265	177	132	88
3.0	6	181	126	95	63	47	32
	7	261	182	137	91	68	46
	8	382	268	201	136	101	67
	9	552	389	292	195	146	97
ClO_2	6–9	63	26	23	19	15	11
Ozone	6–9	2.9	1.9	1.43	0.95	0.72	0.48
Chloramine	6–9	3800	2200	1850	1500	1100	750

Source: Abstracted from Tables E-1 to E-6, E-8, E-10, and E-12 of the USEPA Guidance Manual (USEPA, 1989a).

of viruses at pH 6 through 9 (USEPA, 1989b). The SWTR Guidance Manual did not include *CT* values at pH above 9 due to the limited research results available at the time of rule promulgation. In November 1997, a new set of proposed rules was developed for the higher pH values, up to pH of 11.5 (Federal Register, 1997).

EXAMPLE 1: What are the percentages of inactivation for 2- and 3.4-log removal of *Giardia lamblia*?

Solution:

Using Eq. (6.185)

$$y = 100 - 100/10^x$$

as

$$x = 2$$

$$y = 100 - 100/10^2 = 100 - 1 = 99(\%)$$

as

$$x = 3.4$$

$$y = 100 - 100/10^{3.4} = 100 - 0.04 = 99.96(\%)$$

EXAMPLE 2: A water system of 100,000 gpd (0.0044 m^3/s) slow sand filtration system serves a small town of 1000 persons. The filter effluent turbidity values are 0.4 to 0.6 mg/L and pH is around 7.5. Chlorine is dosed after filtration and prior to the clearwell. The 4 in (10 cm) transmission pipeline to the first customer is 1640 ft (500 m) in distance. The residual chlorine concentrations in

TABLE 6.8 *CT* Values ((mg/L) min) for Achieving 90 Percent (1 log) Inactivation of *Giardia lamblia*

Disinfectant mg/L	pH	Temperature, °C 0.5 or < 1	5	10	15	20	25
Free chlorine ≤ 0.4	6	46	32	24	16	12	8
	7	65	46	35	23	17	12
	8	92	66	50	33	25	17
	9	130	93	70	47	35	23
1.0	6	49	35	26	18	13	9
	7	70	50	37	25	19	12
	8	101	72	54	36	27	18
	9	146	104	78	52	39	26
1.6	6	52	37	28	19	14	9
	7	75	52	40	26	20	13
	8	110	77	58	39	29	19
	9	159	112	84	56	42	28
2.0	6	55	39	29	19	15	10
	7	79	55	41	28	21	14
	8	115	81	61	41	30	20
	9	167	118	88	59	46	29
3.0	6	60	42	32	21	16	11
	7	87	61	46	30	23	15
	8	127	89	67	45	36	22
	9	184	130	97	65	49	32
Chlorine dioxide	6–9	21	8.7	7.7	6.3	5.0	3.7
Ozone	6–9	0.97	0.63	0.48	0.32	0.24	0.16
Chloramine	6–9	1270	735	615	500	370	250

Source: Abstracted from Tables E-1 to E-6, E-8, E-10, and E-12 of Guidance Manual (USEPA, 1989a).

TABLE 6.9 *CT* Values ((mg/L) min) for Achieving Inactivation of Viruses at pH 6 through 9

Disinfectant mg/L	Log inactivation	Temperature, °C ≤ 1	5	10	15	20	25
Free chlorine	2	6	4	3	2	1	1
	3	9	6	4	3	2	1
	4	12	8	6	4	3	2
Chlorine dioxide	2	8.4	5.6	4.2	2.8	2.1	1.4
	3	25.6	17.1	12.8	8.6	6.4	4.3
	4	50.1	33.4	25.1	16.7	12.5	8.4
Ozone	2	0.9	0.6	0.5	0.3	0.25	0.15
	3	1.4	0.9	0.8	0.5	0.4	0.25
	4	1.8	1.2	1.0	0.6	0.5	0.3
Chloramine	2	1243	857	643	428	321	214
	3	2063	1423	1067	712	534	365
	4	2883	1988	1491	994	746	497

Source: Modified from Tables E-7, E-9, E-11, and E-13 of Guidance Manual (USEPA, 1989a).

the clearwell and the distribution main are 1.6 and 1.0 mg/L, respectively. The volume of the clearwell is 70,000 gallons (265 m^3). Determine *Giardia* inactivation at water temperature 10°C at the peak hour flow of 100 gpm.

Solution:

Step 1. An overall inactivation of 3 logs for *Giardia* and 4 logs for viruses is required. The Primacy Agency can credit the slow sand filter process, which produces water with turbidity ranging from 0.6 to 0.8 NTU, with a 2-log *Giardia* and virus inactivation. For this example, the water system meets the turbidity standards. Thus, disinfection must achieve an additional 1-log *Giardia* and 2-log virus removal/inactivation to meet the overall treatment efficiency.

Step 2. Calculate T_{10} at the clearwell (one half of volume used, see Step 7a of Ex. 5)
T_{10} can be determined by the trace study at the peak hour flow or by calculation

$$T_{10} = V_{10}/Q = 0.1 \times (70{,}000 \text{ gal}/2)/100 \text{ gpm}$$
$$= 35 \text{ min}$$

Step 3. Calculate CT_{cal} in the clearwell

$$CT_{cal} = 1.6 \text{ mg/L} \times 35 \text{ min} = 56 \text{ (mg/L) min}$$

Step 4. Calculate $CT_{cal}/CT_{99.9}$
From Table 6.7, for 3-log removal for 1.6 mg/L chlorine residual at 10°C and pH 7.5.

$$CT_{99.9} = 145 \text{ (mg/L) min (by linear proportion between pH 7 and 8)}$$

then

$$CT_{cal}/CT_{99.9} = 56/145 = 0.38$$

Step 5. Calculate contact time at transmission main

$$Q = 100{,}000 \text{ gal/d} \times 1 \text{ ft}^3/7.48 \text{ gal} \times 1 \text{ day}/1440 \text{ min}$$
$$= 9.28 \text{ ft}^3\text{/min}$$
$$A = 3.14 \text{ (4 in/2/12 in/ft)}^2 = 0.0872 \text{ ft}^2$$
$$v = \frac{Q}{A} = \frac{9.28 \text{ ft}^3\text{/min}}{0.0872 \text{ ft}^2}$$
$$= 106 \text{ ft/min}$$
$$T = \text{length}/v = 1640 \text{ ft}/106 \text{ ft/min}$$
$$= 15.5 \text{ min}$$

where A is the cross-sectional area of the 4-in diameter pipe, and v is the flow velocity.

Step 6. Calculate CT_{cal} and $CT_{cal}/CT_{99.9}$ for the pipeline

$$CT_{cal} = 1.0 \text{ mg/L} \times 15.5 \text{ min}$$
$$= 15.5 \text{ (mg/L) min}$$

From Table 6.7, for 3-log removal of 1.0 mg/L chlorine residual at 10°C and pH 7.5

$$CT_{99.9} = 137 \text{ (mg/L) min}$$

then

$$CT_{cal}/CT_{99.9} = 15.5/137 = 0.11 \text{ (log)}$$

Step 7. Sum of $CT_{cal}/CT_{99.9}$ from Steps 4 and 6

$$\text{Total } CT_{cal}/CT_{99.9} = 0.38 + 0.11 = 0.49 \text{ (log)}$$

Since this calculation is based on a 3-log removal, the ratio needs to be multiplied by 3. Then the equivalent *Giardia* inactivation is

$$3 \times \text{total } CT_{cal}/CT_{99.9} = 3 \times 0.49 \text{ log} = 1.47 \text{ log}$$

The 1.47-log *Giardia* inactivation by chlorine disinfection in this system exceeds the 1-log additional inactivation needed to meet the overall treatment objectives. Generally, if the treatment meets *Giardia* inactivation, it also meets the virus inactivation requirement.

EXAMPLE 3: For a 2000 ft long, 12 in transmission main with a flow rate of 600 gpm, what is the credit of inactivation of *Giardia* and viruses using chlorine dioxide, residual = 0.6 mg/L, at 5°C and pH of 8.5.

Solution:

Step 1. Calculate $CT_{cal}/CT_{99.9}$ for transmission pipe

$$\begin{aligned} T &= \pi r^2 L/Q \\ &= 3.14 \text{ (0.5 ft)}^2 \text{ (2000 ft) (7.48 gal/ft}^3\text{)/(600 gal/min)} \\ &= 19.6 \text{ min} \end{aligned}$$

where r is the radius of the pipe and L is the length of the pipe.

$$\begin{aligned} CT_{cal} &= 0.6 \text{ mg/L} \times 19.6 \text{ min} \\ &= 11.7 \text{ (mg/L) min} \end{aligned}$$

At 5°C and pH of 8.5, using ClO_2, from Table 6.7

$$\begin{aligned} CT_{99.9} &= 26 \text{ (mg/L) min} \\ CT_{cal}/CT_{99.9} &= 11.7/26 = 0.45 \text{ (log)} \end{aligned}$$

Step 2. Calculate log inactivation for *Giardia* (X) and viruses (Y)

$$\begin{aligned} X &= 3\ (CT_{cal}/CT_{99.9}) = 3 \times 0.45 \text{ log} \\ &= 1.35 \text{ log} \\ Y &= 4\ (CT_{cal}/CT_{99.9}) = 4 \times 0.45 \text{ log} \\ &= 1.8 \text{ log} \end{aligned}$$

EXAMPLE 4: A water system pumps its raw water from a remote lake. No filtration is required due to good water quality. Chlorine is dosed at the pumping station near the lake. The peak pumping rate is 320 gpm (0.02 m^3/s). The distance from the pumps to the storage reservoir (tank) is 3300 ft (1000 m) with a transmission pipe of 10 in diameter. The chlorine residual at the outlet of the tank is 1.0 mg/L (*C* for the tank). The T_{10} for the tank is 88 min at the peak flow rate determined by a tracer study. Assuming the service connection to the first customer is negligible, determine the minimum chlorine residue required at the inlet of the tank (*C* at the pipe) to meet *Giardia* 3-log removal at 5°C and pH of 7.0.

Solution:

Step 1. Calculate CT_{cal} for the pipe

$$Q = 320 \text{ gpm} = 320 \text{ gal/min} \times 0.1337 \text{ ft}^3/\text{gal}$$
$$= 42.8 \text{ ft}^3/\text{min}$$
$$T = \pi r^2 L/Q = 3.14\,(0.417 \text{ ft})^2 \times 3300 \text{ ft}/(42.8 \text{ ft}^3/\text{min})$$
$$= 42 \text{ min}$$
$$CT_{cal}\text{ (pipe)} = 42\, C_p$$

where C_p is chlorine residual at the end of the pipe.

Step 2. Calculate CT_{cal} for the tank, CT_{cal} (tank)

$$CT_{cal}\text{ (tank)} = 1.0 \text{ mg/L} \times 88 \text{ min} = 88 \text{ (mg/L) min}$$

Step 3. Find $CT_{99.9}$ for *Giardia* removal
At water temperature of 5°C, pH of 7.0, and residual chlorine of 1.0 mg/L, from Table 6.7:

$$CT_{99.9} = 149 \text{ (mg/L) min}$$

Step 4. Calculate chlorine residue required at the end of the pipe (tank inlet), C_p

$$CT_{cal}\text{ (pipe)} + CT_{cal}\text{ (tank)} = CT_{99.9}$$

From Steps 1 to 3, we obtain

$$42 \text{ min} \times C_p \text{ (mg/L)} + 88 \text{ (mg/L) min} = 149 \text{ (mg/L) min}$$
$$C_p = 1.45 \text{ mg/L (minimum required)}$$

EXAMPLE 5: In a 1 MGD water plant (1 MGD = 694 gpm = 0.0438 m^3/s), water is pumped from a lake and prechlorinated with chlorine dioxide at the lake site. The source water is pretreated with chlorine at the intake near the plant. The peak flow is 600 gpm. The worst conditions to be evaluated are at 5°C and pH of 8.0. Gaseous chlorine is added at the plant after filtration (clearwell inlet). The service connection to the first customer is located immediately near the clearwell outlet. The following conditions are also given:

Treatment unit	Volume, gal	Outlet res. Cl_2, mg/L	Baffling conditions
Rapid mixer	240	0.4	Baffled at inlet and outlet
Flocculators	18,000	0.3	Baffled at inlet and outlet and with horizontal paddles
Clarifiers	150,000	0.2	Only baffled at outlet
Filters	6600	0.1	
Clearwell	480,000	0.3 (free) 1.6 (combined)	

Chlorine and ammonia are added at the clearwell inlet. Residual free and combined chlorine measured at the outlet of the clearwell are 0.3 and 1.6 mg/L, respectively. Estimate the inactivation level of *Giardia* and viruses at each treatment unit.

Solution:

Step 1. Calculate CT_{cal} for the rapid mixer

$$\text{HRT} = \frac{V}{Q} = \frac{240 \text{ gal}}{600 \text{ gal/min}}$$

$$= 0.4 \text{ min}$$

Assuming average baffling condition, and reading from Table 6.6 we obtain the correction factor $(T_{10}/T) = 0.5$ for average baffling.

Thus

$$\text{ERT} = \text{HRT } (T_{10}/T) = 0.4 \text{ min} \times 0.5$$

$$= 0.2 \text{ min}$$

Use this ERT as T for CT calculation

$$CT_{cal} = 0.4 \text{ mg/L} \times 0.2 \text{ min}$$

$$= 0.08 \text{ (mg/L) min}$$

Step 2. Determine $CT_{cal}/CT_{99.9}$ for the rapid mixer

Under the following conditions:

$$\text{pH} = 8.0$$

$$T = 5°\text{C}$$

$$\text{Residual chlorine} = 0.4 \text{ mg/L}$$

$$\text{log inactivation} = 3$$

From Table 6.7, we obtain:

$$CT_{99.9} = 198 \text{ (mg/L) min}$$

$$CT_{cal}/CT_{99.9} = (0.08 \text{ (mg/L) min})/(198 \text{ (mg/L) min})$$

$$= 0.0004$$

Step 3. Calculate the log inactivation of *Giardia* for the rapid mixer

$$\text{log removal} = 3\ (CT_{cal}/CT_{99.9})$$

$$= 0.0012$$

Step 4. Similar to Steps 1 to 3, determine log inactivation of *Giardia* for the flocculators

$$\text{HRT} = \frac{V}{Q} = \frac{18{,}000 \text{ gal}}{600 \text{ gpm}}$$

$$= 30 \text{ min}$$

with a superior baffling condition from Table 6.6, the correction $T_{10}/T = 0.7$

$$\begin{aligned} \text{ERT} &= \text{HRT } (T_{10}/T) = 30 \text{ min} \times 0.7 \\ &= 21 \text{ min} \end{aligned}$$

then

$$\begin{aligned} CT_{\text{cal}} &= 0.3 \text{ mg/L} \times 21 \text{ min} \\ &= 6.3 \text{ (mg/L) min} \\ \text{log removal} &= 3 \text{ (6.3 (mg/L) min)/(198 (mg/L) min)} \\ &= 0.095 \end{aligned}$$

Step 5. Similarly, for the clarifiers

$$\text{HRT} = \frac{150{,}000 \text{ gal}}{600 \text{ gpm}} = 250 \text{ min}$$

Baffling condition is considered as poor, thus

$$\begin{aligned} T_{10}/T &= 0.3 \\ \text{ERT} &= 250 \text{ min} \times 0.3 = 75 \text{ min} \\ CT_{\text{cal}} &= 0.2 \text{ mg/L} \times 75 \text{ min} = 15 \text{ (mg/L) min} \\ \text{log removal} &= 3 \times 15/198 \\ &= 0.227 \end{aligned}$$

Step 6. Determine *Giardia* removal in the filters

$$\begin{aligned} \text{HRT} &= \frac{6600 \text{ gal}}{600 \text{ gpm}} = 11 \text{ min} \\ T_{10}/T &= 0.5 \text{ (Teefy and Singer, 1990)} \\ \text{ERT} &= 11 \text{ min} \times 0.5 = 5.5 \text{ min} \\ CT_{\text{cal}} &= 0.1 \text{ mg/L} \times 5.5 \text{ min} \\ &= 0.55 \text{ (mg/L) min} \\ \text{log removal} &= 3 \times 0.55/198 \\ &= 0.008 \end{aligned}$$

Step 7. Determine *Giardia* removal in the clearwell.

There are two types of disinfectants, therefore, similar calculations should be performed for each disinfectant.

(a) Free chlorine inactivation. In practice, the minimum volume available in the clearwell during the peak demand period is approximately one-half of the working volume (Illinois EPA, 1992). Thus, one-half of the clearwell volume (240,000 gal) will be used to calculate the HRT:

$$\begin{aligned} \text{HRT} &= 240{,}000 \text{ gal/600 gpm} \\ &= 400 \text{ min} \end{aligned}$$

The clearwell is considered as a poor baffling condition, since it has no inlet, interior, or outlet baffling.

$$T_{10}/T = 0.3 \text{ (Table 6.6), calculate ERT}$$

$$\begin{aligned} \text{ERT} &= \text{HRT} \times 0.3 = 400 \text{ min} \times 0.3 \\ &= 120 \text{ min} \end{aligned}$$

The level of inactivation associated with free chlorine is

$$\begin{aligned} CT_{\text{cal}} &= 0.3 \text{ mg/L} \times 120 \text{ min} \\ &= 36 \text{ (mg/L) min} \end{aligned}$$

From Table 6.7, $CT_{99.9} = 198$ (mg/L) min
The log inactivation is

$$\begin{aligned} \text{log removal} &= 3 \times CT_{\text{cal}}/CT_{99.9} = 3 \times 36/198 \\ &= 0.545 \end{aligned}$$

(b) Chloramines inactivation:

$$\begin{aligned} CT_{\text{cal}} &= 1.6 \text{ mg/L} \times 120 \text{ min} \\ &= 192 \text{ (mg/L) min} \end{aligned}$$

For chloramines, at pH $= 8$, $T = 5$°C and 3-log inactivation,

$$CT_{99.9} = 2200 \text{ (mg/L) min (from Table 6.7)}$$

Log inactivation associated with chloramines is

$$\begin{aligned} \text{log removal} &= 3 \times 192/2200 \\ &= 0.262 \end{aligned}$$

The log removal at the clearwell is

$$\begin{aligned} \text{log removal} &= 0.545 + 0.262 \\ &= 0.807 \end{aligned}$$

Step 8. Summarize the log *Giardia* inactivation of each unit in the water treatment plant

Unit	Log inactivation
Rapid mixing	0.0012 (step 3)
Flocculators	0.095 (step 4)
Clarifiers	0.227 (step 5)
Filters	0.008 (step 6)
Clearwell	0.807 (step 7)
Total	1.138

The sum of log removal is equal to 1.138, which is greater than 1. Therefore, the water system meets the requirements of providing a 3-log inactivation of *Giardia* cysts. The viruses inactivation is estimated as follows.

Step 9. Estimate viruses inactivation in the rapid mixer
Using CT_{cal} obtained in step 1

$$CT_{\text{cal}} = 0.08 \text{ (mg/L) min}$$

Referring to Table 6.9 ($T = 5°\text{C}$, 4 log),

$$CT_{99.99} = 8 \text{ (mg/L) min}$$

The level of viruses inaction in the rapid mixer is

$$\begin{aligned} \text{log removal} &= 4 \times CT_{\text{cal}}/CT_{99.99} = 4 \times 0.08/8 \\ &= 0.04 \end{aligned}$$

Step 10. Estimate viruses inactivation in the flocculators

$$\begin{aligned} CT_{\text{cal}} &= 6.3 \text{ (mg/L) min (step 4)} \\ \text{log removal} &= 4 \times 6.3/8 \\ &= 3.15 \end{aligned}$$

Step 11. Estimate viruses inactivation in the clarifiers

$$\begin{aligned} CT_{\text{cal}} &= 15 \text{ (mg/L) min (step 5)} \\ \text{log removal} &= 4 \times 15/8 \\ &= 7.5 \end{aligned}$$

Step 12. Estimate viruses inactivation in the filters

$$\begin{aligned} CT_{\text{cal}} &= 0.55 \text{ (mg/L) min (step 6)} \\ \text{log removal} &= 4 \times 0.55/8 \\ &= 0.275 \end{aligned}$$

Step 13. Estimate viruses inactivation in the clearwell
For free chlorine:

$$\begin{aligned} CT_{\text{cal}} &= 36 \text{ (mg/L) min (step 6(a))} \\ \text{log removal} &= 4 \times 36/8 \\ &= 18 \end{aligned}$$

For chloramines:

$$\begin{aligned} CT_{\text{cal}} &= 192 \text{ (mg/L) min (step 6(b))} \\ CT_{99.99} &= 1988 \text{ (mg/L) min (Table 6.7)} \\ \text{log removal} &= 4 \times 192/1988 \\ &= 0.39 \\ \text{Total log removal} &= 18.39 \end{aligned}$$

Step 14. Sum of log inactivation for viruses in the plant is:

$$\begin{aligned} \text{Plant log removal} &= \text{sum of steps } 9-13 \\ &= 0.04 + 3.15 + 7.5 + 0.275 + 18.39 \\ &= 29.36 \end{aligned}$$

Note: Total viruses inactivation is well above 1.
The water system meets the viruses inactivation requirement.

In summary, a brief description of regulations in the United States and determination of *CT* values are presented. Under the SWTR, all surface water supplies and groundwater supplies that are under the influence of surface water must calculate *CT* values daily during peak hourly flow. A minimum of 3-log *Giardia lamblia* and 4-log virus removal and/or inactivation performance must be achieved at all times to comply with the existing SWTR. The *CT* values are used to evaluate the achievement of disinfection and to determine compliance with the SWTR. They also are used to compute the log inactivation of *Giardia* and viruses during water treatment and to construct a disinfection profile.

To calculate the *CT* value, operation data, e.g., disinfectant concentration, water pH, and water temperature, should be measured daily during the peak hourly flow. Contact time (T_{10}), actual calculated *CT* value (CT_{cal}), 3-log *Giardia* inactivation ($CT_{99.9}$ from Table 6.7), and/or 4-log virus inactivation (from Table 6.9), $CT_{cal}/CT_{99.9}$, and $CT_{cal}/CT_{99.99}$ should be calculated for the estimated segment log inactivations. The total plant log inactivation is the sum of all segment log inactivations. The daily log inactivation values can be used to develop a disinfection profile for the treatment system.

19 WATER FLUORIDATION

Fluoride occurs naturally in water at low concentrations. Water fluoridation is the intentional addition of fluoride to drinking water. During the past five decades, water fluoridation has been proven to be both wise and a most cost-effective method to prevent dental decay. The ratio of benefits (reductions in dental bills) to cost of water fluoridation is 50:1 (U.S. Public Health Service, 1984). In order to prevent decay, the optimum fluoride concentration has been established at 1 mg/L (American Dental Association, 1980). The benefits of fluoridation last a lifetime. While it is true that children reap the greatest benefits from fluoridation, adults benefit also. Although there are benefits, the opponents of water fluoridation exist. The charges and the facts are discussed in the manual (U.S. PHS, 1984).

Fluoride in drinking water is regulated under section 1412 of the Safe Drinking Water Act (SDWA). In 1986, USEPA promulgated an enforceable standard, a Maximum Contamination Level (MCL) of 4 mg/L for fluoride. This level is considered as protective of crippling skeletal fluorosis, an adverse health effect. A review of human data led the EPA (1990) to conclude that there was no evidence that fluoride in water presented a cancer risk in humans. A nonenforceable Secondary Maximum Contamination Level (SMCL) of 2 mg/L was set to protect against objectionable dental fluorisis, a cosmetic effect.

Since water fluoridation has been implemented in the United States some communities (organizations) have been against it. There are 22 charges. The major reason of public opposition is due to its carcinogenic and toxic characteristics. Research results prove that fluoridation is safe and efficient, so the controversy does not make any scientific sense (Reeves, 1986, US PHS 1984).

A number of scattered location through out the United States have high fluoride levels (1.6–8.0 mg/L). Both sodium and potassium fluoride are soluble. Many metals and nonmetals form complex fluoro anions. Removal of fluoride or any other trace contaminant from water is generally difficult and costly. The fluoride in water can be removed by activated carbon adsorption, reverse osmosis, electrodialysis, distillation, or ion exchange process.

19.1 Fluoride Chemicals

Fluoride is a pale yellow noxious gaseous halogen. It is the thirteenth most abundant element in the earth's crust and is not found in a free state in nature. The three most commonly used fluoride compounds in the water system are hydrofluosilicic acid (H_2SiF_6), sodium fluoride (NaF), and sodium silicofluoride (Na_2SiF_6). Fluoride chemicals, like chlorine, caustic soda, and many other chemicals used in water treatment can cause a safety hazard. The operators should observe the safe handling of the chemical.

The three commonly used fluoride chemicals have virtually 100 percent dissociation (Reeves, 1986, 1990):

$$NaF \leftrightarrow Na^+ + F^- \tag{6.186}$$

$$Na_2SiF_6 \leftrightarrow 2Na^+ + SiF_6^{2-} \tag{6.187}$$

The SiF_6 radical will be dissociated in two ways: hydrolysis mostly dissociates very slowly

$$SiF_6^{2-} + 2H_2O \leftrightarrow 4H^+ + 6F^- + SiO_2 \downarrow \tag{6.188}$$

and/or

$$SiF_6^{2-} \leftrightarrow 2F^- + SiF_4 \uparrow \tag{6.189}$$

Silicon tetrafluoride (SiO_4) is a gas which will easily volatilize out of water when present in high concentration. It also reacts quickly with water to form silicic acid and silica (SiO_2):

$$SiF_4 + 3H_2O \leftrightarrow 4HF + H_2SiO_3 \tag{6.190}$$

and

$$SiF_4 + 2H_2O \leftrightarrow 4HF + SiO_2 \downarrow \tag{6.191}$$

then

$$HF \leftrightarrow H^+ + F^- \tag{6.192}$$

Hydrofluosilicic acid has a dissociation very similar to sodium silicofluoride:

$$H_2SiF_6 \leftrightarrow 2HF + SiF_4 \uparrow \tag{6.193}$$

then it follows Eqs. (6.190), (6.191), and (6.192).

Hydrofluoric acid (HF) is very volatile and will attack glass and electrical parts and will tend to evaporate in high concentrations.

19.2 Optimal Fluoride Concentrations

The manual (Reeves, 1986) presents the recommended optimal fluoride levels for fluoridated water supply systems based on the average and the maximum daily air temperature in the area of the involved school and community. For community water system in the United States the recommended fluoride levels are from 0.7 mg/L in the south to 1.2 mg/L to the north of the country. The recommended fluoride concentrations for schools are from 4 to 5 times greater than that for communities.

The optimal fluoride concentration can be calculated by the following equation:

$$F(\text{mg/L}) = 0.34/E \tag{6.194}$$

where E is the estimated average daily water consumption for children through age 10, in ounces of water per pound of body weight. The value of E can be computed from

$$E = 0.038 + 0.0062\,T \tag{6.195}$$

where T is the annual average of maximum daily air temperature in degrees Fahrenheit.

EXAMPLE 1: The annual average of maximum daily air temperature of a city is 60.5°F (15.8°C). What should be the recommended optimal fluoride concentration?

Solution:

Step 1. Determine E by Eq. (6.195)

$$E = 0.038 + 0.0062\,T = 0.038 + 0.0062 \times 60.5$$
$$= 0.413$$

Step 2. Calculate recommended F by Eq. (6.194)

$$F = 0.34/E = 0.34/0.413$$
$$= 0.82\ \text{(mg/L)}$$

EXAMPLE 2: Calculate the dosage of fluoride to the water plant in Example 1, if the naturally occurring fluoride concentration is 0.03 mg/L.

Solution:

The dosage can be obtained by subtracting the natural fluoride level in water from the desired concentration:

$$\text{Dosage} = 0.82\ \text{mg/L} - 0.03\ \text{mg/L}$$
$$= 0.79\ \text{mg/L}$$

As for other chemicals used in water treatment, the fluoride chemicals are not 100 percent pure. The purity of a chemical is available from the manufacturers. The information on the molecular weight, purity (p), and the available fluoride ion (AFI) concentration for the three commonly used fluoride chemicals are listed in Table 6.10. The AFI is determined by the weight of fluoride portion divided by the molecular weight.

EXAMPLE 3: What is the percent available fluoride in the commercial hydrofluosilicic acid (purity, $p = 23\%$)?

Solution:

From Table 6.10

$$\%\ \text{available}\ F = p \times \text{AFI} \tag{6.196}$$
$$= 23 \times 0.792$$
$$= 18.2$$

TABLE 6.10 Purity and Available Fluoride Ion (AFI) Concentrations for the Three Commonly Used Chemicals for Water Fluoridation

Chemical	Molecular weight	Purity, %	AFI
Sodium fluoride, NaF	42	98	0.452
Sodium silicofluoride, Na_2SiF_6	188	98.5	0.606
Hydrofluosilicic acid, H_2SiF_6	144	23	0.792

19.3 Fluoride Feed Rate (Dry)

Fluoride can be fed into treated water by (1) dry feeders with a mixing tank (for Na_2SiF_6), (2) direct solution feeder for H_2SiF_6, (3) saturated solution of NF, and (4) unsaturated solution of NF or Na_2SiF_6. The feed rate, FR, can be calculated as follows:

$$\text{FR (lb/d)} = \frac{\text{dosage (mg/L)} \times \text{plant flow (MGD)} \times 8.34 \text{ lb/gal}}{p \times \text{AFI}} \tag{6.197a}$$

or

$$\text{FR (lb/min)} = \frac{\text{(mg/L)} \times \text{gpm} \times 8.34 \text{ lb/gal}}{1{,}000{,}000 \times p \times \text{AFI}} \tag{6.197b}$$

EXAMPLE 1: Calculate the fluoride feed rate in lb/d and g/min, if 1.0 mg/L of fluoride is required using sodium silicofluoride in a 10 MGD (6944 gpm, 0.438 m^3/s) water plant assuming zero natural fluoride content.

Solution:

Using Eq. (6.196) and data from Table 6.10, feed rate FR is

$$\text{FR} = \frac{1.0 \text{ mg/L} \times 10 \text{ MGD} \times 8.34 \text{ lb/gal}}{0.985 \times 0.606}$$

$$= 139.7 \text{ lb/d}$$

or

$$= 139.7 \text{ lb/d} \times 453.6 \text{ g/lb} \times (1 \text{ d}/1440 \text{ min})$$

$$= 44.0 \text{ g/min}$$

EXAMPLE 2: A water plant has a flow of 1600 gpm (0.1 m^3/s). The fluoride concentration of the finished water requires 0.9 mg/L using sodium fluoride in a dry feeder. Determine the feed rate if 0.1 mg/L natural fluoride is in the water.

Solution:

Using Eq. (6.197), and data from Table 6.10

$$\text{FR} = \frac{(0.9 - 0.1) \text{ mg/L} \times 1600 \text{ gpm} \times 8.34 \text{ lb/gal}}{1{,}000{,}000 \times 0.98 \times 0.452}$$

$$= 0.024 \text{ lb/min}$$

or

$$= 10.9 \text{ g/min}$$

EXAMPLE 3: If a water plant treats 3500 gpm (5 MGD, 0.219 m^2/s) of water with a natural fluoride level of 0.1 mg/L. Find the hydrofluosilicic acid feed rate (mL/min) with the desired fluoride level of 1.1 mg/L.

Solution:

Step 1. Calculate FR in gal/min

$$\text{FR} = \frac{(1.1 - 0.1) \text{ mg/L} \times 3500 \text{ gpm}}{1{,}000{,}000 \times 0.23 \times 0.792}$$

$$= 0.0192 \text{ gpm}$$

Step 2. Convert gpm into mL/min

$$\begin{aligned} \text{FR} &= 0.0192 \text{ gpm} \\ &= 0.0192 \text{ gal/min} \times 3785 \text{ mL/gal} \\ &= 72.7 \text{ mL/min} \end{aligned}$$

19.4 Fluoride Feed Rate for Saturator

In practice 40 g of sodium fluoride will dissolve in 1 L of water. It gives 40,000 mg/L (4% solution) of NF. The AFI is 0.45. Thus the concentration of fluoride in the saturator is 18,000 mg/L (40,000 mg/L × 0.45). The sodium fluoride feed rate can be calculated as

$$\text{FR (gpm or L/min)} = \frac{\text{dosage (mg/L)} \times \text{ plant flow (gpm or L/min)}}{18{,}000 \text{ mg/L}} \tag{6.198}$$

EXAMPLE 1: In a 1 MGD plant with 0.2 mg/L natural fluoride, what would the fluoride (NF saturator) feed rate be to maintain 1.2 mg/L of fluoride in the water?

Solution:

By Eq. (6.198)

$$\begin{aligned} \text{FR} &= \frac{(1.2 - 0.2) \text{ mg/L} \times 1{,}000{,}000 \text{ gpd}}{18{,}000 \text{ mg/L}} \\ &= 55.6 \text{ gpd} \end{aligned}$$

Note: Approximately 56 gal of saturated NF solution to treat 1 Mgal of water at 1.0 mg/L dosage.

EXAMPLE 2: Convert the NF feed rate in Example 1 to mL/min.

Solution:

$$\begin{aligned} \text{FR} &= 55.6 \text{ gal/d} \times 3785 \text{ mL/gal} \times (1 \text{ d}/1440 \text{ min}) \\ &= 146 \text{ mL/min} \end{aligned}$$

or

$$= 2.5 \text{ mL/s}$$

Note: Approximately 2.5 mL/s (or 150 mL/min) of saturated sodium fluoride solution to 1 MGD flow to obtain 1.0 mg/L of fluoride concentration.

19.5 Fluoride Dosage

Fluoride dosage can be calculated from Eqs. (6.196) and (6.198) by rearranging those equations as follows:

$$\text{Dosage (mg/L)} = \frac{\text{FR (lb/d)} \times p \times \text{AFI}}{\text{plant flow (MGD)} \times 8.34 \text{ lb/(Mgal} \cdot \text{mg/L)}} \tag{6.199}$$

and for the saturator:

$$\text{Dosage (mg/L)} = \frac{\text{FR (gpm)} \times 18{,}000 \text{ mg/L}}{\text{plant flow (gpm)}} \tag{6.200}$$

For a solution concentration, C, of unsaturated NF solution

$$C = \frac{18{,}000 \text{ mg/L} \times \text{solution strength}(\%)}{4\%} \tag{6.201}$$

EXAMPLE 1: A water plant (1 MGD) feeds 22 pounds per day of sodium fluoride into the water. Calculate the fluoride dosage.

Solution:

Using Eq. (6.199) with Table 6.10:

$$\text{Dosage} = \frac{22 \text{ lb/d} \times 0.98 \times 0.452}{1 \text{ MGD} \times 8.34 \text{ lb/(Mgal} \cdot \text{mg/L)}}$$

$$= 1.17 \text{ mg/L}$$

EXAMPLE 2: A water plant adds 5 gallons of sodium fluoride from its saturator to treat 100,000 gallons of water. Find the dosage of the solution.

Solution:

Using Eq. (6.200)

$$\text{Dosage} = \frac{5 \text{ gal} \times 18{,}000 \text{ mg/L}}{100{,}000 \text{ gal}}$$

$$= 0.9 \text{ mg/L}$$

EXAMPLE 3: A water system uses 400 gpd of a 2.8 percent solution of sodium fluoride in treating 500,000 gpd of water. Calculate the fluoride dosage.

Solution:

Step 1. Find fluoride concentration in NF solution
Using Eq. (6.201)

$$C = \frac{18{,}000 \text{ mg/L} \times 2.8\%}{4\%}$$

$$= 12{,}600 \text{ mg/L}$$

Step 2. Calculate fluoride dosage by Eq. (6.200)

$$\text{Dosage} = \frac{400 \text{ gpd} \times 12{,}600 \text{ mg/L}}{500{,}000 \text{ gpd}}$$

$$= 10.08 \text{ mg/L}$$

EXAMPLE 4: A water system adds 6.5 pounds per day of sodium silicofluoride to fluoridate 0.5 MGD of water. What is the fluoride dosage?

Solution:

Using Eq. (6.199), $p = 0.985$, and AFI $= 0.606$

$$\text{Dosage} = \frac{6.5 \text{ lb/d} \times 0.985 \times 0.606}{0.5 \text{ MGD} \times 8.34 \text{ lb/(Mgal} \cdot \text{mg/L)}}$$

$$= 0.93 \text{ mg/L}$$

EXAMPLE 5: A water plant uses 2.0 pounds of sodium silicofluoride to fluoridate 160,000 gal water. What is the fluoride dosage?

Solution:

$$\text{Dosage} = \frac{2.0 \text{ lb} \times 0.985 \times 0.606 \times 10^6 \text{ mg/L}}{160{,}000 \text{ gal} \times 8.34 \text{ lb/(Mgal} \cdot \text{mg/L)}}$$

$$= 0.9 \text{ mg/L}$$

EXAMPLE 6: A water plant feeds 100 pounds of hydrofluosilicic acid of 23% purity during 5 days to fluoridate 0.437 MGD of water. Calculate the fluoride dosage.

Solution:

$$\text{FR} = 100 \text{ lb/5 days} = 20 \text{ lb/d}$$

$$\text{Dosage} = \frac{\text{FR} \times p \times \text{AFI}}{\text{(MGD)} \times 8.34 \text{ lb/(Mgal} \cdot \text{mg/L)}}$$

$$= \frac{20 \text{ lb/d} \times 0.23 \times 0.792}{0.437 \text{ MGD} \times 8.34 \text{ lb/(Mgal} \cdot \text{mg/L)}} \quad \text{(from Table 6.8, AFI} = 0.792)$$

$$= 1.0 \text{ mg/L}$$

EXAMPLE 7: A water system uses 1200 pounds of 25% hydrofluosilicic acid in treating 26 Mgal. The natural fluoride level in the water is 0.1 mg/L. What is the final fluoride concentration in the finished water.

Solution:

$$\text{Plant dosage} = \frac{1200 \text{ lb} \times 0.25 \times 0.792}{26 \text{ Mgal} \times 8.34 \text{ lb/(Mgal} \cdot \text{mg/L)}}$$

$$= 1.1 \text{ mg/L}$$

$$\text{Final fluoride} = (1.1 + 0.1) \text{ mg/L}$$

$$= 1.2 \text{ mg/L}$$

REFERENCES

Allied Signal. 1970. *Fluid systems, Reverse osmosis principals and applications*. San Diego, California: Allied Signal.

American Dental Association. 1980. *Fluoridation facts*. G21 Chicago, Illinois: American Dental Association.

American Public Health Association, American Water Works Association, Water Environment Federation. 1998. *Standard methods for the examination of water and wastewater*. 20th edn.

American Society of Civil Engineers (ASCE) and American Water Works Association (AWWA). 1990. *Water treatment plant design*. 2nd edn. New York: McGraw-Hill.

American Water Works Association (AWWA). 1978. *Corrosion control by deposition of $CaCO_3$ films*. Denver, Colorado: AWWA.

American Water Works Association and American Society of Civil Engineers. 1998. *Water treatment plant design,* 3rd edn. New York: McGraw-Hill.

Amirtharajah, A. and O'Melia, C. R. 1990. Coagulation processes: Destabilization, mixing, and flocculation. In: AWWA, *Water quality and treatment*. New York: McGraw-Hill.

Anderson, R. E. 1975. Estimation of ion exchange process limits by selectivity calculations. In: AICHE, *Symposium Series* **71**: 152, 236.

Applegate, L. 1984. Membrane separation processes. *Chem. Eng*. June 11, 1984, 64–69.

Benefield, L. D. and Morgan, J. S. 1990. Chemical precipitation. In: AWWA, *Water quality and treatment*. New York: McGraw-Hill.

Benefield, L. D., Judkins, J. F. and Weand, B. L. 1982. *Process chemistry for water and wastewater treatment*. Englewood Cliffs, New Jersey: Prentice-Hall.

Brown, T. L. and LeMay, H. E. Jr. 1981. *Chemistry: the central science,* 2nd edn., Englewood Cliffs, New Jersey: Prentice-Hall.

Calderbrook, P. H. and Moo Young, M. B. 1961. The continuous phase heat and mass transfer properties of dispersions. *Chem. Eng. Sci*. **16**: 39.

Camp, T. R. 1968. Floc volume concentration. *J. Amer. Water Works Assoc*. **60**(6): 656–673.

Camp, T. R. and Stein, P. C. 1943. Velocity gradients and internal work in fluid motion. *J. Boston Soc. Civil Engineers* **30**: 219.

Chellam, S., Jacongelo, J. G., Bonacquisti, T. P. and Schauer, B. A. 1997. Effect of pretreatment on surface water nanofiltration. *J. Amer. Water Works Assoc*. **89**(10): 77–89.

Clark, J. W. and Viessman, W. Jr. 1966. *Water supply and pollution control*. Scranton, Pennsylvania: International Textbook Co.

Clark, J. W., Viessman, W. Jr. and Hammer, M. J. 1977. *Water supply and pollution control*. New York: IEP-A Dun-Donnelley.

Cleasby, J. L. 1990. Filtration. In: AWWA, *Water quality and treatment*. New York: McGraw-Hill.

Cleasby, J. L. and Fan, K. S. 1981. Predicting fluidization and expansion of filter media. *J. Environ. Eng. Div. ASCE* **107**(EE3): 355–471.

Clifford, D. A. 1990. Ion exchange and inorganic adsorption: In: AWWA, *Water quality and treatment*, New York: McGraw-Hill.

Conlon, W. J. 1990. Membrane processes. In: AWWA, *Water quality and treatment*, New York: McGraw-Hill.

Cornwell, D. A. 1990. Airstripping and aeration. In: AWWA, *Water quality and treatment*, New York: McGraw-Hill.

Cotruvo, J. A. and Vogt, C. D. 1990. Rationale for water quality standards and goals. In: AWWA, *Water quality and treatment*. New York: McGraw-Hill.

Dean, J. A. 1979. *Lange's handbook of chemistry*. 12th edn. New York: McGraw-Hill.

Diamond Shamrock Company. 1978. Duolite A104-Data leaflet. Functional Polymers Division, DSC., Cleveland, Ohio.

Downs, A. J. and Adams, C. J. 1973. *The chemistry of chlorine, bromine, iodine and astatine*. Oxford: Pergamon.

Doyle, M. L. and Boyle, W. C. 1986. Translation of clean to dirty water oxygen transfer rates. In: Boyle, W. C. (ed.) *Aeration systems, design, testing, operation, and control*. Park Ridge, New Jersey: Noyes.

Fair, G. M., Geyer, J. C. and Morris, J. C. 1963. *Water supply and waste-water disposal*. New York: John Wiley.

Fair, G. M., Geyer, J. C. and Okun, D. A. 1966. *Water and wastewater engineering*, Vol. 1: Water supply and wastewater removal. New York: John Wiley.

Fair, G. M., Geyer, J. C. and Okun, D. A. 1968. *Water and wastewater engineering*, Vol. 2: Water purification and wastewater treatment and disposal. New York: John Wiley.

Federal Register. 1997. Environmental Protection Agency 40 CFR, Part 141 and 142 National Primary Drinking Water Regulation: Interim Enhanced Surface Water Treatment Rule Notice of Data Availability; Proposed Rule, 62 (212): 59521–59540, Nov. 3, 1997.

Ficek, K. J. 1980. Potassium permanganate for iron and manganese removal. In: Sanks, R. L. (ed.), *Water treatment plant design*. Ann Arbor, Michigan: Ann Arbor Science.

Gregory, R. and Zabel, T. F. 1990. Sedimentation and flotation. In AWWA: *Water quality and treatment*. New York: McGraw-Hill.

Haas, C. N. 1990. Disinfection. In: AWWA, *Water quality and treatment*. New York: McGraw-Hill.

Hammer, M. J. 1975. *Water and waste-water technology*. New York: John Wiley.

Hom, L. W. 1972. Kinetics of chlorine disinfection in an eco-system. *J. Sanitary Eng. Div., Pro. Amer. Soc. Civil Eng*. **98**(1): 183.

Hudson, H. E., Jr. 1972. Density considerations in sedimentation. *J. Amer. Water Works Assoc.* **64**(6): 382−386.

Humphrey, S. D. and Eikleberry, M. A. 1962. Iron and manganese removal using $KMnO_4$. *Water and Sewage Works*, **109**(R. N.): R142−R144.

Illinois Environmental Protection Agency (IEPA). 1992. *An example analysis of CT value*. Springfield, Illinois: IEPA.

Illinois Pollution Control Board. 1990. Rules and regulations. Title 35: Environmental Protection, Subtitle C: Water Pollution, Chapter I, Springfield, Illinois: IEPA.

International Fire Service Training Association. 1993. *Water supplies for fire protection*, 4th edn. Stillwater, Oklahoma: Oklahoma State University.

Inversand Company. 1987. *Manganese greensand.* Clayton, New Jersey: Inversand Company.

Ionic, Inc. 1993. *Ionic public water supply systems.* Bulletin no. 141. Watertown, Massachusetts.

James M. Montgomery Consulting Engineering, Inc. 1985. *Water treatment principles and design.* New York: John Wiley.

Jobin, R. and Ghosh, M. M. 1972. Effect of buffer intensity and organic matter on oxygenation of ferrous iron. *J. Amer. Water Works Assoc.* **64**(9): 590−595.

Jones, B. H., Mersereau-Kempf, J. and Shreve, D. 2000. In: WEF Multiphase multimethod remediation. *Industrial Wastewater* **8**(4): 21–25.

Kreuz, D. F. 1962. Iron oxidation with H_2O_2 in secondary oil recovery produce plant, FMC Corp. Internal report, 4079-R, Philadelphia, PA.

Lide, D. R. 1996. *CRC Handbook of chemistry and physics*, 77th edn. Boca Raton, Florida: CRC Press.

Lin, S. D. and Green, C. D. 1987. *Waste from water treatment plants: Literature plants: Literature review, results of an Illinois survey and effects of alum sludge application to cropland.* ILENR/RE-WR-87/18. Springfield, Illinois: Illinois Department of Energy and Natural Resources.

Mattee-Müller, C., Gujer, W. and Giger, W. 1981. Transfer of volatile substances from water to the atmosphere. *Water Res.* **5**(15): 1271−1279.

McKee, J. E., Brokaw, C. J. and McLaughlin, R. T. 1960. Chemical and colicidal effects of halogens in sewage. *J. Water Pollut. Control Fed.* **32** (8): 795−819.

Morris, J. C. 1966. The acid ionization constant of HOCl from 5°C to 35°C. *J. Phys. Chem.* **70**(12): 3789.

National Board Fire Underwriters (NBFU). 1974. *Standard schedule for grading cities and towns of the United States with reference to their fire defenses and physical conditions.* New York: NBFU, American Insurance Association.

Peroxidation System, Inc. 1990. *Perox-pure-process description.* Tucson, Arizona: Peroxidation System Inc.

Pontius, F. W. 1993. Configuration, operation of system affects C × T value. *Opflow* **19**(8): 7−8.

Reeves, T. G. 1986. *Water fluoridation—A manual for engineers and technicians.* US Public Health Service, 00-4789. Atlanta, Georgia: US PHS.

Reeves, T. G. 1990. Water fluoridation. In: AWWA, *Water quality and treatment.* New York: McGraw-Hill.

Rehfeldt, K. R., Raman, R. K., Lin, S. D. and Broms, R. E. 1992. *Asessment of the proposed discharge of ground water to surface waters of the American Bottoms area of Southwestern Illinois.* Contract Report 539. Champaign, Illinois. Illinois State Water Survey.

Rice, R. G. 1986. Instruments for analysis of ozone in air and water. In: Rice, R. G., Bolly, L. J. and Lacy, W. J. (ed) *Analytic aspects of ozone treatment of water and wastewater.* Chelsea, Michigan: Lewis.

Rose, H. E., 1951. On the resistance coefficient–Reynolds number relationship for fluid flow through a bed of granular material. *Proc. Inst. Mech. Engineers (London)*, 154−160.

Schroeder, E. D. 1977. *Water and wastewater treatment.* New York: McGraw-Hill.

Snoeyink, V. L. 1990. Adsorption of organic compounds. In: AWWA, *Water quality and treatment.* New York: McGraw-Hill.

Snoeyink, V. L. and Jenkins, D. 1980. *Water chemistry.* New York: John Wiley.

Steel, E. W. and McGhee, T. J. 1979. *Water supply and sewerage*, 5th edn. New York: McGraw-Hill.

Tchobanoglous, G. and Schroeder, E. D. 1985. *Water quality.* Reading, Massachusetts: Addison-Wesley.

Teefy, S. M. and Singer, P. C. 1990. Performance of a disinfection scheme with the SWTR. *J. Amer. Water Works Assoc.* **82**(12): 88−98.

Treybal, R. D. 1968. *Mass-transfer operations*. New York: McGraw-Hill.

US Environmental Protection Agency (USEPA). 1976. Translation of reports on special problems of water technology, vol. 9—Adsorption. EPA-600/9-76-030, Cincinnati, Ohio.

US Environmental Protection Agency (USEPA). 1983. Nitrate removal for small public water systems. EPA 570/9-83-009. Washington, D.C.

US Environmental Protection Agency (USEPA). 1986. *Design manual: Municipal wastewater disinfection*. EPA/625/1-86/021. Cincinnati, Ohio: USEPA.

US Environmental Protection Agency (USEPA). 1987. The safe drinking water act–Program summary. Region 5, Washington D.C.

US Environmental Protection Agency (USEPA). 1989a. Guidance manual for compliance with the filtration and disinfection requirements for public water systems using surface water sources. USEPA, Oct. 1989 ed. Washington, D.C.

US Environmental Protection Agency (USEPA). 1989b. National primary drinking water regulations: Filtration and disinfection; turbidity, *Giardia lamblia*, viruses *Legionella*, and heterotrophic bacteria: Proposed Rules, Fed. Reg., 54 FR:124:27486.

US Environmental Protection Agency (USEPA). 1990. Environmental pollution control alternatives: Drinking water treatment for small communities. EPA/625/5-90/025, Cincinnati, Ohio.

US Environmental Protection Agency (USEPA). 1996. Capsule Report—Reverse osmosis process. EPA/625/R-96/009.

US Public Health Service (US PHS). 1984. *Fluoridation engineering manual*. Atlanta, Georgia: US Department of Health and Human Services.

US Public Health Service. 1982. Homestudy course 3017-G, water fluoridation. Atlanta, GA. Center of Disease Control, US PHS.

Viessman, W. Jr. and Hammer, M. J. 1993. *Water supply and pollution control*, 5th edn. New York: Harper Collins.

Voelker, D. C. 1984. Quality of water in the alluvial aquifer, American Bottoms, East St Louis, Illinois. Water Resources Investigation Report 84-4180. Chicago: US Geological Survey.

Water, G. C. 1978. *Disinfection of wastewater and water for reuse*. New York: Van Nostrand Reinhold.

Wen, C. Y. and Yu, Y. H. 1966. Mechanics of fluidization. In: Chem., Eng. Progress Symposium Series, **62**: 100–111.

White, G. C. 1972. *Handbook of chlorination*. New York: Litton Educational.

Wilmarth, W. A. 1968. Removal of iron, manganese, and sulfides. *Water and Wast Eng.* **5**(8): 52−4.

Wong, J. M. 1984. Chlorination-filtration for iron and manganese removal. *J. Amer. Water Works Assoc.* **76**(1): 76−79.

Yao, K. M. 1976. Theoretical study of high-rate sedimentation. *J. Water Pollut. Control Fed.* **42**(2): 218−28.

CHAPTER 1.7
WASTEWATER ENGINEERING

Shun Dar Lin

1 WHAT IS WASTEWATER?

"Wastewater," also known as "sewage," originates from household wastes, human and animal wastes, industrial wastewaters, storm runoff, and groundwater infiltration. Wastewater, basically, is the flow of used water from a community. It is 99.94 percent water by weight (Water Pollution Control Federation 1980). The remaining 0.06 percent is material dissolved or suspended in the water. It is largely the water supply of a community after it has been fouled by various uses.

2 CHARACTERISTICS OF WASTEWATER

An understanding of physical, chemical, and biological characteristics of wastewater is very important in design, operation, and management of collection, treatment, and disposal of wastewater. The nature of wastewater includes physical, chemical, and biological characteristics which depend on the water usage in the community, the industrial and commercial contributions, weather, and infiltration/inflow.

2.1 Physical Properties of Wastewater

When fresh, wastewater is gray in color and has a musty and not unpleasant odor. The color gradually changes with time from gray to black. Foul and unpleasant odors may then develop as a result of septic sewage. The most important physical characteristics of wastewater are its temperature and its solids concentration.

Temperature and solids content in wastewater are very important factors for wastewater treatment processes. Temperature affects chemical reaction and biological activities. Solids, such as total suspended solids (TSS), volatile suspended solids (VSS), and settleable solids, affect the operation and sizing of treatment units.

Solids. Solids comprise matter suspended or dissolved in water and wastewater. Solids are divided into several different fractions and their concentrations provide useful information for characterization of wastewater and control of treatment processes.

Total solids. Total solids (TS) is the sum of total suspended solids (TSS) and total dissolved solids (TDS). Each of these groups can be further divided into volatile and fixed fractions. Total solids is the material left in the evaporation dish after it has dried for at least one hour (h) or overnight (preferably) in an oven at 103°C to 105°C and is calculated according to *Standard Methods* (APHA *et al.* 1995)

$$\text{mg TS/L} = \frac{(A - B) \times 1000}{\text{sample volume, mL}} \tag{7.1}$$

where A = weight of dried residue plus dish, mg
B = weight of dish, mg
1000 = conversion of 1000 mL/L

Note: Sample can be natural waters, wastewaters, even treated waters.

Total suspended solids. Total suspended solids are referred to as nonfilterable residue. The TSS is a very important quality parameter for water and wastewater and is a wastewater treatment effluent standard. The TSS standards for primary and secondary effluents are usually set at 30 and 12 mg/L, respectively. TSS is determined by filtering a well-mixed sample through a 0.2 μm pore size, 24 mm diameter membrane; the membrane filter is placed in a Gooch crucible, and the residue retained on the filter is dried in an oven for at least one hour at a constant weight at 103°C to 105°C. It is calculated as

$$\text{mg TSS/L} = \frac{(C - D) \times 1000}{\text{sample volume, mL}} \tag{7.2}$$

where C = weight of filter and crucible plus dried residue, mg
D = weight of filter and crucible, mg

Total dissolved solids. Dissolved solids are also called filterable residues. Total dissolved solids in raw wastewater are in the range of 250°C to 850 mg/L.

TDS is determined as follows. A well-mixed sample is filtered through a standard glass fiber filter of 2.0 μm normal pore size, and the filtrate is evaporated for at least 1 hour in an oven at 180 ± 2°C. The increase in dish weight represents the total dissolved solids, which is calculated as

$$\text{mg/TDS/L} = \frac{(E - F) \times 1000}{\text{sample volume, mL}} \tag{7.3}$$

where E = weight of dried residue plus dish, mg
F = weight of dish, mg

Fixed and volatile solids. The residue from TS, TSS, or TDS tests is ignited to constant weight at 550°C. The weight lost on ignition is called volatile solids, whereas the remaining solids represent the fixed total, suspended, or dissolved solids. The portions of volatile and fixed solids are computed by

$$\text{mg volatile solids/L} = \frac{(G - H) \times 1000}{\text{sample volume, mL}} \tag{7.4}$$

$$\text{mg fixed solids/L} = \frac{(H - I) \times 1000}{\text{sample volume, mL}} \tag{7.5}$$

where G = weight of residue plus crucible before ignition, mg
H = weight of residue plus crucible or filter after ignition, mg
I = weight of dish or filter, mg

The determination of the volatile portion of solids is useful in controlling wastewater treatment plant operations because it gives a rough estimation of the amount of organic matter present in the solid fraction of wastewater, activated sludge, and in industrial waste.

Determination of volatile and fixed solids does not distinguish precisely between organic and inorganic matter. Because the loss on ignition is not confined only to organic matter, it includes losses due to decomposition or volatilization of some mineral salts. The determination of organic matter can be made by tests for biochemical oxygen demand (BOD), chemical oxygen demand (COD), and total organic carbon (TOC).

Settleable solids. Settleable solids is the term applied to material settling out of suspension within a defined time. It may include floating material, depending on the technique. Settled solids may be expressed on either a volume (mL/L) or a weight (mg/L) basis.

The volumetric method for determining settleable solids is as follows. Fill an Imhoff cone to the one-liter mark with a well-mixed sample. Settle for 45 minutes, gently agitate the sample near the sides of the Imhoff cone with a rod or by spinning, then continue to settle for an additional 15 minutes and record the volume of settleable solids in the cones as mL/L.

Another test to determine settleable solids is the gravimetric method. First, determine total suspended solids as stated above. Second, determine nonsettleable suspended solids from the supernatant of the same sample which has settled for one hour. Then determine TSS (mg/L) of this supernatant liquor; this gives the nonsettleable solids. The settleable solids can be calculated as

$$\text{mg settleable solids/L} = \text{mg TSS/L} - \text{mg nonsettleable solids/L} \tag{7.6}$$

EXAMPLE: A well-mixed 25-mL of raw wastewater is used for TS analyses. A well-mixed 50-mL of raw wastewater is used for suspended solids analyses. Weights (wt.) of evaporating dish with and without the sample either dried, evaporated, or ignited were determined to constant weight according to *Standard Methods* (APHA *et al.* 1998). The laboratory results are

Tare wt. of evaporating dish = 42.4723 g

Wt. of dish plus residue after evaporation at 105°C = 42.4986 g

Wt. of dish plus residue after ignition at 550°C = 42.4863 g

Tare wt. of filter plus Gooch crucible = 21.5308 g

Wt. of residue and filter plus crucible after drying at 105°C = 21.5447 g

Wt. of residue and filter plus crucible after ignition at 550°C = 21.5349 g

Compute the concentrations of total solids, volatile solids, fixed solids, total suspended solids, volatile suspended solids, and fixed suspended solids.

Solution:

Step 1. Determine total solids by Eq. (7.1)

$$\text{Sample size} = 25 \text{ mL}$$

$$A = 42{,}498.6 \text{ mg}$$

$$B = 42{,}472.3 \text{ mg}$$

$$\begin{aligned}\text{TS} &= \frac{(A - B)\text{ mg} \times 1000 \text{ mL/L}}{25 \text{ mL}}\\ &= (42{,}498.6 - 42{,}472.3) \times 40 \text{ mg/L}\\ &= 1052 \text{ mg/L}\end{aligned}$$

Step 2. Determine total volatile solids by Eq. (7.4)

$$G = 42{,}498.6 \text{ mg}$$
$$H = 42{,}486.3 \text{ mg}$$
$$\text{VS} = (42{,}498.6 - 42{,}486.3) \times 1000/25$$
$$= 492 \text{ mg/L}$$

Step 3. Determine total fixed solids

$$\text{FS} = \text{TS} - \text{VS} = (1052 - 492) \text{ mg/L}$$
$$= 560 \text{ mg/L}$$

Step 4. Determine total suspended solids by Eq. (7.2)

$$C = 21{,}544.7 \text{ mg}$$
$$D = 21{,}530.8 \text{ mg}$$
$$\text{Sample size} = 50 \text{ mL}$$
$$\text{TSS} = (C - D) \times 1000/50$$
$$= (21{,}544.7 - 21{,}530.8) \times 20$$
$$= 278 \text{ mg/L}$$

Step 5. Determine volatile suspended solids by Eq. (7.4)

$$G = 21{,}544.7 \text{ mg}$$
$$H = 21{,}534.9 \text{ mg}$$
$$\text{VSS} = (G - H) \times 1000/50$$
$$= (21{,}544.7 - 21{,}534.9) \times 20$$
$$= 196 \text{ mg/L}$$

Step 6. Determine fixed suspended solids by Eq. (7.5)

$$H = 21{,}534.9 \text{ mg}$$
$$I = 21{,}530.8 \text{ mg}$$
$$\text{FSS} = (H - I) \times 1000/50$$
$$= (21{,}534.9 - 21{,}530.8) \times 20$$
$$= 82 \text{ mg/L}$$

or

$$\text{FSS} = \text{TSS} - \text{VSS} = (278 - 196) \text{ mg/L}$$
$$= 82 \text{ mg/L}$$

2.2 Chemical Constituents of Wastewater

The dissolved and suspended solids in wastewater contain organic and inorganic material. Organic matter may include carbohydrates, fats, oils, grease, surfactants, proteins, pesticides and other agricultural chemicals, volatile organic compounds, and other toxic chemicals (household and industrial). Inorganics may include heavy metals, nutrients (nitrogen and phosphorus), pH, alkalinity, chlorides, sulfur, and other inorganic pollutants. Gases such as carbon dioxide, nitrogen, oxygen, hydrogen sulfide, and methane may be present in a wastewater.

Normal ranges of nitrogen levels in domestic raw wastewater are 25 to 85 mg/L for total nitrogen (the sum of ammonia, nitrate, nitrite, and organic nitrogen); 12 to 50 mg/L ammonia nitrogen; and 8 to 35 mg/L organic nitrogen (WEF 1996a). The organic nitrogen concentration is determined by a total kjeldahl nitrogen (TKN) analysis, which measures the sum of organic and ammonia nitrogen. Organic nitrogen is then calculated by subtracting ammonia nitrogen from the TKN measurement.

Typical total phosphorus concentrations of raw wastewater range from 2 to 20 mg/L, which includes 1 to 5 mg/L of organic phosphorus and 1 to 15 mg/L of inorganic phosphorus (WEF 1996a). Both nitrogen and phosphorus in wastewater serve as essential elements for biological growth and reproduction during wastewater treatment processes and in the natural water.

The strength (organic content) of a wastewater is usually measured as 5-days biochemical oxygen demand (BOD_5), chemical oxygen demand, and total organic carbon. The BOD_5 test measures the amount of oxygen required to oxidize the organic matter in the sample during 5 days of biological stabilization at 20°C. This is usually referred to as the first stage of carbonaceous BOD (CBOD), not nitrification (second phase). Secondary wastewater treatment plants are typically designed to remove CBOD, not for nitrogenous BOD (except for advanced treatment). The BOD_5 of raw domestic wastewater in the United States is normally between 100 and 250 mg/L. It is higher in other countries. In this chapter, the term BOD represents 5-days BOD, unless stated otherwise.

The ratio of carbon, nitrogen, and phosphorus in wastewater is very important for biological treatment processes, where there is normally a surplus of nutrients. The commonly accepted BOD/N/P weight ratio for biological treatment is 100/5/1; i.e. 100 mg/L BOD to 5 mg/L nitrogen to 1 mg/L phosphorus. The ratios for raw sanitary wastewater and settled (primary) effluent are 100/17/5 and 100/23/7, respectively.

EXAMPLE 1: Calculate the pounds of BOD_5 and TSS produced per capita per day. Assume average domestic wastewater flow is 100 gallons (378 L) per capita per day (gpcpd), containing BOD_5 and TSS concentrations in wastewater of 200 and 240 mg/L, respectively.

Solution:

Step 1. For BOD

$$\begin{aligned} \text{BOD} &= 200 \text{ mg/L} \times 1 \text{ L}/10^6 \text{ mg} \times 100 \text{ gal/(c} \cdot \text{d)} \times 8.34 \text{ lb/gal} \\ &= 0.17 \text{ lb/(c} \cdot \text{d)} \\ &= 77 \text{ g/(c} \cdot \text{d)} \end{aligned}$$

Step 2. For TSS

$$\begin{aligned} \text{TSS} &= 240 \text{ mg/L} \times 10^{-6} \text{ L/mg} \times 100 \text{ gal/(c} \cdot \text{d)} \times 8.34 \text{ lb/gal} \\ &= 0.20 \text{ lb/(c} \cdot \text{d)} \\ &= 90 \text{ g/(c} \cdot \text{d)} \end{aligned}$$

Note: 100 gal/(c · d) of flow and 0.17 lb BOD_5 per capita per day are commonly used for calculation of the population equivalent of hydraulic and BOD loadings for other wastewaters.

In newly developed communities, generally designed wastewater flow is 120 gpcpd. With the same concentrations, the loading rate for design purpose would be

$$\begin{aligned} \text{BOD} &= 0.20 \text{ lb/(c} \cdot \text{d)} \\ &= 90 \text{ g/(c} \cdot \text{d)} \\ \text{TSS} &= 0.24 \text{ lb/(c} \cdot \text{d)} \\ &= 110 \text{ g/(c} \cdot \text{d)} \end{aligned}$$

EXAMPLE 2: An example of industrial waste has an average flow of 1,230,000 gpd (4656 m^3/d) with BOD_5 of 9850 lb/d (4468 kg/d). Calculate BOD concentration and the equivalent populations of hydraulic and BOD loadings.

Solution:

Step 1. Determine BOD concentration

$$Q = 1.23 \text{ MGD}$$

$$\text{BOD} = \frac{9850 \text{ lb/d} \times 1 \text{ M}}{1.23 \text{ Mgal/d} \times 8.34 \text{ lb/(Mgal} \cdot \text{mg/L)}}$$

$$= 960 \text{ mg/L}$$

Note: M = million

Step 2. Calculate equivalent populations, E. P.

$$\text{BOD E. P.} = \frac{9850 \text{ lb/d}}{0.17 \text{ lb/c} \cdot \text{d}}$$

$$= 57{,}940 \text{ persons}$$

$$\text{Hydraulic E. P.} = \frac{1{,}230{,}000 \text{ gpd}}{100 \text{ g/c} \cdot \text{d}}$$

$$= 12{,}300 \text{ persons}$$

For the SI units calculations:

Step 3. Compute BOD concentration

$$\text{BOD} = \frac{4468 \text{ kg/d} \times 10^6 \text{ mg/kg}}{4656 \text{ m}^3\text{/d} \times 10^3 \text{ L/m}^3}$$

$$= 960 \text{ mg/L}$$

Step 4. Compute E. P., refer to Example 1

$$\text{BOD E. P.} = \frac{4468 \text{ kg/d} \times 1000 \text{ g/kg}}{77 \text{ g/c} \cdot \text{d}}$$

$$\cong 58{,}000 \text{ capitas (persons)}$$

$$\text{Hydraulic E. P.} = \frac{4656 \text{ m}^3\text{/d} \times 1000 \text{ L/m}^3}{378.5\text{/c} \cdot \text{d}}$$

$$= 12{,}300 \text{ persons}$$

Chemical oxygen demand. Concept and measurement of BOD are fully covered in Chapter 1.2. Since the 5-day BOD test is time consuming, chemical oxygen demand is routinely performed for wastewater treatment operations after the relationship between BOD and COD has been developed for a wastewater treatment plant. Many regulatory agencies accept the COD test as a tool of the wastewater treatment operation. The total organic carbon test is also a means for defining the organic content in wastewater.

The chemical oxygen demand is a measurement of the oxygen equivalent of the organic matter content of a sample that is susceptible to oxidation by a strong chemical oxidant, such as potassium dichromate. For a sample, COD can be related empirically to BOD, organic carbon, or organic matter.

After the correlation of BOD_5 and COD has been established, the COD test is useful for controlling and monitoring wastewater treatment processes. The COD test takes 3 to 4 h rather than 5 days for BOD data. The COD results are typically higher than the BOD values. The correlation between COD and BOD varies from plant to plant. The BOD:COD ratio also varies across the plant from influent to process units to effluent. The ratio is typically 0.5:1 for raw wastewater and may drop to as low as 0.1 : 1 for well-stabilized secondary effluent. The normal COD range for raw wastewater is 200 to 600 mg/L (WEF 1996a).

COD test. Most types of organic matter in water or wastewater are oxidized by a boiling mixture of sulfuric and chromic acids with excess of potassium dichromate ($K_2Cr_2O_7$). After 2-h refluxing, the remaining unreduced $K_2Cr_2O_7$ is titrated with ferrous ammonium sulfate to measure the amount of $K_2Cr_2O_7$ consumed and the oxidizable organic matter is calculated in terms of oxygen equivalent. The COD test can be performed by either the open reflux or the closed reflux method. The open reflux method is suitable for a wide range of wastewaters, whereas the closed reflux method is more economical but requires homogenization of samples containing suspended solids to obtain reproducible results.

The COD test procedures (open reflux) are (APHA *et al.* 1998):

- Place appropriate size of sample in a 500-mL refluxing flask
  use 50 mL if COD $<$ 900 mg/L
  use less sample and dilute to 50 mL if COD $>$ 900 mg/L
  use large size of sample if COD $<$ 50 mL
- Add 1 g of $HgSO_4$ and several glass beads
- Slowly add 5 mL sulfuric acid reagent and mix to dissolve $HgSO_4$
- Cool the mixture
- Add 25.0 mL of 0.0417 mole $K_2Cr_2O_7$ solution and mix
- Attach the refluxing flask to condenser and turn on cooling water
- Add 70 mL sulfuric acid reagent through open end of condenser, then swirl and mix
- Cover open end of condenser with a small beaker
- Reflux for 2 h (usually 1 blank with several samples)
- Cool and wash down condenser with distilled water to about twice volume
- Cool to room temperature
- Add 2 to 3 drops ferroin indicator
- Titrate excess $K_2Cr_2O_7$ with standard ferrous ammonium sulfate (FAS)
- Take as the end point of the titration of the first sharp color change from blue-green to reddish brown
- Calculate COD result as (APHA, *et al.* 1998)

$$\text{COD as mg } O_2/\text{L} = \frac{(A - B) \times M \times 8000}{\text{mL sample}} \tag{7.7}$$

where A = mL FAS used for blank
B = mL FAS used for sample
M = molarity of FAS to be determined daily against standard $K_2Cr_2O_7$ solution $\cong$ 0.25 mole

EXAMPLE: The results of a COD test for raw wastewater (50 mL used) are given. Volumes of FAS used for blank and the sample are 24.53 and 12.88 mL, respectively. The molarity of FAS is 0.242. Calculate the COD concentration for the sample.

Solution: Using Eq. (7.7)

$$\text{COD as mg } O_2/\text{L} = \frac{(A - B) \times M \times 8000}{\text{mL sample}}$$

$$= (24.53 - 12.88) \times 0.242 \times 8000/50$$

$$= 451$$

2.3 Biological Characteristics of Wastewater

The principal groups of microorganisms found in wastewater are bacteria, fungi, protozoa, microscopic plants and animals, and viruses. Most microorganisms (bacteria, protozoa) are responsible and are beneficial for biological treatment processes of wastewater. However, some pathogenic bacteria, fungi, protozoa, and viruses found in wastewater are of public concern.

Indicator bacteria. Pathogenic organisms are usually excreted by humans from the gastrointestinal tract and discharged to wastewater. Water-borne diseases include cholera, typhoid, paratyphoid fever, diarrhea, and dysentery. The number of pathogenic organisms in wastewaters is generally low in density and they are difficult to isolate and identify. Therefore, indicator bacteria such as total coliform (TC), fecal coliform (FC), and fecal streptococcus (FS) are used as indicator organisms. The concept of indicator bacteria and enumeration of bacterial density are discussed in Chapter 1.2.

Tests for enumeration of TC, FC, and FS can be performed by multiple-tube fermentation (most probable number, MPN) or membrane filter methods; the test media used are different for these three groups of indicators. Most regulatory agencies have adopted fecal coliform density as an effluent standard because FC is mostly from fecal material.

3 SEWER SYSTEMS

Sewers are underground conduits to convey wastewater and stormwater to a treatment plant or to carry stormwater to the point of disposal. Sewers can be classified into three categories: sanitary, storm, and combined. Community sewer systems, according to their discharging types, can be divided into separated and combined sewer systems.

3.1 Separated Sewer System

Separated sewers consist of sanitary sewers and stormwater sewer networks separately. Sanitary sewers carry a mixture of household and commercial sewage, industrial wastewater, water from groundwater infiltration/inflow, basement and foundation drainage connections, and cross-connections between sanitary sewers and stormwater drainage. Separated sanitary sewers should be free of stormwater, but they seldom are.

Storm sewers are commonly buried pipes that convey storm drainage. They may include open channel elements and culverts, particularly when drainage areas are large.

Storm sewer networks convey mainly surface storm runoff from roofs, streets, parking lots, and other surfaces toward the nearest receiving water body. An urban drainage system with separated sewers is more expensive than a combined sewer system because it uses two parallel conduits. Sanitary and storm sewers are usually designed to operate under gravity flow conditions. Pressure or vacuum sewers are rare.

Storm sewers are dry much of the time. When rainfalls are gentle, the surface runoffs are usually clear and low flows present no serious problem. However, flooding runoffs wash and erode unprotected areas and create siltation.

Illicit connections from roofs, yards, and foundations drain flow to sanitary sewers through manhole covers that are not tight. The flow rates vary and are as high as 70 gal/(c · d) and average 30 gal/(c · d). A rainfall of 1 inch (2.54 cm) per hour on 1200 ft^2 (111 m^2) of roof produces a flow of 12.5 gal/min (0.788 L/s) or 18,000 gal/d. Spreading over an acre (0.4047 ha), 1 in/h equals 1.008 ft^3/s (28.5 L/s) of runoff. Leakage through manhole covers can add 20 to 70 gal/min (1.26 to 4.42 L/s) to the sewer when there is as much as 1 in of water on the streets (Fair *et al.* 1966).

EXAMPLE: A house has a roof of 30 ft by 40 ft (9.14 m × 12.19 m) and is occupied by four people. What is the percentage of rainfall of 1 in/h, using a rate of leakage through manholes of 60 gal/(c · d)?

Solution:

Step 1. Compute stormwater runoff (also proove the above statements)

$$\begin{aligned}\text{Runoff} &= 30\ \text{ft} \times 40\ \text{ft} \times 1\ \text{in/h} \times (1\ \text{ft/12 in}) \times 24\ \text{h/d}) \times 7.48\ \text{gal/ft}^3 \div 4\ \text{c} \\ &\cong 18{,}000\ \text{gal/d} \div 4\ \text{c} \\ &= 4500\ \text{gal/(c} \cdot \text{d)}\end{aligned}$$

Step 2. Determine percent of leakage

$$\begin{aligned}\% &= 60\ \text{gal/(c} \cdot \text{d)} \times 100/4500\ \text{gal/(c} \cdot \text{d)} \\ &= 1.33\end{aligned}$$

3.2 Combined Sewers

Combined sewers are designed for collection and conveyance of both sanitary sewage, including industrial wastewaters, and storm surface runoff in one conduit. Combined sewer systems are common in old U.S. and European urban communities. In the United States, there has been a trend to replace combined sewers by separate sewer systems. The dry-weather flow (sanitary and industrial wastewaters, plus street drainage from washing operations and lawn sprinkling and infiltration) is intercepted to the treatment facility. During storm events, when the combined wastewater and stormwater exceed the capacity of the treatment plant, the overflow is bypassed directly to the receiving water body without treatment or is stored for future treatment. The overflow is the so-called combined sewer overflow (CSO).

The average 5-day BOD in stormwater is approximately 30 mg/L. The average BOD_5 in combined sewer overflows is between 60 and 120 mg/L (Novotny and Chesters 1981). Since the CSO bypass may result in significant pollution, even during high river flows, it will create a hazardous threat to downstream water users. Thus, proper management of CSO is required (e.g. storage and later treatment). One obvious solution is to replace combined sewers by separate sewers; however, costs and legalities of shifting are major concerns.

EXAMPLE: The following information is given:

Separate sanitary sewer flow = 250 L/(c · d)
Population density = 120 persons/ha
BOD_5 of raw wastewater = 188 mg/L
BOD_5 of plant effluent or stormwater = 30 mg/L
A storm intensity = 25 mm/d (1 in/d)
Impervious urban watershed = 72%

Compare flow and BOD_5 pollution potential produced by wastewater after treatment and by the stormwater.

Solution:

Step 1. Compute flow and BOD_5 loads for wastewater

$$\begin{aligned}\text{Flow} &= 250\ \text{L/(c} \cdot \text{d)} \times 120\ \text{c/ha} \\ &= 30{,}000\ \text{L/(d} \cdot \text{ha)} \\ &= 30\ \text{m}^3\text{/(d} \cdot \text{ha)}\end{aligned}$$

$$\begin{aligned}
\text{Raw wastewater BOD}_5 &= 30{,}000\ \text{L/(d} \cdot \text{ha)} \times 188\ \text{mg/L} \\
&= 5{,}640{,}000\ \text{mg/(d} \cdot \text{ha)} \\
&= 5640\ \text{g/(d} \cdot \text{ha)} \\
\text{Effluent BOD}_5 &= 30{,}000\ \text{L/(d} \cdot \text{ha)} \times 30\ \text{mg/L} \\
&= 900\ \text{g/(d} \cdot \text{ha)} \\
\text{BOD}_5\ \text{removed} &= (5640 - 900)\ \text{g/(d} \cdot \text{ha)} \\
&= 4740\ \text{g/(d} \cdot \text{ha) (84\% removal)}
\end{aligned}$$

Step 2. Compute flow and BOD_5 load for stormwater

$$\begin{aligned}
\text{Flow} &= 25\ \text{mm/d} \times 10{,}000\ \text{m}^2\text{/ha} \times 0.72 \times 0.001\ \text{m/mm} \times 1000\ \text{L/m}^3 \\
&= 180{,}000\ \text{L/(d} \cdot \text{ha)} \\
&= 180\ \text{m}^3\text{/(d} \cdot \text{ha)} \\
\text{BOD}_5\ \text{load} &= 180{,}000\ \text{L/(d} \cdot \text{ha)} \times 30\ \text{mg/L} \\
&= 5{,}400{,}000\ \text{mg/(d} \cdot \text{ha)} \\
&= 5400\ \text{g/(d} \cdot \text{ha)}
\end{aligned}$$

Step 3. Compare BOD_5 load

$$\begin{aligned}
\text{Stormwater BOD}_5\text{: effluent BOD}_5 &= 5400 : 900 \\
&= 6 : 1
\end{aligned}$$

4. QUANTITY OF WASTEWATER

The quantity of wastewater produced varies in different communities and countries, depending on a number of factors such as water uses, climate, lifestyle, economics, etc. Metcalf and Eddy, Inc. (1991) lists typical municipal water uses and wastewater flow rates in the United States including domestic, commercial, institutional, and recreational facilities and various institutions.

A typical wastewater flow rate from a residential home in the United States might average 70 gal (265 L) per capita per day (gal/(c · d) or gpcpd). Approximately 60 to 85 percent of the per capita consumption of water becomes wastewater. Commonly used quantities of wastewater flow rates that form miscellaneous types of facilities are listed in Table 7.1 (Illinois EPA 1997).

Municipal wastewater is derived largely from the water supply. Some water uses, such as for street washing, lawn watering, fire fighting, and leakage from water supply pipelines, most likely do not reach the sewers. The volume of wastewater is added to by infiltration and inflows.

Wastewater flow rates for commercial developments normally range from 800 to 1500 gal/(acre · d) (7.5 to 14 $m^3/(ha \cdot d)$), while those for industries are 1000 to 1500 gal/(acre · d) (9.4 to 14 $m^3/(ha \cdot d)$) for light industrial developments and 1500 to 3000 gal/(acre · d) (14 to 28 $m^3/(ha \cdot d)$) for medium industrial developments.

Water entering a sewer system from ground through defective connections, pipes, pipe joints, or manhole wells is called infiltration. The amount of flow into a sewer from groundwater, infiltration, may range from 100 to 10,000 gal/(d · inch · mile) (0.0093 to 0.926 $m^3/(d \cdot cm \cdot km)$) or more (Metcalf and Eddy, Inc. 1991). Construction specifications commonly permit a maximum infiltration rate of 500 gal/(d · mile · in) (0.463 $m^3/(d \cdot km \cdot cm)$) of pipe diameter. The quantity of infiltration may be equal to 3 to 5 percent of the peak hourly domestic wastewater flow, or approximately 10 percent of the average daily flow. With better pipe joint material and tight control of construction methods, infiltration allowance can be as low as 200 gal/(d · mile · in) of pipe diameter (Hammer 1986).

TABLE 7.1 Typical Wastewater Flow Rates for Miscellaneous Facilities

Type of establishment	Gallons per person per day (unless otherwise noted)
Airports (per passenger)	5
Bathhouses and swimming pools	10
Camps:	
Campground with central comfort station	35
With flush toilets, no showers	25
Construction camps (semi-permanent)	50
Day camps (no meals served)	15
Resort camps (night and day) with limited plumbing	50
Luxury camps	100
Cottages and small dwellings with seasonal occupancy	75
Country clubs (per resident member)	100
Country clubs (per nonresident member present)	25
Dwellings:	
Boarding houses	50
(additional for nonresident boarders)	10
Rooming houses	40
Factories (gallons per person, per shift, exclusive of industrial wastes)	35
Hospitals (per bed space)	250
Hotels with laundry (2 persons per room) per room	150
Institutions other than hospitals including nursing homes (per bed space)	125
Laundries—self service (gallons per wash)	30
Motels (per bed) with laundry	50
Picnic parks (toilet wastes only per park user)	5
Picnic parks with bathhouses, showers and flush toilets (per park user)	10
Restaurants (toilet and kitchen wastes per patron)	10
Restaurants (kitchen wastes per meal served)	3
Restaurants (additional for bars and cocktail lounges)	2
Schools:	
Boarding	100
Day, without gyms, cafeterias or showers	15
Day, with gyms, cafeterias and showers	25
Day, with cafeterias, but without gyms or showers	20
Service stations (per vehicle served)	5
Swimming pools and bathhouses	10
Theaters:	
Movie (per auditorium set)	5
Drive-in (per car space)	10
Travel trailer parks without individual water and sewer hook-ups (per space)	50
Travel trailer parks with individual water and sewer hook-ups (per space)	100
Workers:	
Offices, schools and business establishments (per shift)	15

Source: Illinois EPA (1997).

Inflow to a sewer includes steady inflow and direct inflow. Steady inflow is water drained from springs and swampy areas, discharged from foundation, cellar drains, and cooling facilities, etc. Direct inflow is from stormwater runoff direct to the sanitary sewer.

EXAMPLE 1: Convert to the SI units for the construction allowable infiltration rate of 500 gal/(d · mile) per inch of pipe diameter.

Solution:

$$500 \text{ gal/(d} \cdot \text{mile} \cdot \text{in)} = \frac{500 \text{ gal/d} \times 0.003785 \text{ m}^3\text{/gal}}{1 \text{ mile} \times 1.609 \text{ km/mile} \times 1 \text{ in} \times 2.54 \text{ cm/in}}$$

$$= 0.463 \text{ m}^3\text{/(d} \cdot \text{km) per cm of pipe diameter}$$

EXAMPLE 2: The following data are given:

Sewered population = 55,000

Average domestic wastewater flow = 100 gal/(c · d)

Assumed infiltration flow rate = 500 gal/(d · mile) per inch of pipe diameter

Sanitary sewer systems for the city:

4-in house sewers = 66.6 miles

6-in building sewers = 13.2 miles

8-in street laterals = 35.2 miles

12-in submains = 9.8 miles

18-in mains = 7.4 miles

Estimate the infiltration flow rate and its percentage of the average daily and peak hourly domestic wastewater flows.

Solution:

Step 1. Calculate the average daily flow (Q) and peak hourly flow (Q_p)
Assuming $Q_p = 3Q$

$$Q = 100 \text{ gal/(c} \cdot \text{d)} \times 55{,}000 \text{ persons}$$
$$= 5{,}500{,}000 \text{ gal/d}$$
$$Q_p = 5{,}500{,}000 \text{ gal/d} \times 3$$
$$= 16{,}500{,}000 \text{ gal/d}$$

Step 2. Compute total infiltration flow, I

$$I = \text{infiltration rate} \times \text{length} \times \text{diameter}$$
$$= 500 \text{ gal/(d} \cdot \text{mile} \cdot \text{in)} \times (66.6 \times 4 + 13.2 \times 6 + 35.2 \times 8 + 9.8 \times 12 + 7.4 \times 18) \text{ mile} \cdot \text{in}$$
$$= 439{,}000 \text{ gal/d}$$

Step 3. Compute percentages of infiltration to daily average and peak hourly flows

$$I/Q = (439{,}000 \text{ gal/d})/(5{,}500{,}000 \text{ gal/d}) \times 100\%$$
$$= 8.0\%$$
$$I/Q_p = (439{,}000 \text{ gal/d})/(16{,}500{,}000 \text{ gal/d}) \times 100\%$$
$$= 2.66\%$$

4.1 Design Flow Rates

The average daily flow (volume per unit time), maximum daily flow, peak hourly flow, minimum hourly and daily flows, and design peak flow are generally used as the basis of design for sewers, lift stations, sewage (wastewater) treatment plants, treatment units, and other wastewater handling facilities. Definitions and purposes of flow are given as follows.

The design average flow is the average of the daily volumes to be received for a continuous 12-month period of the design year. The average flow may be used to estimate pumping and chemical costs, sludge generation, and organic-loading rates.

The maximum daily flow is the largest volume of flow to be received during a continuous 24-hour period. It is employed in the calculation of retention time for equalization basin and chlorine contact time.

The peak hourly flow is the largest volume received during a one-hour period, based on annual data. It is used for the design of collection and interceptor sewers, wet wells, wastewater pumping stations, wastewater flow measurements, grit chambers, settling basins, chlorine contact tanks, and pipings. The design peak flow is the instantaneous maximum flow rate to be received. The peak hourly flow is commonly assumed as three times the average daily flow.

The minimum daily flow is the smallest volume of flow received during a 24-hour period. The minimum daily flow is important in the sizing of conduits where solids might be deposited at low flow rates.

The minimum hourly flow is the smallest hourly flow rate occurring over a 24-hour period, based on annual data. It is important to the sizing of wastewater flowmeters, chemical-feed systems, and pumping systems.

EXAMPLE: Estimate the average and maximum hourly flow for a community of 10,000 persons.

Step 1. Estimate wastewater daily flow rate
Assume average water consumptiom = 200 L/(c · d)
Assume 80% of water consumption goes to the sewer

$$\begin{aligned}\text{Average wastewater flow} &= 200\ \text{L/(c}\cdot\text{d)} \times 0.80 \times 10{,}000\ \text{persons} \times 0.001\ \text{m}^3\text{/L} \\ &= 1600\ \text{m}^3\text{/d}\end{aligned}$$

Step 2. Compute average hourly flow rate

$$\begin{aligned}\text{Average hourly flow rate} &= 1600\ \text{m}^3\text{/d} \times 1\ \text{d/24 h} \\ &= 66.67\ \text{m}^3\text{/h}\end{aligned}$$

Step 3. Estimate the maximum hourly flow rate
Assume the maximum hourly flow rate is three times the average hourly flow rate, thus

$$\begin{aligned}\text{Maximum hourly flow rate} &= 66.67\ \text{m}^3\text{/h} \times 3 \\ &= 200\ \text{m}^3\text{/h}\end{aligned}$$

5 URBAN STORMWATER MANAGEMENT

In the 1980s stormwater detention or retention basin became one of the most popular and widely used best management practices (BMPs) for quality enhancement of stormwater. In the United States, Congress mandated local governments to research ways to reduce the impact of separate storm sewer systems and CSO discharges on all receiving water bodies. In Europe, most communities use combined sewer systems, with some separate storm sewer systems in newer suburban communities. The CSO problem has received considerable attention in Europe.

Both quality and quantity of stormwater should be considered when protecting water quality and reducing property damage and traffic delay by urban flooding. Stormwater detention is an important measure for both quality and quantity control. Temporarily storing or detaining stormwater is a very effective method. Infiltration practices are the most effective in removing stormwater pollutants (Livingston 1995). When stormwater is retained long enough, its quality will be enhanced.

Several publications (USEPA 1974c, 1983a, Northeastern Illinois Planning Commission 1992, WEF and ASCE 1992, Wanielista and Yousef 1993, Urbonas and Stahre 1993, Pitt and Voorhees 1995,

Shoemaker *et al.* 1995, Terstriep and Lee 1995, Truong and Phua 1995) describe stormwater management plans and design guidelines for control (e.g. detention or storage facilities) in detail. Storage facilities include local disposal, inlet control at source (rooftops, parking lots, commercial or industrial yards, and other surfaces), on-site detention (swales or ditches, dry basins or ponds, wet ponds, concrete basins, underground pipe packages or clusters), in-line detention (concrete basins, excess volume in the sewer system, pipe packages, tunnels, underground caverns, surface ponds), off-line storage (direct to storage), storage at treatment plant, and constructed wetland.

Two of the largest projects for urban stormwater management are given. Since the early 1950s, Toronto, Canada, has expanded with urban development. The city is located at the lower end of a watershed. In 1954, Hurricane Hazel brought a heavy storm (6 inches in one hour). Subsequently, overflow from the Don River flooded the city. Huge damage and losses occurred. Thereafter, an urban management commission was formed and two large flood control reservoirs were constructed. Natural techniques were applied to river basin management.

In Chicago, USA, before the 1930s combined sewers were built, as in other older municipalities. About 100 storms per year caused a combination of raw wastewater and stormwater to discharge into Chicagoland waterways. Approximately 500,000 homes had chronic flooding problems. Based on the need for pollution and flood controls, the mass \$3.66 billion Tunnel and Reservoir Plan (TARP) was formed. Also known as Chicago's Deep Tunnel, this project of intergovernmental efforts has undergone more than 25 years of planning and construction. The two-phased plan was approved by the water district commissions in 1972. Phase 1 aimed to control pollution and included 109 miles (175.4 km) of tunnels and three dewatering pumping stations. Phase 2 was designed for flood control and resulted in construction of three reservoirs with a total capacity of 16 billion gallons (60.6 billion liters) (Carder 1997).

The entire plan consisted of four separate systems—Mainstream, Calumet, Des Plaines, and O'Hare. TARP has a total of 130 miles (209 km) of tunnels, 243 vertical drop shafts, 460 near-surface collecting structures, three pumping stations, and 126,000 acre · ft (155,400,000 m^3) of storage in three reservoirs. The reservoir project began in the middle of the 1990s (F. W. Dodge Profile 1994, Kirk 1997).

The tunnels were excavated through the dolomite rock strata, using tunnel boring machines. The rough, excavated diameter of the tunnel is about 33 ft (10 m). Tunnel depths range from 150 to 360 ft (45.7 to 110 m) below the ground surface and their diameter ranges from 9 to 33 ft (2.7 to 10 m). The construction of the Mainstream tunnel started in 1976 and it went into service in 1985.

The TARP's mainstream pumping station in Hodgkins, Illinois, is the largest pumping station in the United States. It boasts six pumps: four with a combined capacity of 710 Mgal/d (31.1 m^3/s) and two with a combined capacity of 316 Mgal/d (13.8 m^3/s). The stored wastewater is pumped to wastewater treatment plants. The TARP system received the 1986 Outstanding Civil Engineering Achievement award from the American Society of Civil Engineers (Robison 1986) and the 1989 Outstanding Achievement in Water Pollution Control award from the Water Pollution Control Federation (current name, Water Environment Federation). It is now one of the important tour sites in Chicago.

5.1 Urban Drainage Systems

Urban drainage systems or storm drainage systems consist of flood runoff paths, called the major system. Major (total) drainage systems are physical facilities that collect, store, convey, and treat runoff that exceeds the minor systems. It is composed of paths for runoff to flow to a receiving stream. These facilities normally include detention and retention facilities, street storm sewers, open channels, and special structures such as inlets, manholes, and energy dissipators (Metcalf and Eddy, Inc. 1970, WEF and ASCE 1992).

The minor or primary system is the portion of the total drainage system that collects, stores, and conveys frequently occurring runoff, and provides relief from nuisance and inconvenience. The minor system includes streets, sewers, and open channels, either natural or constructed. It has traditionally been carefully planned and constructed, and usually represents the major portion of the urban drainage infrastructure investment. The major drainage system is usually less controlled than the minor system and will function regardless of whether or not it has been deliberately designed and/or protected against encroachment, including when the minor system is blocked or otherwise inoperable (WEF and ASCE 1992). The minor system is traditionally planned and designed to safely convey runoff from storms with a specific recurrence interval, usually 5 to 10 years (Novotny *et al.* 1989). In the United States, flood drainage systems are included in flood insurance studies required by the Office of Insurance and Hazard Mitigation of the Federal Emergency Management Agency.

6 DESIGN OF STORM DRAINAGE SYSTEMS

Urban drainage control facilities have progressed from crude drain ditches to the present complex systems. The systems start from surface runoff management control facilities, such as vegetation covers, detention or storage facilities, sedimentation basin, filtration, to curbs, gutters, inlets, manholes, and underground conduits.

Managing surface runoff in urban areas is a complex and costly task. Hydrology, climate, and physical characteristics of the drainage area should be considered for design purposes. Hydrologic conditions of precipitate–runoff relationship (such as magnitude, frequency, and duration of the various runoffs, maximum events) and local conditions (soil types and moisture, evapotranspiration, size, shape, slope, land-use, etc.) of the drainage (watershed) should be determined.

Clark and Viessman (1966) presented an example of hydraulic design of urban storm drainage systems to carry 10-year storm runoffs from eight inlets. A step-by-step solution of the problem was given.

7 PRECIPITATION AND RUNOFF

Precipitation includes rainfall, snow, hail, and sleet. It is one part of the hydrologic cycle. Precipitation is the primary source of water in springs, streams and rivers, lakes and reservoirs, groundwater, and well waters. The US National Weather Service maintains observation stations throughout the United States. The US Geological Survey operates and maintains a national network of stream-flow gage stations throughout the country's major streams. State agencies, such as the Illinois State Water Survey and the Illinois Geological Survey, also have some stream flow data similar to the national records. In other countries, similar governmental agencies also maintain long-term precipitation, watershed runoff, stream flow data, and groundwater information.

Upon a catchment area (or watershed), some of the precipitation runs off immediately to streams, lakes or lower land areas, while some evaporates from land to air or penetrates the ground (infiltration). Snow remains where it falls, with some evaporation. After snow melts, the water may run off as surface and groundwater.

Rainwater or melting snow that travels as overland flow across the ground surface to the nearest channel is called surface runoff. Some water may infiltrate into the soil and flow laterally in the surface soil to a lower land water body as interflow. A third portion of the water may percolate downward through the soil until it reaches the groundwater. Overland flows and interflow are usually grouped together as direct runoff.

7.1 Rainfall Intensity

The rainfall intensity is dependent on the recurrence interval and the time of concentration. The intensity can be determined from the cumulative rainfall diagram. From records of rainfall one can format the relationship between intensity and frequency as well as duration. Their relationship is expressed as (Fair *et al.* 1966, Wanielista 1990)

$$i = \frac{at^m}{(b + d)^n} \tag{7.8}$$

where i = rainfall intensity, in/h or cm/h
t = frequency of occurrences, yr
d = duration of storm, min
a, b, m, n = constants varying from place to place

By fixing a frequency of occurrence, the intensity–duration of n-year rainstorm events can be drawn. For many year storm events, the intensity–duration–frequency curves (Fig. 7.1) can be formatted for future use. The storm design requires some relationship between expected rainfall intensity and duration.

Let $A = at^m$ and $n = 1$. Eliminating t in Eq. (7.8) the intensity–duration relationship (Talbot parabola formula) is

$$i = \frac{A}{B + d} \tag{7.9}$$

where i = rainfall intensity, mm/h or in/h
d = the duration, min or h
A, B = constants

Rearrange the Talbot equation for a straight line

$$d = \frac{A}{i} - B$$

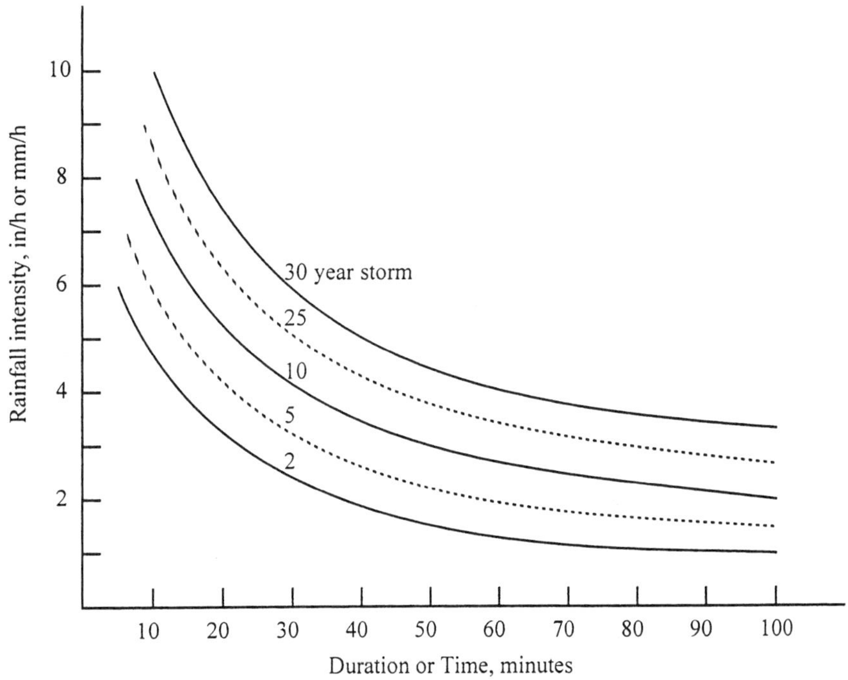

FIGURE 7.1 Intensity—duration—frequency curves.

Use the following two equations to solve constants A and B:

$$\begin{cases} \sum d = A \sum \frac{1}{i} - nB \\ \sum \frac{d}{i} = A \sum \frac{1}{i^2} - nB \sum \frac{1}{i} \end{cases}$$

From the data of rainfall intensity and duration, calculate Σd, $\Sigma(1/i)$, $\Sigma(1/i^2)$, $\Sigma(d/i)$; then A and B will be solved. For example, for a city of 5-year storm, $i(\text{mm/h}) = 5680/(42 + d(\text{min}))$. Similarly, a curve such as that in Fig. 7.1 can be determined.

7.2 Time of Concentration

The time for rainwater to flow overland from the most remote area of the drainage area to an inlet is called the inlet time, t_i. The time for water to flow from this inlet through a branch sewer to the storm sewer inlet is called the time of sewer flow, t_s. The time of concentration for a particular inlet to a storm sewer, t, is the sum of the inlet time and time of sewer flow, i.e.

$$t = t_i + t_s \tag{7.10}$$

The inlet time can be estimated by

$$t_i = C(L/S\, i^2)^{1/3} \tag{7.11}$$

where t_i = time of overland flow, min
L = distance of overland flow, ft
S = slope of land, ft/ft
i = rainfall intensity, in/h
C = coefficient
= 0.5 for paved areas
= 1.0 for bare earth
= 2.5 for turf

The time of concentration is difficult to estimate since the different types of roof, size of housing, population density, land coverage, topography, and slope of land are very complicated. In general, the time of concentration may be between 5 and 10 min for both the mains and the submains.

EXAMPLE: The subdrain sewer line to a storm sewer is 1.620 km (1 mile). The average flow rate is 1 m/s. Inlet time from the surface inlet through the overland is 7 min. Find the time of concentration to the storm sewer.

Solution:

Step 1. Determine the time of sewer flow, t_s

$$t_s = \frac{L}{v} = \frac{1620 \text{ m}}{1 \text{ m/s}} = 1620 \text{ s} = 27 \text{ min}$$

Step 2. Find t

$$\begin{aligned} t = t_i + t_s &= 7 \text{ min} + 27 \text{ min} \\ &= 34 \text{ min} \end{aligned}$$

7.3 Estimation of Runoff

The quantity of stormwater runoff can be estimated by the rational method or by the empirical formula. Each method has its advantages. The rational formula for the determination of the quantity of stormwater runoff is

$$Q = CIA \tag{7.12a}$$

where Q = peak runoff from rainfall, ft^3/s
C = runoff coefficient, dimensionless
I = rainfall intensity, inches rain/h
A = drainage area, acres

The formula in SI units is

$$Q = 0.278\, CIA \tag{7.12b}$$

where Q = peak runoff, m^3/s
C = runoff coefficient, see Table 7.2
I = rainfall intensity, mm/h
A = drainage area, km^2

The coefficient of runoff for a specific area depends upon the character and the slope of the surface, the type and extent of vegetation, and other factors. The approximate values of the runoff coefficient are shown in Table 7.2. There are empirical formulas for the runoff coefficient. However, the values of C in Table 7.2 are most commonly used.

EXAMPLE: In a suburban residential area of 1000 acres (4.047 km^2), the rainfall intensity–duration of a 20-min 25-year storm is 6.1 in/h (155 mm/h). Find the maximum rate of runoff.

Solution: From Table 7.2, $C = 0.50$ (for rolling suburban)

Using Eq. (7.12a)

$$\begin{aligned} Q &= CIA \\ &= 0.5 \times 6.1 \text{ in/h} \times 1000 \text{ acre} \\ &= 3050 \frac{\text{acre} \cdot \text{in}}{\text{h}} \times \frac{1 \text{ h}}{3600 \text{ s}} \times \frac{1 \text{ ft}}{12 \text{ in}} \times \frac{43{,}560 \text{ ft}^2}{1 \text{ acre}} = 3075 \text{ ft}^3/\text{s} \end{aligned}$$

Note: In practice, a factor of 1.0083 conversion is not necessary. The answer could be just 3050 ft^3/s.

TABLE 7.2 Coefficient of Runoff for Various Surfaces

Type of surface	Flat slope < 2%	Rolling slope 2–10%	Hilly slope > 10%
Pavements, roofs	0.90	0.90	0.90
City business areas	0.80	0.85	0.85
Dense residential areas	0.60	0.65	0.70
Suburban residential areas	0.45	0.50	0.55
Earth areas	0.60	0.65	0.70
Grassed areas	0.25	0.30	0.30
Cultivated land			
clay, loam	0.50	0.55	0.60
sand	0.25	0.30	0.35
Meadows and pasture lands	0.25	0.30	0.35
Forests and wooded areas	0.10	0.15	0.20

Source: Perry (1967).

8 STORMWATER QUALITY

The quality and quantity of stormwaters depend on several factors—intensity, duration, and area extent of storms. The time interval between successive storms also has significant effects on both the quantity and quality of stormwater runoff. Land contours, urban location permeability, land uses and developments, population densities, incidence and nature of industries, size and layout of sewer systems, and other factors are also influential.

Since the 1950s, many studies on stormwater quality indicate that runoff quality differs widely in pattern, background conditions, and from location to location. Wanielista and Yousef (1993) summarized runoff quality on city street, lawn surface, rural road, highway, and by land-use categories. Rainwater quality and pollutant loading rates are also presented.

8.1 National Urban Runoff Program

In 1981 and 1982, the USEPA's National Urban Runoff Program (NURP) collected urban stormwater runoff data from 81 sites located in 22 cities throughout the United States. The data covered more than 2300 separate storm events. Data was evaluated for solids, oxygen demand, nutrients, metals, toxic chemicals, and bacteria. The median event mean concentrations (EMC) and coefficients of variance for ten standard parameters for four different land use categories are listed in Table 7.3 (USEPA 1983a). It was found that lead, copper, and zinc are the most significant heavy metals found in urban runoff and showed the highest concentrations. The EMC for a storm is determined by flow-weighted calculation.

8.2 Event Mean Concentration

Individual pollutants are measured on the flow-weighted composite to determine the event mean concentrations. The event mean concentration is defined as the event loading for a specific constituent divided by the event stormwater volume. It is expressed as

$$C = \frac{W}{V} \tag{7.13}$$

TABLE 7.3 Mean EMCS and Coefficient of Variance for all NURP Sites by Land Use Category

	Residential		Mixed		Commercial		Open/Nonurban	
Pollutant	Median	CV	Median	CV	Median	CV	Median	CV
BOD (mg/L)	10.0	0.41	7.8	0.52	9.3	0.31	–	–
COD (mg/L)	73	0.55	65	0.58	57	0.39	40	0.78
TSS (mg/L)	101	0.96	67	1.14	69	0.35	70	2.92
Total lead (μg/L)	144	0.75	114	1.35	104	0.68	30	1.52
Total copper (μ/L)	33	0.99	27	1.32	29	0.81	–	–
Total zinc (μg/L)	135	0.84	154	0.78	226	1.07	195	0.66
Total kjeldahl nitrogen (μg/L)	1900	0.73	1288	0.50	1179	0.43	965	1.00
NO_2-N + NO_3-N (μg/L)	736	0.83	558	0.67	572	0.48	543	0.91
Total P (μg/L)	383	0.69	263	0.75	201	0.67	121	1.66
Soluble P (μg/L)	143	0.46	56	0.75	80	0.71	26	2.11

Notes: EMC = event mean concentration.
NURP = National urban runoff program.
CV = coefficient of variance.
Source: USEPA (1983a).

where C = event mean concentration, mg/L
W = total loading per event, mg
V = volume per event, L

The total loading for a storm event is determined by the sum of the loadings during each sampling period, the loading being the flow rate (volume) multiplied by the concentration. The loading per event is

$$W = \sum_{i=1}^{n} V_i C_i \tag{7.14}$$

where W = loading for a storm event, mg
n = total number of samples taken during a storm event
V_i = volume proportional to flow rate at time i, L
C_i = concentration at time i, mg/L

EXAMPLE: An automatic sampler was installed at the confluence of the main tributary of a lake. The following sampling, time and tributary flow rate, were determined in the field and in the laboratory. The results are shown in Table 7.4. The watershed covers mainly agricultural lands. The flow rate and total phosphorus (TP) concentrations listed are the recorded values subtracted by the dry flow rate and TP concentration during normal flow period. Estimate the EMC and TP concentration for this storm event.

Solution:

Step 1. Calculate sampling interval Δt = time intervals of 2 successive samplings in col. 1 of Table 7.5.

Step 2. Calculate mean flow q (col. 2 of Table 7.5)

$$q = \text{average of 2 successive flows measured in col. 2 of Table 7.4}$$

Step 3. Calculate mean runoff volume, ΔV

In Table 7.5, col. 3 = col. 1 × 60 × col. 2

Step 4. Calculate total volume of the storm

$$V = \Sigma \Delta V = 1{,}920{,}720 \text{ L}$$

TABLE 7.4 Data Collected from Stormwater Sampling

Time of collection (1)	Creek flow, L/s (2)	Total phosphorus, mg/L (3)
19:30	0	0
19:36	44	52
19:46	188	129
19:56	215	179
20:06	367	238
20:16	626	288
20:26	752	302
20:36	643	265
20:46	303	189
20:56	105	89
21:06	50	44

TABLE 7.5 Calculations for Total Phosphorus Loading

Sampling interval, Δt min (1)	Mean flow, L/s (2)	Mean runoff volume ΔV, L (3)	Mean TP, μg/L (4)	Loading ΔW, mg (5)
6	22	7,920	26	206
10	116	69,600	90.5	6,300
10	202	121,200	153.5	18,604
10	291	174,600	288	50,285
10	497	298,200	263	78,427
10	689	413,400	295	121,953
10	638	382,800	283.5	108,524
10	473	283,800	227	64,423
10	204	122,400	139	17,014
10	78	46,800	66.5	2,713
		Σ = 1,920,720		Σ = 468,449

Step 5. Calculate mean total phosphorus, TP (col. 4 of Table 7.5)

$$\text{TP} = \text{average of 2 successive TP values in Table 7.4}$$

Step 6. Calculate loading (col. 5 of Table 7.5)

$$\Delta W = (\Delta V \text{ in L})\ (\text{TP in } \mu\text{g/L})/(1000\ \mu\text{g/mg})$$

Step 7. Calculate total load by Eq. (7.14)

$$W = \Sigma \Delta W = 468{,}449 \text{ mg}$$

Step 8. Calculate EMC

$$C = W/V = 468{,}449 \text{ mg}/1{,}920{,}720 \text{ L}$$
$$= 0.244 \text{ mg/L}$$

Note: If the TP concentrations are not flow weighted, the mathematic mean of TP is 0.183 mg/L which is less than the EMC.

8.3 Street and Road Loading Rate

Rainfall will carry pollutants from the atmosphere and remove deposits on impervious surfaces. The rainfall intensity, duration, and storm runoff affect the type and the amount of pollutants removed. According to the estimation by the USEPA (1974a), the times required for 90 percent particle removal from impervious surfaces are 300, 90, 60, and 30 min respectively for 0.1, 0.33, 0.50, and 1.00 in/h storms. Average loading of certain water quality parameters and some heavy metals from city street, highway, and rural road are given in Table 7.6. Loading rates vary with local conditions. Loads from highways are larger than those from city streets and rural roads. Loading is related to daily traffic volume.

EXAMPLE: A storm sewer drains a section of a downtown city street 260 m (850 ft) and 15 m (49 ft) wide with curbs on both sides. Estimate the BOD_5 level in the stormwater following a 60-min storm of 0.5 in/h. Assume there is no other source of water contribution or water losses; also, it has been 4 days since the last street cleansing.

TABLE 7.6 Average Loads (kg/curb · km · d) of Water Quality Para-meters and Heavy Metals on Streets and Roads

Parameter	Highway	City street	Rural road
BOD_5	0.900	0.850	0.140
COD	10.000	5.000	4.300
PO_4	0.080	0.060	0.150
NO_3	0.015	0.015	0.025
N (total)	0.200	0.150	0.055
Cr	0.067	0.015	0.019
Cu	0.015	0.007	0.003
Fe	7.620	1.360	2.020
Mn	0.134	0.026	0.076
Ni	0.038	0.002	0.009
Pb	0.178	0.093	0.006
Sr	0.018	0.012	0.004
Zn	0.070	0.023	0.006

Source: Wanielista and Youset (1993).

Solution:

Step 1. Determine BOD loading

From Table 7.6, average BOD loading for a city street is 0.850 kg/(curb km · d)

$$= 850{,}000 \text{ mg/(curb km} \cdot \text{d)}$$

$$\text{Total street length} = 0.26 \text{ km} \times 2 \text{ curb} = 0.52 \text{ km} \cdot \text{curb}$$

$$\text{Total BOD loading} = 850{,}000 \text{ mg/(curb km} \cdot \text{d)} \times 0.52 \text{ km} \cdot \text{curb} \times 4 \text{ day}$$

$$= 1{,}768{,}000 \text{ mg}$$

A 0.5 in/h storm of 60 minutes' duration will remove 90 percent of pollutant; thus

$$\text{Runoff BOD mass} = 1{,}768{,}000 \text{ mg} \times 0.9$$

$$= 1{,}591{,}000 \text{ mg}$$

Step 2. Determine the volume of runoff

$$\text{Area of street} = 260 \text{ m} \times 15 \text{ m} = 3900 \text{ m}^2$$

$$\text{Volume} = 0.5 \text{ in/h} \times 1 \text{ h} \times 0.0254 \text{ m/in} \times 3900 \text{ m}^2$$

$$= 49.53 \text{ m}^3$$

$$= 49{,}530 \text{ L}$$

Step 3. Calculate BOD concentration

$$\text{BOD} = 1{,}591{,}000 \text{ mg} \div 49{,}530 \text{ L}$$

$$= 32.1 \text{ mg/L}$$

8.4 Runoff Models

Several mathematical models have been proposed to predict water quality characteristics from storm runoffs. A consortium of research constructions proposed an urban runoff model based on pollutants deposited on street surfaces and washed off by rainfall runoff. The amount of pollutants transported from a watershed (dP) in any time interval (dt) is proportional to the amount of pollutants remaining in the watershed (P). This assumption is a first-order reaction (discussed in Chapter 1.2):

$$-\frac{dP}{dt} = kP \tag{7.15}$$

Integrating Eq. (7.15) to form an equation for mass remaining

$$P = P_0 e^{-kt} \tag{7.16}$$

The amount removed (proposed model) is

$$P_0 - P = P_0(1 - e^{-kt}) \tag{7.17}$$

where P_0 = amount of pollutant on the surface initially present, lb
P = amount of pollutant remaining after time t, lb
$P_0 - P$ = amount of pollutant washed away in time t, lb
k = transport rate constant, per unit time
t = time

The transport rate constant k is a factor proportional to the rate of runoff. For impervious surfaces,

$$k = br$$

where b = constant
r = rainfall excess

To determine b, it was assumed that a uniform runoff of 0.5 in/h would wash 90 percent of the pollutant in one h (as stated above). We can calculate b from Eq. (7.16)

$$P/P_0 = 0.1 = e^{-brt} = e^{-b(0.5\text{ in/h})\,(1\text{h})} = e^{-0.5b}$$

$$2.304 = 0.5b\ (\text{in})$$

$$b = 4.6\text{ in}$$

Substituting b into Eq. 7.17 leads to the equation for impervious surface areas

$$P_0 - P = P_0(1 - e^{-4.6\,rt}) \tag{7.18}$$

Certain modifications to the basic model (Eq. (7.18)) for predicting total suspended solids and BOD have been proposed to refine the agreement between the observed and predicted values for these parameters. The University of Cincinnati Department of Civil Engineering (1970) developed a mathematical model for urban runoff quality based essentially on the same principles and assumptions as in Eq. (7.18). The major difference is that an integral solution was developed by this group instead of the stepwise solution suggested by the consortium. The amount of a pollutant remaining on a runoff surface at a particular time, the rate of runoff at that time, and the general characteristics of the watershed were found to be given by the following relationship

$$P = P_0 e^{-kV_t} \tag{7.19}$$

where P_0 = amount of pollutant initially on the surface, lb
P = amount of pollutant remaining on the surface at time t, lb
V_t = accumulated runoff water volume up to time, t
$= \int_0^t q\delta t$
q = runoff intensity at time t
k = constant characterizing the drainage area

Many computer simulation models for runoff have been proposed and have been discussed by McGhee (1991).

9 SEWER HYDRAULICS

The fundamental concepts of hydraulics can be applied to both sanitary sewers and stormwater drainage systems. Conservation of mass, momentum, energy, and other hydraulic characteristics of conduits are discussed in Chapter 1.5 and elsewhere (Metcalf and Eddy Inc. 1991, WEF and ASCE 1992). WEF and ASCE (1992) also provide the design guidelines for urban stormwater drainage systems, including system layout, sewer inlets, street and intersection, drainage ways (channels, culverts, bridges), erosion control facilities, check dams, energy dissipators, drop shaft, siphons, side-overflow weirs, flow splitters, junctions, flap gates, manholes, pumping, combined sewer systems, evaluation and mitigation of combined sewer overflows, stormwater impoundments, and stormwater management practices for water quality enhancement, etc.

10 SEWER APPURTENANCES

The major appurtenances used for wastewater collection systems include street (stormwater) inlets, catch basins, manholes, building connection, flushing devices, junctions, transitions, inverted siphons, vertical drops, energy dissipators, overflow and diversion structure, regulators, outlets, and pumping stations.

10.1 Street Inlets

Street inlets are structures to transfer stormwater and street cleansing wastewater to the storm sewers. The catch basin is an inlet with a basin which allows debris to settle. There are four major types of inlet, and multiple inlets. Location and design of inlets should consider traffic safety (for both pedestrian and vehicle) and comfort. Gutter inlet is more efficient than curb inlet in capturing gutter flow, but clogging by debris is a problem. Combination inlets are better. Various manufactured inlets and assembled gratings are available.

Street inlets are generally placed near the corner of the street, depending on the street length. The distance between inlets depends on the amount of stormwater, the water depth of the gutter, and the depression to the gutter. The permissible depth of stormwater in most US cities is limited to 6 in (15 cm) on residential streets.

Flow in the street gutter can be calculated by the Manning formula, modified for a triangular gutter cross section (McGhee 1991):

$$Q = K(z/n)s^{1/2}y^{8/3} \tag{7.20}$$

where Q = gutter flow · m^3/s or cfs
K = constant = 22.61m^3/(min · m) = 0.38 m^3/(s · m)
or = 0.56 ft^3/(s · ft)
z = reciprocal of the cross transverse slope of the gutter
n = roughness coefficient
= 0.015 for smooth concrete gutters
s = slope of the gutter
y = water depth in the gutter at the curb

The water depth at the curve can be calculated from the flow and the street cross section and slope. The width over which the water will spread is equal to yz.

EXAMPLE 1: A street has a longitudinal slope of 0.81 percent, a transverse slope of 3.5 percent, a curb height of 15 cm (6 in), and a coefficient of surface roughness (n) of 0.016. The width of the street is 12 m (40 ft). Under storm design conditions, 4 m of street width should be kept clear during the storm. Determine the maximum flow that can be carried by the gutter.

Solution:

Step 1. Calculate the street spread limit, w

$$w = (12 \text{ m} - 4 \text{ m})/2 = 4 \text{ m}$$

Step 2. Calculate the curb depth (d) with the spread limit for a transverse slope of 3.5%

$$d = 4 \text{ m} \times 0.035 = 0.14 \text{ m} = 14 \text{ cm}$$

The street gutter flow is limited by either the curb height (15 cm) or the curb depth with the spread limit. Since

$$d = 14 \text{ cm}$$

a curb height of 14 cm is the limit factor for the gutter flow.

Step 3. Calculate the maximum gutter flow Q by Eq. (7.20)

$$\begin{aligned} z &= 1/0.035 = 28.57 \\ Q &= K(z/n)s^{1/2}y^{8/3} \\ &= 0.38 \text{ m}^3/(\text{s} \cdot \text{m})(28.57/0.016)(0.0081)^{1/2}(0.14 \text{ m})^{8/3} \\ &= 0.323 \text{ m}^3/\text{s} \\ &= 11.4 \text{ ft}^3/\text{s} \end{aligned}$$

EXAMPLE 2: The stormwater flow of a street gutter at an inlet is 0.40 m^3/s (14.1 ft^3/s). The longitudinal slope of the gutter is 0.01 and its cross transverse slope is 0.025. The value of roughness coefficient is 0.016. Estimate the water depth in the gutter at the curb.

Solution:

Step 1. Calculate the value of z

$$z = 1/0.025 = 40$$

Step 2. Find y by Eq. (7.20)

$$\begin{aligned} 0.40 \text{ m}^3/\text{s} &= 0.38 \text{ m}^3/(\text{s} \cdot \text{m})\ (40/0.016)\ (0.01)^{1/2}(y \cdot \text{m})^{8/3} \\ y^{8/3} &= 0.00421 \text{ m} \\ y &= 0.129 \text{ m} \\ &= 5.06 \text{ in} \end{aligned}$$

10.2 Manholes

Manholes provide an access to the sewer for inspection and maintenance operations. They also serve as ventilation, multiple pipe intersections, and pressure relief. Most manholes are cylindrical in shape.

The manhole cover must be secured so that it remains in place and avoids a blow-out during peak flooding period. Leakage from around the edges of the manhole cover should be kept to a minimum.

For small sewers, a minimum inside diameter of 1.2 m (4 ft) at the bottom tapering to a cast-iron frame that provides a clear opening usually specified as 0.6 m (2 ft) has been widely adopted (WEF and ASCE 1992). For sewers larger than 600 mm (24 in), larger manhole bases are needed. Sometimes a platform is provided at one side, or the manhole is simply a vertical shaft over the center of the sewer. Various sizes and shapes of manholes are available locally.

Manholes are commonly located at the junctions of sanitary sewers, at changes in grades or alignment except in curved alignments, and at locations that provide ready access to the sewer for preventive maintenance and emergency service. Manholes are usually installed at street intersections (Parcher 1988).

Manhole spacing varies with available sanitary sewer maintenance methods. Typical manhole spacings range from 90 to 150 m (300 to 500 ft) in straight lines. For sewers larger than 1.5 m (5 ft), spacings of 150 to 300 m (500 to 1000 ft) may be used (ASCE and WPCF 1982).

Where the elevation difference between inflow and outflow sewers exceeds about 0.5 m (1.5 ft), sewer inflow that is dropped to the elevation of the outflow sewer by an inside or outside connection is called a drop manhole (or drop inlet). Its purpose is to protect workers from the splashing of wastewater, objectionable gases, and odors.

10.3 Inverted Siphons (Depressed Sewers)

A sewer that drops below the hydraulic gradient to pass under an obstruction, such as a railroad cut, subway, highway, conduit, or stream, is often called an inverted siphon. More properly, it should be called a depressed sewer. Because a depressed sewer acts as a trap, the velocity of sewer flow should be greater than 0.9 m/s (3 ft/s) or more for domestic wastewater, and 1.25 to 1.5 m/s (4 to 5 ft/s) for stormwater, to prevent deposition of solids (Metcalf and Eddy Inc. 1991). Thus, sometimes, two or more siphons are needed with an inlet splitter box.

In practice, minimum diameters for depressed sewers are usually the same as for ordinary sewers: 150 or 200 mm (6 or 8 in) in sanitary sewers, and about 300 mm (12 in) in storm sewers (Metcalf and Eddy, Inc. 1991).

The determination of the pipe size for depressed sewers is the same as for water and wastewater mains. The size depends upon the maximum wastewater flow and the hydraulic gradient.

Due to high velocities in depressed sewers, several pipes in parallel are commonly used. For example, it may be that a small pipe may be designed large enough to carry the minimum flow; a second pipe carries the difference between the minimum and average flow (or maximum dry-weather flow); and a third pipe carries peak flow above the average flow. Depressed sewers can be constructed of ductile iron, concrete, PVC, or tile.

EXAMPLE: Design a depressed sewer system using the following given conditions:

- Diameter of gravity sewer to be connected by depressed sewer = 910 mm (36 in)
- Slope of incoming sewer, $S = 0.0016$ m/m (ft/ft)
- Minimum flow velocity in depressed sewer = 0.9 m/s (3 ft/s)
- Length of depressed sewer = 100 m (328 ft)
- Maximum sewer deression = 2.44 m (8 ft)
- Design flows:

  minimum flow = 0.079 m^3/s (2.8 ft^3/s)

  average flow = 0.303 m^3/s (10.7 ft^3/s) = max. dry-weather flow

  full (maximum) flow = capacity of gravity sanitary sewer
- Design three inverted siphons from the inlet chamber

  (1) to carry minimum flow

  (2) to carry flows from minimum to average

  (3) to carry all flows above the average flow
- Available fall from invert to invert = 1.0 m (3.3 ft)
- Available head loss at inlet = 125 mm (0.5 ft)
- Available head loss for friction in depressed sewer = 1.0 m (3.3 ft)
- Available hydraulic grade line = 1 m/100 m = 0.01 m/m
- $n = 0.015$ (ductile-iron pipe)

Note: The above information is required to design a depressed sewer system.

Solution:

Step 1. Design the depressed sewer

(a) Calculate velocity and flow of the 910 mm (D) sewer for full flow

by Eq. (5.21)
$$R = D/4 = 910 \text{ mm}/4 = 227.5 \text{ mm}$$
$$= 0.2275 \text{ m}$$

by Eq. (5.22)
$$V = (1/n)R^{2/3}S^{1/2}$$
$$= (1/0.015)\,(0.2275)^{2/3}(0.0016)^{1/2}$$
$$= 0.994 \text{ (m/s)}$$

$$\text{Flow } Q = AV = 3.14\,(0.455 \text{ m})^2\,(0.994 \text{ m/s})$$
$$= 0.646 \text{ m}^3\text{/s}$$

(b) Determine the size of the small depressed sewer pipe to carry the minimum flow (d = diameter of the pipe)

$$Q = \pi(d/2)^2(1/n)(d/4)^{2/3}S^{1/2}$$
$$= (0.3115/n)\,d^{8/3}\,S^{1/2}$$
$$= 0.079 \text{ (m}^3\text{/s)}$$

and

$$S = 0.01 \text{ m/m}$$

then

$$0.079 = (0.3115/0.015)\,d^{8/3}(0.01)^{1/2}$$
$$0.079 = 2.078\,d^{8/3}$$
$$d^{8/3} = 0.038$$
$$d = 0.293 \text{ m}$$
$$\cong 300 \text{ mm}$$
$$= 12 \text{ in}$$

Using a 12-in (300 mm) pipe will just carry the 0.079 m³/s flow
Check velocity

$$V = (0.397/n)\,d^{2/3}S^{1/2}$$
$$= (0.397/0.015)\,(0.304)^{2/3}(0.01)^{1/2}$$
$$= 2.647\,(0.304)^{2/3}$$
$$= 1.20 \text{ (m/s) (verified, > 0.9 m/s)}$$

Note: A nomograph for the Manning equation can be used without calculation.

$$0.397 = 1/4^{2/3}$$

(c) Determine the size of the second depressed sewer pipe for maximum dry-weather flow above the minimum flow

$$Q = (0.303 - 0.079) \text{ m}^3\text{/s}$$
$$= 0.224 \text{ m}^3\text{/s}$$
$$0.224 = 2.077\,d^{8/3} \text{ (refer to Step (b))}$$
$$d^{8/3} = 0.1078$$
$$d = 0.434 \text{ (m)}$$
$$= 17.1 \text{ in}$$

A standard 18-in (460 mm) pipe would be used. Check velocity of 460 mm pipe, from Step 1(b)

$$V = 2.647\ (0.460)^{2/3} = 1.58\ (\text{m/s})$$

The capacity of the 460 mm pipe would be (refer to Step (b))

$$Q = 2.077\ (0.46)^{8/3}$$
$$= 0.262\ (\text{m}^3/\text{s})$$

(d) Determine the size of the third pipe to carry the peak flow, from Steps (a) and (c)
The third pipe must carry (0.646 − 0.079 − 0.262) m^3/s

$$= 0.305\ \text{m}^3/\text{s}$$

The size (d) required would be

$$d^{8/3} = 0.305/2.077$$
$$d^{8/3} = 0.1468$$
$$d = 0.487\ (\text{m})$$

The size of 500-mm (20-in) diameter standard is chosen.
The capacity and velocity of a 500-mm pipe with 0.01 hydraulic slope is

$$Q = 2.077\ (0.50)^{8/3}$$
$$= 0.327\ (\text{m}^3/\text{s})$$
$$V = 2.647\ (0.50)^{2/3}$$
$$= 1.67\ (\text{m/s})$$

(e) Calculate total capacity of the three pipes (300-, 460-, and 500-mm)

$$Q = (0.079 + 0.262 + 0.327)\ \text{m}^3/\text{s}$$
$$= 0.668\ \text{m}^3/\text{s}\ (0.646\ \text{m}^3/\text{s is needed})$$

Step 2. Design the inlet and outlet chambers
These depressed sewer pipes are connected from the inlet chamber and outlet chamber. Weirs (2 m in length) are installed to divide the chamber into three portions. The design detail can be found elsewhere (Metcalf and Eddy, Inc. 1981).

11 PUMPING STATIONS

The pumping station must be able to adjust to the variations of wastewater flows. The smallest capacity pump should be able to pump from the wet well and discharge at a self-cleansing velocity of about 0.6 m/s (2 ft/s). It should be connected to a 100-mm (4-in) force main which would have a capacity of approximately 280 L/min or 74 gal/min. The wet well capacity should contain sufficient wastewater to permit the pump to run for at least 2 min and restart not more than once in 5 min (Steel and McGhee 1979). The pump running time (t_r) and the filling time (t_f) in the wet well are computed as

$$t_r = \frac{V}{D - Q} \tag{7.21}$$

and

$$t_f = \frac{V}{Q} \tag{7.22}$$

where t_r = pump running time, min
V = storage volume of wet well, L or gal
D = pump discharge at peak flow, L/min (m^3/s) or gpm
Q = inflow (daily average), L/min or gpm
t_f = filling time with the pump off, min

and the total cycle time (t) is

$$t = t_r + t_f = \frac{V}{D - Q} + \frac{V}{Q} \tag{7.23}$$

The starting limitations on pump motors usually dictate the minimum size of a well. The wet well should be large enough to prevent pump motors from overheating due to extensive cycling, but small enough to accommodate cycling times that will reduce septicity and odor problems. Typically, submersible pumps can cycle 4 to 10 times per hour. A maximum storage volume for a cycling time should be no more than 30 min.

If the selected pumps have a capacity equal to the peak (maximum) flow rate, the volume of a wet well is calculated as (WEF 1993a)

$$V = TQ/4 \tag{7.24}$$

where V = storage volume of wet well, gal
T = pump cycle time, min
Q = peak flow, gal/min

EXAMPLE 1: A subdivision generates an average daily wastewater flow of 144,000 L/d (38,000 gal/d). The minimum hourly flow rate is 20,000 L/d (5300 gal/d) and the peak flow is 500,000 L/d (132,000 gal/d). Determine the pumping conditions and the size of a wet well.

Solution:

Step 1. Determine pump capacity for peak flow

$$D = 500{,}000 \text{ L/d} \times 1 \text{ d/1440 min}$$
$$= 347 \text{ L/min}$$

Step 2. Calculate the minimum volume (V_1) for 2-min running time

$$V_1 = 347 \text{ L/min} \times 2 \text{ min}$$
$$= 694 \text{ L}$$

Step 3. Calculate volume (V_2) for 5-min cycle using Eq. (7.23)

Average flow Q = 144,000 L/d = 100 L/min

$$t = \frac{V_2}{D - Q} + \frac{V_2}{Q}$$

$$5 \text{ min} = \frac{V_2}{(347 - 100) \text{ L/min}} + \frac{V_2}{100 \text{ L/min}}$$

$$100V_2 + 247V_2 = 5 \times 247 \times 100$$
$$V_2 = 356 \text{ L}$$

Step 4. Determine the control factor

Since $V_1 > V_2$, therefore the pump running time is the control factor. Say V_1 = 700 L for design.

Step 5. Calculate the actual time of the pumping cycle

$$t = \frac{700 \text{ L}}{(347 - 100) \text{ L/min}} + \frac{700 \text{ L}}{100 \text{ L/min}}$$
$$= 9.83 \text{ min}$$

Step 6. Determine size of wet well

A submergence of 0.3 m (1 ft) above the top of the suction pipe is required for an intake velocity of 0.6 m/s (2 ft/s). The depth between the well bottom and the top of

submergence is 0.5 m (1.6 ft). If a 1.20-m (4-ft) diameter of wet well is chosen, surface area is 1.13 m^2 (12.2 ft^2)
For storage

$$V_2 = 700 \text{ L} = 0.7 \text{ m}^3$$

the depth would be

$$d = 0.7/1.13 = 0.62 \text{ (m)}$$

Typically, 0.6 m (2 ft) of freeboard is required

$$\begin{aligned}\text{Thus total depth of the wet well} &= (0.50 + 0.62 + 0.60) \text{ m} \\ &= 1.72 \text{ m} \\ &= 5.6 \text{ ft}\end{aligned}$$

EXAMPLE 2: Wastewater is collected from a subdivision of 98.8 acres (40.0 ha) area that consists of 480 residential units and 2.2 acres (0.89 ha) of commercial center. Each of the two pumps will be cycled, alternately, 4 times per hour. Determine the volume of wet well needed.

Solution:

Step 1. Determine domestic sewer flow q_1
Assume the residential units have 3.5 persons (United States) and each produces 100 gal/d (378 L/d)

$$\begin{aligned}q_1 &= 100 \text{ gal/(c} \cdot \text{d)} \times 3.5 \text{ person/unit} \times 480 \text{ unit} \\ &= 168{,}000 \text{ gal/d}\end{aligned}$$

Step 2. Estimate commercial area contribution q_2
Assume 1500 gal/d · per acre (468 m^3/d)

$$\begin{aligned}q_2 &= 1500 \text{ gal/(d} \cdot \text{acre)} \times 2.2 \text{ acre} \\ &= 3300 \text{ gpd}\end{aligned}$$

Step 3. Estimate infiltration/inflow (I/I) q_3
Assume I/I is 1000 gal/(d · acre) or (9.35 m^3/(d · ha))

$$\begin{aligned}q_3 &= 1000 \text{ gal/(d} \cdot \text{acre)} \times 98.8 \text{ acre} \\ &= 98{,}800 \text{ gal/d}\end{aligned}$$

Step 4. Determine total sewer flow q. This average daily flow

$$\begin{aligned}q &= q_1 + q_2 + q_3 \\ &= (168{,}000 + 3300 + 98{,}800) \text{ gal/d} \\ &= 270{,}100 \text{ gal/d} \\ &= 188 \text{ gal/min}\end{aligned}$$

Step 5. Estimate population equivalent (PE) and peak flow (Q)

$$\begin{aligned}\text{PE} &= 270{,}100 \text{ gal/d/100 gal/(c} \cdot \text{d)} \\ &= 2700 \text{ persons}\end{aligned}$$

Take

$$\begin{aligned}Q &= 3.5q \text{ for the peak flow} \\ Q &= 188 \text{ gal/min} \times 3.5 \\ &= 658 \text{ gal/min}\end{aligned}$$

Note: The selected pump should be able to deliver 658 gal/min. The pipe diameter that will carry the flow and maintain at least 2.0 ft/s (0.6 m/s) of velocity should be selected by the manufacturer's specification. For this example a 10-in (254-mm) pressure class ductile iron pipe will be used.

Step 6. Determine the volume of the wet well, V
The selected pumps can cycle 4 times per hour.
Alternating each pump between starts gives 8 cycles per hour. The time between starts, T, is

$$T = 60 \text{ min}/8 = 7.5 \text{ min}$$

It means that one pump is capable of starting every 7.5 min.
Using Eq. (7.24)

$$\begin{aligned} V &= TQ/4 \\ &= 7.5 \text{ min} \times 658 \text{ (gal/min)}/4 \\ &= 1230 \text{ gal} \\ &= (4660 \text{ L}) \\ &= 164 \text{ ft}^3 \end{aligned}$$

Wet wells are typically available in cylindrical sections of various sizes. In this example, a 6-ft (1.8-m) diameter gives 28.3 ft^2 of surface area.
The depth D of the wet well is

$$\begin{aligned} D &= 164/28.3 \\ &= 5.8 \text{ ft} \end{aligned}$$

Note: One foot of freeboard should be added. Thus the well is 6 ft in diameter and 6.8 ft in depth.

12 SEWER CONSTRUCTION

Conduit material for sewer construction consists of two types: rigid pipe and flexible pipe. Specified rigid materials include asbestos—cement, cast iron, concrete, and vitrified clay. Flexible materials include ductile iron, fabricated steel, corrugated aluminum, thermoset plastic (reinforced plastic mortar and reinforced thermosetting resin), and thermoplastic. Thermoplastic consists of acrylonitrile—butadiene—styrene (ABS), ABS composite, polyethylene (PE), and polyvinyl chloride (PVC). Their advantages, disadvantages, and applications are discussed in detail elsewhere (ASCE and WPCF 1982, WEF and ASCE 1993a).

Nonpressure sewer pipe is commercially available in the size range from 4 to 42 in (102 to 1067 mm) in diameter and 13 ft (4.0 m) in length. Half-length sections of 6.5 ft 2 m are available for smaller size pipes.

12.1 Loads and Buried Sewers

Loads on sewer lines are affected by conditions of flow, groundwater, adjacent earth, and superimposed situation. Loads include hydraulic loads, earth loads, groundwater loads, and superimposed loads (weight and impact of vehicles or other structure). Crushing strength of the sewer material, type of bedding, and backfill load are important factors.

Marston's equation. Figure 7.2 illustrates common cuts used for sewer pipe installations. Marston's equation is widely used to determine the vertical load on buried conduits caused by earth forces in all of the most commonly encountered construction conditions (Marston 1930). The general form of Marston's formula is

$$W = CwB^2 \tag{7.25}$$

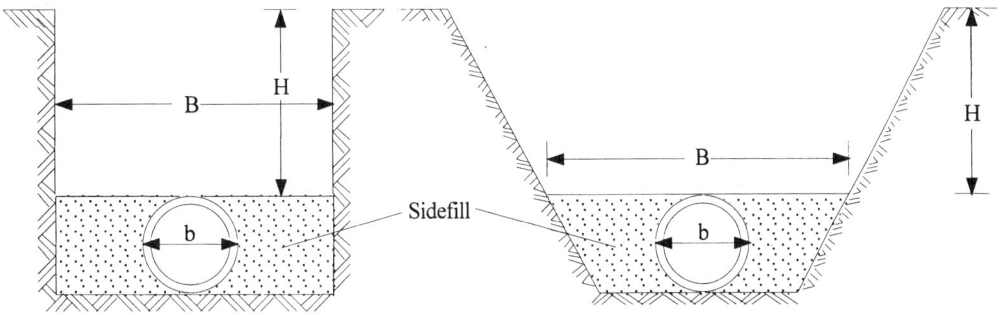

FIGURE 7.2 Common trench cuts for sewer pipes.

where W = vertical load on pipe as a result of backfill, lb/lin ft
C = dimensionless load coefficient based on the backfill and ratio of trench depth to width; a nomograph can be used
w = unit weight of backfill, lb/ft^3
B = width of trench at top of sewer pipe, ft (see Fig. 7.2)

The load coefficient C can be calculated as

$$C = \frac{1 - e^{-2k\mu'(H/B)}}{2k\mu'} \tag{7.26}$$

where e = base of natural logarithms
k = Rankine's ratio of lateral pressure to vertical pressure

$$k = \frac{\sqrt{\mu^2 + 1} - \mu}{\sqrt{\mu^2 + 1} + \mu} = \frac{1 - \sin\Phi}{1 + \sin\Phi} \tag{7.27}$$

μ = tan Φ
= coefficient of internal friction of backfill material
μ' = tan Φ'
= coefficient of friction between backfill material and the sides of the trench $\leq \mu$
H = height of backfill above pipe, ft (see Fig. 7.2)

Load on sewer for trench condition. The load on a sewer conduit for the trench condition is affected directly by the soil backfill. The load varies widely over different soil types, from a minimum of approximately 100 lb/ft^3 (1600 kg/m^3) to a maximum of about 135 lb/ft^3 (2160 kg/m^3) (WEF and ASCE 1992). The unit weight (density) of backfill material is as follows (McGhee 1991):

100 lb/ft^3 (1600 kg/m^3) for dry sand, and sand and damp topsoil;

115 lb/ft^3 (1840 kg/m^3) for saturated topsoil and ordinary sand;

120 lb/ft^3 (1920 kg/m^3) for wet sand and damp clay; and

130 lb/ft^3 (2080 kg/m^3) for saturated clay.

The average maximum unit weight of soil which will constitute the backfill over the sewer pipe may be determined by density measurements in advance of the structural design of the sewer pipe. A design value of not less than 120 or 125 lb/ft^3 (1920 or 2000 kg/m^3) is recommended (WEF and ASCE 1992).

The load on a sewer pipe is also influenced by the coefficient of friction between the backfill and the side of the trench (μ') and by the coefficient of internal friction of the backfill soil (μ). For most cases these two values are considered the same for design purposes. But, if the backfill is sharp sand and the sides of the trench are sheeted with finished lumber, μ may be substantially greater than μ'. Unless specific information to the contrary is available, values of the products $k\mu$ and $k\mu'$ may be assumed to be the same and equal to 0.103. If the backfill soil is slippery clay, $k\mu$ and $k\mu'$ are equal to 0.110 (WEF and ASCE 1992).

The values of the product $k\mu'$ in Eq. (7.26) range from 0.10 to 0.16 for most soils; specifically, 0.110 for saturated clay, 0.130 for clay, 0.150 for saturated top soil, 0.165 for sand and gravel, and 0.192 for cohesionless granular material (McGhee 1991). Graphical solutions of Eq. (7.26) are presented elsewhere (ASCE and WPCF 1982).

EXAMPLE: A 20-in (508-mm) ductile iron pipe is to be installed in an ordinary trench of 10 ft (3.05 m) depth at the top of the pipe and 4 ft (1.22 m) wide. The cut will be filled with damp clay. Determine the load on the sewer pipe.

Solution:

Step 1. Compute load coefficient C by Eq. (7.26)

$$k\mu' = 0.11$$

$$H/B = 3.05 \text{ m}/1.22 \text{ m} = 2.5$$

$$C = \frac{1 - e^{-2k\mu' H/B}}{2k\mu'} = \frac{1 - e^{-2(0.11)(2.5)}}{2(0.11)}$$

$$= 1.92$$

Step 2. Compute the load W by Eq. (7.25)

$$w = 120 \text{ lb/ft}^3 = 1920 \text{ kg/m}^3$$

$$W = CwB^2 = 1.92 \times 1920 \text{ kg/m}^3 \times (1.22 \text{ m})^2$$

$$= 5487 \text{ kg/m}$$

$$= 3687 \text{ lb/ft}$$

13 WASTEWATER TREATMENT SYSTEMS

As discussed in Chapters 1.2 and 1.3, the natural waters in streams, rivers, lakes, and reservoirs have a natural waste assimilative capacity to remove solids, organic matter, even toxic chemicals in the wastewater. However, it is a long process.

Wastewater treatment facilities are designed to speed up the natural purification process that occurs in natural waters and to remove contaminants in wastewater that might otherwise interfere with the natural process in the receiving waters.

Wastewater contains varying quantities of suspended and floating solids, organic matter, and fragments of debris. Conventional wastewater treatment systems are combinations of physical and biological (sometimes with chemical) processes to remove its impurities.

The alternative methods for municipal wastewater treatment are simply classified into three major categories: (1) primary (physical process) treatment, (2) secondary (biological process) treatment, and (3) tertiary (combination of physical, chemical, and biological process) or advanced treatment. As can be seen in Fig. 7.3, each category should include previous treatment devices (preliminary), disinfection, and sludge management (treatment and disposal). The treatment devices shown in the preliminary treatment are not necessarily to be included, depending on the wastewater characteristics and regulatory requirements.

For over a century, environmental engineers and aquatic scientists have developed wastewater treatment technologies. Many patented treatment process and package treatment plants have been

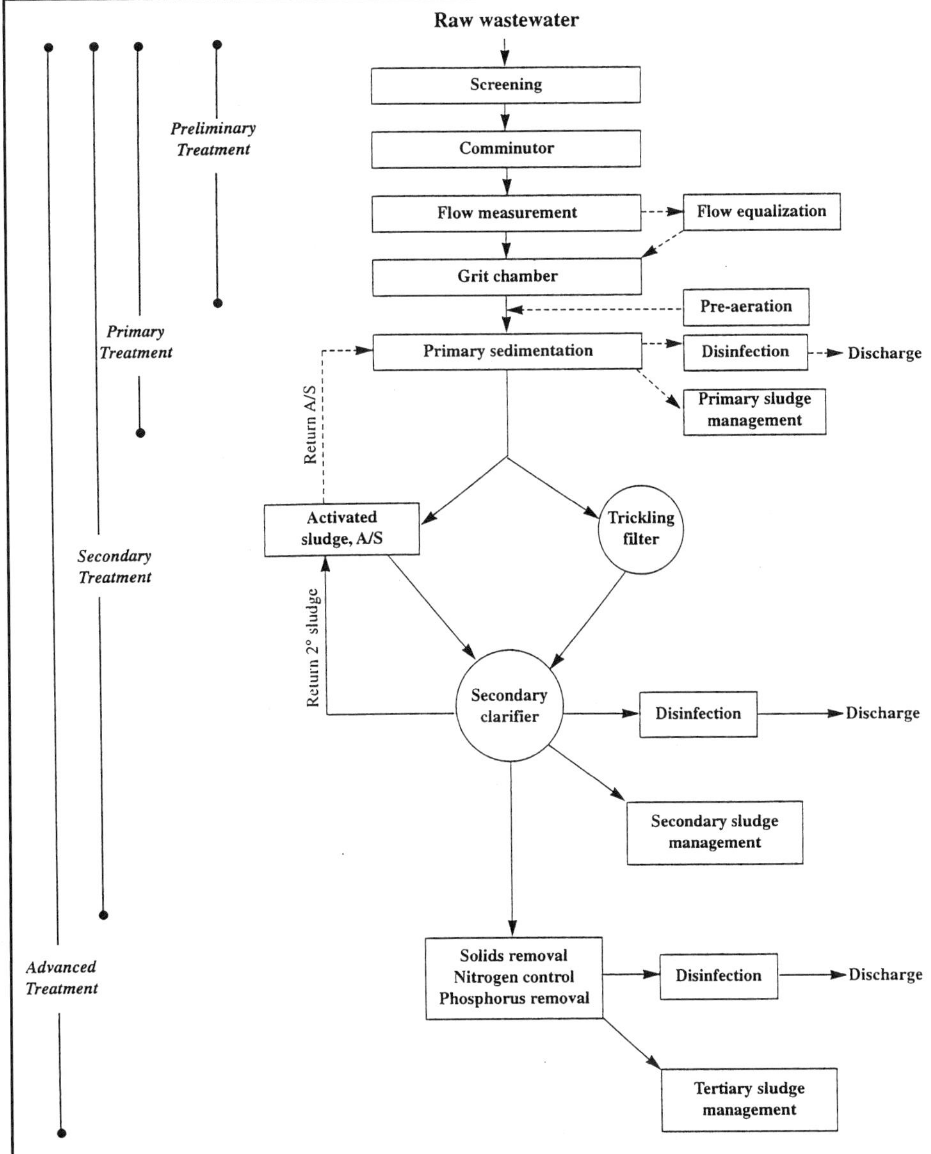

FIGURE 7.3 Flow chart for wastewater treatment processes.

developed and applied. The goal of wastewater treatment processes is to produce clean effluents and to protect public health, natural resources, and the ambient environment.

The Ten States Recommended Standards for Sewage Works, adopted by the Great Lakes–Upper Mississippi River Board (GLUMRB), was revised five times as a model for the design of wastewater treatment plants and for the recommended standards for other regional and state agencies. The original members of the ten states were Illinois, Indiana, Iowa, Michigan, Minnesota, Missouri, New York, Ohio, Pennsylvania, and Wisconsin. Recently, Ontario, Canada was added as a new member.

The new title of the standards is "Recommended Standards for Wastewater Facilities—Policies for design, review, and approval of plans and specifications for wastewater collection and treatment facilities," 1996 edition, by Great Lakes–Upper Mississippi River Board of State and Provincial Public Health and Environmental Managers.

13.1 Preliminary Treatment Systems

Preliminary systems are designed to physically remove or cut up the larger suspended and floating materials, and to remove the heavy inorganic solids and excessive amounts of oil and grease. The purpose of preliminary treatment is to protect pumping equipment and the subsequent treatment units. Preliminary systems consist of flow measurement devices and regulators (flow equalization), racks and screens, comminuting devices (grinders, cutters, and shredders), flow equalization, grit chambers, pre-aeration tanks, and (possibly) chlorination. The quality of wastewater is not substantially improved by preliminary treatment.

13.2 Primary Treatment Systems

The object of primary treatment is to reduce the flow velocity of the wastewater sufficiently to permit suspended solids to settle, i.e. to remove settleable materials. Floating materials are also removed by skimming. Thus, a primary treatment device may be called a settling tank (or basin). Due to variations in design and operation, settling tanks can be divided into four groups: plain sedimentation with mechanical sludge removal, two story tanks (Imhoff tank, and several patented units), upflow clarifiers with mechanical sludge removal, and septic tanks. When chemicals are applied, other auxiliary units are needed. Auxiliary units such as chemical feeders, mixing devices, and flocculators (New York State Department of Health 1950) and sludge (biosolids) management (treatment and dispose of) are required if there is no further treatment.

The physical process of sedimentation in settling tanks removes approximately 50 to 70 percent of total suspended solids from the wastewater. The BOD_5 removal efficiency by primary system is 25 to 35 percent. When certain coagulants are applied in settling tanks, much of the colloidal as well as the settleable solids, or a total of 80 to 90 percent of TSS, is removed. Approximately 10 percent of the phosphorus corresponding insoluble is normally removed by primary settling. During the primary treatment process, biological activity in the wastewater is negligible.

Primary clarification is achieved commonly in large sedimentation basins under relatively quiescent conditions. The settled solids are then collected by mechanical scrapers into a hopper and pumped to a sludge treatment unit. Fats, oils, greases, and other floating matter are skimmed off from the basin surface. The settling basin effluent is discharged over weirs into a collection conduit for further treatment, or to a discharging outfall.

In many cases, especially in developing countries, primary treatment is adequate to permit the wastewater effluent discharge, due to proper receiving water conditions or to the economic situation. Unfortunately, many wastewaters are untreated and discharged in many countries. If primary systems only are used, solids management and disinfection processes should be included.

13.3 Secondary Treatment Systems

After primary treatment the wastewater still contains organic matter in suspended, colloidal, and dissolved states. This matter should be removed before discharging to receiving waters, to avoid interfering with subsequent downstream users.

Secondary treatment is used to remove the soluble and colloidal organic matter which remains after primary treatment. Although the removal of those materials can be effected by physicochemical means providing further removal of suspended solids, secondary treatment is commonly referred to as the biological process.

Biological treatment consists of application of a controlled natural process in which a very large number of microorganisms consume soluble and colloidal organic matter from the wastewater in a relatively small container over a reasonable time. It is comparable to biological reactions that would occur in the zone of recovery during the self-purification of a stream.

Secondary treatment devices may be divided into two groups: attached and suspended growth processes. The attached (film) growth processes are trickling filters, rotating biologic contactors (RBC) and intermittent sand filters. The suspended growth processes include activated sludge and its modifications, such as contact stabilization (aeration) tanks, sequencing batch reactors, aerobic and anaerobic digestors, anaerobic filters, stabilization ponds, and aerated lagoons. Secondary treatment can also be achieved by physical–chemical or land application systems.

Secondary treatment processes may remove more than 85 percent of BOD_5 and TSS. However, they are not effective for the removal of nutrients (N and P), heavy metals, nonbiodegradable organic matter, bacteria, viruses, and other microorganisms. Disinfection is needed to reduce densities of microorganisms. In addition, a secondary clarifier is required to remove solids from the secondary processes. Sludges generated from the primary and secondary clarifiers need to undergo treatment and proper disposal.

13.4 Advanced Treatment Systems

Advanced wastewater treatment is defined as the methods and processes that remove more contaminants from wastewater than the conventional treatment. The term advanced treatment may be applied to any system that follows the secondary, or that modifies or replaces a step in the conventional process. The term tertiary treatment is often used as a synonym; however, the two are not synonymous. A tertiary system is the third treatment step that is used after primary and secondary treatment processes.

Since the early 1970s, the use of advanced wastewater treatment facilities has increased significantly in the United States. Most of their goals are to remove nitrogen, phosphorus, and suspended solids (including BOD_5) and to meet certain regulations for specific conditions. In some areas where water supply sources are limited, reuse of wastewater is becoming more important. Also, there are strict rules and regulations regarding the removal of suspended solids, organic matter, nutrients, specific toxic compounds and refractory organics that cannot be achieved by conventional secondary treatment systems as well as some industrial wastewaters; thus, advanced wastewater treatment processes are needed.

In the U.S. federal standards for secondary effluent are BOD 30 mg/L and TSS 30 mg/L. In Illinois, the standards are more stringent: BOD 20 mg/L and TSS 25 mg/L for secondary effluent. In some areas, 10 to 12 (BOD = 10 mg/L and TSS = 12 mg/L) standards are implied. For ammonia nitrogen standards, very complicated formulas depending on the time of the year and local conditions are used.

In the European Community, the European Community Commission for Environmental Protection has drafted the minimum effluent standards for large wastewater treatment plants. The standards include: $BOD_5 < 25$ mg/L, COD < 125 mg/L, suspended solids < 35 mg/L, total nitrogen < 10 mg/L, and phosphorus < 1 mg/L. Stricter standards are presented in various countries. The new regulations were expected to be ratified in 1998 (Boehnke *et al.* 1997).

TSS concentrations less than 20 mg/L are difficult to achieve by sedimentation through the primary and secondary systems. The purpose of advanced wastewater treatment techniques is specifically to reduce TSS, TDS, BOD, organic nitrogen, ammonia nitrogen, total nitrogen, or phosphorus. Biological nutrient removal processes can eliminate nitrogen or phosphorus, and any combination.

Advanced process use some processes for the drinking water treatment. Those include chemical coagulation of wastewater, wedge-wire screens, granular media filters, diatomaceous earth filters, microscreening, and ultrafiltration and nanofiltration, which are used to remove colloidal and fine-size suspended solids.

For nitrogen control, techniques such as biological assimilation, nitrification (conversion of ammonia to nitrogen and nitrate), and denitrification, ion exchange, breakpoint chlorination, air stripping are used. Soluble phosphorus may be removed from wastewater by chemical precipitation and biological (bacteria and algae) uptake for normal cell growth in a control system. Filtration is required after chemical and biological processes. Physical processes such as reverse osmosis and ultrafiltration also help to achieve phosphorus reduction, but these are primarily for overall dissolved inorganic solids reduction. Oxidation ditch, Bardenpho process, anaerobic/oxidation (A/O) process, and other patented processes are available.

The use of lagoons, aerated lagoons, and natural and constructed wetlands is an effective method for nutrients (N and P) removal.

Removal of some species of groups of toxic compounds and refractory organics can be achieved by activated carbon adsorption, air stripping, activated-sludge, powder activated-carbon processes, and chemical oxidation. Conventional coagulation–sedimentation–filtration and biological treatment (trickling filter, RBC, and activated sludge) processes are also used to remove the priority pollutants and some refractory organic compounds.

13.5 Compliance with Standards

The National Pollutant Discharge Elimination System (NPDES) has promulgated discharge standards to protect and preserve beneficial uses of receiving water bodies based on water quality criteria, or technology-based limits, or both. The receiving water quality criteria are typically established for a 7-day, 10-year, low-flow period.

Table 7.7 shows the national minimum performance standards for secondary treatment and its equivalency for public owned treatment works (POTW) (Federal Register, 1991). The secondary

TABLE 7.7 Minimum National Performance Standards for Public Owned Treatment Works (Secondary Treatment and Its Equivalency)

Parameter	30-day average shall not exceed	7-day average shall not exceed
Conventional secondary treatment processes		
5-day biochemical oxygen demand,* BOD_5		
Effluent, mg/L	30	45
Percent removal†	85	—
5-day carbonaceous biochemical oxygen demand,* $CBOD_5$		
Effluent, mg/L	25	40
Percent removal†	85	—
Suspended solids, SS		
Effluent, mg/L	30	45
Percent removal†	85	—
pH	6.0 to 9.0 at all times	—
Whole effluent toxicity	Site specific	—
Fecal coliform, MPN/100 mL	200	400
Stabilization ponds and other equivalent of secondary treatment		
5-day biochemical oxygen demand,* BOD_5		
Effluent, mg/L	45	65
Percent removal†	65	—
5-day carbonaceous biochemical oxygen demand,* $CBOD_5$		
Effluent, mg/L	40	60
Percent removal†	65	—
Suspended solids, SS		
Effluents, mg/L	45	65
Percent removal†	65	—
pH	6.0 to 9.0 at all times	—
Whole effluent toxicity	Site specific	—
Fecal coliform, MPN/100 mL	200	400

Notes: *Chemical oxygen demand (COD) or total organic carbon (TOC) may be substituted for BOD_5 when a long-term BOD_5 : COD or BOD_5 : TOC correlation has been demonstrated.

†Percent removal may be waived on a case-by-case basis for combined sewer service areas and for separated sewer areas not subject to excessive inflow and infiltration (I/I) where the base flow plus infiltration is ≤ 120 gpd/capita and the base flow plus I/I is ≤ 275 gpd/capita.

Source: Federal Register, 1991 (40 CFR 133; 49 FR 37006, September 20, 1984; revised through July 1, 1991. http/www.access.gpo.gov./nara/cfr/index.htm/, 1999).

treatment regulation was established from 40 Code of Regulation, Part 133, 49 Federal Register 37006, on September 20, 1984; and revised through July 1, 1991.

The values in Table 7.7 are still relevant as far as national standards go; however they really do not have much applicability in Illinois and many other states because state standards for most facilities are more stringent than the national standards. For example, in Illinois, most POTWs must meet monthly averages of 10/12 (means 10 mg/L of BOD and 12 mg/L of TSS) effluent standards if the dilution factor is less than 5 : 1. The daily maximum effluent concentrations are 20/24. For lagoon effluents, the standards are 30/30 if the dilution factor is greater than 5 : 1. The NPDES applies to each facility of interest because most have some water quality based effluent limits, especially for ammonia. The Illinois state effluent standards are in Title 35, Subtile C, Chapter II, Part 370 (IEPA, 1997) which can be found on the Illinois Pollution Control Board's web site at http://www.state.il.us/title35/35conten.htm.

14 SCREENING DEVICES

The wastewater from the sewer system either flows by gravity or is pumped into the treatment plant. Screening is usually the first unit operation at wastewater treatment plants. The screening units include racks, coarse screens, and fine screens. The racks and screens used in preliminary treatment are to remove large objects such as rags, plastics, paper, metals, dead animals, and the like. The purpose is to protect pumps and to prevent solids from fouling subsequent treatment facilities.

14.1 Racks and Screens

Coarse screens are classified as either bar racks (trash racks) or bar screens, depending on the spacing between the bars. Bar racks have clear spacing of 5.08 to 10.16 cm (2.0 to 4.0 in), whereas bar screens typically have clear spacing of 0.64 to 5.08 cm (0.25 to 2.0 in). Both consist of a vertical arrangement of equally spaced parallel bars designed to trap coarse debris. The debris captured on the bar screen depends on the bar spacing and the amount of debris caught on the screen (WEF 1996a).

Clear openings for manually cleaned screens between bars should be from 25 to 44 mm (1 to 1 3/4 in). Manually cleaned screens should be placed on a slope of 30 to 45 degrees to the horizontal. For manually or mechanically raked bar screens, the maximum velocities during peak flow periods should not exceed 0.76 m/s (2.5 ft/s) (Ten States Standards, (GLUMRB) 1996, Illinois EPA 1998).

Hydraulic losses through *bar racks* are a function of approach (upstream) velocity, and the velocity through the bars (downstream), with a discharge coefficient. Referring to Fig. 7.4, Bernoulli's equation can be used to estimate the headloss through bar racks:

$$h_1 + \frac{v^2}{2g} = h_2 + \frac{V^2}{2g} + \Delta h \tag{7.28}$$

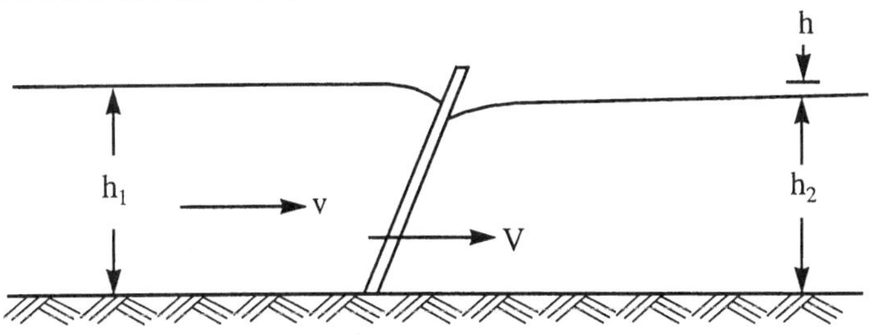

FIGURE 7.4 Profile for wastewater flowing through a bar screen.

and

$$h = h_1 - h_2 = \frac{V^2 - v^2}{2gC^2} \tag{7.29}$$

where h_1 = upstream depth of water, m or ft
h_2 = downstream depth of water, m or ft
h = headloss, m or ft
V = flow velocity through the bar rack, m/s or ft/s
v = approach velocity in upstream channel, m/s or ft/s
g = acceleration due to gravity, 9.81 m/s^2 or 32.2 ft/s^2

The headloss is usually incorporated into a discharge coefficient C; a typical value of $C = 0.84$, thus $C^2 = 0.7$. Eq. (7.29) becomes (for bar racks)

$$h = \frac{1}{0.7}\left(\frac{V^2 - v^2}{2g}\right) \tag{7.30}$$

Kirschmer (1926) proposed the following equation to describe the headloss through racks:

$$H = B\left(\frac{w}{b}\right)^{4/3}\frac{v^2}{2g}\sin\theta \tag{7.31}$$

where H = headloss, m
w = maximum width of the bar facing the flow, m
b = minimum clear spacing of bars, m
v = velocity of flow approaching the rack, m/s
g = gravitational acceleration, 9.81 m/s^2
θ = angle of the rack to the horizontal
B = bar shape factor, as follows

Bar type	B
Sharp-edged rectangular	2.42
Rectangular with semicircular face	1.83
Circular	1.79
Rectangular with semicircular upstream and downstream faces	1.67
Tear shape	0.76

The maximum allowable headloss for a rack is about 0.60 to 0.70 m. Racks should be cleaned when headloss is more than the allowable values.

EXAMPLE 1: Compute the velocity through a rack when the approach velocity is 0.60 m/s (2 ft/s) and the measured headloss is 38 mm (0.15 in)

Solution: Using Eq. (7.30)

$$h = \frac{V^2 - v^2}{0.7(2g)}$$

$$0.038\text{ m} = \frac{V^2 - (0.6\text{ m/s})^2}{0.7(2 \times 9.81\text{ m/s}^2)}$$

$$V^2 = 0.882$$

$$V = 0.94\text{ m/s}$$

$$= 3.08\text{ ft/s}$$

EXAMPLE 2: Design a coarse screen and calculate the headloss through the rack, using the following information:

Peak design wet weather flow = 0.631 m^3/s (10,000 gal/min)
Velocity through rack at peak wet weather flow = 0.90 m/s (3 ft/s)
Velocity through rack at maximum design dry weather flow = 0.6 m/s (2 ft/s)
$\theta = 60°$, with a mechanical cleaning device
Upstream depth of wastewater = 1.12 m (3.67 ft)

Solution:

Step 1. Calculate bar spacing and dimensions

(a) Determine total clear area (A) through the rack

$$A = \frac{\text{peak flow}}{v} = \frac{0.631\ \text{m}^3/\text{s}}{0.90\ \text{m/s}} = 0.70\ \text{m}^2$$

(b) Calculate total width of the opening at the rack, w

$$w = A/d = 0.70\ \text{m}^2/1.12\ \text{m} = 0.625\ \text{m}$$

(c) Choose a 25 mm clear opening

(d) Calculate number of opening, n

$$n = w/\text{opening} = 0.625\ \text{m}/0.025\ \text{m} = 25$$

Note: Use 24 bars with 10 mm (0.01 m) width and 50 mm thick.

(e) Calculate the width (W) of the chamber

$$\text{width} = 0.625\ \text{m} + 0.01\ \text{m} \times 24 = 0.865\ \text{m}$$

(f) Calculate the height of the rack

$$\text{height} = 1.12\ \text{m}/\sin 60° = 1.12\ \text{m}/0.866 = 1.29\ \text{m}$$

Allowing at least 0.6 m of freeboard, a 2-m height is selected.

(g) Determine the efficiency coefficient, EC

$$\text{EC} = \frac{\text{clear opening}}{\text{width of the chamber}} = 0.625\ \text{m}/0.865\ \text{m} = 0.72$$

Note: The efficiency coefficient is available from the manufacturer.

Step 2. Determine headloss of the rack by Eq. (7.31)
Select rectangular bars with semicircular upstream face, thus

$$B = 1.83$$
$$w/b = 1$$
$$\sin\theta = \sin 60° = 0.866$$
$$H = B\left(\frac{w}{b}\right)^{4/3}\frac{v^2}{2g}\sin\theta$$
$$= 1.83 \times 1 \times [(0.9 \text{ m/s})^2/(2 \times 9.81 \text{ m})] \times 0.866$$
$$= 0.065 \text{ m}$$

Note: If we want to calculate the headloss through the rack at 50 percent clogging, many engineers use an approximate method. When the rack becomes half-plugged, the area of the flow is reduced to one-half, and velocity through the rack is doubled. Thus the headloss will be 0.26 m (4 times 0.065 m).

14.2 Fine Screens

Fine screens are used more frequently in wastewater treatment plants for preliminary treatment or preliminary/primary treatment purposes. Fine screens typically consist of wedge-wire, perforated plate, or closely spaced bars with openings 1.5 to 6.4 mm (0.06 to 0.25 in). Fine screens used for preliminary treatment are rotary or stationary-type units (USEPA 1987a).

The clean water headloss through *fine screens* may be obtained from manufacturers' rating tables, or may be computed by means of the common orifice equation

$$h = \frac{1}{2g}\left(\frac{v}{C}\right)^2 = \frac{1}{2}\left(\frac{Q}{CA}\right)^2 \qquad (7.32)$$

where h = headloss, m or ft
v = approach velocity, m/s or ft/s
C = coefficient of discharge for the screen
g = gravitational acceleration, m/s^2 or ft/s^2
Q = discharge through the screen, m^3/s or ft^3/s
A = area of effective opening of submerged screen, m^2 or ft^2

Values of C depend on the size and milling of slots, the diameter and weave of the wire, and the percentage of open area. They must be determined experimentally. A typical value of C for a clean screen is 0.60. The headloss of clean water through a clean screen is relatively less. However, the headloss of wastewater through a fine screen during operation depends on the method and frequency of cleaning, the size and quantity of suspended solids in the wastewater, and the size of the screen opening.

The quantity of screenings generated at wastewater treatment plants varies with the bar opening, type of screen, wastewater flow, characteristics of served communities, and type of collection system. Roughly, 3.5 to 35 L (0.93 to 9.25 gal) of screenings is produced from 1000 m^3 (264,200 gal) wastewater treated. Screenings are normally 10 to 20 percent dry solids, with bulk density of 640 to 1120 kg/m^3 (40 to 70 lb/ft^3) (WEF and ASCE 1991a).

15 COMMINUTORS

As an alternative to racks or screens, a comminutor or shredder cuts and grinds up the coarse solids in the wastewater to about 6 to 10 mm (1/4 to 3/8 in) so that the solids will not harm subsequent treatment equipment. The chopped or ground solids are then removed in primary sedimentation basins. A comminutor consists of a fixed screen and a moving cutter. Comminution can eliminate the messy and offensive screenings for solids handling and disposal. However, rags and large objects cause clogging problems.

Comminutors are installed directly in the wastewater flow channel and are equipped with a bypass so that the unit can be isolated for service maintenance. The size, installation, operation, and maintenance of the comminutors are available by the manufactures.

16 GRIT CHAMBER

Grit originates from domestic wastes, stormwater runoff, industrial wastes, pumpage from excavations, and groundwater seepage. It consists of inert inorganic material such as sand, cinders, rocks, gravel, cigarette filter tips, metal fragments, etc. In addition grit includes bone chips, eggshells, coffee grounds, seeds, and large food wastes (organic particles). These substances can promote excessive wear of mechanical equipment and sludge pumps, and even clog pipes by deposition.

Composition of grit varies widely, with moisture content ranging from 13 to 63 percent, and volatile content ranging from 1 to 56 percent. The specific gravity of clean grit particles may be as high as 2.7 with inert material, and as low as 1.3 when substantial organic matter is agglomerated with inert. The bulk density of grit is about 1600 kg/m^3 or 100 lb/ft^3 (Metcalf and Eddy, Inc. 1991).

Grit chambers should be provided for all wastewater treatment plants, and are used on systems required for plants receiving sewage from combined sewers or from sewer systems receiving a substantial amount of ground garbage or grit. Grit chambers are usually installed ahead of pumps and comminuting devices.

Grit chambers for plants treating wastewater from combined sewers usually have at least two hand cleaned units, or a mechanically cleaned unit with bypass. There are three types of grit settling chamber: hand cleaned, mechanically cleaned, and aerated or vortex-type degritting units. The chambers can be square, rectangular, or circular. A velocity of 0.3 m/s (1 ft/s) is commonly used to separate grit from the organic material. Typically, 0.0005 to 0.00236 m^3/s (1 to 5 ft^3/min) of air per foot of chamber length is required for a proper aerated grit chamber; or 4.6 to 7.7 L/s per meter of length. The transverse velocity at the surface should be 0.6 to 0.8 m/s or 2 to 2.5 ft/s (WEF 1996a).

Grit chambers are commonly constructed as fairly shallow longitudinal channels to catch high specific gravity grit (1.65). The units are designed to maintain a velocity close to 0.3 m/s (1.0 ft/s) and to provide sufficient time for the grit particle to settle to the bottom of the chamber.

EXAMPLE: The designed hourly average flow of a municipal wastewater plant is 0.438 m^3/s (10 Mgal/d). Design an aerated grit chamber where the detention time of the peak flow rate is 4.0 min (generally 3 to 5 min).

Solution:

Step 1. Determine the peak hourly flow Q
Using a peaking factor of 3.0

$$\begin{aligned} Q &= 0.438 \text{ m}^3/\text{s} \times 3 \\ &= 1.314 \text{ m}^3/\text{s} \\ &= 30 \text{ Mgal/d} \end{aligned}$$

Step 2. Calculate the volume of the grit chamber
Two chambers will be used; thus, for each unit

$$\begin{aligned} \text{Volume} &= 1.314 \text{ m}^3/\text{s} \times 4 \text{ min} \times 60 \text{ s/min} \div 2 \\ &= 157.7 \text{ m}^3 \\ &= 5570 \text{ ft}^3 \end{aligned}$$

Step 3. Determine the size of a rectangular chamber
Select the width of 3 m (10 ft), and use a depth-to-width ratio of 1.5 : 1 (typically 1.5 : 1 to 2.0 : 1)

$$\begin{aligned} \text{Depth} &= 3 \text{ m} \times 1.5 = 4.5 \text{ m} \\ &= 15 \text{ ft} \\ \text{Length} &= \text{volume}/(\text{depth} \times \text{width}) = 157.7 \text{ m}^3/(4.5 \text{ m} \times 3 \text{ m}) \\ &= 11.7 \text{ m} \\ &= 36 \text{ ft} \end{aligned}$$

Note: Each of the two chambers has a size of 3 m × 4.5 m × 11.7 m or 10 ft × 15 ft × 36 ft.

Step 4. Compute the air supply needed
Use 5 std ft^3/min (scfm) or (0.00236 m^3/s/ft (0.3 m) length.

$$\begin{aligned} \text{Air needed} &= 0.00236\ \text{m}^3/(\text{s}\cdot\text{ft}) \times 36\ \text{ft} \\ &= 0.085\ \text{m}^3/\text{s} \\ \text{or} &= 5\ \text{ft}^3/\text{min}\cdot\text{ft} \times 36\ \text{ft} \\ &= 180\ \text{ft}^3/\text{min} \end{aligned}$$

Step 5. Estimate the average volume of grit produced
Assume 52.4 mL/m^3 (7 ft^3/Mgal) of grit produced

$$\begin{aligned} \text{Volume of grit} &= 52.4\ \text{mL/m}^3 \times 0.438\ \text{m}^3/\text{s} \times 86{,}400\ \text{s/d} \\ &= 1{,}980{,}000\ \text{mL/d} \\ &= 1.98\ \text{m}^3/\text{d} \\ \text{or} &= 7\ \text{ft}^3/\text{Mgal} \times 10\ \text{Mgal/d} \\ &= 70\ \text{ft}^3/\text{d} \end{aligned}$$

17 FLOW EQUALIZATION

The Parshall flume is commonly used in wastewater treatment plants. Methods of flow measurement are discussed in Chapter 1.5.

The incoming raw wastewater varies with the time of the day, the so-called diurnal variation, ranging from less than one-half to more than 200 percent of the average flow rate. A storm event increases the flow. Flow equalization is used to reduce flow fluctuations in the collection system or in the in-plant storage basins as well as in the water treatment plant. This benefits the performance of the downstream treatment processes and reduces the size and cost of treatment units.

Flow equalization facilities include the temporary storage of flows in existing sewers, the use of in-line or on-line separate flow-equalization facilities or retention basins.

The volume for a flow equalization basin is determined from mass diagrams based on average diurnal flow patterns.

EXAMPLE: Determine a flow equalization basin using the following diurnal flow record:

Time	Flow, m^3/s	Time	Flow, m^3/s
Midnight	0.0492	Noon	0.1033
1	0.0401	1 p.m.	0.0975
2	0.0345	2	0.0889
3	0.0296	3	0.0810
4	0.0288	4	0.0777
5	0.0312	5	0.0755
6	0.0375	6	0.0740
7	0.0545	7	0.0700
8	0.0720	8	0.0688
9	0.0886	9	0.0644
10	0.0972	10	0.0542
11	0.1022	11	0.0513

Solution 1:

Step 1. Compute the average flow rate Q

$$Q = \sum q/24 = 0.0655 \text{ m}^3/\text{s}$$

Step 2. Compare the observed flows and average flow from the data shown above. The first observed flow to exceed Q is at 8 a.m.

Step 3. Construct a table which is arranged in order, beginning at 8 a.m. See Table 7.8. Calculations of col. 3 to 6 are given in the following steps.

Step 4. For col. 3, convert the flows to volume for 1 h time interval.

$$\begin{aligned} \text{Volume} &= 0.072 \text{ m}^3/\text{s} \times 1 \text{ h} \times 3600 \text{ s/h} \\ &= 259.2 \text{ m}^3 \end{aligned}$$

Step 5. For col. 4, for each row: calculate average volume to be treated.

$$\begin{aligned} \text{Volume} &= Q \times 1 \text{ h} \times 3600 \text{ s/h} \\ &= 0.0655 \text{ m}^3/\text{s} \times 1 \text{ h} \times 3600 \text{ s/h} \\ &= 235.8 \text{ m}^3 \end{aligned}$$

TABLE 7.8 Analysis of Flow Equalization

(1) Time	(2) Flow m^3/s	(3) Volume in, m^3	(4) Volume out, m^3	(5) Storage m^3	(6) Σ storage, m^3
8 a.m.	0.072	259.2	235.8	23.4	23.4
9	0.0886	318.96	235.8	83.16	106.56
10	0.0972	349.92	235.8	114.12	220.68
11	0.1022	367.92	235.8	132.12	352.8
12	0.1033	371.88	235.8	136.08	488.88
1 p.m.	0.0975	351	235.8	115.2	604.08
2	0.0889	320.04	235.8	84.24	688.32
3	0.081	291.6	235.8	55.8	744.12
4	0.0777	279.72	235.8	43.92	788.04
5	0.0755	271.8	235.8	36	824.04
6	0.0740	266.4	235.8	30.6	854.64
7	0.0700	252	235.8	16.2	870.84
8	0.0688	247.68	235.8	11.88	882.72
9	0.0644	231.84	235.8	−3.96	878.76
10	0.0542	195.12	235.8	−40.68	838.08
11	0.0513	184.68	235.8	−51.12	786.96
12	0.0492	177.12	235.8	−58.68	728.28
1 a.m.	0.0401	144.36	235.8	−91.44	636.84
2	0.0345	124.2	235.8	−111.6	525.24
3	0.0296	106.56	235.8	−129.24	396
4	0.0288	103.68	235.8	−132.12	263.88
5	0.0312	112.32	235.8	123.48	140.4
6	0.0375	135	235.8	−100.8	39.6
7	0.0545	196.2	235.8	−39.6	0

Step 6. For col. 5: calculate the excess volume needed to be stored.

$$\text{Col. } 5 = \text{col. } 3 - \text{col. } 4$$

$$\text{Example: } 259.2 \text{ m}^3 - 235.8 \text{ m}^3 = 23.4 \text{ m}^3$$

Step 7. For col. 6: calculate the cumulative sum of the difference (col. 5).
Example: For the second time interval, the cumulative storage (cs) is

$$\text{cs} = 23.4 \text{ m}^3 + 83.16 \text{ m}^3 = 106.56 \text{ m}^3$$

Note: The last value for the cumulative storage should be zero. Theoretically, it means that the flow equalization basin is empty and ready to begin the next day's cycle.

Step 8. Find the required volume for the basin.
The required volume for the flow equalization basin for this day is the maximum cumulative storage. In this case, it is 882.72 m^3 at 8 p.m. (col. 6, Table 7.8). However, it is common to provide 20 to 50 percent excess capacity for unexpected flow variations, equipment, and solids deposition. In this case, we provide 35 percent excess capacity. Thus the total storage volume should be

$$\begin{aligned}\text{Total basin volume} &= 882.72 \text{ m}^3 \times 1.35 \\ &= 1192 \text{ m}^3\end{aligned}$$

Solution 2: Graphic method (Fig. 7.5).

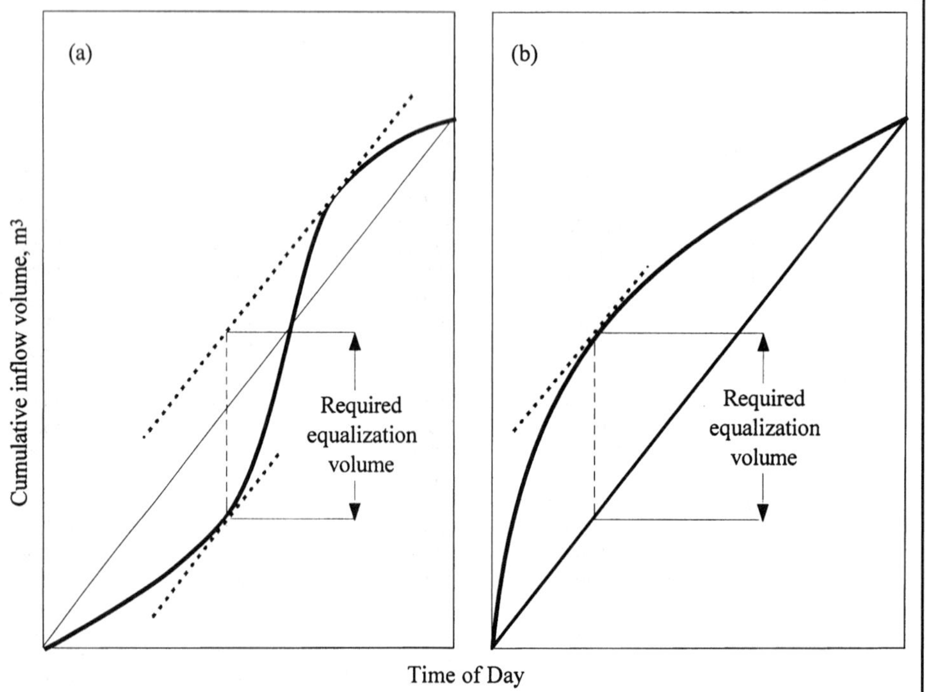

FIGURE 7.5 Mass diagram for determining required equalization volume.

Step 1. Calculate cumulative volumes as in solution 1.

Step 2. Plot time of day at *X*-axis (starting at midnight) versus cumulative volume at *Y*-axis to produce a mass curve.

Step 3. Connect the origin and the final point on the mass curve. This gives the daily average flow rate (m^3/d or Mgal/d).

Step 4. Draw two lines parallel to the average flow rate and tangent to the mass curve at the highest and lowest points.

Step 5. Determine the required volume for the flow equalization basin.
The vertical distance between two parallels drawn in Step 4 is the required basin capacity.
Note: In the above example, the storage starts to fill in at 8 a.m. and it is empty 24 h later. At the highest point of the cumulative volume draw a tangent line (Fig. 7.5b). The distance between this tangent line and the average flow line is the storage volume required.

18 SEDIMENTATION

Sedimentation is the process of removing solid particles heavier than water by gravity settling. It is the oldest and most widely used unit operation in water and wastewater treatments. The terms sedimentation, settling, and clarification are used interchangeably. The unit sedimentation basin may also be referred to as a sedimentation tank, clarifier, settling basin, or settling tank.

In wastewater treatment, sedimentation is used to remove both inorganic and organic materials which are settleable in continuous-flow conditions. It removes grit, particulate matter in the primary settling tank, and chemical flocs from a chemical precipitation unit. Sedimentation is also used for solids concentration in sludge thickeners.

Based on the solids concentration, and the tendency of particle interaction, there are four types of settling which may occur in wastewater settling operations. The four categories are discrete, flocculant, hindered (also called zone), and compression settlings. They are also known as types 1, 2, 3, and 4 sedimentation, respectively. Some discussion of sedimentation is covered in Chapter 1.6. The following describes each type of settling.

18.1 Discrete Particle Sedimentation (Type 1)

The plain sedimentation of a discrete spherical particle, described by Newton's law, can be applied to grit removal in grit chambers and sedimentation tanks. The terminal settling velocity is determined as (also in Chapter 1.6, Eq. (6.60))

$$v_s = \left[\frac{4g(\rho_s - \rho)d}{3C_D\rho}\right]^{1/2} \tag{7.33}$$

where v_s = terminal settling velocity, m/s or ft/s
ρ_s = mass density of particle, kg/m^3 or lb/ft^3
ρ = mass density of fluid, kg/m^3 or lb/ft^3
g = acceleration due to gravitation, 9.81 m/s^2 or 32.2 ft/s^2
d = diameter of particle, mm or in
C_D = dimensionless drag coefficient

The drag coefficient C_D is not constant. It varies with the Reynolds number and the shape of the particle. The Reynolds number $\mathbf{R} = vd\rho/\mu$, where μ is the absolute viscosity of the fluid, and the other terms are defined as above.

C_D varies with the effective resistance area per unit volume and shape of the particle. The relationship between **R** and C_D is as follows

$$1 > \mathbf{R}: \quad C_D = \frac{24}{\mathbf{R}} = \frac{24\mu}{v\rho d} \tag{7.34}$$

$$1 < \mathbf{R} < 1000: \quad C_D = \frac{24}{\mathbf{R}} + \frac{3}{\mathbf{R}^{0.5}} + 0.34 \tag{7.35}$$

$$\text{or} = \frac{18.5}{\mathbf{R}^{0.5}} \tag{7.36}$$

$$\mathbf{R} > 1000: \quad C_D = 0.34 \text{ to } 0.40 \tag{7.37}$$

For small **R** (< 1 or 2) with laminar flow. Eq. (7.34) is applied. Eq. (7.35) or (7.36) is applicable for **R** up to 1000, which includes all situations of water and wastewater treatment processes. For fully developed turbulent settling use $C_D = 0.34$ to 0.40 (Eq. (7.37)).

When the Reynolds number is less than 1, substitution of Eq. (7.34) for C_D in Eq. (7.33) yields Stokes' law (Eq. (6.63))

$$v_s = \frac{g(\rho_s - \rho)d^2}{18\mu} \tag{7.38}$$

Discrete particle settling refers to type 1 sedimentation. Under quiescent conditions, suspended particles in water or wastewater exhibit a natural tendency to agglomerate, or the addition of coagulant chemicals promotes flocculation. The phenomenon is called flocculation–sedimentation or type 2 sedimentation. For flocculated particles the principles of settling are the same as for a discrete particle, but settling merely occurs at a faster rate.

18.2 Scour

The horizontal velocity in grit chambers or in sedimentation tanks must be controlled to a value less than what would carry the particles in traction along the bottom. The horizontal velocity of fluid flow just sufficient to create scour is described as (Camp 1946)

$$V = \left[\frac{8\beta(s-1)gd}{f}\right]^{1/2} \tag{7.39}$$

where V = horizontal velocity, m/s
β = constant for the type of scoured particles
= 0.04 for unigranular material
= 0.06 for sticky interlocking material
s = specific gravity of particle
g = acceleration due to gravity, 9.81 m/s^2
d = diameter of particle, m
f = Darcy–Weisbach friction factor, 0.02–0.03

The f values are a function of the Reynolds number and surface characteristics of the settled solids. The horizontal velocity in most sedimentation tanks is well below that which would cause scour. In grit chambers, scour is an important factor for design.

EXAMPLE: Determine the surface overflow rate and horizontal velocity of a grit chamber to remove the grit without removing organic material. Assume that grit particles have a diameter of 0.2 mm (0.01 in) and a specific gravity of 2.65 (sand, silt, and clay); and organic material has the same diameter and a specific gravity of 1.20. Assume $C_D = 10$.

Solution:

Step 1. Compute the terminal settling velocity, using Eq. (7.33)

$$C_D = 10$$

$$d = 0.2 \text{ mm} = 0.02 \text{ cm}$$

$$v_s = \left[\frac{4g(\rho_s - \rho)d}{3C_D\rho}\right]^{1/2}$$

$$= \left[\frac{4 \times 981 \times (2.65 - 1) \times 0.02}{3 \times 10 \times 1}\right]^{1/2}$$

$$= 2.08 \text{ (cm/s)}$$

Note: This will be the surface overflow rate to settle grit, not organic matter.

Step 2. Compute the horizontal velocity (V_1) just sufficient to cause the grit particles to scour Use $\beta = 0.06$ and $f = 0.03$. Using Eq. (7.39)

$$V_1 = \left[\frac{8\beta(s - 1)gd}{f}\right]^{1/2}$$

$$= \left[\frac{8 \times 0.06 \times (2.65 - 1) \times 981 \times 0.02}{0.03}\right]^{1/2}$$

$$= 22.8 \text{ (cm/s)}$$

Step 3. Compute the scouring velocity V_2 for organic material, using Eq. (7.39)

$$V_2 = \left[\frac{8 \times 0.06 \times (1.20 - 1) \times 981 \times 0.02}{0.03}\right]^{1/2}$$

$$= 7.9 \text{ (cm/s)}$$

Note: The grit chamber is designed to have a surface overflow rate (settling velocity) of 2.1 cm/s and a horizontal velocity less than 22.8 cm/s but greater than 7.9 cm/s. Under these conditions, the grit will be removed and organic matter will not. If the horizontal velocity is close to the scour velocity, the grit will be reasonably clean.

18.3 Sedimentation Tank (Basin) Configuration

Sedimentation tanks can be rectangular, square, or circular. Imhoff tanks perform the dual function of settling and aerobic treatment with two-story chambers; however, the Imhoff tank is old technology and is no longer allowed in the developed countries.

For a continuous flow sedimentation tank, the shape can be either rectangular or circular. Camp (1953) divided the ideal sedimentation tank into four zones which affect settling, namely the inlet zone, theoretical effective settling zone, sludge zone (beneath the settling zone), and outlet zone (Fig. 7.6). The inlet and outlet condition and tank geometry influence short circuiting, which can be minimized in narrow rectangular horizontal flow basins. Short circuiting is a common problem in circular radial flow clarifiers.

Figure 7.6 illustrates an ideal rectangular continuous horizontal flow settling tank. The inlet zone uniformly distributes wastewater flows and solids over the cross-sectional area of the tank in such a manner that flow through the settling zone follows horizontal paths to prevent short circuiting. In the settling zone, a uniform concentration of particles settles at terminal settling velocity to the sludge zone at the bottom of the tank. In the real world, there is no theoretical effective settling zone. Particle settling vectors are difficult to predict. However, it is usually assumed that the flow of wastewater through the settling zone is steady and that the concentration of each sized particle is uniform throughout the cross section normal to the flow direction. The sludge zone is a region for storing the settled sediments below the settling zone. This zone may be neglected for practical purposes, if mechanical equipment continually removes the sediment. In the outlet zone, the supernatant (clarified effluent) is collected through an outlet weir and discharged to further treatment units or to the outfall.

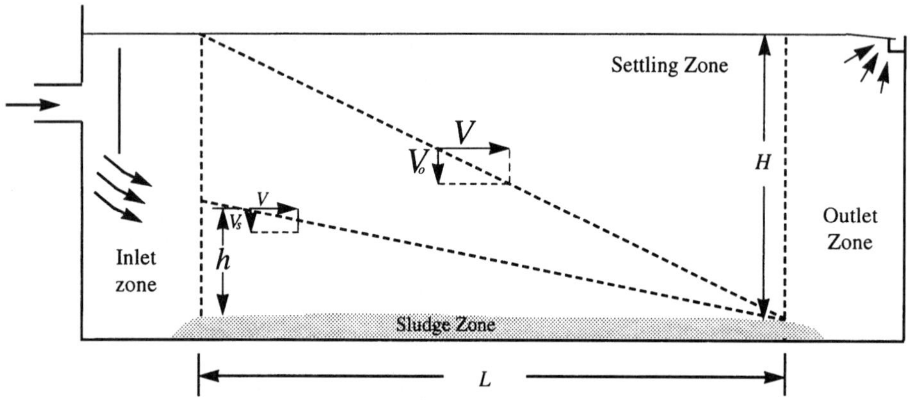

FIGURE 7.6 Sketch of the discrete particle settling in an ideal settling tank.

In the design of clarifiers, a particle terminal velocity V_0 is used as a design overflow settling velocity, which is the settling velocity of the particle which will settle through the total effective depth H of the tank in the theoretical detention time. All particles that have a terminal velocity (V_s) equal to or greater than V_0 will be removed. The surface overflow rate of wastewater is (Stoke's law)

$$V_0 = Q/A = Q/WL \tag{7.40}$$

$$= \frac{g(\rho_s - \rho)d^2}{18\mu} \tag{7.41}$$

where Q = flow, m^3/d or gal/d
A = surface area of the settling zone, m^2 or ft^2
V_0 = overflow rate or surface loading rate, $m^3/(m^2 \cdot d)$ or $gal/(ft^2 \cdot d)$
W, L = width and length of the tank, m or ft

This is called type 1 settling. Flow capacity is independent of the depth of a clarifier. Basin depth H is a product of the design overflow velocity and detention time t

$$H = V_0 t \tag{7.42}$$

The flow through velocity V_f is

$$V_f = Q/HW \tag{7.43}$$

where H is the depth of the settling zone. The retention time t is

$$t = \text{Volume}/Q \tag{7.44}$$

The removal ratio r (or fraction of removal) of particles having a settling velocity equal to V_s will be h/H. Since depth equals the product of the settling velocity and retention time t (Fig. 7.6)

$$r = \frac{h}{H} = \frac{V_s t}{V_0 t} = \frac{V_s}{V_0} \tag{7.45}$$

where r is the fraction of the particles with settling velocity V_s that are removed. This means that in a horizontal flow particles with settling velocity V_s less than V_0 will also be removed if they enter the settling zone at a depth less than H.

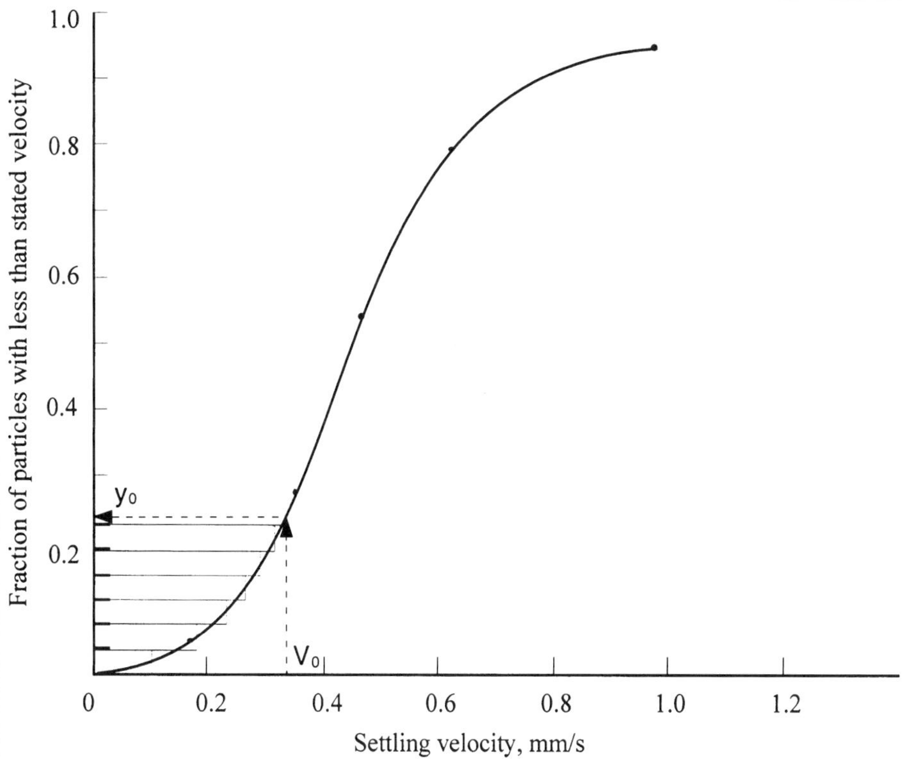

FIGURE 7.7 Cumulative particles removal versus settling velocity curve.

The settling velocity distribution for a suspension sample can be determined from a column settling test. The data obtained from the test can be used to construct a cumulative settling velocity frequency distribution curve, as shown in Fig. 7.7.

For a given clarification flow rate Q, only those particles having settling velocity $\geq V_0$ ($= Q/A$) will be completely removed. Let y_0 represent the portion of particles with a settling velocity $< V_0$; then the percentage removed will be $1 - y_0$. Also, for each size particle with $V_s < V_0$ its proportion of removal, expressed as Eq. (7.45), is equal to $r = V_s/V_0$. When considering various particle sizes in this group, the percentage of removal is

$$\int_0^{y_0} \frac{V_s}{V_0} dy$$

The overall fraction of particles removed, F, would be

$$F = (1 - y_0) + \frac{1}{V_0}\int_0^{y_0} V_s\, dy \tag{7.46}$$

Approximation:

$$F = 1 - y_0 + \frac{V_0 + V_1}{2V_0}(y_0 - y_1) + \frac{V_1 + V_2}{2V_0}(y_1 - y_2) + \cdots + \frac{V_i + V_{i+1}}{2V_0}(y_1 - y_{i+1})$$

$$F = 1 - y_0 + \frac{1}{V_0}\sum V\Delta y \tag{7.47}$$

where y_0 = fraction of particles by weight with $V_s \geq V_0$
i = ith particle

EXAMPLE (type 1): A clarifier is designed to have a surface overflow rate of 28.53 $m^3/(m^2 \cdot d)$ (700 gal/($ft^2 \cdot d$)). Estimate the overall removal with the settling analysis data and particle size distribution in cols. 1 and 2 of Table 7.8. The wastewater temperature is 15°C and the specific gravity of the particles is 1.20.

Solution:

Step 1. Determine settling velocities of particles by Stokes' law, Eq. (7.41)
From Table 5.1a, at 15°C

$$\mu = 0.00113 \text{ N} \cdot \text{s/m}^2 = 0.00113 \text{ kg/(s} \cdot \text{m)}$$

$$\rho = 0.9990$$

$$V = \frac{g(\rho_s - \rho)d^2}{18\mu}$$

$$= \frac{9.81 \text{ m/s}^2(1200 - 999) \text{ kg/m}^3 \times d^2}{18 \times 0.00113 \text{ kg/(s} \cdot \text{m)}}$$

$$= 96{,}942d^2 \text{ m/s}$$

where d is in m

Step 2. Calculate V for each particle size (col. 3 of Table 7.9)
For d = 0.1 mm = 0.0001 m

$$V = 96{,}942\ (0.0001)^2$$
$$= 0.000969 \text{ m/s}$$
$$= 0.968 \text{ mm/s}$$

Similarly, calculate the settling velocities for other particle sizes.

Step 3. Construct the settling velocities versus cumulative distribution curve shown in Fig. 7.7.

Step 4. Calculate designed settling velocity V_0

$$V_0 = 28.53 \text{ m/d}$$
$$= 28{,}530 \text{ mm/d} \times 1 \text{ day/86,400 s}$$
$$= 0.33 \text{ mm/s}$$

TABLE 7.9 Results of Settling Analysis Test and Estimation of Overall Solids Removal

Particle size mm	Weight fraction < size, %	Settling velocity V, mm/s
0.10	12	0.968
0.08	18	0.620
0.07	35	0.475
0.06	72	0.349
0.05	86	0.242
0.04	94	0.155
0.02	99	0.039
0.01	100	0.010

Note: All particles with settling velocities greater than 0.33 mm/s (700 gal/(d · ft^2)) will be removed.

Step 5. Find the fraction $(1 - y_0)$
From Fig. 7.7 we read $y_0 = 0.25$ at $V_0 = 0.33$ mm/s

then $$1 - y_0 = 1 - 0.25 = 0.75$$

Step 6. Graphical determination of $\Sigma V\Delta y$

Referring to Figure 7.7

Δy	0.04	0.04	0.04	0.04	0.04	0.04	0.01
V	0.09	0.17	0.23	0.26	0.28	0.31	0.33
$V\Delta y$	0.0036	0.0068	0.0092	0.0104	0.0112	0.0124	0.0033

$$\sum V\Delta y = 0.0569$$

Step 7. Determine overall removal R
Using Eq. (7.47)

$$\begin{aligned} F &= (1 - y_0) + \frac{1}{V_0}\sum V\Delta y \\ &= 0.75 + 0.0569/0.33 \\ &= 0.92 \\ &= 92 \text{ percent} \end{aligned}$$

18.4 Flocculant Settling (Type 2)

In practice, the actual settling performance cannot be adequately predicted because of unrealistic assumptions on ideal discrete particle settling. Under quiescent conditions, suspended particles in water or wastewater exhibit a natural tendency to agglomerate. Also, suspended solids in wastewater are not discrete particles and vary more than light and small particles, as they contact and agglomerate and grow in size. As coalescence of flocculation occurs, including chemical coagulation and biological flocs, the mass of the particles increases and they settle faster. This phenomenon is called flocculant or type 2 sedimentation.

The flocculation process increases removal efficiency but it cannot be adequately expressed by equations. Settling-column analysis is usually used to determine the settling characteristics of flocculated particles. A column can be of any diameter and equal in length to the proposed clarifier. Satisfactory results can be achieved with 15 cm (6 in) diameter plastic tube 3 m (10 ft) in height (Metcalf and Eddy, Inc. 1991). Sampling ports are uniformly spaced (45 to 60 cm or 1.5 to 2 ft) along the length of the column. The test suspension is placed in the settle-column and allowed to settle in a quiescent manner. The initial suspended solids concentration is measured. Samples are withdrawn from the sampling ports at various selected time intervals from different depths. Analyses of SS are performed for each sample, and the data used to calculate the percentage of removal is plotted as a number against time and depth. Between the plotted points, curves of equal percent removal are drawn. The results of settling-column analyses are presented in Fig. 7.8. Use of the curves in Fig. 7.8 is illustrated in the following example.

EXAMPLE: Using the settling curves of Fig. 7.8, determine the overall removal of solids in a sedimentation basin (type 2 flocculant settling) with a depth equal to the test cylinder and at a detention time of 20 min. The total depth is 2.5 m.

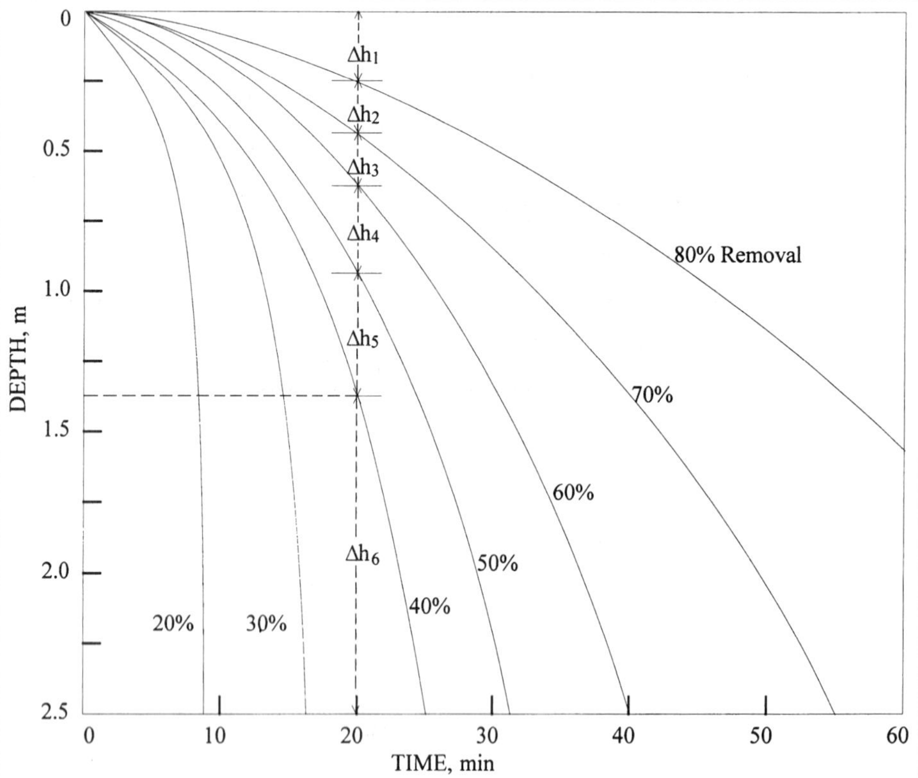

FIGURE 7.8 Settling characteristics for flocculent particles.

Solution:

Step 1. From Fig. 7.8, 40 percent of the particles will have a settling velocity of 0.1 m/min (2.5 m/25 min)

At t = 20 min, the volume of the test cylinder within Δh_6 has 40% removal

Step 2. Determine percent removal of each volume of the tank
In the volume of the tank corresponding to Δh_5 between 50 and 40 percent removal will occur. Similarly, in the tank volume corresponding to Δh_4 between 60 and 50 percent will be removed. In like fashion, this is applied to other tank volumes.

Step 3. Calculate the overall removal
Since $1/h = 1/2.5 = 0.4$

$\Delta h_1 = 0.23$ m

$\Delta h_2 = 0.14$ m

$\Delta h_3 = 0.20$ m

$\Delta h_4 = 0.32$ m

$\Delta h_5 = 0.50$ m

$$F = 40 + \frac{\Delta h_5}{h}\left(\frac{40 + 50}{2}\right) + \frac{\Delta h_4}{h}\left(\frac{50 + 60}{2}\right) + \frac{\Delta h_3}{h}\left(\frac{60 + 70}{2}\right)$$

$$+ \frac{\Delta h_2}{h}\left(\frac{70 + 80}{2}\right) + \frac{\Delta h_1}{h}\left(\frac{80 + 100}{2}\right)$$

$$= 40 + 0.4(0.5 \times 45 + 0.32 \times 55 + 0.20 \times 65 + 0.14 \times 75 + 0.23 \times 90)$$

$$= 73.7\% \text{ removal}$$

18.5 Hindered Sedimentation (Type 3)

In systems with high concentrations of suspended solids, the velocity fields of closely spaced particles are obstructed, causing an upward displacement of the fluid and hindered or zone settling (type 3) and compression settling (type 4). In addition, discrete (free) settling (type 1) and flocculant settling (type 2) occur. This settling phenomenon of concentrated suspensions (such as activated sludge) is illustrated in a graduated cylinder, as shown in Fig. 7.9.

Hindered (zone) settling occurs in sludge thickeners and at the bottom of a secondary clarifier in biological treatment processes. The velocity of hindered settling is estimated by (Steel and McGhee 1979)

$$v_h/v = (1 - C_v)^{4.65} \tag{7.48}$$

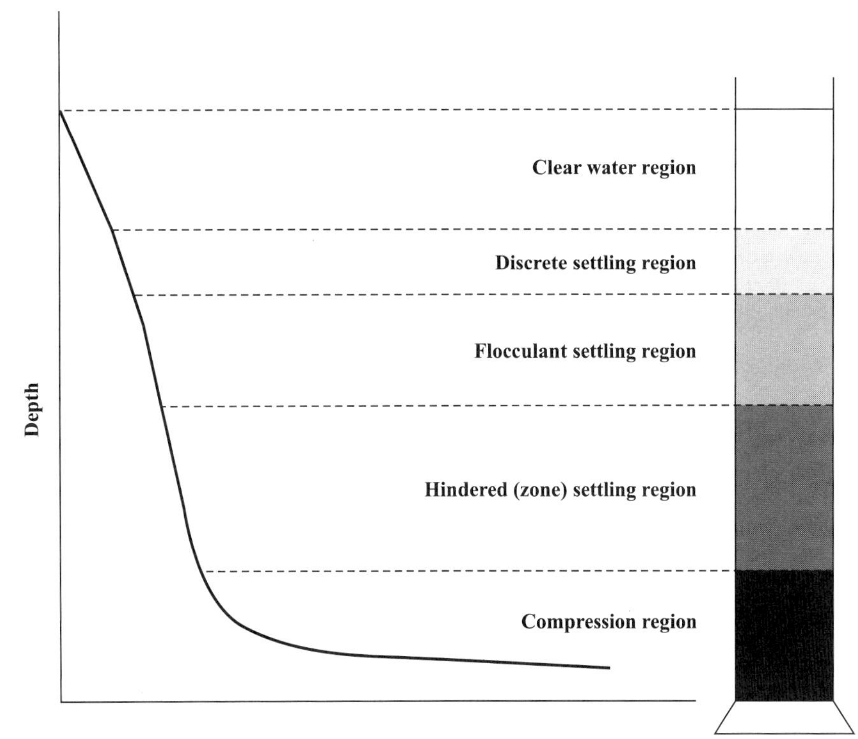

FIGURE 7.9 Schematic drawing of settling regions for concentrated suspensions.

where v_h = hindered settling velocity, m/s or ft/s
v = free settling velocity, calculated by Eq. (7.33) or (7.38)
C_v = volume of particles divided by the volume of the suspension

Eq. (7.48) is valid for Reynolds numbers less than 0.2, which is generally the situation in hindered settling.

A typical curve of interface height versus time for activated sludge is shown in Fig. 7.10. From A to B, there is a hindered settling of the particles and this is called liquid interface. From B to C there is a deceleration marking the transition from hindered settling into the compression zone. From C to D there is a compression zone where settling depends on compression of the sludge blanket.

The system design for handling concentrated suspensions for hindered settling must consider three factors: (1) the area needed for discrete settling of particles at the top of the clarifier; (2) the area needed for thickening (settling of the interface between the discrete and hindered settling zones); and (3) the rate of sludge withdrawal. The settling rate of the interface is usually the controlling factor.

Column settling tests, as previously described, can be used to determine the area needed for hindered settling. The height of the interface is plotted against time, as shown in Fig. 7.10. The area needed for clarification is

$$A = Q/v_s \tag{7.49}$$

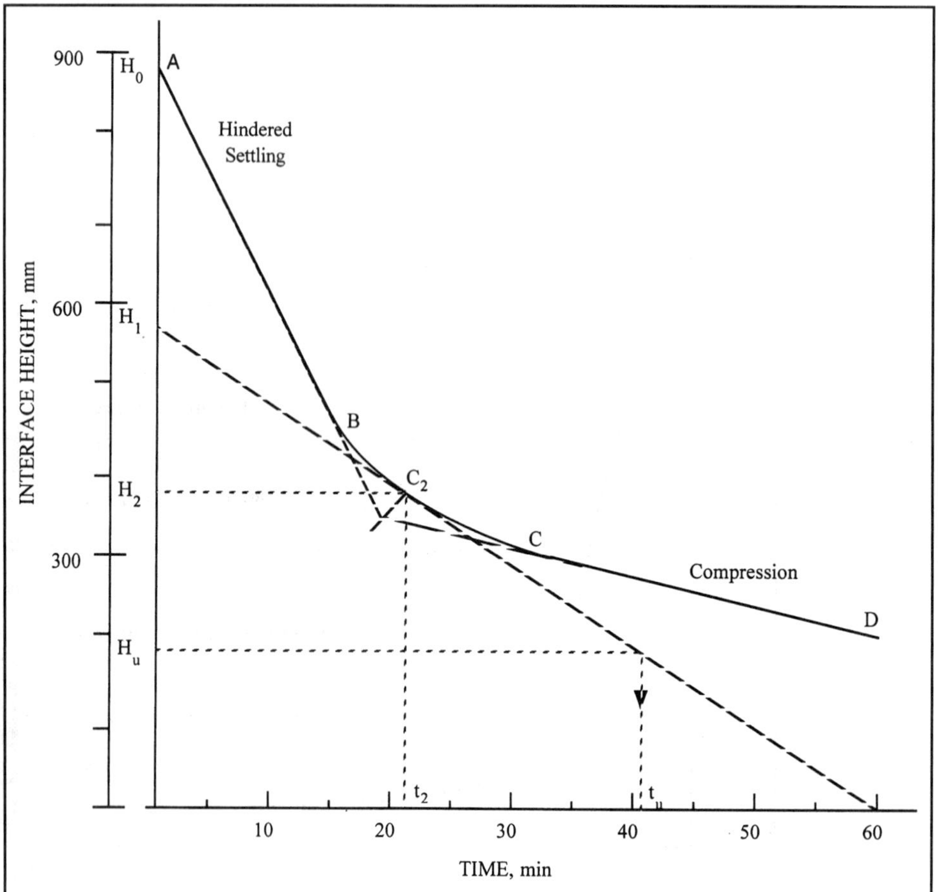

FIGURE 7.10 Graphic analysis of interface.

where A = surface area of the settling zone, m^2 or ft^2
Q = overflow rate, m^3/s or gal/min
v_s = subsidence rate in the zone of hindering settling, mm/s or in/s

A value of v_s is determined from batch settling column test data by computing the slope of the hindered settling portion of the interface height versus time curve (Fig. 7.10). The area needed for thickening is obtained from the batch settling test of a thick suspension. The critical area required for adequate thickening is (Rich 1961)

$$A = \frac{Qt_u}{H_0} \tag{7.50}$$

where A = area needed for sludge thickening, m^2 or ft^2
Q = flow into settling tank, m^3/s or ft^3/s
t_u = time to reach a desired underflow or solids concentration, s
H_0 = depth of the settling column (initial interface height), m or ft

From Fig. 7.10, the critical concentration (C_2) is determined by extending the tangent from the hindered and compression settling lines to their point of intersection and bisecting the angle formed. The bisector intersects the subsidence curve at C_2 which is the critical concentration. The critical concentration controls the sludge-handling capacity of the tank at a height of H_2.

A tangent is drawn to the subsidence curve at C_2 and the intersection of this tangent with depth H_u, required for the desired underflow (or solids concentration C_u), will yield the required retention time t_u. Since the total weight of solids in the system must remain constant, i.e. $C_0H_0A = C_uH_uA$, the height H_u of the particle–liquid interface at the underflow desired concentration C_u is

$$H_u = \frac{C_0H_0}{C_u} \tag{7.51}$$

The time t_u can be determined as:
draw a horizontal line through H_u and draw a tangent to the subsidence settling curve at C_2. Draw a vertical line from the point of intersection of the two lines drawn above to the time axis to find the value of t_u. With this value of t_u, the area needed for thickening can be calculated using Eq. (7.50). The area required for clarification is then determined. The larger of the two calculated areas from Eqs. (7.40) and (7.50) is the controlling factor for design.

EXAMPLE: The batch-settling curve shown in Fig. 7.10 is obtained for an activated sludge with an initial solids concentration C_0 of 3600 mg/L. The initial height of the interface in the settling column is 900 mm. This continuous inflow to the unit is 380 m^3/d (0.10 Mgal/d). Determine the surface area required to yield a thickened sludge of 1.8 percent by weight. Also determine solids and hydraulic loading rate.

Solution:

Step 1. Calculate H_u by Eq. (7.51)

$$C_u = 1.8\% = 18{,}000 \text{ mg/L}$$

$$H_u = \frac{C_0H_0}{C_u} = \frac{3600 \text{ mg/L} \times 900 \text{ mm}}{18{,}000 \text{ mg/L}}$$

$$= 180 \text{ mm}$$

Step 2. Determine t_u
Using the method described above to find the value of t_u

$$t_u = 41 \text{ min} = 41 \text{ min}/1440 \text{ min/d}$$

$$= 0.0285 \text{ day}$$

Step 3. Calculate the area required for the thickening, using Eq. (7.50)

$$A = \frac{Qt_u}{H_0} = \frac{380 \text{ m}^3/\text{d} \times 0.0285 \text{ d}}{0.90 \text{ m}}$$
$$= 12.02 \text{ m}^2$$
$$= 129 \text{ ft}^2$$

Step 4. Calculate the subsidence velocity v_s in the hindered settling portion of the curve
In 10 min

$$v_s = \frac{(900 - 617) \text{ mm}}{10 \text{ min} \times 60 \text{ s/min}}$$
$$= 0.47 \text{ mm/s}$$
$$= 40.6 \text{ m/d}$$

Step 5. Calculate the area required for clarification
Using Eq. (7.49)

$$A = Q/v_s = 380 \text{ m}^3/\text{d} \div 40.6 \text{ m/d}$$
$$= 9.36 \text{ m}^2$$

Step 6. Determine the controlling area
From comparison of areas calculated from Steps 3 and 5, the larger area is the controlling area. Thus the controlling area is the thickening area of 12.02 m² (129 ft²)

Step 7. Calculate the solids loading

$$C_0 = 3600 \text{ mg/L} = 3600 \text{ g/m}^3 = 3.6 \text{ kg/m}^3$$
$$\text{Solids weight} = QC_0 = 380 \text{ m}^3/\text{d} \times 3.6 \text{ kg/m}^3$$
$$= 1368 \text{ kg/d}$$
$$= 3016 \text{ lb/d}$$

$$\text{Solids loading rate} = 1368 \text{ kg/d} \div 12.02 \text{ m}^2$$
$$= 114 \text{ kg/(m}^2 \cdot \text{d)}$$
$$= 23.3 \text{ lb/(ft}^2 \cdot \text{d)}$$

Step 8. Determine the hydraulic (overflow) loading rate

$$\text{Hydraulic loading rate} = 380 \text{ m}^3/\text{d} \div 12.02 \text{ m}^2$$
$$= 31.6 \text{ m}^3/(\text{m}^2 \cdot \text{d}) = 31.6 \text{ m/d}$$
$$\text{or} = 100{,}000 \text{ gal/d} \div 129 \text{ ft}^2$$
$$= 776 \text{ gal/(ft}^2 \cdot \text{d)}$$

18.6 Compression Settling (Type 4)

When the concentration of particles is high enough to bring the particles into physical contact with each other, compression settling will occur. Consolidation of sediment at the bottom of the clarifier is extremely slow. The rate of settlement decreases with time due to increased resistance to flow of the fluid.

The volume needed for the sludge in the compression region (thickening) can also be estimated by settling tests. The rate of consolidation in this region has been found to be approximately proportional to the difference in sludge height H at time t and the final sludge height H_∞ obtained after a long period of time, perhaps 1 day. It is expressed as (Coulson and Richardson 1955)

$$\frac{dH}{dt} = i(H - H_\infty) \tag{7.52}$$

where H = sludge height at time t
i = constant for a given suspension
H_∞ = final sludge height

Integrating Eq. (7.52) between the limits of sludge height H_t at time t and H_1 at time t_1, the resulting expression is

$$H_t - H_\infty = (H_1 - H_\infty)e^{-i(t-t_1)} \tag{7.53}$$

or

$$i(t - t_1) = \ln (H_t - H_\infty) - \ln (H_1 - H_\infty) \tag{7.54}$$

A plot of ln $[(H_t - H_\infty) - \ln (H_1 - \ln H_\infty)]$ versus $(t - t_1)$ is a straight line having the slope $-i$. The final sludge height H_∞ depends on the liquid surface film which adheres to the particles.

It has been found that gentle stirring serves to compact sludge in the compression region by breaking up the floc and permitting water to escape. The use of mechanical rakes with 4 to 5 revolutions per hour will serve this purpose. Dick and Ewing (1967) reported that gentle stirring also helped to improve settling in the hindered settling region.

19 PRIMARY SEDIMENTATION TANKS

Primary treatment has traditionally implied a sedimentation process to separate the readily settleable and floatable solids from the wastewater. The treatment unit used to settle raw wastewater is referred to as the primary sedimentation tank (basin), primary tank (basin) or primary clarifier. Sedimentation is the oldest and most widely used process in the effective treatment of wastewater.

After the wastewater passes the preliminary processes, it enters sedimentation tanks. The suspended solids that are too light to fall out in the grit chamber will settle in the tank over a few hours. The settled sludge is then removed by mechanical scrapers, or pumped. The floatable substances on the tank surface are removed by a surface skimmer device. The effluent flows to the secondary treatment units or is discharged off (not in the United States and some countries).

The primary sedimentation tank is where the flow velocity of the wastewater is reduced by plain sedimentation. The process commonly removes particles with a settling rate of 0.3 to 0.6 mm/s (0.7 to 1.4 in/min). In some cases, chemicals may be added. The benefits of primary sedimentation are reduced suspended solids content, equalization of sidestream flow, and BOD removal. The overflow rate of the primary sedimentation tanks ranges from 24.5 to 49 $m^3/(m^2 \cdot d)$ (600 to 1200 $gal/(d \cdot ft^2)$). The detention time in the tank is usually 1 to 3 h (typically 2 h). Primary tanks (or primary clarifiers) should remove 90 to 95 percent of settleable solids, 50 to 60 percent of total suspended solids, and 25 to 35 percent of the BOD_5 (NY Department of Health 1950).

Settling characteristics in the primary clarifier are generally characterized by type 2 flocculant settling. The Stokes formula for settling velocity cannot be used because the flocculated particles are continuously changing in shape, size, and specific gravity. Since no mathematical equation can describe flocculant settling satisfactorily, laboratory analyses of settling-column tests are commonly used to generate design information.

Some recommended standards for the design of primary clarifiers are as follows (GLUMRB–Ten States Standards 1996, Illinois EPA 1998). Multiple tanks capable of independent operation are desirable and shall be provided in all plants where design average flows exceed 380 m^3/d (100,000 gal/d). The minimum length of flow from inlet to outlet should be 3.0 m (10 ft) unless special provisions are made to prevent short circuiting. The side depth for primary clarifiers shall be as shallow as practicable,

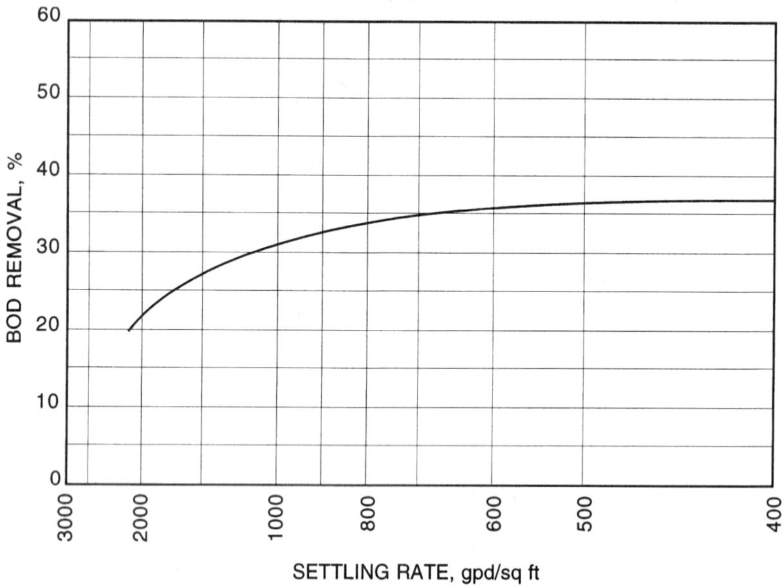

FIGURE 7.11 BOD_5, removal in primary settling tank (*Source: Illinois EPA 1998*).

but not less than 3.0 m (10 ft). Hydraulic surface settling rates (overflow rates) of the clarifier shall be based on the anticipated peak hourly flow. For normal domestic wastewater, the overflow rate, with some indication of BOD removal, can be obtained from Fig. 7.11. If waste activated sludge is returned to the primary clarifier, the design surface settling rate shall not exceed 41 $m^3/(m^2 \cdot d)$ (1000 $gal/(d \cdot ft^2)$). The maximum surfaced settling rate for combined sewer overflow and bypass settling shall not exceed 73.3 $m^3/(m^2 \cdot d)$ (1800 $gal/(d \cdot ft^2)$), based on peak hourly flow. Weir loading rate shall not exceed 250 m^3 per day linear meter (20,000 $gal/(d \cdot ft)$), based on design peak hourly flow for plants having a design average of 3785 m^3/d (1 Mgal/d) or less. Weir loading rates shall not exceed 373 $m^3/(d \cdot m)$ (30,000 $gal/(d \cdot ft)$), based on peak design hourly flow for plants having a design average flow greater than 3785 m^3/d (1.0 Mgal/d). Overflow rates, side water depths, and weir loading rates recommended by various institutions for primary settling tanks are listed elsewhere (WEF and ASCE 1991a).

In cases where a reliable loading–performance relationship is not available, the primary tank design may be based on the overflow rates and side water depths listed in Table 7.10. The design surface settling is selected on the basis of Fig. 7.11 and Table 7.10. The hydraulic detention time t in the primary clarifier can be calculated from Eq. (7.44). The hydraulic detention times for primary clarifier design range from 1.5 to 2.5 h, typically 2 h. Consideration should be made for low flow period to ensure that longer detention times will not cause septic conditions. Septic conditions may cause a potential odor problem, stabilization and loading to the downstream treatment processes. In the cold climatic region, a detention time multiplier should be included when wastewater temperature is below 20°C (68°F). The multiplier can be calculated by the following equation (Water Pollution Control Federation 1985a)

$$M = 1.82e^{-0.03T} \tag{7.55}$$

where M = detention time multiplier
T = temperature of wastewater, °C

TABLE 7.10 Typical Design Parameters for Primary Clarifiers

Type of treatment	Source	Surface settling rate, $m^3/(m^2 \cdot d)$, $(gal/(d \cdot ft^2))$ Average	Peak	Depth, m (ft)
Primary settling followed by secondary treatment	USEPA 1975a	33–49 (800–1200)	81–122 (2000–3000)	3–3.7 (10–12)
	GLUMRB—Ten States Standards and Illinois EPA 1998	600	Figure 7.11	minimum 2.1 (7)
Primary settling with waste activated sludge return	USEPA 1975a	24–33 (600–800)	49–61 (1200–1500)	3.7–4.6 (12–15)
	Ten States Standards, GLUMRB 1996	≤ 41	≤ 61	3.0
		(≤ 1000)	(≤ 1500)	(10) minimum

In practice, the linear flow-through velocity (scour velocity) is limited to 1.2 to 1.5 m/min (4 to 5 ft/min) to avoid resuspension of settled solids in the sludge zone (Theroux and Betz 1959). Camp (1946) suggested that the critical scour velocity can be computed by Eq. (7.39). Scouring velocity may resuspend settled solids and must be avoided with clarifier design. Camp (1953) reported that horizontal velocities up to 18 ft/min (9 cm/s) may not create scouring; but design horizontal velocities should still be designed substantially below 18 ft/min. As long as scouring velocities are not approached, solids removal in the clarifier is independent of the tank depth.

EXAMPLE 1: Determine the detention time multipliers for wastewater temperatures of 12 and 6°C

Solution:

Using Eq. (7.55), for $T = 12°C$

$$\begin{aligned} M &= 1.82e^{-0.03\times 12} = 1.82 \times 0.70 \\ &= 1.27 \end{aligned} \tag{7.56}$$

For $T = 6°C$

$$\begin{aligned} M &= 1.82 \times e^{-0.03\times 6} = 1.82 \times 0.835 \\ &= 1.52 \end{aligned}$$

EXAMPLE 2: Two rectangular settling tanks are each 6 m (20 ft) wide, 24 m (80 ft) long, and 2.7 m (9 ft) deep. Each is used alternately to treat 1900 m^3 (0.50 Mgal) in a 12 h period. Compute the surface overflow (settling) rate, detention time, horizontal velocity, and outlet weir loading rate using H-shaped weir with three times the width.

Solution:

Step 1. Determine the design flow Q

$$\begin{aligned} Q &= \frac{1900 \text{ m}^3}{12 \text{ h}} \times \frac{24 \text{ h}}{1 \text{ day}} \\ &= 3800 \text{ m}^3/\text{d} \end{aligned}$$

Step 2. Compute surface overflow rate v_0

$$v_0 = Q/A = 3800\ \text{m}^3/\text{d} \div (6\ \text{m} \times 24\ \text{m})$$
$$= 26.4\ \text{m}^3/(\text{m}^2 \cdot \text{d})$$
$$= 650\ \text{gal}/(\text{d} \cdot \text{ft}^2)$$

Step 3. Compute detention time t

$$\text{Tank volume } V = 6\ \text{m} \times 24\ \text{m} \times 2.1\ \text{m} \times 2$$
$$= 604.8\ \text{m}^3$$
$$t = V/Q = 604.8\ \text{m}^3/(3800\ \text{m}^3/\text{d})$$
$$= 0.159\ \text{day}$$
$$= 3.8\ \text{h}$$

Step 4. Compute horizontal velocity v_h

$$v_h = \frac{3800\ \text{m}^3/\text{d}}{6\ \text{m} \times\ 2.1\ \text{m}}$$
$$= 301\ \text{m/d}$$
$$= 0.209\ \text{m/min}$$
$$= 0.686\ \text{ft/min}$$

Step 5. Compute outlet weir loading, wl

$$wl = \frac{3800\ \text{m}^3/\text{d}}{6\ \text{m} \times 3}$$
$$= 211\ \text{m}^3/(\text{d} \cdot \text{m})$$
$$= 17{,}000\ \text{gal}/(\text{d} \cdot \text{ft})$$

19.1 Rectangular Basin Design

Multiple units with common walls shall be designed for independent operation. A bypass to the aeration basin shall be provided for emergency conditions. Basin dimensions are to be designed on the basis of surface overflow (settling) rate to determine the required basin surface area. The area required is the flow divided by the selected overflow rate. An overflow rate of 36 $\text{m}^3/(\text{m}^2 \cdot \text{d})$ (884 $\text{gal}/(\text{ft}^2 \cdot \text{d})$) at average design flow is generally acceptable.

Basin surface dimensions, the length (l) to width (w) ratio (l/w), can be increased or decreased without changing the volume of the basin. The greater the l/w ratio, the better the basin conforms to plug flow conditions. Also, for greater l/w ratio, the basin has a proportionally larger effective settling zone and smaller percent inlet and outlet zones. Increased basin length allows the development of a more stable flow. Best conformance to the plug flow model has been reported by a basin with l/w ratio of 3 : 1 or greater (Aqua-Aerobic Systems 1976).

Basin design should be cross-checked for detention time for conformance with recommended standards by the regulatory agencies.

For bean bridge (cross the basin) design, a commercially available economical basin width can be used, such as 1.5 m (5 ft), 3.0 m (10 ft), 5.5 m (18 ft), 8.5 m (28 ft) or 11.6 m (38 ft).

The inlet structures in rectangular clarifiers are placed at one end and are designed to dissipate the inlet velocity to diffuse the flow equally across the entire cross section of the basin and to prevent short circuiting. Typical inlets consist of small pipes with upward ells, perforated baffles, multiple ports discharging against baffles, a single pipe turned back to discharge against the headwall, simple weirs, submerged weirs sloping upward to a horizontal baffle, etc. (Steel and McGhee 1979, McGhee 1991). The inlet structure should be designed not to trap scum or settling solids. The inlet channel should have a velocity of 0.3 m/s (1 ft/s) at one-half design flow (Ten States Standards, GLUMRB 1996).

Baffles are installed 0.6 to 0.9 m (2 to 3 ft) in front of inlets to assist in diffusing the flow and submerged 0.45 to 0.60 m (1.5 to 2 ft) with 5 cm (2 in) water depth above the baffle to permit floating material to pass. Scum baffles are placed ahead of outlet weirs to hold back floating material from the basin effluent and extend 15 to 30 cm (6 to 12 in) below the water surface (Ten States Standards, GLUMRB 1971). Outlets in a rectangular basin consist of weirs located toward the discharge end of the basin. Weir loading rates range from 250 to 373 $m^3/(d \cdot m)$ (20,000 to 30,000 $gal/(d \cdot ft)$) (Ten State Standards, GLUMRB 1996).

Walls of settling tanks should extend at least 150 mm (6 in) above the surrounding ground surface and shall provide not less than 300 mm (12 in) freeboard (GLUMRB 1996).

Mechanical sludge collection and withdrawal facilities are usually designed to assure rapid removal of the settled solids. The minimum slope of the side wall of the sludge hopper is 1.7 vertical to 1 horizontal. The hopper bottom dimension should not exceed 600 mm (2 ft). The sludge withdrawal line is at least 150 mm (6 in) in diameter and has a static head of 760 mm (30 in) or greater with a velocity of 0.9 m/s (3 ft/s) (Ten States Standards, GLUMRB 1996).

EXAMPLE: Design a primary clarification system for a design average wastewater flow of 7570 m^3/d (2.0 Mgal/d) with a peak hourly flow of 18,900 m^3/d (5.0 Mgal/d) and a minimum flow of 4540 m^3/d (1.2 Mgal/d). Design a multiple units system using Ten States Standards for an estimated 35 percent BOD_5 removal at the design flow.

Solution:

Step 1. List Ten States Standards
Referring to Fig. 7.11, for 35% BOD removal:

$$\text{Surface settling rate } \upsilon = 28.5 \text{ m}^3/(\text{m}^2 \cdot \text{d}) \text{ or } (700 \text{ gal}/(\text{d} \cdot \text{ft}^2))$$

$$\text{Minimum depth} = 3.0 \text{ m (10 ft)}$$

$$\text{Maximum weir loading} = 124 \text{ m}^3/(\text{d} \cdot \text{m}) \text{ or } (10{,}000 \text{ gal}/(\text{d} \cdot \text{ft})) \text{ for average daily flow}$$

Step 2. Determine tank dimensions

1. Surface area needed
Use 2 settling tanks, each with design flow of 3785 m^3/d (1.0 Mgal/d)

$$\text{Surface area } A = Q/\upsilon = 3785 \text{ m}^3/\text{d} \div 28.5 \text{ m}^3/(\text{m}^2 \cdot \text{d})$$
$$= 132.8 \text{ m}^2$$

2. Determine length l and width w using l/w ratio of 4/1

$$(w)(4w) = 132.8$$
$$w^2 = 33.2$$
$$w = 5.76 \text{ (m)}$$

Standard economic widths of bean bridge are: 1.52 m (5 ft), 3.05 m (10 ft), 5.49 m (18 ft), 8.53 m (28 ft), and 11.58 m (38 ft).
Select standard 5.49 m (18 ft width); $w = 5.49$ m

$$5.49l = 132.8$$
$$l = 24.19 \text{ m}$$
$$= 79.34 \text{ ft}$$

Say area is 5.5 m × 24 m (18 ft × 79 ft)
Compute tank surface area A

$$A = 132 \text{ m}^2 = 1422 \text{ ft}^2$$

3. Compute volume V with a depth of 3 m

$$V = 132\ \text{m}^2 \times 3\ \text{m} = 396\ \text{m}^3 = 14{,}000\ \text{ft}^3$$

Step 3. Check detention time t
At average design flow

$$t = V/Q = 396\ \text{m}^3/(3785\ \text{m}^3/\text{d}) = 0.105\ \text{day} \times 24\ \text{h/d} = 2.5\ \text{h}$$

At peak flow

$$t = 2.5\ \text{h} \times (2/5) = 1.0\ \text{h}$$

Step 4. Check overflow rate at peak flow

$$v = 28.5\ \text{m}^3/(\text{m}^2 \cdot \text{d}) \times (5/2) = 71.25\ \text{m}^3/(\text{m}^2 \cdot \text{d})$$

Step 5. Determine the length of outlet weir

$$\text{Length} = \text{flow} \div \text{weir loading rate} = 3785\ \text{m}^3/\text{d} \div 124\ \text{m}^3/(\text{d} \cdot \text{m}) = 30.5\ \text{m}$$

It is 5.5 times the width of the tank

19.2 Circular Basin Design

Inlets in circular or square basins are typically at the center and the flow is directed upward through a baffle that channels the wastewater (influent) toward the periphery of the tank. Inlet baffles are 10 to 20 percent of the basin diameter and extend 0.9 to 1.8 m (3 to 6 ft) below the wastewater surface (McGhee 1991).

Circular basins have a higher degree of turbulence than rectangular basins. Thus circular basins are more efficient as flocculators.

A typical depth of sidewall of a circular tank is 3 m (10 ft). As shown in Figure 7.12, the floor slope of the tank is typically 300 mm (12 in) horizontal to 25 mm (1 in) vertical (Aqua-Aerobic Systems 1976).

Outlet weirs extend around the periphery of the tank with baffles extending 200 to 300 mm (8 to 12 in) below the wastewater surface to retain floating material (McGhee 1991). Overflow weirs shall be located to optimum actual hydraulic detention time and minimize short circuiting. Peripheral weirs shall be placed at least 30 cm (1 ft) from the well (Ten States Standards, GLUMRB 1996).

EXAMPLE 1: Design circular clarifiers using US customary units with the same given information as in the example for rectangular clarifiers design.

Solution:

Step 1. Calculate surface area A
Design 2 circular clarifiers, each treating 1.0 Mgal/d with an overflow rate of 700 gal/(d · ft²).

$$A = 1{,}000{,}000\ (\text{gal/d})/700\ \text{gal}/(\text{d} \cdot \text{ft}^2) = 1429\ \text{ft}^2 = 132.8\ \text{m}^2$$

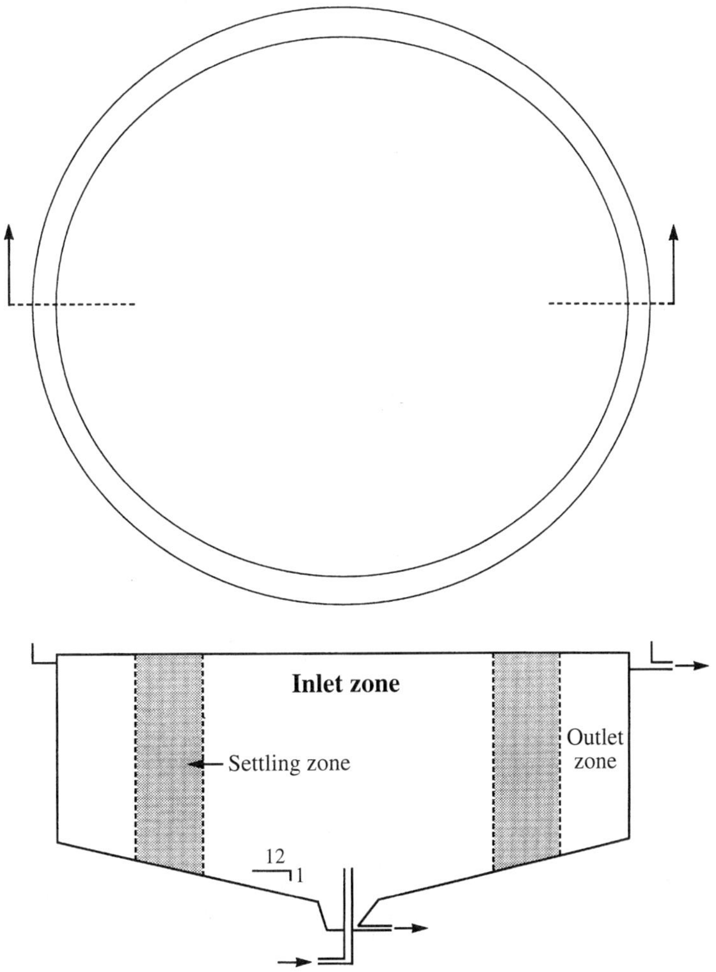

FIGURE 7.12 Typical circular basin design-plan view and cross section.

Step 2. Determine the tank radius r

$$\pi r^2 = 1429$$
$$r^2 = 1429/3.14 = 455$$
$$r = 21.3 \text{ ft}$$

Use $d = 44$ ft (13.4 m) diameter tank
Surface area = 1520 ft^2 (141 m^2)

Step 3. Check overflow rate

$$\begin{aligned}\text{Overflow rate} &= 1{,}000{,}000 \text{ (gal/d)}/1520 \text{ ft}^2 \\ &= 658 \text{ gal/(d} \cdot \text{ft}^2) \\ &= 26.8 \text{ m}^3/(\text{m}^2 \cdot \text{d})\end{aligned}$$

From Fig. 7.11, BOD_5 removal for an overflow rate of 658 gal/(d · ft^2) is 35%.

Step 4. Determine detention time t
Use a wastewater side wall depth of 10 ft and, commonly, plus 2-ft freeboard for wind protection

$$t = \frac{1520\ \text{ft}^2 \times 10\ \text{ft} \times 7.48\ \text{gal/ft}^3}{1{,}000{,}000\ \text{gal}}$$

$$= 0.114\ \text{day}$$

$$= 2.73\ \text{h}$$

Step 5. Calculate weir loading rate
Use an inboard weir with diameter of 40 ft (12.2 m)

$$\text{Length of periphery} = \pi \times 40\ \text{ft} = 125.6\ \text{ft}$$

$$\text{Weir loading} = 1{,}000{,}000\ \text{(gal/d)/125.6 ft}$$

$$= 7960\ \text{gal/(d} \cdot \text{ft)}$$

$$= 98.8\ \text{m}^3/\text{m} \cdot \text{d}$$

Step 6. Calculate the number (n) of V notches
Use 90° standard V notches at rate of 8 in center-to-center of the launders.

$$n = 125.6\ \text{ft} \times 12\ \text{in/ft} \div 8\ \text{in}$$

$$= 188$$

Step 7. Calculate average discharge per notch at average design flow q

$$q = 1{,}000{,}000\ \text{gal/d} \times (1\ \text{day/1440 min}) \div 188$$

$$= 3.7\ \text{gal/min}$$

$$= 14\ \text{L/min}$$

EXAMPLE 2: If the surface overflow rate is 40 m^3/(m^2 · d) [982 gal/(d · ft^2)] and the weir overflow rate is 360 m^3/(d · m) [29,000 gal/(d · ft)], determine the maximum radius for a circular primary clarifier with a single peripheral weir.

Solution:

Step 1. Compute the area (A) required with a flow (Q)

Let r = the radius of the clarifier

$$A = \pi r^2 = Q/40\ \text{m}^3/(\text{m}^2 \cdot \text{d}) \text{ or } Q = 40\pi r^2$$

Step 2. Compute the weir length ($= 2\pi r$)

$$2\pi r = Q/360\ \text{m}^3/(\text{d} \cdot \text{m})$$

or

$$Q = 720\pi r\ \text{m}^3/(\text{d} \cdot \text{m})$$

Step 3. Compute r by solving the equations in Steps 1 and 2

$$\pi r^2 = 720\pi r[\text{m}^3/(\text{d} \cdot \text{m})]/40[\text{m}^3/(\text{m}^2 \cdot \text{d})]$$

$$r = 18\ \text{m (59 ft)}$$

20 BIOLOGICAL (SECONDARY) TREATMENT SYSTEMS

The purpose of primary treatment is to remove suspended solids and floating material. In many situations in some countries, primary treatment with the resulting removal of approximately 40 to 60 percent of the suspended solids and 25 to 35 percent of BOD_5, together with removal of material from the wastewater, is adequate to meet the requirement of the receiving water body. If primary treatment is not sufficient to meet the regulatory effluent standards, secondary treatment using a biological process is mostly used for further treatment due to its greater removal efficiency and less cost than chemical coagulation. Secondary treatment processes are intended to remove the soluble and colloidal organics (BOD) which remain after primary treatment and to achieve further removal of suspended solids and, in some cases, also to remove nutrients such as phosphorus and nitrogen. Biological treatment processes provide similar biological activities to waste assimilation, which would take place in the receiving waters, but in a reasonably shorter time. Secondary treatment may remove more than 85 percent of BOD_5 and suspended matter, but is generally not effective for removing nutrients, nonbiodegradable organics, heavy metals, and microorganisms.

Biological treatment systems are designed to maintain a large active mass and a variety of microorganisms, (principally bacteria and fungi, protozoa, rotifers, algae, etc.), within the confined system under favorable environmental conditions, such as dissolved oxygen, nutrient, etc. Biological treatment processes are generally classified mainly as suspended growth processes (activated sludge, Fig. 7.13), attached (film) growth processes (trickling filter and rotating biological contactor, RBC), and dual-process systems (combined). Other biological wastewater treatment processes include the stabilization pond, aerated lagoon, contaminant pond, oxidation ditch, high-purity oxygen activated sludge, biological nitrification, denitrification, and phosphorus removal units.

In the suspended biological treatment process, under continuous supply of air or oxygen, living aerobic microorganisms are mixed thoroughly with the organics in the wastewater and use the organics as food for their growth. As they grow, they clump or flocculate to form an active mass of microbes. This is so-called biologic floc or activated sludge.

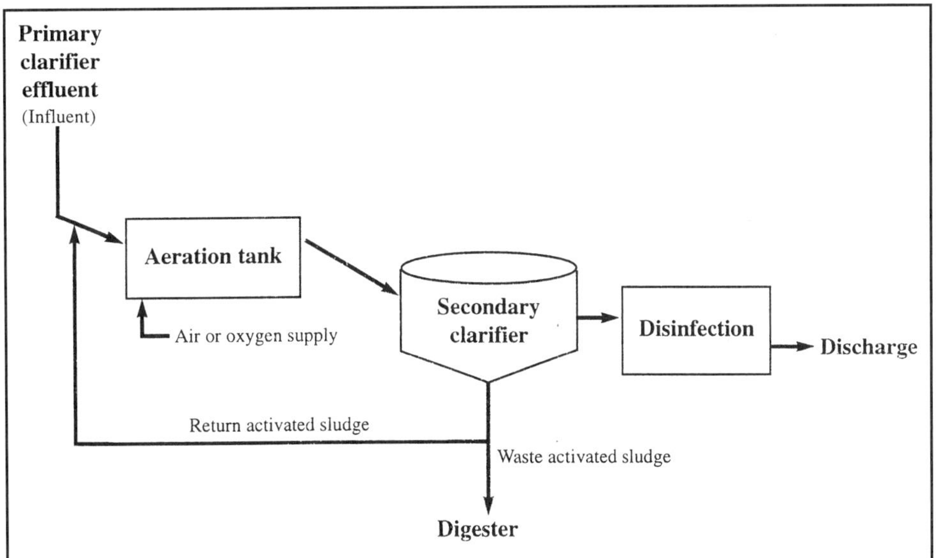

FIGURE 7.13 Conventional activated sludge process.

20.1 Cell Growth

Each unicellular bacterium grows and, after reaching a certain size, divides to produce two complete individuals by binary fission. The period of time required for a growing population to double is called the generation time. With each generation, the total number increases as the power (exponent) of 2, i.e. 2^0, 2^1, 2^2, 2^3 . . . The exponent of 2 corresponding to any given number is the logarithm of that number to the base 2 ($\log_2$). Therefore, in an exponentially growing culture, the $\log_2$ of the number of cells increasing in proportion to time, is often referred to as logarithmic growth.

The growth rate of microorganisms is affected by environmental conditions, such as DO levels, temperature, nutrient levels, pH, microbial community, etc. Exponential growth does not normally continue for a long period of time. A general growth pattern for fission-reproduction bacteria in a batch culture is sketched in Fig. 7.14. They are the lag phase, the exponential (logarithmic) growth phase, the maximum (stationary) phase, and the death phase.

When a small number of bacteria is inoculated into a fixed volume of vessel with culture medium, bacteria generally require time to acclimatize to their environmental condition. For this period of time, called the lag phase, the bacterial density is almost unchanged. Under excess food supply, a rapid increase in number and mass of bacteria occurs in the log growth phase. During this maximum rate of growth a maximum rate of substrate removal occurs. Either some nutrients become exhausted, or toxic metabolic products may accumulate. Subsequently, the growth rate decreases and then ceases. For this period, the cell number remains stationary at the stationary phase. As bacterial density increases and food runs short, cell growth will decline. The total mass of protoplasm exceeds the mass of viable cells, because bacteria form resistant structures such as endospores.

In the endogenous growth period, the microorganisms compete for the limiting substrate even to metabolize their own protoplasm. The rate of metabolism decreases and starvation occurs. The rate of death exceeds the rate of reproduction; cells also become old, die exponentially, and lysis. Lysis

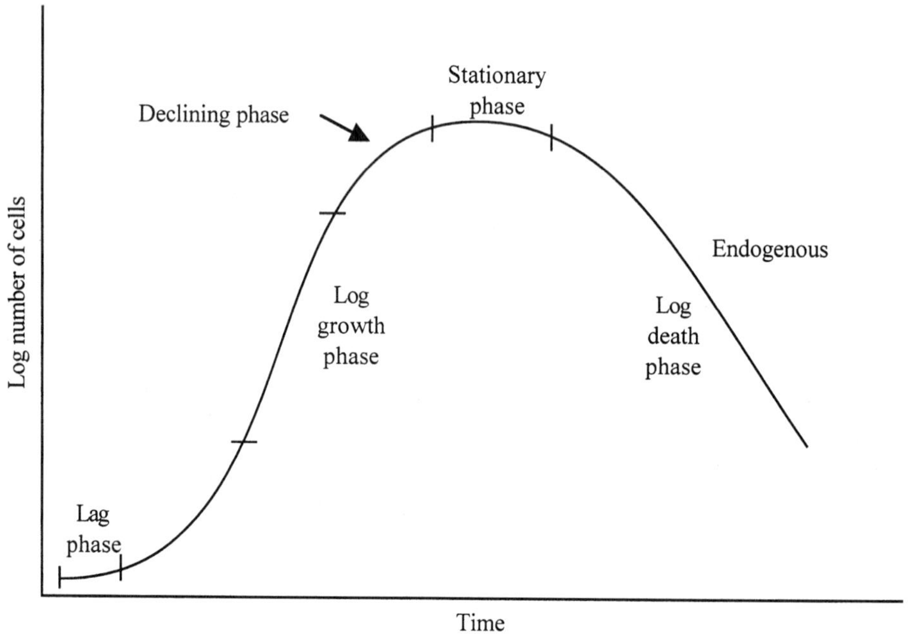

FIGURE 7.14 Bacterial density with growth time.

can occur in which the nutrients from the dead cells diffuse out to furnish the remaining living cells as food. The results of cell lysis decrease both the population and the mass of microorganisms.

In the activated-sludge process, the balance of food to microorganisms is very important. Metabolism of the organic matter results in an increased mass of microorganisms in the process. The excess mass of microorganisms must be removed (waste sludge) from the system to maintain a proper balance between mass of microorganisms and substrate concentration in the aeration basin.

In both batch and continuous culture systems the growth rate of bacterial cells can be expressed as

$$r_g = \mu X \tag{7.57}$$

where r_g = growth rate of bacteria, mg/(L · d)
μ = specific growth rate, per day
X = mass of microorganism, mg/L

since

$$\frac{dX}{dt} = r_g \tag{7.58}$$

Therefore

$$\frac{dX}{dt} = \mu X \tag{7.59}$$

Substrate limited growth. Under substrate limited growth conditions, or for an equilibrium system, the quantity of solids produced is equal to that lost, and specific growth rate (the quantity produced per day) can be expressed as the well-known Monod equation

$$\mu = \mu_m \frac{S}{K_s + S} - k_d \tag{7.60}$$

where μ = specific growth rate, per day
μ_m = maximum specific growth rate, per day
S = concentration for substrate in solution, mg/L
K_s = half velocity constant, substrate concentration at one-half the maximum growth rate, mg/L
k_d = cell decay coefficient reflecting the endogenous burn-up of cell mass, mg/(mg · d)

Substituting the value of μ from Eq. (7.60) into Eq. (7.57), the resulting equation for the cell growth rate is

$$r_g = \frac{\mu_m SX}{K_s + S} - k_d \tag{7.61}$$

21 ACTIVATED-SLUDGE PROCESS

The activated-sludge process was first used in Manchester, England. This process is perhaps the most widely used process for secondary treatment of wastewaters. Recently, the activated-sludge process has been applied as a nitrification and denitrification process, and modified with an anoxic and anaerobic section for phosphorus removal.

A basic suspended-growth (activated-sludge) process is depicted in Fig. 7.13. The wastewater continuously enters an aeration tank where previously developed biological flocs are brought into contact with the organic material of the wastewater. Air or oxygen-enriched air is continuously injected into the aeration tank as an oxygen source to keep the system aerobic and the activated sludge in suspension. Approximately 8 m^3 of air is required for each m^3 of wastewater.

The microbes in the activated sludge consist of a gelatinous matrix of filamentous and unicellular bacteria which are fed on protozoa. The predominant bacteria, such as *Pseudomonas*, utilize carbohydrate and hydrocarbon wastes, whereas *Bacillus, Flavobacterium*, and *Alcaligenes* consume protein wastes. When a new activated-sludge system is first started, it should be seeded with activated sludge from an existing operating plant. If seed sludge is not available, activated sludge can be prepared by simply continuously aerating, settling, and returning settled solids to the wastewater for a few weeks (4 to 6 weeks) (Alessi *et al.* 1978, Cheremisinoff 1995).

The microorganisms utilize the absorbed organic matter as a carbon and energy source for cell growth and convert it to cell tissue, water, and oxidized products (mainly carbon dioxide, CO_2). Some bacteria attack the original complex substance to produce simple compounds as their waste products. Other bacteria then use these waste products to produce simpler compounds until the food is used up.

The mixture of wastewater and activated sludge in the aeration basis is called mixed liquor. The biological mass (biomass) in the mixed liquor is called the mixed liquor suspended solids (MLSS) or mixed liquor volatile suspended solids (MLVSS). The MLSS consists mostly of microorganisms, nonbiodegradable suspended organic matter, and other inert suspended matter. The microorganisms in MLSS are composed of 70 to 90 percent organic and 10 to 30 percent inorganic matter (Okun 1949, WEF and ASCE 1996a). The types of bacterial cell vary, depending on the chemical characteristics of the influent wastewater tank conditions and the specific characteristics of the microorganisms in the flocs. Microbial growth in the mixed liquor is maintained in the declining or endogenous growth phase to insure good settling properties.

After a certain reaction time (4 to 8 h), the mixed liquor is discharged from the aeration tank to a secondary sedimentation basin (settling tank, clarifier) where the suspended solids are settled out from the treated wastewater by gravity. However, in a sequencing batch reactor (SBR), mixing and aeration in the aeration tank are stopped for a time interval to allow MLSS to settle and to decant the treated wastewater; thus a secondary clarifier is not needed in an SBR system. Most concentrated biological settled sludge is recycled back to the aeration tank (so-called return activated sludge, RAS) to maintain a high population of microorganisms to achieve rapid breakdown of the organics in the wastewater. The volume of RAS is typically 20 to 30 percent of the wastewater flow. Usually more activated sludge is produced than return sludge.

The treated wastewater is commonly chlorinated and dechlorinated, then discharged to receiving water or to a tertiary treatment system (Fig. 7.3). The preliminary, primary, and activated-sludge (biological) processes are included in the so-called secondary treatment process.

21.1 Aeration Periods and BOD Loadings

The empirical design of activated sludge is based on BOD loading, food-to-microorganism ratio (F/M), sludge age, and aeration period. Empirical design concepts are still acceptable. The Ten States Standards states that when activated-sludge process design calculations are not submitted, the aeration tank capacities and permissible loadings for the several adoptions of the processes shown in Table 7.11 are used as simple design criteria. Those values apply to plants receiving diurnal load ratios of design peak hourly BOD_5 to design BOD_5 ranging from 2 : 1 to 4 : 1 (GLUMRB 1996).

The aeration period is the retention time of the influent wastewater flow in the aeration basin and is expressed in hours. It is computed from the basin volume divided by the average daily flow excluding the return sludge flow. For normal domestic sewage, the aeration period commonly ranges from 4 to 8 h with an air supply of 0.5 to 2.0 ft^3 or per gallon (3.7 to 15.0 m^3/m^3) of wastewater. The return activated sludge is expressed as a percentage of the wastewater influent of the aeration tank.

The organic (BOD_5) loading on an aeration basin can be computed using the BOD in the influent wastewater without including the return sludge flow. BOD loadings are expressed in terms of lb BOD applied per day per 1000 ft^3 or ($kg/(d \cdot m^3)$) of liquid volume in the aeration basin and in terms of lb BOD applied per day per lb of mixed liquid volatile suspended solids. The latter is called the food-to-microorganism ratio.

BOD loadings per unit volume of aeration basin vary widely from 10 to more than 100 lb/1000 ft^3 (0.16 to 1.6 $kg/(d \cdot m^3)$), while the aeration periods correspondingly vary from 2.5 to 24 h (Clark *et al.* 1977). The relationship between the two parameters is directly related to the BOD concentration in the wastewater.

TABLE 7.11 Permissible Aeration Tank Capacities and Loadings

Process	Organic (BOD_5) loading lb/(d · 1000 ft^3) (kg/d · m^3)*	F/M ratio, lb BOD_5/d per lb MLVSS	MLSS, mg/L‡
Conventional step aeration complete mix	40 (0.64)	0.2–0.5	1000–3000
Contact stabilization†	50† (0.80)	0.2–0.6	1000–3000
Extended aeration single-stage nitrification	15 (0.24)	0.05–0.1	3000–5000

*Loadings are based on the influent organic load to the aeration tank at plant design average BOD_5.
†Total aeration capacity includes both contact and reaeration capacities; normally the contact zone equals 30 to 35 percent of the total aeration capacity.
‡The values of MLSS are dependent on the surface area provided for secondary settling and the rate of sludge return as well as the aeration processes.
Source: GLUMRB (Ten States Standards) (1996).

21.2 F/M Ratio

The F/M ratio is used to express BOD loadings with regard to the microbial mass in the process. The value of the F/M ratio can be calculated by the following equation:

$$\text{F/M} = \frac{\text{BOD, lb/d}}{\text{MLSS, lb}} \tag{7.62}$$

$$= \frac{Q\text{ (Mgal/d)} \times \text{BOD (mg/L)} \times 8.34\text{ (lb/gal)}}{V\text{ (Mgal)} \times \text{MLSS (mg/L)} \times 8.34\text{ (lb/gal)}} \tag{7.63}$$

or

$$\text{F/M} = \frac{\text{BOD, kg/d}}{\text{MLSS, kg}} \tag{7.64}$$

where F/M = food-to-microorganism ratio, kg (lb) of BOD per day per kg (lb) of MLSS
Q = wastewater flow, m^3/d or Mgal/d
BOD = wastewater 5-day BOD, mg/L
V = liquid volume of aeration tank, m^3 or Mgal
MLSS = mixed liquor suspended solids in the aeration tank, mg/L

In Eqs. (7.62) and (7.63), some authors use mixed liquor volatile suspended solids instead of MLSS. The MLVSS is the volatile portion of the MLSS and ranges from 0.75 to 0.85. Typically they are related, for design purposes, by MLVSS = 0.80 × MLSS. The use of MLVSS may more closely approximate the total active biological mass in the process.

The F/M ratio is also called the sludge loading ratio (SLR). The equation for the calculation of the SLR is (Cheremisinoff 1995)

$$\text{SLR} = \frac{24\text{ BOD}}{\text{MLVSS}(t)(1 + R)} \tag{7.65}$$

where SLR = sludge loading, g of BOD/d per g of MLVSS
BOD = wastewater BOD, mg/L
MLVSS = mixed liquor volatile suspended solids, mg/L
t = retention time, d
R = recyle ratio

EXAMPLE 1: An activated-sludge process has a tank influent BOD concentration of 140 mg/L, influent flow of 5.0 Mgal/d (18,900 m^3/d) and 35,500 lb (16,100 kg) of suspended solids under aeration. Calculate the F/M ratio.

Solution:

Step 1. Calculate BOD in lb/d

$$\begin{aligned} \text{BOD} &= Q \times \text{BOD} \times 8.34 \\ &= 5.0\ \text{Mgal/d} \times 140\ \text{mg/L} \times 8.34\ \text{lb/gal} \\ &= 5838\ \text{lb/d} \end{aligned}$$

Step 2. Calculate the volatile SS under aeration
Assume VSS is 80% of TSS

$$\begin{aligned} \text{MLVSS} &= 35{,}500\ \text{lb} \times 0.8 \\ &= 28{,}400\ \text{lb} \end{aligned}$$

Step 3. Calculate F/M ratio
Using Eq. (7.62)

$$\begin{aligned} \text{F/M} &= (5838\ \text{lb/d})/(28{,}400\ \text{lb}) \\ &= 0.206\ \text{lb BOD/d per lb MLVSS} \end{aligned}$$

EXAMPLE 2: Convert the BOD concentration of 160 mg/L in the primary effluent into BOD loading rate in terms of kg/m^3 and lb/1000 ft^3. If this is used for 24 h high rate aeration, what is the rate for 6 h aeration?

Solution:

Step 1. Calculate BOD loading in kg/m^3

$$\begin{aligned} 160\ \text{mg/L} &= \frac{160\ \text{mg} \times (1\ \text{g}/1000\ \text{mg})}{1\ \text{L} \times (1\ \text{m}^3/1000\ \text{L})} \\ &= 160\ \text{g/m}^3 \\ &= 0.16\ \text{kg/m}^3 \end{aligned}$$

Step 2. Calculate BOD loading in lb/1000 ft^3

$$\begin{aligned} 0.16\ \text{kg/m}^3 &= \frac{0.16\ \text{kg} \times 2.205\ \text{lb/kg}}{1\ \text{m}^3 \times 35.3147\ \text{ft}^3/\text{m}^3} \\ &= 0.01\ \text{lb/ft}^3 \\ &= \frac{0.01\ \text{lb} \times 1000}{1000\ \text{ft}^3} \\ &= 10\ \text{lb/1000 ft}^3 \end{aligned}$$

Thus $\quad 1\ \text{lb/1000 ft}^3 = 0.016\ \text{kg/m}^3$

Step 3. Calculate loading for 6-h aeration

$$0.16\ \text{kg/(d} \cdot \text{m}^3) \times \frac{24\ \text{h}}{6\ \text{h}} = 0.64\ \text{kg/(d} \cdot \text{m}^3)$$

and

$$10.0\ \text{lb/d/1000 ft}^3 \times \frac{24\ \text{h}}{6\ \text{h}} = 40\ \text{lb/(d} \cdot \text{m}^3)$$

Note: Refer to Table 7.11 to meet Ten States Standards.
The influent BOD to the conventional activated-sludge process is limited to 160 mg/L for 6-h aeration.

21.3 Biochemical Reactions

The mechanism of removal of biodegradable organic matter in aerobic suspended-growth systems can be expressed by the energy production or respiration equation

$$\underset{\text{(CHONS)}}{\text{Organic matter}} + \underset{\text{(heterotrophic)}}{\text{bacteria}} + O_2 \rightarrow CO_2 + H_2O + NH_4^+ + \underset{\text{(energy)}}{\text{new cells}} \tag{7.66}$$

Further nitrification process can take place by selected autotrophs with oxidation of ammonia to nitrate and protoplasm synthesis

$$NH_4^+ + O_2 + CO_2 + HCO_3 \underset{\text{Energy}}{\overset{\text{Bacteria}}{\rightarrow}} NO_3^- + H_2O + H^+ \quad \underset{\text{(protoplasm)}}{\text{new cells}} \tag{7.67}$$

The oxidation of protoplasm is a metabolic reaction which breaks down the protoplasm into elemental constituents, so that cells die. This is called endogenous respiration or cell maintenance, as follows

$$\text{Protoplasm} + O_2 \rightarrow CO_2 + NH_3 + H_2O + \text{dead cells} \tag{7.68}$$

21.4 Process Design Concepts

The activated-sludge process has been used extensively in its original basic form as well as in many modified forms. The process design considerations include hydraulic retention time (HRT) for reaction kinetics; wastewater characteristics; environmental conditions, such as temperature, pH, and alkalinity; and oxygen transfer.

Single or multiple aeration tanks are typically designed for completed mixed flow, plug flow, or intermediate patterns and sized to provide an HRT in the range of 0.5 to 24 hours or more (WEF and ASCE 1991a).

In the past, designs of activated-sludge processes were generally based on empirical parameters such as BOD_5 (simplified as BOD) loadings and aeration time (hydraulic retention time). In general, short HRTs were used for weak wastewaters and long HRTs for strong wastewaters. Nowadays, the basic theory and design parameters for the activated-sludge process are well developed and generally accepted. The different design approaches were proposed by researchers on the basis of the concepts of BOD_5, mass balance, and microbial growth kinetics (McKinney 1962, Eckenfelder 1966, Jenkins and Garrison 1968, Eckenfelder and Ford 1970, Lawrence and McCarty 1970, Ramanathan and Gaudy 1971, Gaudy and Kincannon 1977, Schroeder 1977, Bidstrup and Grady 1988).

Solution of the theoretical sophisticated design equations and computer models requires knowledge of microbial metabolism and growth kinetics, with pilot studies to obtain design information. Alternatives to such studies are: (1) to assume certain wastewater characteristics and embark on a semi-empirical design; and (2) to use an entirely empirical approach relying on regulatory recommended standards (WEF and ASCE 1991a).

21.5 Process Mathematical Modeling

For almost half a century, numerous design criteria utilizing empirical and rational parameters based on biological kinetic equations have been developed for suspended-growth systems. A survey of major consulting firms in the United States indicates that the basic Lawrence and McCarty (1970) model is most widely used. Details of its development can be obtained in the references (Lawrence and McCarty 1970, Grady and Lim 1980, Qasim 1985, Metcalf and Eddy Inc. 1991). The basic Lawrence and McCarty design equations used for sizing suspended-growth systems are listed below.

Complete mix with recycle. The flow in a reactor is continuously stirred. The contents of the reactor are mixed completely. It is called the complete-mix reactor or continuous flow stirred tank reactor. Ideally, it is uniform throughout the tank. If the mass input rate into the reactor remains constant, the content of the effluent remains constant.

For a complete-mix system, the mean hydraulic retention time (HRT) θ for the aeration tank is

$$\theta = V/Q \tag{7.69}$$

where θ = hydraulic retention time, d
V = volume of aeration tank, m^3
Q = influent wastewater flow, m^3/d

Referring to Fig. 7.15(a), the mean cell residence time θ_c (or sludge age or SRT) in the system is defined as the mass of organisms in the aeration tank divided by the mass of organisms removed from the system per day, and is expressed as

$$\theta_c = \frac{X}{(\Delta X/\Delta t)} \tag{7.70}$$

$$\theta_c = \frac{VX}{Q_{wa}X + Q_e X_e} = \frac{\text{total mass SS in reactor}}{\text{SS wasting rate}} \tag{7.71}$$

where θ_c = mean cell residence time based on solids in the tank, d
X = concentration of MLVSS maintained in the tank, mg/L
$\Delta X/\Delta t$ = growth of biological sludge over time period Δt, mg/(L · d)
Q_{wa} = flow of waste sludge removed from the aeration tank, m^3/d
Q_e = flow of treated effluent, m^3/d
X_e = microorganism concentration (VSS) in effluent, mg/L

For system-drawn waste sludge from the return sludge line (Fig. 7.15b), the mean cell residence time would be

$$\theta_c = \frac{VX}{Q_{wr}X_r + Q_e X_e} \tag{7.72}$$

where Q_{wr} = flow of waste sludge from return sludge line, m^3/d
X_r = microorganism concentration in return sludge line, mg/L

Microorganism and substrate mass balance. Because the term $V \times$ MLSS in Eq. (7.63) is a function of SRT or θ_c and not HRT or return sludge ratio, the F/M ratio is also a function only of SRT. Therefore, operation of an activated-sludge plant at constant SRT will result in operation at a constant F/M ratio.

The mass balance for the microorganisms in the entire activated sludge system is expressed as the rate of accumulation of the microorganisms in the inflow plus net growth, minus that in the outflow. Mathematically, it is expressed as (Metcalf and Eddy Inc. 1991)

$$V\frac{dX}{dt} = QX_0 + V(r'_g) - (Q_{wa}X + Q_e X_e) \tag{7.73}$$

where V = volume of aeration tank, m^3
dX/dt = rate of change of microorganisms concentration (VSS), mg/(L · m^3 · d)
Q = influent flow, m^3/d
X_0 = microorganisms concentration (VSS) in influent, mg/L
X = microorganisms concentration in tank, mg/L
r'_g = net rate of microorganism growth (VSS), mg/(L · d)
Other terms are as in the above equations.

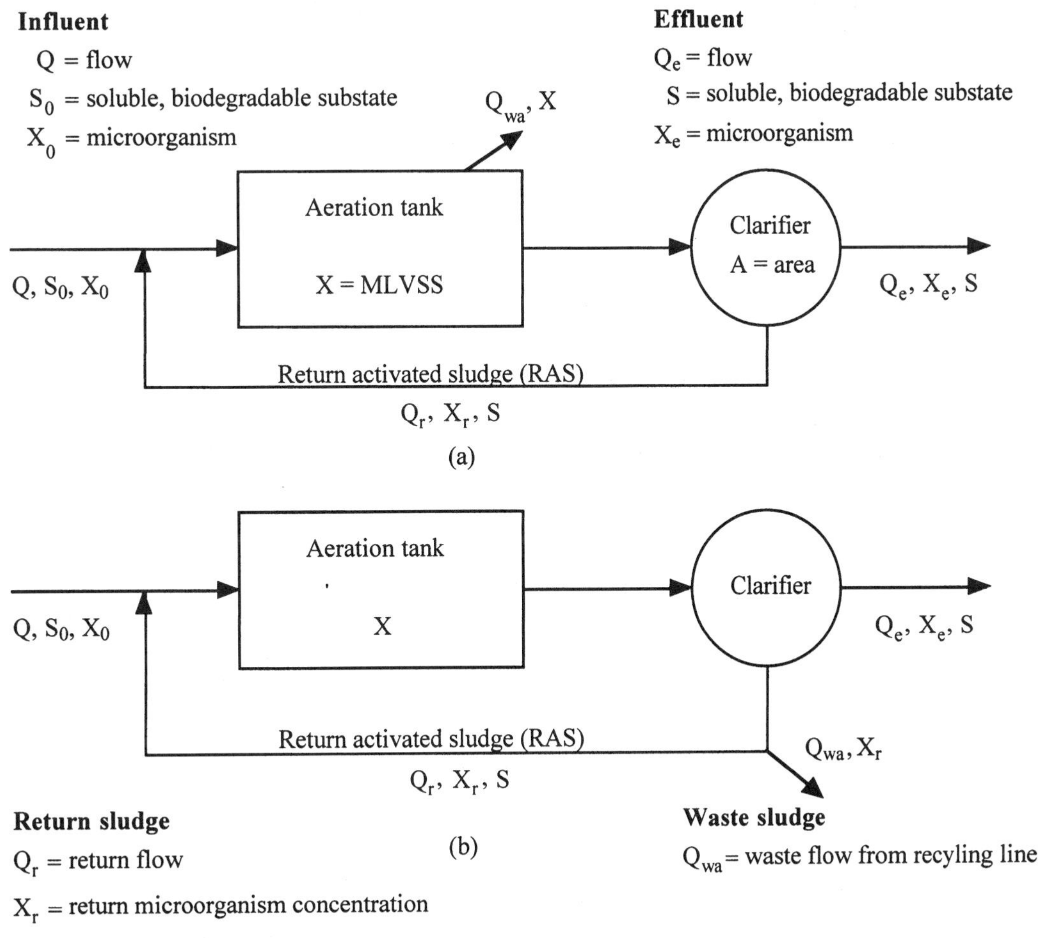

FIGURE 7.15 Schematic chart of complete-mix activated sludge reactor: (a) sludging wasting from the aeration tank; (b) sludge wasting from return sludge line.

The net rate of bacterial growth is expressed as

$$r'_g = Yr_{su} - k_d X \tag{7.74}$$

where Y = maximum yield coefficient over finite period of log growth, mg/mg
r_{su} = substrate utilization rate, mg/(m^3 · d)
k_d = endogenous decay coefficient, per day

Substituting Eq. (7.74) into Eq. (7.73), and assuming the cell concentration in the influent is zero and steady-state conditions, this yields

$$\frac{Q_{wa}X + Q_e X_e}{VX} = -Y\frac{r_{su}}{X} - k_d \tag{7.75}$$

The left-hand side of Eq. (7.75) is the inverse of the mean cell residence time θ_c as defined in Eq. (7.71); thus

$$\frac{1}{\theta_c} = -Y\frac{r_{su}}{X} - k_d \tag{7.76}$$

The term $1/\theta_c$ is the net specific growth rate.

The term r_{su} can be computed from the following equation

$$r_{su} = \frac{Q}{V}(S_0 - S) = \frac{S_0 - S}{\theta} \tag{7.77}$$

where $S_0 - S$ = mass concentration of substrate utilized, mg/L
S_0 = substrate concentration in influent, mg/L
S = substrate concentration in effluent, mg/L
θ = hydraulic retention time (Eq. (7.69)), day

Effluent microorganism and substrate concentrations. The mass concentration of microorganisms X in the aeration tank can be derived by substituting Eq. (7.77) into Eq. (7.76)

$$X = \frac{\theta_c Y(S_0 - S)}{\theta(1 + k_d\theta_c)} = \frac{\mu_m(S_0 - S)}{k(1 + k_d\theta_c)} \tag{7.78}$$

where μ_m = maximum specific growth rate, per day
k = maximum rate of substrate utilization per unit mass of microorganism, per day

Substituting for θ from Eq. (7.69) for (7.78) and solving for the reactor (aeration tank) volume yields

$$V = \frac{\theta_c QY(S_0 - S)}{X(1 + k_d\theta_c)} \tag{7.79}$$

The substrate concentration in effluent S can also be determined from the substrate mass balance by the following equation

$$S = \frac{K_s(1 + \theta_c k_d)}{\theta_c(Yk - k_d) - 1} \tag{7.80}$$

where S = effluent substrate (soluble BOD_5) concentration, mg/L
K_s = half-velocity constant, substrate concentration at one-half the maximum growth rate, mg/L
Other terms are as mentioned in previous equations.

The ranges of typical biological kinetic coefficients for activated-sludge systems for domestic wastewater are given in Table 7.12. When the kinetic coefficients are available, Eqs. (7.78) and (7.80) can be used to predict densities of effluent microorganisms and substrate (soluble BOD_5) concentrations,

TABLE 7.12 Ranges and Typical Biological Kinetic Coefficients for the Activated-Sludge Process for Domestic Wastewater

Coefficient	Range	Typical value
k, per day	11–20	5
k_d, per day	0.025–0.075	0.06
K_s, mg/L BOD_5	25–100	60
mg/L COD	15–70	40
Y, mg VSS/mg BOD_5	0.4–0.8	0.6

Source: Metcalf and Eddy, Inc. (1991), Techobanoglous and Schroeder (1985).

respectively. They do not take into account influent suspended solids concentrations (primary effluent). They can be used to evaluate the effectiveness of various treatment system changes.

Substituting the value of X given by Eq. (7.78) for r'_g in Eq. (7.74), and dividing by the term $S_0 - S$ which corresponds to the value of r_{su} expressed as concentration value, the observed yield in the system with recycle is

$$Y_{obs} = \frac{Y}{1 + K_d \theta_c \text{ or } \theta_{ct}} \tag{7.81}$$

where Y_{obs} = observed yield in the system with recycle, mg/mg
θ_{ct} = mean of all residence times based on solids in the aeration tank and in the secondary clarifier, d

Other terms are defined previously.

Process design and control relationships. To predict the effluent biomass and BOD_5 concentration, the use of Eqs. (7.78) and (7.80) is difficult because many coefficients have to be known. In practice, the relationship between specific substrate utilization rate, mean cell residence time, and the food to microorganism (F/M) ratio is commonly used for activated-sludge process design and process control.

In Eq. (6.76), the term $(-r_{su}/X)$ is called the specific substrate utilization rate (or food to microorganisms ratio), U. Applying r_{su} in Eq. (7.77), the specific substrate utilization rate can be computed by

$$U = -\frac{r_{su}}{X} \tag{7.82}$$

$$U = \frac{Q(S_0 - S)}{VX} = \frac{S_0 - S}{\theta X} \tag{7.83}$$

The term U is substituted for the term $(-r_{su}/X)$ in Eq. (7.76). The resulting equation becomes

$$\frac{1}{\theta_c} = YU - k_d \tag{7.84}$$

The term $1/\theta_c$ is the net specific growth rate and is directly related to U, the specific substrate utilization rate.

In order to determine the specific substrate utilization rate, the substrate utilized and the biomass effective in the utilization must be given. The substrate utilized can be computed from the difference between the influent and the effluent BOD_5 or COD.

In the complete-mix activated-sludge process with recycle, waste sludge (cells) can be withdrawn from the tank or from the recycling line. If waste sludge is withdrawn from the tank and the VSS in the effluent X_e is negligible ($Q_e X_e \approx 0$), Eq. (7.71) will (if X_e is very small) be approximately rewritten as

$$\theta_c \approx \frac{VX}{Q_{wa}X} \tag{7.85}$$

or

$$Q_{wa} \approx \frac{V}{\theta_c} \tag{7.86}$$

The flow rate of waste sludge from the sludge return line will be approximately

$$Q_{wr} = \frac{VX}{\theta_c X_r} \tag{7.87}$$

where X_r is the concentration (in mg/L) of sludge in the sludge return line.

In practice, the food-to-microorganism (F/M) ratio is widely used and is closely related to the specific substrate utilization rate U. The F/M (in per day) ratio is defined as the influent soluble

BOD_5 concentration (S_0) divided by the product of hydraulic retention time θ and MLVSS concentration X in the aeration tank

$$\text{F/M} = \frac{S_0}{\theta X} = \frac{QS_0}{VX} = \frac{\text{mg BOD}_5\text{/d}}{\text{mg MLVSS}} \tag{7.88}$$

F/M and U are related by the efficiency E of the activated-sludge process as follows

$$U = \frac{(F/M)E}{100} \tag{7.89}$$

The value of E is determined by

$$E = \frac{S_0 - S}{S_0} \times 100 \tag{7.90}$$

where E = process efficiency, %
S_0 = influent substrate concentration, mg/L
S = effluent substrate concentration, mg/L

Sludge production. The amount of sludge generated (increased) per day affects the design of the sludge treatment and disposal facilities. It can be calculated by

$$P_x = Y_{obs}Q(S_0 - S) \div (1000 \text{ g/kg}) \qquad \text{(SI units)} \tag{7.91}$$

$$P_x = Y_{obs}Q(S_0 - S)\,(8.34) \qquad \text{(US customary units)} \tag{7.92}$$

where P_x = net waste activated sludge (VSS), kg/d or lb/d
Y_{obs} = observed yield (Eq. (7.81)), g/g or lb/lb
Q = influent wastewater flow, m^3/d or Mgal/d
S_0 = influent soluble BOD_5 concentration, mg/L
S = effluent soluble BOD_5 concentration, mg/L
8.34 = conversion factor, lb/(Mgal · (mg/L))

Oxygen requirements in the process. The theoretical oxygen requirement in the activated sludge is determined from BOD_5 of the wastewater and the amount of microorganisms wasted per day from the process. The biochemical reaction can be expressed as below

$$\begin{array}{lcccl} C_5H_7NO_2 & + & 5O_2 & \rightarrow & 5CO_2 + 2H_2O + NH_3 + \text{energy} \\ 113 & & 5 \times 32 = 160 & & \\ \text{organism cells} & & & & \\ 1 & & 1.42 & & \end{array} \tag{7.93}$$

Equation (7.93) suggests that the BOD_u (ultimate BOD) for one mole of cells requires 1.42 (160/113) mole of oxygen. Thus the theoretical oxygen requirement to remove the carbonaceous organic matter in wastewater for an activated-sludge process is expressed as (Metcalf and Eddy Inc. 1991).

$$\text{Mass of O}_2\text{/d} = \text{total mass of BOD}_u \text{ used} - 1.42 \text{ (mass of organisms wasted, } p_x)$$

$$\text{kg O}_2\text{/d} = \frac{Q(S_0 - S)}{(1000 \text{ g/kg})f} - 1.42\,P_x \qquad \text{(SI units)} \tag{7.94a}$$

$$\text{kg O}_2\text{/d} = \frac{Q(S_0 - S)}{1000 \text{ g/kg}}\left(\frac{1}{f} - 1.42\,Y_{obs}\right) \tag{7.94b}$$

$$\text{lb O}_2\text{/d} = Q(S_0 - S) \times 8.34\left(\frac{1}{f} - 1.42\,Y_{obs}\right) \qquad \text{(US customary units)} \tag{7.95}$$

where Q = influent flow, m^3/d or Mgal/d
S_0 = influent soluble BOD_5 concentration, mg/L
S = effluent soluble BOD_5 concentration, mg/L
f = conversion factor for converting BOD_5 to BOD_u
Y_{obs} = observed yield, g/g or lb/lb
8.34 = conversion factor, lb/(Mgal · (mg/L))

When nitrification is considered, the total oxygen requirement is the mass of oxygen per day for removal of carbonaceous matter and for nitrification. It can be calculated as

$$\text{kg } O_2/\text{d} = \frac{Q(S_0 - S)}{1000 \text{ g/kg}}\left(\frac{1}{f} - 1.42\, Y_{obs}\right) + \frac{Q(N_0 - N)}{1000 \text{ g/kg}} \quad \text{(SI units)} \tag{7.96}$$

$$\text{lb } O_2/\text{d} = 8.34\,[Q(S_0 - S)(1/f - 1.42\, Y_{obs}) + 4.75(N_0 - N)] \quad \text{(US customary units)} \tag{7.97}$$

where N_0 = influent total kjeldahl nitrogen concentration, mg/L
N = effluent total kjeldahl nitrogen concentration, mg/L
4.75 = conversion factor for oxygen requirement for complete oxidation of TKN

Oxygen requirements generally depend on the design peak hourly BOD_5, MLSS, and degree of treatment. Aeration equipment must be able to maintain a minimum of 2.0 mg/L of dissolved oxygen concentration in the mixed liquor at all times and provide adequate mixing. The normal air requirements for all activated-sludge systems, except extended aeration, are 1.1 kg of oxygen (93.5 m^3 of air) per kg BOD_5 or 1.1 lb of oxygen (1500 ft^3 of air) per lb BOD_5, for design peak aeration tank loading. That is 94 m^3 of air per kg of BOD_5 (1500 ft^3/lb BOD_5) at standard conditions of temperature, pressure, and humidity. For the extended aeration process, normal air requirements are 128 m^3/kg BOD_5 or 2050 ft^3/lb BOD_5 (GLUMRB 1996).

For F/M ratios greater than 0.3/day, the air requirements for conventional activated-sludge systems amount to 31 to 56 m^3/kg (500 to 900 ft^3/lb) of BOD_5 removed for coarse bubble (nonporous) diffusers and 25 to 37 m^3/kg (400 to 600 ft^3/lb) BOD_5 removal for fine bubble (porous) diffusers. For lower F/M ratios, endogenous respiration, nitrification, and prolonged aeration increase air use to 75 to 112 m^3/kg (1200 to 1800 ft^3/lb) of BOD_5 removal. In practice, air requirements range from 3.75 to 15.0 m^3 air/m^3 water (0.5 to 2 ft^3/gal) with a typical value of 7.5 m^3/m^3 or 1.0 ft^3/gal (Metcalf and Eddy Inc. 1991).

EXAMPLE 1A: Design a complete-mix activated-sludge system.

Given:

Average design flow	= 0.32 m^3/s (7.30 Mgal/d)
Peak design flow	= 0.80 m^3/s (18.25 Mgal/d)
Raw wastewater BOD_5	= 240 mg/L
Raw wastewater TSS	= 280 mg/L
Effluent BOD_5	≤ 20 mg/L
Effluent TSS	≤ 24 mg/L
Wastewater temperature	= 20°C

Operational parameters and biological kinetic coefficients:

Design mean cell residence time θ_c = 10 day
MLVSS = 2400 mg/L (can be 3600 mg/L)
VSS/TSS = 0.8

$$\text{TSS concentration in RAS} = 9300 \text{ mg/L}$$
$$Y = 0.5 \text{ mg VSS/mg BOD}_5$$
$$k_d = 0.06/\text{d}$$
$$\text{BOD}_5/\text{ultimate BOD}_u = 0.67$$

Assume:

1. BOD (i.e. BOD_5) and TSS removal in the primary clarifiers are 33 and 67 percent, respectively.
2. Specific gravity of the primary sludge is 1.05 and the sludge has 4.4% of solids content.
3. Oxygen consumption is 1.42 mg per mg of cell oxidized.

Solution:

Step 1. Calculate BOD and TSS loading to the plant

$$\begin{aligned} \text{Design flow } Q &= 0.32 \text{ m}^3/\text{s} \times 86{,}400 \text{ s/d} \\ &= 27{,}648 \text{ m}^3/\text{d} \\ \text{Since } 1 \text{ mg/L} &= 1 \text{ g/m}^3 = 0.001 \text{ kg/m}^3 \\ \text{BOD loading} &= 0.24 \text{ kg/m}^3 \times 27{,}648 \text{ m}^3/\text{d} \\ &= 6636 \text{ kg/d} \\ \text{TSS loading} &= 0.28 \text{ kg/m}^3 \times 27{,}648 \text{ m}^3/\text{d} \\ &= 7741 \text{ kg/d} \end{aligned}$$

Step 2. Calculate characteristics of primary sludge

$$\begin{aligned} \text{BOD removed} &= 6636 \text{ kg/d} \times 0.33 = 2190 \text{ kg/d} \\ \text{TSS removed} &= 7741 \text{ kg/d} \times 0.67 = 5186 \text{ kg/d} \\ \text{Specific gravity of sludge} &= 1.05 \\ \text{Solids concentration} &= 4.4\% = 0.044 \text{ kg/kg} \\ \text{Sludge flow rate} &= \frac{5186 \text{ kg/d}}{1.05 \times 1000 \text{ kg/m}^3} \div 0.044 \text{ kg/kg} \\ &= 112 \text{ m}^3/\text{d} \end{aligned}$$

Step 3. Calculate flow, BOD, and TSS in primary effluent (secondary influent)

$$\begin{aligned} \text{Flow} &= \text{design flow, } 27{,}648 \text{ m}^3/\text{d} - 112 \text{ m}^3/\text{d} \\ &= 27{,}536 \text{ m}^3/\text{d} \\ &= Q \text{ for Step 6} \\ \text{BOD} &= 6636 \text{ kg/d} - 2190 \text{ kg/d} \\ &= 4446 \text{ kg/d} \\ &= \frac{4446 \text{ kg/d} \times 1000 \text{ g/kg}}{27{,}536 \text{ m}^3/\text{d}} \\ &= 161.5 \text{ g/m}^3 \\ &= 161.5 \text{ mg/L} = S_0 \\ \text{TSS} &= 7741 \text{ kg/d} - 5186 \text{ kg/d} \\ &= 2555 \text{ kg/d} \end{aligned}$$

$$= \frac{2555 \text{ kg/d} \times 1000 \text{ g/kg}}{27{,}536 \text{ m}^3\text{/d}}$$

$$= 92.8 \text{ g/m}^3$$

$$= 92.8 \text{ mg/L}$$

Step 4. Estimate the soluble BOD_5 escaping treatment, S, in the effluent
Use the following relationship

Effluent BOD = influent soluble BOD escaping treatment,
S + BOD of effluent suspended solids

(a) Determine the BOD_5 of the effluent SS (Assuming 63% biodegradable)

$$\text{Biodegradable effluent solids} = 24 \text{ mg/L} \times 0.63 = 15.1 \text{ mg/L}$$

$$\text{Ultimate BOD}_u \text{ of the biodegradable effluent solids} = 15.1 \text{ mg/L} \times 1.42 \text{ mg O}_2\text{/mg cell}$$

$$= 21.4 \text{ mg/L}$$

$$\text{BOD}_5 = 0.67 \text{ BOD}_u = 0.67 \times 21.4 \text{ mg/L}$$

$$= 14.3 \text{ mg/L}$$

(b) Solve for influent soluble BOD_5 escaping treatment

$$20 \text{ mg/L} = S + 14.3 \text{ mg/L}$$

$$S = 5.7 \text{ mg/L}$$

Step 5. Calculate the treatment efficiency E using Eq. (7.90)
(a) The efficiency of biological treatment based on soluble BOD is

$$E = \frac{S_0 - S}{S_0} \times 100 = \frac{(161.5 - 5.7 \text{ mg/L}) \times 100\%}{161.5 \text{ mg/L}}$$

$$= 96.5\%$$

(b) The overall plant efficiency including primary treatment is

$$E = \frac{(240 - 20)\text{mg/L} \times 100\%}{240 \text{ mg/L}}$$

$$= 91.7\%$$

Step 6. Calculate the reactor volume using Eq. (7.79)

$$V = \frac{\theta_c QY(S_0 - S)}{X(1 + k_d\theta_c)}$$

$\theta_c = 10$ day

$Q = 27.536 \text{ m}^3\text{/d}$ (from Step 3)

$Y = 0.5$ mg/mg

$S_0 = 161.5$ mg/L (from Step 3)

$S = 5.7$ mg/L (from Step 4(b))

$$X = 2400 \text{ mg/L}$$
$$k_d = 0.06 \text{ d}^{-1}$$
$$V = \frac{(10 \text{ day})(27{,}536 \text{ m}^3/\text{d})(0.5)(161.5 - 5.7) \text{ mg/L}}{(2400 \text{ mg/L})(1 + 0.06 \text{ d}^{-1} \times 10 \text{ day})}$$
$$= 5586 \text{ m}^3$$
$$= 1.48 \text{ Mgal}$$

Step 7. Determine the dimensions of the aeration tank
Provide 4 rectangular tanks with common walls.
Use width-to-length ratio of 1 : 2 and water depth of 4.4 m with 0.6 m freeboard

$$w \times 2w \times (4.4 \text{ m}) \times 4 = 5586 \text{ m}^3$$
$$w = 12.6 \text{ m}$$
$$\text{width} = 12.6 \text{ m}$$
$$\text{length} = 25.2 \text{ m}$$
$$\text{water depth} = 4.4 \text{ m (total tank depth} = 5.0 \text{ m)}$$

Note: In the Ten States Standards, liquid depth should be 3 to 9 m (10 to 30 ft). The tank size would be smaller if a higher design value of MLVSS were used.

Step 8. Calculate the sludge wasting flow rate from the aeration tank
Using Eq. (7.71), also $V = 5586 \text{ m}^3$ and VSS = 0.8 SS

$$\theta_c = \frac{VX}{Q_{wa}X + Q_eX_e}$$
$$10 \text{ day} = \frac{(5586 \text{ m}^3)(2400 \text{ mg/L})}{Q_{wa}(3000 \text{ mg/L}) + (27{,}536 \text{ m}^3/\text{d})(24 \text{ mg/L} \times 0.8)}$$
$$Q_{wa} = 270 \text{ m}^3/\text{d}$$
$$= 0.0715 \text{ Mgal/d}$$

Step 9. Estimate the quantity of sludge to be wasted daily
(a) Calculate observed yield
Using Eq. (7.81)

$$Y_{obs} = \frac{Y}{1 + K_d\theta_c} = \frac{0.5}{1 + 0.06 \times 10}$$
$$= 0.3125$$

(b) Calculate the increase in the mass of MLVSS for Eq. (7.91)

$$p_x = Y_{obs}Q(S_0 - S) \div 1000 \text{ g/kg}$$
$$= 0.3125 \times 27{,}536 \text{ m}^3/\text{d} \times (161.5 - 5.7) \text{ g/m}^3 \div 1000 \text{ g/kg}$$
$$= 1341 \text{ kg/d}$$

Note: A factor of 8.34 lb/Mgal is used if Q is in Mgal/d.

(c) Calculate the increase in MLSS (or TSS), p_{ss}

$$p_{ss} = 1341 \text{ kg/d} \div 0.8$$
$$= 1676 \text{ kg/d}$$

(d) Calculate TSS lost in the effluent, p_e

$$p_e = (27{,}536 - 270)\ \text{m}^3/\text{d} \times 24\ \text{g/m}^3 \div 1000\ \text{g/kg}$$
$$= 654\ \text{kg/d}$$

Note: Flow is less sludge wasting rate from Step 8.

(e) Calculate the amount of sludge that must be wasted

$$\text{Wastewater sludge} = p_{ss} - p_e$$
$$= 1676\ \text{kg/d} - 654\ \text{kg/d}$$
$$= 1022\ \text{kg/d}$$

Step 10. Estimate return activated sludge rate
Using a mass balance of VSS, Q and Q_r are the influent and RAS flow rates, respectively.

$$\text{VSS in aerator} = 2400\ \text{mg/L}$$
$$\text{VSS in RAS} = 9300\ \text{mg/L} \times 0.8 = 7440\ \text{mg/L}$$
$$2400\ (Q + Q_r) = 7440\ Q_r$$
$$Q_r/Q = 0.4762$$
$$Q_r = 0.4762 \times 27{,}536\ \text{m}^3/\text{d}$$
$$= 13{,}112\ \text{m}^3/\text{d}$$
$$= 0.152\ \text{m}^3/\text{s}$$

Step 11. Check hydraulic retention time (HRT = θ)

$$\theta = V/Q = 5586\ \text{m}^3/(27{,}536\ \text{m}^3/\text{d})$$
$$= 0.203\ \text{day} \times 24\ \text{h/d}$$
$$= 4.87\ \text{h}$$

Note: The preferred range is of HRT is 5 to 15 h.

Step 12. Check F/M ratio using U in Eq. (7.83)

$$U = \frac{S_0 - S}{\theta X} = \frac{161.5\ \text{mg/L} - 5.7\ \text{mg/L}}{(0.203\ \text{day})(2400\ \text{mg/L})}$$
$$= 0.32\ \text{d}^{-1}$$

Step 13. Check organic loading rate and mass of ultimate BOD_u utilized

$$\text{Loading} = \frac{QS_0}{V} = \frac{27{,}536\ \text{m}^3/\text{d} \times 161.5\ \text{g/m}^3}{5586\ \text{m}^3 \times 1000\ \text{g/kg}} = 0.80\ \text{kg BOD}_5/(\text{m}^3 \cdot \text{d})$$

$$\text{BOD}_5 = 0.67\ \text{BOD}_u\ \text{(given)}$$
$$\text{BOD}_u\ \text{used} = Q(S_0 - S)/0.67$$
$$= \frac{27{,}536\ \text{m}^3/\text{d} \times (161.5 - 5.7)\ \text{g/m}^3}{0.67 \times 1000\ \text{g/kg}}$$
$$= 6403\ \text{kg/d}$$

Step 14. Compute theoretical oxygen requirements
The theoretical oxygen required is calculated from Eq. (7.94a)

$$O_2 = \frac{Q(S_0 - S)}{(1000 \text{ g/kg}) f} - 1.42 p_x$$

$$= 6403 \text{ kg/d (from Step 13)} - 1.42 \times 1341 \text{ kg/d (from Step 9(b))}$$

$$= 4499 \text{ kg/d}$$

Step 15. Compute the volume of air required
Assume that air weighs 1.202 kg/m^3 (0.075 lb/ft^3) and contains 23.2% oxygen by weight; the oxygen transfer efficiency for the aeration equipment is 8%; and a safety factor of 2 is used to determine the actual volume for sizing the blowers.

(a) The theoretical air required is

$$\text{Air} = \frac{4499 \text{ kg/d}}{1.202 \text{ kg/m}^3 \times 0.232 \text{ g } O_2/\text{g air}}$$

$$= 16{,}200 \text{ m}^3/\text{d}$$

(b) The actual air required at an 8% oxygen transfer efficiency

$$\text{Air} = 16{,}200 \text{ m}^3/\text{d} \div 0.08$$

$$= 202{,}000 \text{ m}^3/\text{d}$$

$$= 140 \text{ m}^3/\text{min}$$

$$= 4950 \text{ ft}^3/\text{min}$$

(c) The design air required (with a factor of safety 2) is

$$\text{Air} = 140 \text{ m}^3/\text{min} \times 2$$

$$= 280 \text{ m}^3/\text{min}$$

$$= 9900 \text{ ft}^3/\text{min}$$

$$= 165 \text{ ft}^3/\text{s (cfs)}$$

Step 16. Check the volume of air required per unit mass BOD_5 removed, and per unit volume of wastewater and aeration tank, using the actual value obtained in Step 15b.

(a) Air supplied per kg of BOD_5 removed

$$\text{Air} = \frac{202{,}000 \text{ m}^3/\text{d} \times 1000 \text{ g/kg}}{27{,}536 \text{ m}^3/\text{d} \times (161.5 - 5.7) \text{ g/m}^3}$$

$$= 47.1 \text{ m}^3 \text{ of air/kg } BOD_5$$

$$= 754 \text{ ft}^3/\text{lb}$$

(b) Air supplied per m^3 of wastewater treated

$$\text{Air} = \frac{202{,}000 \text{ m}^3/\text{d}}{27{,}536 \text{ m}^3/\text{d}}$$

$$= 7.34 \text{ m}^3 \text{ air/m}^3 \text{ wastewater}$$

$$= 0.98 \text{ ft}^3/\text{gal}$$

(c) Air supplied per m^3 of aeration tank

$$\text{Air} = \frac{202{,}000\ \text{m}^3/\text{d}}{5586\ \text{m}^3}$$
$$= 36.2\ \text{m}^3/(\text{m}^3 \cdot \text{d})$$
$$= 36.2\ \text{ft}^3/(\text{ft}^3 \cdot \text{d})$$

EXAMPLE 1B: Design secondary clarifiers using the data in Example 1A and the MLSS settling test results. The MLSS setting data is derived from a pilot plant study and is shown below:

MLSS, mg/L	1200	1800	2400	3300	4000	5500	6800	8100
Initial settling velocity, m/h	4.1	3.1	2.1	1.2	0.77	0.26	0.13	0.06

Solution:

Step 1. Plot the MLSS settling curve (Fig. 7.16) on log–log paper from the observed data.

Step 2. Construct the gravity solid-flux curve from Fig. 7.17.
Data in columns 1 and 2 of the following table is adopted from Fig. 7.16. Values in column 3 are determined by col. 1 × col. 2. Plot the solids–flux curve using MLSS concentration and calculate solids flux as shown in Fig. 7.17.

(1) MLSS concentration X, mg/L or g/m^3	(2) Initial settling velocity V_1, m/h	(3) Solids flux XV_1, kg/(m^2 · h)
1000	4.2	4.20
1500	3.7	5.55
2000	2.8	5.60
2500	2.0	5.00
3000	1.5	4.50
4000	0.76	3.04
5000	0.41	2.05
6000	0.22	1.32
7000	0.105	0.74
8000	0.062	0.50
9000	0.033	0.30

Step 3. Determine the limiting solids flux value
From Fig. 7.17, determine the limiting solids flux (SF) for an underflow concentration of 9300 mg/L. This is achieved by drawing a tangent to the solids flux curve from 9300 mg/L (the desired underflow) solids concentration. The limiting solids flux value is

$$1.3\ \text{kg}/(\text{m}^2 \cdot \text{h}) \equiv 31.2\ \text{kg}/(\text{m}^2 \cdot \text{d}) \equiv 6.4\ \text{lb}/(\text{ft}^2 \cdot \text{d})$$

Step 4. Calculate design flow to the secondary clarifiers, Q
From Steps 8 and 10 of Example 1A

$$Q = \text{average design flow} + \text{return sludge flow} - \text{MLSS wasted}$$
$$= (27{,}563 + 13{,}112 - 270)\ \text{m}^3/\text{d}$$
$$= 40{,}405\ \text{m}^3/\text{d}$$
$$= 0.468\ \text{m}^3/\text{s}$$

Use two clarifiers, each one with flow of 20,200 m^3/d.

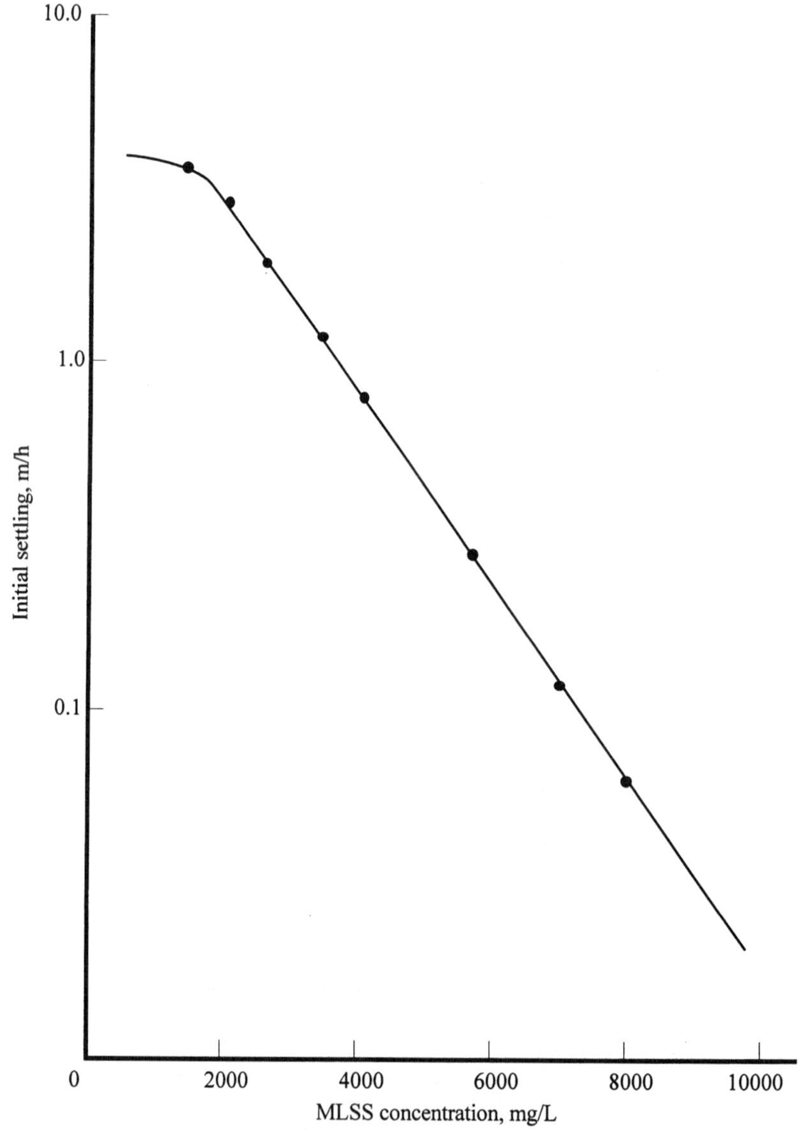

FIGURE 7.16 Solids settling rate: experimental results for design of secondary clarifier.

Step 5. Compute the area A and diameter d of the clarifier

$$A = \frac{QX}{\text{SF}} \tag{7.98}$$

where A = area of the secondary clarifier, m^2 or ft^2
Q = influent flow of the clarifier, m^3/h or gal/h
X = MLSS concentration, kg/m^3 or lb/ft^3
SF = limiting solids flux, $kg/(m^2 \cdot h)$ or $lb/(ft^2 \cdot h)$

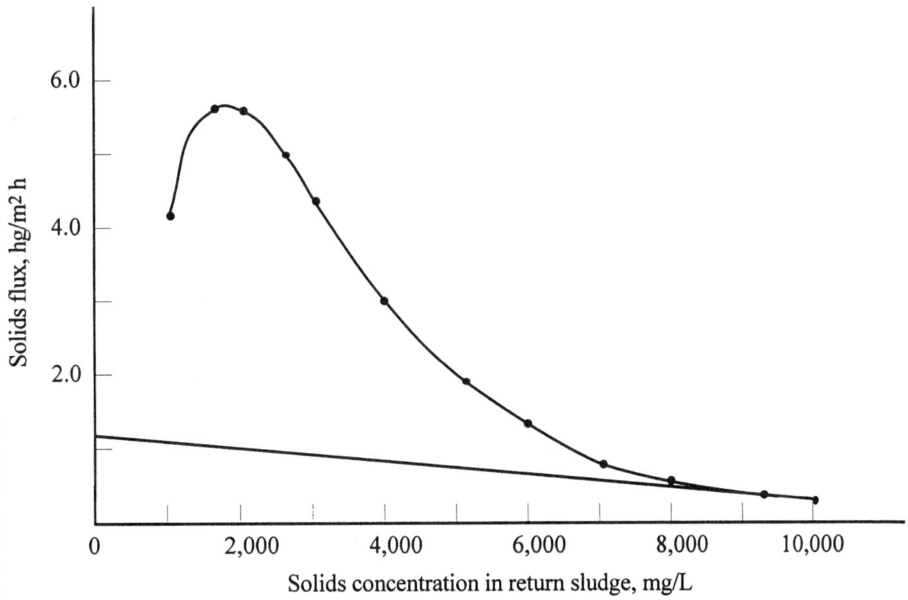

FIGURE 7.17 Solids flux curve.

From Eq. (7.98), for each clarifier

$$Q = 20{,}200 \text{ m}^3\text{/d} = 841.7 \text{ m}^3\text{/h}$$

$$\text{MLSS} = (2400/0.8) \text{ mg/L} = 3.0 \text{ kg/m}^3$$

$$\text{SF} = 1.3 \text{ kg/(m}^2 \cdot \text{h) (from Step 3)}$$

$$\text{Therefore } A = \frac{QX}{\text{SF}} = \frac{841.7 \text{ m}^3\text{/h} \times 3.0 \text{ kg/m}^3}{1.3 \text{ kg/(m}^2 \cdot \text{h)}}$$

$$= 1942 \text{ m}^2$$

$$A = \pi(d/2)^2 = 1942 \text{ m}^2$$

$$d = \sqrt{1942 \text{ m}^2 \times 4 \div 3.14}$$

$$= 49.7 \text{ m} \approx 50 \text{ m}$$

$$= 164 \text{ ft}$$

Step 6. Check the surface overflow rate at average design flow

$$\text{Overflow rate} = \frac{Q}{A} = \frac{20{,}200 \text{ m}^3\text{/d}}{1942 \text{ m}^2}$$

$$= 10.4 \text{ m}^3\text{/(m}^2 \cdot \text{d)}$$

$$= 255 \text{ gal/(d} \cdot \text{ft}^2)$$

Note: This is less than the design criteria of 15 $\text{m}^3/(\text{m}^2 \cdot \text{d})$

Step 7. Check the clarifier's area for clarification requirements
From Step 6, the calculated surface overflow rate

$$Q/A = 10.4 \text{ m}^3\text{/(m}^2 \cdot \text{d)} = 0.433 \text{ m/h}$$

From Fig. 7.16, the MLSS concentration corresponding to a 0.433 m/h of settling rate is 4700 mg/L. The design MLSS is only 2400 mg/L. The area is sufficient.

Step 8. Check the surface overflow rate at peak design flow

$$\text{Peak flow} = 0.80\ \text{m}^3/\text{s} \times 86{,}400\ \text{s/d} + 13{,}112\ \text{m}^3/\text{d}$$
$$= 82{,}232\ \text{m}^3/\text{d}$$
$$\text{Overflow rate} = \frac{82{,}232\ \text{m}^3/\text{d}}{1942\ \text{m}^2 \times 2}$$
$$= 21.2\ \text{m}^3/(\text{m}^2 \cdot \text{d})$$

Step 9. Determine recycle ratio required to maintain MLSS concentration at 3000 mg/L

$$(Q + Q_r) \times 3000 = QX + Q_r X_u$$

where Q = influent flow, m³/d or Mgal/d
Q_r = recycle flow, m³/d or Mgal/d
X = influent SS concentration, mg/L
X_u = underflow SS concentration, mg/L

$$Q(3000 - X) = Q_r (X_u - 3000)$$
$$\frac{Q_r}{Q} = \frac{3000 - X}{X_u - 3000} = \alpha = \text{recycle ratio}$$

when X_u = 9300 mg/L (from Step 3) and

$$X = 93\ \text{mg/L (from Step 3 of Example 1A, 92.8 mg/L)}$$
$$\alpha = \frac{3000\ \text{mg/L} - 93\ \text{mg/L}}{9300\ \text{mg/L} - 3000\ \text{mg/L}}$$
$$= 0.46$$

Step 10. Estimate the depth required for the thickening zone. The total depth of the secondary clarifier is the sum of the required depths of the clear water zone, the solids thickening zone, and the sludge storage zone. In order to estimate the depth of the thickening zone, it is assumed that, under normal conditions, the mass of solids retained in the secondary clarifier is 30 percent of the mass of solids in the aeration basin, and that the average concentration of solids in the sludge zone is 7000 mg/L (Metcalf and Eddy Inc. 1991). The system has 4 aeration tanks and 2 clarifiers.

(a) Compute total mass of solids in each aeration basin

$$\text{MLSS} = 3000\ \text{mg/L} = 3.0\ \text{kg/m}^3$$

Refer to Step 7 of Example 1A

$$\text{Solids in each aeration tank} = 3.0\ \text{kg/m}^3 \times 4.4\ \text{m} \times 12.6\ \text{m} \times 25.2\ \text{m}$$
$$= 4191\ \text{kg}$$

(b) Compute the mass of solids in a clarifier for two days (see Step 11)

$$\text{Solids in each clarifier} = 4191\ \text{kg} \times 0.3 \times 2$$
$$= 2515\ \text{kg}$$

(c) Compute depth of sludge zone

$$\text{Depth} = \frac{\text{mass}}{\text{area} \times \text{concentration}} = \frac{2515 \text{ kg} \times 1000 \text{ g/kg}}{1942 \text{ m}^2 \times 7000 \text{ g/m}^3}$$

$$= 0.19 \text{ m}$$

$$\approx 0.2 \text{ m}$$

Step 11. Estimate the depth of the sludge storage zone. This zone is provided to store excess solids at peak flow conditions or at a period over which the sludge-processing facilities are unable to handle the sludge quantity. Assume storage capacity for 2 days' sustained peak flow (of 2.5 average flow) and for 7 days' sustained peak BOD loading (of 1.5 BOD average).

(a) Calculate total volatile solids generated under sustained BOD loading. Using Eq. (7.91)

$$P_x = Y_{obs} Q(S_0 - S) \div (1000 \text{ g/kg})$$

$$Y_{obs} = 0.3125 \text{ (from Step 9(a) of Example 1A)}$$

$$Q = 2.5\ (0.32 \text{ m}^3\text{/s}) = 0.8 \text{ m}^3\text{/s} = 69{,}120 \text{ m}^3\text{/d}$$

$$S_0 = 161.5 \text{ mg/L} \times 1.5 = 242 \text{ mg/L}$$

$$S = 5.7 \text{ mg/L} \times 1.5 = 9 \text{ mg/L}$$

$$P_x = 0.3125 \times 69{,}120 \text{ m}^3\text{/d} \times (242 - 9) \text{ g/m}^3 \div (1000 \text{ g/kg})$$

$$= 5033 \text{ kg/d}$$

(b) Calculate mass of solid for 2-day storage

$$\text{Total solids stored} = 5033 \text{ kg/d} \times 2 \text{ day} \div 0.8$$

$$= 12{,}580 \text{ kg}$$

(c) Calculate stored solids per clarifier

$$\text{Solids to be stored} = 12{,}580 \text{ kg} \div 2$$

$$= 6290 \text{ kg}$$

(d) Calculate total solids in each secondary clarifier

$$\text{Total} = 6290 \text{ kg} + 2515 \text{ kg (from Step 10(b))}$$

$$= 8805 \text{ kg}$$

(e) Calculate the required depth for sludge storage in the clarifier

$$\text{Depth} = \frac{8505 \text{ kg} \times 1000 \text{ g/kg}}{7000 \text{ g/m}^3 \times 1942 \text{ m}^2}$$

$$= 0.63 \text{ m}$$

Step 12. Calculate total required depth of a clarifier
The depth of clear water and settling zone is commonly 1.5 to 2 m. Provide 2 m for this example.

$$\text{Total required depth of a clarifier} = 2 \text{ m} + 0.2 \text{ m} + 0.63 \text{ m}$$

$$= 2.83 \text{ m}$$

$$\approx 3 \text{ m}$$

With addition of 0.65 m free board,

$$\text{the total water depth of the clarifier} = 3.65\text{ m} = 12\text{ ft}$$

Step 13. Check hydraulic retention time of the secondary clarifier

$$\text{Diameter} = 50\text{ m (from Step 5)}$$

$$\text{Volume} = 3.14\,(50\text{ m}/2)^2 \times 3.0\text{ m} = 5888\text{ m}^3$$

Under average design flow plus recirculation, from Step 4

$$\text{HRT} = \frac{5888\text{ m}^3 \times 24\text{ h/d}}{40{,}405\text{ m}^3\text{/d (from Step 4)}} = 3.5\text{ h}$$

At peak design flow with recirculation, from Step 8

$$\text{HRT} = \frac{5888\text{ m}^3 \times 24\text{ h/d}}{82{,}232\text{ m}^3\text{/d (from Step 8)}} = 1.72\text{ h}$$

Plug-flow with recycle. Conventional and many modified activated-sludge processes are designed approximately like the plug-flow model (Fig. 7.18). In a time plug-flow reactor, all fluid particles entering the reactor stay in the reactor for an equal amount of time and pass through the reactor in sequence. Those that enter first leave first. It is assumed that there is no mixing in the lateral direction. In practice, a time plug-flow reactor is difficult to obtain because of longitudinal dispersion. In a real plug-flow, some particles may make more passes through the reactor as a result of recycling.

The mathematical modeling for the plug-flow process is difficult. The most accepted useful kinetic model of the plug-flow reactor for the activated-sludge process was developed by Lawrence and McCarty (1969). They made two assumptions: (1) The concentration of microorganisms in the influent of the reactor is approximately the same as that in the effluent of the reactor. This

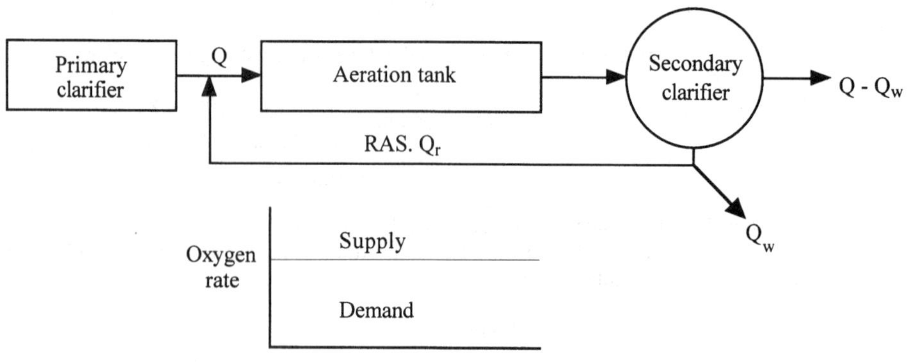

FIGURE 7.18 Conventional activated-sludge process and DO in aeration tank.

assumption is valid only when $\theta_c/\theta > 5$. (2) The rate of substrate utilization r_{su} as wastewater flows through the aeration tank is expressed as

$$r_{su} = \frac{kSX}{K_s + S} \tag{7.99}$$

The reverse of the mean cell residence time is

$$\frac{1}{\theta_c} = \frac{Yk(S_0 - S)}{(S_0 - S) + (1 + \alpha)K_s \ln (S_i/S)} - k_d \tag{7.100}$$

and

$$S_i = \frac{S_0 + \alpha S}{1 + \alpha} \tag{7.101}$$

where r_{su} = substrate utilization rate, mg/L
S = effluent concentration of substrate, mg/L
S_0 = influent concentration of substrate, mg/L
X = average concentration of microorganisms in the reactor, mg/L
k, K_s, Y = kinetic coefficients as defined previously
θ_c = mean cell residence time, d
α = recycle ratio
S_i = influent concentration in reactor after dilution with recycle flow

21.6 Operation and Control of Activated-Sludge Processes

The aeration basin is the heart of the activated-sludge process. Process operation and system controls usually need well trained operators, with operational parameters such as air or oxygen supply, F/M ratio, and mass balance to return activated sludge. However, some non-theoretical indexes such as sludge volume index, sludge density index, sludge settleability, and sludge age are widely used and valuable tools for the daily process of operation and control.

Sludge volume index. The sludge volume index (SVI) is the volume of mL occupied by 1 g of suspension after 30 minutes of settling. Although SVI is not theoretically supported, experience has shown it is useful in routine process control. SVI typically is used to monitor the settling characteristics of activated sludge and can impact on return sludge rate and MLSS.

SVI is calculated from the laboratory test results of the suspended solids concentration of a well mixed sample of the suspension and the 30-min settled sludge volume. The sludge volume is measured by filling a one-liter graduated cylinder to the 1.0 liter mark, allowing settling for 30 minutes, and then reading the volume of settled solids. The mixed liquid suspended solids is determined by filtering, drying, and weighing a sample of the mixed liquor as stated in the previous section in this chapter. The value of SVI can be calculated by the following formula (Standard Methods—APHA *et al.* 1998).

$$\text{SVI} = \frac{\text{SV} \times 1000 \text{ mg/g}}{\text{MLSS}} \tag{7.102}$$

or

$$\text{SVI} = \frac{\text{Wet settled sludge, mL/L}}{\text{Dry sludge solid, mg/L}} \tag{7.102a}$$

where SVI = sludge volume index, mL/g
SV = settled sludge volume, mL/L
MLSS = mixed liquor suspended solids, mg/L
1000 = milligrams per gram, mg/g

Typical values of SVI for domestic activated-sludge plants operating with an MLSS concentration of 2000 to 3500 mg/L range from 80 to 150 mL/g (Davis and Cornwell 1991).

The SVI is an important factor in process design. It limits the tank MLSS concentration and return sludge rate.

Sludge density index. Sludge density index is used in a way similar to the sludge volume index to indicate the settleability of a sludge in a secondary clarifier or effluent. The weight in grams of one milliliter of sludge, after settling for 30 minutes, is calculated as

$$\text{SDI} = 100/\text{SVI} \tag{7.103}$$

where SDI = sludge density index, g/mL
SVI = sludge volume index, mL/g

A sludge with good settling characteristics has an SDI between 1.0 and 2.0, whereas an SDI of 0.5 indicates a bulky or nonsettleable sludge (Cheremisinoff 1995).

Return activated sludge. Return activated-sludge is the settled activated sludge that is collected in the secondary clarifier and returned to the aeration tank to mix with the influent wastewater. The efficiency of the activated-sludge process is measured by BOD removal, which is directly related to the volatile activated-sludge solids in the aeration basin. The purpose of sludge return is to maintain a sufficient concentration of activated sludge in the aeration tank. The RAS makes it possible for the microorganisms to be in the treatment system longer than the flowing wastewater. The RAS flow for a conventional activated-sludge system is usually 20 to 40 percent of the tank influent flow rate. Table 7.13 lists typical ranges of RAS flow rates for some of the activated-sludge processes. Sludge volume index (Eq. (7.102)) is empirical and is used to control the rate of return sludge. The minimum percentage of return activated sludge is related to SVI and the solids concentration in the mixed liquor and is expressed as (Clark and Viessman 1966)

$$\%\text{ of return sludge} = \frac{100}{100/(\text{SVI})P - 1} \tag{7.104}$$

where SVI = sludge volume index, mL/g
P = percentage of solids in the mixed liquor

EXAMPLE 1: Determine the aeration basin dimensions for a town of 20,000 population. Assume the mixed liquor suspended solids is 2600 mg/L; BOD loading rate is 0.48 kg/(d · m^3) or 0.03 lb/(d · ft^3); SVI = 110 mL/g; and P = 2600 mg/L.

TABLE 7.13 Guidelines for Return Activated Sludge Flow Rate

Type of process	Percent of design average flow	
	Minimum	Maximum
Conventional	15	100
Carbonaceous stage of separate-stage nitrification	15	100
Step-feed aeration	15	100
Complete-mix	15	100
Contact stabilization	50	150
Extended aeration	50	150
Nitrification stage of separate stage nitrification	50	200

Source: GLUMRB (Ten States Standards) 1996.

Solution:

Step 1. Calculate total BOD loading (L) on aeration basin
Using average BOD contribution of 0.091 kg/(person · d) or 0.20 lb/(c · d)

$$\text{Total BOD load to the wastewater treatment plant} = 20{,}000 \text{ person} \times 0.091 \text{ kg/(person} \cdot \text{d)}$$
$$= 1820 \text{ kg/d}$$

Assume 30% BOD removal in the primary clarifiers.
The total daily BOD load on aeration basin would be

$$L = 1820 \text{ kg/d} \times (1.0 - 0.3) = 1274 \text{ kg/d}$$

Step 2. Calculate percentage return activated sludge
Using Eq. (7.104)

$$\text{SVI} = 110 \text{ mL/g}$$
$$P = 2600 \text{ mg/L} = 0.26\%$$
$$\%\text{ return} = \frac{100}{100/(\text{SVI})P - 1} = \frac{100}{100/(110 \times 0.26) - 1}$$
$$= 40$$

Step 3. Determine the total BOD loading on the basin
Assume the BOD concentration for both the RAS and the influent are the same. Allowing an additional 40% of return sludge, the total BOD loading would be

$$1274 \text{ kg/d} \times 1.40 = 1784 \text{ kg/d}$$

Step 4. Determine the volume V required for the aeration tank

$$V = \frac{\text{Total BOD loading}}{\text{Allowed loading per m}^3}$$
$$= \frac{1784 \text{ kg/d}}{0.48 \text{ kg/(d} \cdot \text{m}^3)}$$
$$= 3717 \text{ m}^3$$

Step 5. Determine the dimensions of the aeration tank
Select eight tanks, the water depth of 4.4 m, and add 0.6 m for the freeboard, and width of 7 m. The length of a tank is

$$\text{length} = \frac{3717 \text{ m}^3}{8 \times 4.4 \text{ m} \times 7 \text{ m}}$$
$$= 15 \text{ m}$$

Note: Each tank measures 5 m × 7 m × 15 m.

EXAMPLE 2: Determine the aeration tank based on 6 h of aeration period, using the data given in Example 1. Assume 0.53 m^3/(person · d) or 140 gal/(c · d) of waste flow.

Solution:

Step 1. Calculate plant flow rate Q

$$Q = 0.53 \text{ m}^3\text{/(person} \cdot \text{d)} \times 20{,}000 \text{ persons}$$
$$= 10{,}600 \text{ m}^3\text{/d}$$

Step 2. Plus 40% of Q_r

$$Q + Q_r = 1.4 \times 10{,}600\ \text{m}^3/\text{d}$$
$$= 14{,}840\ \text{m}^3/\text{d}$$

Step 3. Calculate the required volume V of a tank

$$V = 14{,}840\ \text{m}^3/\text{d} \times (1\ \text{d}/24\ \text{h}) \times 6\ \text{h}$$
$$= 3710\ \text{m}^3$$

Note: For these two examples, the tank volume determined is based on either organic loading or hydraulic loading; these give almost identical results.

Step 4. Determine tank dimensions
Use eight tanks with depth of 5 m, width of 7 m, and length of 15 m, as in Example 1.

Return activated-sludge flow rate. The RAS, at a constant flow rate, is independent of the tank influent flow rate and changes the MLSS in the tank continuously. The MLSS concentration will be at a maximum during low influent flows and at a minimum during peak influent flows. The secondary clarifier must constantly change the depth of the sludge blanket. To maintain a good F/M ratio, a programmed control device should be installed to maintain a constant percent of RAS flow to the tank influent flow. However, any change in the activated-sludge quality will require a different RAS flow rate because of the changing settling characteristics of the sludge.

The relationship between maximum RAS concentration (X_r) and SVI can be derived from Eq. (7.69) using settled sludge volume SV = 1000 mL/L. The suspended solids concentration in RAS can be calculated by

$$X_r = \left(\frac{\text{SV} \times 1000\ \text{mg/g}}{\text{SVI}}\right)$$
$$= \frac{1000\ \text{mL/L} \times 1000\ \text{mg/g}}{\text{SVI mL/g}} \tag{7.105}$$
$$X_r = \frac{10^6}{\text{SVI}}\ \text{mg/L}$$

Sludge settleability. Another method of calculating RAS flow rate is based on the settleability approach. The settleability is defined as the percentage of volume occupied by the sludge after settling for 30 min. The RAS flow rate is related to settled sludge volume as follows

$$1000Q_r = (\text{SV})\,(Q + Q_r) \tag{7.106}$$

or

$$Q_r = \frac{(\text{SV})Q}{1000 - \text{SV}} \tag{7.107}$$

where 1000 = factor mg/L
Q_r = flow of return activated sludge, m³/s or Mgal/d
Q = flow of tank influent, m³/s or Mgal/d
SV = settled sludge volume (30-min settling), mL/L

EXAMPLE: In practice, the operator checks the VSS concentration in the return activated sludge at least once every shift and makes the appropriate RAS flow rate adjustment. The previous operator recorded that RAS flow was 44 gal/min (240 m³/d) with the VSS in RAS of 5800 mg/L. The on-duty operator determines the VSS in RAS as 5500 mg/L. What should the RAS flow rate be adjusted to?

Solution: Since the VSS in the RAS is decreasing, the sludge wasting flow should be increased proportionally to waste the amount of VSS.

$$Q_{adj} \times 5550 \text{ mg/L} = 44 \text{ gal/min} \times 5800 \text{ mg/L}$$
$$Q_{adj} = 44 \times 5800/5550 \text{ gal/min}$$
$$= 46 \text{ gal/min}$$
$$= 250 \text{ m}^3\text{/d}$$

Aeration tank mass balance. In practice, the maximum RAS pumping capacity is commonly designed as 100 percent of design average flow (use 150 percent for oxidation ditch). The required RAS pumpage can be determined by aeration tank mass balance. Assume new cell growth is negligible and there is no accumulation in the aeration tank. Also, the solids concentration in the tank influent is negligible compared to the MLSS, X, in the tank. Thus mass of inflow is equal to that of outflow. This can be expressed as

$$X_r Q_r = X(Q + Q_r) \tag{7.108}$$

or

$$Q_r = \frac{X}{X_r - X} Q \tag{7.109}$$

$$X_r = \frac{X(Q + Q_r)}{Q_r} = \frac{\text{MLSS}(Q + Q_r)}{Q_r} \tag{7.110}$$

where X_r = return activated-sludge suspended solids, mg/L
Q_r = flow of RAS, m³/s or Mgal/d
X = mixed liquor suspended solids (MLSS), mg/L
Q = flow of secondary influent (or plant flow), m³/s or Mgal/d

The values of X and X_r include both the volatile and non-volatile (inert) fractions. Eqs. (7.106) and (7.110) assume no loss of suspended solids in the basin. Eq. (7.109) is essentially the same as Eq. (7.107).

EXAMPLE 1: Determine the return activated-sludge flow as a percentage of the influent flow 10.0 Mgal/d (37,850 m³/d). The sludge settling volume in 30 min is 255 mL.

Solution:

Step 1. Compute RAS flow Q_r in %
Using Eq. (7.107)

$$Q_r, \% = \frac{\text{SV}}{1000 - \text{SV}} = \frac{255 \text{ mL/L} \times 100\%}{1000 \text{ mL/L} - 255 \text{ mL/L}}$$
$$= 34\%$$

Step 2. Compute RAS flow rate in Mgal/d

$$Q_r = 0.34\,Q = 0.34 \times 10 \text{ Mgal/d}$$
$$= 3.40 \text{ Mgal/d}$$
$$= 12{,}900 \text{ m}^3\text{/d}$$

EXAMPLE 2: The MLSS concentration in the aeration tank is 2800 mg/L. The sludge settleability test showed that the sludge volume, settled for 30 min in a 1-L graduated cylinder, is 285 mL. Calculate the sludge volume index and estimate the SS concentration in the RAS and the required return sludge ratio.

Solution:

Step 1. Calculate SVI
Using Eq. (7.102)

$$\mathrm{SVI} = \frac{\mathrm{SV} \times 1000\ \mathrm{mg/g}}{\mathrm{MLSS}} = \frac{285\ \mathrm{mL/L} \times 1000\ \mathrm{mg/g}}{2800\ \mathrm{mg/L}}$$

$$= 102\ \mathrm{mL/g}$$

This is in the typical range of 80 to 150 mL/g

Step 2. Calculate SS in RAS
Using Eq. (7.105)

$$X_r = \frac{1{,}000{,}000}{\mathrm{SVI}} = \frac{1{,}000{,}000\ (\mathrm{mL/L})\ (\mathrm{mg/g})}{102\ \mathrm{mL/g}}$$

$$= 9804\ \mathrm{mg/L}$$
$$= 0.98\%$$

Step 3. Calculate Q_r/Q
Using Eq. (7.107)

$$\frac{Q_r}{Q} = \frac{\mathrm{SV}}{1000 - \mathrm{SV}} = \frac{285\ \mathrm{mg/L}}{(1000 - 285)\ \mathrm{mg/L}}$$

$$= 0.40$$
$$= 40\%\ \text{return}$$

EXAMPLE 3: Compute the return activated-sludge flow rate in m^3/d and as a percentage of the influent flow of 37,850 m^3/d (10 Mgal/d). The laboratory results show that the SVI is 110 mg/L and the MLSS is 2500 mg/L.

Solution:

Step 1. Compute the suspended solids in RAS based on the SVI
Using Eq. (7.105)

$$X_r\ \text{in RAS} = \frac{1{,}000{,}000\ \mathrm{mg/L}}{\mathrm{SVI}} = \frac{1{,}000{,}000\ \mathrm{mg/L}}{110}$$

$$= 9090\ \mathrm{mg/L}$$

Step 2. Compute RAS flow rate Q_r, based on SVI
Using Eq. (7.109)

$$Q_r = \frac{X}{X_r - X} Q = \frac{2500\ \mathrm{mg/L} \times 37{,}850\ \mathrm{m^3/d}}{9090\ \mathrm{mg/L} - 2500\ \mathrm{mg/L}}$$

$$= 14{,}360\ \mathrm{m^3/d}$$
$$= 3.79\ \mathrm{Mgal/d}$$

Step 3. Compute RAS flow as percentage of influent flow

$$\text{RAS flow, \%} = \frac{Q_r \times 100\%}{Q} = \frac{3.79\ \mathrm{Mgal/d} \times 100\%}{10.0\ \mathrm{Mgal/d}}$$

$$= 37.9\%$$

$$\text{Say} = 38\%\ \text{of influent flow}$$

Waste activated sludge. The excess of activated sludge generated from the secondary clarifier must be wasted, usually from the return sludge line. The waste activated-sludge (WAS) is discharged to either the primary clarifiers or to sludge thickeners. Withdrawing mixed liquor directly from the aeration basin or from the basin effluent line is an alternative method of sludge wasting. The amount of WAS removed affects mixed liquor settleability, oxygen consumption, growth rate of the microorganisms, nutrient quantities, input occurrence of foaming and sludge bulking effluent quality, and possibility of nitrification.

The purpose of WAS is to maintain a given food-to-microorganisms ratio or mean cell residence time of the system. It should remove just the amount of microorganisms that grow in excess of the microorganism death rate.

Sludge wasting can be accomplished on an intermittent or continuous basis. The parameters used for control guidelines are F/M ratio, mean cell residence time, MLVSS, and sludge age.

Secondary clarifier mass balance. Although it assumes the sludge blanket is level in the secondary clarifier, the secondary clarifier mass balance approach is a useful tool to determine the RAS flow rate. The calculations are based on the secondary clarifier (Fig. 7.15). Assuming the sludge blanket in the clarifier is not changed and the effluent suspended solids are negligible, the SS entering the clarifier is equal to the SS leaving. The relationship can be written as

$$(Q + Q_{wr})(\text{MLSS}) = Q_{wa}(\text{WAS}) + Q_{wr}(\text{RAS})$$

$$Q(\text{MLSS}) - Q_{wa}(\text{WAS}) = Q_{wr}(\text{RAS} - \text{MLSS})$$

$$Q_{wr} = \frac{Q(\text{MLSS}) - Q_{wa}(\text{WAS})}{\text{RAS} - \text{MLSS}} \tag{7.111}$$

where Q = flow of tank influent, m^3/d
Q_{wr} = return activated sludge flow, m^3/d
Q_{wa} = waste activated sludge flow rate, m^3/d
MLSS = mixed liquor suspended solids, mg/L
WAS = SS of waste activated sludge, mg/L
RAS = SS of return activated sludge, mg/L

Note: WAS is equal to RAS (Fig. 7.15a)

Sludge age. Sludge age is a measure of the length of time a particle of suspended solids has been retained in the activated-sludge process. It is also referred to as mean cell residence time and is an operational parameter related to the F/M ratio. Sludge age is defined as that of the suspended solids under aeration (kg or lb) divided by the suspended solids added (kg/d or lb/d) and is calculated by the following equation which relates MLSS in the system to the influent suspended solids

$$\text{Sludge age} = \frac{\text{SS under aeration, kg or lb}}{\text{SS added, kg/d or lb/d}}$$

$$\text{Sludge age} = \frac{\text{MLSS} \times V}{\text{SS}_w \times Q_w + \text{SS}_e \times Q_e} \tag{7.112}$$

where Sludge age = mean cell residence time, d
MLSS = mixed liquor suspended solids, mg/L
V = volume of aeration basin, m^3 or Mgal
SS_w = suspended solids in waste sludge, mg/L
Q_w = flow of waste sludge, m^3/d or Mgal/d
SS_e = suspended solids in wastewater effluent, mg/L
Q_e = flow of wastewater effluent, m^3/d or Mgal/d

Note: Eqs. (7.112) and (7.72) are the same.

Sludge age can also be expressed in terms of the volatile portion of suspended solids, which is more representative of biological mass. Solids retention time in an activated-sludge system is measured in days, whereas the liquid aeration period is in hours. In most activated-sludge plants, sludge age ranges from 3 to 8 days (USEPA 1979). The liquid aeration periods vary from 3 to 30 hours (Hammer 1986). Wastewater passes through the aeration tank only once and rather quickly, whereas the resultant biological growths and extracted waste organics are repeatedly recycled from the secondary clarifier back to the aeration basin.

EXAMPLE 1: An activated-sludge system has an influent flow of 22,700 m^3/d (6.0 Mgal/d) with suspended solids of 96 mg/L. Three aeration tanks hold 1500 m^3 (53,000 ft^3) each with MLSS of 2600 mg/L. Calculate the sludge age for the system.

Solution:

Step 1. Calculate the SS under aeration, MLSS $\times$ V

$$\text{MLSS} = 2600 \text{ mg/L} = 2600 \text{ g/m}^3 = 2.6 \text{ kg/m}^3$$

$$\begin{aligned} \text{MLSS} \times V &= 2.6 \text{ kg/m}^3 \times 1500 \text{ m}^3 \times 3 \\ &= 11{,}700 \text{ kg} \\ &= 25{,}800 \text{ lb} \end{aligned}$$

Step 2. Calculate the SS added

$$\begin{aligned} \text{SS added} &= \text{Influent flow} \times \text{SS conc.} \\ &= 22{,}700 \text{ m}^3\text{/d} \times 0.096 \text{ kg/m}^3 \\ &= 2180 \text{ kg/d} \\ &= 4800 \text{ lb/d} \end{aligned}$$

Step 3. Calculate sludge age
Using Eq. (7.112)

$$\begin{aligned} \text{Sludge age} &= \frac{\text{SS under aeration}}{\text{SS added}} = \frac{11{,}700 \text{ kg}}{2180 \text{ kg/d}} \\ &= 5.4 \text{ days} \end{aligned}$$

Note: It is in the typical range of 3 to 8 days.

EXAMPLE 2: In an activated-sludge system, the solids under aeration are 13,000 kg (28,700 lb); the solids added are 2200 kg/d (4850 lb/d); the return activated-sludge SS concentration is 6600 mg/L; the desired sludge age is 5.5 days; and the current waste activated-sludge is 2100 kg/d (4630 lb/d). Calculate the WAS flow rate using the sludge age control technique.

Solution:

Step 1. Calculate the desired SS under aeration for the desired sludge age of 5.5 days

$$\begin{aligned} \text{SS} &= \text{solid added} \times \text{sludge age} \\ &= 2200 \text{ kg/d} \times 5.5 \text{ day} \\ &= 12{,}100 \text{ kg} \\ &= 26{,}800 \text{ lb} \end{aligned}$$

Step 2. Calculate the additional suspended solids removed per day

$$\begin{aligned} \text{SS aerated} - \text{SS desired} &= 13{,}000 \text{ kg} - 12{,}100 \text{ kg} \\ &= 900 \text{ kg} \end{aligned}$$

Step 3. Calculate the additional WAS flow (q) to maintain the desired sludge age

$$q\ (\text{SS in RAS}) = 900\ \text{kg/d}$$
$$q = (900\ \text{kg/d})/(6.6\ \text{kg/m}^3)$$
$$= 136.4\ \text{m}^3/\text{d} = 0.095\ \text{m}^3/\text{min}$$
$$= 25.0\ \text{gal/min}$$

Step 4. Calculate total WAS flow

$$\text{Current flow} = (2100\ \text{kg/d})/(6.6\ \text{kg/m}^3)$$
$$= 318.2\ \text{m}^3/\text{d} = 0.221\ \text{m}^3/\text{min}$$
$$= 58.4\ \text{gal/min}$$
$$\text{Total WAS flow} = (0.095 + 0.221)\ \text{m}^3/\text{min}$$
$$= 0.316\ \text{m}^3/\text{min}$$
$$= 83.4\ \text{gal/min}$$

EXAMPLE 3: The aeration basin volume is 6600 m^3 (1.74 Mgal). The influent flow to the basin is 37,850 m^3/d (10.0 Mgal/d) with BOD of 140 mg/L. The MLSS is 3200 mg/L with 80 percent volatile portion. The SS concentration in the return activated sludge is 6600 mg/L. The current waste activated-sludge flow rate is 340 m^3/d (0.090 Mgal/d). Determine the desired WAS flow rate using the F/M ratio control technique with a desired F/M ratio of 0.32.

Solution:

Step 1. Calculate BOD loading

$$\text{BOD} = 140\ \text{mg/L} = 140\ \text{g/m}^3 = 0.14\ \text{kg/m}^3$$
$$\text{Loading} = \text{BOD} \times \text{flow}$$
$$= 0.14\ \text{kg/m}^3 \times 37{,}850\ \text{m}^3/\text{d}$$
$$= 5299\ \text{kg/d}$$
$$= 11{,}680\ \text{lb/d}$$

Step 2. Calculate the desired MLVSS with the desired F/M = 0.32

$$\text{Desired MLVSS} = \text{BOD loading/(F/M) ratio}$$
$$= (5299\ \text{kg/d})/(0.32\ \text{kg/(d} \cdot \text{kg))}$$
$$= 16{,}560\ \text{kg}$$

Step 3. Calculate the desired MLSS

$$\text{Desired MLSS} = \text{MLVSS}/0.80 = 16{,}560\ \text{kg}/0.80$$
$$= 20{,}700\ \text{kg}$$

Step 4. Calculate actual MLSS under aeration

$$\text{Actual MLSS} = \text{concentration} \times \text{basin volume}$$
$$= 3.2\ \text{kg/m}^3 \times 6600\ \text{m}^3$$
$$= 21{,}120\ \text{kg}$$

Step 5. Calculate the additional solids to be removed daily

$$(\text{actual} - \text{desired})\ \text{MLSS} = 21{,}120\ \text{kg/d} - 20{,}700\ \text{kg/d}$$
$$= 420\ \text{kg/d}$$

Step 6. Calculate the additional WAS flow q required

$$\begin{aligned} q &= \text{additional SS removed/SS in RAS} \\ &= (420 \text{ kg/d})/(6.6 \text{ kg/m}^3) \\ &= 63.6 \text{ m}^3\text{/d} \end{aligned}$$

Step 7. Calculate the total WAS flow Q

$$\begin{aligned} Q &= (340 + 63.6) \text{ m}^3\text{/d} \\ &= 403.6 \text{ m}^3\text{/d} = 0.28 \text{ m}^3\text{/min} \\ &= 74 \text{ gal/min} \end{aligned}$$

EXAMPLE 4: Calculate the waste activated-sludge flow rate using the mean cell residence time (MCRT) method. The given conditions are the same as those in Example 3. In addition, the desired MCRT is 7.5 days and the SS level in the effluent is 12 mg/L.

Solution: (use US customary units)

Step 1. Calculate SS concentration in the aeration tank

$$\begin{aligned} \text{MLSS} &= 3200 \text{ mg/L} \\ V &= 1.74 \text{ Mgal} \\ \text{SS} &= 1.74 \text{ Mgal} \times 3200 \text{ mg/L} \times 8.34 \text{ lb/(Mgal} \cdot \text{mg/L)} \\ &= 46{,}440 \text{ lb} \\ &= 21{,}070 \text{ kg} \end{aligned}$$

Step 2. Determine SS lost in the effluent

$$\begin{aligned} \text{SS}_e &= 12 \text{ mg/L} \\ Q_e &= 10 \text{ Mgal/d} \\ \text{SS}_e \times Q_e &= 10 \text{ Mgal/d} \times 12 \text{ mg/L} \times 8.34 \text{ lb/(Mgal} \cdot \text{mg/L)} \\ &= 1000 \text{ lb/d} = 453 \text{ kg/d} \end{aligned}$$

Step 3. Calculate the desired SS in waste sludge
Using Eq. (7.113)

$$\text{MCRT, d} = \frac{\text{SS in aerator, lb}}{\text{SS wasted, lb/d} + \text{SS in effluent, lb/d}}$$

$$7.5 \text{ day} = \frac{46{,}440 \text{ lb}}{\text{SS wasted, lb/d} + 1000 \text{ lb/d}}$$

$$\begin{aligned} \text{SS wasted} &= (46{,}440 \text{ lb} - 7500 \text{ lb})/7.5 \text{ day} \\ &= 5192 \text{ lb/d} \\ &= 2355 \text{ kg/d} \end{aligned}$$

Step 4. Calculate the WAS flow rate Q_w

$$\begin{aligned} \text{SS}_w &= 6600 \text{ mg/L} \\ \text{SS wasted} &= Q_w \times \text{SS}_w \times 8.34 \end{aligned}$$

$$Q_w = \frac{\text{SS wasted}}{\text{SS}_w \times 8.34} = \frac{5192 \text{ lb/d}}{6600 \text{ mg/L} \times 8.34 \text{ lb/(Mgal} \cdot \text{mg/L)}}$$

$$\begin{aligned} &= 0.0943 \text{ Mgal/d} \\ &= 0.0943 \text{ Mgal/d} \times 694 \text{ (gal/min)/(Mgal/d)} \\ &= 65.4 \text{ gal/min} \\ &= 356 \text{ m}^3\text{/d} \end{aligned}$$

EXAMPLE 5: The given operating conditions are the same as in Example 3. The desired MLVSS is 16,500 kg (36,400 lb). The SS concentration in RAS is 6600 mg/L. Using MLVSS as a control technique, calculate the desired WAS flow rate. The current WAS flow rate is 420 m^3/d (0.11 Mgal/d).

Solution:

Step 1. Calculate actual MLVSS

$$\begin{aligned}\text{Tank volume} &= 6600\ \text{m}^3\\ \text{VSS/TSS} &= 0.80\\ \text{MLSS} &= 3200\ \text{mg/L} = 3.2\ \text{kg/m}^3\\ \text{Actual MLVSS} &= \text{tank volume} \times \text{volatiles} \times \text{MLSS}\\ &= 6600\ \text{m}^3 \times 0.8 \times 3.2\ \text{kg/m}^3\\ &= 16{,}900\ \text{kg}\end{aligned}$$

Step 2. Calculate additional VSS to be wasted daily

$$\begin{aligned}\text{Wasted} &= \text{actual} - \text{desired (Step 3 of Example 3)}\\ &= (16{,}900 - 16{,}500)\ \text{kg/d}\\ &= 400\ \text{kg/d}\end{aligned}$$

Step 3. Calculate additional WAS flow, q

$$\begin{aligned}\text{VSS in RAS} &= 6600\ \text{mg/L} \times 0.8 = 5280\ \text{mg/L}\\ &= 5.28\ \text{kg/m}^3\\ q &= (400\ \text{kg/d})/(5.28\ \text{kg/m}^3)\\ &= 75.8\ \text{m}^3\text{/d}\\ &= 13.9\ \text{gal/min}\end{aligned}$$

Step 4. Calculate the desired WAS flow rate

$$\begin{aligned}Q &= \text{current} + \text{additional flow}\\ &= (420 + 75.8)\ \text{m}^3\text{/d}\\ &= 495.8\ \text{m}^3\text{/d}\\ &= 91\ \text{gal/min}\end{aligned}$$

Sludge bulking. A desirable activated-sludge is one which settles rapidly leaving a clear, odorless, and stable supernatant. The efficiency of treatment achieved in an activated-sludge process depends directly on settleability of the sludge in the secondary clarifier. At times, poorly settling sludge and foam formation cause the most common operational problems of the activated-sludge process. Sludge with poorly flocculated (pin) particles or buoyant filamentous growths increases its volume and does not settle well in the clarifier. Light sludge in the clarifier then spills over the weirs and is carried away in the effluent. The concentrations of BOD and suspended solids increase in the effluent. This phenomenon is called sludge bulking and it frequently occurs unexpectedly.

Sludge bulking is caused by (1) the growth of filamentous organisms, or organisms that can grow in filamentous form under adverse conditions, and (2) adverse environmental conditions such as excessive flow, insufficient aeration, short circulating of aeration tanks, lack of nutrients, septic influent, presence of toxic substances, or overloading. *Nocardia amarae, Microthrix parvicella, N. amarae*-like organisms, *N. pinensis*-like organisms, and type 0092 were most commonly found in foam and bulk sludge. *Nocardia* growth is supported by high sludge age, low F/M ratio, and higher rather than lower wastewater temperature (Droste, 1997). Fungi are mostly filamentous microorganisms. Some bacteria such as *Beggiatoa, Thiotrix,* and *Leucothrix* grow in filamentous sheaths.

There are no certain rules for prevention and control of sludge bulking. If bulking develops, the solution is to determine the cause and then either eliminate or correct it or take compensatory steps in operation control. Some remedial steps may be taken, e.g. changing parameters such as wastewater characteristics, BOD loading, dissolved oxygen concentration in the aeration basin, return sludge pumping rate, microscopic examination of organisms (check for protozoa, rotifers, filamentous bacteria, and nematodes) in clarifiers and other operating units. Chlorination of RAS in the range of 2 to 3 mg/L of chlorine per 1000 mg/L of MLVSS is suggested (Metcalf and Eddy, Inc. 1991). Hydroperoxide also can be used as an oxidant. Reducing the suspended solids in the aeration basin, increasing air supply rate, or increasing the BOD loading (which may depress filamentous growth), and addition of lime to the mixed liquor for pH adjustment are the remedial methods which have been used. The F/M ratio should be maintained at 0.5 to 0.2.

21.7 Modified Activated-Sludge Processes

Numerous modifications (variations) of the conventional activated-sludge plug flow process have been developed and proved effective for removal of BOD and/or nitrification. The modified processes include tapered aeration, step-feed aeration, complete-mix extended aeration, modified aeration, high-rate aeration, contact stabilization, Hatfield process, Kraus process, sequencing batch reactor, high-purity oxygen, oxidation ditch, deep shaft reactor, single-stage nitrification, and separate stage nitrification. Except for a few processes, most modified processes basically differ on the range of the F/M ratio maintained and in the introduction locations for air supply and wastewater. The design parameters and operational characteristics for various activated-sludge processes are presented elsewhere (Metcalf and Eddy, Inc. 1991, WEF and ASCE 1991a). Standard and modified processes are discussed below.

Conventional process. The earliest activated-sludge process was contained in long narrow tanks with air for oxygen supply and mixing through diffusers at the bottom of the aeration tank. A conventional plug-flow process consists of a primary clarifier, an aeration tank with air diffusers for mixing the wastewater and the activated sludge in the presence of dissolved oxygen, a secondary settling basin for solids removal, and a sludge-return line from the clarifier bottom to the influent of the aeration tank (Fig. 7.18). The return activated-sludge is mixed with the incoming wastewater and passes through the aeration tank in a plug-flow fashion. High organics concentration and high microbial solids at the head end of the aeration tank lead to a high oxygen demand. The conventional process is more susceptible to upset from shock loads and toxic materials, and is used for low-strength domestic wastewaters.

The conventional process is designed to treat 0.3 to 0.6 kg BOD_5 applied/($m^3 \cdot d$) (20 to 40 lb/(1000 cu ft · d)) in the United States with MLSS of 1500 to 3000 mg/L. The aeration period ranges between 4 and 8 hours with return activated-sludge flow ratios of 0.27 to 0.75. The cell residence time (θ_c) is 5 to 15 days and the F/M ratio is 0.2 to 0.4 per day. The process generally removed 85 to 95 percent of BOD_5 and produces highly nitrified effluent (WEF and ASCE 1991a).

EXAMPLE: The operational records at a conventional activated-sludge plant are shown as average values as below.

Wastewater flow Q	7570 m^3/d (2.0 Mgal/d)
Wastewater temperature	20°C
Volume of aeration tanks	2260 m^3 (79,800 ft^3)
Influent BOD_5 (say BOD)	143 mg/L
Influent total suspended solids	125 mg/L
Influent total solids	513 mg/L
Effluent total solids	418 mg/L
Effluent TSS	24 mg/L
Effluent BOD	20 mg/L
Return sludge flow Q_r	3180 m^3/d (0.84 Mgal/d)
MLSS	2600 mg/L
TSS in waste sludge	8900 mg/L
Volume of waste sludge	200 m^3/d (0.053 Mgal/d)

Calculate the following operational parameters: (1) volumetric BOD loading rate; (2) F/M ratio; (3) hydraulic retention time θ; (4) cell residence time θ_c; (5) return activated-sludge ratio; and (6) removal efficiencies for BOD, TSS, and total solids.

Solution:

Step 1. Calculate volumetric BOD loading rate

$$\begin{aligned}\text{Influent BOD} &= 143 \text{ mg/L} = 143 \text{ g/m}^3 \\ &= 0.143 \text{ kg/m}^3 \\ \text{BOD load} &= \frac{\text{amount}}{\text{volume}} = \frac{7570 \text{ m}^3\text{/d} \times 0.143 \text{ kg/m}^3}{2260 \text{ m}^3} \\ &= 0.48 \text{ kg/(m}^3 \cdot \text{d)}\end{aligned}$$

Note: It is in the range of 0.3 to 0.6 kg/(m$^3 \cdot$ d).

Step 2. Calculate F/M ratio using Eq. (7.64)

$$\begin{aligned}\text{Assume MLVSS} &= 0.8 \text{ MLSS} = 0.8 \times 2600 \text{ mg/L} \\ &= 2080 \text{ mg/L} \\ \text{F/M} &= \frac{7570 \text{ m}^3\text{/d} \times 143 \text{ mg/L BOD applied}}{2260 \text{ m}^3 \times 2080 \text{ mg/L MLVSS}} \\ &= 0.23 \text{ kg BOD applied/(kg MLVSS} \cdot \text{d)} \\ &= 0.23 \text{ lb BOD applied/(lb MLVSS} \cdot \text{d)}\end{aligned}$$

Note: $0.2 < \text{F/M} < 0.4$.

Step 3. Calculate aeration time HRT

$$\begin{aligned}\text{HRT} &= V/Q = 2260 \text{ m}^3 \div 7570 \text{ m}^3\text{/d} \\ &= 0.30 \text{ day} \\ &= 7.17 \text{ h}\end{aligned}$$

Note: $4 \text{ h} < \text{HRT} < 8 \text{ h}$.

Step 4. Calculate cell residence time θ_c

$$\begin{aligned}\text{TSS in the aeration tank} &= 7570 \text{ m}^3\text{/d} \times 2.6 \text{ kg/m}^3 \\ &= 19{,}682 \text{ kg/d} \\ \text{TSS in sludging wasting} &= 200 \text{ m}^3\text{/d} \times 8.9 \text{ kg/m}^3 \\ &= 1780 \text{ kg/d} \\ \text{TSS in effluent} &= 7570 \text{ m}^3\text{/d} \times 0.024 \text{ kg/m}^3 \\ &= 182 \text{ kg/d} \\ \theta_c &= \frac{19{,}682 \text{ kg/d}}{1780 \text{ kg/d} + 182 \text{ kg/d}} \\ &= 10.0 \text{ days}\end{aligned}$$

Note: $5 \text{ days} < \theta_c < 15 \text{ days}$.

Step 5. Calculate RAS ratio

$$Q_r/Q = 3180 \text{ m}^3/\text{d} \div 7570 \text{ m}^3/\text{d}$$
$$= 0.42$$
$$= 42\%$$

Note: $0.25 < Q_r/Q < 0.75$.

Step 6. Calculate removal efficiencies

$$\text{BOD} = \frac{(143 - 20) \text{ mg/L} \times 100\%}{143 \text{ mg/L}}$$
$$= 86\%$$

$$\text{TSS} = \frac{(125 - 22) \text{ mg/L} \times 100\%}{125 \text{ mg/L}}$$
$$= 82\%$$

$$\text{TS} = \frac{(513 - 418) \text{ mg/L} \times 100\%}{513 \text{ mg/L}}$$
$$= 19\%$$

Tapered aeration. Since oxygen demands decrease along the length of the plug-flow reactor, the tapered aeration process attempts to match the oxygen supply to demand by adding more air at the influent end of the aeration tank than at the effluent end (Fig. 7.19). This is obtained by varying the diffuser spacing. The best results can be achieved by supplying 55 to 75 percent of the total air supply to the first half of the tank (Al-Layla *et al.* 1980). Advantages of tapered aeration are those of reducing below-capacity and operational costs, providing better operational control, and inhibiting nitrification if desired.

Step aeration. In step (step-feed) aeration activated-sludge systems, the incoming wastewater is distributed to the aeration tank at a number (3 to 4) of points along the plug-flow tank, and the return activated sludge is introduced at the head of the aeration tank. The organic load is distributed over

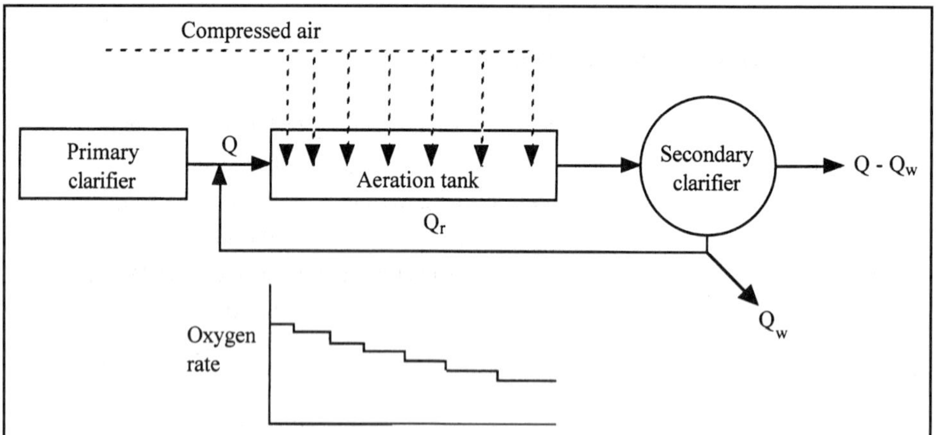

FIGURE 7.19 Conventional process with tapered aeration.

the length of the tank (Fig. 7.20), thus avoiding the locally lengthy oxygen demand encountered in conventional and tapered aeration. This process leads to shorter retention time and lower activated-sludge concentrations in the mixed liquor. The process is used for general application.

The BOD_5 loading rates are higher, 0.6 to 1.0 kg/($m^3 \cdot d$) (40 to 60 lb/(1000 $ft^3 \cdot d$)) and MLSS are also higher, 2000 to 3500 mg/L. The hydraulic retention time is shorter by 3 to 5 hours. The following parameters for the step aeration process are the same as for a conventional plant: $\theta_c = 5$ to 15 days; F/M = 0.2 to 0.4 per day; $Q_r/Q = 0.25$ to 0.75; and BOD removal = 85 to 95 percent.

This process, with treatment efficiency practically equivalent to that of the conventional activated-sludge process, can be carried out in about one-half of the aeration time while maintaining the sludge

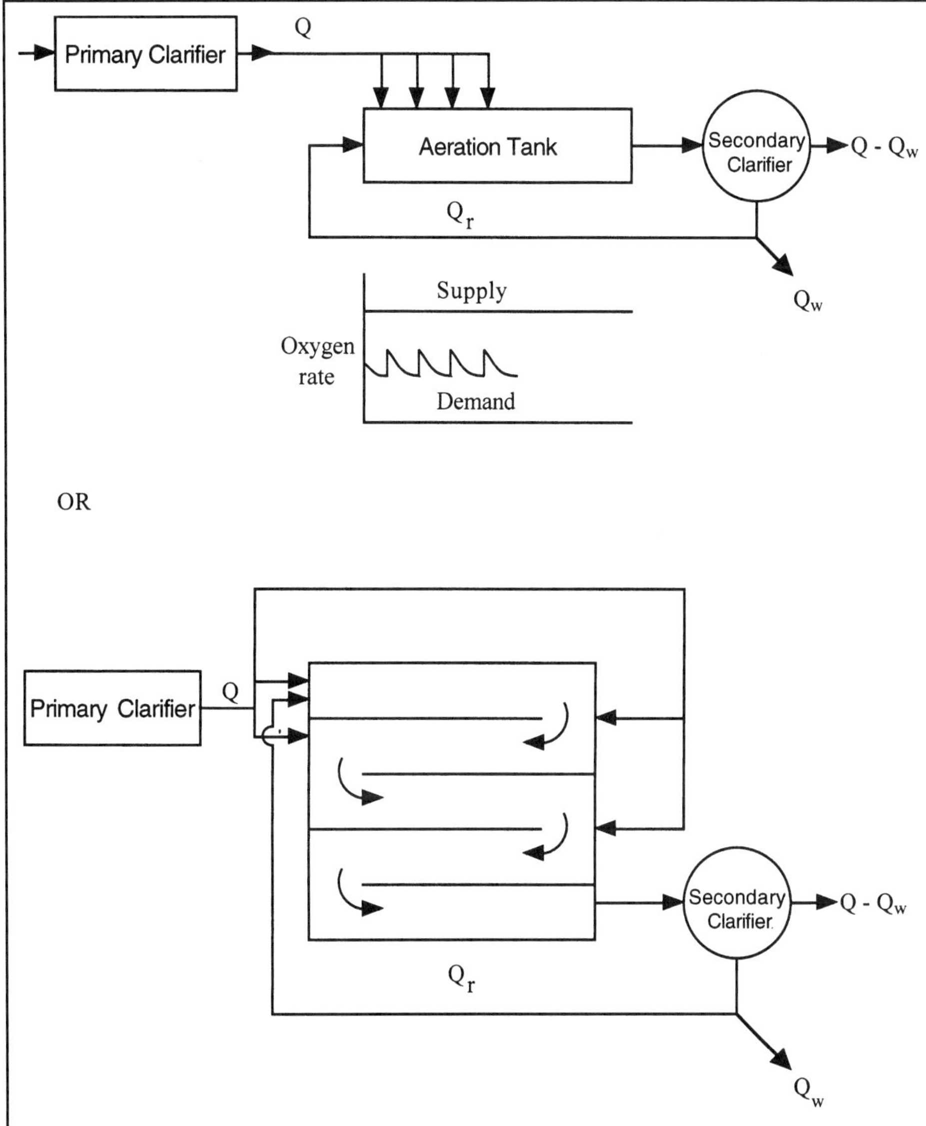

FIGURE 7.20 Step aeration activated-sludge process.

age within proper limits of 3 to 4 days. The construction cost and the area required for step aeration are less than for the conventional process. Operational costs are about the same for both the conventional and step aeration processes.

Complete-mix process. Complete-mix processes disperse the influent wastewater and return activated-sludge uniformly throughout the aeration tank. The shape of the reactor is not important. With careful selection of aeration and mixing equipment, the process should provide practically complete mixing. The oxygen demand is also uniform throughout the tank (Fig. 7.21).

Complete-mix processes protect against hydraulic and organic shock loadings commonly encountered in the process. Toxic materials are usually diluted below their threshold concentration. The treatment efficiency of the complete-mix process is comparable to that of the conventional process (85 to 95 percent BOD removal). The complete-mix process has increased in popularity for the treatment of industrial wastewaters. The process is used for general application, but is susceptible to filamentous growths.

The design criteria for the complete-mix process are: 0.45 to 2.0 kg $BOD_5/(m^3 \cdot d)$ (28 to 125 lb/(1000 $ft^3 \cdot d$)), F/M = 0.2 to 0.6 per day, θ_c = 5 to 15 days, MLSS = 2500 to 6500 mg/L, HRT = 3 to 5 h, Q_r/Q = 0.25 to 1.0 (Metcalf and Eddy, Inc., 1991). A design example for the complete-mix activated-sludge process is given in the section on mathematical modeling.

Extended aeration. The extended aeration process is a complete-mix activated-sludge process operated at a long HRT (θ = 16 to 24 or 36 h) and a high cell residence time (sludge age θ_c = 20 to 30 days). The process may be characterized as having a long aeration time, high MLSS concentration, high RAS pumping rate, and low sludge wastage. Extended aeration is typically used in small plants for plant flows of 3780 m^3/d (1 Mgal/d) or less, such as schools, villages, subdivisions, trailer parks, etc. Many extended aeration plants are prefabricated units or so-called package plants (Guo *et al.* 1981). The process is flexible and is also used where nitrification is required.

The influent wastewater may be only screened and degritted without primary sedimentation. The extended aeration system allows the plant to operate effectively over widely varying flows and organic loadings without upset. Organic loading rates are designed as 0.1 to 0.4 kg $BOD_5/(m^3 \cdot d)$ (16 to 25 lb/(1000 $ft^3 \cdot d$)). The other design parameters are: MLSS = 1500 to 5000 mg/L, and Q_r/Q = 0.5 to 1.50 (WEF and ASCE 1992).

The extended aeration process has a food to microorganism ratio from 0.05 to 0.20 kg BOD_5 per kg MLSS per day. It is developed to minimize waste activated sludge production by providing a large endogenous decay of the sludge mass. The process is so designed that the mass of cells synthesized per day equals the mass of cells endogenous decay per day. Therefore there is theoretically no cell mass production.

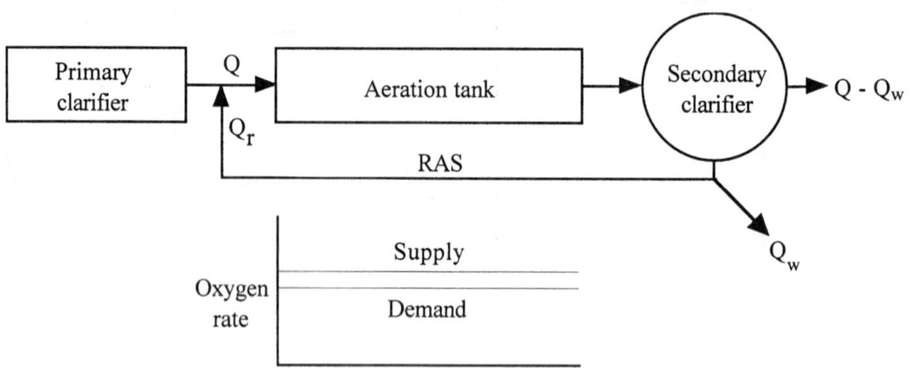

FIGURE 7.21 Complete-mix activated-sludge process.

Secondary clarifiers must be designed to handle the variations in hydraulic loadings and high MLSS concentration associated with the system. The overflow rates range from 8 to 24 $m^3/(m^2 \cdot d)$ (200 to 600 $gal/(d \cdot ft^2)$) and with long retention time. Sludge may be returned to the aeration tank through a slot opening or by an air-lift pump (Hammer 1986). Floating materials and used sludge on the surface of the clarifier can be removed by a skimming device.

Since the process maintains a high concentration of microorganisms in the aeration tank for a long time, endogenous respiration plays a major role in activated sludge quality. More dissolved oxygen (DO) is required. Nitrification may occur in the system. The volatile portion of the sludge remaining is not degraded at the same rate as normal activated sludge and thereby results in a lower BOD exertion rate. The effluent of extended aeration often meets BOD standards (75 to 95 percent removal) but does not meet TSS standards due to continuous loss of pinpoint flocs (rising sludge). To overcome TSS loss, periodic sludge wasting is required. It has been suggested that the average MLSS concentration should not fall below 2000 mg/L (Guo *et al.* 1981). In cold climates, heat lost in the extended aeration system must be controlled. However, Guo *et al.* (1981) pointed out that insufficient staff, inadequate training, or both are the most common causes of poor performance of the extended aeration process, especially in small package plants.

EXAMPLE: An extended aeration activated-sludge plant without sludge wasting has a BOD_5 loading rate of 0.29 $kg/(m^3 \cdot d)$ (18 $lb/(1000\ ft^3 \cdot d)$) with aeration for 24 h. The daily MLSS buildup in the aeration basin is measured as 55 mg/L. Determine the percentage of the influent BOD_5 that is converted to MLSS and retained in the basin. Estimate the time required to increase the MLSS from 1500 mg/L to 5500 mg/L (this range is in the design criteria).

Solution:

Step 1. Calculate influent BOD concentration in mg/L

$$\begin{aligned} \text{BOD} &= 0.29\ \text{kg}\ (\text{m}^3 \cdot \text{d}) \times 1\ \text{day} \\ &= 290\ \text{g/m}^3 \\ &= 290{,}000\ \text{mg/1000 L} \\ &= 290\ \text{mg/L} \end{aligned}$$

Step 2. Determine the percentage

$$\frac{\text{MLSS buildup}}{\text{influent BOD}} = \frac{55\ \text{mg/L} \times 100\%}{290\ \text{mg/L}}$$

$$= 19\%$$

thus 81% of influent BOD retained in the basin

Step 3. Estimate buildup time required, t

$$t = \frac{5500\ \text{mg/L} - 1500\ \text{mg/L}}{55\ \text{mg/(L} \cdot \text{d)}}$$

$$= 72.7\ \text{days}$$

Short-term aeration. Short-term aeration or modified aeration is a plug-flow pretreatment process. The process has extremely high loading rates of 1.2 to 2.4 kg $BOD_5/(m^3 \cdot d)$ (75 to 150 $lb/(1000\ ft^3 \cdot d)$) with an F/M ratio of 1.5 to 5.0 kg BOD/d per kg MLVSS. The volumetric loadings are low, ranging from 200 to 1000 mg/L of MLSS. The HRT (θ) are 1.5 to 3 h; sludge age $\theta_c = 0.2$ to 0.5 days; and $Q_r/Q = 0.05$ to 0.25. Short retention time and low sludge age lead to a poor effluent quality and relatively high solids production. This process can be used as the first stage of a two-stage nitrification process and is used for an intermediate degree of treatment.

The short-term aeration process offers cost savings of construction by increasing BOD loading rates and reducing the required volume of aeration tank. The process produces a relatively large amount of MLVSS which may cause disposal problems. If the sludge is allowed to remain in the system, the BOD_5 removal efficiency ranges from 50 to 75 percent (Chermisinoff 1995).

High rate aeration. For some cases it may not be necessary to treat the wastewater to the high degree of effluent quality achieved by conventional or other improved processes. High rate aeration is an application of the complete-mix activated-sludge process with a short HRT ($\theta = 0.5$ to 2.0 h), a short sludge age ($\theta_c = 5$ to 10 days), a high sludge recycle ratio ($Q_r/Q = 1.0$ to 5.0), and an organic loading rate of 1.6 to 16 kg $BOD_5/m^3 \cdot d$ (100 to 1000 lb/(1000/$ft^3 \cdot d$)). The MLSS concentration in the aeration tanks ranges from 4000 to 10,000 mg/L, and the F/M ratios are 0.4 to 1.5 per day which are higher than those for the conventional process (WEF and ASCE 1991a). The process reduces the cost of construction.

Complete-mix aeration and the hydraulic thickening action of a rapid sludge return are mandatory to offset decreased sludge settleability of the biological flocs (Bruce and Merkens 1973). It is designed to maintain the biomass in the growth phase. Poor performance of a high-rate aeration system is usually due to insufficient aeration capacity to supply adequate dissolved oxygen during peak loading periods. Subsequently, suspended flocs are carried over to the secondary clarifier effluent. Also, inadequate RAS flow rates and high solids flux may cause sludge to be washed out through the clarifier.

Well-operated high-rate aeration processes can produce effluent quality comparable to that of a conventional plant. A BOD_5 removal efficiency of 75 to 90 percent can be achieved. A high-rate single-stage aeration system can be used for general application and to partially remove carbonaceous BOD_5 at the first stage of a two-stage nitrification system.

Contact stabilization. The contact stabilization process or biosorption was developed to take advantage of the adsorptive properties of activated sludge. A schematic flow diagram of the contact stabilization system is shown in Fig. 7.22. Returned sludge which has been aerated (3 to 6 h) in the stabilization basin (44,000 to 10,000 mg/L of MLSS) for stabilization of previously adsorbed organic matter is mixed with influent wastewater for a brief period (20 to 40 min). BOD removal from wastewater takes place on immediate contact with a high concentration (1000 to 4000 mg/L of MLSS in contact tank) of stabilized activated sludge. This adsorbs suspended and colloidal solids quickly, but not dissolved organic matter. Following the contact period, the activated sludge is separated from the mixed liquor in a clarifier. A small portion of the sludge is wasted while the remainder flows to the stabilization tank. During the stabilization period, the stored organic matter is utilized for cell growth and respiration; as a result, it becomes stabilized or activated again, then is recycled.

Contact stabilization was initially used to provide a partial treatment (60 to 75 percent BOD_5 removal) at larger coastal plants. It has been designed for package systems for industrial waste application and is used for expansion of existing systems and package plants.

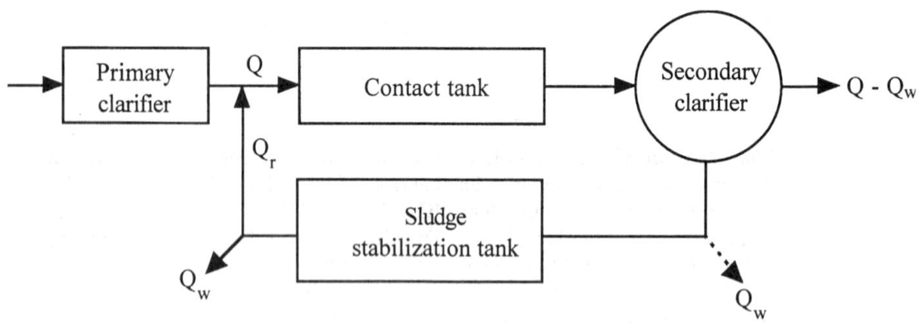

FIGURE 7.22 Schematic flow diagram of contact stabilization process.

The process is a plug-flow system and is characterized by relatively short retention times (0.5 to 3 h) with very high organic loading of 1.0 to 1.2 kg $BOD_5/(m^3 \cdot d)$ (62 to 75 lb/(1000 $ft^3 \cdot d$)). Compared to the conventional activated-sludge process, oxygen requirements are lower but waste sludge quantities are higher. Designs may either include or omit primary treatment. In general, the process has a BOD_5 removal efficiency of 80 to 90 percent (WEF and ASCE 1992). This process can be effectively used as the first stage of a multiple system.

The disadvantages of the process are:

1. sensitive to variations in organic and hydraulic loadings due to its short HRT and low MLSS concentration;
2. neither as economical nor as efficient in BOD removal.

These shortcomings may result in noncompliance with effluent quality standards.

Hatfield process. The Hatfield process (developed by William Hatfield of the Decatur Sanitary District, Illinois) differs from the contact stabilization process. The process aerates anaerobic digester supernatant or sludge (rich in nitrogen), with (all) return sludge from the secondary clarifier then fed back to the aeration tank (Fig. 7.23). The process is used to treat wastewater with low nitrogen levels and high carbonaceous material levels, such as some industrial wastewaters. Supplying an aerobic digester effluent to the aeration tank fortifies the MLSS with amino acids and other nitrogenous substances.

The Hatfield process has the advantage, as in the contact stabilization system, of being able to maintain a large amount of microorganisms under aeration in a relatively small aeration tank. Heavier types of solids are produced in the aeration tank, which can prevent bulking problems. However, the Hatfield process is not widely used in the United States.

Kraus process. The conventional activated-sludge plant at Peoria, Illinois, was found to be operated improperly due to sludge bulking because of a heavy load of carbohydrates from breweries, packing houses, and paper mills. Kraus (1955) improved the process performance by aerating the

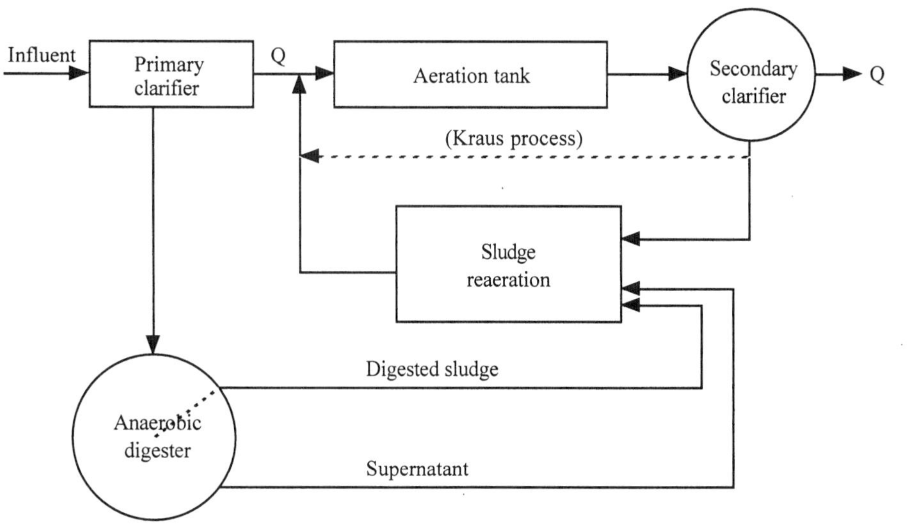

FIGURE 7.23 Schematic flow diagram of Hatfield activated-sludge process and Kraus process.

mixture of anaerobic digester supernatant and a small portion of the return activated sludge in a separate aeration basin (Fig. 7.23). Some portion of the return activated sludge bypasses the sludge aeration basin and is introduced directly to the mixed liquor aeration tank. This is a modification of the Hatfield process and lies between the conventional system and the Hatfield process. The MLSS become highly nitrified materials which are brought into contact with the influent wastewater in the aeration tank. The BOD_5 loading is 1.8 kg/($m^3 \cdot$ d) (112 lb/(1000 $ft^3 \cdot$ d)) with 90 percent removal.

The specific gravity of the digester solids is greater than that of the activated sludge, and the settling characteristics are thus improved.

Since high dissolved oxygen is required for the Kraus process, in order to cope with high oxygen demand, Kraus applied a dual system in the sludge reaeration basin in which there was a combination of coarse-bubble aeration at the top of the basin and fine-bubble aeration at the bottom of the basin. The process is used for high-strength wastewaters with low nitrogen levels.

The design criteria for the Kraus process are: BOD_5 loading = 0.6 to 1.6 kg/($m^3 \cdot$ d) (40 to 100 lb/(1000 $ft^3 \cdot$ d)), F/M = 0.3 to 0.8 per day, MLSS = 2000 to 3000 mg/L, HRT = 4 to 8 days, Q_r/Q = 0.5 to 1.0.

Sequencing batch reactor. A sequencing batch reactor (SBR) is a periodically operated, fill-and-draw activated-sludge system. The unit process used in the SBR and the conventional activated-sludge system are essentially the same. Both have aeration and sedimentation. In the activated-sludge plant, the processes are taking place simultaneously in separate basins, whereas in the SBR the aeration and clarification processes are carried out sequentially in the same basin. The system is used for small communities with limited land.

Each reactor in an SBR system has five discrete periods (steps) in each cycle: fill, react (aeration), settle (sedimentation/clarification), draw (decant), and idle (Herzbrun *et al.* 1985). Biological activities are initiated as the influent wastewater fills the basin. During the fill and react period, the wastewater is aerated in the same manner as in a conventional activated-sludge system. After the reaction step, the mixed liquor suspended solids are allowed to settle in the same basin. The treated supernatant is withdrawn during the draw period. The idle period, the time between the draw and fill, may be zero or some certain period (days). The process is flexible and can remove nitrogen and phosphorus. Since the development of SBR technology in the early 1960s, a number of process modifications have been made to achieve specific treatment objectives.

In an SBR operation, sludge wasting is an important task relating to system performance. It usually takes place during the settle and idle phases. There is no return activated-sludge system in the SBR system.

High-purity oxygen system. The pure oxygen activated-sludge process was first studied by Okun in 1947 (Okun 1949). The process achieved commercial status in the 1970s. Currently a large number of high-purity oxygen activated-sludge plants have been put into operation.

The process has been developed in an attempt to match oxygen supply and oxygen demand and to use a high-rate process through maintenance of high concentrations of biomass. The major components of the process are an oxygen generator, using high-purity oxygen in lieu of air, a specially compartmented aeration basin, a clarifier, pumps for recirculating activated sludge, and sludge disposal facilities. Oxygen is generated by manufacturing liquid oxygen cryogenically on-site for large plants and by pressure swing adsorption for small plants. Standard cryogenic air separation involves liquefaction of air followed by fractional distillation to separate the major components of oxygen and nitrogen.

The aeration basin is divided into compartments by baffles and is covered with a gastight cover. An agitator is included in each compartment (Fig. 7.24). High-purity oxygen is fed concurrently with wastewater flow. Influent wastewater, return activated-sludge, and oxygen gas under slight pressure are introduced to the first compartment. A dissolved oxygen concentration of 4 to 10 mg/L is normally maintained in the mixed liquor. Successive aeration compartments are connected to each other through submerged ports. Exhausted waste gas is a mixture of nitrogen, carbon dioxide, and approximately 10 percent of oxygen supplied and vented from the last stage of the system. Effluent mixed liquor is then settled in a secondary clarifier. Settled activated sludge is recirculated to the aeration

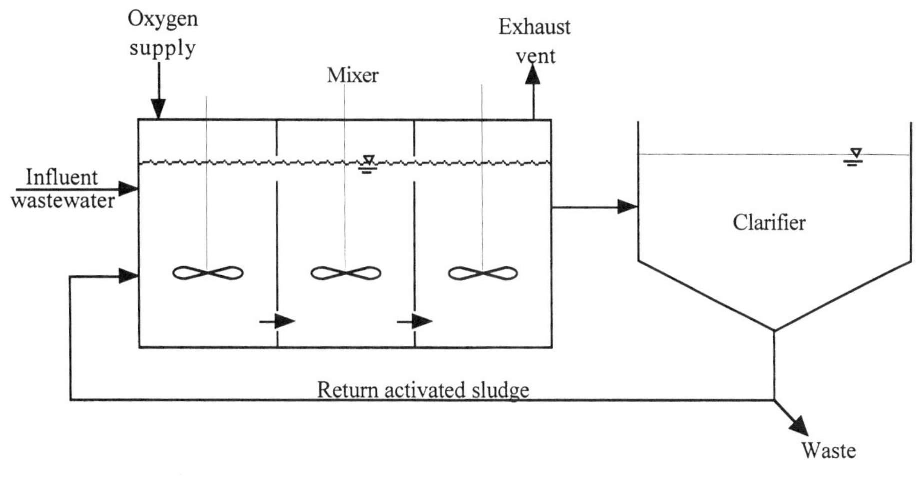

FIGURE 7.24 Schematic flow diagram of a high-purity activated-sludge system.

basin, with some wasted. The process is used for general application with high-strength wastes with limited space.

Design parameters of high-purity oxygen systems are: BOD_5 loading = 1.6 to 3.2 kg/(m^3 · d) (100 to 200 lb/(1000 ft^3 · d)), MLSS = 3000 to 8000 mg/L, F/M = 0.25 to 1.0 per day, HRT = 1 to 3 days, θ_c = 8 to 20 days, and Q_r/Q = 0.25 to 0.5. The BOD_5 removal efficiency for the process is 85 to 95 percent (WEF and ASCE 1992).

Advantages of the high-purity oxygen activated-sludge process include higher oxygen transfer gradient and rates, smaller reaction chambers, improved biokinetics, the ability to treat high-strength soluble wastewater, greater tolerance for peak organic loadings, reduced energy requirements, reduction in bulking problems, a better settling sludge (1 to 2 percent solids), and effective odor control.

In a comparison of air and pure oxygen systems, researchers reported little or no significant difference in carbonaceous BOD_5 removal kinetics with DO concentration about 1 to 2 mg/L. Lower pH values (6.0 to 6.5) occur with high-purity oxygen and loss of alkalinity, with nitrate production. Covered aeration tanks have a potential explosion possibility.

EXAMPLE: The high-purity oxygen activated-sludge process is selected for treating a high-strength industrial wastewater due to limited space. Determine the volume of the aeration tank and check with design parameters, using the following data.

Design average flow	1 Mgal/d (3785 m^3/d)
Influent BOD	280 mg/L
Influent TSS	250 mg/L
F/M	0.70 lb BOD applied/(lb MLVSS · d)
MLSS	5500 mg/L
VSS/TSS	0.8
Maximum volumetric BOD loading	200 lb/(1000 ft^3 · d) = 3.2 kg/(m^3 · d)
Minimum aeration time	2 h
Minimum mean cell residence time	3 days

Solution:

Step 1. Determine the tank volume
Using Eq. (7.88)

$$\text{F/M} = \frac{S_0}{\theta X} = \frac{QS_0}{VX}$$

$$V = \frac{QS_0}{(\text{F/M})X}$$

$$\text{F/M} = 0.7/\text{d}$$

where $X = 5500\text{ mg/L} \times 0.8 = 4400\text{ mg/L}$
$8.34 =$ conversion factor, lb/(Mgal · (mg/L))

$$\text{Therefore } V = \frac{1\text{ Mgal/d} \times 280\text{ mg/L} \times 8.34}{(0.7/\text{d}) \times 4400\text{ mg/L} \times 8.34}$$

$$= 0.0909\text{ Mgal}$$

$$= 344\text{ m}^3$$

$$= 12{,}150\text{ ft}^3$$

Step 2. Check the BOD loading

$$\text{BOD loading} = \frac{1.0\text{ Mgal/d} \times 280\text{ mg/L} \times 8.34\text{ lb/(Mgal} \cdot \text{(mg/L))}}{12.15 \times 1000\text{ ft}^3}$$

$$= 192\text{ lb/(1000 ft}^3 \cdot \text{d)}$$

$$< 200\text{ lb/(1000 ft}^3 \cdot \text{d) (OK)}$$

$$= 3.08\text{ kg/(m}^3 \cdot \text{d)}$$

Step 3. Check the aeration time

$$\text{HRT} = \frac{V}{Q} = \frac{0.0909\text{ day} \times 24\text{ h/d}}{1}$$

$$= 0.0909\text{ day} \times 24\text{ h/d}$$

$$= 2.2\text{ h}$$

$$> 2.0\text{ h (OK)}$$

Step 4. Check mean cell residence time (sludge age)

$$\text{Influent VSS } X_i = 0.8 \times 250\text{ mg/L}$$

$$= 200\text{ mg/L}$$

$$\theta_c = \frac{VX}{QX_i} = \frac{0.0909\text{ Mgal} \times 4400\text{ mg/L}}{1\text{ Mgal/d} \times 200\text{ mg/L}}$$

$$= 2.0\text{ days}$$

$$> 3.0\text{ days (OK)}$$

Oxidation ditch. The oxidation ditch process was developed by Pasveer (1960) in Holland and is a modification of the conventional activated-sludge plug-flow process. The system is especially applicable to small communities needing low-cost treatment. The oxidation ditch is typically an

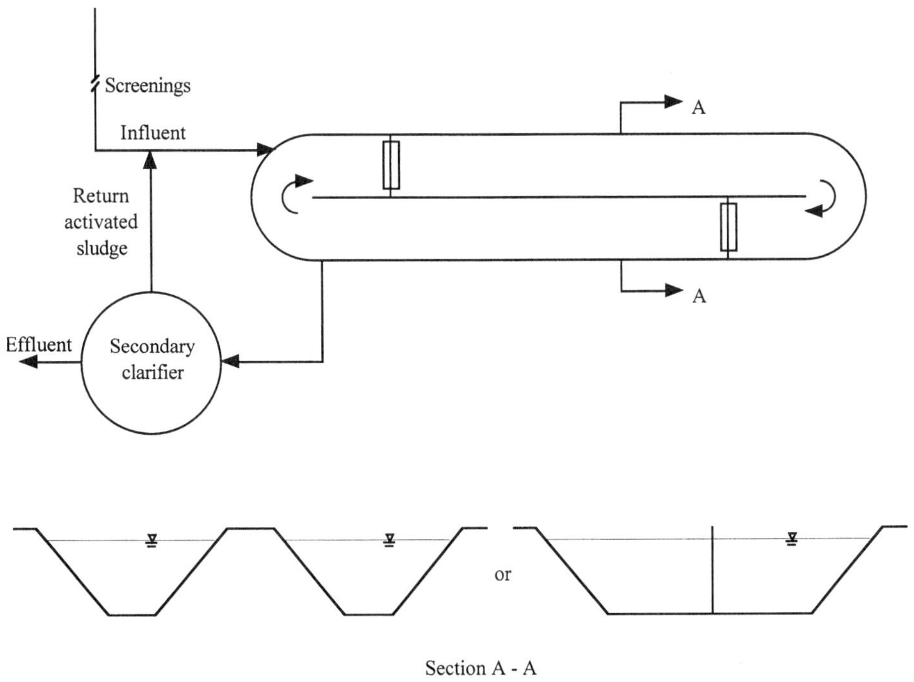

FIGURE 7.25 Schematic diagram of oxidation ditch activated-sludge process.

extended aeration mode. It consists typically of a single or closed-loop elongated oval channel with a liquid depth of 1.2 to 1.8 m (4 to 6 ft) and 45-degree sloping sidewalls (Fig. 7.25).

Wastewater is given preliminary treatment such as screening, comminution, or grit removal (usually without primary treatment). The wastewater is introduced into the ditch and is aerated using mechanical aerator(s) (Kessener brush) which are mounted across the channel for an extended period of time with long HRT of 8 to 36 h (typically 24 h) and SRT solids residence time of 20 to 30 days (WEF and ASCE 1992).

The aerators provide mixing and circulation in the ditch, and oxygen transfer. The mechanical brush operates in the 60 to 110 rev/min range and keeps the liquid in motion at velocities from 0.24 to 0.37 m/s (0.8 to 1.2 ft/s), sufficient to prevent deposition of solids. A large fraction of the required oxygen is supplied by transfer through the free surface of the wastewater rather than by aeration. As the wastewater passes the aerator, the DO concentration rises and then falls while the wastewater traverses the circuit.

The BOD_5 loading in the oxidation ditch is usually as low as 0.08 to 0.48 kg/($m^3 \cdot d$) (5 to 30 lb/(1000 $ft^3 \cdot d$)). The F/M ratios are 0.05 to 0.30 per day. The MLSS are in the range of 1500 to 5000 mg/L. The activated sludge in the oxidation ditch removes the organic matter and converts it to cell protoplasm. This biomass will be degraded for long solids residence time.

Separate or interchannel clarifiers are used for separating and returning MLSS to the ditch. Recommendations for unit size and power requirements are available from equipment manufacturers.

The oxidation ditch is popular because of its high BOD_5 removal efficiency (85 to 95 percent) and easy operation. Nitrification occurs also in the system. The oxidation ditch is a flexible process and is used for small communities where large land area is available. However, the surface aerators may become iced during the winter months. Electrical heating must be installed at the critical area. During cold weather the operation must prevent loss of efficiency, mechanical damage, and danger to the operators.

EXAMPLE 6: Estimate the effluent BOD_5 to be expected of an oxidation ditch treating a municipal wastewater with an influent BOD of 220 mg/L. Use the following typical kinetic coefficients: k = 5/day; K_s = 60 mg/L of BOD; Y = 0.6 mg VSS/mg BOD; k_d = 0.06/d; and θ_c = 20 day.

Solution:

Step 1. Estimate effluent substrate (soluble BOD_5) concentration
From Eq. (7.80)

$$S = \frac{K_s(1 + \theta_c k_d)}{\theta_c(Yk - k_d) - 1}$$

$$= \frac{60 \text{ mg/L}(1 + 20 \text{ day} \times 0.06/\text{d})}{20 \text{ d}(0.6 \times 5/\text{d} - 0.06/\text{d}) - 1}$$

$$= 2.3 \text{ mg/L}$$

Step 2. Estimate TSS in the effluent
Using Eq. (7.78)

$$X = \frac{\theta_c Y(S_0 - S)}{\theta (1 + k_d\theta_c)}$$

$$= \frac{20 \text{ d} \times 0.6(220 - 2.3) \text{ mg/L}}{20 \text{ d}(1 + 0.06/\text{d} \times 20 \text{ d})}$$

$$= 59.4 \text{ mg/L}$$

Step 3. Estimate the effluent BOD
It is commonly used that 0.63 mg BOD will be exerted (used) for each mg of TSS. Then, the effluent total BOD is

$$\text{BOD} = 2.3 \text{ mg/L} + 0.63 \times 59.4 \text{ mg/L}$$

$$= 39.7 \text{ mg/L}$$

Deep shaft reactor. A new variation of the activated-sludge process, developed in England, the deep shaft reactor is used where costs of land are high. It is used for general application with high-strength wastes. There are a number of installations in Japan. It has been marketed in the United States and Canada.

The deep shaft reactor consists of a vertical shaft about 120 to 150 m (400 to 500 ft) deep, utilizing a U-tube aeration system (Fig. 7.26). The shaft replaces the primary clarifiers and aeration tanks. The deep shaft is lined with a steel shell and fitted with a concentric pipe to form an annular reactor. Wastewater, return activated sludge, and air is forced down the center of the shaft and allowed to rise upward through the annulus (riser). The higher pressures obtained in the deep shaft improve the oxygen transfer rate and more oxygen can be dissolved, with beneficial effects on substrate utilization. DO concentration can range from 25 to 60 mg/L (Droste 1997). Microbial reaction rates are not affected by the high pressures. Temperature in the system is fairly constant throughout the year. The flow in the shaft is a plug-flow mode. The F/M ratios are 0.5 to 5.0 per day and the HRT ranges from 0.5 to 5 h.

A flotation tank serves for the separation of the biomass (solids) from the supernatant. Most of the solids are recirculated to the deep shaft reactor, and some are wasted to an aerobic digester. The effluent of the reactor is supersaturated with air when exposed to atmospheric pressure. The release of dissolved air from the mixed liquor forms bubbles that float suspended solids to the surface of the flotation tank.

The advantages of the deep shaft reactor include lower construction and operational costs, less land requirement, capability to handle big organic loads, and suitability for all climatic conditions.

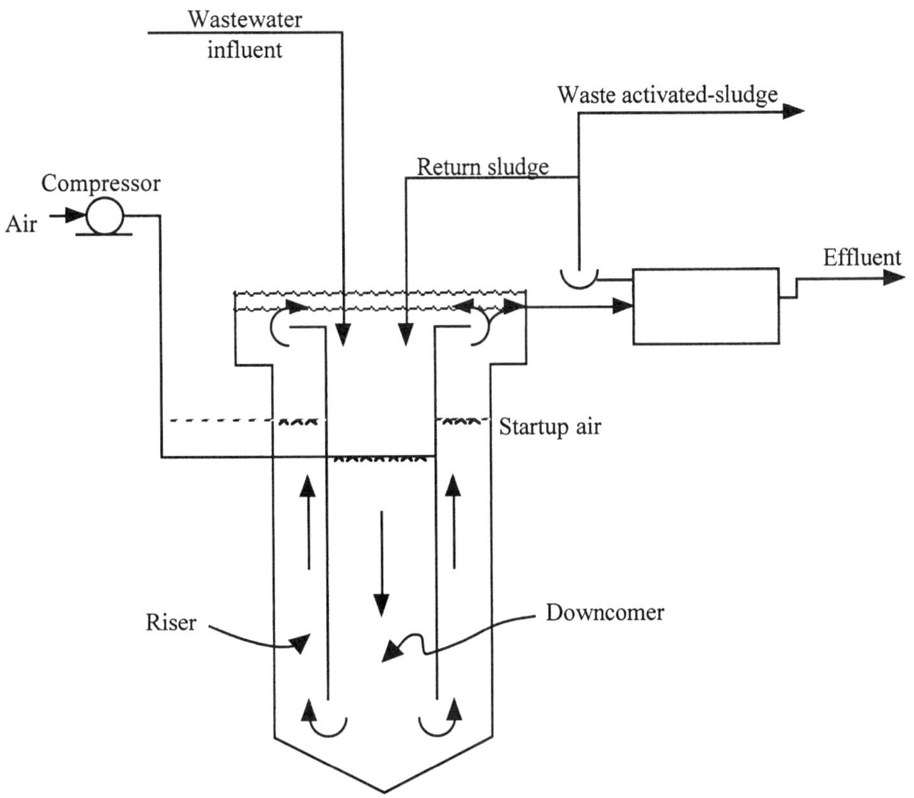

FIGURE 7.26 Schematic diagram of deep shaft activated-sludge system.

Biological nitrification. Conversion of ammonia to nitrite and nitrate can be achieved by biological nitrification. Biological nitrification is performed either in single-stage activated-sludge carbon oxidation and ammonia reduction or in separate continuous-flow stir-tank reactors or plug-flow.

The single-stage nitrification process is used for general application for nitrogen control where inhibitory industrial wastes are not present. The process also removes 85 to 95 percent of BOD.

The design criteria for single-stage activated-sludge nitrification are: volumetric BOD loading = 0.08 to 0.32 kg/(m^3 · d) (5 to 20 lb/(1000 ft^3 · d)); MLSS = 1500 to 3500 mg/L; θ = 6 to 15 h; θ_c = 8 to 12 days; F/M = 0.10 to 0.25 per day; and Q_r/Q = 0.50 to 1.50 (Metcalf and Eddy, Inc. 1991, WEF and ASCE 1991a).

The advantage of the separate-stage suspended-growth nitrification system is its flexibility to be optimized to conform to the nitrification needs. The separate-stage process is used to update existing systems where nitrogen standards are stringent or where inhibitory industrial wastes are present and can not be removed in the previous stages.

The design criteria for the separate-stage nitrification process are: volumetric BOD loading = 0.05 to 0.14 kg/(m^3 · d) (3 to 9 lb/1000 (ft^3 · d)), very low; F/M = 0.05 to 0.2 per day; MLSS = 1500 to 3500 mg/L; HRT = 3 to 6 h; θ_c = 15 to 100 days (very long); and Q_r/Q = 0.50 to 2.00. More details on single-stage and separate-stage biological nitrification are given under advanced treatment for nitrogen removal in a later section.

21.8 Aeration and Mixing Systems

The treatment efficiency of an activated-sludge process depends on the oxygen dissolved in the aeration tank and the transfer of oxygen supply from the air and from pure oxygen. Oxygen is transferred to the mixed liquor in an aeration tank by dispersing compressed air bubbles through submerged diffusers with compressors or by entraining air into the mixed liquor by surface aerators. A combination of the above has also been used. An air diffusion system consists of air filters and other conditioning equipment, blowers, air piping, and diffusers, including diffuser cleaning devices.

A large number of combinations of mixing/aeration equipment and tank configuration have been developed.

Oxygen transfer and utilization. Oxygen transfer is a two-step process. Gaseous oxygen is first dissolved in the mixed liquor by diffusion or mechanical aeration or both. The dissolved oxygen is utilized by the microorganisms in the process of metabolism of the organic matter. When the rate of oxygen utilization exceeds the rate of oxygen dissolution, the dissolved oxygen (DO) in the mixed liquor will be depleted. This should be avoided. DO should be measured periodically, especially during the period of peak loading, and the air supply then adjusted accordingly.

The rate of oxygen transfer from air bubbles into solution is written as (Hammer 1986)

$$r = k(\beta C_s - C_t) \tag{7.113}$$

where

r = rate of oxygen transfer from air to liquid, mg/(L · h)
k = transfer coefficient, per hour
β = oxygen saturation coefficient of the wastewater
= 0.8 to 0.9
C_s = DO concentration at saturation of top water (Table 2.1), mg/L
C_t = DO concentration in mixed liquor, mg/L
$\beta C_s - C_t$ = DO deficit, mg/L

The greater the DO deficit, the higher the rate of oxygen transfer. The oxygen transfer coefficient k depends on wastewater characteristics and, more importantly, on the physical features of the aeration, liquid depth of the aeration tank, mixing turbulence, and tank configuration.

The rate of DO utilization is essentially a function of F/M ratio and liquid temperature. The biological uptake of DO is approximately 30 mg/(L · h) for the conventional activated-sludge process, generally less than 10 mg/(L · h) for extended aeration, and as high as 100 mg/(L · h) for high-rate aeration. Aerobic biological reaction is independent of DO above a minimum critical concentration. However, if DO is below the critical value, the metabolism of microorganisms is limited by reduced oxygen supply. Critical concentrations reported for various activated-sludge systems range from 0.2 to 2.0 mg/L, the most commonly being 0.5 mg/L (Hammer 1986). Ten States Standards stipulates that aeration equipment shall be capable of maintaining a minimum of 2.0 mg/L of DO in mixed liquor at all times and provide thorough mixing of the mixed liquor (GLUMRB 1996).

Oxygen requirements generally depend on maximum diurnal organic loading (design peak hourly BOD_5), degree of treatment, and MLSS. The Ten States Standards (GLUMRB 1996) further recommends that the normal design air requirement for all activated-sludge processes except extended aeration shall provide 1500 ft^3 per pound of BOD_5 loading (94 m^3/kg BOD_5). For extended aeration the value shall be 2050 ft^3 per pound of BOD_5 (128 m^3/kg of BOD_5). The design oxygen requirement for all activated-sludge processes shall be 1.1 lb oxygen/lb design peak hourly BOD_5 applied.

EXAMPLE 1: Compare the rate of oxygen transfer when the existing DOs in the mixed liquor are 3.0 and 0.5 mg/L. Assume the transfer coefficient is 2.8 per hour and oxygen saturation coefficient is 0.88 for wastewater at 15.5°C.

Solution:

Step 1. Find C_s at 15.5°C
From Table 2.1

$$C_s = 9.92 \text{ mg/L}$$

Step 2. Calculate oxygen transfer rates, using Eq. (7.113)

$$\text{At DO} = 3.0 \text{ mg/L}$$

$$r = k(\beta C_s - C_t)$$

$$= (2.8/\text{h})\,(0.88 \times 9.92 - 3.0) \text{ mg/L}$$

$$= 16.0 \text{ mg/(L} \cdot \text{h)}$$

$$\text{At DO} = 0.5 \text{ mg/L}$$

$$r = (2.8/\text{h})\,(0.88 \times 9.92 - 0.5) \text{ mg/L}$$

$$= 23.0 \text{ mg/(L} \cdot \text{h)}$$

Note: The higher the DO deficit, the greater the oxygen transfer rate.

EXAMPLE 2: Determine the oxygen transfer efficiency of a diffused aeration system providing 60 m^3/kg BOD_5 applied (960 ft^3/lb BOD_5). Assume that the aeration system is capable of transferring 1.1 kg of atmospheric oxygen to DO in the mixed liquor per kg BOD_5 applied, and that one m^3 of air at standard conditions (20°C, 760 mm Hg, and 36% relative humidity) contains 0.279 kg of oxygen.

Solution:

Step 1. Calculate oxygen provided per kg BOD_5

$$\text{Oxygen provided} = 60 \text{ m}^3\text{/kg BOD} \times 0.279 \text{ kg/m}^3$$

$$= 16.7 \text{ kg/kg BOD}$$

Step 2. Determine oxygen transfer efficiency e

$$e = \text{oxygen transferred/oxygen provided}$$

$$= (1.1 \text{ kg/kg BOD})/(16.7 \text{ kg/kg BOD})$$

$$= 0.066$$

$$= 6.6\%$$

Diffused air aeration. A diffused air aeration system consists of diffusers submerged in the mixed liquor, header pipes, air mains, blowers (air compressors), and appurtenances through which the air passes. Ceramic diffusers (porous and nonporous) are manufactured as tubes, domes, or plates. Other diffusers, such as jets or U-tubes, may be used. Compressed air forced through the diffusers is released as fine bubbles. The oxygen is then introduced to the liquid.

The oxygen requirement for the aeration can be calculated using Eq. (7.94) or (7.95). In practice, the efficiency of oxygen transfer of a diffusion and sparging device seldom exceeds 8 percent. The oxygen transfer efficiency for various bubble sizes is available from the manufacturer's data. The diffuser system should be capable of providing for 200 percent of the designed average daily oxygen demand (GLUMRB 1996).

Air piping. As compressed air flows through a pipe, its volume changes according to the pressure drop. The total headloss from the blower (or silencer outlet) to the diffuser inlet shall not exceed 0.5 lb/in^2 (3.4 kPa) under average operational conditions (GLUMRB 1996). The pressure drop in a pipeline carrying compressed air can be calculated by the Darcy–Weisbach formula (Eq. (5.17)).

The value of friction factor f can be determined from the Moody diagram. However, as an approximation, the friction factor of a steel pipe carrying compressed air is expressed as (Steel and McGhee 1979)

$$f = 0.029D^{0.027}/Q^{0.148} \quad \text{(SI units)} \tag{7.114}$$

$$f = 0.013D^{0.027}/Q^{0.148} \quad \text{(US customary units)} \tag{7.115}$$

where f = friction factor
D = diameter of pipe, m or ft
Q = flow of air, m^3/min or gal/min

The headloss in a straight pipe is calculated from

$$H = f(L/D)\, h_r \tag{7.116}$$

where H = headloss in pipe, mm or in
L = length of pipe, m or ft
D = diameter of pipe, m or ft
h_r = velocity head, mm or in

Another approximation of the estimation of headloss in a pipe can be expressed as

$$H = 9.82 \times 10^{-8}\, (fLTQ^2)/(PD^5) \tag{7.117}$$

where T = temperature of air, K
P = absolute pressure, atm
Other terms are as given above.

The absolute temperature may be estimated from the pressure rise using the following equation

$$T_2 = T_1(P_2/P_1)^{0.283} \tag{7.118}$$

Fittings in an air distribution system may be converted into the equivalent length of straight pipe as below

$$L = 55.4\, CD^{1.2} \quad \text{(SI units)} \tag{7.119}$$

and

$$L = 230\, \mathrm{CD}^{1.2} \quad \text{(US customary units)} \tag{7.120}$$

where L = equivalent length, m or ft
D = diameter of pipe, m or ft
C = resistance factor (0.25 to 2.0)

EXAMPLE: Determine the headloss in 250 m (820 ft) of 0.30-m (12-in) steel pipeline carrying 42.5 m^3/min (1500 ft^3/min = 25 ft^3/s = 0.7 m^3/s) of air at a pressure of 0.66 atm (gauge). The ambient air temperature is 26°C (78°F).

Solution:

Step 1. Estimate the temperature in the pipe using Eq. (7.118)

$$\begin{aligned} P_1 &= 1 \text{ atm} \\ P_2 &= 1 \text{ atm} + 0.66 \text{ atm} = 1.66 \text{ atm} \\ T_1 &= 26 + 273 = 299 \text{ (K)} \\ T_2 &= T_1(P_2/P_1)^{0.283} \\ &= 299 \text{ K } (1.66/1)^{0.283} \\ &= 345 \text{ K} \end{aligned}$$

Step 2. Estimate the friction factor using Eq. (7.114)

$$\begin{aligned} f &= 0.029\, D^{0.027}/Q^{0.148} \\ &= 0.029\, (0.3)^{0.027}/(42.5)^{0.148} \\ &= 0.0161 \end{aligned}$$

Step 3. Estimate the head loss using Eq. (7.117)

$$\begin{aligned} H &= 9.82 \times 10^{-8}(fLTQ^2)/(PD^5) \\ &= 9.82 \times 10^{-8} \times 0.0161 \times 250 \times 345 \times 42.5^2/(1.66 \times 0.3^5) \\ &= 61 \text{ (mm of water)} \end{aligned}$$

Blower. The blower (compressor) is used to convey air up to 103 kPa (15 lb/in^2). There are two types of blower commonly used for aeration, i.e. rotary position displacement (PD) and centrifugal units. Centrifugal blowers are usually used when the air supply is greater than 85 m^3/min (3000 ft^3/min or cfm) and are popular in Europe. For pressures up to 50 to 70 kPa (7 to 10 lb/in^2) and flows greater than 15 m^3/min (530 ft^3/min), centrifugal blowers are preferable. Rotary PD blowers are used primarily in small plants. Air flow is low when pressure is over 40 to 50 kPa (6 to 7 lb/in^2) and plant flows are less than 15 m^3/min (530 ft^3/min) (Steel and McGhee 1979).

As an advantage, PD blowers are capable of operating over a wide range of discharge pressure, but it is difficult to adjust their air flow rate (only the speed is adjustable), they require more maintenance, and are noisy. In contrast, centrifugal blowers are less noisy in operation and take up less space. Their disadvantages include a limited range of operating pressure and reduced air flow when there is any increase in backpressure because of diffuser clogging. In many plants, single-stage centrifugal blowers can provide the head requirement (6 m or 20 ft of water). If higher pressures are needed, multiple units can provide a pressure of 28 m (90 ft) of water.

The power requirements for a blower may be estimated from the air flow, inlet and discharge pressures, and air temperature by the following equation which is based on the assumption of adiabatic compression conditions.

$$p = \frac{wRT_1}{29.7ne}\left[\left(\frac{P_2}{P_1}\right)^{0.283} - 1\right]$$

or

$$p = \frac{wRT_1}{8.41e}\left[\left(\frac{P_2}{P_1}\right)^{0.283} - 1\right] \quad \text{(SI units)} \tag{7.121}$$

$$p = \frac{wRT_1}{550ne}\left[\left(\frac{P_2}{P_1}\right)^{0.283} - 1\right]$$

or

$$p = \frac{wRT_1}{155.6\,e}\left[\left(\frac{P_2}{P_1}\right)^{0.283} - 1\right] \quad \text{(US customary units)} \tag{7.122}$$

where p = power requried, kW or hp
w = weight of flow of air, kg/s or lb/s
R = gas constant for air
= 8.314 kJ/(kmol · K) for SI units
= 53.3 ft · lb/(lb air · R)
(R = °F + 460 for US customary units)
T_1 = inlet absolute temperature, K or R
P_1 = inlet absolute pressure, atm or lb/in^2
P_2 = outlet absolute pressure, atm or lb/in^2
n = $(k - 1)/k$ = 0.283 for air
k = 1.395 for air
550 = ft · lb/(s · hp)
29.7 = constant for conversion to SI units
e = efficiency of the machine (usually 0.7 to 0.9)

As stated above, for higher discharge pressure application (> 55 kPa, > 8 lb/in^2) and for capacity smaller than 85 m^3/min (3000 ft^3/min) of free air per unit, rotary-lobe PD blowers are generally used.

PD blowers are also used when significant water level variations are expected. The units cannot be throttled. Rugged inlet and discharge silencers are essential (Metcalf and Eddy, Inc. 1991).

EXAMPLE: Estimate the power requirement of a blower providing 890 kg of oxygen per day through a diffused air system with an oxygen transfer efficiency of 7.2 percent. Assume the inlet temperature is 27°C, the discharge pressure is 6 m of water, and the efficiency of the blower is 80 percent.

Solution:

Step 1. Calculate weight of air flow w

$$\text{Oxygen required} = (890 \text{ kg/d})/0.072$$
$$= 12{,}360 \text{ kg/d}$$

In air, 23.2% is oxygen

$$w = (12{,}360 \text{ kg/d})/0.232$$
$$= 53{,}280 \text{ kg/d}$$
$$= 0.617 \text{ kg/s}$$

Step 2. Determine the power requirement for a blower p

$$T_1 = 27 + 273 = 300 \text{ (K)}$$
$$P_1 = 1 \text{ atm}$$
$$P_2 = 1 \text{ atm} + 6 \text{ m}/10.345 \text{ m/atm}$$
$$= 1.58 \text{ atm}$$
$$e = 0.8$$

Using Eq. (7.121)

$$p = \frac{wRT_1}{8.41e}\left[\left(\frac{P_2}{P_1}\right)^{0.283} - 1\right]$$
$$= \frac{0.617 \times 8.314 \times 300}{8.41 \times 0.8}\left[\left(\frac{1.58}{1}\right)^{0.283} - 1\right]$$
$$= 31.6 \text{ (kW)}$$

Mechanically aerated systems. Mechanical aerators consist of electrical motors and propellers mounted on either a floating or a fixed support. The electrically driven propellers throw the bulk liquid through the air and oxygen transfer occurs both at the surface of the droplets and at the surface of the mixed liquor. Mechanical aerators may be mounted on either a horizontal or a vertical axis. Each group is divided into surface or submerged, high-speed (900 to 1800 rev/min) or low-speed (40 to 50 rev/min) (Steel and McGhee 1979). Low-speed aerators are more expensive than high-speed ones but have fewer mechanical problems and are more desirable for biological floc formation. Selection of aerators is based on oxygen transfer efficiency and mixing requirements. Effective mixing is a function of liquid depth, unit design, and power supply.

Mechanical aerators are rated on the basis of oxygen transfer rate, expressed as kg of oxygen per kW · h (or lb/hp · h) under standard conditions, in which tap water with 0.0 mg/L DO (with sodium sulfite added) is tested at the temperature of 20°C. Commercially available surface reaerators range from 1.2 to 2.4 kg O_2/kW · h (2 to 4 lb O_2/hp · h) (Metcalf and Eddy, Inc. 1991).

Mechanical aerator requirements depend on the manufacturer's rating, the wastewater quality, the temperature, the altitude, and the desired DO level. The standard performance data must be adjusted to the anticipated field conditions, using the following equation (Eckenfelder 1966)

$$N = N_0\left(\frac{\beta C_w - C_L}{C_{s20}}\right)1.024^{T-20}\alpha \tag{7.123}$$

where N = oxygen transfer rate under field conditions, kg/kW · h or lb/hp · h
N_0 = oxygen transfer rate provided by manufacturer, kg/kW · h or lb/hp · h
β = salinity surface tension correction factor
= 1 (usually)
C_w = oxygen saturation concentration for tap water at given altitude and temperature, mg/L
C_L = operating DO concentration (2.0 mg/L)
C_{s20} = oxygen saturation concentration in tap water at 20°C, mg/L
= 9.02
T = temperature, °C
α = oxygen transfer correction factor for wastewater
= 0.8 to 0.9

EXAMPLE: An activated-sludge plant is located at an elevation of 210 m. The desired DO level in the aeration tank is 2.0 mg/L. The range of operating temperature is from 8°C to 32°C. The saturated DO values at 8°C, 20°C, and 32°C are 11.84, 9.02, and 7.29 mg/L (from Table 2.2), respectively. The oxygen-transfer correction factor for the wastewater is 0.85. The manufacturer's rating for oxygen-transfer rate of the aerator under standard conditions is 2.0 kg oxygen/kW · h. Determine the power requirement for providing 780 kg oxygen per day to the aeration system.

Solution:

Step 1. Determine DO saturation levels at altitude 210 m
For DO, after altitude correction (zero saturation at 9450 m)

$$C_w = C\left(1 - \frac{\text{altitude, m}}{9450\text{ m}}\right) = C\left(1 - \frac{210\text{ m}}{9450\text{ m}}\right)$$

$$= 0.98\ C$$

At 8°C

$$C_w = 0.98 \times 11.84\text{ mg/L}$$
$$= 11.6\text{ mg/L}$$

At 32°C

$$C_w = 0.98 \times 7.29\text{ mg/L}$$
$$= 7.14\text{ mg/L}$$

Step 2. Calculate oxygen transfer under field conditions
At 8°C, by Eq. (7.123)

N_0 = 2.0 kg oxygen/kW · h
β = 1
C_w = 11.6 mg/L
C_L = 2.0 mg/L
C_{s20} = 9.02 mg/L
T = 8°C
α = 0.85

$$N = N_0\left(\frac{\beta C_w - C_L}{C_{s20}}\right)1.024^{T-20}\alpha$$

$$= (2\text{ kg/kW}\cdot\text{h})\left(\frac{1 \times 11.6\text{ mg/L} - 2.0\text{ mg/L}}{9.02\text{ mg/L}}\right)(1.024^{8-20})0.85$$

$$= 1.36\text{ kg/kW}\cdot\text{h}$$

At 32°C

$$N = 2\left(\frac{7.14 - 2}{9.02}\right)(1.024^{32-20})0.85$$
$$= 1.29\ (\text{kg/kW} \cdot \text{h})$$
$$= 31\ \text{kg/kW} \cdot \text{d}$$

Step 3. Calculate the power required per day, p
At 32°C

$$p = (780\ \text{kg/d})/(31\ \text{kg/kW} \cdot \text{d})$$
$$= 25\ \text{kW}$$

Note: Transfer rate is lower with higher temperature.

Aerated lagoon. An aerated lagoon (pond) is a complete-mix flow-through system with or without solids recycle. Most systems operate without solids recycle. If the solids are returned to the lagoon, the system becomes a modified activated-sludge process.

The lagoons are 3 to 4 m (10 to 13 ft) deep. Solids in the complete-mix aerated pond are kept suspended at all times by aerators. Oxygen is usually supplied by means of surface aerators or diffused air devices. Depending on the hydraulic retention time, the effluent from an aerated pond will contain from one-third to one-half the concentration of the influent BOD in the form of cell tissue (Metcalf and Eddy Inc. 1991). These solids must be removed by settling before the effluent is discharged. Settling can take place at a part of the aerated pond system separated with baffles or in a sedimentation basin.

The design factors for the process include BOD removal, temperature effects, oxygen requirements, mixing requirements, solids separation, and effluent characteristics. BOD removal and the effluent characteristics are generally estimated using a complete-mix hydraulic model and first-order reaction kinetics. The mathematical relationship for BOD removal in a complete-mix aerated lagoon is derived from the following equation

$$QS_0 - QS - kSV = 0 \tag{7.124}$$

Rearranging

$$\frac{S}{S_0} = \frac{1}{1 + k(V/Q)} = \frac{\text{effluent BOD}}{\text{influent BOD}} \tag{7.125}$$

$$= \frac{1}{1 + k\theta} \tag{7.126}$$

where S = effluent BOD_5 concentration, mg/L
S_0 = influent BOD_5 concentration, mg/L
k = overall first-order BOD_5 removal rate, per day
= 0.25 to 1.0, based on e
Q = wastewater flow, m^3/d or Mgal/d
θ = total hydraulic retention time, d

Typical design values of θ for aerated ponds used for treating domestic wastewater vary from 3 to 6 days. The amounts of oxygen required for aerated lagoons range from 0.7 to 1.4 times the amount of BOD_5 removed (Metcalf and Eddy, Inc. 1991)

Mancini and Barnhart (1968) developed the resulting temperature in the aerated lagoon from the influent wastewater temperature, air temperature, surface area, and flow. The equation is

$$T_i - T_w = \frac{(T_w - T_a)fA}{Q} \tag{7.127}$$

where T_i = influent wastewater temperature, °C or °F
T_w = lagoon water temperature, °C or °F
T_a = ambient air temperature, °C or °F
f = proportionality factor
= 12×10^{-6} (for US customary units)
= 0.5 (for SI units)
A = surface area of lagoon, m^2 or ft^2
Q = wastewater flow, m^3/d or Mgal/d

Rearranging Eq. (7.127), the lagoon water temperature is

$$T_w = \frac{AfT_a + QT_i}{Af + Q} \tag{7.128}$$

Aerated lagoons, usually followed by facultative lagoons, are used for first-stage treatment of high-strength domestic wastewaters and for pretreatment of industrial wastewaters. Their BOD removal efficiencies range from 60 to 70 percent. Low efficiency and foul odors may occur for improperly designed or poorly operated plants, especially if the aeration devices are inadequate. Wet weather flow, infiltration, and icing may cause the process to be upset.

EXAMPLE: Design a complete-mix aerated lagoon system using the conditions given below.

Wastewater flow	3000 m^3/d (0.8 Mgal/d)
Influent soluble BOD_5	180 mg/L
Effluent soluble BOD_5	20 mg/L
Soluble BOD_5 first-order k_{20}	2.4 per day
Influent TSS (not biodegraded)	190 mg/L
Final effluent TSS	22 mg/L
MLVSS/MLSS	0.8
Kinetic coefficients:	
k	5 per day
K_s	60 mg/L BOD
Y	0.6 mg/mg
k_d	0.06 per day
Design depth of lagoon	3 m (10 ft)
Design HRT, θ	5 days
Detention time at settling basin	2 days
Temperature coefficient	1.07
Wastewater temperature	15°C (59°F)
Summer mean air temperature	26.5°C (78°F)
Winter mean air temperature	9°C (48°F)
Summer mean wastewater temperature	20.2°C (68°F)
Winter mean wastewater temperature	12.3°C (54°F)
Aeration constant α	0.86
Aeration constant β	1.0
Plant site elevation	210 m (640 ft)

Solution:

Step 1. Determine the surface area of the lagoon with a depth of 3 m

$$\text{Volume } V = Q\theta = 3000 \text{ m}^3/\text{d} \times 5 \text{ days}$$
$$= 15{,}000 \text{ m}^3$$
$$\text{Area } A = V/(3 \text{ m}) = 15{,}000 \text{ m}^3/3 \text{ m}$$
$$= 5000 \text{ m}^2$$
$$= 1.23 \text{ acres}$$

Step 2. Estimate lagoon wastewater temperature in summer and winter
Using Eq. (7.128), in summer (T_a = 26.5°C)

$$T_w = \frac{AfT_a + QT_i}{Af + Q}$$

$$= \frac{5000 \times 0.5 \times 26.5 + 3000 \times 15}{5000 \times 0.5 + 3000}$$

$$= 20.2(°C)$$

$$= 68.4°F$$

In winter (T_a = 9°C)

$$T_w = \frac{5000 \times 0.5 \times 9 + 3000 \times 15}{5000 \times 0.5 + 3000}$$

$$= 12.3(°C)$$

$$= 54.1°F$$

Step 3. Calculate the soluble BOD_5 during the summer
Using Eq. (7.80)

$$S = \frac{K_s(1 + \theta k_d)}{\theta(Yk - k_d) - 1}$$

$$= \frac{60\ \text{mg/L}(1 + 5\ \text{d} \times 0.06/\text{d})}{5\ \text{d}(0.6 \times 5/\text{d} - 0.06/\text{d}) - 1}$$

$$= 5.7\ \text{mg/L}$$

Step 4. Calculate first-order BOD removal rate constant for temperature effects
In summer at the mean wastewater temperature 20.2°C

$$k_t = k_{20}(1.07)^{T-20}$$

$$k_{20.2} = 2.4/\text{d}\ (1.07)^{20.2-20}$$

$$= 2.43/\text{d}$$

In winter at 12.3°C

$$k_{12.3} = 2.4/\text{d}\ (1.07)^{12.3-20}$$

$$= 1.43/\text{d}$$

Step 5. Calculate the effluent BOD_5 using Eq. (7.126)
In summer at 20.2°C

$$\frac{S}{S_0} = \frac{1}{1 + k\theta}$$

$$\frac{S}{180\ \text{mg/L}} = \frac{1}{1 + 2.43 \times 5}$$

$$S = 13.7\ \text{mg/L}$$

In winter at 12.3°C

$$S = \frac{180\ \text{mg/L}}{1 + 1.43 \times 5}$$

$$= 22.1\ \text{mg/L}$$

Ratio of winter to summer = 22.1 : 13.7 = 1.6 : 1

Step 6. Estimate the biological solids produced using Eq. (7.78), $\theta = \theta_c$

$$X = \frac{Y(S_0 - S)}{1 + k_d\theta} = \frac{0.6(180 - 5.7) \text{ mg/L}}{1 + 0.06 \times 5}$$

$$= 80 \text{ mg/L (of VSS)}$$

$$\text{TSS} = 80 \text{ mg/L} \div 0.8$$

$$= 100 \text{ mg/L}$$

Step 7. Calculate TSS in the lagoon effluent before settling

$$\text{TSS} = 190 \text{ mg/L} + 100 \text{ mg/L}$$

$$= 290 \text{ mg/L}$$

Note: With low overflow rate and a long detention time of 2 days, the final effluent can achieve 22 mg/L of TSS.

Step 8. Calculate the amount of biological solids wasted per day
Since $X = 80 \text{ mg/L} = 80 \text{ g/m}^3 = 0.08 \text{ kg/m}^3$ (from Step 6)

$$p_x = 3000 \text{ m}^3\text{/d} \times 0.08 \text{ kg/m}^3$$

$$= 240 \text{ kg/d}$$

Step 9. Calculate the oxygen required using Eq. (7.94a)
Using the conversion factor f for BOD_5 to BOD_L, $f = 0.67$

$$\text{Oxygen} = \frac{Q(S_0 - S)}{(1000 \text{ g/kg}) f} - 1.42\, p_x$$

$$= \frac{3000 \text{ m}^3 \times (180 - 5.7) \text{ g/m}^3}{1000 \text{ g/kg} \times 0.67} - 1.42 \times 240 \text{ kg/d}$$

$$= 440 \text{ kg/d}$$

$$= 970 \text{ lb/d}$$

Step 10. Calculate the ratio of oxygen required to BOD removal

$$\text{BOD removal} = 3000 \text{ m}^3\text{/d} \times (180 - 5.7) \text{ g/m}^3 \div 1000 \text{ g/kg}$$

$$= 522.9 \text{ kg/d}$$

$$= 1153 \text{ lb/d}$$

$$\frac{\text{Oxygen required}}{\text{BOD removed}} = \frac{440 \text{ kg/d}}{522.9 \text{ kg/d}}$$

$$= 0.84$$

Step 11. Determine the field transfer rate for the surface aerators
The example in the mechanical aerator shows that the oxygen transfer rate is lower in summer (higher temperature). From that example, the altitude correction factor is 0.98 for an elevation of 210 m. Select the surface aerator rating as 2.1 kg oxygen/kW · h (3.5 lb/hp · h). The solubility of tap water at 26.5 and 20.0°C is 7.95 and 9.02 mg/L, respectively. Let C_L = 2.0 g/L as GLUMRB (1996) recommended. The power requirements under field conditions can be estimated using Eq. (7.123).

$$
\begin{aligned}
N &= N_0\left(\frac{\beta C_{\mathrm{w}} - C_{\mathrm{L}}}{C_{\mathrm{s20}}}\right)1.024^{T-20}\alpha \\
&= (2.1\ \mathrm{kg/kW \cdot h})\left(\frac{1 \times 0.98 \times 7.95\ \mathrm{mg/L} - 2\ \mathrm{mg/L}}{9.02\ \mathrm{mg/L}}\right)(1.024)^{26.5-20}(0.86) \\
&= 1.35\ \mathrm{kg/kW \cdot h} \\
&= 32.4\ \mathrm{kg/kW \cdot d}
\end{aligned}
$$

Step 12. Determine the power requirements of the surface aerator
From Steps 9 and 11

$$
\text{Power required} = \frac{\text{oxygen required}}{\text{field transfer rate}} = \frac{440\ \mathrm{kg/d}}{32.4\ \mathrm{kg/kW \cdot d}}
$$

$$
= 13.6\ \mathrm{kW}
$$

Note: This is the power required for oxygen transfer.

Step 13. Determine the total energy required for mixing the lagoon
Assume power required for complete-mix flow is 0.015 kW/m^3 (0.57 hp/1000 ft^3)
From Step 1

$$
\begin{aligned}
\text{Volume of the lagoon} &= 15{,}000\ \mathrm{m^3} \\
\text{Total power needed} &= 0.015\ \mathrm{kW/m^3} \times 15{,}000\ \mathrm{m^3} \\
&= 225\ \mathrm{kW}
\end{aligned}
$$

Use 8 of 30 kW (40 hp) surface aerators providing 240 kW.

Note: This is the power required for mixing and is commonly the control factor in sizing the aerators for domestic wastewater treatment. For industrial wastewater treatment, the control factor is usually reversed.

22 TRICKLING FILTER

A trickling filter is actually a unit process for introducing primary effluent into contact with biological growth and is a biological oxidation bed. The word "filter" does not mean any filtering or straining action; nevertheless, it is popularly and universally used.

22.1 Process Description

The trickling filter is the most commonly used unit of the fixed-growth film-flow-type process. A trickling filter consists of (1) a bed of coarse material, such as stone slates or plastic media, over which wastewater from primary effluent is sprayed; (2) an underdrain system; and (3) distributors. The underdrain is used to carry wastewater passing through the biological filter and drain to the subsequent treatment units and to provide ventilation of the filter and maintenance of the aerobic condition. Wastewater from the primary effluent is distributed to the surface of the filter bed by fixed spray nozzles (first developed) or rotary distributors. Sloughs of biomass from the media are settled in the secondary sedimentation tank.

Biological slime occurs on the surface of the support media while oxygen is supplied by air diffusion through the void spaces. It allows wastewater to trickle (usually in an intermittent fashion) downward through the bed media. Organic and inorganic nutrients are extracted from the liquid film by the microorganisms in the slime. The biological slime layer consists of aerobic, anaerobic, and facultative bacteria, algae, fungi, and protozoans. Higher animals such as sludge worms, filter-fly larvae, rotifers, and snails are also present. Facultative bacteria are the predominant microorganisms in the trickling filter. Nitrifying bacteria may occur in the lower part of a deep filter.

The biological activity of the trickling filter process can be described as shown in Fig. 7.27. The microbial layer on the filter is aerobic usually to a depth of only 0.1 to 0.2 mm. Most of the depth of the microbial film is anaerobic. As the wastewater flows over the slime layer, organic matter (nutrient) and dissolved oxygen are transferred to the aerobic zone by diffusion and extracted, and then metabolic end products such as carbon dioxide are released to the water.

Dissolved oxygen in the liquid is replenished by adsorption from air in the voids surrounding the support media. Microorganisms near the surface of the filter bed are in a rapid growth rate due to plenteous food supply, whereas microorganisms in the lower portion of the filter may be in a state of starvation. Overall, a trickling filter operation is considered to be in the endogenous growth phase. When slime becomes thicker and cells die and lyse, the slime layer will slough off and is subsequently removed by secondary settling.

For over a century, the trickling filter has been used as a secondary treatment unit. In a rock-fill conventional trickling filter, the rock size is 25 to 100 mm (1 to 4 in). The depth of the rock bed varies from 0.9 to 2.5 m (3 to 8 ft) with an average of 1.8 m (6 ft). The Ten States Standards recommends a

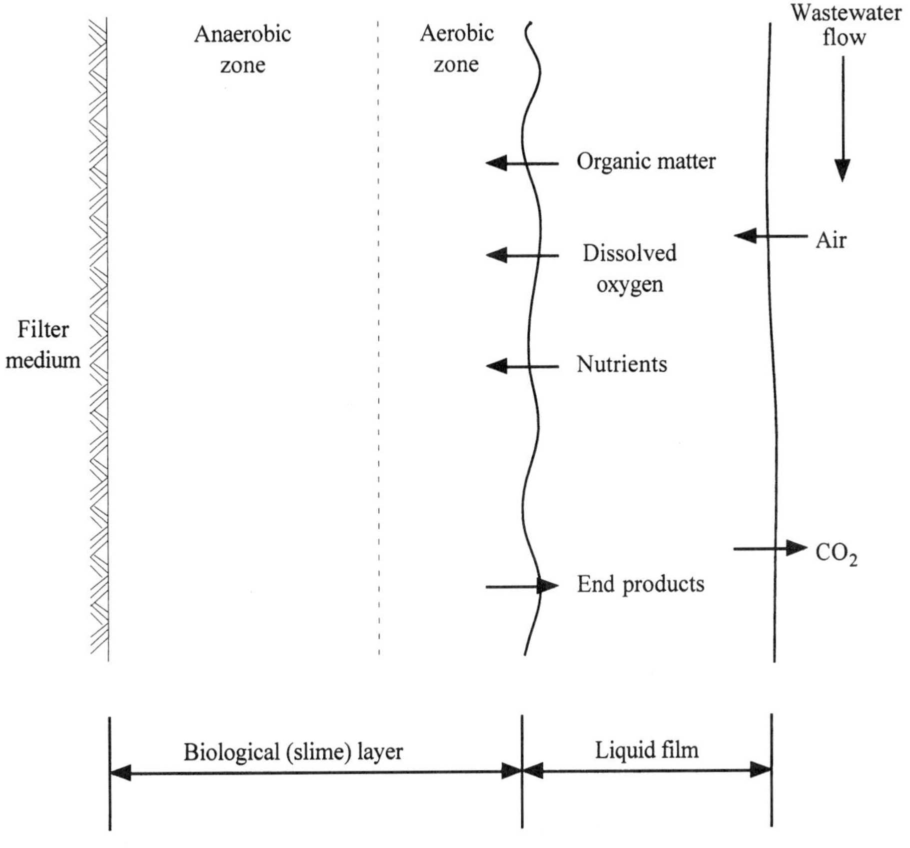

FIGURE 7.27 Schematic diagram of attached-growth process.

minimum depth of trickling filter media of 1.8 m above the underdrains and that a rock and/or slag filter shall not exceed 3 m (10 ft) in depth (GLUMRB 1996).

After declining in use in the late 1960s and early 1970s because of the development of RBC, trickling filters regained popularity in the late 1970s and 1980s, primarily due to new synthetic media. The new high-rate media were generally preferred over rocks because they are lighter, increased the surface area for biological growth, and improved treatment efficiency. High-rate media minimizes many of the common problems with rock media, such as uncontrolled sloughing, plugging, odors, and filter flies. Consequently, almost all trickling filters constructed in the late 1980s have been of high-rate media type (WPCF 1988b).

Plastic media built in square, round, and other modules of corrugated shape have become popular. The depths of these plastic media range from 4 to 12 m (14 to 40 ft) (USEPA 1980, Metcalf and Eddy, Inc. 1991). These materials increase void ratios and air flow. The plastics are 30% lighter than rock. A minimum clearance of 0.3 m (1 ft) between media and distribution arms shall be provided (GLUMRB 1996).

22.2 Filter Classification

Trickling filters are classified according to the applied hydraulic and organic loading rates. The hydraulic loading rate is expressed as the quantity of wastewater applied per day per unit area of bulk filter surface ($m^3/(d \cdot m^2)$, $gal/(d \cdot ft^2)$, or $Mgal/(d \cdot acre)$ or as depth of wastewater applied per unit of time. Organic loading rate is expressed as mass of BOD_5 applied per day per unit of bulk filter volume ($kg/(m^3 \cdot d)$, $lb/(1000\ ft^3 \cdot d)$). Common classifications, are low- or standard-rate, intermediate-rate, high-rate, super-high-rate, and roughing. Two-stage filters are frequently used, in which two trickling filters are connected in series. Various trickling filter classifications are summarized in Table 7.14.

22.3 Recirculation

Recirculation of a portion of the effluent to flow back through the filter is generally practised in modern trickling filter plants. The ratio of the return flow Q_r, to the influent flow Q is called the recirculation ratio r. Techniques of recirculation vary widely, with a variety of configurations. The recirculation ratios range from 0 to 4 (Table 7.14) with usual ratios being 0.5 to 3.0.

The advantages of recirculation include an increase in biological solids in the system with continuous seeding of active biological material; elimination of shock load by diluting strong influent; maintenance of more uniform hydraulic and organic loads; an increase in the DO level of the influent; thinning of the biological slime layer; an improvement of treatment efficiency; reduction of filter clogging; and less nuisance problems.

22.4 Design Formulas

Attempts have been made by numerous investigators to correlate operational data with the design parameters of trickling filters. The design of trickling filter plants is based on empirical, semi-empirical, and mass balance concepts. Mathematical equations have been developed for calculating the BOD_5 removal efficiency of biological filters on the basis of factors such as bed depth, types of media, recirculation, temperature, and loading rates. The design formulations of trickling filters of major interest include the NRC formula (1946), Velz formula (1948), Fairall (1956), Schultz formula (1960), Eckenfelder formula (1963), Galler and Gotaas formulas (1964, 1966), Germain formula (1966), Kincannon and Stover formula (1982), Logan formula (Logan *et al.* 1987a, b), and British multiple regression analysis equation (1988). These are summarized elsewhere (WEF and ASCE 1991a, McGhee 1991).

NRC formula. The NRC (National Research Council) formula for trickling-filter performance is an empirical expression developed by the National Research Council from an extensive study of the operating data of trickling treatment plants at military bases within the United States during World

TABLE 7.14 Typical Design Information for Trickling Filter Categories

Parameter	Low-rate (Standard)	Intermediate-rate	High-rate	Super high-rate	Roughing	Two-stage
Filter media	Rock, slag	Rock, slag	Rock	Plastic	Plastic, redwood	Rock, plastic
Hydraulic loading						
$m^3/(m^2 \cdot d)$	1–3.7	3.7–9.4	9.4–37	14–86	47–187	9.4–37
$gal/(d \cdot ft^2)$	25–90	90–230	230–900	350–2100	1150–4600	230–900
$Mgal/(d \cdot acre)$	1–4	4–10	10–39	15–91	50–200	10–39
BOD_5 loading						
$kg/(m^3 \cdot d)$	0.08–0.32	0.24–0.48	0.48–1.0	0.48–1.6	1.6–8.0	1.0–2.0
$lb/(1000\ ft^3 \cdot d)$	5–20	15–30	30–60	30–100	100–500	60–120
Recirculation rate	0	0–1	1–2	1–2	1–4	0.5–2
Filter flies	many	some	few	few or none	few or none	few or none
Sloughing	intermittent	intermittent	continuous	continuous	continuous	continuous
Depth, m	1.8–2.4	1.8–2.4	0.9–1.8	3–12	4.6–12	1.8–2.4
ft	6–8	6–8	3–6	10–40	15–40	6–8
BOD_5 removal % (includes settling)	80–90	50–70	65–85	65–80	40–65	85–95
Nitrification	well	partial	little	little	no	well
Combined process						
for secondary treatment	no	unlikely	likely	yes	yes	yes
for tertiary treatment	yes	yes	yes	yes	yes	yes
Type normally used*	SC/TF ABF	SC/TF ABF	SC/TF RBC/TF	AS/TF AS/BF	AS/TF –	TF/TF –

*SC/TF = trickling filter after solid contactor
AS = activated sludge
ABF = activated biofilter
RBC = rotating biological contactor
BF = biofilter
Sources: USEPA (1974a), Metcalf and Eddy, Inc. (1991), WEF and ASCE (1991a, 1996a).

War II in the early 1940s (NRC 1946). It may be applied to single-stage and multistage rock filters with varying recirculation ratios. Graphic expressions for BOD removal efficiency are available. The equation for a single-stage or first-stage rock filter is

$$E_1 = \frac{100}{1 + 0.532\sqrt{W/VF}} \quad \text{(SI units)} \tag{7.129}$$

$$E_1 = \frac{100}{1 + 0.0561\sqrt{W/VF}} \quad \text{(US customary units)} \tag{7.130}$$

where E_1 = efficiency of BOD removal for first stage at 20°C including recirculation and sedimentation, %
W = BOD loading to filter, kg/d or lb/d
= flow times influent concentration
V = volume of filter media, m^3 or 1000 ft^3
F = recirculation factor

The recirculation factor is calculated by

$$F = \frac{1 + r}{(1 + 0.1r)^2} \tag{7.131}$$

where r = recirculation ratio, Q_r/Q
Q_r = recirculation flow, m^3/d or Mgal/d
Q = wastewater flow, m^3/d or Mgal/d

The recirculation factor represents the average number of passes of the influent organic matter through the trickling filter. For the second-stage filter, the formula becomes

$$E_2 = \frac{100}{1 + \dfrac{0.0532}{1 - E_1}\sqrt{\dfrac{W'}{VF}}} \quad \text{(SI units)} \tag{7.132}$$

$$E_2 = \frac{100}{1 + \dfrac{0.0561}{1 - E_1}\sqrt{\dfrac{W'}{VF}}} \quad \text{(US customary units)} \tag{7.133}$$

where E_2 = efficiency of BOD_5 removal for second-stage filter, %
W' = BOD loading applied to second-stage filter, kg/d or lb/d

Other terms are as described previously. Overall BOD removal efficiency of a two-stage filter system can be computed by

$$E = 100 - 100\left(1 - \frac{35}{100}\right)\left(1 - \frac{E_1}{100}\right)\left(1 - \frac{E_2}{100}\right) \tag{7.134}$$

where the term 35 means that 35 percent of BOD of raw wastewater is removed by primary settling.

BOD removal efficiency in biological treatment process is significantly influenced by wastewater temperature. The effect of temperature can be calculated as

$$E_T = E_{20}1.035^{T-20} \tag{7.135}$$

EXAMPLE 1: Estimate the BOD removal efficiency and effluent BOD_5 of a two-stage trickling filter using the NRC formula with the following given conditions.

Wastewater temperature	20°C
Plant flow Q	2 Mgal/d (7570 m^3)
BOD_5 in raw waste	300 mg/L
Volume of filter (each)	16,000 ft^3 (453 m^3)
Depth of filter	7 ft (2.13 m)
Recirculation for filter 1	$= 1.5Q$
Recirculation for filter 2	$= 0.8Q$

Solution:

Step 1. Estimate BOD loading at the first stage

$$\text{Influent BOD } C_1 = 300 \text{ mg/L } (1 - 0.35) = 195 \text{ mg/L}$$

$$W = QC_1 = 2 \text{ Mgal/d} \times 195 \text{ mg/L} \times 8.34 \text{ lb/(Mgal} \cdot \text{mg/L)}$$

$$= 3252 \text{ lb/d}$$

Step 2. Calculate BOD removal efficiency of filter 1
Using Eqs. (7.131) and (7.130)

$$F = \frac{1 + r_1}{(1 + 0.1r_1)^2} = \frac{1 + 1.5}{(1 + 0.1 \times 1.5)^2}$$

$$= 1.89$$

$$E_1 = \frac{100}{1 + 0.0561\sqrt{W/VF}}$$

$$= \frac{100}{1 + 0.0561\sqrt{3252/(16 \times 1.89)}}$$

$$= 63.2\%$$

Step 3. Calculate effluent BOD concentration of filter 1

$$C_{1e} = 195 \text{ mg/L } (1 - 0.632)$$

$$= 71.8 \text{ mg/L}$$

Step 4. Calculate BOD removal efficiency of filter 2

$$F' = \frac{1 + 0.8}{(1 + 0.1 \times 0.8)^2} = 1.54$$

$$\text{Mass of influent BOD} = 2 \times 71.8 \times 8.34$$

$$= 1198 \text{ (lb/d)}$$

$$E_2 = \frac{100}{1 + 0.0561\sqrt{1198/(16 \times 1.54)}}$$

$$= 71.9\%$$

Step 5. Calculate effluent BOD concentration of filter 2

$$C_{2e} = 71.8 \text{ mg/L} \times (1 - 0.719)$$

$$= 20.1 \text{ mg/L}$$

Step 6. Calculate the overall efficiency

(a) Using Eq. (7.134)

$$E = 100 - 100[(1 - 0.35)(1 - 0.632)(1 - 0.719)]$$
$$= 93.3\%$$

(b)

$$E = (300 \text{ mg/L} - 20.1 \text{ mg/L})/(300 \text{ mg/L})$$
$$= 0.933$$
$$= 93.3\%$$

Note: At 20°C, no temperature correction is needed.

EXAMPLE 2: Determine the size of a two-stage trickling filter using the NRC equations. Assume both filters have the same efficiency of BOD_5 removal and the same recirculation ratio. Other conditions are as follows.

Wastewater temperature	20°C
Wastewater flow Q	3785 m³/d (1 Mgal/d)
Influent BOD_5	195 mg/L
Design effluent BOD_5	20 mg/L
Depth of each filter	2 m (6.6 ft)
Recirculations for filters 1 and 2, $r_1 = r_2$	1.8

Solution:

Step 1. Determine E_1 and E_2

$$\text{Overall efficiency} = (195 - 20) \text{ mg/L} \times 100\%/195 \text{ mg/L}$$
$$= 89.7\%$$

$$E_1 + E_2(1 - E_1) = 0.897$$
$$E_1 + E_2 - E_1E_2 = 0.897$$

Since $E_1 = E_2$

Thus $E_1^2 - 2E_1 + 0.897 = 0$

$$E_1 = \frac{2 \pm \sqrt{4 - 4 \times 0.897}}{2}$$

$$E_1 = 0.68 = 68\%$$

Step 2. Calculate the recirculation factor F

$$F = \frac{1 + r}{(1 + 0.1r)^2} = \frac{1 + 1.8}{(1 + 0.1 \times 1.8)^2}$$
$$= 2.01$$

Step 3. Calculate mass BOD load to the first-stage filter, W

$$\text{Influent BOD} = 195 \text{ mg/L} = 195 \text{ g/m}^3$$
$$= 0.195 \text{ kg/m}^3$$
$$W = QC_1 = 3785 \text{ m}^3\text{/d} \times 0.195 \text{ kg/m}^3$$
$$= 738 \text{ kg/d}$$

Step 4. Calculate the volume V of the first filter using Eq. (7.129)

$$E_1 = \frac{100}{1 + 0.532\sqrt{W/VF}}$$

From Step 1

$$68 = \frac{100}{1 + 0.532\sqrt{738/2.01V}}$$

$$\sqrt{367.2/V} = 0.885$$

$$V = 469 \text{ (m}^3\text{)}$$

Step 5. Calculate the diameter d of the first filter

$$\text{Area} = V/\text{depth} = 469 \text{ m}^3/2 \text{ m}$$
$$= 234.5 \text{ m}^2$$
$$d^2\pi/4 = 234.5 \text{ m}^2$$
$$d = 17.3 \text{ m}$$

Step 6. Calculate the mass BOD loading W' to the second-stage filter

$$W' = W(1 - E) = 738 \text{ kg/d } (1 - 0.68)$$
$$= 236.2 \text{ kg/d}$$

Step 7. Calculate the volume of the second filter using Eq. (7.132)

$$E_2 = \frac{100}{1 + \dfrac{0.532}{1 - E_1}\sqrt{\dfrac{W'}{VF}}}$$

$$68 = \frac{100}{1 + \dfrac{0.532}{1 - 0.68}\sqrt{\dfrac{236.2}{2.01V}}}$$

$$68 + 1225\sqrt{1/V} = 100$$

$$V = 1467 \text{ (m}^3\text{)}$$

Step 8. Calculate the diameter of the second-stage filter

$$A = 1467 \text{ m}^3/2 \text{ m}$$
$$= 733.5 \text{ m}^2$$
$$d^2 = 733.5 \text{ m}^2 \times 4/\pi$$
$$d = 30.6 \text{ m}$$

Step 9. Check the BOD loading rate to each filter

(a) First-stage filter, from Steps 3 and 4

$$\text{BOD loading rate} = 738 \text{ kg/d} \div 469 \text{ m}^3$$
$$= 1.57 \text{ kg/(m}^3 \cdot \text{d)}$$

Note: The rates are between 1.0 and 2.0 kg/(m^3 · d) (Table 7.14).

(b) Second-stage filter, from Steps 6 and 7

$$\text{BOD loading rate} = 236.2 \text{ kg/d} \div 1467 \text{ m}^3 = 0.161 \text{ kg/(m}^3 \cdot \text{d)}$$

Step 10. Check hydraulic loading rate to each filter

(a) For first-stage filter, recirculation ratio $r_1 = 1.8$ and from Step 5

$$\text{HLR} = \frac{(1 + 1.8) \times 3785 \text{ m}^3/\text{d}}{234.5 \text{ m}^2} = 45.2 \text{ m}^3/(\text{m}^2 \cdot \text{d})$$

(b) For second-stage filter, $r_2 = 1.8$ and from Step 8

$$\text{HLR} = \frac{(1 + 1.8) \times 3785 \text{ m}^3/\text{d}}{733.5 \text{ m}^2} = 14.4 \text{ m}^3/(\text{m}^2 \cdot \text{d})$$

Formulation for plastic media. Numerous investigations have been undertaken to predict the performance of plastic media in the trickling filter process. The Eckenfelder formula (1963) and the Germain (1965) applied Schultz formulation (1960) are the ones most commonly used to describe the performance of plastic media packed trickling filters.

Eckenfelder formula. Eckenfelder (1963) and Eckenfelder and Barnhart (1963) developed an exponential formula based on the rate of waste removal for a pseudo-first-order reaction, as below:

$$S_e/S_i = \exp[-KA_s^{1+m}D/q^n] \tag{7.136}$$

where S_e = effluent soluble BOD_5, mg/L
S_i = influent soluble BOD_5, mg/L
K = observed reaction rate constant, m/d or ft/d
A_s = specific surface area
= surface area/volume, m^2/m^3 or ft^2/ft^3
D = depth of media, m or ft
q = influent volumetric flow rate
= Q/A
Q = influent flow, m^3/d or ft^3/d
A = cross-sectional area of filter, m^2 or ft^2
m, n = empirical constants based on filter media

The mean time of contact t of wastewater with the filter media is related to the filter depth, the hydraulic loading rate, and the nature of the filter packing. The relationship is expressed as

$$\frac{t}{D} = \frac{C}{q^n} = \frac{C}{(Q/A)^n} \tag{7.137}$$

and

$$C \cong 1/D^m \tag{7.138}$$

where t = mean detention time
C, n = constants related to the specific surface and configuration of the packing

Other terms are as in Eq. (7.136).

Eq. (7.136) may be simplified to the following form

$$S_e/S_i = \exp[-kD/q^n] \tag{7.139}$$

where k is a new rate constant, per day.

EXAMPLE: Estimate the effluent soluble BOD_5 from a 6-m (20-ft) plastic packed trickling filter with a diameter of 18 m (60 ft). The influent flow is 4540 m^3/d (1.2 Mgal/d) and the influent BOD is 140 mg/L. Assume that the rate constant $k = 1.95$ per day and $n = 0.68$.

Solution:

Step 1. Calculate the area of the filter A

$$A = \pi(18 \text{ m}/2)^2$$
$$= 254 \text{ m}^2$$

Step 2. Calculate the hydraulic loading rate q

$$q = Q/A = 4540 \text{ m}^3/\text{d} \div 254 \text{ m}^2$$
$$= 17.9 \text{ m}^3/(\text{m}^2 \cdot \text{d})$$
$$= 19.1 \text{ (Mgal/d)/acre}$$

Step 3. Calculate the effluent soluble BOD using Eq. (7.139)

$$S_e = S_i \exp[-kD/q^n]$$
$$= 140 \text{ mg/L} \exp[-1.95 \times 6/(17.9)^{0.68}]$$
$$= 27 \text{ mg/L}$$

Germain formula. In 1966, Germain applied the Schultz (1960) formulation to a plastic media trickling filter and proposed a first-order equation as follows:

$$S_e/S_i = \exp[-k_{20}D/q^n] \tag{7.140}$$

where S_e = total BOD_5 of settled effluent, mg/L
S_i = total BOD_5 of wastewater applied to filter, mg/L
k_{20} = treatability constant corresponding to depth of filter at 20°C
D = depth of filter
q = hydraulic loading rate, $m^3/(m^2 \cdot d)$ or gal/min/ft^2
n = exponent constant of media, usually 0.5

The treatability constant at another depth of the filter must be corrected for depth when the k_{20} value is determined at one depth. The relationship proposed by Albertson and Davis (1984) is

$$k_2 = k_1(D_1/D_2)^x \tag{7.141}$$

where k_2 = treatability constant corresponding to depth D_2 of filter 2
k_1 = treatability constant corresponding to depth D_1 of filter 1
D_1 = depth of filter 1, ft
D_2 = depth of filter 2, ft
x = 0.5 for vertical and rock media filters
= 0.3 for crossflow plastic media filters

The values of k_1 and k_2 are a function of wastewater characteristics, the depth and configuration of the media, surface area of the filter, dosing cycle, and hydraulic loading rate. They are interdependent. Germain (1966) reported that the value of k for a plastic media filter 6.6 m (21.6 ft) deep, treating domestic wastewater, was 0.24 $(L/s)^n/m^2$ and that n is 0.5. This VFC media had a surface area of 88 m^2/m^3 (27 ft^2/ft^3). The ranges of k values in $(L/s)^{0.5}/m^2$ for a 6 m (20 ft) tower trickling filter packed with plastic media at 20°C are 0.18 to 0.27 for domestic wastewater, 0.16 to 0.22 for domestic and food waste, 0.054 to 0.14 for fruit-canning wastes, 0.081 to 0.14 for meat packing wastes, 0.054 to 0.11 for paper mill wastes, 0.095 to 0.14 for potato processing, and 0.054 to 0.19 for a refinery. Multiplying the value in $(L/s)^{0.5}/m^2$ by 0.37 obtains the value in $(gal/min)^{0.5}/ft^2$.

Distributor speed. The dosing rate of BOD_5 is very important for treatment efficiency. The instantaneous dosing rate is a function of the distributor speed or the on–off times for a fixed distributor. The rotational speed of a rotary distributor is expressed as follows

$$n = \frac{0.00044 q_t}{a(\text{DR})} \quad \text{(SI units)} \tag{7.142}$$

$$n = \frac{1.6 q_t}{a(\text{DR})} \quad \text{(US customary units)} \tag{7.143}$$

where n = rotational speed of distributor, rev/min
q_t = total applied hydraulic loading rate, $m^3/(m^2 \cdot d)$ or $(gal/min)/ft^2$
$= q + q_r$
q = influent wastewater hydraulic loading rate, $m^3/(m^2 \cdot d)$ or $(gal/min)/ft^2$
q_r = recycle flow hydraulic loading rate, $m^3/(m^2 \cdot d)$ or $(gal/min)/ft^2$
a = number of arms in rotary distributor assembly
DR = dosing rate, cm or in per pass of distributor arm

The required dosing rates in inches per pass for trickling filters is determined approximately by multiplying the organic loading rate, expressed in lb $BOD_5/1000\ ft^3$, by 0.12. For SI units, the dosing rate in cm/pass can be obtained by multiplying the loading rate in kg/m^3 by 0.30.

EXAMPLE: Design a 8.0 m (26 ft) deep plastic packed trickling filter to treat domestic wastewater and seasonal (summer) food-process wastewater with given conditions as follows:

Average year-round domestic flow Q	5590 m^3/d (1.48 Mgal/d)
Industrial wastewater flow	4160 m^3/d (1.1 Mgal/d)
Influent domestic total BOD_5	240 mg/L
Influent domestic plus industrial BOD_5	520 mg/L
Final effluent BOD_5	≤ 24 mg/L
Value of k at 26°C and at 6 m (from pilot plant study)	0.27 $(L/s)^{0.5}/m^2$ or 0.10 $(gal/min)^{0.5}/ft^2$
Sustained low temperature in summer	20°C
Sustained low temperature in winter	10°C

Solution:

Step 1. Compute k value at 20°C at 6 m

$$\begin{aligned} k_{20@6} &= k_{26@6}\theta^{T-26} \\ &= 0.27\ (L/s)^{0.5}/m^2 \times 1.035^{20-26} \\ &= 0.22\ (L/s)^{0.5}/m^2 \end{aligned}$$

Step 2. Correct the observed k_{20} value for depth of 8 m, using Eq. (7.141)

$$k_2 = k_1(D_1/D_2)^x$$

At 8 m depth

$$\begin{aligned} k_{20@8} &= k_{20@6}(6/8)^{0.5} \\ &= 0.22\ (L/s)^{0.5}/m^2 \times 0.866 \\ &= 0.19\ (L/s)^{0.5}/m^2 \end{aligned}$$

Step 3. Compute the summer total flow

$$\begin{aligned} Q &= (5590 + 4160)\ m^3/d \\ &= 9750\ m^3/d \times 1000\ L/m^3 \div 86{,}400\ s/d \\ &= 112.8\ L/s \end{aligned}$$

Step 4. Determine the surface area required for an 8 m deep filter for summer, using Eq. (7.140)

$$S_e/S_i = \exp[-k_{20}D/q^n]$$

Substituting Q/A for q in Eq. (7.140) and rearranging yields

$$\begin{aligned}\ln S_e/S_i &= -k_{20}D/(Q/A)^n \\ A &= Q[-(\ln S_e/S_i)/k_{20}D]^{1/n} \\ &= 112.8[-(\ln 24/520)/(0.19 \times 8)]^{1/0.5} \\ &= 462\ (\text{m}^2)\end{aligned}$$

Step 5. Similarly, determine the surface area required for an 8 m deep filter during the winter at 10°C to meet the effluent requirements

(a) Determine k_{10} for 6-m filter

$$\begin{aligned}k_{10@6} &= 0.27\ (\text{L/s})^{0.5}/\text{m}^2 \times 1.035^{10-26} \\ &= 0.156\ (\text{L/s})^{0.5}/\text{m}^2\end{aligned}$$

(b) Correct k_{10} value for 8-m filter

$$\begin{aligned}k_{10@8} &= 0.156[(\text{L/s})^{0.5}/\text{m}^2](6/8)^{0.5} \\ &= 0.135\ (\text{L/s})^{0.5}/\text{m}^3\end{aligned}$$

(c) Compute the area required (without industrial wastes in the winter)

$$\begin{aligned}Q &= 5590\ \text{m}^3/\text{d} \times 1000\ \text{L/m}^3 \div 86{,}400\ \text{s/d} \\ &= 64.7\ \text{L/s} \\ A &= 64.7\ [-(\ln 24/240)/(0.135 \times 8)]^2 \\ &= 294\ (\text{m}^2)\end{aligned}$$

Step 6. Select the design area required
The required design area is controlled by the summer condition. Because the area required for the summer condition is larger (see Steps 4 and 5), the design area required is 462 m^2.

Step 7. Compute the hydraulic loading rates, HLR or q_t

(a) For summer

$$\begin{aligned}\text{HLR} &= 9750\ \text{m}^3/\text{d} \div 462\ \text{m}^2 \\ &= 21.1\ \text{m}^3/(\text{m}^2 \cdot \text{d}) \\ &= 518\ \text{gal}/(\text{d} \cdot \text{ft}^2)\end{aligned}$$

(b) For winter

$$\begin{aligned}\text{HLR} &= 5590\ \text{m}^3/\text{d} \div 462\ \text{m}^2 \\ &= 12.1\ \text{m}^3/(\text{m}^2 \cdot \text{d}) \\ &= 297\ \text{gal}/(\text{d} \cdot \text{ft}^2)\end{aligned}$$

Step 8. Check the organic (BOD) loading rates

(a) For summer

$$\begin{aligned}\text{Volume of filter} &= 8\ \text{m} \times 462\ \text{m}^2 \\ &= 3696\ \text{m}^3\end{aligned}$$

$$\text{BOD loading} = \frac{9750 \text{ m}^3/\text{d} \times 520 \text{ g/m}^3}{3696 \text{ m}^3 \times 1000 \text{ g/kg}}$$
$$= 1.37 \text{ kg/(m}^3 \cdot \text{d)}$$
$$= 85.6 \text{ lb/(1000 ft}^3 \cdot \text{d)}$$

(b) For winter

$$\text{BOD loading} = \frac{5590 \text{ m}^3/\text{d} \times 240 \text{ g/m}^3}{3696 \text{ m}^3 \times 1000 \text{ g/kg}}$$
$$= 0.36 \text{ kg/(m}^3 \cdot \text{d)}$$
$$= 22.5 \text{ lb/(1000 ft}^3 \cdot \text{d)}$$

Step 9. Determine rotation speed of rotary distributor, using Eq. (7.142)
The required dosing rates in cm/pass of arm for the trickling filter can be approximately estimated by multiplying the BOD loading rate in kg/(m³ · d) by 0.30.

(a) For summer, dose rate DR is, from Step 8(a)

$$\text{DR} = 1.37 \times 0.3 = 0.41 \text{ (cm/pass)}$$

Using 2 arms in the rotary distributor, $a = 2$; from Step 7, $q_1 = 21.1 \text{ m}^3/(\text{m}^2 \cdot \text{d})$

$$n = \frac{0.00044q_t}{a(\text{DR})} = \frac{0.00044 \times 21.1}{2(0.41)}$$
$$= 0.0113 \text{ (rev/min)}$$

or 1 revolution every 88.5 min

(b) For winter
From Step 8(b)

$$\text{DR} = 0.36 \times 0.3 = 0.108$$

From Step 7(b)

$$q_t = 12.1 \text{ m}^3/(\text{m}^2 \cdot \text{d})$$

Thus

$$n = \frac{0.00044 \times 12.1}{2 \times 0.108}$$
$$= 0.025 \text{ (rev/min)}$$

or one revolution every 40 min

23 ROTATING BIOLOGICAL CONTACTOR

The use of plastic media to develop the rotating biological contactor (RBC) was commercialized in Germany in the late 1960s. The process is a fixed-film (attached growth) either aerobic or anaerobic biological treatment system for removal of carbonaceous and nitrogenous materials from domestic and industrial wastewaters. It was very popular during the late 1970s and early 1980s in the United States.

The RBC process can be used to modify and upgrade existing treatment systems as secondary or tertiary treatment. It has been successfully applied to all three steps of biological treatment, that is BOD_5 removal, nitrification, and denitrification.

The majority of RBC installations in the northern United States have been designed for removal of BOD_5 or ammonia nitrogen (NH_3-N), or both, from domestic wastewater. Currently in the United States there are approximately 600 installations for domestic wastewater treatment and more than 200 for industrial wastewater treatment. Over 1000 installations have been used in Europe, especially in Germany.

23.1 Hardware

The basic elements of the RBC system are media, shaft, bearings, drive, and cover. The RBC hardware consists of a large-diameter and closely spaced circular plastic media which is mounted on a horizontal shaft (Fig. 7.28). The shaft is supported by bearings and is slowly rotated by an electric motor. The plastic media are made of corrugated polyethylene or polystyrene material with various sizes and configurations and with various densities. The configuration designs are based on increasing stiffness and surface area, serving as spaces providing a tortuous wastewater flow path and stimulating air turbulence. As an exception, the Bio-Drum process consists of a drum filled with 38 mm plastic balls.

FIGURE 7.28 (a) RBC units. (b) A view of an 84-unit RBCs for nitrification.

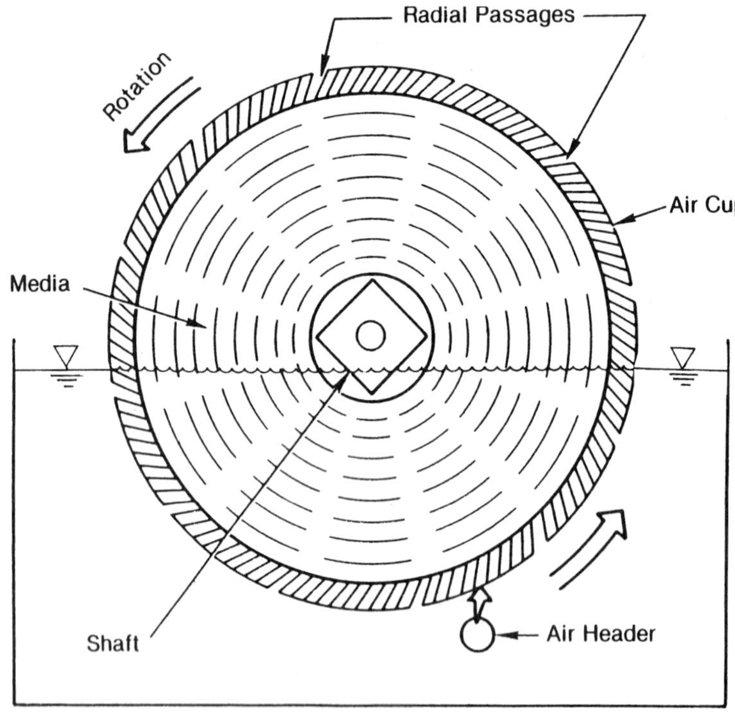

FIGURE 7.29 Schematic diagram of air-drive RBC.

The diameters of the media range from 4 ft (1.22 m) to 12 ft (3.66 m), depending on the treatment capacity. The shaft lengths vary from 5 ft (1.52 m) to 27 ft (8.23 m), depending on the size of the RBC unit (Banerji 1980). Commonly used RBC shafts are generally 25 to 27 ft (7.62 to 8.23 m) in length with a media diameter of 12 ft (3.66 m). Standard density media provide 100,000 ft^2 (9290 m^2) of surface area.

About 40 percent of the media is immersed in the wastewater at any time in trapezoidal, semicircular, or flat-bottom rectangular tanks with intermediate partitions in some cases. The shaft rotates at 1.5 to 1.7 rev/min for mechanical drive and 1.0 to 1.3 rev/min for air drive units (Fig. 7.29). The wastewater flows can be perpendicular to or parallel to the shafts.

23.2 Process Description

In domestic wastewater treatment, the RBC does not require seeding to establish the biological growth. After RBC system startup, microorganisms naturally present in the wastewater begin to adhere to the rotating media surface and propagate until, in about one week, the entire surface area will be covered with an approximately 1- to 4-mm thick layer of biological mass (biomass). The attached biomass contains about 50,000 to 100,000 mg/L suspended solids (Antonie 1978). The microorganisms in the film of biomass (biofilm) on the media remove biodegradable organic matter, nitrogen, and dissolved oxygen in wastewater and convert the pollutants to more benign components (biomass and gaseous by-products).

In the first-stage RBC biofilm, the most commonly observed filamentous bacterium is *Sphaerotilus*, *Beggiatoa* (a sulfur bacterium); *Cladothrix*, *Nocardia*, *Oscillatoria*, and filamentous fungus, *Fusarium*, are also found, though less frequently. Nonfilamentous organisms in the first stage are *Zocloea* and *Zooglear filipendula*, *Aerobacter aerogen*, *Escherichia coli*, unicellular rods, spirilla and spirochaetes, and unicellular algae. The final stages harbor mostly the same forms of biota in addition to *Athrobotrys* and *Streptomyces* reported by investigators (Pretorius 1971 Torpey *et al.* 1971, Pescod and Nair 1972, Sudo *et al.* 1977, Clark *et al.* 1978, Hitdlebaugh and Miller 1980, Hoag *et al.* 1980, Kinner *et al.* 1982).

In continuous rotation, the media carry films of wastewater into the air, which then trickle down through the liquid film surface into the bulk liquid (Figs. 7.30 and 7.31). They also provide the surface area necessary for absorbing oxygen from air. Intimate contact between the wastewater and the biomass creates a constantly moving surface area for the bacteria–substrate–oxygen interface. The renewed liquid layer (wastewater film) on the biomass is rich in DO. Both substrates and DO penetrate the liquid film through mixing and diffusion into the biofilm for biological oxidation. Excess DO in the wastewater film is mixed with bulk wastewater in the tank and results in aeration of the wastewater. The rotating media are used for supporting growth of microorganisms and for providing contact between the microorganisms, the substrates, and DO.

A group of RBC units is usually separated by baffling into stages to avoid short circuiting in the tank. There can be one shaft or more in a stage. The hydraulic detention time in each stage is relatively short, on the order of 20 minutes under normal loading. Each RBC stage tends to operate as a complete-mix reactor. The density and species of microbial population in each stage can vary significantly, depending on wastewater loading conditions. If RBCs are designed for secondary treatment, heavy microbial growth, shaggy and gray in color, develops. Good carbonaceous removal usually occurs in the first and second stages. The succeeding stages can be used for nitrification, if designed, which will exhibit nitrifier growth, brown in color.

The shearing process from rotation exerted on the biomass periodically sloughs off the excess biomass from the media into the wastewater stream. This sloughing action prevents bridging and clogging between adjacent media. The mixing action of the rotating media keeps the sloughed biomass in suspension from settling to the RBC tank. The sloughed-off solids flow from stage to stage and finally into the clarifier following the RBC units. Intermediate clarification and sludge recycle are not necessary for the RBC process.

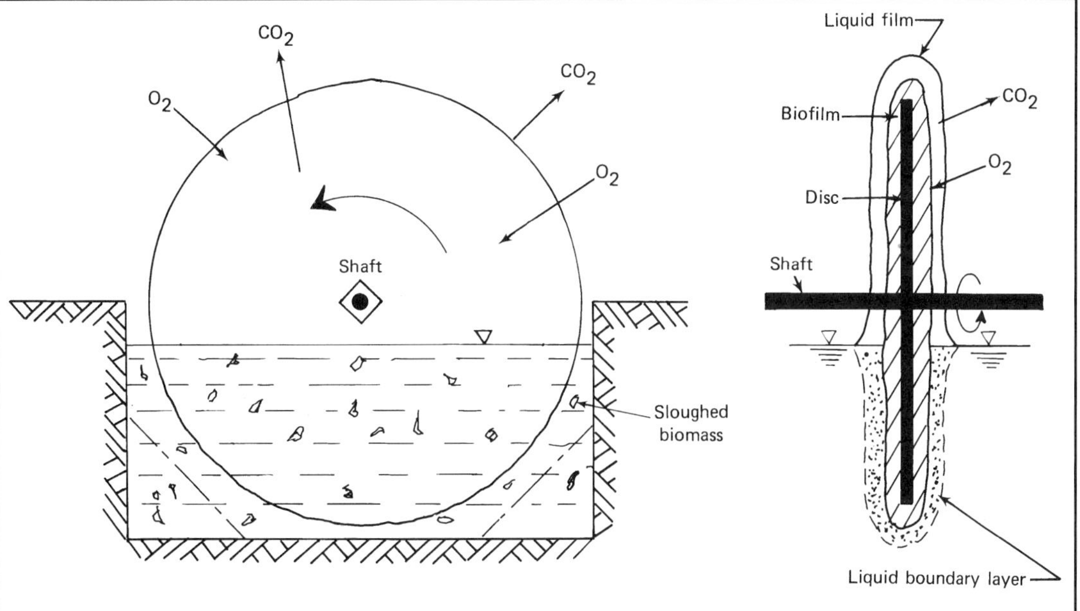

FIGURE 7.30 Mechanism for attached growth media in RBC system.

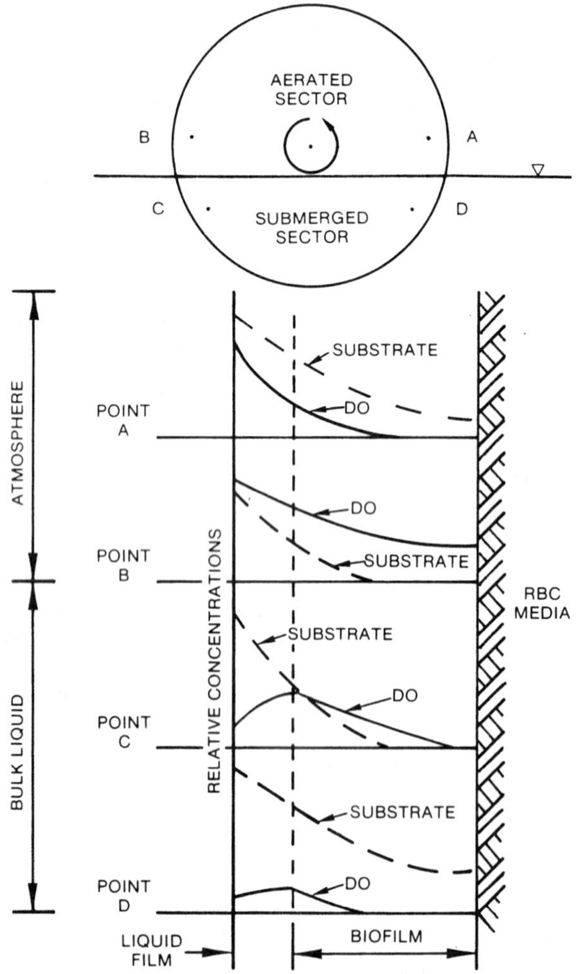

FIGURE 7.31 Relative concentration of substrate and dissolved oxygen for a loading condition and RBC rotation speed as a function of media location (*Source: USEPA 1984*).

In comparison with other biological treatment process, the RBC process differs from the trickling filter process by having substantially longer retention time and dynamic rather than stationary media; and from the activated-sludge process by having attached (fixed) biomass rather than a suspended culture and sludge recycle.

The patented Surfact process was created and developed by the Philadelphia Wastewater Department (Nelson and Guarino 1977). The process uses air-driven RBCs that are partially submerged in the aeration basins of an activated-sludge system. The RBCs provide fixed-film media for biological growth that are present in the recycled activated-sludge in the aeration tanks. The results are more active biological coating on the fixed-film media than that which is found on such media when used as a separate secondary treatment. Surfact combines the advantages of both RBC (fixed-film growth) and activated-sludge systems in a single tank, producing additional biological solids in the system. The results can be either a higher treatment efficiency at the same flow rate or the same level of treatment at a higher flow rate.

23.3 Advantages

Whether used in a small or a large municipal wastewater treatment plant, the RBC process has provided 85 percent or more of BOD_5 and ammonia nitrogen removal from sewage. In addition to high treatment capacity, it gives good sludge separation. It has the advantages of smaller operation and maintenance costs, and of simplicity in operation. It can be retrofitted easily to existing plants.

The RBC process is similar to the trickling filter (fixed-film biological reactor) and to the activated-sludge process (suspended culture in the mixed liquor). However, the RBC has advantages over the trickling filter process, such as longer contact time (8 to 10 times), relatively low land requirement (40 percent less), and less excavation, more surface area renewal for aeration, greater effectiveness for handling shock loadings, effective sloughing off of the excessive biomass, and without the nuisance of "filter flies." The RBC system may use less power than either the mechanical aeration (activated-sludge) or the trickling filter system of an equivalent capacity. It is anticipated that the RBC would exhibit a more consistent treatment efficiency during the winter months.

Over 95 percent of biological solids in the RBC units are attached to the media. These result in lower maintenance and power consumption of the RBC process over the activated-sludge process. In comparison with the activated-sludge process, there is no sludge or effluent recycle with a minimum process control requirement. Less skilled personnel are needed to operate the RBC process. A hydraulic surge or organic overloading will cause activated-sludge units to upset in process operation and thereby cause sludge bulking. This is not the case for the RBC process, which has a better process stability. Other advantages of the RBC process over the activated-sludge process are relatively low land requirement and less excavation, more flexibility for upgrading treatment facilities, less expense for nitrification, and better sludge settling without hindered settling.

23.4 Disadvantages

The problems of the first generation of RBC units are mainly caused by the failure of hardware and equipment. Significant effort has been made to correct these problems by the manufacturers. The second or third generation of RBC units may perform as designed.

Capital and installation costs for the RBC system, including an overhead structure, will be higher than that for an activated-sludge system of equal capacity. The land area requirement for the RBC process is about 30 to 40 percent about that for the activated-sludge process.

If low dissolved oxygen is coupled with available sulfide, the nuisance bacteria *Beggiatoa* may grow on the RBC media (Hitdlebaugh and Miller 1980). The white biomass phenomenon is caused by the *Beggiatoa* propagation. These problems can be corrected by addition of hydrogen peroxide. Some minor disadvantages related to the RBC process as well as other biological treatment processes include that a large land area may be required for a very large facility, additional cost for enclosures, possible foul odor problem, shock loading recovery, extremes of wastewater, pH, *Thiotrix* or *Beggiatoa* growth, overloading, and controversy regarding the technological nature of RBC.

23.5 Soluble BOD_5

It is accepted that soluble BOD_5 (SBOD) is the control parameter for RBC performance. The SBOD can be determined by the BOD test using filtrate from the wastewater samples. For RBC design purposes, it can also be estimated from historical data on total BOD (TBOD) and suspended solids. The SBOD is suspended BOD subtracted from TBOD. Suspended BOD is directly correlated with total suspended solids (TSS). Their relationships are expressed as

$$\text{SBOD} = \text{TBOD} - \text{suspended BOD} \tag{7.144}$$

$$\text{Suspended BOD} = c\,(\text{TSS}) \tag{7.145}$$

$$\text{SBOD} = \text{TBOD} - c\,(\text{TSS}) \tag{7.146}$$

where c = a coefficient
= 0.5 to 0.7 for domestic wastewater
= 0.5 for raw domestic wastewter (TSS > TBOD)
= 0.6 for raw wastewater (TSS ≅ TBOD)
(municipal with commercial and industrial wastewaters)
= 0.6 for primary effluents
= 0.5 for secondary effluents

EXAMPLE: Historical data of the primary effluent show that the average TBOD is 145 mg/L and TSS is 130 mg/L. What is the influent SBOD concentration that can be used for the design of an RBC system? RBC is used as the secondary treatment unit.

Solution:

For the primary effluent (RBC influent)

$$c = 0.6$$

Estimate SBOD concentration of RBC influent using Eq. (7.146)

$$\begin{aligned} \text{SBOD} &= \text{TBOD} - c\,(\text{TSS}) \\ &= 145 \text{ mg/L} - 0.6\,(130 \text{ mg/L}) \\ &= 67 \text{ mg/L} \end{aligned}$$

23.6 RBC Process Design

Many studies employ either the Monod growth kinetics or Michaelis–Menton enzyme kinetics for modeling organic and nitrogenous removals by the RBC process. Under steady-state conditions and a complete-mix chamber, based on the material mass balance expression for RBC stages, zero-, half-, and first-order kinetics, nonlinear second-order differential equations, conceptional models, and empirical models have been proposed. However, the usefulness of these models to predict RBC performance is not well established.

Factors affecting the RBC performance include wastewater temperature, influent substrate concentration, hydraulic retention time, tank volume to media surface area ratio, media rotational speed, and dissolved oxygen (Poon *et al.* 1979).

Manufacturers' empirical design approach. In the United States, the design of an RBC system for most municipal wastewaters is normally based on empirical curves developed by various manufacturers. Unfortunately the empirical approach is often the least rational in its methodologies and omits many important performance parameters. The designers or the users have relied heavily on the manufacturers for planning and design assistance.

The RBC design aspects for organic removal vary considerably among the various manufacturers. For the design loading, Autotrol (Envirex) (1979) and Clow (1980) use applied $SBOD_5$ while Lyco (1982) uses applied $TBOD_5$. All can predict the water quality at intermediate points in the treatment process and at the effluent. However, the predicted performances are quite different among manufacturers.

The manufacturers' design curves are developed on the basis of observed municipal RBC wastewater treatment performances. For example, Figs. 7.32 and 7.33 show organic removal design curves. The manufacturers' empirical curves define effluent concentration (soluble or total BOD_5 or NH_3-N) or percent removal in terms of applied hydraulic loading rate or organic loading rate in conjunction with influent substrate concentration.

In design, starting with a desired effluent concentration or percent removal (① in Figs. 7.32 and 7.33) and given influent concentration (②), the hydraulic loading (③) can be selected. The RBC total media-surface area required is calculated by dividing the design flow rate by the hydraulic loading derived.

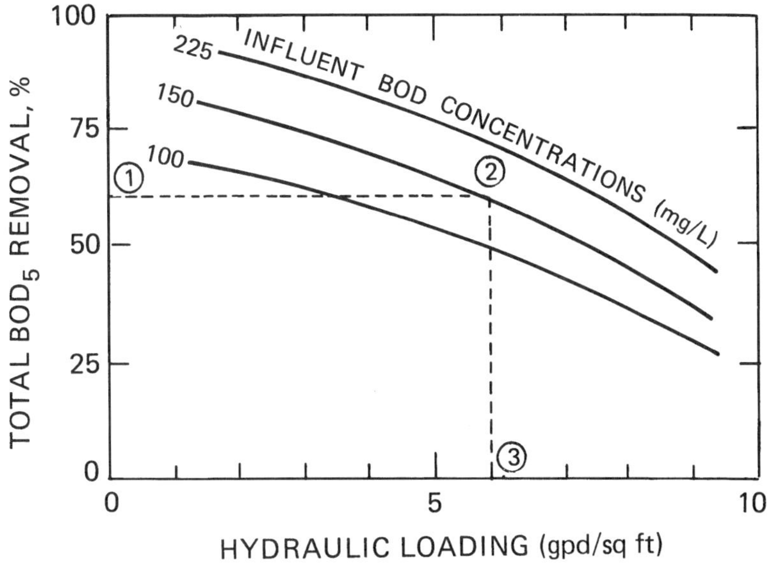

FIGURE 7.32 RBC process (mechanical drive) design curves for percent total BOD removal (temperature > 13°C, 55°F).

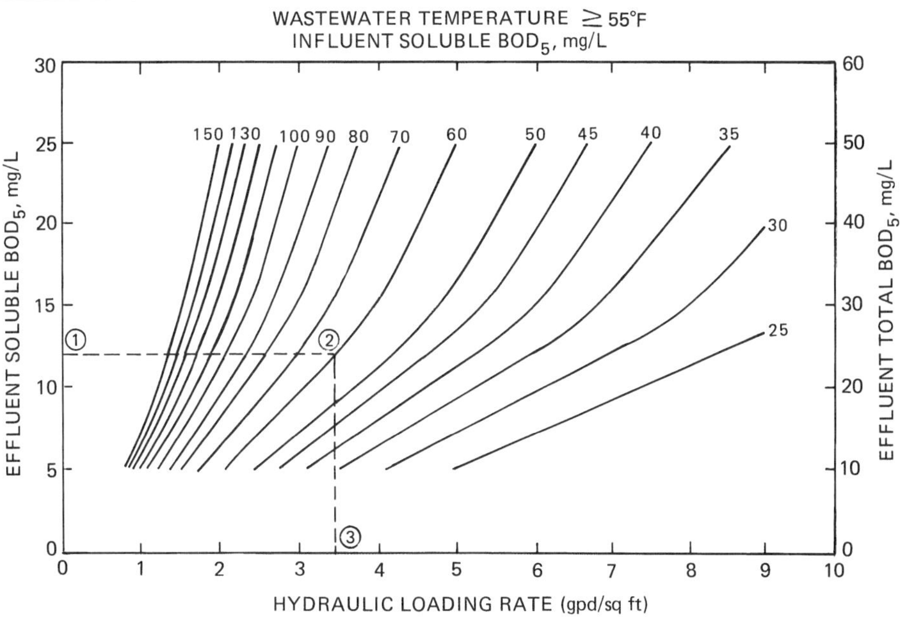

FIGURE 7.33 RBC (mechanical drive) process design curves for total and soluble BOD removal ($T > 13°C$, 55°F).

TABLE 7.15 RBC Stage Recommendations

Autotrol (1979)			Lyco (1980)	
Target effluent $SBOD_5$, mg/L	Min. number of stages	Clow (1980)	Target $TBOD_5$ reduction	Number of stages
> 25	1	At least 4 stages per train	up to 40%	1
15–25	1 or 2		35–65%	2
10–15	2 or 3		60–85%	3
< 10	3 or 4		80–95%	4
			Minimum of four stages recommended for combined BOD_5 and NH_3-N removal	

Work by Antonie (1978) demonstrated that, for any kinetics order higher than zero, overall BOD_5 removal for a given media surface area is enhanced by increasing the number of stages. Table 7.15 gives guidelines for stages recommended by three manufacturers.

One to four stages are recommended in most design manuals. A large number of stages is required to achieve higher removal efficiency. The relationship between design curves and stages has not been established and the reasons for recommendations are not generally given. The stage selection is an integral part of the design procedure and should be used intelligently. Based on substrate loading, the percent surface area for the first stage can be determined from the manufacturer's design curves (not included). Typical design information for RBC used for various treatment levels is summarized in Table 7.16.

To avoid organic overloading and possible growth of nuisance organisms, and shaft overloads, limitation of organic loadings has been recommended as follows (Autotrol 1979):

TABLE 7.16 Typical Design Parameters for Rotating Biological Contactors used for Various Treatment Levels

Parameter	Secondary	Combined nitrification	Separate nitrification
Hydraulic loading			
gal/(d · ft²)	2.0–40	0.75–2.0	1.0–2.5
m³/(m² · d)	0.081–0.163	0.030–0.081	0.041–0.102
Soluble BOD_5 loading			
lb/(1000 ft² · d)	0.75–2.0	0.5–1.5	0.1–0.3
Total (T) BOD_5 loading			
lb/(1000 ft² · d)	2.0–3.5	1.5–3.0	0.2–0.6
Maximum loading on first stage			
lb SBOD/(1000 ft² · d)	4–6	4–6	
lb TBOD/(1000 ft² · d)	8–12	8–12	
NH_3-N loading			
lb/(1000 ft² · d)		0.15–0.3	0.2–0.4
Hydraulic retention time θ, h	0.7–1.5	1.5–4	1.2–2.9
Effluent BOD_5, mg/L	15–30	7–15	7–15
Effluent NH_3-N, mg/L		< 2	1–2

Note: lb/(1000 ft² · d) × 0.00488 = kg/(m² · d).
Source: Metcalf and Eddy Inc. (1991).

Media	Organic ($SBOD_5$) loading, lb/(d · 1000 ft²)	
	Mechanical drive	Air drive
Standard density	4.0	5.0
High density	1.7	2.5

Note: lb/(1000 ft² · d) × 0.00488 = kg/(m² · d).

Then media distribution and choice of configurations can be made with engineering judgment or according to the manufacturer's design manual. Most RBC manufacturers have a standard ratio of tank volume to media surface area at 1.2 gal/ft² (48.9 L/m²).

The manufacturers universally contend that wastewater temperatures above 55°F (12.8°C) do not affect the organic removal rate. Below 55°F varying degrees of decreased biological activity rate are predicted by the manufacturers. Surface area correction factors for wastewater treatment below 55°F are shown in Table 7.17. Most manufacturers require wastewater temperature not to exceed 90°F (32.2°C) which can damage the RBC media. Fortunately, this is normally not the case for municipal wastewaters.

In summary, an RBC system can be designed on the basis of information on total required media surface area, temperature correction, staging, and organic loading limits.

RBC performance histories relative to empirical design have been examined. Operational data indicated that the prediction results from manufacturer's design curves were considerably different from actual results (Opatken 1982, Brenner *et al.* 1984). They usually gave an overestimate of attainable removals. The manufacturer's design curves should be used with caution whenever first-stage loading exceeds the recommended values.

EXAMPLE 1: At the first stage of the RBC system, what is the maximum hydraulic loading if the total BOD_5 for the influent and effluent are 150 and 60 mg/L, respectively?

Solution:

Step 1. Calculate percent TBOD removal

$$\% \text{ removal} = \frac{(150 - 60)\text{ mg/L} \times 100\%}{150 \text{ mg/L}}$$

$$= 60\%$$

TABLE 7.17 Factors for Required Surface Area Correction to Temperature

Temperature		For soluble BOD_5 removal	For ammonia-nitrogen removal
°F	°C		
64	17.8	1.0	0.71
62	16.7	1.0	0.77
60	15.5	1.0	0.82
58	14.4	1.0	0.89
56	13.3	1.0	0.98
55	12.8	1.0	1.00
54	12.2	1.03	1.02
52	11.1	1.09	1.14
50	10.0	1.15	1.28
48	8.9	1.21	1.40
46	7.8	1.27	1.63
45	7.2	1.31	1.75
44	6.7	1.34	1.85
42	5.6	1.42	2.32
40	4.4	1.50	–

Step 2. Determine hydraulic loading limit
From Fig. 7.32, starting with 60 percent TBOD removal at point 1, draw a horizontal line to the intersect with the influent TBOD concentration of 150 mg/L at point 2. Next, starting at point 2, draw a vertical line to intersect the hydraulic loading axis at point 3. Read the value at point 3 as 5.8 gal/(d · ft^2). This is the answer.

EXAMPLE 2: Compute the hydraulic and organic loading rates of an RBC system and of the first stage. Given:

Primary effluent (influent) flow = 1.5 Mgal/d

Influent total BOD_5 = 140 mg/L

Influent $SBOD_5$ = 75 mg/L

Area of each RBC shaft = 100,000 ft^2

Number of RBC shafts = 6

Number of trains = 2 (3 stages for each train)

No recirculation

Solution:

Step 1. Compute the overall system hydraulic loading (HL)

$$\text{Surface area of total system} = 6 \times 100{,}000 \text{ ft}^2$$
$$= 600{,}000 \text{ ft}^2$$
$$\text{System HL} = \frac{1{,}500{,}000 \text{ gal/d}}{600{,}000 \text{ ft}^2}$$
$$= 2.5 \text{ gal/(d} \cdot \text{ft}^2)$$

Note: It is in the range of 2 to 4 gal/(d · ft^2).

Step 2. Compute the organic loadings of the overall RBC system

$$\text{Total BOD loading} = \frac{1.5 \text{ Mgal/d} \times 140 \text{ mg/L} \times 8.34 \text{ lb/(Mgal} \cdot \text{mg/L)}}{600 \times 1000 \text{ ft}^2}$$
$$= 2.92 \text{ lb/(1000 ft}^2 \cdot \text{d)}$$
$$= 14.2 \text{ g/(m}^2 \cdot \text{d)}$$

Note: Design range = 2 to 3.5 lb/1000 ft^2.

$$\text{Soluble BOD loading} = \frac{1.5 \times 75 \times 8.34}{600 \times 1000}$$
$$= 1.56 \text{ lb/(1000 ft}^2 \cdot \text{d)}$$
$$= 7.6 \text{ g/(m}^2 \cdot \text{d)}$$

Note: Design range = 0.75 to 2.0 lb/1000 ft^2

Step 3. Compute the organic loadings of the first stage
Two units are used as the first stage.

$$\text{Total BOD loading} = \frac{1.5 \times 140 \times 8.34}{2 \times 100{,}000}$$
$$= 8.76 \text{ [lb/(1000 ft}^2 \cdot \text{d)]}$$

$$\text{Soluble BOD loading} = \frac{1.5 \times 75 \times 8.34}{200 \times 1000}$$

$$= 4.69 \ [\text{lb}/(1000 \ \text{ft}^2 \cdot \text{d})]$$

EXAMPLE 3: A municipal primary effluent with total and soluble BOD_5 of 140 and 60 mg/L, respectively, is to be treated with an RBC system. The RBC effluent TBOD and SBOD is to be 24 and 12 mg/L, respectively, or less. The temperature of wastewater is 60°F (above 55°F; no temperature correction needed). The design plant flow is 4.0 Mgal/d (15,140 m^3/d). The maximum hourly flow is 10.0 Mgal/d (37,850 m^3/d).

Solution:

Step 1. Determine the allowable hydraulic loading rate
From Fig. 7.33, at effluent

$$\text{SBOD} = 12 \ \text{mg/L}$$

$$\text{Hydraulic loading} = 3.44 \ \text{gal}/(\text{d} \cdot \text{ft}^2)$$

Step 2. Compute the required surface area of the RBC

$$\text{Area} = \frac{4{,}000{,}000 \ \text{gal/d}}{3.44 \ \text{gal}/(\text{d} \cdot \text{ft}^2)}$$

$$= 1{,}160{,}000 \ \text{ft}^2$$

Step 3. Check the design for overall organic loading rate

$$\text{SBOD loading} = \frac{4 \ \text{Mgal/d} \times 60 \ \text{mg/L} \times 8.34 \ \text{lb}/(\text{Mgal} \cdot \text{mg/L})}{1{,}160{,}000 \ \text{ft}^2}$$

$$= 1.73 \ \text{lb}/(1000 \ \text{ft}^2 \cdot \text{d})$$

It is under 2.0 lb/(1000 ft^2 · d), therefore OK.

Step 4. Size the first stage
The size factor for the area required for the first stage is 0.2 × SBOD loading. For this case, the size factor is 0.346 (0.2 × 1.73, or 34.6 percent of total surface area). This value can be obtained from the manufacturer's curves.

$$\text{Surface area for first stage} = 0.346 \times 1{,}160{,}000 \ \text{ft}^2$$

$$\cong 400{,}000 \ \text{ft}^2$$

Step 5. Determine configuration

(a) Choice standard media assemblies each having 100,000 ft^2 for the first stage

$$\text{Unit for first stage} = 400{,}000 \ \text{ft}^2/100{,}000 \ \text{ft}^2$$

$$= 4$$

Use 4 standard media; each is installed at the first stage of 4 trains.

(b) Choice high-density media assemblies have 150,000 ft^2 for the following stages

$$\text{High-density media area} = (1{,}160{,}000 - 400{,}000) \ \text{ft}^2$$

$$= 760{,}000 \ \text{ft}^2$$

$$\text{No. of high-density assemblies} = 760{,}000 \ \text{ft}^2/150{,}000 \ \text{ft}^2$$

$$= 5.1$$

Alternative 1: Use 4 standard media at stages 2 and 3.
Alternative 2: Use 4 high-density media at stages 2 and 4 standard density media at stage 3
Alternative 1 is less costly.
Alternative 2 has an advantage in preventing shock loadings.

(c) Possible configurations are designed using 4 trains with 3 stages each, as shown below:

(1) SSSS + SSSS + SSSS

or

(2) SSSS + HHHH + SSSS

Note: S = standard density media; H = high density media.

Step 6. Determine the required surface area for the secondary settling tank

(a) Compute surface area based on average (design) flow using an overflow rate of 600 gal/(d · ft²)

$$\text{Area} = 4{,}000{,}000\ (\text{gal/d})/600\ (\text{gal/(d} \cdot \text{ft}^2)$$
$$= 6670\ \text{ft}^2$$

(b) Compute surface area based on peak flow using an overflow rate of 1200 gal/(d · ft²)

$$\text{Area} = 10{,}000{,}000\ (\text{gal/d})/1200\ (\text{gal/(d} \cdot \text{ft}^2))$$
$$= 8330\ \text{ft}^2$$

(c) On the basis of the above calculations, the size of the secondary settling tank is controlled by the maximum hourly flow rate.

24 DUAL BIOLOGICAL TREATMENT

Dual biological treatment processes use a fixed film reactor in series with a suspended growth reactor. Dual processes include such as activated biofilter, trickling filter-solids contact, roughing filter-activated-sludge, biofilter-activated-sludge, trickling filter-activated-sludge, roughing filter-RBC, roughing filter-aerated lagoon, roughing filter-facultative lagoon, roughing filter-pure oxygen activated-sludge. Descriptions of these dual processes may be found elsewhere (Metcalf and Eddy, Inc. 1991, WEF and ASCE 1991a).

25 STABILIZATION PONDS

The terms stabilization pond, lagoon, oxidation pond have been used synonymously. This is a relatively shallow earthen basin used as secondary or tertiary wastewater treatment, especially in rural areas. Stabilization ponds have been employed for treatment of wastewater for over 300 years. Ponds have become popular in small communities because of their low construction and operating costs. They are used to treat a variety of wastewaters from domestic and industrial wastes and functions under a wide range of weather conditions. Ponds can be employed alone or in combination with other treatment processes.

Stabilization ponds can be classified as facultative (aerobic–anaerobic), aerated, aerobic, and anaerobic ponds according to the dominant type of biological activity or reactions occurring in the pond. Other classifications can be based on the types of the influent (untreated, screened, settled effluent, or secondary (activated-sludge) effluent), the duration of discharge (nonexistent, intermittent, and continuous), and the method of oxygenation (photosynthesis, atmospheric surface reaeration, and mechanical aeration). Aerated ponds (aerated lagoons) have been discussed previously.

25.1 Facultative Ponds

The most common type of stabilization pond is the facultative pond. It is also called the wastewater lagoon. Facultative ponds are usually 1.2 to 2.5 m (4 to 8 ft) in depth, with an aerobic layer overlying an anaerobic layer, often containing sludge deposits The detention time is usually 5 to 30 days (USEPA 1983b).

The ponds commonly receive no more pretreatment than screening (few with primary effluent). It can also be used to follow trickling filters, aerated ponds, or anaerobic ponds. They then store grit and heavy solids in the first or primary ponds to form an anaerobic layer. The system is a symbiotic relationship between heterotrophic bacteria and algae.

Bacteria found in an aerobic zone of a stabilization pond are primarily of the same type as those found in an activated-sludge process or in the zoogleal mass of a trickling filter. The most frequently isolated bacteria include *Beggiatoa alba, Sphaerotilus natans, Achromobacter, Alcaligenes, Flavobacterium, Pseudomonas*, and *Zoogloea* spp. (Lynch and Poole 1979). These organisms decompose the organic materials present in the aerobic zone into oxidized end products.

Organic matter in wastewater is decomposed by bacterial activities, including both aerobic and anaerobic, which release phophorus, nitrogen, and carbon dioxide. Oxygen in the aerobic layer is supplied by surface reaeration and algal photosynthesis. Algae consume nutrient and carbon dioxide produced by bacteria and release oxygen to water. DO is used by bacteria, thus forming a symbiotic cycle. In the pond bottom, anaerobic breakdown of the solids in the sludge layer produces dissolved organics and gases such as methane, carbon dioxide, and hydrogen sulfide. Between the aerobic and anaerobic zones, there is a zone called the facultative zone. Temperature is a major factor for the biological symbiotic activities.

Organic loading rates on stabilization ponds are expressed in terms of kg BOD_5 applied per hectare of surface area per day (lb BOD/(acre · d)), or sometimes as BOD equivalent population per unit area. Typical organic loading rates are 22 to 67 kg BOD/(ha · d) (20 to 60 lb BOD/(acre · d)). Typical detention times range from 25 to 180 days. Typical dimensions are 1.2 to 2.5 m (4 to 8 ft) deep with 4 to 60 ha (10 to 150 acres) of surface area (USEPA 1983b).

Facultative ponds are commonly designed to reduce BOD to about 30 mg/L; but, in practice, it ranges from 30 to 40 mg/L or more due to algae. Volatile organic removal is between 77 and 96 percent. Nitrogen removal achieves 40 to 95 percent. Less phophorus removal is observed, being less than 40 percent. Effluent TSS levels range from 40 to 100 mg/L, mainly contributed by algae (WEF and ASCE 1991b). The presence of algae in pond effluent is one of the most serious performance problems associated with facultative ponds. The ponds are effective in removal of fecal coliform (FC) due to their dying off. In most cases effluent FC densities are less than the limit of 200 FC/100 ml.

Process design. Several design formulas with operational data were presented in the design manual (USEPA 1974b). Calculations of the size of a facultative pond are also illustrated for the areal loading rate procedure, Gloyna equation, Marais–Shaw equation, plug-flow model, and Wehner–Wilhelm equation. The design for partial-mix aerated lagoon is given elsewhere (WPCF 1990). In this section, design methods of the areal loading rate method and the Wehner and Wilhelm model are discussed.

Areal loading rate method. The design procedure is usually based on organic loading rate and hydraulic residence time. Several empirical and rational models for the design of facultative ponds have been proposed. Several proposed design methods have been discussed elsewhere (USEPA 1983b). The areal loading rate is the most conservative design method and can be adapted to specific standards. The recommended BOD loading rates based on average winter air temperature are given in Table 7.18 (USEPA 1974b).

TABLE 7.18 Recommended BOD_5 Loading Rates for Facultative Ponds

Average winter air temperature, °C	Water depth		BOD loading rate	
	m	ft	kg/(ha · d)	lb/(acre · d)
< 0	1.5–2.1	5–7	11–22	10–20
0–15 (59°F)	1.2–1.8	4–6	22–45	20–40
> 15	1.1	3.6	45–90	40–80

Source: USEPA (1974b).

The surface area required for the facultative pond is determined by dividing the organic (BOD) load by the appropriate BOD loading rate listed in Table 7.18 or from the specific state standards. It can be expressed as the following equation (as used for other processes):

$$A = \frac{Q(\text{BOD})}{(\text{LR})(1000)} \quad \text{(SI units)} \tag{7.147}$$

$$A = \frac{Q(\text{BOD})(8.34)}{(\text{LR})} \quad \text{(US customary units)} \tag{7.148}$$

where A = area required for facultative pond, ha or acre
BOD = BOD concentration in influent, mg/L
Q = flow of influent, m^3/d or Mgal/d
LR = BOD loading rate for average winter air temperature, kg/(ha · d) or lb/(acre · d)
1000 = conversion factor, 1000 g = 1 kg
8.34 = conversion factor, lb/(Mgal · mg/L) = 8.34 lb

The BOD loading rate at the first cell in a series of cells (ponds) should not exceed 100 kg/(ha · d) (90 lb/(acre · d)) for warm climates, average winter air temperature greater than 15°C (59°F), and 40 kg/(ha · d) (36 lb/(acre · d)) for average winter air temperature less than 0°C (32°F).

EXAMPLE 1: The design flow of facultative ponds for a small town is 1100 m^3/d (0.29 Mgal/d). The expected influent BOD is 210 mg/L. The average winter temperature is 10°C (50°F). Design a three-cell system with organic loading less than 80 kg/(ha · d) (72 lb/(acre · d)) in the primary cell. Also estimate the hydraulic detention time when average sludge depth is 0.5 m and there are seepage and evaporation losses of 2.0 mm of water per day.

Solution:

Step 1. Determine area required for the total ponds
From Table 7.18, choose BOD loading rate
LR = 38 kg/(ha · d) (33.9 lb/(acre · d)), because mean air temperature is 10°C.
Using Eq. (7.147)

$$A = \frac{Q(\text{BOD})}{(\text{LR})(1000)} = \frac{1100\ \text{m}^3/\text{d} \times 210\ \text{g/m}^3}{38\ \text{kg/(ha} \cdot \text{d)} \times 1000\ \text{g/kg}}$$

$$= 6.08\ \text{ha (16.6 areas)}$$

$$= 60{,}800\ \text{m}^2$$

Step 2. Determine the area required for the primary cell (use LR = 80 kg/(ha · d))

$$\text{Area} = \frac{1100 \times 210}{80 \times 1000}$$

$$= 2.88\ \text{(ha)}$$

Step 3. Sizing for 3 cells

Referring to Table 7.18, choose water depth of 1.5 m (5 ft) for all cells.

(a) Primary cell:

$$\text{Area} = 2.88 \text{ ha} = 28{,}800 \text{ m}^2$$

Use 100 m (328 ft) wide, 288 m (945 ft) long and 1.5 m (5 ft) deep pond.

(b) Two other cells:

$$\begin{aligned}\text{Area for each} &= (60{,}800 - 28{,}800) \text{ m}^2/2 \\ &= 16{,}000 \text{ m}^2\end{aligned}$$

Choose 144 m in length, then the width = 111 m

The pond arrangements are as shown below.

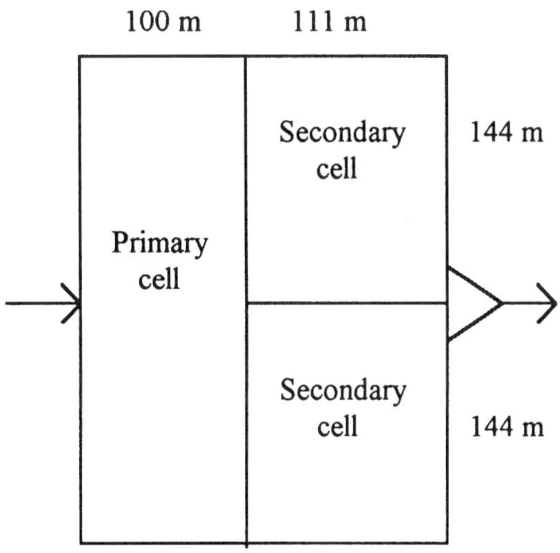

Step 4. Estimate the hydraulic retention time

(a) Calculate the storage volume V (when average sludge depth = 0.5 m)

$$\begin{aligned}V &= (1.5 \text{ m} - 0.5 \text{ m}) \times 60{,}800 \text{ m}^2 \\ &= 60{,}800 \text{ m}^3\end{aligned}$$

(b) Calculate the water loss V'

$$\begin{aligned}V' &= 0.002 \text{ m/d} \times 60{,}800 \text{ m}^2 \\ &= 122 \text{ m}^3\text{/d}\end{aligned}$$

(c) Calculate the HRT storage time

$$\begin{aligned}\text{HRT} &= \frac{V}{Q - V'} = \frac{60{,}800 \text{ m}^2}{1100 \text{ m}^3\text{/d} - 122 \text{ m}^3\text{/d}} \\ &= 62 \text{ days}\end{aligned}$$

EXAMPLE 2: A lagoon system contains two cells in series. Each cell is 110 m × 220 m (360 ft × 720 ft) in size with a maximum operating depth of 1.64 m (5.4 ft) and a minimum operating depth of 0.55 m. (1.8 ft). The wastewater flow is 950 m^3/d (0.25 MGD) with an average BOD_5 of 105 mg/L. Determine the organic loading rate and the detention time of the lagoon. The temperature ranges 10°C to 15°C.

Solution:

Step 1. Calculate the BOD_5 loading rate

$$\begin{aligned} BOD_5 \text{ mass} &= 950 \text{ m}^3/\text{d} \times 105 \text{ mg/L} \times 1000 \text{ L/m}^3 \times 10^{-6} \text{ kg/mg} \\ &= 99.75 \text{ kg/d} \end{aligned}$$

$$\begin{aligned} \text{Area of the primary cell} &= 110 \text{ m} \times 220 \text{ m} \\ &= 24{,}200 \text{ m}^2 \times 10^{-4} \text{ ha/m}^2 \\ &= 2.42 \text{ ha} \end{aligned}$$

$$\begin{aligned} BOD_5 \text{ loading rate} &= (99.75 \text{ kg/d})/(2.42 \text{ ha}) \\ &= 41.2 \text{ kg/(ha} \cdot \text{d)} \end{aligned}$$

This loading rate is acceptable by checking with Table 7.18, at temperature 0°C to 15°C, the recommended BOD_5 loading rate ranges 22 to 45 kg/(ha · d).

Step 2. Calculate the detention time

$$\begin{aligned} \text{The detention time} &= 24{,}000 \text{ m}^2/\text{cell} \times (1.64 \text{ m} - 0.55 \text{ m}) \times 2 \text{ cells} \div 950 \text{ m}^3/\text{d} \\ &= 55.5 \text{ days} \end{aligned}$$

Wehner and Wilhelm equation. Wehner and Wilhelm (1958) used the first-order substrate removal rate equation for a reactor with an arbitrary flow-through pattern, which is between a plug-flow pattern and a complete-mix pattern. Their proposed equation is

$$\frac{C}{C_0} = \frac{4a \exp(1/2D)}{(1 + a)^2 \exp(a/2D) - (1 - a)^2 \exp(-a/2D)} \tag{7.149}$$

where C = effluent substrate concentration, mg/L
C_0 = influent substrate concentration, mg/L
a = $\sqrt{1 + 4ktD}$
k = first-order reaction constant, 1/h
t = detention time, h
D = dispersion factor = H/uL
H = axial dispersion coefficient, m^2/h or ft^2/h
u = fluid velocity, m/h or ft/h
L = length of travel path of a typical particle, m or ft

The Wehner–Wilhelm equation for arbitrary flow was proposed by Thirumurthi (1969) as a method of designing facultative pond systems. Thirumurthi developed a graph, shown in Fig. 7.34, to facilitate the use of the equation. In Fig. 7.34, the term kt is plotted against the percent BOD_5 remaining (C/C_0) in the effluent for dispersion factors varying from zero for an ideal plug-flow reactor to infinity for a complete-mix reactor. Dispersion factors for stabilization ponds range from 0.1 to 2.0, with most values not exceeding 1.0 due to the mixing requirement. Typical values for the overall first-order BOD_5 removal rate constant k vary from 0.05 to 1.0 per day, depending on the operating and hydraulic characteristics of the pond. The use of the arbitrary flow equation is complicated in selecting k and D values. A value of 0.15 per day is recommended for k_{20} (USEPA 1984). Temperature adjustment for k can be determined by

$$k_T = k_{20}(1.09)^{T-20} \tag{7.150}$$

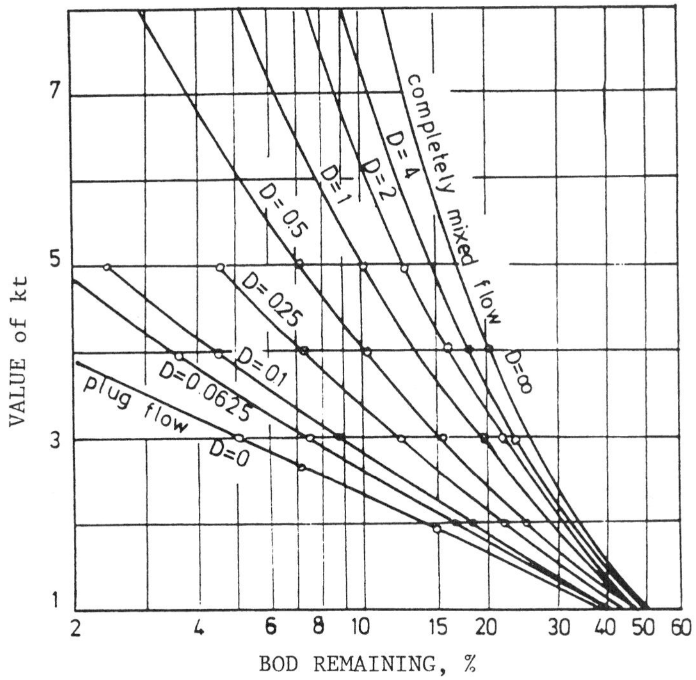

FIGURE 7.34 Relationship between *kt* values and percent BOD remaining for various dispersions factors, by the Wehner–Wilhelm equation.

where k_T = reaction rate at minimum operating water temperature *T*, per day
k_{20} = reaction rate at 20°C, per day
T = minimum operating temperature, °C

EXAMPLE: Design a facultative pond system using the Wehner–Wilhelm model and Thirumurthi application with the following given data.

Design flow rate Q = 1100 m^3/d (0.29 Mgal/d)

Influent TSS = 220 mg/L

Influent BOD_5 = 210 mg/L

Effluent BOD_5 = 30 mg/L

Overall first-order *k* at 20°C = 0.22 per day

Pond dispersion factor *D* = 0.5

Water temperature at critical period = 1°C

Pond depth = 2 m (6.6 ft)

Effective depth = 1.5 m (5 ft)

Solution:

Step 1. Calculate the percentage of BOD remaining in the effluent

$$C/C_0 = 30 \text{ mg/L} \times 100\%/(210 \text{ mg/L})$$
$$= 14.3\%$$

Step 2. Calculate the temperature adjustment for k_{20} using Eq. (7.150)

$$k_T = k_{20}(1.09)^{T-20} = 0.22\,(1.09)^{1-20}$$
$$= 0.043 \text{ (per day)}$$

Step 3. Determine the value of $k_T t$ from Fig. 7.34

$$\text{At } C/C_0 = 14.3\% \text{ and } D = 0.5$$
$$k_T t = 3.1$$

Step 4. Calculate the detention time for the critical period of year

$$t = 3.1/(0.043\text{ d}^{-1})$$
$$= 72 \text{ days}$$

Step 5. Determine the pond volume and surface area requirements

$$\text{Volume} = Qt = 1100\text{ m}^3\text{/d} \times 72\text{ days}$$
$$= 79{,}200\text{ m}^3$$
$$\text{Area} = \text{volume/effective depth} = 79{,}200\text{ m}^3/1.5\text{ m}$$
$$= 52{,}800\text{ m}^2$$
$$= 5.28\text{ ha}$$
$$= 13.0\text{ acres}$$

Step 6. Check BOD_5 loading rate

$$\text{Loading} = \frac{1100\text{ m}^3\text{/d} \times 210\text{ g/m}^3}{5.28\text{ ha} \times 1000\text{ g/kg}}$$
$$= 43.8\text{ kg/(ha} \cdot \text{d)}$$
$$= 40.0\text{ lb/(acre} \cdot \text{d)}$$

Step 7. Determine the power requirement for the surface aerators
Assume that the oxygen transfer capacity of the aerators is twice the value of the BOD applied per day and that a typical aerator has a transfer capacity of 22 kg O_2/(hp · d)

$$O_2\text{ required} = 2 \times 1100\text{ m}^3\text{/d} \times 210(\text{g/m}^3)/(1000\text{ g/kg})$$
$$= 462\text{ kg/d}$$
$$\text{Power} = (462\text{ kg/d})/(22\text{ kg/hp} \cdot \text{d})$$
$$= 21.0\text{ hp}$$
$$= 15.7\text{ kW}$$

Use seven 3-hp units

Step 8: Check the power input to determine the degree of mixing

$$\text{Power input} = 15.7\text{ kW}/79.2 \times 1000\text{ m}^3$$
$$= 0.20\text{ kW}/1000\text{ m}^3$$
$$= 0.0076\text{ hp}/1000\text{ ft}^3$$

Note: In practice, the minimum power input is about 28 to 54 kW/1000 m^3 (1.06 to 2.05 hp/1000 ft^3) for mixing (Metcalf and Eddy, Inc. 1991).

Tertiary ponds. Tertiary ponds, also called polishing or maturation ponds, serve as the third-stage process for effluent from activated-sludge or trickling filter secondary clarifier effluent. They are also used as a second stage following facultative ponds and are aerobic throughout their depth.

The water depth of the tertiary pond is usually 1 to 1.5 m (3 to 4.5 ft). BOD loading rates are less than 17 kg/(ha · d) (15 lb/(acre · d)). Detention times are relatively short and vary from 4 to 15 days.

25.2 Aerobic Ponds

Aerobic ponds, also referred to as high-rate aerobic ponds, are relatively shallow with usual depths ranging from 0.3 to 0.6 m (1 to 2 ft), allowing light to penetrate the full depth. The ponds maintain DO throughout their entire depth. Mixing is often provided to expose all algae to sunlight and to prevent deposition and subsequent anaerobic conditions. DO is supplied by algal photosynthesis and surface reaeration. Aerobic bacteria utilize and stabilize the organic waste. The HRT in the ponds is short—3 to 5 days.

The use of aerobic ponds is limited to warm and sunny climates, especially where a high degree of BOD removal is required but the land area is not limited. However, very little coliform die-off occurs.

25.3 Anaerobic Ponds

Anaerobic ponds are usually deep and are subjected to heavy organic loading. There is no aerobic zone in an anaerobic pond. The depths of anaerobic ponds usually range from 2.5 to 5 m (8 to 16 ft). The detention times are 20 to 50 days (USEPA 1983b).

Anaerobic bacteria decompose organic matter to carbon dioxide and methane. The principal biological reactions are acid formation and methane fermentation. These processes are similar to those of anaerobic digestion of sludge. Odorous compounds, such as organic acids and hydrogen sulfide, are also produced.

Anaerobic ponds are usually used to treat strong industrial and agricultural wastes. They have been used as pretreatment to facultative or aerobic ponds for strong industrial wastewaters and for rural communities with high organic load, such as food processing. They are not in wide application for municipal wastewater treatment.

The advantages of anaerobic ponds compared with an aerobic treatment process are low production of waste biological sludge and no need for aeration equipment. An important disadvantage of the anaerobic pond is the generation of odorous compounds. Its incomplete stabilization of waste requires a second-stage aerobic process. It requires a relatively high temperature for anaerobic decomposition of wastes.

Normal operation, to achieve a BOD removal efficiency of at least 75 percent, entails a loading rate of 0.32 kg BOD/(m^3 · d) (20 lb/(1000 ft^3 · d)), a minimum detention time of four days, and a minimum operating temperature of 24°C (75°F) (Hammer 1986).

26 SECONDARY CLARIFIER

The secondary settling tank is an integral part of both suspended- and attached-growth biological treatment processes. It is vital to the operation and performance of the activated-sludge process. The clarifier separates MLSS from the treated wastewater prior to discharge. It thickens the MLSS before their turn in the aeration tank or their wasting.

Secondary clarifiers are generally classified as Type 3 settling. Types 1 and 2 may also occur. The sloughed-off solids are commonly well-oxidized particles that settle readily. On the other hand, greater viability of activated-sludge results in lighter, more buoyant flocs, with reduced settling velocities. In addition, this is the result of microbial production of gas bubbles that buoy up the tiny biological clusters.

26.1 Basin Sizing for Attached-Growth Biological Treatment Effluent

The Ten States Standards (GLUMRB 1996) recommends that surface overflow rate for the settling tank following trickling filter should not exceed 1200 gal/(d · ft^2) (49 m^3/(m^2 · d)) based on design peak hourly flow. In practice, typical overflow rates are 600 gal/(d · ft^2) (24.4 m^3/(m^2 · d)) for plants smaller than 1 Mgal/d (3785 m^3/d) and 800 gal/(d · ft^2) (33 m^3/(m^2 · d)) for larger plants. The minimum side water depth is 10 ft (3 m) with greater depths for larger diameter basins. The retention time in the secondary settling tank is in the range of 2 to 3 h. The maximum recommended weir loading rate is 20,000 gal/(d · ft) (250 m^3/(d · m)) for plants equal or less than 1 Mgal/d and 30,000 gal/(d · ft) (373 m^3/(d · m)) for plants greater than 1 Mgal/d (3785 m^3/d).

Secondary clarifiers following trickling filters are usually sized on the basis of the hydraulic loading rate. Typical design parameters for secondary clarifiers are listed in Table 7.19 (USEPA 1975a). Referring to the hydraulic loading rates from Table 7.19, sizing should be calculated for both peak and design average flow conditions and the largest value calculated should be used. At the selected hydraulic loading rates, settled effluent quality is limited primarily by the performance of the biological reactor, not of the settling basins. Solids loading limits are not involved in the size of clarifiers following trickling filters. Where further treatment follows the clarifier, cost optimization may be considered in sizing the settling basins.

EXAMPLE 1: Calculate the residual chlorine concentration that must be maintained to achieve a fecal coliform (FC) density equal or less than 200/100 mL, in chlorinated secondary effluent. The influent of the chlorine contact tank contains 4 × 10^6 FC/100 mL. The contact time is 30 min.

Step 1. Calculate the BOD_5 loading to the secondary clarifiers

$$\begin{aligned} BOD_5 &= (1 - 0.32) - (200 \text{ mg/L}) \times 1.0 \text{ Mgal/d} \times 8.34 \text{ lb/[(Mgal} \cdot \text{mg/L)]} \\ &= 1134 \text{ lb/d } (= 515 \text{ kg/d}) \end{aligned}$$

Step 2. Estimate solids (S_2) production of the secondary clarifiers

Note: The biological solids generated in secondary treatment range from 0.4 to 0.5 lb per lb (kg/kg) BOD_5 applied. Use 0.45 lb/lb in this example.

$$\begin{aligned} S_2 &= 0.45 \text{ lb/lb} \times 1134 \text{ lb/d} \\ &= 510 \text{ lb/d } (= 232 \text{ kg/d}) \end{aligned}$$

TABLE 7.19 Typical Design Parameters for Secondary Sedimentation Tanks

	Hydraulic loading gal/(d · ft^2)†		Solids loading,* lb solids/(d · ft^2)†		
Type of treatment	Average	Peak	Average	Peak	Depth, ft
Settling following tracking filtration	400–600	1000–2000	–	–	10–12
Settling following air activated-sludge (excluding extended aeration)	400–800	1000–1200	20–30	50	12–15
Settling following extended aeration	200–400	800	20–30	50	12–15
Settling following oxygen activated-sludge with primary settling	400–800	1000–1200	25–35	50	12–15

*Allowable solids loading are generally governed by sludge thickening characteristics associated with cold weather operations.
†gal/(d · ft^2) × 0.0407 = m^3/(m^2 · d).
lb/(d · ft^2) × 4.883 = kg/(d · m^2).
Source: USEPA (1975a).

Step 3. Estimate solids (S_1) generated from the primary clarifiers

$$S_1 = 0.65 \times (240 \text{ mg/L}) \times 1.0 \text{ Mgal/d} \times 8.34 \text{ lb/(Mgal} \cdot \text{mg/L)}$$
$$= 1300 \text{ lb/d} (= 590 \text{ kg/d})$$

Step 4. Calculate the total daily solids production (S)

$$S = S_1 + S_2 = 1300 \text{ lb/d} + 510 \text{ lb/d}$$
$$= 1810 \text{ lb/d} (= 822 \text{ kg/d})$$

Or solve this problem using SI units

Step 1. Calculate the BOD_5 loading to the secondary clarifiers

$$BOD_5 = (1 - 0.32) \times (200 \text{ mg/L}) \times (3785 \text{ m}^3\text{/d}) \times 0.001 \text{ [(kg/m}^3\text{)/(mg/L)]}$$
$$= 515 \text{ kg/d} (= 1134 \text{ lb/d})$$

Step 2. Estimate solids (S_2) production of the secondary clarifiers

Note: The biological solids generated in secondary treatment range from 0.4 to 0.5 lb per lb (kg/kg) BOD_5 applied. Use 0.45 kg/kg in this example.

$$S_2 = 0.45 \text{ kg/kg} \times 515 \text{ kg/d}$$
$$= 232 \text{ kg/d} (= 510 \text{ lb/d})$$

Step 3. Estimate solids (S_1) generated from the primary clarifiers

$$S_1 = 0.65 \times (240 \text{ mg/L}) \times 3785 \text{ m}^3\text{/d} \times 1000 \text{ kg/m}^3 \times 10^{-6} \text{ L/mg}$$
$$= 590 \text{ kg/d} (= 1300 \text{ lb/d})$$

Step 4. Calculate the total daily solids production (S)

$$S = S_1 + S_2 = 590 \text{ kg/d} + 232 \text{ kg/d}$$
$$= 822 \text{ kg/d} (= 1810 \text{ lb/d})$$

EXAMPLE 2: Design a two-stage trickling filter system with a design average flow of 5680 m^3/d (1.5 Mgal/d) and intermediate and secondary settling tanks under conditions given as follows. The primary effluent (system influent) BOD = 190 mg/L. The design BOD loading rate is 1.5 kg/(m^3 · d) (93.8 lb/(1000 ft^3 · d)).

The recycle ratio of both filters is 0.8. Both clarifiers have 20 percent recycled to the influent.

Solution:

Step 1. Determine volume required for the filters (plastic media)

$$\text{volume} = \frac{5680 \text{ m}^3\text{/d} \times 0.19 \text{ kg/m}^3}{1.5 \text{ kg/(m}^3 \cdot \text{d)}}$$
$$= 719 \text{ m}^3$$

Volume of each filter = 360 m^3

Step 2. Determine surface area of the filter
Use side water depth of 4 m (13 ft) since minimum depth is 3 m.

$$\begin{aligned}\text{Area} &= 360\ \text{m}^3/4\ \text{m} = 90\ \text{m}^2\\ \text{Diameter} &= (4 \times 90\ \text{m}^2/3.14)^{0.5}\\ &= 10.7\ \text{m}\\ &= 35\ \text{ft}\end{aligned}$$

Step 3. Check hydraulic loading rate

$$\begin{aligned}\text{HRT} &= (5680\ \text{m}^3/\text{d})(1 + 0.8)/90\ \text{m}^2\\ &= 113.6\ \text{m}^3/(\text{m}^2 \cdot \text{d}) \quad (\text{OK}, < 375\ \text{m}^3/(\text{m}^2 \cdot \text{d}))\end{aligned}$$

Use 4 m (13 ft) deep filters with 90 m^2 (970 ft^2) area.

Step 4. Sizing for the intermediate clarifier
Use HLR = 41 m^3/(m^2 · d)(1000 gal/(d · ft^2)) and minimum depth of 3 m (10 ft)

$$\begin{aligned}\text{Area required} &= \frac{5680\ \text{m}^3/\text{d} \times 1.2}{41\ \text{m}^3/(\text{m}^2 \cdot \text{d})}\\ &= 166\ \text{m}^2\\ \text{Diameter} &= (4 \times 166\ \text{m}^2/3.14)^{1/2}\\ &= 14.6\ \text{m}\\ &= 48\ \text{ft}\end{aligned}$$

Step 5. Sizing for the secondary clarifier
Using HLR = 31 m^3/(m^2 · d) (760 gal/(d · ft^2)) and minimum depth of 3 m (10 ft)

$$\begin{aligned}\text{Area required} &= \frac{5680\ \text{m}^3/\text{d} \times 1.2}{31\ \text{m}^3/(\text{m}^2 \cdot \text{d})}\\ &= 220\ \text{m}^2\\ \text{Diameter} &= (4 \times 220\ \text{m}^2/3.14)^{1/2}\\ &= 16.7\ \text{m}\\ &= 55\ \text{ft}\end{aligned}$$

26.2 Basin Sizing for Suspended-Growth Biological Treatment

In order to produce the proper concentration of return sludge, activated-sludge settling tanks must be designed to meet thickening as well as separation requirements. Since the rate of recirculation of RAS from the secondary settling tanks to the aeration or reaeration basins is quite high in activated-sludge processes, their surface overflow rate and weir overflow rate should be adjusted for the various modified processes to minimize the problems with sludge loadings, density currents, inlet hydraulic turbulence, and occasional poor sludge settleability. The size of the secondary settling tank must be based on the large surface area determined for surface overflow rate and solids loading rate. Table 7.20 presents the design criteria for secondary clarifiers following activated-sludge processes (GLUMRB 1996). The values given in Tables 7.19 and 7.20 are comparable.

Solids loading rate is of primary importance to ensure adequate function in the secondary settling tanks following aeration basins. In practice, most domestic wastewater plants have values of solids volume index in the range of 100 to 250 mg/L (WEF and ASCE 1991a). Detail discussions of the secondary clarifier can be found in this manual. Most design engineers prefer to keep the maximum

TABLE 7.20 Recommended Design Overflow Rate and Peak Solids Loading Rate for Secondary Settling Tanks Following Activated-Sludge Processes

Treatment process	Surface loading at design peak hourly flow,* gal/(d · ft^2) (m^3/m^2 · d))	Peak solids loading rate,‡ lb/(d · ft^2) (kg/(d · m^2))
Conventional Step aeration Complete-mix Contact stabilization Carbonaceous stage of separate-stage nitrification	1200 (49) or 1000 (41)†	50 (244)
Extended aeration Single-stage nitrification	1000 (41)	35 (171)
Two-stage nitrification	800 (33)	35 (171)
Activated-sludge with chemical addition to mixed liquor for phosphorus removal	900 (37)§	as above

*Based on influent flow only.
†Computed on the basis of design maximum daily flow rate plus design maximum return sludge rate requirement, and the design MLSS under aeration.
‡For plant effluent TSS ≤ 20 mg/L.
§When effluent P concentration of 1.0 mg/L or less is required.
Source: GLUMBRS (1996).

solids loading rates in the range of 4 to 6 kg/(m^2 · h) (20 to 30 lb/(d · ft^2)). Solids loadings rates of 10 kg/(m^2 · h) (50 lb/(d · ft^2) or more have been found in some well-operating plants.

The maximum allowable hydraulic loading rate HLR as a function of the initial settling velocity ISV at the design MLSS concentration was proposed by Wilson and Lee (1982). The equation is expressed as follows

$$\text{HRT} = Q/A = 24 \times \text{ISV/CSF} \tag{7.151}$$

where HRT = hydraulic retention time, h
Q = limiting hydraulic capacity, m^3/d
A = area of the clarifier, m^2
24 = unit conversion factor, 24 h/d
ISV = initial settling velocity at the design MLSS concentration, m/h
CSF = clarifier safety factor, 1.5 to 3, typically 2

The values of ISV change with MLSS concentrations and other conditions. Batch-settling analyses should be performed. The maximum anticipated operational MLSS or the corresponding minimum ISV should be used in Eq. (7.151).

Numerous state regulations limit the maximum allowable weir loading rates to 124 m^3/(d · m) (10,000 gal/(d · ft)) for small plants (less than 3785 m^3/d or 1 Mgal/d) and to 186 m^3/(d · m) (15,000 gal/(d · ft)) for larger treatment plants (WEF and ASCE 1991a). It is a general consensus that substantially high weir loading rates will not impair performance.

The depth of secondary clarifiers is commonly designed as 4 to 5 m (13 to 16 ft). The deeper tanks increase TSS removal and RAS concentration as well as costs. Typically, the larger the tank diameter, the deeper the sidewall depth. The shapes of settling tanks include rectangular, circular, and square. A design example of a secondary clarifier following the activated-sludge process is given previously in the section on the complete-mix process (Example 1B, Sec. 21.5).

EXAMPLE: Determine the size of three identical secondary settling tanks for an activated-sludge process with a recycle rate of 25 percent, an MLSS concentration of 3600 mg/L, an average design flow of 22,710 m^3/d (6 Mgal/d), and an anticipated peak hourly flow of 53,000 m^3/d (14 Mgal/d). Use solids loading rates of 4.0 and 10.0 $kg/(m^2 \cdot h)$ for average and peak flow respectively.

Solution:

Step 1. Compute the peak solids loading

$$3600 \text{ mg/L} = 3600 \text{ g/m}^3 = 3.6 \text{ kg/m}^3$$

$$\begin{aligned} \text{Loading} &= 53{,}000 \text{ m}^3\text{/d} \times 3.6 \text{ kg/m}^3 \times 1.25 \\ &= 238{,}500 \text{ kg/d} \end{aligned}$$

Step 2. Compute the design average solids loading

$$\begin{aligned} \text{Loading} &= 22{,}710 \text{ m}^3\text{/d} \times 3.6 \text{ kg/m}^3 \times 1.25 \\ &= 102{,}200 \text{ kg/d} \end{aligned}$$

Step 3. Compute the surface area required (each of three)

(a) At design flow

$$\begin{aligned} A &= \frac{102{,}200 \text{ kg/d}}{4.0 \text{ kg/(m}^2 \cdot \text{h)} \times 24 \text{ h/d} \times 3 \text{ (units)}} \\ &= 355 \text{ m}^2 \\ &= 3820 \text{ ft}^2 \end{aligned}$$

(b) At peak flow

$$\begin{aligned} A &= 238{,}500/(10.0 \times 24 \times 3) \\ &= 331 \text{ (m}^2\text{)} \\ &= 3563 \text{ ft}^2 \end{aligned}$$

The design flow is the controlling factor; and the required surface area $A = 355 \text{ m}^2$.

Step 3. Determine the diameter of circular clarifiers

$$\begin{aligned} d^2 &= 4A/\pi = 4 \times 355 \text{ m}^2/3.14 \\ d &= 21.3 \text{ m} \\ &= 70 \text{ ft} \end{aligned}$$

Use sidewall depth of 4 m (13 ft) for the tank diameter of 21 to 30 m (70 to 100 ft).

Step 4. Check hydraulic loading rate at the design flow

$$\begin{aligned} \text{HLR} &= (22{,}710 \text{ m}^3\text{/d})/(3 \times 355 \text{ m}^2) \\ &= 21.3 \text{ m}^3/(\text{m}^2 \cdot \text{d}) \end{aligned}$$

Step 5. Check HLR at the peak flow

$$\begin{aligned} \text{HLR} &= (53{,}000 \text{ m}^3\text{/d})/(3 \times 355 \text{ m}^2) \\ &= 49.8 \text{ m}^3/(\text{m}^2 \cdot \text{d}) \end{aligned}$$

(It is slightly over the limit of 49 $m^3/(m^2 \cdot d)$; see Table 7.20.)

Step 6. Compute the weir loading rate

$$\text{Perimeter} = \pi d = 3.14 \times 21.3 \text{ m} = 66.9 \text{ m}$$

At average design flow

$$\text{Weir loading} = (22{,}710 \text{ m}^3/\text{d})/(66.9 \text{ m} \times 3) = 113 \text{ m}^3/(\text{m} \cdot \text{d}) \quad (\text{OK}, < 186 \text{ m}^3/\text{m} \cdot \text{d limit})$$

At the peak flow

$$\text{Weir loading} = (53{,}000 \text{ m}^3/\text{d})/(66.9 \text{ m} \times 3) = 264 \text{ m}^3/(\text{m} \cdot \text{d})$$

27 EFFLUENT DISINFECTION

Effluent disinfection is the last treatment step of a secondary or tertiary treatment process. Disinfection is a chemical treatment method carried out by adding the selected disinfectant to an effluent to destroy or inactivate the disease-causing organisms. The purposes of effluent disinfection are to protect public health by killing or inactivating pathogenic organisms such as enteric bacteria, viruses, and protozoans, and to meet the effluent discharge standards.

The disinfection agents (chemicals) include chlorine, ozone, ultraviolet (UV) radiation, chlorine dioxide, and bromine. Design of UV irradiation can be referred to the manufacturer or elsewhere (WEF and ASCE 1996a).

The chlorination–dechlorination process is currently widely practiced in the United States. Chlorine is added to a secondary effluent for a certain contact time (20 to 45 min for average flow and 15 min at peak flow), then the effluent is dechlorinated before discharge only during warm weather when people use water as primary contact. In the United States most states adopt a coliform limitation of 200 fecal coliform/100 ml.

Chlorination of effluents is usually accomplished with liquid chlorine. Alternative methods use calcium or sodium hypochlorite or chlorine dioxide. Disinfection kinetics and chemistry of chlorination are discussed in Chapter 1.6, numerous literature, and text books. An excellent review of effluent disinfection can be found in Design Manual (USEPA 1986).

27.1 Chlorine Dosage

If a small quantity of chlorine is added to wastewater or effluent, it will react rapidly with reducing substances such as hydrogen sulfide and ferrous iron, and be destroyed. Under these conditions, there are no disinfection effects. If enough chlorine is added to react with all reducing compounds, then a little more added chlorine will react with organic materials present in wastewater and form chlororganic compounds, which have slight disinfection activities. Again, if enough chlorine is introduced to react with all reducing compounds and all organic materials, then a little more chlorine added will react with ammonia or other nitrogenous compounds to produce chloramines or other combined forms of chlorine, which do have disinfection capabilities. Therefore chlorine dosage and residual chlorine are very important factors of disinfection operation. In addition to its disinfection purpose, chlorination is also applied for prevention of wastewater decomposition, prechlorination of primary influent, control of activated-sludge bulking, and reduction of BOD.

Chlorinators are designed to have a capacity adequate to produce an effluent to meet coliform density limits specified by the regulatory agency. Usually, multiple units are installed for adequate capacity and to prevent excessive chlorine residuals in the effluent. Table 7.21 shows the recommended chlorine dosing capacity for treating normal domestic wastewater, based on design average flow (Illinois EPA 1997, GLUMRB 1996).

TABLE 7.21 Recommended Chlorine Dosing Capacity for Various Types of Treatment Based on Design Average Flow

Type of treatment	Illinois EPA dosage, mg/L	GLUMRB dosage, mg/L
Primary settled effluent	20	
Lagoon effluent (unfiltered)	20	
Lagoon effluent (filtered)	10	
Trickling filter plant effluent	10	10
Activated sludge plant effluent	6	8
Activated sludge plant with chemical addition	4	
Nitrified effluent		6
Filtered effluent following mechanical biological treatment	4	6

Sources: Illinois EPA (1997), GLUMRB (1996).

For small applications, 150-lb (68-kg) chlorine cylinders are typically used where chlorine gas consumption is less than 150 lb/day. Chlorine cylinders are stored in an upright position with adequate support brackets and chains at 2/3 of cylinder height for each cylinder. For larger applications where the average daily chlorine gas consumption is greater than 150 lb, one-tone (907 kg) containers are employed. Tank cars, usually accompanied by evaporators, are used for large installations (> 10 Mgal/d, 0.44 m^3/s). In this case, area-wide public safety should be evaluated as part of the design consideration.

EXAMPLE 1: Estimate a monthly supply of liquid chlorine for trickling filter plant effluent disinfection. The design average flow of the plant is 3.0 Mgal/d (11,360 m^3/d).

Solution:

Step 1. Find the recommended dosage
From Table 7.21, the recommended dosage for trickling filter plant effluent is 10 mg/L.

Step 2. Compute the daily consumption

$$\begin{aligned}\text{Chlorine} &= 3.0 \text{ Mgal/d} \times 10 \text{ mg/L} \times 8.34 \text{ lb/(Mgal} \cdot \text{mg/L)} \\ &= 250 \text{ lb/d}\end{aligned}$$

The daily consumption is over 150 lb/d (68 kg/d); thus choose one-ton (2000 lb, 907 kg) containers.

Step 3. Compute the number of one-ton containers required for one month's (M) supply

$$\begin{aligned}\text{Monthly need} &= 250 \text{ lb/d} \times 30 \text{ d/M} \\ &= 7500 \text{ lb/M}\end{aligned}$$

The plant needs 4 one-ton containers, which is enough for one month's consumption.

EXAMPLE 2: Determine the feeding rate in gallons per minute of sodium hypochlorite (NaOCl) solution containing 10 percent available chlorine. The daily chlorine dosage for the plant is 480 kg/d (1060 lb/d).

Solution:

Step 1. Calculate chlorine concentration of the solution

$$10\% = 100{,}000 \text{ mg/L} = 100 \text{ g/L}$$

Step 2. Calculate the solution feed rate

$$\text{Feeding rate} = \frac{480{,}000 \text{ g/d}}{100 \text{ g/L} \times 1440 \text{ min/d}}$$

$$= 3.333 \text{ L/min}$$

$$= 0.88 \text{ gal/min}$$

EXAMPLE 3: A wastewater treatment plant having an average flow of 28,400 m^3/d (7.5 Mgal/d) needs an average chlorine dosage of 8 mg/L. Chlorine is applied daily via 10 percent NaOCl solution. The time of shipment from the vendor to the plant is 2 days. A minimum 10-day supply is required for reserve purposes. Estimate the storage tank capacity required for NaOCl solution with a 0.03 percent per day decay rate.

Solution:

Step 1. Determine daily chlorine requirement

$$\text{Chlorine} = \frac{28{,}400 \text{ m}^3\text{/d} \times 8 \text{ g/m}^3}{1000 \text{ g/kg}}$$

$$= 227.2 \text{ kg/d}$$

Step 2. Calculate daily volume of NaOCl solution used

$$10\% \text{ solution} = 100 \text{ g/L} = 100 \text{ kg/m}^3$$

$$\text{Volume} = \frac{227.2 \text{ kg/d}}{100 \text{ kg/m}^3}$$

$$= 2.27 \text{ m}^3\text{/d}$$

Step 3. Calculate the storage tank for NaOCl solution

$$\text{Tank volume} = (2.27 \text{ m}^3\text{/d})(2 \text{ d} + 10 \text{ d})$$

$$= 27.24 \text{ m}^3$$

Step 4. Calculate the tank volume with NaOCl decay correction.

$$\text{Correction factor} = (0.03\%\text{/d}) \times 12 \text{ d}$$

$$= 0.36\%$$

$$\text{The volume requirement} = \frac{27.24 \text{ m}^3 \times 10\%}{10\% - 0.36\%}$$

$$= 28.3 \text{ m}^3$$

EXAMPLE 4: Calculate the residual chlorine concentration that must be maintained to achieve a fecal coliform (FC) density equal or less than 200/100 mL in the chlorinated secondary effluent. The influent of chlorine contact tank contains 4 × 10^6 FC/100 mL. The contact time is 30 min.

Solution:

Step 1. Calculate the residual chlorine needed during the average flow using equation (White, 1986) modified from Eq. (6.180a)

$$N/N_0 = (1 + 0.23C_t t)^{-3} \quad \text{where } C_t \text{ is chlorine residual at time } t \text{ in mg/L}$$

$$200/(4 \times 10^6) = (1 + 0.23\, C_t t)^{-3}$$

$$5 \times 10^{-5} = (1 + 0.23\, C_t t)^{-3}$$

$$1 + 0.23\, C_t t = (0.2 \times 10^5)^{1/3} = 27.14$$

$$C_t t = 113.7$$

$$\text{When } t = 30 \text{ min, } C_t = 113.7/30 = 3.8 \text{ (mg/L)}$$

Step 2. Calculate the residual chlorine needed during the peak hourly flow

Assume the ratio of the peak hourly flow to the average flow is 3.0

$$C_t = 3.8 \text{ mg/L} \times 3 = 11.4 \text{ mg/L}$$

27.2 Dechlorination

There have been concerns that wastewater disinfection may do more harm than good due to the toxicity of disinfection by-products. Many states have adopted seasonal chlorination (warm weather periods) in addition to dechlorination. Dechlorination is important in cases where chlorinated wastewater comes into contact with fish and other aquatic animals. Sulfur compounds, activated carbon, hydrogen peroxide, and ammonia can be used to reduce the residual chlorine in a disinfected wastewater prior to discharge. The first two are the most widely used. Sulfur compounds include sulfur dioxide (SO_2), sodium metabisulfite (NaS_2O_5), sodium bisulfite ($NaHSO_3$), and sodium sulfite (Na_2SO_3). Dechlorination reactions with those compounds are shown below.

Free chlorine:

$$SO_2 + 2H_2O + Cl_2 \rightarrow H_2SO_4 + 2HCl \tag{7.152}$$

$$SO_2 + H_2O + HOCl \rightarrow 3H^+ + Cl^- + SO_4^{2-} \tag{7.153}$$

$$Na_2S_2O_5 + 2Cl_2 + 3H_2O \rightarrow 2NaHSO_4 + 4HCl \tag{7.154}$$

$$NaHSO_3 + H_2O + Cl_2 \rightarrow NaHSO_4 + 2HCl \tag{7.155}$$

Chloramine:

$$SO_2 + 2H_2O + NH_2Cl \rightarrow NH_4^+ + 2H^+ + Cl^- + SO_4^{2-} \tag{7.156}$$

$$3Na_2S_2O_5 + 9H_2O + 2NH_3 + 6Cl_2 \rightarrow 6NaHSO_4 + 10HCl + 2NH_4Cl \tag{7.157}$$

$$3NaHSO_3 + 3H_2O + NH_3 + 3Cl_2 \rightarrow 3NaHSO_4 + 5HCl + NH_4Cl \tag{7.158}$$

Activated carbon absorption:

$$C^* + 2H_2O + 2Cl_2 \rightarrow 4HCl + C^*O_2 \tag{7.159}$$

$$C^* + H_2O + NH_2Cl \rightarrow NH_4^+ + Cl^- + C^*O \tag{7.160}$$

$$C^*O + 2NH_2Cl \rightarrow N_2 + 2HCl + H_2O + C^* \tag{7.161}$$

$$C^* + H_2O + 2NHCl_2 \rightarrow N_2 + 4HCl + C^*O \tag{7.162}$$

The most common dechlorination chemicals are sulfur compounds, particularly sulfur dioxide gas, or aqueous solutions of bisulfite or sulfite. A pellet dechlorination system can be used for small plants. Liquid sulfur dioxide gas cylinders are in 50 gal (190 L) drums.

Sulfur dioxide is a deadly gas that affects the central nervous system. It is colorless and nonflammable. When sulfur dioxide dissolves in water, it forms sulfurous acid which is a strong reducing chemical. As sulfur dioxide is introduced to chlorinated effluent, it reduces all forms of chlorine to chloride and converts sulfur to sulfate. Both byproducts, chlorides and sulfates, occur commonly in natural waters.

Sodium metabisulfite is a cream-colored powder readily soluble in water at various strengths. Sodium bisulfite is a white powder, or in granular form, and is available in solution with strengths up to 44 percent. Both sodium metabisulfite and sodium bisulfite are safe chemicals. These are most commonly used in small plants and in a few larger plants.

The dosage of dechlorination chemical depends on the chlorine residual in the effluent, the final residual chlorine standards, and the type of dechlorinating chemical used. Theoretical requirements to neutralize 1 mg/L of chlorine for sulfur dioxide (gas), sodium bisulfite, and sodium metabisulfite are 0.90, 1.46, and 1.34 mg/L, respectively. Theoretical values may be employed for initial estimation for sizing dechlorinating equipment under good mixing conditions. An extra 10 percent of dechlorination is designed above theoretical values. However, excess sulfur dioxide may consume dissolved oxygen at a maximum of 0.25 mg DO for every 1 mg of sulfur dioxide. Since the sulfur dioxide reacts very rapidly with the chlorine residual, no additional contact time is necessary. Dechlorination may remove all chlorine-induced toxicity from the effluents.

27.3 Process Design

The design of an ozonation or ultraviolet irradiation system can usually be provided by the manufacturers. Only chlorination is discussed in this section.

Traditional design. The disinfection efficiency of effluent chlorination depends on chlorine dosage, contact time, temperature, pH, and characteristics of wastewater such as TSS, nitrogen concentrations, and type and density of organisms. The first two factors can be designed for the best performance for chlorination of water and wastewater effluent. In general, residual chlorine at 0.5 mg/L after 20 to 30 min of contact time is required. The State of California requires a minimum of 30 min contact time. The chlorine contact tank can be sized on the basis of contact time and plant flow. However, many engineers disagree with this method.

Contact time is related to the configuration of the chlorine contact tank. The drag on the sides of a deep, narrow tank causes relatively poor dispersion characteristics. Sepp (1977) observed that the dispersion number usually decreases with increasing length-to-width (L/W) ratios. The depth-to-width (D/W) ratio should be 1.0 or less in chlorine tanks. Adequate baffling, reduction of side drag, and elimination of dead spaces can provide a reasonably long tank with adequate plug-flow hydraulics (Sepp 1981). Marske and Boyle (1973) claimed that adequate plug-flow tanks can be achieved by L/W ratios of 40 to 70 to 1. In design, in addition to L/W ratios, the usual dispersion parameters and geometric configurations must be considered.

EXAMPLE: Determine the size of a chlorine contact tank for a design average flow of 0.131 m^3/s (3 Mgal/d) and a peak flow of 0.329 m^3/s (7.5 Mgal/d).

Solution:

Step 1. Calculate the tank volume required at peak design flow. Select contact time of 20 min at the peak flow

$$\text{Volume} = 0.329\ \text{m}^3/\text{s} \times 60\ \text{s/min} \times 20\ \text{min}$$
$$= 395\ \text{m}^3$$

Step 2. Calculate the volume required at the average flow
Use contact time of 30 min at the design flow

$$\text{Volume} = 0.131\ \text{m}^3/\text{s} \times 60\ \text{s/min} \times 30\ \text{min}$$
$$= 236\ \text{m}^3$$

Peak flow is the control factor.

Step 3. Determine chlorine contact tank configuration and dimensions
Provide one basin with three-pass-around-the-end baffled arrangement. Select channel width W of 2.2 m (7.1 ft) and water depth D of 1.8 m (5.9 ft) with 0.6 m (2.0 ft) of freeboard; thus

$$D/W = 1.8\ \text{m}/2.2\ \text{m}$$
$$= 0.82\ (\text{OK if} < 1.0)$$
$$\text{Cross-sectional area} = 1.8\ \text{m} \times 2.2\ \text{m}$$
$$= 3.96\ \text{m}^2$$
$$\text{Total length } L = 395\ \text{m}^3/3.96\ \text{m}^2 = 99.7\ \text{m (Say 100 m)}$$

$$\text{Length of each pass} = 100 \text{ m}/3 = 33.3 \text{ m}$$
$$\text{Use each pass} = 33.5 \text{ m (110 ft)}$$
$$\text{Check } L/W = 100 \text{ m}/2.2 \text{ m}$$
$$= 45.5 \text{ (OK if in the range of 40 to 70)}$$
$$\text{Tank size: } D \times W \times L = (1.8 \text{ m} + 0.6 \text{ m}) \times (2.2 \text{ m} \times 3) \times 33.5 \text{ m}$$
$$\text{Actual liquid volume} = 1.8 \text{ m} \times 2.2 \text{ m} \times 100 \text{ m}$$
$$= 396 \text{ m}^3$$

Step 4. Check the contact time t at peak flow

$$t = \frac{396 \text{ m}^3}{0.329 \text{ m}^3/\text{s} \times 60 \text{ s/min}}$$
$$= 20.1 \text{ min}$$

Collins–Selleck model. Traditionally, chlorination has been considered as a first-order reaction that follows the theoretical Chick or Chick–Watson model. However, in practice, it has been observed not to be the case for wastewater chlorination. Collins *et al.* (1971) found that initial mixing of the chlorine solution and wastewater has a profound effect on coliform bacteria reduction. The efficiency of wastewater chlorination, as measured by coliform reduction, was related to the residence time. It was recommended that the chlorine contact tank be designed to approach plug-flow reactors with rapid initial mixing. Collins *et al.* (1971) suggested that amperometric chlorine residuals and contact time could be used to predict the effluent coliform density.

Collins and Selleck (1972) found that the coliform bacteria survival ratio often produces an initial log period and a declining rate of inactivation as a plot of the log survival ratio versus time (Fig. 7.35a). After the initial lag period, there is a straight-line relationship between log survival and log of the product of chlorine residual concentration C and contact time t (Fig. 7.35b).

The rate equation of the Collins and Selleck model is

$$\frac{dN}{dt} = -kN \tag{7.163}$$

where N = bacteria density at time t
t = time
k = rate constant characteristic of type of disinfectant, microorganism, and water quality
= 0 for $Ct < b$
= K for $Ct = b$
= $K/b(Ct)$ for $Ct > b$
b = log time

After integration and application of the boundary conditions, the rate equation for Collins and Selleck becomes

$$\frac{N}{N_0} = 1 \quad \text{for } Ct < b \tag{7.164}$$

$$\frac{N}{N_0} = \left(\frac{Ct}{b}\right)^n \quad \text{for } Ct > b \tag{7.165}$$

where N = bacteria density at time t after chlorine contact
N_0 = bacteria density at time zero in unchlorinated effluent
C = chlorine residual after contact, mg/L
t = theoretical contact time, min
b = an intercept when N/N_0 = 1.0, min · mg/L
= a product of concentration and time
n = slope of the curve

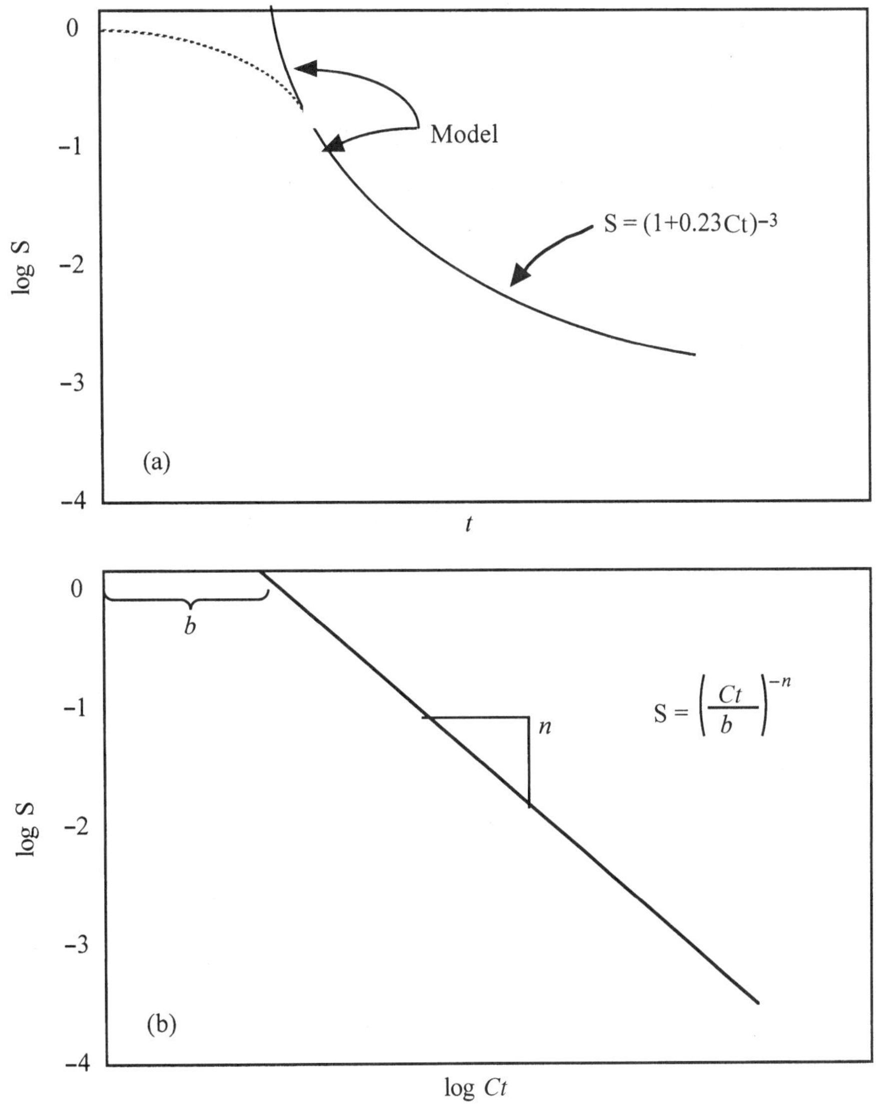

FIGURE 7.35 Form of plots after Collins–Selleck model: (a) log S versus time and (b) log S versus log Ct.

Equation (7.164) applies to the lag period. Equation (7.165) plots as a straight line on log–log paper and expresses the declining rate. The values of b and n, obtained from several researchers for various types of chlorine applied to various stages of wastewaters with different bacteria species, are summarized by J. M. Montgomery Engineers (1985). Wide variations of those values were observed. Collins and Selleck (1972) found that the slope $n = -3$ for ideal plug flow, $n = -1.5$ for complete-mix flow, and b was about 4 in a primary effluent.

Robert *et al.* (1980) stated that the Collins–Selleck model is a useful empirical design tool although it has no rational mechanistic basis in describing chemical disinfection. Sepp (1977) recommended that the design of chlorine disinfection systems should incorporate rapid initial mixing, reliable automatic chlorine residual control, and adequate chlorine contact in well-designed contact tanks.

EXAMPLE: A pilot study was conducted to determine the efficiency of combined chlorine to disinfect a secondary effluent. The results show that when $b = 2.87$ min · mg/L and $n = -3.38$, the geometric mean density of fecal coliform (FC) in the effluent is 14,000 FC/100 mL. Estimate FC density in the chlorinated effluent when contact times are 10, 20, and 30 min with the same chlorine residual of 1.0 mg/L.

Solution:

Step 1. Determine mean FC survival ratios

$$Ct = 1.0 \text{ mg/L} \times 10 \text{ min}$$
$$= 10 \text{ min} \cdot \text{mg/L} \; (> b = 2.87)$$

Using Eq. (7.165)

(1) At $t = 10$ min

$$\frac{N}{N_0} = \left(\frac{1.0 \text{ mg/L} \times 10 \text{ min}}{2.87 \text{ min} \cdot \text{mg/L}}\right)^{-3.38}$$
$$= 0.0147$$

(2) At $t = 20$ min

$$\frac{N}{N_0} = \left(\frac{1.0 \times 20}{2.87}\right)^{-3.38}$$
$$= 0.00141$$

(3) At $t = 30$ min

$$\frac{N}{N_0} = \left(\frac{1.0 \times 30}{2.87}\right)^{-3.38}$$
$$= 0.000359$$

Step 2. Calculate FC densities in chlorinated effluent

(1) At $t = 10$ min

$$N = 0.0147 \times N_0 = 0.0147 \times 14{,}000 \text{ FC/100 mL}$$
$$= 210 \text{ FC/100 mL}$$

(2) At $t = 20$ min

$$N = 0.00141 \times 14{,}000$$
$$= 20 \text{ (FC/100 mL)}$$

(3) At $t = 30$ min

$$N = 0.000359 \times 14{,}000$$
$$= 5 \text{ (FC/100 mL)}$$

28 ADVANCED WASTEWATER TREATMENT

Advanced wastewater treatment (AWT) refers to those additional treatment techniques needed to further reduce suspended and dissolved substances remaining after secondary treatment. It is also called tertiary treatment. Secondary treatment removes 85 to 95 percent of BOD and TSS and minor portions of nitrogen, phosphorus, and heavy metals. The purposes of AWT are to improve the effluent quality to meet stringent effluent standards and to reclaim wastewater for reuse as a valuable water resource.

The targets for removal by the AWT process include suspended solids, organic matter, nutrients, dissolved solids, refractory organics, and specific toxic compounds. More specifically, the common AWT processes are suspended solids removal, nitrogen control, and phosphorus removal. Suspended solids are removed by chemical coagulation and filtration. Phosphorus removal is done to reduce eutrophication of receiving waters from wastewater discharge. Ammonia is oxidized to nitrate to reduce its toxicity to aquatic life and nitrogenous oxygen demand in the receiving water bodies.

28.1 Suspended Solids Removal

Treatment techniques for the reduction of suspended solids in secondary effluents include chemical coagulation followed by gravity sedimentation or dissolved air flotation, and physical straining or filtration such as wedge-wire screens, microscreens, other screening devices, diatomaceous earth filters, ultrafiltration, and granular media filters. These treatment processes are discussed in Chapter 6. More detailed descriptions and practical applications for SS removal are given in the USEPA's manual (1975a). Chemical coagulants used for SS removal include aluminum compounds, iron compounds, soda ash, caustic soda, carbon dioxide, and polymers. Some of the above processes are described in Chapter 6 and in other textbooks.

The use of a filtration process similar to that employed in drinking water treatment plant can remove the residual SS, BOD, and microorganisms from secondary effluent. Conventional sand filters or granular-media filters for water treatment may serve as advanced wastewater treatment and are discussed by Metcalf and Eddy, Inc. (1991). However, the filters may require more frequent backwashing. Design surface loading rates for wastewater filtration usually range from 2.0 to 2.7 $L/(m^2 \cdot s)$ (3 to 4 $gal/(min \cdot ft^2)$). Filtration lasts about 24 h, and the effluent quality expected is 5 to 10 mg/L of suspended solids.

28.2 Phosphorus Removal

The typical forms of phosphorus found in wastewater include the orthophosphates, polyphosphates (molecularly dehydrated phosphates), and organic phosphates. Orthophosphates such as PO_4^{3-}, HPO_4^{2-}, $H_2PO_4^-$, H_3PO_4 are available for biological uptake without further breakdown. The polyphosphates undergo hydrolysis in aqueous solutions and revert to the orthophosphate forms. This hydrolysis is a very slow process. The organic phosphorus is an important constituent of industrial wastes and less important in most domestic wastewaters.

The total domestic phosphorus contribution to wastewater is about 1.6 kg per person per year (3.5 lb per capita per year). The average total phosphorus concentration in domestic raw wastewater is about 10 mg/L, expressed as elemental phosphorus, P (USEPA 1976). Approximately 30 to 50 percent of the phosphorus is from sanitary wastes, while the remaining 70 to 50 percent is from phosphate builders in detergents.

Phosphorus is one of major contributors to eutrophication of receiving waters. Removal of phosphorus is a necessary part of pollution prevention to reduce eutrophication. In most cases, the effluent standards range from 0.1 to 2.0 mg/L as P, with many established at 1.0 mg/L. Percentage reduction requirements range from 80 to 95 percent.

In practice, phosphorus in wastewater is removed by chemical precipitation with metal salt (aluminum, iron) or/and polymer, or lime. The chemical can be dosed before primary sedimentation, in the trickling filter, in the activated-sludge process, and in the secondary effluent.

Chemical precipitation. Phosphorus removal can be achieved by chemical precipitation or a biological method. Chemical precipitation is usually carried out by addition of mineral (aluminum or iron salts) and lime. Aluminum ions can flocculate phosphate ions to form aluminum phosphate which then precipitates:

$$Al^{3+} + H_nPO_4^{3-,n+} \rightarrow AlPO_4 \downarrow + nH^+ \qquad (7.166)$$

Alum, $Al_2(SO_4)_3 \cdot 14H_2O$, is most commonly used as a source of aluminum. The precipitation reactions for phosphorus removal by three chemical additions are as follows:

Alum:

$$Al_2(SO_4)_3 \cdot 18H_2O + 2HPO_4^{2-} \rightarrow 2AlPO_4\downarrow + 2H^+ + 3SO_4^{2-} + 18H_2O \qquad (7.167)$$

Sodium aluminate:

$$Na_2O \cdot Al_2O_3 + 2HPO_4^{2-} + 4H_2O \rightarrow 2AlPO_4\downarrow + 2NaOH + 6OH^- + 2H^+ \qquad (7.168)$$

Ferric chloride, $FeCl_3 \cdot 6H_2O$:

$$FeCl_3 + HPO_4^{2-} \rightarrow FePO_4 \downarrow + H^+ + 3Cl^- \qquad (7.169)$$

Lime, CaO:

$$CaO + H_2O \rightarrow Ca(OH)_2 \qquad (7.170)$$

$$5Ca(OH)_2 + 3HPO_4^{2-} \rightarrow Ca_5(PO_4)_3OH \downarrow + 3H_2O + 6OH^- \qquad (7.171)$$

Alum and ferric chloride decrease the pH while lime increases pH. The optimum pH for alum and ferric chloride is between 5.5 and 7.0. The effective pH for lime is above 10.0. Theoretically, 9.6 g of alum is required to remove 1 g of phosphorus. However, there is an excess alum requirement due to the competing reactions with natural alkalinity. The alkalinity reaction is

$$Al_2(SO_4)_3 \cdot 14H_2O + 6HCO_3^- \rightarrow 2Al(OH)_3 \downarrow + 3SO_4^{2-} + 6CO_2 + 14H_2O \qquad (7.172)$$

As a result, the requirement of alum to remove phosphorus demands weight ratios of approximately 13 : 1, 16 : 1, and 22 : 1 for 75, 85, and 95 percent phosphorus removal, respectively (from Eqs. (7.167) and (7.172)).

The molar ratio of ferric ion to orthophosphate is 1 to 1. Similarly to alum, a greater amount of iron is required to precipitate phosphorus and to react with alkalinity in wastewater. Its competing reaction with natural alkalinity is

$$FeCl_2 \cdot 6H_2O + 3HCO_3^- \rightarrow Fe(OH)_3\downarrow + 3Cl^- + 3CO_2 + 6H_2O \qquad (7.173)$$

The stoichiometric weight ratios of Fe : P is 1.8 : 1. Weight ratio of lime (CaO) to phosphorus is 2.2 : 1. The reaction of Eq. (7.173) is slow. Therefore, lime or some other alkali may be added to raise the pH and supply hydroxyl ion for better coagulation. The reaction is

$$2FeCl_3 \cdot 6H_2O + 3Ca(OH)_2 \rightarrow 2Fe(OH)_3 \downarrow + 3CaCl_2 + 12H_2O \qquad (7.174)$$

Both ferric (Fe^{3+}) and ferrous (Fe^{2+}) ions are used in the precipitation of phosphorus. Ferrous sulfate application for phosphorus removal is similar to that of ferric sulfate.

Calcium ion reacts with phosphate ion in the presence of hydroxyl ion to form hydroxy-apatite. Lime usually comes in a dry form, calcium oxide. It must be mixed with water to form a slurry (calcium hydroxide, $Ca(OH)_2$) in order to be fed to a wastewater. Equipment required for lime precipitation includes a lime-feed device, mixing chamber, settling tank, and pumps and piping.

Phosphorus removal in primary and secondary plants. In conventional primary and secondary treatment of wastewater, phophorus removal is only sparingly undertaken because the majority of

TABLE 7.22 Efficiencies of primary and secondary treatment without and with mineral addition for phosphorus and other constituents removal

	Phosphorus removal %		Suspended solids removal, %		BOD removal, %	
Treatment process	Without	With	Without	With	Without	With
Primary	5–10	70–90	40–70	60–75	25–40	40–50
Secondary						
Activated-sludge	10–20	80–95	85–95	85–95	85–95	85–95
Trickling filter	10–20	80–95	70–92	85–95	80–90	80–95
RBC*	8–12					

*RBC = rotating biological contactor.
Source: USEPA 1976.

phosphorus in wastewater is in soluble form. Primary sedimentation removes only 5 to 10 percent of phosphorus. Secondary biological treatment removes 10 to 20 percent of phosphorus by biological uptake. The effectiveness of primary and secondary treatment without (as in most conventional plants) and with chemical addition for phosphorus removal as well as for SS and BOD removals is presented in Table 7.22. Generally, the total phosphorus concentration of 10 mg/L in the raw wastewater is reduced to about 9 mg/L in the primary effluent and to 8 mg/L in the secondary effluent.

EXAMPLE: Alum is added in the aeration basin of a conventional activated-sludge process for phosphorus removal. The effluent limit for total phosphorus is 0.5 mg/L. The alum is dosed at 140 mg/L. The average total phophorus concentrations are 10.0 and 9.0 mg/L respectively for the influent and effluent (influent of the aeration basin) of the primary clarifier. Determine the molar ratio of aluminum to phosphorus and the weight ratio of alum dosed to the phosphorus content in the wastewater. Estimate the amount of sludge generated, assuming 0.5 mg/L of biological solids are produced by 1 mg/L of BOD reduction and using the following given data for the aeration basin (with alum dosage)

Influent BOD = 148 mg/L

Effluent BOD = 10 mg/L

Influent TSS = 140 mg/L

Effluent TSS = 12 mg/L

Also estimate the increase of sludge production of chemical–biological treatment compared to biological treatment, assuming 30 mg/L for both TSS and BOD in the conventional system.

Solution:

Step 1. Compute the molar ratio of Al to P

$$\begin{aligned} \text{Molecular wt of } Al_2(SO_4)_3 \cdot 18H_2O &= 27 \times 2 + 3(32 + 16 \times 4) + 14(2 + 16) \\ &= 666 \\ \text{Alum dosage} &= 140 \text{ mg/L} \\ \text{Aluminum dosage} &= 140 \text{ mg/L } (2 \times 27/594) \\ &= 12.7 \text{ mg/L} \\ \text{Molar ratio of Al : P} &= 12.7 \text{ mg/L} : 9.0 \text{ mg/L} \\ &= 1.4 : 1 \end{aligned}$$

Step 2. Compute the weight ratio of alum dosed to P in the influent

$$\frac{\text{Alum dosed}}{\text{P in the influent}} = \frac{140 \text{ mg/L}}{9.0 \text{ mg/L}} = \frac{15.6}{1} \tag{7.174}$$

Step 3. Estimate sludge residue for TSS removal

$$\text{TSS removed} = 140 \text{ mg/L} - 12 \text{ mg/L} = 128 \text{ mg/L}$$

Step 4. Estimate biological solids from BOD removal

$$\begin{aligned}\text{Biological solids} &= 0.5\ (148 \text{ mg/L} - 10 \text{ mg/L})\\ &= 69 \text{ mg/L}\end{aligned}$$

Step 5. Estimate organic P in biological solids
Assuming organic P in biological solids removed is 2% by weight,

$$\begin{aligned}\text{P in biological solids} &= 69 \text{ mg/L} \times 0.02\\ &= 1.4 \text{ mg/L}\end{aligned}$$

Step 6. Compute P removed by alum precipitation

$$\begin{aligned}\text{P removed by alum} &= \text{P in (influent} - \text{biological solids} - \text{effluent)}\\ &= (9.0 - 1.4 - 0.5) \text{ mg/L}\\ &= 7.1 \text{ mg/L}\end{aligned}$$

Step 7. Compute $AlPO_4$ precipitate

$$\begin{aligned}AlPO_4 \text{ precipitate} &= \frac{\text{(P removed by alum) (M.wt. of } AlPO_4)}{\text{(M.wt. of P)}}\\ &= \frac{7.1 \text{ mg/L } (27 + 31 + 16 \times 4)}{31}\\ &= 27.9 \text{ mg/L}\end{aligned}$$

Step 8. Compute unused alum
Refer to Eq. (7.167) to find alum used

$$\begin{aligned}\text{Alum used} &= \frac{\text{(P removed)(M.wt. of alum)}}{2 \times \text{M.wt. of P}}\\ &= \frac{7.1 \text{ mg/L} \times 666}{2 \times 31}\\ &= 76.3 \text{ mg/L}\end{aligned}$$

$$\begin{aligned}\text{Unused alum} &= \text{alum dosed} - \text{alum used} = 140 \text{ mg/L} - 76.3 \text{ mg/L}\\ &= 63.7 \text{ mg/L}\end{aligned}$$

Step 9. Compute $Al(OH)_3$ precipitate
Referring to Eq. (7.172), the unused alum reacts with natural alkalinity.

$$\begin{aligned}Al(OH)_3 \text{ precipitate} &= \frac{\text{(unused alum)(2} \times \text{M.wt. of } Al(OH)_3)}{\text{M.wt. of alum}}\\ &= \frac{63.7 \text{ mg/L}(2 \times (27 + 17 \times 3))}{666}\\ &= 14.9 \text{ mg/L}\end{aligned}$$

Step 10. Compute total sludge produced in the secondary clarifier

$$\text{Total sludge} = \text{TSS removed} + \text{biological solids} + AlPO_4 \downarrow + Al(OH)_3 \downarrow$$
$$= (128 + 69 + 27.9 + 14.9)\ \text{mg/L}$$
$$= 239.8\ \text{mg/L}$$

Step 11. Estimate sludge generated from the conventional activated-sludge system

$$\text{TSS removal} = 140\ \text{mg/L} - 30\ \text{mg/L} = 110\ \text{mg/L}$$
$$\text{Biological sludge} = 0.5\ (148\ \text{mg/L} - 30\ \text{mg/L})$$
$$= 59\ \text{mg/L}$$
$$\text{Total sludge} = 110\ \text{mg/L} + 59\ \text{mg/L}$$
$$= 169\ \text{mg/L}$$

Step 12. Increase of sludge production by chemical–biological process over biological process

$$\text{Increase} = 243.8\ \text{mg/L} - 169\ \text{mg/L}$$
$$= 74.8\ \text{mg/L}$$
$$\%\ \text{increase} = 74.8\ \text{mg/L} \times 100\%/169\ \text{mg/L}$$
$$= 44.3\%$$

EXAMPLE 2: A wastewater has a soluble phosphorus concentration of 7.0 mg/L as P. Determine the stoichiometric quantity of ferric chloride required to remove P completely.

Solution:

Step 1. Calculate the gram molecular weights
Equation (6.169) shows that one mole of ferric chloride reacts with 1 mole of orthophosphate to be removed. The pertinent gram molecular weights for $FeCl_3$ and P are:

$$FeCl_3 = 55.845 + 35.453 \times 3 = 162.204\ (\text{g})$$
$$P = 30.975\ \text{g}$$

Step 2. Calculate the stoichiometric amount (Y) of $FeCl_3$ required

$$Y = 7.0\ \text{mg/L}\ (162.204\ \text{g}/30.974\ \text{g})$$
$$= 36.66\ \text{mg/L}$$

Notes: In practice, the actual $FeCl_3$ dose would be 1.5 to 3.0 times the stoichiometric amount. Likewise, the alum dose would be 1.3 to 2.5 times the stoichiometric amount due to side reaction, solubility products limitations, and wastewater quality variations. Jar tests on wastewater are needed for determining the actual amount.

Phosphorus removal by mineral addition to secondary effluent. Generally, an alum dosage of about 200 mg/L is required for phosphorus removal from typical municipal raw wastewater and a dosage of 50 to 100 mg/L is sufficient for secondary effluent. Iron salts have little application because of residual iron remaining in the treated wastewater. An Al : P molar ratio of 1 : 1 to 2 : 1 is required. The optimum pH for alum treatment is near 6.0, whereas that for iron is near 5.0 (Recht and Ghassemi 1970). The pH of high alkalinity waters may be reduced either by using high dosages of alum or by adding supplementary dosages of sulfuric acid. Anionic polyelectrolytes (coagulant aids) can be used to enhance P removal. If higher phosphorus removal is required, a filtration process must be used to achieve 0.1 mg/L of residual P.

The surface overflow rates for settling tanks range from 24 to 58 $m^3/(m^2 \cdot d)$ (580 to 1440 gal/$(d \cdot ft^2)$). Filtration rates are 0.08 to 0.20 $m^3/(m^2 \cdot min)$ (2 to 5 gal/$(min \cdot ft^2)$) (USEPA 1976).

EXAMPLE: Estimate the daily liquid alum requirement for phosphorus removal from the secondary effluent which has an average design flow of 1 Mgal/d (3785 m^3/d) and an average phosphorus concentration of 8 mg/L. Assume that the specific weight of alum is 11.1 lb/gal and alum contains 4.37 percent of aluminum.

Solution:

Step 1. Calculate the mass of the incoming P loading rate

$$\text{Load} = 1 \text{ Mgal/d} \times 8 \text{ mg/L} \times 8.34 \text{ lb/(Mgal} \cdot \text{mg/L)}$$
$$= 66.7 \text{ lb/d}$$

Step 2. Determine aluminum in mol/d required
Atomic weights of phosphorus and aluminum are 31 and 27, respectively.

$$\text{P load} = 66.7 \text{ lb/d} \div 31 \text{ lb/mol}$$
$$= 2.15 \text{ mol/d}$$

Usually a molar ratio of Al : P = 2 : 1 is used, referring to Eq. (7.167)

$$\text{Al required} = 2.15 \text{ mol/d} \times 2$$
$$= 4.3 \text{ mol/d}$$

Step 3. Determine liquid alum required

$$\text{Mass of Al} = 4.3 \text{ mol/d} \times 27 \text{ lb/mol}$$
$$= 116.1 \text{ lb/d}$$

Note: Or this can be solved simply as

$$\frac{\text{Al required}}{\text{P load}} = \frac{2 \times 27}{31}$$
$$\text{Al required} = 66.7 \text{ lb/d} \times 54/31$$
$$= 116.1 \text{ lb/d}$$

Using liquid alum having 4.37% of Al, with 11.1 lb/gal

$$\text{Liquid alum required} = \frac{116.1 \text{ lb/d}}{0.0437(11.1 \text{ lb/gal})}$$
$$= 239 \text{ gal/d} = 0.166 \text{ gal/min}$$
$$= 0.628 \text{ L/min}$$

Phosphorus removal by lime treatment of secondary effluent. For phosphorus removal, lime can be added to the primary sedimentation tank and to the secondary effluent. Lime treatment of wastewater is essentially the same process as is used for lime softening of drinking water, but with a different purpose. While softening may occur, the primary objective is to remove phosphorus by precipitation as hydroxyapatite (Eq. (7.171)).

During phosphorus precipitation, a competing reaction of lime with alkalinity will occur. This reaction results in calcium removal, an action of softening which has a very important effect on the phosphorus removal efficiency of the process. This reaction may occur in two ways:

$$Ca(OH)_2 + Ca(HCO_3)_2 \rightarrow 2\ CaCO_3 \downarrow + 2H_2O \tag{7.175}$$
$$Ca(OH)_2 + NaHCO_3 \rightarrow CaCO_3 \downarrow + NaOH + H_2O \tag{7.176}$$

Another reaction may occur to precipitate magnesium hydroxide:

$$Mg^{2+} + 2OH^- \rightarrow Mg(OH)_2 \downarrow \quad (7.177)$$

In the two-stage lime treatment process, recarbonation after the first stage is required to reduce the pH and to precipitate excess lime as calcium carbonate by adding carbon dioxide. The chemical reaction is

$$Ca^{2+} + CO_2 + 2OH^- \rightarrow CaCO_3 + H_2O \quad (7.178)$$

Lime can be added to secondary effluent by single-stage or two-stage treatment. For the single-stage process (Figure (7.36)), lime is mixed with feed wastewater to raise the pH (10 to 11); this is then followed by flocculation and sedimentation. The clarified water is adjusted to lower pH with carbon dioxide and is filtered through a multimedia filter to prevent post-precipitation of calcium carbonate before discharge. The settled lime sediment may be disposed of to a landfill or may be recalcined for recovery of lime. In recovery, the sediment is thickened, dewatered by centrifuge or vacuum filter, calcined, and then reused.

For a typical two-stage process (Figure (7.37)), in the first-stage clarifier of the two-stage process, sufficient lime is dosed to raise the pH above 11 to precipitate the soluble phosphorus as hydroxyapatite, calcium carbonate, and magnesium hydroxide.

The calcium carbonate precipitate formed in the process acts as a coagulant for suspended solids removal. The excess soluble calcium is removed in the second clarifier as a calcium carbonate precipitate by adding carbon dioxide gas to reduce the pH to about 10. The calcium carbonate settles out in the secondary (or tertiary) clarifier. To remove the residual levels of phosphorus and suspended solids, the secondary (or tertiary) clarified effluent is filtered, then discharged. Usually pH is reduced to about 8.0 and 8.5 in first-stage recarbonation and further reduced to 7.0 in second-stage recarbonation. Fifteen minutes' contact time in the recarbonation basin is recommended.

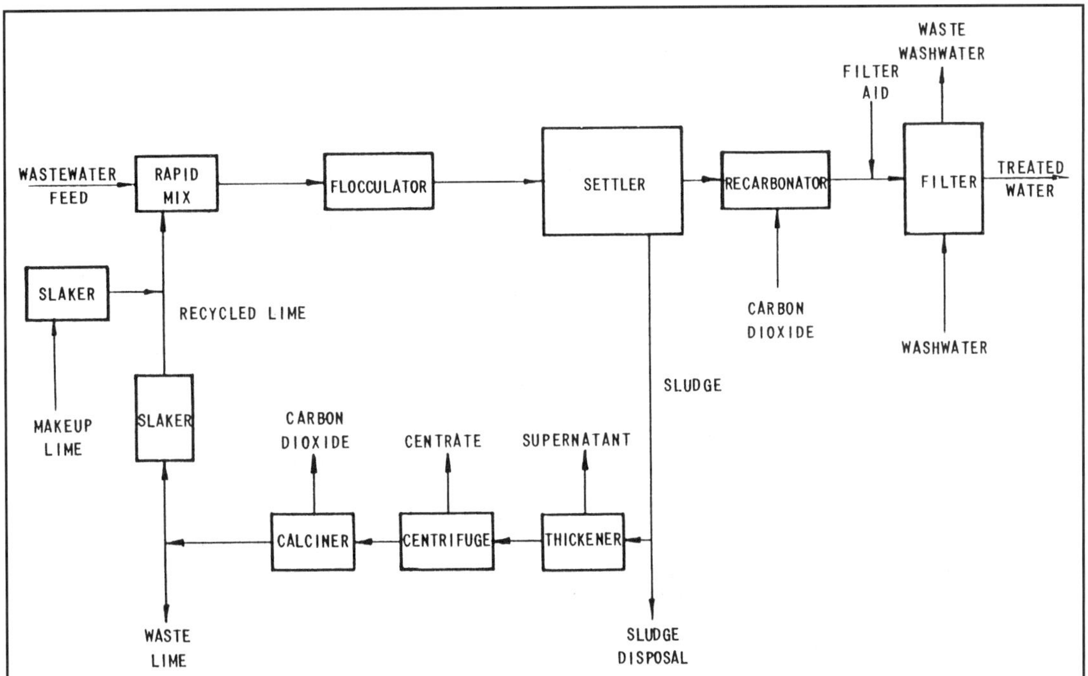

FIGURE 7.36 Single-stage lime treatment process for phosphorus removal (*Source*: *USEPA 1976*).

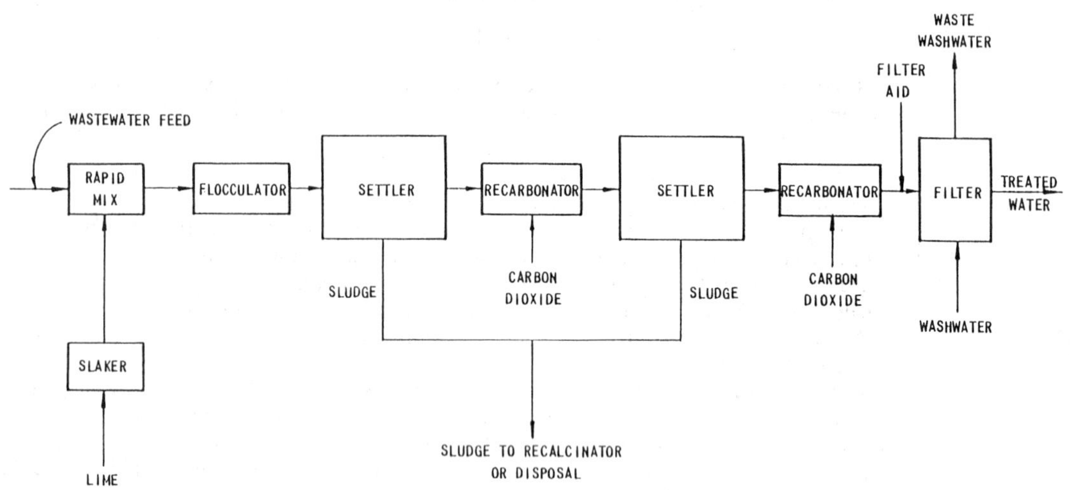

FIGURE 7.37 Two-stage lime treatment process for phosphorus removal (*Source: USEPA 1976*).

(Typical hydraulic loading rates for chemical treatment clarifiers are 32 to 61 $m^3/(m^2 \cdot d)$ (800 to 1500 $gal/(d \cdot ft^2)$).

EXAMPLE 1: From the results of jar tests, calcium concentration drops from 56 to 38 mg/L when the pH is reduced from 10 to 7 by addition of carbon dioxide. Determine the carbon dioxide dose rate required for recarbonation.

Solution:

Step 1. Determine theoretical Ca : CO_2 ratio, using Eq. (7.178)

$$Ca^{2+} + CO_2 + 2OH^- \rightarrow CaCO_3 \downarrow + H_2O$$

then

$$\begin{aligned} CO_2 : Ca &= (12 + 2 \times 16) : 40 \\ &= 1.1 : 1 \end{aligned}$$

Step 2. Calculate CO_2 theoretical requirement

$$\begin{aligned} CO_2 \text{ required} : \text{Ca reduction} &= 1.1 : 1 \\ CO_2 &= (56 - 38) \text{ mg/L} \times 1.1 \\ &= 19.8 \text{ mg/L} \\ &\cong 20 \text{ mg/L} \end{aligned}$$

Step 3. Calculate field requirement of CO_2
A safe factor of 20% should be added to calculated dosage to compensate for inefficiency in absorption.

$$\begin{aligned} CO_2 \text{ required} &= 20 \text{ mg/L} \times 1.2 \\ &= 24 \text{ mg/L} \end{aligned}$$

EXAMPLE 2: A single-stage lime treatment system is used as advanced treatment for phosphorus removal. Estimate the residue mass and volume generated in the clarifier under the following conditions.

Influent design flow = 37,850 m^3/d (10 Mgal/d)
Lime dosage as $Ca(OH)_2$ = 280 mg/L
Influent PO_4^{3-} as P = 8 mg/L
Influent Ca^{2+} = 62 mg/L
Influent Mg^{2+} = 3 mg/L
Effluent PO_4^{3-} as P = 0.6 mg/L
Effluent Ca^{2+} = 20 mg/L
Effluent Mg = 0 mg/L
Influent suspended solids = 20 mg/L
Effluent suspended solids = 2 mg/L
Specific gravity of residue = 1.07
Moisture content of residue = 92%

Solution:

Step 1. Compute the mass of $Ca_5(PO_4)_3(OH)$ formed

(a) Calculate the moles of P removed

$$\text{Mole of P removed} = \frac{8\text{ mg/L} - 0.6\text{ mg/L}}{30.974\text{ g/mol} \times 1000\text{ mg/g}}$$
$$= 0.239 \times 10^{-3}\text{ mol/L}$$
$$= 0.239\text{ mM/L}$$

where mM = millimol

(b) Calculate the moles of $Ca_5(PO_4)_3(OH)$ formed
From Eq. (7.171), 3 moles of P removed will form 1 mole of $Ca_5(PO_4)_3(OH)$

$$\text{mole of } Ca_5(PO_4)_3(OH) \text{ formed} = 1/3 \times 0.239\text{ mM/L}$$
$$= 0.080\text{ mM/L}$$

(c) Calculate the mass of $Ca_5(PO_4)_3(OH)$ formed

$$\text{Mass of 1 mole } Ca_5(PO_4)_3(OH) = 40 \times 5 + 30.97 \times 3 + 16 \times 13 + 1\text{ g/mol}$$
$$= 501.9\text{ g/mol}$$
$$= 501.9\text{ mg/mM}$$
$$\text{Mass of } Ca_5(PO_4)_3(OH) \text{ formed} = 0.08\text{ mM/L} \times 501.9\text{ mg/mM}$$
$$= 40.2\text{ mg/L}$$

Step 2. Calculate the mass of calcium carbonate formed

(a) Calculate the mass of Ca added in the lime dosage

$$\text{Mass of Ca in } Ca(OH)_2 = 280\text{ mg/L} \times 40/(40 + 17 \times 2)$$
$$= 151.4\text{ mg/L}$$

(b) Determine the mass of Ca^{2+} in $Ca_5(PO_4)_3(OH)$

$$\text{Mass of Ca in } Ca_5(PO_4)_3(OH) = 5\ (40\ \text{mg/mM}) \times 0.08\ \text{mM/L} = 16\ \text{mg/L}$$

(c) Calculate the mass of total Ca present in $CaCO_3$

$$\begin{aligned}\text{Ca in } CaCO_3 &= \text{Ca in } Ca(OH)_2 + \text{Ca in influent} - \text{Ca in } Ca_5(PO_4)_3(OH) - \text{Ca in effluent}\\ &= (151.4 + 62 - 16 - 20)\ \text{mg/L}\\ &= 177.4\ \text{mg/L}\end{aligned}$$

(d) Convert the mass of Ca as expressed in $CaCO_3$

$$\begin{aligned}\text{Mass of Ca as } CaCO_3 &= 177.4\ \text{mg/L} \times (40 + 12 + 48)/40\\ &= 443.5\ \text{mg/L}\end{aligned}$$

Step 3. Determine the mass of $Mg(OH)_2$ formed

(a) Calculate the mass of Mg removed

$$\begin{aligned}\text{Mole of } Mg^{2+} \text{ removed} &= 3\ \text{mg/L}/(24.3\ \text{g/mol} \times 1000\ \text{mg/g})\\ &= 0.123 \times 10^{-3}\ \text{mol/L}\\ &= 0.123\ \text{mM/L}\end{aligned}$$

(b) Calculate the mass of $Mg(OH)_2$ formed

$$\begin{aligned}\text{Mass of } Mg(OH)_2 &= 0.123\ \text{mM/L} \times (24.3 + 17 \times 2)\ \text{mg/mM}\\ &= 7.2\ \text{mg/L}\end{aligned}$$

Step 4. Calculate the total mass of residue removed as a result of the lime application

(a) Mass of residue due to lime dosage

$$\begin{aligned}\text{Chemical mass} &= \text{sum of mass from } Ca_5(PO_4)_3(OH),\ CaCO_3, \text{ and Mg } (OH)_2\\ &= (40.2 + 443.5 + 7.2)\ \text{mg/L} \times 37{,}850\ \text{m}^3/\text{d}\\ &= 490.9\ \text{g/m}^3 \times 37{,}850\ \text{m}^3/\text{d}\\ &= 18{,}580{,}000\ \text{g/d}\\ &= 18{,}580\ \text{kg/d}\end{aligned}$$

(b) Calculate the mass of SS removed

$$\begin{aligned}\text{Mass of SS} &= (20 - 2)\ \text{mg/L} \times 37{,}850\ \text{m}^3/\text{d}\\ &= 681\ \text{kg/d}\end{aligned}$$

(c) Calculate the total mass of residues

$$\begin{aligned}\text{Total mass} &= \text{mass of chemical residue} + \text{mass of SS}\\ &= 18{,}580\ \text{kg/d} + 681\ \text{kg/d}\\ &= 19{,}261\ \text{kg/d}\end{aligned}$$

Step 5. Determine total volume V of residue from lime precipitation process
Given: specific gravity of residue = 1.07, moisture content of residue = 92%, then

$$\text{Solid content} = 8\%$$

$$V = \frac{19{,}261 \text{ kg/d}}{1000 \text{ kg/m}^3 \times 1.07 \times 0.08}$$

$$= 225 \text{ m}^3/\text{d}$$

$$= 7945 \text{ ft}^3/\text{d}$$

Phosphorus removal by biological processes. Numerous biological methods of removing phosphorus have been developed. Integrated biological processes for nutrient removal from wastewater use a combination of biological and chemical methods to bring phosphorus and nitrogen concentrations to below the effluent standards. Some patented (17 years) integrated biological processes have been installed in the United States. The A/O process (Fig. 7.38) for mainstream P removal, Phostrip process (Fig. 7.39) for sidestream P removal, sequencing batch reactor (SBR, Fig. 7.40), sidestream fermentation process (OWASA nitrification, Fig. 7.41), and chemical polishing are used for phosphorus removal. Combined removal of phosphorus and nitrogen by biological methods includes the A^2/O process (Fig. 7.42a), the modified (5-stage) Bardenpho process (Fig. 7.42b), the University of Cape Town (UCT) process (Fig. 7.42c), the VIP process (Virginia Institute Plant in Norfolk, Virginia) (Fig. 7.42d), PhoStrip II process (Fig. 7.43), SBR (Fig. 7.44), and phased isolation ditch. For nitrogen control, biological denitrification includes the Wuhrmann process (Fig. 7.45), Ludzack–Ettinger process (Figs. 7.46 and 7.47), Bordenpho (4-stage) process, oxidation ditch, phase isolation ditch, dual-sludge process (Fig. 7.48), triple-stage process (Fig. 7.49), denitrification filter, RBC, and fluidized bed process. The basic theory, stoichiometry, kinetics, design considerations, and practice of integrated systems in the United States are discussed in detail elsewhere (WEF and ASCE 1991b, Metcalf and Eddy, Inc. 1991).

Biological processes for nutrient removal feature the exposure of alternate anaerobic and aerobic conditions to the microorganisms so that their uptake of phosphorus will be above normal levels. Phosphorus is not only utilized for cell synthesis, maintenance, and energy transport, but is also stored for subsequent use by the microorganisms. The sludge produced containing phosphorus is either wasted or removed through a sidestream (return sludge stream). The alternate exposure of the microorganisms to anaerobic and aerobic (oxic) conditions in the main biological treatment is called the mainstream process. A design example (Ex. 15.2) is illustrated by WEF and ASCE (1991b); this uses the activated-sludge kinetic model developed by Lawrence and McCarty (1970), combined with empirical equations for complete denitrification using a four-stage Bardenpho process.

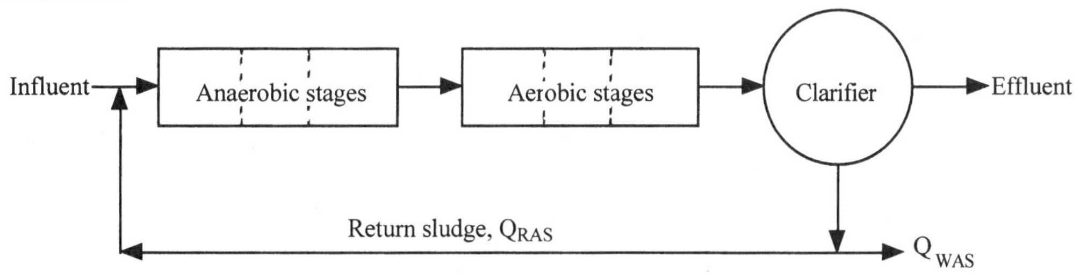

FIGURE 7.38 A/O process.

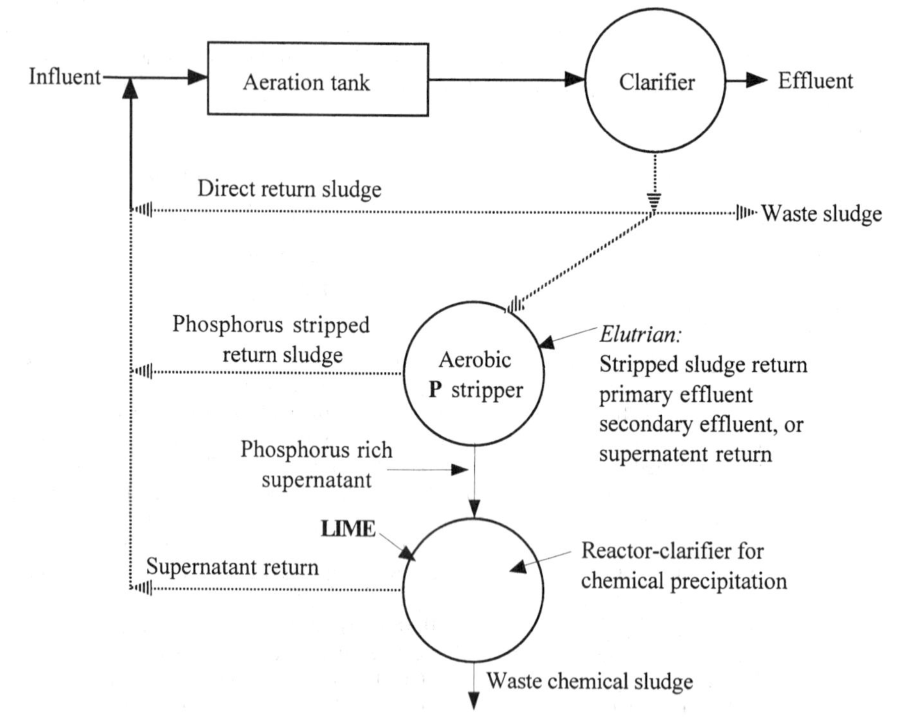

FIGURE 7.39 PhoStrip process.

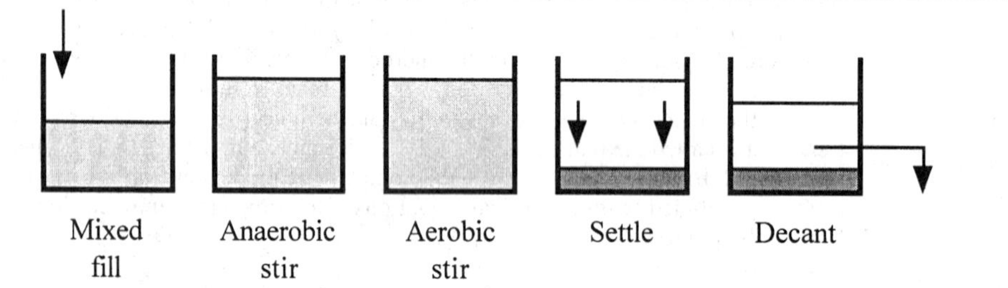

FIGURE 7.40 Sequencing batch reactor for removal of phosphorus and carbonaceous BOD.

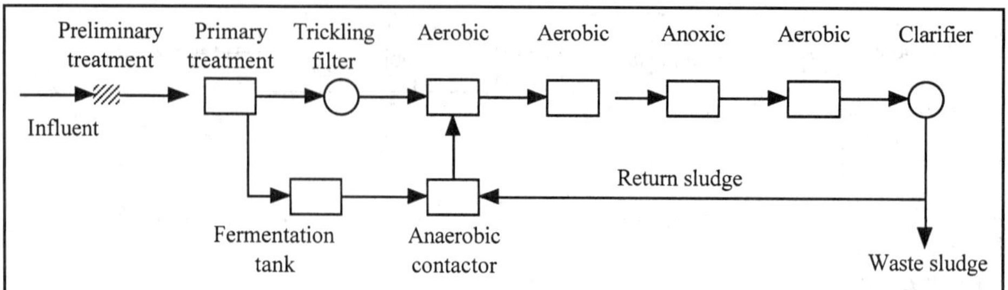

FIGURE 7.41 OWASA nitrification process.

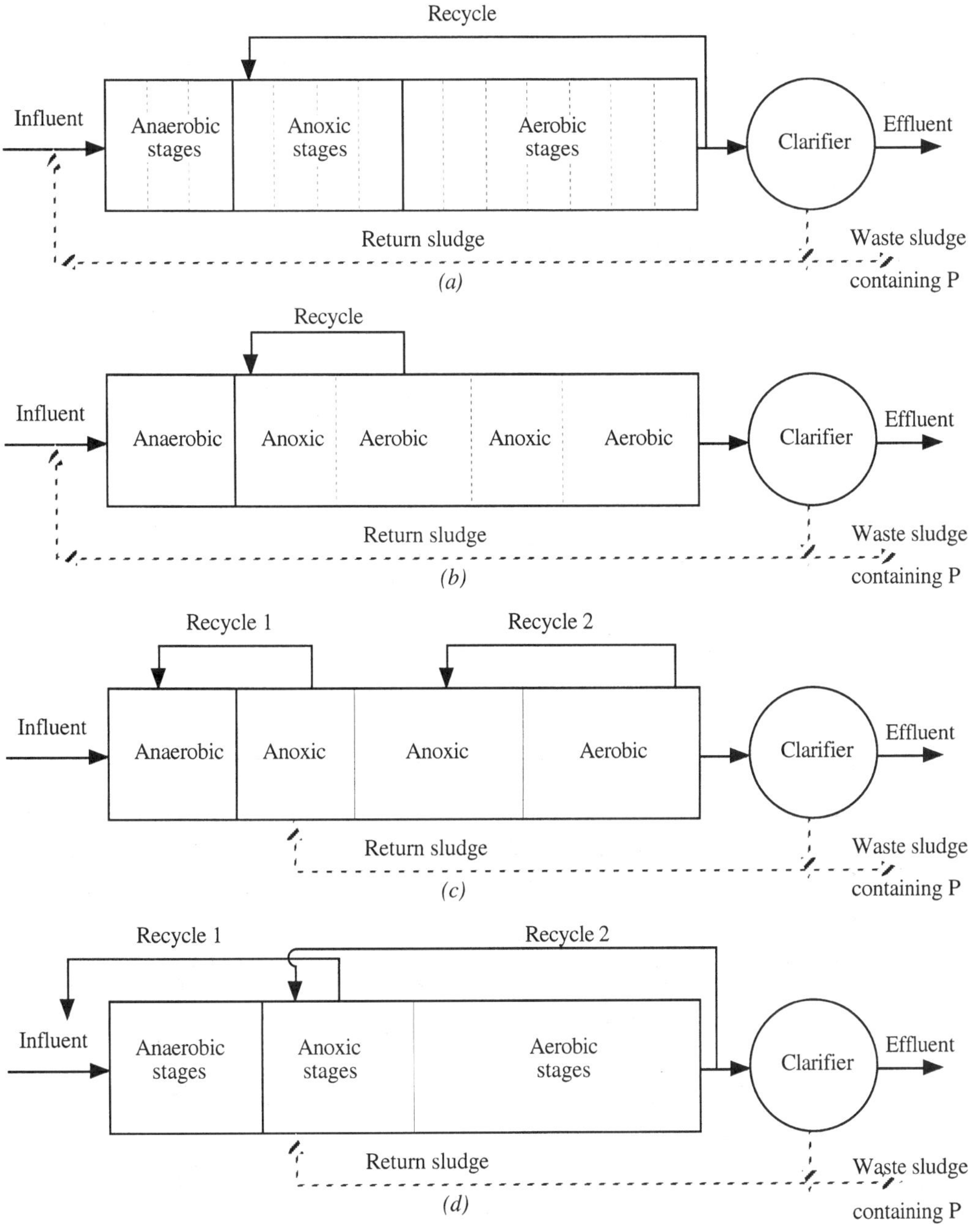

FIGURE 7.42 Combined biological nitrogen and phosphorus removal processes: (a) A^2/O process; (b) five-stage Bardenpho process; (c) UCT process; and (d) VIP process (*Source: Metcalf and Eddy, Inc. 1991*).

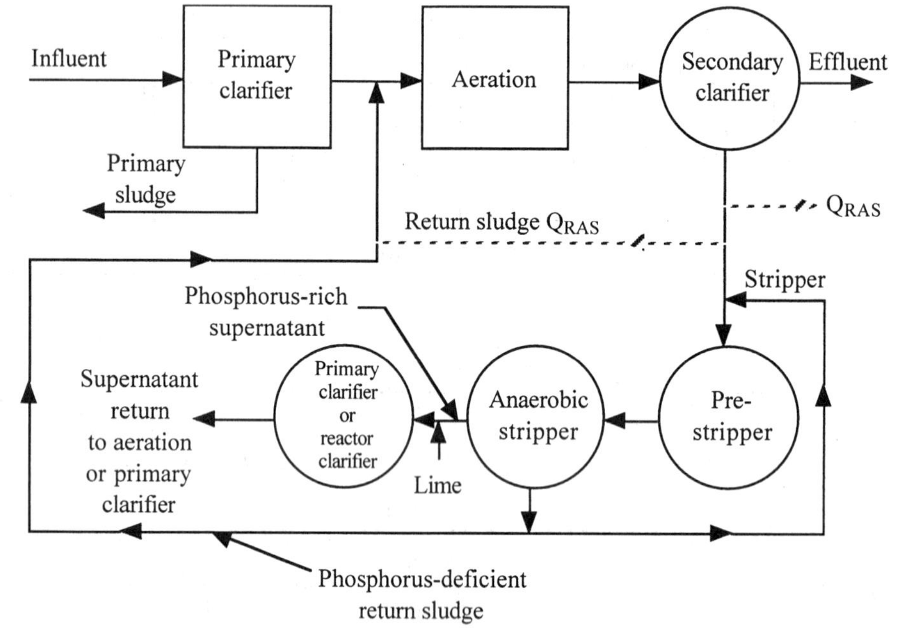

FIGURE 7.43 PhoStrip II process for phosphorus and nitrogen removal.

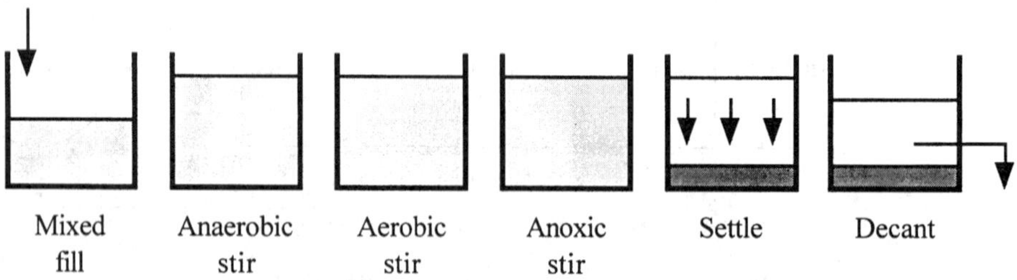

FIGURE 7.44 Sequencing batch reactor for carbon oxidation plus phosphorus and nitrogen removal.

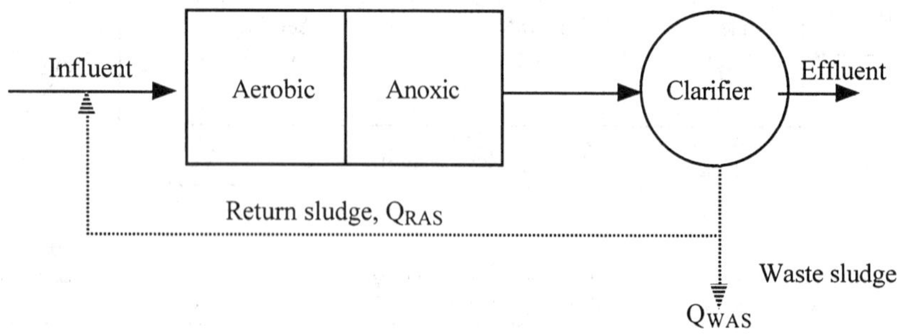

FIGURE 7.45 Wuhrmann process for nitrogen removal.

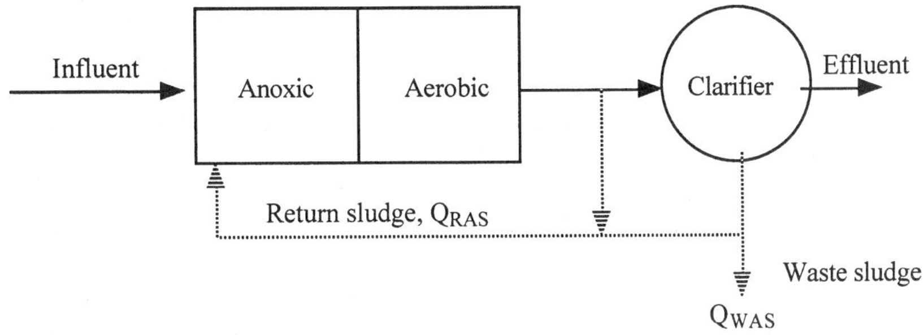

FIGURE 7.46 Ludzack–Ettinger process for nitrogen removal.

28.3 Nitrogen Control

Nitrogen (N) in wastewater exists commonly in the form of organic, ammonia, nitrite, nitrate, and gaseous nitrogen. The organic nitrogen includes both soluble and particulate forms. The soluble organic nitrogen is mainly in the form of urea and amino acids. The main sources are from human excreta, kitchen garbage, and industrial (food processing) wastes. Typical domestic wastewater contains 20 mg/L of organic nitrogen and 15 mg/L of inorganic nitrogen.

Environmental effects. Nitrogen compounds, particularly ammonia, will exert a significant oxygen demand through biological nitrification and may cause eutrophication in receiving waters. Ammonia (unionized) can be toxic to aquatic organisms and it readily reacts with chlorine (affecting disinfection efficiency). A high nitrate (NO_3) level in water supplies has been reported to cause methemoglobinemia in infants. The need for nitrogen control in wastewater effluents has generally been recognized.

Treatment process. Many treatment processes have been developed with the specific purpose of transforming nitrogen compounds or removing nitrogen from the wastewater stream.

Conventional processes. In conventional treatment processes, the concentration of inorganic nitrogen is not affected by primary sedimentation and is increased more than 50 percent to 24 mg/L after secondary (biological) treatment. The overall primary and secondary treatment removes 25 to 75 percent

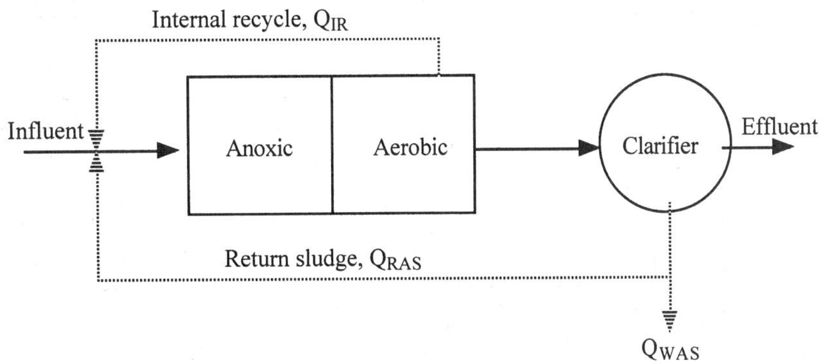

FIGURE 7.47 Modified Ludzack–Ettinger process for nitrogen removal.

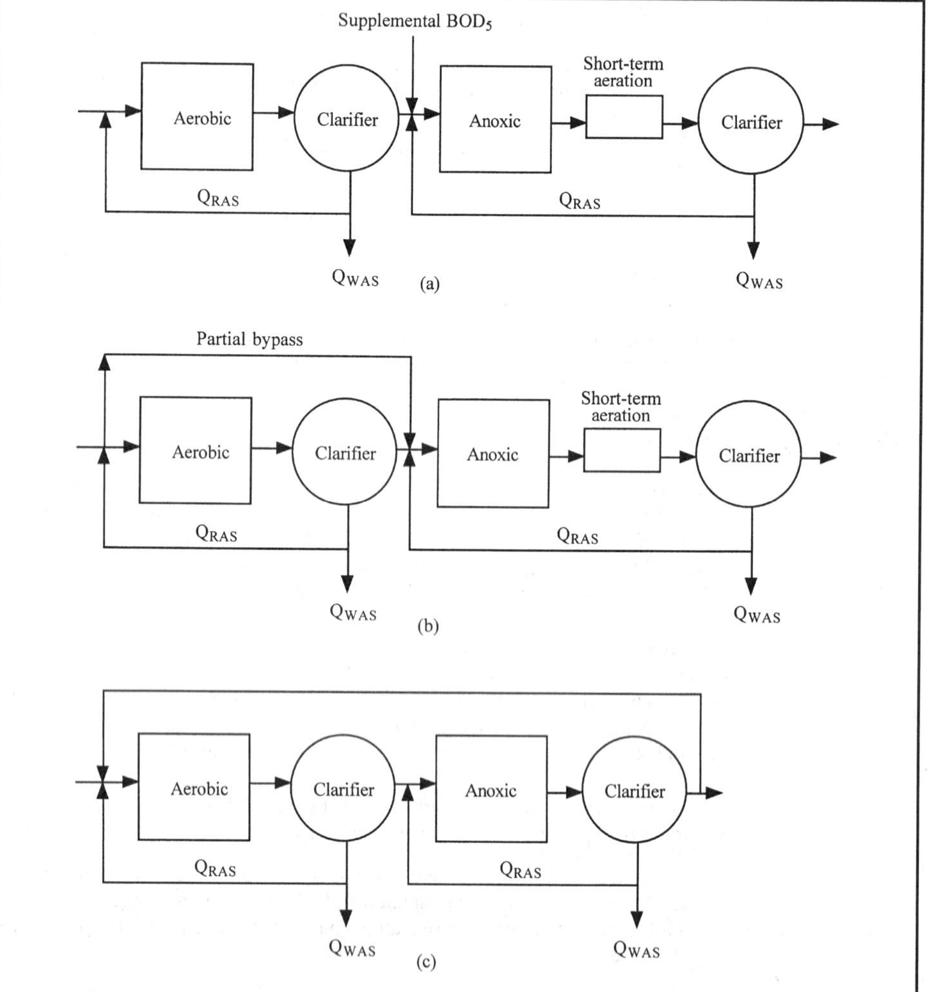

FIGURE 7.48 Dual-sludge processes for nitrogen removal (*Source: Grady and Lim 1980*).

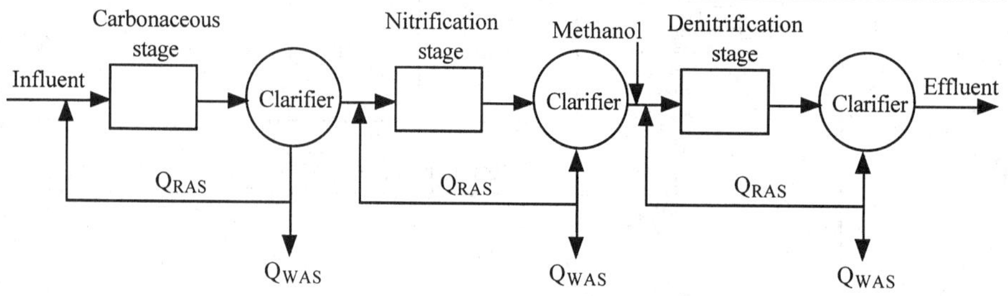

FIGURE 7.49 Triple-sludge process for nitrogen removal.

(5 to 15 mg/L) of organic nitrogen. Typically, about 14 and 26 percent of total nitrogen from raw wastewater are removed by conventional primary and secondary treatment processes, respectively.

Biological processes remove particulate organic nitrogen and transform some to ammonium and other inorganic forms. A fraction of the ammonium present will be assimilated into organic materials of cells. Soluble organic nitrogen is partially transformed to ammonium by microorganisms, with 1 to 3 mg/L of organic nitrogen remaining insoluble in the secondary effluent.

Advanced processes. Advanced wastewater treatment processes designed to remove wastewater constituents other than nitrogen often remove some nitrogen compounds as well. Removal is usually limited to particulate forms, and the removal efficiency is not high.

Tertiary filtration removes the suspended organic nitrogen from the secondary effluent. However, the majority of nitrogen is inorganic (ammonium). Reverse osmosis and electrodialysis can be used as a tertiary process for ammonium removal. Their effectiveness is 80 and 40 percent, respectively. In practice, RO and electrodialysis are not used for wastewater treatment.

Chemical coagulation for phosphorus removal also removes particulate organic nitrogen. The process for nitrogen control may be divided into two categories, i.e. nitrification and nitrification–denitrification, depending on the quality requirements of wastewater effluent. The nitrification process is the oxidation of organic and ammonia nitrogen (NH_3-N) to nitrate, a less objectionable form; it is merely the conversion of nitrogen from one form to another form in the wastewater. Denitrification is the reduction of nitrate to nitrogen gas, constituting a removal of nitrogen from wastewater. Nitrification is used only to control NH_3-N concentration in the wastewater. Nitrification–denitrification is employed to reduce the total level of nitrogen in the effluent.

There are several methods of nitrogen control. They are biological nitrification–denitrification, break-point chlorination, selective ion exchange, and air (ammonia) stripping. This section is mainly devoted to biological nitrification or biological technologies, since most processes have been discussed previously in Chapter 1.6.

Biological nitrification can be achieved in separate stage processes following secondary treatment (most cases) or in combination for carbon oxidation–nitrification and for carbon oxidation–nitrification–denitrification. The secondary effluent with its high ammonia content and low BOD provides greater growth potential for the nitrifiers relative to the heterotrophic bacteria. The nitrification process is operated at an increased sludge age to compensate for lower temperature. After the nitrifiers have oxidized the ammonia in the aeration tank, the activated sludge containing a large fraction of nitrifiers is settled in the final clarifier for return to the aeration tank and for waste.

Nitrification reaction. Biological nitrification is an aerobic autotrophic process in which the energy for bacterial growth is derived from the oxidation of inorganic compounds, primarily ammonia nitrogen. Autotrophic nitrifiers, in contrast to heterotrophs, use inorganic carbon dioxide instead of organic carbon for cell synthesis. The yield of nitrifier cells per unit of substrate metabolized is many times smaller than that for heterotrophic bacteria.

Although a variety of nitrifying bacteria exist in nature, the two genera associated with biological nitrification are *Nitrosomonas* and *Nitrobacter*. The oxidation of ammonia to nitrate is a two-step process requiring both nitrifiers for the conversion. *Nitrosomonas* oxidizes ammonia to nitrite, while *Nitrobacter* subsequently transforms nitrite to nitrate. The respective oxidation reactions are as follows:

Ammonia oxidation:

$$NH_4^+ + 1.5O_2 + 2HCO_3^- \xrightarrow{\textit{Nitrosomonas}} NO_2^- + 2H_2CO_3 + H_2O \tag{7.179}$$

Nitrite oxidation:

$$NO_2^- + 0.5O_2 \xrightarrow{\textit{Nitrobacter}} NO_3^- \tag{7.180}$$

Overall reaction:

$$NH_4^+ + 2O_2 + 2HCO_3^- \xrightarrow{\text{nitrifiers}} NO_3^- + 2H_2CO_3 + H_2O \tag{7.181}$$

Note: The above three equations are essentially the same as Eqs. (2.87), (2.88), and (2.89), respectively.

To oxidize 1 mg/L of NH_3-N, theoretically 4.56 mg/L of oxygen is required when synthesis of nitrifiers is neglected. As ammonia is oxidized and bicarbonate is utilized, nitrate is formed and carbonic acid is produced. The carbonic acid will further depress the pH of the wastewater. Theoretically, alkalinity of 7.14 mg/L $CaCO_3$ is consumed for each mg/L of NH_3-N oxidized. The destruction of alkalinity and increase of carbonic acid results in a drop in pH of the wastewater.

Reviews of the synthesis and energy relationship associated with biological nitrification are available elsewhere (Painter 1970, 1975, Haug and McCarty 1972, USEPA 1975). Overall oxidation and synthesis–oxidation reactions of ammonia are presented below (USEPA 1975c):

$$55NH_4^+ + 76O_2 + 109HCO_3^- \xrightarrow[\textit{Nitrosomonas}]{} C_5H_7NO_2 + 54NO_2^- + 57H_2O + 104H_2CO_3 \quad (7.182)$$

and

$$400NO_2^- + NH_4^+ + 4H_2CO_3 + HCO_3^- + 195O_2 \xrightarrow[\textit{Nitrobacter}]{} C_5H_7NO_2 + 3H_2O + 400NO_3^- \quad (7.183)$$

The overall synthesis and oxidation reaction is

$$NH_4^+ + 1.83O_2 + 1.98HCO_3^- \rightarrow 0.021C_5H_7NO_2 + 1.041H_2O + 0.98NO_3^- + 1.88H_2CO_3 \quad (7.184)$$

Nitrifying biofilm properties. Although the principal genera *Nitrosomonas and Nitrobacter* are responsible for biological nitrification, heterotrophic nitrification can also occur when nitrite and/or nitrate are produced from organic or inorganic compounds by heterotrophic organisms (> 100 species, including fungi). However, the amount of oxidation nitrogen formed by heterotrophic organisms is relatively small (Painter 1970).

The growth rate for nitrifying bacteria is much less than of heterotrophic bacteria. Nitrifying bacteria have a longer generation time of at least 10 to 30 hours (Painter 1970, 1975). They are also much more sensitive to environmental conditions as well as to growth inhibitors. The growth rate for nitrite oxidizers is much greater than that for the ammonia oxidizers.

According to Painter (1970), experimental yield values of *Nitrosomonas* lay between 0.04 and 0.13 lb volatile suspended solids (VSS) grown per pound of ammonia nitrogen oxidized; and for *Nitrobacter* the yield ranged from 0.02 to 0.07 pounds VSS per pound of nitrite nitrogen oxidized. Yield values based on thermodynamic theory are 0.29 and 0.084 respectively for *Nitrosomonas* and *Nitrobacter* (Haug and McCarty 1972). The lower value of the experimental yield is due to the diversion of a portion of the free energy released by oxidation to microorganism maintenance functions. Based on Eqs. (7.182) and (7.183), the yield for *Nitrosomonas* and *Nitrobacter* should be 0.15 mg cells/mg NH_4^+-N and 0.02 mg cells/mg NO_2-N, respectively.

Kinetics of nitrification. The kinetics of biological nitrification, using mathematical expressions for the oxidation of ammonia and nitrite, have been proposed by many investigators. A variety of environmental factors, such as ammonia concentration, pH, temperature, and dissolved oxygen level, affect the kinetics. For the combined carbon oxidation–nitrification processes, BOD to TKN ratio is also an important factor. The reaction is enhanced with higher pH, higher temperature, and higher DO concentration.

Effect of ammonia concentration on kinetics. Numerous investigators have developed growth kinetics and substrate utilization equations for nitrifiers. However, the most popular method is the Monod model (1949) of population dynamics. The growth rate of nitrifiers is a function of ammonia concentration dynamics, as follows

$$\mu = \frac{1}{X}\frac{dS}{dt} = \hat{\mu}\frac{S}{K_S + S} \quad (7.185)$$

where μ = growth rate of nitrifier, per day
$\hat{\mu}$ = maximum specific growth rate of nitrifier, per day
X = concentration of bacteria
S = concentration of substrate (NH_3-N), mg/L
t = time, days
K_S = half velocity constant, i.e. substrate concentration (mg/L) at half of the maximum growth rate

Equation (7.185) assumes no mass transfer or oxygen transfer limitations.

Values of K_S for *Nitrosomonas* and *Nitrobacter* are generally equal to or less than 1 mg/L (0.18 to 1.0 mg/L mostly) at liquid temperatures of 20°C or less (Haug and McCarty 1972). The Monod expression for biological nitrification provides for a continuous transition between first- and zero-order kinetics based on substrate concentration. Since K_S values for nitrification are much less than ammonia concentrations found in wastewater, many investigators (Huang and Hopson 1974, Wild *et al.* 1971, Kiff 1972) found that the growth kinetic model reduces to a zero-order expression

$$\mu = \frac{1}{X}\frac{dS}{dt} = \hat{\mu} \tag{7.186}$$

Equation (7.186) indicates that the nitrification rate is independent of the initial substrate concentration and the mixing regime. This suggests that nitrifiers are reproducing at or near their maximum growth rate. As substrate is removed, and as S approaches the value of K_S or less, further substrate removal will begin to approximate a first-order reaction.

Estimates of the maximum growth rates ($\hat{\mu}$) of *Nitrosomonas* and *Nitrobacter* under various environmental conditions, found by other investigators, are summarized elsewhere (USEPA 1975c, Brenner *et al.* 1984). Both $\hat{\mu}$ and K_S are found to increase with increasing temperature. All nitrification studies were conducted with activated-sludge processes or river waters. Maximum growth rates might be expected to approximate those of suspended growth process in which ammonia mass transfer and DO are not limiting.

The ranges and typical values of kinetic coefficients for suspended growth nitrification processes using pure cultures are summarized in Table 7.23. In practice, the values of those coefficients for nitrifiers in activated-sludge processes will be considerably less than the values listed in the table.

The reported data indicate that the maximum growth rate of *Nitrobacter* is considerably greater than that of *Nitrosomonas.* Therefore the oxidation of ammonia to nitrite is the rate-limiting reaction in nitrification. For this reason, nitrite does not accumulate in large amounts in mature nitrification systems for municipal wastewaters.

TABLE 7.23 Values of Kinetic Coefficients for the Suspended Growth Nitrification Process (Pure Culture Values)

		Value (20°C)	
Kinetic coefficient	Unit	Range	Typical
Nitrosomonas			
μ_m	per day	0.3–2.0	0.7
K_s	NH_4^+-N, mg/L	0.2–2.0	0.6
Nitrobacter			
μ_m	per day	0.4–3.0	1.0
K_s	NO_2^--N, mg/L	0.2–5.0	1.4
Overall			
μ_m	per day	0.3–3.0	1.0
K_s	NH_4^+-N, mg/L	0.2–5.0	1.4
Y	mg VSS/mg NH_4^+-N	0.1–0.3	0.2
k_d	per day	0.03–0.06	0.05

Sources: USEPA 1975b and c, Schroeder 1977.

Oxidation rate. The ammonia oxidation rate is related to the *Nitrosomonas* growth rate. Its maximum oxidation rate can be expressed as

$$r_N = \frac{\mu_N}{Y_N} = \hat{r}_N \frac{N}{K_N + N} \tag{7.187}$$

where r_N = ammonia oxidation rate, lb NH_4^+-N oxidized/(lb VSS · d)
μ_N = *Nitrosomonas* growth rate, d^{-1}
Y_N = organism yield coefficient, lb *Nitrosomonas* grown (VSS)/lb NH_4^+-N removed
$\hat{r}_N = \hat{\mu}_N / Y_N$ = peak ammonia oxidation rate, lb NH_4^+-oxidized/(lb VSS · d)
$\hat{\mu}_N$ = peak *Nitrosomonas* growth rate, d^{-1}
N = NH_4^+-N concentration, mg/L
K_N = half saturation constant, mg/L NH_4^+-N, mg/L

Loading rate. Referring to Eqs. (7.185) and (7.187), the growth rate of nitrifier is proportional to substrate (NH_3-N) concentration. The ammonia loading rates applied to the biological nitrification unit (aeration tank) are 160 to 320 g/(m^3 · d) (10 to 20 lb/(1000 ft^3 · d)) with corresponding wastewater temperature of 10 to 20°C, respectively. The aeration periods are 4 to 6 h for an average wastewater secondary effluent (Hammer 1986).

BOD concentration. The combined process oxidizes a high proportion of the influent organics ($SBOD_5$) relative to the NH_3-N concentration, while less nitrifiers are present in the biofilm. Separate nitrification processes have relatively low $SBOD_5$ values, relative to the NH_3-N content, when a high level of $SBOD_5$ removal is provided prior to the nitrification stages. High populations of nitrifiers occur in the separate-stage nitrification process.

In combined carbon oxidation–nitrification processes the ratio of BOD to TKN is greater than 5, whereas in separate processes the BOD to TKN ratio in the second stage is greater than 1 and less than 3. Reducing the ratio to 3 does not require a high degree of treatment in the first stage (USEPA 1975b, McGhee 1991).

Biological nitrification is sensitive to organic loading, depending on the carbon removal sections, because of the nitrifiers having a much slower growth rate. Nitrification will take place when organic content is reduced to a certain level. There are discrepancies about what is the critical concentration for organic matter. The values cited in the literature are as follows: $TBOD_5$ approaching 30 mg/L (Antonie 1978, Khan and Raman 1980); $TBOD_5$ reaching 20 mg/L (Banerji 1980); $SBOD_5$ less than 20 mg/L (Autotrol Corp 1979); $SBOD_5$ reduced to 15 mg/L; $SBOD_5$ around 10 mg/L (Miller *et al.* 1980).

Temperature. The optimum range for nitrification is 30°C to 36°C (Haug and McCarty 1972, Ford *et al.* 1980). Nitrifiers have been found not to grow at temperatures below 4°C or above 45°C (Ford *et al.* 1980).

In a suspended-growth activated-sludge system, Downing and Hopwood (1964) found that the maximum growth $\hat{\mu}$ and the half-velocity constant K_N for both *Nitrosomonas* and *Nitrobacter* were markedly influenced by temperature. The relationships developed for *Nitrosomonas* are as follows

$$\hat{\mu} = 0.47 e^{0.098(T - 15)}, \text{ per day} \tag{7.188}$$

and

$$K_N = 10^{0.051T - 1.158}, \text{ mg/L as N} \tag{7.189}$$

where T is the wastewater temperature, °C.

For attached-growth systems, different temperature effects have been reported. Attached-growth systems have an advantage in withstanding low temperature (< 15°C) without as severe a loss of nitrification rates as suspended-growth systems.

The following equation describes temperature effects on nitrification rate at a temperature T, °C

$$\mu_T = \mu_{20} \theta^{T - 20} \tag{7.190}$$

where T is wastewater temperature, °C and θ is a temperature correction coefficient. From laboratory and pilot RBC studies, the value of θ is found to be 1.10 (Mueller *et al.* 1980).

Dissolved oxygen. When dissolved oxygen concentration is a limiting factor, the growth rate of nitrifiers can be expressed as following the Monod-type relationships as follows

$$\mu_N = \hat{\mu}_N\left(\frac{DO}{K_{O_2} + DO}\right) \tag{7.191}$$

where K_{O_2} is the half-saturation constant for oxygen in mg/L. Values of K_{O_2} have been reported to vary from 0.15 mg/L at 15°C to 2.0 mg/L at 20°C (USEPA 1975b).

Alkalinity and pH. From stoichiometry Eq. (7.181), 7.14 mg/L of alkalinity as $CaCO_3$ will be destroyed for every one mg/L of NH_3-N oxidized. However, Sherrard (1976) claimed this value to be in error and suggested that it should be less for biological nitrification, because ammonia is incorporated into the biomass resulting in a lesser quantity of NH_3-N available for oxidation to nitrate. The alkalinity destruction in a nitrifying activated-sludge process is a function of solids retention time and influent wastewater BOD_5 : N : P ratio. Alleman and Irving (1980) observed a considerably low value (2.73 mg/L of alkalinity as $CaCO_3$ per mg/L of NH_3-N oxidized) for nitrification in a sequential batch reactor.

With the RBC system at Cadillac, Michigan, alkalinity declined 8.1 mg/L for each mg/L of NH_3-N oxidized (Singhal 1980, Chou *et al.* 1980). At Princeton, Illinois, the alkalinity reduction ratios were respectively 6.8 and 9.1 for stages 3 and 4 in a combined BOD_5-nitrification 5-stage RBC system (Lin *et al.* 1982). Therefore it is important to have sufficient alkalinity buffering capacity for nitrifiers to maintain the acceptable pH range.

Reported data show a wide range of optimum pH of 7.0 to 9.5 with maximum activity at approximately pH 8.5 (Painter 1970, 1975, Haug and McCarty 1972, Antonie 1978). Below pH 7.0, adverse effects on ammonia oxidation become pronounced.

Full-scale kinetic studies conducted at the Autotrol Corporation (1979) showed that the nitrification rate declined from 0.31 lb NH_3-N/(d · 1000 ft^2) (1.52 g NH_3-N/(m^2 · d)) at pH 7.0 to 0.17 lb/(d · 1000 ft^2) (0.83 g/(m^2 · d)) at pH 6.5.

Downing and Hopwood (1964) proposed the following relationship for the growth rate for nitrifiers and pH values up to 7.2 in a combined carbon oxidation and nitrification system.

$$\mu_N = \hat{\mu}_N[1 - 0.833(7.2 - \text{pH})] \tag{7.192}$$

It is assumed that the growth rate is constant for pH levels between 7.2 and 8.0. Eq. (7.192), when applied to separate-stage nitrification systems, is probably conservative.

Combined kinetics expression. As previously presented, the major factors which affect the nitrification rate are ammonia nitrogen concentration, temperature, DO, and pH. In the absence of toxic or inhibitory substance in the wastewater, USEPA (1975b) applied Chen's model (1970) which combined Monod's expression for ammonia-N concentration (Eq. (7.185)), DO (Eq. (7.191)), and pH (Eq. (7.192)) effects on nitrifier growth as follows

$$\mu_N = \hat{\mu}_N\left(\frac{N}{K_N + N}\right)\left(\frac{DO}{K_{O_2} + DO}\right)[1 - 0.833(7.2 - \text{pH})] \tag{7.193a}$$

Substituting the effects of temperature (Eq. (7.188)), Eq. (7.189) for K_N, 1.3 mg/L for K_{O_2}, the following equation is valid for *Nitrosomonas* for PH > 7.2 and wastewater temperature between 8 and 30°C

$$\mu_N = 0.47\left[e^{0.098(T-15)}\right]\left[\frac{N}{10^{(0.051T-1.158)} + N}\right]\left[\frac{DO}{1.3 + DO}\right][1 - 0.833(7.2 - \text{pH})] \tag{7.193b}$$

The terms in the first bracket represent the effect of temperature. The second bracket terms are the Monod expression for the effect of NH_3-N concentration. The terms in the third bracket take into account the effect of DO. The terms in the last bracket account for the effect of pH: it will be unity for pH $\geq$ 7.2. From Eq. (7.193) it can be seen that if any one factor becomes limiting, even if all the others are non-limiting, the nitrification rate will be much lower, perhaps even approaching zero.

EXAMPLE: Design an activated-sludge process for carbon oxidation–nitrification using the following given data. Determine the volume of the aeration tank, daily oxygen requirement, and the mass of organisms removed daily from the system.

Design average flow	3785 m^3/d (1 Mgal/d)
BOD of influent (primary effluent)	160 mg/L
TKN of influent (primary effluent)	30 mg/L
NH_3-N of influent (primary effluent)	15 mg/L
Minimum DO in aeration tank	2.0 mg/L
Temperature (minimum)	16°C
Maximum growth rate, $\hat{\mu}$	1.0 per day
Minimum pH	7.2
Assume a safety factor, SF	2.5
(required due to transient loading conditions)	
MLSS	2500 mg/L
Maximum effluent SS or BOD	15 mg/L
Total alkalinity, as $CaCO_3$	190 mg/L

Solution:

Step 1. Compute the maximum growth rate of nitrifiers under the stated operating conditions

$$T = 16°\text{C}$$

$$\text{DO} = 2 \text{ mg/L}$$

$$\text{pH} \geq 7.2$$

Additional required data for Eqs. (7.193a) and (7.193b)

$$\hat{\mu}_N = 0.47 \text{ d}^{-1}$$

$$K_{O_2} = 1.3 \text{ mg/L}$$

$$K_N = 10^{0.051T-1.158} = 10^{0.051\times 16-1.158}$$

$$= 0.455 \text{ mg/L as N}$$

$$N = 15 \text{ mg/L as N}$$

$$\text{Temperature correction factor} = e^{0.098(16-15)} = 1.10$$

$$\text{pH correction factor} = 1 \text{ for pH} \geq 7.2$$

Eqs. (7.193a) and (7.193b)

$$\mu_N = \hat{\mu}_N\left(\frac{N}{K_N + N}\right)\left(\frac{\text{DO}}{K_{O_2} + \text{DO}}\right)[1 - 0.833(7.2 - \text{pH})]$$

$$= 0.47\text{d}^{-1}\left[e^{0.098(T-15)}\right]\left(\frac{15}{0.455 + 15}\right)\left(\frac{2}{1.3 + 2}\right)(1)$$

$$= (0.47 \text{ d}^{-1})(1.1)(0.97)(0.61)(1)$$

$$= 0.30 \text{ d}^{-1}$$

Step 2. Compute the maximum ammonia oxidation rate under the environmental conditions of temperature, pH, and DO

Using Eq. (7.187)

$$r_N = \frac{\mu_N}{Y_N}$$

Referring to Table 7.23

$$\begin{aligned} Y_N &= 0.2 \text{ mg VSS/mg NH}_4^+\text{-N} \\ \text{Max. } \hat{r}_N &= (0.30 \text{ d}^{-1})/0.2 \\ &= 1.5 \text{ d}^{-1} \end{aligned}$$

Step 3. Compute the minimum cell residence time

Using Eq. (7.76)

$$\begin{aligned} 1/\theta_c &\approx Yk' - k_d \\ Y &= Y_N = 0.2 \quad \text{(from Table 7.23)} \\ k' &= \hat{r}_N = 1.5 \text{ d}^{-1} \quad \text{(from Step 2)} \\ k_d &= 0.05 \quad \text{(from Table 7.23)} \\ 1/\theta_c &= 0.2(1.5 \text{ d}^{-1}) - 0.05 \text{ d}^{-1} \\ &= 0.25 \text{ d}^{-1} \\ \text{Minimum } \theta_{c-\min} &= 1/0.25 \text{ d}^{-1} \\ &= 4 \text{ days} \end{aligned}$$

Note: Another rough estimate of θ_c (solids retention time) can be determined by

$$\theta_c = 1/\mu_N = 1/0.30 \text{ d}^{-1} = 3.33 \text{ days}$$

Step 4. Compute the design cell residence time

$$\begin{aligned} \text{Design } \theta_{c-d} &= \text{SF} \times \text{min. } \theta_{c-\min} = 2.5 \times 4 \text{ days} \\ &= 10 \text{ days} \end{aligned}$$

Step 5. Compute the design specific substrate utilization rate U for ammonia oxidation

Using Eq. (7.84)

$$\frac{1}{\theta_c} = YU - k_d$$

or

$$U = \frac{1}{Y}\left(\frac{1}{\theta_c} + k_d\right)$$

Applying θ_{c-d} as θ_c

$$\begin{aligned} &= \frac{1}{0.2}\left(\frac{1}{10 \text{ days}} + 0.05 \text{ d}^{-1}\right) \\ &= 0.75 \text{ d}^{-1} \end{aligned}$$

Note: $U = r_N$ for Eq. (7.187).

Step 6. Compute the steady-state ammonia concentration of the effluent, N
Using Eq. (7.187)

$$r_{\mathrm{N}} = \hat{r}_{\mathrm{N}} \frac{N}{K_{\mathrm{N}} + N}$$

where $U = 0.75\ \mathrm{d}^{-1}$ (Step 5)
$K_{\mathrm{N}} = 0.455$ mg/L for $\mathrm{NH_4^+}$-N (Step 1)
$\hat{r}_{\mathrm{N}} = 1.50\ \mathrm{d}^{-1}$ = maximum growth rate (Step 2)

$$0.75\ \mathrm{d}^{-1} = \frac{(1.50\ \mathrm{d}^{-1})N}{(0.455\ \mathrm{mg/L}) + N}$$

$$N = 0.455\ \mathrm{mg/L}$$

Step 7. Compute organic (BOD) removal rate U
The design cell residence time $\theta_{\mathrm{c-d}}$ applies to both the nitrifiers and heterotrophic bacteria.

$$\frac{1}{\theta_{\mathrm{c-d}}} = YU - k_{\mathrm{d}}$$

$$\theta_{\mathrm{c-d}} = 10\ \mathrm{days}$$

Referring to Table 7.12, for heterotrophic kinetics

$$Y = 0.6\ \mathrm{kg\ VSS/kg\ BOD_5}$$
$$k_{\mathrm{d}} = 0.06\ \mathrm{d}^{-1}$$

Then

$$\frac{1}{10\ \mathrm{days}} = 0.6U - (0.06\ \mathrm{d}^{-1})$$

$$U = 0.27\ \mathrm{kg\ BOD_5\ removed/(kg\ MLVSS \cdot d)}$$

assuming there is 90 percent $\mathrm{BOD_5}$ removal efficiency.
The food to microorganism ratio is

$$\mathrm{F/M} = 0.27/0.9$$
$$= 0.30\ \mathrm{kg\ BOD_5\ applied/(kg\ MLVSS \cdot d)}$$

Step 8. Compute the hydraulic retention time θ required for organic and ammonia oxidations
Using Eq. (7.83)

$$U = \frac{S_0 - S}{\theta X} \quad \text{or} \quad \theta = \frac{S_0 - S}{UX}$$

(a) For organic oxidation

$$S_0 = 160\ \mathrm{mg/L} \quad \text{(given)}$$
$$S = 15\ \mathrm{mg/L} \quad \text{(regulated)}$$
$$U = 0.27\ \mathrm{d}^{-1} \quad \text{(Step 7)}$$
$$X = \mathrm{MLVSS} = 0.8\ \mathrm{MLSS} = 0.8 \times 2500\ \mathrm{mg/L}$$
$$= 2000\ \mathrm{mg/L}$$

Therefore

$$\theta = \frac{(160 - 15)\ \mathrm{mg/L}}{(0.27\ \mathrm{d}^{-1})(2000\ \mathrm{mg/L})}$$
$$= 0.269\ \mathrm{days}$$
$$= 6.4\ \mathrm{h}$$

(b) For nitrification

$$N_0 = \text{TKN} = 30 \text{ mg/L} \qquad \text{(given)}$$
$$N = 0.45 \text{ mg/L} \qquad \text{(Step 6)}$$
$$U = 0.75 \text{ d}^{-1} \qquad \text{(Step 5)}$$
$$X = 2000 \text{ mg/L} \times 0.08 \qquad \text{(assuming 8\% of VSS is nitrifiers)}$$
$$= 160 \text{ mg/L}$$
$$\theta = \frac{(30 - 0.45) \text{ mg/L}}{(0.75 \text{ d}^{-1})(160 \text{ mg/L})}$$
$$= 0.246 \text{ days}$$
$$= 5.9 \text{ h}$$

Note: Organic oxidation controls HRT; $\theta = 6.4$ h

Step 9. Determine the volume V of aeration tank required, based on organic removal process

$$V = Q\theta = (3785 \text{ m}^3\text{/d}) \text{ } (0.269 \text{ day})$$
$$= 1020 \text{ m}^3$$
$$= 36{,}000 \text{ ft}^3$$

Step 10. Compute BOD_5 loading rate

$$\text{BOD loading} = (3785 \text{ m}^3\text{/d})(160 \text{ g/m}^3)/(1000 \text{ g/kg})$$
$$= 606 \text{ kg/d}$$
$$\text{BOD loading rate} = (606 \text{ kg/d})/(1020 \text{ m}^3)$$
$$= 0.59 \text{ kg/(m}^3 \cdot \text{d)}$$
$$= 37.1 \text{ lb/(1000 ft}^3 \cdot \text{d) (OK)}$$

Note: This rate is below the range of 50 to 120 lb/(1000 ft^3 · d) for the complete-mix activated-sludge process.

Step 11. Estimate the total quantity of oxygen supply needed. The total quantity of oxygen required can be estimated on the basis of oxygen demand for BOD_5 and TKN removal and net mass of volatile solids (cells) produced with conversion factors. A procedure similar to the one illustrated in the section on the activated-sludge process may be used.

Alternatively, a rough estimation can be calculated by the following equation from the influent BOD_5 and TKN concentrations (Metcalf and Eddy, Inc. 1991)

$$O_2, \text{ lb/d} = Q(kS_0 + 4.57 \text{ TKN}) \text{ } 8.34$$

where Q = influent flow, Mgal/d
k = conversion factor for BOD loading on nitrification system
= 1.1 to 1.25
4.57 = conversion factor for complete oxidation of TKN
8.34 = a conversion factor, lb/(Mgal · mg/L)

For this example, let $k = 1.18$ and SF = 2.5. Then

$$\text{Required } O_2, \text{ lb/d} = 1 \text{ Mgal/d } [1.18(160 \text{ mg/L}) + 4.57(30 \text{ mg/L})]$$
$$\times [8.34 \text{ lb/(Mgal} \cdot \text{mg/L)}] \times 2.5$$
$$= 6795 \text{ lb/d}$$
$$= 3082 \text{ kg/d}$$

Step 12. Determine the sludge wasting schedule

Sludge wasting includes solids contained in the effluent from secondary clarifier effluent and sludge waste in the return activated-sludge or mixed liquor.

The sludge to be wasted under steady-state conditions is the denominator of Eq. (7.71)

$$\theta_c = \frac{VX}{Q_{wa}X + Q_e X_e}$$

Total solids wasted per day is

$$\begin{aligned} Q_{wa}X + Q_e X_e &= VX/\theta_c \\ V &= 1020 \text{ m}^3 \\ X &= \text{MLVSS} = 0.8 \times \text{MLSS} = 0.8 \times 2500 \text{ mg/L} \\ &= 2000 \text{ mg/L} \\ VX &= 1020 \text{ m}^3 \times 2000 \text{ (g/m}^3\text{)/(1000 g/kg)} \\ &= 2040 \text{ kg} \\ \theta_c &= 10 \text{ days (use design } \theta_{c-d}\text{, Step 4)} \\ VX/\theta_c &= 2040 \text{ kg/10 days} \\ &= 204 \text{ kg/d} \end{aligned}$$

The solids contained in the clarifier effluent at 3785 m^3/d (1 Mgal/d) are calculated from the VSS in the effluent

$$\begin{aligned} \text{Effluent VSS} &= 0.8 \times 15 \text{ mg/L} = 12 \text{ mg/L} = 12 \text{ g/m}^3 \\ Q_e X_e &= 3785 \text{ m}^3\text{/d} \times 12 \text{(g/m}^3\text{)/(1000 g/kg)} \\ &= 45.4 \text{ kg/d} \end{aligned}$$

The VSS (microorganism concentration) to be wasted from mixed liquor or return sludge is

$$\begin{aligned} Q_{wa}X &= VX/\theta_c - Q_e X_e = (204 - 45) \text{ kg/d} \\ &= 159 \text{ kg/d} \end{aligned}$$

Step 13. Check the buffering capacity of the wastewater, theoretically 7.14 mg/L of alkalinity as $CaCO_3$ is destroyed per mg/L of NM_4^+-N oxidized. The alkalinity remaining after nitrification would be at least

$$\begin{aligned} \text{Alk.} &= 190 \text{ mg/L} - 7.14 \text{ (15 mg/L)} \\ &= 83 \text{ mg/L as CaCO}_3 \end{aligned}$$

This should be sufficient to maintain the pH value in the aeration tanks above 7.2.

Combined carbon oxidation–nitrification in attached growth reactors. Two attached growth reactors, the trickling filter process and the RBC process, can be used for combined carbon oxidation–nitrification. Detailed design procedures for these two processes are described elsewhere (USEPA 1975c) and in the manufacturers' design manual.

In the design of nitrification with trickling filters, the total surface area required is determined on the basis of empirical unit surface area for unit NH_4^+-N oxidized per day under the BOD/TKN and the sustained temperature conditions. The choice of filter media is based on the effluent BOD_5 and SS requirements. A circulation rate of 1 : 1 is usually adequate.

Nitrification with RBC system. For a combined oxidation–nitrification process with the RBC system, a two-step design procedure is needed. Significant nitrification will not occur in the RBC

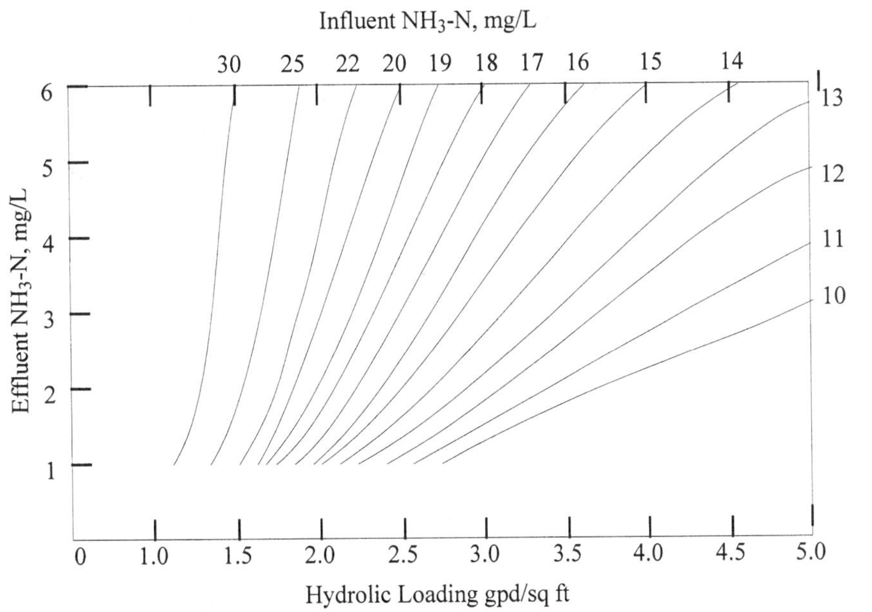

FIGURE 7.50 RBC process design curve for nitrification of domestic wastewater (*Source: Autotrol 1979*).

process until the soluble BOD concentration is reduced to 15 mg/L or less (Autotrol 1979). In the first design the media surface area required to reduce SBOD is determined as shown in the example illustrated in the previous section. For influent ammonia nitrogen concentrations of 15 mg/L or above, it is necessary to reduce SBOD concentrations to less than 15 mg/L, i.e. to about the same value as the ammonia nitrogen concentration. The second design uses the nitrification design curves (Fig. 7.50) to determine the RBC area to reduce the influent ammonia nitrogen level to the required effluent concentration. The sum of the two RBC surface areas is determined as the total surface area required for the combined carbon oxidation–nitrification system. Figure 7.51 shows the design curves for RBC carbon oxidization–nitrification of municipal wastewater. If the wastewater temperature is less than 55°F (12.8°C), a separate temperature correction should be made to each of the surface areas determined.

EXAMPLE: Design an RBC system for organic and nitrogen removal with the following data. The effluent SBOD and NH_3-N are to be 6 and 2 mg/L, respectively. Total BOD = 12 mg/L.

Design inflow	4.0 Mgal/d (15,140 m^3/d)
Influent soluble BOD	72 mg/L
Effluent soluble BOD	6 mg/L
Influent NH_3-N	18 mg/L
Effluent NH_3-N	2 mg/L
Wastewater temperature	≥ 55°F (12.8°C)

Solution:

Step 1. Determine the required surface area

(a) Determine hydraulic loading HL to reduce SBOD from 70 mg/L to 15 mg/L for starting nitrification

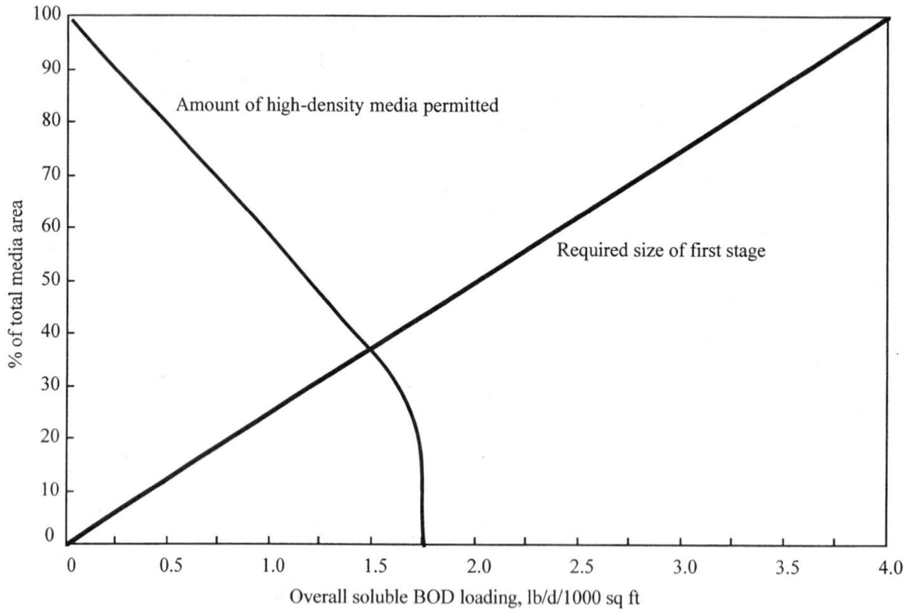

FIGURE 7.51 Design curves for RBC carbon oxidation–nitrification of domestic wastewater.

From Fig. 7.33

$$HL_b = 3.32 \text{ gal/(d} \cdot \text{ft}^2) \text{ for the influent SBOD} = 72 \text{ mg/L}$$

(b) Determine HL to reduce NH_3-N from 18 mg/L to 2 mg/L
From Fig. 7.50

$$HL_n = 2.1 \text{ gal/(d} \cdot \text{ft}^2)$$

(c) Determine overall hydraulic loading HL_o

$$\frac{1}{HL_o} = \frac{1}{HL_b} + \frac{1}{HL_n} = \frac{1}{3.32} + \frac{1}{2.1}$$

$$HL_o = 1.29 \text{ gal/(d} \cdot \text{ft}^2)$$

(d) Determine HL for reducing SBOD to 6 mg/L
From Fig. 7.33

$$HL_b = 1.9 \text{ gal/(d} \cdot \text{ft}^2)$$

This is greater than HL_o; thus, nitrification controls the overall design HL.
Calculate the surface area requirement

$$\text{Area} = \frac{4{,}000{,}000 \text{ gal/d}}{1.29 \text{ gal/(d} \cdot \text{ft}^2)}$$

$$= 3{,}100{,}000 \text{ ft}^2$$

Step 2. Determine the size of first stage

(a) Calculate overall SBOD loading area

$$\text{SBOD loading} = \frac{4\ \text{Mgal/d} \times 70\ \text{mg/L} \times 8.34\ \text{lb/(Mgal} \cdot \text{mg/L)}}{3{,}100{,}000\ \text{ft}^2}$$

$$= 0.75\ \text{lb/1000 ft}^2$$

(b) Sizing of first stage

The size of the first stage can be found from the manufacturer's design curve (Fig. 7.51) or by proportional calculation. An SBOD loading rate of 4.00 lb/(d · 1000 ft^2) uses 100 percent of the total media area. Therefore

$$\frac{X\%}{100\%} = \frac{0.75\ \text{lb/(d} \cdot 1000\ \text{ft}^2)}{4.00\ \text{lb/(d} \cdot 1000\ \text{ft}^2)}$$

$$X = 18.8\% \text{ or } 19\%$$

(c) Calculate surface area of first stage

$$\text{Area} = 0.19\ (3{,}100{,}000\ \text{ft}^2)$$
$$= 589{,}000\ \text{ft}^2$$

Step 3. Determine media distribution

(a) Using high-density media, from Fig. 7.51,

$$\text{maximum \% of total media area} = 67\% \text{ for } 0.75\ \text{lb SBOD/(d} \cdot 1000\ \text{ft}^2)$$

(b) High-density surface area

$$= 0.67\ (3{,}100{,}000\ \text{ft}^2)$$
$$= 2{,}077{,}000\ \text{ft}^2$$

(c) Standard media surface area

$$= (1 - 0.67)(3{,}100{,}000\ \text{ft}^2)$$
$$= 1{,}023{,}000\ \text{ft}^2$$

Step 4. Select configuration

(a) Standard media (100,000 ft^2 per unit) assemblies in first stage

$$= \frac{589{,}000\ \text{ft}^2}{100{,}000\ \text{ft}^2\text{/unit}}$$

$$= 5.9 \text{ units (use 6 units with 3 trains)}$$

(b) Standard media (total)

$$= \frac{1{,}023{,}000}{100{,}000}$$

$$= 10.23 \text{ (use 12 units)}$$

(c) High-density media (150,000 ft^2 per unit) assemblies

$$= \frac{2{,}077{,}000}{150{,}000}$$

$$= 13.8 \text{ (use 15 units)}$$

(d) Choose between three trains of five-stage operation as shown below

3 (SS + SS + HH + HH + H) = 27 (units with 5 stages in 3 trains)

or three trains of four-stage operation

3 (SS + SS + HHH + HH) = 27 (units with 4 stages in 3 trains)

Step 5. No temperature correction is needed, since the wastewater temperature is above 12.8°C (55°F).

Step 6. Determine power consumption
A rough estimation of power consumption is 2.5 kW per shaft

$$\text{Power consumption} = 2.5 \text{ kW/shaft} \times 27 \text{ shafts}$$
$$= 67.5 \text{ kW}$$

Denitrification. Biological denitrification is the conversion of nitrate to gaseous nitrogen species and to cell material by the ubiquitous heterotrophic facultative aerobic bacteria and some fungi. Denitrifiers include a broad group of bacteria such as *Pseudomonas, Micrococcus, Archromobacter,* and *Bacillus* (USEPA 1975b). These groups of organisms can use either nitrate or oxygen as electron acceptor (hydrogen donor) for conversion of nitrate to nitrogen gas. Denitrification occurs in both aerobic and anoxic conditions. An anaerobic condition in the liquid is not necessary for denitrification, and 1 to 2 mg/L of DO does not influence denitrification. The term anoxic is preferred over anaerobic when describing the process of denitrification.

The conversion of nitrate to gaseous end products is a two-step process, called dissimilatory denitrification, with a series of enzymatic reactions. The first step is a conversion of nitrate to nitrite, and the second step converts the nitrite to nitrogen gas. The nitrate dissimilations are expressed as follows (Stanier *et al.* 1963)

$$2NO_3^- + 4e^- + 4H^+ \rightarrow 2NO_2^- + 2H_2O \tag{7.194}$$

$$2NO_2^- + 6e^- + 8H^+ \rightarrow N_2 + 4H_2O \tag{7.195}$$

Overall transformation yields

$$2NO_3^- + 10e^- + 12H^+ \rightarrow N_2 + 6H_2O \tag{7.196}$$

For each molecule of nitrate reduced, five electrons can be accepted. When methanol (CH_3OH) is used on the organic carbon sources, the dissimilatory denitrification can be expressed as follows:

First step

$$6NO_3^- + 2CH_3OH \rightarrow 6NO_2^- + 4H_2O + 2CO_2 \tag{7.197}$$

Second step

$$6NO_2^- + 3CH_3OH \rightarrow 3N_2 + 3H_2O + 3CO_2 + 6OH^- \tag{7.198}$$

Overall transformation

$$6NO_3^- + 5CH_3OH \rightarrow 3N_2 + 7H_2O + 5CO_2 + 6OH^- \tag{7.199}$$

In Eq. (7.199), nitrate serves as the electron acceptor and methanol as the electron donor.

When methanol is used as an organic carbon source in the conversion of nitrate to cell material, a process termed assimilatory (or synthesis) denitrification, the stoichiometric equation is

$$3NO_3^- + 14CH_3OH + 4H_2CO_3 \rightarrow 3C_5H_7O_2N + 20H_2O + 3HCO_3^- \tag{7.200}$$

From Eq. (7.200), neglecting cell synthesis, 1.9 mg of methanol is required for each mg of NO_3-N reduction (M/N ratio). Including synthesis results in an increase in the methanol requirement to 2.47 mg. The total methanol requirement can be calculated from that required for nitrate and nitrite reductions and deoxygenation as follows (McCarty *et al.* 1969)

$$C_m = 2.47\,(NO_3\text{-N}) + 1.53\,(NO_2\text{-N}) + 0.87\,\text{DO} \tag{7.201}$$

where C_m = methanol required, mg/L
NO_3-N = nitrate nitrogen removed, mg/L
NO_2-N = nitrite nitrogen removed, mg/L
DO = dissolved oxygen removed, mg/L

Biomass production can be computed similarly

$$C_b = 0.53\,(NO_3\text{-N}) + 0.32\,(NO_2\text{-N}) + 0.19\,\text{DO} \tag{7.202}$$

where C_b = biomass production, mg/L

In general, an M/N ratio of 2.5 to 3.0 is sufficient for complete denitrification (USEPA 1975c). A commonly used design value for the required methanol dosage is 3 mg/L per mg/L of NO_3-N to be reduced.

EXAMPLE: Determine the methanol dosage requirement, the M/N ratio, and biomass generated for complete denitrification of an influent with a nitrate-N of 24 mg/L, nitrite-N of 0.5 mg/L, and DO of 2.5 mg/L.

Solution:

Step 1. Determine methanol required, using Eq. (7.201)

$$\begin{aligned} C_m &= 2.47\,(NO_3\text{-N}) + 1.53\,(NO_2\text{-N}) + 0.87\,\text{DO} \\ &= 2.47\,(24\text{ mg/L}) + 1.53\,(0.5\text{ mg/L}) + 0.87\,(2.5\text{ mg/L}) \\ &= 62.2\text{ mg/L} \end{aligned}$$

Step 2. Calculate M/N ratio

$$\begin{aligned} \text{M/N} &= (62.2\text{ mg/L})/(24\text{ mg/L}) \\ &= 2.6 \end{aligned}$$

Step 3. Calculate biomass generated, using Eq. (7.202)

$$\begin{aligned} C_b &= 0.53\,(NO_3\text{-N}) + 0.32\,(NO_2\text{-N}) + 0.19\,\text{DO} \\ &= 0.53\,(24\text{ mg/L}) + 0.32\,(0.5\text{ mg/L}) + 0.19\,(2.5\text{ mg/L}) \\ &= 13.4\text{ mg/L} \end{aligned}$$

Kinetics of denitrification. Similarly to nitrification, environmental factors affect the kinetic rate of denitrifier growth and nitrate removal. These factors are mentioned in the previous section.

Effect of nitrate on kinetics. The effect of nitrate on denitrifier growth rate can be expressed by the Monod equation

$$\mu_D = \hat{\mu}_D \frac{D}{K_D + D} \tag{7.203}$$

where μ_D = growth rate of denitrifier, per day
$\hat{\mu}_D$ = maximum growth rate of denitrifier, per day
D = nitrate-N concentration, mg/L
K_D = half saturation constant, mg/L nitrate-N
= 0.08 mg/L NO_3-N, for suspended-growth systems without solids recycle at 20°C
= 0.16 mg/L NO_3-N, for suspended-growth systems with solids recycle at 20°C
= 0.06 mg/L NO_3-N, for attached-growth systems at 25°C

With the effects of wastewater temperature (T, °C) and dissolved oxygen (DO, mg/L), the overall rate of denitrification can be described as follows:

$$\mu_D = \hat{\mu}_D \times \frac{D}{K_D + D} \times 1.09^{(T-20)} \times (1 - \text{DO}) \tag{7.203a}$$

Denitrification rates can be related to denitrifier growth rates by the following relationship (USEPA 1975c)

$$r_D = \mu_D / Y_D \tag{7.204}$$

where r_D = nitrate removal rate, lb (NO_3-N)/lb VSS/d
Y_D = gross yield of denitrifier, lb VSS/lb (NO_3-N)

Similarly, peak denitrification rates are related to maximum denitrifier growth rates as follows

$$\hat{r}_D = \hat{\mu}_D / Y_D \tag{7.205}$$

Solids retention time. Consideration of solids production and solids retention time is an important design consideration for the system. Similar to Eq. (7.84), a mass balance of the biomass in a completely mixed reactor yields the relationship (Lawrence and McCarty 1970)

$$\frac{1}{\theta_c} = Y_D r_D - K_d \tag{7.206}$$

where θ_c = solids retention time, d
K_d = decay coefficient, d^{-1}

Denitrification with RBC process. The RBC process has been applied to biological denitrification by completely submerging the rotating media and by adding an appropriate source of organic carbon. Fig. 7.52 is a schematic process flow diagram of carbon oxidation–nitrification–denitrification. Methanol is added to the denitrification stage, in which the rotational speed is reduced. Methanol requirements are a significant portion of the operating cost. In a completely submerged mode, RBC will remove nitrate nitrogen at a rate of approximately 1 lb/(d · 1000 ft^2) while treating influent nitrate concentrations up to 25 mg/L and producing effluent nitrate nitrogen concentrations below 5 mg/L. The design curves for RBC denitrification of municipal wastewater are presented in Fig. 7.53 (Autotrol 1979).

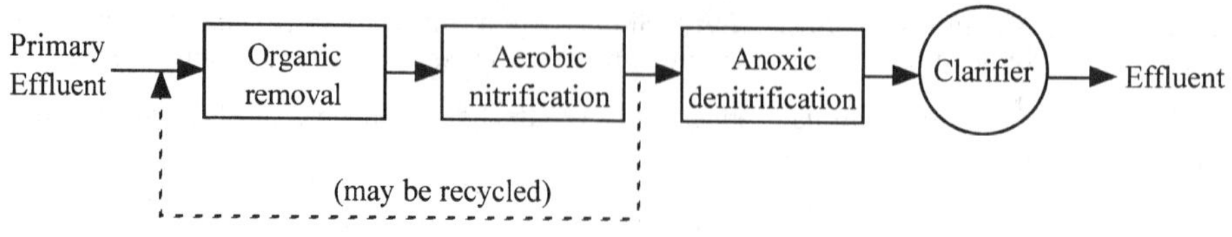

FIGURE 7.52 Schematic RBC process for BOD removal, nitrification, and denitrification.

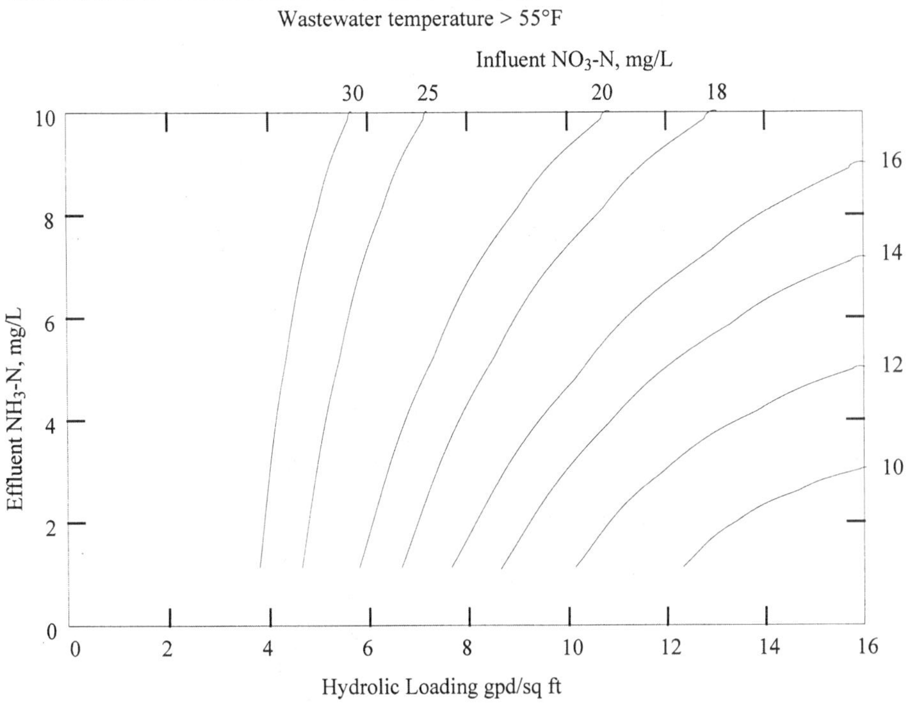

FIGURE 7.53 Design curves for RBC denitrification of domestic wastewater.

Denitrification is a relatively rapid reaction compared to nitrification. It is generally more economical to reduce nitrate nitrogen to as low a level as can be achieved by the RBC process, i.e. $\leq$ 1.0 mg/L.

EXAMPLE: Design a denitrification RBC system following organic oxidation–nitrification, using data for the example in the section on nitrification with the RBC system. The treatment plant is designed to produce a final effluent of 4 mg/L total nitrogen (2 mg/L of NH_3-N, 3 mg/L of TKN, and 1 mg/L of NO_3-N).

Given conditions:

Design flow = 4.0 Mgal/d

Influent NH_3-N (nitrification) = 18 mg/L

Effluent NH_3-N = 2 mg/L

Wastewater temperature $\geq$ 55°F

Solution:

Step 1. Determine surface area for denitrification shafts.

(a) As in the previous RBC example for BOD removal and nitrification, the overall hydraulic loading HL_o = 1.29 gal/(d · ft^2)

$$\text{The surface area required} = 3{,}100{,}000 \text{ ft}^2$$

(b) Influent NO_3-N of denitrification

$$= (\text{influent} - \text{effluent})\ NH_3\text{-N}$$
$$= (18 - 2)\ \text{mg/L}$$
$$= 16\ \text{mg/L}$$

Note: Assume that all NH_3-N removed is stoichiometrically converted to NO_3-N, even though field practice often shows a loss of NO_3-N.

(c) Referring to Fig. 7.53, the hydraulic loading rate to reduce NO_3-N from 16 mg/L to 1 mg/L

$$\text{HL} = 7.45\ \text{gal/(d} \cdot \text{ft}^2)$$

(d) Calculate the surface area for denitrification

$$\text{Surface area} = \frac{4{,}000{,}000\ \text{gal/d}}{7.45\ \text{gal/(d} \cdot \text{ft}^2)}$$
$$= 540{,}000\ \text{ft}^2$$

Step 2. Select configuration

(a) Use standard media (100,000 ft²) for denitrification
(b) Number of shafts

$$= \frac{540{,}000\ \text{ft}^2}{100{,}000\ \text{ft}^2}$$
$$= 5.4\ \text{(use 6 shafts)}$$

(c) A two-stage operation with three trains is recommended. Install two assemblies in a separate basin with baffles between adjacent shafts.

29 SLUDGE (RESIDUALS) TREATMENT AND MANAGEMENT

Residuals is a term currently used to refer to "sludge." The bulk of residuals (sludge) generated from wastewater by physical primary and biological (secondary) and advanced (tertiary) treatment processes must be treated and properly disposed of. The higher the degree of wastewater treatment, the larger the quantity of sludge to be treated and handled. With the advent of strict rules and regulations involving the handling and disposal of sludge, the need for reducing the volume of sludge has become increasingly important in order to reduce the operating costs (approximately 50 percent of the plant costs) of wastewater treatment plants. Sludge treatment and disposal is a complex problem facing wastewater treatment professionals. A properly designed and efficiently operated sludge processing and disposal system is essential to the overall success of the wastewater treatment effort.

29.1 Quantity and Characteristics of Sludge

The quantity and characteristics of the sludge produced depend on the character of raw wastewater and the wastewater treatment processes. In the United States, approximately 12,750 public owned treatment works (POTW) generate 5.4 million dry metric tons of sludge annually, or 21 kg (47 lb) of dry sewage sludge (biosolids) per person (Federal Register 1993).

Some estimates of the amounts of solids generated by the treatment unit processes may be inferred from previous examples. Grit collected from the preliminary treatment units is not biodegradable. It is usually transported to a sanitary landfill without further treatment.

TABLE 7.24 Solids Concentrations and Other Characteristics of Various Types of Sludge

Wastewater treatment	Primary, gravity	Secondary, biological	Advanced (tertiary), chemical precipitation, filtration
Sludge			
Amounts generated, L/m^3 of wastewater	2.5–3.5	15–20	25–30
Solids content, %	3–7	0.5–2	0.2–1.5
Organic content, %	60–80	50–60	35–50
Treatability, relative	easy	difficult	difficult
Dewatered by belt filter			
Feed solids, %	3–7	3–6	
Cake solids, %	28–44	20–35	

Sources: WPCF (1988a), WEF and ASCE (1991b), USEPA (1991).

Characteristics of wastewater sludges, including total solids and volatile solids contents, pH, nutrients, organic matter, pathogens, metals, organic chemicals, and hazardous pollutants, are discussed in detail elsewhere (Federal Register 1993, USEPA 1995).

Sludge from primary settling tanks contains from 3 to 7 percent solids which are approximately 60 to 80 percent organic (Davis and Cornwell 1991, Federal Register 1993). Primary sludge solids are usually gray in color, slimy, fairly coarse, and with highly obnoxious odors. Primary sludge is readily digested under suitable operational conditions (organics are decomposed by bacteria). Table 7.24 provides the solids concentrations of primary sludge and sludge produced in different biological treatment systems.

Sludge from secondary settling tanks commonly has a brownish, flocculant appearance and an earthy odor. If the color is dark, the sludge may be approaching a septic condition. Secondary sludge consists mainly of microorganisms (75 to 90 percent organic) and inert materials. The organic matter may be assumed to have a specific gravity of 1.01 to 1.06, depending on its source, whereas the inorganic particles have a specific gravity of 2.5 (McGhee 1991).

In general, secondary sludges are more flocculant than primary sludge solids, less fibrous. Waste activated-sludge usually contains 0.5 to 2 percent solids, whereas trickling filter sludge has 2 to 5 percent solids (Davis and Cornwell 1991, Hammer 1986). Activated-sludge and trickling filter sludge can be digested readily, either alone or when mixed with primary sludge.

Sludge from chemical (metal salts) precipitation is generally dark in color or red (with iron) and slimy. Lime sludge is grayish brown. Chemical sludge may create odor. Decomposition of chemical sludge occurs at a slower rate.

The nature of sludge from the tertiary (advanced) treatment process depends on the unit process. Chemical sludge from phosphorus removal is difficult to handle and treat. Tertiary sludge combined with biological nitrification and denitrification is similar to waste activated-sludge.

EXAMPLE: Estimate the solids generated in the primary and secondary clarifiers at a secondary (activated-sludge) treatment plant. Assume that the primary settling tank removes 65 percent of the TSS and 33 percent of the BOD_5. Also determine the volume of each sludge, assuming 6 and 1.2 percent of solids in the primary and secondary effluents, respectively.

Average plant flow = 3785 m^3/d (1 Mgal/d)

Primary influent TSS = 240 mg/L

Primary influent BOD = 200 mg/L

Secondary effluent BOD = 30 mg/L

Secondary effluent TSS = 24 mg/L

Bacteria growth rate $Y = 0.23$ kg (0.5 lb) sludge solids per kg (lb) BOD removed

Solution:

Step 1. Calculate the quantity of dry primary solids produced daily

$$1 \text{ mg/L} = 1 \text{ g/m}^3$$
$$\text{Primary sludge} = 3785 \text{ m}^3\text{/d} \times 240 \text{ g/m}^3 \times 0.65/(1000 \text{ g/kg})$$
$$= 590 \text{ kg/d}$$
$$= 1300 \text{ lb/d}$$

Step 2. Calculate the primary effluent TSS and BOD concentrations

$$\text{TSS} = 240 \text{ mg/L} \times (1 - 0.65) = 84 \text{ mg/L}$$
$$\text{BOD} = 200 \text{ mg/L} \times (1 - 0.33) = 134 \text{ mg/L}$$

Step 3. Calculate TSS removed in the secondary clarifier

$$\text{Secondary (TSS) solids} = 3785 \text{ m}^3\text{/d} \times (84 - 24) \text{ g/m}^3/(1000 \text{ g/kg})$$
$$= 227 \text{ kg/d}$$
$$= 500 \text{ lb/d}$$

Step 4. Calculate biological solids produced due to BOD removal

$$\text{BOD removed} = 3785 \text{ m}^3\text{/d } (134 - 30) \text{ g/m}^3/(1000 \text{ g/kg})$$
$$= 394 \text{ kg/d}$$
$$= 869 \text{ lb/d}$$
$$\text{Biological solids} = 394 \text{ kg/d} \times Y = 394 \text{ kg/d} \times 0.23 \text{ kg/kg}$$
$$= 90 \text{ kg/d}$$
$$= 198 \text{ lb/d}$$

Step 5. Calculate total amount of solids produced from the secondary clarifier and from the whole plant (from Steps 3 and 4)

$$\text{Secondary solids} = (227 + 90) \text{ kg/d}$$
$$= 317 \text{ kg/d}$$
$$= 698 \text{ lb/d}$$
$$\text{Solids of the plant} = \text{Step 1} + \text{Step 5}$$
$$= (590 + 317) \text{ kg/d}$$
$$= 907 \text{ kg/d}$$
$$= 1998 \text{ lb/d}$$

Step 6. Determine the volume of each type of sludge
Assuming sp gr of sludge = 1.0

$$\text{Primary sludge volume } V_1 = \frac{590 \text{ kg/d}}{0.06 \times 1000 \text{ kg/m}^3}$$
$$= 9.8 \text{ m}^3\text{/d}$$
$$= 2590 \text{ gal/d}$$

$$\text{Secondary sludge volume } V_2 = \frac{317 \text{ kg/d}}{0.012 \times 1000 \text{ kg/m}^3}$$
$$= 26.4 \text{ m}^3\text{/d}$$
$$= 6980 \text{ gal/d}$$

Mass–volume relation. The mass of solids in a slurry is related to volatile and fixed suspended solids contents. The sp gr of a slurry is

$$S_s = \frac{m_w + m_v + m_f}{V_s} \tag{7.207}$$

where S_s = sp gr of slurry, g/cm^3 or lb/ft^3
m_w = mass of water, kg or lb
m_v = mass of VSS, kg or lb
m_f = mass of FSS, kg or lb
V_s = volume of sludge slurry, m^3 or ft^3

Since

$$V_s = V_w + V_v + V_f \tag{7.208}$$

where

$$V_w, V_v, V_f = \text{volume of water, VSS, FSS, m}^3 \text{ or ft}^3$$

then

$$\frac{m_s}{S_s} = \frac{m_w}{S_w} + \frac{m_v}{S_v} + \frac{m_f}{S_f} \tag{7.209}$$

where m_s = mass of slurry, kg or lb

Moisture content. The moisture (water) ρ_w or total solids ρ_s content of a sludge, expressed on a percentage basis, can be computed as

$$\rho_w = \frac{100m_w}{m_w + m_s} = \frac{100m_w}{m_w + m_v + m_f} \tag{7.210}$$

$$\rho_s = 100 - \rho_w \tag{7.211}$$

The volume of a sludge, related to its total solids content, is

$$V_s = \frac{m_s}{(\rho_s/100)S_s} \tag{7.212}$$

The sp gr of organic matter (VSS) is close to that of water (1.00). The sp gr of activated-sludge is 1.01 to 1.10, the sp gr of FSS is 2.5, and that for chemical sludge ranges from 1.5 to 2.5 (Droste 1997).

EXAMPLE: Determine the sp gr of waste activated-sludge that contains 80 percent VSS and has a solids concentration of 1.5 percent. Also, determine the volume of 1 kg of sludge.

Solution:

Step 1. Calculate mass of VSS, FSS, and water in 1000 g of sludge

$$1.5\% \text{ sludge} = 15{,}000 \text{ mg/L of solids}$$
$$= 15 \text{ g/L}$$
$$\text{VSS} = 15 \text{ g/L} \times 0.8 = 12 \text{ g/L} = m_v$$
$$\text{FSS} = 15 \text{ g/L} \times 0.2 = 3 \text{ g/L} = m_f$$

If $$m_s = 1000 \text{ g}$$

Then $$m_w = (1000 - 12 - 3) \text{ g} = 985 \text{ g}$$

Step 2. Calculate sp gr of the sludge, using Eq. (7.209)

$$\frac{m_s}{S_s} = \frac{m_w}{S_w} + \frac{m_v}{S_v} + \frac{m_f}{S_f}$$

$$\frac{1000}{S_s} = \frac{985}{1.0} + \frac{12}{1.0} + \frac{3}{2.5}$$

$$S_s = \frac{1}{0.9982} = 1.0018 \text{ (g/cm}^3\text{)}$$

Step 3. Calculate the volume per kg of sludge

$$m_s = 1000 \text{ g} \times 0.015 = 15 \text{ g}$$

Using Eq. (7.212)

$$V_s = \frac{m_s}{(\rho_s/100)S_s} = \frac{15 \text{ g}}{(1.5/100)\, S_s}$$

$$= \frac{15 \text{ g}}{(1.5/100)(1.0018 \text{ g/cm}^3)(1000 \text{ cm}^3\text{/L})}$$

$$= 0.998 \text{ L}$$

29.2 Sludge Treatment Alternatives

A variety of treatment processes and overall sludge management options can be established, depending on the type and quantity of sludge generated. Figure 7.54 illustrates schematically the wastewater sludge treatment alternatives. The basic processes for sludge treatment include thickening, stabilization, conditioning, dewatering, and volume reduction. More details of sludge treatments are available from manufacturers' manuals and elsewhere (USEPA 1979, 1991, Metcalf and Eddy Inc. 1991, WEF and ASCE 1991b, 1996b).

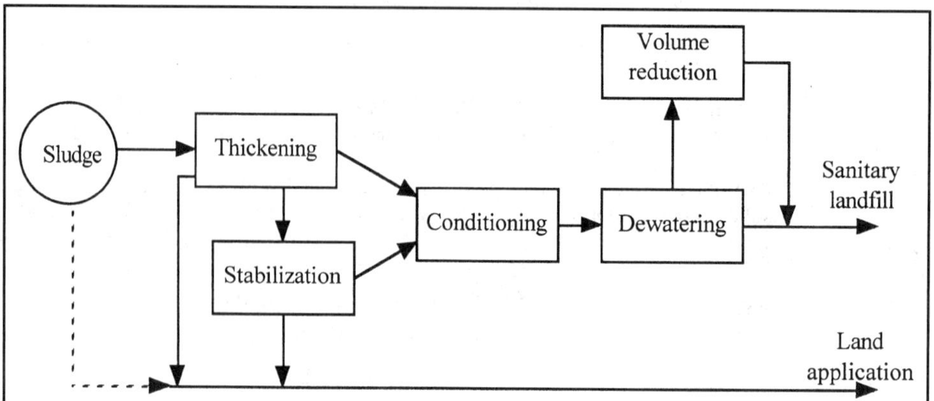

FIGURE 7.54 Sludge processing alternatives.

Sludge thickening. As stated previously, all types of sludge contain a large volume of water (Table 7.24). The purposes of sludge thickening are to reduce the sludge volume to be handled in the subsequent sludge processing units (pump, digester, dewatering equipment) and to reduce the construction and operating costs of subsequent processes. Sludge thickening is a procedure used to remove water and increase the solids content. For example, if waste activated-sludge with 0.6 percent solids is thickened to a content of 3.0 percent solids, a fivefold decrease in sludge volume is achieved.

Thickening a sludge with 3 to 8 percent solids may reduce its volume by 50 percent. Sludge thickening mainly involves physical processes such as gravity settling, flotation, centrifugation, and gravity belts.

EXAMPLE: Estimate the sludge volume reduction when the sludge is thickened from 4 percent to 7 percent solids concentration. The daily sludge production is 100 m^3 (26,420 gal).

Solution:

Step 1. Calculate amount of dry sludge produced

$$\begin{aligned}\text{Dry solids} &= 100\ \text{m}^3 \cdot \text{d} \times 1000\ \text{kg/m}^3 \times 0.04 \\ &= 4000\ \text{kg/d}\end{aligned}$$

Step 2. Calculate volume in 7% solids content

$$\begin{aligned}\text{Volume} &= (4000\ \text{kg/d})/[0.07(1000\ \text{kg/m}^3)] \\ &= 57.1\ \text{m}^3/\text{d} \\ &= 15{,}100\ \text{gal/d}\end{aligned}$$

Note: Steps 1 and 2 can be solved by the mass balance method:

$$\begin{aligned}\text{Volume} &= 100\ \text{m}^3/\text{d} \times 4\%/7\% \\ &= 57.1\ \text{m}^3/\text{d}\end{aligned}$$

Step 3. Calculate percentage sludge volume reduction

$$\begin{aligned}\text{Percent volume reduction} &= \frac{(100 - 57.1)\text{m}^3 \times 100\%}{100\ \text{m}^3} \\ &= 42.9\%\end{aligned}$$

Gravity thickening. Gravity thickening (Fig. 7.55) uses gravity forces to separate solids from the sludge. The equipment is similar in design to a conventional sedimentation basin. Sludge withdrawn from primary clarifiers or sludge blending tanks is applied to the gravity thickener through a central inlet wall. The normal solids loading rates range from 30 to 60 kg solids per m^2 of tank bottom per day (6 to 12 $lb/(ft^2 \cdot d)$) (Hammer 1986). Coagulant is sometimes added for improving the settling.

Typical hydraulic loading rates are from 16 to 32 $m^3/(m^2 \cdot d)$ (390 to 785 $gal/(d \cdot ft^2)$). For waste activated-sludge or for a very thin mixture, hydraulic loading rates range from 4 to 8 $m^3/(m^2 \cdot d)$ (100 to 200 $gal/(d \cdot ft^2)$) for secondary sludges, and 16 to 32 $m^3/(m^2 \cdot d)$ (390 to 785 $gal/(d \cdot ft^2)$) for primary sludges (USEPA 1979). For activated-sludge, residence time in the thickener needs to be more than 18 h to reduce gas production and other undesirable effects (WEF and ASCE 1991b). A typical design is a circular tank with a side depth of 3 to 4 m and a floor sloping at 1 : 4 to 1 : 6.

EXAMPLE 1: A residual with 4 percent solids is thickened to a 9 percent solids content. What is the concentration factor?

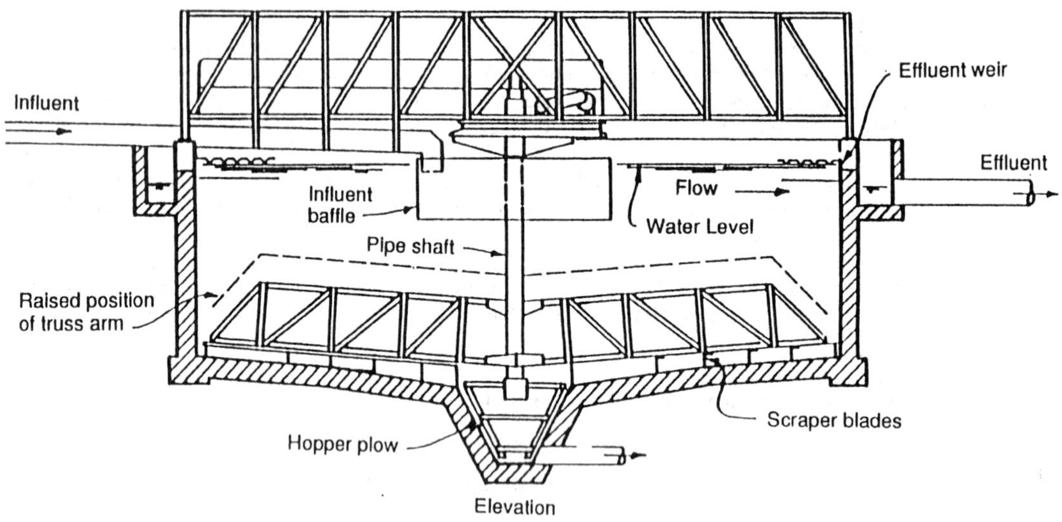

FIGURE 7.55 Schematic of gravity thickener (*Source: USEPA 1991*).

Solution:

$$\text{Concentration factor} = \frac{\text{thickened solids concentration, \%}}{\text{solids content in influent, \%}}$$
$$= \frac{9\%}{4\%}$$
$$= 2.25$$

Note: Concentration factor should be 2 or more for primary sludges, and 3 or more for secondary sludges.

EXAMPLE 2: A 9 m (30 ft) diameter by 3 m (10 ft) side wall depth gravity thickener is concentrating a primary sludge. The sludge flow is 303 L/min (80 gal/min) with average solids content of 4.4 percent. The thickened sludge is withdrawn at 125 L/min (33 gal/min) with 6.8 percent solids. The sludge blanket is 1 m (3.3 ft) thick. The effluent of the thickener has a TSS level of 660 mg/L. Determine (1) the sludge detention time; (2) whether the blanket will increase or decrease in depth under the stated conditions.

Solution:

Step 1. Determine the sludge detention time

(a) Compute the volume of the sludge blanket

$$\text{Volume} = \pi\,(9\text{ m}/2)^2 \times 1\text{ m}$$
$$= 63.6\text{ m}^3$$
$$= 63{,}600\text{ L}$$

(b) Compute the daily sludge pumped

$$\text{Pumpage} = 125\text{ L/min} \times 1440\text{ min/d}$$
$$= 180{,}000\text{ L/d}$$

(c) Compute the sludge detention time (DT)

$$\begin{aligned} \text{DT} &= \text{sludge volume/sludge pumpage} \\ &= 63{,}600 \text{ L/(180,000 L/d)} \\ &= 0.353 \text{ day} \\ &= 8.48 \text{ h} \end{aligned}$$

Step 2. Determine sludge blanket increase or otherwise

(a) Estimate the amount of solids entering the thickener

$$\begin{aligned} \text{Assume sludge} &= 1 \text{ kg/L} \\ \text{Solids in} &= 303 \text{ L/min} \times 1440 \text{ min/d} \times 1 \text{ kg/L} \times 0.044 \\ &= 19{,}200 \text{ kg/d} \end{aligned}$$

(b) Compute the withdrawal rate

$$\begin{aligned} \text{Withdrawal rate} &= 125 \text{ L/min} \times 1440 \text{ min/d} \times 1 \text{ kg/L} \times 0.068 \\ &= 12{,}240 \text{ kg/d} \end{aligned}$$

(c) Compute solids discharge in the effluent of the thickener

$$\begin{aligned} \text{Effluent TSS} &= 660 \text{ mg/L} = 0.066\% \\ \text{Solids lost} &= 303 \times 1440 \times 1 \times (0.066\%/100\%) \\ &= 288 \text{ (kg/d)} \\ \text{say} &= 290 \text{ kg/d} \end{aligned}$$

(d) Compute total solids out

$$\begin{aligned} \text{Solids out} &= \text{(Step 2b)} + \text{(Step 2c)} = (12{,}240 + 290) \text{ kg/d} \\ &= 12{,}530 \text{ kg/d} \end{aligned}$$

(e) Compare Steps 2a and 2d

$$\text{Solids in } 19{,}200 \text{ kg/d} > \text{solids out, } 12{,}530 \text{ kg/d}$$

Answer: The sludge blanket will increase in depth.

Note: The withdraw rate can use per minute instead of per day.

Dissolved air flotation thickening. Flotation thickeners include dissolved air flotation (DAF), vacuum flotation, and depressed-air flotation. Only DAF is used for wastewater sludge thickening in the US. It offers significant advantages in thickening light sludges such as activated-sludge. The DAF thickener (Fig. 7.56) separates solids from the liquid phase in an upward direction by attaching fine air bubbles (60 to 100 μm) to particles of suspended solids which then float. The influent stream at the tank bottom is saturated with air, pressurized (280 to 550 kPa), then released to the inlet distributor. The retention tank is maintained at a pressure of 3.2 to 4.9 kg/cm^2 (45 to 70 lb/in^2) (USEPA 1987b).

The ratio of the quantity of air supplied and dissolved into the recycle or waste stream to that of solids (the air-to-solids, A/S ratio) is probably the most important factor affecting the performance of the flotation thickener. Normal loading rates for waste activated-sludge range from 10 to 20 kg solids/(m^2 · d) (2 to 4 lb/(ft^2 · d)). DAF thickening produces about 4 percent solids with a solids recovery of 85 percent (Hammer 1986). The sludge volume index, SVI, is also an important factor for DAF operation.

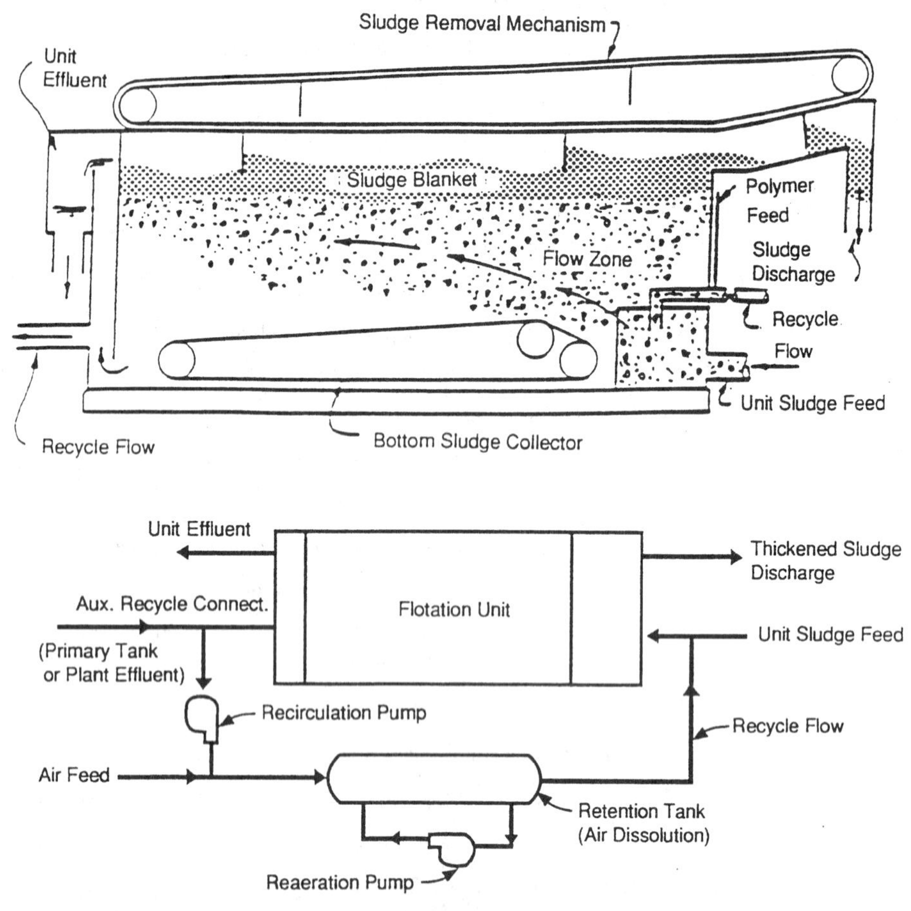

FIGURE 7.56 Dissolved air flotation thickener (*Source: USEPA 1991*).

EXAMPLE 1: Determine hydraulic and solids loading rates for a DAF thickener. The thickener, of 9 m (30 ft) diameter, treats 303 L/min (80 gal/min) of waste activated-sludge with a TSS concentration of 7800 mg/L.

Solution:

Step 1. Calculate liquid surface area

$$\text{Area} = \pi\ (9\ \text{m}/2)^2 = 63.6\ \text{m}^2$$

$$\text{Hydraulic loading} = (303\ \text{L/min})/63.6\ \text{m}^2$$

$$= 4.76\ \text{L/(min} \cdot \text{m}^2)$$

Step 2. Calculate solids loading rate

$$\text{TSS in WAS} = 7800\ \text{mg/L} = 0.78\%$$

$$\text{Solids loaded} = 303\ \text{L/min} \times 60\ \text{min/h} \times 1\ \text{kg/L} \times (0.78\%/100\%)$$

$$= 141.8\ \text{kg/h}$$

$$\begin{aligned}\text{Solids loading rate} &= (141.8\ \text{kg/h})/(63.6\ \text{m}^2)\\ &= 2.23\ \text{kg/(m}^2\cdot\text{h)}\\ &= 53.3\ \text{kg/(m}^2\cdot\text{d)}\\ &= 10.9\ \text{lb/(ft}^2\cdot\text{h)}\end{aligned}$$

EXAMPLE 2: A DAF thickener treats 303 L/min (80 ft/min) of waste-activated-sludge at 8600 mg/L. Air is added at a rate of 170 L/min (6.0 ft^3/min). Determine the air-to-solids ratio. Use 1.2 g of air per liter of air (0.075 lb/ft^3) under the plant conditions.

Solution:

Step 1. Compute mass of air added

$$\begin{aligned}\text{Air} &= 170\ \text{L/min} \times 1.2\ \text{g/L}\\ &= 204\ \text{g/min}\end{aligned}$$

Step 2. Compute mass of solids treated

$$\begin{aligned}\text{Solids} &= 8.6\ \text{g/L} \times 303\ \text{L/min}\\ &= 2606\ \text{g/min}\end{aligned}$$

Step 3. Compute A/S ratio

$$\begin{aligned}\text{A/S} &= (204\ \text{g/min})/(2606\ \text{g/min})\\ &= 0.078\ \text{g air/g solids}\\ \text{or} \quad &= 0.078\ \text{lb air/lb solids}\end{aligned}$$

EXAMPLE 3: Determine the concentration factor and the solids removal efficiency of a DAF thickener with conditions the same as in Example 2. The thickened sludge or float has 3.7 percent solids and the effluent TSS concentration is 166 mg/L.

Solution:

Step 1. Compute the concentration factor CF
From Example 2,

$$\begin{aligned}\text{The influent sludge solids content} &= 8600\ \text{mg/L}\\ &= 0.86\%\end{aligned}$$

$$\begin{aligned}\text{CF} &= \frac{\text{float solids content}}{\text{influent solids content}} = \frac{3.7\%}{0.86\%}\\ &= 4.3\end{aligned}$$

Step 2. Compute the solids removal efficiency

$$\begin{aligned}\text{Efficiency} &= \frac{\text{influent TSS} - \text{effluent TSS}}{\text{influent TSS}}\\ &= \frac{(8600 - 166)\ \text{mg/L} \times 100\%}{8600\ \text{mg/L}}\\ &= 98.1\%\end{aligned}$$

Centrifuge thickening. A centrifuge acts both to thicken and to dewater sludge. The centrifuge process separates liquid and solids by the influence of centrifugal force which is typically 50 to 300 times that of gravity (WEF and ASCE 1996b). Centrifuges may be used for thickening waste activated-sludge or as dewatering devices for digested or conditioned sludges. Three basic types (solid bowl, imperforate basket, and disc-nozzle) are commonly installed to thicken or dewater wastewater sludge.

The solid bowl scroll centrifuge (Fig. 7.57) is the most widely used type. It rotates along a horizontal axis and operates in a continuous-feed manner. It consists of a rotating bowl having a cylindrical-conical shape and a screw conveyer. Sludge is introduced into the rotating bowl through a stationary feed pipe continuously and the solids concentrate on the periphery. The gravitational force causes the solids to settle out on the inner surface of the rotating bowl. A helical scroll, spinning at a slightly different speed, moves the settled solids toward the tapered end (outlet ports) and then discharges them. The light liquid pools above the sludge layer and flows towards the centrate outlet ports. The units has a low cost/capacity ratio.

The basket centrifuge, also called the imperforate bowl (Fig. 7.58), is a knife-discharge type and operates on a batch basis. Liquid sludge is transported by a pipe through the top and is fed to the bottom of a vertically mounted spinning bowl. Solids accumulate against the wall of the bowl by centrifugal force and the centrate is decanted. The duration of the feed time is controlled by a preset timer or a centrate monitor (usually 60 to 85 percent of maximum depth). When the feed is stopped, the bowl begins to decelerate. As a certain point a nozzle skimmer (plow or knife) enters the bowl to remove the retained solids. The solids fall through the bottom of the bowl into a hopper. The plow retracts and the bowl accelerates, starting a new cycle. The units needs a skilled operator.

The disc-nozzle centrifuge (Fig. 7.59) rotates along a vertical axis and operates in a continuous manner. The liquid sludge is fed normally through the top of the unit (bottom feed is also possible) and flows through a feedwell in the center of the rotor and to a set of some 50 conical discs. An impeller within the rotor accelerates and distributes the feed slurry, filling the rotor interior. The centrifugal force is applied to the relatively thin film of liquid and solid between the discs. The force throws the denser solid materials to the wall of the rotor bowl, where it is subjected to additional

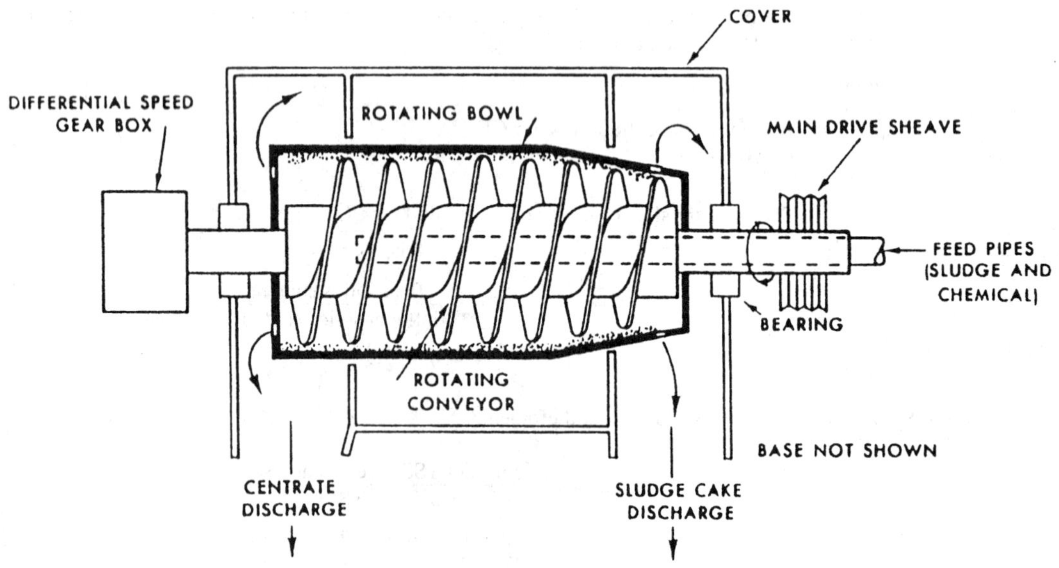

FIGURE 7.57 Solid bowl scroll centrifuge (*Source: USEPA 1991*).

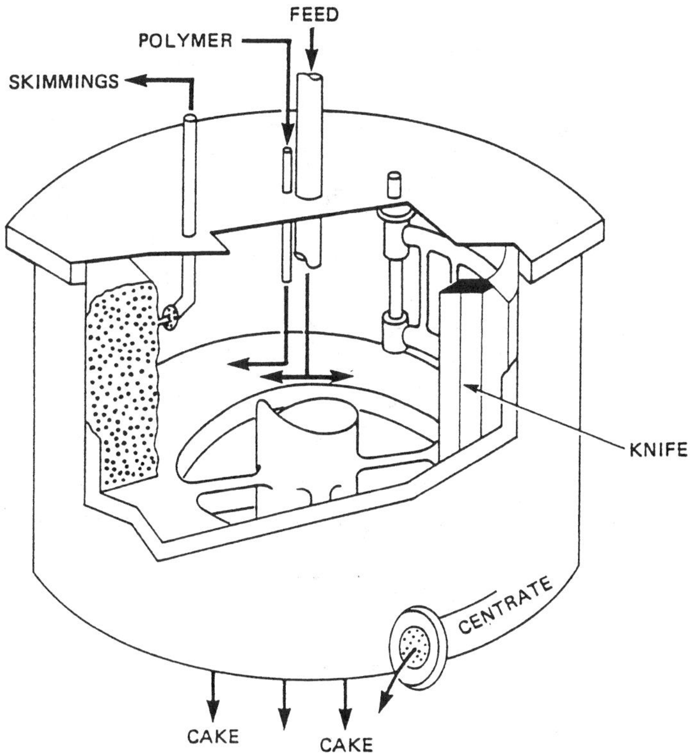

FIGURE 7.58 Schematic of imperforate basket centrifuge (*Source: USEPA 1991*).

centrifugal force and concentrated before it is discharged through nozzles located on the periphery. The clarified liquid passes on through the disc stack into the weir at the top of the bowl, and is then discharged.

The performance of a centrifuge is usually determined by the percentage of capture. It can be calculated as follows

$$\text{Percent capture} = \left[1 - \frac{C_r(C_c - C_s)}{C_s(C_c - C_r)}\right] \times 100 \qquad (7.213)$$

where C_r = concentration of solids in rejected wastewater (centrate), mg/L, %
C_c = concentration of solids in sludge cake, mg/L, %
C_s = concentration of solids in sludge feed, mg/L, %

EXAMPLE: A basket centrifuge is applied with 378 L/min (100 gal/min) of waste activated-sludge at a solids content of 0.77 percent. The basket run time for 80 percent depth is 18 min with a skimming operating time of 2 min. The average solids content in the thickened sludge is 7.2 percent. The solids concentration in the effluent is 950 mg/L. Determine hourly hydraulic loading and solids loading, dry solids produced per cycle and per day, and the efficiency of solids capture.

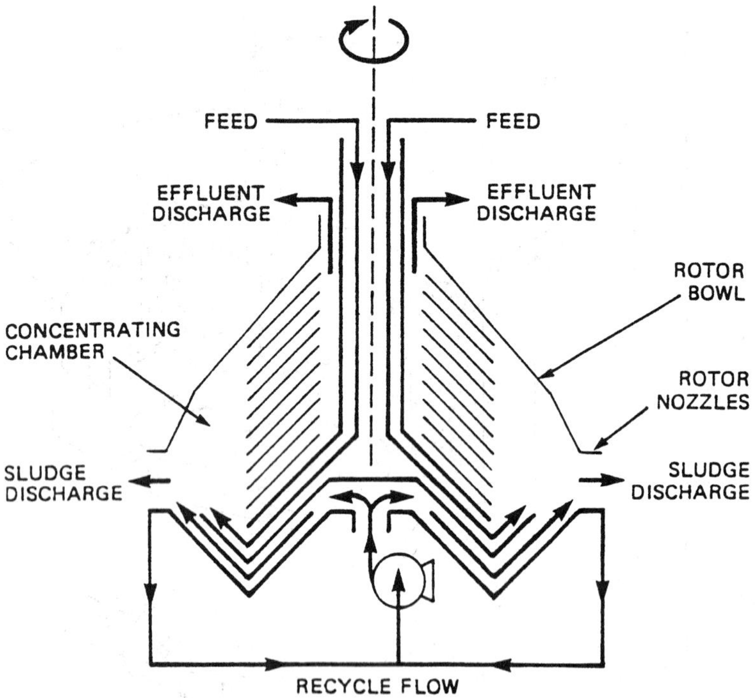

FIGURE 7.59 Schematic of a disc-nozzle centrifuge (*Source: USEPA 1991*).

Solution:

Step 1. Calculate hydraulic loading (HL) rate

$$\begin{aligned} HL &= 378 \text{ L/min} \times 60 \text{ min/h} \\ &= 22{,}680 \text{ L/h } (22.7 \text{ m}^3\text{/h}) \\ &= 5992 \text{ gal/h} \end{aligned}$$

Step 2. Calculate solids loading SL

$$\begin{aligned} SL &= 378 \text{ L/min} \times 60 \text{ min/h} \times (0.77\%/100\%) \times 1 \text{ kg/L} \\ &= 175 \text{ kg/h} \\ &= 385 \text{ lb/h} \end{aligned}$$

Step 3. Calculate the solids production per cycle and per day

$$\text{Solids produced per cycle} = \frac{175 \text{ kg/h} \times 18 \text{ min/cycle}}{60 \text{ min/h}}$$

$$= 52.5 \text{ kg/cycle}$$

$$\text{Total time for each cycle} = 18 \text{ min} + 2 \text{ min} = 20 \text{ min/cycle}$$

$$\text{Solids produced per day} = \frac{52.5 \text{ kg/cycle} \times 1440 \text{ min/d}}{20 \text{ min/cycle}}$$

$$= 3780 \text{ kg/d}$$

$$= 8330 \text{ lb/d}$$

Step 4. Calculate the efficiency of solids capture

Giving:

$$C_r = 950 \text{ mg/L} = 0.095\%$$
$$C_c = 7.2\%$$
$$C_s = 0.77\%$$

Using Eq. (7.213)

$$\text{Efficiency} = \left[1 - \frac{C_r(C_c - C_s)}{C_s(C_c - C_r)}\right] \times 100\%$$

$$= \left[1 - \frac{0.095(7.2 - 0.77)}{0.77(7.2 - 0.095)}\right] \times 100\%$$

$$= 88.8\%$$

Gravity belt thickening. The gravity belt thickener is a relatively new sludge thickening technique, stemming from the application of belt presses for sludge dewatering. It is effective for raw sludge and digested sludge with less than 2 percent solids content. A gravity belt moves over rollers driven by a variable-speed drive unit. The sludge is conditioned with polymer and applied into a feed/distribution box at one end. The sludge is distributed evenly across the width of the moving belt as the liquid drains through, and the solids are carried toward the discharge end of the thickener and removed. The belt travels through a wash cycle. Additional material on gravity belt thickeners and rotary drum thickeners may be found in WEF and ASCE (1991b, 1996b).

EXAMPLE 1: An operator adds 2 pounds (0.91 kg) of a dry polymer to 50 gal (189 L) of water. What is the polymer strength?

Solution:

Step 1. Calculate the mass of water plus polymer, W

$$W = 50 \text{ gal} \times 8.34 \text{ lb/gal} + 2 \text{ lb}$$

$$= 419 \text{ lb}$$

Step 2. Calculate the strength of polymer solution

$$\text{Polymer strength} = \frac{\text{mass of polymer}}{\text{total mass}} = \frac{2 \text{ lb}}{419 \text{ lb}} \times 100\% = 0.48\%$$

EXAMPLE 2: Determine the amount of dry polymer needed to prepare a 0.12% solution in a 1000-gal (3.78 m^3) tank. If 8 gal/min of this 0.12% polymer solution is added to a waste activated-sludge flow of 88 gal/min (333 L/min) with 8600 mg/L solids concentration, compute the dosage of polymer per ton of sludge.

Solution:

Step 1. Compute dry polymer required for 1000-gal solution

$$\text{Polymer} = 1000 \text{ gal} \times 8.34 \text{ lb/gal} \times (0.12\%/100\%)$$

$$= 10.0 \text{ lb}$$

Step 2. Compute the dosage in lb/ton of sludge

$$\text{Solids at 8600 mg/L} = 0.86\%$$

$$\text{Dosage} = \frac{\text{lb polymer added}}{\text{lb dry solid}} \times \frac{2000 \text{ lb}}{\text{ton}}$$

$$= \frac{8 \text{ gal/min} \times 8.34 \text{ lb/gal} \times 0.12\% \times 2000 \text{ lb/ton}}{88 \text{ gal/min} \times 8.34 \text{ lb/gal} \times 0.86\%}$$

$$= 25.4 \text{ lb polymer/ton dry sludge solid}$$

Sludge stabilization. After sludge has thickened, it requires stabilization to convert the organic solids to a more refractory or inert form. Thus the sludge can be handled or used as a soil conditioner without causing a nuisance or health hazard. The purposes of sludge stabilization are to reduce pathogens, eliminate odor-causing materials, and to inhibit, reduce, and eliminate the potential for putrefaction.

Treatment processes commonly used for stabilization of wastewater sludges include anaerobic digestion, aerobic digestion, and chemical (lime, disinfectants) stabilization and composting.

Anaerobic digestion. Anaerobic sludge digestion is the biochemical degradation (oxidation) of complex organic substances in the absence of free oxygen. The process is one of the oldest and most widely used methods. During anaerobic digestion, energy is released, and much of the volatile organic matter is converted to methane, carbon dioxide, and water. Thus little carbon and energy are available to sustain further biological activity, and the remaining residuals are rendered stable.

Anaerobic digestion involves three basic successive phases of fermentation: hydrolysis, acid formation, and methane formation (WEF and ASCE 1996b, USEPA 1991). In the first phase of digestion, extracellular enzymes (enzymes operating outside the cells) break down complex organic substances (proteins, cellulose, lignins, lipids) into soluble organic fatty acids, alcohols, carbon dioxide, and ammonia.

In the second phase, acid-forming bacteria, including facultative bacteria, convert the products of the first stage into short-chain organic acids such as acetic acid, propionic acid, other low molecular weight organic acids, carbon dioxide, and hydrogen. These volatile organic acids tend to reduce the pH, although alkalinity buffering materials are also produced. Organic matter is converted into a form suitable for breakdown by the second group of bacteria.

The third phase, strictly anaerobic, involves two groups of methane-forming bacteria (methanogens). One group converts carbon and hydrogen to methane. The other group converts acetate to methane, carbon dioxide, and other trace gases. Both groups of bacteria are anaerobic. Closed digesters are used for anaerobic digestion. It should be noted that many authors consider only two phases (excluding phase 1 above).

The most important factors affecting the performance of anaerobic digesters are solids residence time, hydraulic residence time, temperature, pH, and toxic materials. Methanogens are very sensitive to environmental conditions. Anaerobic digesters are usually heated to maintain a temperature of 34°C to 36°C (94°F to 97°F). The methane bacteria are active in the mesophilic range (27°C to 43°C, 80°F to 110°F). They have a slower growth rate than the acid formers and are very specific in food supply requirements. The anaerobic digester (two-stage) usually provides 10 to 20 days' detention of sludge (WEF and ASCE 1996b). Optimum methane production typically occurs when the pH is maintained between 6.8 and 7.2. If the temperature falls below the operating range and/or the digestion times falls below 15 days, the digester may become upset and require close monitoring and attention.

If concentrations of certain materials such as ammonia, sulfide, light metal cations, and heavy metals increase significantly in anaerobic digesters, they can inhibit or upset the process performance.

The key parameter for digester sizing is solids residence time. For digester systems without recycle, there is no difference between SRT and HRT. Volatile solids loading rate is also used frequently as a basis for design. Typically, design SRT values are from 30 to 60 days for low-rate digesters and 10 to 20 days for high-rate digesters (WEF and ASCE 1991b).

Volatile solids loading criteria are generally based on sustained loading conditions (typically peak month or peak week solids production), with provisions for avoiding excessive loading during shorter time periods. A typical design sustained peak volatile solids loading rate is 1.9 to 2.5 kg VS/($m^3 \cdot d$) (0.12 to 0.16 lb VS/($ft^3 \cdot d$)). A maximum limit of 3.2 kg VS/($m^3 \cdot d$) (0.20 lb VS/($ft^3 \cdot d$)) is often used (WEF and ASCE 1991b).

The configuration of an anaerobic digester is typically a single-stage or two-stage process (Figure 7.60). For the low-rate, single-stage digester, three separate layers (scum, supernatant, and sludge layers) form as decomposition occurs. The stabilized (digested) sludge settles at the bottom of the digester. The supernatant is usually returned to the plant influent. In a single-stage high-rate process, the digester is heated and mixed, and supernatant is not withdrawn. In a two-stage system,

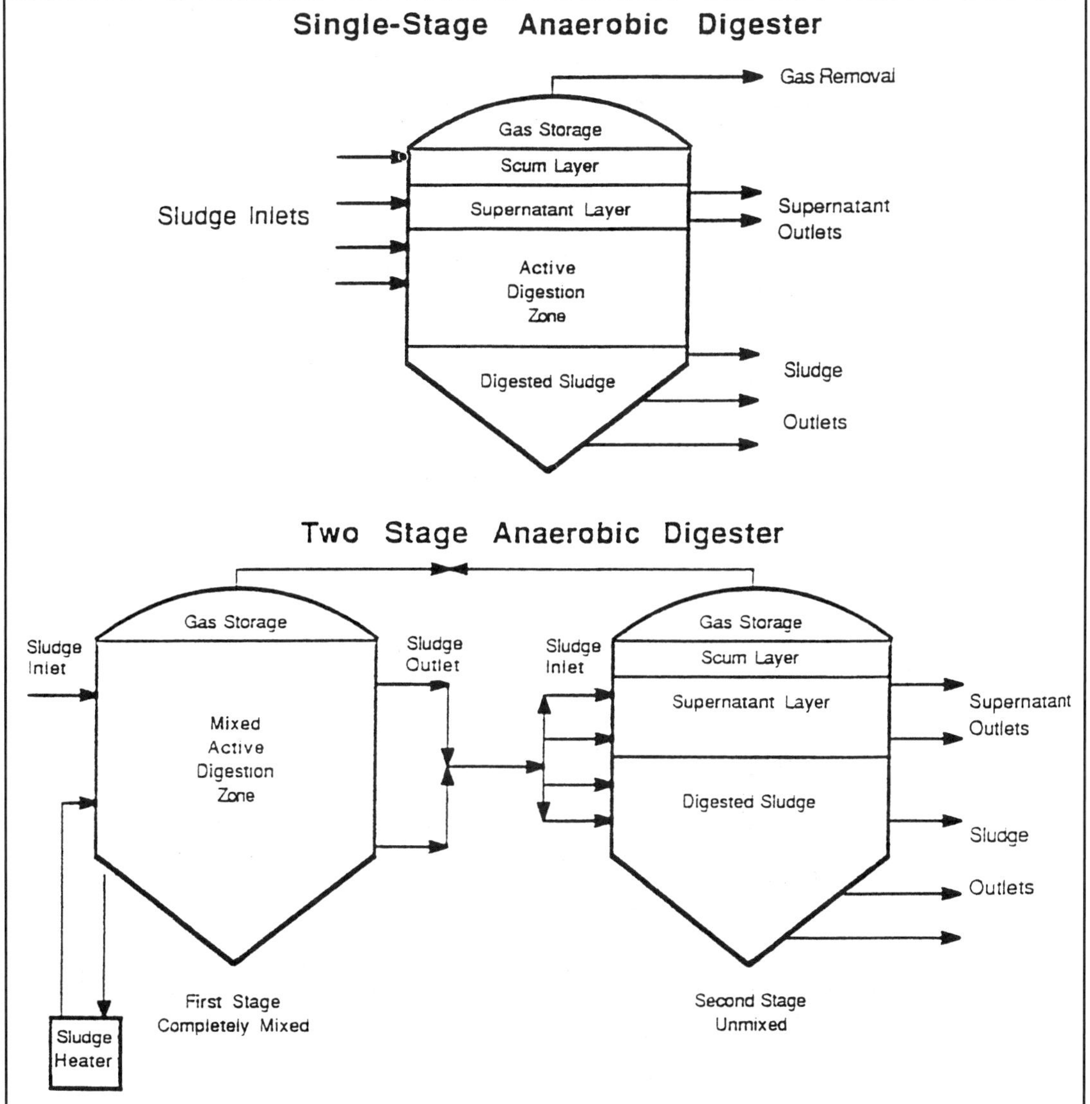

FIGURE 7.60 Configuration of (a) single-stage and (b) two-stage anaerobic digesters (*Source: USEPA 1991*).

sludge is stabilized in the first stage, whereas the second stage provides settling and thickening. The digester is heated to 34°C to 36°C.

An anaerobic digester is designed to provide warm, oxygen-free, and well-mixed conditions to digest organic matter and to reduce pathogenic organisms. The hardware include covers, heaters, and mixer. Operation of the process demands control of food supply, temperature (31°C to 36°C, 88°F to 97°F), pH (6.8 to 7.2), alkalinity (2000 to 3500 mg/L as $CaCO_3$), and detention time (60 days at 20°C to 15 days at 35°C).

Gas production. Gas production is one of the important parameters for measuring the performance of the digester. Typically, gas production ranges from 810 to 1120 L of digester gas per kg volatile solids (13 to 18 ft^3 gas/lb VS) destroyed. Gas produced from a properly operated digester contains approximately 65 to 69 percent methane and 31 to 35 percent carbon dioxide. If more than 35 percent of gas is carbon dioxide, there is probably something wrong with the digestion system (USEPA 1991). The quantity of methane gas produced can be computed by the following equation (McCarty 1964)

$$V = 350[Q(S_0 - S)/(1000) - 1.42\,P_x] \quad \text{(SI units)} \tag{7.214}$$

$$V = 5.62\,[Q(S_0 - S)8.34 - 1.42\,P_x] \quad \text{(US customary units)} \tag{7.215}$$

where V = volume of methane produced at standard conditions (0°C, 32°F and 1 atm), L/d or ft^3/d
350, 5.62 = theoretical conversion factor for the amount of methane produced per kg (lb) of ultimate BOD oxidized (see Example 1), 350 L/kg or 5.62 ft^3/lb
1000 = 1000 g/kg
Q = flow rate, m^3/d or Mgal/d
S_0 = influent ultimate BOD, mg/L
S = effluent ultimate BOD, mg/L
8.34 = conversion factor, lb/(Mgal · mg/L)
P_x = net mass of cell tissue produced, kg/d or lb/d

For a complete-mix high-rate two-stage anaerobic digester without recycle, the mass of biological solids synthesized daily, P_x, can be estimated by the equation below

$$P_x = \frac{Y[Q(S_0 - S)]}{1 + k_d\theta_c} \quad \text{(SI units)} \tag{7.216}$$

$$P_x = \frac{Y[Q(S_0 - S)8.34]}{1 + k_d\theta_c} \quad \text{(US customary units)} \tag{7.217}$$

where Y = yield coefficient, kg/kg or lb/lb
k_d = endogenous coefficient, per day
θ_c = mean cell residence time, d
Other terms are as defined previously.

EXAMPLE 1: Determine the amount of methane generated per kg of ultimate BOD stabilized. Use glucose, $C_6H_{12}O_6$, as BOD.

Solution:

Step 1. Use the balanced equation for converting glucose to methane and carbon dioxide under anaerobic conditions

	$C_6H_{12}O_6$	$\rightarrow$	$3CH_4$	+	$3CO_2$
MW	180		48		
	1		x		

$x = 48/180 = 0.267$

Step 2. Determine oxygen requirement for methane
Using a balanced equation of oxidation of methane to carbon dioxide and water

$$\begin{array}{lcccccccc} & 3CH_4 & + & 6O_2 & \rightarrow & 3CO_2 & + & 6H_2O \\ \text{MW} & 48 & & 192 & & & & \end{array}$$

$$\begin{aligned} \text{Ultimate BOD per kg of glucose} &= (192/180)\ \text{kg} \\ &= 1.07\ \text{kg} \end{aligned}$$

Step 3. Calculate the rate of the amount of methane generated per kg of BOD_L converted

$$\frac{\text{kg } CH_4}{\text{kg } BOD_L} = \frac{0.267}{1.07} = \frac{0.25}{1.0}$$

or simply 48/192 = 0.25/1.0

Note: 0.25 kg of methane is produced by each kg of BOD_L stabilized.

Step 4. Calculate the volume equivalent of 0.25 kg of methane at the standard conditions (0°C and 1 atm)

$$\begin{aligned} \text{Volume} &= (0.25\ \text{kg} \times 1000\ \text{g/kg}) \left(\frac{1\ \text{mol}}{16\ \text{g}}\right)\left(\frac{22.4\ \text{L}}{\text{mol}}\right) \\ &= 350\ \text{L} \\ &= 12.36\ \text{ft}^3 \end{aligned}$$

Note: 350 L of methane is produced per kg of ultimate BOD stabilized, or 5.60 ft^3 of methane is produced per lb of ultimate BOD stabilized (L/kg × 0.016 = ft^3/lb).

EXAMPLE 2: Determine the size of anaerobic digester required to treat primary sludge under the conditions given below, using a complete-mix reactor. Also check loading rate and estimate methane and total gas production.

Average design flow	3785 m^3/d	(1 Mgal/d)
Dry solid removed	0.15 kg/m^3	(1250 lb/Mgal)
Ultimate BOD_L removed	0.14 kg/m^3	(1170 lb/Mgal)
Sludge solids content	5%	(95% moisture)
Specific gravity of solids	1.01	
θ_c	15 day	
Temperature	35°C	
Y	0.5 kg cells/kg BOD_L	
K_d	0.04 per day	
Efficiency of waste utilization	0.66	

Solution:

Step 1. Calculate the sludge volume produced per day

$$\text{Sludge volume per day} = \frac{(3785\ \text{m}^3/\text{d})\ (0.15\ \text{kg/m}^3)}{1.01\ (1000\ \text{kg/m}^3)\ (0.05\ \text{kg/kg})}$$

$$\begin{aligned} Q &= 11.24\ \text{m3/d} \\ &= 397\ \text{ft}^3/\text{d} \end{aligned}$$

Step 2. Calculate the total BOD_L loading rate

$$BOD_L \text{ loading} = 0.14 \text{ kg/m}^3 \times 3785 \text{ m}^3\text{/d}$$
$$= 530 \text{ kg/d}$$
$$= 1.170 \text{ lb/d}$$

Step 3. Calculate the volume of the digester
From Step 1

$$Q = 11.24 \text{ m}^3\text{/d}$$
$$V = Q\theta_c = (11.24 \text{ m}^3\text{/d})\ (15 \text{ day})$$
$$= 169 \text{ m}^3$$
$$= 5968 \text{ ft}^3$$

Use the same volume for both the first and second stages.

Step 4. Check BOD_L volumetric loading rate

$$BOD_L = \frac{530 \text{ kg/d}}{169 \text{ m}^3}$$
$$= 3.14 \text{ kg/(m}^3 \cdot \text{d)}$$

Step 5. Estimate the mass of volatile solids produced per day, using Eq. (7.216)

From Step 2: $QS_0 = 530$ kg/d

$$QS = 530 \text{ kg/d } (1 - 0.66) = 180 \text{ kg/d}$$
$$P_x = \frac{YQ(S_0 - S)}{1 + k_d\theta_c}$$
$$= \frac{0.05(530 - 180) \text{ kg/d}}{1 + (0.04 \text{ d}^{-1})\ 15 \text{ day}}$$
$$= 10.9 \text{ kg/d}$$

Step 6. Calculate the percentage of stabilization

$$\% \text{ stabilization} = \frac{[(530 - 180) - 1.42(10.9)] \text{ kg/d} \times 100\%}{530 \text{ kg/d}}$$
$$= 63.1\%$$

Step 7: Calculate the volume of methane produced per day, using Eq. (7.214)

$$V = 350 \text{ L/kg } [(530 - 180) - 1.42\ (10.9)] \text{ kg/d}$$
$$= 117{,}000 \text{ L/d}$$
$$= 117 \text{ m}^3\text{/d}$$
$$= 4130 \text{ ft}^3\text{/d}$$

Step 8. Estimate total gas ($CH_4 + CO_2$) production
Normally 65 to 69 percent of gas is methane. Use 67 percent.

$$\text{Total gas volume} = (117 \text{ m}^3\text{/d})/0.67$$
$$= 175 \text{ m}^3\text{/d}$$
$$= 6170 \text{ ft}^3\text{/d}$$

EXAMPLE 3: Estimate the gas produced in a anaerobic digester treating the secondary sludge from 1.0 MGD (3785 m^3/d) of wastewater containing 240 mg/L of total suspended solids (TSS).

Solution:

Step 1. Estimate TSS removed, assuming the over-all secondary treatment process has TSS removal of 90 percent

$$\text{TSS removed} = 240\ \text{mg/L} \times 0.9 = 216\ \text{mg/L}$$

Step 2. Estimate volatile suspended solids (VSS) in raw sewage removed, assuming VSS/TSS = 0.75

$$\text{VSS removed} = 216\ \text{mg/L} \times 0.75 = 162\ \text{mg/L}$$

Step 3. Estimate volatile suspended solids (VSS) in the sludge reduced 65 percent

$$\text{VSS reduced} = 162\ \text{mg/L} \times 0.65 = 105.3\ \text{mg/L}$$

Step 4. Calculate volatile solids (VS) reduced in 1 Mgal/d of sewage

$$\text{VS reduced} = (1\ \text{Mgal/d}) \times 8.34\ \text{lb/(Mgal} \cdot \text{mg/L)} \times 105.3\ \text{mg/L} = 878\ \text{lb/d}$$

Step 5. Estimate daily gas production for 1 MGD of sewage, assuming 15.0 ft^3/lb of VS

$$\begin{aligned}\text{Gas produced} &= 15.0\ \text{ft}^3/\text{lb} \times 878\ \text{lb/d}\\ &= 13{,}170\ \text{ft}^3/\text{d}\end{aligned}$$

EXAMPLE 4: The desired digester temperature is 91°F (32.8°C). The average raw sludge temperature is 45°F (7.2°C). The volume of raw sludge is 1800 gal/d (6810 L/d) with a specific weight of 1.20. Determine the quantity of heat needed for the digest, if SRT = 50 days and radiation heat loss of the inner surface is 7 Btu/h/ft^2.

Solution:

Step 1. Calculate the heat required to raise the temperature of the sludge

$$\begin{aligned}\text{Heat} &= 1800\ \text{gal/d} \times 8.34\ \text{lb/gal} \times (91 - 45)\ \text{F}\\ &= 690{,}600\ \text{Btu/d (lb} \cdot \text{F/d)}\end{aligned}$$

Step 2. Determine the capacity of the digester with SRT = 50 days

$$\begin{aligned}\text{Digester} &= 1800\ \text{gal/d} \times 0.1337\ \text{ft}^3/\text{gal} \times 50\ \text{day}\\ &= 12{,}030\ \text{ft}^3\end{aligned}$$

Use one digester 15 ft deep, thus the diameter (D) is

$$\begin{aligned}15 \times (3.14/4)D^2 &= 12{,}030\\ D &= 32.0\ \text{ft}\end{aligned}$$

Step 3. Calculate radiation heat loss, assuming 7 Btu/h/ft^2 from the inner surface

Area of the tank floor or roof = 3.14 × 16 ft × 16 ft = 804 ft^2

Area of the wall = 32 ft × 3.14 × 15 ft = 1507 ft^2

Total area = 804 ft^2 + 804 ft^2 + 1507 ft^2 = 3115 ft^2

Radiation loss = 7 Btu/h/ft^2 × 24 h/d × 3115 ft^2 = 523,300 Btu/d

Step 4. Calculate the total heat required

$$\text{Sum of heat to raise sludge and radiation loss} = 690{,}600 \text{ Btu/d} + 523{,}300 \text{ Btu/d} = 1{,}213{,}900 \text{ Btu/d}$$

Adding extra heat (25%) for other losses

$$\text{Total heat required} = 1{,}213{,}900 \text{ Btu/d} \times 1.25 = 1{,}517{,}400 \text{ Btu/d}$$

Egg-shaped digesters. The egg-shaped digester looks similar to an upright egg and is different from the conventional American digester and the conventional German digester. The American digester typically is a relatively shallow cylindrical vessel with moderate floor and roof slopes. The typical conventional German digester is a deep cylindrical tank with steeply sloped top and bottom cones. The egg-shaped digester was first installed in Germany in the 1950s.

The advantages of the egg-shaped digester over conventional digesters include the elimination of abrupt change in the vessel geometry, reduced inactive volume of the digester, more cost effectiveness and less energy costs. The egg shape of the vessel provides unique advantages. Its mechanisms are as follows. Heavy solids cannot settle out over a large unmanageable bottom area; they must concentrate in the steep-sided bottom cone. Light solids do not accumulate across a large diameter surface area; they concentrate at the top of a steep-sided converging cone. The settleable materials can be pumped to the top of the vessel to sink light material into the digesting main mass. Likewise, light materials can be pumped to the bottom zone to stir the heavier materials. Small amounts of energy are required to manage the top and bottom zones. Moving materials through the main digesting mass to assure homogeneous digester conditions can be done more effectively (CBI Walker 1998).

The egg-shaped digester is currently constructed of steel. It has become popular in the United States and other countries. More detailed description and design information may be found elsewhere (CBI Walker 1998, Stukenberg *et al.* 1990).

Aerobic digestion. Aerobic digestion is used to stabilize primary sludge, secondary sludge, or a combination of these by long-term aeration. The process converts organic sludge solids to carbon dioxide, ammonia, and water by aerobic bacteria with reduction of volatile solids, pathogens, and offensive odor.

In a conventional aerobic digester, concentration of influent VSS must be no more than 3 percent for retention times of 15 to 20 days. High-purity oxygen may be used for oxygen supply.

Sludge is introduced to the aerobic digester on a batch (mostly), semibatch, or continuous basis. Aerobic digesters are typically a single-stage open tank like the activated-sludge aeration tank. In the batch basis, the digester is filled with raw sludge and aerated for 2 to 3 weeks, then stopped. The supernatant is decanted and the settled solids are removed (USEPA 1991). For the semibatch basis, raw sludge is added every couple of days; the supernatant is decanted periodically, and the settled solids are held in the digester for a long time before being removed.

The design standard for aerobic digesters varies with sludge characteristics and method of ultimate sludge disposal. Sizing of the digester is commonly based on volatile solids loading of the digester or the volatile solids loading rate. It is generally determined by pilot and/or full-scale study. In general, volatile suspended solids loading rates for aerobic digesters vary from 1.1 to 3.2 kg VSS/($m^3 \cdot d$) (0.07 to 0.20 lb VSS/($ft^3 \cdot d$)), depending on the temperature and type of sludge. Solids retention time is 15 to 20 days for activated-sludge only, and that for primary plus activated-sludge is 20 to 25 days (USEPA 1991). An aeration period ranges from 200 to 300 degree-days which is computed by multiplying the digester's temperature in °C by the sludge age. According to Rule 503 (Federal Register 1993), if the sludge is for land application, the residence time requirements range from 60 days at 15°C ($900° \cdot d$) to 40 days at 20°C ($800° \cdot d$). Desirable aerobic digestion temperatures are approximately 18°C to 27°C (65°F to 80°F). A properly operated aerobic digester is capable of achieving a volatile solids reduction of 40 to 50 percent and a pathogens reduction of 90 percent.

GLUMRB (1996) recommended volume requirements based on population equivalent (PE). Table 7.25 presents the recommended digestion tank capacity based on a solids content of 2 percent with supernatant separation performed in a separate tank. If supernatant separation is performed in

TABLE 7.25 Recommended Volume Required for Aerobic Digester

Type of sludge	Volume per population equivalent	
	m^3	ft^3
Waste activated-sludge—no primary settling	0.13*	4.5
Primary plus activated sludge	0.11*	4.0
Waste activated sludge exclusive of primary sludge	0.06*	2.0
Extended aeration activated sludge	0.09	3.0
Primary plus fixed film reactor sludge	0.09	3.0

*These volumes also apply to waste activated sludge from single-stage nitrification facilities with less than 24 h detention time based on design average flow.
Source: Greater Lakes Upper Mississippi River Board 1996.

the digestion tank, a minimum of 25 percent additional volume is required. If high solids content (> 2%) is applied, the digestion tank volume may be reduced proportionally. A digestion temperature of 15°C (59°F) and a solids retention time of 27 days (405° · d) is recommended. Cover and heating are required for cold temperature climate areas.

The air requirements of aerobic digestion are based on DO (1 to 2 mg/L) and mixing to keep solids in suspension. If DO in the digestion tank falls below 1.0 mg/L, the aerobic digestion process will be negatively impacted. The diffused air requirements for waste activated-sludge only is 20 to 35 ft^3/min per 1000 ft^3 (20 to 35 L/(min · m^3)) (USEPA 1991). The oxygen requirement is 2.3 kg oxygen per kg VSS destroyed (Metcalf and Eddy Inc. 1991).

The aerobic tank volume can be computed by the following equation (WPCF 1985b)

$$V = \frac{Q_i(X_i + YS_i)}{X(K_d P_v + 1/\theta_c)} \tag{7.218}$$

where V = volume of aerobic digester, ft^3
Q_i = influent average flow rate to digester, ft^3/d
X_i = influent suspended solids concentration, mg/L
Y = fraction of the influent BOD_5 consisting of raw primary sludge, in decimals
S_i = influent BOD_5, mg/L
X = digester suspended solids concentration, mg/L
K_d = reaction-rate constant, d^{-1}
P_v = volatile fraction of digester suspended solids, in decimals
θ = solids retention time, d

The term YS_i can be eliminated if no primary sludge is included in the influent of the aerobic digester. Eq. (7.218) is not to be used for a system where significant nitrification will occur.

EXAMPLE 1: The pH of an aerobic digester is found to have declined to 6.3. How much sodium hydroxide must be added to raise the pH to 7.0? The volume of the digester is 378 m^3 (0.1 Mgal). Results from jar tests show that 36 mg of caustic soda will raise the pH to 7.0 in a 2-L jar.

Solution:

$$\begin{aligned} \text{NaOH required per m}^3 &= 36 \text{ mg/2 L} = 18 \text{ mg/L} \\ &= 18 \text{ g/m}^3 \\ \text{NaOH to be added} &= 18 \text{ g/m}^3 \times 378 \text{ m}^3 \\ &= 6800 \text{ g} \\ &= 6.8 \text{ kg} = 15 \text{ lb} \end{aligned}$$

EXAMPLE 2: A 50 ft (15 m) diameter aerobic digester with 10 ft (3 m) sidewall depth treats 13,000 gal/d (49.2 m^3/d) of thickened secondary sludge. The sludge has 3.0 percent of solids content and is 80 percent volatile matter. Determine the hydraulic digestion time and volatile solids loading rate.

Solution:

Step 1. Compute the effective volume (V) of the digester

$$V = 3.14\ (50\ \text{ft}/2)^2 \times 10\ \text{ft} \times 7.48\ \text{gal/ft}^3$$
$$= 147{,}000\ \text{gal}$$
$$= 556\ \text{m}^3$$

Step 2. Compute the hydraulic digestion time

$$\text{Digestion time} = V/Q = 147{,}000\ \text{gal}/13{,}000\ \text{gal/d}$$
$$= 11.3\ \text{day}$$

Step 3. Compute the volatile solids loading rate

$$\text{Solids} = 3.0\% = 30{,}000\ \text{mg/L}$$
$$\text{VSS} = 30{,}000\ \text{mg/L} \times 0.8 = 24{,}000\ \text{mg/L}$$
$$\text{Volume} = 147{,}000\ \text{gal}/(7.48\ \text{ft}^3/\text{gal}) = 19{,}650\ \text{ft}^3$$
$$\text{VSS loading} = \frac{\text{VSS applied, lb/d}}{\text{Volume of digester, ft}^3}$$
$$= \frac{0.013\ \text{Mgal/d} \times 24{,}000\ \text{mg/L} \times 8.34\ \text{lb/(Mgal/d)(mg/L)}}{19{,}650\ \text{ft}^3}$$
$$= 0.132\ \text{lb}/(\text{d} \cdot \text{ft}^3)$$
$$= 2.11\ \text{kg}/(\text{d} \cdot \text{m}^3)$$

EXAMPLE 3: Design an aerobic digester to receive thickened waste activated-sludge at 1500 kg/d (3300 lb/d) with 2.5 percent solids content and a specific gravity of 1.02. Assume that the following conditions apply:

Minimum winter temperature	15°C
Maximum summer temperature	25°C
Residence time (land application by Rule 503)	900° · d at 15°C 700° · d at 25°C
VSS/TSS	0.80
SS concentration in digester	75% of influent SS
K_d at 15°C	0.05 d^{-1}
Oxygen supply and mixing	diffused air

Solution:

Step 1. Calculate the quantity of sludge to be treated per day, Q_i

$$Q_i = \frac{1500\ \text{kg/d}}{1000\ \text{kg/m}^3(1.02)(0.025)}$$
$$= 58.8\ \text{m}^3/\text{d}$$

Step 2. Calculate sludge age required for winter and summer conditions
During the winter:

$$\text{Sludge age} = 900°\text{d}/15° = 60 \text{ day}$$

During the summer:

$$\text{Sludge age} = 700°\text{d}/25° = 28 \text{ day}$$

The winter period is the controlling factor.
Under these conditions, assume the volatile solids reduction is 45 percent (normally 40 to 50 percent).
Note: A curve for determining volatile solids reduction in an aerobic digester as a function of digester liquid temperature and sludge age is available (WPCF 1985b). It can be used to compare winter and summer conditions.

Step 3. Calculate volatile solids reduction

$$\begin{aligned}\text{Total mass of VSS} &= 1500 \text{ kg/d} \times 0.8 \\ &= 1200 \text{ kg/d} \\ \text{VSS reduction} &= 1200 \text{ kg/d} \times 0.45 \\ &= 540 \text{ kg/d}\end{aligned}$$

Step 4. Calculate the oxygen required
Using 2.3 kg O_2/kg VSS reduction

$$\begin{aligned}O_2 &= 540 \text{ kg/d} \times 2.3 \\ &= 1242 \text{ kg/d}\end{aligned}$$

Convert to the volume of air required under standard conditions, taking the mass of air as 1.2 kg/m^3 (0.75 lb/ft^3) with 23.2 percent of oxygen.

$$\begin{aligned}\text{Air} &= \frac{1242 \text{ kg/d}}{1.2 \text{ kg/m}^3 \times 0.232} \\ &= 4460 \text{ m}^3/\text{d}\end{aligned}$$

Assume an oxygen transfer efficiency of 9 percent. Total air required is

$$\begin{aligned}\text{Air} &= (4460 \text{ m}^3/\text{d})/0.09 \\ &= 49{,}600 \text{ m}^3/\text{d} \\ &= 34.4 \text{ m}^3/\text{min} \\ &= 1210 \text{ ft}^3/\text{min}\end{aligned}$$

Step 5. Calculate the volume of the digester under winter conditions
Using Eq. (7.218)

$$\begin{aligned}X_i &= 2.5\% = 25{,}000 \text{ mg/L} \\ X &= 25{,}000 \text{ mg/L} \times 0.75 = 18{,}750 \text{ mg/L} \\ K_d &= 0.05 \text{ d}^{-1} \\ P_v &= 0.8 \\ \theta_c &= 60 \text{ day}\end{aligned}$$

The terms YS_i is neglected, since no primary sludge is applied.

$$V = \frac{Q_i(X_i + YS_i)}{X(K_d P_v + 1/\theta_c)}$$

$$= \frac{58.8 \text{ m}^3\text{/d} \times 25{,}000 \text{ mg/L}}{18{,}750 \text{ mg/L}(0.05 \text{ d}^{-1} \times 0.8 + 1/60 \text{ day})}$$

$$= 1383 \text{ m}^3$$

$$= 48{,}860 \text{ ft}^3$$

Step 6. Check the diffused air requirement per m^3 of digester volume, from Step 4

$$\text{Air} = \frac{34.4 \text{ m}^3\text{/min}}{1383 \text{ m}^3}$$

$$= 0.025 \text{ m}^3/(\text{min} \cdot \text{m}^3)$$

$$= 25 \text{ L/min} \cdot \text{m}^3$$

$$= 25 \text{ ft}^3\text{/min/1000 ft}^3$$

This value is at the lowest range of regulatory requirement, 25 to 35 L/min · m^3.

EXAMPLE 4: Determine the size of an aerobic digester to treat the primary sludge of 590 kg/d (1300 lb/d) plus activated-sludge of 380 kg/d (840 kg/d). Assume the mixed sludge has a solids content of 3.8 percent and a retention time of 18 days.

Solution:

Step 1. Calculate the mass of mixed sludge

$$\text{Mass} = (590 \text{ kg/d} + 380 \text{ kg/d})/0.038$$

$$= 25{,}530 \text{ kg/d} \cong 25.5 \text{ m}^3\text{/d}$$

Step 2. Calculate the volume (V) of the aerobic digester

The volume of daily sludge production is approximately 25.5 m^3/d, thus

$$V = 25.5 \text{ m}^3\text{/d} \times 18 \text{ day}$$

$$= 459 \text{ m}^3$$

Step 3. Check the solids loading rate assuming the solids are 80 percent volatile

$$\text{VSS} = (590 + 380) \text{ kg/d} \times 0.8 = 776 \text{ kg/d}$$

$$\text{Loading rate} = (776 \text{ kg VSS/d})/459 \text{ m}^3$$

$$= 1.69 \text{ kg VSS(m}^3 \cdot \text{d)[in the range of 1.1 to 3.2 kg VSS/(m}^3 \cdot \text{d)]}$$

Lime stabilization. Chemical stabilization of wastewater sludge is an alternative to biological stabilization. It involves chemical oxidation (commonly using chlorine) and pH adjustment under basic conditions, which is achieved by addition of lime or lime-containing matter. Chemical oxidation is achieved by dosing the sludge with chlorine (or ozone, hydrogen peroxide). The sludge is deodorized and microbiological activities slowed down. The sludge can then be dewatered and disposed of.

In the lime stabilization process, lime is added to sludge in sufficient quantity to raise the pH to 12 or higher for a minimum of 2 h contact. The highly alkaline environment will inactivate biological growth and destroy pathogens. The sludge does not putrefy, create odors, or pose a health hazard. However, if the pH drops below 11, renewed bacteria and pathogen growth can reoccur. Since the addition of lime does not reduce the volatile organics, the sludge must be further treated and disposed of before the organic matter starts to putrefy again.

With 40 CFR Part 503 regulations (Federal Register 1993) to reduce pathogens, a supplemental heat system can be used to reduce the quantity of lime required to obtain the proper temperature. Additional heating reduces the lime dosage and operating costs.

Composting. Composting is an aerobic biological decomposition of organic material to a stable end product at an elevated temperature. Some facilities have utilized anaerobic composting in a sanitary, nuisance-free environment to create a stable humus-like material suitable for plant growth. Approximately 20 to 30 percent of volatile solids are converted to carbon dioxide and water at temperatures in the pasteurization range of 50°C to 70°C (122°F to 158°F). At these temperatures most enteric pathogens are destroyed. Thermophilic bacteria are responsible for decomposing organic matter. The finished compost has a moisture content of 40 to 50 percent and a volatile solids content of 40 percent or less (USEPA 1991).

The optimum moisture content for composting is 50 to 60 percent water. Dewatered wastewater sludges are generally too wet to meet optimum composing conditions. If the moisture content is over 60 percent water, proper structural integrity must be obtained. The dewatered sludge and bulking agent must be uniformly mixed. Sufficient air must be supplied to the composting pile, either by forced aeration or windrow turning to maintain oxygen levels between 5 and 15 percent. Odor control may be required on some sites.

In order to comply with the "process to significantly reduce pathogens" (PSRP) requirements of 40 CFR Part 257 (Federal Register 1993), the compost pile must be maintained at a minimum operating temperature of 40°C (104°F) for at least 5 days. During this time, the temperature must be allowed to increase above 55°C (131°F) for at least 4 h to ensure pathogen destruction. In order to comply with the "process to further reduce pathogens" (PFRP) requirements of 40 CFR Part 257 (Federal Register 1993), the in-vessel and static aerated compost piles must be maintained at a minimum operating temperature of 55°C for at least 3 days. The windrow pile must be maintained at a minimum operating temperature of 55°C for 15 days. In addition, there must be at least three turnings of the compost pile during this period.

Sludge conditioning. Mechanically concentrated (thickened) sludges and biologically and chemically stabilized sludges still require some conditioning steps. Sludge conditioning can involve chemical and/or physical treatment to enhance water removal. Sludge conditioning is undertaken before sludge dewatering.

Details of process chemistry, design considerations, mechanical components, system layout, operation, and costs of conditional methods are presented in *Sludge Conditioning Manual* (WPCF 1988a). Some additional sludge conditioning processes disinfect sludge, control odors, alter the nature of solids, provide limited solids destruction, and increase solids recovery.

Chemical conditioning. Chemical conditioning can reduce the 90 to 99 percent incoming sludge moisture content to 65 to 80 percent, depending on the nature of the sludge to be treated (WPCF 1988a). Chemical conditioning results in coagulation of the solids and release of the absorbed water. Chemicals used for sludge conditioning include inorganic compounds such as lime, pebble quicklime, ferric chloride, alum, and organic polymers (polyelectrolytes). Addition of conditioning chemicals may increase the dry solids of the sludge. Inorganic chemicals can increase the dry solids by 20 to 30 percent, but polymers do not increase the dry solids significantly.

Physical conditioning. The physical process includes using hot and cold temperatures to change sludge characteristics. The commonly used physical conditioning methods are thermal conditioning and elutriation. Less commonly used methods include freeze–thaw, solvent extraction, irradiation, and ultrasonic vibration.

The thermal conditioning (heat treatment) process involves heating the sludge to a temperature of 177°C to 240°C (350°F to 464°F) in a reaction vessel under pressure of 1720 to 2760 kN/m^2 (250 to 400 lb/in^2 [psig]) for a period of 15 to 40 minutes (USEPA 1991). One modification of the process involves the addition of a small amount of air. Heat coagulates solids, breaks down the structure of microbial cells in waste activated-sludge, and releases the water bound in the cell. The heat-treated sludge is sterilized and practically deodorized. It has excellent dewatering characteristics and does not normally require chemical conditioning to dewater well on mechanical equipment. yielding cake solids concentrations of 40 to 50 percent.

Heat treatment is suitable for many types of sludge that cannot be stabilized biologically due to the presence of toxic materials and is relatively insensitive to change in sludge composition. However, the process produces liquid sidestreams with high concentrations of organics, ammonia nitrogen, and color. Significant odorous off-gases are also generated which must be collected and treated before release. The process has high capital cost because of its mechanical complexity and the use of corrosion-resistant material. It also requires close supervision, skilled operators, and O and M programs. More detailed description of thermal processing of sludge can be found elsewhere (WPCF 1988a, WEF and ASCE 1991b).

A physical conditioning method used in the past is elutriation. In elutriation, a washing process, sludge is mixed with a liquid for the purpose of transferring certain soluble organic or inorganic components to the liquid. Wastewater effluent is usually used for elutriation. The volume of wastewater is 2 to 6 times the volume of sludge. The elutriation of the sludge produces large volumes of liquid that contain a high concentration of suspended solids. This liquid, when returned to the treatment plant, increases the solids and organic loadings. Elutriation tanks are designed to act as gravity thickeners with solids loading rates of 39 to 49 kg/($m^2 \cdot d$) (8 to 10 lb/($ft^2 \cdot d$)) (Qasim 1985).

Sludge dewatering. Following stabilization and/or conditioning, wastewater sludge can be ultimately disposed of, or can be dewatered prior to further treatment and/or ultimate disposal. It is generally more economical to dewater before disposal. The primary objective of dewatering is to reduce sludge moisture. Subsequently, it reduces the costs of pumping and hauling to the disposal site. Dewatered sludge is easier to handle than thickened or liquid sludge. Dewatering is also required before composting, prior to sludge incineration, and prior to landfill. An advantage of dewatering is that it makes the sludge odorless and nonputrescible.

Sludge can be dewatered by slow natural evaporation and percolation (drying beds, drying lagoons) or by mechanical devices, mechanically assisted physical means, such as vacuum filtration, pressure filtration, centrifugation, or recessed plate filtration.

Vacuum filtration. Vacuum filtration has been used for wastewater sludge dewatering for almost seven decades. Its use has declined owing to competition from the belt press. A rotary vacuum filter consists of a cylindrical drum covered with cloth of natural or synthetic fabric, coil springs, or woven stainless steel mesh. The drum is partly submerged (20 to 40 percent) in a vat containing the sludge to be dewatered.

The filter drum is divided into compartments. In sequence, each compartment is subject to vacuums ranging from 38 to 75 cm (18 to 30 in) of mercury. As it slowly rotates, vacuum is applied immediately under the mat formation or sludge pick-up zone. Suction continues to dewater the solids adhering to the filter medium as it rotates out of the liquid. This is called the drying zone of the cycle. The cake drying zone represents from 40 to 60 percent of the drum surface. The vacuum is then stopped, and the solids cake is removed to a sludge hopper or a conveyor. The filter medium is washed by water sprays before reentering the vat.

For small plants, 35 h operation per week is often designed. This allows daily 7 h operation with 1 h for start-up and wash-down. For larger plants, vacuum filters are often operated for a period of 16 to 20 h/d (Hammer 1986).

EXAMPLE: A 3 m (10 ft) diameter by 4.5 m (15 ft) long vacuum filter dewaters 5400 kg/d primary plus secondary sludge solids. The filter is operated 7 h/d with a drum cycle time of 4 min. The dewatered sludge has 25 percent solids content. The filter yield is 17.3 kg/($m^2 \cdot h$). Determine the filter loading rate and percent solids recovery. What is the daily operating hours required if the dewatered sludge is 30 percent solids with the same percentage of solids recovery and a filter yield of 9.8 kg/($m^2 \cdot h$)?

Solution:

Step 1. Compute the surface area of the filter

$$\text{Area} = \pi(3\text{ m})(4.5\text{ m}) = 42.4\text{ m}^2$$

Step 2. Compute the filter loading rate

$$\text{Loading} = \frac{\text{solids loading}}{\text{area} \times \text{operating time}} = \frac{5400\text{ kg/d}}{42.4\text{ m}^2 \times 7\text{ h/d}}$$

$$= 18.2\text{ kg/(m}^2 \cdot \text{h)}$$

$$= 3.7\text{ lb/(ft}^2 \cdot \text{h)}$$

Step 3. Compute the percent solids recovery

$$\% = \frac{(17.3\text{ kg/(m}^2 \cdot \text{h)}) \times 100\%}{18.2\text{ kg/(m}^2 \cdot \text{h)}}$$

$$= 95\%$$

Step 4. Compute the daily filter operation required with 30 percent solids
Since filter yield is determined by

$$\text{Filter yield} = \frac{\text{solids loading} \times \text{factor of recovery}}{\text{h/d operation} \times \text{filter area}}$$

rearranging the above:

$$\text{Operation, h/d} = \frac{\text{solids loading} \times \text{factor of recovery}}{\text{filter yield} \times \text{filter area}}$$

$$= \frac{5400\text{ kg/d} \times 0.95}{9.8\text{ kg/(m}^2 \cdot \text{h)} \times 42.4\text{ m}^2}$$

$$= 12.3\text{ h/d}$$

Pressure filtration. There are two types of pressure filtration process used for sludge dewatering. Normally, they are the belt filter press and the plate and frame filter press. The operating mechanics of these two filter press types are completely different. The belt filter press is popular due to availability of smaller sizes and uses polymer for chemical flocculation of the sludge. It consumes much less energy than the vacuum filter. The plate and frame filter press is used primarily to dewater chemical sludges. It is a large machine and uses lime and ferric chloride for conditioning of organic sludge prior to dewatering. The filtered cake is very compact and dry.

BELT FILTER PRESS A belt filter press consists of two endless, tensioned, porous belts. The belts travel continuously over a series of rollers of varied diameter that squeeze out the water from the sludge and produce a dried cake that can be easily removed from the belts. Variations in belt filter press design are available from different manufacturers.

Chemical conditioning of sludge is vital to the efficiency of a belt filter press. Polymers are used as a conditioning agent, especially cationic polymers. The operation of the belt filter press comprises three zones, i.e. gravity drainage, low-pressure, and high-pressure zones. Conditioned sludge is first introduced uniformly to the gravity drainage zone and onto the moving belt where it is allowed to thicken. The majority of free water is removed from the sludge by gravity. Free water readily separates from the slurry and is recycled back through the treatment system. The efficiency of gravity drainage depends on the type of sludge, the quality of the sludge, sludge conditioning, the belt screen mesh, and design of the drainage section. Typically, gravity drainage occurs on a flat or slightly inclined belt for a period of 1 to 2 min. A 5 to 10 percent increase in solids content is achieved in this zone; that is, 1 to 5 percent sludge feed produces 6 to 15 percent solids prior to compression (USEPA 1991).

Following the gravity drainage zone, pressure is applied in the low-pressure (wedge) zone. The pressure can come from the compression of the sludge between two belts or from the application of a vacuum on the lower belt. In the wedge zone, the thickened sludge is subjected to low pressure to further remove water from the sludge matrix. This is to prepare an even firmer sludge cake that can withstand the shear forces to which it is subjected by rollers.

On some units, the low-pressure zone is followed by a high-pressure zone, where the sludge is sandwiched between porous belts and subjected to shearing forces as the belts pass through a series of rollers of decreasing diameter. The roller arrangement progressively increases the pressure. The squeezing and shearing forces induce the further release of additional amounts of water from the sludge cake.

The final dewatered filter cake is removed from the belts by scraper blades into a hopper or conveyor belt for transfer to the sludge management area. After the cake is removed, a spray of water is applied to wash and rinse the belts. The spray rinse water, together with the filtrate, is recycled back to primary or secondary treatment.

Belt filter presses are commercially available in metric sizes from 0.5 to 3.5 m in belt width, with the most common size between 1.0 and 2.5 m. Hydraulic loading based on belt width varies from 25 to 100 gal/(min · m) (1.6 to 6.3 L/(m · s)). Sludge loading rates range from 200 to 1500 lb/(m · h) (90 to 680 kg/(m · h)) (Metcalf and Eddy, Inc. 1991). Typical operating parameters for belt filter press dewatering of polymer-conditioned wastewater sludges are presented in Table 7.26.

Odor is often a problem with the belt filter press. Adequate ventilation to remove hydrogen sulfide or other gases is also a safety consideration. Hydrogen peroxide or potassium permanganate can be used to oxidize the odor-causing chemical (H_2S). Potassium permanganate can improve the dewatering efficiency of the sludge and can reduce the quantity of polymer required.

EXAMPLE: A belt filter press (BFP) is designed to dewater anaerobically digested primary plus waste activated-sludge for seven hours per day and five days per week. The effective belt width is 2.0 m (commonly used). Estimate hydraulic and solids loading rates, polymer dosage, and solids capture (recovery) with the following given data.

TABLE 7.26 Typical Operating Parameters for Belt Filter Press Dewatering of Polymer Conditioned Sludge

Type of sludge	Feed solids, percent	Hydraulic loading, gal/(min · m*)	Solids loading, lb/(m* · h)	Cake solids, percent	Polymer dosage, lb/ton
Anaerobically digested					
Primary	4–7	40–50	1000–1600	25–44	3–6
Primary plus WAS	2–6	40–50	500–1000	15–35	6–12
Aerobically digested without primary	1–3	30–45	200–500	12–20	8–14
Raw primary and waste activated	3–6	40–50	800–1200	20–35	4–10
Thickened waste activated	3–5	40–50	800–1000	14–20	6–8
Extended aeration waste activated	1–3	30–45	200–500	11–22	8–14
Heat-treated primary plus waste activated	4–8	35–50	1000–1800	25–50	1–2

Notes: 1.0 gal/(min · m) = 0.225 m^3/(m · h).
1.0 lb/(m · h) = 0.454 kg/(m · h).
1.1 lb/ton = 0.500 kg/tonne.
*Loading per meter belt width.
Sources: WPCF 1983, Viessman and Hammer 1993, WEF 1996b.

Sludge production	20 gal/min
Total solids in sludge feed	3.5%
Total solids in cake	25%
Sp gr of sludge feed	1.03
Sp gr of dewatered cake	1.08
Sp gr of filtrate	1.01
Polymer dose (0.2% by weight)	7.2 gal/min
TSS in wastewater of BFP	1900 mg/L
Wash water	40 gal/min

Solution:

Step 1. Calculate the average weekly sludge production

$$\begin{aligned}\text{Wet sludge} &= 20\ \text{gal/min} \times 1440\ \text{min/d} \times 7\ \text{d/wk} \times 8.34\ \text{lb/gal} \times 1.03\\ &= 1{,}731{,}780\ \text{lb/wk}\\ \text{Dry solids} &= 1{,}731{,}780\ \text{lb/wk} \times 0.035\\ &= 60{,}610\ \text{lb/wk}\end{aligned}$$

Step 2. Calculate daily and hourly dry solids dewatering requirement based on the designed operation schedule: 7 h/d and 5 d/wk

$$\begin{aligned}\text{Daily rate} &= (60{,}610\ \text{lb/wk})/(5\ \text{day/wk})\\ &= 12{,}120\ \text{lb/d}\\ \text{Hourly rate} &= (12{,}120\ \text{lb/d})/(7\ \text{h/d})\\ &= 1730\ \text{lb/h}\end{aligned}$$

Step 3. Select belt filter press size and calculate solids loading rate
Use one 2.0 m belt (most common size) and one more identical size for standby.
Calculate solids loading rate

$$\begin{aligned}\text{Solids loading} &= (1730\ \text{lb/h})/2\ \text{m}\\ &= 865\ \text{lb/(h}\cdot\text{m)}\ \text{(OK, within 500 to 1000 lb/(h}\cdot\text{m) range)}\end{aligned}$$

Step 4. Calculate hydraulic loading rate

$$\begin{aligned}\text{Flow to the press} &= 20\ \text{gal/min} \times (7\ \text{days/5 days})(24\ \text{h/7 h})\\ &= 96\ \text{gal/min}\\ \text{HL} &= 96\ \text{(gal/min)/2 m}\\ &= 48\ \text{gal/(min}\cdot\text{m)}\ \text{(OK, within 40 to 50 gal/(min}\cdot\text{m) range)}\end{aligned}$$

Step 5. Calculate dosage of polymer (0.2% by weight) with a 7.2 gal/min rate

$$\begin{aligned}\text{Dosage} &= \frac{(7.2\ \text{gal/min} \times 60\ \text{min/h} \times 0.002) \times 8.34\ \text{lb/gal}}{2.0\ \text{m} \times 865\ \text{lb/(h}\cdot\text{m)} \times (1\ \text{ton/2000 lb})}\\ &= 8.33\ \text{lb/ton}\\ &= 4.16\ \text{kg/metric ton}\end{aligned}$$

Step 6. Calculate solids recovery rate

(a) Estimate the volumetric flow of cake, q (from Step 4)

$$q = \frac{96 \text{ gal/min} \times 0.035}{(25\%/100\%) \times 1.08}$$
$$= 12.4 \text{ gal/min}$$

(b) Calculate flow rate of filtrate

$$\text{Flow of filtrate} = \text{sludge feed flow} - \text{cake flow}$$
$$= 96 \text{ gal/min} - 12.4 \text{ gal/min}$$
$$= 83.6 \text{ gal/min}$$

(c) Calculate total solids in the filtrate of wastewater from the press

$$\text{Solids} = \frac{83.6 \text{ gal/min} \times 60 \text{ min/h} \times 1900 \text{ mg/L} \times 8.34 \text{ lb/gal}}{2.0 \text{ m} \times 1{,}000{,}000 \text{ mg/L}}$$
$$= 40 \text{ lb/(h} \cdot \text{m)}$$

(d) Calculate solids capture (recovery)

$$\text{Solids capture} = \frac{\text{solids in feed} - \text{solids in filtrate}}{\text{solids in feed}} \times 100\%$$
$$= \frac{(865 - 40) \text{ lb/(h} \cdot \text{m)} \times 100\%}{865 \text{ lb/(h} \cdot \text{m)}}$$
$$= 95.4\%$$

PLATE AND FRAME FILTER PRESS There are several types of filter press available. The most common type, the plate and frame filter press, consists of vertical plates that are held rigidly in a frame and pressed together between fixed and moving ends (Figure 7.61a). As shown in Figure 7.61b, each chamber is formed by paired recessed plates. A series of individual chambers have a common feed port and two or more common filtrate channels. Fabric filter media are mounted on the face of each individual plate, thus providing initial solids capture.

The filter press operates in batch manner to dewater sludge. Despite its name, the filter press does not press or squeeze sludge. Instead, when the filter is closed, the recessed faces of adjacent plates form a pressure-tight chamber; the sludge is pumped into the press at pressure up to 225 lb/in^2 (15 atm, 1550 kN/m^2) and maintained for less than 2 h (CPC Eng. 1980). The water passes through the filter cloth, while the solids are retained and form a cake on the cloth. During the initial phase of operation the fabric cloth retains the solids which then become the filter media. This action results in high solids capture and a filtrate of maximum clarity. At the end of a filtration cycle, the feed pump system is stopped and the chamber plates are opened, allowing the cake to fall into appropriate hoppers or conveyors. Filter cloth usually requires a precoat (ash or diatomaceous earth) to aid retention of solids on the cloth and release of cake. Drainage ports are provided at both the top and the bottom of each chamber (Figure 7.61b). This figure shows a fixed volume, recessed plate filter press. Another type, the variable volume recessed plate filter press, is also used for sludge dewatering.

The advantages of the plate filter press include production of higher solids cake, 35 to 50 percent versus 18 percent for both belt press (USEPA 1991) and vacuum filters, higher clarity filtrate, lower energy consumption, lower O & M costs, longer equipment service life, and relatively simple operation. However, sufficient washing and air drying time between cycles is required.

EXAMPLE: A plate and frame filter press has a plate surface of 400 ft^2 and is used to dewater conditioned waste activated-sludge with a solids concentration of 2.5 percent. The filtration time is 110 min at 225 lb/in^2 (gage) and the time to discharge the sludge and restart the feed is 10 min. The total volume of sludge dewatered is 3450 gal. (1) Determine the solids loading rate and the net filter yield

(a)

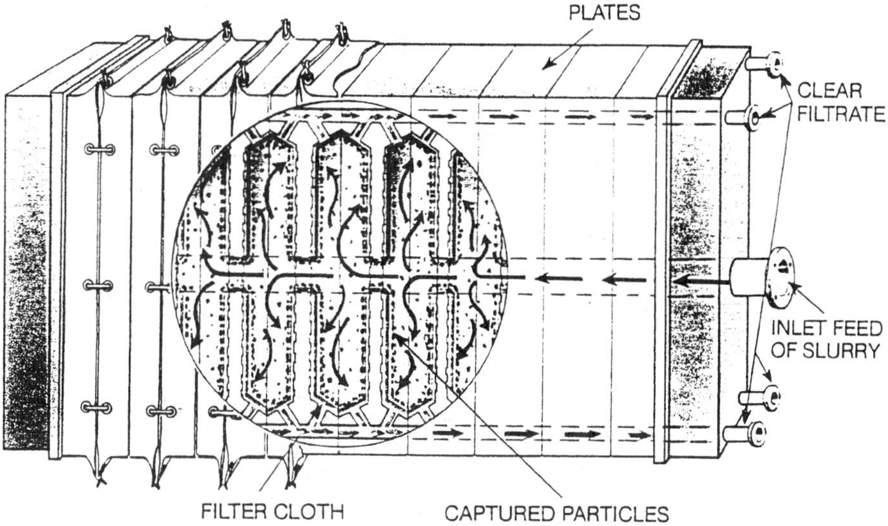

(b)

FIGURE 7.61 Plate and frame filter press: (a) side view of a filter press; (b) schematic cross section of chamber area during fill cycle.

in lb/(h · ft^2). (2) What will be the time of filtration required to produce the same net yield if the feed solids concentration is reduced to 2 percent?

Solution:

Step 1. Compute solids loading rate

$$\text{Solids loading} = \frac{\text{mass of dry solids applied}}{\text{filtration time} \times \text{plate area}}$$
$$= \frac{3450 \text{ gal} \times 8.34 \text{ lb/gal} \times 0.025}{110 \text{ min/(60 min/h)} \times 400 \text{ ft}^2}$$
$$= 0.98 \text{ lb/(h} \cdot \text{ft}^2)$$

Step 2. Compute the net filter yield

$$\text{Net yield} = \text{solids loading} \times \frac{\text{filtration time}}{\text{time per cycle}}$$
$$= 0.98 \text{ lb/(h} \cdot \text{ft}^2) \times \frac{110 \text{ min}}{(110 + 10) \text{ min}}$$
$$= 0.90 \text{ lb/(h} \cdot \text{ft}^2)$$

Step 3. Determine the filtration time required at 2% solids feed

$$\text{Time} = 110 \text{ min } (2.5\%/2\%)$$
$$= 137.5 \text{ min}$$

Centrifugation. The centrifugation process is also used for dewatering raw, digested, and waste activated-sludge. The process is discussed in more detail in the section on sludge thickening. Typically, a centrifuge produces a cake with 15 to 30 percent dry solids. The solids capture ranges from 50 to 80 percent and 80 to 95 percent, without and with proper chemical sludge conditioning respectively (USEPA 1987b).

Sludge drying beds. Sludge drying beds remove moisture by natural evaporation, gravity, and/or induced drainage, and are the most widely used method of dewatering municipal sludge in the United States. They are usually used for dewatering well-digested sludge. Drying beds include conventional sand beds, paved beds, unpaved beds, wedge-wire beds, and vacuum-assisted beds.

Drying beds are less complex, easier to operate, require less operating energy, and produce higher solids cake than mechanical dewatering systems. However, more land is needed for them. Drying beds are usually used for small- and mid-size wastewater treatment plants with design flow of less than 7500 m^3/d (2 Mgal/d), due to land restrictions (WEF 1996b).

SAND DRYING BEDS. Sand drying beds generally consist of 10 to 23 cm (4 to 9 in) of sand placed over a 20 to 50 cm (8 to 20 in) layer of gravel. The diameters of sand and gravel range from 0.3 to 1.22 mm and from 0.3 to 2.5 cm, respectively. The water drains to an underdrain system that consists of perforated pipe at least 10 cm (4 in) in diameter and spaced 2.4 to 6 m (8 to 20 ft) apart. Drying beds usually consist of a 0.3 to 1 m (1 to 3 ft) high retaining wall enclosing process drainage media (USEPA 1989, 1991). The drying area is partitioned into individual beds, 6 m wide by 6 to 30 m long (20 ft wide by 20 to 100 ft long), or of a convenient size so that one or two beds will be filled in a normal cycle (Metcalf and Eddy Inc. 1991).

In a typical sand drying bed digested and/or conditioned sludge is discharged (at least 0.75 m/s velocity) on the bed in a 30 to 45 cm (12 to 18 in) layer and allowed to dewater by drainage through the sludge mass and supporting sand and by evaporation from the surface exposed to air. Dissolved gases are released and rise to the surface. The water drains through the sand and is collected in the underdrain system and generally returned to the plant for further treatment.

Design area requirements for open sludge drying beds are based on sludge loading rate, which is calculated on a per capita basis or on a unit load of mass of dry solids per unit area per year. On a per capita basis, surface areas range from 1.0 to 2.5 ft^2/person (0.09 to 0.23 m^2/person), depending on sludge type. Sludge loading rates are between 12 and 23 lb dry solids/($ft^2 \cdot y$) (59 and 113 kg/($m^2 \cdot y$) (Metcalf and Eddy Inc. 1991).

The design, use, and performance of the drying bed are affected by the type of sludge, sludge conditioning, sludge application rates and depth, dewatered sludge removal techniques, and climatic conditions.

EXAMPLE: A sand drying bed, 6 m by 30 m (20 by 100 ft), receives conditioned sludge to a depth of 30 cm (12 in). The sludge feed contains 3 percent solids. The sand bed requires 29 days to dry and one day to remove the sludge from the bed for another application. Determine the amount of sludge applied per application and the annual solids loading rate. The sp gr of sludge feed is 1.02.

Solution:

Step 1. Compute the volume of sludge applied per application, V

$$\begin{aligned} V &= 6 \text{ m} \times 30 \text{ m} \times 0.3 \text{ m/app.} \\ &= 54 \text{ m}^3\text{/app.} \\ &= 1907 \text{ ft}^3\text{/app.} \\ &= 14{,}300 \text{ gal/app.} \end{aligned}$$

Step 2. Compute the solids applied per application

$$\begin{aligned} \text{Solids} &= 54 \text{ m}^3\text{/app.} \times 1000 \text{ kg/m}^3 \times 1.02 \times 0.03 \\ &= 1650 \text{ kg/app.} \\ &= 3640 \text{ lb/app.} \end{aligned}$$

Step 3. Compute annual solids loading rate

$$\text{Surface area of the bed} = 6 \text{ m} \times 30 \text{ m} = 180 \text{ m}^2$$

$$\begin{aligned} \text{Loading rate} &= \frac{\text{solids applied} \times 365 \text{ d/y}}{\text{surface area} \times \text{cycle}} \\ &= \frac{1650 \text{ kg/app.} \times 365 \text{ d/y}}{180 \text{ m}^2 \times 30 \text{ d/app.}} \\ &= 111 \text{ kg/(m}^2 \cdot \text{y) (OK within 59–113 kg/(m}^2 \cdot \text{y) range)} \\ &= 22.8 \text{ lb/(ft}^2 \cdot \text{y)} \end{aligned}$$

PAVED BEDS. Paved beds consist of a concrete or asphalt pavement above a porous gravel subbase with a slope of at least 1.5 percent. There are two types of drainage, i.e. a drainage type and a decant type. Unpaved areas, constructed as sand drains, are placed around the perimeter or along the center of the bed to collect and convey drainage water. The advantage of the paved bed is that sludge can be removed by a front-end loader. The bed areas are larger and typically rectangular in shape, being 6 to 15 m (20 to 50 ft) wide by 21 to 46 m (70 to 150 ft) long with vertical side walls as sand beds. Cake solids contents as much as 50 percent can be achieved in an arid climate.

Well designed paved beds may remove about 20 to 30 percent of the water with good settling solids. Solids concentration may range from 40 to 50 percent for a 30 to 40 days drying time in an arid climate for a 30 cm (1 ft) sludge layer (Metcalf and Eddy, Inc. 1991).

UNPAVED BEDS. Unpaved beds may be used in warm and dry climate areas where groundwater is not a concern. The beds are similar to paved beds with decanting. Sufficient storage area is required during a wet weather period, because access to the beds is restricted during wet weather.

VACUUM-ASSISTED DRYING BEDS. The vacuum-assisted drying bed consists of a reinforced concrete ground slab, a layer of supporting aggregate, and a rigid porous media plate on top. The space between the concrete slab and the multimedia plate is the vacuum chamber, connected to a vacuum pump. This is to accelerate sludge dewatering and drying.

Polymer-preconditioned sludge is applied to the surface of the media plates until it is entirely covered. The vacuum is then applied to remove the free water from the sludge. The sludge is air dried for 24 to 48 hours. Essentially all of the solids remain on top of the media plates form a cake of fairly uniform thickness. The cake has solids concentration from 9 to 35 percent, depending on sludge feed (WEF 1996b). The dewatered sludge is removed by a front-end loader. The porous multimedia plates are washed with a high-pressure hose to remove the remaining sludge residue before another application.

The vacuum-assisted drying bed has a short cycle time and needs less footage. However, it is labor intensive and is expensive to operate. Drying beds are sometimes enclosed in a greenhouse-type glass structure to increase drying efficiency in wet or colder climates. The enclosure also serves for odor and insect control, and improves the overall appearance of the treatment plant. Good ventilation is required to allow moisture to escape.

WEDGE-WIRE DRYING BEDS. The wedge-wire (wedgewater) drying bed was developed in England and there exist a few installations in the United States. The material for the drying bed consists of stainless steel wedge wire or high-density polyurethane. It is a physical process similar to the vacuum-assisted drying bed.

The bed consists of a shallow rectangular watertight basin fitted with a false floor of artificial media, wedge-wire panels (Figure 7.62). The panels (wedge-wire septum) have wedge-shaped slotted openings of 0.25 mm (0.01 in). The false floor is made watertight with caulking where the panels abut the walls. An outlet valve to control the drainage rate is located underneath the false floor. Water or plant effluent is introduced to the bed from beneath the wedge-wire septum up to a depth of 2.5 cm (1 in) over the septum. This water serves as a cushion and prevents compression or other disturbance of the colloidal particles. The sludge is introduced slowly onto a horizontal water level to float without causing upward or downward pressure across the wedge-wire surface. After the bed is filled with sludge, the initially separate water layer and drainage water are allowed to percolate away at a controlled flow rate through the outlet valve. After the free water has been drained, the sludge further concentrates by drainage and evaporation until there is a requirement for sludge removal (USEPA 1987b).

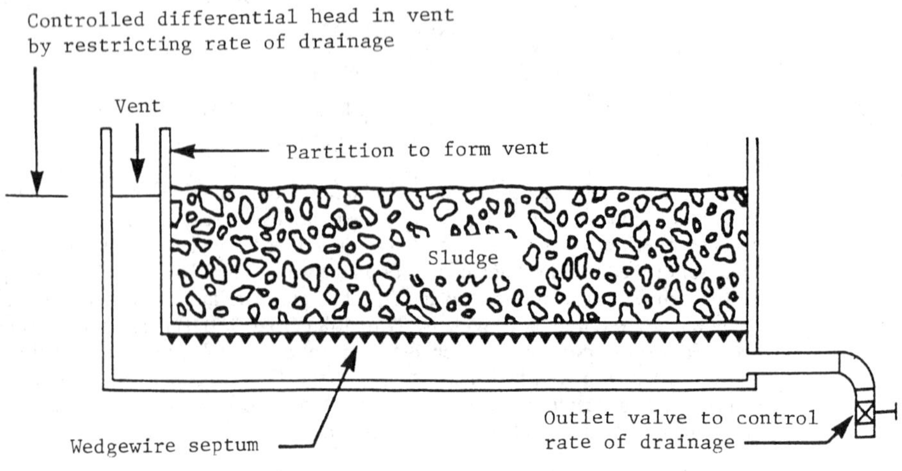

FIGURE 7.62 Crosssection of a wedgewire drying bed (*Source*: *USEPA 1991*).

The advantages of this process include no clogging in the wedge-wire, treatability of aerobically digested sludge, constant and rapid drainage, higher throughput than the sand bed, and ease of operation and maintenance. However, the capital cost is higher than that of other drying beds.

SLUDGE DRYING LAGOONS. A sludge drying lagoon is similar to a sand bed in that the digested sludge is introduced to a lagoon and removed after a period of drying. Unlike the sand drying bed, the lagoon does not have an underdrain system to drain water. Sludge drying lagoons are not suitable for dewatering untreated sludges, lime treated sludges, or sludge with a high strength of supernatant, due to odor and nuisance potential.

Sludge drying lagoons are operated by periodically decanting the supernatant back to the plant and by evaporation. They are periodically dredged to remove sludge. Unconditioned digested sludge is introduced to the lagoon to a depth of 0.75 to 1.25 m (2.5 to 4 ft). It will dry mainly by evaporation in 3 to 5 months, depending on climate. Dried sludge is removed mechanically at a solids content of 20 to 30 percent. If sludge is to be used for soil conditioning, it needs to be stored for further drying. A 3-year cycle may be applied: lagooning for 1 year, drying for 18 months, and cleaning and resting for 6 months.

Solids loading rates are 36 to 39 kg/(m^3 · y) (2.2 to 2.4 lb/(ft^3 · y)) of lagoon capacity (USEPA 1987b). The lagoons are operated in parallel; at least two units are essential. Very little process control is needed. Sludge lagoons are the most basic treatment units.

OTHER SLUDGE VOLUME REDUCTION PROCESSES. Processes such as heat treatment and thermal combustion can reduce the moisture content and volume of the sludge. Detailed discussions are presented in the section on solids treatment.

EXAMPLE: A POTW generates 1800 kg/d (3970 lb/d) of dewatered sludge at 33 percent solids concentration. The moisture content of composted material and compost mixture are 35 and 55 percent, respectively. What is the mass of composted material that must be blended daily with the dewatered sludge?

Solution:

Step 1. Compute moisture of dewatered sludge

$$\begin{aligned}\text{Sludge moisture} &= 100\% - \text{solids content} = 100\% - 33\% \\ &= 67\%\end{aligned}$$

Step 2. Compute composted material required daily
Let X and Y be the mass of dewatered sludge and composted material, respectively
Then $X = 1800$ kg/d

$$\begin{aligned}(X + Y)\,(\text{mixture moisture}) &= X\,(\text{sl. mois.}) + Y\,(\text{comp. mois.}) \\ (X + Y)\,55\% &= X\,(67\%) + Y\,(35\%) \\ Y\,(55 - 35) &= X\,(67 - 55) \\ Y &= (1800\ \text{kg/d})\,(12/20) \\ &= 1080\ \text{kg/d} \\ &= 2380\ \text{lb/d}\end{aligned}$$

29.3 Sewage Sludge Biosolids

Sewage sludge processing, (i.e., thickening, stabilization, conditioning, and dewatering) produces a volume reduction, and decreases the capital and operating costs. Digestion or composting of the sludge reduces the level of pathogens and odors. The degree of sludge treatment process is very important in order to eliminate pathogens when considering land application of the sludge, when distributing and marketing it, and when placing it in monofills or on a surface disposal site.

The end products of wastewater sludge treatment processes are referred to as "biosolids." Webster's Collegiate Dictionary (1997) defines biosolids as solid organic matter recovered from a sewage treatment process and used especially as fertilizer for plants. The McGraw-Hill Dictionary of Scientific and Technical Terms (5th ed., 1994) defines biosolid as a recyclable, primarily organic solid material produced by wastewater treatment processes.

The term "biosolids" has gained recently in popularity as a synonym for sewage sludge because it perhaps has more reuse potential than the term "treated wastewater sludge." The name was chosen by the Water Environmental Federation (WEF) in early 1990s.

29.4 Use and Disposal of Sewage Sludge Biosolids

Biosolids are commonly used and disposed of in many ways. The beneficial uses of biosolids include land application to agricultural lands, land application to non-agricultural lands, and sales or give-away of biosolids for use on home gardens. Non-agricultural areas may include compost, forests, public contact (parks, highways, recreational areas, golf courses etc.). Case histories of beneficial use programs for biosolids use and management can be found elsewhere (WEF 1994).

Disposal methods of biosolids include disposal in municipal landfills, disposal on delicate sites, surface disposal, and incineration. Surface disposal includes piles of biosolids left on the land surface and land application to dedicated non-agricultural land, and disposal in sludge-only landfills (monofills).

29.5 Regulatory Requirements

In 1993, the USEPA promulgated "the standard for the use and disposal of sewage sludge, final rule, title 40 of the Code of Federation Regulations (CFR), Parts 257, 403, and 503" (Federal Register 1993). All the above three parts plus "Phase-in submission of sewage sludge permit application: final rule, revisions to 40 CFR Parts 122, 123, and 501" are reprinted by the Water Environmental Federation, stock #P0100 (1993b). "The Part 503 rule" or "Part 503" was developed to protect public health and the environment from any reasonably anticipated adverse effects of using or disposing of sewage sludge biosolids. Two easy-to-read versions of the Part 503 rule were published by USEPA (1994, 1995).

The CFR 40 Part 503 includes five subparts. They are general provisions, requirement for land application, surface disposal, pathogen and vector attraction (flies, mosquitoes, and other potential disease-carrying organisms) reduction, and incineration. For each of the regulated use or disposal methods, the rule covers general requirements, pollutant limits, operational standards (pathogen and vector attraction reduction for land application and surface disposal; total hydrocarbons or carbon dioxide for incineration), management practices, and requirement for the frequency of monitoring, record keeping, and reporting. The requirements of the Part 503 are self-implementing and must be followed even without the insurance of a permit. State regulatory agencies may have their own rules governing the use or disposal of sewage sludge biosolids or domestic septage.

Part 503 applies to any person who applies biosolids to the land, or burns the biosolids in an incinerator, and to the operator/owner of a surface disposal site, or to any person preparing to use, dispose of, and incinerate biosolids. A person is defined as an individual, association, partnership, corporation, municipality, state, or federal agency, or an agent or employee thereof.

A person must apply for a permit covering biosolids use or disposal standards if they own or operate a treatment works treating domestic sewage (TWTDS). In most cases, Part 503 rule requirements will be incorporated over a time period into National Pollutant Discharge Elimination System (NPDES) permits issued to public owned treatment works (POTWs) and TWTDSs. Application for a federal biosolids permit must be submitted to the appropriate USEPA Regional Office, not the state. Until the biosolids management programs of individual states are approved by the USEPA, the USEPA will remain the permitting authority.

Most sewage sludge biosolids currently generated by POTWs in the US meet the minimum pollutant limits and pathogen and vector attraction reduction requirements set forth in the Part 503 rule. Some biosolids already meet the most stringent Part 503 pollutant limit standards and requirements of pathogen and vector attraction reduction.

Pathogen reduction requirements. The Part 503 pathogen reduction requirements for land application of biosolids are divided into two categories: Class A and Class B biosolids. In addition to meeting the requirement in one of the six treatment alternatives, the Class A requirement is to reduce the pathogens in biosolids (fecal coliform, FC, or *Salmonella* sp. bacteria, enteric viruses, parasites, and viable helminth ova) to below detectable levels. When this goal is reached, Class A biosolids can be applied without any pathogen-related restriction on the site. The six treatment alternatives for biosolids include alternative 1, thermally treated; alternative 2, high pH–high temperature treatment; alternative 3, other processes treatment; alternative 4, unknown processes; alternative 5, use of the processes to further reduce pathogens (PFRP); and alternative 6, use of a process equivalent to PFRP.

Class A biosolids. The pathogen reduction requirements must be met for all six alternatives (503.32a) and vector attraction reduction (503.33) for Class A with respect to pathogens. Regardless of the alternative chosen, either the FC density in the biosolids must be less than 1000, the most probable number (MPN) per gram dry total solids; or the density of *Salmonella* sp. in the biosolids must be less than 3 MPN per 4 grams of dry total solids. Either of these requirements must be met at one of the following times: (1) when the biosolids are used or disposed of; (2) when the biosolids are prepared for sale or giveaway in a bag or other container for land application; or (3) when the biosolids or derived materials are prepared to meet the requirements for excellent quality biosolids.

Table 7.27 lists the four time-temperature regimes for Class A pathogen reduction under alternative 1 (time and temperature). Alternative 2 (pH and time) raises and maintains sludge pH above 12 for 72 h and keeps it at 52°C for 12 h. Alternative 3 is an analysis of the effective the treatment processes. Enteric viruses and helminth ova which shall be determined prior to and after treatment. The final density shall be less than one plaque-forming unit per 4 g of dry solids. Alternative 4 (analysis with unidentified treatment process) analyzes for fecal coliform, *Salmonella*, enteric viruses, and helminth ova at the time of use or disposal. Alternative 5 (PFRP process) treats sludge with one of composting and Class B alternative 2 processes (aerobic digestion, air drying, anaerobic

TABLE 7.27 The Four Time-Temperature Regimes for Class A Pathogen Reduction Under Alternative 1

Regime	Applies to	Requirement	Time-temperature relationships*
A	Biosolids with 7 percent solids or greater (except those covered by Regime B)	Temperature of biosolids must be 50°C or higher for 20 min or longer	$D = \dfrac{131{,}700{,}000}{10^{0.14T}}$ (Eq. 3 of Section 503.32)
B	Biosolids with 7 percent solids or greater in the form of small particles and heated by contact with either warmed gases or an immiscible liquid	Temperature of biosolids must be 50°C or higher for 15 s or longer	$D = \dfrac{131{,}700{,}000}{10^{0.14T}}$
C	Biosolids with less than 7 percent solids	Heated for at least 15 s but less than 30 min	$D = \dfrac{131{,}700{,}000}{10^{0.14T}}$
D	Biosolids with less than 7 percent solids	Temperature of sludge is 50°C or higher with at least 30 min or longer contact time	$D = \dfrac{50{,}070{,}000}{10^{0.14T}}$ (Equation 4 of Section 503.32)

*D = time in days; T = temperature in degrees Celsius.
Source: USEPA (1994, 1995).

digestion, heat drying, heat treatment, thermophilic aerobic digestion, beta ray irradiation, gamma ray irradiation, and pasteurization). Alternative 6 (PFRP equivalent process) allows the permitting authority to approve a process not currently identified as a PFRP process through a review by the Pathogen Equivalency Committee (Federal Register 1993).

EXAMPLE 1: What is the required minimum time to achieve Class A pathogen requirement when a biosolid with 15 percent solids content is heated at 60°C (140°F)?

Solution:

Step 1. Select the regime under the given condition
Referring to Table 7.27, Regime A is chosen

Step 2. Compute minimum time required

$$D = \frac{131{,}700{,}000}{10^{0.14T}} = \frac{131{,}700{,}000}{10^{0.14(60)}}$$
$$= 0.5243 \text{ day}$$
$$= 12.58 \text{ h}$$

EXAMPLE 2: Determine the required minimum temperature to treat a biosolid containing 15 percent solids with a heating time of 30 min.

Solution:

$$D = 30 \text{ min} = 0.0208 \text{ day}$$
$$D(10^{0.14T}) = 131{,}700{,}000$$
$$10^{0.14T} = 131{,}700{,}000/0.0208 = 6{,}331{,}700{,}000$$
$$0.14T = 9.80$$
$$T = 70.0\ (^\circ\text{C})$$

Class B biosolids. Class B requirements ensure that pathogens have been reduced to levels that are unlikely to pose a threat to public health and the environment under specific conditions of use. Class B biosolids also must meet one of the three alternatives: (1) monitoring of indicator organisms (fecal coliform) density; (2) biosolids are treated in one of the processes to significantly reduce pathogens (PSRP); and (3) use of the processes equivalent to PSRP.

Land application. Land application is the application of biosolids to land, either to condition the soil or to fertilize crops or other vegetation grown in the soil. Biosolids can be applied to land in bulk or sold or given away in bags or other containers. The application sites may be categorized as non-public contact sites (agricultural lands, forests, reclamation sites) and public contact sites (public parks, roadsides, golf courses, nurseries, lawns, and home gardens). In the United States, about half of the biosolids produced are ultimately disposed of through land application.

Approximately one-third of the 5.4 million dry metric tons of wastewater sludge produced annually in the United States at POTWs is used for land application. Of that, about two-thirds is applied on agricultural lands (USEPA 1995). Land application of wastewater sludge has been practiced for centuries in many countries, because the nutrients (nitrogen and phosphorus) and organic matter in sludge can be beneficially utilized to grow crops and vegetation. However, microorganisms (bacteria, viruses, protozoa, and other pathogens), heavy metals, and toxic organic chemicals are major public health concerns for land application of biosolids. Proper management of biosolids utilization is required.

The application rate of biosolids applied to agricultural land must be equal to or less than the "agronomic rate." The agronomic rate is defined in the Part 503 as the rate designed to provide the amount of nitrogen needed by the crop or vegetation while minimizing the nitrogen in the biosolids passing below the root zone of the crop or vegetation and flowing to the ground water.

Biosolids may be sprayed or spread on the soil surface and left on the surface (pasture, range land, lawn, forest), tilled into the soil after being applied, or injected directly below the surface. Land application of biosolids must meet risk-based pollutant limits specified in the Part 503. Operation standards are to control pathogens and to reduce the attraction of vectors. In addition, the application must meet the general requirements; management practices; and the monitoring, record keeping, and reporting requirements.

All land application of biosolids must meet the ceiling concentration limits for ten heavy metals, listed in the second column of Table 7.28. If a limit for any one of the pollutants is exceeded, the biosolids cannot be applied to the land until such time that the ceiling concentration limits are no longer exceeded.

Biosolids applied to the land also must meet either pollutant concentration (PC) limits, cumulative pollutant loading rate (CPLR) limits, or annual pollutant loading rate limits for these 10 heavy metals. Either Class A or Class B pathogen requirements and site restrictions must be met. Finally, one of 10 options for vector attraction reduction must also be met.

Annual whole sludge application rate. The annual whole sludge application rate (AWSAR) for biosolids sold or given away in a bag or other container for application to land can be determined by dividing the annual pollutant loading rate (APLR) by pollutant concentration (PC). It is expressed as

$$\text{AWSAR} = \frac{\text{APLR}}{0.001\ \text{PC}} \tag{7.219}$$

where AWSAR = annual whole sludge application rate, metric ton/(ha · y)
APLR = annual pollutant loading rate, kg/(ha · y) (Table 7.27)
0.001 = a conversion factor, kg/metric ton or mg/kg
PC = pollutant concentration, mg/kg, dry weight

EXAMPLE: Ten heavy metals listed in Table 7.27 were analyzed in biosolids to be sold. The chromium concentration is 240 mg/kg of biosolids. Determine the AWSAR of chromium for the biosolids.

TABLE 7.28 Pollutant Limits for Land Application of Sewage Biosolids

Pollutant	Ceiling concentration limits for all biosolids applied to land, mg/kg*	Pollutant concentration (PC) limits, mg/kg*†	Cumulative pollutant loading rate limits (CPLR), kg/ha	Annual pollutant loading rate limits (APLR), kg/ha over 365-day period
Arsenic	75	41	41	2.0
Cadmium	85	39	39	1.9
Chromium	3000	1200	3000	150
Copper	4300	1500	1500	75
Lead	840	300	300	15
Mercury	57	17	17	0.85
Molybdenum†	75	—‡	—‡	—‡
Nickel	420	420	420	21
Selenium	100	36	100	5.0
Zinc	7500	2800	2800	140
Applies to:	All biosolids that are land applied	Bulk biosolids and bagged biosolids§	Bulk biosolids	Bagged biosolids§
From Part 503, Section 503.13	Table 1	Table 3	Table 2	Table 4

*Dry-weight basis.
†Monthly average: also include exceptional quality (EQ) biosolids.
‡EPA is re-examining these limits.
§Bagged biosolids are sold or given away in a bag or other container.
Source: USEPA 1994, 1995.

Solution:

Step 1. Find APLR from Table 7.27

$$\text{APLR} = 150 \text{ kg/(ha} \cdot \text{y)}$$

Step 2. Compute AWSAR for chromium, using Eq. (7.219)

$$\text{PC} = 1200 \text{ mg/kg}$$

$$\begin{aligned}
\text{AWSAR} &= \frac{\text{APLR}}{0.001 \text{ PC}} \\
&= \frac{150 \text{ kg/(ha} \cdot \text{y)}}{0.001 \text{ kg/(metric ton} \cdot \text{(mg/kg))} \times 1200 \text{ mg/kg}} \\
&= 125 \text{ metric ton/(ha} \cdot \text{y)} \\
&= \frac{(125 \text{ metric ton/(ha} \cdot \text{y)} \times (2205 \text{ lb/metric ton})}{107{,}600 \text{ ft}^2\text{/ha}} \\
&= 2560 \text{ lb/1000 ft}^2\text{/y} \\
&= 1.28 \text{ ton/1000 ft}^2\text{/y}
\end{aligned}$$

Note: 1 lb/1000 ft^2/y = 48.83 kg/ha/y; 1 ton/1000 ft^2/y = 97.65 metric ton/ha/y

Site evaluation and selection. Details of site evaluation and the selection process are discussed by Federal Register (1993) and by the USEPA (1995). The discussion covers Part 503 requirements, preliminary planning, phases 1 and 2, site evaluation, and site screening.

Calculation of annual biosolids application rate on agricultural land. Sewage sludge has being applied to agricultural land for a long time. The use of biosolids in agricultural land can partially replace costly commercial fertilizers. Generally, the intention is to optimize crop yields with application of biosolids and, if required, supplemental fertilizers. The annual application rates of biosolids should not exceed the nutrients (N and P) requirements of the crop grown on an agricultural soil. In addition, land application of biosolids must meet the Part 503 requirements stated above.

Calculation based on nitrogen. In order to prevent groundwater contamination by ammonia nitrogen due to excess land application of biosolids, Part 503 requires that bulk biosolids be applied to the site at a rate equal to or less than the agronomic rate for nitrogen at the site.

Nitrogen may be present in biosolids in inorganic forms such as ammonium (NH_4) or nitrate (NO_3), or in organic forms. NO_3-N is the most water-soluble form of N, and is of most concern for groundwater contamination. NH_4-N can readily be volatilized as ammonia. Inorganic N is the plant available nitrogen (PAN). Not all the N in biosolids is immediately available for crop use, because some N is present as difficulty decomposable (refractory) organic N (Org-N), Org-N = total N − (NO_3-N) − (NH_4-N) which is in microbial cell tissue and other organic compounds. Organic N must be decomposed into mineral or inorganic forms, such as NH_4-N and NO_3, before it can be used by plants. Thus, the availability of Org-N for crops depends on the microbial breakdown of organic materials (biosolids, manure, crop residuals, soil organics, etc.) in soil.

When calculating the agronomic N rate for biosolids, the residual N from previously applied biosolids that will be mineralized and released as PAN must be accounted for as part of the overall budget for the total PAN. The residual N credit can be estimated for some sites undergoing soil nitrate tests, but the PAN credit is commonly estimated by multiplying a mineralization factor (K_{min}) by the amount of biosolids Org-N still remaining in the soil one or two years after biosolids application. Mineralization factors for different types of biosolids may be found in Table 7.29.

EXAMPLE 1: An aerobic digested biosolid with 2.8 percent of organic nitrogen was applied at a rate of 4.5 dry ton per acre for the 1997 growing season. No other biosolids were applied to the same land in 1998. Determine the amount of plant-available nitrogen that will be mineralized from the biosolids Org-N in 1997 for the 1999 growing season.

TABLE 7.29 Estimated Mineralization Rate for Various Type of Sewage Sludge

Year of growing season	Unstabilized primary and WAS	Aerobically digested	Anaerobically digested	Composted
0–1 (year of application)	0.40	0.30	0.20	0.1
1–2	0.20	0.15	0.10	0.05
2–3	0.10	0.08	0.05	–
3–4	0.05	0.04	–	–

Source: Sommers *et al.* (1981).

Solution:

Step 1. Calculate the biosolids Org-N applied in 1997

$$\text{Org-N} = 4.5 \text{ ton/acre} \times (2.8\%/100\%) \times 2000 \text{ lb/ton}$$
$$= 252 \text{ lb/acre}$$

Step 2. Determine Org-N released as PAN in 1999 growing season. Construct a PAN credit for the application year, one year, and two years later for growing seasons as follows:

Year of growing season (1)	Mineralization rate (2)	Org-N, lb/acre Starting (3)	Mineralized (4)	Remaining (5)
0–1 (1997 application)	0.30	252	76	176
1–2 (1998)	0.15	176	26	150
2–3 (1999)	0.08	150	**12**	138

Notes: 1. Values of col. 2 are from Table 7.29.
2. Col. 4 = col. 2 × col. 3.
3. Col. 5 = col. 3 − col. 4.
4. 1 lb/acre = 1.12 kg/ha.

Answer: Org-N released as PAN in 1999 = 12 lb/acre (as bold shown above).

The NH_4-N and NO_3-N added by biosolids is considered to be available for plant use. Plant-available nitrogen comprises NH_4-N and NO_3-N provided by biosolids and by fertilizer salts or other sources of these mineral forms of N.

When biosolids or animal manure are applied on a soil surface, the amount of PAN is reduced by the amount of NH_4-N lost by volatilization of ammonia. The volatilization factor, K_{vol} is used for estimating the amount of NH_4-N.

The PAN of biosolids for the first year of application may be determined as

$$\text{PAN} = (NO_3\text{-N}) + K_{vol}(NH_4\text{-N}) + K_{min}(\text{Org-N}) \quad (7.220)$$

where PAN = plant available N in biosolids, lb/dry ton
NO_3-N = nitrate N content in biosolids, lb/dry ton
K_{vol} = volatilization factor, or fraction of NH_4-N not lost as NH_3 gas
= 0.5 for liquid and surface applied
= 1.0 for liquid and injection into soil
= 0.5 for dewatered and surface applied
NH_4-N = ammonium N content in biosolids, lb/dry ton
K_{min} = mineralization factor, or fraction of Org-N converted to PAN, Table 7.29
Org-N = organic N content in biosolids, lb/dry ton
= Total N − (NO_3-N) − (NH_4-N)

EXAMPLE 2: An anaerobically digested and dewatered biosolid is to be surface applied on an agricultural land. The biosolids analysis of N content is: $NO_3 - N = 1500$ mg/kg, NH_4-N = 1.2%, and total N = 3.7%, based on the dry weight. The solids content of the biosolids is 4.3 percent. Determine PAN.

Solution:

Step 1. Convert N forms into lb/dry ton

$$\begin{aligned} NO_3\text{-N} &= 1500 \text{ mg/kg} = 1.500 \times 10^{-3} \text{ kg/kg} \\ &= 1.5 \times 10^{-3} \text{ lb/lb} \\ &= 3 \text{ lb/ton} \\ NH_4\text{-N} &= 2000 \text{ lb/ton} \times 1.2\%/100\% \\ &= 24 \text{ lb/ton} \\ \text{Total N} &= 2000 \text{ lb/ton} \times 0.037 = 74 \text{ lb/ton} \\ \text{Org-N} &= \text{Total N} - (NO_3\text{-N}) - (NH_4\text{-N}) \\ &= (74 - 3 - 24) \text{ lb/ton} \\ &= 47 \text{ lb/ton} \end{aligned}$$

Step 2. Compute PAN
For dewatered biosolids and surface applied, $K_{vol} = 0.5$
Referring to Table 7.29, $K_{min} = 0.20$ for the first year.
Using Eq. (7.220),

$$\begin{aligned} \text{PAN} &= NO_3\text{-N} + K_{vol}(NH_4\text{-N}) + K_{min}(\text{Org-N}) \\ &= 3 \text{ lb/ton} + 0.5 \text{ (24 lb/ton)} + 0.2 \text{ (47 lb/ton)} \\ &= 24.4 \text{ lb/ton} \end{aligned}$$

Note: 1 lb/ton = 0.5 kg/metric ton = 0.5 mg/kg

It is necessary to determine the adjusted fertilizer N rate by subtracting "total N available from existing, anticipated, and planned sources" from the total N requirement. The procedures to compute the adjusted fertilizer N requirement for the crop to be grown may be summarized as follows (USEPA 1995).

1. Determine the total N requirement (lb/ton) of crop to be grown. This value can be obtained from the Cooperative Extension Service of agricultural agents or universities, USDA–Natural Resources Conservation Service, or other agronomy professionals.
2. Determine N provided from other N sources added or mineralized in the soil
   a. N from a previous legume crop (legume credit) or green manure crop
   b. N from supplemental fertilizers already added or expected to be added
   c. N that will be added by irrigation water
   d. Estimate of available N from previous biosolids application (Example 1)
   e. Estimate of available N from previous manure application
   f. Soil nitrate test of available N present in soil [this quantity can be substituted in place of (a + d + e) if test is conducted properly; do not use this test value if estimates for a, d, and e are used]
   Total N available from existing, expected, and planned sources of N = a + b + c + d + e
   or = b + c + f
3. Determine loss of available N by denitrification, immobilization, or NH_4 fixation
   Check with state regulatory agency for approval before using this site-specific factor.

4. Compute the adjusted fertilizer N required for the crop to be grown

$$\text{N required} = \text{Step 1} - \text{Step 2} + \text{Step 3}$$

Finally, the agronomic N application rate of biosolids can be calculated by dividing the adjusted fertilizer N required (step 4 above) by PAN (using Eq. (7.220)).

EXAMPLE 3: Determine the agronomic N rate from Example 2, assuming the adjusted fertilizer N rate is 61 lb/acre (6.83 kg/ha). The solids content of the biosolids is 4.3 percent.

Solution:

Step 1. Compute agronomic N rate in dry ton/acre

$$\begin{aligned}\text{Rate} &= (\text{adj. fert. N rate}) \div \text{PAN} \\ &= (61 \text{ lb/acre}) \div (24.4 \text{ lb/ton}) \\ &= 2.5 \text{ dry ton/acre}\end{aligned}$$

Step 2. Convert agronomic N rate in dry ton/acre to wet gal/acre

$$\begin{aligned}\text{Rate} &= (2.5 \text{ dry ton/acre}) \div (4.3 \text{ dry ton/100 wet ton}) \\ &= 58 \text{ wet ton/acre} \\ &= (58 \text{ wet ton/acre})\,(2000 \text{ lb/wet ton})\,(1 \text{ gal/8.34 lb}) \\ &= 13{,}900 \text{ gal/acre}\end{aligned}$$

Note: 1 gal/acre = 9.353 L/ha

Calculation based on phosphorus. The majority of phosphorus (P) in biosolids is present as inorganic forms, resulting from mineralization of biosolids organic matter. Generally, the P concentration in biosolids is considered to be about 50 percent available for plant uptake, as is the P normally applied to soils with commercial fertilizers. The P fertilizer required for the crop to be grown is determined from the soil fertility test for available P and the crop yield. The agronomic P rate of biosolids for land application is expressed as (USEPA 1995)

$$\text{Agronomic P rate} = P_{req} \div (\text{available } P_2O_5\text{/dry ton}) \qquad (7.221)$$

where P_{req} = P fertilizer recommended for harvested crop, or the quantity of P removed by the crop

$$\text{Available } P_2O_5 = 0.5\,(\text{total } P_2O_5\text{/dry ton}) \qquad (7.222)$$

$$\text{Total } P_2O_5\text{/dry ton} = \%\text{ of P in biosolids} \times 20 \times 2.3 \qquad (7.223)$$

where 20 = 0.01 (= 1%) × 2000 lb/ton
2.3 = M.W. ratio of $P_2O_5 : P_2$ = 142 : 62 = 2.3 : 1

If biosolids application rates are based on the crop's P requirement, supplemental N fertilization is needed to optimize crop yield for nearly all biosolids applications.

Calculation based on pollutant limitation. Most biosolids are likely to contain heavy metals concentrations that do not exceed the Part 503 pollutant concentration limits. Therefore, pollutant loading limits are not a limiting factor for determining annual biosolids application rates on agricultural lands. Biosolids meeting pollutant concentration limits, as well as certain pathogen and vector attraction reduction requirements, generally are subject to meet the CPLRs requirement. A CPLR is the maximum amount of a pollutant that can be applied to a site by all bulk biosolids applications after July 20, 1993. When the maximum CPLR is reached at the application site for any one of the 10 metals regulated by Part 503 (molybdenum was deleted with effect from February 25, 1994), no more additional biosolids are allowed to be applied to the site.

TABLE 7.30 Part 503 Rule Cumulative Pollutant Loading Rate Limits

Pollutant	CPLR limits	
	kg/ha	lb/acre
Arsenic	41	37
Cadmium	39	35
Chromium*	3000	2700
Copper	1500	1300
Lead	300	270
Mercury	17	15
Molybdenum†	–	–
Nickel	420	380
Selenium	100	90
Zinc	2800	2500

*The chromium limit will most likely be deleted from the Part 503 rule.
†The CRLR for Mo was deleted from Part 503 with effect from February 25, 1994.

For some biosolids with one or more of the pollutant concentrations exceeding the Part 503 pollutant concentration limits, the CPLRs as shown in Table 7.30 must be met. In these cases, the CPLRs could eventually be the limiting factor for annual biosolids applications rather than the agronomic (N or P) rate of application.

For biosolids meeting the Part 503 CPLRs, the maximum total quantity of biosolids allowed to be applied to a site can be calculated on the basis of the CPLR and the pollutant concentration in the biosolids as follows

$$\text{Maximum allowed in dry ton/acre} = (\text{CPLR in lb/acre}) \div [0.002\ (\text{mg pollutant/kg dry biosolids})] \tag{7.224}$$

After computing for each of the 8 or 9 pollutants regulated by the Part 503 rule, the lowest total biosolids value will be used as the maximum quantity of biosolids permitted to be applied to the site. The individual pollutant loading applied by each biosolid application can be calculated by

$$\text{lb of pollutant/acre} = \text{biosolids application rate in dry ton/acre} \times (0.002\ \text{mg/kg}) \tag{7.225}$$

The pollutant loading for each individual biosolids application must be calculated and recorded to keep a cumulative summation of the total quantity of each pollutant that has been applied to each site receiving the biosolids.

Supplemental K fertilizer. Once the agronomic application rate of biosolids has been determined, the amounts of plant available N, P, and K added by the biosolids must be computed and compared to the fertilizer recommendation for the crop grown at a specified yield level. If one or more of these three nutrients provided by the biosolids is less than the amount recommended, then supplemental fertilizers are needed to achieve crop yield.

Potassium, K, is a soluble nutrient. Most of the concentration of K in raw wastewater is discharged with the treatment plant effluents. Generally, biosolids contain low concentrations of potassium, which is one of the major plant nutrients. Fertilizer potash (K_2O) or other sources of K are usually needed to supplement the amounts of K_2O added by biosolids application.

Because K is readily soluble, all the K in biosolids is considered to be available for crop growth. The quantity of K_2O provided (credited) by biosolids application can be calculated as

$$K_2O \text{ added by biosolids} = \text{applied rate in dry ton/acre} \times \text{available } K_2O \text{ in lb/dry ton} \tag{7.226}$$

where

$$\text{Available K}_2\text{O} = \% \text{ of K in biosolids} \times 20 \times 1.2 \tag{7.227}$$

where $20 = 2000$ lb/dry ton · %
$1.2 = K_2O/K_2 = 94/78$

EXAMPLE 4: A K fertilizer recommendation for a soybean field is 150 lb K_2O/acre. The agronomic N rate of biosolids application is 1.5 dry ton/acre. The biosolid contains 0.5% K. Compute the supplemental (additional) K_2O required.

Solution:

Step 1. Compute the available K_2O, using Eq. (7.227)

$$\begin{aligned}\text{Available K}_2\text{O} &= 0.5 \times 20 \text{ lb/dry ton} \times 1.2\\ &= 12 \text{ lb/dry ton}\end{aligned}$$

Step 2. Compute K_2O added by biosolids, using Eq. (7.226)

$$\begin{aligned}\text{Biosolids K}_2\text{O applied} &= \text{biosolids rate} \times \text{available K}_2\text{O}\\ &= 1.5 \text{ dry ton/acre} \times 12 \text{ lb/dry ton}\\ &= 18 \text{ lb/acre}\end{aligned}$$

Step 3. Compute supplemental K_2O needed

$$\begin{aligned}\text{Additional K}_2\text{O needed} &= \text{recommended} - \text{biosolids added}\\ &= 150 \text{ lb/acre} - 18 \text{ lb/acre}\\ &= 132 \text{ lb/acre}\end{aligned}$$

Note: 1 lb/acre = 1.121 kg/ha

EXAMPLE 5: (This example combines Examples 1 to 4 and more) Determine the biosolid annual application for the 1999 growing season for an agricultural site on the basis of the agronomic N rate, P rate, and long-term pollutant limitations required by the Part 503 rule. Also determine the supplemental nutrients to be added. An anaerobically digested liquid biosolid is designed to be applied to farm land at 10 dry ton/acre. The results of laboratory analyses for the biosolids are shown below:

	Heavy metals, mg/kg
Total solids = 5.5%	As = 10
Total N = 4.3%	Cd = 7
NH_4-N = 1.1%	Cr = 120
NO_3-N = 500 mg/kg	Cu = 1100
Total P = 2.2%	Pb = 140
Total K = 0.55%	Hg = 5
pH = 7.0	Mo = 10
	Ni = 120
	Se = 6
	Zn = 3400

If plant nutrients added by biosolids application are not sufficient, supplemental fertilizer nutrients must be provided. Routine soil fertility tests, monitoring, and records must meet the Part 503 and state agency requirements. The field is divided into two portions for rotating cropping of corn, soybeans, and wheat. The same biosolids have been applied to the field during the previous 2 years. The biosolids application rate and its Org-N content are as follows:

	Application rate, ton/acre		Org-N in biosolids, %	
Crop	1997	1998	1997	1998
Corn	3.5	4.4	3.0	3.3
Wheat	5.0	0	3.1	–

According to the County Farm Bureau for the 1999 growing season (personal communication), the crop yield and required nutrients in lb/(acre-y) are given below:

		lb/(acre · y)		
Crop	Yield	N	P_2O_5	K_2O
Corn	160	150	70	140
Wheat	70	70	80	125
Soybeans	45	0	60	150

For the 1999 growing season, one-half of the field will grow wheat in which liquid biosolids will be injected to the soil in the fall (1998) after soybeans are harvested and before the winter wheat is planted in the fall of 1998. For the corn fields, biosolids will be surface applied and tilled in the spring of 1999 before corn is planted. There will be no other source of N for the corn field, except for residual N from 1997 and 1998 biosolids applications.

The wheat field will have 23 lb of N per acre from the preceding soybean crop; and a residual N credit for the 1997 biosolids application. No manure will be applied and no irrigation will be made to either field.

Solution:

Step 1. Compute agronomic rate for each field

(a) Convert N from percent and mg/kg to lb/ton

$$\text{Total N} = 2000 \text{ lb/ton} \times 0.043$$
$$= 86 \text{ lb/ton}$$
$$NH_4\text{-N} = 2000 \text{ lb/ton} \times 0.011 = 22 \text{ lb/ton}$$
$$NO_3\text{-N} = 500 \text{ mg/kg} = 500 \times \frac{1 \text{ lb}}{10^6 \text{ lb}} = 500 \times \frac{1 \text{ lb}}{500 \text{ ton}}$$
$$= 500 \times 0.002 \text{ lb/ton}$$
$$= 1 \text{ lb/ton}$$

(b) Compute Org-N

$$\text{Org-N} = (86 - 22 - 1) \text{ lb/ton}$$
$$= 63 \text{ lb/ton}$$

(c) Compute PAN for surface applied biosolids, using Eq. (7.220)

From Table 7.29

$$K_{\min} = 0.20 \text{ for first year}$$
$$= 0.10 \text{ for second year}$$
$$= 0.05 \text{ for third year}$$

$$K_{vol} = 0.7 \text{ for the corn field, surface applied and tilled}$$
$$= 1.0 \text{ for the wheat field, injected}$$

For the corn field

$$\begin{aligned} \text{PAN} &= (\text{NO}_3\text{-N}) + K_{vol}(\text{NH}_4\text{-N}) + K_{min}(\text{Org-N}) \\ &= 1 \text{ lb/ton} + 0.7\ (22 \text{ lb/ton}) + 0.2\ (63 \text{ lb/ton}) \\ &= 29 \text{ lb/ton} \end{aligned}$$

For the wheat field

$$\begin{aligned} \text{PAN} &= 1 \text{ lb/ton} + 1.0\ (22 \text{ lb/ton}) + 0.2\ (63 \text{ lb/ton}) \\ &= 36 \text{ lb/ton} \end{aligned}$$

(d) Compute the Org-N applied in previous years using the given data
For the corn field, Org-N applied in 1997 and 1998 was

$$\begin{aligned} \text{In 1997: Org-N} &= 3.5 \text{ ton/acre} \times 2000 \text{ lb/ton} \times 3\%/100\% \\ &= 210 \text{ lb/acre} \\ \text{In 1998: Org-N} &= 4.4 \text{ ton/acre} \times 2000 \text{ lb/ton} \times 0.033 \\ &= 290 \text{ lb/acre} \end{aligned}$$

For the wheat field, Org-N originally applied in 1997 was

$$\begin{aligned} \text{In 1997: Org-N} &= 5.0 \text{ ton/acre} \times 2000 \text{ lb/ton} \times 0.031 \\ &= 310 \text{ lb/acre} \end{aligned}$$

(e) Compute the residual N mineralized from previous years' biosolids applications Construct PAN credits for the application year, one year, and two years later than the growing season, as in Example 1. Take values of starting Org-N from Step 1(d).

Year of growing season	Mineralization rate K_{min}	Org-N, lb/acre Starting	Mineralized	Remaining
Corn field: 1997 application				
0–1 (1997)	0.20	210	42	168
1–2 (1998)	0.10	168	17	151
2–3 (1999)	0.05	151	**8**	143
Corn field: 1998 application				
0–1 (1998)	0.20	290	58	232
1–2 (1999)	0.10	232	**23**	209
2–3 (2000)	0.05	209	10	199
Wheat field: 1997 application				
0–1 (1997)	0.20	310	62	248
1–2 (1998)	0.10	248	25	223
2–3 (1999)	0.05	223	**11**	212

The PAN credit for the 1999 growing season on the corn field due to biosolids applications in 1997 and 1998 is **8 + 23** = 31 lb of N per acre, while that on the wheat field due to 1997 application is **11** lb of N per acre.

(f) Determine agronomic N rate for both fields

For the corn field

1.	Total N required for corn grown (given)	= 150 lb/acre
2.	N provided from other sources	
2a.	Estimate of available N from previous applications (from Step 1(e))	= 31 lb/acre
3.	Loss of available N	= 0
4.	Adjusted fertilizer N required = (1) − (2a) − (3)	= 119 lb/acre
5.	The PAN dry ton biosolid to be applied, from Step 1(c)	= 29 lb/ton
6.	The agronomic N rate of biosolids, (4)/(5)	= 4.1 dry ton/acre
7.	Convert biosolids rate into wet ton/acre, 4.1/0.055	= 74.6 wet ton/acre
8.	Convert biosolids rate into gal/acre (74.6 wet ton/acre × 2000 lb/ton ÷ 8.34 lb/gal)	= 17,900 gal/acre

For the wheat field

1.	Total N required for growing wheat	= 70 lb/acre
2.	N provided from other sources	
2a.	N from previous legume credit (given)	= 23 lb/acre
2b.	N from previous biosolids applications, from Step 1(e)	= 11 lb/acre
3.	Loss of available N	= 0
4.	Adjusted fertilizer N required for wheat (1) − (2a) − (2b) − (3)	= 36 lb/acre
5.	PAN/dry ton biosolids to be applied, from Step 1(c)	= 36 lb/ton
6.	Agronomic N rate of biosolids, (4)/(5)	= 1 dry ton/acre
7.	Convert biosolids rate into wet ton/acre	= 18 wet ton/acre
8.	Convert biosolids rate into gal/acre	= 4320 gal/acre

Step 2. Determine cumulative pollution loadings and maximum permitted application rate

(a) Comparing the pollutants concentration in biosolids with the Part 503 pollutant concentration limits, and comparing the concentrations of the ten heavy metals in the biosolids with the Part 503 "pollutant concentration limits" (Table 7.26), the results show that all heavy metals concentrations in biosolids are less than the limits, except for zinc; therefore, the CPLR limits must be met for the biosolids.

(b) Compute the maximum permitted biosolids application rate for each pollutant. The maximum total amount of biosolids permitted to be applied to soil can be calculated from the CPLR limits from Table 7.29 and the concentrations of pollutants, using Eq. (7.224).

The maximum biosolids application rate allows for

As = 37 lb/acre ÷ [0.002 (10 mg/kg)] = 1850 dry ton/acre
Cd = 35 lb/acre ÷ [0.002 (7 mg/kg)] = 2550 dry ton/acre
Cr will be deleted from Part 503
Cu = 1300 lb/acre ÷ [0.002 (1100 mg/kg)] = 590 dry ton/acre
Pb = 270 lb/acre ÷ [0.002 (140 mg/kg)] = 964 dry ton/acre
Hg = 15 lb/acre ÷ [0.002 (5 mg/kg)] = 1500 dry ton/acre
Mo deleted
Ni = 380 lb/acre ÷ [0.002 (120 mg/kg)] = 1583 dry ton/acre
Se = 90 lb/acre ÷ [0.002 (6 mg/kg)] = 7500 dry ton/acre
Zn = 2500 lb/acre ÷ [0.002 (3400 mg/kg)] = 367 dry ton/acre

Note: Zn is the lowest allowed application rate which is much higher than the agronomic N rate. Thus heavy metal pollutants are not the limit factor.

(c) Compute the amount of each pollutant in the 1999 growing season
According to Part the 503, the pollutant loading for each individual biosolids application must be calculated and recorded to keep a cumulative summation of the total

amount of each pollution. The amount of each pollutant can be calculated, using Eq. (7.225), from the given pollutant concentrations for corn and wheat fields. The results are shown below.

		Amount applied, lb/acre	
Pollutant	Concentration in biosolids, mg/kg	Corn field (1.4 ton/acre)[†]	Wheat field (1.0 ton/acre)
As	10	0.028[*]	0.020
Cd	7	0.020	0.014
Cr (deleted)	–	–	–
Cu	1100	3.08	2.2
Pb	140	0.39	0.28
Hg	5	0.014	0.010
Mo (deleted)	–	–	–
Ni	120	0.34	0.24
Se	6	0.017	0.012
Zn	2400	6.72	6.8

[†]see Step 4.
[*]lb pollutant/acre: 1.4 ton/acre × 0.002 (10 mg/kg) = 0.028 lb/acre.

(d) Estimate the number of years biosolids can be applied.
Assume that biosolids continue to have the same quality over time, and zinc would continue to be the limiting pollutant (see Step 2(b), 367 dry ton/acre is the maximum biosolids application rate for zinc). For this example, it is assumed that an average rate of 1.4 dry ton (acre · year) would be applied over time. The number of years biosolids could be applied before reaching the CPLR can be computed from maximum biosolids allowed divided by average annual application rate, as follows:

$$\text{Number of years} = 367 \text{ dry ton/acre} \div 1.4 \text{ dry ton/(acre} \cdot \text{year)}$$
$$= 262 \text{ years}$$

Step 3. Compute agronomic P rate for each field

(a) The total P_2O_5 can be computed by Eq. (7.223). Given that anaerobic biosolids have 2.2% total P

$$\text{Total } P_2O_5 = 2.2\% \times 20 \times 2.3 \text{ lb/(ton} \cdot \%)$$
$$= 101 \text{ lb/dry ton}$$

(b) The plant available P_2O_5 is computed by Eq. (7.222)

$$\text{Available } P_2O_5 = 0.5 \text{ (101 lb/dry ton)}$$
$$= 50 \text{ lb/dry ton}$$

(c) For the corn field

$$P_{req} = 70 \text{ lb } P_2O_5\text{/acre (given)}$$

For the wheat field

$$P_{req} = 80 \text{ lb } P_2O_5\text{/acre (given)}$$

(d) Calculate the agronomic P rate using Eq. (7.221)

For the corn field

$$\begin{aligned}\text{Agronomic P rate} &= P_{reg} \div (\text{available } P_2O_5/\text{dry ton})\\ &= (70 \text{ lb/acre}) \div (50 \text{ lb/dry ton})\\ &= 1.4 \text{ dry ton/acre}\end{aligned}$$

For the wheat field

$$\begin{aligned}\text{Agronomic P rate} &= 80 \div 50\\ &= 1.6 \text{ dry ton/acre}\end{aligned}$$

Step 4. Determine the application rate of biosolids on corn field

Since the agronomic P rate for the corn field (1.4 dry ton/acre) is less than the agronomic N rate (4.1 dry ton/acre), this rate, 1.4 dry ton/acre, is selected to be used for the 1999 growing season. A supplemental N fertilizer must be added to fulfill the remaining N needs for corn not supplied by biosolids.

Step 5. Determine the supplemental N fertilizer for the corn field

The amount of additional N needed for the corn can be computed by multiplying the PAN in lb/dry ton by the rate of biosolids application, then subtracting this PAN in lb/acre from the adjusted fertilizer N requirement. Referring to Step 1(c) and Step 1 (f-4), supplemental N can be determined as

$$\begin{aligned}\text{Biosolids N credit} &= 29 \text{ lb PAN/dry ton} \times 1.4 \text{ dry ton/acre}\\ &= 41 \text{ lb PAN/acre}\\ \text{Supplemental N} &= 119 \text{ lb/acre} - 41 \text{ lb/acre}\\ &= 78 \text{ lb/acre}\end{aligned}$$

Step 6. Determine the supplemental K fertilizer for the corn field

(a) Calculate the available K_2O added by biosolids, using Eq. (7.227)

Given: % K in biosolids = 0.55

$$\begin{aligned}\text{Available } K_2O &= \%\text{ K in biosolids} \times 20 \text{ lb/dry ton} \cdot \% \times 1.2\\ &= 0.55 \times 24 \text{ lb/dry ton}\\ &= 13 \text{ lb/dry ton}\end{aligned}$$

(b) Compute the amount of biosolids K_2O applied, using Eq. (7.226)

$$\begin{aligned}\text{Biosolids } K_2O \text{ added} &= 1.4 \text{ dry ton/acre (see Step 4)} \times 13 \text{ lb/dry ton}\\ &= 18.2 \text{ lb/acre (say 18 lb/acre)}\end{aligned}$$

(c) Compute supplemental K_2O needed

$$\begin{aligned}\text{Additional } K_2O \text{ needed} &= \text{required (given)} - \text{biosolids } K_2O\\ &= 140 \text{ lb/acre} - 18 \text{ lb/acre}\\ &= 122 \text{ lb/acre}\end{aligned}$$

Step 7. Determine the application rate of biosolids for the wheat field

For the wheat field, the agronomic N rate of 1.0 dry ton/acre of biosolids application is the choice since it is less than the agronomic P rate (1.6 dry ton/acre).

Step 8. Determine the supplemental P for the wheat field, referring to Steps 3 and 5

$$\text{Biosolids P credit} = 50 \text{ lb } P_2O_5/\text{dry ton} \times 1.0 \text{ dry ton/acre}$$
$$= 50 \text{ lb } P_2O_5/\text{acre}$$
$$\text{Additional P needed} = (80 - 50) \text{ lb } P_2O_5/\text{acre}$$
$$= 30 \text{ lb } P_2O_5/\text{acre}$$

Step 9. Determine the supplemental K for the wheat field

(a) Compute the biosolid K_2O added
From Step 6(a)

$$\text{available } K_2O = 13 \text{ lb/dry ton (from Step 6(a))}$$
$$\text{Biosolid } K_2O \text{ added} = 1.0 \text{ dry ton/acre} \times 13 \text{ lb/dry ton}$$
$$= 13 \text{ lb/acre}$$

(b) Compute additional K_2O needed

$$= 125 \text{ lb/acre} - 13 \text{ lb/acre}$$
$$= 112 \text{ lb/acre}$$

REFERENCES

Albertson, O. E. and Davis, G. 1984. Analysis of process factors controlling performance of plastic biomedia. Presented at the 57th Annual Meeting of the Water Pollution Control Federation, October 1984, New Orleans, Louisiana.

Alessi, C. J. *et al.* 1978. Design and operation of the activated-sludge process. SUNY/BUFFALO-WREE-7802. Buffalo: State University of New York at Buffalo.

Al-Layla, M. A., Ahmad, S. and Middlebrooks, E. J. 1980. *Handbook of wastewater collection and treatment: principles and practice.* New York: Garland STPM Press.

Alleman, J. E. and Irving, R. L. 1980. Nitrification in the sequencing batch biological reactor. *J. Water Pollut. Control Fed.*, **52**(11): 2747–2754.

American Society of Civil Engineers (ASCE) and Water Pollution Control Federation (WPCF). 1982. *Design and construction of urban stormwater management systems.* New York: ASCE and WPCF.

American Society of Civil Engineers (ASCE) and Water Environment Federation (WEF). 1992. *Design and construction of urban stormwater management systems.* New York: ASCE & WEF.

Antonie, R. L. 1978 *Fixed biological surfaces: wastewater treatment.* West Palm Beach, Florida: CRC Press.

APHA, AWWA and WEF. 1998. *Standard methods for the examination of water and wastewater*, 20th edn. Washington, DC: APHA.

Aqua-Aerobic Systems. 1976. *Clarifiers design.* Rockford, Illinois: Aqua-Aerobic Systems.

Autotrol Corporation. 1979. *Wastewater treatment systems: design manual.* Milwaukee, Wisconsin: Autotrol Corp.

Banerji, S. K. 1980. ASCE Water Pollution Management Task Committee report on "Rotating biological contactor for secondary treatment." In: *Proc. First National Symposium/Workshop on Rotating Biological Contactor Technology (FNSWRBCT),* Smith, E. D., Miller, R. D. and Wu, Y. C. (eds.), vol. I, p. 31. Pittsburgh: University of Pittsburgh.

Bidstrup, S. M. and Grady, Jr., C. P. L. 1988. SSSP: simulation of single-sludge processes. *J. Water Pollut. Control Fed.,* **60**(3): 351–361.

Boehnke, B., Diering, B. and Zuckut, S. W. 1997. Cost-effective wastewater treatment process for removal of organics and nutrients. *Water Eng. Mgmt.,* **145**(2): 30–35.

Bruce, A. M. and Merkens, J. C. 1973. Further study of partial treatment of sewage by high-rate biological filtration. *J. Inst. Water Pollut. Control*, **72**(3): 499–527.

Brenner, R. C. *et al.* 1984. Design information on rotation biological contactors. USEPA-600/2-84-106, Cincinnati, Ohio, Chaps. 2, 3, and 5.

Camp, T. R. 1946. Corrosiveness of water to metals. *J. Net Engl. Water Works Assoc.* **60**(1): 188.

Camp, T. R. 1953. Studies of sedimentation basin design. Presented at 1952 Annual Meeting, Pennsylvania Sewage and Industrial Wastes Association, Stage College Pennsylvania, August 27–29, 1952. Washington, DC: Sewage and Industrial Wastes.

Carder, C. 1997. Chicago's Deep Tunnel. *Compressed Air Mag*.

CBI Walker, Inc. 1998. *Small egg shaped digester facilities*. Plainfield, Illinois: CBI Walker, Inc.

Cheremisinoff, P. N. 1995. *Handbook of water and wastewater treatment technology*. New York: Marcel Dekker.

Chou, C. C., Hynek, R. J. and Sullivan, R. A. 1980. Comparison of full scale RBC performance with design criteria. *Proc. FNSDWRBCT*, vol. II, p. 1101.

Clark, J. H., Moseng, E. M. and Asano, T. 1978. Performance of a rotating biological contactor under varying wastewater flow. *J. Water Pollut. Control Fed.*, **50**: 896.

Clark, J. W. and W. Viessman, Jr. 1966. *Water supply and pollution control*. New York: A Dun-Donnelley.

Clark, J. W., Viessman, Jr., W. and Hammer, M. J. 1977. *Water supply and pollution control*. New York: IEP-A Dun-Donnelley Publisher.

Clow Corporation. 1980. *Clow envirodisc rotating biological contactor system*. Florence, Kentucky: Clow Corp.

Collins, H. F. and Selleck, R. E. 1972. Process kinetics of wastewater chlorination. SERL Report 72–75. Berkeley: University of California.

Collins, H. F., Selleck, R. E. and White, G. C. 1971. Problems in obtaining adequate sewage disinfection. *J. Sanitary Eng. Div., Proc. ASCE*, **87**(SA5): 549–562.

Coulson, J. M. and Richardson, J. F. 1955. *Chemical engineering* vol. II. New York: McGraw-Hill.

Davis, M. L. and Cornwell, D. A. 1991. *Introduction to environmental engineering*, 2nd edn. New York: McGraw-Hill.

Dick, R. I. and Ewing, B. B. 1967. Evaluation of activated-sludge thickening theories. *J. Sanitary Eng. Div., Proc. ASCE*, **93**(SA4): 9–29.

Downing, A. L. and Hopwood, A. P. 1964. Some observations on the kinetics of nitrifying activated-sludge plants. *Schweizerische Zeitschrift für Hydrologie*, **26**: 271.

Droste, R. L. 1997. *Theory and practice of water and wastewater treatment*. New York: John Wiley.

Eckenfelder, W. W. 1963. Trickling filter design and performance. *Trans. Am. Soc. Civil. Eng.*, **128** (part III): 371–384.

Eckenfelder, W. W., Jr. 1966. *Industrial water pollution control*. New York: McGraw-Hill.

Eckenfelder, W. W. and Barnhart, W. 1963. Performance of a high-rate trickling filter using selected media. *J. Water Pollut. Control Fed.*, **35**(12): 1535–1551.

Eckenfelder, W. W. and Ford, D. L. 1970. *Water pollution control*. Austin and New York: Jenkins.

F. W. Dodge Profile. 1994. *Perini goes underground with Chicago's Deep Tunnel project*. New York: McGraw-Hill.

Fair, G. M., Geyer, J. C. and Okun, D. A. 1966. Water and wastewater engineering, vol. 1. *Water supply and wastewater removal*. New York: John Wiley.

Fairall, J. M. 1956. Correlation of trickling filter data. *Sewage and Industrial Wastes*, **28**(9): 1069–1074.

Federal Register. 1991. Secondary treatment regulation. 40CFR Part 133, July 1, 1991, Washington, DC.

Federal Register. 1993. Standards for the use or disposal of sewage sludge; final rules. 40CFR Part 257 *et al.* Part II, EPA, *Federal Register* **58**(32): 9248–9415. Friday, February 19, 1993. Washington, DC.

Galler, W. S. and Gotaas, H. G. 1964. Analysis of biological filter variables. *J. Sanitary Eng. Div., Proc. ASCE*, **90**(SA6): 59–79.

Galler, W. S. and Gotaas, H. G. 1966. Optimization analysis for biological filter design. *J. Sanitary Eng. Div., Proc. ASCE*, **92**(SA1): 163–182.

Gaudy, A. F. and Kincannon, D. F. (1977). Comparing design models for activated-sludge. *Water and Sewage Works*, **123**(7): 66–77.

Germain, J. E. 1966. Economical treatment of domestic waste by plastic medium trickling filters. *J. Water Pollut. Control Fed.*, **38**(2): 192–203.

Grady, C. P. L. and Lim, H. C. 1980. *Biological wastewater treatment: theory and application.* New York: Marcel Dekker.

Great Lakes–Upper Mississippi River Board (GLUMRB) of State Sanitary Engineers, Health Education Service. 1971. *Recommended (Ten States) standards for sewage works.* Albany, New York: Health Research, Inc.

Great Lakes–Upper Mississippi River Board of State Public Health and Environmental Managers. 1971 *Recommended standards for wastewater facilities.* Albany, New York: Health Research, Inc.

Great Lakes–Upper Mississippi River Board of State and Provincial Public Health and Environmental Managers. 1996. *Revision 5 (draft edition). Recommended (Ten States) standards for wastewater facilities: policies for the design, review, and approval of plans and specifications for wastewater collection and treatment facilities.* Albany, New York: Health Research, Inc.

Guo, P. H. M., Thirumurthi, D. and Jank, B. E. 1981. Evaluation of extended aeration activated-sludge package plants. *J. Water Pollut. Control Fed.,* **53**(1): 33–42.

Hammer, M. J. 1986. *Water and wastewater technology.* New York: John Wiley.

Haug, R. T. and McCarty, P. L. 1972. Nitrification with submerged filters. *J. Water Pollut. Control Fed.*, **44**(11): 2086–2102.

Herzbrun, R. A., Irvine, R. L. and Malinowski, K. C. 1985. Biological treatment of hazardous waste in sequencing batch reactors. *J. Water Pollut. Control Fed.*, **57**(12): 1163–1167.

Hitdlebaugh, J. A. and Miller, R. D. 1980. Full-scale rotating biological contactor for secondary treatment and nitrification. *Proc. FNSWRBCT,* vol. I, p. 269.

Hoag, G., Widmer, W. and Hovey, W. 1980. Microfauna and RBC performance: laboratory and full-scale system. *Proc. FNSWRBCT.,* vol. I, p. 167.

Huang, C.-S. and Hopson, N. 1974. Nitrification rate in biological processes. *J. Environ. Eng. Div., Proc. ASCE,* **100**(EE2): 409.

Illinois Environmental Protection Agency. 1998. *Recommended standards for sewage work.* Part 370 of Chapter II, EPA, Subtitle C: Water Pollution, Title 35: Environmental Protection, Springfield, Illinois: IEPA.

Illinois Environmental Protection Agency. 1998. *Illinois recommended standards for sewage works.* State of Illinois Rules and Regulations Title 35, Subtitle C, Chapter II, parts 370. Springfield: Illinois EPA.

Institute of Water and Environmental Management (IWEM). 1988. *Unit process biological filtration: manuals of British practice in water pollution.* London: IWEM.

Jenkins, D. and Garrison, W. E. 1968. Control of activated-sludge by mean cell residence time. *J. Water Pollut. Control Fed.*, **40**(11): 1905–1919.

James A. Montgomery, Consulting Engineers, Inc. 1985. *Water Treatment Principles and Design.* New York: John Wiley.

Khan, A. N. and Raman, V. 1980. Rotating biological contactor for the treatment of wastewater in India. *Proc. FNSWRBCT,* vol. I: p. 235.

Kiff, R. J. 1972. The ecology of nitrification/denitrification systems in activated-sludge. *Water Pollut. Control,* **71**: 475.

Kincannon, D. F. and Stover, E. L. 1982. Design methodology for fixed film reactors, RBCs and trickling filters. *Civil Eng. Practicing and Design Eng.*, **2**: 107.

Kink, B. 1997. Touring the deep tunnel. *The Regional News*, Jan. 9.

Kinner, N. E., Balkwill, D. L. and Bishop, P. L. 1982. The microbiology of rotating biological contactor films. *Proc. FICFFBP*, vol. I, p. 184.

Kirschmer, O. 1926. *Untersuchungen uber den Gefallsverlust an rechen.* Trans. Hydraulic Inst. 21, Munich: R. Oldenbourg.

Kraus, L. S. 1955. Dual aeration as rugged activated-sludge process. *Sewage and Industrial Wastes,* **27**(12): 1347–1355.

Lawrence, A. W. and P. L. McCarty. 1969. Kinetics of methane fermentation in anaerobic treatment. *J. Water Pollut. Control Fed.*, **41**(2): R1–R17.

Lawrence, A. W. and McCarty, P. L. 1970. Unified basis for biological treatment design and operation. *J. Sanitary Eng. Div., Proc. ASCE,* **96**(SA3): 757–778.

Lin, S. D., Evans, R. L. and Dawson, W. 1982. RBC for BOD and ammonia nitrogen removals at Princeton wastewater treatment plant. *Proc. FICFFBP,* vol. I, p. 590.

Livingston, E. H. 1995. Infiltration practices: The good, the bed, and the ugly. In: *National Conference on Urban Runoff Management: Enhancing urban watershed management at the local, county, and state levels,* March 30–April 2, Chicago, Illinois, EPA.625/R-95/003, pp. 352–362. Cincinnati: USEPA.

Logan, B. E., Hermanowicz, S. W. and Parker, D. S. 1987a. Engineering implication of a new trickling filter model. *J. Water Pollut. Control Fed.*, **59**(12): 1017–1028.

Logan, B. E., Hermanowicz, S. W. and Parker, D. S. 1987b. A fundamental model for trickling filter process design. *J. Water Pollut. Control Fed.*, **59**(12): 1029–1042.

Lyco Division of Remsco Assoc. 1982. *Lyco Wastewater Products—RBC Systems.* Marlbor, New Jersey: Lyco Corp.

Lynch, J. M. and Poole, N. J. 1979. *Microbial ecology: a conceptual approach.* New York: John Wiley.

Mancini, J. L. and Barnhart, E. L. 1968. Industrial waste treatment in aerated lagoons. In: Gloyna, E. F. and Eckenfelder, Jr., W. W. (eds.), *Advances in water quality improvement.* Austin: University of Texas Press.

Marske, D. M. and Boyle, J. D. 1973. Chlorine contact chamber design: a field evaluation, *Water and Sewage Works,* **120**(1): 70–77.

Marston, A. 1930. The theory of external loads on closed conduits in the light of latest experiments. Bulletin 96, Iowa Engineering Experimental Station, Iowa City.

McCarty, P. L. 1964. Anaerobic waste treatment fundamentals. *Public Works* **95**(9): 107–112.

McCarty, P. L., Beck, L. and St. Amant, P. 1969. Biological denitrification of wastewaters by addition of organic materials. In: *Proc. of the 24th Industrial Waste Conference,* May 6–8, 1969. Lafayette, Indiana: Purdue University.

McGhee, T. J. 1991. *Water supply and sewerage,* 6th edn. New York: McGraw-Hill.

McKinney, R. E. 1962. Mathematics of complete mixing activated-sludge. *J. Sanitary Eng. Div., Proc. ASCE,* **88**(SA3).

Metcalf and Eddy, Inc. 1970. *Stormwater management model,* vol. I: *Final report.* Water Pollution Control Research Series 1124 D0C07/71, USEPA.

Metcalf and Eddy, Inc. 1981. *Wastewater engineering: collection and pumping.* New York: McGraw-Hill.

Metcalf and Eddy, Inc. 1991. *Wastewater engineering treatment, disposal, and reuse.* New York: McGraw-Hill.

Miller, R. D. *et al.* 1980. Rotating biological contactor process for secondary treatment and nitrification following a trickling filter. *Proc. FNSWRBCT,* vol. II, p. 1035.

Monod, J. 1949. The growth of bacterial cultures. *Ann. Rev. Microbiol.,* **3**: 371.

Mueller, J. A., Paquin, P. L. and Famularo, J. 1980. Nitrification in rotating biological contactor. *J. Water Pollut. Control Fed.*, **52**(4): 688–710.

National Research Council. 1946. Sewage treatment in military installations, Chapter V: Trickling filter. *Sewage Works J.,* **18**(5): 897–982.

Nelson, M. D. and Guarino, C. F. 1977. New "Philadelphia story" being written by Pollution Control Division. *Water and Waste Eng.*, **14**, 9–22.

New York State Department of Health. 1950. *Manual of instruction for sewage treatment plant operators.* Albany, New York: NY State Department of Health.

Northeastern Illinois Planning Commission. (1992). *Stormwater detention for water quality benefits.* Chicago: NIPC.

Novotny, V. and Chesters, G. 1981. *Handbook of nonpoint pollution: sources and management.* New York: Van Nostrand-Reinhold.

Novotny, V. *et al.* 1989. *Handbook of urban drainage and wastewater disposal.* New York: John Wiley.

Okun, D. A. 1949. A system of bioprecipitation of organic matter from activated-sludge. *Sewage Works J.,* **21**(5): 763–792.

Opatken, E. J. 1982. Rotating biological contactors: second order kinetics. *Proc. FICFFBP,* vol. I, p. 210.

Painter, H. A. 1970. A review of literature on inorganic nitrogen metabolism. *Water Res.*, **4**: 393.

Painter, H. A. 1975. Microbial transformations of inorganic nitrogen. In: *Proc. Conf. on Nitrogen as Wastewater Pollutant,* Copenhagen, Denmark.

Parcher, M. J. 1988. *Wastewater collection system maintenance.* Lancaster, Pennsylvania: Technomic.

Pasveer, A. 1960. New developments in the application of Kesener brushes (aeration rotors) in the activated-sludge treatment of trade waste waters. In: *Water treatment: Proc. Second Symposium on the Treatment of Waste Waters,* ed. P. C. G. Issac. New York: Pergamon Press.

Perry, R. H. 1967. *Engineering manual: A practical reference of data and methods in architectural, chemical, civil, electrical, mechanical, and nuclear engineering.* New York: McGraw-Hill.

Pescod, M. B. and Nair, J. V. 1972. Biological disc filtration for tropical waste treatment, experimental studies. *Water Res.*, **61**, 1509.

Pitt, R. and Voorhees, J. 1995. Source loading and management model (SLAMM). In: *National Conference on Urban Runoff Management*, pp. 225–243, EPA/625/R-95/003. Cincinnati: USEPA.

Poon, C. P. C., Chao, Y. L. and Mikucki, W. J. 1979. Factors controlling rotating biological contactor performance. *J. Water Pollut. Control Fed.*, **51**, 601.

Pretorius, W. A. 1971. Some operating characteristics of a bacteria disc unit. *Water Res.*, **5**, 1141.

Qasim, S. R. 1985. *Wastewater treatment plant—plan, design, and operation.* New York: Holt Rinehart & Winston.

Ramanathan, M. and Gaudy, A. F. 1971. Steady state model for activated-sludge with constant recycle sludge concentration. *Biotechnol. Bioeng*, **13**: 125.

Recht, H. L. and Ghassemi, M. 1970. Kinetics and mechanism of precipitation and nature of the precipitate obtained in phosphate removal from wastewater using aluminum(III) and iron(III) salts. Water Pollution Control Series 17010 EKI, Contract 14-12-158. Washington, DC: US Department of the Interior.

Rich, L. G. 1961. *Unit operations of sanitary engineering.* New York: John Wiley.

Roberts, P. V. *et al.* 1980. Chlorine dioxide for wastewater disinfection: a feasibility evaluation, Technical Report 251. San Jose: Stanford University.

Robison, R. 1986. The tunnel that cleans up Chicago. *Civil Eng.,* **56**(7): 34–37.

Schroeder, E. D. 1977. *Water and wastewater treatment.* New York: McGraw-Hill.

Schultz, K. L. 1960. Load and efficiency of trickling filters. *J. Water Pollut. Control Fed.,* **32**(3): 245–261.

Sepp, E. 1977. *Tracer evaluation of chlorine contact tanks.* Berkeley: California State Department of Health.

Sepp, E. 1981. Optimization of chlorine disinfection efficiency. *J. Environ. Eng. Div., Proc. ASCE,* **107**(EE1): 139–153.

Sherard, J. H. 1976. Destruction of alkalinity in aerobic biological wastewater treatment. *J. Water Pollut. Control Fed.*, **48**(7): 1834–1839.

Shoemaker, L. L. *et al.* 1995. Watershed screening and targeting tool (WSTT). In: *National Conference on Urban Runoff Management*, pp. 250–258. EPA/625/R-95/003. Cincinnati: USEPA.

Singhal, A. K. 1980. Phosphorus and nitrogen removal at Cadillac, Michigan. *J. Water Pollut. Control Fed.*, **52**(11): 2761–2770.

Sommers, L., Parker C. and Meyers, G. 1981. Volatilization, plant uptake and mineralization of nitrogen in soils treated with sewage sludge. Technical Report 133, Water Resources Research Center. West Lafayette, Indiana: Purdue University.

Stanier, R. Y., Doudoroff, M. and Adelberg, E. A. 1963. *The Microbial World.* 2nd edn. Englewood Cliffs, New Jersey: Prentice-Hall.

Steel, E. W. and McGhee, T. J. 1979. *Water supply and sewerage*, 5th edn. New York: McGraw-Hill.

Stukenberg, J. R. *et al.* 1990. Egg-shaped digester: from Germany to the United States. Presented at the 63rd Annual Conference of Water Pollution Control Federation, Washington, DC, October 7–11, 1990.

Sudo, R., Okad, M. and Mori, T. 1977. Rotating biological contactor microbial control in RBC. *J. Water and Waste,* **19**: 1.

Techobanoglous, G. and Schroeder, E. D. 1985. *Water quality*. Reading, Massachusetts: Addison-Wesley.

Terstriep, M. L. and Lee, M. T. 1995. AUTO-QI: an urban runoff quality I quantity model with a GIS interface. In: *National Conference on Urban Runoff Management*, pp. 213–224. EPA/625/R-95/003. Cincinnati: USEPA.

Theroux, R. J. and Betz, J. M. 1959. Sedimentation and preparation experiments in Los Angeles. *Sewage and Industrial Wastes,* **31**(11): 1259–1266.

Thirumurthi, D. 1969. Design principles of waste stabilization ponds. *J. Sanitary Eng. Div., Proc. ASCE,* **95**(SA2): 311–330.

Torpey, W. N. 1971. Rotating disks with biological growths prepare wastewater for disposal as reuse. *J. Water Pollut. Control Fed.*, **43**(11): 2181–2188.

Truong, H. V. and Phua, M. S. 1995. Application of the Washington, DC, sand filter for urban runoff control. In: *National Conference on Urban Runoff Management*, pp. 375–383, EPA/625/R-95/003. Cincinnati: USEPA.

US Environmental Protection Agency. 1974a. *Process design manual for upgrading existing wastewater treatment plants.* Technology Transfer, EPA 625/1-71-004a. Washington, DC: USEPA.

US Environmental Protection Agency. 1974b. *Wastewater treatment ponds.* EPA-430/9-74-001, MCD-14. Washington, DC: USEPA.

US Environmental Protection Agency. 1974c. *Water quality management planning for urban runoff.* EPA 440/9-75-004. Washington, DC: USEPA.

US Environmental Protection Agency. 1975a. *Process design manual for suspended solids removal.* EPA 625/1-75-003a. Washington, DC: USEPA.

US Environmental Protection Agency. 1975b. *Process design manual for nitrogen control.* Office of Technical Transfer. Washington, DC: USEPA.

US Environmental Protection Agency. 1975c. *Process design manual for nitrogen control.* EPA-625/1-75-007, Center for Environmental Research. Cincinnati: USEPA.

US Environmental Protection Agency. 1976. *Process design manual for phosphorus removal.* EPA 625/1-76-00/a. Washington, DC: USEPA.

USEPA. 1979. *Process design manual—Sludge treatment and disposal.* EPA 625/1-79-011, Cincinnati, Ohio: USEPA.

US Environmental Protection Agency. 1980. *Converting rock trickling filters to plastic media: design and performance.* EPA-600/2-80-120, Cincinnati: USEPA.

US Environmental Protection Agency. 1983a. *National urban runoff program*, vol. I. NTIS PB 84-185552. Washington, DC: USEPA.

US Environmental Protection Agency. 1983b. *Municipal wastewater stabilization ponds: design manual.* EPA-625/1-83-015. Cincinnati: USEPA.

US Environmental Protection Agency. 1984. *Design information on rotating biological contactors.* EPA-600/2-84-106. Cincinnati: USEPA.

US Environmental Protection Agency. 1987a. *Preliminary treatment facilities, design and operational considerations.* EPA-430/09-87-007. Washington, DC: USEPA.

US Environmental Protection Agency. 1987b. *Design manual for dewatering municipal wastewater sludges.* Washington, DC: USEPA.

USEPA. 1989. *Design manual—Dewatering municipal wastewater sludges,* EPA 625/1-79-011, Cincinnati, Ohio: USEPA.

US Environmental Protection Agency. 1991. *Evaluating sludge treatment processes.* Washington, DC: USEPA.

US Environmental Protection Agency. 1993. *Nitrogen control.* EPA/625/R-93/010. Washington, DC: USEPA.

US Environmental Protection Agency. 1994. *A plain English guide to the EPA part 503 biosolids rule.* EPA/832/R-93/003. Washington, DC: USEPA.

US Environmental Protection Agency. 1995. *Process design manual: land application of sewage sludge and domestic septage.* EPA/625/R-95/0001. Washington, DC: USEPA.

University of Cincinnati, Department of Civil Engineering. 1970. *Urban runoff characteristics.* Water Pollution Control Research Series 11024DQU10/70. Cincinnati: USEPA.

Urbonas, B. and Stahre, P. 1993. *Stormwater: best management practices and detention for water quality, drainage, and CSO management.* Englewood Cliffs, New Jersey: Prentice-Hall.

Velz, C. J. 1948. A basic law for the performance of biological filters. *Sewage Works J.* **20**(4): 607–617.

Viessman, Jr., W. and Hammer, M. J. 1993. *Water supply and pollution control*, 5th edn. New York: Harper Collins.

Wanielista, M. 1990. *Hydrology and water quality control.* New York: John Wiley.

Wanielista, M. 1992. Stormwater reuse: an alternative method of infiltration. In: *National Conference on Urban Runoff Management*, pp. 363–371, EPA/625/R-95/003. Cincinnati: USEPA.

Wanielista, M. P. and Yousef, Y. A. 1993. *Stormwater management.* New York: John Wiley.

Water Environment Federation (WEF) and American Society of Civil Engineers (ASCE). 1991a. *Design of municipal wastewater treatment plants*, vol. I. Alexandria, Virginia: WEF.

Water Environment Federation (WEF) and American Society of Civil Engineers (ASCE). 1991b. *Design of municipal wastewater treatment plants*, vol. II. Alexandria, Virginia: WEF.

Water Environment Federation (WEF) and American Society of Civil Engineers (ASCE). 1992. *Design and construction of urban stormwater management systems.* Alexandria, Virginia: WEF

Water Environment Federation (WEF). 1993a. *Design of wastewater and stormwater pumping stations.* Alexandria, Virginia: WEF.

Water Environment Federation (WEF). 1993b. *Standards for the use and disposal of sewage sludge (40CFR Part 257, 403, and 503) final rule and phase-in submission of sewage sludge permit application (Revisions to 40CFR Parts 122, 123, and 501) final rule.* Alexandria, Virginia: WEF.

Water Environment Federation (WEF). 1994. *Beneficial use programs for biosolids management.* Alexandria, Virginia: WEF.

Water Environment Federation (WEF) and American Society of Civil Engineers (ASCE). 1996a. *Operation of municipal wastewater treatment plants*, 5th edn., vol. 2. Alexandria, Virginia: WEF.

Water Environment Federation (WEF) and American Society of Civil Engineers (ASCE). 1996b. *Operation of municipal wastewater treatment plants*, 5th edn., vol. 3. Alexandria, Virginia: WEF.

Water Pollution Control Federation (WPCF). 1980. *Clean water for today: what is wastewater treatment.* Washington, DC: WPCF.

Water Pollution Control Federation (WPCF). 1983. *Sludge dewatering,* Manual of Practice 20. Alexandria, Virginia: WPCF.

Water Pollution Control Federation (WPCF). 1985a. *Clarifier design.* Manual of Practice FD-8. Alexandria, Virginia: WPCF.

Water Pollution Control Federation (WPCF). 1985b. *Sludge stabilization.* Manual of Practice FD-9. Alexandria, Virginia: WPCF.

Water Pollution Control Federation (WPCF). 1988a. *Sludge conditioning.* Manual of Practice FD-14. Alexandria, Virginia: WPCF.

Water Pollution Control Federation (WPCF). 1988b. *Operation and maintenance of trickling filters, RBCs and related processes.* Manual of Practice OM-10. Alexandria, Virginia: WPCF.

Water Pollution Control Federation. 1990. *Natural systems for wastewater treatment.* Manual of Practice FD-16. Alexandria, Virginia: WPCF.

Wehner, J. F. and Wilhelm, R. H. 1958. Boundary conditions of flow reactor. *Chem. Eng. Sci.*, **6**(1): 89–93.

Wild, A. E., Sawyer, C. E. and McMahon, T. C. 1971. Factors affecting nitrification kinetics. *J. Water Pollut. Control Fed.*, **43**(9): 1845–1854.

Wilson, T. E. and Lee. J. S. 1982. Comparison of final clarifier design techniques. *J. Water Pollut. Control Fed.*, **54**(10): 1376–1381.

APPENDIX A

ILLINOIS ENVIRONMENTAL PROTECTION AGENCY'S MACROINVERTEBRATE TOLERANCE LIST

Macroinvertebrate	Tolerance value
PLATYHELMINTHES	
TURBELLARIA	6
ANNELIDA	
OLIGOCHAETA	10
HIRUDINEA	8
Rhynchobdellida	
Glossiphoniidae	8
Piscicolidae	7
Gnathobdellida	
Hirudinidae	7
Pharyngobdellida	
Erpobdellidae	8
ARTHROPODA	
CRUSTACEA	
ISOPODA	
Asellidae	6
Caecidotea	6
brevicauda	6
intermedia	6
Lirceus	4
AMPHIPODA	
Hyalellidae	
Hyalella	
azteca	5
Gammaridae	
Bactrurus	1
Crangonyx	4
Gammarus	3
DECAPODA	
Cambaridae	5
Palaemonidae	
Palaemonetes	4
INSECTA	
EPHEMEROPTERA	
Siphlonuridae	
Ameletus	0
Siphlonurus	2
Oligoneuriidae	
Heptageniidae	
Arthroplea	3
Epeorus	1
vitreus	0
Heptagenia	3
diabasia	4
flavescens	2
hebe	3
lucidipennis	3
maculipennis	3
marginalis	1
perfida	1
pulla	0
Rhithrogena	0
Stenacron	4
candidum	1
gildersleevei	1
interpunctatum	4
minnetonka	4
Stenonema	4
annexum	4
ares	3
exiguum	5
femoratum	7
integrum	4
luteum	1
mediopunctatum	2
modestum	3
nepotellum	5
pudicum	2
pulchellum	3
quinquespinum	5
rubromaculatum	2
scitulum	1
terminatum	4
vicarium	3
Ephemerellidae	
Attenella	2
Danella	2

Macroinvertebrate	Tolerance value
Isonychia	3
Metretopodidae	
Siphloplecton	2
Baetidae	
Baetis	4
brunneicolor	4
flavistriga	4
frondalis	4
intercalaris	7
longipalpus	6
macdunnoughi	4
propinquus	4
pygmaeus	4
tricaudatus	1
Callibaetis	4
fluctuans	4
Centroptilum	2
Cloeon	3
Pseudocloeon	4
dubium	4
parvulum	4
punctiventris	4
Hexagenia	6
limbata	5
munda	7
Palingeniidae	
Pentagenia	4
vittigera	4
Polymitarcyidae	
Ephoron	2
Tortopus	4
ODONATA	
ANISOPTERA	
Cordulegasteridae	
Cordulegaster	2
Gomphidae	
Dromogomphus	4
Gomphus	7
Hagenius	3
Lanthus	6
Ophiogomphus	2
Progomphus	5
Aeshnidae	
Aeshna	4
Anax	5
Basiaeschna	2
Boyeria	3
Epiaeschna	1
Nasiaeschna	2
Macromiidae	
Didymops	4
Macromia	3
Drunella	1
Ephemerella	2
Eurylophella	4
Seratella	1
Tricorythidae	
Tricorythodes	5
Caenidae	
Brachycercus	3
Caenis	6
Baetiscidae	
Baetisca	3
Leptophlebiidae	
Choroterpes	2
Habrophlebiodes	2
americana	2
Leptophlebia	3
Paraleptophlebia	2
Potamanthidae	
Potamanthus	4
Ephemeridae	
Ephemera	3
simulans	3
PLECOPTERA	
Pteronarcyidae	
Pteronarcys	2
Taeniopterygidae	
Taeniopteryx	2
Nemoundae	
Nemoura	1
Leuctridae	
Leuctra	1
Capniidae	
Allocapnia	2
Capnia	1
Perlidae	
Acroneuria	1
Atoperia	1
Neoperia	1
Perlesta	4
placida	4
Perlinella	2
Periodidae	
Hydroperia	1
Isoperia	2
Chloroperiidae	
Chloroperia	3
MEGALOPTERA	
Sialidae	
Sialis	4
Corydalidae	
Chauliodes	4
Corydalus	3

Macroinvertebrate	Tolerance value	Macroinvertebrate	Tolerance value
Corduliidae		*Nigronia*	2
Cordulia	2	NEUROPTERA	
Epitheca	4	Sisyridae	1
Helocordulia	2	TRICHOPTERA	
Neurocordulia	3	Hydropsychidae	
Somatochlora	1	*Cheumatopsyche*	6
Libellulidae		*Diplectrona*	2
Celithemis	2	*Hydropsyche*	5
Erythemis	5	*arinale*	5
Erythrodiplax	5	*betteni*	5
Libellula	8	*bidens*	5
Pachydiplax	8	*cuanis*	5
Pantala	7	*frisoni*	5
Perithemis	4	*orris*	4
Plathemis	3	*phalerata*	2
Sympetrum	4	*placoda*	4
Tramea	4	*simulans*	5
ZYGOPTERA		*Macronema*	2
Calopterygidae		*Potamyia*	4
Calopteryx	4	*Symphitopsyche*	4
Hetaerina	3	Philopotamidae	
Lestidae		*Chimarra*	3
Archilestes	1	*Dolophilodes*	0
Lestes	6	Polycentropodidae	
Coenagrionidae		*Cyrnellus*	5
Amphiagrion	5	*Neureclipsis*	3
Argia	5	*Nyctiophylax*	1
moesta	5	*Polycentropus*	3
tibialis	5	Psychomyiidae	
Enallagma	6	*Psychomyia*	2
signatum	6	Glossosomatidae	
Ischnura	6	*Agapetus*	2
Nehalennia	7	*Protoptila*	1
Hydroptilidae		DIPTERA	
Agraylea	2	Blephariceridae	0
Hydroptila	2	Tipulidae	4
Ithytrichia	1	*Antocha*	5
Leucotrichia	3	*Dicranota*	4
Mayatrichia	1	*Eriocera*	7
Neotrichia	4	*Helius*	5
Ochrotrichia	4	*Hesperoconopa*	2
Orthotrichia	1	*Hexatoma*	4
Oxyethira	2	*Limnophila*	4
Rhyacophilidae		*Limonia*	3
Rhyacophila	1	*Liriope*	7
Brachycentridae		*Pedicia*	4
Brachycentrus	1	*Pilaria*	4
Lepidostomatidae		*Polymeda*	2
Lepidostoma	3	*Pseudolimnophila*	2
Limnephilidae		*Tipula*	4
Hydatophylax	2	Chaoboridae	8
Limnephilus	3	Culicidae	8
Neophylax	3	*Aedes*	8

Macroinvertebrate	Tolerance value	Macroinvertebrate	Tolerance value
Platycentropus	3	*Anopheles*	6
Pycnopsyche	3	*Culex*	8
Phryganeidae		Psychodidae	11
Agrypnia	3	Ceratopogonidae	5
Banksiola	2	*Atrichopogon*	2
Phryganea	3	*Palpomyia*	6
Ptilostomis	3	Simuliidae	
Helicopsychidae		*Cnephia*	4
Helicopsyche	2	*Prosimulium*	2
Leptoceridae		*Simulium*	6
Ceraclea	3	*clarkei*	4
Leptocerus	3	*corbis*	0
Mystacides	2	*decorum*	4
Nectopsyche	3	*jenningsi*	4
Oecetis	5	*luggeri*	2
Triaenodes	3	*meridionale*	1
COLEOPTERA		*tuberosum*	4
Gyrinidae (larvae only)		*venustum*	6
Dineutus	4	*verecundum*	6
Gyrinus	4	*vittatum*	8
Psephenidae (larvae only)	4	Chironomidae	
Psephenus	4	Tanypodinae	
herricki	4	*Ablabesmyia*	6
Eubriidae	4	*mallochi*	6
Ectopria	4	*parajanta*	6
thoracica	4	*peleensis*	6
Dryopidae	4	*Clinotanypus*	6
Helichus	4	*pinguis*	6
lithophilus	4	*Coelotanypus*	4
Helodidae (larvae only)	7	*Labrundinia*	4
Elmidae		*Larsia*	6
Ancyronyx	2	*Macropelopia*	7
variegatus	2	*Natarsia*	6
Dubiraphia	5	*Pentaneura*	3
bivittata	2	*Procladius*	8
quadrinotata	7	*Psectrotanypus*	8
vittata	7	*Tanypus*	8
Macronychus	2	*Thienemannimyia group*	6
glabratus	2	*Zavrelimyia*	8
Microcylloepus	2	Diamesinae	
Optioservus	4	*Diamesa*	1
ovalis	4	*Pseudodiamesa*	1
Stenelmis	7		
crenata	7		
vittipennis	6		
Orthocladiinae		Syrphidae	11
Cardiocladius	6	Ephydridae	8
Chaetocladius	6	Sciomyzidae	10
Corynoneura	2	Muscidae	8
Cricotopus	8	Athencidae	
bicinctus	10	*Atherix*	4
trifasciatus	6	MOLLUSCA	
Eukiefferiella	4	GASTROPODA	

Macroinvertebrate	Tolerance value
Hydrobaenus	2
Nanocladius	3
Orthocladius	4
Parametriocnemus	4
Prodiamesa	3
Psectrocladius	5
Rheocricotopus	6
Thienemaniella	2
xena	2
Chironominae	
Chironomus	11
attenuatus	10
riparius	11
Cryptochironomus	8
Cryptotendipes	6
Dicrotendipes	6
modestus	6
neomodestus	6
nervosus	6
Einfeldia	10
Endochironomus	6
Glyptotendipes	10
Harnischia	6
Kiefferulus	7
Microtendipes	6
Parachironomus	8
Paracladopelma	4
Paralauterborniella	6
Paratendipes	3
Phaenopsectra	4
Polypedilum	6
fallax	6
halterale	4
illinoense	5
scalaenum	6
Pseudochironomus	5
Stenochironomus	3
Stictochironomus	5
Tribelos	5
Xenochironomus	4
Tanytarsini	
Cladotanytarsus	7
Micropsectra	4
Rheotanytarsus	6
Tanytarsus	7
Ptychopteridae	8
Tabanidae	7
Chrysops	7
Tabanus	7
Dolichopodidae	5
Empididae	6
Hemerodromia	6
Viviparidae	
Campeloma	7
Lioplax	7
Viviparus	1
Valvatidae	
Valvata	2
Bulimidae	
Amnicola	4
Pleurocandae	
Goniobasis	5
Pleurocera	7
Physidae	
Aplexa	7
Physa	9
Lymnaeidae	
Lymnara	7
Stagnicola	7
Planorbidae	
Gyraulus	6
Helisoma	7
Planorbula	7
Ancylidae	
Ferrissia	7
PELECYPODA	
Unionidae	
Actinonaias	
carinata	1
Alasmidonta	
marginata	1
triangulata	0
Anodonta	3
Carunculina	7
Elliptio	2
Fusronaia	1
Lampsilis	1
Ligumia	1
Margaritifera	1
Micromya	1
Obliquaria	1
Proplera	1
Strophitus	4
Tritagonia	1
Truncilla	1
Utterbackia	1
Sphaeridae	
Musculium	5
Pisidium	5
Sphaeriuni	5
Cyrenidae	
Corbicula	4

Source: Illinois Environmental Protection Agency, 1987.

APPENDIX B
WELL FUNCTION FOR CONFINED AQUIFERS

10^{-10} well functions							
u	$W(u)$	u	$W(u)$	u	$W(u)$	u	$W(u)$
1.0E – 10	22.45	3.3E – 10	21.25	5.6E – 10	20.73	7.9E – 10	20.38
1.1E – 10	22.35	3.4E – 10	21.22	5.7E – 10	20.71	8.0E – 10	20.37
1.2E – 10	22.27	3.5E – 10	21.20	5.8E – 10	20.69	8.1E – 10	20.36
1.3E – 10	22.19	3.6E – 10	21.17	5.9E – 10	20.67	8.2E – 10	20.34
1.4E – 10	22.11	3.7E – 10	21.14	6.0E – 10	20.66	8.3E – 10	20.33
1.5E – 10	22.04	3.8E – 10	21.11	6.1E – 10	20.64	8.4E – 10	20.32
1.6E – 10	21.98	3.9E – 10	21.09	6.2E – 10	20.62	8.5E – 10	20.31
1.7E – 10	21.92	4.0E – 10	21.06	6.3E – 10	20.61	8.6E – 10	20.30
1.8E – 10	21.86	4.1E – 10	21.04	6.4E – 10	20.59	8.7E – 10	20.29
1.9E – 10	21.81	4.2E – 10	21.01	6.5E – 10	20.58	8.8E – 10	20.27
2.0E – 10	21.76	4.3E – 10	20.99	6.6E – 10	20.56	8.9E – 10	20.26
2.1E – 10	21.71	4.4E – 10	20.97	6.7E – 10	20.55	9.0E – 10	20.25
2.2E – 10	21.66	4.5E – 10	20.94	6.8E – 10	20.53	9.1E – 10	20.24
2.3E – 10	21.62	4.6E – 10	20.92	6.9E – 10	20.52	9.2E – 10	20.23
2.4E – 10	21.57	4.7E – 10	20.90	7.0E – 10	20.50	9.3E – 10	20.22
2.5E – 10	21.53	4.8E – 10	20.88	7.1E – 10	20.49	9.4E – 10	20.21
2.6E – 10	21.49	4.9E – 10	20.86	7.2E – 10	20.47	9.5E – 10	20.20
2.7E – 10	21.46	5.0E – 10	20.84	7.3E – 10	20.46	9.6E – 10	20.19
2.8E – 10	21.42	5.1E – 10	20.82	7.4E – 10	20.45	9.7E – 10	20.18
2.9E – 10	21.38	5.2E – 10	20.80	7.5E – 10	20.43	9.8E – 10	20.17
3.0E – 10	21.35	5.3E – 10	20.78	7.6E – 10	20.42	9.9E – 10	20.16
3.1E – 10	21.32	5.4E – 10	20.76	7.7E – 10	20.41		
3.2E – 10	21.29	5.5E – 10	20.74	7.8E – 10	20.39		

10⁻⁹ well functions							
u	$W(u)$	u	$W(u)$	u	$W(u)$	u	$W(u)$
1.0E – 09	20.15	3.3E – 09	18.95	5.6E – 09	18.42	7.9E – 09	18.08
1.1E – 09	20.05	3.4E – 09	18.92	5.7E – 09	18.41	8.0E – 09	18.07
1.2E – 09	19.96	3.5E – 09	18.89	5.8E – 09	18.39	8.1E – 09	18.05
1.3E – 09	19.88	3.6E – 09	18.87	5.9E – 09	18.37	8.2E – 09	18.04
1.4E – 09	19.81	3.7E – 09	18.84	6.0E – 09	18.35	8.3E – 09	18.03
1.5E – 09	19.74	3.8E – 09	18.81	6.1E – 09	18.34	8.4E – 09	18.02
1.6E – 09	19.68	3.9E – 09	18.79	6.2E – 09	18.32	8.5E – 09	18.01
1.7E – 09	19.62	4.0E – 09	18.76	6.3E – 09	18.31	8.6E – 09	17.99
1.8E – 09	19.56	4.1E – 09	18.74	6.4E – 09	18.29	8.7E – 09	17.98
1.9E – 09	19.50	4.2E – 09	18.71	6.5E – 09	18.27	8.8E – 09	17.97
2.0E – 09	19.45	4.3E – 09	18.69	6.6E – 09	18.26	8.9E – 09	17.96
2.1E – 09	19.40	4.4E – 09	18.66	6.7E – 09	18.24	9.0E – 09	17.95
2.2E – 09	19.36	4.5E – 09	18.64	6.8E – 09	18.23	9.1E – 09	17.94
2.3E – 09	19.31	4.6E – 09	18.62	6.9E – 09	18.21	9.2E – 09	17.93
2.4E – 09	19.27	4.7E – 09	18.60	7.0E – 09	18.20	9.3E – 09	17.92
2.5E – 09	19.23	4.8E – 09	18.58	7.1E – 09	18.19	9.4E – 09	17.91
2.6E – 09	19.19	4.9E – 09	18.56	7.2E – 09	18.17	9.5E – 09	17.89
2.7E – 09	19.15	5.0E – 09	18.54	7.3E – 09	18.16	9.6E – 09	17.88
2.8E – 09	19.12	5.1E – 09	18.52	7.4E – 09	18.14	9.7E – 09	17.87
2.9E – 09	19.08	5.2E – 09	18.50	7.5E – 09	18.13	9.8E – 09	17.86
3.0E – 09	19.05	5.3E – 09	18.48	7.6E – 09	18.12	9.9E – 09	17.85
3.1E – 09	19.01	5.4E – 09	18.46	7.7E – 09	18.10		
3.2E – 09	18.98	5.5E – 09	18.44	7.8E – 09	18.09		

10⁻⁸ well functions							
u	$W(u)$	u	$W(u)$	u	$W(u)$	u	$W(u)$
1.0E – 08	17.84	3.3E – 08	16.65	5.6E – 08	16.12	7.9E – 08	15.78
1.1E – 08	17.75	3.4E – 08	16.62	5.7E – 08	16.10	8.0E – 08	15.76
1.2E – 08	17.66	3.5E – 08	16.59	5.8E – 08	16.09	8.1E – 08	15.75
1.3E – 08	17.58	3.6E – 08	16.56	5.9E – 08	16.07	8.2E – 08	15.74
1.4E – 08	17.51	3.7E – 08	16.54	6.0E – 08	16.05	8.3E – 08	15.73
1.5E – 08	17.44	3.8E – 08	16.51	6.1E – 08	16.04	8.4E – 08	15.72
1.6E – 08	17.37	3.9E – 08	16.48	6.2E – 08	16.02	8.5E – 08	15.70
1.7E – 08	17.31	4.0E – 08	16.46	6.3E – 08	16.00	8.6E – 08	15.69
1.8E – 08	17.26	4.1E – 08	16.43	6.4E – 08	15.99	8.7E – 08	15.68
1.9E – 08	17.20	4.2E – 08	16.41	6.5E – 08	15.97	8.8E – 08	15.67
2.0E – 08	17.15	4.3E – 08	16.38	6.6E – 08	15.96	8.9E – 08	15.66
2.1E – 08	17.10	4.4E – 08	16.36	6.7E – 08	15.94	9.0E – 08	15.65
2.2E – 08	17.06	4.5E – 08	16.34	6.8E – 08	15.93	9.1E – 08	15.64
2.3E – 08	17.01	4.6E – 08	16.32	6.9E – 08	15.91	9.2E – 08	15.62
2.4E – 08	16.97	4.7E – 08	16.30	7.0E – 08	15.90	9.3E – 08	15.61
2.5E – 08	16.93	4.8E – 08	16.27	7.1E – 08	15.88	9.4E – 08	15.60
2.6E – 08	16.89	4.9E – 08	16.25	7.2E – 08	15.87	9.5E – 08	15.59
2.7E – 08	16.85	5.0E – 08	16.23	7.3E – 08	15.86	9.6E – 08	15.58
2.8E – 08	16.81	5.1E – 08	16.21	7.4E – 08	15.84	9.7E – 08	15.57
2.9E – 08	16.78	5.2E – 08	16.19	7.5E – 08	15.83	9.8E – 08	15.56
3.0E – 08	16.74	5.3E – 08	16.18	7.6E – 08	15.82	9.9E – 08	15.55
3.1E – 08	16.71	5.4E – 08	16.16	7.7E – 08	15.80		
3.2E – 08	16.68	5.5E – 08	16.14	7.8E – 08	15.79		

10^{-7} well functions							
u	$W(u)$	u	$W(u)$	u	$W(u)$	u	$W(u)$
1.0E – 07	15.54	3.3E – 07	14.35	5.6E – 07	13.82	7.9E – 07	13.47
1.1E – 07	15.45	3.4E – 07	14.32	5.7E – 07	13.80	8.0E – 07	13.46
1.2E – 07	15.36	3.5E – 07	14.29	5.8E – 07	13.78	8.1E – 07	13.45
1.3E – 07	15.28	3.6E – 07	14.26	5.9E – 07	13.77	8.2E – 07	13.44
1.4E – 07	15.20	3.7E – 07	14.23	6.0E – 07	13.75	8.3E – 07	13.42
1.5E – 07	15.14	3.8E – 07	14.21	6.1E – 07	13.73	8.4E – 07	13.41
1.6E – 07	15.07	3.9E – 07	14.18	6.2E – 07	13.72	8.5E – 07	13.40
1.7E – 07	15.01	4.0E – 07	14.15	6.3E – 07	13.70	8.6E – 07	13.39
1.8E – 07	14.95	4.1E – 07	14.13	6.4E – 07	13.68	8.7E – 07	13.38
1.9E – 07	14.90	4.2E – 07	14.11	6.5E – 07	13.67	8.8E – 07	13.37
2.0E – 07	14.85	4.3E – 07	14.08	6.6E – 07	13.65	8.9E – 07	13.35
2.1E – 07	14.80	4.4E – 07	14.06	6.7E – 07	13.64	9.0E – 07	13.34
2.2E – 07	14.75	4.5E – 07	14.04	6.8E – 07	13.62	9.1E – 07	13.33
2.3E – 07	14.71	4.6E – 07	14.01	6.9E – 07	13.61	9.2E – 07	13.32
2.4E – 07	14.67	4.7E – 07	13.99	7.0E – 07	13.59	9.3E – 07	13.31
2.5E – 07	14.62	4.8E – 07	13.97	7.1E – 07	13.58	9.4E – 07	13.30
2.6E – 07	14.59	4.9E – 07	13.95	7.2E – 07	13.57	9.5E – 07	13.29
2.7E – 07	14.55	5.0E – 07	13.93	7.3E – 07	13.55	9.6E – 07	13.28
2.8E – 07	14.51	5.1E – 07	13.91	7.4E – 07	13.54	9.7E – 07	13.27
2.9E – 07	14.48	5.2E – 07	13.89	7.5E – 07	13.53	9.8E – 07	13.26
3.0E – 07	14.44	5.3E – 07	13.87	7.6E – 07	13.51	9.9E – 07	13.25
3.1E – 07	14.41	5.4E – 07	13.85	7.7E – 07	13.50		
3.2E – 07	14.38	5.5E – 07	13.84	7.8E – 07	13.49		

10^{-6} well functions							
u	$W(u)$	u	$W(u)$	u	$W(u)$	u	$W(u)$
1.0E – 06	13.24	3.3E – 06	12.04	5.6E – 06	11.52	7.9E – 06	11.17
1.1E – 06	13.14	3.4E – 06	12.01	5.7E – 06	11.50	8.0E – 06	11.16
1.2E – 06	13.06	3.5E – 06	11.99	5.8E – 06	11.48	8.1E – 06	11.15
1.3E – 06	12.98	3.6E – 06	11.96	5.9E – 06	11.46	8.2E – 06	11.13
1.4E – 06	12.90	3.7E – 06	11.93	6.0E – 06	11.45	8.3E – 06	11.12
1.5E – 06	12.83	3.8E – 06	11.90	6.1E – 06	11.43	8.4E – 06	11.11
1.6E – 06	12.77	3.9E – 06	11.88	6.2E – 06	11.41	8.5E – 06	11.10
1.7E – 06	12.71	4.0E – 06	11.85	6.3E – 06	11.40	8.6E – 06	11.09
1.8E – 06	12.65	4.1E – 06	11.83	6.4E – 06	11.38	8.7E – 06	11.07
1.9E – 06	12.60	4.2E – 06	11.80	6.5E – 06	11.37	8.8E – 06	11.06
2.0E – 06	12.55	4.3E – 06	11.78	6.6E – 06	11.35	8.9E – 06	11.05
2.1E – 06	12.50	4.4E – 06	11.76	6.7E – 06	11.34	9.0E – 06	11.04
2.2E – 06	12.45	4.5E – 06	11.73	6.8E – 06	11.32	9.1E – 06	11.03
2.3E – 06	12.41	4.6E – 06	11.71	6.9E – 06	11.31	9.2E – 06	11.02
2.4E – 06	12.36	4.7E – 06	11.69	7.0E – 06	11.29	9.3E – 06	11.01
2.5E – 06	12.32	4.8E – 06	11.67	7.1E – 06	11.28	9.4E – 06	11.00
2.6E – 06	12.28	4.9E – 06	11.65	7.2E – 06	11.26	9.5E – 06	10.99
2.7E – 06	12.25	5.0E – 06	11.63	7.3E – 06	11.25	9.6E – 06	10.98
2.8E – 06	12.21	5.1E – 06	11.61	7.4E – 06	11.24	9.7E – 06	10.97
2.9E – 06	12.17	5.2E – 06	11.59	7.5E – 06	11.22	9.8E – 06	10.96
3.0E – 06	12.14	5.3E – 06	11.57	7.6E – 06	11.21	9.9E – 06	10.95
3.1E – 06	12.11	5.4E – 06	11.55	7.7E – 06	11.20		
3.2E – 06	12.08	5.5E – 06	11.53	7.8E – 06	11.18		

10^{-5} well functions							
u	$W(u)$	u	$W(u)$	u	$W(u)$	u	$W(u)$
1.0E – 05	10.94	3.3E – 05	9.74	5.6E – 05	9.21	7.9E – 05	8.87
1.1E – 05	10.84	3.4E – 05	9.71	5.7E – 05	9.20	8.0E – 05	8.86
1.2E – 05	10.75	3.5E – 05	9.68	5.8E – 05	9.18	8.1E – 05	8.84
1.3E – 05	10.67	3.6E – 05	9.65	5.9E – 05	9.16	8.2E – 05	8.83
1.4E – 05	10.60	3.7E – 05	9.63	6.0E – 05	9.14	8.3E – 05	8.82
1.5E – 05	10.53	3.8E – 05	9.60	6.1E – 05	9.13	8.4E – 05	8.81
1.6E – 05	10.47	3.9E – 05	9.57	6.2E – 05	9.11	8.5E – 05	8.80
1.7E – 05	10.41	4.0E – 05	9.55	6.3E – 05	9.10	8.6E – 05	8.78
1.8E – 05	10.35	4.1E – 05	9.52	6.4E – 05	9.08	8.7E – 05	8.77
1.9E – 05	10.29	4.2E – 05	9.50	6.5E – 05	9.06	8.8E – 05	8.76
2.0E – 05	10.24	4.3E – 05	9.48	6.6E – 05	9.05	8.9E – 05	8.75
2.1E – 05	10.19	4.4E – 05	9.45	6.7E – 05	9.03	9.0E – 05	8.74
2.2E – 05	10.15	4.5E – 05	9.43	6.8E – 05	9.02	9.1E – 05	8.73
2.3E – 05	10.10	4.6E – 05	9.41	6.9E – 05	9.00	9.2E – 05	8.72
2.4E – 05	10.06	4.7E – 05	9.39	7.0E – 05	8.99	9.3E – 05	8.71
2.5E – 05	10.02	4.8E – 05	9.37	7.1E – 05	8.98	9.4E – 05	8.70
2.6E – 05	9.98	4.9E – 05	9.35	7.2E – 05	8.96	9.5E – 05	8.68
2.7E – 05	9.94	5.0E – 05	9.33	7.3E – 05	8.95	9.6E – 05	8.67
2.8E – 05	9.91	5.1E – 05	9.31	7.4E – 05	8.93	9.7E – 05	8.66
2.9E – 05	9.87	5.2E – 05	9.29	7.5E – 05	8.92	9.8E – 05	8.65
3.0E – 05	9.84	5.3E – 05	9.27	7.6E – 05	8.91	9.9E – 05	8.64
3.1E – 05	9.80	5.4E – 05	9.25	7.7E – 05	8.89		
3.2E – 05	9.77	5.5E – 05	9.23	7.8E – 05	8.88		

10^{-4} well functions							
u	$W(u)$	u	$W(u)$	u	$W(u)$	u	$W(u)$
1.0E – 04	8.63	3.3E – 04	7.44	5.6E – 04	6.91	7.9E – 04	6.57
1.1E – 04	8.54	3.4E – 04	7.41	5.7E – 04	6.89	8.0E – 04	6.55
1.2E – 04	8.45	3.5E – 04	7.38	5.8E – 04	6.88	8.1E – 04	6.54
1.3E – 04	8.37	3.6E – 04	7.35	5.9E – 04	6.86	8.2E – 04	6.53
1.4E – 04	8.30	3.7E – 04	7.33	6.0E – 04	6.84	8.3E – 04	6.52
1.5E – 04	8.23	3.8E – 04	7.30	6.1E – 04	6.83	8.4E – 04	6.51
1.6E – 04	8.16	3.9E – 04	7.27	6.2E – 04	6.81	8.5E – 04	6.49
1.7E – 04	8.10	4.0E – 04	7.25	6.3E – 04	6.79	8.6E – 04	6.48
1.8E – 04	8.05	4.1E – 04	7.22	6.4E – 04	6.78	8.7E – 04	6.47
1.9E – 04	7.99	4.2E – 04	7.20	6.5E – 04	6.76	8.8E – 04	6.46
2.0E – 04	7.94	4.3E – 04	7.17	6.6E – 04	6.75	8.9E – 04	6.45
2.1E – 04	7.89	4.4E – 04	7.15	6.7E – 04	6.73	9.0E – 04	6.44
2.2E – 04	7.84	4.5E – 04	7.13	6.8E – 04	6.72	9.1E – 04	6.43
2.3E – 04	7.80	4.6E – 04	7.11	6.9E – 04	6.70	9.2E – 04	6.41
2.4E – 04	7.76	4.7E – 04	7.09	7.0E – 04	6.69	9.3E – 04	6.40
2.5E – 04	7.72	4.8E – 04	7.06	7.1E – 04	6.67	9.4E – 04	6.39
2.6E – 04	7.68	4.9E – 04	7.04	7.2E – 04	6.66	9.5E – 04	6.38
2.7E – 04	7.64	5.0E – 04	7.02	7.3E – 04	6.65	9.6E – 04	6.37
2.8E – 04	7.60	5.1E – 04	7.00	7.4E – 04	6.63	9.7E – 04	6.36
2.9E – 04	7.57	5.2E – 04	6.98	7.5E – 04	6.62	9.8E – 04	6.35
3.0E – 04	7.53	5.3E – 04	6.97	7.6E – 04	6.61	9.9E – 04	6.34
3.1E – 04	7.50	5.4E – 04	6.95	7.7E – 04	6.59		
3.2E – 04	7.47	5.5E – 04	6.93	7.8E – 04	6.58		

10^{-3} well functions							
u	$W(u)$	u	$W(u)$	u	$W(u)$	u	$W(u)$
1.0E – 03	6.33	3.3E – 03	5.14	5.6E – 03	4.61	7.9E – 03	4.27
1.1E – 03	6.24	3.4E – 03	5.11	5.7E – 03	4.60	8.0E – 03	4.26
1.2E – 03	6.15	3.5E – 03	5.08	5.8E – 03	4.58	8.1E – 03	4.25
1.3E – 03	6.07	3.6E – 03	5.05	5.9E – 03	4.56	8.2E – 03	4.23
1.4E – 03	6.00	3.7E – 03	5.03	6.0E – 03	4.54	8.3E – 03	4.22
1.5E – 03	5.93	3.8E – 03	5.00	6.1E – 03	4.53	8.4E – 03	4.21
1.6E – 03	5.86	3.9E – 03	4.97	6.2E – 03	4.51	8.5E – 03	4.20
1.7E – 03	5.80	4.0E – 03	4.95	6.3E – 03	4.50	8.6E – 03	4.19
1.8E – 03	5.74	4.1E – 03	4.92	6.4E – 03	4.48	8.7E – 03	4.18
1.9E – 03	5.69	4.2E – 03	4.90	6.5E – 03	4.47	8.8E – 03	4.16
2.0E – 03	5.64	4.3E – 03	4.88	6.6E – 03	4.45	8.9E – 03	4.15
2.1E – 03	5.59	4.4E – 03	4.85	6.7E – 03	4.44	9.0E – 03	4.14
2.2E – 03	5.54	4.5E – 03	4.83	6.8E – 03	4.42	9.1E – 03	4.13
2.3E – 03	5.50	4.6E – 03	4.81	6.9E – 03	4.41	9.2E – 03	4.12
2.4E – 03	5.46	4.7E – 03	4.79	7.0E – 03	4.39	9.3E – 03	4.11
2.5E – 03	5.42	4.8E – 03	4.77	7.1E – 03	4.38	9.4E – 03	4.10
2.6E – 03	5.38	4.9E – 03	4.75	7.2E – 03	4.36	9.5E – 03	4.09
2.7E – 03	5.34	5.0E – 03	4.73	7.3E – 03	4.35	9.6E – 03	4.08
2.8E – 03	5.30	5.1E – 03	4.71	7.4E – 03	4.34	9.7E – 03	4.07
2.9E – 03	5.27	5.2E – 03	4.69	7.5E – 03	4.32	9.8E – 03	4.06
3.0E – 03	5.23	5.3E – 03	4.67	7.6E – 03	4.31	9.9E – 03	4.05
3.1E – 03	5.20	5.4E – 03	4.65	7.7E – 03	4.30		
3.2E – 03	5.17	5.5E – 03	4.63	7.8E – 03	4.28		

10^{-2} well functions							
u	$W(u)$	u	$W(u)$	u	$W(u)$	u	$W(u)$
1.0E – 02	4.04	3.3E – 02	2.87	5.6E – 02	2.36	7.9E – 02	2.04
1.1E – 02	3.94	3.4E – 02	2.84	5.7E – 02	2.34	8.0E – 02	2.03
1.2E – 02	3.86	3.5E – 02	2.81	5.8E – 02	2.33	8.1E – 02	2.02
1.3E – 02	3.78	3.6E – 02	2.78	5.9E – 02	2.31	8.2E – 02	2.00
1.4E – 02	3.71	3.7E – 02	2.76	6.0E – 02	2.30	8.3E – 02	1.993
1.5E – 02	3.64	3.8E – 02	2.73	6.1E – 02	2.28	8.4E – 02	1.982
1.6E – 02	3.57	3.9E – 02	2.71	6.2E – 02	2.26	8.5E – 02	1.971
1.7E – 02	3.51	4.0E – 02	2.68	6.3E – 02	2.25	8.6E – 02	1.960
1.8E – 02	3.46	4.1E – 02	2.66	6.4E – 02	2.23	8.7E – 02	1.950
1.9E – 02	3.41	4.2E – 02	2.63	6.5E – 02	2.22	8.8E – 02	1.939
2.0E – 02	3.35	4.3E – 02	2.61	6.6E – 02	2.21	8.9E – 02	1.929
2.1E – 02	3.31	4.4E – 02	2.59	6.7E – 02	2.19	9.0E – 02	1.919
2.2E – 02	3.26	4.5E – 02	2.57	6.8E – 02	2.18	9.1E – 02	1.909
2.3E – 02	3.22	4.6E – 02	2.55	6.9E – 02	2.16	9.2E – 02	1.899
2.4E – 02	3.18	4.7E – 02	2.53	7.0E – 02	2.15	9.3E – 02	1.889
2.5E – 02	3.14	4.8E – 02	2.51	7.1E – 02	2.14	9.4E – 02	1.879
2.6E – 02	3.10	4.9E – 02	2.49	7.2E – 02	2.12	9.5E – 02	1.869
2.7E – 02	3.06	5.0E – 02	2.47	7.3E – 02	2.11	9.6E – 02	1.860
2.8E – 02	3.03	5.1E – 02	2.45	7.4E – 02	2.10	9.7E – 02	1.851
2.9E – 02	2.99	5.2E – 02	2.43	7.5E – 02	2.09	9.8E – 02	1.841
3.0E – 02	2.96	5.3E – 02	2.41	7.6E – 02	2.07	9.9E – 02	1.832
3.1E – 02	2.93	5.4E – 02	2.39	7.7E – 02	2.06		
3.2E – 02	2.90	5.5E – 02	2.38	7.8E – 02	2.05		

10^{-1} well functions							
u	$W(u)$	u	$W(u)$	u	$W(u)$	u	$W(u)$
1.0E – 01	1.823	3.3E – 01	0.836	5.6E – 01	0.493	7.9E – 01	0.316
1.1E – 01	1.737	3.4E – 01	0.815	5.7E – 01	0.483	8.0E – 01	0.311
1.2E – 01	1.660	3.5E – 01	0.794	5.8E – 01	0.473	8.1E – 01	0.305
1.3E – 01	1.589	3.6E – 01	0.774	5.9E – 01	0.464	8.2E – 01	0.300
1.4E – 01	1.524	3.7E – 01	0.755	6.0E – 01	0.454	8.3E – 01	0.294
1.5E – 01	1.464	3.8E – 01	0.737	6.1E – 01	0.445	8.4E – 01	0.289
1.6E – 01	1.409	3.9E – 01	0.719	6.2E – 01	0.437	8.5E – 01	0.284
1.7E – 01	1.358	4.0E – 01	0.702	6.3E – 01	0.428	8.6E – 01	0.279
1.8E – 01	1.310	4.1E – 01	0.686	6.4E – 01	0.420	8.7E – 01	0.274
1.9E – 01	1.265	4.2E – 01	0.670	6.5E – 01	0.412	8.8E – 01	0.265
2.0E – 01	1.223	4.3E – 01	0.655	6.6E – 01	0.404	8.9E – 01	0.265
2.1E – 01	1.183	4.4E – 01	0.640	6.7E – 01	0.396	9.0E – 01	0.260
2.2E – 01	1.145	4.5E – 01	0.625	6.8E – 01	0.388	9.1E – 01	0.256
2.3E – 01	1.110	4.6E – 01	0.611	6.9E – 01	0.381	9.2E – 01	0.251
2.4E – 01	1.076	4.7E – 01	0.598	7.0E – 01	0.374	9.3E – 01	0.247
2.5E – 01	1.044	4.8E – 01	0.585	7.1E – 01	0.367	9.4E – 01	0.243
2.6E – 01	1.014	4.9E – 01	0.572	7.2E – 01	0.360	9.5E – 01	0.239
2.7E – 01	0.985	5.0E – 01	0.560	7.3E – 01	0.353	9.6E – 01	0.235
2.8E – 01	0.957	5.1E – 01	0.548	7.4E – 01	0.347	9.7E – 01	0.231
2.9E – 01	0.931	5.2E – 01	0.536	7.5E – 01	0.340	9.8E – 01	0.227
3.0E – 01	0.906	5.3E – 01	0.525	7.6E – 01	0.334	9.9E – 01	0.223
3.1E – 01	0.882	5.4E – 01	0.514	7.7E – 01	0.328		
3.2E – 01	0.858	5.5E – 01	0.503	7.8E – 01	0.322		

0 well functions							
u	$W(u)$	u	$W(u)$	u	$W(u)$	u	$W(u)$
1.0E + 00	0.219	2.1E + 00	0.043	3.2E + 00	0.010	4.3E + 00	0.003
1.1E + 00	0.186	2.2E + 00	0.037	3.3E + 00	0.009	4.4E + 00	0.002
1.2E + 00	0.158	2.3E + 00	0.033	3.4E + 00	0.008	4.5E + 00	0.002
1.3E + 00	0.135	2.4E + 00	0.028	3.5E + 00	0.007	4.6E + 00	0.002
1.4E + 00	0.116	2.5E + 00	0.025	3.6E + 00	0.006	4.7E + 00	0.002
1.5E + 00	0.100	2.6E + 00	0.022	3.7E + 00	0.005	4.8E + 00	0.001
1.6E + 00	0.086	2.7E + 00	0.019	3.8E + 00	0.005	4.9E + 00	0.001
1.7E + 00	0.075	2.8E + 00	0.017	3.9E + 00	0.004	5.0E + 00	0.001
1.8E + 00	0.065	2.9E + 00	0.015	4.0E + 00	0.004		
1.9E + 00	0.056	3.0E + 00	0.013	4.1E + 00	0.003		
2.0E + 00	0.049	3.1E + 00	0.011	4.2E + 00	0.003		

Source: Illinois Environmental Protection Agency (1990).

APPENDIX C

SOLUBILITY PRODUCT CONSTANTS FOR SOLUTION AT OR NEAR ROOM TEMPERATURE

Substance	Formula	$K_{sp}^{\dagger}$
Aluminum hydroxide	$Al(OH)_3$	2×10^{-32}
Barium arsenate	$Ba_3(AsO_4)_2$	7.7×10^{-51}
Barium carbonate	$BaCO_3$	8.1×10^{-9}
Barium chromate	$BaCrO_4$	2.4×10^{-10}
Barium fluoride	BaF_2	1.7×10^{-6}
Barium iodate	$Ba(IO_3)_2 \cdot 2H_2O$	1.5×10^{-9}
Barium oxalate	$BaC_2O_4 \cdot H_2O$	2.3×10^{-8}
Barium sulfate	$BaSO_4$	1.08×10^{-10}
Beryllium hydroxide	$Be(OH)_2$	7×10^{-22}
Bismuth iodide	BiI_3	8.1×10^{-19}
Bismuth phosphate	$BiPO_4$	1.3×10^{-23}
Bismuth sulfide	Bi_2S_3	1×10^{-97}
Cadmium arsenate	$Cd_3(AsO_4)_2$	2.2×10^{-33}
Cadmium hydroxide	$Cd(OH)_2$	5.9×10^{-15}
Cadmium oxalate	$CdC_2O_4 \cdot 3H_2O$	1.5×10^{-8}
Cadmium sulfide	CdS	7.8×10^{-27}
Calcium arsenate	$Ca_3(AsO_4)_2$	6.8×10^{-19}
Calcium carbonate	$CaCO_3$	8.7×10^{-9}
Calcium fluoride	CaF_2	4.0×10^{-11}
Calcium hydroxide	$Ca(OH)_2$	5.5×10^{-6}
Calcium iodate	$Ca(IO_3)_2 \cdot 6H_2O$	6.4×10^{-7}
Calcium oxalate	$CaC_2O_4 \cdot H_2O$	2.6×10^{-9}
Calcium phosphate	$Ca_3(PO_4)_2$	2.0×10^{-29}
Calcium sulfate	$CaSO_4$	1.9×10^{-4}
Cerium(III) hydroxide	$Ce(OH)_3$	2×10^{-20}
Cerium(III) iodate	$Ce(IO_3)_3$	3.2×10^{-10}
Cerium(III) oxalate	$Ce_2(C_2O_4)_3 \cdot 9H_2O$	3×10^{-29}
Chromium(II) hydroxide	$Cr(OH)_2$	1.0×10^{-17}
Chromium(III) hydroxide	$Cr(OH)_3$	6×10^{-31}
Cobalt(II) hydroxide	$Co(OH)_2$	2×10^{-16}
Cobalt(III) hydroxide	$Co(OH)_3$	1×10^{-43}
Copper(II) arsenate	$Cu_3(AsO_4)_2$	7.6×10^{-76}
Copper(I) bromide	$CuBr$	5.2×10^{-9}
Copper(I) chloride	$CuCl$	1.2×10^{-6}
Copper(I) iodide	CuI	5.1×10^{-12}
Copper(II) iodate	$Cu(IO_3)_2$	7.4×10^{-8}
Copper(I) sulfide	Cu_2S	2×10^{-47}
Copper(II) sulfide	CuS	9×10^{-36}
Copper(I) thiocyanate	$CuSCN$	4.8×10^{-15}

Substance	Formula	$K_{sp}^{\dagger}$
Iron(III) arsenate	$FeAsO_4$	5.7×10^{-21}
Iron(II) carbonate	$FeCO_3$	3.5×10^{-11}
Iron(II) hydroxide	$Fe(OH)_2$	8×10^{-16}
Iron(III) hydroxide	$Fe(OH)_3$	4×10^{-38}
Lead arsenate	$Pb_3(AsO_4)_2$	4.1×10^{-36}
Lead bromide	$PbBr_2$	3.9×10^{-5}
Lead carbonate	$PbCO_3$	3.3×10^{-14}
Lead chloride	$PbCl_2$	1.6×10^{-5}
Lead chromate	$PbCrO_4$	1.8×10^{-14}
Lead fluoride	PbF_2	3.7×10^{-8}
Lead iodate	$Pb(IO_3)_2$	2.6×10^{-13}
Lead iodide	PbI_2	7.1×10^{-9}
Lead oxalate	PbC_2O_4	4.8×10^{-10}
Lead sulfate	$PbSO_4$	1.6×10^{-8}
Lead sulfide	PbS	8×10^{-28}
Magnesium ammonium phosphate	$MgNH_4PO_4$	2.5×10^{-13}
Magnesium arsenate	$Mg_3(AsO_4)_2$	2.1×10^{-20}
Magnesium carbonate	$MgCO_3 \cdot 3H_2O$	1×10^{-5}
Magnesium fluoride	MgF_2	6.5×10^{-9}
Magnesium hydroxide	$Mg(OH)_2$	1.2×10^{-11}
Magnesium oxalate	$MgC_2O_4 \cdot 2H_2O$	1×10^{-8}
Manganese(II) hydroxide	$Mn(OH)_2$	1.9×10^{-13}
Mercury(I) bromide	Hg_2Br_2	5.8×10^{-23}
Mercury(I) chloride	Hg_2Cl_2	1.3×10^{-18}
Mercury(I) iodide	Hg_2I_2	4.5×10^{-29}
Mercury(I) sulfate	Hg_2SO_4	7.4×10^{-7}
Mercury(II) sulfide	HgS	4×10^{-53}
Mercury(I) thiocyanate	$Hg_2(SCN)_2$	3.0×10^{-20}
Nickel arsenate	$Ni_3(AsO_4)_2$	3.1×10^{-26}
Nickel carbonate	$NiCO_3$	6.6×10^{-9}
Nickel hydroxide	$Ni(OH)_2$	6.5×10^{-18}
Nickel sulfide	NiS	3×10^{-19}
Silver arsenate	Ag_3AsO_4	1×10^{-22}
Silver bromate	$AgBrO_3$	5.77×10^{-5}
Silver bromide	$AgBr$	5.25×10^{-13}
Silver carbonate	Ag_2CO_3	8.1×10^{-12}
Silver chloride	$AgCl$	1.78×10^{-10}
Silver chromate	Ag_2CrO_4	2.45×10^{-12}
Silver cyanide	$Ag[Ag(CN)_2]$	5.0×10^{-12}
Silver iodate	$AgIO_3$	3.02×10^{-8}
Silver iodide	AgI	8.31×10^{-17}
Silver oxalate	$Ag_2C_2O_4$	3.5×10^{-11}
Silver oxide	Ag_2O	2.6×10^{-8}
Silver phosphate	Ag_3PO_4	1.3×10^{-20}
Silver sulfate	Ag_2SO_4	1.6×10^{-5}
Silver sulfide	Ag_2S	2×10^{-49}
Silver thiocyanate	$AgSCN$	1.00×10^{-12}
Strontium carbonate	$SrCO_3$	1.1×10^{-10}
Strontium chromate	$SrCrO_4$	3.6×10^{-5}
Strontium fluoride	SrF_2	2.8×10^{-9}
Strontium iodate	$Sr(IO_3)_2$	3.3×10^{-7}
Strontium oxalate	$SrC_2O_4 \cdot H_2O$	1.6×10^{-7}
Strontium sulfate	$SrSO_4$	3.8×10^{-7}

Substance	Formula	$K_{sp}^{\dagger}$
Thallium(I) bromate	$TlBrO_3$	8.5×10^{-5}
Thallium(I) bromide	$TlBr$	3.4×10^{-6}
Thallium(I) chloride	$TlCl$	1.7×10^{-4}
Thallium(I) chromate	Tl_2CrO_4	9.8×10^{-13}
Thallium(I) iodate	$TlIO_3$	3.1×10^{-6}
Thallium(I) iodide	TlI	6.5×10^{-8}
Thallium(I) sulfide	Tl_2S	5×10^{-21}
Tin(II) sulfide	SnS	1×10^{-25}
Titanium(III) hydroxide	$Ti(OH)_3$	1×10^{-40}
Zinc arsenate	$Zn_3(AsO_4)_2$	1.3×10^{-28}
Zinc carbonate	$ZnCO_3$	1.4×10^{-11}
Zinc ferrocyanide	$Zn_2Fe(CN)_6$	4.1×10^{-16}
Zinc hydroxide	$Zn(OH)_2$	1.2×10^{-17}
Zinc oxalate	$ZnC_2O_4 \cdot 2H_2O$	2.8×10^{-8}
Zinc phosphate	$Zn_3(PO_4)_2$	9.1×10^{-33}
Zinc sulfide	ZnS	1×10^{-21}

†The solubility of many metals is altered by carbonate complexation. Solubility predictions without consideration for complexation can be highly inaccurate.

Source: Benefield, L. D. and Morgan J. S. 1990. Chemical precipitation. In: AWWA, *Water Quality and Treatment.* New York: McGraw-Hill. Reprinted with permission of the McGraw-Hill Co.

APPENDIX D

FREUNDLICH ADSORPTION ISOTHERM CONSTANTS FOR TOXIC ORGANIC COMPOUNDS

Compound	$K(\text{mg/g})(\text{L/mg})^{1/n}$	$1/n$
PCB	14,100	1.03
Bis(2-ethylhexyl phthalate)	11,300	1.5
Heptachlor	9,320	0.92
Heptachlor epoxide	2,120	0.75
Butylbenzyl phthalate	1,520	1.26
Toxaphene	950	0.74
Endosulfan sulfate	686	0.81
Endrin	666	0.80
Fluoranthene	664	0.61
Aldrin	651	0.92
PCB-1232	630	0.73
β-Endosulfan	615	0.83
Dieldrin	606	0.51
Alachlor	479	0.26
Hexachlorobenzene	450	0.60
Pentachlorophenol	436	0.34
Anthracene	376	0.70
4-Nitrobiphenyl	370	0.27
Fluorene	330	0.28
Styrene	327	0.48
DDT	322	0.50
2-Acetylaminofluorene	318	0.12
α-BHC	303	0.43
Anethole	300	0.42
3,3-Dichlorobenzidine	300	0.20
γ-BHC (lindane)	285	0.43
2-Chloronaphthalene	280	0.46
Phenylmercuric acetate	270	0.44
Carbofuran	266	0.41
1,2-Dichlorobenzene	263	0.38
Hexachlorobutadiene	258	0.45
p-Nonylphenol	250	0.37
4-Dimethylaminoazobenzene	249	0.24
PCB-1221	242	0.70
DDE	232	0.37
m-Xylene	230	0.75
Acridine yellow	230	0.12
Dibromochloropropane (DBCP)	224	0.51
Benzidine dihydrochloride	220	0.37
β-BHC	220	0.49
n-Butylphthalate	220	0.45
n-Nitrosodiphenylamine	220	0.37
Silvex	215	0.38

Compound	$K(\text{mg/g})(\text{L/mg})^{1/n}$	$1/n$
Phenanthrene	215	0.44
Dimethylphenylcarbinol	210	0.34
4-Aminobiphenyl	200	0.26
β-Naphthol	200	0.26
p-Xylene	200	0.42
α-Endosulfan	194	0.50
Chlordane	190	0.33
Acenaphthene	190	0.36
4,4′-Methylene-bis-(2-chloroaniline)	190	0.64
Benzo[*k*]fluoranthene	181	0.57
Acridine orange	180	0.29
α-Naphthol	180	0.32
Ethylbenzene	175	0.53
o-Xylene	174	0.47
4,6-Dinitro-*o*-cresol	169	0.27
α-Naphthylamine	160	0.34
2,4-Dichlorophenol	157	0.15
1,2,4-Trichlorobenzene	157	0.31
2,4,6-Trichlorophenol	155	0.40
β-Naphthylamine	150	0.30
2,4-Dinitrotoluene	146	0.31
2,6-Dinitrotoluene	145	0.32
4-Bromophenyl phenyl ether	144	0.68
p-Nitroaniline	140	0.27
1,1-Diphenylhydrazine	135	0.16
Naphthalene	132	0.42
Aldicarb	132	0.40
1-Chloro-2-nitrobenzene	130	0.46
p-Chlorometacresol	124	0.16
1,4-Dichlorobenzene	121	0.47
Benzothiazole	120	0.27
Diphenylamine	120	0.31
Guanine	120	0.40
1,3-Dichlorobenzene	118	0.45
Acenaphthylene	115	0.37
Methoxychlor	115	0.36
4-Chlorophenyl phenyl ether	111	0.26
Diethyl phthalate	110	0.27
Chlorobenzene	100	0.35
Toluene	100	0.45
2-Nitrophenol	99	0.34
Dimethyl phthalate	97	0.41
Hexachloroethane	97	0.38
2,4-Dimethylphenol	78	0.44
4-Nitrophenol	76	0.25
Acetophenone	74	0.44
1,2,3,4-Tetrahydronaphthalene	74	0.81
Adenine	71	0.38
Dibenzo[*a,h*]anthracene	69	0.75
Nitrobenzene	68	0.43
2,4-D	67	0.27
3,4-Benzofluoranthene	57	0.37
2-Chlorophenol	51	0.41
Tetrachloroethylene	51	0.56
o-Anisidine	50	0.34

Compound	$K(\text{mg/g})(\text{L/mg})^{1/n}$	$1/n$
5-Bromouracil	44	0.47
Benzo[*a*]pyrene	34	0.44
2,4-Dinitrophenol	33	0.61
Isophorone	32	0.39
Trichloroethylene	28	0.62
Thymine	27	0.51
5-Chlorouracil	25	0.58
N-Nitrosodi-*n*-propylamine	24	0.26
Bis(2-Chloroisopropyl) ether	24	0.57
1,2-Dibromoethene (EBD)	22	0.46
Phenol	21	0.54
Bromoform	20	0.52
1,2-Dichloropropane	19	0.59
1,2-*trans*-Dichloroethylene	14	0.45
cis-1,2-Dichloroethylene	12	0.59
Carbon tetrachloride	11	0.83
Bis(2-Chloroethyoxy) methane	11	0.65
Uracil	11	0.63
Benzo[*g*,*h*,*i*]perylene	11	0.37
1,1,2,2-Tetrachloroethane	11	0.37
1,2-Dichloropropene	8.2	0.46
Dichlorobromomethane	7.9	0.61
Cyclohexanone	6.2	0.75
1,1,2-Trichloroethane	5.8	0.60
Trichlorofluoromethane	5.6	0.24
5-Fluorouracil	5.5	1.0
1,1-Dichloroethylene	4.9	0.54
Dibromochloromethane	4.8	0.34
2-Chloroethyl vinyl ether	3.9	0.80
1,2-Dichloroethane	3.6	0.83
Chloroform	2.6	0.73
1,1,1-Trichloroethane	2.5	0.34
1,1-Dichloroethane	1.8	0.53
Acrylonitrile	1.4	0.51
Methylene chloride	1.3	1.16
Acrolein	1.2	0.65
Cytosine	1.1	1.6
Benzene	1.0	1.6
Ethylenediaminetetraacetic acid	0.86	1.5
Benzoic acid	0.76	1.8
Chloroethane	0.59	0.95
N-Dimethylnitrosamine	6.8×10^{-5}	6.6

[†]The isotherms are for the compounds in distilled water, with different activated carbons. The values of K and $1/n$ should be used only as rough estimates of the values that will be obtained using other types of water and other activated carbon.

Source: Snoeyink, V. L. 1990. Adsorption of organic compounds. In: AWWA, *Water Quality and Treatment*. New York: McGraw-Hill. Reprinted with permission of the McGraw-Hill Co.

P • A • R • T • 2

SOLID WASTE CALCULATIONS

CHAPTER 2.1
THERMODYNAMICS USED IN ENVIRONMENTAL ENGINEERING

C. C. Lee and G. L. Huffman
U.S. Environmental Protection Agency, Cincinnati, Ohio

1 INTRODUCTION

Thermodynamics is a science in which energy transformations are studied as well as their relationships to the changes in the chemical properties of a system. It is the fundamental basis of many engineering fields. The profession of environmental engineering is no exception. In particular, the design and operation of many types of pollution control equipment must be based on the principles of thermodynamics. The purpose of this chapter is to provide information on how thermodynamics is applied to selected waste treatment processes. The authors use many calculational examples to illustrate the application of the thermodynamic concept to incineration. Some of the examples provided herein were originated by the authors, and some of them were edited from or quoted from previously published EPA examples.

2 THERMODYNAMIC TERMS AND CALCULATIONS

This section provides key thermodynamic terms, their definitions, and their related calculations as they apply to the environmental field. For the convenience of information search, the terms are provided in alphabetical order.

2.1 Air

Air is the natural source of oxygen for incineration and is considered as an ideal gas in many combustion and incineration calculations. Some air properties are as follows:

Molar value (volumetric value) of dry air

- Oxygen: 21 percent of air or 0.21 mole, or 0.21 lb-mole, when air is assumed at 1 mole.
- Nitrogen: 79 percent of air or 0.79 mole, or 0.79 lb-mole, when air is assumed at 1 mole. Thus,
- 0.21 moles of $O_2 + 0.79$ moles of $N_2 = 1$ mole of air, or
- 1 mole of $O_2 + 3.76$ moles of $N_2 = 4.76$ moles of air.

This equation shows that for each volume (mole) of oxygen, 3.76 volumes (moles) of nitrogen or 4.76 volumes (moles) of air are involved.

Gravimetric value of dry air

- Air molecular weight $= 0.21(32) + 0.79(28) = 28.84$ lb/lb-mole
- Oxygen mass fraction $= 0.21(32/28.84) = 23\%$
- Nitrogen mass fraction $= 0.79(28/28.84) = 77\%$
- Thus, one pound of air can be expressed by:
- 0.23 lb $O_2 + 0.77$ lb $N_2 = 1$ lb air, or
- 1 lb $O_2 + 3.35$ lb $N_2 = 4.35$ lb air

This equation shows that for each 1 lb of oxygen, 3.35 lbs of nitrogen or 4.35 lbs of air are involved.

Air standard volume. Any ideal gas at standard conditions has the same volume. Air is considered as an ideal gas. At standard conditions, the standard volume of air is about

387 scf (standard cubic feet)/lb-mole or about 24 scm (standard cubic meters)/kg-mole. These values will be derived in the two sections that follow.

Air standard volume in British units. Standard conditions in the British unit system are at pressure = 1 atm and temperature = 70°F.

Applying the ideal gas equation, $PV = nR_u T$.

Air standard volume:

$$\begin{aligned} V &= nR_u T/P \\ &= 1(1545)(460 + 70)/[14.6959(144)] \\ &= 386.94 \text{ scf/lb-mole} \end{aligned}$$

Air standard volume in metric units. Standard conditions in the metric unit system are at pressure = 1 atm and temperature = 20°C.

Air standard volume:

$$\begin{aligned} V &= nR_u T/P \\ &= 1(0.0821)(273 + 20)/1 \\ &= 24.04 \text{ scm/kg-mole} \end{aligned}$$

where P = pressure
V = volume
n = number of moles
R_u = universal gas constant
scf = standard cubic feet
scm = standard cubic meters

Air density. Air density (ρ) is defined as the ratio of air mass to air volume. At standard conditions, pressure = 1 atm, and temperature = 70°F

Air density (ρ) = (air molecular weight)/(standard volume)

$$\begin{aligned} D &= 28.84/386.94 \\ &= 0.0745 \text{ lb/scf} \end{aligned}$$

2.2 Air–Fuel Ratio

The air–fuel ratio is defined as follows:

(1) Air–fuel ratio by mole

$$AF_m = (\text{air/fuel}) = n(1 + 3.76)/n_f \tag{1.1}$$

(2) Air–fuel ratio by weight

$$AF_w = (\text{air/fuel}) = n(1 + 3.76)29/(n_f \times M_f) \tag{1.2}$$

where n = moles of oxygen
n_f = moles of fuel
M_f = molecular weight of fuel
29 = molecular weight of air (actually 28.84)

EXAMPLE: air–fuel ratio
Assume that the fuel octane (C_8H_{18}) is under theoretical air combustion. Determine the theoretical air–fuel ratio.

Solution:

1. Write the octane combustion equation.

$$C_8H_{18} + 12.5(O_2 + 3.76N_2) \rightarrow 8CO_2 + 9H_2O + 47.0N_2$$

2. Calculate the theoretical air–fuel ratio on a mole basis (AF_m) by applying Eq. (1.1).

$$AF_m = 12.5(1 + 3.76)/1 = 59.5 \text{ mole-air/mole-fuel}$$

3. Calculate the theoretical air–fuel ratio on a mass basis (AF_w) by applying Eq. (1.2).

$$AF_w = 59.5(28.84)/114.0 = 15 \text{ lb air/lb fuel}$$

2.3 Boyle's Law

Boyle's law states that when the temperature is held constant, the volume of a given ideal gas is inversely proportional to the pressure. The relationship of the volume and the pressure of the ideal gas can be expressed as

$$V_2/V_1 = P_1/P_2$$

where P_1 = initial pressure
P_2 = final pressure
V_1 = initial volume
V_2 = final volume

2.4 Charles' Law

Charles' law is also known as the Charles–Gay Lussac law. It states that when the pressure is held constant, the volume of a given mass of an ideal gas is directly proportional to the absolute temperature. The relationship of the volume and the absolute temperature of the ideal gas can be expressed as

$$V_2/V_1 = T_2/T_1$$

where V_1 = initial volume
V_2 = final volume
T_1 = initial absolute temperature
T_2 = final absolute temperature

References used in this section: EPA-80/02, p. 2-6; EPA-86/03, p. 13.

EXAMPLE: actual flow rate
Assume that:

(1) The volumetric flow rate within an incinerator is 5000 scfm (standard cubic feet/minute) (70°F, 1 atm);
(2) The operating temperature and pressure of the unit are 1200°F and 1 atm, respectively.

Determine the actual flow rate in acfm (actual cubic feet/minute).

Solution:
Because the pressure remains constant, calculate the acfm using the Charles–Gay Lussac law.

$$V_2/V_1 = T_2/T_1$$

or

$$\begin{aligned} V_2 &= V_1(T_2/T_1) \\ &= 5000(460 + 1200)/(460 + 70) \\ &= 15{,}660 \text{ acfm} \end{aligned}$$

2.5 Chemical Equilibrium

The following is a qualitative discussion of the chemical reaction equilibrium and kinetics in an attempt to clarify the roles which concentrations and temperature play in combustion processes. Chemical reactions are considered to be reversible to some extent. How far a reaction proceeds depends on the relative rates of the forward and reverse reactions. Consider a reaction where reactants A and B form products C and D:

$$\mathrm{A} + \mathrm{B} \leftrightarrow \mathrm{C} + \mathrm{D} \tag{1.3}$$

From the law of mass action, the rates of reactions are proportional to the concentrations of the reactants. Hence, the forward rate r_f is

$$r_\mathrm{f} = k_\mathrm{f}[\mathrm{A}][\mathrm{B}] \tag{1.4}$$

and the reverse rate is

$$r_\mathrm{r} = k_\mathrm{r}[\mathrm{C}][\mathrm{D}] \tag{1.5}$$

where the k's represents the reaction velocity constants, and the square brackets the concentrations of the respective species.

At equilibrium, the forward and reverse rates are necessarily equal. Thus

$$k_\mathrm{f}[\mathrm{A}][\mathrm{B}] = k_\mathrm{r}[\mathrm{C}][\mathrm{D}] \text{ or}$$

$$k_\mathrm{f}/k_\mathrm{r} = ([\mathrm{C}][\mathrm{D}])/([\mathrm{A}[\mathrm{B}]) \tag{1.6}$$

It is now convenient to define an equilibrium constant K:

$$K = k_\mathrm{f}/k_\mathrm{r} = ([\mathrm{C}][\mathrm{D}])/([\mathrm{A}][\mathrm{B}]) \tag{1.7}$$

The equilibrium constant K is a function of temperature because of the temperature effect on the reaction velocity constants k_f and k_r. Note that if it were desired to reduce the concentration of one of the reactants, say reactant A for example, this could be accomplished by increasing the concentration of B. This is exactly the rationale for using excess air to assure complete combustion of the fuel.

It is common knowledge that some reactions proceed faster than others. The reaction rates depend on the chemical bonding in the materials. Enough energy must be supplied to break the chemical bonds in the fuel and in the molecular oxygen before new bonds can be formed. It is convenient to think of this energy as elevating the reactants to a new higher energy state, called the transition state, where an activated but unstable complex is formed from the reactants.

References used in this section: EPA-80/02, p. 2-10.

EXAMPLE: chemical equilibrium
Carbon monoxide and oxygen are in equilibrium at 1 atm, expressed as: $CO + 0.5O_2 \rightarrow CO_2$. The partial pressure for oxygen is 0.004 atm and for carbon monoxide is 0.008 atm. The chemical equilibrium constant is assumed to be 0.122. Determine:

(1) partial presssure of CO_2;
(2) ppm of CO_2.

Given conditions: the data in question are summarized below.

- Total pressure, $p = 1$ atm
- Partial pressure of oxygen, $p(O_2) = 0.004$ atm
- Partial pressure of CO, $p(CO) = 0.008$ atm
- Equilibrium constant, $K = 0.122$

Solution:
Solution for question (1)

1.1. Write the chemical equilibrium constant equation.

$$K = p(CO_2)/(p(CO) \times p(O_2)^{0.5})$$

$$p(CO_2) = K \times p(CO) \times p(O_2)^{0.5}$$

1.2. Calculate the partial pressure of CO_2.

$$\begin{aligned} p(CO_2) &= 0.122(0.008)(0.004)^{0.5} \\ &= 0.122(0.008)(0.063) \\ &= 0.0000615 \text{ atm} \\ &= 6.15 \times 10^{-5} \text{ atm} \end{aligned}$$

Solution for question (2)
The result in step 1.2 can be re-written as

$$p(CO_2) = 61.5 \times 10^{-6}, \text{ therefore}$$

$$\text{ppm of } CO_2 = 61.5 \text{ ppm}$$

2.6 Chemical Kinetics

In general, many chemical reactions can be satisfactorily described by the first-order reaction equation

$$dW/dt = -kW$$

where W = concentration of the material undergoing reaction
k = reaction rate constant = $A \exp(-E/R_u T)$ (Arrhenius equation)
t = time
A = frequency factor, units of reciprocal time or s^{-1}
E = activation energy, cal/g-mole or Btu/lb-mole

Some A and E values can be obtained from a study by the University of Dayton Research Institute (UDRI) for the US EPA and are provided in Attachment 1 of this chapter.

Integration of this equation results in

$$\ln(W_0/W_i) = -(t_0 - t_i)k \tag{1.8}$$

$$\ln(W_0/W_i) = -(t_0 - t_i)(A)\exp(-E/R_u T) \tag{1.9}$$

For a 99.99% DRE (destruction and removal efficiency),

$$\text{DRE} = (1 - W_0/W_i) = 0.9999$$

$$W_0/W_i = 0.0001 = 10^{-4} = 1/10{,}000$$

$$\ln(10^{-4}) = -(t_0 - t_i)(A)\exp(-E/R_u T)$$

$$-9.21 = -(t_0 - t_i)(A)\exp(-E/R_u T) \tag{1.10}$$

where W_0 and W_i are the concentrations of the compound at time $t = t_0$ and $t = t_i$, respectively
T = absolute temperature in K or °R
R_u = 1.987 cal/g-mole-K or Btu/lb-mole-°R

EXAMPLE 1: chemical kinetics
It is assumed that the combustion rate of chlorobenzene (C_6H_5Cl) is governed by the first-order reaction equation. Determine the minimum temperature needed to thermally destroy chlorobenzene to a DRE of 99.99%, i.e., to "4-9's."

Given conditions: the data in question are summarized below [see Attachment 1 and (UDRI-90)].

- Universal gas constant, $R_u = 1.987$ cal/g-mole-K
- Frequency factor, $A = 1.34 \times 10^{17}$ (1/s)
- Activation energy, $E = 76600$ cal/g-mole

Solution:
Realize that the incineration of hazardous waste involves many intermediate reaction processes. From a chemical kinetic point of view, the basic equation that governs the destruction of a chemical compound (assuming a first-order chemical reaction mechanism) is as follows:

1. Rewrite Eq. (1.10)

$$-9.21 = -(t_0 - t_i)(A)\exp(-E/R_u T)$$

$$\ln[(t_0 - t_i)(A)/9.21] = E/R_u T$$

$$T = (E)/[(R_u)\ln(0.109(A)(t_0 - t_i))]$$

$$T = 0.503(E)/[\ln(0.109(A)(t_0 - t_i))]$$

2. Calculate reaction temperature.
Assume that the residence is 2 s (i.e., $t_0 - t_i = 2$). The destruction temperature for chlorobenzene can be roughly estimated by the above-derived equation.

$$
\begin{aligned}
T &= 0.503(E)/[\ln(0.109(A)(t_0 - t_i))] \\
&= 0.503(76600)/[\ln(0.109)(1.34 \times 10^{17})(2))] \\
&= 38{,}529.8/[\ln(0.109)(1.34 \times 10^{17})(2))] \\
&= 38{,}529.8/[\ln(0.29212 \times 10^{17})] \\
&= 38{,}529.8/[\ln(0.29212) + 17\ln(10)] \\
&= 38{,}529.8/37.869 \\
&= 1017.0\,\mathrm{K} \\
&= 744^\circ\mathrm{C} \\
&= 1371^\circ\mathrm{F}
\end{aligned}
$$

This calculated result compares favorably with experimental values gotten by the University of Dayton Research Institute (for EPA) for chlorobenzene's "4-nines/2 seconds" destruction temperature (UDRI-90).

EXAMPLE 2: chemical kinetics
Xylene is under thermal destruction with the following given conditions. Determine the minimum required temperature.

Given conditions: the data in question are summarized below.

- Destruction efficiency, $DE = 0.999999 = 6\text{-}9\text{'s}$; $t = 2\,\mathrm{s}$
- Arrhenius constant (activation energy) for xylene, $A = 5.00 \times 10^{13}$ (1/s)
- Arrhenius constant (frequency factor) for xylene, $E = 76{,}200\,\mathrm{cal/g\text{-}mole}$ (Arrhenius constants from *Chemical Engineers' Handbook* by Perry and Chilton, 5th edition, p. 4-37)

Solution:

1. Write the destruction efficiency equation.

$$DE = (1 - W_0/W_i)$$

2. Calculate the ratio W_0/W_i for 99.9999% DE.

$$0.999999 = 1 - W_0/W_i$$
$$W_0/W_i = 0.000001 = 10^{-6}$$

3. Apply Eq. (1.9).

$$
\begin{aligned}
\ln(10^{-6}) &= -(A)(t)\exp(-E/R_\mathrm{u}T) \\
(-6)\ln(10) &= -(A)(t)\exp(-E/R_\mathrm{u}T) \\
(-6)(2.303) &= -(A)(t)\exp(-E/R_\mathrm{u}T) \\
13.82 &= (A)(t)\exp(-E/R_\mathrm{u}T) \\
13.82/((A)(t)) &= \exp(-E/R_\mathrm{u}t) \\
(A)(t)/13.82 &= \exp(+E/R_\mathrm{u}T) \\
(2)(5.0 \times 10^{13})/13.82 &= \exp(76{,}200/((1.987)T)) \\
7.24 \times 10^{12} &= \exp(38{,}349.27/T) \\
\ln(7.2 \times 10^{12}) &= 38{,}349.27/T
\end{aligned}
$$

$$\ln(7.2) + \ln(10^{12}) = 38{,}349.27/T$$

$$1.9741 + (12)\ln(10) = 38{,}349.27/T$$

$$1.9741 + (12)(2.303) = 38.349.27/T$$

$$29.5741 = 38{,}349.27/T$$

$$T = 1297\,\text{K}$$

$$T = 1024^\circ\text{C}$$
$$= 1875^\circ\text{F}$$

EXAMPLE 3: chemical kinetics
Toluene is incinerated in an incinerator with the following given conditions. Calculate the residence time needed to reach 99.99% destruction efficiency.

Given conditions: the data in question are summarized below.

- Incineration temperature, $T = 1000^\circ$C
- Destruction efficiency, $DE = 0.9999$
- Arrhenius constant (activation energy) for toluene, $A = 2.28 \times 10^{13}$ (1/s)
- Arrhenius constant (frequency factor) for toluene, $E = 56{,}500$ cal/g-mole [Arrhenius constants from Attachment 1 and (UDRI-90)]

Solution:

1. Write the destruction efficiency equation.

$$DE = (1 - W_0/W_i)$$

2. Calculate the ratio W_0/W_i for 99.99% DE.

$$0.9999 = 1 - W_0/W_i$$

$$W_0/W_i = 0.0001$$

3. Apply Eq. (1.10).

$$\ln(0.0001) = -(A)(t)\exp(-E/R_\text{u}T)$$

$$-9.21 = -(A)(t)\exp(-E/R_\text{u}t)$$

$$9.21 = (A)(t)\exp(-E/R_\text{u}T)$$

$$9.21/((A)(t)) = \exp(-E/R_\text{u}T)$$

$$(t)(A)/9.21 = \exp(E/R_\text{u}T)$$

$$(t)(2.28 \times 10^{13})/9.21 = \exp(56{,}500/(1.987(273 + 1000))$$

$$2.4756 \times 10^{12}(t) = \exp(22.33686)$$

$$\ln(t) = -6.20544$$

$$t = 0.002\,\text{s}$$

EXAMPLE 4: chemical kinetics
Methane gas is used to thermally react with residual waste hydrocarbons in a reactor operating at 1750°F. The waste destruction efficiency is 99.96%. The feeding rate of fuel and waste residual hydrocarbon is 3500 scfm at the standard condition, $T = 60°\text{F}$ and $P = 1\,\text{atm}$. The velocity of the flow is 15 ft/s. The reaction rate constant (k) is assumed to be 6.5/s. Determine the reactor diameter and length.

Given conditions: the data in question are summarized below.

- Initial temperature, $T_1 = 60°\text{F}$
- Pressure, $P = 1\,\text{atm}$
- Initial flow rate, $Q_1 = 3500\,\text{scfm}$
- Velocity of the flow, $v = 15\,\text{ft/s}$
- Reaction rate constant, $k = 6.5\,(1/\text{s})$
- Reaction temperature, $T_2 = 1750°\text{F}$

Solution:

1. Determine the flow rate at the reactor operating temperature.

$$\begin{aligned} Q_2 &= Q_1(T_2/T_1) \\ &= 3500(460 + 1750)/(460 + 60) \\ &= 14{,}875\,\text{acfm} \end{aligned}$$

2. Determine the reactor cross-sectional area A.

$$\begin{aligned} A &= (\text{flow rate in acfm})/\text{velocity} \\ &= 14{,}875/((15)(60)) \\ &= 16.53\,\text{ft}^2 \end{aligned}$$

3. Calculate the reactor diameter D.

$$\begin{aligned} D &= (4A/\pi)^{0.5} \\ &= (4(16.53)/\pi)^{0.5} \\ &= 4.59\,\text{ft} \end{aligned}$$

4. Write the differential equations for determining the reactor length L.

$$\mathrm{d}W/\mathrm{d}t = -kW \tag{1.11}$$

$$\mathrm{d}z = v\,\mathrm{d}t \tag{1.12}$$

Combining Eqs. (1.11) and (1.12) gives: $v(\mathrm{d}W/\mathrm{d}z) = -kW$
Solving the differential equation gives: $v(\ln W) = -kz +$ "constant"
Determine the "constant" by using the conditions

$$z = 0,\ W = W_i$$

$$z = L,\ W = W_0$$

Solving for the "constant" gives:

$$L = -(v/k)\ln(W_0/W_i)$$
$$= -(v/k)\ln(1 - \text{destruction efficiency})$$

5. Calculate the reactor length L.

$$L = -(15/6.5)\ln(1 - 0.9996)$$
$$= -(15/6.5)(-7.824)$$
$$= 18.06\,\text{ft}$$

2.7 Collection Efficiency

Collection efficiency is a measure of the degree of performance of a control device; it specifically refers to the degree of removal of a pollutant. *Loading* refers to the concentration of a pollutant, usually in grains of pollutant per cubic feet of contaminated gas stream. Mathematically, the collection efficiency is defined as

$$E = [(\text{inlet loading} - \text{outlet loading})/(\text{inlet loading})](100)$$

From the above equation, the collected amount of a pollutant by a control unit is the product of collection efficiency (E) and inlet loading. The amount discharged to the atmosphere is given by the inlet loading minus the amount collected.

Another term used to describe the performance of control devices is *penetration* P_t. It is given by:

$$\text{Penetration, } P_t = 1 - E/100;\ \text{fraction basis}$$
$$\text{Penetration, } P_t = 100 - E;\ \text{percent basis (EPA-84/09, p. 33)}$$

EXAMPLE 1: air pollution control equipment efficiency
Given the inlet loading and outlet loading of a control unit, determine the collection efficiency of the unit.

Given conditions: the data in question are summarized below.

- Inlet loading, $W_i = 5\,\text{grains/ft}^3$
- Outlet loading, $W_0 = 0.2\,\text{grains/ft}^3$

Solution:

1. Write an equation describing collection efficiency E.

$$E = [(\text{inlet loading} - \text{outlet loading})/(\text{inlet loading})](100)$$
$$= (1 - W_0/W_i)(100)$$

2. Calculate the collection efficiency of the control unit in percent.

$$E = (1 - 0.2/5)(100)$$
$$= 96\%$$

EXAMPLE 2: air pollution control equipment efficiency
A combustion gas carried 200 ppm of particulates. Two particulate collectors are connected in a sequential manner for particulate removal. The first collector is a cyclone which has an efficiency of 70%. The second collector is a bagfilter. The required emission level is 0.05 ppm. Determine:

(1) Overall efficiency;
(2) The particulate loading exiting both the cyclone and the bagfilter;
(3) The minimum efficiency of the bagfilter.

Given conditions: the data in question are summarized below.

- Particulate concentration, $W_{i1} = 200\,\text{ppm}$
- Particulate exit concentration, $W_{02} = 0.05\,\text{ppm}$
- Cyclone efficiency, $E_1 = 70\%$

Solution:
Solution to question (1)

1.1. Write overall collection efficiency equation E_{12}.

$$E_{12} = (1 - W_{02}/W_{i1})(100)$$

1.2. Calculate the overall collection efficiency.

$$\begin{aligned} E_{12} &= (1 - 0.05/200)(100) \\ &= 99.975\% \end{aligned}$$

Solution to question (2)

2.1. Cyclone particulate loading, W_{i1}.

$$W_{i1} = 200\,\text{ppm}$$

2.2. Calculate the mass of particulates leaving the cyclone W_{01}.

$$E_1 = (1 - W_{01}/W_{i1})100, \text{ or}$$

$$\begin{aligned} W_{01} &= (1 - (E_1/100))W_{i1} \\ &= (1 - 0.7)(200) \\ &= 60\,\text{ppm} \end{aligned}$$

2.3. Bagfilter particulate loading W_{i2}.

$$\begin{aligned} W_{i2} &= W_{01} \\ &= 60\,\text{ppm} \end{aligned}$$

Solution to question (3)

3.1. Write bagfilter collection efficiency equation E_2.

$$E_2 = (1 - W_{02}/W_{i2})(100)$$

3.2. Calculate the bagfilter collection efficiency E_2.

$$E_2 = (1 - 0.05/60)(100)$$
$$= 99.92\%$$

2.8 Combustion and Incineration

Combustion or incineration basically refers to the burning of substances during an extremely rapid chemical oxidation process. In contrast, rusting is a very slow chemical oxidation. When oxidation is rapid, the temperature of the material rises rapidly due to its inability to transfer heat to the surroundings as rapidly as it is produced by the oxidation reaction. As a result, the material emits visible radiation, which is referred to as a flame.

Scientifically, the terms combustion and incineration have the same definition. Both of these terms have been used interchangeably in waste incineration documents. Combustion, however, is generally used more often in the area of fossil fuel burning for steam or power generation, and incineration is used more often when referring to waste destruction. Nevertheless, incineration uses many of the terminologies which were originally defined in the field of conventional combustion. For more information related to combustion and incineration, readers are encouraged to see later chapters.

2.9 Density

Density (ρ) is defined as the mass of a substance in the system divided by its volume, or the mass per unit volume. It can be expressed as

$$\rho = (\text{mass})/(\text{volume}) = m/V$$

Water at 4°C has the maximum density. In general, water at 4°C is used as the reference substance for solids and liquids (Schaum-66, p. 9).

$$\text{Density of water at } 4^\circ\text{C} = 62.4\,\text{lb/ft}^3 \text{ in British units}$$
$$= 1.0\ \text{g/cm}^3 \text{ in cgs units}$$
$$= 1.0\ \text{g/cc (cubic centimeter)}$$
$$= 1.0\ \text{g/ml (milliliter)}$$

EXAMPLE 1: density
The specific gravity of a substance is 0.95. Determine the density of the substance.

Given conditions

- Specific gravity of the substance = 0.95
- Density of reference substance (water density at 60°F), $\rho_w = 62.4\,\text{lb/ft}^3$

Solution:

1. Write the specific gravity equation.

$$\text{Specific gravity} = (\text{density of substance})/\rho_w$$
$$\text{Density of substance} = (\text{specific gravity})(\rho_w)$$
$$= 0.95(62.4)$$
$$= 59.28\,\text{lb/ft}^3$$

EXAMPLE 2: density
Calculate the density and specific gravity of a body that weighs 520 g and has a volume of 80 cc.

Given conditions

- Body weight, $m = 520\,\text{g}$
- Volume, $V = 80\,\text{cc}$
- Reference substance density (water at 4°C), $\rho_w = 1\,\text{g/cc}$

Solution:

1. Write the definition of density equation.

$$\rho = m/V$$

where ρ = density
m = mass
V = volume

2. Calculate the density.

$$\begin{aligned}\rho &= m/V \\ &= 520/80 \\ &= 6.50\,\text{g/cc}\end{aligned}$$

3. Write the definition of specific gravity, ρ_g

$$\rho_g = (\text{density of a substance})/(\rho_w)$$

4. Calculate the specific gravity.

$$\begin{aligned}\rho_g &= 6.50 \\ &= 6.50\end{aligned}$$

EXAMPLE 3: density
(1) Calculate the volume (in ft^3) of 1.0 lb-mole of any ideal gas at 60°F and 14.7 psia.
(2) Calculate the density of a gas ($M = 29$) in g/cm^3 at 20°C and 1.2 atm using the ideal gas law.

Given conditions

- Number of moles, $n = 1.0$ lb-mole
- Universal gas constant (British units), $R_u = 1545\,(\text{lb}_f\text{-ft})/(\text{lb-mole-°R})$
- Universal gas constant (metric units), $R_u = 82.0575\,(\text{atm-cm}^3)/(\text{g-mole-K})$
- Pressure, $P = 14.7$ psia
- Temperature, $T = 60°\text{F}$
- Molecular weight, $M = 29$

Solution:

1. Solve the ideal gas law for V and calculate the volume.

$$\begin{aligned} V &= nR_u T/P \\ &= 1(1545)(460+60)/(14.7(144)) \\ &= 380\,\text{ft}^3 \end{aligned}$$

2. Calculate the density of the gas using the ideal gas law.

$$\begin{aligned} \text{Density} &= m/V \\ &= PM/R_u T \\ &= 1.2(29)/[(82.0575)(20+273)] \\ &= 0.00145\,\text{g/cm}^3 \end{aligned}$$

2.10 Destruction and Removal Efficiency

Destruction and removal efficiency (DRE) is a legal term defined in 40CFR264.34(a)(1) as

$$\begin{aligned} \text{DRE} &= ((W_i - W_0)/W_i))100\%, \text{ or} \\ &= (1 - W_0/W_i)100\% \end{aligned}$$

where W_i = mass feed rate of principal organic hazardous constituents (POHCs)
W_0 = mass emission rate of POHC in flue gas (downstream of all air pollution control equipment)

DRE is a measure of the amount of hazardous constituents listed in 40CFR261 Appendix VIII, or other organic hazardous constituents that may be emitted to the air from an operating incinerator. The definition does not include the organic constituent remaining in the ash nor that captured by the air pollution control equipment (APCE) as part of the W_0 term. Any organic constituent remaining with the ash increases the DRE, as does any captured by the air pollution control device. The DRE requirement of 99.99% [(99.9999% for PCB (polychlorinated biphenyls) and dioxin-listed wastes] is an emission standard, i.e., no more than 0.01% of POHC or 0.0001% of PCB and dioxins fed to the combustor may be emitted to the atmosphere).

EXAMPLE 1: destruction and removal efficiency
A POHC (principal organic hazardous constituent) of a waste stream is fed into an incinerator at the rate of 200 lb/h. The waste rates leaving the incinerator for the different cases are shown below. Calculate the destruction and removal efficiency for each case shown

1. 20 lb/h
2. 2
3. 0.2
4. 0.02
5. 0.002
6. 0.0002

Solution:

1. Write the destruction and removal efficiency (DRE) equation.

$$DRE = (1 - W_0/W_i)100\%$$

where DRE = destruction and removal efficiency
W_i = mass feed rate of POHC
W_0 = mass emission rate of the POHC

2. Calculate DRE for case 1.

$$\begin{aligned} DRE &= (1 - 20/200)(100) \\ &= 90\% \end{aligned}$$

3. Calculate DRE for other cases. The calculated DRE results are given in Table 1.1.

TABLE 1.1 DRE Calculated for Cases 1–6

Case	W_i (lb/h)	W_0 (lb/h)	DRE (%)
1	200	20	90
2	200	2	99
3	200	0.2	99.9
4	200	0.02	99.99
5	200	0.002	99.999
6	200	0.0002	99.9999

EXAMPLE 2: destruction and removal efficiency
This example is for the determination of POHC compliance. Methane gas (CH_4) is used to incinerate the POHC compounds, chloroform ($CHCl_3$) and dichlorobenzene ($C_6H_4Cl_2$), in an incinerator. The inlet and outlet of feeding streams are shown in Table 1.2. Determine if each feeding stream is in compliance with regulatory requirements.

TABLE 1.2 Inlet and Outlet of Feeding Streams

	W_i (lb/h)	W_0 (lb/h)
Methane (CH_4)	1000	0.155
Chloroform ($CHCl_3$)	950	0.088
Dichlorobenzene ($C_6H_4Cl_2$)	850	0.056

Solution:

1. Write the destruction and removal efficiency (DRE) equation.

$$DRE = (1 - W_0/W_i)100\%$$

where DRE = destruction and removal efficiency
W_i = mass feed rate of POHC
W_0 = mass emission rate of POHC

2. Calculate the DRE for each case.

$$\text{DRE for methane} = (1 - 0.155/1000) = 99.98\%$$

$$\text{DRE for chloroform} = (1 - 0.088/950) = 99.99\%$$

$$\text{DRE for dichlorobenzene} = (1 - 0.056/850) = 99.99\%$$

The above data show that:

1. although methane does not meet 99.99% DRE, methane is not a POHC and is therefore not subject to the DRE requirement;
2. both chloroform and dichlorobenzene are in compliance.

EXAMPLE 3: destruction and removal efficiency
This example is for the determination of HCl compliance. Assume that all the chlorine in the previous example is converted into HCl. Determine if the incinerator needs a scrubber and at what efficiency.

Solution:

1. Write the chemical balance equation.

For chloroform:

$$2CHCl_3 + CH_4 + 3O_2 \rightarrow 3CO_2 + 6HCl$$

For dichlorobenzene:

$$C_6H_4Cl_2 + CH_4 + 8.5O_2 \rightarrow 7CO_2 + 3H_2O + 2HCl$$

2. Calculate the amount of HCl produced from chloroform incineration.

$$2CHCl_3 + CH_4 + 3O_2 \rightarrow 3CO_2 + 6HCl$$

$2CHCl_3$	$6HCl$
239	219
1	0.9163
950	870.50

This means that 950 lb/h of chloroform incinerated will produce 863.28 lb/h of HCl.

3. Calculate the amount of HCl produced from dichlorobenzene incineration.

$$C_6H_4Cl_2 + CH_4 + 8.5O_2 \rightarrow 7CO_2 + 3H_2O + 2HCl$$

$C_6H_4Cl_2$	$2HCl$
147	73
1	0.4966
850	422.11

This means that 850 lb/h of chloroform incinerated will produce 422.11 lb/h of HCl.

4. Calculate the total amount of HCl produced.

The total amount of HCl produced from the incineration of chloroform and dichlorobenzene is the sum of 870.50 and 422.11, which is equal to 1292.61 lb/h.

5. Write the regulatory requirement of HCl.

The RCRA standards require that the emission of HCl be less than 4 lb/h (40CFR264.343).

Obviously, the 1285.39 lb/h of HCl is much greater than the 4 lb/h regulatory requirement. Therefore, a scrubber is needed to remove the HCl prior to its emission to the air. The minimum collection efficiency (*CE*) of the scrubber is calculated below:

$$CE = (1 - W_0/W_i)100$$
$$= (1 - 4/1292.61)100$$
$$= 99.69\%$$

EXAMPLE 4: destruction and removal efficiency
This example is for the determination of particulate compliance. Assume that the particulate emission rate is 6.55 lb/h and the stack gas flow rate is 12,350 dry standard cubic feet per minute (dscfm). Determine the particulate loading and determine if the particulate emission meets the regulatory requirement.

Given conditions

- Particulate emission rate, $W_0 = 6.55\,\text{lb/h}$
- Stack gas flow rate, $Q = 12{,}350$ dscfm

Solution:

1. Calculate the particulate loading (PL).

$$PL = (\text{particulate flow rate})/(\text{flue gas flow rate})$$
$$= W_0(7000\ \text{grains/lb})/((\text{flow rate})(60))$$
$$= 6.55(7000)/(12{,}350(60))$$
$$= 0.0619\ \text{grains/dscf}$$

2. Write the particulate regulatory requirement.

The RCRA standards require that the emission of particulates be less than 0.08 grains/dscf (40CFR264.343). The particulate emission therefore meets the regulatory requirements.

2.11 Dew Point

The dew point is defined as the saturation temperature of an air–water mixture. It is the temperature at which condensation begins if the mixture is cooled at a constant pressure.

EXAMPLE 1: dew point
This example is for the determination of dew point temperature. Assume that atmospheric air at 14.696 psia, 90°F, has a relative humidity of 70%. Determine the dew point.

Given conditions

- Atmospheric pressure, $P = 14.696$ psia
- Atmospheric temperature, $T = 90°\text{F}$
- Relative humidity, $RH = 70\%$

Solution:

1. Draw a figure to help solve the problem.

The phenomena of the example is graphically shown in the temperature–specific volume figure (Fig. 1.1).

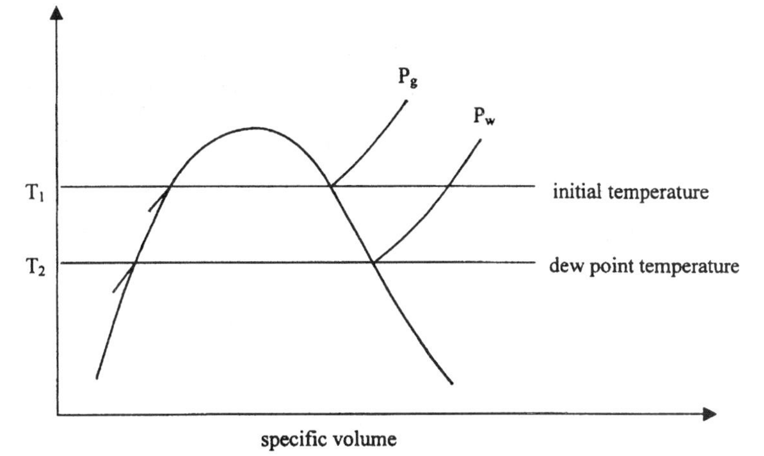

FIGURE 1.1 Dew point temperature in a T–v diagram.

2. Determine the saturation pressure for water vapor at 90°F.

From steam tables (Table A.1.1, p. 649, Wylen-1972), the saturation pressure for water vapor at 90°F is 0.6982 psia.

The partial pressure of the water vapor in the mixture is then

$$\begin{aligned} P_w &= RH(\text{Pg}) \\ &= 0.7(0.6982) \\ &= 0.489\ \text{psia} \end{aligned}$$

3. Determine the dew point.

The saturation temperature corresponding to this water vapor partial pressure is 78.76°F. Therefore the dew point is 78.76°F, which was calculated from the following format:

T_1	70	0.3631	P_1
	T	0.4890	P, given data
T_2	80	0.5069	P_2

The values of T_1, T_2, P_1, and P_2 are found from the steam table (Holman-74, p. 551)

$$(T - T_1)/(T_2 - T_1) = (P - P_1)/(P_2 - P_1)$$

$$T = T_1 + (T_2 - T_1)(P - P_1)/(P_2 - P_1)$$

$$T = 78.76^\circ\text{F}$$

EXAMPLE 2: dew point

Assume that trichloroethylene (C_2HCl_3) is incinerated with methane gas in theoretical air conditions. Determine:

(1) The mole analysis of the combustion products;
(2) The dew point of the products for a total pressure of 14.696 psia.

Solution:

1. Determine the theoretical air combustion equation.

For incineration in theoretical air conditions, the reaction equation can be written as follows:

$$C_2HCl_3 + CH_4 + 3.5(O_2 + 3.76N_2) \rightarrow 3CO_2 + 3HCl + H_2O + 13.16N_2$$

2. Determine mole fraction of each species in the products.

The total number of moles in the product is

$$\begin{aligned} n &= 3 + 3 + 1 + 13.16 \\ &= 20.16 \text{ moles} \end{aligned}$$

The mole fraction of the products is

$$\begin{aligned} y(CO_2) &= 3/20.16 = 14.88\% \\ y(HCl) &= 3/20.16 = 14.88\% \\ y(H_2O) &= 1/20.16 = 4.96\% \\ y(N_2) &= 13.16/20.16 = 65.28\% \end{aligned}$$

3. Determine the dew point.

The water dew point of the products is the temperature at which the vapor is saturated with water. It corresponds to the partial pressure of the water vapor, or

$$\begin{aligned} P(H_2O) &= y(H_2O) \times P \\ &= 0.0496(14.6959) \\ &= 0.7289 \text{ psia} \end{aligned}$$

The saturation temperature corresponding to 0.7289 psia is about 91.22°F. Therefore: dew point = 91.22°F.

T_1	90	0.6982	P_1
	T	0.7289	P, given data
T_2	100	0.9492	P_2

The values of T_1, T_2, P_1, and P_2 are found from the steam table (Holman-74, p. 551)

$$(T - T_1)/(T_2 - T_1) = (P - P_1)/(P_2 - P_1)$$

$$T = T_1 + (T_2 - T_1)(P - P_1)/(P_2 - P_1)$$

$$T = 91.22°\text{F}$$

This example illustrates three facts.

1. Nitrogen is about 65 percent of the combustion product gas. This means that most of the fuel is used to heat the nitrogen in the combustion air to the incineration temperature.
2. The dew point is 91.22°F. The moisture in the combustion gas would condense if the stack gas temperature is cooled below 91.22°F. Because the condensed moisture may contain HCl, and HCl is corrosive to the stack or other parts in the incinerator that are downstream of the condensation point, when designing this incinerator stack, the combustion gas should be maintained above 91.22°F

3. The dew point of organics is much lower than that of water. For example, the dew point of freon 12 (dichlorodifluoromethane) is −100°F at pressure 1.4280 psia (Holman-74, p. 567). Therefore, if temperature can be maintained above the water dew point, other organics will not condense before the water vapor does.

2.12 Dry Versus Wet Analysis

A mass balance can be performed on either a molar or weight basis but it is necessary to define all streams the same way. Typically, the simplest method of making such conversions is to reduce the ultimate analysis (wet basis) or the numbers of atoms in the compound to a mass or molar flow rate, respectively. It is extremely important to know the basis (wet or dry) and the way the values were determined when examining analytical data (see more related examples in the sections Fuel and Waste Characterization).

Conversion between wet and dry basis analyses is made by the following equation:

$$P_{\text{wet}} = P_{\text{dry}}(1 - H_2O)$$

where P_{dry} = fraction of a constituent on a dry basis
P_{wet} = fraction of a constituent on a wet basis
H_2O = fraction of water (or moisture) in an original sample. Note that this value must be given on a decimal or fractional basis rather than as a percent

P_{wet} must always be smaller than P_{dry}.

Table 1.3 shows the analysis of dry and wet basis data. Consider carbon data as an example, its wet base fraction was obtained by the following calculation:

$$P_{\text{wet}} = 0.6300(1 - 0.3632) = 0.4012$$

TABLE 1.3 Analysis of Dry and Wet Basis Data

Composition	Dry basis	Wet basis
C	0.63	0.4012
H_2	0.0937	0.0597
Cl	0.047	0.0299
O_2	0.0747	0.0476
N_2	0.0012	0.0008
S	0	0
H_2O	0	0.3632
Ash	0.1534	0.0977
Total	1	1

2.13 Duct Diameter

EXAMPLE: Flue gas at a temperature of 1800°F is introduced to a scrubber through a pipe which has an inside diameter of 4.0 ft. The inlet velocity to and the outlet velocity from the scrubber are 25 ft/s and 20 ft/s, respectively. The scrubber cools the flue gas to 550°F. Determine the duct size required at the outlet of the unit.

Solution:

1. Calculate the inlet cross-sectional area A_i.

$$A_i = \pi[(D_i)^2]/4$$
$$= \pi(4^2)/4$$
$$= 12.57\,\text{ft}^2$$

2. Calculate the inlet volumetric flow rate Q_i.

$$Q_i = (\text{velocity}) \times (\text{cross-sectional area})$$
$$= 25(12.57)$$
$$= 314.16\,\text{ft}^3/\text{s}$$

3. Calculate the outlet volumetric flow rate using Charles' law.

$$Q_o = Q_i(T_o/T_i)$$
$$= 314.16(460 + 550)/(460 + 1800)$$
$$= 140.40\,\text{ft}^3/\text{s}$$

4. Calculate the outlet cross-sectional area.

$$A_o = Q_o/v_o$$
$$= 140.40/20$$
$$= 7.02\,\text{ft}^2$$

5. Calculate the outlet duct diameter.

$$(D_o)^2 = 4(A_o)/\pi$$
$$= 4(7.02)/\pi$$
$$D_o = (4(7.02)/\pi)^{0.5}$$
$$= 2.99\,\text{ft}$$

2.14 Duct Flow

EXAMPLE: Air flows through a duct with the following given conditions. Calculate density, absolute viscosity, and Reynolds number.

Given conditions

- Duct inside diameter, $D = 5\,\text{m}$
- Pressure, $P = 1\,\text{atm}$
- Temperature, $T = 25°\text{C}$
- Universal gas constant, $R_u = 0.082\,\text{atm-L/g-mole-K}$
- K_v (kinematic viscosity) $= 1.1 \times 10^{-5}\,\text{m}^2/\text{s} = 1.1 \times 10^{-1}\,\text{cm}^2/\text{s}$
- Air velocity, $v = 0.8\,\text{m/s} = 80\,\text{cm/s}$
- Air molecular weight $= 29\,\text{g/(g-mole)}$

Solution:

1. Calculate the density d using the ideal gas law.

$$\begin{aligned} PV &= nR_uT \\ &= (m/M)R_uT \end{aligned}$$

$$\begin{aligned} \rho &= m/V \\ &= PM/R_uT \\ &= 1(29)/(0.082(25+273)) \\ &= 1.19\ \mathrm{g/L} \\ &= 1.19 \times 10^{-3}\ \mathrm{g/cm^3} \end{aligned}$$

2. Write the absolute viscosity (A_v) equation.

$$A_v = K_v(\rho)$$

where A_v = absolute viscosity
K_v = kinematic viscosity
ρ = density

3. Calculate the absolute viscosity (A_v).

$$\begin{aligned} A_v &= K_v(\rho) \\ &= (1.1 \times 10^{-1})(1.19 \times 10^{-3}) \\ &= 1.3 \times 10^{-4}\ \text{g-cm/s} \end{aligned}$$

4. Write the Reynolds number equation Re.

$$Re = vD/K_v$$

where Re = Reynolds number
v = velocity
D = diameter
K_v = kinematic viscosity

5. Calculate the Reynolds number Re.

$$\begin{aligned} Re &= 80(500)/0.11 \\ &= 3.64 \times 10^5 \end{aligned}$$

2.15 Enthalpy for Air–Water Vapor Mixture Application

During the energy balance calculation within an incinerator, the enthalpy information for the air–water vapor mixture is very important. The mixture enthalpy h in units of kJ/kg or Btu/lb is defined as the total energy of a substance at one temperature relative to the energy of the same substance at another temperature. It is a relative property, and whenever it is used it must be related to a base or datum point. Table 1.4 shows some enthalpy values of several commonly used compounds. Figures 1.2 and 1.3 (OME-88/12) present the enthalpy of water vapor and air, respectively, as a function of temperature. The enthalpies are related to a base of zero degrees Celsius.

TABLE 1.4 Enthalpy Values of Combustion Gas, in Btu/lb

Temperature (°F)	N_2	Air MW 28.7	CO_2	H_2O
32	0	0	0	0
60	194.9	194.6	243.1	224.2
77	312.2	312.7	392.2	360.5
100	473.3	472.7	597.9	545.3
200	1170	1170	1527	1353
300	1868	1870	2509	2171
400	2570	2576	3537	3001
500	3277	3289	4607	3842
600	3991	4010	5714	4700
700	4713	4740	6855	5572
800	5443	5479	8026	6460
900	6182	6227	9224	7364
1000	6929	6984	10447	8284
1200	9452	8524	12960	10176
1500	10799	10895	16860	13140
2000	14840	14970	23630	18380
2500	19020	19170	30620	23950
3000	23280	23460	37750	29750

Source: EPA-84/09, p. 93.

2.16 Enthalpy for Ideal Gas Application

Because the combination of $U + PV$ occurs frequently in many thermodynamic applications, the combination has been given the name *enthalpy*. Mathematically, enthalpy is defined as $H = U + PV$, where H is enthalpy, U is internal energy, P is pressure, and V is volume. Enthalpy is an arbitrary combination of other properties and therefore it is a property. Enthalpy has units of Btu/lbm (Jones-60, p. 68). Mathematically, enthalpy is related by the following equations:

(1) Enthalpy definition:

$$H = U + PV$$

where H = enthalpy
U = internal energy
P = pressure
V = volume

(2) Enthalpy and the first thermodynamic law equation

In engineering practice, enthalpy is a term of one of the thermodynamic energy equations. Under steady state, it is the first thermodynamic law equation without considering work, potential energy, and kinetic energy.

$$Q = H_2 - H_1$$

where Q = heat
H_1 = enthalpy value at state 1
H_2 = enthalpy value at state 2

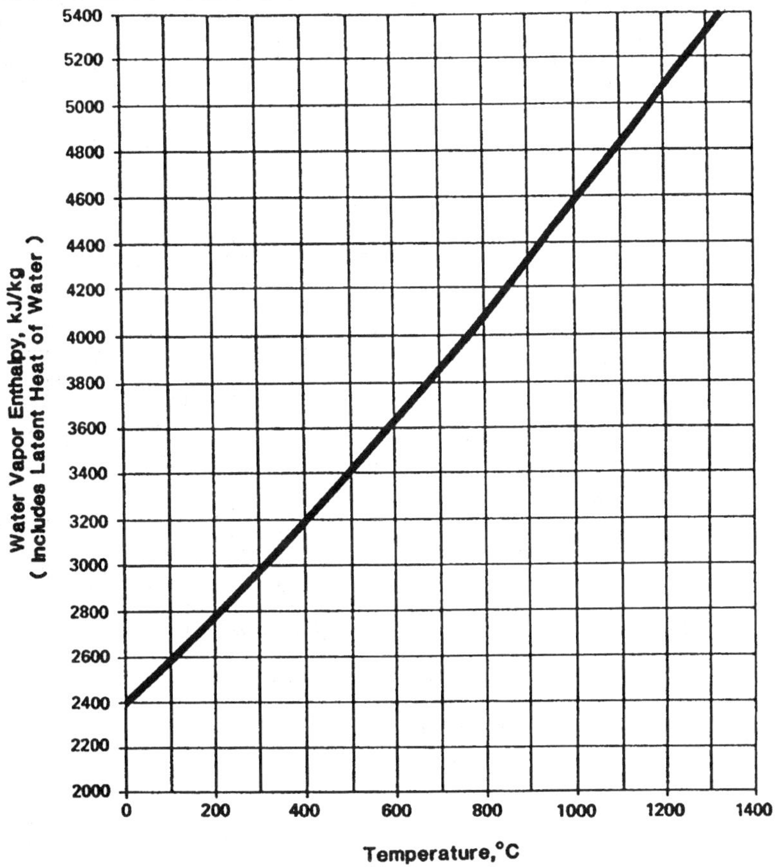

FIGURE 1.2 Enthalpy, water vapor relative to 0°C (*OME-88/12, p. C1-9*).

(3) Enthalpy and specific heat

$$H_2 - H_1 = mC_p(T_2 - T_1)$$

where H_1 = enthalpy value at state 1
H_2 = enthalpy value at state 2
m = mass
C_p = constant pressure specific heat
T_1 = temperature at state 1
T_2 = temperature at state 2

Combining the equations from (2) and (3) above results in

$$Q = mC_p(T_2 - T_1)$$

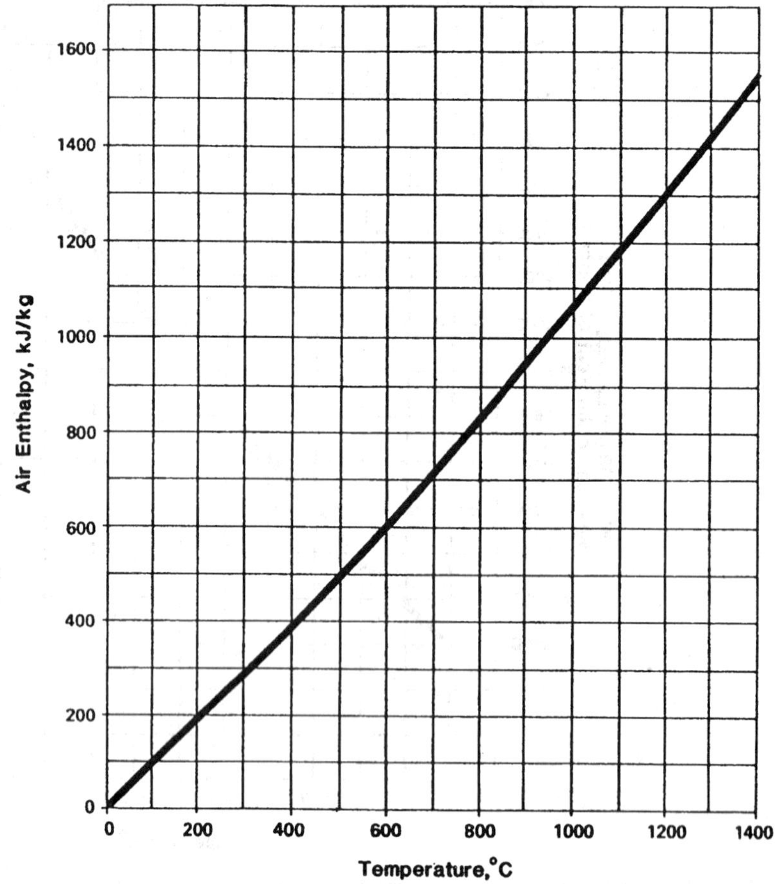

FIGURE 1.3 Enthalpy, dry air relative to 0°C (*OME-88/12, p. C1-10*).

2.17 Entropy

In thermodynamics, entropy is the ratio of the heat added to a system to the absolute temperature at which it was added. For a reversible process, entropy is defined as

$$dS = dQ/T$$

where dS = differential amount of entropy
dQ = differential amount of heat
T = absolute temperature

Entropy is a measure of the unavailable energy in a system. An increase in entropy is accompanied by a decrease in available energy (Holman-74, p. 163; Jones-60, p. 266).

2.18 Flammability of Gases and Vapors

The flammability of gases and vapors is also known as the combustion limits of gases and vapors. Not all mixtures of fuel and air are able to support combustion. The flammable or explosive limits for a mixture are the maximum and minimum concentrations of fuel in air that will support combustion. The upper explosive limit (UEL) is defined as the concentration of fuel which produces a nonburning mixture due to a lack of oxygen. The lower explosive limit (LEL) is defined as the concentration of fuel below which combustion will not be self-sustaining (EPA-81/12, p. 3-6).

At concentrations below LEL the localized heat release rate of the oxidation reaction at the ignition source is lower than the rate at which heat is dissipated to the surroundings, and therefore it is not possible to maintain a high enough temperature for flame propagation or sustained combustion. Above the upper flammability limit, there is less than the necessary amount of oxygen, with the result that the flame does not propagate due to the local depletion of oxygen, thus causing the temperature, and hence the oxidation rate, to drop below the levels required for sustained combustion.

The rate of flame propagation in combustible mixtures covers a wide range as it depends on a number of factors, including the nature of the combustible substances, mixture composition, temperature, and pressure. For a given substance the flame propagation rate is maximum at or near the stoichiometric mixture composition, and drops off to zero at the upper and lower explosive limits.

Figure 1.4 (EPA-80/02, p. 2-17) is typical of the effect of temperature on the limits of flammability. Here T_L is defined as the lowest temperature at which a liquid combustible has vapor pressure high enough to produce a vapor–air mixture within the flammability

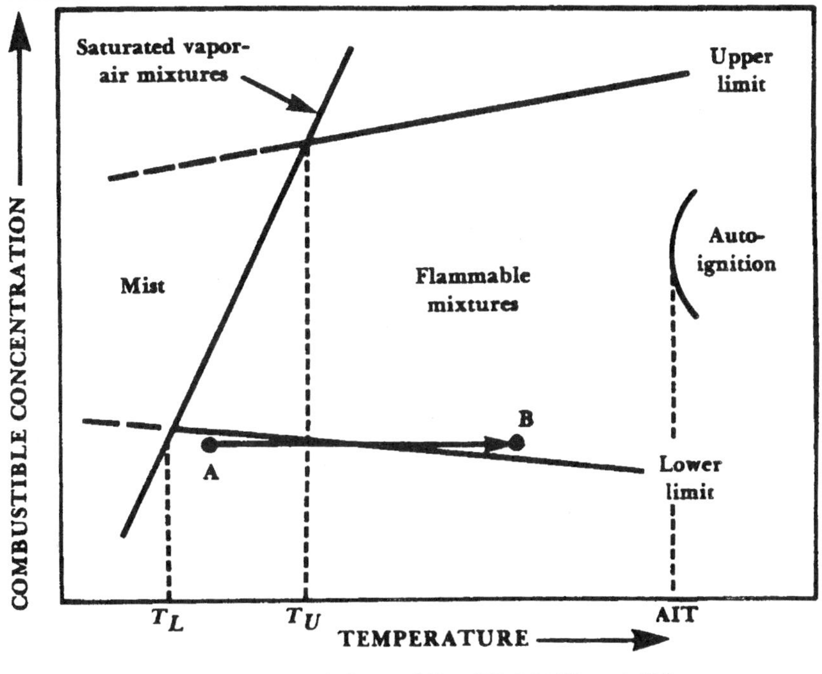

FIGURE 1.4 Temperature effect on air flammability (*EPA-80/02, p. 2-170*).

range (at LEL). The autoignition temperature (AIT), on the other hand, is the lowest temperature at which a uniformly heated mixture will ignite spontaneously.

2.19 Fuel

Fuel is a material which is capable of releasing energy or power by combustion or other chemical or physical reaction (40CFR79.2-91). In many cases, if the heating value of a waste is not high enough to sustain the combustion by itself, supplemental fuels are needed. If supplemental fuels are required, the ultimate analysis information of the fuels is needed for calculating the energy balance in an incinerator. Table 1.5 provides the ultimate analysis of major fuels. Table 1.6 shows the conversion of a typical fuel between the wet basis and dry basis. The conversion equation is $P_{dry} = P_{wet}/(1 -$ moisture fraction in wet basis analysis) (See Dry versus Wet Analysis section in this chapter).

TABLE 1.5 Ultimate Analysis of Fuels

	Natural gas*	Residual fuel oil (No. 6)†	Distillate fuel oil (No. 2)*	Coal‡
C	0.693	0.8852	0.872	76.9
H_2	0.227	0.1087	0.123	5.1
O_2	0	0.0006	0	6.9
N_2	0.08	0.001	0	1.5
S	0	0.004	0.005	2.4
Ash	0	0.0005	0	7.2
Heating value	22,027 (Btu/lb)§ 1,037 (Btu/ft^3)§	17,500 (Btu/lb)¶ 150,000 (Btu/gal)¶	16,450 (Btu/lb)¶ 141,000 (Btu/gal)¶	13,000 (Btu/lb)

* = (EPA-81/09, p. 4-87); † = (EPA-86/03, p. 73); ‡ = (EPA-80/02, p. 3-17); § = (EPA-80/02, p. 3-7); ¶ = (EPA-80/02, p. 3-10).

TABLE 1.6 An Example Analysis of a Typical Fuel

Proximate analysis		Ultimate analysis (wet basis or as received)		Ultimate analysis (dry basis)	
Component	Weight (%)	Component	Weight (%)	Component	Weight (%)
Moisture (free)	2	Moisture (free)	2	Carbon	71.43
Volatile matter	38	Carbon	70	Hydrogen	10.2
Fixed carbon	53	Hydrogen	10	Sulfur	2.35
Ash	7	Sulfur	2.3	Nitrogen	2.04
Total	100	Nitrogen	2	Oxygen	6.84
Heating value, Btu/lb	13,000	Oxygen	6.7	Ash	7.14
		Ash	7	Total	100
		Total	100		

Source: (EPA-80/02, p. 3-17).

2.20 Heat (or Heat Load)

Heat is also known as heat load. It is an interactive flow of energy between a system and its surroundings which is caused by a temperature difference between the system and the surroundings. Typical heat units include the Btu and the calorie. The conventional signs of heat are as follows: Heat added to the system is positive, and heat liberated from the system is negative (cf. work). An adiabatic process is a process in which no heat is transferred between a system and its surroundings. Applying the first thermodynamic law under steady state, heat can be obtained:

$$Q = mC_p(T_2 - T_1)$$

where m = mass flow
C_p = constant pressure specific heat
T_1 = temperature at state 1
T_2 = temperature at state 2

EXAMPLE 1: heat load
Determine the heat transfer rate by using the following given conditions.

Given conditions

- Mass flow rate = 2000 lb/min
- Specific heat, $C_p = 0.24$ Btu/lb-°F
- Initial temperature, $T_1 = 200$°F
- Final temeprature, $T_2 = 1800$°F

Solution:

1. Write the energy equation.

$$Q = mC_p(T_2 - T_1)$$

2. Calculate heat transfer.

$$Q = 2000(0.24)(1800 - 200)$$
$$= 7.68 \times 10^5 \text{ Btu/min}$$

EXAMPLE 2: heat load
Heat is transferred from the flue gas of an incinerator to the environment with the following given conditions. Calculate the exit temperature of the flue gas at the stack.

Given conditions

- Heat transferred, $Q = -30 \times 10^6$ Btu/h (negative sign means that heat is transferred from the system to its surroundings)
- Average heat capacity of the gas, $C_p = 0.24$ Btu/lb-°F
- Gas mass flow rate, $m = 80{,}000$ lb/h
- Initial gas temperature from the incinerator, $T_1 = 1800$°F

Solution:

1. Write the energy equation.

$$Q = mC_p(T_2 - T_1) \text{ or}$$

$$T_2 = T_1 + Q/mC_p$$

where m = mass flow
C_p = constant pressure specific heat
T_1 = temperature at state 1
T_2 = temperature at state 2

2. Calculate the gas exit temperature T_2.

$$\begin{aligned} T_2 &= T_1 + Q/mC_p \\ &= 1800 - 30 \times 10^6/(80{,}000(0.24)) \\ &= 237.5^\circ\text{F} \end{aligned}$$

2.21 Heat Capacity and Specific Heat

Heat capacity is defined as the amount of heat required to raise the temperature of one mole of a substance by one degree (molar heat capacity) or the amount of heat required to raise the temperature of one unit mass of a substance by one degree. *Specific heat* of a substance is defined as the ratio of heat capacity of a substance to the heat capacity of a reference substance at a specified temperature. Numerically, it can be expressed as

$$\text{Specific heat} = \frac{\text{(heat capacity of a substance)}}{\text{(heat capacity of a reference substance at a specified temperature)}}$$

The reference substance is normally water at 17°C, at which temperature the heat capacity is 1.0 Btu/lb-°F. Because the difference between heat capacity and specific heat is the factor of water heat capacity, which is 1.0 Btu/lb-°F or 1.0 cal/g-°C, the heat capacity and specific heat are interchangeable in common practice.

Heat capacity includes: (1) constant pressure heat capacity (C_p) and (2) constant volume heat capacity (C_v). Because heat capacity and specific heat are interchangeable, C_p is also known as constant pressure specific heat, and C_v constant volume specific heat. Typical units of heat capacity include:

- Btu/lb-°F (Btu/lb-°R)
- Btu/(lb-mole-°F)
- cal/g-°C
- cal/(g-mole-°C)

Some frequently used C_p and C_v values are shown in Table 1.7.

References used in this section: EPA-84/09; EPA-86/03; Marks-67, p. 4-10; Holman-74, p. 49; Jones-60, p. 124.

EXAMPLE: heat capacity and specific heat
Assume that the specific heat of a liquid is 4.0. Calculate the heat capacity in Btu/lb-°F and cal/g-°C.

TABLE 1.7 Some C_p and C_v Values

Gas	C_p (Btu/lb-°R)	C_v (Btu/lb-°R)
Air	0.24	0.171
CO	0.249	0.178
CO_2	0.203	0.158
H_2O	0.445	0.335
N_2	0.248	0.177
O_2	0.219	0.157

Solution:

1. By definition, the heat capacity of water (C_{pw}) at 17°C is

$$C_{pw} = 1.0 \text{ Btu/lb-°F}$$
$$= 1.0 \text{ cal/g-°C}$$

Note: by definition, water is used as the reference substance for the specific heat calculation.

2. Calculate the heat capacity of the liquid (C_{pl}).

$$C_{pl} = (\text{specific heat}) \times (C_{pw})$$
$$= 4.0 \text{ Btu/lb-°F}$$
$$= 4.0 \text{ cal/g-°C}$$

2.22 Heat of Combustion

Heat of combustion is also called enthalpy of combustion, heat of reaction, or heating value. It is defined as the energy liberated when a compound experiences complete combustion with oxygen with both the reactants starting, and the products ending, at the same conditions, usually 25°C or 60°F and 1 atm. Since all combustion processes result in a decrease of enthalpy of the system, energy in the form of heat is released during the reaction. The heating value can be determined experimentally by calorimeters in which the products of combustion are cooled to the initial temperature and the heat adsorbed by the cooling medium is measured. The heating value of a waste is a measure of the energy released when the waste is burned. The heat transfer of the heating value during a combustion reaction can be computed by using the thermodynamic first law (Lee-92/06). Table 1.8 gives the heat of combustion of several typical compounds.

The heating value of a waste is a measure of the energy released when the waste is incinerated. It is measured in units of Btu/lb (J/kg). A heating value of about 5000 Btu/lb (11.6×10^6 J/kg) or greater is needed to sustain combustion. Wastes with lower heating values can be burned, but they will not maintain adequate temperature without the addition of auxiliary fuel. The heating value of the waste also is needed to calculate heat input to the incinerator, where

$$\text{Heat input (Btu/h)} = \text{Feed rate (lb/h)} \times \text{Heating value (Btu/lb)}$$

Moisture is evaporated from the waste as the temperature of the waste is raised in the combustion chamber; it passes through the incinerator, unchanged, as water vapor. This evaporation of moisture uses energy and reduces the temperature in the combustion chamber. The

TABLE 1.8 Heat of Combustion at 25°C (77°F)

Substance		lb/ft^3	ft^3/lb	Btu/ft^3 Gross (high)	Btu/ft^3 Net (low)	Btu/lb Gross (high)	Btu/lb Net (low)
Carbon	C					14.093	14.093
Hydrogen	H_2	0.0053	187.723	325	275	61100	51623
Oxygen	O_2	0.0846	11.819				
Nitrogen	N_2	0.0744	13.443				
Carbon monoxide	CO	0.074	13.506	322	322	4.347	3.347
Carbon dioxide	CO_2	0.117	8.543				
Paraffin series							
Methane	CH_4	0.0424	23.565	1013	913	23879	21520
Ethane	C_2H_6	0.0803	12.455	1792	1641	22320	20432
Propane	C_3H_8	0.1196	8.365	2590	2385	21661	19944
n-Butane	C_4H_{10}	0.1582	6.321	3370	3113	21308	19680
Isobutane	C_4H_{10}	0.1582	6.321	3363	3105	21257	19629
n-Pentane	C_5H_{12}	0.1904	5.252	4016	3709	21091	19517
Isopentane	C_5H_{12}	0.1904	5.252	4008	3716	21052	19478
Neopentane	C_5H_{12}	0.1904	5.252	3993	3693	20970	19596
n-Hexane	C_6H_{14}	0.2274	4.398	4762	4412	20940	19403
Olefin series							
Ethylene	C_2H_4	0.0746	13.412	1614	1513	21644	20295
Propylene	C_3H_6	0.111	9.007	2336	2186	21041	19691
n-Butene	C_4H_6	0.148	6.756	3084	2885	20840	19496
Isobutene	C_4H_6	0.148	6.756	3068	2869	20730	19382
n-Pentene	C_5H_{10}	0.1852	5.4	3836	3586	20712	19363
Aromatic series							
Benzene	C_6H_6	0.206	4.852	3751	3601	18210	17480
Toluene	C_7H_8	0.2431	4.113	4484	4284	18440	17620
Xylene	C_8H_{10}	0.2803	3.567	5230	4980	18650	17760

Source: (EPA-84/09, p. 37).

water vapor also increases the combustion gas flow rate, which reduces combustion gas residence time.

For example, dichloromethane would release 1.7 kcal/g during combustion, as shown in the following equation:

$$CH_2Cl_2 + O_2 + 3.76N_2 \rightarrow CO_2 + 2HCl + 3.76N_2 + 1.70\,\text{kcal/g}$$

The heat of combustion is normally written with a negative sign. This is a thermodynamic convention and indicates that energy flows out of the system. The heat of combustion of any compound at standard conditions can be calculated from the standard heats of formation of the compound and of the oxidation products.

If a fuel is only partially oxidized, the entire heat of combustion is not released; some of this heat remains bound up as potential chemical energy in the bonds of the partial oxidation species. Therefore, the final or peak temperature produced will be lower than if the fuel were completely oxidized.

Heat of combustion was initially proposed as one of the criteria to determine the ranking of hazardous waste incinerability. The rationale was that if a compound has a higher heat of combustion or can release more heat than other compounds during combustion, the compound would be easier to be incinerated. Under an EPA study, a list of heat of combustion values has been developed and is shown in Attachment 2 of this chapter for RCRAs Appendix VIII compounds (those then in Appendix VIII).

2.23 Heat of Formation

Heat of formation has also been called the enthalpy of reaction, the enthalpy of combustion, the enthalpy of formation, or the enthalpy of hydration. It is defined as the quantity of heat transferred during the formation of a compound from its elements at standard conditions (temperature = 25°C (77°F) and pressure = 1 atm) where the energy level of all elements (reactants, in this case) is assigned to be zero. Heat of formation can be determined from the change in enthalpy resulting when a compound is formed from its elements at constant temeprature and pressure conditions.

EXAMPLE 1: heat of combustion and heat of formation
Assume that at standard conditions, methane formation is expressed as $C + 2H_2 \rightarrow CH_4$. Determine heat of formation (H_f) for methane.

Solution:

1. Express the methane formation equation below.

$$C + 2H_2 \rightarrow CH_4$$

2. Apply the first law to this process.

$$\begin{aligned} H_f &= H_2 - H_1 \\ &= H_f(CH_4) - H(C) - H(2H_2) \\ &= -17.90 \text{ kcal/g-mole}) - 0 - 0 \\ &= -17.90 \text{ kcal/g-mole} \end{aligned}$$

Note: See Table 12.3 (Wylen-72, p. 502) for more information on heat of formation.

Thus, the measurement of the heat transferred actually provides the enthalpy difference between the products and the reactants. If zero value is assigned to the enthalpy of all elements at standard conditions (25°C and 1 atm pressure), then the enthalpy of the reactants in this case is zero. The enthalpy of CH_4 at 25°C and 1 atm pressure (relative to this base in which the enthalpy of its elements is assigned to be zero) is called its heat of formation.

The enthalpy of CH_4 at any other state would be found by adding the change of enthalpy between 25°C and 1 atm and the given state to the heat of formation. That is, the enthalpy at any temperature and pressure is

$$\begin{aligned} H(\text{t, p}) = H_f + &\text{enthalpy change from the conditions of 25°C} \\ &\text{and 1 atm and the given conditions.} \end{aligned}$$

EXAMPLE 2: heat of combustion and heat of formation
Calculate the enthalpy of combustion of butane (C_4H_{10}) at 77°F on both a lb-mole and a lb basis under the following conditions:

(1) Liquid butane with liquid H_2O in the products
(2) Liquid butane with gaseous H_2O in the products
(3) Gaseous butane with liquid H_2O in the products
(4) Gaseous butane with gaseous H_2O in the products

This example is designed to show how the enthalpy of combustion can be determined from the enthalpies of formation.

Given conditions

- Temperature, $T = 77°F$
- Pressure, $P = 1$ atm
- Enthalpy of formation (H_f) for the following compounds (Wylen-72, p. 502)
  $H_f(C_4H_{10}$ gas$) = -54{,}270$ Btu/(lb-mole)
  $H_f(CO_2) = -169{,}297$
  $H_f(H_2O$, liquid$) = -122{,}971$
  $H_f(H_2O$, gas$) = -104{,}036$

Solution to question (1)

1.1. Write butane reaction equation.

$$C_4H_{10} + 6.5O_2 \rightarrow 4CO_2 + 5H_2O$$

1.2. Find heating value (H_v) of C_4H_{10} at a liquid and gas state (Wylen-72, p. 511).

$$H_v(C_4H_{10}, \text{liquid}) = -21{,}134 \text{ Btu/lb}$$
$$H_v(C_4H_{10}, \text{gas}) = -21{,}293$$

1.3. Calculate the enthalpy of evaporation of butane H_e.

$$\begin{aligned} H_e &= -21{,}293 - (-21{,}134) \\ &= -159 \text{ Btu/lb} \\ &= -159(58) \text{ Btu/(lb-mole)} \\ &= -9222 \text{ Btu/(lb-mole)} \end{aligned}$$

1.4. Calculate enthalpy of formation for liquid C_4H_{10} from gas C_4H_{10}.

$$\begin{aligned} H_f \text{ of } C_4H_{10}(\text{liquid}) &= -54{,}270 + \text{enthalpy of evaporation} \\ &= -54{,}270 - 9222 \\ &= -63{,}492 \text{ Btu/(lb-mole)} \end{aligned}$$

1.5. Calculate enthalpy of combustion for liquid butane.
Liquid butane with liquid H_2O in the combustion products.

$$\begin{aligned} H_c &= 4H_f(CO_2) + 5H_f(H_2O, \text{liquid}) - H_f(C_4H_{10}, \text{liquid}) \\ &= 4(-169{,}297) + 5(-122{,}971) - (-63{,}492) \\ &= -1{,}228{,}551 \text{ Btu/(lb-mole)} \\ &= -1{,}228{,}551/58 \\ &= -21{,}182 \text{ Btu/lb} \end{aligned}$$

This is the higher heating value of liquid butane.

Solution to question (2)
Liquid butane with gaseous H_2O in the combustion products.

$$\begin{aligned} H_c &= 4H_f(CO_2) + 5H_f(H_2O,\ \text{gas}) - H_f(C_4H_{10},\ \text{liquid}) \\ &= 4(-169{,}297) + 5(-104{,}036) - (-63{,}492) \\ &= -1{,}133{,}876\ \text{Btu/(lb-mole)} \\ &= -1{,}133{,}876/58\ \text{Btu/lb} \\ &= -19{,}550\ \text{Btu/lb} \end{aligned}$$

This is the lower heating value of liquid butane.

Solution to question (3)
Gaseous butane with liquid H_2O in the combustion products.

$$\begin{aligned} H_c &= 4H_f(CO_2) + 5H_f(H_2O,\ \text{liquid}) - H_f(C_4H_{10},\ \text{gas}) \\ &= 4(-169{,}297) + 5(-122{,}971) - (-54{,}270) \\ &= -1{,}237{,}773\ \text{Btu/(lb-mole)} \\ &= -1{,}237{,}773/58\ \text{Btu/lb} \\ &= -21{,}341\ \text{Btu/lb} \end{aligned}$$

This is the higher heating value of gaseous butane.

Solution to question (4)
Gaseous butane with gaseous H_2O in the combustion products.

$$\begin{aligned} H_c &= 4H_f(CO_2) + 5H_f(H_2O,\ \text{gas}) - H_f(C_4H_{10},\ \text{gas}) \\ &= 4(-169{,}297) + 5(-104{,}036) - (-54{,}270) \\ &= -1{,}143{,}098\ \text{Btu/(lb-mole)} \\ &= -1{,}143{,}098/58\ \text{Btu/lb} \\ &= -19{,}708.6\ \text{Btu/lb} \end{aligned}$$

This is the lower heating value of gaseous butane.

2.24 Heat Transfer

Heat transfer is defined as the transfer of heat from a higher temperature point to a lower temperature point. This means that the temperature levels in a body are reduced by the heat flowing from regions of higher temperature to those of lower temperature. This process takes place in all substances (in solids, liquids, and gases). In general, in environmental engineering practice, only the conditions of steady state are considered. Steady state is defined as when the temperature at any point is constant with respect to time (Jones-60, p. 661).

There are three modes of heat transfer. In an industrial application, all three modes often participate simultaneously in the transmission of heat. The three modes of heat transfer are briefly described below.

1. *Conduction*: In a solid body, the flow of heat results from the transfer of internal energy from one molecule to another. Conduction can also be interpreted as the transfer of heat from one part of a body to another part or to another body by short-range interaction of molecules and/or electrons.

2. *Convection*: Convection is akin to mass transfer. In liquids and gases, convection is the flow of heat resulting from the transport of internal energy of the flowing medium from one point to another. Convection can also be interpreted as the transfer of heat by the combined mechanisms of fluid mixing and conduction. It can be natural or forced convection.
3. *Radiation*: The emission of energy in the form of electromagnetic waves. All bodies above absolute zero temperature radiate. Radiation incident on a body may be absorbed, reflected, and transmitted.

Conduction in a single-layer wall. Under the condition of steady state and one-dimensional heat flow, the basic conduction equation for a flat wall is expressed as

$$Q = -kA(\mathrm{d}T/\mathrm{d}x)$$

where Q = heat transfer rate
A = cross-sectional area normal to the direction of heat transfer
x = distance in the direction of heat transfer
T = temperature
k = thermal conductivity

The minus sign in the above equation indicates that heat is transferred in the direction of decreasing temperature. Integration of the equation results in

$$Q = -kA(T_2 - T_1)/(x_2 - x_1) \qquad \text{or}$$

$$T_1 - T_2 = (a/kA)Q$$

where T_2 and T_1 are temperatures at locations x_2 and x_1, respectively
$a = x_2 - x_1$ = distance in the direction of heat transfer

The expression a/kA is known as the *thermal resistance*. The amount of heat penetrating a unit area of the surface per unit of time is called the *specific rate of heat flow*.

Conduction in multilayer walls. Assume that a plane wall consists of three layers of materials with different conductivities in different layers. The equations that govern the heat transfer process are as follows:

For layer 1: $T_1 - T_{12} = (a_1/k_1A)Q$
For layer 2: $T_{12} - T_{23} = (a_2/k_2A)Q$
For layer 3: $T_{23} - T_3 = (a_3/k_3A)Q$

Adding all the equations, we obtain the equation

$$T_1 - T_3 = ((a_1/k_1A) + (a_2/k_2A) + (a_3/k_3A))Q$$

The value of $(a_1/k_1A) + (a_2/k_2A) + (a_3/k_3A)$ is the overall thermal resistance

where k_1, k_2, and k_3 are thermal conductivities for layers 1, 2, and 3, respectively
a_1, a_2, and a_3 are the thickness of layers 1, 2, and 3, respectively
T_1 = temperature at the beginning of the wall
T_{12} = temperature at the location between layer 1 and layer 2
T_{23} = temperature at the location between layer 2 and layer 3
T_3 = temperature at the other side of the wall

Conduction in a single cylindrical wall. For a steady heat flow in a cylindrical wall, the heat transfer is also considered to be one dimensional, because the temperature is only a function of its radius. The basic heat conduction equation for a cylindrical wall is expressed as

$$T_1 - T_2 = (Q/pk)(\ln(r_2/r_1))$$

where Q = heat transfer rate
p = constant = $2(\pi)(L)$
L = axial length of the cylinder
T = temperature
k = thermal conductivity
T_1 and T_2 are temperatures at the radius locations r_1 and r_2, respectively

Conduction in multilayer cylindrical walls. Assume that the wall of a pipe consists of two layers of materials with different conductivities in different layers. The equations that govern the heat transfer process are as follows:

$$\text{For layer 1:} \quad T_1 - T_{12} = ((Q/p)(1/k_1))(\ln(r_2/r_1))$$
$$\text{For layer 2:} \quad T_{12} - T_3 = ((Q/p)(1/k_2))(\ln(r_3/r_2))$$

Adding the above two equations results in

$$T_1 - T_3 = (Q/p)((1/k_1))\ln(r_2/r_1) + (1/k_2)\ln(r_3/r_2))$$

The expression $(1/p)((1/k_1)\ln(r_2/r_1) + (1/k_2)\ln(r_3/r_2))$ is called the *overall thermal resistance*

where k_1 and k_2 are thermal conductivities for layers 1 and 2, respectively
p = constant = $2(\pi)(L)$
L = axial length of the cylinder
r_1 = inner radius of layer 1
r_2 = outer radius of layer 1 = inner radius of layer 2
r_3 = outer radius of layer 2
T_1 = temperature at the inner surface of layer 1
T_{12} = temperature at the location between layer 1 and layer 2
T_3 = temperature at the outer surface of layer 2

Convection over one side of a solid wall. The basic equation for convective heat transfer between a fluid flow (liquid or gas) and a solid surface is expressed as

$$Q = hA(T_1 - T_2)$$

where T_1 = temperature in a fluid flow outside the convective film
T_2 = temperature at a solid surface
A = the cross-sectional area normal to the direction of heat transfer
h = convective coefficient

Convection over two sides of a solid wall. For this case, the overall heat transfer involves two processes of convective heat transfer and one process of conduction heat transfer. Their governing equations are as follows:

(1) For convection between fluid flow 1 and a solid surface,

$$T_1 - T_2 = (Q/A)(1/h_1)$$

(2) For conduction within the solid surface,

$$T_2 - T_3 = (Q/A)(a/k)$$

(3) For convection between a solid surface and fluid flow 2,

$$T_3 - T_4 = (Q/A)(1/h_2)$$

Adding all the equations in items (1), (2), and (3), we obtain the equation

$$T_1 - T_4 = (1/h_1 + a/k + 1/h_2)(Q/A)$$

The expression $(1/h_1 + a/k + 1/h_2)$ is called the *overall heat transfer coefficient* and the expression $(1/h_1 + a/k + 1/h_2)/A$ is called the *overall thermal resistance*

where T_1 = temperature in a fluid flow 1 outside the convective film
T_2 = temperature at surface 1 of a solid wall
T_3 = temperature at surface 2 of a solid wall
T_4 = temperature in a fluid flow 2 outside the convective film
A = the cross-sectional area normal to the direction of heat transfer
h_1 and h_2 = convective coefficient at fluid flow 1 and fluid flow 2
k = thermal conductivity within the solid wall

Convection within a tubular heat exchanger. Heat exchangers are used in many industrial fields. An example of a heat exchanger is that a smaller diameter tube is installed inside a large diameter tube. A hot fluid is flowing inside the inner tube and a cold fluid is flowing between the inner and outer tubes. Heat is transferred from the hot fluid to a metal wall by convection, through the wall by conduction, and to the cold fluid by convection. The equations that govern the heat transfer process are as follows:

(1) For convection between hot fluid and the inner tube solid surface,

$$T_1 - T_2 = (Q/A_1)(1/h_1)$$

(2) For conduction within the inner tube,

$$T_2 - T_3 = ((Q/pk)(\ln(r_2/r_1))$$

(3) For convection between the outer surface of the small tube and the cold fluid,

$$T_3 - T_4 = (Q/A_2)(1/h_2)$$

Adding the above equations results in

$$T_1 - T_4 = Q(1/h_1A_1 + (\ln(r_2/r_1))/(pk) + 1/h_2A_2))$$

where A_1 = inner surface area of the small tube = $(2\pi) \times (r_1) \times (L)$
A_2 = outer surface area of the small tube = $2(\pi) \times (r_2) \times (L)$
p = constant = $2(\pi)(L)$
L = axial length of the small tube
Q = heat transfer rate
r_1, r_2 are the inner and outer radii of the small tube
T_1 = temperature of hot fluid inside the small tube
T_2 = temperature at the inner surface of the small tube
T_3 = temperature at the outer surface of the small tube
T_4 = temperature of cold fluid between the small and large tubes

EXAMPLE 1: heat transfer
The temperatures of the inside and outside of a boiler's refractory wall are 1800°F and 200°F, respectively. The heat conductivity of the wall is $k = 0.02$ Btu/h-ft-°F). The area of the refractory wall is $A = 50\,\text{ft}^2$ and the thickness of the refractory wall is 6 inches. Calculate the heat transfer rate.

Given conditions

- Conductivity, $k = 0.02$ Btu/(h-ft-°F)
- Area of refractory wall, $A = 50\,\text{ft}^2$

Solution:

1. Write the basic heat conduction equation.

$$Q = -kA(T_2 - T_1)/(x_2 - x_1)$$

where
Q = heat transfer rate
k = conductivity
A = area
T_2, T_1 = temperatures at location at x_2 and x_1, respectively

2. Calculate the heat transfer rate.

$$\begin{aligned} Q &= -0.02(50)(200 - 1800)/(6/12) \\ &= 3200\,\text{Btu/h} \end{aligned}$$

EXAMPLE 2: heat transfer
A flat steel plate, 0.4 in thick, and a copper plate, 0.2 in thick, are laid one upon the other, leaving an air space of 0.0008 in between them. Calculate the thermal resistance of the composite wall.

Given conditions

- Steel conductivity, $k_1 = 30$ Btu/(h-ft-°F)
- Air gap conductivity, $k_2 = 0.015$ Btu/(h-ft-°F)
- Copper conductivity, $k_3 = 210$ Btu/(h-ft-°F)
- Thickness of steel, $a_1 = 0.4$ inches
- Thickness of air gap, $a_2 = 0.0008$ inches
- Thickness of copper, $a_3 = 0.2$ inches
- Area of heat transfer, $A = 5\,\text{ft}^2$

Solution:

1. Write the steel wall conduction equation.

$$T_1 - T_2 = Q(a_1)/(Ak_1)$$

2. Write the air gap conduction equation.

$$T_2 - T_3 = Q(a_2)/(Ak_2)$$

3. Write the copper conduction equation.

$$T_3 - T_4 = Q(a_3)/(Ak_3)$$

4. Combine equations in items 1, 2, and 3.

$$T_1 - T_4 = (a_1/k_1 + a_2/k_2 + a_3/k_3)(Q/A)$$

5. Write the thermal resistance equation (R).

$$R = (a_1/k_1 + a_2/k_2 + a_3/k_3)(1/A)$$

6. Calculate the value of R.

$$\begin{aligned} R &= (0.4/(12(30)) + 0.0008/(12(0.015)) + 0.2/(12(210))/5 \\ &= (1/900 + 1/225 + 1/12{,}600)/5 \\ &= 0.001127 \text{ Btu/(h-ft}^2\text{-°F)} \end{aligned}$$

The thermal resistance of the composite wall above is therefore influenced essentially by the air space, which, however small, cannot be avoided even by the most careful construction. The resistance without an air space would be

$$\begin{aligned} R &= (1/900 + 1/12{,}600)/5 \text{ Btu/(h-ft-°F)} \\ &= 0.000238 \end{aligned}$$

This value is less than one-quarter of that in step 6.

EXAMPLE 3: heat transfer

A boiler wall, 10 in thick and 20 ft^2 area, has thermal conductivity k and an average thermal conductivity 0.9 Btu/(h-ft-°F). The inner surface is exposed to a hot gas at an average temperature of 1000°F, and the outer surface is in contact with air at an average temperature of 150°F. If the gas and air film coefficients are 18 and 1.5 Btu/(h-ft^2-°F), respectively, compute:

(1) Overall heat transfer coefficient
(2) Overall thermal resistance
(3) Heat transfer rate

Given conditions

- Thermal conductivity, $k = 0.9$ Btu/(h-ft-°F)
- Inner boiler surface temperature, $T_{w1} = 1000$°F
- Outer boiler surface temperature, $T_{w2} = 150$°F
- Film coefficient at side 1, $h_1 = 18$ Btu/(h-ft^2-°F)
- Film coefficient at side 2, $h_2 = 1.5$ Btu/(h-ft^2-°F)
- Wall area, $A = 20$ ft^2

Solution to question (1)

1.1. Write side 1 convection equation.

$$T_1 - T_{w1} = Q/(h_1 A)$$

1.2. Write the boiler wall conduction equation.

$$T_{w1} - T_{w2} = aQ/(kA)$$

1.3. Write side 2 convection equation.

$$T_{w2} - T_2 = Q/(h_2 A)$$

where Q = heat transfer rate
T_1 = hot fluid temperature
T_{w1} = inner boiler surface temperature
T_{w2} = outer boiler surface temperature
T_2 = ambient temperature
k = thermal conductivity
h_1 = film coefficient at side 1
h_2 = film coefficient at side 2
A = boiler wall area

1.4. Combine equations.

$$T_1 - T_2 = (1/h_1 + a/k + 1/h_2)(Q/A)$$

1.5. Write the overall heat transfer coefficient equation (U).

$$1/U = (1/h_1 + a/k + 1/h_2)$$

1.6. Calculate the value of U.

$$\begin{aligned} 1/U &= (1/18 + 10/(12(0.9)) + 1/1.5) \\ &= 1.0482 \\ U &= 0.95\ \text{Btu/(h-ft}^2\text{-°F)} \end{aligned}$$

Solution to question (2)

2.1. Write the overall thermal resistance equation (R).

$$R = 1/AU$$

2.2. Calculate the R value.

$$\begin{aligned} R &= 1/(20(0.95)) \\ &= 0.05\ \text{(h-°F)/Btu} \end{aligned}$$

Solution to question (3)

3.1. Write the overall heat transfer equation.

$$Q = AU(T_{w1} - T_{w2})$$

3.2. Calculate the heat transfer rate.

$$\begin{aligned} Q &= 20(0.95)(1000 - 150) \\ &= 16{,}150\ \text{Btu/h} \end{aligned}$$

Radiation. The key phenomenon of radiation is that radiation incident on a body may be absorbed, reflected, or transmitted. The relationship of absorptivity (α), reflectivity (r), and transmissivity (t) of radiation is

$$\alpha + r + t = 1$$

The basic radiation equation for radiative energy is expressed as

$$P = \sigma \varepsilon T^4$$

where P = blackbody radiative power radiated per unit area, W/cm^2
σ = Stefan–Boltzmann constant, $5.68 \times 10^{-12}\ W/(cm^2 \cdot K^4)$
ε = emissivity of the substance, $0 < \varepsilon < 1$
T = absolute temperature of the substance, K

EXAMPLE: radiative heat transfer

A plasma arc is used to illustrate how radiative heat transfer is applied to waste treatment. The most common method of plasma generation is electrical discharge through a gas. The gas used is relatively unimportant in creating the discharge, but will ultimately affect the products formed. In passing through the gas, electrical energy is converted to thermal energy and is absorbed by gas molecules, which are activated into ionized atomic states, losing electrons in the process. Arc temperatures up to 10,000°C may be achieved along the centerline of the recirculation vortex. Radiation is emitted when molecules or atoms relax from highly activated states to lower energy levels.

The plasma, when applied to waste disposal, can be best understood by thinking of it as an energy-conversion and energy-transfer device. As the activated components of the plasma decay, their energy is transferred to waste materials exposed to the plasma. The wastes are then atomized, ionized, pyrolyzed, and finally destroyed as they interact with the decaying plasma species. Theoretically speaking, the destruction of wastes should result in simple molecules or atoms such as hydrogen, carbon monoxide, carbon, and hydrochloric acid. The off-gases from the plasma system are typically scrubbed to remove hydrochloric acid and then flared.

One of the key factors in waste destruction by plasma arc is radiative heat transfer. The following is an example to demonstrate how rapidly a carbon particle can increase its temperature from an atmospheric temperature to 2200°C under the radiative heat-transfer environment.

Once again, the radiative heat-transfer equation can be expressed as (Lee-83/07)

$$P = \sigma \varepsilon T^4$$

where P = blackbody radiative power radiated per unit area, W/cm^2
σ = Stefan–Boltzmann constant, $5.68 \times 10^{-12}\ W/(cm^2 \cdot K^4)$
ε = emissivity of the substance, $0 < \varepsilon < 1$
T = absolute temperature of the substance, K

This function is plotted in Fig. 1.5. Inspection of Fig. 1.5 shows that radiation as a heat-transfer method does not become significant until temperatures of the order of 1650°C are reached.

The spectrum of the blackbody radiator changes, of course, with temperature. Figure 1.6 shows the relative spectral distribution of the blackbody.

In the near-infrared, many materials exhibit very strong absorption coefficients. Coal, for example, has an absorption coefficient of the order of $10^5\ cm^{-1}$, which means (to a first approximation) that most of the incident energy is absorbed in a surface layer of the order of 10^{-5} cm thickness.

The rate of temperature rise on the surface of a coal particle can be estimated by the equation

$$T/t = P/MC \text{ and } M = \rho/\alpha$$

where T = temperature rise
t = unit time
P = radiative power
C = specific heat
M = mass
ρ = density
α = absorption coefficient

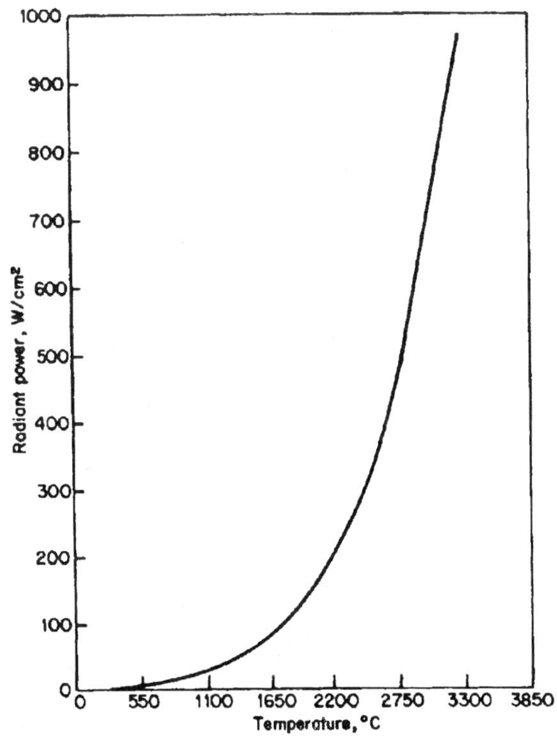

FIGURE 1.5 Blackbody radiant emission vs. temperature.

At 2200°C, the blackbody radiative power is of the order of 200 W/cm^2. The specific heat of coal is 0.9 J/(g · °F) and its density is 1.4 g/cm^2; therefore the rate of temperature rise on the surface of a coal particle is

$$
\begin{aligned}
T/t &= (200 \times 10^5)/(1.4 \times 0.9) \\
&= 1.58 \times 10^7 \ {}^\circ\mathrm{F/s}
\end{aligned}
$$

where 1 W = 1 J/s.

Thus, the time needed to raise the temperature to 2200°C (4000°F) on the surface of an absorbing coal particle is

$$
\begin{aligned}
t &= 4000/(1.58 \times 10^7) \\
&= 2.5 \times 10^{-4}\ \mathrm{s}
\end{aligned}
$$

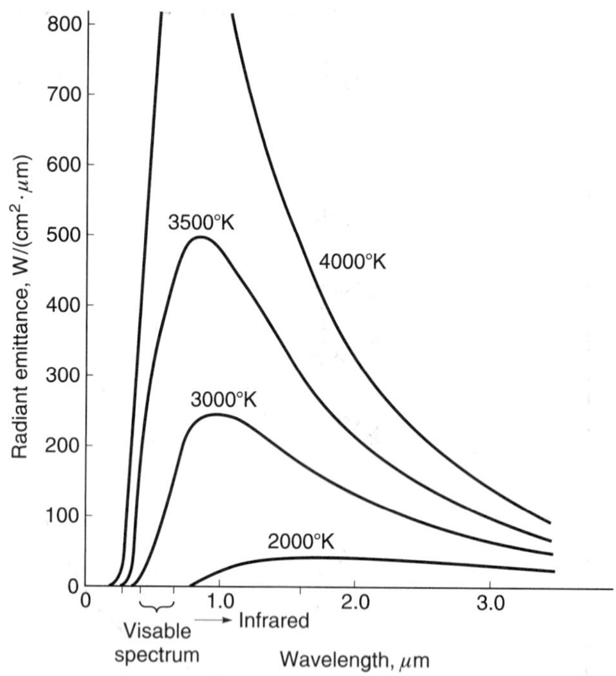

FIGURE 1.6 Spectral output of blackbody radiators at various temperatures.

2.25 Heating Value

This section covers:

- high heating value (HHV)
- higher heating value of combustible elements
- low heat value (LHV)

High heating value. The HHV represents the enthalpy change or heat released when a chemical compound is stoichiometrically combusted and the products of combustion are cooled to the initial temperature of the compound. Stoichiometric combustion requires that no oxygen be present in the flue gas following the complete combustion of the hydrocarbons.

Any fuel containing hydrogen yields water as one product of combustion. At atmospheric pressure, the partial pressure of the water vapor in the resulting combustion gas mixture is usually sufficiently high to cause water to condense out if the temperature is allowed to fall below 120–140°F. This causes liberation of the heat of vaporization of any water condensed. Thus, HHV is the heat transfer with liquid water in the products. HHV is also referred to as *higher heating value* or *gross heating value*. It is usually expressed as Btu/lb fuel or Btu/scf fuel (Lee-92/06).

Higher heating value of combustible elements. The analysis of fuels and wastes is determined by ASTM laboratory procedures. Samples of solid wastes may be shredded in the field or in the laboratory to reduce the size of the particles and hence of the sample needed for laboratory analysis. Liquids can be analyzed as is in the laboratory (Hasselriis-98/12).

The laboratory analysis will typically provide a proximate and ultimate analysis which determines the combustible components of fuels and wastes, carbon, hydrogen, sulfur and nitrogen, as well as the oxygen content, and the noncombustible components, moisture, and ash residues. The heating value may be obtained by bomb calorimeter or by use of empirical equations such as the Du Long equation below.

The heating value of the basic combustible elements carbon, hydrogen, sulfur, and nitrogen, determined by calorimeter, are given in Table 1.9.

When there is oxygen in the fuel or waste, account must be taken of the combination of oxygen with hydrogen to form water. Since 1 lb hydrogen reacts with 8 lb of oxygen, the energy which would be available from the oxidation of hydrogen would have to be reduced by 1/8, or 61,004/8 = 7400 Btu/lb. In real wastes which contain inorganic oxides, some of the oxygen is already consumed, so this coefficient needs to be revised.

Analyses of various empirical equations which aim to correlate calorimeter data with formulas based on ultimate analysis have given the empirical equations of Du Long, and of Boie and Vondracek, which result in the best match for municipal wastes.

Du Long:

$$\text{HHV [Btu/lb]} = 14{,}455[\text{C}] + 62{,}028[\text{H}] + 4050[\text{S}] + 2700[\text{N}] - 7753[\text{O}]$$

Boie and Vondracek:

$$\text{HHV [Btu/lb]} = 15{,}122[\text{C}] + 50{,}000[\text{H}] + 4500[\text{S}] + 2700[\text{N}] - 4771[\text{O}]$$

The Du Long formula is generally used in the utility field, and the Boie formula is often used in the municipal waste field. They give slightly different results.

TABLE 1.9 Heating Value of Combustible Components

Element	Btu/lb
Carbon	14,455
Hydrogen	61,004
Sulfur	4,044
Nitrogen	2,538
Oil	18,400
Natural gas	21,800

EXAMPLES

Fuel oil [#2]: $[\text{C}] = 0.86 \quad [\text{H}] = 0.14;$

$$\text{HHV} = 14{,}455(0.86) + 61{,}004(0.14) = 20.972\ \text{Btu/lb}$$

Methane $[\text{C}] = 0.75, \quad [\text{H}] = 0.25, \quad \text{HHV} = 26{,}092\ \text{Btu/lb}$

Heating value of cellulose. To calculate the heating value of cellulose, $C_6H_{10}O_5$, we need the weight fractions of the C, H, and O (Hasselriis-98/12).

Carbon: 6 moles at 12 = 72/162 = 0.4444
Hydrogen: 10 moles at 1 = 10/162 = 0.0617
Oxygen: 5 moles at 16 = 80/162 = 0.4938

The weight formula is then $C_{0.444}H_{0.062}O_{0.494}$, and the heating value is

$$\text{HHV [Btu/lb]} = 15{,}122[\text{C}] + 50{,}000[\text{H}] - 4771[\text{O}]$$
$$\text{HHV [Btu/lb]} = 15{,}122[0.444] + 50{,}000[0.062] - 4771[0.4938]$$
$$= 6714 + 3100 - 2356 = 7458\ \text{Btu/lb}$$

Low heat value. The LHV is similar to the HHV except that the water produced by the combustion is not condensed but retained as vapor. It is also referred to as the *lower heating value* or *net heating value* (EPA-84/09).

The higher heating value is based on the heat liberated in a calorimeter. When the products of combustion contain moisture, this moisture will be condensed in the calorimeter, liberating the heat of condensation of the water vapor. In actual combustion systems, this water vapor is not condensed. The lower heating value represents the heat released when the vapor is not condensed. The equation for the lower heating value (Hasselriis-98/12) is

$$\text{LHV} = \text{HHV} - 1050[\text{H}_2\text{O} + 9(\text{H} - \text{Cl}/35.5 - \text{F}/19)]$$

Lower heating value is the energy which can be captured by heat recovery devices such as boilers, taking into account the fact that the vapors formed by combustion cannot economically be condensed to recover the heat of vaporization.

Net heating value is essentially the same as lower heating value. Net heating value is the term used when there is moisture in the fuel as well as the moisture produced by the combustion of hydrogen.

Consider first the commonly used (and often misused) term heat of combustion H_c. It represents the enthalpy change due to the chemical transformation of the reactive species referenced to a standard condition; 1 atmosphere and 20°C or 70°F are used here for metric and British units, respectively. The values for the wastes and auxiliary fuels are usually supplied by the applicant and are frequently published for specific compounds. They can also be determined by assuming the heats of formation H_f of the compounds. Suffice it to say for now that the heat of combustion is normally given for a fuel or waste as either the lower heating value (LHV) or the higher heating value (HHV). The two are best explained by considering the combustion of hydrogen by the following equation:

$$H_2 + (1/2)O_2 \rightarrow H_2O$$

The hydrogen and oxygen enter the system as gases. The water forms as a gas, but if the system is allowed to return to the standard temperature it will condense and release its latent heat of vaporization H_v. If the heat of combustion is measured before the water is condensed, the value is termed the LHV, and after the water is condensed, the HHV. Hence, for this system, the LHV and HHV are related by the following equation:

$$\text{HHV} = \text{LHV} + H_v$$

The same equation holds when the materials being burned are more complicated, but one must then determine the latent heat of vaporization of the water entering the combustion chamber as the compound and the water formed by the combustion. In summary, the net, or lower, heating value (LHV) is calculated by using the heat of formation for water vapor rather than condensed liquid water. It is recommended that the LHV be used for all thermal energy and heat transfer calculations in combustion applications because the water exits in the combustion chamber as a vapor.

EXAMPLE 1: heating value
Consider the combustion of methane gas, and determine its heat of combustion.

$$CH_4 + 2O_2 \rightarrow CO_2 + 2H_2O + H_c$$

$$H_c = H_f(CO_2) + 2H_f(H_2O) - H_f(CH_4) - 2H_f(O_2)$$

where H_c = the standard heat of combustion of CH_4 (kcal/g-mole) at 25°C, 1 atm
H_f = the standard heat of formation of the individual compound in the reaction (kcal/g-mole)

The LHV of methane is calculated from the standard heat of formation for CH_4, CO_2, and H_2O.

$$\begin{aligned} H_f &= 94.051 + (2 \times 68.315) - 17.88 - 0 \\ &= 212.8 \text{ kcal/g-mole} \\ &= 383{,}040 \text{ Btu/lb-mole} \end{aligned}$$

(Note: the standard heat of formation of a basic element is zero by definition.)

EXAMPLE 2: heating value
Given a gas mixture, determine its gross (or higher) heating value (HHV) (EPA-84/09, p. 36).

Given conditions

- Composition (mole fraction) of the gas mixture shown in Table 1.10.

Solution:

1. Determine the gross heating value of each component HHV from Table 1.11 (EPA-84/09, p. 37).

2. Calculate the gross heating value of the gas mixture HHV in Btu/scf.

$$\begin{aligned} \text{HHV} &= (0.0515)(0) + (0.811)(1013) + (0.0967)(1792) + (0.0351)(2590) + (0.0056)(3370) \\ &= 1104.71 \text{ Btu/scf} \end{aligned}$$

TABLE 1.10 Composition of Gas Mixtures

Composition	Mole fraction
N_2	0.0515
CH_2	0.8111
C_2H_6	0.0967
C_3H_8	0.0351
C_4H_{10}	0.0056
Total	1

TABLE 1.11 Gross Heating Value

Composition	Mole fraction	HHV_i (Btu/scf)	Gross heating value
N_2	0.0515	0	0.00
CH_4	0.8111	1013	821.64
C_2H_6	0.0967	1792	173.29
C_3H_8	0.0351	2590	90.91
C_4H_{10}	0.0056	3370	18.87
Total			1104.71

EXAMPLE 3: heating value
This example is to show the relationship between HHV and LHV. A waste stream has a higher heating value (HHV) of 12,000 Btu/lb at 70°F. The moisture content and the mass fraction of the waste stream are given below.

$$H_2O = 0.12 \text{ lb/lb-waste}$$
$$H = 0.095$$
$$Cl = 0.005$$
$$F = 0.002$$

Note: The sum of the mass fraction is not equal to one, because only these elements affect the LHV calculation.

Solution:

1. Write the net heating value equation in terms of HHV.

$$LHV = HHV - 1050(H_2O + 9(H - Cl/35.5 - F/19))$$

(This equation was given in EPA-81/09, p. 4-86, and EPA-86/03, p. 80.)

2. Calculate the LHV.

$$\begin{aligned} LHV &= 12{,}000 - 1050(0.12 + 9(0.095 - 0.005/35.5 - 0.002/19)) \\ &= 10{,}979 \text{ Btu/lb} \end{aligned}$$

Notes

- 1050 is the heat of vaporization of water at 25°C
- The H_2O term accounts for the energy to vaporize the water
- The H_2 term accounts for the water formed due to the hydrogen present in the waste
- Other terms account for the hydrogen reacting with the halogens present
- There are 9 lb of water formed for every 1 lb of H formed or reacted

The following equations are recommended by the author of EPA-86/03, Dr. Louis Theodore, for use in estimating heating values of wastes if experimental data are not available (EPA-86/03, p. 83). For this calculation, the equations are termed the Theodore approximation.

$$\text{HHV, Btu/lb} = 14{,}000(\text{C}) + 54{,}500(\text{H}) - 0.125(\text{O}) + 150(\text{Cl}) + 4500(\text{S})$$

$$\text{LHV, Btu/lb} = 14{,}000(\text{C}) + 4500(\text{H}) - 0.125(\text{O}) + 760(\text{Cl}) + 4500(\text{S})$$

HHV can also be calculated by the Du Long approximation (Brunner-84, p. 145).

$$\text{HHV, Btu/lb} = 14{,}455(\text{C}) + 62{,}028(\text{H}) - 7753.5(\text{O}) + 4050(\text{S})$$

where C, H, O, and S are the mass fractions of carbon, hydrogen, oxygen, and sulfur, respectively
O represents the weight fraction of O and not O_2
The heating value of fuels is usually available. Typical values for residual oil (No. 6), distillate (No. 2) oil, and natural gas are 17,500 Btu/lb, 18,300 Btu/lb, and 19,700 Btu/lb (1000 Btu/scf), respectively. These values assume complete combustion of the fuel.

EXAMPLE 4: heating value
The analyses of three fuels are provided in Table 1.12. Estimate the HHV and LHV.

TABLE 1.12 Analyses of Three Fuels

	No. 6 oil*	No. 2 oil†	Natural gas‡
C	0.8852	0.872	0.693
H	0.1087	0.123	0.227
O	0.0006	0	0
N	0.001	0	0.08
S	0.004	0.005	0
Ash	0.0005	0	0
HHV‡	17,500	18,300	19,700

No. 6 oil = residual fuel oil.
No. 2 oil = distillate fuel oil.
HHV = high heating value (Btu/lb).
*from EPA-86/03, p. 73.
†from EPA-81/09, p. 4-87.

Solution:

1. Calculate HHV.

1.1 Estimate HHV for No. 6 oil by the Theodore approximation.

$$\begin{aligned}\text{HHV} &= 14{,}000(\text{C}) + 54{,}500(\text{H}) - 0.125(\text{O}) + 150(\text{Cl}) + 4500(\text{S}) \\ &= 14{,}000(0.8852) + 54{,}500(0.1087) - 0.125(0.0006) + 150(0) + 4500(0.0040) \\ &= 18{,}335 \text{ Btu/lb}\end{aligned}$$

1.2. Estimate HHV for No. 6 oil by the Du Long approximation.

$$\begin{aligned}\text{HHV} &= 14{,}455(\text{C}) + 62{,}028(\text{H}) - 7753.5(\text{O}) + 4050(\text{S}) \\ &= 14{,}455(0.8852) + 62{,}028(0.1087) - 7753.5(0.0006) + 4050(0.0040) \\ &= 19{,}550 \text{ Btu/lb}\end{aligned}$$

1.3. Compare the estimated HHV values with the reported HHV value.
Deviation from the Theodore approximation

$$= (18{,}335 - 17{,}500)/17{,}500$$
$$= 0.0477$$
$$= 4.77\%$$

Deviation from the Du Long approximation

$$= (19{,}550 - 17{,}500)/17{,}500$$
$$= 0.1171$$
$$= 11.71\%$$

2. Calculate LHV.

2.1. Estimate LHV for No. 6 oil by the Theodore approximation.

$$\begin{aligned} \text{LHV} &= 14{,}000(\text{C}) + 4500(\text{H}) - 0.125(\text{O}) + 760(\text{Cl}) + 4500(\text{S}) \\ &= 14{,}000(0.8852) + 4500(0.1087) - 0.125(0.0006) + 760(0) + 4500(0.0040) \\ &= 12{,}906 \text{ Btu/lb} \end{aligned}$$

2.2. Estimate LHV for No. 6 oil by the equation given in EPA-81/09, p. 4-86, and EPA-86/03, p. 80.

$$\begin{aligned} \text{LHV} &= \text{HHV} - 1050(\text{H}_2\text{O} + 9(\text{H} - \text{Cl}/35.5 - \text{F}/19)) \\ &= 17{,}500 - 1050(0 + 9(0.1087 - 0/35.5 - 0/19)) \\ &= 16{,}473 \text{ Btu/lb} \end{aligned}$$

EXAMPLE 5: heating value
Ciba–Geigy conducted a trial burn on one of their incinerators during November 12–17, 1984. The POHCs (Principal Organic Hazardous Constituents) and other parameters are provided in the following given conditions.

1. Calculate the heating value of the POHCs by heat of combustion.
2. Calculate the heating value of the POHCs by Du Long approximation.
3. Compare the calculated heating values with Ciba–Geigy's reported result.

Given conditions

- The flow rate of POHCs, $m = 1200$ lb per hour
- POHCs, their mass fraction (x_i) and their heat combustion (H_c) are provided in Table 1.13.

Solution:
Solution 1: by actual calculation.

1. Calculate the heat content of each POHC species by multiplying the mass fraction by the heat of combustion (Table 1.14).
2. Calculate the heating value.

The sum of the heat content column is the heating value per pound of POHC. This means that 14,707 Btu/lb is the heating value of the combined POHC compounds.

3. Compare with Ciba–Geigy's reported value.

The heating value reported by Ciba–Geigy was 15,200 Btu/lb. The deviation of the heating value calculated from this example and Ciba–Geigy's reported value is

$$(15{,}200 - 14{,}707)/15{,}200 = 3.24\%$$

4. Calculate the total heat load of the POHCs.

$$\begin{aligned}\text{Total heat content} &= 14{,}707(1200)\\ &= 17.65 \times 10^6 \text{ Btu/h}\end{aligned}$$

Solution 2: by the Du Long approximation.

1. Calculate the flow rate of POHC species (Table 1.15).

2. Analyze the chemical composition.

2.1. Calculate the ultimate composition of hexachloroethane (C_2Cl_6) (Table 1.16).

where x_i represents the element of the compound
x_i/M = mass fraction
M means the molecular weight of the compound. This is the sum of the molecular weights of each element in Table 1.16.

2.2. Calculate the ultimate composition of tetrachloroethene (C_2Cl_4) (Table 1.17).

2.3. Calculate the ultimate composition of chlorobenzene (C_6H_5Cl) (Table 1.18).

2.4. Calculate the ultimate composition of toluene (C_7H_8) (Table 1.19).

2.5. Calculate the ultimate composition of POHCs by combining items 2.1, 2.2, 2.3, and 2.4 (Table 1.20).

3. The Du Long approximation is expressed as shown below (Marks-67, p. 7-11).

The value of this Du Long approximation represents the high heating value of the substances.

$$\begin{aligned}\text{HHV} &= 14{,}455\text{ C} + 62{,}028\text{ H}_2 - 7753.5\text{ O}_2 + 4050\text{ S}\\ &= (14{,}455)(0.75) + (62{,}028)(0.07)\\ &= 10{,}841 + 4342\\ &= 15{,}183 \text{ Btu/lb (high heating value of volatile substances)}\end{aligned}$$

TABLE 1.13 POHCs Parameters

POHCs	POHC x_i	H_c (Btu/lb)
Hexachloroethane, C_2Cl_6	0.0487	828
Tetrachloroethene, C_2Cl_4	0.0503	2141
Chlorobenzene, C_6H_5Cl	0.2952	11,876
Toluene, C_7H_8	0.6058	18,246
Total	1	

TABLE 1.14

POHC	POHC x_i	H_c (Btu/lb)	Heat content (Btu/lb)
Hexachloroethane	0.0487	828	40
Tetrachloroethene	0.0503	2141	108
Chlorobenzene (C_6H_5Cl)	0.2952	11,876	3506
Toluene (C_7H_8)	0.6058	18,246	11,053
Total	1		14,707

TABLE 1.15 Flow Rate of POHC Species

POHC	POHC x_i	1200 lb/h
Hexachloroethane (C_2Cl_6)	0.0487	58.44
Tetrachloroethene (C_2Cl_4)	0.0503	60.36
Chlorobenzene (C_6H_5Cl)	0.2952	354.24
Toluene (C_7H_8)	0.6058	726.96
Total	1	1200.00

TABLE 1.16 Composition of Hexachloroethane

x_i	C	H	Cl	O	M
C_2Cl_6	24	0	213	0	237.00
x_i/M	0.10	0.00	0.90	0.00	
58.44 lb	5.84	0.00	52.60	0.00	

TABLE 1.17 Composition of Tetrachloroethene

x_i	C	H	Cl	O	M
C_2Cl_4	24	0	142	0	166.00
x_i/M	0.14	0.00	0.86	0.00	
60.36 lb	8.45	0.00	51.91	0.00	

TABLE 1.18 Composition of Chlorobenzene

x_i	C	H	Cl	O	M
C_6H_5Cl	72	5	35.5	0	112.50
x_i/M	0.64	0.04	0.32	0.00	
354.24 lb	226.71	14.17	113.36	0.00	

TABLE 1.19 Composition of Toluene

x_i	C	H	Cl	O	M
C_7H_8	84	8	0	0	92.00
x_i/M	0.91	0.09	0.00	0.00	
726.96 lb	661.53	65.43	0.00	0.00	

TABLE 1.20 Composition of POHCs

x_i	C	H	Cl	O	M
C_2Cl_6	5.84	0	52.6	0	58.44
C_2Cl_4	8.45	0	51.91	0	60.36
C_6H_5Cl	226.71	14.17	113.36	0	354.24
C_7H_8	661.53	65.43	0	0	726.96
Sum(x_i)	902.53	79.60	217.87	0.00	1200.00
x_i/sum(x_i)	0.75	0.07	0.18	0.00	1.00

4. Compare with Ciba–Geigy's reported value.

The heating value reported by Ciba–Geigy was 15,200 Btu/lb. The deviation of the heating value calculated from this example and Ciba–Geigy's reported value is

$$(15{,}200 - 15{,}183)/15{,}200 = 0.11\%$$

5. Calculate the deviation between the two methods.

$$\begin{aligned}\text{Deviation} &= (\text{solution 2} - \text{solution 1})/(\text{solution 1})\\ &= (15{,}183 - 14{,}707)/14{,}707\\ &= 3.24\%\end{aligned}$$

6. Calculate the total heat content.

$$\begin{aligned}\text{Total heat content} &= 15{,}183(1200)\\ &= 18.22 \times 10^6\ \text{Btu/h}\end{aligned}$$

7. Low heating value is expressed as

$$\begin{aligned}\text{Low heating value (LHV)} &= \text{HHV} - [(92.7) \times \text{hydrogen fraction of the substance}]\\ &= 15{,}181 - (92.7(0.07))\\ &= 15{,}175\ \text{Btu/lb}\end{aligned}$$

Comment: The weight fraction of all elements in the reaction such as C, H_2, O_2, S_2, N_2, Cl, ash, etc., should add to make 100%. However, only the C, H_2, O_2, and S_2 fractions are used for the Du Long approximation.

2.26 Henry's Law

Henry's law states that the partial pressure of a solute in equilibrium in a solution is proportional to its mole fraction in the limit of zero concentration (dilute solution). In air pollution applications, the *solute* refers to the pollutant (EPA-81/12, p. 4-5).

For dilute solutions where the components do not interact, the resulting partial pressure (p) of a component "A" in equilibrium with other components in a solution can be expressed as: $p = x_A H$

where p = equilibrium partial pressure of component A over a solution
x_A = mole fraction or concentration of A in the liquid phase, g-mole/cm^3
H = Henry's law constant (atm-cm^3)/(g-mole) of pure A at the same temperature and pressure as the solution

Unlike Henry's law, Raoult's law is for concentrated solutions (EPA-84/09, p. 39). For more information on H values, see 40CFR265.10-84 and Appendix VI to 40CFR265.

EXAMPLE: Henry's law
Given Henry's law constant and the partial pressure of a solute, determine the maximum mole fraction (concentration) of a solute that can be dissolved in solution (EPA-84/09, p. 39).

Given conditions

- Partial pressure of hydrogen sulfide, $H_2S = 0.01$ atm
- Total pressure = 1 atm
- Henry's law constant = 483 atm/mole fraction

Solution:

1. Write the equation describing Henry's law.

$$p(H_2S) = xH$$

where $p(H_2S)$ = partial pressure of H_2S, atm
H = Henry's law constant, atm/mole fraction
x = mole fraction of H_2S in solution

For an ideal gas, the partial pressure of a component in a gas mixture is given by

$$p(H_2S) = y(H_2S)P$$

where P = total pressure.

2. Calculate the maximum mole fraction of H_2S that can be dissolved in solution.

$$x(H_2S) = p(H_2S)/H$$
$$= 0.01/483$$
$$= 2.07 \times 10^{-5}$$

2.27 Ideal Gas

An ideal gas is an imaginary (or hypothetical) gas or vapor which obeys the ideal gas law at pressure approaching to zero (0) (very low density). No real gas obeys the ideal gas law exactly over all ranges of temperature and pressure. Although the lighter gases (hydrogen, oxygen, air, etc.) at ambient conditions approach ideal gas law behavior, the heavier gases such as sulfur dioxide and hydrocarbons, particularly at high pressures and low temperatures, deviate considerably from the ideal gas law. Despite these deviations, the ideal gas law is routinely used in all air pollution calculations (EPA-84/09). In virtually all but the most detailed calculations, the gases for a combustion process can be assumed to obey the ideal gas law.

The *ideal gas law* is a law or equation describing the relationship among pressure, volume, and temperature of an ideal gas. All gases which follow the ideal gas law will have the same molar volume at a specified temperature and pressure, such as at standard temperature and pressure (STP). Equations applicable to the ideal gas law are referred to as the ideal gas equations, which can generally be expressed in the following two forms (for more ideal gas equation applications, see the section on "Standard Condition"):

$$PV = nR_uT = mRT, \qquad n = m/M, \qquad R = R_uM \tag{1.13}$$

$$P = \rho RT \tag{1.14}$$

where
P = absolute pressure (psi, atm, kPa, dynes/m^2)
V = volume (ft^3)
T = absolute temperature (°R or K)
m = mass (lb or g)
ρ = density (g/cm^3, g/m^3, lb/ft^3)
M = molecular weight (lb/lb-mole, or g/g-mole)
n = number of moles (lb-moles, g-moles or kg-moles)
R = gas constant (different gases have different R values)
R_u = universal gas constant (this constant is applicable for all gases)
= 82.0575 (atm-cm^3)/(g-mole-K)
= 0.0821 (atm-m^3)/(kg-mole-K)
= 0.7302 (atm-ft^3)/(lb-mole-°R)
= 0.0821 (atm-L)/(g-mole-K)
= 83.144 × 10^6 (g-cm^2)/(s^2-g-mole-K)
= 8.3144 × 10^4 (kg-m^2)/(s^2-kg-mole-K)
= 1544 (lb$_f$-ft)/(lb-mole-°R)
= 4.9686 × 10^4 (lb$_m$-ft^2)/(s^2-lb-mole-°R)
= 8.3144 × 10^6 (Pa-cm)/(s-g-mole-K)
= 8.3144 × 10^3 (kPa-cm)/(s-g-mole-K)
= 8.3144 × 10^5 (kPa-m)/(s-g-mole-K)
= 0.1724 (psi-ft^3)/(lb-mole-°R)
= 10.73 (psia-ft^3)/(lb-mole-°R)
= 21.83 (in-Hg)(ft^3)/(lb-mole-°R)
= 62.4 (mm-Hg-L)/(g-mole-K)
= 555 (mm-Hg)(ft^3)/(lb-mole-°R)
= 1.9872 Btu/(lb-mole-°R)
= 1.9872 cal/(g-mole-K)
= 8.3144 J/(g-mole-K)
= 8.3144 (N-m)/(g-mole-K)

For an ideal gas in states 1 and 2, Eqs. (1.13) and (1.14) can be modified to become:

$$\text{Volume equation: } V_2/V_1 = (P_1/P_2)(T_2/T_1) \tag{1.15}$$

$$\text{Density equation: } \rho_2/\rho_1 = (P_2/P_1)(T_1/T_2) \tag{1.16}$$

$$\text{Concentration equation: } C_2/C_1 = (P_2/P_1)(T_1/T_2) \tag{1.17}$$

where V_1 = volume at condition 1
V_2 = volume at condition 2
P_1 = pressure at condition 1
P_2 = pressure at condition 2
T_1 = absolute temperature at condition 1
T_2 = absolute temperature at condition 2
ρ_1 = density at condition 1
ρ_2 = density at condition 2
C_1 = concentration at condition 1
C_2 = concentration at condition 2
The subscript 1 denotes the observed conditions and the subscript 2 the standard conditions, or vice versa

References used in this section: Holman-74, p. 24; Jones-60, p. 151; Wark-66, p. 42.

EXAMPLE 1: ideal gas law
Determine the density of an ideal gas with the following given conditions.

Given conditions

- Pressure: 1.5 atm
- Temperature: 70°F
- Molecular weight: 29
- Universal gas constant: 0.7302 (atm-ft^3)/(lb-mole-°R)

Solution:

1. Write the ideal gas equation.

$$PV = nR_u T$$

where P = absolute pressure
V = volume
T = absolute temperature
n = number of moles
R_u = universal gas constant

2. Rewrite the ideal gas equation in terms of density p.

Remember that the density is defined as the mass divided by the volume.

$$\begin{aligned}\rho &= m/V\\ &= nM/V\\ &= PM/(R_u T)\end{aligned}$$

3. Calculate the density of the gas using the appropriate value of R_u.

$$\begin{aligned}\rho &= PM/(R_u T)\\ &= (1.5)(29)/(0.7302(460 + 70))\\ &= 0.1124\,\text{lb/ft}^3\end{aligned}$$

EXAMPLE 2: ideal gas law
Determine the pressure of 10 lb of air at 70°F occupying a volume of 30 ft^3.

Given conditions

- Air weight, $m = 10\,\text{lb}$
- Air molecular weight, $M = 29\,\text{lb/lb-mole}$
- Temperature, $T = 70^\circ\text{F}$
- Volume, $V = 30\,\text{ft}^3$
- Universal gas constant, $R_u = 1545\,(\text{lb-ft})/(\text{lb-mole-}^\circ\text{R})$

Solution:

1. Assume that nitrogen is an ideal gas.
2. Write the ideal gas equation.

$$PV = nR_u T$$

where
P = absolute pressure
V = volume
T = absolute temperature
n = number of moles
R_u = universal gas constant

3. Rewrite the ideal gas equation in terms of pressure.

$$\begin{aligned} P &= nR_u T/V \\ &= (m/M)R_u T/V \end{aligned}$$

4. Calculate the pressure.

$$\begin{aligned} P &= 10(1545)(460 + 70)/(30(29)) \\ &= 9412\,\text{psfa} \\ &= 65.36\,\text{psia} \end{aligned}$$

2.28 Mass and Weight

EXAMPLE: A person in a spaceship has 300 newtons weight. The gravitational acceleration at this particular location is $15.0\,\text{ft/s}^2$. Calculate the mass of the man and his weight at sea level on Earth.

Given conditions: mass and weight

- Man's weight: 300 newtons
- $g = 15\,\text{ft/s}^2 = 4.572\,\text{m/s}^2$
- $1\,\text{ft} = 0.3048\,\text{m}$
- $1\,\text{lb}_\text{m} = 0.4536\,\text{kg}$
- $1\,\text{lb}_\text{f} = 4.4481\,\text{newtons}$

Solution:

$$\text{Apply equation } W = mg/g_c$$
$$300 \text{ newtons} = m(4.572\,\text{m/s}^2)/(1.0\,\text{newton-m/kg-s}^2)$$
$$m = 65.62\,\text{kg}$$
$$= 144.66\,\text{lb}_\text{m}$$

$$\text{At sea level, } g = 32.174\,\text{ft/s}^2 = 9.8066\,\text{m/s}^2$$
$$W = mg/g_c = (65.6168\,\text{kg})(9.8066\,\text{m/s}^2)/(1\,\text{newton-m/kg-s}^2)$$
$$= 643.48\,\text{newtons}$$
$$= 144.66\,\text{lb}_\text{f}$$

Note: This example shows that 1 lb_m weighs 1 lb_f at sea level.

2.29 Molar Fraction

A molar fraction, or molecular fraction, is a ratio employed in expressing concentrations of solutions and mixtures. The mole fraction of any component of a mixture or solution is defined as the number of moles of that component divided by the sum of the number of moles of all components (EPA-84/09).

EXAMPLE 1: molar fraction
A mixture contains 40 pounds of O_2, 25 pounds of SO_2, and 30 pounds of SO_3. Determine the weight and mole fractions of each component.

Given conditions

- $O_2 = 40$ lb
- $SO_2 = 25$ lb
- $SO_3 = 30$ lb

By definition:

- Weight fraction = (weight of a component)/(total weight)
- Moles of a component = (weight of a component)/(molecular weight of the component)
- Mole fraction = (moles of a component)/(total moles)

Solution:

1. Determine the molecular weight of each component.

- Molecular weight of $O_2 = 32$
- Molecular weight of $SO_2 = 64$
- Molecular weight of $SO_3 = 80$

The atomic weight is needed to determine the molecular weight, e.g., the atomic weight of the oxygen atom (O) is 16, and the molecular weight of the oxygen molecule (O_2) is 32. The atomic weight of sulfur is 32.

2. Calculate the weight fraction of each component (Table 1.21).

3. Calculate the molar fraction of each component (Table 1.22).

TABLE 1.21 Weight Fractions

Compound	Weight	Weight fraction
O_2	40	0.42
SO_2	25	0.26
SO_3	30	0.32
Total	95	1.00

TABLE 1.22 Molar Fractions

Compound	(1) Weight	(2) Weight fraction	(3) Molar weight	(4) Mole	(5) Molar fraction
O_2	40	0.42	32	1.25	0.62
SO_2	25	0.26	64	0.39	0.19
SO_3	30	0.32	80	0.38	0.19
Total		1.00		2.02	1.00

Notes

- Column (1): given data
- Colume (2): weight fraction
- Column (3): given data
- Column (4): column (1) divided by column (3); mole = (weight)/(molar weight)
- Column (5): moles of each component divided by the total moles in the mixture, i.e., 2.02

The procedure can be extended to calculate the average molecular weight of a mixture.

EXAMPLE 2: molar fraction
A mixture of gases and the mole percent of each gas in the mixture are given in Table 1.23. Determine the average molecular weight.

Solution:

1. Determine the molar weight of each component.

The molar weight of each component is provided in Table 1.24.

2. Multiply the molecular weight by its mole fraction.

The results are provided in Table 1.25.

3. Calculate the average molecular weight of the gas mixture.

$$\begin{aligned}\text{Average molecular weight} &= 22.12 + 6.40 + 0.31 + 0.08 \\ &= 28.91 \text{ lb/lb–mole}\end{aligned}$$

The concentration of gases in a stack or mixture of gases is typically determined by an Orsat analysis (CO, O, and CO are measured and N_2 is calculated). The concentrations are expressed on a mole (or volume) percent basis. These values are on a dry basis, i.e., the amount of water is not considered.

TABLE 1.23 Mole Fractions

Gas composition	Mole fraction
N_2	0.79
O_2	0.2
CO_2	0.007
CO	0.003
Total	1

TABLE 1.24 Molar Weights

(1) Gas composition	(2) Mole fraction	(3) Molar weight
N_2	0.79	28
O_2	0.2	32
CO_2	0.007	44
CO	0.003	28
Total	1	

TABLE 1.25 Molar Fractions

Gas composition	Mole fraction	Molar weight	Average molecular weight (lb/lb-mole)
N_2	0.79	28	22.12
O_2	0.2	32	6.40
CO_2	0.007	44	0.31
CO	0.003	28	0.08
Total	1		28.91

2.30 Molar Volume

The molar volumes of ideal gases at standard conditions are provided in Table 1.26 (EPA-80/02, p. 2-16).

TABLE 1.26 Molar Volume at Standard Conditions

	Universal scientific	Natural gas industry
Temperature	0°C (273.15 K)	60°F (530 °R)
Pressure	1 atm (1.013×10^5 Pa)	30 in. Hg
Molar volume*	22.4 L/g-mole 22.4 m^3/kg-mole 359 ft^3/lb-mole	379 ft^3/lb-mole

*See standard condition section for molar volume calculations.

2.31 Molar Weight of Common Combustion Compounds

During the analysis of combustion gases, the atomic weight and the molar (molecular) weight of several compounds are important and they are frequently used. This information is provided in Table 1.27.

TABLE 1.27 Atomic and Molecular Weights

Compounds	Atomic weight	Molecular weight
Air	–	29
C	12	12
Cl_2	35.5	71
CO	–	28
CO_2	–	44
H_2	1	2
H_2O	–	18
N_2	14	28
NO	–	30
NO_2	–	46
O_2	16	32
P	31	31
P_2O_5	–	142
S	32	32
SO_2	–	64
SO_3	–	80

2.32 Mole and Avogadro's Number

A mole is defined as the number of atoms or molecules which produce a weight equal to the atomic (or molecular) weight of the substance. The term mole is technically incorrect for elements; the recognized designation for an element is g-atom or lb-atom but the term mole is commonly used to specify both elements and compounds and there is no need to differentiate between the two.

The mole has units which must be specified, i.e., g-mole or lb-mole. A g-mole is an accumulation of (approximately) 6.023×10^{23} atoms or molecules. This is known as *Avogadro's number*. In English units a pound mole (lb-mole) is 2.73×10^{26} atoms or molecules. To illustrate, 1 lb-mole of carbon (atomic weight 12) is 12 lb, 1 g-mole weighs 12 g. One lb-mole of oxygen gas (molecular weight 32) weighs 32 lb, 1 g-mole weighs 32 g.

The measure applies to mixtures of gases, e.g., a pound mole of air is approximately 29 lb. A pound mole of the gases just mentioned or any gas or mixture occupies the same volume under standard conditions at 60°F and 1 atm, i.e., 379 cu ft. This important fact, based on Avogadro's principle, establishes the basis for the calculation of gas flow rates and other factors which are a part of control equipment design (EPA-84/09). The conversion between gram and pound moles is equivalent to that between grams and pounds, i.e., 454 g-moles equals one lb-mole.

The number of moles can be calculated by $n = m/M$, where m = weight of a substance and M = molecular weight of the substance. For example: for 64 lb of O_2, $n = 64\,\text{lb}/(32\ \text{lb}/(\text{lb-mole})) = 2\,\text{lb-mole}$.

The actual number of atoms or molecules in a pound or gram mole (Avogadro's number) does not have to be known when performing a mass balance. One simply performs the calculations on the basis of moles taking care that the units (lb-mole or g-mole) are consistent throughout the calculations. The molecular weight of other compounds can be determined by adding the atomic weights of the constituent atoms.

EXAMPLE: number of moles
A container contains 10 lb HCl.

(1) How many lb-moles of HCl does it contain?
(2) How many g-moles of HCl does it contain?

Given conditions

- HCl weight, $m = 10\,\text{lb}$
- Molecular weight of hydrogen $= 1\,\text{lb/(lb-mole)}$
- Molecular weight of chlorine $= 35.5\,\text{lb/(lb-mole)}$
- $1\,\text{lb} = 453.5\,\text{g}$

Solution to question (1)

1.1. Calculate the molecular weight of HCl, M.

$$M = 1 + 35.5 \\ = 36.5$$

1.2. Write the lb-mole equation.

$$n = m/M$$

where n = number of moles
m = weight
M = molecular weight

1.3. Calculate the lb-moles of HCl.

$$n = 10/36.5 \\ = 0.27\,\text{lb-mole}$$

Solution to question (2)

2.1. Convert 10 lb to grams.

$$m = 10(453.5) = 4535\,\text{g}$$

2.2. Calculate the g-mole.

$$n = 4535/36.5 \\ = 124.2\,\text{g-mole}$$

2.33 Orsat Analysis

An orsat analyzer is an apparatus used to volumetrically analyze O_2, CO, and CO_2 by passing the mixture gases through various solvents that absorb them. A gas sample, on a dry basis, is first contained in an Orsat analyzer at a known temperature and pressure. A liquid which can absorb O_2 is then brought into contact with the sample gas. Because the temperature and pressure are maintained constant, the volume of the gas decreases as O_2 is absorbed. The reduced volume is recorded. Similarly, different liquids which can absorb CO_2 and CO selectively are sequentially brought into contact with the sample gas. For each test, the change in volume is recorded. Based on the property of a gas mixture, the change in volume is the measurement of the volumetric fraction of each gas species. Also based on the property of a gas mixture, the volumetric fraction is equal to the mole fraction of each gas species. The remaining gas in the apparatus is assumed to be nitrogen.

EXAMPLE 1: Orsat analysis
An Orsat analysis of a flue gas from an unknown hydrocarbon fuel combustion shows the following results:

$$CO_2 = 9.00\%$$
$$O_2 = 2.00\%$$
$$CO = 1.00\%$$

Determine:

1. The chemical equation for the actual reaction
2. The composition of the fuel
3. The air-fuel ratio used during the test.

Solution to question (1)

1.1. Calculate the amount of analyzed component.

$$\begin{aligned} CO_2 &= 9.000 \\ O_2 &= 2.000 \\ CO &= \underline{1.000} \\ & \ 12.000 \end{aligned}$$

Based on the Orsat definition, percent nitrogen in the Orsat = 100 – 12 = 88%

1.2. Determine moles of oxygen in the reaction.

In air, 1 mole of O_2 is associated with 3.76 moles of N_2. Therefore, for 88% of nitrogen, the moles of $O_2 = 88/3.76 = 23$

1.3. Write the possible reaction equation for the unknown hydrocarbon.

$$C_xH_y + 23O_2 + 88N_2 \rightarrow 9CO_2 + 2O_2 + CO + 88N_2 + (a)H_2O$$

1.4. Write the mass balance equation.

$$\text{For carbon: } x = 9 + 1$$
$$\text{For hydrogen: } y = 2a$$
$$\text{For oxgyen: } 2(23) = 2(9) + 2(2) + 1 + a$$

1.5. Solve the above equations.

$$x = 10$$
$$y = 46$$
$$a = 23$$

1.6. The reaction equation can be written as

$$C_{10}H_{46} + 23O_2 + 88N_2 \rightarrow 9CO_2 + 2O_2 + CO + 88N_2 + 23H_2O$$

Solution to question (3)

3.1. Calculate air mass (m_a) and fuel mass (m_f).

$$m_a = 23(32) + 88(28) = 3200$$
$$m_f = 12(10) + 1(46) = 166$$

3.2. Calculate the air-fuel ratio (AF).

$$AF = 3200/166$$
$$= 19.28$$

These calculations are summarized in Table 1.28.

TABLE 1.28 Orsat Analyses Results

	Moles in	Moles out	Mole fraction
N_2	88	88	0.7154
O_2	23	2	0.0163
CO_2	0	9	0.0732
CO	0	1	0.0081
H_2O	0	23	0.1870
Total	111	123	1

EXAMPLE 2: Orsat analysis
An Orsat analyzer was used to analyze methane combustion gas. The mole fractions of the dry products are as follows:

$$CO_2 = 10\%$$
$$O_2 = 1.0\%$$
$$CO = 1.5\%$$

Determine (1) the mole balance, and (2) the percent theoretical air used during the test; (3) calculate the mass flows and the air-to-fuel ratio.

Solution to question (1)

1.1. Calculate the amount of analyzed component.

$$\begin{aligned} CO_2 &= 10.000 \\ O_2 &= 1.000 \\ CO &= \underline{1.500} \\ & \ 12.500 \end{aligned}$$

Percent nitrogen $= 100 - 12.5 = 87.5$, say 87.

1.2. Determine moles of oxygen in the reaction.

Because 1 mole of O_2 is associated with 3.76 moles of N_2, then moles of $O_2 = 87/3.76 = 23.1$, say 23.

1.3. Write the possible reaction equation for the unknown hydrocarbon.

$$(a)CH_4 + 23O_2 + 87N_2 \rightarrow 10CO_2 + O_2 + 1.5CO + 87N_2 + (b)H_2O$$

1.4. Write the mass balance equation.

$$\begin{aligned} \text{For carbon: } a &= 10 + 1.5 = 11.5 \\ \text{For hydrogen: } 4a &= 2b \\ b &= 23.00 \\ \text{For oxygen: } 2(23) &= 2(10) + 2(1) + 1.5 + b \\ b &= 22.5 \end{aligned}$$

The b values calculated from the hydrogen balance and the oxygen balance are close enough.

1.5. The reaction equation can be written as.

$$11.5CH_4 + 23.1O_2 + 87.5N_2 \rightarrow 10CO_2 + O_2 + 1.5CO + 87.5N_2 + 23H_2O$$

1.6. These mole balance calculations are summarized in Table 1.29.

Solution to question (2)

2.1. Write a theoretical combustion equation for CH_4.

$$CH_4 + 2O_2 + 2(3.76)N_2 \rightarrow CO_2 + 2(3.76)N_2 + 2H_2O$$

TABLE 1.29 Orsat Analyses Results

	Moles in	Moles out	Mole fraction
N_2	87.5	87.5	0.7114
O_2	23.10	1	0.0081
CO_2	0	10	0.0813
CO	0	1.5	0.0122
H_2O	0	23	0.187
Total	110.6	123	1

2.2. Compare the equations in steps 1.6. and 2.1.

- Oxygen for theoretical combustion = 2 moles
- Oxygen for actual combustion = 23.1/11.5 = 2.01 moles
- The oxygen need for actual combustion is more than the theoretical air combustion, and therefore this is an excess-air combustion (even if it is very slight).

2.3. Calculate percent excess air (EA).

$$EA = (2.01 - 2)/2 = 0.5\%$$

2.4. Calculate percent theoretical air.

$$\text{Percent theoretical air} = 2.01(100)/2 = 100.5\%$$

Solution to question (3)

3.1. Calculate air mass (m_a) and fuel mass (m_f).

$$\begin{aligned} m_a &= 23.3(32) + 87.5(28) \\ &= 3195 \\ m_f &= 11.5(12 + 4) \\ &= 184 \end{aligned}$$

3.2. Calculate air-fuel ratio (AF).

$$\begin{aligned} AF &= 3213/168 \\ &= 17.36 \end{aligned}$$

2.34 Oxygen Amount for Elementary Combustion

Oxygen is an essential element for combustion to occur. To achieve complete combustion of a compound, a sufficient supply of oxygen must be present to convert all of the carbon to CO_2 and all hydrogen to H_2O. This quantity of oxygen is referred to as the stoichiometric or theoretical amount. The stoichiometric amount of oxygen is determined from a balanced chemical equation summarizing the oxidation reactions. For example, the following equation shows the reaction of methane (CH_4) with oxygen. It takes 1 mole of methane to react with 2 moles of oxygen for complete combustion (EPA-81/12, p. 3-4).

$$CH_4 + 2O_2 \rightarrow CO_2 + 2H_2O$$

The air amount associated with $2O_2$ is

$$\text{Air} = 2(O_2 + 3.76N_2)$$

If an insufficient amount of oxygen is supplied, the mixture is referred to as a *rich combustion*. Under this condition, there is not enough oxygen to react with all elements in the fuel, and thus incomplete combustion occurs. This condition results in black smoke being exhausted. If more than the stoichiometric amount of oxygen is supplied, the mixture is referred to as a *lean combustion*. The extra oxygen plays no part in the oxidation reaction and passes through the incinerator.

Oxygen for combustion processes is supplied by using air. Since air is essentially 79% nitrogen and 21% oxygen, a larger volume of air is required than if pure oxygen were used.

To balance the above methane combustion equation, 9.52 moles of air would be required to completely combust the 1 mole of methane. The 9.52 was calculated from 2 moles of O_2 + 2(3.76) moles of N_2.

In industrial applications, more than the stoichiometric amount of air is used to ensure complete combustion. This extra volume is referred to as excess air. If ideal mixing were achievable, no excess air would be necessary. However, most combustion devices are not capable of achieving ideal mixing of the fuel and air streams. The amount of excess air is held to a minimum in order to reduce heat losses. Excess air takes no part in the reaction but does absorb some of the heat produced. To raise the excess air to the combustion temperature, additional fuel must be used to make up for this loss of heat. Operating at a high volume of excess air can be very costly in terms of the added fuel required.

2.35 Oxygen Enrichment

Oxygen enrichment has been considered as one of the options to increase the effectiveness of waste incineration. It has long been recognized that operational benefits from oxygen enrichment are possible for special applications if the combustion air is enriched with pure oxygen. In recent years, incinerators have become more sophisticated and the cost of oxygen relative to the overall cost of incineration has declined. As a result, oxygen enrichment, where a portion of the combustion air is replaced by pure oxygen, is gaining usage.

One of the advantages of oxygen enrichment is evident from the management of nitrogen in air. Nearly 66% of the gas being moved through the incinerator is nitrogen, which does not participate constructively in the incineration process. In fact, the nitrogen reacts to form undesirable pollutants, such as nitrogen oxides (NO_x). Furthermore, the incinerator has to heat the nitrogen, increasing the demand for auxiliary fuel when the wastes have low heating values. The nitrogen increases the amount of gases that need to be moved. The incinerator must have larger combustion chambers, ducts, and fans. Oxygen enrichment reduces the amount of nitrogen carried with the oxygen. In new incinerators, its use can allow the use of smaller equipment to achieve the desired operating throughput. In existing units (if air enrichment is mechanically practical) the technique can result in greater waste throughput.

That is not to say that oxygen enrichment is desirable in all cases. The pure oxygen could cause wastes with high heating values to burn at temperatures beyond the equipment's design limit or the fusion temperature of the ash. Combustion chambers often rely on the nitrogen in the air to cool the combustion gases to the ranges that the refractory and ductwork can withstand. The nitrogen gas also creates higher gas velocities in the combustion chamber, with a corresponding increase in turbulence and improved mixing. Reducing the turbulence may result in lower destruction efficiencies because of poorer mixing of the gases. Finally, the cost of oxygen is significant and must be weighed against the resultant equipment savings or increases in throughput.

Oxygen enrichment also impacts on the correction factors used for particulate and CO emission limits. RCRA (and some other) regulations require the reporting of pollutant concentrations in the flue gas corrected to 7 percent oxygen (for oxygen no enrichment combustion). EPA published in the Federal Register (Vol. 55, No. 82, Friday, April 27, 1990, p. 17,918) a revised correction method for oxygen enrichment combustion. It is illustrated for CO, but applies equally well to particulates. The revised correction factor is

$$CO_c = CO_m \times 14/(E - Y)$$

where CO_c = corrected CO concentration
CO_m = measured CO concentration
E = enrichment factor, percentage of oxygen used in the combustion air (21% for no enrichment).
Y = measured oxygen concentration in the stack (by Orsat) in %

2.36 Oxygen Residual in Flue Gas

The residual oxygen (% O_2) in flue gas can be calculated by the following equation:

$$\text{Excess air (\%)} = \%\ O_2/(21\% - \%\ O_2), \text{ or} \tag{1.18}$$

$$\text{Fraction of excess air } (EA) = (O_2)/(0.21 - O_2), \text{ or}$$

$$O_2 = 0.21(EA)/(1 + EA) \tag{1.19}$$

where O_2 = residual O_2 fraction in flue gas.

In this equation, excess air is a given condition which is the ratio of excess air to the theoretical air amount. Thus, the residual O_2 after combustion can be determined. For example, a 50% excess air combustion of propane is

$$C_3H_8 + 5(1.5)(O_2 + 3.76N_2) \rightarrow 3CO_2 + 4H_2O + 2.5O_2 + 28.2N_2$$

Applying Eq. (1.19):

$$O_2 = 0.21(0.5)/(1 + 0.5) = 0.07$$

$$\%\ O_2 = 7\%$$

This value can be checked by actual calculation of the O_2 fraction in the flue gas, i.e., $O_2 = 2.5/(3 + 4 + 2.5 + 28.2) = 6.6\%$. The discrepancy between the 7% and 6.6% values is probably due to rounding off certain values; in any case, the values are very close.

2.37 Pressure

Pressure is defined as the total load or force per unit area acting on a surface, or the normal component of force per unit area (40CFR146.3-91). The terms that are related to pressure include those listed below.

Atmospheric pressure (atm). The pressure of the air and the atmosphere at sea level. Atmospheric pressure is also known as *barometric pressure*. In engineering practice, atmospheric pressure is expressed in different units as follows:

1 atm	= 14.6959 pounds/in^2 absolute (psia)
	= 2116 pounds/ft^2 absolute (psfa)
	= 406.9 in water
	= 33.91 ft water
	= 29.92 in mercury (in Hg) at 32°F
	= 760 mm mercury (mm Hg)
	= 760 Torr
	= 1.013×10^6 dynes/cm^2
	= 1.013×10^5 newtons/m^2
1 in Hg	= 0.491 psi
1 micron Hg	= 1 μ = 10^{-6} m = 10^{-3} mm Hg
	= 1.933×10^{-5} psi
1 microbar	= 1 dyne/cm^2
1 Torr	= 1 mm Hg

Many pressure terms in air pollution are expressed in inches of H_2O. Correspondingly, pressure drops (or pressure changes) across control equipment are also expressed in inches of H_2O.

Absolute pressure (psia). The total pressure exerted on a surface. This is a measure of pressure referred to a complete vacuum or zero pressure.

Gauge pressure (psig). A measure of pressure expressed as a quantity above atmospheric pressure or some other reference pressure. Gauge pressure is also known as gage pressure.

Vacuum pressure. A measure of pressure expressed as a quantity below atmospheric pressure or some other reference pressure.

The relationships of atmospheric pressure, absolute pressure, gage pressure, and vacuum pressure are shown in the following equations and in Fig. 1.7.

- Absolute pressure (P_{abs}) = atmospheric pressure (P_{atm}) + gage pressure (P_g), or

$$P_{abs} = P_{atm} + P_g$$

- Vacuum pressure = negative gage pressure
- In engineering practice, gas pressure is usually measured by Bourdon tube gages and fluid pressure by manometers
- A manometer measures a pressure difference by using a measurable length of a fluid column. For a fluid in static equilibrium, the relationship between the pressure and elevation within the fluid is expressed as:

$$\text{pressure difference} = (\text{density of measured fluid}) \times (\text{height of the fluid column})$$

- The unit is usually denoted by psig (pound force per square inch gage).

Figure 1.7 graphically shows the relationship among absolute pressure, gage pressure, and negative pressure (EPA-81/12, p. 2-4). Gage pressure P_{g1} is above the gage pressure zero (the atmospheric pressure), and hence is expressed as a positive value; gage pressure P_{g2} is below the gage pressure zero, and therefore is expressed as a negative value.

EXAMPLE 1: pressure
If an atmospheric pressure measured by a barometer is 29.92 in. Hg, and if a gage pressure reads a vacuum pressure 10 in Hg, determine the absolute pressure in psia.

Given conditions

- Barometer pressure, atm = 29.92 in Hg
- Vacuum pressure, psig = −10 in Hg
- 1 in Hg = 0.491 psi

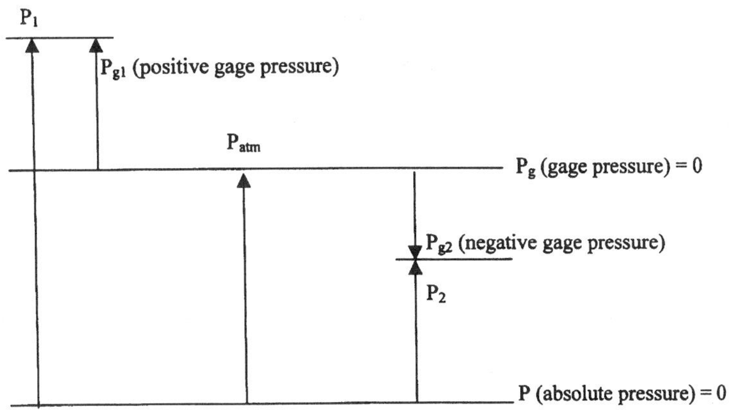

FIGURE 1.7 Pressure relationship.

Solution

1. Write the absolute pressure equation P_{abs}.

$$P_{abs} = P_{atm} + P_g$$

where P_{atm} = atmospheric pressure
P_g = gage pressure

2. Calculate the absolute pressure P_{abs}.

$$\begin{aligned} P_{abs} &= 29.92 - 10 \\ &= 19.92 \text{ in Hg} \\ &= 19.92 \text{ (in Hg)(0.491 psi/(in Hg))} \\ &= 9.78 \text{ psia} \end{aligned}$$

EXAMPLE 2: pressure
If a pressure gage reads 30.0 psi and a barometer reads 29.92 in. Hg, determine the absolute pressure of the system (psia).

Given conditions

- Atmospheric pressure measured by a barometer = 29.92 in Hg
- Pressure gage = 30 psi
- Mercury density, $D_m = 0.491 \text{ lb/in}^3$

Solution

1. Write atmospheric pressure equation P_{atm}.

$$P_{atm} = D_m(H)$$

where D_m = mercury density
H = height of mercury column

2. Calculate the atmospheric pressure.

$$\begin{aligned} P_{atm} &= 0.491(29.92) \\ &= 14.691 \text{ lb/in}^2 \\ &= 14.691 \text{ psi} \end{aligned}$$

3. Write the absolute pressure equation P_{abs}.

$$P_{abs} = P_{atm} + P_g$$

4. Calculate the absolute pressure P_{abs}.

$$\begin{aligned} P_{abs} &= 14.691 + 30 \\ &= 44.691 \text{ psia} \end{aligned}$$

EXAMPLE 3: pressure
Determine the pressure difference when it is recorded as (1) 12 ft of water, and (2) 2 in of mercury (EPA-84/09, p. 45).

Given conditions

- Pressure difference measured by water = 12 ft water
- Pressure difference measured by mercury = 2 in Hg
- Density of water $\rho_w = 62.4\,\mathrm{lb_f/(ft^3)}$
- Specific gravity of mercury $\gamma_m = 13.6$

Solution to question (1)

1.1. Write gauge pressure in terms of column height and liquid density.

$$p_g = g\rho_w h/g_c$$

where
p_g = gauge pressure or pressure difference, $\mathrm{lb_f/lb}$
g = acceleration of gravity
ρ_w = density of water, $\mathrm{lb/ft^3}$
h = height of the fluid column, ft
g_c = conversion factor
g/g_c = $1\,\mathrm{lb_f/lb}$ in engineering units

1.2. Calculate the pressure difference.

$$\begin{aligned} p_g &= 62.4(12)/144 \\ &= 5.20\,\mathrm{lb/in^2} \\ &= 5.20\,\mathrm{psi} \end{aligned}$$

Solution for mercury (2)

2.1. Write mercury density equation ρ_m.

$$\rho_m = \gamma_m \times \rho_w$$

where
ρ_m = mercury density
γ_m = mercury specific gravity
ρ_w = water density

2.2. Calculate mercury density ρ_m.

$$\begin{aligned} \rho_m &= 13.6(62.4) \\ &= 848.64\,\mathrm{lb/ft^3} \\ &= 848.64/(1728) \\ &= 0.49\,\mathrm{lb/in^3} \end{aligned}$$

2.3. Write the pressure difference equation p_d.

$$p_g = g\rho_m h/g_c$$

2.4. Calculate the pressure difference.

$$
\begin{aligned}
p_d &= (0.49)(2) \\
&= 0.98\,\mathrm{lb/in^2} \\
&= 0.98\,\mathrm{psi}
\end{aligned}
$$

Partial pressure

EXAMPLE: partial pressure

An analysis of an incinerator flue gas shows a CO_2 concentration with a partial pressure of 0.25 mm Hg. How many ppm is this?

Solution:

1. Write the mole fraction equation.

$$y = p/P$$

where y = mole fraction
p = partial pressure
P = total pressure = 760 mm Hg

2. Calculate the mole fraction.

$$
\begin{aligned}
y &= p/P \\
&= (0.25)/(760) \\
&= 0.000329 \\
&= 3.29(10^{-4})
\end{aligned}
$$

3. Calculate ppm by multiplying the mole fraction by 10^6.

$$
\begin{aligned}
\mathrm{ppm} &= y(10^6) \\
&= 329
\end{aligned}
$$

2.38 Principal Organic Hazardous Constituents (POHCs)

Appendix VIII to 40CFR261 provides a list of hazardous constituents. Since it is impossible to measure and consider all potential organics that may be fed to an incinerator during a trial burn, the regulations (40CFR264.342(b)(1)) allow the test to be conducted on a selected number of them. In general, incinerator operators and EPA permit writers to select the compounds from Appendix VIII for conducting trial burns. The compounds selected for testing are called the principal organic hazardous constituents (POHCs).

The POHCs are selected on the basis of their high concentration in the waste, their higher toxicity, environmental impact, and/or stability (difficulty of destruction) in an incinerator. The POHCs only have meaning for the trial burn and as reference organics which may be used as a permit condition to restrict the combustion of certain organic compounds. Because they are selected as being the most difficult to destroy of all of the organics that the incinerator may burn, they are considered to be representative of the Destruction and Removal Efficiency (DRE) of the incinerator for all hazardous organic constituents, and the permit may specify no organics more difficult to destroy than those burned during the trial burn. The permit may also specify the incinerability ranking procedure which would be used to determine the difficulty of destruction.

2.39 Property

A thermodynamic property (or properties) is a parameter which is used to define the state, or the condition, of a substance in a system. In engineering practice, except for heat and work, almost all other thermodynamic parameters are properties. Examples of properties include density, specific volume, temperature, pressure, mass, volume, etc. The terms that are related to the property include those listed below.

Extensive property. Also known as the extrinsic property. This is a property whose value for a system equals the sum of its values for the various parts of the system. Examples of extensive properties include mass, volume, weight, energy, enthalpy, entropy, etc.

Intensive property. Also known as the intrinsic property. This is a property whose value is the same for any part of a homogeneous system as it is for the whole system. Examples of intensive properties include pressure, temperature, density, etc.

Specific property. The value of an extensive property divided by the mass of the system. The value becomes an intensive property and is called a specific property.

References used in this section: Holman-74, p. 5; Jones-60, p. 10; Wark-66, p. 4.

2.40 Property of Gas Mixture

The terms and equations that are related to the properties of a gas mixture include those listed below.

Mass analysis or gravimetric analysis. The mass of a mixture is equal to the sum of the masses of its components, namely

$$m = m_1 + m_2 + \cdots m_i + \cdots$$

where m = total mass of the mixture
m_i = mass of component i

Mass fraction. The mass of component i divided by the total mass of the system. Mathematically, mass fraction (x_i) is defined as

$$x_i = m_i/m$$

$$x_1 + x_2 + x_3 + \cdots = 1$$

where x_i = mass fraction of component i
m_i = mass of component i
m = total mass of the mixture
x_1, x_2, x_3 = mass fraction of components 1, 2, and 3, respectively

Molar analysis. The total number of moles of a mixture is equal to the sum of the number of moles of its components, namely

$$n = n_1 + n_2 + \cdots n_i + \cdots$$

where n = total number of moles of the mixture
n_i = moles of component i

Mole fraction. The moles of component i divided by the total number of moles. Mathematically, mole fraction (y_i) is defined as

$$y_i = n_i/n$$

$$y_1 + y_2 + y_3 + \cdots = 1$$

where y_i = mole fraction
n_i = moles of component i
n = total number of moles
y_1, y_2, y_3 = mole fraction of components 1, 2, and 3, respectively

By the definition of the molar mass or molecular weight (M_i), the number of moles, the mass, and the molecular weight of a substance are related by

$$m_i = n_i M_i$$

where m_i = mass of component i
n_i = number of moles of component i
M_i = molecular weight of component i

Molecular weight (apparent molecular weight or average molecular weight). The molecular weight of an ideal gas mixture is the sum of all the components of the mole fraction times the molecular weight. Molecular weight is also known as apparent molecular weight or average molecular weight. Mathematically, it can be expressed as

$$M_{\mathrm{m}} = y_1 M_1 + y_2 M_2 + y_3 M_3 + \cdots$$

The above equation can be derived from the mass equation

$$m_{\mathrm{m}} = m_1 + m_2 + m_3 + \cdots$$

because $m_i = n_i M_i$, the above equation becomes

$$n_{\mathrm{m}} M_{\mathrm{m}} = n_1 M_1 + n_2 M_2 + n_3 M_3 + \cdots$$
$$M_{\mathrm{m}} = (n_1/n_{\mathrm{m}})M_1 + (n_2/n_{\mathrm{m}})M_2 + (n_3/n_{\mathrm{m}})M_3 + \cdots$$
$$M_{\mathrm{m}} = y_1 M_1 + y_2 M_2 + y_3 M_3 + \cdots$$

where m_i = mass of component i
n_i = number of moles of component i
M_i = molecular weight
M_{m} = molecular weight of the mixture
n_{m} = number of moles of the mixture
n_1, n_2, n_3 = number of moles of components 1, 2, and 3, respectively
M_1, M_2, M_3 = molecular weight of components 1, 2, and 3, respectively
y_1, y_2, y_3 = mole fraction of components 1, 2, and 3, respectively

Pressure. The total pressure of a mixture is the sum of the partial pressures, namely

$$p = p_1 + p_2 + p_3 + \cdots$$

where p = total pressure
p_1, p_2, p_3 = partial pressure of components 1, 2, and 3, respectively

Partial pressure and mole fraction. Partial pressure is the pressure that is exerted by one component of a mixture in a system. The relationship between partial pressure and mole fraction is

$$p_i = y_i p$$

where p_i = partial pressure of component i
y_i = mole fraction
p = total pressure of the system

For this case, the partial pressure of component i is equal to the mole fraction of component i in the mixture, i.e.,

$$y_i = n_i/n = p_i/p$$

References used in this section: Holman-74, p. 312; Jones-60, p. 391; Wark-66, p. 322.

2.41 Property of Ideal Gas Mixture

A mixture of ideal gases is also an ideal gas. All the terms and equations under the property of the gas mixture are applicable to the property of an ideal gas mixture. This section summarizes all the terms and equations.

Constant pressure specific heat. The constant pressure specific heat of an ideal gas mixture is the sum of each component.

$$\begin{aligned} m_\mathrm{m} C_\mathrm{pm} &= m_1 C_\mathrm{p1} + m_2 C_\mathrm{p2} + m_3 C_\mathrm{p3} + \cdots \\ C_\mathrm{pm} &= (m_1/m_\mathrm{m}) C_\mathrm{p1} + (m_2/m_\mathrm{m}) C_\mathrm{p2} + (m_3/m_\mathrm{m}) C_\mathrm{p3} + \cdots \\ C_\mathrm{pm} &= x_1 C_\mathrm{p1} + x_2 C_\mathrm{p2} + x_3 C_\mathrm{p3} + \cdots \end{aligned}$$

where C_pm = constant pressure specific heat of the mixture
$C_\mathrm{p1}, C_\mathrm{p2}, C_\mathrm{p3}$ = constant pressure specific heat of components, 1, 2, and 3, respectively
m_m = total mass of the mixture
m_1, m_2, m_3 = mass of components 1, 2, and 3, respectively
$C_\mathrm{p1}, C_\mathrm{p2}, C_\mathrm{p3}$ = constant pressure specific heat of components 1, 2, and 3, respectively
x_1, x_2, x_3 = mass fraction of components 1, 2, and 3, respectively

Constant volume specific heat. The constant volume specific heat of an ideal gas mixture is the sum of each component.

$$\begin{aligned} m_\mathrm{m} C_\mathrm{vm} &= m_1 C_\mathrm{v1} + m_2 C_\mathrm{v2} + m_3 C_\mathrm{v3} + \cdots \\ C_\mathrm{vm} &= (m_1/m_\mathrm{m}) C_\mathrm{v1} + (m_2/m_\mathrm{m}) C_\mathrm{v2} + (m_3/m_\mathrm{m}) C_\mathrm{v3} + \cdots \\ C_\mathrm{vm} &= x_1 C_\mathrm{v1} + x_2 C_\mathrm{v2} + x_3 C_\mathrm{v3} + \cdots \end{aligned}$$

where C_vm = constant volume specific heat of the mixture
$C_\mathrm{v1}, C_\mathrm{v2}, C_\mathrm{v3}$ = constant volume specific heat of components 1, 2, and 3, respectively
m_m = total mass of the mixture
m_1, m_2, m_3 = mass of components 1, 2, and 3, respectively
$C_\mathrm{v1}, C_\mathrm{v2}, C_\mathrm{v3}$ = constant volume specific heat of components 1, 2, and 3, respectively
x_1, x_2, x_3 = mass fraction of components 1, 2, and 3, respectively

Enthalpy. The enthalpy of an ideal gas mixture is the sum of each component.

$$\begin{aligned} H_\mathrm{m} &= H_1 + H_2 + H_3 + \cdots \\ m_\mathrm{m} h_\mathrm{m} &= m_1 h_1 + m_2 h_2 + m_3 h_3 + \cdots \\ h_\mathrm{m} &= (m_1/m_\mathrm{m}) h_1 + (m_2/m_\mathrm{m}) h_2 + (m_3/m_\mathrm{m}) h_3 + \cdots \\ h_\mathrm{m} &= x_1 h_1 + x_2 h_2 + x_3 h_3 + \cdots \end{aligned}$$

where H_m = enthalpy of the mixture
H_1, H_2, H_3 = enthalpy of components 1, 2, and 3, respectively
m_m = total mass of the mixture
m_1, m_2, m_3 = mass of components 1, 2, and 3, respectively
h_1, h_2, h_3 = specific enthalpy of components 1, 2, and 3, respectively
x_1, x_2, x_3 = mass fraction of components 1, 2, and 3, respectively

Entropy. The entropy of an ideal gas mixture is the sum of each component.

$$S_m = S_1 + S_2 + S_3 + \cdots$$
$$m_m s_m = m_1 s_1 + m_s s_2 + m_3 s_3 + \cdots$$
$$s_m = (m_1/m_m)s_1 + (m_2/m_m)s_2 + (m_3/m_m)s_3 + \cdots$$
$$s_m = x_1 s_1 + x_2 s_2 + x_3 s_3 + \cdots$$

where S_m = entropy of the mixture
S_1, S_2, S_3 = entropy of components 1, 2, and 3, respectively
m_m = total mass of the mixture
m_1, m_2, m_3 = mass of components 1, 2, and 3, respectively
s_1, s_2, s_3 = specific entropy of components 1, 2, and 3, respectively
x_1, x_2, x_3 = mass fraction of components 1, 2, and 3, respectively

Gas constant (or apparent gas constant). The gas constant of an ideal gas mixture is the sum of each component.

$$m_m R_m = m_1 R_1 + m_2 R_2 + m_3 R_3 + \cdots$$
$$R_m = (m_1/m_m)R_1 + (m_2/m_m)R_2 + (m_3/m_m)R_3 + \cdots$$
$$R_m = x_1 R_1 + x_2 R_2 + x_3 R_3 + \cdots$$

where R_m = gas constant of the mixture
R_1, R_2, R_3 = gas constant of components 1, 2, and 3, respectively
m_m = total mass of the mixture
m_1, m_2, m_3 = mass of components 1, 2, and 3, respectively
x_1, x_2, x_3 = mass fraction of components 1, 2, and 3, respectively

The universal gas constant (R_u) can be expressed as

$$R_u = M_m R_m$$

where R_u = universal gas constant
R_m = gas constant of the mixture
M_m = molecular weight of the mixture

Internal energy. The internal energy of an ideal gas mixture is the sum of each component.

$$U_m = U_1 + U_2 + U_3 + \cdots$$
$$m_m u_m = m_1 u_1 + m_2 u_2 + m_3 u_3 + \cdots$$
$$u_m = (m_1/m_m)u_1 + (m_2/m_m)u_2 + (m_3/m_m)u_3 + \cdots$$
$$u_m = x_1 u_1 + x_2 u_2 + x_3 u_3 + \cdots$$

where U_m = internal energy of the mixture
U_1, U_2, U_3 = internal energy of components 1, 2, and 3, respectively
m_m = total mass of the mixture
m_1, m_2, m_3 = mass of components 1, 2, and 3, respectively
u_1, u_2, u_3 = specific internal energy of components 1, 2, and 3, respectively
x_1, x_2, x_3 = mass fraction of components 1, 2, and 3, respectively

Mass (or apparent mass). Under gravimetric analysis, the mass of an ideal gas mixture is the sum of each component. This definition has been given under the property of a gas mixture. Equations are given here again for easy reference.

$$m_m = m_1 + m_2 + m_3 + \cdots$$

where m_m = total mass of the mixture
m_1, m_2, m_3 = mass of components 1, 2, and 3, respectively

Mass fraction. The mass of component of an ideal gas mixture i divided by the total mass of the system. This definition has been given under the property of a gas mixture. Equations are given here again for easy reference.

$$x_i = m_i/m$$

$$x_1 + x_2 + x_3 + \cdots = 1$$

where x_i = mass fraction of component i
m_i = mass of component i
m = total mass of the mixture
x_1, x_2, x_3 = mass fraction of components 1, 2, and 3, respectively

Mole (or apparent molar mass). Under the molar analysis, the number of moles of an ideal gas mixture is the sum of each component. This definition has been given under the property of a gas mixture. Equations are given here again for easy reference.

$$n_m = n_1 + n_2 + n_3 + \cdots$$

where n_m = number of moles of the mixture
n_1, n_2, n_3 = number of moles of components 1, 2, and 3, respectively

Mole fraction. The moles of component i divided by the total number of moles. This definition has been given under the property of a gas mixture. Equations are given here again for easy reference.

$$y_i = n_i/n$$

$$y_1 + y_2 + y_3 + \cdots = 1$$

where y_i = mole fraction
n_i = moles of component i
n = total number of moles
y_1, y_2, y_3 = mole fraction of components 1, 2, and 3, respectively

By the definition of the molar mass or molecular weight (M_i), the number of moles, the mass, and the molecular weight of a substance are related by

$$m_i = n_i M_i$$

where m_i = mass of component i
n_i = number of moles of component i
M_i = molecular weight of component i

Molecular weight (apparent molecular weight or average molecular weight). The molecular weight of an ideal gas mixture is the sum of all the components of the mole fraction times the molecular weight. Molecular weight is also known as apparent molecular weight or average molecular weight. This definition has been given under the property of a gas mixture. Equations are given here again for easy reference.

$$M_m = y_1 M_1 + y_2 M_2 + y_3 M_3 + \cdots$$

The above equation can be derived from the mass equation.

$$m_m = m_1 + m_2 + m_3 + \cdots$$

because $m_i = n_i M_i$, the above equation becomes

$$
\begin{aligned}
n_m M_m &= n_1 M_1 + n_2 M_2 + n_3 M_3 + \cdots \\
M_m &= (n_1/n_m)M_1 + (n_2/n_m)M_2 + (n_3/n_m)M_3 + \cdots \\
M_m &= y_1 M_1 + y_2 M_2 + y_3 M_3 + \cdots
\end{aligned}
$$

where m_i = mass of component i
n_i = number of moles of component i
M_i = molecular weight
M_m = molecular weight of mixture
n_m = number of moles of the mixture
n_1, n_2, n_3 = number of moles of components 1, 2, and 3, respectively
M_1, M_2, M_3 = molecular weight of components 1, 2, and 3, respectively
y_1, y_2, y_3 = mole fraction of components 1, 2, and 3, respectively

Pressure. Under the Dalton law or the law of additive pressures, the pressure of a mixture of ideal gases equals the sum of the pressures of its components as if each component existed alone at the temperature and volume of the mixture (Jones-60, p. 393). Mathematically, it can be derived from the ideal gas law equation as follows:

$$
\begin{aligned}
p_m &= n_m R_u T_m / V_m \\
&= (n_1 + n_2 + n_3 + \cdots) R_u T_m / V_m \\
&= n_1 R_u T_m / V_m + n_2 R_u T_m / V_m + n_3 R_u T_m / V_m + \cdots \\
p_m &= p_1(T_m, V_m) + p_2(T_m, V_m) + p_3(T_m, V_m) + \cdots
\end{aligned}
$$

where p_m = pressure of the mixture
R_u = universal gas constant
T_m = temperature of the mixture
V_m = volume of the mixture
n_m = number of moles of the mixture
n_1, n_2, n_3 = number of moles of components 1, 2, and 3, respectively
$p_1(T_m, V_m), p_2(T_m, V_m), p_3(T_m, V_m)$ = pressure of components 1, 2, and 3 existing at the temperature T_m and the volume V_m

Pressure fraction. Pressure fraction is equal to mole fraction (y_i). Using component 1 as an example, the mole fraction for component 1 is

$$
\begin{aligned}
y_1 &= n_1/n_m \\
&= [(p_1 V_m)/(R_u T_m)][(R_u T_m)/p_m V_m)] \\
&= p_1/p_m
\end{aligned}
$$

where y_1 = mole fraction of component 1
p_1/p_m = definition of pressure fraction
V_m = volume of mixture
n_m = number of moles of the mixture
n_1 = number of moles of component 1
p_1 = pressure of component 1 existing at the temperature T_m and the pressure V_m

Temperature. The temperature of an ideal gas mixture is the same for each component and for the mixture.

$$
T_m = T_1 = T_2 = T_3 = \cdots
$$

where T_m = temperature of the mxiture
T_1, T_2, T_3 = temperature of components 1, 2, and 3, respectively

Volume. Under the Amagat law, Leduc law, or the law of additive volumes, the volume of a mixture of ideal gases equals the sum of the volumes of its components as if each component existed alone at the temperature and pressure of the mixture (Jones-60, p. 394). Mathematically, this can be derived from the ideal gas law equation as follows:

$$\begin{aligned} V_m &= n_m R_u T_m / p_m \\ &= (n_1 + n_2 + n_3 + \cdots) R_u T_m / p_m \\ &= n_1 R_u T_m / p_m + n_2 R_u T_m / p_m + n_3 R_u T_m / p_m + \cdots \\ V_m &= V_1(T_m, p_m) + V_2(T_m, p_m) + V_3(T_m, p_m) + \cdots \end{aligned}$$

where p_m = pressure of the mixture
R_u = universal gas constant
T_m = temperature of the mixture
V_m = volume of the mixture
n_m = number of moles of the mixture
n_1, n_2, n_3 = number of moles of components 1, 2, and 3, respectively
$V_1(T_m, p_m), V_2(T_m, p_m), V_3(T_m, p_m)$ = volume of components 1, 2, and 3, respectively, existing at temperature T_m and pressure p_m

Volume fraction. Volume fraction is equal to mole fraction (y_i). Using component 1 as an example, the mole fraction for component 1 is

$$\begin{aligned} y_1 &= n_1 / n_m \\ &= [(p_m V_1)/(R_u T_m)][(R_u T_m)/p_m V_m)] \\ &= V_1 / V_m \end{aligned}$$

where y_1 = mole fraction of component 1
V_1/V_m = definition of volume fraction
V_m = volume of mixture
n_m = number of moles of the mixture
n_1 = number of moles of component 1
V_1 = volume of component 1 existing at temperature T_m and pressure p_m

References used in this section: Holman-74, p. 312; Jones-60, p. 391; Wark-66, p. 322.

EXAMPLE 1: property
A gaseous fuel consists of 60 percent CH_4, 25 percent C_2H_6, and 15 percent C_3H_8 by volume. The following given conditions summarize the data for calculation. Determine:

(1) The gravimetric analysis of the fuel
(2) The apparent molecular weight of the mixture
(3) The apparent gas constant for the mixture

Given conditions

CH_4: 0.60
C_2H_6: 0.25
C_3H_8: 0.15

Solution to question (1)

1.1. Determine the gravimetric analysis of fuel (see Table 1.30).

TABLE 1.30 Gravimetric Analysis

	(1) Volume fraction	(2) Mole fraction, mole/mole-mixture	(3) Molar mass, lb/mole	(4) lb/(mole-mixture)	(5) lb/(lb-mixture)
CH_4	0.6	0.6	16	9.6	0.40
C_2H_6	0.25	0.25	30	7.5	0.32
C_3H_8	0.15	0.15	44	6.6	0.28
Total	1	1		23.7	1

Notes

- Column (1) = given conditions
- Column (2) = column (1) because the volume fraction equals the mole fraction
- Column (3) = given conditions
- Column (4) = column (2) times column (3)
- Column (5) = individual weight divided by the total weight 23.70

Solution to question (2)

The apparent molecular weight of the fuel is 23.70 (column 4).

Solution to question (3)

3.1. Write the apparent gas constant equation.

$$Rm = R_u/Mm$$

where Rm = apparent gas constant
R_u = universal gas constant
Mm = apparent molecular weight

3.2. Calculate the apparent gas constant.

$$Rm = 1545/23.70$$
$$= 65.19\,(\text{ft-lbf})/(\text{lbm-°R})$$

EXAMPLE 2: property
A combustion gas consists of 15 percent hydrogen, 45 percent oxygen, and 40 percent carbon monoxide by weight. Determine:

(1) The volumetric analysis, in percent
(2) The apparent molar mass.

Given conditions

H_2: 0.15
O_2: 0.45
CO: 0.40

1. Solution to question (1)

1.1. Determine the volumetric analysis of fuel (see Table 1.31).

TABLE 1.31 Volumetric Analysis

	(1) Mass fraction, lb/lb-mixture	(2) Molar mass, lb/mole	(3) Mole/ lb-mixture	(4) Mole fraction	(5) Volume fraction, mole/ mole-mixture
H_2	0.15	2	0.075	0.7257	0.7257
O_2	0.45	32	0.0141	0.1361	0.1361
CO	0.4	28	0.0142	0.1382	0.1382
Total	1		0.1033	1	1

Notes

- Column (1) = given conditions
- Column (2) = given conditions
- Column (3) = column (1) divided by column (2)
- Column (4) = individual moles divided by the total moles, 0.1033
- Column (5) = volumetric analysis results, because mole fraction equals the volume fraction

2. *Solution to question (2)*

2.1. Write the mole, mass, and molar mass equation.

$$n = m/M \text{ or } M = m/n$$

where n = number of moles
m = apparent mass in column 1
M = apparent molar mass

2.2. Calculate the apparent molar mass.

$$M = 1/0.1033$$
$$= 9.68$$

EXAMPLE 3: property
The mass analysis of a combustion gas is 16 percent hydrogen, 25 percent nitrogen, 15 percent carbon monoxide, and 44 percent carbon dioxide. For the mixture at 14.696 psia, 70°F, compare

1. the partial pressure of the constituents;
2. the constant volume specific heat;
3. the internal energy;
4. the enthalpy.

Assume that $u_m = 0$, $h_m = 0$, at $T = 0\,°R$. Approximate specific heats of the constituents are given in Table 1.32.

Given conditions

- Mixture pressure, $p_m = 14.696$ psia
- Mixture temperature, $T = 70°F$

TABLE 1.32 Mole Fraction

	(1) Mass analysis	(2) Molar mass	(3) Mole	(4) Mole fraction
H_2	0.16	2	0.08	0.7671
N_2	0.25	28	0.0089	0.0856
CO	0.15	28	0.0054	0.0514
CO_2	0.44	44	0.01	0.0959
Total	1		0.1043	1

Notes

- Column (1) = given conditions
- Column (2) = given conditions
- Column (3) = column (1) divided by column (2)
- Column (4) = individual moles divided by the total moles 0.1043

- Initial mixture internal energy, $u_m = 0$ at $T = 0$
- Initial mixture enthalpy, $h_m = 0$ at $T = 0$
- Mass analysis of each component in the mixture is as follows:
  Hydrogen, $H_2 = 0.16$
  Nitrogen, $N_2 = 0.25$
  Carbon monoxide, $CO = 0.15$
  Carbon dioxide, $CO_2 = 0.44$
- Heat capacity (Btu/(lb-°R)) of each component in the mixture
  Hydrogen, $C_p = 3.41$; $C_v = 2.42$
  Nitrogen, $C_p = 0.248$; $C_v = 0.177$
  Carbon monoxide, $C_p = 0.248$; $C_v = 0.177$
  Carbon dioxide, $C_p = 0.195$; $C_v = 0.150$

Solution:

1. Find the mole fraction for calculating the partial pressure.

2. Write the partial pressure equation.

$$\text{Partial pressure } p_i = (y_i)(p_m)$$
$$\text{Partial pressure of } H_2 = (0.7671)(14.696) = 11.2736 \text{ psia}$$
$$\text{Partial pressure of } N_2 = (0.0856)(14.696) = 1.2582 \text{ psia}$$
$$\text{Partial pressure of CO} = (0.0514)(14.696) = 0.7549 \text{ psia}$$
$$\text{Partial pressure of } CO_2 = (0.0959)(14.696) = 1.4092 \text{ psia}$$

3. Write the constant volume specific heat equation.

$$\begin{aligned} C_{vm} &= (m_1/m_m)C_{v1} + (m_2/m_m)C_{v2} + (m_3/m_m)C_{v3} + (m_4/m_m)C_{v4} \\ &(0.16)(2.42) + 0.25(0.177) + (0.15)(0.177) + (0.44)(0.150) \\ &= 0.524 \text{ Btu/(lb-°R)} \end{aligned}$$

where $m_m = 1$ in column (1).

4. Find the internal energy.

For an ideal gas, $u_{2m} - u_{1m} = C_{vm}(T_2 - T_1)$
From the given conditions, $u_{1m} = 0$ at $T_1 = 0$

$$\begin{aligned} U_{2m} &= C_{vm}(T_2) \\ &= 0.524(460 + 70) \\ &= 278 \text{ Btu/lb} \end{aligned}$$

5. Method 1 for calculating enthalpy.

5.1. Write the constant pressure specific heat equation.

$$\begin{aligned} C_{pm} &= (m_1/m_m)C_{p1} + (m_2/m_m)C_{p2} + (m_3/m_m)C_{p3} + (m_4/m_m)C_{p4} \\ &= (0.16)(3.41) + 0.25(0.248) + (0.15)(0.248) + (0.44)(0.195) \\ &= 0.7108 \text{ Btu/(lb-}^\circ\text{R)} \end{aligned}$$

5.2. Calculate the enthalpy.

For an ideal gas, $h_{2m} - h_{1m} = C_{pm}(T_1 - T_1)$
From the given conditions, $h_{1m} = 0$ at $T_1 = 0$

$$\begin{aligned} h_{2m} &= C_{pm}(T_2) \\ &= 0.7108(460 + 70) \\ &= 377 \text{ Btu/lb} \end{aligned}$$

6. Method 2 for calculating enthalpy.

6.1. Write the enthalpy definition equation.

$$\begin{aligned} h &= u + pv \\ &= u + RmT \\ &= u + (R_u/M)T \\ &= u + (R_u)(n_m/m_m)T \\ &= 278 + 1544(0.1043/1)(460 + 70)/\left(\frac{778 \text{ ft-lb}_f}{\text{Btu}}\right) \\ &= 387 \text{ Btu/lb} \end{aligned}$$

The results from steps 5 and 6 are similar. The slight difference is caused by the rounding up during the calculation.

EXAMPLE 4: property
For an oxygen enrichment combustion, oxygen is contained in a rigid tank. The rigid tank originally contains 0.2 lb of nitrogen at 150 psia and 70°F. A sufficient quantity of oxygen is added to the tank such that the pressure increases to 380 psia while the temperature remains constant at 70°F. Determine:

(1) The mass of oxygen added
(2) The volumes which would be occupied by the oxygen and nitrogen if each gas were contained separately at the temperature and pressure of the mixture, i.e., 70°F and 380 psia.

Given conditions

- Pressure, $p_1 = 150\,\text{psia}$
- Pressure, $p_2 = 380\,\text{psia}$
- Temperature, $T_1 = 70°\text{F}$
- Nitrogen mass, $m_{N2} = 0.2\,\text{lb}$
- Nitrogen molar mass, $N_2 = 28\,\text{lb/lb-mole}$
- Oxygen molar mass, $O_2 = 32\,\text{lb/lb-mole}$

Solution to part (1)

1.1. Calculate the number of moles of nitrogen.

In this process, the number of moles of nitrogen remains constant.

1.2. Write the number of moles equation.

$$n = m/M$$

where n = number of moles
m = apparent mass
M = apparent molar mass

1.3. Calculate the number of moles of nitrogen.

$$\begin{aligned} n &= 0.2/28 \\ &= 0.0071 \end{aligned}$$

1.4. Calculate the oxygen partial pressure.

After the oxygen is added, the partial pressure of the nitrogen is 150 psia, because the volume does not change. Therefore, the partial pressure for oxygen is

$$\begin{aligned} p(O_2) &= 380 - 150 \\ &= 230\,\text{psia} \end{aligned}$$

1.5. Write the ideal gas equation for oxygen and nitrogen.

$$\begin{aligned} p(O_2) &= n(O_2)(R_u)(T)/V \\ p(N_2) &= n(N_2)(R_u)(T)/V \end{aligned}$$

Ratio of the above two equations:

$$n(O_2)/n(N_2) = p(O_2)/p(N_2) \text{ or}$$

$$n(O_2) = n(N_2)p(O_2)/p(N_2)$$

1.6. Calculate the number of moles of oxygen.

$$\begin{aligned} n(O_2) &= (0.0071)(230)/(150) \\ &= 0.0110 \end{aligned}$$

1.7. Calculate the mass of oxygen added.

Apply the equation $m = nM$

$$m(O_2) = (0.0110)(32)$$
$$= 0.352\,\text{lb}$$

Solution to part (2)

2.1. Write the ideal gas equation for nitrogen.

$$V(N_2) = n(N_2)(R_u)(T)/p(N_2)$$

2.2. Calculate the volume of nitrogen.

$$V(N_2) = (0.0071)(1545)(460 + 70)/(380)(144)$$
$$= 0.1062\,\text{ft}^3$$

2.3. Calculate the volume of oxygen.

$$V(O_2) = (0.0110)(1545)(460 + 70)/(380)(144)$$
$$= 0.1646\,\text{ft}^3$$

2.4. Calculate the total volume of the mixture.

$$V = V(N_2) + V(O_2)$$
$$= 0.1062 + 0.1646$$
$$= 0.2708\,\text{ft}^3$$

2.42 Property of Vapor Mixture

Properties of vapour mixtures can be shown on property diagrams such as the P–v (pressure–specific volume) diagram (Fig. 1.8) and the T–v (temperature–specific volume) diagram (Fig. 1.9) (Lee-90/05). Steam properties are probably the most widely used set of thermodynamic properties applied to engineering applications. Steam properties can be found in many thermodynamic text books.

where afcgb is the saturated curve
afc is the saturated liquid curve
cgb is the saturated vapor curve
df is the liquid zone
ge is the vapor zone
fg is the liquid–vapor mixture zone
c is the critical point
hfgi is the constant temperature line in the P–v diagram
hfgi is the constant pressure line in the T–v diagram

The properties at point y are

- Specific volume $= v_y = v_f + x(v_g - v_f)$
- Internal energy $= u_y = u_f + x(u_g - u_f)$
- Enthalpy $= h_y = h_f + x(h_g - h_f)$

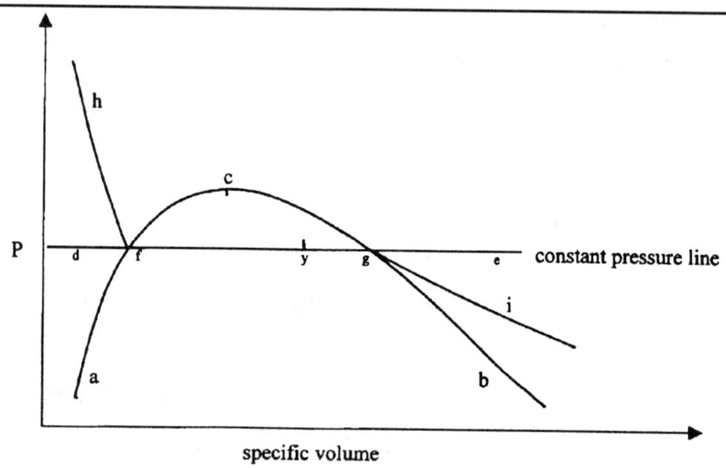

FIGURE 1.8 Water and air mixture properties in a *P–v* diagram.

where quality (x) = (vapor mass)/(liquid mass + vapor mass).

The values of v and h along the saturated liquid and vapor curves can be obtained from a steam table in many thermodynamic text books.

The terms that are related to a pure substance include those listed below.

Compressed liquid or subcooled liquid. The liquid at a temperature lower than the saturation temperature at a given pressure (Jones-60, p. 143).

Critical point or critical state. The conditions where the states (temperature, pressure, and specific volume) of saturated-liquid and saturated-vapor are identical. At the critical point, the difference of the following properties between the saturated liquid and the saturated vapor is zero:

- Enthalpy: $h_{fg} = h_g - h_f = 0$
- Entropy: $s_{fg} = 0$

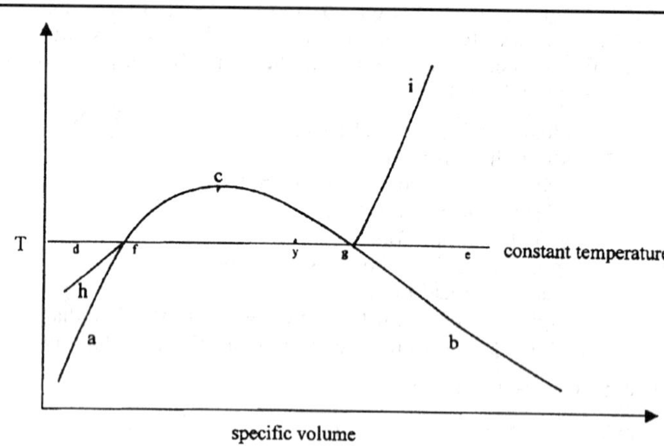

FIGURE 1.9 Water and air mixture properties in a *T–v* diagram.

- Internal energy: $u_{fg} = 0$
- Specific volume: $v_{fg} = 0$

Critical pressure. The pressure of the critical point where the states (temperature, pressure, and specific volume) of saturated-liquid and saturated-vapor are identical.

Critical temperature. The temperature of the critical point where the states (temperature, pressure, and specific volume) of saturated-liquid and saturated-vapor are identical.

Fusion. The change of a pure substance from a solid phase to its liquid phase. Fusion is an endothermic process, that is, energy must be added to a solid substance (solid phase) to convert it to a liquid (liquid phase). This energy is commonly referred to as the latent heat of fusion.

Liquid saturation line. The curved line which separates the liquid region from the liquid–vapor region (Wark-66, p. 56).

Phase. In physical chemistry, the uniform appearance of substances. Three phases, gas, liquid, and solid, can usually be identified from their appearance. A substance may exist in several phases:

- A pure solid phase (e.g. ice)
- A pure liquid phase
- A pure vapor phase (e.g. steam)
- An equilibrium mixture of liquid and vapor phases
- An equilibrium mixture of liquid and solid phases
- An equilibrium mixture of solid and vapor phases

Quality (x). The fraction of vapor mass in a saturated liquid–vapor mixture. Quality-related equations include:

- By definition
  $x = m_g/(m_f + m_g)$, where
  m_g = vapor mass in a mixture
  m_f = liquid mass in a mixture
- Enthalpy (h_x), (Btu/lb), of a saturated liquid–vapor mixture at a given temperature or pressure
  $h_x = h_f + x(h_g - h_f) = h_f + xh_{fg}$, where
  h_f = enthalpy of saturated liquid
  h_g = enthalpy of saturated vapor
  h_{fg} = enthalpy difference between saturated vapor and saturated liquid; h_{fg} is also known as latent heat of vaporization
  The values of h_f, h_g, and h_{fg} can be obtained from a steam table.
- Entropy (s_x), (Btu/(lb-°R)), of a saturated liquid–vapor mixture
  $s_x = s_f + x(s_g - s_f) = s_f + xs_{fg}$, where
  s_f = entropy of saturated liquid
  s_g = entropy of saturated vapor
  s_{fg} = entropy difference between saturated vapor and saturated liquid
  The values of s_f, s_g, and s_{fg} can be obtained from a steam table.
- Internal energy (u_x), (Btu/lb), of a saturated liquid–vapor mixture
  $u_x = u_f + x(u_g - u_f) = u_f + xu_{fg}$, where
  u_f = internal energy of saturated liquid
  u_g = internal energy of saturated vapor
  u_{fg} = internal energy difference between saturated vapor and saturated liquid
  The values of s_f, s_g, and s_{fg} can be obtained from a steam table.

- Specific volume (v_x), (ft^3/lb), of a saturated liquid–vapor mixture
  $v_x = v_f + x(v_g - v_f) = v_f + xv_{fg}$, where
  v_f = specific volume of saturated liquid
  v_g = specific volume of saturated vapor
  v_{fg} = specifice volume difference between saturated vapor and saturated liquid

The values of v_f, v_g, and v_{fg} can be obtained from a steam table (Holman-74, p. 66; Jones-60, p. 120; Wark-66, p. 66).

Saturated liquid state. Any state represented by a point on the liquid saturation line (Wark-66, p. 56).

Saturated liquid. The liquid which is present under the saturation curve.

Saturated phase. Any phase of a pure substance under the saturation state or saturation conditions (Jones-60, p. 111).

Saturated vapor state. Any state represented by a point on the vapor saturation line (Wark-66, p. 56).

Saturated vapor. The vapor which is present under the saturation curve.

Saturation. The condition for coexistence in stable equilibrium of the vapor and liquid phases or the vapor and solid phases of the same substance.

Saturation curve. The combination of the liquid saturation line and the vapor saturation line.

Saturation pressure or saturated pressure. The pressure at which two phases of a pure substance can coexist in equilibrium at a given temperature (Jones-60, p. 143).

Saturation state or saturation conditions. The conditions under which two or more phases of a pure substance can coexist in equilibrium. At the saturation state, a change of phase may occur without a change of pressure or temperature (Jones-60, p. 111; Wark-66, p. 56).

Saturation temperature or saturated temperature. The temperature at which two phases of a pure substance can coexist in equilibrium at a given pressure (Jones-60, p. 143).

Sublimation. A change of state directly from a solid state to a gas state without the appearance of the liquid state.

Supercritical state. The state where the pressure is greater than the critical pressure (Wark-66, p. 57).

Superheated steam, superheated vapor, or superheated gas. Steam, vapor, or gas at a temperature higher than the saturation temperature at a given pressure (Jones-60, p. 143).

Triple point or triple phase point. The location where solid, liquid, and vapor phases can coexist in equilibrium.

Vapor saturation line. The curved line which separates the vapor region from the liquid–vapor region (Wark-66, p. 56).

Vaporization. The change of a pure substance from a liquid phase to its vapor phase. Vaporization is an endothermic process. That is, energy must be added to a liquid substance (liquid phase) to convert it to a vapor phase (gas phase). This energy is commonly referred to as the latent heat of vaporization.

References used in this section: Holman-74, p. 63; Jones-60, p. 106; Wark-66, p. 53.

EXAMPLE 1: property of vapor
Fifty pounds of steam at 14.696 psia has a volume of 400 ft^3. Determine the temperature, the enthalpy per pound, and the internal energy.

Given conditions

- Steam mass, $m = 50\,lb$
- Volume, $V = 400\,ft^3$
- Pressure, $P = 14.696\,psia$

Solution:

1. Apply the specific volume equation.

$$v = V/m$$

where v = specific volume
V = volume
m = mass

2. Calculate the specific volume.

$$v = 400/50$$
$$= 8\ \text{ft}^3/\text{lb}$$

3. Find data from a steam table.

From a steam table (Wark-66, p. 596), at 1 psia, the corresponding saturated liquid specific volume (v_f) and saturated vapor specific volume (v_g) are

$$v_f = 0.01672\ \text{ft}^3/\text{lb}$$
$$v_g = 26.8\ \text{ft}^3/\text{lb}$$

These two values show that the steam is a wet vapor, i.e., ($0 < x < 100$ percent) at a temperature of 212°F, which corresponds to the saturation pressure of 14.696 psia.

4. Apply quality equation.

By definition, $v = v_f + x(v_g - v_f)$, where x = quality.

5. Solve for x.

$$x = (v - v_f)/(v_g - v_f)$$
$$= (8 - 0.01672)/(26.8 - 0.01672)$$
$$= 0.30 \text{ or } 30\%$$

6. Calculate the enthalpy per pound.

$$h = h_f + x(h_g - h_f)$$

7. Find data from a steam table.

From a steam table (Wark-66, p. 597), at 14.696 psia, the corresponding saturated liquid enthalpy (h_f) and saturated vapor enthalpy (h_g) are

$$h_f = 180.07\ \text{Btu/lb}$$
$$h_g = 1150.4\ \text{Btu/lb}$$

8. Calculate the enthalpy.

$$h = 180.07 + 0.3(1150.4 - 180.07)$$
$$= 471.2\ \text{Btu/lb}$$

9. Method 1 for finding the internal energy.

The internal energy per pound can be found by

$$u = u_f + x(u_g - u_f)$$

10. Find data from a steam table.

From a steam table (Wark-66, p. 597), at 14.696 psia, the corresponding saturated liquid internal energy (u_f) and saturated vapor internal energy (u_g) are

$$u_f = 180.02\ \text{Btu/lb}$$
$$u_g = 1077.5\ \text{Btu/lb}$$

11. Calculate the internal energy.

$$u = 180.02 + 0.3(1077.5 - 180.02)$$
$$= 449.3\ \text{Btu/lb}$$

12. Method 2 for finding the internal energy by the equation

$$h = u + pv \text{ or}$$
$$u = h - pv$$

13. Calculate the internal energy.

$$u = 471.2 - (14.696)(144)(8)/(778)$$
$$= 449.4\ \text{Btu/lb}$$

The u values obtained from steps 11 and 13 are essentially the same.

EXAMPLE 2: property of vapor
Determine the enthalpy, specific volume, and internal energy of steam at 1400 psia and 900°F.

Given conditions

- Pressure: 1400 psia
- Temperature: 900°F

Solution:

1. Find data from a steam table.

From a steam table (Wark-66, p. 604), the saturation temperature for a pressure of 1400 psia is 587.10°F. The temperature of the steam under consideration is greater than the saturation temperature; hence it is superheated.

2. Find the enthalpy required and the specific volume.

Referring to the same steam table under conditions of pressure 1400 psia and temperature 900°F, the enthalpy and specific volume of the steam can be found as

$$h = 1433.1\ \text{Btu/lb}$$
$$v = 0.5281\ \text{ft}^3/\text{lb}$$

3. Find the internal energy from the enthalpy equation.

Because internal energy values are not given in the superheated steam table, the u value has to be calculated from the following equations:

$$h = u + pv \text{ or}$$
$$u = h - pv$$

4. Calculate the internal energy.

$$\begin{aligned} u &= 1433.1 - 1400(144)(0.5281)/(778) \\ &= 1296.3 \text{ Btu/lb} \end{aligned}$$

EXAMPLE 3: property of vapor
Steam is compressed reversibly and adiabatically at a steady rate of 50 lb/min from 60 psia, 80 percent quality, to 500 psia. Determine the power input to the steam.

Given conditions

- Entropy is constant because of the reversible and adiabatic process
- Steam flow rate, $Q = 50$ lb/min
- Initial pressure, $P_1 = 60$ psia
- Initial quality, $x_1 = 0.8$
- Exit pressure, $P_2 = 500$ psia
- 1 hp = 42.4 Btu/min

Solution:

1. Write the first law equation under steady-state conditions.

$$-w = h_2 - h_1$$

2. Calculate h_1 by the equation.

$$h_1 = h_f + x_1(h_g - h_f)$$

3. Find data from a steam table at pressure 60 psia (Wark-66, p. 598).

$$\begin{aligned} h_f &= 262.09 \text{ Btu/lb} \\ h_g &= 1177.60 \text{ Btu/lb} \end{aligned}$$

4. Calculate h_1.

$$\begin{aligned} h_1 &= 262.09 + 0.8(1177.60 - 262.09) \\ &= 994.50 \text{ Btu/lb} \end{aligned}$$

5. Determine the state of the exit conditions.

The exit conditions could be in the superheated region or the wet region. Because of the reversible and adiabatic process, $s_2 = s_1$, s_1 can be found from the equation

$$s_1 = s_f + x_1(s_g - s_f)$$

6. Find data from a steam table at pressure 60 psia.

$$\begin{aligned} s_f &= 0.427 \text{ Btu/(lb-°R)} \\ s_g &= 1.6438 \text{ Btu/(lb-°R)} \end{aligned}$$

7. Calculate s_1.

$$\begin{aligned} s_1 &= 0.427 + 0.8(1.6438 - 0.427) \\ &= 1.40044\ \text{Btu/(lb-°R)} \end{aligned}$$

8. Find data from a steam table at pressure 500 psia.

$$\begin{aligned} s_{f2} &= 0.6487\ \text{Btu/(lb-°R)} \\ s_{g2} &= 1.4634\ \text{Btu/(lb-°R)} \end{aligned}$$

From step 7, s_1 value is less than s_{g2}. Hence, the discharge is wet.

9. Determine the quality x_2, at the exit condition.

$$s_1 = s_2 = s_{f2} + x_2(s_{g2} - s_{f2})$$

10. Calculate x_2.

$$\begin{aligned} 1.4004 &= 0.6487 + x_2(1.4634 - 0.6487) \\ x_2 &= 0.9227 \end{aligned}$$

11. Find data from a steam table at pressure 500 psia.

$$\begin{aligned} h_f &= 449.4\ \text{Btu/lb} \\ h_g &= 1204.4\ \text{Btu/lb} \end{aligned}$$

12. Calculate h_2.

$$\begin{aligned} h_2 &= 449.4 + 0.9227(1204.4 - 449.4) \\ &= 1146.05\ \text{Btu/lb} \end{aligned}$$

13. Calculate the work.

$$\begin{aligned} -w &= 1146.05 - 994.50 \\ -w &= 151.55 \\ -W &= mw = 50(151.55)/42.4 \\ -W &= 178.7\ \text{hp} \\ W &= -178.7\ \text{hp} \end{aligned}$$

A negative sign means that work is done on the system.

2.43 Property of a Vapor and Ideal Gas Mixture

This section covers the properties of mixtures which involve ideal gases, condensable vapors, and water. This class of mixture is encountered in many applications such as the air we breathe and the heating and air conditioning units that are installed in many houses. Furthermore, calculations of combustion air within incinerators and air pollution control devices also depend on the understanding of the behavior of gas–vapor–water mixtures. The terms that are related to gas–vapor–water mixtures include those listed below.

Adiabatic saturation temperature. The temperature which results from adiabatically adding water to a gas–vapor–water mixture in a steady flow until it becomes saturated, the water being supplied at the final temperature of the mixture.

Dew point temperature. The saturation temperature of the water corresponding to its partial pressure in the mixture. It is thus the temperature at which condensation begins if the mixture is cooled at constant pressure.

Dry bulb temperature. The temperature measured by a thermometer whose bulb (mercury holder at the bottom of a thermometer) is dry (exposed to air). It is the temperature indicated by an ordinary thermometer placed in the mixture.

Humidity ratio (HR) (absolute humidity or specific humidity)

$$\begin{aligned} HR &= [\text{water vapor mass } (m_v) \text{ in air–vapor mixture}]/[\text{air mass } (m_a) \text{ in air–vapor mixture}] \\ &= [\text{vapor molecular weight } (M_v) \times \text{vapor partial pressure } (p_v)]/[\text{air molecular weight } (M_a) \\ &\quad \times \text{air partial pressure } (p_a)] \\ &= 0.622 p_v/p_a = 0.622 \times RH \times p_g/p_a \end{aligned}$$

where RH = relative humidity
p_g = saturation vapor pressure

Psychrometric chart. A chart that graphically displays the relationship between air–water–vapor mixtures and their properties (EPA-82/11f). An important conversion factor used with the chart is 1 lb_m = 7000 grains.

Relative humidity (RH). The ratio of the actual amount of water vapor present in the air to the amount which could exist at saturation. When a radio or TV announcer talks about percent humidity, he or she means percent relative humidity (% RH) (EPA-84/09). RH can be expressed in the following equation formats:

$$RH = [\text{actual vapor mass } (m_v)]/[\text{vapor mass required to produce a saturated mixture } (m_g)] \quad (1.20)$$

$$= [\text{vapor partial pressure } (p_v) \text{ in air–vapor mixture}]/[\text{saturation vapor pressure } (p_g) \text{ at mixture temperature}] \quad (1.21)$$

$$= [\text{saturation specific volume } (v_g) \text{ of air–water vapor mixture}]/[\text{specific vapor volume } (v_v) \text{ in the air–water vapor mixture}] \quad (1.22)$$

Equations (1.21) and (1.22) can be obtained by applying the ideal gas equation to Eq. (1.20).

Saturated mixture. A gas–vapor mixture is saturated when a reduction in temperature would cause part of the vapor to condense.

Wet bulb temperature. The temperature measured by a thermometer whose bulb (mercury holder at the bottom of a thermometer) is covered with a cotton wick which is saturated with water (wet). It is indicated by a wet bulb psychrometer.

References used in this section: Holman-74, p. 319; Jones-60, p. 409; Wark-66, p. 334.

In performing the calculations in this section, two key factors were followed.

1. For an ideal gas: properties of ideal gas components are calculated from the application of the ideal gas equation.
2. For water vapor: properties of water vapors are obtained from various steam tables which are published in many thermodynamics text books.

EXAMPLE 1: property of a vapor and ideal gas mixture

A wet scrubber contains 0.5 lb of saturated water vapor and 10 lb of air at a temperature of 300°F. Compute:

1. The volume of the water vapor
2. The total pressure of the scrubber
3. The relative humidity

Given conditions

- Water vapor, $m_g = 0.5\,\text{lb}$
- Air, $m_a = 10\,\text{lb}$
- Temperature: 300°F
- Air molecular weight: 29 lb/(lb-mole)
- Universal gas constant: 1544 (ft-lbf)/(mole-°R)

Solution to question (1)

1.1. Find data from a steam table for a temperature of 300°F (Wark-66, p. 594).

Saturation pressure, $p_v = 67.013\,\text{psia}$
Saturation specific volume, $v_g = 6.466\,\text{ft}^3/\text{lb}$

1.2. Calculate the volume of water vapor.

$$\begin{aligned} V &= (m_g)(v_g) \\ &= (0.5)(6.466) \\ &= 3.23\,\text{ft}^3 \end{aligned}$$

Solution to question (2)

2.1. Write the ideal gas equation.

$$\begin{aligned} pV &= nR_uT \text{ or} \\ p &= mR_uT/VM \end{aligned}$$

where
p = pressure
V = volume
n = number of moles
R_u = universal gas constant
M = molecular weight

2.2. Calculate the air pressure.

$$\begin{aligned} p_a &= 10(1544)(460 + 300)/(3.23(29)(144)) \\ &= 869.96\,\text{psia} \end{aligned}$$

2.3. Calculate the total pressure by applying the equation.

$$\begin{aligned} p &= p_a + p_v \\ p &= 869.96 + 67.01 \\ &= 936.97\,\text{psia} \end{aligned}$$

Solution to question (3)

1. Write the definition of the relative humidity equation.

$$RH = p_v/p_g$$

Because the property of the mixture is at the saturated vapor state, $p_v = p_g$, the relative humidity for this mixture is 100 percent.

EXAMPLE 2: property of a vapor and ideal gas mixture
A wet scrubber contains 10 lb of wet steam with a quality of 30 percent and 25 lb of air at a temperature of 200°F. Compute:

1. The total pressure of the mixture
2. The mass of air per pound of liquid water

Given conditions

- Wet steam weight, $m = 10\,\text{lb}$
- Air $= 25\,\text{lb}$
- Temperature, $T = 200°\text{F}$
- Wet vapor quality, $x = 30$ percent
- Air molecular weight, $M = 29\,\text{lb/(lb-mole)}$
- Universal gas constant, $R_u = 1544\,\text{(ft-lbf)/(mole-°R)}$

Solution to question (1)

1.1. Find data from a steam table for a temperature of 200°F (Wark-66, p. 594).

Saturation pressure, $p_v = 11.5260$ psia
Saturated liquid specific volume, $v_f = 0.0166\,\text{ft}^3/\text{lb}$
Saturation specific volume, $v_g = 33.6400\,\text{ft}^3/\text{lb}$

1.2. Write the equation for wet vapor specific volume.

$$v = v_f + x(v_g - v_f)$$

1.3. Calculate v.

$$\begin{aligned} v &= 0.0166 + (0.3)(33.64 - 0.0166) \\ &= 10.1036\,\text{ft}^3/\text{lb} \end{aligned}$$

1.4. Calculate the volume of wet vapor.

$$\begin{aligned} V &= (m)(v) \\ &= (10)(10.1036) \\ &= 101.04\,\text{ft}^3 \end{aligned}$$

Solution to question (2)

2.1. Write the ideal gas equation.

$$\begin{aligned} pV &= mR_uT/M \text{ or} \\ p &= mR_uT/VM \end{aligned}$$

2.2. Calculate the air pressure.

$$\begin{aligned} p_a &= 25(1544)(460 + 200)/(101.04(144)(29)) \\ &= 60.38\,\text{psia} \end{aligned}$$

2.3. Calculate the total pressure by applying the equation

$$p = p_a + p_v$$
$$p = 11.5260 + 60.38$$
$$= 71.91 \text{ psia}$$

2.4. Calculate the water mass.

Because the quality is 30 percent, this means that 70 percent of wet vapor is liquid. The mass of the liquid is

$$m_f = (\text{wet vapor mass})\ (0.7)$$
$$= 10(0.7)$$
$$= 7 \text{ lb-water}$$

2.5. Find the air mass per pound of liquid water.

$$m_a/m_f = 25/7$$
$$= 3.57 \text{ (lb-air)/(lb-water)}$$

EXAMPLE 3: property of a vapor and ideal gas mixture
A wet scrubber contains 2 lb water vapor and 0.5 lb air at a temperature of 400°F. The scrubber has a volume of 12.44 ft^3. Compute:

1. The pressure of the mixture
2. The relative humidity.

Given conditions

- Water vapor, $m = 2$ lb
- Air: 0.5 lb
- Temperature, $T = 400$°F
- Volume, $V = 12.44$ ft^3
- Air molecular weight, $M = 29$ lb/(lb-mole)

Solution to question (1)

1.1. Write the equation for the specific water vapor volume.

$$v = V/m$$

1.2. Calculate the specific volume.

$$v = 12.44/2$$
$$= 6.22 \text{ ft}^3/\text{lb}$$

1.3. Determine the state of the water vapor.

Based on the temperature and the specific volume conditions, the water vapor is located in the superheated region, and its pressure is roughly

$$p_v = 80 \text{ psia (from steam tables)}$$

1.4. Write the ideal gas equation.

$$p = mR_u T/VM$$

1.5. Calculate the air pressure.

$$\begin{aligned} p_a &= 0.5(1544)(400 + 460)/((12.44)(29)(144)) \\ &= 12.78 \text{ psia} \end{aligned}$$

1.6. Calculate the total pressure by applying the equation

$$\begin{aligned} p &= p_a + p_v \\ p &= 12.78 + 80 \\ &= 92.78 \text{ psia} \end{aligned}$$

Solution to question (2)

2.1. Find the saturation pressure from a steam table (Wark-66, p. 594).

From a steam table, the saturation pressure at a temperature of 400°F is

$$p_g = 247.31 \text{ psia}$$

2.2. Write the relative humidity equation.

$$RH = p_v/p_g$$

2.3. Calculate the relative humidity.

$$\begin{aligned} RH &= 80/247.31 \\ &= 0.3235 \text{ or } 32.35 \text{ percent} \end{aligned}$$

EXAMPLE 4: property of a vapor and ideal gas mixture
A wet scrubber contains a gas vapor mixture at 35 psia, 160°F, and 65 percent relative humidity. Calculate:

1. Humidity ratio and water vapor mass fraction
2. Dew point
3. Amount of water added or removed per pound dry air if the mixture undergoes a process during which its state is changed to 60°F and 40 percent relative humidity
4. The quantity of water condensed if the mixture is simply cooled at constant pressure to a final temperature of 50°F.

Given conditions

- Atmospheric pressure, $p = 14.696$ psia
- State 1 temperature, $T_1 = 160°F$
- State 2 temperature, $T_2 = 60°F$
- State 3 temperature, $T_3 = 50°F$
- State 1 relative humidity, $RH_1 = 65$ percent
- State 2 relative humidity, $RH_2 = 40$ percent
- 1 lb-water = 7000 grains water

Solution to question (1)

1.1. Find the saturation pressure from a steam table for a temperature of 160°F.

$$p_g = 4.731 \text{ psia}$$

1.2. Write the relative humidity equation.

$$RH = p_v/p_g \text{ or } p_v = (RH)(p_g)$$

1.3. Calculate the vapor pressure of the water.

$$\begin{aligned} p_v &= (0.65)(4.731) \\ &= 3.0752 \text{ psia} \end{aligned}$$

1.4. Write the air partial pressure related equation.

$$p_a = p - p_v$$

where p_a = air partial pressure
p = total air/vapor mixture pressure
p_v = vapor pressure of the water

1.5. Calculate the air partial pressure.

$$\begin{aligned} p_{a1} &= 14.696 - 3.0752 \\ &= 11.6208 \text{ psia} \end{aligned}$$

1.6. Write the humidity ratio equation.

$$HR = 0.622(p_v)/(p_a)$$

1.7. Calculate the humidity ratio.

$$\begin{aligned} HR_1 &= (0.622)(3.0752)/11.6208 \\ &= 0.1646 \text{ (lb-water)/(lb-air)} \end{aligned}$$

1.8. Write the equation of the water vapor mass fraction.

$$\begin{aligned} x &= m_v/(m_v + m_v) \\ &= HR/(1 + HR) \end{aligned}$$

1.9. Calculate the water vapor mass fraction.

$$\begin{aligned} x &= 0.1646/(1 + 0.1646) \\ &= 0.1413 \text{(lb-water)/(lb-air)} \end{aligned}$$

Solution to question (2)
By definition, the dew point is the temperature at which the vapor becomes saturated. From a steam table (Wark-66, p. 594), at a vapor pressure of 3.0752 psia the saturated temperature is roughly 145.15°F. Thus 145.15°F is the dew point temperature.

Solution to question (3)
The amount of water added or removed can be calculated from the difference between the state 2 and state 1 humidity ratios.

3.1. Find the saturation pressure from a steam table for a temperature of 60°F.

$$p_{g2} = 0.256\,\text{psia}$$

3.2. Write the relative humidity equation.

$$RH = p_v/p_g \text{ or } p_v = (RH)(p_g)$$

3.3. Calculate the vapor pressure.

$$\begin{aligned} p_{v2} &= (0.4)(0.256) \\ &= 0.1024\,\text{psia} \end{aligned}$$

3.4. Write the air partial pressure related equation.

$$p_a = p - p_v$$

3.5. Calculate the air partial pressure.

$$\begin{aligned} p_{a2} &= 14.696 - 0.1024 \\ &= 14.5936\,\text{psia} \end{aligned}$$

3.6. Write the humidity ratio equation.

$$HR = 0.622(p_v)/(p_a)$$

3.7. Calculate the final humidity ratio (HR_2).

$$\begin{aligned} HR_2 &= (0.622)(0.1024)/14.5936 \\ &= 0.0044\,\text{(lb-water)/(lb-dry-air) or} \\ &= (0.0044)(7000) \\ &= 30.55\,\text{(grains-water)/(lb-dry-air)} \end{aligned}$$

3.8. Calculate water added or removed.

$$\begin{aligned} \text{Water} &= HR_2 - HR_1 \\ &= 0.0044 - 0.1646 \\ &= -0.1602\,\text{lb-water/(lb-dry-air)} \end{aligned}$$

Negative means that the water is removed.

Solution to question (4)
Because the dew point temperature at state 1 is roughly 145.15°F, water must condense if the mixture is cooled to 50°F.

4.1. Find the saturation pressure at 50°F from a steam table.

$$p_{g3} = 0.178\,\text{psia}$$

4.2. Find the relative humidity.

Because the temperature is lower than the dew point, the mixture must be saturated, and the relative humidity is 1. Therefore, $p_{v3} = p_{g3} = 0.178$ psia

4.3. Calculate the humidity ratio.

$$HR_3 = 0.622(0.178)/(14.696 - 0.178)$$
$$= 0.0076$$

4.4. Calculate the water removed.

$$\text{Water} = HR_3 - HR_1$$
$$= 0.0076 - 0.1646$$
$$= -0.1570\,\text{(lb-water)/(lb-dry-air)}$$

Negative means that the water is removed.

EXAMPLE 5: property of a vapor and ideal gas mixture
For air at a pressure of 14.7 psia, a dry bulb temperature of 82°F, and a wet bulb temperature of 70°F, determine:

1. The humidity ratio
2. The relative humidity
3. The vapor pressure in psia
4. The dew point
5. The enthalpy
6. The specific volume per pound of dry air

Given conditions

- Dry bulb temperature, $T_d = 82°F$
- Wet bulb temperature, $T_w = 70°F$
- Total pressure, $p = 14.696$ psia

Solution:
A psychrometric chart (Wark-66, p. 636) was used to obtain the answers to the questions. Actually, a psychrometric chart in any thermodynamic text book should contain the information used for this example.

Solution to question (1)
The humidity ratio can be read from the intersection of the dry bulb temperature line and the wet bulb temperature line. The value is roughly more than 92 grains/(lb-dry-air) or 0.0133 (lb-water)/(lb-air).

Solution to question (2)
At the intersection point, the relative humidity is roughly 58%.

Solution to question (3)
The vapor pressure is roughly 0.108 psia, which is shown on the right-hand side of the psychrometric chart in Wark.

Solution to question (4)
The dew point is the temperature at which condensation begins if the vapor is cooled at constant pressure. Therefore, the dew point can be found by moving horizontally to the left from the initial point until the saturation line is reached. The dry bulb temperature at the intersection of the constant humidity ratio and the saturation line is found to be roughly 65°F. This temperature is the dew point temperature.

Solution to question (5)
The enthalpy is obtained by following the wet bulb temperature from the initial point to the enthalpy line in the upper left of the chart. The enthalpy is read roughly as 34.2 Btu/(lb-dry-air).

Solution to question (6)
The specific volume is roughly 13.9 ft^3/(lb-dry-air).

EXAMPLE 6: property of a vapor and ideal gas mixture
An air flow at 95°F and 60 percent relative humidity is passed through a chiller unit which cools the air flow to 75°F and 50 percent relative humidity. Determine:

1. The amount of water removed per pound of dry air passing through the unit
2. The amount of heat removed (Btu/lb dry air)

Given conditions

- Temperature at state 1, $T_1 = 95°F$
- Temperature at state 2, $T_2 = 75°F$
- Relative humidity at state 1, $RH_1 = 60$ percent
- Relative humidity at state 2, $RH_2 = 50$ percent

Solution:
A psychrometric chart (Jones-60) was used to obtain the answers to the questions. A psychrometric chart in any thermodynamic text book should contain the information used for this example.

Solution to question (1)
The amount of water removed can be calculated from the difference of humidity ratios betwen state 1 and state 2. From the psychrometric chart, the humidity ratio at 95°F and 60 percent relative humidity is roughly

$$HR_1 = 0.0211\,\text{(lb-water)/(lb-air)}$$

The humidity ratio at 75°F and 50 percent relative humidity is roughly

$$HR_2 = 0.0094\,\text{(lb-water)/(lb-air)}$$

The water removed is $HR_2 - HR_1 = -0.0117$ (lb-water)/(lb-air). The negative sign shows that water is removed from the system.

Solution to question (2)
The amount of heat transferred can be calculated from the enthalpy difference between state 1 and state 2. From the psychrometric chart, the enthalpy at 95°F and 60 percent relative humidity is roughly

$$h_1 = 46.8\,\text{Btu/(lb-air)}$$

The enthalpy at 75°F and 50 percent relative humidity is roughly

$$h_2 = 28.5\,\text{Btu/(lb-air)}$$

The water removed is $h_2 - h_1 = -18.3$ Btu/(lb-air). The negative sign shows that water is removed from the system.

EXAMPLE 7: property of a vapor and ideal gas mixture
An air flow is heated from a relative humidity of 40 percent and a dry bulb temperature of 40°F to a temperature of 75°F. Determine the amount of heat added per pound of dry air. Assume that atmospheric pressure equals 14.696 psia.

Given conditions

- Temperature at state 1, $T_1 = 40°F$
- Relative humidity at state 1, $RH_1 = 40\%$
- Temperature at state 2, $T_2 = 75°F$

Solution
The psychrometric chart (Wark-66, p. 636) was used to obtain the answers to the questions. A psychrometric chart in any thermodynamic text book should contain the information used for this example.

1. Determine the humidity ratio.

From the psychrometric chart, the humidity ratio at 40°F and 40 percent relative humidity is roughly

$$HR = 0.0016\,(\text{lb-water})/(\text{lb-air})$$

2. Write the first law energy equation under steady-state conditions.

$$q = h_2 - h_1$$

where q = heat transferred
h_1 and h_2 = enthalpy at state 1 and 2, respectively

From the psychrometric chart, the enthalpy at 40°F and 40 percent relative humidity is roughly

$$h_1 = 11.7\,\text{Btu/(lb-air)}$$

Because the water vapor mass remains constant, HR = constant from state 1 to state 2. From the psychrometric chart, the enthalpy at 75°F and humidity ratio at 0.0016 (lb-water)/(lb-air) is

$$h_2 = 20.1\,\text{Btu/(lb-air)}$$

3. Calculate the heat transferred.

$$\begin{aligned} q &= 20.1 - 11.7 \\ &= 8.4\,\text{Btu/(lb-air)} \end{aligned}$$

Positive means that the heat is added to the air flow.

2.44 Property of Waste Mixture

This example is to illustrate how properties of a waste mixture can be found. This example was derived from an actual case of a "Superfund" site waste analysis. The waste was analyzed for a proposed incinerator trial burn. Because this example is used only to illustrate how thermodynamic properties are applied to "real-world" wastes, the actual site name and the detailed waste descriptions are not included.

EXAMPLE: property of waste mixture
With the following given conditions, determine:

1. Total waste quantity to be fed to the incinerator
2. Composite waste heat content per pound
3. Composite waste moisture content
4. Composite waste ash content

For these calculations, the density of the water in the feed is assumed to be 1785 lb/yd^3.

Given conditions

The results of the waste analysis are given in Table 1.33.

TABLE 1.33 Waste Analysis Results

	S_g	V	Q	% H_2O	% ash
Waste 1	1.01	5,200	11,000	0.46	0.11
Waste 2	2.5	91,000	6,000	0.25	0.32
Waste 3	2.69	19,000	3,700	0.54	0.49
Waste 4	3.65	9,000	1,500	0.25	0.9

S_g = specific gravity. Water density is used as the reference substance to determine the waste specific gravity. The water density in question was assumed to be 1785 lb/yd^3.
V = volume (yard^3).
Q = heat content of waste streams (Btu/lb).

Solution to question (1)
For waste stream 1

1.1.1. Calculate the density of waste stream 1, D_1.

Write the definition equation for specific gravity S_g

$$S_g \text{ (waste only)} = \text{(density of a substance)/(density of water), or}$$

$$D_1 = S_g \text{ (density of water)}$$
$$D_1 = 1.01(1785)$$
$$= 1803 \text{ lb/yd}^3$$

1.1.2. Calculate the mixture density of waste stream 1, D_{m1}.

$$D_{m1} = 1785(0.46) + 1803(1 - 0.46)$$
$$= 1795 \text{ lb/yd}^3$$

1.1.3. Calculate the mixture weight of waste stream 1, W_1.

Write the definition equation of mixture weight W_1

$$W_1 = \text{(density)(volume)}$$
$$W_1 = 5200(1795)$$
$$= 9.33 \times 10^6 \text{ lb}$$

For waste stream 2

1.2.1. Calculate the density of waste stream 2, D_2.

$$\begin{aligned} D_2 &= 2.5(1785) \\ &= 4463 \end{aligned}$$

1.2.2. Calculate the mixture density of waste stream 2, D_{m2}.

$$\begin{aligned} D_{m2} &= 1785(0.25) + 4463(1 - 0.25) \\ &= 3793\,\text{lb/yd}^3 \end{aligned}$$

1.2.3. Calculate the mixture weight of waste stream 2, W_2.

$$\begin{aligned} W_2 &= 91{,}000(3793) \\ &= 345 \times 10^6\ \text{lb} \end{aligned}$$

For waste stream 3

1.3.1. Calculate the density of waste stream 3, D_3.

$$\begin{aligned} D_3 &= 2.69(1785) \\ &= 4802 \end{aligned}$$

1.3.2. Calculate the mixture density of waste stream 3, D_{m3}.

$$\begin{aligned} D_{m3} &= 1785(0.54) + 4802(1 - 0.54) \\ &= 3173\,\text{lb/yd}^3 \end{aligned}$$

1.3.3. Calculate the mixture weight of waste stream 3, W_3.

$$\begin{aligned} W_3 &= 19{,}000(3173) \\ &= 60.3 \times 10^6\,\text{lb} \end{aligned}$$

For waste stream 4

1.4.1. Calculate the density of waste stream 4, D_4.

$$\begin{aligned} D_4 &= 3.65(1785) \\ &= 6515\,\text{lb/yd}^3 \end{aligned}$$

1.4.2. Calculate the mixture density of waste stream 4, D_{m4}.

$$\begin{aligned} D_{m4} &= (1785\,\text{lb/yd}^3)(0.25) + (6515\,\text{lb/yd}^3)(1 - 0.25) \\ &= 5333\,\text{lb/yd}^3 \end{aligned}$$

1.4.3. Calculate the mixture weight of waste stream 4, W_4.

$$\begin{aligned} W_4 &= 9000(5333) \\ &= 48 \times 10^6\,\text{lb} \end{aligned}$$

1.4.4. Calculate the total waste to be fed to the incinerator W_t.

Assuming that all four waste streams are simultaneously fed to the incinerator,

$$
\begin{aligned}
W_t &= W_1 + W_2 + W_3 + W_4 \\
&= (9.33 + 345 + 60.3 + 48) \times 10^6 \\
&= 463 \times 10^6
\end{aligned}
$$

Solution to question (2)

2.1. Calculate the weight fractions of mixture waste 1, x_1.

Write the weight fraction equation for mixture waste stream 1, x_1

$$
\begin{aligned}
x_1 &= W_1/W_t \\
&= 9.33 \times 10^6/463 \times 10^6 \\
&= 0.0202
\end{aligned}
$$

2.2. Calculate the weight fractions of mixture waste 2, x_2.

$$
\begin{aligned}
x_2 &= 345 \times 10^6/463 \times 10^6 \\
&= 0.7451
\end{aligned}
$$

2.3. Calculate the weight fractions of mixture waste 3, x_3.

$$
\begin{aligned}
x_3 &= 60.3 \times 10^6/463 \times 10^6 \\
&= 0.1302
\end{aligned}
$$

2.4. Calculate the weight fractions of mixture waste 4, x_4.

$$
\begin{aligned}
x_4 &= 48 \times 10^6/463 \times 10^6 \\
&= 0.1037
\end{aligned}
$$

2.5. Calculate the waste heat content Q.

Write the total heat content equation.

$$
\begin{aligned}
Q &= Q_1(x_1) + Q_2(x_2) + Q_3(x_3) + Q_4(x_4) \\
&= 11{,}000(0.0202) + 6000(0.7451) + 3700(0.1302) + 1500(0.1037) \\
&= 5330 \text{ Btu/lb}
\end{aligned}
$$

Solution to question (3)

Write the total moisture content equation

$$
\begin{aligned}
H_2O &= (\%H_2O)_1(x_1) + (\%H_2O)_2(x_2) + (\%H_2O)_3(x_3) + (\%H_2O)_4(x_4) \\
&= 0.46(0.0202) + 0.25(0.7451) + 0.54(0.1302) + 0.250(0.1037) \\
&= 0.2920\%
\end{aligned}
$$

Solution to question (4)

Write the total ash content

$$
\begin{aligned}
\text{Ash} &= (\%\ \text{ash})_1(x_1) + (\%\ \text{ash})_2(x_2) + (\%\ \text{ash})_3(x_3) + (\%\ \text{ash})_4(x_4) \\
&= 0.11(0.0202) + 0.32(0.7451) + 0.49(0.1302) + 0.90(0.1037) \\
&= 0.3978\%
\end{aligned}
$$

2.45 Psychrometric Chart

A psychrometric chart is a chart that graphically displays the relationship between air–water–vapor mixtures and their properties (EPA-82/11f). An important conversion factor used with the chart is 1 lb_m = 7000 grains. A schematic diagram of a psychrometric chart is shown in Fig. 1.10 (Wark-66, p. 342).

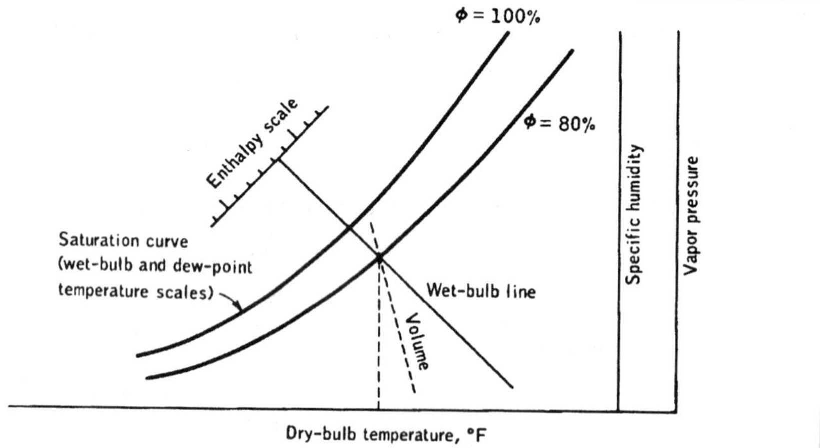

FIGURE 1.10 Chart: outline of a psychrometric chart.

2.46 Raoult's Law

Raoult's law. For concentrated solutions where the components do not interact, the resulting vapor pressure (p) of component "a" in equilibrium with other solutions can be expressed as

$$p = x_a P_a$$

where p = resulting vapor pressure
x = mole fraction of component "a" in solution
P_a = vapor pressure of pure "a" at the same temperature and pressure as the solution.

Comparing Raoult's law and Henry's law, Raoult's law is applicable to highly concentrated solutions and Henry's law for dilute solutions.

EXAMPLE: Raoult's law
Ethylene glycol ($C_2H_6O_2$) is used as an antifreeze compound. The compound not only lowers the freezing temperature, but also increases the boiling temperature of the water in a radiator. For this example, consider a solution which contains 70 percent ethylene glycol. Determine the boiling temperature of the solution.

Given conditions

- Ethylene glycol mass fraction = 70%
- Water mass fraction = 0.30
- Pure water vapor pressure at 100°C, p_w = 760 mm Hg

Solution:

1. Determine the molecular weight of ethylene glycol and water.

The chemical formula of ethylene glycol is $C_2H_6O_2$; its molecular weight is therefore equal to 62. Similarly, water's chemical formula is H_2O, and its molecular weight is 18.

2. Determine the moles of ethylene glycol and water.

The equation for mole calculation is

$$\text{mole} = (\text{compound weight})/(\text{molecular weight})$$

Moles of ethylene glycol (assuming 1 lb of solution),

$$\begin{aligned} n &= 0.7/62 \\ &= 0.0113 \end{aligned}$$

Moles of water,

$$\begin{aligned} n &= 0.3/18 \\ &= 0.0167 \end{aligned}$$

3. Determine the mole fraction of water.
The mole fraction of water is

$$\begin{aligned} x &= 0.0167/(0.0167 + 0.0113) \\ &= 0.5962 \end{aligned}$$

4. Write Raoult's law equation.

$$p = x_{\mathrm{w}} P_{\mathrm{w}}$$

where p = resulting vapor pressure
x_{w} = water mole fraction
P_{w} = vapor pressure of water at the same temperature and pressure as the solution.

5. Calculate the resulting vapor pressure of the solution.

$$\begin{aligned} p &= 0.5961(760) \\ &= 453.08\,\text{mm Hg} \end{aligned}$$

6. Determine the boiling point of water.

Remember that for pure water to boil, its vapor pressure needs to be at $p = 760$ mm Hg and $T = 212°\text{F}$ (100°C). Now because of the component ethylene glycol, the water vapor pressure drops from 760 mm Hg to 453.08 mm Hg. For water to boil at this condition, the temperature needs to be increased to the level so that the water partial pressure will be at 760 mm Hg. This pressure (p_2) can be obtained by

$$p_2 = 760/0.5962, \text{ which is equal to } 1274.84 \text{ mm Hg}$$

7. Convert p_2 into pounds per square inch units.

$$\begin{aligned} p_2 &= 1274.84\,\text{mm Hg} \\ &= 50.19\,\text{in Hg} \\ &= 24.64\ \text{psi (pound per square inch)} \end{aligned}$$

Note: 1in Hg = 0.491 psi

8. Determine the boiling temperature.

At pressure $p_2 = 24.64$ psi, the corresponding saturated temperature is roughly 242°F. This means that the solution of ethylene glycol and water will boil at a temperature of 242°F.

2.47 Reynolds Number

The Reynolds number provides information on flow behavior. In general, a laminar flow has a Reynolds number below 2100 in a tube, but it can be up to Reynolds numbers of several thousand under special conditions. A turbulent flow has a Reynolds number above about 4000. Between 2100 and 4000 a transition region is found where the type of flow may be either laminar or turbulent. By definition, the Reynolds number is dimensionless. For most air pollution applications, the gas flow is a turbulent (high number) flow.

EXAMPLE 1: Reynolds number
Assume that the second chamber of an incinerator has an internal diameter of 2 m. The gas flow is at a temperature of 1000°C, and at a velocity of 10 m/s (OME-86/10, p. 53).

Given conditions

- Secondary chamber diameter = 2 m
- Velocity through the duct = 10 m/s
- Kinematic viscosity = 140×10^{-6}
- Temperature = 1000°C

Solution:

1. Write the definition of Reynolds number.

$$Re = VD/K$$

where Re = Reynolds number
V = velocity
D = diameter
K = kinematic viscosity = (absolute viscosity)/density

2. Calculate the Reynolds number of the gas stream.

$$\begin{aligned} Re &= VD/K \\ &= (10)(2)/(140 \times 10^{-6}) \\ &= 1.43 \times 10^5 \end{aligned}$$

This is greater than the critical *Re* (4000) and turbulence can be assumed to be adequate.

EXAMPLE 2: Reynolds number
A gas with viscosity of 1.16×10^{-5} lb/ft-s and density of 0.075 lb/ft^3 flows through a 1.5-ft-diameter pipe at a velocity of 25 ft/s. Calculate the Reynolds number and determine whether or not the flow is a turbulent flow (EPA-84/09, p. 43).

Given conditions

- Velocity through the duct = 25 ft/s
- Duct diameter = 1.5 ft
- Gas viscosity = 1.16×10^{-5} lb/ft-s
- Gas density = 0.075 lb/ft^3

Solution:

1. Write the definition of Reynolds number.

$$Re = Dvd/\mu$$

where D = duct diameter
v = velocity
d = gas density
μ = gas viscosity

2. Calculate the Reynolds number of the gas stream.

$$\begin{aligned} Re &= Dvd/\mu \\ &= (1.5)(25)(0.075)/(1.16 \times 10^{-5}) \\ &= 2.42 \times 10^{5} \end{aligned}$$

The flow is a turbulent flow.

2.48 Sampling and Analytical Methods

Sampling and analytical methods are summarized below.

1. Sample: Method 5 (M5) train
   - Filter, probe rinse
     —Analytical parameter: particulate
     —Analytical method: EPA method 5
   - Water impinger and caustic impinger
     —Analytical parameter: Cl^-
     —Analytical method: ion chromatography or EPA 352.2
2. Multiple metals train
   - Analytical parameter: metals
   - Analytical method: (see Table 13 in EPA-89/06)
3. Semi-volatile organic sampling (VOST)
   - Filter, probe rinse
     —Analytical parameter: semi-volatile POHC (SV–POHC)
     —Analytical method: (see appendix A in EPA-89/06)

- XAD-2 (Sorbent Resin)
  —Analytical parameter: SV–POHC
  —Analytical method: GC/MS (Gas Chromatograph/Mass Spectrometry)
- Condensate
  —Analytical parameter: SV–POHC
  —Analytical method: not specified

4. VOST
- Analytical parameter: V–POHC
- Analytical method: GC/MS per EPA SW-846, method 5040

5. Mylar gas bag
- Analytical parameter: CO_2, O_2
- Analytical method: EPA method 3

6. Tedlar gas bag
- Analytical parameter: V–POHC
- Analytical method: transfer to Tenax trap and GC/MS per
- SW-846, method 5040

EXAMPLE 1: sampling and analytical methods
Estimate POHC concentration in stack gas at a DRE (see Section 2.10) of 99.99% (EPA-85/11, p. 27).

Given conditions

- Waste feed flow rate approximately = 4 gpm
- POHC concentration in waste feed is near minimum significant level of 200 ppm (200 ppm = 0.000200 g-POHC/g-feed)
- Stack gas flow rate unknown, but total heat input to incinerator is approximately 30×10^6 Btu/h with 100% excess air.

Solution:
General assumption (or rule of thumb): each 100 Btu of heat input produces about 1 dscf of flue gas at 0% excess air, or 2 dscf (dry standard cubic foot) of dry flue gas at 100% excess air. This assumption applies in most, but not all, cases.

1. Calculate flue gas flow rate Q.

$$
\begin{aligned}
Q &= 30 \times 10^6\,(\text{Btu/h})(\text{h}/60\,\text{min})(2\,\text{dscf}/100\,\text{Btu}) \\
&= 10{,}000\,\text{dscf/min} \\
&= 10{,}000\,(\text{dscf/min})(1\,\text{m}^3/35.3\,\text{ft}^3) \\
&= 283\,\text{m}^3/\text{min}
\end{aligned}
$$

2. Calculate waste feed rate V.

$$
\begin{aligned}
V &= (4\,\text{gal/min})(9\,\text{lb/gal})(454\,\text{g/lb}) \\
&= 16{,}344\,\text{g/min}
\end{aligned}
$$

3. Calculate POHC input rate.

$$\text{POHC} = (16{,}344\ \text{g-feed/min})(0.000200\ \text{g-POHC/g-feed})$$
$$= 3.27\ \text{g-POHC/min} = W_i$$

4. Calculate POHC stack output rate at 99.99% DRE.

$$\text{DRE} = (1 - W_0/W_i)100$$
$$0.9999 = 1 - W_0/W_i$$
$$W_0 = (1 - 0.9999)W_i$$
$$= 0.000327\ \text{g/min}$$

5. Calculate POHC concentration in stack gas at 99.99 DRE.

$$\text{POHC concentration} = 0.000326/283$$
$$= 1.15 \times 10^{-6}\ \text{g/m}^3$$
$$= 1.15\ \mu\text{g/m}^3$$
$$= 1.15\ \text{ng/L}$$

EXAMPLE 2: sampling and analytical method
The following is a sample calculation showing the method used to convert the analytical results to DREs for trichloroethylene (TCE) in Run 2 using the VOST sample (EPA-85/11).

Given conditions

- Consider two waste streams:
  —Waste stream 1 is an organic waste with flow rate = 4010 g/min
  —Waste stream 2 is an aqueous waste with flow rate = 5380 g/min
- TCE is a POHC which is contained in both waste streams
  —TCE concentration in waste stream 1 = 5500 μ/g
  —TCE concentration in waste stream 2 = less than 1 μ/g

Solution:

1. Calculate total TCE feed W_i

W_i = (waste stream flow rate 1) × (TCE concentration in waste stream 1) + (waste stream flow rate 2) × (TCE concentration in waste stream 2)

$$= 4010(5500) + 5380(1)$$
$$= 22{,}060{,}380\ \mu/\text{min}$$
$$= 22.06038\ \text{g/min, or}$$
$$= 22\ \text{g/min}$$

2. Calculate output rate W_0.

It is assumed that:

- Stack flow rate (measured): 76 Nm^3/min
- There are 3 VOST tests. The results of each test show TCE concentration as follows:
  —test 1: 20 ng/L
  —test 2: 1.8 ng/L
  —test 3: 1ng/L

Solution:

2.1. Calculate the average of the test results.

$$\text{avg} = (20 + 1.8 + 1)/3 = 7.6\,\text{ng/L}$$

2.2. Determine blank correction.

- VOST blank correction values for TCE =< 1 ng/sample (not blank corrected)
- VOST sample volumes (dry standard liters) = 18.5 L/sample
- VOST =< 1/18.5 =< 0.05 ng/L
- Blank corrected value = 7.6– < 0.05 =< 7.6 ng/L =< 7.6 μ/m^3

2.3. VOST output rate.

$$\begin{aligned} W_0 &= (< 7.6\,\mu/\text{m}^3)(76\,\text{Nm}^3)(1 \times 10^{-6}\,\text{g}/\mu) \\ &=< 577.6 \times 10^{-6}\ \text{g/min} \end{aligned}$$

2.4. Calculate DRE.

$$\begin{aligned} \text{DRE} &= (1 - W_0/W_i)100 \\ &= (1- < 0.0005776/22) \\ &= > 99.9974\% \end{aligned}$$

2.49 Specific Gravity

The specific gravity is defined as the ratio of the density of a substance to the density of a reference substance. Water at 4°C has its maximum density. In general, water at a temperature of 4°C is used as the reference substance for solids and liquids (Schaum-66, p. 9).

$$\begin{aligned} \text{Density of water at } 4^\circ\text{C} &= 62.4\,\text{lb/ft}^3 \text{ in British units} \\ &= 1.0\,\text{g/cm}^3 \text{ in cgs units} \\ &= 1.0\,\text{g/cc (cubic centimeter)} \\ &= 1.0\,\text{g/ml (milliliter)} \end{aligned}$$

EXAMPLE: specific gravity
The specific gravity of a liquid is given as 0.755. Calculate its density at 4°C in lb/ft^3 and g/cm^3.

Solution:

1. Write the specific gravity γ definition equation.

$$\gamma = \text{(density of a substance)/(density of a reference substance)}$$

2. Rearrange the above equation.

$$\text{Density of the liquid} = (\gamma) \times (\text{density of water at } 4^\circ\text{C})$$

3. Calculate the density of the liquid.

$$\text{Density of the liquid} = 0.755(62.4)$$
$$= 47.11\ \text{lb/ft}^3, \text{ or}$$
$$\text{Density of the liquid} = 0.755(1)$$
$$= 0.755\ \text{g/cc}$$

2.50 Specific Heat (see Section 2.21 Heat Capacity and Specific Heat)

2.51 Specific Volume

Specific volume v is defined as the volume of a system divided by the mass of substances in the system. It is the reciprocal of density ρ or the volume per unit mass. It can be expressed as

$$v = (\text{volume})/(\text{mass}) = V/m = 1/\rho$$

2.52 Specific weight

Specific weight w is defined as the weight of a substance divided by its volume, or the weight per unit volume. It can be expressed as (Holman-74, p. 18)

$$w = (\text{weight})/(\text{volume})$$
$$= (\text{force})/(\text{volume})$$
$$= (mg/g_c)/V = \rho g/g_c$$

where m = mass
g = local acceleration of gravity
V = volume
ρ = density
g_c = gravitational acceleration

The example values and units of g_c are

$$g_c = 32.17\ (\text{lbm-ft})/(\text{lbf-s}^2)$$
$$= 1\ (\text{slug-ft})/(\text{lbf-s}^2)$$
$$= 1\ (\text{kg mass-m})/(\text{newton-s}^2)$$
$$= 1\ (\text{g mass-cm})/(\text{dyne-s}^2)$$

At sea level, $g = 32.17\,(\text{ft})/(\text{s}^2) = 9.8066\,(\text{m})/(\text{s}^2)$, and therefore $g = g_c$, which means 1 lbm weighs 1 lbf.

2.53 Stack Cross-sectional Area

The stack cross-sectional area is the area where the exhaust gas from a process is discharged to the atmosphere. Actual volumetric flow rates are always used to calculate the stack cross-sectional area. The equation used to calculate the stack cross-sectional area is

$$A = Q/v$$

where A = stack cross-sectional area
Q = actual volumetric gas flow rate
v = gas velocity

EXAMPLE: stack cross-sectional area
The actual exhaust gas flow rate from a facility is 2200 acfm with a discharge velocity of 20 ft/s. Calculate the stack cross-sectional area.

Solution:

1. Write the stack cross-sectional area equation.

$$A = Q/v$$

2. Calculate the stack cross-sectional area equation.

$$\begin{aligned} A &= 2200/20 \\ &= 110\,\text{ft}^2 \end{aligned}$$

2.54 Stack Discharge Velocity

The stack discharge velocity is the velocity at which the exhaust gas from a process is discharged to the atmosphere. Actual volumetric flow rates are always used to calculate the stack discharge velocity.

The velocity calculated here is the bulk or average velocity. The average velocity can be simply calculated by dividing the actual volumetric gas flow rate by the cross-sectional area through which the gas flows.

EXAMPLE: stack discharge velocity
The exhaust gas flow rate from a facility is 1000 scfm. All of the gas is vented through a small stack which has a diameter of 1.2 ft. The exhaust gas temperature is 300°F. What is the velocity of the gas through the stack inlet in feet per second. Assume standard conditions to be 60°F and 1.0 atmosphere (EPA-84/09, p. 41).

Given conditions

- Standard volumetric flow rate of exhaust gas, $Q_1 = 1000$ scfm
- Standard conditions, $T_1 = 60°$F and 1.0 atm
- Actual temperature of exhaust gas, $T_2 = 300°$F
- Diameter of the stack = 1.2 ft

Solution:

1. Write the ideal gas equation for actual volumetric flow rate (acfm) calculation.

$$Q_2 = Q_1(T_2/T_1)(P_1/P_2)$$

where Q_2 = actual volumetric flow rate
Q_1 = standard volumetric flow rate
T_2 = actual operating temperature, °R or K
T_1 = standard temperature, °R or K
P_2 = actual operating pressure
P_1 = standard absolute pressure

Note that (P_1/P_2) is usually equal to 1.0, since the exhaust gas is discharged to the atmosphere.

2. Calculate the actual volumetric flow rate Q_2.

$$\begin{aligned} Q_2 &= 1000(460 + 300)/(460 + 60) \\ &= 1462\,\text{acfm} \end{aligned}$$

3. Write the equation for the stack cross-sectional area calculation A.

$$A = \pi(D^2)/4$$

where A = stack cross-sectional area
D = stack diameter

4. Calculate the cross-sectional area of the stack.

$$\begin{aligned} A &= \pi(D^2)/4 \\ &= \pi(1.2)^2/(4) \\ &= 1.131\,\text{ft}^2 \end{aligned}$$

5. Calculate the discharge velocity v.

$$\begin{aligned} v &= Q_2/A \\ &= (1461.5)/(1.131) \\ &= 1292.2\,\text{ft/min} \\ &= 21.5\,\text{ft/s} \end{aligned}$$

2.55 Standard Condition

Thermodynamic and gas volume calculations require the definition of a reference condition or standard temperature and pressure (STP) against which all values will be calculated. In general, the following definitions are used as the STP.

Standard conditions for British units. In general, the standard conditions for British units are

- Temperature = 70°F
- Pressure = 1 atm (or 14.696 lb/in^2)
- Volume = 387 ft^3

The standard volume is calculated from the ideal gas equation as follows:

For air at 1 mole, 14.6959 psia, and 70°F,

$$\begin{aligned} V &= nR_uT/P \\ &= 1(1544)(460 + 70)/(14.696(144)) \\ &= 387\,\text{ft}^3 \end{aligned}$$

where P = absolute pressure
V = volume, ft^3
T = absolute temperature, °R or K
n = number of moles (lb-moles, g-moles or kg-moles)
R_u = universal gas constant (see Section 2.27)

Standard conditions for metric units. In general, the standard conditions for metric units are

- Temperatutre = 0°C
- Pressure = 1 atm (or 101.3 kPa)
- Volume = 22.40 m^3

The standard volume is calculated from the ideal gas equation as follows:

For air at 1 kg-mole, 1 atm, 0°C,

$$\begin{aligned} V &= nR_u T/P \\ &= 1(0.08205)(273 + 0)/1 \\ &= 22.40\ \text{m}^3 \end{aligned}$$

Commonly used standard temperatures. The following shows the volume occupied by one mole of an ideal gas at STP for British and metric units and at different commonly used standard temperatures:

- 359 ft^3 at 32°F (0°C), 1 atm (for 1 lb-mole)
- 380 ft^3 at 60°F, 1 atm (for 1 lb-mole)
- 387 ft^3 at 70°F, 1 atm (for 1 lb-mole)
- 22.4 liter at 0°C, 1 atm (for 1 g-mole)
- 0.0224 m^3 at 0°C, 1 atm (for 1 g-mole)

To illustrate its use for temperature at 32°F or 0°C,

- 32 pounds or one lb-mole of oxygen (O_2, MW = 32) occupies 359 ft^3 at STP
- 32 grams, or one g-mole of oxygen occupies 24.40 L
- 44 lb of CO_2 occupies the same 359 ft^3 and 44 g occupies 22.40 L

While the standard conditions for metric and British units are slightly different, the difference is inconsequential for the calculations discussed herein. Other temperatures, such as 60°F, 68°F, 32°F and 0°C are sometimes used in the literature. It is important to know the STP whenever thermodynamic data are obtained from the literature. It is rarely necessary to correct for them when performing combustion calculations because the high temperatures of combustion make the differences between the various standard temperatures negligible. The impact may, however, be significant when evaluating gases at lower temperatures as in a quench, scrubber, or stack.

Gaseous elements under normal conditions exist only as two-atom molecules. For example, hydrogen gas exists as H_2, oxygen and nitrogen as O_2 and N_2. This is important when performing gas volume calculations, since a mole of O_2, for example, will occupy one half the volume of a mole of the element O, and if this is not taken into account, it will lead to errors in the gas flow calculations.

Standard flow rate and actual flow rate. The volume or volumetric flow rate (V) of a gas stream at any temperature (T) and pressure (P) can be calculated using the ideal gas law if the quantity is known at the standard temperature and pressure. The equation is

$$V_2 = V_1(T_2/T_1)(P_1/P_2) \tag{1.23}$$

The absolute temperature (K or °R) must be used in the equation, and in all thermodynamic and kinetic calculations. Any absolute units of pressure may be used as long as the base or zero pressure is a vacuum. For example, the use of inches of mercury of vacuum where 1 atmosphere is equal to zero is not acceptable because it is not an absolute pressure scale. Volume will not be proportional to pressure if such a scale is used. Both temperatures and both pressures must be given in the same units.

Since the pressures involved in most incinerators are close to 1 atmospheric pressure, Eq. (1.23) can be simplified to the equation

$$V_2 = V_1(T_2/T_1) \tag{1.24}$$

Equation (1.24) can be generalized to relate the molar volume or density of gases to their temperature at any two states by

$$V_1/V_2 = T_1/T_2 \text{ or} \tag{1.25}$$

$$\rho_2/\rho_1 = T_1/T_2 \tag{1.26}$$

where 1 and 2 are two different temperatures for the same gas. Equation (1.25) states that the volume or volumetric flow rate of a unit of a gas is proportional to temperature at constant pressure.

Ideal gas calculations can also be conducted on the basis of mass. In this case, the ideal gas law is written as

$$P = \rho RT \tag{1.27}$$

where P = the pressure of the gas (psi, atm, kPa, dynes/m^2)
ρ = the density of the gas (g/cm^3, g/m^3, lb/ft^3)
T = the absolute temperature (K or °R)
R = R_u/MW where MW is the molecular weight of the gas

The effect of pressure, temperature and molecular weight on density can be obtained directly from the ideal gas law equation. Increasing the pressure and molecular weight increases the density; increasing the temperature decreases the density.

EXAMPLE: standard flow rate and actual flow rate
Two hundred (200) lb per hour of methane gas, CH_4, is used to incinerate a waste at 70°F standard condition and at 2000°F temperature. Determine the methane gas flow rate at standard and at 2000°F conditions.

Solution:

1. Write the ideal gas equation.

$$V = nR_uT/P = (m/M)R_uT/P$$

where P = absolute pressure
V = volume, ft^3
T = absolute temperature, °R or K
n = number of moles (lb-moles, g-moles or kg-moles)
R_u = universal gas constant
m = mass flow rate
M = molecular weight

2. Calculate flow rate (V) at the standard condition $T = 70°\text{F}$.

$$
\begin{aligned}
V &= (200/16)(1545)(460 + 70)/(14.696(144)) \\
&= 4837\ \text{ft}^3/\text{h} \\
&= 4837\ \text{scfh} \\
&= 81\ \text{scfm} \\
&= 1.34\ \text{scfs}
\end{aligned}
$$

where scfh = standard cubic feet per hour
scfm = standard cubic feet per minute
scfs = standard cubic feet per second

3. Calculate flow rate (V) at the actual condition $T = 2000\,°\text{F}$

$$
\begin{aligned}
V &= (200/16)(1545)(460 + 2000)/[14.696(144)] \\
&= 22{,}450\ \text{ft}^3/\text{h} \\
&= 22{,}450\ \text{acfh} \\
&= 374\ \text{acfm} \\
&= 6.24\ \text{acfs}
\end{aligned}
$$

where acfh = actual cubic feet per hour
acfm = actual cubic feet per minute
acfs = actual cubic feet per second

2.56 Stoichiometry

Stoichiometry is defined as the material balances involving chemical reactions. *Stoichiometric air* is the air that is required to assure the complete combustion of chemical compounds such as a fuel, a waste, or a combination of a fuel and a waste. Chemical reaction equations are usually needed in order to calculate the material balances involving chemical reactions. The chemical reaction equation also provides a variety of qualitative and quantitative information essential for the determination of the weight of reactants reacted and products formed in a chemical process.

For a fuel which contains only carbon, hydrogen, sulfur, and oxygen, the stoichiometric coefficients can be obtained from the following generalized equation (EPA-80/02, p. 2-2; EPA-84/09, p. 34):

$$C_xH_yS_zO_w + (x + y/4 + z - w/2)O_2 \rightarrow xCO_2 + (y/2)H_2O + zSO_2 + Q \tag{1.28}$$

where Q represents the heat of combustion.

The amount of air needed for the above reaction can be expressed as

$$
\begin{aligned}
\text{air} &= (x + y/4 + z - w/2)O_2 + (0.79/0.21)(x + y/4 + z - w/2)N_2 \\
&= (x + y/4 + z - w/2)O_2 + 3.76(x + y/4 + z - w/2)N_2
\end{aligned} \tag{1.29}
$$

The above reaction assumes that

- Air consists of 21% by volume of oxygen, with the remaining 79% made up of nitrogen and other inert gases.
- Combined oxygen in fuel is available for combustion, thus reducing air requirements.

- Fuel contains no combined nitrogen, so no "fuel NO_x" is produced.
- Thermal NO_x via nitrogen fixation is small, so that it is neglected in stoichiometric air calculations.
- Sulfur in fuel is oxidized to SO_2 with negligible S_3 formation.

Equation (1.28) relates the reactants on a molar basis. One gram-mole of a substance is the mass of that substance equal to its molecular weight in grams. A gram-mole of any substance contains Avogadro's number of molecules of that substance, i.e., there are 6.02×10^{23} molecules/g-mole. Pound-moles (lb-mole) are also in common use. Since one pound-mole is equivalent to the molecular weight of the substance in pounds, it contains 454 times as many molecules as a gram-mole.

The generalized combustion equation Eq. (1.28) can be converted to a mass basis simply by multiplying the number of moles of each substance by its respective molecular weight.

Avogadro's law states that equal volumes of different gases at the same pressure and temperature contain equal numbers of molecules.

Thus it follows that the volumes of gaseous reactants in Eq. (1.28) are in the same ratios as their respective numbers of moles.

EXAMPLE: stoichiometry
One mole of methane, CH_4, is combusted in a combustion chamber. Determine:

(1) moles of CO_2 that are formed from the combustion process;
(2) the amount of air that is required.

Solution to question (1)

1.1. Apply the generalized equation.

Comparing CH_4 with the above generalized equation, it is found that

$$x = 1; \qquad y = 4; \qquad z = 0; \qquad w = 0$$

1.2. Balance the CH_4 reaction equation.

$$CH_4 + 2O_2 + 2(3.76)N_2 \rightarrow CO_2 + 2H_2O + 2(3.76)N_2$$

The balanced chemical equation must have the same number of atoms of each type in the reactants and products. Therefore the balanced equation shows that

- Number of Cs in the reactants = number of Cs in the products = 1
- Number of Os in the reactants = number of Os in the products = 4
- Number of Hs in the reactants = number of Hs in the products = 4
- Number of Ns in the reactants = number of Ns in the products = 15

Remember that the chemical equation tells us in terms of moles (not mass) the ratios among reactants and products.

1.3. Determine number of moles of CO_2 formed.

$$\text{Moles of } CO_2 = 1$$

Solution to question (2)

2.1. Determine the number of moles of oxygen required.

From the above chemical balance equation, 2 moles of oxygen are needed.

2.2. Determine the number of moles of nitrogen required.

From the above chemical balance equation, 7.52 moles of nitrogen are needed.

2.3. Determine the number of moles of air required.

Because every mole of oxygen is associated with 3.76 moles of nitrogen in the air, the air needed is equal to the sum of moles of oxygen and moles of nitrogen, and is equal to 9.52 moles.

2.57 Temperature

Temperature is an indicator of the thermal state of matter. There are two systems of temperature scales: the Celsius (°C) and Fahrenheit (°F) scales. Their relationship is as follows: $°F = 1.8°C + 32$. On many occasions, absolute temperatures are used for engineering calculations. Absolute temperature is the temperature measured on the thermodynamic scale, designated as degree Kelvin (K) or Rankine (°R). It is measured from absolute zero (−273.15°C or −459.76°F) and has the following relationship with °C and °F (EPA-83/06):

- Kelvin scale (K): $K = °C + 273.15$
- Rankine scale (°R): $°R = °F + 459.76$; and $°R = 1.8\,K$
- $(T_2/T_1)_{\text{Rankine}} = (T_2/T_1)_{\text{Kelvin}}$

The rate at which a combustible compound is oxidized is greatly affected by temperature. The higher the temperature, the faster the oxidation reaction will proceed. The chemical reactions involved in the combination of a fuel and oxygen can occur even at room temperature, but very slowly. For this reason, a pile of oily rags can be a fire hazard. Small amounts of heat are liberated by the slow oxidation of the oils. This in turn raises the temperature of the rags and increases the oxidation rate, liberating more heat. Eventually a full-fledged fire can break out (EPA-81/12, p. 3-2).

For combustion processes, ignition is accomplished by adding heat to speed up the oxidation process. Heat is needed to combust any mixture of air and fuel until the mixture ignition temperature is reached. By gradually heating a mixture of fuel and air, the rate of reaction and energy released will gradually increase until the reaction no longer depends on the outside heat source. More heat is being generated than is lost to the surroundings. The ignition temperature must be reached or exceeded to ensure complete combustion.

The ignition temperature of various fuels and compounds can be found in combustion handbooks such as the *North American Combustion Handbook* (1965). These temperatures are dependent on combustion conditions and therefore should be used only as a guide. Ignition depends on

- Concentration of combustibles in the waste stream
- Inlet temperature of the waste stream
- Rate of heat lost from combustion chamber
- Residence time and flow pattern of the waste stream
- Combustion chamber geometry and materials of construction

Most incinerators operate at a higher temperature than the ignition temepratur e, which is a minimum temperature. Thermal destruction of most organic compounds occurs between 590 and 650°C (1100 and 1200°F). However, most incinerators are operated at 700–820°C (1300–1500°F) to convert CO to CO_2, which occurs only at these higher temperatures.

Figure 1.11 graphically shows the relationships of various temperature scales (EPA-81/12, p. 2-2).

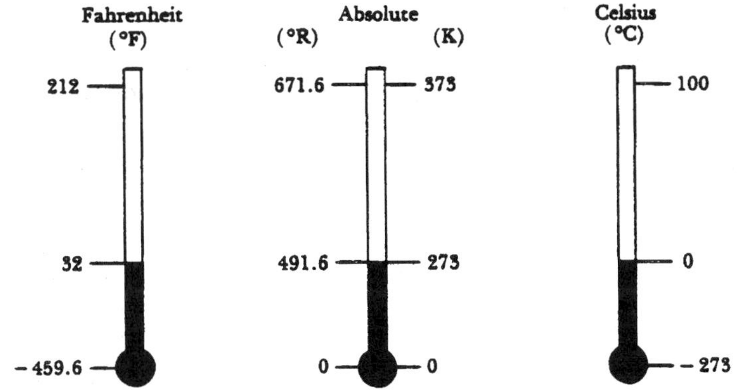

FIGURE 1.11 Relationship of various temperatures scales.

2.58 Thermochemical Relation

Combustion reaction, with its release of heat and light, is referred to as an *exothermic reaction*. Energy, which is released as the result of rearranging chemical bonds, can be utilized for power generation, space heating, drying, or for air pollution abatement, just to mention a few applications. Thermochemical calculations are concerned with the heat effects associated with combustion. These calculations permit determination of the energy released by burning a specific fuel. Only a part of this heat will be available for useful work, however (EPA-80/02, p. 2-7).

Each combustion installation has heat losses, some of which can be controlled to a certain extent, and others over which there is little or no control. The avoidable heat losses are those which can be minimized by good design and careful operation. The efficiency of a combustion installation reflects how well the designer succeeded in this respect. The percent efficiency is defined as 100 minus the sum of all losses, expressed as a percent of the energy input from the fuel.

In order to make efficiency as well as other thermochemical calculations, one needs to be able to determine the fuel heating values, heat contents of entering and leaving streams, and any other heat losses. Since rather specialized terminology is involved, a definition of terms is provided to avoid confusion and ambiguities later.

Heat of combustion. The heat energy evolved from the union of a combustible substance with oxygen to form CO_2, H_2O (and SO_2) as the end products, with both the reactants starting and the products ending at the same conditions, usually 25°C and 1 atm.

Gross or higher heating value (GHV or HHV). The quantity of heat evolved as determined by a calorimeter where the combustion products are cooled to 60°F and all water vapor condensed to liquid. Usually expressed in terms of Btu/lb or Btu/scf. Net or lower heating value (NHV or LHV) is similar to the higher heating value except that the water produced by the combustion is not condensed but retained as vapor at 60°F. Expressed in the same units as the gross heating value.

Enthalpy or heat content. The total heat content, expressed in Btu/lb, above a standard reference condition.

Sensible heat. The heat, the addition or removal or which results in a change of temperature.

Latent heat. The heat effect associated with a change of phase, e.g., from liquid to vapor (vaporization), or from liquid to solid (fusion), etc., without a change in temperature. Expressed usually as Btu/lb. Available heat means the quantity of heat available for intended (useful) purposes (the difference between the gross heat input to a combustion chamber and all the losses).

Accordining to a heat balance, energy outflow from a system and accumulation within the system equals the energy input to the system. For steady-state operations, the accumulation term is zero. Therefore,

$$\text{Heat in (sensible} + \text{HHV)} = \text{Heat out (sensible} + \text{latent} + \text{available)}$$

EXAMPLE: thermochemical relation
Given a gas mixture, determine its gross (or higher) heating value (GHV).

Given conditions

- Composition (molar fraction) of the gas mixture is shown in Table 1.34.
- Heat of combustion of each compound is provided in Table 1.35.

Solution:

1. Determine the gross heating value of each component GHV from Table 1.36.
2. Calculate the gross heating value of the gas mixture GHV in Btu/ft^3.

$$\begin{aligned} \text{GHV} &= (0.0515)(0) + (0.811)(1013) + (0.0967)(1792) + (0.0351)(2590) + (0.0056)(3370) \\ &= 1104.71\ \text{Btu/ft}^3 \end{aligned}$$

References used in this section: EPA-80/02, p. 2-7; EPA-84/09, p. 36.

TABLE 1.34 Composition of Gas Mixture

Component	Molar fraction
Nitrogen, N_2	0.0515
Methane, CH_4	0.8111
Ethane, C_2H_6	0.0967
Propane, C_3H_8	0.0351
n-Butane, C_4H_{10}	0.0056
Total	1

TABLE 1.35 Heat of Combustion at 25°C (77°F)

			Btu/ft^3		Btu/lb	
Substance	lb/ft^3	ft^3/lb	Gross (high)	Net (low)	Gross (high)	Net (low)
Nitrogen, N_2	0.0744	13.443				
Methane, CH_4	0.0424	23.565	1013	913	23,879	21,520
Ethane, C_2H_6	0.0803	12.455	1792	1641	22,320	20,432
Propane, C_3H_8	0.1196	8.365	2590	2385	21,661	19,944
n-Butane, C_4H_{10}	0.1582	6.231	3370	3113	21,308	19.68

Source: (EPA-84/09, p. 37).

TABLE 1.36 Composition of Gas Mixture

Component	Molar fraction	GHV of each compound (Btu/scf)	Mixture GHV
Nitrogen, N_2	0.0515	0	0.00
Methane, CH_4	0.8111	1013	821.64
Ethane, C_2H_6	0.0967	1792	173.29
Propane, C_3H_8	0.0351	2590	90.91
n-Butane, C_4H_{10}	0.0056	3370	18.87
Total	1.00		1104.71

2.59 Thermodynamic First Law

The thermodynamic first law is the law of energy conservation. This means that energy can be neither created nor destroyed, although it can be stored in various forms and can be transferred from one system to another as heat or work. The law also means that the energy of an isolated system remains constant.

For a steady-state, steady-flow situation between the entrance and exit of an incineration system, the first law of thermodynamics can be expressed as

$$\text{Heat } (Q) - \text{work } (W) = \text{enthalpy change} + \text{kinetic energy change} + \text{potential energy change}$$

In general, because the velocity of a combustion gas and the difference in height between the feed point and the stack gas exit point of an incinerator are relatively small, the kinetic energy change and potential energy change are negligible. Therefore, the first law of thermodynamics can be simplified to

$$Q - W = \text{enthalpy change} = m(h_2 - h_1)$$

where h_1 = enthalpy value at state 1
h_2 = enthalpy value at state 2
m = mass or mass flow rate

Because most incinerators do not involve "work" per se, the above equation can be further simplified to

$$Q = m(h_2 - h_1)$$
$$= mC_p(T_2 - T_1) \text{ for an ideal gas}$$

where m = the mass or mass flow rate
C_p = the specific heat of the gas (in Btu/lb-°R)
T_1 = temperature at state 1
T_2 = temperature at state 2

Applying the first law of thermodynamics to steady state conditions, heat can be calculated according to, $Q = mC_p(T_2 - T_1)$, where m is mass flow, C_p is specific heat, and T_2 and T_1 are temperatures at different states.

References used in this section: Holman-74, p. 35; Jones-60, p. 54; Wark-66, p. 98.

EXAMPLE 1: thermodynamic first law
Hot air at 1200 lb/min flows from a temperature of 1200°F to 200°F. Determine the heat transfer.

Given conditions

- Mass flow rate, $Q = 1200$ lb/min
- Air specific heat, $C_p = 0.26$ Btu/lb-°F
- Initial temperatures, $T_1 = 1200$°F
- Final temperature, $T_2 = 200$°F

Solution:

$$\begin{aligned} Q &= mC_p(T_2 - T_1) \\ &= 1200(0.26)(200 - 1200) \\ &= -3.12 \times 10^5 \text{ Btu/min} \end{aligned}$$

The negative sign shows that heat is liberated from the system.

EXAMPLE 2: thermodynamic first law
A mixture of air and water vapor with an enthalpy of 60 Btu/lb enters the demister section of an air pollution control system at a rate of 1500 lb/h. Liquid water drains out of the demister with an enthalpy of 25 Btu/lb at a rate of 500 lb/h. An air–vapor mixture leaves with an enthalpy of 20.1 Btu/lb. Determine the rate of heat removal from the fluids passing through the dehumidifier.

Given conditions

- Enthalpy of air and water entering demister = 60 Btu/lb
- Rate of air and water entering demister = 1500 lb/h
- Enthalpy of water leaving demister = 25 Btu/lb
- Rate of water leaving demister = 500 lb/h
- Enthalpy of air and steam leaving demister = 20.1 Btu/lb

Solution:

1. Write the thermodynamic law for a steady flow.

$$Q = H_{21} + H_{22} - H_1$$

where Q = heat transfer
H_1 = enthalpy of air and water at entrance section of demister
H_{21} = enthalpy of water at exit section
H_{22} = enthalpy of air and steam at exit section

2. Rewrite the equation above.

$$\begin{aligned} Q &= (m_{21})(h_{21}) + (m_{22})(h_{22}) - (m_1)(h_1) \\ &= (m_{21})(h_{21}) + (m_1 - m_{21})(h_{22}) - (m_1)(h_1) \end{aligned}$$

where m_1 = flow rate of air and water at entrance section
m_{21} = flow rate of water at exit section
m_{22} = flow rate of air and steam at exit section

h_1 = enthalpy per pound of air and water at entrance section
h_{21} = enthalpy per pound of water at exit section
h_{22} = enthalpy per pound of air and steam at exit section

3. Calculate the heat transfer.

$$Q = (500)(25) + (1500 - 500)(20.1) - (1500)(60)$$
$$= -57{,}400 \text{ Btu/lb}$$

Negative sign means that heat is removed from the system.

2.60 Thermodynamic State

A thermodynamic state is set by the conditions specified by the values of its parmeters. Terms that are related to the state include

- Equilibrium state: A system is in an equilibrium state or is in equilibrium if no change occurs in the system without the aid of external stimulus.
- Process or thermodynamic process: The transformation of a system from one equilibrium state to another over a period of time can be called a thermodynamic process.
- Cycle or cyclic process: A process or a series of processes which returns the system to the state of the original conditions. The conditions of a thermodynamic system change during the execution of a cyclic process. At the completion of the cyclic process, all conditions return to their initial values.
- Path function: A thermodynamic function whose results depend on the path taken between the two states. Heat and work are path functions. They can be calculated only when the path of the process is known (Jones-60, p. 50; Wark-66, p. 5).
- Point function: A thermodynamic function whose results depend only on the values of two end states. For example, temperature is a point function. Therefore, the temperature change from state 1 (T_1) to state 2 (T_2) can be expressed as $T_2 - T_1$ regardless what processes the temperature change follows between the two states (Wark-66, p. 6).

Additional references used in this section: Holman-74, p. 5; Jones-60, p. 10; Wark-66, p. 4.

2.61 Thermodynamic System

A thermodynamic system is a three-dimensional space bounded by arbitrary geometric surfaces. The bounding surfaces can be real or imaginary and can be at rest or in motion. The terms that are related to the thermodynamic system include

- Closed system: A thermodynamic system in which no mass crosses its boundaries. In a closed system, energy is allowed to cross its boundaries in various forms as the system changes from one state to another.
- Control mass: The quantity of mass within a thermodynamic system.
- Isolated system: A thermodynamic system in which neither mass nor energy interacts with its surroundings (or crosses its boundaries). An isolated system must be a closed system.
- Open system: A thermodynamic system that allows the transfer of mass across its boundaries. An open system is also called a control volume, and its boundary is called a control surface.
- Surroundings: The state or matter outside of a system.

References used in this section: Holman-74, p. 3; Jones-60, p. 7; Wark-66, p. 3.

2.62 Turbulence

Proper mixing is important in combustion processes for two reasons. First, for complete combustion to occur, every particle of fuel must come in contact with air (oxygen). If not, unreacted fuel will be exhausted from the stack. Second, not all of the fuel or waste gas stream is able to be in direct contact with the burner flame. In most incinerators a portion of the waste stream bypasses the flame and is mixed at some point downstream of the burner with the hot products of combustion. If the two streams are not completely mixed, a portion of the waste stream will not react at the required temperature and incomplete combustion will occur.

A number of methods are available to improve mixing the air and combustion streams. Some of these include the use of refractory baffles, swirl-fired burners, and baffle plates. The problem of obtaining complete mixing is not easily solved. Unless properly designed, many of these mixing devices may create dead spots and reduce operating temperatures. Merely inserting obstructions to increase turbulence is not the answer. According to one study of afterburner systems, the process of mixing the flame and fume stream to obtain a uniform temperature for decomposition of pollutants is the most difficult part in the design of the afterburner.

References used in this section: EPA-81/12, p. 3-4.

2.63 Viscosity

Newton's viscosity law. In fluid mechanics, the applied shear stress (S_s) is proportional to the rate of deformation or to the velocity gradient normal to the velocity. Mathematically, it can be expressed as

$$S_s = \mu(dv/dy)$$

where S_s = shear stress
dv = velocity in differential amount
dy = distance normal to the velocity direction in differential amount
μ = absolute or dynamic viscosity

Figure 1.12 graphically shows the concept of viscosity (EPA-81/12, p. 2-5).

Viscosity is a measure of a fluid's resistance to flow. Viscosities vary greatly, from materials like heavy lubricating oils to water. Viscosity is also a strong function of temperature, increasing with increasing temperature for gases and decreasing with increasing temperature for liquids.

Viscosity is the result of two phenomena:

1. Intermolecular cohesive forces;
2. Momentum transfer between flowing strata caused by molecular agitation perpendicular to the direction of motion. Between adjacent strata of a flowing fluid, a shearing stress results that is directly proportional to the velocity gradient. Viscosity is often defined as resistance to flow.

Viscosity has the following unit systems:

1. For dynamic viscosity (D_v)
   - In the cgs system, the unit of dynamic viscosity is a poise, with the dimensions, g/(s-cm) or dyne-s/cm^2

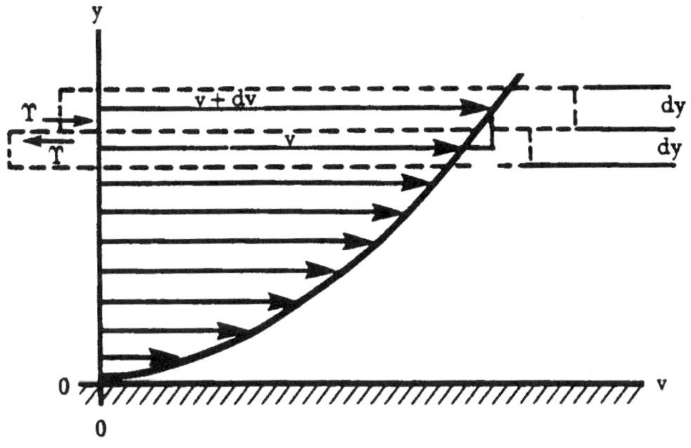

FIGURE 1.12 Shearing stress in a moving fluid.

- 1 centipoise = 1 poise/10^2 = 0.01 poise = 0.01 g/(s-cm)
- In the British system, the unit of dynamic viscosity is lb-s/ft^2
- Dynamic viscosity of water = 1 centipoise at 68.4°F and atmospheric pressure

2. For kinematic viscosity (K_v)
   - Kinematic viscosity = (dynamic viscosity)/density
   - In the cgs system, the unit of kinematic viscosity is a stoke (cm^2/s)
   - 1 centistoke = 1 stoke/10^2
   - In the British system, the unit of kinematic viscosity is ft^2/s
   - 1 centistoke = $1.076 \times 10^{-5} ft^2/s$
3. For absolute viscosity
   - In the cgs system, the unit of absolute viscosity is a Pascal second (Pa-s)
   - In the British system, the unit of absolute viscosity is lbm/ft-s

Viscosity measurement systems

1. A Saybolt Universal viscometer is commonly used for petroleum products and lubricating oils.
2. A Saybolt Furol viscometer is used for heavy oils.

See Fig. 1.13 for the viscosity of air at different temperatures and Fig. 1.14 for the viscosity of various gases.

References used in this section: EPA-81/10, p. 2-9; EPA-81/12; EPA-84/09; EPA-86/03; Marks-67, p. 3-49.

EXAMPLE: viscosity
The density and viscosity of a flow are provided in the following given conditions. Calculate the kinematic viscosity in cm^2/s.

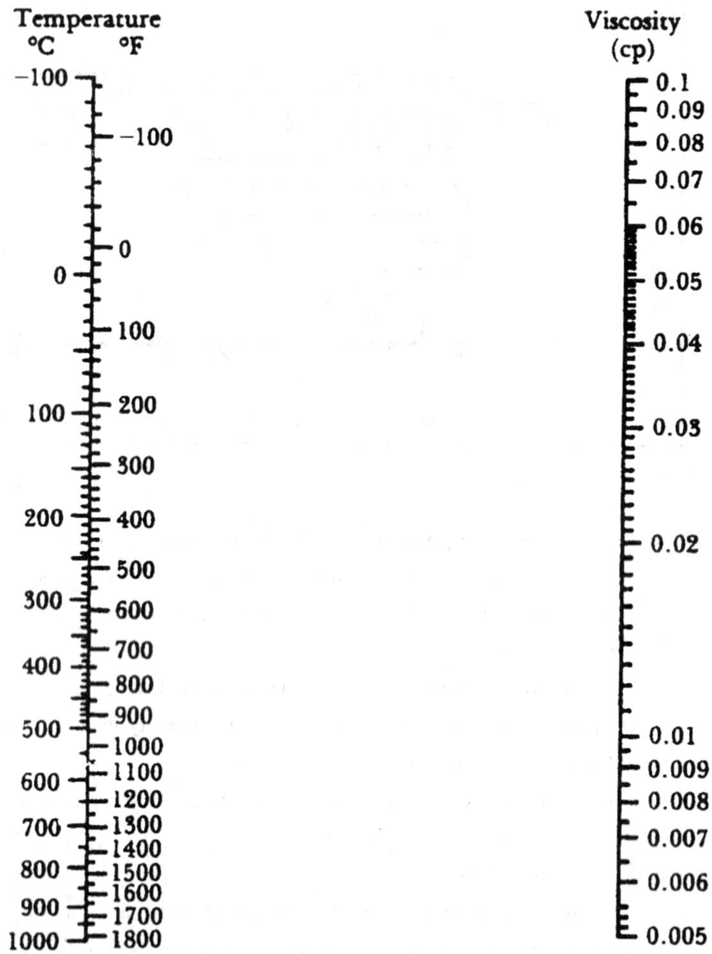

FIGURE 1.13 Viscosity of air at 1.0 atmosphere (*EPA-84/09, p. 167*).

Given conditions

- Density of the flow, $\rho = 0.768\,\text{g/cm}^3$
- Dynamic viscosity of the flow $D_v = 0.022$ cP (centipoise)

Solution:

1. Convert centipoise (cP) to units of g/cm-s.

By definition,

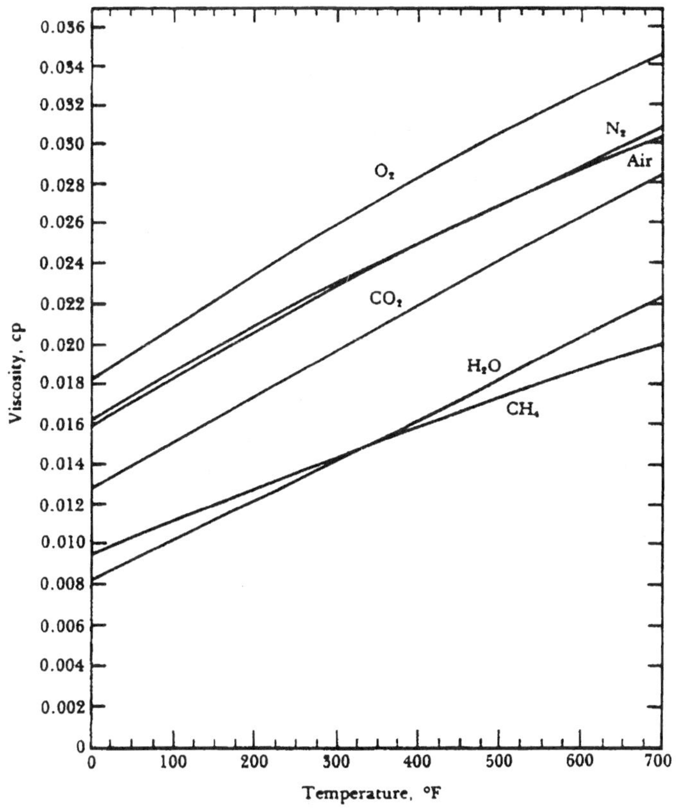

FIGURE 1.14 Viscosity of gases at 1.0 atmosphere (*EPA-84/09, p. 168*).

$$
\begin{aligned}
1 \text{ poise} = 100 \text{ cP} &= 1 \text{ g/s-cm} \\
1 \text{ cP} &= 0.01 \text{ g/s-cm} \\
0.022 \text{ cP} &= 0.00022 \text{ g/s-cm}
\end{aligned}
$$

2. Calculate the kinematic viscosity.

By definition,

$$
\begin{aligned}
\text{kinematic viscosity} &= (\text{dynamic viscosity})/\text{density} \\
&= 0.00022/0.768 \\
&= 2.86 \times 10^{-4} \text{ cm}^2/\text{s}
\end{aligned}
$$

2.64 VOST (Volatile Organic Sampling Train)

VOST is a sampling method, developed by EPA, to capture volatile organic molecules from combustion stack gases. Sorbent traps capture the organics, which then are taken to ovens to desorb the material into an analyzer. Generally, VOST uses two traps to capture volatile organics. One trap contains tenax and the other tenax plus charcoal (Lee-92/06).

EXAMPLE 1: VOST
With the following given conditions, this example is to determine whether VOST sample size is sufficient to measure 99.99% DRE for the POHC, CCl_4 (carbon tetrachloride) (EPA-89/06, p. 6).

Given conditions

- Waste feed rate $Q = 15.2\,\text{kg/min}$ (2000 lb/h)
- POHC = CCl_4 (carbon tetrachloride)
- POHC feed concentration $Q_p = 500\,\text{ppm}$ ($500\,\text{ppm} = 500\,\text{kg}/10^6\,\text{kg} = 0.5\,\text{g/kg}$)
- Stack gas flow rate $V = 4500\,\text{scfm}$ ($127.4\,\text{m}^3/\text{min}$)
- $1\,\text{m}^3 = 35.31\,\text{ft}^3$
- Lower detection limit = 2 ng per trap (ng = 10^{-9} g)

Solution:
VOST: three trap pairs (measured at different times), 500 mL/min (0.5 L/min) flow rate per pair; 20 L sample/pair.

1. POHC input rate W_i.

$$\begin{aligned} W_i &= (\text{waste feed rate}) \times (\text{POHC concentration}) \\ &= 15.2\,(\text{kg/min}) \times 0.5\,(\text{g/kg}) \\ &= 7.6\,\text{g/min} \end{aligned}$$

2. POHC stack output rate at 99.99% DRE, W_0.

$$\begin{aligned} W_0 &= W_i(1 - \text{DRE}/100) \\ W_0 &= 7.6(1 - 0.9999) \\ &= 0.00076\,\text{g/min} \end{aligned}$$

3. POHC concentration (P_c) in stack gas at 99.99% DRE.

$$\begin{aligned} P_c &= W_0/(\text{stack gas flow rate}) \\ &= 0.00076/127.4 \\ &= 0.0000060\,\text{g/m}^3 \text{ (stack gas flow rate is a given condition)} \\ &= 0.0000060\,\text{g/m}^3 \times (10^9\,\text{ng/g}) \times (10^{-3}\,\text{m}^3/\text{L}) \\ &= 6.0\,\text{ng/L} \end{aligned}$$

$$1\,\text{g} = (10^9) \times (10^{-9})\text{g} = (10^9\,\text{ng})$$

$$1\,\text{l} = 1000\,\text{cc} = 10^3\,\text{cm}^3\,[\text{m}^3/(10^6\,\text{cm}^3)] = 10^{-3}\,\text{m}^3$$

4. Sample amount collected in one pair of traps = $20\,(\text{L}) \times 6.0\,(\text{ng/L}) = 120\,\text{ng}$.
5. Conclusion

Since the VOST lower detection limit for CCl_4 is 2 ng, the sample is sufficient to detect CCl_4 to determine a DRE of 99.99% or lower. A margin of safety above the detection limit is desirable. This calculation assumes both traps in a pair are combined for analysis. If they are analyzed separately, the distribution of mass on each trap must be considered.

EXAMPLE 2: VOST
This example is to calculate the sample volume required to show 99.99% DRE with the following given conditions (EPA-81/09, p. 5-123).

Given conditions

- POHC designated by the permit writer = hexachlorobenzene
- Concentration of POHC in waste feed = 1.0%
- Waste feed rate = 1000 lb/h
- Stack gas flow = 85,382 scf/min

Solution:

1. Computation of maximum W_0 to satisfy 99.99% DRE.

$$\begin{aligned} W_\mathrm{i} &= \text{(concentration of POHC in waste)(waste feed rate)} \\ &= (0.01)(1000\,\mathrm{lb/h}) \\ &= 10\,\mathrm{lb/h} \\ W_0 &= W_\mathrm{i}(1 - \mathrm{DRE}) \\ &= (10)(1 - 0.9999) \\ &= 0.001\,\mathrm{lb/h} \end{aligned}$$

2. Compute the minimum weight of POHC sample that can be collected.

Given conditions

- Detection limit of hexachlorobenzene in the analytical sample extract as injected in the GC/MS: 1 ng/μL or 1 μ/mL
- Average extraction efficiency 60%

Because the extracted sample is concentrated via evaporation before injection into the GC/MS, then the minimum weight of collected hexachlorobenzene is independent of extract liquid volume. The minimum detectable total weight of POHC collected as obtained from laboratory analysis is

$$\begin{aligned} W_\mathrm{sample} &= \text{(detection limit)/(extraction efficiency)} \\ &= (1\,\mu/\mathrm{mL})/(0.60) \\ &= 1.667\mu \end{aligned}$$

3. Compute the POHC stack gas loading.

It was given that the stack gas volume flow rate at standard conditions $Q = 85{,}382\,\mathrm{scf/min}$.

$$\begin{aligned} \text{POHC concentration } (P_c) &= \text{(total weight of POHC in sample)/} \\ &\quad \text{(volume of sample at standard conditions)} \\ &= (W_0/Q)(1\,\text{h}/60\,\text{min}) \\ &= [(0.001\,\text{lb/h})/(85{,}382\,\text{scf/min})][1\,\text{h}/60\text{min}] \\ &= 1.95 \times 10^{-10}\,\text{lb/scf} \\ &= 8.85 \times 10^{-8}\,\text{g/scf} \end{aligned}$$

Note: This computation assumes 100% collection of the POHC on the filter, resin, and impingers.

4. Compute the minimum stack gas sample volume (V_{std}).

$$\begin{aligned} V_{std} &= W_{sample}/P_c \\ &= (1.667 \times 10^{-6}\,\text{g})/(8.85 \times 10^{-8}\,\text{g/scf}) \\ &= 18.84\,\text{scf} \end{aligned}$$

2.65 Waste Characterization

EXAMPLE: waste characterization
This example is to show the conversion from a pure compound basis (theoretical chemistry) to an ash and moisture basis.

Given conditions

- Cellulose ($C_6H_{10}O_5$) contains 82% organics, 15% moisture, 3% ash
- Cellulose H_c (HHV) = 7500 Btu/lb
- Polyethylene (PE) (C_2H_4) contains 96.5% organics, 2% moisture, 1.5% ash
- Polyethylene H_c (HHV) = 20,000 Btu/lb

Solution:

1. Analyze the chemical composition.

1.1. Calculate the ultimate composition of $C_6H_{10}O_5$ (Table 1.37).

TABLE 1.37 Composition of $C_6H_{10}O_5$

x_i	C	H	Cl	O	M
$C_6H_{10}O_5$	72	10	0	80	162.00
x_i/M	0.4444	0.0617	0.0000	0.4938	1.00
$0.82\ x_i/M$	0.3644	0.0506	0.0000	0.4049	

1.2. Calculate the ultimate composition of C_2H_4 (Table 1.38).

The heats of combustion indicated are for 100% pure materials. That means ash-free and moisture-free in the case of cellulose and PE. To illustrate the conversion technique, it is assumed that the actual as-received conditions of the waste streams are such that the following moisture and ash contents are applicable.

TABLE 1.38 Composition of C_2H_4

x_i	C	H	Cl	O	M
C_2H_4	24	4	0	0	28.00
x_i/M	0.8571	0.1429	0.0000	0.0000	
$0.965x_i/M$	0.8271	0.1379	0.0000	0.0000	

2. Determine the composition of 100% pure materials.

2.1. Cellulose: 82% organics; 15% moisture; 3% ash (Table 1.39).

TABLE 1.39 Cellulose

Cellulose (%)	(1)	(2)
C=C/M	0.4444	0.3644
H	0.0617	0.0506
Cl	0	0
O	0.4938	0.4049
N	0	0
Moisture		0.15
Ash		0.03
Total	1	1
H_c(HHV)	7500	6150

Notes

- Column (1) contains 100% organic material
- Column (2) contains organics, moisture, and ash
- For the conversion between dry and wet analyses, see Dry versus Wet Analysis in this book

2.2. Polyethylene: 96.5% organics; 2% moisture; 1.5% ash (Table 1.40).

TABLE 1.40 Polyethylene

PE content (%)	(1)	(2)
C=C/M	0.8571	0.8271
H	0.1429	0.1379
Cl	0	0
O	0	0
N	0	0
Moisture		0.02
Ash		0.015
Total	1	1
H_c(HHV)	20,000	19,300

Notes

- Column (1) contains 100% organic
- Column (2) contains organics, moisture, and ash

3. Determine the composition of a composite of both compounds.

It is assumed that the sample waste consists of 40% and 60% of cellulose and polyethylene components, respectively. Now that a consistent basis of ash, moisture, and theoretical component chemistries is available, calculate the chemistry and heat of combustion as described below.

3.1. Calculate 40% of cellulose (Table 1.41).

3.2. Calculate 60% of polyethylene (Table 1.42).

3.3. Combine 40% of cellulose and 60% of polyethylene (Table 1.43).

TABLE 1.41 Cellulose

Cellulose (%)	(1)	(2)	(3)
C=C/M	0.4444	0.3644	0.1458
H	0.0167	0.0506	0.0202
Cl	0	0.0000	0.0000
O	0.4938	0.4049	0.1620
N	0	0.0000	0.0000
Moisture		0.1500	0.0600
Ash		0.0300	0.0120
Total	1	1.0000	
H_c(HHV)	7500	6150	2460

Notes

- Column (1): cellulose contains 100% organic material
- Column (2): cellulose contains organics, moisture, and ash
- Column (3): 40% of column (2)
- Column (4): polyethylene contains 100% organic material
- Column (5): polyethylene contains organics, moisture, and ash
- Column (6): 60% of column (5)
- Column (7) = column (3)
- Column (8) = column (6)
- Column (9) = column (3) + column (6)

TABLE 1.42 Polyethylene

PE content (%)	(4)	(5)	(6)
C=C/M	0.8571	0.8271	0.4963
H	0.1429	0.1379	0.0827
Cl	0	0.0000	0.0000
O	0	0.0000	0.0000
N	0	0.0000	0.0000
Moisture		0.0200	0.0120
Ash		0.0150	0.0090
Total	1	1.0000	
H_c(HHV)	20,000	19,300	11,580

TABLE 1.43 Cellulose and Polyethylene

Cellulose (%)	(7)	(8)	(9)
C=C/M	0.1458	0.4963	0.6420
H	0.0202	0.0827	0.1030
Cl	0.0000	0.0000	0.0000
O	0.1620	0.0000	0.1620
N	0.0000	0.0000	0.0000
Moisture	0.0600	0.0120	0.0720
Ash	0.0120	0.0090	0.0210
Total			1.0000
H_c(HHV)	2460	11,580	14,040

2.66 Work

Work can be defined as either

1. Work = (force) × (distance), or
2. Work = (pressure) × (volume change)

This definition shows that work is an interaction between a system and its surroundings which is caused by a force displacing the boundary between the system and the surroundings. Both heat and work are path functions and thus, to evaluate their magnitude, the entire process must be considered. Typical work units are ft-lb_f.

where ft = feet
lb_f = pound force

The conventional signs of work are

- Work done on the system is negative
- Work done by the system is positive

Heat and work conversion factors

- 1 Btu = 778 ft-lb_f
- 1 hp = 42.4 Btu/min
- 1 hp = 33,000 ft-lb_f/min
- 1 hp = 0.746 kW
- 1 calorie = 4.184 joules
- 1 Btu/[(lb-mole)(°F)] = 1 cal/[(g-mole)(°C)]
- 1 Btu/[(lb)(°F)] = 1 cal/[(g)(°C)]

REFERENCES

40CFRxxx-yy: 40 Code of Federal Regulations (CFR), Parts xxx, published in yy year. For example, 40CFR260.10-91 means 40 Code of Federal Reguation (CFR), Parts 260.10, published in 1991.

(Brunner-84), "Incineration Systems – Selection and Design," Calvin R. Brunner, Van Nostrand Reinhold, 1984.

(EPA-80/02), "Combustion Evaluation—Student Manual, Course 427," EPA Air Pollution Training Institute (APTI), EPA450-2-80-063, February 1980.

(EPA-80/02a), "Combustion Evaluation—Instructor Manual, Course 427," EPA Air Pollution Training Institute (APTI), EPA450-2-80-065, February 1980.

(EPA-81/09), "Engineering Handbook for Hazardous Waste Incineration," September 1981, SW-889.

(EPA-81/10), "Control of Particulate Emissions, Course 413," EPA Air Pollution Training Institute (APTI), EPA450-2-80-063, February 1980.

(EPA-81/12), "Control of Gaseous Emissions, Course 415." EPA Air Pollution Training Institute (APTI), EPA450-2-81-005, December 1981.

(EPA-82/11f), "Development Document for Effluent Limitations Guidelines and Standards and Pretreatment Standards for the Steam Electric, Point Source Category," EPA440-1-82-029, November 1982.

(EPA-83/06), "Atmospheric Sampling," Course 435, EPA450-2-80-004, June 1983.

(EPA-84/09), "Control of Gaseous and Particulate Emissions, Course SI:412D," EPA Air Pollution Training Institute (APTI), EPA450-2-84-007, September 1984.

(EPA-85/11), "Practical Guide—Trial Burns for Hazardous Waste Incinerators," EPA600-2-86-050, November 1985, NTIS: PB86-190246.

(EPA-86/03), "Course 502 Hazardous Waste Incineration," Prepared by Louis Theodore for the EPA Air Pollution Training Institute, March 7, 1986.

(EPA-86/12), "Prevention References Manual—Chemical Specific, Vol. f: Carbon Tetrachloride," EPA Contract No. 68-02-3889, Work Assignment 98, December 1986.

(EPA-89/03), "Hospital Incinerator Operator Training Course: Vol. II, Presentation Slides," EPA450-3-89-003, March 1989.

(EPA-89/06), "Hazardous Waste Incinceration Measurement Guidance Manual, Vol. III of the Hazardous Waste Incineration Guidance Series," EPA625-6-89/021, June 1989.

(EPA-90/06), "State of the Art Assessment of Medical Waste Thermal Treatment," Final Draft Prepared by EER Corporation for the EPA Risk Reduction Engineering Laboratory.

(Hasselriis-98/12), The information was provided by Mr. Floyd Hasselriis, P. E. December 1998.

(Hesketh-79), "Air Pollution Control," Howard E. Hesketh, Ann Arbor Science, 1979.

(Holman-74), "Thermodynamics," J. P. Holman, McGraw-Hill, 1974.

(Jones-60), "Engineering Thermodynamics," J. B. Jones and G. A. Hawkins, Wiley, 1960.

(Lee-83/07), "Proceedings of the First Annual Hazardous Materials Conference", Philadelphia, July 12–14, 1983, pp. 278–303.

(Lee-84/11), "An Overview of Who is Doing What in Laboratory- and Bench-Scale Hazardous Waste Incineration Research," C. C. Lee and G. L. Huffman. Proceedings from the Fifth National Conference on Management of Uncontrolled Hazardous Waste Sites, held in Washington, DC, November 7–9, 1984.

(Lee-89/08), "Incineration of Solid Waste," C. C. Lee and George L. Huffman, Environmental Progress, 8(3), August 1989.

(Lee-89/08a), "Innovative Thermal Destruction Technologies," C. C. Lee and George L. Huffman, Environmental Progress, 8(3), August 1989.

(Lee-90/05), "Thermodynamic Fundamentals Used in Hazardous Waste Incineration," by C. C. Lee and G. L. Huffman. Presented at the 1990 Incineration Conference, San Diego, California, May 14–18, 1990, EPA Report, EPA600/D-90/087, and for publication in Hazardous Materials Control, 1994.

(Lee-90/11), "Medical Waste Incineration Handbook," C. C. Lee, Published by the Government Institutes, November 1990.

(Lee-92/06). "Environmental Engineering Dictionary," C.C. Lee, Government Institutes, June 1992.

(Lee-93/08), "Innovative Thermal Destruction Technologies," C. C. Lee and G. L. Huffman, 1993 Environmental and Development Workshop, Beijing, August 1993.

(Marks-67), "Standard Handbook for Mechanical Engineers," Lionel S. Marks, Editor, 1916–1951, McGraw-Hill, 1967.

(OME-86/10), "Incinerator Design and Operating Criteria, Vol. II, Biomedical Waste Incinerators," Ontario Ministry of the Environment (OME), October 1986.

(OME-88/12), "Guidance for Incinerator Design and Operation—General, Vol. I," Ontario Ministry of the Environment (OME), December 1988.

(Oppelt-87/05), "Incineration of Hazardous Waste: A Critical Review," E. Timothy Oppelt, Journal of Air Pollution Control Association, 37(5), 558–586, May 1987.

(Schaum-66), "Theory and Problems of College Chemistry," 5th edn, Schaum's Outline Series, McGraw-Hill, 1966.

(UDRI-90), "Development of a Thermal Stability Based Ranking of Hazardous Organic Compound Incinerability," P. H. Taylor and B. Dellinger, University of Dayton Research Institute (UDRI), and C. C. Lee, Environmental Protection Agency, Environmental Science and Technology, 24(3), 1990.

(Wark-66), "Thermodynamics," K. Wark, McGraw-Hill, 1966.

(Wylen-72), "Fundamentals of Classical Thermodynamics," Gordon J. Van Wylen and Richard E. Sonntag, John Wiley, 1972.

Attachment 1: Thermal oxidation parameters

Compound	A (1/s)	E (cal/mole)
Acrolein	3.30×10^{10}	35,900
Acrylonitrile	2.13×10^{12}	52,100
Allyl alcohol	1.75×10^{6}	21,400
Allyl chloride	3.89×10^{7}	29,100
Benzene	7.43×10^{21}	95,900
Butene-1	3.74×10^{14}	58,200
Chlorobenzene	1.34×10^{17}	76,600
1-2 Dichloroethane	4.82×10^{11}	45,600
Ethane	5.65×10^{14}	63,600
Ethanol	5.37×10^{11}	48,100
Ethyl acrylate	2.19×10^{12}	46,000
Ethylene	1.37×10^{12}	50,800
Ethyl formate	4.39×10^{11}	47,700
Ethyl mercaptan	5.20×10^{5}	14,700
Methane	1.68×10^{11}	52,100
Methyl chloride	7.34×10^{8}	40,900
Methyl ethyl ketone	1.45×10^{14}	58,400
Propane	5.25×10^{19}	85,200
Propylene	4.63×10^{8}	24,300
Toluene	2.28×10^{13}	56,500
Triethylamine	8.10×10^{11}	43,200
Vinyl acetate	2.54×10^{9}	35,900
Vinyl chloride	3.57×10^{14}	63,300

Attachment 2: Ranking of incinerability of organic hazardous constituents from Appendix VIII, Part 261, in heat of combustion order and in alphabetical order (Lee-84/11)

Incinerability ranking by heat of combustion (H_c) (kcal/g) and in alphabetical order

H_c	H_c order	H_c	Alphabetic order
0.11	Trichloromonofluoromethane	7.37	Acetonitrile
0.13	Tribromomethane	7	Acetonylbenzyl(3-(alpha-))-4-hydroxycoumarin and salts
0.22	Dichlorodifluoromethane	8.26	Acetophenone
0.24	Tetrachloromethane	2.77	Acetyl chloride
0.41	Tetranitromethane	4.55	Acetyl-2-thioreau(1-)
0.46	Hexachloroethane	7.82	Acetylaminofluorine(2-)
0.5	Dibromomethane	6.96	Acrolein
0.53	Pentachloroethane	5.75	Acrylamide
0.7	Hexachloropropene	7.83	Acrylonitrile
0.75	Chloroform	5.73	Aflatoxins
0.8	Chloral	3.75	Aldrin
0.81	Cyanogen bromide	7.75	Allyl alcohol
0.84	Trichloromethanethiol	9	Amino biphenyl(4-)
1.12	Hexachlorocyclohexane	4.78	Aminomethyl-3-isoxazolol(5-)
1.19	Tetrachloroethane	7.37	Aminopyridine(4-)
1.29	Cyanogen chloride	4.01	Amitrole
1.32	Formic acid	8.73	Aniline
1.34	Iodomethane	7.69	Auramine
1.39	Tetrachloroethane, NOS	3.21	Azaserine
1.39	Tetrachloroethane(1,1,1,2-)	10.03	Benzene
1.39	Tetrachloroethane(1,1,2,2-)	3.4	Benzenearsonic acid
1.43	Dibromomethane(1,1-)	8.43	Benzenethiol
1.48	Dibromo-3-chloropropane(1,2-)	9.18	Benzidine
1.62	Pentachloronitrobenzene	6.07	Benzoquinone
1.7	Bromomethane	3.9	Benzotrichloride
1.7	Dichloromethane	9.25	Benzo[a]pyrene
1.74	Trichloroethane	9.25	Benzo[b]fluoranthene
1.79	Hexachlorobenzene	9.25	Benzo[j]fluoranthene
1.97	Bis(chloromethyl) ether	6.18	Benzyl chloride
1.99	Trichloroethane(1,1,1-)	9.39	Benz[a]anthracene
1.99	Trichloroethane(1,1,2-)	8.92	Benz[c]acridine
2.05	Pentachlorobenzene	4.6	Bis(2-chloroethoxy) methane
2.09	Pentachlorophenol	3.38	Bis(2-chloroethyl) ether
2.1	Hexachlorocyclopentadiene	6.64	Bis(2-chloroethyl)-2-naphthylamine(n,n-)
2.12	Hexachlorobutadiene	4.93	Bis(2-chloroisopropyl)ether
2.15	Kepone	8.42	Bis(2-ethylhexyl) phthalate
2.23	Tetrachlorophenol(2,3,4,6-)	1.97	Bis(chloromethyl) ether
2.31	Decachlorobiphenyl	2.66	Bromoacetone
2.31	Dichlorophenylarsine	1.7	Bromomethane
2.33	Endosulfan	5.84	Bromophenyl phenyl ether(4-)
2.5	Nonachlorobiphenyl	6.96	Butanone peroxide(2-)
2.5	Toxaphene	8.29	Butyl benzyl phthalate
2.61	Tetrachlorobenzene(1,2,4,5-)	5.46	Butyl-4,6-dinitrophenol(2-sec-)
2.66	Bromoacetone	0.8	Chloral
2.7	Dichloroethylene, NOS	2.71	Chlordane
2.7	Dichloroethylene(1,1-)	5.08	Chloro-m-cresol(p-)
2.71	Chlordane	5.93	Chloroambucil
2.71	Heptachlor epoxide	6.14	Chloroaniline(p-)
2.71	Phenylmercury acetate	6.6	Chlorobenzene

Attachment 2: Ranking of incinerability of organic hazardous constituents from Appendix VIII, Part 261, in heat of combustion order and in alphabetical order (Lee-84/11) (*contd.*)

Incinerability ranking by heat of combustion (H_c) (kcal/g) and in alphabetical order

H_c	H_c order	H_c	Alphabetic order
2.72	Octachlorobiphenyl	5.5	Chlorobenzilate
2.77	Acetyl chloride	0.75	Chloroform
2.81	Trichloropropane, NOS	3.25	Chloromethane
2.81	Trichloropropane(1,2,3-)	3.48	Chloromethyl methyl ether
2.84	Dichloropropanol, NOS	7.37	Chloronaphthalene(2-)
2.86	Dimethyl sulfate	6.89	Chlorophenol(2-)
2.86	T(2,4,5-)	5.3	Chlorophenyl(1-(o-)) thiourea
2.88	Trichlorophenol(2,4,5-)	4.5	Chloropropionitrile(3-)
2.88	Trichlorophenol(2,4,6-)	8.18	Cresol
2.89	Nitroso(n-)-n-methylurea	8.09	Cresylic acid
2.98	Heptachlorobiphenyl	7.73	Crotonaldehyde
3	Dichloroethane(1,1-)	6.79	Cyanogen
3	Dichloroethane(1,2-)	0.81	Cyanogen bromide
3	Dichloroethane(*trans*-1,2-)	1.29	Cyanogen chloride
3.12	Phenyl dichloroarsine	3.92	Cycasin
3.19	Nitrosoarcosine(n-)	3.97	Cyclophosphamide
3.21	Azaserine	3.62	D(2,4-)
3.24	Fluoroacetamide(2-)	5.7	Daunomycin
3.25	Chloromethane	5.14	DDD
3.28	Hexachlorobiphenyl	5.05	DDE
3.38	Bis(2-chloroethyl) ether	4.51	DDT
3.38	Hexachloro(1,2,3,4,10,10-)-1,4,4a,5,7,8a,hexahydro-1,4,:5,8-endo,endo-dimethanonaphthalene	2.31	Decachlorobiphenyl
3.4	Benzenearsonic acid	7.34	Di-n-butyl phthalate
3.4	Maleic anhydride	7.83	Di-n-propylnitrosoamine
3.4	Trichlorobenzene(1,2,4-)	5.62	Diallate
3.43	TCDD	8.9	Dibenzo[c,g]carbazole(7h-)
3.44	Dichloropropene, NOS	9.33	Dibenzo[a,e]pyrene
3.44	Dichloropropene(1,3-)	9.33	Dibenzo[a,h]pyrene
3.46	Endrin	9.33	Dibenzo[a,i]pyrene
3.48	Chloromethyl methyl ether	9.53	Dibenz[a,h]acridine
3.52	Dinitrophenol(2,4-)	9.4	Dibenz[a,h]anthracene
3.56	Nitrogen mustard n-oxide and hydrochloride salt		
3.61	Parathion	9.53	Dibenz[a,j]acridine
3.62	D(2,4-)	1.48	Dibromo-3-chloropropane(1,2-)
3.66	Pentachlorobiphenyl	0.5	Dibromomethane
3.67	Propane sultone(1,3-)	1.43	Dibromomethane(1,1-)
3.74	Methyl methanesulfonate	4.27	Dichloro-2-butene(1,4-)
3.75	Aldrin	4.57	Dichlorobenzene, NOS
3.79	Nitroglycerine	5.72	Dichlorobenzidine(3,3′-)
3.81	Dichlorophenol(2,4-)	6.36	Dichlorobiphenyl
3.81	Dichlorophenol(2,6-)	0.22	Dichlorodifluoromethane
3.82	Hexachlorophene	3	Dichloroethane(1,1-)
3.84	Trypan blue	3	Dichloroethane(1,2-)
3.9	Benzotrichloride	3	Dichloroethane(*trans*-1,2-)
3.92	Cycasin	2.7	Dichloroethylene, NOS
3.92	Nitroso-n-ethylurea(n-)	2.7	Dichloroethylene(1,1-)
3.97	Cyclophosphamide	1.7	Dichloromethane

Attachment 2: Ranking of incinerability of organic hazardous constituents from Appendix VIII, Part 261, in heat of combustion order and in alphabetical order (Lee-84/11) (*contd.*)

Incinerability ranking by heat of combustion (H_c) (kcal/g) and in alphabetical order

H_c	H_c order	H_c	Alphabetic order
3.99	Dichloropropane, NOS	5.09	Dichloromethylbenzene
3.99	Dichloropropane(1,2-)	3.81	Dichlorophenol(2,4-)
4	Methylparathion	3.81	Dichlorophenol(2,6-)
4	Uracil mustard	2.31	Dichlorphenylarsine
4.01	Amitrole	3.99	Dichloropropane, NOS
4.02	Dimethoate	3.99	Dichloropropane(1,2-)
4.04	Tetraethyl lead	2.84	Dichloropropanol, NOS
4.06	Dinitro-o-cresol and salts(4,6-)	3.44	Dichloropropene, NOS
4.06	Methyl-n′-nitro-n-nitrosoguanidine(n-)	3.44	Dichloropropene(1,3-)
4.06	Mustard gas	5.56	Dieldrin
4.1	Maleic hydrazide	5.74	Diepoxybutane
4.15	Dinitrobenzene, NOS	6.39	Diethyl phthalate
4.18	Nitroso-n-methylurethane(n-)	5.25	Diethylarsine
4.27	Dichloro-2-butene(1,4-)	8.68	Diethylhydrazine(1,2-)
4.28	Nitrogen mustard and hydrochloride salt	8.54	Diethylstilbestrol
4.29	Tetrachloribiphenyl	7.66	Dihydrosafrole
4.44	Hydrazine	6.05	Dihydroxy-alpha-(methylamino)methyl benzyl alcohol(3,4-)
4.45	Vinyl chloride	4.02	Dimethoate
4.47	Formaldehyde	7.36	Dimethoxybenzidine(3,3′-)
4.49	Saccharin	5.74	Dimethyl phthalate
4.5	Chloropropionitrile(3-)	2.86	Dimethyl sulfate
4.51	DDT	5.82	Dimethyl(3,3-)-1-(methylthio)-2-butanone-o-(methylamino)carbonyl oxime
4.51	Thiourea	6.97	Dimethylaminoazobenzene(p-)
4.55	Acetyl-2-thiourea(1-)	9.61	Dimethylbenz[a] anthracene(7,12-)
4.55	Thiosemicarbazide	5.08	Dimethylcarbamoyl chloride
4.57	Dichlorobenzene, NOS	7.87	Dimethylhydrazine(1,1-)
4.6	Bis(2-chloroethoxy) methane	7.87	Dimethylhydrazine(1,2-)
4.68	Dinitrotoluene(2,4-)	5.14	Dimethylnitrosamine
4.69	Isocyanic acid, methyl ester 7-oxabicyclo(2.2.1) heptane-2,4.70 3-dicarboxylic acid	9.54	Dimethylphenethylamine(alpha, alpha-)
4.73	Ethyl carbamate	8.51	Dimethylphenol(2,4-)
4.78	Aminomethyl-3-isoxazolol(5-)	4.06	Dinitro-o-cresol and salts(4,6-)
4.79	Methylthiouracil	4.15	Dinitrobenzene, NOS
4.84	Methylene-bis-(2-chloroaniline(4,4′-)	3.52	Dinitrophenol(2,4-)
4.93	Bis(2-chloroisopropyl)ether	5.74	Dinitrophenol(4,6-)
4.95	Nitrophenol(4-)	4.68	Dinitrotoluene(2,4-)
5.05	DDE	6.67	Dinitrotoluene di-n-octyl phthalate(2,6-)
5.08	Chloro(p-)-m-cresol	6.41	Dioxane
5.08	Dimethylcarbamoyl chloride	9.09	Diphenylamine
5.09	Dichloromethylbenzene	8.73	Diphenylhydrazine(1,2-)
5.1	Trichlorobiphenyl	5.73	Disulfoton
5.14	DDD	2.33	Endosulfan
5.14	Dimethylnitrosamine	3.46	Endrin
5.14	Nitrosodimethylamine(n-)	4.73	Ethyl carbamate
5.25	Diethylarsine	8.32	Ethyl cyanide
5.29	Phthalic anhydride	7.27	Ethyl methacrylate
5.3	Chlorophenyl(1-(o-)) thiourea	6.86	Ethylene oxide

Attachment 2: Ranking of incinerability of organic hazardous constituents from Appendix VIII, Part 261, in heat of combustion order and in alphabetical order (Lee-84/11) (*contd.*)

Incinerability ranking by heat of combustion (H_c) (kcal/g) and in alphabetical order

H_c	H_c order	H_c	Alphabetic order
5.34	Methyl(1-)-2-(methylthio)propionaldehyde-o-(methlcarbonyl) oxime	5.7	Ethylenebisdithiocarbamate
5.46	Butyl-4,6 dinitrophenol(2-sec-)	7.86	Ethyleneimine
5.5	Chlorobenzilate	5.98	Ethylenethiourea
5.5	Nitroananiline(p-)	9.35	Fluoranthene
5.56	Dieldrin	3.24	Fluoroacetamide(2-)
5.58	TP(2,4,5-)	4.47	Formaldehyde
5.59	Methoxychlor	1.32	Formic acid
5.59	Nitroquinoline-1-oxide(4-)	5.74	Glycidylaldehyde
5.62	Diallate	2.71	Heptachlor epoxide
5.7	Daunomycin	2.98	Heptachlorobiphenyl
5.7	Ethylenebisdithiocarbamate	3.38	Hexachloro(1,2,3,4,10,10-)-1,4,4a,5,7,8a,hexahydro-1,4:5,8-endo,endo-dimethanonaphthalene
5.72	Dichlorobenzidine(3,3′-)	1.79	Hexachlorobenzene
5.72	Pronamide	3.28	Hexachlorobiphenyl
5.73	Aflatoxins	2.12	Hexachlorobutadiene
5.73	Disulfoton	1.12	Hexachlorocyclohexane
5.74	Diepoxybutane	2.1	Hexachlorocyclopentadiene
5.74	Dimethyl phthalate	0.46	Hexachloroethane
5.74	Dinitrophenol(4,6-)	3.82	Hexachlorophene
5.74	Glycidylaldehyde	0.7	Hexachloropropene
5.75	Acrylamide	4.44	Hydrazine
5.82	Dimethyl(3,3-)-1-(methylthio)-2-butanone-o-(methylamino)carbonyl oxime	5.82	Indenol pyrene(1,2,3-c,d)
5.82	Indenol pyrene(1,2,3-c,d)	1.34	Iodomethane
5.84	Bromophenyl phenyl ether(4-)	8.62	Isobutyl alcohol
5.85	Thiuram	4.69	Isocyanic acid, methyl ester 7-oxabicyclo(2.2.1)heptane-2,4.70 3-dicarboxylic acid
5.91	Methanethiol	7.62	Isosafrole
5.92	Tolylene diisocyanate	2.15	Kepone
5.93	Chloroambucil	3.4	Maleic anhydride
5.95	Thioacetamide	4.1	Maleic hydrazide
5.98	Ethylenethiourea	5.98	Malononitrile
5.98	Malononitrile	5.91	Methanethiol
5.98	Nitro-o-toluidine(5-)	7.93	Methapyrilene
6.01	Nitrobenzene	5.59	Methoxychlor
6.05	Dihydroxy-alpha-(methylamino)-methyl benzyl alcohol(3,4-)	8.07	Methyl ethyl ketone (MEK)
6.07	Benzoquinone	6.78	Methyl hydrazine
6.13	Nitrosomethylethylamine(n-)	6.52	Methyl methacrylate
6.14	Chloroaniline(p-)	3.74	Methyl methanesulfonate
6.18	Benzyl chloride	5.34	Methyl(1-)-2-(methylthio)propionaldehyde-o-(methylcarbonyl) oxime
6.19	Resorcinol	4.06	Methyl-n′-nitro-n-nitrosoguanidine(n-)
6.28	Propylthiouracil	8.55	Methylacrylonitrile
6.3	Paraldehyde	9.09	Methylaziridine(2-)
6.36	Dichlorobiphenyl	9.57	Methylcholanthrene(3-)

Attachment 2: Ranking of incinerability of organic hazardous constituents from Appendix VIII, Part 261, in heat of combustion order and in alphabetical order (Lee-84/11) (*contd.*)

Incinerability ranking by heat of combustion (H_c) (kcal/g) and in alphabetical order

H_c	H_c order	H_c	Alphabetic order
6.39	Diethyl phthalate	4.84	Methylene-bis-(2-chloroaniline)(4,4′-)
6.41	Dioxane	6.43	Methyllactonitrile(2-)
6.43	Methyllactonitrile(2-)	4	Methylparathion
6.43	Nitrosopyrrolidone(n-)	4.79	Methylthiouracil
6.52	Methyl methacrylate	7.75	Monochlorobiphenyl
6.6	Chlorobenzene	4.06	Mustard gas
6.63	Toluidine hydrochloride(o-)	9.62	Naphthalene
6.64	Bis(2-chloroethyl)-2-naphthylamine (n,n-)	6.97	Naphthoquinone(1,4-)
6.67	Dinitrotoluene di-n-octyl phthalate (2,6-)	7.5	Naphthyl-2-thiourea(1-)
6.7	Reserpine	8.54	Naphthylamine(1-)
6.78	Methyl hydrazine	8.54	Naphthylamine(2-)
6.79	Cyanogen	8.92	Nicotine and salts
6.86	Ethylene oxide	5.98	Nitro-o-toluidine(5-)
6.86	Nitrosodiethylamine(n-)	5.5	Nitroananiline(p-)
6.89	Chlorophenol(2-)	6.01	Nitrobenzene
6.93	Phenylthiourea(n-)	4.28	Nitrogen mustard and hydrochloride salt
6.96	Acrolein	3.56	Nitrogen mustard n-oxide and hydrochloride salt
6.96	Butanone peroxide(2-)	3.79	Nitroglycerine
6.97	Dimethylaminoazobenzene(p-)	4.95	Nitrophenol(4-)
6.97	Naphthoquinone(1,4-)	5.59	Nitroquinoline-1-oxide(4-)
7	Acetonylbenzyl(3-(alpha-))-4-hydroxycoumarin and salts	3.92	Nitroso(n-)-n-ethylurea
7	Nitrosonornicotine(n-)	4.18	Nitroso(n-)-n-methylurethane
7.02	Nitrosodiethanolamine(n-)	2.89	Nitroso(n-)-n-methylurea
7.04	Nitrosopiperidine(n-)	3.19	Nitrosoarcosine(n-)
7.17	Phenacetin	8.46	Nitrosodi-n-butylamine(n-)
7.27	Ethyl methacrylate	7.02	Nitrosodiethanolamine(n-)
7.34	Di-n-butyl phthalate	6.86	Nitrosodiethylamine(n-)
7.36	Dimethoxybenzidine(3,3′-)	5.14	Nitrosodimethylamine(n-)
7.37	Acetonitrile	6.13	Nitrosomethylethylamine(n-)
7.37	Aminopyridine(4-)	7.91	Nitrosomethylvinylamine(n-)
7.37	Chloronaphthalene(2-)	7	Nitrosonornicotine(n-)
7.43	Propyn-1-ol(2-)	7.04	Nitrosopiperidine(n-)
7.5	Naphthyl-2-thiourea(1-)	6.43	Nitrosopyrrolidone(n-)
7.62	Isosafrole	2.5	Nonachlorobiphenyl
7.66	Dihydrosafrole	2.72	Octachlorobiphenyl
7.68	Safrole	6.3	Paraldehyde
7.69	Auramine	3.61	Parathion
7.73	Crotonaldehyde	2.05	Pentachlorobenzene
7.75	Allyl alcohol	3.66	Pentachlorobiphenyl
7.75	Monochlorobiphenyl	0.53	Pentachloroethane
7.78	Phenol	1.62	Pentachloronitrobenzene
7.81	Phenylenediamine	2.09	Pentachlorophenol
7.82	Acetylaminofluorine(2-)	7.17	Phenacetin
7.83	Acrylonitrile	7.78	Phenol
7.83	Di-n-propylnitrosoamine	3.12	Phenyl dichloroarsine
7.86	Ethyleneimine	7.81	Phenylenediamine
7.87	Dimethylhydrazine(1,1-)	2.71	Phenylmercury acetate
		6.93	Phenylthiourea(n-)

Attachment 2: Ranking of incinerability of organic hazardous constituents from Appendix VIII, Part 261, in heat of combustion order and in alphabetical order (Lee-84/11) (*contd.*)

Incinerability ranking by heat of combustion (H_c) (kcal/g) and in alphabetical order

H_c	H_c order	H_c	Alphabetic order
7.87	Dimethylhydrazine(1,2-)	5.29	Phthalic anhydride
7.91	Nitrosomethylvinylamine(n-)	8.72	Picoline(1-)
7.93	Methapyrilene	5.72	Pronamide
7.93	Pyridine	3.67	Propane sultone(1,3-)
8.03	Strychnine and salts	9.58	Propylamine(n-)
8.07	Methyl ethyl ketone (MEK)	6.28	Propylthiouracil
8.09	Cresylic acid	7.43	Propyn-1-ol(2-)
8.18	Cresol	7.93	Pyridine
8.24	Toluene diamine	6.7	Reserpine
8.26	Acetophenone	6.19	Resorcinol
8.29	Butyl benzyl phthalate	4.49	Saccharin
8.32	Ethyl cyanide	7.68	Safrole
8.42	Bis(2-ethylhexyl) phthalate	8.03	Strychnine and salts
8.43	Benzenethiol	2.86	T(2,4,5-)
8.46	Nitrosodi-n-butylamine(n-)	3.43	TCDD
8.51	Dimethylphenol(2,4-)	4.29	Tetrachloribiphenyl
8.54	Diethylstilbestrol	2.61	Tetrachlorobenzene(1,2,4,5-)
8.54	Naphthylamine(1-)	1.19	Tetrachloroethane
8.54	Naphthylamine(2-)	1.39	Tetrachloroethane, NOS
8.55	Methylacrylonitrile	1.39	Tetrachloroethane((1,1,2,2-)
8.62	Isobutyl alcohol	1.39	Tetrachloroethane(1,1,1,2-)
8.68	Diethylhydrazine(1,2-)	0.24	Tetrachloromethane (carbon tetrachloride)
8.72	Picoline(1-)	2.23	Tetrachlorophenol(2,3,4,6-)
8.73	Aniline	4.04	Tetraethyl lead
8.73	Diphenylhydrazine(1,2-)	0.41	Tetranitromethane
8.9	Dibenzo(7h-)[c,g]carbazole	5.95	Thioacetamide
8.92	Benz[c]acridine	4.55	Thiosemicarbazide
8.92	Nicotine and salts	4.51	Thiourea
9	Amino biphenyl(4-)	5.85	Thiuram
9.09	Diphenylamine	10.14	Toluene
9.09	Methylaziridine(2-)	8.24	Toluene diamine
9.18	Benzidine	6.63	Toluidine hydrochloride(o-)
9.25	Benzo[a]pyrene	5.92	Tolylene diisocyanate
9.25	Benzo[b]fluoranthene	2.5	Toxaphene
9.25	Benzo[j]fluoranthene	5.58	TP(2,4,5-)
9.33	Dibenzo[a,e]pyrene	0.13	Tribromomethane
9.33	Dibenzo[a,h]pyrene	3.4	Trichlorobenzene(1,2,4-)
9.33	Dibenzo[a,i]pyrene	5.1	Trichlorobiphenyl
9.35	Fluoranthene	1.74	Trichloroethane
9.39	Benz[a]anthracene	1.99	Trichloroethane(1,1,1-)
9.4	Dibenz[a,h]anthracene (dibenzo (a,h) anthracene)	1.99	Trichloroethane(1,1,2-)
9.53	Deibenz[a,h]acridine	0.84	Trichloromethanetiol
9.53	Dibenz[a,j]acridine	0.11	Trichloromonofluoromethane
9.54	Dimethylphenethylamine(alpha, alpha-)	2.88	Trichlorophenol(2,4,5-)
9.57	Methylcholanthrene(3-)	2.88	Trichlorophenol(2,4,6-)
9.58	Propylamine(n-)	2.81	Trichloropropane, NOS
9.61	Dimethylbenz[a] anthracene(7,12-)	2.81	Trichloropropane(1,2,3-)
9.62	Naphthalene	3.84	Trypan blue
10.03	Benzene	4	Uracil mustard
10.14	Toluene	4.45	Vinyl chloride

CHAPTER 2.2

BASIC COMBUSTION AND INCINERATION

C. C. Lee and G. L. Huffman
U.S. Environmental Protection Agency, Cincinnati, Ohio

Thomas C. Ho
Department of Chemical Engineering, Lamar University, Beaumont, Texas

Floyd Hasselriis
Consulting Engineer, Forest Hills, New York

1 INTRODUCTION

The basic principle of solid waste incineration is similar to that of conventional fossil fuel combustion. Fossil fuels may include coal, oil, and natural gas, and solid wastes may include medical waste, municipal waste, hazardous waste, and sludge from the treatment of domestic and industrial wastewater. Conventional combustion has been a subject of research for many decades, but incineration research for solid waste destruction had been fairly minimal until the mid-1970s, when incineration was found to be effective for the disposal of various wastes.

2 BASIC COMBUSTION PRINCIPLES

Combustion or incineration is a very complicated subject. It is so complicated that many consider it an art rather than a science. Understanding of the combustion process, however, is essential in order to effectively design, operate, and regulate incinerators. Key areas requiring understanding of the combustion process include (Lee-89/08):

1. Mass balance (this determines the amounts of the products formed by the reactants);
2. Energy balance (this determines the energy transferred within a combustion system or how much auxiliary fuel is needed for an incinerator to reach a certain temperature);
3. Thermodynamic analysis (this reveals information about the changes in the chemical components of a combustion system; however, it does not reveal how rapidly these changes will occur);
4. Kinetic analysis (this provides information on how quickly changes can occur, but does not predict the extent of change that is ultimately possible);
5. Heat transfer (this determines the temperature distribution within a combustion system);
6. Turbulent mixing (this determines whether the waste compounds are effectively put in contact with oxygen for reaction);
7. Residence time (this determines the volumetric size of a combustor).

The last three items, i.e., temperature, turbulence, and time, are called "The three t's" of waste incineration.

Basic combustion covers two major subjects, namely, fuel combustion and waste incineration.

2.1 Fuel Combustion

Combustion involves heat and light. It is a process of burning resulting from the rapid oxidation of organic (fuel) compounds. The process can be expressed in terms of fundamental chemical reaction equations. Consider the oxidation of carbon and the oxidation of hydrogen:

Reactants		Products
$C + O_2$	$\rightarrow$	CO_2
$2H_2 + O_2$	$\rightarrow$	$2H_2O$

These equations state that one mole of carbon reacts with one mole of oxygen to form one mole of carbon dioxide, and two moles of hydrogen react with one mole of oxygen to form two moles of water. This also means that 12 lbs of carbon react with 32 lbs of oxygen to form 44 lbs of carbon dioxide, and 4 lbs of hydrogen react with 32 lbs of oxygen to form 36 lbs of water. A schematic of these reactions are shown in Fig. 2.1. All feed substances that undergo

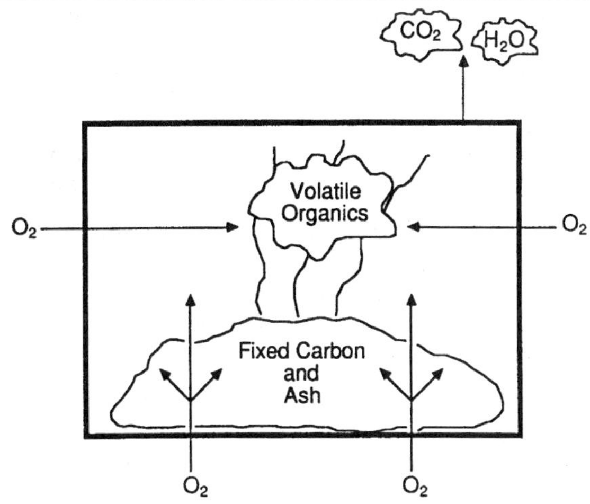

FIGURE 2.1 Oxygen reaction (*EPA-89/03*).

the combustion process are called the *reactants*, and the substances that result from the combustion process are called the *products*.

Obviously, combustion has to follow the law of mass conservation and the law of energy conservation. Therefore, during combustion, chemical elements can react with each other but the mass and energy level of the entire combustion system must remain the same. Mass and energy balance calculations are two key ways to define a combustion system. The mass balance determines the levels of products formed by the reactants, and the energy balance determines the amount of energy transfer within a combustion system.

2.2 Waste Incineration

A thermal treatment process is defined as the treatment of wastes in a device which uses elevated temperatures as the primary means to change the chemical, physical, or biological character or composition of the hazardous waste (40CFR260.10-91). Of the thermal processes, incineration is the most used technology worldwide. For example, it has been adopted by several laws in the United States as a proven technology to dispose of:

1. Hazardous waste, medical waste, municipal waste and mixed waste (hazardous waste mixed with radioactive components) regulated under the Resource Conservation and Recovery Act (RCRA);
2. Toxic substances under the Toxic Substances Control Act (TSCA);
3. Sludge waste under the Clean Water Act (CWA);
4. Superfund waste or remediation substances under the Comprehensive Environmental Response, Compensation, and Liability Act (CERCLA);
5. Extremely hazardous substances under the Superfund Amendments and Reauthorization Act (SARA);
6. Pesticides under the Federal Insecticide, Fungicide and Rodenticide Act (FIFRA).

The disposal of solid wastes is an environmental issue and it will continue to be as long as waste is continuously produced. During the late 1970s to early 1980s, research on solid waste disposal, particularly hazardous waste disposal, was dominated by the RCRA requirements and was very much focused on stationary incineration processes. The incineration activities during that period were later summarized by Oppelt in his critical review paper entitled, "Incineration of Hazardous Waste—A Critical Review" (Oppelt-87/05). In this paper, he concluded that EPAs research data and industry's operating experience indicate that incineration, when compared with the alternative technologies, has the highest overall degree of destruction and control for the broadest range of waste streams.

Although there are many potential treatment technologies, none are as universally applicable as incineration in treating the many types of waste governed by the many different Federal laws and State regulations. Compared with other treatment technologies, incineration has several major advantages:

1. Volume reduction: the reduction rate depends on the ash content of the waste incinerated.
2. Detoxification: incineration can achieve almost 100% destruction of any pathogen, toxic, or hazardous substance contained in the waste.
3. No long-term liability: once a waste is incinerated, the problem will never re-surface again as it does in a landfill.
4. Effectiveness: it takes only a few seconds to destroy what may take years to decompose in landfills.
5. Waste heat recovery potential: waste combustion may feature the beneficial conversion of waste to energy for heating or for electrical energy applications.

However, waste incinerators may produce unwanted toxic combustion byproducts (TCBs) or pollutants of environmental concern if they are not well designed and operated. TCBs (as depicted in Fig. 2.2) include those listed below.

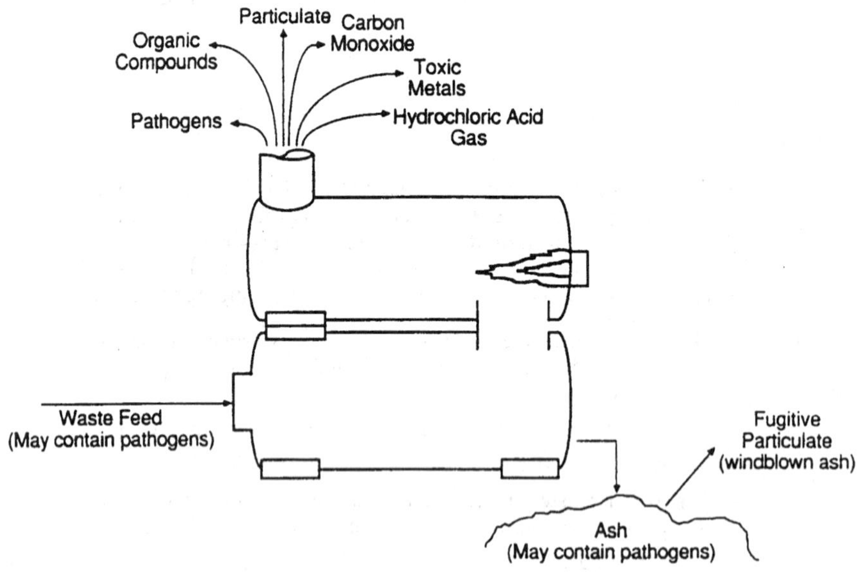

FIGURE 2.2 Toxic combustion byproducts (*EPA-89/03*).

1. Gaseous emissions.
   - Acid gas: including hydrogen chloride (HCl), nitrogen oxide (NO_x), and sulfur oxide (SO_2).
   - Organics or hydrocarbons: including dioxins and furans (PCDDs and PCDFs) [This category is generally referred to as the products of incomplete combustion (PICs)].
2. Particulate matter.
   - Trace metals (conventional metals and radioactive metals).
   - Fly ash.
   - Soots.
3. Contaminants in ash.
4. Contaminants in spent wastewater.

The issue of TCBs has been one of the major technical and sociological issues surrounding the implementation of incineration as a waste treatment alternative. The current RCRA regulation on "destruction and removal efficiency" has led to the unfortunate public misconception of incineration as a "landfill in the sky." As a result, the public has developed the so-called NIMBY (not in my back yard) attitude, which makes siting an incineration facility extremely difficult.

Incineration is an engineered process using controlled flame and postflame combustion conditions to thermally degrade waste materials. Theoretically speaking, the incineration process follows the same principles as the combustion process. However, it involves three simultaneous chemical reaction modes: (1) strong oxidation; (2) weak pyrolysis; (3) weak radical attack (Lee-89/08).

OXIDATION: The oxidation of waste is shown in the following example in which dichloromethane is oxidized to produce harmless products:

$$CH_2Cl_2 + O_2 \rightarrow CO_2 + 2HCl$$

PYROLYSIS: Pyrolysis is a thermal degradation process wherein carbonaceous materials are destroyed or chemically rearranged in the absence or near absence of oxygen or air. It uses heat to break the bonds of the elements contained in a compound.

Although incineration requires about 50–150% excess air to ensure enough oxygen in the combustion chamber to effectively contact with the waste, some small fraction of the waste still may not have a chance to contact the oxygen. These small waste fractions that remain in the high-temperature environment may undergo pyrolysis. For example, the pyrolysis of cellulose and polychlorinated biphenyl (PCB) would be

$$\text{Cellulose: } C_6H_{10}O_5 \rightarrow 2CO + CH_4 + 3H_2O + 3C$$

$$\text{PCB: } C_{12}H_7Cl_3 \rightarrow 12C + 3HCl + 2H_2$$

The pyrolyzed compounds will generally produce simpler compounds such as CO, CH_4, and H_2O, which will be in the gaseous phase, and carbon (C) or char which will be in the solid phase. These simpler compounds generally come into contact with the excess oxygen contained in the incinerator to react fully to form water and carbon dioxide.

RADICAL ATTACK: During incineration, flames are characterized by temperatures usually in the neighborhood of 1000°C and a radical-rich gas flow. This gas flow consists primarily of atomic hydrogen (H), atomic oxygen (O), atomic chlorine (Cl), hydroxyl radicals (OH·), possibly methyl radicals (CH_3·) in carbon–hydrogen–oxygen systems, and chloroxy radicals (ClO·) in chlorine-containing systems. When these radicals contact the waste organics they react with them quickly, thus accelerating the decomposition of the waste.

Also during incineration there are at least three possible situations.

1. The majority of the waste can be easily oxidized and totally destroyed in the primary combustion chamber. Or, if a small fraction of the waste is pyrolyzed in the primary chamber, the pyrolysis products are totally destroyed in the secondary combustion chamber, the afterburner.
2. A very small amount of waste, for some reason, may escape the incineration process and is not destroyed or is only partially destroyed. In this case, emissions of POHCs (principal organic hazardous constituents) will generally be too high.
3. Waste constituents may produce intermediate compounds and may result in the emission of unwanted products of incomplete combustion (PICs), which may be more hazardous than the parent compounds but which are at *much* lower concentrations (ppm levels versus percent levels).

2.3 Complete Combustion

Complete combustion is also known as *stoichiometric combustion* or *theoretical combustion*. Complete combustion is defined as a combustion process in which all carbon and hydrogen elements in the reactants are converted into only carbon dioxide and water in the products. For complete combustion, the following are valid (Hasselriis-98/12):

- no fuel and/or waste remains in the products;
- no free oxygen remains in the products of combustion (flue gas);
- all carbon has been combusted to CO_2, not CO;
- sulfur is oxidized to SO_2, not SO_3.

An example of complete combustion of a propane gas is

$$C_3H_8 + 5(1.0)(O_2 + 3.76N_2) \rightarrow 3CO_2 + 4H_2O + 18.8N_2$$

The chemical equation for the complete combustion of a fuel is called the *stoichiometric equation*. To obtain the optimum temperature during combustion, it is necessary to convert all the chemical energy stored in the reactants into thermal energy. To reach this goal, all carbon and hydrogen elements in a combustion system must be fully oxidized and become only carbon dioxide and water.

Oxygen is necessary for combustion. The amount of oxygen required for complete combustion is known as the stoichiometric or theoretical oxygen, and is determined by the nature and, of course, the quantity of the combustible material to be burned. With the exception of some exotic fuels, combustion oxygen is usually obtained from atmospheric air. Figure 2.3 depicts the concept of complete combustion. It shows that the reaction of oxygen and organics produces only CO_2 and H_2O (EPA-89/03).

Consider a generalized fuel with a chemical formula $C_xH_yS_zO_w$ where the indices x, y, z, and w represent the relative number of atoms of carbon, hydrogen, sulfur, and oxygen respectively. Balancing the chemical reaction for the complete oxidation (combustion) of this fuel with oxygen from air gives (EPA-80/02, p. 2):

$$C_xH_yS_zO_w + (x + y/4 + z - w/2)O_2 + (0.79/0.21)(x + y/4 + z - w/2)N_2 \rightarrow$$
$$xCO_2 + (y/2(H_2O + zSO_2 + (0.79/0.21)(x + y/4 + z - w/2)N_2 + Q \qquad (2.1)$$

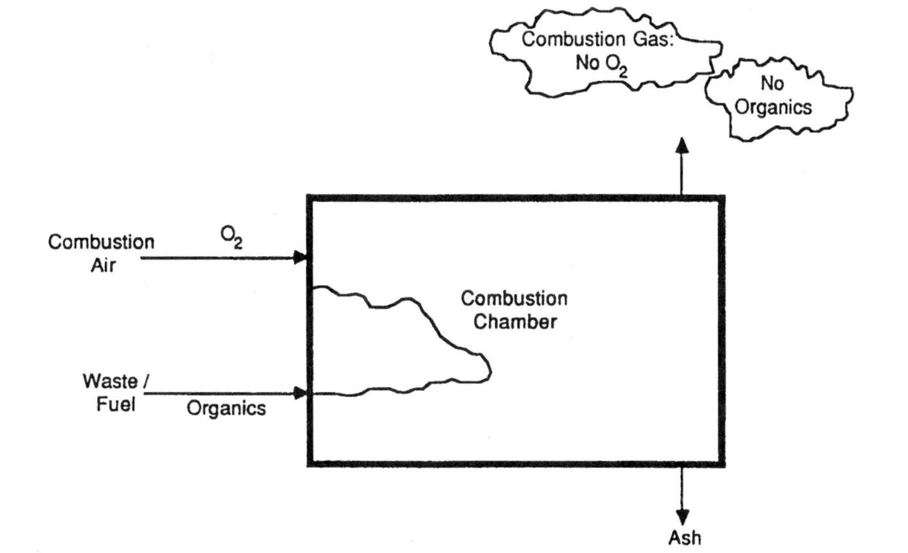

FIGURE 2.3 Complete combustion.

where Q represents the heat of combustion.

The above reaction assumes that:

1. air consists of 21% by volume of oxygen with the remaining 79% made up of nitrogen and other inert gases;
2. bound oxygen in the fuel is available for combustion, thus reducing air requirements;
3. fuel contains no bound nitrogen, so no "fuel NO_x" is produced;
4. "thermal NO_x" via nitrogen fixation is small, so that it is neglected in stoichiometric air calculations;
5. sulfur in fuel is oxidized to SO_2 with negligible SO_3 formation.

Equation (2.1) relates the reactants on a molar basis. One gram-mole of a substance is the mass of that substance equal to its molecular weight in grams. A gram-mole of any substance contains Avogadro's number of molecules of that substance, i.e., there are 6.02×10^{23} molecules/g-mole. Pound-moles (lb-mole) are also in common use. Since one pound-mole is equivalent to the molecular weight of the substance in pounds, it contains 454 times as many molecules as a gram-mole.

The generalized combustion equation, Eq. (2.1), can be converted to a mass basis simply by multiplying the number of moles of each substance by its respective molecular weight.

Referring to Eq. (2.1), for CH_4: $x = 1$; $y = 4$; $z = w = 0$. Thus balancing the combustion equation gives

$$CH_4 + 2O_2 + 2(3.76)N_2 \rightarrow CO_2 + 2H_2O + 7.52N_2 \tag{2.2}$$

Mole balance (volume balance):

$$1 + 2 + 7.52 \rightarrow 1 + 2 + 7.52 \tag{2.3}$$

$= 10.52$ moles for both reactants and products.

Mass balance:

$$16 + 64 + 211 \rightarrow 44 + 36 + 211 \quad (2.4)$$

= 291 lb for both reactants and products.

$$\text{Mass/combustibles} = (\text{mass of each component})/(CH_4 \text{ mass}): \quad (2.5)$$

$$1 + 4 + 13.19 \rightarrow 2.75 + 2.25 + 13.19$$

The above expression gives not only the theoretical air requirements in terms of moles or volume in Eq. (2.3), and mass in Eqs. (2.4) and (2.5), but it also permits the determination of the resulting combustion products which the flue needs to handle.

Chemical reactions take place as a function of the molecular weights of the reactants. Based on the molecular weights of elements, consider the oxygen needed for the following five cases (Hasselriis-98/12).

1. One mole of carbon [12 lb] reacts with one half mole of oxygen (gas) [32/2 lb] to form one mole of carbon monoxide [28 lb].

$$\begin{array}{ccccc} C & + & 1/2(O_2) & \rightarrow & CO \\ 12\,\text{lb carbon} & + & 16\,\text{lb oxygen} & \rightarrow & 28\,\text{lb carbon monoxide} \\ 1\,\text{lb carbon} & + & 1.333\,\text{lb oxygen} & \rightarrow & 2.333\,\text{lb carbon monoxide} \end{array}$$

2. One mole of carbon reacts with one mole of oxygen (gas) to form one mole of carbon dioxide.

$$\begin{array}{ccccc} C & + & O_2 & \rightarrow & CO_2 \\ 12\,\text{lb carbon} & + & 32\,\text{lb oxygen} & \rightarrow & 44\,\text{lb carbon dioxide} \\ 1\,\text{lb carbon} & + & 2.667\,\text{lb oxygen} & \rightarrow & 3.667\,\text{lb carbon dioxide} \end{array}$$

3. Two moles hydrogen + one mole oxygen (gas) → two moles water.

$$\begin{array}{ccccc} 2H_2 & + & O_2 & \rightarrow & 2H_2O \\ 4\,\text{lb hydrogen} & + & 32\,\text{lb oxygen} & \rightarrow & 36\,\text{lb } H_2O \\ 1\,\text{lb hydrogen} & + & 8\,\text{lb oxygen} & \rightarrow & 9\,\text{lb } H_2O \end{array}$$

4. One mole of sulfur reacts with one mole of oxygen (gas) to form one mole of sulfur dioxide.

$$\begin{array}{ccccc} S & + & O_2 & \rightarrow & SO_2 \\ 12\,\text{lb sulfur} & + & 32\,\text{lb oxygen} & \rightarrow & 44\,\text{lb sulfur dioxide} \\ 1\,\text{lb sulfur} & + & 2.667\,\text{lb oxygen} & \rightarrow & 3.667\,\text{lb sulfur dioxide} \end{array}$$

5. One mole of oxygen plus one half mole of nitrogen (gas) forms one mole of nitrogen dioxide.

$$\begin{array}{ccccc} N & + & O_2 & \rightarrow & NO_2 \\ 14\,\text{lb nitrogen} & + & 32\,\text{lb oxygen} & \rightarrow & 46\,\text{lb nitrogen dioxide} \\ 1\,\text{lb nitrogen} & + & 2.286\,\text{lb oxygen} & \rightarrow & 3.386\,\text{lb nitrogen dioxide} \end{array}$$

EXAMPLE 1 combustion of hydrogen
Calculate the weight of oxygen required for combustion of methane CH_4.

Step 1. Determine the C–H combustion balance equation. The same number of carbons and hydrogens must appear on the left and right sides of the equation. One mole of CO_2 and two moles of H_2O are produced. Two moles of oxygen are needed to balance the equation.

$$\begin{array}{lllll} \text{Methane} + \text{oxygen} & \rightarrow & \text{carbon dioxide} & + & \text{water} \\ CH_4 \quad + \quad 2O_2 & \rightarrow & CO_2 & + & 2H_2O \\ 1\ \text{mol} \quad + 2\ \text{moles} & \rightarrow & 1\ \text{mole} & + & 2\ \text{moles} \end{array}$$

Step 2. Using the molecular weights of carbon (12), hydrogen (1), and oxygen gas (32), write the weight equation and determine the weights of the reactants.

$$\begin{array}{lllllll} CH_4 & + & 2O_2 & \rightarrow & CO_2 & + & 2H_2O \\ 1[12+4]CH_4 & + & 2[32]O_2 & \rightarrow & 1[12+32]CO_2 & + & 2[2+16]H_2O \\ 16\,\text{lb}\,CH_4 & + & 64O_2 & \rightarrow & 44\,\text{lb}\,CO_2 & + & 36\,\text{lb}\,H_2O \\ 1\,\text{lb}\,CH_4 & + & 4O_2 & \rightarrow & 2.75\,\text{lb}\,CO_2 & + & 2.25\,\text{lb}\,H_2O \end{array}$$

EXAMPLE 2 combustion of hydrogen

Calculate how many pounds of oxygen are theoretically required to completely burn (oxidize) a pound of Number 2 fuel oil (C_6H_{10}).

Step 1. Determine the C–H combustion balance equation.

$$\begin{array}{rllllll} C_6H_{10} & + & \text{oxygen} & \rightarrow & \text{carbon dioxide} & + & \text{water} \\ \text{Balance the equation: } C_6H_{10} & + & \frac{17}{2}O_2 & \rightarrow & 6CO_2 & + & 5H_2O \\ \text{Mol equation: } 1\ \text{mol} & + & 8.5\ \text{moles} & \rightarrow & 6\ \text{moles} & + & 5\ \text{moles} \end{array}$$

Five moles of H_2O are produced and 8.5 moles of oxygen are needed to balance the equation.

Step 2. Determine the weights of the reactants.

Use the molecular weights of carbon (12), hydrogen (1), oxygen (16), and oxygen gas (32) to balance and write the weight equation.

$$\begin{array}{lllllll} C_6H_{10} & + & 6O_2 & \rightarrow & 6CO_2 & + & 5H_2O \\ (6[12]+1[10])C_6H_{10} & + & 17/2[32]O_2 & \rightarrow & 6[12+32]CO_2 & + & 5[2+16]H_2O \\ 82\,\text{lb}\,C_6H_{10} & + & 272\,\text{lb}\,O_2 & \rightarrow & 264\,\text{lb}\,CO_2 & + & 90\,\text{lb}\,H_2O \end{array}$$

Step 3. Divide by weight of fuel.

$$\begin{array}{lllllll} 1\,\text{lb}\,C_6H_{10} & + & 272/82\,\text{lb}\,O_2 & \rightarrow & 264/82\,\text{lb}\,CO_2 & + & 90/82\,\text{lb}\,H_2O \\ 1\,\text{lb}\,C_6H_{10} & + & 3.317\,\text{lb}\,O_2 & \rightarrow & 3.22\,\text{lb}\,CO_2 & + & 1.097\,\text{lb}\,H_2O \end{array}$$

EXAMPLE 3 compounds containing hydrogen, carbon, and oxygen

Calculate how many pounds of oxygen are theoretically required to completely burn a pound of cellulose ($C_6H_{10}O_5$).

Step 1. Determine the C-H-O combustion balance equation.

$$\begin{array}{lllllll} \text{Cellulose} & + & \text{oxygen} & \rightarrow & \text{carbon dioxide} & + & \text{water} \\ C_6H_{10}O_5 & + & 6O_2 & \rightarrow & 6CO_2 & + & 5H_2O \\ 1\ \text{mole} & + & 6\ \text{moles} & \rightarrow & 6\ \text{moles} & + & 5\ \text{moles} \end{array}$$

Six moles of oxygen produce 6 moles CO_2 and five moles of H_2O.

Step 2. Determine the weights of the reactants.

$$
\begin{array}{lcccccccc}
C_6H_{10}O_5 & + & 6O_2 & \rightarrow & 6CO_2 & + & 5H_2O \\
(6[12] + 1[10] + 5[16])C_6H_{10}O_5 & + & 6[32]O_2 & \rightarrow & 6[12 + 32]CO_2 & + & 5[2 + 16]H_2O \\
162\,\text{lb}\, C_6H_{10}O_5 & + & 192\,\text{lb}\, O_2 & \rightarrow & 264\,\text{lb}\, CO_2 & + & 90\,\text{lb}\, H_2O
\end{array}
$$

Step 3. Divide by weight of fuel.

$$1\,\text{lb}\, C_6H_{10}O_5 + 192/162\,\text{lb}\, O_2 \rightarrow 264/162\,\text{lb}\, CO_2 + 90/162\,\text{lb}\, H_2O$$
$$1\,\text{lb}\, C_6H_{10}O_5 + 1.185\,\text{lb}\, O_2 \rightarrow 1.6300\,\text{lb}\, CO_2 + 0.555\,\text{lb}\, H_2O$$

2.4 Complete Incineration

Parallel to complete combustion, complete incineration is a process where the elements in the waste are generally assumed to follow the reaction patterns listed in Table 2.1. The ultimate goal of incinerating a waste is to convert the waste materials into harmless combustion products so that they can be safely emitted to the environment.

However, complete incineration is solely a theoretical concept. In actual practice, ppm levels of partially oxidized products of incomplete combustion (PICs) are formed. These PICs may include carbon monoxide (CO), soot, and a whole myriad of other organic products. It is always possible to over-design an incinerator or to use extra fuel for higher flame temperatures to ensure sufficiently complete combustion. However, either of these corrective measures increases the cost of incineration.

The organic materials that enter an incinerator with the waste and fuel are primarily made up of carbon, hydrogen, and oxygen. Ideally, these organic materials react with oxygen in the combustion gas to form carbon dioxide (CO_2) and water vapor (H_2O). The chemical reaction for this ideal situation is

$$\text{Organics (C, H, O)} + O_2 \rightarrow CO_2 + H_2O + \text{heat}$$

This ideal reaction represents complete incineration or complete combustion, and is depicted in Fig. 2.4. However, this ideal reaction may or may not occur in actual waste combustion systems. For example, when chlorinated compounds are in the waste, HCl is emitted from the incinerator combustion chamber.

TABLE 2.1 Complete Incineration Reaction Pattern

Waste elements	Conversion products
Hydrogen, H	H_2O
Carbon, C	CO_2
Chloride, Cl	HCl or Cl_2
Fluoride, F	HF or F_2
Sulfur, S	SO_2
Nitrogen, N	N_2
Alkali metals	Carbonate
Sodium, Na	Na_2CO_3
Potassium, K	KOH
Non-alkali metals	Oxides
Copper, Cu	CuO
Iron, Fe	Fe_2O

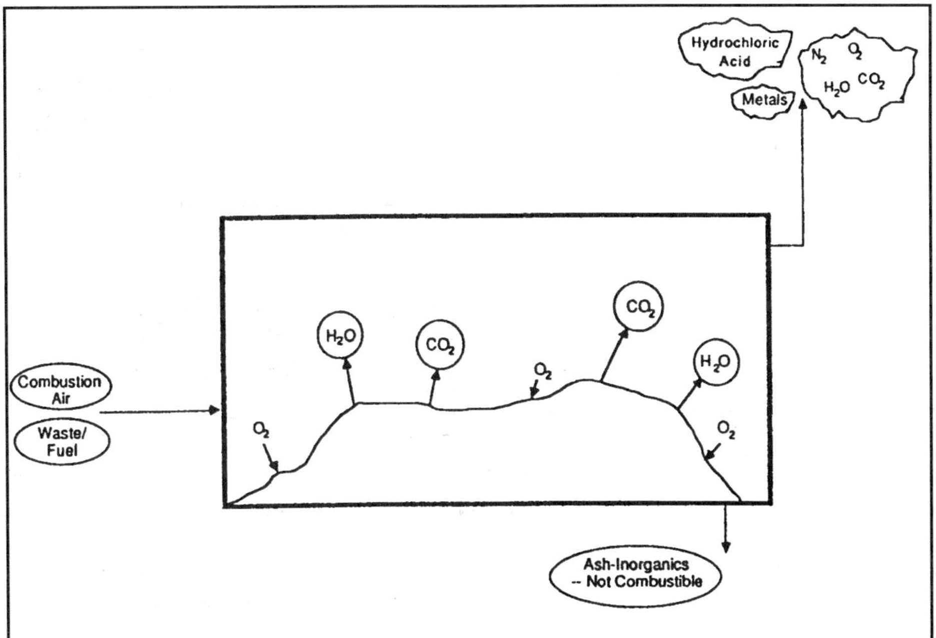

FIGURE 2.4 Complete incineration (*EPA-89/03*).

2.5 Incomplete Combustion and Incomplete Incineration

Incomplete combustion means that the combustion process has not undergone complete combustion. This is depicted in Fig. 2.5. During the incineration of waste, many factors can cause incomplete combustion. Factors that lead to incomplete combustion include poor mixing, too much or too little combustion air, low temperatures, etc. Under these conditions, undesirable toxic combustion byproducts (TCBs) may be discharged with ash or emitted with stack gas. The gaseous TCBs which are emitted with the stack gas are also known as the products of incomplete combustion (PICs). The most common PIC emitted is CO. Concentrations of CO in the stack gas generally increase with any of the poor combustion conditions described above. Soot is an another TCB that often is emitted under poor operating conditions. Soot particles are an elemental carbon which is in general very fine and can result in high opacity at the combustion stack. Other PICs that cause concern because of their health impacts are benzene, dioxins, furans, and other hazardous organic compounds.

The main products of waste incineration under either complete or incomplete incineration are combustion gases, solid residue (ash), and energy. The primary objectives of designing or operating an incinerator are to ensure that:

1. the combustion gases contain minimum PICs (products of incomplete combustion);
2. ash or solid residue is free of organic compounds;
3. the energy generated is recovered.

EXAMPLE: complete combustion
Methanol (CH_4O) undergoes incomplete combustion and produces two intermediate compounds, formaldehyde (CH_2O) and formic acid (CH_2O_2). Calculate the volume of air required to combust one pound of methanol at 60°F and 14.6959 psia.

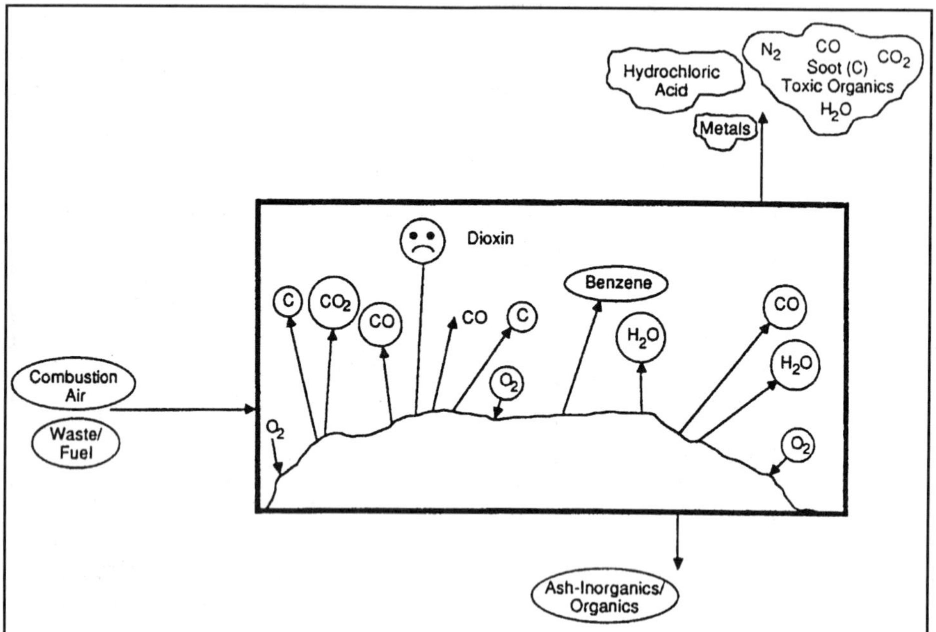

FIGURE 2.5 Incomplete incineration (*EPA-89/03*).

Solution:

1. Complete the following combustion equations.

$$\text{(eq. a): } CH_4O + 0.5O_2 \rightarrow CH_2O + H_2O$$
$$\text{(eq. b): } CH_4O + O_2 \rightarrow CH_2O_2 + H_2O$$
$$\text{(eq. c): } CH_4O + 1.5O_2 \rightarrow CO_2 + 2H_2O$$

2. Calculate the number of moles of methanol n in one pound of methanol.

$$n = m/M = 1/[12 + 1(4) + 16]$$
$$= 0.0313 \text{ lb-mole}$$

$$\text{(eq. a): } (0.0313)CH_4O + (0.0156)O_2 \rightarrow (0.0313)CH_2O + (0.0313)H_2O$$
$$\text{(eq. b): } (0.0313)CH_4O + (0.0313)O_2 \rightarrow (0.0313)CH_2O_2 + (0.0313)H_2O$$
$$\text{(eq. c): } (0.0313)CH_4O + (0.0469)O_2 \rightarrow (0.0313)CO_2 + (0.0626)H_2O$$

The air needed for each of the three cases above is

$$\text{(eq. a): } (0.0156)(O_2 + 3.76N_2)$$
$$\text{(eq. b): } (0.0313)(O_2 + 3.76N_2)$$
$$\text{(eq. c): } (0.0469)(O_2 + 3.76N_2)$$

3. Calculate the air needed.
The standard volume of one lb-mole of air at $T = 60°F$, and $p = 14.6959$ psia is

$$V = nR_u T/P$$
$$= 1(1545)(460 + 60)/(14.6959(144))$$
$$= 380 \text{ scf}$$

4. Calculate the volume of air needed for each of the three cases above.

$$\text{For eq. a, air} = 380(0.0156)(1 + 3.76) = 28.218 \text{ scf}$$
$$\text{For eq. b, air} = 380(0.0313)(1 + 3.76) = 56.615 \text{ scf}$$
$$\text{For eq. c, air} = 380(0.0469)(1 + 3.76) = 84.833 \text{ scf}$$

Notes: As shown above, it is possible to create a worse pollution problem if the combustion is incomplete. Incomplete combustion can arise because of:

- low temperatures;
- insufficient air; and
- incomplete mixing.

2.6 Combustion Air

The above discussion on basic combustion indicates that oxygen is the essential item in both fuel combustion and waste incineration. It is well known that oxygen can be supplied from the air. Figure 2.6 shows the fate of combustion air. The oxygen component in the air reacts with

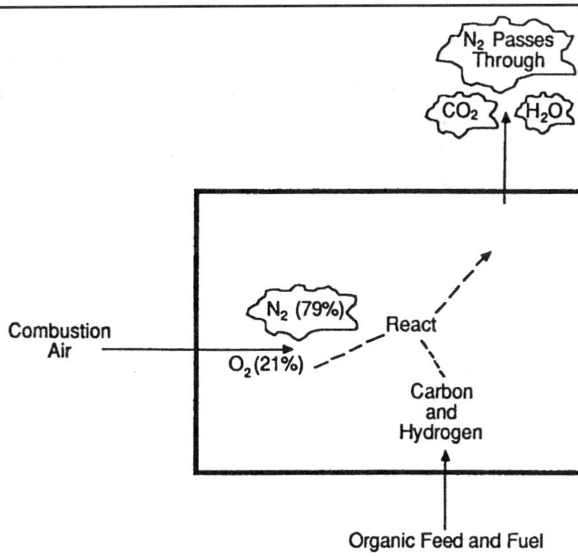

FIGURE 2.6 Fate of combustion air (*EPA-89/03*).

organics and produces CO_2 and H_2O. The nitrogen component in the air passes through the combustion chamber and emits to the atmosphere.

Theoretical air combustion. The following procedures provide the methods for calculating the amounts of oxygen, nitrogen, and air needed for the theoretical combustion of a hydrocarbon compound. It is assumed that a hydrocarbon compound contains elements of carbon, hydrogen, chlorine, oxygen, nitrogen, and sulfur. The theoretical amount of air needed and the combustion products produced can be derived in terms of the mass of input elements.

1. For carbon combustion: $C + O_2 \rightarrow CO_2$
   - Oxygen needs, $O_2 = (32/12)(C) = 2.67(C)$ lb/lb-carbon
   - Combustion products, $CO_2 = (44/12)(C) = 3.67(C)$ lb/lb-carbon
2. For hydrogen combustion: hydrogen reacts with chlorine first; the left-over hydrogen then reacts with oxygen. The oxygen needed for this combustion is
   - $H + Cl \rightarrow HCl$, $H_2 =$ H left over after Cl reaction; $H_2 = H–Cl/35.5$
   - $H_2 + 0.5O_2 = H_2O$
   - Oxygen needs, $O_2 = H_2(0.5)(32/2) = 8(H–Cl/35.5)$
   - Combustion products, $H_2O = (18/2)H_2 = 9(H_2) = 9(H–Cl/35.5)$
3. For sulfur combustion: $S + O_2 \rightarrow SO_2$
   - Oxygen, $O_2 = (32/32)(S) = S$
   - Combustion products, $SO_2 = (64/32)S = 2(S)$
4. For bound oxygen (B_o): B_o is the oxygen associated with (bound to) fuel or waste, not with air. It is assumed that B_o will react with reactant species such as carbon or hydrogen during combustion. Because of the presence of B_o, less air is needed. B_o will replace the amount of oxygen in the air needed for combustion. Therefore, the net oxygen quantity needed for combustion will be the theoretical oxygen quantity less the B_o quantity. B_o is used to represent the amount of bound oxygen.
5. For bound nitrogen (B_n): B_n is the nitrogen associated with (bound to) fuel or waste, not with air. It is assumed that B_n will not react with reactant species during combustion. It stays as molecular nitrogen and emits to the atmosphere as part of the combustion products. B_n is used to represent the amount of bound nitrogen.

Summing up the five items above results in:

Oxygen amount

$$O_2 = 2.67(C) + 8(H–Cl/35.5) + S - B_o \tag{2.6}$$

Nitrogen amount

$$\begin{aligned} N_2 &= O_2(0.79/0.21)(\text{MW of } N_2/\text{MW of } O_2) = O_2(3.76)(28)/32 \\ &= 3.29(O_2) \end{aligned} \tag{2.7}$$

Air needed for theoretical combustion A_t

$$\begin{aligned} A_t &= O_2 + N_2 \\ &= O_2 + 3.29(O_2) \\ &= 4.29(O_2) \end{aligned} \tag{2.8}$$

Combustion products from theoretical combustion P_t

$$\begin{aligned} P_t &= CO_2 + H_2O + SO_2 + N_2 + B_n \\ &= 3.67(C) + 9(H{-}Cl/35.5) + 2(S) + 3.29(O_2) + B_n \end{aligned} \quad (2.9)$$

The above results are also summarized in Tables 2.2 and 2.3. Table 2.2 can be used to calculate air requirements during theoretical or complete combustion.

Table 2.3 can be used to calculate combustion products during complete combustion.

TABLE 2.2 Computing the Reactants of Theoretical Air Combustion (R_t)

Basic reaction	Oxygen, nitrogen, and air requirements
$C + O_2 \rightarrow CO_2$	$O_2 = C(32/12) = 2.67(C)$
$H_2 + 0.5O_2 \rightarrow H_2O$	H_2 = H left over after Cl reaction; $H_2 = H{-}Cl/35.5$ $O_2 = H_2(0.5)(32/2) = 8(H{-}Cl/35.5)$
$S + O_2 \rightarrow SO_2$	$O_2 = S(32/32) = S$
B_o = bound oxygen	B_o is the oxygen associated with waste or fuel. It reacts with reactants and is a negative value if the air amount is positive
O_2	$O_2 = 2.67(C) + 8(H{-}Cl/35.5) + S - B_o$
N_2	$N_2 = (O_2)(3.76 \times 28/32)$
R_t	$R_t = O_2 + N_2$

TABLE 2.3 Computing the Products of Theoretical Air Combustion (P_t)

Basic reaction	Combustion products
$C + O_2 \rightarrow CO_2$	$CO_2 = (C)(44/12)$
$H_2 + 0.5O_2 \rightarrow H_2O$	H_2 = H left over after Cl reaction; $H_2 = H{-}Cl/35.5$ $H_2O = H_2(18/2) = 9(H{-}Cl/35.5)$
$S + O_2 \rightarrow SO_2$	$SO_2 = S(64/32) = 2(S)$
$N_2 = 3.29(O_2)$	N_2 associated with air, $N_2 = (O_2)(3.76 \times 28/32)$
B_n = bound nitrogen	N_2 associated with waste or fuel
P_t	$P_t = 3.67(C) + 9(H{-}Cl/35.5) + 2(S) + 3.29(O_2) + B_n$

EXAMPLE 1 air for theoretical combustion

Calculate the amounts of O_2 and N_2 needed in the air and the combustion products produced for the complete combustion of 200 pounds CH_4.

Given conditions:

- $CH_4 = 200$ lb
- $Cl = 0$
- $S = 0$
- $B_o = 0$
- $B_n = 0$

Solution 1 by chemical analysis

1.1. Calculate the ultimate composition of CH_4 in 200 lb (Table 2.4)

where

- x_i = compound element
- M = molecular weight of CH_4
- x_i/M = mass per pound fuel

TABLE 2.4 200 lb CH_4

x_i	C	H	Cl	O	M
CH_4	12	4	0	0	16.00
x_i/M	0.7500	0.2500	0.0000	0.0000	
200 lb	150.0000	50.0000	0.0000	0.0000	

1.2. Apply results from 1.1 to calculate the oxygen amount.

$$\begin{aligned} O_2 &= 2.67(C) + 8(H\text{–}Cl/35.5) + S - B_o \\ &= 2.67(150) + 8(50 - 0) + 0 - 0 \\ &= 401 + 400 \\ &= 801 \text{ lb} \end{aligned}$$

1.3. Apply results from 1.2 to calculate the nitrogen amount.

$$\begin{aligned} N_2 &= O_2(3.76)(28)/32 \\ &= 3.29(O_2) \\ &= 3.29(801) \\ &= 2635 \text{ lb} \end{aligned}$$

1.4. Apply results from 1.3 to calculate the theoretical air A_t.

$$\begin{aligned} A_t &= O_2 + N_2 \\ &= 801 + 2635 \\ &= 3436 \text{ lb} \end{aligned}$$

1.5. Calculate the total reactants.

$$\begin{aligned} \text{Reactants} &= CH_4 + \text{air} \\ &= 200 + 3436 \\ &= 3636 \end{aligned}$$

1.6. Apply results of 1.4 to calculate the combustion products P_t.

$$\begin{aligned} P_t &= CO_2 + H_2O + SO_2 + N_2 + B_n \\ &= 3.67(C) + 9(H\text{–}Cl/35.5) + 2(S) + 3.29(O_2) + B_n \\ &= 3.67(150) + 9(50 - 0/35.5) + 0 + 3.29(801) + 0 \\ &= 551 + 450 + 2635 \\ &= 3636 \text{ lb} \end{aligned}$$

1.7. Compare the results of reactant and product calculations.
Comparing the values of reactants and products shows that the two numbers are identical. This means that the mass input and mass output are balanced and the calculation is correct.

Solution 2 by tabulation
2.1. Calculate the theoretical combustion air (A_t) by the procedures in Table 2.5.

TABLE 2.5 Computing the Reactants of Theoretical Air Combustion (R_t)

Basic reaction	Oxygen, nitrogen, and air requirements	Mass
$C + O_2 \rightarrow CO_2$	$O_2 = C(32/12) = 2.67(C) = 2.67(150)$	401
$H_2 + 0.5O_2 \rightarrow H_2O$	H_2 = H left over after Cl reaction; H_2 = H–Cl/35.5 $O_2 = H_2(0.5)(32/2) = 8(H–Cl/35.5) = 8(50 - 0)$	400
$S + O_2 \rightarrow SO_2$	$O_2 = S(32/32) = S$	0
B_o = bound oxygen	B_o is the oxygen associated with waste or fuel. It reacts with reactants and is a negative value if the air amount is positive	0
O_2	$O_2 = 2.67(C) + 8(H–Cl/35.5) + S - B_o$	801
N_2	$N_2 = (O_2)(3.76 \times 28/32)$	2635
R_t	$R_t = O_2 + N_2$	3436

2.2. Calculate the theoretical combustion products (G_t) by the procedures in Table 2.6.

TABLE 2.6 Computing the Products of Theoretical Air Combustion (P_t)

Basic reaction	Combustion products	Mass
$C + O_2 \rightarrow CO_2$	$CO_2 = (C)(44/12)$	551
$H_2 + 0.5O_2 \rightarrow H_2O$	H_2 = H left over after Cl reaction; H_2 = H–Cl/35.5 $H_2O = H_2(18/2) = 9(H–Cl/35.5)$	450
$S + O_2 \rightarrow SO_2$	$SO_2 = S(64/32) = 2(S)$	0
$N_2 = 3.29(O_2)$	N_2 associated with air, $N_2 = (O_2)(3.76 \times 28/32)$	2635
B_n = bound nitrogen	N_2 associated with waste or fuel	0
P_t	$P_t = 3.67(C) + 9(H–Cl/35.5) + 2(S) + 3.29(O_2) + B_n$	3636

2.3. Check the mass input and mass output results.
The results shows that the mass input and mass output are identical.

EXAMPLE 2 weight of air for theoretical combustion
Calculate how many pounds of air are required to completely burn a pound of cellulose ($C_6H_{10}O_5$) [see the previous example in this chapter] (Hasselriis-98/12).

Step 1. Add the nitrogen carried with the theoretical oxygen to both sides.
For 1.18 lb oxygen there will be 1.185(3.31)* or 3.92 lb N_2:

$$1\,\text{lb}\,C_6H_{10}O_5 + 1.185\,\text{lb}\,O_2 + 3.92\,\text{lb}\,N_2 \rightarrow 1.6296\,\text{lb}\,CO_2 + 0.555\,\text{lb}\,H_2O + 3.92\,\text{lb}\,N_2$$

$$1\,\text{lb cellulose} \quad + \quad 5.1\,\text{lb air} \rightarrow \quad 6.1\,\text{lb products}$$

* This value is derived elsewhere in this chapter as 3.29, however 3.31 will be used in Examples 2 and 3.

EXAMPLE 3 weight of air for theoretical combustion
Calculate the products of combustion when 50% excess air is used to burn cellulose.
Add 50% more oxygen and nitrogen to both sides of the equation:

$$1\,\text{lb}\,C_6H_{10}O_5 + 1.185\,\text{lb}\,O_2 + 3.92\,\text{lb}\,N_2 + 0.5\,(1.185)\,\text{lb}\,O_2 + 0.5\,(3.92)\,\text{lb}\,N_2 \rightarrow$$
$$1.63\,\text{lb}\,CO_2 + 0.555\,\text{lb}\,H_2O + 3.92\,\text{lb}\,N_2 + 0.5\,(1.185)\,\text{lb}\,O_2 + 0.5\,(3.92)\,\text{lb}\,N_2$$

$$1\,\text{lb}\,C_6H_{10}O_5 + 1.78\,\text{lb}\,O_2 + 5.88\,\text{lb}\,N_2 \rightarrow 1.63\,\text{lb}\,CO_2 + 0.56\,\text{lb}\,H_2O + 1.78\,\text{lb}\,O_2 + 5.88\,\text{lb}\,N_2$$
$$1\,\text{lb cellulose} + 7.66\,\text{lb air} \rightarrow 8.66\,\text{lb products}$$

This can be written as:

Stoichiometric air:	1 lb cellulose +	5.10 lb air →	6.10 lb products
With 50% excess air:	1 lb cellulose +	7.66 lb air →	8.66 lb products
With 100% excess air:	1 lb cellulose +	10.21 lb air →	11.21 lb products

EXAMPLE 4 volume of air for theoretical combustion
Calculate the volume of air required for complete combustion of methane.

Step 1. Determine and balance the CH_4 and O_2 combustion equation.

$$1\text{ mol} + 2\text{ mol} \rightarrow 1\text{ mol} + 2\text{ mol}$$
$$CH_4 + 2O_2 \rightarrow CO_2 + 2H_2O$$

The equation shows that 1 mole of CH_4 requires 2 moles of O_2. It also shows that 1 scf of CH_4 requires 2 scf of O_2.

Step 2. Calculate the volume of air required.
The oxygen required for combustion carries nitrogen with it. The components of air can be expressed in terms of either moles or standard cubic feet (scf) as follows:

$$\text{Moles:}\quad 0.21O_2 + 0.79N_2 = 1\text{ mole of air, or}$$
$$O_2 + 3.76N_2 = 4.76\text{ mole of air}$$
$$0.21O_2 + 0.79N_2 = 1\text{ scf of air, or}$$
$$O_2 + 3.76N_2 = 4.76\text{ scf of air}$$

For 2 scf of O_2, $2 \times 4.76 = 9.52$ scf of air is needed per scf of CH_4.

EXAMPLE 5 calculate the average molecular weight of the products
Calculate the average molecular weight of the *wet* products of combustion of cellulose at 0% excess air.

Wt. fraction	× Mol. wt.	= weight
CO_2 1.63	× 44	= 71.72
H_2O 0.555	× 18	= 9.99
N_2 3.92	× 28	= 109.76
Sum: 6.105		191.47

Average: sum/wt. = 191.47/6.105 = 31.363 lb/mole

EXAMPLE 6 calculate the average molecular weight of the products
Calculate the average molecular weight of the *dry* products from combustion of cellulose at 0% excess air; subtract the moisture (H_2O).

Wt. fraction × Mol. wt. = weight

CO_2 1.63 × 44 = 71.72
H_2O (0.555) × 18 = –
N_2 3.92 × 28 = 109.76

Total: 5.55 181.48
Average: sum/wt. = 181.48/5.55 = 32.70 lb/mole

EXAMPLE 7 calculate the average molecular weight of the products
Calculate the average molecular weight of the dry products from combustion of cellulose at 50% excess air.

Wt. fraction × Mol. wt.

CO_2 1.63 × 44 = 71.72
H_2O (0.555) × 18 = –
O_2 1.78 × 32 = 56.96
N_2 5.88 × 28 = 164.64

Wt: 9.21 Average: sum/wt. = 293.32/9.21 = 31.85 lb/mole

Deficient (starved) air combustion. Deficient air combustion is also known as starved air combustion. Deficient air combustion results when the supplied combustion air is less than that required for theoretical combustion. Insufficient air will result in incomplete combustion, with emissions of pollutants such as carbon monoxide, solid carbon particulates in the form of smoke or soot, and unburned and/or partially oxidized hydrocarbons. Figure 2.7 shows the phenomenon of starved air combustion in which the organics in the waste are not completely destroyed and are emitted to the atmosphere (EPA-89/03).

Burning carbon with insufficient oxygen can produce CO.

$$C + 0.5O_2 \rightarrow CO \quad (2.10)$$

With additional oxygen the carbon monoxide can be converted to CO_2.

$$CO + 0.5O_2 \rightarrow CO_2 \quad (2.11)$$

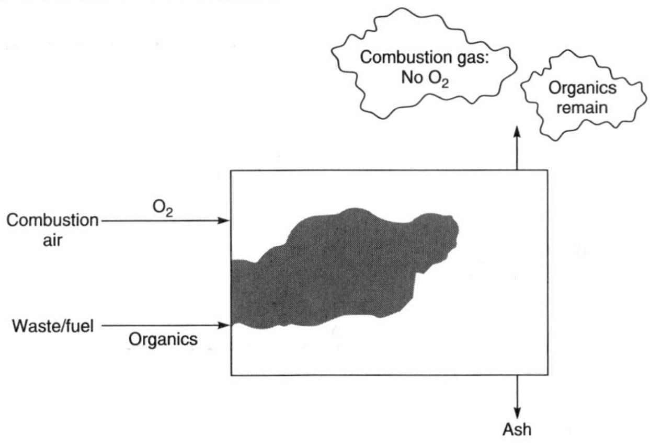

FIGURE 2.7 Starved air combustion (*EPA-89/03*).

Even clean-burning gaseous fuels, such as methane, could produce pollutants when burned with too little oxygen.

$$CH_4 + O_2 \rightarrow C(\text{solid}) + 2H_2O \tag{2.12}$$

The solid carbon particles can agglomerate resulting in smoke and soot. Somewhat more oxygen, but still less than theoretically necessary, could lead to carbon monoxide formation by the following reaction:

$$CH_4 + (1.5)O_2 \rightarrow CO + 2H_2O \tag{2.13}$$

Reactions similar to those represented by Eqs (2.10) and (2.13) can occur in the presence of adequate air if (a) the oxygen is not readily available for the burning process, as a result of inadequate mixing or turbulence, (b) the flame is quenched too rapidly, and/or (c) the residence time is too short. These "three T's of combustion" (i.e., turbulence, temperature, and time) are all interrelated and need to be considered carefully in order to achieve efficient combustion with a minimum of pollutant emissions.

Excess air combustion. Excess combustion air or excess air (EA) is the air supplied in excess of that necessary to burn a compound completely and it appears in the products of combustion. The amount of excess air is normally expressed as a percentage of the theoretical (stoichiometric) air required for complete combustion of the compound. Figure 2.8 shows the concept of excess air combustion in that the excess amount of oxygen passes through the combustion chamber and emits to the atmosphere (EPA-80/02, p. 51-4; EPA-89/03).

The excess air (A_e) (fractional basis) is defined by two methods.

METHOD 1

Theoretical air (stoichiometric air) (A_t) = (actual air entering process (A_a))/(theoretical air (A_t))

$$\%A_t = (A_a/A_t)(100) \tag{2.14}$$

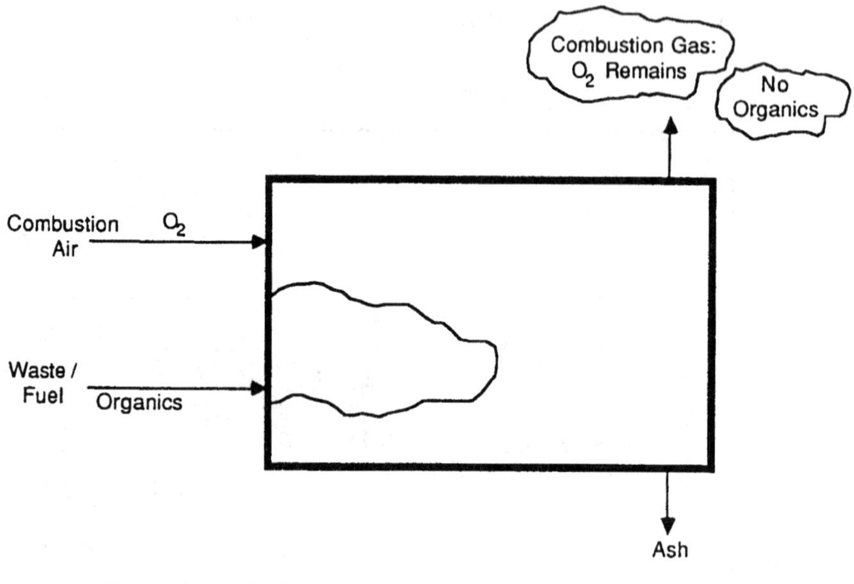

FIGURE 2.8 Excess air combustion (*EPA-89/03*).

Excess air (A_e) = (actual air entering process (A_a) – theoretical air (A_t))/(theoretical air (A_t))

$$\%A_e = [(A_a - A_t)/A_t](100) \tag{2.15}$$

where

- A_a = moles of actual air used in the combustion process
- A_t = moles of theoretical air (stoichiometric air) used at 100% of stoichiometric combustion
- $\%A_t$ = theoretical air
- $\%A_e$ = excess air

METHOD 2
For example, for the combustion of methane, CH_4:

$$CH_4 + 2(1+e)(O_2 + 3.76N_2) \rightarrow CO_2 + 2H_2O + 2(e)O_2 + 2(1+e)(3.76)N_2 \tag{2.16}$$

where

- e = the number of moles of excess O_2 in the excess air

Calculations of reactants and products from excess air combustion. Tables 2.7 and 2.8 were derived from the tables in Section 2.6. The basic steps needed to calculate the quantities of reactants and products are provided. The method can also be used to calculate the reactants and products of theoretical air combustion.

The basic steps needed to perform the calculation of combustion products are provided in Table 2.8, where the mole is the mass divided by its corresponding molecular weight.

TABLE 2.7 Combustion Reactant Table

Basic reaction	Oxygen, nitrogen, and air requirements
$C + O_2 \rightarrow CO_2$	(1) $O_2 = C(32/12)$
$H_2 + 0.5O_2 \rightarrow H_2O$	(2) $O_2 = 8(H\text{–}Cl/35.5)$*
$S + O_2 \rightarrow SO_2$	(3) $O_2 = S(32/32) = S$
B_o = bound oxygen	(4) Data from waste analysis
O_{2t} = theoretical O_2	$O_{2t} = (1) + (2) + (3) - (4)$
N_{2t} = theoretical N_2	$N_{2t} = (O_{2t}(3.76 \times 28/32) = 3.29O_{2t}$
A_t = theoretical air	$A_t = O_{2t} + N_{2t}$
$\%A_e$ = % excess air	Given data
A_e = excess air	$A_e = (\%A_e)(A_t)$
A_{ada} = actual dry air	$A_{ada} = A_t + A_e$
H_2O from air†	$0.0127(A_{ada})$
A_a = actual air	$A_a = A_{ada} + H_2O$
Waste feed	Data from waste analysis
R_a = actual reactants	$R_a = A_a$ + waste feed + H_2O from air

* If H_2 is the hydrogen left over after Cl reaction, i.e., $H_2 = H\text{–}Cl/35.5$, the O_2 needed to react with H_2 will be equal to $H_2(0.5)(32/2) = 8(H\text{–}Cl/35.5)$.
† In general, at room temperature, the humidity ratio is 0.0127 (lb-H_2O)/(lb-dry-air) or 0.0127 (kg-H_2O)/(kg-dry-air).

TABLE 2.8 Combustion Product Table

Component	Combustion product
CO_2	$CO_2 = (C)(44/12)$
SO_2	$SO_2 = S(64/32)$
HCl	$HCl = Cl(36.5/35.5)$
O_{2e}	$(\%A_e)O_{2t}$, O_{2t} from Table 2.7
N_{2e}	$(\%A_e)N_{2t}$, N_{2t} from Table 2.7
N_2	$N_{2e} + N_{2t} +$ from waste
H_2O from comb.	H_2O from combustion $= 9(H–Cl/35.5)$
H_2O	H_2O from combustion + from air + from waste
$G_a =$ actual flue gas	$G_a = CO_2 + SO_2 + HCl + O_{2e} + N_2 + H_2O$
Ash	Fly ash + bottom ash
$P_a =$ actual product*	$G_a +$ ash

* In performing the mass balance calculation, the values of reactants (R_a) and products (P_a) should be equal, i.e., if $R_a = P_a$, the mass balance calculation is correct.

EXAMPLE 1 excess air combustion
Consider the following combustion cases:

	Case
$CH_4 + (2)O_2$	1
$CH_4 + (3)O_2$	2
$CH_4 + (4)O_2$	3

Determine the percent theoretical air and percent excess air for these combustion cases.

Solution:
1. Complete the balance of chemical equations.

$$CH_4 + 2O_2 \rightarrow CO_2 + 2H_2O \tag{2.17}$$

$$CH_4 + 3O_2 \rightarrow CO_2 + 2H_2O + O_2 \tag{2.18}$$

$$CH_4 + 4O_2 \rightarrow CO_2 + 2H_2O + 2O_2 \tag{2.19}$$

The above equations can also be expressed as

$$CH_4 + 2(1)O_2 \rightarrow CO_2 + 2H_2O \tag{2.20}$$

$$CH_4 + 2(1.5)O_2 \rightarrow CO_2 + 2H_2O + O_2 \tag{2.21}$$

$$CH_4 + 2(2)O_2 \rightarrow CO_2 + 2H_2O + 2O_2 \tag{2.22}$$

2. Apply Eq. (2.14), $\%A_t = (A_a/A_t)(100)$.

- $\%A_t$ for case $(1) = 2/2 = 100\%$ theoretical air
- $\%A_t$ for case $(2) = 3/2 = 150\%$ theoretical air
- $\%A_t$ for case $(3) = 4/2 = 200\%$ theoretical air

The above cases can further be expressed as

$$CH_4 + (2)(1 + 0)O_2 \rightarrow CO_2 + 2H_2O \tag{2.23}$$

$$CH_4 + (2)(1 + 0.5)O_2 \rightarrow CO_2 + 2H_2O + O_2 \tag{2.24}$$

$$CH_4 + (2)(1 + 1)O_2 \rightarrow CO_2 + 2H_2O + 2O_2 \tag{2.25}$$

3. Apply Eq. (2.15) to Eqs. (2.23)–(2.25).

$\%A_e = ((A_a - A_t)/A_t)(100)$

Eq. (2.23): "e" $= (2 - 2)/2 = 0\%$ excess air
Eq. (2.24): "e" $= (3 - 2)/2 = 50\%$ excess air
Eq. (2.25): "e" $= (4 - 2)/2 = 100\%$ excess air

4. In summary, apply Eqs. (2.14)–(2.15) to each case.

Eq. (2.17): $CH_4 + (2)O_2 \rightarrow CO_2 + 2H_2O$
$\%A_t = 2/2 = 1 = 100\%$
$\%A_e = (2 - 2)/2 = 0$

Eq. (2.18): $CH_4 + (3)O_2 \rightarrow CO_2 + 2H_2O + O_2$
$\%A_t = 3/2 = 1.5$ or 150%
$\%A_e = (3 - 2)/2 = 0.5$ or 50%
$\%A_e$ can also be expressed by $2(1 + e) = 3$, $e = 0.5$ or 50%

Eq. (2.19): $CH_4 + (4)O_2 \rightarrow CO_2 + 2H_2O + 2O_2$
$\%A_t = 4/2 = 2$ or 200%
$\%A_e = (4 - 2)/2 = 1$ or 100%
$\%A_e$ can also be expressed by $2(1 + e) = 4$, $e = 1$ or 100%

EXAMPLE 2 excess air combustion
Consider the following combustion cases:

$$CH_4 + (2)O_2$$
$$CH_4 + (3)O_2$$

Determine the theoretical air, excess air, and actual combustion air.

Solution:
1. Write the chemical balance equations.

$$CH_4 + 2O_2 + 2(3.76)N_2 \rightarrow CO_2 + 2H_2O + 2(3.76)N_2 \tag{2.26}$$

$$CH_4 + 3O_2 + 3(3.76)N_2 \rightarrow CO_2 + 2H_2O + 3(3.76)N_2 + O_2 \tag{2.27}$$

2. Determine theoretical combustion air A_t.
Equation (2.26) is a theoretical combustion. The air needed for this combustion is the theoretical combustion air A_t.

$$\begin{aligned} A_t &= 2 + 2(3.76) \\ &= 9.52 \text{ mole, or} \end{aligned}$$

$$\begin{aligned} A_t &= 2(32) + 2(3.76)(28) \\ &= 64 + 210.56 \\ &= 274.56 \text{ lb} \end{aligned}$$

3. Determine the excess combustion air A_e.
For Eq. (2.26), $A_e = 0$
Equation (2.27) is an excess air combustion. The % excess air, % A_e, is

$$\begin{aligned}\%A_e &= (A_a - A_t)/A_t \\ &= (3 - 2)/2 \\ &= 50\%\end{aligned}$$

$$\begin{aligned}A_e &= \%A_e(A_t) \\ &= 0.5(9.52) \\ &= 4.76 \text{ mole, or}\end{aligned}$$

$$\begin{aligned}A_e &= 0.50(274.56) \\ &= 137.28 \text{ lb}\end{aligned}$$

4. Determine actual combustion air A_a.
For Eq. (2.26), actual combustion air is equal to its theoretical combustion air. Hence, $A_a = A_t$.
For Eq. (2.27), actual combustion air is

$$\begin{aligned}A_a &= A_t + A_e \\ &= 9.52 + 4.76 \\ &= 14.28 \text{ mole, or}\end{aligned}$$

$$\begin{aligned}A_a &= 274.56 + 137.28 \\ &= 411.84 \text{ lb}\end{aligned}$$

EXAMPLE 3 more excess air combustion
Consider propane gas (C_3H_8) combustion at different levels of air supply.
At 100% theoretical air or 0% excess air combustion,

$$C_3H_8 + 5(1)(O_2 + 3.76N_2) \rightarrow 3CO_2 + 4H_2O + 18.8N_2$$

At 80% theoretical air or 20% deficient air combustion,

$$C_3H_8 + 5(0.8)(O_2 + 3.76N_2) \rightarrow 2CO + CO_2 + 4H_2O + 15.04N_2$$

At 150% theoretical air or 50% excess air combustion,

$$C_3H_8 + 5(1.5)(O_2 + 3.76N_2) \rightarrow 3CO_2 + 4H_2O + 2.5O_2 + 28.2N_2$$

Consequently, 150% of theoretical air is equivalent to 50% excess air. The above reactions show that at 100% A_t (theoretical air) of propane (C_3H_8) combustion, the reaction is a complete combustion, and at both 80% and 150% A_t, the reactions are an incomplete combustion which results in compounds other than H_2O, CO_2 and N_2 in their products.

2.7 Combustion Components and Incineration Components

Combustion process

Fundamentally, for a combustion system, there are four major combustion components. They are:

- (1) fuel
- (2) oxidizer
- (3) heat
- (4) diluent

Incineration process

For an incineration system, there are five major components. They are:

- (1) fuel
- (2) oxidizer
- (3) heat
- (4) diluent
- (5) organic material

Description

FUEL: A fuel is a mixture of hydrocarbons containing energy-rich bonds such as the carbon–carbon and carbon–hydrogen bonds. These hydrocarbons are a common source of chemical potential energy. Some important families of hydrocarbons are summarized in Table 2.9. During combustion, fuels are oxidized and their chemical potential energy is transformed into thermal energy. Relative to their chemical structures, fuels can be composed of the following items.

1. Unsaturated hydrocarbons. Unsaturated hydrocarbons are those compounds which have two or more adjacent carbon atoms joined by a double or triple bond. For example, the molecular structure of H–C=C–H is an unsaturated hydrocarbon.
2. Saturated hydrocarbons. Saturated hydrocarbons are those compounds whose carbon atoms are each joined by a single bond. For example, the molecular structure of H–C–C–H is a saturated hydrocarbon.
3. Isomers. Two hydrocarbons with the same number of carbon and hydrogen atoms but different structures are called isomers. For example, there are several different octanes, the C_8H_{18} series, each having 8 carbon atoms and 18 hydrogen atoms, but each has a different structure; these different structures are called the isomers of octane.

TABLE 2.9 Characteristics of Some of the Hydrocarbon Families

Family	Formula	Structure	Saturated?
Paraffin	C_nH_{2n+2}	Chain	Yes
Olefin	C_nH_{2n}	Chain	No
Diolefin	C_nH_{2n-2}	Chain	No
Naphthene	C_nH_{2n}	Ring	Yes
Aromatic benzene	C_nH_{2n-6}	Ring	No
Aromatic naphthalene	C_nH_{2n-12}	Ring	No

OXIDIZER: An oxidizer is the chemical species that reacts with the fuel or waste compounds during incineration. Its function is to transform the chemical potential energy stored in the fuel into thermal energy or to convert heavy-molecule waste compounds into light, simple compounds such as CO_2, H_2O and HCl. Most commonly, the oxidizer is molecular oxygen, a constituent of air.

The oxygen needed for the combustion reaction is supplied by the ambient combustion air. Combustion air is supplied to the combustion chambers through air ports by a forced draft fan, by an induced draft fan, or by natural draft. In general, this air contains about 21 percent oxygen (O_2) and 79 percent nitrogen (N_2), so about 21 percent of the total combustion air fed to the incinerator is oxygen that is available to react with the organic material in the waste and fuel.

The combustion reaction between the organic material and oxygen that causes the organics to burn occurs only after the temperature of the organic material is raised to the point that combustion can begin. Each specific organic compound has its own temperature at which the reaction occurs, but temperatures in the range of 1000 to 1800°F generally are considered to provide "good combustion conditions." Energy in the form of heat is required to raise the temperatures of the incinerator chamber and organic material and O_2. Initially this energy usually is supplied by the pilot and auxiliary fuel burners. After the system is in full operation, the energy released from the burning waste is often sufficient to maintain these high temperatures.

HEAT: Heat provides the elevated temperatures needed to start and to maintain incineration.

DILUENT: A diluent is a substance that does not participate chemically in the combustion reaction either as a fuel substance or as an oxidizer. It is physically present and often does influence the combustion process. For example, diluents have heat capacity, and while they do not make a positive contribution to the total energy released, they do act as a thermal sink and limit the temperature rise achieved by combustion. A diluent can be thought of as a substance that participates principally in the physical aspects of the combustion process. There are several possible diluents in an incineration system, and some are listed below.

- Nitrogen. Nitrogen, which comprises almost 79% of air, is the most common diluent.
- Excess amount of oxygen. Incineration normally takes place at about 150–200% of the amount of theoretical air needed for combustion. The excess oxygen (the amount over 100% of theoretical air) will act as a diluent. That is, its participation in the combustion process will be physical but not chemical.
- Water contents. Water contained either in combustion air or in the waste feed (or, for that matter, the amount formed during the incineration process).
- Inorganic ash compounds. Inorganic ash compounds such as trace heavy metals present in the waste or in the fuel. The waste feed generally includes inorganic materials which are not involved in the combustion reaction. The inorganic materials in the waste feed are either retained in the ash or are emitted as particulate matter in the combustion gas. Air velocities in the combustion bed are controlled to reduce the amount of inorganic material entrained (picked up) by the combustion gas and emitted with the effluent gas.

ORGANIC MATERIAL: The waste streams that are incinerable include hazardous waste, municipal waste, toxic substances (PCB), medical waste, spent pesticides, sludges from both municipal and industrial wastewater treatment processes, and other unclassified waste such as nonhazardous military waste. The organic material used in the reaction comes from two sources, waste and auxiliary fuel. Some organic material is contained in most medical waste. Depending on the fraction of organics and the specific organic composition, the waste may be adequate to sustain combustion. The other source of organic material is auxiliary fuel. Auxiliary fuel is always used to preheat the incinerator and to start combustion;

auxiliary fuel may be used to maintain combustion if the waste material does not contain enough organic material to maintain high temperatures.

In general, solid waste contains two types of organic materials. They are:

- volatile matter
- fixed carbon

These two types of materials are involved in distinct types of combustion reactions, and the operating variables that control the two types of reaction are different.

VOLATILE MATTER: Volatile matter is the portion of the waste that is vaporized (or evaporated) when the waste is heated. Combustion occurs after the material becomes a gas. The combustion variables that influence this reaction are gas temperature, residence time, and mixing. A minimum temperature is needed to start and sustain the chemical reaction. Residence time is the length of time, generally measured in seconds, that the combustion gas spends in the high temperature combustion chamber. The residence time must be long enough for the reaction to be completed before it leaves the high temperature zone. Turbulent mixing of the volatile matter and combustion air is required to ensure that the organic material and oxygen are well mixed.

FIXED CARBON: Fixed carbon is the nonvolatile organic portion of the waste. For fixed carbon, the combustion reaction is a solid-phase reaction that occurs primarily in the waste bed (although some materials may burn in suspension). Key operating parameters are bed temperature, solids retention time, and mechanical turbulence in the bed. The solids retention time is the length of time that the waste bed remains in the primary chamber. Mechanical turbulence of the bed is needed to expose all the solid waste to oxygen for complete burnout. Without mechanical turbulence, the ash liberated during combustion can cover the unburned waste and prevent the oxygen necessary for combustion from contacting the waste.

2.8 Combustion Constants

The American Gas Association (AGA) constructed a table which was included on page 2-14 of reference EPA-80/02. The table contains various constants of basic combustion. Examples of the constants include the amount of theoretical air and the amount of 150% excess air needed for the combustion of one mole or one pound of methane. The following two calculations show how some of the constants are obtained. Case 1 is for theoretical combustion and case 2 for 150% theoretical air combustion. It is understood that 150% of theoretical combustion air is equivalent to 50% excess air.

Combustion constants: theoretical air combustion

$$\begin{array}{ccccccccccc} CH_4 & + & 2O_2 & + & 2(3.76)N_2 & \rightarrow & CO_2 & + & 2H_2O & + & 2(3.76)N_2 \\ CH_4 & + & 2O_2 & + & 7.52N_2 & \rightarrow & CO_2 & + & 2H_2O & + & 7.52N_2 \\ 16 & & 64 & & 210.56 & & 44 & & 36 & & 210.56 \\ 1 & & 4 & & 13.16 & & 2.75 & & 2.25 & & 13.16 \end{array}$$

(1) Molar analysis (mole/mole-CH_4)

$$CH_4 + 2O_2 + 7.52N_2 \rightarrow CO_2 + 2H_2O + 7.52N_2$$

$CH_4 = 1$; $O_2 = 2$; $N_2 = 7.52$; $CO_2 = 1$; $H_2O = 2$; $N_2 = 7.52$; Air $= 9.52$

(2) Mass analysis (lb/lb-CH_4)

$$\begin{array}{cccccc} CH_4 + 2O_2 + 7.52N_2 & \rightarrow & CO_2 + 2H_2O + 7.52N_2 \end{array}$$

CH_4	$2O_2$	$7.52N_2$	CO_2	$2H_2O$	$7.52N_2$
16	64	210.56	44	36	210.56
1	4	13.16	2.75	2.25	13.16

$CH_4 = 1$; $O_2 = 4$; $N_2 = 13.16$; $CO_2 = 2.75$; $H_2O = 2.25$; $N_2 = 13.16$; Air $= 17.16$

Combustion constants: excess air combustion

Write the chemical balance equation with 150% theoretical combustion air.

$$CH_4 + 2(1.50)(O_2 + 3.76N_2) \rightarrow CO_2 + 2H_2O + 11.28N_2 + O_2$$

(1) Molar analysis (mole/mole-CH_4)

$$CH_4 + 3O_2 + 11.28N_2 \rightarrow CO_2 + 2H_2O + 11.28N_2 + O_2$$

$CH_4 = 1$; $O_2 = 3$; $N_2 = 11.28$; $CO_2 = 1$; $H_2O = 2$; $N_2 = 11.28$; Residual $O_2 = 1$; Air $= 14.28$

(2) Mass analysis (lb/lb-CH_4)

$$CH_4 + 3O_2 + 11.28N_2 \rightarrow CO_2 + 2H_2O + 11.28N_2 + O_2$$

CH_4	$3O_2$	$11.28N_2$	CO_2	$2H_2O$	$11.28N_2$	O_2
16	96	315.84	44	36	315.84	32
1	6	19.74	2.75	2.25	19.74	2

$CH_4 = 1$; $O_2 = 6$; $N_2 = 19.74$; $CO_2 = 2.75$; $H_2 = 2.25$; $N_2 = 19.74$; Residual $O_2 = 2$; Air $= 25.74$

The same approach can be used to calculate the data from the combustion of the following compounds:

Paraffin series

- ethane, C_2H_6
- propane, C_3H_8
- *n*-butane, C_4H_{10}
- isobutane, C_4H_{10}
- *n*-pentane, C_5H_{12}
- isopentane, C_5H_{12}
- neopentane, C_5H_{12}
- *n*-hexane, C_6H_{14}

Olefin series

- ethylene, C_2H_4
- propylene, C_3H_6
- *n*-butene, C_4H_8
- isobutene, C_4H_8
- *n*-pentene, C_5H_{10}

Aromatic series

- benzene, C_6H_6
- toluene, C_7H_8
- xylene, C_8H_{10}

2.9 Combustion Correction Factor

For comparison purposes, it is sometimes necessary to conect a measured value of a compound's concentration at the stack to a certain desired concentration. The way to make such a correction follows.

The correction factor (CF) for oxygen is defined as

$$\text{CF} = (21 - \text{desired } O_2)/(21 - \text{measured } O_2)$$

EXAMPLE 1 combustion correction factor
Consider the following combustion cases:

$$CH_4 + 2(O_2 + 3.76N_2) \rightarrow CO_2 + 2H_2O + 7.52N_2 \tag{2.28}$$

$$CH_4 + 2(1.5)(O_2 + 3.76N_2) \rightarrow CO_2 + 2H_2O + O_2 + 11.28N_2 \tag{2.29}$$

Equation (2.28) depicts combustion at 0% excess air, and Eq. (2.29) shows 150% of theoretical air combustion.

Determine how CO_2 measured at 150% of theoretical air combustion (50% excess air) is converted to 0% excess air combustion.

Solution:
O_2 measured at the wet condition (with H_2O) of Eq. (2.29) $= 1/(1 + 2 + 1 + 11.28) = 6.5\%$.
O_2 measured at the dry condition (without H_2O) of Eq. (2.29) $= 1/(1 + 1 + 11.28) = 7.5\%$.
To correct the measured (dry) O_2 value to a desired level (0% O_2 in products):

$$\begin{aligned}\text{CF} &= (21 - 0)/(21 - 7.5)\\ &= 1.55\end{aligned}$$

CO_2 measured at the dry condition of Eq. (2.29) $= 1/(1 + 1 + 11.28) = 7.5\%$
To correct the measured CO_2 to a desired O_2 level of 0% excess air (i.e., at the theoretical air condition):

$$C = \text{CF} \times C_m$$

where C = corrected parameter value at the desired O_2 level
C_m = measured parameters such as CO or CO_2 which are required to be corrected to a desired O_2 level

For this example,

$$\begin{aligned}CO_2 \text{ (at 0\% } O_2 \text{ level)} &= \text{CF} \times (CO_2)_m\\ &= 1.55 \times 7.5\%\\ &= 11.625\%\end{aligned}$$

This number can be checked with the dry CO_2 level from the theoretical air combustion of CH_4 from Eq. (2.28). That is,

$$CO_2 = 1/(1 + 7.52) = 11.73\%.$$

The discrepancy between the 11.625% and 11.73% values is probably due to rounding off certain values; in any case, the values are very close.

EXAMPLE 2 combustion correction factor

This example is to show the conversion between two correction factors. Assume that CO concentration at 7% O_2 is 150 ppm. Determine CO concentration at 10% O_2 level.

Solution:
CO concentration at 7% O_2 is

$$(CO)_7 = ((21 - 7)/(21 - O_m))(CO)_m \tag{2.30}$$

where subscript m refers to the measured value.
CO concentration at 10% O_2 is

$$(CO)_{10} = ((21 - 10)/(21 - O_m))(CO)_m \tag{2.31}$$

From the ratio of these two equations:

$$\begin{aligned}(CO)_{10} &= (11/14) \times (CO)_7 \\ &= 0.7857(150) \\ &= 117.86 \text{ ppm}\end{aligned}$$

EXAMPLE 3 combustion correction factor
Consider the following two equations. Equation (2.32) is a theoretical air combustion of carbon tetrachloride (CCl_4) with methane (CH_4), and Eq. (2.33) is a 150% of theoretical air combustion of the same carbon tetrachloride with methane.

$$CH_4 + CCl_4 + 2(O_2 + 3.76N_2) \rightarrow 2CO_2 + 4HCl + 7.52N_2 \tag{2.32}$$

$$CH_4 + CCl_4 + 2(1.5)(O_2 + 3.76N_2) \rightarrow 2CO_2 + 4HCl + O_2 + 11.28N_2 \tag{2.33}$$

Determine:

1. Correction factor
2. Corrected CO_2 value
3. Prove that the corrected CO_2 value under 150% theoretical air combustion in Eq. (2.33) is equal to the CO_2 molar fraction in Eq. (2.32) for a theoretical air combustion

Solution for question (1)

1.1. Determine O_2 molar fraction ($(O_2)_m$) in Eq. (2.33)

$$\begin{aligned}(O_2)_m &= 1/(2 + 4 + 1 + 11.28) \\ &= 0.0547\end{aligned}$$

1.2. Calculate correction factor (CF)

CF for correcting the measured O_2 in Eq. (2.33) to a 0% excess air is

$$\begin{aligned}CF &= (0.21 - 0)/(0.21 - 0.0547) \\ &= 1.35\end{aligned}$$

Solution for question (2)

2.1. Determine CO_2 molar fraction $((CO_2)_m)$ in Eq. (2.33)

$$(CO_2)_m = 2/(2 + 4 + 1 + 11.28) = 0.11$$

2.2. Calculate the corrected CO_2

$$\begin{aligned} CO_2 \text{ (at 0\% } O_2 \text{ level)} &= CF \times (CO_2)_m \\ &= 1.35 \times 0.11 \\ &= 0.15 \end{aligned} \tag{2.34}$$

Solution for question (3)

3.1. Determine CO_2 molar fraction $((CO_2)_m)$ in Eq. (2.32)

$$\begin{aligned} (CO_2)_m &= 2/(2 + 4 + 7.52) \\ &= 0.15 \end{aligned} \tag{2.35}$$

The results in Eqs. (2.34) and (2.35) are identical. This means that the corrected CO_2 value under 50% excess air combustion in Eq. (2.33) is equal to the CO_2 molar fraction in Eq. (2.32) for a theoretical air combustion.

2.10 Combustion Efficiency

Definition

In fossil fuel combustion or solid waste incineration, carbon monoxide (CO) represents one of the most stable products of incomplete combustion. It is therefore used to indicate the extent of incomplete combustion in the combustion system. Combustion efficiency (CE) is defined as

$$CE = CO_2/(CO_2 + CO)$$

where CO_2 = carbon dioxide
CO = carbon monoxide

Equation (2.32) shows that if $CO = 0$, the combustion efficiency would be equal to 100% combustion. Thus, the less CO that is produced, the better the combustion efficiency is. In general, the concentration of CO and CO_2 is obtained from a continuous emission monitor (CEM). The concentration often has a unit of "parts per million by volume, or ppm by weight."

EXAMPLE combustion efficiency

Consider that propane gas (C_3H_8) is under combustion at 80% theoretical air, or 20% deficient air supply.

$$C_3H_8 + 5(0.8)(O_2 + 3.76N_2) \rightarrow 2CO + CO_2 + 4H_2O + 15.04N_2$$

Determine the combustion efficiency

Solution 1: By the volume of product species

Because the volume of product species is proportional to the coefficient of each species in the chemical reaction equation, the combustion efficiency of the propane gas can be expressed as

$$\begin{aligned} \text{CE} &= (\text{coefficient of } CO_2)/(\text{coefficient of } CO_2 + \text{coefficient of CO}) \\ &= 1/(1+2) \\ &= 33\% \text{ (by volume)} \end{aligned}$$

Solution 2: By the mass of product species

If the mass of product species is used to calculate combustion efficiency, CE can be expressed as

$$\begin{aligned} \text{CE} &= (\text{mass of } CO_2)/(\text{mass of } CO_2 + \text{mass of CO}) \\ &= (12+32)/((12+32)+2(12+16)) \\ &= 44/(44+56) \\ &= 44/100 \\ &= 44\% \text{ (by mass)} \end{aligned}$$

Factors affecting combustion performance

The factors that may affect the combustion performance include:

- combustion air
- combustion temperature
- waste characteristics

COMBUSTION AIR: Combustion air is usually the ambient air associated with the oxygen needed for burning a fuel or waste materials. The two key questions about combustion air are: (1) How much combustion air is needed to sustain the combustion reaction? and (2) What happens if there is too much or too little combustion air?

Ambient air is a typical source of oxygen for combustion, and is considered as an ideal gas in many incineration calculations. It is, of course, a mixture of oxygen, nitrogen, and small amounts of water vapor, carbon dioxide, argon, and other elements. For the purposes of combustion calculations, the last four items are usually included with the nitrogen.

In a chemical reaction between organic materials and oxygen, the amount of oxygen required under ideal or "perfect" conditions to burn all of the organic materials into CO_2 and H_2O with no oxygen left over is called the *stoichiometric* (or *theoretical*) *oxygen level* and the process is called *complete combustion*. The amount of combustion air associated with that oxygen level is called the stoichiometric air level. Air flows greater than those required at stoichiometric levels are called the *excess air levels*, and airflows less than those required at stoichiometric levels are called the *deficient air* or *substoichiometric starved air levels*. Typically, an incinerator operates with an overall 150–200 percent of the stoichiometric air level. That is, the incinerator operates with about 1.5-2 times more air than is required at stoichiometric levels. Excess air is used to assure that enough oxygen is available for complete combustion.

Computation of exact stoichiometric air requirements for a waste incinerator is difficult because it depends on the chemical composition of the waste and fuel. However, stoichiometric air requirements can be estimated on the basis of the energy input (or heat input) to the incinerator. Heat input is a measure of the energy released when the waste and fuel are burned. It is measured in British thermal units (Btu's). Generally about 1 standard cubic foot of combustion air is required per 100 Btu's of heat input to the incinerator.

Maximum combustion temperatures are always attained at stoichiometric conditions. As the amount of excess air is increased above the stoichiometric point, the temperature in the incinerator drops because energy is used to heat the extra combustion air. If the amount of combustion air is too great, the temperature can drop below "good combustion temperature," and undesirable combustion products are generated as a result of incomplete combustion. As the amount of excess air is decreased, the combustion temperature increases until it becomes maximum at the stoichiometric point. Below the stoichiometric point, the temperature decreases because complete combustion has not occurred. A graphical representation of the relationship between combustion temperature and excess air level is shown in Fig. 2.9 (EPA-89/03). At air levels below the stoichiometric point, some of the organic compounds are not reacted, and pollutants are emitted as a result of incomplete combustion.

COMBUSTION TEMPERATURE: Temperature plays an important role in the combustion of solid waste. Temperatures should be maintained at the level specified by combustor manufacturers to sustain the combustion reaction. However, temperatures that are too high may cause problems. Continuous operation at high temperatures is generally not desirable because it can cause the ash to fuse and cause thermal damage to the refractory.

WASTE CHARACTERISTICS: The primary characteristics of the waste that affects the combustion reaction are:

- heating value
- chlorine content
- ultimate composition
- moisture content

(1) Heating value

The heating value of a waste is a measure of the energy released when the waste is incinerated. It is measured in units of Btu/lb (1 Btu/lb = 2.324 J/g). In general, a heating value of about 5000 Btu/lb (11,620 J/g) or greater is needed to sustain combustion. Wastes with lower heating values can be burned, but they will not maintain adequate temperature without the addition of auxiliary fuel. The heating value of the waste is needed to calculate total heat input to the incinerator, where

$$\text{Heat input (Btu/h)} = \text{Feed rate (lb/h)} \times \text{Heating value (Btu/lb)}$$

The heating value of waste varies from nearly zero for some pathological wastes and other materials with very high moisture contents to around 15,000 Btu/lb (35,000 J/g) for plastic packaging materials.

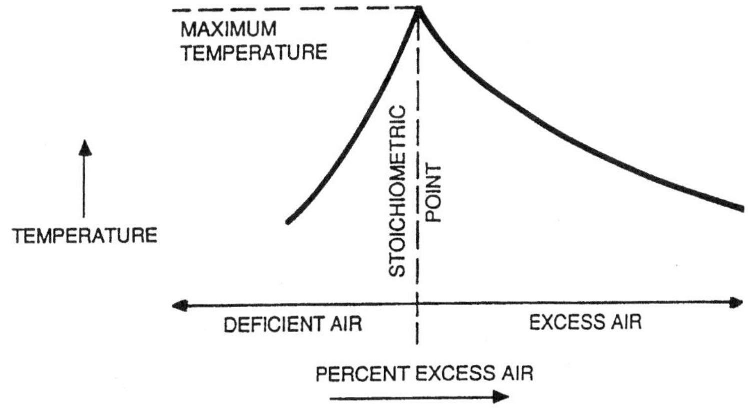

FIGURE 2.9 Relationship of combustion temperature and excess air.

High heating value (HHV)

HHV is the heating value produced from the combustion of a waste under the condition that the water vapor in the combustion gas has condensed into the liquid state. Any fuel containing hydrogen yields water as one product of combustion. At atmospheric pressure, the partial pressure of the water vapor in the resulting combustion gas mixture is usually sufficiently high to cause water to condense out if the temperature of the exhaust gas is allowed to fall below 120–140°F. This causes liberation of the heat of vaporization of any water condensed. Thus, HHV is the heat transfer with liquid water in the products. HHV is also referred to as higher heating value or gross heating value. It is usually expressed as Btu/lb fuel or Btu/scf fuel (Lee-89/09).

Low heating value (LHV)

Similar to HHV, LHV is the heating value produced from the combustion of a waste under the condition that the water vapor in the combustion gas is not condensed into the liquid state. It is also referred to as the lower heating value or net heating value (Lee-89/09).

(2) Chlorine content

Most chlorine in plastics or solvents in the waste feed react with hydrogen during incineration to form hydrochloric acid (HCl), which mostly requires a downstream scrubbing system to remove it. This HCl is an emission problem which can create corrosion problems downstream from the incinerator.

(3) Ultimate composition

The ultimate composition of a compound is the chemical elements in the compound such as C, O, H, Cl, S, etc. The ultimate composition provides information relative to (1) the quantity of combustion air needed during incineration, and (2) the quantity of ash to be produced after incineration.

(4) Moisture content

Moisture is evaporated from the waste as the temperature of the waste is raised in the combustion chamber; it passes through the incinerator, unchanged, as water vapor. This evaporation of moisture uses energy and reduces the temperature in the combustion chamber. The water vapor also increases the combustion gas flow rate, which reduces combustion gas residence time.

2.11 Combustion of Chlorinated Compounds

Compounds containing carbon and chlorine rather than hydrogen require a source of hydrogen in order to attain complete destruction (hydrogen is used to react with chlorine for chemical balance). For example, methane gas (CH_4) can be used to completely destroy (incinerate) carbon tetrachloride (CCl_4) in an incinerator.

Consider the combustion of a mixture of methane and carbon tetrachloride in air. The balanced chemical reaction for the combustion is

$$\begin{array}{llllll} CH_4 + & CCl_4 + & 2(O_2 + 3.76N_2) \rightarrow & 2CO_2 + & 4HCl + & 7.52N_2 + \text{heat of combustion} \\ 16 & 154 & 275 & 88 & 146 & 211 \\ 0.10 & 1 & 1.78 & 0.57 & 0.95 & 1.37 \end{array}$$

The result shows that 1 lb CCl_4 requires 0.1 lb CH_4, and 1.78 lb air for complete combustion. In other words, one mole of methane reacts with one mole of carbon tetrachloride and two moles of oxygen to form two moles of carbon dioxide and four moles of HCl. The nitrogen does not participate in the main combustion reaction. It is carried along with the oxygen. The coefficient 7.52 for nitrogen is obtained by multiplying the number of moles of oxygen participating in the chemical reaction by the volumetric ratio of nitrogen and oxygen in air (79/21).

Under standard temperature and pressure (STP) conditions (°C and 1 atm), 1 lb-mole occupies a standard volume of 359 ft^3 and 1 g-mole 0.0224 m^3.

- The combustion requires 2 moles of air. If the calculations are performed in lb-moles, the combustion will require 2(359) = 718 scf (standard cubic feet).
- The combustion produces (2 + 4 + 7.52) = 13.52 moles of flue gas. If the calculations are performed in lb-moles, the combustor will produce 13.52(359) = 4854 scf of flue gas.
- If the calculation were performed in metric units and the equations were read as g-moles, the 13.52 g-moles of flue gas would occupy 0.303 m^3 at STP.

The air volume can also be computed by applying the ideal gas law. For this example, the above equation shows that 2 moles of air are needed for the complete incineration of CCl_4.

Assume that the standard conditions are $P = 1$ atm and $T = 32°$F; the volume of air can be obtained by applying the ideal gas equation:

$$PV = nR_u T$$

where
P = pressure
V = volume
n = moles
R_u = universal gas constant
T = absolute temperature

$$\begin{aligned} V &= nR_u T/P \\ &= 2(1544)(32 + 460)/[14.6959(144)] \\ &= 718 \text{ scf (standard cubic feet)} \end{aligned}$$

The air volume (718 scf) that was calculated from the ideal gas equation and the air volume (718 scf) that was calculated from the standard volume are identical.

Mass balances on more complex systems are performed in the same way. No matter how complex the system is, the calculations can be based on the elemental feed rates of the streams.

In this example, chlorine consumes the hydrogen preventing its reaction with oxygen to form water. The thermodynamics of HCl formation will always result in chlorine consuming the hydrogen to form HC1 before any of the hydrogen forms water. In addition, the only time that more than trace quantities of chlorine gas (Cl_2) will be formed is when insufficient hydrogen of any form is present in the combustion chamber—a virtual impossibility in an actual combustor.

2.12 More Examples of Basic Combustion

EXAMPLE 1 basic combustion
Propane (C_3H_8) is burned with 100 percent theoretical air. Calculate:

(1) air/fuel ratio
(2) molar analysis of the products of combustion
(3) dew point of the products for a total pressure of 14.696 psia
(4) volume of air required to burn gaseous propane to completion, if the ambient air temperature is 80°F
(5) volume of air per pound of fuel on the basis of per pound of fuel
(6) volume of air per pound of fuel on the basis of per mole of fuel

Given conditions
The data in question are:

- fuel: propane
- total pressure: 14.696 psia
- temperature: 80°F

Solution:
Solution for question (1)
1.1. Write the theoretical air combustion equation.

$$C_3H_8 + 5O_2 + 5(3.76)N_2 \rightarrow 3CO_2 + 4H_2O + 5(3.76)N_2$$

1.2. Write the equation for mass calculation.

$$m = nM$$

where m = mass
n = number of moles
M = molecular weight

1.3. Calculate air mass (m_a) and fuel mass (m_f).

$$m_f = 1(44) = 44$$
$$m_a = 5(32 + (3.76)(28)) = 686.4$$

1.4. Calculate the air/fuel ratio (mass base).

$$AF = 686.4/44 = 15.6$$

1.5. Calculate the air/fuel ratio (mole base).

$$AF = 5/1 = 5$$

Solution for question (2)
2.1. Calculate the total number of moles of products.

$$n = 3 + 4 + 5(3.76)$$
$$= 25.8 \text{ mole/(mole-fuel)}$$

2.2. Calculate the mole fraction of product components.

$$y(CO_2) = 3/25.8 = 0.1163$$
$$y(H_2O) = 4/25.8 = 0.1550$$
$$y(N_2) = (5)(3.76)/25.8 = 0.7287$$

The sum of y values = 1.

Solution for question (3)
3.1. Write the partial pressure equation for H_2O.

$$p(H_2O) = y(H_2O)p(\text{total})$$
$$= (0.1550)(14.696)$$
$$= 2.28 \text{ psia}$$

3.2. Find the saturation temperature (from a steam table).

- The temperature corresponding to 2.28 psia is the dew point temperature, which is roughly 138°F. Thus moisture would condense if the products were cooled below 138°F.
- This explains, of course, why water drips from the tailpipe of an automobile on a cold morning before the exhaust system has warmed up.

Solution for question (4)

4.1. Write the ideal gas equation.

$$V = nR_u T/p$$
$$= (m/M)R_u T/p$$
$$v = V/m$$
$$= R_u T(pM)$$

4.2. Calculate the specific volume of air.

$$v = (1544)(460 + 80)/((14.696)(144)(29))$$
$$= 13.59 \text{ ft}^3/(\text{lb-air})$$

Solution for question (5)

Because the air/fuel ratio (mass basis) is 15.6, the volume of air required per pound of fuel is

$$V = (13.59)(15.6)$$
$$= (212.00 \text{ ft}^3/(\text{lb-fuel})$$

Solution for question (6)

$$V = (212.00)(\text{ft}^3/(\text{lb-fuel}))(44)(\text{lb-fuel})/(\text{mole-fuel})$$
$$= 9328 \text{ ft}^3/(\text{mole-fuel})$$

EXAMPLE 2 basic combustion

Propane (C_3H_8) is burned with 150 percent theoretical air. Calculate:

(1) the air/fuel ratio
(2) molal analysis of the products of combustion
(3) the dew point of the products for a total pressure of 14.696 psia
(4) the volume of air required to burn gaseous propane to completion, if the ambient air temperature is 80°F
(5) volume of air per pound of fuel on the basis of per pound of fuel
(6) volume of air per pound of fuel on the basis of per mole of fuel

Given conditions

The data in question are:

- fuel: propane
- total pressure: 14.696 psia
- temperature: 80°F
- excess air: 50 percent

Solution:

Solution for question (1)

1.1. Write the theoretical air combustion equation.

$$C_3H_8 + 5O_2 + 5(3.76)N_2 \rightarrow 3CO_2 + 4H_2O + 5(3.76)N_2$$

1.2. Write the 150% theoretical air combustion equation.

$$C_3H_8 + 1.5(5)O_2 + 1.5(5)(3.76)N_2 \rightarrow 3CO_2 + 4H_2O + 1.5(5)(3.76)N_2 + 0.5(5)O_2$$

1.3. Write the equation for mass calculation.

$$m = nM$$

where m = mass
n = number of moles
M = molecular weight

1.4. Calculate air mass (m_a) and fuel mass (m_f).

$$m_f = 1(44) = 44$$

$$m_a = 1.5(5)(32 + (3.76)(28)) = 1029.6$$

1.5. Calculate the air/fuel ratio (mass base).

$$\mathrm{AF} = 1029.6/44 = 23.4$$

1.6. Calculate the air/fuel ratio (mole base).

$$\mathrm{AF} = 1.5(5)/1 = 7.5$$

Solution for question (2)

2.1. Calculate the total number of moles of products.

$$n = 3 + 4 + (1.5)(5)(3.76) + 0.5(5) = 37.7 \text{ mole/(mole-fuel)}$$

2.2. Calculate the mole fraction of product components.

$$\begin{aligned} y(CO_2) &= 3/37.7 \\ &= 0.0796 \end{aligned}$$

$$\begin{aligned} y(H_2O) &= 4/37.7 \\ &= 0.1061 \end{aligned}$$

$$\begin{aligned} y(O_2) &= (0.5)(5)/37.7 \\ &= 0.0663 \end{aligned}$$

$$\begin{aligned} y(N_2) &= (1.5)(5)(3.76)/37.7 \\ &= 0.7480 \end{aligned}$$

The sum of mole fractions equals 1, i.e., $y(CO_2) + y(H_2O) + y(O_2) + y(N_2) = 1$

Solution for question (3)

3.1. Write the partial pressure equation for H_2O.

$$\begin{aligned} p(H_2O) &= y(H_2O)p(\text{total}) \\ &= 0.1061(14.696) \\ &= 1.56 \text{ psia} \end{aligned}$$

3.2. Find the saturation temperature corresponding to 1.56 psia (from a steam table). Dew point temperature is roughly 117°F

Solution for question (4)

4.1. Write the ideal gas equation.

$$\begin{aligned} V &= nR_u T/p \\ &= (m/M)R_u T/p \end{aligned}$$

$$\begin{aligned} v &= V/m \\ &= R_u T/(pM) \end{aligned}$$

4.2. Calculate the specific volume of air.

$$\begin{aligned} v &= 1544(460 + 80)/((14.696)(144)(29)) \\ &= 13.59 \text{ ft}^3/(\text{lb-air}) \end{aligned}$$

Solution for question (5)

Because the air/fuel ratio (mass basis) is 23.4, the volume of air required per pound of fuel is

$$\begin{aligned} V &= 13.59(23.4) \\ &= 318.01 \text{ ft}^3/(\text{lb-fuel}) \end{aligned}$$

Solution for question (6)

$$\begin{aligned} V &= 318.01(\text{ft}^3/(\text{lb-fuel}))(44)(\text{lb-fuel})/(\text{mole-fuel}) \\ &= 13{,}992 \text{ ft}^3/(\text{mole-fuel}) \end{aligned}$$

EXAMPLE 3 basic combustion
Propane (C_3H_8) is burned with 80 percent theoretical air. Calculate:

(1) the air/fuel ratio
(2) molal analysis of the products of combustion
(3) the dew point of the products for a total pressure of 14.696 psia.
(4) the volume of air required to burn gaseous propane to completion, if the air temperature is 80°F
(5) volume of air per pound of fuel on the basis of per pound of fuel
(6) volume of air per pound of fuel on the basis of per mole of fuel

Given conditions
The data in question are:

- fuel: propane
- total pressure: 14 696 psia
- temperature: 80°F
- theoretical air: 0.8

Solution:
Solution for question (1)
1.1. Write the theoretical air combustion equation.

$$C_3H_8 + 5O_2 + 5(3.76)N_2 \rightarrow 3CO_2 + 4H_2O + 5(3.76)N_2$$

1.2. Write the 80% of theoretical air combustion equation.

$$C_3H_8 + 0.8(5)O_2 + 0.8(5)(3.76)N_2 \rightarrow (1)CO_2 + (2)CO + 4H_2O + 0.8(5)(3.76)N_2$$

1.3. Write the mass balance equation for each component.

For carbon: $3 = a + b$
For oxygen: $8 = 2a + b + 4$

The solution to these two equations is:

$a = 1$
$b = 2$

1.4. Write the 80% theoretical air combustion equation.

$$C_3H_8 + 0.8(5)O_2 + 0.8(5)(3.76)N_2 \rightarrow CO_2 + 2CO + 4H_2O + 0.8(5)(3.76)N_2$$

1.5. Write the equation for mass calculation.

$$m = nM$$

where m = mass
n = number of moles
M = molecular weight

1.6. Calculate air mass (m_a) and fuel mass (m_f).

$$\begin{aligned} m_f &= 1(44) \\ &= 44 \\ m_a &= 0.8(5)(32 + (3.76)(28)) \\ &= 549.12 \end{aligned}$$

1.7. Calculate the air/fuel ratio (mass base).

$$\begin{aligned} AF &= 549.12/44 \\ &= 12.48 \end{aligned}$$

1.8. Calculate the air/fuel ratio (mole base).

$$\begin{aligned} AF &= (0.8)(5)/1 \\ &= 4 \end{aligned}$$

Solution for question (2)

2.1. Calculate the total number of moles of products.

$$\begin{aligned} n &= 1 + 2 + 4 + (0.8)(5)(3.76) \\ &= 22.04 \text{ mole/(mole-fuel)} \end{aligned}$$

2.2. Calculate the mole fraction of product components.

$$\begin{aligned} y(CO_2) &= 1/22.04 \\ &= 0.0454 \\ y(CO) &= 2/22.04 \\ &= 0.0907 \\ y(H_2O) &= 4/22.04 \\ &= 0.1815 \\ y(N_2) &= 0.8(5)(3.76)/22.04 \\ &= 0.6824 \end{aligned}$$

The sum of mole fractions equals 1, i.e., $y(CO_2) + y(CO) + y(H_2O) + y(N_2) = 1$

Solution for question (3)

3.1. Write the partial pressure equation for H_2O.

$$\begin{aligned} p(H_2O) &= y(H_2O)p(\text{total}) \\ &= 0.1815(14.696) \\ &= 2.67 \text{ psia} \end{aligned}$$

3.2. Find the saturation temperature corresponding to 2.67 psia (from a steam table). Dew point temperature is roughly 129.41°F.

Solution for question (4)

4.1. Write the ideal gas equation.

$$\begin{aligned} V &= nR_u T/p \\ &= (m/M)R_u T/p \\ v &= V/m \\ &= R_u T/(pM) \end{aligned}$$

4.2. Calculate the specific volume of air.

$$\begin{aligned} v &= 1544(460 + 80)/((14.696)(144)(29)) \\ &= 13.59 \text{ ft}^3/(\text{lb-air}) \end{aligned}$$

Solution for question (5)

Because the air/fuel ratio (mass basis) is 12.48, the volume of air required per pound of fuel is

$$\begin{aligned} V &= 13.59(12.48) \\ &= 169.60 \text{ ft}^3/(\text{lb-fuel}) \end{aligned}$$

Solution for question (6)

$$\begin{aligned} V &= (169.60)[\text{ft}^3/(\text{lb-fuel})](44)(\text{lb-fuel})/(\text{mole-fuel}) \\ &= 7462.4 \text{ ft}^3/(\text{mole-fuel}) \end{aligned}$$

EXAMPLE 4 basic combustion

Propane (C_3H_8) is burned with 150 percent theoretical air. Air has an initial relative humidity of 60%. Calculate:

(1) the air/fuel ratio

(2) molal analysis of the products of combustion

(3) the dew point of the products for a total pressure of 14.696 psia

(4) the volume of air required to burn gaseous propane to completion, if the air temperature is 80°F

(5) volume of air per pound of fuel on the basis of per pound of fuel

(6) volume of air per pound of fuel on the basis of per mole of fuel

Given conditions

The data in question are:

- fuel: propane
- total pressure: 14.696 psia
- temperature: 80°F
- excess air: 50 percent
- relative humidity: 60%

Solution:

Solution for question (1)

1.1. Write the theoretical air combustion equation.

$$C_3H_8 + 5O_2 + 5(3.76)N_2 \rightarrow 3CO_2 + 4H_2O + 5(3.76)N_2 + (HR)(\text{air mass})^*$$

where

- humidity ratio (HR) at temperature = 80°F
- relative humidity = 60% from psychrometric chart, HR = 0.01315 (lb-water)/(lb-dry-air)*

1.2. Write the 150% of theoretical air combustion equation.

$$C_3H_8 + 1.5(5)O_2 + 1.5(5)(3.76)N_2 \rightarrow 3CO_2 + 4H_2O + 1.5(5)(3.76)N_2 + 0.5(5)O_2 + (0.01315)(\text{air mass})$$

1.3. Write the equation for mass calculation.

$$m = nM$$

where m = mass
n = number of moles
M = molecular weight

1.4. Calculate air mass (m_a) and fuel mass (m_f).

$$m_f = 1(44) = 44$$

$$\text{Dry air mass} = 1.5(5)(32 + (3.76)(28)) = 1029.6 \text{ lb}$$

$$\text{Water vapor associated with dry air} = 0.01315(1029.6) = 13.54(\text{lb-water})$$

$$\text{Moles of water} = 13.54/18 = 0.7522(\text{mole-water})$$

Actual air mass, m_a

$$m_a = 1029.6 + 13.54 = 1043.14$$

1.5. Calculate the air/fuel ratio (mass base).

$$AF = 1043.14/44 = 23.71$$

1.6. Calculate the air/fuel ratio (mole base).

$$AF = 1.5(5)/1 = 7.5$$

*Theoretically, the (HR) (air mass) term should appear on both sides of the equation.

Solution for question (2)
2.1. Calculate the total number of moles of products.

$$n = 3 + 4 + (1.5)(5)(3.76) + 0.5(5) + 0.7522$$
$$= 38.45 \text{ mole/(mole-fuel)}$$

2.2. Calculate the mole fraction of product components.

$$y(CO_2) = 3/38.45$$
$$= 0.0780$$
$$y(H_2O) = (4 + 0.7522)/38.45$$
$$= 0.1236$$
$$y(O_2) = 0.5(5)/38.45$$
$$= 0.0650$$
$$y(N_2) = (1.5)(5)(3.76)/38.45$$
$$= 0.7334$$

The sum of mole fractions equals 1, i.e., $y(CO_2) + y(H_2O) + y(O_2) + y(N_2) = 1$

Solution for question (3)
3.1. Write the partial pressure equation for H_2O.

$$p(H_2O) = y(H_2O)p(\text{total})$$
$$= 0.1236(14.696)$$
$$= 1.82 \text{ psia}$$

3.2. Find the saturation temperature corresponding to 1.82 psia (from a steam table). Dew point temperature is roughly 129°F.

Solution for question (4)
4.1. Write the ideal gas equation.

$$V = nR_u T/p$$
$$= (m/M)R_u T/p$$
$$v = V/m$$
$$= R_u T/(pM)$$

4.2. Calculate the specific volume of air.

$$v = 1544(460 + 80)/((14.696)(144)(29))$$
$$= 13.59 \text{ ft}^3/(\text{lb-air})$$

Solution for question (5)
Because the air fuel ratio (mass basis) is 23.71, the volume of air required per pound of fuel is

$$V = 13.59(23.71)$$
$$= 322.21 \text{ ft}^3/(\text{lb-fuel})$$

Solution for question (6)

$$V = 322.21[\text{ft}^3/(\text{lb-fuel})](44)(\text{lb-fuel})/(\text{mole-fuel})$$
$$= 14{,}177 \text{ ft}^3/(\text{lb-fuel})$$

EXAMPLE 5 basic combustion
Assume that 200 lb of methane gas (CH_4) is used to incinerate 100 lb of carbon tetrachloride (CCl_4) in an incinerator. Determine the quantity of theoretical air needed for complete incineration.

Solution 1: By chemical analysis
1. Analyze chemical composition.
1.1 Calculate the ultimate composition of CCl_4 on a 100 lb basis (Table 2.10)

where x_I = compound element
M = the molecular weight of the compound
x_i/M = mass per pound fuel

1.2. Calculate the ultimate composition of CH_4 on a 200 lb basis (Table 2.11).
1.3. Calculate the ultimate composition of compounds by combining items 1.1 and 1.2 (Table 2.12).

2. Determine theoretical combustion air.
2.1. Calculate oxygen needs for theoretical combustion.
Apply the method in Section 2.6 (Table 2.13).

TABLE 2.10 Composition of CCl_4

x_i	C	H	Cl	O	M
CCl_4	12	0	142	0	154.00
x_i/M	0.0779	0.0000	0.9221	0.0000	
100 lb	7.7900	0.0000	92.2100	0.0000	

TABLE 2.11 Composition of CH_4

x_i	C	H	Cl	O	M
CH_4	12	4	0	0	16.00
x_i/M	0.7500	0.2500	0.0000	0.0000	
200 lb	150.0000	50.0000	0.0000	0.0000	

TABLE 2.12 Composition of Compounds

x_i	C	H	Cl	O	Weight
CCl_4	7.79	0	92.21	0	100.00
CH_4	150	50	0	0	200.00
Total	157.79	50.00	92.21	0.00	300.00

TABLE 2.13 Combustion Reactant Table

Basic reaction	Oxygen, nitrogen, and air requirements	Mass (lb/h)	Mole (mol/h)
$C + O_2 \rightarrow CO_2$	(1) $O_2 = C(32/12)$	420.77	13.15
$H_2 + 0.5O_2 \rightarrow H_2O$	(2) $O_2 = 8(H\text{–}Cl/35.5)$*	379.22	11.85
$S + O_2 \rightarrow SO_2$	(3) $O_2 = S(32/32) = S$	0.00	0.00
B_o = bound oxygen	(4) Data from waste analysis	0.00	0.00
O_{2t} = theoretical O_2	$O_{2t} = (1) + (2) + (3) - (4)$	799.99	25.00
N_{2t} = theoretical N_2	$N_{2t} = (O_{2t})(3.76 \times 28/32)$	2631.98	94.00
A_t = theoretical air	$A_t = O_{2t} + N_{2t}$	3431.97	119.00
$\%A_e$ = % excess air	Given data	0	
A_e = excess air	$A_e = (\%A_e)(A_t)$	0.00	0.00
A_{ada} = actual dry air	$A_{ada} = A_t + A_e$	3431.97	119.00
H_2O from air†	$0.0127(A_{ada})$	43.59	2.42
A_a = actual air	$A_a = A_{ada} + H_2O$	3475.56	121.42
Waste feed	Data from waste analysis	300.00	
R_a = actual reactant	$R_a = A_a$ + waste feed + H_2O from air	3775.56	

* If H_2 is the hydrogen left over after Cl reaction, i.e., $H_2 = H\text{–}Cl/35.5$, the O_2 needed to react with H_2 will be equal to $H_2(0.5)(32/2) = 8(H\text{–}Cl/35.5)$.

† In general, at room temperature, the humidity ratio is 0.0127 (lb-H_2O)/(lb-dry-air) or 0.0127 (kg-H_2O)/(kg-dry-air).

Table 2.13 shows that the oxygen, nitrogen and air needed for the complete combustion of 200 lb of CH_4 and 100 lb of CCl_4 are:

- oxygen = 799.99 lb
- nitrogen = 2631.98 lb
- water produced = 43.59 lb
- combustion air = 3475.56 lb
- total reactants = 3775.56 lb

Solution 2: By chemical balance

1. Determine CH_4 and air needed for CCl_4 incineration.

CH_4 +	CCl_4 +	$2O_2$ →	$2CO_2$ +	$4HCl$
16	154	64	88	146
0.10	1	0.42	0.57	0.95
10	100	42	57	95

$$\text{Theoretical nitrogen, } N_2 = O_2(3.76)(28)/32$$

$$= 42(3.76)(28)/32$$

$$= 138$$

$$\text{Theoretical dry air} = O_2 + N_2$$

$$= 42 + 138$$

$$= 180$$

This calculation shows that 100 lb CCl_4 requires 10 lb CH_4 and 180 lb air for complete incineration.

2. Determine air needed to react with unreacted CH_4.

$$\begin{array}{cccc} CH_4 + & 2O_2 \rightarrow & CO_2 + & 2H_2O \\ 16 & 64 & 44 & 36 \\ 1 & 4 & 2.75 & 2.25 \\ 190 & 760 & 523 & 427 \end{array}$$

$$\begin{aligned} \text{Theoretical nitrogen, } N_2 &= O_2(3.76)(28)/32 \\ &= 760(3.76)(28)/32 \\ &= 2500 \end{aligned}$$

$$\begin{aligned} \text{Theoretical dry air} &= O_2 + N_2 \\ &= 760 + 2500 \\ &= 3260 \end{aligned}$$

This calculation shows that the unreacted 190 lb CH_4 needs 3260 lb air.
3. Determine the total theoretical air needed.
The total air needed is the sum of air found in 1 and 2 above, which is equal to $180 + 3260 = 3440$ lb. This amount is approximately equal to the amount obtained by method 1, which is 3476 lb. The numerical difference between the two methods is due to rounding off certain values.

EXAMPLE 6 basic combustion
Assume that a waste fuel consists of several components with the following volume fraction composition:

$$N_2 = 0.05$$

$$CH_4 = 0.81$$

$$C_2H_6 = 0.10$$

$$C_3H_8 = 0.04$$

Determine:

(1) The volumetric flow rate of air required for complete combustion of 1.0 lb-mole/h of the above fuel with 100% theoretical air. Assume that the standard volume at the standard conditions is 387 scf (standard cubic feet) per lb-mole.

(2) The volumetric flow rates of the combustion products (EPA-86/03, p. 55).

First, the nitrogen in the feed is (0.05)(28) = 1.40 lb.

Solution 1 for question (1)
1.1. Analyze chemical composition.
1.1.1. Calculate the ultimate composition of CH_4 (Table 2.14).

TABLE 2.14 Composition of CH_4

x_i	C	H	Cl	O	M
CH_4	12	4	0	0	16.00
x_i/M	0.7500	0.2500	0.0000	0.0000	
$(0.81) \times (M) = 12.96$	9.7200	3.2400	0.0000	0.0000	

1.1.2. Calculate the ultimate composition of C_2H_6 (Table 2.15).
1.1.3. Calculate the ultimate composition of C_3H_8 (Table 2.16).
1.1.4. Calculate the ultimate composition by combining items 1.1.1, 1.1.2, and 1.1.3 (Table 2.17).
1.2. Determine the theoretical combustion air (Table 2.18).
Apply the method in Section 2.6.
Table 2.18 shows that the oxygen, nitrogen, and air needed for this combustion are:

- oxygen = 69.44 lb/(lb-waste)
- nitrogen = 228.46 lb/(lb-waste)
- combustion air = 297.90 lb/(lb-waste)
- total reactants = 315.62 lb

1.3. Determine the air volumetric flow rate.
From Table 2.18, the air flow rate A_a is 10.33 mol/h, and from the standard volume given in the question the volumetric flow rate V_r is:

$$V_r = 387(10.33)$$
$$= 3997.71 \text{ scfh}$$

TABLE 2.15 Composition of C_2H_6

x_i	C	H	Cl	O	M
C_2H_6	24	6	0	0	30.00
x_i/M	0.8000	0.2000	0.0000	0.0000	
$(0.10) \times (M) = 3.00$	2.4000	0.6000	0.0000	0.0000	

TABLE 2.16 Composition of C_3H_8

x_i	C	H	Cl	O	M
C_3H_8	36	8	0	0	44.00
x_i/M	0.8182	0.1818	0.0000	0.0000	
$(0.04) \times (M) = 1.76$	1.4400	0.3200	0.0000	0.0000	

TABLE 2.17 Composition of Organic Compounds

x_i	C	H	Cl	O	Weight
CH_4	9.72	3.24	0	0	12.96
C_2H_6	2.4	0.6	0	0	3.00
C_3H_8	1.44	0.32	0	0	1.76
Total	13.56	4.16	0.00	0.00	17.72

TABLE 2.18 Combustion Reactant Table

Basic reaction	Oxygen, nitrogen, and air requirements	Mass (lb/h)	Mole (mol/h)
$C + O_2 \rightarrow CO_2$	(1) $O_2 = C(32/12)$	36.16	1.13
$H_2 + 0.5O_2 \rightarrow H_2O$	(2) $O_2 = 8(H–Cl/35.5)$*	33.28	1.04
$S + O_2 \rightarrow SO_2$	(3) $O_2 = S(32/32) = S$	0.00	0.00
B_o = bound oxygen	(4) Data from waste analysis	0.00	0.00
O_{2t} = theoretical O_2	$O_{2t} = (1) + (2) + (3) - (4)$	69.44	2.17
N_{2t} = theoretical N_2	$N_{2t} = (O_{2t})(3.76 \times 28/32)$	228.46	8.16
A_t = theoretical air	$A_t = O_{2t} + N_{2t}$	297.90	10.33
$\%A_e$ = % excess air	Given data	0	
A_e = excess air	$A_e = (\%A_e)(A_t)$	0.00	0.00
A_{ada} = actual dry air	$A_{ada} = A_t + A_e$	297.90	10.33
H_2O from air†	$0.0127(A_{ada})$, not considered for this case	0.00	0.00
A_a = actual air	$A_a = A_{ada} + H_2O$	297.90	10.33
Waste feed	Data from waste analysis (17.72 + 1.40)	19.12	
R_a = actual reactant	$R_a = A_a$ + waste feed + H_2O from air	317.02	

* If H_2 is the hydrogen left over after Cl reaction, i.e., $H_2 = H–Cl/35.5$, the O_2 needed to react with H_2 will be equal to $H_2(0.5)(32/2) = 8(H–Cl/35.5)$.
† In general, at room temperature, the humidity ratio is 0.0127 (lb-H_2O)/(lb-dry-air) or 0.0127 (kg-H_2O)/(kg-dry-air).

Solution 1 for question (2)

2.1. Analyze the combustion products (Table 2.19).
Apply the method in Section 2.6.

2.2. Calculate the volumetric flow rate of combustion products.
Table 2.19 shows that the combustion flue gas G_a = 11.37 mol/h. The volumetric flow rate of the combustion products is

$$V_p = 387(11.37)$$
$$= 4400.19 \text{ scfh}$$

Solution 2: by the chemical balance of each individual compound

1. Determine the mass of each compound by multiplying the molecular weight with its corresponding compound.

TABLE 2.19 Combustion Product Table

Component	Combustion product	Mass (lb/h)	Mole (mol/h)
CO_2	$CO_2 = (C)(44/12)$	49.72	1.13
SO_2	$SO_2 = S(64/32)$	0.00	0.00
HCl	$HCl = Cl(36.5/35.5)$	0.00	0.00
O_{2e}	$(\%A_e)O_{2t}$, O_{2t} from Table 2.18	0.00	0.00
N_{2e}	$(\%A_e)N_{2t}$, N_{2t} from Table 2.18	0.00	
N_2	$N_{2e} + N_{2t}$ + from waste (0 + 228.46 + 1.40)	229.86	8.21
H_2O from combustion	H_2O from combustion = 9(H–Cl/35.5)	37.44	
H_2O	H_2O from combustion + from air + from waste	37.44	2.08
G_a = actual flue gas	$G_a = CO_2 + SO_2 + HCl + O_{2e} + N_2 + H_2O$	315.62	11.37
Ash	Fly ash + bottom ash	0.00	
P_a = actual product*	G_a + ash	317.02	

* In performing the mass balance calculation, the values of reactants (R_a) and products (P_a) should be equal, i.e., if $R_a = P_a$, the mass balance calculation is correct.

$$\text{Mass of } N_2 = 0.05(28) = 1.40 \text{ lb/h}$$
$$\text{Mass of } CH_4 = 0.81(16) = 12.96 \text{ lb/h}$$
$$\text{Mass of } C_2H_6 = 0.10(30) = 3.00 \text{ lb/h}$$
$$\text{Mass of } C_3H_8 = 0.04(44) = 1.76 \text{ lb/h}$$

2. Determine the chemical reaction balance equation for each compound.

CH_4 +	$2O_2$	$\rightarrow CO_2 + 2H_2O$
16.00	64.00	
1.00	4.00	
12.96	51.84	

C_2H_6 +	$3.5O_2$	$\rightarrow 2CO_2 + 3H_2O$
30.00	112.00	
1.00	3.73	
3.00	11.20	

C_3H_8 +	$5O_2$	$\rightarrow 3CO_2 + 4H_2O$
44.00	160.00	
1.00	3.64	
1.76	6.40	

These calculations show that the total oxygen O_2 required is:

$$O_2 = 51.84 + 11.20 + 6.40$$
$$= 69.44 \text{ lb/h}$$

3. Calculate the theoretical air amount (m_{air}).

$$\text{Theoretical nitrogen, } N_2 = O_2(3.76)(28)/32$$
$$= 69.44(3.76)(28)/32$$
$$= 228.46 \text{ lb}$$

$$\text{Theoretical dry air} = O_2 + N_2$$
$$= 69.44 + 228.46$$
$$= 297.90 \text{ lb}$$

This air mass flow rate is identical to that in Solution 1.

Following the same procedures as in Solution 1, the volumetric flow rate of combustion air and the volumetric flow rate of combustion products can be found.

Solution 3: by total chemical balance

1. Write the balanced chemical equation for complete combustion.

$$0.81CH_4 + 0.10C_2H_6 + 0.04C_3H_8 + 0.05N_2 + u(O_2 + 3.76N_2) \rightarrow$$
$$vCO_2 + wH_2O + u(3.76)N_2 + 0.05N_2 \quad (2.36)$$

2. Determine the chemical balance.
For carbon balance

$$0.81(12) + 0.1(12)(2) + 0.04(12)(3) = 12v$$
$$13.56 = 12v$$
$$v = 1.13 \text{ lb-mole/h} \tag{2.37}$$

For hydrogen balance

$$0.81(4) + 0.1(6) + 0.04(8) = 2w$$
$$4.16 = 2w$$
$$w = 2.08 \text{ lb-mole/h} \tag{2.38}$$

For oxygen balance

$$32u = 32v + 16w$$
$$32u = 32(1.13) + 16(2.08)$$
$$u = 2.17 \text{ lb-mole/h} \tag{2.39}$$

3. Summarize the calculated coefficients from Eqs. (2.37)–(2.39).

$$u = 2.17 \tag{2.40}$$
$$v = 1.13 \tag{2.41}$$
$$w = 2.08 \tag{2.42}$$

4. Write the complete balance reaction for Eq. (2.36).

$$0.81CH_4 + 0.10C_2H_6 + 0.04C_3H_8 + 0.05N_2 + 2.17(O_2 + 3.76N_2)$$
$$\rightarrow 1.13CO_2 + 2.08H_2O + 2.17(3.76)N_2 + 0.05N_2 \tag{2.43}$$

5. Calculate the volumetric flow rate of air required.
From Eq. (2.43), the air component is

$$\text{Air} = 2.17(O_2 + 3.76N_2)$$

Air flow rate is the sum of the coefficients from O_2 and N_2.
Therefore, air flow rate $= 2.17(1 + 3.76) = 10.33$ lb-mole/h

$$\text{Air mass flow rate} = 2.17(32 + 3.76(28)) = 297.9 \text{ lb/h}$$

Again this air mass flow rate is identical to that in Solution 1.
Following the same procedures as in Solution 1, the volumetric flow rate of combustion air and the volumetric flow rate of combustion products can be found.

EXAMPLE 7 basic combustion
Given conditions

- A waste is incinerated in an incinerator. The waste feeding rate is 300 lb/h with the following components:

$$C = 219.00 \text{ lb/h}$$
$$H_2 = 50.00$$
$$Cl = 2.00$$
$$O_2 = 5.00$$
$$N_2 = 16.00$$
$$S = 2.00$$
$$H_2O = 3.00$$
$$\text{bottom ash} = 2.00$$
$$\text{fly ash} = 1.00$$

- Air humidity ratio = 0.0127 (lb-H_2O/lb-air)
- Excess air = 150% = 250% theoretical air = 1.5 (as a fraction)

The waste consists of several components, as shown below.
Determine:

(1) The air flow rate required for 150% excess air combustion.
(2) The flow rate of combustion products.

1. Solution for question (1)
1.1. Calculate the combustion air flow rate.
Apply the method in Section 2.6 (Table 2.20).
From Table 2.20, the air flow rate for 150% excess air combustion is 10,649.95 lb/h.

TABLE 2.20 Combustion Reactant Table

Basic reaction	Oxygen, nitrogen, and air requirements	Mass (lb/h)	Mole (mol/h)
$C + O_2 \rightarrow CO_2$	(1) $O_2 = C(32/12)$	584.00	18.25
$H_2 + 0.5O_2 \rightarrow H_2O$	(2) $O_2 = 8(H\text{–}Cl/35.5)$*	399.55	12.49
$S + O_2 \rightarrow SO_2$	(3) $O_2 = S(32/32) = S$	2.00	0.06
B_o = bound oxygen	(4) Data from waste analysis	5.00	0.16
O_{2t} = theoretical O_2	$O_{2t} = (1) + (2) + (3) - (4)$	980.55	30.64
N_{2t} = theoretical N_2	$N_{2t} = (O_{2t})(3.76 \times 28/32)$	3226.01	115.21
A_t = theoretical air	$A_t = O_{2t} + N_{2t}$	4206.56	145.86
$\%A_e$ = % excess air	Given data (use % A_e as a fraction)	1.5	
A_e = excess air	$A_e = (\%A_e)(A_t)$	6309.83	218.79
A_{ada} = actual dry air	$A_{ada} = A_t + A_e$	10,516.39	364.65
H_2O from air†	$0.0127(A_{ada})$	133.56	7.42
A_a = actual air	$A_a = A_{ada} + H_2O$	10,649.95	372.07
Waste feed	Data from waste analysis	300.00	
R_a = actual reactant	$R_a = A_a$ + waste feed + H_2O from air	10,949.95	

* If H_2 is the hydrogen left over after Cl reaction, i.e., $H_2 = H\text{–}Cl/35.5$, the O_2 needed to react with H_2 will be equal to $H_2(0.5)(32/2) = 8(H\text{–}Cl/35.5)$.
† In general, at room temperature, the humidity ratio is 0.0127 (lb-H_2O)/(lb-dry-air) or 0.0127 (kg-H_2O)/(kg-dry-air).

2. Solution for question (2)
Calculate the combustion products flow rate by applying the method in Section 2.6 (Table 2.21).

From Table 2.21 the flow rate of combustion products for 150% excess air combustion is 10,949.96 lb/h.

TABLE 2.21 Combustion Product Table

Component	Combustion product	Mass (lb/h)	Mole (mol/h)
CO_2	$CO_2 = (C)(44/12)$	803.00	18.25
SO_2	$SO_2 = S(64/32)$	4.00	0.06
HCl	$HCl = Cl(36.5/35.5)$	2.06	0.06
O_{2e}	$(\%A_e)O_{2t}$, O_{2t} from Table 2.20	1470.83	45.96
N_{2e}	$(\%A_e)N_{2t}$, N_{2t} from Table 2.20	4839.02	172.82
N_2	$N_{2e} + N_{2t}$ + from waste	8081.03	288.61
H_2O from combustion	H_2O from combustion = 9(H–Cl/35.5)	449.49	24.97
H_2O	H_2O from combustion + from air + from waste	586.05	32.56
G_a = actual flue gas	$G_a = CO_2 + SO_2 + HCl + O_{2e} + N_2 + H_2O$	10,946.96	385.50
Ash	Fly ash + bottom ash	3.00	
P_a = actual product*	G_a + ash	10,949.96	

* In performing the mass balance calculation, the values of reactants (R_a) and products (P_a) should be equal, i.e., if $R_a = P_a$, the mass balance calculation is correct.

3 BASIC MASS AND ENERGY BALANCE CALCULATIONS

The purpose of this section is to provide readers with a universally applicable, step-by-step, and easy-to-follow procedure to perform mass and energy balance calculations on thermal treatment operations. Examples will be included to illustrate the procedure in different applications (Ho-98/12).

3.1 Solid Waste Thermal Treatment

There are a variety of solid wastes which can be managed through thermal treatment processes. These wastes may include medical waste, municipal waste, hazardous waste, and sludge waste from the treatment of domestic and industrial wastewater. Although these different types of waste may have different chemical compositions, energy levels, water contents, and so on, and they may be burned in different types of combustors, e.g., rotary kiln, fixed hearth, liquid injection, fluidized bed, etc., the procedures of performing the mass and energy balance calculations are essentially identical.

Thermal treatment (or combustion/incineration) is an effective method for treating solid wastes. Under normal operations, a properly designed thermal treatment unit is capable of destroying more than 99.99% of the incoming organic wastes. Depending on the heating values of the wastes, the treatment process may or may not require auxiliary fuels. Energy recovery during and after the treatment process is also possible if the unit is equipped with waste heat recovery systems. The amounts of auxiliary fuels required and potential heat recovery can theoretically be estimated through mass and energy balance calculations.

Combustion fundamentals. The combustion/incineration process is the burning of waste with air or pure oxygen. Theoretically, it is the conversion of elemental constituents in the organic waste, e.g., carbon (C), hydrogen (H), oxygen (O), nitrogen (N), chlorine (Cl), and sulfur (S), into both nontoxic combustion products such as carbon dioxide (CO_2), water (H_2O), oxygen (O_2) and nitrogen (N_2), and potentially toxic products such as carbon monoxide (CO), hydrochloric acid (HCl), chlorine (Cl_2), nitrogen oxides (NO_x, or NO and NO_2), and sulfur oxides (SO_x, or SO_2 and SO_3). These combustion phenomena can be represented by the following simplified stoichiometric equations:

A. $$1C + 1O_2 \rightarrow 1CO_2$$
B. $$1C + 1/2O_2 \rightarrow 1CO$$
C. $$1H + 1Cl \rightarrow 1HCl$$
D. $$2H + 1/2O_2 \rightarrow 1H_2O$$
E. $$1Cl + 1Cl \rightarrow 1Cl_2$$
F. $$1S + 1O_2 \rightarrow 1SO_2$$
G. $$1S + 3/2O_2 \rightarrow 1SO_3$$
H. $$1N + 1/2O_2 \rightarrow 1NO$$
I. $$1N + 1O_2 \rightarrow 1NO_2$$
J. $$1N + 1N \rightarrow 1N_2$$
K. $$1O + 1O \rightarrow 1O_2$$

where C, H, O, N, Cl, and S in the above equations represent the elemental constituents in the waste, and O_2 represents the supplied oxygen from air or other oxygen sources. Note that reactions A, C, D, and F represent the major combustion reactions and are commonly the only reactions considered in the material and energy balance calculations.

Theoretical oxygen requirement. One of the important steps in performing material and energy balance calculations is to determine the *theoretical oxygen requirement* (n_{s,O_2}) for the waste to be combusted, i.e., the minimum amount of oxygen (O_2) which is required to be supplied to completely combust the waste according to combustion stoichiometry. This theoretical oxygen requirement can be calculated by the equation

$$n_{s,O_2} = n_C + (n_H - n_{Cl})/4 + n_S - n_O/2 \tag{2.44}$$

where n_C, n_H, n_{Cl}, n_S, and n_O represent the number of pound-moles of carbon, hydrogen, chlorine, sulfur, and oxygen atoms in the waste. Note that this equation reflects the combustion stoichiometry that each pound-mole of carbon and sulfur atoms requires one pound-mole of oxygen gas (reactions A and F) for complete combustion; and each two pound-moles of available hydrogen atoms ($n_H - n_{Cl}$) requires one-half pound-mole of oxygen gas (reaction D). It also reflects the reaction that each pound-mole of chlorine atoms will consume one pound-mole of hydrogen atoms to form one pound-mole of hydrochloric acid (reaction C), which reduces the amount of hydrogen available for reacting with oxygen. The existence of oxygen atoms in the waste also reduces the theoretical oxygen requirement as reflected in the equation.

Excess oxygen supply. In combustion/incineration practice, the actual amount of oxygen supply is normally greater than the theoretical oxygen requirement to ensure sufficient oxygen for combustion. In such practice, the amount of oxygen supply in excess of that of the theoretical requirement is called the excess oxygen, and the fraction in excess is called the fractional excess. The fractional excess (f) is thus defined as

$$f = (n_{O_2} - n_{s,O_2})/n_{s,O_2} \tag{2.45}$$

where n_{O_2} represents the actual amount of oxygen supply in pound-moles and n_{s,O_2} is the theoretical amount of oxygen requirement in pound-moles as calculated in Eq. (2.44). In cases where the fractional excess is to be fixed, Eq. (2.45) can be rearranged to determine the actual amount of oxygen supply, i.e.,

$$n_{O_2} = (1 + f)n_{s,O_2} \tag{2.46}$$

Theoretical and excess air supply. When air is used as the oxygen source, the following two equations can be derived based on a common assumption that air is composed of 21 mole% oxygen and 79 mole% nitrogen, i.e.,

$$n_{O_2} = 0.21n_{air} \quad \text{or} \quad n_{air} = n_{O_2}/0.21 \tag{2.47}$$

and

$$n_{s,O_2} = 0.21n_{s,air} \quad \text{or} \quad n_{s,air} = n_{s,O_2}/0.21 \tag{2.48}$$

where n_{air} represents the actual amount of air supply in pound-moles and $n_{s,air}$ is the theoretical amount of air requirement in pound-moles. Substituting Eqs. (2.47) and (2.48) into Eqs. (2.45) and (2.46), the corresponding two equations in terms of n_{air} and $n_{s,air}$ will have the following forms:

$$f = (n_{air} - n_{s,air})/n_{s,air} \tag{2.49}$$

and

$$n_{air} = (1 + f)n_{s,air} \tag{2.50}$$

All the above equations are commonly involved in the material and energy balance calculations.

3.2 Mass and Energy Balance Principle

Material balance principle. The material balance principle for a combustion system is simply that, under steady state operations, the inlet molar flow rate of any elemental species must be equal to the total molar flow rate of that particular element in the outlet, which may appear in one or more flue gas species. Note that the inlet molar flow rate of a particular elemental species is the sum of that species from all inlet streams. Normally, an inlet stream is considered to be composed of moisture (H_2O), combustible substances, and noncombustible (ash) materials. In the calculations, the moisture and noncombustible materials are considered to pass through the unit with only physical changes, i.e., temperature changes and the vaporization of water. For the combustible portion, the elemental species commonly considered are carbon (C), hydrogen (H), chlorine (Cl), sulfur (S), oxygen (O), and nitrogen (N). The following sequence of assumptions are generally involved in determining the outlet species from the inlet combustible substances.

Carbon element. All the incoming carbon (C) is either converted to carbon dioxide (CO_2) according to reaction A, i.e.,

(A) $$1C + 1O_2 \rightarrow 1CO_2$$

or remains unburned and exits as solid elemental carbon. Note that the conversion of carbon to carbon monoxide (CO) is generally not considered in material balance calculations because carbon monoxide is not expected to be present in any significant amount in a well-designed and well-operated combustion system.

Hydrogen element. All the incoming hydrogen (H) will either react with chlorine to form hydrochloric acid (HCl) according to reaction C, i.e.,

(C) $$1H + 1Cl \rightarrow 1HCl$$

or is converted to water (H_2O) according to reaction D, i.e.,

(D) $$2H + 1/2O_2 \rightarrow 1H_2O$$

Chlorine element. Essentially all of the incoming chlorine (Cl) is converted to hydrochloric acid (HCl) according to reaction C.

Sulfur element. Essentially all of the incoming sulfur (S) is converted to sulfur dioxide (SO_2) according to reaction F, i.e.,

(F) $$1S + 1O_2 \rightarrow 1SO_2$$

Oxygen element. All the incoming oxygen, including that in the inlet combustible materials and that from the air or oxygen supply, will be in either carbon dioxide (CO_2), water (H_2O), sulfur dioxide (SO_2), or oxygen gas (O_2).

Nitrogen element. All the incoming nitrogen, including that in the inlet combustible materials and that from the air supply, will be in nitrogen gas (N_2). Note again that the formation of nitrogen oxides (NO_x) is generally in small amounts and is not considered in the material and energy balance calculations.

In summary, the outlet species considered in the material balance calculations are carbon dioxide (CO_2), water (H_2O), hydrochloric acid (HCl), sulfur dioxide (SO_2), oxygen gas (O_2), nitrogen gas (N_2), and solid ash, including noncombustible materials and unburned carbon.

Based on the above assumptions, the following material balance equations are derived to determine the molar flow rate of each outlet species:

$$n_{\text{carbon dioxide}} = n_C \tag{2.51}$$

$$n_{\text{water}} = (n_H - n_{Cl})/2 + n_{\text{moisture}} \tag{2.52}$$

$$n_{\text{hydrochloric acid}} = n_{Cl} \tag{2.53}$$

$$n_{\text{sulfur dioxide}} = n_S \tag{2.54}$$

$$n_{\text{oxygen}} = n_O/2 + n_{O_2} - n_C - n_S - (n_H - n_{Cl})/4 \tag{2.55}$$

$$n_{\text{nitrogen}} = n_N/2 + n_{N_2} \tag{2.56}$$

where

$$n_{N_2} = 0.79 n_{\text{air}} \tag{2.57}$$

Equations (2.49)–(2.57) completely describe the material balance calculations for waste combustion processes.

Energy balance principle. The energy balance principle for a combustion system is simply that, under steady state operations, the difference in total energy content between the outlet and inlet streams must be equal to the amount of heat transfer. In formula, it is expressed as

$$Q = (H_{\text{total}})_{\text{outlet}} - (H_{\text{total}})_{\text{inlet}} \tag{2.58}$$

Note that the $(H_{\text{total}})_{\text{outlet}}$ is generally less than $(H_{\text{total}})_{\text{inlet}}$ in combustion processes, and therefore the amount of heat transfer (Q) appearing in the equation is always negative, indicating that energy is transfering out of the combustion process. This heat transfer can be in the forms of energy recovery through generating useful steam or energy lost to the surroundings.

In performing energy balance calculations, the following equation, derived from Eq. (2.58), is usually involved, i.e.,

$$Q = -(\mathrm{HHV})_{\mathrm{total}} + (H_{\mathrm{sensible}})_{\mathrm{outlet}} - (H_{\mathrm{sensible}})_{\mathrm{inlet}} \tag{2.59}$$

where $(\mathrm{HHV})_{\mathrm{total}}$ is the sum of the higher heating values from all the inlet streams, $(H_{\mathrm{sensible}})_{\mathrm{outlet}}$ is the sum of the sensible heat from all the outlet flue gas species (plus the heat of vaporization of water), and $(H_{\mathrm{sensible}})_{\mathrm{inlet}}$ is the sum of the sensible heat from all the inlet streams. The three terms in Eq. (2.59) can be calculated by the following equations, respectively:

$$(\mathrm{HHV})_{\mathrm{total}} = \sum \{w(\mathrm{HHV})\}_i, \qquad i = \text{each inlet stream} \tag{2.60}$$

$$(H_{\mathrm{sensible}})_{\mathrm{outlet}} = \sum \{nC_{\mathrm{pm}}(T - 77)\}_i + n_{\mathrm{water}}(\Delta H_{\mathrm{v}})_{\mathrm{water}}, \qquad i = \text{each flue gas species} \tag{2.61}$$

and

$$(H_{\mathrm{sensible}})_{\mathrm{inlet}} = \sum \{wC_{\mathrm{pm}}(T - 77)\}_i, \qquad i = \text{each inlet stream} \tag{2.62}$$

where w is the mass or molar flow rate of each inlet stream, HHV is the higher heating value of each inlet stream on either a per mass or per mole basis, n is the molar flow rate of each flue gas species, n_{water} is the molar flow rate of water in the flue gas, $(\Delta H_{\mathrm{v}})_{\mathrm{water}}$ is the heat of vaporization of water at 77°F (= 1050 Btu/lbm), C_{pm} is the mean specific heat again on either a per mass or per mole basis, and T is temperature in degrees Fahrenheit.

In the application, the HHV for each inlet stream can either be looked up from thermodynamic tables, measured experimentally, or estimated using the following equation:

$$\mathrm{HHV} = 14{,}545X_{\mathrm{C}} + 62{,}031(X_{\mathrm{H}} - 0.125X_{\mathrm{O}}) - 760X_{\mathrm{Cl}} + 4500X_{\mathrm{S}} \tag{2.63}$$

where HHV is in the unit of Btu/lbm and X_{C}, X_{H}, X_{O}, X_{Cl} and X_{S} are mass fraction of carbon, hydrogen, oxygen, chlorine, and sulfur, respectively, in the combustible portion of an inlet stream. For the mean specific heat appearing in Eqs (2.61) and (2.62), Table 2.22 lists the corresponding values at different temperatures.

3.3 Mass and Energy Balance Calculational Procedure

Calculational procedure. The step-by-step procedure for performing material and energy balance calculations is described below.

Step 1. Break down each inlet stream into moisture, noncombustible material and various elemental species; then add up all the inlet streams to determine the total flow rate of moisture, noncombustible material, and each elemental species in terms of carbon, hydrogen, chlorine, sulfur, hydrogen, and nitrogen.

Step 2. Determine the theoretical oxygen requirement based on Eq. (2.44) and the results from step 1.

Step 3. Determine the actual oxygen (or air) supply under a given value of fractional excess (f) based on either Eq. (2.46) or Eq. (2.50).

Step 4. Perform the material balance calculation by identifying the flue gas species and determine the corresponding molar flow rate based on Eqs. (2.51)–(2.56).

Step 5. Perform the energy balance calculation based on Eqs. (2.59)–(2.62).

TABLE 2.22 Mean Specific Heat (C_{pm}) of Various Flue Gas Species (Reference Temperature 77°F)

	C_{pm} (Btu/lb-mole°F)						
T(°F)	Air	O_2	N_2	CO_2	HCl	SO_2	H_2O(g)
77	6.94	7.02	6.94	8.88	6.95	9.97	8.04
100	6.95	7.04	6.95	8.94	6.95	10.06	8.05
200	6.98	7.11	6.97	9.2	6.96	10.44	8.1
300	7.01	7.18	6.99	9.45	6.98	10.78	8.15
400	7.04	7.25	7.01	9.69	7.01	11.07	8.21
500	7.08	7.32	7.04	9.91	7.05	11.33	8.27
600	7.11	7.38	7.06	10.12	7.1	11.56	8.33
700	7.15	7.44	7.09	10.32	7.16	11.76	8.4
800	7.18	7.49	7.12	10.51	7.22	11.92	8.47
900	7.22	7.55	7.16	10.68	7.3	12.07	8.54
1000	7.25	7.6	7.19	10.85	7.34	12.2	8.61
1100	7.29	7.65	7.23	11.01	7.4	12.3	8.68
1200	7.33	7.7	7.26	11.16	7.46	12.39	8.75
1300	7.37	7.74	7.3	11.3	7.52	12.48	8.83
1400	7.41	7.79	7.33	11.43	7.57	12.55	8.91
1500	7.44	7.83	7.37	11.55	7.62	12.61	8.98
1600	7.48	7.87	7.41	11.67	7.66	12.68	9.06
1700	7.52	7.91	7.44	11.78	7.68	12.74	9.14
1800	7.55	7.95	7.48	11.88	7.7	12.81	9.22
1900	7.59	7.98	7.52	11.98	7.7	12.89	9.29
2000	7.63	8.02	7.55	12.07	7.7	12.97	9.37

Source: R.M. Felder and W.R. Ronald, *Elementary Principles of Chemical Processes*, 2nd edn. Wiley, 1986.

3.4 Example Calculation

EXAMPLE 1 combustion of pure methane gas (CH_4)
Determine the amount of heat transfer (Q) = ? Btu/h.

Given conditions

- methane flow rate = 5000 ft^3/h @ 1 atm, 77°F
- fractional excess air = 0.15 @77°F
- flue gas temperature = 1800°F
- higher heating value of methane = 383,109 Btu/lb-mole

Solution:

Step 1. Determine the molar flow rate of methane and the corresponding molar flow rates of C and H based on the ideal gas law and the following equations:

$$\begin{aligned} n_{CH_4} &= (PV)/(RT) \\ &= (14.7 \times 5000)/(10.73 \times 537) \\ &= 12.76 \text{ (lb-mole/h)} \end{aligned}$$

Therefore,

$$n_C = 1 \times n_{CH_4} = 12.76 \text{ lb-mole/h}$$

and

$$n_H = 4 \times n_{CH_4} = 51.04 \text{ lb-mole/h}$$

Step 2. Determine the theoretical air requirement based on Eq. (2.44), i.e.,

$$\begin{aligned} n_{s,O_2} &= n_C + (n_H - n_{Cl})/4 + n_S - n_O/2 \\ &= 12.76 + (51.04 - 0)/4 + 0 - 0 \\ &= 25.52 \text{ (lb-mole/h)} \end{aligned}$$

Therefore,

$$n_{s,air} = 25.52/0.21 = 121.52 \text{ (lb-mole/h)}$$

Step 3. Determine the actual air supply based on the given value of fractional excess according to Eq. (2.50), i.e.,

$$\begin{aligned} n_{air} &= (1 + f)n_{s,air} \\ &= (1 + 0.15)(121.52) \\ &= 139.75 \text{ (lb-mole/h)} \end{aligned}$$

Therefore

$$\begin{aligned} n_{O_2} &= 0.21 n_{air} \\ &= 29.35 \text{ (lb-mole/h)} \end{aligned}$$

and

$$\begin{aligned} n_{N_2} &= 0.79 n_{air} \\ &= 110.40 \text{ (lb-mole/h)} \end{aligned}$$

Step 4. Determine the molar flow rate of flue gas based on Eqs. (2.50)–(2.56), i.e.,

$$\begin{aligned} n_{\text{carbon dioxide}} &= n_C \\ &= 12.76 \text{ lb-mole/h} \\ n_{\text{water}} &= (n_H - n_{Cl})/2 + n_{\text{moisture}} \\ &= (51.04 - 0)/2 + 0 \\ &= 25.52 \text{ lb-mole/h} \\ n_{\text{hydrochloric acid}} &= n_{Cl} \\ &= 0.0 \text{ lb-mole/h} \\ n_{\text{sulfur dioxide}} &= n_S \\ &= 0.0 \text{ lb-mole/h} \\ n_{\text{oxygen}} &= n_O/2 + n_{O_2} - n_C - n_S - (n_H - n_{Cl})/4 \\ &= 0 + 29.35 - 12.76 - 0 - (51.04 - 0)/4 \\ &= 3.83 \text{ lb-mole/h} \\ n_{\text{nitrogen}} &= n_N/2 + n_{N_2} \\ &= 0 + 110.40 \\ &= 110.40 \text{ lb-mole/h} \end{aligned}$$

Step 5. Determine Q based on the given $(\mathrm{HHV})_{\mathrm{methane}}$ and Eq. (2.60), i.e.,

$$Q = -(\mathrm{HHV})_{\mathrm{total}} + (H_{\mathrm{sensible}})_{\mathrm{outlet}} - (H_{\mathrm{sensible}})_{\mathrm{inlet}}$$

(a) For $(\mathrm{HHV})_{\mathrm{total}}$:

$$\begin{aligned}(\mathrm{HHV})_{\mathrm{total}} &= \sum \{w(\mathrm{HHV})\}_i, \qquad i = \text{each inlet stream} \\ &= (12.76 \times 383{,}109)_{\mathrm{methane}} \\ &= 4.89 \times 10^6\ (\mathrm{Btu/h})\end{aligned}$$

(b) For $(H_{\mathrm{sensible}})_{\mathrm{outlet}}$ at 1800°F based on Table 2.22:

$$\begin{aligned}C_{\mathrm{pm,CO_2,1800^\circ F}} &= 11.88\ (\mathrm{Btu/lb\text{-}mole^\circ F}) \\ C_{\mathrm{pm,H_2O,1800^\circ F}} &= 9.22\ (\mathrm{Btu/lb\text{-}mole^\circ F}) \\ C_{\mathrm{pm,O_2,1800^\circ F}} &= 7.95\ (\mathrm{Btu/lb\text{-}mole^\circ F}) \\ C_{\mathrm{pm,N_2,1800^\circ F}} &= 7.48\ (\mathrm{Btu/lb\text{-}mole^\circ F})\end{aligned}$$

Then

$$\begin{aligned}(H_{\mathrm{sensible}})_{\mathrm{outlet}} &= \sum \{nC_{\mathrm{pm}}(T-77)\}_i + n_{\mathrm{water}}(\Delta H_{\mathrm{v}})_{\mathrm{water}} \\ &= \{nC_{\mathrm{pm}}(T-77)\}_{\mathrm{CO_2}} + \{nC_{\mathrm{pm}}(T-77)\}_{\mathrm{H_2O}} + \{nC_{\mathrm{pm}}(T-77\}_{\mathrm{O_2}} \\ &\quad + \{nC_{\mathrm{pm}}(T-77)\}_{\mathrm{N_2}} + n_{\mathrm{water}}(\Delta H_{\mathrm{v}})_{\mathrm{water}} \\ &= \{12.76C_{\mathrm{pm}}(1800-77)\}_{\mathrm{CO_2}} + \{25.52C_{\mathrm{pm}}(1800-77)\}_{\mathrm{H_2O}} \\ &\quad + \{3.83C_{\mathrm{pm}}(1800-77)\}_{\mathrm{O_2}} + \{110.40C_{\mathrm{pm}}(1800-77)\}_{\mathrm{N_2}} \\ &\quad + 25.52 \times 18 \times 1050 \\ &= 2.62 \times 10^6\ (\mathrm{Btu/h})\end{aligned}$$

(c) For $(H_{\mathrm{sensible}})_{\mathrm{inlet}}$ at 77°F:

$$\begin{aligned}(H_{\mathrm{sensible}})_{\mathrm{inlet}} &= \sum \{nC_{\mathrm{pm}}(T-77)\}_i \qquad i = \text{each flue gas species} \\ &= \{nC_{\mathrm{pm}}(T-77)\}_{\mathrm{methane}} + \{nC_{\mathrm{pm}}(T-77)\}_{\mathrm{air}} \\ &= \{12.76C_{\mathrm{pm}}(77-77)\}_{\mathrm{methane}} + \{139.75C_{\mathrm{pm}}(77-77)\}_{\mathrm{air}} \\ &= 0\ (\mathrm{Btu/h})\end{aligned}$$

Therefore,

$$\begin{aligned}Q &= -(\mathrm{HHV})_{\mathrm{total}} + (H_{\mathrm{sensible}})_{\mathrm{outlet}} - (H_{\mathrm{sensible}})_{\mathrm{inlet}} \\ &= -4.89 \times 10^6 + 2.62 \times 10^6 - 0 \\ &= -2.27 \times 10^6\ (\mathrm{Btu/h})\end{aligned}$$

EXAMPLE 2 co-combustion of liquid benzene and No. 2 oil
Determine the amount of heat transfer $(Q) = ?$ Btu/h.

Given conditions

- Stream 1: liquid benzene (C_6H_6)

$$\text{benzene flow rate} = 100 \text{ lbm/h at } 77^\circ\text{F}$$
$$\text{HHV (benzene)} = 18{,}026 \text{ Btu/lbm}$$

- Stream 2: No. 2 oil (C: 0.872; H: 0.123; S: 0.005 by mass)

$$\text{oil flow rate} = 300 \text{ lbm/h at } 77^\circ\text{F}$$
$$\text{fractional excess air} = 0.1 \text{ at } 77^\circ\text{F}$$
$$\text{flue gas temperature} = 1800^\circ\text{F}$$

Solution:

Step 1. Determine the combined molar flow rate of C, H, and S.
(a) Stream 1 (benzene)

$$M_{\text{benzene}}(\text{benzene molecular weight}) = 12 \times 6 + 1 \times 6 = 78$$
$$n_{\text{benzene}} = 100/M_{\text{benzene}} = 100/78 = 1.28 \text{ (lb-mole/h)}$$

Therefore

$$n_{\text{C}} = 6 \times n_{\text{benzene}} = 7.68 \text{ lb-mole/h}$$
$$n_{\text{H}} = 6 \times n_{\text{benzene}} = 7.68 \text{ lb-mole/h}$$

(b) Stream 2 (No. 2 oil)

$$n_{\text{C}} = (300 \times 0.872)/12 = 21.8 \text{ lb-mole/h}$$
$$n_{\text{H}} = (300 \times 0.123)/1 = 36.9 \text{ lb-mole/h}$$
$$n_{\text{S}} = (300 \times 0.005)/32 = 0.047 \text{ lb-mole/h}$$

(c) Total input (benzene and No. 2 oil)

$$n_{\text{C}} = 7.68 + 21.8 = 29.48 \text{ lb-mole/h}$$
$$n_{\text{H}} = 7.68 + 36.9 = 44.58 \text{ lb-mole/h}$$
$$n_{\text{S}} = 0 + 0.047 = 0.047 \text{ lb-mole/h}$$

Step 2. Determine the theoretical air requirement based on Eq. (2.44), i.e.,

$$\begin{aligned} n_{\text{s,O}_2} &= n_{\text{C}} + (n_{\text{H}} - n_{\text{Cl}})/4 + n_{\text{S}} - n_{\text{O}}/2 \\ &= 29.48 + (44.58 - 0)/4 + 0.047 - 0 \\ &= 40.672 \text{ (lb-mole/h)} \end{aligned}$$

Therefore

$$n_{\text{s,air}} = 40.672/0.21 = 193.68 \text{ (lb-mole/h)}$$

Step 3. Determine the actual air supply based on the given value of fractional excess according to Eq. (2.50), i.e.,

$$\begin{aligned} n_{\text{air}} &= (1 + f)n_{\text{s,air}} \\ &= (1 + 0.1)(193.68) \\ &= 213.05 \text{ (lb-mole/h)} \end{aligned}$$

Therefore,

$$\begin{aligned} n_{O_2} &= 0.21 n_{\text{air}} \\ &= 44.74 \text{ (lb-mole/h)} \end{aligned}$$

and

$$\begin{aligned} n_{N_2} &= 0.79 n_{\text{air}} \\ &= 168.31 \text{ (lb-mole/h)} \end{aligned}$$

Step 4. Determine the molar flow rate of flue gas based on Eqs. (2.50)–(2.56), i.e.,

$$\begin{aligned} n_{\text{carbon dioxide}} &= n_C \\ &= 29.48 \text{ lb-mole/h} \\ n_{\text{water}} &= (n_H - n_{Cl})/2 + n_{\text{moisture}} \\ &= (44.58 - 0)/2 + 0 \\ &= 22.29 \text{ lb-mole/h} \\ n_{\text{hydrochloric acid}} &= n_{Cl} \\ &= 0.0 \text{ lb-mole/h} \\ n_{\text{sulfur dioxide}} &= n_S \\ &= 0.047 \text{ lb-mole/h} \\ n_{\text{oxygen}} &= n_O/2 + n_{O_2} - n_C - n_S - (n_H - n_{Cl})/4 \\ &= 0 + 44.74 - 29.48 - 0.047 - (44.58 - 0)/4 \\ &= 4.07 \text{ lb-mole/h} \\ n_{\text{nitrogen}} &= n_N/2 + n_{N_2} \\ &= 0 + 168.31 \\ &= 168.31 \text{ lb-mole/h} \end{aligned}$$

Step 5. Determine Q based on the given $(\text{HHV})_{\text{benzene}}$, the estimated $(\text{HHV})_{\text{oil}}$, and Eq. (2.60), i.e.,

$$Q = -(\text{HHV})_{\text{total}} + (H_{\text{sensible}})_{\text{outlet}} - (H_{\text{sensible}})_{\text{inlet}}$$

(a) For $(\text{HHV})_{\text{total}}$:

(1) Estimated $(\text{HHV})_{\text{oil}}$ based on Eq. (2.63), i.e.,

$$\begin{aligned} (\text{HHV})_{\text{oil}} &= 14{,}545 X_C + 62{,}031(X_H - 0.125 X_O) - 760 X_{Cl} + 4500 X_S \\ &= 14{,}545(0.872) + 62{,}031(0.123 - 0) - 0 + 4500(0.005) \\ &= 20{,}335.6 \text{ (Btu/lbm)} \end{aligned}$$

(2) Determine $(\mathrm{HHV})_{\mathrm{total}}$

$$\begin{aligned}(\mathrm{HHV})_{\mathrm{total}} &= \sum \{w(\mathrm{HHV})\}_i, \qquad i = \text{each inlet stream}\\ &= (100 \times 18{,}026)_{\mathrm{benzene}} + (300 \times 20{,}335.6)_{\mathrm{oil}}\\ &= 7.90 \times 10^6 \ (\mathrm{Btu/h})\end{aligned}$$

(b) For $(H_{\mathrm{sensible}})_{\mathrm{outlet}}$ at 1800°F, based on Table 2.22:

$$\begin{aligned}C_{\mathrm{pm,CO_2,1800^\circ F}} &= 11.88 \ (\mathrm{Btu/lb\text{-}mole^\circ F})\\ C_{\mathrm{pm,steam,1800^\circ F}} &= 9.22 \ (\mathrm{Btu/lb\text{-}mole^\circ F})\\ C_{\mathrm{pm,O_2,1800^\circ F}} &= 7.95 \ (\mathrm{Btu/lb\text{-}mole^\circ F})\\ C_{\mathrm{pm,N_2,1800^\circ F}} &= 7.48 \ (\mathrm{Btu/lb\text{-}mole^\circ F})\\ C_{\mathrm{pm,SO_2,1800^\circ F}} &= 12.81 \ (\mathrm{Btu/lb\text{-}mole^\circ F})\end{aligned}$$

Then

$$\begin{aligned}(H_{\mathrm{sensible}})_{\mathrm{outlet}} &= \sum \{nC_{\mathrm{pm}}(T-77)\}_i + n_{\mathrm{water}}(\Delta H_{\mathrm{v}})_{\mathrm{water}}\\ &= \{nC_{\mathrm{pm}}(T-77)\}_{\mathrm{CO_2}} + \{nC_{\mathrm{pm}}(T-77)\}_{\mathrm{H_2O}} + \{nC_{\mathrm{pm}}(T-77)\}_{\mathrm{O_2}}\\ &\quad + \{nC_{\mathrm{pm}}(T-77)\}_{\mathrm{N_2}} + n_{\mathrm{water}}(\Delta H_{\mathrm{v}})_{\mathrm{water}}\\ &= \{29.48C_{\mathrm{pm}}(1800-77)\}_{\mathrm{CO_2}} + \{22.29C_{\mathrm{pm}}(1800-77)\}_{\mathrm{H_2O}}\\ &\quad + \{4.07C_{\mathrm{pm}}(1800-77)\}_{\mathrm{O_2}} + \{168.31C_{\mathrm{pm}}(1800-77)\}_{\mathrm{N_2}}\\ &\quad + \{0.047C_{\mathrm{pm}}(1800-77)\}_{\mathrm{SO_2}} + 22.29 \times 18 \times 1050\\ &= 3.60 \times 10^6 \ (\mathrm{Btu/h})\end{aligned}$$

(c) For $(H_{\mathrm{sensible}})_{\mathrm{inlet}}$ at 77°F:

$$\begin{aligned}(H_{\mathrm{sensible}})_{\mathrm{inlet}} &= \sum \{nC_{\mathrm{pm}}(T-77)\}_i, \qquad i = \text{each flue gas species}\\ &= \{nC_{\mathrm{pm}}(T-77)\}_{\mathrm{benzene}} + \{nC_{\mathrm{pm}}(T-77)\}_{\mathrm{oil}} + \{nC_{\mathrm{pm}}(T-77)\}_{\mathrm{air}}\\ &= \{100C_{\mathrm{pm}}(77-77)\}_{\mathrm{benzene}} + \{300C_{\mathrm{pm}}(77-77)\}_{\mathrm{oil}}\\ &\quad + \{213.05C_{\mathrm{pm}}(77-77)\}_{\mathrm{air}}\\ &= 0 \ (\mathrm{Btu/h})\end{aligned}$$

Therefore,

$$\begin{aligned}Q &= -(\mathrm{HHV})_{\mathrm{total}} + (H_{\mathrm{sensible}})_{\mathrm{outlet}} - (H_{\mathrm{sensible}})_{\mathrm{inlet}}\\ &= -7.90 \times 10^6 + 3.60 \times 10^6 - 0\\ &= -4.30 \times 10^6 \ (\mathrm{Btu/h})\end{aligned}$$

EXAMPLE 3 combustion of MSW with No. 2 oil
Determine the required oil flow rate = ? (lbm/hr)

Given conditions

- Stream 1: municipal solid waste (MSW) at MSW flow rate = 500 lbm/h at 77°F
  proximate analysis: moisture 32%
  noncombustible 44%
  combustible 24%
  ultimate analysis (combustible portion):
  C: 55 wt%
  H: 20 wt%
  Cl: 5 wt%
  O: 10 wt%
  N: 5 wt%
  S: 5 wt%
  higher heating value (HHV) = 4756 Btu/lbm MSW
- Stream 2: No. 2 oil (C: 0.872; H: 0.123; S: 0.005 by mass, HHV = 20,335.6 Btu/lbm)
  Fractional excess air = 1.2 (at 77°F)
  Flue gas temperature = 2000°F

Solution
The solution to this problem requires a trial and error approach. First, it will be assumed that no auxiliary fuel is required and the standard procedure will be followed to determine Q. If Q is less than or equal to 0, then the assumption is correct. If not, then the assumption is not correct and an oil flow rate will be guessed to repeat the trial and error procedure.

(A) Assume no auxiliary fuel is required, i.e., oil flow rate = 0 lbm/h.

Step 1. Determine the flow rate of moisture and noncombustible materials (inert), and molar flow rate of C, H, Cl, O, N, and S.

$$n_{\text{moisture}} = 500 \times 0.32/18 = 8.89\ \text{lb-mole/h}$$
$$n_{\text{inert}} = 500 \times 0.44 = 220\ \text{lbm/h}$$
$$n_{\text{C}} = (500 \times 0.24 \times 0.55)/12 = 5.5\ \text{lb-mole/h}$$
$$n_{\text{H}} = (500 \times 0.24 \times 0.2)/1 = 24.0\ \text{lb-mole/h}$$
$$n_{\text{Cl}} = (500 \times 0.24 \times 0.05)/36.5 = 0.16\ \text{lb-mole/h}$$
$$n_{\text{O}} = (500 \times 0.24 \times 0.1)/16 = 0.75\ \text{lb-mole/h}$$
$$n_{\text{N}} = (500 \times 0.24 \times 0.05)/14 = 0.43\ \text{lb-mole/h}$$
$$n_{\text{S}} = (500 \times 0.24 \times 0.05)/32 = 0.19\ \text{lb-mole/h}$$

Step 2. Determine the theoretical air requirement based on Eq. (2.44), i.e.,

$$\begin{aligned} n_{\text{s,O}_2} &= n_{\text{C}} + (n_{\text{H}} - n_{\text{Cl}})/4 + n_{\text{S}} - n_{\text{O}}/2 \\ &= 5.5 + (24.0 - 0.16)/4 + 0.19 - 0.75/2 \\ &= 11.275\ (\text{lb-mole/h}) \end{aligned}$$

Therefore,

$$n_{\text{s,air}} = 11.275/0.21 = 53.69\ (\text{lb-mole/h})$$

Step 3. Determine the actual air supply based on the given value of fractional excess according to Eq. (2.50), i.e.,

$$\begin{aligned} n_{\text{air}} &= (1 + f)n_{\text{s,air}} \\ &= (1 + 1.2)(53.69) \\ &= 118.12 \text{ (lb-mole/h)} \end{aligned}$$

Therefore,

$$\begin{aligned} n_{O_2} &= 0.21 n_{\text{air}} \\ &= 24.81 \text{ (lb-mole/h)} \end{aligned}$$

and

$$\begin{aligned} n_{N_2} &= 0.79 n_{\text{air}} \\ &= 93.31 \text{ (lb-mole/h)} \end{aligned}$$

Step 4. Determine the molar flow rate of flue gas based on Eqs. (2.50)–(2.56), i.e.,

$$\begin{aligned} n_{\text{carbon dioxide}} &= n_{\text{C}} \\ &= 5.5 \text{ lb-mole/h} \\ n_{\text{water}} &= (n_{\text{H}} - n_{\text{Cl}})/2 + n_{\text{moisture}} \\ &= (24.0 - 0.16)/2 + 8.89 \\ &= 20.81 \text{ lb-mole/h} \\ n_{\text{hydrochloric acid}} &= n_{\text{Cl}} \\ &= 0.16 \text{ lb-mole/h} \\ n_{\text{sulfur dioxide}} &= n_{\text{S}} \\ &= 0.19 \text{ lb-mole/h} \\ n_{\text{oxygen}} &= n_{\text{O}}/2 + n_{O_2} - n_{\text{C}} - n_{\text{S}} - (n_{\text{H}} - n_{\text{Cl}})/4 \\ &= 0.75/2 + 24.81 - 5.5 - 0.19 - (24.0 - 0.16)/4 \\ &= 13.54 \text{ lb-mole/h} \\ n_{\text{nitrogen}} &= n_{\text{N}}/2 + n_{N_2} \\ &= 0.43/2 + 93.31 \\ &= 93.53 \text{ lb-mole/h} \end{aligned}$$

Step 5. Determine Q based on the given $(\text{HHV})_{\text{MSW}}$ and Eq. (2.60), i.e.,

$$Q = -(\text{HHV})_{\text{total}} + (H_{\text{sensible}})_{\text{outlet}} - (H_{\text{sensible}})_{\text{inlet}}$$

(a) For $(\text{HHV})_{\text{total}}$:

$$\begin{aligned} (\text{HHV})_{\text{total}} &= \sum \{w(\text{HHV})\}_i, \qquad i = \text{each inlet stream} \\ &= (500 \times 4756)_{\text{MSW}} \\ &= 2.38 \times 10^6 \text{ (Btu/h)} \end{aligned}$$

(b) For $(H_{sensible})_{outlet}$ at 2000°F, based on Table 2.22:

$$C_{pm,CO_2,2000°F} = 12.07 \text{ (Btu/lb-mole°F)}$$
$$C_{pm,H_2O,2000°F} = 9.37 \text{ (Btu/lb-mole°F)}$$
$$C_{pm,HCl,2000°F} = 7.70 \text{ (Btu/lb-mole°F)}$$
$$C_{pm,SO_2,2000°F} = 12.97 \text{ (Btu/lb-mole°F)}$$
$$C_{pm,O_2,2000°F} = 8.02 \text{ (Btu/lb-mole°F)}$$
$$C_{pm,N_2,2000°F} = 7.55 \text{ (Btu/lb-mole°F)}$$

and, assuming

$$C_{pm,inert,2000°F} = 0.24 \text{ (Btu/lbm°F)}$$

then

$$\begin{aligned}(H_{sensible})_{outlet} &= \sum \{nC_{pm}(T-77)\}_i + n_{water}(\Delta H_v)_{water} \\ &= \{nC_{pm}(T-77)\}_{CO_2} + \{nC_{pm}(T-77)\}_{H_2O} + \{nC_{pm}(T-77)\}_{HCl} \\ &\quad + \{nC_{pm}(T-77)\}_{SO_2} + \{nC_{pm}(T-77)\}_{O_2} + \{nC_{pm}(T-77)\}_{N_2} \\ &\quad + \{nC_{pm}(T-77)\}_{inert} + n_{water}(\Delta H_v)_{water} \\ &= \{5.5C_{pm}(2000-77)\}_{CO_2} + \{20.81C_{pm}(2000-77)\}_{H_2O} \\ &\quad + \{0.16C_{pm}(2000-77)\}_{HCl} + \{0.19C_{pm}(2000-77)\}_{SO_2} \\ &\quad + \{13.54C_{pm}(2000-77)\}_{O_2} + \{93.53C_{pm}(2000-77)\}_{N_2} \\ &\quad + \{220C_{pm}(2000-77)\}_{inert} + 20.81 \times 18 \times 1050 \\ &= 2.18 \times 10^6 + 0.39 \times 10^6 \\ &= 2.57 \times 10^6 \text{ (Btu/h)}\end{aligned}$$

(c) For $(H_{sensible})_{inlet}$ at 77°F:

$$\begin{aligned}(H_{sensible})_{inlet} &= \sum \{nC_{pm}(T-77)\}_i, \qquad i = \text{each inlet stream} \\ &= \{nC_{pm}(T-77)\}_{MSW} + \{nC_{pm}(T-77)\}_{air} \\ &= \{500C_{pm}(77-77)\}_{MSW} + \{118.12C_{pm}(77-77)\}_{air} \\ &= 0 \text{ (Btu/h)}\end{aligned}$$

Therefore,

$$\begin{aligned}Q &= -(\text{HHV})_{total} + (H_{sensible})_{outlet} - (H_{sensible})_{inlet} \\ &= -2.38 \times 10^6 + 2.57 \times 10^6 - 0 \\ &= 0.19 \times 10^6 \text{ (Btu/h)}\end{aligned}$$

Discussion: since Q is positive, the assumption that no auxiliary fuel is required is not correct. To continue the trial and error procedure, assume oil flow rate = 100 lbm/h.

(B) Assume oil flow rate = 100 lbm/h

Step 1. Determine the combined flow rate of moisture and noncombustible materials (inert), and molar flow rates of C, H, Cl, O, N, and S.

(a) MSW

$$
\begin{aligned}
n_{\text{moisture}} &= 500 \times 0.32/18 = 8.89 \text{ lb-mole/h} \\
n_{\text{inert}} &= 500 \times 0.44 = 220 \text{ lbm/h} \\
n_{\text{C}} &= (500 \times 0.24 \times 0.55)/12 = 5.5 \text{ lb-mole/h} \\
n_{\text{H}} &= (500 \times 0.24 \times 0.2)/1 = 24.0 \text{ lb-mole/h} \\
n_{\text{Cl}} &= (500 \times 0.24 \times 0.05)/36.5 = 0.16 \text{ lb-mole/h} \\
n_{\text{S}} &= (500 \times 0.24 \times 0.05)/32 = 0.19 \text{ lb-mole/h} \\
n_{\text{O}} &= (500 \times 0.24 \times 0.1)/16 = 0.75 \text{ lb-mole/h} \\
n_{\text{N}} &= (500 \times 0.24 \times 0.05)/14 = 0.43 \text{ lb-mole/h}
\end{aligned}
$$

(b) No. 2 oil

$$
\begin{aligned}
n_{\text{C}} &= (100 \times 0.872)/12 = 7.27 \text{ lb-mole/h} \\
n_{\text{H}} &= (100 \times 0.123)/1 = 12.3 \text{ lb-mole/h} \\
n_{\text{S}} &= (100 \times 0.005)/32 = 0.016 \text{ lb-mole/h}
\end{aligned}
$$

(c) Combined (MSW + oil)

$$
\begin{aligned}
n_{\text{moisture}} &= 8.89 + 0 = 8.89 \text{ lb-mole/h} \\
n_{\text{inert}} &= 220 + 0 = 220 \text{ lbm/h} \\
n_{\text{C}} &= 5.5 + 7.27 = 12.77 \text{ lb-mole/h} \\
n_{\text{H}} &= 24.0 + 12.3 = 36.30 \text{ lb-mole/h} \\
n_{\text{Cl}} &= 0.16 + 0 = 0.16 \text{ lb-mole/h} \\
n_{\text{S}} &= 0.19 + 0.016 = 0.21 \text{ lb-mole/h} \\
n_{\text{O}} &= 0.75 + 0 = 0.75 \text{ lb-mole/h} \\
n_{\text{N}} &= 0.43 + 0 = 0.43 \text{ lb-mole/h}
\end{aligned}
$$

Step 2. Determine the theoretical air requirement based on Eq. (2.44), i.e.,

$$
\begin{aligned}
n_{\text{s},\text{O}_2} &= n_{\text{C}} + (n_{\text{H}} - n_{\text{Cl}})/4 + n_{\text{S}} - n_{\text{O}}/2 \\
&= 12.77 + (36.3 - 0.16)/4 + 0.21 - 0.75/2 \\
&= 21.64 \text{ (lb-mole/h)}
\end{aligned}
$$

Therefore

$$
n_{\text{s,air}} = 21.64/0.21 = 103.05 \text{ (lb-mole/h)}
$$

Step 3. Determine the actual air supply based on the given value of fractional excess according to Eq. (2.50), i.e.,

$$
\begin{aligned}
n_{\text{air}} &= (1 + f)n_{\text{s,air}} \\
&= (1 + 1.2)(103.05) \\
&= 226.71 \text{ (lb-mole/h)}
\end{aligned}
$$

Therefore,

$$\begin{aligned} n_{O_2} &= 0.21 n_{\text{air}} \\ &= 47.61 \text{ (lb-mole/h)} \end{aligned}$$

and

$$\begin{aligned} n_{N_2} &= 0.79 n_{\text{air}} \\ &= 179.10 \text{ (lb-mole/h)} \end{aligned}$$

Step 4. Determine the molar flow rate of flue gas based on Eqs. (2.50)–(2.56), i.e.,

$$\begin{aligned} n_{\text{carbon dioxide}} &= n_{C} \\ &= 12.77 \text{ lb-mole/h} \\ n_{\text{water}} &= (n_{H} - n_{Cl})/2 + n_{\text{moisture}} \\ &= (36.3 - 0.16)/2 + 8.89 \\ &= 26.96 \text{ lb-mole/h} \\ n_{\text{hydrochloric acid}} &= n_{Cl} \\ &= 0.16 \text{ lb-mole/h} \\ n_{\text{sulfur dioxide}} &= n_{S} \\ &= 0.21 \text{ lb-mole/h} \\ n_{\text{oxygen}} &= n_{O}/2 + n_{O_2} - n_{C} - n_{S} - (n_{H} - n_{Cl})/4 \\ &= 0.75/2 + 47.61 - 12.77 - 0.21 - (36.3 - 0.16)/4 \\ &= 25.97 \text{ lb-mole/h} \\ n_{\text{nitrogen}} &= n_{N}/2 + n_{N_2} \\ &= 0.43/2 + 179.1 \\ &= 179.32 \text{ lb-mole/h} \end{aligned}$$

Step 5. Determine Q based on the given $(\text{HHV})_{\text{MSW}}$ and Eq. (2.60), i.e.,

$$Q = -(\text{HHV})_{\text{total}} + (H_{\text{sensible}})_{\text{outlet}} - (H_{\text{sensible}})_{\text{inlet}}$$

(a) For $(\text{HHV})_{\text{total}}$:

$$\begin{aligned} (\text{HHV})_{\text{total}} &= \sum \{w(\text{HHV})\}_i, \qquad i = \text{each inlet stream} \\ &= (500 \times 4756)_{\text{MSW}} + (100 \times 20{,}335.6)_{\text{oil}} \\ &= 4.41 \times 10^6 \text{ (Btu/h)} \end{aligned}$$

(b) For $(H_{\text{sensible}})_{\text{outlet}}$ at 2000°F, based on Table 2.22:

$$C_{pm,CO_2,2000°F} = 12.07 \text{ (Btu/lb-mole°F)}$$
$$C_{pm,H_2O,2000°F} = 9.37 \text{ (Btu/lb-mole°F)}$$
$$C_{pm,HCl,2000°F} = 7.70 \text{ (Btu/lb-mole°F)}$$
$$C_{pm,SO_2,2000°F} = 12.97 \text{ (Btu/lb-mole°F)}$$
$$C_{pm,O_2,2000°F} = 8.02 \text{ (Btu/lb-mole°F)}$$
$$C_{pm,N_2,2000°F} = 7.55 \text{ (Btu/lb-mole°F)}$$

and, assuming

$$C_{pm,inert,2000°F} = 0.24 \text{ (Btu/lbm°F)}$$

Then

$$\begin{aligned}(H_{sensible})_{outlet} &= \sum \{nC_{pm}(T-77)\}_i + n_{water}(\Delta H_v)_{water} \\ &= \{nC_{pm}(T-77)\}_{CO_2} + \{nC_{pm}(T-77)\}_{H_2O} + \{nC_{pm}(T-77)\}_{HCl} \\ &\quad + \{nC_{pm}(T-77)\}_{SO_2} + \{nC_{pm}(T-77)\}_{O_2} + \{nC_{pm}(T-77)\}_{N_2} \\ &\quad + \{nC_{pm}(T-77)\}_{inert} + n_{water}(\Delta H_v)_{water} \\ &= \{12.77C_{pm}(2000-77)\}_{CO_2} + \{26.96C_{pm}(2000-77)\}_{H_2O} \\ &\quad + \{0.16C_{pm}(2000-77)\}_{HCl} + \{0.21C_{pm}(2000-77)\}_{SO_2} \\ &\quad + \{25.97C_{pm}(2000-77)\}_{O_2} + \{179.32C_{pm}(2000-77)\}_{N_2} \\ &\quad + \{220C_{pm}(2000-77)\}_{inert} + 26.96 \times 18 \times 1050 \\ &= 3.90 \times 10^6 + 0.51 \times 10^6 \\ &= 4.41 \times 10^6 \text{ (Btu/h)}\end{aligned}$$

(c) For $(H_{sensible})_{inlet}$ at 77°F

$$\begin{aligned}(H_{sensible})_{inlet} &= \sum \{nC_{pm}(T-77)\}_i, \qquad i = \text{each inlet stream} \\ &= \{nC_{pm}(T-77)\}_{MSW} + \{nC_{pm}(T-77)\}_{air} \\ &= \{500C_{pm}(77-77)\}_{MSW} + \{226.71C_{pm}(77-77)\}_{air} \\ &= 0 \text{ (Btu/h)}\end{aligned}$$

Therefore

$$\begin{aligned}Q &= -(\text{HHV})_{total} + (H_{sensible})_{outlet} - (H_{sensible})_{inlet} \\ &= -4.41 \times 10^6 + 4.41 \times 10^6 - 0 \\ &= 0 \text{ (Btu/h)}\end{aligned}$$

Discussion: this calculation result indicates that the guessed oil flow rate, i.e., 100 lbm/h, is exact for the combustion process. Note that if there is heat loss involved in the process, then the oil flow rate must be increased correspondingly.

EXAMPLE 4 combustion of MSW with preheated air
Determine the required oil flow rate = ? (lbm/h).

Given conditions

- Stream 1: municipal solid waste (MSW)
  MSW flow rate = 500 lbm/h at 77°F
  proximate analysis: moisture 32%
  noncombustible 44%
  combustible 24%
  ultimate analysis (combustible portion):
  C: 55 wt%
  H: 20 wt%
  Cl: 5 wt%
  O: 10 wt%
  N: 5 wt%
  S: 5 wt%
  higher heating value (HHV) = 4756 Btu/lbm MSW
- Stream 2: No. 2 oil (C: 0.872; H: 0.123; S: 0.005 by mass, HHV = 20, 355.6 Btu/lbm)
  Fractional excess air = 1.2 (preheated to 300°F)
  Flue gas temperature = 2000°F

Solution:
This problem is exactly identical to Example 3 except that the supplied air is preheated to 300°F instead of 77°F. Similar to that for Example 3, the solution to this problem requires a trial and error approach. First, it will be assumed that no auxiliary fuel is required.

(A) Assume no auxiliary fuel is required, i.e., oil flow rate = 0 lbm/h.

Step 1. Determine the flow rate of moisture and noncombustible materials (inert), and molar flow rate of C, H, Cl, O, N, and S.

$$\begin{aligned} n_{\text{moisture}} &= 500 \times 0.32/18 = 8.89 \text{ lb-mole/h} \\ n_{\text{inert}} &= 500 \times 0.44 = 220 \text{ lbm/h} \\ n_{\text{C}} &= (500 \times 0.24 \times 0.55)/12 = 5.5 \text{ lb-mole/h} \\ n_{\text{H}} &= (500 \times 0.24 \times 0.2)/1 = 24.0 \text{ lb-mole/h} \\ n_{\text{Cl}} &= (500 \times 0.24 \times 0.05)/36.5 = 0.16 \text{ lb-mole/h} \\ n_{\text{O}} &= (500 \times 0.24 \times 0.1)/16 = 0.75 \text{ lb-mole/h} \\ n_{\text{N}} &= (500 \times 0.24 \times 0.05)/14 = 0.43 \text{ lb-mole/h} \\ n_{\text{S}} &= (500 \times 0.24 \times 0.05)/32 = 0.19 \text{ lb-mole/h} \end{aligned}$$

Step 2. Determine the theoretical air requirement based on Eq. (2.44), i.e.,

$$\begin{aligned} n_{\text{s,O}_2} &= n_{\text{C}} + (n_{\text{H}} - n_{\text{Cl}})/4 + n_{\text{S}} - n_{\text{O}}/2 \\ &= 5.5 + (24.0 - 0.16)/4 + 0.19 - 0.75/2 \\ &= 11.275 \text{ (lb-mole/h)} \end{aligned}$$

Therefore,

$$n_{\text{s,air}} = 11.275/0.21 = 53.69 \text{ (lb-mole/h)}$$

Step 3. Determine the actual air supply based on the given value of fractional excess according to Eq. (2.50), i.e.,

$$\begin{aligned} n_{\text{air}} &= (1+f)n_{\text{s,air}} \\ &= (1+1.2)(53.69) \\ &= 118.12 \text{ (lb-mole/h)} \end{aligned}$$

Therefore,

$$\begin{aligned} n_{\text{O}_2} &= 0.21 n_{\text{air}} \\ &= 24.81 \text{ (lb-mole/h)} \end{aligned}$$

and

$$\begin{aligned} n_{\text{N}_2} &= 0.79 n_{\text{air}} \\ &= 93.31 \text{ (lb-mole/h)} \end{aligned}$$

Step 4. Determine the molar flow rate of flue gas based on Eqs. (2.50)–(2.56), i.e.,

$$\begin{aligned} n_{\text{carbon dioxide}} &= n_{\text{C}} \\ &= 5.5 \text{ lb-mole/h} \\ n_{\text{water}} &= (n_{\text{H}} - n_{\text{Cl}})/2 + n_{\text{moisture}} \\ &= (24.0 - 0.16)/2 + 8.89 \\ &= 20.81 \text{ lb-mole/h} \\ n_{\text{hydrochloric acid}} &= n_{\text{Cl}} \\ &= 0.16 \text{ lb-mole/h} \\ n_{\text{sulfur dioxide}} &= n_{\text{S}} \\ &= 0.19 \text{ lb-mole/h} \\ n_{\text{oxygen}} &= n_{\text{O}}/2 + n_{\text{O}_2} - n_{\text{C}} - n_{\text{S}} - (n_{\text{H}} - n_{\text{Cl}})/4 \\ &= 0.75/2 + 24.81 - 5.5 - 0.19 - (24.0 - 0.16)/4 \\ &= 13.54 \text{ lb-mole/h} \\ n_{\text{nitrogen}} &= n_{\text{N}}/2 + n_{\text{N}_2} \\ &= 0.43/2 + 93.31 \\ &= 93.53 \text{ lb-mole/h} \end{aligned}$$

Step 5. Determine Q based on the given $(\text{HHV})_{\text{MSW}}$ and Eq. (2.60), i.e.,

$$Q = -(\text{HHV})_{\text{total}} + (H_{\text{sensible}})_{\text{outlet}} - (H_{\text{sensible}})_{\text{inlet}}$$

(a) For $(\text{HHV})_{\text{total}}$:

$$\begin{aligned} (\text{HHV})_{\text{total}} &= \sum \{w(\text{HHV})\}_i, \qquad i = \text{each inlet stream} \\ &= (500 \times 4756)_{\text{MSW}} \\ &= 2.38 \times 10^6 \text{ (Btu/h)} \end{aligned}$$

(b) For $(H_{\text{sensible}})_{\text{outlet}}$ at 2000°F, based on Table 2.22:

$$C_{\text{pm,CO}_2,2000^\circ\text{F}} = 12.07 \text{ (Btu/lb-mole}^\circ\text{F)}$$
$$C_{\text{pm,H}_2\text{O},2000^\circ\text{F}} = 9.37 \text{ (Btu/lb-mole}^\circ\text{F)}$$
$$C_{\text{pm,HCl},2000^\circ\text{F}} = 7.70 \text{ (Btu/lb-mole}^\circ\text{F)}$$
$$C_{\text{pm,SO}_2,2000^\circ\text{F}} = 12.97 \text{ (Btu/lb-mole}^\circ\text{F)}$$
$$C_{\text{pm,O}_2,2000^\circ\text{F}} = 8.02 \text{ (Btu/lb-mole}^\circ\text{F)}$$
$$C_{\text{pm,N}_2,2000^\circ\text{F}} = 7.55 \text{ (Btu/lb-mole}^\circ\text{F)}$$

and assuming

$$C_{\text{pm,inert},2000^\circ\text{F}} = 0.24 \text{ (Btu/lbm}^\circ\text{F)}$$

then

$$\begin{aligned}(H_{\text{sensible}})_{\text{outlet}} &= \sum \{nC_{\text{pm}}(T-77)\}_i + n_{\text{water}}(\Delta H_{\text{v}})_{\text{water}} \\ &= \{nC_{\text{pm}}(T-77)\}_{\text{CO}_2} + \{nC_{\text{pm}}(T-77)\}_{\text{H}_2\text{O}} + \{nC_{\text{pm}}(T-77)\}_{\text{HCl}} \\ &\quad + \{nC_{\text{pm}}(T-77)\}_{\text{SO}_2} + \{nC_{\text{pm}}(T-77)\}_{\text{O}_2} + \{nC_{\text{pm}}(T-77)\}_{\text{N}_2} \\ &\quad + \{nC_{\text{pm}}(T-77)\}_{\text{inert}} + n_{\text{water}}(\Delta H_{\text{v}})_{\text{water}} \\ &= \{5.5C_{\text{pm}}(2000-77)\}_{\text{CO}_2} + \{20.81C_{\text{pm}}(2000-77)\}_{\text{H}_2\text{O}} \\ &\quad + \{0.16C_{\text{pm}}(2000-77)\}_{\text{HCl}} + \{0.19C_{\text{pm}}(2000-77)\}_{\text{SO}_2} \\ &\quad + \{13.54C_{\text{pm}}(2000-77)\}_{\text{O}_2} + \{93.53C_{\text{pm}}(2000-77)\}_{\text{N}_2} \\ &\quad + \{220C_{\text{pm}}(2000-77)\}_{\text{inert}} + 20.81 \times 18 \times 1050 \\ &= 2.18 \times 10^6 + 0.39 \times 10^6 \\ &= 2.57 \times 10^6 \text{ (Btu/h)}\end{aligned}$$

(c) For $(H_{\text{sensible}})_{\text{inlet}}$ at 77°F and 300°F:

$$\begin{aligned}(H_{\text{sensible}})_{\text{inlet}} &= \sum \{nC_{\text{pm}}(T-77)\}_i, \qquad i = \text{each inlet stream} \\ &= \{nC_{\text{pm}}(T-77)\}_{\text{MSW}} + \{nC_{\text{pm}}(T-77)\}_{\text{air}} \\ &= \{500C_{\text{pm}}(77-77)\}_{\text{MSW}} + \{118.12C_{\text{pm}}(300-77)\}_{\text{air}} \\ &= 0 + 118.12 \times 7.04 \times 223 \\ &= 0.19 \times 10^6 \text{ (Btu/h)}\end{aligned}$$

Therefore,

$$\begin{aligned}Q &= -(\text{HHV})_{\text{total}} + (H_{\text{sensible}})_{\text{outlet}} - (H_{\text{sensible}})_{\text{inlet}} \\ &= -2.38 \times 10^6 + 2.57 \times 10^6 - 0.19 \times 10^6 \\ &= 0 \text{ (Btu/h)}\end{aligned}$$

Discussion: this calculation result indicates that no auxiliary fuel is required if the air is preheated to 300°F. Note that its effectiveness is equivalent to adding 100 lbm/h of the No. 2 oil as auxiliary fuel, as calculated in Example 3.

Nomenclature for this example

- C_{pm,CO_2} = heat capacity of CO_2(g), Btu/lb-mole°F
- $C_{pm,HCl}$ = heat capacity of HCl(g), Btu/lb-mole°F
- C_{pm,H_2O} = heat capacity of H_2O(g), Btu/lb-mole°F
- $C_{pm,inert}$ = heat capacity of inert materials, Btu/lbm°F
- C_{pm,N_2} = heat capacity of N_2(g), Btu/lb-mole°F
- C_{pm,O_2} = heat capacity of O_2(g), Btu/lb-mole°F
- C_{pm,SO_2} = heat capacity of SO_2(g), Btu/lb-mole°F
- f = fractional excess of oxygen or air supply
- $H_{sensible}$ = enthalpy relative to process species at 77°F, Btu
- $(H_{sensible})_{inlet}$ = total inlet rate of $H_{sensible}$ defined in Eq. (2.62), Btu/h
- $(H_{sensible})_{outlet}$ = total outlet rate of $H_{sensible}$ defined in Eq. (2.61), Btu/h
- H_{total} = enthalpy relative to atomic species at 77°F, Btu
- $(H_{total})_{inlet}$ = total inlet rate of H_{total}, Btu/h
- $(H_{total})_{outlet}$ = total outlet rate of H_{total}, Btu/h
- HHV = higher heating value, Btu/lbm or Btu/lb-mole
- $(\text{HHV})_{methane}$ = HHV of methane, Btu/lb-mole
- $(\text{HHV})_{total}$ = total inlet rate of HHV, Btu/h
- M = molecular weight, lbm/lb-mole
- $M_{benzene}$ = molecular weight of benzene, 78 lbm/lb-mole
- n_{air} = molar flow rate of air supply, lb-mole/h
- n_C = molar flow rate of C in an inlet stream, lb-mole/h
- n_{CH_4} = molar flow rate of methane, lb-mole/h
- n_{Cl} = molar flow rate of Cl in an inlet stream, lb-mole/h
- $n_{carbon\ dioxide}$ = molar flow rate of CO_2(g) in the outlet stream, lb-mole/h
- n_H = molar flow rate of H in an inlet stream, lb-mole/h
- $n_{hydrochloric\ acid}$ = molar flow rate of HCl(g) in the outlet stream, lb-mole/h
- n_{inert} = molar flow rate of inert, lbm/h
- $n_{moisture}$ = molar flow rate of water in an inlet stream, lb-mole/h
- $n_{nitrogen}$ = molar flow rate of N_2(g) in the outlet stream lb-mole/h
- n_N = molar flow rate of N in an inlet stream, lb-mole/h
- n_{N_2} = molar flow rate of N_2 in air supply, lb-mole/h
- n_O = molar flow rate of O_2 in an inlet stream, lb-mole/h
- n_{O_2} = molar flow rate of O_2 in air supply, lb-mole/h
- n_{oxygen} = molar flow rate of O_2(g) in the outlet stream, lb-mole/h
- n_S = molar flow rate of S in an inlet stream, lb-mole/h
- $n_{s,air}$ = theoretical air required for complete combustion, lb-mole/h
- n_{s,O_2} = theoretical oxygen required for complete combustion, lb-mole/h
- $n_{sulfur\ dioxide}$ = molar flow rate of SO_2(g) in the outlet stream, lb-mole/h
- n_{water} = molar flow rate of H_2O(g) in the outlet stream, lb-mole/h
- P = pressure, psia

- Q = rate of heat transfer, Btu/h
- R = gas constant, 10.73 (psia-ft^3)/lb-mole-°R
- T = temperature, °F or °R
- V = volume flow rate, ft^3/h
- X_C = mass fraction of carbon in an inlet stream
- X_{Cl} = mass fraction of chlorine in an inlet stream
- X_H = mass fraction of hydrogen in an inlet stream
- X_O = mass fraction of oxygen in an inlet stream
- X_S = mass fraction of sulfur in an inlet stream
- w = mass or mole flow rate of an inlet stream, lbm/h or lb-mole/h
- $(\Delta H_v)_{water}$ = heat of vaporization of water at 77°F, Btu/lb-mole

4 BASIC INCINERATOR DESIGN

This section describes the key items needed for the design of waste incinerators. These include:

- adiabatic temperature
- waste properties
- auxiliary fuel
- incinerator design calculation
- system design feature
- refractory selection
- exhaust gas monitoring
- acid gas corrosion

4.1 Adiabatic Temperature

The temperature at which a waste will burn without considering process losses is its adiabatic temperature. These process losses would include radiation losses from the furnace walls and heat lost in the ash discharge. The adiabatic temperature is a function of the waste combustible heating value, waste combustible content, moisture content and excess air requirement. As a first-order approximation of adiabatic temperature for any combustible material, the as-received waste heating value can be used. Figure 2.10 is a graph of adiabatic temperature versus gross (as-received or as-fired) heating value for various values of excess air. It is applicable for any material, liquid, solid, or gas. Excess air values for typical furnaces are given in Table 2.23 (OME-88/12, p. 4-12).

Using Fig. 2.10, for example, a waste with an as-received heating value of 8000 kJ/kg (3440 Btu/lb) is burned in a fluid bed incinerator using 60% excess air. As an example, the combustion of this waste is assumed to require 1000°C (1830°F) for effective destruction. From the graph, the adiabatic temperature is approximately 900°C (1650°F), and therefore supplemental fuel is required to achieve a combustion temperature of 1000°C (1830°F).

This graph gives an approximate temperature of combustion and provides an indication of whether or not to proceed with a detailed analysis of incinerator parameters.

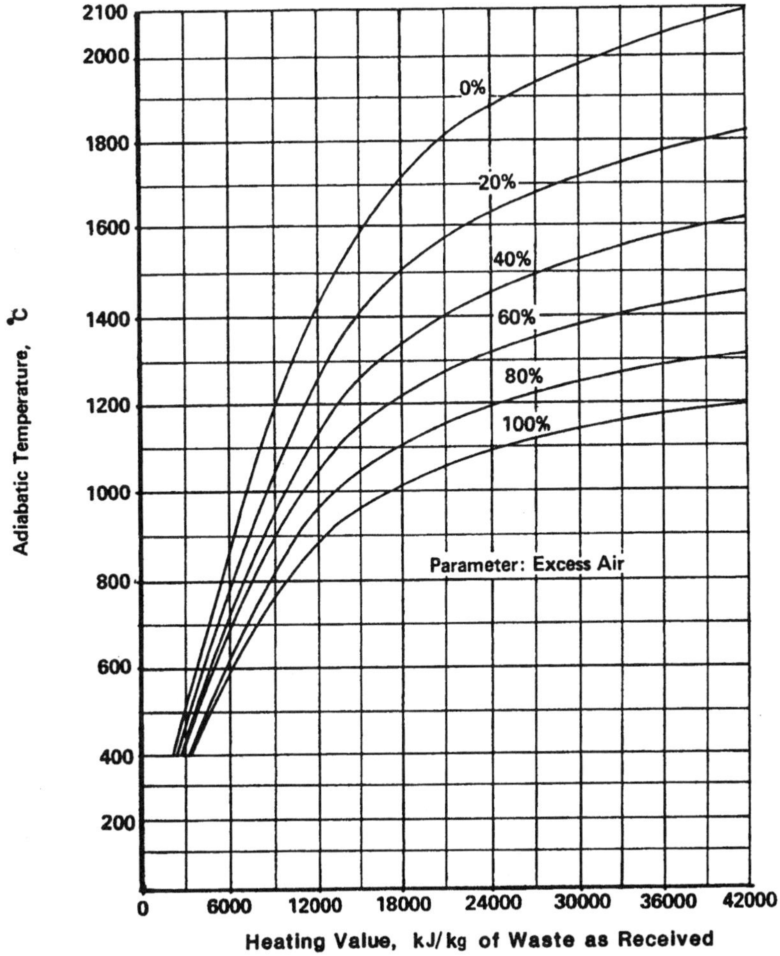

FIGURE 2.10 Adiabatic temperature vs. excess air (*OME-88/12, p. 4-11*).

TABLE 2.23 Typical Excess Air Values

Incinerator type	Excess air range
Fluid bed	40–60
Gaseous waste	10–15
Liquid waste	15–30
Multiple chamber (retort, in-line)	100–200
Multiple hearth	100–125
Rotary kiln	100
Waterwall	65–100

4.2 Incinerator Step-by-step Calculation

There are a number of incinerator parameters that are normally of interest, including:

- furnace temperature
- supplemental fuel requirement
- combustion air requirement
- flue gas discharge quantity
- residence time
- turbulence/mixing

From these parameters, the sizes of equipment associated with the incinerator system can be calculated.

Table 2.24 (OME-88/12, p. 4-14) represents a step-by-step method of analysis for an incinerator system (see sludge incineration sample calculations in Chapter 2.3). It is a simplified mass and heat balance calculation. The feed quantity, moisture fraction, noncombustible fraction, and combustible heating value must be known. An excess air and humidity figure must be selected, as well as a reasonable radiation loss figure (from 1% to 5% of the total heat release) and an ash heat content (as will normally represent a heat loss in the range 160–340 kJ/kg [69–146 Btu/lb] of ash). The required gas temperature for destruction of the waste stream must be selected.

This analysis is a detailed heat balance based on the principle that the total heat leaving (exiting) a system is equal to the heat entering that system. Heat entering a system includes the potential heat release of a waste, and/or fuel, that is fired within that system.

One important assumption made in this analysis is that flue gas from an incinerator is composed of moisture plus dry air. Actual incinerator off-gas contains more carbon dioxide and less oxygen than is found in air. The assumption that dry flue gas has the properties of air, however, greatly simplifies calculations while introducing a relatively small error (less than 3%) in calculated temperatures and heat requirements.

The air quantities required to support combustion must be determined to perform these calculations. If the exact waste composition is known, the air requirements can be determined by calculation from the equilibrium equations, as follows:

- for combustion of carbon, $C + O_2 \rightarrow CO_2$
- for combustion of hydrogen, $H_2 + 0.5O_2 \rightarrow H_2O$
- for sulfur, $S + O_2 \rightarrow SO_2$

With each kilogram of oxygen, nitrogen is present in air:

- 1 kg air = 0.2315 kg O_2 + 0.7685 kg N_2
- 1 kg oxygen carries 3.3197 kg nitrogen from air

Usually there is some uncertainty about the range of waste composition and properties. The curves in Figs. 2.11 and 2.12 were generated based on the DuLong equation. Using this equation, a relationship between the composition and heating value was established and plotted. Figure 2.11 relates the stoichiometric air requirements to heating value; Fig. 2.12 relates moisture produced from combustion to heating value. These values are approximate, recognizing that the DuLong equation represents only an estimate of heating value.

Figures 2.13 and 2.14 and Table 2.25 indicate the net heat available from supplemental fuel combustion in an incineration process. As the temperature of the process (the incinerator exit temperature) increases, the heat required to raise the temperature of the products of combustion of supplemental fuel to the incinerator exit temperature increases. The heat available for process heating (the net heating value) is the heat of combustion of the fuel less heat lost to the

TABLE 2.24 Incinerator Analysis

Step	Description	Calculation	Unit
1	Waste feed rate (given)		kg/h
2	Water fraction in waste (given)		
3	Water quantity in waste (step 1 × 2)		kg/h
4	Solid quantity (step 1 − step 3)		kg/h
5	Noncombustible fraction (given)		
6	Noncombustible quantity (step 4 × 5)		kg/h
7	Combustible quantity (step 4 − step 6)		kg/h
8	Heating value of combustibles (given)		kJ/kg
9	Total heat generated (step 7 × 8)		kJ/h
10	Theoretical (stoichiometric) air ratio (Fig. 2.11)		
11	Theoretical air (step 7 × 10)		kg/h
12	Excess air fraction (given)		
13	Excess air required (step 11 × 12)		kg/h
14	Total air required (steps 11 + 13)		kg/h
15	Water formed due to combustion of combustible waste (Fig. 2.12)		kg-water/ kg-combustible
16	Total water due to combustion (step 7 × 15)		kg/h
17	Humidity in air (given)		kg-water/kg-air
18	Water due to humidity (step 14 × 17)		kg/h
19	Total water in flue gas (steps 3 + 16 + 18)		kg/h
20	Dry flue gas (steps 7 + 14–16)		kg/h
21	Radiation loss fraction (given)		
22	Radiation heat loss (step 9 × 21)		kJ/h
23	Ash heating value (given)		kJ/kg
24	Ash heat loss (step 6 × 23)		kJ/kg
25	Humidity correction due to latent heat (step18) × 2186		kJ/h
26	Miscellaneous heat loss (given)		kJ/h
27	Heat in flue gas (steps 9 − 22 − 24 + 25 − 26)		kJ/h
28	Estimate flue gas temperature*		°C
29	Required gas temperature (given)		°C
30	Required enthalpy of flue gas†		kJ/h
31	Net heat required (step 30 − step 27)		kJ/h
32	Fuel excess air (given)		
33	Fuel heating value at required temperature (Fig. 2.13 or 2.14)		kJ/L
34	Fuel required (step 31/33)		L/h
35	Air for fuel (Table 2.25)		kg air/L
36	Total air required for fuel (step 35 × 34)		kg/h
37	Moisture from fuel (Table 2.25)		kg/L
38	Moisture generated (step 37 × 34)		kg/h
39	Dry flue gas from fuel combustion (Table 2.25)		kg/L
40	Dry gas generated (step 39 × 34)		kg/h
41	Total dry gas flow (steps 40 + 20)		kg/h
42	Total moisture flow (steps 38 + 19)		kg/h
43	Fuel gross heating value (Table 2.25)		kJ/L
44	Heat generated from fuel (step 43 × 34)		kJ/h
45	Total heat at exit (steps 44 + 27)		kJ/h

* Calculation for step 28. The flue gas temperature can be calculated from the heat in the flue and Figs. 2.15 and 2.16. The procedure involves estimating a temperature, calculating the flue gas enthalpy, and comparing the calculated enthalpy with the enthalpy in step 27. Repeat these steps until the calculated enthalpy and the enthalpy in step 27 are nearly equal. Actual calculations are as in the next footnote.

† Calculation for step 30. The required enthalpy of the flue gas can be calculated from the required temperature, 760°C, and Figs. 2.15 and 2.16. The procedures are similar to those in the previous footnote.

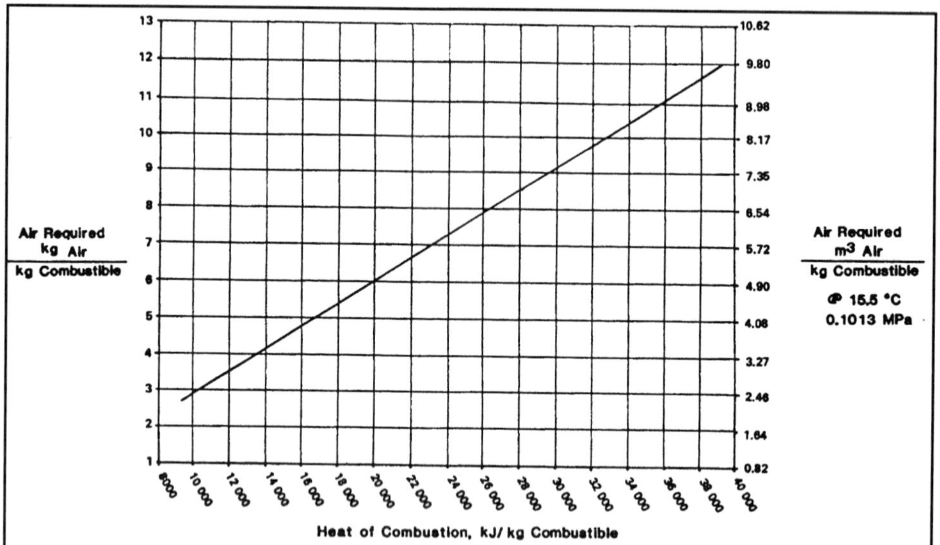

FIGURE 2.11 Air required from theoretical combustion (*OME-88/12, p. 4-16*).

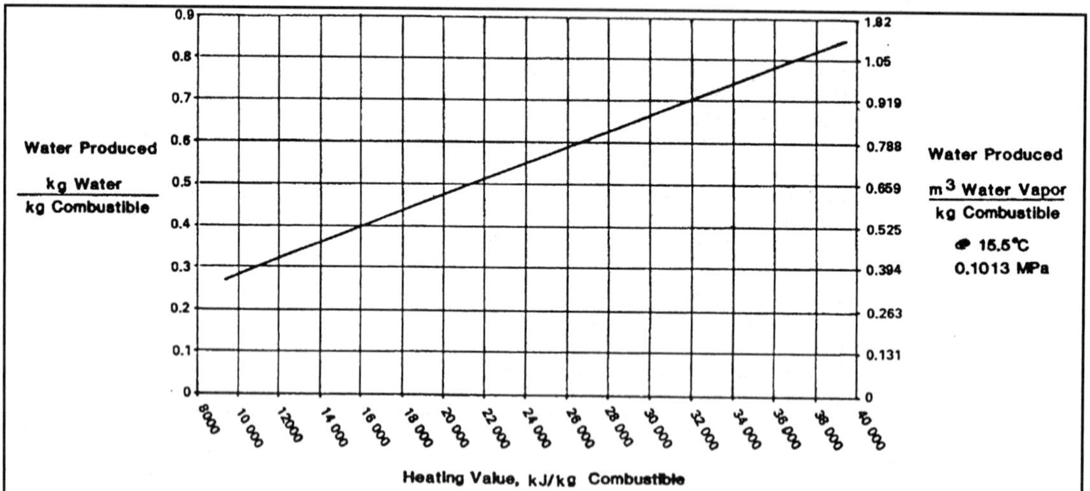

FIGURE 2.12 Water produced from combustion (zero humidity) (*OME-88/12, p. 4-17*).

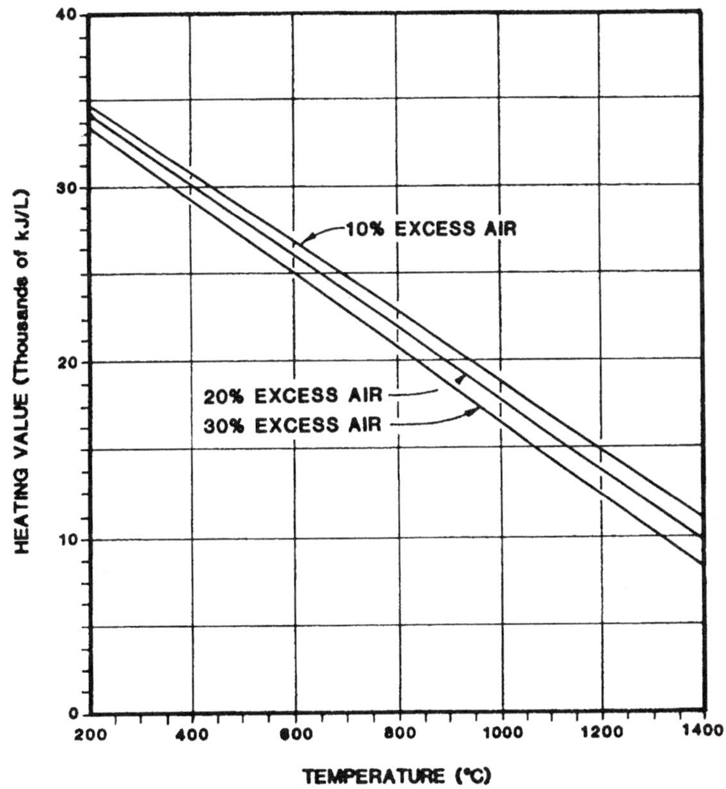

FIGURE 2.13 Net heat available from No. 2 fuel (*OME-88/12, p. 4-18*).

TABLE 2.25 Supplemental Fuel Combustion

No. 2 fuel oil, heat content 39,100 kJ/L, density 0.91 kg/L			
Percent excess air	10	20	30
kg air/L	13.78	15.03	16.29
kg dry gas/L*	13.84	15.09	16.34
kg water/L*	1.04	1.05	1.07
Natural gas, heat content 37,256 kJ/m^3, density 0.80 kg/m^3			
Percent excess air	5	10	15
kg-air/m^3	12.09	12.66	13.24
kg-dry-air/m^3*	11.4	11.97	12.55
kg-water/m^3*	1.65	1.65	1.66

* Products of combustion.

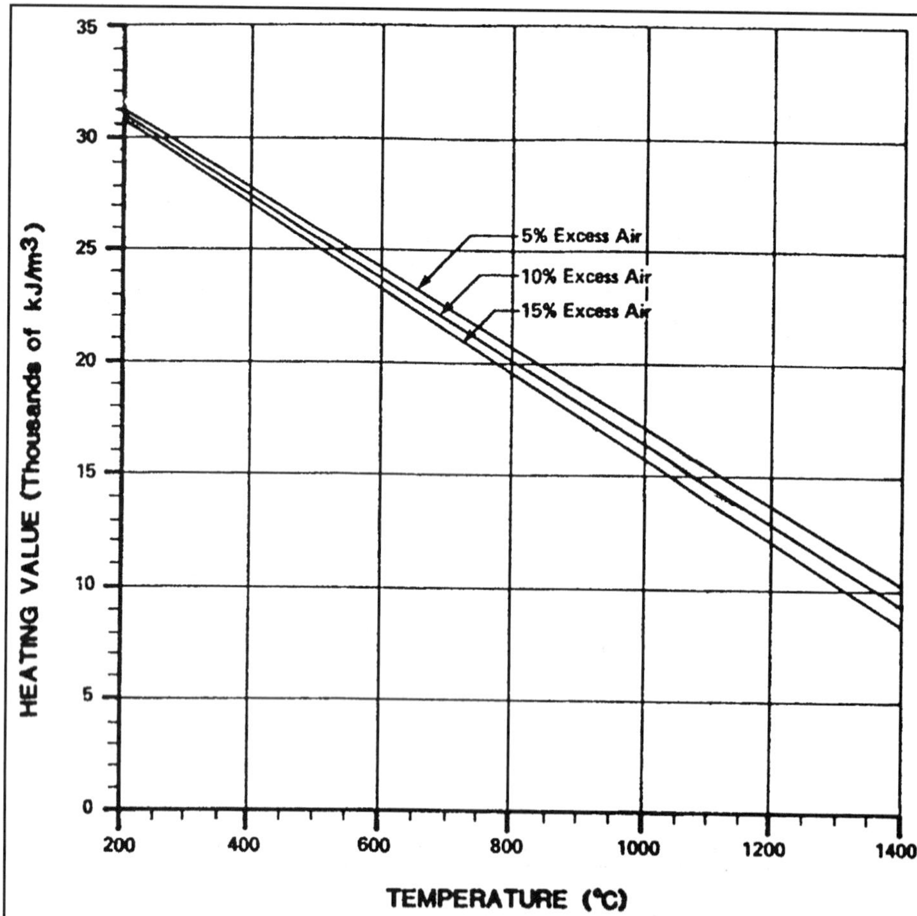

FIGURE 2.14 Net heat available from natural gas (*OME-88/12, p. 4-19*).

products of combustion of supplemental fuel. The available heat, therefore, decreases with increasing temperature, as shown in Figs. 2.13 and 2.14 for No. 2 fuel oil and natural gas, respectively. The net heating value must be used in incinerator calculations.

4.3 Incinerator Residence Time

Residence time means the length of time that the combustion gas is exposed to the combustion temperature in an incinerator. It can be expressed as:

$$t = V/q$$

where t = residence time
V = combustion chamber volume
q = combustion gas flow rate

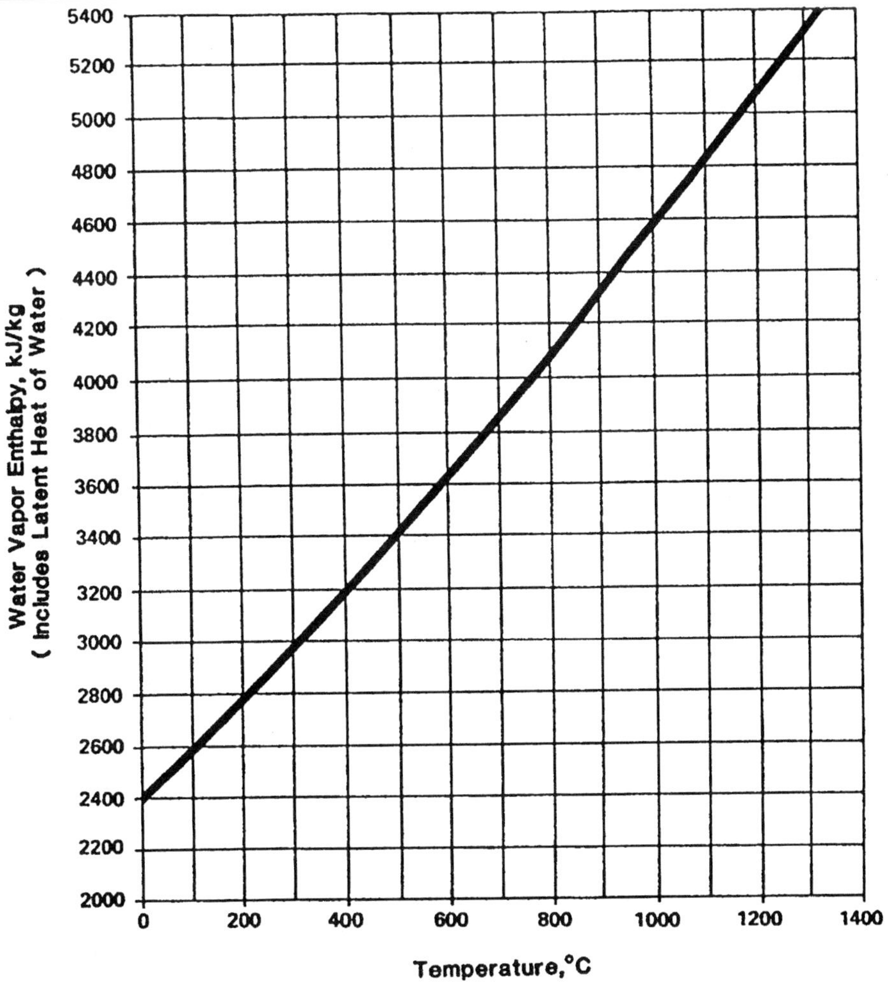

FIGURE 2.15 Enthalpy, water vapor relative to 0°C (*OME-88/12, p. C1-9*).

Time and temperature affect combustion in much the same manner as temperature and pressure affect the volume of a gas. When one variable is increased, the other may be decreased with the same end result. With a higher temperature, a shorter residence time can achieve the same degree of oxidation. The reverse is also true; a higher residence time allows the use of a lower temperature. In describing an incinerator operation, these two terms are always mentioned together. One has little meaning without specifying the other.

The choice between higher temperature or longer residence time is based on economic considerations. Increasing residence time involves using a larger combustion chamber, resulting in a higher capital cost. Raising the operating temperature increases fuel usage which also adds to the operating costs. Fuel costs are the major operating expense for most incinerators. Within certain limits, lowering the temperature and adding volume to increase residence time can be a cost-effective alternative method of operation.

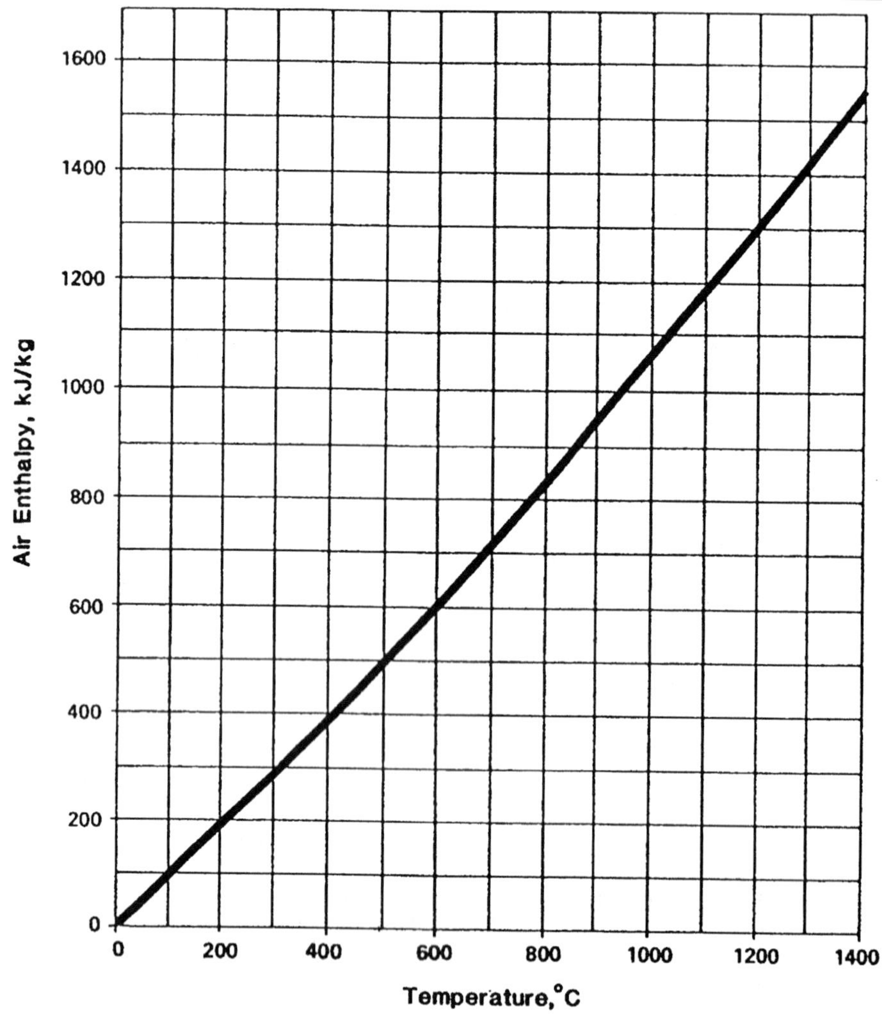

FIGURE 2.16 Enthalpy, dry air relative to 0°C (*OME-88/12, p. C1-10*).

Once again, the residence time of gases in the combustion chamber may be calculated from

$$t = V/Q$$

where t = residence time
V = volume of incinerator chamber
Q = gas volumetric flow rate at combustion conditions

Q is the total flow of hot gases in the combustion chamber. Adjustments to the flow rate must include any outside air added for combustion.

References used in this section: EPA-81/09, p. 4-98; EPA-81/12, p. 3-7.

EXAMPLE 1 residence time
Calculate the maximum achievable gas residence time in an incinerator after the desired operating temperature has been achieved.

Case 1. Use the following given conditions to calculate:

(1) the maximum achievable gas residence time after the desired operating temperature has been achieved;
(2) the required volume of the incinerator chamber.

Case 2. If the operating temperature $T_2 = 2000°F$, answer the same questions.
Case 3. If the incinerator requires a minimum residence time of 2 s, calculate the required volume of the incinerator for cases 1 and 2.

Given conditions
The data in question are summarized below:

- standard volumetric flow rate in the combustion chamber, $Q_1 = 1000$ scfm
- standard conditions $T_1 = 60°F$ and $P_1 = 1$ atm
- combustion chamber temperature $T_2 = 1500°F$
- volume of incinerator chamber $V = 120\ ft^3$

Solution for case 1
1.1 Write the ideal gas equation for the actual volumetric flow rate (acfm) calculation.

$$Q_2 = Q_1(T_2/T_1)(P_1/P_2)$$

where
Q_2 = actual volumetric flow rate
Q_1 = standard volumetric flow rate
T_2 = actual operating temperature, °R or K
T_1 = standard temperature, °R or K
P_2 = actual operating pressure
P_1 = standard pressure

Note: in general, $P_1 = P_2$ for an incinerator.

1.2. Calculate the actual volumetric flow rate, Q_2.

$$\begin{aligned} Q_2 &= (1000)[(460 + 1500)/(460 + 60)] \\ &= 3769 \text{ acfm (actual cubic feet per minute)} \\ &= 62.82 \text{ acfs (actual cubic feet per second)} \end{aligned}$$

1.3. Write the residence time equation.

$$\begin{aligned} t &= V/Q_2 \\ &= 120/62.82 \\ &= 1.91 \text{ s} \end{aligned}$$

Solution for case 2
2.1. Write the ideal gas equation for the actual volumetric flow rate (acfm) calculation.

$$Q_2 = Q_1(T_2/T_1)(P_1/P_2)$$

2.2. Calculate the actual volumetric flow rate Q_2.

$$\begin{aligned} Q_2 &= (1000)[(460 + 2000)/(460 + 60)] \\ &= 4731 \text{ acfm} \\ &= 78.85 \text{ acfs} \end{aligned}$$

2.3. Write the residence time equation.

$$\begin{aligned} t &= V/Q_2 \\ &= 120/78.85 \\ &= 1.52 \text{ s} \end{aligned}$$

Note: the results of cases 1 and 2 show that the higher the operating temperature, the shorter the residence time.

Solution for case 3

3.1. Write the volume equation from either sub-section 1.3 or 2.3.

$$V = Q_2 t$$

3.2. Calculate the incinerator volume for case 1.

$$\begin{aligned} V &= 62.82(2) \\ &= 126 \text{ ft}^3 \end{aligned}$$

3.3. Calculate the incinerator volume for case 2.

$$\begin{aligned} V &= 78.85(2) \\ &= 158 \text{ ft}^3 \end{aligned}$$

4.4 Incinerator Retention Time

Retention time means the length of time that solid materials remain in the primary combustion chamber during incineration. It can be expressed as

$$t = 0.19(L)/(DSN)$$

where
t = retention time in minutes
L = kiln length in ft
D = kiln diameter in ft
S = slope of kiln ft/ft
N = rotation velocity in rpm

References used in this section: EPA-81/09, p. 4-104; EPA-86/03, p. 98.

EXAMPLE retention time

A rotary kiln is used to incinerate solid waste. Calculate:

(1) retention time
(2) if the solid waste requires at least 60 min of retention time, determine the rotational velocity with kiln slope remaining unchanged, or
(3) determine the kiln slope with rotational velocity remaining unchanged.

Given conditions

The data in question are summarized below:

- kiln length, $L = 20$ ft
- kiln diameter, $D = 6$ ft
- kiln slope, $S = 0.02$ ft/ft
- rotation velocity, $N = 1.5$ rpm

Solution for question (1)

1.1. Write the retention time equation.

$$t = 0.19(L)/(DSN)$$

1.2. Calculate the retention time.

$$\begin{aligned} t &= 0.19(20)/[6(0.02)(1.5)] \\ &= 21.11 \text{ min} \end{aligned}$$

Solution for question (2)

2.1. Write the rotational velocity equation.

$$N = 0.19(L)/(DSt)$$

2.2. Calculate the rotational velocity.

$$\begin{aligned} N &= 0.19(20)/(6(0.02)(60)) \\ &= 0.53 \text{ rpm} \end{aligned}$$

Solution for question (3)

3.1. Write the rotational velocity equation.

$$S = 0.19(L)/(DNt)$$

3.2. Calculate the rotational velocity.

$$\begin{aligned} S &= 0.19(20)/[6(1.5)(60)] \\ &\cong 0.01 \text{ rpm} \end{aligned}$$

4.5 Duct Sizing

Ducts convey air and are normally sized to provide no more than 12 m/s (40 ft/s) air flow velocity. Higher velocities result in excessive noise and excessive pressure drop.

4.6 Flue Sizing

Flues carry gaseous products of combustion (flue gas) from an incinerator. They are generally sized to allow no more than 9 m/s (30 ft/s) gas flow. At higher velocities the particulate normally present in the exhaust will tend to accelerate flue erosion.

4.7 Chamber Sizing

Chamber sizing is based on heat release. There is a limit to the quantity of heat that can be released in a particular furnace chamber. Heat release is that amount of heat generated when combustible material burns. When combustible gases burn or when liquids burn in suspension, the volume of the furnace chamber will limit the total amount of heat released within that chamber. The furnace volume must be large enough to allow release of the heat generated by the anticipated waste and the supplemental fuel.

When a solid or sludge waste is fired, the heat release (heat generated per chamber volume or hearth area per hour) of that waste is characterized by the area of the surface on which it is placed, i.e., the hearth or grate.

Table 2.26 (OME-88/12, p. 1-22) lists typical heat release values for some common incinerator systems. The calculated heat release is determined by dividing the value in step 45 of Table 2.24 (which includes heat released from the waste plus the heat released from supplemental fuel) by the hearth or grate area (or chamber volume).

As an example of the use of this information, a reasonable size of a rotary kiln used for destruction of 2000 kg/h (4410 lb/h) waste with a heat content of 17,000 kJ/kg (7309 Btu/lb) is calculated as described below.

$$(2000\ \text{kg/h})(17{,}000\ \text{kJ/kg}) = 34{,}000{,}000\ \text{kJ/h}$$

Using a 1,000,000 kJ/m^3-h heat release rate,

$$(34{,}000{,}000\ \text{kJ/h})/(1{,}000{,}000\ \text{kJ/m}^3\text{-h}) = 34\ \text{m}^3(1200\ \text{ft}^3)\ \text{volume}$$

For a kiln with a 2.5 m internal diameter (unit cross sectional area = 4.91 m^2), the required length is 34 m^3/4.91 m^2 = 6.93 m (22.7 ft).

The kiln for this application would have an internal diameter of 2.5 m (8.2 ft) and a length of 7.0 m (23 ft).

In general, an incinerator volume can be calculated by two approaches.

Approach 1: by incinerator dimensions. For example, the volume of a rotary kiln can be calculated by the equation

$$V = \pi(r^2)(L)$$

where r = radius of the kiln
L = length of the kiln

Approach 2: by heat release rate.

TABLE 2.26 Typical Heat Release Rates

Incinerator	Heat release
Fluid bed*	350,000–500,000 kJ/m^2-h (bed-area)
Multiple chamber	300,000–400,000 kJ/m^3-h
Multiple hearth	300,000–400,000 kJ/m^3-h
Multiple hearth heat release*	250,000–350,000 kJ/m^2-h (hearth area)
Gaseous waste incinerator	3,000,000–10,000,000 kJ/m^3-h
Liquid waste incinerator	1,000,000–3,000,000 kJ/m^3-h
Rotary kiln	500,000–1,500,000 kJ/m^3-h
Solid waste grate*	150,000–300,000 kJ/m^2-h (grate area)

* These heat release values are based on surface area, not volume.

EXAMPLE 1 incinerator volume
An incinerator is used to incinerate waste with the following given conditions. Calculate the maximum waste burning rate.

Given conditions

- incinerator energy release rate $Q_r = 30{,}000$ Btu/(h-ft^3)
- combustion chamber volume $V = 800$ ft^3
- waste heating value $H = 9500$ Btu/lb

Solution:
1. Calculate the incinerator maximum heat capacity Q.

$$\begin{aligned} Q &= Q_r V \\ &= 30{,}000(800) \\ &= 2.40 \times 10^7 \text{ Btu/h} \end{aligned}$$

2. Calculate the maximum waste burning rate W.

$$\begin{aligned} W &= Q/H \\ &= 2.40 \times 10^7/9500 \\ &= 2526 \text{ lb/h} \end{aligned}$$

EXAMPLE 2 incinerator volume
A rotary kiln was designed to have an average energy release rate of 28,500 Btu/h-ft^3 in the combustion chamber. It is anticipated that the kiln will be used to incinerate 4000 lb/h of solid waste which has heating value as high as 9500 Btu/lb. Assume that the length-to-diameter ratio of the kiln is 3.5. Determine (1) the volume of the kiln, and (2) the diameter of the kiln.

Given conditions

- energy release rate: $Q_r = 28{,}500$ Btu/h-ft^3
- waste feed rate $W = 4000$ lb/h
- heating value of the waste $H = 9500$ Btu/lb
- length-to-diameter ratio $L/D = 3.5$

Solution:
1. Calculate the amount of heat released by the waste Q.

$$\begin{aligned} Q &= (4000 \text{ lb/h})(9500 \text{ Btu/lb}) \\ &= 3.80 \times 10^7 \text{ Btu/h} \end{aligned}$$

2. Calculate the combustion chamber volume.

$$\begin{aligned} V &= Q/Q_r \\ &= 3.80 \times 10^7/(28{,}500 \text{ Btu/h-ft}^3) \\ &= 1333 \text{ ft}^3 \end{aligned}$$

3. Calculate the diameter of the kiln.

$$V = \pi(D/2)^2(L)$$
$$D^3 = 1333(4)/[\pi(3.5)]$$
$$= 485 \text{ ft}^3$$
$$D = 7.86 \text{ ft}$$

4. Calculate the length of the kiln.

$$L = 3.5D$$
$$= 3.5(7.86)$$
$$= 27.51 \text{ ft}$$

4.8 Turbulence and Mixing

In order to achieve high combustion efficiency in incinerators, it is particularly important to achieve good mixing between the primary combustion products (primarily CO and organics) and a stoichiometric excess secondary combustion air. This mixing can be promoted by a range of physical parameters as well as by promoting highly turbulent flow of the gases. Physical parameters which are used to promote mixing include:

- location and direction of secondary air jets
- volume and pressure of secondary air addition
- changes in flow direction(s)
- other baffling techniques

As well as the provision of overall mixing using the above methods, the eddies formed by turbulent flow promote local mixing of the combustible gases and air. The degree of turbulence is typically assessed by use of the Reynolds number *Re*:

$$Re = VD/K$$

where V = average velocity, m/s
D = diameter (or equivalent diameter) of flow stream, m
K = kinematic viscosity, m^2/s.

Figure 2.17 shows the relationship between the kinematic viscosity and temperature. When the Reynolds number is below 2300 (defined as the critical Reynolds number), flow is generally considered laminar with no significant turbulence.

As a guide, the gas flow within an incinerator chamber or flue should have a Reynolds number well above 10,000. As an example of this calculation, assume a flue exiting a furnace has an internal diameter of 2 m (6.6 ft). The gas flow is at a temperature of 1000°C (1832°F), and is at a velocity of 10 m/s (33 ft/s). From the figure, $K = 140 \times 10^{-6}$ m^2/s at 1000°C. Therefore,

$$Re = VD/K = 10(2)/140 \times 10^{-6} = 143{,}000$$

A Reynolds number of 143,000, in combination with good mixing, should be ample to provide high combustion efficiency and burnout.

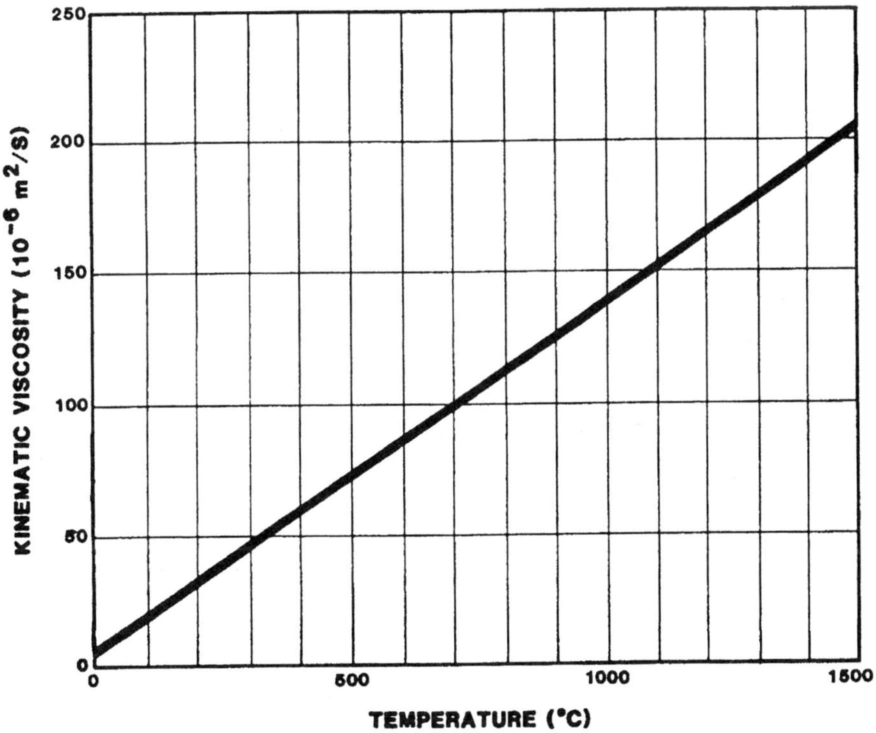

FIGURE 2.17 Kinematic viscosity of air vs. temperature (*OME-88/12, p. 4-25*).

For a rectangular or other shape of flue, the equivalent diameter (D_e) should be used. For example, for a rectangle, D_e is equal to four times the cross-sectional area divided by the wetted perimeter, i.e.,

$$D_e = 2ab/(a+b)$$

where a and b are the dimensions of the sides of the rectangle. For example, $a = 2$ m and $b = 3$ m. Therefore, $D_e = 2(2)(3)/(2+3) = 2.4$ m. This value of D_e would be used in lieu of D when applied to a noncircular cross section.

4.9 Fan Sizing

The fan power requirement is a function of air or gas flow rate (cubic meter per second), the pressure developed across the fan (kilopascals), and the fan efficiency. The efficiency of fans located downstream of air pollution control systems (clean gas) is normally in the range 80–85%; fans located upstream of control devices (dirty gas) require specific designs which may be limited to the range 60–75% efficiency. Efficiency will also drop as the fan becomes dirty, or when fan material erodes with wear. The fan power consumption is calculated as follows:

$$S = QWp/h$$

where Q = flow rate, m^3/s
Wp = pressure across fan, kPa
h = fan efficiency
S = fan power consumption, kW

The flow rate of air or gas is determined from the incinerator calculations procedure described above. Typical pressure drops for fans associated with incinerator systems are listed in Table 2.27 (OME-88/12, p. 4-28). These pressure values are values developed across the fan and should be used as Wp in the above equation.

4.10 Induced Fan Sizing

An induced draft fan (ID fan) is the fan at the discharge of an incinerator that draws flue gas from the incinerator and discharges the gas through a stack to the atmosphere. Many incinerators such as multiple hearth furnaces, grate type systems, and rotary kilns, use ID fans.

Hot exhaust gases leaving incinerators may first be cooled by passage through a heat recovery section and/or passage through a water quench. Normally, sufficient water will be added to the hot gas stream to saturate that gas stream before entering the ID fan. When a fabric filter or an electrostatic precipitator is included within the system, only a limited amount of water is added; the gas flow would not be saturated and less water would be required.

Figure 2.18 relates the gas flow rate of saturated gas to dry air flow as a function of temperature. Determination of this ratio is necessary to allow a determination of the fan size. As noted previously, the assumption that flue gas has a dry component with the properties of dry air is used in these calculations.

As an example, if an ID fan is placed downstream of a wet scrubber which passes 18,000 kg/h (39,683 lb/h) of dry gas at a temperature of 50°C (122°F) entering the ID fan, the flow rate is calculated as follows.

From Fig. 2.18, at 50°C (122°F), 1 kg of dry air will have a volume of 1.03 m^3 (36 ft^3) when saturated with water. The 18,000 kg/h dry gas will therefore produce (18,000 kg/h) (1.03 m^3/kg), or 18,540 m^3/h of flue gas entering the ID fan. The ID fan can be sized on the basis of this flow.

4.11 Electric Power Requirements

As a general rule, when an ID fan is used in an incineration system, the total electric power requirements of that system can be approximated as double the ID fan rated kW. If an ID fan is not used, the power rating of each of the major items of equipment must be determined and totaled to obtain the facility power requirement.

TABLE 2.27 Typical Fan Pressure Drop Values

Fan type	Pressure drop
Forced draft	1.0–1.5 kPa
Heated gas recirculation	1.0–2.0 kPa
Induced draft	5.0–15.0 kPa
Supplemental fuel combustion air	0.2–0.5 kPa
Fluidizing air blower	1.2–5.0 kPa

Note: The above values are typical of the pressure drop across a particular fan usage that should be used in determining fan power requirements.

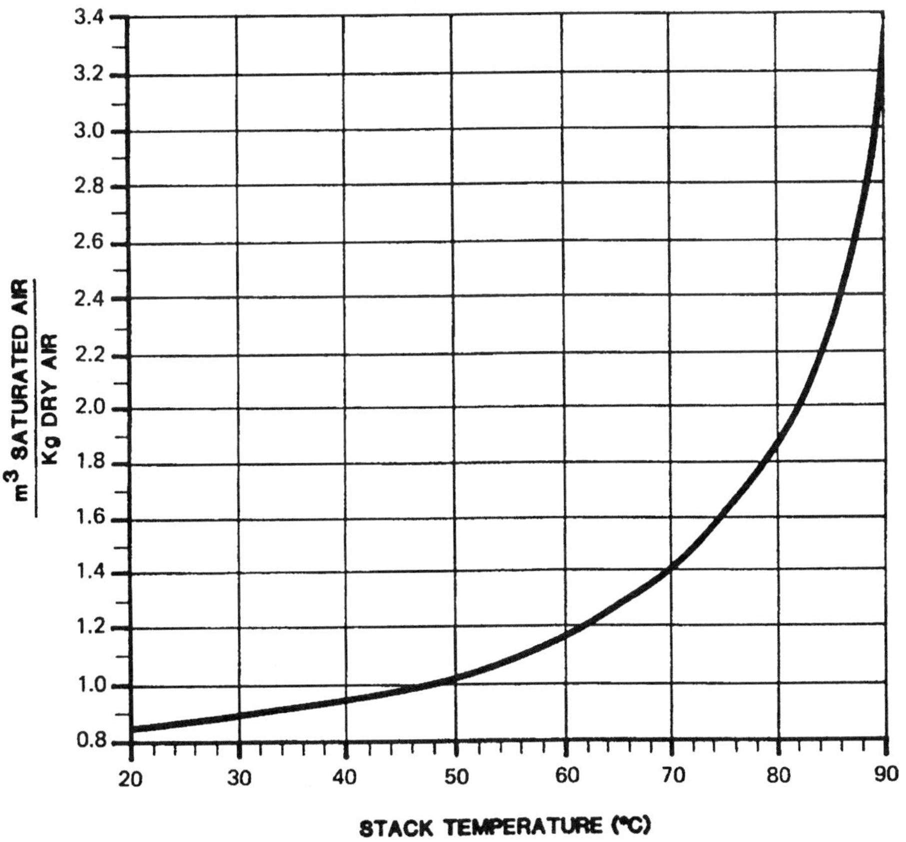

FIGURE 2.18 Incinerator induced draft fan (*OME-88/12, p. 4-30*).

4.12 Refractory Selection

Refractory selection involves the following factors:

- refractory parameters
- castables (refractory concrete)
- firebrick
- incinerator refractory selection

Refractory parameters. There are a wide variety of refractory and insulation products available. Their significant properties include those listed below.

- Abrasion: the washing away or physical destruction of material under forces due to physical contact with the gaseous, liquid, or solid materials on the hot face.
- Resistance to slagging or corrosion (chemical processes that destroy the refractory bond or the chemical integrity of the insulation).

- Mechanical shock or cold crushing strength: the ability to withstand handling and shipping without damage and impact strength at low temperature operations.
- Modulus of rupture: a standard measure of structural strength provided by load testing (ASTM Methods C16 and C216), providing another measure of mechanical shock resistance.
- Spalling: a deterioration of the surface of the refractory by flaking caused by abrasion, corrosion, or mechanical or thermal shock.
- Porosity: the susceptibility to penetration by slags or gases, indicative of fly ash adherence.
- Maintaining strength under high temperature conditions.
- Insulating value: the ability to provide resistance to the flow of heat.
- Reheat (ASTM Method C113): a measure of irreversible changes in linear dimensions under repeated heating and cooling.
- Specific heat: the amount of heat required to raise the temperature of a refractory material.
- Thermal expansion: the reversible change in linear dimensions under heating and cooling. Where the refractory is used as a liner in a flue, for instance, the expansion of the refractory must be evaluated with respect to thermal expansion of the flue; if the expansions do not match, provision for material expansion must be considered.
- Resistance to the operating environment: of particular interest is the performance of refractory in an oxidizing or reducing atmosphere. A reducing atmosphere, one deficient in oxygen, will tend to degrade refractory material containing iron or silicon components.

Castables (refractory concrete). These materials are supplied dry and are to be mixed with water before installation. They are installed by either pouring (casting in place), troweling, pneumatic gunning (as with gunned fireproofing), or ramming. A castable refractory should provide a smooth, continuous, monolithic mass. Castable materials are normally placed in an area with pine, mesh, or other anchor devices to hold the refractory in place during placement and curing. Mesh, grid, studs, or needles may also be used to enhance the strength of the refractory installation.

Castable refractories are classified as dense or lightweight (insulating). Dense castables have excellent mechanical strength and low permeability. Their insulating properties, however, are relatively poor. As dense materials, over 1600 kg/m^3 (100 lb/ft^3) specific weight, they offer good resistance in wet service such as quencher linings.

Lightweight castables are excellent insulators. Castable materials have the advantage of relatively easy installation in irregular areas such as furnace transitions, burner openings, etc. Lightweight castable service temperatures are generally limited to below 1650°C (3000°F), whereas other castable refractories can sustain higher temperatures.

Firebrick. Conventional firebrick is kiln-baked to uniform, controlled consistency and quality. The term firebrick refers to dense brick, over 1600 kg/m^3 (100 lb/ft^3), containing up to 44% alumina, normally placed in direct contact with the hot gas stream. Conventional firebrick has relatively poor insulating qualities.

Insulating firebrick (IFB) is a lightweight, porous brick, normally less than 800 kg/m^3 (50 lb/ft^3), which can be placed in direct contact with the gas stream and which provides good insulating characteristics. IFB is machined to its final shape, providing excellent dimensional control as compared with conventional firebrick which is used as cast.

IFB is lower in structural strength than firebrick and, because of its porosity, is a soft material not suitable for erosive gas streams, i.e., gas streams with high particulate components. This low abrasion resistance limits the maximum velocities allowed adjacent to it.

Where refractory brick is required and high abrasive resistance is necessary, firebrick will often be provided with an insulation block as backup between the firebrick and the furnace/flue wall.

Silicon carbide refractory has relatively good thermal conductivity, good mechanical resistance, and abrasion resistance.

There are numerous types and grades of firebrick such as super-duty (high alumina content, improved strength and stability properties at high temperature), high-duty (good thermal shock resistance), low-duty (applicable for lower temperature and low abrasion service), and silicon carbide (excellent resistance to chemical attack), etc.

Incinerator refractory selection. Table 2.28 (OME-88/12, p. 4-35) describes typical refractory applications for various incinerator sections.

4.13 Exhaust Gas Monitoring

Permanently mounted instrumentation is required for the measurement of temperatures, total hydrocarbons (or carbon monoxide), and opacity. Additionally, oxygen, carbon dioxide, and nitrogen oxides monitoring may be required.

As a minimum, temperature should be monitored and recorded for the gas leaving the furnace and the exhaust gas exiting the stack. Gaseous monitors should be located as close as practicable to the furnace exhaust point.

Special care must be exercised in locating opacity meters on systems which incorporate wet gas cleaning devices (such as a venturi scrubber). Entrained and/or condensing moisture downstream will appear and will affect the opacity reading. An upstream location will provide an indication of particulate and other carryover from the incinerator.

4.14 Acid Gas Corrosion

The exhaust gases from incinerators usually contain HCl, SO_2, and other acidic gases. Special design considerations may be necessary where:

- gas temperatures will be below 150°C (300°F; potential for acid gas condensation), and/or
- heat recovery media (e.g., boiler tubes) will be above 370°C (700°F; potential for acid gas corrosion).

5 SYSTEM CALCULATIONS

5.1 Sludge Incineration

A stream of dewatered sludge is fed to a multiple hearth incinerator (Fig. 2.19) at the rate of 5000 kg/h. The sludge is 75% water and has 12.5% noncombustible (ash). The heating value of the combustible fraction has been determined to be 23,000 kJ/kg. This is to be burned using 100% excess air with fuel oil (20% excess air) as auxiliary fuel. Standard air is available at 15.5°C and 60% relative humidity, which is equivalent to a humidity of 0.00657 kg moisture per kg of dry air. Radiant heat losses are estimated at 1% of the heat released from the waste. The ash is estimated to remove 200 kJ/kg ash from the incinerator. The exhaust gas temperature is to be maintained at 760°C (OME-88/12, p. C.3-1).

The solution to this problem is provided below by using the information (tables and figures) in Section 4.2 and Table 2.29.

TABLE 2.28 Refractory Selection for Incinerators

Incinerator part	Temperature range (°C)	Abrasion	Slagging	Mechanical shock	Spalling	Fly ash adherence	Recommended refractory
Charging gate	20–1400	Severe, very important	Slight	Severe	Severe	None	Super-duty
Furnace walls, grate to 2 m above	20–1400	Severe	Severe, very important	Severe	Severe	None	Silicon carbide or super-duty
Furnace walls, upper portion	20–1400	Slight	Severe	Moderate	Severe	None	Super-duty
Stoking doors	20–1400	Severe, very important	Severe	Severe	Severe	None	Super-duty
Furnace ceiling	20–1400	Slight	Moderate	Slight	Severe	Moderate	Super-duty
Flue from combustion chamber	650–1400	Slight	Severe, very important	None	Moderate	Moderate	Silicon carbide or super-duty
Combustion chamber	650–1400	Slight	Moderate	None	Moderate	Moderate	Super-duty
Combustion chamber ceiling	650–1400	Slight	Moderate	None	Moderate	Moderate	Super-duty
Breeching walls	650–1650	Slight	Slight	None	Moderate	Moderate	Super-duty
Breeching ceiling	650–1650	Slight	Slight	None	Moderate	Moderate	Super-duty
Subsidence chamber walls	650–1650	Slight	Slight	None	Slight	Moderate	Firebrick or super-duty
Subsidence chamber ceiling	650–1650	Slight	Slight	None	Slight	Moderate	Firebrick or super-duty
Stack	260–540	Slight	None	None	Slight	Slight	Firebrick or super-duty

Source: OME-88/12.

TABLE 2.29 Sludge Incinerator Analysis

Step	Description	Calculation	Unit
1	Waste feed rate (given)	5000	kg/h
2	Water fraction in waste (given)	0.75	
3	Water quantity in waste (step 1 × 2)	3750.00	kg/h
4	Solid quantity (step 1 − step 3)	1250.00	kg/h
5	Noncombustible fraction (given)	0.125	
6	Noncombustible quantity (step 4 × 1)	625.00	kg/h
7	Combustible quantity (step 4 − step 6)	625.00	kg/h
8	Heating value of combustibles (given)	23,000	kJ/kg
9	Total heat generated (step 7 × 8)	14,375,000.00	kJ/h
10	Theoretical (stoichiometric) air ratio (see Fig. 2.11)*	7	
11	Theoretical air (step 7 × 10)	4375.00	kg/h
12	Excess air fraction (given)	1	
13	Excess air required (step 11 × 12)	4375.00	kg/h
14	Total air required (steps 11 + 13)	8750.00	kg/h
15	Water formed due to combustion of combustible waste (see Fig. 2.12)*	0.52	kg-water/ kg-combustible
16	Total water due to combustion (step 7 × 15)	325.00	kg/h
17	Humidity in air (given)	0.00657	kg-water/kg-air
18	Water due to humidity (steps 14 × 17)	57.49	kg/h
19	Total water in flue gas (steps 3 + 16 + 18)	4132.49	kg/h
20	Dry flue gas (steps 7 + 14 − 16)	9050.00	kg/h
21	Radiation loss fraction (given)	0.01	
22	Radiation heat loss (step 9 × 21)	143,750.00	kJ/h
23	Ash heating value (given)	200	kJ/kg
24	Ash heat loss (step 6 × 23)	125,000.00	kJ/h
25	Humidity correction due to latent heat (step18) ×2186	125,673.14	kJ/h
26	Miscellaneous heat loss (given)	0	kJ/h
27	Heat in flue gas (steps 9 − 22 − 24 + 25 − 26)	14,231,923.14	kJ/h
28	Estimate flue gas temperature (see note 1 calculation below)	258	°C
29	Required gas temperature (given)	760	°C
30	Required enthalpy of flue gas (see note 2 calculation below)	23,547,625	kJ/h
31	Net heat required (step 30 − step 27)	9,315,701.99	kJ/h
32	Fuel excess air (20% excess air, given)	0.2	
33	Fuel heating value at required temperature (Figs. 2.13 or 2.14)	22,000	kJ/L
34	Fuel required (step 31/33)	423.44	L/h
35	Air for fuel (see Table 2.25)*	15.03	kg air/L
36	Total air required for fuel (step 35 × 34)	6364.30	kg/h
37	Moisture from fuel (see Table 2.25)*	1.05	kg/L
38	Moisture generated (step 37 × 34)	444.61	kg/h
39	Dry flue gas from fuel combustion (see Table 2.25)*	15.09	kg/L
40	Dry gas generated (step 39 × 34)	6389.71	kg/h
41	Total dry gas flow (steps 40 + 20)	15,439.71	kg/h
42	Total moisture flow (steps 38 + 19)	4577.10	kg/h
43	Fuel gross heating value (Table 2.25)	39,100	kJ/L
44	Heat generated from fuel (step 43 × 34)	16,556,504.00	kJ/h
45	Total heat at exit (steps 44 + 27)	30,788,427.14	kJ/h

* See Section 4.2.

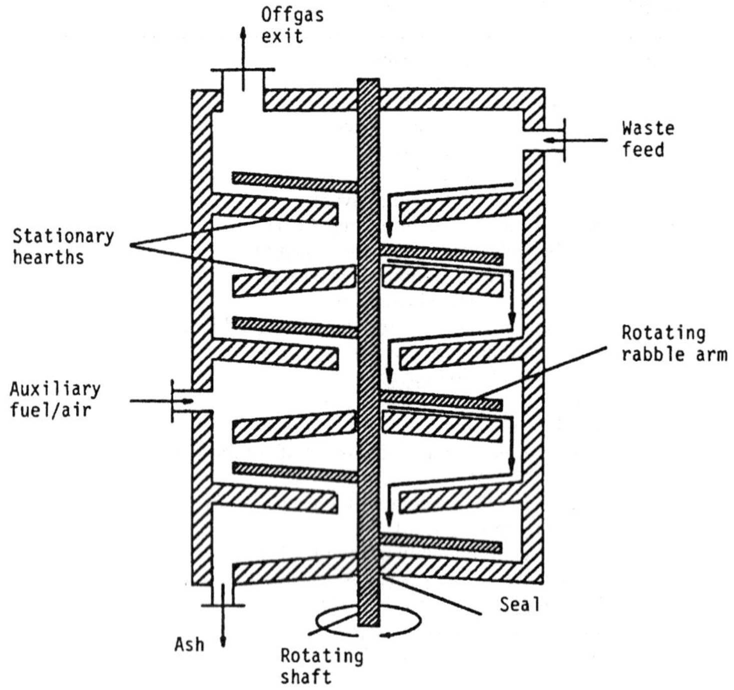

FIGURE 2.19 Multiple hearth incinerator (*DOE-93/03, p. C-5*).

NOTE 1: Calculation for step 28

The flue gas temperature can be calculated from the heat content (enthalpy) in the flue gas (step 27) and the enthalpy in Figs. 2.15 and 2.16. The procedure involves estimating a temperature and calculating the flue gas enthalpy by using the enthalpy values obtained from the figures and the water mass in step 19 and dry flue gas in step 20. The calculated gas enthalpy is then compared with the enthalpy in step 27. Repeat these steps until the calculated enthalpy and the enthalpy in step 27 are nearly equal. Actual calculations are given in detail below.

Assume that the flue gas exit temperature is 200°C.

- From Fig. 2.15, moisture enthalpy = 2790 kJ/kg
- From Fig. 2.16, dry air enthalpy = 190 kJ/kg
- Total enthalpy = 190(9050) + 2790(4132) = 13,249,140.13 kJ/h. This value is less than the value of 14,231,923.14 in step 27.

Assume that the exit temperature is 400°C.

- From Fig. 2.15, moisture enthalpy = 3200 kJ/kg
- From Fig. 2.16, dry air enthalpy = 380 kJ/kg
- Total enthalpy = 380(9050) + 3200(4132) = 16,662,960 kJ/h. This value is greater than the value of 14,231,923.14 in step 27.

The kiln temperature (x) is estimated by the following interpolation method.

x_1	200°C	$1.32E+07$ kJ/h	y_1
Estimate	x	$1.42E+07$	y
x_2	400°C	$1.67E+07$	y_2

$$(x - x_1)/(y - y_1) = (x_2 - x_1)/(y_2 - y_1)$$
$$x = x_1 + (y - y_1)(x_2 - x_1)/(y_2 - y_1)$$
$$x = 257.58°\text{C}$$

NOTE 2: Calculation for step 30

The required enthalpy of flue gas can be calculated from required temperature 760°C and Figs. 2.15 and 2.16.

- From Fig. 2.15, moisture enthalpy = 3990 kJ/kg
- From Fig. 2.16, dry air enthalpy = 780 kJ/kg
- Total enthalpy = 780(9050) + 3990(4132) = 23,547,625.13 kJ/h

5.2 Refuse Incineration

Refuse at 50 tonnes per day (2083 kg/h) is to be incinerated. It has a moisture content of 24%, a heating value of 11,000 kJ/kg as received and an ash (noncombustible) content of 11% as received. The incinerator is of grate design, and requires 85% excess air for complete combustion. A minimum temperature of 980°C is required for effective burnout. Assume an ash content of 250 kJ/kg, 2% radiation heat loss, and a humidity of 0.00657 kg/kg dry air. Calculations follow and are summarized in Table 2.30, this refuse incinerator analysis (OME-88/12, p. 3-7).

The solution to this problem is provided below by using the information (tables and figures) in Section 4.2.

NOTE 1: Calculation for step 28

The flue gas temperature can be calculated from the heat content (enthalpy) in the flue gas (step 27) and the enthalpy in Figs. 2.15 and 2.16. The procedure involves estimating a temperature and calculating the flue gas enthalpy by using the enthalpy values obtained from these figures and the water mass in step 19 and dry flue gas in step 20. The calculated gas enthalpy is then compared with the enthalpy in step 27. Repeat these steps until the calculated enthalpy and the enthalpy in step 27 are nearly equal. Actual calculations are given in detail below.

Assume that the flue gas exit temperature is 1000°C.

- From Fig. 2.15, moisture enthalpy = 4590 kJ/kg
- From Fig. 2.16, dry air enthalpy = 1070 kJ/kg
- Total enthalpy = 1165.84(4590) + 13,542.06(1070) = 19,841,209 kJ/h. This value is less than the value of 22,580,760.35 in step 27.

Assume that the exit temperature is 1200°C.

- From Fig. 2.15, moisture enthalpy = 5060 kJ/kg
- From Fig. 2.16, dry air enthalpy: 1300 kJ/kg
- Total enthalpy = 1165.84(5060) + 13,542.06(1300) = 23,503,828.40 kJ/h. This value is less than the value of 22,580,760.35 in step 27.

TABLE 2.30 Refuse Incinerator Analysis

Step	Description	Calculation	Unit
1	Waste feed rate (given)	2083	kg/h
2	Water fraction in waste (given)	0.24	
3	Water quantity in waste (step 1 × 2)	499.92	kg/h
4	Solid quantity (step 1 − step 3)	1583.08	kg/h
5	Dry base of noncombustible fraction = (wet fraction)/(1 − water fraction) = 0.11/(1 − 0.24)	0.145	
6	Noncombustible quantity (step 4 × 5)	229.55	kg/h
7	Combustible quantity (step 4 − step 6)	1353.53	kg/h
8	Heating value of combustibles = 11,000(2083/1353.53)	16,928.33	kJ/kg
9	Total heat generated (step 7 × 8)	22,913,002.50	kJ/h
10	Theoretical (stoichiometric) air ratio (Fig. 2.11)*	5.1	
11	Theoretical air (step 7 × 10)	6903.00	kg/h
12	Excess air fraction (given)	0.85	
13	Excess air required (step 11 × 12)	5867.55	kg/h
14	Total air required (steps 11 + 13)	12,770.55	kg/h
15	Water formed due to combustion of combustible waste (see Fig. 2.12)*	0.43	kg-water/kg-combustible
16	Total water due to combustion (step 7 × 15)	582.02	kg/h
17	Humidity in air (given)	0.00657	kg-water/kg-air
18	Water due to humidity (step 14 × 17)	83.90	kg/h
19	Total water in flue gas (steps 3 + 16 + 18)	1165.84	kg/h
20	Dry flue gas (steps 7 + 14 − 16)	13,542.06	kg/h
21	Radiation loss fraction (given)	0.02	
22	Radiation heat loss (step 9 × 21)	458,260.05	kJ/h
23	Ash heating value (given)	250	kJ/kg
24	Ash heat loss (step 6 × 23)	57,387.50	kJ/kg
25	Humidity correction due to latent heat (step18) ×2186	183,405.40	kJ/h
26	Miscellaneous heat loss (given)	0	kJ/h
27	Heat in flue gas (steps 9 − 22 − 24 + 25 − 26)	22,580,760.35	kJ/h
28	Estimate flue gas temperature (see note 1 calculation below)	1150	°C
29	Required gas temperature (given)	980	°C
30	Required enthalpy of flue gas (sufficient heat exists)	–	kJ/h
31	Net heat required (step 30 − step 27)	–	kJ/h
32	Fuel excess air (given)	–	
33	Fuel heating value at required temperature (see Figs. 2.13 or 2.14)	–	kJ/L
34	Fuel required (step 31/33)	–	L/h
35	Air for fuel (see Table 2.25)*	–	kg air/L
36	Total air required for fuel (step 35 × 34)	–	kg/h
37	Moisture from fuel (see Table 2.25)*	–	kg/L
38	Moisture generated (step 37 × 34)	–	kg/h
39	Dry flue gas from fuel combustion (see Table 2.25)*	–	kg/L
40	Dry gas generated (step 39 × 34)	–	kg/h
41	Total dry gas flow (steps 40 + 20)	13,542.06	kg/h
42	Total moisture flow (steps 38 + 19)	1165.84	kg/h
43	Fuel gross heating value (Table 2.25)*	–	kJ/L
44	Heat generated from fuel (step 43 × 34)	0.00	kJ/h
45	Total heat at exit (steps 44 + 27)	22,580,760.35	kJ/h

* See Section 4.2.

The furnace temperature (x) is estimated by the following interpolation method.

x_1	1000°C	1.98 $E+07$ kJ/h	y_1
Estimate	x	2.26 $E+07$	y
x_2	1200	2.35 $E+07$	y_2

$$(x - x_1)/(y - y_1) = (x_2 - x_1)/(y_2 - y_1)$$

$$x = x_1 + (y - y_1)(x_2 - x_1)/(y_2 - y_1)$$

$$x = 1149.60°\text{C}$$

5.3 Starved Air Incineration

For the previous example (heating value = 11,000 kJ/kg), it was assumed that the waste was incinerated in a starved air unit consisting of two chambers, and the total gas flow and heat release from the system would be the same. Conditions at the exit of the primary chamber could be calculated as follows, assuming 60% of the stoichiometric air requirement was provided in the primary chamber.

The heating value of the combustible content would not be fully realized in the primary chamber because only 60% of the stoichiometric air requirement is provided. From Fig. 2.20, at 60% of stoichiometric, 72% of the total heat capacity of the combustibles is released, which is 0.72(16,928.33) kJ/kg = 12,188.40 kJ/kg. The air requirement is based on 16,928.33 kJ/kg, while moisture generation is based on 12,188.40 kJ/kg actual heat release. Calculations follow, and are summarized in Table 2.31 for the analysis of a starved air refuse incinerator.

The solution to this problem is provided by using the information (tables and figures) in Section 4.2.

NOTE 1: Calculation for step 28

The flue gas temperature can be calculated from the heat content (enthalpy) in the flue gas (step 27) and the enthalpy in Figs. 2.15 and 2.16. The procedure involves estimating a temperature and calculating the flue gas enthalpy by using the enthalpy values obtained from these figures and the water mass in step 19 and dry flue gas in step 20. The calculated gas enthalpy is then compared with the enthalpy in step 27. Repeat these steps until the calculated enthalpy and the enthalpy in step 27 are nearly equal. Actual calculations are given in detail below.

Assume that the exit temperature is 1200°C.

- From Fig. 2.15, moisture enthalpy = 5060 kJ/kg
- From Fig. 2.16, dry air enthalpy: 1300 kJ/kg
- Total enthalpy = 960.26(5060) + 5062.20(1300) = 11,439,775.60 kJ/h. This value is less than the value of 16,169,511.31 in step 27. The temperature exiting the primary chamber is therefore in excess of 1200°C.

5.4 Furnace

The combustion chamber is a volume where the fuel and air mixture (in proper proportions) is exposed to an ignition source and burned. The residence time needed to achieve complete oxidation of the fuel depends on the temperature maintained in the combustion chamber, commonly referred to as the *furnace*. From the temperature effect on the reaction rate, the higher the furnace temperature, the faster the oxidation reaction and hence the smaller the furnace would need to be. This size reduction, however, is limited by Charles' Law.

Adiabatic flame temperatures, which are the highest temperatures which may theoretically be attained in the furnace, are for most fuels considerably higher than the commonly used

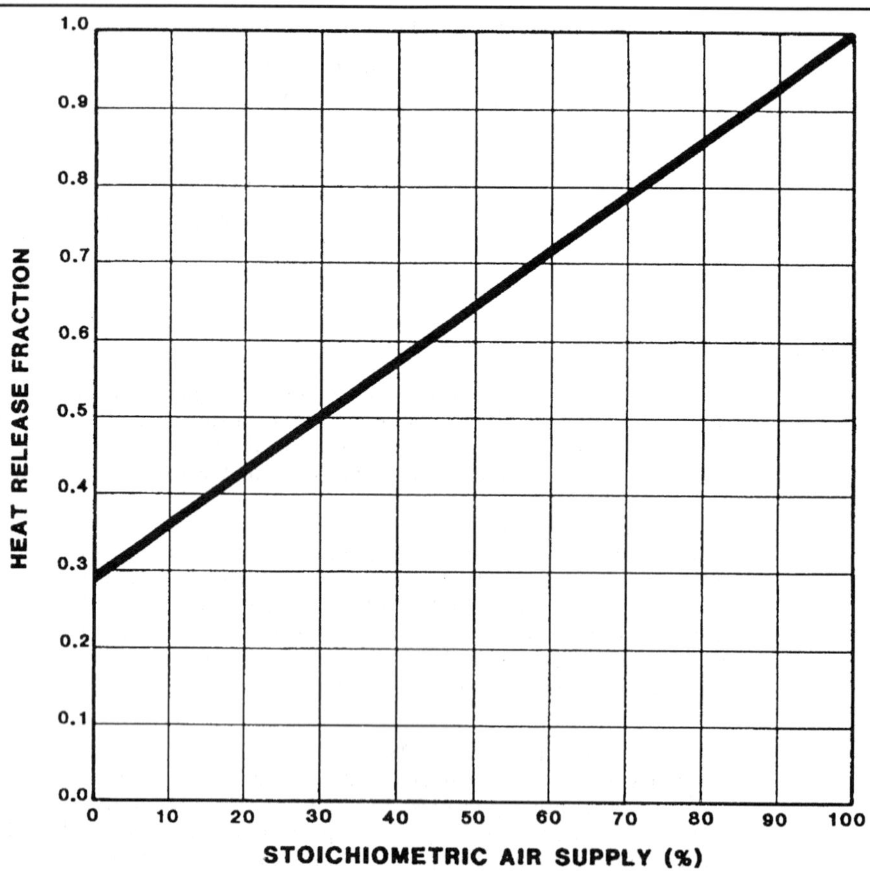

FIGURE 2.20 Heat release fraction vs. stoichiometric air supply for starved air combustion (*OME-88/12, p. C3-1*).

furnace materials can tolerate. Uncooled furnace walls constructed of refractory materials normally require the furnace gas temperatures not to exceed 1800–2200°F. Furnace temperature control, therefore, takes on primary importance. This can be accomplished by:

1. using excess air in quantities sufficient to produce the desired temperature;
2. heat removal across heat transfer surfaces;
3. some combination of (1) and (2).

EXAMPLE 1 furnace temperature
Consider a furnace burning No. 6 fuel oil having a specific gravity of 0.986, a HHV of 18,640 Btu/lb, and an ultimate analysis of 85.7% C, 10.5% H_2, 0.92% O_2, 2.8% S, 0.08% ash, and a net heating value, NHV, of 17,620 Btu/lb (EPA-80/02, p. 4-3).

TABLE 2.31 Starved Air Incinerator Analysis

Step	Description	Calculation	Unit
1	Waste feed rate (given)	2083	kg/h
2	Water fraction in waste (given)	0.24	
3	Water quantity in waste (step 1 × 2)	499.92	kg/h
4	Solid quantity (step 1 − step 3)	1583.08	kg/h
5	Dry base of noncombustible fraction = (wet fraction)/(1-water fraction) = 0.11/(1 − 0.24)	0.145	
6	Noncombustible quantity (step 4 × 5)	229.55	kg/h
7	Combustible quantity (step 4 − step 6)	1353.53	kg/h
8	Heating value of combustibles = 11,000(2083/1353.53) = 16,928.33. From Fig. 2.11, at 60% of stoichiometric air, 72% of the total heat capacity is released, i.e., 0.72(16,928.33) = 12,188.40	12,188.4	kJ/kg
9	Total heat generated (step 7 × 8)	16,497,365.05	kJ/h
10	Theoretical (stoichiometric) air ratio (based on 16,928.33 from Fig. 2.11)*	5.1	kg-air/kg-combustible
11	Theoretical air (step 7 × 10)	6903.00	kg/h
12	Excess air fraction (60% stoichiometric air = −40% excess air, given)	−0.4	
13	Excess air required (step 11 × 12)	−2761.20	kg/h
14	Total air required (steps 11 + 13)	4141.80	kg/h
15	Water formed due to combustion of combustible waste (based on 12,188.40 from Fig. 2.12)*	0.32	kg-water/kg-combustible
16	Total water due to combustion (step 7 × 15)	433.13	kg/h
17	Humidity in air (given)	0.00657	kg-water/kg-air
18	Water due to humidity (step 14 × 17)	27.21	kg/h
19	Total water in flue gas (steps 3 + 16 + 18)	960.26	kg/h
20	Dry flue gas (steps 7 + 14 − 16)	5062.20	kg/h
21	Radiation loss fraction (given)	0.02	
22	Radiation heat loss (step 9 × 21)	329,947.30	kJ/h
23	Ash heating value (given)	250	kJ/kg
24	Ash heat loss (step 6 × 23)	57,387.50	kJ/kg
25	Humidity correction due to latent heat (step 18) × 2186	59,481.06	kJ/h
26	Miscellaneous heat loss (given)	0	kJ/h
27	Heat in flue gas (steps 9 − 22 − 24 + 25 − 26)	16,169,511.31	kJ/h
28	Estimate flue gas temperature (see note 1 calculation below)	>1200	°C
29	Required gas temperature (given)	–	°C
30	Required enthalpy of flue gas	–	kJ/h
31	Net heat required (step 30 − step 27)	–	kJ/h
32	Fuel excess air (20% excess air, given)	–	
33	Fuel heating value at required temperature (Figs. 2.13 or 2.14)*	–	kJ/L
34	Fuel required (step 31/33)	None	L/h
35	Air for fuel (see Table 2.25)*	–	kg air/L
36	Total air required for fuel (step 35 × 34)	–	kg/h
37	Moisture from fuel (see Table 2.25)*	–	kg/L
38	Moisture generated (step 37 × 34)	–	kg/h
39	Dry flue gas from fuel combustion (see Table 2.25)*	–	kg/L
40	Dry gas generated (step 39 × 34)	–	kg/h
41	Total dry gas flow (steps 40 + 20)	5062.20	kg/h
42	Total moisture flow (steps 38 + 19)	960.26	kg/h
43	Fuel gross heating value (Table 2.25)*	–	kJ/L
44	Heat generated from fuel (step 43 × 34)	–	kJ/h
45	Total heat at exit (steps 44 + 27)	16,169,511.31	kJ/h

* See Section 4.2.

Determine:

1. The furnace gas temperature with the following system design alternatives.

   Case 1. Adiabatic combustion (no loss or useful heat transfer) with stoichiometric air.

   Case 2. Stoichiometric air, and 5% energy loss from the furnace to the surroundings.

2. Excess air or heat transfer necessary to achieve 2200°F furnace temperature.

   Case 3. Excess air but no heat transfer other than 5% energy loss.

   Case 4. Excess air limited to 10%, 5% energy loss, and heat transfer is needed to limit the temperature to 2200°F.

Given conditions
The data in question are summarized below.

- specific gravity of No. 6 oil $S_g = 0.986$
- high heating value (HHV) = 18,640 Btu/lb
- net heating value (NHV) = 17,620 Btu/lb
- ultimate analysis of No. 6 oil:

$$C = 0.857$$
$$H_2 = 0.105$$
$$O_2 = 0.0092$$
$$S = 0.028$$
$$\text{Ash} = 0.0008$$

- sum of the ultimate analysis = 1.00

Solution for Case 1
First we need to determine the amount of stoichiometric (theoretical) air required for complete combustion (Table 2.32). The amount of combustion air can be found by applying the method in Section 2.6. Assume 1.0 lb/h of No. 6 fuel oil fed.

For the No. 6 fuel oil given here, air required for complete combustion from the calculation contained in Table 2.32 is

$$A_t = 13.4883 \text{ lb-air/lb-oil}$$

When a fuel is burned, mass must be conserved. It is possible then to predict the mass of combustion gas from the air required and the combustible matter actually burned (Table 2.33). The mass of flue gas produced is calculated by applying the method in Section 2.6.

CHECK ON MASS BALANCE: The above calculations show that the flow rate of total reactants (R_a) and total products (P_a) are considered to be equal (14.5, accurate to one decimal point). Also from the above calculations, the actual combustion flue gas G_a becomes the theoretical combustion flue gas because there are no excess air components in the calculation. Let G_t be the theoretical combustion flue gas, so that

$$G_t = G_a$$

The noncombustibles here are either the ash in the fuel or the ash together with the unburned combustibles in solid form. Gaseous unburned components would remain part of the flue gas. With one pound of fuel as a basis, G_t for the No. 6 oil specified here becomes

$$G_t = 14.5 \text{ lb-air/lb-oil}$$

The mass of each individual gas in the product can be calculated, and an average or effective specific heat (C_p) for the mixture can be computed. A value applicable to oil combustion gas temperatures is approximately 0.25 Btu/lb-°F. With this value, one can estimate the adiabatic flame temperature t_{ad} from

$$H = G_t C_p (t_{ad} - t_a) \tag{2.64}$$

where H = net heating value
G_t = theoretical combustion flue gas
t_{ad} = adiabatic combustion temperature
t_a = combustion air intake temperature, initial temperature (100°F)

Substitute the proper numbers into the above equation and solve for t_{ad}.

TABLE 2.32 Combustion Reactant Table

Basic reaction	Oxygen, nitrogen, and air requirements	Mass (lb/h)	Mole (mol/h)
$C + O_2 \rightarrow CO_2$	(1) $O_2 = C(32/12)$	2.2853	0.0714
$H_2 + 0.5O_2 \rightarrow H_2O$	(2) $O_2 = 8(H\text{–}Cl/35.5)$*	0.8400	0.0263
$S + O_2 \rightarrow SO_2$	(3) $O_2 = S(32/32) = S$	0.0280	0.0009
B_o = bound oxygen	(4) Data from waste analysis	0.0092	0.0003
O_{2t} = theoretical O_2	$O_{2t} = (1) + (2) + (3) - (4)$	3.1441	0.0983
N_{2t} = theoretical N_2	$N_{2t} = (O_{2t})(3.76 \times 28/32)$	10.3442	0.3694
A_t = theoretical air	$A_t = O_{2t} + N_{2t}$	13.4883	0.4677
$\%A_e$ = % excess air	Given data	0.0000	
A_e = excess air	$A_e = (\%A_e)(A_t)$	0.0000	0.0000
A_{ada} = actual dry air	$A_{ada} = A_t + A_e$	13.4883	0.4677
H_2O from air†	$0.0127(A_{ada})$ = N/A	0.0000	0.0000
A_a = actual air	$A_a = A_{ada} + H_2O$	13.4883	0.4677
Waste feed	Data from waste analysis	1.0000	
R_a = actual reactant	$R_a = A_a$ + waste feed + H_2O from air	14.4883	

* If H_2 is the hydrogen left over after Cl reaction, i.e., $H_2 = H\text{–}Cl/35.5$, the O_2 needed to react with H_2 will be equal to $H_2(0.5)(32/2) = 8(H\text{–}Cl/35.5)$.
† In general, at room temperature, the humidity ratio is 0.0127 (lb-H_2O)/(lb-dry-air) or 0.0127 (kg-H_2O)/(kg-dry-air).

TABLE 2.33 Combustion Product Table

Component	Combustion product	Mass (lb/h)	Mole (mol/h)
CO_2	$CO_2 = C(44/12)$	3.1423	0.0714
SO_2	$SO_2 = S(64/32)$	0.0560	0.0009
HCl	$HCl = Cl(36.5/35.5)$	0.0000	0.0000
O_{2e}	$(\%A_e)O_{2t}$, O_{2t} from Table 2.32	0.0000	0.0000
N_{2e}	$(\%A_e)N_{2t}$, N_{2t} from Table 2.32	0.0000	
N_2	$N_{2e} + N_{2t}$ + from waste	10.3400	0.3693
H_2O from combustion	H_2O from combustion = $9(H\text{–}Cl/35.5)$	0.9450	
H_2O	H_2O from combustion + from air + from waste	0.9450	0.0525
G_a = actual flue gas	$G_a = CO_2 + SO_2 + HCl + O_{2e} + N_2 + H_2O$	14.4833	0.4941
Ash	Fly ash + bottom ash	0.0008	
P_a = actual product*	$P_a = G_a$ + ash	14.4841	

* In performing the mass balance calculation, the values of reactants (R_a) and products (P_a) should be equal, i.e., if $R_a = P_a$, the mass balance calculation is correct.

$$17{,}620 = 14.5(0.25)(t_{ad} - 100)$$
$$t_{ad} = 100 + 17{,}620/(14.51(0.25))$$
$$= 4957°F$$

Note that this temperature is considerably greater than the furnace materials of construction can tolerate. Therefore, case 1 is not a viable option.

Solution for Case 2

A second approach involves predicting the gas temperature when the system has heat transfer losses to the structure and surroundings. Equation (2.64) must be modified by the loss term Q_L to yield the nonadiabatic furnace temperature t_f as given by

$$H - Q_L = G_t C_p(t_f - t_a) \tag{2.65}$$

Here, with $Q_L = 0.05H$, the furnace temperature is

$$H - 0.05H = G_t C_p(t_f - t_a)$$
$$0.95H = G_t C_p(t_f - t_a)$$
$$t_f = 100 + 0.95(17{,}620)/(14.51(0.25))$$
$$= 4714°F$$

This gas temperature, while lower than that calculated for the adiabatic situation (case 1), is still too high to be practical.

Solution for Case 3

The third alternative proposes imposing a limit to the furnace temperature, with a 5% energy loss and no other heat transfer. This can be realized only through the use of excess air (A_e). The quantity of excess air needed is determined by a calculation of the mass of combustion product gas G_f required to absorb the net heating value of the fuel H with the gases leaving the furnace at the specified temperature (2200°F). The gas per pound of fuel is

$$G_f = (A_e + G_t)$$

Equation (2.65) is modified as given by

$$H - Q_L = G_f C_p(t_f - t_a) \tag{2.66}$$

Now if the $t_f = 2200°F$ condition is imposed on the system and assuming $C_p = 0.25$ Btu/lb°F as before, G_f can be calculated from

$$H - 0.05H = G_f(0.25)(2200 - 100)$$
$$0.95(17{,}620) = G_f(525)$$
$$16{,}739 = G_f(525)$$
$$G_f = 31.88 \text{ lb}$$

The excess air needed to reduce the temperature is then

$$A_e = G_f - G_t$$
$$= 31.88 - 14.51$$
$$= 17.37 \text{ lb-air/lb-fuel}$$

Percent excess air (A_e/A_t):

$$A_e/A_t = 17.37/13.51$$
$$= 128\%$$

This is substantially greater than the excess air normally found necessary for proper combustion of No. 6 oil.

Solution to Case 4

The logical next alternative is to limit the temperature by transferring energy to some useful purpose while limiting the excess air to the amount required for complete combustion. The energy equation for this case becomes

$$H = G_f C_p(t_f - t_a) + Q_L + Q_u \tag{2.67}$$

where Q_u is the energy to be transferred in order to maintain the furnace temperature at t_f. Rearranging Eq. (2.67) gives

$$Q_u = H - Q_L - G_f C_p(t_f - t_a) \tag{2.68}$$

Recall that case 4 prescribes 10% excess air. For this case, Eq. (2.69) becomes

$$Q_u = H - Q_L - G_{f10} C_p(t_f - t_a) \tag{2.69}$$

$$A_{e10} = 0.10(13.51)$$
$$= 1.35 \text{ lb-air/lb-fuel}$$
$$G_{f10} = G_t + A_{e10}$$
$$= 14.51 + 1.35$$
$$= 15.86 \text{ lb-air/lb-fuel}$$

where G_{f10} = flue gas for combustion with 10% excess air (lb/lb-fuel)
A_{e10} = 10% excess air (lb/lb-fuel)

Substituting the appropriate numerical values into Eq. (2.69) gives

$$Q_u = 17{,}620 - 0.05(17{,}620) - (15.86)(0.25)(2200 - 100)$$
$$= 16{,}739 - 8327$$
$$= 8412 \text{ Btu/lb fuel}$$

Percent available energy for recovery:

$$Q_u/H = 8412/17{,}620$$
$$= 47.74\%$$

Here Q_u represents 47.74% of the net heating value of the fuel. Useful application of this energy obviously depends upon the primary purpose of the combustion system. Steam generation would dictate water walls in the furnace to absorb this energy. Other systems would have to utilize this energy in some other appropriate manner with the heat transfer surface and medium compatible with the intended end use.

Summarizing the design process to this point, the primary alternatives for controlling the furnace temperature are to use a great deal of excess air, or to use some appropriate heat transfer surface to remove sufficient energy from the combustion gas to effect a control of temperature. The use of excess air alone as a control is wasteful of energy and should be avoided whenever possible.

This potentially wasteful aspect is also evident when considering the utilization of the energy remaining in the combustion products after they leave the furnace.

Nomenclature for this example

- A_e = excess air (lb/lb-fuel)
- A_{e10} = 10% excess air (lb/lb-fuel)
- A_t = theoretical (stoichiometric) air (lb/lb-fuel)
- C_p = constant pressure specific heat (Btu/lb-°F)
- G_t = theoretical combustion flue gas (lb/lb-fuel)
- G_f = flue gas for combustion with excess air (lb/lb-fuel)
- G_{f10} = flue gas for combustion with 10% excess air (lb/lb-fuel)
- H = net heating value (Btu/lb-fuel) = HHV – Q_v
- HHV = high heating value (Btu/lb-fuel)
- Q_L = energy transfer losses to structure and surrounding (Btu/lb-fuel)
- Q_u = useful energy transferred in the furnace (Btu/lb-fuel)
- t_a = combustion air temperature (°F)
- t_{ad} = adiabatic flame temperature (°F)
- t_f = furnace temperature (°F)

5.5 Energy Utilization in the Nonfurnace Region

Further utilization of energy, represented by the elevated temperatures of gases leaving a furnace, has a significant impact on the overall combustion system thermal efficiency E. The efficiency and its related equations are provided below (EPA-80/02, p. 4-6).

$$E = Q_s/Q_h \tag{2.70}$$

$$Q_h = m_f(\text{HHV}) \tag{2.71}$$

$$Q_s = m_f(q_s) = Q_h - \text{sum}(Q_{loss}) \tag{2.72}$$

where
E = overall combustion efficiency
HHV = high heating value
m_f = fuel firing rate
Q_s = total energy transferred for a useful purpose
q_s = energy to useful purpose per lb of fuel
Q_h = total energy input to the system

Losses identified earlier were limited to the energy transferred to the structure and the surroundings in the furnace Q_L. Additional losses occur in the regions through which the gas must flow upon leaving the furnace. A major loss is due to the heat content of flue gas leaving the system. This loss Q_{fg} arises from the fact that the flue gas stack temperature t_{fg} is higher than ambient temperature, and is expressed as

$$Q_{fg} = G_f(C_p)(t_{fg} - t_{amb}) \tag{2.73}$$

Equation (2.73) indicates that Q_{fg} is directly proportional to the total mass of the flue gases G_f, the specific heat of the gas, and the difference between the flue gas and the ambient. Increasing excess air beyond that which is required to ensure proper combustion increases G_f, which tends to increase the flue losses. The desirability of reducing the flue gas temperature t_{fg} is also apparent. In almost all combustion energy utilization devices, it is impractical to reduce t_{fg} to t_{amb}. Design, material, and economic factors prevent this and, in fact, dictate limits for various cases. Flue gas temperatures in steam boilers are limited to a low of about 250–300°F because of the potential dew-point and SO_x-associated corrosion problems which can develop at lower temperatures. Achieving even these flue gas exit temperatures requires considerable energy recovery equipment such as economizers and air preheaters.

The overall energy utilization pattern is summarized in Fig. 2.21, and by the following terms of the energy balance relationship (see the Nomenclature at the end of this sub-section).

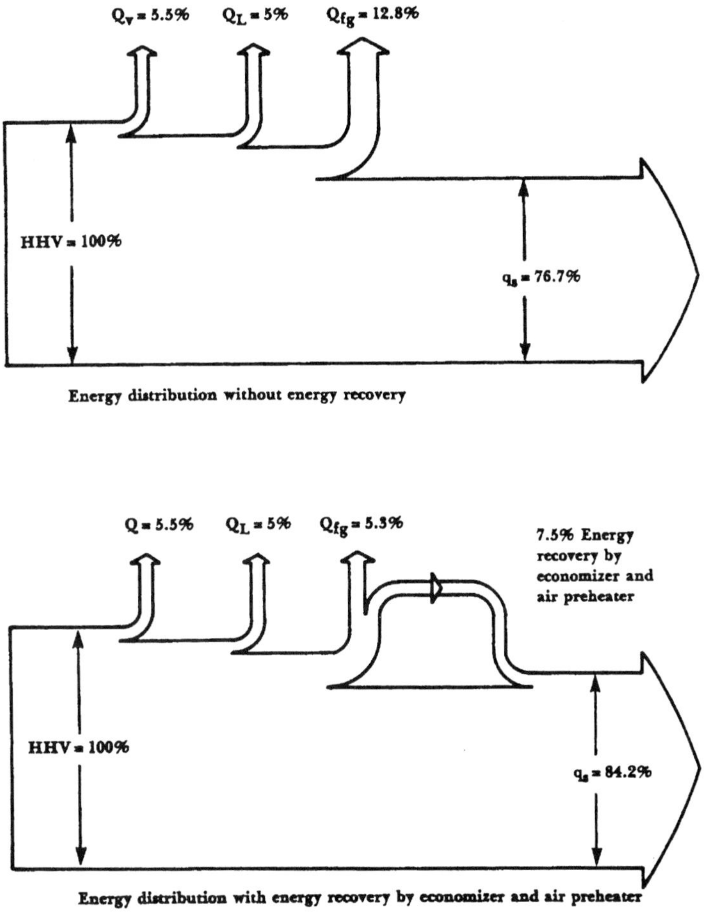

FIGURE 2.21 Energy distribution (*EPA-80/02, p. 4-12*).

Input: HHV

Losses: $\text{sum}(Q_{\text{loss}}) = Q_L + Q_{fg} + Q_v$

Available (utilized) energy: $Q_s = Q_u + Q_{nf}$

$Q_s = \text{HHV} - \text{sum}(Q_{\text{loss}})$

Note that in terms of the net heating value of the fuel H, the energy balance would become

$$H = \text{HHV} - Q_v$$
$$= Q_L + Q_{fg} + Q_s$$

The interaction of these several energy quantities is illustrated by the next example, which presumes a steam boiler where the fuel is already identified.

EXAMPLE furnace system thermal efficiency
A steam generator is to be designed for firing the No. 6 fuel oil from the previous example. Its rated output is to be 60,000 lb/h, output steam at $p = 650$ psia, $t = 800°\text{F}$, with the feedwater at 320°F (EPA-80/02, p. 4-8). No. 6 fuel oil has a specific gravity of 0.986, a HHV of 18,640 Btu/lb, and an ultimate analysis of 85.7% C, 10.5% H_2, 0.92% O_2, 2.8% S, 0.08% ash, and a net heating value (NHV) of 17,620 Btu/lb (EPA-80/02, p. 4-3).

Determine the distribution of the available energy utilization in this steam generator.

Solution:
The design begins with a determination of Q_s for this unit. This is done by accounting for the energy which is added to the working fluid (water) as it passes through the unit. Figure 2.22 shows the energy distribution in question (EPA-80/02, p. 4-8).

Letting m_s represent the steaming rate, Q_s becomes

$$Q_s = m_s(h_2 - h_1)$$

where
h_1 = enthalpy of the entering water
h_2 = enthalpy of the output steam
m_s = steam rate
Q_s = total energy transferred for a useful purpose

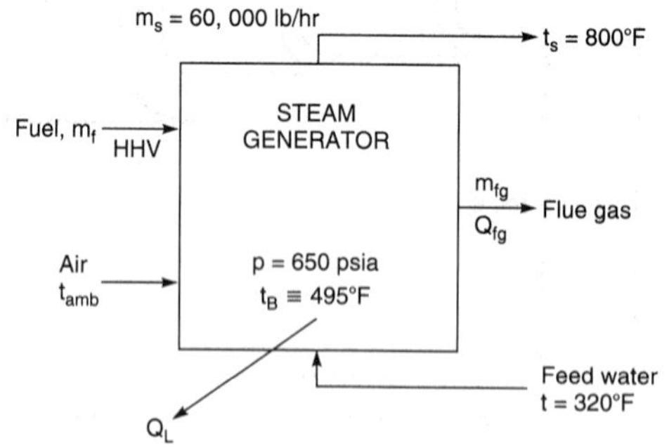

FIGURE 2.22 Energy distribution (*EPA-80/02, p. 4-8*).

$$h_1 = 290.3 \text{ Btu/lb at } p = 650 \text{ psia, } T = 320°\text{F (data from a steam table)}$$
$$h_2 = 1406.0 \text{ Btu/lb at } p = 650 \text{ psia, } T = 800°\text{F (data from a steam table)}$$
$$Q_s = 60{,}000(1406.0 - 290.3)$$
$$= 66.9 \times 10^6 \text{ Btu/h}$$

This is the available useful energy represented by $m_f(Q_u + Q_{nf})$.

The fuel supply rate needed to provide this energy depends on the overall efficiency E, which in turn depends on the energy recovery devices incorporated into the design.

$$\begin{aligned} q_s &= Q_s/m_f \\ &= \text{HHV} - Q_v - Q_L - Q_{fg} \\ &= H - Q_L - Q_{fg} \\ &= 17{,}620 - Q_L - Q_{fg} \end{aligned} \tag{2.74}$$

where H = net heating value (Btu/lb-fuel)
HHV = high heating value (Btu/lb-fuel)
Q_{fg} = energy loss as sensible heat in flue gas (Btu/lb-fuel)
Q_L = energy transfer losses to structure and surroundings (Btu/lb-fuel)
Q_v = energy loss due to latent heat of the water vapor formed by combustion (Btu/lb-fuel)

Suppose that Q_L can be limited to a maximum of 5% of HHV. Before the remaining loss term Q_{fg} can be determined, it is in order to consider some of the temperatures in the system.

- Gas leaves the furnace at $t_f = 2200°$F, while steam leaves the
- Steam superheater at $t_s = 800°$F
- Steam boiler temperature $t_b = 495°$F (saturation temperature at 650 psia)

The reason for listing these temperatures is to emphasize the limitations imposed by thermodynamic and heat transfer considerations. Energy exchange by heat transfer requires a temperature difference between the energy source and the heated medium. The superheater, if located in the convection zone, might reduce the gas temperature typically from 2200°F to say 1000°F, which will still allow a 200°F temperature difference for heat transfer requirements. The boiler operating at the 495°F boiling temperature can remove enough energy to bring the gas temperature to about 700°F. These temperatures are practical values, that is, they recognize the need for a finite temperature difference for heat exchange at realistic rates. In any event, temperatures lower than 800°F for the superheater outlet and 495°F for the boiler cannot be realized even with infinite heat transfer areas.

If the steam generator design does not include either an economizer or an air preheater, the gas temperature leaving the system would be approximately 700°F. For this case, the energy loss in the flue gas is given by

$$\begin{aligned} Q_{fg} &= G_{f10}(C_p)(t_{fg} - t_{amb}) \\ &= 15.86(0.25)(700 - 100) \\ &= 2379 \text{ Btu/lb-fuel} \end{aligned} \tag{2.75}$$

$$\begin{aligned} Q_L &= 0.05(\text{HHV}) \\ &= 0.05(18{,}640) \\ &= 930 \text{ Btu/lb} \end{aligned}$$

The useful energy per pound of fuel q_s is calculated by solving Eq. (2.74).

$$q_s = 17{,}620 - Q_L - Q_{fg}$$
$$= 17{,}620 - 930 - 2379$$
$$= 14{,}300 \text{ Btu/lb-oil}$$

The efficiency E, with Q_s and Q_h each based on one pound of fuel, is

$$E = Q_s/Q_h$$

where E = overall combustion efficiency
Q_s = total energy transferred for a useful purpose
Q_h = total energy input to the system

Substituting the appropriate numerical values into the above equation gives

$$E = 14{,}300/18{,}640$$
$$= 76.7\%$$

The fuel firing rate can now be determined noting that the total useful energy Q_s is 66.9×10^6 Btu/h and solving for m_f from equation

$$m_f = Q_s/q_s$$
$$= (66.9 \times 10^6)/14{,}300$$
$$= 4680 \text{ lb-oil/h}$$

The specific gravity of this No. 6 fuel oil was specified to be 0.986, and therefore a required fuel flow of approximately 569 gal/h is indicated.

The efficiency obtainable with a unit which extracts useful energy only in the furnace water walls, superheater, and boiler is not as high as could be realized. Continuing the design process, one would seek means to reduce the flue gas temperature still further, thereby reducing the flue losses and increasing the thermal efficiency. Recall that the feedwater temperature was specified to be 320°F. This is 175°F lower than the boiler temperature of 495°F. It would therefore appear to be possible to insert a heat exchange surface in the flue gas stream to extract energy by transferring energy to the colder feedwater. Such an exchange surface is called the *economizer*, and with temperatures as hypothesized here, flue gas temperature could be reduced to 500°F. With this lower flue gas temperature, the flue loss Q_{fg} would be reduced to 1590 Btu/lb, q_s would increase to 15,100 Btu/lb, and the effciency would increase to 81.0%.

Continuing the design analysis, one would note that the flue gas leaves the economizer at 500°F and that the ambient air enters at 100°F. Why not preheat combustion air? A decision to do so or not should, at least in part, be based upon economics. The additional hardware would have a higher first-cost and operating cost, which would have to be balanced against the value of the energy saved. An air preheater could certainly be expected to reduce flue gas temperatures to 350°F.

At 350°F flue gas temperature, Eq. (2.75) is used to calculate the loss Q_{fg}:

$$Q_{fg} = G_{f10}(C_p)(t_{fg} - t_{amb})$$
$$= 15.86(0.25)(350 - 100)$$
$$= 991 \text{ Btu/lb-fuel}$$

Now, from Eq. (2.74)

$$q_s = 17{,}620 - 932 - 991$$
$$= 15{,}697 \text{ Btu/lb fuel}$$

The efficiency E would be

$$E = 15{,}697/18{,}640$$
$$= 84.21\%$$

The fuel firing rate m_f would be

$$m_f = Q_s/q_s$$
$$= (66.9 \times 10^6)/15{,}697$$
$$= 4262 \text{ lb-oil/h}$$

Nomenclature for this example

- A_a = actual combustion air (lb/lb-fuel)
- A_e = excess air (lb/lb-fuel)
- A_{e10} = 10% excess air (lb/lb-fuel)
- A_t = theoretical (stoichiometric) air (lb/lb-fuel)
- C_p = constant pressure specific heat (Btu/lb-°F)
- G_t = theoretical combustion flue gas (lb/lb-fuel)
- G_f = flue gas for combustion with excess air (lb/lb-fuel)
- G_{f10} = flue gas for combustion with 10% excess air (lb/lb-fuel)
- h = specific enthalpy (Btu/lb-fuel)
- h_1 = enthalpy of the entering water (Btu/lb-fuel)
- h_2 = enthalpy of the output steam (Btu/lb-fuel)
- H = net heating value (Btu/lb-fuel) = HHV – Q_v
- HHV = high heating value (Btu/lb-fuel)
- m_f = fuel firing rate (lb/h)
- m_{nc} = noncombustibles in fuel (lb/h)
- m_s = steam rate (lb/h)
- Q_{fg} = energy loss as sensible heat in flue gas (Btu/lb-fuel)
- Q_h = total energy input to the system (Btu/h)
- Q_L = energy transfer losses to structure and surroundings (Btu/lb-fuel)
- Q_{nf} = useful energy (Btu/lb-fuel), transferred in nonfurnace region
- q_s = total energy transferred for a useful purpose (Btu/lb)
- Q_s = total energy transferred for a useful purpose (Btu/h)
- Q_u = useful energy transferred in the furnace (Btu/lb-fuel)
- Q_v = energy loss due to latent heat of the water vapor formed by combustion (Btu/lb-fuel)
- t_a = combustion air temperature (°F)
- t_b = steam boiler temperature (°F)
- t_{ad} = adiabatic flame temperature (°F)
- t_{amb} = ambient air temperature (°F)
- t_f = furnace temperature (°F)
- t_{fg} = flue gas temperature (°F)

5.6 Rotary and Afterburner System

A computer program was developed by the Energy and Environmental Research Corporation (EER) under a contract to the US EPA to perform energy and mass balance calculations on rotary kiln incinerators for hazardous waste incineration (EPA Contract 68-CO-0094, 92/10). A typical rotary kiln incineration system is shown in Fig. 2.23. The calculations are performed for a single-zone incinerator or for an incinerator system composed of single zones in series. The composition of gases and solids in a zone is assumed to be the same as the exit composition—that of complete reaction, the temperature in a zone is assumed to be the same as the exit temperature, and the residence time within a zone is assumed to be the mean residence time. Products of one zone enter the next zone at the exit temperature of the first zone. In cases where there is a significant heat loss between the incinerator units, a transition zone is included between the units to allow the products to enter the second unit at a lower temperature than they exit the first.

Inputs to the energy and mass balance calculation include:

- feed rate, temperature, heating value, heat capacity, heat of vaporization, and composition of all input streams to each unit including wastes, fuels, water, air, and oxygen;
- incinerator design specifications including the thickness and conductivity of the refractory, the volume of the unit, the area of the refractory and any cooled surfaces, and the outer shell temperature;
- the air pollution control devices (APCD) design specifications including gas volumetric capacity, acid gas capacity, quench water capacity and temperature, and the temperature to which the gas must be quenched.

Tables 2.34–2.37 show typical blank input forms.

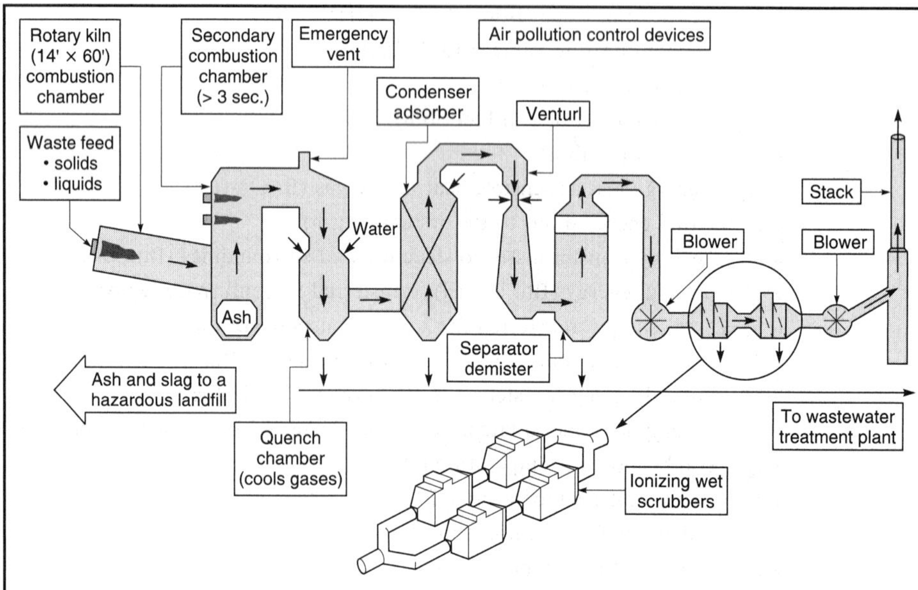

FIGURE 2.23 Schematic of incineration process.

TABLE 2.34 PCC Energy and Mass Balance Input Data*

	Proximate analysis (as received)							Elemental analysis (dry)						
	Feed rate (lb/h)	Preheat (°F)	Fixed carbon (%)	Volatiles (%)	Ash (%)	Moisture (%)	HHV (Btu/lb)	C (%)	H (%)	N (%)	S (%)	O (%)	Ash (%)	Cl (%)
Waste stream														
Fuel														
Air														
Water														
Oxygen														
Ash dropout														

* PCC = primary combustion chamber.

TABLE 2.35 SCC Energy and Mass Balance Input Data*

	Proximate analysis (as received)							Elemental analysis (dry)						
	Feed rate (lb/h)	Preheat (°F)	Fixed carbon (%)	Volatiles (%)	Ash (%)	Moisture (%)	HHV (Btu/lb)	C (%)	H (%)	N (%)	S (%)	O (%)	Ash (%)	Cl (%)
Waste stream														
Fuel														
Air														
Water														
Oxygen														
Ash dropout														

* SCC = secondary combustion chamber = afterburner.

TABLE 2.36 Incinerator Design Specifications

Unit volume (ft^3)
Refractory thickness (in)
Refractory conductivity (Btu-in/(h-ft^2-°F))
Refractory surface area (ft^2)
Cooled surface area (ft^2)
Others

TABLE 2.37 Air Pollution Control Device Design Specifications

Quench water temperature (°F)
Stack temperature (°F)
Others

Feed rates are in the form of mass flows. Preheat denotes the temperature at which a stream enters the incinerator. Proximate analysis is a standard analytical procedure used to characterize fuel/waste consisting of mass percentages of fixed carbon, volatiles, ash, and moisture. Elemental analysis consists of the dry mass percentages of carbon, hydrogen, nitrogen, sulfur, ash, oxygen, and chlorine. Halogens other than chlorine should be treated as chlorine by multiplying their mass percentage by the ratio of atomic weight of chlorine to the atomic weight of the halogen and renormalizing so that the total mass percentage is 100.

Heating value is input as the higher or gross heating value (HHV) the way it is typically measured and reported. The higher heating value is defined as the heat of complete combustion of the fuel/waste at 76.4°F (298 K) with all product water in liquid form and all product chlorine in the form of HCl. If the heating value is unknown, it can be estimated from the fuel/waste composition by the following equation (EPA-92/10):

$$\begin{aligned} HHV = {} & 1.8(100 - \text{ash} - \text{moisture})/100\{83.2(C) + 275.15(H) + 25.0(S) \\ & + 15.0(N) - 25.8(O) - 568.4[1 - \exp(-0.582(Cl/C))]\}\ \text{Btu/lb} \end{aligned}$$

EXAMPLE energy and mass balance calculation for a rotary kiln and afterburner system
A rotary kiln burns 3000 lb/h of typical medical waste with a higher heating value of 8724 Btu/lb. The waste has about 36% moisture on a wet basis, and about 15% ash and 4% chlorine on a dry basis. No auxiliary fuel is required for the rotary kiln. Almost 50,000 lb/h of air is used to burn the waste. Heat loss from the kiln is a little more than 5%. There is no preheat of either the waste or the air. The hot combustion products from the rotary kiln pass directly into the afterburner except that 98% of the ash drops out in the kiln.

The afterburner burns about 550 lb/h of No. 2 fuel oil in almost 8000 lb/h of additional air. Heat loss from the afterburner is almost 6%.

In this example, the rotary kiln is also referred to as the primary combustion chamber (PCC) and the afterburner as the secondary combustion chamber (SCC). More detailed inputs for the example are provided in Tables 2.38 and 2.39.

Solution:
1. Calculate mass balance on PCC
1.1. Convert each input stream to composition on a wet, as-fired basis
From the proximate analysis, the waste consists of 36.32% moisture by mass and 100–36.32 = 63.68% dry components by mass.

TABLE 2.38 PCC Energy and Mass Balance Input Data*

			Proximate analysis (as received)					Elemental analysis (dry)						
	Feed rate (lb/h)	Preheat (°F)	Fixed carbon (%)	Volatiles (%)	Ash (%)	Moisture (%)	HHV (Btu/lb)	C (%)	H (%)	N (%)	S (%)	O (%)	Ash (%)	Cl (%)
Waste stream	3000	76.4	0	53.65	10.03	36.32	8724	63	9.37	0.12	0	7.47	15.76	4.27
Fuel														
Air	49,439	76.4												
Water	0													
Oxygen	0													
Ash dropout	98%													

* PCC = primary combustion chamber; PCC heat loss = 5.34%; PCC unit volume = 1525 ft^3.

TABLE 2.39 SCC Energy and Mass Balance Input Data*

			Proximate analysis (as received)					Elemental analysis (dry)						
	Feed rate (lb/h)	Preheat (°F)	Fixed carbon (%)	Volatiles (%)	Ash (%)	Moisture (%)	HHV (Btu/lb)	C (%)	H (%)	N (%)	S (%)	O (%)	Ash (%)	Cl (%)
Waste stream														
Fuel (No. 2 oil)	554	76.4	0	100	0	0	25,296	87.13	12.6	0.01	0.22	0.04	0	0
Air	7,932	76.4												
Water	0													
Oxygen	0													
Ash dropout	0%													

* SCC = secondary combustion chamber; SCC heat loss = 5.96%; SCC unit volume = 2113 ft^3.

Converting the elemental analysis from a dry mass to a wet mass basis, the dry components are

$$C = (63.00/100)(63.68) = 40.12\%$$
$$H = (9.37/100)(63.68) = 5.97\%$$
$$N = (0.12/100)(63.68) = 0.08\%$$
$$S = (0.00/100)(63.68) = 0.00\%$$
$$O = (7.47/100)(63.68) = 4.76\%$$
$$\text{Ash} = (15.76/100)(63.68) = 10.04\%$$
$$Cl = (4.27/100)(63.68) = 2.72\%$$

1.2. Calculate individual component flowrates

$$\text{mass}_C = 3000(40.12)/100 = 1203.60 \text{ lb/h}$$
$$\text{mass}_H = 3000(5.97)/100 = 179.10 \text{ lb/h}$$
$$\text{mass}_N = 3000(0.08)/100 = 2.40 \text{ lb/h}$$
$$\text{mass}_S = 3000(0.00)/100 = 0.00 \text{ lb/h}$$
$$\text{mass}_O = 3000(4.76)/100 = 142.80 \text{ lb/h}$$
$$\text{mass}_{ash} = 3000(10.04)/100 = 301.20 \text{ lb/h}$$
$$\text{mass}_{Cl} = 3000(2.72)/100 = 81.60 \text{ lb/h}$$
$$\text{mass}_{H_2O} = 3000(36.32)/100 = 1089.60 \text{ lb/h}$$

1.3. Calculate the O_2 and N_2 amount from the given air flowrate

$$\text{mass}_O = 49{,}439(23.2)/100 = 11{,}470 \text{ lb/h}$$
$$\text{mass}_N = 49{,}439(76.8)/100 = 37{,}969 \text{ lb/h}$$

1.4. Calculate combustion air
Calculate the combustion air by applying the method in Section 2.6 and using Table 2.40.

TABLE 2.40 Combustion Air in PCC

Basic reaction	Oxygen, nitrogen, and air requirements	Mass (lb/h)	Mole (mol/h)
$C + O_2 \rightarrow CO_2$	(1) $O_2 = C(32/12)$	3209.60	100.30
$H_2 + 0.5O_2 \rightarrow H_2O$	(2) $O_2 = 8(H–Cl/35.5)$*	1414.41	44.20
$S + O_2 \rightarrow SO_2$	(3) $O_2 = S(32/32) = S$	0.00	0.00
B_o = bound oxygen	(4) Data from waste analysis	142.80	4.46
O_{2t} = theoretical O_2	$O_{2t} = (1) + (2) + (3) - (4)$	4481.21	140.04
N_{2t} = theoretical N_2	$N_{2t} = (O_{2t})(3.76 \times 28/32)$	14,743.19	526.54
A_t = theoretical air	$A_t = O_{2t} + N_{2t}$	19,224.40	666.58
$\%A_e$ = % excess air	See step 1.5	1.5717	(as a fraction)
A_e = excess air	$A_e = (\%A_e)(A_t)$	30,214.98	1047.66
A_{ada} = actual dry air	$A_{ada} = A_t + A_e$	49,439.38	1714.24
H_2O from air†	$0.0127(A_{ada})$ = N/A	0.00	0.00
A_a = actual air	$A_a = A_{ada} + H_2O$	49,439.38	1714.24
Waste feed	Data from waste analysis	3000.00	
R_a = actual reactant	$R_a = A_a$ + waste feed + H_2O from air	52,439.38	

* If H_2 is the hydrogen left over after Cl reaction, i.e., $H_2 = H–Cl/35.5$, the O_2 needed to react with H_2 will be equal to $H_2(0.5)(32/2) = 8(H–Cl/35.5)$.

† In general, at room temperature, the humidity ratio is 0.0127 (lb-H_2O)/(lb-dry-air) or 0.0127 (kg-H_2O)/(kg-dry-air).

1.5. Calculate excess air amount $\%A_e$
For this example, A_a = 49,439 lb/h (given condition).

The excess air amount is

$$A_e = A_a - A_t$$
$$= 49{,}439 - 19{,}224$$
$$= 30{,}215 \text{ lb/h}$$
$$\%A_e = A_e/A_t(100\%)$$
$$= 30{,}215/19{,}224(100\%)$$
$$= 157.17\%$$

1.6. Calculate actual combustion products P_a
P_a can be calculated from the method in Section 2.6 and using Table 2.41.
1.7 Check on mass balance
The above calculations show that the flow rate of reactants and products are checked. They are both equal to 52,439 lb/h.

2. Calculate mass balance on secondary combustion chamber
The mass balance on the secondary combustion chamber is performed first on a local basis (i.e., ignoring the primary effluent), and then on an overall basis (i.e., combining secondary with primary combustion gas products).

2.1. Convert each input stream to composition on a wet, as-fired basis
From the proximate analysis, No. 2 fuel oil consists of 0% moisture by mass and 100 – 0 = 100% dry components by mass.

Converting the elemental analysis from a dry mass to a wet mass basis, the dry components are

TABLE 2.41 Combustion Product in PCC

Component	Combustion product	Mass (lb/h)	Mole (mol/h)
CO_2	$CO_2 = (C)(44/12)$	4413.20	100.30
SO_2	$SO_2 = S(64/32)$	0.00	0.00
HCl	$HCl = Cl(36.5/35.5)$	83.90	2.30
O_{2e}	$(\%A_e)O_{2t}$, O_{2t} from Table 2.40	7043.12	220.10
N_{2e}	$(\%A_e)N_{2t}$, N_{2t} from Table 2.40	23,171.87	
N_2	$N_{2e} + N_{2t}$ + from waste	37,917.46	1354.20
H_2O from combustion	H_2O from combustion $= 9(H–Cl/35.5)$	1591.21	
H_2O	H_2O from combustion + from air + from waste	2680.81	148.93
G_a = actual flue gas	$G_a = CO_2 + SO_2 + HCl + O_{2e} + N_2 + H_2O$	52,138.49	1825.83
Ash	Fly ash + bottom ash	301.20	
P_a = actual product*	G_a + ash	52,439.69	

* In performing the mass balance calculation, the values of reactants (R_a) and products (P_a) should be equal, i.e., if $R_a = P_a$, the mass balance calculation is correct.

$$C = (87.13/100)(100) = 87.13\%$$
$$H = (12.60/100)(100) = 12.60\%$$
$$N = (0.01/100)(100) = 0.01\%$$
$$S = (0.22/100)(100) = 0.22\%$$
$$O = (0.04/100)(100) = 0.04\%$$
$$\text{Ash} = (0.00/100)(100) = 0.00\%$$
$$Cl = (0.00/100)(100) = 0.00\%$$

2.2. Calculate individual component flowrates

$$\text{mass}_C = 554(87.13)/100 = 482.70 \text{ lb/h}$$
$$\text{mass}_H = 554(12.60)/100 = 69.80 \text{ lb/h}$$
$$\text{mass}_N = 554(0.01)/100 = 0.0554 \text{ lb/h}$$
$$\text{mass}_S = 554(0.22)/100 = 1.22 \text{ lb/h}$$
$$\text{mass}_O = 554(0.04)/100 = 0.22 \text{ lb/h}$$
$$\text{mass}_{ash} = 554(0.00)/100 = 0.00 \text{ lb/h}$$
$$\text{mass}_{Cl} = 554(0.00)/100 = 0.00 \text{ lb/h}$$
$$\text{mass}_{H_2O} = 554(0.00)/100 = 0.00 \text{ lb/h}$$

2.3. Calculate the O_2 and N_2 amount from the given air flowrate

$$\text{mass}_O = 7932(23.2)/100 = 1840 \text{ lb/h}$$
$$\text{mass}_N = 7932(76.8)/100 = 6092$$

2.4. Calculate theoretical combustion air (A_t)
Calculate the combustion air by applying the method in Section 2.6 and using Table 2.42.

TABLE 2.42 Combustion Air in SCC

Basic reaction	Oxygen, nitrogen, and air requirements	Mass (lb/h)	Mole (mol/h)
$C + O_2 \rightarrow CO_2$	(1) $O_2 = C(32/12)$	1287.20	40.23
$H_2 + 0.5O_2 \rightarrow H_2O$	(2) $O_2 = 8(H\text{–}Cl/35.5)$*	558.72	17.46
$S + O_2 \rightarrow SO_2$	(3) $O_2 = S(32/32) = S$	1.22	0.04
B_o = bound oxygen	(4) Data from waste analysis	0.22	0.01
O_{2t} = theoretical O_2	$O_{2t} = (1) + (2) + (3) - (4)$	1846.92	57.72
N_{2t} = theoretical N_2	$N_{2t} = (O_{2t})(3.76 \times 28/32)$	6076.37	217.01
A_t = theoretical air	$A_t = O_{2t} + N_{2t}$	7923.29	274.73
$\%A_e$ = % excess air	See step 2.5	0.0011	
A_e = excess air	$A_e = (\%A_e)(A_t)$	8.72	0.30
A_{ada} = actual dry air	$A_{ada} = A_t + A_e$	7932.00	275.03
H_2O from air†	$0.0127(A_{ada})$ = N/A	0.00	0.00
A_a = actual air	$A_a = A_{ada} + H_2O$	7932.00	275.03
Waste feed	Data from waste analysis	554.00	
R_a = actual reactant	$R_a = A_a$ + waste feed + H_2O from air	8486.00	

* If H_2 is the hydrogen left over after Cl reaction, i.e., $H_2 = H\text{–}Cl/35.5$, the O_2 needed to react with H_2 will be equal to $H_2(0.5)(32/2) = 8(H\text{–}Cl/35.5)$.
† In general, at room temperature, the humidity ratio is 0.0127 (lb-H_2O)/(lb-dry-air) or 0.0127 (kg-H_2O)/(kg-dry-air).

2.5. Calculate excess air amount $\%A_e$
For this example, $A_a = 7932$ lb/h (given condition)

$$\begin{aligned} A_e &= A_a - A_t \\ &= 7932 - 7923 \\ &= 9 \text{ lb/h} \\ \%A_e &= A_e/A_t(100\%) \\ &= 9/7923(100\%) \\ &= 0.11\% \end{aligned}$$

2.6. Calculate actual combustion products P_a
P_a can be calculated from the method in Section 2.6 and using Table 2.43.

TABLE 2.43 Combustion Product in SCC

Component	Combustion product	Mass (lb/h)	Mole (mol/h)
CO_2	$CO_2 = (C)(44/12)$	1769.90	40.23
SO_2	$SO_2 = S(64/32)$	2.44	0.04
HCl	$HCl = Cl(36.5/35.5)$	0.00	0.00
O_{2e}	$(\%A_e)O_{2t}$, O_{2t} from Table 2.42	2.03	0.06
N_{2e}	$(\%A_e)N_{2t}$, N_{2t} from Table 2.42	6.68	
N_2	$N_{2e} + N_{2t}$ + from waste	6083.11	217.25
H_2O from combustion	H_2O from combustion = $9(H\text{–}Cl/35.5)$	628.20	
H_2O	H_2O from combustion + from air + from waste	628.20	34.90
G_a = actual flue gas	$G_a = CO_2 + SO_2 + HCl + O_{2e} + N_2 + H_2O$	8485.68	292.48
Ash	Fly ash + bottom ash	0.00	
P_a = actual product*	$P_a = G_a$ + ash	8485.68	

* In performing the mass balance calculation, the values of reactants (R_a) and products (P_a) should be equal, i.e., if $R_a = P_a$, the mass balance calculation is correct.

2.7. Check on mass balance
The above calculations show that the flow rate of reactants and products are close enough.

3. Combine combustion products from both primary and secondary combustion chambers
3.1. Calculate the composite combustion gas from the PCC and the SCC (Table 2.44)

3.2. Calculate the overall excess oxygen (Table 2.45)

4. Energy balance on PCC (rotary kiln) (EPA-92/10)
The energy balance equates the energy released in combustion to the energy required to vaporize all moisture, the energy required to heat the combustion products to the exit temperature, and the energy lost in the process. The energy balance can be expressed as

$$\text{(heat of combustion)} - \text{(heat of vaporization)} - \text{(sensible heat)} - \text{(heat loss)} = 0$$

or

$$\text{(heat of combustion)} = \text{(heat of vaporization)} + \text{(sensible heat)} + \text{(heat loss)}$$

4.1. Calculate heat of combustion
The heat of combustion is the chemical heat released as the reactants are converted to products at the reference state of 76.4°F with the product H_2O in liquid form. It is calculated by the following equation:

$$\text{heat of combustion} = \text{HHV(mass flow)} - \text{(heat of solution)}$$

where HHV is the higher heating value of the fuel/waste, and the heat of solution is the energy released when gaseous HCl dissolves in water. The heat of solution is calculated by the following equation:

$$\text{heat of solution} = \text{(mass flow)(Cl)}(1 - \text{moisture})(887.36)(1 - \exp((-\text{H}_2\text{O/HCl})^{0.77}/1.92))$$

where H_2O/HCl is the molar ratio of H_2O to HCl produced in the oxidation of the fuel/waste. The molar values of H_2O and HCl can be calculated from the following equations:

$$\text{Molar value of H}_2\text{O} = 1590/18 = 88.33$$
$$\text{Molar value of HCl} = 84/36.5 = 2.30$$

where the values of 1590 and 84 are obtained from Table 2.41, and the values of 18 and 36.5 are the molecular weight of H_2O and HCl, respectively.

TABLE 2.44 PCC and SCC Composite Combustion Products (lb/h)

Combustion products	PCC (lb/h)	SCC (lb/h)	PCC + SCC (lb/h)	Molar weight	PCC + SCC (mole/h)
CO_2	4413	1769.9	6182.90	44	140.52
SO_2	0	2.44	2.44	64	0.04
HCl	84	0	84.00	36.5	2.30
O_{2e}	7043	2.03	7045.03	32	220.16
N_2	37,917	6083.11	44,000.11	28	1571.43
H_2O	2681	628.2	3309.20	18	183.84
Fly ash	6	0	6.00	–	–
	52,144.00	8485.68	60,629.68		2118.29

TABLE 2.45 PCC and SCC Composite Oxygen Flowrate (lb/h)

Combustion products	PCC	SCC	PCC + SCC
O_{2t}	4481	1847	6328.00
O_{2e}	7043	2.03	7045.03
O_{2a}	11,524	1849.03	13,373.03
Excess $\%O_2 = (7045/6328)100\% = 111.33\%$			

Thus, the molar ratio (H_2O/HCl) = 88.33/2.30 = 38.40 and the heat of solution is

$$\text{Heat of solution} = 3000(4.27/100)(1 - 36.32/100)(887.38)(1 - \exp((-38.40)^{0.77}/1.92))$$
$$= 72{,}000 \text{ Btu/h}$$

Thus, the heat of combustion = 8724(3000) − 72,000 = 26.10 $E + 06$ Btu/h where the values of 3000, 4.27, 36.32, and 8724 were given conditions provided in Table 2.38.

4.2. Calculate heat of vaporization
As shown in the preceding section, the higher heating value is used for the heat of combustion. This is based on the assumption that the product water is in liquid form. However, in combustion, the product water is in vapor form. Therefore, the heat required to vaporize that water must be factored into the energy balance.

The heat of vaporization is the heat required to vaporize all moisture and combustion water. It is calculated according to the equation

$$\begin{aligned}\text{Heat of vaporization} &= (\text{water flowrate})(\text{heat of vaporization})\\ &= 2681(1050.54)\\ &= 2.82\ E + 06 \text{ Btu/h}\end{aligned}$$

where the value of 2681 lb/h was from Table 2.41 and 1050.54 Btu/lb is the water latent heat, that is, the heat needed to vaporize one pound of water to vapor form.

4.3. Calculate heat loss
The heat loss can be rigorously calculated based on the radiative and convective heat transfer, or it can be prescribed. For this calculation, assume that the heat loss is 5.34% of the heat entering the kiln. Since there is no preheat, the heat entering the kiln is the heat of combustion.

$$\begin{aligned}\text{Heat loss} &= 26.10\ E + 06(5.34/100)\\ &= 1.39\ E + 06 \text{ Btu/h}\end{aligned}$$

4.4. Calculate sensible heat
The sensible heat is the energy required to heat the combustion gas from the reference temperature to the combustion chamber exit temperature. It is the sum of the sensible heat from the components of combustion gases.

$$\text{Sensible heat} = \text{sum}(m_i(Cp_i)(T - T_{\text{ref}}))$$

where
- m_i = mass flowrate of component i of combustion gases
- Cp_i = heat capacity of component i of combustion gases
- T = combustion chamber exit temperature (this is the unknown variable which the energy balance is being used to calculate)
- T_{ref} = reference temperature = 76.4°F (298 K)

Cp_i is the mean heat capacity of each component of the combustion gas. For gaseous species, the mean heat capacity is calculated according to the equation below.

$$Cp_i = (a_i(T - T_{\text{ref}}) + b_i(T^2 - T_{\text{ref}}^2)/2 + c_i(T^3 - T_{\text{ref}}^3)/3 + d_i(T^4 - T_{\text{ref}}^4)/4(T - T_{\text{ref}})$$

where T is the combustion chamber exit temperature in degrees Kelvin. This is the unknown variable which the energy balance is being used to calculate. It is solved by iteration, calculating the sensible heats and performing the energy balance until it closes. For simplification, in this example we will fortuitously choose the correct combustion exit temperature of 1600°F (1144.4 K) as a first guess. Thus, only one iteration will be required. T_{ref} is the reference temperature of 76.4°F (298 K).

The constants a, b, c, and d for the gaseous species are provided in Table 2.46 (EPA-92/10).

Using the above equation, the mean heat capacities for the gaseous species are provided in Table 2.47.

For ash, heat capacity is calculated from an integration of a formula from Perry (1973):

$$\begin{aligned} Cp_{\text{ash}} &= (0.18(T - T_{\text{ref}}) + 0.00003(T^2 - T_{\text{ref}}^2))/(T - T_{\text{ref}}) \\ &= 0.2303 \text{ Btu/(lb-°F)} \end{aligned}$$

Thus, the sensible heat of the various species is expressed in Table 2.48

where the values of column "mass flow" were calculated in Table 2.41 of this example
temperature difference = combustion chamber exit temperature − reference temperature = 1600°F − 76.4°F = 1523.6°F

4.5. Calculate the total energy balance for a PCC (rotary kiln)
The energy balance terms are combined according to the equation

$$\text{Heat of combustion} - \text{heat of vaporization} - \text{heat loss} - \text{sensible heat}$$

$$26.10\ E+06 - 2.82\ E+06 - 1.39\ E+06 - 21.89\ E+06 = 0 \text{ Btu/h}$$

Since the sum of the heats is zero, the energy balance is closed. If the sum had come out positive, it would have been an indication that the sensible heat was too low. In such a case, the guess for the combustion chamber exit temperature should be revised upward, and the energy balance should be iterated until closure is achieved.

5. Energy balance on a SCC (afterburner)
The energy balance equates the energy released in combustion with the energy required to vaporize all moisture, the energy required to heat the combustion products to the exit temperature, and the energy lost in the process. The energy balance can be expressed as

(Heat passed on from kiln) + (heat of combustion) − (heat of vaporization) − (heat loss) − (sensible heat) = 0

TABLE 2.46 Constants for Heat Capacity Calculation

Species	a	b	c	d
CO_2	5.316	1.4284×10^{12}	-8.362×10^{-6}	1.784×10^{-9}
SO_2	6.157	1.3840×10^{-2}	-9.103×10^{-6}	2.057×10^{-9}
HCl	7.244	-1.8200×10^{-3}	3.170×10^{-6}	-1.036×10^{-9}
O_2	6.085	3.6310×10^{-3}	-1.709×10^{-6}	3.133×10^{-10}
N_2	6.903	-3.7530×10^{-4}	1.930×10^{-6}	-6.861×10^{-10}
H_2O	7.7	4.5940×10^{-4}	2.521×10^{-6}	-8.587×10^{-10}

TABLE 2.47 Heat Capacity

Species	Mole base Cp (Btu/(mole-°F))	Molar weight (M)	Mass base Cp (Btu/(lb-°F)) = Cp (Btu/(mole-F))/M
CO_2	11.67	44	0.2652
SO_2	11.9	64	0.1859
HCl	7.25	36.5	0.1986
O_2	7.87	32	0.2459
N_2	7.41	28	0.2646
H_2O	9.06	18	0.5033

or

(Heat passed on from kiln) + (heat of combustion) = (heat of vaporization) + (heat loss) + (sensible heat)

The heat passed on from the kiln includes the sensible heat and the heat of vaporization of the combustion gas entering the afterburner.

The sensible heat entering the afterburner is the same as the sensible heat exiting the rotary kiln except that a portion of that sensible heat is lost due to the ash which drops out in the kiln:

Sensible heat entering the afterburner = sensible heat exiting kiln − sensible heat of ash drop out

Since 98% of the ash drops out in the kiln, 98% of the sensible heat of the ash also drop out:

$$\text{Sensible heat of ash dropout} = 105{,}616(0.98) = 103{,}504 \text{ Btu/lb}$$

$$\text{Sensible heat entering the afterburner} = 21{,}890{,}000 \text{ (from Table 2.48)} - 103{,}504 = 21.78\ E+06 \text{ Btu/lb}$$

The heat of vaporization of the combustion gas entering the afterburner is the same as the heat of vaporization of the gas exiting the kiln as calculated in step 4.2 of this example.

TABLE 2.48 Sensible Heat Exiting PCC

Species	Mass flow	Cp (Btu/(lb-°F))	Temperature difference	Sensible heat of species (Btu/h)
CO_2	4413	0.2652	1523.6	1,783,111
SO_2	0	0.1859	1523.6	0
HCl	84	0.1986	1523.6	25,417
O_{2e}	7043	0.2459	1523.6	2,638,683
N_2	37,917	0.2646	1523.6	15,286,032
H_2O	2681	0.5033	1523.6	2,055,866
Ash	301	0.2303	1523.6	105,616
Sum of sensible heat is roughly equal to 21.89 $E+06$				21,894,725

$$\begin{aligned}\text{Heat passed on from kiln} &= \text{sensible heat entering afterburner}\\ &\quad + \text{heat of vaporization entering afterburner}\\ &= 21.78\ E+06 + 2.82\ E+06\\ &= 24.60\ E+06\ \text{Btu/h}\end{aligned}$$

5.1. Calculate heat of combustion
The heat of combustion is the chemical heat released as the reactants are converted to products at the reference state of 76.4°F with the product H_2O in liquid form. It is calculated by the following equation:

$$\text{Heat of combustion} = \text{HHV(mass flow)} - \text{(heat of solution)}$$

where HHV is the higher heating value of the fuel/waste, and heat of solution is the energy released when gaseous HCl dissolves in water. The heat of solution is calculated by the following equation:

$$\text{Heat of solution} = \text{(mass flow)(Cl)}(1 - \text{moisture})887.36(1 - \exp((-H_2O/HCl)^{0.77}/1.92))$$

Since the No. 2 fuel oil charged to the afterburner has no chlorine, Cl = 0, the heat of solution is 0.

$$\begin{aligned}\text{Thus, heat of combustion} &= 25{,}296(554) - 0\\ &= 14.01\ E+06\ \text{Btu/lb}\end{aligned}$$

where the values of 25,296 and 554 were given in Table 2.39.

5.2. Calculate heat of vaporization
As shown in the preceding section, the higher heating value is used for the heat of combustion. This is based on the assumption that the product water is in liquid form. However, in combustion, the product water is in vapor form. Therefore, the heat required to vaporize that water must be factored into the energy balance.

The heat of vaporization is the heat required to vaporize all moisture and combustion water. It is calculated according the equation

$$\begin{aligned}\text{Heat of vaporization} &= (H_2O \text{ in kiln} + H_2O \text{ in afterburner})(\text{heat of vaporization})\\ &= 3309(1050.54)\\ &= 3.48\ E+06\ \text{Btu/h}\end{aligned}$$

where the value of 3308 is from Table 2.44 of this example.

5.3. Calculate heat loss
The heat loss can be rigorously calculated based on the radiative and convective heat transfer, or it can be prescribed. For this calculation, assume that the heat loss is 5.96% of the heat entering the afterburner. The heat load to the afterburner includes the heat of combustion and the heat passed on from the kiln.

$$\begin{aligned}\text{Heat loss} &= (14.01\ E+06 + 24.60\ E+06)(5.96/100)\\ &= 2.30\ E+06\ \text{Btu/h}\end{aligned}$$

5.4. Calculate sensible heat
The sensible heat exiting the afterburner is the energy required to heat the combustion gas (including both kiln and afterburner combustion products) from the reference temperature to

the combustion chamber exit temperature. It is the sum of the sensible heats from the components of combustion gases:

$$\text{Sensible heat} = \text{sum}(m_i(Cp_i)(T - T_{ref}))$$

where
m_i = mass flowrate of component i of combustion gases
Cp_i = heat capacity of component i of combustion gases
T = combustion chamber exit temperature (this is the unknown variable which the energy balance is being used to calculate)
T_{ref} = reference temperature = 76.4°F (298 K)

Cp_i is the mean heat capacity of each component of the combustion gas. For gaseous species, the mean heat capacity is calculated according to the equation below.

$$Cp_i = (a_i(T - T_{ref}) + b_i(T^2 - T_{ref}^2)/2 + c_i(T^3 - T_{ref}^3)/3 + d_i(T^4 - T_{ref}^4)/4)/(T - T_{ref})$$

where T is the combustion chamber exit temperature in degrees Kelvin. This is the unknown variable which the energy is being used to calculate. It is solved by iteration, calculating the sensible heats, and performing the energy balance until it closes. For simplification, in this example we will fortuitously choose the correct combustion exit temperature of 2000°F (1366.7 K) as a first guess. Thus, only one iteration will be required. T_{ref} is the reference temperature of 76.4°F (298 K).

The constants a, b, c, and d for the gaseous species are given in Table 2.46 (EPA-92/10).

Using the above equation, the mean heat capacities for the gaseous species are provided in Table 2.49.

For ash, heat capacity is calculated from an integration of a formula from Perry (1973):

$$Cp_{ash} = (0.18(T - T_{ref}) + 0.00003(T^2 - T_{ref}^2))/(T - T_{ref})$$
$$= 0.2423 \text{ Btu/(lb-°F)}$$

Thus, the sensible heat of the species is given in Table 2.50.

where the values of the column "mass flow" were calculated in Table 2.41 of this example.

Temperature difference = combustion chamber exit temperature − reference temperature = 2000 − 76.4 = 1923.6°F

5.5. Calculate the total energy balance for the SCC (afterburner)

The energy balance terms are combined according to the equation

TABLE 2.49 Heat Capacity

Species	Mole base Cp (Btu/(mole-°F))	Molar weight (M)	Mass base Cp (Btu/(lb-°F)) = Cp (Btu/(mole-F))/M
CO_2	12.07	44	0.2743
SO_2	12.18	64	0.1903
HCl	7.38	36.5	0.2022
O_2	8.02	32	0.2506
N_2	7.55	28	0.2696
H_2O	9.37	18	0.5206

TABLE 2.50 Sensible Heat Exiting Afterburner

Species	Mass flow	Cp (Btu/(lb-°F))	Temperature difference	Sensible heat of species (Btu/h)
CO_2	6183	0.2743	1923.6	3,262,420
SO_2	2.44	0.1903	1923.6	893
HCl	84	0.2022	1923.6	32,672
O_{2e}	7045	0.2506	1923.6	3,396,072
N_{2a}	44,000	0.2696	1923.6	22,818,513
H_2O	3309	0.5206	1923.6	3,313,719
Ash	6	0.2423	1923.6	2797
Sum of sensible heat is roughly equal to 32.83 $E+06$				32,827,086

$$\text{Heat passed on from PCC} + \text{heat of combustion} - \text{heat of vaporization} - \text{heat loss}$$
$$-\text{sensible heat} = 0$$

$$24.60\ E+06 + 14.01\ E+06 - 3.48\ E+06 - 2.30\ E+06 - 32.83\ E+06 = 0$$

Since the sum of the heats is zero, the energy balance is closed. If the sum had come out positive, it would have been an indication that the sensible heat was too low. In such a case, the guess for the combustion chamber exit temperature should be revised upward, and the energy balance should be iterated until closure is achieved.

6. Calculate residence time
6.1. Calculate the standard condition flow volumetric flowrate
Mean residence time in a combustion chamber is calculated from the temperature, volume, and volumetric flow of the combustion chamber.

$$\text{Volumetric flow at } 70°\text{F} = (\text{molar flow})(386.7\ \text{sft}^3/\text{mole})$$

where the value of 386.7 sft^3 is the gas volume at standard conditions.

$$\begin{aligned}\text{Volumetric flow (at } 70°\text{F})_{\text{PCC}} &= (\text{molar flow in Table 2.41})(386.7\ \text{sft}^3/\text{mole})\\ &= 1825.83(386.7)\\ &= 706{,}048.46\ \text{sft}^3/\text{h (standard cubic feet per hour)}\\ &= 196.12\ \text{sft}^3/\text{s}\end{aligned}$$

$$\begin{aligned}\text{Volumetric flow (at } 70°\text{F})_{\text{SCC}} &= (\text{molar flow in Table 2.44})(386.7\ \text{sft}^3/\text{mole})\\ &= 2118.29(386.7)\\ &= 819{,}142.74\ \text{sft}^3/\text{h}\\ &= 227.54\ \text{sft}^3/\text{s}\end{aligned}$$

6.2. Calculate the actual condition flow volumetric flowrate
By the ideal gas law:

$$\text{Volumetric flow (at } T) = \text{volumetric flow (at 70)}(T + 460)/(70 + 460)$$

$$\text{Volumetric flow (at } T)_{\text{PCC}} = 196.12(1600 + 460)/(70 + 460) = 762.27 \text{ ft}^3/\text{s}$$

$$\text{Volumetric flow (at } T)_{\text{SCC}} = 227.54(2000 + 460)/(70 + 460) = 1056.13 \text{ ft}^3/\text{s}$$

6.3. Calculate the residence time

$$\text{Residence time} = \text{volume}/(\text{volumetric flow})$$

$$\text{Residence time}_{\text{PCC}} = 1525/762.27 = 2.00 \text{ s}$$

$$\text{Residence time}_{\text{SCC}} = 2113/1056.13 = 2.00 \text{ s}$$

5.7 Stationary Hearth System

This section provides the application of a stationary hearth to the incineration of medical waste (OME-86/10, p. 28). This type of incinerator is also known as a starved air incinerator or a controlled air incinerator. The purpose of this section is to compare the difference between a stationary hearth and a rotary kiln provided in Section 5.6.

Figure 2.24 shows the operating principle of a stationary hearth (or controlled air) incinerator.
Figure 2.25 shows the major components of a stationary hearth incinerator.
Figure 2.26 shows a typical feeding system for a stationary hearth incinerator.
Figure 2.27 shows a typical staged hearth and automatic ash removal system for a stationary hearth incinerator.
Figure 2.28 shows a waste heat recovery boiler and a bypass stack for emergency use.

The design criteria for this type of incinerator are briefly discussed under the following headings:

- Feed system
- Primary chamber
- Secondary chamber
- Turbulence
- Combustion air
- Burners
- Process monitoring/control
- Quenching
- Ash disposal
- Operating procedure
- Incinerator stack

FEED SYSTEM

- All feed systems should be designed to prevent leakage of liquids that may be contained in the waste, and should be sloped towards the opening of the incinerator.
- Provision should be made to disinfect the hopper sides and containers when mechanical feed systems are used. Cleaning fluids should be drained to a controlled area or container and properly disposed of.

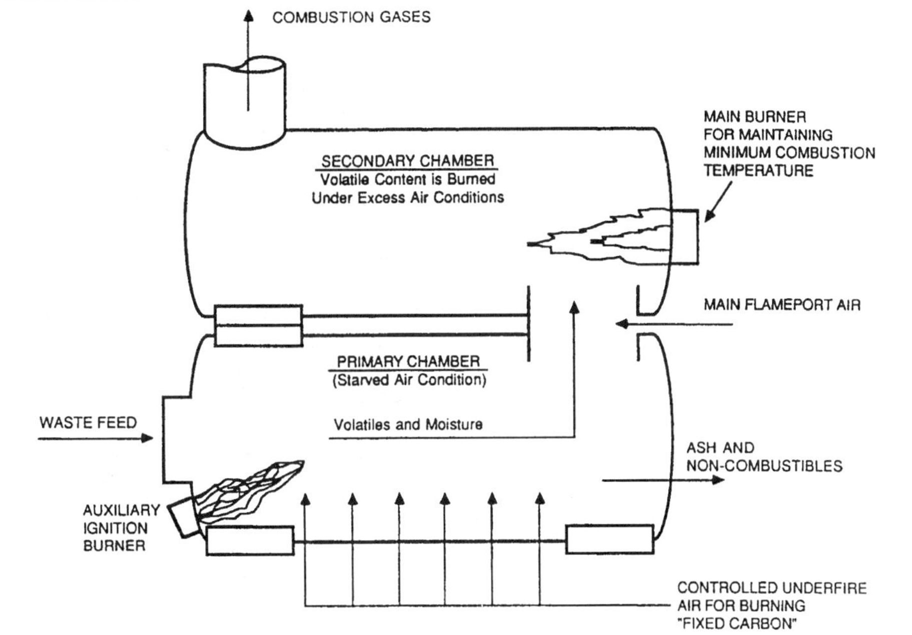

FIGURE 2.24 Principle of controlled air incinerator (*EPA-89/03*).

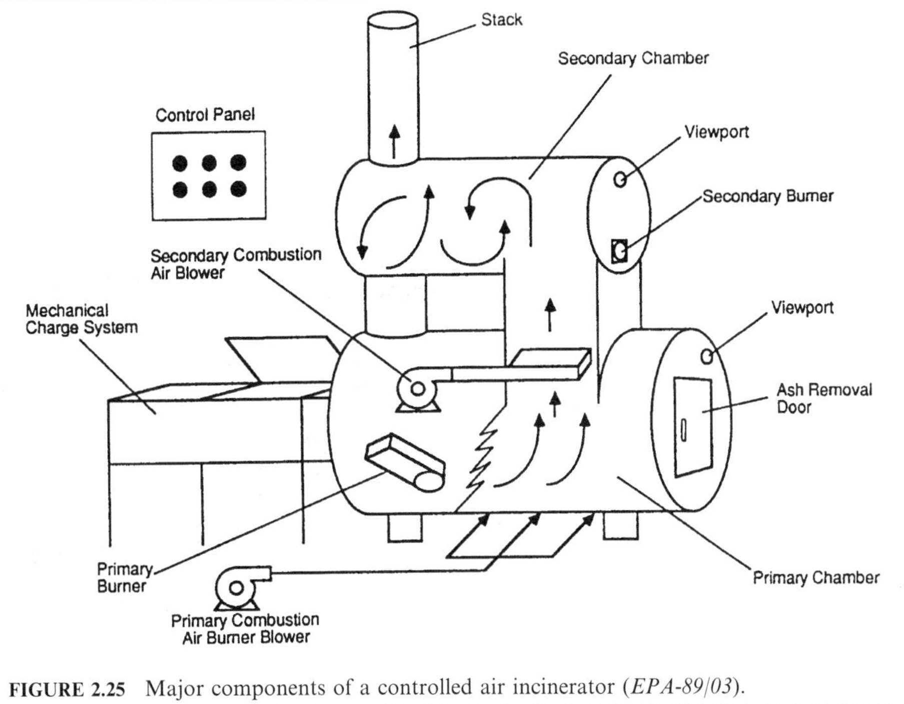

FIGURE 2.25 Major components of a controlled air incinerator (*EPA-89/03*).

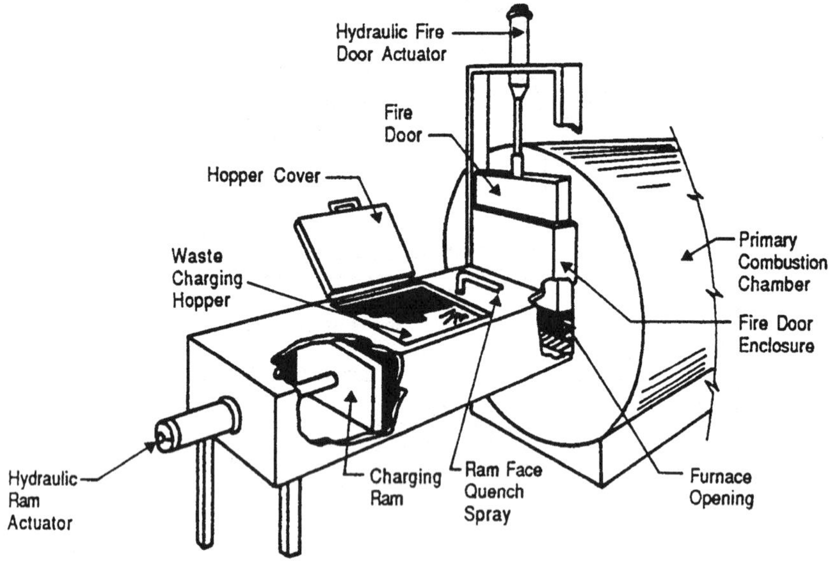

FIGURE 2.26 Hopper ram assembly (*EPA-89/03*).

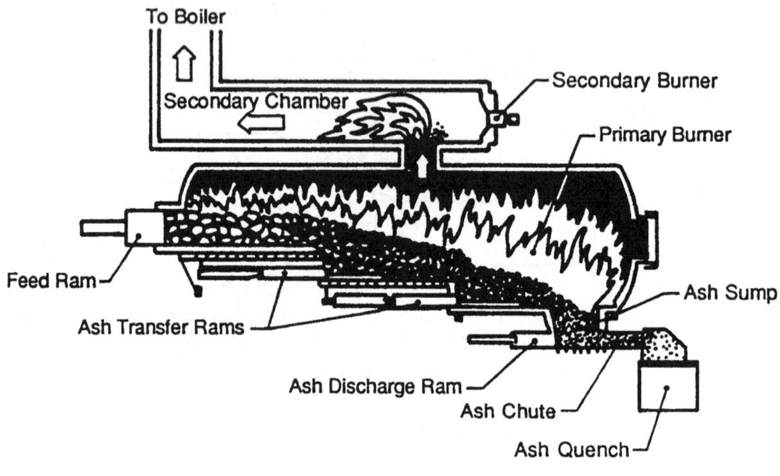

FIGURE 2.27 Staged hearth and automatic ash removal (*EPA-89/03*).

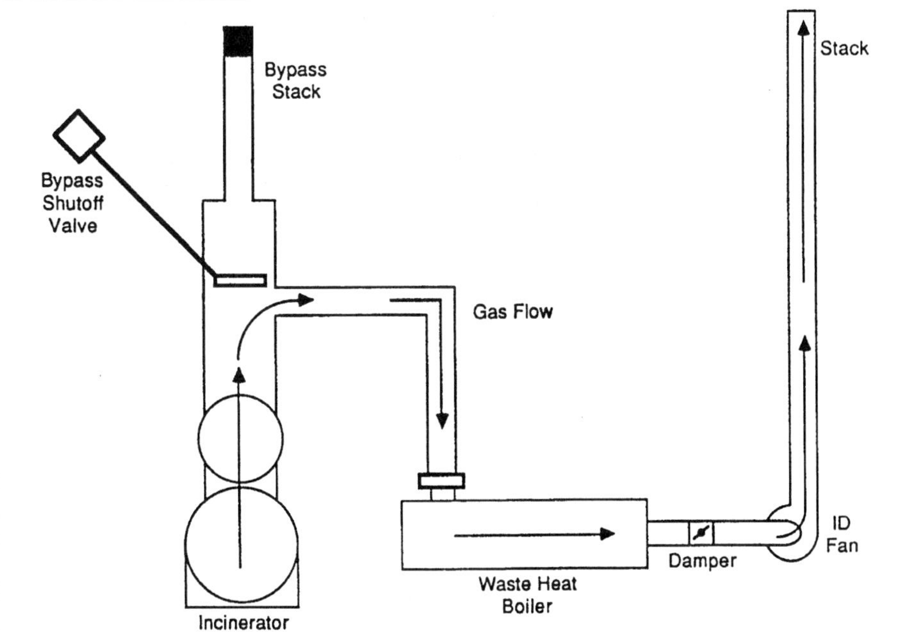

FIGURE 2.28 Incinerator with waste heat boiler and bypass stack (*EPA-89/03*).

- Mechanical feed systems should be interlocked with the charging mechanism to facilitate lock-out of the waste feed if the temperature in the secondary chamber falls below 1000°C (1830°F).
- Automated mechanical feed systems are desirable to maintain steady operation.

PRIMARY CHAMBER

- The rate at which waste is fed into the incinerator is critical to the successful operation of the unit because the feed rate determines the heat release rate of the medical waste and the size of the primary chamber. Medical waste contains low-density, high heating value wastes (e.g., plastics) as well as high-density, low heating value wastes (e.g., tissue, bones). Therefore, the primary chamber should be sized to accommodate the variation in the waste composition.
- The volume of the primary chamber should be designed to allow for a total heat release rate of between 445,000 $kJ/(h\,m^3)$ (12,000 $Btu/(h\,ft^3)$) and 930,000 $kJ/(h\,m^3)$ (25,000 $Btu/(h\,ft^3)$). A heat release rate of over 930,000 $kJ/(h\,m^3)$ (25,000 $Btu/(h\,ft^3)$) is not recommended for use in the design as this may lead to uncontrolled conditions, resulting in high particulate emissions, poor micro-organism destruction, and the discharge of incomplete combustion products to the environment.
- The hearth area should be designed to allow a burning rate of 58.5–68.5 $kg/(h\,m^2)$ (12–14 $lb/(h\,ft^2)$).
- The primary chamber should be designed to operate continuously under negative pressure to prevent fugitive emissions.

- The temperature in the primary chamber should be maintained in the range 400–760°C (750–1400°F). Avoiding temperature peaks above 760°C will minimize excursions in gas velocities, thereby reducing ash carry-over and particulate emissions.
- The external casing of the chamber should be designed to maintain a maximum temperature of 70–90°C (160–195°F). This can be accomplished with the use of refractory and/or insulation materials. Where appropriate, an expanded metal shield or other suitable means of shielding should be installed for the protection of personnel.
- The primary chamber hearth should have an adequate grease containment lip at the opening to prevent escape or leakage of fluids from the chamber. The lip should be a minimum of 5 cm (2 inches) deep.
- A "hot hearth" is recommended for use in all primary chambers where economically feasible.

Secondary chamber

- The temperature in the secondary chamber should be designed for a minimum of 1100°C (2010°F) with an operating temperature of not less than 1000°C (1830°F) at all times.
- The incinerator should be designed to provide no less than 6% residual oxygen in the flue gas exhaust from the secondary chamber.
- The secondary chamber should be designed for a gas residence time of not less than 1 s at 1000°C (1830°F). This residence time is to be based on the volume of the secondary chamber from the flame front to the location of the temperature sensing device.
- The temperature in the secondary chamber should be controlled by a thermocouple or other temperature sensor located at a point representing 1-s retention time at the exit of the secondary chamber or at the breaching. The thermocouple should be connected to a system to provide automatic temperature control, and it should also regulate the modulating secondary chamber burner.
- The refractory surface of the secondary chamber should be heated over a minimum period of half an hour prior to feeding waste into the incinerator to ensure optimum conditions for the destruction of micro-organisms.
- The external casing of the secondary chamber should be designed to maintain a maximum temperature of 70–90°C (160–195°F) by means of insulation and refractory. For protection of personnel, an expanded metal shield or other suitable means of protection should also be installed on the casing.

Turbulence

- Gas turbulence is an important parameter in the design of incinerators and can be achieved by high combustion gas velocity, tangential air injection, abrupt changes in flow direction, and the installation of combustion gas restrictions (e.g., orifices, checkerwork, or baffles).
- Turbulence is difficult to quantify; however, use of the Reynolds number (*Re*) has been suggested to provide an indication of the gas-phase turbulence in the incinerator.
- The calculated Reynolds number in the secondary chamber should be over 10,000 to ensure turbulent flow.

Combustion air

- For starved air incinerators, air into the primary chamber should be supplied at 30–80% of that required for stoichiometric combustion.
- The air supply in the secondary chamber of all incinerators should be able to provide excess air at 40–250% of that theoretically required during the peak burning rate.

- The combustion air supply should be automatically adjustable with a temperature recorder control system to maintain the set temperatures in the primary and secondary chambers of the incinerator.

BURNERS

- The burners must be able to maintain a stable flame throughout the range of pressures, input rates, and fuel/air ratios experienced in the primary and secondary chambers.
- The burners should be designed to supply a minimum of 80% of the total heat input of the incinerator design capacity. The burners must also be capable of modulating down to 15% of the total heat input requirement.
- The burner(s) in the primary chamber should be:
  - — located at a downward angle to provide maximum impingement of the flame onto the wastes. The alignment of the burner(s) should not allow the flame to impinge on the refractory walls or on other burner(s)
  - — set to maintain a temperature of 400–760°C (750–1400°F) in the primary chamber once the burn cycle is initiated
  - — constructed with a sealed casing to eliminate the flow of tramp air into the chamber
- The burner(s) in the secondary chamber should be:
  - — mounted to promote thorough mixing throughout the whole chamber. The alignment of the burner(s) should not allow the flame to impinge on the refractory walls or on other burner(s)
  - — designed to automatically lock out the primary chamber charging mechanism in the event of burner(s) failure
  - — set to maintain a temperature of not less than 1000°C (1830°F) in the secondary chamber at all times
  - — fully modulating with a low "hold fire" setting to ensure a flame throughout the incineration cycle

PROCESS MONITORING/CONTROL

- One, or preferably two, viewports should be installed in the primary chamber immediately behind the burners to facilitate visual inspection of the burn. The location should be selected to reduce particulate impingement, so that the viewport will remain relatively clean.
- A temperature recorder/controller (TRC) should be used to control primary and secondary temperatures by:
  - — turning off or reducing heat input from the burners
  - — turning off, throttling back, or increasing the air supply
  - — turning on the quench water system where used

The TRCs should provide a graphical recording of the temperature variations and feedback of the operating fluctuations. A TRC should be used to monitor the temperature at the exit of the primary chamber and also control the air supply to the primary chamber and its auxiliary burners. A second TRC should be installed to monitor the temperature at the exit of the secondary chamber or at the base of the stack, and also to control the secondary air supply and its auxiliary burners.

- All incinerators should be equipped with continuous total hydrocarbon or carbon monoxide monitoring equipment. An opacity meter should also be provided in the incinerator stack.

QUENCHING

- A water quench system should be provided in the primary chamber of the incinerator to prevent the temperature from developing into runaway conditions and/or to reduce flue gas temperatures at the exit of the secondary chamber.
- The quench system should be activated by the primary chamber TRC and sized to reduce the temperature in the primary chamber by 2000°C (3630°F) within 60 s.

ASH DISPOSAL

- Ash resulting from the incineration of medical waste may contain significant quantities of sharps, needles, and glass; therefore, care should be exercised in the removal and disposal of incinerator ash.
- The incinerator ash should be wetted prior to handling to minimize the potential for generating airborne dust.
- All personnel handling the ash should wear or use dust masks, gloves, and protective clothing as a safety precaution.
- The incinerator ash should be stored in enclosed containers and transported to an approved landfill site for disposal.

OPERATING PROCEDURE

- Waste should not be charged into the incinerator during the start-up period until the refractory surface of the secondary chamber has been heated to the operating temperature.
- The initial charges to the incinerator should be noninfectious waste; infectious waste should be fed later in the incineration cycle.
- The waste should be weighed and logged prior to charging to ensure that the design feed rate is not exceeded and to maintain a record of the quantities of waste processed.
- Incinerator operators should be properly trained and be familiar with all the manufacturer's operating procedures for the unit.
- The ash in the primary chamber should be discharged on a batch basis at the end of each incineration cycle to ensure complete destruction of the micro-organisms.

INCINERATOR STACK

- For natural draft systems, calculations for stack design should be based on a gas temperature of 1000°C (1830°F). If substantial heat losses through the stack are expected, such losses should be taken into account in determining the average stack temperature and the available draft.
- The stack height should be calculated to provide a minimum available draft of 6.3 mm (0.25 in.) water gauge (WG) at the breaching. The latter is an absolute minimum draft provision for all natural draft medical waste incinerators and must result in a draft of at least 2.5 mm (0.1 in.) WG at the burner air inlets. Perry's Chemical Engineers' Handbook outlines procedures for calculating stack draft.

EXAMPLE stationary hearth system

An energy and material balance is an important part of designing and/or evaluating incinerators. The following example entails a mathematical evaluation of the input and output conditions of a medical waste incinerator. It determines the combustion air and auxiliary fuel requirements for incinerating a given medical waste, and the limitations of an existing incinerator when charged with a known waste.

An incinerator is to be designed to incinerate a mixture of 30% red bag and 70% yellow bag (with a PVC content of 4%) medical waste. Throughput is to be 100 kilogram per hour (kg/h). The auxiliary fuel is natural gas, the waste has been ignited, and the secondary burner is modulating. Design requirements are summarized as follows:

- secondary chamber temperature: 1100°C
- flue gas residence time at 1000°C: 1 s
- residual oxygen in flue gas: 6% minimum (OME-86/10, p. 42)

For comparison purposes, this example provides both the metric system and the British system in the process of solution calculation.

Given conditions

- waste throughput = 100 kg/h or 220 lb/h
- input temperature of waste, fuel, and air is 15.5°C or 60°F
- air contains 23% by weight O_2 and 77% by weight N_2
- air contains 0.0132 kg-H_2O/kg-dry-air at 60% relative humidity, and 26.7°C dry bulb temperature, or 0.0132 lb-H_2O/kg-dry-air at 60% relative humidity and 80°F dry bulb temperature
- for any ideal gas, 1 kg-mole is equal to 22.4 m^3 at 0°C and 101.3 kPa
- latent heat of vaporization of water is 2460.3 kJ/kg at 15.5°C or 1057.35 Btu/lb at 60°F
- 1 Btu = 1.0551 kJ
- 1 Btu/lb = 2.3269 kJ/kg = 0.5557 kcal/kg
- 1 Btu/lb-°F = 4.1883 kJ/kg-°C
- 1 lb = 453.5 g = 0.4535 kg
- 1 ft^3 = 0.02832 m^3

Solution:
1. Analyze waste throughput

Waste throughput =100.00 kg/h or 220.51 lb/h
Red bag throughput = 30%(100) = 30 kg/h or 66.15 lb/h
Yellow bag throughput = 70%(100) = 70 kg/h or 154.36 lb/h

1.1. Analyze red bag waste components for 30 kg/h throughput

	%	kg/h	lb/h
Tissue	0.15	4.50	9.92
Water	0.80	24.00	52.92
Ash	0.05	1.50	3.31
	1.00	30	66.15

1.2. Analyze yellow bag waste components for 70 kg/h throughput

	%	kg/h	lb/h
Polyethylene	0.35	24.50	54.02
PVC	0.04	2.80	6.17
Cellulose	0.51	35.70	15.44
Ash	0.10	7.00	15.44
	1.00	70	154.36

2. Calculate theoretical combustion components
2.1. Write chemical reaction equations for combustible organics
For tissue ($C_5H_{10}O_3$) incineration

$C_5H_{10}O_3 + 6O_2$		$\rightarrow 5CO_2 + 5H_2O$	
118	192	220	90
1	1.63	1.86	0.76
4.5	7.32	8.39	3.43 for kg base
9.92	16.15	18.50	7.57 for lb lb base

For polyethylene (C_2H_4) incineration

$C_2H_4 + 3O_2$		$\rightarrow 2CO_2 + 2H_2O$	
28	96	88	36
1	3.43	3.14	1.29
24.5	84.00	77.0	31.50 for kg base
54.02	185.23	169.79	69.46 for lb base

Polyvinyl chloride (C_2H_3Cl) incineration

$2(C_2H_3Cl) + 5O_2$		$\rightarrow 4CO_2 + 2H_2O + 2HCl$		
125	160	176	36	73
1	1.28	1.41	0.29	0.58
2.80	3.58	3.94	0.81	1.64 for kg base
6.17	7.90	8.69	1.78	3.61 for lb base

Cellulose ($C_6H_{10}O_5$) incineration

$C_6H_{10}O_5 + 6O_2$		$\rightarrow 6CO_2 + 5H_2O$	
162	192	264	90
1	1.19	1.63	0.56
35.7	42.31	58.18	19.99 for kg base
78.72	93.30	128.29	44.08 for lb base

2.2. Calculate theoretical combustion air

Theoretical oxygen (kg base) $= 7.32 + 84.00 + 3.58 + 42.31$
$= 137.22$ kg/h
Theoretical oxygen (lb base) $= 16.15 + 185.23 + 7.90 + 93.30$
$= 302.57$ lb/h
Theoretical nitrogen (kg base) $= 3.29(137.22)$
$= 451.44$ kg/h
Theoretical nitrogen (lb base) $= 3.29(302.57)$
$= 995.47$ lb/h
Theoretical air (kg base) = theoretical oxygen + theoretical nitrogen
Theoretical air (kg base) $= 137.22 + 451.44 = 588.66$ kg/h
Theoretical air (lb base) $= 302.57 + 995.47 = 1298.04$ lb/h

2.3. Calculate CO_2 formed

CO_2 products (kg base) $= 8.39 + 77.00 + 3.94 + 58.18 = 147.51$ kg/h
CO_2 products (lb base) $= 18.50 + 169.79 + 8.69 + 128.29 = 325.27$ lb/h

2.4. Calculate moisture (H_2O) formed

H_2O products (kg base) = 3.43 + 31.50 + 0.81 + 19.99 = 55.73 kg/h
H_2O products (lb base) = 7.57 + 69.46 + 1.78 + 44.08 = 122.89 lb/h

2.5. Calculate HCl formed

HCl products (kg base) = 1.64 kg/h
HCl products (kg base) = 3.61 lb/h

2.6. Calculate ash dropout

Ash from waste feed (kg base) = 8.50 kg/h
Ash from waste feed (lb base) = 18.75 lb/h

3. Calculate actual combustion components
3.1. Calculate actual dry combustion air

By given conditions at the 150% excess air combustion,

Excess air amount (kg base) = 588.66(1.5) = 882.99 kg/h
Excess air amount (lb base) = 1298.04(1.5) = 1947.06 lb/h
Actual air amount = theoretical oxygen + excess oxygen
Actual air amount (kg base) = theoretical oxygen + excess oxygen
= 588.66 + 882.99 = 1471.65 kg/h
The actual air amount (lb base) = 1298.04 + 1947.06 = 3245.10 lb/h

3.2. Calculate the moisture from actual dry combustion air

By given conditions, the humidity = 0.0132 kg-H_2O/kg-dry-air, or 0.0132 lb-H_2O/lb-dry-air

Moisture = humidity (dry air)
Moisture (kg base) = 0.0132(1471.65) =19.43 kg/h
Moisture (lb base) = 0.0132(3245.10) = 42.84 lb/h

4. Calculate material balance
4.1. Calculate mass in

Mass in = waste feed rate + actual dry combustion air + moisture from combustion air
Mass in (kg base) = 100.00 + 1471.65 + 19.43 = 1591.08 kg/h
Mass in (lb base) = 220.51 + 3245.10 + 42.84 = 3508.44 lb/h

4.2. Calculate dry combustion products

Dry combustion products = excess air + theoretical nitrogen + CO_2 from combustion
Dry combustion products (kg base) = 882.99 + 451.44 + 147.51 = 1481.94 kg/h
Dry combustion products (lb base) = 1947.06 + 995.47 + 325.27 = 3267.80 lb/h

4.3. Calculate H_2O in products

H_2O in products = H_2O from waste + H_2O from combustion reaction
+ H_2O from actual combustion air

H_2O in products (kg base) = 24.00 + 55.73 + 19.43 = 99.16 kg/h
H_2O in products (lb base) = 52.92 + 122.89 + 42.84 = 218.65 lb/h

4.4. Calculate flue gas flowrate (W_{pcc})

Flue gas flowrate (W_{pcc}) = dry combustion products (step 4.2) + moisture in products (step 4.3)

$W_{pcc} = 1481.94 + 99.16 = 1581.10$ kg/h in kg base
$W_{pcc} = 3267.80 + 218.65 = 3486.45$ lb/h in lb base

4.5. Calculate total mass out
Total mass out = the sum of (dry combustion products (step 4.2) + H_2O in products (step 4.3) + HCl (step 2.5) + ash (step 2.6))

$$\text{Total mass out (kg base)} = 1481.94 + 99.16 + 1.64 + 8.50 = 1591.24 \text{ kg/h}$$
$$\text{Total mass out (lb base)} = 3267.80 + 218.65 + 3.61 + 18.75 = 3508.81 \text{ lb/h}$$

4.6. Check on material balance
Comparing data in steps 4.1 and 4.4, the results should be identical if the combustion materials are balanced. However, for this example, the discrepancy between the values of 1591.08 kg/h and 1591.24 kg/h or between the values of 3508.44 lb/h and 3508.81 lb/h is probably due to rounding off of certain values; in any case, the values are considered to be very close.
5. Calculate heat balance
5.1. Calculate heat input (Q_i)

Combustible components	Waste (kg/h)	Hc (kJ/kg)	Heat input (kJ/h)
Tissue	4.50	20,471.00	92,119.50
Polyethylene	24.50	46,304.00	1,134,448.00
PVC	2.80	22,630.00	63,364.00
Cellulose	35.70	18,568.00	662,877.60
	67.50		1,952,809.10 kJ/h or 1,850,828.45 Btu/h

$$Q_i = 1{,}952{,}809.10 \text{ kJ/h or } 1{,}850{,}828.45 \text{ Btu/h}$$

where Hc = HHV (high heating value)
1 Btu = 1.0551 kJ

5.2. Calculate total heat out based on an equilibrium temperature of 1100°C (Q_o)
5.2.1. Calculate radiation loss (Q_r)

$$\begin{aligned} Q_r &= 5\% \text{ of total heat available} \\ &= 0.05(1{,}952{,}809.10) \\ &= 97{,}640.46 \text{ kJ/h or } 92541.42 \text{ Btu/h} \end{aligned}$$

5.2.2. Calculate heat loss to ash (Q_a)

$$\begin{aligned} Q_a &= mCpdT \\ &= 8.5(0.831)(1100 - 15.5) \\ &= 8.5(0.831)(1084.5) \\ &= 7660.37 \text{ kJ/h or } 7260.32 \text{ Btu/h} \end{aligned}$$

where m = weight of ash = 8.5 kg/h
Cp = heat capacity = 0.831 kJ/kg-°C
ΔT = 1100 − 15.5 = 1084.5°C

5.2.3. Calculate heat loss to dry combustion products (Q_g)

$$\begin{aligned} Q_g &= mCpdT \\ &= 1481.94(1.086)(1100 - 15.5) \\ &= 1481.94(1.086)(1084.5) \\ &= 1{,}745{,}380.03 \text{ kJ/h or } 1{,}654{,}231.85 \text{ Btu/h} \end{aligned}$$

where m = weight of combustion products = 1481.94 kg/h
Cp = heat capacity of combustion products = 1.086 kJ/kg-°C
ΔT = 1100 − 15.5 = 1084.5°C

5.2.4. Calculate heat loss to moisture (Q_m)

$$\begin{aligned} Q_m &= mCpdT + mH_v \\ &= 99.16(2.347)(1100 - 15.5) + 99.16(2460.3) \\ &= 99.16(2.347)(1084.5) + 99.16(2460.3) \\ &= 252{,}394.08 + 243{,}963.35 \\ &= 496{,}357.43 \text{ kJ/h or } 470{,}436.39 \text{ Btu/h} \end{aligned}$$

where m = weight of water = 99.16 kg/h
Cp = heat capacity of water = 2.347 kJ/kg-°C
ΔT = 1100 − 15.5 = 1084.5°C
H_v = latent heat of water vaporization = 2460.3 kJ/kg

5.2.5. Calculate total heat out (Q_o)

$$\begin{aligned} Q_o &= \text{sum of } (5.2.1 + 5.2.2 + 5.2.3 + 5.2.4) \\ &= 2{,}347{,}038.29 \text{ kJ/h or } 2{,}224{,}469.98 \text{ Btu/h} \end{aligned}$$

5.2.6. Calculate net heat balance (Q_n)

$$\begin{aligned} Q_n &= Q_i - Q_o \\ &= 1{,}952.809.10 - 2{,}347{,}038.29 \\ &= -394{,}229.19 \text{ kJ/h (deficiency)} \end{aligned}$$

The negative sign indicates that auxiliary fuel must be supplied to achieve the design temperature of 1100°C.

6. Calculate auxiliary fuel required to reach 1100°C
6.1. Calculate total heat required from fuel (Q_f)

$$\begin{aligned} Q_f &= 394{,}229.19 + 5\% \text{ radiation loss} \\ &= 413{,}940.65 \text{ kJ/h or } 392{,}323.62 \text{ Btu/h} \end{aligned}$$

6.2. Calculate natural gas requirements
Available heat (net) from natural gas at 1100°C and 20% excess air = 15,805.2 kJ/m^3 (Gordon-79/09)

Natural gas required: 413,940.65/15,805.2

$$= 26.19 \text{ m}^3\text{/h or } 26.19/0.02832 = 924.79 \text{ ft}^3\text{/h}$$

7. Calculate the products of combustion from auxiliary fuel
Dry products density from fuel at 20% excess air = 16.0 kg/m^3-fuel (Gordon-79/09)
Dry products from fuel at 20% excess air = 16.0(26.19) = 419.04 kg/h
Moisture quantity from fuel at 20% excess air = 1.59 kg/m^3-fuel (Gordon-79/09)
Moisture from fuel at 20% excess air = 1.59(26.19) = 41.64 kg/h

8. Calculate the secondary combustion chamber volume required to achieve 1 s residence time at 1000°C
8.1. Calculate the total dry products from waste and fuel
Total dry products from waste (step 4.2) and fuel (step 7)

$$= 1481.94 + 419.04$$
$$= 1900.98 \text{ kg/h}$$

Assuming that dry products have the properties of air and using the ideal gas law, the volumetric flow rate (V_p) of dry products at 1000°C can be calculated as follows:

$$V_p = 1900.98[(22.4)/29][1273/273][1/3600] = 1.9019 \text{ m}^3/\text{s}$$

where 22.4 m^3 = standard volume at standard conditions
29 kg = air molar weight
1273 K = 1000°C + 273
273 K = 0°C + 273
3600 s = 1 h

8.2. Calculate the moisture from waste and fuel that is in the products
Total water in products (step 4.3) and from the fuel (step 7)

$$= 99.16 + 41.64 = 140.80 \text{ kg/h}$$

Using the ideal gas law, the volumetric flow rate (V_m) of moisture at 1000°C can be calculated as follows:

$$V_m = 140.80[(22.4)/18][1273/273][1/3600] = 0.2270 \text{ m}^3/\text{s}$$

where 22.4 m^3 = standard volume at standard conditions
18 kg = moisture molar weight
1273 K = 1000°C + 273
273 K = 0°C + 273
3600 s = 1 h

8.3. Calculate the volumetric flow rate

$$\text{Total volumetric flow rate} = \sum \text{ of steps 8.1 and 8.2}$$
$$= 1.9019 + 0.2270 = 2.1289 \text{ m}^3/\text{s}$$

Therefore, the active chamber volume required to achieve 1 s retention is 2.1289 m^3 (void areas with retention volume). This should not be included in the secondary chamber to meet the 1 s retention time required; the length of chamber should be calculated from the flame front to the location of the temperature-sensing device.

9. Calculate oxygen in the flue gas
The residual oxygen ($\%O_2$) can be determined using the following equation:

$$EA \text{ (excess air)} = \%O_2/(21\% - \%O_2)$$
$$1.50 = \%O_2/(21\% - \%O_2)$$
$$\%O_2 = 12.60\%$$

10. Calculate gas velocity in the rotary kiln (primary combustion chamber)
At standard conditions, i.e., temperature $T_1 = 0°C$, the flue gas flow rate (step 4.4)

$$W_{pcc} = 1481.94 + 99.16 = 1581.10 \text{ kg/h in kg base}$$

Assuming that the flue gas has a molecular weight similar to air molecular weight, which is equal to 29 kg/kg-mole, the molar flow rate $(n) = W_{pcc}/29 = 54.52$ kg-mole/h.

Because the volume at standard conditions per kg-mole is 22.4 m^3, the volumetric flow rate $(V_{std}) = 54.52(22.4) = 1221.26$ m^3/h.

The actual flue gas volumetric flow rate (V_{act}) can be calculated from the equations below.

$$\begin{aligned} V_{act} &= T_{act}(V_{std}/T_{std}) \\ &= (1000 + 273)(1221.26/(0 + 273)) \\ &= (1273)(1221.26/273) \\ &= 5694.74 \text{ m}^3/\text{h} \\ &= 1.582 \text{ m}^3/\text{s} \end{aligned}$$

11. Calculate flue gas velocity (v)
Assuming that the internal cross-sectional area (A_i) of the kiln = 4 m^2, the internal diameter (D_i) of the kiln can be calculated from

$$\begin{aligned} D &= (4A_i/3.1416)^{0.5} \\ &= (4(4)/3.1416)^{0.5} \\ &= 2.26 \text{ m} \end{aligned}$$

The flue gas velocity (v) can be calculated from

$$\begin{aligned} v &= V_{act}/A_i \\ &= 1.582/4 \\ &= 0.3955 \text{ m/s} \end{aligned}$$

12. Calculate the residence time in the kiln (last half only)
Assuming that the kiln length = 2 m
Use half the length for residence time calculation, i.e., = 1 m

$$\begin{aligned} \text{Residence time } (t) &= \text{length/velocity} \\ t &= 1/0.3955 \\ &= 2.53 \text{ s} \end{aligned}$$

13. Calculate the kiln volume

$$\text{Kiln useful volume} = \text{(useful length)(cross-sectional area)} = 1(4) = 4\ \text{m}^3$$

$$\text{Total kiln volume} = \text{(kiln length)(cross-sectional area)} = 2(4) = 8\ \text{m}^3$$

14. Gas flow turbulence

$$\text{Reynolds number} = Re = VD/ki$$

where V = average velocity, m/s
D = diameter of flow
Ki = kinematic viscosity = 0.00014 m^2/s

$$Re = 0.3955(2.26)/0.00014 = 6385$$

i.e., probably sufficiently turbulent because the *Re* is above 5000.

In accordance with Perry's *Chemical Engineers' Handbook*, p. 5-4, the critical Reynolds number corresponds to the transition from turbulent to laminar flow as the velocity is reduced. Its value is in the range 2000–3000 for a circular pipe.

15. Calculate the specific heat release

$$\text{Kiln volume} = 8\ \text{m}^3$$

$$\text{Heat input} = 1{,}952{,}809.10\ \text{kJ/h (step 5.1)}$$

$$\text{Specific heat release} = \text{(heat input)/(kiln volume)} = 1{,}952{,}809.10/8 = 244{,}101.13\ \text{kJ/m}^3\text{-h}$$

5.8 Evaluation of Trial Burn Data

The operation of an incinerator is subject to the regulations developed under the Resource Conservation and Recovery Act and the Clean Air Act. Federal and State permit writers are responsible for implementing the regulations. One of the major problems that the permit writers often encounter is the uncertainty of determining whether data submitted are adequate or accurate. For example, if an applicant's data show that his incinerator can reach a certain temperature by burning certain wastes at certain combustion air levels, the question is, "Are the claimed data dependable?"

To help answer this question, the authors developed a simple calculational computer model on a Lotus 1-2-3 program in the mid-1990s to calculate an energy and mass balance for a rotary kiln incinerator (Lee-98). The main purpose of the model is to assist the permit writers in evaluating the adequacy of the data submitted by applicants seeking incinerator permits. Key parameters that the model can calculate include theoretical combustion air, excess air needed for actual combustion cases, flue gas rate, and exit temperature.

Because the model is considered to be of interest to the readers, the calculational procedure has been simplified and provided here for the readers' reference.

EXAMPLE application of energy and mass balance model to evaluate Ciba–Geigy trial burn data

Statement of the problem. The Ciba–Geigy Corporation conducted a trial burn on their rotary kiln incinerator on November 12–17, 1984. The measured data were later summarized in an EPA report (EPA-86/09, p, B-11) and key aspects of it appear below. A schematic of the Ciba–Geigy incinerator is shown in Fig. 2.29.

EQUIPMENT INFORMATION

- Type of unit: private incinerator—rotary kiln with secondary chamber
- Capacity: 50 tpd (tons per day) with 10% excess capacity (30×10^6 Btu/h for each burner)
- Pollution control system: quench tower, polygon venturi scrubber (25-in. pressure drop), and packed tower scrubber
- Waste feed system:
  - — Liquid: Hauck Model 780 wide-range burners (kiln and secondary burners)
  - — Solid: ram feed
- Residence time: 5.05 s (kiln); 3.09 s (secondary chamber)

TEST CONDITIONS

- Waste feed data: hazardous liquid and nonhazardous solid wastes usually burned; for this run only, synthetic hazardous liquid waste was tested
- Length of burn: 6–9 h (2-h sampling time)
- Total amount of waste burned: 480 gal (liquid) and 0 lb (solid)
- Waste feed rate: 4 gpm (liquid); 0 lb/h (solid)
- POHCs (principal organic hazardous constituents) selected and percent concentration in waste feed:
  - — hexachloroethane 4.87
  - — tetrachloroethene 5.03
  - — chlorobenzene 29.52
  - — toluene 60.58

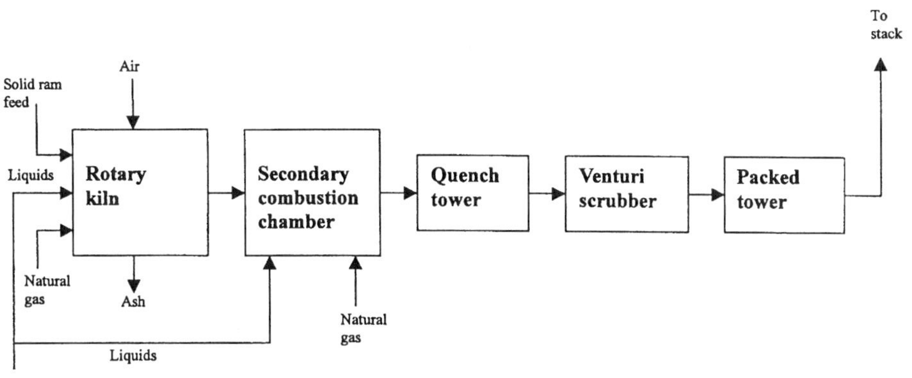

FIGURE 2.29 Ciba–Geigy incinerator schematic (*EPA-86/09, p. B-12*).

- Btu content: 15,200 Btu/lb
- ash content: not measured
- chlorine content: 20.8% (calculated)
- moisture content: not measured

OPERATING CONDITIONS

- Temperature: range 1750–1850°F (kiln); 1950-2050°F (secondary chamber)
- Average 1800°F (kiln); 2000°F (secondary chamber)
- Auxiliary fuel used: natural gas (1200 scfh to the kiln)
- Airflow:
  — primary air to kiln 2200 cfm
  — secondary air to kiln 1400 cfm
- Flue gas oxygen content: 10.3%

Energy and mass balance calculations for primary chamber (kiln) (Table 2.51)
A. GIVEN CONDITIONS: the data in question are

a1. Waste feed rate (gpm): 4 gpm
Assume that 1 gal = 5 lb
Waste feed rate in lb/h: 1200 lb/h
a2. Fly ash (% of waste feed): 0 (assumed)
a3. % of ash due to unburned carbon: 0 (assumed)
a4. Ash quench temperature: undefined
a5. Exit temperature: unspecified
a6. Reference temperature: 70°F
a7. Radiation loss (assumed): 0.05 (5%)
a8. Excess air rate (EAR) (assumed), by weight: 0.885 (i.e., 88.5% excess air)
a9. Humidity at 60% relative humidity (RH), 80°F: 0.0127 kg-H_2O/kg-dry-air; 0.0127 lb-H_2O/lb-dry-air
a10. Standard volume: 24.04 scm/kg-mol; 386.9 scf/lb-mol
a11. Water latent heat: 2460 kJ/kg; 1054 Btu/lb
a12. Ash specific heat: 0.83 kJ/kg-°C; 0.25 Btu/lb-°F
a13. Flue gas specific heat: 1.09 kJ/kg-°C; 0.26 Btu/lb-°F
a14. Water specific heat: 2.35 kJ/kg-°C; 0.49 Btu/lb-°F
a15. 1 kcal/g = 4187 kJ/kg = 1799 Btu/lb; 2.33 kJ/kg = 1 Btu/lb; 1 kJ/kg = 0.43 Btu/lb; 1.06 kJ = 1 Btu; 1 m = 3.28 ft
a16. Natural gas (heat of combustion): 13.3 kcal/g = 23,932 Btu/lb
a17. Waste analysis

TABLE 2.51 Primary chamber

POHC name	POHC ratio	Waste feed (lb/h)	ΔH (kcal/g)	ΔH (Btu/lb)	Mixture (Btu/h)
Hexachloroethane, C_2Cl_6	0.0487	58.44	0.46	828	4.84 $E+04$
Tetrachloroethene, C_2Cl_4	0.0503	60.36	1.19	2141	1.29 $E+05$
Chlorobenzene, C_6H_5Cl	0.2952	354.24	6.6	11873	4.21 $E+06$
Toluene, C_7H_8	0.6058	726.96	10.14	18242	1.33 $E+07$
Total	1	1200	18.39	33084	1.76 $E+07$
Therefore, the heat value of the POHC mixture is (in Btu/lb)					1.47 $E+04$

a18. Chemical analysis
a18.1. Calculate the ultimate composition of hexachloroethane, C_2Cl_6 (Table 2.52).
a18.2. Calculate the ultimate composition of tetrachloroethene, C_2Cl_4 (Table 2.53).
a18.3. Calculate the ultimate composition of chlorobenzene, C_6H_5Cl (Table 2.54)
a18.4. Calculate the ultimate composition of toluene, C_7H_8 (Table 2.55).
a18.5. Calculate the ultimate composition of fuel (natural gas), CH_4 (Table 2.56).

- Fuel (natural gas, CH_4) to kiln: 1200 scf/h
- Fuel density = molecular wt/std volume (a10) = 16/386.9 = 0.04135 lb/scf
- Fuel weight flow rate = fuel density × fuel volume flow rate = 0.04135(1200) = 49.62 lb/h

a18.6. Calculate the ultimate composition of compounds by combining items a18.1–a18.5 (Table 2.57).
a19. Fuel heat input = weight rate × HHV = (49.62)(23932) = 1.188 $E + 06$ Btu/h
a20. Total heat in = waste input (a17) + fuel input (a19): 1.88 $E + 07$ Btu/h

Total average heating value = a20/(waste + fuel): 15,045 Btu/lb

Because the test data gave 15,200 Btu/lb, it shows that the calculated results and test data are consistent.

TABLE 2.52 Composition of C_2Cl_6

x_i	C	H	Cl	O	M
C_2Cl_6	24	0	213	0	237.00
x_i/M	0.1013	0.0000	0.8987	0.0000	
58.44	5.9200	0.0000	52.5200	0.0000	

TABLE 2.53 Composition of C_2Cl_4

x_i	C	H	Cl	O	M
C_2Cl_4	24	0	142	0	166.00
x_i/M	0.1446	0.0000	0.8554	0.0000	
60.36	8.7281	0.0000	51.6319	0.0000	

TABLE 2.54 Composition of C_6H_5Cl

x_i	C	H	Cl	O	M
C_6H_5Cl	72	5	35.5	0	112.50
x_i/M	0.6400	0.0444	0.3156	0.0000	
354.24	226.7136	15.7238	111.7981	0.0000	

TABLE 2.55 Composition of C_7H_8

x_i	C	H	Cl	O	M
C_7H_8	84	8	0	0	92.00
x_i/M	0.9130	0.0870	0.0000	0.0000	
726.96	663.7145	63.2455	0.0000	0.0000	

TABLE 2.56 Composition of CH_4

x_i	C	H	Cl	O	M
CH_4	12	4	0	0	16.00
x_i/M	0.7500	0.2500	0.0000	0.0000	
49.62	37.2150	12.4050	0.0000	0.0000	

TABLE 2.57 Composition of Compounds

x_i	C	H	Cl	O	Weight
C_2Cl_6	5.92	0	52.52	0	58.44
C_2Cl_4	8.73	0	51.63	0	60.36
C_6H_5Cl	226.71	15.73	111.8	0	354.24
C_7H_8	663.71	63.25	0.00	0.00	726.96
CH_4	37.22	12.40	0.00	0.00	49.62
Total	942.29	91.38	215.95	0.00	1249.62

B. COMBUSTION AIR

The combustion air with 88.5% excess air needed to incinerate the following compounds is shown in Table 2.58 (88.5% excess air is an assumed value).

- C = 942.29 lb/h
- H = 91.38
- Cl = 215.95

The format of Table 2.58 was given in Section 2.6.

Calculate the combustion products flow rate (Table 2.59).
Apply the method in Section 2.6.

Check on mass balance.
The above calculations show that the mass of reactants and the mass of products are balanced because $R_a = P_a = 27{,}415.81$ lb/h.

C. COMBUSTION AIR FLOW RATE AND OXYGEN CONTENT

C1. Combustion air flow rate
From Table 2.58, the actual combustion air is

TABLE 2.58 Combustion Reactant Table

Basic reaction	Oxygen, nitrogen, and air requirements	Mass (lb/h)	Mole (mol/h)
$C + O_2 \rightarrow CO_2$	(1) $O_2 = C(32/12)$	2512.77	78.52
$H_2 + 0.5O_2 \rightarrow H_2O$	(2) $O_2 = 8(H\text{–}Cl/35.5)$*	682.38	21.32
$S + O_2 \rightarrow SO_2$	(3) $O_2 = S(32/32) = S$	0.00	0.00
B_o = bound oxygen	(4) Data from waste analysis	0.00	0.00
O_{2t} = theoretical O_2	$O_{2t} = (1) + (2) + (3) - (4)$	3195.15	99.85
N_{2t} = theoretical N_2	$N_{2t} = (O_{2t})(3.76 \times 28/32)$	10,512.04	375.43
A_t = theoretical air	$A_t = O_{2t} + N_{2t}$	13,707.19	475.28
$\%A_e$ = % excess air	Given data (as a fraction)	0.885	
A_e = excess air	$A_e = (\%A_e)(A_t)$	12,130.86	420.62
A_{ada} = actual dry air	$A_{ada} = A_t + A_e$	25,838.05	895.90
H_2O from air†	$0.0127(A_{ada})$	328.14	18.23
A_a = actual air	$A_a = A_{ada} + H_2O$	26,166.19	914.13
Waste feed	Data from waste analysis	1249.62	
R_a = actual reactant	$R_a = A_a$ + waste feed + H_2O from air	27,415.81	

* If H_2 is the hydrogen left over after Cl reaction, i.e., $H_2 = H\text{–}Cl/35.5$, the O_2 needed to react with H_2 will be equal to $H_2(0.5)(32/2) = 8(H\text{–}Cl/35.5)$.

† In general, at room temperature, the humidity ratio is 0.0127 (lb-H_2O)/(lb-dry-air) or 0.0127 (kg-H_2O)/(kg-dry-air).

$$A_a = 26166.19 \text{ lb/h} = 914.13 \text{ mol/h}$$

$$\begin{aligned}\text{Air flow rate} &= \text{std. volume(a10)(mol/h)}\\ &= 386.9(914.13)\\ &= 353{,}676.90 \text{ scfh}\\ &= 5894.61 \text{ scfm}\end{aligned}$$

The test result was 3600 scfm with primary air 2200 scfm and secondary air 1400 scfm. When this test result is compared with the calculated result, there is a considerable difference. See the discussion of results that follows.

TABLE 2.59 Combustion Product Table

Component	Combustion product	Mass (lb/h)	Mole (mol/h)
CO_2	$CO_2 = (C)(44/12)$	3455.06	78.52
SO_2	$SO_2 = S(64/32)$	0.00	0.00
HCl	$HCl = Cl(36.5/35.5)$	222.03	6.08
O_{2e}	$(\%A_e)O_{2t}$, O_{2t} from Table 2.58	2827.71	88.37
N_{2e}	$(\%A_e)N_{2t}$, N_{2t} from Table 2.58	9303.16	
N_2	$N_{2e} + N_{2t}$ + from waste	19,815.20	707.69
H_2O from combustion	H_2O from combustion $= 9(H\text{–}Cl/35.5)$	767.67	
H_2O	H_2O from combustion + from air + from waste	1095.81	60.88
G_a = actual flue gas	$G_a = CO_2 + SO_2 + HCl + O_{2e} + N_2 + H_2O$	27,415.81	941.54
Ash	Fly ash + bottom ash	0.00	
P_a = actual product*	G_a + ash	27,415.81	

* In performing the mass balance calculation, the values of reactants (R_a) and products (P_a) should be equal, i.e., if $R_a = P_a$, the mass balance calculation is correct.

c2. Calculate oxygen content in flue gas
Table 2.59 shows that the excess oxygen $O_{2e} = 2827.71$ lb/h and the product flue gas, $P_a =$ 27,416 lb/h

$$\text{Therefore, the oxygen content in the flue gas} = 2827.71/27{,}416$$
$$= 10.31\%$$

The test result was 10.3%. This shows that the test result and the calculated result are consistent (and that the calculated air rate is more dependable than the measured one).
c3. Total water vapor in flue gas from Table 2.59: 1095.8 lb/h
c4. Total dry flue gas $= P_a - H_2O = 27{,}416 - 1096 = 26{,}320$ lb/h

D. CALCULATION OF EXIT TEMPERATURE FROM KILN

d1. Total heat in = waste heat input (a18) + fuel heat input (a19) = a20: 1.88 $E + 07$ Btu/h
d2. Overall heat loss (assumed as 5%, see a7): 0.0940 $E + 07$
Unburned carbon = 0 (not measured)
d3. Unreleased heat (due to unburned carbon): 0
Trial 1
Assumed exit temperature: 1500°F
Reference temperature: 70°F
d4. Temperature difference (ΔT): 1430°F
d5. Heat in dry flue gas $= mCp\Delta T$
$= [\text{c4} \times \text{a13} \times \text{d4}]$
$= 26{,}320(0.26)(1430)$
$= 9.79\ E + 06$
d6. Heat in water $= mCp\Delta T$
$= (\text{c3} \times \text{a14} \times \text{d4})$
$= 1096(0.49)(1430)$
$= 0.0768\ E + 07$
d7. Total latent heat $= (\text{c3} \times \text{a11})$
$= 1096(1054)$
$= 0.1158\ E + 07$
d8. Heat in ash $= mCp\Delta T$
$= (\text{a2} \times \text{a12} \times \text{d4})$
$= 0.0000$
d9. Total heat accounted for $= (\text{d2} + \text{d3} + \text{d5} + \text{d6} + \text{d7} + \text{d8})$
$= 1.2660\ E + 07$
d10. Net heat balance $= (\text{d1} - \text{d9})$
$= 0.6140\ E + 07$ Btu/h
Trial 2
Assumed exit temperature: 2500°F
Reference temperature: 70°F
d11. Temperature difference (ΔT): 2430°F
d12. Heat in dry flue gas $= mCp\Delta T$
$= [\text{c4} \times \text{a13} \times \text{d4}]$
$= 26{,}320(0.26)(2430)$
$= 1.6643\ E + 07$
d13. Heat in water $= mCp\Delta T$
$= (\text{c3} \times \text{a14} \times \text{d4})$
$= 1096(0.49)(2430)$
$= 0.1305\ E + 07$

d14. Total latent heat $= (\text{c3} \times \text{a11})$
$= 1096(1054)$
$= 0.1158 \quad E + 07$

d15. Heat in ash $= mCp\Delta t$
$= (\text{a2} \times \text{a12} \times \text{d11})$
$= 0.0000$

d16. Total heat accounted for $= (\text{d2} + \text{d12} + \text{d13} + \text{d14} + \text{d15})$
$= 2.0046 \quad E + 07$ Btu/h

d17. Net heat balance $= (\text{d1} - \text{d16})$
$= -0.1245 \text{ E} + 07$ Btu/h

d18. Using the interpolation method to estimate kiln temperature, we have:

x_1	1500°F	$0.6140\ E + 07$ Btu/h	y_1
x		0.00	$y(y = 0)$
x_2	2500	$-0.1246\ E + 07$	y_2

$$(x - x_1)/(0 - y_1) = (x_2 - x_1)/(y_2 - y_1)$$

$$x = x_1 - y_1(x_2 - x_1)/(y_2 - y_1)$$

d19. $x = 2331$°F
Test data were at 1800°F (average kiln exit temperature).

Summary/discussion of results. Based on the calculations contained here and information provided by the trial burn results, a summary of key data is provided in Table 2.60.

Table 2.60 shows that the differences between the calculated and the measured results are small, with the exception of the kiln exit temperature and the air flow rate. The calculated value of the air flow rate is about 63% greater than that of the trial burn (measured) value. The difference is due to the fact that the measured (upstream-from-the-combustor) air rate values neglected to account for the amount of air in-leakage which has to occur in any actual (negative pressure) kiln combustion operation. The measured data relative to oxygen content (taken downstream of the combustor) show that the calculated air in the system (the 5855 scfm amount) *is* reasonable because the oxygen content measured downstream of the kiln matches the calculated oxygen concentration (the 10.3%). The calculation therefore confirms that the air needed is much more than the 3600 cftn measured value (which, of course, proves that the air in-leakage phenomenon does occur). The fact that the measured kiln exit temperature is also much lower (about 530°F lower) than the calculated kiln exit temperature indicates that the assumed amount of heat loss (the 5%) is probably too low.

TABLE 2.60 Summary of Calculated and Trial Burn Results

	Calculated results	Measured results
O_2 content in flue gas	10.31%	10.3%
Heating value, Btu/lb	15,045	15,200
Exit kiln temperature	2331°F	1800°F (average)
Air flow rate	5855 scfm	3600 cfm

6 EVALUATION OF PERMIT APPLICATION DATA

The permit writer may receive incinerator performance data requiring detailed evaluation from the results of a trial burn conducted in accordance with an approved plan. The purpose of this example is to provide numerical methods for interpretation and validation of these data. Sample calculations are provided for:

- computation of destruction and removal efficiencies (DRE)
- scrubber efficiencies
- particulate loadings in the stack gas

The purpose of these calculations is to ensure compliance with the regulatory performance requirements. Guidance is presented to evaluate the routine process monitoring, safety systems inspections, and incinerator effluent management described in a permit application. Finally, the permit writer is presented with guidance for the specification of the regulatory operating requirements under 40CFR Parts 264.345 which are incorporated into an operating permit (EPA-80/12, p. 87).

6.1 Evaluation of Compliance with Performance Standards

An owner or operator of a hazardous waste incinerator must submit a computation of the destruction and removal efficiency (DRE), particulate emissions, and scrubber efficiency, if applicable, derived from data accompanying a permit application. The permit writer must ensure that these values are computed correctly and are within the regulatory performance standards specified under 40CFR Parts 264.343. These calculations must be provided for each different set of data or for each trial test conducted. Only in those situations during which the waste feed composition and incinerator operating conditions were exactly the same may these calculated values be averaged. Otherwise, each set of data will be judged independently and only those demonstrating compliance with the performance standards may be used to establish incinerator operating conditions in a permit. A permit application may be recommended for rejection if any of the performance standards are not attained.

Destruction and removal efficiency

EXAMPLE calculation of destruction and removal efficiency (DRE)

Incinerators burning hazardous waste must achieve a destruction and removal efficiency (DRE) of 99.99% for each principal organic hazardous constituent (POHC) in the waste feed as required under 4OCFR Parts 264.343(a). The DRE is determined from the following equation:

$$\text{DRE} = [(W_{\text{in}} - W_{\text{out}})/W_{\text{in}}](100)$$

where DRE = destruction and removal efficiency

W_{in} = mass feed rate of the principal organic hazardous constituent (POHC) in the waste stream feeding the incinerator

W_{out} = mass emission rate of the principal organic hazardous constituent (POHC) present in exhaust emissions prior to release to the atmosphere

W_{in} is calculated using the formula

$$W_{\text{in}} = (\text{concentration of POHC in waste})(\text{waste feed rate})$$

POHCs are given as concentrations and must be expressed in percentages when applying this formula. The waste feed rate is expressed in mass per unit time and must be consistent with the units used to express W_{out}. If a waste is cofired with auxiliary fuel, the auxiliary fuel feed rate does not affect the calculation of W_{in}.

W_{out} is calculated from stack sampling data and involves three steps:

- computation of stack gas sample volume
- computation of POHC concentration in stack sample
- computation of stack gas volume flow rate

Stack gas sample volume and stack gas volume flow rate may be determined by EPA methods 2 and 5 or ASTM method D2928. Stack emissions of POHCs may be determined using a modified EPA method 5 apparatus, and includes collection and analysis of particulate matter, gas phase organics, and water present in the stack gas.

Methods of laboratory analysis of a stack sample for POHCs are presented in *Test Methods for Evaluating Solid Waste*, EPA, SW-846, 1980. Sampling and analytical data should be included with a permit application. The following sample calculation of DRE will identify the necessary data and provide methods of computation allowing the permit writer to verify a DRE claimed by an applicant.

Given conditions

- POHC = hexachlorobenzene
- concentration of POHC in waste feed = 14.5%
- waste feed rate = 1000 lbs/h

Calculation:
Step 1. Computation of W_{in}

$$\begin{aligned} W_{in} &= \text{(concentration of POHC in waste)(waste feed rate)} \\ &= 0.145(1000\ \text{lb/h}) \\ &= 145\ \text{lb/h} \end{aligned}$$

Step 2. Computation of stack gas sample volume
This sample volume must include water collected during sampling and be expressed under standard conditions (293 K, 760 mm Hg or 528 °R, 29.92 in. Hg). The following formula from EPA method 5 may be used to compute the corrected dry gas volume V_m (std):

$$V_{m(std)} = K_1 V_m Y[P + (H/13.6)]/\dot{T}_m$$

where
$V_{m(std)}$ = sample volume under standard conditions
K_1 = 0.3858 K/mm Hg = 17.64 °R/in Hg
V_m = volume of gas measured by dry gas meter, corrected if necessary
Y = dry gas meter calibration factor
P = barometric pressure
H = average pressure differential across sampler orifice meter
T_m = absolute average dry gas meter temperature

Given conditions

- $K_l = 0.3858$ K/mm Hg $= 17.64$ °R/in. Hg
- $V_m = 31.153$ ft^3
- $Y = 1.12$
- $P = 29.82$ in. Hg
- $H = 0.705$ in. H_2O
- $T_m = 554$ °R

Calculation:

$$\begin{aligned} V_{m(std)} &= K_l V_m Y[P + (H/13.6)]/T_m \\ &= (17.64\ ^\circ\text{R/in. Hg})(31.153\ \text{ft}^3)(1.12)[29.82\ \text{in. Hg} + 0.705\ \text{in. H}_2\text{O}/13.6]/(554\ ^\circ\text{R}) \\ &= 33.19\ \text{scf} \end{aligned}$$

This volume must be corrected for the volume of water collected during sampling using the formula

$$V_{W(GAS)} = 0.0472(V_w - \text{grams SO}_2 - \text{grams H}_2\text{S})$$

where $V_{W(GAS)}$ = volume of water vapor at standard conditions, scf
V_w = volume of water collected in impingers and silica gel, ml

Given conditions

- $V_w = 90.4$ ml
- grams SO_2 = grams H_2S = 0

Calculation:

$$\begin{aligned} V_{W(GAS)} &= 0.0472(90.4) \\ &= 4.267\ \text{scf} \end{aligned}$$

The volume of water vapor is added to $V_{m(std)}$ to obtain the sample volume.

$$\begin{aligned} \text{Sample volume} &= V_{W(GAS)} + V_{m(std)} \\ &= 4.27 + 33.19 \\ &= 37.46\ \text{scf} \end{aligned}$$

Step 3. Computation of POHC concentration in stack sample
The concentration of the POHC in the stack gas (C_g) is determined from the following equation:

$$C_g = (\text{total weight of POHC in sample})/(\text{volume of sample at standard conditions})$$

The total weight of POHC is obtained from laboratory analysis.

Given conditions

- 3.5 μg hexachlorobenzene extracted from water
- 12.3 μg hexachlorobenzene extracted from particulate matter
- 26.0 μg hexachlorobenzene extracted from gas-phase trap

Total 41.8 μg hexachlorobenzene
or 4.18×10^{-5} grams hexachlorobenzene

Calculation:

$$\begin{aligned} C_g &= (4.18 \times 10^{-5}\ \text{g})/(37.46\ \text{scf}) \\ &= 1.12 \times 10^{-6}\ \text{g/scf} \\ &= 2.46 \times 10^{-9}\ \text{lb/scf} \end{aligned}$$

Step 4. Computation of stack gas volume flow rate
This value is determined by a pitot tube which measures the difference between the total and static pressures in a flue. Gas velocity determinations are made at several locations during sampling and the values are averaged. EPA method 2 or ASTM method D2928 may be used to obtain data. The gas velocity V_g in feet per minute may be calculated using the formula

$$V_g = 174(C)[(29.92/P_s)(1.00/G_s)(h)(T_s)]^{0.5}$$

where V_g = gas velocity
C = Pitot tube correction factor (usually 0.85 for Type S and 1.00 for others)
P_s = absolute pressure in flue, inches of mercury
h = velocity pressure at sampling point, inches of water. If the velocities differ greatly from one sampling point to another, the averages of the square roots of the velocity pressures must be used (see ASTM D2928 Section 5.6)
T_s = absolute temperature of stack gas, °R
G_s = specific gravity of flue gas with respect to air

G_s can be computed by the formulas

$$\begin{aligned} G_s &= M_s/(387)(0.0749) \\ M_s &= M_d[(100 - W)/100] + 0.18(W) \\ M_d &= 0.44[\%(CO_2)] + 0.28[9.00(CO)] + 0.32[9.00(O_2) + 0.28[\%/N_2] \end{aligned}$$

where W = water content of flue gas

Given conditions

- $C = 0.835$
- $P_s = 29.81$ in. Hg
- $G_s = 0.850$
- $h = 0.202$ in. Hg
- $T_s = 582$ °R

Calculation:

$$\begin{aligned} V_g &= 174(0.835)[(29.92/29.81)(1.00/0.85)(0.202)(582)]^{0.5} \\ &= 1700\ \text{ft/min} \end{aligned}$$

The stack gas volume flow rate at standard conditions Q is determined by the formula

$$Q = V_g A_s(530/T_s)(P_s/29.92)$$

where A_s = cross-sectional area of stack, ft^2

Given conditions

- $A_s = 55.0\ \text{ft}^2$

Calculation:

$$\begin{aligned} Q &= V_g A_s(530/T_s)(P_s/29.92) \\ &= (1700\ \text{ft/min})(55.0\ \text{ft}^2)(530/582(29.81/29.92)) \\ &= 84{,}833\ \text{scf/min} \end{aligned}$$

Step 5. Computation of W_{out}
W_{out} may be calculated using the formula

$$\begin{aligned} W_{out} &= C_g Q \\ &= (2.46 \times 10^{-9}\ \text{lb/scf})(84.833\ \text{scf/min})(60\ \text{min/h}) \\ &= 0.0125\ \text{lb/h} \end{aligned}$$

Step 6. Computation of DRE
The DRE is computed from the equation

$$\begin{aligned} \text{DRE} &= [(W_{in} - W_{out})/W_{in}](100\%) \\ &= \{[(145\ \text{lb/h}) - (0.0125\ \text{lb/h})]/(145\ \text{lb/h})\}(100\%) \\ &= 99.9914\% \end{aligned}$$

Note: the expression of the DRE to 5 or 6 decimal places is justified because an error by as much as 25% in the W_{out} would often affect only the fifth decimal place.

If a DRE of 99.99% is not demonstrated, the permit writer must recommend rejection of a permit application for this reason. If the DRE is 99.99% or better, the permit writer must evaluate the scrubber efficiency and particulate emission values submitted by the applicant.

Hydrogen chloride control efficiency. An incinerator destroying hazardous waste having an organically bound chlorine concentration greater than 0.5 percent must be equipped with emission control equipment capable of removing at least 99 percent of hydrogen chloride from the exhaust gases, as required under 40CFR Parts 264.343(b). Waste and stack gas sample analyses are usually generated in terms of chloride (Cl^-) concentrations, and the scrubber efficiency (SE) may be defined as

$$\text{SE} = [Cl_{in} - Cl_{out})/Cl_{in}](100\%)$$

where SE = scrubber efficiency
Cl_{in} = mass feed rate of organically bound chlorides entering the scrubber system
Cl_{out} = mass emission rate of chlorides in the scrubber exhaust gas prior to emission to the atmosphere

In the above definition for scrubber efficiency the words "scrubber system" are used, as opposed to the word "scrubber." The former term is used on purpose, since many incinerator facilities do indeed have a scrubber system, and rarely have only a single scrubber. A scrubber system is usually composed of a prescrubber quench section (to precool the hot gases, and prevent excess evaporation of scrubber water), and one or more scrubbers installed in a series-flow arrangement. In some installations, the first scrubber is a venturi scrubber which eliminates most particulates and some of the acid gases. Subsequent scrubbers remove acid gases as their primary function. Residual water from the quench section is generally recycled to the quench or to one of the scrubbers. Thus, where the term "scrubber efficiency" is used, the acid gas removal efficiency of the whole scrubber system is usually the parameter being considered. The efficiency of individual scrubber units can be determined, but a high efficiency for the total scrubber system is the fundamental desired parameter.

Since the gases exiting the scrubber are generally cool (180°F), sampling and analysis of this gas is comparatively easy and safe. Sampling of the hot incinerator exhaust gases is not simple and it may be a somewhat hazardous procedure. A more common basis for estimating Cl_{in} is to base this quantity on the waste feed rate to the incinerator and the average organic chlorine analysis of this feed. If this method for estimating Cl_{in} overstates the "true" chlorine content of the unscrubbed exhaust gas (and it may), then the calculated value for scrubber efficiency may also be slightly overstated. Two sources of error, which may be considered negligible for calculation of the scrubber efficiency, derived from using a calculated Cl_{in} value are as follows:

- Some chlorine may react with alkaline metal components of the feed and end up in any slag or fly-ash.
- Some chlorine may exit the combustion chamber in elemental form (Cl_2). This will not dissolve in scrubber water to any large extent, but some will react with alkaline components of scrubber liquid to form hypochlorites. The remainder will exit the system via the stack.

A sample calculation is presented below demonstrating the use of this equation for scrubber efficiency.

EXAMPLE hydrogen chloride emissions

Step 1. Cl_{in} is calculated from the information below

Given conditions

- Organic (free) chlorine content of incinerator feed = 25.0%. Chlorine content must be expressed as chloride (Cl^-). If it is expressed as free chlorine (Cl_2), $Cl^- = Cl_2/2$.
- Incinerator feed rate = 2000 lb/h hazardous waste

Calculation:

$$\begin{aligned} Cl_{in} &= \text{(concentration of chlorine in feed)(feed rate)} \\ &= (0.250)(2000 \text{ lb/h}) \\ &= 500 \text{ lb/h} \\ &= 8.33 \text{ lb/min} \end{aligned}$$

Cl_{out} is computed from stack monitoring data. Necessary data include:

- volume of the stack gas sample at standard conditions
- total chlorides (Cl^-) collected during sampling
- stack gas volume flow rate at standard conditions

The method for computing the volume of the stack gas sample ($V_{m(std)}$) employed for the DRE calculation may be used to compute the scrubber efficiency. Assume sampling conditions are the same for this sample calculation as for the DRE example (immediately preceding).

Step 2. Total chlorides are determined by laboratory analysis of the impinger solution using the formula

$$\text{mgCl}^- = 35.45ANV_I/V_A$$

where
A = ml of titrant for sample
N = normality of mercuric nitrate titrant
V_I = volume of impinger solution
V_A = volume of sample aliquot

Given conditions

- $A = 3.12$ ml
- $N = 0.01$
- $V_I = 40$ ml
- $V_A = 10$ ml

Calculation:

$$\begin{aligned} \text{mgCl}^- &= 35.45(3.12)(0.01)(40)/10 \\ &= 4.43 \\ &= 9.77 \times 10^{-6} \text{ lb} \end{aligned}$$

Step 3. The concentration of chlorides (C_{Cl}) in the stack gas is

$$\begin{aligned} C_{Cl} &= \text{lb of Cl}^-/(\text{sample volume}) \\ &= 9.77 \times 10^{-6} \text{ lb}/37.46 \text{ scf} \\ &= 2.61 \times 10^{-7} \text{ lb/scf} \end{aligned}$$

The stack gas volume flow rate (Q) may be calculated by the method used to compute the DRE. The same conditions exist as for the DRE example, $Q = 84{,}833$ scf/min.

Step 4. Cl_{out} may be computed by the formula

$$\begin{aligned} Cl_{out} &= Q(C_{Cl}) \\ &= (84{,}833 \text{ scf/min})(2.61 \times 10^{-7} \text{ lb/scf}) \\ &= 0.0221 \text{ lb/min} \end{aligned}$$

Step 5. The scrubber efficiency (SE) may be computed from the equation

$$\begin{aligned} \text{SE} &= [(Cl_{in} - Cl_{out})/Cl_{in}](100\%) \\ &= [(4.17 \text{ lb/min} - 0.0221 \text{ lb/min})/4.17 \text{ lb/min}](100\%) \\ &= 99.47\% \end{aligned}$$

If the scrubber efficiency is less than 99 percent, the permit writer must recommend rejection of a permit application for this reason. If the scrubber efficiency is 99 percent or better, the permit writer must evaluate incinerator particulate emissions.

Particle emission control efficiency. As of 1991, incinerators destroying hazardous wastes must not emit particulate matter at concentrations greater than 180 milligrams of particulates per dry standard cubic meter of stack gas (0.08 grains per dry standard cubic foot) when the stack gas is corrected to a 12 percent carbon dioxide concentration as required under 40CFR Parts 264.343(c). Two stack sampling methods for particulate emissions are the EPA method 5 and ASTM method D2928. A sample calculation of particulate matter concentration in the stack gas is presented using the Subpart E, 40CFR Parts 60.50 methods referenced in the regulations. The calculation involves the following steps:

- determination of the stack gas sample volume
- determination of the weight of collected particulate matter
- calculation of the particulate concentration in the stack gas
- determination of the carbon dioxide concentration in the stack gas
- correction of the particulate concentration to 12% carbon dioxide

EXAMPLE particulate emissions
Step 1. Calculation of sample volume
The stack gas sample volume may be computed in a manner similar to that in EPA method 5 or ASTM method D2928. The method for computing the volume of the stack gas sample ($V_{m(std)}$) employed for the DRE calculation may be used for this example. Assume that the sampling conditions are the same for this sample calculation as for the DRE example. For this example, the same sample volume will be used, $V_{m(std)} = 33.19$ scf.

Step 2. Calculation of sample weight
The weight of the collected particulate matter is determined gravimetrically. For this example, a particulate matter weight of 137.0 milligrams (2.113 grains) will be assumed.

Step 3. Calculation of sample concentration
The concentration of particulate matter (P) in the stack gas is obtained by dividing the weight of the collected particulate matter by the stack gas sample volume.

$$P = 2.113 \text{ grains}/33.19 \text{ scf} = 0.0637 \text{ grains/scf}$$

Step 4. Calculation of carbon dioxide concentration
The carbon dioxide concentration in the stack gas may be determined by the method presented in 40CFR Parts 60.521(c). This method involves a CO_2 determination by EPA method 3 and a measurement of the stack gas volumetric flow rate by EPA method 2. The stack gas volume flow rate (Q) may be calculated by the method used to compute the DRE.

Alternatively, continuous monitoring equipment may be used to measure CO_2 concentrations. If an incinerator is equipped with a scrubber, volumetric flow rates and CO_2 concentrations must be determined in the flue gas prior to a scrubber as well as in the stack after the scrubber in order to allow for CO_2 absorption in the scrubber. A sample calculation of the corrected CO_2 concentration $(CO_2)_{adj}$ in the stack gas of an incinerator equipped with a scrubber is presented below. The following equation is used:

$$(CO_2)_{adj} = (CO_2)_b(Q_b/Q_a)$$

where $(CO_2)_{adj}$ = adjusted (corrected) CO_2 concentration
$(CO_2)_b$ = CO_2 concentration in the stack gas measured before the scrubber
Q_b = volumetric gas flow rate before the scrubber, dry basis
Q_a = volumetric gas flow rate after the scrubber, dry basis

Given conditions

- $(CO_2)_b = 17.10\%$
- $Q_b = 80{,}259$ scfm
- $Q_a = 84{,}833$ scfm

Calculation:

$$\begin{aligned}(CO_2)_{adj} &= 17.10(80{,}259/84{,}833)\\ &= 16.18\%\end{aligned}$$

Step 5. Calculation of particulate matter concentration adjusted to 12% CO_2
The particulate matter concentration is adjusted to 12% CO_2 using the formula

$$P_{12} = 12P/(CO_2)_{adj}$$

where P_{12} = particulate matter concentration in the stack gas adjusted to 12% CO_2

$$P_{12} = 12(0.0637)/16.18$$
$$= 0.047 \text{ gr/dscf}$$

Rather than follow the procedure provided in this example, an applicant may submit particulate emission monitoring data from an in-stack instrument in order to demonstrate compliance with the performance requirements. The permit writer must ensure that such instruments are properly calibrated and are functioning properly. The permit writer may request instrument calibration data from the applicant for this purpose. The permit writer should be aware that continuous particle emission monitoring equipment must operate near its sensitivity limit in order to detect 180 mg/dscm. Equipment response is dependent upon particle size distribution and particle color, and requires calibration for each different waste fed to an incinerator.

If the particulate emissions are greater than 180 mg/dscm (0.08 gr/dscf), the permit writer may recommend rejection of a permit application for this reason. If the particulate emissions are less than 180 mg/dscm, the permit writer must evaluate a permit application for compliance with regulatory monitoring and inspection requirements. [Be aware that these particulate emissions limits became much more stringent in 1999.]

6.2 Evaluation of Compliance with Monitoring and Inspection

The routine incinerator monitoring and inspection requirements are defined in 40CFR Parts 264.346 and 264.347 of the regulation. The owners and operators of hazardous waste incinerators must include routine monitoring and inspection data in the operating record required under 40CFR Parts 264.73 and 264.74. These records must be furnished upon request and made available for inspection by officers, employees, or representatives of EPA who are duly designated by the Administrator. The following sections provide the permit writer guidance to evaluate information on monitoring, inspection, safety systems, and effluent management included in a permit application.

Required process monitoring. An owner or operator of a hazardous waste incinerator must continuously monitor

- combustion temperature
- waste feed rate
- air feed rate
- carbon monoxide concentration in the stack gas

Instrument capabilities are described in the *Engineering Handbook for Hazardous Waste Incineration* (EPA-81/09, p. 5-102). It is recommended that these instruments should be equipped with continuous recorders in order to ease the record-keeping burden and determination of compliance with regulatory operating requirements. The permit writer must ensure that provisions to continuously monitor these parameters are included in the facility description.

Supplementary process monitoring. An applicant may monitor parameters other than those required under the regulations, and the permit writer may wish to evaluate this supplementary process monitoring. Pressure drops across major incinerator unit components are frequently monitored. Monitoring of the pressure drop is important because it is sensitive to changes in the gas flow rate, liquid flow rate, and clogging phenomena in the system. During the design phase, a proper pressure drop value or range to maintain design removal efficiency is specified.

Monitoring this parameter provides a continuous check on the normal operation of the scrubber. A change in the pressure drop or gage pressure is an indication that other measured parameters in the system need to be observed immediately to find the cause of any malfunction in order to take corrective action.

Many kinds of pressure measurement devices are commercially available to measure pressure drop across a device; however, a differential pressure gage calibrated in inches of water is usually recommended for this purpose. In selecting a pressure measuring device, the following items are considered:

- pressure range
- temperature sensitivity
- corrosivity of the fluid
- durability and ease of maintenance
- frequency response

Another parameter that may be monitored is the current draw of blower motors. Rapid fluctuations in the current draw indicates upset operating conditions of the incinerator unit. Pressure drop and current monitoring instruments may be connected to the incinerator safety system, as described in the next section.

Stack gases may be continuously monitored for a variety of parameters including sulfur dioxide, nitrogen oxides, nitrogen dioxide, unburned hydrocarbons, carbon dioxide, and oxygen. Capabilities of instruments available for monitoring of these parameters are discussed in the *Engineering Handbook for Hazardous Waste Incineration* (EPA-81/09, p. 5-102). Composition of the waste feed to an incinerator, as well as any state or local laws, will govern supplemental stack monitoring. Incinerators destroying wastes with a high sulfur content may monitor the stack for sulfur dioxide, and those destroying wastes with a high nitrogen content may monitor nitrogen oxides.

Safety system. The objective of an incinerator safety system is to stop waste feed when normal operating conditions are upset. If an incinerator is not equipped with a safety system, hazardous waste could be emitted to the environment and the combustion chamber may become filled with an explosive mixture. Every hazardous waste incinerator must be equipped with a safety shut-off system under 40CFR Parts 264.346(e).

A pilot does not offer sufficient protection and may not be considered a safety system. It may be extinguished or be unable to relight the main flame. Automatic feed shut-off valves are recommended. Automatic shut-off valves should close upon the failure of the:

- combustion or atomizing air blower
- elements of input control systems
- response from the flame detector
- response from other safety devices
- electrical power to the facility

Often, two shut-off valves are placed in series because a leak or failure of one valve may be extremely dangerous. Shut-off valves are often connected to flame detectors. Several types of flame detectors are available. Only ultraviolet detectors are suitable for use in hazardous waste incinerators. Unacceptable types include:

- thermopiles and bimetal warping devices — used in low-input heating applications
- photocells (cadmium sulfide and lead sulfide) — respond to light sources other than the flame and are not suitable for hazardous waste incinerators
- flame electrodes — suitable only for clean gas flames

Use of flame supervising systems is strongly recommended on each operating burner on any hazardous waste incinerator operating below 1400°F (760°C). Even when such combustion chambers are normally operated above 1400°F (760°C), it must be remembered that they are not above this temperature level during start-up and warm-up, which is a particularly hazardous time. The 1400°F level is a generally agreed upon temperature at which accidental fuel input would be ignited by the hot furnace interior before dangerous accumulation could occur. For installations that run continuously above 1400°F for long periods, it may be desirable to electrically bypass flame supervising equipment after the furnace is above 1400°F.

Additional potential incinerator malfunctions are identified and appropriate safety system responses are summarized in Table 2.61 (EPA-80/12, p. 105).

Evaluation of incinerator effluent management. All effluents, except for air emissions, from a hazardous waste incinerator must be managed as hazardous wastes under RCRA, unless analyzed by the methods specified in Subpart C of 40CFR 261 and found to be nonhazardous. Major incinerator effluents will most likely be ash and APCD (air pollution control device) wastewater and sludges. Management of incinerator effluents is regulated under 40CFR Parts 264, and specific restrictions are imposed for the operation of tanks, surface impoundments, waste piles, land treatment, landfills, chemical, physical, and biological treatment, and underground injection.

Inspection. An incinerator facility inspection plan should be included with a permit application. The plan must include, at a minimum as required under 40CFR Parts 264.347(d), daily inspection of the complete incinerator and associated equipment (pumps, valves, conveyors, pipes, etc.) for leaks, spills, and fugitive emissions, and checks of all emergency shut-down controls and system alarms to ensure proper operation. The results of these daily inspections must be included in the operating record maintained at a facility. Because facility diagrams vary, the inspection plans may not be similar. However, the permit writer must ensure that inspection of all facility components shown on a facility diagram are included in an inspection plan.

6.3 Specification of Operating Requirements

The permit writer must specify incinerator operating requirements based upon acceptable performance data submitted with an application. These operating requirements are incorporated with the facility operating permit recommendation, and an incinerator must operate in accordance with the requirement when permitted, as specified in 40CFR Parts 264.345. In this section the permit writer will find guidance to check the validity of incinerator operating data and to specify the operating requirements for carbon monoxide level in the stack exhaust gas, waste feed rate, combustion temperature, air feed rate and allowable variations in waste composition and incinerator operating conditions.

Carbon monoxide level in the stack gas. The measurement of CO has been selected as a combustion control parameter for the following reasons:

- well developed, reliable, and rugged instrumentation exists that can measure the concentration of CO in exhaust gases in a range from a very few parts per million to a 100% basis;
- response time of the equipment is rapid (1–90 s, depending on the instrument);
- measurement of CO maintains its meaningfulness as the excess air is lowered toward stoichiometric, or as combustion temperatures are lowered;
- instrumentation does not need to be mounted inside the gas stream, and a nonsampling noncontacting methodology for measuring the CO content of a flowing gas stream can be utilized; and

- the cost of the instrumentation (~$1500–$3000) is nominal considering the capital required to build a complete hazardous waste incinerator.

The amount of CO present in any combustion exhaust gas is a function of many factors, including:

- (a) combustion temperatures,
- (b) residence time of the combusting gases at the combustion temperature
- (c) degree of mixing of fuel(s) and air,
- (d) amount of air used in excess of stoichiometric requirements.

To some extent, these factors are interdependent. However, residence time and the degree of mixing of air and fuel(s) are largely fixed after the designs of the combustion chambers and the burners are selected. Therefore in any given facility, any changes of CO concentration will primarily reflect changes in excess air usage and in combustion temperatures. CO concentration does increase as combustion temperatures are decreased. With other factors constant, CO concentration will be fairly uniform over a wide range of excess air levels, but with continued decrease in excess air there will be a point where the CO content will very rapidly increase with further decreases in excess air. Thus, the monitoring of CO concentration, with temperature monitoring, is an excellent basis to gauge the proper operating range of combustion temperatures and excess air usage.

Some consideration has been given to the concept of a correlation between CO concentration of an exhaust gas and the destruction and removal efficiency (DRE) of a POHC. Correlations have been attempted with no significant measure of success. In general, time and temperature considerations are primary for a high DRE, but temperature and excess air considerations are more important for a low CO value.

Although monitoring CO in an exhaust gas is more convenient in an exhaust stack (where temperatures are low), measurements of CO at other points within the system (for example, in the take-off ducting immediately after the combustion chamber, or any afterburner) are also acceptable.

An upper limit of CO concentration in the exhaust gas has not been specified, and for the present, the matter will be left to the discretion of the EPA Regional Administrator. As a guide, CO levels for a number of circumstances are given in Table 2.62.

The maximum CO concentration monitored in the stack during a trial burn in compliance with the regulatory performance standards should be designated as the operating requirement for the maximum CO concentration allowable.

Waste feed rate. In order to achieve stable operating conditions, the waste feed should remain relatively constant during a trial test or during a trial burn. The permit writer should specify as an operating requirement that the waste feed rate should not vary more than ten percent from the average value established during a trial test which complies with the regulatory performance standards (unless incinerator operating data demonstrate that greater excursions still allow compliance).

Combustion temperature. The permit writer must verify the operating combustion zone temperature included in a permit application prior to specifying an operating requirement. The following information is necessary for this calculation:

- mass feed rates of waste and any auxiliary fuel to the incinerator,
- chemical composition of the waste and fuel, plus the proportion of any free or bound water,
- gross heat of combustion of the waste and the fuel, and
- amount of excess air to be used for incineration.

TABLE 2.61 Hazardous Waste Incinerator Malfunctions, and Remedial or Emergency Responses

No.	Malfunction	Type*	Malfunction indication	Response
1	Partial or complete stoppage of liquid waste feed delivery to all liquid burners	L, C	(a) Flow meter reading out of specified range (b) Pressure build-up in feed lines (c) Change in combustion zone temperature (d) Feed pump failure, zero amps	Halt waste feed, start trouble-shooting and maintenance in affected system. Reinitiate or increase auxiliary fuel feed to maintain combustion zone temperatures; continue operation of air pollution control devices (APCD)
2	Partial or complete stoppage of liquid waste to only one burner	L, C	As in (a), (b), and (c), above	Halt waste feed to affected burner only
3	Partial or complete stoppage of solid wastes feed to rotary kiln	RK, C	(a) Drop in RK combustion temperature (b) Power loss in waste feed conveyor or other feed system	As in 1, above
4	Puffing, or sudden occurrence of fugitive emissions from RK due to thermal instability or excessive feed rate of wastes to RK or failure of seals	RK, C	(a) Pressure surge in kiln (rapid change in manometer level) (b) Visible emission from air seals at either end of kiln	(a) Halt feeding of any solid waste to kiln for 10–30 min, but continue combustion (b) Evacuate unneeded personnel from immediate vicinity of kiln (c) Re-evaluate waste prior to further incineration
5	Failure of forced air supply to liquid waste feed or fuel burners	L, RK, C	(a) Flowmeter reading for air supply off scale (b) Automatic flame detector alarm activated (c) Zero amps or excessive current draw on blower motor(s)	(a) Halt waste and fuel feed immediately (b) Start trouble-shooting immediately and restart as possible (c) Continue operation of APCDs but reduce air flow at induced draft fan by "damping accessory"
6	Combustion temperature too high	L, RK, C	(a) Temperature indicator(s) at instrument control panel (b) Annunciator, or other alarm sounded	(a) Check fuel or waste feed flow rates; reduce if necessary (b) Check temperature sensors (c) Check other indicators in combustor if multiple sensors used (d) Automatic or manual activation of combustion chamber vent (sometimes called an "emergency stack cap")
7	Combustion temperature too low	L, RK, C	(a) as above; (b) as above	(a) Check other indicators in combustor if multiple sensors are used (b) Check fuel or waste feed flow rates; increase if necessary (c) Check sensor accuracy
8	Sudden loss of integrity of refractory lining	L, RK, C	(a) Sudden loud noise (b) Partial stoppage of air drawn into combustor, resulting in decreasing combustion temperatures, increased particulate emissions, and development of hot spots on external combustor shell	Shut down facility as quickly as possible

9	Excess opacity of stack plume	L, RK, C	Visual or instrument opacity readings which are above maximum allowable operating point	(a) Check combustion conditions, especially temperatures and O_2 (excess air) and CO monitor (b) Check APCD operation (c) Check nature and feed rates of wastes being burned (d) Check ESP rapping interval, cycle duration, and intensity
10	CO in exhaust gas in excess of 100 ppm, or in excess of normal CO values	L	CO indicator	Check and adjust combustion conditions, especially temperature and excess air (O_2 in stack gas), and adjust accordingly
11	Indication of, or actual failure of, induced draft fan	L, RK, C	(a) Motor overheating (b) Excessive or zero current (amps) (c) Total stoppage of fan (d) Δp drop across blower inlet and outlet	(a) Switch to stand-by fan, if available (b) If two induced draft fans are used in series, reduce operational levels immediately, stop the failing unit, and operate at reduced rate on one fan only until maintenance can be completed (c) If there is only one fan, and the fan failure appears serious, shift into an emergency shutdown mode for entire incinerator
12	Increase in gas temperature after quench zone, affecting scrubber operation	L, RK, C	(a) Partial or total loss of water supply to quench zone (b) Increase of combustion temperatures	(a) Check water flow to quench zone. Prepare for limited operation rate until water suply is restored (b) Check combustion conditions, especially temperature
13	Partial or complete stoppage of water or caustic solution to scrubber(s)	L, RK	(a) Decrease in Δp across scrubber, as indicated by manometers or other instruments (b) Zero or increased amps on water or solution pumps (c) Flowmeter readings out of specified range (d) Large increase in acid components in stack gas as detected by NDIR or other type instruments	(a) Halt waste feed, start trouble-shooting and maintenance in affected system (b) Start-up redundant pumps, if available (c) Check recycle water or solution tank levels (d) If using alkaline solution, switch to water supply if available (e) Check for deposition of solids from recycled liquors in pump lines (f) Use emergency (stand-by) water supply which will feed water by gravity until the whole system can be shut down
14	Deposition of solids in scrubber from recycled wastes or caustic solution, or from excess solids emissions from combustor	L, RK, C	(a) Build-up of Δp across scrubber as indicated by manometers or other instruments (b) Increased hold-up of liquor in packed or tray towers, up to and including flooded condition. This can also be detected by liquid level indicators	This requires a shut-down to clean out the tower and internals. The shut-down can be scheduled if the deposit build-up is gradual and is monitored
15	pH of recycled scrubber liquor not within specifications	L, RK, C	(a) Continuous, or spot-checking pH indicator shows actual pH to be outside of desired operating range (b) Drop in scrubber efficiency with excess acid gas in stack gas	(a) Check for adequate supply and metering of alkaline agent (b) Check accuracy of pH meter and alkaline solution metering pump associated with recycling of scrubber liquor
16	Failure of demister operation	L, RK, C	Increased Δp, as measured by manometer, due to solids accumulation in demister element	Back-wash element

* L, liquid injection; RK, rotary kiln; C, combination liquid injection and rotary kiln.

TABLE 2.62 CO Levels in Exhaust Gas

CO level (ppm) in exhaust gas	Source
30–40	Most coal-burning utility boilers
>1	Commercial hazardous waste incinerators
5–50	Trial burns of miscellaneous hazardous wastes at several test facilities sponsored by EPA

In any theoretical system, heat in = heat out. However, heat is lost from the incinerator by radiation and conductive/convective means, and this loss is estimated to be about 5 percent of the heat input. Assuming that all remaining heat input is available to increase product gas temperatures, the permit writer may use the equation

$$(0.95)(\Delta H \text{ gross}) = \sum (W_n)(Cp_n)(T_c - 60)$$

where ΔH gross = sum of the total Btu's from the combustion of the waste materials and any auxiliary fuel, per unit time,

$\sum$ = sum of the products of $(W)(Cp)(T_c - 60)$ for each individual gas in the combustion (flue) gases,

W_n = weight, in pounds, of each gaseous product, per unit time,

Cp_n = specific heat (at constant pressure) of each gaseous product, in Btu/lb-°F. (For approximation purposes, the Cp of water vapor is 0.49, and 0.26 for most other combustion gases including O_2, N_2, CO_2, and HCl.)

T_c = combustion temperature in °F, which is the parameter to be calculated.

If the permit writer is not prepared to perform this calculation, the simple method of estimation presented at the end of this section may be used to verify the combustion zone temperature.

A sample calculation for the determination of combustion temperatures is given in the following paragraphs.

Given conditions

- waste heat content = 6000 Btu/lb
- fuel heat content = 17,500 Btu/lb
- waste feed rate = 1000 lb/h-waste
- fuel feed rate = 1000 lb/h-fuel
- nominal combustion zone temperature = 2200°F
- incinerator diameter = 5 ft
- incinerator length = 26 ft
- excess air = 30%
- composition of waste and fuel:

	Waste	Fuel
%C	30	92
%H	7	8
%Cl_2	63	–
%H_2O	–	–

Calculation:
Step 1. Analyze waste and fuel (Table 2.63).

Step 2. Calculate the pound-moles of each element per hour.

$$C: 1220/12 = 101.7 \text{ lb-mole/h}$$
$$H_2: 150/2 = 75$$
$$Cl_2: 630/(2)(35.45) = 8.9$$

Step 3. Calculate total heat input per hour.

Heat from waste: (1000 lb-waste/h)(6000 Btu/lb)= 6×10^6 Btu/h
Heat from fuel: (1000 lb-fuel/h)(17,500 Btu/lb)= 17.5×10^6 Btu/h
Total heat input = 6×10^6 Btu/h + 17.5×10^6 Btu/h = 23.5×10^6 Btu/h
Heat available after loss = $(0.95)(23.5 \times 10^6) = 22.3 \times 10^6$ Btu/h

Step 4. Calculate oxygen needed for complete combustion.
The combustion equation for stoichiometric air is (in lb-moles/h)

$$101.7C + 75H_2 + 8.9Cl_2 + 134.8O_2 + (134.8)(3.76)N_2 \rightarrow$$
$$101.7CO_2 + 17.8HCl + 66.1H_2O + (134.8)(3.76)N_2$$

O_2 requirements are 134.8 moles, of which 101.7 moles are for $C \rightarrow CO_2$ and 33.05 moles are for $H \rightarrow H_2O$; 33.05 moles was calculated from $(75 - 8.9)/2$.

With 30% excess air, we obtain the equation

$$101.7C + 75H_2 + 8.9Cl_2 + 175.2O_2 + 658.9N_2 \rightarrow$$
$$101.7CO_2 + 17.8HCl + 66.1H_2O + 40.4O_2 + 658.9N_2$$

The heat output (heat absorbed by the gaseous products of combustion) is the sum of the equations given below (Table 2.64).

TABLE 2.63 Analysis of Waste and Fuel

Item	Rate (lb/h)	Elemental composition	%, by wt.	lb element/h
Waste	1000	C	30	300
Fuel	1000	C	92	920
Total C				1220
Waste	1000	H	7	70
Fuel	1000	H	8	80
Total H				150
Waste	1000	Cl	63	630
Total				2000

TABLE 2.64 Heat Output

	lb-mol	lb/mol	Sp. heat	ΔT (°F)
CO_2	101.7	44	0.26	$(T_c - 60)$
HCl	17.8	36.46	0.26	$(T_c - 60)$
H_2O	66.1	18	0.49	$(T_c - 60)$
O_2	40.4	32	0.26	$(T_c - 60)$
N_2	658.9	28	0.26	$(T_c - 60)$
Total flue gas	884.9			

For heat input = heat output,

$$\begin{aligned} 22.3 \times 10^6 &= (101.7)(44)(0.26)(T_c - 60) + (17.8)(36.46)(0.26)(T_c - 60) \\ &\quad + (66.1)(18)(0.49)(T_c - 60) + (40.4)(32)(0.26)(T_c - 60) \\ &\quad + (658.9)(28)(0.26)(T_c - 60) \end{aligned}$$

$$\begin{aligned} 22.3 \times 10^6 &= 7048(T_c - 60) \\ T_c &= 3224°\text{F} \end{aligned}$$

The calculated combustion temperature of 3224°F is considerably higher than the proposed temperature of 2200°F, and therefore it appears that there is excessive use of fuel. If the same total combined flow of waste and fuel is maintained, then the proportion of waste can possibly be increased. By so doing, residence time and turbulent flow in the combustion chamber would remain essentially unchanged, but one should not allow the load imposed on the downstream scrubber (to remove HCl) to exceed the scrubber's capacity.

An internal check of these calculations can be made at this point. A total of 885 lb-moles of flue gas have been produced. The volumetric flow rate at 2200°F is calculated as follows:

$$\begin{aligned} \text{Standard volumetric flow rate} &= (885\ \text{lb-mol/h})(359\ \text{ft}^3/\text{lb-mol}) \\ &= 317{,}715\ \text{ft}^3/\text{h} \\ &= 88.25\ \text{ft}^3/\text{s} \\ \text{Actual volumetric flow rate} &= 88.25[(460 + 2200)/(460 + 60)] \\ &= 451\ \text{ft}^3/\text{s} \end{aligned}$$

An alternative useful set of data, illustrating the relationship between the gross heating value of wastes, percent excess air, and adiabatic temperature attained in the combustion gases, is shown in Figs. 2.30 and 2.31. These figures apply to liquid wastes (in Btu/lb) and to gaseous wastes (in Btu/ft^3). The curves in these figures were based on theoretical considerations of the average combustion temperatures attained in combustion gases burned under adiabatic (no heat loss) conditions. Thus, these calculated temperatures would be somewhat higher than would be encountered in actual practice.

The use of these figures may be demonstrated by the following example. Using the same values presented in the previous example, the heating value of the incinerator feed is 11,750 Btu/lb and the excess air is 30%. Locating these values on Fig. 2.30, the corresponding adiabatic temperature is about 2900°F. This value agrees fairly well with the value of 3224°F obtained from stoichiometric calculations.

After the permit writer has ensured that the combustion temperature is consistent with the other incinerator operating data, the operating requirements must be specified. The minimum allowable temperature should be the minimum average temperature attained during incinera-

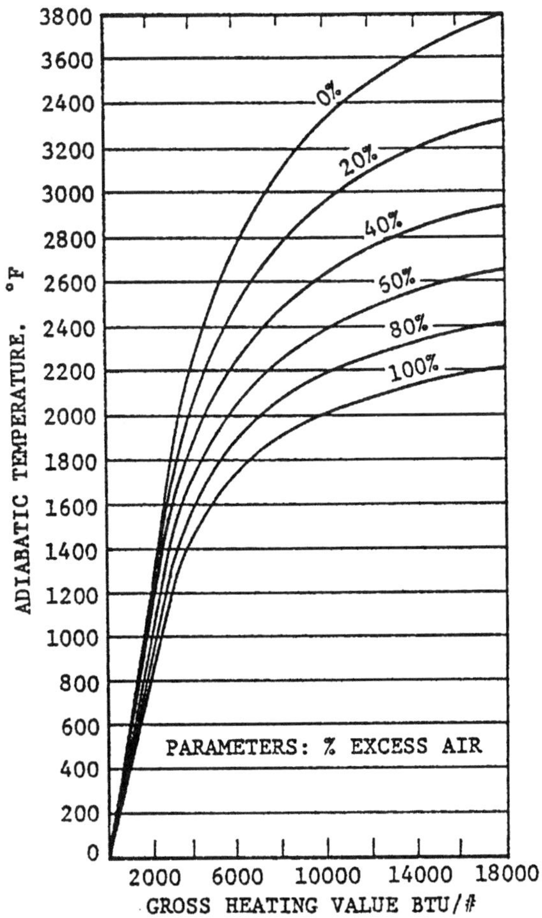

FIGURE 2.30 Adiabatic combustion temperature from a typical liquid (*EPA-80/12, p. 120*).

tion occuring within the regulatory performance standards. If only one set of data or test burn results are in compliance, the minimum operating temperature allowable is the average temperature of that run determined from continuous monitoring. An incinerator may be operated at temperatures higher than this minimum operating temperature at the discretion of an owner or operator. The maximum operating temperature is limited by the refractory lining material in the combustion chamber.

Air feed rate. It is suggested that the permit writer consider four factors when evaluating air feed rates prior to specifying an operating requirement:

- agreement of the air feed rate with excess air values and other operating data,
- residence time in the combustion chamber
- capacities and ratings of air moving equipment, and
- turbulence or mixing in the combustion chamber.

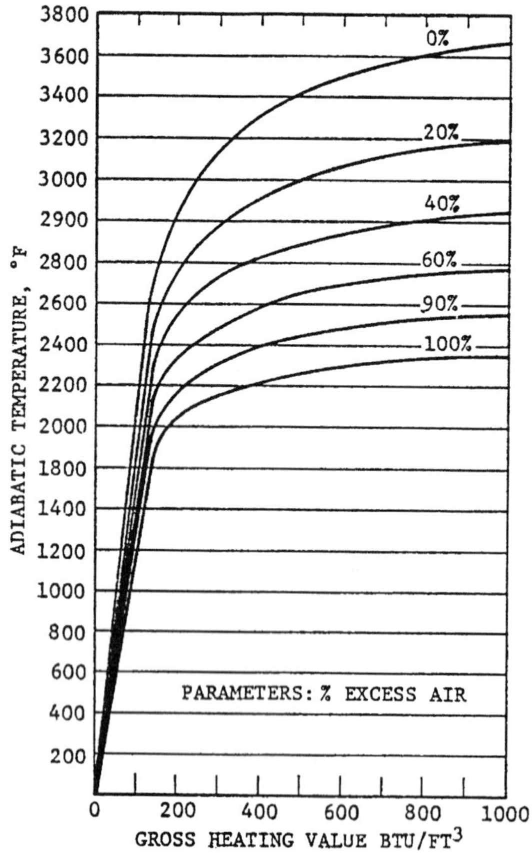

FIGURE 2.31 Adiabatic combustion temperature from a typical gas (*EPA-80/12, p. 121*).

The amount of excess air used in a given application determines the degree of air/waste mixing achieved in the primary combustion zone, process-dependent secondary combustion requirements, and the desired degree of combustion gas cooling. Since excess air acts as a diluent in the combustion process, it reduces the temperature in the incinerator (e.g., maximum theoretical temperatures are achieved at zero percent excess air). This temperature reduction is desirable to limit refractory degradation when readily combustible, high heating value wastes are burned. When highly aqueous or other low heating value waste is being burned, however excess air should be minimized to keep the system temperature as high as possible. Even with combustible waste, it is desirable to limit excess air to some extent so that the combustion chamber volume and the downstream air pollution control system capacities can be minimized.

The percentage of excess air used during incineration may be estimated from the stack gas composition under standard conditions, waste and auxiliary feed rates, and net heating value of the feed. Excess air is defined as that air supplied in addition to the quantity required for stoichiometric (perfect) combustion and may be expressed as

$$\% \text{ excess air} = [(\text{air feed rate per pound of feed})/(\text{stoichiometric air feed rate per pound of feed})] \times 100$$

Excess air may be determined from stack monitoring data by EPA method 3B, Appendix A, 40CFR Part 60. The following equation is used for that computation:

$$\%\text{ excess air} = [(O_2 - 0.5CO)/(0.264N_2 - O_2 + 0.5CO)] \times 100$$

where O_2 = percent oxygen in the stack gas by volume, dry basis
N_2 = percent nitrogen by volume, dry basis
CO = percent carbon monoxide by volume, dry basis

The stoichiometric air (SA) volume may be estimated from the approximation that 1 scf of air is required for each 100 Btu of incinerator feed material. This estimate is true only for stoichiometric combustion (zero excess air). This relation may be expressed as

$$(\text{scf of SA}) = (\text{gross heating value of feed, Btu})/100$$

If auxiliary fuel is co-fired with the waste, the heating value (HV) of the feed may be expressed as

$$\text{HV(feed)} = \text{HV(waste)} \times (\text{fraction of waste in feed}) + \text{HV(fuel)} \times (\text{fraction of fuel in feed})$$

It should be noted that this approximation is valid for estimating the volume of flue gas generated during combustion (see Section 6.1).

The volume of stack gas measured under operating conditions, or actual cubic feet (acf), may be converted to the volume under standard conditions, or standard cubic feet (scf), by use of the ideal gas law. Since both conditions are essentially at atmospheric pressure, this law simplifies to

$$\begin{aligned}(\text{acf})/(\text{scf}) &= (\text{stack gas temperature in }^\circ\text{R})/492 \\ &= (\text{stack gas temperature in K})/273\end{aligned}$$

Another useful and approximate relationship converts the volume of combustion gas obtained with stoichiometric air to the total volume of gas generated with a known percentage of excess air. The calculation is as follows:

$$(\text{stoichiometric gas volume}) \times [1 + (\%\text{ excess air})/100] = (\text{total gas volume})$$

This approximation is adequate for most fuels or waste/fuel mixtures and for excess air values to 200%, with differences from theoretical calculations seldom exceeding ±10%. The utility of these approximations is presented in the following sample calculation.

EXAMPLE standard volumetric flow rate and actual volumetric flow rate
An applicant claims that the volume flow rate of stack gas is 5525 scf/min when using 22% excess air. The permit writer finds the following data to check these values (Table 2.65).

TABLE 2.65 Gross Heating Values of Several Fuels

	Btu/lb	Btu/scf	Btu/gal
Natural gas	21,830	1020	–
No. 2 distillate oil	18,993	–	137,000
No. 6 residual oil	18,126	–	153,000

Given conditions

- waste feed rate = 17 lb/min
- heating value of waste = 5000 Btu/lb
- auxiliary fuel feed rate (No. 6 residual oil) = 17 lb/min
- heating value of fuel = 18,126 Btu/lb

Calculation:
Calculate the flue gas volume flow rate for stoichiometric air.

$$[(17\ \text{lb/min})(5000\ \text{Btu/lb}) + (17\ \text{lb/min})(18{,}126\ \text{Btu/lb})]/(100\ \text{Btu/scf}) = 3931\ \text{scf/min}$$

Note: 1 scf of air is required for each 100 Btu of waste incinerated at standard conditions (see previous statement).

Calculate the flue volume gas flow rate including excess air.

$$(3931\ \text{scf/min})(1 + 22/100) = 4796\ \text{scf/min}$$

Because this calculated value differs by more than 10% from the volume flow rate claimed by the applicant, the permit writer should check the applicant's calculation of the excess air value and volume flow rate.

In liquid injection incinerators, two air rates must be considered: (1) the air present in the primary combustion air mixed with the feed in the burner, and (2) the total air, which includes secondary combustion air brought in around the burner.

Normally, 10 percent to 20 percent excess air (e.g. 1.1–1.2 times the stoichiometric requirement) must be supplied to the burner to prevent smoke formation in the flame zone. When relatively homogeneous wastes are being burned in high efficiency burners, 5 percent excess air may be adequate. Too much excess air through the burner is undesirable, since this can extinguish the flame. Burner manufacturer specifications are the best source of information for case-by-case analysis.

In general, the total excess air rate in liquid injection incinerator units should exceed 20 percent to 25 percent to ensure adequate waste/air contact in the secondary combustion zone. In rotary kiln/afterburner incineration units, three air levels must be considered: (1) air present in the primary combustion air introduced through liquid waste burners in the kiln, (2) total air fed to the kiln, and (3) the air percentage maintained in the afterburner.

High excess air rates are needed in rotary kilns because the efficacy of air/solids contact is less than that for air and atomized liquid droplets. Typical excess air rates range from 140 percent to 210 percent or greater, depending on the desired operating temperature and the heating value of the waste. When high aqueous wastes are being burned, lower excess air rates may be needed. However, less than 100 percent excess air in the kiln may not provide adequate air/solids contact.

The residence time in the combustion zone may be estimated if the volume of the combustion gas is known. The residence time is approximated by dividing the volume of the combustion chamber by the flue gas volume flow rate. A sample calculation is provided below.

EXAMPLE residence time
The combustion zone is cylindrical with a diameter of 5 feet and a length of 26 feet.

$$\begin{aligned}\text{Volume of combustion} &= \pi d^2 L/4\\ &= 3.14(5^2)(26)/4\\ &= 510\ \text{ft}^3\end{aligned}$$

The volume flow rate is the same as was calculated in the previous example: 4796 scf/min. This volume flow rate must be corrected to combustion zone conditions using the ideal gas law, $V_1/T_1 = V_2/T_2$. If the combustion zone is 2200°F,

$$
\begin{aligned}
(\text{acf flue gas})/(460 + 2200) &= (4796\ \text{scf})/(460 + 32) \\
\text{acf flue gas} &= 25{,}930\ \text{acf/min} \\
\text{Residence time} &= (510\ \text{ft}^3)/(25{,}930\ \text{acf/min}) \\
&= 0.0197\ \text{min} \\
&= 1.18\ \text{s}
\end{aligned}
$$

In rotary kiln/afterburner incinerators, residence time is dependent on the physical state of the waste. Finely-divided solids may incinerate within a fraction of a second; dense, bulky materials may require up to an hour. Waste liquids with a high heating value may be co-incinerated with solids, slurries, or tarry materials, but will be injected into a space above the kiln bed. Residence times for these injected materials would generally be much shorter than for the denser materials.

REFERENCES

40CFR260.10-91: 40 Code of Federal Regulation (CFR), Parts 260.10, Published in 1991.

(DOE-93/03), Summary of Thermal Treatment Technologies Within The Department, of Energy Nuclear Weapons Complex, A Department of Energy (DOE) Report Prepared by Ralph A. Koenig, Merlin Co., Order Number: DE-AP01-92EW30054-A000; Reference Number: 01-92EW30054.000, March 31, 1993.

(EPA-80/02), Combustion Evaluation — Student Manual Course 427, EPA Air Pollution Training Institute (APTI), EPA450-2-80-063, February 1980.

(EPA-80/02a), Combustion Evaluation — Instructor Manual Course 427, EPA Air Pollution Training Institute (APTI), EPA450-2-80-065, February 1980.

(EPA080/12), Guidance Manual for Evaluating Permit Applications for the Operation of Incinerator Units. A Draft Report Prepared by The Mitre Corporation for EPA, Under Contract No. 68-01-6092, December 31, 1980.

(EPA-81/09), Engineering Handbook for Hazardous Waste Incineration, September 1981, SW-889.

(EPA-81/12), Control of Gaseous Emissions, Course 415, EPA Air Pollution Training Institute (APTI), EPA450-2-81-005, December 1981.

(EPA-86/03), Hazardous Waste Incineration, Course 502, EPA Air Pollution Training Institute (APTI), March 7, 1986.

(EPA-86/09), Permit Writer's Guide Test Burn Data: Hazardous Waste Incineration, EPA625-6-86-012, September 1986.

(EPA-89/03), Hospital Incinerator Operator Training Course: Volume II, Presentation Slides, EPA450-3-89-003, March 1989.

(EPA-92/10), Engineering Analysis of Medical Waste Incinerator Emissions Data, An EPA Draft Report, Prepared by the Energy and Environmental Research Corporation, October 1992.

(Gordon-79/09), Disposal of Hospital Waste Containing Pathologic Organisms, Judith Gordon, Neal Zank, et al., NTIS Report AD-A084-913, September 1979.

(Hasselriis-98/12), Information provided by Mr. Floyd Hasselriis, P.E. December 1998.

(Ho-98/12), Information provided by Dr. Thomas C. Ho, December 1998.

(Lee-89/08), Incineration of Solid Waste, C. C. Lee and George L. Huffman, *Environmental Progress*, **8**(3) August 1989.

(Lee-89/09), *Environmental Engineering Dictionary*, C. C. Lee, Published by Government Institutes, Inc., 4 Research Place, Suite 200, Rockville MD 20850, Phone: 301-921-2300; Fax: 301-921-0373.

(Lee-98), Energy and Mass Balance Calculations for Incinerators, C.C. Lee and G.L. Huffman (Editors), *Energy Sources* Taylor and Francis, 1998.

(OME-86/10), Incinerator Design and Operating Criteria – Medical Waste Incinerator, Volume II, Ontario Ministry of the Environment (OME), October 1986.

(OME-88/12), Guidance for Incinerator Design and Operation, Volume I, Ontario Ministry of the Environment (OME), December 1988.

(Oppelt-87/05), Incineration of Hazardous Waste, A Critical Review, E. Timothy Oppelt, *Journal of Air Pollution Control Association*, **37**(5), 558–586, May 1987.

(Perry), *Perry's Chemical Engineers' Handbook*, Robert H. Perry, McGraw-Hill, 1973.

CHAPTER 2.3

PRACTICAL DESIGN OF WASTE INCINERATION

Floyd Hasselriis

Consulting Engineer, Forest Hills, New York

1 INTRODUCTION

Waste combustion systems may be classified in accordance with the nature of the wastes which they are designed to burn, i.e., municipal refuse, commercial or industrial waste, medical waste or biomedical waste, and the wide range of wastes classified as hazardous. The form of the wastes, such as gaseous, liquid or solid, and the size of the combustion system are also important factors.

2 COMBUSTION PROCESS CALCULATIONS

2.1 Heat Balance

As the heat of combustion is released, the temperature of the air supplied for combustion combined with the gaseous products of combustion rises. The temperature achievable when only the stoichiometric or "ideal" air is supplied is above that which the refractory materials used to form combustion chambers can withstand without slagging and rapid deterioration. From this standpoint alone, at least 50% and up to 150% excess air must be provided in the combustion process, resulting in "total air" of 150% to 250% of ideal air required for complete combustion. The amount of excess air which must be supplied is determined by the desired final temperature, usually in the range from 1800°F to 2200°F, and can be calculated by performing a heat balance between the heat released and the quantity of gases generated by the combustion process, taking into account the specific heats of the gases generated throughout the temperature range between supply and exhaust of the gases.

The following is the heat balance equation:

$$Q = WC_p\Delta t$$

where Q = heat transferred, Btu per hour (Btu/h) or calories/h
W = mass flow of gas, pounds per hour (lbg/h) or mol/h
C_p = specific heat of gas, Btu per pound per degree F (Btu/(lb·°F)) or calories per degree per mole (cal/(deg·mol))
Δt = temperature difference

2.2 Specific Heat and Enthalpy

Heat balance calculations can be made by using the mean specific heat of each component of the products of combustion over the range of temperatures; alternatively, the enthalpy (Btu per pound) of the gases at the temperatures involved may be used.

The specific heats of the products of combustion vary with temperature. The overall specific heat over the range of temperatures is somewhat difficult to calculate because the variation with temperature is not linear. Table 3.1 gives specific heats at constant pressure (C_p) of gases in the products of combustion versus temperature. In addition, the average specific heat is given for ranges of temperatures, assuming combustion air at 70°F is used.

TABLE 3.1 Specific Heats of Gases versus Temperature

T (°F)	H_2O	CO_2	N_2	O_2	T (K)*
0	0.440	0.249	0.241	0.171	255
70	0.445	0.256	0.244	0.206	294
500	0.478	0.286	0.259	0.254	533
1000	0.517	0.317	0.277	0.265	811
1500	0.555	0.349	0.295	0.271	1089
1800	0.578	0.367	0.306	0.274	1255
2000	0.593	0.380	0.313	0.275	1366
2500	0.632	0.411	0.331	0.280	1644
3000	0.670	0.442	0.348	0.284	1921
Avg to 1800°F	0.515	0.315	0.276	0.254	
Avg to 2000	0.528	0.326	0.282	0.257	
Avg to 2500	0.543	0.338	0.289	0.261	
Avg to 3000	0.559	0.351	0.296	0.264	

*T(K) = T(°C) + 273

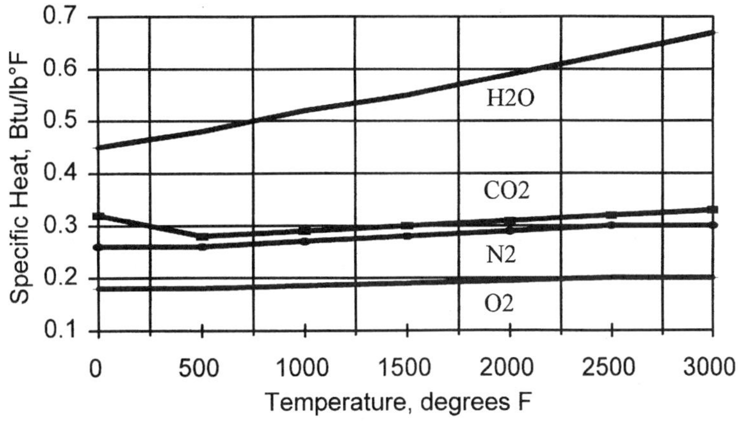

FIGURE 3.1 Specific heat of common gases versus temperature.

From *Chemical Engineers' Handbook*, McGraw Hill:

$$C_p \text{ of } CO_2 = 10.34 + 0.00274T - 19{,}550/T^2 \text{ cal/(deg·mol)}$$

$$O_2 = 8.27 + 0.000258T - 187{,}700/T^2 \text{ cal/(deg·mol)}$$

$$N_2 = 6.50 + 0.00100T \text{ cal/(deg·mol)}$$

Figure 3.1 graphs the specific heats of the main components of products of combustion and air. The substantial range of the specific heat of water vapor (H_2O) makes it important to consider this highly variable factor, especially when there is a wide range of temperature.

Table 3.2 shows the enthalpy H, the product of specified heat and temperature ($C_p \times T$) of the combustion products versus temperature, assuming theoretical (stoichiometric) air, using the average specific heats of the products.

2.3 Temperature Calculations

The heat released by combustion (Q) raises the temperature of the flow stream (W) of the products of combustion. The temperature rise (ΔT) of the products is dependent on the specific heat (C_p) of the mixture of gases.

TABLE 3.2 Average Enthalpy ($C_p \times T$) of Products vs. Temperature

T (°F)	cal/g	Btu/(lb · mol)	Btu/scf	cal/(g · mol)	cal/scm	T (°C)
32	−5.1	−265	−0.69	−143	−5941	0
70	0	0	0	11	439	21
150	12	572	1.48	338	14,071	66
212	21	1,025	2.65	596	24,800	100
400	50	2,437	6.3	1,390	57,960	205
1000	145	7,201	18.61	4,060	168,845	538
1500	228	11,400	29.46	6,400	266,216	816
1800	280	14,014	36.21	7,853	326,681	982
2000	315	15,792	40.81	8,843	367,829	1,094
2500	406	20,364	52.62	11,386	473,608	1,371

$$Q = W \times C_p \times \Delta T = W \times C_p \times (T_2 - T_1)$$

where Q = heat released, Btu/h
W = flow rate, lb/h
C_p = specific heat, Btu/(lb·°F)
T_1, T_2 = temperature of the gases entering and leaving the heat exchange zone
$T_2 = T_1 + Q/(W \times C_p)$

The heat capacity of the gases varies with temperature, as is seen in Fig. 3.1. Over the range from 70°F to 2000°F the C_p of nitrogen averages 0.282 Btu/(lb · °F) (see Table 3.1), for carbon dioxide it is 0.282, for water vapor 0.528, and for oxygen 0.257 Btu/(lb · °F). For a mixture of combustion gases from municipal waste, the specific heat is typically about 0.30 Btu/(lb · °F), dominated by the major component, nitrogen, but substantially influenced by the water vapor. This typical value of C_p can be used to make a fairly accurate estimate of the theoretical flame temperature and the final temperature of the gases after adding excess air.

Note that the heat released from combustion is the *lower* heating value (LHV), because the moisture content is not condensed in order to absorb the latent heat of condensation.

EXAMPLE: Calculate theoretical flame temperature of municipal waste. Assume that the heat input is the LHV of 4600 Btu/lbf, and the weight of gaseous products is 4.6 pounds of products per pound of fuel (lbp/lbf).

The theoretical flame temperature

$$T_f = 70\ (°\text{F}) + 4600\ (\text{Btu/lbf})/(0.30\ (\text{Btu/(lb} \cdot °\text{F)}) \times 4.6\ (\text{lbg/lbf})) = 3403°\text{F}$$

Theoretical air required for complete combustion. When the ultimate analysis is known, the following equation may be used to determine the theoretical or "stoichiometric" air required for complete combustion, using the weight composition of the waste. Note that moisture and ash are not included, so the component total is less than 100%.

$$\text{TA (lba/lbf)} = 11.50 \times (\%\text{C}) + 34.5 \times ((\%\text{H}) - (\%\text{O}_2/10) - (\%\text{Cl}/35.5) - (\%\text{F}/19)) + 4.32 \times (\%\text{S})$$

The air required for the complete combustion of fuels and wastes depends upon the heating value, since this determines the amount of oxygen required and hence the corresponding nitrogen and air. The weight of air to weight of fuel ranges from as little as 3.31 lb air per lb fuel for municipal refuse to almost 16 lb/lb for natural gas. On the other hand,

> *the volume of air required for complete combustion, per unit of heat, 9450 to 10,200 std ft^3/(MBtu), or 1062 to 1150 std m^3/(Mcal) is little affected by the fuel type, and thus may be used when no other information is available.*

Addition of excess air to control furnace exit temperature. The temperature resulting from combustion with stoichiometric air is high enough to cause damage to the chamber in which the combustion takes place. Additional air, called excess air, is therefore required in order to bring the final gas temperature down to acceptable levels. Excess air is also required to assure complete combustion.

The temperature of the mixture of gases at the theoretical maximum temperature with excess air is calculated by a heat and mass balance. For example, assume that the oxygen demand for complete combustion was found to be 0.882 lb/lbf. Adding the 3.31 lb N_2 associated with 1 lb oxygen, we get 4.192 lb air per lb fuel, plus 1 lb fuel to get 5.192 lbp per lbf. Adding 100% excess air, we add another 4.192 lb/lbf for a total of 9.384 lbp/lbf. Assume an average specific heat of 0.30 Btu/(lb · °F). The temperature of the mixture is thus

$$T_{mix} = 70°F + 4378\ (Btu/lbf)/(9.384 \times 0.3) = 70 + 1555 = 1625°F$$

For other amounts of excess air, we can include a factor n = fraction excess air (%xsa/100), as follows

$$T_{mix} = T_1 + Q/(W_s + (n \times W_s) \times 0.3) = T_1 + \frac{Q}{W_s \times (1+n) \times 0.3}$$

where W_s = lb stoichiometric air per lb of fuel

This can be solved for n, given $T_{mix} = 2000°F$ and $T_1 = 70°F$:

$$n = \left(\frac{Q}{(T_{mix} - T_1) \times W_s \times 0.3}\right) - 1$$

$$= (4378/((2000 - 70) \times 4.6 \times 0.3)) - 1 = 1.64 - 1 = 0.64 = 64\%$$

After obtaining T_{mix} using the above approximate values of C_p, we can refine these estimates and obtain a more exact calculation of T_{mix}, or n. Then

$$T_{mix} = \frac{(W_h \times T_h + W_c \times T_c)}{(W_h + W_c)}$$

where W_h = mass flow of hot fluid
W_c = mass flow of cold fluid
T_h = temperature of hot fluid
T_c = temperature of cold fluid
T_{mix} = temperature of the mixture

Mass balance to achieve 1800°F gas temperature. A simple way to estimate the amount of excess air needed to achieve a gas temperature such as 1800°F is to calculate how much cold combustion air would have to be blended with the gases at the flame temperature theoretically achieved under stoichiometric conditions.

Mixing calculation:

Input (hot) + input (cold) = output (mixture)

$$(1 - X) \times (T_h)$$

$$\downarrow$$

$$X \times (T_c) \rightarrow [MIX] \rightarrow (1) \times (T_m)$$

or

$$X \times T_h + (1 - X)T_c = 1 \times T_m$$

where X = fraction of cold gas
$1 - X$ = fraction of hot gas
T_h = hot temperature
T_c = cold temperature
T_m = mixed temperature

For $T_h = 3600°F$
$T_c = 100°F$
$X = 0.4$

$$T_m = X \times T_c + (1 - X) \times T_h = 0.4 \times 100(°F) + (1 - 0.4) \times 3600(°F) = 2200°F$$

How much excess air would be required to control the mixed temperature to $T_m = 2000°F$?

$$X = (T_h - T_m)/(T_h - T_c)(3600 - 2000)/(3600 - 100) = 0.457 = 45.7\%$$

To take into account heat losses from the furnace, we can subtract the losses from Q, the net heating value (NHV), or apply a factor $Q = \text{NHV} - \text{losses}$. Losses are probably less than 5%, so we can safely use $Q = 0.95 \times \text{NHV}$ to take the losses into account. In addition, it must not be forgotten that a supplementary fuel burner may have to remain ignited at a minimum low fire position which may contribute about 5%, canceling out the heat losses.

Figure 3.2 shows the percentage of excess air and excess oxygen needed to obtain a given gas temperature, based on combustion of municipal waste having a heating value of 5500 Btu/lb. Note that excess oxygen is linear with the temperature, but excess air is not.

Estimating heat losses. Heat losses from the casings of furnaces can be estimated by assuming that the outer surfaces of the primary and secondary furnaces are at 170°F, and assuming a heat transfer coefficient of 5 Btu/(h · ft^2 · °F). Then each square foot of surface will have a loss of 5 × (170 – 70) or 500 Btu/(h · ft^2).

EXAMPLE: An incinerator has a primary chamber 10 ft diameter by 30 ft long, and a secondary chamber 10 ft diameter by 20 ft long. The outside surface is approximately 1500 ft^2, so the losses will be about 750,000 Btu/h. If this incinerator had a capacity of 2000 lb/h of 6000 Btu/lb waste, or 12 million Btu/h, the heat loss would be 0.75/12=6% of the HHV.

Many refractory-lined furnaces are constructed with double-wall jackets, and the combustion air is passed through this annular volume to recover heat. Typically, the air is heated from room temperature to about 140°F. This rise of about 70°F represents about 4% of the 1800°F final temperature. Together with the reduced skin temperature, typically about 140°F, the surface differential may be reduced from 100°F to 70°F, the outside loss to 750,000 × 70/100 = 500,000 Btu/h, or about 4%. Thus must of the heat is recovered, and any loss may be neglected.

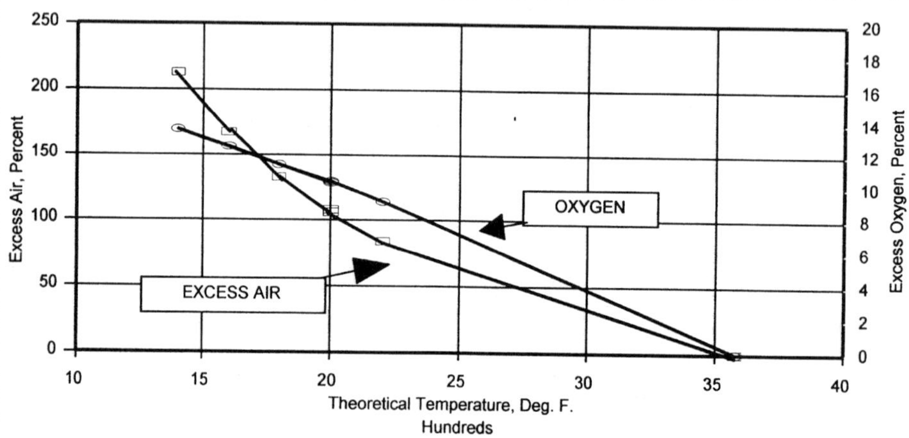

FIGURE 3.2 Theoretical gas temperature as a function of excess air and excess oxygen.

2.4 Heat Balance of Waste Combustion

Combustion of waste of known composition. Given the approximate analysis of a waste, a heat balance calculation can be carried out.

Step 1. For the given fuel composition, as shown in Table 3.3, determine the oxygen required for complete combustion of the carbon and hydrogen (stoichiometric oxygen), $(O_2)_{st}$, to be 0.816 lb per lb fuel, and the products of combustion at zero excess air to be 4.314 lbp/lbf. The final temperature is set at 1800°F.

Step 2. Calculate additional N_2 and O_2 due to excess air (EA), where EA is *assumed* to be 1.4 times stoichiometric air:

$$(O_2)_{ea} = EA \times (O_2)_{st} = 1.4 \times 0.816 = 1.14 \text{ lb } O_2/\text{lbf}$$

$$(N_2)_{ea} = 4.31 \times (O_2)_{st} = 4.31 \times 0.816 = 3.76 \text{ lb } N_2/\text{lbf}$$

$$\text{Stoichiometric products} = \underline{\qquad 4.31 \text{ lbp/lbf}}$$

$$\text{Total product with excess air} = 9.21 \text{ lb/lbf}$$

Step 3. Calculate the heating value of the fuel, using the modified DuLong formula:

$$\text{HHV} = 144.55[\text{C}] + 610\,[\text{H} - \text{Cl}/35.5 - \text{F}/19 - \text{O}/10.5] + 40[\text{S}] + 25[\text{N}]$$

$$= 5003 \text{ Btu/lbf}$$

Step 4. Calculate the *net* heating value (NHV)* by subtracting 1050 × ("already formed compounds"):

$$\text{NHV} = \text{HHV} - [1050 \times (H_2O + 9(\text{H} - \text{Cl}/35 - \text{F}/19)]$$

$$4336 \text{ Btu/lbf}$$

*The quantity of heat released in the combustion chamber(s) is derived from the net heating value (NHV). The higher heating value (HHV) includes the heat of condensation of the water present in the waste *plus* the heat formed in the combustion of hydrogen. The lower heating value (LHV) represents the heat formed in the combustion reaction; the net heating value (NHV) is the higher heating value minus the energy necessary to vaporize any moisture present and react with Cl and F, as noted in the heating value equation.

TABLE 3.3 Oxygen Required for Complete Combustion of MSW

Fuel composition	lb/lbf	Mult. factor	Stoichiometric oxygen, lb/lbf	Products	Mult. factor	Stoichiometric products, lbp/lbf
Carbon	0.2633	2.67	0.703	CO_2	3.67	0.966
Hydrogen	0.0372	8	0.297	H_2O	9	0.334
Oxygen	0.186	−1	−0.186	O_2		0
Nitrogen	0.0038					2.704
Chlorine	0.0031	−0.225	−0.0007	HCl	1.03	0.003
Fluorine	0.001	−0.421	−0.0003	HF	1.05	0.001
Sulfur	0.0019	1	0.002	SO_2	2	0.004
Moisture	0.3			H_2O	1	0.302
Ash	0.2					
Totals	1		$(O_2)_{st} = 0.816$		lbp/lbf=	4.314

Step 5. Calculate input enthalpy from preheated air and from the net heating value:

$$\Delta H_1 = \text{enthalpy of combustion air above } 77°\text{F} = 0 \text{ Btu/lbf}$$

$$\Delta H_2 = \text{NHV} = 4336 \text{ Btu/lbf}$$

Step 6. Calculate the change in enthalpy for the products of combustion at the specified final temperature of 1800°F, using appropriate average specific heats:

$$0.315 \times (CO_2 = 0.974) \times (T - 77) = 528.6 \text{ Btu/lbf}$$

$$0.276 \times (N_2 = 2.704) \times (T - 77) = 1285.9 \text{ Btu/lbf}$$

$$0.515 \times (H_2O = 0.337 + 0.303) \times (T - 77) = 567.9$$

$$0.20 \times (SO_2 + HCl = 0.007) \times (T - 77) = \underline{2.4}$$

$$\text{Total } \Delta H_3 = 2384.8 \text{ Btu/lbf}$$

Step 7. Calculate the enthalpy change of the excess air supplied to control the temperature. Note that the oxygen plus nitrogen in air is 4.31 times the oxygen:

$$\Delta H_4 = 0.26 \times 4.31 \text{ (lba/lb oxygen)} \times (T - 77) \times (O_2)_{st} = 1575 \text{ Btu/lbf}$$

Step 8. Determine the additional air needed to provide a heat balance between input enthalpy and the enthalpy change of the excess air plus products, expressed as a percentage of the stoichiometric air (note: Q = heat lost from the furnace):

$$\text{EA} = 100 \times (\Delta H_1 + \Delta H_2 - Q - \Delta H_3)/(\Delta H_4 - \Delta H_1) = 143\%$$

Note that the excess air determined here is the same at that assumed in Step 2. This is a trial-and-error calculation, which is solved in a spreadsheet calculation by a simple loop in which the answer of Step 8 is supplied as the input to Step 2.

Step 9. Calculate the products of combustion on a volumetric basis. As seen in Table 3.4, conversion from weight (mass) basis to volumetric basis is accomplished by dividing the weight of each compound by the appropriate molecular weight. The factor 386/MW converts the result to standard cubic feet of gaseous product per pound of fuel

TABLE 3.4 Products of Combustion of MSW with Excess Air for 1800°F

Products at excess air	lbp/lbf	Molecular weight	Factor 386/MW	Products (scfg/lbf)	Vol.% wet	Vol.% dry
CO_2	0.969	44	8.77	8.54	7.26	8.3
H_2O (from H_2)	0.335	18	21.44	7.27	6.14	0
O_2	1.148	28	12.06	79.26	67.35	77.2
N_2	6.488	32	13.79	16.10	13.68	15.7
HCl	0.0032	36	10.58	0.034	0.03	0.03
HF	0	19	19.3	0.002	0.002	0.002
SO_2	0.004	44	6.03	0.024	0.02	0.023
H_2O (moisture)	0.301	18	21.44	6.47	5.52	0
Totals	9.25			117.68	100%	100%

(scfg/lbf). The percent volume (wet) is converted to (dry) by subtracting the volume of moisture from combustion of H_2 and from moisture in the fuel.

Step 10. Determine parts per million by dry volume:

$$HCl = (0.03/100) \times 1{,}000{,}000 = 300 \text{ ppmdv}$$
$$SO_2 = (0.023/100) \times 1{,}000{,}000 = 230 \text{ ppmdv}$$

Combustion of a mixture of aqueous waste and auxiliary fuel. Aqueous waste is water contaminated with combustible materials which cannot be removed readily. To burn aqueous wastes, an auxiliary fuel having a substantial heating value is needed. To calculate a heat and mass balance for a mixture of wastes or, specifically, for a mixture of fuel and an aqueous waste which does not support combustion, the first step determines the composition of a mixture which can be burned with sufficient excess air to assure complete combustion, generally no less than 50% excess air.

A few heat and mass balance trials found that 13.1 lb of fuel oil per 100 lb of aqueous waste and fuel, as shown in Table 3.5, resulted in a furnace temperature of 1800°F at 50% excess air. Table 3.6 determines the products of combustion, and Table 3.7 shows the products of combustion on a mass and volume basis.

Step 1. Calculate additional N_2 and O_2 due to excess air

$$(O_2)_{EA} = EA \times (O_2)_{EA} = 0.34 \text{ lb } O_2/\text{lbf}$$
$$(N_2)_{EA} = 4.31 \times (O_2)_{EA} = 1.13 \text{ lb } N_2/\text{lb}$$
$$\text{Stoichiometric products} = 3.95$$
$$\text{Total} = 5.42 \text{ lb/lbf}$$

Step 2. Calculate heating values

$$HHV = 4164$$
$$NHV = HHV - [1050 \times (H_2O + 9(H - Cl/35 - F/19))]$$
$$= 2988 \text{ Btu/lbf}$$

TABLE 3.5 Composition of Mixture of Auxiliary Fuel Oil and Aqueous Waste

	Composition		Fraction composition		
	Fuel oil	Aqueous waste	Fuel oil	Aqueous waste	Mixture
Carbon	88.6	8.98	11.61	7.80	19.41
Hydrogen	9.54	1.19	1.26	1.03	2.28
Oxygen	0.58	1.57	0.08	1.36	1.44
Nitrogen	0.14	2.48	0.02	2.15	2.17
Chlorine	0	0	0.00	0.00	0.00
Fluorine	0	0	0.00	0.00	0.00
Sulfur	1.14	0.39	0.16	0.34	0.49
Moisture	0	81.48	0.00	70.81	70.81
Ash/inerts	0	3.92	0.00	3.40	3.40
Totals	100	100	13.1	86.9	100

Step 3. Calculate enthalpy of products

$$\Delta H_1 = \text{enthalpy of products above } 77^\circ\text{F} = 0 \text{ Btu/lbf}$$
$$\Delta H_2 = \text{NHV} = 4164 \text{ Btu/lbf}$$

Step 4. Calculate enthalpy of combustion products at 1800°F

$$\Delta H = C_p \times W_p \times (T_{out} - T_{in})$$
$$0.26 \times (CO_2 + N_2) \times (T - 77) = 1354 \text{ Btu/lb}$$
$$0.49 \times (H_2O) \times (T - 77) = 771$$
$$0.20 \times (SO_2 + HCl)(T - 77) = 3$$
$$\text{Total } \Delta H_3 = 2129$$

TABLE 3.6 Oxygen Required for Complete Combustion of Mixture of Fuel Oil and Aqueous Waste

Fuel composition	lb/lbf	Multiplying factor	Oxygen, lb/lbf	Products	Multiplying factor	Products, lbp/lbf
Carbon	0.194	2.67	0.518	CO_2	3.67	0.712
Hydrogen	0.023	8	0.183	H_2O	9	0.206
Oxygen	0.014	−1	−0.014	O_2		0
Nitrogen	0.022				$O_2 * 3.31 + N_2$	2.311
Chlorine	0	−0.225	0	HCl	1.03	0
Fluorine	0	−0.421	0	HF	1.05	0
Sulfur	0.005	1	0.005	SO_2	2	0.01
Moisture	0.708			H_2O	1	0.708
Ash	0.034					
Totals	1		0.692			3.946

TABLE 3.7 Products of Combustion Mixture with Excess Air for 1800°F

Products at excess air	Molecular weight	Factor 386/MW	Products (scfg/lbf)	Vol.% wet	Vol.% dry
CO_2	44	8.77	6.25	8.07	10.79
H_2O (from H_2)	18	21.44	19.56	25.27	0
N_2	28	12.06	47.38	61.21	74.73
O_2	32	13.79	4.13	5.34	7.15
HCl	36	10.58	0	0	0
HF	19	19.3	0	0	0
SO_2	44	6.03	0.09	0.116	0.16
H_2O (moisture)	18	21.44			0
Totals			77.4	100%	100%

Step 5. Calculate enthalpy of excess air

$$\Delta H_4 = 0.26 \times 4.31 \times (T - 77) \times (O_2)_{xs} = 1315 \text{ Btu/lbf}$$

Step 6. Calculate excess air percentage

$$EA = 100 \times (\Delta H_1 + \Delta H_2 - Q - \Delta H_3)/(\Delta H_4 - \Delta H_1) = 50\%$$

2.5 Calculation of *F*-Factor for Stoichiometric Combustion

When converting stack test data from a volumetric basis (as measured) to a mass basis, it is useful, indeed necessary, to use a factor relating stack gas flow to the heat released in combustion. This factor, expressed as standard cubic feet per million Btu, depends upon the composition of the fuel. The US EPA has used a factor of 9750 scf per million Btu for conversions when no information was available about the composition of the waste. The method which can be used to determine the *F*-factor from the chemical composition of the waste is presented below, using a composition typical of municipal solid waste (MSW). The factor determined in Table 3.8 is 10,968 scf/(MBtu), which is about 12% higher than the standard EPA factor, thus showing that it is not exact.

2.6 Air Required for Combustion

Starved-air combustion. The principle of "starved-air" or "controlled-air" combustion is used in incinerators having refractory primary chambers as a means of keeping the temperature in this chamber below the temperatures at which glassy materials melt and cause slagging. By starving the air, that is, keeping the stoichiometric ratio below one, the flame temperature is reduced accordingly, as CO is produced rather than CO_2.

Figure 3.2 showed the relationship between excess air or oxygen and the theoretical flame temperature in the excess-air region.

Figure 3.3 shows both the starved-air and excess-air regions. In the starved-air region, where less than half the air required to achieve the maximum flame temperature is supplied, the temperature is less than half the flame temperature. In spite of typical maximum flame temperatures as high as 3600°F, target temperatures of 1400 to 1600°F can be maintained, to minimize slagging. The air needed to complete combustion is provided in the secondary chamber where temperatures in the 1800°F to 2200°F range are achieved, and are controlled by modulating the quantity of combustion air added to the secondary chamber.

The air requirement for a given temperature depends upon the heating value of the waste, such as Types 0, 1 and 2, and typical municipal wastes, which have heating values ranging from 4300 to 8500 Btu/lb. The properties of these wastes are shown in Table 3.9. The oval in Fig. 3.3 marked "primary chamber" shows the starved-air conditions at temperatures from 1300°F to 1500°F, at about 45 to 65% of theoretical air. The oval marked "secondary chamber" shows the temperatures resulting from 125 to 200% of theoretical air (25 to 100% excess air), which range from 1750° to 2000°F. A heat loss of 10% has been assumed. Under these conditions, the peak flame temperature ranges from 2200 to 3000°F.

EXAMPLE: Estimate the combustion air required to maintain the primary chamber at 1500°F and the secondary chamber at 1800°F, assuming the theoretical flame temperature is 3600°F.

Step 1. Calculate combustion air needed to maintain starved-air conditions and 1500°F in the primary chamber. This is done by restricting combustion air (and oxygen) so that, instead of heating the 100°F combustion air to 3600°F, a rise of

TABLE 3.8 Calculation of *F*-Factor for Stoichiometric Combustion of Municipal Waste

MSW	Wt	Wt.%	× mult. factor	Stoichiometric O_2 (lbp/lbf)	Prod.	Mult. factor	Stoichiometric product (lb/lbf)	Mol. wt	Wet products (scfw/lbf)	Dry products (scfd/lbf)	Volume percent wet	Volume percent dry
Carbon	0.263	26.33	2.67	0.703	CO_2	3.67	0.966	44	8.48	8.48	15.44%	20.58%
Hydrogen	0.037	3.72	8	0.298	H_2O	9	0.335	18	7.18		13.08%	
Oxygen	0.186	18.60	−1	−0.186	O_2			28				
Nitrogen	0.004	0.38			$O_2 \times 3.31 + N_2$		2.706	32	32.64	32.64	59.45%	79.22%
Chlorine	0.003	0.31	−0.0225	−0.000	HCl	1.03	0.003	36	0.03	0.03	0.06%	0.08%
Fluorine	0.001	0.06	−0.421	−0.000	HF	1.05	0.001	19	0.01	0.01	0.02%	0.03%
Sulfur	0.002	0.19	1	0.002	SO_2	2	0.004	44	0.03	0.03	0.06%	0.08%
Moisture	0.304	30.40		0.000	H_2O	1	0.304	18	6.52		11.88%	
Ash	0.200	20.00										
Totals	1.000	100.00	Stoic. O_2	0.816		Stoic. prod.	4.319		54.90	41.20	100%	100%

Notes HHV = 144.55[C] + 610[H − Cl/35.5 − F/19 − O/10.5] + 40[S] + 25[N]
= 5005 Btu/lb
lbf/million Btu = 1,000,000 Btu/HHV
F-factor (scfd/million Btu) = 41.2 × 1,000,000/5005 = 8231 standard cubic feet per million Btu, dry
F-factor (scfw/million Btu) = 54.9 × 1,000,000/5005 = 10, 968 standard cubic feet per million Btu, wet.

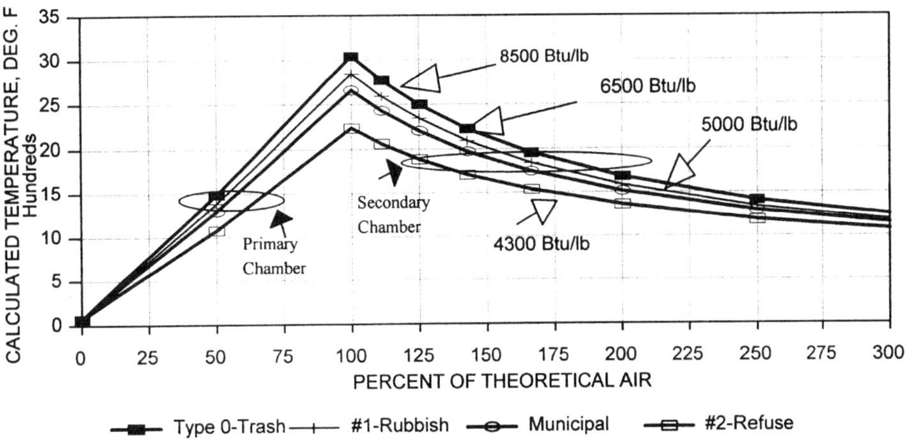

FIGURE 3.3 Temperature versus theoretical air in starved-air and excess-air regions.

$3600 - 100 = 3500°F$, the rise is only $1500 - 100 = 1400°F$. The air needed is then $1400/3500 = 0.4 = 40\%$ of theoretical air.

Step 2. Calculate the total combustion air needed to maintain 1800°F in the secondary chamber. This is done by supplying the air needed to complete combustion ($100 - 40\% = 60\%$ of SA) plus excess cold air at 100°F with gases at the theoretical flame temperature to reduce the temperature to 1800°F.

$$X = (T_h - T_m)/(T_h - T_c)$$
$$= (3600 - 1800)/(3600 - 100)$$
$$= 0.51 = 51\% \text{ excess air}$$
$$\text{Total air} = \text{primary air plus secondary air}$$
$$= (0.40 + 0.60 + 0.51)$$
$$= 1.51 \times \text{stoichiometric air}$$

TABLE 3.9 Properties of Types of Medical Waste and Trash

Type	HHV	Excess air at 1800°F	Theoretical flame temp. (°F)	$\%O_2$ in dry gas
0	8500	168	4060	13.2
1 + 0	7500	133	3510	12
1	6500	146	3690	12.5
2	4300	90	2800	9.5
MSW	5000	139	3490	12.2

Heat release in primary furnace. The primary furnace is used to "release" the heat, and the secondary chamber is used to complete the combustion process. In order to minimize slagging of refractory-lined furnace walls, experience indicates that heat release rates should be 15,000 Btu/(h·ft^3) or less.

EXAMPLE: What would be the maximum amount of heat which would be released within a rectangular primary furnace, 9 ft high, 9 ft wide and 30 ft long, based on a limit of 15,000 Btu/(h · ft^3) for a refractory-lined chamber?

$$\text{Volume} = (9 \times 9 \times 30) = 2430\ \text{ft}^3$$

$$\text{Heat release} = 15{,}000\ (\text{Btu}/(\text{h} \cdot \text{ft}^3)) \times 2430\ (\text{ft}^3) = 37\ \text{million Btu/h}$$

Secondary chamber volume and retention or residence time. An essential criterion in the design of combustion systems, especially those designed to burn difficult-to-burn wastes, is the time provided for the combustion process to be completed while the gases are maintained, calculated at the typically mandated 1800° or 2000°F. This time period is variously called "retention time" and "residence time," and is typically measured from the location where the last overfire air ports are placed.

Retention time is determined by the velocity of the gases and the distance they travel through the combustion chamber. By multiplying the velocity and distance by the cross-section of the chamber we obtain the volume flow rate of the gases and the volume of the chamber. Retention time is thus determined by the equation:

$$t = (\text{chamber length, ft})/(\text{gas velocity, ft/s}) = \text{s}$$

or

$$t = (\text{chamber volume, ft}^3)/(\text{gas volume, ft}^3/\text{s}) = \text{s}$$

EXAMPLE: A chamber having a volume of 1000 cubic feet passes 60,000 cubic feet per minute of gas. The retention time is then 1000/(60,000/60) = 1000/1000 = 1 second.

The gas volume must be calculated at the required temperature (i.e. 1800°F).

EXAMPLE: Calculate the secondary furnace volume required for burning 10,000 pounds per hour of the waste at a temperature of 1800°F.

Solution:

Based on a heat and mass balance calculation, the volume of gas produced from burning 1 pound of this waste is 13.6 cubic feet, at standard conditions of 70°F and atmospheric pressure

$$\text{Cubic feet per second at standard conditions} = 13.6 \times 10{,}000\ (\text{lb/h})/3600(\text{s/h}) = 38\ \text{ft}^3/\text{s}$$

Correcting to 1800°F:

$$\text{Actual CFS} = 38 \times [1800 + 460]/[460 + 70] = 38 \times 4.26 = 160\ \text{ft}^3/\text{s}$$

For a retention time of one second, a chamber of 160 cubic feet would be required.

Retention time and induced-draft fan capacity. The capacity of the waste combustor is thus determined by the volume of the secondary chamber or afterburner needed to provide one second retention time. The gas flow which provides 1 s retention time thus becomes the design gas flow for the system.

Since it is necessary to heat up the combustion chambers to the required temperature of 1600°F to 1800°F before introducing wastes, whether liquid or solid, clean waste or auxiliary fuel must be used for this purpose. Hence the calculation of the design capacity of the system should be based on the startup operation using clean fuel. The example which follows uses distillate #2 oil as the basis. A similar calculation based on natural gas was found to require less chamber volume and fan capacity, hence the oil basis was selected as being more conservative.

The capacity of the induced-draft fan which will provide the gas flow associated with the retention time is determined by whether water injection or a boiler is used to reduce the temperature of the gases from 1800°F. If a boiler is used, the fan will carry the same mass flow, but the fan capacity will be based on the temperature of the gases leaving the boiler. Typically, the boiler cools the gases to about 450°F. A water quench or a spray-dry scrubber could then reduce the gas temperature to about 300°F, or alternatively, if a wet scrubber is used, to about 130°F. If a wet scrubber is used, without a boiler, typically the gases are cooled to the adiabatic saturation temperature of about 180°F. In any case, the water so evaporated must be added to the products of combustion.

The following calculation determines the gas flow associated with a chamber temperature of 1800°F and a retention time of one second, at a heat release of one million Btu per hour. Then the induced-draft fan capacity is determined on the basis of use of a water quench and scrubber which reduces the gas temperature to 180°F. The basic steps in this calculation are as follows:

(a) Perform a heat balance using the auxiliary fuel properties to determine the excess air required to achieve a furnace temperature of 1800°F.

(b) Calculate the volume of the products of combustion. Use this volume and the excess air obtained to determine the volume of the furnace per million Btu of heat release.

(c) Calculate the quantity and volume of water evaporated in reducing the gas temperature from 1800°F to 180°F.

(d) The capacity of the fan is the sum of the combustion products and the water evaporated. The chamber volume and fan capacity, based on one million Btu/h, is multiplied by the desired heat release to determine the design capacity of the furnace and fan.

(e) The capacity of the furnace and fan to burn wastes having different compositions can be estimated closely by using the heat release rate. However, if the waste, or one of several wastes fed to the furnace, contains a large amount of water, a complete calculation will have to be made.

EXAMPLE: Calculate secondary chamber volume, and fan capacity for startup.

Step 1. Identify auxiliary fuel and operating conditions:

Auxiliary fuel type	distillate #2 oil
Fuel heating value, NHV	18,300 Btu/lb
Desired operating temperature, T_{out}	1800°F
Stoichiometric oxygen required for fuel	3.32 lb/lbf

Step 2. Identify stoichiometric oxygen requirements and combustion product yields for auxiliary fuel

	Combustion products, lbp/lbf				
	(O_2) stoic(f), lb/lbf	CO_2	H_2O	N_2	Net heat value
Residual fuel #6	3.16	3.18	0.92	10.5	17,500
Distillate fuel #2	3.32	3.2	1.11	11	18,300
Natural gas	3.67	2.54	2.04	12.2	19,700

Step 3. Calculate enthalpy of the fuel combustion gases:

$$\Delta H_1 = [0.26(CO_2 + N_2) + 0.49\ H_2O]\ (T_{out} - 77)$$
$$= [0.26(3.2 + 11) + 0.49 \times 1.11](1800 - 77)$$
$$= 7298\ \text{Btu/lbf}$$

Step 4. Calculate the enthalpy of excess air leaving combustion:

$$\Delta H_2 = (4.31 \times 0.26) \times (O_2)\text{stoic(f)} \times (T_{out} - 77)$$
$$= (4.31 \times 0.26) \times (3.32) \times (1800 - 77)$$
$$= 6410\ \text{Btu/lbf}$$

Step 5. Calculate the heat of fuel combustion, less 5% heat loss through walls:

$$\Delta H_3 = 0.95 \times 18{,}300 = 17{,}385\ \text{Btu/lbf}$$

Step 6. Calculate heat loss through chamber walls Q:

$$Q = 5\%\ \text{loss} = 0.05 \times 18{,}300\ \text{Btu/lbf} = 915\ \text{Btu/lbf}$$

Step 7. Calculate excess air percentage:

$$\text{EA} = 100 \times (\Delta H_3 - Q - \Delta H_1)/(\Delta H_2)$$
$$\text{EA} = 100 \times (17{,}385 - 915 - 7298)/(6410) = 143\ \text{percent}$$

Step 8. Calculate products with excess air:

		lb/lbf	Molecular weight	(lb/lbf) × 387/MW = ft^3/lbf
CO_2		2.54	44	22.34
H_2O		2.04	18	43.86
N_2		12.20	28	168.62
O_2	1.43 × (O_2)stoic	5.25	32	63.47
N_2	1.43 × N_2 =	17.45	28	241.13
Total		39.47		539.42

Step 9. Correct volume to 1800°F:

$$539.42\ (\text{lbm/lbf}) \times (1800 + 460)/(77 + 460) = 2270\ \text{ft}^3/\text{lbf}$$

Step 10. Calculate chamber volume needed for one second retention time:

For 1 million Btu/h, we need 1000,000/18,300 = 54.64 lbf/h

For 1 second, 2270 (ft^3/lbf)/3600 (s/h) = 0.63 ft^3/(MBtu/h)

For 40 million Btu/h, we need 40 × 54.645 × 0.631 = 1378 ft^3/s
Select 8 ft diam. × 27 ft: Volume = 1358 ft^3

Step 11. Calculate water needed to cool gases to 180°F at 1 million Btu/h:

$$@\ 1 \text{ million Btu/h}: \text{ products} = 54.645 \text{ (lbf/h)} \times 39.47 \text{ (lbp/lbf)} = 2157 \text{ lbg/h}$$

$$\text{Heat to be removed, } Q = 2157 \text{ (lb/h)} \times 0.26 \times (1800 - 180) = 908{,}551 \text{ Btu/h}$$

$$\text{At 1000 Btu/lb water, water added} = 908{,}000/1000 = 909 \text{ lbw/h}$$

$$\text{Volume of water at STP} = 386/18 = 21.44 \text{ ft}^3\text{/lbw}$$

$$\text{Volume of water at 180°F} = 21.44 \times (460 + 180)/(460 + 70) = 25.89 \text{ ft}^3\text{/lbw}$$

Step 12. Calculate volume at fan:

$$\text{Specific volume of products at STP} = 386/30 = 12.87 \text{ ft}^3\text{/lbp}$$

$$\text{Volume of products at STP} = 2157 \times (386/30)/60 = 463 \text{ ft}^3\text{/min}$$

$$\text{Volume of products @ 180°F} = 462.57 \times (460 + 180)/(460 + 70) = 525 \text{ ft}^3\text{/min}$$

$$\text{Add volume of water vapor} = 909 \times 21.44/60 = \underline{325 \text{ ft}^3\text{/min}}$$

$$\text{Fan capacity} = \text{Total} = 850 \text{ ft}^3\text{/min}$$

$$\text{Fan capacity at 40 million Btu/h} = 850 \times 40 \text{ million Btu/h} = 34{,}000 \text{ ft}^3\text{/min}$$

$$\text{Fan capacity per million Btu/h} = 850 \text{ ft}^3\text{/(min} \cdot \text{MBtu)}$$

3 WASTE COMBUSTION SYSTEMS

Systems designed to burn fuels and wastes must feed the fuels and wastes to the combustion chambers, supply combustion air where and as needed, agitate or mix the fuel with this air, remove the ash and flyash residues, measure and control combustion temperatures, and cool the combustion products. Finally, the system must process and clean the combustion gases before discharge through the stack to the atmosphere. The functions of a complete system are controlled by computer systems which measure, record and display operating temperatures and other data and control grate operation, and by fans and dampers which adjust combustion air and furnace draft, among other variables.

Combustion systems must be designed to accommodate the special characteristics of the wastes to be burned. While the principles are common, evolution of the designs has found that certain designs and configurations are most suitable for specific fuels and wastes. Given a design and configuration, there are limitations as to what types, quantities and mixtures of fuels and wastes can be burned satisfactorily.

Combustion of wastes generally requires these stages:

- Preparation of fuel or waste for feeding
- Feeding fuel or waste under control
- Drying: removing moisture from outer surface of waste
- Evaporating liquids to volatile gases
- Volatilizing solids to gases
- Raising volatile gases to ignition temperature
- Supplying combustion air sufficient to establish flame
- Adding "underfire" combustion air as combustion process continues
- Adding "overfire" combustion air to continue combustion process
- Adding "secondary" air to maintain secondary chamber temperatures.

The combustion process is controlled so that, in the progression of fuel and air through the chambers, the optimum amount of air is supplied to combustibles (or, alternatively, the right amount of waste is fed to match the air) at all times and all places: how well this is accomplished is the measure of success. If too much air is supplied too soon, ignition may not take place where desired, and the flame temperature can be too low to assist rapid combustion. Too little air can result in insufficient combustion in the primary chamber, or in excessively high temperatures which can cause slagging and other damage.

Emission control devices are an integral part of a complete system. The induced-draft fan at the stack end of the process must be controlled by a damper and/or speed control to maintain the draft in the combustion chambers so that hot gases do not leak into the atmosphere, and so that too much cold air is not sucked into the chambers.

3.1 Types of Wastes to be Burned

Liquid and gaseous wastes can be burned in relatively simple refractory combustion chambers provided with a burner suitable for burning conventional fuels (gas and oil) and one or more burners designed to atomize and burn the liquid, or to atomize essentially non-combustible liquids, which will be destroyed at the temperatures provided by high-energy fuels or wastes. Heating values of liquid wastes can range from negative to 23,000 Btu per pound. The combustion process is easy to control because the fuel has consistent properties and the combustion air can be controlled in direct proportion to the combustible feed, or vice versa.

Solid wastes are more difficult to burn due to the highly variable nature of the waste material. Solid wastes vary widely in composition and physical characteristics, spanning a wide range of bulk densities and moisture contents, resulting in heating values ranging from negative to as high as 22,000 Btu/lb. Solid materials require complex handling, feeding, firing, and residue removal equipment. Municipal and medical wastes, while highly variable in composition and heating value, have a fairly consistent range in their properties.

Hazardous waste incinerators may be dedicated to the wastes of a specific facility, or to wastes from a wide range of generators. They are generally capable of processing wastes that include aqueous liquids with negative heating value, high energy liquids, sludges which are difficult to pump, solids which are slow to ignite and burn, and in general wastes containing potentially hazardous chemicals and compounds.

Note: the term "incinerator" has been associated with small waste combustors which often lacked emission controls. To avoid this connotation, the term "waste combustor" is preferred.

3.2 Types of Combustion System

Municipal, commercial and medical waste combustion systems may be classified by size (throughput capacity) and application:

Small refractory-lined incinerators have capacities from 20 to 750 lb/h:

- Single-chamber incinerators, batch feed, 8 to 16-hour operation
- Multiple-chamber incinerators, batch or intermittent feed, 8 to 16-hour operation.

Medium-size refractory-lined two-chamber combustors with capacities from 500 to 2000 lb/h (50 tpd), capable of 24-hour continuous operation.

For municipal and commercial wastes:

Small mass-burn combustors with capacities from 50 to 250 tpd:

- Refractory-wall, continuous feed
- Water-cooled wall, continuous feed

Large-scale combustors, water-wall, continuous feed, with capacities from 250 to 1000 tpd:

- Mass-burn stoker
- Suspension-fired RDF stoker
- Fluidized-bed RDF

For hazardous wastes:

- Refractory-lined furnaces
- Rotary kiln refractory furnaces
- Cement kilns
- Lightweight aggregate kilns
- Boilers and industrial furnaces

Mass burn combustors. Mass burn combustion systems burn solid wastes such as municipal solid waste (MSW) and biomedical medical wastes (BMW) without preprocessing other than removal of items too large to go through the feed system.

Small refractory-wall combustors. Figure 3.4 shows a multiple-chamber rectangular refractory-wall combustor, representing units which have been installed in many hospitals and crematoriums, commonly called incinerators. These are used for capacities in the range of 500 to 2000 lb/h. They may be charged manually through the charging door and, after ignition of the waste, allowed to burn out before recharging. Normally, however, they are fitted with

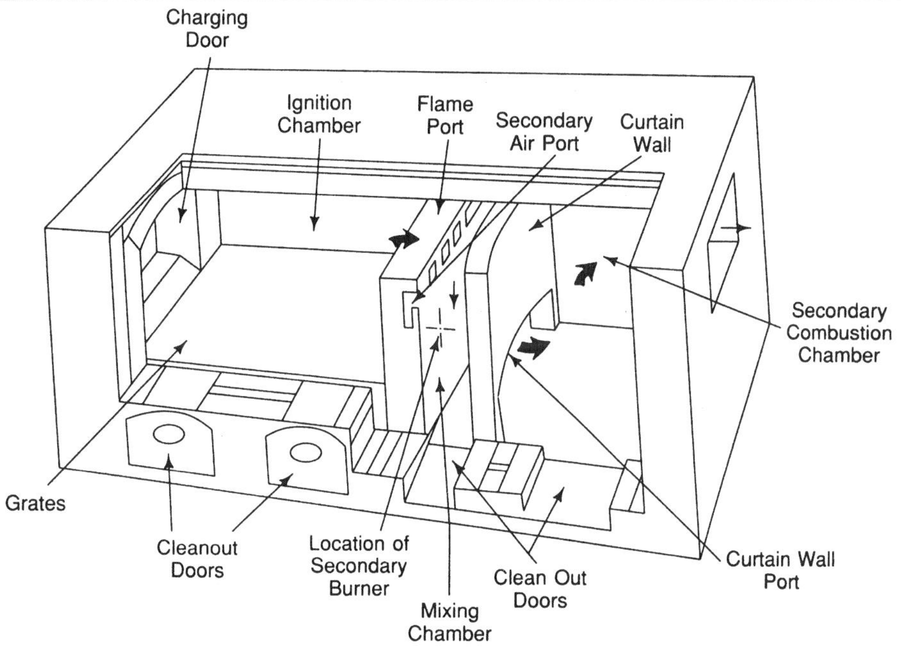

FIGURE 3.4 Small multiple-chamber refractory combustor.

feed hoppers with tight-closing doors and ram feeders so that they can receive additional charges during the course of the 8 to 12-hour working day, while maintaining the air seal needed to keep the combustion under control and avoiding flash-backs when the charging door is opened.

Waste is fed through the charging door onto the stationary grate, which may be perforated so that combustion air can be supplied from under the waste as it burns in the ignition chamber; alternatively, combustion air may be injected from nozzles in the chamber walls. An ignition burner is provided to ignite the waste and maintain a minimum temperature (such as 1200°F) in the chamber. Flames and hot gases exit through the flame port to the mixing chamber where mixing with secondary air occurs. A secondary burner maintains the temperature when the wastes do not provide enough energy. The gases pass under the curtain wall into the secondary combustion chamber where combustion is completed before exiting to the stack. Not shown is a barometric damper which controls the chamber draft by adding cold air before the gases enter the stack. Air ports are provided in the secondary combustion chamber and in the mixing chamber. A supplemental fuel burner is provided in the primary chamber for ignition and heating up of this chamber. Another burner in the secondary chamber is used for heat up, to maintain the flame, and to assure that the required chamber temperature is maintained at all times that waste is being fed, regardless of the energy supplied by the waste. Ash is removed only after the waste has burned out overnight.

Modular refractory combustors. Small two-chamber refractory-wall combustors having capacities ranging roughly from 20 to 750 lb/h are illustrated in Fig. 3.5. A ram feeder pushes the waste from the feed hopper into the lower chamber. The waste is agitated by the internal ram, progressing it toward the ash quench pit, into which ash residues fall and are pulled out by an ash hoe, in this case, into an ash cart. A primary burner (not shown) is used to preheat the primary chamber and maintain temperature, while primary combustion air is introduced through ports in the chamber walls. Secondary combustion air is introduced in the throat between the lower and upper chambers. The secondary burner maintains the temperature in the upper chamber, in which combustion is completed before the gases leave through the stack.

With intermittent feeding, each charge produces an upset in the combustion conditions. The size and frequency of charging determine the degree of upset, and challenge the ability of the control system to adjust the combustion air to maintain essentially constant furnace temperatures. The size of the feed hopper determines the maximum volume of the charge, but neither its weight nor its thermal value; hence it is the duty of the operator to keep weight and quality in mind while charging the hopper.

The time between charges may be automatic, with the operator option to extend it. Generally charges should not follow more closely than 10–15 minutes apart, in order not to exceed the capacity of the combustion air fans during the spike in heat release which occurs about one minute after the charge enters.

The incinerator shown in Fig. 3.5 has a hydraulically operated feed hopper with cover, guillotine shutoff door, charging ram, and reciprocating grates for agitating the waste and moving the burning waste to the ash discharge. The ash discharges into a water-quench tank. An ash conveyor retrieves the ash and places it in containers.

Continuous feed two-chamber waste incinerator with heat recovery boiler. Figure 3.6 shows a complete incinerator for continuous feeding of medical waste. A waste cart with tipping mechanism is used to load the feeder while the feeder lid is open and the shrouded feeder door is closed. After loading, the feeder lid closes, the feeder door opens, and the ram feeder pushes the waste into the primary chamber. A refractory ram in the primary chamber pushes the ash to the ash quench pit, where it is removed and discharged into ash carts. The inducted draft fan draws gases from the secondary chamber through the breeching to the heat recovery boiler, generating steam or hot water, and discharges the cooled gases to the boiler exhaust stack.

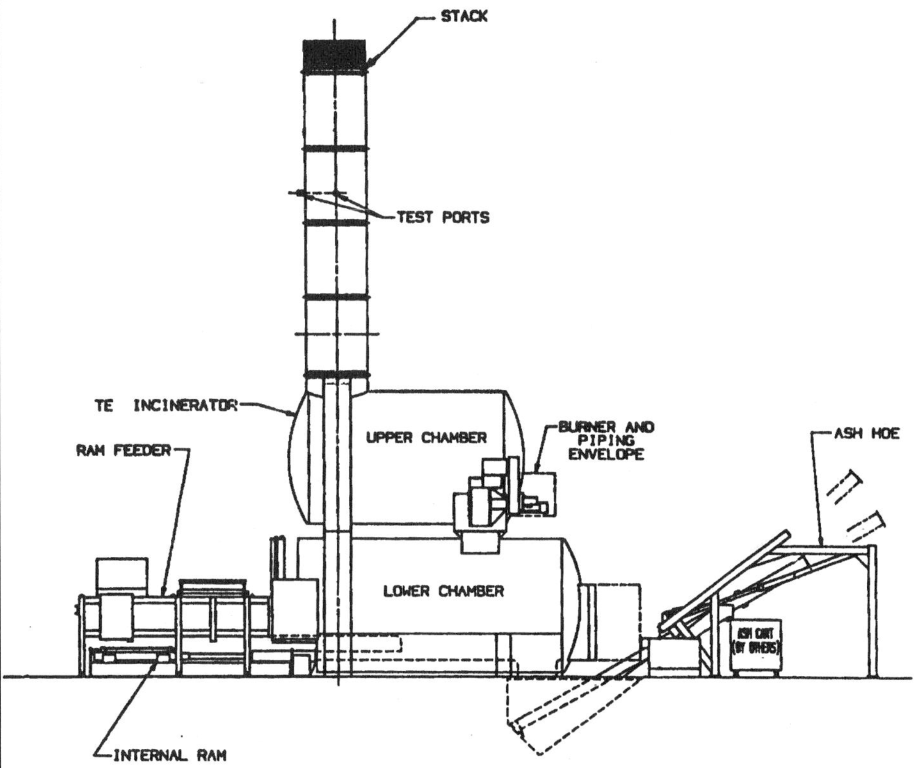

FIGURE 3.5 Small two-chamber continuous feed refractory combustor.

Pathological and crematory waste incinerators. These combustors are specially designed to burn high-moisture wastes which do not support combustion; hence they rely entirely on the supplemental fuel burners to raise and maintain the furnace temperature until the charge has been totally burned to ash. They are generally fed in batches which are allowed to burn down before a new charge is introduced.

Larger two-chamber modular refractory combustors. Large two-chamber combustors, shop fabricated for the capacity range of 5–120 tpd of MSW throughput, are illustrated in Fig. 3.7. They were developed to serve the need for smaller, less costly refractory systems which could be shipped in modules for assembly on the site. The refractory is brick or castable, depending on the application. They burn waste that has not been preprocessed other than to remove oversize objects. They are usually fed from a tipping floor by front-loaders.

Multiple hearths are provided, each having a transfer ram to progress the waste down the hearths and into the residue sump and into the water pit, from which they are removed by conveyors. The hot gases from the secondary chamber are generally cooled by a waste heat boiler before discharge through an emission control system to the stack.

Modular starved air combustors. These units are designed to operate in starved air (also called controlled air) mode. Air is supplied at levels below that required for complete combustion (sub-stoichiometric level) in order to assist combustion while maintaining temperatures

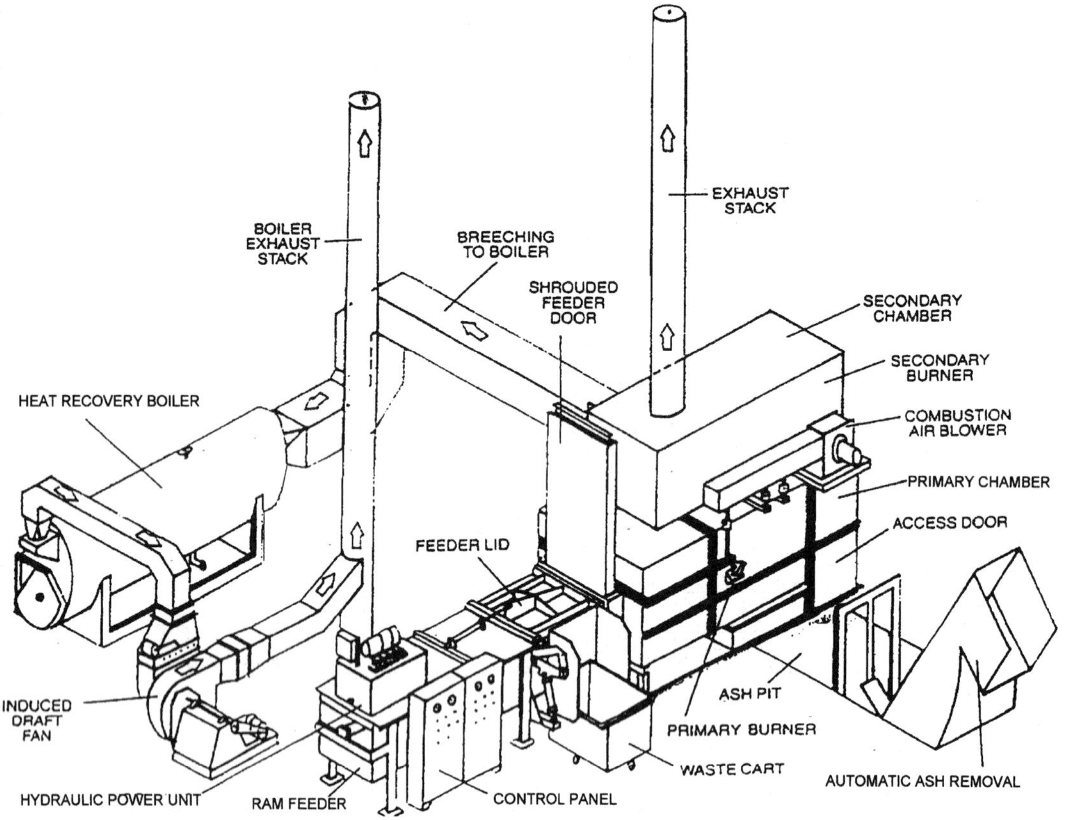

FIGURE 3.6 Two-chamber modular incinerator with feeder, ash removal and heat recovery (*Simonds Mfg Corp., Auburndale, Florida*).

below the levels which cause slagging. Refractory or water-cooled reciprocating rams agitate the waste and move it along the refractory floor to the ash discharge. The incompletely oxidized combustion products (CO and organic compounds) pass into the secondary combustion chamber, where additional air is added to control temperature while providing sufficient retention time for completion of combustion.

Modular excess air combustors. A few modular refractory combustors, limited to a capacity of 120 tpd per unit, have been designed to operate in the excess air mode, using recirculating flue gas to control the temperatures in the primary and secondary combustion chambers. Waste is batch-fed by hydraulically operated feed rams to the refractory-lined primary chamber, typically on a 6–10 minute cycle between charges, adjustable to compensate for waste composition. The waste is moved through the primary chamber by hydraulically operated refractory hearths. The ash residue is discharged to a wet quench pit. The refractory secondary combustion chamber provides residence time for burnout.

Medical waste combustion systems. Medical wastes have been burned in systems similar to those shown in Figs. 3.4–3.6 without emission controls, discharging the hot gases directly to

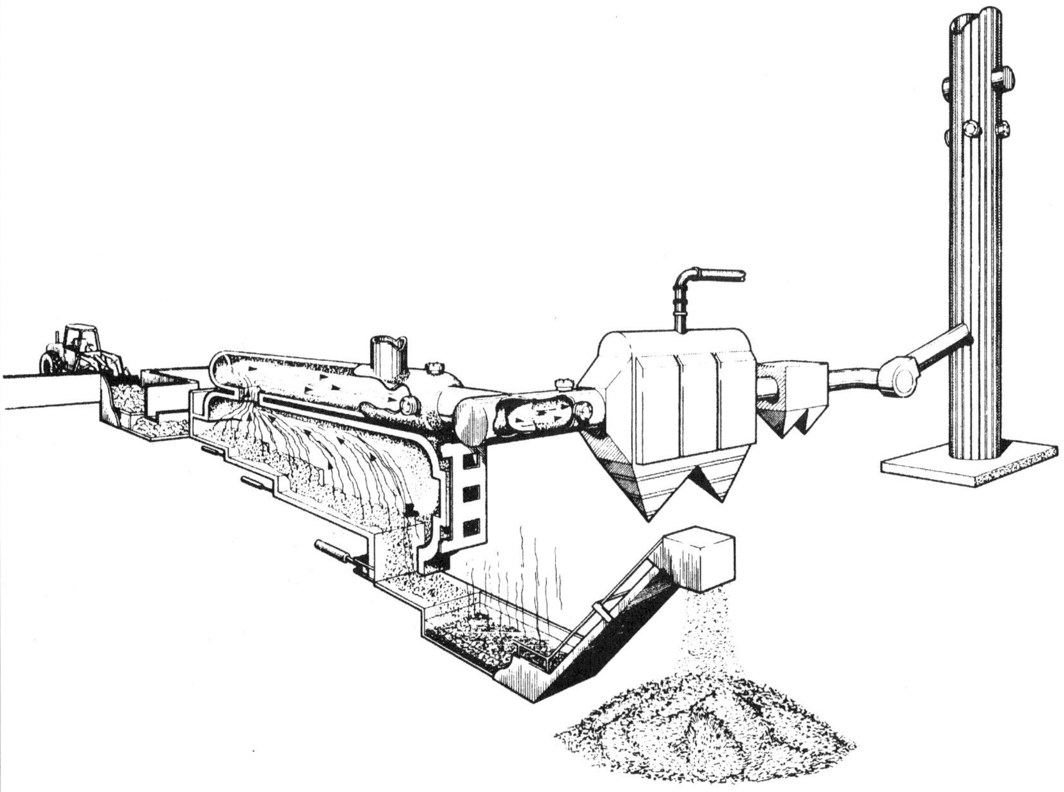

FIGURE 3.7 Large two-chamber refractory combustor.

the stack, or passing them through a boiler for heat recovery before discharge to the stack. More recently, these systems have been provided with wet scrubbers or baghouses for emission control.

More stringent regulations have required that medical waste combustors be provided with emission controls to remove not only particulate matter but acid gases. Figure 3.8 shows a complete system including a waste hopper (1), an auger feeder (2), a rotating primary chamber (3), a secondary chamber (4), an emergency bypass stack (5), a heat recovery boiler (6) to reduce the temperature of the gases, a baghouse (7) to remove particulates, a wet scrubber (8) to remove acid gases, an induced-draft fan (9) to provide system draft requirements and maintain furnace draft, and a stack (10) discharging to the atmosphere. Some systems of this type have been provided with dry activated carbon injection systems to remove mercury and dioxins.

Large refractory-wall combustion systems. Early designs of mass-burn combustors for municipal solid waste built in unit sizes ranging from 90 to 270 Mg/d (100–300 tpd) had refractory chambers which were constructed in the field from refractory brick, generally anchored to structural steel frames, and often air cooled. They were generally fed from waste storage pits by cranes using grapples to pick up and discharge the waste into feed hoppers.

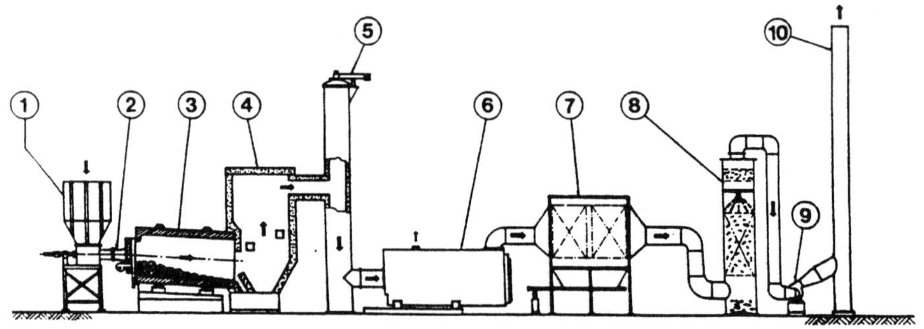

FIGURE 3.8 Complete medical waste combustion system with baghouse and wet scrubber (*Enercon, Elyria, Ohio*).

To permit continuous duty, large refractory-chamber combustors were provided with rocking or reciprocating grates which stir and aerate the waste bed as it advances through the combustion chamber, thereby improving contact between the waste and combustion air and increasing the burnout of combustibles. The ash is discharged at the end of the grate into a water quench pit for collection and disposal in a landfill. Most of these plants, that were built in the 1970s and 1980s, used water quench towers in lieu of heat recovery boilers to reduce the gas temperatures and electrostatic precipitators (ESPs) to reduce PM emissions.

Because refractory-walled combustors do not generally have waterwalls to remove heat from the primary combustion chamber, they need to use relatively large amounts of combustion air, typically 150–300 percent more than that required to maintain the desired furnace temperature range of 1600°–2000°F. These excess air levels result in relatively high carryover of PM from the combustion chamber and through the boiler passes before entering the air pollution control device (APCD).

Mass burn waterwall combustors. These systems range in capacity from 46 to 900 Mg/d (50–1000 tpd) of MSW throughput per unit. They are usually erected at the site (as opposed to being prefabricated at another location), and employ from one to three units per plant. This category includes mass burn waterwall (MB/WW), mass burn rotary waterwall combustor (MB/RC) and mass burn refractory wall (MB/REF) designs. These capacities are possible due to the cooling effect of the water-cooled walls of the combustion chamber.

As seen in Fig. 3.9, the waste is discharged from trucks from the tipping floor (2) to a pit (4) where a crane (5) transfers it to the feed hopper (6) from which it is injected onto the stoker grate in the primary combustion chamber (7). The ash residues pass through a quench tank (8) and into an ash hopper (9). In some plants these residues are passed under a magnet to separate ferrous metals, and screened to producing an aggregate for beneficial use. Otherwise the ash residues are sent to landfill.

The hot gases leaving the combustion chamber pass through a waterwall furnace (10) and through a boiler, superheater and economizer (11) and (12). In the conditioning tower (13) the gases are cooled before entering the lime reactor (14). The gases then pass through the fabric filter (15) and the induced draft fan which discharges them to the stack (16). Facilities with spray-dry scrubbers use the vessel (14) to cool the gases and evaporate the lime slurry before they enter the fabric filter. The flyash removed from the boiler hoppers is combined with the bottom ash residue. The flyash removed from the baghouse may be either mixed with the bottom ash or disposed of separately. The flyash is stored for use as a cement supplement, or mixed with the quench tank ash for discharge to the landfill.

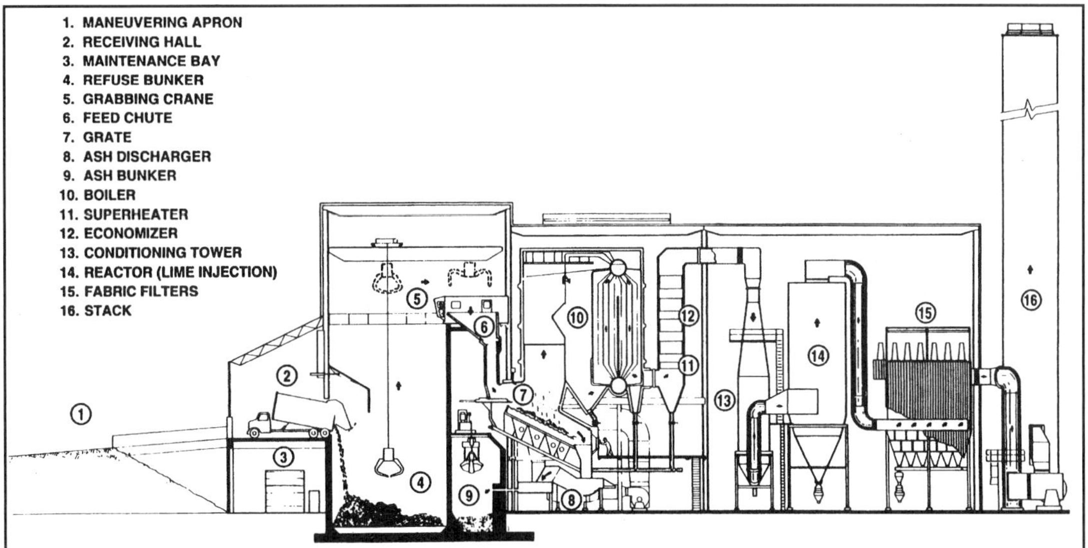

FIGURE 3.9 Complete mass-burn waterwall combustion system with scrubber/baghouse (Martin combustor, Flakt emission controls).

The combustor walls are constructed of steel tubes that contain circulating pressurized water used to recover heat from the combustion chamber. In the lower, actively burning region of the chamber, where corrosive conditions may exist, the walls are generally lined with refractory. Heat continues to be recovered in the convective sections, superheater and economizer of the boiler.

Modern MB/WW facilities utilize hydraulically controlled reciprocating stoker grates or roller grates to move the waste through the combustion chamber. The grates typically include three or more sections which can be controlled separately. On the drying grate the moisture content of the waste is reduced prior to ignition. The next set of grates, referred to as the burning grates, is where the majority of active burning takes place. The third grate section, referred to as the burnout grates, is where remaining combustibles in the waste are burned. Bottom ash is discharged from the burnout grate into a water-filled ash quench pit. From there, the wet ash is discharged by a ram or drag conveyor to a conveyor system which transports the ash to a load-out or storage area prior to disposal. Dry ash systems, which have been used in some designs, require a suitable dust collection system to prevent discharge of dust to the environment.

The stoker grates transport the refuse through the furnace and, at the same time, promote combustion by adequate agitation and good mixing with combustion air, while being cooled by the underfire air. Abrupt tumbling caused by the dropping of burning solid waste from one tier to another promotes combustion. This action, however, also contributes to carryover of particulate matter with the exiting flue gases.

Combustion is initiated by radiation from the active combustion areas onto the incoming waste, resulting in drying the exposed surface and raising it to the ignition temperature. Combustion air is added from beneath the grate sections through underfire air plenums, usually one per grate section, to permit proportioning of the air supply as needed to control burning and heat release from the waste bed. Air injected below the grates is also needed to cool the grates. The grates may be divided into multiple parallel sections, each with a controllable air supply. Underfire air may be preheated to compensate for moisture in the MSW. The tendency for slagging waste components (ash and glass) causes fouling and deterioration of the furnace refractory walls. The fusion and melting temperatures of these slags limit the operating temperatures above the grates. With simple refractory walls, flame temperatures may have to be limited to as low as 1800°F. Air-cooled or water-cooled refractory walls permit operation at higher temperatures, up to 2600°, with manageable slagging.

Overfire air is injected through rows of high-pressure nozzles located on the side walls of the chamber to complete the oxidation of the fuel-rich gases evolved from the bed and reduce the temperature of the gases below the theoretical flame temperature. Secondary air is added in the secondary combustion chamber. Properly designed and operated overfire air systems are essential for good mixing and burnout of organics in the flue gas. Typically, 80–100 percent excess air is needed to control the temperature of the gases leaving the combustion chamber.

The flue gas leaving the primary combustion chamber passes through the waterwall-cooled furnace section which cools the gases, from temperatures from 2200°F to 2600°F, generally to about 1500°F before they enter the superheater and/or convection sections, to minimize the formation of slag on the tubes. The superheater and convection sections are followed by economizer sections, altogether cooling the gases to about 450–550°F before they enter the air pollution control device (APCD).

Mass burn rotary waterwall combustors. Mass burn rotary combustor (MB/RC) facilities are similar to the units shown in Fig. 3.9 except that the primary chamber is a water-cooled rotary cylinder. These plants have been built in a range of unit capacities from 160 to 800 Mg/d (176–880 tpd), with typically 2 or 3 units per plant. Following removal of objects too large to feed into the combustor, the waste is ram-fed to the inclined cylindrical combustion chamber, which rotates slowly, causing the waste to advance and tumble as it burns. Underfire air is injected through the waste bed, and overfire air is provided above the waste bed. Bottom ash is discharged from the rotary combustor to an afterburner grate and then into a wet quench pit. The wet ash is conveyed to an ash load-out or storage area prior to disposal.

Combustion air is provided along the rotary combustion chamber length, with most of the air provided in the first half of the chamber. The rest of the combustion air is supplied to the afterburner grate and above the rotary combustor outlet in the boiler. The rotary waterwall system normally operates with about 50 percent excess air, compared with 80–100 percent for typical waterwall firing systems, since, due to the effective cooling in the combustor, less cooling air is needed to maintain the desired final combustion temperature of about 1800°–2000°F. Water circulated through the tubes in the rotary chamber recovers heat from combustion. Additional heat recovery occurs in the boiler waterwall, superheater, convector and economizer sections. From the economizer, the flue gas passes to the APCD at a temperature of about 450°F.

Refuse-derived fuel combustors. Refuse-derived fuel (RDF) combustors burn processed waste suitable for combustion in stoker-fired boilers, similar to those used to burn granular coal in utility boilers. Combustor sizes range from 290 to 1300 Mg/d (320–1400 tpd). A typical RDF combustion furnace is shown in Fig. 3.10. The waste processing system (not shown) consists of a tipping floor where front-loaders move to the waste after it is dumped from trucks into piles or to the feed conveyor leading to shredding magnetic separation and screening operations. Objectionable items from the waste are placed on a picking floor. Additional steps may remove glass and other non-combustibles, and multiple-stage shredding may provide a more uniform fuel to facilitate feeding to the furnace. Systems doing a high degree of processing to remove glass also produce roughly 20% residue which is normally landfilled, which increases the heating value of the remaining RDF product.

Figure 3.10 shows an RDF-firing system with associated emission controls. The RDF is fed by vibrating feeders and a speed-controlled screw to the air-swept spouts through which it is injected over the traveling-grate spreader-stoker. Underfire air passes through perforations in the stoker grate, and overfire nozzles supply secondary air. The forced-draft fan is followed by

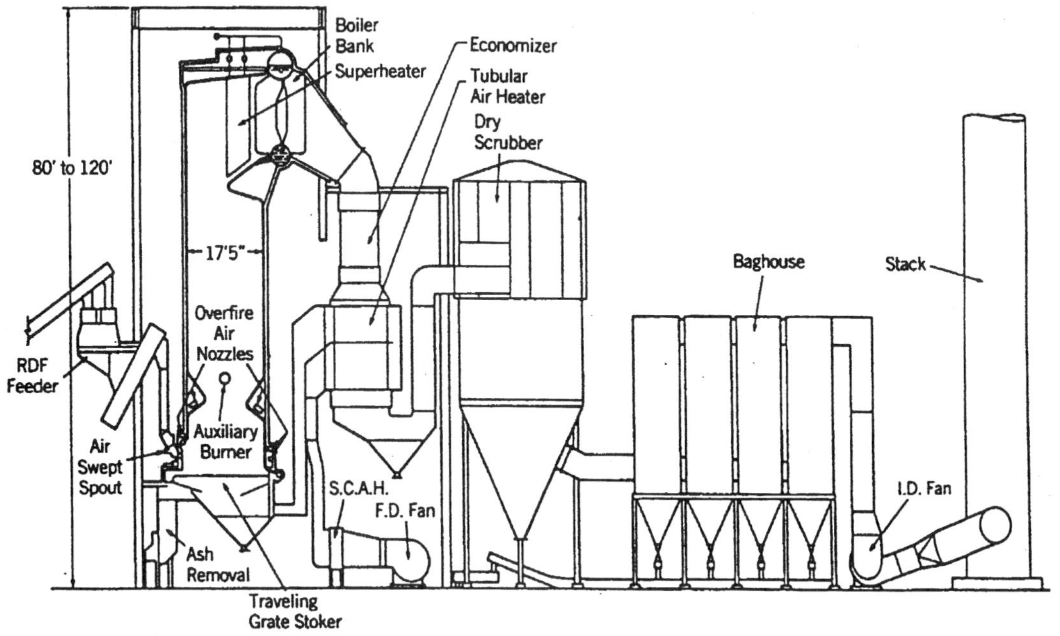

FIGURE 3.10 RDF-firing system with associated emission controls (Palm Beach, Florida).

a steam heater to preheat the combustion air if needed. The RDF is fired in a semi-suspension mode, using an air-swept distributor, which allows a portion of the RDF to burn in suspension and the remainder to be burned out after falling on a horizontal traveling grate. Because the traveling grate moves from the rear to the front of the furnace, distributor settings are adjusted so that most of the waste lands on the rear two-thirds of the grate. This allows more time for combustion to be completed on the grate. Bottom ash drops into a water-filled quench chamber. The speed of the traveling grate can be manually adjusted to accommodate variations in burning conditions. Underfire air is normally preheated and introduced beneath the grate by a single plenum. Overfire air is injected through rows of high-pressures nozzles, providing a zone for mixing and completion of the combustion process.

The combustion gases lose temperature while passing through the waterwall-cooled chamber to the superheater, boiler bank, economizer and tubular air heater which heats the combustion air, thus generating superheated steam for use in the steam turbine-driven generator. Particulate matter (PM) concentrations at the inlet to the pollution control device are approximately double those of mass burn systems and more than an order of magnitude higher than in combustors using starved-air, two-stage combustion.

RDF-fired boilers can totally avoid the use of refractory by extending the water-cooled walls down to the grate level. Corrosion-resistant alloy cladding is used on the waterwalls to provide long life. The boilers have tall furnace chambers, providing retention time to burn out the RDF particles, followed by the superheater, convector and economizer sections, which reduce the gas temperatures to roughly 450°F. Excess air provided is typically 40–80%, similar to that of mass burn waterwall units.

Fluidized bed combustors. Fluidized bed furnaces can burn municipal solid waste as well as wood and other suitable wastes. It is necessary to process the waste, reducing particle size to facilitate feeding to the fluid bed, and preferable to remove glass and metals from the waste by preprocessing. This technology requires that the furnace temperature be kept below the melting point of glass and metals such as aluminum and zinc, which are removed from the circulating sand at the bottom.

These furnaces have the ability to add limestone (or other alkali) to the bed to capture halogens (chlorides and fluorides) and other compound, significantly reducing the discharge of acid gases. The ash residues are of a high quality which permits their beneficial use, reducing the costs of ash residue disposal.

In an FBC, illustrated in Fig. 3.11, shredded MSW (RDF) is fed to a turbulent bed of noncombustible material such as limestone, sand or silica, where combustion air is supplied through a gas distribution plate and underfire air windbox. The combustion bed, which lies over the gas distribution plate, is suspended or "fluidized" through the introduction of underfire air at a high flow rate. Other wastes and supplemental fuel may be blended with the RDF outside the combustor or added into the combustor through separate openings. Overfire air is used to complete the combustion process and maintain the furnace temperature below the 1600°F range.

Bubbling bed combustors maintain most of the solids near the bottom of the combustor by using relatively low air fluidization velocities to reduce the entrainment of solids from the bed into the flue gas, minimizing recirculation or re-injection of bed particles.

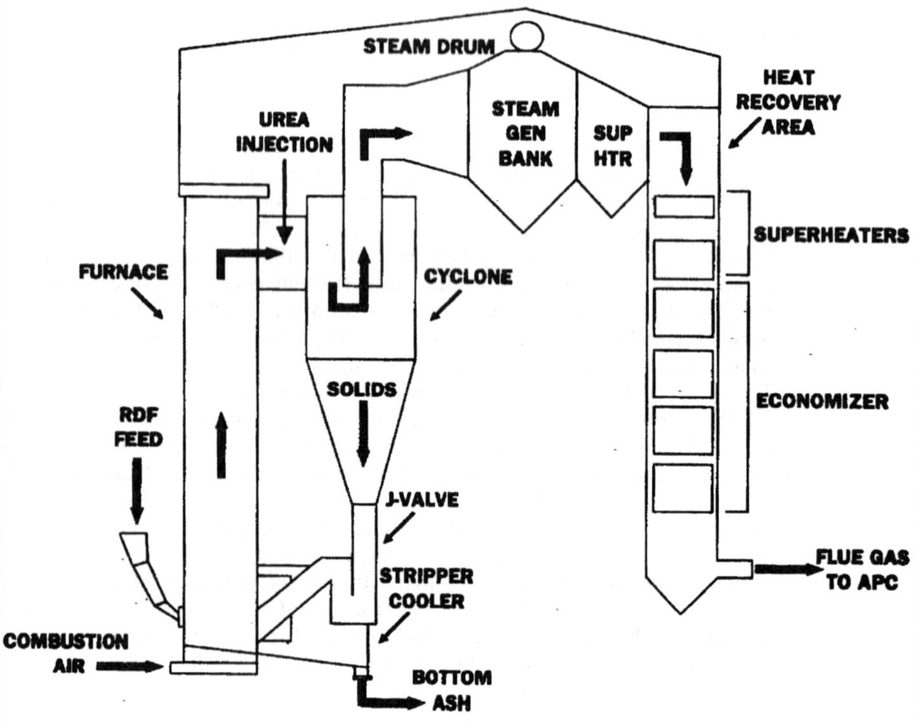

FIGURE 3.11 Fluidized bed combustor with boiler (*Foster Wheeler, Robbins, Illinois*).

Circulating bed combustors operate at higher fluidization velocities to promote carryover of solids into the upper section of the combustor. A cyclone is used to return the entrained bed material to the bed. Combustion occurs in both the bed and the upper section of the combustor. A fraction of the bed material is entrained in the combustion gas and enters a cyclone separator which recycles unburned waste and inert particles to the lower bed.

Good mixing is inherent in the FBC design. Fluidized bed combustors have very uniform gas temperatures in both the bed and the upper region of the combustor. This allows the FBCs to operate at lower excess air and temperature levels than conventional combustion systems. Waste-fired FBCs typically operate at excess air levels between 30 and 100 percent and at bed temperatures around 815°C (1500°F). These low temperatures are necessary for waste-firing FBCs because higher temperatures lead to bed agglomeration. FBCs employ conventional heat recovery boilers including superheaters, convection banks and economizers, and air preheaters to heat the combustion air.

Other waste combustion systems. *Sewage sludge* can be burned in a sludge incinerator specially designed for this purpose. Sludge normally has a moisture content in the range of 70–80 percent (conventionally reported as about 50% solids). Municipal solid waste has a moisture content in the range of 20–30 percent. The only effective method of firing these two waste streams together has been by a reduction of the sludge moisture content to that of the MSW.

Cement kilns are used to burn a variety of wastes, usually a very small fraction of the weight of material processed by the kiln. The wastes are used as a source of heat, supplementing fossil fuels such as coal and oil, while serving as an efficient way to dispose of these wastes. Generally, organic wastes or components are, essentially, completely destroyed provided the quantities are not excessive. The inorganic components, especially the regulated toxic metals, determine the limitations on the quantity of waste which may be fed to these furnaces.

Lightweight aggregate kilns, like cement kilns, are used to burn a variety of wastes, usually a small fraction of the weight of material processed by the kiln. In addition to providing an economical source of heat, they provide an efficient method of disposal of wastes. The difficulty in establishing numerical limits for their emission when burning wastes is that the natural materials being processed into lightweight aggregate also contain naturally occurring organics and inorganic matter, including heavy metals. Generally, organic wastes or components are, essentially, completely destroyed provided the quantities are not excessive. The inorganic components, especially the regulated toxic metals, determine the limitations on the quantity of waste which may be fed to these furnaces.

Boilers and industrial furnaces have provided an effective and convenient means of disposal of wastes, especially liquid wastes which could be injected through conventional oil burners. Generally, organic wastes or components are, essentially, completely destroyed provided the quantities are not excessive. The inorganic components, especially the regulated toxic metals, determine the limitations on the quantity of waste which may be fed to these furnaces.

Hazardous waste combustion systems. Hazardous waste combustors range from those designed only to burn pumpable liquids to those capable of burning solids as well as liquids. Rotary kilns are usually used for burning solids due to their tolerance of a wide range of solid, liquid and viscous materials. They have thermal capacities up to 100 million Btu/h.

Where hazardous regulations must be complied with, the incinerator design and operation are subject to the Resource Conservation and Recovery Act (RCRA) regulations for handling and disposal. Incineration regulations under RCRA require an extensive analytical compliance process. To obtain construction and operating permits, operating conditions are determined by trial burn tests, which restrict operation. A detailed operator training program must be implemented.

Rotary kiln technology. A rotary kiln provides a versatile primary chamber design for combustion of solid wastes, due to the tumbling action which exposes the waste to high temperatures for ignition and continuously exposes the waste to oxygen to continue the combustion process. A secondary chamber is used to complete the combustion of the gases produced in the kiln. Rotary kilns can be used for the simultaneous combustion of a wide variety of solid and sludge wastes as well as for the incineration of liquid and gaseous waste. Solid and liquid wastes and sludges are burned in the kiln, whereas liquid wastes, especially aqueous wastes, are burned in the secondary chamber.

The rotary kiln system used for waste incineration, as shown in Fig. 3.12, includes provisions for feeding, air supply, kiln, secondary chamber (afterburner), ash collection system, waste-heat boiler and emission control system. The induced-draft (ID) or exhaust fan draws the gases through the system and discharges them through the stack to the atmosphere. The emission control system employs alkali injection for acid gas control. Dry scrubbing may be used in lieu of the wet scrubbing system shown.

The rotary kiln has a horizontal cylinder, lined with refractory, which turns about its horizontal axis. Waste is deposited in the kiln at one end, and burns out to an ash by the time it reaches the other end. The rotation of the kiln is achieved by trunnion rollers, driven by a gear drive around the kiln periphery, or through a chain driving a large sprocket around the body of the kiln. Supplemental fuel is normally injected into the kiln through a conventional burner, to bring the kiln up to operating temperature and to maintain its temperature during incineration of the waste feed and during standby and shutdown periods.

Kilns designed to burn hazardous wastes may be operated in either a non-slagging or a slagging mode. Some kiln designs have a zero or slightly negative rake, with lips at the input and discharge ends so that they can be operated in the slagging mode with the internal kiln geometry designed to maintain a pool of molten slag between the kiln lips. The temperature in a slagging kiln must be sufficiently high to maintain the ash as a molten slag. Temperatures as high as 2600–2800°F are not uncommon. A non-slagging kiln will normally operate at a temperature below 2000°F.

Solid waste retention time in a kiln can be varied as a function of kiln geometry and kiln speed, as shown in the following equation:

$$t = \frac{2.28\ L/D}{SN}$$

where t = mean residence time, min
L/D = internal length-to-diameter ratio
S = kiln rake slope, in/ft of length
N = rotational speed, r/min

For a given L/D ratio and rake, the solids residence time within the unit is inversely proportional to the kiln speed. Doubling the speed halves the residence time. An example of this calculation is as follows.

EXAMPLE: Calculate the residence time for a kiln rotating at $N = 0.75$ r/min with a 1 percent slope ($S = 0.12$ in/ft of length), with a 4 ft inside diameter and 12 ft length L.

$$t = \frac{2.28(12/4)}{0.12 \times 0.75} = 76\,\text{min}$$

Doubling of the rotation N would halve the retention time, and halving the slope rake S would double the retention time.

3.3 Operation of Waste Combustors

Temperature and combustion control. Figure 3.13 shows that temperature is directly related to the excess air provided; hence temperature can be used to control fuel supply and/or airflow in both primary and secondary chambers. By providing the primary chamber with less than

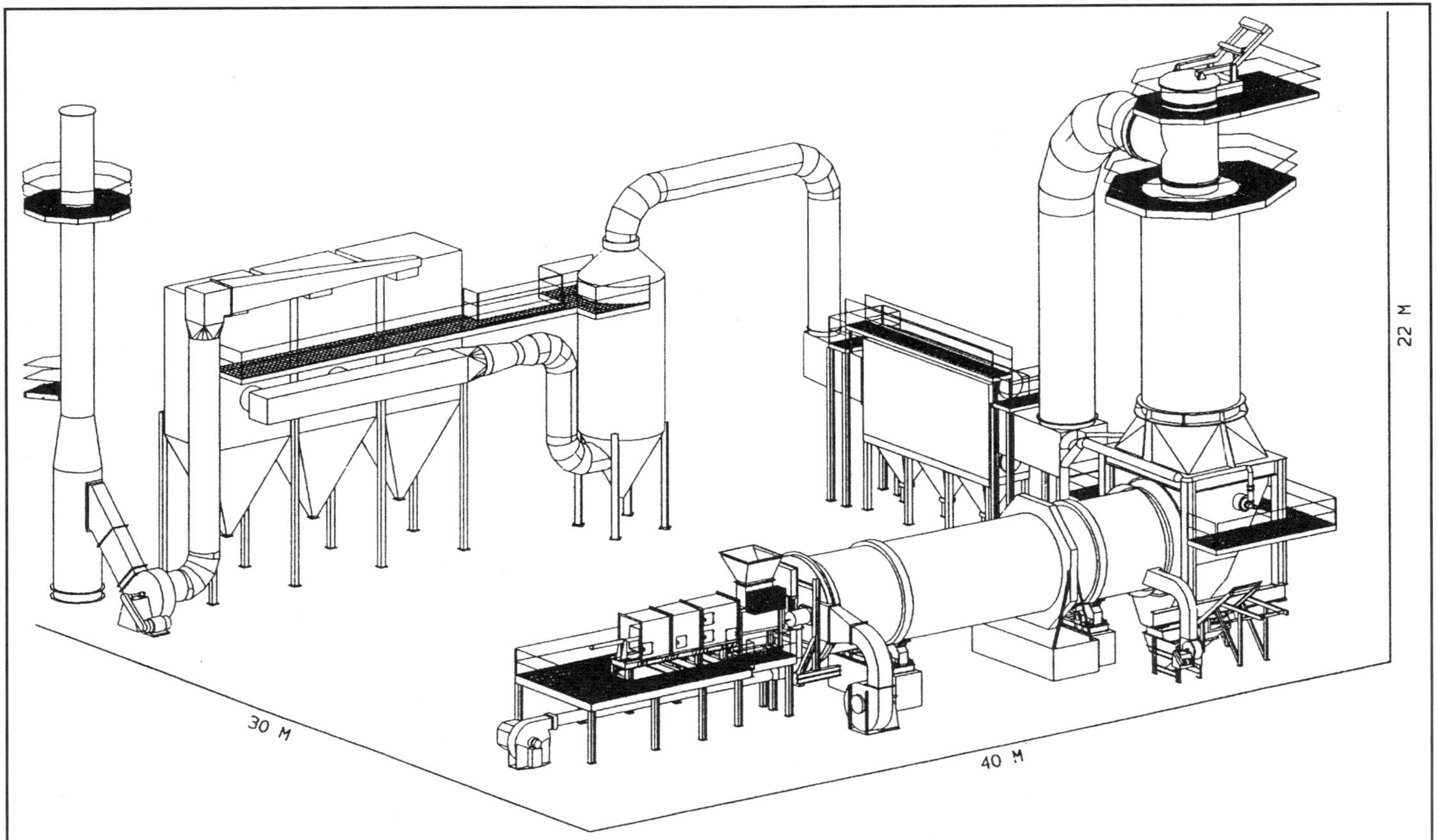

FIGURE 3.12 Complete rotary kiln waste combustion system.

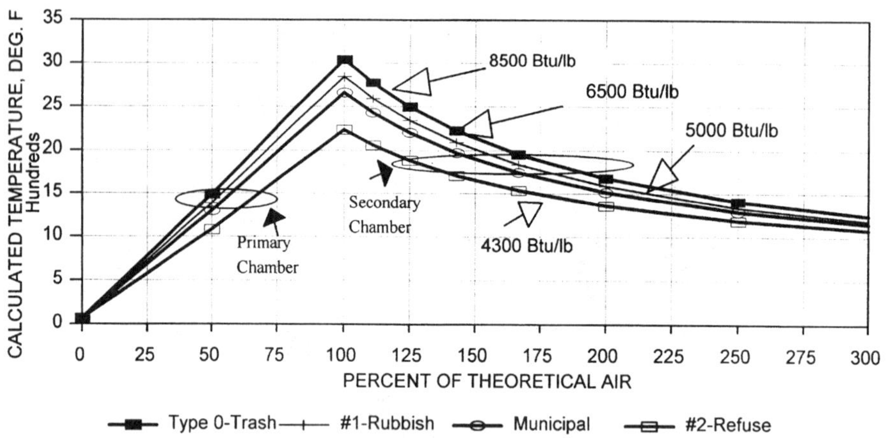

FIGURE 3.13 Relationship between temperature and excess air for various waste types.

stoichiometric air (starved-air conditions), carbon is converted mainly to CO, releasing roughly half of the heat of combustion of CO_2, so that the process must be completed in the secondary chamber. The temperature can be maintained at about 1400°F, about half of the theoretical flame temperature, by this method, protecting the furnace refractory from excessive temperatures and restraining the rate of combustion. Setting the primary air supply to a fixed level serves to stabilize combustion in spite of the highly variable heat release rates of some of the components of the waste.

The primary chamber behaves this way with constant air supply:

- To obtain a higher temperature, decrease the solid feed rate.
- To lower the temperature, increase the solid feed rate.

The secondary chamber completes the combustion, using greater than stoichiometric air supplied to control the temperature, by adjusting the secondary combustion air damper and the auxiliary burner as needed. This control behaves this way:

- More air is added to reduce the temperature.
- Air is reduced to increase the temperature.
- Auxiliary fuel is added if needed to keep the temperature up when the waste feed rate or heating value is insufficient.

Combustion controls. It is the responsibility of the operator to feed waste in quantities and in mixtures which are within the ability of the combustion air controls to adjust, while maintaining the required primary and secondary furnace temperatures. When the waste has a low heating value, more weight can be fed, and when it has a large quantity of plastics, the weight fed must be reduced. The operator can use the furnace temperatures resulting from each charge to gauge when and how much waste of what type should be fed in the next cycle. Many incinerator controls do this automatically, delaying the next load when the present one indicates a high content, i.e., heating value times weight.

Since much of the waste charged into an incinerator is in opaque bags or boxes, so that the waste cannot be identified, the operator and the controls can only respond after the charge is ignited and starts to burn. The timing and content of the next charge can be varied to compensate for low or high heat release. The impact of each load on the performance of the system is especially great when the charge weight is large compared with the incinerator capacity, since the size of the waste containers and hence the feed hopper cannot be scaled down to the small units.

While similar in requirements, the combustion controls of the three basic systems described above differ in important details.

Small incinerators are controlled almost entirely by the furnace temperatures, which modulate the supplemental fuel burners, and the combustion air. Whether or not the furnace is charged, the controls maintain the required temperatures.

Starved-air combustors operate best under steady conditions whereby the combustion air fans and induced-draft fans operate at essentially constant capacity while the feed is varied as needed to maintain the necessary heat input. Steam flow may be used to adjust the air supply.

Municipal waste combustors including waste-to-energy systems operate best under steady conditions, usually maintaining constant steam flow from the boilers in order to keep the electric generator operating at full capacity. Steam flow adjusts combustion air flow, the time cycle of the waste charging ram, and perhaps the grate cycle, whereas furnace temperature and oxygen are somewhat loosely controlled. The operator watches the grate burnout and may make further adjustments to accommodate the type of waste being burned.

Hazardous waste combustors also operate best under steady gas flow conditions so that the emission control system operates at best flow rate. The temperature controls then maintain the primary and secondary chamber temperatures by adjusting the feed of wastes and supplemental fuel. The operators schedule the burning of different types of waste in different locations to optimize the burning of wastes while maintaining the required conditions in the system. Since aqueous wastes do not contribute enough energy to burn, it is necessary to provide sufficient high-energy waste or auxiliary fuel to maintain the required temperatures. Burning of containerized wastes is another special problem since there can be a significant surge in energy release from each container which the controls must be able to accommodate without straying out of the temperature envelope. While this can be a complex process to control, the net result is that the temperatures can be maintained at all times by adjusting the various fuel and waste inputs.

Draft control. It is essential for waste combustion systems that excessive air in-leakage not occur, potentially defeating the efforts of the combustion control system to maintain the desired furnace temperatures. The feeding ports as well as the dry ash collection systems must have some clearances to allow for temperature variations. Water-sealed ash quench systems are used to avoid leakage. To ensure that the leakage is into the system, a negative draft must be maintained at all times. Induced-draft fans having damper and/or speed controls actuated by a sensitive draft control are needed to maintain a slight negative pressure in the furnaces so that leakage is always into, not out of, the furnace system, and so that no hot, dirty gases leak out of the furnaces into the surrounding areas, especially into the feed hopper where ignition can take place.

Ash disposal. With smaller medical waste units, after burnout the chamber is opened and ash residue is manually raked out. With continuous operating units such as those shown in Figures 3.5 and 3.6, ash is continually discharged, normally into a wet well, then transferred to a container or truck by means of a drag conveyor. It is necessary to quench the dry ash with water to eliminate possible burning embers so that it can be disposed of safely.

Heat recovery boilers. Boilers are frequently installed, especially with the larger units, as seen in Fig. 3.7, where there is a use for the heat as steam or hot water and as a means for reducing the combustion gas temperatures to 450°F or less, as is needed before the gas enters the stack or emission control system. If heat recovery is not used, the gases must be discharged at 1800°F through refractory stacks directly to the atmosphere. Dump stacks, as seen in Fig. 3.7, are generally provided for discharging hot gas during emergency conditions where it is necessary to bypass the boiler or air pollution control device.

3.4 Calculation of a Waste-to-Energy Plant

Assume: 500 ton/d (500/24 = 20.83 ton/h= 41,666 lb/h MSW)

Higher heating value = 5000 Btu/lb
The plant produces 700 kW·h per ton of MSW (14.6 MW)
Ash residue is 20% of MSW, dry
Ash leaving plant has 50% moisture content
Combustion products are 10 pounds per pound of fuel (lbp/lbf)
The specific heat of combustion products is 0.28 Btu/(lb · °F)
Furnace temperature is 1800°F
Boiler outlet temperature is 450°F
Boiler efficiency is 75%
Water sprays are used to cool the product gases from 450°F to 300°F
The turbine generator produces 14.6 MW of power (equal to 50 MBtu/h)
The steam leaving the turbine is condensed by cooling tower water
The cooling tower removes 106 MBtu/h by evaporating water
The boiler blowdown is 10% of the steam generated by the boiler
The cooling tower blowdown is 10% of the water evaporated by the tower

Calculate:

1. Btu/h produced by burning the waste, using higher heating value

$$= 500 \times 5000 \times 2000/24 = 208{,}000{,}000 \text{ Btu/h} = 208 \text{ MBtu/h}$$

2. Pounds per hour of combustion products at 10 lb/lb fuel

$$= 41{,}666 \times 10 = 416{,}660 \text{ lb/h}$$

3. Heat delivered to steam based on boiler efficiency of 75%

$$= 208 \times 0.75 = 156 \text{ MBtu/h}$$

4. Pounds per hour of steam produced at 1200 Btu/lb of steam

$$= 156{,}000{,}000/1200 = 130{,}000 \text{ lb/h of steam}$$

5. Heat equivalent of 14.6 MW of power at 3416 Btu/(kW · h)

$$= 14{,}600 \text{ kW} \times 3416 \text{ Btu/(kW} \cdot \text{h)} = 50 \text{ MBtu/h}$$

6. Heat removed to condense steam leaving the turbine

$$= 156 - 50 = 106 \text{ MBtu/h}$$

7. Water needed to cool gases from 450°F to 300°F

$$= 416{,}000\ \text{lb/h} \times (450 - 300) \times 0.28\ \text{Btu/(lb}\cdot{}^\circ\text{F)}/1000\ \text{(Btu/lb)} = 17{,}500\ \text{lb/h}$$

8. Water needed for boiler makeup if blowdown is 10% of steam produced

$$= 130{,}000 \times 0.1 = 13{,}000\ \text{lb/h}$$

9. Water needed to cool ash from 1600 to 200°F at 1000 Btu/lb water

$$= 41{,}600\ \text{lb/h} \times 0.20 \times 0.1 \times (1600 - 200)/1000\ \text{Btu/(lb}\cdot{}^\circ\text{F)} = 1164\ \text{lb/h}$$

10. Water needed to wet ash from dry to 50% moisture content

$$= 41{,}600\ \text{lb/h} \times 0.20 \times (1\ \text{lb water})/(\text{lb dry ash}) = 8320\ \text{lb/h}$$

11. Water evaporated in cooling tower at 1000 Btu/lb water

$$= 106{,}000{,}000\ \text{(Btu/h)}/(1000\ \text{Btu/lb water}) = 106{,}000\ \text{lb/h}$$

12. Water needed to supply 10% blowdown from the cooling tower

$$= 106{,}000 \times 0.10 = 10{,}600\ \text{lb/h}$$

13. Total amount of water needed for all of the above uses

$$= 17{,}500 + 13{,}000 + 1165 + 8320 + 106{,}000 + 10{,}600$$
$$= 156{,}585\ \text{lb/h} = 313\ \text{gal/min}$$

4 CONTROL OF EMISSIONS FROM COMBUSTION

Emission control systems are provided to control the various pollutants contained in the products of combustion, including particulate matter, SO_2, HCl, NO_x, metals and organics. Each pollutant is subject to a different mechanism of control, however, and a given emission control technique has a different impact on individual pollutants.

4.1 Pollutants of Concern

The pollutants of concern which may be found in the combustion products, prior to the air pollution control device (APCD) or in the stack emissions, along with the controls used to reduce their discharge to the atmosphere, are discussed below.

Particulate matter. The quantity and concentration of particulate matter (PM) exiting the furnace of a waste combustion system depends on the waste characteristics and the design and operation of the combustion system. While most of the inorganic, non-combustible fraction of MSW and BMW will be discharged as bottom ash, a substantial fraction will be formed from combustion and released into the flue gas. Generally 99% or more of this particulate is captured by the APCD and is not emitted to the atmosphere.

Particulate matter varies in particle diameter from less than 1 micron (μm) to hundreds of microns. Fine particulates, having diameters less than 10 μm (known as PM10), and particulates smaller than 2.5 μm are of concern because of the greater potential for inhalation and passage of these fine particles into the pulmonary region of the lungs. Also, acid gases, metals and toxic organics preferentially adsorb onto particulates in this size range and can therefore be absorbed within the lungs.

The physical properties of the waste being fed, the method of feeding and the distribution of underfire air influence PM concentrations in the flue gas. The concentration of PM emissions at the inlet of the APCD will depend on the combustor design, air distribution and waste characteristics. The higher the underfire/overfire air ratio or the excess air levels, the greater the entrainment of PM in the flue gases, and the higher the PM levels at the APCD inlet. Combustors with boilers that change the direction of the flue gas flow may remove a significant portion of the PM prior to the APCD. For instance, RDF furnaces typically have higher PM carryover due to the suspension firing of the RDF, and starved-air furnaces have substantially lower carryover than excess-air furnaces due to the low gas velocities in the primary furnace.

Metals. Metals are present throughout the components of municipal and medical wastes. The metals emitted as components of PM (e.g. arsenic [As], cadmium [Cd], chromium [Cr] and lead [Pb]) and as vapors, such as mercury [Hg], are highly variable and are essentially independent of the combustor type. Most of the metals are vaporized during combustion and condense onto particulates in the flue gas as its temperature is reduced; hence the metal can be effectively removed by the PM control device, at efficiencies greater than 99 percent. Mercury, on the other hand, still has a high vapor pressure at typical APCD operating temperatures, and capture by the PM control device is highly variable. A high level of carbon in the fly ash, or the injection of activated carbon, greatly enhances Hg adsorption onto the particles removed by the PM control device.

Acid gases. Combustion of wastes produces HCl and SO_2, as well as hydrogen fluoride (HF), hydrogen bromide (HBr) and sulfur trioxide (SO_3) at much lower concentrations. Concentrations of HCl and SO_2 in flue gases are related directly to the chlorine and sulfur content in the waste, which vary considerably with seasonal and local waste variations. Emissions of SO_2 and HCl depend partly on the chemical form of sulfur and chlorine in the waste, as well as on the availability of alkali materials in combustion-generated fly ash that act as sorbents, and the type of emission control system used. The major sources of chlorine in MSW are paper, food and plastics. The presence of PVC plastics in BMW results in relatively high HCl emissions. Sulfur in MSW may derive from asphalt shingles, gypsum wallboard, and tires.

Carbon monoxide. Carbon monoxide emissions result when not all of the carbon in the waste is oxidized to carbon dioxide (CO_2). High levels of CO indicate that the combustion gases were not held at a sufficiently high temperature in the presence of sufficient oxygen (O_2) for a long enough time to convert CO to CO_2. In the first stages of combustion in a fuel bed, waste first releases CO, hydrogen (H_2), and unburned hydrocarbons, then additional air converts these gases to CO_2 and H_2O. Adding too much air to the combustion zone can lower the local gas temperature and quench (or retard) the oxidation reactions, whereas if too little air is added the probability of incomplete mixing increases, allowing greater quantities of unburned hydrocarbons to escape the furnace, thus increasing CO emissions. Because O_2 levels, air distribution and the effectiveness of mixing vary among combustor types, CO levels also vary substantially. For example, semi-suspension-fired RDF units generally have higher CO levels than mass burn units, due to the effects of carryover of incompletely combusted materials into low temperature portions of the furnace and, in some cases, due to instabilities that result from fuel feed characteristics and distribution over the fuel bed. Likewise, two-chamber starved-air systems usually have very low CO emissions due to the inherently effective mixing which they can achieve.

Carbon monoxide concentration is a direct indicator of combustion efficiency, and is an important indicator of instabilities and non-uniformities in the combustion process. During unstable combustion conditions when more carbonaceous material is available, serving as precursors, higher CDD/CDF and organic hazardous air pollutant levels are likely to occur. The relationship between emissions of CDD/CDF and CO indicates that high levels of CO (several hundred parts per million by volume [ppmv]), corresponding to poor combustion conditions, frequently correlate with high CDD/CDF emissions. When CO levels are low, however, correlations between CO and CDDs/CDFs may not be found because many mechanisms contribute to CDD/CDF formation, though CDD/CDF emissions are generally lower.

Nitrogen oxides. Nitrogen oxides are produced by all combustion processes using air as a source of oxygen. Nitric oxide (NO) is the primary component, but nitrogen dioxide (NO_2) and nitrous oxide (N_2O) are also formed in smaller amounts. The combination of these compounds is referred to as NO_x. Nitrogen oxides are formed during combustion through oxidation at relatively low temperatures (less than 1090°C [2000°F]), and fixation of atmospheric nitrogen occurs at higher temperatures. Because of the relatively low temperatures at which municipal, medical and hazardous waste furnaces operate, 70–80 percent of NO_x is associated with nitrogen in the waste. Acrylic plastics are a major source of nitrogen in MSW and BMW. Hazardous wastes may contain nitrogen in many forms.

Organic compounds. Organic compounds, including chlorinated dibenzodioxins and chlorinated dibenzofurans (CDDs/CDFs), chlorobenzenes (CB), polychlorinated biphenyls (PCBs), chlorophenols (CPs) and polyaromatic hydrocarbons (PAHs), are present in MSW and BMW or can be formed during the combustion and post-combination processes. Organics in the flue gas exist in the vapor phase or may be condensed or absorbed on fine particulates. Organics are controlled by proper design and operation of both the combustor and the APCDs. Activated carbon injection has been found to be effective in adsorbing CDDs/CDFs and probably other trace organic compounds.

Due to their toxicity levels, emphasis is placed on levels of CDDs/CDFs in the tetra- through octa- homolog groups and specific isomers within those groups that have chlorine substituted in the 2, 3, 7 and 9 positions. The NSPS and emission guidelines for MWCs and BMCs regulate the total tetra- through octa-CDDs/CDFs.

4.2 Emission Control Devices

Technologies which are currently available or in use will be discussed below:

- Cyclone separators
- Quench venturi
- Wet venturi scrubber (WVS)
- Packed-bed scrubber (PBS)
- Plate scrubber (PS)
- Electrostatic precipitator (ESP)
- Wet electrostatic precipitator (WESP)
- Dry venturi (DV)
- Dry sorbent injection (DSI)
- Reactor tower (RT)
- Conditioning tower
- Spray-dry reactor (SDR)
- Fabric filter (FF)

Cyclone separators use the centrifugal forces attained by forcing the gases through a tangential entry to a cylinder to cause solid particles to follow the walls and drop out of the gas stream, while the gas stream is diverted away without those collected particles. Cyclones are useful for collecting particles larger than 10 μm in diameter, and are still often used to knock out the large particles in gas streams entering scrubbing towers and ESPs, but since they are not effective on smaller particles we will concentrate on the devices which actually collect the fine PM which is of environmental concern.

Venturi scrubber/packed tower scrubber. Medical waste and hazardous waste combustors are often served by venturi scrubbers followed by packed tower scrubbers, due to their effectiveness in removing acid gases and organic vapors and where solid, inorganic particulate matter is not a major contaminant in the flue gases (Hesketh, 1991).

In single-stage scrubbers, the flue gas reacts with an alkaline scrubber liquid to simultaneously remove HCl and SO_2. In two-stage scrubbers, a low-pH water scrubber for HCl removal is installed upstream of the alkaline SO_2 scrubber. The alkaline solution, typically containing calcium hydroxide ($Ca[OH]_2$), reacts with the acid gas to form salts which are generally insoluble and may be removed by sequential clarifying, thickening, and vacuum filtering. The dewatered salts or sludges must then be disposed of in suitable landfills.

Quench sections are used to rapidly cool combustion products down to the temperatures at which emission control devices must operate in order to be effective. Water sprays are used for this process, evaporating the water and cooling the gases down to close to the wet bulb temperature. Gases entering at a secondary chamber exit temperature of 1800°F are cooled to about 180°F, whereas those which have been cooled by a heat recovery boiler or heat exchanger will be cooled to about 130°F. Scrubbers which follow a quench section cool the gases closer to the wet bulb temperature by providing more time and surface contact.

Wet venturi scrubbers use a converging duct section followed by a diverging section to accelerate and then decelerate the gas stream, while at the same time spraying water into the converging section of the scrubber. As the gases pass through the diverging section, most of the pressure drop lost in the converging section is recovered, resulting in a permanent pressure loss which must be overcome by the fan which moves the gases through the system.

The water droplets, moving at a slower velocity than the gases, take significant time to pass through the venturi, while becoming targets for the particulate matter carried by the gases. The droplets thus absorb fine particulate, at the same time absorbing some of the acid gases such as HCl and HF. As the gases pass through the diverging section, slowing up, the accelerated droplets of water continue to capture dust particles, while agglomerating into larger water droplets which can be dropped out and collected beyond the venturi, typically by cyclonic action.

The effectiveness of wet venturi scrubbers in collecting particulate matter depends mainly on the difference between the entering and the throat gas velocity, which in turn depends upon the relative cross-sectional areas of the duct and the throat. The pressure drop measured between entrance and throat is directly dependent upon these areas and the corresponding velocities achieved.

Complete hazardous waste incineration system. A complete hazardous waste incineration system is illustrated in Fig. 3.14. A solids shredder shreds drums and other large objects which are conveyed to the inlet of the rotary kiln. Liquid wastes, atomized by steam, are also fed to the kiln, where auxiliary fuel burners provide heat-up and temperature maintaining functions. The ash residues leaving the kiln at the end of the kiln are conveyed out to ash storage where they are analyzed to determine suitable disposal procedures. The hot gases at 1000°C go to a secondary chamber, or afterburner, where burners maintain the temperatures at 1200°C for 4 s. The off-gas system consists of a quench tower, used to cool the gases before they enter the venturi scrubber, a packed tower scrubber, and an ionizing wet scrubber which removes the mist. The induced draft fan discharges the cleaned gases to the stack. Water enters

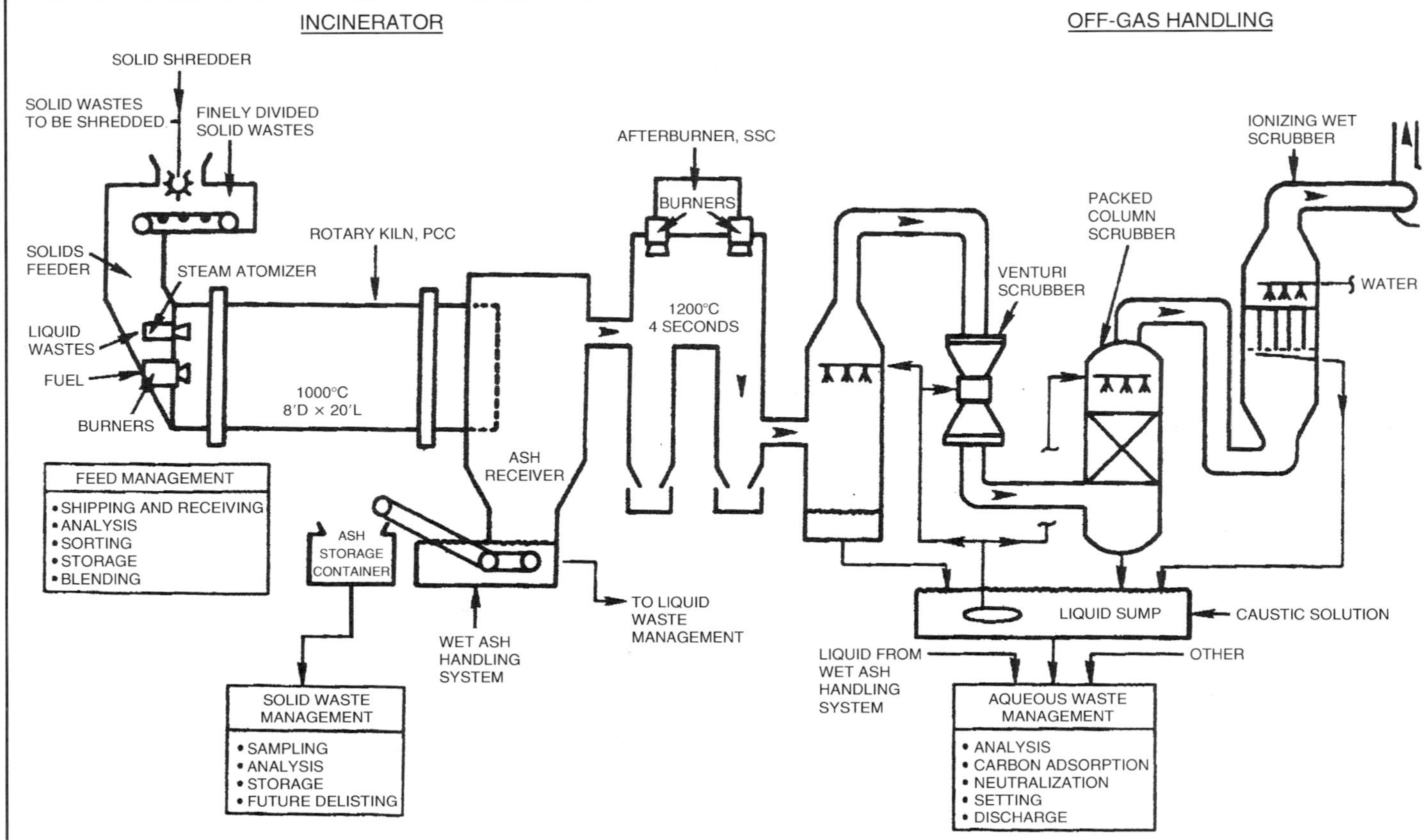

FIGURE 3.14 Complete hazardous waste combustion system with wet scrubber.

the wet scrubber and is drained to the liquid sump, where caustic soda is used to neutralize the acids absorbed by the system. The sump water is recirculated for use in the quench, venturi and column sprays. The liquids from the wet ash system and other wastes are then analyzed, treated, and discharged properly.

Particle size analysis. The efficiency of cyclones and venturi scrubbers varies with the size of the particles to be collected. To obtain the overall collection efficiency and the actual quantity and size distribution of the emissions, it is necessary to calculate the efficiency for each size group and obtain the quantities of each size in the emissions.

Sample calculation: Pressure drop determination for typical venturi scrubber application

Step 1. Plot the data given in Table 3.10 on a log–log graph to obtain the fractions in each size range for the particle diameter ranges listed.

Step 2. Determine the required collection efficiency:

(a) Inlet loading:

1.83 gr/dscf × 183,000 dscfm × (1/7000 grains) × 60 (min/h) = 2870 lb/h

(b) Collection efficiency:

$$[2870\ (\text{lb/h}) - 48.79\ (\text{lb/h})] \times 100/2870 = 98.3\%$$

(c) Allowable penetration:

$$100.0 - 98.3 = 1.7\% = 0.017\ \text{penetration}$$

Step 3. Plot the penetration from Table 3.11 as a function of pressure drop, and read 36.5″ W.G. for 1.7% penetration.

Step 4. To meet the regulation of 0.03 gr/dscf,

$$0.03\ \text{gr/dscf} \times 183{,}000\ \text{dscfm} \times 60/7000 = 47.057\ \text{lb/h}$$

Select $\Delta p = 40''$ W.G.

Collection efficiency. A venturi scrubber can remove 92 to 99% of particulate in the size range from 0.02 to 5 μm, which makes it effective for smoke and fumes. Since the efficiency is dependent on particle size, it is necessary to obtain the particle size distribution of the PM and perform a fractional efficiency analysis in order to determine the overall efficiency of the venturi for the specific size distribution. Penetration, the fraction passing through, or 1 minus the efficiency/100, is plotted in Fig. 3.15, which plots fractional overall penetration (one minus fractional overall efficiency) against the ratio of the aerodynamic cut diameter d_{50}

TABLE 3.10 Data Supplied for Pressure Drop Determination

Inlet dust loading to scrubber	1.83 gr/dscf
Inlet gas flow to scrubber	183,000 dscfm
Allowable emission	48.79 lb/h = 0.03 gr/dscfm
Particle diameter (μm)	Weight less than (%)
10.0	99.0
5.2	89.5
2.0	43.5
1.0	9.7
0.5	1.0

TABLE 3.11 Determination of Penetration and PM at Outlet of Venturi Scrubber

Particle diameter (μm)	Fraction in size range		Fractional efficiency = Pt (Dp)		
			$\Delta p = 10$ in.	$\Delta p = 20$ in.	$\Delta p = 40$ in.
6	0.063	×	0 = 0	0 = 0	0 = 0
5–6	0.042	×	0.01 = 0.00042	0 = 0	0 = 0
4–5	0.077	×	0.03 = 0.00231	0 = 0	0 = 0
3–4	0.138	×	0.08 = 0.1104	0.02 = 0.00276	0 = 0
2–3	0.245	×	0.20 = 0.049	0.06 = 0.0174	0 = 0
1.5–2	0.165	×	0.35 = 0.05775	0.10 = 0.0165	0.005 = 0.0008
1.0–1.5	0.173	×	0.45 = 0.07785	0.16 = 0.02768	0.025 = 0.0043
0.5–1.0	0.087	×	0.60 = 0.0522	0.27 = 0.02349	0.05 = 0.00435
0–0.5	0.0100–0.5	×	0.87 = 0.0087	0.50 = 0.005	0.20 = 0.002
Sum Pt × (Dp) × dPp = total penetration =			0.25927	0.09013	0.0115
1 – total penetration = collection =			0.91	0.91	0.989

to aerodynamic mass mean diameter d_g. The parameter on this plot is the particle standard geometric deviation σ_g.

Pressure drop through venturi scrubber. The pressure drop of a venturi scrubber can be calculated as (Hesketh, 1991)

$$P = v^2 d_g A^{0.122} L^{0.78} / 3870$$

where
P = venturi scrubber pressure drop, cm H_2O
v = throat velocity of the gas and particles, cm/s
A = throat cross-sectional area, cm^2
L = liquid/gas ratio, l/m^3
d_g = gas density, g/cm^3

or

$$P = v^2 d_g A^{0.122} L^{0.78} / 1270$$

where
P = venturi scrubber pressure drop, in. H_2O
v = throat velocity of the gas and particles, ft/s
A = throat cross-sectional area, ft^2
L = liquid/gas ratio, gallons per 1000 actual cubic feet gas
d_g = gas density, lb/ft^3

TABLE 3.12 Wet Scrubber Design Parameters

	Venturi stage	Absorber stage
Gas velocity, ft/s	90–150	6–10
Pressure drop, in. w.c.	40–70	4–8
L/G, gal/(1000 acfm)	10–20	20–40
Scrubbing medium	Water	Caustic
Solution pH	< 1–2	6.5–9
Materials of construction	High-alloy steel	FRP Lined carbon steel

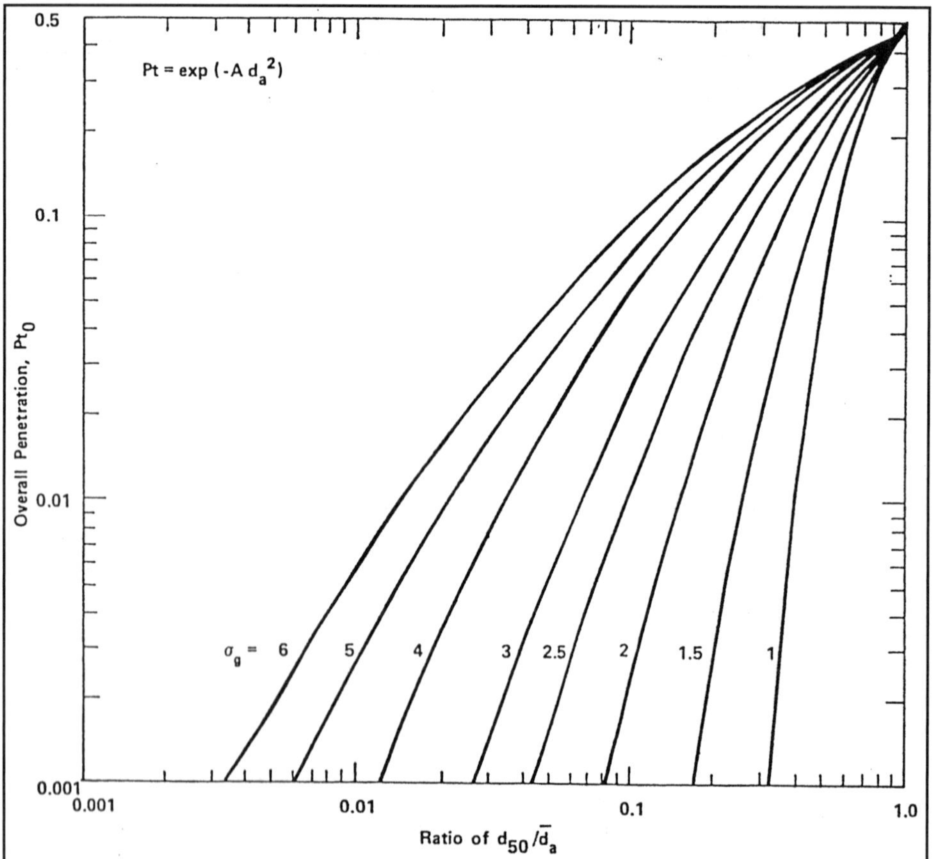

FIGURE 3.15 Overall penetration versus ratio of aerodynamic cut diameter to aerodynamic mass mean diameter.

Packed-bed scrubbers employ a vertical bed of surface-generating shapes through which the gases to be scrubbed pass vertically, while water or recirculated solution enters from the top and contacts the gases. These scrubbers are highly effective in absorbing HCl and SO_2 from the gas stream. Their effectiveness depends upon the temperatures at which the scrubber is operated (usually close to the saturation or wet bulb temperature of the solution), the gas velocity, liquid to gas ratio, and packing height.

Calculations of performance of packed-bed scrubbers are based on tests which determine the performance factors. These are best obtained from the manufacturer who performed the tests and who will guarantee the performance.

Given a specific scrubber, with a set of performance conditions, it is possible to extrapolate its performance under different conditions of operation. For instance, when the number of transfer units (NTU) built into the scrubber is known versus removal efficiency, the efficiency under different conditions can be calculated, or the change in efficiency which would result from an increase in the NTU.

A scrubber used to remove SO_2 from a given concentration of 300 mg/cm to 30 mg/cm has an NTU as follows:

$$\mathrm{NTU} = \ln\,(Y_1(\mathrm{inlet})/Y_2(\mathrm{outlet})) = \ln(300/30) = 2.3$$

The NTU required to reduce the SO_2 from 3000 to 30 mg/cm would be 4.6.

Plate scrubbers operate in a similar fashion to packed-bed scrubbers, and are likewise tested and guaranteed by the manufacturer to perform specific functions.

Wet scrubbers (Fig. 3.16) perform as a function of the *difference* in concentration between the entering gas and leaving absorbing liquid (Y_1) and the leaving gas and the entering absorbing liquid (Y_2). This means that the concentration in the liquid is an important factor: the liquid absorbs the acid, accumulating it and increasing its concentration. To control concentration, a portion of the liquid must be "blown down" and discharged from the system. Achieving a high removal efficiency is obtained at the cost of blowing down and probably

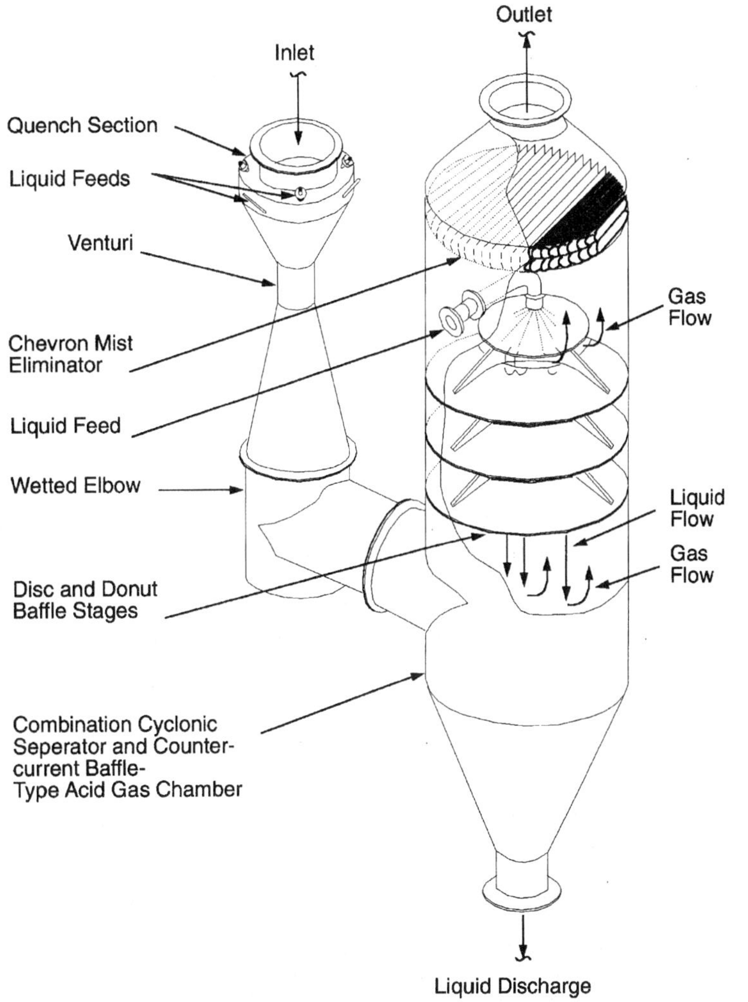

FIGURE 3.16 Wet scrubber and tray tower wet scrubbing arrangement.

having to treat the blowdown to make it suitable for dischargee to the sewer or publicly operated treatment works (POTW). The ideal method would be to remove the products in a dry benign form, rather than diluted with water, which increases the cost or limitations of disposal.

EXAMPLE: Calculate the blowdown rate required to maintain stable concentrations.

Makeup $= x =$ vapor + blowdown

Blowdown $= y$

Makeup concentration = 250 ppm

Blowdown concentration = 1000 ppm

If vapor = 1 gallons per minute (gpm), $x = 1 + y$

Concentration balance

$$x(250) = 1(0) + y(1000)$$

hence

$$x = 4y, \text{ or } y = x/4$$

Then

$$\text{makeup} = x = 1 + y = 1 + 0.25x$$

$$x - 0.25x = 1 = (0.75)x = 1$$

Then:

$$x = 1/0.75 = 1.33 \text{ gpm}$$

$$y = 1.33/4 = 0.33 \text{ gpm}$$

Electrostatic precipitators. Electrostatic precipitators are used to remove dry dust particles from gas streams by employing a series of high-voltage discharge electrodes to charge particulate matter, and attracting the charged particles to grounded metal plates parallel to the direction of gas flow where they are collected and periodically cleaned by rapping the plates, causing the particles to fall off. Although most of the dust is removed in the first stage, upon rapping, some of the collected PM becomes re-entrained in the flue gas, so it is necessary to have two or more sets of charging wires and plates in series so that the next set can pick up the re-entrained particles. In order to meet present standards for PM emissions, ESPs are provided with four or more sets in series. The voltage which may be applied to each bank of the ESP is limited, since sparking, which occurs at some point, varies with the amount and nature of the particulate on the plates. The time intervals between rapping cycles and the voltages which can be maintained for optimum performance are variables which greatly affect the overall collection efficiency of the ESP.

Small particles have lower migration velocities than large particles and are therefore more difficult to collect. As compared to pulverized coal fired combustors, in which only 1–3 percent of the fly ash is generally smaller than 1 μm, 20–70 percent of the fly ash at the inlet of the PM control device for MWCs is reported to be smaller than 1 μm, requiring greater collection areas and lower flue gas velocities than are required for many other combustion types.

The specific collection area (SCA) of an ESP is used as an indicator of collection efficiency. The greater the collection plate area, the greater the ESP's PM collection efficiency. The SCA is expressed as square meters per cubic meter per minute (square feet per cubic feet per minute)

TABLE 3.13 Electrostatic Precipitator Design Parameters

Particulate load, gr/acf		0.5–9
Required efficiency, %		98–99.9
Number of fields		3–4
SCA, ft^2/1000 acfm		400–550
Average secondary voltage, kV		35–55
Average secondary current, mA/1000 ft^3		30–50
Gas velocity, ft/s		3.0–3.5
	Particulate	Acid-gas control
Flue gas temperature, °F	350–450	230–300
Flue gas moisture, % vol.	8–16	12–20
Ash resistivity, ohm · cm	10^9–10^{12}	10^8–10^9

of flue gas, in effect giving the gas velocity. Most recent ESPs have SCAs in the range of 400 to 600 ft^2/(1000 acfm).

ESP efficiency. The efficiency of an ESP can be estimated by the Deutsch–Anderson equation

$$\eta = 1 - \exp(-Aw/Q)$$

where
η = fractional collection efficiency
A = area of the collection plates, m^2
w = drift velocity of the charged particles, m/s
Q = flow rate of the gas stream, m^3/s

The drift velocity is the velocity at which a particle approaches the collection surface. It can be estimated from the equation

$$w = ad_p$$

where
w = drift velocity, m/s
a = constant, a function of the charging field, the carrier gas properties, and the ability of the particle to accept an electrical charge, s^{-1}
d_p = particle size, µm

Drift velocities commonly range from 0.02 to 0.2 m/s and can be determined experimentally.

EXAMPLE: An electrostatic precipitator is to remove flyash particles of 0.5 µm diameter from a gas stream flowing at 0.7 m^3/s. Determine the plate area needed to collect these particles at 98% efficiency. The value of the constant a has been found to be 0.24×10^6/s.

The drift velocity will be

$$w = 0.24 \times 10^6/\text{s} \times 0.5 \times 10^{-6}\text{m} = 0.12 \text{ m/s}$$

The removal efficiency is then

$$0.98 = 1 - \exp -\left[\left(0.12 \text{ (m/s)}/0.7 \text{ (m}^3\text{/s)}\right) \times A\right]$$

Then

$$A = 23\ \text{m}^2$$

Considering that both sides of the plate are available for collecting particles, the plate area required is $23/2 = 11.5\ \text{m}^2$. A square plate measuring 3.4 m × 3.4 m would suffice.

Wet electrostatic precipitator. The fine mist which escapes a wet scrubber contains the particulate which has been collected by the scrubber; hence, if allowed to escape to the stack, the same particulate matter which was to be collected will be emitted. In order to collect these droplets of mist, a wet electrostatic precipitator can be used. As seen in Fig. 3.17, the fine droplets of mist are absorbed by a fine spray of water, then the droplets are charged by electrodes kept under high voltage so that they are collected by grounded plates. These units are widely used in hazardous waste incinerators.

Fabric filters. Fabric filters (FF), also known as "baghouses," are widely used for PM, metals and acid gas control. They remove particulate matter by passing the flue gas through a large number of porous cylindrical fabric bags hanging vertically from a tube sheet. Particulate matter is collected on one of the bag surfaces, from which it is periodically removed and

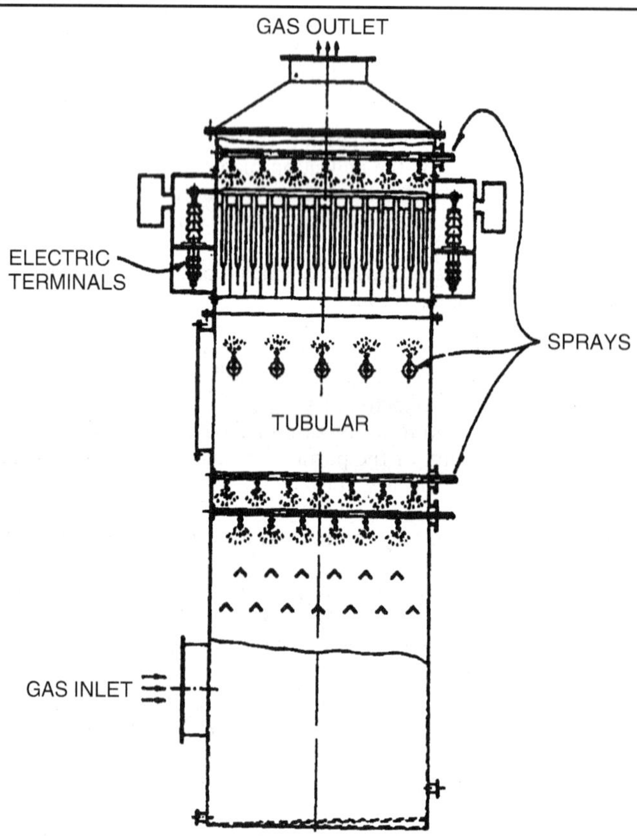

FIGURE 3.17 Wet electrostatic precipitator (*Beltran Associates, Inc., Brooklyn, NY*).

collected in hoppers. The cleaning process may use either reverse air, with the bags "off line," or pulse cleaning, with the bags either on or off line.

A fabric filter consists of 4 to 16 individual compartments that can be independently operated and taken off-line for maintenance and/or cleaning. The collected particulate builds up on the bag, forming a filter cake. As the thickness of the filter cake increases, the pressure drop across the bag also increases. Once the pressure drop across the bags in a given compartment reaches a set limit, that compartment is subjected to cleaning, either on-line (pulse cleaning) or off-line.

Figure 3.18 shows a fabric filter (baghouse). Dirty gases enter the enclosure containing the suspended bags and flow through the bags to exit as clean gases, leaving the dust on the outside of the bags. The dust is dropped from the bags after being loosened by pulses of air applied internally to the bags.

In *reverse-air* FFs, flue gas flows through the filter bags, leaving the particulate on the bag. Once the preset pressure drop across the filter cake is reached, air is blown through the filter in the opposite direction, the filter bag collapses, and the filter cake falls off and is collected in a hopper below.

In a *pulse-jet* FF, compressed air is used, pulsed through the inside of the filter bag, to remove the particulate filter cake; the filter bag expands and collapses to its pre-pulsed shape, and the filter cake falls off and is collected in the hopper.

The bags are usually 8 in. in diameter, with lengths ranging from 10 ft to 25 ft. Baghouses are generally provided with a weather-protection house above the tube sheet so that the bags can be removed and replaced.

Filtering area. The total filtering area required for a given application is

$$A_f = Q/v_f$$

where A_f = total filtering area, m^2
Q = volume of gas stream, m^3/min
v_f = filtering velocity (air-to-cloth ratio), m/min

For cylindrical bags,

A_b = filtering area for each bag, m^2
π = 3.1416
d = diameter of bag, m
h = length of bag, m

TABLE 3.14 Fabric Filter Design Parameters

	Type of fabric filter	
	Reverse air	Pulse jet
Operating temperature, °F	230–450	
Type of fabric	Woven fiberglass	
Fabric coating	10% Teflon B or acid resistant	
Fabric weight, oz/yd^2	9.5	16 or 22
Bag diameter, in.	8	6
Net air-to-cloth ratio	1.5–2.0 : 1	3.5–4.0 : 1
Minimum number of compartments	6	4
Overall pressure drop, in. w.g.	4–6	8–10
Estimated bag life, yr	3–4	1.5–2

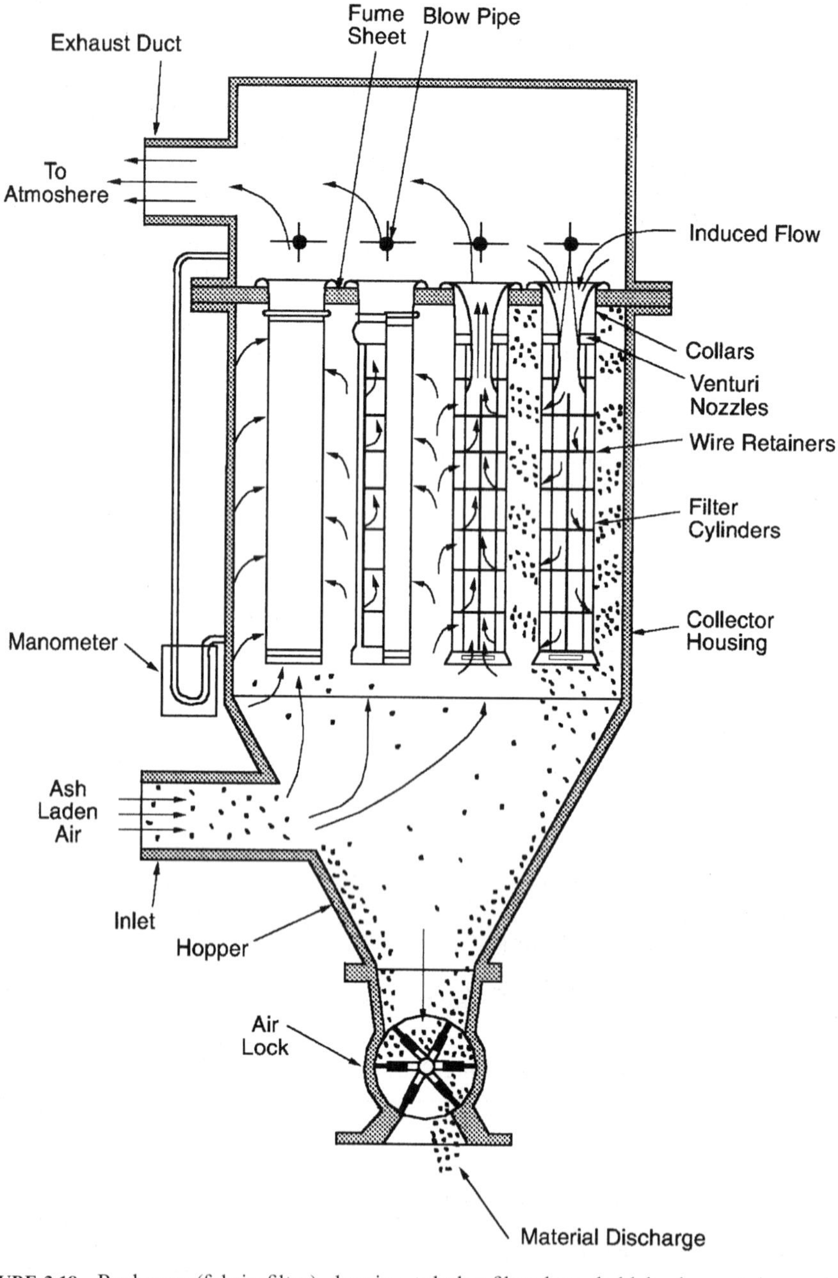

FIGURE 3.18 Baghouse (fabric filter) showing tubular filter bags held by bag retainers, dust hopper, and rotary valve through which the dust is discharged.

The number of bags required is

$$N = A_f / A_b$$

Pressure drop. The pressure drop for a baghouse filter can be estimated from the equation

$$\Delta P = S_e v_f + k_2 c v_f^2 t$$

where ΔP = pressure drop
S_e = effective residual drag
v_f = gas velocity at filter surface
k_2 = specific resistance coefficient of collected particles
c = concentration of particulate in gas stream
t = filtration time, min

Values for S_e and k_2 can be found in the McGraw-Hill *Chemical Engineer's Handbook* (Perry, 1984). Below is an equation which includes values for S_e and k_2 which may be suitable. Obviously they will be dependent on the nature of the filter material.

$$\Delta P = 1.6 v_f + 8.52 c v_f^2 t$$

where ΔP = pressure drop, cm
v_f = filtering velocity, m/min
c = particle concentration in the air stream, kg/m^3
t = time required between cleanings, min

EXAMPLE: Determine the filter cloth area and the required frequency of cleaning on the basis of the following data:

$$\text{Volume of contaminated air stream } Q = 400 \text{ m}^3/\text{min}$$

$$\text{Filtering velocity } v_f = 0.8 \text{ m/min}$$

$$\text{Particulate concentration in stream } c = 0.03 \text{ kg/m}^3$$

$$\text{Diameter of filter bag } d = 0.2 \text{ m}$$

$$\text{Length of filter bag } h = 5.0 \text{ m}$$

Cleaning occurs when ΔP reaches 9 cm water.

$$\text{Total filter area } A_f = 400 \text{ (m}^3\text{/min)}/0.8 \text{ (m/min)} = 500 \text{ m}^2$$

$$\text{Bag filtering area } A_b = 3.14\ (0.2)\ (5.0) = 3.14 \text{ m}^2 \text{ per bag}$$

$$\text{Number of bags } N = 500 \text{ m}^2/3.14 \text{ m}^2 = 160 \text{ bags}$$

At a ΔP of 9 cm, the frequency of cleaning is determined as follows:

$$9 \text{ cm} = 1.6\ (0.8 \text{ m/min}) + 8.52\ (0.03 \text{ kg/m}^3)\ (0.80 \text{ m/min})^2 t$$

$$t = 47 \text{ min}$$

Spray dryers. Spray dryers (SD) are frequently used as the acid gas control technology for waste combustors. When used in combination with an ESP or FF, the system can control CDD/CDF, PM (and metals), SO_2 and HCl emissions. Spray dryer/fabric filter systems have become favored over SD/ESP systems due to their more efficient metals removal. In the spray drying process, lime slurry is injected into the SD through either a rotary atomizer or dual-fluid nozzles, using steam or air for atomization. The water in the slurry evaporates to cool the flue gas, but, before evaporating, the droplets absorb the acid gases where the acids react with the lime to form calcium salts that can be removed by the PM control device. The SD is designed to provide sufficient contact and residence time to produce a dry product before it leaves the SD adsorber vessel. The residence time in the adsorber vessel is typically 10–15 s, resulting in very large vessel diameters. The particulate leaving the SD contains fly ash plus calcium salts, water, and unreacted hydrated lime.

The SD outlet temperature and lime-to-acid gas stoichiometric ratio are the key design and operating parameters that significantly affect SD performance. The outlet temperature must be high enough to ensure that the slurry and reaction products are adequately dried prior to collection in the PM control device. For MWC flue gas containing significant chlorine, the SD outlet temperature must be higher than about 115°C (240°F) to control agglomeration of PM and sorbent by calcium chloride. To provide a necessary safety margin, the outlet gas temperature from the SD is usually kept around 140°C (285°F). This temperature is controlled by the quantity of water sprayed into the gases. The acid gas concentrations are controlled by the quantity of lime added to the slurry.

A measure of performance of spray-dry scrubbers is the stoichiometric ratio, the molar ratio of calcium in the lime slurry fed to the SD divided by the theoretical amount of calcium required to completely react with the inlet HCl and SO_2. At a ratio of 1.0, the moles of calcium are equal to the moles of incoming HCl and SO_2. More than the theoretical amount of lime is generally fed to the SD because of mass transfer limitations, incomplete mixing, and differing rates of reaction (SO_2 reacts more slowly than HCl). The stoichiometric ratio used in SD systems varies depending on the level of acid gas reduction required, the temperature of the flue gas at the SD exit and the type of PM control device used. Lime is fed in quantities sufficient to react with the peak acid gas concentrations expected without severely decreasing performance. The lime content in the slurry is generally about 10 percent by weight, but cannot exceed approximately 30 percent by weight without clogging of the lime slurry feed system and spray nozzles.

Dry sorbent injection. Dry sorbent injection is effective in the control of acid gas as well as CDD/CDF and PM emissions from MWCs.

Duct sorbent injection (DSI) involves injecting dry alkali sorbents into flue gas downstream of the combustor boiler outlet and upstream of the PM control device.

In DSI, powdered sorbent is pneumatically injected into either a separate reaction vessel or a section of flue gas duct located downstream of the boiler's economizer, or quench tower if no boiler is present. Alkali in the sorbent (generally calcium hydroxide, sodium hydroxide, or sodium bicarbonate) reacts with HCl, HF and SO_2 to form alkali salts (e.g., calcium chloride [$CaCl_2$], calcium fluoride [CaF_2] and calcium sulfite [$CaSO_3$]). By lowering the acid content of the flue gas and not evaporating water into the gas stream, downstream equipment can be operated at reduced temperatures while minimizing the potential for acid corrosion of equipment. Solid reaction products, fly ash and unreacted sorbent are collected with either an ESP or an FF.

Acid gas removal efficiency with DSI depends on the method of sorbent injection, flue gas temperature, sorbent type and feed rate, and the extent of sorbent mixing with the flue gas. Flue gas temperature at the point of sorbent injection can range from about 150 to 320°C (300–600°F) depending on the sorbent being used, the means of cooling the gases, and other aspects of the process. Sorbents that have been successfully used include hydrated lime ($Ca[OH]_2$), soda ash (Na_2CO_3) and sodium bicarbonate ($NaHCO_3$). DSI systems can achieve removal efficiencies comparable to those of SD systems. Recirculation towers which increase residence time may be required to achieve comparable acid gas and sorbent efficiencies. Flue gas cooling by dry heat exchange or water injection, combined with DSI, makes it possible to

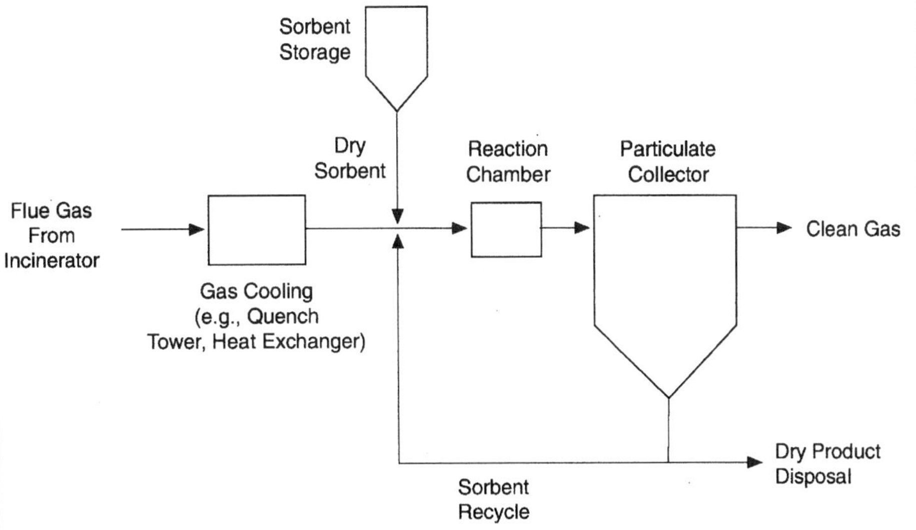

FIGURE 3.19 Dry scrubbing sorbent injection process schematic.

increase CDD/CDF and acid gas removal through a combination of vapor condensation and adsorption onto the sorbent surface.

Furnace injection has been employed to achieve a degree of acid gas control. The basic chemistry of FSI is similar to DSI. Both use a reaction of sorbent with acid gases to form alkali salts. By injecting sorbent directly into the furnace at temperatures of 870–1200°C (1600–2200°F), limestone can be calcined in the combustor to form the more reactive lime, thereby allowing use of less expensive limestone as a sorbent. At these temperatures SO_2 and lime react in the combustor, thus providing a mechanism for effective removal of SO_2. By injecting sorbent into the furnace rather than into a downstream duct, additional time is available for mixing and reaction between the sorbent and acid gases. Removing a significant portion of the HCl before the glue gas exits the combustor can reduce the formation of CDD/CDF in later sections of the flue gas ducting.

Acid gas control with alkaline reagents. Various lime and related alkaline products are used in spray dry and dry injection scrubbers to react with the acid gases, HCl, SO_2 and HF, to convert them into salts which can be collected by the particulate filter.

Calcium oxide (CaO) is called "pebble lime" or "quick lime." Hydrated lime [$Ca(OH)_2$] is made from CaO by adding 32% by weight of water in a hydrator. $Ca(OH)_2$ is a power having a mean particle size of 5 μm, and is highly reactive.

CaO is not very reactive with acid gases at the temperatures and conditions that exist in waste combustion facilities, and has to be converted to the hydrate form to be reactive in scrubbing systems. CaO has been demonstrated to remove acid gases in high temperature applications such as furnace injection.

CaO converts to $Ca(OH)_2$ in the slaking process, in which four parts of water are added to one part of CaO to form $Ca(OH)_2$ in a slurry that is about 25% solids. This conversion requires two phases and takes place in a slaker. The first phase is to convert the hydrate by mixing 3.96 lb of free water with one part of hydrate (1.32 lb), resulting in a 25% slurry (5.28 lb).

The reactions are as follows:

$$\begin{array}{ccccccccc} CaO & + & H_2O & \Rightarrow & Ca(OH)_2 & + & heat \\ 56 & & 18 & & 74 & & \\ Ca(OH)_2 & + & SO_2 & \Rightarrow & CaSO_3 & + & H_2O \\ 74 & & 64 & & 120 & & 18 \\ Ca(OH)_2 & + & 2HCl & \Rightarrow & CaCl_2 & + & 2H_2O \\ 74 & & 73 & & 111 & & 36 \end{array}$$

The capture ratio is $74/73 = 1.104$.

1 lb of CaO yields 1.32 lb of $Ca(OH)_2$

1.156 lb $Ca(OH)_2$ captures 1.0 lb SO_2

1.04 lb $Ca(OH)_2$ captures 1.0 lb HCl

EXAMPLE: The following emission factors are listed for a typical MWC facility:

$$SO_2 = 5.03 \text{ lb/ton } (212 \text{ ppmv @ } 7\% \ O_2)$$
$$HCl = 7.03 \text{ lb/ton } (532 \text{ ppmv @ } 7\% \ O_2)$$

The characteristic stoichiometric reaction of $Ca(OH)_2$ is

$$5.03 \text{ lb } SO_2/\text{ton MSW} \times 1.156 \text{ lb } Ca(OH)_2/\text{lb } SO_2 = 5.815 \text{ lb } Ca(OH)_2/\text{ton MSW}$$
$$7.03 \text{ lb HCl/ton MSW} \times 1.104 \text{ lb } Ca(OH)_2/\text{lb HCl} = 7.761 \text{ lb } Ca(OH)_2/\text{ton MSW}$$
$$\text{Total} = 13.575 \text{ lb/ton}$$

Pebble lime:

$$\begin{array}{ccccc} CaO & + & H_2O & \Rightarrow & Ca(OH)_2 \\ 56 & & 18 & & 74 \end{array}$$

The characteristic stoichiometric reaction of CaO is

$$5.03 \text{ lb } SO_2/\text{ton MSW} \times 0.875 \text{ lb CaO/lb } SO_2 = 4.401 \text{ lb CaO/ton MSW}$$
$$7.03 \text{ lb HCl/ton MSW} \times 0.767 \text{ lb CaO/lb HCl} = 5.392 \text{ lb CaO/ton MSW}$$
$$\text{Total} = 9.793 \text{ lb/ton}$$

Since CaO contains about 7% unreactive material and inerts that are lost in the slaking process, the usage is adjusted to compensate: the adjustment is

$$1.07 \times 9.793 = 10.5$$

Due to the inability to provide absolute contact between the lime and the acid gases, more lime is required in the process. In addition, several unwanted chemical reactions take place that also use some undefined portion of the lime. For example, lime will react with carbon dioxide in the flue gas as follows:

$$Ca(OH)_2 + CO_2 = CaCO_3 + H_2O$$

The ratio of the actual amount of lime used to the theoretical amount required is called the stoichiometric ratio. A typical MWC equipped with a spray dryer and an ESP will require about 35 lb pebble lime per ton of MSW, while an MWC with a spray dry baghouse will require about 20 lb per ton, to meet the NSPS standards of 25 ppm VI of HCl (a 93.5% reduction) and 30 ppm of SO_2 (an 85.8% reduction). The stoichiometric ratio for a plant with an ESP would be

$$35 + 10.5 = 45.5 \text{ lb per ton MSW}$$

Tests of dry sodium bicarbonate injection after a municipal waste combustor. The refuse combustion unit tested consisted of two identical lines working in parallel, merging before the common electrofilter. Each line consisted of a combustor, a tubular reactor (residence time 5 s) and a cyclone. The bicarbonate was pulverized and then distributed equally into the two lines, being injected into the exit of a venturi tube placed at the entrance to the reactor. The solid residues were trapped by the electrofilter.

The waste combustion plant treating 5 t/h of refuse produced 85 m^3/h water at 180°C under a 15 bar pressure, corresponding to a power of 8.6 MW. Flue gas flow was 28,000 Nm^3/h.

The heterogeneous nature of the fuel prevented determination in advance of the exact bicarbonate requirements. The injection rate varied between 50 and 150 kg/h.

Stoichiometric ratios. For removal of SO, and HCl, the following global reactions apply:

$$NaHCO_3 + HCl = NaCl + H_2O + CO_2$$

$$2NaHCO_3 + SO_2 + \tfrac{1}{2}O_2 = Na_2SO_4 + H_2O + 2CO_2$$

In other words, $84/36.5 = 2.30$ kg of $NaHCO_3$ are required in order to remove 1 kg of HCl and $2 \times 84/64 = 2.625$ kg of $NaHCO_3$ are required in order to remove 1 kg of SO_2.

Testing. The following average values for the two lines together were measured at the bicarbonate injection points at a flue gas temperature of 225°C:

HCL content: 880 mg/Nm^3 (dry) ± 28%

SO_2 content: 143 mg/Nm^3 (dry) ± 34%

HF content: 6 mg/Nm^3 (dry) ± 36%

During testing the composition of the flue gases to be purified fluctuated widely. A range of stoichiometric ratios between 0.9 and 2.03 was covered. On leaving the reactors and cyclones, the gases were already 75% purified compared with levels at the electrofilter outlet.

Nitrogen oxides control techniques. Nitrogen oxides emissions derive from two sources: the fuel, and conversion of nitrogen in the air. The conversion from nitrogen takes place in the flame, and depends upon the flame temperature.

The control of NO_x emissions can be accomplished through either control of the combustion process, injection of reactants, or the use of add-on controls. Combustion controls include use of refractory furnaces (without waterwall cooling), staged combustion, low excess air (LEA), and flue gas recirculation (FGR). Add-on controls which have been used on MWCs include selective noncatalytic reduction (SNCR), selective catalytic reduction (SCR), and natural gas reburning.

Combustion controls involve the control of temperature or O_2 to reduce NO_x formation. With LEA, less air is supplied, which lowers the supply of O_2 that is available to react with N_2 in the combustion air. In staged combustion, the amount of underfire air is reduced, which generates a starved-air region.

In FGR, cooled flue gas and ambient air are mixed to become the combustion air. This mixing reduces the O_2 content of the combustion air supply and lowers combustion temperatures. Due to the lower combustion temperatures present in MWCs, especially in refractory furnaces, most NO_x is produced from the oxidation of nitrogen present in the fuel.

SNCR, also known as thermal deNOx, is the most common method used for reduction of NO_x. Injection of ammonia or urea into the furnace in a region having the optimum gas temperature achieves the optimum reduction. With SNCR, ammonia (NH_3) or urea is injected into the furnace along with chemical additives to reduce NO_x to NO_2 without the use of catalysts. MWCs equipped with SNCR have achieved NO_x reductions of about 45 percent.

With SCR, NH_3 is injected into the flue gas *downstream* of the boiler where it mixes with NO_x in the flue gas and passes through a catalyst bed, where NO_x is reduced to NO_2 by a reaction with the NH_3. Reductions of up to 80 percent have been observed, but problems with catalyst poisoning and deactivation reduce performance over time.

Natural gas reburning involves limiting combustion air to produce an LEA zone. Recirculated flue gas and natural gas are then added to this LEA zone to produce a fuel-rich zone that inhibits NO_x formation and promotes reduction of NO_x to NO_2. Natural gas reburning has achieved NO_x reductions of 50–60 percent.

Table 3.15 compares the NO_x emissions from three California MSW combustion systems equipped with thermal deNOx.

TABLE 3.15 NO_x Emissions from MWCs with Thermal DeNOx

Emissions (ppm @ 7% oxygen)	Commerce	Stanislaus		SERRF		
		Unit 1	Unit 2	Unit 1	Unit 2	Unit 3
Uncontrolled NO_x	128–217	298	318	–	210	259
Controlled NO_x	104	93	112	49	72	54
Ammonia slip	~ 2	3.7	5	–	–	35

Source: McDonald *et al.*, 1996.

At Commerce, a study evaluated the two injection levels installed, carrier air injection pressure and ammonia injection rate. The study concluded that optimum performance was achieved by injection of an NH_3-to-NO_x molar ratio of about 1.5 through the upper elevation of nozzles. Even when there was substantial ammonia slip at the economizer exit, the level at the stack due to the spray dryer-baghouse was held to less than 5 ppm. The lower levels achieved by SERRF are due to a higher rate of ammonia injection as compared with Commerce and Stanislaus. The higher ammonia injection rate also explains the higher ammonia slip numbers, which become the limiting factor in NO_x control by this method.

Mercury and dioxins control. Injection of dry activated carbon (DAC) has been proven to be effective in removing dioxins and mercury from the flue gases produced in municipal waste and medical waste combustion systems. Test data gives an indication of the quantity of DAC required in relation to the concentrations of mercury and/or dioxins present upstream of the point of injection. This quantity is affected by the temperature of the gases and the retention time before the gases enter the particulate collection device, such as ESP or fabric filter (Rigo and Chandler, 1996).

Sodium sulfide has been used for mercury control, using DSI/FF as the APC. Aqueous sodium sulfide will react with mercury to form solid HgS that can be collected in the PM control device. In this process, a dilute Na_2S solution is injected into cooled gas (< 400°F) prior to injection of hydrated lime used to control the acid gases. Mercury initially in the gas is

TABLE 3.16 Control of Mercury Emissions—Davis County Tests

Reagent feed		Mercury measured in stack (μg/dscm @ 7% oxygen) Stack temperature (°F)			
Trona (lb/h)	PAC (lb/h)	420	350	320	300
150	0	155	89		104
		92	202		258
		247			
150	15		20	22	
			25	17	
150	30	60	20	17	22
		43	24	15	
Average		116	53	18	128

Source: A Rigo and Chandler, 1996.

TABLE 3.17 Control of Dioxin Emissions—Davis County Tests

Reagent feed		Total dioxins measured in stack (μg/dscm @ 7% oxygen) Stack temperature (°F)			
Trona (lb/h)	PAC (lb/h)	420	350	320	300
150	0	116	83		151
			250		127
		91			
		87			
150	15		55	12	
				11	
150	30	37	20	33	6
		15	17		
			73		
Average		80	49	7.4	95

Source: Rigo and Chandler, 1996.

absorbed by the solution droplet. As the droplet evaporates, HgS and sodium salts precipitate. Feed rates of Na_2S vary from 0.05 to 0.5 kg/Mg (0.1 to 1 lb/ton) of MSW, depending on site-specific conditions such as the amount of mercury in the flue gas, the level of control required, and the level of carbon in flyash (which also absorbs mercury).

Tests of three WTE facilities in California provide comprehensive data, obtained when the input and output concentrations were measured simultaneously along with operating conditions. Such data are normally available only from government-sponsored tests, such as those carried out by the US EPA.

Uncontrolled mercury levels prior to APCDs range from roughly 200 to 1400 micrograms per dry standard cubic meter (μg/dscm), and average roughly 650 μg/dscm. Some typical data for various types of APC are shown in Table 3.18. The high variability of mercury levels entering and leaving the APC makes it difficult to obtain anything better than a range of removal efficiency.

Note that the emissions limits for new HWCs were met, except for one PM test and one mercury test. The average of the two PM tests, 0.0282 gr/dscf, does meet the limit of 0.030 gr/dscf. This points out the importance of performing a number of tests and using the average.

TABLE 3.18 Summary of APC Performance in Removing Mercury

Facility	Carbon addition rate (mg/dscm)	Operating temperature (°F)	Inlet Hg level (μg/dscm)	Outlet Hg level (μg/dscm)	Removal efficiency (%)
Stanislaus MWC (SD/FF)	183,672	285	400–700	300–550	20–40
			500–650	100–200	65–85
			350–1200	40–140	70–90
			500–1200	30–60	92–96
	Sodium sulfide injection rate (kg/h)				
Burnaby MWC (DSI/FF)	236	400	670	84	87
			1200–1500	470–750	50–60
			660–780	90–105	85–90
Fergus Falls, MN—MWC, wet scrubber			600	6–50	92–99
Basel, Switzerland—MWC, wet scrubber	Unit 1		170–510	15–20	90–96
	Unit 2		75–360	< 15–30	88–94

Source: McDonald *et al.*, 1996.

TABLE 3.19 Performance of Hybrid Air Pollution Control System

Pollutant	Test	Spray dryer inlet*	Fabric filter outlet	Control efficiency (%)	Scrubber outlet	Control efficiency (%)
Particulate	1	2.43	0.008	99.64	0.0054	99.78
(gr/dscf)	2	2.08	0.047	97.71	0.0075	99.63
HCl (ppmv)	1		1.86		1.15	38.3
Metals (μg/dscf):						
Arsenic	1	14,750	2.36	99.984	1.58	99.9895
	2	1770	0.389	99.978	0.113	99.994
Cadmium	1	820	3.36	99.059	2.98	99.165
	2	127	5.14	95.953	1.74	98.630
Chromium	1	2168	12.1	99.442	6.78	99.687
	2	920	15.0	98.37	4.08	99.557
Lead	1	90,333	5.42	99.994	4.44	99.995
	2	4,000,000			27.8	99.9993
Mercury	1	2.36×10^6			20,815.3	99.120
	2					99.936
Nickel	1	536	9.73	98.185	3.54	98.185
	2	902	21.1	97.663	8.65	99.042
Zinc	1	179,000	129,340	99.928	44.7	99.975
	2	33,800	150	99.562	0.338	99.999

Source: Radian Corp., 1989.
Notes: Test condition 1 : 487 lb/h contaminated soil, 2730 lb/h capacitors, 2432 lb/h PCB liquid.
Test condition 2: 682 lb/h contaminated soil and latex paint, 1054 lb/h capacitors, 4223 lb/h PCB liquid.
*Inlet = outlet/(1-control efficiency/100).

The high mercury level in one test, in spite of almost equal input, points out the difficulty in controlling mercury emissions. Carbon injection may be indicated.

Achievable emission limits and averaging times. Due to the high variability of mercury in the waste and hence in the uncontrolled and controlled emissions, it is necessary to analyze the data statistically in order to make estimates of the future performance of the control system, and to decide on the amount of carbon which should be injected on a continuous basis. On the basis of extensive tests at Stanislaus, it appears that the question of whether to assess compliance based on the average of multiple one-hour tests or on one multiple-hour test appears to be arbitrary from a statistical viewpoint (White, 1993b).

5 CONTROLLED AND UNCONTROLLED EMISSION FACTORS

Emission factors (EF), defined as pounds of a pollutant per ton of waste, pounds per million tons, grams/tonne, pounds per million Btu, etc., provide convenient reference numbers for tabulating stack test data, and hence for use in predicting emissions from facilities having similar combustion and emission control configurations.

Table 3.20 is a compilation published by the US EPA in AP-42 (EPA, 1993), providing EFs for uncontrolled emissions and for emissions controlled by ESPs, spray-dryer + ESP, dry sorbent injection + FF, and spray-dry scrubber + FF. AP-42 also expressed emission factors alternatively as μg/standard cubic meter. The conversion factor is 8.06 lb/million tons per μg/scm, based on the assumption that the heating value of the waste is 4500 Btu/lb. The value in μg/ scm is presumably based on actual stack measurements, converted to 7% oxygen or 12% CO_2, from which the weight units are calculated. The numbers in this table have been calculated by assuming uncontrolled emissions considered to be typical and controlled emissions from selected facilities. It is important to note that the uncontrolled emissions have not been measured at the same facility as the controlled emissions. While the efficiencies determined this way may be useful for general guidance, data from facilities where input and output of the emission controls were measured at the same time should be more reliable. Tables 3.18 and 3.19 present measurements and efficiency calculations which were obtained at the same facility. Table 3.21 shows, for general guidance, uncontrolled emission factors obtained by averaging a number of EPA tests on medical waste incinerators (Walker and Cooper, 1992).

TABLE 3.20 Particulate Matter, Metals and Acid Gas Emission Factors

	Control	ESP Control		SD/ESP Control		DSI/FF Control		SD/FF Control	
	E.F.*	E.F.	Effy. (%)†	E.F.	Effy. (%)	E.F.	Effy. (%)	E.F.	Effy.(%)
PM	25,100,000	210,000	99.16	70,300	99.72	17,900	99.93	62,000	99.75
As	4,370	21.7	99.50	13.7	99.69	10.3	99.76	4.2	99.90
Cd	10.000	646	93.54	75.1	99.25	23.4	99.77	27.1	99.73
Cr	8,970	113	98.74	259	97.11	200	97.77	30	99.67
Hg	4,790	6,620	−38.20!	3,260	31.94	2,200	54.07	2,200	54.07
Ni	7,850	112	98.57	270	96.56	143	98.18	52	99.34
Pb	213,000	3,000	98.59	915	99.57	297	99.86	261	99.88

Source: EPA, 1993.
*E.F. = emission factor, pounds per million tons of MSW.
†Control efficiencies calculated from uncontrolled emissions of various other WTE facilities.

TABLE 3.21 Uncontrolled Emission Factors for General Medical Waste

Pollutant	Uncontrolled emission factors (μg or mg/kg waste)
μg/kg waste:	
Dioxin/furan	32
Cd	2,000
Pb	28,600
Hg	25,500
Cr (total)	422
Cr (VI)	32
Ni	< 124
Fe	4,780
Mn	245
As	118
mg/kg waste:	
CO	2,500
NO_x	1,350
SO_2	566
PM	3,000
HCl	11,000
Benzene	1,300

Source: Walker and Cooper, 1992.

5.1 Tracing Metals from Waste to Emissions

It is important to have a clear perspective regarding the fate of air pollutants, especially the heavy metals, starting from concentrations in the waste, in the combustion process, before and after the air pollution control systems, and as concentrations in the stack gases, as they are diluted in the atmosphere before the metals reach the ground. From such analyses the metals concentrations in the stack may be compared with those of regulatory standards, to see whether they comply with the standards, and with what safety factor. Finally, after dilution before they reach ground level, as estimated by modeling studies, comparisons may be made with health standards and with "acceptable ground level concentrations" established by health authorities and amended by state regulatory authorities.

5.2 Reduction and Partitioning of Metals

In the combustion of wastes, a substantial fraction of metals is not volatilized but remains in the ash residues, depending upon the volatility of the metals. In the process of emissions control, nearly all of the metals remain in the captured flyash.

Only a few comprehensive studies have been made to measure the "partitioning" of metals from the waste to the various streams leaving the process. Such studies are costly and difficult to run, especially since the samples taken for analysis should be taken as near simultaneously as possible. Three studies are cited below, the first for a European facility having an ESP followed by a scrubbing tower, the second for a Canadian facility having a dry-sorbent injection system and baghouse, and a third for a US facility having a spray-dry scrubber and fabric filter. These facilities represent the major types of air pollution control system in common use at this time.

TABLE 3.22 Distribution of Emissions from MSW Combustion System

Sampling point	Annual emissions (kg/yr)		
	Lead	Cadmium	Mercury
Waste feed	38,060	434	139
Bottom ash	35,950	272	9
ESP flyash	1,871	104	3
Scrubber filter cake	146	28	121
Stack gas	92	33	7
Drain water	0.80	0.02	0.80
Stack concentration	0.162 mg/m^3	0.057 mg/m^3	0.06 mg/m^3
Background conc.*	0.1 μg/m^3	0.01 μg/m^3	
Ratio: stack/backgnd	1,620	5,700	

Source: Sorum *et al.*, 1997.
*Based on ground level concentrations measured in the New York/New Jersey metropolitan area, 1986 (NYSDEC, 1990).

EXAMPLE 1: A study by Sorum *et al.* (1997) analyzed the MSW, bottom ash, ESP ash, scrubber filter cake and drain water, as well as the flue gas leaving the stack. This facility has an ESP followed by a washing tower (wet scrubber). Table 3.22 summarizes the results for cadmium, mercury and lead.

The bottom ash contained 39,950/38,060 = 94.4% of the lead. The emissions were 0.2% of the MSW, or, in other words, the lead in the MSW was reduced by 99.76%, or by a factor of 413. Likewise, 62.6% of the cadmium stayed in the bottom ash, 24% was removed in the ESP ash, 6% was in the filter cake, and 7.6% was emitted from the stack. Most of the mercury (87%) was collected in the scrubber filter cake, and only 5% was emitted.

The schematic of the facility with annual lead emissions shown is as follows:

38,000 kg/year [MSW] Lead → [combustor] → [ESP] → [scrubber] → stack: 92kg/yr emissions

[combustor] ↓ 39,950 ash residue

[ESP] ↓ 1871 flyash

[scrubber] ↓ 146 kg/yr filter cake

The stack emissions for lead are therefore 92/38,000 or 1/413 of the input, or, stated another way, the removal efficiency is 99.76%.

EXAMPLE 2: In another mass-balance test, of a WTE facility, having a dry lime injection scrubber and baghouse (Burnaby), the mass-balance given in Table 3.23 was determined.

Here the ash residues contained (326,000 – 1871)/326,000 = 99.4% of the lead, a reduction by a factor of 174. The reduction from waste to stack was 326,000/363 or a factor of 898.

Table 3.23 gives a numerical demonstration of the factors by which the metals are reduced, and the efficiency of capture by the emission controls, as measured in extensive tests of the Burnaby facility. With the exception of the highly volatile mercury, of which over 50% reached the stack, the other metals were reduced by factors from 20 times for vanadium to as much as 5000 times for lead. Overall reduction of metals from 4700,000 to only 4720 constitutes a 99.4% reduction. Individual APC control efficiencies for the critical metals, lead and cadmium, were 98.3 and 99.7% respectively.

TABLE 3.23 Reduction and Partitioning of Metals in Municipal Waste

Metal	Metal in waste (lb/Mton)	Boiler out (lb/Mton)	Boiler reduction waste/ boiler out	Stack emissions (lb/Mton)	Overall reduction waste/stack	APC reduction boiler out/ stack	APC control efficiency (%)
Mercury	3,630	3,630	1	1934	2	2	46.7
Boron	222,000	7,496	30	1370	162	5	81.7
Zinc	3746,000	249,860	15	725	5167	345	99.7
Lead	326,000	21,681	15	363	898	60	98.3
Nickel	33,000	1,612	20	105	314	15	93.5
Chromium	185,000	2,821	66	97	1907	29	96.6
Tin	98,000	1,120	88	31	3161	36	97.2
Cadmium	27,000	5,723	5	18	1500	318	99.7
Arsenic	15,800	1,048	15	11	1436	95	99.0
Selenium	9,600	81	119	10	960	8	87.7
Vanadium	40	40	1	2	20	20	95.0
Copper	28,400	14,508	2	54	526	269	99.6
Totals	4694,470	309,620		4720			98.5

Source: Rigo and Chandler, 1994.

TABLE 3.24 Heavy Metals Collected and Emitted by the Commerce Resource-to-Energy Facility

Metal	Boiler emissions (μg/Nm³)	Stack emissions (μg/Nm³)	Control effy- (%)	Collected lb/Mton MSW	Emitted lb/Mton MSW	Range of AP-42 lb/Mton MSW
Magnesium	89,933	270	> 99.70%	89,663	< 2160	
Barium	4695	117	97.51%	4578	936	
Silicon	1860	66	96.45%	1794	528	
Calcium	193,000	56	99.97%	192,944	448	
Copper	8818	54	99.39%	8764	< 432	9–153
Iron	84,167	54	99.94%	84,113	< 432	
Mercury	475	41	91.28%	434	331	113–3460
Zinc	90,933	38	99.96%	90,895	308	90–420
Aluminum	178,000	16	> 99.99%	177,984	< 130	
Molybdenum	522	12	> 97.61%	510	< 100	
Nickel	4240	6	99.85%	4234	50	2–258
Selenium	84	2.7	> 96.76%	81	< 22	1–8
Chromium	3620	2.3	99.94%	3618	19	1–210
Tin	800	2	> 99.75%	798	< 16	
Cadmium	1680	2	99.88%	1678	16	3–145
Lead	18,133	2	99.99%	18,131	16	8–230
Manganese	3235	1	99.97%	3234	8	4–129
Cobalt	111	0.3	99.69%	111	3	
Antimony	822	0.3	> 99.96%	822	< 2	1–23
Beryllium	7	0.2	> 97.24%	7	< 2	0.01–4
Bismuth	31	0.16	> 99.49%	31	< 1	
Arsenic	78	0.16	> 99.79%	78	< 1	
Vanadium	257	0.09	99.96%	257	1	
Totals	685,501	745.2	99.89%	684,756	5962	

Source: Teller, 1994.

EXAMPLE 3: The Commerce facility, having a spray-dry scrubber and fabric filter (baghouse) provided the data for Table 3.24. Stack emissions are compared with boiler outlet concentrations to obtain control efficiencies ranging from 91% for mercury to 99.99% for lead. The stack emissions for lead are 1/1133 of the stack emissions. The emission factors, in lb per million tons, are also shown. The overall collection efficiency is found to be (684,756 − 5962)/684,756 = 99.1%. The input is thus 115 times the stack emissions.

5.3 Emissions from WTE versus Fossil Fuels

Table 3.25 gives a comparison between the emissions from waste-to-energy facilities and those from fossil-fuel-fired utility boilers. Note that, based on equivalent electric power generation, WTE facilities generally have much lower emissions than those of the fossil fuels (not including distillate oil and gas). This calculation is, of course, for illustration purposes only. Individual comparisons, while more accurate, would show the same general tendencies.

5.4 Recycling and Pollution Prevention

Table 3.26 lists the components of MSW analyzed at Burnaby, and the concentrations of the metals in each component.

It is notable that the highest concentrations of mercury were found in certain paper fractions, plastic film, lawn waste, unfinished wood, textiles, food, fiberglass, and, as expected, in batteries. The use of mercury as an anti-fungal agent in corrugated cardboard has essentially ceased. Mercury in printing inks has also been eliminated by use of organic colors. Mercury in batteries has been phased out. It has been found that the levels measured at Burnaby were declining at the time of the tests, and have declined further. Therefore these emission factors are no longer valid. In any case, concentrating on elimination of batteries would not have resulted in a substantial reduction: the total of 0.73 parts per million of MSW would be reduced by only 0.03 (4%) to 0.70 ppm if the battery fraction were eliminated, assuming that there was a direct relationship. It is not known whether the mercury in the batteries reports to the ash residues or to the stack gases. It *is* known that most of the lead goes to the ash residues (Rigo *et al.*, 1993).

TABLE 3.25 Comparison of Emissions from WTE Facilities with those from Fossil Fuels (lb/1000 MWh)

Emission	Residual oil	Bituminous coal (pulverized)	Lignite coal (pulverized)	Waste-to-energy (mass burn/refuse-derived fuel)
Arsenic (As)	0.22	0.46	0.91	< 0.033
Beryllium (Be)	0.06	0.03	0.06	< 0.017
Cadmium (Cd)	0.18	0.10	0.11	0.063
Chromium (Cr)	0.24	4.56	570	< 0.19
Copper (Cu)	3.19	2.28	3.42	0.43
Mercury (Hg)	0.04	0.23	0.23	0.17
Nickel (Ni)	1436	3.42	3.42	0.84
Lead (Pb)	0.34	0.87	0.11	0.44
Selenium (Se)	NR	0.29	0.29	< 0.022
Vanadium (V)	3.4	4.0	4.0	0.025
Zinc (Zn)	0.47	8.0	8.0	1.23
Particulate	1030	440	440	150

Source: Getz, 1993.

TABLE 3.26 Contribution of Components of MSW to Metals

Component			Percent in MSW	Parts per million parts of MSW			
				Cd	Cr	Hg	Pb
Paper	fine		2.09	0.002	0.07	0.006	0.09
	books		0.24	0.001	0.02	0.000	0.00
	magazines	glued	0.88	0.000	0.15	0.003	0.00
		not glued	0.93	0.003	0.05	0.003	0.05
	laminates	wax/plastic	1.66	0.005	0.05	0.002	0.12
		foil	0.30	0.000	0.13	0.000	0.28
	newsprint	glued	0.29	0.000	0.00	0.000	0.01
		not – b&w	4.55	0.005	0.17	0.014	0.33
		color	1.32	0.001	2.84	0.038	0.08
	browns	corrugate	9.19	0.009	0.17	0.028	0.35
		kraft	1.86	0.002	0.09	0.002	0.17
		box	1.68	0.003	0.09	0.008	0.20
	mixed paper		13.52	0.230	4.46	0.027	30.96
Plastic	film	color	3.13	0.207	3.60	0.013	11.33
		flexible	2.51	0.070	2.16	0.005	7.00
		rigid	0.3	0.112	0.36	0.001	0.10
	food	pete	0.015	0.001	0.00	0.000	0.01
		hdpe	0.182	0.005	0.03	0.000	0.11
		pvc	0.001	0.000	0.00	0.000	0.02
		dpe	0.001	0.000	0.00	0.000	0.00
		pp	0.026	0.000	0.01	0.000	0.02
		ps	0.006	0.000	0.00	0.000	0.00
		misc	0.684	0.542	0.30	0.003	1.08
	housewares	clear	0.064	0.001	0.00	0.000	0.04
		white	0.262	0.007	1.56	0.001	0.11
		blue	0.039	0.113	0.00	0.000	0.03
		yellow	0.049	0.001	0.63	0.000	1.21
		other	0.663	0.670	2.38	0.002	4.29
	toys etc		0.257	0.195	0.59	0.000	0.00
	video tape		0.001	0.022	0.00	0.000	0.01
Organics	yard	lawn	10.87	0.652	10.98	0.152	16.74
		branches	2.46	0.027	0.59	0.010	1.53
	food	organic	6.76	0.066	0.75	0.010	2.39
	wood	finished	3.29	0.036	3.72	0.007	18.52
		unfinished	6.06	0.002	3.51	0.024	19.63
	textiles		4.4	0.123	19.36	0.048	5.63
	footware		0.65	0.077	11.90	0.001	0.87
Metals	ferrous	beer cans	0.015	0.009	0.05	0.005	0.03
		soft drinks	0.012	0.007	0.04	0.004	0.03
		food	1.26	0.543	3.64	0.071	4.33
		band	0.06	0.009	0.30	0.000	0.36
	non-ferrous	beer	0.058	0.002	0.55	0.000	0.04
		soft drink	0.182	0.011	0.16	0.001	0.06
		food	0.016	0.000	0.03	0.000	0.02
		manufactured	0.40	0.022	5.42	0.001	0.38
		foil	0.326	0.166	0.44	0.003	0.00
		other	0.001	0.000	0.00	0.000	0.00

TABLE 3.26 *(continued)*

Component			Percent in MSW	Parts per million parts of MSW			
				Cd	Cr	Hg	Pb
Glass	combined	clear	1.52	0.073	0.43	0.003	1.67
		green	0.12	0.000	1.13	0.000	0.02
		brown	0.13	0.002	0.06	0.001	0.13
		other	0.02	0.001	0.02	0.000	0.02
Inorganic light		dirt, rock	0.60	0.120	1.12	0.002	9.27
construction		drywall	0.09	0.002	0.01	0.000	0.03
		fiberglass	0	0.050	14.10	1.100	40.80
		other	0.87	0.400	34.00	0.100	30.10
Small appliances		plastic	0.15	0.005	0.38	0.000	0.99
Household batteries		carbon	0.011	0.003	0.00	0.002	0.00
		Ni-Cad	0.007	8.400	0.00	0.000	0.01
		alkaline	0.012	0.233	0.01	0.029	0.02
Fines			7.6	0.334	8.74	0.106	19.68
Total percent:			93.24				
Total parts per million:				13.5	93.5	0.73	163.40

Source: Rigo and Chandler, 1994.

6 CONVERSIONS AND CORRECTIONS

6.1 Correction of Emission Factors for Heating Value

The EPA has listed emission factors for MSW in AP-42, using a higher heating value (HHV) of 4500 Btu/lb as a reference value. When the heating value is different than 4500 Btu/lb, a correction must be made, as follows:

EXAMPLE: AP-42 lists emissions of nitrogen oxides (NO_x) as 2.86 kg/mg of fuel, and as 358 ppmdv at 4500 Btu/lb.

Correct these emissions to the actual higher heating value of 5000 Btu/lb:

Calculate lb/ton and lb/million Btu:

multiply 2.86 kg/mg *by* 5000/4500 *to get* 3.18 kg/mg at 5000 Btu/lb

multiply 3.18 kg/mg *by* 2 *to get* 6.36 lb/ton

Convert from lb/ton to lb/million Btu:

One ton at 4500 Btu/lb × 2000 lb/ton = 9 million Btu/ton

2.86 kg/mg @ 4500 Btu/lb *multiplied by* 2 = 5.72 lb/ton

5.72 lb/ton divided by 9 MBtu = 0.64 lb/MBtu

One ton at 5000 Btu/lb contains 2000 lb = 10 million Btu/ton

6.36 lb/ton divided by 10 Mbtu/ton = 0.64 lb/MBtu

Correct ppmv at 4500 Btu/lb to 5000 Btu/lb:

multiply 358 ppmdv @4500 Btu/lb *by* 5000/4500 *to get* 398 ppmdv

6.2 Conversions from Volumetric to Other Bases

The US EPA has developed the volumetric emission factors, in grams per cubic meter based on a standardized "F" factor of 9570 ft^3/MBtu for stoichiometric combustion (zero excess air or zero oxygen). This F-factor is reasonably accurate for combustion of wastes and fuels, but a more precise number may be calculated from the actual composition of the waste, as shown in Section 2.5.

Convert μg/dscm to (lb pollutant)/(MBtu of waste)

$$1\left[\frac{\mu g}{m^3}\right] @\ 7\%\ O_2 \times \left[\frac{21-0}{21-7}\right] \times \left[\frac{m^3}{35.3\ ft^3}\right] \times \left[\frac{g}{10^5\ \mu g}\right] \times \left[\frac{9570\ ft^3}{10^6\ Btu}\right] = 0.9\left[\frac{lb}{10^6\ Btu}\right]$$

Convert ppmv to lb/MBtu

$$1\ [ppmv]\ @\ 7\%\ O_2 \times MW\left[\frac{lb}{[lb \cdot mol]}\right] \times \left[\frac{21-0}{21-7}\right] \times \frac{1}{35.3}\left[\frac{lb \cdot mol}{ft^3}\right]$$

$$\times\ 9570\left[\frac{ft^3}{10^6\ Btu}\right] = \frac{1}{26{,}820}\left[\frac{lb}{10^6\ Btu}\right]$$

Convert ppmv to mg/m³

$$385/(454 \times 35.3) = 0.24\left[\frac{mg}{m^3}\right] \times \left[\frac{1000 \times 0.024}{MW}\right] = ppmv$$

or

$$[mg/m^3] \times 24/MW = [ppmv]$$

TABLE 3.27 Conversion of ppmv to lb/MBtu and mg/m³

Gas	Molecular weight MW	lb/MBtu 100 ppmv =	mg/m³ 100 ppmv =
CO_2	28	0.104	116.5
SO_2	44	0.163	183.1
NO_x	46	0.172	191.3
HCl	36.5	0.136	151.8
HF	20	0.746	83.2
CH_4	14	0.060	66.6
NH_3	32	0.063	112.3

6.3 Corrections for Excess Air

Emissions tests performed in the stack of combustion devices are measured at the actual gas flow, but must be reported at "reference" conditions such as 7% oxygen or 12% CO_2 in order to standardize all reported data and relate it to regulated emission standards. Reference standards vary from 3% oxygen, used in California, to 11% oxygen, used in Europe.

Calculate excess air:

$$\%\text{EA} = \frac{\text{O}_2 - 0.5\ \text{CO}}{0.266\ (\text{N}_2 - 0.5\ N_\text{f}) - \text{O}_2} \times 100\%$$

where

O_2, CO_2 and CO are the molar or volume fractions of the gases in the flue gas as determined by an Orsat or equivalent analysis

N_f is the mole fraction of fuel nitrogen in the combined waste and fuel feeds to the combustor, determined from the ultimate analysis. This value is usually negligible

Normally N_f and CO can be neglected, yielding this simpler expression:

$$\%\text{EA} = \frac{\text{O}_2}{0.266\ (\text{N}_2) - \text{O}_2} \times 100\%$$

Correction from (concentration)$_{\text{actual}}$ *to (concentration)*$_{\text{standard}}$:

Since the sum of O_2 plus CO_2 in the *dry* products of combustion is about 20.5, corrections for excess air are usually approximated by this expression:

$$\text{ratio} = \frac{20.5 - (\text{O}_2)_{\text{actual}}}{20.5 - (\text{O}_2)_{\text{standard}}} \times 100\%$$

6.4 Calculation of Destruction and Removal Efficiency

The regulations for emissions from hazardous waste incinerators were based on destruction and removal efficiency (DRE). Essentially, it was required that 99.99% of the principal organic hazardous components (POHC) be removed, and that for critical chlorinated compounds 99.999% removal was required, during trial burn tests. The operating conditions under which these efficiencies were demonstrated were to be the upper limits of operating parameters, such as waste feed rates and furnace temperatures.

In order to calculate DREs, it is necessary to determine the feed rate of the organics or metals, and measure the stack emissions. To predict the necessary feed rate it is necessary to estimate the partitioning of the metals from ash to flyash and the efficiency of the air pollution control system (APC). For organics, the efficiency of the APC is applied to obtain the estimate of stack emissions.

Since the stack emission levels would be extremely low, it is important that they be high enough to measure. Hence the detection limit becomes a controlling factor. The expected emissions must be sufficiently higher than the detection limit to permit demonstration of the required DRE. The procedure for making this calculation is as follows.

Destruction and removal efficiency for an incinerator/air pollution control system is defined by the following formula:

$$\text{DRE} = \frac{W_{\text{in}} - W_{\text{out}}}{W_{\text{in}}}(100)$$

where DRE = destruction and removal efficiency, %

W_{in} = mass feed rate of the principal organic hazardous constituents to the incinerator

W_{out} = mass emission rate of the principal organic hazardous constituent(s) to the atmosphere as measured in the stack prior to discharge

DRE calculations are based on the combined efficiencies of destruction in the incinerator and removal from the gas stream in the air pollution control system.

USEPA Part 264, Subpart 0 regulations for hazardous waste incineration require a DRE of 99.99% for all principal organic hazardous components of a waste unless it can be demonstrated that a higher or lower DRE is more appropriate on the basis of human health criteria.

Destruction and removal efficiencies are normally measured only during trial burns and occasional compliance tests, and are used as a basis for determining whether or not the operating conditions of the incinerator/air pollution control system are adequate.

Sample calculation. A liquid injection incinerator equipped with a quench tower, venturi scrubber and packed bed caustic scrubber has been constructed to burn a mixture of waste oils and chlorinated solvents with the following empirical composition:

73.0 wt% carbon

16.5 wt% chlorine

10.5 wt% hydrogen

The principal organic hazardous components are trichloroethylene, 1,1,1-trichloroethane, methylene chloride and perchloroethylene. Each of these compounds constitutes about 5% of the total waste feed to the incinerator.

During a trial burn, the incinerator was operated at a waste feed rate of 5000 lb/h and 50% excess air. The gas flow rate measured in the stack was 19,200 dscm. Under these conditions, the measured concentrations of the principal organic hazardous components were:

trichloroethylene	4.9 pg/dscf
1,1,1-trichloroethane	1.0 pg/dscf
methylene chloride	49 pg/dscf
perchloroethylene	490 pg/dscf

In order to calculate destruction and removal efficiency for each of these compounds using the equation

$$\mathrm{DRE} = \frac{W_{\mathrm{in}} - W_{\mathrm{out}}}{W_{\mathrm{in}}}(100)$$

it is necessary to calculate the mass flow of each component entering and exiting the system. Because each hazardous component constitutes about 5% of the waste and the total waste feed rate is 5000 lb/h, W_{in} for each component is

$$W_{\mathrm{in}} = 0.05\ (5000\ \mathrm{lb/h}) = 250\ \mathrm{lb/h}$$

The mass flow rate of each component exiting the stack is then calculated by the following equation

$$W_{\mathrm{out}} = C_i \times \frac{(19{,}200\ \mathrm{dscfm})\ (60\ \mathrm{min/h})}{4.54 \times 10^8\ \mu\mathrm{g/lb}}$$

where W_{out} = mass flow rate of component i exiting the stack, lb/h
C_i = concentration of component i in the stack gas, µg/dscf

Using this equation to calculate W_{out} for each component and the previously cited equation for destruction and removal efficiency, the following results are obtained:

Component	W_{out} (lb/h)	DRE (%)
Trichloroethylene	0.0124	99.995
1,1,-trichloroethane	0.00254	99.999
Methylene chloride	0.124	99.95
Perchloroethylene	1.24	99.5

These results indicate that the required 99.99% destruction and removal efficiency was achieved in the trial burn for trichloroethylene and 1,1,1-trichloroethane, but not for methylene chloride and perchloroethylene.

6.5 Useful Relationships

The stoichiometric combustion air needed per million Btu can be calculated:

$$\text{lb air/MBtu} = \frac{1{,}000{,}000\ [\text{Btu}]\ [\text{lbf}]}{7457\ [\text{Btu}]} \times \frac{5.105\ [\text{lba}]}{[\text{lbf}]} = 684.6\ \text{lba/MBtu}$$

The pounds of stoichiometric products per million Btu would be

$$\text{lb products/MBtu} = \frac{1{,}000{,}000\ [\text{Btu}]\ [\text{lbf}]}{7457\ [\text{Btu}]} \times \frac{6.105\ [\text{lba}]}{[\text{lbf}]} = 819\ \text{lbp/MBtu}$$

With 50% excess air, we get pounds combustion air per million Btu:

$$\text{lb air/MBtu} = \frac{1{,}000{,}000\ [\text{Btu}]\ [\text{lbf}]}{7457\ [\text{Btu}]} \times \frac{7.66\ [\text{lba}]}{[\text{lbf}]} = 713\ \text{lba/MBtu}$$

and the pounds of products at 50% excess air would be

$$\text{lb products/MBtu} = \frac{1{,}000{,}000\ [\text{Btu}]\ [\text{lbf}]}{7457\ [\text{Btu}]} \times \frac{8.66\ [\text{lba}]}{[\text{lbf}]} = 1161\ \text{lba/MBtu}$$

Weight and volume of products per million Btu as standard temperature and pressure:

Volume of 1 mole of ideal gas = 387 $\text{ft}^3/\text{lb}\cdot\text{mol}$ @ 70°F (20°C), 1 atmosphere

1 lb · mol of dry products (see above) weighs 29.51 lb

1 lb · mol of products weights about 29.75 lb at *zero excess air*

Therefore the volume of wet products *corrected* to 70°F (70 + 460 = 530°R) is

$$\begin{aligned} V_{wp} &= 387/29.75 = 13.0\ \text{ft}^3/\text{lbf} \\ &= 6.105\ [\text{lbp/lbf}] \times 13\ \text{ft}^3/\text{lbf} = 79.416\ \text{ft}^3/\text{lbf} \end{aligned}$$

The volume of dry products is calculated when reporting emissions. To get the volume of dry products we subtract 0.555 [H_2O] from 6.105 to obtain

$$V_{dp} = (6.105 - 0.555)\ [\text{lbdp/lbf}] \times 13.0\ \text{ft}^3/\text{lbf} = 72.15\ \text{ft}^3/\text{lbf}$$

6.6 Common Conversion Factors

To convert from	*To*	*Multiply by*
Milligrams/m^3	Micrograms/m^3	1000
	Micrograms/liter	1.0
	ppm by volume (20°C)	(24.04/M)
	ppm by weight	0.8347
	lb/ft^3	62.43×10^{-9}
Micrograms/m^3	Milligrams/m^3	0.001
	Micrograms/liter	0.001
	ppm by volume (20°C)	(0.02404/M)
	ppm by weight	834.7×10^{-6}
	lb/ft^3	62.43×10^{-12}
Micrograms/liter	Milligrams/m^3	1.0
	Micrograms/m^3	1,000
	ppm by volume (20°C)	(24.04/M)
	ppm by weight	0.8347
	lb/ft^3	62.43×10^{-9}
ppm by volume (20°C)	Milligrams/m^3	(M/24.04)
	Micrograms/m^3	(M/0.02404)
	Micrograms/liter	(M/24.04)
	ppm by weight	(M/28.8)
	lb/ft^3	$(M/385.1 \times 10^6)$
ppm by weight	Milligrams/m^3	1.198
	Micrograms/m^3	1.198×10^{-3}
	Micrograms/liter	1.198
	ppm by volume (20°C)	(28.8/M)
	lb/ft^3	7.48×10^{-6}
lb/ft^3	Milligrams/m^3	16.018×10^6
	Micrograms/m^3	16.018×10^9
	Micrograms/liter	16.018×10
	ppm by volume (20°C)	$(385.1 \times 10^6/M)$
	ppm by weight	133.7×10^3

Note: cm = 0.0328 ft
gal (US) = 0.1337 ft^3
Liter = 0.03532 ft^3 = 0.001 m^3
Microgram = 0.000001 g
Micron = 0.0000394 in. = 0.001 mm
Milligram = 0.001 g
lb = 7,000 grains = 453.6 g
M = Molecular weight

BIBLIOGRAPHY

Andersson, C., and B. Weimer, "Mercury Emission Control—Sodium Sulfide Dosing at the Hogdalen Plant in Stockholm," *Proceedings of the 2nd Annual Conference on Municipal Waste Combustion*, Air and Waste Management Association, VIP-19, Tampa, FL, April 1991, pp. 664–674.

Balfour, Raymond L., "Mercury, Batteries and MSW," *8th Annual Meeting of A&WMA*, Denver, CO, June 1993.

Brunner, C. R., *Medical Waste Disposal Handbook*, Incinerator Consultants Inc., Reston, VA, 1998.

Buonicore, A. J., and W. T. Davis (editors), *Air Pollution Control Engineering Manual*, A&WMA, Van Nostrand Reinhold, New York, 1992.

Clarke, Marjorie J., "The Development of New Jersey's Mercury Emissions Standards for Municipal Waste Combustors," *Municipal Waste Combustion*, VIP-32, A&WMA, 1993, 966–982.

Clement, "Understanding the Sources, Trends and Impacts of Mercury in the Environment," Clement Risk Assessment Division of ICF Kaiser Engineers, Fairfax, VA, 1992.

Comans, Rob, Hans van der Sloot and Petra Bonouvie, "Geochemical Reactions Controlling the Solubility of Major and Trace Elements During Leaching of Municipal Solid Waste Incinerator Residues," *Municipal Waste Combustion*, VIP-32, A&WMA, 1993, 667–679.

EPA, "Test Methods for Evaluating Solid Waste," *Report SW-846*, 1982.

EPA, "Combustion Emissions Technical Resource Document (CETRED)," *Report EPA530-R-94-014*, May 1994.

EPA, "Draft Technical Support Document for MWC MACT Standards," US EPA OSWER, Washington, DC, February 1996.

EPA, "Compilation of Air Pollution Emission Factors," *Report AP-42*, US EPA Office of Air Quality, PB93, May 1993.

Franklin Associates Ltd, "Characterization of Products Containing Mercury in Municipal Solid Waste in the United States, 1970 to 2000," *Report EPA530-R-92-013*, US EPA, Washington, DC, 1992.

Getz, Norman, "How does Waste-to-Energy 'Stack' up?" *Municipal Waste Combustion*, VIP-32, A&WMA, 1993, 951–965.

Gleiser, Rick, K. Nielsen and K. Felsvang, "Control of Mercury from MSW Combustors by Spray Dryer Absorption Systems and Activated Carbon Injection," *Municipal Waste Combustion*, VIP-32, A&WMA, 1993, 106–120.

Hasselriis, Floyd, "Variability of Composition of Municipal Solid Waste and Emissions from its Combustion," *ASME Solid Waste Conference*, Miami, FL, 1982.

Hasselriis, Floyd, "Relationship Between Waste Composition and Environmental Impact," Paper No. 90-38.2, *83rd A&WMA Conference*, Pittsburgh, PA, June 1990.

Hasselriis, Floyd, "Relationship between Input and Output," Chapter 5 of *Medical Waste Incineration and Pollution Prevention*, Van Nostrand Reinhold, 1992, pp. 97–126.

Hasselriis, Floyd, "Analysis of Data Obtained from an Historic Ash Residue Leaching Investigation," *16th ASME Biennial National Waste Processing Conference*, Boston, MA, June 1994.

Hasselriis, F., "Variability of Metals and Dioxins in Stack Emissions of Three Types of Municipal Waste Combustors over Four Year Period," *Paper #95-RP147B.03*, A&WMA, San Antonio, TX, June 1995.

Hesketh, H. E., *Air Pollution Control for Traditional and Hazardous Pollutants*, Technomic, Lancaster, PA, 1991.

Kilgroe, James D., T. D. Brna *et al.*, "Camden County MWC Carbon Injection Test Results," *Municipal Waste Combustion*, VIP-32, A&WMA, 1993, 123–142.

Licata, Anthony, Manyam Babu and Lutz-Peter Nethe, "Acid Gases, Mercury and Dioxin from MWCs," *1994 ASME National Waste Processing Conference*, Boston, MA, pp. 39–48.

Licata, Anthony, Manyam Babu and Lutz-Peter Nethe, "An Economic Alternative to Controlling Acid Gases, Mercury and Dioxin from MWCs," *1994 A&WMA Conference*, Cincinnati, OH, Paper 94-MP17.06.

Linak, William P., Ravi K. Srivastava and Jost Wendt, "Metal Aerosol Formation in a Laboratory Swirl Flame Incinerator," *Municipal Waste Combustion*, VIP-32, A&WMA, 1993, 644–663.

McDonald, Barry, G. Fields and M. McDannel, *Selective Non-Catalytic Reduction (SNCR) Performance of Three California WTE Facilities*, Carnot, Tustin, CA, 1996.

McKenna, J. D., and J. H. Turner, *Fabric Filter-Baghouses I: Theory, Design and Selection*, ETS International Inc., 1989.

Malcolm Pirnie, Inc., RTP Environment Association, Inc. and Clement International Corp., "A Report on the Mercury Control System for the Lee County Resource Recovery Facility," prepared for the Florida Department of Utilities, Lee County, FL, April 1992.

NITEP, "The National Incinerator Testing and Evaluation Program: Two-stage Combustion (Prince Edward Island)," *Report EPS 3/UP1*, Environment Canada, 1985.

NITEP, "The National Incinerator Testing and Evaluation Program: Environmental Characterization of Mass Burning Incinerator Technology in Quebec City," *Report EPS 3/UP/5*, Environment Canada, 1988.

NYSDEC, "Incinerator 2000," New York State Department of Environmental Conservation, 1990.

Perry's Chemical Engineer's Handbook, McGraw-Hill, New York, 1984.

Radian Corp., Draft Test Report, *DCN 89-232-011-034-06,* OSW, US EPA, Oct. 1989.

Richman, D., and Jeff Hahn, "Mercury Removal Studies at a Municipal Waste Combustor in Marion County, Oregon," *Municipal Waste Combustion*, VIP-32, A&WMA, 1993, 918–932.

Ricket, William S., and Murray Kaiserman, "Levels of Lead, Cadmium and Mercury in Canadian Cigarette Tobacco as Indicators of Environmental Change: Results from a 21-year Study (1968–1988)," *Environ. Sci. Technol.,* **28**(5), 1994.

Rigo, H. G. and J. Chandler, "Metals in MSW – Where are They and Where Do They Go in an Incinerator?" *1994 National Waste Processing Conference*. ASME, New York, 1994.

Rigo, G. and J. Chandler, "Retrofit of Waste-to-Energy Facilities Equipped with Electrostatic Precipitators," *CRTD,* **39**, ASME, April 1996.

Rigo, H. Gregor, John Chandler and Steven Sawell, "Impact of Lead Acid Batteries and Cadmium Stabilizers on Incinerator Emissions," *Municipal Waste Combustion*, VIP-32, A&WMA, 1993, 628–643.

Rigo, H. Gregor, John Chandler and Steven Sawell, "Debunking some myths about metals," *Municipal Waste Combustion*, VIP-32, A&WMA, 1993, 609–627

Rood, Mark J., and R. E. Simek, "Modeling the Combustion of Municipal Solid Waste," *85th Annual Meeting of A&WMA*, Kansas City, June 1992.

Sheppard, S., "Operating Experience with the Ionizing Wet Scrubber on Hazwaste Incinerators," *Proceedings of the 1992 Incineration Conference*, Albuquerque, NM, pp. 24–29.

Sigg, Alfred, "Combustion Process Control in Rotary Kiln Incinerators," *Paper #90-169.4*, A&WMA, Pittsburgh, PA, June 1990.

Sorum, Lars, M. Fossum, E. Evesen and J. E. Hustad, "Heavy Metal Partitioning in a Municipal Solid Waste Incinerator," *Proceedings of the North American Waste-to-Energy Conference*, 1997.

Srinivasachar, S., *et al.*, "Heavy Metal Transformations and Capture during Incineration," *85th Annual Meeting of A&WMA*, Kansas City, June 1992.

Teller, Aaron J., "Heavy Metal Emissions and Control," *Municipal Waste Combustion*, VIP-32, A&WMA, 1993, 217–237.

Teller, Aaron, "Emission Control," p. 11.159, *Handbook of Solid Waste Management*, McGraw-Hill, New York, 1994.

Walker, B. L. and C. D. Cooper, "Air Pollutant Emission Factors for Medical Waste Incinerators," *J. Air Waste Management Assoc.,* **42**(6), June 1992, 784–791.

WASTE, "Waste Analysis, Testing and Evaulation: The Fate and Behavior of Metals in Mass Burn Incineration (Burnaby, BC)," A. J. Chandler Associates Ltd. *et al*, Willodale, Ontario, April, 1993.

White, David M., and Anne M. Jackson, "The Potential of Materials Separation as a Control Technique for Compliance with Mercury Emission Limits," *Municipal Waste Combustion*, Williamsburg, VA, March 1993a, A&WMA VIP-32, 797–808.

White, David M., and Anne M. Jackson, Technical Work Paper on "Mercury Emissions from Waste Combustors," Minnesota Pollution Control Agency, St Paul, MN, 1993b.

Wood, Roy, and F. Hasselriis, "Relationship between Short-term Emissions from Waste Combustion over Four Year Period," *Paper #97-A350*, A&WMA, Toronto, Canada, June, 1997.

2.4 CALCULATIONS FOR PERMITTING AND COMPLIANCE

Floyd Hasselriis
Consulting Engineer, Forest Hills, New York

1 INTRODUCTION

This chapter addresses the procedures which can or must be followed in order to determine that the emissions from the stack (chimney) of a combustion device burning municipal, medical or hazardous wastes can or are expected to comply with regulations permitting operation of the facility, as well as to evaluate the environmental and health impact of these emissions.

The procedures promulgated by the US EPA for performing stack tests are outlined. The emissions determined by these procedures are compared with permit requirements to confirm compliance with the regulations.

The emissions measured by stack tests can be used to determine their impact at ground level or in sensitive areas by the use of modeling methods which are outlined in this chapter, and their environmental and health risk impact compared with health risk standards.

2 CALCULATIONS OF EMISSIONS FROM THE STACK

Stack testing is performed to measure the discharges to the atmosphere from the combustion system and its emission control equipment. This testing is performed in accordance with the US EPA methods and procedures, which are briefly summarized below.

The US Environmental Protection Agency (EPA) has published sampling and analytical procedures in the Code of Federal Regulations, Title 40, Part 60 (40 CFR 60), Appendix A, and Title 40, Part 61 (40 CFR 61), Appendix B, among others. The following are the "basic specific methods:"

- EPA Method 1 for determination of sampling and traverse points.
- EPA Method 2 for calculating stack gas velocity, volumetric flow rate, density and mass flow rate.
- EPA Method 3 for sampling of flue gas composition and molecular weight determinations. Dry oxygen and dry volume are calculated using the measured percent moisture. The specific volume at measured temperature is calculated as measured on a wet basis and used to determine the actual cubic feet per minute (acfm) and the dry standard cubic feet per minute (scfm) of flue gas, as well as the percent excess air.
- EPA Method 3/4 for flue gas moisture content. Stack gas moisture is determined by condensation of water from the extracted sample. Orsat analysis of an extracted gas sample is used to determine the percent oxygen and CO_2 of the gases in their wet condition.
- EPA Method 5 for determination of total particulate emissions. Particulate matter (PM) and metals mass rate and composition, measured by sampling train, are shown in Fig. 4.1.
- EPA Method 23 is used for determination of the emission of polychlorinated dibenzodioxins (PCDDs), polychlorinated dibenzofurans (PCDFs), and poly-aromated hydrocarbons (PAHs).
- EPA Method 29 is used for determination of metals emissions (arsenic, barium, beryllium, cadmium, chromium, lead, nickel, selenium, vanadium, zinc and mercury).

Oxygen is continuously measured in the stack by instruments which can measure it in the dry or wet condition.

CO is measured by a continuous monitor located in the stack, or by continuous extraction of a gas sample.

SO_2 is measured by the sampling train of EPA Method 6, or by continuous monitoring.

HCl is measured by the sampling train of EPA Method 26A, or by continuous monitoring.

All gas flow measurements are corrected from actual temperature and pressure to standard conditions of one standard atmosphere, 29.92 inches mercury and a temperature of 68°F, and by use of the measured oxygen and moisture.

For analysis of total hydrocarbons and other volatiles, gas samples can be collected in Tedlar bags and taken to the laboratory for analysis.

2.1 EPA Method of Calculating Stack Gas

Stack gas flow volume is determined by measuring velocity, temperature and barometric pressure. Velocity is measured by a Pitot tube traverse of equal areas which determine the

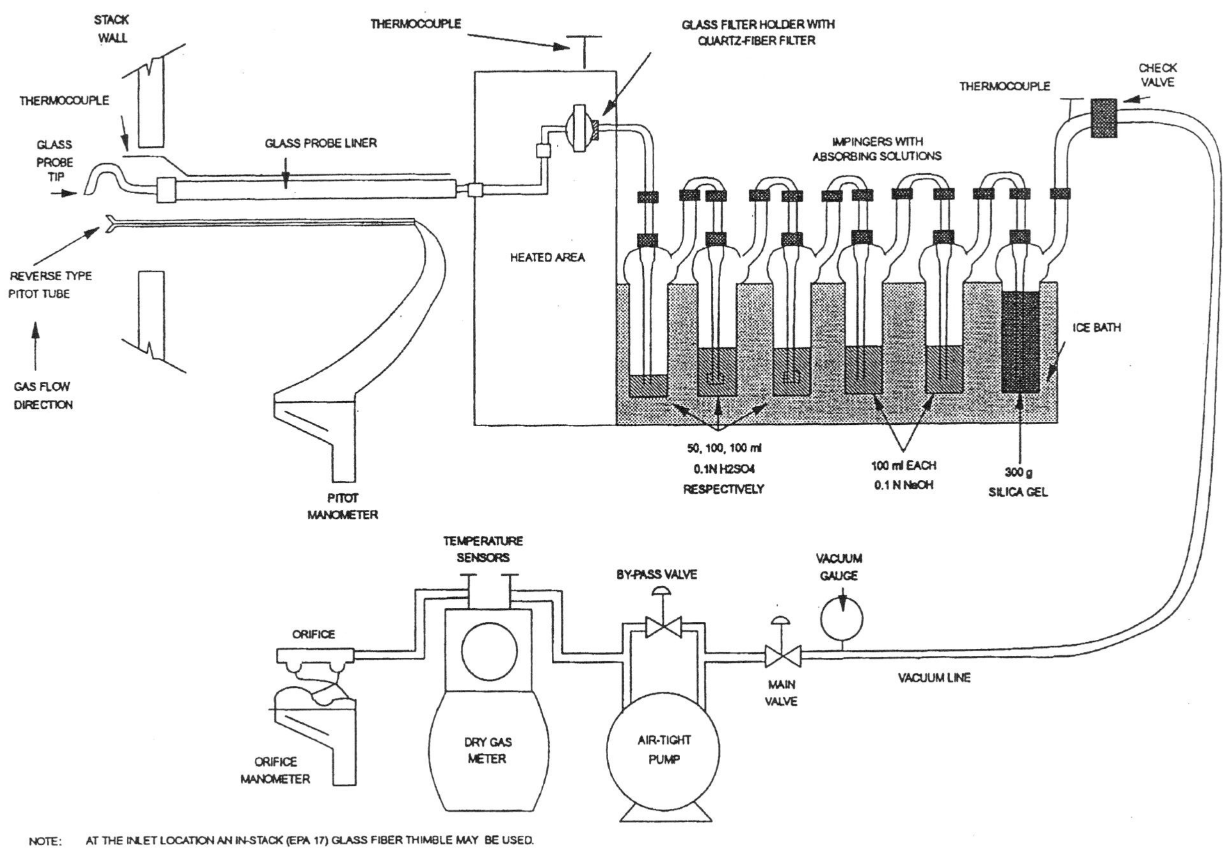

FIGURE 4.1 Combined EPA Method 5 and 26A particulate/anion sampling train.

average velocity. Flow volume is calculated by use of the diameter of the stack at the outlet, and mass flow is determined after calculating the actual gas density.

Step 1. Determine the percentages by volume of oxygen, carbon dioxide and nitrogen, measured on a dry basis by Orsat analysis.

For instance:

$\%CO_2 =$	8.6%
$\%O_2 =$	10.30%
$\%N_2 =$	81.1%
Total =	100.00% dry volume

Step 2. Determine stack gas moisture by Method 4: a sample of stack gas is drawn from the stack through a condenser in the sampling train from which a volume of water condensate (V_{1c} in ml) is collected and weighed, while the volume of gas extracted (V_m in ft^3) is measured by gas meter.

The sample volume is corrected to standard conditions:

$$V_{std} = 17.647 \times V_m \times Y \times (P_b + \Delta H/13.6)/T_m$$

The volume of water vapor sampled is

$$V_w = 0.04707 \times V_{1c}$$

The proportion of water vapor in sampled gas is

$$B_{ws} = V_w/(V_w + V_{std})$$

where B_{ws} = water vapor in gas stream (proportion by volume)
ΔH = orifice meter differential presure (inches water column)
P_b = barometric pressure (inches Hg)
T_m = sampling train meter temperature (°R)
V_{1c} = total volume of liquid collected in sampling train (ml)
V_m = volume of gas sample measured by gas meter (ft^3 or cf)
V_{std} = gas volume corrected to standard conditions (dscf)
V_w = volume of water vapor in gas sample (scf)
Y = dry gas meter calibration coefficient

For this example, assume that the moisture content has been determined to be

$$\text{moisture content} = B_{ws} \times 100 = 17.3\%$$

Step 3. Determine molecular weight of the stack gas.

In order to calculate the molecular weight of the actual wet gas, M_w, it is necessary to have the gas composition, measured on a dry basis by an Orsat, and a measurement of the moisture content of the flue gas, obtained by condensing the water vapor in a measured volume of sample gas extracted from the stack.

The *molecular weight* of the *dry gas* is calculated as follows:

$$M_d = (0.44 \times \%CO_2) + (0.32 \times \%O_2) + (0.28 \times (\%N_2 + \%CO))$$
$$= (0.44 \times 8.6) + (0.32 \times 10.30) + (0.28 \times 81.1) = 29.788\ \text{lb/(lb}\cdot\text{mol)}$$

The *molecular weight* of the *wet gas* is calculated from M_d as follows:

$$M_s = M_d \times (1 - B_{ws}) + (18 \times B_{ws})$$
$$= 29.788 \times (1 - 0.1743) + (18 \times 0.1743) = 27.733\ \text{lb/(lb}\cdot\text{mol)}$$

Step 4. Calculate stack gas linear velocity.

The *stack gas velocity* is measured by a Pitot tube, which reads the velocity head of impact as the gases impact on the Pitot tube mouth perpendicular to the direction of flow.

A number of readings of velocity pressure ΔP are taken over equal areas to represent the total cross-section of the duct. The average of the square roots of these readings $\sqrt{\Delta P}$ represents the average velocity through the duct. The stack gas velocity V_s is calculated by a constant 85.49 multiplied by the Pitot calibration coefficient C_p, the average velocity $\sqrt{\Delta P}$ times the square root of the flue gas temperature in degrees Rankine T_s, the static pressure in the duct P_s and the molecular weight of the flue gas in its wet condition M_w, as follows:

$$V_s = 85.49 \times C_p \times \sqrt{\Delta P} \times \sqrt{(T_s/P_s \times M_s)}$$
$$= 85.49 \times 0.835 \times 0.7843 \times \sqrt{(1004/(28.31 \times 27.733))} = 63.31\ \text{ft/s}$$

where T_s = stack temperature = 544°F + 460 = 1004°R
P_s = barometric pressure + absolute pressure in duct
= 28.35 − 0.51 = 28.31 in. H_2O.

Step 5. Calculate volumetric flow rates acfm, scfm and dscfm.
The duct area (A) for a diameter of 41.5 in. = $(41.5/12)^2 \times 0.786 = 9.4\ \text{ft}^2$

Actual volumetric gas flow:

$$Q = 60 \times V_s \times A = 60 \times 63.31 \times 9.4 = 35{,}707\ \text{actual ft}^3/\text{min (acfm)}$$

Correct the volumetric stack flow (Q) to standard conditions, 68°F and 29.92 in. Hg:

$$Q_s = Q \times (528/T_s) \times (P_s/29.92) = Q \times 17.647 \times (P_s/T_s)$$
$$= 35{,}707 \times 17.647 \times (28.31/1004) = 17{,}768\ \text{std. ft}^3/\text{min (scfm)}$$

Correct Q_s to dry basis:

$$Q_{sd} = Q_s \times (1 - B_{ws}) = 17{,}768 \times (1 - 0.1743) = 14{,}676\ \text{dry std. ft}^3/\text{min (dscfm)}$$

where A = cross-sectional area of duct at sample point (ft^2) = 9.4 ft^2
B_{ws} = water vapor in gas stream (proportion by volume) = 17.43%
C_p = Pitot tube calibration coefficient = 0.843
G_d = flue gas specific gravity relative to air, dimensionless = $\rho/0.075$
m_g = mass flow rate of wet flue gas
M_s = molecular weight of wet flue gas
M_d = molecular weight of dry flue gas
P_s = absolute gas pressure of duct (in. Hg) = 28.31
ΔP = velocity pressure measured by Pitot tube (in. WC) = 0.615
Q = actual flue gas volumetric flow rate (acfm)

Q_s = volumetric gas flow at standard conditions (scfm)
Q_{sd} = dry standard volumetric gas flow rate (dscfm) $= Q_s \times (1 - B_{ws})$
T_s = flue gas temperature (°R) $= 544 + 460 = 1004°R$
V_s = flue gas linear velocity (ft/s) $= 63.9$
ρ = actual flue gas density (lb/ft^3)
V_s = stack gas linear velocity, ft/s $= 85.49 \times C_p \times \sqrt{(\Delta P)} \times \sqrt{(T/(P_s \times M_s))}$

Step 6. Calculate actual gas density and mass flow rate of wet flue gas.

The *actual gas density* in the *stack* ρ is calculated at the absolute pressure in the duct P_s, the actual temperature T_s and the molecular weight M_s, as follows:

$$\rho = 0.04585 \times P_s \times M_s/T_s = 0.03585 \text{ lb/ft}^3$$

$$= 0.075(460 + 68)(28.31)/(544 + 460)(29.96) = 0.03721 \text{ lb/ft}^3$$

where P_s = 28.35 in. Hg
T_s = $544 + 460 = 1004°R$

The *mass flow rate* in the *stack* m_g is calculated from the volumetric flow Q and the gas density ρ:

$$m_g = Q \times 60 \times \rho = 35{,}707 \times 60 \times 0.03721 = 79{,}719 \text{ pounds per hour (lb/h)}$$

or

$$m_g = 4.995 \times Q_{sd} \times G_d/(1 - B_{ws})$$

where P_s = 28.35 in. Hg
T_s = 1004°R

Step 7. Calculate percent excess air

$$\%EA = 100 \times \left\{\frac{\%O_2 - (0.05 \times \%CO)}{(0.264 \times \%N_2) - \%O_2 + (0.5 \times \%CO)}\right\}$$

or, neglecting CO (usually very small):

$$\%EA = 100 \times (\%O_2)/(0.264 \times \%N_2 - \%O_2)$$

$$= 100 \times (10.3)/(0.264 \times 91.1 - 10.3) = 74.9\% \text{ excess air}$$

where B_{ws} = water vapor in gas stream (proportion by volume)
%CO = carbon monoxide in gas stream (percent)
$\%CO_2$ = carbon dioxide in gas stream (percent)
%EA = excess air for combustion (percent)
F_0 = correction factor for results comparison
M_d = molecular weight of dry flue gas (lb/(lb · mol))
M_s = molecular weight of wet flue gas (lb/(lb · mol))
$\%N_2$ = nitrogen in gas stream (percent)
$\%O_2$ = oxygen in gas stream (percent)

Step 8. Calculate correction factor, for comparison with standard O_2 or CO_2.

$$F_0 = (20.9 - \%O_2)/\%CO_2 = (20.9 - \%O_2)/(20.9 - \%O_2)$$

2.2 EPA Method 5 Calculations for Particulate Matter (PM)

A sample is extracted from the stack using the sampling train shown in Fig. 4.1.

Sample volume, correcting to standard conditions

$$V_{\mathrm{std}} = 17.647 \times V_{\mathrm{m}} \times Y \times (P_{\mathrm{b}} + \Delta H/13.6)/T_{\mathrm{m}}$$

Isokinetic variation

$$I = 0.09450 \times (T_{\mathrm{s}} \times V_{\mathrm{m}})/[P_{\mathrm{s}} \times V_{\mathrm{s}} \times A_{\mathrm{n}} \times \theta \times (1 - B_{\mathrm{ws}})]$$

Particulate concentration

$$C_{\mathrm{s}} = 15.432 \times [m_{\mathrm{n}}/V_{\mathrm{std}}]$$

Particulate mass rate

$$m_{\mathrm{p}} = 0.008571 \times C_{\mathrm{s}} \times Q_{\mathrm{sd}}$$

where A_{n} = cross-sectional area of nozzle opening (ft^2)
B_{ws} = water vapor in gas stream (proportion by volume)
C_{s} = particulate concentration of gas stream (gr/dscf)
ΔH = orifice meter differential pressure (in. WC)
I = isokinetic variation of sampling rate (percent)
m_{n} = total particulate collected in sampling train (g)
m_{p} = particulate mass flow rate (lb/h)
P_{b} = barometric pressure (in. Hg)
P_{s} = absolute gas pressure of duct (in. Hg)
Q_{sd} = dry standard volumetric gas flow rate (dscfm)
T_{m} = sampling train meter temperature (°R)
T_{s} = flue gas temperature (°R)
V_{m} = volume of gas sample measured by gas meter (ft^3)
V_{std} = gas volume corrected to standard conditions (dscf)
V_{s} = flue gas linear velocity (ft/s)
Y = dry gas meter calibration coefficient
θ = total sampling time of run (min).

2.3 Calculation for Metals

The sample of particulate matter collected on the filter is analyzed for metals. The following calculation is used to relate this data to the concentration and emission rate of the metals:

Step 1. Concentration ($\mu\mathrm{g/Nm}^3$)

$$\mu\mathrm{g/Nm}^3 = \frac{35.3\ \mathrm{ft}^3}{\mathrm{m}^3} \times \frac{\mu\mathrm{g\ in\ sample}}{V_{\mathrm{m}}\ (\mathrm{dscf})}$$

where V_{m} = exhaust gas volume through meter (dscf)

Step 2. Emission rate (lb/h)

$$\begin{aligned}(\mathrm{lb/h}) &= \frac{\mu\mathrm{g}}{\mathrm{Nm}^3} \times \frac{\mathrm{lb}}{454} \times \frac{\mathrm{g}}{1000\ \mathrm{mg}} \times \frac{m_{\mathrm{g}}}{1000\ \mu\mathrm{g}} \times \frac{\mathrm{dscf}}{\mathrm{min}} \times \frac{60\ \mathrm{min}}{\mathrm{h}} \times \frac{0.0283\ \mathrm{Nm}^3}{\mathrm{dscf}} \\ &= (3.74 \times 10^{-9}) \times \mathrm{dscfm} \times \mu\mathrm{g/Nm}^3\end{aligned}$$

where dscfm = volumetric flow rate in the source (stack)

EXAMPLE:

$$\text{Sum of metals} = 284\ \mu g/Nm^3$$

$$\text{dscfm} = 14{,}676\ Nm^3$$

$$\text{lb/h} = 3.74 \times 10^{-9} \times 14{,}676\ (Nm^3) \times 284\ (\mu g/Nm^3)$$

$$= 0.015\ \text{lb/h}$$

2.4 Typical Calculation Summary for MSW Incinerator ESP Stack

Parameter	Run 1	Run 2	Run 3
Date of run	6/14/95	6/14/95	6/14/95
Sample duration (min)	60	60	60
Barometric pressure (in. Hg)	28.35	28.35	28.35
Static pressure of duct (in. H_2O)	−0.51	−0.51	−0.51
Absolute pressure of duct (in. Hg)	28.31	28.31	28.31
Meter coefficient	1.0019	1.0019	1.0019
Pitot tube coefficient	0.843	0.843	0.843
Nozzle diameter (in.)	0.31	0.31	0.31
Area of nozzle opening (ft^2)	0.000524	0.000524	0.000524
Average ΔP	0.7843	0.7418	0.7846
Average ΔH	2.45	2.52	2.85
Average stack temperature (°R)	1004.00	979.88	981.67
Average meter temperature (°R)	569.38	572.88	583.17
Meter volume (ft^3)	51.59	52.5	56.45
Dry standard sample volume (ft^3)	45.70	46.24	48.88
Collected condensate volume (ml)	205.0	222.0	183.0
Volume of water vapor (std. ft^3)	9.65	10.45	8.61
Moisture content of flue gas (% v/v)	17.43	18.43	14.98
Sum of metals collected (μg)	148.2	208.4	180.2
Sum of metals concentrations (μg/dscm)	284.0	390.1	344.6
Sum of metals mass rates (lb/h)	0.006	0.008	0.007
Source gas velocity (ft/s)	63.9	59.8	63.0
Isokinetic variation (%)	92.2	98.5	95.1

3 REGULATORY EMISSION STANDARDS AND GUIDELINES

3.1 Technology-based Standards versus Risk-based Standards

The trend of regulations has been from risk-based standards toward technology-based standards. Before the present federal regulations were in place for municipal waste combustion, many states took the initiative and wrote standards based on the health risk associated with emissions at ground level, and required risk assessments to be prepared on the basis of modeling.

The Clean Air Act Amendments of 1990 required that Maximum Available Control Technology (MACT) standards be developed which would take into account the "best performing" combustion systems, requiring that new facilities meet these more stringent emission standards and that existing facilities be upgraded to higher standards.

In spite of having to comply with the new federal standards and more stringent state standards, permits may continue to be subject to public scrutiny on a one-by-one basis, which means that environmental impact will still be considered for pollutants which are not numerically regulated by federal and state regulations.

New Source Performance Standards (NSPS) are firm for new facilities, but less stringent guidelines are set for existing units. Note that "unit" means a single line of equipment, not the entire facility. All stack emission data is reported after correction to 7% oxygen.

3.2 Proposed Guidelines

In addition to stack emission requirements, guidelines have been issued regarding design and operation of waste combustion systems. Guidelines are used for review, but need not necessarily be followed if the performance and compliance tests can be met.

Combustion temperatures are to be maintained at 1800°F for one (or two) seconds after the last injection of secondary combustion air.

To confirm the probability of a given system being able to comply with these regulations, a number of calculations may be made. Some of these will be described and carried out for typical systems in the following sections.

3.3 Regulatory Standards

The following summaries greatly simplify the requirements, but are presented in order to point out the differences in general approaches to the various types of systems. More detailed summaries are presented in Tables 4.1–4.5.

Municipal Waste Combustor (MWC) emission limits have been established by the US EPA for existing and new units, and for two size categories: between 38.6 tons per day (t/d) and 248 t/d, and greater than 248 t/d, as shown in Table 4.1.

MWCs are required to demonstrate the following in periodic compliance tests:

Particulate matter (PM) emissions are limited to 0.015 grains/dscf at 7% oxygen.

Carbon monoxide (CO) emissions are not to exceed 50 ppm on a rolling average, with exemptions of 150 ppm for refuse-derived fuel (RDF) systems.

Hydrogen chloride (HCl) is limited to 30 ppm at 7% oxygen or 95% control.

Sulfur dioxide (SO_2) is limited to 50 ppm at 7% oxygen or 80% control.

Nitrogen oxides (NO_x) are limited to 150 ppm at 7% oxygen.

Mercury, lead and cadmium emissions are limited quantitatively.

The system is permitted to operate at a maximum feed rate, usually specified as tons per day of defined wastes having a specified reference heating value.

Biomedical Waste Combustor (BWC) emission limits have been established by the US EPA for existing and for new units for three size categories: small units, < 200 pounds per hour (pph); medium units, 200–500 pph; and large units, > 500 pph. Special consideration has been made for MWCs in rural areas, which have small environmental impact and also cannot support the costs of elaborate controls from an economic point of view. Emission limits for existing BWCs are shown in Table 4.2, and for new BWCs in Table 4.3.

BWCs are required to demonstrate the following:

TABLE 4.1 US EPA Emission Standards for Municipal Waste Combustors (MWCs)

Pollutant (test method)	Units	Emission limits at 7% oxygen			
		Existing units		New units	
		> 38.6 t/d	> 248 t/d	> 38.6 t.d	> 348 t/d
Particulates (EPA Method 5 or 29)	mg/dscm (gr/dscf)	70 (0.03)	27 (0.012)	24 (0.010)	24 (0.010)
Opacity	6 min. avg.	10%	10%	10%	10%
CO (EPA Method 10 or 108)	ppmv	100	40	20	20
Dioxins/furans (EPA Method 23)	ng/dscm Total	125	60 (ESP) 30 (FF)	13	13
HCl (EPA Method 26)	ppmv or % reduction	250 or 50%	31 or 95%	25 or 95%	25 or 95%
SO_2	ppmv or % reduction	80 or 50%	31 or 75%	30 or 80%	30 or 80%
NO_x	ppmv	N/A	200	N/A	150
Lead (EPA Method 29)	µg/dscm	1600	490	200	200
Cadmium (EPA Method 29)	µg/dscm	100	40	20	20
Mercury (EPA Method 29)	µg/dscm or % reduction	80 or 85%	80 or 85%	80 or 85%	80 or 85%

Source: EPA, December 1995.

TABLE 4.2 Emission Limits for Existing Medical Waste Incinerators

Pollutant (test method)	Small units (< 200 pph)	Medium units (200–500 pph)	Large units (> 500 pph)
Particulates (EPA Method 5 or 29)	115 mg/dscm (0.05 gr/dscf)	69 mg/dscm (0.03 gr/dscf)	34 mg/dscm (0.015 gr/dscf)
CO (EPA Method 10 or 108)	40 ppmv	40 ppmv	40 ppmv
Dioxins/furans (EPA Method 23)	125 ng/dscm total CDD/CDF (2.3 ng/dscm TEQ)	125 ng/dscm total CDD/CDF (2.3 ng/dscm TEQ)	125 ng/dscm total CDD/CDF (2.3 ng/dscm TEQ)
HCl (EPA Method 26)	100 ppmv or 93% reduction	100 ppmv or 93% reduction	100 ppmv or 93% reduction
SO_2 (testing not required)	55 ppmv	55 ppmv	55 ppmv
NO_x (testing not required)	250 ppmv	250 ppmv	250 ppmv
Lead (EPA Method 29)	1.2 mg/dscm or 70% reduction	1.2 mg/dscm or 70% reduction	1.2 mg/dscm or 70% reduction
Cadmium (EPA Method 29)	0.16 mg/dscm or 65% reduction	0.16 mg/dscm or 65% reduction	0.16 mg/dscm or 65% reduction
Mercury (EPA Method 29)	0.55 mg/dscm or 85% reduction	0.55 mg/dscm or 85% reduction	0.55 mg/dscm or 85% reduction

Source: EPA, July 1999.

TABLE 4.3 Emission Limits for New Medical Waste Incinerators

Pollutant (test method)	Small units (< 200 pph)	Medium units (200–500 pph)	Large units (> 500 pph)
Particulates (EPA Method 5 or 29)	69 mg/dscm (0.03 gr/dscf)	34 mg/dscm (0.015 gr/dscf)	34 mg/dscm (0.015 gr/dscf)
CO (EPA Method 10 or 10B)	40 ppmv	40 ppmv	40 ppmv
Dioxins/furans (EPA Method 23)	125 ng/dscm total CDD/CDF (2.3 ng/dscm TEQ)	25 ng/dscm total CDD/CDF (0.6 ng/dscm TEQ)	25 ng/dscm total CDD/CDF (0.6 ng/dscm TEQ)
HCl (EPA Method 26)	15 ppmv or 99% reduction	15 ppmv or 99% reduction	15 ppmv or 99% reduction
SO_2 (testing not required)	55 ppmv	55 ppmv	55 ppmv
NO_x (testing not required)	250 ppmv	250 ppmv	250 ppmv
Lead (EPA Method 29)	1.2 mg/dscm or 70% reduction	0.07 mg/dscm or 98% reduction	0.07 mg/dscm or 98% reduction
Cadmium (EPA Method 29)	0.16 mg/dscm or 65% reduction	0.04 mg/dscm or 90% reduction	0.04 mg/dscm or 90% reduction
Mercury (EPA Method 29)	0.55 mg/dscm or 85% reduction	0.55 mg/dscm or 85% reduction	0.55 mg/dscm or 85% reduction

Source: EPA, July 1999.

Particulate matter (PM) emissions are limited to 0.015 grains/dscf at 7% oxygen.

Carbon monoxide (CO) emissions are not to exceed 100 ppm on a rolling average.

Hydrogen chloride (HCl) is limited to 30 ppm at 7% oxygen or 95% control.

Sulfur dioxide (SO_2) is limited to 50 ppm at 7% oxygen or 80% control.

Nitrogen oxides (NO_x) are limited to 150 ppm at 7% oxygen.

Mercury, lead and cadmium emissions are limited quantitatively.

The system is permitted to operate at a maximum feed rate, usually specified as pounds per hour or tons per day of defined wastes.

Hazardous Waste Combustor (HWC) emission limits have been established by the US EPA, as shown in Table 4.4, for existing and new units, with specific limits for hazardous waste incinerators (HWIs), cement kilns (CKs) and lightweight aggregate kilns (LWAKs).

The metals are divided into Hg on the one hand, and semi- and low-volatile groups on the other. Averaging times are different for different pollutants. All data is corrected to 7% oxygen.

HWIs must demonstrate the following in trial burns and periodic compliance tests:

POHCs are required to demonstrate a 99.99% destruction and removal efficiency (DRE).

Hydrogen chloride (HCl) emissions are to be less than 5 lb/h or 99% removed.

Particulate matter (PM) emissions are limited to 0.03 grains/dscf at 7% oxygen.

Carbon monoxide (CO) emissions are not to exceed 150 ppm on a rolling average.

Mercury, lead and cadmium emissions are limited quantitatively.

TABLE 4.4 US EPA Emission Limits for Hazardous Waste Combustors, Final Rule, Sept. 30, 1999

			Emission limits					
			Incinerator		Cement kiln		Lightweight aggregate kiln	
Pollutant	Averaging time	Units	Existing	New	Existing	New	Existing	New
PM	CEM 2 h	mg/dscm (gr/dscf)	34.5 (0.015)	34.5 (0.015)			57 (0.025)	57 (0.025)
		kg/Mg			0.15	0.15		
CDD/DF	Stack	ng/dscm	0.2 or 0.4[1]	0.2	0.2 or 0.4[1]	0.2	0.2 or 0.4[1]	0.20
HC	CEM	ppmdv	10	10	10	10	20	20
CO	CEM	ppmdv	100	100	100	100	100	100
$HCl + C_2$	CEM 10 h	ppmdv	77	21	130	86	230	41
Hg	CEM 10 h	μg/dscm	130	45	120	56	47	33
Semivol. metals Pb, Cd, (sum)	CEM 10 h	μg/dscm	240	24	240	180	250	43
Low vol. metals As, Be, Cr Sb (sum)	Stack or CEM 10 h	μg/dscm	97	97	56	56	110	110

Note: All emission rates corrected to 7% oxygen.
Note: If the gas temperature at the inlet of the initial particulate control device is maintained below 400°F.

Critical operating parameters are to be maintained within limits established by the trial burn tests, and may require waste feed shutoffs when these limits are exceeded.

Waste Incinerator Emission Guidelines for some European countries are outlined in Table 4.5. They are notably different from the US EPA limits in the averaging times, i.e., 24-h average, maximum $\frac{1}{2}$ h, weekly mean and 24-h maximum. Also notable is the division of heavy metals into three classes, each class being the sum of a group of like-behaving metals. Note that the Federal Republic of Germany requires correction of the data only when the oxygen level exceeds 11% oxygen, not otherwise, whereas the European Community Directive calls for correcting all data to 11% oxygen.

3.4 Good Combustion Practice

US EPA Standards for Good Combustion Practice (GCP) have been developed which apply to all waste combustors: essentially, they require that no waste be fed until the furnace temperature is at least 1600°F, and that the gaseous products of combustion be retained for at least one second at 1800°F or higher for most wastes, or for two seconds at 2000°F for wastes containing large quantities of halogenated compounds which are more difficult to destroy. Hazardous waste incinerators have been subjected to stringent regulation due to the potentially toxic emissions from the wide range of organic and inorganic compounds which may be in the wastes, and which define them as hazardous. These regulations require that trial burns be performed as a permit condition, during which it is demonstrated that the system can achieve 99.99% destruction and removal efficiency (DRE) while burning the principal organic hazardous compounds (POHC) and heavy metals. The assumption behind

TABLE 4.5 Waste Incinerator Emission Guidelines for Some European Countries

Measurement	Federal Republic of Germany 17.BImSch V, Nov. 1990 All waste plants mg/Nm^3 dry at > 11% oxygen		European Community Directive for new facilities mg/Nm^3 at 11% O_2	
	24-h avg.	max. $\frac{1}{2}$ h	max. $\frac{1}{2}$ h	24-h max.
HCl	10	60	10	5
SO ($SO_2 + SO_3$)	50	200	50	25
HF	1	4	2	1
$NO_x(NO_2)$	200	400	–	–
CO	50	(h) 100	–	50
C (organic)	10	20	10	5
Particulates	10	30	10	5
Heavy metals:				
Class I	Cd + Tl $\Sigma = 0.05$ (> 0.5 h)		Cd + Hg = 0.2	
Class II	Hg = 0.05 (> 0.5 h)		Hg = 0.05 (> 0.5 h)	
Class III	Sb, As, Pb, Co, Cr, Cu, Mn, V, Sn, Ni $\Sigma = 0.5(> \frac{1}{2}\ h < 2\ h)$		Sb, As, Pb, Co, Cu, Mn, V, Sn, Ni $\Sigma = 0.5(> 0.5\ h)$	
PCDD/PCDF	0.1 ng/Nm^3 I-TEQ > 500 min		0.1 ng/Nm^3 I-TEQ (8-h avg.)	
Combustion temperature	1200°C > 2 s		850°C at > 6% O_2 > 2 s	

Source: Licata *et al.*, 1999.

these tests was that the emissions would be related to the feed materials; hence the feed must be analyzed in normal operation and be anticipated as the basis for the trial burns. In view of the fact that modern combustion and emission control systems which can meet present regulatory standards actually destroy or remove all but traces of the target substances, the tendency has been to focus on the actual emissions and assure that they will not exceed the new, stringent standards.

3.5 Stack Testing and Monitoring

An essential part of this focus on actual stack emissions, which applies to all waste combustors, is the requirement that monitoring instruments be installed and operated, providing a continuous record of compliance. Continuous measurement of carbon monoxide (CO) and/or hydrocarbons (HC) and oxygen provides assurance of good combustion. These are useful surrogates for trace organic compounds including dioxins and furans. Measurement of HCl, SO_2 and NO_x may be required, where applicable. Opacity measurements provide continuous supervision of the combustion process as well as the emission control system.

3.6 Ash Residue Management

Ash residues from combustion of wastes are generally required to be tested for toxicity and managed appropriately. If found to fail the toxic characteristic leaching procedure (TCLP) test, they may have to be treated, or disposed in suitable landfills. The source of toxicity is generally the heavy metals, specifically lead and cadmium, which are not destroyed by the combustion process and hence end up in the flyash or bottom ash residues. The use of alkaline reagents for control of the acid gases adds another component to the disposal problem, since excess alkalinity can increase the solubility of these metals in the ash residues.

3.7 Operator Certification and Training

In addition to requiring good combustion practice and regulating emissions, an essential component of the newly promulgated regulations is the requirement for operator training and certification, applied appropriately to municipal, medical and hazardous waste combustion systems.

4 CALCULATIONS TO CONFIRM COMPLIANCE WITH STANDARDS

Any combustion/emission control system consists of a succession of components which must be designed to operate as a unified system while burning the specified wastes, achieving specified furnace temperatures and meeting the regulatory emission limits, while being fired at a specified heat input, or waste feed rate(s).

Calculations made regarding such unified systems must show that the components can all perform at the required rates, specifically at the rates allowed by the permits. Compliance tests are required to confirm emissions which, in principle, are dependent on the performance of the system.

For example, the gas flows provided by the combustion air fans, passing through the primary and secondary chambers, boilers and, taking into consideration infiltration, the emission control systems and induced draft fan, should be consistent, and provide the desired secondary furnace retention time at the required furnace temperature. In addition, the flow rates of quench water, scrubber blowdown and reagent feed must be sufficient to control the specified temperatures and quantities of acid gases on which the design is based. In order to establish these quantities, a complete mass flow calculation is needed, based on a heat and mass-flow analysis.

In the following sections, example calculations are provided, encompassing the complete system from feed to stack.

The proposer of a waste combustion system must make an application to regulatory authorities which predicts that the system will comply with local, state and federal requirements. The permit writer of the regulatory authority must review this submission and make a judgement as to whether the system will probably comply. Subsequently, the system will have to be tested in order to confirm compliance, and to obtain a permit to operate.

This section will review the basic requirements which must be met by three basic types of combustion system, each of which is subject to substantially different regulatory requirements.

4.1 Municipal Waste Combustion System

The following is an approximate calculation for a 1000 ton per day (tpd) municipal waste combustion system with heat and power recovery. A single unit is considered. In most cases, plant capacity is divided into two or even three units. The emission consequences of each of the multiple units would be essentially the same as those of this single unit.

Waste composition—Typical composition of municipal solid waste:

Component	Percent in MSW
Carbon	26.4
Hydrogen	3.74
Oxygen	18.19
Nitrogen	0.40
Chlorine	0.31
Fluorine	0.01
Sulfur	0.20
Moisture	30.0
Ash/inert	20.5

$$\text{Higher heating value (HHV)} = 5000 \text{ Btu/lb}$$

$$\text{Moisture and ash-free HHV (MAFHHV)} = 10{,}000 \text{ Btu/lb}$$

$$\text{Waste feed rate at 1000 tpd} = 41.7 \text{ tph} = 83{,}000 \text{ lb/h}$$

$$\begin{aligned}\text{Heat released by waste} &= 83{,}000 \text{ (lb/h)} \times 5000 \text{ (Btu/lb)}\\ &= 416{,}600{,}000 \text{ Btu/h}\end{aligned}$$

The heat and mass balance determines that a furnace temperature of 2200°F requires the use of 88% excess air, assuming zero heat loss from the primary furnace.

The heat and mass balance determines the flow rate of gaseous products for the 2200°F furnace temperature to be 7.48 lbp/lbf.

The mass flow of product gases = 83,000 (lb/h) × 7.48(lbp/lbf) = 620,840 lb/h

For a typical stoker heat release rate of 350,000 Btu/(h · ft^2), the grate area would be:

$$416{,}000{,}000 \text{ (Btu/h)}/350{,}000 \text{ (Btu/(h} \cdot \text{ft}^2\text{))} = 1190 \text{ ft}^2$$

At 25 ft long, width = 46ft

For a furnace volumetric heat release of 8000 Btu/(h · ft^3), the furnace volume would be:

$$416{,}600{,}000 \text{ (Btu/h)}/8000 \text{ (Btu/(h} \cdot \text{ft}^3\text{))} = 52{,}000 \text{ ft}^3$$

For a furnace plan area of 25 ft × 46 ft = 1150 ft^2

$$\text{furnace height} = 52{,}000/1150 = 45 \text{ ft}$$

Determine the volume of combustion chamber equivalent to 1 s retention time:

$$\begin{aligned}\text{Volume of gases at 2200°F} &= 620{,}840 \text{ (lbp/h)} \times (460 + 2200) \times 13.5 \text{ (ft}^3\text{/lb)}/(530 \times 3600)\\ &= 11{,}670 \text{ ft}^3\text{/s}\end{aligned}$$

For a retention time of 1 s

$$\text{volume is } 11{,}670 \text{ ft}^3$$

For a plan area of 25 × 46 = 1150 ft^2

$$\text{height} = 11{,}670/1150 = 10 \text{ ft}$$

For actual height of 45 ft

$$\text{retention time} = 4.4 \text{ s}$$

Determine heat recovered by the boiler, Q_b, assuming the boiler exit temperature is 450°F.

Assuming average specific heat of gases is 0.30 Btu/(lb · °F), the heat absorbed by the boiler, Q_b is

$$Q_b = wc(T_{in} - T_{out}) = 620{,}840 \text{ [lbp/h]} \times 0.30 \text{ [Btu/(lb} \cdot {}^\circ\text{F)]} \times (2200 - 450) \text{ [}^\circ\text{F]}$$

$$= 325{,}940{,}000 \text{ Btu/h}$$

Calculate boiler efficiency = heat absorbed by boiler/heat supplied in fuel:

$$\eta_b = 325{,}940{,}000/416{,}600{,}000 = 78.5\%$$

Calculate steam generation and power production for typical boiler conditions:

Steam generation at 1200 Btu/lb steam

$$= 325{,}940{,}000/1200 \text{ [Btu/lb]} = 27{,}162 \text{ lb/h}$$

Power generation at 11,000 Btu/(kW · h)

$$= 325{,}940{,}000 \text{ [Btu/h]}/11{,}000 \text{ [Btu/(kW} \cdot \text{h)]} = 29{,}631 \text{ kW} = 29.6 \text{ MW}$$

Calculate water spray needed to cool gases to 300°F:

$$620{,}840 \text{ [lbp/h]} \times 0.3 \text{ [Btu/(lb} \cdot {}^\circ\text{F)]} \times (450 - 300) \text{ [}^\circ\text{F]} = 27{,}900{,}000 \text{ Btu/h}$$

$$\text{Water evaporated} = 27{,}900{,}000 \text{ [Btu/h]}/1000 \text{ [Btu/lb]}$$

$$= 27{,}900 \text{ lb/h} = 56.5 \text{ gal/min}$$

Gas flow entering fabric filter and handled by the induced-draft fan:

$$620{,}900 + 27{,}900 = 648{,}000 \text{ lb/h}$$

The specific volume of the gases is calculated to be 13.59 std. ft^3/lb at standard conditions.

$$\text{At } 300^\circ\text{F, the actual flow volume} = 13.59 \text{ [std. ft}^3\text{/lb]} \times [(460 + 300)/530]$$

$$\times\ 648{,}000 \text{ [lb/h]}/60 \text{ [min/h]} = 210{,}470 \text{ acfm}$$

Calculate the volumetric composition of the stack gases:

Gas	Wet basis (%)	Dry basis (%)	
CO_2	8.47	9.80	($CO_2 + O_2 = 19.73\%$)
O_2	8.58	9.93	
N_2	69.30	80.18	
H_2O	13.58	0	
	100%	100%	
HCl = 396 ppmv (parts per million by volume)			
HF = 24.5 ppmv			
SO_2 = 406 ppmv			

Calculate factor for correction to composition of stack gases to 7% oxygen, dry. Note the difference between the actual sum of $CO_2 + O_2$ and the "standard" factor:

$$\text{Actual factor} = [CO_2 + O_2 - O_2\%]/[CO_2 + O_2 - 7\%]$$
$$= [19.73 - 9.93]/[19.7 - 7] = 0.77$$

"Regulatory" factor =

$$\text{ratio} = \frac{20.5 - (O_2)_{\text{actual}}}{20.5 - (O_2)_{\text{standard}}} = \frac{20.5 - 9.93}{20.5 - 7} = 10.58/13.5$$
$$= 0.78 = 1/1.276$$

Using the "standard" factor to correct from the higher oxygen content (more excess air) to the "reference" 7% oxygen increases the concentration:

$$(HCl)_{\text{corrected}} = 396 \times 1.276 = 504 \text{ ppm}$$
$$(SO_2)_{\text{corrected}} = 406 \times 1.276 = 517 \text{ ppm}$$

To comply with 25 ppmv HCl, the removal efficiency must be

$$(500 - 25)/500 = 95\%$$

To comply with 30 ppmv SO_2, the removal efficiency must be

$$(517 - 30)/640 = 94\%$$

The alternative is 80% removal, reducing 640 to

$$(517 - (0.85 \times 517)) = 103 \text{ ppmv}$$

Estimate alkaline reagent needed to control SO_2 and HCl to regulatory limits.

Calculate stoichiometric requirements:

HCl input is 0.31 lb/100 lb, or

$$83{,}000 \text{ [lbw/h]} \times 0.0031 \text{ [Cl]} = 257 \text{ lb/h} \times [35 + 1]/35 = 264 \text{ lb/h}$$

From Table 4.6, at 300°F, the $Ca(OH)_2$ required would be

$$1.6 \times 264 = 422 \text{ lb/h}$$

Sulfur input is 0.20 lb/100 lb, or

$$83{,}000 \times 0.002 = 166 \text{ lb/h [S]} \times [44/12] = 609 \text{ lb/h}$$

From Table 4.6, at 300°F, the $Ca(OH)_2$ required would be

$$3.7 \times 609 = 2253 \text{ lb/h}$$

In practice, the lime which must be added is greater than the stoichiometric quantity, depending upon the percent removal required. The above numbers should probably be increased by a factor of at least 50%, depending upon the effectiveness of the APC system itself.

Calculate metals in stack gases after emission controls of MSW facility

Using the data from the tests of the 650 tpd combustor at Burnaby, a facility with dry lime injection and with reactor tower and fabric filter:

TABLE 4.6 Calculated Theoretical Reagent Requirements per Tonne of Gaseous Pollutants

Pollutant	Gas temperature °C	Gas temperature °F	Stoichiometric ratio $Ca(OH)_2$	Consumption in tonnes (94%)
SO_2	120–160	250–320	3.0	3.7
	160–220	320–428	3.5	4.3
	220–280	428–536	5.0	6.2
HCl	120–160	250–320	1.5	1.6
	160–220	320–428	1.8	2.0
	220–280	428–536	2.0	2.2
HF	120–280	250–536	1.0	2.1

Source: Benson and Licata, 1998.

$$\text{Waste flow} = 650 \times 2000/24 = 54{,}170\ \text{lb/h} = 27\ \text{t/h}$$

$$\text{Stack gas flow} = 830\ \text{dscm/min} \times 60 = 49{,}800\ \text{dry std m}^3\text{/h}$$

Cadmium emissions:

Cadmium in the waste at Burnaby was estimated to be 27,000 lb/million tons (Mt). (This is 13.5 parts per million parts (ppm) weight of MSW. Assuming ash is 20% of MSW, this is 13.5/0.2 = 67.5 ppm in the ash if all goes to ash, or, if 80% goes to ash, 54 ppm cadmium in the dry ash.) The cadmium leaving the boiler was measured to be 5723 lb/Mt, a reduction factor of 5. This means that 80% of the cadmium reported to the ash residues and 20% to the gases entering the emission controls. From the measured stack emissions of 18 lb/Mt, the control efficiency is calculated to be 99.7%.

Cadmium in the stack was

$$18\ [\text{lb/Mt}] \times 27\ [\text{t/h}] \times 454\ [\text{g/lb}]/1{,}000{,}000\ [\text{lb/Mt}] = 0.221\ \text{g/h}$$

Cadmium concentration in stack

$$= 0.221\ [\text{g/h}]/49{,}800\ [\text{dry std. m}^3\text{/h}] = 4.43\ \mu\text{g/dscm}$$

This is 22% of the US EPA standard of 20 μg/dscm for new MWCs. If the control efficiency had been 99.00 instead of 99.70, the result would have been 15 μg/dscm.

European guideline:	200 μg/dscm corrected to 11% for Hg + Cd
US EPA Cd standard:	
for small existing MSW units:	100 μg/dscm corrected to 7% oxygen
for large existing MSW units:	40 μg/dscm corrected to 7% oxygen
for new MSW units:	20 μg/dscm corrected to 7% oxygen

Lead emissions:

Lead in the stack was

$$363\ [\text{lb/Mt}] \times 27\ [\text{t/h}] \times 454\ [\text{g/lb}]/1{,}000{,}000\ [\text{lb/Mt}] = 4.45\ \text{g/h}$$

Lead concentration in stack

$$= 4.45\ [\text{g/h}]/49{,}800\ [\text{m}^3\text{/h}] = 89.4\ \mu\text{g/dscm}$$

This is 44% of the US EPA standard of 200 μg/dscm for new MWCs. If the control efficiency had been 97.00 instead of 98.3, this would have resulted in 158 μg/dscm.

Mercury emissions:

Mercury entering the APC was

$$3630\ [\text{lb/Mt}] \times 27\ [\text{t/h}] \times 454\ [\text{g/lb}]/1{,}000{,}000\ [\text{lb/Mt}] = 44.5\ \text{g/h}$$

Mercury concentration in stack

$$= 44.5\ [\text{g/h}]/49{,}800\ [\text{m}^3/\text{h}] = 894\ \mu\text{g/dscm}$$

This exceeds the standard of 20 μg/dscm for new units and 100 μg/dscm for small existing units. Carbon injection would be required, with a control efficiency of 90%.

4.2 Biomedical Waste Combustion System

The following is an approximate calculation for a 1000 tpd municipal waste combustion system with heat and power recovery. A refractory chamber with 10% heat loss is assumed.

Composition of typical biomedical waste:

Component	Weight percent
Carbon	42.19
Hydrogen	4.20
Oxygen	19.41
Nitrogen	0.74
Chlorine	0.60
Fluorine	0.10
Sulfur	0.36
Moisture	20.76
Ash/inert	9.00

$$\text{HHV} = 7000\ \text{Btu/lb}$$

$$\text{MAF HHV} = 10{,}000\ \text{Btu/lb}$$

Heat released by waste

$$= 1000\ [\text{lb/h}] \times 7500\ [\text{Btu/lbw}] = 7{,}500{,}000\ \text{Btu/h}$$

Determine by heat balance the excess air required to maintain this temperature:

$$141\%$$

By heat and mass balance,

$$\text{product gas for } 1800^\circ\text{F primary furnace temperature is: } 9.36\ \text{lbp/lbw}$$

$$\text{Product gas flow, } W = 1000\ [\text{lbw/h}] \times 9.36\ [\text{lbp/lbw}] = 9360\ \text{lbp/h}$$

Grate area for heat release of 7,500,000 Btu/h at 150,000 Btu/(h · ft^2):

$$\text{Grate area} = 7{,}500{,}000\ [\text{Btu/h}]/150{,}000\ [\text{Btu/(h}\cdot\text{ft}^2)] = 50\ \text{ft}^2$$

For length = 12 ft, width = 4.2 ft

Volume of primary chamber for heat release of 25,000 Btu/(h · ft^3)

$$= 7{,}500{,}000/25{,}000 = 300 \text{ ft}^3$$

For diameter of 5ft,

$$\text{area} = 25 \times 0.786 = 19.65 \text{ ft}^2$$

$$\text{length} = 300 \text{ ft}^3/19.65 \text{ ft}^2 = 15.27 \text{ ft}$$

Determine volume of secondary combustion chamber equivalent to 1 s retention time:

Volume of gases at 1800°F

$$= 9360 \text{ [lb/h]} \times [460 + 1800] \times 13.5 \text{ [ft}^3\text{/lb]}/(530 \times 3600) = 150 \text{ ft}^3\text{/s}$$

For retention time of 2 s,

$$\text{volume is } 300 \text{ ft}^3$$

For diameter of 4.5 ft,

$$\text{area} = 15.92 \text{ ft}^2$$

$$\text{length} = 300/15.92 = 18.8 \text{ ft}$$

Secondary chamber is about 4.5 ft diam. × 20 ft long.

Determine heat exchange to boiler, assuming boiler exit temperature is 450°F.

Assuming average specific heat, c, of gases is 0.30 Btu/(lb · °F),

$$Q = wc(T_{\text{in}} - T_{\text{out}}) = 9360 \text{ [lb/h]} \times 0.30 \text{ [Btu/(lb} \cdot \text{°F)]} \times (1800 - 450)$$
$$= 3{,}790{,}000 \text{ Btu/h}$$

Boiler efficiency

$$= 3{,}790{,}000 \text{ [Btu/h output]}/7{,}500{,}000 \text{ [Btu/h input]} = 50.5\%$$

Calculate steam generation and power production for typical boiler conditions.

Steam generation at 1000 Btu/lb steam

$$= 3{,}790{,}000/1000 = 3790 \text{ lb/h at 200 psig}$$

Power generation at 15,000 Btu/(kW · h)

$$= 3790/15{,}000 = 2530 \text{ kW} = 2.53 \text{ MW}$$

Calculate water spray needed to cool gases to 300°F

$$9360 \text{ [lbp/h]} \times 0.3 \text{ [Btu/(lb} \cdot \text{°F)]} \times (450 - 300) \text{ [°F]} = 421{,}200 \text{ Btu/h}$$

Water evaporated

$$= 421{,}200 \text{ [Btu/h]}/1000 \text{ [Btu/lb]} = 421 \text{ [lb/h]}/8.6 \text{ [lb/gal]} \times 60 \text{ [min/h]}$$
$$= 0.82 \text{ gal/min}$$

Gas flow entering fabric filter and handled by the induced-draft fan:

$$9360 + 421 = 9781 \text{ lb/h}$$

The specific volume is calculated to be 13.59 std. ft^3/lb, at standard conditions.

The actual flow volume

$$= 13.59 \text{ [std. ft}^3\text{/lb]} \times [(460 + 300)/530] \times 9781 \text{ [lb/h]}/60 \text{ [min/h]}$$

$$= 3206 \text{ ft}^3\text{/min.}$$

Calculate volumetric composition of stack gases:

Gas	Wet basis (%)	Dry basis (%)	
CO_2	6.73	7.54	($CO_2 + O_2 = 7.54 + 12.37 = 19.91\%$)
O_2	5.69	12.37	
N_2	71.39	80.04	
H_2O	10.80	0.0	
	100%	100%	
HCl = 305 ppmv			
HF = 19 ppmv			
SO_2 = 313 ppmv			

Calculate composition of stack gases corrected to 7% oxygen, dry:

$$\text{Actual factor} = [CO_2 + O_2 - O_2\%]/[CO_2 + O_2 - 7\%] = [19.91 - 11.03]/[19.91 - 7]$$

$$= 0.688 = 1/1.454$$

$$\text{Regulatory ratio} = \frac{20.5 - (O_2)_{\text{actual}}}{20.5 - (O_2)_{\text{standard}}} = \frac{20.5 - 11.03}{20.5 - 7} = 10.58/13.5 = 0.70 = 1/1.276$$

Correcting from the higher oxygen content (more excess air) to 7% oxygen increases the concentration:

$$(HCl)_{\text{corrected}} = 305 \times 1.276 = 389 \text{ ppm}$$

$$(SO_2)_{\text{corrected}} = 313 \times 1.276 = 400 \text{ ppm}$$

To comply with 15 ppm HCl, the removal efficiency must be

$$(389 - 15)/389 = 96\%$$

Alternatively, 99% removal

$$= 389 \times (1 - 0.99) = 3.9 \text{ ppmv}$$

To comply with 55 ppm SO_2, the removal efficiency must be

$$(400 - 55)/400 = 86.3\%$$

Using the alternative of 85% removal would reduce 400 ppm to

$$(400 - (0.85 \times 400)) = 60 \text{ ppm}$$

Estimate alkaline reagent needed to control SO_2 and HCl to regulatory limits.

HCl input is 0.31 lb/100 lb, or

$$83{,}000\ [\text{lbw/h}] \times 0.0031\ [\text{Cl}] = 257\ \text{lb/h} \times [35 + 1]/35$$
$$= 264\ \text{lb/h}$$

Sulfur input is 0.20 lb/100 lb, or

$$83{,}000 \times 0.002 = 166\ \text{lb/h}\ [\text{S}] \times [44/12] = 609\ \text{lb/h}$$

To remove all of the sulfur,

$$\text{Ca(OH)}_2 = 609\ [\text{lb/h}] \times 3.7\ [\text{lb/lb}] = 2253\ \text{lb/h}$$

To remove all of the HCl,

$$\text{Ca(OH)}_2 = 264\ [\text{lb/h}] \times 1.6\ [\text{lb/lb}] = 422\ \text{lb/h}$$

The effectiveness of the emission control system is less than 100%; hence more reagent will actually be required.

4.3 Liquid Waste Combustion System

The following is a calculation for a 40 million Btu/h rotary kiln hazardous waste combustion system with heat recovery. A refractory chamber heat loss of 10% is assumed.

Liquid waste composition:

Component	Weight %
Carbon	57.30
Hydrogen	7.56
Oxygen	6.00
Nitrogen	22.52
Chlorine	1.47
Fluorine	0.01
Sulfur	0.01
Moisture	4.21
Ash/inert	0.92

$$\text{HHV} = 12{,}975\ \text{Btu/lb}$$
$$\text{MAF HHV} = 13{,}676\ \text{Btu/lbf}$$

Calculate waste feed rate for input of 40 million Btu/h:

$$40{,}000{,}000\ [\text{Btu/h}]/13{,}676 = 2925\ \text{lbw/h}$$

Determine excess air required to maintain 2200°F furnace temperature:

141%

Calculate product gas flow rate for 2200°F secondary furnace temperature:

19.33 lbg/lbw

Product gas flow

$$= 3083 \text{ lbw/h} \times 19.33 \text{ lbp/lbwf}$$
$$= 59{,}592 \text{ lbp/h}$$

Determine kiln volume for heat release rate of 35,000 Btu/(h · ft^3):

$$40{,}000{,}000 \text{ [Btu/h]}/35{,}000 \text{ [Btu/(h} \cdot \text{ft}^3\text{)]} = 11{,}428 \text{ ft}^3$$

For diameter of 10 ft,

$$\text{area} = 78.6 \text{ ft}^2$$
$$\text{length} = 1142/78.6 = 14.5 \text{ ft}$$

Determine volume of combustion chamber equivalent to 1 s retention time:

Volume of gases at 2200°F

$$= 59{,}592 \text{ [lbp/h]} \times (2200 + 460) \text{ [°F]} \times 13.3 \text{ [ft}^3\text{/lbp]}/(530 \times 3600)$$
$$= 1105 \text{ ft}^3\text{/s}$$

For retention time of 2 s,

$$\text{volume is } 2243 \text{ ft}^3$$

For diameter of 10 ft

$$\text{area} = 78.6 \text{ ft}^2$$
$$\text{length} = 2243/78.6 = 28.5 \text{ ft (OK)}$$

Heat release

$$= 40{,}000{,}000/2243 = 17{,}833 \text{ Btu/(h} \cdot \text{ft}^3\text{)}$$

Determine heat exchange to boiler, assuming boiler exit temperature is 450°F.

Assuming average specific heat, c, of gases is 0.30 Btu/(lb · °F),

$$Q = wc(T_{\text{in}} - T_{\text{out}}) = 59{,}592 \text{ [lbp/h]} \times 0.30 \text{ [Btu/(lb} \cdot \text{°F)]} \times (2200 - 450) \text{ [°F]}$$
$$= 31{,}286{,}000 \text{ Btu/h}$$

Boiler efficiency

$$= \text{output/input} = 31{,}286{,}000/40{,}000{,}000$$
$$= 78\%$$

Calculate steam generation and power production for typical boiler conditions:

Steam generation at 1000 Btu/lb steam

$$= 31{,}286{,}000/1000 = 31{,}286 \text{ lb/h at 200 psig}$$

Power generation at 13,000 Btu/(kW · h)

$$= 31{,}286{,}000/13{,}000 = 2407 \text{ kW} = 2.4 \text{ MW}$$

Calculate water spray needed to cool gases to 300°F:

$$59{,}592\ [\text{lb/h}] \times 0.3\ [\text{Btu/(lb}\cdot{}^{\circ}\text{F)}] \times (450 - 300)\ [{}^{\circ}\text{F}] = 2{,}681{,}600\ \text{Btu/h}$$

Water evaporated

$$= 2{,}681{,}600\ [\text{Btu/h}]/1000 = 2681\ \text{lb/h}$$
$$= 2681/8.6\ [\text{lb/gal}] \times 60\ [\text{min/h}]$$
$$= 5.24\ \text{gal/min}$$

Gas flow entering fabric filter and handled by the induced-draft fan:

$$59{,}592 + 2681 = 62{,}273\ \text{lb/h}$$

The specific volume is calculated to be 13.3 scf/lb at standard conditions.

$$\text{At } 300^{\circ}\text{F, the actual flow volume} = 62{,}273\ [\text{lb/h}] \times 13.3\ [\text{std. ft}^3/\text{lb}] \times [(460 + 300)/530]/60\ [\text{min/h}] = 19{,}794\ \text{acfm}$$

Calculate composition of actual stack gases:

Gas	Wet basis	Dry basis	
CO_2	7.17	7.63	$(CO_2 + O_2 = 18.51)$
O_2	10.23	10.88	
N_2	76.55	81.42	
H_2O	5.99	0	
	100%	100%	
HCl = 672 ppmv			
HF = 7.9 ppmv			
SO_2 = 7.2 ppmv			

Calculate composition of stack gases corrected to 7% oxygen, dry:

$$\text{Actual factor} = [CO_2 + O_2 - O_2\%]/[CO_2 + O_2 - 7\%] = [18.51 - 10.88]/[18.51 - 7] = 0.663 = 1/1.51$$

$$\text{Regulatory ratio} = \frac{20.5 - (O_2)_{\text{actual}}}{20.5 - (O_2)_{\text{standard}}} = \frac{20.5 - 10.88}{20.5 - 7} = 9.62/13.5 = 0.7126 = 1/1.403$$

Correcting from the higher oxygen content (more excess air) to 7% oxygen increases the concentration:

$$(\text{HCl})_{\text{corrected}} = 672 \times 1.4 = 941\ \text{ppmv}$$
$$(SO_2)_{\text{corrected}} = 7.2 \times 1.4 = 10.1\ \text{ppmv}$$

To comply with 99% control, HCl emissions would be

$$(1 - 0.99) \times 941 = 10\ \text{ppmv}$$

There is no requirement for SO_2 control, but it would absorb reagent, which must be considered.

Estimate alkaline reagent needed to control SO_2 and HCl to regulatory limits.

HCl input is 1.47 lb/100 lb, or

$$0.0147 \times 3083\ [\text{lbw/h}] = 45.3\ \text{lb/h} \times [35 + 1]/35$$
$$= 46.6\ \text{lb/h}$$

The 99% control standard would reduce the emissions to $(1 - 0.99) \times 46.6 = 0.5$ ppmv. The standard of 5 lb/h could be met with $(45.3 - 9)/45.3 = 90\%$ control.

In order to remove the HCl with lime, more than 1.5×46.6 lb/h = 70 lb/h $Ca(OH)_2$ would be needed, depending on the effectiveness of the acid gas control system.

4.4 Aqueous Waste Combustion System

The following is a calculation for a 40 million Btu/h rotary kiln hazardous waste combustion system with heat recovery. A heat loss of 10% is assumed for this refractory system.

Liquid waste composition: Fuel oil must be burned in order to destroy this aqueous waste. The quantity of oil and the composition of the mixture which would achieve a temperature of 1800°F with 50% excess air has been determined by heat balance and is shown in Table 3.7.

$$\text{HHV} = 4194\ \text{Btu/lb}$$
$$\text{MAF HHV} = 16{,}392\ \text{Btu/lbf (combustible heating value)}$$

Calculate waste feed rate for input of 40 million Btu/h:

$$40{,}000{,}000\ [\text{Btu/h}]/4194 = 9537\ \text{lbf/h}$$

Determine excess air required to maintain this temperature:

$$51\%$$

Calculate product gas flow rate for 1800°F secondary furnace temperature:

$$5.50\ \text{lbg/lbf}$$

Product gas flow

$$= 9537\ [\text{lbw/h}] \times 5.50\ [\text{lbp/lbw}] = 52{,}456\ \text{lbp/h}$$

Determine kiln volume for a volumetric heat release rate of 35,000 Btu/(h · ft^3):

$$40{,}000{,}000\ [\text{Btu/h}]/35{,}000\ [\text{Btu/(h} \cdot \text{ft}^3)] = 11{,}428\ \text{ft}^3$$

For diameter of 10 ft

$$\text{area} = 78.6\ \text{ft}^2$$
$$\text{length} = 1142/78.6 = 14.5\ \text{ft}$$

Determine volume of combustion chamber equivalent to 1 s retention time:

Volume at 1800°F

$$= 52{,}456\ [\text{lbp/h}] \times [1800 + 460] \times 14.26\ [\text{ft}^3/\text{lb}]/(530 \times 3600)$$
$$= 886\ \text{ft}^3/\text{s}$$

For retention time of 2 s

$$\text{volume is } 2 \times 839 = 1678 \text{ ft}^3$$

For diameter of 10 ft

$$\text{area} = 78.6 \text{ ft}^2$$

$$\text{length} = 1678/78.6 = 21.3 \text{ ft (OK)}$$

Heat release

$$= 40{,}000{,}000/1678 = 23{,}838 \text{ Btu/(h} \cdot \text{ft}^3)$$

Determine heat transferred to boiler, assuming boiler exit temperature is 450°F.

Assuming average specific heat of gases is 0.30 Btu/(lb · °F),

$$Q = wc(T_{\text{in}} - T_{\text{out}}) = 52{,}456 \text{ [lbp/h]} \times 0.30 \text{ [Btu/(lb} \cdot {}^\circ\text{F)]} \times (1800 - 450) \text{ [}^\circ\text{F]}$$
$$= 21{,}245{,}000 \text{ Btu/h}$$

Boiler efficiency

$$= \text{output/input} = 21{,}245{,}000/40{,}000{,}000$$
$$= 53\%$$

Calculate steam generation and power production for typical boiler conditions:

Steam generation at 1000 Btu/lb steam

$$= 21{,}245{,}000/1000 = 21{,}245 \text{ lb/h at 200 psig.}$$

Power generation at 13,000 Btu/(kW · h)

$$= 21{,}245{,}000/13{,}000 = 1634 \text{ kW} = 1.63 \text{ MW}$$

Calculate water spray needed to cool gases to 300°F:

$$52{,}456 \text{ [lbp/h]} \times 0.3 \text{ [Btu/(lb} \cdot {}^\circ\text{F)]} \times (450 - 300) \text{ [}^\circ\text{F]} = 2{,}360{,}500 \text{ Btu/h}$$

Water evaporated

$$= 2{,}360{,}500 \text{ [Btu/h]}/1000 \text{ [Btu/lb]}$$
$$= 2360 \text{ [lb/h]}/8.6 \text{ [lb/gal]} \times 60 \text{ [min/h]}$$
$$= 4.57 \text{ gal/min}$$

Gas flow entering fabric filter and handled by the induced-draft fan:

$$52{,}456 + 2360 = 54{,}816 \text{ lb/h}$$

The specific volume is calculated to be 14.26 ft^3/lb at standard conditions.

$$\text{The actual volume at } 300^\circ\text{F} = 54{,}816 \text{ [lb/h]} \times 14.26 \text{ [std. ft}^3\text{/lb]} \times [(460 + 300)/530]/60 \text{ [min/h]}$$
$$= 18{,}682 \text{ acfm}$$

Calculate composition of actual stack gases (see Section 2.4):

Gas	Wet basis (vol %)	Dry basis (vol. %)	
CO_2	7.99	10.65	($CO_2 + O_2 = 17.94$)
O_2	5.47	7.29	
N_2	61.44	81.91	
H_2O	24.99	0	
	100.00	100.00	
HCl = 38 ppmv			
HF = 33 ppmv			
SO_2 = 1465 ppmv			

Calculate composition of stack gases corrected to 7% oxygen, dry:

$$\text{Actual factor} = [CO_2 + O_2 - O_2\%]/[CO_2 + O_2 - 7\%] = [17.94 - 7.29]/[17.94 - 7]$$
$$= 0.973 = 1/1.027$$

$$\text{Regulatory ratio} = \frac{20.5 - (O_2)_{\text{actual}}}{20.5 - (O_2)_{\text{standard}}} = \frac{20.5 - 10.88}{20.5 - 7} = 9.62/13.5 = 0.9785 = 1/1.022$$

Correcting from the higher oxygen content (more excess air) to 7% oxygen increases the concentration:

$$(HCl)_{\text{corrected}} = 38 \times 1.022 = 39 \text{ ppm}$$
$$(SO_2)_{\text{corrected}} = 1465 \times 1.022 = 1497 \text{ ppm}$$

The HCl limit of 67 ppmv is complied with.

There is no SO_2 removal requirement. However, to meet 30 ppmv, (1497 – 30)/1497 = 98% control efficiency would be needed.

Calculate metals in stack gases after emission controls of HWC:

Assume the waste has been analyzed for cadmium, and the ash residue has been found to contain 5% lead. The input of cadmium is thus 3.4% ash, of which 5% or 0.17% is lead. The input of cadmium is then 2360 lbf/h (waste)×0.17/100 (lead) = 0.401 lb/h cadmium.

Assume all of the cadmium enters the APC, and that the APC efficiency for cadmium is 99.9%.

Cadmium emissions

$$= (1 - 0.999) \times 0.401 \text{ lb/h} = 4.1 \times 10^{-4} \text{ lb/h}$$

The actual wet volume of gases containing this cadmium is 19,948 acfm at 300°F. Correcting to dry gases, take out the volume of moisture: 24.99%. The dry volume is thus

$$19{,}948 \text{ acfm} \times (100 - 24.99)/100 = 14{,}963 \text{ acfm}$$

Correcting to 70°F :

$$\text{Standard dry volume} = 14{,}963 \text{ acfm} \times (70 + 460)/(300 + 460)$$
$$= 10{,}434 \text{ scfm}$$

The concentration in the stack gases, corrected for temperature, is

$$4.1 \times 10^{-4} \text{ [lb/h]} \times 434 \text{ [g/lb]} \times [35.3 \text{ ft}^3/\text{m}^3]/(10{,}434 \text{ [ft}^3/\text{min]} \times 60 \text{ [min/h]})$$
$$= 10\,\mu\text{g/m}^3$$

This must also be corrected from 7.29% to 7% oxygen using the factor derived above, 1.022:

$$\text{Corrected concentration of stack gas} = 1.022 \times 10$$
$$= 10.22\ \mu\text{g/dscm, corrected to 7\% oxygen}$$

This concentration can be compared with:

European guideline:	200 μg/dscm corrected to 11% for Hg + Cd
US EPA standard:	
for small existing MSW units:	100 μg/dscm corrected to 7% oxygen
for large existing MSW units:	40 μg/dscm corrected to 7% oxygen
for new MSW units:	20 μg/dscm corrected to 7% oxygen

5 ENVIRONMENTAL IMPACT OF STACK EMISSIONS

The environmental impact of incinerator stack emissions depends upon many factors which will be discussed in this section. Concentrations of pollutants emitted by the stack are reduced by dispersion in the atmosphere before the gases reach ground level or elevated receptors of concern. The resulting concentrations are evaluated by comparison with health risk standards which have established "acceptable ground level concentrations" (AGLC).

5.1 Stack Emissions Concentrations

Volume of gases leaving the stack. The volume of stack gases discharged to the atmosphere depends upon whether or not a heat recovery boiler, heat exchanger, dilution air or water evaporation is used to cool the gases. Hence, for a given mass emission rate, the concentration of the pollutants in the gases leaving the stack will depend upon the actual volume of gases discharged.

Figure 4.2 shows the major alternative emission control systems and the stack temperatures characteristic of each system, ranging from 150°F to 390°F.

Table 4.7 shows the range of actual cubic feet per minute (acfm) discharged *per million Btu* of heat released by the waste, and the corresponding actual cubic meters per second (acms).

At a heat release rate of one million Btu per hour, an incinerator with *no* emission control device would discharge 1510 acfm of gases at 1800°F directly to the atmosphere.

- If the gases are diluted by a draft control device such as a barometric damper, reducing their temperature to 1300°F, the quantity increases to 1730 acfm.
- If a wet scrubber is applied to the 1800°F gases, evaporation of the water will reduce the temperature of the gases to about 180°F and the quantity will be reduced to 660 acfm.
- If a boiler is used to cool the gases, and the gases are then cooled by water to 350°F and passed through a baghouse, the gas flow will be reduced to 520 acfm. If a wet scrubber is used, the gases will be cooled to about 145°F and the quantity reduced to 450 acfm.

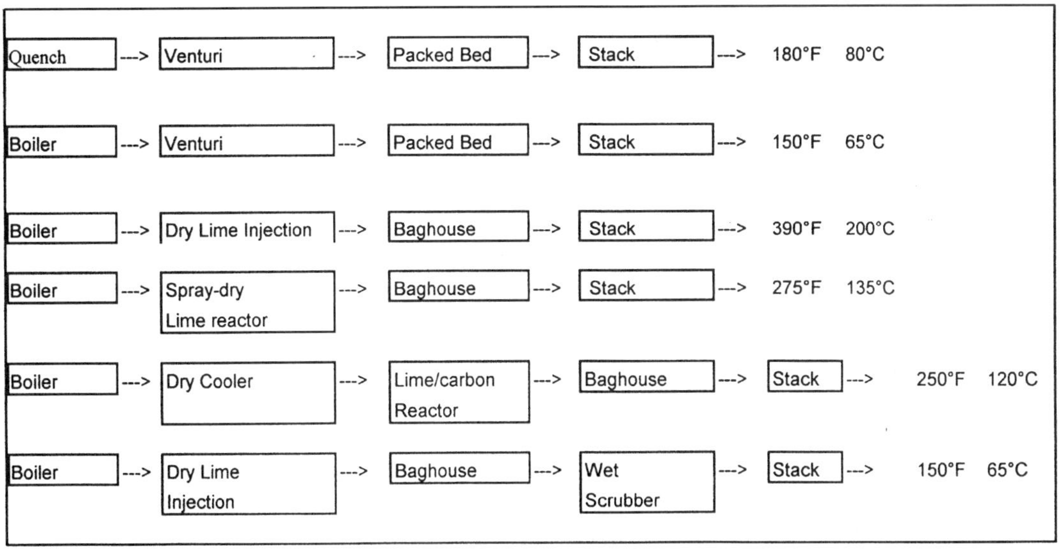

FIGURE 4.2 Alternative emission control systems, with corresponding stack temperatures.

Note that an MWI having *no* boiler or emission controls discharges direct to the stack over *three* times the gas volume of one with a scrubber. For the same pollutant emission *rate*, this results in a reduction in stack pollutant *concentration* by a factor of three.

The actual volume discharged by a typical oil, gas or coal-fired boiler with a stack temperature of 450°F is about 400 acfm per million Btu/h.

Emission limits are regulated on the basis of grains per dry standard cubic foot (gr/sdcf), corrected to 7% oxygen (or its equivalent, 12% CO_2). This represents about 50% excess air, which is typical for operation of fossil fuel-fired boilers. This correction is made in order to use the same regulatory basis for all facilities, eliminating variations in temperature and excess air.

Note that making this correction brings the stack gas flow to 236 *standard* cubic feet per minute (scfm), whereas in reality the various systems actually discharge 450–1500 acfm at temperatures ranging from 145°F to 1800°F. Thus a dilution of 2–6 times actually takes place

TABLE 4.7 Stack Gas Flow at Various Stack Temperatures (per million Btu of heat released in the furnace)

Emission control	Stack temp. (°F)	ACFM (ft^3/min)	ACMS (m^3/s)
Direct to stack*	1800	1510	0.71
Diluted with air*	1300	1730	0.82
Boiler only*	450	610	0.29
Boiler and filter*	350	520	0.24
Boiler, spray-dry + filter*	300	490	0.23
Wet scrubber*	180	660	0.31
Boiler and wet scrubber*	145	450	0.21
Fossil-fuel boiler*	450	400	0.19
Reference gas flow, 12% CO_2†	68	236	0.11

Source: Hasselriis, 1990.
*Based on medical waste at 7500 Btu/lb.
† Based on municipal waste at 5000 Btu/lb/

prior to discharge by the stack, relative to the corrected dry scfm. Making the correction does not change the actual emissions.

If combustors with emission controls are allowed to emit 0.02 gr/dscf of particulate matter, a direct-discharge incinerator logically should be allowed $1729/450 \times 0.02$ or 0.08 gr/dscf, corrected to 12% CO_2 or 7% O_2, since they achieve more dilution before the gases are discharged. The original US EPA Federal Standard was 0.08 gr/dscf for all incinerators.

Likewise, if the particulate emissions from an incinerator are the concern, a direct-discharge incinerator without boiler or scrubber with emissions of 0.045 gr/dscf, corrected, would have emissions equivalent to 0.015 gr/dscf for systems with scrubbers, and the stack concentration would actually be the same.

5.2 Dispersion in the Atmosphere

Dispersion of stack gases by the atmosphere. After the gases leave the stack, dispersion takes place, resulting in substantial dilution of the concentration of pollutants in the plume of gases as it spreads out into the environment by mixing with the ambient air.

Before the plume of gases reaches critical receptors in the environment at ground level or at elevated windows or buildings, the gases have been dispersed and diluted substantially. The environmental and health risk arising from the emission of pollutants from MWIs depends upon how effectively this dispersion process takes place.

The relationship between stack emissions and ground level concentrations (GLC) is determined by computer modeling. The burning capacity of the incinerator, stack diameter, velocity, temperature and height, meteorological (weather) data, the configuration of buildings and the general terrain are taken into account.

Dispersion models provide calculated concentrations of pollutants at ground and elevated levels. Simple models, such as the EPA SCREEN model, produce the average hourly concentration. Concentrations averaged over other time periods can be adjusted by factors roughly like those in Table 4.8. More complex models take into account average meteorological data obtained in the region over a period of years, as well as the effect of local terrain and nearby buildings, arriving at concentrations at various averaging times up to one year. Hourly averages are used to evaluate short-term exposures. The annual average applies to the region of maximum concentration and the most affected individual. Note that the hourly average may be over 20 times the annual average, sometimes over 40 times (Hasselriis, 1990).

TABLE 4.8 Relative Factors for Various Times of Exposure

Annual average	1
Quarterly average	1.7
24-h average	10
8-h average	15
3-h average	20
1-h average	30
30-min average	36

Source: Wood and Hasselriis, 1997.

Modeling calculations. Modeling calculations are based on the Gaussian plume model, illustrated in Fig. 4.3, showing the stack height, height to plume axis, and two cross sections along the path of the plume.

Plume rise. The plume of gases leaving the stack rises, due to its velocity and buoyancy. The distance above the stack outlet that a plume will rise into the atmosphere before leveling off is called the plume rise, Δh. The plume rise is affected not only by the temperature but also by wind speed, physical stack height, its inside diameter, and the ambient pressure.

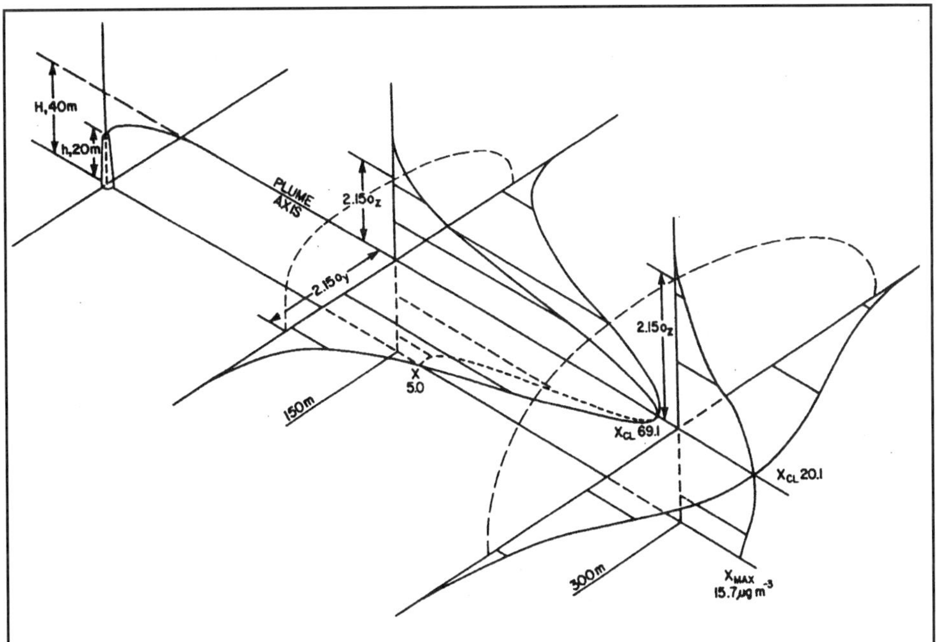

FIGURE 4.3 Gaussian plume model, showing stack height, plume height, and two cross sections through the plume (*Zanetti, 1990*).

The emission height used in the dispersion models is the effective stack height H which comprises the physical stack height h and the plume rise Δh.

EXAMPLE: Determine the effective stack height, given the Holland equation and the following data:

$$H = h + \Delta h$$

$$\Delta h = \frac{v_s d}{u}\left\{1.5 + 2.68(10)^{-3} p\left[\frac{T_s - T_a}{T_a}\right]d\right\}$$

where
H = effective stack height, m
h = physical stack height, m
Δh = rise of the plume above the stack
v_s = stack gas velocity, m/s
u = mean wind speed at stack height, m/s
d = stack inner diameter, m
p = atmospheric pressure mbar
T_s = stack temperature, K
T_a = atmospheric temperature, K

This equation is valid for neutral stability conditions; however, Holland suggests that the plume rise be adjusted by a factor of 1.1–1.2 for unstable conditions such as stability types A and B, and 0.8–0.9 for stability conditions E and F (see Table 4.9).

To estimate the mean wind speed (u) at the top of the stack, an empirical formula can be used, as follows:

TABLE 4.9 Pasquill Stability Types

Surface wind speed (m/s)	Day: Incoming solar radiation (sunshine) — Strong	Moderate	Slight	Night: Thinly overcast or =4/8 low cloud	=3/8 low cloud
< 2	A	A–B	B		
2	A–B	B	C	E	F
4	B	B–C	C	D	E
6	C	C–D	D	D	D
> 6	C	D	D	D	D

Note: A = extremely unstable, B = moderately unstable, C = slightly unstable, D = neutral, E = slightly stable, F = moderately stable. Neutral class D should be assumed for overcast conditions during day or night.

$$\frac{v}{v_0} = \left(\frac{z}{z_0}\right)^k$$

where v = wind speed at height z, m/s
v_0 = wind speed at anemometer level z_0, m/s
k = exponent or coefficient, dimensionless

The exponent k has been taken generally as 1/7 but may actually be related to the stability class, ranging from 0.141 for A to 0.414 for G (very stable).

EXAMPLE: Determine the effective stack height, given the following data:

Physical stack height $h = 183$ m, inside diameter $d = 6$ m
Wind velocity at anemometer level (2 m above ground) = 5 m/s
Air temperature = 10°C
Atmospheric pressure = 1000 mbar
Stack gas velocity = 16 m/s
Stack gas temperature = 135°C
Class B Pasquill stability type

Determine mean wind speed at the top of the stack:

$$u = (5\ \text{m/s})\ [183/(2\ \text{m})^{0.176} = 11\ \text{m/s}$$

Convert temperatures to K:

$$T_a = 273 + 10 = 283\ \text{K}$$

$$T_s = 273 + 135 = 408\ \text{K}$$

Substitute the given and calculated values in Holland's equation to determine plume rise Δh:

$$\Delta h = \frac{(16\ \text{m/s})(6\ \text{m})}{11\ \text{m/s}} \left\{1.5 + 2.68(10^{-3}\ (1000\ \text{mbar}))\left[\frac{408\ \text{K} - 283\ \text{K}}{408\ \text{K}}\right] 6\ \text{m}\right\} = 56\ \text{m}$$

Then

$$H = 183 \text{ m} + 56 = 239 \text{ m}$$

Gaussian dispersion model. The Gaussian plume model is used to estimate concentrations of pollutants in the plume leaving the stack and in the process of dispersion before it reaches the ground or defined objects. Figure 4.4 shows the concentrations at 150 m and 300 m downwind of a stack having a height of 20 m and a plume rise of 20 m, for a plume axis of 40 m. The concentrations have the shape of the probability curve in the horizontal and vertical directions, but the plume has already intersected the ground prior to the 150 m distance.

The concentration χ of a gas or aerosol (< 20 μm) calculated at ground level for a distance downwind x is expressed as follows:

$$\chi(x, y) = \frac{Q}{\pi \sigma_y \sigma_z \bar{u}} \exp\left\{-\frac{1}{2}\left(\frac{y}{\sigma_y}\right)^2\right\} \exp\left\{-\frac{1}{2}\left(\frac{H}{\sigma_z}\right)^2\right\}$$

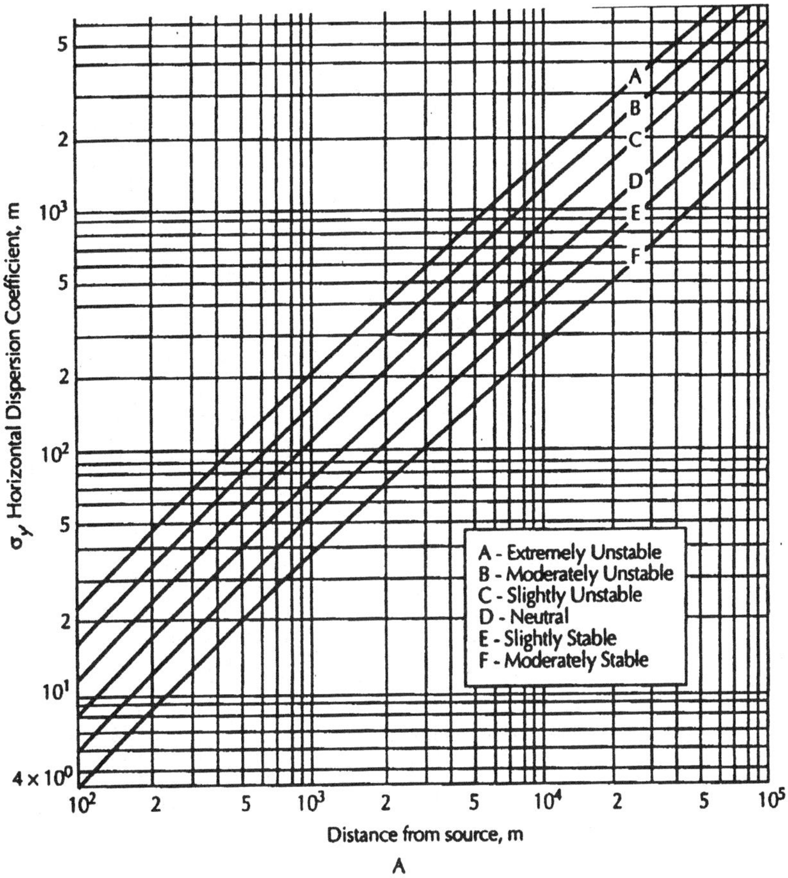

FIGURE 4.4 Horizontal dispersion coefficient σ_y versus downwind distance from the source.

where (x, y) = receptor coordinates, m
χ = ground level concentration, g/m^3
Q = emission rate, g/s
H = effective stack height
$\bar{u}$ = mean wind speed, m/s
σ_y, σ_x = dispersion coefficients, m
π = 3.14159
exp = base of natural logs, 2.7182818

The dispersion coefficients can be obtained as a function of the distance from the source and the stability classes by use of the graphs, Figs. 4.4 and 4.5.

This equation can be simplified if only the ground level downwind concentration along the center line of the plume is needed. In this case, $y = 0$ and the equation becomes

$$\chi(x, y) = \frac{Q}{\pi\sigma_y\sigma_z\bar{u}}\exp\left\{-\frac{1}{2}\left(\frac{H}{\sigma_z}\right)^2\right\}$$

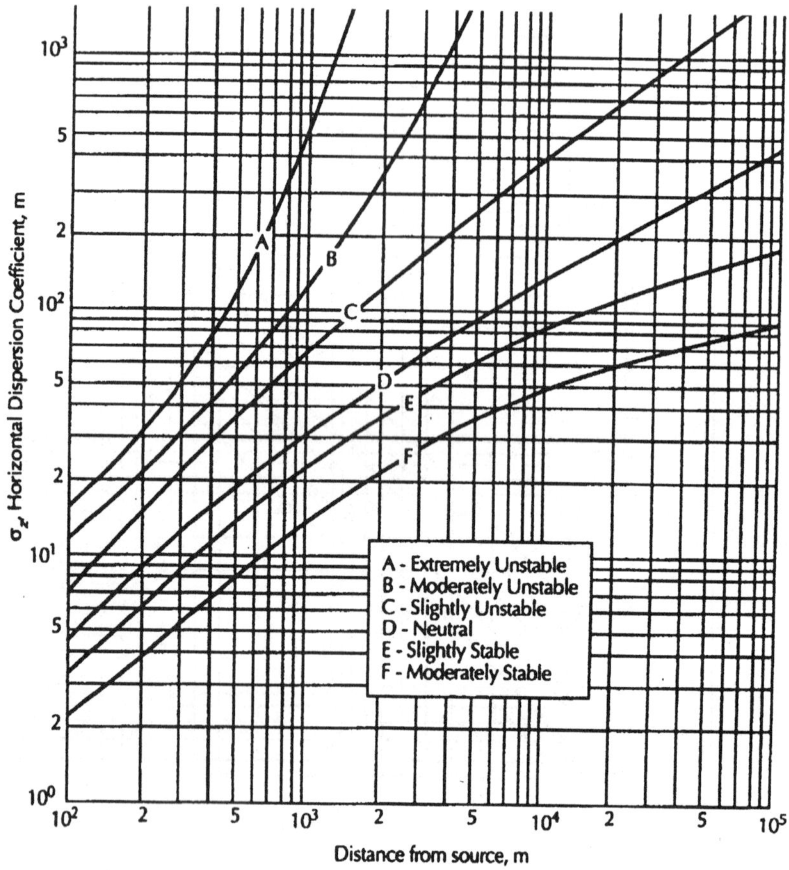

FIGURE 4.5 Vertical dispersion coefficient σ_x versus downwind distance from the source.

The maximum downwind ground level concentration of a contaminant occurs on the center line of the plume. This concentration can be estimated by taking the differential of a modified version of the equation, resulting in the following expression:

$$\sigma_z = 0.70\ H$$

Wake effects. Modeling calculations are carried out in accordance with US EPA established procedures. If there is a building within the area of influence of an MWI stack, any such structure must be evaluated for wake effects. The area of influence is five times the lesser of the height or width of the building(s). This building may justify the use of a good engineering practice (GEP) stack height.* The structure justifying the greatest GEP stack height is then used in the wake effect calculations. It makes no difference whether the stack is upwind, downwind or to the side of a structure. As long as it is within the area of influence, wake effects must be considered (Turner, 1970).

In evaluating wake effects, three zones are considered:

- The cavity zone is considered to be three times the building height H_b, and is between the stack and the building under consideration. Not much dilution takes place in this zone.
- The wake zone extends beyond the building such that the cavity plus the wake zone is 10 H_b. In this zone dispersion takes place as a result of the building geometry.
- In the reattachment zone stability is regained and the gases gradually become further diluted.

In the case of a short stack in a cavity zone, the area of influence is considered to be five times the lesser of the height or width of the building. The structure justifying the greatest GEP stack height is then used in the wake effect calculations.

If the cavity is entirely on plant property, EPA considers that it may not be necessary to include the concentrations in the cavity region. However, state regulations may require analysis of these on-site conditions, to make certain that the ground level conditions do not exceed safe concentrations of pollutants. The MASC (maximum allowable stack concentration) is calculated using modeling equations. The MASC factor is actually a dilution ratio, which we shall call the dilution factor (DF) below.

The concentration in the cavity is calculated on the basis that the emissions are diluted in a volume equal to the area of the building facing the wind, multiplied by the wind speed and a factor of 1.5. Cavity concentrations are used to determine the one-hour concentrations of pollutants, in accordance with health standards.

The wake region may be three to ten times the lesser of the building height or width. Wake effect may be ignored if the plume elevation at two times H_b downwind is greater than 1.2 H_b. After the plume passes the wake region it proceeds as if there were no disturbance by the building.

Any stack release height will result in a concentration profile that is zero at the source, increases to some maximum value, and decreases slowly with distance. The decay is logarithmic.

Urban conditions reduce the axial concentrations of the plume and increase the GLC due to the turbulence produced by the buildings and uneven ground conditions. Rural conditions produce the opposite effect. These effects are taken into account in modeling.

**Note:* A good engineering practice (GEP) stack has a height $H + 1.5L$, where H is the height of the worst case building, and L is the lesser of H or projected width of the building.

5.3 Modeling of Specific Facilities

Height of the plume. Figure 4.6 shows, for a typical medical waste combustor, the plume height as a function of the temperature of the gases as they leave the stack. If the gases leave the stack at the temperature of the secondary chamber, 1800°F, the rise in this illustration is

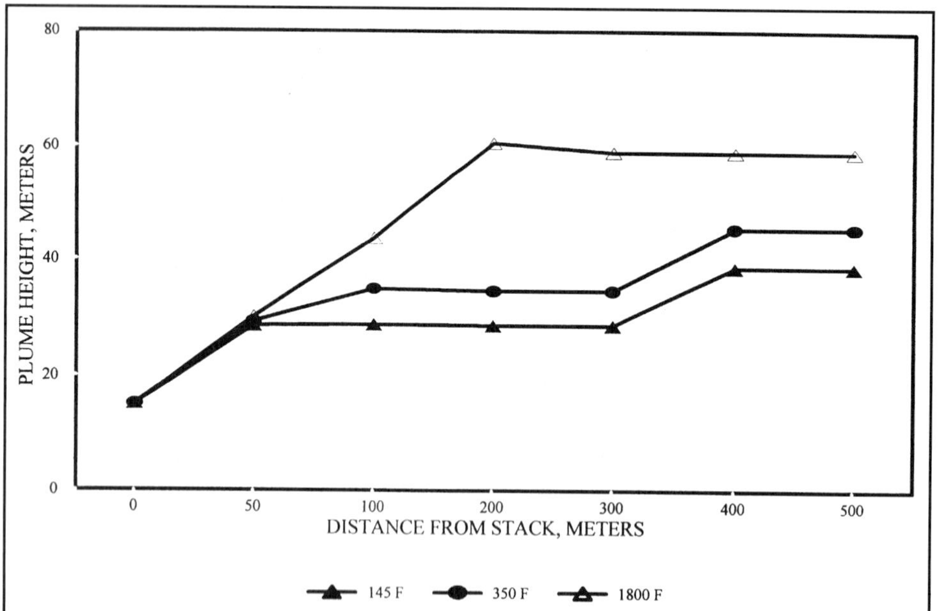

FIGURE 4.6 Height of plume center of 200 lb/h incinerator with 145°F stack temperature with wet scrubber, 350°F stack temperature with boiler and baghouse, and 1800°F with no emission control.

about twice as high as when emission controls are used, necessarily reducing the stack gas temperature to 145°F for a wet scrubber, or to 350°F when a boiler and spray-dry scrubber are used.

Plume dimensions. Figure 4.7 shows the height of upper limit, centerline and bottom of the plume which is achieved, as a function of the size of the incinerator, and the temperature of the gases (resulting from the emission control used). This stack is shorter than a GEP stack, resulting in high concentrations downwind of the building. These graphs, as well as Figs. 4.8 and 4.9, have been calculated by using a US EPA model called SCREEN, which is considered to be quite conservative (Hasselriis, 1990; OAQPS, 1992).

Ground level concentrations. Ground level concentrations are calculated from the grams per second (g/s) emission rate, the stack height, diameter and velocity, the temperature of the stack gases, and meteorological data such as temperature, with speed, direction and stability conditions.

The modeling calculations produce the ground or elevated level concentrations in grams of pollutant per cubic meter of air (g/m^3) resulting from an emission rate of one g/s of pollutant. These "unit" concentrations can be multiplied by the emissions of specific pollutants in g/s to obtain the ground level concentrations of these pollutants.

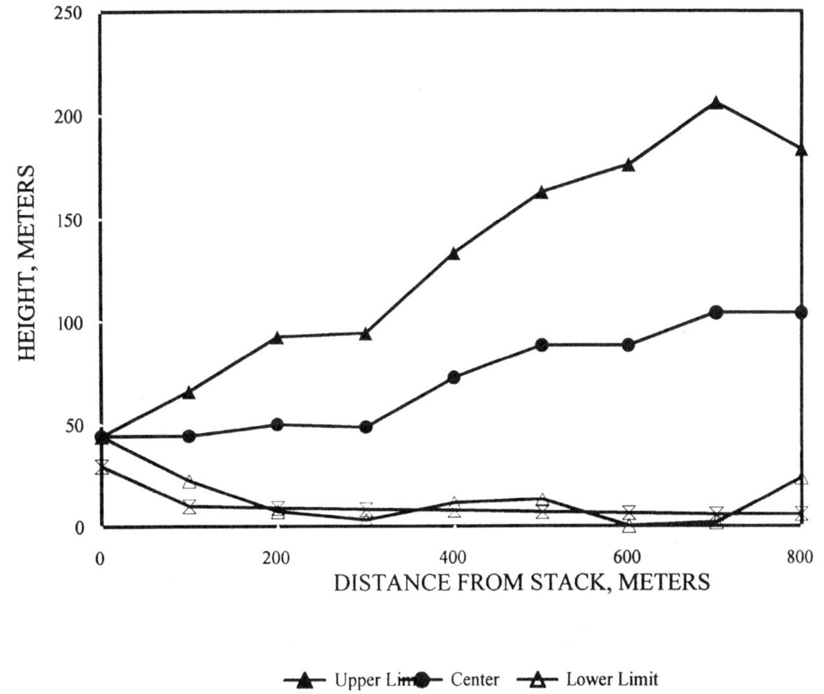

FIGURE 4.7 Plume dimensions for 2000 lb/h medical waste combustor without emission controls, with 1800°F stack temperature. Upper and lower lines represent the 10% point, where the concentration is 10% of the maximum (centerline) value.

Ground level concentrations resulting from emissions from a 2000 lb/h incinerator, per g/s of pollutant, are shown in Fig. 4.8. In this case, the concentrations from a stack temperature of 1800°F, without an emission control system, range from a maximum of 10 μg/m^3 down to 2 μg/m^2, compared with a peak of 100–200 μg/m^3 at 100 m down to about 30 μg/m^3 at 1000 m from the stack.

The impact of particulate emissions of 0.10 gr/dscf of particulate emitted by the uncontrolled emissions would be substantially less than the impact of incinerators with emission controls, in fact less than 1/10th as much, or roughly 0.010 gr/m^3. This finding helps to justify the application of less stringent emission limits to small rural medical waste incinerators and, specifically, waiving of the need for applying scrubbers.

Dilution factor. The factor by which the pollutant concentration leaving the stack is diluted by dispersion can be calculated by comparing the stack concentration (in g/m^3) at 1 g/s of pollutant with the GLC (in g/m^3) resulting from the same 1 g/s emission rate. This factor may be defined and called the dilution factor (DF), which can conveniently be used to evaluate the phenomenon of dilution *per se* resulting from dispersion (Hasselriis, 1991).

The dilution factor is calculated for a given facility in accordance with these equations:

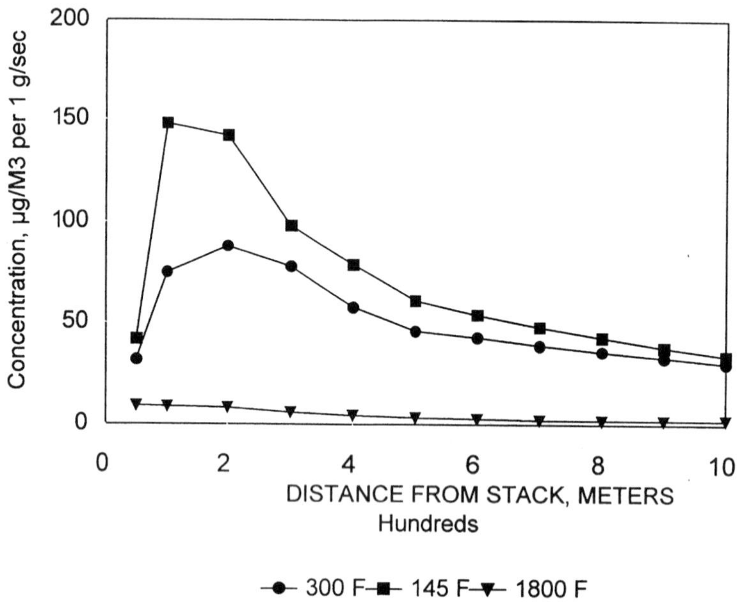

FIGURE 4.8 Ground level concentrations calculated per g/s of pollutant, using the SCREEN model, for a 2000 lb/h incinerator, showing the effect of stack temperature (*OAQPS, 1992*).

$$\mathrm{DF} = \frac{\text{concentration in the stack (g/m}^3)}{\text{maximum ground level concentration (g/m}^3)}$$

$$= \frac{(\mathrm{g/s})/(\mathrm{m^3/s})}{(\mathrm{g/m^3})/(\mathrm{g/s})}$$

where $\mathrm{m^3/s}$ = actual stack flow volume

= (meters/second)×(square meters stack area)

$\dfrac{\mathrm{g/m^3}}{\mathrm{g/s}}$ = grams per cubic meter per g/s, by modeling.

The actual cubic meters per second (acms) discharged by the stack depends upon the burning capacity of the incinerator, the composition of the fuel, the excess air used, the type of emission control (if any) used, and the leaving gas temperature, as listed in Table 4.7. The increase in dilution factor with distance from the stack is shown in Fig. 4.9, calculated for a 200 lb/h incinerator having a 45 m stack, discharging 1800°F gases directly (and vertically) from the stack, using the SCREEN model, and ranges from 2000 to 8000.

Dilution factors which have been calculated from modeling data performed for a wide range of combustion facilities are plotted, in Fig. 4.10, against the stack height upon which the modeling was based. The stack concentrations in µg/m^3, corresponding to a unit emission rate of 1 g/s, have been divided by the maximum annual average ground level concentrations in µg/m^3 in order to obtain the dilution factor for each facility. Most of these modeling studies were performed for the California Air Resources Board (CARB) as part of the board's cadmium and dioxin studies (Fry *et al.*, 1990).

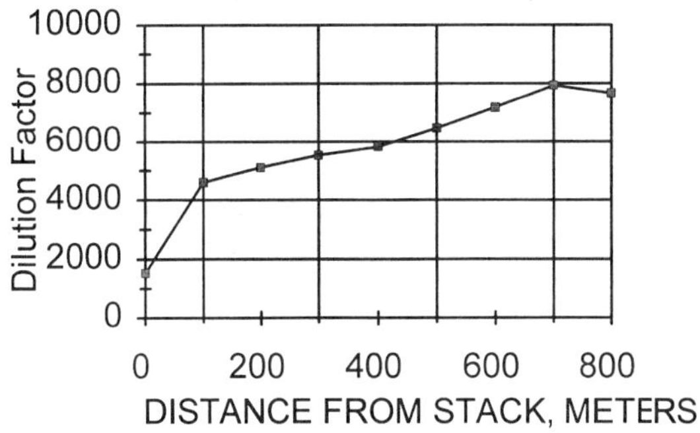

FIGURE 4.9 Dilution factors versus distance from the stack, calculated for a 200 lb/h incinerator having direct discharge of gases at 1800°F.

It is interesting to note that, in spite of the fact that this data set includes medical and industrial waste incinerators, asphalt heaters burning waste oil, municipal sludge incinerators, biomass (wood) burners, and municipal waste incinerators, the range of DFs for each category is remarkably similar. Most of the DFs range from 10,000 to 100,000. The small MWIs with short stacks produced DFs which were similar to those of the other categories, even the municipal incinerators with high stacks (Hasselriis, 1990).

Effect of emission controls on ground level concentration. Incinerators with greater burning capacities usually have higher stacks. Building heights may vary from as little as 5 m for small incinerators to 30 m for large waste-to-energy power plants. Good engineering practice (GEP)

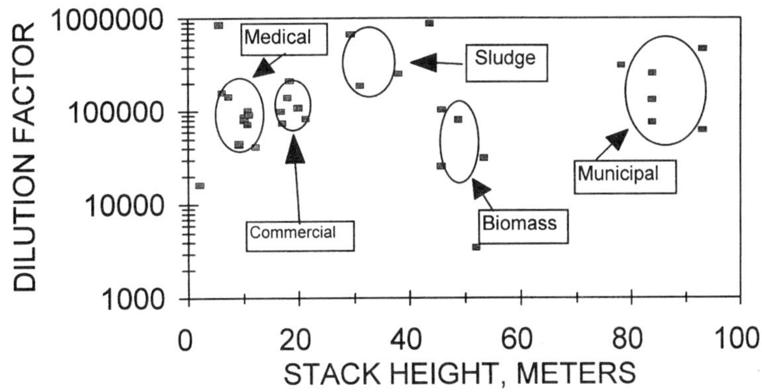

FIGURE 4.10 Dilution factors achieved by various combustors, versus stack height.

usually results in a stack height of 2.5 times the height of the building. Large municipal waste-to-energy facilities have stack heights of 90 m or greater.

It appears from Fig. 4.10 that in spite of the fact that the larger MSW incinerators have capacities 50–100 times greater than those of MWIs, their higher stacks result in achieving the same DFs as the small MWIs.

The type of emission control system and the burning capacity of the incinerator affect the ground level concentration for several reasons: the amount of pollutant removed (if controls are used), the dilution before discharge, and the effect of the stack temperature on dispersion. In the following, only the effects of quantity and temperature are discussed.

Effect of burning capacity and stack temperature on plume rise. The higher the temperature of the stack gases, and the higher the stack exit velocity, the higher the plume of gases will rise as they blend in with the atmospheric air and become dispersed. Higher temperatures therefore result in lower pollutant concentrations before the gases reach critical "receptors" at ground or elevated levels.

Table 4.10 shows the range of unit dispersion factors (per g/s) and annual average dilution factors resulting from three specific combustion sources, obtained by detailed modeling of the sites. The facilities selected illustrate the fact that two similar MWIs exhibited radically different dispersion and dilution factors. On the other hand, one MWI showed a dilution factor higher than that of a large municipal solid waste (MSW) incinerator (Hasselriis, 1990).

Combustors of a wide range of sizes and types show similar ranges of dilution factors, as stack heights vary with facility burning capacity and size.

5.4 Acceptable Ground Level Concentrations

Acceptable ground level concentrations of many pollutants have been determined by various health agencies, such as OSHA. Exposure is expected to take place during 8-hour working days. Some of these standards are listed in Table 4.11. Standards based on health effects use annual averages. Corrections for other exposure times are made in accordance with Table 4.8 above.

Health risks of cancer due to exposure to dioxins and furans, as well as to toxic metals such as cadmium, are based on the annual average, since cancer-based risk involves exposure of the most affected individual (MED) for 70 years at that location.

The health risk impact of incinerator emissions should include not only inhalation but also other pathways. Some studies have concluded that total risk is five times greater than that of inhalation alone (Hahn and Sofaer, 1990). It should not be overlooked that on-site MWIs are generally operated 8 h per day or less, thus contributing only 1/3 the emissions that their capacity implies. The shorter operating time of on-site MWIs reduces the total risk to a factor of two times inhalation alone.

TABLE 4.10 Dispersion and Dilution Factors for Specific Sources

Facility	Burn rate (lb/h)	Unit dispersion ($\mu g/m^3$ per g/s)	Dilution factor
Cedars Sinai * (MWI, 2 m stack)	980	27	16,000
Kaiser Permanente * (MWI, 7.3 m stack)	980	5	145,000
St Stanislaus† (Municipal waste)	57,000	0.117	133,500

*Fry *et al.*, 1990.
†Hahn and Sofaer, 1990.

TABLE 4.11 Acceptable Workplace Concentrations and Unit Risk Values for Pollutants

Pollutant	Ground level concentration* ($\mu g/m^3$)	Unit risk value for inhalation (cases per million†‡ at 1 $\mu g/m^3$)
Total suspended particulate:		
24-h	150	
annual	7	
Hydrochloric acid:		
3-min	140	
annual	50	
Arsenic	0.00023	0.004
Beryllium	0.00042	0.003
Cadmium	0.00056	0.002
Chromium VI	0.000083	0.012
Lead	0.150–1.5	
Mercury	0.012–0.80	
Nickel	0.0033	0.0005
Benzene	0.00042–0.000120	
Dioxin equivalent	3.0×10^{-8}	33 @ 1 pg/m^3‡

*New York State DEC "Air Guide 1," also regulations of Pennsylvania and North Carolina.
† Health Risk Assessment, St Lawrence County, NY, Ram Trac Corp.
‡ Unit risk is posed by 70-year exposure to an airborne concentration of 1 $\mu g/m^3$ of each substance except for dioxin equivalent 2,3,7,8-TCDD, for which the applicable unit is picograms per cubic meter of air (pg/m^3), where 1 pg = 10^{-6} μg.

Health risks due to cancer are established on the basis of "unit risks." The unit risk is the risk number of additional cancer cases per million caused by exposure to an airborne concentration of 1 $\mu g/m^3$. The unit risk factors for the major cancer causing or promoting pollutants are listed in Table 4.11. For dioxins, a concentration of 3×10^{-8} $\mu g/m^3$, or 30 femtograms per cubic meter, has been adopted by some states.

Comparison of predicted with acceptable GLC. Once the hierarchy has been established as above, it is possible to trace through, from waste composition to stack to ground, what happens to individual pollutants, and how they compare with acceptable levels of air contaminants, and what the resulting health effect may be. The influence of alternative emission control devices can be discerned, as well as those of other variables such as burning capacity, stack height and gas velocity and temperature.

One remedy would be to use an emission control capable of removing 90–95% of the HCl. Another might be to raise the stack to increase the DF from the poor level of 500 to 10,000, thus achieving a 20-fold improvement or the equivalent of 95% control. If a DF of 50,000 were achieved, it would be the equivalent of 99% control by the scrubber.

Finally, a reduction in chlorine content from 1% (400 ppm emission) to 0.1% (40 ppm) would achieve a 90% reduction, perhaps eliminating the need for a scrubber.

Table 4.12 shows the effect of stack temperature and type of emission control equipment on the average annual ground level concentration, for the 2000 lb/h MWI with 25 m stack, compared with the acceptable air quality limits established by New York State.

TABLE 4.12 Comparison of GLCs with Acceptable Air Quality Limits (2000 lb/h MWI with 25 m stack). Units are micrograms per cubic meter ($\mu g/m^3$)

Pollutant	Temperature leaving stack: Direct discharge* 1800°F	Boiler, no APC 400°F	Wet scrubber† 145°F	Acceptable air quality limit
Particulate	0.1	0.4	0.006	50
HCl	1.0	3.0	0.6	7
Cadmium	3×10^{-4}	10×10^{-4}	2×10^{-4}	6×10^{-4}
Lead	2.5×10^{-3}	8.8×10^{-3}	2×10^{-3}	1.5
Dioxin T.E.	2.0×10^{-8}	7.4×10^{-8}	1×10^{-8}	3×10^{-8}

*Uncontrolled emissions assumed in this example have been derived from California Air Resources Board data from tests of the St. Agnes MWI (Fry *et al.*, 1990).
† For simplicity, the wet scrubber is assumed to have 90% control efficiency for all listed pollutants, although it is actually much greater.

Particulate concentrations are less than 1% of AAQL *without* emission controls. HCl concentrations range from 15 to 40% of AAQL without emission controls. Lead emissions are insignificant relative to the limit.

Cadmium concentrations in this example range from 50 to 150% of the AAQL, without the use of emission controls. A reduction in cadmium content of plastics in the waste could significantly reduce these percentages.

The TCDD (dioxin) toxic equivalent concentrations are significant in this example. Improvements in combustion could bring these concentrations down. Other MWIs tested by CARB showed TE concentrations were about one-half of those found at St Agnes.

Dioxin emissions and ground level concentrations. Dioxins are found in the emissions of wood stoves, automobiles, coal-fired boilers, and medical and municipal waste combustors. Finding that background levels of dioxins in the Los Angeles area occasionally exceeded standards, the California Air Resources Board (CARB) developed guidelines for combustion of medical waste which mandated the use of scrubbers for MWI.

CARB obtained dioxin data from old incinerators which were equipped with no emission controls, or only with the simple devices employed at that time (Fry *et al.*, 1990). Due to the installation of scrubbers, data is now becoming available regarding the effectiveness of modern scrubbers in removing not only heavy metals, but also dioxins, from incinerator emissions.

Table 4.13 lists the toxic equivalent TCDD (dioxin) emissions of *un*controlled as well as controlled MWI, based on US EPA data. These are expressed as nanograms per standard cubic meter as well as in pounds per million tons. Controlled TCDD, based on tests of state-of-the-art baghouse and scrubber emission control systems, show the range of emissions measured.

The limit of 3×10^{-8} μg/g (30 femtograms/gram or 30×10^{-15} g/g) is recognized internationally, and by the US EPA, as a very conservative estimate of a risk of one additional cancer case per million affected persons. Risks at this level are not considered to be significant.

Sample calculation:

Assume an uncontrolled MWI has emissions of 300 ng/dscm, and the dilution factor is 10,000. What is the ground level concentration?

$$\text{Ground level concentration} = 300 \times 10^{-9}/10{,}000 = 3 \times 10^{-11} \text{ g/dscm}$$

TABLE 4.13 Uncontrolled and Controlled Toxic Equivalent TCDD Emissions for MWI

	Range of TEQ emissions	
Type of control	(ng/m^3)	(pounds per million tons)
Uncontrolled	3.24–540	0.04–7.0
Spray-dry + baghouse	0.06	0.0008
Baghouse + wet scrubber	0.14	0.0019
Baghouse	0.15–115	0.002–1.5
Wet scrubber	2.6–41	0.04–0.53

Note: conversion factors
1 g/m^3 = 8 lb/ton at 4500 Btu/lb
1 g/m^3 = 13 lb/ton at 7500 Btu/lb
1 ng/m^3 = 0.013 lb/million tons at 7500 Btu/lb
Source: Walker and Cooper, 1992.

Compare with 3×10^{-8} $\mu g/m^3$:

$$\text{Safety factor} = 3 \times 10^{-8}\ [\mu g/m^3]/3 \times 10^{-11}\ [g/dscm] = 1000 \times \text{limit}$$

Source substitution versus emission controls. The primary emissions from a medical waste incinerator without emission controls which are likely to fail to meet acceptable ground level pollutant standards are the HCl, heavy metal and dioxin emissions. Hospital purchasing agents can potentially reduce the presence of chlorinated plastics and plastics containing critical metals such as cadmium in the purchased materials.

Overall effect of various factors on GLC. A series of factors influence dispersion. Emission controls may achieve a reduction of 10–100 times; dilution before the stack may contribute a factor of 3–5. A change in stack height may produce as much as 3–10-fold reductions if downwash effects are avoided: dispersion factors may range from as low as 500 in the case of high nearby buildings to well over 100,000 with a GEP stack. Finally, removal of pollutants from the waste may result in reductions of perhaps 10–20 times. All of these factors may be considered in planning new or upgraded installations.

Some points to consider:

- Modeling studies show that small refractory-lined incinerators, burning commercial or industrial wastes and medical wastes, have pollutant impacts at ground level which have similar ranges to those of other waste burning incinerators, including municipal waste incinerators. The shorter stacks are compensated for by their relatively low mass emissions and more diluted stack discharges.
- The high stack temperatures of small *uncontrolled* incinerators cause the plume to rise higher than plumes of incinerators equipped with heat recovery boilers and scrubbers, resulting in relatively low ground level concentrations.
- When flue gases are bypassed directly to the atmosphere, the rapid rise and height of the hot gas plume results in very high dilution factors, producing relatively low ground level concentrations in spite of bypassing of emission controls. This assumes that the bypass stack is of equal height. Shorter stacks may still achieve acceptable ground level or elevated level concentrations, at least over the short term of the bypass condition.
- Dilution factors are a useful parameter for relating stack emission concentrations to ground level concentrations. Dilution factors may range from as low as 500 for MWI with short incinerator-mounted stacks and nearby high buildings to well over 100,000 when stacks are

TABLE 4.14 Uncontrolled and Controlled Emission Factors

Pollutant	Uncontrolled emissions (lb/ton)	Control efficiency (percent)	Controlled emissions (lb/ton or lb/million tons)
Solid matter:			
Particulate	7.35	99	0.02 lb/ton
Arsenic	0.00018	99	2.0 lb/million tons
Cadmium	0.000035	99	3.0 lb/million tons
Chromium	0.00072	99	7.2 lb/million tons
Lead	0.00007	99	0.7 lb/million tons
Manganese	0.0371	99	371 lb/million tons
Mercury	0.00054	50	270 lb/million tons
Nickel	0.000255	99	2 lb/million tons
Zinc	0.0019	99	20 lb/million tons
Acid gases:			
Sulfur dioxide	1.85	90	0.185 lb/ton
Nitrogen oxides	6.02	50	3.0 lb/ton
Carbon monoxide	1.2	0	1.2 lb/ton
Hydrogen chloride	38.3	95	2 lb/ton
Trace organics:			
Benzene	0.41	95	0.02 lb/ton
TCDD toxic equiv.	1.37×10^{-6}	95	0.07 lb/million tons
Formaldehyde	0.02	95	0.001 lb/ton
Polyaromatic hydrocarbons (PAHC)	0.001	95	50 lb/million tons

Source: Doucet and Mainka, 1991.

GEP and/or higher than adjoining buildings. Municipal incinerators showed DFs ranging from 60,000 to 300,000.

- Detailed modeling helps to obtain the most effective and economic technology for environmental controls. The type of emission control and stack height are essential variables.

6 ENVIRONMENTAL RISK ASSESSMENT

6.1 Health Risk Assessment

Health risk assessments of waste combustion systems emissions are carried out using air dispersion models and exposure assessment methodologies for estimating atmospheric dispersion to determine the concentrations of pollutants to which the exposed population would hypothetically be exposed. EPA recommends the Industrial Source Complex–Long Term (ISC–LT) model be run, using site-specific meteorology data from local weather stations (Hahn and Sofaer, 1990).

A variety of health risk estimators are used to help quantify the health risks of toxic substances. Past studies have shown that metals pose potentially significant health risks and may be of greater health concern than organics.

Estimates of carcinogen potency factors developed by the US EPA Carcinogen Assessment Group (CAG) are used to calculate excess risk of developing cancer to individuals and populations as a result of exposure to carcinogens detected in the stack emissions (arsenic, beryllium, cadmium, chromium, nickel and PCBs).

Guidelines for evaluating non-carcinogenic effects can use upper limits of "safe" (no observed adverse effect) exposures to derive "acceptable" continuous exposure limits

(CELs), which may reflect realistic exposure conditions as well as additional uncertainty factors which allow for the increased susceptibility of sensitive populations to disease.

Major assumptions and limitations:

- Air dispersion modeling is a source of error, perhaps by a factor of two.
- The flat and urban models used may not account for actual variances in elevation in the area.
- Health effects due to chronic exposures, particularly carcinogenesis, are considered rather than the effects from acute exposures.
- The facility is conservatively assumed to be in operation for 74 years, the average length of a human lifetime, to be consistent with other health risk estimates, although 20 years may be more realistic.
- Emission rates are assumed to be constant and continuous, with the facility running under normal conditions 90% of the year.
- Atmospheric dispersion of stack emissions is assumed to be the primary transport mechanism of pollutants. In fact, the food chain has been shown to be a potentially significant pathway of exposure.
- The combined health effects of various substances are assumed to be additive, in accordance with EPA guidelines, since it is difficult if not impossible to evaluate the synergistic and antagonistic reactions between elements.
- A large margin of safety (perhaps three orders of magnitude) is incorporated into extrapolating the results of animal toxicity testing to prediction of human health effects because of the large degree of uncertainty in predicting these correlations, and in order to protect public health to the greatest extent possible.

6.2 Risk Analysis of a Hazardous Waste Facility

The data and calculations provided below show how risk analysis can be calculated on the basis of data on waste composition, uncontrolled emissions of pollutants, controlled emissions, and modeled ground level concentrations obtained by computer modeling, based on a hazardous waste incineration facility (EPA, 1997).

Annual capacity:	52,000–77,000 tons of liquid, solid and semi-solid RCRA wastes.
Types of waste:	Bulk solid wastes, sampled upon arrival; bulk liquid wastes, sampled on arrival; and drummed wastes, at least one out of 10 drums of each waste stream normally sampled. Non-pumpable liquids and sludges are usually mixed with pumpable wastes.
Description:	Rotary kiln, waste feed mechanisms, secondary combustion chamber, heat recovery boiler, air pollution control devices, stack, solid residue removal equipment, and computerized process control and instrumentation systems.
Rotary kiln:	Refractory-lined shell, 15 ft in diameter and 43 ft long, at temperatures of 1800–2200°F.
Secondary chamber:	61 ft high, 21 ft long, 22 ft wide, at 1350–1500°F.
Stack:	150 ft high (45.7 m), 1.83 m inside diameter, exit temperature 201°F.

This facility was subjected to extensive modeling, using the US EPA ISC–COMPDEP atmospheric dispersion model, which incorporates the ISC–ST short-term model and the COMPLEX I screening model for complex terrain, using modeling receptors between stack top elevation and plume height with both the simple and complex terrain models, and using the larger of the two concentration estimates as being the more representative.

The result of these modeling studies was that the maximum annual concentrations in the air were predicted to occur at a location approximately 1 km to the east of the stack. The maximum annual concentration was estimated to be between 0.9108 and 0.9223 μg/m^3 per 1 g/s of emissions.

Taking into account the stack velocity and area, the stack flow was 46.7 m^3/s. Dividing 1 g/s by this flow, the stack concentration was 0.0214 g/m^3. The dilution factor is then calculated as follows:

$$\text{DF} = 0.0214/(0.102 \times 10^{-6}) = 209{,}952$$

6.3 Comparison with EPA "Screen" Model

The US EPA screening model, SCREEN, is useful for making a first estimate of ground level concentrations on the basis of site data. This model is shown in Table 4.15.

This model produces the concentration for a one-hour exposure. Using the one-hour concentration of 4.781 μg/m^3, compared with the 0.101 μg/m^3 calculated with the detailed model indicates a conversion factor of 47 from one-hour to annual. Typically this ratio ranges from 30 to 50, depending upon local conditions. In this case this appears to be a good match with SCREEN.

EXAMPLE: Calculate the dilution factor (DF) for one-hour exposure:

The stack concentration for a flow rate of 1 g/s is calculated using the stack diameter and velocity:

$$\text{Stack area} = D^2 \times (\pi/4) = 1.83^2 \times 0.786 = 2.63\text{m}^2$$

$$\text{Stack volume} = (\text{velocity}) \times (\text{area}) = 17.74\ [\text{m/s}] \times 2.63\ [\text{m}^2] = 46.7\ \text{m}^3/\text{s}$$

$$\text{Stack concentration} = 1\ [\text{g/s}]/46.7\ [\text{m}^3/\text{s}] = 0.0214\ [\text{g/m}^3]\ \text{per 1 g/s}$$

$$\text{GLC} = 4.781\ [\text{g/m}^3]\ \text{per 1 g/s}$$

$$\text{One-hour DF} = (\text{stack concentration})/(\text{GLC}) = 0.0214\ [\text{g/m}^3]/4.781\ [\mu\text{g/m}^3] = 4480$$

Using the factor of 47 to correct to annual, we get an annual DF of 4480 × 47 = 210,500.

Comments regarding Table 4.15:

- SCREEN tests each of the stability classes and selects the highest concentration. Hence, as distance from the stack increases, different stability classes are maxima. Stability classes are described in Table 4.9.
- The mixing height varies with the stability class. At distances up to 1200 m the mixing height is as low as 360 at stability 1, but beyond that it is fixed at 5000 m. This illustrates the importance of searching for the maximum at each distance.
- The critical wind velocity at 10 m ranges from 10 m/s at 100 m down to 1 m/s, illustrating its importance in determining ground level concentrations closer to the stack.
- The plume height is highest with stability 1 (260 m), but beyond 1200 m stability 5 results in a plume height of only 124.5 m, bringing the plume closer to the ground.
- σ_y, the plume width, is maximum with stability 1 at 900 m, but at stability 5 it increases from 110 m to 165.5 m. At the maximum point it is 161.8 m.
- There is no downwash component for these site conditions, due to the high stack.

TABLE 4.15 Screen Model Run

```
Simple Terrain Inputs:
  Source                    =    POINT
  Emission rate (g/s)       =      1.000
  Stack ht. (m)             =     45.70
  Stack diameter (m)        =      1.83
  Stack velocity (m/s)      =     17.74
  Stack gas temp. (K)       =    367.00
  Ambient air temp. (K)     =    293.00
  Receptor height (m)       =      0.00
  Iopt (1=URB, 2=RUR)       =      1
```

Buoy. flux = 29.37 m^4/s^3;
Mom. flux = 210.35 m^4/s^2.
Full Meteorology
Terrain height of 0 m above stack base use for following distances:

Distance (m)	Conc. (μg/m^3)	Stab.	U10m (m/s)	USTK (m/s)	Mix Ht. (m)	Plume ht. (m)	σ_y (m)	σ_x (m)	D_{wash}
10	0.0000	0	0.0	0.0	0.0	0.0	0.0	0.0	
100	0.2948	3	10.0	13.6	3200.0	64.9	21.7	20.1	NO
200	3.812	2	5.0	6.3	1600.0	88.7	62.1	53.1	NO
300	4.543	3	5.0	6.8	1600.0	85.6	63.1	60.7	NO
400	4.476	3	3.0	4.1	960.0	112.2	83.8	82.2	NO
500	4.249	4	4.0	5.8	1280.0	91.9	74.2	66.6	NO
600	4.163	4	3.0	4.4	960.0	107.3	88.0	79.3	NO
700	4.127	1	1.0	1.3	320.0	260.9	207.3	227.5	NO
800	4.050	1	1.0	1.3	320.0	260.9	231.1	264.8	NO
900	3.824	1	1.0	1.3	320.0	260.9	254.5	304.0	NO
1000	3.858	3	1.0	1.4	320.0	245.1	194.5	208.0	NO
1100	3.849	3	1.0	1.4	320.0	245.1	209.6	227.3	NO
1200	3.843	5	1.0	1.6	5000.0	124.5	110.8	61.6	NO
1300	4.130	5	1.0	1.6	5000.0	124.5	118.2	64.6	NO
1400	4.352	5	1.0	1.6	5000.0	124.5	125.3	67.5	NO
1500	4.518	5	1.0	1.6	5000.0	124.5	132.4	70.3	NO
1600	4.635	5	1.0	1.6	5000.0	124.5	139.3	73.0	NO
1700	4.714	5	1.0	1.6	5000.0	124.5	146.0	75.6	NO
1800	4.760	5	1.0	1.6	5000.0	124.5	152.6	78.2	NO
1900	4.780	5	1.0	1.6	5000.0	124.5	159.1	80.7	NO
2000	4.778	5	1.0	1.6	5000.0	124.5	165.5	83.1	NO
Maximum 1-h concentration at or beyond 10 m:									
1942	4.781	5	1.0	1.6	5000.0	124.5	161.8	81.7	NO

Summary of screen model results:

Calculation procedure	Max conc. μg/m^3	Dist. to max. (m)	Terrain ht. (m)
Simple terrain	4.781	1942	0

The next step in the risk analysis is to determine the relationship between pollutants in the stack emissions, their ground level concentrations, and the comparison with "acceptable" concentrations at ground level.

Table 4.16 lists the stack concentrations of the critical toxic metals and the critical trace organic, toxic equivalent dioxin (TEQ). These were determined from trial burn tests and "normal" emissions, selecting the maximum. These are translated to ground level conditions using the dispersion factor (0.9108 $\mu g/m^3$ per 1 g/s) or the dilution factor (209,952) derived from it, to permit comparison with health risk concentrations.

TABLE 4.16 Ratio of Acceptable GLC to Annual GLC. Hazardous Waste Incinerator

Pollutant	Feed rate (g/s)	Stack emissions (g/s)	System removal efficiency (%)	Annual average GLC	NY State acceptable GLC	Ratio: AGLC/GLC
Arsenic	0.16	37×10^{-6}	99.98	40×10^{-6}	230×10^{-6}	5.8
Cadmium	0.12	16×10^{-6}	99.99	17.6×10^{-6}	560×10^{-6}	31.8
Chromium	0.10	0.71×10^{-6}	99.999	0.8×10^{-6}	1.20	126,600
Lead	0.44	43×10^{-6}	99.99	47×10^{-6}	1.50	21,276
Mercury	0.0014	210×10^{-6}	85.0	230×10^{-6}	0.12	600

Source: Cowley *et al.*, 1994.

6.4 Human Health Risk Assessment

The risk assessment process consists of the following components:

- *hazard identification*: identifying the chemical substances of concern and compiling, reviewing and evaluating data relevant to toxic properties of these substances;
- *dose–response evaluation*: assessing the relationship between dose and response for each chemical of potential concern;
- *exposure assessment*: identification of potential exposure pathways, the fate and transport of chemicals in the environment (including dispersion modeling), and estimation of the magnitude of chemical exposure for the potential exposure pathways;
- *risk characterization*: calculation of numerical estimates of risks for each substance through each route of exposure, using the dose–response information and the exposure estimates.

The general approach used by the EPA guidance provides estimates of:

- individual risks based on exposure within defined subareas surrounding the facility, expressed both as averages across the subareas and at the location of maximum chemical concentrations within each subarea;
- risks to potentially more highly exposed or susceptible subgroups, such as young children, within the general population;
- risks associated with specific activities that may result in elevated exposures, such as subsistence fishing;
- individual risks based on "high end" exposure to subgroups of the population that are believed to be potentially more highly exposed;

- cumulative risks to the population in the vicinity of the incinerator as a result of stack emissions.

This approach allows for the estimation of risk to specific segments of the population, taking into account site-specific activity patterns, the numbers of individuals in each subgroup, and actual locations of individuals within these subgroups.

Toxicity assessment. Consistent with US EPA guidance, potential carcinogenic and non-carcinogenic effects are evaluated separately, assuming potentially carcinogenic substances to pose a finite cancer risk at all exposure levels; therefore a "no-threshold" assumption, based on a 70-year lifetime exposure. The US EPA uses a linearized multistage model to develop the cancer slope factor (SF), which is generally believed to overpredict the true potency of a chemical.

Non-cancer effects assume that a minimum threshold level of exposure must be reached before the effect will occur. The estimated level of daily human exposure below which it is unlikely that adverse effects will result is known as the reference dose (RfD). The reference concentration (RfC) reflects the non-carcinogenic effects of certain chemicals, typically derived from experimental animal studies, incorporating uncertainty factors to extrapolate from the high dose exposures in the animal experiments to the low doses likely to be received by humans from environmental sources, and taking into account individuals who are likely to be more susceptible than the general population to the chemical.

For incinerator emissions, exposure to individuals living and working in the vicinity of a facility is evaluated for both inhalation and indirect, multipathway routes of exposure, specifically:

- inhalation of air;
- ingestion of and dermal contact with soil;
- consumption of meat, dairy products and eggs from locally raised livestock;
- consumption of locally grown vegetables;
- ingestion of and dermal contact with surface water during swimming.

Fate and transport modeling. The ISC–COMPDEP model is used to estimate chemical concentrations in air associated with the routine emissions from the facility. The results of this modeling can be used directly to assess inhalation exposures, and at the starting point for evaluating exposures through indirect pathways, including the result of wet and dry deposition of particulate matter and vapor onto soil and vegetation, followed by ingestion by livestock, and ingestion of vegetables or livestock grown or raised locally.

Estimated average cancer risks and health indices. The final results of the complex analysis performed for the WTI facility are listed in Table 4.17. Cancer risks range from 1×10^{-6} for the theoretical subsistence farmer to 6×10^{-7} for the farmer adult or child. *The US EPA considers these risk levels to be of no significance*. The hazard indices were all less than 0.1.

Sample calculation. The following calculation shows the relationship between the ground level exposure for Cd with the inhalation exposure and risk for a rotary kiln with a stack volume of 50,000 m^3/h and a 33 m stack:

TABLE 4.17 Estimated Area Average Cancer Risks and Hazard Indices Due to Exposure from Direct and Indirect Pathways. WTI Hazardous Waste Incineration Facility

Population subgroup	Cancer risk (per million)	Hazard index
Resident adult	0.2	0.01
Resident child	0.4	0.05
Farmer adult	0.6	0.01
Farmer child	0.6	0.06
School-age child	0.3	0.03
Subsistence farmer adult	1	0.02
Subsistence farmer child	1	0.07

Source: EPA, 1997.

Step 1. Cd ground level exposure concentration

$$= \text{max. exposure concentration} \times \frac{\text{Cd emission rate}}{\text{Modeled emission rate}}$$

$$= 0.373\ \mu\text{g/m}^3 \times \frac{0.0008\ \text{g/s}}{1\ \text{g/s}} = 0.00030\ \mu\text{g/m}^3$$

Step 2.

Inhalation exposure =

$$\frac{\text{contaminant concentration} \times \text{inhalation rate} \times \text{exposure duration} \times \text{absorption fraction}}{\text{body weight} \times \text{lifetime}}$$

$$= \frac{3.0 \times 10^{-7}\ \text{mg} \times \frac{23\ \text{m}^3}{\text{d}}\ 74\ \text{y} \times 1}{70\ \text{kg} \times 74\ \text{y}}$$

$$= 9.9 \times 10^{-8}\ \text{mg/(kg} \cdot \text{d)}$$

Step 3.

$$\text{risk} = \text{potency} \times \text{exposure}$$

$$= \frac{7.8\ \text{kg} \cdot \text{d}}{\text{mg}} \times \frac{9.9 \times 10^{-8}\text{mg}}{\text{kg} \cdot \text{d}}$$

$= 7.7 \times 10^{-7}$ increased lifetime risk of developing cancer to the maximum exposed individual, or one chance in 1.3 million (worst case)

6.5 Risk Assessment with Failure of Emission Controls

Public concern has been expressed regarding the short-term effects associated with failure of the emission control system and/or bypass of the incinerator emissions directly to the atmosphere. In order to address this issue, short-term health effect risks have been evaluated, and compared with typical uncontrolled stack emissions.

TABLE 4.18 IDLH versus One-half Hour GLC with APC Failure

Pollutant	Controlled stack emissions (g/s)	Stack emissions if APC fails (g/s)	One-half hour GLC (μg/dscm)	IDLH (μg/dscm)	Ratio: IDLH/$\frac{1}{2}$ h GLC)
Arsenic	37×10^{-6}	31,268	313	230×10^{-6}	29
Cadmium	16×10^{-6}	13,523	135	560×10^{-6}	37
Chromium	0.71×10^{-6}	600	6	1.20	16,663
Lead	43×10^{-6}	36,343	363	1.50	41
Mercury	210×10^{-6}	150,882	1775	0.12	5.6

Source: Wood and Hasselriis, 1997).

The IDLH is the level considered to be "Immediately Dangerous to Life and Health." Table 4.18 shows the results of this investigation. Modeled GLCs are compared with the IDLH for a half-hour exposure considered possible until the incinerator is brought back under control. The ratio of IDLH to GLC ranges from 5.6 for mercury to 29 to 41 for the other metals except chromium at 16,663. This ratio may be considered to be the safety factor, or even the number of facilities of this size which could fail at one time before the IDLH would be exceeded.

REFERENCES

Benson, Lew and A. Licata, "The Application of Lime Sorbents in Municipal Waste Combustors," 6th Annual North American Waste-to-Energy Conference, Miami Beach, FL, May 1998.

Cowley, R., B. Gallagher and B. Nee, "Development and Execution of a Metals Pretest Program for a Hazardous Waste Incinerator," *Hazardous Waste & Hazardous Materials*, Vol. 11, No. 1, 1994.

Doucet, Lawrence, "Air Quality Modeling and Risk Assessment Analysis for Peninsula Hospital Center," Doucet & Mainka, P.C., Peekskill, NY, February, 1991.

EPA, "Background Information Document for the Development of Regulations to Control the Burning of Hazardous Waste in Boilers and Industrial Furnaces," US EPA, FA035/83D, 1987.

EPA, "Standards for Performance for New Stationary Sources and Emission Guidelines for Existing Sources: Medical Waste Incinerators," 40 CFR Part 60, February 27, 1995.

EPA, "New Source Performance Standards and Emission Guidelines for New and Existing Medical Waste Incinerators," February 27, 1995.

EPA, "Standards for Performance for Stationary Sources and Emission Guidelines for Existing Sources—Municipal Waste Combustors," 40 CFR Part 60, December 19, 1995.

EPA, "Risk Assessment for the WTI Hazardous Waste Incineration Facility," *Report EPA-905-R97-002*a, USEPA Region 5, Chicago, IL 60604, May, 1997.

EPA, "Hazardous Waste Combustors; Revised Standards; Final Rule," 40 CFR Part 60, June 19, 1998.

EPA, "Hospital/Medical/Infectious Waste Incinerators New Source Performance Standards, Final Rule," 40 CFR Part 62, July 6, 1999.

EPA, "New Source Performance Standards for New Small Municipal Waste Combustors," 40 CFR Part 60, August 30, 1999.

Fry, Barbara, *et al.*, "Technical Support Document to Proposed Dioxins and Cadmium Control Measure for Medical Waste Incinerators," California Air Resources Board, Sacramento, CA, May 1990.

Hahn, Jeffrey, and Donna Sofaer, "A Comparison of Health Risk Assessments for Three Ogden Martin Systems Resource Recovery Facilities Using Estimated (Permitted) and Actual Emission Levels," *Proceedings of the ASME Solid Waste Processing Conference*, Los Angeles, CA, June 1990.

Hasselriis, Floyd, "Relationship between Waste Composition and Environmental Impact," *Paper 90-38.2*, A&WMA, June 1990.

Hasselriis, Floyd, "Environmental and Health Risk Analysis of Medical Waste Incinerators Employing State of the Art Emission Controls," 84th A&WMA, Paper 91-30.3, 1991.

Levin, Arlene, D. Fratt *et al.*, "Comparative Analysis of Health Risk Assessments for Municipal Waste Combustors," *J. Air & Waste Management Assoc.*, **41**(1), Jan. 1991.

Licata, Anthony, H. Hartenstein and L. Terracciano, "Status of US EPA and European Emission Standards for Combustion and Incineration Technologies," Third Sorbalit Symposium, New Orleans, LA, November 2–5, 1999.

OAQPS, "SCREEN2 Model User's Guide," 450/4-92-006, US EPA, Research Triangle Park, NC, 1992.

Turner, D.B., "Workbook of Atmospheric Dispersion Estimates," US EPA Ref. AP-26 (NTIS PB 191–482), 1970.

Walker, B.L. and C.D. Cooper, "Air Pollution Emission Factors for Medical Waste Incinerators," *J. Air & Waste Management Assoc.*, **42**(6) June 1992, 784–791.

Wood, Roy and F. Hasselriis, "Relationship Between Short-term Emissions from Waste Combustion Systems and Potential Health Effects," Paper Number 97-FA85.04, *Air and Waste Management Association 90th Annual Meeting*, Toronto, Ontario, 1997.

Zanetti, P., *Air Pollution Modeling*, Van Nostrand Reinhold, New York, 1990.

CHAPTER 2.5
CALCULATIONAL PROCEDURES FOR ASH STABILIZATION AND SOLIDIFICATION

SECTION 2.5.1

CALCULATIONAL PROCEDURES FOR ASH STABILIZATION

Carl F. Isonhart

Mixer Systems, Inc., Pewaukee, Wisconsin

1 OVERVIEW OF PROCESSING AND STABILIZING FLY ASH

1.1 Introduction

What is coal ash? Coal ash is the noncombustible part of the coal which has been burned to produce electrical power or steam. Coal ash also includes flue gas desulfurization residue, and breaks down into the following constituents.

1. *Bottom ash*—This is a coarser particle and is collected at the bottom of the pulverized coal (PC) and stoker-fired boilers.
2. *Fly ash*—This is a fine material collected in a Baghouse or precipitator.
3. *Boiler slag*—This is a coarse ash particle collected at the bottom of cyclone-fired boilers.

Flue gas desulfurization (FGD) develops from the injection of calcium materials into the flue gas for the removal of sulfur.

As stated above, coal ash is divided into three parts. This chapter examines five specific cases.

- The first three examples are for processing fly ash with water only.
- The fourth example is for processing FBC fly and bottom ash with water.
- The fifth example covers hazardous medical waste from an incinerator, stabilizing both fly and bottom ash by chemistry using the silicate treatment.

However, before proceeding let us list some of the variables affecting fly ash:

- type of coal, i.e., bituminous, sub-bituminous, lignite, anthracite, petroleum, coke, etc.;

- co-gen-fired oil, wood, waste materials, mill and industrial sludges, etc.;
- boiler type, i.e., PC, stoker, FBC;
- boiler efficiencies;
- emission controls;
- collection systems.

It is clear that the above variables will have a profound effect upon the type of fly ash the equipment must process to meet legal requirements.

In order to perform in this type of hostile atmosphere, you need properly designed equipment that will wet all the ash particles with water or, if needed, chemical reagents to condition or stabilize the fly ash for disposal. It is imperative to coat all the particles with the proper chemical reagents to maximize the use of the chemicals and at the same time minimize costs without supersaturating the batch. The proper equipment will give you the capabilities to reach this goal. We have developed a large number of applications for our DustMASTER equipment. It is not feasible in this chapter to describe all the applications that are in service. However, we have selected five different applications that will give an idea of the flexibility of this equipment.

The objective of this chapter is to familiarize those who are interested in processing or stabilizing fly and bottom ash with:

- equipment for processing or stabilizing ash;
- how the equipment is controlled and how it operates;
- applications;
- terminologies often used in processing ash.

1.2 Understanding the Proper Equipment

Standard fly ash conditioning system specifications

Mixer. The "low profile" Turbin Mixer is the mixer utilized. The size used varies with the production requirement. The production rates are based on material weighing fifty (50) pounds per cubic foot and a minimum systems cycle time (charge–mix–discharge) of sixty (60) to ninety (90) seconds (Table 5.1).

The standard mixer is equipped with one (1) swivel discharge door (maximum three (3) discharge doors optionally available). The door is hydraulically operated. The opening and closing of the door is controlled through a single solenoid valve.

The mixer has a dual voltage 230/460 drive motor, 60 Hz, 3 Ph, and is totally enclosed and fan-cooled for long life.

TABLE 5.1 DustMASTER Capacity

DustMASTER	Turbin	Tons per hour capacity*
Model 20 TPH	Model 50	20
Model 40 TPH	Model 100	40
Model 60 TPH	Model 150	60
Model 80 TPH	Model 200	80
Model 120 TPH	Model 450	120
Model 160 TPH	Model 450	160
Model 200 TPH	Model 500	200

* Dry material weight.

The mixer cover is a movable, self-supporting assembly that has hinged doors for access to the mixing area inside the mixer. These are used for inspection and clean-out. A heavy rubber seal is fitted around the access door edges as well as the cover to provide a dust-tight seal. Quick release and latch clamps secure the doors to the cover assembly.

Mixing Process. The Turbin is a high-intensity mixer. The mixing is produced by a series of paddles which move through the mixer at the rate of six hundred (600) feet per minute.

There are from four (4) to ten (10) mixing arms, depending on the model, with replaceable cast ni-hard paddles plus inner and outer wall scrapers mounted on the rotating mixing assembly. These cast ni-hard wall scrapers are 8 in. wide and provide additional mixing action. The cast ni-hard paddles are right- or left-handed to provide the most effective braiding, mixing action. They have locked mounts with 3/4 in. of vertical wear adjustment.

The mixer is lined with 3/8 in. thick steel segmented liners on the floor and the inner and outer wall. They are bolted in place and can be reversed for longer life.

The mixer discharge door is comprised of a circular segment flush with the floor, pivoted on an external shaft, mounted in a pair of shaft bearings which are sealed and self-lubricated for longer life. The edge of the door matches the floor opening and rides on a machined steel rail to ensure door and floor alignment. The door surface is covered with a bolt in a replaceable 3/8-in.-thick abrasion-resistant liner. The door is externally sealed with a reversible, adjustable steel wiper bar. The door is actuated by a hydraulic cylinder to provide an angular opening of 95°.

The transmission is totally sealed in a lubricant. This prevents foreign materials from entering the transmission causing damage to gears and bearing.

Air system. All air cylinders have corrosion-resistant barrels, hard chrome-plated rods, are cushioned at both ends of the stroke, and have repackable piston and nose pieces. The 115-V solenoid valves include a lockable manual operator. The air cylinders, solenoid, and vibrator are piped to a common air supply header which includes an airline filter, pressure regulator, and airline lubricator.

Hopper for series I. The weigh-batching hopper is suspended from load cells in a self-supporting frame and is made from all-welded steel. It has a rolling blade-type fill gate and gas-tight butterfly valve discharge gates. All gates are air-cylinder operated through a 115-V, four-way valve. All gates have limit switches to indicate the closed position. The load cells are used for weight measurements. Although we have standard-size hoppers, they can be altered to fit specific space requirements.

Customer-furnished venting is required for each hopper. A 12-in.-diameter vent outlet must be provided.

The batching hopper has an access hatch on top for inspection or repair. The hopper utilizes externally replaceable aeration pads to condition the fly ash for uniform flow upon discharge, along with an air vibrator that is activated only when the final empty signal exists.

The hopper includes a flanged vent on top for connecting back to the silo bag house for a fugitive dust collecting system. A cross vent between the weigh hopper and the mixer allows the air in the mixer to be displaced by the incoming fly ash to flow back to the hopper without pressurizing the mixer.

Water system. Water volume is measured by weight or a meter which allows the water to pass through a rotary water system into the mixer. A meter controls the water volume per batch. The ash volume is controlled by load cells. The load cells and programmable controller are used to automatically control the batching, mixing, and truck loading. The batching controller is in a NEMA 4x enclosure that serves as the main junction box for all control connections to the DustMASTER.

Programmable controller. The principal job of the controller is to obtain a consistent weight of ash per batch. In addition, it provides interlocks, and contains programs designed to allow production in the event of mechanical failure of gates, limit switches, water valves, level probes, and even a breakdown of the master programmable controller.

This controller is housed in a wall-mounted NEMA 4x enclosure with terminal strip wiring. The programmable controller is an industrial modular type complete with a plug-in memory module that requires no battery back-up. It has a data access and display module. With this arrangement, the program is protected while allowing access and control of all timer and counter set points used in the process. The touch panel interface on the door indicates key functions and proved visual–manual control. With our manual control, it is possible to reach production rates equal to the automatic mode.

Since the number of loads to be cycled into a truck can be pre-set, it is possible to have this function at an optional truck driver start station and the main control. The automatic mode requires someone to start the system. It batches, mixes, and counts the loads into a truck, and then shuts down to a ready condition.

How the equipment is controlled and works. The DustMASTER is a batch system and is controlled by a programmable controller. All the weights of the ingredients are controlled by load cells; the water is controlled by a meter or weight. Any reagents can be weighed cumulatively. When the automatic process is activated by the operator, the following sequence of events takes place.

DustMASTER system operation:

- the mixer starts;
- the ash batch gates are open, charging ash into the volumetric or weigh batcher;
- the ash batch gates close when the full signal in the volumetric ash batcher is activated;
- the batcher gates open, discharging fly ash into the mixer;
- the mixer door opens (the timer starts when the batcher gate opens);
- the water start timers begin when the batcher gate opens;
- water starts metering into the mixer when the water start timers expire;
- when the ash in the batcher discharges to the empty level, a batcher discharge gate timer is activated and the ash batcher vibrator is started;
- when another batch is to follow immediately (not the last batch), the ash batcher refills;
- water flow ends when the set volume has been metered into the mixer;
- when the mixer timer expires, the mixer discharge door opens;
- a dishcarge door-open timer is started;
- when the mixer discharge door-open timer expires (if this is not the last batch), the door closes (if this is the last batch the door stays open for a longer set clean-out time);
- the operator can initiate the stop or last-batch signal or it can be set for automatic shut down when a pre-set number of batches have been loaded onto the haul vehicle;
- when the door closes the volumetric ash batcher gate opens and the mix cycle is repeated unless this is the last batch.

The system will operate in this manner until either it is shut down, or the fly ash storage silo is empty. There are three (3) methods of cleaning the mixer.

1. Hand-scrape the floor, side walls, and paddles (requires two men and approximately 15 min).
2. Pressure water wash (requires high-pressure washer, commerically available).
3. Scour the mixer with small aggregate, fill the mixer, and run it for about 10 min., or as long as required to clean the floor and walls. Discharge aggregate into a container for re-use.

As a result of our batch system we are able to introduce both liquid and powder reagents automatically for stabilizing and fixation during the mixing cycle. The weights of the ingredients are controlled by load cells, and the water addition by time or weight. Neither weight per cubic foot nor particle size affects the mixing ability of the unit. We have mixed materials from 0#/c.f. up to 287#/c.f. With the programmable controller it is easy to change the cycle times within the sequence of events, giving complete flexibility in the system. In addition, we provide 100% manual override of the control in the event of automatic mode failure.

The design of our system is an accumulation tank process for stabilization. We process and stabilize cupola and electric furnace dust, general foundry sands, sludges, and many other types of hazardous and nonhazardous materials.

2 CALCULATIONS OF PROCESSING AND STABILIZING FLY ASH

2.1 Applications

We have a wide variety and range of applications, including both hazardous and nonhazardous materials. During the mixing cycle, the unit has a 99% distribution level. What this statement really means is that all dust particles are coated with moisture, as well as all reagents used in the processing of waste materials, both hazardous and nonhazardous.

It is clear that this type of performance is vital to the conditioning and stabilizing of all waste materials, regardless of application or final use. This is why DustMASTER Enviro Systems issues a money-back performance guarantee on systems designed, engineered, and installed by our company. We have the knowledge and expertise to meet your environmental needs.

2.2 Stabilization and Fixation

As discussed in the districts' proposal for the modification of ash from CREF dated 1989 (CSDLAC, 1989), silicate treatment is by far the largest category of treatment process and has demonstrated the greatest effectiveness in achieving treatment goals. Silicate treatment includes processes which add sodium and potassium silicates as well as those which add calcium silicate (cement). Both published research and research performed by the districts indicate that silicate-based treatment provides the most effective and economically feasible treatment of ash.

As a result of studies and research on alternative treatment technologies, it was ultimately concluded that silicate treatment presented the best opportunity for successful treatment of CREF ash. The districts proposed to employ calcium silicate (cement amendment) as a treatment method.

SILICATE TREATMENT TECHNOLOGY: Silicate treatment systems have been found to be effective in reducing the soluble metals levels of a variety of wastes, including incinerator ash. Trezek (1987) has conducted tests on a wide variety of metals and confirmed the efficacy of silicate treatment. Silicate treatment technology utilizes commercially available sodium, potassium, and calcium silicates. Sodium and potassium silicates are normally applied in addition to a calcium silicate in the form of cement, of which calcium silicate is a major constituent. Other cementitious materials may be used which contain, or react to produce, calcium silicates. Cementitious materials include cement kiln dust and natural pozzolans (materials which upon reacting with lime produce calcium silicate). Natural pozzolans include volcanic ash and certain diatamaceous earths. Synthetic pozzalans include flyash and burnt clays. Treatment is achieved by mixing the silicate materials and allowing the mixture to cure.

Several chemical mechanisms are reported to bring about the treatment of heavy metals using silicate treatment. In applications of proprietary silicate blends, there is clear evidence that metasilicate formation stabilizes heavy metals (Trezek, 1987). For example, in the case of lead, an insoluble lead metasilicate ($PbSiO_3$) precipitate is formed (Zirschy and Pizner, 1988). It is suggested that other insoluble metal complexes are formed by various chemical mechanisms, including absorption by cement hydrates, or substitution and solid solution in the hydrate structure (Poon et al., 1983). Physical mechanisms such as binding and coating of the metal wastes in the cement matrix are also reported to contribute to the stabilization of contaminants.

However, it appears that chemical stabilization resulting in the formation of insoluble, stable complexes which incorporate metals into a crystalline silicate structure serves as the dominant treatment mechanism. The specific chemical reactions involving calcium silicate fixation of metallic ions are discussed below.

The application of portland cement is a form of silicate treatment. Published research demonstrates that calcium silicate, which comprises the majority of cement, behaves similarly to sodium and potassium silicates in chemically stabilizing heavy metals. Zirschky and Pizner (1988), Benson et al. (1985), and Shively et al. (1986) showed that portland cement alone may be effectively applied to wastes as a form of silicate treatment. Results of research conducted by the districts support the efficacy of calcium silicate treatment, and indicate that in the instance of CREF ash, silicate treatment with portland cement achieves the desired treatment objectives more cost-effectively than treatment with proprietary silicate blends.

Portland cement is a hydraulic cement which, when fully cured, achieves stone-like strength. The cement is manufactured by making cement clinker from the raw materials lime (CaO), silica (SiO_2), and alumina (Al_2O_3) in a cement kiln. The two major components of the clinker are tricalcium silicate ($3CaO–SiO_2$) and dicalcium silicate ($2CaO–SiO_2$), which comprise approximately 70–80% of the portland cement. Tricalcium silicate is the major cementitious phase of most portland cements (Bhatty, 1987). All the constituents of the cement clinker are anhydrous. However, when mixed with water, the anhydrous constituents chemically react and begin to form a dense structure of new, hydrated compounds. This process is most significant during the first 90 days after mixing, but can continue for a year or more. The hydration reaction is usually represented by the following chemical equation (Lea, 1971):

$3CaO–SiO_2$ + water	$\Rightarrow$	$x CaO y SiO_2 z H_2O$ +	$Ca(OH)_2$
(calcium silicate)		(calcium silicate hydrate)	(calcium hydroxide)

The calcium silicate hydrate comprises approximately 70% of the total amount of the cement hydrated phases, and is believed to play a major role in the fixation of metallic ions [7]. It has been shown that a metallic ion such as sodium can be retained in the structure of calcium silicate hydrates (Poon et al., 1985).

Bhatty (1987) presented four probable mechanisms involved in the fixation of metallic ions in solids formed with portland cement: (1) fixation by addition, (2) fixation by substitution reaction, (3) fixation by formation of a new compound or compounds, and (4) fixation by a combination of the above mechanisms. These mechanisms are discussed below.

FIXATION BY ADDITION: In this mechanism, a metallic ion is incorporated into the structure of calcium silicate hydrate by an addition reaction, as follows:

$x CaO y SiO_2 z H_2O$ +	M	$\Rightarrow$	$M x CaO y SiO_2 z H_2O$
(calcium silicate hydrate)	(metallic ion)		(metallic calcium silicate hydrate)

It is likely that calcium silicate hydrates containing low calcium/silicate (C/S) ratios favor fixation of metallic ions by this mechanism.

FIXATION BY SUBSTITUTION: In this mechanism, a calcium ion in the calcium silicate hydrate is substituted with a metallic ion in a reaction as follows:

$x CaO y SiO_2 z H_2O$ +	M	$\Rightarrow$	$M(x-1) CaO y SiO_2 z H_2O$ +	Ca^{2+}
(calcium silicate hydrate)	(metallic ion)		(metallic calcium silicate hydrate)	(calcium ion)

Substitution of calcium ion from the structure of calcium silicate hydrate is generally found to occur with high C/S ratios.

FIXATION BY FORMATION OF NEW COMPOUNDS: Insoluble compounds containing metallic ions such as metasilicates may also be formed during hydration of portland cement. These compounds are probably the products of a substitution reaction between the calcium ions in the calcium silicate hydrate and metallic ions.

FIXATION BY MULTIPLE MECHANISMS: In a complex system such as cement and ash waste, it is likely that a combination of the above mechanisms operates to fixate metallic ions. These treatment mechanisms begin upon mixing, and continue as the cement cures through a period of months.

Several published studies show that portland cement or cement-based materials can be used to stabilize wastes containing heavy metals. Zirschky and Pizner (1988) shows that contaminated foundry sand having a soluble lead concentration of 79.8 mg/l could be stabilized to meet EP-toxicity criteria of 5.0 mg/l using portland cement. The advantages associated with using cement included achivement of stabilization of hazardous foundry sands at a low cost, and the ease of mixing cement with the waste. Benson et al. (1985) shows that sand-blasting residue containing approximately 0.4 and 9.0 mg of cadmium and lead, respectively, per gram of waste was stabilized when added (up to 15%) as an admixture for concrete. The concrete mixtures were found to retain the heavy metals so that none was detected following the EP-toxicity extraction procedure. Shively et al. (1986) shows that the leaching of industrial sludge containing heavy metals was substantially reduced by solidification with Type II portland cement. All samples met EP-toxicity criteria for nonhazardous wastes. During these studies, the treated wastes were cured for varying periods of time (Appendix A).

Physical characteristics of ash. The CREF produces combined ash, which is a mixture of bottom and fly ashes. The bottom ash consists of two fractions, the siftings and the grate ash, of which the latter comprises the greater fraction. Siftings ash consists of the material, generally less than one inch in size, that passes through moving grates on the floor of the furnace. The material that remains on the grates after combustion constitutes the grate ash. The grate ash is very heterogeneous, containing all size fractions from the very fine up to about four feet in length. The larger items may be inerts such as steel or masonry. The fly ashes consist of three distinct fractions: the boiler tube residue, the spray dyer ash, and the baghouse ash. Each of these, though unique in character, is of a relatively uniformly fine particle size. In all ashes, the finer fractions were found to have a greater concentration of the metals cadmium, copper, lead, and zinc. Ash finer than one inch constitutes about 75% of the total waste stream. The balance consists of ordinary uncombusted and inert refuse.

The large items present in the bottom ash (the ash that passes through the furnace) are easily recognizable as nonhazardous and inert material (CCR, 1983). Such items include, but are not limited to, rocks and pebbles, wood and plant pieces, and fragments of manufactured items of metal, plastic, rubber, and glass (CCR, 1983). Large metal and concrete items also occur in the bottom ash. The material larger than one inch in size may be separated by screening. Removal of large items improves the homogeneity of the ash and cement mixture. Treatment becomes more efficient through targeting the ash fraction, which requires treatment and removing large extraneous inert materials.

As we mentioned in the beginning, this chapter will deal with the processing of fly and bottom ash. We have selected the following applications to review.

2.3 Scopes and Designs

1. Utility

Conditioning fly ash with water only.

Fly ash from Western Coal (powder river basin).

Fly ash contained 26% calcium oxide class C chemical analysis (see Appendix B).

Scope and design

We supplied four 160 TPH Series I DustMASTERS including volumetric hoppers, all valves, 125 hp electric motors, connection chutes, and controls with software.

Batch size 8000 lbs.

Water content 15–18% by weight.

Cycle time 90 s.

The units operate 8 h per day, 5 days per week; at the end of the day shift, two workers clean the units by hand – time required 15 min. These units have been in operation since 1985, and the maintenance cost is $0.01 per ton. This system was installed in 1985.

2. Utility

Conditioning fly ash with water only.

Fly ash type class F (see Appendix C), physical and chemical properties.

Scope and design
One 40 TPH Series I DustMASTER including weigh hopper, all valves, 40 hp electric motor, connection chutes, and controls with software and steel support structure. This was a turn-key project.

Batch size 2000 lbs.

Water content 8–10% by weight.

Cycle time 90 s, 40 batches per hour.

The system operates approximately 2 days per week; at the end of the shift the mixer is washed down with water. This system was installed in 1990.

3. Utility

Conditioning fly ash with water only.

Fly ash from western coal PC boiler.

Scope and design
Two 120 TPH Series II DustMASTERS including mixer mounted on load cells, all valves, 100 hp electric motor, connection chutes, and controls with software.

Batch size 6000 lbs.

Water content 15–18% by weight.

Cycle time 90 s, 40 batches per hour, 5 days per week.

At the end of the shift, the mixer is washed down with water.

Maintenance cost $0.01 per ton.

This system was installed in 1994.

4. Utility

Conditioning fly and bottom ash with water only.

Partially hydrated FBC fly ash (see Appendix D) for physical and chemical properties.

Scope and design
One 60 TPH Series II DustMASTER including mixer mounted on load cells, all valves, 50 hp electric motor, connection chutes, and controls with software.

Batch size 3000 lbs.

Water content 20–25% by weight.

Cycle time 90 s, 40 batches per hour, 5 days per week.

At the end of each day, one man used a power wash to clean the mixer, requiring approximately 20 min. This system was installed in 1989.

5. Medical waste incinerator

Stabilizing both fly and bottom ash with the silicate treatment.

Fly and bottom ash from incinerated hazardous medical waste (see Appendices E and F).

Scope and design turn key
One 20 TPH Series II DustMASTER including mixer mounted on load cells, all valves, 20 hp electric motor, connection chutes, and controls with software.

One 535 ft^3 insulated and heat-traced silo.

One dust collector.

One 55-foot-long drag conveyor, heat-traced and insulated.

One cement reagent hopper with delivery screw feeder.

One all-steel support structure.

One recorder and print-out, time-dated by batch and ingredients.

One complete installation of all equipment.

Batch size 800 lbs.

Water content 15% by weight.

Reagent (cement) 10%.

Cycle time 90 s, 40 batches per hour.

This system was installed in 1994.

Flexibility is the norm with our DustMASTER approach to processing all types of waste. With our programmable controller we can adjust all times within the sequence of events, and also change the batch weights. Therefore, if there is a change in the chemical composition of the fly ash requiring more or less process water, the change can be excepted at once via the programmable controller. Furthermore, if at a later date the waste material was declared hazardous, the DustMASTER base equipment would remain the same and a few components would be added to meet the new requirements.

The five examples described above were used as a cross seciton of the types of fly ash generated from various types of coal-fired boilers in the US, Asia and Europe. The most significant point is that the DustMASTER was able to process with water, or stabilize with chemical reagents, all types of fly ash and dust. Furthermore, our equipment has not encountered a material that could not be processed, up to and including nuclear waste. We do not infer that our unit is a panacea, but to date we have been able to condition all materials submitted to us for processing. In some cases we are conditioning the waste material for recycling to add to anotehr product or application. In some cases we are able to feed it back into the process, eliminating the need for land filling. The ultimate goal for all is recycling. However, in some cases the technology is not in place at this time to accomplish complete recycling of all types of waste material.

3 REFERENCES

Benson, R. E., Chandler, H. W., and Chacey, K. A. (1985) Hazardous waste disposal as concrete admixture. *J. Environ. Eng.*, III(4), 442–447.

Bhatty, M. S. Y. (1987) Fixation of metallic ions in portland cement. Portland Cement Association, Construction Technologies Laboratories, Skokie, IL.

CCR, Title 22, Division 4, Chapter 30, Initial statement of reasons. 1983.

CSDLAC, Incinerator ash modification proposal for the Puente Hills landfill, July 25, 1989.

Lea, F. M. (1971) *The chemistry of cement and contrete.* Chemical Publishing Company, New York, 4th ed., 727 pp.

Poon, C. S., et al. (1983) Use of stabilization processes in the control of toxic wastes. *Effluent Water Treat. J.*, 23, 451–459.

Poon, C. S., et al. (1985) Mechanisms of metal stabilization by cement-based fixation processes. *Sci. Total Environ.*, 41, 55–71.

Shively, W., et al. (1986) Leaching tests of heavy metals stabilized with portland cement. *J. Water Pollut. Control Fed.*, 58(3), 234–241.

Trezek, G. J. (1987) Application of the polysilicate technology to heavy metal waste streams. Submitted to State of California Department of Health Services, Toxic Substances Control Division, Alternative Technology Section, Grant Number 85-00180.

Zirschky, J. and Pizner, M. (1988) Cement stabilization of foundry sands. *J. Environ. Eng.*, 114(3), 175–719.

APPENDIX A MATERIAL SAFETY DATA SHEET (DATE LAST REVISED: SEPTEMBER 15, 1988)

I. General information

Chemical name and synonyms Fly ash	*Trade name and synonyms* N/A
Chemical family N/A	*Formula* Mixture
Proper DOT shipping name N/A	*DOT hazard classification* ORM-E
Manufacturer North Dakota Utilities	*Manufacturer's phone number*
Manufacturer's address	*Chemtrec phone number*

II. Ingredients

Principal hazardous components	*Percent*	*Threshold limit value (units)*
Calcium oxide	18–25%	5 mg/m^3

III. Physical data

Boiling point (°F) None	*Specific gravity ($H_2O = 1$)* 2.3–2.8
Vapor pressure (mm Hg) None	*Percent volatile by volume (%)* N/A
Vapor density (air = 1) N/A	*Evaporation rate (_____=1)* N/A
Solubility in water None	*pH*
Appearance and odor Brown to tan powder, no peculiar odor	

IV. Fire and explosion hazard data

Flash point (test method) None		*Auto ignition temperature* None
Flammable limits	*LEL* None	*UEL* None
Extinguishing media N/A		
Special fire-fighting procedures N/A		
Unusual fire and explosion hazards N/A		

V. Health hazard data

OSHA permissible exposure limit TWA 5 mg/m^3	*ACGIH threshold limit value* TWA = 2 mg m^3
Carcinogen—NTP program N/A	*Carcinogen—IARC program* N/A
Symptoms of exposure Irritation of eyes, nose, throat, and skin	
Medical conditions aggravated by exposure	
Primary route(s) of entry Breathing dust/hand contact	
Emergency first aid Eyes—wash with large amounts of water, seek medical attention Breathing—move to fresh air Skin—flush with large amounts of water	

VI. Reactivity data

Stability	XX	*Unstable* *Stable*	*Conditions to avoid* None
Incompatibility		*Materials to avoid* None	
Hazardous polymerization	XX	*May occur* *Will not occur*	*Conditions to avoid*
Hazardous decomposition products			

VII. Environmental protection procedures

Spill response Ventilate area. Collect spilled material.
Waste disposal method Dispose of in approved landfill
Other protection

VIII. Special protection information

Eye protection Dust-resistant goggles	*Skin protection* Gloves
Respiratory protection (specific type) Dust mask	*Ventilation recommended*

IX. Special precautions

Hygienic practices in handling and storage None
Precautions for repair and maintenance of contaminated equipment None
Other precautions None

APPENDIX B MATERIAL SAFETY DATA SHEET. FLY ASH

I. Product identification

Processor's name:	Minnesota Utility
Telephone:	
Address:	
Date prepared:	
Chemical name and synonyms:	Fly ash
Trade name and synonyms:	N/A

II. Ingredients

Typical analysis (wt. %):	Normal	Design	Upset
CaO	16.5%	20.5	24
$CaSO_4$	8.5	11	15
SiO_2	–	45–50	–
Al_2O_3	–	20–25	–
Fe_2O_3	–	10–17	–

III. Physical data

Density for volumetric design:	$38\,lb/ft^3$
Density for structural design:	$80\,lb/ft^3$
Normal size:	200 microns × 0 for fly ash
Temperature:	300°F

IV. Fire and explosion hazard data

Flash point:	–
Flammable limits:	–
Extingushing media:	–
Special fire-fighting procedures	–
Unusual fire and explosion hazards:	–

V. Health hazard data

Route(s)	Inhalation? Skin? Ingestion?	– – –
Health hazards (acute and chronic):		–
Carcinogenicity: Monographs?	NPT? IARC OSHA regulated?	– – –
Signs and symptoms of exposure:		–
Medical conditions generally aggravated by exposure:		–
Emergency and first aid procedures:		–

VI. Reactivity data

Stability:	–
Conditions to avoid	–
Incompatibility (materials to avoid):	–
Hazardous decomposition products:	–
Hazardous polymerization:	–
Conditions to avoid:	–

VII. Special or lead procedures

Steps to be taken in case material is released or spilled	–
Precautions to be taken in handling and storing	–
Other precautions:	–

VIII. Special protection information

IX. Emergency and first aid procedures

APPENDIX C MATERIAL SAFETY DATA SHEET. MIXTURES

I. Product identification

Processor's name:	Wisconsin Utility
Telephone:	
Address:	
Date prepared:	
Chemical name and synonyms:	Mixture
Trade name and synonyms:	Bituminous and sub-bituminous ash, coal ash, bottom ash, bottom slag, fly ash
Container:	Varies
Use:	Varies

II. Ingredients

Chemical	%	CAS	TLV/PEL	STEL	CEIL
Amorphous silica	20–50	7631–86–9	6 mg/m^3	–	–
Aluminum oxide	10–30	1344–28–1	10 mg/m^3	–	–
Iron oxide	5–30	1309–37–1	10 mg/m^3	–	–
Crystalline silica**	5–10	14,808–60–7	0.1 mg/m^3	–	–
Calcium oxide	1–30	1305–78–8	5 mg/m^3	–	–
Many trace metals	All <0.1	N/A	N/A	–	–

Section 313. Supplier notification

This product contains the following toxic chemicals subject to the report requirements of Section 313 of the Emergency Planning and Community Right-To-Know Act of 1986 (40 CFR 372)

III. Physical data

Boiling point:	N/A °F N/A °C
Melting point:	N/A °F N/A °C
Vapor density (air = 1)	N/A
Vapor pressure (mm Hg):	N/A
Bulk density:	N/A °F N/A
Solubility in water:	Slight
Appearance and odor:	Earth-tone powder and granular material
Specific gravity (H_2O = 1)	2–3
Percent volatile by volume:	N/A
pH/concentration:	N/A N/A
Corrosivity on metals:	No

IV. Health hazard data

Inhalation: Skin contact: Eye contact: Ingestion:	Little hazard produced by normal operation in open or well-ventilated areas May be abrasive and/or irritating May be irritating and abrasive Mild irritation of throat and GI tract
Health hazards (acute and chronic): .	Lung damage. Specifically silicosis** Also see section IX**
Warning properties:	Eye, nose, or throat irritation
Signs and symptoms of exposure:	Skin redness or burning
Emergency and first aid procedures:	
Inhalation	Remove to fresh air. Seek medical help promptly. Use artificial respiration if necessary
Eye contact:	Flush with large amounts of water for at least 15 min. Seek medical attention
Skin contact:	Wash with mild soap and water
Ingestion:	Give milk or water. Do not induce vomiting Seek medical attention promptly

V. Reactivity data

Stability:	N/A
Conditions to avoid:	N/A
Incompatibility (materials to avoid):	N/A
Hazardous decomposition products:	None
Hazardous polymerization:	Will not occur
Conditions to avoid:	None

VI. Environmental/handling/storage

Spill/disposal procedures:	Collect and dispose per local, state, and federal regulations
Storage requirements:	Contain as appropriate to minimize dusting **Keep all containers labeled**
Is this product listed or does it contain any chemical listed for the following:	
1. Disposal of product or any residue a hazardous waste? Hazardous waste code:	No N/A
2. An extremely hazardous substance under emergency planning and community right-to-know?	No
3. An EPA hazardous substance requiring spill reporting?	No
4. An OSHA hazardous chemical? Chemical name:	 Amorphous silica Aluminum oxide Crystalline silica
5. Does it contain any materials regulated as a hazardous material or hazardous substance by the Department of Transportation? Proper shipping name: Hazard class: Identification No.: Label required: Shipping paper required: Quantity required for placarding: Packing GRP:	 N/A N/A N/A N/A N/A N/A N/A

VII. Special protection information

Ventilation requirements: Consider on a case-by-case basis/usually not required		
Personal protective equipment	Respiratory: Eyes: Skin: Special clothing:	NIOSH-approved dust respirator based on OSHA PEL Safety glasses with dust goggles recommended Gloves recommended if irritating to skin Cover skin areas as needed/fly ash can be irritating

VIII. Other

** Listed by IARC as a possible human carcinogen based on laboratory animal test data/no evidence of human carcinogenicity.

Notice

This information was based upon current scientific literature. Information may be developed from time to time which may render this document incorrect. Therefore, the company makes any warranties to its agents, employees, or contractors as to the applicability of this data to the user's intended purpose.

APPENDIX D MATERIAL SAFETY DATA SHEET. CFBC ASH

I. Product identification

Processor's name:	Midwestern University
Telephone:	
Address:	
Date prepared:	
Chemical name and synonyms:	Partially hydrated CFBC ash
Trade name and synonyms:	Partially hydrated circulating fluidized bed combustor ash
Description:	Inorganic mineral ash
NFPA ratings:	Health, 1; flammability, 1; reactivity, 1

II. Ingredients

Hazardous ingredients	CAS	%
Calcium oxide	1305–78–8	10–25
Calcium sulfate	7778–18–9	40–60
Crystalline silica	14808–60–7	< 2.5
Magnesium oxide	1309–48–4	0.4–2.0

Section 313. Supplier notification

The remaining components are not known to be hazardous as defined by OSHA's Hazard Communication Standard, 29 CFR 1910.1200

III. Physical data

Boiling point:	N/A
Freezing point:	N/A
Solubility in water	Insoluble
Appearance and odor	Gray to white, no odor
Form:	Dust
Bulk density:	Approximately 75 lb/ft^3
pH (1 : 1):	12.3 approximately

IV. Fire and explosion hazard data

Not a fire or explosion hazard under normal conditions

V. Health hazard data

Route(s) of entry: Inhalation? Skin? Ingestion?	Yes No No
Signs and symptoms of exposure to:	
Calcium oxide:	Calcium oxide is a strong alkali, is corrosive, and is dangerous when improperly handled. Marked corrosive action results from contact with all tissues of the body, and it can cause severe burns to the eyes, resulting in blindness. Since signs and symptoms of irritation are frequently not evident immediately after contact with calcium oxide, injury may result before one realizes that the chemical is in contact with the body. Therefore, adequate protection against such exposure should be provided for all parts of the body
Crystalline silica:	Eye contact may initially include irritation with discomfort, tears, or blurring of vision. Short exposures to very high concentrations of crystalline silica may be lethal. The predominant effect of overexposure to airborne crystalline silica in humans is silicosis. Silicosis is a chronic disease characterized by pulmonary fibrosis and the formation of silica-containing scar tissue in the lungs, with symptoms of coughing, dyspnea, wheezing, and nonspecific respiratory ailments. Individuals with pre-existing diseases of the lungs may have increased susceptibility to the toxicity of excessive exposure to crystalline silica
Carcinogenicity:	Crystalline silica is listed by IARC and NTP as a carcinogen
Applicable exposure limits:	
Calcium oxide:	TLV (ACGIH): 2 mg/m^3, 8 h TWA PEL (OSHA): 5 mg/m^3, 8 h TWA
Calcium sulfate:	TLV (ACGIH): 10 mg/m^3 of total dust containing no asbestos and $<1\%$ free silica, 8 h TWA PEL (OSHA): 15 mg/m^3 of total dust, 8 h TWA 5 mg/m^3 of respirable dust, 8 h TWA
Silicon dioxide (crystalline silica):	TLV (ACGIH): 0.1 mg/m^3, or respirable dust 8 h TWA PEL (OSHA): 0.1 mg/m^3, or respirable dust 8 h TWA
Magnesium oxide:	TLV (ACGIH): 10 mg/m^3, 8 h TWA PEL (OSHA): 10 mg/m^3, 8 h TWA

VI. Reactivity data

Stability:	Stable. Reacts exothermically when water is added.
Incompatibility (materials to avoid):	None known
Hazardous decomposition products:	None
Hazardous polymerization:	Will not occur

VII. Safe handling and storage

Do not take internally. Excessive inhalation of fine particles may result in respiratory disorders. Treat as a nuisance dust. Clean up with shovel or vacuum, taking steps to minimize fugitive dust. **CAUTION:** Material is exothermic upon exposure to water; avoid direct contact.	
Shelf life limitations:	N/A
Incompatible materials for packaging:	Non-waterproofed
Incompatible materials for storage or transport:	Non-waterproofed

VIII. Special protection information

Eye protection:	Close-fitting goggles
Skin protection:	Minimize skin exposure by use of sufficient protective clothing, including gloves. Wash contaminated clothing after use. Wash skin with soap and water to avoid prolonged exposure
Respiratory protection:	Under normal use, use any NIOSH-approved fine particulate filtering mask. Extreme conditions may require the use of a self-contained breathing apparatus
Ventilation to be used:	Local exhaust

IX. Emergency and first aid procedures

Eye contact:	Flush with water for 15 min or more. Call a physician
Skin contact:	Thoroughly wash skin with soap and water Wash contaminated clothing before re-use. Obtain medical attention if irritation persists
Inhalation:	Remove to area with fresh air. Give oxygen or CPR if necessary. Seek medical attention if needed
Ingestion:	Not considered to be a potential route of exposure

X. Waste disposal

Dispose of in accordance with all local, state, and federal regulations

Notice

All statements, information, and data provided in this material safety data sheet are believed to be accurate and reliable, but are presented without guarantee, representation, warranty, or responsibility of any kind, expressed or implied. Any and all representations and/or warranties of merchantability or fitness for a particular purpose are specifically disclaimed. Users should make their own investigations to determine the suitability of the information or products of their particular purpose. Nothing contained herein is intended as permission, inducement, or recommendations to violate any laws or to practice any invention covered by existing patents, copyrights, or inventions.

APPENDIX E MEDICAL WASTE INCINERATOR. BOTTOM ASH

Parameter	Method	TDL	Concentration (mg/kg), dry weight basis 1	2	3	4	Average	Std. Dev.
(Log number)			(2528-01A)	(2528-01B)	(2528-01C)	(2528-01D)		
Physical (%)								
Moisture	ASTM D3173	0.05	66.1	63.6	64	NR	64.57	1.1
Combustibles	ASTM D3174	0.05	16.7	17.2	17.8	NR	17.23	0.45
Composition								
pH	9045		9.4	10.4	9.9	10.8	10.1	0.53
Aluminum	6010	5	59,000	54,000	65,000	45,000	55,750	7327.18
Arsenic	7060	2	< 6	< 6	< 6	< 6	5.9*	0
Cadmium	6010	0.8	3	3	6	3	3.8	1.3
Lead	6010	5	340	110	110	110	168	99.59
Manganese	6010	0.6	310	270	280	280	285	15
Mercury	7470	0.02	< 0.06	0.06	< 0.06	< 0.06	0.06*	0
Nickel	6010	0.4	68	42	100	48	65	22.64
Selenium	7740	2	< 6	< 6	< 6	< 6	5.9*	0
Zinc	6010	0.8	1400	1500	1300	1700	1475	147.9

* MPC.

APPENDIX F MEDICAL WASTE INCINERATOR. FLY ASH

			Concentration (mg/kg), dry weight basis				
Parameter	Method	TDL	1	2	3	Average	Std. Dev.
(Log number)			(2528-02A)	(2528-02B)	(2528-02C)		
Physical (%)							
Moisture	ASTM D3173	0.05	1.84	1.93	1.5	1.76	0.19
Combustibles	ASTM D3174	0.05	16.7	18.4	18.2	17.8	0.76
Composition							
pH	9045		11.1	11.2	NR	11.2	0.05
Aluminum	6010	5	3000	3000	NR	3000	0
Arsenic	7060	2	< 2	< 2	NR	1.9*	0
Cadmium	6010	0.8	100	110	NR	105	5
Lead	6010	5	620	640	NR	630	10
Manganese	6010	0.6	41	41	NR	41	0
Mercury	7470	2.5	58	63	NR	61	2.5
Nickel	6010	0.4	4	3	NR	3.5	0.05
Selenium	7740	2	< 2	< 2	NR	1.9*	0
Zinc	6010	0.8	5600	5600	NR	5600	0

* MPC.

SECTION 2.5.2

CATALYTIC EXTRACTION PROCESSING: CALCULATING PROCEDURES FOR ASSESSING LOW GRADE MATERIAL PROCESSING POTENTIAL

Christopher J. Nagel

Quantum Catalytics, Fall River, Massachusetts

3 OVERVIEW OF FUNDAMENTAL CATALYTIC EXTRACTION PROCESSING

3.1 Introduction

Wastes are typically heterogeneous materials that are an economic burden because value-recovery mechanisms are either unknown or uneconomical (Schung and Realff, 1997). If a technology can view wastes as the elements which comprise the original material, efficient separation and recovery of material from diverse feeds can occur (Nagel, 1990). Catalytic Extraction Processing (CEP) is a technology that uses thermodynamic and solution chemistry principles to achieve the above goal. It provides a means for manufacturing commercial products (i.e., industrial gases, alloys, and ceramics) from heterogeneous low grade materials (including hazardous, radioactive, and mixed waste) using a molten metal bath as both a catalyst for elemental dissociation and a solution for reaction engineering Nagel et al., 1996).

CEP is able to manufacture products from a wide range of solid, liquid, and gaseous feeds due to its ability to dissociate feeds and dissolve those associated elements into predictable intermediates. Applications that are currently commercially deployed or under development include chlorinated and nonchlorinated organic liquids (Mather et al., 1995), incineration flyash, radioactive ion exchange resins (Castagnacci et al., 1997), mixed wastes from nuclear power plants (e.g., filters, grit blast media, personal protection equipment (PPE), paint, etc.)

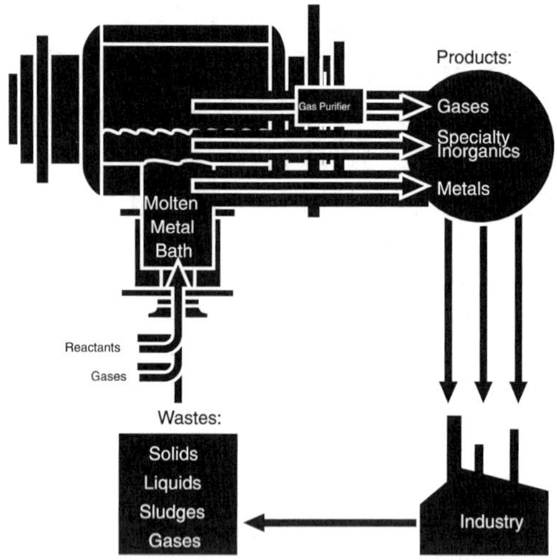

FIGURE 5.1 CEP recycling overview.

(Evans et al., 1997; Centec-XXI, 1997), radioactive nuclear wastes (e.g., supernate sodium nitrate), and chemical weapons (e.g., nerve agent) (Battelle Memorial Institute, 1996). Products that are generated typically include synthesis gas (i.e., carbon monoxide and hydrogen), ceramics (e.g., abrasives, cementitious media, filters, etc.), and metals (e.g., master alloys of iron, nickel, copper, etc.) (ENCORE Technical Resources, Inc., 1997).

This chapter will cover

1. Catalytic Extraction Processing (CEP) fundamentals
2. standard engineering calculations to predict CEP performances
3. actual waste processing applications.

3.2 Understanding Catalytic Extraction Processing

At the core of CEP is a metal bath, operated at temperatures above its liquidus point. The synthesis route to final product involves two simple steps (see also Fig. 5.2):

Elemental intermediate formation. The liquid metal contained in the reaction zone, under standard operating conditions, causes feed materials to catalytically dissociate into their basic building blocks (e.g., elements). These constituents form single primary bonded elemental intermediates (e.g., $Fe_{(3+\gamma)}C$) (Bach et al., 1996, Glukhovstev et al., 1997) as a result of the favorable dissolution driving forces governing the metal–element interaction. The parent atom bound to the surrounding metal media (i.e., the dissolved intermediate) exchanges with any given metal atom(s) at diffusion controlled rates. Hence, the formation of the elemental intermediates is affected only by the elemental composition of the feed, not the physical form or molecular structure of the arranged atoms contained therein; and ensures that the resultant products are uncontaminated with their progenitors. This solution property of the CEP system also allows complete conversion of feed constituents and predictability in subsequent product generation.

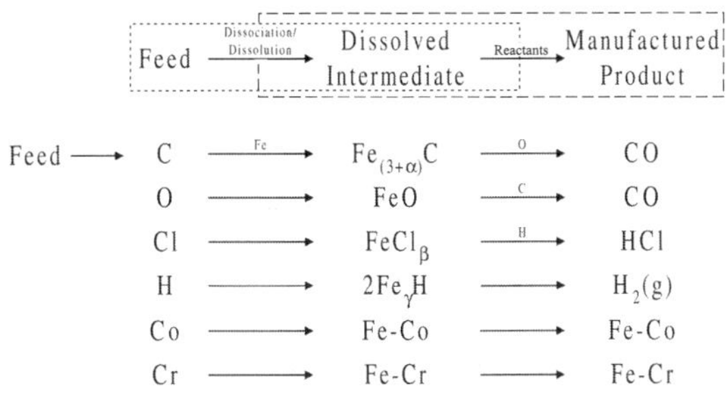

FIGURE 5.2 Product synthesis via dissolved intermediates.

Product formation and recovery. Thermodynamic driving forces govern product formation favorability under reaction conditions. Through the addition of select co-reactants and/or controlling operating conditions, the dissolved elemental intermediate(s) are reacted to form targeted chemical products with commercial utility. Product recovery is facilitated by partitioning the desired species between the metallic, ceramic, and gaseous product streams. Product partitioning between recoverable streams is governing by solution equilibria.

Engineering science calculations which elucidate the utilization of the above reaction pathway are:

1. *process chemistry* basics with attending solution requirements to enable elemental intermediate formation
2. *thermodynamic* basics including Gibbs free energy calculations, elemental partitioning prediction, and cofeed determination
3. *engineering calculations* including mass balance, energy balance, and reactor sizing.

3.3 Process Chemistry

With the appropriate choice of operating conditions (e.g., temperature (T), pressure (P), and moles of species (N_i)), organic compounds and reducible materials including metals and certain radionuclides can be dissolved in the metal bath. By virtue of the catalytic effect and solution chemistry of the bath, the metal actively participates in the formation of discrete elemental intermediates, creating a chemical-physical separation of feed from products (Glukhovtsev et al., 1997). The change in the Gibbs free energy of solution shows the thermodynamic favorability of elemental intermediate formation as contrasted to dissociation energy requirements in the absence of a solvent (Table 5.2). The relationship between the solubility of elements in molten iron and bond dissociation (Elliot and Gleiser, 1963) is further confirmed by experimental results (Fig. 5.3). The solubility of the element in the metal bath and the bulk concentration of that element in the metal establish dissolution driving forces and conversion performances at constant conditions (Fig. 5.4).

3.4 Thermodynamics

Elemental partitioning. The reaction of dissolved intermediates creates metallic (alloys), ceramic (inorganic oxides, halides, sulfides), and gaseous (industrial gases) products. Density differentials and immiscibility characteristics enable these products to be separated and recovered in distinct phases. The number of product phases and the final disposition of products are

TABLE 5.2 Comparison of Dissolution/Dissociation Energetics in the Presence and Absence of a Metal Solvent

	Liquid solvent (iron)			Absence of solvent	
Elemental intermediate formation	Gibbs free energy of solution*,† (J/g atom)	Gibbs free energy of solution at 1500°C (J/g atom)	Dissociated element	Gibbs free energy of dissociation (J/g atom)‡	Gibbs free energy of dissociation at 1500°C (J/g atom)
$C(gr) \rightarrow \underline{C}$§	22, 594 – 42.256 T(K)	−52, 336	$C(gr) \rightarrow C$	718, 567 – 157.69 T(K)	438, 959
$\frac{1}{2}H_2(g) \rightarrow \underline{H}$	36, 485 + 30.460 T(K)	90, 495	$\frac{1}{2}H_2(g) \rightarrow H$	222, 132 – 57.47 T(K)	120, 229
$\frac{1}{2}O_2(g) \rightarrow \underline{O}$	−117, 152 – 2.887 T(K)	−122, 271	$\frac{1}{2}O_2(g) \rightarrow O$	252, 358 – 65.11 T(K)	136, 908
$\frac{1}{2}N_2(g) \rightarrow \underline{N}$	3, 598 + 23.891 T(K)	45, 960	$\frac{1}{2}N_2(g) \rightarrow N$	476, 310 – 64.81 (TK)	361, 392
$\frac{1}{2}S_2(g) \rightarrow \underline{S}$	−135, 060 + 23.430 T(K)	−93, 519	$\frac{1}{2}S_2(g) \rightarrow S$	218, 650 – 59.93 T(K)	112, 385
$\frac{1}{2}P_2(g) \rightarrow \underline{P}$	−122, 173 – 19.246 T(K)	−156, 299	$\frac{1}{2}P_2(g) \rightarrow P$	246, 671 – 58.46 T(K)	143, 013

* In liquid iron.
† Extracted from Rao, 1985.
‡ Extracted from *HSC Chemistry Version 2.0*, Copyright Outokumpu Research Oy, Pori, Finland, A. Roine.
[d] $\underline{C}$ is the dissolved elemental C in the iron bath.

determined by thermodynamics and solution equilibria. The relative likelihood of a reaction can be obtained by comparing the negativity of the Gibbs free energy (ΔG) change upon reaction for the given conditions (Figs. 5.5 and 5.6) (United States Steel, 1985). Targeted product formation can be achieved through manipulation of thermodynamic space (T, P, N_i) (Fig. 5.7). This is accomplished by changing bath composition (via modification of feed material addition and bath metal selection) and, to a much lesser extent, pressure of operation.

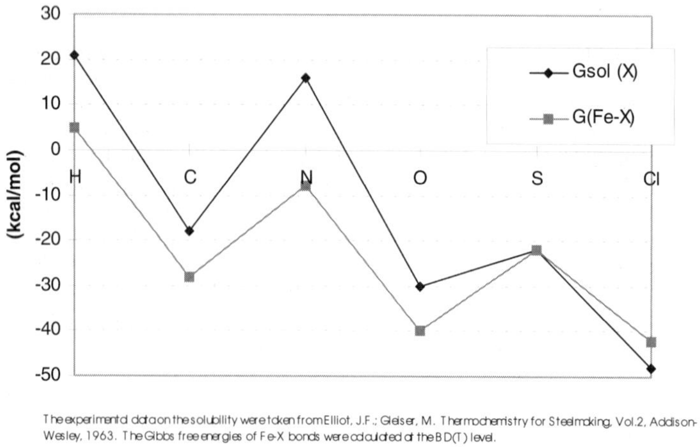

FIGURE 5.3 The relationship between the solubility of elements in molten iron (ΔG_{sol}(X)) and their free energy of bond dissociation (ΔG (Fe–X)). (The experimental data on the solubility were taken from Elliot and Gleiser, 1963. The Gibbs free energies of Fe–X bonds were calculated at the BD(T) level.)

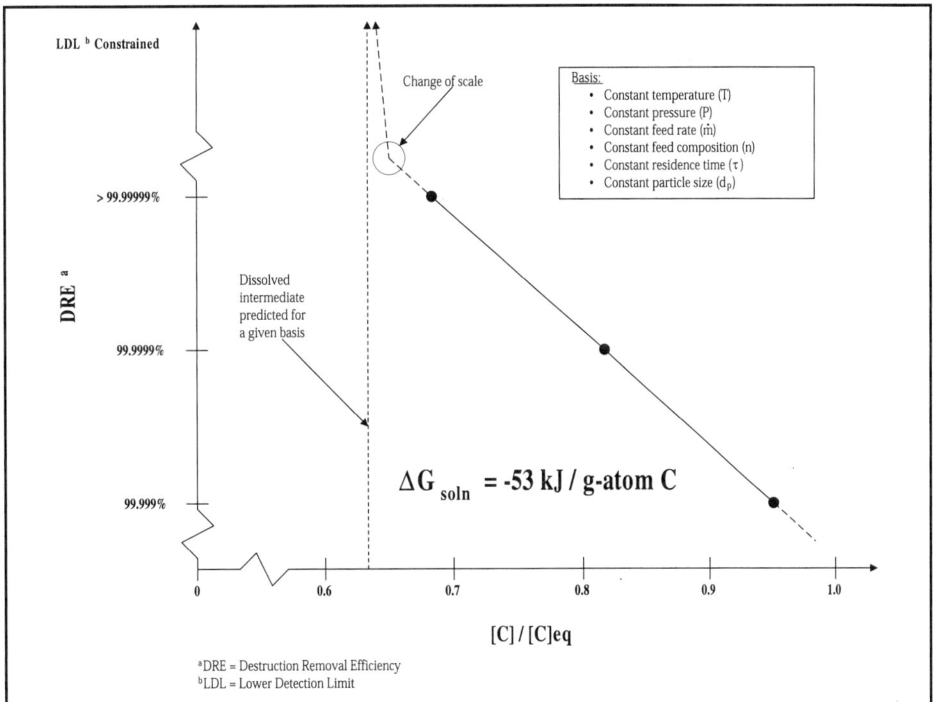

FIGURE 5.4 Effect of carbon concentration on conversion performance.

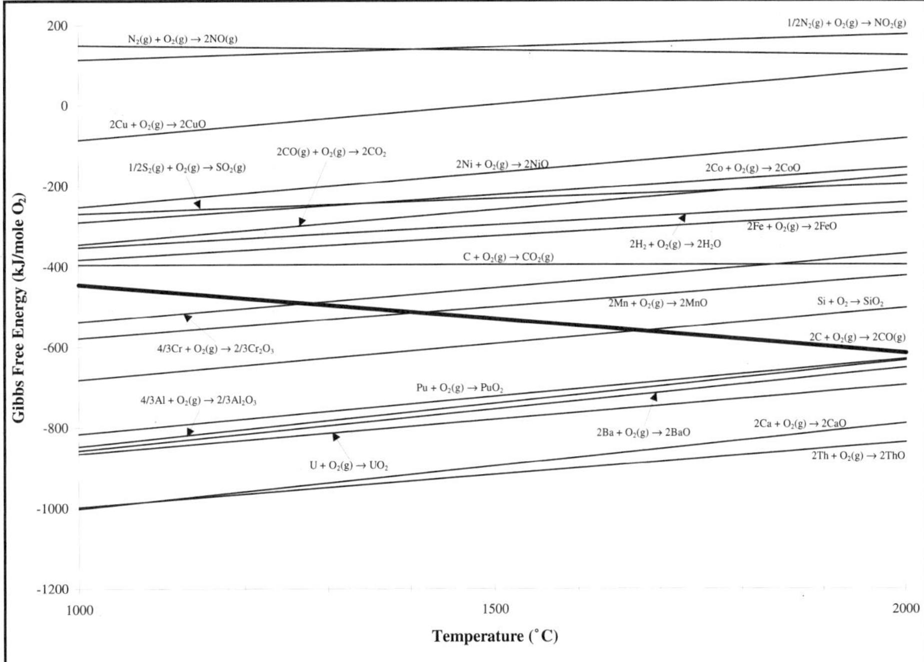

FIGURE 5.5 Gibbs free energy of oxidation (*J. J. Moore, 1990*).

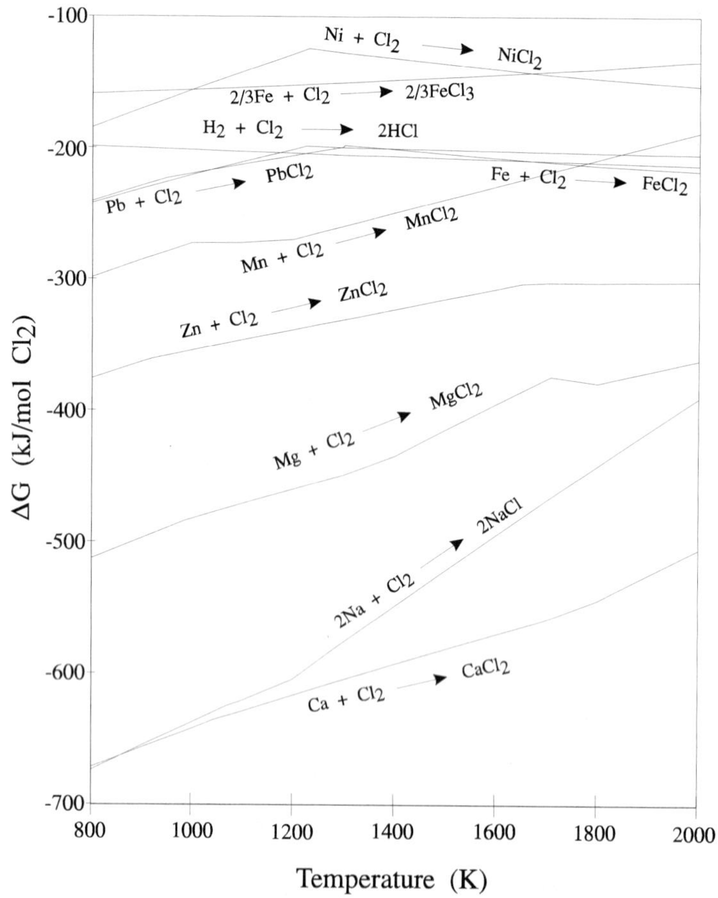

FIGURE 5.6 Gibbs free energy for chlorination (*T. Barin, 1989*).

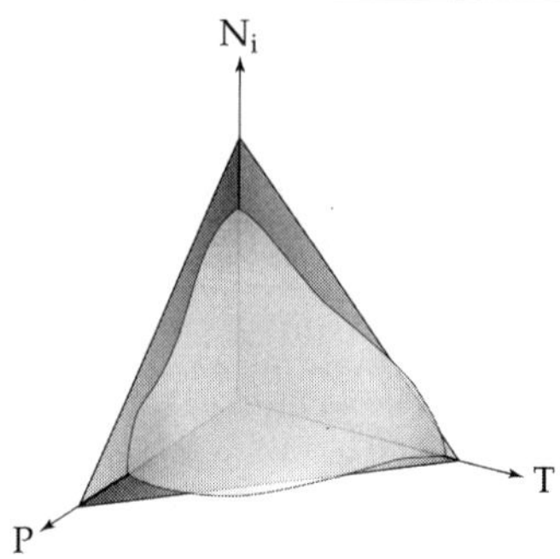

FIGURE 5.7 Thermodynamic space.

Oxygen thermochemistry. Synthesis gas is a mixture of carbon monoxide and hydrogen which is widely used in the manufacture of chemicals (e.g., methanol, methyl tertbutyl ether (MTBE), etc.). The end use product determines the syngas specification, including H_2/CO ratio and the type and level of acceptable contaminants (e.g., metals, $\cdot N_2$, CO_2, etc.). In CEP, synthesis gas can be generated from any source of C, H, and O, provided the feed rates are controlled to achieve the targeted product specification. Since the bath carbon activity sets the bath partial pressure of oxygen which in turn bounds the CO/CO_2 and H_2/H_2O ratio, the desired product composition can be tailored by modifying the equilibrium position in the defining product forming reaction equations. The reaction of oxygen with carbon monoxide to form carbon dioxide is illustrated by the following reaction:

$$2CO + O_2 \longleftrightarrow 2CO_2$$

This reaction equilibrium coupled with assumed ideal gas behavior can be used to derive a very simple expression relating the partial pressure of oxygen to the ratio of the carbon monoxide concentration and carbon dioxide concentration as shown by the expression:

$$P_{O_2} \cong \left(\frac{y_{CO}}{y_{CO_2}}\right)^{-2} K_{eq}^{-1}$$

where the equilibrium constant, K_{eq}, is determined directly using gas phase thermochemical data. It is related to the Gibbs free energy of reaction by the relation:

$$\ln K_{eq} \equiv \frac{-\Delta G}{RT}$$

where R is the ideal gas constant and T is the absolute temperature. The CO/CO_2 ratio as a function of oxygen partial pressure at 1600°C is illustrated in Fig. 5.8. The vertical dashed lines have been added to highlight those conditions consistent with the oxidation of pure iron (Fe), pure nickel (Ni), and pure copper (Cu).

Similarly, the overall oxygen distribution between the various species within the product gas is given by the water-gas-shift reaction as indicated below:

$$H_2 + CO_2 \leftrightarrow H_2O + CO$$

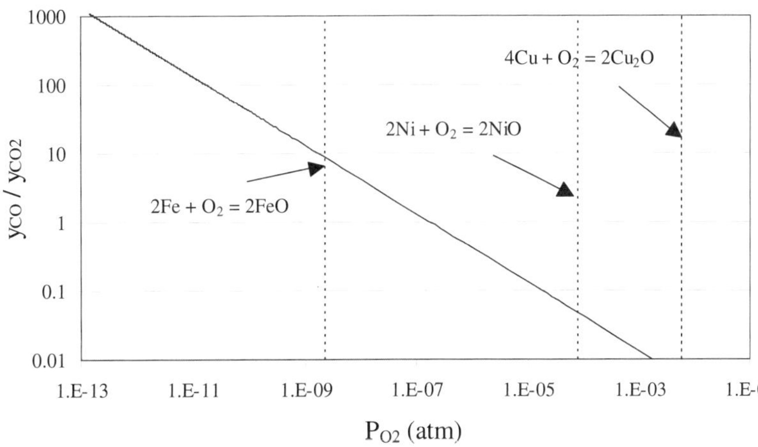

FIGURE 5.8 CO/CO_2 ratio as a function of oxygen partial pressure at 1600°C.

Hence the overall oxygen requirement, represented as a normalized O/C molar ratio, is a function of the CO/CO_2 ratio. In chlorinated organic processing applications, the combined feed (H – Cl)/C molar ratio is assumed to be fixed at a value, for example 2. The overall oxygen requirement for various fixed combined feed (H – Cl)/C molar ratios at a temperature of 1600°C is shown as a function of the CO/CO_2 ratio in Fig. 5.9. These equilibrium data can be used to determine the synthesis gas quality under various operating conditions (e.g., temperature, CO/CO_2 ratio).

3.5 Engineering

Based on the above reaction pathways, organic feeds with elemental constituents carbon, hydrogen, oxygen, chlorine, and sulfur will form carbon monoxide, hydrogen, hydrogen chloride, and hydrogen sulfide. The amount of carbon dioxide inerts will be determined by the oxygen potential. Any metallic feeds with elemental constituents cobalt, nickel, copper, etc., will dissolve in the bath metal and form a metal alloy. Inorganic feeds with elemental constituents aluminum, calcium, magnesium, etc., will form an oxide layer based on their affinity to oxygen. Hence, once the elemental composition of feed material is known, the final destination of the elements and the final composition of the three phase products (gas, ceramic, metal) can be determined. A mass balance can be generated once component partitioning is calculated and the cofeed strategy is determined. An energy balance can be determined based on the heat of reaction and heat loss around the reactor. The mass balance, energy balance, throughput required, and targeted conversion performance will be used as baseline information for specifying the reactor sizing: reactor diameter, bath height requirement, headspace requirement, induction coil requirement, and refractory design.

Product development can be engineered through process chemistry, operating conditions optimization, and equipment modifications (shown in Fig. 5.10). For example, synthesis gas quality and contaminates (e.g., CO_2 levels) can be controlled by the oxygen potential and

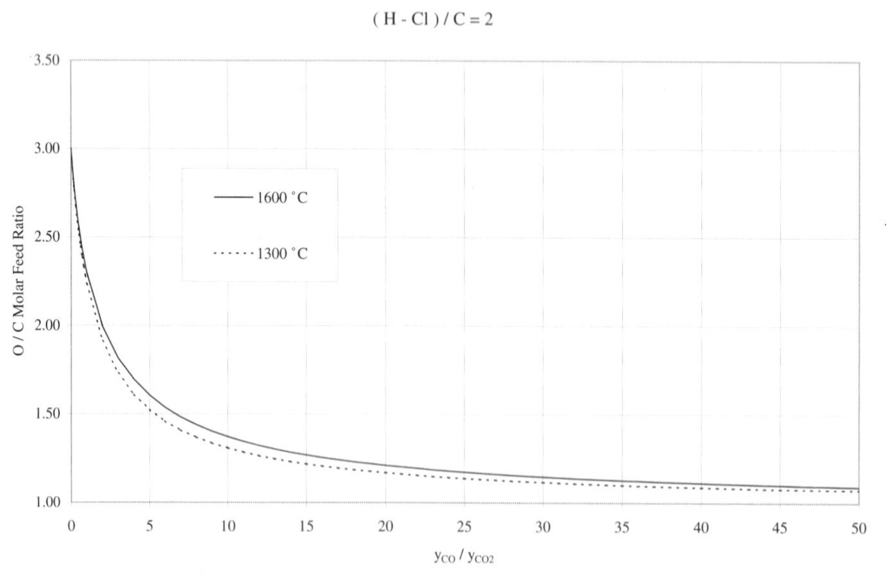

FIGURE 5.9 Oxygen requirement temperature sensitivity.

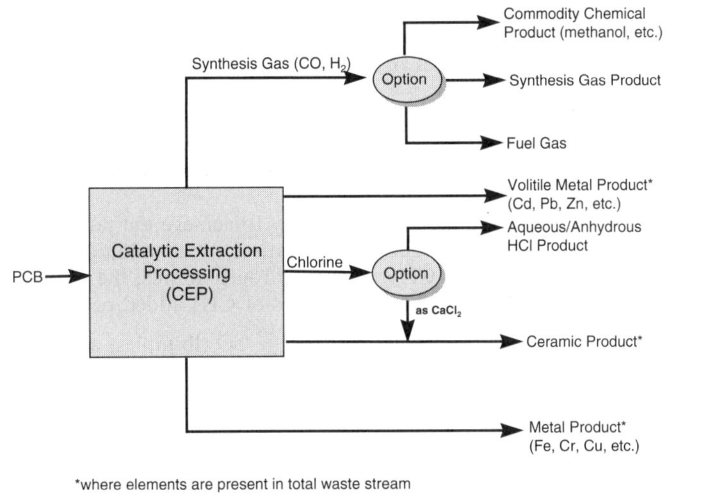

FIGURE 5.10 CEP product recovery scenarios.

pressure in the system with subsequent conditioning for end use application. Similarly, ceramic product quality is typically determined by chemical composition and post-process operation procedures; hence by predetermining cofeed addition strategies and post-process product handling mechanisms, targeted ceramic product quality can be achieved.

4 ENGINEERING CALCULATIONS OF CATALYTIC EXTRACTION PROCESSING

4.1 Process Chemistry

PROBLEM 1: Determination of Cofeed Strategies for Processing Chlorinated Organic Wastes

A highly chlorinated organic waste is requested to process in CEP to generate high quality synthesis gas with a CO : H_2 ratio ration 1 : 1 (vol.), and maximum HCl generation. What is the cofeed consumption to achieve the target product distribution? The elemental composition for the chlorinated organic waste is C 25%, H 5%, Cl 70%.

Goal: To complete the stoichiometric consumption of oxygen, and hydrogen cofeeds to achieve targeted product distribution.

Solution:

Step 1: To determine the chemical equations of the targeted reactions
The targeted chemical equation is

$$2C + O_2 = 2CO\ (g)$$

$$2H = H_2\ (g)$$

$$H + Cl = HCl\ (g)$$

Step 2: To determine the cofeed requirement to complete the reactions
Assuming 100 lb of feed materials, the number of lb moles of C, H, and Cl are: $25/12 = 2.08$ lb moles of C, $5/1 = 5$ lb moles of H, and $70/35.5 = 1.97$ lb moles of Cl.

Based on the above equations: 2.08 lb moles of C require $2.08/2 = 1.04$ lb moles of O_2 to form 2.08 lb moles of CO (g), 1.97 lb moles of Cl require 1.97 lb moles of H to form 1.97 lb moles of HCl (g), $5 - 1.97 = 3.03$ lb moles of H form $3.03/2 = 1.52$ lb moles of H_2 (g).

Hence, without fine-tuning the CO : H_2 ratio, the final product (per 100 lb of feed) has 2.08 lb moles of CO (g), 1.52 lb moles of H_2 (g), 1.97 lb moles of HCl (g), and require 1.04 lb moles of O_2 (g) to complete the reaction.

Step 3: To determine the cofeed strategy to achieve the target product composition

To achieve the target CO : H_2 ratio 1 : 1, an additional hydrogen source is required. Using methane (CH_4) as the hydrogen source, the following equation is used.

Let X be the number of moles of CH_4 added per 100 lb feed.

CO (lb moles) original + CO (lb moles) added by CH_4 = H_2 (lb moles) original + H_2 (lb moles) added by CH_4

$$2.08 + X = 1.52 + 2X$$

$$X = 2.08 - 1.52$$

$$= 0.56$$

Hence, 0.56 lb moles of CH_4 and 0.56 lb moles of additional O_2 is required.

To summarize, for 100 lb of feed, the cofeed required is 0.56 lb moles (8.96 lb) CH_4, $0.56 + 1.47 = 2.03$ lb moles (64.96 lb) O_2, with a final product of 2.64 lb moles (73.92 lb) CO, 2.64 lb moles (2.64 lb) H_2, and 1.97 lb moles (71.91 lb) HCl.

Alternatively, using a per lb feed basis, the results can be summarized as:

Components	lb/lb feed
O_2 (cofeed)	0.65
CH_4 (cofeed)	0.09
CO (final product)	0.74
H_2 (final product)	0.03
HCl (final product)	0.72

PROBLEM 2: Determination of Product Phase for Processing Inorganic Materials

An inorganic-based waste has a composition as shown in the following table:

Component	wt%
Al_2O_3	6
CaO	45
Fe_2O_3	4
MgO	6
MnO	7
NiO	5
SiO_2	20
ZnO	2
C	5
Total	100

The generation rate of this waste stream is 500 tons per year. What is the annual product generation rate:

Goal: To determine the partitioning performance of the elements and calculate the various product phase generation.

Solution:

Step 1: To determine the partitioning performance of individual elements
Using the Gibbs free energy diagram or the calculation procedure as shown in Problem 1 (Section 4.2), the resultant partitioning phases for the elements are:

Component	Partitioning phase	Partitioning form
Al_2O_3	Ceramic	Al_2O_3
CaO	Ceramic	CaO
Fe_2O_3	Metal	Fe
MgO	Ceramic	MgO
MnO	Metal	Mn
NiO	Metal	Ni
SiO_2	Ceramic	SiO_2
ZnO	Gas	Zn (g)
C	Gas	CO

Step 2: To determine the mass balance for product generation

Ceramic phase:

$$Al_2O_3 + CaO + MgO + SiO_2 = (500/100) \times (6 + 45 + 6 + 20)\ \text{ton} = 385\ \text{ton}$$

Metal phase:

$$\begin{aligned} Fe + Mn + Ni &= (500/100) \times (4/(56 \times 2 + 16 \times 3) \times 56 \times 2 + 7/(55 + 16) \\ &\quad \times 55 + 5/(59 + 16) \times 59)\ \text{ton} \\ &= (500/100) \times (0.025 \times 56 \times 2 + 0.098 \times 55 + 0.067 \times 59)\ \text{ton} \\ &= (500/100) \times 12\ \text{ton} = 61\ \text{ton} \end{aligned}$$

Gas metal phase:

$$\begin{aligned} Zn &= (500/100) \times 2/(65 + 16) \times 65\ \text{ton} = (500/100) \times 0.025 \times 65\ \text{ton} \\ &= (500/100) \times 1.6\ \text{ton} = 8\ \text{ton} \end{aligned}$$

Gas phase:

$$CO = (500/100) \times (5/12) \times (12 + 16) = 58\ \text{ton CO}$$

Step 3: To determine oxygen balance and cofeed strategy
Total oxygen input in excess of that used in ceramic phase generation:

$$\begin{aligned} \text{From } Fe_2O_3 + \text{From MnO} + \text{From NiO} + \text{From ZnO} &= (500/100) \times 16(0.025/2 \times 3 \\ &\quad + 0.098 + 0.067 + 0.025) \\ &= 21\ \text{ton oxygen} \end{aligned}$$

Total oxygen output used in CO formation:

$$58 \text{ ton}/(12 + 16) \times 16 = 33 \text{ ton oxygen}$$

Hence $33 - 21 = 12$ ton oxygen is required to add to the system to complete the carbon conversion.

The results can be summarized as

Inputs	
Feed	500 ton
Oxygen cofeed	12 ton
Input total	**512 ton**
Outputs	
CO	58
Ceramic product	385
Metal product	61
Gaseous metal	8
Output Total	**512 ton**

CHALLENGE 1: Determination of the cofeed strategy for an inorganic feed material

Based on the feed composition in Problem 2, what will be the cofeed strategy if the targeted ceramic product requires a composition of CaO 45%, SiO_2 30%, Al_2O_3 20%, MgO 5%?

CHALLENGE 2: Determination of the intermediate structures

Why is the Gibbs free energy of solution in Table 5.2 positive for hydrogen and nitrogen but negative for chlorine or oxygen? Explain why $G(\text{Fe} - \text{X})$ is negative, and $G_{sol}(\text{X})$ is positive for $\text{X} = \text{N}$ in Fig. 5.3? Using Table 5.2 and Fig. 5.3, what is an average energy in a metal parent atom bond? What are the likely chemical structures resulting from the dissolution of carbon, oxygen, and chlorine in iron?

4.2 Thermodynamics

PROBLEM 1: Determination of the partitioning results for various feeds

Consider a feed containing elements carbon, nitrogen, cobalt, nickel, calcium, and uranium-238. Determine where these elements will partition in a CEP system.

Goal: To determine how the listed elements partition between gas, ceramic, and metal phase.

Solution:

Step 1: To calculate the Gibbs free energy of oxide formation for these elements
The Gibbs free energy for oxide formation is defined as

$$\Delta G = \Delta H - T\Delta S$$

where ΔG is the standard Gibbs free energy change, ΔH is the standard enthalpy change, ΔS is the standard entropy change, and T is the temperature.
The corresponding equations for these elements to be evaluated are:

$$2C + O_2\ (g) = 2CO\ (g)$$

$$N + O_2\ (g) = NO_2\ (g)$$

$$2Co + O_2\ (g) = 2CoO\ (l)$$

$$2Ni + O_2\ (g) = 2NiO\ (l)$$

$$2Ca + O_2\ (g) = 2CaO\ (l)$$

$$U + O_2\ (g) = UO_2\ (l)$$

Based on the thermodynamic equilibrium data from the literature (*Perry's Engineers' Handbook*, 1997), the Gibbs free energy for the above equations at the typical CEP operating temperature (1500°C) are:

$$2C + O_2\ (g) = 2CO\ (g) \quad \Delta G_f = -530.9\ \text{kJ/mol of oxygen}$$

$$N + O_2\ (g) = NO_2\ (g) \quad \Delta G_f = +180\ \text{kJ/mol of oxygen}$$

$$2Co + O_2\ (g) = 2CoO\ (l) \quad \Delta G_f = -226\ \text{kJ/mol of oxygen}$$

$$2Ni + O_2\ (g) = 2NiO\ (l) \quad \Delta G_f = -167\ \text{kJ/mol of oxygen}$$

$$U + O_2\ (g) = UO_2\ (l) \quad \Delta G_f = -810\ \text{kJ/mol of oxygen}$$

Step 2: Comparing the Gibbs free energy values
Given that carbon is used as the reducing agent, oxidation reactions with a Gibbs free energy change more negative than that for CO formation will be favorable. Hence, Ca and U-238 will partition to the ceramic phase as metal oxides. The other metals, Co and Ni will remain in the metal and be recovered as ferroalloy. Oxidation byproducts like NO_2 formation is thermodynamically unfavorable under CEP operating conditions. The above evaluation can also be determined based on the standard Gibbs free energy diagram.

Note: The partitioning results indicated that radioactive materials can be concentrated into the ceramic phase to achieve significant volume reduction for radioactive final form disposal.

PROBLEM 2: Determination of chlorine partitioning

Where does chlorine partition in CEP systems?

Goal: To determine the Gibbs free energy of chlorine formation in the presence of various components: primarily with and without a ceramic phase.

Solution: Refer to Fig. 5.6: Gibbs free energy of chloride formation at the typical CEP operating temperature 1500°C. In the presence of a ceramic phase (alumina–calcia–silicate based) calcium chloride is highly favorable. Hence chlorine is expected to partition to the ceramic phase. In the absence of a ceramic phase, i.e., absence of Ca, Na, Ca, and other metals, hydrogen chloride will be the favorable product. Hence, hydrogen chloride is the expected product in CEP systems.

Note: This manipulation of elemental partitioning based on cofeed strategy is used if the partitioning phase of certain elements is required by the customer.

PROBLEM 3: Determination of sodium partitioning

If sodium is present in the inorganic materials as shown in Problem 2 in Section 4.1, what is the expected partitioning of sodium with/without a ceramic phase? What is the expected partitioning of sodium in the presence of chlorine and sulfur?

Goal: To determine the effect of various cofeeds and pressure on sodium partitioning.

Solution:

Step 1: To determine the standard sodium partitioning determination
The melting point of pure sodium is 700°C (Lide, 1975); hence in the absence of a ceramic phase, sodium will vaporize and be recovered in the gas phase.

In the presence of a ceramic phase, the Gibbs free energy diagram (Fig. 5.2) for oxide formation is used. The diagram indicated that sodium oxide formation is favorable in CEP conditions. The melting point of sodium oxide is 1500°C; hence in the presence of a ceramic phase, sodium will stay in the bath as sodium oxide.

Step 2: To determine sodium partitioning in the presence of chlorine and sulfur
In the presence of other anions, alternative reactions are considered:

$$\mathrm{Na} + \mathrm{Cl} = \mathrm{NaCl}$$

$$2\mathrm{Na} + \mathrm{S} = \mathrm{Na_2S}$$

To determine the competing reactions between oxygen, chlorine, and sulfur for sodium, the following equations are considered:

$$\mathrm{Na_2O} + 2\mathrm{HCl} = 2\mathrm{NaCl} + \mathrm{H_2O}$$

$$\mathrm{Na_2O} + \mathrm{H_2S} = \mathrm{Na_2S} + \mathrm{H_2O}$$

Based on the Gibbs free energy calculations of the above equations, the product distribution is determined by the thermochemistry equilibrium, and NaCl, Na_2S are the predominant products. Hence in the presence of Cl and S, Na will partition to the gas phase as NaCl and ceramic phase as Na_2S, respectively.

PROBLEM 4: Determination of bath metal for chlorinated organic wastes processing

The R&D scientists are considering the different options of bath metal in processing a highly chlorinated organic feed. The goal is to maximize the generation of hydrogen chloride which will be used as a product. The scientists are considering using a iron, copper, or nickel bath. Based on the Gibbs free energy determination, which bath metal should be used?

Goal: To determine the formation of chlorides in various metal baths and identify the one with minimal chloride formation.

Solution: Based on a diagram similar to Fig. 5.6 or Gibbs free energy calculation of the following equations:

$$\mathrm{Ni} + \mathrm{Cl_2} = \mathrm{NiCl_2} \quad \Delta G_f = -147\ \mathrm{kJ/mol\ Cl_2}$$

$$\mathrm{Fe} + \mathrm{Cl_2} = \mathrm{FeCl_2} \quad \Delta G_f = -200\ \mathrm{kJ/mol\ Cl_2}$$

$$\mathrm{Cu} + \mathrm{Cl_2} = \mathrm{CuCl_2} \quad \Delta G_f = -82\ \mathrm{kJ/mol\ Cl_2}$$

By comparing the respective ΔG_f, it is apparent that a copper system will minimize metal chloride formation and hence will maximize HCl generation.

CHALLENGE 3: Sulfur partitioning

How will sulfur partition in an iron bath system? How will partitioning be affected upon saturation of sulfur in the bath? How will the partitioning be affected if the system is operated at 6 bar? Repeat the previous question for an iron bath system that contains sulfur and sodium.

CHALLENGE 4: HCl formation

Which metal systems are reasonable for the formation for anhydrous HCl? Which metal system yields: the highest CO/CO_2 ratio; the lowest operating temperature; the lowest energy costs, least hazardous maintenance; the highest carbon solubility; the highest hydrogen solubility, the highest chlorine solubility? Why can the HCl/C ratio be assumed to be fixed for chlorinated organic processing applications?

CHALLENGE 5: UO_2 partitioning

What is the solubility of UO_2 in molten iron, nickel, and copper? If the targeted decontamination potential limit of the metal is 10^{-8}, how would the UO_2 be extracted to achieve this level and how should the thermodynamic space be changed to achieve the targeted equilibrium?

4.3 Engineering/Product Applications

PROBLEM 1: Energy balance for chlorinated organic wastes processing

Determine the energy balance when processing a chlorinated organic waste stream with the following composition:

Element	wt.%
C	72.5%
H	9.76%
O	0.1%
Cl	17.63%

Solution:

Consider the reactions that will take place:

$$C + O = CO$$
$$2H = H_2$$
$$H + Cl = HCl$$

The resultant energy balance is:

$$\text{Heat duty} + \text{heating value of feed} - H(CO \rightarrow CO_2,\ 298\ \text{K}) - H(H_2 \rightarrow H_2O,\ 298\ \text{K})$$
$$= H(CO,\ 298\ \text{K} \rightarrow 1800\ \text{K}) + H(H_2, 298\ \text{K} \rightarrow 1800\ \text{K}) + H(HCl,\ 298\ \text{K} \rightarrow 1800\ \text{K})$$

Hence

$$\begin{aligned}\text{Heat duty} = {} & H(CO,\ 298\ \text{K} \rightarrow 1800\ \text{K}) + H(H_2,\ 298\ \text{K} \rightarrow 1800\ \text{K}) \\ & + H(HCl,\ 298\ \text{K} \rightarrow 1800\ \text{K}) + H(CO \rightarrow CO_2,\ 298\ \text{K}) \\ & + H(H_2 \rightarrow H_2O,\ 298\ \text{K}) - \text{heating value of feed}\end{aligned}$$

The final energy balance can be determined based on feed heating value (9000 Btu/lb), heat of combustion, heat capacity of these products, as shown in the following table:

	Heat of combustion		Enthalpy		Heat capacity polynomial regression constants		
Compound	Btu/lb mol	Btu/lb	Btu/lb mol	Btu/lb	a	b	c
CO (gas)	20.873	745	−47, 555	−1, 698	2.37E-6	2.8E-2	−8.6588
H_2 (gas)	19.477	9642	0	0	1.70E-6	2.71E-2	−8.15575
CO_2 (gas)	–	–	−169, 286	−3, 847			
H_2O (liq)	–	–	−122, 964	−6, 828			
HCl (gas)	20.184	554	−39, 713	−1, 089	2.29E-6	2.70E-2	−8.24483

Based on the above equation, the reactor heat duty is −1156 Btu/lb (exothermic reaction), i.e., the energy required to release out of the system is 1156 Btu/lb.

PROBLEM 2: Impact of operating conditions on energy requirements

Products resulting from the processing of 10 pounds per minute of allyl chloride (C_3H_5Cl) in a copper bath at 1350°C are given below with attending thermodynamic data at operating conditions. Determine the energy requirements if the stoichiometric ratio selected, under conditions of operation, promote the formation of carbon monoxide and hydrogen exclusively. What are the energy requirements if the stoichiometric ratio selected, under conditions of operation, promote only the formation of carbon dioxide and water? What is the percent of oxidation under substoichiometric conditions? How does the percent of oxidation change with varying CO/CO_2 ratio?

$$\begin{aligned} K_{eq} &= y_{CO}y_{H_2O}/y_{CO_2}y_{H_2} \\ &= 2.96 \text{ (at } 1350^\circ\text{C)} \end{aligned}$$

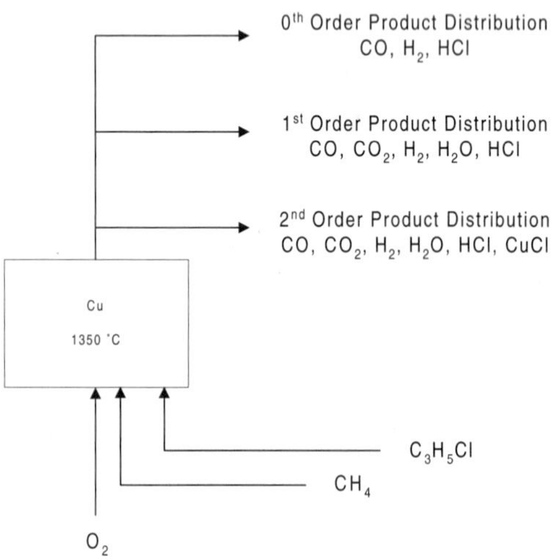

	ΔH_{25} (kJ/mol)	ΔH_{1350} (kJ/mol)
CH_4	−74.873	
C_3H_5Cl	−0.628	
CO		−67.73
CO_2		−324.674
H_2		39.53
H_2O		−187.82
HCl		−52.659

Solution:

Product distribution is set by the stoichiometric conditions specified. Since carbon and hydrogen are the only species reactive with oxygen at the conditions specified, stoichiometric conditions favoring the formation of carbon monoxide and hydrogen result in a product distribution of:

C_3H_5Cl	CH_4	O_2	CO	CO_2	H_2	H_2O	HCl
1.000	1.000	2.000	4.000	0.000	4.000	0.000	1.000

Similarly, stoichiometric conditions favoring the formation of carbon dioxide and water result in a product distribution of:

C_3H_5Cl	CH_4	O_2	CO	CO_2	H_2	H_2O	HCl
1.000	1.000	6.000	0.000	4.000	0.000	4.000	1.000

The energy requirements are determined by establishing an enthalpy balance on the species formed under the conditions of interest. Using the data provided above, this balance becomes:

$$\begin{aligned}
\Delta H &= \Delta H_{\text{out}} - \Delta H_{\text{in}} \text{ (kJ)} \\
&= \big\{\big[\alpha CO \times \Delta H_{CO,1350^\circ C} + \beta H_2 \times \Delta H_{H_2 1350^\circ C} + \chi HCl \times \Delta H_{HCl,1350^\circ C}\big] \\
&\quad -\big[\delta C_3H_5Cl \times \Delta H_{CO,25^\circ C} + \epsilon CH_4 \times \Delta H_{CH_4,25^\circ C} + \phi O_2 \times \Delta H_{O_2 25^\circ C}\big]\big\} \\
&= [(4 \times -67.73) + (4 \times 39.53) + (1 \times -52.659)] - [(1 \times -74.873) + (1 \times -0.628)] \\
&= -89.958 \text{ kJ}
\end{aligned}$$

Likewise, at stoichiometric conditions favoring the formation of carbon dioxide and water (e.g., the higher oxides) the energy requirement becomes:

$$\begin{aligned}
\Delta H &= \Delta H_{\text{out}} - \Delta H_{\text{in}} \text{ (kJ)} \\
&= \big\{\big[\gamma CO_2 \times \Delta H_{CO_2,1350^\circ C} + \eta H_2O \times \Delta H_{H_2O,1350^\circ C} + \kappa HCl \times \Delta H_{HCl,1350^\circ C}\big] \\
&\quad -\big[\lambda C_3H_5Cl \times \Delta H_{CO,25^\circ C} + \nu CH_4 \times \Delta H_{CH_4 25^\circ C} + n O_2 \times \Delta H_{O_2 25^\circ C}\big]\big\} \\
&= [(4 \times -324.674) + (4 \times -187.82) + (1 \times -52.659)] - [1 \times -74.873) + (1 \times -0.628)] \\
&= -2027.134 \text{ kJ}
\end{aligned}$$

The percent of oxidation is determined by calculating the enthalpic ratio for the condition of interest and comparing it to the condition of full oxidation:

$$\begin{aligned}
\text{Percent oxidation} &= \Delta H_{\text{partial}}/\Delta H_{\text{full}} \\
&= (-89.958/(-2027.134)) \times 100 \\
&= 4.44\%
\end{aligned}$$

Now to calculate the percent of oxidation at varying CO/CO_2 ratios. By using the targeted CO/CO_2 ratio in conjunction with the equilibrium relationship provided in the problem statement, the moles of each species at each ratio specified can be calculated. The product distribution at varying CO/CO_2 ratios is:

CO/CO_2	C_3H_5Cl	CH_4	O_2	CO	CO_2	H_2	H_2O	HCl
1000	0.098788	1.000	1.987	3.960	0.004	3.970	0.006	0.0988
100	1.000	1.000	2.049	3.960	0.040	3.942	0.058	1.000
10	1	1	2.440	3.636	0.364	3.484	0.516	1.000
1	1	1	4.194	2.000	2.000	1.613	2.387	1.000
0.1	1	1	5.692	0.364	3.636	0.253	3.747	1.000
0.01	1	1	5.967	0.040	3.960	0.027	3.973	1.000
0.001	1	1	5.997	0.004	3.996	0.003	3.997	1.000
0.0001	1	1	6.000	0.000	4.000	0.000	4.000	1.000
0.00001	1	1	6.000	0.000	4.000	0.000	4.000	1.000

By calculating the energy requirements for each CO/CO_2 ratio and comparing it to the conditions at full oxidation, the percent oxidation at each condition can be determined:

Targeted CO/CO_2 ratio	Energy requirements (kJ)	Percent oxidation (%)
1000	−90.1757	4.45
100	−113.3968	5.59
10	−300.6319	14.83
1	−1146.5525	56.56
0.1	−1876.1429	92.55
0.01	−2010.8546	99.20
0.001	−2025.4932	99.92
0.0001	−2026.9698	99.99
0.00001	−2027.1176	100.00

CHALLENGE 6: Methane consumption

What is the methane requirement at the balanced energy point ($\Delta H = 0$) for the processing of allyl chloride (C_3H_5Cl) at 1350°C? Assume that the allyl chloride feed rate is 10 lb/min. If the natural heat loss rate is 100 kW, what impact does the target CO/CO_2 ratio have on the methane requirement?

CHALLENGE 7: Heat duty

What is the resultant heat duty if a CO/CO_2 ratio of 8 is required for the above reaction; what if the required CO/CO_2 ratio is 100? At what CO/CO_2 ratio is the heat duty 0?

CHALLENGE 8: Product purity

How does synthesis gas yield and purity, resulting from the CEP processing of hydrocarbons, compare with other alternative approaches? What if the feed contains a substantial amount of chlorine (e.g., a chlorinated organic) or sulfur or phosphorous?

PROBLEM 3: Product quality of CEP synthesis gas

The composition of the synthesis gas generated from processing organic liquid sludge is:

Component	vol %
Carbon monoxide	40.0
Hydrogen	41.8
Carbon dioxide	1.0
Nitrogen	8.2
Argon	9

What is the corresponding heating value of the synthesis gas? To meet the synthesis gas product specifications, the targeted CO_2 should be $< 1\%$. What are the potential options to achieve the low CO_2 content? The product is delivered in pressure. What is the impact of pressure on throughput?

Solution:

Step 1: To determine the heating value

	Heat of combustion data	MW
Hydrogen	−68.317 kcal/g mol	2
Carbon monoxide	−67.636 kcal/g mol	28

The heating value of the synthesis gas is:

Heating of combustion of $H_2 \times H_2$ vol% + heating of combustion of

$CO \times CO$ vol% = 263.8 Btu/SCF

Step 2: To determine the operating conditions
The product specification for synthesis gas is

Criterion	Specifications
$H_2 : CO$	1 : 1
Total inerts (N_2, CO_2, Ar, CH_4)	< 2–4 vol%
HCN	< 0.1 ppmv
NH_3	< 0.1 ppmv
Total nitrogen	< 1 ppmv
Total sulfur	< 0.1 ppmv
Oxygen	< 1 ppmv
Total halogen	< 1 ppmv
Total metals	< 1 ppmv
Pressure	60 psig

To consider the partial pressure of oxygen in various metal baths, the CO/CO_2 ratio has to be at least > 8 to achieve CO_2 less than 1 vol%. Given the wider range of operable CO/CO_2 ratio for a nickel and copper bath when compared with an iron bath, Ni and Cu are the preferred bath metal to be used. The other specifications can be achieved through standard purification process design.

Step 3: To evaluate the effect of pressure on product generation
The synthesis gas is required to be compressed to 60 psig for delivery. By increasing the pressure, the throughput is increased given the maximum off-gas flow rate is kept constant (the off-gas flow rate is determined by the gas handling train design).

PROBLEM 4: Environmental performance for processing nerve agent (Battelle Memorial Institute, 1996)

Processing of chemical weapons mustard gas at the bench-scale CEP unit provided the following data:

	Time	Feed weight, g	Product gas, ft^3
Start	17:50	204.57	834.97
Stop	18:37	172.70	843.77
Difference	47 min	31.87 g	8.8 ft^3

The XAD resin trap for testing the outlet concentration of mustard shows non-detection to the limit of 0.2 μg/ml. What is the Destruction Removal Efficiency (DRE) of CEP in processing mustard gas?

Solution:

$$\text{DRE} = (\text{In} - \text{Out})/\text{In} \times 100\%$$

$$\begin{aligned}\text{Input} &= \text{agent feed rate/time}\\ &= 31.87\ \text{g}/47\ \text{min}\\ &= 0.678\ \text{g/min}\end{aligned}$$

$$\begin{aligned}\text{Output} &= (\text{detection limit/gas volume})\times(\text{gas volume/time})\\ &\le (0.2\text{E} - 6\ \text{g}/8.8\ \text{ft}^3) \times (8.8\ \text{ft}^3/47\ \text{min})\\ &\le 4\text{E} - 9\ \text{g/min}\end{aligned}$$

Hence,

$$\begin{aligned}\text{DRE} &\ge (0.678 - 4\text{E} - 9)/0.678 \times 100\%\\ &\ge 99.9999993\%\end{aligned}$$

CHALLENGE 9: VX processing performance

What is the DRE for VX based on the same above operating conditions, with a non-detection limit of 0.05 μg/ml for the XAD trap?

CHALLENGE 10: VX processing optimization

What is the chemical structure of VX? What is the metal system of choice for processing VX and why? What pressure should the system be operated at and why?

CHALLENGE 11: DRE determination of organics processing

The processing of chlorinated organics with an off-gas flowrate of 110 SCFM yielded a TO-14 analysis of:

Compound	Reporting limit (ppb)	Results (ppb)
Chloromethane	4.9	ND
Vinyl chloride	3.9	ND
Chloroethane	3.8	ND
Bromomethane	2.6	ND
Acetone	4.2	ND
Trichlorofluoromethane	1.8	ND
1,1-Dichloroethene	2.5	ND
Methylene chloride	2.9	ND
Carbon disulfide	3.2	ND
Trichlorotrifluoroethane	1.3	ND
trans-1,2-Dichloroethene	2.5	ND
cis-1,2-Dichloroethene	2.5	ND
1,1-Dichloroethane	2.5	ND
Methyl tert-butyl ether	2.8	ND
Vinyl acetate	2.8	ND
2-Butanone	3.4	ND
Chloroform	2.1	ND
1,2-Dichloroethane	2.5	ND
1,1,1-Trichloroethane	1.9	ND
Benzene	3.1	ND
Carbon tetrachloride	1.6	ND
1,2-Dichloropropane	2.2	ND
Bromodichloromethane	1.5	ND
Trichloroethene	1.9	ND
cis-1,3-Dichloropropene	2.2	ND
4-Methyl-2-pentanone	2.4	ND
trans-1,3-Dichloropropene	2.2	ND
1,1,2-Trichloroethane	1.9	ND
Toluene	2.7	ND
Dibromochloromethane	1.2	ND
2-Hexanone	2.4	ND
1,2-Dibromoethane	1.3	ND
Tetrachloroethene	1.5	ND
Chlorobenzene	2.2	ND
Ethylbenzene	2.3	ND
Bromoform	0.98	ND
Styrene	2.4	ND
m,p-Xylenes	2.3	ND
o-Xylene	2.3	ND
1,1,2,2-Tetrachloroethane	1.5	ND
1,3-Dichlorobenzene	1.7	ND
1,4-Dichlorobenzene	1.7	ND
1,2-Dichlorobenzene	1.7	ND

What is the appropriate Principal Organic Hazardous Constituent (POHC)(s) and what is the DRE based on EPA BDAT requirements? Feed was injected at a rate of 95 lb/min and the feed composition is shown as:

Volatile organics	μg/L
1,1,2-Trichloroethane	18,000,000
Tetrachloroethane	9,700,000
Chlorobenzene	12,000,000
1,2-Dichloroethane	6,800,000
1,1,2,2-Tetrachloroethane	11,000,000
Trichloroethene	24,000,000
Xylenes	610,000
1,1,1,2-Tetrachloroethane	1,000,000
sec-Butylbenzene	540,000
tert-Butylbenzene	2,000,000
Hexachlorobutadiene	3,600,000
p-Isopropyltoluene	870,000
Naphthalene	900,000
1,2,4-Trimethylbenzene	1,100,000

PROBLEM 5: Operating strategies for processing ion exchange resins

Ion exchange resins are processed in a batch mode. Typical throughput per batch is 50,000 lb of resins. The final crucible is disposed in a 42 ft^3 stainless-steel container. The typical composition, radioactivity, and volume of ion exchange resins is shown below. What is the product distribution of ion exchange resins? What is the distribution of radioactivity? What is the volume reduction?

Element	Composition, wt%, dry basis
C	54.25
H	6.0
N	2.7
S	8.7
O	25.0
Other elements, including Fe, Na, Ca, Mg, B, Al, Si, K and Cl	3.35
Total	100

Radionuclides	Activity, mCi	% of total
Ni-63	100.34	80.3%
Fe-55	8.53	6.8%
Cs-137	5.48	4.4%
Co-60	4.73	3.8%
Cs-134	2.04	1.6%
H-3	0.94	0.7%
Co-58	0.93	0.7%
Mn-54	0.60	0.5%
Ce-144	0.19	0.2%
Other*	1.25	1.0%
Total	125.02	100%

* Includes Fe-59, Sb-124, C-14, Nb-95 and Zr-95.

	Preprocessed volume of ion exchange resins, ft^3
Utility 1	1335
Utility 2	400
Utility 3	570
Utility 4	360
Utility 5	67

Solution:

Step 1: To determine the partitioning of bulk elements
The organic constituents (C, H, and O) in the IER feed are partitioned into the gas phase primarily as synthesis gas, which is a mixture of carbon monoxide and hydrogen. Nitrogen partitions to the gas phase as N_2. Sulfur partitions primarily to the gas phase as H_2S. The syngas, after purification by removing H_2S in the GHT, is either ignited in a thermal oxidizer or used as fuel in a steam boiler. About 97 wt% (dry basis) of the IER feed was C, H, N, S, and O. Trace elements, such as Fe, Na, Ca, Mg, B, K, Al and Si which represented about 3 wt%, were captured in the metal bath. Particulates in the reactor off-gases includes carryover from the bath, and volatile metals and compounds. After separation from the off-gases, particulates can be recycled to the CPU. The mass balance for processing ion exchange resin is as shown below:

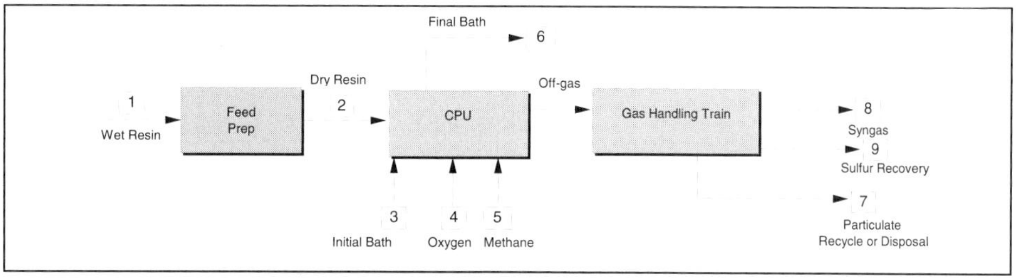

Stream	Weight, lb	Components/comments
1 Wet resin	50,000	
2 Dry resin	25,000	Composition shown
6 Final bath	1200 + Fe in initial charge	Bath growth due to partitioning of elements from resin feed such as Fe, Na, Ca, Mg, B, Al, and Si
7 Particulates	Recycled or purged as needed	Includes bath carryover and volatile metals
8 Syngas	35,500	CO, H_2, and inerts
9 Sulfur recovery	1250	Sulfur compounds from caustic scrubbing

Step 2: To determine the distribution of radionuclide
For the radionuclide as shown above, the nonvolatile metals, including Ni-63, Fe-55, Co-60, Co-58, Mn-54 and Ce-144, partitioning to the metal bath and disposed in the crucible; and volatile metals, such as Cs-137 and Cs-134, and H-3 partitioning to off-gases. Cs-137 and Cs-134 were removed from the off-gases in the GHT and disposed together with the crucible. Hence all the radionuclides are disposed in the final form. The distribution of radioactivity of final exit are shown below:

Radionuclides	Activity, mCi	% of total	% in bath	% in off-gas and collected for disposal in the final form	% in off-gas
Ni-63	100.34	80.3%	80.3%		
Fe-55	8.53	6.8%	6.8%		
Cs-137	5.48	4.4%		4.4%	
Co-60	4.73	3.8%	3.8%		
Cs-134	2.04	1.6%		1.6%	
H-3	0.94	0.7%			0.7%
Co-58	0.93	0.7%	0.7%		
Mn-54	0.60	0.5%	0.5%		
Ce-144	0.19	0.2%	0.2%		
Other*	1.25	1.0%			1.0%
Total		100.0%	92.3%	6%	1.7%

Step 3: To determine the volume reduction
The total volume of wastes processed is per batch

$$= 1335 + 400 + 570 + 360 + 67 = 2732 \text{ ft}^2$$

The final disposal volume is the total ingot = 42 ft^3
Hence, the volume reduction is:

$$2732 \text{ ft}^3/42 \text{ ft}^3 = 65$$

CHALLENGE 12: Weapon components processing

Devise a flowsheet and operations concept for the manufacture of a copper master alloy resulting from the catalytic extraction processing of 20,000 tons of weapon components (see composition below); repeat for the manufacture of an iron master alloy.

Component	+10 Mesh	−10 Mesh	Generator*
Ag, %	NA†	NA	0.16
Al, %	34.65	8.72	32.30
As, %	0.07	0.07	NA
Ba, %	< 0.01	< 0.01	0.06
Be, %	0.08	0.10	NA
Ca, %	0.45	1.20	NA
Cd, %	0.03	0.05	0.10
Co, %	0.09	0.19	0.04
Cr, %	0.93	0.61	1.12
Cu, %	9.13	14.84	7.02
Fe, %	8.00	5.88	8.12
Mg, %	0.05	0.06	0.19
Mn, %	0.09	0.10	0.11
Mo, %	2.79	0.05	0.04
Ni, %	4.34	1.06	0.93
Pb, %	0.32	0.48	0.39
Se, %	< 0.01	< 0.01	0.02
Sn, %	0.39	0.28	0.42
V, %	< 0.07	< 0.07	0.02
Zn, %	0.56	0.37	0.66

* Analysis provided by generator of the SMC.
† NA = not available

Ultimate analysis of source material	As received	Dry	Air dried
Moisture, %	0.40	0.00	0.40
Carbon, %	47.09	47.28	47.09
Hydrogen, %	5.50	5.52	5.50
Nitrogen, %	1.67	1.68	1.67
Sulfur, %	0.08	0.08	0.08

CHALLENGE 13: UF_6. recycling

Devise a flowsheet and operations concept for the manufacture of anhydrous hydrogen fluoride (HF) and uranium oxide resulting from the catalytic extraction processing of 20,000 tons of uranium hexafluoride (UF_6). If UO_2 is not wetted by the metal, how might the process flowsheet change?

CHALLENGE 14: Biosolids recycling

Devise a flowsheet and operations concept for the manufacture of synthesis gas ($CO/H_2 = 1 \pm 0.1$, $CO/CO_2 \geq 18$) from 20,000 tons of biosolids (see composition below)

Element	Average composition, wt%
Carbon	45.5
Hydrogen	6.6
Oxygen	26.4
Nitrogen	9.0
Sulfur	1.2
Chlorine	0.9
Ash	10.4
Total	100.0

REFERENCES

Bach, R. D., Shobe, D. S., and Schlegel, H. B. (1996). "Thermochemistry of iron chlorides and their positive and negative ions," *Journal of Physical Chemistry*, 100(21), 8770–8776.

Barin, T (1989) *Thermochemical Data of Pure Substances*.

Battelle Memory Institute (1996) "Alternative Technologies for Chemical Demilitarization," submitted to U.S. Army PMCD.

Castagnacci, A., Kaczmarsky, M., and Sills, R. (1997) "Quantum-CEP® Processing Spent Ion Exchange Resins from Nuclear Power Stations", presented at Waste Management 97.

Centec-XXI. (Under Review 1997) "Advanced Mixed Waste Treatment, Results of Mixed Waste Treatment at the M-4 Facility," EPRI TR-107974.

Elliot, J. F., and Gleiser, M. (1963). Thermochemistry for Steelmaking, Vol. 2, Addison-Wesley.

ENCORE Technical Resources, Inc. (Under Review 1997) "Mixed Waste Advanced Treatment Technology: Waste Treatment Products and Their Recycling Applications," EPRI TR-107990.

Evans, L., Pernsteiner, A., and Wong, E. (1997) "Applications of Quantum-CEP® technology to utility mixed wastes," *Journal of The Franklin Institute*, 334a(1).

Glukhovtsev, M. N., Bach, R. D., and Nagel, C. J. (1997) "Performance of the B3LYP/ECP DFT calculations of iron-containing compounds," *Journal of Physical Chemistry*, 101(3), 316–323.

Lide, D. (1975). *CRC Handbook of Chemistry & Physics*.

Mather, R., Steckler, D., Kimmel, S., and Tanner, A. (1995). "Integrated Recycling of Industrial Waste Using Catalytic Extraction Processing at a Chemical Manufacturing Site," presented at the Spring National Meeting of The American Institute of Chemical Engineers.

Moore, J. J. (1990) *Chemical Metallurgy*, 2d edn.

Nagel, C. J. (1996) "Novel Chemical Processing of Hazardous Waste," presented at Annual Symposium on Frontiers of Engineering, National Academy of Engineering.

Nagel, C. J., Chanenchuk, C. A., Wong, E. W., and Bach, R. (1996). "Catalytic extraction processing: An elemental recycling process," *Environmental Science and Technology*, 30(7), 2155–2167.

Schug, B. W., and Realff, M. J. (1997) "Analysis of waste vitrification product-process systems," submitted to *Computers & Chemical Engineering*.

Perry, R. H., and Green, D. W. (1997) *Perry's Chemical Engineers' Handbook*, 7th edn., McGraw-Hill, New York.

Rao, Y. K. (1985) *Stoichiometry and Thermodynamics of Metallurgical Process*, Cambridge University Press.

United States Steel (1985) *The Making, Shaping and Treating of Steel.*

GLOSSARY

Catalytic Extraction Processing (CEP) is a recycling technology which utilizes the catalytic and solvation effect of molten metal to convert low grade materials to high value products.

BDAT (Best Demonstrated Available Technology) is the designation given by EPA for hazardous wastes processing. The regulation is shown in 41 Part CFR.

CHAPTER 2.6

INCINERATION TECHNOLOGIES AND FACILITY REQUIREMENTS

C. C. Lee and G. L. Huffmann
U.S. Environmental Protection Agency, Cincinnati, Ohio

1 INTRODUCTION

Incineration is a controlled combustion process used in the treatment of unusable wastes. Careful selection of equipment and processes for the incineration of wastes is essential to ensure that the basic obligations of safe handling and proper ultimate disposal are met in a satisfactory manner. Because of past research and development there are many incineration technologies that are available on the market. For obvious reasons, different waste may require different treatment technology. In other words, technology A, which is suitable for incinerating waste A, may not be suitable for incinerating waste B. Therefore, proper selection of an adequate technology becomes a very important factor in designing a treatment facility.

This chapter describes a collection of thermal treatment technologies. The technology variation ranges from a very simple technology such as a liquid injection incinerator to a very complicated technology such as the recently developed vitrification process. The intended purpose is for readers, particularly beginners, to be aware of the breadth of technological options. The calculational methods which we provided in previous chapters are generally applicable to the technologies given in this chapter, because every technology has to meet the mass and energy conservation principle.

Construction of an effective system that will satisfy regulatory requirements and meet with the community approval is the ultimate goal of understanding the incineration processes. Accordingly, in the later part of this chapter, key elements that are needed for the design and operation of an incineration facility are provided for reference.

2 INCINERATION TECHNOLOGY

Because the technologies described in this section are "stand-alone" technologies, they are given in alphabetical order for easy search.

2.1 Catalytic Incinerator

PROCESS DESCRIPTION: Catalytic incineration is for the destruction of gaseous waste. The objective of catalytic incinerators is to destroy the organic components of a gas stream at lower temperatures than those required by direct combustion, thus effecting significant energy savings for wastes containing low concentrations of combustibles. The catalyst bed (or matrix) in commercial units is typically a metal mesh mat, ceramic honeycomb, or other ceramic matrix structure with a surface deposit or coating of finely divided platinum or other platinum family metals. The finely divided metal deposit is the catalyst for the oxidation reactions, while the matrix serves to support the catalyst on a high geometric surface area and promote good contact between waste stream and catalyst. Figure 6.1 shows a schematic of a catalytic process for fume incineration.

The basic steps of catalytic incineration include:

- gas phase heat and mass transfer of reactants to the surface film on the catalyst. The gas is normally preheated with fired burners and then passed over a catalyst bed where the oxidation reactions proceed;
- diffusion of adsorbed reactants within the surface film;
- oxidation reaction between reactants;
- diffusion and desorption of reaction products;
- gas phase heat and mass transfer of products leaving the catalyst.

At the appropriate temperature and residence time with sufficient oxygen present, 85–95% destruction of hydrocarbons can be achieved with catalytic incinerators.

Catalytic incinerators normally operate at temperatures below 600°C (1110°F) and residence times of approximately 1 s. The catalyst section is normally sized to provide a pressure drop no greater than 1 kPa (4 in WC, water column). It should not be used unless the gas stream constituents are consistent in analysis and contain neither particulate matter, halogens, nor other contaminants listed under Technology Application below.

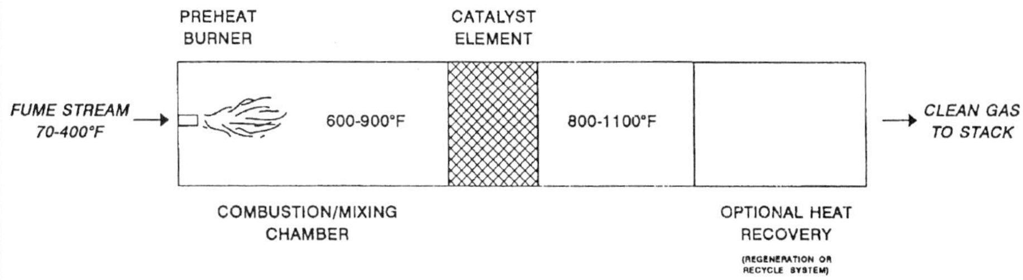

FIGURE 6.1 Catalytic incinerator (*EPA-72/08, p56*).

The gas stream entering the catalyst shall have a minimum temperature of 315°C (600°F) and a maximum inhibitor concentration of 25 ppmv.

Flares or direct flame incinerators can be used with gases above the upper explosive limit (UEL) or below the lower explosive limit (LEL). Catalytic incinerators operate most efficiently when the gas stream concentration is no more than 25% of the LEL. The LEL is the lowest concentration of a gas in air that will sustain combustion. The UEL is that concentration above which there is insufficient air present to sustain combustion. Table 6.1 (OME-88/12, p. 4-92) lists LEL and UEL values for common compounds.

TECHNOLOGY APPLICATION: Catalytic incineration is most applicable to the destruction of gaseous wastes with low concentrations of combustibles. The waste stream must be free of particulates and several other contaminants (e.g., heavy metals, phosphates, arsenic compounds, halogens, sulfur compounds, alumina and silica dusts, iron oxides and silicones) which can foul the catalyst materials. Table 6.2 (OME-88/12, p. 91) lists a number of these catalyst inhibitors and their effect on the process. Although commercially available, catalytic incinerators are not widely used for waste gas destruction.

TECHNOLOGY ADVANTAGE: The technological advantage is the feature of treating waste containing low concentrations of waste constituents. The efficiency of most catalytic reactors in destroying products of incomplete combustion make catalytic incinerators valuable as secondary afterburners in some applications (DOE-93/03 p. C-9).

TECHNOLOGY DISADVANTAGE: The technological disadvantages include:

- the system is only suitable for treating gaseous streams; and
- catalysts will usually degrade over time by mechanical blinding of the active surface by particles and soot, or by deactivation (poisoning) of the surface by certain metals in the waste gas stream, or by internal changes in the catalyst resulting from inadvertent overheating of the catalyst; common poisons for catalysts are copper, phosphorus, and zinc (DOE-93/03, p. C-10).

2.2 Controlled Air Incinerator

PROCESS DESCRIPTION: Controlled air incinerators are also known as fixed hearth or starved air incinerators. They typically contain two furnace chambers: a primary and a secondary chamber. Typical fixed hearth systems are shown in Figs. 6.2 and 6.3. Solid and liquid wastes may be charged into the primary chamber. Small units are normally batch-fed, while larger units may be continuously fed with a screw feeder or moving grate, or semicontinuously fed with a raw pusher. In some designs, there may be two or three step hearths on which the ash and waste are pushed with rams through the system. In other designs, rotating rabble arms stir the solid waste material on the grate. A controlled flow of "underfire" combustion air is

TABLE 6.1. Combustibility Characteristics of Pure Gases and Vapors in Air

Gas or vapour	LEL (% by volume)	UEL (% by volume)
Acetaldehyde	4	57
Acetone	2.5	12.8
Acetylene	2.5	80
Allyl alcohol	2.5	–
Ammonia	15.5	26.6
Amyl acetate	1	7.5
Amylene	1.6	77
Benzene (bensol)	1.3	68
Benzyl chloride	1.1	–
Butene	1.8	8.4
Butyl acetate	1.4	15
Butyl alcohol	1.7	–
Butyl cellosolve	–	–
Carbon disulfide	1.2	50
Carbon monoxide	12.5	74.2
Chlorobenzene	1.3	7.1
Cottonseed oil	–	–
Cresol, m- or p-	1.1	–
Crotonaldehyde	2.1	15.5
Cyclohexane	1.3	8.4
Cyclohexanone	1.1	–
Cyclopropane	2.4	10.5
Cymene	0.7	–
Dichlorobenzene	2.2	9.2
Dichloroethylene (1,2)	9.7	12.8
Diethyl selenide	2.5	–
Dimethyl formamide	2.2	–
Dioxane	2	22.2
Ethane	3.1	15.5
Ether (diethyl)	1.8	36.5
Ethyl acetate	2.2	11.5
Ethyl alcohol	3.3	19
Ethyl bromide	6.7	11.3
Ethyl cellosolve	2.6	15.7
Ethyl chloride	4	14.8
Ethyl ether	1.9	48
Ethyl lactate	1.5	–
Ethylene	2.7	28.6
Ethylene dichloride	6.2	15.9
Ethyl formate	2.7	16.5
Ethyl nitride	3	50
Ethylene oxide	3	80
Furfural	2.1	–
Gasoline (variable)	1.4–1.5	7.4–7.6
Heptane	1	60
Hexane	1.2	69
Hydrogen cyanide	5.6	40
Hydrogen	4	74.2
Hydrogen sulfide	4.3	45.5
Illuminating gas (coal gas)	5.3	33
Isobutyl alcohol	1.7	–
Isopentane	1.3	–

TABLE 6.1 (*Continued*)

Gas or vapor	LEL (% by volume)	UEL (% by volume)
Isopropyl acetate	1.8	7.8
Isopropyl alcohol	2.0	–
Kerosene	0.7	5
Linseed oil	–	–
Methane	5	15
Methyl acetate	3.1	15.5
Methyl alcohol	6.7	36.5
Methyl bromide	13.5	14.5
Methyl butyl ketone	1.2	8
Methyl chloride	8.2	18.7
Methyl cyclohexane	1.1	–
Methyl ether	3.4	18
Methyl ethyl ether	2	10.1
Methyl ethyl ketone	1.8	9.5
Methyl formate	5	22.7
Methyl propyl ketone	1.5	8.2
Mineral spirits No. 10	0.8	–
Naphthalene	0.9	–
Nitrobenzene	1.8	–
Nitroethane	4	–
Nitromethane	7.3	–
Nonane	0.83	2.9
Octane	0.95	32
Paraldehyde	1.3	–
Paraffin oil	–	–
Pentane	1.4	7.8
Propane	2.1	10.1
Propyl acetate	1.8	8
Propyl alcohol	2.1	13.5
Propylene	2	11.1
Propylene dichloride	3	14.5
Propylene oxide	2	22
Pyridine	1.8	12.4
Rosin oil	–	–
Toluene (toluol)	1.3	7
Turpentine	0.8	–
Vinyl ether	1.7	27
Vinyl chloride	4	21.7
Water gas (variable)	6	6
Xylene (xylol)	1	6

LEL, lower explosive limits: UEL, upper explosive limit; –, not determined.

introduced, usually up through the hearth on which the waste sits. In some designs, combustion air may also be provided from the wall over the waste bed (EPA-94/05, p. 32).

The term starved-air is derived from the principle of combustion. The combustion air to the primary combustion chamber (PCC) into which the waste is fed is strictly controlled, so that the amount of air present is less than that needed for complete combustion, i.e., the chamber is "starved" for air. This type of incinerator is often referred to as a controlled-air incinerator because the amount and distribution of air to each combustion chamber is controlled to meet the design requirements.

TABLE 6.2 Catalyst Inhibitors

Type of inhibitor	Effect
Fast-acting inhibitors: phosphorus, bismuth, lead, arsenic, antimony, mercury	Irreversible reduction of catalyst activity at a rate dependent on concentration and temperature
Slow-acting inhibitors: iron, tin, silicon	Irreversible reduction of catalyst activity. Higher concentrations than those of fast-acting catalyst inhibitors may be tolerated
Reversible inhibitors: sulfur, halogens, zinc	Reversible surface coating of active catalyst area. Removed by increasing catalyst temperature
Surface eroders and maskers: inert particulates	Surface coating of active catalyst area. Also erosion of catalyst surface at a rate dependent on particulate size, grain loading, and gas stream velocity

The temperature in the primary chamber is usually maintained in the range 760–870°C (1400–1600°F), whereas the secondary chamber may operate at temperatures as high as 1100°C (2010°F). Typically, 70–80% of the stoichiometric air requirement is introduced into the primary chamber. Approximately 140–200% of the primary chamber's off-gas stoichiometric requirement is fed to the secondary chamber. The net air flow to the system is in the range 100–200% of the stoichiometric requirement.

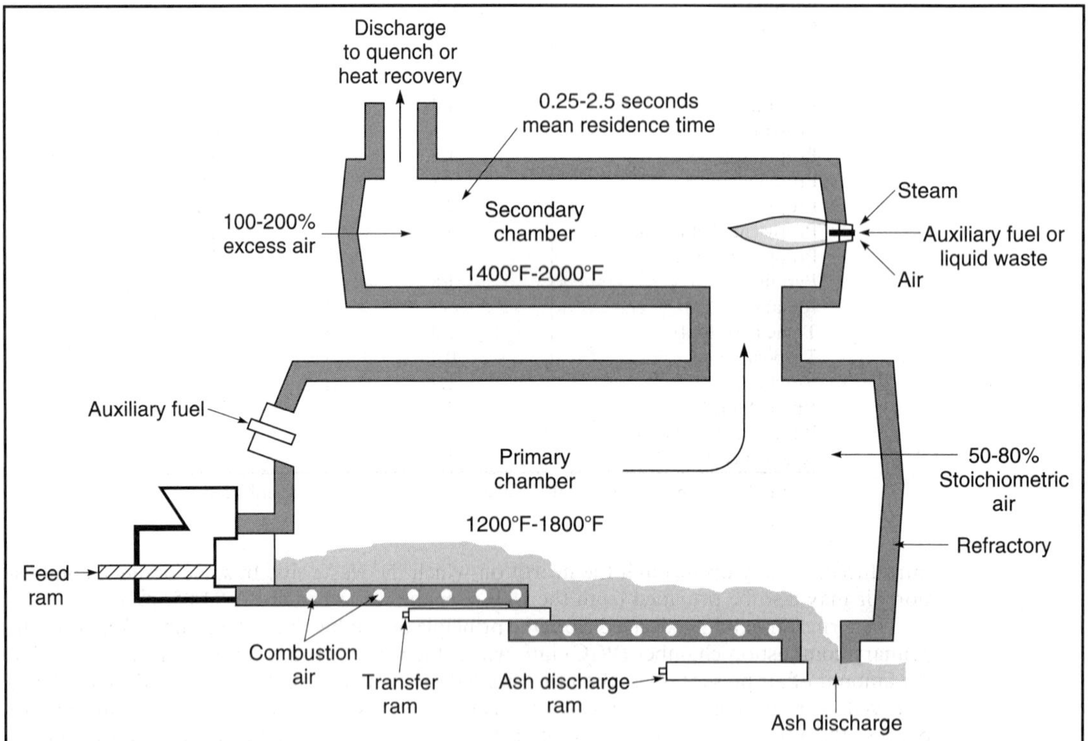

FIGURE 6.2 Fixed hearth incinerators (*EPA-94/05, p. 2-33*).

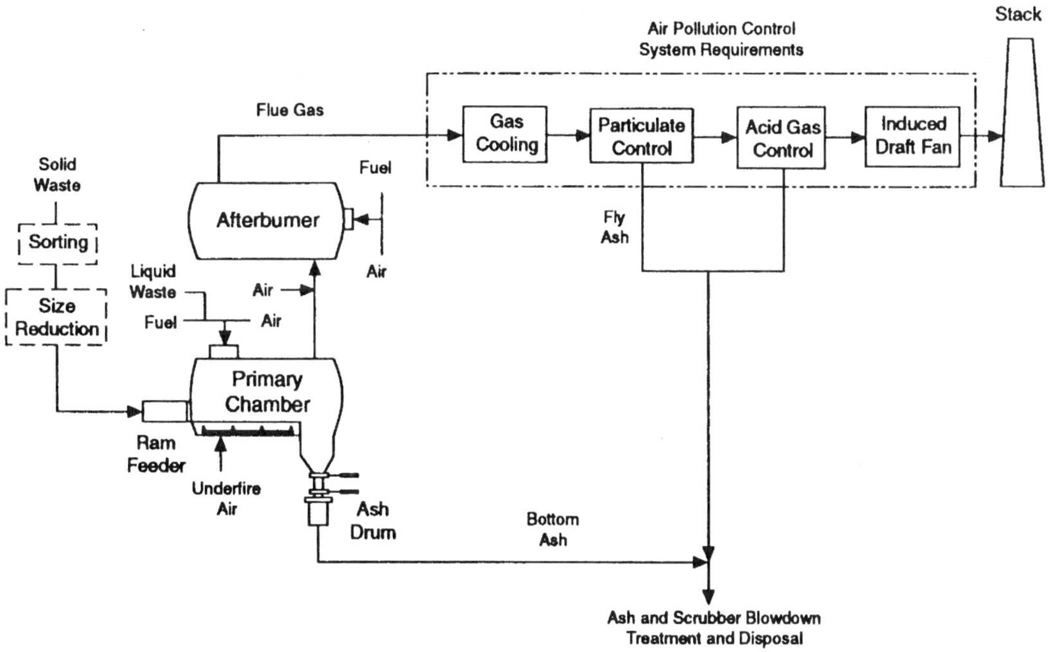

FIGURE 6.3 Starved-air incinerator (*EPA-94/09, p. 5-17*).

The air feed rate to each chamber is controlled by temperature. Below the stoichiometric air requirement (in the primary chamber) temperatures increase with increasing airflow, whereas in the secondary chamber, where excess air is provided, increasing airflow decreases the operating temperature. Automatic fan damper positioning is normally used to control the airflow, and thus the temperature, in each chamber.

Starved-air incinerators come in all sizes and shapes. Incinerators are available with design capacities ranging from 50 lb/h (23 kg/h) to 4000 lb/h (1800 kg/h). Some are manually controlled and others are automatically controlled. Some use manual waste loading and ash removal, and others are fully automated.

Critical fixed hearth operating parameters include primary chamber and afterburner temperature. Primary chamber temperature is controlled by the underfire air supply. Afterburner temperature can be controlled by auxiliary fuel firing and primary waste feed rate adjustments. Note that the capacity of the afterburner or secondary chamber limits the primary chamber burning rate. The afterburner must have adequate volume to accept and oxidize all volatile gases generated in the primary chamber. Additionally, the afterburner must be kept operating at excess air conditions to ensure complete volatile burnout.

Due to the starved air combustion in the primary chamber, the gas and entrained particulate matter flowing to the afterburner will contain unburned hydrocarbons (including hazardous organics) and high levels of CO and H_2. In the afterburner, where 140–200 percent of the stoichiometric air required is supplied, these compounds are oxidized to CO_2 and H_2O. Complete hazardous organic destruction will occur if sufficient time and high enough temperatures are provided in this oxidizing environment. Normally afterburners are sized to provide a minimum of 2 s of residence time in the afterburner at temperatures as high as 2200°F, depending upon the waste composition. An auxiliary natural gas burner in the secondary chamber is used to maintain temperature at the set point.

TECHNOLOGY APPLICATION: Controlled air incinerators are capable of treating solid, sludge, liquid and mixed wastes. For example, at DOE's Los Alamos National Laboratory (LANL), all solid waste feed for their starved air incinerator is prepackaged in 1 ft × 1 ft × 2 ft cardboard boxes. In the staging glovebox, which can hold 24-h worth of waste feed, a state-of-the-art multiple energy gamma assay system (MEGAS) and a micro-dose X-ray system are used to screen for large noncombustible waste items and to assay for radioisotopes. Boxes of acceptable waste pass out of the staging glovebox on a conveyor and are individually loaded into the hydraulic feed ram for charging into the primary chamber. For liquid wastes, a high intensity vortex liquid burner was installed on the side of the primary chamber. This burner is also used for heat-up and temperature modulation using natural gas or fuel oil.

TECHNOLOGY ADVANTAGE: The technological advantages include:

- the small size of these units, which makes them favorable for onsite treatment of small quantities of solid waste;
- the primary chamber can be operated with a wide variety of wastes with different chemical properties;
- the secondary chamber can be operated with conditions necessary to achieve high destruction efficiencies for organic materials;
- maintenance costs are typically low because there are no moving parts inside the incineration chamber;
- generally, the low combustion air input volume (starved air) in the primary chamber maintains a quiescent environment, resulting in lowered entrained ash or particulate matter in the combustion gases entering the secondary combustion chamber.

TECHNOLOGY DISADVANTAGE: The technological disadvantage lies primarily in limitations on acceptance of certain materials which require turbulence for effective combustion. Such materials include powdered carbon, pulp wastes, sludges, and highly viscous wastes.

2.3 Direct Flame Incinerator

Direct flame incineration is the most common method of gas incineration. Catalytic incineration uses catalysts to reduce the temperature required for waste destruction. Direct flame incineration includes:

- flare incineration
- fume incineration
- gas combustion
- afterburner.

FLARE INCINERATOR: Flaring is the simplest means of disposing of relatively large quantities of waste gas with combustible components. The gas is ignited and discharged to the atmosphere without heat recovery. The process is particularly suited to the disposal of intermittent gas discharges.

Supplemental fuel for flares is required when the gas heating value is less than 5600 kJ/Nm^3 (150 Btu/scf). Radiative heat from flares should be controlled to less than 34,000 kJ/(m^2-h) [3000 Btu/(ft^2-h)] for equipment, 5000 kJ/(m^2-h) [440 Btu/(ft^2-h)] for personnel on a continuous basis, and 17,000 kJ/(m^2-h) [1500 Btu/(ft^2-h)] for personnel on an intermittent basis. Provisions should be included for noise control in flare design.

As illustrated in Fig. 6.4 (OME-88/12, p. 4-87), a flare is basically a stack discharging a combustible gas to the atmosphere. A pilot flame at the end of the stack ignites the gas, the combustion of which is supported by the surrounding atmosphere. Steam is frequently intro-

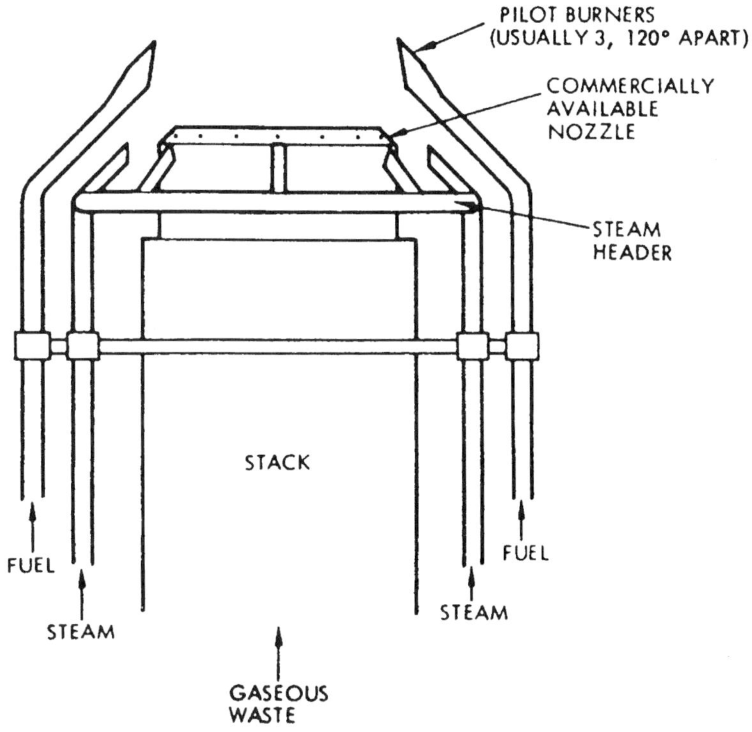

FIGURE 6.4 Flare burner (*OME-88/12, p. 4-87*).

duced into the flame to generate turbulence and promote mixing with the surrounding air, to aid in cracking complex molecules, and to help oxidize carbon.

FUME INCINERATOR: A fume incinerator is one of the gas incineration devices. It is used to destroy gaseous or fume wastes. The combustion chambers are comparable with those of liquid injection incinerators in that they are usually single-chamber units, are vertical or horizontal in configuration, and use nozzles to inject the wastes into the unit for combustion. Wastes are injected by pressure or atomization through the burner nozzles. Using the waste in this manner to maintain combustion requirements reduces secondary fuel requirements. Wastes may be combusted solely by thermal or catalytic oxidation (EPA-86/09 p. 2-7).

Castable and brick refractories are used in the combustion chamber of a fume incinerator. The type used depends on the temperature required to incinerate the waste. For some units, the combustion chamber temperature is maintained at 650–980°C (1200–1800°F), with a fume retention time of 0.3–1.0 s to achieve maximum conversion to carbon dioxide and water. Use of a catalyst such as alumina coated with noble metals (e.g., platinum, palladium, or rhodium) and other materials (e.g., copper chromate, or oxides of copper, chromium, or manganese) can lower the required temperature to 260–480°C (500–900°F) and can also decrease retention time.

Exhaust gas from the incinerator can be passed through a heat exchanger before discharge to recover heat energy for a variety of uses. Fume incinerators may be equipped with air pollution control devices for removing SO_x or Cl^- gases, depending on the composition of the waste gases. Particulate controls and ash collection equipment are seldom needed because gaseous wastes yield very little ash when completely incinerated.

TECHNOLOGY APPLICATION: Direct flame incineration is best applied to waste gas streams rich in combustible constituents. Probably the most common application of flares is the combustion of waste gases and vent gases from petroleum refining operations. Fume incinerators have been successfully applied to control smoke and solvent emissions from surface coating operations and odorous emissions from food facilities, as well as combustible fumes from a wide variety of industries such as printing, food processing, and chemical processing.

TECHNOLOGY ADVANTAGE: The technological advantages include:

- flue incinerators can incinerate a wide range of gaseous wastes;
- continuous ash removal and particulate control systems are usually not required; and
- these incinerators have virtually no moving parts.

TECHNOLOGY DISADVANTAGE: The technological disadvantages include:

- if the heat content of the burned waste is not adequate to maintain ignition and incineration temperatures, a supplemental fuel must be provided; and
- if a catalyst is used in a fume incinerator, it must be replaced periodically, because of deactivation.

2.4 Fluidized Bed Incinerator

PROCESS DESCRIPTION: A fluidized bed is essentially a vertical cylinder containing a bed of granular material at the bottom. Combustion air is introduced at the bottom of the cylinder and flows up through the bed material, suspending the granular particles. Solid, liquid, and gaseous waste fuels may be injected into the bed, where they mix with the combustion air and burn. This section provides a process description and a discussion of the process operating parameters (EPA-94/05, p. 2-29; EPA-86/10, p. 4-41; OME-88/12, p. 4-73).

A fluidized bed incinerator consists of a fluidized bed reactor, fluidizing air blower, waste feed system, auxiliary fuel feed system, and an air pollution control device system. One such reactor, as shown in Fig. 6.5, has an inside diameter of 26 ft (8 m) and elevation of 33 ft (10 m). Silica beds are commonly used and have a depth of 3 ft (1 m) at rest and extending up to 6.5 ft (2 m) in height when fluidizing air is passed through the bed. The waste is put in direct contact with the bed media, causing heat transfer from the bed particles. At the proper temperature, waste ignition and combustion occur. The bed media acts to scrub the waste particles, exposing a fresh surface by the abrasion process, which encourages rapid combustion of the waste. Waste and auxiliary fuels are injected radially into the bed and react at temperatures from 840°F to 1500°F (450°C to 810°C). Further reaction occurs in the volume above the bed at temperatures up to 1800°F (980°C). An auxiliary burner is located above the bed to provide heat for start-up, reheat, and maintenance of bed temperature.

Generally, fluidized bed incinerators have two separate waste preparation/feed systems, one for solids and one for liquids. In some cases, four feed systems are employed: wet solids, dry solids, viscous fluids, and nonviscous fluids. Solid wastes are usually fed into a coarse shredder. The coarsely shredded waste falls into a classifier which separates the light and dense particles. The lighter particles are transferred to a secondary shredder, and from there they are conveyed to the hopper to feed the fluidized bed. Liquid waste is pumped into a larger holding tank. To ensure that the mixture is as homogeneous as possible, the liquid waste is continuously pumped through a recirculating loop from the bottom of the tank to the top. A metering pump draws the waste fuel to be burned from the tank to the primary reactor. Nozzles are used to atomize and distribute the liquid waste within the bed.

Critical parameters for optimum performance include fluidized bed temperature, oxygen level in the bed, solids residence time, bed fluidization, and combustion gas residence time.

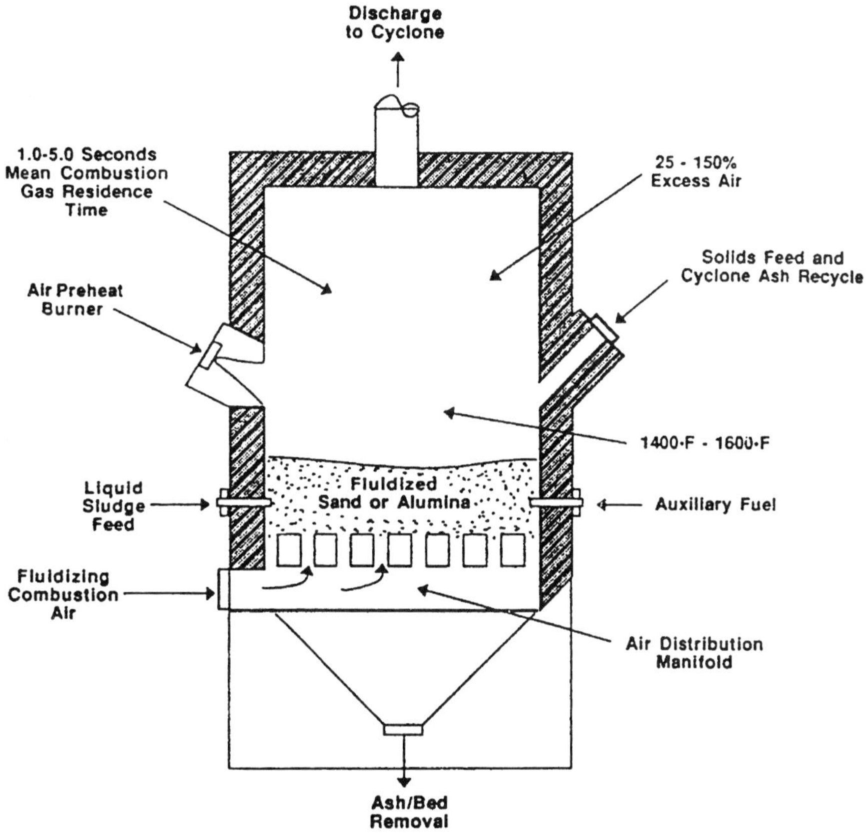

FIGURE 6.5 Fluidized bed incinerator (*EPA-94/05, p. 2-30*).

Specifically:

- The fluidized bed temperature must be monitored and controlled by a waste feed and auxiliary burner to ensure that the temperature remains above an established minimum. Operating temperatures are normally maintained in the 1400–1600°F range.
- The oxygen level in the bed must be maintained to ensure the potential for complete combustion. Monitoring the carbon monoxide and total hydrocarbon concentration in the flue gas is also an indication of complete combustion.
- Solids retention time in the bed is a measure of the thorough treatment of the waste. Retention time is especially important for relatively noncombustible wastes.
- Uniform bed fluidization must be maintained to properly treat the waste.
- The combustion gas residence time must be monitored to ensure that the combustion gas has been exposed to a volatiles destruction temperature for a sufficient period of time.

The use of a control system could aid in maintaining fluidized bed temperature and flue gas residence time. The control system would involve the adjustment of waste and auxiliary fuel feed rates, as well as the ratio of oxidation/inert fluidization gases.

TECHNOLOGY APPLICATION: Fluid bed furnaces have been used for combustion of many materials, including coal, coke, wood, oil sands, and several other fuels. In incineration applications, fluidized bed systems have been used principally for the combustion of sludges, particularly sewage sludge. The first fluid bed used for incineration of sewage sludge was in 1962. Since that time it has been gaining popularity in North America. A prime requisite for the successful combustion of solid waste is that it be adequately prepared by shredding and sorting.

TECHNOLOGY ADVANTAGE: The technological advantages include:

- fluidized bed incinerators are generally applicable for the disposal of combustible solids, liquid, and gaseous wastes;
- the design concept is simple, and no moving parts are required in the combustion zone;
- because of the compact design, a fluidized bed can have a high heating rate per unit volume (100,000–200,000 Btu/h-ft^3);
- fluctuations in the feed rate and composition are easily tolerated because of the large quantities of heat stored in the bed;
- because of its intimate contact with heated bed particles, waste can be combusted at lower temperatures than those of conventional incinerators;
- relatively low gas temperatures and excess air requirements tend to minimize nitrogen oxide formation and contribute to smaller, lower-cost emission control systems; temperatures in the vessel are reportedly high enough to destroy wastes, but low enough to prevent formation of significant amounts of N_{Ox}.
- the large active surface area resulting from the fluidizing action increases the combustion efficiency;
- the bed material acts as a scrubber to capture acid gas from the process, reportedly creating a nontoxic solid residue, and selection of proper bed material suppresses acid gas formation, thereby reducing emission control requirements;
- there is the potential for some metals capture in the bed, thereby lowering the emissions to the environment; and
- the fluid bed furnace system is compact, of airtight construction, and with most manufacturers it is maintained as a positive pressure system. By maintaining the entire system under positive pressure, the furnace must be designed to be airtight. This feature is useful in applications where the furnace is called upon to operate on a noncontinuous basis. If the furnace is to operate 5 days per week, for instance, its airtight construction allows it to be sealed fairly effectively. Dampers on its inlet and outlet will hold in its heat, and the refractory and sand within the unit provide significant thermal inertia. As a result of its construction the fluid bed furnace will lose as little as 5°C (9°F) per hour after a shutdown. It can be shut down on a Friday evening and will require only a few hours of heat-up on a Monday morning to be available for waste feeding.

TECHNOLOGY DISADVANTAGE: The technological disadvantages include:

- the presence of heavy, noncombustible materials in the waste may cause the bed volume to increase with time, necessitating tapping to prevent excessive agglomeration;
- feed must be selected to avoid bed degradation caused by corrosion or reaction, and waste feed particle size should be controlled to maintain a uniform feed rate;
- removal of the residual materials from the bed is a potential problem area. As waste combustion proceeds, noncombustible ash accumulates in the bed;
- beds with higher than 20 wt% ash are undesirable, since at ash levels of this magnitude, defluidization can occur;
- disposal of the supposedly inert residual bed material may present a problem (if the removed bed materials are considered to be hazardous, they would have to be disposed of at a secure landfill site);

- relatively large amounts of fine particulate matter entrained in the exhaust gases may require elaborate pollution control devices;
- accurate control is needed to ensure that retention time in the bed is sufficient for complete combustion, and that radical increases in the waste's heat value will not drastically boost bed temperatures and adversely affect bed operations;
- sophisticated analytical methodologies for the analysis of hazardous residues in the spent bed materials may be required. The analysis is normally required to determine how much unburned waste remains in the residues and to examine how much waste residue is leachable to the environment. It is quite possible that milligram amounts of hazardous materials may remain in the spent bed materials undetected. Disposal of the spent materials could then recycle the hazardous material back into the environment. This situation would be especially dangerous in the case of chemical warfare agents;
- other problems include corrosion of the combustor and mechanical devices for feeding wastes to the combustor;
- the fluidized bed incineration technique is not well suited for irregular, bulky solids, tarry solids, or wastes with a high fusible ash content. Formation of eutectics (compounds with low melting or fusion temperatures) can result in bed fouling. Problems caused by wastes with low ash fusion temperatures can be avoided by keeping operating temperatures below the ash fusion level or by using chemical additives to raise the ash fusion temperature. Waste containing bulky or irregular solids may require pretreatment in the form of drying, shredding, and sorting prior to entering the reactor (EPA-86/10); and
- labor utilization is high, since regular preparation and maintenance of the fluid bed must be performed. These costs can increase dramatically if it becomes difficult to remove residual materials from the bed (EPA-86/10).

2.5 Liquid Injection Incinerator

PROCESS DESCRIPTION: For purposes of incinerator design, a material is considered a liquid if it can be pumped to a burner, atomized, and fired in suspension. Liquid injection incinerators are either horizontally or vertically oriented, cylindrical, refractory-lined chambers fitted with nozzles firing axially or tangentially into the furnace. Vertical units may be upfired (i.e., the burner is on the lower end and fires upward), and combustion gases exit at the top of the combustion chamber. Downfired units are equipped with a wet quench at the combustion chamber exits at the bottom of the unit; this feature is especially important when wastes have a high salt content. Figures 6.6 and 6.7 show a horizontal and a vertical liquid injection incinerator, respectively (EPA-94/05, p. 2-27; EPA-86/10, p. 4-32; OME-88/12, p. 4-79).

The typical liquid injection incinerator includes a waste burner system, an auxiliary fuel system, an air supply system, a combustion chamber, and an air pollution control system. Liquid wastes are fed and atomized into the combustion chamber through the waste burner nozzles. These nozzles atomize the waste and mix it with air. Atomization is usually achieved either by mechanical methods such as a rotary cup or pressure atomization systems, or by twin-fluid nozzles which use high-pressure air or steam. With a relatively large surface area, the atomized particles vaporize quickly, forming a highly combustible mix of waste fumes and combustion air. This mixture ignites and burns in the combustion chamber. Typical combustion chamber residence time and temperature ranges are 0.5–2 s and 1300–3000°F (700–1650°C), respectively. Typical liquid feed rates are as high as 200 cubic feet per hour (5600 L/h).

Liquid injection can be used to incinerate virtually any combustible liquid waste, including slurries and sludges with a viscosity of up to 10,000 Saybolt Second Units (SSU). This viscosity represents the upper limit at which atomization can be used to expedite the conversion of liquid waste to a gas before combustion. Atomization is accomplished by

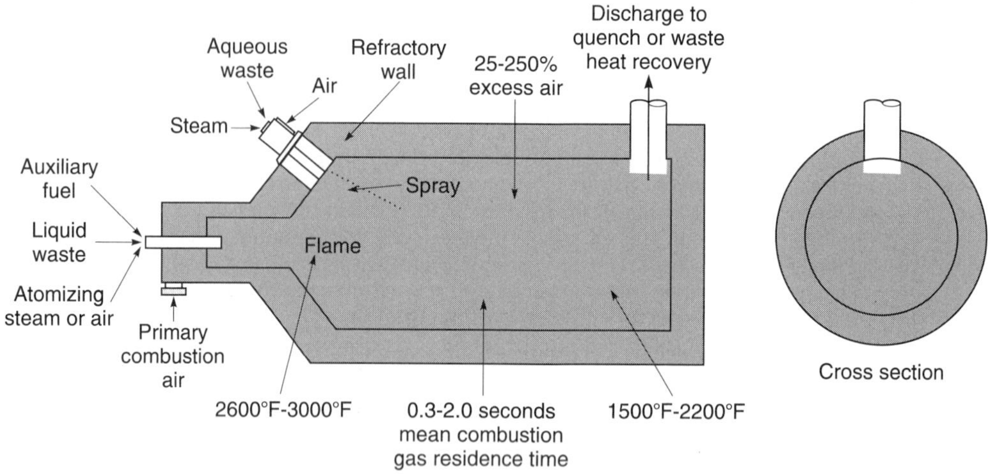

FIGURE 6.6 Liquid injection incinerator (*EPA-94/05, p. 2-28*).

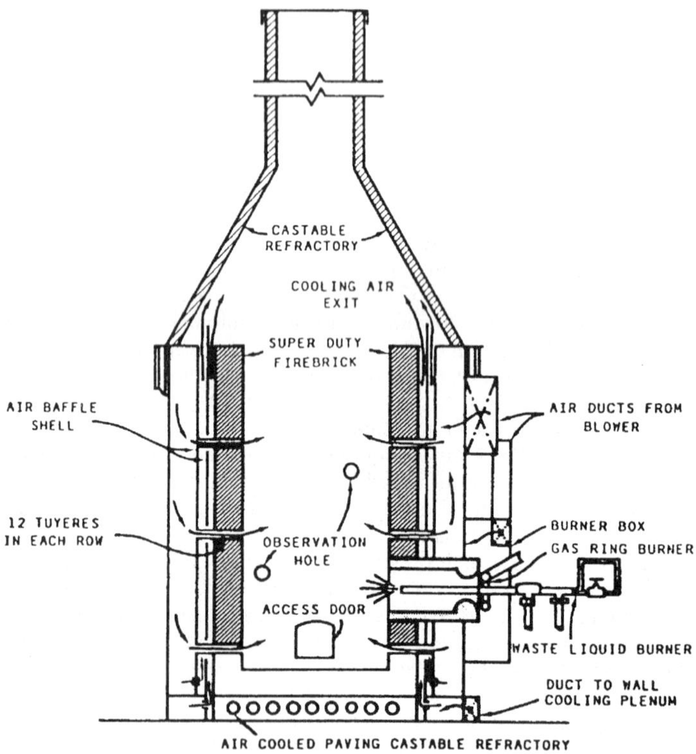

FIGURE 6.7 Vertical liquid injection incinerator (*OME-88/12, p. 4-81*).

the use of gas-fluid nozzles with high-pressure air or steam. The waste must be atomized to small droplets of 40 μm or less. Efficient destruction of liquid waste results from minimizing unevaporated droplets and unreacted vapors. Wastes with high solids are filtered prior to entering the feed tank. The liquid waste fuel system transfers waste from drums into a feed tank.

The tank is pressurized with nitrogen, and waste is fed to the incinerator using a control valve and a flow meter. The fuel supply line is purged with nitrogen after use. A recirculation system is used to mix the tank contents.

One of the very important components for liquid injection incinerators is the nozzle. Types of nozzles for liquid injection incinerators include:

- mechanical atomizing nozzle
- rotary cup burner
- external low pressure air atomizing burner
- external high pressure two-fluid burner
- internal mix nozzle
- sonic nozzle

MECHANICAL ATOMIZING NOZZLE: These are the most common types of burner nozzles in current use. Fuel is pumped into the nozzle at pressures of 515–1030 kPa (70–140 psig) through a small fixed-orifice discharge. The fuel is given a strong cyclonic or whirling velocity before it is released through the orifice. Combustion air is provided around the periphery of the conical spray of fuel produced. The combination of combustion air introduced tangentially into the burner and the action of the swirling fuel produce effective atomization. Normal turndown ratios are in the range 2.5 : 1 to 3.5 : 1. By utilizing a return flow line for fuel oil, the turndown ratio can be increased to as high as 10 : 1. A major disadvantage of this type of burner/nozzle is its susceptibility to erosion and pluggage from solids components of the fluid stream. Flames tend to be short, bushy, or low velocity, and this results in slower combustion, requiring relatively large combustion chamber volumes. This burner is applicable for fluids with relatively low kinematic viscosity, under $0.22 \times 10^{-4}\ m^2/s$ (100 SSU).

ROTARY CUP BURNER: Atomization is provided by throwing fuel centrifugally from a rotating cup or plate. Oil is thrown from the lip of the cup in the form of conical sheets which break up into droplets by the effect of surface tension. No air is mixed with fuel prior to atomization. Instead, it is introduced through an annular space around the rotary cup. Normally a common motor drives the oil pump, rotating cup, and combustion air blower. The liquid pressure required for this burner is relatively low, since atomization is a function of cup rotation and combustion air injection, not fuel pressure. This low pressure requirement and the relatively large openings within the burner fuel path allow passage of fluids with relatively high solids contents, as high as 20% by weight. They have a turndown ratio of approximately 5:1 and can fire liquids with kinematic viscosity up to $0.65 \times 10^{-4}\ m^2/s$ (300 SSU). Rotary cup burners are sensitive to combustion air flow adjustment. Insufficient air flow will result in fuel impingement on furnace walls, while excessive combustion air will cause a flame-out.

EXTERNAL LOW PRESSURE AIR ATOMIZING BURNER: The major portion of the combustion air requirement is provided at 7–35 kPa (1–5 psig) near the burner tip. Air is injected externally to the fuel nozzle and is directed to the liquid stream to produce high turbulence and effective atomization. The liquid pressure necessary for operation is only enough for positive delivery, normally less than 10 kPa (1.4 psig). Secondary combustion air is provided around the periphery of the atomized liquid mixture. The flame is relatively short because of the high amount of air provided at the burner (atomization and secondary combustion air). The short flame allows the design of smaller combustion chambers. These burners normally operate with liquids in the range $0.43–3.24 \times 10^{-4}\ m^2/s$ (200–1500 SSU) and can handle solids concentrations in the liquid of up to 30%. A small quantity of the air flow passes around the fuel discharge to aid in optimization of the fuel flow pattern.

EXTERNAL HIGH PRESSURE TWO-FLUID BURNER: The atomizing fluid, air or steam (or nitrogen or other gas), impinges the fuel stream at high velocity to generate small particles that encourage quick vaporization and effective atomization of tars and other heavy liquids. The required atomization pressure varies from 200 to 1030 kPa (30 to 140 psig). Turndown is in the range 3:1–4:1. The flame produced is relatively long, requiring appropriately constructed combustion chambers. The fuel viscosity normally handled by these burners ranges from 0.3 to 10.8 m^2/s (150–5000 SSU) for either air or steam atomization. A solids content of up to 70% can be accommodated by these burners.

INTERNAL MIX NOZZLE: Air or steam is introduced within the nozzle to provide impingement of atomization fluid on the fuel stream prior to spraying. Atomization air is provided at pressures less than 200 kPa (30 psig), and steam is normally introduced at 620–1030 kPa (85–140 psig). The turndown ratio for this type of burner is from 3:1 to 4:1. These nozzles cannot tolerate a significant solids content and can handle only low-viscosity fuels, under 0.22 m^2/s (100 SSU). This burner is used for clean, low-viscosity liquids. Its advantage is in its low cost compared with other burners.

SONIC NOZZLE: These nozzles utilize a compressed gas such as air or steam to create high frequency sound waves which are directed at the fuel stream. This acoustic energy is transferred to the liquid stream and creates an atomizing force, breaking the stream into minute particles. The fuel nozzle diameter is relatively large, allowing passage of solid particulate streams such as slurries and sludges with high particulate content. Little fuel pressurization is required. The spray pattern is not well defined, with finely atomized, uniformly distributed droplets travelling at low velocities. These nozzles are difficult to adjust, have low turndown, and generate an extremely high noise level during operation.

TECHNOLOGY APPLICATION: Liquid injection incinerators can be used to dispose of almost any combustible liquid waste, including lubrication oils, paints, solvents, and pesticides. With appropriate nozzle design, liquids with kinematic viscosities of up to 1.62×10^{-4} m^2/s (750 SSU) can be fired in suspension. A wide variety of liquid injection incinerators are marketed today, and these represent probably the most common incinerator type for waste disposal.

TECHNOLOGY ADVANTAGE: The technological advantages include (EPA-86/09, p. 2-6):

- they can incinerate a wide range of liquid wastes;
- they are capable of a fairly high turndown ratio;
- they have virtually no moving parts.

TECHNOLOGY DISADVANTAGE: The technological disadvantages include:

- they are generally limited to wastes that can be atomized through a burner;
- the burners are susceptible to plugging (burners are designed to accept a certain particle size; thus the particle size of any solids contained in the liquid waste feed is a critical parameter for successful operation);
- burners may not be able to accept a material that dries and cakes as it passes through the nozzles.

2.6 Mass Burn Incinerator

PROCESS DESCRIPTION: A mass burning incinerator is an incineration process which burns unprocessed, mixed municipal solid waste (MSW) in a single combustion chamber under conditions of excess air, as illustrated in Fig. 6.8 (OME-88/12, p. 4–54).

MSW, which is usually bulky and heterogeneous in nature, is stored in a pit and fed into the incinerator by a crane, which can also remove oversized items. It is burned in a sloping, moving grate. The movement (e.g., vibrating, reciprocating, or pulsing) helps agitate the MSW

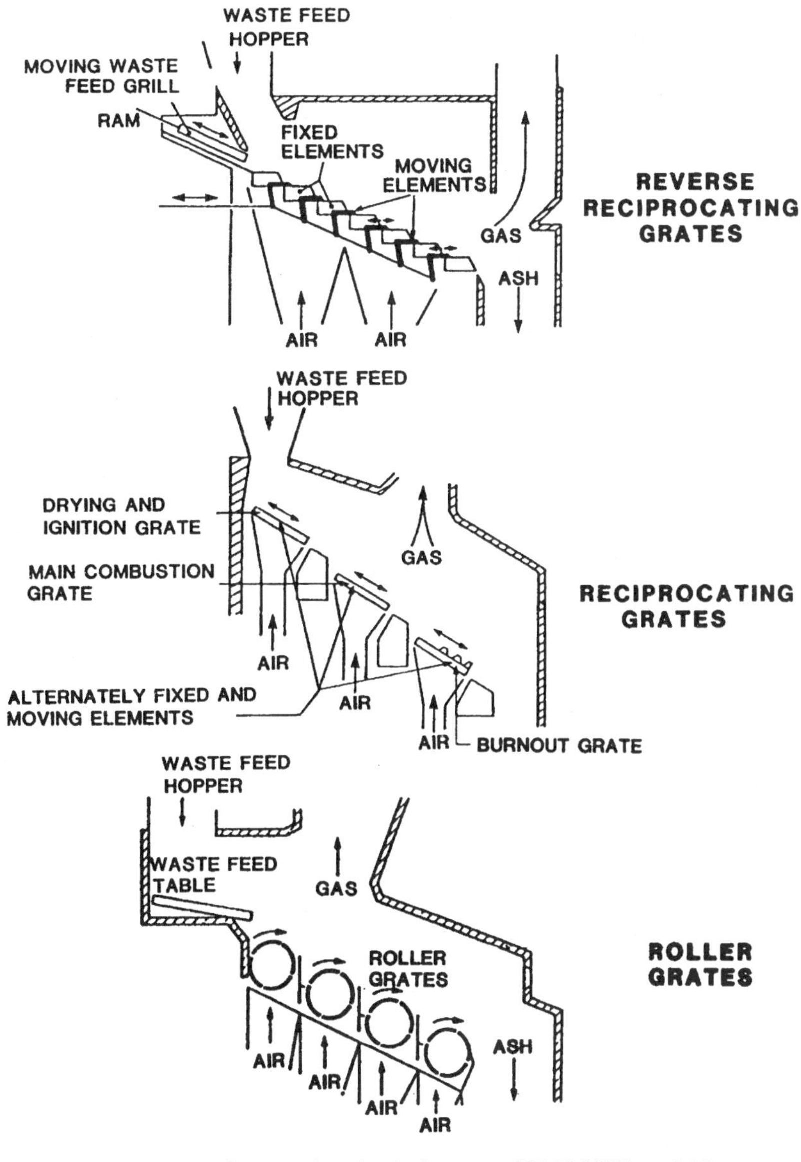

FIGURE 6.8 Moving grates for mass burning incinerators (*OME-88/12, p. 4-56*).

and mix it with air, and causes it to tumble down the slope. Many proprietary grates have been designed. Some systems use a rotating (or rotary) kiln rather than grates to agitate the waste and mix it with air. Many new mass burn incinerators use computer systems to precisely control grate movement, underfire air, and overfire air (OTA-89/10).

An MSW incinerator is normally (in excess of 50 t/d [55 T/d]) a continuously operated installation equipped with heat recovery equipment. Waste is burned in these incinerators without preprocessing.

The features of large central mass burning incinerators are distinguished by the design of the grate system. The grate must transport the MSW through the furnace and promote combustion by providing adequate agitation without contributing to excessive particulate emissions.

As the waste moves progressively through the furnace, it is dried, burned, and combusted to ash. Approximately 40–60% of the total air entering the furnace is provided as underfire air to cool the grates and prevent ash slagging. The balance is supplied as overfire air to completely combust the flue gas and particulates rising from the grates.

Many central waste incineration systems are built with waterwall construction in addition to boiler tubes within the flue gas stream (convection sections) to maximize energy recovery from the incinerator. Refractory-lined combustion chambers with a separate downstream boiler section may also be used in lieu of waterwall construction. If refractory walls are used, higher excess air is required to control the operating temperature.

Operating temperatures in mass burning central waste incinerators are normally maintained in the order of 1000°C (1832°F), and refuse residence time on the grate ranges from 20 to 45 min. Refractory wall systems normally require 100–150% excess air to maintain operating temperatures, whereas waterwall systems require only about 80% excess air. This offers the advantage of a smaller furnace volume and reduced NO_x formation with the latter system, due to the lower airflow. Waterwalls extract heat from the burning waste. Without waterwalls, where the furnace chamber is lined with refractory, the furnace temperature must be controlled by the injection of cool air. In refractory or waterwall furnaces, the maximum temperatures should be below 1100°C (2010°F), the temperature at which slagging problems will begin to occur.

Underfire air is provided beneath the grates to prevent overheating of the grate system and to supply part of the waste combustion air requirement. Air is also provided above the grates (overfire air) to burn off the products of combustion of the waste and to properly direct flue gas flow within the furnace. Underfire air will comprise from 40% to 60% of the air flow to the furnace, with the overfire air flow inserting the balance of the air required for incineration.

Bulky goods (stoves, refrigerators, etc.) must be removed from the waste feed. With some systems there is a limited ability to handle waste tires, and they must be removed or distributed throughout the daily feed.

There are two types of corrosion mechanisms normally associated with waste burning facilities. One is dew point corrosion and the other is high temperature corrosion. Dew point corrosion occurs when the flue gas stream is reduced in temperature to less than 150°C (302°F). Refuse will contain plastics which, in turn, have a chloride component. In the presence of water vapor, which is always present when burning refuse, the chloride will convert to hydrogen chloride gas. When the gas temperature drops below 150°C (302°F), the hydrogen chloride will begin to condense to a liquid hydrochloric acid, which will produce severe corrosion to steel and other metallic surfaces.

High temperature corrosion of steel occurs when hydrogen chloride is present in the flue gas. At temperatures in excess of 370°C (700°F), ferric chloride will form; this compound is friable, i.e., it will form a series of flakes on the steel surface and the surface will "waste" away. As the temperature increases, this type of corrosion will increase significantly.

To reduce dew point corrosion, the exiting flue gas must never be allowed to drop too low in temperature. This is particularly important when an economizer or an air preheater is used to try to extract relatively low-level heat from the flue gas.

Control of high temperature corrosion is more difficult. By generating steam at a temperature less than 370°C (700°F), the incidence of high temperature corrosion will be greatly reduced (the temperature of the steel boiler tubes is essentially equal to the temperature of the generated steam). The concentration of hydrogen chloride in flue gas is highest just above the grate. To further reduce the corrosion potential of the flue gas, the lower section of a waterwall is normally coated with a refractory material with heat conduction properties, such as carbide brick or cement.

When superheated steam is desired, with temperatures in excess of 370°C (700°C), superheater sections should be placed as far as possible from the burning waste, downstream or immediately upstream of the boiler convection section. By the time the flue gas reaches this area of the incinerator, there should have been sufficient mixing by virtue of the turbulent upstream conditions to reduce the occurrence of concentrated pockets of hydrogen chloride. The hydrogen chloride should be fairly evenly distributed through the gas stream, resulting in a relatively lower concentration.

TECHNOLOGY APPLICATION: Central waste incineration systems are used primarily for refuse incineration. The practice has been more prevalent and more long-standing in Europe than in North America.

Mass burning of MSW in central incineration facilities is a commercially demonstrated technology. Due to the relatively large size of such units and unique features of each system with respect to waste handling, energy recovery, etc., these facilities are normally designed and built to meet each customer's specific needs.

2.7 Microwave Discharge Reactor

PROCESS DESCRIPTION: A microwave discharge, also known as microwave plasma, in the treatment of wastes is a special application in the general field of plasma chemistry. A microwave is an electromagnetic wave that has a wavelength between about 0.3 and 30 cm, corresponding to a frequency of 1–100 GHz. The technology uses microwave energy to excite the molecules of a carrier gas such as helium or air, thus raising electron energy levels and essentially forming very reactive free radicals. The gas in this high energy condition is called plasma. The excited electrons transfer this energy to break the chemical bonds of materials. Carbon–carbon bonds are among those most susceptible. Thus, theoretically, any organic waste (liquid, solid, or gas) placed into the plasma can be degraded to intermediate or ultimate products, perhaps destroying their toxic properties. Residence time within the plasma varies from 0.1 to 1.0 s. The temperature for plasma destruction can be as low as 150°C (300°F).

TECHNOLOGY APPLICATION: Lockheed, under the auspices of the US EPA, investigated the microwave plasma decomposition of wastes. The basic elements of Lockheed's microwave discharge reactor were: (a) a microwave generator or source of power; (b) a waveguide which leads the microwave power from the generator to a resonant cavity, which surrounds the reactor; (c) a reaction chamber, which must be made of silica or quartz to minimize dielectric losses; (d) a system for generating or feeding vapors, or reactants or condensed phases into the reaction chamber; (e) a system for controlling pressure and flow; (f) a system for collecting reaction products (EPA-76/11).

2.8 Molten Metal

PROCESS DESCRIPTION: The Molten Metal Technology Company is currently developing the Quantum Catalytic Extraction Process™ (CEP) for treatment of mixed wastes. The CEP is shown schematically in Fig. 6.9 (EPA-94/09, p. 5-100).

Wastes are fed into a molten metal bath where they absorb heat, mix with oxygen, and decompose. The waste organics are partially oxidized. Waste metals are dissolved and assimilated into the molten metal bath. Waste inorganic oxides which are insoluble in the metal bath rise to the top of the molten metal bath and form a distinct vitrified slag layer. The slag layer is tapped and disposed of, theoretically without further processing such as cementation. The metal layer is tapped as required. An afterburner is employed to complete combustion of the flue gas emitted from the surface of the melt. The main subsystems of the CEP are the waste preparation and feeding, the melting chamber, and the air pollution control systems.

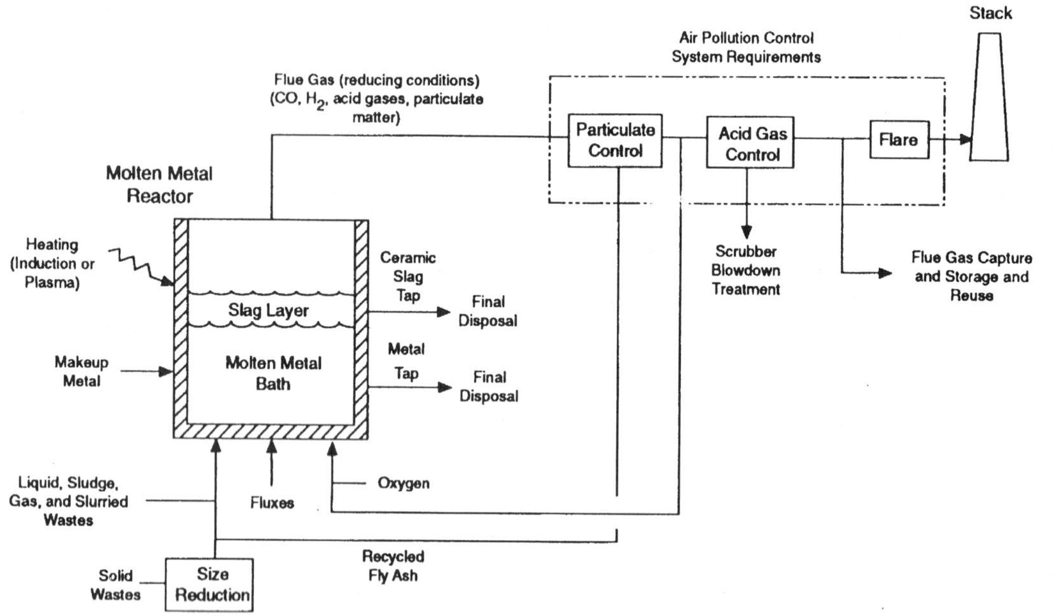

FIGURE 6.9 Molten metal technology (*EPA-94/09, p. 5-105*).

The CEP is capable of treating liquid, shredded solid, slurry, and sludge wastes. These waste forms are introduced at the bottom of the furnace, beneath the molten metal surface. Methods to feed bulk solid wastes are shown in Fig. 6.10.

It may also be most effective to separate organics and nonmetal wastes from metal wastes. Additionally, it may be desirable to segregate metal wastes based upon metal chemistry so they are compatible with thermodynamic considerations, and to segregate them based upon radionuclide contamination. Work on melt refining of uranium-contaminated metal demonstrated the necessity of selecting proper slag and furnace melting procedures based upon the specific metal or alloy. For example, decontamination of aluminum requires a fluoride-type slag; borosilicate and oxidizing slags are most effective for nickel, ferrous metals, and copper; sodium-based fluxes should be used for zinc, lead, and tin.

The heart of the CEP is the bath of molten metal contained in a refractory-lined, enclosed vessel. The bath temperature is maintained between 2500 and 3500°F. Induction heating coils that circle the furnace are used to maintain bath temperature. Alternating current flowing though the coil produces a magnetic field which induces an electric current in the molten metal. The resistance of the metal to the current generates heat. Plasma heating, as described in a later section, may also be used. Partial combustion of waste organics with oxygen (not air) serves as an additional heat source. The molten bath remains well mixed due to natural convection currents and oxygen injection up through the bath.

When the waste comes into contact with the hot molten bath, it disassociates into elements. The catalytic and solvent nature of the molten metal bath assists in the conversion process. Oxygen is added to the bath to gasify waste organics and strip carbon from the molten metal. Oxygen is bubbled up through the bath through three "tuyeres" located at the bottom of the vessel. As in the steelmaking industry, the tuyeres are cooled using methane gas. The tuyeres consist of two concentric tubes; oxygen is transported in the inner tube and methane in the

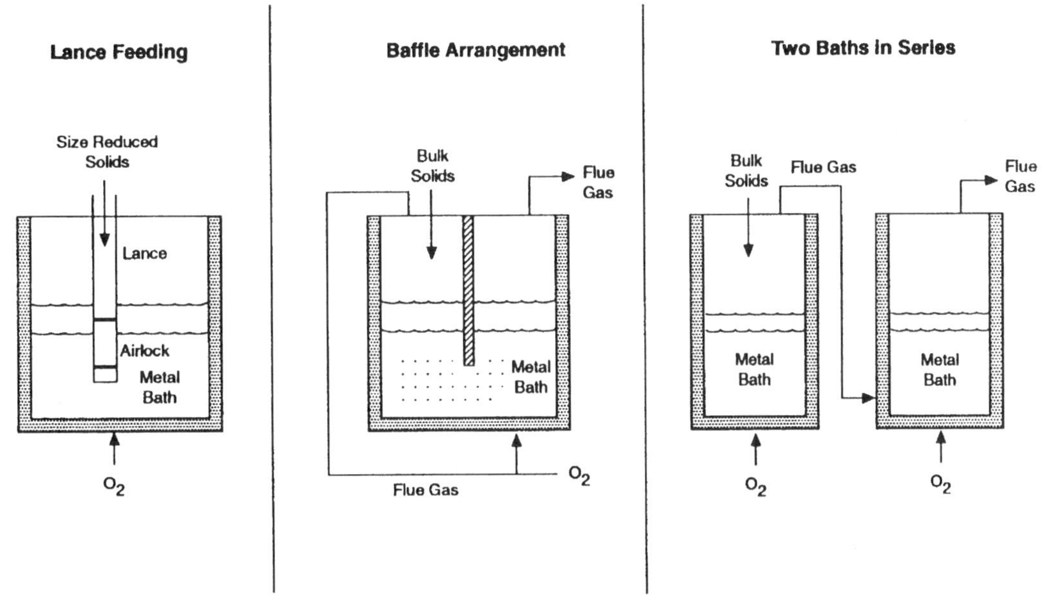

FIGURE 6.10 Waste feeding techniques for MMT (*EPA-94/09, p. 5-106*).

outer tube. The methane is endothermically cracked, absorbing heat generated from the oxygen reacting with molten metal. A limited supply of oxygen is provided to create a reducing atmosphere within the melt. The flue gas leaving the melt contains carbon monoxide, hydrogen gas, and hydrocarbons.

The inorganic oxides in the waste which are insoluble in the metal bath rise to the top of the molten metal and form a distinct, vitrified slag layer. The slag layer may act as an insulator to keep the bath hot; it may also prevent emission of some volatilized metals and radionuclides. The slag typically contains silica oxides, iron oxides, magnesium oxides, and sulfur, chlorine, and phosphorus compounds. Controlling the properties of the slag is critical to the success of this process. The slag chemistry, viscosity, melting temperature, and final-form properties such as durability and leachability are controlled by adding different fluxes to the bath. Fluxes may include calcium, silica, alumina, magnesium, sodium, and potassium in the form of raw materials such as limestone, fluorospar, soda ash, borax, and dolomite. For example, borate or calcium fluoride can be added to provide increased slag fluidity, and ferric oxide may be added to increase its oxidative potential.

The accumulated slag and molten metal are to be tapped (removed from the furnace) separately. The slag and metals are poured into containers and allowed to cool and solidify. If the proper chemicals are added, the final slag product is ready for final disposal without further processing. The tapped metal layer may contain radionuclides and leachable metals—it may require stabilization prior to disposal. It is worth noting that because the melt stoichiometry is reducing, the metals do not mix with the slag. This is different from the plasma process, in which the oxidizing hearth stoichiometry allows the metals and other inorganics to be incorporated into one slag material, which is reportedly nonleachable.

Technology application: The emerging CEP is derived from a metal separation process which has been used in the steelmaking industry for decades. The CEP has treated chlorinated organics, soils, and used electronics.

The CEP has been used to treat homogeneous wastes of known composition. It has not been demonstrated to treat heterogeneous wastes or wastes of unknown composition because the slag and bath properties become more difficult to control. Wastes with high organic content may not be appropriate for the metal melt refining process due to the importance of control of the thermochemistry and chemical make-up of the feed. Of particular concern are the presence of chemicals which cause explosive energy releases or disrupt refining processes when placed in the metal bath. Wet waste materials may cause steam explosions in the molten metal bath.

2.9 Molten Salt Oxidation

PROCESS DESCRIPTION: The molten salt oxidation (MSO) process is shown schematically in Fig. 6.11. The molten salt bath provides a variety of functions: (1) it provides a thermally stable medium in which the waste and oxygen may mix, (2) it provides a catalyst for the waste oxidation reactions and accelerates destruction of organic materials, (3) it retains soot, char, and ash generated during waste decomposition in the melt, and (4) it acts as an in-situ acid gas scrubber for neutralization and removal of acid gases such as HCl, HF, and SO_x formed during waste decomposition. The five main subsystems of the MSO process are the waste pretreatment, molten bath, salt recycle, air pollution control, and ash disposal systems. Each is described below (EPA-94/09 p. 5–7).

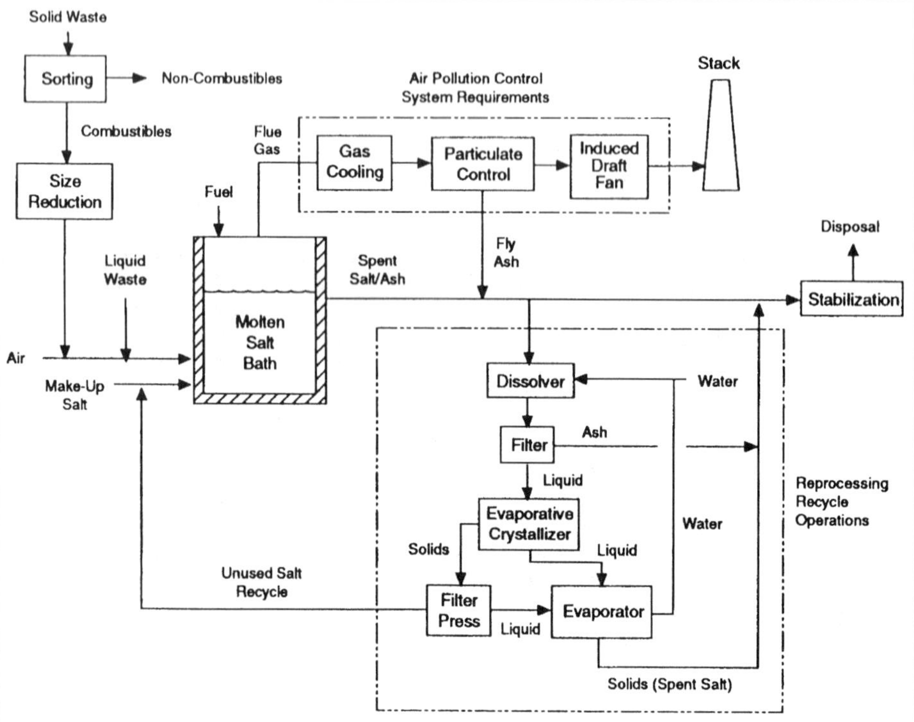

FIGURE 6.11 Molten salt oxidation (*EPA-94/09, p. 5-7*).

The heart of the MSO system is the refractory-lined vessel containing the molten salt bath. The temperature range of the bath is maintained at 900–1000°C, which is above the nominal salt melting temperature of 650°C. The molten bath is primarily composed of sodium carbonate, but may be supplemented with smaller quantities of other salts such as potassium carbonate, lithium carbonate, and/or sodium sulfate. The addition of lithium carbonate acts to lower the temperature of the melt. Addition of sodium sulfate acts to increase oxygen availability for more efficient waste organics destruction.

The liquid and size-reduced solid wastes are fed via pneumatic or screw-type feeders into the reactor vessel, near the bottom of the molten salt bath. The air may either be utilized as transport fluid for the wastes, or is bubbled up through the molten bath through a dedicated injector. The wastes come up to temperature and oxidize in a uniform fashion as they contact the air in the bubbling molten material. The heat content of the waste acts to maintain melt temperature; auxiliary fuel is used for temperature control. Chlorine and other halides react with the salts to achieve in-situ acid gas capture.

Molten salts near the surface are continuously removed from the reaction vessel and replaced with fresh salts near the bottom. The removed material consists of spent salts, unused salts, and ash. The salts become nonuseful once they have reacted to neutralize waste constituents such as chloride, fluoride, sulfate, or phosphate. The residual ash from thermal treatment of waste material may consist of inorganic metal oxides, aluminates, or silicate oxides, which are present as either dissolved solids or as distinct layers in the melt. For example, lead and aluminum may form separate metallic phases in the salt melt.

A higher melt ash content results in increased bath viscosity, preventing efficient mixing between the waste and the oxygen. Also, a more viscous melt is difficult to remove from the reactor. The rate at which the melt must be replaced depends upon the halide and ash contents of the wastes being treated. Specifically, the replacement rate should be adjusted to maintain an ash to salt ratio of less than 20 percent by weight, or to keep dissolved impurities low enough so that they do not have a negative effect on acid gas retention or other operating parameters. If equal amounts of silicon and aluminum oxides are present in the molten bath, an ash to salt ratio of less than 5 percent by weight must be maintained.

An additional problem is the formation of solid alumina, silica, and iron oxides, which will deposit and remain on the reactor walls until the unit is shut down for cleaning. To avoid this problem the Lawrence Livermore National Laboratory is developing a two-stage molten salt treatment process. The first bath is operated under reducing conditions to limit the formation of oxides. This allows the first bath to accept wastes with higher ash content. The second stage, which has minimal ash buildup due to ash removal in the primary bath, is operated under oxidizing conditions.

Technology application: Rockwell International developed this technology and has conducted numerous pilot-scale tests. Although all testing results showed excellent destruction and removal efficiency, the technology has not been used commercially (EPA-86/10).

2.10 Multiple Chamber Incinerator

Process description: To overcome the problem of incomplete combustion associated with single-chamber incinerators, multiple-chamber units were developed. A primary chamber is used for combustion of the solid waste, while the secondary chamber provides the residence time, temperature, and supplementary fuel for combustion of the unburned products carried over from the primary chamber.

There are two basic types of multiple chamber incinerators:

- retort incinerator;
- in-line incinerator.

Retort incinerator: This unit is a compact incinerator in the form of a cube with multiple internal baffles. The baffles are positioned to guide the combustion gases through 90-degree turns in both lateral (horizontal) and vertical directions. At each turn, ash (soot) drops out of the flue gas flow. The primary chamber has elevated grates for burning of the waste and an ash pit for collection of ash residue. Cut-away views of a typical retort incinerator are shown in Figs. 6.12 and 6.13 (EPA-73/05, p. 438).

In-line incinerator: This is a larger unit than the retort incinerator. The flow of combustion gases is straight through the incinerator axially, with abrupt changes in direction, as shown in Fig. 6.14 (EPA-73/05, p. 439). Waste is charged on the grate, which can be either stationary or moving. As with the retort type, changes in the flow path and flow restrictions in an in-line incinerator provide settling out of larger airborne particles and increased turbulence for more efficient burning.

With both types of systems, supplemental fuel burners are provided in both the primary and secondary chambers. Depending upon the nature of the waste burned, the fuel supply in the primary chamber may be unnecessary after start-up although the secondary chamber normally requires a continuous fuel supply (at least a pilot flame). Overfire and underfire air to the primary chamber is normally aspirated through ports in the wall of the furnace, or is supplied at a controlled rate by forced draft fans. Additional air is added to the secondary chamber to ensure complete combustion.

Dimensional data for typical retort and in-line incinerators of varying capacity can be obtained from OME-88/12, p. 4-44 and p. 4-45. Although they represent the product of a single manufacturer, the dimensional data can be considered generic. They are typically operated at a temperature of 760–1000°C (1400–1832°F) in the secondary chamber, and normally use approximately 200% excess air. Approximately half of the required air enters as leakage.

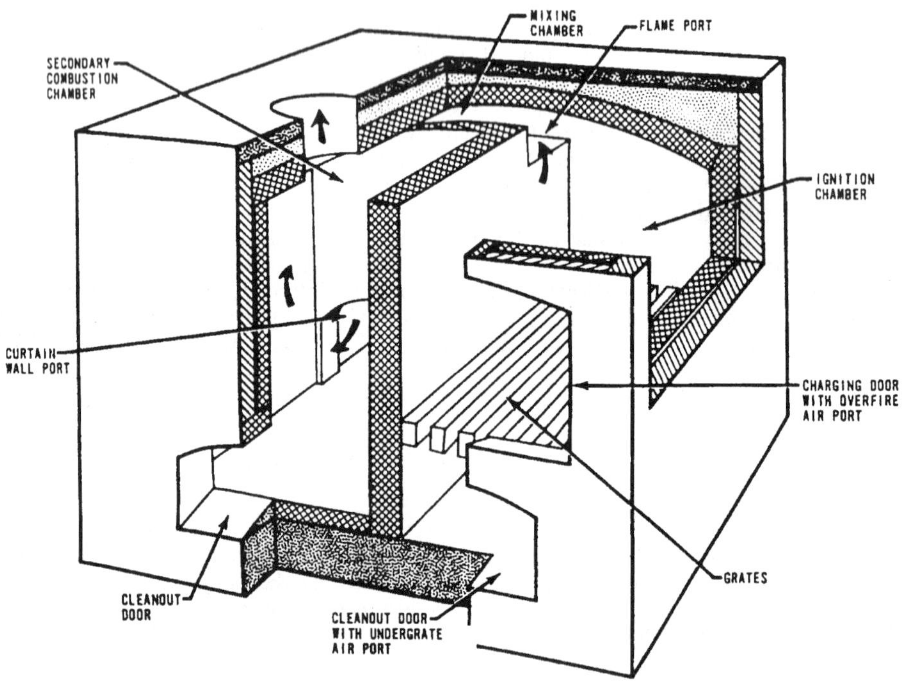

FIGURE 6.12 Retort incinerator (*EPA-73/05, p. 438*).

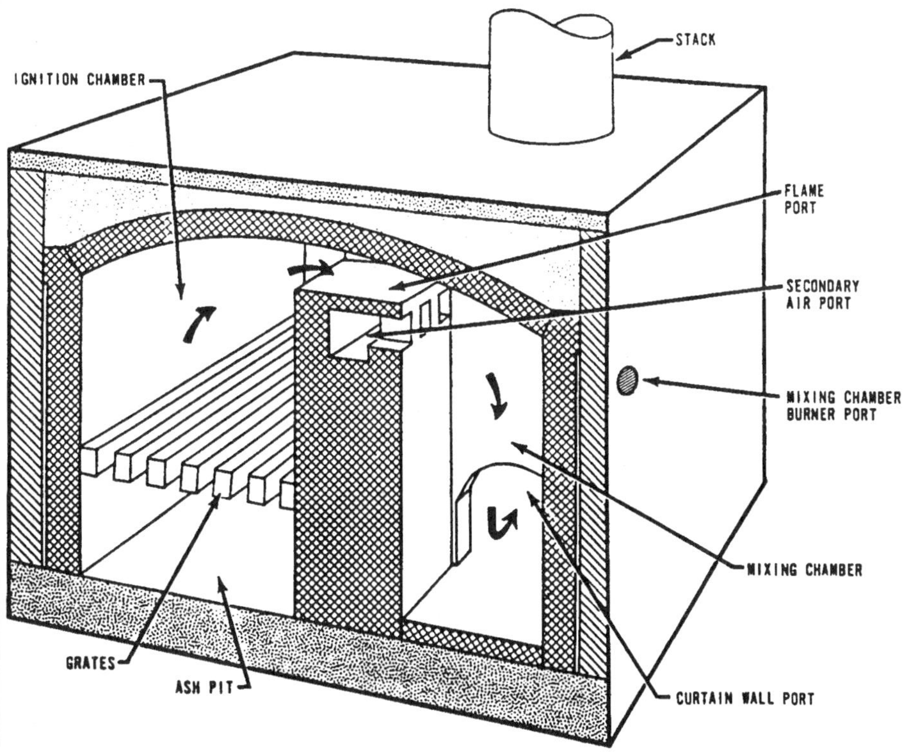

FIGURE 6.13 Retort incinerator (*EPA-73/05, p. 438*).

Of the balance, 70% should be provided in the primary combustion chamber as overfire air, 10% as underfire air, and 20% in the mixing or secondary chamber.

They are applicable for refuse, trash, rubbish, and biomedical waste, but it is generally difficult to incorporate the air emission control equipment required to meet regulatory emission standards due to space and economic limitations.

TECHNOLOGY APPLICATION: Multiple chamber incinerators are commonly used for all types of solid waste as well as for specialty applications such as incineration of medical wastes, crematory incinerators, and metal drum reclamation.

Multiple chamber incinerators are commercially available, many as packaged systems. Retort incinerators are normally used for batch or semicontinuous operation in the capacity range 10–350 kg/h (22–772 lb/h). In-line incinerators are normally provided in the 225–900 kg/h (496–1984 lb/h) range, and may be equipped with automatic charging and/or ash removal equipment on units above 20 kg/h (44 lb/h) capacity.

2.11 Multiple Hearth System

PROCESS DESCRIPTION: As the name implies, a multiple hearth incinerator is designed with several combustion hearths. It is a vertical cylindrical structure, refractory-lined, with a series of refractory hearths positioned one beneath the other, as shown in Fig. 6.15. The

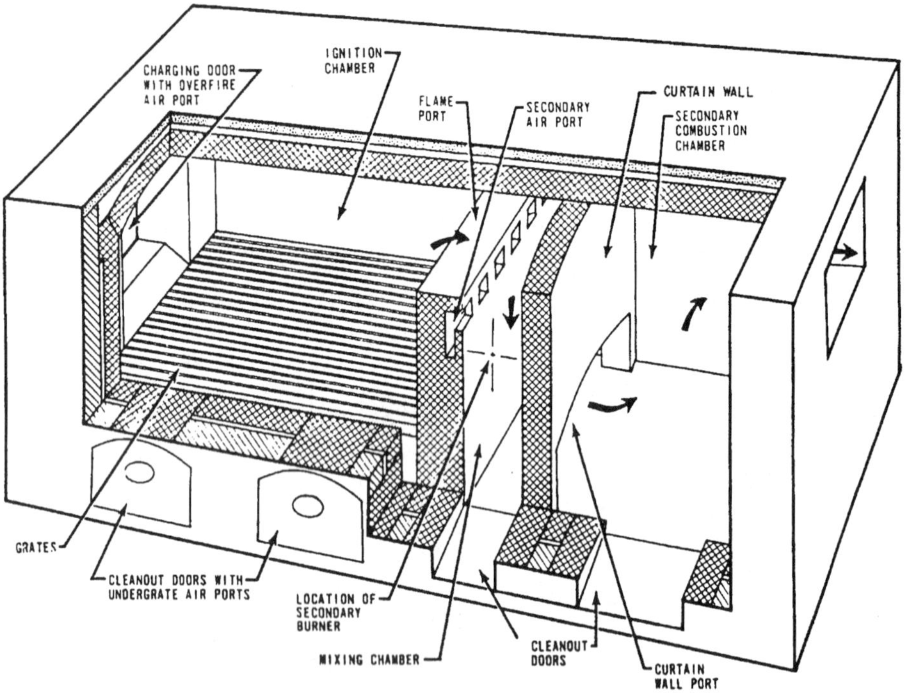

FIGURE 6.14 In-line incinerator (*EPA-73/05, p. 439*).

transport of the waste over the hearths is by rabble-arms that stir the waste pile and eventually move the waste to edges or ports in the hearth where the waste drops to the next hearth (DOE-93/03, p. C-4).

The temperature over any hearth can be controlled by direct-fired or indirect-fired auxiliary heat burners. The rabble arms and supporting control shaft are often air-cooled. The air that is heated in this manner can be used as preheated combustion air. The center shaft, which is hollow to allow the passage of cooling air through it, rotates within the incinerator at approximately one revolution per minute, carrying the rabble arms with it sweeping waste across the hearths.

The incinerator is particularly well suited to the combustion of sludge. The upper hearths drive volatiles from the incoming waste, the middle hearths combust the waste, and the lower stages cool the ash before discharge. When burning wastes with an appreciable volatile content, an afterburner chamber is usually included after the multiple hearth incinerator to ensure complete combustion of hazardous components in the exhaust from the "rotary" hearth.

Multiple hearth furnaces range from 2 to 8 m (6.6 to 26.2 ft) in diameter and from 4 to 23 m (13 to 76 ft) in height. The number of hearths is dependent upon the waste feed and processing requirements, but generally varies between 5 and 12 hearths. Waste retention time is controlled by the rabble tooth pattern and the rotational speed of the central shaft.

The upper hearths of the furnace comprise the drying zone where moisture is evaporated from the sludge. Flue gas exits this upper zone of the incinerator at 430–650°C (806–1202°F). Cooling air exiting the top of the center shaft is approximately 90 to 230°C (194–446°F), and is often recirculated to the furnace as preheated combustion air.

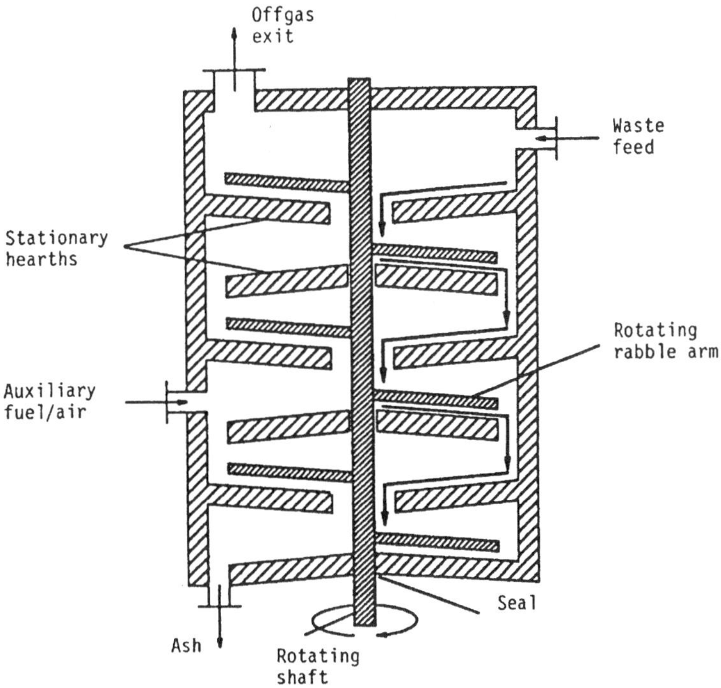

FIGURE 6.15 Multiple hearth incinerator (*DOE-93/03, p. C-5*).

In the center hearths, the combustion zone, temperatures are normally in the 760–980°C (1400–1796°F) range. The lower hearths cool the ash before discharge. Ash exits this system dry from beneath the furnace. It can be collected and disposed of dry, or it can be dropped into a wet hopper, mixed with water, and pumped to a lagoon for dewatering and ultimate disposal.

Sludge or other wastes to be incinerated in the hearth of a multiple hearth furnace should have a solids content of 15–40% for proper movement and rabbling through the furnace. The temperature above at least two hearths should be maintained at approximately 870°C (1600°F) at all times when burning sludge cake.

When burning wastes with an appreciable volatile content, an afterburner chamber is usually included after the multiple hearth incinerator to ensure complete combustion of hazardous components in the exhaust from the rotary hearth, i.e., elimination of smoke and odor.

Feeding a multiple hearth is relatively simple. Feed can be dropped onto the top hearth by gravity. It can also be deposited on the top hearth or a lower hearth by means of a screw conveyor. Attempts to feed sludge to a multiple hearth furnace pneumatically have met with failure. This system necessarily injects air into the furnace in strong discrete bursts, which makes maintenance of constant furnace draft difficult, if not impossible.

Fuel or gas can be used as supplemental fuel. Generally solid fuels, such as coal or wood chips, should not be placed on a hearth and used as supplemental fuel. They are relatively dry and will start burning on the top hearth, encouraging premature release of volatiles from the waste stream and inadequate burnout can result. Similarly, grease (scum) should not be added to sludge feed. If grease is to be incinerated in the multiple hearth furnace, it should be added at a lower hearth (a burning hearth) through a separate nozzle(s). Grease will volatilize easily, and introducing it too high in the furnace where the temperatures may be below 760°C (1400°F) will not provide effective burnout.

The off-gas temperature can range from 425°C to 760°C (800°F to 1400°F). The oxygen content of the flue gas exiting the furnace should be continuously measured at the breeching leaving the top hearth. A sample should be continuously extracted from the gas stream, passed through a water bath to clean the sample and to reduce its moisture content, and then measured for oxygen content. The oxygen content should be in the range 6–10% by volume.

Excess air of 100–125% must be provided to ensure complete burnout of the sludge to ash. Since approximately 20% of the ash can be entrained in the flue gas, extensive gas cleaning equipment must be provided for its capture.

The multiple hearth furnace should always be run at a negative pressure (draft) to prevent external leakage of hot flue gas. If a separate afterburner is needed and if an emergency discharge stack is provided, the stack should be located downstream of the afterburner.

TECHNOLOGY APPLICATION: Multiple hearth incinerators are designed to burn wastes with low heating value such as sewage sludge and other high-moisture-content wastes. Its design includes drying as well as burning sections. Materials with less moisture content, such as coal or solid waste, will start to burn too high in the furnace. There would be insufficient residual time in these cases for effective burnout.

This incinerator system is relatively complex, with fans, burners, shaft speed, and feed charging all within the operator's control, and all affecting combustion efficiency and burnout. Particular attention must be paid to operator training in the proper control of these various equipment items and parameters.

This is the most prevalent incineration system for sewage sludge destruction in North America. Many multiple hearth incinerators have been built in the United States and Canada.

TECHNOLOGY ADVANTAGE: The primary technological advantage is the hearth's ability to handle wastes with high moisture and ash content, such as evaporator bottoms and wastewater treatment sludges. This incinerator can even handle sludges in the range 50–85% moisture. Hearths also are energy-efficient because of provisions for reclaiming heat from the ash and the ease of preheating combustion air.

TECHNOLOGY DISADVANTAGE: The technological disadvantages of hearths include:

- they are generally not applicable to the incineration of solid materials;
- the designs of the hearths are mechanically intensive and the maximum feed component particle size is limited;
- generally, the multiple hearth furnace cannot accommodate a temperature in excess of 1000°C (1830°F) without damage. If higher temperatures are required, an afterburner must be provided, which represents higher capital costs and higher operating (fuel) costs;
- it is virtually impossible to maintain heat in a multiple hearth furnace without firing supplemental fuel. This furnace, as noted above, has many areas of leakage, and therefore heat cannot be as effectively maintained within the units as in, for example, a fluid bed furnace.

2.12 Plasma Arc System

PROCESS DESCRIPTION: Plasma arc technology was originally developed for the United States space program. The evaluation of heat shields that protect space vehicles on re-entry required an intense heat source with plasma characteristics.

Plasmas have been referred to as the fourth state of matter since they do not always behave as a solid, liquid, or gas. A plasma may be defined as a conductive gas flow consisting of charged and neutral particles, having an overall charge of approximately zero, and all exhibiting collective behavior.

The plasma, when applied to waste disposal, can best be understood by thinking of it as an energy conversion and energy transfer device. The electrical energy input is transformed into a plasma with a temperature equivalent of up to 10,000°C at the centerline of the reactor. As the activated components of the plasma decay, their energy is transferred to waste materials exposed to the plasma. The wastes are then broken down into atoms, ionized, pyrolyzed, and finally destroyed as they interact with the decaying plasma species. The heart of this technology is that the breakdown of the wastes into atoms occurs virtually instantaneously and no large molecular intermediary compounds are produced during the kinetic recombination.

The most common method of plasma generation is electrical discharge through a gas. A low pressure gas is used as a medium through which an electrical current is passed. The type of gas used is relatively unimportant in creating the discharge, but will ultimately affect the products formed. In passing through the gas, electrical energy is converted to thermal energy by the absorption of gas molecules which are activated into ionized atomic states, losing electrons in the process. Arc temperatures up to 10,000°C may be achieved along the centerline recirculation vortex. Ultraviolet radiation is emitted when molecules or atoms relax from the highly activated states to lower energy levels.

A flow diagram of the plasma pyrolysis system is shown in Fig. 6.16. The plasma device is horizontally mounted in a refractory-lined pyrolysis chamber. Liquid wastes are injected through the colinear electrodes of the plasma device where the waste molecules dissociate into their atomic elements. These elements then enter the pyrolysis chamber, which serves as a mixing zone where the atoms recombine to form hydrogen, carbon monoxide, hydrogen chloride, and particulate carbon. The approximate residence times in the atomization zone and the recombination zone are 500 microseconds and 1 second, respectively. The temperature in the recombination zone is normally maintained at 900–1200°C.

After the pyrolysis chamber, the product gases are typically scrubbed with water and caustic soda to remove hydrochloric acid and particulate matter. The remaining gases, a high percentage of which are combustible, are drawn by an induction fan to the flare stack where they are electrically ignited. In the event of a power failure, the product gases are vectored through an activated carbon filter to remove any undestroyed toxic material.

TECHNOLOGY APPLICATION: Since an EPA/New York State pilot plant was tested in 1986 (Kolak-86), the plasma technology has made considerable progress. Full-scale plasma has

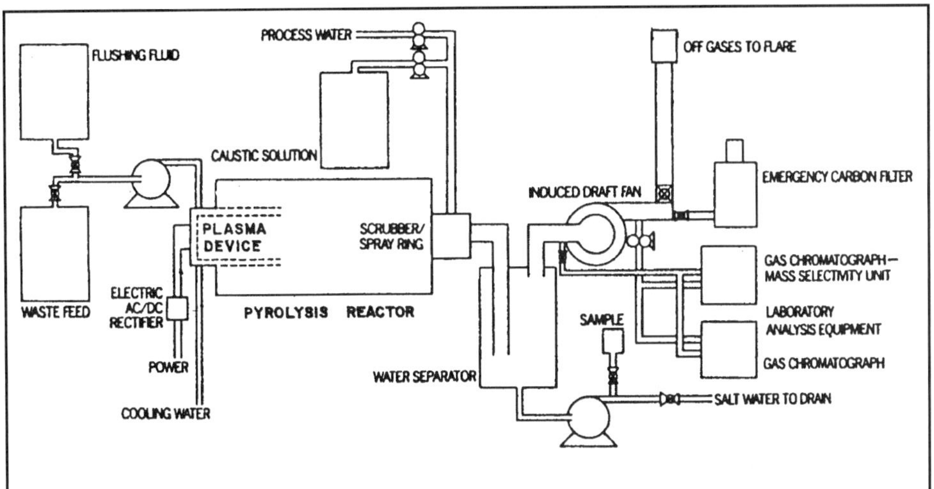

FIGURE 6.16 Plasma arc system (*EPA-86/10, p. 4-66*).

been installed to treat both liquid and solid waste. The authors call this technology the second generation of thermal treatment systems.

TECHNOLOGY ADVANTAGE: The technological advantages include:

- because radiative heat transfer proceeds as a function of the fourth power of temperature, a plasma system has very intense radiative power and therefore is capable of transferring its heat much faster than a conventional flame;
- organic chlorides are known to dehydrogenate when excited by ultraviolet radiation, which is abundant from thermal plasmas;
- because the plasma arc for waste destruction is a pyrolytic process, it requires virtually no oxygen. Compared with conventional incinerators, which normally require about 150% excess air to ensure proper combustion, the plasma arc will save the energy required to heat the excess air to the combustion temperatures, and will also produce significantly less gaseous by-product to be treated downstream;
- the process has a very short on–off cycle; and
- because of its compactness, a plasma arc system has potential for use as a mobile treatment system; the system would be housed in a trailer and moved from site to site.

TECHNOLOGY DISADVANTAGE: The technological disadvantages include:

- because the temperatures are so high (about 10,000°C at the arc centerline), the durability of the arc and the refractory materials could be a potential problem (from a corrosion or embrittlement standpoint); and
- because the arc is very sensitive to many factors such as sudden drops in voltage, the operation of the system requires highly trained professionals.

2.13 Pyrolysis System

PROCESS DESCRIPTION: Pyrolysis is the destruction of carbonaceous material in the presence of heat and substoichiometric oxygen levels. In practice, the heat is provided by combustion of a portion of the waste. Under ideal conditions, the pyrolysis of pure cellulose will produce combustible off-gases (methane and carbon monoxide), water (steam), and a residual char (carbon). The ideal pyrolysis reaction, of which there are many variations, is as follows (DOE-93/03, p. C-1) (Fig. 6.17):

$$C_6H_{10}O_5 \rightarrow CH_4 + 2CO + 3H_2O + 3C$$

In the actual destruction of waste, the off-gases are a mixture of many simple and complex volatile organic compounds, and the char is often a liquid containing residual carbon or tars and ash. In general, volatile gas production increases with increasing temperature, whereas the generation of tars and other liquors is greatest at lower temperatures.

Pyrolysis reactors typically operate at temperatures of between 500°C and 900°C (930°F and 1650°F), with waste residence times of 12–15 min. Systems designed to generate heat internally by combustion use less than the stoichiometric oxygen requirement. Depending on the waste characteristics and process design features, off-gas will typically have a heating value of between 5000 and 15,000 kJ/m^3 (134–403 Btu/ft^3) which can represent as high as 80% of the heat energy originally contained within the waste. The solid residue normally represents less than 5% of the original waste volume, although pyrolysis systems will have a higher fixed carbon content than excess air incinerators. Pyrolysis units have been found to be applicable to the destruction of organic industrial sludges.

TECHNOLOGY APPLICATION: Pyrolysis has been used as an industrial process for many years for the production of charcoal from wood chips, coke and coke gas from coal, fuel gas

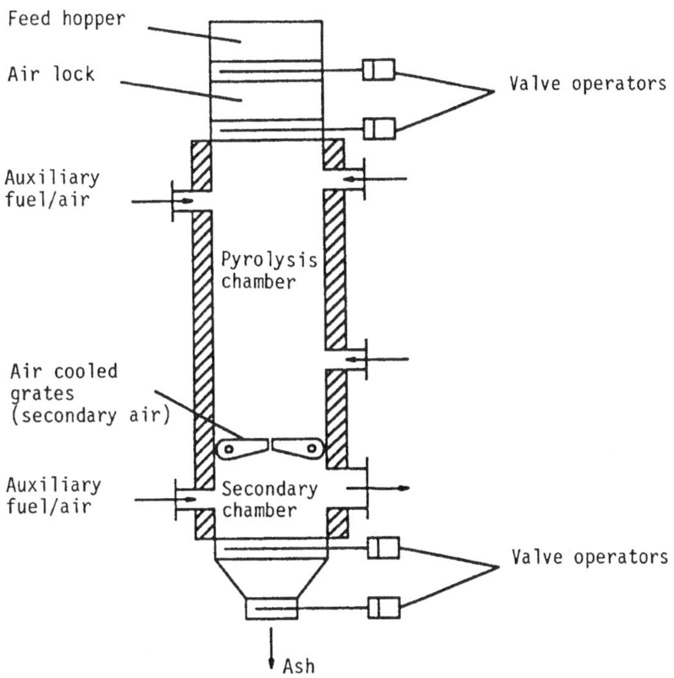

FIGURE 6.17 Pyrolysis reactor (*DOE-93/03, p. C-3*).

and pitch from heavy hydrocarbon still bottoms, etc. The process has been promoted for waste destruction only since the late 1960s. The only commercial pyrolysis system that has been found to operate successfully in North America is one that handles industrial semiliquid and sludge wastes at a number of pharmaceutical plants in the United States. The use of pyrolysis for the destruction of municipal sludge and solid wastes has not been successfully demonstrated to date on a commercial scale.

2.14 Refuse-Derived-Fuel Incinerator

PROCESS DESCRIPTION: The refuse-derived-fuel (RDF) incinerator is also known as the semi-suspension incinerator. RDF incineration differs from mass burning in that the waste is pretreated to produce a finer, more homogeneous matter, commonly referred to as refuse-derived-fuel or RDF. The pretreatment steps normally include shredding and ferrous metal recovery by magnetic separation, and can include air classification for removal of other noncombustible material (e.g., nonferrous metal and glass).

RDF can be coarse, fluff, powder, and densified pellets, briquettes, or similar forms and can be burned in two types of boilers. It can be used as the sole or primary fuel in dedicated boilers, or it can be cofired with conventional fossil fuels (e.g., coal and oil) or even wood in existing industrial or utility boilers. Boilers using RDF can recover energy. In addition, materials such as steel and glass recovered during the initial processing can be sold (OTA-89/10).

The RDF is blown into the furnace through a pneumatic charging system. Primary combustion (of approximately 70% of the waste's combustible content) occurs in suspension. Ash and unburned material drop onto a travelling grate where final burnout occurs. The grate discharges the ash to a hopper for removal to final disposal.

The high degree of turbulence in the furnace contributes to high ash carryover in the flue gas stream, necessitating the installation of higher efficiency particulate removal equipment with this type of incinerator than with mass burn units.

These RDF incineration systems require waste preprocessing which, at a minimum, includes shredders and magnetic separators. Although the heating value of the RDF generated by these processes can be 20% higher than the specific heating value of municipal solid waste (MSW), production of this fuel requires a significant investment in capital equipment and high operating costs. Likewise, system reliability is generally lower than for mass burn systems because of the greater amount of process equipment.

TECHNOLOGY APPLICATION: Although RDF incineration systems are commercially available, they have not gained the widespread acceptance that mass burning systems currently enjoy, largely due to the additional complexity and cost associated with waste preparation. They do, however, allow the reclamation of salvageable materials from the waste stream (OME-88/12, p. 4-61).

2.15 Rollins Rotary Reactor

PROCESS DESCRIPTION: The Rollins rotary reactor (RRR) is essentially a cross between a rotary kiln and a fluidized bed. The primary reactor contains a bed of inert media which is effectively fluidized by the rotating action. Shredded solids, liquid, and sludge wastes are injected directly into the inert material and burned. The off-gas flows to an afterburner which ensures complete destruction of the hazardous organics. The RRR provides the capability for in-situ acid gas control by using an alkali agent in the media bed (EPA-94/09, p. 5-33).

The RRR incineration system consists of four major subsystems: waste preparation, primary reactor, afterburner, and air pollution control systems. The RRR system is shown in Fig. 6.18 (EPA-94/09, p. 5-38).

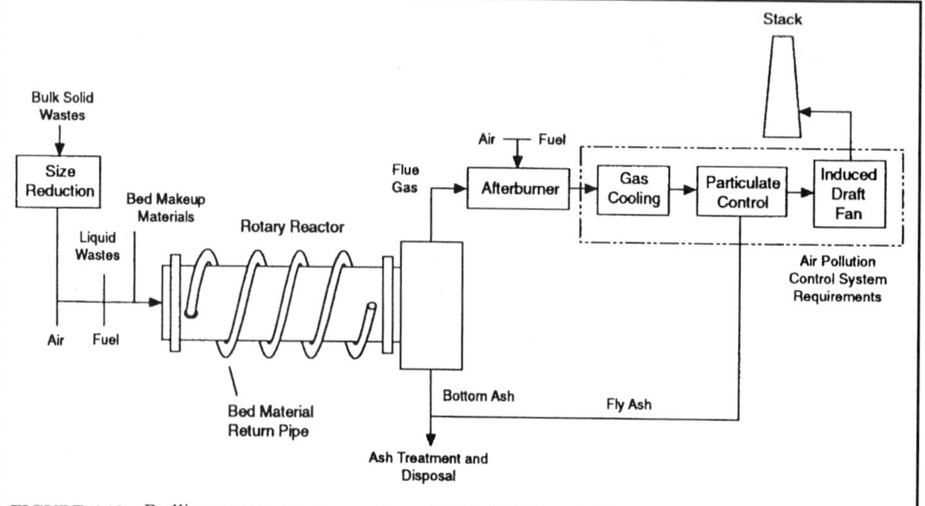

FIGURE 6.18 Rollins rotary reactor system (*EPA-94/09, p. 5-38*).

The RRR is a rotating horizontal cylindrical combustion chamber which rotates at a rate of five revolutions per minute—approximately ten times faster than a conventional rotary kiln. Externally, the cylinder looks like a rotary kiln with the exception of the helixes attached to the outer shell. Internally, the RRR incinerator is refractory-lined, and equipped with a series of equally spaced lifter assemblies around the circumference. Heat transfer media (such as sand) reside in the combustion chamber and are thoroughly mixed by the mechanical energy imparted by the combination of high speed rotation and the internal mixing devices.

Waste materials are charged through the front wall directly into the noncombustible media, which rapidly heats the waste to combustion conditions. To further enhance mixing, the inert media are recirculated from the exit to the inlet of the combustion chamber. This is accomplished through the use of external helical coils wrapped around the RRR shell which convey media as the unit rotates. Axial movement of the inert solids is driven by gravity due to a slight downward slope and the rapid rotation of the RRR shell. The external helixes are conduits for recirculating the inert bed material from the discharge end of the combustion chamber to the inlet end.

The inert bed is effectively fluidized by the rotation of the incinerator shell. In contrast to stationary fluidized beds, where the combustion air must be supplemented with additional air or nitrogen to effectively fluidize the bed, the quantity of combustion air required by the RRR is set according to the stoichiometric requirements of the waste material being burned without regard to maintaining a minimum gas velocity. This feature allows optimal balancing of combustion air/fuel ratios in the RRR, providing maximum thermal efficiencies. Moreover, the fluidization activity within the RRR is less subject to upsets and channeling caused by improperly sized or sticky feed materials. Conventional pneumatic fluidized beds must be operated with carefully graded and conditioned feed materials to prevent short circuiting and channeling in the bed, which may produce poor destruction efficiencies and operating economies.

The primary combustor operates under excess air conditions, at slightly subatmospheric pressures, and within a temperature range of 1200–1600°F. An alkali reagent is added to the bed material at approximately 1.5 times the stoichiometrically required amount. In the event that the heating value of the waste is not sufficient to allow combustion at these low temperatures, natural gas or fuel oil is fired in a conventional burner, while wood chips are mixed with contaminated soils and fed through the shredder/screw conveyor system. Temperature is monitored at several locations in the primary reactor and a feedback control loop modulates solid, liquid, or auxiliary fuel flows to maintain the temperature at the set point. Excess oxygen is also monitored at the primary chamber exit. This measurement is used to control air flow. If excess O_2 drops below 2 percent, all feeds to the primary chamber are stopped.

At the exit end of the primary combustor, there is a separation zone in which the velocity is reduced so that the larger particles drop to the bottom of the chamber and the flue gas exits through the top. The ash particles are drawn from the bottom in a screw conveyor for final disposal. The flue gas travels from the top of the separation chamber to the afterburner.

A refractory-lined afterburner chamber is sized to provide a 2-s residence time at a minimum temperature of 2000°F. The temperature and oxygen are measured at the secondary chamber exit. Auxiliary fuel is fired to ensure that the temperature is maintained. If oxygen drops below 3 percent, feed to the primary reactor is shut off.

Technology application: Rollins Environmental tested the RRR in a pilot program using a prototype reactor in the mid-1980s. Since that time, a commercial unit was built and has been operating at their plant in Deer Park, Texas, for more than 4 years. This particular unit is operated in conjunction with a 4.4-m rotary kiln incinerator, both use a common afterburner and wet scrubber system. The reason for this is that the rotary kiln and scrubber were already in place at the plant when the RRR was installed. Because the RRR is capable of in-situ capture of HCl in a stand-alone system, a wet scrubber with its associated waste stream would not be required.

2.16 Rotary Kiln System

PROCESS DESCRIPTION: Rotary kiln systems used for waste incineration typically consist of two incineration chambers: the kiln and an afterbumer. The waste is fed into the rotary kiln chamber and burned. The ash is removed at the end of the chamber. The combustion gases travel to the afterburner and are maintained at a specified temperature for a specified residence time to complete destruction of any remaining organic. A typical rotary kiln incinerator system is shown in Fig. 6.19 (EPA-94/09, p. 5-27).

The rotary kiln itself is a inclined horizontal steel cylinder with an outside diameter typically less than 15 ft, so the kiln can be shipped by rail or truck. On the inside, the kiln is lined with approximately 2–4 in. of insulating refractory covered with 6–10 in. of temperature- and erosion-resistant refractory. The kiln length is generally 2–5 times the diameter dimension. The shell is supported by two or more steel "trundles" that ride on rollers allowing the kiln to rotate about its axis. The inside surface is usually smooth, although recent designs have included internal vanes or paddles to promote mixing of the waste with the combustion air. Sometimes, internal dams are utilized to retard solid waste movement through the kiln.

The kiln is oriented on a slight incline from the horizontal, known as the rake. The rake is normally 2–4 degrees, and the waste is charged on the high side of the kiln. The kiln rotates at rates ranging from 0.5–2 rpm. The incline facilitates ash and slag removal. The rotation of the shell provides transportation of the waste through the kiln and enhances mixing of the waste with combustion air. The rotational speed is used to control waste residence time and mixing. Although increasing the rotation rate can enhance the mixing of waste and air mixing, this also acts to move the solids through the kiln more rapidly, thus reducing their residence time.

Two types of rotary kilns are currently being manufactured: cocurrent and counter-current. In cocurrent rotary kilns, the burner is located at the front end where the waste is fed; in counter-current rotary kilns, the burner is located at the end opposite the feed.

In the kiln, wastes are heated by the primary flame, bulk gases, and refractory walls. Through a series of volatilization and partial combustion reactions, combustible fractions of the wastes are gasified. The solids continue to heat and burn as they travel down the kiln. The solids retention time in the kiln is 0.5–1.5 h, while gas residence time through the kiln is roughly 2 s. Waste feed to the kiln is controlled so that the waste occupies no more than 20 percent of the kiln volume. The typical bulk temperature in the kiln ranges from 1500 to 1900°F.

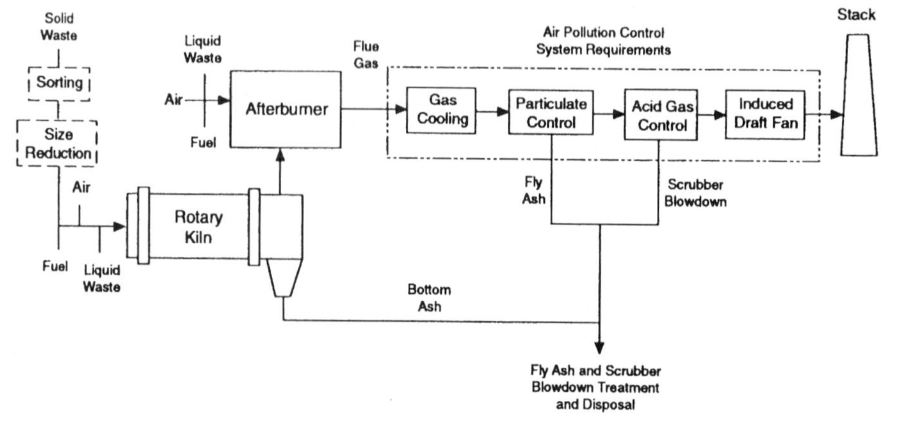

FIGURE 6.19 Rotary kiln system (*EPA-94/09, p. 5-27*).

An auxiliary burner is located at the feeding end of the kiln, and is used for start-up and to maintain the desired kiln temperature when sufficient heat input is not available from the waste. Wastes with an average heating value of 4500 Btu/lb are adequate to sustain combustion at a kiln temperature between 1600 and 1800°F. Combustion air is provided through ports on the face of the kiln, as well as through rotary seal leaks. The kiln typically operates at 50–300 percent excess air. Kiln operating pressure is maintained slightly below atmospheric (0.5–2 in water) through the use of an induced-draft fan located downstream of the air pollution control system. Operation at negative pressure minimizes fugitive emissions of partially combusted waste gases to the atmosphere through gaps in the rotary seals and any small cracks or holes that may be present. To avoid over-pressurization caused by feeding too much highly combustible waste, an emergency relief vent is typically provided between the afterburner and the air pollution control system to protect the equipment. Use of this valve constitutes a serious release of hazardous and, potentially, radioactive compounds to the environment.

Inorganics, ash, slag, and other noncombustible items that remain when the waste reaches the end of the kiln are removed by gravity into an ash pit. A typical rotary kiln is equipped with a water-filled sump with a drag chain to remove the ash from the sump and discharge it into a container. At the exit of the kiln, there is a separation chamber which serves to reduce the flue gas velocity and separate the gases from the particulate matter.

Rotary kilns which burn waste generally employ an afterburner downstream of the kiln to ensure complete destruction of organic compounds. An afterburner is simply a refractory-lined chamber sized to provide a residence time of 4 s at the maximum firing rate. An auxiliary burner on the front face of the afterburner heats the flue gas to the set-point temperature. Liquid wastes may also be fired directly into the afterburner via a lance or atomizer. A typical afterburner combustion temperature ranges from 1800 to 2200°F. The afterburner supplies an oxidizing environment to ensure complete destruction of hazardous organics.

TECHNOLOGY APPLICATION: The rotary kiln system is highly versatile as it is capable of burning solids, sludges, liquids, and gases. Solids are typically ram-fed or dropped by conveyor onto the kiln hearth, while sludges, liquids, and gases are injected through nozzles located in the front or rear face of the kiln or in the afterburner. They are particularly attractive for the destruction of toxic wastes due to their ability to operate at temperatures in excess of 1400°C (2550°F) when fitted with an afterburner.

TECHNOLOGY ADVANTAGE: The technological advantages include those listed below.

- The greatest advantage of a rotary kiln incineration system is its ability to retain and tumble the wastes to provide high turbulence and air exposure to solid wastes, thus achieving better combustion efficiency. This ability is especially important when high ash waste is involved.
- The rotary kiln incinerator will incinerate a wide variety of liquid and solid wastes. It is capable of incinerating materials passing through a melt phase. Liquids and solids can be received independently or in combination.
- The rotary kiln incinerator is adaptable to a wide variety of feed mechanism designs. Drums and bulk containers can even be accepted in the feed.
- It incorporates continuous ash removal which does not interfere with the waste oxidation.
- There are no moving parts inside the kiln (except when chains are added to facilitate heat transfer or to enhance mixing).
- The retention time of the waste can be controlled by adjusting the rotational speed.

TECHNOLOGY DISADVANTAGE: The technological disadvantages includes those listed below.

- Capital cost for installation is high. The systems are mechanically intensive. They have precise drive mechanisms and large rotating seals. The seals at either end of the kiln cylinder are susceptible to leakage at the interface between the rotating cylinder and the stationary portions of the system. Seal design is challenging, having to accommodate linear and radial expansion and the imperfect roundness of the heavy, large diameter reactor tube.

- The rotation action of a rotary kiln tends to cause a high degree of solids turbulence resulting in significant particle loadings in the exhaust gas, which must be considered in the design of the off-gas cleaning system.
- Operating care is necessary to prevent refractory damage; thermal shock is a particularly damaging event.
- Spherical or cylindrical items may roll through the kiln before complete combustion.
- Under certain conditions (e.g. temperature, rotation speed, waste feed rate, and composition), molten solids can form and accumulate on the walls of the kiln, forming layers or rings which can restrict the flow of wastes or interfere with the overall operation of the unit.
- Airborne particles may be carried out of the kiln before combustion is complete.
- Problems in maintaining seals at either end of the kiln can result in operating difficulties. Also, the induced-draft fan and air pollution control equipment must be oversized to handle extra flue-gas flow resulting from infiltration of gas through leaking seals.

2.17 Shaft-type Incinerator

PROCESS DESCRIPTION: The main combustion chamber is a vertically oriented, cylindrical furnace (Fig. 6.20). Waste is fed into the lower third of the furnace tangentially with the combustion air, and burns as it flows up. The flue gas exits the top of the furnace and flows through the air pollution control system, which includes an afterburner. The main

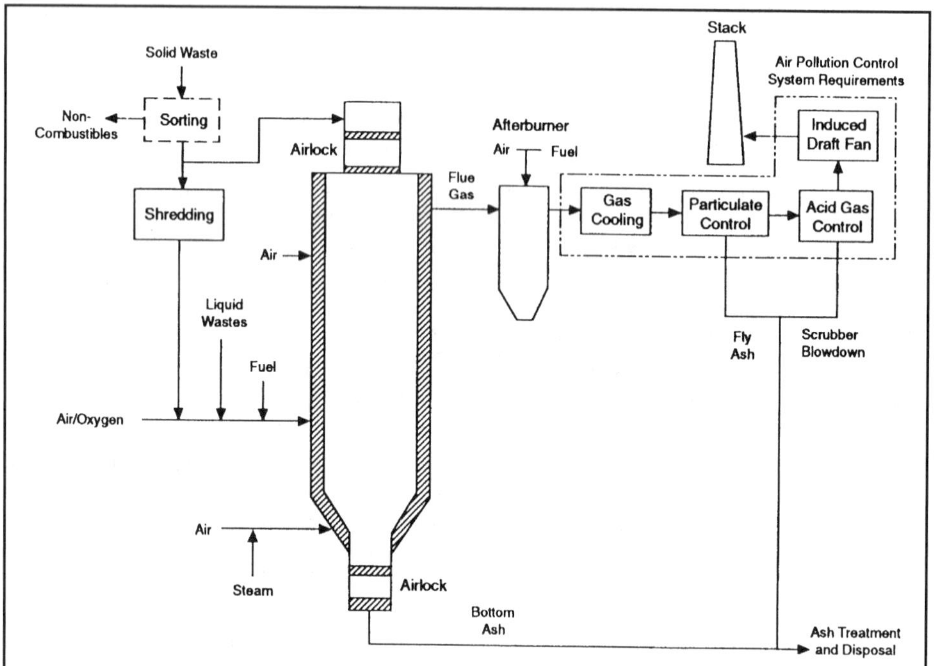

FIGURE 6.20 Shaft-type incinerator (*EPA-94/09, p. 5-48*).

subsystems of this thermal treatment system are the waste pretreatment, shaft furnace, and off-gas clean-up systems. Each is described below (EPA-94/09).

Combustible liquids, sludge and solids are suitable for treatment in the excess air shaft-type incinerator. The solids are sorted to remove large noncombustibles. Size reduction of solid waste is not strictly required, but, as is true of most incinerators, performance is enhanced with shredding. Shredded waste is pneumatically fed through nozzles into the lower portion of the furnace. If solids are not shredded, they must be batch-fed through an airlock at the top of the reactor. Liquid and sludge wastes are simply injected through nozzles near the bottom of the furnace.

The shaft furnace is a refractory-lined cylinder with a typical height to inside diameter ratio of approximately six. Shredded solid and liquid wastes are introduced near the bottom of the furnace. The primary combustion air is injected tangentially into the reactor at the same elevation as the shredded solid and liquid wastes. Two lesser combustion air flows are also added just above the bottom air lock and just below the top of the furnace. Sufficient combustion air is provided to maintain a set excess oxygen level at the primary reactor exit. The shaft furnace is maintained at a moderate temperature of about 850°C (1562°F) through use of supplemental fuel and solid waste feed rate control.

An induced-draft fan located just upstream of the stack maintains negative pressure in the furnace and allows the shredded waste to burn in suspension as it flows to the top of the furnace. The bottom of the furnace tapers into a small cylindrical volume which serves as a receptacle for the ash and noncombustibles. Steam and air are added at the bottom cone section to prevent the ash from slagging and from adhering to the refractory walls; the temperature of the ash bed is maintained below 800°C. The residual materials that accumulate at the bottom of the reactor are periodically removed in batches through a double enclosure airlock to prevent the release of contamination.

TECHNOLOGY APPLICATION: The excess air shaft-type incinerator was originally designed at the Karlsruhe Nuclear Research Center in Germany, and is sometimes referred to as the "Karlsruhe" incinerator. The Karlsruhe system is similar to a starved-air incinerator except that the solids burn in suspension rather than on a grate.

TECHNOLOGY ADVANTAGE: The technological advantage is its simplicity. It has been reported that the original Karlsruhe system has treated thousands of tons of low-level mixed waste with only minor interruptions for maintenance of the combustion chamber (DOE-93/03, p. C-6).

TECHNOLOGY DISADVANTAGE: The technological disadvantage is the need to sort most metals and glass from the waste feed, and the need to shred the waste in advance.

2.18 Supercritical Fluid (SCF)

PROCESS DESCRIPTION: SCF is characterized as a form of matter in which the liquid and gaseous states are indistinguishable from one another. It is formed when both temperatures and pressures to which the fluid is subjected exceed the critical point (T_c and P_c), as shown in Fig. 6.21. Under these supercritical states the character of the fluid becomes very unusual compared with that under subcritical conditions. For example, if water is under supercritical conditions (pressures greater than 218 atmospheres combined with temperatures above 374°C), the density, dielectric constant, hydrogen bonding and certain other physical properties are so altered that water behaves much as a moderately polar organic liquid. Thus, *n*-heptane or benzene could become miscible in all proportions with SCW (supercritical water), which cannot happen with water under subcritical conditions. Even some types of wood fully dissolve in SCW. On the other hand, the solubility of sodium chloride (NaCl) could be as low as 100 ppm, and that of calcium chloride can be less than 10 ppm. This is the reverse of the solubilities in water that are encountered under subcritical conditions—under which the two salt solubilities are about 37 wt% and up to 70 wt%, respectively. Thus, organics become

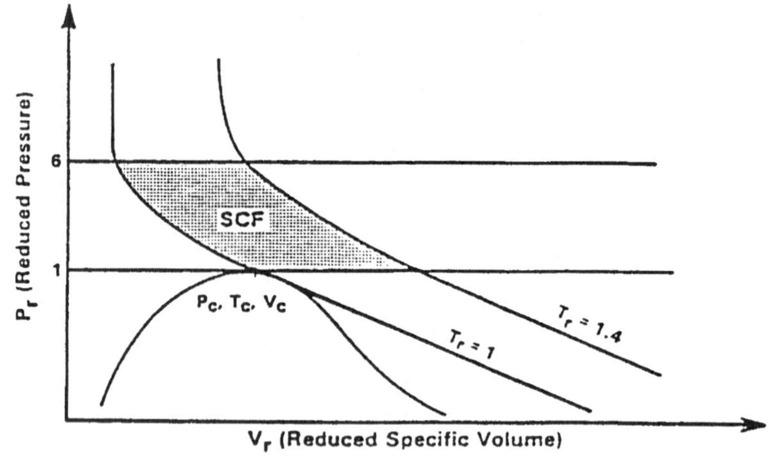

FIGURE 6.21 SCF operating zone (*Lee-83/07*).

almost completely soluble and inorganic salts become only sparingly soluble and tend to precipitate as shown in Fig. 6.22 (EPA-86/10).

The SCF region can best be visualized with the aid of a reduced pressure and volume diagram, as in Fig. 6.21. The reduced pressure, temperature, and volume are defined as

$$P_r = P/P_c, \qquad T_r = T/T_c, \qquad V_r = V/V_c$$

where the subscripts r and c denote a reduced property and the property at the critical point, respectively.

In a study of the use of supercritical CO_2 for the regeneration of activated carbon, Arthur D. Little, Inc., suggested that the SCF region be in the range of 1–1.4 of reduced temperatures and 1–6 of reduced pressures, as shown in Fig. 6.21 (for CO_2, $P_c = 1071$ psia and $T_c = 548°R$).

The steps involved in the SCF oxidation process are as follows. Initially, the waste (in the form of an aqueous solution or slurry) is pressurized and heated to supercritical conditions by mixing it with recycled reactor effluent. Compressed air is also mixed with the feed to serve as a source of oxygen for the reactions. Oxygen and air are miscible with water under supercritical conditions, thereby enabling the homogeneous operation of the process. The homogenized mixture is then pumped to the oxidizer where organics are rapidly (residence times average 1 min) oxidized. Oxidation is achieved under homogeneous conditions (single-phase supercritical fluid), and therefore higher effective oxygen concentrations and destruction efficiencies can be achieved with shorter residence times than with other similar processes (i.e., the wet oxidation process) (EPA-86/10).

The release of combustion heat from the oxidation reactions causes temperatures in the oxidizer reactor to rise to 1112–1202°F. The reactor effluent then enters a cyclone (solids separator) where inorganic salts are precipitated out (at temperatures above 450°C). The fluid effluent of the solid separator consists of superheated supercritical water, nitrogen, and carbon dioxide. A portion of the superheated supercritical water is directed to an eductor so that it can be recycled to heat the incoming waste feed (initial step in the process).

The supercritical oxidation process results in conversion of carbon and hydrogen compounds from the organic compound to CO_2 and H_2O. Chlorine atoms are converted to chloride ions and can be precipitated as sodium chloride with the addition of basic materials to the feed. Gaseous emissions consist primarily of carbon dioxide with smaller amounts of oxygen and nitrogen gas, which do not require auxiliary treatment for off gases. Solid

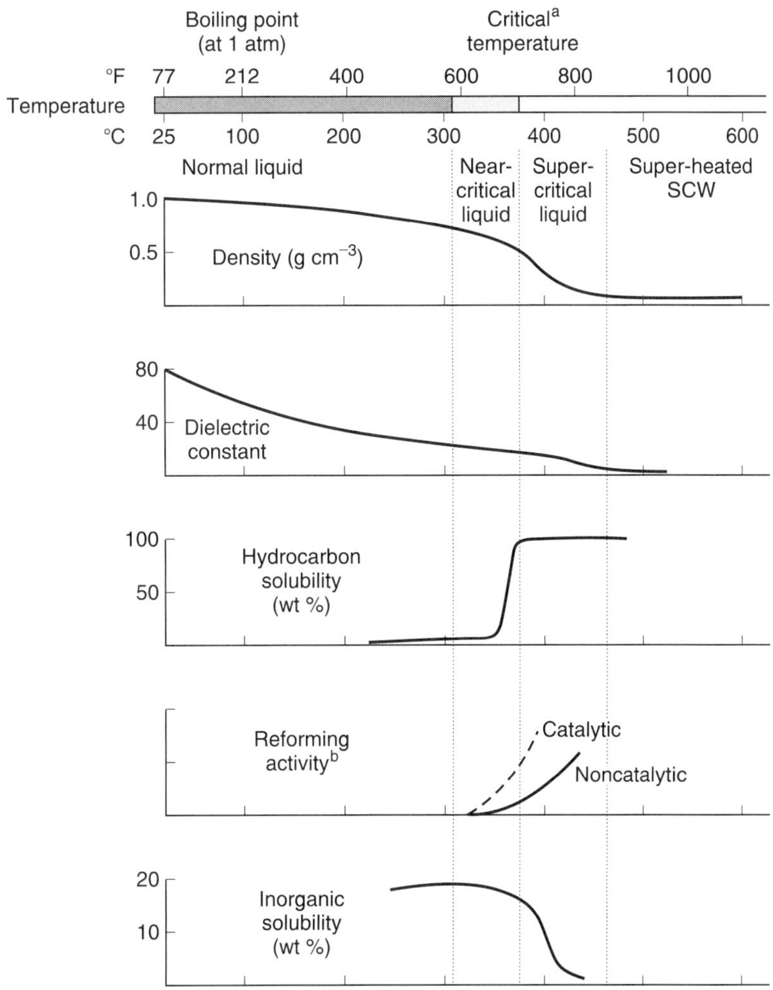

a For pressure range 218.3-300 atm
b The ability to reform (break down) organic molecules
Source: Michael Modell, Modar Inc.

FIGURE 6.22 Water property at supercritical condition.

emissions consist of precipitated inorganic salts (chlorine produces chloride salts, nitro-compounds precipitate as nitrates, sulfur compounds as sulfates, and phosphorus compounds as phosphates). The liquid effluent consists of a purified water stream, which can be used for process water.

Typical operating parameters for a SCF system are:

Waste form	Aqueous solution or slurry of organics
Temperature	450–650°C
Pressure	220–250 atmospheres
Average residence time	Less than 1 min

TECHNOLOGY APPLICATION: At the superitical condition (for water, $P_c = 3206$ psia and $T_c = 1165°R$), water acts as a solvent. Therefore it can oxidize many hazardous compounds without extra energy. Similar to the wet air oxidation process, one technological advantage is that since oxidation takes place in the liquid state, it is not necessary to evaporate the water content of the waste. The process can treat the wastes which are too dilute to incinerate economically yet too toxic to treat biologically. This is another advantage of the technology. A technological disadvantage is that at elevated temperatures and pressure, SCF may require the use of high-alloy materials such as titanium to prevent corrosion. This use of high alloys increases capital costs.

2.19 Thermal Desorption Process

PROCESS DESCRIPTION: Volatile and semivolatile organics are removed from the soil by thermal treatment. Indirect heating of the soil in a rotating chamber volatilizes organic contaminants along with any moisture present in the soil. The soil passes through the chamber and is collected as a dry solid. Condensing of the volatilized organics and water generates separate liquid phases. The organic phase is decanted and removed for disposal. The contaminated aqueous phase is treated by passing it through activated carbon and removing soluble organics before combining it with the thermally treated soil (EPA-96/12, p. 288).

Inorganic contaminants are removed by three physical and chemical separation techniques: (1) gravity separation of high density particles, (2) chemical precipitation of soluble metals, and (3) chelant extraction of chemically bound metals. Gravity separation will be used to separate higher density particles from common soil. Radionuclide contaminants are typically found in this fraction. Selection of the gravity separation device—shaker table, jig, cone or spiral—is based on the distribution of contaminants and the physical properties of the thermally treated soil.

Many of the pollutants (such as radionuclide and other heavy metals) are in a form that makes them partially soluble or suspended in the aqueous media used for separation. These contaminants are separated from the soils and are precipitated. A potassium ferrate formulation is used to precipitate radionuclides. The resulting microcrystalline precipitant is removed, allowing recycling of the aqueous stream, during the process cycle.

Some of the radionuclides that are not in soluble form remain with the soil through the gravity separation process. These radionuclides are removed from the soil via extraction with a chelant. The chelant solution then passes through an ion exchange resin to remove the radionuclides. The chelant solution is recycled back to the soil extraction step.

The contaminants are collected as concentrates from all waste process streams for recovery or off-site disposal at commercial waste or radiological waste facilities. The decontaminated soil is then returned to the site as clean fill.

TECHNOLOGY APPLICATION: It was reported that thermal desorption is capable of separating polychlorinated biphenyls (PCB) from soil contaminated with uranium and technetium (EPA-96/12).

2.20 Vitrification by Joule Heating

PROCESS DESCRIPTION: In a vitrification process, high temperatures are used to convert waste materials into a dense glass which is ready for final disposal. The high temperatures may be achieved through the use of electricity, plasma, microwave, or fossil energy. When the energy source is electricity, the process is referred to as joule heating. A joule-heated vitrification furnace (also known as "melter") system is shown schematically in Fig. 6.23 (EPA-94/09).

Solid and liquid wastes are fed into a bath of molten glass. Oxidation air is bubbled up through the bottom of the melter. Glass-making additives are also added as required to maintain operations and provide for suitable glass formation. Electrodes immersed in the

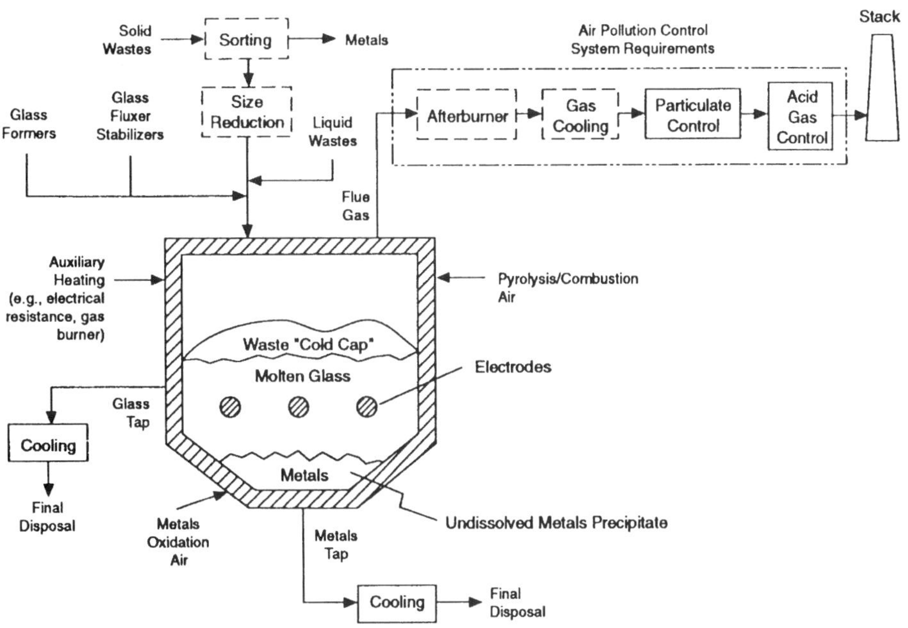

FIGURE 6.23 Joule-heated vitrification (*EPA-94/09, p. 5-68*).

glass bath provide the melt energy through joule heating. The waste melts and oxidizes when exposed to the air in the high-temperature environment. Waste organics are destroyed, while many of the inorganics are incorporated into the glass melt. The main subsystems of the joule-heated vitrification system are the waste preparation and feeding, melter, glass removal, flue gas clean-up, and automatic control systems. Each is discussed below.

Vitrification furnaces come in a variety of designs, depending on the waste type and the desired operating temperature. Typically, they are rectangular-shaped vessels, lined with high temperature refractory (such as chromium/alumina oxide). Some designs are long tunnels with relatively shallow flat bottoms, while others may have deep or sloped bottoms.

Solid, liquid, slurry, and sludge-type wastes may be fed either semicontinuously with a ram feeder and/or continuously with a screw feeder directly onto the surface of the glass melt. Alternatively, wastes may be introduced below the surface of the glass melt with the use of a drop-tube lance. In some operations a "cold-cap" or "batch-blanket" is maintained over the glass surface. In cold-cap operations, waste is fed at a rate such that a blanket layer over the entire surface of the glass is formed. The waste layer acts as a filter to capture inorganic volatiles released during waste heating and melting and allows for possible reincorporation back into the melt.

Wastes are ignited by the radiant heat of the glass pool and continue to partially oxidize and melt within the pool. Combustion air/oxygen is provided above the glass pool or into the cold cap layer to ensure complete combustion of the partially oxidized flue gas exiting the melt. In some designs, excess oxygen is maintained above the bath. In other designs, reducing conditions are maintained above the bath and an afterburner is utilized to complete combustion of the organics. Auxiliary fuel-fired burners or electric space heaters may also be used to ensure adequate destruction of volatiles.

The melt temperature is typically maintained between 1040°C (1900°F) and 1200°C (2220°F). Temperature is maintained by immersing metal electrodes in the melt, either vertically from the top or horizontally from the sides of the furnace. The electrodes make use of the joule effect, in which alternating electrical current passing between the electrodes and the glass is dissipated in the form of thermal heat due to the glass resistivity. Because the glass conductivity is ionic, alternating current must be used to avoid the risk of electrolysis or anodization of electrodes. Glass has decreased resistivity in its high temperature molten state, making it amenable to conducting and dissipating electricity. Natural convection currents created around and in the region between the electrodes act to stir and mix the molten glass, providing relative homogeneity of the glass melt. Because the burning waste is also a heat source, the electrode voltage is adjusted as necessary to control melt temperature. At start-up, a glass charge is initially melted with auxiliary heating, either by a fuel-fired burner or electrical resistance heating. When the bath reaches a temperature of about 730°C (1350°F), the auxiliary heating is stopped, and heat is provided by the immersed electrodes and the waste.

The electrodes must be designed to withstand corrosion from the glass bath, maintain strength at high temperature, and have low resistivity. Inconel™ is commonly selected for electrodes when the design melt temperature is below 1150°C; at higher temperatures, molybdenum electrodes are commonly used. In some designs, the electrodes are internally air-cooled.

In a unique furnace design, a rotating impeller paddle–mixer is submerged in the glass pool and used to stir the glass bath. The stirrer draws unmelted feed into the molten glass, acting to enhance the mixing between the waste and molten glass. This allows for a substantial increase in the melter waste throughput/volume ratio (about eight times higher than that of a conventional melter), while also providing the ability to operate at a lower average melt temperature, The stir-melting furnaces come in two designs depending on the desired operating temperature. In the low temperature design [1050–1150°C (1900–2100°F)], the paddle and furnace are constructed of steel alloy and act as the electrodes to provide the joule heating. In the high temperature design [1350–1500°C (2460–2730°F)], the impeller and electrodes are molybdenum, and the furnace is refractory-lined.

In another unique design used to treat ion-exchange resins, the furnace operates at a high temperature without refractory lining. This is desirable since liquid metals can corrode refractories under reducing conditions. Initially, a thick solidified "skull" of sintered sand is formed on the inside of the water-cooled, steel-lined furnace. The skull crust layer acts as a protective barrier between the molten glass and the steel furnace lining. The furnace can operate at 980–1500°C, and as high as 1650°C for short periods of time.

During operation, the glass is sampled and checked for composition. Depending on the waste composition, additives are fed with the waste to maintain the proper glass viscosity, melt temperature, electrical properties, and residual slag properties. The glass viscosity must be low enough to achieve sufficient convective current mixing in the furnace, allow the required throughput to be met, and facilitate glass pouring operations. Alternatively, the viscosity must be high enough to avoid significant refractory corrosion. The bath temperature must be maintained above the melting point to avoid formation of crystals and deposits in the melter. Electrical resistance must be high enough to promote generation of heat, but low enough to avoid prohibitively high voltage requirements to dissipate current. Finally, the leachability, durability, and stability of the final glass product is ensured by adding the proper compounds to the melt.

Additives may be classified into two main groups based on their impact on the glass physical properties: "network formers" and "network modifiers." Large amounts of network formers such as silicates, phosphates, and borates are generally required since they are the backbone of the glass structure. Network formers provide the glass with durability, mechanical strength, and resistance to leachability, but tend to increase glass viscosity. Network modifiers include stabilizers and fluxes. Stabilizers such as limestone, dolomite, and feldspar (providing alkali earth elements such as calcium, magnesium, and aluminum) also act to improve glass durability and maintain low melting temperature, but increase melt viscosity and increase the potential for the glass to devitrify when cooled. Fluxes such as soda ash (those providing alkali metals sodium, potassium, or lithium) are added to decrease melt viscosity and melting temperature; they are incorporated into the glass but break interconnected network bonds. Their addition adversely affects glass leachability and durability and lowers glass resistivity.

For the final glass product to have acceptable durability and resistance to leachability, the glass must have the right proportion of fluxes, formers, and stabilizers. Note that a pure glass has an "amorphous", unordered structure. On the other hand, a ceramic has a crystalline structure, with fused grains, and a highly ordered structure. So-called "slags" are a mixture of both amorphous and crystalline structures. The ultimate glass formation depends on a variety of factors such as melting temperature, glass composition, and glass cooling rate.

The three most common types of glass for waste processing are borosilicate-, aluminosilicate, and sodiumsilicate-based glasses. Borosilicate is almost always chosen for treating high-level radioactive wastes. The presence of boron allows for increased solubility of waste components; additionally, the glass is stable, durable, and has a moderate melt temperature [1150°C (2100°F)]. Aluminosilicate, although the most stable type, is not as popular due to the high melt temperature [1350°C (2500°F)].

Inorganics such as metals, silica, halides, and sulfur which are contained in the waste are either encapsulated (surrounded by glass), melted and incorporated into the glass network, or contained in the flue gas as volatilized vapors or entrained solid particulates. The fate of waste metals in the glass melt depends on factors such as operating temperature, metal type and form, and glass composition. Metal objects will either soften or melt, depending upon the operating temperature and reducing/oxidizing melt conditions. For designs operating at lower temperatures, metals will sink and accumulate at the bottom of the furnace. For higher temperature designs, metals may melt and form a molten layer at the bottom of the melt. In some arrangements, multizone heating is used by positioning electrodes at various depths within the glass melt and powering the electrodes at different levels; higher temperatures are used near the bottom of the melter to accommodate for metals melting.

If the melt operates under reducing conditions, metallic pools will collect at the bottom. However, if oxygen is available, or provided in the melt, a portion of the metal oxides may be absorbed within the glass structure as glass formers or modifiers. In a unique design, an air bubbling and stirring lance is inserted into the glass melt near the floor. Injected air creates a curtain of bubbles between the electrodes, which act to mix the molten bath. The air also provides an oxidizing environment for the oxidation of metals and prevents them from forming a detrimental metallic layer on the bottom of the melter. The bubbler also disrupts the formation of secondary-phase layers on the top of the molten glass, such as sulfides. Glass additives may also enhance the ability to accommodate additional metals in the glass network.

When high level radioactive wastes are treated, the melt temperature is low to minimize the volatilization of radionuclides. To maintain the viscosity at low enough levels to allow processing, the waste loading in the glass is limited to about 30 percent by weight. However, if the waste composition is correct (contains the right amounts of glass formers and fluxes), waste loadings in the glass above 90 percent have been demonstrated, leading to very high waste volume reduction.

Chlorine and sulfur have a low solubility in glass; these constituents are contained primarily in the furnace flue gas. Fluorine has a higher retention in glass than chlorine or sulfur. Any halogens that are incorporated in glass reduce the glass durability and increase glass porosity. Alkali and sulfate salts may also accumulate and form a layer on top of the glass.

The presence of metals in the furnace, as either unmelted or separate melted phases, may lead to degradation of the joule heating due to changes in the conductivity of the metal layer and/or short-circuiting of the electrodes. One solution is to design the furnace with a deep glass bath or slanted bottom. The metals are allowed to accumulate, and periodically, the temperature of the bath is increased enough to melt the metals; the molten metal is then poured from the furnace. Electrodes must be placed a good distance above the metal layer. Other alternatives include sorting metals from the waste feed, or adding metals into the furnace at a slow enough rate to allow all the metal to be incorporated into the glass melt.

The molten glass is drawn off as required to maintain proper glass pool height. Glass may be removed from the furnace in a number of different methods. In most high-level radioactive waste furnace designs, glass is removed through a "teapot spout" or "bottom takeoff overflow weir." An air lift is used to control glass draining, and ensures consistent, uniform removal of glass from the system. Air is injected into the glass as it rises to the overflow level, acting to control the glass flowrate. In other designs, a tapping hole is located in the side of the furnace; the flow of glass can be controlled with the use of "freeze" valves, which are used to intermittently remove glass. The valve is heated (typically with induction, resistance, or electrode joule heating) to initiate the flow of glass. To stop the flow, the heating is cut off and the glass is solidified. In this arrangement, a slide valve is included for emergency shutoff situations. A bottom drain may also be included for drainage of the entire furnace and removal of accumulated metals. Additionally some furnace designs may include an overflow drain, which is used to ensure a minimum glass level and aid in the removal of unwanted layers that form on the glass surface (such as sulfide salts).

Once removed, the molten glass can be fed directly into canisters or made into small marbles or "gems" depending on storage and disposal requirements. The marble form may be desirable to allow for convenient handling, sampling, repackaging, and/or reprocessing. The furnace may also have an additional separate tap for liquid metals collected at the bottom of the furnace. Tap and drain lines must be heated to ensure proper flow, typically with resistance, induction, or electrode heating.

2.21 Vitrification by Plasma Arc

Process description: Plasma arc vitrification occurs in a plasma centrifugal furnace by a thermal treatment process where heat from a transferred arc plasma creates a molten bath that detoxifies the feed material. Organic contaminants are vaporized and react at temperatures of 2000–2500 degrees Fahrenheit (°F) to form innocuous products. Solids melt and are vitrified in the molten bath at 2800–3000°F. Metals are retained in this phase. When cooled, this phase is a nonleachable, glassy residue which meets the toxicity characteristic leachate procedure (TCLP) criteria (EPA-94/09, p. 5-78).

Contaminated soils enter the sealed furnace through the bulk feeder. The reactor well rotates during waste processing. Centrifugal force created by this rotation prevents material from falling out of the bottom and helps to evenly transfer heat and electrical energy throughout the molten phase. Periodically, a fraction of the molten slag is tapped, falling into the slag chamber to solidify.

The term plasma is used to refer to a highly ionized and therefore electrically conductive gas. The types of plasma include dc or ac arc-generated thermal plasmas. The plasma torch utilizes a flowing gas such as nitrogen to maintain an electric discharge or arc between two high voltage electrodes. The first electrode is contained within the torch, and the second electrode is typically the solid material being treated. The electrical resistance of the plasma gas leads to joule heating, which is the conversion of electrical energy into thermal energy. The result is very high plasma gas temperatures and direct heating of the waste.

The primary chamber is a rotating, refractory-lined, and externally water-cooled tub with a central orifice, or copper throat. The copper throat at the bottom of the primary chamber is used to initially strike the arc of the plasma torch. This torch is capable of using air or nitrogen as the stabilizing plasma gas. Once the arc has been struck, the torch is moved slowly up and down the sides of the primary chamber to heat it up. The operator moves the torch remotely, utilizing a video camera to see the furnace condition. Feeding of the waste material begins when the primary chamber's temperature is greater than 2000°F and the secondary chamber's temperature is greater than 1800°F.

When the loading of waste begins, the operator moves the torch and monitors the condition and location of the slag in the furnace. The arc heats the waste material by convective heat transfer from the plasma plume, and to a lesser extent joule heating of the waste material itself, until it is completely molten and at least partially oxidized. The primary chamber may operate under reducing or excess air conditions depending upon the amount of organics in the waste charged. However, the air flow is sufficient to allow oxidation of most of the organics in the primary chamber. The tub spins at approximately 40 rpm to keep the newly fed waste on the sides of the tub, preventing it from rolling down the throat. The slag along the walls of the tub is at approximately 3000°F and actually serves to heat and volatilize the fresh waste as it is fed.

When a sufficient amount of waste has been fed to the tub, the molten material begins to flow down the throat. At this point, the operator halts waste feeding and allows time for destruction of the organics in the latest batch of waste fed to the tub. The rotation of the reaction chamber is then slowed to between 5 and 10 revolutions per minute, allowing gravity to pull the treated glass through the center hole (thimble), falling through the gas/slag separation chamber to be collected in a pig mold below. The slag then pours out through the center opening into a slag collector, where it cools and solidifies. The gases in the separation chamber flow to the second combustion chamber for further treatment.

The primary chamber bulk temperature ranges from 2100°F to 2400°F, and the pressure is slightly below atmospheric (−22 mbar). Typical feedrates range from 45 to 130 kg/h. Typical torch power is 500–750 kW, and the average primary chamber gas residence time is 10–15 s. It is important to note that the centrifuge is a double–walled, water–cooled chamber. As a result, the seals between the various furnace elements can be close to room temperature, and thus they seal very effectively. Also, the system is designed with a relief system and a surge tank, so that even if an explosion occurred, causing a pressure surge of as much as one bar, no untreated material would be released assuming that the seals are truly tight.

The slag separation chamber is sealed off from the receptacles with a simple locking system to prevent untreated material from entering the slag containers. During operation, a refractory-lined collection thimble is sealed to the bottom of the slag separation system by means of a pressure-regulated hydraulic lift in the bottom of the slag lock chamber. Any untreated material will drop into the collection thimble, which can periodically be removed for emptying and recycling of the untreated debris. Prior to slag tapping, all feeding of material is stopped. The slag thimble is lowered and moved from below the slag separation chamber. A slag container is moved into place and sealed to the slag collection system prior to casting. Upon completion of the casting cycle, the slag container is lowered and transferred to its original storage position. The collection thimble is then repositioned and sealed to the slag separation chamber. After the molten slag (in a 48-gallon drum) has cooled sufficiently to be transported to the cooling chamber, the slag lock chamber is purged with air and the slag drum is moved through the slag lock door to the cooling chamber. After 48 h, it is sealed inside a 55-gallon drum and brought to the outside.

TECHNOLOGY APPLICATION: Plasma furnaces originated in the metals industry, where high temperatures and controlled environments were required to ensure metal purity. EPA demonstrated its interest in a particular plasma technology for the treatment of waste when it sponsored tests to evaluate Retech's plasma arc centrifugal treatment (PACT) process through the Superfund Innovative Technology Evaluation (SITE) program. Interest in the PACT was also demonstrated by DOE's funding of a test series to evaluate the performance on simulated radioactive and RCRA wastes. The PACT process is shown in Figs. 6.24 and 6.25.

2.22 Wet Air Oxidation (WAO)

PROCESS DESCRIPTION: A temperature–volume diagram is shown in Fig. 6.26. The figure shows the operating zone of WAO. The water-waste mixture is maintained within the saturated zone within EBCF in the figure. For the particular case shown, total pressure is maintained at 2000 psi, and the temperature varies from 175°C to 320°C.

Wet air oxidation refers to the aqueous-phase oxidation of dissolved or suspended organic substances at elevated temperatures and pressures. Water, which represents the aqueous phase, serves to catalyze the oxidation reactions so that they proceed at relatively low tem-

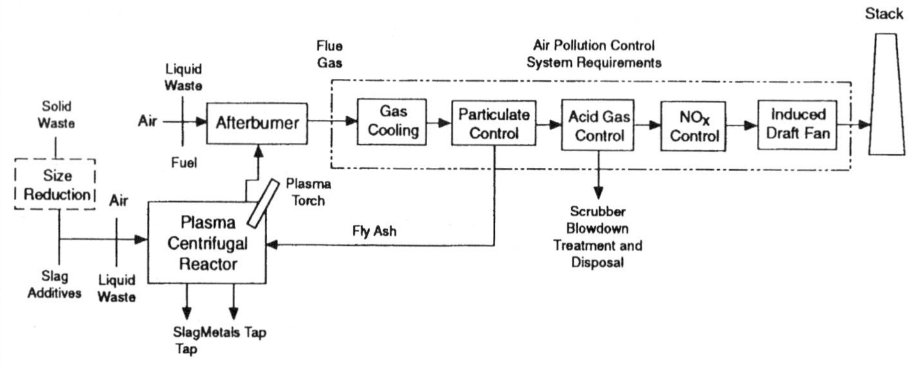

FIGURE 6.24 Plasma arc centrifugal treatment (PACT) system (*EPA-94/09, p. 5-82*).

peratures (175–345°C), and at the same time serves to moderate the oxidation rates by removing excess heat by evaporation. Water also provides an excellent heat transfer medium which enables the wet air oxidation process to be thermally self-sustaining with relatively low organic feed concentration.

The oxygen required by the wet air oxidation reactions is provided by an oxygen-containing gas, usually air, bubbled through the liquid phase in a reactor used to contain the process; thus the commonly used term "wet air oxidation." The process pressure is maintained at a level high enough to prevent excessive evaporation of the liquid phase, generally between 200 and 3000 psi.

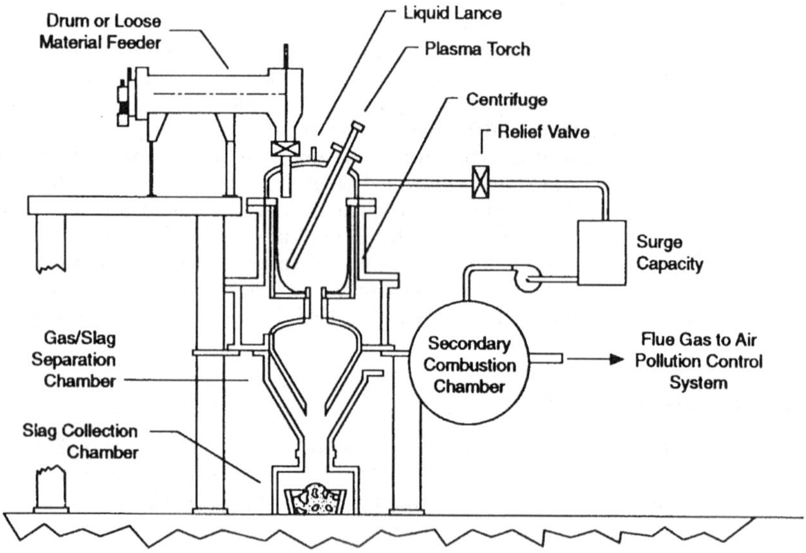

FIGURE 6.25 Plasma arc centrifugal treatment (PACT) reactor (*EPA-94/09, p. 5-83*).

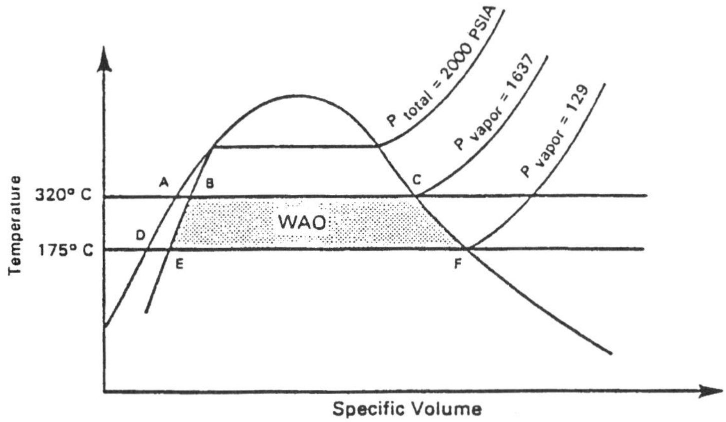

FIGURE 6.26 WAO operating zone (*Lee-83/07*).

TECHNOLOGY APPLICATION: WAO has been used worldwide to treat both industrial and municipal waste streams for the last 30–40 years. Operating designs vary from low oxidations, for sludge conditioning, to high oxidations, for chemical and power recovery and for carbon regeneration.

TECHNOLOGY ADVANTAGE: The technological advantages include:

- since oxidation takes place in the liquid state, it is not necessary to evaporate the water content of the waste. The process is therefore most useful for wastes which are too dilute to incinerate economically yet too toxic to treat biologically; and
- because wet air oxidation reactions take place in the liquid phase, the process can operate without consumption of heat for vaporizing water and organics. The sensible heat required to heat the waste to reaction temperature is recovered from the effluent by heat exchangers.

TECHNOLOGY DISADVANTAGE: The principal technological disadvantage is that at elevated temperatures and pressure, WAO may require the use of high alloy materials such as titanium to prevent corrosion. This use of high alloys increases capital costs.

3 INCINERATION TECHNOLOGY SUMMARY

Table 6.3 (OME-88/12, p. 4-95) summarizes the qualitative features of major incineration technologies.

4 RESOURCE RECOVERY SYSTEM

Combustion systems have been in use for many years burning wastes for energy and material recovery. The systems include:

- Cement kilns for resource recovery. For example, a significant fraction of the waste tires in the United States are fed to cement kilns for disposal and energy and material recovery.

TABLE 6.3 Incineration System Summary

Incinerator type	Waste Feed				Feed preparation	Agitation of waste	Hazardous waste	Medical waste	Ash discharge		Supplemental fuel			Excess air required	Waste water discharge	Particulate emission
	Solid	Liquid	Sludge	Gas					Wet	Dry	Liquid	Gas	Solid			
• Catalytic	No	No	No	Yes	None	N/A	Yes	N/A	N/A	N/A	Yes	Yes	No	Low	None	Low
• Fluidized bed	Limited	Yes	Yes	Limited	High	High	Yes	No	Yes	Yes	Yes	Limited	Yes	Low	Yes	Moderate
• Gas incineration	No	No	No	Yes	None	N/A	Yes	N/A	N/A	N/A	Yes	Yes	No	Low	None	Low
• Liquid injection	No	Yes	No	Limited	Limited	N/A	Yes	N/A	N/A	N/A	Yes	Yes	No	Low	None	Low
• Mass burn	Yes	No	Limited	No	None	Limited	No	No	Yes	Yes	Yes	Yes	Yes	Moderate	Ash pit	Moderate
• Multiple chamber	Yes	Limited	Limited	Limited	Limited	Limited	Limited	Yes	Yes	Yes	Yes	Yes	Yes	High	None	High
• Multiple hearth	Limited	Limited	Yes	Limited	Limited	High	Limited	Limited	Yes	Yes	Yes	Yes	No	High	Yes	Moderate
• Pyrolysis	Yes	Limited	Yes	Limited	High	None	Limited	No	Yes	No	Yes	Yes	Yes	None	Yes	High
• RDF	Yes	No	No	No	High	High	No	No	Yes	No	Yes	Yes	Yes	High	Ash pit	Moderate
• Rotary with afterburner	Yes	Yes	Yes	Yes	Limited	High	Yes	Yes	Yes	Yes	Yes	Yes	Yes	High	Yes	Moderate
• Starved air	Yes	Limited	Limited	Limited	Limited	Limited	Limited	Limited	Yes	Yes	Yes	Yes	No	Low	Ash pit	Moderate

Spent solvents and waste oils are used as fuel in the production of cement and lightweight aggregates.

- Industrial boilers for resource recovery. For example, it is a common practice to fire waste oils in industrial boilers to recover heat for power generation, and to fire spent pulping liquors in alkaline sulfate (kraft) and acid sulfite pulping to recover heat and chemicals in the manufacture of paper and wood cellulose products (EPA-94/05; OME-88/12).

4.1 Cement Kilns for Resource Recovery

In general, cement manufacturing processes that can be used to recover waste energy and materials include:

- cement kilns
- lime kilns
- lightweight aggregate kilns

Cement kilns. Cement is made from a carefully proportioned mixture of raw materials containing calcium (typically limestone), silica, and alumina (typically clay, shale, slate, and/or sand), and iron (typically steel mill scale or iron ore). These materials are ground to a fine powder (80% passing 200 mesh), homogenized, and heated to a very high temperature to produce a cement "clinker" product. The raw feed material, known as "meal," is heated in a kiln which is a large, inclined, rotating cylindrical steel furnace lined with refractory materials. The kilns are operated in a "counter-current" configuration; the gases and solids flow in opposite directions through the kiln, providing for more efficient heat transfer compared with "cocurrent" operations. The raw meal is fed at the upper, "cold" end; the slope (3–6°) and rotation (50–70 revolutions per hour) cause the meal to move toward the "hot" lower end. The kiln is fired at the "hot" end, usually with coal or petroleum coke as the primary fuel; natural gas or fuel oil may also be used as a supplemental fuel. As the meal moves through the kiln and is heated, it undergoes drying and pyroprocessing reactions to form the clinker. The reactions can be categorized into three major stages (EPA-94/05, p. 2-1).

DRYING AND PREHEATING ZONE: Residual water is evaporated from the raw meal feed, and clay materials begin to decompose and are dehydrated (removal of bound water) in a temperature range of 70–1100°F.

CALCINING ZONE: Material is "calcined"; that is, calcium carbonate in the limestone is dissociated, producing calcium oxide ("burnt lime") and carbon dioxide in the temperature range 1100–1650°F.

BURNING ZONE:: In the "burning" zone, also known as the "clinkering" or "sintering" zone, calcium oxide reacts with silicates, iron, and aluminum to form "clinker." The clinker is a chemically complex mixture of calcium silicates, aluminates, and aluminoferrites. A minimum meal temperature of 2700°F is necessary in the burning zone of the kiln to produce the clinker.

The clinker is removed from the kiln at the hot end. After it passes through the burning zone and by the kiln flame, it enters a short cooling area where the clinker melt begins to solidify; the cooling rate from the burning zone to the kiln exit is important since it determines the microstructure of the clinker. The clinker leaves the kiln at about 2000°F and falls into a clinker cooler. The cooler is typically a moving grate on which the clinker sits. Cooling air is blown through the clinker bed. The cooled clinker consists of grey-colored nodules of variable diameters, typically up to 2 inches. The clinker is blended with gypsum and ground in a ball mill to produce the final product, cement. Hot exhaust air produced from the clinker cooler is either directed to the kiln, where it is used as combustion air, or used to pre-dry the raw feed material in the case of dry-process kilns.

Kiln exhaust flue gases contain significant amounts of entrained particulate matter due to the turbulence in the kiln from the rotary action and from the use of finely ground feed material. The entrained particulate matter, known as "cement kiln dust" (CKD), is removed from the flue gas by some type of air pollution control device. Many plants return a portion of the CKD to the raw feed materials. However, in most cases, some CKD must be removed from the kiln system entirely to lower the build-up of alkali salts. CKD can be used in other industries as neutralizers or additives; however, the excess CKD is usually land-disposed.

In summary, the production of cement involves four steps:

- quarrying and crushing the raw materials;
- grinding and blending these materials into feed in the proper proportions;
- calcining the raw materials at extremely high temperatures to form clinker (an interim product);
- finish-grinding of the clinker, blending the clinker with gypsum, and packaging the finished product.

In the United States, some 200 process kilns are currently in operation across the country. Typical kilns range in size from 18 m (60 ft) long and 1.8 m (6 ft) in diameter to 230 m (760 ft) long and 7.6 m (25 ft) in diameter. These kilns are often larger than those used to incinerate wastes. About 2.9 metric tons (3.2 tons) of raw material (limestone, alumina, silica, and iron) and 6.1 million Btu are required to produce 1 ton of cement. About 90% of the energy is supplied by coal.

Cement can be produced in three major different types of arrangement—wet, dry, or semidry processes—as described below (EPA-94/05, p. 2-2; EPA-86/09, p. 2-12):

- wet process;
- dry process;
- semidry process.

Wet process. A schematic of the wet cement-making process is shown in Fig. 6.27. Ground raw materials are mixed with water (about 30% by weight) to form a slurried meal. The slurry is fed to the kiln through a flow-metered pump. Solids in the slurry typically occupy no more than 15–20% of the internal kiln volume. Wet kilns typically have a length-to-diameter ratio of about 30 : 1 to 40 : 1, and rotate at from 70 to 80 revolutions per hour. Kiln rotational speed is adjusted to maintain clinker quality and heat transfer. The wet cement making process is the older process, characterized in part by handling, mixing, and blending of the raw materials in the slurry form and lower emissions of kiln dust. However, because all water must be evaporated out of the slurry mixture, wet process kilns require greater energy input than other types of cement kilns; typically, from 5 to 7 MMBtu/ton of clinker product is required (where MM = million = 10^6).

Dry process. In the dry process, the moisture content is reduced to less than 1% before or during grinding, and the ground raw material is pneumatically transported to the kiln, or to a preheater if used. The dry process can be as much as twice as energy-efficient (3.4–4.5 MMBtu/ton of clinker) as the wet process because there is little water to evaporate from the feed. Kiln exhaust flue gases or hot clinker cooler air are typically used for drying the raw materials. There are three different dry process configurations (EPA-94/05, p. 2-7; EPA-86/09):

- long dry kilns
- preheater dry kiln;
- preheater/precalciner kiln.

LONG DRY KILN: A long dry process kiln schematic is shown in Fig. 6.27. It has the same flow path as the "dry process"; the raw meal is fed in a dry form. Similar to wet kilns, long dry

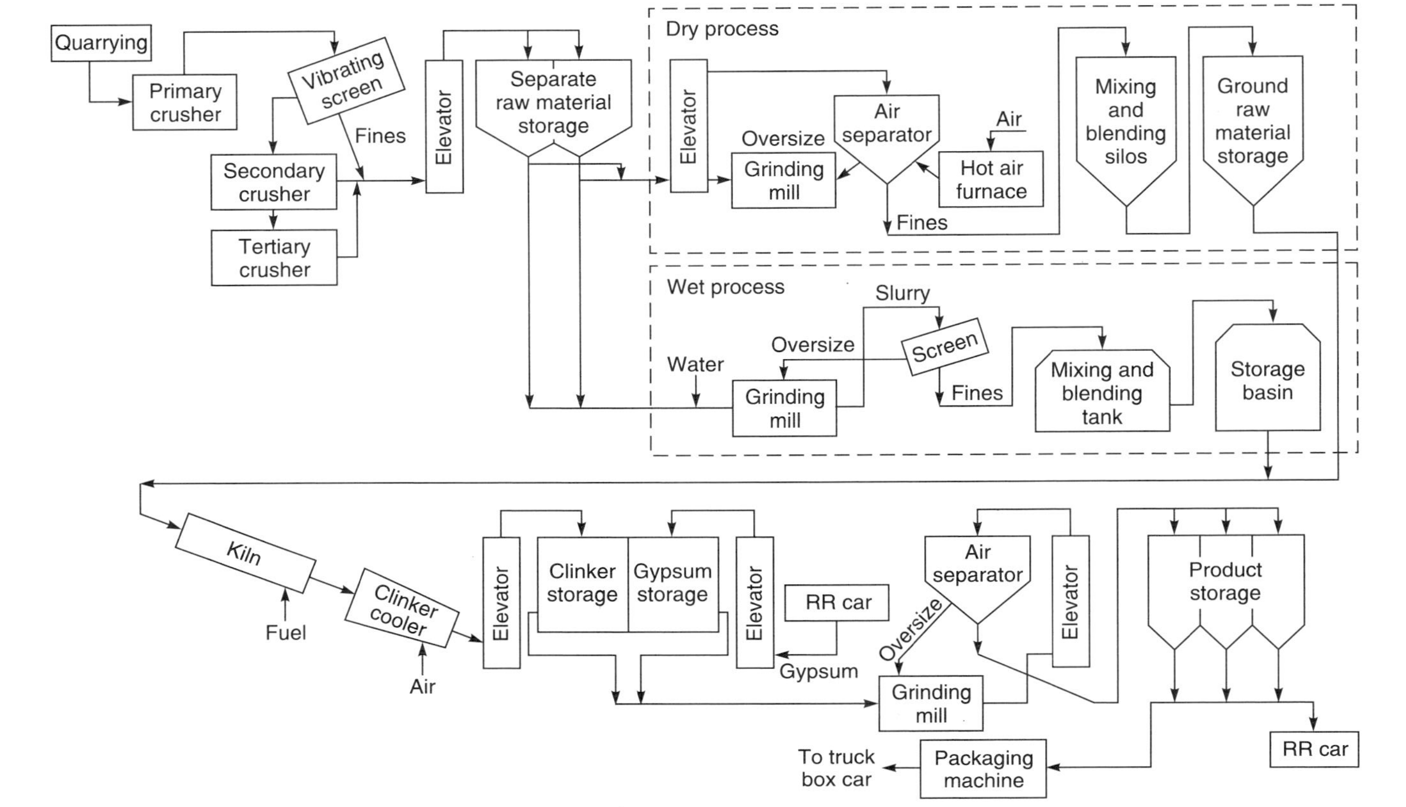

FIGURE 6.27 Long kiln (dry and wet) schematic (*EPA-86/09, p. 2-12*).

kilns have typical length-to-diameter ratios of about 30 : 1. Note that in comparison to wet kilns, the raw meal feed is heated more rapidly since the energy required for the evaporation of water is small. Kiln gas and solids residence times are similar to those of wet kilns (of order of 10 s and 2–3 h, respectively). Internal chains are also used in dry process long kilns to increase energy efficiency.

PREHEATER DRY KILN: A preheater arrangement process schematic is shown in Fig. 6.28. Preheaters are used to further increase the thermal efficiency of the cement-making process. A suspension preheater ("Humboldt" design) consists of a vertical tower containing a series of cyclone-type vessels (typically containing four stages). Raw meal is introduced at the top of the tower. Hot kiln exhaust flue gases pass counter-current through the downward moving meal to heat it prior to introduction into the kiln. The meal is separated from the kiln flue gases in the cyclone, and then dropped into the next stage. Because the meal enters the kiln at a higher temperature than that of the conventional long dry kilns, the length of the preheater kiln is shorter; kilns with preheaters typically have length-to-diameter ratios of about 15 : 1. The kiln has a gas residence time of about 6 s and a solids residence time of about 30 min. The gas residence time through the preheater cyclones (typically four stages) is about 5.5 s.

With preheater systems, it is often necessary to utilize an "alkali" bypass in which a portion of the kiln flue gases are routed away from the preheater tower at a location between the feed end of the rotary kiln and the preheater tower. The bypass is used to remove undesirable components, such as certain alkali constituents, that may accumulate in the kiln due to an internal circulation loop caused by volatilization at high temperatures in the kiln and condensation in the lower temperatures of the preheater. Accumulated alkali salts may cause preheater operating problems such as clogging of the cyclones and an increase in fine NaCl

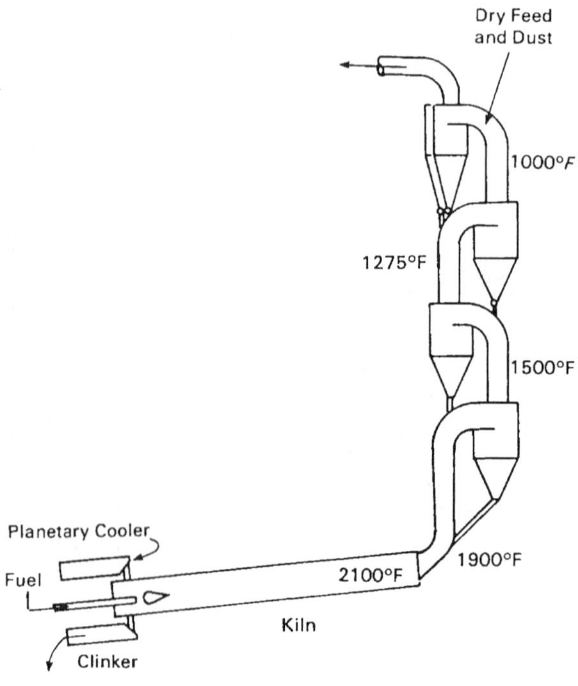

FIGURE 6.28 Four-stage preheater dry kiln (*EPA-86/09, p. 2-12*).

or KCl fumes in the emission gas. Typically 10–15% of the flue gas is routed through the bypass. Systems without bypasses are limited with respect to raw meal and waste concentrations of alkali metals, chloride (greater than 0.015% Cl by weight), and sulfur that can be tolerated in the raw materials.

The internal circulation of alkali components is greater in systems with preheaters compared to systems without preheaters due to the filtering effect of feed material flow in the preheater cyclones. Systems without preheaters have a kiln dust with a high content of alkali salts, which can be removed from the internal cycle when caught in the air pollution control device. However, for a preheater, a bypass is required to reduce the alkali build-up.

PREHEATER / PRECALCINER KILN: A preheater/precalciner process schematic shown in Fig. 6.29. A preheater/precalciner is similar to the preheater arrangement described above, with the addition of an auxiliary firing system to further increase the raw materials temperature prior to introduction into the kiln. An additional precalciner combustion vessel is added to the bottom of the preheater tower. Typical systems use 30–60% of the kiln fuel in the precalciner to release up to 95% of the CO_2 from the raw material. Precalciner air can be supplied either directly with the precalciner fuel, or at the hot end of the kiln. In another arrangement, the kiln flue gas may be routed around the calciner directly to the preheater. Kilns with preheater/precalciners can be even shorter than those with preheaters only (length-to-diameter ratio of 10 : 1).

The primary advantage of using the precalciner is that it increases the production capacity of the kiln, since only the clinker burning is performed there. The use of the precalciner also increases the kiln refractory lifetime due to reduced thermal load on the burning zone. These configurations also require a bypass system for alkali control.

Semi-dry process. In the semidry process, the ground feed material is pelletized with 12–14% water. The pellets are put on a moving "Lepol" grate on which they are dried and

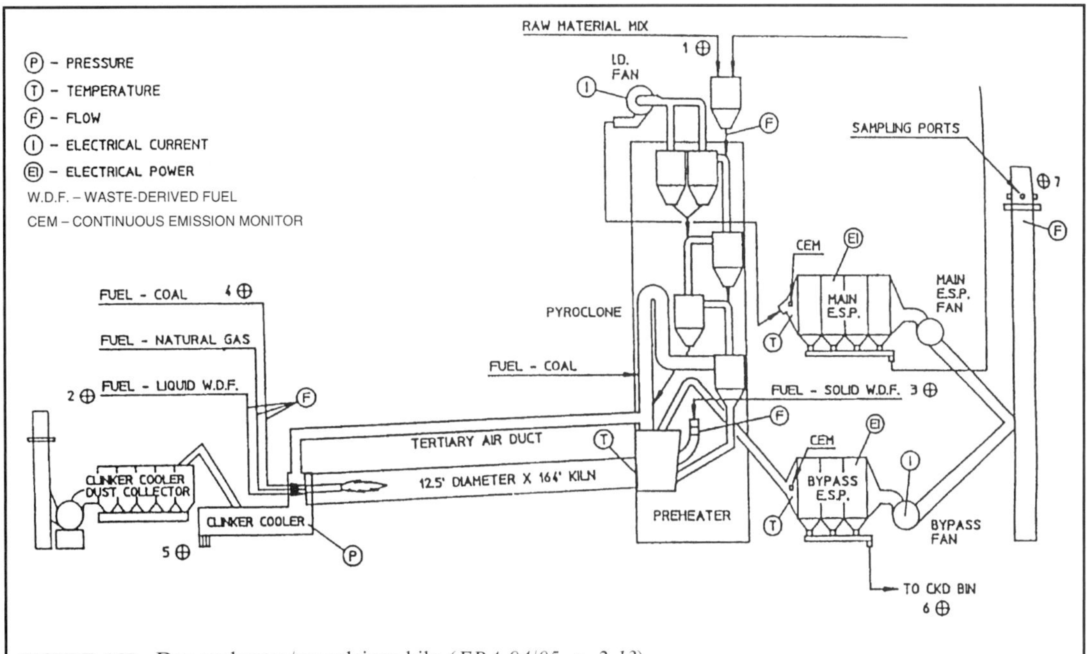

FIGURE 6.29 Dry preheater/precalciner kiln (*EPA-94/05, p. 2-13*).

partially calcined by hot kiln exhaust gases before being fed to the rotary kiln. A semidry process schematic is shown in Fig. 6.30.

Lime kiln. The United States is the second largest producer of lime in the world. In 1984, lime producers at 137 plants in 38 states sold or used 14.6 million metric tons (16.1 million tons) of lime. The term "lime" is a general term that includes the various chemical and physical forms of quicklime and hydrated lime, the two types generally produced. About 6.7 million Btu of energy is required for each 0.91 metric ton (1 ton) of quicklime produced. The cost of this high energy requirement has led to increased energy efficiency in the industry and to the use of more readily available and lower-cost fuels, especially coal or high-energy content industrial waste. Recent new plant installations and modernization projects have incorporated pulverized-coal-burning systems and energy-saving preheater systems. Figure 6.31 is a schematic of a lime kiln (EPA-86/09, p. 2-13).

The lime manufacturing process is similar to that of cement in that the raw material (usually limestone or dolomite) is quarried, crushed, and sized, and calcined in a kiln at 1093°C (2000°F). Although a variety of kiln types can be used, about 85% of the US producers use the rotary kiln. Kiln sizes vary. The largest is 152 m (500 ft) long and 5.2 m (17 ft) in diameter, and is capable of producing more than 1090 metric tons (1200 tons) of quicklime per day.

The calcining drives off nearly half the limestone's weight as carbon dioxide (CO_2) and leaves a soft, porous, highly reactive lime known as quicklime (CaO). Heating beyond this

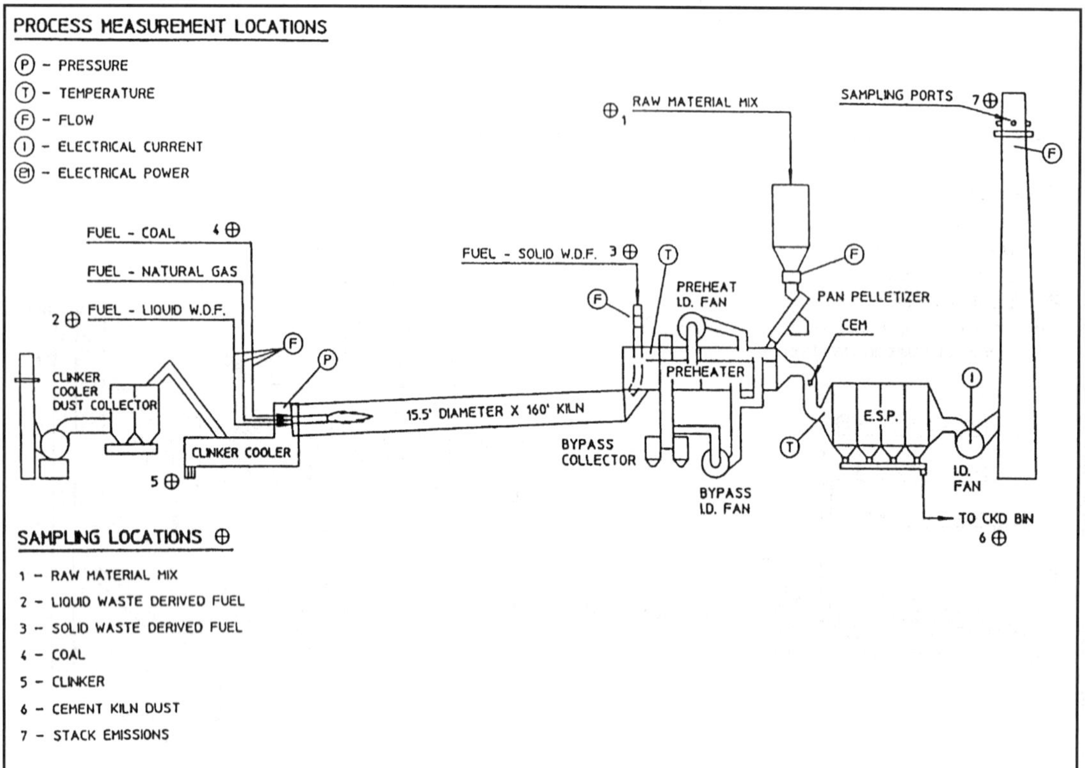

FIGURE 6.30 Semi-dry cement kiln (*EPA-94/05, p. 2-16*).

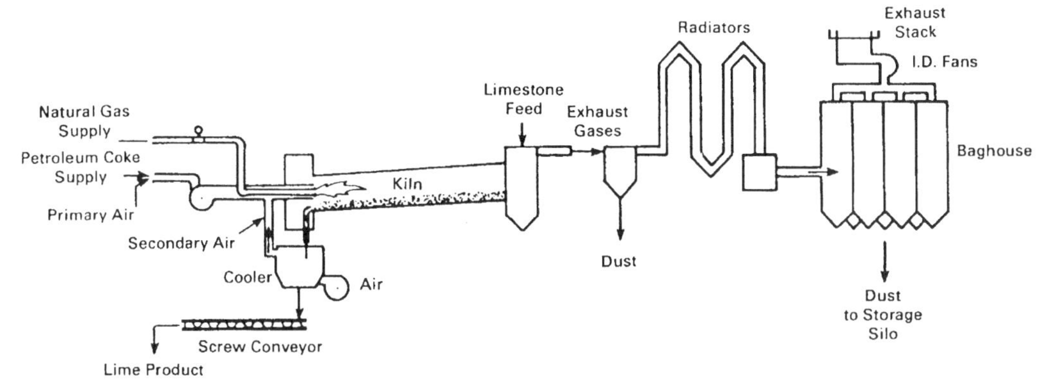

FIGURE 6.31 Lime kiln (*EPA-86/09, p. 2-13*).

stage can result in lumps of inert, semivitrified material (known as overburned or dead-burned lime) that is often used in the manufacture of refractory materials. The quicklime is discharged at the lower end of the kiln into the cooling system, where it is air-cooled and then stored in silos. A portion of the quicklime is hydrated before storage. Hydrated lime is produced by combining quicklime with sufficient water to cause the formation of a dry, white powder.

Lightweight aggregate kiln. Similar to the cement kiln and lime kiln processes, a lightweight aggregate kiln (Fig. 6.32) is an energy-intensive process to produce concrete products either for structural purposes or for thermal insulation purposes. Because of intensive energy needs, industrial waste is used to supplement the energy requirements (EPA-94/05, p. 2-19).

A lightweight aggregate plant includes a wide variety of raw materials which, combined with cement, form concrete products. The lightweight aggregate concrete is produced either for structural purposes or for thermal insulation purposes. A lightweight aggregate plant is composed of a quarry, a raw material preparation area, a kiln, a cooler, and a product storage area. From there, the material is fed into a rotary kiln.

A rotary kiln consists of a long steel cylinder, lined internally with refractory bricks, which is capable of rotating about its axis and is inclined at an angle of about 5 degrees to the horizontal. The length of the kiln depends in part upon the composition of the raw material to

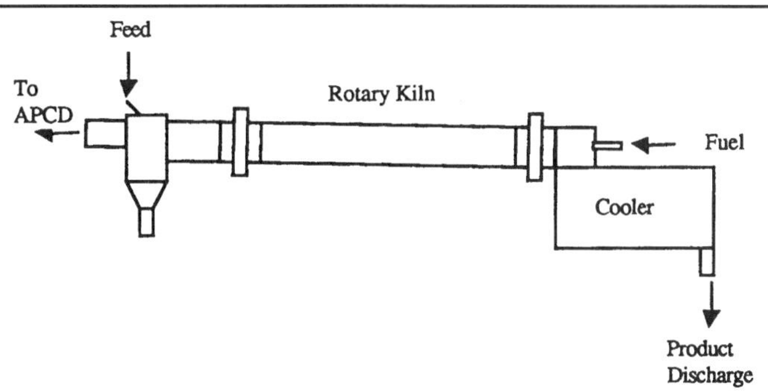

FIGURE 6.32 Lightweight aggregate kiln (*EPA-94/05, p. 2-20*).

be processed, but is usually 30–60 meters. The prepared raw material is fed into the kiln at the higher end, while firing takes place at the lower end. The dry raw material fed into the kiln is initially preheated by hot combustion gases.

Once the material is preheated, it passes into a second furnace zone where it melts to a semiplastic state and begins to generate gases which serve as the bloating or expanding agent. In this zone, specific compounds begin to decompose and form gases such as SO_2, CO_2, SO_3 and O_2 that eventually trigger the desired bloating action within the material. As temperatures reach their maximum (approximately 2100°F), the semiplastic raw material becomes viscous and entraps the expanding gases. This bloating action produces small, unconnected gas cells, which remain in the material after it cools and solidifies. The product exits the kiln and enters the cooler section of the process, where it is cooled with cold air and then conveyed to the discharge point.

Waste feed. The method of feeding waste into a kiln is a primary factor in determining the destruction of the waste, which is dependent on temperature, residence time, and turbulence. The method of introducing the waste also depends on its physical state, including (EPA-94/05, p. 2-14):

- liquid;
- solid;
- containerized solid waste;
- powdered waste.

LIQUID WASTE: Liquid wastes are either blended directly with conventional fuels provided at the hot end of the kiln, or they are injected through a separate burner/atomizer into the primary kiln flame.

SOLID WASTE:

- For long-type wet and dry kilns and dry kilns with preheaters, a method has been patented by Benoit and Hansen under US Patent No. 4,850,290 for charging solid wastes directly through a hatch on the rotating kiln wall at an intermediate location within the calcining zone. At each rotation, the hatch is opened, and containerized waste solids are fed down a drop tube that is inserted through the hatch and into the rotating kiln. The drop tube prevents hot mineral material from escaping through the port or contacting the enclosure. It is important that the volume of the volatile components do not exceed the capacity for their complete combustion in the gas stream. Thus, wastes are containerized to minimize the potential for overloading the combustion capacity (creating local reducing conditions).
- For preheater or precalcining kilns, the solid waste may be injected directly into the precalciner vessel or preheater inlet.

CONTAINERIZED SOLID WASTE: Containerized solid wastes may be injected at the hot end of the kiln at a high enough velocity so that they are projected into the calcining zone. An "air cannon," which is mounted to the kiln hood, is used to propel the waste containers.

POWDERED WASTE: If the waste is in powdered form, it may be injected directly into the primary burning-zone coal flame of the cement kiln.

4.2 Industrial Boilers for Resource Recovery

In contrast to incinerators, whose main objective is to destroy wastes, boilers are constructed to produce steam for electrical generation (utility boilers) or for onsite process needs (industrial boilers). Also, wastes compose the primary feed to incinerators, whereas they are usually a supplementary fuel for boilers. Fuel inputs to industrial boilers vary with process requirements, which may fluctuate considerably more than the waste feed to a waste incinerator.

Before chlorinated wastes can be fired to boilers, the incompatibility with materials of construction and air pollution control equipment must be considered so as to minimize corrosion problems and hydrogen chloride emissions (EPA-94/05, p. 2-34; EPA-86/09, p. 2-9).

Reportedly there are approximately 2600 fossil-fuel-fired utility boilers and more than 23,000 fossil-fuel-fired industrial boilers (9800 with capacities greater than 50×10^6 Btu/h) in the United States. Coal is the primary fuel in both boiler sectors, but oil and gas are also used. The concept of disposing of wastes in boilers has centered around industrial boilers because (1) their operation is more flexible than utility boilers, (2) they offer the potential of destroying wastes generated onsite, and (3) the storage and handling facilities for wastes generated onsite generally already exist.

When burning wastes in boilers, blending and operation plans should be developed to address (OME-88/12, p. 4-99):

- additional gas cleaning requirements necessary to handle the ash content of waste fuels;
- control of hydrogen chloride or other acid gas resulting from the burning of waste fuels;
- increased maintenance of boilers and air-cleaning equipment resulting from the use of waste fuels; and
- installation of ash discharge equipment at the bottom of the boiler to catch and discharge accumulated bottom ash.

In general, criteria for waste firing in a boiler include:

- gross waste heating value > 18,000 kJ/kg, as fired;
- ash content < 1.5%, as fired;
- chloride content of the organic matter < 0.5%, as fired;
- minimum operation temperature 870°C (1600°F).

By definition, a boiler is a closed vessel in which water under pressure is transformed into steam by the application of heat. More specifically, EPA has defined the following characteristics of a boiler for regulatory purposes under the Resource Conservation and Recovery Act (RCRA):

- the combustion chamber and primary energy recovery section must be of integral design;
- thermal recovery efficiency must be at least 60%;
- at least 75% of the recovered energy must be "exported" (i.e., not for internal boiler uses).

Boilers come in a variety of sizes, configurations, and designs. The three common boiler design categories which burn hazardous waste are

- fire tube boilers
- water tube boilers
- stoker-fired boilers

Fire tube boilers. The fire tube boiler can be described in simple terms as a water-filled cylinder with tubes running through it which provide the escape path for the combustion gases or flue gas (as shown in Fig. 6.33). As the flue gas passes through the tubes, the hot gases heat the tubes which then heat the water to produce steam. The fire tube boiler is primarily used in industrial applications.

Fire tube boilers are compact, low in initial cost, and easy to modularize based on plant requirements. However, they are also slow to respond to changes in demand for steam (load) compared with water tube boilers, and circulation is slower. Also, stresses are greater in boilers because of their rigid design and subsequent inability to expand and contract easily. Fire tube boilers usually range in size from less than 2 to 50 million Btu per hour.

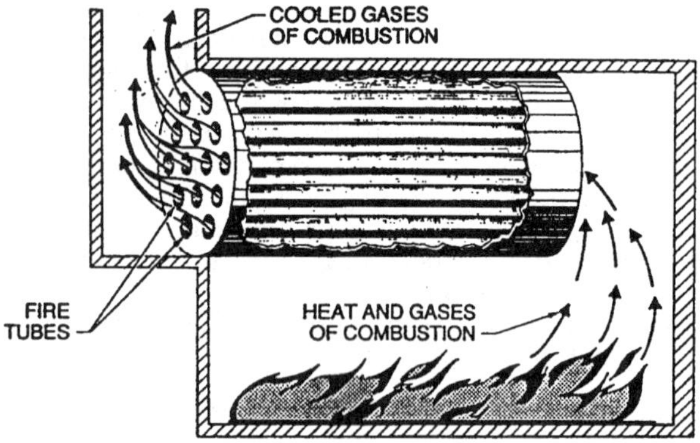

FIGURE 6.33 Fire tube boiler (*EPA-94/05, p. 2-35*).

Water tube boiler. The basic design of a water tube boiler (Figs 6.34 and 6.35) is that the water is circulated through tubes, with the hot combustion gases passing over the outside surfaces of the tubes. Generally, the boiler can be physically divided into two sections, the furnace and the convection pass. Furnaces (fireboxes, combustion chambers) will vary in configuration and size, but their function is to contain the flaming combustion gases and transfer the heat energy to the water-cooled walls. The convection pass contains the superheaters, reheater, economizer and air preheater heat exchangers, where the heat of the combustion flue gases is used to increase the temperature of the steam, water, and combustion air. The superheaters and reheaters are designed to increase the temperature of the steam generated within the tubes of the furnace walls. Steam flows inside the tube, and flue gas passes along the outside surface of the tubes.

The economizer is a counter flow heat exchanger designed to recover energy from the flue gas after the superheater and the reheater. The boiler economizer is a tube bank-type, hot-gas-to-water heat exchanger. It increases the temperature of the water entering the steam drum. The tube bundle is typically an arrangement of parallel horizontal serpentine tubes with the water flowing inside but in an opposite direction to the flue gas. Tube spacing is as small as possible to promote heat transfer while still permitting adequate tube surface cleaning and limiting flue gas side pressure loss. By design, steam is not usually generated inside these tubes.

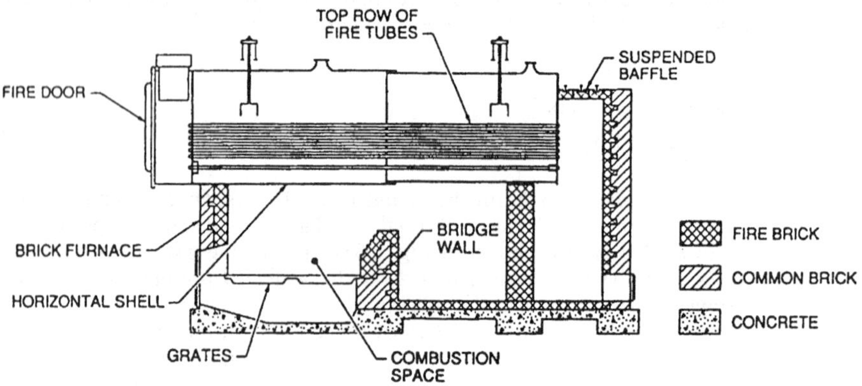

FIGURE 6.34 Industrial water tube boiler (*EPA-94/05, p. 2-35*).

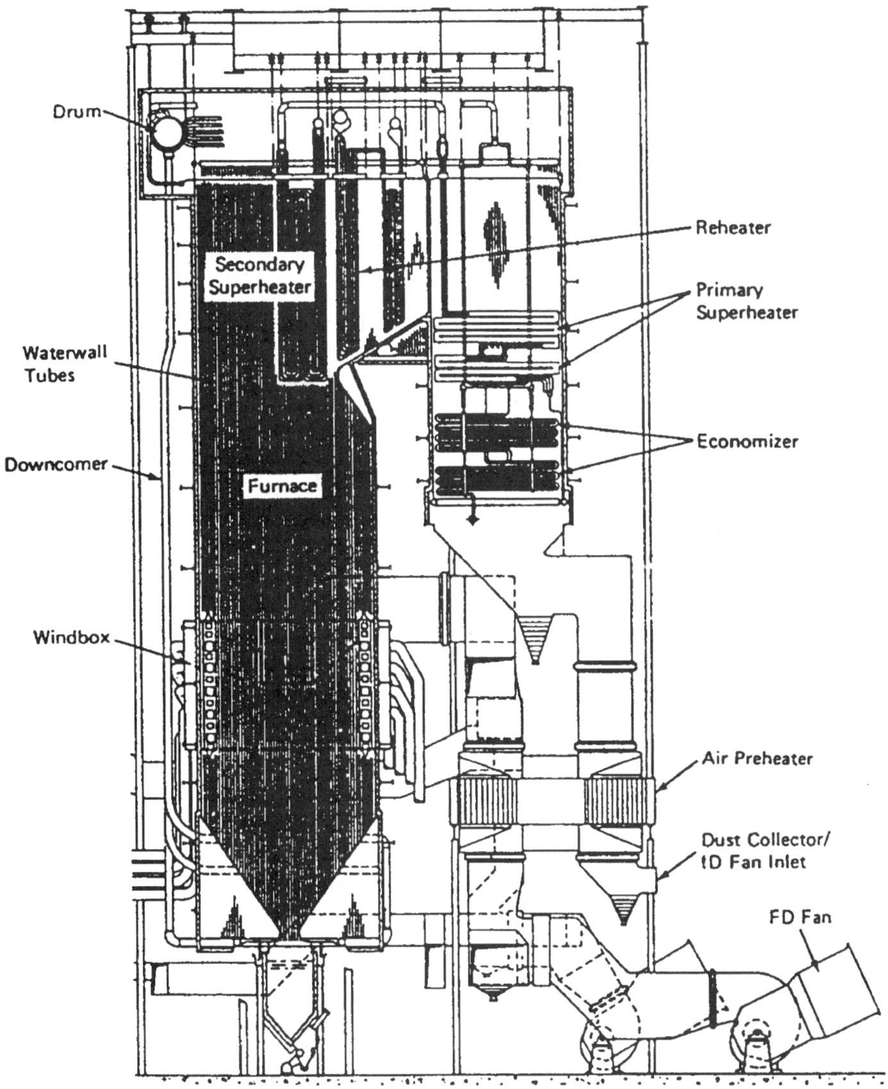

FIGURE 6.35 Utility water tube boiler (*EPA-94/05, p. 2-41*).

The air preheater is not a portion of the steam–water circuit, but serves a key role in the steam generator system to provide heat transfer and efficiency. In many cases, especially in a high pressure boiler, the temperature of the flue gas leaving the economizer is still quite high. The air heater recovers much of this energy and adds it to the combustion air. Heating the combustion air prior to its entrance to the furnace reduces fuel usage.

Water tube boilers are rapid steamers and respond quickly to changes in demand for steam due to improved water circulation. They can withstand much higher operating pressures and temperatures than fire tube boilers. In addition, the water tube boiler design is safer. They can also burn a wide variety of fuels and have the ability to expand and contract more easily than fire tube boilers. The major drawback is that water tube boilers are more expensive to install. They also require more complicated furnaces and repair techniques.

Stoker boiler

STOKER: Stokers are mechanical devices that feed solid fuels such as coal, wood wastes, and bagasse (as well as residential and commercial refuse) onto a grate at the bottom of the furnace and remove the ash residue after combustion. Stokers are designed to permit continuous or intermittent fuel feed, fuel ignition, air supply for combustion, free passage for the resulting gaseous products, and disposal of noncombustible materials. A stoker firing system typically consists of a fuel supply system, a stationary or moving grate assembly which supports the burning mass of fuel and admits most of the combustion air to the fuel, an overfire air system to complete combustion, and an ash or residual discharge system. Figure 6.36 shows a cross section of an overfeed mass burning traveling grate stoker.

FUEL FEEDER: There are three types of fuel feeders utilized in spreader stokers: (1) reciprocating feeder, (2) chain feeder, and (3) drum feeder. In general, fuel feeders have a device which meters the fuel to the combustion control system and delivers it to the built-in rotor.

GRATE: The purpose of the grate includes providing

- a floor on which the fuel can burn
- a means of distributing air evenly through the grates
- a method of discharging the ashes that accumulate on the grate from the consumed fuel.

The various grate designs include stationary and dumping grates, reciprocating grates, vibrating grates, traveling grates, and vibrating, water-cooled grates.

5 FACILITY DESIGN FEATURE

The facility design and operation of waste incinerators is significantly influenced by the category of waste involved, i.e., solids, liquids, or sludges. The systematic approach to facility design and operation therefore requires investigation of the composition of each class of waste to define the equipment and operating procedures for each of the following elements. The

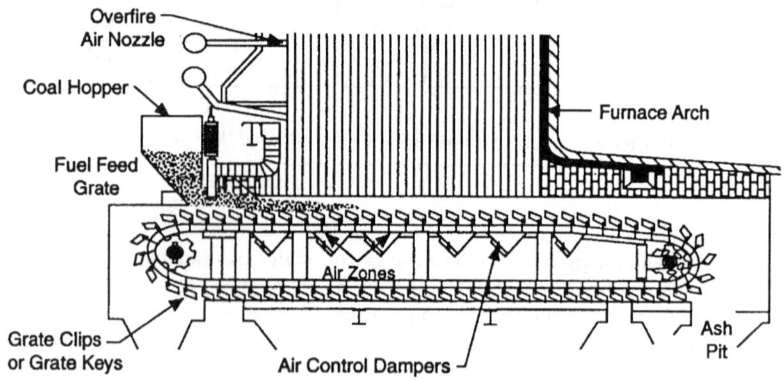

FIGURE 6.36 Overfeed traveling grate stoker (*EPA-94/05, p. 2-35*).

overall success of an incinerator facility depends upon the successful integration of these elements.

- Compliance with regulatory requirements.
- Waste characteristics.
- Auxiliary fuel.
- Factors affecting hazardous waste incinerator facility design.
- Site selection.
- Waste receiving area.
- Waste segregation and storage.
- Waste blending and/or processing before incineration.
- Combustion process monitoring.
- Preplanning of facility operation.
- Safety (toxicity, fire explosion).
- Transportation and unloading.
- Fugitive emission control.
- Scrubber/quench water treatment.
- Residue handling and disposal.
- Secondary problems (e.g., stream pollution, runoff, groundwater contamination).
- Permit procedure.

Some of the above items are further described below.

5.1 Compliance with Regulatory Requirements

A major consideration in the design and operation of waste combustion systems is the regulatory requirements under which an incineration facility will be permitted and operated. Emissions from combustion of municipal, medical, and hazardous wastes are regulated under entirely different sets of environmental laws, even though many waste types can generally be burned in a given unit, separately or mixed. Key components for the compliance with regulatory requirements include (Hasselriis-98/12):

- environmental laws and environmental regulations;
- pollutants of concern;
- stack emissions and the environment;
- emission standards and guidelines;
- good combustion practice;
- stack testing and monitoring;
- ash residue management;
- operator certification and training.

Environmental laws and environmental regulations: Based on the environmental laws enacted by Congress, US EPA develops detailed environmental regulations and codifies them in 40 Code of Federal Regulations (40CFR). Table 6.4 provides the relationship between the environmental laws and environmental regulations. For the compliance of air emissions, readers are encouraged to refer to the air program in 40 CFR Parts 50 to 90 and for solid waste treatment, the readers should refer to the solid waste program in 40 CFR Parts 240 to 299.

TABLE 6.4 Major Environmental Laws and Environmental Regulations

Environmental laws	Environmental regulations (CFR cites)
EPA's purpose and functions	Subchapter A—General; 40CFR1-29
Government Performance and Results Act (GPRA) of 1993 (Public Law 103-62)	Web address: http://www.epa.gov/Indicator/gpra.htm#law
EPA's regulatory authorities	Subchapter B—Grants and Other Federal Assistance; 40CFR30-47
Clean Air Act (CAA) of 1970	Subchapter C—Air Programs; 40CFR50-99
Federal Water Pollution Control Act (FWPCA) of 1972	Subchapter D–Water Programs; 40CFR100-140 (oil discharge and prevention related regulations)
Safe Drinking Water Act (SDWA) of 1974	Subchapter D—Water Programs; 40CFR141-149 (drinking water regulations)
Federal Insecticide, Fungicide, and Rodenticide Act (FIFRA) of 1947	Subchapter E—Pesticide Programs; 40CFR150-189
Atomic Energy Act (AEA) of 1954	Subchapter F—Radiation Protection Programs; 40CFR190-192
Noise Control Act (NCA) of 1972	Subchapter G—Noise Abatement Programs; 40CFR201-211
Marine Protection, Research, and Sanctuaries Act (MPRSA) of 1972	Subchapter H—Ocean Dumping; 40CFR220-238
Resource Conservation Recovery Act (RCRA) of 1976	Subchapter I—Solid Wastes; 40CFR240-259 (municipal waste, land disposal, and resource recovery regulations)
Hazardous and Solid Waste Act (HSWA) of 1984	Subchapter I—Solid Wastes; 40CFR260-299 (hazardous waste)
Comprehensive Environmental Response, Compensation, and Liability Act of 1980 (CERCLA)	Subchapter J—Superfund, Emergency Planning, and Community Right-To-Know Programs; 40CFR300-399
Superfund Amendments and Reauthorization Act (SARA) of 1986	
Clean Water Act (CWA) of 1997	Subchapter N—Effluent Guidelines and Standards; 40CFR400-599
Motor Vehicle Information and Cost Savings Act (MVICSA) (15USC1901)	Subchapter P (reserved)
	Subchapter Q—Energy Policy; 40CFR600-699
Toxic Substances Control Act (TSCA) of 1976	Subchapter R—Toxic Substances Control Act; 40CFR700-799
National Environmental Policy Act (NEPA) of 1969	Subchapter V–Council on Environmental Quality; 40CFR1500-1517

POLLUTANTS OF CONCERN: The pollutants which are regulated by air programs are particulate matter (PM), carbon monoxide (CO), sulfur dioxide (SO_2), nitrogen oxides (NO_x), hydrogen chloride (HCl), lead, cadmium, and mercury. Other pollutants, not expressly regulated, which may need to be investigated in environmental impact studies, may be evaluated by the same calculation procedures.

Permits for hazardous waste combustors (HWCs) have been granted on the basis of demonstrating 99.99% destruction and removal efficiency (DRE) for the principal organic hazardous components (POHC), and for specific metals in the feed, as well as CO limitations. These tests are designed so that the anticipated concentrations of these components are fed to the HWC during the trial burn test while stack emissions are measured. In order to obtain stack emissions high enough to measure with accuracy, sufficiently above the detection limit of the analytical procedures, it is often necessary to feed substantially higher quantities of the target substance than are likely to be encountered in reality. Hence, the measured emissions may not be appropriate for estimating environmental and health effects resulting from normal operation of the HWC.

The composition of the fuel or the waste determines the potential for emissions of pollutants from the stack, although the chemical forms of the emissions will generally change as a result of the process, and a substantial fraction may be discharged as ash residue. Additives used to alter or control emissions will also affect the composition of the stack gases. These may include water used for cooling of the products of combustion, alkaline reagents to control acid gases, urea or ammonia to control NO_x, and activated carbon used to adsorb mercury and dioxins.

STACK EMISSIONS AND THE ENVIRONMENT: It is important to understand that the regulated emissions measured at the stack through which they are discharged are expressed as concentrations, weight per unit volume, referred to standard atmospheric temperature and pressure. The regulations express the allowable emissions on a dry basis (water vapor not included), and with an arbitrary standard representing approximately 50% excess air, usually exactly 7% oxygen, and approximately 11% carbon dioxide. The European Union requires correction to 3% oxygen.

These corrected values are used to report and compile emission numbers on a common basis to make valid comparisons of data. The data base of these emission factors, maintained by the US EPA and others, is useful in predicting emissions from facilities yet to be constructed and tested. They do not represent actual emission concentrations which take place at stack temperatures, including the water vapor and excess air. Actual emission concentrations of the pollutants in the actual gaseous emission volumes are used to calculate ground level or other local concentrations resulting from the dilution and dispersion of the stack gases in the ambient atmosphere. These concentrations are appropriate for the estimation of health effects and other environmental impacts.

EMISSION STANDARDS AND GUIDELINES: The reader should consult Federal, State, and local emission standards for an incinerator to be designed or operated. The trend of regulations has been from risk-based standards toward technology-based standards. Before present federal regulations were in place for municipal waste combustion, many states took the initiative and wrote standards based on the health risk associated with the emissions at ground level, and required risk assessments to be prepared on the basis of modeling.

The Clean Air Act Amendments of 1990 required that Maximum Available Control Technology (MACT) standards be developed which would take into account the "best performing" combustion systems, requiring that new facilities meet these more stringent emission standards, and that existing facilities be upgraded to higher standards.

In spite of having to comply with the new federal standards, and more stringent state standards, permits may continue to be subject to public scrutiny on a one-by-one basis, which means that environmental impact will still be considered for pollutants which are not numerically regulated by federal and state regulations.

GOOD COMBUSTION PRACTICE: Standards for Good Combustion Practice (GCP) have been developed, which apply to all waste combustors: essentially they require that no waste be fed until the furnace temperature is at least 1600°F, and that the gaseous products of combustion be retained for at least 1 s at 1800°F or higher for most wastes, or for 2 s at 2000°F for wastes containing large quantities of halogenated compounds, which are more diffcult to destroy. Hazardous waste incinerators have been subjected to stringent regulation due to the potentially toxic emissions from the wide range of organic and inorganic compounds which may be in the wastes, and which define them as hazardous. These regulations require that trial burns be performed as a permit condition, during which it is demonstrated that the system can achieve 99.99% destruction and removal efficiency (DRE) while burning the principal organic hazardous compounds (POHC) and heavy metals. The assumption behind these tests was that the emissions would be related to the feed materials; hence the feed must be analyzed in normal operation, and be anticipated as the basis for the trial burns. In view of the fact that modern combustion and emission control systems which can meet present regulatory standards actually destroy or remove all but traces of the target substances, the tendency has been to focus on the actual emissions and ensure that they will not exceed the new, stringent standards.

STACK TESTING AND MONITORING: An essential part of the focus on actual stack emissions, which applies to all waste combustors, is the requirement that monitoring instruments be installed and operated, providing a continuous record of compliance. Continuous measurement of carbon monoxide (CO) and/or hydrocarbons (HC) and oxygen provides assurance of good combustion. These are useful surrogates for trace organic compounds, including dioxins and furans. Measurement of HCl, SO_2, and NO_x may be required, where applicable. Opacity measurements provide continuous supervision of the combustion process as well as the emission control system.

ASH RESIDUE MANAGEMENT: Ash residues from combustion of wastes are generally required to be tested for toxicity and managed appropriately. If found to fail the Extraction Procedure Toxicity (TCLP) test, they may have to be treated, or disposed in suitable landfills. The source of toxicity is generally the heavy metals, specifically lead, cadmium, and other metals, which are not destroyed by the combustion process, and hence end up in the flyash or bottom ash residues. The use of alkaline reagents for control of the acid gases adds another component to the disposal problem, since excess alkalinity can increase the solubility of these metals in the ash residues, and carbon, which absorbs dioxins and mercury.

OPERATOR CERTIFICATION AND TRAINING: In addition to requiring good combustion practice and regulating emissions, an essential component of the newly promulgated regulations is the requirement for operator training and certification, applied appropriately to municipal, medical, and hazardous waste combustion systems.

5.2 Waste Characteristics

Wastes are, by definition, nonuniform in quality. In order to proceed with equipment selection and design, however, waste characteristics must be determined or developed. The objective of undertaking waste characterization is

- to obtain the heating value,
- to obtain the range of the moisture content,
- to determine recyclables,
- to determine ash quantity for disposal,
- to determine the fate of metals,
- to determine the sources of chlorine,
- to determine composition of medical and/or infectious wastes.

A waste material considered as an incineration candidate will normally have a combustible component and a noncombustible component, and may contain moisture. Generally, heating value is stated relative to the waste combustible component. Heating value can be noted as a gross number applied to the total waste quantity, but the combustible, noncombustible, and moisture components must be identified in order to determine the heat content of a material. Key topics related to waste characteristics include (OME-88/12, p. 4-2):

- proximate analysis;
- ultimate analysis;
- calculated heating value.

PROXIMATE ANALYSIS: Proximate analysis and ultimate analysis are two important methods for the determination of whether a given waste is suitable for incineration, and how the incineration should take place (OME-88/12, p. 4-2).

Proximate analysis as per ASTM D 3172 is a relatively quick and inexpensive laboratory determination of the percentages of moisture, volatile matter, fixed carbon, and ash. The analytic procedure is as follows:

- heat a sample for 1 h at 105–110°C (221–230°F) and report the weight loss fraction as percent moisture;
- raise the temperature of the dried sample in a covered crucible to 725°C (1337°F) and hold it at this temperature for 7 min, and then report the sample weight loss fraction as volatile matter percentage;
- ignite the remaining sample in an open crucible at 950°C (1742°F) and allow it to burn to a constant weight, then report the sample weight loss as a percent fixed carbon.

The sample residue is to be reported as percent ash. The sum of moisture, volatiles, fixed carbon, and ash should equal 100%.

ULTIMATE ANALYSIS: Ultimate analysis is a standard procedure used for a determination of the quantities of elemental components present in a sample (ASTM D 3176). It is required in order to determine the products of combustion of a material, its combustion air requirement, and the nature of the off-gas or combustion products.

In this procedure the following elemental percentages are normally determined:

- carbon
- hydrogen
- sulfur
- oxygen
- nitrogen
- halogens (chlorine, fluorine, etc.)
- heavy metals (mercury, lead, etc.)
- other elements that can affect the combustion process

In addition to these components, analyses may be performed under the heading of ultimate analysis for the presence of certain compounds that may be in the waste, such as chlorobenzene, chlorophenols, PCBs (polychlorinated biphenyls), dioxins, etc.

It must be recognized that wastes are often extremely heterogeneous in nature. The proper selection of a representative sample is at least as important as performing the test itself. Most commonly used methods involve dumping a truckload of material in an enclosed area, and then reducing the truckload to an approximately 100-kg sample by a process of repeatedly quartering the waste and discarding three-fourths. Generally, landfill equipment (a bulldozer or wheeled loader) is used for quartering. Alternative methods include selecting a random sample using a grid method, or treating the entire truckload as the sample.

After sample selection is completed, the material can be sorted into standard components (food wastes, paper, cardboard, plastics, textiles, rubber, leather, garden trimmings, wood, glass, nonferrous metals, ferrous metals, dirt, ashes, etc.) so that each component can be analyzed.

CALCULATED HEATING VALUE: If the fractions of the elemental composition such as carbon, hydrogen, oxygen, and sulfur of a waste are given, the DuLong equation can be used to calculate the approximate heating value of the waste. This equation is as follows:

$$Q = 33{,}829\,\mathrm{C} + 144{,}277\,(\mathrm{H_2} - 0.125\,\mathrm{O_2}) + 9420\,\mathrm{S}$$

where Q is in kJ/kg and C, H_2, O_2, and S are the weight fractions of carbon, hydrogen, oxygen, and sulfur, respectively. The weight fractions should add to 100% unless an inert material such as nitrogen is present. With the presence of nitrogen, for instance, the weight fraction (%) of C, H_2, O_2, and S will total 100% less the fraction of nitrogen present.

WASTE AND FUEL DATA: Tables 6.5–6.10 give a collection of waste data pertinent for incineration references.

Table 6.5 (OME-88/12, p. 4-3) lists properties of various municipal waste materials relative to incineration. Included are heating value, density, and typical ash and moisture components.

Table 6.6 (OME-88/12, p. 4-4) identifies common waste designations.

Table 6.7 (OME 88/12, p. 4-5) lists municipal solid waste composition in the US and Canada on a national average basis.

Table 6.8 (OME-88/12, p. 4-6) lists combustion characteristics of those wastes generated in Hamilton, Canada.

Table 6.9 (Hasselriis-98/12) lists the moisture, ash, heating value, and density of the standard types of municipal, commercial, and medical wastes according to waste types established by the Incinerator Institute of America. The wastes have been classified as Types 0–4 as a simplified definition.

Table 6.10 (Hasselriis-98/12) gives the approximate ultimate analysis of hazardous waste in different combustion chambers.

TABLE 6.5 Characteristics of Selected Municipal Waste*

Waste	Heating value as fired (kJ/kg)	Density (kg/m^3)	% Ash (by weight)	% Moisture
Newspaper	18,600	110	1.5	6
Brown paper	16,900	110	1	6
Magazines	12,200	560	22.5	5
Corrugated paper	16,400	110	5	5
Plastic-coated paper	17,100	110	2.6	5
Coated milk cartons	26,400	80	1	3.5
Citrus rinds	3950	640	0.75	75
Shoe leather	16,800	320	21	7.5
Butyl sole composition	25,400	400	30	1
Polyethylene	46,500	960	0	0
Polyurethane (foamed)	30,200	32	0	0
Latex	23,000	720	0	0
Rubber waste	23,300	1500	25	0
Wax (paraffin)	43,300	890	0	0
1/3 tar–2/3 paper	26,700	140	3	1
Tar or asphalt	39,500	960	1	0
Wood sawdust (pine)	22,300	180	3	10
Wood sawdust	19,000	180	3	10
Wood bark (fir)	22,100	260	3	10
Wood bark	19,200	260	3	10
Corn cobs	18,600	200	3	5
Rags (silk or wool)	21,000	200	2	5
Rags (linen or cotton)	16,700	200	2	5
Animal fats	39,500	880	–	0
Cotton seed hulls	20,000	440	2	10
Coffee grounds	23,200	440	20	20
Linoleum scraps	25,600	1400	25	1

* This table shows the various heating values of materials commonly encountered in incinerator designs. The values given are approximate and may vary based in their exact characteristics or moisture content. The heating value is the higher heating value.

TABLE 6.6 Ranges of Heating Values and Other Physical Characteristics of Various Waste Types

	Type and description	% moisture (wet basis)	HHV (kJ/kg) (wet basis)	HHV (kJ/kg) (dry basis)	Bulk density (kg/m^3 as fired)
A	Cellulosic solids with up to 15% moisture	0–15	7900–27,900	9300–27,900	80–961
B	Cellulosic solids with 10–50% moisture	10–50	4600–20,900	9300–27,900	48–961
C	Cellulosic solids with over 40% moisture	40–80	2300–16,700	9300–27,900	48–993
D	Plastics and asphaltic solids, non-halogenated	0	17,400–46,400	18,600–46,400	32–1298
E	Plastics and asphaltic solids, halogenated	0	15,800–30,600	20,400–30,600	80–2307
F	Rubber	0	20,200–36,200	23,600–36,200	144–2003
G	Animal materials	5–85	2300–22,000	23,200	16–1298
H	Animal and human wastes	30–85	2300–20,200	15,600–28,800	481–1282
I	Noncombustible solids	0	0	0	80–4486
J	Pathological materials	36–97	10,400–18,800	11,600–20,900	80–1041
K	Pathological remains	85	2300	23,200	801–1202
L	Cadavers, coffin-encased	40	13,700	24,100	641–1282
M	Organic liquids with under 30% water	0–30	70–39,500	70–39,500	721–1202
N	Organic liquids with over 30% water	30–80	700-27,900	700–39,500	641–1121
O	Fumes	0–80	1900–141,600	9300–141,600	0.64–5.45
P	Particulates, gas-borne	0	0–1200	0–1200	0.64–1.60
Q	Radioactive materials	0–50	0–46,400	0–46,400	32.0–8010
R	Special wastes	0–80	0–41,800	0–41,800	32–4486

Source: This table is from Incinerator Performance, CSA Standard Z103-1976

5.3 Auxiliary Fuel

Table 6.11 provides the proximate and ultimate analysis and higher heating values of the most commonly used auxiliary fuels, No. 2 light distillate fuel oil, No. 6 heavy residual fuel oil, and natural gas.

5.4 Factors Affecting Hazardous Waste Incinerator Facility Design

The overall facility design of hazardous waste incinerators is significantly influenced by the category of waste involved; i.e., solids, liquids, or sludges. The systematic approach to facility design therefore requires investigation of the composition of each class of waste to define the equipment and operating procedures for each of the following elements (EPA-81/09, p. 5-2):

1. safety (toxicity, fire explosion);
2. transportation and unloading;
3. segregation of wastes during storage;
4. storage;
5. handling and feeding;
6. monitoring;

TABLE 6.7 Composition of Municipal Solid Waste from US and Canada (percent wet weight)

	US national average	Canada national average
Combustibles		
Paper	33.5	36.5
Food waste	17	27.6
Yard waste	17.5	6.1
Plastics	3.6	5.2*
Rubber, leather	2.6	–
Textiles	2	4.3
Wood	3.2	4.2
Miscellaneous organics	–	1.1
Total	79.4	85
Noncombustibles		
Glass, ceramics	9.9	8.4
Metals	9.2	6.6
ferrous	–8	
aluminum	–0.9	
Other	–0.3	
Fines	–	
Miscellaneous organics	1.5	
Total	20.6	15

* Includes rubber.

TABLE 6.8 Combustion Characteristics of Municipal Solid Waste

	Percent by weight
Ultimate analysis	
Carbon	18.5
Hydrogen	2.9
Nitrogen	0.5
Oxygen	23
Sulfur	0.1
Moisture	25
Ash	30
Total	100
Higher heating value (kJ/kg)	11,165
Proximate analysis	
Volatile matter	38.5
Fixed carbon	6.5
Moisture	25
Ash	30
Total	100

TABLE 6.9 Standard Types of Municipal, Commercial and Medical Wastes

Waste type		Moisture (%)	Ash (%)	Heating value (Btu/lb)	Density (lb/ft^3)
Type 0	Trash	10	5	8500	8–10
Highly combustible waste, paper, wood cardboard cartons, including up to 10% treated papers, plastic or rubber scraps; commercial and industrial sources					
Type 1	Rubbish	25	10	6500	8–10
Combustible waste, paper, cartons, rags, wood scraps, combustible floor sweepings; domestic, commercial, and industrial sources					
Type 2	Refuse	50	7	4300	15–20
Rubbish and garbage; residential sources					
Type 3	Garbage	70	5	2500	30–35
Animal and vegetable wastes; restaurants, hotels, markets; institutional; commercial and club sources					
Type 4	Organic waste	85	5	1000	45–55
Carcasses, organs, solid organic wastes; hospitals, laboratories, abattoirs, animal pounds, and similar sources					

7. fugitive emission control;
8. scrubber/quench water treatment;
9. residue handling and disposal; and
10. secondary problems (e.g., stream pollution, runoff, groundwater contamination).

The overall success of an incinerator facility depends upon the successful integration of storage, feeding, and firing equipment; often these are areas which do not receive as much attention as is necessary. In the case of hazardous waste incineration, it is crucial that these areas receive special attention.

Figure 6.37 is a block diagram of a typical incinerator facility layout. In an overall facility evaluation the key areas are the facilities and equipment before and after the combustors, i.e., waste receiving, waste storage, waste blending, transfer between these areas, equipment feeding waste to the incinerator, handling and treatment of quench and scrubber waters, and ash disposal.

5.5 Site Selection

The selection of a site for a waste incineration facility is a phased decision process which occurs prior to making a permit application. Site screening is the process of identifying and evaluating a parcel of land for its suitability as a waste disposal site. Specific site screening criteria which the permit applicant must address include the following aspects (EPA-81/09, p. 5-2):

- geological
- topographic
- climatic
- ecological
- cultural
- public

TABLE 6.10 Typical Hazardous Waste Stream Analysis

Component/property	% by weight	Range
Primary combustion chamber (pumpable sludge waste)		
Carbon	24.35	20.52–41.40
Hydrogen	3.22	1.61–5.87
Nitrogen	0	0.00–3.47
Oxygen	0	0.00–4.73
Chlorine	0	1.17–29.48
Fluorine	0	0.00–1.00
Sulfur	0	0.07–1.00
Water	62.43	4.00–70.00
Ash	10	12.63–25.00
HHV (Btu/lb)	5400	3200–9500
Viscosity	500 cp	200–1000
Specific gravity	1.2	1.0–1.71
Primary combustion chamber (solid waste)		
Carbon	13.8	3.39–20.70
Hydrogen	1.95	0.43–2.97
Nitrogen	3.25	0.00–7.80
Oxygen	24	0.00–7.62
Chlorine	15	0.76–16.00
Fluorine	0	0.00–1.00
Sulfur	1	0.07–1.00
Water	16	10.00–22.00
Ash	25	48.33–58.00
HHV (Btu/lb)	3000	700–4500
Secondary combustion chamber (liquid waste)		
Carbon	57.3	48.61–75.90
Hydrogen	7.56	6.42–10.88
Nitrogen	22.52	0.58–28.61
Oxygen	6	7.48–10.26
Chlorine	1.47	1.10–4.68
Fluorine	0.01	0.00–1.00
Sulfur	0.01	0.00–1.00
Water	4.21	0.00–5.02
Ash	0.92	0.54–1.00
HHV	12,000	10,000–20,000
Viscosity	15.2 cp	1.0–20.00
Specific gravity	0.95	1.0–1.10

While many sites may exist which meet technical, economic, and ecological criteria, public acceptance or rejection may ultimately decide the fate of the facility.

GEOLOGICAL: The main geological constraints that can render a site unsuitable for a hazardous waste incinerator facility are historical or predicted seismic activity, landslide potential, and volcanic or hot spring activities.

TOPOGRAPHIC: The main topographic constraints are susceptibility to flooding, erosion, and offsite drainage runoff. The site will need sufficient area for the construction of a runoff holding pond (or diversion to an existing holding pond) to retain surface runoff which may contain hazardous substances in solution. Because of the holding pond and flood protection criteria, siting in flood plains is not normally acceptable.

TABLE 6.11 Common Auxiliary Fuels

Component	Residual oil (No. 6)*	Distillate fuel (No. 2)**	Natural gas**
C	0.8852	0.872	0.693
H	0.1087	0.123	0.227
O	0.0006	0	0
N	0.001	0	0.08
S	0.004	0.005	0
Ash	0.0005	0	0
HHV (Btu/lb)***	18,300	19,430	23,170

* ***Source*** (EPA-86/03, p. 73)
** ***Source*** (EPA-81/09, p. 4-87).
*** Hasselriis-98/12.

CLIMATIC: The primary climatic features which can adversely affect an incineration site are the amount of annual or seasonal precipitation and the incidence of severe storms. Copious precipitation will cause surface runoff and water infiltration through the soil. Runoff, i.e., that amount of rainfall that does not infiltrate the soil, depends on such factors as the intensity and duration of the precipitation, the soil moisture content, vegetation cover, permeability of the soil, and slope of the site. Normally, the runoff from a 10-year storm (recurrence interval of only once in 10 years) or annual spring thaw, whichever is greater, is containable by the site's natural topography. If not, berms, dikes, and other runoff control measures must be constructed to modify the site.

ECOLOGICAL: Ecological site features are those elements determined through earlier studies and environmental impact statements (EIS) which determine whether ecosystems at

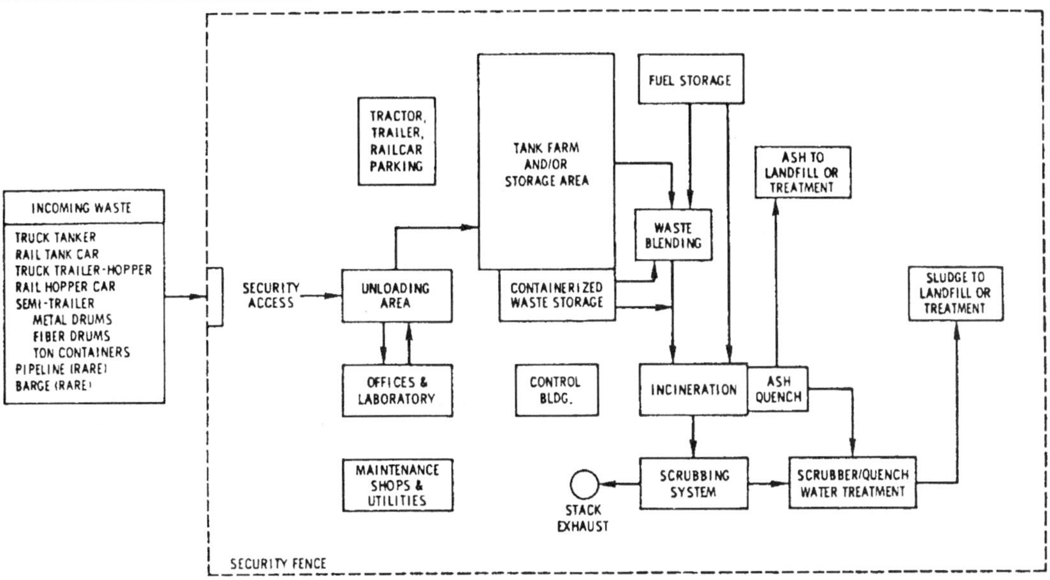

FIGURE 6.37 Typical incinerator facility layout (*EPA-81/09, p. 5-3*).

the site are in a delicate balance. Whether a site is a habitat for rare and endangered species, or used seasonally by migratory wildlife is also a factor in final site selection.

CULTURAL: Cultural site features are those elements that are a direct result of human activities which modify and affect the site's desirability as a hazardous waste incineration facility such as access, land use, and aesthetics. Land areas zoned for nonresidential uses and adequate buffer zones are generally preferred for siting a hazardous waste management facility. The site ideally needs to contain sufficient land area to provide a concentric ring of unoccupied space as a buffer zone between active storage, treatment, and disposal areas, and the nearest area of human activity. Vegetation, topography, distance, and artificial barriers are all potential means to screen facility activities from line of sight observations from commercial, residential, or recreational areas.

PUBLIC: One of the most difficult problems faced by a hazardous waste incineration facility applicant has been that of gaining public approval from a community for construction of the facility. No matter how thoroughly the above parameters have been examined in the facility site selection, public acceptance or rejection probably decides the fate of the facility. Public awareness of the planned facility, early planning input, and active participation by political leaders, public officials, and environmental groups, as well as other public interest groups and adjacent industry have led to successful facility sitings in the past.

5.6 Waste Receiving Area

The type and nature of waste received at an incinerator facility will dictate the design and equipment of the waste receiving area. The physical types of waste which may be received are:

- liquid,
- containerized materials, liquid and solids,
- dry solid materials, and
- wet solid materials:
  (a) pumpable, and
  (b) nonpumpable.

The types of receiving equipment for unloading can be divided into three general areas:

- pumpable liquid transfer,
- container transfer, and
- bulk solids transfer.

Careful consideration must be given to the layout, safety, and recordkeeping arrangements of the waste receiving area. Unloading material offers one of the greatest spill or toxic exposure potentials at a waste facility. For recordkeeping, the waste receiving area poses the first interface with the transporter and manifest system. Proper operation of waste receiving areas includes:

- typical operation and layouts,
- laboratory for waste verification and/or characterization,
- liquid unloading, including safety/emergency provisions, and spill and runoff containment,
- container unloading, including drums, barrels, and special bulk units, and
- bulk solids unloading, including mechanical conveyors and pneumatic conveyors.

5.7 Waste Segregation and Storage

The manner in which a waste is handled onsite is dependent on:

- the nature of the waste (corrosivity, explosivity, etc.),
- the plant storage facilities, and
- the heat content of the fuel.

Wastes received for incineration at a disposal facility are either incinerated directly (in some cases via pumping directly from the tank truck), or stored until they can be handled more conveniently. A plant operator may want to segregate and store some of the incoming wastes with higher heating values to possibly blend with other wastes which have heating values too low to support combustion alone. At some plants, waste blending occurs prior to storage.

Storage capacity is based on:

- seasonal inventory build-up,
- redundancy or excess incinerator capacity,
- maintenance schedules and downtime,
- operating schedules (i.e., number of shifts vs. in-shipment rates), and
- amounts and nature of waste blending to be done.

Depending on the type of incinerator installation, storage facilities may be required to hold both liquid and solid wastes. If an incinerator cannot burn solids, facilities for solids storage are obviously not necessary.

A waste storage area is designed to address three problem areas:

- segregation of incompatible corrosive and reactive waste types;
- fire hazards due to flammable liquids and solids; and
- toxic hazards to prevent human exposure during storage, transfer, and spill possibilities.

Other factors for consideration for waste storage include:

- types of storage, including liquid storage, bulk solids storage, container storage, and tank cars,
- segregation of waste during storage, and
- safety provisions for storage areas, including fire safety and spill/toxicity safety.

5.8 Waste Blending and/or Processing Before Incineration

The methods by which wastes are removed from storage, prepared for incineration, and fed to the incinerator are dependent on the nature of the waste and the type of incinerator. Careful design consideration is given to:

- the layout for liquid waste blending, pumping, and associated pipework; and
- the handling and feeding arrangements for nonpumpable sludges, solids, and containerized wastes, where applicable.

Operating experience has shown that these are areas that do not receive as much attention as is necessary; the overall success of an incineration facility depends upon the successful integration of storage, feeding, and firing equipment.

Other factors for consideration for waste blending and/or processing before incineration include:

- waste compatibilities,
- liquid feed and blending equipment,
- pumps and piping, including positive-displacement pumps, centrifugal pumps, pump emission control, and pump and piping safety,
- valving and controls, including on/off service, throttling service, pressure control, etc.,
- valving and control safety considerations, including safety shutoffs, gages, meters, and gage glasses, operating controls,
- solids feeding equipment, including shredders, explosion suppression and safety consideration for shredders, feeders, container feeding equipment, and
- waste processing instrumentation.

5.9 Combustion Process Monitoring

Before incineration process conditions can be controlled automatically they must be measured with precision and reliability. Instrumentation for an incineration process is essential because of the variability of the many factors involved in attaining good combustion. For example, as the heat content of the solid waste rises, changes in the combustion process become necessary. Instrumentation indicates these variations so that automatic or manual control adjustments can be made.

The uses of instrumentation and controls include a means of process control, protection of the environment, protection of the equipment, and data collection. A control system must have four basic elements:

- standard of desired performance;
- sensor (instrument) to determine actual performance;
- capability to compare actual versus desired performance (error); and
- control device to effect a corrective change.

The four major factors governing incineration efficiency for a given waste feed are:

- temperature,
- residence time,
- oxygen concentration, and
- turbulence achieved.

The temperature in the incinerator can be measured directly. Instrumentation is also available to directly monitor CO, CO_2, and oxygen concentration in the combustion gas to ensure that excess air levels are maintained. Residence time and mixing efficiency cannot be directly measured, however, so other parameters indicative of these conditions need to be measured instead.

Gas residence time in the combustion zone depends upon the volume of the combustion chamber and the volumetric flow rate. Since the volume of the chamber is fixed for a given unit, residence time is directly related to combustion gas volumetric flow rate. Therefore, measuring this flow rate is equivalent to residence time measurement for a given incinerator.

Mixing in liquid waste incinerators or afterburners is a function of burner configuration, gas flow patterns, and turbulence. Burner configuration and gas flow pattern are a function of the incinerator design and will not vary from baseline conditions. Turbulence is determined by gas velocity in the combustion chamber, which is proportional to gas volumetric flow rate. Therefore, combustion gas flow rate is an indicator of mixing as well as residence time in liquid injection incinerators.

In incinerators burning solid wastes, other factors need to be considered to determine solids retention time and degree of agitation. These factors vary from one type of incinerator to another.

TEMPERATURE MONITORING: Incinerator temperature is monitored on a continuous basis to ensure that the minimum acceptable temperature for waste destruction is maintained.

This requires one or more temperature sensors in the hot zone and a strip chart recorder or equivalent recording device.

Generally, wall temperatures and/or gas stream temperatures are determined using shielded thermocouples as sensors. Thermocouples are the most commonly used contact sensors for measuring temperatures above 1000°F. Specifically, thermocouples can measure the following thermal parameters:

- Average gas temperature—accomplished using a shielded thermocouple with a relatively large thermal capacity anchored to a relatively large mass. The metering circuit is provided with a 30-s time constant to further smooth and average the readings.
- Instantaneous gas temperature—accomplished using a shielded thermocouple with a very small thermal capacity with the output metered by a circuit with a 1-s time constant. (Nominally, the reaction rates within the hot gas stream are strongly temperature-dependent; they should thus depend on the highest temperature to which the constituents are exposed.)
- Open flame temperature–obtained using an unshielded low thermal mass thermocouple with the output metered by an amplifier with a 30-s time constant.
- Average wall temperature—obtained using a shielded thermocouple imbedded in the refractory wall. (Here, the averaging is accomplished by the thermal inertia of the refractory material.)

Optical pyrometers are not recommended for these measurements owing to spectral bias factors present in the combustion area which can cause unacceptable measurement error.

The location at which temperature measurements are taken is important, owing to possible variations from one point to another in the combustion chamber. Temperatures are highest in the flame and lowest in the refractory wall or at a point of significant air infiltration. Ideally, temperatures are measured in the bulk gas flow at a point after which the gas has traversed the combustion chamber volume that provides the specified residence time for the unit. Generally, temperature measurement at a point of flame impingement or at a point directly in sight of radiation from the flame is not recommended. Figures 6.38 and 6.39 show typical monitoring locations for liquid injection and rotary kiln incinerators, respectively.

Table 6.12 (EPA-81/09, p. 5-80) provides the types of thermocouples used, including J, K, E, R, S, T, and B. The letter symbols identifying the thermocouple types are those defined in ANSI Standard C96.1. These symbols are in common use throughout industry. The table also lists the limits of error for the common thermocouple types; most manufacturers supply thermocouples and thermocouple wire to these limits of error or better.

Since the thermocouple element in a thermocouple assembly is usually expendable, conformance to established emf–temperature relationships is necessary to permit interchangeability. Calibration of a thermocouple consists of the determination of its emf at a sufficient number of known temperatures so that with some accepted means of interpolation, its emf will be known over the entire range in which it is to be used. The process requires a standard thermometer with high-level calibration to indicate temperatures on a standard scale, a means for measuring the emf of the thermocouple, and a controlled environment in which the thermocouple and standard can be brought to the same temperature.

Thermocouples use one of three different types of measuring junctions—grounded, ungrounded, and exposed. The grounded junction is the most popular. The ungrounded junction is the most rugged, but its speed of response is slower than that of the grounded

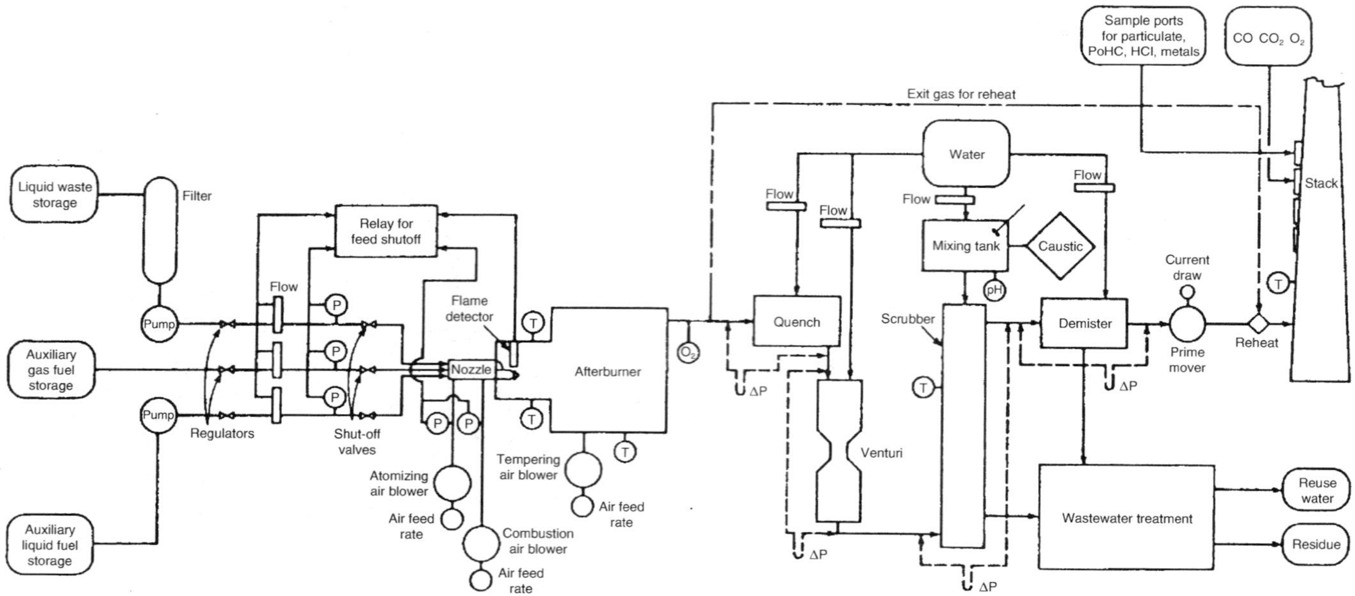

FIGURE 6.38 Typical monitoring locations for a liquid injection incinerator (*EPA-81/09, p. 5-78*).

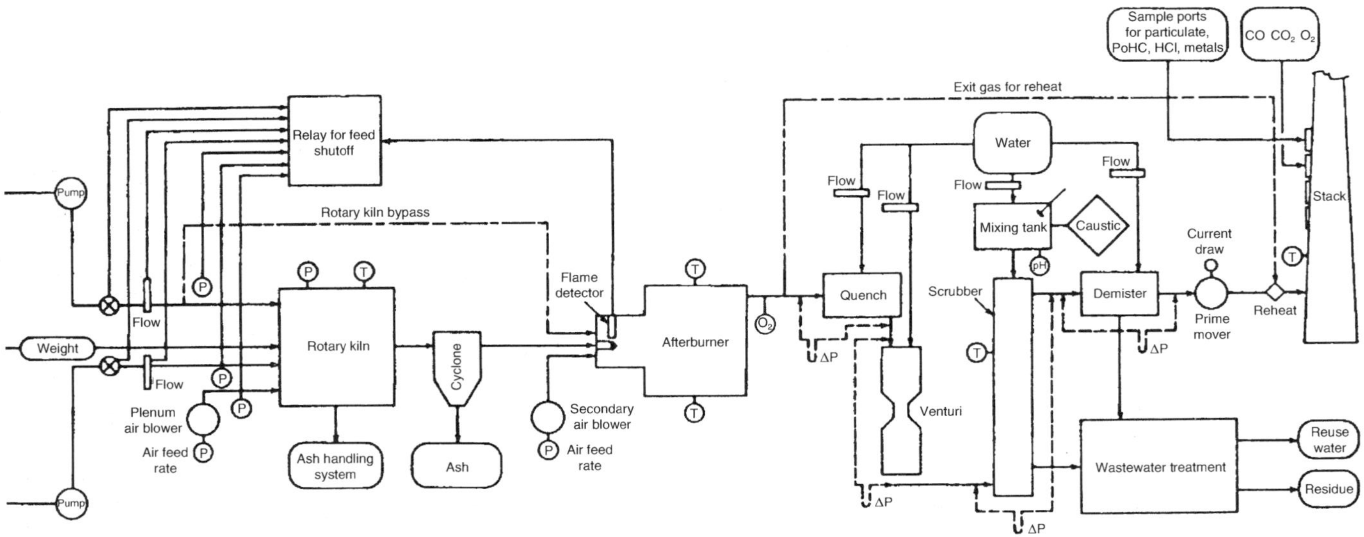

FIGURE 6.39 Typical monitoring locations for a rotary kiln incinerator (*EPA-81/09, p. 5-79*).

TABLE 6.12 Types and Limits of Error for Thermocouples

Type	Temperature range	Limits of error: Standard	Limits of error: Special
Type J—iron versus constantan (modified 1913 calibration)	32–530	±4°F	±2°F
	53–1400	±3/4%	±3/8%
Type K—originally chromel-P versus alumel	32–530	±4°F	2°F
	530–2300	±3/4%	±3/8%
Type R—platinum 13% rhodium versus platinum	32–1000	±5°F	±2.5°F
Type S—platinum 10% rhodium versus platinum	1000–2700	±1/2%	±1/4%
Type T—copper versus constantan	−300 to −75	–	±1%
	−150 to −75	±2%	±1%
	−75 to +200	±1.5°F	±3/4°F
	200–700	±3/4%	±3/8%
Type E—originally chromel-P versus constantan	32–600	±3°F	±2.25°F
	600–1600	±1/2%	±3/8%
Type B—platinum 30% rhodium versus platinum 6% rhodium	1600–3100	±1/2%	–

type. The unprotected exposed junction responds the fastest but is more vulnerable to corrosion and mechanical damage.

A complete thermocouple assembly consists of:

- a sensing element assembly, including in its most basic form two dissimilar wires joined at one end and separated by an electrical insulator;
- a protection tube, either ceramic or metal, or a thermowell. In some cases, both primary and secondary protection tubes are used;
- a thermocouple head or connector; and
- miscellaneous-type hardware such as pipe nipples or adaptors to join the protection tube to the head and thermocouple glands for mounting and pressure sealing.

Protection tubes and thermowells serve the double purpose of guarding the thermocouple against mechanical damage and shielding it from corrosive atmospheres. The choice of the proper material for the protection tube or thermowell is governed by the conditions of use and by the tolerable life of the thermocouple. There may be times when the strength of the protection tube is more important than the long-term stability of the thermocouple. On the other hand, gas tightness, resistance to thermal shock, or chemical compatibility of the protection tube with the process may be the deciding factors.

OXYGEN MONITORING: Oxygen concentration in the combustion gas is usually measured at a point of high turbulence after the gas has traversed the full length of the combustion chamber. A good location for measurement is at the inlet to the duct leading from the combustion chamber to the quench zone, immediately after the gas has gone through a 90° turn. Figures 6.38 and 6.39 show such a location.

Oxygen measurements are made on a continuous basis. Commercially available instruments are discussed in Section 5.9. Whichever type of sensor is used, it is typically equipped with a gas conditioning system specified by the manufacturer for the gas environment in which the instrument is used.

When measuring oxygen concentration directly in a high temperature flow, some difficulty can be experienced because of molten slag impingement on the probe. Trial-and-error solutions of location and probe length have minimized this problem. A redundant system for scheduled maintenance is desirable.

GAS FLOW MEASUREMENT: Gas flow rates can be measured or approximated in several ways: by insertion in the flue gas duct of an air pressure measuring element (e.g., pitot tube) or by measuring the drop in pressure across a restriction to the gas flow (e.g., baffle plate, venturi section, or orifice) downstream of the combustor. Exhaust gas flow, however, is the most difficult flow measurement application in the incinerator for the reasons given below.

- Because the gas is dusty, moist, and corrosive, pressure taps will tend to plug. For this reason it is extremely important that the connection to the duct be made sufficiently large and with clean-out provision.
- If the two pressure-sensing points are at widely different temperatures, the resulting difference in density of the gas in the connecting lines to the instrument will create an error in measurement. For this reason, avoid measurement across spray chambers or other locations where gas temperature changes radically.
- If taken across a restriction to gas flow, the fouling tendencies of the dirty gas will cause the restriction to increase with time, thereby changing the differential measurement for a given rate of flow.

For the reasons stated above, the usefulness of this measurement as an indication of quantitative flow is limited, and care should be taken in this application.

Flow measurements are performed at either of two locations: (1) in the duct between the combustion chamber and quench zone, or (2) in the stack (Figs. 6.38 and 6.39). Both locations have their advantages and disadvantages. In the combustion chamber outlet duct, a sufficiently long length of duct may not be available for flow pattem development. Access to this location can also be a problem when the incinerator is vertically oriented and because of the necessity to breech the duct at a high temperature point. High temperatures at this location may require special construction materials (e.g., inconel) for measurement elements.

The advantages of flow rate measurement in the stack are relief of the problems associated with high temperature gas flow measurement, increased accessibility to the gas flow, and increased likelihood of having a proper section of duct for the flow measurement. One minor disadvantage associated with this position is the increased possibility that ambient air leaking into the system upstream of the draft fan could bias the flow measurement.

Of the instruments available to measure gas flow in closed conduits, pressure or velocity head meters are among the oldest and most common. The principal shortcomings are the need for elements to be inserted directly into the flow paths (in contact with the gas stream), making them susceptible to corrosion, erosion, and fouling, the requirement for seals, the likelihood that the conduit may have to be opened for inspection or service, and permanent pressure losses caused by restrictions placed in the channels.

Head-type flow meters incorporate primary elements, which interact directly with the streams to induce velocity changes, and secondary elements, which sense the resulting pressure perturbations. The flow rate of interest is a function of the differential pressures which can be detected.

5.10 Preplanning of Facility Operation

Preplanning the proper operation of a waste incineration facility is necessary for protection and to prevent adverse effects of the facility on the public health or the environment. Proper facility operation, on a day-to-day basis, includes:

- operation plans including waste classification, waste storage, and disposal procedures, incinerator monitoring, and administrative procedures such as hours of operation per day and days per week;
- operation manuals, including equipment manuals, spare parts, etc.;

- safety at the site, including emergency manual or handbook, leak detection and repair plan, chemical spill handling, facility security, operator practices and training, and a loss prevention program;
- monitoring of operating parameters;
- monitoring to ensure protection of the environment; and
- operator training.

These plans are developed within the operating company (and corporate structure) and are done in cooperation with other neighboring or similar organizations and with governmental agencies. It may not always be possible for all of them to fully cooperate or participate, but through planned action each organization is made aware of certain available assistance.

6 SUMMARY OF INCINERATION SYSTEM

This section is a summary of all discussions about solid waste treatment, and is self-explanatory.

Figure 6.40 depicts the hardware and software requirements in operating a treatment facility and in getting a permit to operate one.

Figure 6.41 describes the possible configuration of an incineration system. The PCC can be (1) a rotary kiln for mostly industrial waste combustion, (2) a mass burning combustor for

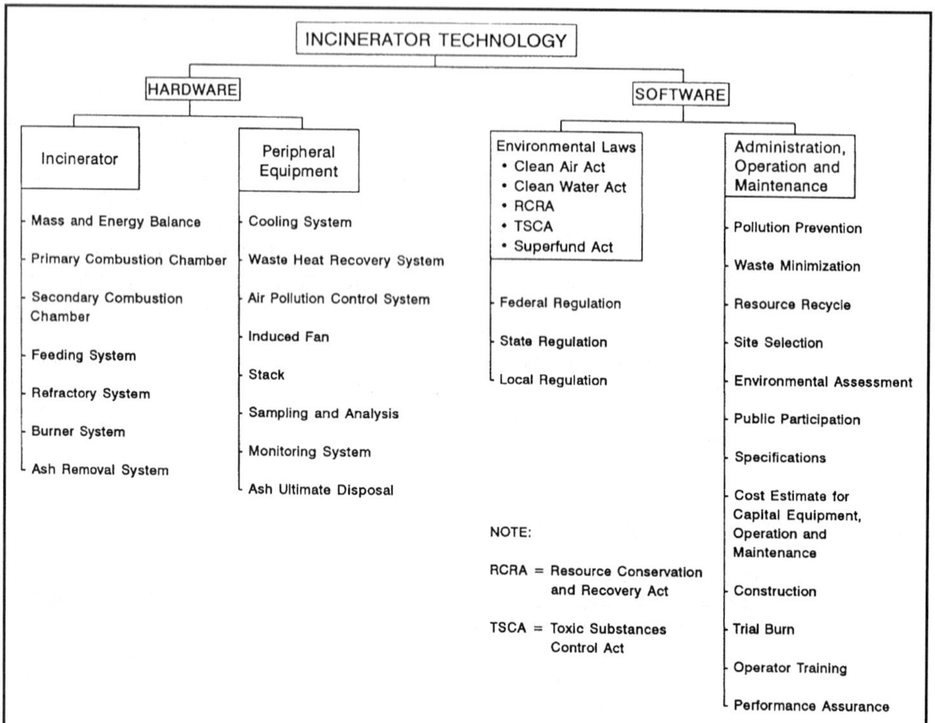

FIGURE 6.40 Key information requirements for permitting an incinerator.

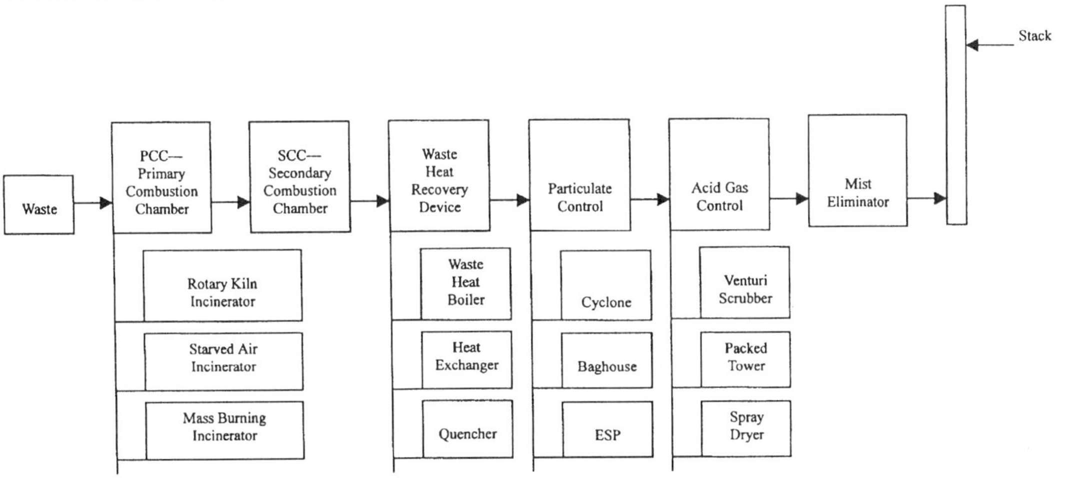

FIGURE 6.41 Potential configurations of an incineration system. The PCC can be a rotary kiln, starved-air, or mass burning incinerator depending on the type of wastes. The waste heat recovery device can be a waste heat boiler, heat exchanger, or quencher. Particulate control can be a cyclone, baghouse, or ESP (electrostatic precipitator). Acid gas control can be a venturi scrubber, packed tower, or spray dryer.

municipal waste incineration, or (3) a starved-air combustor for mostly medical waste incineration. Other types of PCC may include multihearth combustors and fluidized beds for mostly sludge incineration. The waste heat system is to extract heat from combustion gas for potential reuse and to reduce the combustion gas temperature for air pollution control devices which follow. The particulate control system can be a cyclone, a baghouse, or an ESP (electrostatic precipitator). The acid gas control system can be a venturi scrubber, a packed tower, or a spray dryer. The particulate control system and the acid gas control system are sometimes in reverse order. In other words, after the waste heat recovery device, the system sometimes has acid gas control and then particulate control.

Figure 6.42 shows some typical temperature distributions in different configurations of a rotary kiln incineration system.

The following information summarizes the typical ranges for temperature and residence time in six popular incinerators (EPA-81/09, p. 2-2):

- rotary kiln 1500–2900°F
- liquid injection 1500–2900
- fluidized bed 840–1800
- coincineration 300–2900
- starved air 900–1500
- multiple hearth
  - drying zone 600–1000
  - incinerating zone 1400–1800

In general, for gas or liquid waste, the residence time is less than 2 s. For solid waste, it is in the range of hours, depending on the solids size and nature. This general rule applies to all types of incinerators.

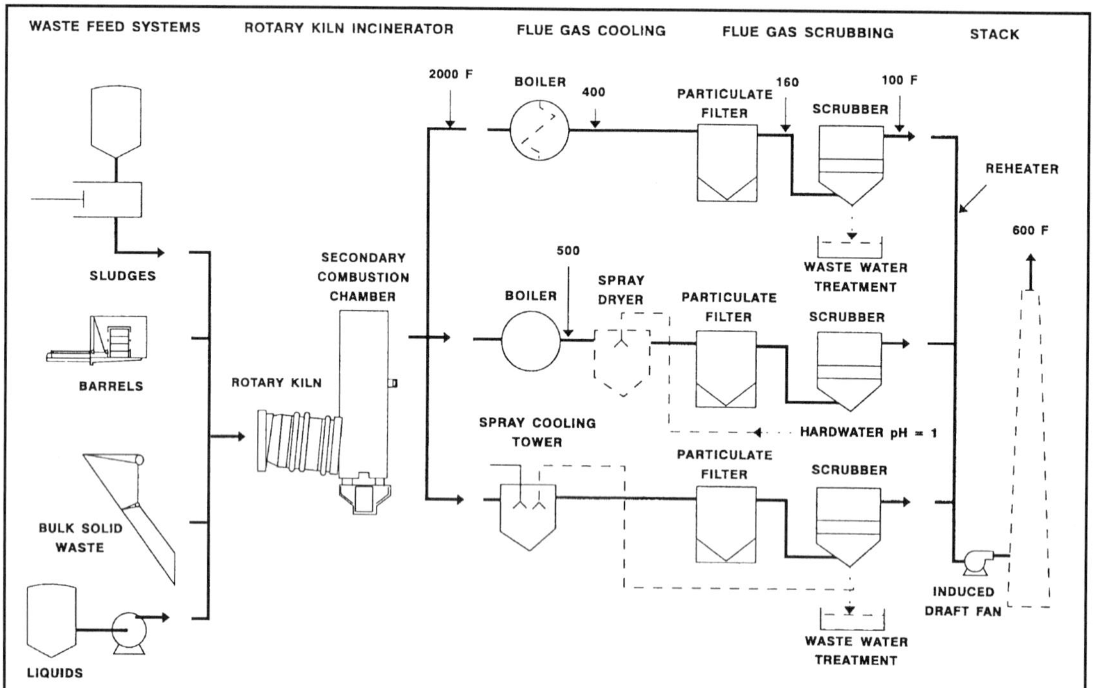

FIGURE 6.42 Different configurations of a rotary kiln system.

REFERENCES

(Dempsey-93/01), Incineration of hazardous waste: A critical review update, by C. R. Dempsey and E. T. Oppelt. An article published in Air and Waste, Vol. 43, January 1993.

(DOE-93/03), Summary of thermal treatment technologies within the Department of Energy Nuclear Weapons Complex, A Department of Energy (DOE) report prepared by Ralph A. Koenig, Merlin Co., Order number: DE-AP01-92 EW30054-A000; Reference number: 01-92EW30054.000, March 31, 1993.

(EPA-72/08), Afterburner system study, EPA R2-72-062, NTIS PB-212-560, August 1972.

(EPA-73/05), Air pollution engineering manual, 2nd edn., AP-40, Air Pollution Control District County of Los Angeles, May 1973.

(EPA-76/11), Physical, chemical, and biological treatment techniques for industrial wastes. An EPA report prepared by Arthur D. Little, Inc., NTIS PB-275-054 (Vol. I), and PB-275-287 (Vol. II), November 1976.

(EPA-81/09), Engineering handbook for hazardous waste incineration, SW889, September 1981.

(EPA-86/01), RCRA orientation manual, EPA530-SW-86-001, January 1986.

(EPA-86/03), Hazardous waste incineration – Course 502, EPA Air Pollution Training Institute (APTI), March 7, 1986.

(EPA-86/09), Permit writer's guide to test burn data, EPA625-6-86-012, September 1986.

(EPA-86/10), Technical resource document: Treatment technologies for dioxin-containing wastes, EPA600-2-86-096, October 1986.

(EPA-91/11), The superfund innovative technology evaluation: Technology profiles, 4th edn, EPA540-R-91-008, November 1991.

(EPA-94/05), Combustion emissions technical resource document (CETRED), EPA530-R-94-014, May 1994.

(EPA-94/09), Development and application of thermal treatment technology selection criteria for DOE/Rocky Flats mixed waste, An EPA report prepared by the Energy and Environmental Research Corporation, September 1994.

(EPA-96/03), Environmental fact-sheet, EPA530-F-96-003, March 1996.

(EPA-96/12), Superfund innovative technology evaluation program, 9th edn, EPA540-R-97-502, December 1996.

(Gordon-79/09), Disposal of hospital waste containing pathologic organisms, Judith Gordon, Neal Zank, et al. NTIS report AD-A084-913, September 1979.

(Hasselriis-98/12), Private communication materials, Floyd Hasselriis, December 1998.

(HWHM-91), Minimization of combustion by-products: Regulatory framework, by C. C. Lee and G. L. Huffman. An article published in Hazardous Waste and Hazardous Materials, Vol. 8, No. 4, 1991 (EPA Report #EPA/600/J-94/452, NTIS #PB95-133138).

(Kolak-86), Trial burns—plasma arc technology, N. P. Kolak, T. G. Barton, C. C. Lee, and E. F. Peduto, Proceedings of the Twelfth Annual Research Symposium, EPA/600-9-86/022, August 1986.

(Lee-83/07), A comparison of innovative technology for thermal destruction of hazardous waste, C. C. Lee, Proceedings of the First Annual Hazardous Materials Management Conference, Philadelphia, PA, July 12–14, 1983.

(Lee-90/11), Regulatory framework for combustion by-products from incineration sources, C. C. Lee and G. L. Huffman, Presented at the 1990 Pacific Basin Conference on Hazardous Waste, Honolulu, Hawaii, November 12–16, 1990.

(Lee-91), Minimization of combustion by-products: Regulatory framework, by C. C. Lee and G. L. Huffman. An article published in Hazardous Waste and Hazardous Materials, Vol. 8, No. 4, 1991 (EPA Report #EPA/600/J-94/452, NTIS #PB95-133138).

(Lee-91/02), Minimization of combustion by-products: Characteristics of hazardous waste, by C. C. Lee and G. L. Huffman. Presented at the National Research and Development Conference on the Control of Hazardous Materials, held in Anaheim, California, on February 20–22, 1991 (EPA Report, EPA600/D-90/223, NTIS PB91-162396/AS).

(Lee-98), Energy and mass balance calculations for incinerators, C. C. Lee and G. L. Huffman, Energy Sources, Taylor & Francis, 1998.

(Lee-98/05), Environmental engineering dictionary, 3rd edn. C. C. Lee. Published by Government Institutes, Inc., May 1998.

(OME-86/10), Incinerator design and operating criteria—medical waste incinerator, Vol. II, Ontario Ministry of the Environment (OME), October 1986.

(OME-88/12), Guidance for incinerator design and operation, Vol. I, Ontario Ministry of the Environment (OME). December 1988.

(Oppelt-87/05, Incineration of hazardous waste: A critical review, E. T. Oppelt, Journal of the Air Pollution Control Association (JAPCA), 37, 5, 1987.

(OTA-89/10), Facing America's trash—what next for municipal solid waste?, Congress of the United States, Office of Technology Assessment, OTA-O-424, October 1989.

P • A • R • T • 3

AIR POLLUTION CONTROL CALCULATIONS

CHAPTER 3.1
AIR EMISSION CONTROL

C. C. Lee and G. L. Huffman
U.S. Environmental Protection Agency, Cincinnati, Ohio

J. C. S. Chang
U.S. Environmental Protection Agency, Research Triangle Park, North Carolina

1 INTRODUCTION

The US EPA considers air pollution control as one of its major research topics and has published a significant amount of air pollution control educational materials and other related reports in the past. This chapter and chapters 3.2 and 3.3 have been heavily excerpted from US EPA publications including those listed in the References. The excerpted materials were then rearranged and edited to make the materials more compact for beginners to understand the basic concepts and basic equipment of air pollution control processes.

Air emissions can result from many sources. The major air emission sources include fossil fuel combustion, waste incineration, and industrial processes. The controls of air emissions include many processes and equipment. The main objective of installing air pollution control equipment (APCE) is to scrub pollutants from contaminated gases to meet regulatory requirements. Many types of equipment are available for use in controlling air emissions from incinerators. Generally speaking, air pollution control processes include:

- Gaseous emission control process: description of this process is provided in this chapter.
- Particulate emission control process: description of this process is provided in later chapters.

Most often, however, one control technique is used more frequently than others for a given source–pollutant combination. For example, absorption is commonly used to remove SO_2 from boiler flue gas. A summary of air emission control is provided in Table 1.1.

Many factors influence the choice of a control device used to reduce air emissions. For example, for particulate matter control, the composition of the particles in terms of concentration, size, and chemical and physical characteristics must be considered. If emitted materials can be re-used in the process, dry collection may be desirable. If the pollutant has little economic value, collection should be accomplished with the safe and economical disposal of the collected pollutants.

The industrial process and potential control devices must both be carefully reviewed. It is very important to not convert an air pollution problem into a water or land pollution problem because of air pollution control activities.

The stringency of air pollution control requirements might also include other methods in addition to the traditional add-on control devices. This calls for the consideration and application of alternative production procedures, modifications of processes, and control techniques that result in minimum emissions from a source.

For example, waste generators who send their waste to industrial furnaces such as cement kilns that act as commercial waste management facilities will have an incentive to reduce the generation of metal- and chlorine-bearing wastes because waste management fees are likely to increase for such waste given that the burner has a fixed metal and chlorine feed rate allotment (due to prescribed feed rates and facility operating conditions). Wastes with extremely high metals content may no longer be acceptable for burning in many cases unless the waste generator reduces the metals content of the waste. Any alternative for the disposal of such wastes may be unavailable or the costs of such treatment may be high enough to create the incentive to reduce waste generation rates at the source. This is a typical scenario for pollution prevention measures to be undertaken by waste generators.

Similarly, generators who burn their wastes onsite also have the incentive to reduce the generation of metal- and chlorine-bearing wastes given that the regulatory rule that they operate under will often require a fixed feed rate allotment of these materials for their boiler or industrial furnace.

2 AIR POLLUTION DEFINITION

An air pollutant is also known as an air contaminant. It is a substance in air which could, if in high enough concentration, harm human health or the environment. Pollutants may include almost any natural or artificial composition of matter capable of being airborne. Pollutants may be in the form of solid particles, liquid droplets, gases, or in combinations of these forms. Generally, they fall into two main groups: (1) those emitted directly from identifiable sources and (2) those produced in the air by interaction between two or more primary pollutants, or by reaction with normal atmospheric constituents, with or without photoactivation. Table 1.2 shows typical air pollutants, their sources, and their health effects (EPA-73/05).

The common air pollutants discharged from various sources, particularly from waste incineration, include:

- particulate matter including metals
- inorganic gases
- organic gases

TABLE 1.1 Air Pollution Control

Pollutant types	Control process	Control equipment	Subset of control equipment
1. Gaseous emission*	1.1. Absorption	1.1.1. Wet absorption scrubber	1.1.1.1. Spray tower or chamber 1.1.1.2. Venturi 1.1.1.3. Ejector 1.1.1.4. Packed column 1.1.1.5. Plate tower 1.1.1.6. Mobile packed bed
		1.1.2. Dry absorption scrubber	1.1.2.1. Spray dryer 1.1.2.2. Dry injection
	1.2. Adsorption	1.2.1. Non-regenerable adsorption system	
		1.2.2. Regenerable adsorption system	1.2.2.1. Fixed bed adsorber 1.2.2.2. Moving bed adsorber 1.2.2.3. Fluidized bed adsorber
	1.3. Incineration	1.3.1. Direct combustor or flare 1.3.2. Incinerator (after-burner) 1.3.3. Catalytic oxidizer 1.3.4. Process boiler used as incinerator	
	1.4. Condensation	Condenser	
	1.5. NO_x emission control	1.5.1. Fuel denitrogenation (before burning) 1.5.2. Combustion modification (during burning) 1.5.3. Flue gas treatment (after burning)	
	1.6. SO_x emission control	1.6.1. Fuel desulfurization 1.6.2. Combustion of coal and limestone mixture 1.6.3. Coal gasification 1.6.4. Coal liquefaction 1.6.5. Flue gas desulfurization (FGD)	
2. Particulate emission	2.1. Dry process†	2.1.1. Gravity force	Gravity settling chamber
		2.1.2. Centrifugal force	Cyclone
		2.1.3. Impaction, interception, diffusion	Baghouse
		2.1.4. Electrostatic attraction	Electrostatic precipitator (ESP)
	2.2. Wet process‡	3.1. Impaction, interception, diffusion	Wet scrubber

* For details, see this chapter.
† For details, see Chapter 3.2.
‡ For details, see Chapter 3.3.

TABLE 1.2 Air Pollution Sources versus Health Effects

Pollutants and sources	Health effects* A	B	C	D
1. Organic gases [hydrocarbons (HC)]				
• Paraffins: processing and transfer of petroleum products; use of solvents; motor vehicles, etc.		×	×	
• Olefins: processing and transfer of gasoline; motor vehicles, etc.	×	×	×	
• Aromatics: same as for paraffins	×	×	×	odor
• Others:				
—oxygenated HC† from use of solvents and motor vehicles		×	×	odor
—halogenated HC‡ from use of solvents		×	×	odor
2. Inorganic gases				
• Oxides of nitrogen: combustion of fuels; motor vehicles, etc.	×	×	×	×
• Oxides of sulfur: combustion of fuels; chemical industry, etc.	×	×	×	×
• Carbon monoxide: motor vehicles; petroleum and metals industry, etc.				×
3. Particles or aerosols				
• Solid particles (carbon or soot): combustion of fuels; motor vehicles, etc.			×	×
• Metal oxides and salts: catalyst dusts from refineries; motor vehicles; combustion of fuel oil, etc.			×	
• Silicates, mineral dusts and metallic fumes: metals industry, etc.			×	

* Health effects: A, plant damage; B, eye irritation; C, visibility reduction; and, D, other.
† For example, aldehydes, ketones and alcohols.
‡ For example, carbon tetrachloride.

- acid gas
- odor
- noise

2.1 Particulate Matter

The terms ‘particulate matter’ and ‘particle’ have been used interchangeably in many air pollution control documents. Particulate matter is defined as any airborne finely divided solid or liquid material with an aerodynamic diameter smaller than 100 micrometers (micro $= 10^{-6}$) (40CFR51.100-90). Some examples of particulates are trace metals, smoke, dusts, fumes, mists, and sprays. Control of particulate matter is necessary to maintain or improve air quality in the atmosphere.

The most visible discharge from a burning process is smoke. Smoke is a suspension of solid or liquid particulate matter in a gaseous discharge. The particles range from fractions of a micron to over 50 microns in diameter. One micron (μm) is one-millionth of a meter and is commonly used to describe particulate size. The visibility of smoke is related to the quantity of particles present rather than to the weight of the particulate matter. The weight of particulate emissions is, therefore, not necessarily indicative of the density of the emission. Neither is the color of a discharge related to opacity or smoke density. Smoke can be black or can appear non-black; in the latter case, it is termed white smoke. These two types of smoke are discussed below:

White smoke: The formation of white or other opaque, non-black smoke is usually due to insufficient furnace temperatures when burning carbonaceous materials. Hydrocarbons will be heated to a level where evaporation and/or cracking will occur within the furnace when white smoke is produced. The temperatures will not be high enough to produce complete combustion of these hydrocarbons. With a stack temperature in the range of 150–260°C (300–500°F), many of these hydrocarbons will condense to liquid aerosols; with solid particulate present, these will appear as non-black smoke.

A method for controlling white smoke is to increase the furnace/stack temperatures and increase turbulence to help ensure uniformity of this temperature within the off-gas flow.

Excessive airflow may result in excessive cooling. An evaluation of reducing white smoke discharges should include investigating the air quantity introduced into the furnace. Inorganics in the exit gas may also produce a non-black smoke discharge. For instance, sulfur and sulfur compounds will appear yellow in a discharge; calcium and silicon oxides in the discharge will appear light to dark brown.

Black smoke: When burned in an oxygen-deficient atmosphere, hydrocarbons will not completely destruct, and carbon particles will be found in the off-gas. Causes of local oxygen deficiency include poor atomization, inadequate turbulence (or mixing), and poor air distribution within a furnace chamber. Each of these factors will generate carbon particles that produce dark, black smoke in the off-gas.

A common method of reducing or eliminating black smoke has been steam injection into the furnace. The carbon present is converted to methane and carbon monoxide as follows:

$$3C \text{ (smoke)} + 2H_2O \rightarrow CH_4 + 2CO$$

Hydrocarbons react similarly, and the methane and carbon monoxide produced burn clean in the heat of the furnace, eliminating the black carbonaceous smoke that would have been produced without steam injection:

$$CH_4 + 2O_2 \rightarrow CO_2 + 2H_2O \text{ (smokeless)}$$

$$2CO + O_2 \rightarrow 2CO_2 \text{ (smokeless)}$$

A number of devices can be used to control the emissions of particulate matter. The devices include:

- mechanical collector
- baghouse
- electrostatic precipitator (ESP)

2.2 Inorganic Gas

Inorganic gases produced from the burning process normally include water vapor, carbon dioxide, carbon monoxide, oxides of nitrogen, and, when sulfur is present, oxides of sulfur.

Nitrogen is a reactive substance at the high temperatures of combustion that forms nitrogen oxide (NO) and nitrogen dioxide (NO_2) from the combustion process.

Sulfur is released into the atmosphere from burning processes in the form of sulfur dioxide, SO_2, and sulfur trioxide, SO_3.

2.3 Organic Gas

Of the many organic discharges from incinerators, the more significant ones are:

- Oxygenated hydrocarbons, which include aldehydes, ketones, alcohols, and acids. In sufficient quantity they will produce eye irritation, reduce visibility in the atmosphere, and react with other components of the atmosphere to form additional pollutants. Many of these compounds are also odorous.

- Halogenated hydrocarbons, such as carbon tetrachloride, perchloroethylene, etc. These may contribute to atmospheric clarity problems and may also generate odor.
- Polychlorinated dibenzo-*p*-dioxins (PCDD) and polychlorinated dibenzo-furans (PCDF). These compounds may be formed when materials containing chlorine are incinerated. These compounds are of considerable concern due to their toxicity in trace quantities to laboratory animals.
- Olefins, which are a group of unsaturated hydrocarbon compounds that readily react with many other chemical compounds. Olefins take part in photochemical reactions in the presence of nitrogen oxides and several other pollutants.
- Aromatics, including benzene, toluene, xylene, and polycyclic aromatic hydrocarbons (PAH). One of the most toxic of these occurring in industrial processes is benzo(a)pyrene (also written as 3,4,benzopyrene). This compound is relatively simple to detect. A primary source of these compounds is the incomplete combustion of organic materials.

The most apparent effect of organic gas discharges is their effect on the atmosphere. They react with other elements of the atmosphere to form photochemical smog, which is as much a danger to health as it is to the aesthetic quality of an area.

2.4 Acid Gas

Industrial waste will often have either a halogen or sulfide component. Municipal wastes have relatively high proportions of plastics, and many of these materials have a significant amount of chlorides. Chlorides and/or sulfides in an incinerator charge will produce hydrogen chloride and/or chlorine, and sulfur dioxide/trioxide gas in the exhaust stream. Acid gas neutralization systems (including water scrubbing followed by water neutralization) should be provided when significant quantities of hydrogen chloride are present in an exhaust. Control of these emissions is often necessary for the protection of downstream equipment (OTA-84/06).

2.5 Odor

Odors associated with incineration are normally organic and result from the incomplete combustion of organic matter in the waste feed. The most effective means of dealing with the problem of odor generation is alteration of the burning process to increase burning efficiency, thereby decreasing the occurrence of odor-causing compounds in the exhaust stream.

Since waste streams vary in chemistry or quality, the efficiency of the waste burning process can also change, and incineration parameters must be adjusted to reduce the possibility of odor generation.

Secondary odors may originate from equipment and systems within the incinerator facility such as storage tanks, waste storage pits, etc. Where practical, ventilation systems should be designed to contain such odors and to bring the odorous air to the incinerator for destruction. Other mechanisms of odor control may be required, such as fume incineration, use of a packed tower (absorber), catalytic oxidation system, adsorption (using carbon on silicon beds), dilution, or masking.

2.6 Noise

Noises generated by an incineration facility are generally those resulting from the movement of air or gas. Fans and blowers will create the greatest noise levels; of secondary concern is the noise generated by passage of air or gas through ducts, flues, and nozzles.

Local noise can usually be controlled by initial equipment selection, use of sound-absorbing equipment, and provision of sound-deadening or sound-absorbing enclosures around noisy equipment. Exterior noise, however, can be broadcast hundreds of meters, perhaps kilometers. The usual cause of broadcast noise is the pulsation of the ID fan.

Sound control equipment sizing is proportional to the wavelengths of the sound generated. The lower the frequency, the longer the wavelength and the larger the sound control equipment. In controlling ID fan broadcast noise, therefore, it is first necessary to try to increase the beat frequency. If possible, the fan speed should be increased, or if this is impractical, the number of blades should be increased. The ID fan should be close-coupled to the stack. Ductwork with appropriate expansion joints will tend to absorb some of the potential broadcast noise.

Two types of insertion sound control devices are used to reduce broadcast noise. The first is an exhaust silencer, which is placed within the stack or between the ID fan and the stack. Exhaust silencers are resonating devices that tend to balance the generated beat by producing (by static geometry) a frequency that is 180 degrees out of phase from the fan beat. The net result, ideally, is cancellation of the beat frequency plus harmonics. The second type of sound control device is a baffle, usually perforated, that absorbs sound energy with its acoustical fill: fiberglass, mineral wool, or another absorbent.

3 POLLUTION EMISSION CALCULATION

3.1 Units of Emission Standards

Combustion sources constitute a significant air quality control problem because of the gaseous and particulate emissions which can be produced. With a variety of combustion systems devised for a multitude of end uses, the US EPA has been developing emission standards for the combustion or incineration industry to follow. Accordingly, the emission standards usually establish the maximum allowable limit for the discharge of specific pollutants. These limits are usually based upon volume or mass flows at specified conditions of temperature and pressure. Actual field measurements of gas flow likely would not be made with gas at standard conditions. It is therefore necessary to adjust the observed volume flow to account for difference in pressure and temperature (EPA-80/02, p. 5-1; EPA-89/03).

Emissions can be measured in terms of the concentration of pollutant per volume or mass of flue (stack) gas, the pollutant mass rates, or a rate applicable to a given process. Standards therefore fall into the following general categories:

- pollutant mass rate standards
- process rate standards
- concentration standards
- ambient concentration standards
- reduction standards
- opacity standards

Pollutant mass rate standards. Under this general category, the standards are based on the fixed rate of emissions: for example, the mass of pollutant emitted per unit time, lb/h or kg/h.

Process rate standards. Process rate standards usually establish the allowable emission in terms of either the input energy or the raw material feed of a process. New source standards for fossil fuel-fired steam power plants are an example of an energy basis standard. Allowable emissions for such operations as acid plants are based upon the mass of acid produced, while a portland cement plant emission standard is in terms of the number of tons of material fed into the kiln.

Where combustion sources are involved, a standard may include not only the allowable concentration but also may specify the quantity of excess air the system may use while achieving this concentration. The standard for solid waste incinerators of 50 T/day (tons/day) or greater is an example of this type of standard (for a calculational example, see Sec. 3.5).

Concentration standards. The most common type of emission limit is the concentration standards, which limit either the mass (weight) or volume of the pollutants in the gas exiting the stack. This type of emission limit is expressed as follows and is shown in Figs. 1.1 and 1.2:

- 1 grain per dry standard cubic foot (1 gr/dscf) at 7 percent oxygen
- Milligrams per dry standard cubic meter (mg/dscm)
- Parts per million parts (ppm)

One grain per dry standard cubic foot (1 gr/dscf) at 7 percent oxygen: This is a British (Imperial) concentration unit to express the limit of particulate matter (by weight) per unit of stack gas volume. This unit means that no more than 1 grain of particulate matter is allowed to be contained in each cubic foot of gas leaving the stack corrected to 7% oxygen and to standard conditions (20°C and 1 atm) (1 pound = 7000 grains).

Milligrams per dry standard cubic meter (mg/dscm): This is a metric concentration unit to express the limit of particulate matter per unit stack gas volume. The conversion factor is

$$1\ \text{gr/dscf} = 2300\ \text{mg/dscm}$$

Parts per million parts (ppm): This unit expresses the limit of pollutants (by volume) per stack gas volume. The pollutants under this limit include (1) carbon monoxide; (2) sulfur dioxide; (3) nitrogen oxides; and (4) hydrogen chloride. This unit is applicable to both British and metric systems. By definition, ppm can be expressed as

$$1\ \text{ppm} = (1\ \text{mole of pollutant})/(10^6\ \text{moles of flue gas})$$

For example, a 100 ppm standard means that no more than 100 cubic feet (cubic meters) of pollutants are allowed to be contained in 1 million cubic feet (cubic meters) of gas leaving the stack.

Because a concentration standard limits the amount of pollutants in a certain amount of stack gas, someone having a problem meeting the standard might be tempted to increase the

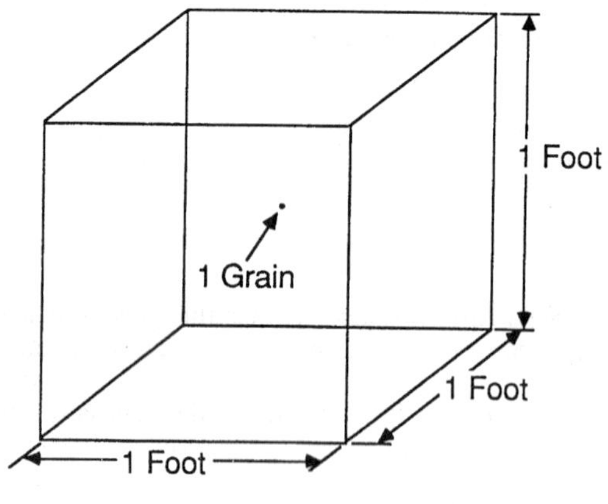

FIGURE 1.1 *Concentration standards (EPA-89/03).*

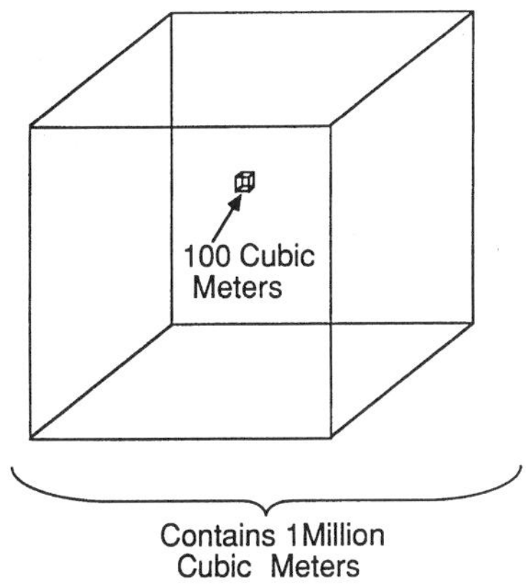

FIGURE 1.2 Concept of 100 ppm in concentration standards (*EPA-89/03*).

amount of dilution air to dilute the concentration of the pollutant in the exhaust gas. As air is added, the oxygen concentration in the gas increases because the air contains 21% oxygen. To keep this from happening, regulations usually either forbid the addition of dilution air or require that the concentration be 'corrected' to a standard level of oxygen, usually 7%, or a standard level of carbon dioxide, usually 12%. Figure 1.3 illustrates this concept.

Emission limits are often given for standard conditions, e.g. 0.1 grain/dry standard cubic foot. Standard temperature is often at 68°F (20°C) and standard pressure is 29.92 in. w.c. (760

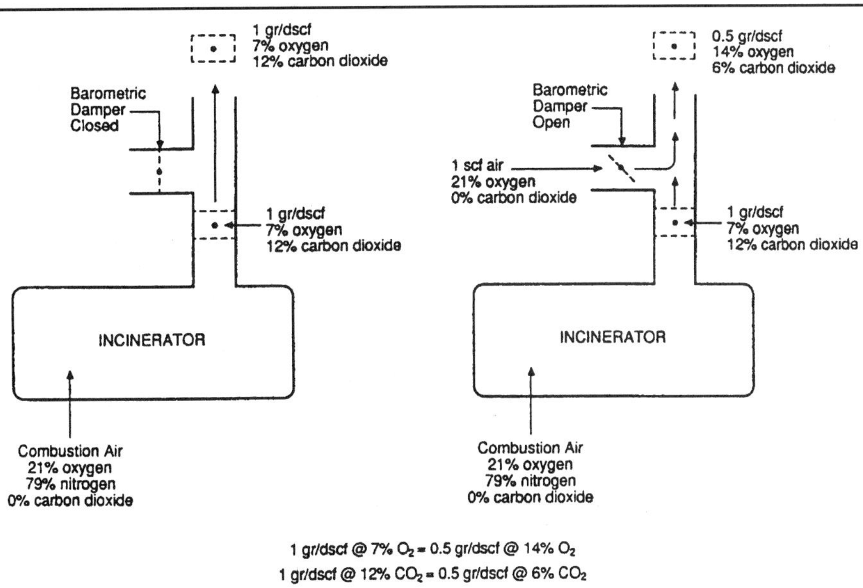

FIGURE 1.3 Concept of correction for dilution (*EPA-89/03*).

millimeters of mercury). A cubic foot measured at this temperature and pressure is known as a standard cubic foot. When a stack test is performed to check the level of emissions from an incinerator, both temperature and pressure are measured during the test in addition to the pollutant of interest. The test results are then converted to standard conditions (grain/dry standard cubic foot) using the temperature and pressure measured. In this way, all test results of all sources including incinerators can be compared on the same basis; i.e. all results are reduced to standard conditions.

Ambient concentration standards. Microgram per cubic meter ($1\,\mu g/m^3$) is the unit for this standard category. It is a metric concentration unit to express the limit of pollutant (by weight) per unit ambient air volume. The pollutants under this limit include (1) toxic metals (arsenic, beryllium, cadmium, chromium, nickel, lead, mercury); (2) organics (dioxins/furans, ethylene, propylene); and (3) hydrogen chloride. For example, $1\,\mu g/m^3$ means that no more than 1 microgram of pollutant is allowed to be contained in each cubic meter of ambient air (there are 1 million micrograms in 1 gram).

This unit is sometimes found in state regulations. It limits the amount of pollutants that collect at ground level in areas surrounding the emission source. Usually, the regulations require that the pollutant be measured as it leaves the stack. This measurement information is then used by a computer to calculate the amount of the pollutants at various locations near the source.

Reduction standards. Sometimes the emission limit is expressed as a percent reduction of the pollutants. In other words the pollution control device must operate at or above a specified efficiency level (such as 90% removal) to reduce the pollutant emissions. This type of standard frequently is used for acid gases such as HCl. For example, if the emission standard requires at least 90% reduction of HCl, and the HCl in the combustion gas is entering the scrubber inlet at a rate of 20 lb/h (9.1 kg/h), then the allowed emission rate is 2 lb/h (0.9 kg/h), which is 10% of the amount entering the scrubber.

Opacity standards. Opacity standards are almost always included in regulations. Opacity standards express the limit on the degree to which the stack emissions are visible and block the visibility of objects in the background. Stack emissions of 100% opacity would totally block the view of background objects and indicate high pollutant levels. Zero percent opacity would provide a clear view of the background and indicate no detectable particulate matter emissions. Opacity may be estimated by taking 'readings' every 15 seconds and averaging the readings over a specified time period. The 'reader' must be a certified opacity reader. The US EPA Reference Method 9 '*Visual determination of the opacity of emissions*' establishes the procedures and criteria for taking opacity readings and for certification. Additionally, opacity may be estimated by comparing the opacity of the smoke to the six sections of a Ringelmann Smoke Chart (Fig. 1.4). The six sections are numbered from 0 to 5 with No. 0 being completely white and No. 5 completely black. Sections 1 through 4 correspond to opacities of 20% (No. 1), 40% (No. 2), 60% (No. 3), and 80% (No. 4). Opacity is estimated by choosing the section which most closely resembles the opacity of the exhaust gas. Opacity may also be measured by an instrument called a transmissometer that is installed in the stack. The following further illustrates an opacity standard: for example, 10% opacity (6-minute average) means that the opacity of the particulate matter emissions cannot average more than 10% for any 6-minute period.

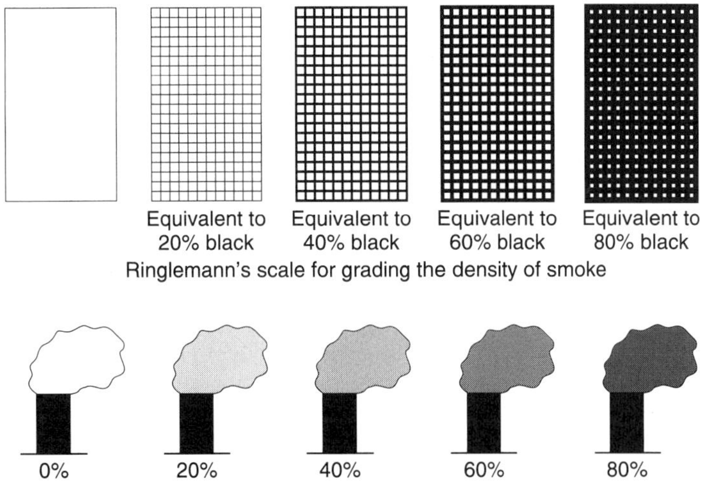

FIGURE 1.4 Opacity and the Ringelmann chart (*EPA-89/03*).

3.2 Volume Correction

Because combustion devices always produce flue gas that is at a higher temperature and pressure than those of the standards, corrections for the difference must be made. Consider one cubic foot of gas at some specified condition, say 14.7 psia and 70°F. Does this volume increase or decrease if one raises the gas temperature? Ask a similar question regarding the effect of a pressure increase. What volume would the gas occupy if both pressure and temperature were raised? The answers to these questions can be developed using the ideal gas equation (EPA-80/02, p. 5-1).

Consider the combustion gas as an ideal gas. For an ideal gas at states 1 and 2, the following three equations are valid:

$$\textit{Volume equation: } V_2/V_1 = (P_1/P_2)(T_2/T_1) \tag{1.1}$$

$$\textit{Density equation: } \rho_2/\rho_1 = (P_2/P_1)(T_1/T_2) \tag{1.2}$$

$$\textit{Concentration equation: } C_2/C_1 = (P_2/P_1)(T_1/T_2) \tag{1.3}$$

where V_1 = volume at condition 1
V_2 = volume at condition 2
P_1 = pressure at condition 1
P_2 = pressure at condition 2
T_1 = absolute temperature at condition 1
T_2 = absolute temperature at condition 2
ρ_1 = density at condition 1
ρ_2 = density at condition 2
C_1 = concentration at condition 1
C_2 = concentration at condition 2
The subscript 1 denotes the observed conditions and the subscript 2 the standard conditions or vice versa

EXAMPLE 1: volume correction

A volume of $20\,m^3$ was drawn from a spirometer at 20°C and 700 mm Hg. What was the standard volume drawn?

Solution: Using the ideal gas equation in state 1 and state 2 conditions, the equation $V_2 = V_1(P_1T_2/P_2T_1)$ can be applied to this case.

where V_2 = volume at condition 2 (standard condition to be determined)
V_1 = volume at condition 1 = $20\,m^3$
P_1 = pressure at condition 1 = 700 mmHg
P_1 = pressure at condition 2 = 760 mmHg
T_1 = temperature at condition 1 = 20°C + 273 = 293 K
T_2 = temperature at condition 2 = 25°C + 273 = 298 K

$$V_2 = (20 \times 700 \times 298)/(760 \times 293) = 18.7\,m^3$$

Conversion of ppm into $\mu g/m^3$

As an applied example, consider using the ideal gas equation to help develop a conversion factor with which ppm can be converted into $\mu g/m^3$. Begin with the definition:

$$1\text{ ppm} = (\text{moles of product})/(10^6\text{ moles of air})$$

Note: This is basically a volume measure; the definition is based on $T = 25°C$ and $P = 760$ mmHg. Recall here that a mole of any gas will occupy a volume of 22.4 liters when $P = 760$ mmHg and $T = 0°C$. The definition of ppm is based on $T = 25°C$; therefore, one must calculate the new volume using Eq. (1.1)

$$V_2/V_1 = (P_1/P_2)(T_2/T_1)$$

Normally, a constant pressure is assumed between conditions 1 and 2, such that

$$\begin{aligned} V_2 &= V_1(T_2/T_1) \\ &= 22.4(273 + 25)/273 \\ &= 24.5\text{ liter} \end{aligned}$$

In turn, there are $10^{-3}\,m^3$/liter and the mass of the moles of product = molecular weight (MW), g/mole.

Combining these conversions:

$$\begin{aligned} 1\text{ ppm} &= (10^{-6}/24.5)[(\text{moles per mole air})/(\text{liters per mole air})][\text{MW}/10^{-3}][\text{g/mole}]/[\text{m}^3/\text{liter}] \\ &= (10^{-3}/24.5)(\text{MW})[(10^6)(\mu\text{g/g})] \\ &= 40.8[\text{MW}][\text{this term has the units of } \mu\text{g/m}^3] \end{aligned}$$

$$1\text{ ppm} = 40.8[\text{MW}], \quad \mu\text{g/m}^3 \tag{1.4}$$

EXAMPLE 2: units of ppm and $\mu g/m^3$ for SO_2

Apply Eq. (1.4)

$$\begin{aligned} 1\text{ ppm } SO_2 &= 40.8(64) \\ &= 2611\ \mu\text{g/m}^3 \end{aligned}$$

3.3 Determination of Excess Air

Another type of calculation often necessary involves combustion equipment stack gas samples obtained by Orsat analysis. Before outlining the fundamental basis of corrections here, it would be well to note several aspects of the problem. The stack sampling is directed to determine the pollutants emitted by equipment and compared with standards. The raw gas leaving a combustion device contains certain levels of pollutants, which can be made to appear smaller if the total gas quantity is increased by adding nonpollutant gas to the stream. For example, consider the ideal combustion of carbon monoxide with air (EPA-80/02, p. 5-4).

$$CO + 0.5(O_2 + 3.76N_2) \rightarrow CO_2 + 1.88N_2 \tag{1.5}$$

Here, the percentage of CO_2 in the flue gas is

$$\begin{aligned}\%CO_2 &= (\text{mole of } CO_2)/(\text{mole of } CO_2 + \text{mole of } N_2)\\ &= 1/(1 + 1.88)\\ &= 34.8\% \text{ by volume}\end{aligned}$$

Suppose the same mole of CO were burned with 100% excess air? The combustion reaction now is given by

$$CO + O_2 + 3.76N_2 \rightarrow CO_2 + 0.5O_2 + 3.76N_2 \tag{1.6}$$

Now the total moles of product is given by

$$1 \text{ mole } CO_2 + 0.5 \text{ mole } O_2 + 3.76 \text{ mole } N_2 = 5.26 \text{ moles}$$

$$\%CO_2 = 1/5.26 = 19.0\% \text{ by volume}$$

Here the volume fraction of CO_2 was reduced by adding more air—in effect a dilution of the products by additional air. The original 2.88 moles of flue gas also could have been diluted through the addition of steam, a practice which is fundamentally possible since flue gas temperatures are normally higher than dew point temperatures. Suppose one added two moles of steam to the flue gas of Eq. (1.5):

$$CO_2 + 1.88N_2 + 2 \text{ moles steam} \tag{1.7}$$

Now there are 4.88 moles of product and the CO_2 percentage would be

$$\begin{aligned}\%CO_2 &= 1/4.88\\ &= 20.5\% \text{ by volume}\end{aligned}$$

Clearly, the volume fraction of any gas present in the flue gas can be reduced by dilution, either by adding air or steam. It is for this reason that combustion equipment emission standards are written with a specified amount of excess air and based on dry flue gas. Flue gases which indicate combustion occurred with excess air different from 50% require correction of observed concentration to that which would have been realized with 50% excess air.

Stack gas measurements are usually made with the Orsat apparatus, an absorption device with separate chambers to remove CO_2, CO, and O_2 from the flue gas in a manner permitting measurement of percentage of each present on a volume basis. The device is designed so that a dry basis measurement is realized. Excess air can be determined from the Orsat readings by computation as follows:

Consider the complete combustion of carbon with air:

$$C + O_2 + 3.76N_2 \rightarrow CO_2 + 3.76N_2 \tag{1.8}$$

Here the product contains only $CO_2 + N_2$. With excess air, the reaction becomes

$$C + (1 + e)O_2 + (1 + e)3.76N_2 \rightarrow CO_2 + eO_2 + (1 + e)3.76N_2 \tag{1.9}$$

where e is the number of moles of excess O_2 in the excess air. By definition, the percent of excess air is

$$\%A_e = [(A_a - A_t)/A_t](100) \tag{1.10}$$

where A_a = moles of actual air used in the combustion process
A_t = moles of theoretical air (stoichiometric air) used at 100% of stoichiometric combustion
A_e = excess air

The theoretical air is $O_2 + 3.76N_2$ from Eq. (1.8) with the actual air $(1 + e)O_2 + (1 + e)3.76N_2$ as given by Eq. (1.9). Combining Eqs. (1.8), (1.9) and (1.10):

$$\%A_e = [(eO_2 + e3.76N_2)/(O_2 + 3.76N_2)]100\% \tag{1.11}$$

Equation (1.11) requires knowledge of the excess oxygen e in order to compute the excess air. Actually, the Orsat analysis contains the information to accomplish the same result based on knowledge of the product composition alone.

Note that oxygen can only appear in the product if excess air is present, assuming complete combustion. Noting product with a subscript p:

$$C + (1 + e)O_2 + (1 + e)3.76N_2 \rightarrow CO_{2p} + O_{2p} + N_{2p} \tag{1.12}$$

where O_{2p} = eO_2, the excess air required
N_{2p} = $(1 + e)3.76N_2$, the nitrogen which was part of the total air supplied
e = moles of excess air

Now the nitrogen present in the product came from the combustion air (unless fuel contained significant nitrogen). Therefore, the actual O_2 supplied can be determined by computing the moles O_2 which were associated with N_{2p}. In air, 1 mole of O_2 is associated with 3.76 moles of N_2 (by volume); therefore, for N_{2p} moles of nitrogen, the oxygen supplied is given by

$$O_2 \text{ supplied} = N_{2p}/3.76 = 0.264N_{2p} \tag{1.13}$$

$$\text{The theoretical } O_2 \text{ is } 0.264N_{2p} - O_{2p} \tag{1.14}$$

From Eq. (1.10)

$$\begin{aligned}\%A_e &= [(A_a - A_t)/A_t](100)\\ &= \{[0.264N_{2p} - (0.264N_{2p} - O_{2p})]/(0.264N_{2p} - O_{2p})\}100\%\\ &= [O_{2p}/(0.264N_{2p} - O_{2p})](100)\end{aligned} \tag{1.15}$$

If the combustion produced both CO and CO_2 (case of incomplete combustion), the O_{2p} measured must be reduced by the amount of oxygen which would have combined with CO to form CO_2.

Then:

$$\%A_e = \{[O_{2p} - 0.5CO_p]/[0.264N_{2p} - (O_{2p} - 0.5CO_p)]\}100\% \tag{1.16}$$

In each case, the quantity introduced is the percentage of each constituent as measured by the Orsat analyzer.

EXAMPLE: combustion air analysis by Orsat method

Find the percent excess air by using the following Orsat analysis data:

$$CO_2 = 10\%$$
$$O_2 = 4\%$$
$$CO = 1\%$$

By difference:

$$N_2 = 100 - (10 + 4 + 1) = 85\%$$

Find $\%A_e$ from Eq. (1.16):

$$\%A_e = \{[4 - 0.5(1)]/[0.264(85) - (4 - 0.5(1)]\}100\% = 18.5\%$$

One caution must be mentioned regarding the CO_2 measurement as determined by an Orsat analyzer. The chemical, caustic potash, employed to absorb CO_2 also absorbs SO_2. Therefore, SO_2 must be measured separately from CO_2 and the percentage SO_2 determined must be subtracted from the observed CO_2 reading. Also, the cuprous chloride solution used to absorb CO also absorbs O_2; therefore, a sample which is not correctly analyzed could erroneously indicate O_2 for CO.

3.4 Excess Air Correction

Correction of concentrations where excess air is different from 50% is accomplished by adjusting the gas volume to that which would have been present if 50% excess air had been used. Equation (1.16) and correction factors for 50% excess air, 12% CO_2, and 6% O_2 are presented (EPA-80/02, p. 5-20) below:

Determination of excess air

$$\%A_e = \{[0.264N_{2p} - (0.264N_{2p} - O_{2p})]/(0.264N_{2p} - O_{2p})\}100\% \quad (1.17)$$

Factors for correction to 50% excess air

$$F_{50v} = 1 - [1.5O_{2p} - 0.133N_{2p} - 0.75CO_p]/0.21 \quad (1.18)$$
$$C_{50v} = C_{vs}/F_{50v} \quad (1.19)$$
$$F_{50m} = 1 - (29/M_e)[1.5O_{2p} - 0.133N_{2p} - 0.75CO_p]/0.21 \quad (1.20)$$
$$C_{50m} = C_{ms}/F_{50m} \quad (1.21)$$

Factor for correction to 12% CO_2

$$F_{12v} = CO_{2p}/0.12 \quad (1.22)$$
$$C_{12v} = C_{vs}/F_{12v} \quad (1.23)$$
$$F_{12m} = 1 - (29/M_e)[1.5O_{2p} - CO_{2p}/0.12] \quad (1.24)$$
$$C_{12m} = C_{ms}/F_{12m} \quad (1.25)$$

Factor for correction to 6% O_2

$$F_{6v} = (0.21 - O_{2p})/0.15 \tag{1.26}$$

$$C_{6v} = C_{vs}/F_{6v} \tag{1.27}$$

$$F_{6m} = 1 - (29/M_e)[(1.5O_{2p} - 0.06)/0.15] \tag{1.28}$$

$$C_{6m} = C_{ms}/F_{6m} \tag{1.29}$$

Note: If CO is present in the product gas, the value ($O_{2p} - 0.5CO_p$) must be substituted for O_{2p} in Eqs (1.26) and (1.28).

EXAMPLE: excess air correction
Given conditions

power plant steam generator data
stack gas temperature = 756°R
pressure = 28.49 in. Hg
wet gas flow = Q_o = 367,000 acfm, 6.25% moisture by volume
apparent molecular weight of gas is 29.29
Orsat analysis is CO_2 = 10.7%; O_2 = 8.2%; CO = 0
pollutant mass rate (PMR) is 103 lb/min

With these data, find the following:

(A) Pollutant mass rate, tons/day
(B) Mass and volume basis concentration at the standards: $T_S = 530$ R; $P_S = 29.92$ inches Hg; $\rho_s = 0.0732$ lb/ft^3
(C) Percent excess air in effluent
(D) Concentrations found in B corrected to 50% EA
(E) Concentrations corrected to 12% CO_2
(F) Concentrations corrected to 6% O_2

Solution:
(A) Pollutant mass rate (PMR), tons/day:

$$\text{PMR} = (103\,\text{lb/min})(60\,\text{min/h})(24\,\text{h/day})(\text{ton}/2000\,\text{lb}) = 74.2\,\text{ton/day}$$

(B) Concentration-mass and volume basis

Dry volume at the observed condition, V_o dry = 367,000(1 − 0.0625) = 344,062 acfm

Concentration by volume at the observed condition, $C_{vo} = \text{PMR}/V_o$
$= 103/344{,}062$

Using Eq. (1.3), concentration by volume at the standard condition

$$C_{vs} = (103/344{,}062)(29.92/28.49)(756/530)$$
$$= 4.48 \times 10^{-4}\,\text{lb/dscf or } 3.14\,\text{grain/dscf}$$

Concentration by mass at the standard condition

$$\begin{aligned} C_{ms} &= (C_{vs}\,\text{lb/ft}^3)(\text{ft}^3/0.0732\,\text{lb})(1000/1000) \\ &= (6.12\,\text{lb})/(1000\,\text{lb}) \end{aligned}$$

(C) % Excess air in effluent using Eq. (1.17) is

$$\begin{aligned} \%A_e &= \{[O_{2p} - 0.5CO_p]/[0.264N_{2p} - (O_{2p} - 0.5CO_p)]\}100\% \\ &= (8.2 - 0)(100)/[(0.264)(81.1) - 8.2] \\ &= 62.1\% \end{aligned}$$

(D) Concentration corrected to 50% EA is accomplished using Eqs. (1.18) and (1.19) for the volume basis, and Eqs. (1.20) and (1.21) for the mass basis concentrations

$$\begin{aligned} F_{50v} &= 1 - [1.5O_{2p} - 0.133N_{2p} - 0.75CO_p]/0.21 \\ &= 1 - [1.5(0.082) - 0.133(0.811)]/0.21 \\ &= 0.928 \end{aligned}$$

Concentration by volume corrected to 50% excess air

$$\begin{aligned} C_{50v} &= C_{vs}/F_{50v} \\ &= 3.14/0.928 \\ &= 3.38\ \text{grain/scf} \end{aligned}$$

$$\begin{aligned} F_{50m} &= 1 - (29/M_e)[1.5O_{2p} - 0.133N_{2p} - 0.75CO_p]/0.21 \\ &= 0.930 \end{aligned}$$

Concentration by mass corrected to 50% excess air

$$\begin{aligned} C_{50m} &= C_{ms}/F_{50m} \\ &= 6.12/0.930 \\ &= 6.56\,\text{lb}/1000\,\text{lb dry} \end{aligned}$$

(E) Correction to 12% CO_2 is accomplished with Eqs. (1.22) and (1.23)

$$\begin{aligned} F_{12v} &= CO_{2p}/0.12 \\ C_{12v} &= C_{vs}/F_{12v} \\ &= C_{vs}(0.12)/CO_{2p} \\ &= 3.14[0.12/0.107] \\ &= 3.52\,\text{grain/dscf} \end{aligned}$$

(F) Correction for 6% O_2 is

$$F_{6v} = (0.21 - O_{2p})/0.15$$
$$= (0.21 - 0.082)/0.15$$
$$= 0.85$$

$$C_{6v} = C_{vs}/F_{6v}$$
$$= 3.14/0.85$$
$$= 3.69 \text{ grain/dscf}$$

This example clearly illustrates how one applies corrections for temperature, pressure, and excess air. The emissions in this sample were expressed as concentrations given by Orsat.

3.5 Process Rate Factor

Process rates are normally based on either energy or material input to a process, and the next example illustrates application of a process rate standard applied to a combustion source. Figure 1.5 is a process rate standard for particulates taken from the State of Virginia air quality control regulations (EPA-80/02, p. 5-10).

EXAMPLE

Determine whether this plant meets the standard imposed by the Virginia code.

Given conditions

pollutant mass rate $(PMR)_{particulate} = 1800$ g/s

fuel: coal @ 23 tons/hour, high heating value (HHV) = 12,500 Btu/lb

proposed abatement uses an electrostatic precipitator with 99% rated collection efficiency

Solution:

(A) Find the process energy rate, H

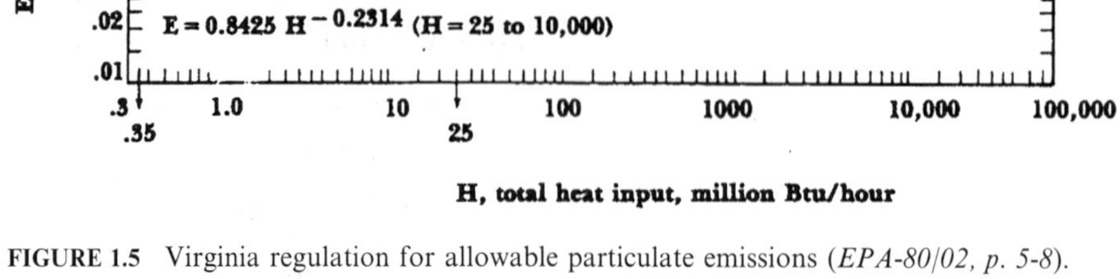

FIGURE 1.5 Virginia regulation for allowable particulate emissions (*EPA-80/02, p. 5-8*).

$$\begin{aligned} H &= \text{(mass of coal)(energy value per unit mass)} \\ &= (23 \text{ ton/h})(12{,}500 \text{ Btu/lb})(2000 \text{ lb/ton}) \\ &= 575 \times 10^6 \text{ Btu/h} \end{aligned}$$

(B) Find the allowable emission rate from Fig. 1.5.
From graph at $H = 575 \times 10^6$ Btu/h

$$E = 0.19 \text{ lb}/10^6 \text{ Btu}$$

or calculate from the equation in the graph

$$\begin{aligned} E &= 0.8425(575)^{-0.2314} \\ &= 0.194 \text{ lb}/10^6 \text{ Btu} \end{aligned}$$

(C) Now find the actual particulate emission rate

$$\begin{aligned} E_{\text{actual}} &= [(1800 \text{ g/s})(\text{lb}/454 \text{ g})(3600 \text{ s/h})(1 - 0.99)]/(575 \times 10^6 \text{ Btu/h}) \\ &= 0.25 \text{ lb}/10^6 \text{ Btu} \end{aligned}$$

$0.25 > 0.19$; therefore, this unit does not conform to the Virginia regulations.

3.6 Use of Emission Factor

US EPA publication AP-42 is a compilation of emission factors which have been gathered from various references. These factors, while quite valuable when calculations of gross inventory for a large number of sources are involved, are not necessarily valid for a specific single source (EPA-80/02, p. 5-14).

While more precise emission information is needed in order to pinpoint actual emissions, factors such as those presented in AP-42 can be used to form estimates of the control required.

The next example (using Table 1.3) indicates that the particulate loading a spreader stoker might produce (in lb/ton of coal) is thirteen times the coal ash weight percentage (EPA-80/02, p. 5-27). This factor tells us that a large number of spreader stoker fired units operating without control would produce, on the average, 13 lb of particulates for each 1 percent of ash in the coal burned. Any given unit might produce this amount at some operating capacity but not at all operating levels. At light loads, for example, gas flows are reduced compared to design capacity, and particulate entrainment is reduced because of lower gas velocity.

The emission factors are essentially process emission rate values expressed in terms of mass fired (lb/ton). These values are convertible to pollutant mass rate, PMR, by knowing the firing rate in lb/h.

EXAMPLE 1: emission factor
If one burns 6 tons/h of coal with $A = 10\%$ (i.e., % ash = 10) and a heating value HHV of 12,500 Btu/lb in a spreader stoker fired boiler, the uncontrolled emission rate is (EPA-80/02, p. 5-14):

From Table 1.3, the particulate emissions are 13A

$$\begin{aligned} E &= (13 \text{ lb/ton})(10) \\ &= 130 \text{ lb/ton} \end{aligned}$$

and the pollutant mass rate (PMR) is

TABLE 1.3 Emission Factors for Bituminous Coal Combustion without Control Equipment Emission Factor Rating *A*

Furnace size, 10^6 Btu/hr heat input[a]	Particulates[b]		Sulfur oxides[c]		Carbon monoxide		Hydrocarbons[d]		Nitrogen oxides		Aldehydes	
	lb/ton coal burned	kg/MT coal burned	lb/ton coal burned	kg/MT coal burned	lb/ton coal burned	kg/MT coal burned	lb/ton coal burned	kg/MT coal burned	lb/ton coal burned	kg/MT coal burned	lb/ton coal burned	kg/MT coal burned
Greater than 100[e] (utility and large industrial boilers)												
Pulverized general	16A	8A	38S	19S	1	0.5	0.3	0.15	18	9	0.005	0.0025
Pulverized wet bottom	13A[f]	6.5A	38S	19S	1	0.5	0.3	0.15	30	15	0.005	0.0025
Pulverized dry bottom	17A	8.5A	38S	19S	1	0.5	0.3	0.15	18	9	0.005	0.0025
Cyclone	2A	1A	38S	19S	1	0.5	0.3	0.15	55	27.5	0.005	0.0025
10 to 100[g] (large commercial and general industrial boilers)												
Spreader stoker[h]	13A[i]	6.5A	38S	19S	2	1	1	0.5	15	7.5	0.005	0.0025
Less than 10[j] (commercial and domestic furnaces)												
Under feed stoker	2A	1A	38S	19S	10	5	3	1.5	6	3	0.005	0.0025
Hand-fired units	20	10	38S	19S	90	45	20	10	3	1.5	0.005	0.0025

[a] 1 Btu/h = 0.252 kcal/h.
[b] The letter A on all units other than hand-fired equipment indicates that the weight percentage of ash in the coal should be multiplied by the value given. Example: If the factor is 16 and the ash content is 10 percent, the particulate emissions before the control equipment would be 10 times 16, or 160 pounds of particulate per ton of coal (10 × 8, or 80 kg of particulates per MT of coal).
[c] S equals the sulfur content (see footnote *b* above).
[d] expressed as methane.
[e] References 1 and 3 through 7 in (EPA-80/02, p. 5-28).
[f] Without fly-ash reinjection.
[g] References 1, 4, and 7 through 9 in (EPA-80/02, p. 5-28).
[h] For all other stokers use 5A for particulate emission factor.
[i] Without fly-ash reinjection. With fly-ash reinjection use 20 A. This value is not an emission factor but represents loading reaching the control equipment.'
[j] References 7, 9, and 10 in (EPA-80/02, p. 5-28).

$$\begin{aligned} \text{PMR} &= (130\,\text{lb/ton})(6\,\text{ton/h}) \\ &= 780\,\text{lb/h} \end{aligned}$$

Conversion of the emission rate from lb/ton to lb per million Btu is as follows:

$$\begin{aligned} \text{HHV} &= 12{,}500\ \text{Btu/lb} \\ &= (12{,}500\,\text{Btu/lb})(2000\,\text{lb/ton}) \\ &= 25 \times 10^6\ \text{Btu/ton} \end{aligned}$$

Therefore,

$$\begin{aligned} E &= (130\,\text{lb/ton})(1/25 \times 10^6\ \text{Btu/ton}) \\ &= 5.2\,\text{lb}/10^6\ \text{Btu} \end{aligned}$$

The degree of control required for a source performance standard of $0.1\,\text{lb}/10^6$ Btu would be determined as follows:

$$\begin{aligned}\eta &= (\text{collected}/\text{input})100\% \\ &= [(\text{input} - \text{allowable})/(\text{input})]100\% \\ &= [(5.2 - 0.1)/5.2]100\% \\ &= 98.1\%\end{aligned}$$

This would be an estimate only. More precise emission data for a specific unit would be desirable.

The SO_2 factor is more nearly representative of an actual case since the sulfur in the fuel is measurable. The factor 38S in Table 1.3 assumes 4% of the sulfur in the fuel does not appear as SO_2. This difference is greater if the system has a high percentage of unburned fuel in the ash. Where unburned combustible in the ash is a specified value, the SO_2 reduction is calculable, again provided the sulfur appearing as SO_3 can be predicted. The 38S emission factor is a valid first approximation of the uncontrolled SO_2 to be expected. Using the coal in Example 1 above with 1.3% sulfur, the following can be seen.

EXAMPLE 2: emission factor
Compute SO_2 emission per 10^6 Btu for the coal in Example 1.

$$\begin{aligned}E(SO_2) &= 38(1.3) = 49.4\,\text{lb/ton} \\ (\text{PMR}) &= (49.4\,\text{lb/ton})(1/25 \times 10^6\,\text{Btu/ton}) \\ &= 1.98\,\text{lb}/10^6\,\text{Btu}\end{aligned}$$

If one assumes that the new source performance standard for SO_2 is $1.2\,\text{lb}/10^6$ Btu, this would require a

$$(1.98 - 1.2)/1.98 = 39.3\%$$

reduction of SO_2 in the flue gas.

Similar calculations of uncontrolled emissions are possible using factors for hydrocarbons and NO_x.

4 GASEOUS EMISSION CONTROL TECHNIQUES

Gaseous pollutants are emitted from a variety of processes. Table 1.4 lists typical gaseous pollutants and their sources (EPA-81/12, p. 1-4).

The main techniques used to control gaseous emissions are

- absorption
- adsorption
- incineration
- condensation

The applicability of a given technique depends on the physical and chemical properties of the pollutant and the exhaust stream. More than one technique may be capable of controlling emissions from a given source. For example, vapors generated from loading gasoline into tank trucks at large bulk terminals are controlled by using any of these four techniques. Most often, however, one control technique is used more frequently than others for a given source–pollutant combination. For example, absorption is commonly used to remove SO_2, from boiler flue gas.

TABLE 1.4 Typical Gaseous Pollutants and Their Sources

Key element	Pollutant	Source
S	SO_2	Boiler flue gas
	SO_3	Sulfuric acid manufacturing
	H_2SO_4 vapor	Sulfuric acid manufacturing, pickling operations
	H_2S	Natural gas processing, pulp and paper mills, sewage treatment
	R-SH (mercaptans)	Petroleum refining, pulp and paper mills
N	NO, NO_2	Nitric acid manufacturing, boiler flue gas
	HNO_3 vapors	Nitric acid manufacturing, pickling operations
	NH_3	Ammonia manufacturing
	Other N compounds (i.e. amines, pyridines)	Sewage, rendering, solvent processes
C	Inorganic CO	Incomplete combustion
	Organics: volatile organic compounds (VOC)	
	VOC hydrocarbons • paraffins • olefins • aromatics	Solvent uses, gasoline marketing, petrochemical plants
	VOC oxygenated hydrocarbons • aldehydes • ketones • alcohols • phenols	Surface-coating operations, petroleum processing, plastics manufacturing
	VOC chlorinated solvents	Dry cleaning, degreasing
Halogen—F	HF	Phosphate fertilizer plant, aluminum plant
	SiF_4	Ceramics, fertilizer plant
Halogen—Cl	HCl	HCl manufacturing, PVC combustion
	Cl_2	Chlorine manufacturing

4.1 Absorption for Gaseous Emission Control

Absorption is a mass transfer operation in which a gas is dissolved in a liquid. It is the transfer of a gaseous component from the gas phase to a liquid phase. In air pollution control, absorption involves the removal of objectionable gaseous contaminants from a process stream by dissolving them in a liquid; i.e., a contaminant (pollutant) exhaust stream contacts a liquid, and the contaminant diffuses from the gas phase into the liquid phase. The absorption rate is enhanced by high diffusion rates of the contaminant in both the liquid and gas phase, high solubility of the contaminant, large liquid–gas contact area, and good mixing between liquid and gas phases (turbulence) (EPA-81/12, p. 4-1).

Some common terms used when discussing the absorption process are:

- *absorbent*: the liquid, usually water mixed with neutralizing agents, into which the contaminant is absorbed
- *absorbate or solute*: the gaseous contaminant being absorbed, such as SO_2, H_2S, etc
- *carrier gas*: the inert portion of the gas stream, usually flue gas, from which the contaminant is to be removed
- *interface*: the area where the gas phase and the absorbent contact each other

- *solubility*: the capability of a gas to be dissolved in a liquid

Absorption is a mass transfer operation. The mass transfer can be compared to heat transfer in that both occur because a system is trying to reach equilibrium conditions. For example, in heat transfer, if a hot slab of metal is placed on top of a cold slab, heat will be transferred from the hot slab to the cold slab until both are at the same temperature (equilibrium). In heat transfer, the process continues as long as a temperature differential or gradient exists. In absorption, mass instead of heat is transferred, and instead of occurring due to a temperature difference, absorption occurs as a result of a concentration difference. Absorption continues as long as a concentration differential exists between the absorbent and the gas from which the contaminant is being removed. In absorption, equilibrium is not as easily defined as in heat transfer, since the concentration difference depends on the solubility of the solute.

Absorption devices used to remove gaseous contaminants are referred to as *absorbers* or *wet scrubbers*. Wet scrubbers are also used to remove particulate matter from gas streams. Wet scrubbers usually cannot be operated to optimize simultaneous removal of both gases and particulate matter. In designing absorbers for gaseous emissions, optimum mass transfer can be accomplished by

- providing a large interfacial contact area
- providing good mixing between gas and liquid phases
- allowing sufficient residence or contact time between the phases
- ensuring a high degree of solubility of the contaminant in the absorbent

Mechanism of absorption. To remove a gaseous contaminant by absorption, the contaminant-laden exhaust stream must be passed through (contacted with) a liquid with the following steps (EPA-81/12, p. 4-2):

- Step 1: The solute diffuses from the bulk area of the gas phase to the gas–liquid interface
- Step 2: The solute transfers across the interface to the liquid phase
- Step 3: The solute diffuses into the bulk area of the liquid, making room for additional gas molecules to be absorbed

Figure 1.6 illustrates the steps involved in the absorption process. The purpose of analyzing these three steps is to determine which variables control the process. The most efficient system can be designed by knowing these variables. It is assumed that once the solute arrives at the interface area, transfer across it occurs instantaneously. This second step in the absorption mechanism is extremely rapid; therefore, it does not need to be considered when deriving absorption efficiency equations. The rate of mass transfer (absorption) is dependent upon the diffusion rate in either the gas phase or the liquid phase.

Two terms used to describe the mass transfer rate are gas-phase controlled absorption and liquid-phase controlled absorption. Each mechanism depends on the rate of diffusion in both phases and upon the solubility of the pollutant in the liquid.

The diffusion rate of a gaseous pollutant molecule through a gas is always faster than its diffusion rate through a liquid because gas molecules in a gas are further apart than liquid molecules in a liquid. The pollutant molecule will move faster through the gas phase than through the liquid phase since, in a gas, obstructions such as other molecules are farther apart than in a liquid. However, the mass transfer rate depends primarily upon the solubility of the pollutant in the liquid.

Consider two gaseous pollutants A and B that are approximately equal in size. Assume that A is not very soluble in the liquid while B is very soluble in the liquid. Assume also that both A and B will move through the gas phase at the same rate. The rates at which the pollutants will be absorbed in the liquid phase are different because B is very soluble while A is not. A is absorbed very slowly and B is absorbed very quickly.

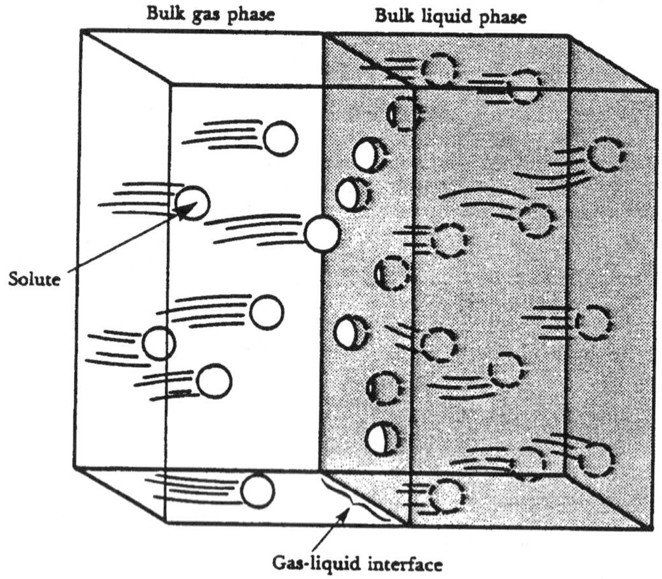

FIGURE 1.6 Gas collection by absorption (*EPA-81/12, p. 4-2*).

The transfer rate of A (slowly absorbed) is greatly affected by its absorption and diffusion in the liquid phase and is therefore referred to as liquid-phase controlled. The transfer rate of B (quickly absorbed) is also greatly affected by its rate of movement through the liquid phase but is absorbed so rapidly that it doesn't influence the overall transfer rate. The mass transfer rate for B depends only on how fast it diffuses through the gas phase towards the gas–liquid interface and is referred to as gas-phase controlled.

The gas-phase controlled systems absorb pollutants more readily than do the liquid-phase controlled systems. Thus, absorption systems used in the field of air pollution control are usually designed to be gas-phase controlled.

The previously mentioned three-step mechanism of absorption occurs on two levels: a micro or molecular level and a macro or bulk level. Molecular diffusion of individual molecules occurs due to a concentration difference because all systems try to achieve an equilibrium state. Molecules migrate from areas of high concentrations to areas of low concentrations. Macro or bulk diffusion occurs as a mass of liquid or gas moves to or from the interface. In an absorber, bulk diffusion is accomplished by turbulent mixing of the gas and liquid phases. To be efficient, an absorber must provide both turbulent mixing of the gas and liquid phases and sufficient residence time to allow pollutant molecules to be absorbed.

Solubility. A very important factor affecting the amount of a contaminant that can be absorbed is the solubility of the contaminant. Solubility is a function of both the temperature and to a lesser extent the pressure of the system. As temperature increases, the amount of gas that can be absorbed by a liquid decreases. From the ideal gas law: as temperature increases, the volume of a gas also increases; therefore, at the higher temperature less gas is absorbed due to the increased volume it occupies. Pressure affects the solubility of a gas in the opposite manner. By increasing the pressure of a system the amount of gas absorbed generally increases (EPA-81/12, p. 4-3).

Under certain conditions, Henry's law may also be used to express equilibrium solubility of gas–liquid systems. Henry's law states that for dilute solutions, where the components do not interact, the resulting partial pressure (p) of a component A in equilibrium with other components in a solution can be expressed as

$$p = Hx_A$$

where p = equilibrium partial pressure of component A over a solution
x_A = mole fraction or concentration of A in liquid phase, g-mole/cm^3
H = Henry's law constant (atm-cm^3)/(g-mole) of pure A at the same temperature and pressure as the solution

This equation is the equation of a straight line, where the slope (m) is equal to H. Henry's law can be used to predict solubility only when the equilibrium line is straight. Equilibrium lines are usually straight when the solute concentrations are very dilute. In air pollution control applications this is usually the case. For example, an exhaust stream that contains a 1000 ppm SO_2 concentration corresponds to a mole fraction of SO_2 in the gas phase of only 0.001.

Another restriction on using Henry's law is that it does not hold true for gases that react or dissociate upon dissolution. If this happens, the gas no longer exists as a simple molecule. For example, scrubbing HF or HCl gases with water causes both compounds to dissociate in solution. In these cases, the equilibrium lines are curved rather than straight. Data on systems that exhibit curved equilibrium lines must be obtained from experiments.

The units of Henry's law constants are atm/(mole fraction). The smaller the constant, the more soluble the gas. The following example illustrates how to develop an equilibrium diagram from solubility data.

EXAMPLE 1: Henry's law
Given the data shown in Table 1.5 for the solubility of SO_2 in pure water at 303K (30°C) and 101.3 kPa (760 mm Hg), plot the equilibrium diagram and determine if Henry's law applies (EPA-81/12, p. 4-7).

Solution:
1. The data must first be converted to mole fraction units.
The mole fraction in the gas phase y is obtained by dividing the partial pressure of SO_2 by the total pressure of the system:

$$y = p/P = 6\,\text{kPa}/(101.3\,\text{kPa}) = 0.06$$

The mole fraction in the liquid phase x is obtained by dividing the moles of SO_2 by the total moles of liquid (basis = 100g of H_2O):

TABLE 1.5 Solubility of SO_2 in Pure Water

Equilibrium data	
C (g of SO_2 per 100 g/H_2O)	p (partial pressure of SO_2)
0.5	6 kPa (42 mmHg)
1.0	11.6 kPa (85 mmHg)
1.5	18.3 kPa (129 mmHg)
2.0	24.3 kPa (176 mmHg)
2.5	30.0 kPa (224 mmHg)
3.0	36.4 kPa (273 mmHg)

x = (moles SO_2 in solution)/(moles SO_2 in solution + moles of H_2O)

moles of SO_2 in solution = (concentration of SO_2)/(64 g of SO_2 per mole)

$= 0.5/64$

$= 0.0078$

moles of H_2O = (100 g of H_2O)/(18 g H_2O per mole) = 5.55

$x = 0.0078/(0.0078 + 5.55)$

$= 0.0014$

2. Table 1.6 is completed. The data from Table 1.6 is plotted in Fig. 1.7. Henry's law applies in the given concentration range with Henry's law constant equal to 42.7 mole fraction SO_2 in air/mole fraction SO_2 in water.

TABLE 1.6 Solubility Data for SO_2

C = (g of SO_2)/(100 g H_2O)	p (kPa)	$y = p/101.3$	$x = (C/64)/(C/64 + 5.55)$
0.5	6	0.06	0.0014
1.0	11.6	0.115	0.0028
1.5	18.3	0.18	0.0042
2.0	24.3	0.239	0.0056
2.5	30	0.298	0.007
3.0	36.4	0.359	0.0084

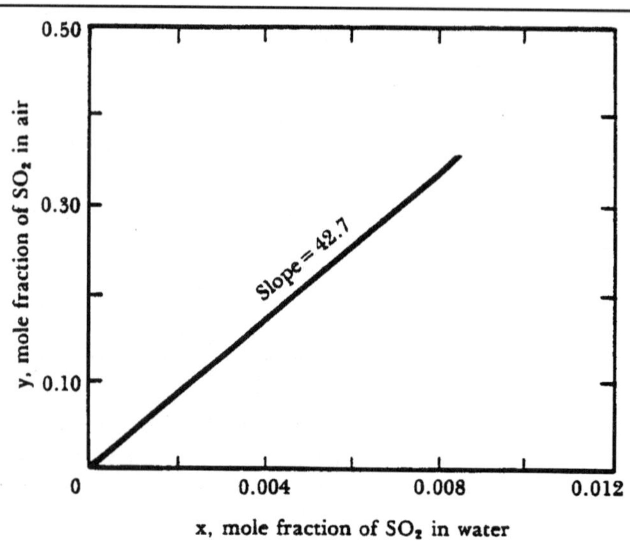

FIGURE 1.7 Equilibrium diagram for SO_2–H_2O system (*EPA-81/12, p. 4-8*).

Material balance. In designing or reviewing the design of an absorption control system, the first task is to determine the flow rates and composition of each stream entering the system. From the law of conservation of mass, the material entering a process must either accumulate or exit. In other words, 'what comes in must go out'. A material balance is used to help determine flow rates and compositions of individual streams. Figure 1.8 illustrates a typical countercurrent flow absorber in which a material balance is drawn. The solute is used as the 'material' in the material balance (EPA-81/12, p. 4-14).

For material balance, the following equation can be obtained:

$$Y_1 - Y_2 = (L_\mathrm{m}/G_\mathrm{m})(X_1 - X_2)$$

where Y_1 = inlet solute concentration
Y_2 = outlet solute concentration
X_2 = inlet composition of scrubbing liquid
X_1 = outlet composition of scrubbing liquid
L_m = liquid flow rate, g-mole/h
G_m = gas flow rate, g-mole/h

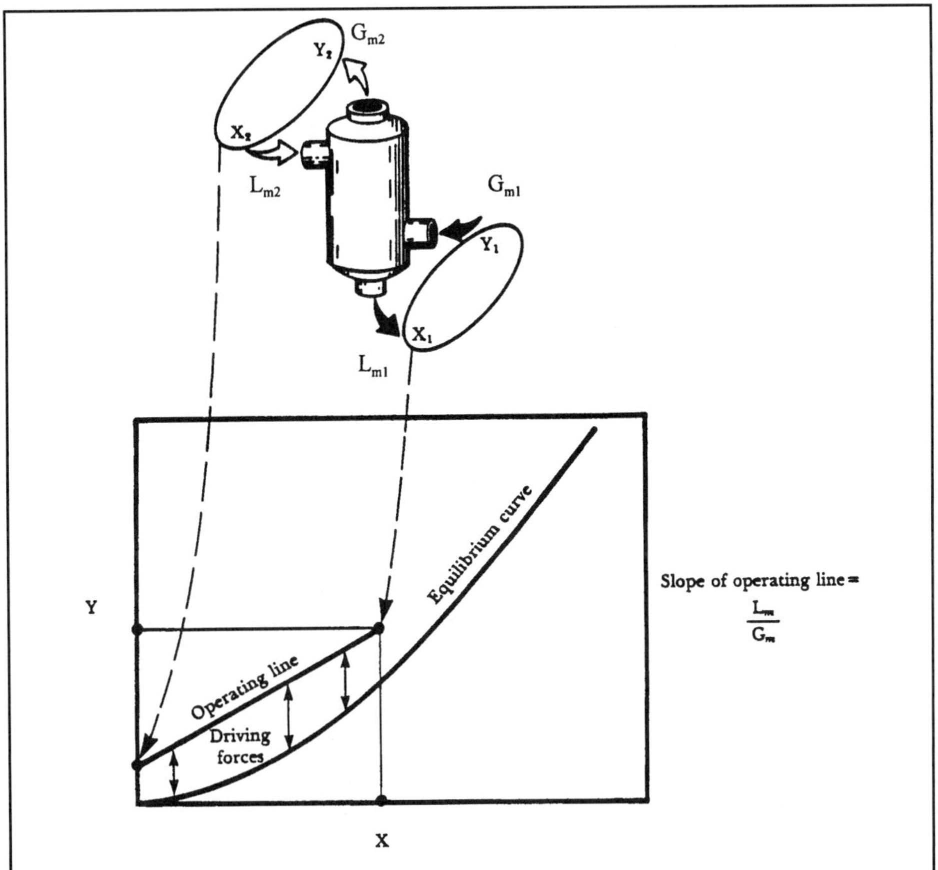

FIGURE 1.8 Operating line for a countercurrent flow absorber (*EPA-81/12, p. 4-17*).

The above equation is the equation of a straight line. When this line is plotted on an equilibrium diagram it is referred to as an *operating line* (Fig. 1.8). This line defines operating conditions within the absorber: what is going in and what is coming out. The slope of the operating line is the liquid mass flow rate divided by the gas mass flow rate, which is the liquid-to-gas ratio or (L_m/G_m). The liquid-to-gas ratio is used extensively when describing or comparing absorption systems.

EXAMPLE 2: Henry's law
Using the data and results from Example 1, compute the minimum liquid rate of pure water required to remove 90% of the SO_2 from a gas stream of $84.9\,m^3$/min (3000 acfm) containing 3% SO_2 by volume. The temperature is 293 K and the pressure is 101.3 kPa (EPA-81/12, p. 4-20).

Given conditions

Inlet gas solute concentration $(Y_1) = 0.03$

Minimum acceptable standards (outlet solute concentration) $(Y_2) = 0.003$

Composition of the liquid into the absorber $(X_2) = 0$

Gas flow rate $(Q) = 84.9\,m^3/min$

Outlet liquid concentration $(X_1) = ?$

Liquid flow rate $(L) = ?$

Solution:

1. First, sketch and label a drawing of the system (Fig. 1.9).

$$Y_1 = 3\% \text{ by volume} = 0.03$$

$$Y_2 = 90\% \text{ reduction from } Y_1 \text{ or only } 10\% \text{ of } Y_1; \text{ therefore}$$

$$Y_2 = (0.10)(0.03) = 0.003$$

2. At the minimum liquid rate, Y_1 and X_1 will be in equilibrium; the liquid will be saturated with SO_2.
At equilibrium: $Y_1 = H'X_1$
From Fig. 1.7,

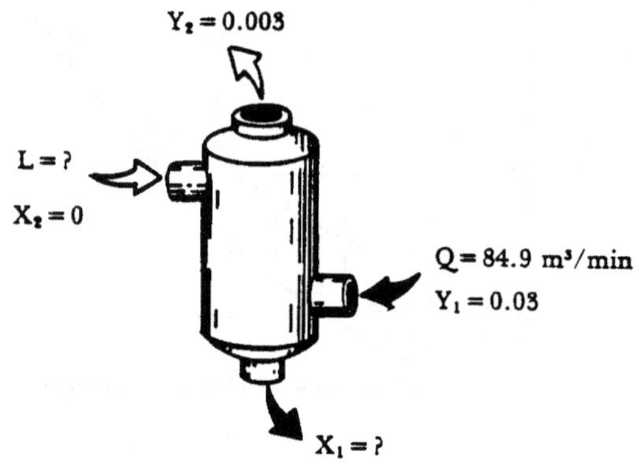

FIGURE 1.9 Material balance for absorber (*EPA-81/12, p. 4-20*).

$$H' = 42.7 \text{ (mole fraction SO}_2\text{ in air)/(mole fraction SO}_2\text{ in water)}$$
$$0.03 = 42.7X_1$$
$$X_1 = 0.000703 \text{ mole fraction}$$

3. The minimum liquid-to-gas ratio is

$$Y_1 - Y_2 = (L_m/G_m)(X_1 - X_2)$$
$$(L_m/G_m) = (Y_1 - Y_2)/(X_1 - X_2)$$
$$= (0.03 - 0.003)/(0.000703 - 0)$$
$$= 38.4 \text{ (g mole water)/(g mole of air)}$$

4. Compute the minimum required liquid flow rate:
First, convert m^3 of air to g moles.
At 0°C (or 273 K) and 101.3 kPa, there are 0.0224 m^3/g mole of an ideal gas.
At 20°C (293 K): 0.0224(293/273) = 0.024 m^3/g mole

$$G_m = 84.9 \text{ (m}^3\text{/min)(g-mole air/0.024 m}^3\text{)}$$
$$= 3538 \text{ g-mole air/min}$$
$$L_m/G_m = 38.4 \text{ (g-mole water)/(g-mole of air) at minimum conditions}$$
$$L_m = 38.4(3538)$$
$$= 136.0 \text{ kg-mole water min}$$

In mass units:

$$L = 136 \text{ kg-mole/min (18 kg/kg mole)}$$
$$= 2448 \text{ kg/min}$$

5. Figure 1.10 illustrates the graphical solution to this problem.
The slope of the minimum operating line × 1.5 = the slope of the actual operating line (line AC).

$$38.4 \times 1.5 = 57.6$$

Sizing packed column diameter of an absorber. The main parameter that affects the size of a packed column is the gas velocity at which liquid droplets become entrained in the exiting gas stream. Consider a packed column operating at set gas and liquid flow rates. By decreasing the diameter of the column, the gas flow rate (m/s or ft/s) through the column will increase. If the gas flow rate through the column is gradually increased (by using smaller and smaller diameter columns) a point will be reached where the liquid flowing down over the packing begins to be held in the void spaces between the packing. This gas-to-liquid flow ratio is termed the *loading point*. The pressure drop over the column begins to increase and the degree of mixing between the phases decreases. A further increase in gas velocity will cause the liquid to completely fill the void spaces in the packing. The liquid forms a layer over the top of the packing and no more liquid can flow down through the tower. The pressure drop increases substantially and mixing between the phases is minimal. This condition is referred to as *flooding* and the gas velocity at which it occurs is the *flooding velocity*. Using an extremely large diameter tower would eliminate this problem. However, as the diameter increases the cost of the tower increases (EPA-81/12, p. 4-22).

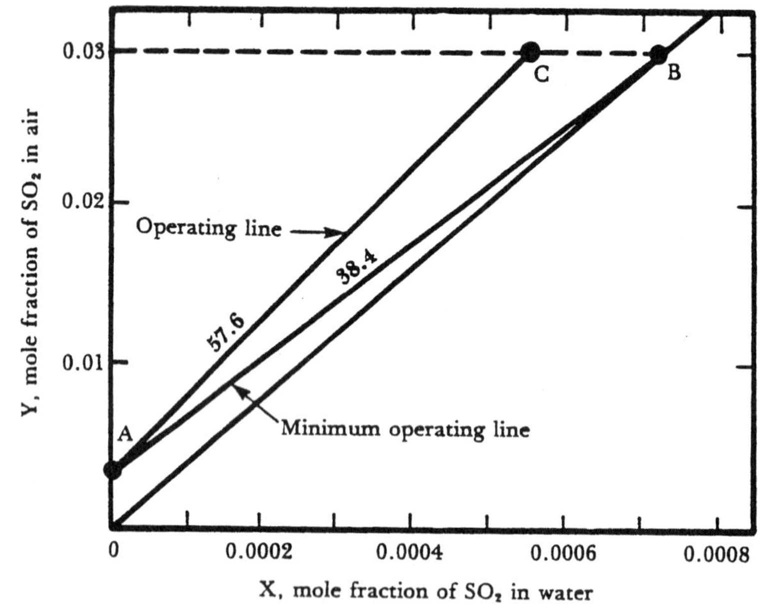

FIGURE 1.10 Solution to Example 2 (*EPA-81/12, p. 4-21*).

Normal practice is to size a packed column diameter to operate at a certain percent of the flooding velocity. A typical operating range for the gas velocity through the columns is 50–75 percent of the flooding velocity. It is assumed that by operating in this range the gas velocity will also be below the loading point.

A common and relatively simple procedure to estimate the flooding velocity (thus setting a minimum column diameter) is to use a generalized flooding and pressure drop correlation. One version of the flooding and pressure drop relationship in a packed tower is shown in Fig. 1.11. This correlation is based on the physical properties of the gas and liquid streams and tower packing characteristics. The procedure to determine the tower diameter is

1. Calculate the value of the abscissa.

$$(L/G)(\rho_g/\rho_l)^{0.5} \tag{1.30}$$

where

L and G = mass flow rates: any consistent set of units may be used as long as the abscissa is dimensionless

ρ_g = density of the gas stream

ρ_l = density of the absorbing liquid

2. From the point calculated in Eq. (1.30) proceed up the graph to the flooding line and read the ordinate ε.

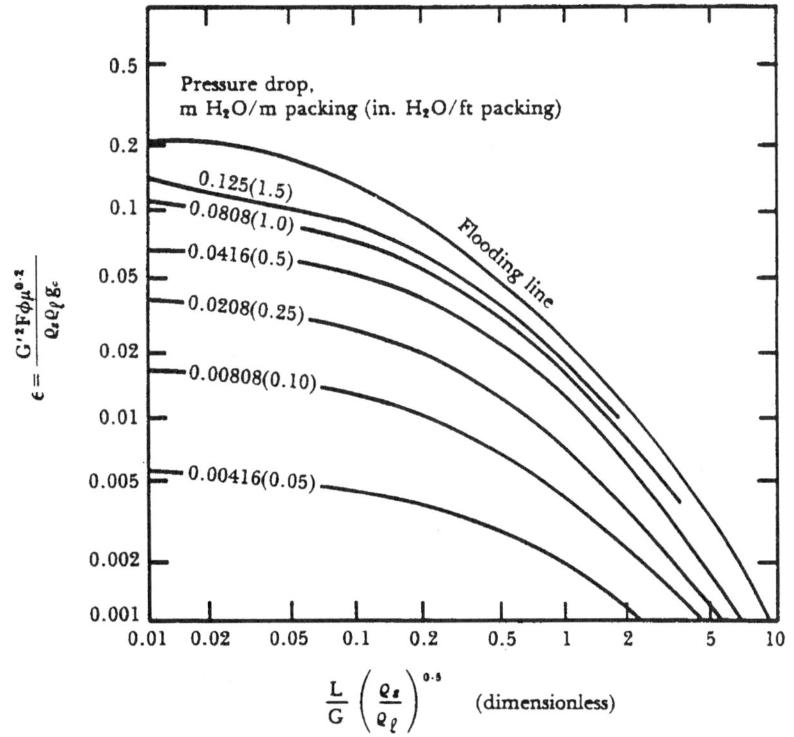

FIGURE 1.11 Generalized flooding and pressure drop correlation (*EPA-81/12, p. 4-23*).

3. Rearrange the equation of the ordinate and solve for G'.

$$G' = \{[(\varepsilon)(\rho_g)(\rho_1)(g_c)]/[F\phi(\mu_1)^{0.2}]\}^{0.5} \tag{1.31}$$

where

G' = mass flow rate of gas per unit cross-sectional area of column at flooding, g/s-m^2 (lb/s-ft^2)

ε = ordinate value in Fig. 1.11

ρ_g = density of the gas stream, kg/m^3 (lb/ft^3)

ρ_1 = density of the absorbing liquid, kg/m^3 (lb/ft^3)

g_c = gravitational constant = 9.82 m/s^2 (32.2 ft/s^2)

F = packing factor

ϕ = ratio of specific gravity of the scrubbing liquid to that of water

μ_1 = viscosity of liquid (for water = 0.8 centipoise = 0.0008 Pa-s) [use Pa-s in this equation]

4. G' at operating conditions is a fraction of G' at flooding:

$$G'_{\text{operating}} = f\,G'_{\text{flooding}} \tag{1.32}$$

5. The cross-sectional area of column A is calculated from

$$A = G/G'_{\text{operating}} \tag{1.33}$$

6. The diameter of the column is obtained from

$$d = (4A/\pi)^{0.5} \tag{1.34}$$

EXAMPLE 3: column sizing
For the scrubber in Example 2, determine the column diameter if the operating liquid rate is 1.5 times the minimum. The gas velocity should be no greater than 75% of the flooding velocity and the packing material is 2-inch ceramic Intalox saddles.

Solution:
From Example 2:

$$G_{\text{m}} = 3538\,\text{g-mole/min}$$
$$L_{\text{m}} = 2448\,\text{kg/min}$$

Convert gas molar flow to a mass flow, assuming molecular weight of the gas to be 29 kg/mole.

$$G = 3.538\ (\text{kg-mole/min})(29)(\text{kg/mole}) = 102.6\,\text{kg/min}$$

Adjusting the liquid flow to 1.5 times the minimum:

$$L = 1.5(2448) = 3672\,\text{kg/min}$$

The densities of water and air at 20°C are:

$$\rho_{\text{l}} = 1000\,\text{kg/m}^3$$
$$\rho_{\text{g}} = 1.17\,\text{kg/m}^3$$

1. Compute the abscissa from Eq. (1.30).

$$(L/G)(\rho_{\text{g}}/\rho_{\text{l}})^{0.5} = (3672/102.6)(1.17/1000)^{0.5} = 1.22$$

2. From Fig. 1.11, proceed up to the flooding line from 1.22. The ordinate ε is 0.019.
3. From Eq. (1.31), calculate G':

$$G' = \{[(\varepsilon)(\rho_{\text{g}})(\rho_{\text{l}})(g_{\text{c}})]/[F\phi(\mu_1)^{0.2}]\}^{0.5}$$

where

G' = mass flow rate of gas per unit cross-sectional area of column at flooding, g/s-m^2 (lb/s-ft^2)
ε = ordinate value in Fig. 1.11
ρ_{g} = density of the gas stream, kg/m^3 (lb/ft^3)
ρ_{l} = density of the absorbing liquid, kg/m^3 (lb/ft^3)
g_{c} = gravitational constant = 9.82 m/s^2 (32.2 ft/s^2)
F = packing factor. For this example, the factor is assumed to be 40 ft^2/ft^3 or 131 m^2/m^3 for 2-inch Intalox saddles

ϕ = ratio of specific gravity of the scrubbing liquid to that of water

μ_1 = viscosity of liquid (for water = 0.8 centipoise = 0.0008 Pa-s)

For water, $\phi = 1.0$ and $\mu_1 = 0.0008$ Pa-s
For 2-inch Intalox saddles, $F = 40\,\mathrm{ft^2/ft^3}$ or $131\,\mathrm{m^2/m^3}$.

$$G' = \{(0.019)(1.17)(1000)(9.82)/[(1)(131)(0.0008)^{0.2}]\}^{0.5}$$
$$= 2.63\,\mathrm{kg/m^2\text{-}s} \text{ at flooding}$$

4. The superficial gas velocity at operating is obtained from Eq. (1.32).

$$G'_{\mathrm{operating}} = f\,G'_{\mathrm{flooding}} = 0.75 \times 2.63 = 1.97\,\mathrm{kg/(m^2\text{-}s)}$$

5. From Eq. (1.33), the cross-sectional area of the tower is

$$A = G/G'_{\mathrm{operating}}$$
$$= (102.6\,\mathrm{kg/min})(\mathrm{min}/60\ \mathrm{s})/(1.97\,\mathrm{kg/m^2\text{-}s})$$
$$= 0.87\,\mathrm{m^2}$$

6. From Eq. (1.34), the diameter of the tower is

$$d = (4A/\pi)^{0.5}$$
$$= [(4)(0.87\,\mathrm{m^2})]^{0.5}$$
$$= 1.05\,\mathrm{m} \text{ or at least } 1\ \mathrm{m}\ (3.5\,\mathrm{ft})$$

Figure 1.11 may also be used to estimate the pressure drop of the tower once $G'_{\mathrm{operating}}$ is set. This is done by plugging G' back into the equation to compute ε

$$\varepsilon = [(G')^2 F\phi(\mu_1)^{0.2}]/[(\rho_g)(\rho_1)(g_c)]$$
$$= (1.97)^2(1)(131)(0.0008)^{0.2}/[(1.17)(1000)(9.82)]$$
$$= 0.0106$$

The abscissa remains unchanged and equals 1.22. The pressure drop through the column is the point at which these two lines cross. From Fig. 1.11

$$\Delta p = (0.0416 \text{ meter of } H_2O)/(\text{meter of packing})(0.5 \text{ inches of } H_2O)/(\text{foot of packing})$$

Sizing the packed column height of an absorber. The height of a packed column refers to the depth of packing material needed to accomplish the required removal efficiency. The more difficult the separation, the larger the packing height required. For example, a much larger packing height would be required to remove SO_2 than to remove Cl_2 from an exhaust stream using water as the absorbent. This is because Cl_2 is more soluble in water than SO_2. Determining the proper height of packing is important since it affects both the rate and efficiency of absorption (EPA-81/12, p. 4-26).

The height of a packed column can be expressed as

$$Z = \mathrm{HTU} \times \mathrm{NTU} \tag{1.35}$$

where Z = height of packed column

HTU = height of a transfer unit

NTU = number of transfer units

The concept of a transfer unit comes from the operation of plate columns. Discrete stages (plates) of separation occur in plate columns. These stages can be visualized as a transfer unit with the number and height of each giving the total tower height. Although a packed column operates as one continuous separation process, in design terminology it is treated as if it were broken into discrete sections (height of a transfer unit). The number and the height of a transfer unit are based on either the gas or liquid phase. Equation (1.35) now becomes

$$Z = N_{OG}H_{OG} = N_{OL}H_{OL} \tag{1.36}$$

where Z = height of packing, m
N_{OG} = number of transfer units based on overall gas film coefficient
H_{OG} = height of a transfer unit based on overall gas film coefficient, m
N_{OL} = number of transfer units based on overall liquid film coefficient
H_{OL} = height of a transfer unit based on overall liquid film coefficient, m

Values for the height of a transfer unit used in designing absorption systems are usually obtained from experimental data. To ensure greatest accuracy, vendors of absorption equipment normally perform pilot plant studies to determine the height of a transfer unit.

When no experimental data are available or if only a preliminary estimate of absorber efficiency is needed, there are generalized correlations available to predict the height of a transfer unit. The correlations for predicting the H_{OG}, or the H_{OL}, are empirical in nature and are a function of

Type of packing
Liquid and gas flow rates
Concentration and solubility of the contaminant
Liquid properties
System temperature

For most applications, the height of a transfer unit ranges between 0.3 and 1.2 m (1 and 4 ft). As a rough estimate, 0.6 m (2.0 ft) can be used.

The number of transfer units, NTU, can be obtained experimentally or calculated from a variety of methods. For the case where the solute concentration is very low and the equilibrium line is straight, the following equation can be used to determine the number of transfer units (N_{OG}) based on the gas phase resistance.

$$N_{OG} = \ln[(Y_1 - mX_2)/(Y_2 - mX_2)(1 - mG_m/L_m) + mG_m/L_m]/(1 - mG_m/L_m) \tag{1.37}$$

where N_{OG} = transfer units
Y_1 = mole fraction of solute in entering gas
X_2 = mole fraction of solute entering the column
Y_2 = mole fraction of solute in exiting gas
m = slope of equilibrium line
G_m = molar flow rate of gas, kg-mole/h
L_m = molar flow rate of liquid, kg-mole/h

Equation (1.37) may be solved directly or graphically by using the Colburn diagram which is presented in Fig. 1.12. The Colburn diagram is a plot of the N_{OG} versus $\ln[Y_1 - mX_2/(Y_2 - mX_2)]$ at various values of mG_m/L_m. The term mG_m/L_m is referred to as the *absorption factor*. Figure 1.12 is used by first computing the value of $\ln[Y_1 - mX_2/(Y_2 - mX_2)]$, reading up the graph to the line corresponding to (mG_m/L_m), and then reading across to obtain the N_{OG}.

Equation (1.37) can be further simplified for situations where a chemical reaction occurs or if the solute is extremely soluble. In these cases the solute exhibits almost no partial pressure

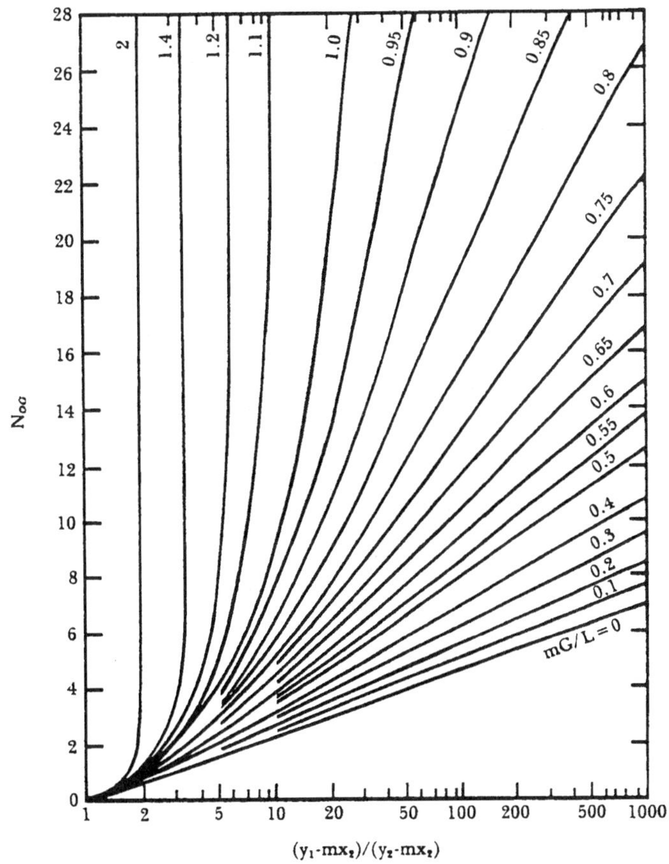

FIGURE 1.12 Colburn diagram (*EPA-84/03, p. 104; EPA-81/12, p. 4-30*).

and therefore the slope of the equilibrium line approaches zero ($m - 0$). For either of these cases, the equation is reduced to

$$N_{OG} = \ln(Y_1/Y_2) \tag{1.38}$$

This means that the number of transfer units only depends on the inlet and outlet concentration of the solute (contaminant). For example, if the conditions of Eq. (1.38) are met, to achieve 90% removal of any pollutant requires 2.3 transfer units. The equation applies only when the equilibrium line is straight and approaches zero (for very soluble or reactive gases). Example 4 illustrates a procedure to calculate the packed column height.

EXAMPLE 4: column sizing

From pilot plant studies of the absorption system in Example 2, it was determined that the H_{OG} for the SO_2–water system is 0.829 m (2.72 ft). Calculate the total height of packing required to achieve 90% removal. The following data were taken from the previous examples:

Given conditions

$H_{OG} = 0.829\,\text{m}$

$m = 42.7$ (kg-mole SO_2 in air)/(kg-mole SO_2 in water)

$G_m = 3.5$ kg-mole/min

$L_m = 3672$ (kg/min)(kg-mole)/18 kg $= 204$ kg-mole/min

$X_2 = 0$ (no recycle liquid)

$Y_1 = 0.03$

$Y_2 = 0.003$

Solution:

1. Compute the N_{OG} from Eq. (1.37)

$$N_{OG} = \ln[(Y_1 - mX_2)/(Y_2 - mX_2)(1 - mG_m/L_m) + mG_m/L_m]/(1 - mG_m/L_m)$$

where N_{OG} = transfer units
Y_1 = mole fraction of solute in entering gas
X_2 = mole fraction of solute entering the column
Y_2 = mole fraction of solute in exiting gas
m = slope of equilibrium line
G_m = molar flow rate of gas, kg-mole/h
L_m = molar flow rate of liquid, kg-mole/h

$$\begin{aligned} N_{OG} &= \ln[(Y_1 - mX_2)/(Y_2 - mX_2)(1 - mG_m/L_m) + mG_m/L_m]/(1 - mG_m/L_m) \\ &= \ln\{(0.03/0.003)[1 - 42.7(3.5)/204] + 42.7(3.5)/204\}/[1 - 42.7(3.5)/204] \\ &= 4.58 \end{aligned}$$

2. Calculate the total packing height.

$$\begin{aligned} Z &= H_{OG}(N_{OG}) \\ &= 0.829 \times 4.58 \\ &= 3.79\,\text{m of packing height} \end{aligned}$$

Sizing the plate tower. In a plate tower the scrubbing liquid enters at the top of the tower, passes over the top plate, and then down over each lower plate until the liquid reaches the bottom. Absorption occurs as the gas, which enters at the bottom, passes up through the plate and contacts the liquid. In a plate tower, absorption occurs in a stepwise or stage process. The operation of plate towers is discussed in greater detail in EPA-81/12, p. 4-32.

The minimum diameter of a single-pass plate tower is determined by using the gas velocity through the tower. If the gas velocity is too fast, liquid droplets are entrained, causing a condition known as priming. Priming occurs when the gas velocity through the tower is so fast that it causes liquid on one tray to foam and then rise to the tray above. Priming reduces absorber efficiency by inhibiting gas and liquid contact. For the purpose of determining tower diameter, priming in a plate tower is analogous to the flooding point in a packed tower. It determines the minimum acceptable diameter. The actual diameter should be larger.

The smallest allowable diameter for a plate tower is expressed by

$$d = \Psi[Q(\rho_g)^{0.5}]^{0.5} \tag{1.39}$$

where d = plate tower diameter
Ψ = empirical correlation, $m^{0.25}(h)^{0.25}/(kg)^{0.25}$
Q = volumetric gas flow, m^3/h
ρ_g = gas density, kg/m^3

The term Ψ is an empirical correlation and is a function of both the tray spacing and the densities of the gas and liquid streams. Values for Ψ shown in Table 1.7 are for a tray spacing of 61 cm (24 in) and a liquid specific gravity of 1.05. If the specific gravity of a liquid varies significantly from 1.05, the values for Ψ in the table cannot be used.

Depending on operating conditions, trays are spaced at a minimum distance between plates to allow the gas and liquid phases to separate before reaching the plate above. Trays should be spaced to allow for easy maintenance and cleaning. Trays are normally spaced 45 to 70 cm (18 to 28 in) apart. In using the information in Table 1.7 for a tray spacing different from 61 cm, a correction factor must be used. Figure 1.13 is used to determine the correction factor which is multiplied by the estimated diameter. Example 5 illustrates how to estimate the minimum diameter of a plate tower.

TABLE 1.7 Empirical Constants for Eq. (1.39)

Tray	Metric Ψ*	Engineering Ψ†
Bubble cap	0.0162	0.1386
Sieve	0.014	0.1198
Valve	0.0125	0.1069

* Metric Ψ is expressed in $m^{0.25}hr^{0.5}/kg^{0.25}$, for use with Q expressed in m^3/h, and ρ_g expressed in kg/m^3.
† Engineering Ψ is expressed in $ft^{0.25}min^{0.5}/lb^{0.25}$, for use with Q expressed in cfm, and ρ_g expressed in lb/ft^3.

EXAMPLE 5: plate tower diameter
For the condition described in Example 2, determine the minimum acceptable diameter if the scrubber is a bubble cap tray tower. The trays are spaced 0.53 m (21 in) apart (EPA-81/12, p. 4-34).

Solution:
From Example 2 and Example 3, the following information is obtained:

$$\text{Gas flow rate} = Q = 84.9\,\text{m}^3/\text{min}$$

$$\text{Density} = \rho_g = 1.17\,\text{kg/m}^3$$

From Table 1.7 for a bubble cap tray:

$$\Psi = 0.0162\,\text{m}^{0.25}(\text{h})^{0.25}/(\text{kg})^{0.25}$$

Before Eq. (1.39) can be used, Q must be converted to m^3/h.

$$Q = 84.9(\text{m}^3/\text{min})(60)(\text{min/h})$$
$$= 5094\,\text{m}^3/\text{h}$$

Substituting these values into Eq. (1.39) for a minimum d:

$$d = \Psi[Q(\rho_g)^{0.5}]^{0.5}$$

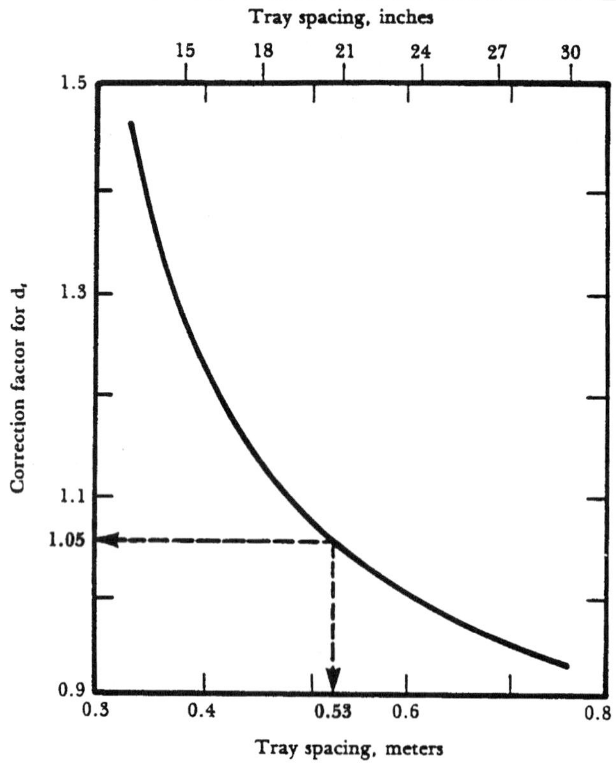

FIGURE 1.13 Tray spacing correction factor (*EPA-81/12, p. 4-33*).

where d = plate tower diameter
Ψ = empirical correlation, $m^{0.25}(h)^{0.25}/(kg)^{0.25}$
Q = volumetric gas flow, m^3/h
ρ_g = gas density, kg/m^3

$$d = \Psi[Q(\rho_g)^{0.5}]^{0.5}$$
$$= (0.0162)[5094(1.17)^{0.5}]^{0.5} = 1.2\,\text{m}$$

Correct this diameter for a tray spacing of 0.53 m. From Fig. 1.13 read the correction factor as 1.05. Therefore, the minimum diameter is

$$d = 1.2(1.05) = 1.26\,\text{m}\ (4.13\,\text{ft})$$

Note: this estimated diameter is a minimum acceptable diameter based on actual conditions. In practice, a larger diameter based on maintenance or economic considerations is usually chosen.

Number of theoretical plates or trays. There are several methods used to determine the number of ideal plates or trays that are required for a given removal efficiency. These methods, however, can become quite complicated. One method used is a graphical technique (EPA-81/12, p. 4-34).

The number of ideal plates is obtained by drawing 'steps' on an operating diagram. This procedure is illustrated in Fig. 1.14. This method can be rather time-consuming and inaccuracies can result at both ends of the graph. The following is a simplified method of estimating the number of plates. It can only be used if both the equilibrium and operating lines for the system are straight. This is a valid assumption for most air pollution control systems.

$$N_p = \ln[(Y_1 - mX_2)/(Y_2 - mX_2)(1 - mG_m/L_m) + mG_m/L_m]/\ln(L_m/mG_m) \qquad (1.40)$$

where N_p = number of theoretical plates
Y_1 = mole fraction of solute in entering gas
X_2 = mole fraction of solute entering the column
Y_2 = mole fraction of solute in exiting gas
m = slope of equilibrium line
G_m = molar flow rate of gas, kg-mole/h
L_m = molar flow rate of liquid, kg-mole/h

Equation (1.40) is used to predict the *number of theoretical plates* (N_p) required to achieve a given removal efficiency. The operating conditions for a theoretical plate assume that the gas and liquid stream leaving the plate are in equilibrium with each other. This ideal condition is never achieved in practice. A larger number of actual trays are required to compensate for this decreased tray efficiency.

Three types of efficiencies are used to describe absorption efficiency for a plate tower:

An overall efficiency, which is concerned with the entire column

Murphree efficiency, which is applicable with a single plate

Local efficiency, which pertains to a specific location on a plate

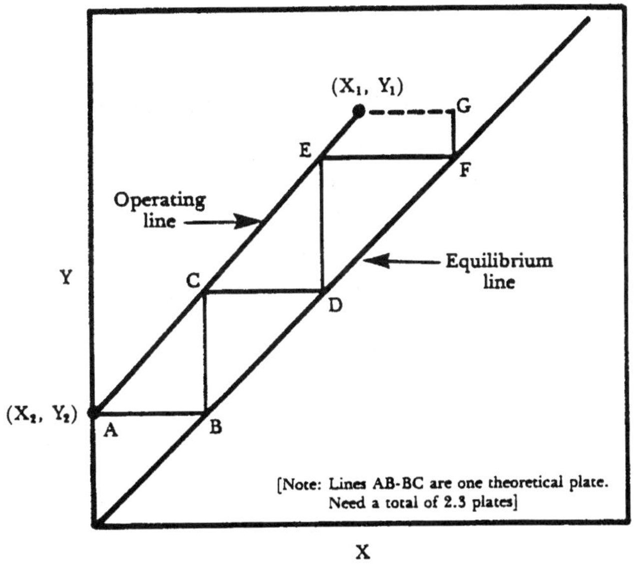

FIGURE 1.14 Graphic determination of the number of theoretical plates (*EPA-81/12, p. 4-35*).

The simplest of the tray efficiency concepts, the overall efficiency, is the ratio of the number of theoretical plates to the number of actual plates. Since overall tray efficiency is an over-simplification of the process, reliable values are difficult to obtain. For a rough estimate, overall tray efficiencies for absorbers operating with low-viscosity liquid normally fall in a 65–80% range. Example 6 shows a procedure to calculate the number of theoretical plates.

EXAMPLE 6: number of theoretical plates
Calculate the number of theoretical plates required for the scrubber in Example 5, using the same conditions as in Example 4. Estimate the total height of the column if the trays are spaced at 0.53 m intervals and assume an overall tray efficiency of 70%.

Solution:
From Example 5 and the previous examples the following data are obtained:

Slope of equilibrium line, $m = 42.7$ (kg-mole SO_2 in air)/(kg-mole SO_2 in water)
$G_m = 3.5$ kg-mole/min
$L_m = 3672$ (kg/min)(kg-mole)/18 kg $= 204$ kg-mole/min
$X_2 = 0$ (no recycle liquid)
$Y_1 = 0.03$
$Y_2 = 0.003$

The number of theoretical plates from Eq. (1.40) is

$$N_p = \ln[(Y_1 - mX_2)/(Y_2 - mX_2)(1 - mG_m/L_m) + mG_m/L_m]/\ln(L_m/mG_m)$$

where N_p = number of theoretical plates
Y_1 = mole fraction of solute in entering gas
X_2 = mole fraction of solute entering the column
Y_2 = mole fraction of solute in exiting gas
m = slope of equilibrium line
G_m = molar flow rate of gas, kg-mole/h
L_m = molar flow rate of liquid, kg-mole/h

$$\begin{aligned} N_p &= \ln[Y_1 - mX_2)/(Y_2 - mX_2)(1 - mG_m/L_m) + mG_m/L_m]/\ln(L_m/mG_m) \\ &= \ln\{(0.03 - 0)/(0.003 - 0)[1 - 42.7(3.5)/204] + 42.7(3.5)/204\}/\ln[204/(42.7)(3.5)] \\ &= 3.94 \text{ theoretical plates} \end{aligned}$$

Assuming that the overall plate efficiency is 70%,

$$\begin{aligned} \text{Actual number of plates} &= 3.94/0.7 \\ &= 5.6 \text{ or } 6 \text{ plates} \end{aligned}$$

The height of the tower is given by

$$Z = N_p \text{ (tray spacing)} + \text{top height}$$

The top height is the distance over the top plate which allows the gas–vapor mixture to separate. This distance is usually the same as the tray spacing.

$$\begin{aligned} Z &= 6 \text{ plates } (0.53 \text{ m}) + 0.53 \text{ m} \\ &= 3.71 \text{ m} \end{aligned}$$

Note: this height is approximately the same as that predicted for the packed tower in Example 4. This seems logical since both the packed and plate towers are efficient gas absorption devices. However, due to the many assumptions, no concrete generalization can be made.

Factors affecting absorption performance. The physical and chemical characteristics of the exhaust gas stream play an important role in both the selection and proper operation of an absorption system. Factors affecting the absorption performance include:

1. *Absorbent*: selection of the proper absorbing liquid (absorbent) is based on the efficiency required and the liquid cost. Water is the usual choice because many gaseous contaminants are soluble in it, it is readily available, and relatively inexpensive. The following properties must also be kept in mind when selecting a liquid:

   - *Gas solubility*: high solubility increases the absorption rate and minimizes the quantity of liquid needed. The solubility of the gaseous contaminant is the first characteristic to evaluate. If the gaseous contaminant is very soluble, then high removal efficiencies can be achieved by almost any absorption device. For a relatively insoluble contaminant only certain systems may be able to achieve the required removal efficiency. In some cases a chemical reagent may have to be added to the absorbing liquid to increase the solubility of the contaminant. These reagents may increase the physical solubility of the contaminant (i.e. sodium citrate added to absorb SO_2) or can chemically react with the contaminant (i.e. lime scrubbing of SO_2). If a precipitate is formed by a chemical reaction when a reagent is used, plugging or corrosion problems may arise.
   - *Volatility*: low volatility of the liquid will reduce the amount of vapor that is lost in the exiting gas stream.
   - *Viscosity*: low viscosity promotes rapid absorption rates, improves flooding characteristics, and lowers the pressure drop.
   - *Chemical stability*: the absorbent should not degrade but remain effective throughout its useful lifetime.
   - *Flammability*: if at all possible, the liquid should be nonflammable, non-corrosive, non-toxic, and inexpensive.

2. *Temperature*: the temperature of the exhaust stream is another important characteristic which affects absorption. The solubility of a gas decreases with an increase in operating temperature. As the temperature increases, so does the kinetic energy of the gas molecules in solution. At these higher states of energy the gas molecules will come out of solution. This loss of solubility at higher temperatures necessitates that some gas streams be cooled before effective absorption occurs. For example, in certain flue gas desulfurization (FGD) systems the exhaust gases from the boiler must be cooled from 150°C (300°F) to approximately 50°C (125°F) to achieve the desired SO_2 removal efficiency. This is accomplished by adding inlet sprays or adjusting the liquid flow rate to the absorber. The temperature of the exhaust stream also affects the size of the absorption system. Decreasing the temperature decreases the volume of gases which must be handled. This decreases the size of the absorption system.
3. *Construction materials*: an important factor in the successful operation of any absorber is to initially select the proper construction materials. Quenching hot gases to their saturation temperatures forms corrosive acids. Depending on the substances present in the exhaust stream, sulfuric, hydrochloric, or hydrofluoric acids may be formed. The presence of these acids can cause severe corrosion problems unless special materials of construction are used.

Absorption equipment. Absorption equipment includes two major categories:

Wet absorption scrubber (or wet scrubber)

Dry absorption scrubber (or dry scrubber)

The description of the absorption equipment is provided in Chapter 3.3.

4.2 Adsorption for Gaseous Emission Control

During adsorption, one or more gaseous components are removed from an effluent gas stream by adhering to the surface of a solid. The gas molecules being removed are referred to as the *adsorbate*, while the solid doing the adsorbing is called the *adsorbent*. Adsorbents are highly porous particles. Adsorption occurs on the internal surfaces of the particles (EPA-81/12, p. 5-1).

The attractive forces which hold the gas to the surface of the solid are the same that cause vapors to condense (van der Waals' forces). All gas–solid interfaces exhibit this attraction, some more than others. Adsorption systems use materials that are highly attracted to each other to separate these gases from the non-adsorbing components of a gas stream. For air pollution control purposes, adsorption is not a final control process. The contaminant gas is merely stored on the surface of the adsorbent. After it becomes saturated with adsorbate, the adsorbent must either be disposed of and replaced, or the vapors must be desorbed. Desorbed vapors are highly concentrated and may be recovered more easily and more economically than before the adsorption step.

Traditionally, adsorption has been used for air purification and solvent recovery. Air purification processes are those in which the contaminant is present in trace quantities (less than 1.0 ppm) but is highly odorous or toxic. Systems used for air purification are small thin bed adsorbers. When the bed becomes saturated with contaminant, it is taken out and replaced. Solvent recovery processes require much larger systems and are usually designed to control organic emissions whose concentrations are greater than 1000 ppm. This has been the point where the recovery value of the solvent could justify the expense of the large adsorption–desorption system. Currently, adsorption is used as a method of recovering valuable organic vapors from flue gases at all concentration levels. This is due to present regulations limiting volatile organic emissions and the higher costs of solvents.

Mechanism of adsorption. Adsorption occurs by a series of three steps (EPA-81/12, p. 5-2):

- Step 1: The contaminant diffuses from the major body of the air stream to the external surface of the adsorbent particle.
- Step 2: The contaminant molecule migrates from the relatively small area of the external surface (a few m^2/g) to the pores within each adsorbent particle. The bulk of adsorption occurs in these pores because the majority of available surface area is there (hundreds of m^2/g).
- Step 3: The contaminant molecule adheres to the surface in the pore.

Figure 1.15 illustrates this overall diffusion and adsorption process. The purpose of analyzing the mechanism of adsorption is to determine which step controls the overall process. By analyzing each step, adsorber performance can be predicted from physical data. The actual adsorption of a molecule, step 3, proceeds relatively quickly compared to steps 1 or 2. Therefore, step 3 can be ignored when developing design equations. Steps 1 and 2 are both diffusional processes. They involve the transport of the adsorbate through a carrier gas phase to an adsorption site.

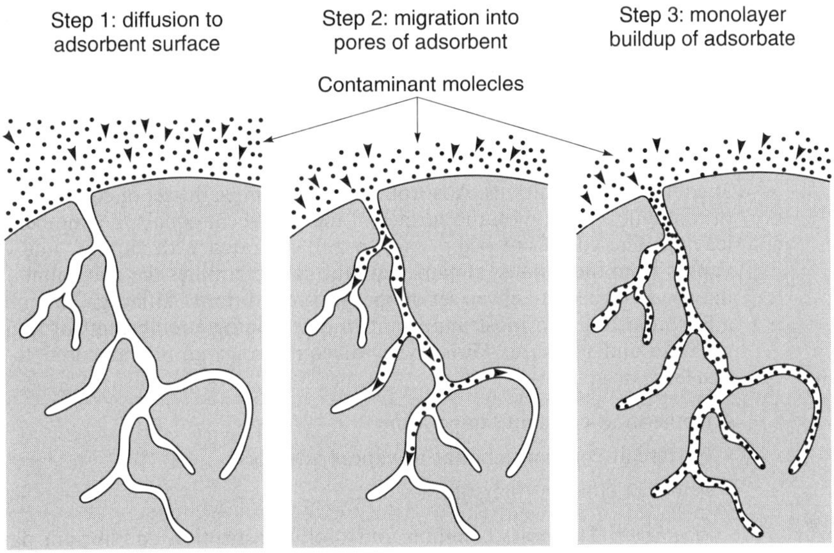

FIGURE 1.15 Gas collection by adsorption (*EPA-81/12, p. 5-2*).

Adsorption forces—physical and chemical. The adsorption process is classified as either *physical* or *chemical*. The basic difference between physical and chemical adsorption is the manner in which the gas molecule is bonded to the adsorbent. In physical adsorption, the gas molecule is bonded to the solid surface by weak forces of intermolecular cohesion. The chemical nature of the adsorbed gas remains unchanged; therefore, physical adsorption is a readily reversible process. In chemical adsorption a much stronger bond is formed between the gas molecule and adsorbent. A sharing or exchange of electrons takes place—as happens in a chemical bond. Chemical adsorption is not easily reversible.

Physical adsorption: The forces active in physical adsorption are electrostatic in nature. These forces are present in all states of matter: gas, liquid, and solid. They are the same forces of attraction which cause gases to condense and real gases to deviate from ideal behavior. This electrostatic force can be measured by the constant a in van der Waals' equation describing non-ideal gas behavior. Physical adsorption is sometimes also referred to as *van der Waals' adsorption*. The electrostatic effect which produces the van der Waals' forces depends on the polarity of both the gas and solid molecules. Molecules in any state are either polar or nonpolar depending on their chemical structure. *Polar substances* are those which exhibit a separation of positive and negative charges within the compound. This separation of positive and negative charges is referred to as a *permanent dipole*. Water is a prime example of a polar substance. Nonpolar substances have both their positive and negative charges in one center so they have no permanent dipole. Most organic compounds, because of their symmetry, are nonpolar.

Chemical adsorption: Chemical adsorption or chemisorption results from the chemical interaction between the gas and the solid. The gas is held to the surface of the adsorbent by the formation of a chemical bond. Adsorbents used in chemisorption can be either pure substances or chemicals deposited on an inert carrier material. One example is using pure iron oxide chips to adsorb H_2S gases. Another example is using activated carbon which has been impregnated with sulfur to remove mercury vapors.

All adsorption processes are exothermic, whether adsorption occurs from chemical or physical forces. In adsorption, molecules are transferred from the gas to the surface of a solid. The fast-moving gas molecules lose their kinetic energy of motion to the adsorbent in the form of heat.

Adsorption equilibrium relationship. Most available data on adsorption systems are determined at equilibrium conditions. Adsorption equilibrium is the set of conditions at which the number of molecules arriving on the surface of the adsorbent equals the number of molecules that are leaving. The adsorbent bed is said to be "saturated with vapors" and can remove no more vapors from the exhaust stream. Equilibrium determines the maximum amount of vapor that may be adsorbed at a given set of operating conditions. Although a number of variables affect adsorption, the two most important in determining equilibrium for a given system are temperature and pressure. Three types of equilibrium graphs are used to describe adsorption systems:

- Isotherm at constant temperature
- Isostere at constant amount of vapors adsorbed
- Isobar at constant pressure

Isotherm: The most common and useful adsorption equilibrium data is the adsorption isotherm. The isotherm is a plot of the adsorbent capacity versus the partial pressure of the adsorbate at a constant temperature. Adsorbent capacity is usually given in weight percent expressed as gram of adsorbate per 100 grams of adsorbent. Figure 1.16 is a typical example of an adsorption isotherm for carbon tetrachloride on activated carbon. Graphs of this type are used to estimate the size of adsorption systems as illustrated in Example 1 to follow.

Isostere: Two additional adsorption equilibrium relationships are the isostere and the isobar. The isostere is a plot of the $\ln(p)$ versus $1/T$ at a constant amount of vapor adsorbed. Adsorption isostere lines are usually straight for most adsorbate–adsorbent systems. Figure 1.17 is an adsorption isostere graph for the adsorption of H_2S gas onto molecular sieves. The isostere is important in that the slope of the isostere corresponds to the heat of adsorption.

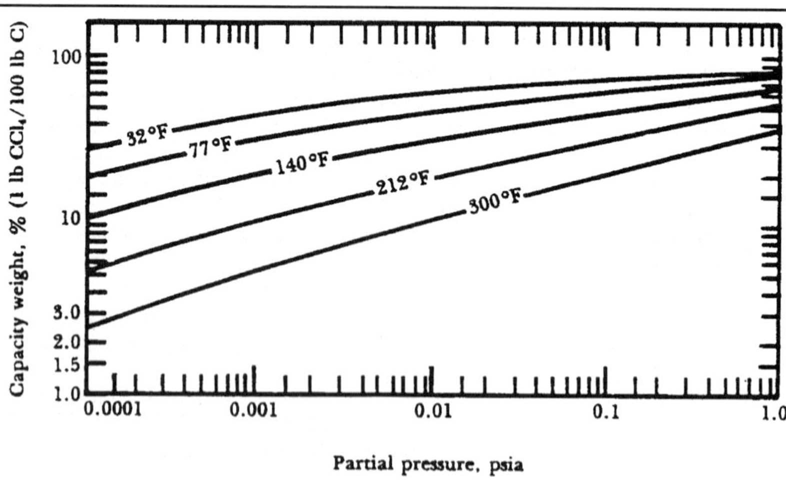

FIGURE 1.16 Adsorption isotherm for carbon tetrachloride on activated carbon (*EPA-81/12, p. 5-8*).

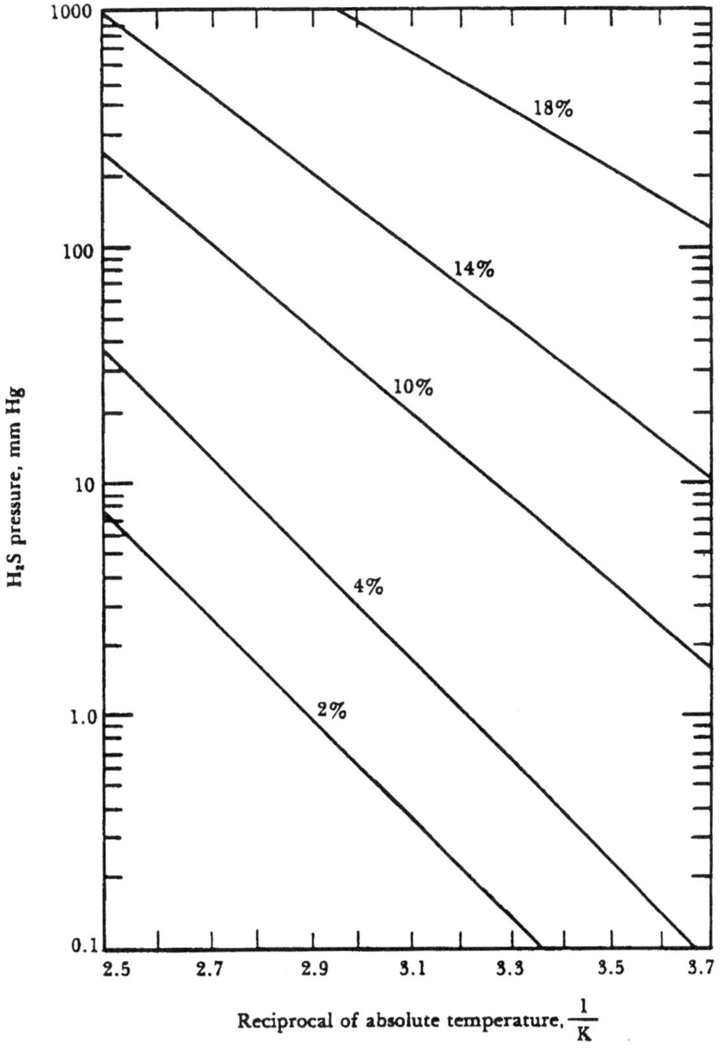

FIGURE 1.17 Adsorption isosteres of H_2S on 13X molecular sieve [loading in % weight] (*EPA-81/12, p. 5-11*).

Isobar: The isobar is a plot of the amount of vapors adsorbed versus temperature at a constant pressure. Figure 1.18 shows an isobar line for the adsorption of benzene vapors on activated carbon. Note that the amount adsorbed decreases with increasing temperature, which is always the case for physical adsorption.

Since these three relationships were developed at equilibrium conditions, they depend on each other. By determining one, such as the isotherm, the other two relationships can be determined for a given system. In the design of a pollution control system, the adsorption isotherm is by far the most commonly used equilibrium relationship.

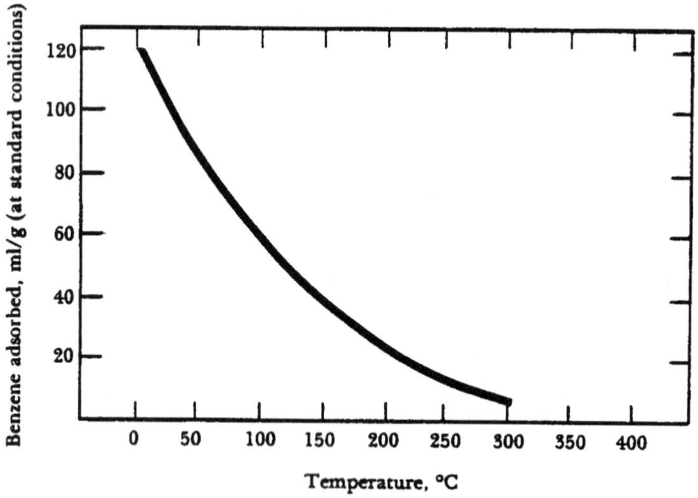

FIGURE 1.18 Adsorption isobar for benzene on carbon (benzene at 10.0 mm Hg) (*EPA-81/12, p. 5-8*).

EXAMPLE 1: adsorption
A dry cleaning process exhausts a 15,000 scfm air stream containing 680 ppm carbon tetrachloride. Using Fig. 1.16 and assuming the exhaust stream is at approximately 140°F and 14.7 psia, determine the saturation capacity of the carbon (EPA-81/12, p. 5-8).

Solution:
In the gas phase, the mole fraction Y is equal to the percent by volume:

$$\begin{aligned} Y &= \%\text{volume} = 680 \text{ ppm} \\ &= 680/(10^6) \\ &= 0.00068 \end{aligned}$$

Obtain the partial pressure:

$$\begin{aligned} p &= YP \\ &= (0.00068)(14.7 \text{ psia}) \\ &= 0.01 \text{ psia} \end{aligned}$$

From Fig. 1.16, at a partial pressure of 0.01 psia and a temperature of 140°F, the carbon capacity is read as 30%. This means that at saturation, 30 lb of vapor are removed per 100 lb of carbon in the adsorber (30 kg/100 kg).

EXAMPLE 2: adsorption
Assume the same conditions as stated above in Example 1. Estimate the amount of carbon that would be required if the adsorber were to operate on a 4-hour cycle. The molecular weight of CCl_4 is 154 lb/lb-mole (EPA-81/12, p. 5-18).

Solution:
From Example 1 we know that the carbon used will remove 30 lb of vapor for every 100 lb of carbon at saturation conditions.

First compute the flow rate of CCl_4:

$$Q(CCl_4) = 15{,}000\,\text{scfm} \times 0.00068 = 10.2\,\text{scfm}\ CCl_4$$

Converting to pounds per hour:

$$10.2\ (\text{ft}^3/\text{min}) \times (\text{lb-mole})/(359\ \text{ft}^3) \times (154\,\text{lb})/(\text{lb-mole}) \times (60\,\text{min})/\text{h} = 262.5\,\text{lb-}CCl_4/\text{h}$$

The amount of carbon (at saturation) required (assuming that the working charge is twice the saturation capacity):

$$2(3500) = 7000\,\text{lb}\ (3182\,\text{kg})\ \text{carbon per 4-hour cycle per adsorber}$$

Note: This gives only a rough estimate of the amount of carbon needed.

Adsorbent materials. Adsorbents are characterized by their chemical nature, extent of their surface area, pore size distribution, and particle size. In physical adsorption, the most important characteristic in distinguishing between adsorbents is their surface polarity. The surface polarity determines the type of vapors a particular adsorbent will have the greatest affinity for (EPA-81/12, p. 5-12).

Several materials are used effectively as adsorbing agents. The most common adsorbents used industrially are

- activated carbon
- silica gel
- molecular sieves
- activated alumina (aluminum oxide)

Of the above absorbents, activated carbon is the primary nonpolar adsorbent. It is possible to manufacture other adsorbing material having nonpolar surfaces. Since their surface area is much less than that of activated carbon, they are not used commercially. Polar adsorbents will preferentially adsorb any water vapor that may be present in a gas stream. Since moisture is present in most polluted gas streams, the use of polar adsorbents is severely limited for an air pollution system.

Activated carbon: Activated carbon can be produced from a variety of feedstocks such as wood, coal, coconut, nutshells, and petroleum-based products. The activation process takes place in two steps.

- First, the feedstock is carbonized. Carbonization involves heating (usually in the absence of air) the material to a temperature high enough (600°C) to drive off all volatile material. Thus, carbon is essentially all that is left. To increase the surface area, the carbon is then activated by using steam, air, or carbon dioxide at higher temperatures. These gases attack the carbon and increase the pore structure. The temperature involved, the amount of oxygen present, and the type of feedstock, all greatly affect the adsorption qualities of the carbon.
- Because of its nonpolar surface, activated carbon is used to control emission of organic solvents, odors, toxic gases, and gasoline vapors. Carbons used in gas phase adsorption systems are manufactured in granular form, usually between 4×6 to 4×20 mesh in size. Bulk density of the packed bed can range from 0.08 to 0.5 g/cm^3 (5 to 30 lb/ft^3) depending on the internal porosity of the carbon. Surface area of the carbon can range from 600 to 1600 m^2/g (2.9×10^6 to 7.8×10^6 ft^2/lb).

Silica gel: Silica gels are made from sodium silicate. Sodium silicate is mixed with sulfuric acid, resulting in a jelly-like precipitant from where the gel name comes. The precipitant is then dried and roasted. Depending on the processes used in manufacturing the gel, different grades varying in activity can be produced. Silica gels have surface areas of approximately 750 m^2/g (3.7×10^6 ft^2/lb). Silica gels are used primarily to remove moisture from exhaust streams, but are ineffective at temperatures above 260°C (500°F).

Molecular sieves: Unlike the other adsorbents, which are amorphous (not crystalline) in nature, molecular sieves have a crystalline structure. The pores are, therefore, uniform in diameter. Molecular sieves can be used to capture or separate gases on the basis of molecular size and shape. An example of this are refining processes which sometimes use molecular sieves to separate straight-chained paraffins from branched and cyclic compounds. However, the main use of molecular sieves is in the removal of moisture from exhaust streams. The surface area of molecular sieves range from 600 to 700 m^2/g (2.9×10^6 to 3.4×10^6 ft^2/lb).

Aluminum oxide (activated alumina): Aluminum oxides are manufactured by thermally activating alumina or bauxite. This is accomplished by heating the alumina in an inert atmosphere to produce a porous aluminum oxide pellet. Aluminum oxides are not commonly used in air pollution applications. They are primarily used for the drying of gases, especially under high pressures, and as support material in catalytic reactions. A prime example is the impregnating of the alumina with platinum or palladium for use in catalytic incineration. Activated alumina's surface areas can range from 200 to 300 m^2/g (0.98×10^6 to 1.5×10^6 ft^2/lb).

Factors affecting adsorption performance. A number of factors or system variables influence the performance of an adsorption system. These variables and their effects on the adsorption process are discussed in turn (EPA-81/12, p. 5-18).

- Temperature
- Pressure
- Gas velocity
- Bed depth
- Humidity
- Contaminants

Temperature: For physical adsorption processes, the capacity of an adsorbent decreases as the temperature of the system increases. As the temperature increases the vapor pressure of the adsorbate increases, raising the energy level of the adsorbed molecules. Adsorbed molecules now have sufficient energy to overcome the van der Waals' attraction and migrate back to the gas phase. Molecules already in the gas phase tend to stay there due to their high vapor pressure. As a general rule, adsorber temperatures are kept below 55°C (130°F) to ensure adequate bed capacities. Temperatures above this limit can be avoided by cooling the exhaust stream that is to be treated.

Pressure: Adsorption capacity increases with an increase in the partial pressure of the vapor. The partial pressure of a vapor is proportional to the total pressure of the system. Any increase in pressure will increase the adsorption capacity of a system. The increase in capacity occurs because of a decrease in the mean free path of vapors at higher pressures: simply, the molecules are packed more tightly together. More molecules have a chance to hit the available adsorption sites, increasing the number of molecules adsorbed.

Gas velocity: The contact or residence time between the contaminant stream and adsorbent is determined by the gas velocity through the adsorber. The residence time directly affects capture efficiency. The slower the contaminant stream flows through the adsorbent bed, the greater the probability of a contaminant molecule hitting an available site. Once a molecule has been captured it will stay on the surface until the physical conditions of the system are changed. To achieve 90% + capture efficiency, most carbon adsorption stems are designed for a maximum gas flow velocity of 30 m/min (100 ft/min) through the adsorber. A lower limit of at least 6 m/min (20 ft/min) is maintained to avoid flow distribution problems, such as channeling.

Bed depth: Providing a sufficient depth of adsorbent is very important in achieving efficient gas removal. If the adsorber bed depth is shorter than the required mass transfer zone, breakthrough will immediately occur rendering the system ineffective. Computing the length of the mass transfer zone (MTZ) is very difficult since it depends upon six factors:

- adsorbent particle size
- gas velocity
- adsorbate concentration
- fluid properties of the gas stream
- temperature
- pressure

The relationship between breakthrough capacity and MTZ is

$$C_B = [0.5C_s(\mathrm{MTZ}) + C_s(D - \mathrm{MTZ})]/D$$

where C_B = breakthrough capacity
C_s = saturation capacity
MTZ = mass transfer zone length
D = adsorption bed depth

The above equation is used mainly as a check to ensure that the proposed bed depth is longer than the MTZ. Actual bed depths are usually many times longer than the length of the MTZ. The additional bed depth allows for adequate cycle times.

The total amount of adsorbent required is usually determined from the adsorption isotherm, as illustrated in Example 2. Once this has been set, the bed depth can then be estimated by knowing the tower diameter and density of the adsorbent. Example 3 illustrates how this is done. Generally, the adsorbent bed is sized to the maximum length allowed by the pressure drop across the bed.

Humidity: As stated previously, activated carbon will preferentially adsorb nonpolar hydrocarbons over polar water vapor. The water vapor molecules in the exhaust stream exhibit strong attractions for each other rather than the adsorbent. At high relative humidity, over 50%, the number of water molecules increases such that they begin to compete with the hydrocarbon molecules for active adsorption sites. This reduces the capacity and the efficiency of the adsorption system. Exhaust streams with humidity greater than 50% may require installation of additional equipment to remove some of the moisture. Coolers to remove the water are one solution. Dilution air with significantly less moisture in it than the process stream has also been used. Also, the contaminant stream may be heated to reduce the humidity as long as the increase in temperature does not greatly affect adsorption efficiency.

Contaminants: In addition to humidity, particulate matter, entrained liquid droplets, and organic compounds that have high boiling points can also reduce adsorber efficiency if present in the air stream. Any micron-sized particle of dust or lint which is not filtered can cover the surface of the adsorbent. This greatly reduces the surface area of the adsorbent available to the gas molecule for adsorption. Covering of active adsorption sites by an inert material is referred to as *blinding or deactivation*. To avoid this situation almost all industrial adsorption systems are equipped with some type of particulate removal device.

Entrained liquid droplets can also cause operational problems. Liquid droplets that are non-adsorbing act the same as particulate matter. The liquid covers the surface, blinding the bed. If the liquid is the same as the adsorbate, high heats of adsorption occur. This is especially a problem in activated carbon systems where liquid organic droplets carried over from the process can cause bed fires from the heat released. Some type of entrainment separator may be required when liquid droplets are present.

Adsorption equipment. Adsorption control systems are summarized in Table 1.8.

Non-regenerable adsorption system. Non-regenerable systems are normally used to control exhaust streams with low pollutant concentrations, below 1.0 ppm. Generally these pollutants are highly odorous or to some degree toxic. When these systems reach the breakthrough point the bed is taken off-stream and replaced with a fresh bed. The used carbon can then be sent back to the manufacturer for reactivation (EPA-81/12, p. 5-26).

Non-regenerable adsorption systems are manufactured in a variety of configurations. Bed areas are sized to control the gas flow through them at between 6.0 and 18 m/min (20 and 60 ft/min). They usually consist of thin adsorbent bed depth, ranging in thickness from 1.25 to 10.0 cm (0.5 to 4.0 in). These thin beds have a low pressure drop, normally below 62 Pa (0.25 in H_2O), dependent on the bed thickness, gas velocity, and particle size of the adsorbent. Service time for these units can range from 6 months for heavy odor concentrations to up to 2 years for trace concentrations or intermittent operations. They are used mainly as air-purification devices for small air flows in offices, laboratory exhausts, and other small exhaust streams (EPA-81/12, p. 5-26).

The shapes of these thin bed adsorbers are

- Flat adsorber
- Pleated absorber
- Cylindrical adsorber
- Canister adsorber

The granules of activated carbon are retained by porous support materials, usually perforated sheet metals. An adsorber system usually consists of a number of retainers or panels placed in one frame.

Flat adsorber: Figure 1.19 shows a nine-cell thin bed adsorber. The panels are similar to home air filters except that instead of containing steel wool they contain activated carbon as the filter. Flat panel beds are sized to handle higher exhaust flow rates, approximately 9.4 m^3/s (20,000 cfm).

Pleated cell adsorber: Figure 1.20 illustrates a pleated cell adsorber. The pleated cell is one continuous retainer of activated carbon, rather than individual panels. Panel and pleated beds are dimensionally about the same size, normally 0.6 m^2 (2.5 ft × 2.5 ft) with the carbon depth ranging from 0.2 to 0.6 m (8 in to 2 ft). Pleated beds are limited to flow rates of approximately 4.7 m^3/s (10,000 cfm).

Cylindrical adsorber: Figure 1.21 illustrates a cylindrical adsorber. It is usually a small unit designed to handle low flow rates of approximately 0.12 m^3/s (250 cfm). Cylindrical adsorbers

TABLE 1.8 Gaseous Adsorption Control Systems

Adsorption process	Control equipment
1. Non-regenerable adsorption system	1.1 Flat adsorber
	1.2 Cylindrical adsorber
	1.3 Pleated adsorber
	1.4 Canister adsorber
2. Regenerable adsorption system	2.1 Fixed bed adsorber
	2.2 Moving bed adsorber
	2.3 Fluidized bed adsorber

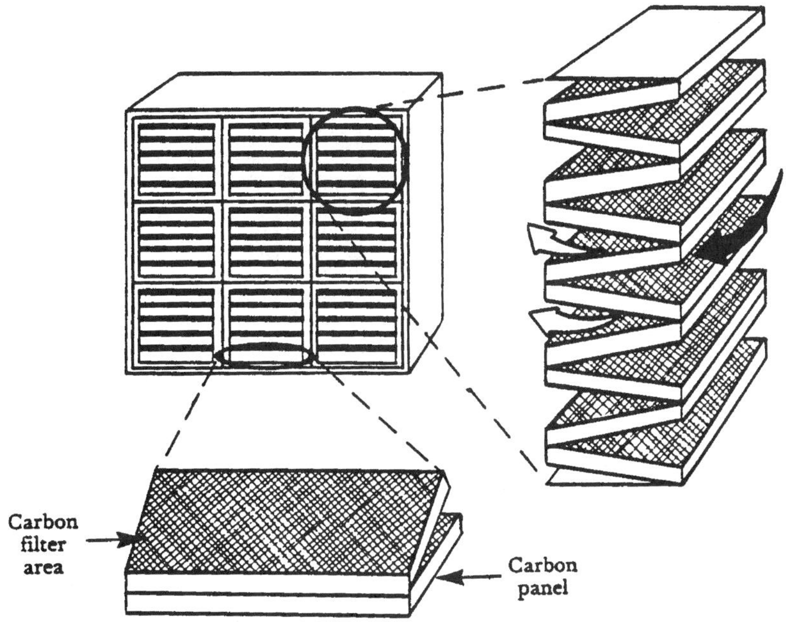

FIGURE 1.19 Non-regenerable flat adsorber (*EPA-81/12, p. 5-27*).

are made of the same materials as the panel and pleated adsorbers except their shape is round rather than square.

Canister absorber: Figure 1.22 illustrates a canister absorber. The unit that can be used is essentially just a 55-gallon drum. The bottom of the drum is filled with gravel to support a bed of activated carbon weighing approximately 150 kg (330 lb). These units are used to treat small flow rates (0.5 m^3/s or 100 cfm) from laboratory hoods, chemical storage tank vents, and chemical reactors.

Regenerable adsorption system. Regenerable systems are used for higher pollutant concentrations, such as in solvent recovery operations. Once the bed reaches the breakthrough

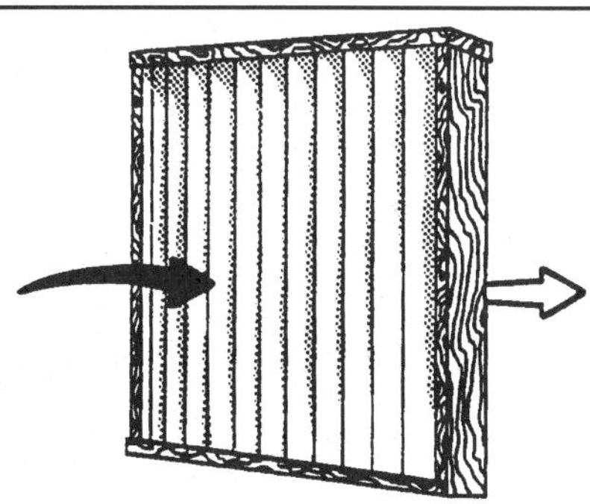

FIGURE 1.20 Non-regenerable pleated thin adsorber (*EPA-81/12, p. 5-28*).

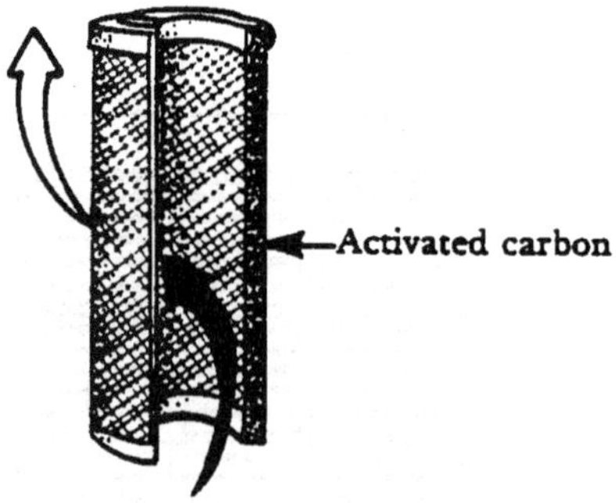

FIGURE 1.21 Non-regenerable cylindrical adsorber (*EPA-81/12, p. 5-28*).

point in a regenerable system, the pollutant vapors are directed to a second bed while the first has the vapors desorbed (EPA-81/12, p. 5-26).

A large regenerable adsorption system can be categorized as

- Fixed bed system
- Moving bed system
- Fluidized bed system

The name refers to the manner in which the vapor stream and adsorbent are brought into contact. The choice of a particular system depends on the pollutants to be controlled and the recovery requirements. The most common adsorption system for controlling air pollutants is the fixed carbon bed. These systems are used to control a variety of organic vapors and are

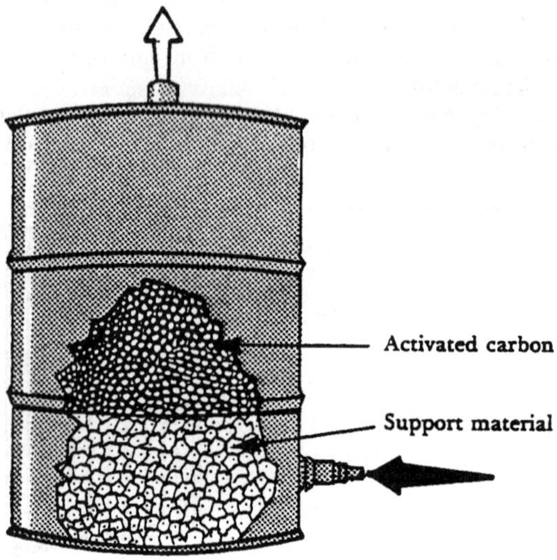

FIGURE 1.22 Non-regenerable canister adsorber (*EPA-81/12, p. 5-29*).

usually regenerated by direct steaming of the bed. The organic compounds may be recovered by condensing the exhaust from the regeneration step and separating out the water and solvent.

Figure 1.23 shows a typical three fixed bed regenerable adsorber. Fixed bed adsorption systems generally involve multiple beds. One or more beds treat the process exhaust while the other beds are either being regenerated or cooled. The solvent-laden air stream is first pre-treated to remove any solid or liquid particles that could blind the carbon bed and decrease its efficiency. The solvent-laden air stream then usually passes down through the fixed carbon bed. Upward flow through the bed is usually avoided (unless flow rates are low (< 500 cfm) to eliminate the risk of entraining carbon particles in the exhaust stream).

After a predetermined length of time, referred to as the *cycle time*, the solvent-laden air stream is directed to the second adsorber by a series of valves. Steam is then injected into the first bed to remove the adsorbed vapors. The steam and desorbed vapors are then usually sent to a recovery system. If the solvents are immiscible in water, they can be separated by condensing the exhaust and decanting off the solvent. If the solvents are miscible in water, distillation may be required. Before the first adsorber is returned to service, cooling and drying of the carbon should be provided. This will ensure against immediate breakthrough occurring from the 'hot, wet' carbon bed. This can be accomplished by venting the solvent-laden air stream through the hot, wet adsorber, then to the on-line adsorber to maintain a high removal efficiency.

Regenerable fixed carbon beds are usually between 0.3 and 1.2 m (1 and 4 ft) thick. The maximum adsorbent depth of 1.2 m is based on pressure drop considerations. Superficial gas velocities through the adsorber range from 6.0 to 30.0 m/min (20 to 100 ft/min) with 30.0 m/min being a maximum permissible flow rate. Pressure drops normally range from 750 to 3730 Pa (3 to 15 in H_2O) depending on the gas velocity, bed depth and carbon particle size.

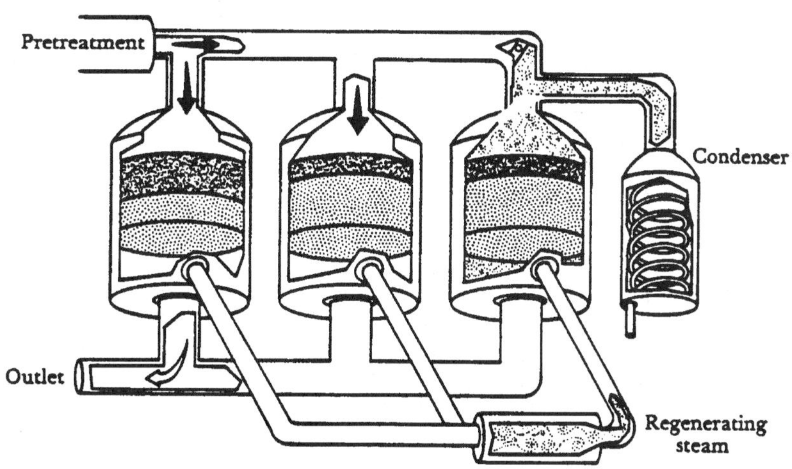

FIGURE 1.23 Regenerable three fixed bed adsorber (*EPA-81/12, p. 5-30*).

EXAMPLE 3: sizing adsorber
A solvent degreaser is designed to recover toluene from a 3.78 m^3/s (8000 acfm) air stream at 25°C (77°F) and atmospheric pressure. The company is planning to use a two-bed carbon adsorption system with a cycle time of 4 hours. The maximum concentration of toluene is kept below 50% of the lower explosive limit (LEL) for safety purposes. Using Fig. 1.24, the adsorption isotherm for toluene, and the additional operational data, estimate (EPA-81/12, p. 5-38):

1. The amount of carbon required for a 4-hour cycle
2. Square feet of surface area required based on a 0.508 m/s (100 fpm) maximum velocity
3. Depth of the carbon bed

Given conditions

LEL for toluene = 1.2%

molecular weight of toluene = 92.1 kg/kg-mol

carbon density = 480 kg/m^3 (30 lb/ft^3)

Solution:
1. To compute amount of carbon required, first, calculate the toluene flow rate.

$$(3.78\,m^3/s)(50\%)(1.2\%) = 0.023\,m^3/s\,\text{toluene}$$

To determine the saturation capacity of the carbon, calculate the partial pressure of toluene at the adsorption conditions.

$$\begin{aligned} p &= YP \\ &= (0.023/3.78)(14.7\,\text{psia}) \\ &= 0.089\,\text{psia} \end{aligned}$$

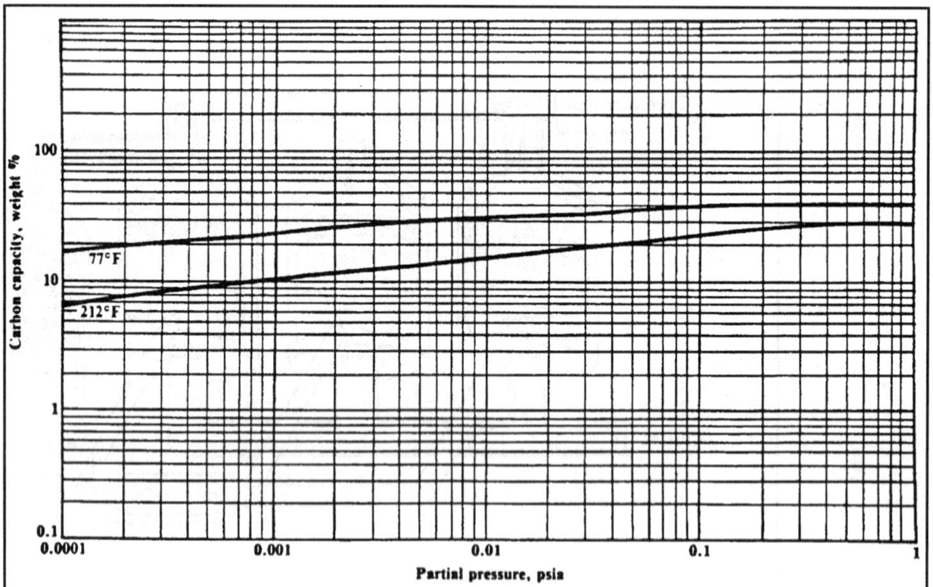

FIGURE 1.24 Adsorption isotherm for toluene (*EPA-81/05, p. 18-9*).

At 0.089 psia and 77°F, from Fig. 1.24 the saturation capacity of the carbon is 40% or 40 kg toluene per 100 kg of carbon.

The flow rate of toluene is

$$(0.023\,\text{m}^3/\text{s}) \times (\text{kg-mol}/22.4\,\text{m}^3) \times (273\,\text{K}/350\,\text{K}) \times (92.1\,\text{kg/kg-mol})$$
$$= 0.074\,\text{kg/s of toluene}$$

The amount of carbon at saturation for a 4-hour cycle is:

$$(0.074\,\text{kg/s toluene}) \times (100\,\text{kg carbon}/40\,\text{kg toluene}) \times (3600\,\text{s/h}) \times (4\,\text{h})$$
$$= 2664\,\text{kg of carbon}$$

The working charge of carbon can be estimated by doubling the saturation capacity. Therefore,

$$\text{working charge} = (2)(2664\,\text{kg of carbon}) = 5328\,\text{kg of carbon}$$

2. The square feet of superficial surface area is the surface area set by the maximum velocity of 0.508 m/s (100 fpm) through the adsorber.

The required surface area is

$$\begin{aligned} A &= Q/(\text{maximum velocity}) \\ &= (3.78\,\text{m}^3/\text{s})/(0.508\,\text{m/s}) \\ &= 7.44\,\text{m}^2 \end{aligned}$$

For a horizontal flow adsorber this would correspond to a vessel approximately 2 m (6.6 ft) in width and 4 m (13.1 ft) in length to give 8.0 m^2 (87 ft^2) surface area. This would supply more than the required area.

The flow rate is too high to be handled by a single vertical flow adsorber. An alternative would be to use three vessels, two adsorbing while one is being regenerated. Each vessel then must be sized to handle 1.89 m^3/s (4000 acfm). The area required for a limiting velocity of 0.508 m/s is

$$\text{Area to handle half flow } A = (1.89\,\text{m}^3/\text{s})/(0.508\,\text{m/s}) = 3.72\,\text{m}^2$$

This cross-sectional area corresponds to a vessel diameter of

$$\begin{aligned} d &= (4A/\pi)^{0.5} \\ &= [4(3.72)/\pi]^{0.5} \\ &= 2.18\,\text{m}\ (7\,\text{ft}) \end{aligned}$$

3. The volume that the carbon would occupy in the horizontal bed system is

$$\begin{aligned} \text{Volume of carbon} &= \text{weight/density} \\ &= 5328\,\text{kg} \times (\text{m}^3/480\,\text{kg}) \\ &= 11.1\,\text{m}^3 \end{aligned}$$

Note: for the three-bed vertical system, the volume of each bed would be half this or 5.55 m^3.

4. The depth of carbon for the horizontal bed is given by

$$\begin{aligned}\text{Depth of carbon} &= (\text{volume of carbon})/(\text{cross-sectional area of adsorber})\\ &= (11.1\ \text{m}^3)/(7.44\ \text{m}^2)\\ &= 1.49\,\text{m}\end{aligned}$$

Note: the depth for the three-bed vertical system is the same, since both the volume and area are halved.

The above solution is based on a number of adsorber design maximum and minimum rules of thumb. It is intended as a guide to illustrate how to 'red flag' any parameters which may be greatly exceeded.

EXAMPLE 4: adsorption working capacity
You are to calculate the working capacity of an adsorption bed given the saturation (equilibrium) capacity, mass transfer zone, and heel (EPA-84/09, p. 110).
Given conditions

depth of absorption bed = 3 ft

saturation capacity = 39%

MTZ = 4 in

heel = 2.5% = amount remaining on the adsorbent following regeneration

Solution:
1. Calculate the breakthrough capacity.
The breakthrough capacity is the capacity when traces of the contaminant first begin to appear in the exit gas stream from an adsorption bed.

$$C_B = [0.5C_s(\text{MTZ}) + C_s(D - \text{MTZ})]/D$$

where C_B = breakthrough capacity
C_s = saturation capacity
MTZ = mass transfer zone length
D = adsorption bed depth

Breakthrough capacity,

$$\begin{aligned}C_B &= [0.5C_s(\text{MTZ}) + C_s(D - \text{MTZ})]/D\\ &= [(0.5)(0.39)(4) + (0.39)(36 - 4)]/36\\ &= 0.368\\ &= 36.8\%\end{aligned}$$

2. Calculate the working capacity.
The working capacity is lower than the saturation capacity or the breakthrough capacity. It results from unrecoverable solvent in the carbon and occasionally from a lower packing factor than laboratory specifications.

$$\text{Working capacity} = (\text{breakthrough capacity}) - (\text{heel}) - (\text{packing factor})$$

Neglect the packing factor for this calculation.

$$\text{Working capacity} = 36.8 - 2.5 = 34.3\%$$

EXAMPLE 5: adsorption–degreaser ventilation clean up
You are asked to determine the required height of adsorbent for an adsorber which treats a degreaser ventilation stream contaminated with trichloroethylene (TCE) given the following operating and design data (EPA-84/09, p. 112).
Given conditions

volumetric flow rate of contaminated air stream = 10,000 scfm

standard conditions = 60°F, 1 atm

operating temperature = 70°F

operating pressure = 20 psia

adsorbent = activated carbon

bulk density of activated carbon = $36\,\mathrm{lb/ft^3}$

working capacity of activated carbon = 28 lb TCE/100 lb carbon

inlet concentration of TCE = 2000 ppm (by volume)

molecular weight of TCE = 131.5

the adsorption column cycle is set at 4 hours in the adsorption mode, 2 hours in heating and desorbing, 1 hour in cooling, and 1 hour in stand-by

the adsorber recovers 99.5% by weight of TCE

a horizontal unit with a width of 6 ft and length of 15 ft is used.

Solution:
1. What is the volume of activated carbon required to treat the contaminated gas stream?
Since the dimensions of the adsorber are known, you only need the volume of activated carbon to calculate the height of adsorbent (activated carbon).
1.a. What is the mass of TCE to be adsorbed during a 4 hour period?
1.a.1. Calculate the actual volumetric flow rate of contaminated gas stream in acfh?

$$Q_a/Q_s = T_a/T_s$$

where Q_a = actual volumetric flow rate
Q_s = standard volumetric flow rate
T_a = actual operating temperature in °R or K
T_a = standard operating temperature in °R or K

$$\begin{aligned} Q_a &= Q_s(T_a/T_s)(P_s/P_a) \\ &= 10{,}000[(70+460)/(60+460)][(14.7)/(20)] \\ &= 7491\ \mathrm{acfm} \\ &= 4.5\times 10^5\ \mathrm{acfh} \end{aligned}$$

1.a.2. Calculate the volumetric flow rate of TCE in acfh?

$$Q_{TCE} = (y_{TCE})(Q_a)$$

where Q_{TCE} = volumetric flow rate of TCE
y_{TCE} = inlet concentration of TCE
Q_a = actual volumetric flow rate

$$\begin{aligned} Q_{TCE} &= (y_{TCE})(Q_a) \\ &= (2000\times 10^{-6})(4.5\times 10^5) \\ &= 900\ \mathrm{acfh} \end{aligned}$$

1.a.3. Calculate the mass flow rate of TCE m_r, in lb/hour?
From the ideal gas equation,

$$\rho = PM/R_u T$$
$$m_r = Q\rho = Q(PM/R_u T)$$

where
ρ = density
P = pressure
M = molecular weight
R_u = universal gas constant = 10.73 (psia-ft^3)/(lb-mole-°R)
T = absolute temperature
Q = volumetric flow rate (ft^3/h)
m_r = mass flow rate (lb/h)

$$\begin{aligned} m_r &= Q\rho = Q(PM/R_u T) \\ &= (900)(20)(131.5)/(10.73)(70 + 460) \\ &= 416.2\,\text{lb/h} \end{aligned}$$

1.a.4. Calculate the mass of TCE to be adsorbed during the 4 hour period?

$$\begin{aligned} \text{TCE adsorbed} &= m_r(0.995)(4) \\ &= (416.2)(0.995)(4) \\ &= 1656.6\,\text{lb} \end{aligned}$$

1.b. What is the volume of activated carbon required in ft^3?

$$\begin{aligned} \text{Activated carbon required} &= \text{(TCE to be adsorbed)/[(bulk density)(ads cap. of carbon)]} \\ &= \text{(TCE to be adsorbed)/} \\ &\quad \text{[((28 lb TCE)/(100 lb carbon))(bulk density)]} \\ &= (1656.6)(100/28)/(36) \\ &= 164\,\text{ft}^3 \end{aligned}$$

2. What is the height of adsorbent in ft?

$$\begin{aligned} \text{Height actual of adsorbent} &= \text{(actual carbon volume)/(cross-sectional area)} \\ &= (164)/(6)(15) \\ &= 1.83\,\text{ft} \end{aligned}$$

EXAMPLE 6: adsorption—plan review
The permit to construct a solvent recovery plant which recovers acetone from a contaminated air stream with three adsorbers has been applied for (Fig. 1.25) and you are asked to review the plans for the adsorption system. In reviewing these plans you should check the time to run each adsorber to breakthrough and the amount of steam required to regenerate the adsorbent (EPA-84/09, p. 114).

Information given in the permit application

1. Volumetric flow rate of contaminated air stream = 30,000 acfm at 1 atm
2. Temperature of contaminated air stream = 20°C
3. Molecular weight of acetone = 58

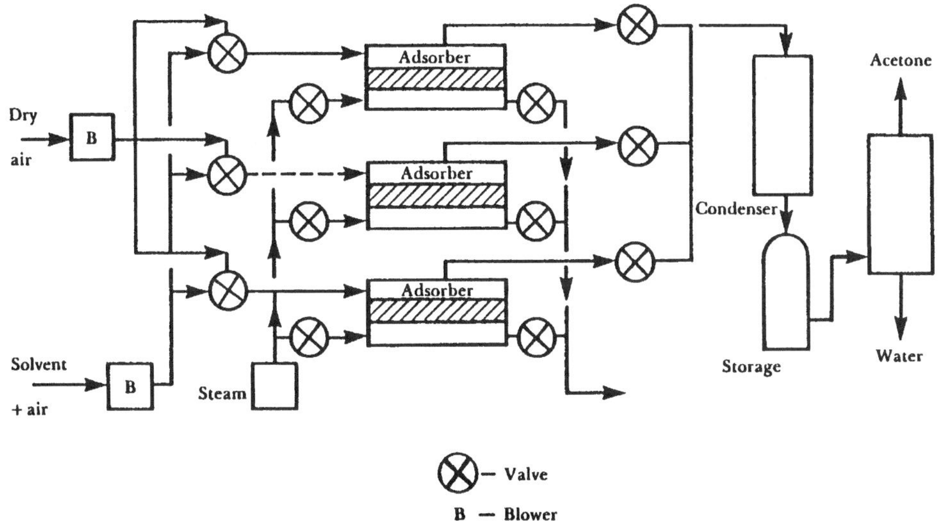

FIGURE 1.25 Graphical representation (*EPA-84/09, p. 115*).

4. Composition of acetone in air stream = 0.15 mole%
5. Carbon type; charge per adsorber = B, 4 × 6 mesh; 15,000 lb
6. Heel = 2 wt %
7. Packing factor (capacity correction due to different packing) = 3 wt %
8. Adsorber dimensions = 10 ft by 24 ft
9. Adsorption bed depth = 1.96 ft
10. Vapor pressure of acetone at 20°C = 170 mmHg
11. MTZ length = 2 inches
12. Figure 1.26 (% relative saturation)
13. Figure 1.27 (steam requirement)
14. Two of the three adsorbers are in the adsorption phase at all times. The plant operates 24 hours a day and 365 days a year. Plant steam is available at 5 psig.

Solution:

1. What is the adsorption time per adsorber bed?

1.a. What is the working capacity of the activated carbon (carbon B)?

The working capacity of the carbon is lower than the saturation capacity or the breakthrough capacity and results from unrecoverable solvent in the carbon and from a lower packing factor than laboratory specifications. Based on experience, the unrecoverable solvent, or heel, is 2 wt % and the lower capacity due to the lower packing factor is estimated to be 3 wt %.

1.a.1. Calculate the relative saturation of acetone in the air stream.

Percent relative saturation = (partial pressure of acetone in air)/
(vapor pressure of acetone at 20°C)

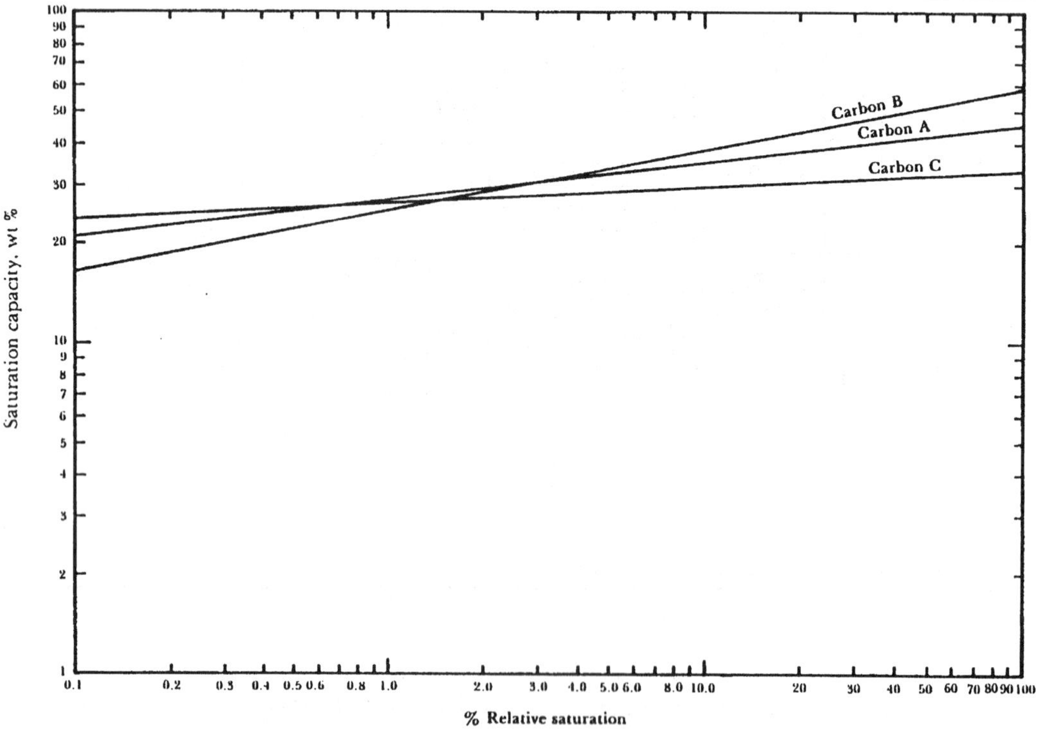

FIGURE 1.26 Saturation capacity vs. relative saturation (*EPA-84/09, p. 116*).

The partial pressure of acetone in air is equal to the product of acetone mole fraction and atmospheric pressure.

$$\begin{aligned}\text{Percent relative saturation} &= [(\text{mole fraction})(\text{air pressure})/\\ &\qquad (\text{vapor pressure of acetone at } 20^\circ\text{C})](100)\\ &= [(0.0015)(760)/(170)](100)\\ &= 0.67\%\end{aligned}$$

1.a.2. Determine the saturation capacity of carbon B and C_s for acetone from Fig. 1.26 using the percent relative saturation calculated above.

From Fig. 1.26, saturation capacity = 23 wt %

1.a.3. Calculate the breakthrough capacity in wt %.

$$C_B = [0.5C_s(\text{MTZ}) + C_s(D - \text{MTZ})]/D$$

where C_B = breakthrough capacity
C_s = saturation capacity
MTZ = mass transfer zone length
D = adsorption bed depth

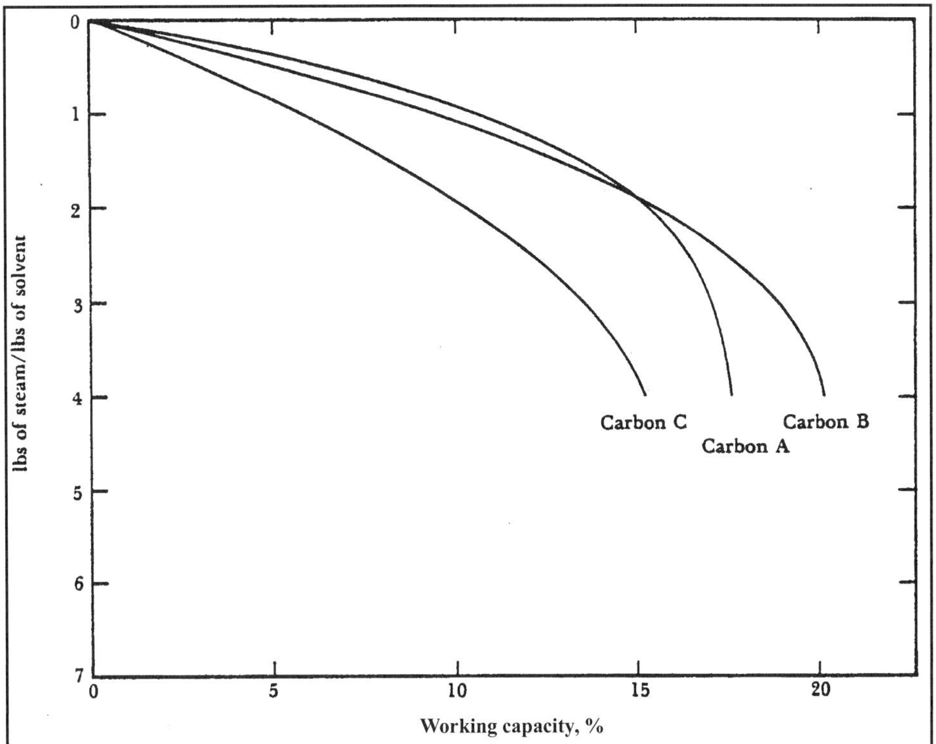

FIGURE 1.27 (lbs of steam)/(lbs of solvent) vs. working capacity (*EPA-84/09, p. 117*).

The breakthrough capacity is the capacity when traces of acetone begin to appear in the exit gas stream.

$$\begin{aligned}\text{Breakthrough capacity} &= [(0.5)(0.23)(2/12) + (0.23)(1.96 - (2/12))]/1.96\\ &= 0.22\\ &= 22\,\text{wt}\,\%\end{aligned}$$

1.a.4. Calculate the working capacity of the carbon B in wt %.

$$\begin{aligned}\text{Working capacity} &= (\text{breakthrough capacity}) - \text{heel} - (\text{packing factor})\\ &= 22 - 2 - 3\\ &= 17\,\text{wt}\,\%\end{aligned}$$

1.b. Calculate the mass flow rate of acetone m_r.
1.b.1. Calculate the volumetric flow rate of acetone, Q, in acfm.

$$\begin{aligned}Q &= (\text{mole fraction of acetone})(\text{volumetric flow rate of air})\\ &= (0.0015)(30{,}000)\\ &= 45\,\text{acfm}\end{aligned}$$

1.b.2. Calculate the mass flow rate, m_r, in lb/min.
From the ideal gas equation,

$$\rho = PM/R_u T$$
$$m_r = Q\rho = Q(PM/R_u T)$$

where
ρ = density
P = pressure
M = molecular weight
R_u = universal gas constant = 0.73 (atm-ft^3)/(lb-mole-°R)
T = absolute temperature
Q = volumetric flow rate (ft^3/h)
m_r = mass flow rate (lb/h)

$$\begin{aligned} m_r &= Q(PM/R_u T) \\ &= (45)(1)(58)/(0.73)(68 + 460) \\ &= 6.77\ \text{lb/min} \end{aligned}$$

1.c. Calculate the adsorption time in minutes for each adsorption bed (time for breakthrough).

Adsorption time = (carbon charge)(working capacity)/[0.5(mass flow rate, m_r)]

Note: two adsorbers are on stream with each cleaning one-half of the total gas stream.

$$\begin{aligned} \text{Adsorption time} &= \text{(carbon charge)(working capacity)}/(0.5m_r) \\ &= (15{,}000)(0.17)/(0.5)(6.77) \\ &= 753\ \text{min} \end{aligned}$$

or

$$= 12\ \text{h}\ 33\ \text{min}$$

1.d. Calculate the regeneration time in hours to complete the total cycle.
The regeneration time should be less than one-half of the time for breakthrough.

Regeneration time = 6 h 16.5 min

2. What is the amount of steam required?
2.a. Determine the steam requirement for carbon B in lb steam/lb solvent removed using the working capacity calculated above and Fig. 1.27.
This is commonly referred to as the steam-to-solvent ratio.
From Fig. 1.27,

Steam requirement = 2.4 lb steam/lb solvent removed

2.b. Calculate the lbs of steam required for each adsorber during regeneration.

$$\begin{aligned} \text{Mass of steam} &= \text{(steam requirement)(working capacity)(carbon charge)} \\ &= (2.4)(0.17)(15{,}000) \\ &= 6120\ \text{lb steam} \end{aligned}$$

EXAMPLE 7: rotogravure printing adsorber
A printing company must reduce the amount of toluene they emit from their printing operation. The company comes in with some preliminary information on installing a carbon adsorption system. You are given the following information (EPA-81/05, p. 18-5):

- air flow is 20,000 cfm
- they operate at 25% of LEL for toluene in the exit air
- LEL for toluene is 1.2%
- toluene MW is 92.1 lb/lb mole
- carbon density is 30 lb/ft^3
- working charge (or working capacity) is 30% of saturation capacity
- regeneration is just under 1 hour
- temperature is 77°F
- maximum velocity through adsorber is 100 fpm
- refer to Fig. 1.24.

Determine the minimum size of adsorber that you would approve:

1. Diameter of the adsorber
2. Square feet of carbon face area
3. Depth of bed

Solution:
(*Note*: all amounts are based on 1 hour because desorption takes only 1 hour)
1. Amount of carbon needed.
First, calculate amount of toluene emitted per hour.

$$(20{,}000\,\text{cfm})(25\%)(1.2\%) = 60\,\text{cfm of toluene}$$
$$(60\,\text{scf}) \times (\text{lb mole})/(359\,\text{scf}) \times (492^\circ\text{R}/537^\circ\text{R}) \times (92.1\,\text{lb/lb mole}) \times (60\,\text{min/h})$$
$$= 846\,\text{lb/h}$$

In order to compute saturation capacity from Fig. 1.24, we need the partial pressure of toluene.

$$p = (60\,\text{cfm}/20{,}000\,\text{cfm})(14.7\,\text{psia})$$
$$= 0.044\,\text{psia}$$
$$\text{saturation capacity} = 38\%\text{ from Fig 1.24}$$
$$\text{working capacity is 30\% of saturation}$$
$$\text{w.c.} = (38\%)(30\%)$$
$$= 11.4\%\text{ or 11.4 lb toluene per 100 lb carbon}$$

Amount of carbon needed for 1-hour cycle =

$$(846\,\text{lb toluene})/(\text{h}) \times (100\,\text{lb carbon}/11.4\,\text{lb toluene}) \times (1\,\text{h})$$
$$= 7421\,\text{lb carbon}$$

Amount of carbon = 7421 lb
2. Volume occupied by the carbon.

$$(7421\,\text{lb carbon}) \times (\text{ft}^3/30\,\text{lb}) = 247\,\text{ft}^3$$

3. Cross-sectional area of bed.

$$20{,}000\ (\text{scf/m}) \times (\text{m}/100\,\text{ft})$$
$$= 200\,\text{ft}^2$$

based on 100 fpm limiting
Note: this requirement could be met by a horizontal flow bed 10 ft wide and 20 ft long.
4. Depth of carbon bed.

$$D = (\text{volume of carbon})/(\text{cross-sectional area})$$
$$= 247\,\text{ft}^3/(200\,\text{ft}^2)$$
$$= 1.24\,\text{ft}$$
$$\text{Depth of carbon bed} = 1.24\,\text{ft}$$

Note: these calculations are based on minimum acceptable design conditions using only two beds. In practical situations, we may opt for three-bed systems, each handling only 10,000 cfm. This will decrease the size (cost) of each individual unit and allow for more carbon length to ensure against breakthrough.

4.3 Incineration for Gaseous Emission Control

The process of incineration is most often used to control the emissions of volatile organic compounds from process industries. At a sufficiently high temperature and adequate residence time, any organic vapor can be incinerated (oxidized) to produce energy and incineration gases which may include carbon dioxide and water. Solid waste can also be treated by the incineration process. Figure 1.28 shows the concept of incineration. Incinerators are capable of

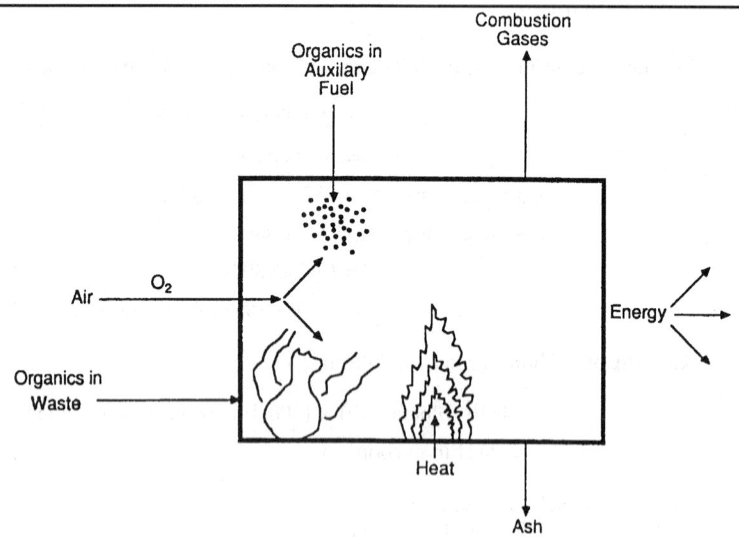

FIGURE 1.28 Waste (gas and solid) treatment by incineration (*EPA-89/03*).

achieving very high removal efficiencies. They consist of burners, which ignite the fuel and organic vapors and a chamber, which provides appropriate residence time for the oxidation process. Due to the high cost and decreasing supply of fuels, incineration systems are designed to include some type of heat recovery. If heat recovery can be used, incineration can be a very effective control technique. For example, pollutant emissions from paint bake ovens can be reduced by 99.9+% using incineration while heat recovered from the incinerator flue gases can be fed back to the oven. Incineration can also be used for serious emission problems that require high destruction efficiencies, such as odor problems or the emission of toxic gases (EPA-81/12, p. 3-1).

There are, however, some problems that may occur when using incineration to control gaseous pollutants. Incomplete incineration of many organic compounds results in the formation of aldehydes and organic acids which may create an additional pollution problem. Oxidizing organic compounds containing sulfur or halogens produce unwanted pollutants such as sulfur dioxide, hydrochloric acid, hydrofluoric acid or phosgene. If present, these pollutants would require a scrubber to remove them prior to release into the atmosphere.

Incineration is a chemical process resulting from the rapid combination of oxygen with various elements or chemical compounds, a process which releases heat. Incineration is also referred to as oxidation. Most fuels used for combustion are composed essentially of carbon and hydrogen, but can include other elements such as sulfur. Simplified reactions for the oxidation of carbon and hydrogen are given as (EPA-81/12, p. 3-1)

$$C + O_2 \rightarrow CO_2 + \text{energy} \tag{1.41}$$

$$2H_2 + O_2 \rightarrow 2H_2O + \text{energy} \tag{1.42}$$

Equations (1.41) and (1.42) show that the final major products of combustion from the burning of an organic fuel are carbon dioxide and water vapor. Although combustion seems to be a very simple process that is well understood, in reality it is not. The exact manner in which a fuel is oxidized does not occur exactly as given in the above equations, but rather in a series of complex, free-radical chain reactions. The precise set of reactions by which combustion occurs is termed the mechanism of combustion. By analyzing the mechanism of combustion, the rate at which the reaction proceeds and the variables affecting the rate can be predicted. For most combustion devices, the rate of reaction proceeds extremely fast. Maintaining efficient and complete combustion is somewhat of an art rather than a science, as anyone who has built a campfire can attest to. Therefore, this chapter focuses on the factors which influence the completeness of combustion, rather than analyzing the mechanisms involved.

To achieve complete combustion once the air (oxygen) and fuel have been brought into contact, the following conditions must be provided: a temperature high enough to ignite the air and fuel mixture; turbulent mixing of the air and fuel; and sufficient residence time for the reaction to occur. These three conditions are referred to as the three T's of combustion: time, temperature, and turbulence. They govern the speed and completeness of reaction. They are not independent variables since changing one affects the other two.

EXAMPLE: combustion air requirements
You are requested to calculate the amount of air required to completely combust a given volumetric flow rate of a particular type of natural gas and also determine the volumetric flow rate of combustion products (EPA-84/09, p. 90).

Given conditions

The composition of this particular "natural gas" is shown in Table 1.9

Volumetric flow rate of natural gas = 1.0 scfm

Solution:

1. Write the balanced chemical equations for complete combustion.

$$CH_4 + 2O_2 \rightarrow CO_2 + 2H_2O$$

$$C_2H_6 + (7/2)O_2 \rightarrow 2CO_2 + 3H_2O$$

$$C_3H_8 + 5O_2 \rightarrow 3CO_2 + 4H_2O$$

Combustion with 100% of theoretical (stoichiometric) air insures that no oxygen will remain in the flue gas with complete combustion. Remember that the combustion products are CO_2, H_2O, and inert N_2 (from both the air and this particular natural gas).
2. Determine the scfm of O_2 required from the above balanced chemical equations.
A chemically balanced equation tells us, in terms of moles (or volume), the ratios among reactants and products. From the mole fraction information given in Table 1.9,

$$0.81CH_4 + 1.62O_2 \rightarrow 0.81CO_2 + 1.62H_2O$$

$$0.1C_2H_6 + 0.35O_2 \rightarrow 0.2CO_2 + 0.3H_2O$$

$$0.04C_3H_8 + 0.2O_2 \rightarrow 0.12CO_2 + 0.16H_2O$$

$$\text{scfm of } O_2 = 1.62 + 0.35 + 0.2 = 2.17 \text{ scfm}$$

3. Calculate scfm of air required.

$$\text{Air components: } 0.21O_2 + 0.79N_2 = 1 \text{ air (by volume or by moles)}$$

$$1 \text{ mole } O_2 + 3.76 \text{ moles } N_2 = 4.76 \text{ moles of air}$$

$$1 \text{ lb } O_2 + 3.31 \text{ lb } N_2 = 4.31 \text{ lb of air}$$

$$\begin{aligned}\text{scfm of air} &= (\text{scfm of } O_2)(1.0 \text{ scfm of air}/0.21 \text{ scfm of } O_2)\\ &= 2.17(1/0.21)\\ &= 10.33 \text{ scfm}\end{aligned}$$

4. Calculate total scfm of N_2 in the flue gas.
Remember that the flue gas nitrogen is the sum of the N_2 in the natural gas and the N_2 in air accompanying the required O_2 for combustion.

$$\begin{aligned}\text{scfm of } N_2 &= 0.05 + (\text{scfm of air})(0.79 \text{ scfm of } N_2/1.0 \text{ scfm of air})\\ &= 0.05 + (10.33)(0.79)\\ &= 8.21 \text{ scfm}\end{aligned}$$

5. Calculate scfm of CO_2 and H_2O in the flue gas.

TABLE 1.9 Composition of Natural Gas

Component		Mole fraction
Nitrogen	N_2	0.05
Methane	CH_4	0.81
Ethane	C_2H_6	0.1
Propane	C_3H_8	0.04

$$CO_2 = 0.81 + 0.2 + 0.12 = 1.13 \text{ scfm}$$

$$H_2O = 1.62 + 0.3 + 0.16 = 2.08 \text{ scfm}$$

6. Calculate the total scfm of flue gas.
The total products of combustion (flue gas) include CO_2, H_2O, and N_2.

$$\text{Total scfm of combustion products} = 1.13 + 2.08 + 8.21$$
$$= 11.42 \text{ scfm}$$

Factors affecting incineration for emission control. Factors affecting combustion for emission control include:

- Temperature
- Time
- Turbulence
- Oxygen requirement
- Combustion limit
- Flame combustion
- Heat

Temperature: The rate at which a combustible compound is oxidized is greatly affected by temperature. The higher the temperature, the faster the oxidation reaction will proceed. The chemical reactions involved in the combination of a fuel and oxygen can occur even at room temperature, but very slowly. For this reason, a pile of oily rags can be a fire hazard. Small amounts of heat are liberated by the slow oxidation of the oils. This in turn raises the temperature of the rags and increases the oxidation rate, liberating more heat. Eventually a fully engulfed fire can break out (EPA-81/12, p. 3-2).

For combustion processes, ignition is accomplished by adding heat to speed up the oxidation process. Heat is needed to combust any mixture of air and fuel until the ignition temperature mixture is reached. By gradually heating a mixture of fuel and air, the rate of reaction and energy released will gradually increase until the reaction no longer depends on the outside heat source. More heat is being generated than is lost to the surroundings. The ignition temperature must be reached or exceeded to ensure combustion. Ignition depends on:

1. Concentration of combustibles in the waste stream
2. Inlet temperature of the waste stream
3. Rate of heat lost from combustion chamber
4. Residence time and flow pattern of the waste stream
5. Combustion chamber geometry and materials of construction

Most incinerators operate at higher temperature than the ignition temperature which is a minimum temperature. Thermal destruction of most organic compounds occurs between 590 and 650°C (1100 and 1200°F). However, most incinerators are operated at 700 to 820°C (1300 to 1500°F) to convert CO to CO_2, which occurs only at these higher temperatures.

Time: Time and temperature affect combustion in much the same manner as temperature and pressure affect the volume of a gas. When one variable is increased, the other may be decreased with the same end result. With a higher temperature, a shorter residence time can achieve the same degree of oxidation. The reverse is also true—a higher residence time allows the use of a lower temperature. In describing incinerator operation, these two terms are always mentioned together. One has little meaning without specifying the other.

The choice between higher temperature or longer residence time is based on economic considerations. Increasing residence time involves using a larger combustion chamber resulting in a higher capital cost. Raising the operating temperature increases fuel usage which also adds to the operating costs. Fuel costs are the major operating expense for most incinerators. Within certain limits, lowering the temperature and adding volume to increase residence time can be a cost-effective alternative method of operation.

The residence time of gases in the combustion chamber may be calculated from

$$t = V/Q \tag{1.43}$$

where: t = residence time, seconds
V = chamber volume, m^3
Q = gas volumetric flow rate at combustion conditions, m^3/s.

Q is the total flow of hot gases in the combustion chamber. Adjustments to the flow rate must include any outside air added for combustion. Example 1 below shows the determination of residence time from the volumetric flow rate of gases.

Turbulence: Proper mixing is important in combustion processes for two reasons. First, for complete combustion to occur, every particle of fuel must come into contact with air (oxygen). If not, unreacted fuel will be exhausted from the stack. Second, not all of the fuel or waste gas stream is able to be in direct contact with the burner flame. In most incinerators a portion of the waste stream bypasses the flame and is mixed at some point downstream of the burner with the hot products of combustion. If the two streams are not completely mixed, a portion of the waste stream will not react at the required temperature and incomplete combustion will occur.

A number of methods are available to improve mixing the air and combustion streams. Some of these include the use of refractory baffles, swirl-fired burners, and baffle plates. These devices are discussed in more detail in the equipment section of this book. The problem of obtaining complete mixing is not easily solved. Unless properly designed, many of these mixing devices may create dead spots and reduce operating temperatures. Merely inserting obstructions to increase turbulence is not always the answer. According to one study of afterburner systems, the process of mixing the flame and the fume streams to obtain a uniform temperature for decomposition of pollutants is the most difficult part in the design of the afterburner (EPA-73).

Oxygen requirement: Oxygen is necessary for combustion to occur. To achieve complete combustion of a compound, a sufficient supply of oxygen must be present to convert all of the carbon to CO_2. This quantity of oxygen is referred to as the stoichiometric or theoretical amount. The stoichiometric amount of oxygen is determined from a balanced chemical equation summarizing the oxidation reactions. For example, from Eq. (1.44), 1 mole of methane requires 2 moles of oxygen for complete combustion (EPA-81/12, p. 3-4):

$$CH_4 + 2O_2 \rightarrow CO_2 + 2H_2O \tag{1.44}$$

If an insufficient amount of oxygen is supplied, the mixture is referred to as rich. There is not enough oxygen to combine with all the fuel so that incomplete combustion occurs. This condition results in black smoke being exhausted. If more than the stoichiometric amount of oxygen is supplied, the mixture is referred to as lean. The added oxygen (i.e., the amount over stoichiometric) plays no part in the oxidation reaction and passes through the incinerator.

Oxygen for combustion processes is supplied by using air. Since air is essentially 79% nitrogen and 21% oxygen, a larger volume of air is required than if pure oxygen were used. To balance Eq. (1.44) 9.53 moles of air would be required to completely combust the 1 mole of methane.

In industrial applications, more than the stoichiometric amount of air is used to ensure complete combustion. This extra volume is referred to as excess air. If ideal mixing were achievable, no excess air would be necessary. However, most combustion devices are not capable of achieving ideal mixing of the fuel and air streams. The amount of excess air is held to a minimum in order to reduce heat losses. Excess air takes no part in the reaction but does absorb some of the heat produced. To raise the excess air to the combustion temperature, additional fuel must be used to make up for this loss of heat.

Combustion limit: Not all mixtures of fuel and air are able to support combustion. The flammable or explosive limits for a mixture are the maximum and minimum concentrations of fuel in air that will support combustion. The *upper explosive limit (UEL)* is defined as the concentration of fuel which produces a non-burning mixture due to a lack of oxygen. The *lower explosive limit (LEL)* is defined as the concentration of fuel below which combustion will not be self-sustaining.

Flame combustion: When mixing fuel and air, two different mechanisms of combustion can occur. A *luminous (yellow) flame* results when air and fuel flowing through separate ports are ignited at the burner nozzle. The yellow flame results from thermal cracking of the fuel. Cracking occurs when hydrocarbons are intensely heated before they have a chance to combine with oxygen. The cracking releases both hydrogen and carbon, which diffuse into the flame to form CO_2 and H_2O. The carbon particles give the flame the yellow appearance. If incomplete combustion occurs from flame temperature cooling or if there is insufficient oxygen, soot and black smoke will form (EPA-81/12, p. 3-7).

Blue flame combustion occurs when the fuel and air are premixed in front of the burner nozzle. This produces a short, intense, blue flame. The reason for the different flame is that the fuel–air mixture is gradually heated. The hydrocarbon molecules are slowly oxidized, going from aldehydes and ketones to CO_2 and H_2O. No cracking occurs and no carbon particles are formed. Incomplete combustion results in the release of the intermediate, partially oxidized compounds. Blue haze and odors are emitted from the stack.

Heat: A main area of concern in vapor incineration deals with the amount of fuel required to raise the temperature of the waste stream to the temperature required for complete oxidation. The first step in computing the heat required is to perform a heat balance around the oxidation system.

From the law of conservation of energy:

$$\text{Heat in} = \text{heat out} + \text{heat lost} \tag{1.45}$$

Heat is a relative term which is compared at a reference temperature. The heat content of a substance is arbitrarily taken as zero at a specified reference temperature. In the gas industry (natural gas), the reference temperature is normally 16°C (60°F). Thus, the heat content can be computed from Eq. (1.45):

$$H = C_p(T - T_0) \tag{1.46}$$

where H = enthalpy, J/kg or Btu/lb
C_p = specific heat at temperature T, J/kg-°C or Btu/lb-°F
T = temperature of the substance, °C or °F
T_0 = reference temperature, °C or °F

Subtracting the heat going in from the heat leaving the system gives the heat which must be supplied by the fuel. This is referred to as a *change in enthalpy or heat content*. Using Eq. (1.46), the enthalpy going in (T_1) is subtracted from the enthalpy leaving (T_2), giving

$$q = m\Delta H = mC_p(T_2 - T_1) \tag{1.47}$$

where q = heat rate, Btu/h
m = mass flow rate, lb/h
ΔH = enthalpy change
C_p = specific heat, J/kg-°C or Btu/lb-°F
T_2 = exit temperature of the substance, °C or °F
T_1 = initial temperature, °C or °F

EXAMPLE 1: oxidizer (solution in the metric system)
Emissions from a paint baking oven are controlled by a thermal oxidizer. The unit has a diameter of 1.5 m (5 ft) and is 3.5 m (11.5 ft) long. The exhaust from the oven is 3.8 m^3/s (8050 scfm). The oxidizer uses 0.14 m^3/s (300 scfm) of natural gas and operates at a temperature of 760°C (1400°F). If all the oxygen necessary for combustion is supplied from the process stream (no outside air added), what is the residence time of the gases in the chamber (EPA-81/12, p. 3-7)?

Solution:
To solve the problem one can use the approximation that 11.5 m^3 of combustion products are formed for every 1.0 m^3 of natural gas burned at standard conditions (16°C and 101.3 kPa). Also, 10.33 m^3 of theoretical air is required to combust 1 m^3 of natural gas at standard conditions.

1. Determine the volume of combustion products from burning the natural gas.

$$(0.14\ \text{m}^3/\text{s}) \times (11.5\ \text{m}^3\ \text{of product})/(1.0\ \text{m}^3\ \text{of gas}) = 1.61\ \text{m}^3/\text{s}$$

2. Determine the air required for combustion.

$$(0.14\ \text{m}^3/\text{s}) \times (10.33\ \text{m}^3\ \text{of air})/(1\ \text{m}^3\ \text{of gas}) = 1.45\ \text{m}^3/\text{s}$$

3. Sum up the volumes.

flow from paint bake oven = 3.8

products from combustion = 1.61

minus the air from exhaust used on combustion = −1.45

total volume = 3.8 + 1.61 − 1.45 = 3.96 m^3/s

4. Convert the m^3/s calculated under the standard conditions to actual conditions.

$$(3.96\ \text{m}^3/\text{s}) \times (273^\circ\text{C} + 760^\circ\text{C})/(273^\circ\text{C}) = 14.98\ \text{m}^3/\text{s}$$

5. Determine the chamber volume V.

$$\begin{aligned} V &= \pi r^2 L \\ &= 3.14(0.75\ \text{m})^2(3.5\ \text{m}) \\ &= 6.18\ \text{m}^3 \end{aligned}$$

6. The residence time from Eq. 1.43 is

$$t = V/Q$$

where t = residence time
V = chamber volume
Q = gas volumetric flow rate at combustion conditions, m^3/s.

$$\begin{aligned} t &= V/Q \\ &= (6.18\ \text{m}^3)/(14.98\ \text{m}^3/\text{s}) \\ &= 0.41\ \text{s} \end{aligned}$$

EXAMPLE 2: oxidizer (solution in the British system)
Emissions from a paint baking oven are controlled by a thermal oxidizer. The unit has a diameter of 5 ft and is 11.5 ft long. The exhaust from the oven is 134 ft^3/s. The oxidizer uses

4.94 ft^3/s of natural gas and operates at a temperature of 1400°F. If all the oxygen necessary for combustion is supplied from the process stream (no outside air added), what is the residence time of the gases in the chamber?

Given conditions

reactor diameter $D = 5\,\text{ft}$

reactor length $L = 11.5\,\text{ft}$

initial flue gas flow rate $= 134\,\text{ft}^3/\text{s}$

natural gas used $= 4.94\,\text{ft}^3/\text{s}$

combustion temperature $T = 1400°\text{F}$

Solution:

General assumptions (or rules of thumb):

- 1 m^3 of natural gas burned produces 11.5 m^3 of combustion products under standard conditions, $T = 16°\text{C}$ and $P = 1\,\text{atm}$.
- 1 m^3 of natural gas requires 10.33 m^3 of theoretical air for complete combustion.

Because $1\,\text{m}^3 = 35.31\,\text{ft}^3$, these two general assumptions become:

- 35.31 ft^3 of natural gas burned produces 406.07 ft^3 of combustion products.
- 35.31 ft^3 of natural gas requires 364.75 ft^3 of theoretical air for complete combustion.

1. Determine the volume Q of combustion products.

Because 35.31 ft^3 of natural gas burned produces 406.07 ft^3 of combustion products, 4.94 ft^3/s of natural gas will produce the combustion products Q as follows:

$$Q = 4.94(406.07)/35.31$$
$$= 56.81\ \text{ft}^3/\text{s}$$

2. Determine the air required for combustion.

$$\text{Air} = 4.94(364.75)/35.31$$
$$= 51.03\ \text{ft}^3/\text{s}$$

3. Sum up the volume Q_1 at the standard condition at $T_1 = 60°\text{F}$.

$$Q_1 = \text{flow from paint bake oven } (134\ \text{ft}^3/\text{s}) + \text{products of combustion } (56.81\ \text{ft}^3/\text{s})$$
$$- \text{the air from exhaust used in combustion } (51.03\ \text{ft}^3/\text{s})^{**}$$
$$= 139.78\ \text{ft}^3/\text{s}$$

**All the oxygen necessary for combustion products is supplied from the process stream (no outside air added).

4. Calculate actual flue gas flow rate Q_2 at $T_2 = 1400°\text{F}$.

$$Q_2 = Q_1(T_2/T_1)(P_1/P_2)$$
$$= 139.78(460 + 1400)/(460 + 60)$$
$$= 500\ \text{acfs}$$

5. Calculate the chamber volume.

$$\begin{aligned} V &= (\pi)(\text{diameter}/2)^2(\text{length}) \\ &= \pi(2.5)^2(11.5) \\ &= 226\,\text{ft}^3 \end{aligned}$$

6. Calculate residence time t.

residence time = (chamber volume)/(gas volumetric flow rate at combustion condition)

$$\begin{aligned} t &= V/Q_2 \\ &= 226/500 \\ &= 0.45\ \text{s} \end{aligned}$$

EXAMPLE 3: heat load
Assume that the exhaust from a meat smokehouse contains obnoxious odors and fumes. The company plans to incinerate the 5000 acfm exhaust stream at the initial temperature of 90°F. The gross heating value of natural gas is 1059 Btu/scf and assume no heat losses. Determine what quantity of natural gas is required to raise the waste gas stream from a temperature of 90°F to the required temperature of 1200°F (EPA-81/12, p. 3-14).

Given conditions

standard condition (state 1), $T_1 = 60°\text{F}$
exhaust gas flow rate $V_2 = 5000$ acfm at temperature $T = 90°\text{F}$
exhaust gas initial temperature $T_2 = 90°\text{F}$
natural gas gross heating value = 1059 Btu/scf (standard cubic foot)
combustion temperature $T_3 = 1200°\text{F}$

Solution:
1. Correct the actual waste gas volume (V_a) to standard condition volume.
The correction equation is

$$\begin{aligned} V_1/T_1 &= V_2/T_2 \\ V_1 &= V_2(T_1/T_2) \end{aligned}$$

$$\begin{aligned} V_1 &= (5000)(460+60)/(460+90) \\ &= 4727\ \text{scfm (standard cubic feet per minute)} \\ &= 283{,}620\ \text{scfh (standard cubic feet per hour)} \end{aligned}$$

2. Convert the volumetric flow rate to a mass flow rate by multiplying by the density.

Mass flow rate = (volume rate) × (density)

Standard volume of an ideal gas at 60°F = 379.64 ft^3/(lb-mole)

Assume that the waste gas molecular weight is the same as air: 29 lb/(lb-mole)

$$\begin{aligned} \text{Density} &= (\text{molecular weight})/\text{volume} \\ &= 29/379.64 \\ &= 0.076388\ \text{lb/ft}^3 \end{aligned}$$

$$\text{Mass flow rate} = 4727(0.076388)$$
$$= 361 \text{ lb/min}$$

3. Calculate the heat rate by using the ideal gas equation.

$$Q = mC_p(T_3 - T_2)$$
$$= 361(0.26)(1200 - 90)$$
$$= 104{,}185 \text{ Btu/min}$$

4. Determine the heating value of natural gas (for the present, ignoring the distinction between "gross" and "available" heating value which will be treated in the next Example).

For natural gas, 1 scf contains 1059 Btu (given)

5. Determine the natural gas quantity, W.

$$W = 104{,}185/1059$$
$$= 98 \text{ scfm}$$

Incineration equipment. Equipment used to control waste gases by incineration can be divided into four categories:

- Direct combustor or flare
- Thermal oxidizer or incinerator or afterburner
- Catalytic oxidizer
- Process boiler

Although these devices are physically similar, the parameters under which they operate are markedly different. Choosing the proper device depends on many factors, including concentration of combustibles in the gas stream, process flow rate, control requirements, presence of contaminants in the waste stream, and an economic evaluation.

A direct combustor or flare is a device in which air and all the combustible waste gases react at the burner. Complete combustion must occur instantaneously because there is no holdup chamber. Therefore, the flame temperature is the most important variable in flaring waste gases. In contrast, in thermal oxidation or incineration, the combustible waste gases pass over or around or through a burner flame into a holdup chamber where oxidation of the waste gases is completed. Catalytic oxidation is very similar to thermal oxidation. The main difference is that after passing through the flame area, the gases pass over a catalyst bed which promotes oxidation at a lower temperature than does thermal oxidation (EPA-81/12, p. 3-15).

Direct combustor or flare: Direct combustors or flares are used for the disposal of intermittent or emergency emissions of combustible gases from industrial sources. Safety and health hazards at or near the plant can be eliminated by using flares to prevent the direct venting of these emissions. Flares have been used mainly at oil refineries and chemical plants which handle large volumes of combustible gases.

Flares are simply burners that have been designed to handle varying rates of fuel while burning smokelessly. In general, flares can be classified as elevated flares and ground-level flares.

The reason for elevating a flare (Fig. 1.29) is to eliminate any potential fire hazard at ground level. Ground-level flares must be completely enclosed to conceal the flame. Either type of flare must be capable of operating over a wide range of flow rates in order to handle all plant emergencies. The range of waste gas flows within which a flare can operate and still burn efficiently is referred to as the *turndown ratio*. Flares are expected to handle turndown ratios of 1000 : 1, while most industrial boilers seldom handle more than a 10 : 1 turndown ratio. For

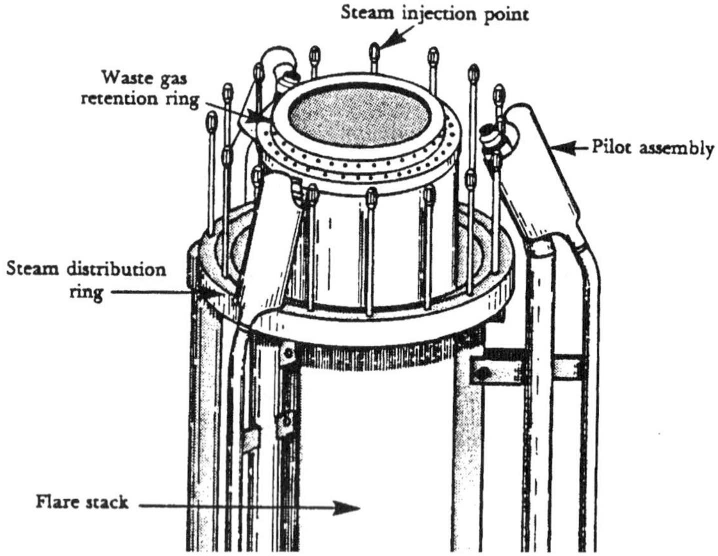

FIGURE 1.29 Flare tip (elevated flare) (*EPA-81/12, p. 3-17*).

example, a flare should be capable of maintaining complete combustion for waste gas flow rates ranging from 20,000 m^3/h to 20 m^3/h.

Incinerator (afterburner): Thermal oxidizers or organic vapor incinerators refer to any device that uses a flame (temperature) combined with a chamber (time and turbulence) to convert combustible materials to carbon dioxide and water. An incinerator usually consists of a refractory-lined chamber that is equipped with one or more sets of burners. A typical incinerator is depicted in Fig. 1.30. The contaminant-laden stream is passed through the burners, where it is heated above its ignition temperature. The hot gases then pass through one or more combustion chambers, where they are held for a certain length of time to ensure complete combustion. Depending on the particular needs of the system, additional fuel and/or excess air can be added through the burners. Also, since the flue gases are discharged at elevated temperatures, a system to recover the heat may be included.

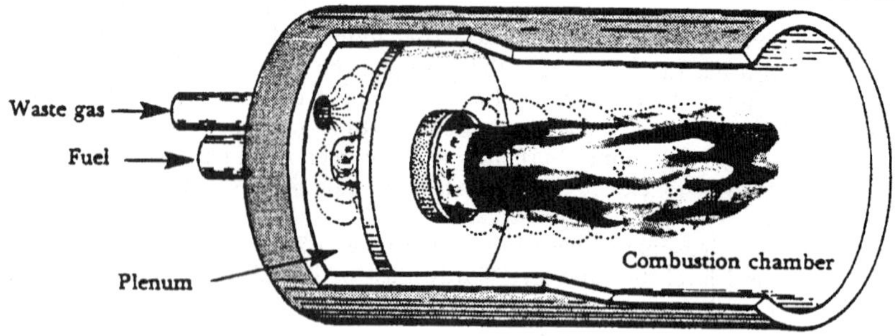

FIGURE 1.30 Thermal incinerator (*EPA-81/12, p. 3-19*).

With the rising costs of fuel and solvents, heat-recovery devices are becoming an integral part of many incineration systems (EPA-81/12, p. 3-19).

EXAMPLE 1: combustion: thermal afterburner design
As an air pollution control engineer, you have been requested to evaluate the gross heating value of a natural gas of a given composition. You are also to determine the available heat of the natural gas at a given temperature, the rate of auxiliary fuel (natural gas) required to heat a known amount of contaminated air to a given temperature, the dimensions of an afterburner treating the contaminated air stream, and the residence time (EPA-84/09, p. 92).
Given conditions

natural gas composition (mole or volume fraction) as shown in Table 1.10

linear gas velocity = 20 ft/s

length-to-diameter ratio of the afterburner = 2.0

temperature of dry natural gas = 60°F

volumetric flow rate of contaminated air $Q = 5000$ scfm (60°F, 1 atm)

it is required to heat the contaminated air from 200°F to 1200°F

TABLE 1.10 Composition of Natural Gas

Component		Mole fraction
Nitrogen	N_2	0.0515
Methane	CH_4	0.8111
Ethane	C_2H_6	0.0967
Propane	C_3H_8	0.0351
n-Butane	C_4H_{10}	0.0056
Total		1

Solution:
1. Determine the gross heating value of the natural gas.
From Table 1.11, the gross heating value of the natural gas is

$$\begin{aligned}\mathrm{HV_G} &= \Sigma n_i \mathrm{HV}_{Gi} \\ &= (0.0515)(0) + (0.8111)(1013) + (0.0967)(1792) + (0.0351)(2590) + (0.0056)(3370) \\ &= 1105 \text{ Btu/scf of natural gas}\end{aligned}$$

2. What is the available heat of this natural gas at 1200°F?
The available heat ($\mathrm{HA_t}$) at any temperature is the gross heating value ($\mathrm{HV_G}$) minus the amount of heat required to take products of combustion to that temperature ($\Sigma\Delta H$).
2.a. Write the balanced chemical combustion equations for each of the four components of this natural gas. Use 1 scf of natural gas as a basis.
For 1 mole of each species:

$$\begin{aligned} CH_4 + 2O_2 &\rightarrow CO_2 + 2H_2O \\ C_2H_6 + (7/2)O_2 &\rightarrow 2CO_2 + 3H_2O \\ C_3H_9 + 5O_2 &\rightarrow 3CO_2 + 4H_2O \\ C_4H_{10} + (13/2)O_2 &\rightarrow 4CO_2 + 5H_2O \end{aligned}$$

TABLE 1.11 Heat of Combustion at 25°C (77°F)

Substance		lb/ft^3	ft^3/lb	Btu/ft^3 Gross (high)	Btu/ft^3 Net (low)	Btu/lb Gross (high)	Btu/lb Net (low)
Nitrogen	N_2	0.0744	13.443				
Methane	CH_4	0.0424	23.565	1013	913	23,879	21,520
Ethane	C_2H_6	0.0803	12.455	1792	1641	22,320	20,432
Propane	C_3H_8	0.1196	8.365	2590	2385	21,661	19,944
n-Butane	C_4H_{10}	0.1582	6.321	3370	3113	21,308	19,680

Source: EPA-84/09, p. 37.

For one mole of the natural gas in question:

$$0.8111CH_4 + 1.622O_2 \rightarrow 0.8111CO_2 + 1.622H_2O$$
$$0.0967C_2H_6 + 0.3385O_2 \rightarrow 0.1934CO_2 + 0.2901H_2O$$
$$0.0351C_3H_9 + 0.1755O_2 \rightarrow 0.1053CO_2 + 0.1404H_2O$$
$$0.0056C_4H_{10} + 0.0364O_2 \rightarrow 0.0224CO_2 + 0.028H_2O$$

2.b. Determine the number of standard cubic feet for each of the following components of combustion.

For O_2: $1.622 + 0.3385 + 0.1755 + 0.0364 = 2.172$ scf/scf of natural gas

For CO_2: $0.8111 + 0.1934 + 0.1053 + 0.0224 = 1.132$ scf/scf of natural gas

For H_2O: $1.6222 + 0.2901 + 0.1404 + 0.028 = 2.081$ scf/scf of natural gas

For N_2: $0.0515 + (79/21)(2.172) = 8.222$ scf/scf of natural gas

The calculation for N_2 assumes that the natural gas is burned with air (21% O_2 and 79% N_2), the total amount of nitrogen in the products of combustion is the sum of N_2 in the natural gas (0.0515 scf) and that amount accompanied by the required O_2 (79/21 × scf of O_2).
2.c. How many cubic feet of products are there per scf of natural gas burned, assuming that a stoichiometric (theoretical) quantity of air (21% O_2 + 79% N_2) is used?

Total cubic feeet of products per scf of natural gas are the sum of the scfs of the product species.

Total cubic feet of combustion products $= 1.132 + 2.081 + 8.222 = 11.435$ scf of products/scf of natural gas

2.d. Determine the amount of heat required to take the products of combustion from 60°F to 1200°F.
The enthalpies of combustion gases are given in Table 1.12 and the latent heat of vaporization of water (ΔH) at 60°F is 1060 Btu/lb. The changes in enthalpy in Btu/scf of natural gas can be calculated by multiplying the change in enthalpy in Btu/lb-mole by scf of gas/scf of natural gas and divided by 379 scf/lb-mole.
2.d.1. From Table 1.12, the following values of enthalpies at 60°F and 1200°F are obtained.

TABLE 1.12 Enthalpies of Combustion Gases (Btu/lb-mole)

°F	N_2	Air MW 28.97	CO_2	H_2O
32	0	0	0	0
60	194.9	194.6	243.1	224.2
77	312.2	312.7	392.2	360.5
100	473.3	472.7	597.9	545.3
200	1170	1170	1527	1353
300	1868	1870	2509	2171
400	2570	2576	3537	3001
500	3277	3289	4607	3842
600	3991	4010	5714	4700
700	4713	4740	6855	5572
800	5443	5479	8026	6460
900	6182	6227	9224	7364
1000	6929	6984	10,447	8284
1200	8452	8524	12,960	10,176
1500	10,799	10,895	16,860	13,140
2000	14,840	14,970	23,630	18,380
2500	19,020	19,170	30,620	23,950
3000	23,280	23,460	37,750	29,780

Source: EPA-84/09, p. 93.

$$\text{For } CO_2\text{: } \Delta H = H \text{ at } 1200°\text{F} - H \text{ at } 60°\text{F} = 12{,}960 - 243.1 = 12{,}716.9 \text{ Btu/lb-mole}$$
$$= (12{,}716.9)(1.132)/(379)$$
$$= 38.0 \text{ Btu/scf of natural gas}$$

This calculation uses the fact that there are 379 scf per lb mole of any ideal gas, and there are 1.132 scf of CO_2/scf of natural gas. Similarly, for N_2:

$$\Delta H = (8452 - 194.9)(8.222)/(379) = 179.1 \text{ Btu/scf of natural gas}$$

for $H_2O(g)$:

$$\Delta H = (10{,}176 - 224.2)(2.081)/(379) = 54.6 \text{ Btu/scf of natural gas}$$

Since the water present in the combustion product is vapor, the latent heat of vaporization should be included in the heat balance calculation.

$$\Delta H \text{ for water latent heat} = (1060 \text{ Btu/lb})(18)\text{lb/lb-mole})$$
$$\times (2.081 \text{ scf of } H_2O\text{/scf of natural gas})/(379 \text{ scf/lb mole})$$
$$= 104.8 \text{ Btu/scf of natural gas}$$

2.d.2. Calculate the heat needed.

$$\Sigma\Delta H = \Delta H \text{ for } CO_2 + \Delta H \text{ for } N_2 + \Delta H \text{ for } H_2O + \Delta H \text{ for water latent heat}$$
$$= 38.0 + 179.1 + 54.6 + 104.8$$
$$= 376.5 \text{ Btu/scf of natural gas}$$

2.e. Calculate the available heat of natural gas at 1200°F in Btu/scf of natural gas.

$$\begin{aligned} \text{HA} &= \text{HV}_G - \Sigma\Delta H \\ &= 1105 - 376.5 \\ &= 728.5 \text{ Btu/scf of natural gas} \end{aligned}$$

3. Calculate the rate of auxiliary fuel required to heat 5000 scfm of contaminated air from 200°F to 1200°F.
To calculate the rate of natural gas required, you need both the enthalpy change of air and the available heat.
3.a. Determine the enthalpy change of air going from 200°F to 1200°F.
From Table 1.12,

$$\Delta H \text{ for air} = H \text{ at } 1200°\text{F} - H \text{ at } 200°\text{F} = 8524 - 1170 = 7354 \text{ Btu/lb-mole}$$

3.b. Calculate the heat rate q required to heat 5000 scfm of the contaminated air from 200°F to 1200°F in Btu/min.

$$\begin{aligned} q &= (5000 \text{ scfm})(7354 \text{ Btu/lb-mole})/(379 \text{ scf/lb-mole}) \\ &= 97{,}018 \text{ Btu/min} \end{aligned}$$

3.c. Calculate the rate of natural gas required to heat 5000 scfm of air from 200°F to 1200°F in scfm.

$$\begin{aligned} \text{Rate of natural gas required} &= q/\text{HA} \\ &= 97{,}018/728.5 \\ &= 133.2 \text{ scfm} \end{aligned}$$

4. What are the dimensions of an afterburner treating 5000 scfm of contaminated air stream?
To determine the dimensions of the afterburner, you need the total flue gas flow rate at 1200°F. The total flue gas flow rate is the sum of the contaminated gas stream and the flow rate of the combustion products. The volumetric flow rate of the products of combustion is obtained by multiplying the total cubic feet of products per scf of natural gas (previously determined) by the natural gas rate (also previously determined).
4.a. What is the total flue gas flow rate?
4.a.1. Calculate the volumetric flow rate of products of combustion in scfm.

$$\begin{aligned} Q \text{ of combustion products} &= (11.435 \text{ scf of products/scf of natural gas}) \\ &\quad \times (133.2 \text{ scfm of natural gas}) \\ &= 1523 \text{ scfm} \end{aligned}$$

4.a.2. Calculate the total flue gas flow rate in scfm.
This includes both the products of combustion (flue) and the process gas stream.

$$\begin{aligned} \text{Total flue gas flow rate} &= 5000 + 1523 \\ &= 6523 \text{ scfm} \end{aligned}$$

4.a.3. Determine the total flue gas flow rate at 1200°F in acfm.

$$\begin{aligned} \text{Actual total flue gas flow rate} &= 6523\,(1200 + 460)/(60 + 460) \\ &= 20{,}823 \text{ acfm} \\ &= 347 \text{ acfs} \end{aligned}$$

4.b. Determine the diameter of the afterburner in ft.

$$S = \pi D^2/4 = Q/v$$

where S = cross-sectional area of the afterburner
D = diameter of the afterburner
Q = total flue gas flow rate
v = linear gas velocity

$$D = (4Q/v\pi)^{1/2}$$
$$= [(4)(347)/(20)(\pi)]^{1/2}$$
$$= 4.7 \text{ ft}$$

4.c. Determine the length of the afterburner in ft.
Since the length-to-diameter ratio is 2.0,

$$L = (2.0)(4.7)$$
$$= 9.4 \text{ ft}$$

4.d. Determine the residence time for the gases in the afterburner in seconds.

$$\text{Residence time} = \text{length}/v$$

Typical residence times in an afterburner are in the 0.1–0.5 second range.

$$t = L/v$$
$$= (9.4)/(20)$$
$$= 0.47 \text{ s}$$

EXAMPLE 2: combustion: plan review of a direct-flame afterburner
You must review plans for a permit to construct a direct flame afterburner serving a lithographer. Review is for the purpose of judging whether the proposed system, when operating as it is designed to operate, will meet emission standards. The permit application provides operating and design data. Agency experience has established design criteria which, if met in an operating system, typically ensure compliance with standards (EPA-84/09, p. 96).
Given conditions: operating and design data from permit application

application = lithography

effluent exhaust volumetric flow rate = 7000 scfm

exhaust temperature = 300°F

hydrocarbons in effluent air to afterburner (assume hydrocarbons to be toluene) = 30 lb/h

afterburner entry temperature of effluent from preheater = 738°F

afterburner exit temperature = 1400°F

afterburner heat loss = 10%

afterburner dimensions = 4.2 ft in diameter, 14 ft in length

Agency design criteria

afterburner temperature = 1300–1500°F

residence time = 0.3–0.5 s

afterburner velocity = 20–40 ft/s

Standard data

gross heating value of natural gas = 1059 Btu/scf of natural gas
combustion products per cubic ft of natural gas burned = 11.5 scf
available heat of natural gas at 1400°F = 600 Btu/scf of natural gas
molecular weight of toluene = 92
specific heat of effluent gases at 738°F (above 0°F) = 7.12 Btu/lb-mole-°F
specific heat of effluent gases at 1400°F (above 0°F) = 7.38 Btu/lb-mole-°F
volume of air required to combust natural gas = 10.33 scf air/scf natural gas

Solution:
1. Does this application meet any of the agency design criteria?
The afterburner exit temperature is 1400°F, which is already within agency criteria. Therefore, you need be concerned only with whether, under the conditions given, the residence time and the afterburner velocity will be within agency design criteria.
2. Determine the fuel requirement for the afterburner.
2.a. Calculate the total heat load (heating rate) required to raise 7000 scfm of the effluent stream from 738°F to 1400°F (in Btu/min).
First, calculate the molar flow rate of the effluent.

$$n = (7000\,\text{scfm})/(379\,\text{scf/lb-mole}) = 18.47\,\text{lb-mole/min}$$

The given conditions shows that the combustion gas at different temperature ranges has different specific heat values. The heat load can now be calculated:

$$\begin{aligned} q &= n[C_{p2}(T_2 - T_0) - C_{p1}(T_1 - T_0)] \\ &= 18.47[(7.38)(1400 - 0) - (7.12)(738 - 0)] \\ &= 93{,}772\,\text{Btu/min} \end{aligned}$$

2.b. Calculate the actual heat load required, accounting for a 10% heat loss, in Btu/min.

$$\begin{aligned} \text{Actual heat load} &= q + 10\%(q) \\ &= (1.1)(93{,}772) \\ &= 103{,}149\,\text{Btu/min} \end{aligned}$$

2.c. Calculate the rate of natural gas required to supply the actual heat required to heat 7000 scfm of effluent from 738°F to 1400°F in scfm.

$$\begin{aligned} \text{Rate of natural gas} &= (\text{actual heat})/(\text{available heat}) \\ &= (103{,}149)/(600) \\ &= 171.9\,\text{scfm} \end{aligned}$$

3. What is the total volumetric flow rate through the afterburner Q_t?
Once you determine the volumetric flow rate, you can calculate the afterburner velocity and residence time, since the dimensions of the afterburner are given. The total volumetric flow rate is the sum of the process effluent from the lithographer plus the combustion products of the natural gas.
3.a. Calculate the volumetric flow rate of the combustion products of the natural gas, Q_1, in scfm.
To calculate the volumetric flow rate of combustion products of the natural gas, you use the amount of natural gas just calculated:

$$\begin{aligned} Q_1 &= \text{(rate of natural gas)(11.5 scf of combustion products/scf of natural gas)} \\ &= (171.9)(11.5) \\ &= 1976 \text{ scfm} \end{aligned}$$

3.b. What is the volumetric flow rate of the effluent stream from the lithographer in scfm?

$$\text{Volumetric flow rate of the effluent} = 7000 \text{ scfm}$$

3.c. Calculate the volumetric flow rate of air required to combust the natural gas, %, in scfm.

$$\begin{aligned} Q_2 &= \text{(rate of natural gas)(10.33 scf of air/scf of natural gas)} \\ &= 1776 \text{ scfm} \end{aligned}$$

3.d. Calculate the total volumetric flow rate through the afterburner, Q_t, in scfm.

$$\begin{aligned} Q_t &= 7000 + Q_1 \\ &= 7000 + 1976 \\ &= 8976 \text{ scfm} \end{aligned}$$

Note: Combustion air for the natural gas is already in the "11.5" figure.
3.e. Calculate Q_t in acfm.

$$\begin{aligned} Q_t &= (8976)(1400 + 460)/(60 + 460) \\ &= 32{,}106 \text{ acfm} \end{aligned}$$

4. Does the afterburner velocity meet the criteria?
To calculate the afterburner velocity you need the cross-sectional area of the afterburner and actual volumetric flow rate through the afterburner.
4.a. Calculate the cross-sectional area of the afterburner S in ft^2.

$$\begin{aligned} S &= \pi D^2/4 \\ &= (\pi)(4.2)^2/4 \\ &= 13.85 \text{ ft}^2 \end{aligned}$$

4.b. Calculate the afterburner velocity v in ft/s.

$$\begin{aligned} v &= Q_t/S \\ &= (32{,}106)/(13.85) \\ &= 2317 \text{ ft/min} \\ &= 38.6 \text{ ft/s} \end{aligned}$$

4.c. Is the afterburner velocity within agency criteria?
Yes.
5. Does the residence time meet agency criteria?
5.a. Calculate the residence time t in s.

$$t = L/v$$

where L = afterburner length
v = velocity

$$t = L/v$$
$$= 14/38.6$$
$$= 0.363 \text{ s}$$

5.b. Is the residence time within agency criteria?
Yes.
Note: The above calculations have assumed combustion of the natural gas to take place with primary air; i.e. air supplied external to the process effluent. In addition, the energy released and volume of gases formed by the combustion of the pollutant (toluene) have been neglected. This is common practice in combustion calculations. If secondary air (from process effluent) is employed for combustion of the natural gas, a trial-and-error calculation is required for a rigorous solution to this problem.

Catalytic oxidizer: A catalyst is a substance which causes or speeds a chemical reaction without itself undergoing a change. In catalytic incineration, a waste gas is passed through a layer of catalyst known as the catalyst bed. The catalyst causes the oxidation reaction to proceed at a faster rate and lower temperature than is possible in thermal oxidation. Catalytic incinerators operating in a 370–480°C (700–900°F) range can achieve the same efficiency as a thermal incinerator operating between 700 and 820°C (1300 and 1500°F). This can result in 40–60% fuel savings.

Catalytic reactions can be classified as either *homogeneous* or *heterogeneous*. Homogeneous reactions occur throughout the bulk of the catalyst, while heterogeneous reactions occur only on the surface of the catalytic material. In air pollution control applications, all reactions are heterogeneous. The oxidation reaction of the organic vapors occurs only on the surface of the catalyst. It should be noted that catalytic reactions produce the same end products (CO_2 and H_2O) and liberate the same heat of combustion as does thermal incineration.

A heterogeneous catalytic reaction proceeds through a series of five basic steps:

1. Organic compounds in the waste gas must first diffuse from the bulk of the vapor to the surface of the catalyst.
2. Organic compounds then adsorb onto the surface of the catalyst.
3. Organic compounds then react (oxidize).
4. New compounds then desorb after reacting.
5. New compounds then diffuse and mix back into the bulk of the exhaust air stream.

The most effective and commonly used catalysts for oxidation reactions come from the noble metals group. Platinum either alone or in combination with other noble metals is by far the most commonly used. Desirable characteristics of platinum are that it gives a high oxidation activity at low temperatures, is stable at high temperatures, and is chemically inert. Palladium is another noble metal which exhibits these properties and is sometimes used in catalytic incinerators.

Since catalytic oxidation is a surface reaction, the noble metal is coated onto the surface of a cheaper support material. The support material can be made of a ceramic or a metal such as alumina, silica-alumina or nickel-chromium. The support material is arranged in a matrix shape to provide high geometric surface area, low pressure drop, uniform flow of the waste gas through the catalyst bed, and a structurally stable surface. Structures which provide these characteristics are pellets, a honeycomb matrix, or a mesh matrix.

A schematic of a catalytic incinerator is shown in Fig. 1.31. Catalytic incinerators consist of a preheat section (burner area), where part of the waste gas is raised to operating temperature. The burners are the same as those used for thermal incineration. The remaining portion of the waste gas is mixed with the hot products of combustion before passing over the catalyst bed. This ensures a homogenous waste gas and a uniform-temperature mixture as it passes over the bed. After passing over the bed, the hot flue gases may be sent to a heat recovery system.

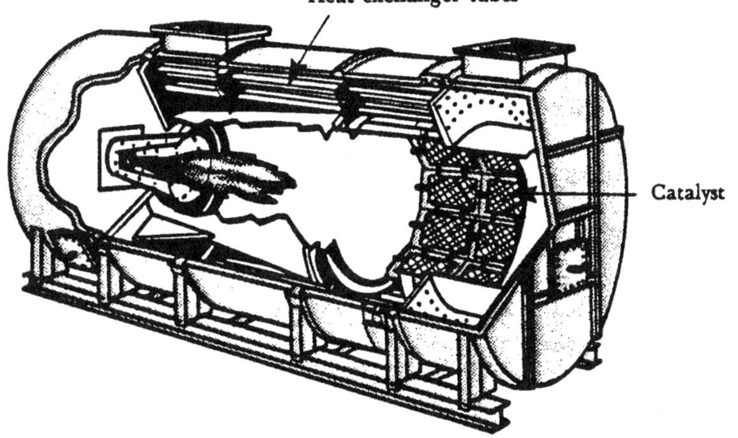

FIGURE 1.31 Catalytic incinerator (*EPA-81/12, p. 3-29*).

Process boiler: An alternative to installing a thermal or catalytic incinerator would be to combust the waste gases in an existing plant or process boiler. This would avoid the capital cost of new equipment and may help to reduce present fuel consumption. Process and plant boilers are normally designed to operate in excess of 980°C (1800°F) with a flue gas residence time of 0.5 to 3.0 seconds. These conditions exceed those recommended for thermal incinerators. However, a number of additional conditions must be satisfied before waste gases can be properly disposed of in this manner.

The following criteria must be considered before a process boiler can be used as an incinerator:

1. The waste gases must be almost completely combustible. If solid particles are present in the waste gases, or formed by incomplete combustion, they can foul heat exchanger surfaces in the boiler, thus reducing boiler efficiency. Particulate matter may also cause boiler emissions to exceed applicable emission regulations. The costs of increased maintenance of the boiler and/or control of particulate matter may well exceed the purchase price of an incinerator.
2. The waste gas should, preferably, constitute only a small fraction of the air requirements of the boiler. If the volume of the waste gas is large, special attention must be paid to the oxygen balance, mixing, and continuation of the air flow in the boiler when the process is shut down.
3. The oxygen concentration of the contaminated gas stream should be close to that of air to avoid incomplete combustion. Incomplete combustion produces tars that coat heat exchanger surfaces, reducing boiler efficiency.
4. The boiler must operate at all times when incineration is required.
5. The waste gas must be free of compounds, such as halogenated hydrocarbons, that accelerate corrosion of the boiler.
6. Baffling may be required in the combustion chamber to ensure adequate mixing and combustion of the waste gas.
7. If the boiler-firing rate varies greatly, it may be necessary to install a small auxiliary boiler that will operate under steady load conditions.

To date, not many industries have been successful in using plant or process boilers as incinerators. Petroleum refineries, which have numerous waste gas streams and process boilers, are one of the few industries which have been able to incinerate waste gases in process boilers.

4.4 Condensation for Gaseous Emission Control

Condensation is the process of reducing a gas or vapor to a liquid. Any gas can be reduced to a liquid by lowering its temperature and/or increasing its pressure. The most common approach is to reduce the temperature of the gas stream, since increasing the pressure of a gas is very costly (EPA-81/12, p. 6-1).

Condensers are simple, relatively inexpensive devices that normally use water or air to cool and condense a vapor stream. Since these devices are usually not capable of reaching low temperatures (below 80°F), high removal efficiencies of most gaseous pollutants are not obtained unless the vapors will condense at high temperatures. Condensers are typically used as pretreatment devices. They are used ahead of incinerators, absorbers, or adsorbers to reduce the total gas volume to be treated by more expensive control equipment. Used in this manner, they help reduce the overall cost of the control system.

Condensation basics. When a hot vapor stream contacts a cooler surface, heat is transferred from the hot gases to the cooler surface. As the temperature of the vapor stream is cooled, the average kinetic energy of the gas molecules is reduced. Also, the volume that these vapors occupy is reduced. Ultimately the gas molecules are slowed down and crowded together so closely that the attractive forces (van de Waals' forces) between the molecules cause them to condense to a liquid.

The two conditions which aid condensation are low temperatures, so that the kinetic energy of the gas molecules are low, and high pressures so that the molecules are brought close together.

The actual conditions at which a particular gas molecule will condense depends on its physical and chemical properties. Condensation occurs when the partial pressure of the pollutant in the gas stream equals its vapor pressure as a pure substance at operating conditions.

Condensation of a gas can occur in three ways:

1. At a given temperature, the system pressure is increased (compressing the gas volume) until the partial pressure of the gas equals its vapor pressure.
2. At a fixed pressure, the gas is cooled until the partial pressure equals its vapor pressure.
3. By using a combination of compression and cooling of the gas until its partial pressure equals its vapor pressure.

These processes are illustrated in Fig. 1.32, a typical vapor pressure diagram for a pure substance. In Fig. 1.32, point I is the initial temperature and pressure of a gas. The dotted lines indicate the paths a quantity of gas would follow to reach the vapor pressure curve. Points on the vapor pressure curve are also referred to as dew points. The *dew point* is defined as the condition at which gas is ready to condense into the first drop of liquid.

Also in Fig. 1.32, the critical point is plotted. Each substance has a critical temperature and critical pressure. The critical temperature is important in that it is the maximum temperature above which the gas will not condense, no matter how great a pressure is applied. The pressure required to liquefy a gas at its critical temperature is the critical pressure.

Once the gas conditions (temperature, pressure, and volume) equal those on the vapor pressure line, liquid begins to condense. From this point on, the gas–liquid mixture follows the vapor pressure line. If the mixture is cooled continuously, the partial pressure of the remaining gas will always equal the vapor pressure. This is important since even though the contaminated gas is being condensed, it still has a certain partial pressure indicating that uncontrolled

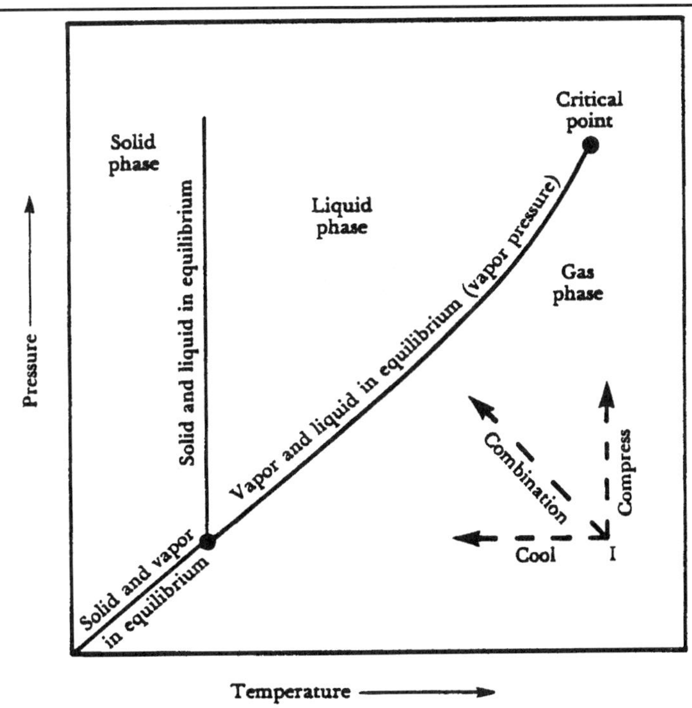

FIGURE 1.32 Typical vapor pressure diagram (*EPA-81/12, p. 6-2*).

vapors are being emitted from the condenser. For most practical applications, the vapor-liquid equilibrium restricts the use of condensers as primary air pollution control devices. Unless very low temperatures or high pressures are attained, condensers are generally not capable of reducing the pollutant concentration to within acceptable emission limits.

Practically, temperature is the only process variable which governs the effectiveness of a condenser. In industrial applications, increasing the system pressure is very costly and therefore rarely used for condensation. At the operating pressure of the system, the outlet temperature from the condenser determines the maximum removal efficiency. Therefore, condensers cannot be used in the same manner as other gaseous pollutant control devices. For example, condensers cannot be used in series, like adsorbers or absorbers, to further reduce outlet concentration unless the outlet temperature of the second condenser is lower than the previous one. Increasing gas residence time or decreasing flow rates in the condenser does not add to the theoretically achievable efficiency as these operations do in incinerators, adsorbers, and absorbers.

Condensation equipment—condenser. Condensers fall into two basic categories (EPA-81/12, p. 6-3):

- Contact condenser
- Surface condenser

Contact condenser: In a contact condenser (Fig. 1.33) the coolant and vapor stream are physically mixed. They leave the condenser as a single exhaust stream. Contact condensers

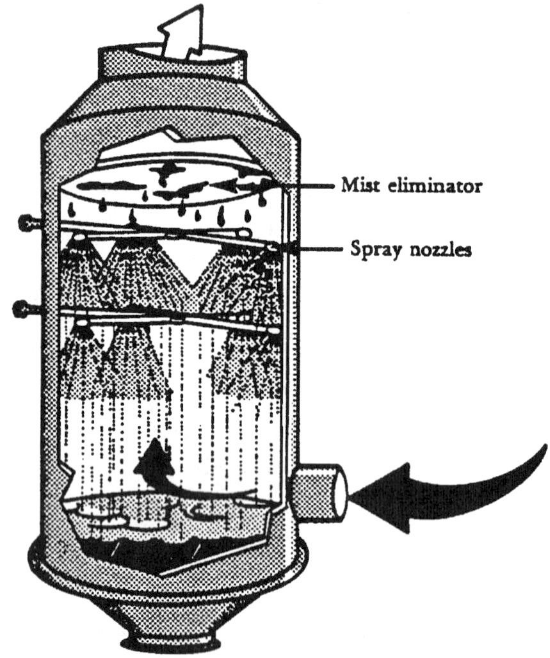

FIGURE 1.33 Direct contact condenser (*EPA-81/12, p. 6-4*).

are simple devices such as spray towers, steam or water jet ejectors, and barometric condensers. These devices bring the coolant, usually water, into direct contact with the vapors. The liquid stream leaving the condenser contains the coolant plus the condensed vapors. If the vapor is soluble in the coolant then absorption also occurs. Absorption increases the amount of contaminant that can be removed at the given conditions.

Condensers reduce the temperature (heat content) of a hot vapor stream by contacting it with a cooler liquid or air stream. Heat is being transferred from a hot to a colder fluid; thus, the process is termed a heat transfer or heat exchange operation. Condensers are designed using the same basic principles and empirical relationships used for heat exchangers.

Simplified heat balance calculations are used to estimate the important parameters. The following equations are not intended to be used as a design method, but only as rough estimates for evaluation purposes.

The first step in analyzing any heat transfer process is to set up a heat balance relationship. For a condensation system, the heat balance can be expressed as

Heat in = Heat out

(Heat required to reduce vapors to the dew point) + (heat required to condense vapors)
= (heat needed to be removed by the coolant)

This heat balance is written in equation form as

$$q = mC_p(T_{G1} - T_{\text{dew point}}) + mH_V = LC_p(T_{L2} - T_{L1}) \tag{1.48}$$

where q = heat transfer rate, Btu/h
m = mass flow rate of vapor, lb/h
C_p = average specific heat of a gas or liquid, Btu/lb-°F
T = temperature of the streams; G for gas and L for liquid coolant
H_v = heat of condensation or vaporization, Btu/lb
L = mass flow rate of liquid coolant, lb/h

In Eq. (1.48) the mass flow rate m and inlet temperature T_{G1} of the vapor stream are set by the process exhaust stream. The temperature of the coolant entering the condenser T_{L1} is also set. The average specific heats C_p of both streams, the heat of condensation H_v, and the dew point temperature can be obtained from chemistry handbooks. Therefore, only the amount of coolant L and its outlet temperature are left to be determined. If either one of these terms are set by process restrictions (i.e. only x pounds an hour of coolant are available or the outlet temperature of coolant must be below a set temperature), then the other term can be solved for directly.

Equation (1.48) is applicable for direct contact condensers and should be used only to obtain rough estimates. The equation has a number of limitations:

1. The specific heat C_p of a substance is dependent on temperature, and the temperature throughout the condenser is constantly changing.
2. The dew point of a substance is dependent on its concentration in the gas phase, and since the mass flow rate is constantly changing (vapors being condensed) the dew point temperature is constantly changing.
3. No provision is made for cooling the vapors below their dew point. An additional term would have to be added to the left side of the equation to account for this amount of cooling.

Surface condenser: In a surface condenser (Fig. 1.34), the coolant is separated from the vapors by tubular heat transfer surfaces. The coolant and condensed vapors leave the device by separate exits. Surface condensers are commonly called shell-and-tube heat *exchangers*. The temperature of the coolant is increased, so these devices also act as heaters (EPA-81/12, p. 6-3).

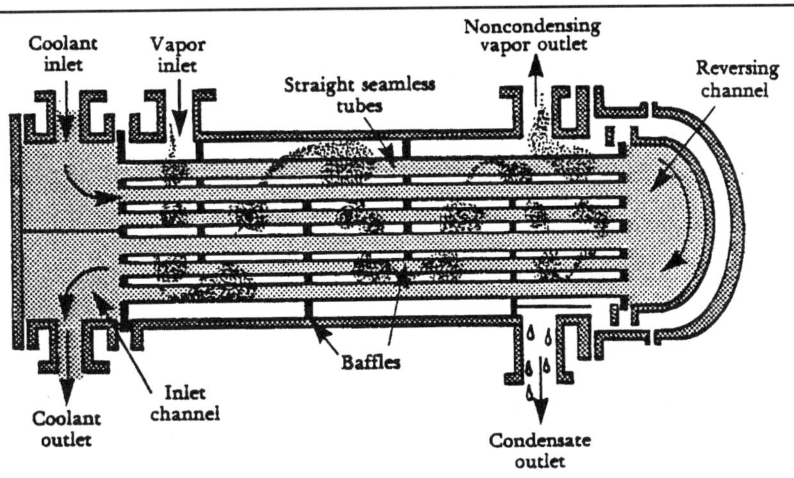

FIGURE 1.34 Surface condenser (*EPA-81/12, p. 6-6*).

A surface condenser consists of a circular or oval cylindrical shell into which the vapor stream flows. Inside the shell are numerous small tubes through which the coolant flows. Vapors contact the cool surface of the tubes, condense, and are collected, while noncondensed vapors are sent for further treatment.

In a surface condenser or heat exchanger, heat is transferred from the vapor stream to the coolant through a heat exchange surface. The rate of heat transfer depends upon three factors:

- Total cooling surface available
- Resistance to heat transfer
- Mean temperature difference between condensing vapor and coolant.

This can be expressed mathematically by

$$q = UA\Delta T_{\mathrm{m}} \tag{1.49}$$

where q = heat transfer rate, Btu/h
U = overall heat transfer coefficient Btu/°F-ft^2-h
ΔT_{m} = mean temperature difference, °F

The overall heat transfer coefficient U is a measure of the total resistance heat experiences while being transferred from a hot body to a cold body. In a shell-and-tube condenser cold water flows through the tubes, causing vapor to condense on the outside surface of the tube wall. Heat is transferred from the vapor to the coolant. The ideal situation for heat transfer would be where heat is transferred from the vapor to the coolant without any heat loss (heat resistance). However, in an engineering application, every time heat moves through a different medium, it encounters a different and additional heat resistance. These heat resistances are throughout the condensate; through any scale or dirt on the outside of the tube (fouling); through the tube itself; and through the film on the inside of the tube (fouling). Each of these resistances are individual heat transfer coefficients and must be added together to obtain an overall heat transfer coefficient.

A number of correlations exist to determine the individual heat transfer coefficient. These correlations are usually presented in terms of dimensionless numbers, such as the Reynolds' number. These correlations consider variables including geometry of the condenser, gas and liquid density and viscosity, and the mechanism of heat transfer (convection, conduction, and radiation). Calculation of individual heat transfer coeffcients require numerous data from the system. An estimate of an overall heat transfer coefficient can be used for preliminary calculations. Table 1.13 lists some typical values for overall heat transfer coefficients. Table 1.13 should be used only for preliminary estimating purposes.

TABLE 1.13 Heat Transfer Coefficients in Tubular Heat Exchangers

Condensing vapor (shell side)	Cooling liquid (tube side)	U (Btu/°F-ft^2-h)
Alcohol vapor	Water	100–200
High-boiling hydrocarbons (vacuum)	Water	20–50
Low-boiling hydrocarbons	Water	80–200
Organic solvents	Water	100–200
Organic solvents with high percentage of non-condensables present	Water or brine	20–60
Naphtha	Water	50–75
Stabilizer reflux vapors	Water	80–120
Sulfur dioxide	Water	150–200
Tall oil derivatives, vegetable oil vapors	Water	20–50
Steam	Feedwater	400–1000

Note: For a water–water (liquid–liquid) heat exchanger (no phase change), the values for U range between 200 and 250.

In a surface heat exchanger the temperature difference between the hot vapor and the coolant usually varies throughout the length of the exchanger. Therefore, a mean temperature difference (ΔT_m) must be used. For the special cases where the flow of both streams is completely co-current, the flow of both streams is completely countercurrent, or the temperature of one of the fluids remains constant (as is the case in condensing a pure liquid), the log mean temperature difference can be used. The temperature profiles for these three conditions are illustrated in Fig. 1.35. The log mean temperature for countercurrent flow can be expressed as

$$\Delta T_m = \Delta T_{lm} = (\Delta T_1 - \Delta T_2)/\ln(\Delta T_1/\Delta T_2) \tag{1.50}$$

where ΔT_{lm} = log mean temperature.

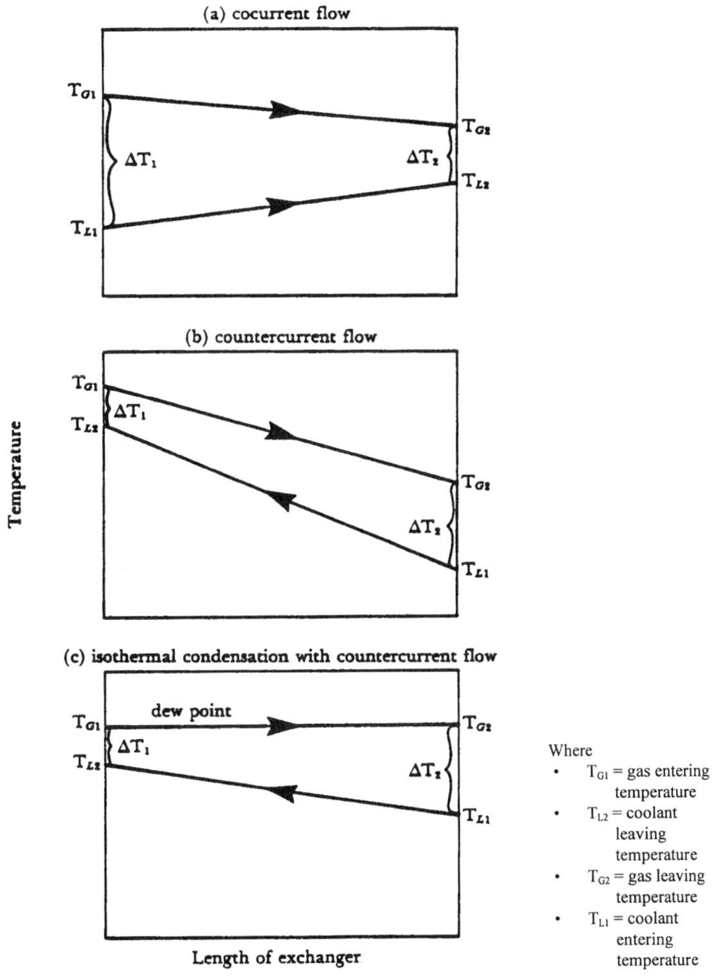

FIGURE 1.35 Temperature profile in heat exchanger (*EPA-81/12, p. 6-13*).

The value calculated from Eq. (1.50) is used for single-pass heat exchangers or condensers. For multiple-pass exchangers a correction factor to the log mean temperature must be included. However, for the special case of isothermal condensation (no change in temperature) of a single-component vapor, $T_{G1} = T_{G2}$ and is equal to the dew point temperature. No correction factor is needed.

In order to size a condenser, Eq. (1.49) must be rearranged to solve for the surface area:

$$A = q/U\Delta T_{lm} \tag{1.51}$$

where A = surface area of a shell-and-tube condenser, ft^2
q = heat transfer rate, Btu/lb
U = overall heat transfer coefficient, Btu/°F-ft^2-h
ΔT_{lm} = log mean temperature difference, °F

For the exact determination of the surface area, U in Eq. (1.51) should be calculated from the individual resistances (coefficients). For a rough estimate, the values in Table 1.13 can be used.

Equation (1.51) is strictly valid only for isothermal condensation of a single component. This implies that the pollutant is a pure vapor stream comprised only of one specific hydrocarbon such as benzene, not a mixture of hydrocarbons. Nearly all air pollution applications involve multicomponents, since air and any other gas are two components. This complicates the design procedure for a condenser.

For preliminary rough estimates of condenser size, the procedure in Example 1 (which follows) for single-component condensation can be used for condensing multicomponents. In choosing a heat transfer coefficient, the smallest value should be used to allow as much overdesign as possible.

Comparison of contact and surface condensers: Since in contact condensers, coolant is merely sprayed on the vapors, these systems are simpler in design, less expensive, and more flexible in application than surface condensers. However, contact condensers require more coolant, and due to direct mixing, produce 10 to 20 times the amount of wastewater (condensate) than surface condensers. Since the wastewater from a contact condenser is contaminated with vapors, it cannot be reused, which poses a water disposal problem. If the condensed vapors have a recovery value, surface condensers are usually used since the condensate can be recovered directly.

EXAMPLE 1: condenser

In a rendering plant, tallow is obtained by removing the moisture from animal matter in a cooker. Exhaust gases from the cookers contain essentially steam; however, the entrained vapors are highly odorous and must be controlled. Condensers are normally used to remove most of the moisture prior to incineration, scrubbing, or carbon adsorption (EPA-81/12, p. 6-15).

The exhaust flow rate from the continuous rendering cooker is 20,000 acfm at 250°F. The exhaust gases are 95% moisture, with the remaining portion consisting of air and obnoxious organic vapors. The exhaust stream is sent first to a shell-and-tube condenser to remove the moisture and then to a carbon adsorption unit. If the coolant water enters at 60°F and leaves at 120°F, estimate the required surface area of the condenser. The condenser is a horizontal, countercurrent flow system with the bottom few tubes flooded to provide subcooling.

Solution:

1. First, compute the pounds of steam condensed per minute.

$$(20{,}000 \text{ acfm}) \times (0.95) = 19{,}000 \text{ acfm steam}$$

From the ideal gas law:

$$PV = nRT$$
$$n = PV/RT$$
$$= (1 \text{ atm})(19{,}000 \text{ acfm})/[(0.73 \text{ atm-ft}^3/\text{lb-mole-}^\circ\text{R})(250 + 460)^\circ\text{R}]$$
$$= 36.66 \text{ lb-mole/min}$$

$$m = (36.66 \text{ lb-mole/min})(18 \text{ lb/lb-mole})$$
$$= 660 \text{ lb/min of steam to be condensed}$$

2. Solve the heat balance to determine q for cooling the superheated steam and condensing only.

$$q = (\text{heat needed to cool steam to condensation temperature}) + (\text{heat of condensation})$$
$$= mC_p\Delta T + mH_v$$

The average specific heat (C_p) of steam at 250°F is roughly 0.45 Btu/lb-°F. The heat of vaporization of steam at 212°F = 970.3 Btu/lb. Substituting into the equation:

$$q = (660 \text{ lb/min})(0.45 \text{ Btu/lb-}^\circ\text{F})(250 - 212^\circ\text{F}) + (660 \text{ lb/min})(970.3 \text{ Btu/lb})$$
$$= 11{,}286 \text{ Btu/min} + 640{,}398 \text{ Btu/min}$$
$$= 651{,}700 \text{ Btu/min}$$

3. Now using Eq. (1.51) to estimate the surface area for this part of the condenser.

$$A = q/U\Delta T_{\text{lm}}$$

where A = surface area of a shell-and-tube condenser, ft^2
q = heat transfer rate, Btu/lb
U = overall heat transfer coefficient. Btu/°F-ft^2-h
ΔT_{lm} = log mean temperature difference, °F

For a countercurrent condenser, the log mean temperature is given by

$$\Delta T_{\text{lm}} = [(T_{G1} - T_{L2}) - (T_{G2} - T_{L1})]/\ln[(T_{G1} - T_{L2})/(T_{G1} - T_{L1})]$$

where ΔT_{lm} = log mean temperature difference
T_{G1} = gas entering temperature
T_{L2} = coolant leaving temperature
T_{G2} = gas leaving temperature
T_{L1} = coolant entering temperature

Remember that the desuperheater–condenser section is designed using the saturation temperature to calculate the log mean temperature difference.

gas entering temperature $T_{G1} = 212^\circ\text{F}$
coolant leaving temperature $T_{L2} = 120^\circ\text{F}$
gas leaving temperature $T_{G2} = 212^\circ\text{F}$
coolant entering temperature $T_{L1} = 60^\circ\text{F}$

$$\Delta T_{\text{lm}} = [(212 - 120) - (212 - 60)]/\ln[(212 - 120)/(212 - 60)]$$
$$= 119.5^\circ\text{F}$$

The overall heat transfer coefficient U is assumed to be 100 Btu/°F-ft^2-h.

Substituting the appropriate values into Eq. (1.51):

$$A = [(651{,}700\ \text{Btu/min})(60\ \text{min/h})]/[(100\ \text{Btu/°F-ft}^2\text{-h})(119.5°\text{F})]$$
$$= 3272\ \text{ft}^2$$

4. To estimate the total size of the condenser, we may need to allow for subcooling of the condensed water (212°–160°F). 160°F is a safe margin.

Re-figuring the heat balance for cooling the water:

$$q = UA\Delta T_{\text{m}}$$

where q = heat transfer rate, Btu/h
U = overall heat transfer coefficient Btu/°F-ft^2-h
ΔT_{m} = mean temperature difference, °F
m = 660 lb/min (assuming all the steam is condensed)
C_p for water = 1 Btu/lb-°F

$$q = (660\ \text{lb/min})(1\ \text{Btu/lb-°F})(212 - 160)°\text{F}$$
$$= 34{,}320\ \text{Btu/min}$$

The new log mean temperature can be calculate from

gas entering temperature $T_{\text{G1}} = 212°\text{F}$
coolant leaving temperature $T_{\text{L2}} = 120°\text{F}$
gas leaving temperature, $T_{\text{G2}} = 160°\text{F}$
coolant entering temperature $T_{\text{L1}} = 60°\text{F}$

$$\Delta T_{\text{lm}} = [(212 - 120) - (160 - 60)]/\ln[(212 - 120)/(160 - 60)]$$
$$= 96°\text{F}$$

$$A = q/U\Delta T_{\text{lm}}$$

where A = surface area of a shell-and-tube condenser, ft^2
q = heat transfer rate, Btu/lb
U = overall heat transfer coefficient, Btu/°F-ft^2-h
ΔT_{m} = mean temperature difference, °F

For cooling water with a water coolant, U is assumed to be 200 Btu/°F-ft^2-h.

$$A = (34{,}320\ \text{Btu/min})/[(200\ \text{Btu/°F-ft}^2\text{-h})(96°\text{F})$$
$$= 1.79\ \text{ft}^2 \text{ or } 2\ \text{ft}^2$$

5. The total area needed is

$$A = 3272 + 2$$
$$= 3274\ \text{ft}^2$$

As illustrated by this example, the area for subcooling is usually very small compared with the area required for condensing.

EXAMPLE 2: contact condenser

In an oil refinery, a stream of light hydrocarbons is to be condensed by a direct contact condenser. The light hydrocarbon stream is essentially benzene. From Perry's *Chemical*

Engineer's Handbook (Perry-73) you found that for benzene: the boiling point is 175°F, the latent heat of vaporization (H_v) is 160 Btu/lb, and the specific heat (C_p) is 0.45 Btu/lb-°F. Water is used as the coolant at 60°F and $C_p = 1.0$ (EPA-81/05, p. 19-1).

Questions
(a) For a benzene mass flow rate (m) of 10,000 lb/h, how much coolant (w) is required if the condensate temperature (T_c) can be no higher than 100°F.
(b) How much benzene is lost in the water if the solubility of benzene at 100°F is 0.05 lb/100 lb water.

Solution:
1. To calculate the amount of water, we must set up a heat balance:

Heat input = heat output

Heat required to condense vapors

+ heat required to cool vapors to outlet temperatures = heat supplied by water

$$mH_v + mC_p(T_{IN} - T_c) = wC_p(T_c - T_{IN})$$

$$(10{,}000\ \text{lb/h}) \times (160\ \text{Btu/lb}) + (10{,}000\ \text{lb/h}) \times (0.45\ \text{Btu/lb-°F}) \times (175 - 100°\text{F}) = w(1.0\ \text{Btu/lb-°F}) \times (100 - 60)$$

$$160 \times 10^4\ \text{Btu/h} + 33.8 \times 10^4\ \text{Btu/h} = w(40\ \text{Btu/lb})$$

$$w = 48{,}450\ \text{lb/h} = 97\ \text{gal/min}\ (8.34\ \text{lb} = 1\ \text{gal}\ H_2O)$$

Amount of coolant w required = 97 gal/min
2. Benzene lost in water

$$\text{benzene lost} = [(0.05\ \text{lb benzene})/(100\ \text{lb-water})] \times (48{,}450\ \text{lb-water/h}) = 24\ \text{lb/h}$$

EXAMPLE 3: surface condenser
A surface condenser is used to condense the hydrocarbon vapors for the same conditions as in Example 2 above (EPA-81/05, p. 19-1), except that the gas outlet temperature is to be 130°F. The overall heat transfer coefficient U from Perry's *Chemical Engineer's Handbook* (Perry-73) is 110 Btu/°F-ft^2-h.

Question
What is the surface of the tubes required for the surface condenser?

Solution:
To calculate the surface area of tubes, use the equation

$$q = UA\Delta T_m$$

where q = heat transfer rate, Btu/h
U = overall heat transfer coefficient, Btu/°F-ft^2-h
ΔT_{lm} = log mean temperature difference, °F

1. We must first calculate the mean temperature change

$$\Delta T_{lm} = [(T_{G1} - T_{L2}) - (T_{G2} - T_{L1})]/\ln[(T_{G1} - T_{L2})/(T_{G1} - T_{L1})]$$

where ΔT_{lm} = log mean temperature difference
T_{G1} = gas entering temperature
T_{L2} = coolant leaving temperature
T_{G2} = gas leaving temperature
T_{L1} = coolant entering temperature

$$\Delta T_{\mathrm{lm}} = [(T_{\mathrm{G1}} - T_{\mathrm{L2}}) - (T_{\mathrm{G2}} - T_{\mathrm{L1}})]/\ln[(T_{\mathrm{G1}} - T_{\mathrm{L2}})/(T_{\mathrm{G1}} - T_{\mathrm{L1}})]$$
$$= [(175 - 60) - (130 - 100)]/\ln[(175 - 60)/(130 - 100)]$$
$$= 63.3°\mathrm{F}$$

2. The heat required is

$$q = \text{heat required to condense vapors} + \text{heat required to subcool}$$
$$q = mH_{\mathrm{v}} + mC_p(T_{\mathrm{G1}} - T_{\mathrm{G2}})$$
$$q = (10{,}000\ \mathrm{lb/h}) \times (160\ \mathrm{Btu/lb}) + (10{,}000\ \mathrm{lb/h}) \times (0.45\ \mathrm{Btu/lb\text{-}°F}) \times (175 - 130°\mathrm{F})$$
$$q = 160 \times 10^4\ \mathrm{Btu/h} + 20.25 \times 10^4\ \mathrm{Btu/h}$$
$$= 180.25 \times 10^4\ \mathrm{Btu/h}$$

3. Surface area

$$A = q/U\Delta T_{\mathrm{lm}}$$

where A = surface area of a shell-and-tube condenser, ft^2
q = heat transfer rate, Btu/lb
U = overall heat transfer coefficient, Btu/°F-ft^2-h
ΔT_{lm} = log mean temperature difference, °F

$$A = (180.25 \times 10^4\ \mathrm{Btu/h})/[(110\ \mathrm{Btu/°F\text{-}ft^2\text{-}h})(63.3°\mathrm{F})$$
$$= 263\ \mathrm{ft}^2$$

4.5 NO_x Emission Control

Approximately one-half of the total nitrogen oxide emissions come from motor vehicles. The other half comes from fossil fuel combustion by utility and industrial boilers, other industrial furnaces and processes, and waste incinerators. Nitrogen oxide emissions that come from fossil fuel combustion result mainly from operating conditions in combustion chambers of furnaces and incinerators. The amount of nitrogen in the combustion air and the amount of nitrogen in the fuel are the two major components that form the nitrogen oxides during combustion or incineration (EPA-81/12, p. 7-1).

Formation of nitrogen oxide. When fossil fuels are burned with air in a boiler or a furnace, some of the oxygen (O_2) and nitrogen (N_2) present combine to form nitrogen oxides (NO_x). Most of the oxides form according to the following reaction:

$$N_2 + O_2 \leftrightarrow 2NO$$

Once the NO forms, the rate of decomposition is very slow and NO does not dissociate back into N_2 and O_2 in any appreciable amounts. The NO formed can react with more oxygen to form NO_2.

$$NO + \tfrac{1}{2}O_2 \leftrightarrow NO_2$$

In large combustion furnaces, the majority of NO_x formed, approximately 95%, is in the form of NO. The main factors involved in NO_x formation include:

- Flame temperature
- Length of time combustion gases are maintained at that temperature

- Amount of excess air present in the flame

Flame temperature in a utility boiler furnace is approximately 1650°C (3000°F). At this high temperature, NO is formed in great abundance (sometimes greater than 1000 ppm). The residence time usually available in combustion equipment is too short for an appreciable fraction (usually less than 5%) of the NO to be oxidized to NO_2. The bulk of the NO_2 is formed in the atmosphere after release from the stack rather than in the confines of the combustion furnace. In this chapter, NO_x refers to the total NO and NO_2 being emitted from major combustion sources.

In small residential heating units, furnace temperatures are usually less than 1090°C (2000°F). These relatively low-temperature operating units produce an exhaust gas containing low NO_x emissions; usually less than 10 ppm.

Thermal NO_x and fuel NO_x. Nitrogen oxide emissions are formed by two chemical processes occurring during combustion: the thermal NO_x process and the fuel NO_x process.

Thermal NO_x: Thermal NO_x results from intense heat during combustion, causing a fraction of the nitrogen content of the combustion air to be oxidized. Its rate of formation is highly sensitive to the flame temperature and to a lesser extent to the local concentration of oxygen at the flame. Virtually all thermal NO_x is formed in the region of the flame which is at the highest temperature. In theory, the formation of thermal NO_x can be reduced in four ways:

- Reduce the nitrogen level (of air) at peak temperature
- Reduce the oxygen level at peak temperature
- Reduce the peak flame temperature
- Reduce the time of exposure at peak temperature

Since the concentration of nitrogen in air–fuel mixtures is relatively fixed, this tactic is not really applicable. Therefore, thermal NO_x is reduced in field practice by reducing oxygen levels, peak flame temperatures, and residence time in the NO_x-producing section of the furnace. This is accomplished by various combustion modification techniques, such as the use of low excess air, staged combustion, reduced air preheat, or flue gas recirculation.

Fuel NO_x: Fuel NO_x occurs as the nitrogen contained in the fuel is oxidized. Therefore, it is dependent on the nitrogen content of the fuel. In fuels such as coal and heavy oil that are relatively high in nitrogen content, approximately 20 to 60% of the fuel-bound nitrogen is oxidized. Its rate of formation is strongly affected by the local oxygen concentration present in the flame and also by the mixing rate of the fuel and air. Thus, like thermal NO_x, fuel NO_x is dominated by the local combustion conditions. One way to reduce fuel NO_x emissions is to reduce the nitrogen content in the fuel. This is not always possible and therefore combustion modification techniques are generally used to reduce NO_x emissions. These include:

- Use of low excess air firing
- Optimum burner designs
- Two-stage combustion or high air preheat (secondary air preheat).

NO_x emission control techniques. The control techniques for nitrogen oxide emissions are summarized in Table 1.14.

Fuel denitrogenation. One control technology for reducing NO_x emissions is to remove the nitrogen contained in the fuel. Amounts of nitrogen vary in a fossil fuel. Coal, shale, and residual fuel oil contain a larger amount of nitrogen than either distillate oil or natural gas. The nitrogen in the fuel can be emitted as NO_x when the fossil fuel is burned in the furnace.

Nitrogen is removed from coal, shale, or heavy fuel oil by liquefying the fuels and mixing with hydrogen gas. The mixture is heated and a catalyst is used to cause the nitrogen in the fuel and the hydrogen to unite. This reaction produces two products: ammonia and a cleaner fuel.

TABLE 1.14 Techniques for the Control of Nitrogen Oxide Emissions

1. Fuel denitrogenation (before burning)		
2. Combustion modification (during burning)	2.1. Low excess air	
	2.2. Staged combustion	
	2.3. Flue gas recirculation	
	2.4. Low NO_x burner	
	2.5. Other techniques	2.5.1. Reduced air preheater
		2.5.2 Load reduction
		2.5.3. Steam and water injection
		2.5.4. Catalytic combustion
3. Flue gas treatment (after burning)	3.1. Exxon thermal $DeNO_x$	
	3.2. Selective catalytic reduction (SCR)	
	3.3. Shell UOP	
	3.4. Wet simultaneous NO_x and SO_2 reduction	

Researchers are developing better catalysts and finding ways to reduce the deposition of carbon on the catalyst surface. Carbon deposits reduce the effectiveness of the catalyst life. This technology can reduce the nitrogen content in both natural fuels and synthetic liquid or gaseous fuel (made from shale and coal). This could become an increasingly important technology with the development and use of synthetic fuels in the future.

Combustion modification. Combustion conditions in the combustion chamber can be modified to reduce NO_x formation. Techniques of combustion modifications to reduce the formation of nitrogen oxides include:

- Low excess air
- Staged combustion
- Flue gas recirculation
- Low NO_x burner
- Other techniques

Low excess air: In a combustion system, a certain amount of excess air is required to ensure complete combustion of the fuel. The more efficient the burners are for air and fuel mixing, the less amount of excess air is required for complete combustion. The minimum amount of excess air is limited by the production of smoke and unburned fuel leaving the furnace.

The level of excess air in an industrial or utility boiler usually ranges from 5% to as high as 50 or 100%. However, most large boilers today operate with excess air less than 5%. Small residential heating furnaces, because of their unsophisticated design, usually operate with approximately 80 to 100% excess air. Gas turbines operate at very high excess air conditions, 300 to 400%. Most of this excess air is added as secondary air at lower temperatures, resulting in low NO_x emissions (EPA-81/12, p. 7-7).

NO_x emissions are reduced from many combustion furnaces by low excess air firing. The local flame zone concentration of oxygen is reduced, thus reducing both thermal and fuel NO. This method is easy to implement and actually increases the efficiency of the furnace slightly. However, there are problems with this combustion modification. Low excess air firing can produce smoke and high CO emissions. Also, fouling and slagging of boiler tube surfaces can occur if various coal and residual oils are burned.

NO_x reductions averaging between 16 and 20% are achieved on gas- and oil-fired utility boilers when the excess air is reduced to a level between 2 and 7%. NO_x reductions averaging around 20% can be achieved on coal-fired utility boilers if the excess air is reduced to the 20% level or lower. For most utility boiler applications, operating at low excess air conditions is considered a routine procedure.

Staged combustion: During staged combustion, air and fuel mixtures are combusted in two separate zones. In one zone, the fuel is fired with less than a stoichiometric amount of air. This creates a fuel-rich local zone in the regions of the primary flame. The second zone is an air-rich zone where the remainder of the combustion air is introduced to complete the combustion of the fuel. The heat in the primary flame zone is not as intense as with normal firing because combustion is incomplete. The air mixed with the fuel is sub-stoichiometric in the NO_x-forming region of the flame, thus creating a low NO_x condition. This modification is also referred to as "off-stoichiometric" combustion (EPA-81/12, p. 7-8).

Staged combustion reduces NO_x emissions by a combination of several factors:

- A lack of available oxygen for NO_x formation in the fuel-rich stage is due to off-stoichiometric firing.
- The flame temperature may be lower in the first stage than with single-stage combustion.
- The peak temperature in the second stage (air rich) is lower.

Staged combustion is an effective technique for controlling both thermal and fuel NO_x due to its ability to control the mixing of fuel with combustion air. The NO_x reduction effectiveness depends on good burner operation to prevent convective boiler tube fouling, unburned hydrocarbon emissions, and poor ignition characteristics which occasionally occur at an excessively fuel-rich boiler operation. Fire-side boiler tube corrosion may occur when burning some coals or heavy oils under staged combustion conditions.

In staged combustion the flame is long, yellow, and slightly smokey as opposed to the short and intense flame observed on normal firing. Fuel combustion (second stage) also extends further into the furnace, sometimes causing excessive temperatures in the convective and superheater sections of the boiler.

Staged combustion is accomplished by using modifications to the boiler such as over fire air (OFA) port and burners out of service (BOOS).

Over fire air port: Over fire air ports are separate ports located above the burner. The burners are operated fuel rich with the remainder of the combustion air coming in through the over fire air ports. These are sometimes referred to as NO_x ports. In some boilers, a number of the burners are operated fuel rich, others air rich in a staggered configuration. This is called *biased firing*. In the case where some burners are operated on air only, this modification is called *burners out of service*.

On existing boilers, a steam load reduction will result in burners out of service if the active fuel burners do not have the capacity to supply fuel for a full load. Most utility boilers installed since 1971 have been designed with over fire air ports so that all fuel burners are active during the staged combustion operation. Using staged combustion modifications on oil- and gas-fired boilers reduces NO_x emissions by approximately 30 to 40%. Modifying existing coal boilers has reduced NO_x emissions 30 to 50%.

Staged combustion can also be accomplished by careful control of air and fuel mixing in the burner.

Flue gas recirculation: Flue gas recirculation (FGR) has been used to reduce thermal NO_x emissions from large oil- and gas-fired boilers. A portion (10 to 30%) of the flue gas exhaust is recycled back into the main combustion chamber by removing it from the stack breeching and mixing it with the secondary air windbox. In order for FGR to be effective in reducing NO_x emissions, the gas must enter directly into the combustion zone. This recirculated gas lowers the flame temperature and dilutes the oxygen content of the combustion air, thus lowering thermal NO_x emissions (EPA-81/12, p. 7-10).

Some operational problems can occur using flue gas recirculation. Possible flame instability, loss of heat exchanger efficiency and, for small package boilers, condensation on intemal heat transfer surfaces, limit the usefulness of gas recirculation.

Flue gas recirculation requires greater capital expenditures than low excess air and staged combustion modifications. High temperature fans (forced or induced draft), ducts, and large space are required for recirculating the gas. However, the costs can be reasonable if the boiler is already equipped with gas recirculation for steam temperature control.

NO_x reduction of approximately 40 to 50% is possible with recirculation of 20 to 30% of the exhaust gas in gas- and oil-fired boilers. Since FGR is used to reduce thermal NO_x, it is not effective for reducing NO_x emissions generated from burning a high nitrogen content fuel such as coal. At high rates of recirculation (greater or equal to 30%) the flame can become unstable, thus increasing carbon monoxide and hydrocarbon emissions.

This combustion modification technique (FGR) should not be confused with flue gas recirculation used by utilities to control boiler tube (superheat) temperature. In this utility practice, flue gas is taken from the stack breeching and mixed into the hopper bottom section of the boiler that is located below the combustion chamber. Since flue gas is not recirculated into the main combustion chamber, this utility practice is not useful in reducing NO_x emissions.

Low NO_x burner: Low NO_x burners have been developed by several manufacturers to reduce NO_x emissions. Burners control mixing of fuel and air in a pattern to keep flame temperature low and to dissipate the heat quickly. Some burners are designed to control the flame shape to minimize the reaction of nitrogen and oxygen at peak flame temperatures. Other designs have fuel-rich and air-rich regions to reduce flame temperature and oxygen availability.

Other combustion modification techniques: Other combustion modifications can also reduce NO_x emissions from combustion sources. These include: reduced air preheater, load reduction, steam and water injection, and catalytic combustion. Reduced air preheat and load reduction are used sparingly in large boilers because of the energy penalty involved and the relatively low emission reduction occurring. Steam and water injection are used mainly for NO_x emission reduction in gas turbines and internal combustion engines. Steam or water is injected into the combustion area to lower the peak flame temperature and thus reduce thermal NO_x emissions. In catalytic combustion, a catalyst is used to achieve oxidation of the fuel rather than using high flame temperatures. These systems have been used in gas turbines to reduce NO_x emissions well below 10 ppm (EPA-81/12).

Flue gas treatment. Nitrogen oxide emissions can be reduced by treating the flue gas after it leaves the combustion zone. Flue gas treatment processes include:

- Exxon Thermal $DeNO_x$
- Selective catalytic reduction (SCR)
- Shell UOP
- Wet simultaneous NO_x and SO_x reduction.

Exxon Thermal $DeNO_x$ process: Thermal $DeNO_x$ is a process that uses ammonia injection to reduce NO_x emissions. The process was developed by Exxon Research and Engineering Co. Ammonia is injected into the post-combustion zone of the boiler (EPA-81/12, p. 7-13).

The ammonia reacts with NO_x (which is 95% NO) to reduce the oxides to molecular nitrogen and water:

$$4NH_3 + 4NO + O_2 \rightarrow 4N_2 + 6H_2O$$

The above reaction is extremely temperature dependent. In a boiler, this reaction successfully takes place at approximately 950°C (1740°F). At higher temperatures (above 1090°C) the ammonia is oxidized, forming additional NO_x. At lower temperatures (below 850°C) the ammonia passes through the boiler unreacted.

Selective catalytic reduction: Selective catalytic reduction (SCR) is a dry process used to reduce NO_x emissions from fossil-fuel-fired boilers. This process is based on the preferential reaction of NH_3 with NO_x rather than with SO_2 in the flue gas. The reactions are expressed as

$$4NH_3 + 4NO + O_2 \rightarrow 4N_2 + 6H_2O$$
$$4NH_3 + 2NO_2 + O_2 \rightarrow 3N_2 + 6H_2O$$

The first equation represents the predominant reaction occurring, since 95% of NO_x emissions in combustion flue gas are in the form of NO.

This process involves injecting NH_3 into the flue gas and passing this mixture through a catalytic reactor. NO_x emissions are reduced to harmless molecular nitrogen (N_2) and water vapor (H_2O). Ammonia is injected on an NH_3 : NO mole ratio of 1 : 1, attaining a 90% NO_x emission reduction with less than 20 ppm NH_3 leaving the reactor.

Shell UOP process: The Shell UOP is a dry process that simultaneously removes both NO_x and SO_x emissions. This process can also be designed to remove either compound separately. The process uses a copper oxide (CuO) catalyst supported on alumina. These catalysts are located in two or more parallel passage reactors (EPA-81/12, p. 7-20).

Flue gas containing both NO_x and SO_x is introduced into the reactor where the SO_x reacts with copper oxide to form copper sulfate ($CuSO_4$). At the same time ammonia is being injected; this reacts with the NO_x. The copper sulfate, and to a less extent the copper oxide, act as catalysts for the NO_x–NH_3 reaction.

The following reactions occur in the reactor (SO_2 and NO_x reduction):

$$CuO + \tfrac{1}{2}O_2 + SO_2 \rightarrow CuSO_4$$
$$4NO + 4NH_3 + O_2 \rightarrow 4N_2 + 6H_2O$$

When the reactor catalyst is saturated with $CuSO_4$, the flue gas is redirected to a fresh reactor and the spent catalyst is regenerated. Hydrogen is used to regenerate the catalyst by reducing the $CuSO_4$ to copper and producing a concentrated SO_2 gas stream. The SO_2 gas is then used to produce sulfuric acid or elemental sulfur for commercial sale. The copper in the reactor is oxidized to CuO and the process is ready to be put on line again. The reactions that take place in the reactor during catalyst regeneration are

$$CuSO_4 + 2H_2 \rightarrow Cu + SO_2 + 2H_2O$$
$$Cu + \tfrac{1}{2}O_2 \rightarrow CuO$$

The Shell UOP process can be operated as a NO_x emission reduction process by eliminating the regeneration cycle. The process can be operated as an SO_x emission reduction process by eliminating the ammonia injection.

Wet NO_x and SO_x process: Wet processes use absorbers to reduce both SO_x and NO_x emissions simultaneously. These will not be discussed in detail.

4.6. SO_x Emission Control

Flue gas desulfurization (FGD) is the generic name of air pollution control technologies that are used to reduce the emissions of sulfur dioxide (SO_2)—a major acid rain precursor (EPA-79/11). Fossil fuels, especially coal, usually contain up to 7% of sulfur. Through the combustion process, almost all of the sulfur in the fuel is oxidized into gaseous SO_2. After the SO_2 is emitted into the atmosphere, it is further oxidized and incorporated into rain, fog, and snow as sulfuric acid. The 'acid rain' or 'acid precipitation' can result in damages to aquatic ecosystems, soil systems, vegetation and man-made objects (Chang-98/07).

The techniques to reduce the sulfur oxide emissions from fossil fuel combustion sources can be grouped into Table 1.15 (EPA-84/03, p. 8-1; EPA-81/12, p. 8-1).

Fuel desulfurization. Fuel desulfurization removes or reduces the sulfur content of coal before it is burned. Coal contains sulfur in two forms:

- Mineral sulfur in the form of inorganic pyrite: mineral sulfur can be removed by physical coal cleaning.
- Organic sulfur which is chemically bound to the coal: organic sulfur requires chemical cleaning.

One of the most straightforward ways to reduce SO_2 emissions from combustion sources is by burning fuel containing less sulfur. This might involve using low sulfur coal, low sulfur fuel oil, or natural gas instead of a high sulfur coal. The use of low sulfur coal (usually supplied from the western United States) can reduce SO_2 emissions. In some cases, low sulfur coal has been used to meet state and federal air pollution regulations. Low sulfur coal generally contains less than 1% sulfur; high sulfur coal contains between 3 and 5% sulfur. Low sulfur coal is more expensive and less available (in the eastern United States) than high sulfur coal. However, the sulfur content can be reduced by a process called fuel desulfurization which includes two major processes (EPA-81/12, p. 8-4):

- Physical coal cleaning
- Chemical coal cleaning

Physical coal cleaning: Physical coal cleaning is important, because it can reduce the ash and sulfur content of coal and hence the potential SO_2 emissions which result from coal combustion. Physical coal cleaning depends on the differences in density of both coal and the impurities. Coal is crushed, washed, and then separated by settling processes using cyclones, air classifiers or magnetic separators. Approximately 40 to 90% of the pyritic sulfur content can be removed by physical coal cleaning (EPA-80/08). Its effectiveness depends on the size of pyritic sulfur particles and the amount of pyritic sulfur contained in the coal.

Chemical coal cleaning: Chemical coal cleaning methods that reduce the organic-bound sulfur include two technologies: (1) microwave desulfurization and (2) hydrothermal desulfurization.

- *Microwave desulfurization*: In microwave desulfurization, coal is crushed, then heated for 30 to 60 seconds by exposure to microwaves. Mineral sulfur selectively absorbs this radiation forming hydrogen sulfide gas (H_2S). The H_2S is usually reduced to elemental sulfur by the Claus process. Another microwave process adds calcium hydroxide [$Ca(OH)_2$] to crushed coal. The organic sulfur converts to calcium sulfite ($CaSO_3$) when exposed to this radiation. The coal is washed with water to remove the $CaSO_3$ and other impurities. As much as 70% of the sulfur can be removed by the microwave process (EPA-80/08).
- *Hydrothermal desulfurization*: In hydrothermal desulfurization, coal is crushed and mixed with a solution of sodium and calcium hydroxides [$NaOH$ and $Ca(OH)_2$]. When this mixture is heated to 275°C in a pressurized vessel, most of the pyritic sulfur and 20 to 50% of the organic sulfur is converted to sodium and calcium sulfites (Na_2SO_3 and $CaSO_3$)(EPA-80/08). The coal is rinsed to remove the sulfites and the water is processed to recycle the sodium and calcium hydroxides. This process is an expensive but effective method for removing sulfur from coal.

Combustion of coal and limestone mixture. Sulfur oxides can be removed by burning coal and limestone mixtures in a boiler. Two promising burning technologies under development are (EPA-81/12, p. 8-5):

- Fluidized bed combustion
- Limestone coal pellets as fuel

TABLE 1.15 Techniques for the Control of Sulfur Oxide Emissions

1. Fuel desulfurization	1.1. Physical coal cleaning		
	1.2. Chemical coal cleaning	1.2.1. Microwave desulfurization	
		1.2.2. Hydrothermal desulfurization	
2. Combustion of coal and limestone mixtures	2.1. Fluidized bed combustion process		
	2.2. Limestone coal pellets as fuel		
3. Coal gasification	3.1. Coal pretreatment		
	3.2. Gasification		
	3.3. Gas cleaning		
4. Coal liquefaction			
5. Flue gas desulfurization (FGD)	5.1. Wet scrubbing process	5.1.1. Throwaway (nonregenerable) process	5.1.1.1. Lime process
			5.1.1.2. Limestone process
			5.1.1.3. Double-alkali process
			5.1.1.4. Sodium-based throwaway process
		5.1.2. Regenerable	5.1.2.1. Wellman–Lord process
			5.1.2.2. Magnesium oxide process
			5.1.2.3. Citrate process
	5.2. Dry scrubbing process		

Fluidized bed combustion process: In the fluidized bed combustion process, a grid supports a bed of coal and limestone (or dolomite) in the firebox of the boiler. Combustion air is forced upward through the grid suspending the coal and limestone bed in a fluid-like motion. Natural gas is used to ignite the pulverized coal. Once the coal is ignited, the gas is turned off. The sulfur in the coal is oxidized to SO_2 and subsequently combined with the limestone to form calcium sulfate ($CaSO_4$). The $CaSO_4$ and flyash particulate matter are usually collected in a baghouse or electrostatic precipitator (ESP).

Limestone coal pellets as fuel: Another way to reduce SO_2 emissions from combustion processes is by burning pellets made of limestone and coal. Pellets are made by pulverizing coal and limestone and adding a binder forming small, cylindrical pellets which are about half the size of a charcoal briquette. These consist of approximately two-thirds coal, and one-third limestone. As the pellet burns, the calcium in the limestone absorbs the SO_2 generated by coal combustion and forms calcium sulfate. Calcium sulfate emissions are subsequently collected in a baghouse or electrostatic precipitator.

Coal gas gasification. Over 70 different processes have been developed for producing a combustible gas from coal. Three basic steps are common to all coal gasification processes:

1. Coal pretreatment
2. Gasification

3. Gas cleaning.

Coal pretreatment: Coal pretreatment involves coal pulverizing and washing.

Gasification: The pulverized coal is gasified in a reactor with limited oxygen. Gasification produces either a low, medium, or high Btu gas by applying heat and pressure or by using a catalyst to break down the components of coal (EPA-80/08). The gas produced contains carbon monoxide (CO), hydrogen (H_2), carbon dioxide (CO_2), water (H_2O), methane (CH_4), and contaminants such as hydrogen sulfide and char. Low and medium Btu gas contains more CO and H than high Btu gas, which contains a higher C content. Methane gas produces more heat when burned. The sulfur in the coal is converted to H_2S during gasification.

Gas cleaning: H_2S is removed during the gas cleaning step generally by a scrubbing process. Hydrogen sulfide is converted to elemental sulfur by partial oxidation and catalytic conversion (EPA-80/08). The synthetic gas produced is sulfur-free and can be burned without releasing harmful pollutants.

Coal liquefaction. *Coal liquefaction* is a process for changing coal into synthetic oil, and is similar to coal gasification. Two basic approaches for liquefaction are used: using a gasifier to convert coal to carbon monoxide, hydrogen, and methane, followed by condensation to convert the gases to oils; using a solvent or slurry to liquefy pulverized coal and then processing this liquid into a heavy fuel oil. Some processes produce both a synthetic gas and a synthetic oil.

Hydrogen is used to convert sulfur in the coal to hydrogen sulfide gas. Hydrogen sulfide is partially oxidized to form elemental sulfur and water. More than 85% of the sulfur is removed from coal by liquefaction (EPA-80/08).

Flue gas desulfurization (FGD). FGD refers to the removal of SO_2 from the process exhaust stream. FGD scrubbing processes can either be wet or dry and can be further grouped into:

- Wet nonregenerable scrubbing processes
- Wet regenerable scrubbing processes
- Dry scrubbing processes

FGD systems can be operated wet or dry. In wet scrubbing systems, liquid absorbs SO_2 in the exhaust stream. The scrubbing liquid contains an alkali reagent to enhance SO_2 absorption. More than a dozen different reagents have been used, with lime and limestone being the most popular for utility boilers, and sodium-based reagents the most popular for industrial boilers. Sodium-based solutions (sometimes referred to as clear solutions) provide better SO_2 solubility and less scaling problems than lime or limestone. However, sodium reagents are much more expensive. Wet FGD scrubbers can further be classified as nonregenerable or regenerable. Nonregenerable processes, sometimes called *throwaway processes*, produce a sludge waste that must be disposed of properly. Most regenerable processes produce a product that may be sold to partially offset the cost of operating the FGD system. Regenerated products include elemental sulfur and sulfuric acid. The throwaway processes are simpler and presently more economical than those used to recover and sell products. As a result, 95% of the FGD processes utilized are nonregenerable, or throwaway (EPA-84/03, p. 8-3).

Most FGD systems employ two stages: one for fly ash removal and the other for SO_2 removal. Attempts have been made to remove both the fly ash and SO_2 in one scrubbing vessel; however, these systems experienced severe maintenance problems and low simultaneous removal efficiencies. The flue gas normally passes first through a fly ash removal device, either an electrostatic precipitator or a wet scrubber, and then into the SO_2 absorber.

Many different types of absorbers have been used in FGD systems, including spray towers. venturis, plate towers, and mobile packed beds. Because of scale buildup, plugging, or erosion, which affect FGD dependability and absorber efficiency, the trend is to use simple scrubbers such as spray towers instead of more complicated ones. The configuration of the tower may be vertical or horizontal, and flue gas can flow co-currently, counter-currently, or cross-currently to the liquid. The chief drawback of spray towers is that they have a higher liquid-to-gas ratio requirement (for equivalent SO_2 removal) than other absorber designs.

FGD is the most popular technology used for controlling sulfur oxide emissions from combustion sources. The majority of FGD systems have been applied to combustion sources such as utility and some industrial coal-fired boilers. FGD systems are also used to reduce SO_2 emissions from some industrial plants such as smelters, acid plants, refineries, and pulp and paper mills. FGD scrubbing processes can either be wet or dry and can be grouped as shown in Table 1.16.

Wet nonregenerable FGD scrubber. Wet nonregenerable FGD processes, also known as wet throwaway FGD processes, generate non-salable (sludge or waste) products. The sludge must be disposed of properly in a pond or landfill. The most common nonregenerable processes used on utility boilers include:

- Lime scrubber
- Limestone scrubber
- Double-alkali scrubber
- Sodium-based throwaway scrubber

LIME SCRUBBER: A lime scrubber is a wet nonregenerable FGD process. It uses an alkaline slurry made by adding lime (CaO), usually 90% pure, to water. The alkaline slurry is sprayed in the absorber and reacts with the SO_2 in the flue gas. Insoluble calcium sulfite ($CaSO_3$) and calcium sulfate ($CaSO_4$) salts are formed in the chemical reaction that occurs in the scrubber and are removed as sludge. The sludge produced can be stabilized to produce an inert landfill material or can be stored in sludge ponds (EPA-84/03, p. 8-6; EPA-81/12, p. 8-7).

Process chemistry

A number of reactions take place in the absorber. Before the calcium can react with the SO_2, both must be broken down into their respective ions.

SO_2 dissociation: SO_2 is absorbed in the water to form sulfite $(SO_3)^-$ and sulfate $(SO_4)^-$ ions.

$$SO_2(\text{gaseous}) \rightarrow SO_2(\text{aqueous})$$
$$SO_2 + H_2O \rightarrow H_2SO_3 \text{ or}$$
$$SO_2 + H_2O + 0.5O_2 \rightarrow H_2SO_4, \text{ if there is excess oxygen}$$
$$H_2SO_3 \rightarrow 2H^+ + (SO_3)^{2-} \text{ or}$$
$$H_2SO_4 \rightarrow 2H^+ + (SO_4)^{2-}$$

TABLE 1.16 Flue Gas Desulfurization (FGD) Techniques

Wet scrubbing process	Throwaway (nonregenerable) process	Lime scrubber
		Limestone scrubber
		Double-alkali scrubber
		Sodium-based throwaway scrubber
	Regenerable process	Wellman–Lord scrubber
		Magnesium oxide scrubber
		Citrate scrubber
Dry scrubbing process		

Lime CaO dissolution: This is accomplished by slaking (dissolving) the lime in water to produce a calcium slurry of CaO and H_2O or calcium hydroxide [$Ca(OH)_2$] and then spraying the slurry into the flue gas to dissolve the SO_2. The calcium hydroxide–water mixture is a solution containing calcium ions (Ca^{2+}) and hydroxide ions (OH^-).

$$CaO(solid) + H_2O \rightarrow Ca(OH)_2(aqueous)$$

$$Ca(OH)_2 \rightarrow Ca^{2+} + 2OH^-$$

Now that the SO_2 and lime are broken into their ions, namely, $(SO_3)^{2-}$ and Ca^{2+}, the following reaction occurs:

$$Ca^{2+} + (SO_3)^{2-} + 2H^+ + 2OH^- \rightarrow CaSO_3(solid) + 2H_2O \text{ or}$$

$$Ca^{2+} + (SO_4)^{2-} + 2H^+ + 2OH^- \rightarrow CaSO_4(solid) + 2H_2O$$

Combining all reactions above, the basic reactions occurring are

$$SO_2 + Ca(OH)_2 + H_2O \rightarrow CaSO_3 + 2H_2O$$

$$SO_2 + Ca(OH)_2 + H_2O + 0.5O_2 \rightarrow CaSO_4 + 2H_2O$$

From the above relationships and assuming that the lime is 90% pure, it will take 1.1 moles of lime to remove 1 mole of SO_2 gas.

System description

The equipment necessary for SO_2 emission reduction is used in four operations:

- *Scrubbing or absorption*: accomplished with scrubbers, holding tanks, liquid-spray nozzles, and circulation pumps.
- *Lime handling and slurry preparation*: Accomplished with lime unloading and storage equipment, lime processing and slurry preparation equipment.
- *Sludge processing*: accomplished with sludge clarifiers for dewatering, sludge pumps and handling equipment, and sludge solidifying equipment.
- *Flue gas handling*: accomplished with inlet and outlet ductwork, dampers, fans, and stack gas reheaters.

Figure 1.36 is a schematic of a typical lime FGD system. Flue gas from the boiler first passes through a particulate emission removal device then into the absorber where the SO_2 is removed. The gas then passes through the entrainment separator to a reheater and is finally exhausted out of the stack. Individual FGD systems vary considerably, depending on the FGD vendor and the plant layout. ESPs or scrubbers can be used for particle removal with the various absorbers used for SO_2 removal.

A slurry of spent scrubbing liquid and sludge from the absorber then goes to a recirculation tank. From this tank, a fixed amount of the slurry is bled off to process the sludge, and, at the same time, an equal amount of fresh lime is added to the recirculation tank. Sludge is sent to a clarifier, where a large portion of water is removed from the sludge and sent to a holding tank. Makeup water is added to the process-water holding tank, and this liquid is returned to the recirculation tank. The partially dewatered sludge from the clarifier is sent to a vacuum filter, where most of the water is removed (and sent to the process-water holding tank) and the sludge is sent to a lined settling pond.

LIMESTONE SCRUBBER: A limestone scrubber is very similar to a lime scrubber. It is a wet nonregenerable process. Limestone scrubbing uses an alkaline slurry from limestone ($CaCO_3$) in an absorber to react with SO_2 in the flue gas. Calcium sulfite ($CaSO_4$) and calcium sulfate ($CaSO_4$) salts are formed in the reaction and are removed as sludge. Two major equipment differences between lime and limestone scrubbing are (EPA-84/03, p. 8-11; EPA-81/12, p. 8-11) lime's use of feed preparation equipment and limestone's use of higher liquid-to-gas

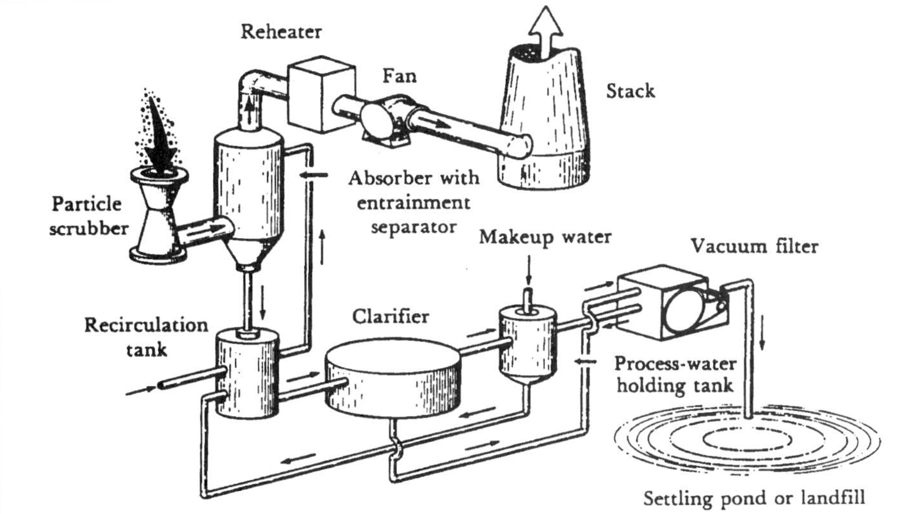

FIGURE 1.36 Process flow for a lime or limestone FGD process (*EPA-84/03, p. 8-7*).

ratios (because limestone is less reactive than lime). Even with these differences, the processes are so similar that an FGD system can be set up to use either lime or limestone in the scrubbing liquid.

The basic chemical reactions occurring in the limestone process are very similar to those in the lime-scrubbing process. Limestone ($CaCO_3$) is slaked with water to form aqueous $CaCO_3$ and is sprayed into the absorber. Sulfite and sulfate ions are produced as the SO_2 gas contacts the water. These ions combine with calcium ions to produce calcium sulfite and calcium sulfate sludge. The basic reactions are

$$SO_2 + CaCO_3 + H_2O + O_2 \rightarrow CaSO_3 + H_2O + CO_2 + O_2$$
$$SO_2 + CaCO_3 + H_2O + 0.5O_2 \rightarrow CaSO_4 + H_2O + CO_2$$

The only difference is in the dissolution reaction that generates the calcium ion. When limestone is mixed with water, the following reaction occurs:

$$CaCO_3(\text{solid}) + H_2O \rightarrow Ca^{2+} + (HCO_3)^- + OH^-$$

The equipment necessary for SO_2 absorption is the same as that for lime scrubbing, except in the slurry preparation. The limestone feed (rock) is reduced in size by crushing it in a ball mill. Limestone is then sent to a size classifier. Pieces larger than 200 mesh are sent back to the ball mill for recrushing. Limestone is mixed with water in a slurry supply tank; it is generally a little cheaper than lime, making it more popular for use in large FGD systems.

DOUBLE-ALKALI SCRUBBER: Dual- or double-alkali scrubbing is a wet nonregenerable or throwaway FGD process that uses a sodium-based alkali solution to remove SO_2, from combustion exhaust gas. Although the double-alkali process regenerates the scrubbing reagent, it is classified as throwaway since it does not produce a salable product and generates solids that must be disposed of in a landfill (EPA-84/03, p. 8-15; EPA-81/12, p. 8-15).

The sodium alkali solution absorbs SO_2 and the spent absorbing liquor is regenerated with lime or limestone. Calcium sulfites and sulfates are precipitated and discarded as sludge. The regenerated sodium scrubbing solution is returned to the absorber loop. The double-alkali process has reduced plugging and scaling problems in the absorber because sodium scrubbing compounds are very soluble. Double-alkali systems are capable of 95% SO_2 reduction.

Particulate matter is removed prior to SO_2 scrubbing by an electrostatic precipitator or a venturi scrubber. This is done to prevent fly ash erosion of the absorber internals and to prevent any appreciable oxidation of the sodium solution in the absorber due to catalytic elements in the fly ash.

Process chemistry

The sodium alkali solution is usually a mixture of three components: sodium carbonate (Na_2CO_3), also called soda ash; sodium sulfite (Na_2SO_3); and sodium hydroxide (NaOH), also called caustic soda.

The SO_2 reacts with the alkaline components to primarily form two salts: sodium sulfite (Na_2SO_3) and sodium bisulfite ($NaHSO_3$) as indicated in the following main absorption reactions:

$$2NaOH + SO_2 \rightarrow Na_2SO_3 + H_2O \tag{1.52}$$

$$NaOH + SO_2 \rightarrow NaHSO_3 \tag{1.53}$$

$$Na_2CO_3 + 2SO_2 + H_2O \leftrightarrow 2NaHSO_3 + CO_2 \tag{1.54}$$

$$Na_2CO_3 + SO_2 \rightarrow 2NaSO_3 + CO_2 \tag{1.55}$$

$$Na_2SO_3 + SO_2 + H_2O \rightarrow 2NaHSO_3 \tag{1.56}$$

In addition to the above reactions, any SO_3 present will react with alkaline components to produce sodium sulfate. For example,

$$2NaOH + SO_3 \rightarrow Na_2SO_4 + H_2O \tag{1.57}$$

Throughout the system, some sodium sulfite is oxidized to sulfate by

$$2Na_2SO_3 + O_2 \rightarrow 2Na_2SO_4 \tag{1.58}$$

After reaction in the absorber, spent scrubbing liquor is bled to a reactor tank for regeneration. Sodium bisulfite and sodium sulfate are inactive salts and do not absorb any SO_2. Actually, it is the hydroxide ion (OH^-), sulfite ion $(SO_3)^{2-}$ and carbonate ion $(CO_3)^{2-}$ that absorb SO_2 gas.

Sodium bisulfite and sodium sulfate are reacted with lime or limestone to produce a calcium sludge and a regenerated sodium solution. The solution is the mixture of sodium sulfite (Na_2SO_3) and sodium hydroxide (NaOH).

$$CaO + H_2O \rightarrow Ca(OH)_2 \text{ (see lime scrubber)}$$

$$2NaHSO_3 + \underset{\text{(lime with water)}}{Ca(OH)_2} \rightarrow Na_2SO_3 + \underset{\text{(sludge)}}{CaSO_3 \cdot 0.5H_2O} + 3/2H_2O \tag{1.59}$$

$$Na_2SO_3 + \underset{\text{(lime with water)}}{Ca(OH)_2} + 0.5H_2O \rightarrow 2NaOH + \underset{\text{(sludge)}}{CaSO_3 \cdot 0.5H_2O} \tag{1.60}$$

$$Na_2SO_4 + \underset{\text{(lime with water)}}{Ca(OH)_2} \rightarrow 2NaOH + \underset{\text{(sludge)}}{CaSO_4} \tag{1.61}$$

At the present time, lime regeneration is the only regeneration process that has been used on commercial double-alkali installations.

Process description

The double-alkali process uses two loops: absorption and regeneration.

Absorption loop: In the absorption loop the sodium solution contacts the flue gas in the absorber to remove SO_2, as shown in Eqs. (1.52)–(1.56). The main products of the absorption loop are two salts: sodium sulfite (Na_2SO_3) and sodium bisulfite ($NaHSO_3$).

Regeneration loop: The products from the absorption loop are mixed with lime (CaO) or limestone ($CaCO_3$) to produce a calcium sludge and a regenerated sodium solution as shown in Eqs. (1.59)–(1.61). In this example, the regenerated sodium solution is a mixture of sodium sulfite (Na_2SO_3) and sodium hydroxide (NaOH). The third component of the sodium solution, sodium carbonate (Na_2CO_3) (also called soda ash), is not regenerated and therefore new soda ash is needed for the system. This need is indicated at the mixing tank shown in Fig. 1.37.

As shown in Fig. 1.37 the scrubbing liquor from the bottom of the absorber is used mixed with regenerated solution and sprayed in at the top of the absorber. A bleed stream from the recirculating liquid is sent to the reactor tank in the regeneration loop. The bleed stream is mixed with a lime slurry in a reactor tank, where insoluble calcium salts are formed and the absorbent is regenerated. The sludge from the reactor is then sent to a clarifier, or thickener, where the calcium sludge is drawn off the bottom, filtered, and washed with water. From the filter, the sodium solution is recycled to the thickener, and the sludge is discarded. From the thickener, the regenerated sodium solution is sent to a holding tank, where the sodium compounds and makeup water are added.

Some sodium sulfate solution is unreacted in the regeneration step. Additional sodium is added to the regenerated solution in the form of soda ash or caustic soda. This regenerated absorbent is now ready to be used again.

SODIUM-BASED SCRUBBER: A sodium-based scrubber is a wet nonregenerable (throwaway or once-through) FGD process. It is the overwhelming choice for FGD systems installed on industrial boilers. This system uses a clear liquid absorbent of either sodium carbonate (Na_2CO_3), sodium hydroxide (NaOH), or sodium bicarbonate ($NaHCO_3$).

Advantages of the sodium-based systems include:

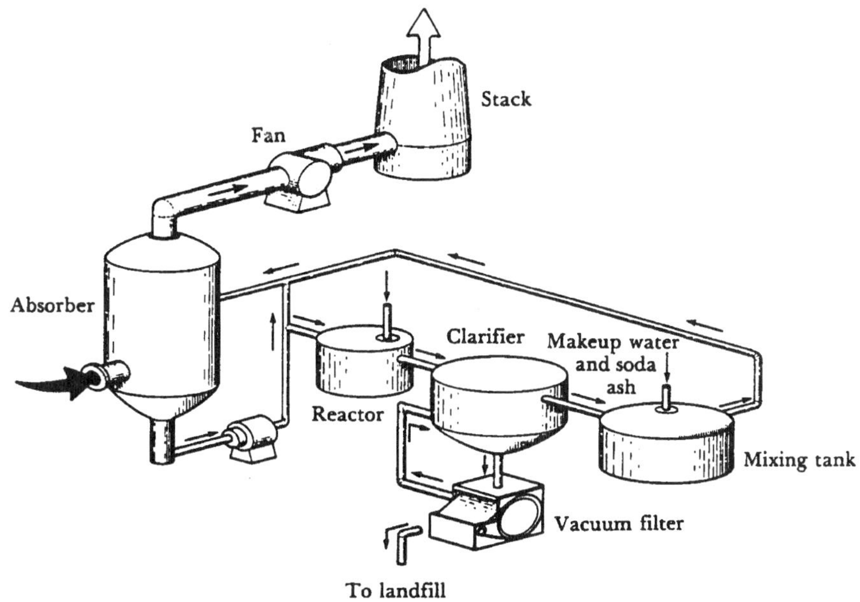

FIGURE 1.37 Typical process flow for a double-alkali system (*EPA-84/03, p. 8-17*).

- Sodium alkali is the most efficient of the commercial reagents in removing SO_2 and the chemistry is relatively simple.
- They are soluble systems—as opposed to slurry systems—making for scale-free operation and fewer components.
- Such systems can handle the wider variations in flue gas composition resulting from the burning of many different fuels by industry.
- The systems are often smaller, and operating costs are a small percentage of total plant costs.
- In some cases, these plants have a waste caustic stream or soda ash available for use as the absorbent.

Disadvantages of the sodium-based systems include:

- The process consumes a premium chemical (NaOH or Na_2CO_3) that is much more costly per pound than calcium-based reagents.
- The liquid wastes contain highly soluble sodium salt compounds. Therefore, the huge quantities of liquid wastes generated by large utilities would have to be sent to ponds to allow the water to evaporate.

The process chemistry is very similar to that of the double-alkali process, except the absorbent is not regenerated.

Figure 1.38 illustrates a basic sodium-based throwaway FGD system. Exhaust gas from the boiler may first pass through an ESP or baghouse to remove particulate matter. Sodium chemicals are mixed with water and sprayed into the absorber. The solution reacts with the

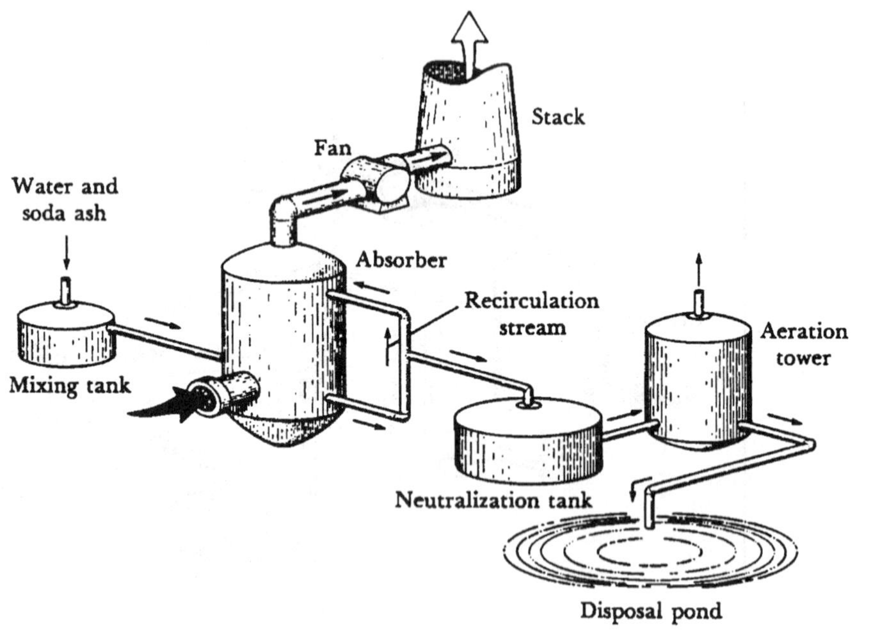

FIGURE 1.38 Sodium-based throwaway FGD system (*EPA-84/03, p. 8-21*).

SO_2 in the flue gas to form sodium sulfite, sodium bisulfite, and a very small amount of sodium sulfate. A bleed stream is taken from the scrubbing liquor recirculation stream at a rate equal to the amount of SO_2 that is being absorbed. The bleed stream is sent to a neutralization tank and aeration tower before being sent to a lined disposal pond.

Some coal-fired units use ESPs or baghouses to remove fly ash before the gas enters the scrubber. In these cases, the absorber can be a plate tower or spray tower that provides good scrubbing efficiency at low pressure drops. For simultaneous SO_2 and fly ash removal, venturi scrubbers can be used. In fact, many of the industrial sodium-based throwaway systems are venturi scrubbers originally designed to remove particulate matter. These units were slightly modified to inject a sodium-based scrubbing liquor. Although removal of both particles and SO_2 in one vessel can be economically attractive, the problems of high pressure drops and using a scrubbing medium to remove fly ash must be considered. However, in cases where the particle concentration is low, such as from oil-fired units, simultaneous particulate and SO_2 emission reduction can be effective.

Wet regenerable scrubbing process. Wet regenerable FGD processes remove SO_2 from the flue gas and generate salable products. Regenerable products include elemental sulfur, sulfuric acid, or, in the case of lime or limestone scrubbing, gypsum (used for wallboard). Regenerable processes do not produce a sludge, thereby eliminating the sludge disposal problem. Most regenerable processes also:

- Have the potential for consistently obtaining a high SO_2 removal efficiency, usually exceeding 90%.
- Utilize the scrubbing reagent more efficiently than nonregenerable processes
- Use scrubbing liquors that do not cause scaling and plugging problems in the scrubber.

The major drawback of using these processes is that these systems are usually more complicated in design and are more expensive to install and operate.

Regenerable processes include:

- Wellman–Lord scrubber
- Magnesium oxide scrubber
- Citrate scrubber

WELLMAN–LORD SCRUBBER: A Wellman–Lord scrubber is the heart of a wet regenerable FGD process. It is a process used to reduce SO_2 emissions from utility and industrial boilers and produces a usable or salable product. This process is sometimes referred to as the Wellman–Lord/Allied Chemical process: Allied Chemical referring to the regeneration step (EPA-84/03, p. 8-26; EPA-81/12, p. 8-18).

Process chemistry

In the Wellman–Lord scrubber, the SO_2 is absorbed by an aqueous sodium sulfite solution which forms a sodium bisulfite solution according to the following equation:

$$SO_2 + Na_2SO_3 + H_2O \rightarrow 2Na_2HSO_3$$

Some oxidation occurs in the absorber to form sodium sulfate, which is unreactive with SO_2 gas.

$$Na_2SO_3 + \tfrac{1}{2}O_2 \rightarrow Na_2SO_4$$

The formation of sodium sulfate depletes the supply of sodium sulfite available for scrubbing. This can be made up by adding sodium carbonate to the scrubbing slurry to combine with sodium bisulfite according to the following chemical reaction:

$$Na_2CO_3 + 2NaHSO_3 \rightarrow 2Na_2SO_3 + CO_2 + H_2O.$$

The absorbent is then regenerated by evaporating the water from the bisulfite solution.

$$2NaHSO_3 \rightarrow Na_2SO_3 + H_2O + SO_2 \text{ (concentrated gas)}$$

The concentrated SO_2 produced in the regeneration step is then sent to the Allied process for conversion to elemental sulfur or sulfuric acid.

Process description

Figure 1.39 shows a typical process flow of a Wellman–Lord system. The process equipment includes an electrostatic precipitator for removing particulate matter; a venturi scrubber for cooling flue gas and removing SO_2 and chlorides; an SO_2 absorber; an evaporator-crystallizer for regenerating the absorbent; and the Allied Chemical process for reducing concentrated SO_2 gas into elemental sulfur or sulfuric acid. The absorber is a plate tower. SO_2 gas is scrubbed with a sodium sulfite solution at each plate. A mist eliminator removes entrained liquid droplets from gas exiting the absorber. There is a direct-fired natural gas reheating system in the absorber stack to reheat cleaned gas for good dispersion of the steam plume.

The solution (sodium bisulfite), collected at the bottom of the absorber, overflows into an absorber surge tank. This solution is pumped through a filter to remove any collected particulate matter. A small side-stream is sent to a purge treatment system, where sodium sulfate is removed. The solution is then pumped to the evaporator for regeneration of the sodium sulfite solution.

The evaporator is a forced-circulation vacuum evaporator. Solution is recirculated in the evaporator, where low-pressure steam evaporates water from the sodium bisulfite solution. When suffcient water is removed, sodium sulfite crystals form and precipitate. Concentrated so, gas (95% by volume) is removed by the steam. The sodium sulfite crystals form a slurry that is withdrawn continuously and sent to a dissolving tank, where condensate from the evaporator is used to dissolve the sodium sulfite crystals into a solution. This solution is pumped back into the top stage of the absorber. The water vapor is removed from the evaporator's overhead SO_2/H_2O vapors by water-cooled condensers.

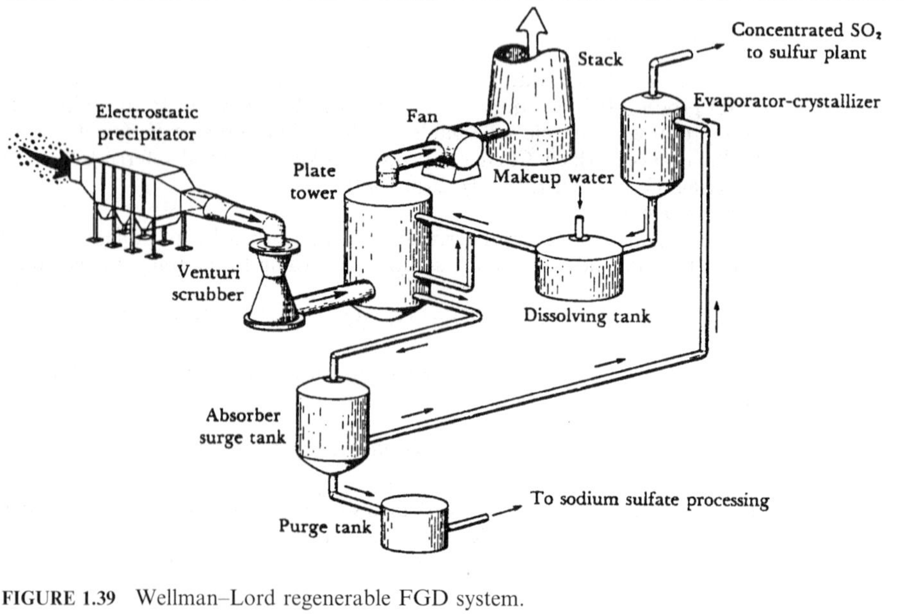

FIGURE 1.39 Wellman–Lord regenerable FGD system.

MAGNESIUM OXIDE SCRUBBER: A magnesium oxide scrubber is a regenerable FGD process used to remove SO_2 from combustion exhaust gas. Magnesium oxide (MgO) slurry absorbs SO_2 and forms magnesium sulfite. Magnesium sulfite solids are separated by centrifugation and dried to remove moisture. The mixture is calcined to regenerate magnesium oxide and produce concentrated SO_2 gas for production of sulfuric acid or elemental sulfur (EPA-81/12, p. 8-21).

Particulate matter is removed from boiler exhaust by a precipitator or wet scrubber prior to entering the absorber. Magnesium oxide slurry is sprayed and absorbs SO_2 according to the following simplified reactions:

$$Mg(OH)_2 + 5H_2O + SO_2 \rightarrow MgSO_3 \cdot 6H_2O$$

$$MgSO_3 \cdot 6H_2O + SO_2 \rightarrow Mg(HSO_3)_2 + 5H_2O$$

$$Mg(HSO_3)_2 + MgO \rightarrow 2MgSO_3 + H_2O$$

$$2MgSO_3 + O_2 + 7H_2O \rightarrow 2MgSO_4 \cdot 7H_2O$$

The aqueous slurry used for scrubbing contains the hydrated crystals of MgO, $MgSO_3$, and $MgSO_4$. A continuous side-stream of this recycled slurry is sent to a centrifuge where partial dewatering produces a moist cake. The liquor removed from the crystals is returned to the main slurry stream. The moist cake is dried at 350 to 450°F in a direct contact or rotary bed dryer. The dried cake is then sent to a calciner, where coke is burned at very high temperatures (1250 to 1340°F) to regenerate magnesium oxide crystals according to the following reactions:
Cake dryer

$$MgSO_3 \cdot 6H_2O \underset{\text{heat}}{\rightarrow} MgSO_3 + 6H_2O \text{ (vapor)}$$

$$MgSO_4 \cdot 7H_2O \underset{\text{heat}}{\rightarrow} MgSO_4 + 7H_2O \text{ (vapor)}$$

MgO regeneration in calciner

$$MgSO_3 \rightarrow MgO + SO_2 \text{ (concentrated gas)}$$

$$C + \tfrac{1}{2}O_2 \underset{\text{heat}}{\rightarrow} CO \text{ (gas)}$$

$$CO + MgSO_4 \rightarrow CO_2 + MgO + SO_2 \text{ (concentrated gas)}$$

CITRATE SCRUBBER: A citrate scrubber is a wet regenerable FGD process that was developed by the US Bureau of Mines. The citrate process uses sodium citrate and citric acid as buffering agents to attain a higher solubility of the SO_2 in an aqueous absorbent solution. The chemistry of this process is very complex. The absorption of SO_2 is pH-dependent, increasing with higher pH. SO_2 forms H_2SO_3 when absorbed by water, resulting in decreasing pH values. This creates a more acidic solution that inhibits additional absorption of SO_2 gas. By using a buffering agent to prevent a pH drop, a substantially higher amount of SO_2 can be absorbed (EPA-81/12, p. 8-22).

One commercial installation of this process was on a 60 MW industrial boiler at St. Joe Minerals in Monaca, Pennsylvania. Particulate matter is removed from the flue gas by an electrostatic precipitator. Chlorides and sulfuric acid mist are removed from the flue gas by a small venturi scrubber before entering the packed tower. SO_2 is absorbed by a solution of sodium citrate, citric acid, and sodium thiosulfate to produce sodium sulfite. SO_2 removal efficiency is approximately 90%. The solution (containing absorbed SO_2) is reacted with hydrogen sulfide gas (H_2S) in a closed vessel to precipitate elemental sulfur and regenerate citrate solution. The elemental sulfur precipitate is concentrated by air flotation into a sulfur slurry which is separated from the regenerated solution. Sulfur slurry is heated to form liquid sulfur and the solution is decanted. Hydrogen sulfide gas (used in regeneration) is obtained either as a by-product of petroleum refining or produced onsite by reacting recovered sulfur with natural gas and steam.

The citrate process has some drawbacks. It is expensive compared with lime and limestone scrubbing. Production of H_2S from methane and sulfur can be a problem whenerer natural gas is not available. Since the size of the absorption, regeneration, and flotation equipment is large, boiler retrofitting is sometimes difficult.

Dry SO_2 scrubber

Dry scrubbing processes use a dry spray to absorb SO_2 gas and form dry particles. These particles are collected in a baghouse or electrostatic precipitator. In dry FGD, the flue gas containing SO_2 is contacted with an alkaline material to produce a dry waste product for disposal.

This technology includes (EPA-81/12, p. 8-23; EPA-89/03, p. 4-14; EPA-92/01, p. 95):

- Injection of an alkaline slurry in a spray dryer with collection of dry particles in a baghouse or electrostatic precipitator (ESP).
- Dry injection of alkaline material into the flue gas stream with collection of dry particles in a baghouse or ESP.
- Addition of alkaline material to the fuel prior to combustion.

The dry FGD has two major types: spray dryer system and dry injection system.

SPRAY DRYER WITH A BAGHOUSE OR ESP: Spray dryers are vessels where hot flue gases are contacted with a finely atomized wet alkaline spray. The high temperatures of the flue gas, 250 to 400°F, evaporate the moisture from the wet alkaline sprays, leaving a dry powdered product. The dry product is collected in a baghouse or ESP (EPA-81/12, p. 8-24).

Flue gas enters the top of the spray dryer and is swirled by a fixed vane ring to cause intimate contact with the slurry spray. The slurry is atomized into extremely fine droplets by rotary atomizers. The turbulent mixing of the flue gas with the fine droplets results in rapid SO_2 absorption and evaporation of the moisture. A small portion of the hot flue gas is added to the spray dryer discharge duct to maintain the temperature of the gas above the dew point. Reheat prevents condensation and corrosion in the duct. Reheat also prevents bags in the baghouse from becoming plugged or caked with moist particles.

Sodium carbonate solutions and lime slurries are the most common absorbents used. A sodium carbonate solution will generally achieve a higher level of SO_2 removal than lime slurries. When sodium carbonate is used, SO_2 removal efficiencies are approximately 75 to 90%, lime removal efficiencies are 70 to 85%. However, vendors of dry scrubbing systems claim that their units are capable of achieving 90% SO_2 reduction using a lime slurry in a spray dryer. Lime is very popular for two reasons:

- Lime is less expensive than sodium carbonate
- Sodium carbonate and SO_2 form sodium sulfite and sodium sulfate which are very soluble causing leaching problems when landfilled.

Some of the evaporated alkaline spray will fall into the bottom of the spray dryer and be recycled. The majority of the spray reacts with SO_2 in the flue gas to form powdered sulfates and sulfites. The particles, along with fly ash in the flue gas, are then collected in a baghouse or electrostatic precipitator. Baghouses have an advantage because unreacted alkaline material collected on the bags can react with any remaining SO_2 in the flue gas.

The major differences between wet absorption SO_2 scrubbers and spray dryer systems is in the scrubbing method and the amount of moisture during the scrubbing action. In a wet scrubber, flue gas is saturated with liquid sprays (usually alkaline). SO_2 is absorbed by the water and also reacts with the chemical. Absorption increases as temperature decreases. Flue gas is cooled and saturated with the scrubbing liquid to remove SO_2. At optimum operating temperatures efficiency is usually $> 90\%$.

DRY INJECTION: In dry injection systems, a dry alkaline material is injected into a flue gas stream. This is accomplished by pneumatically injecting the dry sorbent into a flue gas duct, or by precoating or continuously feeding sorbent onto a fabric filter surface. Most dry injection systems use pneumatic injection of dry alkaline material in the boiler furnace area or in the duct that precedes the ESP or baghouse. Sodium-based sorbents are used more frequently than lime. Many dry injection systems have used nahcolite, a naturally occurring mineral which is 80% sodium bicarbonate found in large reserves in Colorado. Sodium carbonate (soda ash) is also used but is not as reactive as sodium bicarbonate. The major problem of using nahcolite is that it is not presently being mined on a commercial scale (EPA-81/12, p. 8-27).

SO_2 control calculations. This section was authored by Dr John Chang. Dr Chang is an Engineer with the US Environmental Protection Agency in Research Triangle Park, North Carolina.

SO_2 CONTROL REQUIREMENTS: Government standards requiring control of SO_2 emissions are of two types. Ambient air quality standards established by the federal government define the maximum or average permissible level of SO_2 concentration at ground level (40CFR50). The federal government also established emission standards for new and existing combustion sources to avoid any degradation of air quality. For example, the 1979 New Source Performance Standards (NSPS) included SO_2 emission regulations for coal-fired electric utility steam generating units for which construction was commenced after September 18, 1978 (40CFR60). The standards require a minimum of 70% removal when SO_2 emissions are less than 0.6 lb/million Btu heat input. In addition, no gases containing in excess of 1.2 lb/million Btu heat input and 10% of the potential combustion concentration (90% reduction) shall be discharged to the atmosphere (Chang-98/07).

Though it may appear that there are two options in the NSPS for new coal-fired power plants, only one of these sets of requirements applies for a given coal (Molburg-80/02). The new standard can conveniently be regarded as a coal-specific standard. For a coal of specified sulfur content and heating value, the standard determines a unique maximum allowable emission rate and a unique minimum reduction of potential emission.

EXAMPLE 1: SO_2 standards

Consider a 3% sulfur coal with a heating value of 12,000 Btu/lb. Determine the SO_2 removal requirement based on the NSPS.

1. Determine the SO_2 emission without control.

Upon combustion, all the sulfur in the coal is oxidized by air and forms sulfur dioxide:

$$S + O_2 \rightarrow SO_2$$

One mole of sulfur generates 1 mole of SO_2. Since the molecular weight of SO_2 (64) is twice that of S (32), each lb of S generates 2 lb of SO_2. For one million Btu heat input, the amount of coal and SO_2 generated are estimated as

$$\begin{aligned} \text{Coal needed} &= 1{,}000{,}000/12{,}000 \\ &= 83.33 \text{ lb/million Btu} \\ \text{The } SO_2 \text{ generated} &= 3\% \times 83.33 \times 2 \\ &= 5 \text{ lb/million Btu} \end{aligned}$$

2. Determine the NSPS SO_2 removal requirement

First, let us determine the SO_2 emissions with the two removal requirements.

$$\begin{aligned} SO_2 \text{ emission with 70\% removal} &= 5 \times (100 - 70\%) \\ &= 1.5 \text{ lb/million Btu} \\ SO_2 \text{ emission with 90\% removal} &= 5 \times (100 - 90\%) \\ &= 0.5 \text{ lb/million Btu} \end{aligned}$$

The SO_2 emission with 70% removal exceeds 1.2 lb/million Btu, which means greater than 70% removal is required. On the other hand, the SO_2 emission with 90% removal is less than 0.6 lb/million Btu, which means less than 90% removal is required for this particular coal to meet the NSPS (the standards require a minimum of 70% removal when SO_2 emissions are less than 0.6 lb/million Btu heat input).

$$\begin{aligned} SO_2 \text{ removal requirement based on NSPS} &= (1 - 0.6/5) \times 100\% \\ &= 88\% \end{aligned}$$

With 88% removal, the SO_2 emission meets all the NSPS.

EXAMPLE 2: SO_2 standards
Consider three different types of coal: 0.25% sulfur with 9000 Btu/lb heating value, 0.5% sulfur with 9500 Btu/lb heating value, and 4.5% sulfur with 10,500 Btu/lb heating value. Determine the NSPS SO_2 removal requirement for each coal.
1. Coal with 0.25% sulfur and 9000 Btu/lb heating value
The amount of coal needed and SO_2 generated for one million Btu heat input is

$$\begin{aligned} \text{Coal needed} &= 1{,}000{,}000/9000 \\ &= 111.11 \text{ lb/million Btu} \\ SO_2 \text{ generated} &= 0.25\% \times 111.11 \times 2 \\ &= 0.55 \text{ lb/million Btu} \end{aligned}$$

The SO_2 generated is less than 0.6 lb/million Btu and therefore 70% SO_2 removal is required based on the NSPS.
2. Coal with 0.5% sulfur and 9500 Btu/lb heating value

$$\begin{aligned} SO_2 \text{ generated} &= 0.5\% \times (1{,}000{,}000/9500) \times 2 \\ &= 1.05 \text{ lb/million Btu} \\ SO_2 \text{ emission with 70\% removal} &= 1.05 \times (100 - 70\%) \\ &= 0.315 \text{ lb/million Btu} \end{aligned}$$

The SO_2 emission with 70% removal is less than 0.6 lb/million Btu; therefore, 70% removal is the NSPS requirement.
3. Coal with 4.5% sulfur and 10,500 Btu/lb heating value

$$\begin{aligned} SO_2 \text{ generated} &= 4.5\% \times (1{,}000{,}000/10{,}500) \times 2 \\ &= 8.57 \text{ lb/million Btu} \\ SO_2 \text{ emission with 70\% removal} &= 8.57 \times (100 - 70\%) \\ &= 2.57 \text{ lb/million Btu} \end{aligned}$$

The SO_2 emission with 70% removal exceeds 0.6 lb/million Btu, which means greater than 70% removal is required.

$$SO_2 \text{ emission with } 90\% \text{ removal} = 8.57 \times (100 - 90\%)$$
$$= 0.857 \text{ lb/million Btu}$$

The SO_2 emission with 90% removal is greater than 0.6 lb/million Btu but less than 1.2 lb/million Btu. Therefore, the NSPS requirement of this particular coal is 90% SO_2 removal.

LIMESTONE FLUE GAS DESULFURIZATION PROCESS: The primary source of SO_2 emissions in the United States is fuel combustion in utility and industrial boilers, usually coal-fired. An example of such a source are given in Table 1.17.

Note that the flue gas contains not only SO_2 but also O_2, CO_2, HCl, and fly ash, which complicate the application of FGD technologies. Combustion gas from the boiler is usually not cooled below about 150°C because of the condensation of sulfuric acid, which corrodes any of the more economical materials of construction for heat exchangers.

The limestone wet scrubbing process is the most widely used FGD technology in the United States (and the world) for utility and industrial boiler SO_2 emission control (EPA-81/08). This process is based on the chemistry of acid–base titration, buffer capacity, and neutralization (Chang-82/03). The basic idea is to use alkaline water as a medium to absorb the acid gas, water-soluble SO_2, from a flue gas. The acidified water is neutralized by the alkali, $CaCO_3$ and $MgCO_3$, derived from finely ground limestone. Since HCl is very water soluble, almost all the HCl in the flue gas is also absorbed by the scrubbing liquor and eventually neutralized by the limestone.

In brief, the FGD process involves:

- Flue gas from the boiler enters the scrubber, where it is contacted by a recirculating slurry containing the fine ground limestone.
- The cleaned gas leaves the scrubber and flows into the atmosphere via a stack. While passing through the scrubber the gas is cooled and saturated with water vapor.
- During the absorption process, some of the suspended limestone particles in the recirculating slurry dissolve to react with the absorbed SO_2 and precipitate calcium sulfite and sulfate as solid waste. The slurry from the scrubber drains to a hold tank where reactions go to completion. Makeup limestone is added to the tank and part of the slurry recycles back to the scrubber. Typically the continuously recycled scrubbing liquor may contain from 5 to 15% suspended solids, consisting of fresh alkali, reacted waste products, and fly ash.
- The workhorse of the whole system is the scrubber which provides intimate gas/liquid contacting to facilitate SO_2 absorption. To promote maximum gas–liquid surface area and liquor holdup a number of devices are used. Common methods include stages of moving

TABLE 1.17 SO_2 Emission Source

Flue gas feature	Quantity
Flue gas flow rate, 1000 acfm	970
Flue gas temperature, °F	290
SO_2 concentration, ppm	3000
HCl concentration, ppm	95
CO_2 concentration, mol %	12
H_2O concentration, mol %	8.5
Fly ash, g/scf	0.1
O_2 concentration, mol %	4.5
N_2 concentration, mol %	74.7

spheres supported on grids, towers with staged spray headers, towers with closely spaced rods or grids, and perforated trays. Fixed packings are generally not used in slurry service because of a tendency to plug or to scale.

- To remove the solid waste products from the system, a bleed stream is withdrawn from the recycle loop and routed to the solids dewatering equipment. The bleed is routed to a clarifier (also called a thickener or settling tank). The suspended solids settle to the bottom of the tank, and clear liquor is drawn off the top and returned to the scrubber loop. At the bottom of the clarifier, the solids may be concentrated to 20 to 40 %. This underflow then goes to a filter for further thickening, where the separated liquor is also returned to the scrubber loop. The filter cake, consisting of 60% or more solids, is discharged to a waste disposal area.
- Intentional oxidation of the absorbed SO_2 can be conducted by sparging air into the hold tank or a separated oxidation tank. The purposes are to produce marketable gypsum (EPA-78/10) for a solid waste product which occupies less volume and is more suitable for landfill disposal (EPA-79/11). Intentional oxidation also reduces the costs of solids dewatering. Without intentional oxidation the solid product, mostly calcium sulfite, precipitates in the scrubber system as small, thin platelets or rosettes less than 5 microns in the largest dimension. On the other hand, gypsum crystallizes as blocky, large crystals up to 50 microns in size (Chang-86/11). As a result, sulfite sludges and filter cakes retain more water and are also more thixotropic. Without intentional oxidation, the oxygen in flue gas usually can cause 20 to 30% oxidation of the absorbed SO_2. The solid waste product from the final dewatering device contains 30 to 60% interstitial water. With greater than 95% oxidation by intentional air sparging, the water content of the final solid product can be reduced to less than 10%.

EXAMPLE 3: SO_2 control
Consider a wet limestone scrubber for controlling SO_2 emissions with the flue gas features listed in Table 1.17. Determine how much limestone is required per year. The limestone supplier reported that analytical data showed that the limestone contains 94.5 wt % $CaCO_3$, 1.5 wt%$MgCO_3$ and 4.0 wt % of inert material. The actual limestone consumption rate is usually 110% of the theoretical requirement.
1. Determine the overall reaction balance between SO_2, HCl, $CaCO_3$, and $MgCO_3$

$$SO_2 + CaCO_3 + 0.5H_2O \rightarrow CaSO_3 \cdot 0.5H_2O + CO_2$$
$$SO_2 + MgCO_3 \rightarrow MgSO_3 + CO_2$$
$$2HCl + CaCO_3 \rightarrow CaCl_2 + H_2O + CO_2$$
$$2HCl + MgCO_3 \rightarrow MgCl_2 + H_2O + CO_2$$

The equation shows that 1 mole of SO_2 requires 1 mole of $CaCO_3$ and 0.5 mole of H_2O or 1 mole of $MgCO_3$. The equation also shows that one mole of HCl requires 0.5 mole of $CaCO_3$ or $MgCO_3$.
2. Determine the amount of SO_2 and HCl at the scrubber inlet
Assume that the flue gas pressure is about 1 atm and the ideal gas law can be applied to the scrubber inlet flue gas:

$$\begin{aligned} V/n &= RT/P \\ &= 0.73(290 + 460)/1 \\ &= 547.5 \text{ acf/lb-mol} \end{aligned}$$

$$\text{The scrubber inlet flue gas flow rate} = 970{,}000 \times 60/547.5$$
$$= 106{,}301 \text{ lb-mol/h}$$

$$\text{The total amount of } SO_2 \text{ at the scrubber inlet} = 3000 \times 106{,}301/1{,}000{,}000$$
$$= 318.9 \text{ lb-mol/h}$$

$$\text{The total amount of HCl at the scrubber inlet} = 95 \times 106{,}301/1{,}000{,}000$$
$$= 10.1 \text{ lb-mol/h}$$

3. Determine the SO_2 and HCl removal requirement
Under current NSPS regulation, the scrubber system must remove 90% of the inlet SO_2 for this high sulfur coal case.

$$\text{The amount of } SO_2 \text{ to be removed by the scrubber system} = 318.9 \times 90\%$$
$$= 287.01 \text{ lb-mol/h}$$

Due to the high solubility, all of the HCl from the flue gas is removed.
4. Calculate the amount of limestone required
The theoretical limestone requirement depends on the amount of SO_2 and HCl to be removed. Since 1 mole of SO_2 requires 1 mole of alkalinity (as $CaCO_3$ and $MgCO_3$), and 1 mole of HCl requires 0.5 mole alkalinity:

$$\text{The theoretical alkalinity requirement} = 287.01 + 0.5 \times 10.1$$
$$= 292.06 \text{ lb-mol/h}$$

$$\text{The actual alkalinity consumption rate} = 292.06 \times 110\%$$
$$= 321.27 \text{ lb-mol/h}$$

The limestone supplied contains 94.5 wt % $CaCO_3$ (molecular weight 100.09) and 1.5 wt% $MgCO_3$ (molecular weight 84).

$$\text{The alkalinity in each 100 lb of limestone} = 100 \times 94.5\%/100 + 100 \times 1.5\%/84$$
$$= 0.962 \text{ lb-mol}$$

$$\text{The total amount of limestone consumed} = (321.27/0.962) \times 100$$
$$= 33{,}396.05 \text{ lb/h}$$

EXAMPLE 4: SO_2 control
Assume that the scrubber exit flue gas is adiabatically saturated with water. Determine the water required to humidify the flue gas in the scrubber and the flue gas composition at the scrubber exit.
1. Humidity of the scrubber inlet flue gas
The definition of humidity is the mass of water vapor associated with unit mass of dry gas (Bennett-82/01). For the scrubber inlet flue gas, 8.5 mol % of it is water and 91.5% is dry gas.

$$\text{The mass of water in each mole of flue gas} = 18 \times 0.085$$
$$= 1.53 \text{ lb}$$

The average molecular weight of the dry gas can be estimated from the composition data given in Table 1.17.

Average molecular weight of dry gas

$$= (28 \times 0.747 + 32 \times 0.045 + 44 \times 0.12 + 64 \times 0.00300 + 36.5 \times 0.000095)/0.915095$$
$$= 30.42$$

$$\text{The mass of dry gas in each mole of flue gas} = 30.42 \times 0.915$$
$$= 27.83$$

$$\text{Thus, the humidity of the flue gas at } 290°\text{F} = 1.53/27.83$$
$$= 0.055 \text{ lb of water per lb of dry gas}$$

2. Temperature and humidity of the wet flue gas at the scrubber exit
Use of a psychrometric chart (which can be found from any thermodynamic text book) permits rapid estimation of the humidity and temperature of the wet flue gas leaving a scrubber.

When unsaturated hot flue gas is introduced into a scrubber, water evaporates into the flue gas immediately to humidify the flue gas until it is saturated with water vapor. Due to the short residence time (usually only a few seconds) and insignificant pressure change of the flue gas inside the scrubber, the humidification process can be considered as under adiabatic conditions (heat loss through scrubber wall is negligible) at constant pressure. As the flue gas is humidified heat is transferred from the flue gas to the water for evaporation, causing the flue gas temperature to decrease following an adiabatic cooling line in the psychrometric chart. But the wet-bulb temperature (dew point) remains constant throughout the period of vaporization because of the adiabatic conditions.

When evaporatfon continues until the flue gas is 100% saturated with water vapor, the final temperature of the gas will be the same as its initial wet-bulb temperature. For air–water vapor mixtures, the wet-bulb temperature and adiabatic cooling lines are practically the same.

As the vaporization takes place, the humidity of the gas is increased and the dry-bulb temperature must correspondingly decrease along the wet-bulb temperature line (adiabatic cooling line). For use with the psychrometric chart, the humidity of the inlet flue gas must be corrected for air. The average molecular weight of air is 29 (actually 28.97): therefore,

$$\text{The mass of dry air in each mole of gas} = 29 \times 0.915$$
$$= 26.54$$
$$\text{Equivalent humidity of air} = 1.53/26.54$$
$$= 0.0576 \text{ lb of } H_2O/\text{lb of dry air}$$

From the psychrometric chart, point A corresponds to humidity 0.0576 lb H_2O/lb dry air at 290°F. Point B is the intersection of the adiabatic cooling line with the 100% saturation line, which is 0.10 lb H_2O/lb dry air at 127°F. This air–water system humidity has to be corrected to obtain the flue gas humidity.

$$\text{The value for saturation humidity of flue gas} = 0.10 \times 28.97/30.42$$
$$= 0.0952 \text{ lb } H_2O/\text{lb dry gas at } 127°\text{F}$$

Note: It was assumed the molecular weight of the scrubber outlet dry flue gas is the same as that at the scrubber inlet. This approximation is acceptable because the difference is usually within 1%.

3. The amount of water required to saturate the flue gas
The amount of water required to saturate the flue gas can be estimated by the difference between the inlet and outlet of the scrubber.

$$\text{The amount of water vapor in the inlet gas} = 0.085 \times 18 \times 106{,}031 = 162{,}227.43 \text{ lb/h}$$

$$\text{The amount of water vapor in the outlet gas} = 0.0952 \times 106{,}031 \times 0.915 \times 30.42 = 280{,}963.63 \text{ lb/h}$$

$$\text{Water required for humidification} = 280{,}963.63 - 162{,}227.43 = 118{,}736.2 \text{ lb/h or } 237.5 \text{ gpm}$$

4. Flue gas composition at the scrubber exit

To calculate the flue gas composition at the scrubber outlet, the CO_2 flow rate needs to be estimated. Chemical reaction balance showed that 1 mole of CO_2 is produced for each mole of SO_2 removed. Therefore

$$CO_2 \text{ in the scrubber outlet flue gas} = 287.01 + 106{,}031 \times 0.12 = 13{,}010.7 \text{ lb-mol/h}$$

The composition of the cleaned flue gas leaving the scrubber is summarized in Table 1.18 (remember that the N_2 and the O_2 remain unchanged from the inlet).

EXAMPLE 5: SO_2 control

Assume that 20% of the $CaSO_3$ formed was oxidized to $CaSO_4$ in the scrubber and precipitated as $CaSO_3 \cdot CaSO_4 \cdot 0.5H_2O$ solid solution crystal. The waste sludge from the dewatering device contains 60% solids. Determine the waste sludge composition, its production rate, and the total water consumption rate.

1. Solids from the excess limestone supplied

The overall effect of the scrubbing system is that 90% of the incoming SO_2 is removed from the flue gas and transferred to the effluent sludge. The moles of SO_2 removed from the flue gas equal the moles of sulfur in the sludge. The HCl is assumed to be removed as $MgCl_2$ and $CaCl_2$. The remainder of the $MgCO_3$ reacts with SO_2 to form $MgSO_3$. Some of the $MgSO_3$ may oxidize to $MgSO_4$ but for the sake of simplicity the formation of $MgSO_4$ is neglected here.

TABLE 1.18 Composition of Cleaned Flue Gas Leaving the Scrubber

Flue gas	Flow rate, lb-mol/h	Composition, mol %
N_2	79,406.8	70.37
O_2	4,783.5	4.24
CO_2	13,010.7	11.53
SO_2	31.9	0.028
H_2O	15,609.1	13.832
Total	112,842.0	100

$$\begin{aligned}\text{Excess limestone supplied} &= 33{,}396.05 \times 0.1/1.1\\ &= 3036.0 \text{ lb/h}\\ \text{Excess } CaCO_3 \text{ supplied} &= 3036.0 \times 94.5\%\\ &= 2869.0 \text{ lb/h}\\ \text{Excess } MgCO_3 \text{ supplied} &= 3036.0 \times 1.5\%\\ &= 45.54 \text{ lb/h}\\ \text{Inert supplied} &= 33{,}396.05 \times 4\%\\ &= 1335.84 \text{ lb/h}\\ \text{Total solids from the excess limestone supplied} &= 2869.0 + 45.54 + 1335.84\\ &= 4250.38 \text{ lb/h}\end{aligned}$$

2. Solids from the crystals formed
The SO_2 removed in the scrubber either reacts with $CaCO_3$ to precipitate as solids or associates with $MgCO_3$ and remains in the scrubbing liquor.

$$\begin{aligned}\text{Percent of } SO_2 \text{ precipitated} &= 287.01 \times (94.5/100)/0.962\\ &= 281.94 \text{ lb-mol/h}\end{aligned}$$

Each mole of solid solution crystal contains 2 moles of sulfur. Therefore, when 20% of the $CaSO_3$ is oxidized to $CaSO_4$, actually 60% of the sulfur removed is precipitated as $CaSO_3 \cdot 0.5H_2O$.

$$\begin{aligned}\text{The amount of sulfur in the } CaSO_3 \cdot 0.5H_2O &= 281.94 \times 60\%\\ &= 169.16 \text{ lb-mol/h}\\ \text{The } CaSO_3 \cdot 0.5H_2O \text{ formed by the precipitation reaction} &= 169.16 \times 129\\ &= 21{,}821.64 \text{ lb/h}\\ \text{The number of moles of } CaSO_3 \cdot CaSO_4 \cdot 0.5H_2O \text{ formed} &= 281.94 \times 0.2\\ &= 56.39 \text{ lb-mol/h}\\ \text{The } CaSO_3 \cdot CaSO_4 \cdot 0.5H_2O \text{ formed by the precipitation reaction} &= 56.39 \times 265\\ &= 14{,}943.35 \text{ lb/h}\end{aligned}$$

3. Total waste sludge production rate
The total solids production rate is the sum of excess $CaCO_3$, excess $MgCO_3$, inert, $CaSO_3 \cdot 0.5H_2O$ and $CaSO_3 \cdot CaSO_4 \cdot 0.5H_2O$.

$$\begin{aligned}\text{The total solids production rate} &= 2869.0 + 45.54 + 1335.84 + 21{,}821.64 + 14{,}943.35\\ &= 41{,}015.37 \text{ lb/h}\\ \text{The waste sludge production rate} &= 41{,}015.37/0.6\\ &= 68{,}358.95 \text{ lb/h}\end{aligned}$$

4. Waste sludge composition
The waste sludge composition is shown in Table 1.19.

TABLE 1.19 Waste Sludge Composition

Material	Mass flow rate, lb/h	Composition, wt %
$CaCO_3$	2,869.0	4.6
$MgCO_3$	45.54	0.073
$CaSO_3 \cdot 0.5H_2O$	21,821.64	31.56
$CaSO_3 \cdot CaSO_4 \cdot 0.5H_2O$	14,943.35	21.61
Inert	1,335.84	2.157
Water	27,343.58	40
Total	68,358.95	100

5. Total water consumption rate
The total water consumption rate includes the water required for humidification and the water associated with the solid crystals and contained in the waste sludge. There is 0.5 lb-mole of H_2O associated with each lb-mole of $CaSO_3 \cdot 0.5H_2O$ and $CaSO_3 \cdot CaSO_4 \cdot 0.5H_2O$ solid crystals.

$$\text{The water associated with the solid crystals} = 18 \times 0.5 \times (169.16 + 56.39) = 2029.95$$

$$\text{The total water consumption rate} = 118{,}736.2 + 27{,}652.9 + 2029.95 = 148{,}419.05 \text{ lb/h}$$

6. Distribution of water consumption
As shown in Table 1.20, the majority of the water consumed in the scrubber system was for the humidification of the flue gas.

EXAMPLE 6: SO_2 control
Assume all the chlorides leave the scrubber system with the interstitial water of the dewatered sludge. Determine the steady-state chloride concentration in the scrubbing liquor.
1. The amount of chloride removed in the scrubber
It is assumed that 100% of the HCl in the flue gas is absorbed by the scrubbing liquor in the scrubber.

$$\text{The amount of chloride removed in the scrubber} = 35.5 \times 10.1 = 358.55 \text{ lb/h}$$

2. The steady-state chloride concentration in the scrubbing liquor
At steady state, there is no accumulation of chloride in the scrubber system and the mass balance equation requires that

$$\text{Chloride removed in the scrubber} = \text{Chloride leaving the scrubber system}$$

TABLE 1.20 Distribution of Water Consumption

	Consumption rate, lb/h	Distribution, %
Flue gas humidification	118,736.2	80.17
Waste sludge interstitial water	27,343.58	18.46
Solid crystal	2,029.95	1.37
Total	148,109.73	100

When the steady-state chloride concentration in the scrubbing liquor is C_{Cl} and all the chlorides leave the scrubber system with the interstitial water of the dewatered sludge, the chloride balance equation becomes

$$\text{Chloride leaving the scrubber system} = C_{Cl} \times \text{water leaving with the sludge}$$
$$C_{Cl} = 358.55/27{,}343.58$$
$$= 0.013 \text{ or } 1.3\%(13{,}000 \text{ ppm})$$

EXAMPLE 7: SO_2 control
Assume that air is injected to the scrubber hold tank to oxidize 95% of the SO_2 removed to form sulfate ions $(SO_4)^{2-}$ and precipitate as gypsum ($CaSO_4 \cdot 2H_2O$). The waste sludge from the dewatering device contains 90% solids. Determine the theoretical air consumption rate, the waste sludge production rate, the water consumption rate, and the steady-state chloride concentration.

Solution:
1. Determine overall reaction balance with oxidation

$$SO_2 + CaCO_3 + 0.5O_2 + 2H_2O \rightarrow CaSO_4 \cdot 2H_2O + CO_2$$
$$SO_2 + MgCO_3 + 0.5O_2 \rightarrow MgSO_4 + CO_2$$

The equations show that each mole of SO_2 requires 0.5 mole of O_2 for oxidation and each mole of gypsum needs 2 moles of water.
2. Theoretical air consumption rate

$$\text{Theoretical oxygen consumption rate} = 0.5 \times 281.94 \times 0.95$$
$$= 133.92 \text{ lb-mol/h}$$
$$\text{Theoretical air consumption rate} = 133.92/0.21$$
$$= 637.71 \text{ lb-mol/h or } 18493.59 \text{ lb/h}$$

3. Solids from precipitation reactions
When 95% of the SO_2 removed is oxidized by air to form gypsum,

$$CaSO_4 \cdot 2H_2O \text{ production rate} = 172 \times 281.94 \times 0.95$$
$$= 267.84 \text{ lb-mol/h or } 46{,}069 \text{ lb/h}$$
$$CaSO_3 \cdot 0.5H_2O \text{ production rate} = 129 \times 281.94 \times (1 - 0.95)$$
$$= 14.1 \text{ lb-mol/h or } 1818.51 \text{ lb/h}$$

4. Total waste sludge production rate
The total solids production rate is the sum of excess $CaCO_3$, excess $MgCO_3$, inert, $CaSO_3 \cdot 0.5H_2O$ and $CaSO_4 \cdot 2H_2O$.

$$\text{Total solids production rate} = 2869.0 + 45.54 + 1335.84 + 1818.51 + 46{,}069$$
$$= 52{,}137.89 \text{ lb/h}$$
$$\text{The waste sludge production rate} = 52{,}137.89/0.9$$
$$= 57{,}930.99 \text{ lb/h}$$

5. Waste sludge composition
The waste sludge composition is shown in Table 1.21.

6. Total water consumption rate
The total water consumption rate includes the water required for humidification and the water associated with the solid crystals and contained in the waste sludge. There is 0.5 lb-moles of H_2O associated with each lb-mole of $CaSO_3 \cdot 0.5H_2O$ and 2 lb-mole of H_2O with each lb-mole of $CaSO_4 \cdot 2H_2O$ solids.

$$\text{The water associated with the solids} = 18 \times 0.5 \times 14.1 + 18 \times 2 \times 267.84$$
$$= 9769.14$$
$$\text{The total water consumption rate} = 118{,}736.2 + 5793.10 + 9769.14$$
$$= 134{,}298.44 \text{ lb/h}$$

7. The steady-state chloride concentration in the scrubbing liquor

$$\text{Chloride leaving the scrubber system} = C_{\mathrm{Cl}} \times \text{water leaving with the sludge}$$
$$C_{\mathrm{Cl}} = 358.55/5793.10$$
$$= 0.012 \text{ or } 6.2\% \text{ (62,000 ppm)}$$

TABLE 1.21 Waste Sludge Composition

Component	Mass flow rate, lb/h	Composition, wt %
$CaCO_3$	2,869.0	4.952
$MgCO_3$	45.54	0.079
$CaSO_3 \cdot 0.5H_2O$	1,818.51	3.139
$CaSO_4 \cdot 2H_2O$	46,069	79.524
Inert	1,335.84	2.306
Water	5,793.10	10
Total	57,930.99	100

5. MANAGEMENT OF INCINERATION RESIDUES

5.1 Solid Residue

Quantities/characteristics. Ash produced from incineration is primarily inorganic and can be classified into the categories of fly ash and bottom ash. Fly ash is entrained in exhaust gases leaving the incinerator and is usually captured in air pollution control equipment. Bottom ash is that portion that remains in the combustion chamber after incineration and is normally associated with inerts (OME-88/12, p. 6-50).

Ash composition varies greatly and depends on the composition of the waste being incinerated. In municipal refuse incineration, solid materials not susceptible to oxidation (e.g. ceramic or glass) constitute most of the ash; however, depending on the type of incinerator and its efficiency, the ash might also contain unburned waste feed.

The relative proportion of fly ash to bottom ash depends on the waste composition and on the incinerator design and operation. When liquid or gaseous wastes are incinerated, no bottom ash and relatively little fly ash result. However, more ash is produced by liquids derived from complex chemical processes containing inerts or from blending procedures that produce inerts.

When other solid wastes are incinerated, both bottom ash and fly ash can also result. The bottom ash from the incineration of sewage sludge is approximately from 15 to 20% of the total mass feed. Refuse incineration can produce ash in varying amounts depending on the waste material burned, but ash content normally ranges from 5 to 25% of the total mass feed.

Wet and dry residue. Ash can be removed by either wet or dry methods. Fly ash, along with other gaseous pollutants, should be removed by an air pollution control system. If a wet system is used, fly ash and gases will form a sludge if allowed to concentrate in the scrubber. Normally, however, scrubber blowdown will control the solids so that sludges do not form. Again, the composition of the scrubber wastewater stream depends directly on the wastes being incinerated. Blowdown is usually mixed with other liquid and undergoes further treatment and disposal.

If a baghouse or an electrostatic precipitator removes fly ash, a dry solid residue results.

Incineration residue handling. Bottom ash is discharged continuously or periodically. In a continuous system, the ash discharges into a dry ash bin or wet well. From this point, it is transferred to a container or truck, usually by means of drag conveyors. Lighter ash is collected at the outlet of the incinerator and is normally directed into a closed system.

Fly ash collected in a wet scrubber leaves the system with the scrubber blowdown and is further treated in a liquid waste treatment system.

Figure 1.40 (OME-88/12, p. 6-54) is a schematic of a solid waste incineration process including bottom ash collection and a fly ash scrubbing system.

Incineration operations may generate sludge from water/wastewater treatment and ash sludge if a wet ash removal system is employed. Sludges are normally dewatered to reduce the volume before landfilling. Sludge dewatering is usually accomplished by gravity.

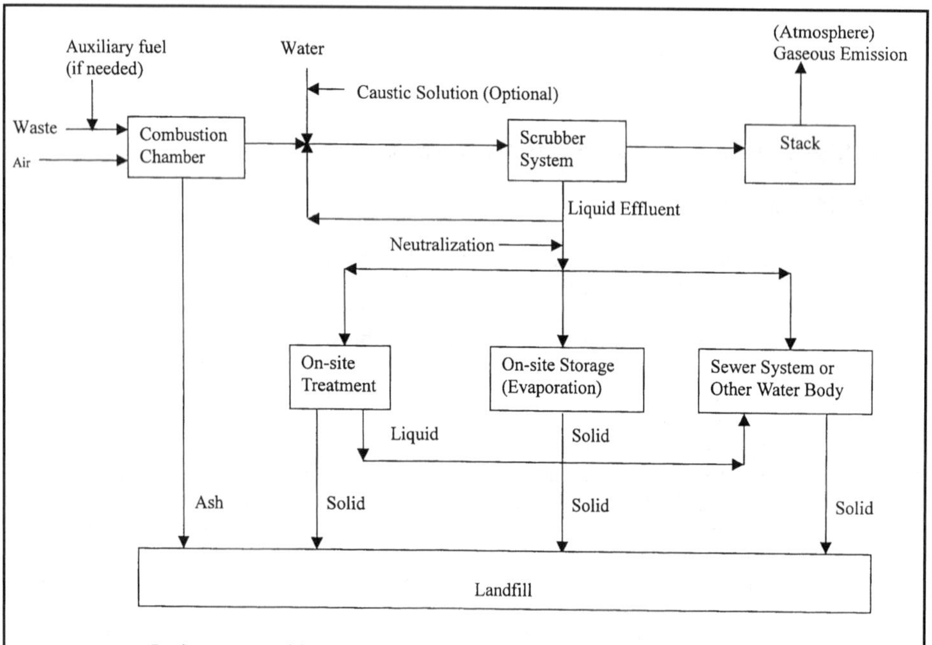

FIGURE 1.40 Incinerator residue treatement and disposal option (*OME-88/12, p. 6-54*).

5.2 Liquid Residue

Gaseous pollutants and particulates resulting from incineration can be captured by one of the following wet methods: (1) water in a quench tower, (2) water in a scrubber, or (3) alkaline liquid in a scrubber. The wet bottom quench tower, used mainly to lower the flue gas temperature, also functions as a scrubber by removing a small amount of particulate and gaseous pollutants from the incinerator exhaust. Water scrubbers are used for capturing the majority of gaseous pollutants in the combustion gas prior to atmospheric discharge. An alkaline scrubber is used to further reduce gaseous components that were not removed in the quench tower or water scrubber.

A water quench tower is used when a waste heat recovery system is not part of the incinerator design package. In addition to lowering the flue gas temperature, it operates like a single-pass scrubber. The quench tower effluent characteristics are highly variable, depending on incinerator operation, waste material being incinerated, quench tower feed rates, and scrubbing efficiency of the tower.

A wet scrubber with water as a medium is normally used downstream of the waste heat recovery system or water quench tower. Water scrubber effluents are quite different from quench tower effluents. Normally, scrubbers are recirculated systems with blowdown that controls the amount of total dissolved solids (TDS). The blowdown rate is variable, depending on the water feed rate and on the amount of acidic gases in the incinerated waste stream. Generally, blowdown occurs when the TDS content reaches approximately 3%. Water scrubber effluents are usually acidic in nature and may contain trace quantities of incinerated waste constituents and/or organics. Acidic effluents may be further processed for acid recovery or may be treated as a liquid waste. As a waste liquid, the stream may or may not be neutralized prior to mixing with other waste streams.

Alkaline (caustic) scrubbers are often used in conjunction with quench towers and/or water scrubbers. This type of scrubber would be used to remove acidic gases and other residual combustion products not captured in the water scrubbers. Alkaline scrubbers are usually recirculated systems with blowdown. Scrubber effluents are generally alkaline, since the scrubbing medium is used in excess to ensure that all noxious substances in the exhaust gas are removed prior to being discharged to the atmosphere. Alkaline scrubber effluents may or may not be neutralized prior to mixing with other liquid waste streams.

Other wastewater sources include waste heat recovery boiler blowdown, raw water treatment wastewaters, cooling tower blowdown, sludge dewatering liquid wastes, site runoff, and drainage from waste storage areas. All of these wastewaters are typically rich in suspended and dissolved solids and are of similar chemical makeup. Some streams may contain hazardous compounds and/or organics. Depending on the overall wastewater treatment program, it may be desirable to (1) separate some streams from the total combination of streams in order to treat them separately prior to disposal or (2) combine them with other wastewater streams.

6 ACCESSORY EOUIPMENT FOR AIR POLLUTION CONTROL

6.1 Duct

Ducts, or ductwork, transport exhaust gas to and from the scrubber. Ducts are carefully designed to keep pressure losses at a minimum. In general, this requires sizing the duct properly and minimizing the number of bends, expansions, and contractions. Sizing the duct to suit the exhaust stream velocity will generally reduce the amount of dust that settles in the ductwork. Bends, expansions, and contractions cause pressure loss in the system and, consequently, increase operating costs (EPA-84/03, p. 7-3; EPA-81/12, p. 9-14).

Abrasion and corrosion are common problems of ductwork. Abrasion is generally more severe on ductwork leading into the scrubber, while corrosion affects ductwork leaving the scrubber. Using proper construction materials or linings greatly reduces corrosion or abrasion. For example, ductwork can be lined partially or fully with brick (especially at elbows) to prevent erosion due to abrasion. For ductwork exiting the scrubber, special alloys resistant to acid attack should be used. Also, ductwork can be insulated to prevent acids in the flue gas from condensing.

In describing air flow through a system, a number of pressure terms are used. The three most important are

- Static pressure
- Velocity pressure
- Total pressure

Air flowing in a pipe is acted on by static and velocity pressure, which when added together give the total pressure (Fig. 1.41).

Any gas within a confined enclosure will have a static pressure, whether or not the gas is in motion. *Static pressure* p_s acts in all directions and is independent of the velocity of the gas. Therefore, static pressure is measured at right angles to the direction of air flow to avoid influence from the air velocity. In duct design, the friction losses in a pipe can sometimes be measured by the difference in static pressure (no change in direction or velocity). Therefore, the static pressure is sometimes called the *frictional or resistance pressure*. Static pressure can be positive or negative, compared with the local atmosphere.

The velocity pressure p_v is the pressure created by a fluid traveling at a specific velocity. Velocity pressure acts only in the direction of flow; therefore, it is always positive. The velocity pressure is obtained by measuring the total pressure and subtracting out the static pressure. There is a basic relationship between the velocity of a gas and the velocity pressure. For an air stream with a density of 0.075 lb/ft^3 (standard conditions) the air velocity can be computed from the velocity pressure by applying Bernoulli's equation:

$$v = (85.486 \text{ ft/s})[T_s \Delta p/(M_s P_s)]^{0.5}$$

where v = gas velocity, ft/s
T_s = temperature of the stack, °R
Δp = velocity pressure, in. H_2O
M_s = molecular weight of the stack gas, lb/lb-mol
P_s = absolute pressure of the stack, in. Hg

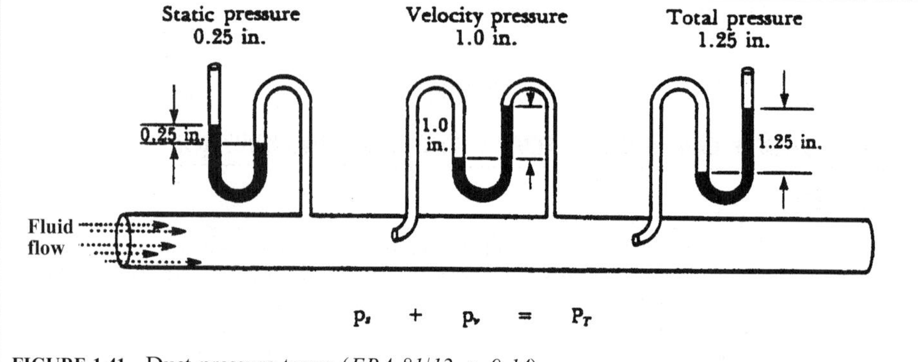

FIGURE 1.41 Duct pressure terms (*EPA-81/12, p. 9-14*).

Note: In stack sampling, the velocity pressure is symbolized by Δp, while in industrial ventilation it is p_v

The sum of the static and the velocity pressures at a given point in the duct is the total pressure P_T. Total pressure is measured in the direction of the flow. Therefore, the total pressure is sometimes referred to as the impact or dynamic pressure. The total pressure should not be confused with the absolute pressure. The absolute pressure is the sum of the atmospheric pressure plus the pressure of the system. The system pressure can be either positive or negative.

Pressure losses from air flowing in a pipe are due to friction and turbulence. Static or friction pressure losses are caused by air abrading or rubbing against the sides of the duct. Turbulent losses are due to rapid changes in direction or velocity. A change in duct cross-sectional area will change the velocity of the gas in the duct. The sum of the friction and turbulent losses over a specific length of pipe is termed the *pressure drop*. The pressure drop of a system is determined by measuring the difference in total pressure at two points in the system.

The friction and dynamic losses can be further subdivided into five categories. These losses are:

- *Inertial losses*: Energy required to accelerate a volume of air from rest. Inertia losses are essentially the same as the velocity pressure.
- *Orifice losses*: Pressure losses at a hood or duct entrance due to turbulence. Orifice losses are dependent on the shape of the opening and are measured by the entry coefficient C_e.
- *Straight duct friction losses*: Pressure losses due to air rubbing along the sides of the duct.
- *Elbow and branch entry losses*: Pressure losses due to change in the streamlines of an air stream moving through an elbow or entering a branched connection.
- *Contraction and expansion losses*: For gradual contractions in the cross-sectional area of a duct, the pressure losses are small. Abrupt contractions are rare in well-designed exhaust systems except for outlets from chambers. For duct expansions, abrupt enlargements result in a greater pressure loss.

6.2 Fan

Fan type. Fans transport exhaust gas through ducts to and from the scrubber, while pumps transport liquids through pipes (EPA-84/03, p. 7-2; EPA-81/12, p. 9-25).

In controlling air pollution, it is necessary to move the contaminants from their point of generation to the control device. Air and gas moving devices using rotary motion are normally termed *fans*. Fans are classified as *axial* or *centrifugal*, depending on the direction of air flow through the impeller. Two other terms used to describe air movers are *blowers* or *compressors*.

Distinguishing between air movers and fans is somewhat arbitrary and is based on the static pressure they are capable of producing. The term 'fan' is generally used to describe air movers with pressure rises up to about 2 psig. The term 'blower' refers to a centrifugal fan or positive displacement compressor operating with static pressures in the 2–15 psig range. For pressures higher than 15 psig, the term 'compressor' is used. Unlike fans, some blowers and compressors operate by piston movement (positive displacement) rather than rotation (EPA-81/12, p. 9-25).

Axial fan: In axial flow fans, air moves directly forward through the axis of rotation of the fan blades. A home window or room fan is an example of an axial flow fan. For industrial use, axial fans are best suited for moving large volumes of clean air with low static pressures.

Centrifugal fan: In centrifugal fans, the air is introduced into the center of a revolving wheel or rotor and exits at right angles to the rotation of the blades. Centrifugal fans are more widely used than axial fans, since they can handle dirtier air streams with higher system resistances.

Fans used in wet scrubbing systems are usually centrifugal. In centrifugal fans, exhaust gas is introduced into the center of a revolving wheel, or rotor, and exits at a right angle (90°) to the rotation of the blades (Fig. 1.42). Centrifugal fans are classified by the type and shape of blades used in the fan. They include:

- Forward-curved fan
- Backward-curved fan
- Radial fan
- Airfoil fan

Forward-curved fan: Forward-curved fans use blades that are curved toward the direction of the wheel rotation. The blades are small and spaced closer together than are the blades in other centrifugal fans. These fans are not usually used if the flue gas contains dust or sticky materials. They have been used for heating, ventilating, and air-conditioning applications in industrial plants.

Backward-curved fan: Backward-curved fans use blades that are curved away from the direction of wheel rotation. The blades will clog when the fan is used to move flue gas containing dust and sticky fumes. They may be used on the clean-air discharge of air pollution control devices or to provide clean combustion air for boilers.

Radial fan: Radial fans use straight blades that are attached to the wheel of the rotor. These fans are built for high mechanical strength and can be easily repaired. Fan blades may be constructed of alloys or coated steel to help prevent deterioration when handling abrasive and corrosive exhaust gas. Radial fans are used most frequently for air pollution control applications; however, backward-curved fans are also used on wet scrubbing systems.

Airfoil fan: Airfoil fans use thick teardrop-shaped blades that are curved away from the wheel rotation. Airfoil fans can clog when handling dust or sticky materials.

Fans used for wet scrubbing systems can be located before or after the scrubber. When located before the scrubber, they are referred to as *forced draft*, *positive pressure*, or *dirty-side fans*. These fans normally move dry air, but can move moist air depending on process conditions. They are subject to abrasion and solids buildup when dust concentration is high. Abrasion on the fan can be reduced by using special wear-resistant alloys, by using replaceable liners on the wheel, or by reducing fan speed (using a large fan that moves slower). The solids buildup can sometimes be controlled by using a spray wash to periodically clean the wheel. If dirty-side fans are used, a cyclone or knockout chamber can reduce dust concentration.

Fans located after the scrubber are always operated wet and are called *induced draft*, *negative pressure*, or *clean-side fans*. These fans are subject to corrosion and solids buildup from mist escaping from the entrainment separator. Corrosion problems can result when the exhaust gas contains acid-forming or soluble electrolytic compounds, especially if the temperature of the gas stream falls below the dew point of these compounds. Corrosion can be reduced by using proper construction materials and careful pH control in the scrubbing system. Solids buildup can occur when the mist escaping from the entrainment separator contains dissolved or settleable solids. As the mist enters the fan, evaporation occurs and some solids deposit on the wheel. Keeping entrainment separators operating efficiently or using clean water sprays on the fan blades will help reduce solids buildup problems.

Fan power. In specifying the proper size fan, it is desirable to know the horsepower requirement for the system. Two terms are used when describing horsepower: *air horsepower (AHP)* and *brake horsepower (BHP)*. Air horsepower is the power output of a fan if the fan were 100% efficient. Since no fan is 100% efficient, the more important term is the brake horse-

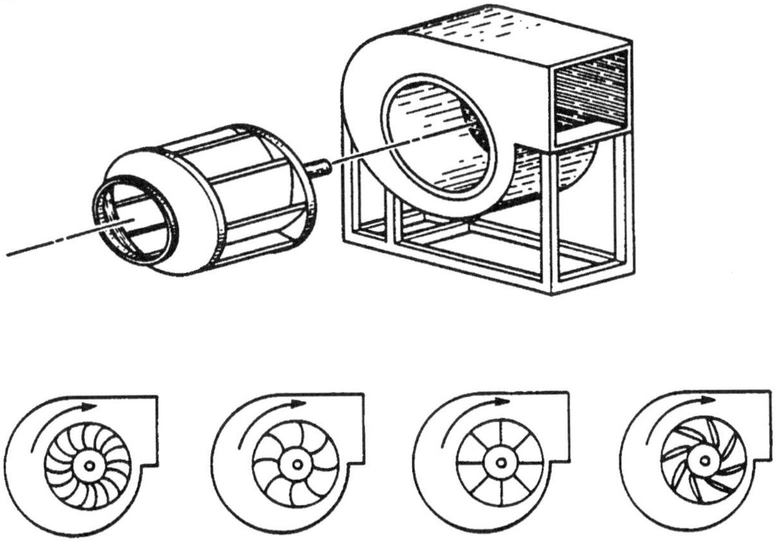

FIGURE 1.42 Centrifugal fan (*EPA-84/03, p. 7-2*).

power. Brake horsepower equals the air horsepower times the efficiency of the fan. Knowing the efficiency of the fan, brake horsepower can be calculated from

$$\text{BHP} = QP_T/(6356E)$$

where BHP = brake horsepower
Q = air flow, cfm
P_T = total pressure of the system, in. H_2O
E = mechanical efficiency of the fan, %

Fan laws. Fan laws are used to predict how a change in one of the performance variables (size, speed, air density, volume, system pressure, power, efficiency and sound level) affects the others. Fan laws are based on the premise that if two fans are geometrically similar (homologous), their performance curves have the same shape. Therefore, if the two fans operate at the same point of rating, their efficiencies are equal and the ratio of all other variables are interrelated.

The fan laws can be written in as many as 10 different ways, depending on which variables change. The following relationships are some of the more useful of the fan laws.

For a change in fan speed ($\text{rpm}_1 \rightarrow \text{rpm}_2$):

- the flow rate (Q) varies directly with the fan speed ratio

  $Q_1/Q_2 = \text{rpm}_1/\text{rpm}_2$

- the power (P) varies with the cube of the fan speed ratio

  $P_1/P_2 = (\text{rpm}_1/\text{rpm}_2)^3$

- the static pressure (p_s) varies with the square of the fan speed ratio

$p_{s1}/p_{s2} = (\text{rpm}_1/\text{rpm}_2)^2$

For a change in fan size (wheel diameter $d_{w1} \rightarrow d_{w2}$):

- the flow rate (Q) varies with the cube of the wheel diameter

  $Q_1/Q_2 = (d_{w1}/d_{w2})^3$

- the power (P) varies with the fifth power of the wheel diameter

  $P_1/P_2 = (d_{w1}/d_{w2})^5$

- the horsepower changes with the fifth power of the fan wheel diameter and the cube of the fan speed

  $P_1/P_2 = (d_1/d_2)^5(\text{rpm}_1/\text{rpm}_2)^3$

- the static pressure (p_s) varies with the square of the wheel diameter

  $p_{s1}/p_{s2} = (d_{w1}/d_{w2})^2$

For a change in gas density ($\rho_1 - \rho_2$)

- the flow rate is constant

  $\rho_1/\rho_2 = Q$

- the power varies directly with the density ratio

  $P_1/P_2 = \rho_1/\rho_2$

- the static pressure (p_s) varies with the density ratio

  $p_{s1}/p_{s2} = \rho_1/\rho_2$

The fan laws can be used to construct performance curves or to determine new operating conditions as process variables change. It is important to note that these fan laws are based on the restriction that the fan at the new conditions will operate at the same point of rating (mechanical efficiency). Therefore, the complete set of operating conditions should be calculated for any change.

EXAMPLE 1: fan laws

A fan operating at a speed of 1474 rpm delivers 10,200 cfm at 4 in. static pressure and requires 8.85 brake horsepower (BHP).

What will be the new operating conditions if the fan is speeded up to 2000 rpm?

Solution:

The new flow is:

$$\begin{aligned} Q_1/Q_2 &= \text{rpm}_1/\text{rpm}_2 \\ Q_2 &= (10{,}200\,\text{cfm})[(2000\text{rpm})/(1474\,\text{rpm})] \\ &= 13{,}840\,\text{cfm} \end{aligned}$$

The new static pressure of the system is:

$$\begin{aligned} p_{s1}/p_{s2} &= (\text{rpm}_1/\text{rpm}_2)^2 \\ p_{s2} &= (4\text{ in.})(2000/1474)^2 \\ &= 7.4\text{ in.} \end{aligned}$$

The horsepower required is:

$$\begin{aligned} P_1/P_2 &= (\text{rpm}_1/\text{rpm}_2)^3 \\ P_2 &= 8.85\,\text{hp}\ (2000/1474)^3 \\ &= 22.1\,\text{hp} \end{aligned}$$

Note the drastic increase in horsepower for increasing the fan speed.

EXAMPLE 2: fan laws

The exhaust system for a rotary dryer is designed to deliver 12,000 cfm of air at 600°F and 4 in. H_2O static pressure. Fan speed is 630 rpm and requires 13 horsepower. If the dryer has been down for repairs, compute the horsepower required to pull the same amount of air at ambient conditions.

Given conditions

density of air at 70°F = 0.075 lb/ft^3
density of air at 600°F = 0.0375 lb/ft^3

Solution:

The horsepower is:

$$P_1/P_2 = \rho_1/\rho_2$$
$$P_2 = 13\,\text{hp}\,(0.075/0.0375)$$
$$= 26\,\text{hp}$$

Note: this would require a motor twice the size as needed for normal operation. An alternative would be to install dampers until the system reaches operating temperature.

EXAMPLE 3: fan (effect of fan wheel and speed)

You are requested to detemine the power requirement of fan "B" which is from the same homologous series as fan "A." The speed (rpm), blade diameter (D), capacity (acfm), and the power (bhp) of fan "A" are given. Fan "B" delivers a gas having the same density as fan "A." (EPA-84/09, p. 120).

Given conditions

Fan "A"

Speed = 1622 rpm
Blade diameter = 46 in.
Gas delivered = 15,120 acfm
Power (brake) = 45.9 bhp

Fan "B"

Speed = 1590 rpm
Blade diameter = 50 in.

Solution:

1. Calculate the power requirement of fan "B"

There are many fan laws. One that relates horsepower with fan wheel (diameter) and speed (rpm) is given below

$$P_B/P_A = (d_B/d_A)^5(N_B/N_A)^3$$

where P_A, P_B = horsepower of fan "A" and "B"
d_A, d_B = blade diameter of fan "A" and "B"
N_A, N_B = speed (rpm) of fan "A" and "B"

Remember that the brake horsepower represents the power the user pays for.

$$P_B/P_A = (d_B/d_A)^5(N_B/N_A)^3$$
$$= (50/46)^5(1590/1622)^3$$
$$= 1.429$$

$$P_B = (1.429)(45.9)$$
$$= 65.6 \text{ bhp}$$

EXAMPLE 4: fan (brake horsepower requirement)
You are requested to calculate the horsepower required for a fan treating a gas stream through a packed tower. The pressure drop across the packing, pressure loss for the duct work, elbows, valves, etc., and overall fan-motor efficiency are given (EPA-84/09, p. 121).

Given conditions

Volumetric flow rate of gas stream = 6000 acfm
Pressure drop across the packing = 6.0 in. H_2O
Pressure loss for the duct work, elbows, valves, etc., and expansion/contraction = 4.0 in. H_2O
Overall fan-motor efficiency = 63%

Solution:

1. Calculate the total pressure drop in inches of H_2O
The total pressure drop is the sum of the pressure drop across the packing and pressure loss due to the duct work, elbows, valves, expansion/contraction, etc.

$$\text{Total pressure drop} = 6.0 + 4.0$$
$$= 10.0 \text{ in. } H_2O$$

2. Calculate the brake horsepower (bhp) required to treat the gas stream in bhp.
The brake horsepower represents the power the user pays for, i.e. it is the power to operate the fan. The term gas or air horsepower is used to describe the power delivered to the gas.

$$\text{Brake horsepower} = (Q)(\Delta p)(1.573 \times 10^{-4})/\eta_f$$

where Q = volumetric flow rate of gas stream, acfm
Δp = pressure drop, in. H_2O
η_f = fan efficiency, usually in the 0.5–0.65 range

$$\text{Brake horsepower} = (6000)(10.0)(1.573 \times 10^{-4})/(0.63)$$
$$= 15 \text{ bhp}$$

6.3 Hood

Control of many air pollutants begins at their point of capture, the hood. The overall control efficiency that can be achieved is a function of the capture efficiency of the hood. For example, if a poorly designed hood captures only 70% of the emissions generated, then the overall efficiency of the system can never be greater than 70% even if the downstream control device is 100% efficient. Any pollutant can be captured effectively with an appropriately sized hood and a fan with enough power to supply the suction. Hoods are traditionally designed, however, to achieve the maximum pollutant capture efficiency with a minimum of air flow and power consumption (EPA-81/12, p. 9-1).

Hoods are generally classified according to their shape and the manner in which they capture the contaminant. Hoods are referred to as the enclosure and non-enclosure types.

Some common hood shapes are shown in Fig. 1.43; they include slot, plain opening, booth, and canopy.

Hood	Description	Entry coefficient (C_e)
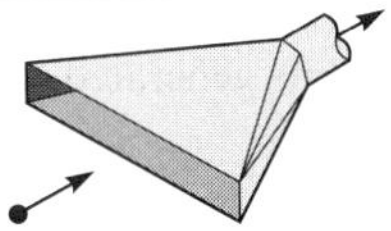	Slot	0.60
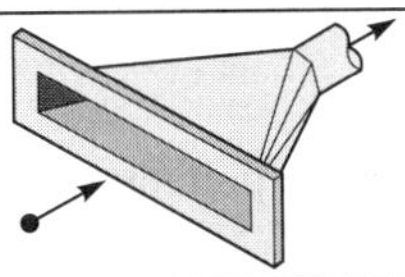	Flanged slot	0.82
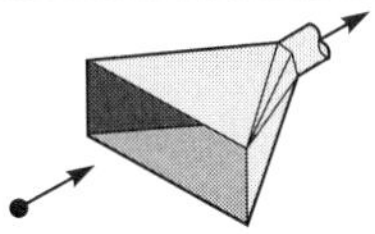	Plain opening	0.72
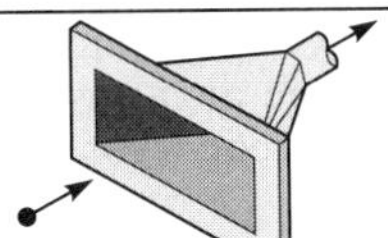	Flanged opening	0.82
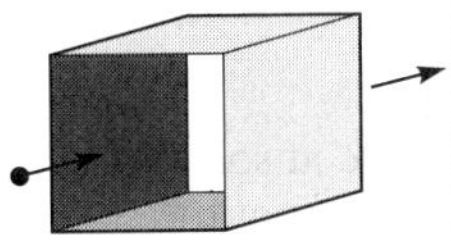	Booth	0.82
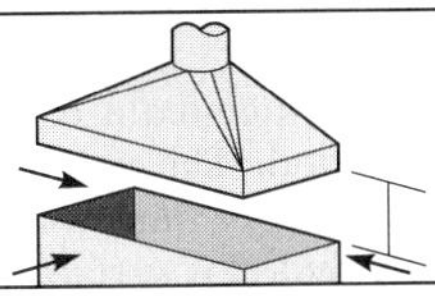	Canopy	0.82

FIGURE 1.43 Hoods and their entry coefficients (*EPA-81/12, p. 9-1*).

The air volume or suction required for a hood depends on hood shape, hood type, hood size, required capture velocity, distance of the hood from the source of pollutants, and temperature of the contaminant exhaust stream. Due to the numerous variables, equations for predicting the required exhaust volume are based on empirical data for a particular type of hood.

When air flows into a hood or duct, pressure losses occur, which result in a decreased flow rate. These losses are due to turbulence and are dependent on the shape of the hood or duct. The coefficient of entry C_e indicates the extent of these losses. For example, in a theoretically perfect hood with no turbulence loss, $C_e = 1.0$. Figure 1.43 lists the entry coefficients for some typical hoods. The coefficient of entry is used to determine the actual flow rate and pressure after the air enters the hood (EPA-81/12, p. 9-14).

Enclosure hood. Enclosure hoods may totally or only partially enclose the source of emissions. The term 'enclosure' is somewhat of a misnomer since some of these hoods can be located close to the point of emissions so that capture occurs before dispersion of the contaminant can take place. For example in many paint-spraying operations, a booth is used to capture the emissions generated. Occasionally, the painting is done inside the booth or it may be done a few feet in front of the booth. Both of these operations would be an example of an enclosure hood (EPA-81/12, p. 9-1).

For any totally enclosed hood or for emissions generated inside a booth, the air flow into the hood can be computed by

$$Q = vA \tag{1.62}$$

where Qair flow (into hood face), cfm

v = air velocity required to capture the pollutant, fpm
A = total open area of the hood, ft^2

The capture velocity is the velocity necessary (at any point) to overcome opposing air currents and to capture the air contaminants.

Non-enclosure hood. Non-enclosure hoods refer to hoods that are usually placed around the perimeter of the contaminant generating source. The most common non-enclosure hood used to capture gaseous emissions is the lip or rim exhaust located on open surface tanks.

For a free-standing or unbounded hood where air flows into a plain, circular duct or a rectangular hood, Eq. (1.62) can be modified to account for the distance a source is from the hood:

$$Q = v_x(10x^2 + A) \tag{1.63}$$

where Q = air flow (into hood face), cfm
v_x = air velocity required for capture at some distance (x) from hood opening, fpm
x = distance of contaminant source from the center of the hood face, ft
A = area of hood face, ft^2

Equation (1.63) shows the rapid increase in air volume required for capture (due to decrease in velocity) as the source gets farther from the hood. The volume increases in relation to the distance squared. It is applied to a free-standing or unobstructed hood. For a square or rectangular hood that is bounded on one side by a flat surface (the floor or wall), Eq. (1.63) then becomes

$$Q = v_x(10x^2 + A)/2 \tag{1.64}$$

EXAMPLE: non-enclosure hood

Determine the total volume of air entering a paint spray booth that is 10 ft wide and 7 ft high. Some objects are painted outside the booth but never at a distance of more than 3 ft. Room air currents are kept low (EPA-81/12, p. 9-6).

Solution:

The paint spray booth would be a plain opening bounded on one side by a flat surface (the floor). Therefore, Eq. (1.64) applies.

$$\begin{aligned} Q &= v_x(10x^2 + A)/2 \\ &= 100[10(3^2) + 2(7)(10)]/2 \\ &= 11{,}500 \text{ cfm} \end{aligned}$$

An important point to note is that, in order to capture the pollutant, the velocity at 3 ft from the hood face must be at least 100 fpm. If painting was done farther away from the hood face, the velocity still must be the same. The volume of air would then greatly increase. For example, if painting is done at 5 ft instead of 3 ft from the hood, then this raises the required air volume pulled into the system to 19,500 cfm.

A common way of expressing control volume is based on the square feet of hood opening. For example, assuming that the hood dimensions are the same as the paint booth, we have

$$
\begin{aligned}
Q/A &= (11{,}500 \text{ cfm})/[(7 \text{ ft})(10 \text{ ft})] \\
&= 164.3 \text{ cfm of air per square foot of hood opening}
\end{aligned}
$$

Non-enclosure lip hood. One of the most common types of non-enclosure hoods for controlling vapors from open surface tanks is the lip hood pictured in Fig. 1.44 (these hoods are sometimes called *slot hoods*). These hoods are rarely used with hot gases. The lips are designed for a velocity of approximately 200 fpm through the lip face at the required ventilation rate (EPA-73).

For a tank with two parallel lip hoods, the ventilation rate required and the lip width may be estimated from Fig. 1.45. If a lip hood is used on only one side of a tank and the opposite side of the tank is bounded by a vertical wall, Fig. 1.45 can be used to estimate the ventilation rate. The procedure used below is to double the tank width and read a ventilation rate from Fig. 1.45. The actual ventilation rate is estimated as one-half the value read from the graph. The following example illustrates the use of Fig. 1.45.

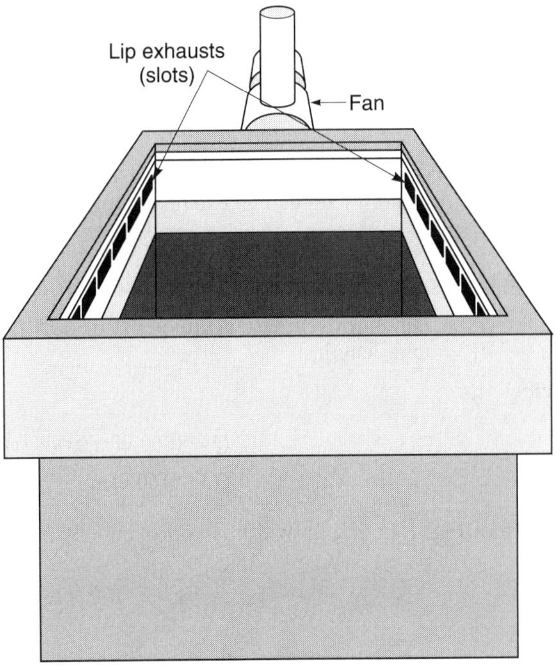

FIGURE 1.44 Open top tank degreaser with lip exhausts (*EPA-81/12, p. 9-12*).

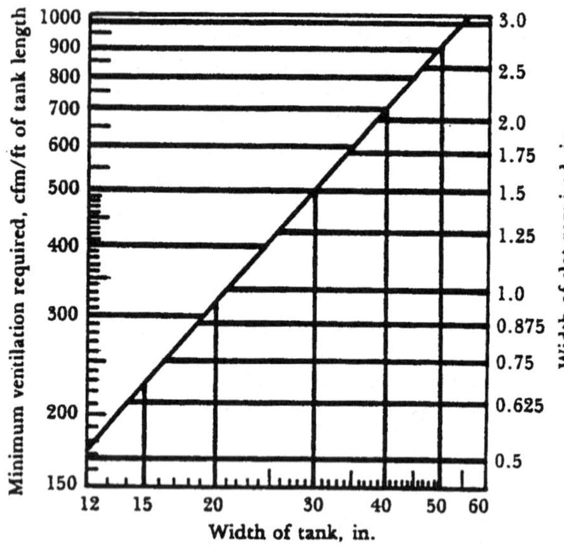

FIGURE 1.45 Minimum ventilation rates for lip exhaust hoods (*EPA-81/12, p. 9-13*).

EXAMPLE: non-enclosure lip hood
A degreasing tank 6 ft long and 4 ft wide (width) is controlled by using parallel lip hoods along each of the 6 ft lengths. Determine the total exhaust rate required and the width of the lips (EPA-81/12, p. 9-13).

Solution:
From Fig. 1.45 and the given tank width, 4 ft (48 in.), the ventilation rate can be read as 860 cfm per ft of tank length. The exhaust volume is

$$Q = Q_v l_T$$

where Q = exhaust volume, cfm
Q_v = minimum ventilation rate per ft of tank length, cfm/ft
l_T = tank length, ft

Then

$$Q = (860 \text{ cfm/ft})(6 \text{ ft})$$

$$Q = 5160 \text{ cfm}$$

Also, from Fig. 1.45, the lip width is approximately 2.6 inches.

6.4 Pipe

Pipes transport liquid to and from the scrubber. As with ducts, pipes are susceptible to abrasion, corrosion, and plugging. A wide variety of materials can be used to make pipes to reduce these problems (EPA-84/03, p. 7-4).

6.5 Pump

A wide variety of pumps are used to transport both the scrubbing liquid and the sludge. The proper choice of a pump depends on flow rate, pressure, temperature, and material being pumped. Electric-motor-driven centrifugal pumps are the pumps most frequently used in wet scrubbing systems. The rotating impeller produces a reduction in pressure at the eye (center) of the impeller, causing liquid to flow into the impeller from the suction pipe. The liquid is then forced outward along the blades and discharged (EPA-84/03, p. 7-3).

As with fans, abrasion and corrosion are the major maintenance problems associated with pumps in scrubbing systems. The impeller housing, and seals are subject to potential corrosion and abrasion problems. Abrasion is caused by solids buildup in the scrubbing liquid. Bleeding this liquid and removing the solids before recycling it back through the pump (or scrubber) will reduce pump wear. Most vendors suggest that the solids content be less than 15%. Special alloys or rubber linings can also be used to help reduce abrasion and corrosion.

7 MONITORING EOUIPMENT

Having adequate equipment is imperative when monitoring the performance of a scrubber. Instrumentation on a wet scrubber can provide three distinct services:

- Obtaining operational information by recording daily data to help detect any problems or misoperation that may occur.
- Providing operating input for other devices to automatically operate some parts of the system.
- Providing for safety by sounding alarms and/or releasing interlocks to protect both the operators and equipment.

A monitoring system must be properly installed and maintained to provide reliable data. Monitors should be installed, operated, and calibrated according to the manufacturer's instructions. This is essential in obtaining reliable information. Because every scrubbing system is unique, the instrumentation and variables measured will vary from source to source. Table 1.22 lists monitors that are typically used in wet scrubbing systems.

For any of these monitors, high and/or low settings can be chosen so that if the set value is exceeded, an alarm sounds, a bypass is opened, or an emergency system is activated. For example, sources scrubbing hot gases normally have a high-temperature alarm and/or an interlock system to automatically introduce emergency water or to bypass the scrubber if the high-temperature setting is exceeded.

TABLE 1.22 Monitoring Equipment for Wet Scrubbing Systems

Monitor	Measurements
Thermometer or thermocouple	Measures inlet and outlet temperatures of gas to and from scrubber. Measures inlet and outlet temperatures of liquid to and from scrubber
Flowmeter	Measures liquid flow rate to scrubber. Measures the amount of recycled liquid and bleed stream. Measures flow rate of fresh makeup liquid to scrubber
Manometer	Measures pressure drop (inlet and outlet static pressure) across fan, scrubber vessel, and entrainment separator
pH meter	Measures pH level in chemical feed stream, scrubbing liquid, recycle liquor, and bleed stream
Ammeter	Monitors the current of the fans and pumps

8 RECORD KEEPING

Keeping good records of instrument readings and operating practices is important because it (EPA-89/03):

- Helps an operator detect problems and develop corrective actions
- Provides proof that an operator is properly operating and maintaining equipment
- Allows an operator to prepare accurate annual (or more frequent) reports that may be required by state regulations

Certain types of records are commonly required by state regulations or operating permits. Most of them are listed below and involve recording the levels indicated on automatic monitoring devices periodically or require recording the parameters continuously. Example record-keeping items are provided below.

Incinerator record keeping includes:

- Temperature of incinerator chamber(s)
- Oxygen concentration of exhaust gas
- Temperature at inlet and/or outlet of control device
- Continuous emission monitoring records (carbon monoxide or opacity)
- Weight of waste charged to incinerator

Air pollution control equipment record keeping includes:

- Scrubber: (1) pressure drop and (2) liquid flow rate
- Fabric filter: pressure drop

REFERENCES

40CFR50, *National primary and secondary ambient air quality standards*, Title 40 Part 50 of Code of Federal Regulations (CFR).

40CFR51.100-90, *Definition*, Title 40 Parts 51.100 of Code of Federal Regulations published 1990.

40CFR60, *Standard for sulfur dioxide*, Title 40 Parts 60.43a of Code of Federal Regulations.

Bennett-82/01, *Momentum, heat, and mass transfer*, 3rd edn, Bennett, C. O. and Myers, J. E. (Eds). New York: McGraw-Hill Book Company, pp. 638, 1982.

Chang-82/03, Effects of organic acid additives on SO_2 absorption into $CaO/CaCO_3$ slurries, Chang, J. C. S. and Rochelle, G. T., *AIChE J.*, **28**(2), 261–66, 1982.

Chang-86/11, Gypsum crystallization for limestone FGD process, Chang, J. C. S. and Brna, T. G., *Chem. Engng Prog.*, **82**(11), 57–62, 1986.

Chang-98/07, Private communication materials, Chang, J. C. S., July 1998.

EPA-73, *Air pollution engineering manual*, Danielson, J. A., Research Triangle Park, NC, US Environmental Protection Agency.

EPA-73/05, *Air pollution engineering manual*, 2nd edn, AP-40, Air Pollution Control, District County of Los Angeles, May 1973.

EPA-78/10, *Feasibility of producing and marketing byproduct gypsum from SO_2 emission control at fossil-fuel-fired power plants*, EPA600-7-78-192, October 1978.

EPA-79/11, *EPA alkali scrubbing test facility: advanced program, fourth progress report*, EPA-600-7-79-244, November 1979.

EPA-79/09, *Acid rain*, EPA600-9-79-036, July 1980.

EPA-80/02, *Combustion evaluation—Student manual, course 427*, EPA Air Pollution Training Institute (APTI), EPA450-2-80-063, February 1980.

EPA-80/08, *Research summary, controlling sulfur oxides*, EPA600-8-80-029, August 1980.

EPA-81/05, *Control of gaseous emissions, course 415*, US EPA Air Pollution Training Institute (APTI), EPA450-2-81-006, May 1981.

EPA-81/08, *Limestone FGD scrubbers: user's handbook*, Henzel, D. S., Laseke, B. A., Smith, E. O., and Swenson, D. O., EPA-600/8-81-017, August 1981.

EPA-81/10, *Control of particulate emissions, course 413*, US EPA Air Pollution Training Institute (APTI), EPA450-2-80-066, October 1981.

EPA-81/12, *Control of gaseous emissions, course 415*, US EPA Air Pollution Training Institute (APTI), EPA450-2-81-005, December 1981.

EPA-84/03, *Wet scrubber plan review, course SI:412C*, US EPA Air Pollution Training Institute (APTI), EPA450-2-82-020, March 1984.

EPA-84/09, *Control of gaseous and particulate emissions, course Si:412D*, US EPA Air Pollution Training Institute (APTI), EPA450-2-84-007, September 1984.

EPA-89/03, *Hospital incinerator operator training course: volume I, student handbook*, EPA-450/3-89-003, March 1989.

EPA-92/01, *Survey of medical waste incinerators and emissions control*, Final draft report prepared by EER Corporation for California Air Resources Board and U.S. EPA, January 1992.

Molburg-80/02, A graphical representation of the new NSPS for sulfur dioxide, Molburg, J., *J. Air Pollution Control Assoc.*, **30**(2), 172, 1980.

OME-88/12, *Guidance for incinerator design and operation, volume I*, Ontario Ministry of the Environment (OME), December 1988.

OTA-84/06, *Acid rain and transported air pollutants—Implications for public policy*, Congress of the United States, Office of Technology Assessment (OTA), June 1984.

Perry-73, *Perry's chemical engineer's handbook*, Perry, R. H. New York: McGraw-Hill.

CHAPTER 3.2
PARTICULATE EMISSION CONTROL

C. C. Lee and G. L. Huffman
U.S. Environmental Protection Agency, Cincinnati, Ohio

1 INTRODUCTION

Particle or particulate matter is defined as any finely divided solid or liquid material, other than uncombined water, emitted to the ambient air as measured by applicable reference methods, or an equivalent or alternative method, or by a test method specified in 40CFR50.

2 BASICS OF PARTICULATE EMISSION CONTROL

2.1 Collection Forces

The basic collection forces which govern the control of particulate emissions include the following mechanisms:

- Collection by gravity
- Collection by centrifugal force

- Collection by impaction, interception, and diffusion
- Collection by electrostatic force

Particle collection can occur from an individual effect or a combined effect of the mechanisms mentioned above. In addition, particles can agglomerate or grow in size by cooling, increasing humidity, or from electrostatic effects. Agglomerated particles thus have a larger aerodynamic diameter and can be collected more easily by impaction, interception, or gravitational forces.

Collection by gravity: All particulate emission control devices collect particles by mechanisms involving an applied force, the simplest being gravity. Large particles moving slowly enough in a gas stream can be overcome by gravity and be collected. Such a phenomenon is shown in Fig. 2.1. Gravity is responsible for particle collection in the simplest devices, such as settling chambers.

Collection by centrifugal force: Centrifugal force is another collection mechanism used for particle capture. The shape or curvature of the collector causes the gas stream to rotate in a spiral motion. Larger particles move toward the outside of the wall by virtue of their momentum. Such a phenomenon is shown in Fig. 2.2. The particles lose kinetic energy there and are separated from the gas stream. Particles are then overcome by gravitational force and are collected. Centrifugal and gravitational forces are both responsible for particle collection in a cyclone.

Collection by impaction, interception, and diffusion: In both fabric filters and wet collectors, three separate forces are responsible for particle collection: impaction, direct interception, and diffusion. In a fabric filter, the target object for particle capture is a stationary fiber. In a wet collector, the target object is a water droplet that is introduced into the gas stream, generally through a nozzle.

Consider the case of an individual fiber in a fabric filter. Impaction occurs when the particle is so large that it cannot follow the gas streamlines around the stationary fiber and impacts on the fiber, as shown in Fig. 2.3. Direct interception is a special case of the impaction mechanism. The center of a particle may follow the streamlines formed around the fibers. A collision will occur if the distance between the particle center and the collection surface is less than the particle radius, as shown in Fig. 2.4. Particles below 0.1 μm (micrometer) in aerodynamic diameter undergo Brownian motion, randomly moving or diffusing throughout the gas volume (1 μm = 1 micron = 10^{-6} m). The mechanism of diffusion is responsible for the collection of particles that are so small that they become affected by collisions of molecules in the gas stream. The randomly moving particles then move or diffuse through the gas to impact on the fiber and are collected as shown in Fig. 2.5.

Collection by electrostatic force: The other primary particle collection mechanism involves electrostatic forces. The particles can be naturally charged, or, as in most cases involving electrostatic attraction, be charged by subjecting the particle to a strong electric field. The charged particles migrate to an oppositely charged collection surface, as shown in Fig. 2.6. This is the collection mechanism responsible for particle capture in both electrostatic

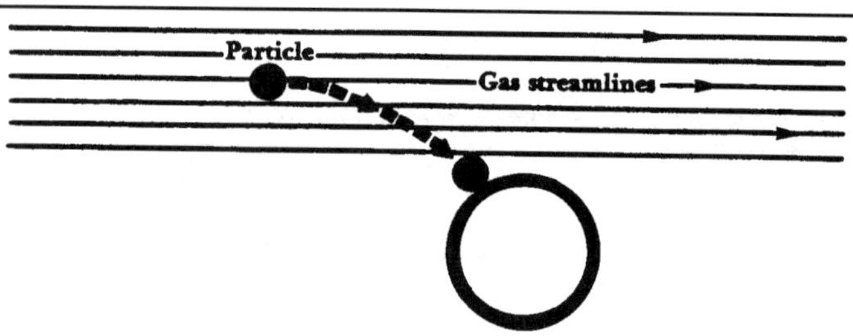

FIGURE 2.1 Particle collection by gravity (*EPA-81/10, p. 1-3*).

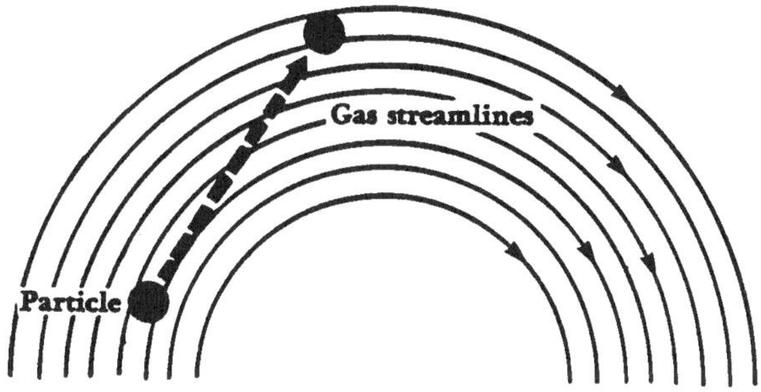

FIGURE 2.2 Particle collection by centrifugal force (*EPA-81/10, p. 1-3*).

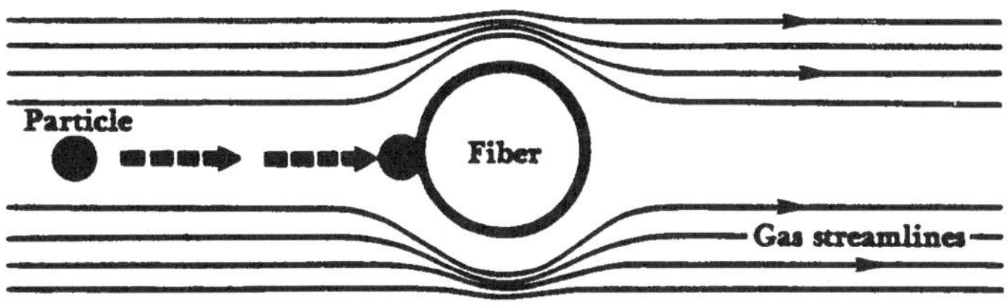

FIGURE 2.3 Particle collection by impaction (*EPA-84/03, p. 1-4*).

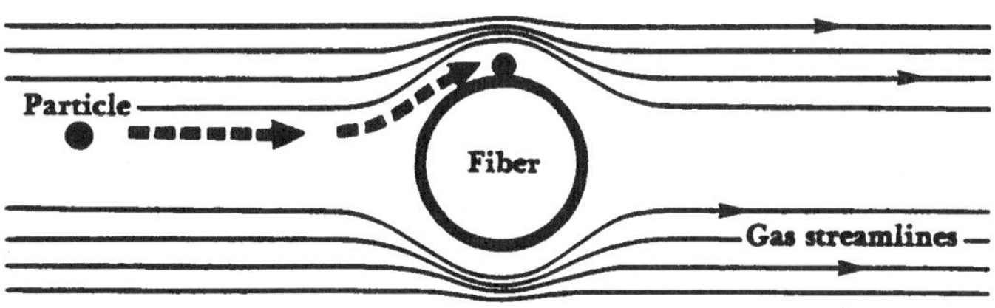

FIGURE 2.4 Particle collection by interception (*EPA-81/10, p. 1-4*).

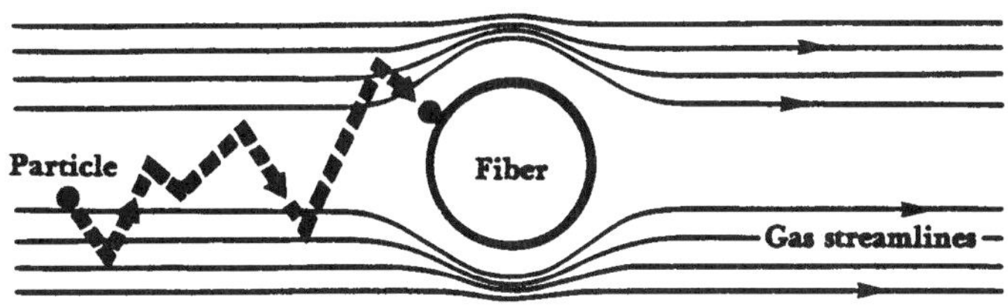

FIGURE 2.5 Particle collection by diffusion (*EPA-84/03, p. 1-5*).

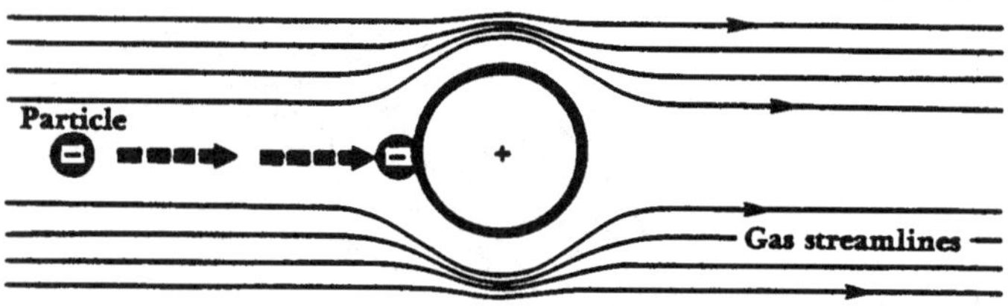

FIGURE 2.6 Particle collection by electrostatic attraction (*EPA-81/10, p. 1-5*).

precipitators (ESPs) and charged droplet scrubbers. In an ESP, particle collection occurs due to electrostatic forces only. In a charged droplet scrubber, particle removal occurs by the combined effects of impaction, direct interception, diffusion, and electrostatic attraction. Particles are charged in these scrubbers to enhance both diffusion and direct interception.

2.2 Particle Dynamics

Collection of solid or liquid particles in an air pollution control device is based upon the movement of a particle in the gas (fluid) stream. For a particle to be captured, the particle must be subjected to external forces large enough to separate it from the gas stream. Forces acting on a particle include three major forces and other forces. They are (EPA-81/10, p. 3-1):

- Gravitational force
- Buoyant force
- Drag force
- Other forces, including magnetic, inertial, electrostatic, and thermal forces

The consequence of acting forces on a particle results in the terminal velocity for a particle to settle. The terminal velocity (also known as the settling velocity) is a constant value of velocity reached when all forces (gravity, drag, buoyancy, etc.) acting on a body are balanced. The sum of all the forces is then equal to zero (no acceleration). To solve for an unknown particle settling velocity, the flow regime of particle motion must be determined. Once the flow regime has been determined, the settling velocity of a particle can be calculated.

The flow regime can be determined by the following equation (EPA-81/10, p. 3-10):

$$K = d_{\rm p}(g\rho_{\rm p}\rho_{\rm a}/\mu^2)^{0.33}$$

where K = a dimensionless constant which determines the range of the fluid-particle dynamic laws
$d_{\rm p}$ = particle diameter, cm or ft
g = gravity force, cm/s^2 or ft/s^2
$\rho_{\rm p}$ = particle density, g/cm^3 or lb/ft^3
$\rho_{\rm a}$ = fluid (gas) density, g/cm^3 or lb/ft^3
μ = fluid (gas) viscosity, g/cm-s or lb/ft-s

The K values corresponding to different flow regimes appear below (EPA-81/10, p. 3-10):

- Laminar regime (also known as Stokes' law range): $K < 3.3$
- Transition regime (also known as intermediate law range): $3.3 < K < 43.6$
- Turbulent regime (also known as Newton's law range): $K > 43.6$

The K value determines the appropriate range of the fluid-particle dynamic laws which apply.

For a laminar regime (Stokes' law range), the terminal velocity is (EPA-81/10, p3-10):

$$v = g\rho_p(d_p)^2/(18\mu)$$

For a transition regime (intermediate law range), the terminal velocity is (EPA-81/10, p. 3-10):

$$v = 0.153g^{0.71}(d_p)^{1.14}(\rho_p)^{0.71}/[\mu^{0.43}(\rho_a)^{0.29}]$$

For a turbulent regime (Newton's law range), the terminal velocity is (EPA-81/10, p. 3-10):

$$v = 1.74(gd_p\rho_p/\rho_a)^{0.5}$$

When particles approach sizes comparable with the mean free path of the fluid molecules, the medium can no longer be regarded as continuous, since particles can fall between the molecules at a faster rate than predicted by aerodynamic theory. To allow for this slip, Cunningham's correction factor is introduced into Stoke's law (EPA-84/09, p. 58):

$$v = g\rho_p(d_p)^2C_f/(18\mu)$$

where C_f = Cunningham correction factor $= 1 + (2A\lambda/d_p)$
A = $1.257 + 0.40e^{-1.10}d_p/2\lambda$
λ = free path of the fluid molecules (6.53×10^{-6} cm for ambient air)

EXAMPLE 1: particle terminal velocity. Calculate the settling velocity of a particle moving in a gas stream. Assume the following information (EPA-81/10, p. 3-11):

Given conditions

d_p = particle diameter = 45 μm (45 microns)
g = gravity force = 980 cm/s^2
ρ_p = particle density = 0.899 g/cm^3
ρ_a = fluid (gas) density = 0.012 g/cm^3
μ = fluid (gas) viscosity = 1.82×10^{-4} g/cm-s
C_f = 1.0 (if applicable)

Solution:

1. Calculate the K parameter to determine the proper flow regime.

$$\begin{aligned} K &= d_p(g\rho_p\rho_a/\mu^2)^{0.33} \\ &= 45 \times 10^{-4}[980(0.899)(0.012)/(1.82 \times 10^{-4})^2]^{0.33} \\ &= 3.07 \end{aligned}$$

Therefore, the flow regime is laminar.

2. The settling velocity is calculated using the laminar regime equation

$$\begin{aligned} v &= g\rho_p(d_p)^2C_f/18\mu \\ &= 980(0.899)(45 \times 10^{-4})^2(1)/[18(1.82 \times 10^{-4})] \\ &= 5.38\ \text{cm/s} \end{aligned}$$

EXAMPLE 2: particle terminal velocity. Three different-sized fly ash particles settle through air. The question is to calculate the particle terminal velocity and determine how far each will fall in 30 seconds. Assume the particles are spherical (EPA-84/09, p. 57).

Given conditions

Fly ash particle diameters = 0.4, 40, 400 microns (μm)

Air temperature and pressure = 238°F, 1 atm

Specific gravity of fly ash = 2.31

Check how the Cunningham correction factor affects the terminal settling velocity for the 0.4 micron particle. The Cunningham correction factor is usually applied to particles equal to or smaller than 1 micron.

Solution:

1. Determine the value for K for each fly ash particle size settling in air.
   1.1. Calculate the particle density using the specific gravity given.

$$\begin{aligned}\rho_p &= \text{particle density} = (\text{specific gravity of fly ash})\,(\text{density of water}) \\ &= 2.31(62.4) \\ &= 144.14\,\text{lb/ft}^3\end{aligned}$$

   1.2. Calculate the density of air.

$$\begin{aligned}\rho &= \text{air density} = PM/RT \\ &= (1)(29)/(0.7302)(238+460) = 0.0569\,\text{lb/ft}^3\end{aligned}$$

$$\mu = \text{air viscosity} = 0.021\ \text{cp} = 1.41\times10^{-5}\,\text{lb/ft-s (EPA-84/09, p. 167)}$$

   1.3. Determine the flow regime (K).

$$K = d_p(g\rho_p\rho_a/\mu^2)^{0.33}$$

For $d_p = 0.4$ micron

$$K = [(0.4)/(25{,}400)(12)][(32.2)(144.14)(0.0569)/(1.41\times10^{-5})^2]^{0.33} = 0.0144$$

where

$$1\ \text{ft} = 25{,}400(12)\,\mu\text{m (EPA-84/09, p. 183)}$$

For $d_p = 40$ microns

$$K = [(40)/(25{,}400)(12)](32.2)(144.14)(0.0569)/(1.41\times10^{-5})^2]^{0.33} = 1.44$$

For $d_p = 400$ microns

$$K = [(400)/(25,400)(12)][(32.2)(144.14)(0.0569)/(1.41\times10^{-5})^2]^{0.33} = 14.4$$

2. Select appropriate law.
   The numerical value of K determines the appropriate law.

   $K < 3.3$; Stokes' law range

   $3.3 < K < 43.6$; intermediate law range

   $43.6 < K < 2360$; Newton's law range

   For $d_p = 0.4$ micron, the flow regime is laminar (EPA-81/10, p. 3-10)

For $d_p = 40$ microns, the flow regime is also laminar
For $d_p = 400$ microns, the flow regime is the transition regime

3. Calculate terminal velocity.
   For $d_p = 0.4$ micron

$$\begin{aligned} v &= g\rho_p(d_p)^2/18\mu \\ &= (32.2)[(0.4)/(25{,}400)(12)]^2(144.14)/(18)(1.41 \times 10^{-5}) \\ &= 3.15 \times 10^{-5}\ \text{ft/s} \end{aligned}$$

For $d_p = 40$ microns

$$\begin{aligned} v &= g\rho_p(d_p)^2/18\mu \\ &= (32.2)[(40)/(25,400)(12)]^2(144.14)/(18)(1.41 \times 10^{-5}) \\ &= 0.315\ \text{ft/s} \end{aligned}$$

For $d_p = 400$ microns, use the transition regime equation

$$\begin{aligned} v &= 0.153g^{0.71}(d_p)^{1.14}(\rho_p)^{0.71}/(\mu^{0.43}\rho^{0.29}) \\ &= 0.153(32.2)^{0.71}[(400)/(25{,}400)(12)]^{1.14}(144.14)^{0.71}/[(1.41 \times 10^{-5})^{0.43}(0.0569)^{0.29}] \\ &= 8.90\ \text{ft/s} \end{aligned}$$

4. Calculate distance.
   For $d_p = 40$ microns, distance = (time)(velocity)

$$\text{Distance} = 30(0.315) = 9.45\ \text{ft}$$

For $d_p = 400$ microns, distance = (time)(velocity)

$$\text{Distance} = 30(8.90) = 267\ \text{ft}$$

For $d_p = 0.4$ micron without Cunningham correction factor, distance = (time)(velocity)

$$\text{Distance} = 30(3.15 \times 10^{-5}) = 94.5 \times 10^{-5}\ \text{ft}$$

For $d_p = 0.4$ micron with Cunningham correction factor, the velocity term needs to be corrected. The Cunningham correction factor (C_f) can be found in Appendix A. The data given in the example are

Particle diameter = 0.4 micron
air temperature = 238°F

For use of the data in Appendix A, use particle diameter = 0.5 micron and temperature = 212°*F* to find the C_f value. Therefore, C_f is approximately equal to 1.446.

$$\text{The corrected velocity} = vC_f = 3.15 \times 10^{-5}(1.446) = 4.55 \times 10^{-5}\ \text{ft/s}$$
$$\text{Distance} = 30(4.55 \times 10^{-5}) = 1.365 \times 10^{-3}\ \text{ft}$$

EXAMPLE 3: fluid particle dynamics—stack application. Determine the minimum distance downstream from a cement dust emitting source that will be free of cement deposit. The source is equipped with a cyclone (EPA-84/09, p. 59).

Given conditions

particle size range of cement dust = 2.5–50.0 microns

specific gravity of the cement dust = 1.96

wind speed = 3.0 miles/h

the cyclone is located 150 ft above ground level. Assume ambient conditions are at 60°F and 1 atm. Neglect meteorological aspects.

μ = air viscosity at 60°F = 1.22×10^{-5} lb/ft-s (EPA-84/09, p. 167)

μm (micrometer or micron) = 3.048×10^{5} ft (EPA-84/09, p. 183)

Note: 1 μm = 1 micron = 10^{-6} m.

Solution:

1. Determine the particle size that should be considered for this question.
   A particle diameter of 2.5 microns is used to calculate the minimum distance downstream free of dust since the smallest particle will travel the greatest horizontal distance.

2. Determine the value of K for the appropriate size of the dust.
   This calculation is similar to that presented in the previous problem.

   2.1. Calculate the particle density (ρ_p) using the specific gravity given.

$$\begin{aligned}\rho_p &= \text{(specific gravity of fly ash)(density of water)} \\ &= 1.96(62.4) \\ &= 122.3\,\text{lb/ft}^3\end{aligned}$$

   2.2. Calculate the air density (ρ).
   Rewrite the ideal gas equation, $PV = nR_uT = (m/M)R_uT$

$$\begin{aligned}\rho &= \text{(mass)/(volume)} \\ &= PM/R_uT \\ &= (1)(29)/[0.73(60+460)] = 0.0764\,\text{lb/ft}^3\end{aligned}$$

   2.3. Determine the flow regime (K).

$$K = d_p(g\rho_p\rho_a/\mu^2)^{0.33}$$

For d_p = 2.5 microns

$$K = [(2.5)/(25{,}400)(12)][(32.2)(122.3)(0.0764)/(1.22 \times 10^{-5})^2]^{0.33} = 0.104$$

where 1 ft = 25,400(12) μm = 304,800 μm (EPA-84/09, p. 183)

3. Determine which fluid-particle dynamic law applies for the above value of K.
   Compare the K value of 0.104 with the following range:

   $K < 3.3$; Stokes' law range

   $3.3 < K < 43.6$; intermediate law range

   $43.6 < K < 2360.0$; Newton's law range

The flow is in the Stokes' law range, therefore it is laminar.

4. Calculate the terminal settling velocity in ft/s.
   For Stokes' law range, the velocity is

$$\begin{aligned} v &= g\rho_p(d_p)^2/18\mu \\ &= (32.2)[(2.5)/(25{,}400)(12)]^2(122.3)/(18)(1.22\times10^{-5}) \\ &= 1.21\times10^{-3}\ \text{ft/s} \end{aligned}$$

5. Calculate the time for descent (settling).

$$\begin{aligned} t &= (\text{outlet height})/(\text{terminal velocity}) \\ &= 150/1.21\times10^{-3} \\ &= 1.24\times10^{5}\ \text{s} = 34.4\ \text{h} \end{aligned}$$

6. Calculate the horizontal distance travelled.

$$\begin{aligned} \text{distance} &= (\text{time for descent})(\text{wind speed}) \\ &= (1.24\times10^{5})(3.0/3600) \\ &= 103.3\ \text{miles} \end{aligned}$$

2.3 Particle Sizing Devices

Particle sizing data is a very important consideration in the design of particulate control equipment. Appropriate design is directly dependent on good particle size data. An ideal particle measuring device should have the following features:

- Measure the exact size of each particle
- Report data instantaneously without averaging data over some specified time interval
- Determine the complete composition of each particle, including shape, density, chemical nature, etc.

It is an extremely difficult task to produce such an instrument. Current devices incorporate only one or two of these ideal functions. Several methods are used to obtain particle size data from industrial sources. They include:

- Microscopy
- Optical counter
- Electrical aerosol analyzer
- Bahco micro-particle classifier
- Impactor

Microscopy. The microscope is an instrument in common use for particle size analysis. It measures the geometric diameter of each individual particle. The determination of particle size analysis is carried out by measuring the size of a number of particles. Particles are sized as they are traversed past the eyepiece micrometer. Each particle, presented in a fixed area of the eyepiece, is sized and tallied into one of a number of size classes. The number of particles sized may range from 100 to several thousand depending on the accuracy desired. This method can be time consuming and extremely tedious.

The particles are collected by deposition on a glass slide or on a filter by using an EPA Method 5 sampling train, and the glass slide or filter is analyzed subsequently by a microscope in a laboratory. The analysis of size distribution of particles collected in the field and transported to the laboratory must be viewed with great caution. It is difficult to collect a representative sample in the first place, and it is almost impossible to maintain the original size distribution under laboratory conditions. For example, laboratory measurements cannot determine whether some of the particles existed in the process stream as agglomerates of smaller particles. In spite of the limitations of the microscopic method, this method is useful in the determination of some properties of interest.

The optical microscope can measure particles from about 0.5 μm to about 100 μm in diameter. Electron microscopes can measure particles with diameters as small as 0.001 μm, which is useful for examining extremely minute particles.

Optical counter. Optical particle counters work on the principle of light scattering. Each particle in a continuously flowing sample stream is passed through a small illuminated viewing chamber. Light scattered by the particle is sensed by a photodetector during the time the particle is in the viewing chamber. The intensity of the scattered light is a function of particle size, shape, and index of refraction. Optical counters give reliable particle size information if only one particle is in the viewing chamber at a single time. The simultaneous presence of more than one particle can be interpreted by the photodetector as a large-sized particle. This error can be avoided by maintaining sample dilutions less than 300 particles per cubic centimeter.

Optical particle counters have not been widely used for particle sizing because they cannot be directly applied to the stack exhaust gas stream. The sample must be extracted, cooled, and diluted before entering the counter. This procedure must be done with extreme care to avoid introducing serious errors in the sample. The major advantage of the counter is its capability of observing emission (particle) fluctuations on an instantaneous level. One can size particles as small as 0.3 μm with the optical counter.

A disadvantage of the optical counter is the dependence of the calibration instrument upon the index of refraction and shape of the particle. Errors in counting can also occur from the presence of high concentrations of very small particles which are sensitive to the light wavelength used.

Electrical aerosol analyzer. An electrical aerosol analyzer (EAA) is an aerosol size distribution measuring device. The EAA uses an electric field (which is set at an intensity dependent upon the size and mass of the particle) to measure the mobility of a charged aerosol. The analyzer operates by first placing a unipolar charge on the aerosol being measured, and measuring the resulting mobility distribution of the charged particles by means of a mobility analyzer.

The EAA has been used for source analysis by pulling a sample from the stack into the chamber and introducing the gas stream into the analyzer. The instrument requires that enough particles pass through the chamber so that a charge can be detected. The concentration range for most efficient operation of the EAA is from 1 to 1000 $\mu g/m^3$. Since stack gas concentrations usually exceed 1000 $\mu g/m^3$, sample dilution with clean air is required. No information on the chemical composition of the particles is possible since the particles are not collected. The major advantage of the EAA is that the instrument can measure particles from 0.003 to 1.0 μm in diameter.

Bahco micro-particle classifier. A Bahco is a versatile particle classifier used for measuring powders, dust, and other finely divided solid materials. The Bahco's working range is approximately 1 to 60 μm. The Bahco uses a combination of elutriation and centrifugation to separate particles in an air stream. Particles can be collected onto a filter by using an EPA Method 5 sampling train. The collected particles are subsequently analyzed in the laboratory.

A weighed sample, usually 5 grams, is introduced into a spiral-shaped air current to separate the particle fractions. The larger particles overcome the viscous forces of the fluid and migrate to the wall of the chamber, while the smaller particles remain suspended. After the two size fractions are separated, one of them is reintroduced into the device and is fractionated further. A different spin speed is used to give a slightly different centrifugal force. This is repeated as many times as desired to give an adequate size distribution. The measurements are grouped into discrete size ranges (i.e. 40–60 μm, 20–40 μm, etc.).

The Bahco provides information on the aerodynamic size of particles. This data can be translated into settling velocity information useful in the design of emission control devices. Several hours are required to complete the fractionation analysis. Once the particles have been fractionated into the discrete range, a chemical analysis can be done on the collected particles.

Impactor. An impactor is a sampling device that employs the principle of impaction (impingement) to collect successively smaller sizes of particles. It is commonly used to determine the particle size distribution of exhaust streams from industrial sources. It can be directly attached to an EPA Method 5 sampling train and easily inserted into the stack of an industrial source.

Typical impactors consist of a series of stacked stages and collection surfaces. Depending on the calibration requirements, each stage contains from one to as many as 400 precisely drilled jet orifices, identical in diameter in each stage but decreasing in diameter in each succeeding stage. Adhesive, electrostatic, and van der Waals' forces hold the particles to each other and to the collection surfaces. Moreover, the particles are not blown off the collecting plate by the jets of air because the jets follow laminar flow paths so that no turbulent areas exist. This results in complete dead air spaces over and around the samples.

Particles are collected on preweighed individual stages, usually filters made of glass fiber or thin metal foil. Once the sample is complete, the collection filters are weighed again, yielding particle size distribution data for the various collection stages. Occasionally, there are some dusts that are very difficult to collect, and require grease on the collection filter for adequate particle capture. Once the particles have been fractionated into discrete ranges, a chemical analysis can be performed on the collected particles.

The effective range for measuring the aerodynamic diameter is generally between 0.3 and 20 μm. Some vendors have claimed size fractionation as small as 0.02 μm with the use of 20 or more stages. Impactors are one of the most useful devices for determining particle size. This is because of the impactor's compact arrangement, mechanical stability, and its ability to draw a sample directly from a stack. In addition, the impactor measures the aerodynamic diameter of particles, which describes the movement of the particles in a gas stream. Particle movement information is extremely useful in designing air pollution control equipment, especially mechanical collectors which depend on aerodynamic drag forces for particle collection.

2.4 Particle Size Presentation

There are various ways to analyze or reduce the data generated from a particle sizing device. The most common methods of expressing particle size data include (EPA-81/10, p. 4-12):

- Frequency distribution curve (histograms)
- Cumulative distribution curve
- Log normal distribution

All forms of graphical presentation employ one axis to represent particle size and the other axis to represent particle amount. The particle amount can be expressed by either the mass of particles or the number of particles.

Frequency distribution curve: This curve is usually plotted on regular coordinate (linear) paper. The curve describes the amount of material (particles) falling within each size range. When gas-borne particles produced in industrial operations are measured, the data have a tendency to show a preferential particle size. A plot of percent mass versus particle size on a linear scale gives a curve with a peak at the preferential size. An example of such a curve is shown in Fig. 2.7

Figure 2.7 shows a normal probability distribution that is symmetrical about the preferential size. This curve is encountered for dusts consisting of large particles. This curve may be

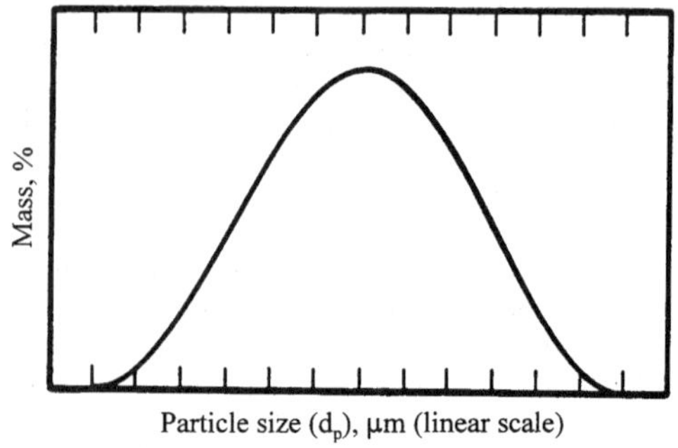

FIGURE 2.7 Particle size distribution with one preferential size (*EPA-81/10, p. 4-12*).

found for particles such as fumes formed by vapor phase reaction and condensation or for tar and acid mists. For very fine particles, the curve will be off-center or skewed about the preferential size. If a particle size distribution shows two peaks, the curve simply shows that the data have two preferential sizes instead of one.

The normal probability of the distribution can also be plotted on a log scale. The representation of the distribution can be seen in Fig. 2.8. The percent mass (concentration) is plotted on a linear scale as the y-axis. The particle size d_p is plotted on a log scale as the x-axis. If the curve takes on a bell-shape, the distribution appears normal.

Cumulative distribution: Particle size data can also be plotted as a cumulative plot. Particle size of each size range is plotted on the ordinate (y-axis). The cumulative percent by weight (frequency) is plotted on the abscissa (x-axis). The cumulative percent by weight can be given as cumulative percent less than stated particle size or cumulative percent larger than a stated particle size. The cumulative percent by weight can be plotted on either a linear percentage or a probability percentage scale. The particle size range (ordinate) is usually a logarithmic scale (EPA-81/10, p. 4-15). If the particle size d_p of each size range is plotted versus the cumulative percent larger than d_p (linear scale) one could get a distribution curve as shown in Fig. 2.9. In Fig. 2.9, the distribution approaches the 0% and 100% values asymptotically. It is evident that the cumulative distribution in this figure is not a straight line for the entire range of particle size in the sample. The majority of the size ranges occur toward the 0% size and the 100% size.

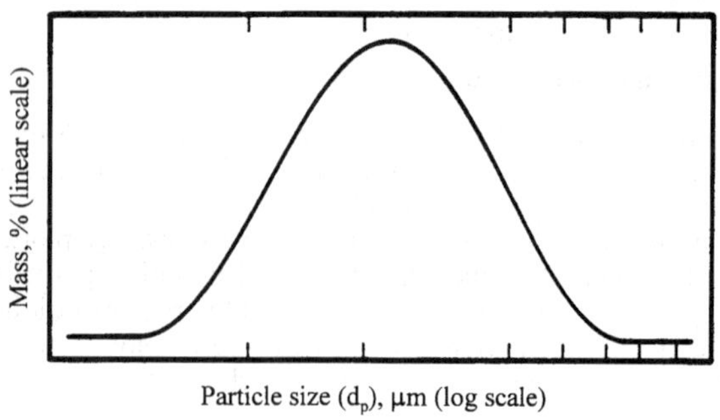

FIGURE 2.8 Particle size distribution plotted on a log scale (*EPA-81/10, p. 4-12*).

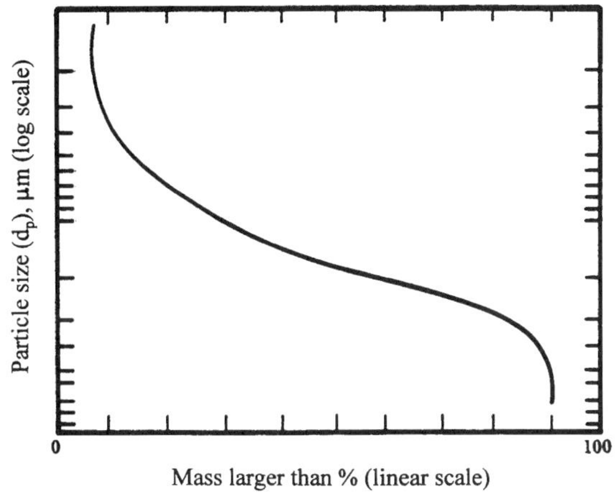

FIGURE 2.9 Cumulative particle size distribution plotted on a linear scale (*EPA-81/10, p. 4-15*).

More frequently, the cumulative distribution is plotted on special coordinate paper called log *probability paper*. The particle size of each size range is plotted on the logarithmic ordinate. The percent by weight larger than d_p is plotted on the probability scale as the abscissa (Fig. 2.10). This allows one to expand the cumulative distribution axis near 0% and near 100%. The cumulative distribution plot in Fig. 2.10 is identical to that in Fig. 2.9, except that the percentage scale is expanded near 0% and 100%. It should be noted that one can just as easily plot percent mass less than on the abscissa.

Log-normal distribution: When measuring dusts from industrial sources, the graph of the particle size distribution often displays the logarithmic variation of the normal distribution. If the distribution follows the log-normal relationship, then the plot will result in a straight line. The linearity of the relationship allows one to describe the distribution statistically with a minimum of individual observations. The distribution is completely specified by two parameters (EPA-81/10, p. 4-16):

- *Geometric mean*: The geometric mean value of log-normal distribution can be read directly from a plot (Fig. 2.11). The geometric mean size is the 50% size on the plot.
- *Geometric standard deviation*: The geometric standard deviation is a good measure of the dispersion or spread of a distribution. The geometric standard deviation is the root-mean square deviation about the mean value. Its derivation and application in significance testing and setting of confidence levels can be found in most statistics textbooks. The geometric standard deviation is identical for specifying the size distribution of a log-normal distribution, whether by particle number, surface, mass, or any other quantity of the form $k(d_p)^n$ where k is a parameter common to all particles and d_p is the diameter. Plots of cumulative distribution on log-probability paper are then parallel straight lines for number, mass, or surface which leads to a greater simplification and easy graphical technique. The geometric standard deviation can be read directly from a plot such as that shown in Fig. 2.11. For a log-normal distribution (that plots d_p maximum versus percent mass larger than d_p), the geometric standard deviation (σ_{gm}) is given by

$$\sigma_{gm} = (50\%\ \text{size})/(84.13\%\ \text{size});\ \text{or}$$

$$\sigma_{gm} = (15.87\%\ \text{size})/(50\%\ \text{size})$$

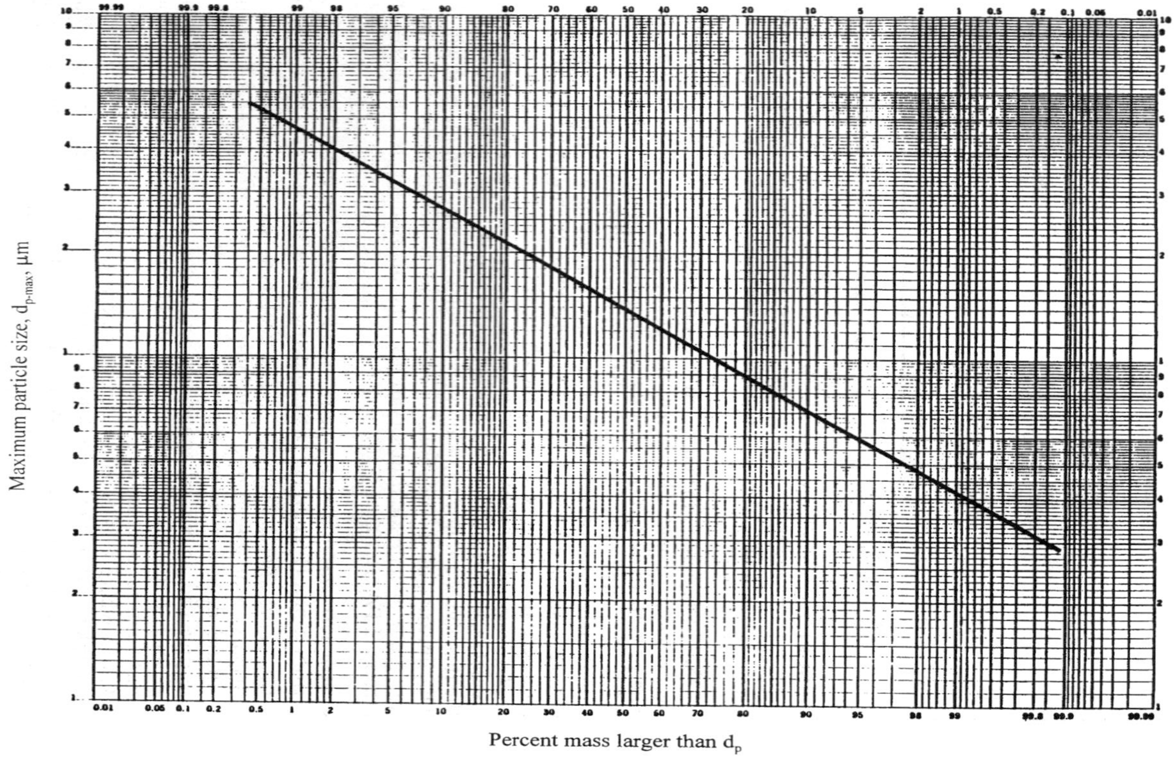

FIGURE 2.10 Cumulative particle size distribution plotted on log probability paper.

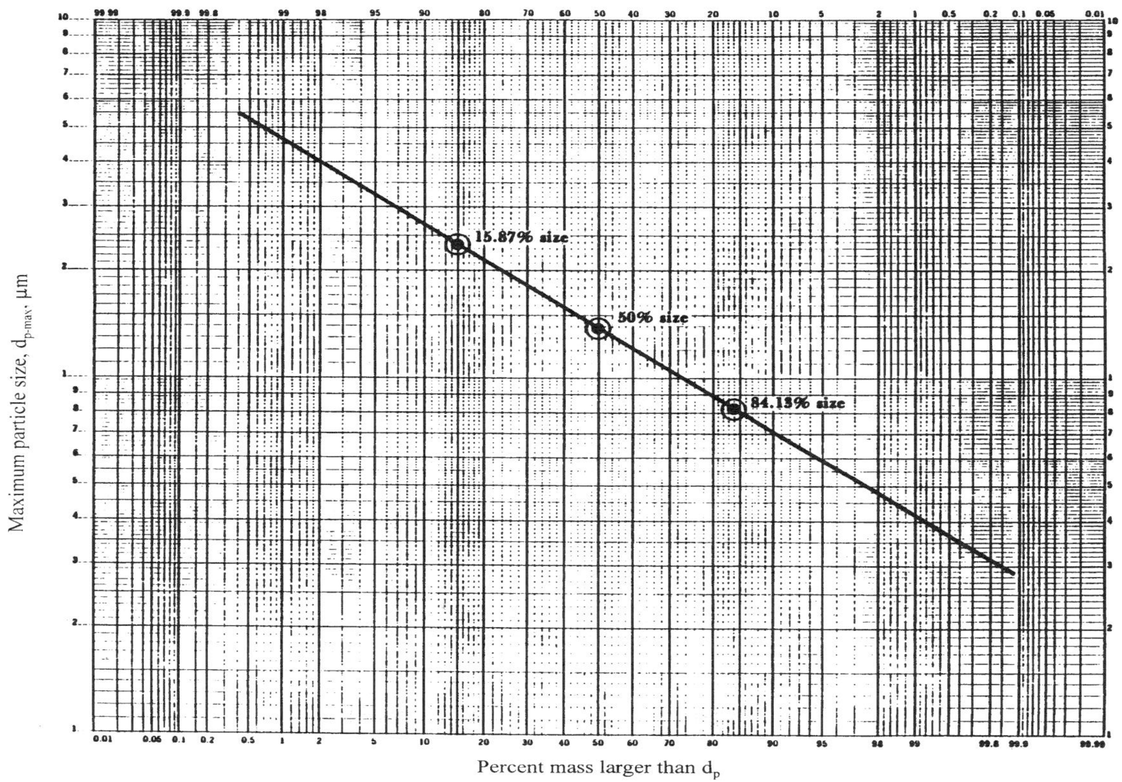

FIGURE 2.11 Geometric mean and standard deviation from a log normal distribution plot.

TABLE 2.1 Typical Particle Size Data

Size range d_p, μm	Concentration (μg/m^3)	Concentration/Δd_p	Concentration/$\Delta \log d_p$
1–2	0.8	0.8/(2.0 − 1.0) = 0.8	0.8/[log 2 − log 1] = 2.66
2–4	12.2	6.1	40.53
4–6	25	12.5	142
6–10	56	14	252.3
10–20	76	7.6	253.2
20–40	27	1.35	89.7
> 40	3	0.075	10

EXAMPLE 1: typical particle size data reduction. Consider the data listed in Table 2.1 (EPA-81/10, p. 4-19). Suppose we were asked to determine if the distribution was log normal, and if so, determine the geometric mean diameter (d_p) and the geometric standard deviation (d_{gm}). How would one approach this problem?

First we could plot the mass concentration (concentration/Δd_p) versus particle diameter d_p on a linear scale (Fig. 2.12). The data appear to be skewed and would therefore lead one to believe that the distribution could be log normal.

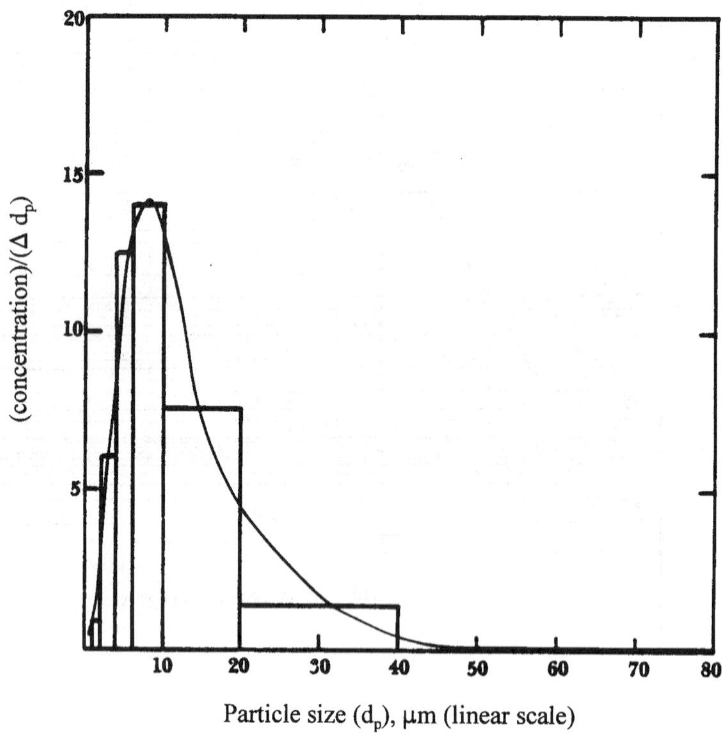

FIGURE 2.12 Mass concentration versus particle diameter on a linear scale (*EPA-81/10, p. 4-20*).

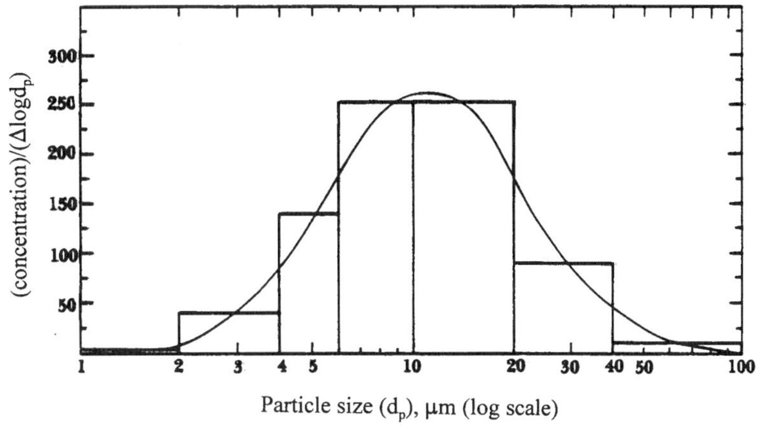

FIGURE 2.13 Mass concentration versus particle diameter on a log scale (*EPA-81/10, p. 4-21*).

We can now plot mass concentration (concentration)/$\Delta \log d_p$ on a log scale (Fig. 2.13). The particle size d_p (x-axis) is plotted on a log scale. The area under the smooth curve between two different values of d_p represents the total particle mass concentration between the two values of d_p. If the y-axis is not 'concentration divided by $\Delta \log d_p$' then the distribution shape will be partly determined by the size ranges of the sampling device (i.e. impactor stages) rather than the true particle distribution shape.

This plot yields an appropriately shaped bell curve that appears to indicate that this is a log-normal distribution. However, it is extremely important to plot the data on log-probability paper to see if a straight line results. If the distribution plots a straight line on log-probability paper, then the distribution is log normal.

First we must calculate the percent in each size range and then the cumulative percent larger than d_p maximum (Table 2.2).

The plot (Fig. 2.14) on log-probability paper yields a straight line. Therefore the distribution is log normal. To determine the geometric mean diameter one can read the 50% size from Fig. 2.14.

$$d_{gm} = 10.5 \text{ micrometers}$$

TABLE 2.2 Cumulative Particle Size Data

Size range d_p, μm	Concentration	Percent weight in size range	Cumulative percent larger than d_p-max
0–2	0.8	0.4	99.6
2–4	12.2	6.1	93.5
4–6	25	12.5	81
6–10	56	28	53
10–20	76	38	15
20–40	27	13.5	1.5
> 40	3	1.5	—
Total	200		

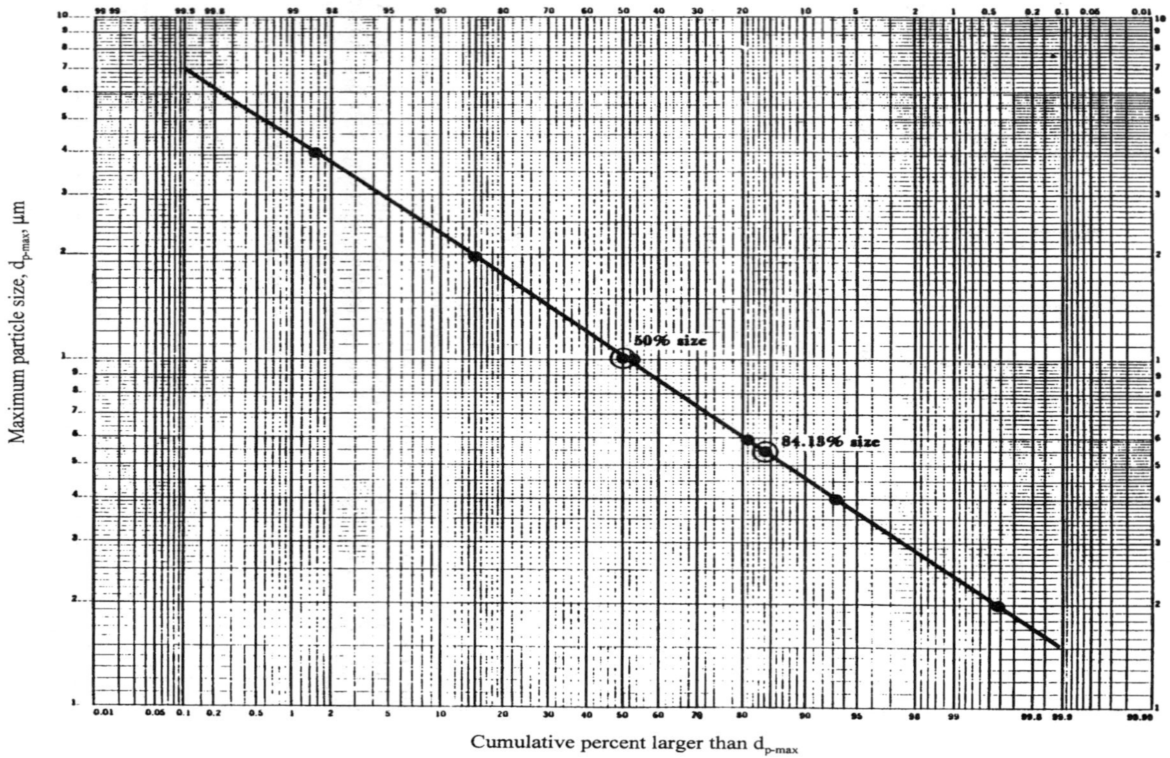

FIGURE 2.14 Log probability distribution.

The geometric standard deviation is

$$\sigma_{gm} = (50\%\ \text{size})/(84.13\%\ \text{size}) \\ = 10.5/5.5 \\ = 1.9$$

EXAMPLE 2: log-normal distribution. Determine if a particle size distribution is log normal, based on the data given in Table 2.3 (EPA-84/09, p. 49).

Solution:

1. Generate a cumulative distribution curve.

Cumulative distribution plots are generated by plotting particle diameter versus cumulative percent. For log-normal distributions, plots of particle diameter versus either percent less than stated size (% LTSS) or percent greater than stated size (% GTSS) produce straight lines on log-probability (normal) coordinates.

1.1. Complete Table 2.4 (given below), calculating the percent of the total mass and the cumulative percent GTSS for each particle size range.

The cumulative percent, based on % GTSS, is the total percent of the mass of particles greater than the upper size for the range in question.

1.2. Plot the cumulative distribution curve on a blank log-probability graph paper as shown in Fig. 2.15.

The cumulative distribution points are plotted using the upper-size particle diameter of each range versus cumulative % GTSS.

2. Is the distribution log normal?

Since a straight line is obtained on log-normal coordinates, the particle size distribution is log normal or a linear (straight line) relationship on log-normal coordinates.

EXAMPLE 3: Andersen 2000 sampler data analysis. Given Andersen 2000 sampler data, plot a cumulative distribution curve on log-probability paper and determine the mean particle diameter and geometric standard deviation (EPA-84/09, p. 52).

Given conditions

volumetric flow rate $Q = 0.5$ cfm

TABLE 2.3 Particle Size Distribution

	(1)* Particle size range d_p, μm	(2)† Distribution, $\mu g/m^3$
	< 0.62	25.50
	0.62–1.0	33.15
	1.0–1.2	17.85
	1.2–3.0	102.00
	3.0–8.0	63.75
	8.0–10.0	5.10
	> 10.0	7.65
Total		255.00

Notes: *Column (1): given data.
†Column (2): given data.

TABLE 2.4 Particle Size Distribution

(1) Particle size range d_p, μm	(2) Distribution, μg/m^3	(3)* Percent of total	(4)† Cumulative % GTSS
< 0.62	25.50	10.00	90.00
0.62–1.0	33.15	13.00	77.00
1.0–1.2	17.85	7.00	70.00
1.2–3.0	102.00	40.00	30.00
3.0–8.0	63.75	25.00	5.00
8.0–10.0	5.10	2.00	3.00
> 10.0	7.65	3.00	0.00
Total	255.00		

Notes: *Column (3): percent of total = [mass/(total mass)](100) or {[column (2)]/255}(100); e.g. (25.5/255)(100) = 10.00.
†Column (4): 100 − cumulative % of column (3); e.g. 100−10 = 90; 100−10−13 = 77.

Fig. 2.16 gives aerodynamic diameter versus flow rate through an Andersen sampler for an impaction efficiency of 95%.

Andersen 2000 sampler data (Table 2.5)

Solution:

1. Generate a cumulative distribution curve.

1.1. Complete Table 2.6 (given below) by calculating the net weight, % of total weight, and cumulative percent for each plate. Use % LTSS.

1.2. Determine the 95% aerodynamic diameter at $Q = 0.5$ cfm for each plate from Fig. 2.16. These values are listed in Table 2.7.

The aerodynamic diameter of a particle is defined as the diameter of a sphere having the same resistance to motion as the particle. The 95% aerodynamic diameter assigned to each plate may be thought of as that particle size that will not pass through the sampler opening.

TABLE 2.5 Andersen 2000 Sampler Data

(1)* Plate No. (Stage No. in Fig. 2.16)	(2)† Tare weight, g	(3)‡ Final weight, g
0	20.48484	20.48628
1	21.38338	21.38394
2	21.92025	21.92066
3	21.55775	21.55817
4	11.40815	11.40854
5	11.61862	11.61961
6	11.7654	11.76664
7	20.99617	20.99737
Back-up filter	0.2081	0.21156

Notes: *Column (1): given data.
†Column (2): given data.
‡Column (3): given data.

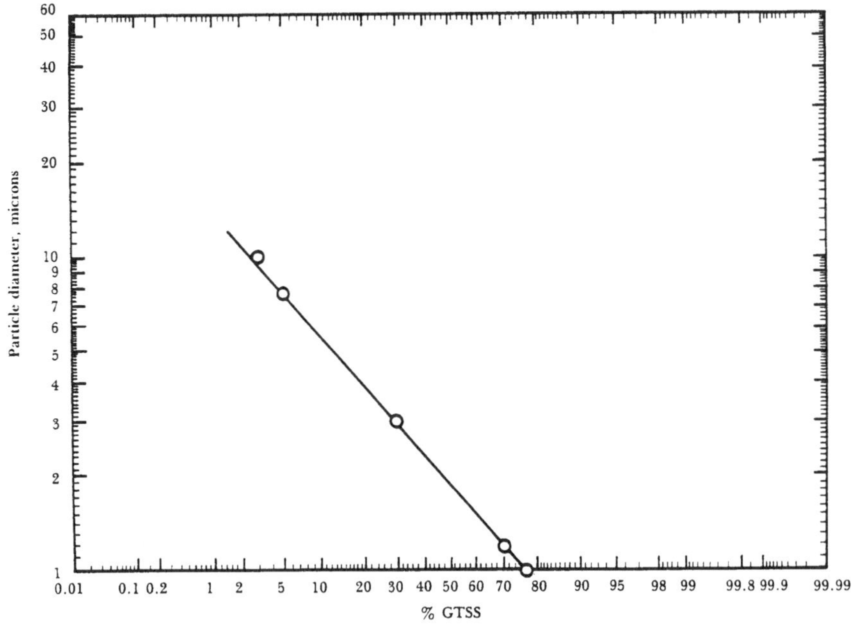

FIGURE 2.15 Cumulative distribution curve (*EPA-84/09, p. 127*).

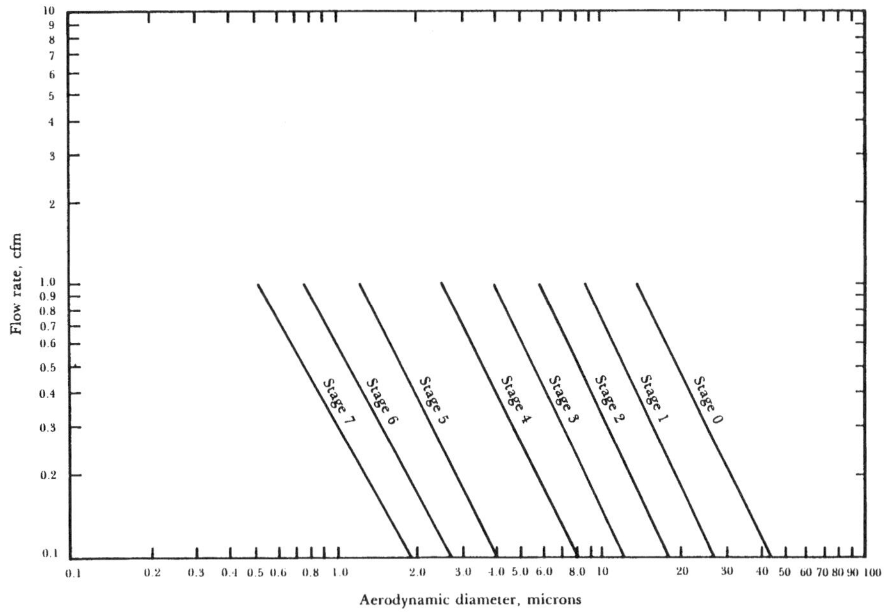

FIGURE 2.16 Aerodynamic diameter vs. flow rate (*EPA-84/09, p. 53*).

TABLE 2.6A Andersen 2000 Sampler Data

(1) Plate No.	(2) Tare weight, g	(3) Final weight, g	(4)* Net wt, g	(5)† Percent of total wt
0	20.48484	20.48628	0.00144	14.24
1	21.38338	21.38394	0.00056	5.54
2	21.92025	21.92066	0.00041	4.06
3	21.55775	21.55817	0.00042	4.15
4	11.40815	11.40854	0.00039	3.86
5	11.61862	11.61961	0.00099	9.79
6	11.7654	11.76664	0.00124	12.27
7	20.99617	20.99737	0.0012	11.87
Back-up filter	0.2081	0.21156	0.00346	34.22
Total	141.34266	141.35277	0.01011	100.00

TABLE 2.6B Andersen 2000 Sampler Data

(1) Plate No.	(2) Tare wt, g	(3) Final wt, g	(4)* Net wt, g	(5)† Percent of total wt	(6)‡ Cumulative %
0	20.48484	20.48628	0.00144	14.24	100.00
1	21.38338	21.38394	0.00056	5.54	85.76
2	21.92025	21.92066	0.00041	4.06	80.22
3	21.55775	21.55817	0.00042	4.15	76.16
4	11.40815	11.40854	0.00039	3.86	72.01
5	11.61862	11.61961	0.00099	9.79	68.15
6	11.7654	11.76664	0.00124	12.27	58.36
7	20.99617	20.99737	0.0012	11.87	46.09
Back-up filter	0.2081	0.21156	0.00346	34.22	34.22
Total	141.34266	141.35277	0.01011	100.00	

Notes for Tables 2.6A and 2.6B
*Column (4) = column (3) − column (2), i.e. net weight = final weight − tare weight.
†Column (5) = ((column (4))/total) (100); where total = 0.01011 g.
‡Column (6) = 100% − cumulative %; e.g. for plate 1, 100% − 14.24% = 85.76%; for plate 2, 85.76% − 5.54% = 80.22%. As indicated in the previous example, cumulative distribution plots are generated by plotting particle diameter versus cumulative percent. For log-normal distributions, % LTSS are usually plotted on the probability normal coordinates. Cumulative % represents the percent less than stated size (% LTSS) associated with each plate. Thus, the calculated cumulative percent is that quantity of particles, in percent, captured by the appropriate plate and all other captured smaller particles.

1.3. Plot the cumulative distribution curve on a blank log-probability sheet as shown in Fig. 2.17.

The cumulative distribution curve is plotted as diameter versus cumulative percent less than stated size (% LTSS).

2. From the cumulative distribution curve, determine the mean particle diameter in microns (μm).

Since the mean particle diameter is the particle diameter corresponding to a cumulative percent of 50%, read off the particle diameter at a cumulative percent of 50% from Fig. 2.17.

$$\text{Mean particle diameter} = Y_{50} = 1.2\,\text{microns}$$

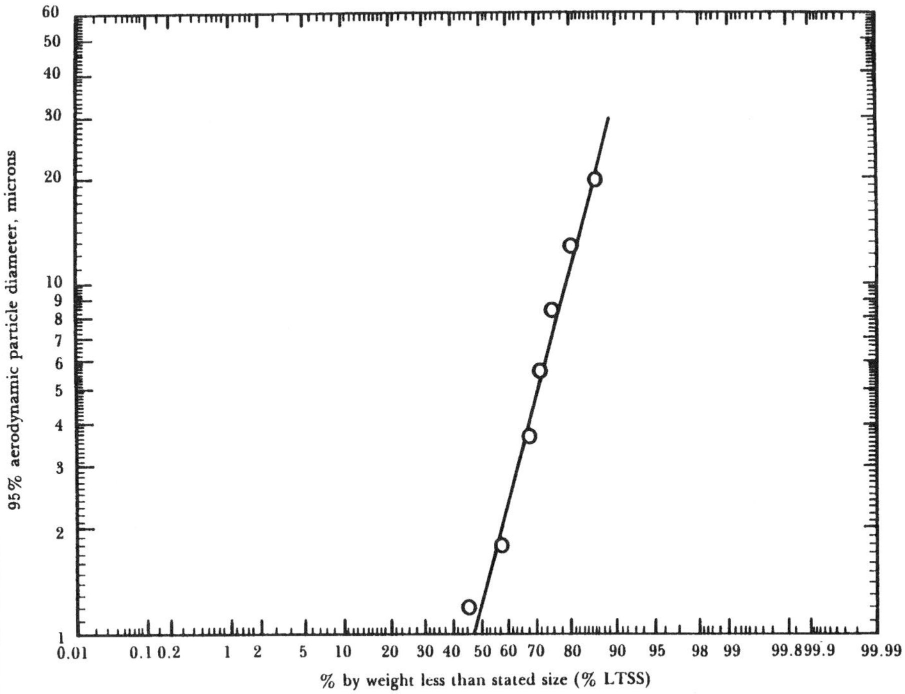

FIGURE 2.17 Cumulative distribution curve (*EPA-84/09, p. 130*).

3. Determine the geometric standard deviation σ_{gm}, if the distribution is approximately log normal.

Geometric standard deviation is defined as (EPA-81/10, p. 4-19):

$$\sigma_{gm} = (50\%\ \text{size})/(84.13\%\ \text{size})$$
$$\sigma_{gm} = (15.87\%\ \text{size})/(50\%\ \text{size})$$

3.1. The distribution approaches log-normal behavior. Read off the particle diameter at 84.13 cumulative% from Fig. 2.17, the particle diameter at 84.13 cumulative%.

TABLE 2.7 Aerodynamic Diameters for Plate Numbers

Plate No. (stage No.)	95% aerodynamic diameter, μm
0	20
1	13
2	8.5
3	5.7
4	3.7
5	1.8
6	1.2
7	0.78

$$Y_{84.13} = 17.0\,\text{microns}$$

3.2. For a log-normal distribution:

$$\sigma_{gm} = Y_{50}/Y_{84.13}$$
$$= 1.2/17.0 = 0.071$$

3 FACTORS AFFECTING THE SELECTION OF PARTICULATE CONTROL EQUIPMENT

Many factors influence the choice of a control device used to reduce particulate emissions. The composition of the particles in terms of concentration, size, and chemical and physical characteristics must be considered. If emitted material can be used in the process, dry collection should be used. If the pollutant has little economic value, collection should be accomplished and the pollutant disposed of safely and economically. The industrial process and potential control devices must both be carefully reviewed. The conversion of an air pollution problem into a water pollution problem can create a more difficult disposal problem. It is necessary to consider many factors (EPA-81/10, p. 1-5), which are illustrated below.

3.1 Particle Concentration or Grain Loading

One selection factor to be considered is the concentration or grain loading of particulate pollutants in the process exhaust stream. Pollutant concentration is typically expressed in terms of pounds per cubic foot (lb/ft^3), grains per cubic foot (gr/ft^3), and grams per cubic meter (g/m^3). Both the level and fluctuation of grain loading are very important. Some control devices are not affected by high levels or great fluctuations in particle concentration such as in fabric filters. Others, such as electrostatic precipitators, do not function effectively with a large fluctuating concentration level. Another related problem can occur when the exhaust gas velocity changes rapidly. Some control devices are designated to operate at a specific exhaust gas velocity; large changes can drastically affect the collection efficiency of the unit.

3.2 Particle Characteristics

Particle characteristics such as size, shape, and density must also be considered. Particle size is usually expressed in terms of aerodynamic diameter. The *aerodynamic diameter* describes how the particle moves in a gas stream. The larger particles ($> 60\,\mu$m aerodynamic diameter) can be collected in simple devices such as settling or baffle chambers. Particles greater than $5\,\mu$m can be collected in cyclones or multicyclones. Smaller particles ($< 5\,\mu$m) must be collected in more sophisticated devices, such as scrubbers, baghouses, or electrostatic precipitators. Particle size thus plays a large role in the collection efficiency of a specific control device.

3.3 Chemical and Physical Properties

Chemical and physical properties of particulate emissions greatly affect the selection of control devices. Electrical properties of the particle can be both a hindrance and an aid to collection. Static electricity can create solid buildup in both inertial and baghouse collectors. Cakes can form on the bag filters, as a result of the static electric forces, and can be difficult to dislodge.

In an electrostatic precipitator, on the other hand, the collection of particulate matter depends on the ability of the precipitator to charge the particle. Particles passing through an electric field are charged and, consequently, migrate to an oppositely charged collection plate. Although there are many factors that govern the ease of charging the particles, the primary factor governing adequate particle collection is particle dust resistivity (a term that describes the resistance of the collected dust layer to the flow of electric current).

If the particles in the gas stream are explosive, electrostatic precipitators cannot be used. Fabric filters might be used, but only if no static electric effects exist in the baghouse. The logical choice of control would be a scrubber in which water is used as the scrubbing liquid. The water has a dampening effect on the explosive dust.

3.4 Hygroscopicity

Hygroscopicity is the tendency of a material to absorb water. It is a physical characteristic of some particles which causes changes in the crystal structure as water is added. Some particles such as sodium sulfate can absorb up to 12 moles of water per mole of anhydrous salt in high humidity conditions. Other particles from processes such as alfalfa dehydration and pelleting, and cotton ginning, are hygroscopic to varying degrees. The hygroscopic nature of the particle affects the performance of mechanical collectors by causing dust deposits to build up on their internal surfaces. This may cause internal plugging or unpredictable dust cake discharges into collection hoppers at various times. Hygroscopic particles also affect the choice of cleaning in a baghouse in that they form cakes on the bags that are difficult to remove.

3.5 Particle Toxicity

Particle toxicity influences the location of the control device and the air moving system (fan). Highly toxic materials require the use of a negative pressure system, so that leaks can be contained within the collector. A positive pressure system could cause fugitive emissions, creating an occupational health and safety problem. In a negative pressure system, the fan is located downstream of the air pollution control devices. The volume of gas to be handled may increase slightly by air leakage into the collector, but little or no contaminant leakage from the collector should occur.

3.6 Carrier Gas Stream

The behavior of the carrier gas stream is also important in the design phases of air pollution control systems. Gas stream temperature affects a number of variables in the design stages of the control device. The size and thus the cost of the unit depends on the temperature of the exhaust gas stream being treated. The volume of gas to be cleaned would be larger at high temperatures than that at correspondingly lower gas temperatures. Reducing the temperature reduces the volume of exhaust gas to be handled; however, this could create some additional problems. The gas stream temperature must be maintained above the dew point of the gas to prevent water and acid from condensing in the collector. Water and acid mists could cause corrosion and complete deterioration of the structural material of the collector.

High gas stream temperatures can also cause equipment failure to components of a fabric filtration system. At exhaust gas temperatures greater than 300°C most fabric materials deteriorate. Gas temperature can also affect conditions such as particle resistivity. By changing the temperature of the exhaust stream in an electrostatic precipitator, one can also change the resistivity of the particles, and thus the collection efficiency of the unit.

3.7 Efficiency/Cost Trade-off

In air pollution work, the control equipment should be designed to meet emission limitations at minimum cost with maximum reliability. The basic trade-offs involve decisions between collection efficiency, pressure drop, installation cost, and operating costs. Of these, the principal one is the trade-off between collection efficiency and pressure drop (which can be translated into power requirements) across the control device.

Collection efficiency (by weight) can be defined by the following formula:

$$\text{Collection efficiency} = [(\text{inlet loading} - \text{outlet loading})/(\text{inlet loading})] \times 100$$

Emission limits are usually set by existing air pollution regulations. The control to be achieved is dependent on the allowed outlet concentration and the quantity of emissions generated from the process.

Equipment should be operated at the pressure drop specified by the design. Pressure drop describes the pressure loss between the inlet and outlet sections of the control device. Collectors with large pressure drops would require larger fans (and greater power requirements) to either push or pull the exhaust gas through the system. An increase in pressure drop means that there is a larger pressure loss in the system. Some control devices such as venturi scrubbers are designed to operate at high pressure drops (as great as 60 in. H_2O; 14.9 kPa). On the other hand, electrostatic precipitators are designed to operate at much lower pressure drops (usually less than 10 in. H_2O; 2.49 kPa) for similar collection efficiencies. It may be advantageous, however, to choose the scrubber as the control device in some cases, especially if the dust to be collected is explosive.

Other conditions such as space limitations for installation may affect the ultimate decision for control equipment. Scrubbers generally require less installation space than either baghouses or electrostatic precipitators. There are, of course, water disposal problems with wet collectors that could sway the choice toward some other type of control. Installed cost and operating costs vary for each type of collector. For a given process, electrostatic precipitators may be more expensive to install than baghouses. However, baghouses may be more expensive to operate and maintain on an annual basis. The cost trade-offs must be examined carefully in making the final choice in control equipment.

4 PARTICULATE EMISSION CONTROL EQUIPMENT

4.1 Gravity Settling Chamber

Gravity settling chambers have long been used by industry for removing solids or liquid particulate matter from gaseous streams. The technology has the advantages of simple construction, low initial cost and maintenance, low pressure losses, and simple disposal of collected materials.

The technology was one of the first devices used to control particulate emissions. It is an expansion chamber in which particle velocity is reduced, thus allowing the particle to settle out under the action of gravity. One primary feature of this device is that the external force causing separation of particles from the gas stream is provided free by nature. This chamber's use in industry, however, is generally limited to the removal of larger sized particles, 40–60 μm in diameter (EPA-81/10, p. 5-1).

Gravity settler type. There are basically three types of gravity settlers (EPA-81/10, p. 5-1):

- Simple expansion chamber
- Multiple-tray settling chamber
- Baffle chamber

Simple expansion chamber: A typical horizontal flow (simple expansion) gravity settling chamber is presented in Fig. 2.18. The unit is constructed in the form of a long horizontal box with an inlet, outlet, and dust collection hoppers. These units depend on gravity for collection of the particles. The particle-laden gas stream enters the unit at the gas inlet. The gas stream then enters the expansion section of the duct. Expansion of the gas stream causes the gas velocity to be reduced. All particles in the gas stream are subject to the force of gravity. However, at reduced gas velocities (in the range of 0.305 to 3.05 m/s) the larger particles can be overcome by gravity and fall into the dust hoppers. Theoretically, a settling chamber of infinite length could collect even the very small particles ($< 10\,\mu$m).

The collection hoppers located at the bottom of the settler are usually designed with positive seal valves and must be emptied as dust buildup occurs. Dust buildup will vary depending on the concentration levels of particulate matter in the gas streams, especially for heavy concentrations of particles greater than 60 μm in diameter.

Multiple-tray settling chamber: The multiple-tray settling chamber, also known as the Howard settling chamber, is shown in Fig. 2.19. Several horizontal collection plates are introduced to shorten the settling path of the particle and to improve the collection efficiency of small particles (as small as 15 μm in diameter). Each shelf or tray in the unit can collect dust that settles out by gravitational force. Since the vertical distance that a particle must fall to be captured is less than the distance in a horizontal settling unit, the overall collection efficiency of the Howard settling chamber can be greater than the horizontal chamber. The gas must be uniformly distributed as it passes through each tray throughout the chamber. Uniform distribution is usually achieved by the use of gradual transition, guide vane, distributor screens, and perforated plates. The particles will settle on the individual trays, which must be cleaned periodically. The vertical distance between trays may be as little as 1 inch, making cleaning much more difficult than with the horizontal settling chamber. Other disadvantages include the tendency of trays to warp during high-temperature operation and the inability of the unit to handle dust concentrations exceeding approximately 1 grain/ft^3 (2.29 g/m^3). For these reasons, the Howard settling chamber is not used very frequently for particulate emission control.

Baffle chamber: A variation of the gravity settling chamber is a baffle chamber, sometimes referred to as an inertial separator. These units have baffles within the chamber to enhance particle separation and collection. This arises by changing the direction of the gas velocity and imparting a downward motion to the particle. This induced motion is superimposed on the motion due to gravity. Thus, particle collection is accomplished by gravity and an inertial or momentum effect. Particles as small as 20 to 40 μm can be collected. An example of this device

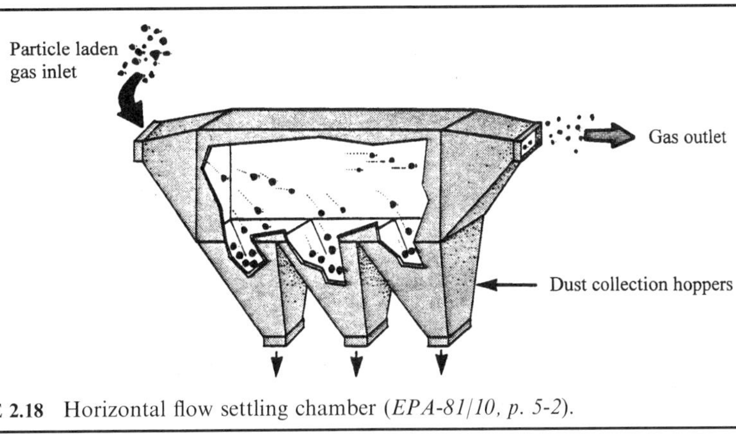

FIGURE 2.18 Horizontal flow settling chamber (*EPA-81/10, p. 5-2*).

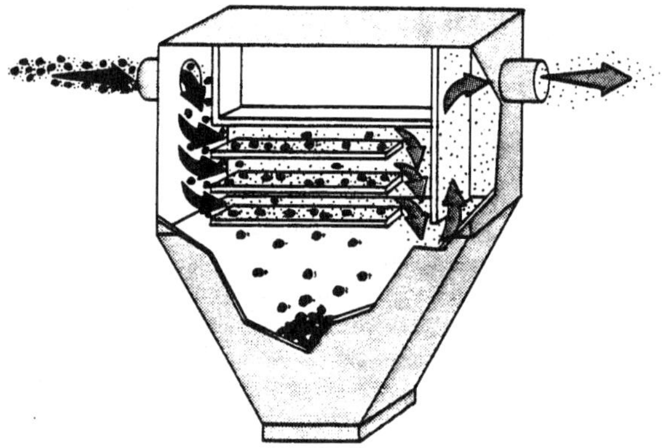

FIGURE 2.19 Multiple tray settling chamber (*EPA-81/10, p. 5-3*).

is shown in Fig. 2.20. These units are more compact and require less space than gravity settling chambers. The pressure drops are slightly higher, ranging from 0.1 to 1.0 in. of H_2O (0.25 to 2.5 cm H_2O).

Design parameter. Understanding the principles governing particle collection in a gravity settler begins by examining the behavior of a single spherical particle in the chamber, as shown in Fig. 2.21. The bulk flow air velocity profile in this case is assumed to be uniform throughout the chamber. The particle flows with the same velocity as the gas stream, and the horizontal velocity is given as v_x. The particle also has a vertical velocity, v_y. The term v_y is also called the terminal settling velocity v_t. The length, width, and height of the chamber are L, B, and H, respectively (EPA-81/10, p. 5-4).

Suppose a particle enters the chamber at a height h_p. The particle must fall this distance (h_p) before it travels the length of the chamber if the particle is to be collected. In other words, the

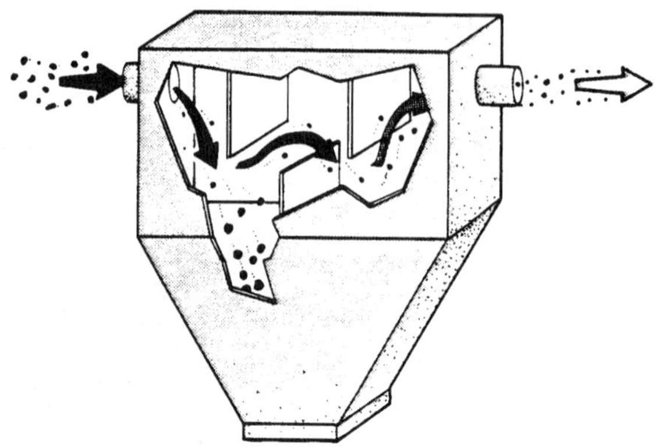

FIGURE 2.20 Baffle chamber (*EPA-81/10, p. 5-3*).

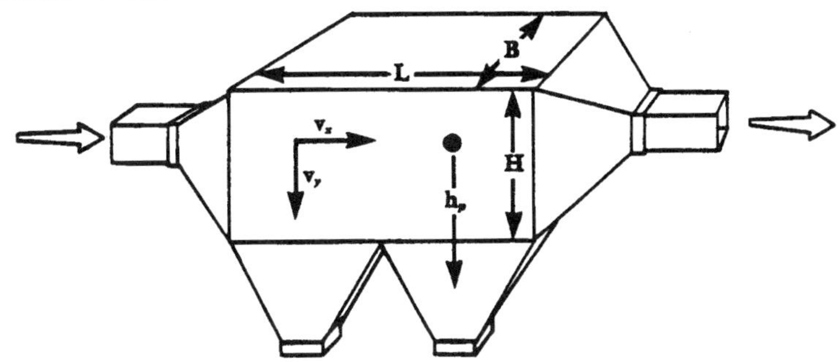

FIGURE 2.21 Settling chamber dimensions (*EPA-81/10, p. 5-4*).

particle will settle if the time required for the particle to settle is less than the time that the particle resides in the chamber.

The theoretical collection efficiency of the settling chamber is given by the expression:

$$\eta = (v_y L)/(v_x H) \tag{2.1}$$

where
η = fractional efficiency of particle size d_p (one size)
v_y = vertical settling velocity
v_x = horizontal gas velocity
L = chamber length
H = chamber height (greatest distance a particle must fall to be collected)

The settling velocity can be calculated from Stokes' law, as previously discussed. As a rule of thumb, Stokes' law applies when the particle size d_p is less than 100 μm in size. The settling velocity is

$$v_t = [g(d_p)^2(\rho_p - \rho_a)]/(18\,\mu) \tag{2.2}$$

where
v_t = settling velocity in Stokes' law range, m/s (ft/s)
g = acceleration due to gravity, 9.8 m/s^2 (32.1 ft/s^2)
d_p = diameter of the particle, μm
ρ_p = particle density, kg/m^3 (lb/ft^3)
ρ_a = gas density (usually air), kg/m^3 (lb/ft^3)
μ = gas viscosity, Pa-sec (lb/ft-s)
Pa = N/m^2
N = kg-m/s^2

Thus, Eq. (2.2) can be rearranged to determine the minimum particle size that can be collected in the unit with 100% efficiency. The minimum particle size d_p^* (in μm) is given as

$$(d_p)^* = \{v_t(18\mu)/[g(\rho_p - \rho_a)]\}^{0.5} \tag{2.3}$$

The density of the particle ρ_p is usually much greater than the density of gas ρ_a. Therefore, the quantity $\rho_p - \rho_a$ reduces to ρ_p. The velocity can be written as

$$v = Q/BL$$

where
Q = volumetric flow rate
B = chamber width
L = chamber length

Equation (2.3) is reduced to

$$(d_p)^* = \{(18\mu Q)/[g\rho_p BL]\}^{0.5} \tag{2.4}$$

The efficiency equation can also be expressed as

$$\eta = [(g\rho_p BLN_c)/(18\mu Q)](d_p)^2 \tag{2.5}$$

where N_c = number of parallel chambers: 1, for simple setting chamber and N trays +1, for a Howard settling chamber.

In Eq. (2.5), ρ_p (particle density) was assumed to be much greater than ρ_a (gas density); hence $\rho_p - \rho_a = \rho_p$. The term in the brackets in Eq. (2.5) is often multiplied by a dimensionless empirical factor to correlate theoretical efficiencies with experimental data. If no information is available, it is suggested that 0.5 be used. Thus, Eq. (2.5) can be written as

$$\eta = 0.5[(g\rho_p BLN_e)/(18\mu Q)](d_p)^2 \tag{2.6}$$

Equations (2.1) and (2.5) give the theoretical collection efficiencies of a settling chamber for a single-sized particle. Since the gas stream entering a unit consists of a distribution of particles of various sizes, a fractional efficiency curve must be used to determine the overall collection efficiency.

The overall efficiency can be calculated using

$$\eta_{TOT} = \Sigma \eta_i w_i \tag{2.7}$$

where η_{TOT} = overall collection efficiency
η_i = fractional efficiency of specific size particle
w_i = weight fraction of specific size particle

Equations (2.1) and (2.2) were developed with the assumption that gas flow through the settling chamber is laminar. This assumption is usually incorrect. The gas flow is usually turbulent. The equation for determining efficiency when the flow is turbulent is

$$\eta = \exp[-(Lv_y/Hv_x)] \tag{2.8}$$

One must be careful when using Eqs. (2.1)–(2.6). For example, Eq. (2.4) is used to find the minimum particle size collected with 100% efficiency. This equation assumes that Stokes' law describes particle settling. However, Stokes' law does not work for particles greater than 100 μm; in some cases, it gives results that are gravely in error.

Process variables. The process design variables for a settling chamber consist of length (L), width (B), and height (H). The parameters are usually chosen by a chamber manufacturer to remove all particles above a specified size. The chamber design must provide conditions for sufficient particle residence time to capture the desired particle size range. This can be accomplished by keeping the velocity of the exhaust gas through the chamber as low as possible. If the velocity is too high dust re-entrainment will occur. However, the design velocity must not be so low as to cause the design of the chamber volume to be exorbitant. Consequently, the units are designed for a gas velocity in the range from 1 to 10 ft/s (0.305 to 3.05 m/s). Errors in estimating settling velocity from equations such as Eq. (2.2) can occur due to agglomeration and electrostatic effects. Therefore, in actual practice, the terminal settling velocities used for design purposes are based upon experience and tests under normal process conditions (EPA-81/10, p. 5-7).

In settling chamber designs, the velocity at which the gas moves through the chamber is usually called the *throughput velocity*. The velocity at which settled materials (particles) become re-entrained is called the *pickup velocity*. In order to avoid re-entrainment of collected dust, the throughput velocity must not exceed the pickup velocity. Experimental data or equipment supplier data such as that presented in Table 2.8 should be used to estimate the pickup velocity. As can be seen from this table, the pickup velocity can exceed 10 ft/s. If no data for determining the pickup velocity is available, the pickup velocity should be assumed to be 10 ft/s. In this case, the velocity of gas through the settling chamber (throughput velocity) must be less than 10 ft/s.

TABLE 2.8 Pickup Velocities of Various Materials

Material	Density (g/cm^3)	Median size (μm)	Pickup velocity (ft/s)
Aluminum chips	2.72	335	14.2
Asbestos	2.20	261	17.0
Nonferrous foundry dust	3.02	117	18.8
Lead oxide	8.26	14.7	25.0
Limestone	2.78	71	21.0
Starch	1.27	64	5.8
Steel shot	6.85	96	15.2
Wood chips	1.18	1370	13.0
Wood sawdust	–	1400	22.3

In terms of overall design considerations for gravity settlers, advantages include:

- low cost of construction and operation
- few maintenance problems
- relatively low operating pressure drops in the range of approximately 0.2–0.5 in. (0.51–1.27 cm) of water
- temperature and pressure limitations imposed only by the materials of construction used
- dry disposal of solid particulates

The disadvantages include:

- large space requirements
- relatively low overall collection efficiency (typically ranging from 20 to 60%)

In general, most gravity settlers are precleaners removing the relatively large particles (greater than 60 μm) before the gas stream enters a more efficient particulate control device such as a multicyclone, baghouse, electrostatic precipitator, or scrubber.

EXAMPLE 1: gravity settler—minimum particle size. A hydrochloric acid mist in air at 25°C is to be collected in a gravity settler. You are requested to calculate the smallest mist droplet (spherical in shape) that will be entirely collected by the settler. Assume the acid concentration to be uniform through the inlet cross-section of the unit and Stokes' law applies (EPA-84/09, p. 61).

Given conditions

dimensions of gravity settler = 30 ft wide, 20 ft high, 50 ft long

actual volumetric flow rate of acid gas in air = 50 ft^3/s

specific gravity of acid = 1.6

viscosity of air = 0.0185 cp = 1.243×10^{-5} lb/ft-s
density of air = 0.076 lb/ft^3

Solution:

1. Calculate the density of the acid mist using the specific gravity given.

$$\rho_p = \text{particle density} = (\text{specific gravity of fly ash})(\text{density of water})$$
$$= 1.6(62.4) = 99.84\ \text{lb/ft}^3$$

2. Calculate the minimum particle diameter both in feet and microns (μm) assuming Stokes' law applies.
For Stokes' law range:

$$\text{Minimum } d_p = (18\mu Q/g\rho_p BL)^{0.5}$$

where μ = gas (fluid) viscosity
Q = volumetric flow rate
g = gravity force
ρ_p = particle or mist density
B = width of settler
L = length of settler

Note that the above equation can be derived by substituting the terminal velocity for Stokes' law into the gravity settler equation $v = Q/BL$.

$$\text{Minimum } d_p = [(18)(1.243 \times 10^{-5})(50)/(32.2)(99.84)(30)(50)]^{0.5}$$
$$= 4.82 \times 10^{-5}\ \text{ft}$$
$$= (4.82 \times 10^{-5}\ \text{ft})(3.048 \times 10^{5}\ \mu\text{m/ft})$$
$$= 14.7\ \mu\text{m}$$

EXAMPLE 2: gravity settler—traveling grate stoker. A settling chamber is installed in a small heat plant which uses a traveling grate stoker. You are requested to determine the overall collection efficiency of the settling chamber given the operating conditions, chamber dimensions, and particle size distribution data (EPA-84/09, p. 62).

Given conditions

chamber width = 10.8 ft
chamber height = 2.46 ft
chamber length = 15.0 ft
volumetric flow rate of contaminated air stream = 70.6 scfs
flue gas temperature = 446°F
flue gas pressure = 1 atm
particle concentration = 0.23 grains/scf
particle specific gravity = 2.65
standard conditions = 32°F, 1 atm
particle size distribution data of the inlet dust from the traveling grate stoker are shown in Table 2.9.

TABLE 2.9 Particle Size Distribution Data

Particle size range, μm	Average particle diameter, μm	Inlet	
		grains/scf	wt %
0–20	10	0.0062	2.7
20–30	25	0.0159	6.9
30–40	35	0.0216	9.4
40–50	45	0.0242	10.5
50–60	55	0.0242	10.5
60–70	65	0.0218	9.5
70–80	75	0.0161	7
80–94	85	0.0218	9.5
94	94	0.0782	34
Total		0.23	100

Assume that the actual terminal settling velocity is one-half of the Stokes' law velocity.

Solution:

1. Plot the size efficiency curve for the settling chamber.

You need the size efficiency curve to calculate the outlet concentration for each particle size (range). These outlet concentrations are then used to calculate the overall collection efficiency of the settling chamber. The collection efficiency for a settling chamber can be expressed in terms of the terminal settling velocity, volumetric flow rate of contaminated stream, and chamber dimensions:

$$\eta = vBL/Q = [g\rho_p(d_p)^2/18\mu](BL/Q)$$

where
η = fractional collection efficiency
v = terminal settling velocity
B = chamber width
L = chamber length
Q = volumetric flow rate of the stream

1.a. Express the collection efficiency in terms of the particle diameter d_p.

1.a.1. Replace the terminal settling velocity in the above equation with Stokes' law.

Since the actual terminal settling velocity is assumed to be one-half of the Stokes' law velocity (according to the problem statement), the velocity equation becomes

$$v = g(d_p)^2\rho_p/36\mu$$
$$\eta = [g\rho_p(d_p)^2/36\mu](BL/Q)$$

1.a.2. Determine the viscosity of the air in lb/ft-s.

$$\text{Viscosity of air at } 446°\text{F} = 1.75 \times 10^{-5}\ \text{lb}_f/\text{ft-s}$$

1.a.3. Determine the particle density in lb/ft^3

$$\rho_p = 2.65(62.4) = 165.4\ \text{lb/ft}^3$$

1.a.4. Determine the actual flow rate in acfs.

In order to calculate the collection efficiency of the system at the operating conditions, the standard volumetric flow rate of contaminated air of 70.6 scfs is converted to actual volumetric flow of 130 acfs.

$$\begin{aligned} Q_a &= Q_s(T_a/T_s) \\ &= 70.6(446 + 460)/(32 + 460) \\ &= 130\,\text{acfs} \end{aligned}$$

1.a.5. Express the collection efficiency in terms of d_p, with d_p in ft. Also express the collection efficiency in terms of d_p, with d_p in microns.

Use the equation developed in 1.a.1. above, substitute values for ρ_p, g, B, L, μ, and Q in consistent units. Use the conversion factor for ft to microns. To convert d_p from ft^2 to μ^2, d_p is divided by $(304{,}800)^2$.

$$\begin{aligned} \eta &= [g\rho_p(d_p)^2/36\mu](BL/Q) \\ &= (32.2)(165.4)(10.8)(15)(d_p)^2/[(36)(1.75 \times 10^{-5})(130)(304{,}800)^2] \\ &= 1.134 \times 10^{-4}(d_p)^2 \end{aligned}$$

where d_p is in μm (microns).

1.b. Calculate the collection efficiency for each particle size.

For a particle diameter of 10 microns:

$$\eta = (1.134 \times 10^{-4})(d_p)^2 = (1.134 \times 10^{-4})(10)^2 = 1.1 \times 10^{-2} = 1.1\%$$

Table 2.10 provides the collection efficiency for each particle size.

1.c. The size efficiency curve for the settling chamber is shown in Fig. 2.22.

2. Read off the collection efficiency of each particle size from Fig. 2.22.

3. Calculate the overall collection efficiency (Table 2.11)

$$\begin{aligned} \eta &= \Sigma w_i \eta_i \\ &= (0.027)(1.1) + (0.069)(7.1) + (0.094)(14.0) + (0.105)(23.0) + (0.105)(34.0) \\ &\quad + (0.095)(48.0) + (0.070)(64.0) + (0.095)(83.0) + (0.340)(100.0) \\ &= 59.0\% \end{aligned}$$

TABLE 2.10 Collection Efficiency for Each Particle Size

d_p, μm	η, %
94	100
90	92
80	73
60	41
40	18.2
20	4.6
10	1.11

TABLE 2.11 Data for Calculation of Overall Collection Efficiency

d_p, μm	Weight fraction w_i	η_i
10	0.027	1.1
25	0.069	7.1
35	0.094	14
45	0.105	23
55	0.105	34
65	0.095	48
75	0.07	64
85	0.095	83
94	0.34	100
Total	1	

4.2 Cyclone

Cyclones provide a relatively low-cost method of removing particulate matter from exhaust gas streams. Cyclones are somewhat more complicated in design than simple gravity settling systems and their removal efficiency is accordingly much better than that of settling chambers. However, cyclones are not as efficient as baghouses and electrostatic precipitators, but are often installed as precleaners before these more effective devices (EPA-81/10, p. 6-1).

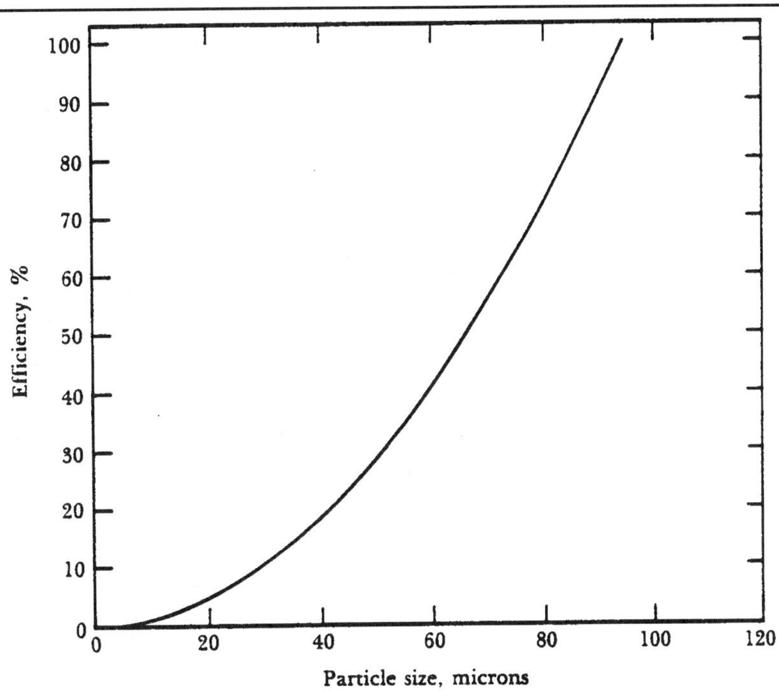

FIGURE 2.22 Size efficiency curve for settling chamber (*EPA-84/09, p. 136*).

Cyclones come in many sizes and shapes and have no moving parts. From the 1- and 2-cm diameter source sampling cyclones used for particle size analysis to the large 5-m diameter cyclone separators used after wet scrubbers, the basic separation principle remains the same. Particles enter the device with the flowing gas, as shown in Fig. 2.23; the gas stream is forced to turn, but the larger particles have too much momentum and cannot turn with the gas. These larger particles impact and fall down the cyclone wall, then are collected in a hopper. The gas stream actually turns a number of times in a helical pattern, much like the funnel of a tornado. The repeated turnings provide many opportunities for particles to break through the streamlines, thus hitting the cyclone wall.

The range of particle sizes collected in a cyclone is dependent upon the overall diameter and relative dimensions of the device. Various refinements, such as the use of skimmers, turning vanes, and water sprays, can in some cases improve efficiency. Stacking cyclones in series or in parallel can provide further alternatives for improving overall collection efficiency.

Cyclone type. Three types of cyclones are shown in Fig. 2.24. Figure 2.24a shows a typical tangential entry cyclone arrangement. These cyclones have a distinctive and easily recognized form and can be found in almost any industrial area of a town or a city at lumber companies, feed mills, cement plants, power plants, smelters, and at many other processing industrial sites. Since top inlet type cyclones are so widely used, most of this chapter will be devoted to their operational characteristics.

In axial entry cyclones (Fig. 2.24b), the gas inlet is parallel to the axis of the cyclone body. The exhaust process gases enter from the top and are directed into a vortex pattern by the vanes attached to the central tube. Axial entry cyclones are commonly used in multicyclone configurations.

The large cyclonic type separator, shown in Fig. 2.24c, is often used after wet scrubbers to collect particulate matter entrained in water droplets. The gas enters tangentially at the bottom of the drum, forming a vortex. The large water droplets are forced against the walls and are removed from the gas stream.

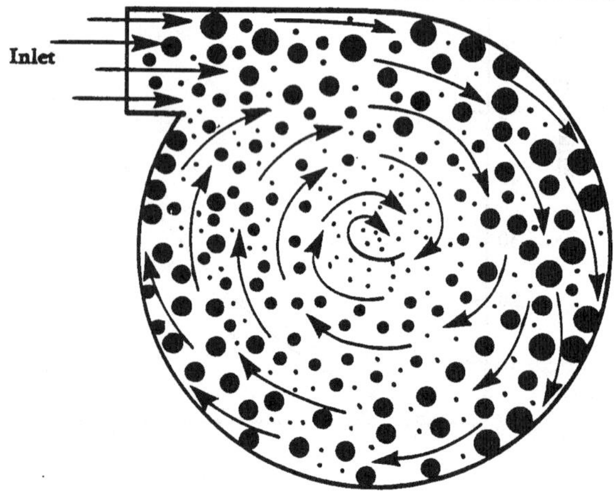

FIGURE 2.23 Particle collection mechanism (*EPA-81/10, p. 6-1*).

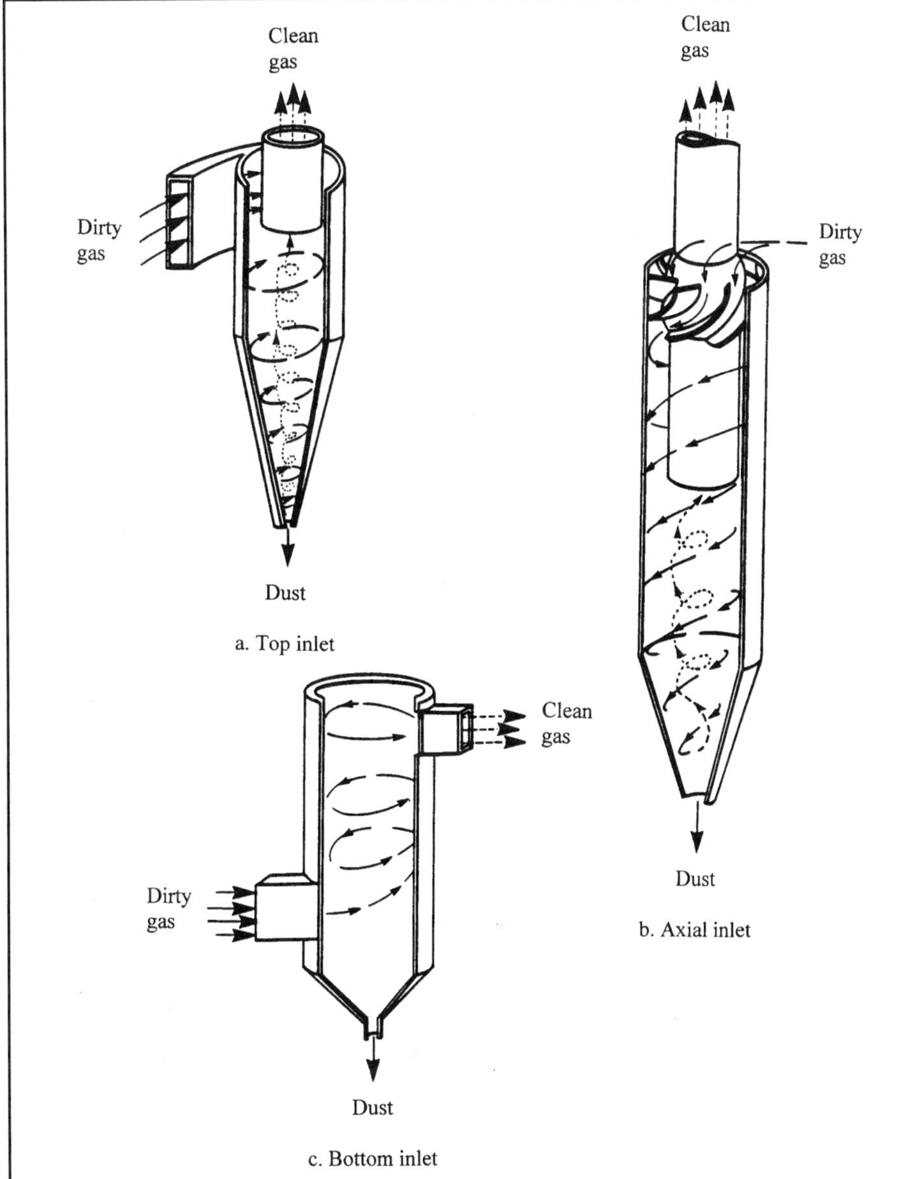

FIGURE 2.24 Types of cyclones (*EPA-81/10, p. 6-3*).

Cyclone design factors. The common cyclone has four major design areas:

- Inlet
- Cyclone body
- Dust discharge system
- Outlet

Each part of the cyclone is important to its overall effectiveness.

Inlet: The inlet directs gas into the cyclone body and is important in forming the vortex. In the cylindrical section, the particulate matter is forced to the wall. The gas spirals down to the cone section, where the body is tapered to give the gas enough rotational velocity to keep the dust against the wall. The particulate matter is collected in the hopper or vortex arrestor, and is continuously or periodically removed. The gas vortex changes from the downward to the upward direction at the bottom of the cone.

A tube extends from the top of the cyclone into the cylinder of the body. This tube extension is sometimes called a *vortex finder*. The ascending vortex enters the tube extension and then exits. A helical exhaust duct or a drum may be placed on the exit tube to straighten the gas flow before it flows through further ductwork. The length and width of the inlet are important, since the smaller the inlet, the greater the inlet velocity becomes. A greater inlet velocity gives greater efficiency but also increases the pressure drop.

Cyclone body and cone: The removal efficiency of a cyclone for a given size particle is very dependent upon the cyclone dimensions. The particle drop at a given volumetric flow rate is most affected by the diameter. The overall length determines the number of turns of the vortex. The greater the number of turns, the greater the efficiency.

The cone primarily serves as a mechanism to remove particulate matter down the walls to the hopper, and to provide greater tangential velocities near the bottom for removal of smaller particles. The vortex formed in a cyclone is, however, eccentric. Just as a tornado moves at an angle to the ground, the cyclone's main vortex can deviate from the vertical axis. For this reason, it has been found that the bottom of the cone should have a diameter of at least 1/4 of the cylinder diameter; otherwise, the outer vortex could touch the cone wall and entrain already separated particulate matter into the ascending vortex.

Even if optimum dimensions are selected, problems can occur within the cyclone which can reduce the efficiency. Rough walls will slow down the gas, decreasing the velocity of the vortex, thus causing small particles to be lost. Particles can bounce off the walls into the inner vortex and also be lost. Secondary flow patterns or eddies can form in the annular region at the inlet. The recirculation and turbulence here can prevent all particles from entering the vortex and instead be drawn into the outlet tube. The lower pressure in the inner vortex caused by its faster velocity can cause particles to drift in from the main vortex. These problems limit the efficiency of a cyclone.

Dust discharge system (hopper): The vortex will extend into the discharge bin if the bin is immediately below the cone and nothing is added at the bottom of the cone to arrest the vortex. Since the static pressure in the vortex core is slightly negative, dust can be re-entrained from the hopper into the inner vortex. Also, if leaks exist in the bin, dust can be sucked back up into the cyclone.

Cyclone gas outlet: The exit tube is an important consideration in the design of any cyclone. Its length must extend beyond the inlet so that the eddies created in the annulus between the tube and the walls do not mix particles up and into the exit tube.

Factors affecting cyclone performance. The factors that affect cyclone performance include:

- Centrifugal force
- Cut diameter
- Pressure drop
- Collection efficiency
- Summary of performance characteristics

Centrifugal force: Objects moving in circular paths tend to move away from the center of their motion. The object moves outward as if a force were pushing it out. This force is known as the *centrifugal force*. The whirling motion of the gas in a cyclone causes particulate matter in the gas to feel this force and move out to the walls. An expression for this phenomenon is as follows:

$$F = [\rho_p(d_p)^3(v_p)^2]/r \tag{2.9}$$

where F = centrifugal force
ρ_p = particles density, lb/ft^3 (kg/m^3)
d_p = particle diameter, ft (μm)
v_p = particle tangential velocity, ft/s (m/s)
r = radius of the circular path, ft (m)

F is the force that the particulate matter views as acting on it. This expression explains several of the cyclone characteristics. For example, $\rho_p(d_p)^3$ is merely proportional to the mass of the particle. The larger the mass, the greater the force. The tendency to move toward the walls is consequently increased and larger particles are more easily collected. The reason all of the particles do not move to the wall is because of the drag resistance of the air. The buffeting molecules in the gas resist the outward motion and act like an opposing force. Particles move to the wall when the centrifugal force is greater than the opposing drag force.

Note also from Eq. (2.9) that as r (the radius of the circular path) decreases, the force again increases. This is why smaller cyclones are more efficient for the collection of smaller-sized particles than are large cyclones.

These types of considerations, in conjunction with considerations of cyclone geometry and vortex formation, have led to the development of numerous performance equations. There are three important parameters which can be used to characterize cyclone performance. These are cut diameter $[d_p]$, pressure drop Δp, and overall collection efficiency η.

Cut diameter: The cut diameter is defined as the size (diameter) of particles collected with 50% efficiency. It is a convenient way of defining efficiency for a control device since it gives an idea of the effectiveness for a particle size range. A frequently used expression for cut diameter is given below:

$$[d_p]_{cut} = \{9\,\mu B/[2\pi n_t v_i(\rho_p - \rho_g)]\}^{0.5} \tag{2.10}$$

where $[d_p]_{cut}$ = cut diameter, ft (μm, microns)
μ = viscosity, lb/s-ft (Pa-s) (or kg/s-m)
B = inlet width, ft (m)
n_t = effective number of turns (5 to 10 for common cyclones)
v_i = inlet gas velocity, ft/s (m/s)
ρ_p = particle density, lb/ft^3 (kg/m^3)
ρ_g = gas density, lb/ft^3(kg/m^3)

The cut diameter, $[d_p]_{cut}$, is a characteristic of the control device and should not be confused with the geometric mean particle diameter d_{gm} of the size distribution.

The expression for the cut diameter (Eq. 2.10) has been found to agree within 4 μm for some experimental data. However, other experimental work has shown limitations to its application. A high-efficiency cyclone will have a cut diameter of typically 5–10 μm. Equation (2.10) is typical of most of those devised for determining the cut or critical diameter. Note that an increase in the number of turns, inlet velocity, or the particle density will decrease the cut size as one would expect. A decrease in viscosity will decrease the drag force opposing the centrifugal force and therefore also reduce the cut size (i.e. smaller-size particles will be collected).

Pressure drop, Δp: The pressure drop across a cyclone is an important parameter to the purchaser of such equipment. Increased pressure drop means greater costs for power to move exhaust gas through the control device. With cyclones, an increase in pressure drop usually means that there will be an improvement in collection efficiency (one exception to this is the use of pressure recovery devices attached to the exit tube; these reduce the pressure drop but do not adversely affect collection efficiency). For these reasons, there have been many attempts to predict pressure drops from design variables. The idea is that having such an equation, one could work backwards and optimize the design of new cyclones.

Collection efficiency: A number of formulations have been developed for determining the fractional cyclone efficiency η_i for a given size particle. *Fractional efficiency* is defined as the fraction of particles of a given size collected in the cyclone, compared to those of that size going into the cyclone.

Summary of performance characteristics: Efficiency, pressure drop, and costs are intimately related in cyclones, as they are with most other particulate control equipment. Many factors affect efficiency and pressure drop. A number of factors can be taken into account in the theoretical formulation; others cannot. A thorough evaluation of cyclone design will depend on previous experience or empirical information derived from experiments on similar cyclones. A summary of changes in performance characteristics is provided in Table 2.12.

Cyclone arrangement. It is apparent from the above discussion that small cyclones are more efficient than large cyclones. Small cyclones, however, have a higher pressure drop and are limited with respect to volumetric flow rates. Smaller cyclones can be arranged either in series or in parallel to substantially increase efficiency at lower pressure drops. These gains are somewhat offset, however, by increased maintenance problems. Multicyclone arrangements tend to plug more easily. When common hoppers are used in such arrangements, different flows through cyclones can lead to re-entrainment problems.

EXAMPLE 1: Cyclone—cut diameter and overall collection efficiency. You are requested to determine the cut size diameter and overall collection efficiency of a cyclone given the particle size distribution of a dust from a cement kiln (EPA-84/09, p. 66).

Given conditions

gas viscosity $\mu = 0.02$ cp $= 0.02\ (6.72 \times 10^{-4})$ lb/ft-s

specific gravity of the particle $= 2.9$

inlet gas velocity to cyclone $= 50$ ft/s

effective number of turns within cyclone $= 5$

cyclone diameter $= 10$ ft

cyclone inlet width $= 2.5$ ft

particle size distribution data is shown in Table 2.13.

TABLE 2.12 Changes in Performance Characteristics

Cyclone and process design changes*	Pressure drop	Efficiency	Cost
Increase cyclone size D	Decreases	Decreases	Increases
Lengthen cylinder L	Decreases slightly	Decreases	Increases
Lengthen cone Z	Decreases slightly	Increases	Increases
Increase exit tube diameter E	Decreases	Decreases	Increases
Increase inlet area (maintaining velocity)	Increases	Decreases	Decreases
Increase velocity	Increases	Increases	Operating costs higher
Increase temperature (maintaining velocity)	Decreases	Decreases	No change
Increase dust concentration	Decrease for large increases	Increases	No change
Increase particle size and/or density	No change	Increases	No change

*See Fig. 2.25.

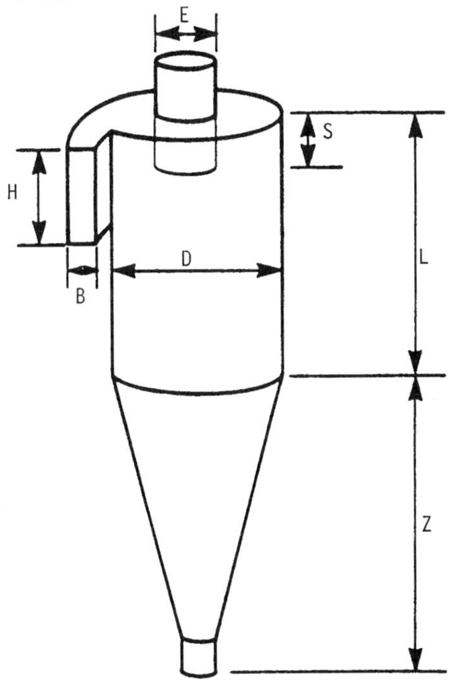

FIGURE 2.25 Nomenclature for a tangential entry cyclone (*EPA-81/10, p. 6-7*).

TABLE 2.13 Particle Size Distribution Data

Average particle size in range d_p, μm	% wt
1	3
5	20
10	15
20	20
30	16
40	10
50	6
60	3
> 60	7

Solution:

1. Calculate the cut diameter $[d_p]_{cut}$.
   The cut diameter is the particle diameter collected at 50% efficiency. For cyclones:

$$[d_p]_{cut} = [9\mu B_c/2\pi n_t v_i(\rho_p - \rho)]^{0.5}$$

where μ = gas viscosity, lb/ft-s
B_c = cyclone inlet width, ft

n_t = number of turns
v_i = inlet gas velocity, ft/s
ρ_p = particle density, lb/ft^3
ρ = gas density, lb/ft^3

1.a. Determine the value of $\rho_p - \rho$.
Since the particle density is much greater than the gas density, $\rho_p - \rho$ can be assumed to be ρ_p.

$$\rho_p - \rho = \rho_p = 2.9(62.4) = 180.96\,\text{lb/ft}^3$$

1.b. Calculate the cut diameter using the equation given.

$$\begin{aligned}[d_p]_{cut} &= [(9)(0.02)(6.72 \times 10^{-4})(2.5)/(2\pi)5(50)(180.96)]^{0.5} \\ &= 3.26 \times 10^{-5}\,\text{ft} \\ &= 9.94\,\mu\text{m}\end{aligned}$$

2. Complete size efficiency table (Table 2.14) using Lapple's method.
Lapple's method provides the collection efficiency as a function of the ratio of particle diameter to cut diameter. Use the equation

$$\eta = 1 - (1.0)/[1.0 + (d_p/[d_p]_{cut})^2]$$

3. Determine overall collection efficiency.

$$\begin{aligned}\Sigma w_i \eta_i(\%) &= 0 + 4 + 7.5 + 16 + 14.4 + 9.3 + 5.7 + 2.94 + 7 \\ &= 66.84\%\end{aligned}$$

TABLE 2.14 Size Efficiency Table

d_p, μm	w_i	$d_p/[d_p]_{cut}$	η_i, %	$w_i\eta_i$, %
1	0.03	0.1	0	0
5	0.2	0.5	20	4
10	0.15	1	50	7.5
20	0.2	2	80	16
30	0.16	3	90	14.4
40	0.1	4	93	9.3
50	0.06	5	95	5.7
60	0.03	6	98	2.94
> 60	0.07	—	100	7

EXAMPLE 2: cyclone—plan review. As an air pollution control officer you have been asked to evaluate a permit application to operate a cyclone as the only device on the ABC Stoneworks plant's gravel drier (EPA-84/09, p. 68).

Given conditions—design and operating data from permit application

average particle diameter = 7.5 microns (μm)
total inlet loading to cyclone = 0.5 grains/ft^3

cyclone diameter = 2.0 ft

inlet velocity = 50 ft/s

specific gravity of the particle = 2.75

number of turns = 4.5 turns

operating temperature = 70°F

viscosity of air at operating temperature = 1.21×10^{-5} lb/ft-s

the cyclone is a conventional one.

Air Pollution Control Agency criteria

maximum total outlet loading = 0.1 grains/ft^3

cyclone efficiency as a function of particle size ratio is provided in Fig. 2.26 (Lapple's curve).

Solution:

1. Determine the collection efficiency of the cyclone.

Utilize Lapple's method which provides collection efficiency values from a graph relating efficiency to the ratio of average particle diameter to the cut diameter. The cut diameter is the particle diameter collected at 50% efficiency (see Fig. 2.26).

1.a. Calculate the cut diameter

The equation for determining cut size diameter is

$$[d_\text{p}]_\text{cut} = [9\mu B_\text{c}/2\pi n_\text{t} v_\text{i}(\rho_\text{p} - \rho)]^{0.5}$$

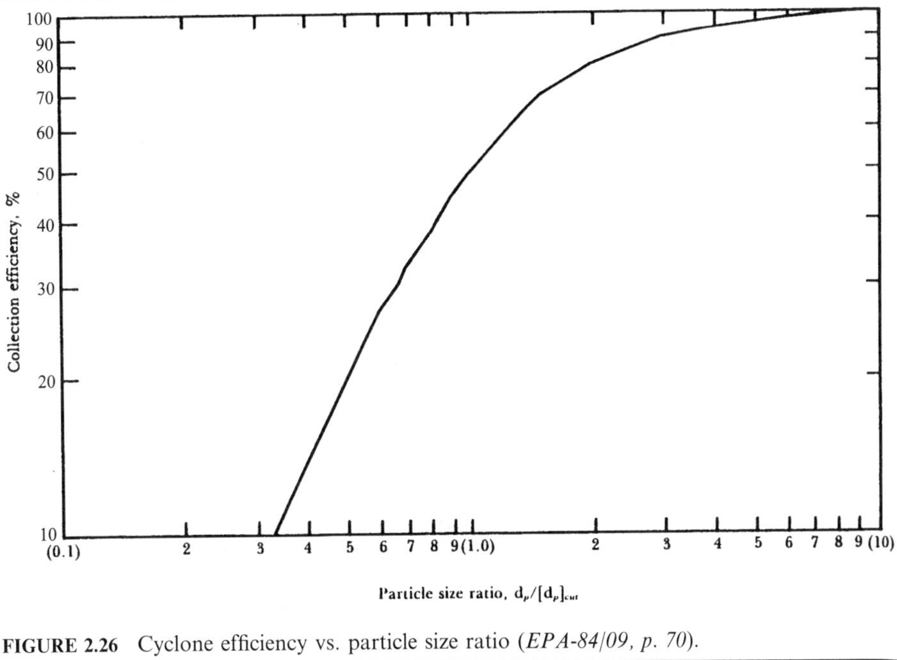

FIGURE 2.26 Cyclone efficiency vs. particle size ratio (*EPA-84/09, p. 70*).

where
μ = gas viscosity, lb/ft-s
B_c = cyclone inlet width, ft
n_t = number of turns
v_i = inlet gas velocity, ft/s
ρ_p = particle density, lb/ft^3
ρ = gas density, lb/ft^3

1.a.1. Determine the inlet width of the cyclone, B_c.

The permit application has established this cyclone as conventional. The inlet width of a conventional cyclone is 1/4 the cyclone diameter.

$$B_c = \text{cyclone diameter}/4 = 2.0/4 = 0.5\,\text{ft}$$

1.a.2. Determine the value of $\rho_p - \rho$.

Since the particle density is much greater than the gas density, $\rho_p - \rho$ can be assumed to be ρ_p.

$$\rho_p - \rho = \rho_p = 2.75(62.4) = 171.6\,\text{lb/ft}^3$$

1.a.3. Calculate the cut diameter using the equation given.

$$\begin{aligned}[d_p]_{cut} &= [9\mu B_c/2\pi n_t v_i(\rho_p - \rho)]^{0.5} \\ &= [(9)(1.21 \times 10^{-5})(0.5)/(2\pi)4.5(50)(171.6)]^{0.5} \\ &= 1.5 \times 10^{-5}\,\text{ft} \\ &= 4.57\,\text{microns}\end{aligned}$$

1.b. Calculate the ratio of average particle diameter to the cut diameter.

$$d_p/[d_p]_{cut} = 7.5/4.57 = 1.64$$

1.c. Determine the collection efficiency utilizing Lapple's curve (see Fig. 2.26).

$$\eta = 72\%$$

2. Calculate the required collection efficiency for the approval of the permit.

$$\begin{aligned}\eta &= [(\text{inlet loading} - \text{outlet loading})/(\text{inlet loading})](100) \\ &= [(0.5 - 0.1)/(0.5)](100) \\ &= 80\%\end{aligned}$$

3. Would you approve the permit?

Since the collection efficiency of the cyclone is lower than the collection efficiency required by the agency, the permit should not be approved.

4.3 Electrostatic Precipitator (ESP)

An electrostatic precipitator (ESP) is an effective device for controlling particle emissions from cement kiln, pulp and paper plants, acid plants, sintering operations, and other industrial sources. The method is extensively used where dust emissions are less than 10–20 μm in size with a predominant portion in the submicron range (EPA-81/10, p. 7-1).

In general, an electrostatic precipitator is comprised of four essential components, as shown in Fig. 2.27. The essential components are

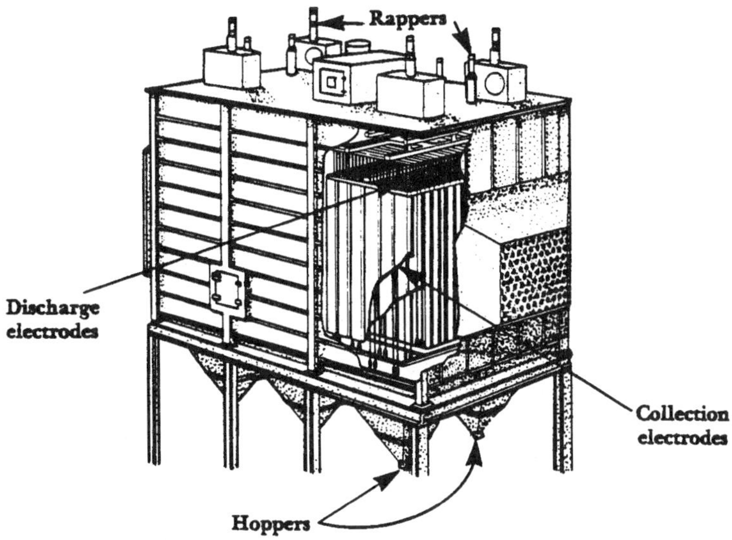

FIGURE 2.27 Typical plate and wire single-stage ESP (*EPA-81/10, p. 7-1*).

- discharge electrodes
- collection electrodes
- rappers
- hoppers

The discharge electrode is normally a wire where a corona discharge occurs. This electrode is used to ionize the gas (which charges the particles) and create an electric field. The collection electrode consists of either a tube or flat plate which is oppositely charged (relative to the discharge electrode) and is the surface where the charged particles are collected. The rapper is a device used to impart a vibration or shock to dislodge the deposited dust on the electrodes. Rappers are used to remove dust accumulated on both the collection electrodes and discharge electrodes. Hoppers are located at the bottom of the precipitator and are used to collect and store the dust removed by the rapping process.

ESP type. The types of electrostatic precipitators include (EPA-81/10, p. 7-2):

- Low voltage two-stage precipitators
- High voltage single-stage precipitators
  Tubular high voltage single-stage precipitators;
  Plate high voltage single-stage precipitators

Low voltage two-stage ESPs: Low voltage two-stage precipitators are limited almost exclusively to the collection of liquid aerosols discharged from sources such as meat smokehouses, pipe-coating machines, asphalt paper saturators, and high-speed grinding machines. The precipitators were originally designed for air purification in conjunction with air conditioning systems (they are also referred to as electronic air filters). Two-stage ESPs have been used primarily for the control of finely divided liquid particles. Controlling solid or sticky materials is usually difficult, and the collector becomes ineffective for dust loadings greater than 0.4

grains per standard cubic foot (0.916 g/m^3). Therefore, two-stage precipitators have limited use for particulate emission control.

High voltage single-stage precipitators: The high voltage single-stage precipitator is the more popular type and has been used successfully to collect both solid and liquid particulate matter in industrial facilities such as smelters, steel furnaces, cement kilns, municipal incinerators, and utility boilers. There are two major types of high voltage single-stage ESP configuration. Particles are both charged and collected in a single stage.

- *Tubular precipitators*: Tubular precipitators consist of cylindrical collection electrodes with discharge electrodes located in the center of the cylinders. Dirty gas flows into the cylinder, where precipitation occurs. The negatively charged particles migrate to and are collected on grounded collecting tubes. The collected dust or liquid is removed by washing the tubes with water sprays located directly above the tubes. These precipitators are generally referred to as water-washed ESPs. Tubular precipitators are generally used for collecting mists or fogs. Tube diameters typically vary from 0.5 to 1 ft (0.15 to 0.31 m), with length usually ranging from 6 to 15 ft (1.85 to 4.6 m).
- *Plate precipitators*: Plate electrostatic precipitators are used more often than tubular ESPs in industrial applications. High voltage is used to subject the particles in the gas stream to an intense electric field. Dirty gas flows into a chamber consisting of a series of discharge electrodes (wires) spaced along the center line of adjacent plates, as shown in Fig. 2.28. Charged particles migrate to and are collected at oppositely charged collection plates. Collected particles are usually removed by rapping (dry precipitator) or by a liquid film (wet precipitator). Particles fall by force of gravity into hoppers, where they are stored prior to removal and final disposal.

Collection efficiency. ESP collection efficiency can be expressed by the following two equations (EPA-81/10, p. 7-9):

- Migration velocity equation
- Deutsch–Anderson equation

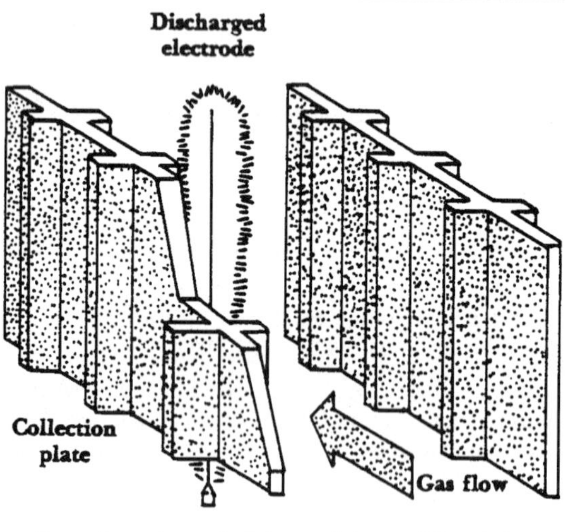

FIGURE 2.28 Gas flow through a wire and plate precipitator (*EPA-81/10, p. 7-4*).

Migration velocity: Once the particle is charged, it migrates toward the grounded collection electrode. An indicator of particle movement toward the collection electrode is denoted by the symbol w and is called the *particle migration velocity* or *drift velocity*. The migration velocity parameter represents the collectability of the particle within the confines of a specific collector. The migration velocity can be expressed in terms of

$$w = d_p E_o E_p / (4\pi\mu) \tag{2.11}$$

where w = migration velocity
d_p = diameter of the particle, μm
E_o = strength of field in which particles are charged, volts per meter (represented by peak voltage)
E_p = strength of field in which particles are collected, volts per meter (normally the field close to the collecting plates)
μ = viscosity of gas, Pa-s

Migration velocity is quite sensitive to the voltage, since the electric field appears twice in Eq. (2.11). Therefore, the precipitator must be designed using the maximum electric field for maximum collection efficiency. The migration velocity is also dependent on particle size; larger particles are collected more easily than smaller ones.

Particle migration velocity can also be determined by the following equation:

$$w = qE_p / (4\pi\mu r) \tag{2.12}$$

where w = migration velocity
q = particle charge (charges)
E_p = strength of field in which particles are collected, volts per meter (normally the field close to the collecting plates)
μ = viscosity of gas, Pa-s
r = radius of the particle, μm

Deutsch–Anderson equation: This equation has been used to determine the collection efficiency of the precipitator under ideal conditions. The simplest form of the equation is

$$\eta = 1 - \exp(-wA/Q) \tag{2.13}$$

where η = fractional collection efficiency
A = collection surface area of the plates
Q = gas volumetric flow rate
w = drift velocity

This equation has been used extensively for many years for theoretical collection efficiency calculations. Unfortunately, while the equation is scientifically valid, there are a number of operating parameters that can cause the results to be in error by a factor of two or more. The Deutsch–Anderson equation neglects three significant process variables.

1. It completely ignores the fact that dust re-entrainment may occur during the rapping process.
2. It assumes that the particle size and, consequently, the migration velocity is uniform for all particles in the gas stream.
3. It assumes that the gas flow rate is uniform everywhere across the precipitator and that particle sneakage through the hopper section does not occur.

Therefore, this equation should be used only for making preliminary estimates of precipitation collection efficiency.

Design parameters. Many parameters must be taken into consideration in the design and specification of electrostatic precipitators. The typical design parameters include (EPA-81/10, p. 7-11):

- Resistivity
- Specific collection area
- Aspect ratio
- Gas flow distribution
- Electrical sectionalization

Resistivity: Particle resistivity is a condition of the particle in the gas stream that can alter the actual collection efficiency of an ESP design. Resistivity is a term that describes the resistance of the collected dust layer to the flow of electrical current. By definition, *resistivity* is the electrical resistance of a dust sample 1.0 cm^2 in cross-sectional area, 1.0 cm thick; it is recorded in units of ohm-cm. It can also be described as the resistance to charge transfer by the dust. Dust resistivity values can be classified roughly into three groups:

- between 10^4 and 10^7 ohm-cm (low resistivity)
- between 10^7 and 10^{10} ohm-cm (normal resistivity)
- above 10^{10} ohm-cm (high resistivity)

Specific collection area (SCA): The specific collection area is defined as the ratio of collection surface area to the gas flow rate into the collector. The importance of this term is that it represents the A/Q relationship in the Deutsch–Anderson equation.

$$\text{SCA} = (\text{Total collection surface, ft}^2)/[\text{flow rate (1000 acfm)}]$$

$$= \text{m}^2/(1000\,\text{m}^3/\text{h}) \text{ in metric units}$$

Increases in the SCA of a precipitator design will in most cases increase the collection efficiency of the precipitator. Most conservative designs call for an SCA of 350 to 400 ft^2 per 1000 acfm (19 and 22 m^2 per 1000 m^3/h) to achieve 99.5% particle removal. The general range of SCA is between 200 and 800 ft^2 per 1000 acfm (11 and 45 m^2 per 1000 m^3/h), depending on precipitator design conditions and desired collection efficiency.

Aspect ratio: The aspect ratio is the ratio of the total length to height of collector surface. The aspect ratio can be calculated by

$$\text{AR} = (\text{effective length})/(\text{effective height})$$

Having a precipitator chamber many times larger in length than in height would be ideal. However, space limitations and cost could be prohibitive. The aspect ratio for ESPs can range from 0.5 to 2.0. For 99.5% collection efficiency, the precipitator design should have an aspect ratio of greater than 1.0.

Gas flow distribution: Gas flow through the ESP chamber should be slow and evenly distributed throughout the unit. The gas velocities in the duct ahead of the ESP are generally between 20 and 80 ft/s (6 and 24 m/s). The gas velocity into the ESP must be reduced for adequate particle collection. This is achieved by using an expansion inlet plenum.

The inlet plenum contains diffuser-perforated plate openings to evenly distribute the gas flow throughout the precipitator. Typical gas velocities in the ESP chamber range from 2 to 8 ft/s (0.6 to 2.4 m/s). With aspect ratios of 1.5, the optimum gas velocity is generally between 5 and 6 ft/s (1.5 and 1.8 m/s).

Electrical sectionalization: Precipitator performance is dependent on the number of individual sections or fields installed. The maximum voltage at which a given field can be maintained depends on the properties of the gas and dust being collected. These parameters may vary from one point to another in the unit. To keep each section of the precipitator working at high efficiency, a high degree of sectionalization is recommended. Multiple fields or stages are used to provide electrical sectionalization. Each field has separate power supplies and controls to adjust for varying gas conditions in the unit.

In general, precipitators have voltage control devices that automatically limit precipitator power input. A well-designed automatic control system keeps the voltage level at approximately the value needed for optimum particle charging by the discharge electrodes.

The voltage control devices operate in the following manner: increases in voltage cause a greater spark rate between the discharge and collection electrodes. Occurrence of a spark counteracts high ESP performance, since it causes an immediate, short-term collapse of the precipitator field. Consequently, less useful power is applied to capture particles. There is, however, an optimal sparking rate where the gains in particle charging are just offset by corona current losses from sparkover.

Measurements on commercial precipitators have determined that the optimal sparking rate is between 50 and 150 sparks per minute per electrical section. The objective in power control is to maintain corona power input at this optimal sparking rate. This can be accomplished by momentarily reducing precipitator power whenever excessive sparking occurs.

The need for separate fields arises mainly because power input requirements differ at various locations in a precipitator. The particulate matter concentration is generally high at the inlet sections of the precipitator. High dust concentrations tend to suppress corona current. Therefore, a great deal of power is needed to generate corona discharge for optimal particle charging at the inlet.

In the downstream fields of a precipitator, the dust loading is usually lighter. Consequently, corona current flows freer in downstream fields. Particle charging will more likely be limited by excessive sparking in downstream fields than in the inlet fields. The power to the outlet sections must still be high in order to collect small particles, particularly if they exhibit high resistivity.

If the precipitator had only one power set, the excessive sparking would limit the power input to the entire precipitator. This would result in a reduction of overall collection efficiency.

Review of ESP design plan. The first step in reviewing design plans for air pollution permits is to read the vendor literature and specifications of the precipitator design. The design specifications should include at least (EPA-81/10, p. 7-26):

- Exhaust gas flow rate and temperature
- Inlet dust concentration
- Specific collection area (SCA)
- Gas velocity in the precipitator
- Distance between the plates
- Aspect ratio
- Number and size of transformer-rectifier (T-R) sets
- Number of fields
- Design migration velocity
- Corona power/1000 m^3/min
- Corona current/ft^2 plate area
- Design collection efficiency
- Outlet dust concentration

The next step is to review the outlet concentration from the ESP. The concentration must meet the grain-loading requirements of air pollution regulations. The design reviewer can determine if the calculated outlet values, using the Deutsch–Anderson equation, are within the regulation limits. In addition, requiring the source to perform a source test to verify the designed collection efficiency of the ESP would be extremely useful.

EXAMPLE 1: electrostatic precipitator—process change. A horizontal parallel-plate electrostatic precipitator consists of a single duct 24 ft high and 20 ft deep with an 11 inch plate-to-plate spacing. Given a collection efficiency at a gas flow rate of 4200 acfm, you are required to determine the bulk velocity of the gas, outlet loading, and drift velocity of this electrostatic precipitator. You are also requested to calculate a revised collection efficiency if the flow rate and the plate spacing are changed (EPA-84/09, p. 71).

Given conditions

inlet loading $= 2.82\,\text{grains/ft}^3$

collection efficiency at 4200 acfm = 88.2%

increased (new) flow rate = 5400 acfm

new plate spacing = 9 in.

Solution:

1. Calculate the bulk flow (throughput) velocity v.

The equation for calculating throughput velocity is

$$V = Q/S$$

where Q = gas volumetric flow rate
S = cross-sectional area through which the gas passes

$$\begin{aligned} V &= Q/S \\ &= (4200)/[(11/12)(24)] \\ &= 191\,\text{ft/min} \\ &= 3.2\,\text{ft/s} \end{aligned}$$

2. Calculate outlet loading

Remember that

$$\eta\ (\text{fractional}) = (\text{inlet loading} - \text{outlet loading})/(\text{inlet loading})$$

Therefore

$$\begin{aligned} \text{Outlet loading} &= (\text{inlet loading})(1 - \eta) \\ &= (2.82)(1 - 0.882) \\ &= 0.333\,\text{grains/ft}^3 \end{aligned}$$

3. Calculate the drift velocity.

The drift velocity is the velocity at which the particle migrates toward the collection electrode within the electrostatic precipitator.

The Deutsch–Anderson equation describing the collection efficiency of an electrostatic precipitator is

$$\eta = 1 - \exp(-wA/Q)$$

where η = fractional collection efficiency
A = collection surface area of the plates
Q = gas volumetric flow rate
w = drift velocity

3.a. Calculate the collection surface area A.

Remember that the particles will be collected on both sides of the plate.

$$A = (2)(24)(20) = 960\,\text{ft}^2$$

3.b. Calculate the drift velocity w.

Since the collection efficiency, gas flow rate, and collection surface area are now known, the drift velocity can easily be found from the Deutsch–Anderson equation:

$$\eta = 1 - \exp(-wA/Q)$$
$$0.882 = 1 - \exp[-(960)(w)/(4200)]$$

Solving for w

$$w = 9.36\,\text{ft/min}$$

4. Calculate the revised collection efficiency when the gas volumetric flow rate is increased to 5400 acfm.

Assume the drift velocity remains the same.

$$\begin{aligned}\eta &= 1 - \exp(-wA/Q)\\ &= 1 - \exp(-(960)(9.36)/(5400)]\\ &= 0.812\\ &= 81.2\%\end{aligned}$$

5. Does the collection efficiency change with changed plate spacing?

No. Note that the Deutsch–Anderson equation does not contain a plate-spacing term.

EXAMPLE 2: electrostatic precipitator—collection efficiency. You have been requested to calculate the collection efficiency of an electrostatic precipitator containing three ducts with plates of a given size, assuming a uniform distribution of particles. Also, determine the collection efficiency if one duct is fed 50% of the gas and the other passages 25% each (EPA-84/09, p. 73).

Given conditions

volumetric flow rate of contaminated gas = 4000 acfm

operating temperature and pressure = 200°C and 1 atm

drift velocity = 0.40 ft/s

size of the plate = 12 ft long and 12 ft high

plate-to-plate spacing = 8 in.

Solution:

1. What is the collection efficiency of the electrostatic precipitator with a uniform volumetric flow rate to each duct?

The Deutsch–Anderson equation describing the collection efficiency of an electrostatic precipitator is

$$\eta = 1 - \exp(-wA/Q)$$

where η = fractional collection efficiency
A = collection surface area of the plates

Q = gas volumetric flow rate
w = drift velocity

1.a. Calculate the collection surface area per duct, A.
Considering both sides of the plate,

$$A = (2)(12)(12) = 288\,\text{ft}^2$$

1.b. Calculate the collection efficiency of the electrostatic precipitator using the Deutsch–Anderson equation.

The volumetric flow rate (Q) through a passage is one-third of the total volumetric flow rate,

$$\begin{aligned} Q &= (4000)/(3)(60) \\ &= 22.22\,\text{acfs} \\ \eta &= 1 - \exp(-wA/Q) \\ &= 1 - \exp[-(288)(0.4)/(22.22)] \\ &= 0.9944 \\ &= 99.44\% \end{aligned}$$

2. What is the collection efficiency of the electrostatic precipitator if one duct is fed 50% of gas and the others 25% each. The collection surface area per duct remains the same.

2.a. What is the collection efficiency of the duct with 50% of gas, η_1?

2.a.1. Calculate the volumetric flow rate of gas through the duct in acfts.

$$Q = (4000)/(2)(60) = 33.33\,\text{acfs}$$

2.a.2. Calculate the collection efficiency of the duct with 50% of gas.

$$\begin{aligned} \eta_1 &= 1 - \exp[-(288)(0.4)/(33.33)] \\ &= 0.9684 \\ &= 96.84\% \end{aligned}$$

2.b. What is the collection efficiency (η_2) of the ducts with 25% of gas flow in each?
2.b.1. Calculate the volumetric flow rate of gas through the duct in acfs.

$$Q = (4000)/(4)(60) = 16.67\,\text{acfs}$$

2.b.2. Calculate the collection efficiency (η_2) of the duct with 25% of gas.

$$\begin{aligned} \eta_2 &= 1 - \exp[-(288)(0.4)/(16.67)] \\ &= 0.9990 \\ &= 99.90\% \end{aligned}$$

2.c. Calculate the new overall collection efficiency.
The key equation becomes:

$$\begin{aligned} \eta_t &= (0.5)(\eta_1) + (2)(0.25)(\eta_2) \\ &= (0.5)(96.84) + (2)(0.25)(99.90) \\ &= 98.37\% \end{aligned}$$

EXAMPLE 3: electrostatic precipitator—plan review. Fractional efficiency curves describing the performance of a specific model of an electrostatic precipitator have been compiled by a vendor. Although you do not possess these curves, the cut diameter is known. The vendor claims that this particular model will perform with a given efficiency under your operating condition. You are asked to verify this claim and to make certain that the effluent loading does not exceed the standard set by EPA (EPA-84/09, p. 75).

Given conditions

plate-to-plate spacing = 10 in.

cut diameter = 0.9 μm (microns)

collection efficiency claimed by the vendor = 98%

inlet loading = 14 grains/ft^3

EPA standard for the outlet loading = 0.2 grains/ft^3 (maximum)

The particle size distribution is given in Table 2.15.
A Deutsch–Anderson type of equation describing the collection efficiency of an electrostatic precipitator is

$$\eta = 1 - \exp(-Kd_p)$$

where η = fractional collection efficiency
K = empirical constant
d_p = particle diameter

TABLE 2.15 Particle Size Distribution

Weight range	Average particle size d_p, μm
0–20%	3.5
20–40%	8
40–60%	13
60–80%	19
80–100%	45

Solution:

1. Is the overall efficiency of the electrostatic precipitator equal to or greater than 98%? Since the weight fractions are given, collection efficiencies of each particle size are needed to calculate the overall collection efficiency.

1.a. Determine the value of K by using the given cut diameter.

Remember that the cut diameter is the particle diameter collected at 50% efficiency. Since the cut diameter is known, you can solve the Deutsch–Anderson type equation directly for K.

$$\eta = 1 - \exp(-Kd_p)$$
$$0.5 = 1 - \exp[-K(0.9)]$$

Solving for K,

$$K = 0.77$$

1.b. Calculate the collection efficiency using the Deutsch–Anderson equation.
Use the Deutsch–Anderson equation to calculate the collection efficiency. For $d_p = 3.5$

$$\begin{aligned}\eta &= 1 - \exp[(-0.77)(3.5)]\\ &= 0.9325\end{aligned}$$

Table 2.16 shows the collection efficiency for each particle size.

TABLE 2.16 Collection Efficiency for Each Particle Size

Weight fraction w_i	Average particle size d_p, μm	η_i
0.2	3.5	0.9325
0.2	8	0.9979
0.2	13	0.9999
0.2	19	0.9999
0.2	45	0.9999

1.c. Calculate the overall collection efficiency.

$$\begin{aligned}\eta &= \Sigma w_i \eta_i\\ &= (0.2)(0.9325) + (0.2)(0.9979) + (0.2)(0.9999) + (0.2)(0.9999) + (0.2)(0.9999)\\ &= 0.9861\\ &= 98.61\%\end{aligned}$$

where η = overall collection efficiency
w_i = weight fraction of ith particle size
η_i = collection efficiency of ith particle size

1.d. Is the overall collection efficiency greater than 98%?
Yes

2. Does the outlet loading meet EPA's standard?

2.a. Calculate the outlet loading in grains/ft^3.

$$\text{Outlet loading} = (1.0 - \eta)(\text{inlet loading})$$

where η is the fractional efficiency for the above equation.

$$\begin{aligned}\text{Outlet loading} &= (1.0 - 0.9861)(14)\\ &= 0.195\ \text{grains/ft}^3\end{aligned}$$

2.b. Is the outlet loading less than 0.2 grains/ft^3.
Yes

3. Is the vendor's claim verified?
Yes.

4.4 Fabric Filtration

Fabric filtration is one of the most common techniques used to collect particulate matter (EPA-81/10, p. 8-1).

Filter type. There are two basic types of filters:

- Disposable filters
- Nondisposable filters

Disposable filters: Disposable filters are similar to those used in a home heating or air conditioning system. Disposable filters can be constructed as mats or as deep beds (12 in. or more). Mat filters are usually made using fiber glass mats with a thin metal plate on the outside of the filter used for structural reinforcement. Depth filters are generally constructed using fiber glass fibers, glass fiber paper, or some other inert material such as fine grade steel to form a deep mesh. The filters are very efficient (99.97%) for the collection of 0.3 μm and larger particles but must be replaced when they become loaded with particulate matter (when the pressure drop across the filter becomes excessive). Depth filters are widely useful for the collection of toxic dust materials.

Nondisposable filters: Nondisposable fabric filters consist of some type of fabric material (nylon, wool, or others) that is commonly used to clean dirty exhaust gas streams from industrial processes. The particles are retained on the fabric material, while the cleaned gas passes through the material. The collected particles are then removed from the filter by a cleaning mechanism: by shaking or using blasts of air. The removed particles are stored in a collection hopper until they are disposed of or are reused in the process.

Collection mechanisms. Particles are collected on a filter by a combination of the mechanisms described earlier. The most important here are

- Impaction
- Direct interception
- Diffusion
- Others

Impaction: In collection by impaction, the particles in the gas stream have too much inertia to follow the gas streamlines around the fiber and are impacted on the fiber surface.

Direct interception: In the case of direct interception, the particles have less inertia and barely follow the gas streamlines around the fiber. If the distance between the center of the particle and the outside of the fiber is less than the particle radius, the particle will graze or hit the fiber and be 'intercepted.'

Impaction and direct interception mechanisms account for 99% of the collection of particles greater than 1 μm aerodynamic diameter in fabric filter systems.

Diffusion: The third collection mechanism is that of diffusion. In diffusion, small particles are affected by collisions on a molecular level. Particles less than 0.1 μm aerodynamic diameter have individual or random motion. The particles do not necessarily follow the gas streamlines, but move randomly throughout the fluid. This is known as Brownian motion. The particles may have a different velocity than the fluid and at some point could come in contact with the fiber and be collected.

Others: Other collection mechanisms such as gravitational settling, agglomeration, and electrostatic attraction may contribute slightly to particle collection. Particles can agglomerate or grow in size and then be more easily collected by the fibers. Some particles have a small electrostatic charge and can be attracted to a material of opposite charge. Electrostatic charges could, on the other hand, have a bad effect if the charges of the particles and fiber are the same. Electrostatic charges can be particularly useful for the capture of particles in the submicron range. The use of a selected fiber material or a specially coated material may enhance particle capture. Different materials will develop electrostatic charges of varying degree and sign.

4.5 Baghouse

Nondisposable fabric filter systems were developed for industrial application as were baghouse systems (EPA-81/10, p. 8-4).

The particle collection surface is composed of the filtering material and usually some type of support structure. Most US baghouse designs employ long cylindrical tubes that contain felted fabric or woven cloth as the filtering medium. The cloth can be supported at the top and bottom of the bag by metal rings or by a cage that completely supports the entire bag, as shown in Fig. 2.29.

Baghouses are usually constructed using many cylindrical bags that hang vertically in the baghouse. The number of bags can vary from a few hundred to a thousand or more depending on the size of the baghouse. When dust layers have built up to a sufficient thickness, the bag is cleaned, causing the dust particles to fall into a collection hopper. Bag cleaning can be done by a number of methods. Particles are stored in the hopper and are usually removed by a pneumatic or screw conveyer. The baghouse is enclosed by sheet metal to contain the collected dust and to protect the bags from atmospheric environmental conditions.

Dirty gas is either pushed or pulled through the baghouse by a fan. When the dust-laden gas is pushed through the baghouse, the collector is called a *positive-pressure baghouse*. When the fan is on the downstream side of the baghouse, the dirty gas is pulled through the baghouse and the collector is called a *negative-pressure baghouse*. The structure of the negative-pressure baghouse must be reinforced because of the suction on the baghouse shell. Vendors can construct positive-pressure baghouses with weaker support structure since the positive pressure will counterbalance the atmospheric pressure on the baghouse shell. Limitations, however, do exist since the fan is located on the dirty side of the system. Premature deterioration of fan blades and bearings can occur in this configuration.

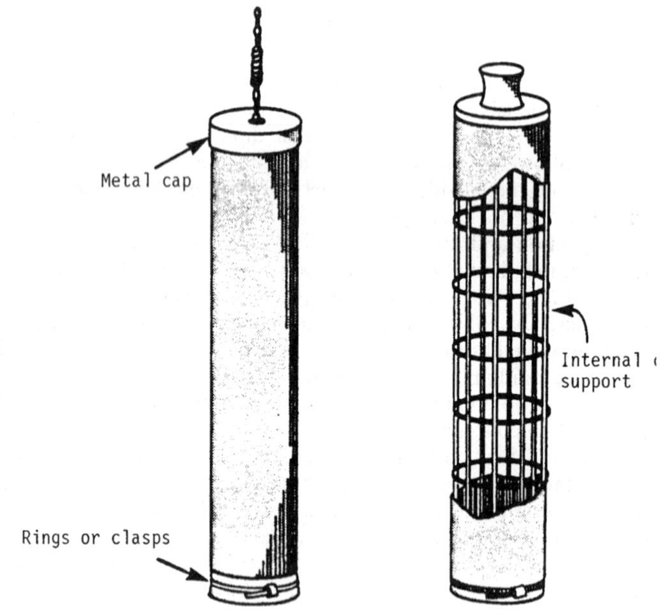

FIGURE 2.29 Bags and support (*EPA-81/10, p. 8-4*).

Baghouse filtration type. There are two filtration designs used in baghouses: interior filtration and exterior filtration.

Interior filtration: In interior filtration, particles are collected on the inside of the bag. The dust-laden gas enters either through the top or bottom of the collector and is directed inside the bag by diffuser vanes and a cell plate. The cell plate is a thin metal sheet surrounding the bag openings. The cell plate separates the clean gas section from the baghouse inlet. The particles are filtered by the bag and clean air exits through the outside of the bag.

Exterior filtration: In exterior filtration, dust is collected on the outside of the bags. The filtering process goes from the outside of the bag to the inside with clean gas exiting through the inside of the bag. Consequently, some type of bag support is necessary, such as an internal bag cage or rings sewn into the bag fabric.

The dust-laden gas inlet position for both filtration systems often depends on the baghouse model and manufacturer. If the gas enters the top of the unit, a downwash of gas occurs which tends to clean the bags somewhat while the bags are filtering. This usually allows slightly higher gas volumes to be filtered through the baghouse before cleaning is required. If the gas enters the bottom of the unit, the inlet is positioned at the very top part of the dust hopper, as shown in Fig. 2.30. Bottom or hopper inlets are easier to design and manufacture structurally than are the top inlets. However, when using hopper inlets, vendors must carefully design gas flows to avoid dust re-entrainment from the hopper.

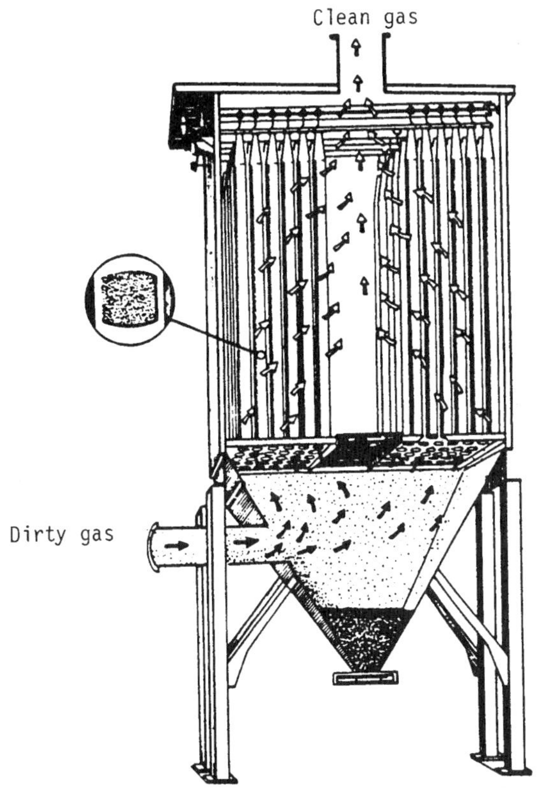

FIGURE 2.30 Dust inlet to the baghouse (*EPA-81/10, p. 8-10*).

Bag. Tubular bags vary in both length and diameter, depending on baghouse design and manufacturer. The length ranges from 10 to 40 feet and the diameter is usually between 6 and 18 inches. Bags are usually hung vertically in the baghouse and are attached at the top and bottom by some type of ring, cap, clamp, or clasp (EPA-81/10, p. 8-11).

Baghouses are constructed as single units or compartmental units. The single unit is generally used on all processes that are not in continuous operation, such as grinding and paint-spraying processes. Compartmental units consist of more than one baghouse compartment and are used in continuous operating processes with large exhaust volumes, such as electric melt steel furnaces and industrial boilers. In both cases, the bags are housed in a shell made of a rigid metal material. Occasionally, it is necessary to include insulation with the shell when treating high-temperature flue gas. This is done to prevent moisture or acid mist from condensing in the unit, causing corrosion and rapid deterioration of the baghouse.

Hopper. Hoppers are used to store the collected dust before it is disposed in a landfill or reused in the process. They are designed usually with a 60° slope to allow dust to flow freely from the top of the hopper to the bottom discharge opening. Some manufacturers add devices to the hopper to promote easy and quick discharge. These devices include strike plates, poke holes, vibrators, and rappers. Strike plates are simply pieces of flat steel which are bolted or welded to the center of the hopper wall. If dust becomes stuck in the hopper, rapping the strike plate several times with a mallet will free this material. Hopper designs also usually include access doors or ports. Access ports provide for easier cleaning, inspection, and maintenance of the hopper.

Some type of discharge device is necessary for emptying the hopper. Discharge devices can be manual or automatic. The simplest manual discharge device is the slide gate, a plate held in place by a frame and sealed with gaskets. When the hopper needs to be emptied, the plate is removed and the material discharges. Other manual discharge devices include hinged doors or drawers. The collector must be shut down before opening any manual discharge device. Thus, manual discharge devices are used on baghouses that operate on a periodic basis.

Automatic continuous discharge devices are installed on baghouses that are used in continuous operation: devices include trickle valves, rotary air lock valves, screw conveyors and pneumatic conveyers. Trickle valve discharge devices are shown in Fig. 2.31. As dust collects in the hopper, the weight of the dust pushes down on the counterweight of the top flap and dust discharges downward. The top flap then closes, the bottom flap opens, and the material falls out. This type of valve is available in gravity-operated and motorized versions.

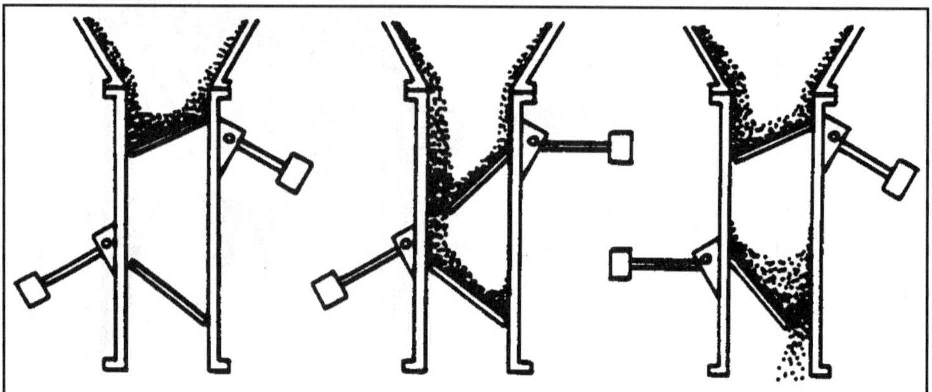

FIGURE 2.31 Trickle valve discharge devices (*EPA-81/10, p. 8-13*).

Rotary airlock valves (Fig. 2.32) are used on medium or large-sized baghouses. The valve is designed with a paddle wheel which is shaft-mounted and driven by a motor. The rotary valve is similar to a revolving door: the paddles or blades form an airtight seal with the housing; the motor slowly moves the blades to allow the dust to discharge from the hopper.

Other automatic dust discharge devices include screw and pneumatic conveyers. Screw conveyers employ a revolving screw feeder located at the bottom of the hopper to remove the dust from the bin. Pneumatic conveyers use compressed air to blow dust from the hopper.

Fabric filter material. Types of fabric filter materials include:

- Woven material
- Felted material

Woven and felted materials are used to make bag filters. Woven filters are made of yarn with a definite repeated pattern. Felted filters are composed of randomly placed fibers compressed into a mat and attached to some loosely woven backing material. Woven filters are used with low-energy cleaning methods, such as shaking and reverse air. Felted fabrics are usually used with higher-energy cleaning systems, such as pulse jet cleaning.

Woven filters: Woven filters are made of yarn with a definite repeated pattern and are used with low-energy cleaning methods, such as shaking and reverse air. Woven filters have open spaces around the fibers. The type of weave used depends on the design and the actual intended application of the woven filter. The simple woven weave is the plain weave. The yarn is woven over and under to form a checkerboard pattern. This weave is not frequently used. Other weaves include the twill and sateen (satin). In the twill weave, yarn is woven over two and under one but in one direction only (Fig. 2.33). This weave is tighter and more durable than the simple weave. Sateen weave goes one over and four under in both directions. Sateen weaves are very tight and allow the use of very fine yarns. Different weaving patterns

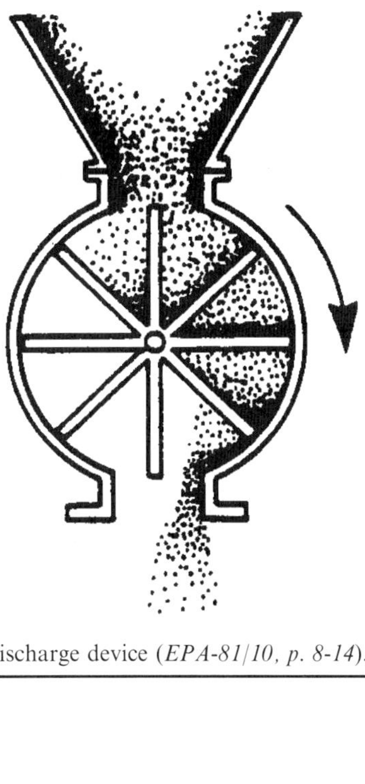

FIGURE 2.32 Rotary airlock discharge device (*EPA-81/10, p. 8-14*).

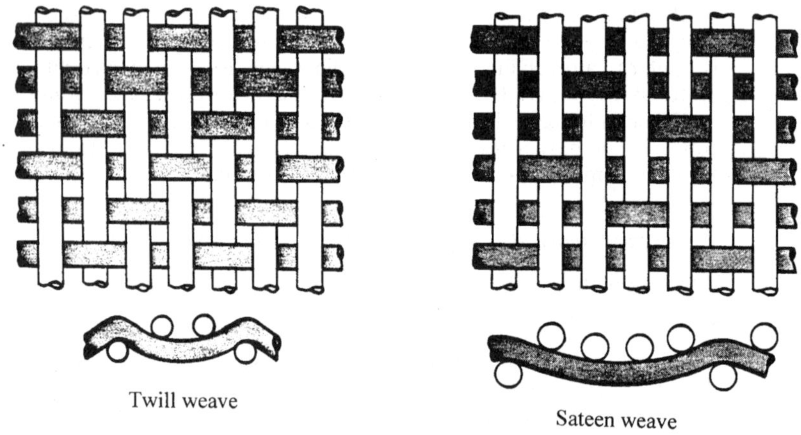

FIGURE 2.33 Twill weave and sateen weave (*EPA-81/10, p. 8-17*).

increase or decrease the open spaces between the fibers, which affects both fabric strength and permeability. Fabric permeability affects the amount of air passing through the filter at a specified pressure drop. A tight weave, for instance, has low permeability and is better for the capture of small particles at the cost of increased pressure drop. The true filtering surface for the woven filter is not the bag itself, but the dust layer or filter cake. The bag simply provides the surface for capture of larger particles. Particles are collected by impaction or interception and the open areas in the weave are closed. This process is referred to as sieving (Fig. 2.34). Some particles escape through the filter until the cake is formed. Once the cake builds up, effective filtering will occur until the bag becomes plugged and cleaning is required. At this point the pressure drop will be exceedingly high and filtering will no longer be cost-effective. The effective filtering time will vary from approximately 15–20 minutes to as long as a number of hours, depending on the concentration of particulate matter in the gas stream.

Felted filters: Felted filters are made by needle punching fibers onto a woven backing called a scrim. The fibers are randomly placed as opposed to the definite repeated pattern of the woven filter. The felts are attached to the scrim by chemical, heat, resin, and stitch-bonding methods. To collect fine particles, the felted filters depend to a lesser degree on the initial dust deposits than do woven filters. The felted filters are generally two to three times thicker than

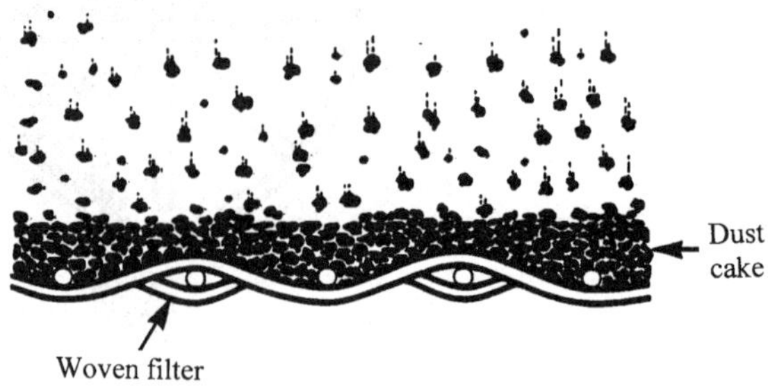

FIGURE 2.34 Sieving collection of particulate matter (*EPA-81/10, p. 8-18*).

woven filters. Each individual randomly-oriented fiber acts as a target for particle capture by impaction and interception. Small particles can be collected on the outer surface of the filter (Fig. 2.35). Felted filters are usually used in pulse jet baghouses. A pulse jet baghouse generally filters more air per cloth area (higher air-to-cloth ratio) than a shaker or reverse air unit. Felted bags should not be used in high-humidity situations, especially if the particles are hydroscopic. Clogging and blinding could result in such situations.

Fiber. The fibers used for fabric filters vary, depending on the industrial application to be controlled. Some filters are made from natural fibers such as cotton or wool. These fibers are relatively inexpensive but have temperature limitations ($< 100^\circ C$) and only average abrasion resistance. Synthetic fibers such as nylon, Orlon, and polyester have slightly higher temperature limitations and chemical resistance. Synthetic fibers are more expensive than natural fibers. Nomex® is a registered trademark of fibers made by DuPont. DuPont makes the fibers, not filter fabrics or bags. Nomex® is widely used, because of its relatively high temperature resistance and its resistance to abrasion. Other fibers such as Teflon® and Fiber glass® can be used in very high temperature situations (230 to 260°C). Both materials have good resistance to acid attack, but are generally more expensive than other fibers.

Fabrics are usually pretreated to improve their mechanical and dimensional stability. They can be treated with silicone to give them better cake-release properties. Natural fabrics (wool and cotton) are usually preshrunk to eliminate bag shrinkage during operation. Both synthetic and natural fabrics usually undergo processes such as heat setting, flame retardation, and napping; these processes increase fabric life, improve dimensional stability and permeability, and ease the work of bag cleaning.

Bag failure mechanism. There are three failure mechanisms that shorten the operating life of a bag. The mechanisms are related to abrasion, thermal durability, and chemical attack. The chief design variable is the upper temperature limit of the fabric. The process exhaust temperature will determine which fabric material should be used for dust collection. Exhaust gas cooling may be feasible, but one must be careful to keep the exhaust gas hot enough to prevent moisture or acid from condensing on the bags.

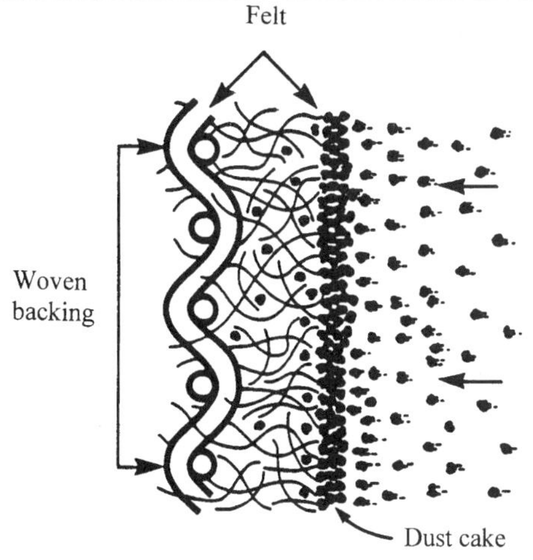

FIGURE 2.35 Felted fabric filter (*EPA-81/10, p. 8-18*).

Another problem frequently encountered in baghouse operation is that of abrasion. Bag abrasion can result from bags rubbing against each other or from the type of bag cleaning employed in the baghouse. For instance, in a shaker baghouse, vigorous shaking may cause premature bag deterioration, particularly at the points where the bags are attached. In pulse jet units, the continual, slight motion of the bags against the supporting cages can also seriously affect bag life. As a result, a 25% per year bag replacement rate is usually encountered. This is the single biggest maintenance problem associated with baghouses.

Bag failure can also occur by chemical attack to the fabric. Changes in dust composition and exhaust gas temperatures from industrial processes can greatly affect the bag material. If the exhaust gas stream is lowered to its dew point or a new chemical species is created, the design of the baghouse (fabric choice) may be completely inadequate. Proper fabric selection and good process operating practices can help eliminate bag deterioration caused by chemical attack.

Gas conditioning. Occasionally, it is necessary to cool the process gas stream before the gas goes to the baghouse. Since there is an upper temperature limit on the fabrics used for bags, gas cooling is sometimes necessary to preserve bag life. This can be accomplished by a number of cooling methods.

Dilution of the exhaust gas stream by air is the easiest and cheapest method, especially at very high temperature. However, air dilution requires the use of a larger baghouse to handle the increased volume of air. Other problems can arise due to the difficulty of controlling the intake of ambient moisture and other contaminants from the dilution air intake.

Radiation cooling can also be used to lower the process exhaust gas temperature. Radiation cooling involves the use of long uninsulated ducts that allow the gas stream to cool as heat radiates from the duct walls. Ducts can be designed in U-shapes to allow more duct surface area to be exposed for radiation cooling. Radiation cooling would not normally be very effective to cool gas temperatures below 300°C. This would require substantial surface area, lengthy duct runs, and increased fan horsepower. Precise temperature control is difficult to maintain and there is a possibility of the ducts becoming plugged due to particle sedimentation.

Evaporative cooling is also used to reduce exhaust gas stream temperature. Evaporative cooling is accomplished by injecting fine water droplets into the gas stream. The water droplets absorb heat from the gas stream as they evaporate. Spray nozzles are located in a quench chamber or somewhere in the duct preceding the baghouse. Evaporative cooling gives a great amount of controlled cooling at a relatively low installation cost. Temperature control can be flexible and accurate. However, this cooling method increases the exhaust volume to the baghouse. The biggest problem with evaporative cooling is keeping the gas temperature above the dewpoint of the gas (SO_2, NO_2, HCl, etc.); otherwise, gases may condense on the bags causing rapid bag deterioration. In addition, all moisture injected into the gas must be evaporated to prevent corrosion of metal parts, and blinding or plugging of caked dust on the bags.

Bag cleaning

Cleaning sequences. Three basic sequences are used for bag cleaning:

- Intermittent cleaning
- Periodic cleaning
- Continuous cleaning

Intermittent cleaning: Intermittent cleaning is done on a demand basis. In intermittently cleaned baghouses, an entire compartment (or baghouse) is bypassed and the bags are cleaned either row by row or simultaneously. Intermittent baghouses are used for batch processes that can be shut down for bag cleaning.

Periodic cleaning: Periodic cleaning is performed on a timed or scheduled basis. Periodically cleaned baghouses consist of a number of compartments or sections. One compartment at a time is removed from service and cleaned on a regular rotation basis. The dirty gas stream is diverted from the compartment being cleaned to the other compartments in the baghouse, so it is not necessary to shut down the process.

Continuous cleaning: Continuously cleaned baghouses are fully automatic and can constantly remain on-line for filtering. The filtering process is momentarily interrupted by a blast of compressed air that cleans the bag; this is called pulse jet cleaning. In continuous cleaning, there is always a row of bags which are being cleaned somewhere in the baghouse. The advantage of continuous cleaning is that it is not necessary to take the baghouse out of service. Large continuous cleaning baghouses are built with compartments to help prevent total baghouse shutdown for bag maintenance and failures to the compressed air cleaning system or hopper conveyers. This allows the baghouse operator to take one compartment off-line to perform necessary maintenance.

Types of bag cleaning. A number of cleaning mechanisms are used to remove caked particles from bags. The three most common are

- Shaking
- Reverse air
- Pulse jet

Shaking: Shaking can be done manually, but is usually performed mechanically in an industrial-scale baghouse. It is a low-energy process that gently shakes the bags to remove deposited particles. The shaking motion and speed depends upon the vendors' design and the composition of dust deposited onto the bag. The shaking motion can be either in a horizontal or vertical direction. The tops of the bags in shaker baghouses are sealed or closed and supported by some type of hook or clasp. Bags are open at the bottom and attached to a cell plate. The cell plate is shaken as a unit, causing the bag to ripple and releasing the dust (Fig. 2.36). In a few systems, shaking is accomplished by sonic vibration (Fig. 2.36). A sound generator is used

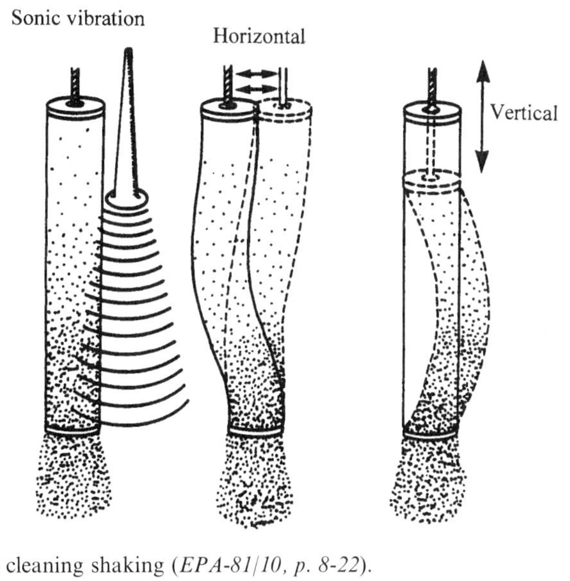

FIGURE 2.36 Bag cleaning shaking (*EPA-81/10, p. 8-22*).

to produce a low-frequency sound that causes the bags to vibrate. The noise level produced by the generator is barely discernible outside the baghouse.

Reverse air: Reverse air, the simplest cleaning mechanism, is accomplished by stopping the flow of dirty gas into the compartment and backwashing the compartment with a low-pressure flow of air. Dust is removed by merely allowing the bags to collapse, thus causing the dust cake to break and fall into the hopper. The cleaning action is very gentle and allows the use of less abrasion-resistant fabrics such as Fiberglass® (Fig. 2.37).

Reverse air cleaning baghouses are usually compartmentalized to permit a section to be off-line for cleaning. Dust can be collected on either the inside or outside of the bag. If collected on the outside, some type of support is needed to prevent bag collapse during the filtering process. Bags can be supported by small steel rings sewn to the inside of the bag.

Reverse air cleaning baghouses generally have very low air-to-cloth (A/C) ratios. Air-to-cloth ratios describe how much dirty gas passes through a given surface area of filter in a given time. A high air-to-cloth ratio means a large volume of air passes through the fabric area. A low air-to-cloth ratio means a small volume of air passes through the fabric. The A/C ratios are usually expressed in units of $(ft^3/min)/ft^2$ of cloth $[(cm^3/s)/cm^2$ of cloth]. The A/C ratio can be used interchangeably with a term called filtration velocity. The units for filtration velocity are ft/min (cm/s). When using the A/C ratios for comparison purposes one should use the units $(ft^3/min)/ft^2$ or $(cm^3/s)/cm^2$. Likewise, when using filtration velocities one should use the units ft/min or cm/s.

For reverse air baghouses, the filtering velocity (filtration velocity) range is usually between 1 and 3 ft/min (0.51 and 1.52 cm/s). For shaker baghouses, the filtering velocity range is between 2 and 6 ft/min (1.02 and 3.05 cm/s). More cloth is generally needed for a given flow rate in a reverse air baghouse than in a shaker baghouse. Hence, reverse air baghouses tend to be larger in size.

Occasionally, baghouse cleaning is accomplished by two methods in combination. Many baghouses have been designed with both reverse air and gentle shaking to remove the dust cake from the bag.

Pulse jet: The third bag cleaning mechanism most commonly used is the pulse jet or pressure jet cleaning (Fig. 2.38). The pulse jet cleaning mechanism uses a high pressure jet

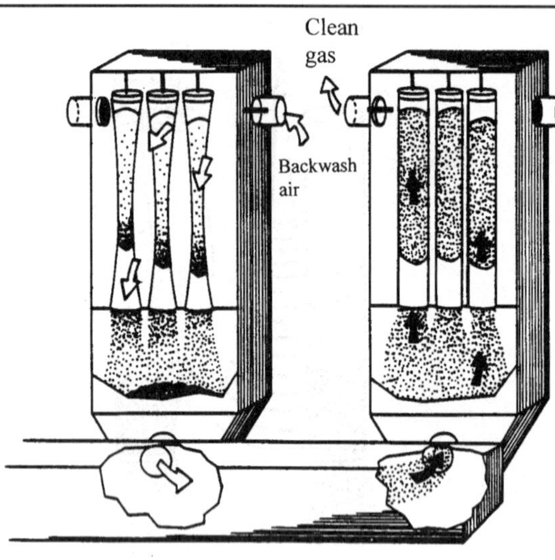

FIGURE 2.37 Reverse air cleaning (*EPA-81/10, p. 8-23*).

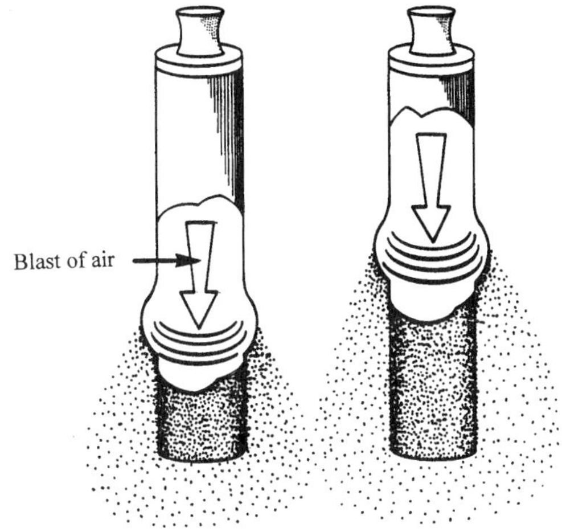

FIGURE 2.38 Pulse jet cleaning (*EPA-81/10, p. 8-25*).

of air to remove the dust from the bag. Bags in the baghouse compartment are supported internally by rings or cages. Bags are held firmly in place at the top by clasps and have an enclosed bottom. Dust-laden gas is filtered through the bag, depositing dust on the outside surface of the bag.

The dust cake is removed from the bag by a blast of compressed air injected into the top of the bag tube. The blast of high-pressure air stops the normal flow of air through the filter. The air blast develops into a standing wave that causes the bag to expand as the bubble travels down the bag tube. As the bag flexes, the cake fractures and deposited particles are discharged from the bag. The air bubble travels down and back up the tube in approximately 0.5 seconds.

The blast of compressed air must be strong enough to travel the length of the bag and shatter or crack the dust cake. Pulse jet units use air supplies from a common header above each bag. In most baghouse designs, a venturi sealed at the top of each bag is used to create a large enough pulse to travel down and up the bag. This occurs in approximately 0.3 to 0.5 s. The pressures involved are commonly between 60 and 100 psig (414 kPa and 689 kPa).

Most pulse jet baghouses use bag tubes that are 4 to 6 in. (10.2 to 15.2 cm) in diameter. The length of the bag is usually around 10 to 12 ft (3.05 to 3.66 m), but can be as long as 25 ft (7.6 m). The shaker and reverse air baghouses use larger bags than the pulse jet units. The bags in these units are 6 to 18 in. (15.2 to 45.7 cm) in diameter and up to 40 ft (12.2 m) in length.

Pulse jet baghouses are designed with filtering velocities between 5 to 15 ft/min (2.5 to 7.5 cm/s); therefore, these units usually use felted fabrics as bag material. Felted material holds up very well under the high filtering rate and vigorous pulse jet cleaning. Pulse jet cleaning methods have the advantage of having no moving parts within the compartment. In addition, pulse jet units can clean bags on a continuous basis without isolating a compartment from service. The duration of the cleaning time is short (< 1.0 s) when compared to the time length between cleaning intervals (approximately 20 minutes to several hours). The major disadvantage of high-pressure cleaning methods is that the bags are subjected to more mechanical stress. Fabric with higher dimensional stability and high tensile strength are required for these units.

Baghouse design variables. Baghouses are designed by considering a number of variables:

- Pressure drop
- Filter drag
- Air-to-cloth ratio
- Collection efficiency
- Gas conditioning

Pressure drop: Pressure drop, a very important baghouse design variable, describes the resistance to air flow across the baghouse. Pressure drop is usually expressed in millimeters of mercury (mmHg) or inches of water (in. H_2O). It can be related to the size of the fan that would be necessary to either push or pull the exhaust gas through the baghouse. A baghouse with a high pressure drop would need a larger fan and more energy to move the exhaust gas through the baghouse.

Filter drag: Filter drag is the filter resistance across the fabric dust layer. It is a function of the quantity of dust accumulated on the filter. As previously mentioned, the true filtering surface is not the bag itself, but the dust layer. Dust bridges the pores or openings in the weave, increasing the drag rapidly. The filter drag increases exponentially up to a constant rate of increase. Following cleaning by shaking, there is a period of cake repair and initial cake buildup. Effective filtration takes place while the filter drag increases at a constant rate. When the total pressure drop reaches a value set by the system design, bag cleaning is initiated. At this point, the pressure drop decreases (almost vertically on the performance curve) to the initial point. Cake repair begins when the cleaning cycle stops and the cycle repeats.

Filtration velocity and air-to–cloth ratio: As previously mentioned, the terms filtration velocity and air-to-cloth ratio can be used interchangeably. The formula used to express filtration velocity is

$$v_f = Q/A_c$$

where v_f = filtration velocity, ft/min (cm/s)
Q = volumetric air flow rate, ft^3/min (cm^3/s)
A_c = area of cloth filter, ft^2 (cm^2)

Air-to-cloth ratio is defined as the ratio of gas filtered in cubic feet per minute to the area of filtering media in square feet. Typical units used to express the A/C ratio are

$$(ft^3/min)/ft^2 \text{ or } (cm^3/s)/cm^2$$

These A/C ratio units essentially reduce to velocity units.

The A/C ratio (filtration velocity) varies for various baghouse designs. Shaker and reverse air baghouses generally have small A/C ratios (shaker units $< 3:1$ $(cm^3/s)/cm^2$ and reverse air units $< 1.5:1$ $(cm^3/s)/cm^2$). On the other hand, pulse jet units usually operate at A/C ratios between 2.5 and $10:1$ $(cm^3/s)/cm^2$. For a given flow rate, pulse jet units can be smaller in size (fewer bags) than the shaker and reverse air baghouse.

The A/C ratio (filtering velocity) is a very important factor used in the design and operation of a baghouse. Improper ratios can cause the baghouse to be in violation of air pollution regulations. Operating at an A/C ratio that is too high may lead to a number of problems. Very high ratios can cause compaction of dust on the bag, resulting in excessive pressure drops. In addition, breakdown of the dust cake could also occur, which in turn results in reduced collection efficiency. The major problem of a baghouse using a very low A/C ratio is that the baghouse will be larger in size.

During a permit review for baghouse installations the reviewer should check the A/C ratio. Typical A/C ratios for shakers, reverse air and pulse jet baghouses are listed in Table 2.17.

TABLE 2.17 Typical Air-to-Cloth Ratios

Baghouse cleaning method	Air-to-cloth ranges	
	$(cm^3/s)/cm^2$	$(ft^3/min)/ft^2$
Shaking	1–3	2–6
Reverse air	0.5–1.5	1–3
Pulse jet	2.5–7.5	5–15

Note: Air-to-cloth ratios are occasionally given as 2.0:1 instead of 2.0 $(cm^3/s)/cm^2$

Baghouses should be operated within a reasonable design A/C ratio range. For example, assume a permit was submitted indicating the use of a reverse air cleaning baghouse using woven Fiberglass® bags for reducing particulate emissions from a small foundry furnace. If the information supplied indicated that the baghouse would operate with an A/C ratio of 6 $(cm^3/s)/cm^2$ of fabric material, one should question this information. Reverse air units should be operated with a much lower A/C ratio. The fabric would probably not be able to withstand the stress from such high filtering rates and could cause premature bag deterioration. Too high an A/C ratio results in excessive pressure drops, reduced collection efficiency, blinding, and rapid wear. In this case a better design might include reducing the A/C ratio within the acceptable range, thus adding more bags. Another alternative would be to use a pulse jet baghouse with the original design A/C ratio of 6 $(cm^3/s)/cm^2$ and use felted bags made of Nomex® fibers. Either alternative would be more acceptable than the original permit submission.

Collection efficiency: Extremely small particles can be efficiently collected in a baghouse. Baghouse units designed with collection efficiencies of 99.99% are common. Exhaust air from a baghouse can even be recirculated back into the plant for heating purposes, as long as the particles collected are not toxic.

Baghouses are not normally designed with the use of fractional efficiency curves as are some of the other particulate emission control devices. Vendors design and size the units strictly on experience. The baghouse units are designed to meet particulate emission outlet loading and opacity regulations. There is no one formula that can determine the collection efficiency of a specific baghouse. Some theoretical formulas for determining collection efficiency have been suggested, but these formulas contain numerous (three to four) experimentally determined coefficients in the equations. Therefore, these efficiency equations give at best only an estimate of baghouse performance.

Summary of design criteria. The principal design criterion is the gas flow rate to the baghouse, measured in cubic meters (cubic feet) per minute. The gas volume to be treated is set by the process exhaust, but the filtration velocity or A/C ratio is determined by the baghouse vendor's design. The A/C ratio depends on a number of variables. A thorough review of baghouse design plans should consider the following factors.

1. Type, shape, and density of the dust; average and maximum concentrations; chemical properties such as abrasiveness, explosiveness, electrostatic charge, and agglomerating tendencies.
2. Gas flow rate: average and maximum flow rate, temperature, moisture content, chemical properties such as dew point, corrosiveness, and combustibility.
3. Fabric construction: woven or felt filters, filter thickness, fiber size, fiber density, filter treatments such as napping, resin and heat setting, and special coatings.
4. Fiber type: natural, synthetic, etc.

5. **Cleaning methods**: low-energy methods, which are shaker and reverse air cleaning; or a high-energy method, which is pulse jet cleaning.
6. **Cleaning time**: ratio of filtering time to cleaning time is the measure of the percent of time the filters are performing; this should be at least 10 : 1 or greater.
7. **Cleaning and filtering stress**: amount of flexing and creasing of the fabric; reverse air is the gentlest, shaking and pulse jet have the most vigorous stress on the fabric.
8. **Bag spacing**: bags must be properly spaced to eliminate rubbing against each other; bags must be accessible for inspection and maintenance service.
9. **Compartment design**: allowance for proper cleaning of bags; design should include an extra compartment to allow for reserve capacity and off-line cleaning, and inspection and maintenance of broken bags.
10. **Space and cost requirements**: baghouses require a good deal of installation space; initial costs, and operating and maintenance costs can be high.
11. **Emission requirements**: efficiency in terms of opacity and grain-loading regulations.
12. **Proper A/C ratio**: reverse air lowest, shakers next, pulse jet baghouses allow the highest A/C ratio.

EXAMPLE 1: baghouse—simple cloth size check. Baghouse sizing is done by the manufacturer. A simple check or estimate of the amount of baghouse cloth needed for a given process flow rate can be computed by using the A/C ratio equation (EPA-81/10, p. 8-34):

$$v_f = Q/A_c \text{ or } A_c = Q/v_f$$

where v_f = filtration velocity, ft/min (cm/s)
Q = volumetric air flow rate, ft^3/min (cm^3/s)
A_c = area of cloth filter, ft^2 (cm^2)

For example, if the process gas exhaust rate is given as 4.72×10^6 cm^3/s (10,000 ft^3/min) and the filtration velocity is 4 cm/s (A/C is 4:1 (cm^3/s)/cm^2), the cloth area would be

$$\begin{aligned} A_c &= 4.72 \times 10^6/4 \\ &= 118\,\text{m}^2 \text{ (cloth required)} \end{aligned}$$

To determine the number of bags required in the baghouse, one would simply use the formula:

$$A_b = \pi dh$$

where A_b = area of bag, m (ft)
d = bag diameter, m (ft)
h = bag height, m (ft)

If the bag diameter is 0.203 m (8 in.) and the bag height is 3.66 m (12 ft), the area of each bag is

$$\begin{aligned} A_b &= 3.14(0.203)(3.66) \\ &= 2.33\,\text{m}^2 \end{aligned}$$

The calculated number of bags in the baghouse is

$$\begin{aligned} \text{Number of bags} &= 118/2.33 \\ &= 51 \text{ bags} \end{aligned}$$

EXAMPLE 2: baghouse—bag selection. It is proposed to install a pulse-jet fabric filter system to clean an air stream containing particulate pollutants. You are asked to select the most appropriate filter bag considering the performance and cost (EPA-84/09, p. 84).

Given conditions

Volumetric flow rate of polluted air stream = 10,000 scfm (60°F, 1 atm)

Operating temperature = 250°F

Concentration of pollutants = 4 grains/ft^3

Average air-to-cloth ratio = 2.5 cfm/ft^2 cloth

Collection efficiency requirement = 99%

Information given by filter bag manufacturers is listed in Table 2.18. *Note*: No bag has an advantage from the standpoint of durability under the operating conditions for which the bag is to be designed.

TABLE 2.18 Filter Bag Properties

	Filter bag			
Property	A	B	C	D
Tensile strength	excellent	above average	fair	excellent
Recommended maximum temperature, °F	260	275	260	220
Resistance factor	0.9	1	0.5	0.9
Cost per bag, $	26.00	38.00	10.00	20.00
Standard size	8 in. × 16 ft	10 in. × 16 ft	1 ft × 16 ft	1 ft × 20 ft

Solution:

1. Eliminate from consideration bags which, on the basis of given characteristics, are unsatisfactory.

Considering operating temperature and bag tensile strength required for a pulse jet system:

- Bag D is eliminated since its recommended maximum temperature (220°F) is below the operating temperature of 250°F.
- Bag C is also eliminated since a pulse jet fabric filter system requires the tensile strength of the bag to be at least above average.

2. Determine comparative costs of the remaining bags.

Total cost for each bag type is the number of bags × cost per bag. No one type of bag is more durable than the other.

2.a. Establish the cost per bag.

From the information given by filter bag manufacturers (Table 2.18), the cost per bag is $26.00 for A bag and $38.00 for B bag.

2.b. Determine number of bags, N, for each type.

The number of bags required, N, is the total filtering area required divided by the filtering area per bag.

2.b.1. Calculate the total filtering area A_t.

2.b.1.a. Convert given flow rate to acfm, Q_a.

$$Q_a = (10{,}000)(250 + 460)/(60 + 460) = 13{,}654 \text{ acfm}$$

2.b.1.b. Establish filtering velocity v_f.

This is given. The A/C ratio, expressed in cfm/ft^2, is the same as the filtering velocity which is given above as 2.5 cfm/ft^2 cloth. From the information given in Table 2.18, the filtering velocity is:

$$v_f = 2.5\,\text{cfm/ft}^2$$
$$= 2.5\,\text{ft/min}$$

2.b.1.c. Calculate the total filtering cloth area A_c from the acfm and filtering velocity determined above.

$$A_c = Q_a/v_f = 13{,}654/2.5 = 5461.6\,\text{ft}^2$$

2.b.2. Calculate the filtering area per bag.

Bags are assumed to be cylindrical; the bag area is $A = \pi Dh$, where D = bag diameter and h = bag length.

$$\text{For bag A: } A = \pi Dh = \pi(8/12)(16) = 33.5\,\text{ft}^2$$
$$\text{For bag B: } A = \pi Dh = \pi(10/12)(16) = 41.9\,\text{ft}^2$$

2.b.3. Determine the number of bags required, N.

N = (filtering cloth area of each bag A_c)/(bag area A)

$$\text{For bag A: } N = A_c/A = 5461.6/33.5 = 163$$
$$\text{For bag B: } N = 5461.6/41.9 = 130$$

2.c. Determine the total cost for each bag.

$$\text{For bag A: total cost} = (N)(\text{cost per bag}) = (163)(26.00) = \$4238$$
$$\text{For bag B: total cost} = (130)(38.00) = \$4940$$

3. Select the most appropriate filter bag considering the performance and cost.

Since the total cost for bag A is less than bag B, select bag A.

EXAMPLE 3: baghouse cleaning frequency. You are requested to determine the number of filtering bags required and cleaning frequency for a plant equipped with a fabric filter system. Operating and design data are given below (EPA-84/09, p. 86).

Given conditions

volumetric flow rate of the gas stream = 50,000 acfm

dust concentration = 5.0 grains/ft^3

efficiency of the fabric filter system = 98.0%

filtration velocity = 10 ft/min

diameter of filtering bag = 1.0 ft

length of filtering bag = 15 ft

the system is designed to begin cleaning when the pressure drop reaches 8.0 inches of water.

the pressure drop is given by $\Delta p = 0.2v_f + 5c(v_f)^2 t$, where

Δp = pressure drop, in. H_2O

v_f = filtration velocity, ft/min

c = dust concentration, lb/ft^3

t = time since the bags were cleaned, min

Solution:

1. What are the number of bags N needed?

To calculate N, you need the total required surface area of the bags and the surface area of each bag.

1.a. Calculate the total required surface area of the bags A_c in ft^2

$$A_c = Q/v_f$$

where A_c = total surface area of the bags
Q = volumetric flow rate
v_f = filtering velocity

$$\begin{aligned} A_c &= Q/v_f \\ &= 50{,}000/10 \\ &= 5000\,\text{ft}^2 \end{aligned}$$

1.b. Calculate the surface area of each bag A, in ft^2

$$A = \pi D h$$

where A = surface area of a bag
D = diameter of the bag
h = length of the bag

$$\begin{aligned} A &= \pi D h \\ &= \pi(1.0)(15) \\ &= 47.12\,\text{ft}^2 \end{aligned}$$

1.c. Calculate the number of bags N required.

$$\begin{aligned} N &= A_c/A \\ &= 5000/47.12 \\ &= 106 \end{aligned}$$

2. Calculate the required frequency of cleaning.

$$\Delta p = 0.2 v_f + 5c(v_f)^2 t$$

Since Δp is given as 8.0 in. H_2O, the time since the bags were cleaned is calculated by solving the above equation.

Since 5.0 grains/ft^3 = 0.0007143 lb/ft^3 and $\Delta p = 0.2 v_f + 5c(v_f)^2 t$

$$8.0 = (0.2)(10) + (5)(0.0007143)(10)^2 t$$

Solving for t,

$$t = 16.8\ \text{min}$$

EXAMPLE 4: baghouse—bag failure. An installed baghouse is presently treating a contaminated gas stream. Suddenly some of the bags are broken. You are now requested to estimate the new outlet loading of this baghouse system (EPA-84/09, p. 88).

Given conditions

operation conditions of the system = 60°F, 1 atm

inlet loading = 4.0 grains/acf

outlet loading before bag failure = 0.02 grains/acf

volumetric flow rate of contaminated gas = 50,000 acfm

number of compartments = 6

number of bags per compartment = 100

bag diameter = 6 inches

pressure drop across the system = 6 in. H_2O

number of broken bags = 2 bags

assume that all the contaminated gas emitted through the broken bags is the same as that passing through the tube sheet thimble.

Solution:

1. Calculate the collection efficiency and penetration before the bag failure(s).

Collection efficiency is a measure of degree of performance of a control device; it specifically refers to degree of removal of pollutants. *Loading* refers to the concentration of pollutants, usually in grains of pollutants per cubic feet of contaminated gas streams. Mathematically, the collection efficiency is defined as

$$\eta = [(\text{inlet loading} - \text{outlet loading})/(\text{inlet loading})](100)$$

From the above equation, the collected amount of pollutants by a control unit is the product of collection efficiency η and inlet loading. The amount discharged to the atmosphere is given by the inlet loading minus the amount collected.

Another term used to describe the performance or collection efficiency of control devices is penetration P_t. It is given by (EPA-84/09, p. 3.3):

$$P_t = 1 - \eta/100 \qquad \text{(fractional basis)}$$
$$P_t = 100 - \eta \qquad \text{(percent basis)}$$

The effect of bag failure on baghouse efficiency is described by the following equation:

$$P_t^* = P_t + P_{tc}$$
$$P_{tc} = 0.582(\Delta p)^{0.5}/\phi$$
$$\phi = Q/(LD^2(T + 460)^{0.5})$$

where P_t^* = penetration after bag failure
P_t = penetration before bag failure
P_{tc} = penetration correction term, contribution of broken bags to P_t^*
Δp = pressure drop, in. H_2O
ϕ = dimensionless parameter
Q = volumetric flow rate of contaminated gas, acfm
L = number of broken bags
D = bag diameter, inches
T = temperature, °F

Collection efficiency η is

$$\begin{aligned}\eta &= (\text{inlet loading} - \text{outlet loading})/(\text{inlet loading})\\ &= (4.0 - 0.02)/(4.0)\\ &= 0.995\\ &= 99.5\%\end{aligned}$$

Penetration is:

$$\begin{aligned}P_t &= 1.0 - \eta\\ &= 0.005\end{aligned}$$

2. Calculate the bag failure parameter ϕ. It is a dimensionless number

$$\begin{aligned}\phi &= Q/(LD^2(T + 460)^{0.5})\\ &= 50{,}000/(2)(6)^2(60 + 460)^{0.5}\\ &= 30.45\end{aligned}$$

3. Calculate the penetration correction P_{tc}.
   This is to determine penetration due to bag failure.

$$\begin{aligned}P_{tc} &= 0.582(\Delta p)^{0.5}/\phi\\ &= (0.582)(6)^{0.5}/30.45\\ &= 0.0468\end{aligned}$$

4. Calculate the penetration and efficiency after the two bag failures.
   Use the results of steps 1 and 3 to calculate P_t^*

$$\begin{aligned}P_t^* &= P_t + P_{tc}\\ &= 0.005 + 0.0468\\ &= 0.0518\end{aligned}$$

$$\begin{aligned}\eta^* &= 1 - 0.0518\\ &= 0.948\end{aligned}$$

5. Calculate the new outlet loading after the bag failures.
   Relate inlet loading and new outlet loading to the revised efficiency or penetration.

$$\begin{aligned}\text{New outlet loading} &= (\text{inlet loading})P_t^*\\ &= (4.0)(0.0518)\\ &= 0.207\ \text{grains/acf}\end{aligned}$$

6. What other system and/or operating variables will affect the new outlet loading (or what other variables will affect penetration after bag failures)? Discuss qualitatively.

Note that

$$P_t^* = P_t + P_{tc}$$
$$P_{tc} = 0.582(\Delta p)^{0.5}/\phi$$
$$\phi = Q/(LD^2(T + 460)^{0.5})$$
$$P_t^* = P_t + 0.582(\Delta p)^{0.5}(LD^2(T + 460)^{0.5})/Q$$

Decreasing either the operating temperature and/or pressure drop will decrease P_t^* (and therefore lower the new outlet loading) assuming L, D, and Q remain the same. Note that increasing Q will also decrease P_t^*, since the fraction of gas emitted through the broken bags is reduced.

5 PARTICULATE CONCENTRATION CALCULATION

EXAMPLE 1: particulate concentration. An analysis of a stack sample indicated that the concentration of particulates from a coal-burning boiler is 2.5 grains per cubic foot. Determine:

1. What is the concentration in the units of lb/ft^3, g/m^3, and μg/m^3?
2. If 350 ft^3 of flue gas are produced for every pound of coal burned, and if all ash in coal becomes particulate in flue gas, what is the maximum ash content in the coal?
3. If the air standard is 75 μg/m^3, calculate the dilution factor.
4. Calculate the particulate collection efficiency, if the air standard is to be met.

Given conditions

particulate concentration, $P_c = 2.5$ grains/ft^3
flue gas volume produced per pound coal burned, $V = 350\,\text{ft}^3/\text{lb}$
air standard level on particulate $P_s = 75\,\mu\text{g/m}^3$
1 lb = 7000 grains
1 lb = 454 grams
1 ft^3 = 0.0283 m^3

Solution for question (1):

1.1 Convert the particulate concentration P_c from grains/ft^3 to lb/ft^3.

$$P_c = 2.5/7000$$
$$= 0.000357\ \text{(lb-particulate)/ft}^3$$

1.2. Convert the particulate concentration P_c from lb/ft^3 to g/m^3.

$$P_c = 0.000357(454)/(0.0283)$$
$$= 5.73\ \text{(g-particulate)/m}^3$$

1.3. Convert the particulate concentration P_c from g/m^3 to μg/m^3.

$$P_c = 5.73(1000000)$$
$$= 5.73 \times 10^6 \ (\mu\text{g-particulate})/\text{m}^3$$

Solution for question (2):

2.1. Calculate the particulate loading P_{lo} in flue gas.

Because it is assumed that all ash becomes particulates in the flue gas, the particulate loading in the flue gas can be calculated from P_c in step 1.1, by multiplying by the flue gas volume rate.

$$P_{lo} = (\text{particulate concentration}) \times (\text{flue gas volume per pound of coal burned})$$
$$= 0.000357(350)$$
$$= 0.12495 \ (\text{lb-particulate/lb-coal})$$
$$= 12.50\%$$

2.2. Determine the ash content in coal.

The ash content in coal is equal to the particulate loading in step 2.1; that is, 12.50%.

Solution for question (3):

Dilution factor (DF) means the amount of additional air that is needed to dilute the flue gas so that the particulate concentration can meet the air standard requirement. To maintain the consistency of the units in this calculation, the P_c value in step 1.3 is used for this calculation.

$$\text{DF} = (\text{particulate loading})/(\text{air standard})$$
$$= (5.73 \times 10^6)/75$$
$$= 7.64 \times 10^4$$

Solution for question (4):

1. Write the collection efficiency equation C_e.

$$C_e = (1 - P_s/P_c)$$

where P_s = air standard level provided in given conditions
P_c = particulate concentration from step 1.3

2. Calculate the collection efficiency.

$$C_e = (1 - 75/5.73 \times 10^6)$$
$$= 0.999987$$
$$= 99.9987\%$$

EXAMPLE 2: particulate concentration. (1) A boiler is burning 8% coal. If 250 ft^3 of flue gas are produced for every pound of coal burned, what is the maximum effluent particulate loading in grains/ft^3? (2) If the 250 ft^3 of flue gas contains 12% oxygen, correct the particulate concentration to the standards requirement of 7% oxygen concentration.

Given conditions

coal ash content = 8%

flue gas volume produced per pound coal burned, $V = 250\,\text{ft}^3/\text{lb}$

flue gas oxygen concentration, $O_2 = 12\%$

flue gas standards oxygen concentration, std $O_2 = 7\%$

1 lb = 7000 grains

Solution for question (1):

1.1. Calculate particulate amount produced.
Assume that all 8% of ash becomes particulates in the flue gas. For 1 lb of coal burned, the amount of particulates emitted is 0.08 lb.

1.2. Calculate the maximum particulate loading P_c.

$$\begin{aligned} P_c &= 0.08/250 \\ &= 0.00032\,\text{lb/ft}^3 \\ &= 2.24\,\text{grains/ft}^3 \end{aligned}$$

Solution for question (2):

2.1. Write the correction equation below from 40CFR264.343.

$$P_c = P_m \times (21 - \text{std}\,O_2)/(21 - Y)$$

where P_c = corrected particulate concentration
P_m = measured particulate concentration (at 12% O_2 for this case)
Y = % O_2 in discharge, dry basis
std O_2 = 7% for this case

2.2. Calculate the corrected particulate concentration

$$\begin{aligned} P_c &= 2.24(21 - 7)/(21 - 12) \\ &= 3.48\,\text{grains/ft}^3 \end{aligned}$$

6 PARTICULATE EMISSION CONTROL COST

6.1 Breakeven Operation

EXAMPLE 1: control costs. A plant emits 50,000 acfm of gas containing a dust loading of 2.0 grains/ft^3. A particulate control device is employed for particle capture and the dust captured from the unit is worth \$0.01/lb of dust. You are requested to determine at what collection efficiency is the cost of power equal to the value of the recovered material. Also determine the pressure drop in inches of H_2O at this condition (EPA-84/09, p. 122).

Given conditions

Overall fan efficiency = 55%

Electric power cost = \$0.06/kW-h

For this control device, assume that the collection efficiency is related to the system pressure drop Δp through the following equation.

$$\eta = \Delta p/(\Delta p + 5.0)$$

where Δp = pressure drop, lb_f/ft^2
η = fractional collection efficiency

Solution:

1. Express the value of the dust collected in terms of collection efficiency η.

$$\text{Amount of dust collected} = (Q)(\text{inlet loading})(\eta)$$

Remember there are 7000 grains per pound.

$$\begin{aligned}\text{The value of dust collected} &= 50{,}000(\text{ft}^3/\text{min})2(\text{grains}/\text{ft}^3)(1/7000)(\text{lb/grain}) \times 0.01(\$/\text{lb})\eta \\ &= 0.143\eta\ \$/\text{min}\end{aligned}$$

2. Express the value of the dust collected in terms of pressure drop Δp.
   Remember that $\eta = \Delta p/(\Delta p + 5.0)$

$$\text{The value of dust collected} = 0.143[\Delta p/(\Delta p + 5.0)]\ \$/\text{min}$$

3. Express the cost of power in terms of pressure drop Δp.

$$\text{bhp} = Q\Delta p/\eta' = \text{brake horsepower}$$

where η' = fan efficiency
Δp = pressure drop, lb_f/ft^2
Q = volumetric flow rate

$$\begin{aligned}\text{Cost of power} &= \Delta p(\text{lb}_f/\text{ft}^2)(50{,}000)[(\text{ft}^3/\text{min})(1/44{,}200)(\text{kW-min/ft-lb}_f)(1/0.55) \\ &\quad \times (0.06)(\$/\text{kW-h})(1/60)(\text{h/min})] \\ &= 0.002\Delta p\ \$/\text{min}\end{aligned}$$

4. Set the cost of power equal to the value of dust collected and solve for Δp in lb_f/ft^2. This represents breakeven operation. Then, convert this pressure drop to in. H_2O.
   To convert from lb_f/ft^2 to in. H_2O, divide by 5.2.

$$(0.143)\Delta p/(\Delta p + 5) = 0.002\Delta p$$

Solving for Δp

$$\Delta p = 66.5\,\text{lb}_f/\text{ft}^2 = 12.8\ \text{in. H}_2\text{O}$$

5. Calculate the collection efficiency using the value of Δp calculated above.
   Use the equation for η in Solution (2):

$$\eta = 66.5/(66.5 + 5) = 0.93 = 93.0\%$$

6.2 Annualized Installed, Operation, and Maintenance Costs

EXAMPLE 2: control costs. You are requested to determine capital, operating, and maintenance costs on an annualized basis for a textile dye and finishing plant (with two coal-fired stoker boilers) where a baghouse is employed for particulate control. Operating, design, and economics factors are given (EPA-84/09, p. 123).

Given conditions

exhaust volumetric flow from two boilers = 70,000 acfm
overall fan efficiency = 60%
operating time = 6240 h/year
surface area of each bag = 12.0 ft^2
bag type = Teflon® felt
air-to-cloth ratio = 5.81 $acfm/ft^2$
total pressure drop across the system = 17.16 lb_f/ft^2
cost of each bag = \$75.00
installed capital costs = \$2.536/acfm
cost of electrical energy = \$0.03/kW-h
yearly maintenance cost = \$5000 plus yearly cost to replace 25% of the bags
salvage value = 0
interest rate (i) = 8%
lifetime of baghouse (m) = 15 yr
annual installed capital cost (AICC) = (installed capital cost) $\{i(1+i)^m/[(1+i)^m - 1]\}$

Solution:

1. What is the annual maintenance cost?
   1.a. Calculate the number of bags N.

$$N = Q/(\text{air-to-cloth ratio})(A)$$

where Q = total exhaust volumetric flow rate
A = surface area of a bag

$$\begin{aligned} N &= Q/(\text{air-to-cloth ratio})(A) \\ &= (70{,}000)/(5.81)(12) \\ &= 1004 \text{ bags} \end{aligned}$$

1.b. Calculate the annual maintenance cost in dollars/year.

$$\begin{aligned} \text{Annual maintenance cost} &= \$5000/\text{year} + \text{cost of replacing 25\% of the bags each year} \\ &= \$5000 + (0.25)(1004)(75.00) \\ &= \$23{,}825/\text{year} \end{aligned}$$

2. What is the annualized installed cost (AICC)?

   2.a. Calculate the installed capital cost in dollars.

$$\begin{aligned} \text{Installed capital cost} &= (Q)(\$2.536/\text{acfm}) \\ &= (70{,}000)(2.536) \\ &= \$177{,}520 \end{aligned}$$

2.b. Calculate the AICC using the equation given above.

$$\begin{aligned}\text{AICC} &= (\text{installed capital cost})\{i(1+i)^m/[(1+i)^m - 1]\} \\ &= (177{,}520)\{0.08(1+0.08)^{15}/[(1+0.08)^{15} - 1)\} \\ &= \$20{,}740/\text{year}\end{aligned}$$

3. Calculate the operating cost in dollars per year.

$$\text{Operating cost} = Q\Delta p(\text{operating time})(0.03/\text{kW-h}/E)$$

Since 1 ft-lb/s = 0.0013558 kW,

$$\begin{aligned}\text{Operating cost} &= (70{,}000/60)(17.16)(6240)(0.03)(0.0013558)/0.6 \\ &= \$8470/\text{year}\end{aligned}$$

4. Calculate the total annualized cost in dollars per year.

$$\begin{aligned}\text{Total annualized cost} &= (\text{maintenance cost}) + \text{AICC} + (\text{operating cost}) \\ &= 23{,}825 + 20{,}740 + 8470 \\ &= \$53{,}035/\text{year}\end{aligned}$$

REFERENCES

40CFR50, *National primary and secondary ambient air quality standards*, 40 Code of Federal Regulations Parts 50.

40CFR264.343, *Performance standards*, 40 Code of Federal Regulations Parts 264.343.

EPA-81/10, *Control of particulate emissions, course 413*, US EPA Air Pollution Training Institute (APTI), EPA450-2-80-066, October 1981.

EPA-84/03, *Wet scrubber plan review, course SI:412C*, US EPA Air Pollution Training Institute (APTI), EPA-450-2-82-020, March 1984.

EPA-84/09, *Control of gaseous and particulate emissions, course SI:412D*, US EPA Air Pollution Training Institute (APTI), EPA450-2-84-007, September 1984.

Lapple-51, *Fluid and particle mechanics*, Lapple, C.E., Newark, Delaware: University of Delaware, 1951.

APPENDIX A

Values of C_f (for air at atmospheric pressure)

Particle diameter (μm)	Values of C_f at temperature of:		
	70°F	212°F	500°F
0.1	2.88	3.61	5.14
0.25	1.682	1.952	2.528
0.5	1.325	1.446	1.711
1.0	1.160	1.217	1.338
2.5	1.064	1.087	1.133
5.0	1.032	1.043	1.067
10.0	1.016	1.022	1.033

Source: Lapple-51.

CHAPTER 3.3
WET AND DRY SCRUBBERS FOR EMISSION CONTROL

C. C. Lee and G. L. Huffman
U.S. Environmental Protection Agency, Cincinnati, Ohio

1 INTRODUCTION

Generally speaking, absorption equipment includes two major categories:

- Wet absorption scrubbers (or wet scrubbers)
- Dry absorption scrubbers (or dry scrubbers)

Wet scrubbers: As the name implies, wet scrubbers (also known as wet collectors) are devices which use a liquid for the removal of source emissions from effluent gas streams (EPA-84/03, p. 1-1; EPA-81/10, p. 9-1). These air pollution control devices can remove both particulate matter and gaseous pollutants.

The dirty exhaust stream is brought into contact with the liquid by spraying it into the liquid; by forcing it through a pool of liquid; or by some other contact method. The advantages or disadvantages of using a wet scrubber instead of some other control device depend on the pollutant (gas or particle) to be controlled. The types of wet scrubber equipment are shown in Table 3.1.

Particulate matter removal: When wet scrubbers are used for removing particles, the particles are captured by and incorporated into liquid droplets. The droplet must then be

TABLE 3.1 Types of Wet Scrubbers

1. Gas-phase contacting scrubber	1.1. Venturi scrubber	1.1.1. Venturi with a wetted throat
		1.1.2. Venturi with throat spray 1.1.3. Venturi with rectangular throat 1.1.4. Adjustable-throat venturi with plunger 1.1.5. Adjustable throat venturi with movable plate 1.1.6. Venturi with a cyclonic separator
	1.2. Plate tower	
	1.3. Orifice scrubber	
2. Liquid-phase contacting scrubber	2.1. Spray tower	
	2.2. Ejector venturi	
3. Combination of liquid-phase and gas-phase contacting scrubber	3.1. Cyclonic spray scrubber	
	3.2. Mobile bed scrubber	
	3.3. Baffle spray scrubber	
	3.4. Mechanically aided scrubber	

separated from the clean exhaust stream. Wet collectors, baghouses, or electrostatic precipitators (ESPs) can be used when collecting small particles at a high efficiency (> 95%) is necessary. When only large particles are to be removed, either a low-energy scrubber or a cyclone can be used. Choosing the 'best' collection system depends on many factors (EPA-84/03, p. 1-1).

Gaseous removal: When wet scrubbers are used for removing gases, the gases are dissolved or absorbed by the liquid. For gaseous pollutant removal, the choice of the control device depends mainly on the type of gaseous pollutant to be controlled. In choosing a system to control organic vapors, the choice of control is among wet scrubbers, adsorbers, thermal oxidizers (incinerators), or condensers; to control most inorganic gases (HCl, H_2S, HF, and SO_2), a wet scrubber is usually the primary control device (EPA-84/03, p. 1-2).

If the exhaust stream contains both particles and gases, wet scrubbers are generally the only air pollution control device used to remove both pollutants. One exception is using a baghouse or an ESP with a spray dryer in a dry SO_2 scrubbing system.

Wet scrubbers can achieve high removal efficiencies for either particles or gases and, in some instances, can achieve a high removal efficiency for both pollutants in the same system. However, in many cases, the best operating conditions for particle collection are the poorest for gas removal. In general, obtaining high simultaneous gas and particle removal efficiencies requires that one of them be easily collected (i.e., that the gases are very soluble in the liquid or that the particles are large and readily captured). Wet scrubbers have been used in a variety of industries such as acid plants, fertilizer plants, steel mills, asphalt plants, and large power plants.

The advantages and disadvantages of using a wet scrubber to remove particulate and gaseous emissions are summarized below:

Advantages

- *Small space requirement*: Scrubbers reduce the temperature and sometimes can reduce the volume of the unsaturated exhaust stream. Therefore, vessel sizes, including fans and ducts downstream, are smaller than those of other control devices. Smaller sizes result in lower capital costs and more flexibility in the site location of the scrubber.

- *No secondary dust sources*: Once particles are collected, they cannot escape from hoppers or during transport.
- *Handle high-temperature, high-humidity gas streams*: No temperature limit or condensation problems can occur as in baghouses or ESPs.
- *Minimal fire and explosion hazard*: Various dry dusts are flammable. Using water eliminates the possibility of explosions.
- *Ability to collect both gases and particles.*

Disadvantages

- *Corrosion problems*: Water and dissolved pollutants can form highly corrosive acidic solutions. Proper construction materials are very important. Also, wet–dry interface areas can result in corrosion.
- *High power requirement*: High collection efficiencies for particles are attainable only at high pressure drops, resulting in high operating costs.
- *Water disposal problems*: Settling ponds or sludge clarifiers may be needed to meet wastewater regulations.
- *Difficult product recovery*: Dewatering and drying of scrubber sludge make recovery of any dust for reuse very expensive and difficult.
- *Meteorological problems*: The saturated exhaust gases can produce a wet, visible steam plume. Fog and precipitation from the plume may cause local meteorological problems.

2 WET ABSORPTION FOR PARTICULATE EMISSION CONTROL

2.1 Collection Mechanism and Efficiency

Wet absorption processes use wet scrubbers to capture relatively small dust particles with large liquid droplets. Droplets are produced by injecting liquid at high pressure through specially designed nozzles, by aspirating the particle-laden gas stream through a liquid pool, or by submerging a whirling rotor in a liquid pool. These droplets collect particles by using one or more of several collection mechanisms. These mechanisms are summarized in Table 3.2 (EPA-84/03, p. 1-4; EPA-81/10, p. 9-5).

Each of these effects can be characterized by a mathematical expression called the separation number. The *separation numbers* are dimensionless groups of parameters—the higher the value of the separation number, the more effective the mechanism. From these basic expressions, theories have been derived to describe scrubber performance (EPA-81/10, p. 9-5)

The design of a scrubber involves many factors including space restriction, pollutant collection efficiency, pressure drop (gas-side), particle size, exhaust gas flow rate, liquid-to-gas ratio, and many construction details such as using corrosion-resistant materials, baffles, nozzles, venturi throats, water sprays, packing, plates, orifices, entrainment separators, inlets, and outlets. These have been discussed in detail in earlier chapters. Officers who review scrubber design plans for air pollution control agencies should consider these factors during the review process (EPA-84/03, p. 9-1).

Because all scrubbers can be used to collect both particulate and gaseous pollutants, the choice of the most appropriate type can sometimes be difficult. Therefore this chapter will point out those scrubber design features that are important when choosing a scrubber to remove particulate and gaseous pollutants, and will also look at a few equations that can be used to estimate pressure drop and collection efficiency.

TABLE 3.2 Particle Collection Mechanisms for Wet Scrubbing Systems

Mechanism*	Explanation
Impaction	Particles too large to follow gas streamlines around a droplet collide with it.
Diffusion	Very tiny particles move randomly, colliding with droplets because they are confined in a limited space.
Direct interception	An extension of the impaction mechanism. The center of a particle follows the streamlines around the droplet, but a collision occurs if the distance between the particle and droplet is less than the radius of the particle.
Electrostatic attraction	Particles and droplets become oppositely charged and attract each other.
Condensation	When hot gas cools rapidly, particles in the gas stream can act as condensation nuclei and, as a result, become larger.
Centrifugal force	The shape or curvature of a collector causes the gas stream to rotate in a spiral motion, throwing larger particles toward the wall.
Gravity	Large particles moving slowly enough will fall from the gas stream and be collected.

*Impaction, direct interception, and diffusion are the three primary collection mechanisms.

Venturi scrubbers are the most popular scrubbers used to remove particulate matter. Other scrubbers used include cyclonic, orifice, mechanically aided, and spray towers. Typical gaseous pressure drops and L/G ratios for these devices are given in Table 3.3.

Wet scrubbers remove particles from an exhaust stream by contacting the particles with liquid, usually water. A number of factors affect particle removal efficiency including:

- particle-size distribution
- liquid flow rate
- exhaust gas flow rate
- method of contacting
- pressure drop across the scrubber

As with gaseous pollutant removal (absorption), efficient particle removal requires contact between the exhaust stream (containing particles) and the scrubbing liquid. However, particle removal occurs instantaneously upon contact with the liquid, whereas efficient absorption requires a long contact time. Therefore, efficient particle removal requires high relative velocities (gas versus liquid velocity).

TABLE 3.3 Ranges of Pressure Drops and Liquid-to-Gas (L/G) Ratios for Various Wet Scrubbers

	Pressure drop, Δp		Liquid-to-gas ratio*	
Scrubber	kPa	in. H_2O	L/m^3	gal/1000 ft^3
Venturi	1.5–25.0	6.0–100.0	0.4–5.0	3.0–40.0
Spray tower	0.12–0.75	0.5–3.0	0.7–2.7	5.0–20.0
Cyclonic spray	0.4–4.0	1.5–10.0	0.3–1.3	2.0–10.0
Moving bed (good for removing particulate and gaseous pollutants	0.5–6.0	2.0–24.0	0.4–8.0	3.0–60.0
Orifice (self-induced spray)	0.5–4.0	2.0–10.0	0.07–0.7	0.5–5.0
Mechanically aided (fan)	1.0–2.0	4.0–8.0	0.07–0.5	0.5–4.0

*Higher L/G reflects those used for gas absorption.

A number of theories have been developed from basic particle movement principles to explain the action of wet scrubbing systems. Many of these start from firm scientific concepts, but yield only qualitative results when predicting collection efficiencies or pressure drops.

Collection efficiency is frequently expressed in terms of penetration. Penetration is defined as the fraction of particles (in the exhaust stream) that passes through the scrubber uncollected. Penetration is the opposite of the fraction of particles collected, and is expressed as (EPA-84/03, p. 9-3):

$$P_t = 1 - \eta$$

where P_t = penetration
η = collection efficiency (expressed as a fraction)

Wet scrubbers usually have an efficiency curve that fits the relationship of:

$$\eta = 1 - e^{\{-[f(\text{system})]\}}$$

where η = collection efficiency
e = exponential function
$f(\text{system})$ = some function of the scrubbing system variables

By substituting for efficiency, penetration can be expressed as

$$\begin{aligned} P_t &= 1 - \eta \\ &= 1 - [1 - e^{\{-[f(\text{system})]\}}] \\ &= e^{-[f(\text{system})]} \end{aligned}$$

In testing the design of a specific scrubber, the vendor can measure operating variables and the collection efficiency of the unit. These data can then be used to evaluate the efficiency of the system. An equation for the scrubbing system variables, $f(\text{system})$, can be developed for that particular design.

Impaction. In a wet scrubbing system, dust particles will tend to follow the streamlines of the exhaust stream. However, when liquid droplets are introduced into the exhaust stream, particles cannot always follow these streamlines as they diverge around the droplet (Fig. 3.1). The

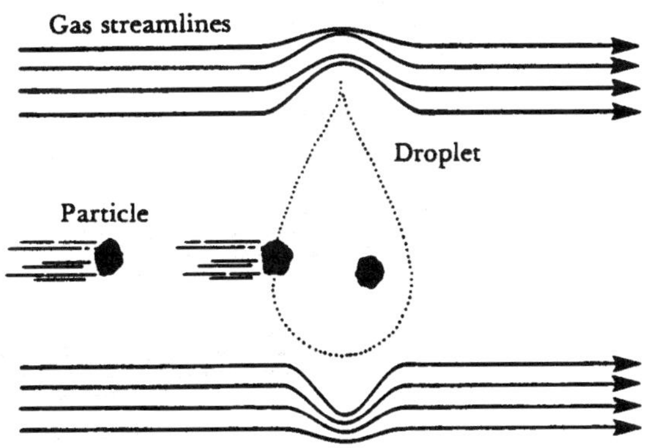

FIGURE 3.1 Particle collection by impaction (*EPA-84/03, p. 1-4*).

particle's mass causes it to break away from the streamlines and impact on the droplet. Impaction is the predominant collection mechanism for scrubbers having gas stream velocities greater than 0.3 m/s (1 ft/s). Most scrubbers do operate with gas stream velocities well above 0.3 m/s. Therefore, at these velocities, particles having diameters greater than 1.0 μm are collected by this mechanism (EPA-84/03, p. 1-4).

As the velocity of the particles in the exhaust stream increases relative to the liquid droplet's velocity, impaction increases. Impaction also increases as the size of the liquid droplet decreases. This is because there will be more droplets (for the same amount of liquid) within the vessel; consequently, this increases the likelihood that the particles will impact on the droplets.

The separation number for impaction is obtained by balancing the force of the moving particle against the resistance of the gas stream to its motion. The separation number for impaction is known as the impaction parameter ψ and is expressed by (EPA-81/10, p. 9-5)

$$\psi = C_f \rho_p v (d_p)^2 / 18 d_d \mu \tag{3.1}$$

where ρ_p = particle density
v = gas velocity at venturi throat, ft/s
d_p = particle diameter, ft
d_d = droplet diameter, ft
μ = gas viscosity, lb/ft-s
C_f = Cunningham correction factor

The collection efficiency associated with this impaction effect is expressed as (EPA-81/10, p. 9-7):

$$\eta_{\text{impaction}} = f(\psi)$$

Interception. The center of a particle follows the streamlines around the droplet, but a collision occurs if the distance between the particle and droplet is less than the radius of the particle (Fig. 3.2).

The separation number, which characterizes this effect, is the ratio of the particle diameter to the droplet diameter and is expressed as (EPA-81/10, p. 9-8)

$$d_p / d_d$$

where d_p = particle diameter, ft
d_d = droplet diameter, ft

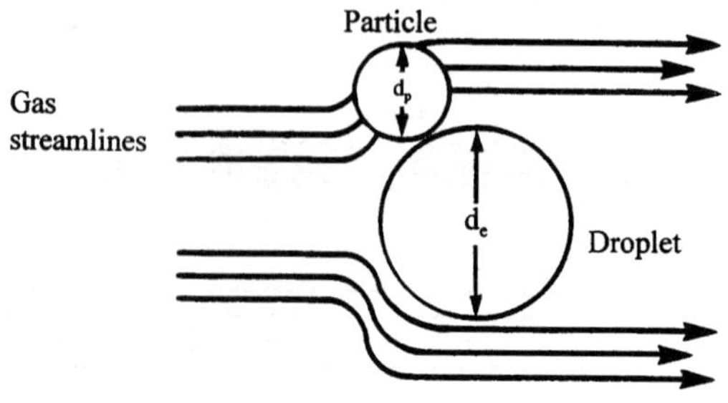

FIGURE 3.2 Particle collection by interception (*EPA-81/10, p. 9-8*).

The collection efficiency associated with this interception effect is a function of d_p and d_d or

$$\eta_{\text{interception}} = f(d_p/d_d)$$

Diffusion. Very small particles (less than 0.1 μm in diameter) experience random (Brownian) movement in an exhaust stream. These particles are so tiny that they are bumped by gas molecules as they move in the exhaust stream. This bumping, or bombardment, causes them to first move one way and then another in a random manner, or diffuse, through the gas. This irregular motion can cause the particles to collide with a droplet and be collected (Fig. 3.3). Because of this, in certain scrubbers, the removal efficiency of particles smaller than 0.1 μm can actually increase.

The rate of diffusion depends on relative velocity, particle diameter, and liquid droplet diameter. As with impaction, collection increases with an increase in relative velocity (liquid- or gas-pressure input) and a decrease in liquid droplet size. However, collection by diffusion increases as particle size decreases. This mechanism enables certain scrubbers to effectively remove the very tiny particles. In the particle size range of approximately 0.1 to 1.0 μm, neither of these two dominates. Particles in this size range are not collected as efficiently as are either larger particles collected by impaction or smaller particles collected by diffusion (EPA-84/03, p. 1-5).

The Brownian diffusion process leading to particle capture is most often described by a parameter called the Péclet number Pe (EPA-81/10, p. 9-8):

$$\text{Pe} = 3\pi\mu v d_p d_d/(C_f k_B T)$$

where
Pe = Péclet number
μ = gas viscosity
v = gas stream velocity
d_p = particle diameter
d_d = liquid droplet diameter
C_f = Cunningham correction factor
k_B = Boltzmann's constant
T = temperature of gas stream

The above equation shows that, as the temperature increases, Pe decreases. An increase in temperature means that gas molecules will move around faster than at lower temperatures.

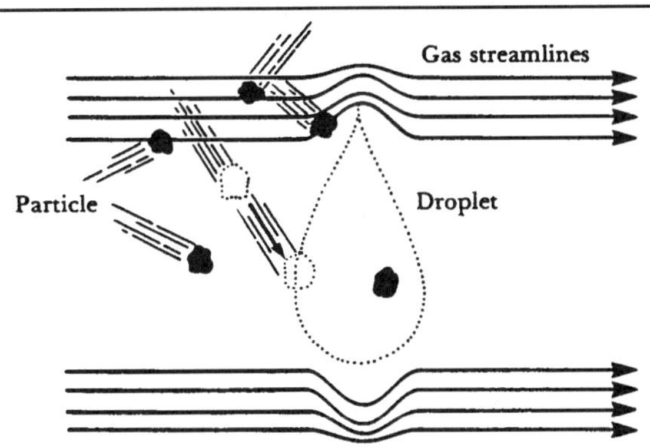

FIGURE 3.3 Particle collection by diffusion (*EPA-84/03, p. 1-5*).

This will lead to increased bombardment of the small particles, increased random motion, and increased collection efficiency by this mechanism. Expressions for collection efficiency by the diffusion process are generally in the form:

$$\eta_{\text{diffusion}} = f(1/\text{Pe})$$

This equation shows that as the Péclet number decreases, collection efficiency by diffusion increases.

2.2 Combined Efficiency

When attempting to collect 100 particles using a collection mechanism having 80% efficiency, 80 particles would be collected and 20 would escape. If another mechanism with 60% efficiency was used to collect the remaining 20 particles, 8 would escape this time. Ninety-two particles would therefore be collected, giving a total efficiency of 92/100 = 0.92 or 92%. This can be expressed as (EPA-81/10, p. 9-9):

$$\begin{aligned}
&100 - \{[100 - 100(0.80)] - [100 - 100(0.80)]0.60\} \\
&\quad = 100 - \{[100 - 100(0.80)](1 - 0.60)\} \\
&\quad = 100 - 100\{(1 - 0.8)(1 - 0.6)\} \\
&\quad = 100 - 8 \\
&\quad = 92
\end{aligned}$$

Generalizing in terms of the number of mechanisms which may be involved in wet collection devices, the combined efficiency can be expressed as

$$\eta_{\text{combined}} = 1 - (1 - \eta_{\text{impaction}})(1 - \eta_{\text{interception}})(1 - \eta_{\text{diffusion}})$$

Depending on flow conditions (laminar, turbulent, etc.), other combinations of these parameters have been used to provide expressions for the efficiency of wet collectors. The number of variables are so great and the interaction between varying size distributions of both particulate matter and liquid droplets is so complex, that this type of theoretical approach is difficult to apply in practice.

This academic approach, although useful in understanding basic phenomena, is commonly sidestepped for more empirical approaches. Correction factors obtained from actual operating systems or pilot plant data are used to correct for complex interactions. A newer concept expressing efficiency in terms of power expended has led to several empirical techniques for evaluating scrubber performance. The subsequent sections of this chapter will describe some of these empirical methods.

2.3 Collection Efficiency Models for Venturi Scrubber

Several models have been used for the calculation of venturi collection efficiency. These models include (EPA-84/03, p. 9-1; EPA-81/10, p. 9-1):

- Johnstone equation
- Infinite throat model
- Cut power method
- Contact power theory
- Pressure drop

Johnstone equation. Johnstone in 1951 developed an expression for the calculation of venturi scrubber efficiency by considering only the predominant mechanism of inertial impaction (EPA-81/10, p. 9-11).

The Johnstone equation is given as

$$\eta = 1 - \exp[-k(Q_L/Q_G)\sqrt{\psi}]$$

where
η = fractional collection efficiency
k = correlation coefficient whose value depends on the system geometry and operating conditions, typically 0.1–0.2, 1000 acf/gal
Q_L/Q_G = liquid-to-gas ratio, gal/1000 acf
ψ = $C_f\rho_p v(d_p)^2/18d_d\mu$ = inertial impaction parameter
ρ_p = particle density
v = gas velocity at venturi throat, ft/s
d_p = particle diameter, ft
d_d = droplet diameter, ft
μ = gas viscosity, lb/ft-s
C_f = Cunningham correction factor

Infinite throat model. One method used to predict particle collection efficiency in a venturi scrubber is called the infinite throat model (EPA-77). The equations presented in the infinite throat model assume that all particles are captured by the liquid in the throat section of the venturi.

The equations listed in the model can be used to predict the penetration (P_t) for one particle size (diameter) or for the overall penetration (P_t^*). The equations are provided below (EPA-84/03, p. 9-4):

$$\ln P_t(d_p) = -B\{[4K_{po} + 4.2 - 5.02(K_{po})^{0.5}(1 + 0.7/K_{po})\tan^{-1}(K_{po}/0.7)^{0.5}]/(K_{po} + 0.7)\} \quad (3.2)$$

where
$P_t(d_p)$ = penetration for one particle size
B = parameter characterizing the liquid-to-gas ratio, dimensionless
K_{po} = inertial parameter at throat entrance, dimensionless

Note: Equation (3.2) was developed assuming that the venturi has an infinite-sized throat length (l). This is valid only when l is greater than 2.0.

$$l = (3l_\tau C_D\rho_g)/(2d_d\rho_l)$$

where
l = throat length parameter, dimensionless
l_τ = venturi throat length, cm
C_D = drag coefficient for the liquid at the throat entrance, dimensionless
ρ_g = gas density, g/cm^3
ρ_l = liquid density, g/cm^3
d_d = droplet diameter, cm

The Nukiyama and Tanasawa equation

$$d_d = 50/v_{g\tau} + 91.8(L/G)^{1.5} \quad (3.3)$$

where
d_d = droplet diameter, cm
$v_{g\tau}$ = gas velocity in the throat, cm/s
L/G = liquid-to-gas ratio, dimensionless

$$B = (L/G)\rho_l/(\rho_g C_D) \quad (3.4)$$

where B = parameter characterizing liquid-to-gas ratio, dimensionless
L/G = liquid-to-gas ratio, dimensionless
ρ_g = gas density, g/cm^3
ρ_l = liquid density, g/cm^3
C_D = drag coefficient for the liquid at the throat entrance, dimensionless

$$K_{po} = (d_p)^2 v_{g\tau}/(9\mu_g d_d) \tag{3.5}$$

where K_{po} = inertial parameter at throat entrance, dimensionless
d_p = particle aerodynamic resistance diameter, cm
$v_{g\tau}$ = gas velocity in the throat, cm/s
μ_g = gas viscosity, g/s-cm
d_d = droplet diameter, cm

$$K_{pg} = (d_{pg})^2 v_{g\tau}/(9\mu_g d_d) \tag{3.6}$$

where K_{pg} = inertial parameter for mass-median diameter, dimensionless
d_{pg} = particle aerodynamic geometric mean diameter, cm
$v_{g\tau}$ = gas velocity in the throat, cm/s
μ_g = gas viscosity, g/s-cm
d_d = droplet diameter, cm

$$C_D = 0.22 + (24/N_{Reo})[1 + 0.15(N_{Reo})^{0.6}] \tag{3.7}$$

where C_D = drag coefficient for the liquid at the throat entrance, dimensionless
N_{Reo} = Reynolds' number for the liquid droplet at the throat inlet, dimensionless

$$N_{Reo} = v_{g\tau} d_d/\nu_g \tag{3.8}$$

where N_{Reo} = Reynolds' number for the liquid droplet at the throat inlet, dimensionless
$v_{g\tau}$ = gas velocity in the throat, cm/s
d_d = droplet diameter, cm
ν_g = gas kinematic viscosity, cm^2/s

$$d_{pg} = d_{ps}(C_f \times \rho_p)^{0.5} \tag{3.9}$$

where d_{pg} = particle aerodynamic geometric mean diameter, μmA [where A [=] (g/cm^3)$^{0.5}$, where [=] means "has the units of"]
d_{ps} = particle physical, or Stokes' diameter, μm
C_f = Cunningham slip correction factor, dimensionless
ρ_p = particle density, g/cm^3

$$C_f = 1 + [(6.21 \times 10^{-4})T]/d_{ps} \tag{3.10}$$

where C_f = Cunningham slip correction factor, dimensionless
T = absolute temperature, K
d_{ps} = particle physical, or Stokes' diameter, μm

The following example illustrates how to use the infinite throat model to predict the performance of a venturi scrubber. When using the equations given in the model, make sure that the units for each equation are consistent.

EXAMPLE: infinite throat model for venturi collection efficiency calculation. Cheeps Disposal Inc. is planning to install a hazardous-waste incinerator that will burn both liquid and solid waste materials. The exhaust gas from the incinerator will pass through a quench spray and then into a venturi scrubber. Caustic will be added to the scrubbing liquor to remove any HCl from the flue gas and to control the pH of the scrubbing liquor. The uncontrolled particulate emissions leaving the incinerator are estimated to be 1100 kg/h (maximum average). The local air pollution regulation states that particulate emissions must not exceed 10 kg/h. Using the following data, estimate the overall collection efficiency of the required scrubbing system (EPA-84/03, p. 9-8).

Given conditions

mass-median particle size (physical) $d_{ps} = 9.0\,\mu m$
geometric standard deviation $\sigma_{gm} = 2.5$
particle density $\rho_p = 1.9\,g/cm^3$
gas viscosity $\mu_g = 2.0 \times 10^{-4}\,g/cm\text{-}s$
gas kinematic viscosity $\nu_g = 0.2\,cm^2/s$
gas density $\rho_g = 1.0\,kg/m^3$
gas flow rate $Q_G = 15\,m^3/s$
gas velocity in venturi throat $v_{g\tau} = 9000\,cm/s$
gas temperature (in venturi) $T_g = 80°C$
water temperature $T_l = 30°C$
liquid density $\rho_l = 1000\,kg/m^3$
liquid flow rate $Q_L = 0.014\,m^3/s$
liquid-to-gas ratio $L/G = 0.0009\,L/m^3$

Solution:

1. Calculate the Cunningham slip correction factor.
The mass-median particle size (physical) d_{ps} is $9.0\,\mu m$. Since the particle aerodynamic geometric mean diameter d_{pg} is not known, you must use Eq. (3.9) to calculate d_{pg}, and Eq. (3.10) to calculate the Cunningham slip correction factor C_f.
From Eq. (3.10)

$$\begin{aligned} C_f &= 1 + [(6.21 \times 10^{-4})T]/d_{ps} \\ &= 1 + [(6.21 \times 10^{-4})(273 + 80)]/9 \\ &= 1.024 \end{aligned}$$

From Eq. (3.9)

$$\begin{aligned} d_{pg} &= d_{ps}(C_f \times \rho_p)^{0.5} \\ &= 9\,\mu m(1.024 \times 1.9\,g/cm^3)^{0.5} \\ &= 12.6\,\mu mA \\ &= 12.6 \times 10^{-4}\,cmA \end{aligned}$$

where $A[=](g/cm^3)^{0.5}$.

Note: This step would not have been required if the particle diameter had been given as the aerodynamic geometric mean diameter d_{pg} and expressed in units of μmA.

2. Calculate the droplet diameter d_d from Eq. (3.3) (Nukiyama and Tanasawa equation).
From Eq. (3.3)

$$d_d = 50/v_{g\tau} + 91.8(L/G)^{1.5}$$

where d_d = droplet diameter, cm
$v_{g\tau}$ = gas velocity in the throat, cm/s
L/G = liquid-to-gas ratio, dimensionless

$$d_d = 50/(9000\,\text{cm/s}) + 91.8(0.0009)^{1.5}$$
$$= 0.0080\,\text{cm}$$

3. Calculate the inertial parameter for the mass-median diameter K_{pg}, using Eq. (3.6).
From Eq. (3.6)

$$K_{pg} = (d_{pg})^2 v_{g\tau}/(9\mu_g d_d)$$

where K_{pg} = inertial parameter for mass-median diameter, dimensionless
d_{pg} = particle aerodynamic geometric mean diameter, cmA
$v_{g\tau}$ = gas velocity in the throat, cm/s
μ_g = gas viscosity, g/s-cm
d_d = droplet diameter, cm

$$K_{pg} = (12.6 \times 10^{-4}\,\text{cm})^2(9000\,\text{cm/s})/\{[9(2.0 \times 10^{-4}\,(\text{g/cm-s})(0.008\,\text{cm})]\}$$
$$= 992$$

4. Calculate the Reynolds' number N_{Reo}, using Eq. (3.8).
From Eq. (3.8)

$$N_{Reo} = v_{g\tau} d_d / \nu_g$$

where N_{Reo} = Reynolds' number for the liquid droplet at the throat inlet, dimensionless
$v_{g\tau}$ = gas velocity in the throat, cm/s
d_d = droplet diameter, cm
ν_g = gas kinematic viscosity, cm^2/s

$$N_{Reo} = v_{g\tau} d_d / \nu_g$$
$$= (9000\,\text{cm/s})(0.008\,\text{cm})/(0.2\,\text{cm}^2/\text{s})$$
$$= 360$$

5. Calculate the drag coefficient for the liquid at the throat entrance C_D using Eq. (3.7).
From Eq. (3.7)

$$C_D = 0.22 + (24/N_{Reo})[1 + 0.15(N_{Reo})^{0.6}]$$

where C_D = drag coefficient for the liquid at the throat entrance, dimensionless
N_{Reo} = Reynolds' number for the liquid droplet at the throat inlet, dimensionless

$$C_D = 0.22 + (24/N_{Reo})[1 + 0.15(N_{Reo})^{0.6}]$$
$$= 0.22 + (24/360)[(1 + 0.15(360)^{0.6}]$$
$$= 0.628$$

6. Now calculate the parameter characterizing the liquid-to-gas ratio B, using Eq. (3.4).
From Eq. (3.4)

$$B = (\text{L/G})\rho_l/(\rho_g C_D)$$

where B = parameter characterizing liquid-to-gas ratio, dimensionless
L/G = liquid-to-gas ratio, dimensionless

ρ_g = gas density, g/cm^3
ρ_l = liquid density, g/cm^3
C_D = drag coefficient for the liquid at the throat entrance, dimensionless

$$B = (L/G)\rho_l/(\rho_g C_D)$$
$$= (0.0009)(1000\ \mathrm{kg/m^3})/(1.0\ \mathrm{kg/m^3})(0.628)$$
$$= 1.43$$

7. The geometric standard deviation σ_{gm} is 2.5. The overall penetration P_t can be found from Fig. 3.4.

$$\sigma_{gm} = 2.5$$
$$B = 1.43$$
$$K_{pg} = 992$$
$$\text{Read } P_t^* = 0.008 \text{ from Fig. 3.4}$$

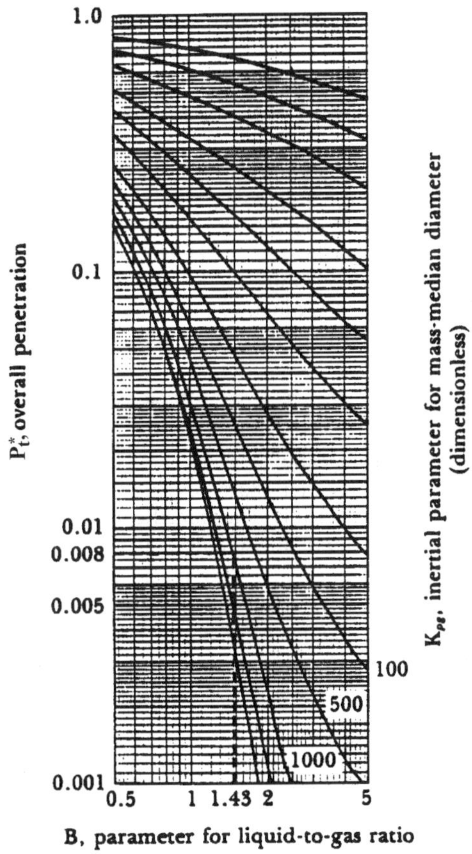

FIGURE 3.4 Overall penetration, P_t*, for σ_{gm} = 2.5.

8. The collection efficiency can be calculated using the equation

$$\begin{aligned}\eta &= 1 - P_t^* \\ &= 1 - 0.008 \\ &= 0.992 \\ &= 99.2\%\end{aligned}$$

9. Compare with regulatory standards
   The local regulations state that the particulate emissions cannot exceed 10 kg/h. The required collection efficiency can be calculated by using

$$\eta_{required} = (\text{dust}_{in} - \text{dust}_{out})/\text{dust}_{in}$$

where dust_{in} = dust concentration leading into the venturi
dust_{out} = dust concentration leaving the venturi

$$\begin{aligned}\eta_{required} &= (1100\,\text{kg/h} - 10\,\text{kg/h})/1100\,\text{kg/h} \\ &= 0.991\end{aligned}$$

The estimated efficiency of the venturi scrubber is slightly higher than the required efficiency.

Cut power method. One empirical correlation that has been used to predict the collection efficiency of a scrubber is the cut power method. In this method, penetration is a function of the cut diameter of the particles to be collected by the scrubber. The cut diameter is the diameter of the particles that are collected by the scrubber with at least 50% efficiency. Since scrubbers have limits to the size of particles they can collect, knowledge of the cut diameter is useful in evaluating the scrubbing system (EPA-84/03, p. 9-11).

In the cut power method, penetration is a function of the particle diameter and is given as

$$P_t = \exp[-A_{cut} d_p (B_{cut})] \tag{3.11}$$

where P_t = penetration
A_{cut} = parameter characterizing the particle size distribution
B_{cut} = empirically determined constant, depending on the scrubber design
d_p = aerodynamic diameter of the particle

Penetration, calculated by Eq. (3.11), is given for only one particle size (d_p). To obtain the overall penetration, the equation can be integrated over the log-normal particle size distribution. By mathematically integrating P_t over a log-normal distribution of particles and by varying the geometric standard deviation σ_{gm} and the geometric mean particle diameter d_{pg}, the overall penetration P_t^* can be obtained.

EXAMPLE 1: cut power method. Given similar conditions used in the example in the infinite throat model section, estimate the cut diameter for a venturi scrubber. The data below are approximate (EPA-84/03, p. 9-12).

geometric standard deviation $\sigma_{gm} = 2.5$

particle aerodynamic geometric mean diameter $d_{pg} = 12.6\,\mu\text{mA}$

required efficiency $\eta = 99.1\%$ or 0.991

Solution:

1. For an efficiency of 99.1%, the overall penetration can be calculated from

$$\begin{aligned} P_t^* &= 1 - \eta \\ &= 1 - 0.991 \\ &= 0.009 \end{aligned}$$

2. The overall penetration is 0.009, and the geometric standard deviation is 2.5. Use the figures on EPA-84/03, p. 9-12 to read $(d_p)_{cut}/d_{pg}$.

$$\begin{aligned} P_t^* &= 0.009 \\ \sigma_{gm} &= 2.5 \\ (d_p)_{cut}/d_{pg} &= 0.09 \end{aligned}$$

3. The cut diameter $(d_p)_{cut}$ is calculated from

$$\begin{aligned} (d_p)_{cut}/d_{pg} &= 0.09 \\ (d_p)_{cut} &= (0.09)d_{pg} \\ (d_p)_{cut} &= 0.09(12.6\,\mu mA) = 1.134\,\mu mA \end{aligned}$$

EXAMPLE 2: cut diameter. A particle size analysis indicated that (EPA-81/10, p. 9-14)

d_{gm} (geometric mean particle diameter) $= 12\,\mu m$
σ_{gm} (standard deviation of the distribution) $= 3.0$
η (wet collector efficiency) $= 99\%$

If a collection efficiency of 99% is required to meet emission standards, what would the cut diameter of the scrubber have to be?

Solution:

1. Write the penetration (P_t) equation

$$\begin{aligned} P_t^* &= 1 - \eta \\ &= 1 - 0.99 \\ &= 0.01 \end{aligned}$$

Use the figures on EPA-84/03, p. 9-12 to read $(d_p)_{cut}/d_{gm}$, for $P_t^* = 0.01$ and $\sigma_{gm} = 3.0$; $[d_p]_{cut}/d_{gm}$ equals 0.063. Since $d_{gm} = 12\,\mu m$, then the scrubber must be able to collect particles of size $0.063 \times 12 = 0.76\,\mu m$ with at least 50% efficiency to achieve an overall scrubber efficiency of 99%.

Contact power theory. Theoretical approaches based on the dynamic interactions between particles and water droplets are limited in applicability, since actual scrubber systems are very complex. For example, the Johnstone and cut power methods assume that the water droplets are uniformly distributed over the cross-section of the scrubber. The actual turbulence and eddies existing in the contact zone are not considered (EPA-81/10, p. 9-15).

A more general theory which avoids the details of how particles and droplets hit each other, is the contact power theory. This theory indicates that collection efficiency is a function of how much power is used, and not the details of design. This has a number of implications in the evaluation and selection of wet collectors. Once it is realized that a certain amount of power (or pressure loss) is needed for a required collection efficiency, the claims about specially located nozzles, baffles, etc., can be evaluated more objectively. The choice between two different scrubbers with the same power requirements may depend primarily on ease of maintenance (EPA-81/10, p. 9-16; EPA-84/03, p. 9-13).

The total pressure loss is expressed in terms of the power expended to inject the liquid into the scrubber plus the power needed to move the process gas through the system (EPA-84/03, p. 9-13):

$$\begin{aligned} P_T &= P_G + P_L \\ P_G &= 0.157\Delta p \\ P_L &= 0.583 p_L (Q_L/Q_G) \end{aligned} \tag{3.12}$$

where P_T = total contacting power (total pressure loss), kWh/100 m^3(hp/1000 acfm)
P_G = power input from gas stream, kWh/100 m^3 (hp/1000 acfm)
P_L = contacting power from liquid injection, kWh/100 m^3 (hp/1000 acfm)

Note: The total pressure loss P_T should not be confused with penetration P_t, which is defined in the previous section.

The power expended in moving the gas through the system P_G is expressed in terms of the scrubber pressure drop:

$$P_G = 2.724 \times 10^{-4} \Delta p, \ \text{kWh/1000 m}^3 \text{ (metric units)} \tag{3.13}$$

or

$$P_G = 0.1575\Delta p, \ \text{hp/1000 acfm (British/US Customary units)}$$

where Δp = pressure drop, kPa (in. H_2O)

The power expended in the liquid stream P_L is expressed as

$$P_L = 0.28 p_L (Q_L/Q_G), \ \text{kWh/1000 m}^3 \text{ (metric units)} \tag{3.14}$$

or

$$P_L = 0.583 p_L (Q_L/Q_G), \ \text{hp/1000 acfm (British/US customary units)}$$

where p_L = liquid inlet pressure, 100 kPa (lb/in^2)
Q_L = liquid feed rate, m^3/h (gal/min)
Q_G = gas flow rate, m^3/h (ft^3/ min)

The constants given in the expressions for P_G and P_L incorporate conversion factors to put the terms on a consistent basis. The total power can therefore be expressed as

$$\begin{aligned} P_T &= P_G + P_L \\ &= 2.724 \times 10^{-4} \Delta p + 0.28 p_L (Q_L/Q_G), \ \text{kWh/1000 m}^3 \text{ (metric units)} \end{aligned} \tag{3.15}$$

or

$$P_T = 0.1575\Delta p + 0.583 p_L (Q_L/Q_G), \ \text{hp/1000 acfm (British/US customary units)}$$

The problem now is to correlate this with scrubber efficiency.

An equation given in Sec. 2.1 shows that efficiency is an exponential function of the system variables for most types of collectors:

$$\eta = 1 - \exp[-f(\text{system})] \tag{3.16}$$

where $f(\text{system})$ is defined as

$$f(\text{system}) = N_t = \alpha(P_T)^\beta \tag{3.17}$$

where N_t = number of transfer units
P_T = total contacting power
α and β = empirical constants that are determined from experiment and depend on the characteristics of the particles

The efficiency then becomes

$$\eta = 1 - \exp[-\alpha(P_T)^\beta] \tag{3.18}$$

The values of α and β which can be used in either the metric or British units can be found from the table in EPA-84/03, p. 9-15.

The scrubber collection efficiency is also expressed as the number of transfer units (EPA-81/10, p. 9-17):

$$N_t = \alpha(P_T)^\beta = \ln[1/(1-\eta)] \tag{3.19}$$

where N_t = number of transfer units
η = fractional collection efficiency
α, β = characteristic parameters for the type of particulates being collected

The contact power theory cannot predict efficiency from a given particle size distribution as can the cut power and Johnstone theories. The contact power theory gives a relationship which is independent of the size of the scrubber. With this observation, a small pilot scrubber could first be used to determine the pressure drop needed for the required collection efficiency. The full-scale scrubber design could then be scaled up from the pilot information. As an example, consider the following problem.

EXAMPLE: contact power theory. A wet scrubber is to be used to control particulate emissions from a foundry cupola. Stack test results reveal that the particulate emissions must be reduced by 85% to meet emission standards. If a 100 acfm pilot unit is operated with a water flow rate of 0.5 gal/min at a water pressure of 80 psi, what pressure drop (Δp) would be needed across a 10,000 acfm scrubber unit (EPA-84/03, p. 9-15; EPA-81/10, p. 9-18))?

Solution:

1. From the table in EPA-84/03, p. 9-15, read the α and β parameters for foundry cupola dust.

$$\alpha = 1.35$$
$$\beta = 0.621$$

2. Calculate the number of transfer units N_t using Eq. (3.16).

$$\begin{aligned}\eta &= 1 - \exp(-N_t)\\ N_t &= \ln[1/(1-\eta)]\\ &= \ln[1/(1-0.85)]\\ &= 1.896\end{aligned}$$

3. Calculate the total contacting power P_T using Eq. (3.17).

$$\begin{aligned} N_t &= \alpha(P_T)^\beta \\ 1.896 &= 1.35(P_T)^{0.621} \\ 1.404 &= (P_T)^{0.621} \\ \ln 1.404 &= 0.621(\ln P_T) \\ 0.3393 &= 0.621(\ln P_T) \\ 0.5464 &= \ln P_T \\ P_T &= 1.73 \text{ hp/1000 acfm} \end{aligned}$$

4. Calculate the pressure drop Δp using Eq. (3.15)

$$\begin{aligned} P_T &= 0.1575\Delta p + 0.583p_L(Q_L/Q_G) \\ 1.73 &= 0.1575\Delta p + 0.583(80)(0.5/100) \\ \Delta p &= 9.5 \text{ in. } H_2O \end{aligned}$$

Pressure drop. As discussed earlier, a number of factors affect particle capture in a scrubber. One of the most important, especially for the contact power theory, is pressure drop. Pressure drop is the difference in pressure between the inlet and the outlet of the scrubber. The pressure drop represents the energy expended in the scrubbing process. From the contact power theory, the higher the pressure drop, the more efficient the scrubber. However, the higher the pressure drop, the higher the operating costs. Most scrubbing systems operate at pressure drops just high enough to ensure adequate collection of particles (EPA-84/03, p. 9-17).

The following factors affect the pressure drop in a scrubber:

- scrubber design and geometry
- gas velocity
- liquid-to-gas ratio

As with calculating collection efficiency, no one equation can predict the pressure drop for all scrubbing systems.

Many theoretical and empirical relationships are available for estimating the pressure drop across a scrubber. Generally, the most accurate are those developed by scrubber manufacturers for their particular scrubbing systems. Due to the lack of validated models, it is recommended that users consult the vendor's literature to estimate pressure drop for the particular scrubbing device of concern.

One expression was developed for venturis and is as follows:

$$\Delta p = 8.24 \times 10^{-4}(v_{gt})^2(L/G) \qquad \text{(metric units)} \tag{3.20}$$

or

$$\Delta p = 4.0 \times 10^{-5}(v_{gt})^2(L/G) \qquad \text{(British units)} \tag{3.21}$$

where Δp = pressure drop, cm H_2O (in. H_2O)
v_{gt} = velocity of gas in the venturi throat, cm/s (ft/s)
L/G = liquid-to-gas ratio, dimensionless [actually in L/m^3 (gal/1000 ft^3)]

Using Eq. (3.20) for the conditions given in the example in the infinite throat model section, we obtain

$$v_{g\tau} = 9000\,\text{cm/s}$$
$$\text{L/G} = 0.0009\,\text{L/m}^3$$
$$\Delta p = 8.24 \times 10^{-4}(9000)^2(0.0009)$$
$$= 60\,\text{cm H}_2\text{O}$$

3 WET ABSORPTION FOR GASEOUS EMISSION CONTROL

Similar to the wet absorption process for particulate emission control, absorption devices used to remove gaseous contaminants are referred to as *absorbers* or *wet scrubbers*. The process of dissolving gaseous pollutants in a liquid is referred to as *absorption*. Absorption is a mass transfer operation. Mass transfer can be compared to heat transfer in that both occur because a system is trying to reach equilibrium conditions. For example, in heat transfer, if a hot slab of metal is placed on top of a cold slab, heat energy will be transferred from the hot slab to the cold slab until both are at the same temperature (equilibrium). In absorption, mass instead of heat is transferred as a result of a concentration difference, rather than a heat energy difference. Absorption continues as long as a concentration differential exists between the liquid and the gas from which the contaminant is being removed. In absorption, equilibrium depends on the solubility of the pollutant in the liquid (EPA-84/03, p. 1-7).

To remove a gaseous pollutant by absorption, the exhaust stream must be passed through (brought into contact with) a liquid (Fig. 3.5). The process involves three steps.

- In the first step, the gaseous pollutant diffuses from the bulk area of the gas phase to the gas–liquid interface.
- In the second step, the gas moves (transfers) across the interface to the liquid phase. This step occurs extremely rapidly once the gas molecules (pollutants) arrive at the interface area.
- In the third step, the gas diffuses into the bulk area of the liquid, thus making room for additional gas molecules to be absorbed. The rate of absorption (mass transfer of the

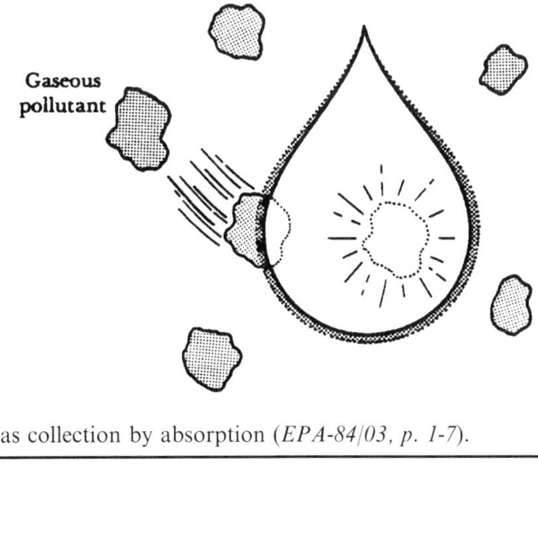

FIGURE 3.5 Gas collection by absorption (*EPA-84/03, p. 1-7*).

pollutant from the gas phase to the liquid phase) depends on the diffusion rates of the pollutant in the gas phase (first step) and in the liquid phase (third step).

To enhance gas diffusion and, therefore, absorption, the steps include:

- Providing a large interfacial contact area between the gas and liquid phases.
- Providing good mixing of the gas and liquid phases (turbulence).
- Allowing sufficient residence, or contact, time between the phases for absorption to occur.

Two of these three gas-collection mechanisms—large contact area and good mixing—are also important for particle collection. The third factor, sufficient residence time, works in direct opposition to efficient particle collection. To increase residence time, the relative velocity of the gas and liquid streams must be reduced. Therefore, achieving a high removal efficiency for both gaseous and particulate pollutants at the same time in the same scrubber is extremely difficult unless the gaseous pollutant is very soluble in the liquid.

As previously mentioned, a very important factor affecting the amount of a pollutant that can be absorbed is its solubility. Solubility governs the amount of liquid (liquid-to-gas ratio) required and the necessary contact time. More soluble gases require less liquid. Also, more soluble gases will be absorbed faster. Solubility is a function of both the temperature and, to a lesser extent, the pressure of the system. As temperature increases, the amount of gas that can be absorbed by a liquid decreases. From the ideal gas law: as temperature increases, the volume of a gas also increases; therefore, at a higher temperature, gas volume increases and less gas is absorbed. For this reason, some absorption systems use inlet quench sprays to cool the incoming exhaust stream, thereby increasing absorption efficiency. Pressure affects the solubility of a gas in the opposite manner. When the pressure of a system is increased, the amount of gas absorbed generally increases.

4 WET SCRUBBERS FOR PARTICULATE AND GASEOUS EMISSION CONTROL

Although many unique wet collector systems are available, commercially designed wet collectors use only a few basic components. Some of these components are (EPA-84/03, p. 2-1; EPA-81/10, p. 9-22):

- Spray nozzles
- Venturi constrictions
- Impingement surfaces
  - Plates
  - Baffles
  - Bubble caps
  - Packing
- Cyclonic openings
- Spray-inducing orifices
- Mechanically-driven rotors

The many possible combinations of these basic pieces of hardware has led to the development of numerous types of wet collectors. This is an advantage for the process engineer, who can specify a system in terms of size, cost, and collection efficiency. However, the profusion of devices can also be a disadvantage, especially when confronted with advertisements for such equipment. In this competitive market, the various claims and counterclaims can be confusing.

Scrubbers can be categorized in terms of:

- Power requirements for bringing the pollutant gas stream into contact with the liquid
- Pressure drop

Power requirement: Although wet collection systems may vary greatly in complexity, most of them follow a general rule: collection efficiency increases with increased power input. Power is defined as energy applied per unit time. Energy from the gas stream, energy from the liquid stream, or energy from a mechanically-driven rotor is used to bring gas stream particles into contact with the scrubbing liquor. The total energy applied per unit time is the *contacting power*. Scrubbers can be categorized in terms of how this power is applied to the system. The categories are given in Table 3.4.

Pressure drop: Scrubbers can also be classed by gas-phase pressure drop. Pressure drop, or gas-side pressure drop, refers to the pressure difference that occurs as the exhaust gas is pushed or pulled through the scrubber, disregarding the pressure that would be used for pumping or spraying the liquid into the scrubber. Pressure drop and gas-side pressure drop are used interchangeably. There are three categories of pressure drops:

- Low-energy scrubbers, having a pressure drop of less than 5 in. (12.7 cm) of water
- Medium-energy scrubbers, having pressure drops between 5 and 15 in. (12.7 to 38.1 cm) of water
- High-energy scrubbers, having a pressure drop greater than 15 in. (38.1 cm) of water

TABLE 3.4 Categories of Scrubbers and Energy Input

Wet collector type	Scrubbers using energy from
Gas-phase contacting scrubbers	Gas stream
Liquid-phase contacting scrubbers	Liquid stream
Liquid-phase/gas-phase contacting scrubbers	Gas and liquid streams
Mechanically-aided scrubbers	Mechanically-driven rotor

4.1 Gas-Phase Contacting Scrubber

Scrubbers using the process gas stream to provide the energy for particle–liquid contact are known as *gas-phase contacting scrubbers*. By moving the gas across or through a liquid surface, the liquid is sheared to form small droplets. Particulate matter in the gas stream impacts on the larger droplets, which in turn are collected by cyclonic action or other means. A number of methods are used to develop this shearing action in a scrubber. The gas can be forced through cascades of liquid falling over flat plates. Holes can be punched in the plates and the gas can aspirate the water flowing over the plate, or the gas can be forced through constricted passages wetted with liquid such as in orifice and venturi scrubbers. Three collectors which work primarily by this action are

- Venturi scrubber
- Plate tower
- Orifice scrubber

Venturi scrubber. A venturi scrubber is designed to effectively use the energy from the exhaust stream to atomize the scrubbing liquid. Venturi devices have been used for over

100 years to measure fluid flow (venturi tubes derived their name from G. B. Venturi, an Italian physicist). In 1949, Johnstone and other researchers found that they could effectively use the venturi configuration to remove particles from an exhaust stream. Figure 3.6 illustrates a venturi configuration (EPA-84/09, p. 3-2; EPA-81/12, p. 4-42; EPA-81/10, p. 9-29).

A venturi scrubber consists of a converging section, a throat section, and a diverging section. The exhaust stream enters the converging section and, as the area decreases the gas velocity increases. The venturi is designed to force gas in and out of a constriction. Since the volumetric flow rate of gas Q must be the same throughout the system, the velocity of the gas must increase at the throat of the venturi.

$$Q_{\text{entrance}} = Q_{\text{throat}}$$
$$V_{\text{entrance}} A_{\text{entrance}} = V_{\text{throat}} A_{\text{throat}}$$

where A = area and V = velocity.

The above equation shows that the velocity at the throat must increase in order to make up for the decrease in area in the throat. Velocities at such a constriction can range from 200 to 800 ft/s (61 to 244 m/s). Now, if water is introduced into the throat, the gas forced to move at high velocity will shear the water into droplets. Particles in the gas stream then impact onto the droplets produced. Moving a large volume of gas through a small constriction gives a high velocity flow, but also a large pressure drop across the system. Collection efficiency for small particles increases with increased velocities (and corresponding increased pressure drops) since the water is sheared into more and smaller droplets than at lower velocities. The large number

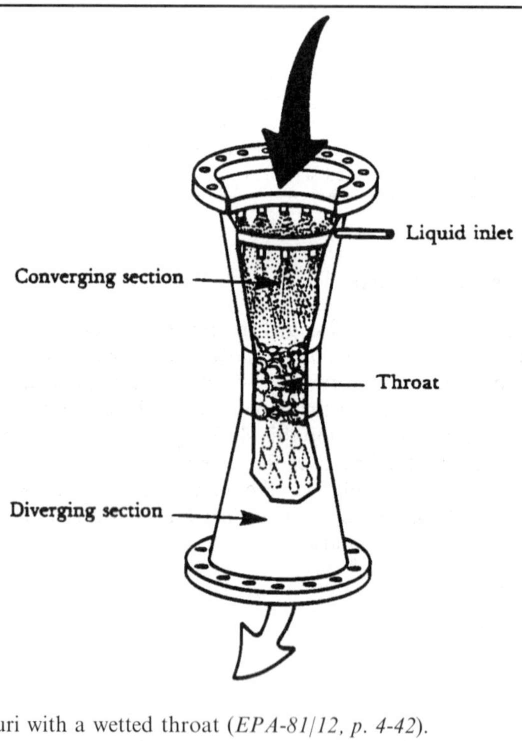

FIGURE 3.6 Venturi with a wetted throat (*EPA-81/12, p. 4-42*).

of small droplets combined with the turbulence in the throat section provides numerous impaction targets for particle collection (EPA-81/10, p. 9-30).

Particle and gas removal occur in the throat section as the exhaust stream mixes with the fog of tiny liquid droplets. The exhaust stream then exits through the diverging section, where it is forced to slow down. Venturis can be used to collect both particulate and gaseous pollutants, but they are more effective in removing particles than in removing gaseous pollutants.

Liquid can be injected at the converging section or at the throat. Figure 3.6 also shows liquid injected at the converging section; thus, the liquid coats the venturi throat. This venturi is very effective for handling hot, dry exhaust gas that contains dust. The dust would have a tendency to cake on or abrade a dry throat. These venturis are sometimes referred to as having a wetted approach.

Figure 3.7 shows liquid injected at the venturi throat. Since it is sprayed at or just before the throat, it does not actually coat the throat surface. These throats are susceptible to solids buildup when the throat is dry. They are also susceptible to abrasion by dust particles. These venturis are best used when the exhaust stream is cool and moist. In this venturi, the relative particle-to-liquid velocity is the highest of any of the venturis; therefore, the smallest particles can be collected efficiently. These venturis are referred to as having a *non-wetted approach*.

Manufacturers have developed other modifications to the basic venturi design to maintain scrubber efficiency by changing the pressure drop for varying exhaust gas rates. Certain types of orifices that create more turbulence than a true venturi were found to be equally efficient for a given unit of energy consumed. Results of these findings led to the development of the annular-orifice, or adjustable throat, venturi scrubber (Fig. 3.8). The throat area is varied by moving a plunger, or adjustable disk, up or down in the throat, decreasing or increasing the

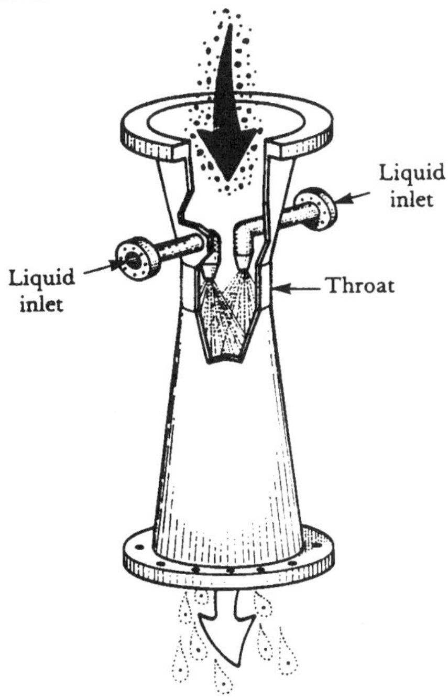

FIGURE 3.7 Venturi with throat sprays (*EPA-84/03, p. 3-3*).

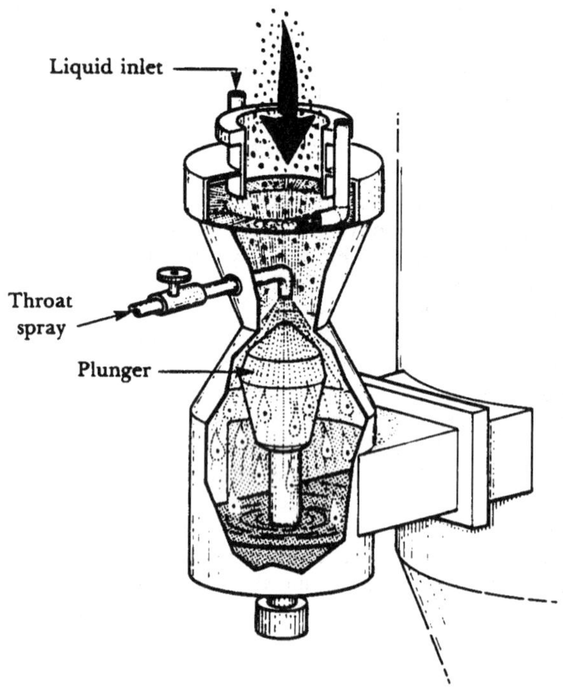

FIGURE 3.8 Adjustable throat venturi with movable plunger (*EPA-84/03, p. 3-5*).

annular opening. Gas flows through the annular opening and atomizes liquid that is sprayed onto the plunger or swirled in from the top.

Venturis with round throats (Figs. 3.6 and 3.7) can handle exhaust flows as large as 68,000 m^3/h (40,000 cfm). At exhaust flow rates greater than this, achieving uniform liquid distribution is difficult, unless additional weirs or baffles are used. To handle large exhaust flows, scrubbers designed with long, narrow, rectangular throats have been used. Figure 3.9 shows an adjustable throat rectangular venturi. In this scrubber, the throat area is varied by using a movable plate. A water-wash spray is used to continually wash collected material from the plate.

All venturi scrubbers require an entrainment separator because the high velocity of gas through the scrubber will have a tendency to exhaust the droplets. Cyclonic, mesh-pad, and blade separators are all used. Cyclonic separators, the most popular, are connected to the venturi vessel by a flooded elbow (Fig. 3.9). The liquid reduces abrasion of the elbow as the exhaust gas passes at high velocities from the venturi to the separator.

Particle collection. Venturis are the most commonly used scrubber for particle collection and are capable of achieving the highest particle collection efficiency of any wet scrubbing system. As the exhaust stream enters the throat, its velocity increases greatly, atomizing and turbulently mixing with any liquid present. The atomized liquid provides an enormous number of tiny droplets for the dust particles to impact on. These liquid droplets incorporating the particles must then be removed from the scrubber exhaust stream, generally by cyclonic separators.

Particle removal efficiency increases with increasing pressure drop (resulting in high gas velocity and turbulence). Venturis can be operated with pressure drops ranging from 12 to 250

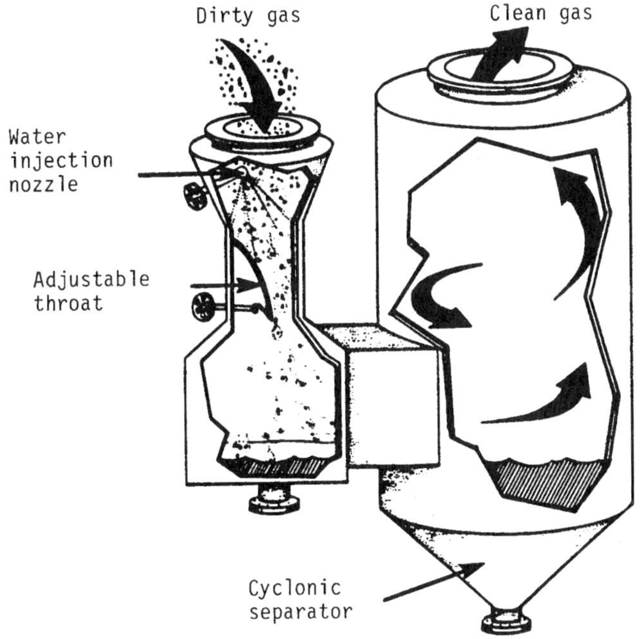

FIGURE 3.9 Adjustable rectangular throat venturi with movable plate and cyclonic separator (*EPA-81/10, p. 9-34*).

cm (5 to 100 in.) of water. Most venturis normally operate with pressure drops in the range of 50 to 150 cm (20 to 60 in.) of water. At these pressure drops, the gas velocity in the throat section is usually between 30 and 120 m/s (100 and 400 ft/s), or approximately 270 mph at the high end. These high pressure drops result in high operating costs.

The liquid-injection rate, or liquid-to-gas ratio (L/G), also affects particle collection. The liquid-injection rate depends on the temperature (evaporation losses) of the incoming exhaust stream and the particle concentration. Most venturi systems operate with an L/G ratio of 0.4 to 1.3 L/m^3 (3 to 10 gal/1000 ft^3). L/G ratios less than 0.4 L/m^3 (3 gal/1000 ft^3) are usually not sufficient to cover the throat, and adding more than 1.3 L/m^3 (10 gal/1000 ft^3) does not usually significantly improve particle collection efficiency.

Gas collection. Venturi scrubbers can be used for removing gaseous pollutants; however, they are not used when removal of gaseous pollutants is the only concern. The high exhaust gas velocities in a venturi result in a very short contact time between the liquid and gas phases. This short contact time limits gas absorption. However, venturi scrubbers are very useful for simultaneous gaseous and particulate pollutant removal, especially when:

- Scaling could be a problem.
- A high concentration of dust is in the exhaust stream.
- The dust is sticky or has a tendency to plug openings.
- The gaseous contaminant is very soluble or chemically reactive with the liquid.

To maximize absorption of gases, venturis operate at a set of conditions different from those used to collect particles. Lower gas velocities and higher liquid-to-gas ratios are necessary for efficient absorption. These L/G values should be approximately 2.7 to 5.3 L/m^3 (20 to 40 gal/1000 ft^3). At high liquid-to-gas ratios, the gas velocity in the venturi throat is reduced (for a given pressure drop). The reduction in gas velocity allows for a longer contact time between phases.

In practice, a venturi scrubber for particle collection is usually connected with an absorber for acid gas scrubbing. Figure 3.10 shows the schematic of a venturi particle scrubber and an acid gas packed bed absorber.

Venturi scrubber/acid gas absorber. The system configurations for venturi particulate scrubbers and acid gas absorbers are illustrated schematically in Fig. 3.10. The basic configurations, with modifications, are used by other vendors as well. As illustrated, hot flue gas from the incinerator or waste heat boiler is directed to the venturi scrubber, which both quenches the hot gas and removes particulate matter. The venturi scrubber also provides partial collection of acid gases and other condensable flue gas constituents. Particulate collection occurs through a variety of fundamental mechanisms, but the predominant phenomenon is inertial impaction of particulate matter with water droplets sprayed into the venturi inlet. High collection efficiency of submicron particulate requires that the flue gas and particulate be accelerated to high velocity in the venturi throat region. The basic venturi configuration is used to minimize pressure loss associated with flow acceleration, but a venturi pressure drop on the order of 30 in. w.c. (7.5 kPa) is required to achieve 0.03 gr/dscf (0.07 g/dscm) particulate matter control and an even higher pressure drop is needed to achieve a 0.015 gr/dscf (0.03 g/dscm) limit (EPA-92/01, p. 77).

Acid gases such as HCl and SO_2 are only partially absorbed by the water spray in the venturi scrubber. High-level acid gas control is achieved in a counterflow acid gas absorber

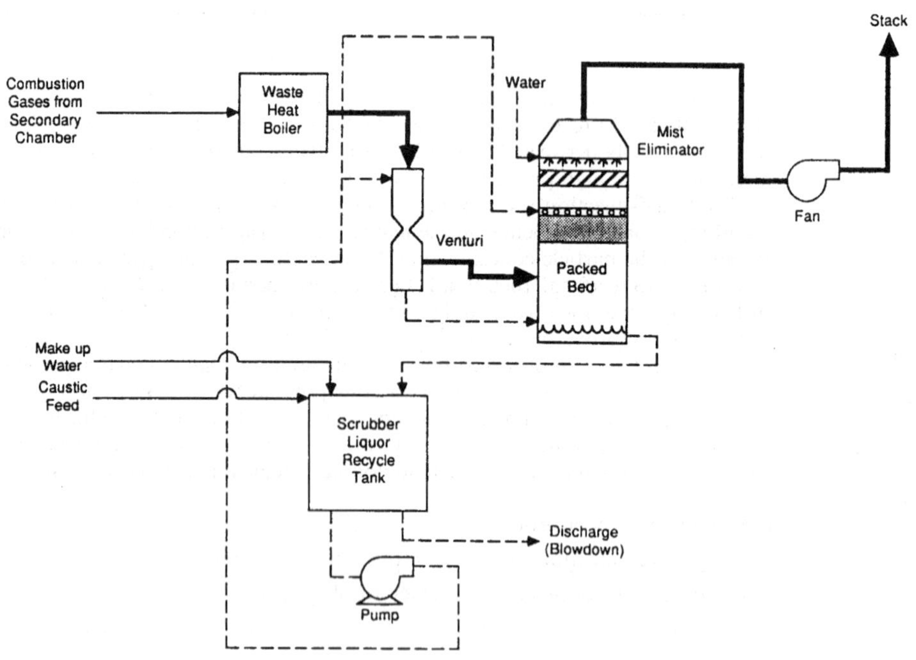

FIGURE 3.10 Venturi particle scrubber with acid gas packed bed (*EPA-89/03*).

downstream of the venturi. As shown in Fig. 3.10, collection of acid gas tends to drop the pH of the scrubber liquor and thus an alkaline buffering reagent is provided for pH control. Either lime or sodium hydroxide may be used for this purpose with NaOH being the typical selection since it can be purchased and stored as a liquid. Common neutralization chemical reactions for sodium hydroxide and absorbed acid gases include

$$HCl + NaOH \rightarrow NaCl + H_2O$$

$$Cl_2 + 2NaOH \rightarrow NaCl + NaOCl + H_2O$$

$$HF + NaOH \rightarrow NaF + H_2O$$

$$SO_2 + 2NaOH \rightarrow Na_2SO_3 + H_2O$$

For lime-based systems, the neutralization reactions include

$$2HCl + Ca(OH)_2 \rightarrow CaCl_2 + 2H_2O$$
$$2Cl_2 + 2Ca(OH)_2 \rightarrow CaCl_2 + Ca(OCl)_2 + 2H_2O$$
$$2HF + Ca(OH)_2 \rightarrow CaF_2 + 2H_2O$$
$$SO_2 + Ca(OH)_2 \rightarrow CaSO_3 + H_2O$$

The rate of buffer reagent addition is controlled to maintain the scrubber liquor pH between 6.5 and 7.0. This is essentially stoichiometric for the above reactions. For sodium-based buffering, the resultant salts are water soluble and in many areas, scrubber blowdown can be discharged directly to the facility sewer system. This blowdown will, however, also contain insoluble material, including particulate matter captured in the venturi scrubber.

Since particulate matter is captured in the venturi by impaction with water droplets, any liquid escaping from the absorption tower contributes particulate to the facility exhaust. For this reason, one or more stages of mist elimination are included at the top of the absorption tower.

The total pressure drop across the absorption tower is typically small [4 to 6 in. w.c. (1.0–1.5 kPa)] in comparison with the pressure drop across the venturi scrubber. The total head loss across the air pollution control equipment plus that of the incinerator is provided by the induced draft fan. Flue gases enter the fan saturated but become unsaturated by the fan's pressure increase before exiting through the stack.

EXAMPLE 1: scrubber design of a venturi scrubber. You are required to calculate the throat area of a venturi scrubber to operate at a specified collection efficiency (EPA-84/09, p. 77).

Given conditions

volumetric flow rate of process gas stream = 11,040 acfm (at 68°F)

density of dust = $187\,\text{lb/ft}^3$

liquid-to-gas ratio = $2\,\text{gal}/1000\,\text{ft}^3$

average particle size = 3.2 microns or μm (1.05×10^{-5} ft)

water droplet size = 48 microns (1.575×10^{-4} ft)

scrubber coefficient $k = 0.14$

required collection efficiency = 98%

viscosity of gas = 1.23×10^{-5} lb/ft-s

Cunningham correction factor = 1.0

Solution:

1. Calculate the inertial impaction parameter, ψ, from Johnstone's equation.
   Johnstone's equation describes the collection efficiency of a venturi scrubber (EPA-81/10, p. 9-11).

$$\eta = 1 - \exp[-k(Q_L/Q_G)\sqrt{\psi}]$$

where
η = fractional collection efficiency
k = correlation coefficient whose value depends on the system geometry and operating conditions, typically 0.1–0.2, 1000 acf/gal
Q_L/Q_G = liquid-to-gas ratio, gal/1000 acf
ψ = $C_f \rho_p v (d_p)^2 / 18 d_d \mu$ = inertial impaction parameter
ρ_p = particle density
v = gas velocity at venturi throat, ft/s
d_p = particle diameter, ft
d_d = droplet diameter, ft
μ = gas viscosity, lb/ft-s
C_f = Cunningham correction factor

2. From the calculated value of ψ (inertial impaction parameter) above, back calculate the gas velocity at the venturi throat v.

2.a. Calculate ψ.

$$\eta = 1 - \exp[-k(Q_L/Q_G)\sqrt{\psi}]$$
$$0.98 = 1 - \exp[-0.14(2)\sqrt{\psi}]$$

Solving for ψ

$$\psi = 195.2$$

2.b. Calculate v.

$$\begin{aligned} \psi &= C_f \rho_p v (d_p)^2 / 18 d_d \mu \\ v &= 18 \psi d_d \mu / C_f \rho_p (d_p)^2 \\ &= (18)(195.2)(1.575 \times 10^{-4})(1.23 \times 10^{-5})/(1)(187)(1.05 \times 10^{-5})^2 \\ &= 330.2\ \text{ft/s} \end{aligned}$$

3. Calculate the throat area S, using gas velocity at the venturi throat v.

$$\begin{aligned} S &= (\text{volumetric flowrate})/(\text{velocity}) \\ &= 11{,}040/(60)(330.2) \\ &= 0.557\ \text{ft}^2 \end{aligned}$$

EXAMPLE 2: scrubber overall collection efficiency. You are requested to calculate the overall collection efficiency of a venturi scrubber which cleans a fly ash-laden gas stream, given the liquid-to-gas ratio, throat velocity, and particle size distribution (EPA-84/09, p. 79).

Given conditions

liquid-to-gas ratio = 8.5 gal/1000 ft^3
throat velocity = 227 ft/s
particle density of fly ash = 43.7 lb/ft^3
gas viscosity = 1.5×10^{-5} lb/ft-s

The particle size distribution data are given in Table 3.5.

Use Johnstone's equation, with a k value of 0.2, to calculate the collection efficiency. Neglect the Cunningham correction factor effect.

TABLE 3.5 Particle Size Distribution Data

d_p, microns	Weight, %
< 0.1	0.01
0.1–0.5	0.21
0.5–1.0	0.78
1.0–5.0	13
5.0–10.0	16
10.0–15.0	12
15.0–20.0	8
> 20.0	50

Solution:

1. What are the parameters used in Johnstone's equation?
   Johnstone's equation is

$$\eta = 1 - \exp[-k(Q_L/Q_G)\sqrt{\psi}]$$

where
η = fractional collection efficiency
k = correlation coefficient whose value depends on the system geometry and operating conditions, typically 0.1–0.2, 1000 acf/gal
Q_L/Q_G = liquid-to-gas ratio, gal/1000 acf
ψ = $C_f \rho_p v(d_p)^2/18 d_d \mu$ = inertial impaction parameter
ρ_p = particle density
v = gas velocity at venturi throat, ft/s
d_p = particle diameter, ft
d_d = droplet diameter, ft
μ = gas viscosity, lb/ft-s
C_f = Cunningham correction factor

1.a. Calculate the average droplet diameter in ft.
The average droplet diameter may be calculated using the equation:

$$d_d = (16{,}400/v) + 1.45(Q_L/Q_G)^{1.5}$$

where d_d = droplet diameter, microns

$$\begin{aligned} d_d &= (16{,}400/272) + 1.45(8.5)^{1.5} \\ &= 96.23 \text{ microns} \\ &= 3.156 \times 10^{-4} \text{ ft} \end{aligned}$$

1.b. Express the inertial impaction parameter in terms of d_p (ft).

$$\begin{aligned} \psi &= C_f \rho_p v(d_p)^2/18 d_d \mu \\ &= (1)(43.7)(272)(d_p)^2/18(3.156 \times 10^{-4})(1.5 \times 10^{-5}) \\ &= 1.3945 \times 10^{11}(d_p)^2 \end{aligned}$$

1.c. Express the fractional collection efficiency η_i, in terms of d_{pi} (d_p in ft).

$$\eta = 1 - \exp[-k(Q_L/Q_G)\sqrt{\psi}]$$
$$\eta_i = 1 - \exp[-(0.2)(8.5)(1.3945 \times 10^{11}(d_{pi})^2)^{0.5}$$
$$= 1 - \exp[-6.348 \times 10^5 d_{pi}]$$

2. Calculate the collection efficiency for each particle size appearing in Table 3.5. For $d_p = 0.05$ micron (1.64×10^{-7} ft), for example:

$$\eta_i = 1 - \exp[-6.348 \times 10^5 d_{pi}]$$
$$= 1 - \exp[-6.348 \times 10^5(1.64 \times 10^{-7})]$$
$$= 0.0989$$
$$w_i\eta_i = 0.01(0.0989) = 9.89 \times 10^{-4}$$

Table 3.6 shows the results of the above calculations for each particle size.

3. Calculate the overall collection efficiency.

$$\eta = \Sigma w_i \eta_i$$
$$= 9.89 \times 10^{-4} + 0.0975 + 0.6325 + 12.980 + 16.00 + 12.00 + 8.00 + 50.00$$
$$= 99.71\%$$

TABLE 3.6 Particle Size Data

d_p, ft*	w_i, %	η_i	$w_i\eta_i$, %
1.64×10^{-7}	0.01	0.0989	9.89×10^{-4}
9.84×10^{-7}	0.21	0.4645	0.0975
2.62×10^{-6}	0.78	0.8109	0.6325
9.84×10^{-6}	13	0.9981	12.98
2.62×10^{-5}	16	1	16
4.27×10^{-5}	12	1	12
5.91×10^{-5}	8	1	8
6.56×10^{-5}	50	1	50

*The average particle diameter has been rounded to the nearest significant figure.

EXAMPLE 3: scrubber plan review. A vendor proposes to use a spray tower on a lime kiln operation to reduce the discharge of solids to the atmosphere. The inlet loading is to be reduced in order to meet state regulations. The vendor's design calls for a certain water pressure drop and gas pressure drop across the tower. You are requested to determine whether this spray tower will meet state regulations. If the spray tower does not meet state regulations, propose a set of operating conditions that will meet the regulations (EPA-84/09, p. 81).

Given conditions

gas flow rate = 10,000 acfm

water rate = 50 gal/min

inlet loading = 5.0 grains/ft^3

maximum gas pressure drop across the unit = 15 in. H_2O

maximum water pressure drop across the unit = 100 psi

water pressure drop = 80 psi

gas pressure drop across the tower = 5.0 in. H_2O

The state regulations require a maximum outlet loading of 0.05 grains/ft^3. Assume that the contact power theory applies (EPA-81/10, p. 9-15).

Solution:

1. Calculate the collection efficiency based on the design data given by the vendor.

The contact power theory is an empirical approach relating particulate collection efficiency and pressure drop in wet scrubber systems. It assumes that particulate collection efficiency is a sole function of the total pressure loss for the unit.

$$P_T = P_G + P_L$$
$$P_G = 0.157\,\Delta p$$
$$P_L = 0.583 p_L (Q_L/Q_G)$$

where P_T = total pressure loss, hp/1000 acfm
P_G = contacting power based on gas stream energy input, hp/1000 acfm
Δp = pressure drop across the scrubber, in. H_2O
P_L = contacting power based on liquid stream energy input, hp/1000 acfm
p_L = liquid inlet pressure, psi
Q_L = liquid feed rate, gal/min
Q_G = gas flow rate, ft^3/min

The scrubber collection efficiency is also expressed as the number of transfer units.

$$N_t = \alpha (P_T)^\beta = \ln[1/(1-\eta)]$$

where N_t = number of transfer units
η = fractional collection efficiency
α, β = characteristic parameters for the type of particulates being collected.

1.a. Calculate the total pressure loss P_T

To calculate the total pressure loss, you need the contacting power for both the gas stream energy input and liquid stream energy input.

1.a.1. Calculate the contacting power based on the gas stream energy input P_G in hp/1000 acfm.

Since the pressure drop across the scrubber is given by the vendor, you can calculate P_G,

$$\begin{aligned} P_G &= 0.157\,\Delta p \\ &= (0.157)(5.0) \\ &= 0.785\ \text{hp}/1000\ \text{acfm} \end{aligned}$$

1.a.2. Calculate the contacting power based on the liquid stream energy input P_L, in hp/1000 acfm.

Since the liquid inlet pressure and liquid-to-gas ratio are given, you can calculate P_L

$$\begin{aligned} P_L &= 0.583 p_L (Q_L/Q_G) \\ &= 0.583(80)(50/10{,}000) \\ &= 0.233 \text{ hp}/1000 \text{ acfm} \end{aligned}$$

1.a.3. Calculate the total pressure loss P_T, in hp/1000 acfm.

$$\begin{aligned} P_T &= P_G + P_L \\ &= 0.785 + 0.233 \\ &= 1.018 \text{ hp}/1000 \text{ acfm} \end{aligned}$$

1.b. Calculate the number of transfer units N_t.

$$N_t = \alpha(P_T)^{\beta}$$

The values of α and β for a lime kiln operation are 1.47 and 1.05, respectively. These coefficients have been previously obtained from field test data. Therefore,

$$\begin{aligned} N_t &= \alpha(P_T)^{\beta} \\ &= (1.47)(1.018)^{1.05} \\ &= 1.50 \end{aligned}$$

1.c. Calculate the collection efficiency based on the design data given by the vendor.

$$\begin{aligned} N_t &= \ln[1/(1-\eta)] \\ 1.50 &= \ln[1/(1-\eta)] \end{aligned}$$

Solving for η

$$\eta = 0.777 = 77.7\%$$

2. Calculate collection efficiency required by state regulations.

Since the inlet loading is known and the outlet loading is set by the regulations, the collection efficiency can be calculated readily.

$$\begin{aligned} \text{Collection efficiency} &= [(\text{inlet loading} - \text{outlet loading})/(\text{inlet loading})](100) \\ &= [(5.0 - 0.05)/(5.0)](100) \\ &= 99.0\% \end{aligned}$$

3. Does the spray tower meet the regulations?

No. The collection efficiency based on the design data given by the vendor should be higher than the collection efficiency required by the state regulations.

4. Assuming the spray tower does not meet the regulations, propose a set of operating conditions that will meet the regulations.

Note that the calculational procedure is now reversed.

4.a. Calculate the total pressure loss P_T, using the collection efficiency required by the regulations in hp/1000 acfm.

4.a.1. Calculate the number of transfer units for the efficiency required by the regulations.

$$
\begin{aligned}
N_t &= \ln[1/(1-\eta)] \\
&= \ln[1(1-0.99)] \\
&= 4.605
\end{aligned}
$$

4.a.2. Calculate the total pressure loss P_T, in hp/1000 acfm.

$$
\begin{aligned}
N_t &= \alpha(P_T)^{\beta} \\
4.605 &= 1.47(P_T)^{1.05}
\end{aligned}
$$

Solving for P_T

$$P_T = 2.96\,\text{hp}/1000\,\text{acfm}$$

4.b. Calculate the contacting power based on the gas stream energy input P_G, using a Δp of 15 in. H_2O.

A pressure drop Δp of 15 in. H_2O is the maximum value allowed by the design

$$
\begin{aligned}
P_G &= 0.157\,\Delta p \\
&= (0.157)(15) \\
&= 2.355\,\text{hp}/1000\,\text{acfm}
\end{aligned}
$$

4.c. Calculate the contacting power based on the liquid stream energy input P_L.

$$
\begin{aligned}
P_L &= P_T - P_G \\
&= 2.96 - 2.355 \\
&= 0.605\,\text{hp}/1000\,\text{acfm}
\end{aligned}
$$

4.d. Calculate Q_L/Q_G in gal/acf, using a p_L of 100 psi.

$$
\begin{aligned}
P_L &= 0.583 p_L(Q_L/Q_G) \\
Q_L/Q_G &= P_L/0.583 p_L \\
&= 0.605/(0.583)(100) \\
&= 0.0104
\end{aligned}
$$

4.e. Determine the new water flow rate Q_L', in gal/min.

$$
\begin{aligned}
(Q_L)' &= (Q_L/Q_G)(10{,}000\,\text{acfm}) \\
&= 0.0104(10{,}000\,\text{acfm}) \\
&= 104\,\text{gal/min}
\end{aligned}
$$

4.f. What are the new set of operating conditions that will meet the regulations?

$$
\begin{aligned}
Q_L' &= 104\,\text{gal/min} \\
P_T &= 2.96\,\text{hp}/1000\,\text{acfm}
\end{aligned}
$$

Plate tower. A plate tower is a vertical column with one or more plates (trays) mounted horizontally inside. As shown in Fig. 3.11, the exhaust stream enters at the bottom and flows upward, passing through openings in the plates. Liquid enters at the top of the tower, traveling across each plate, and reaches either the next plate below or the bottom of the tower. Pollutant

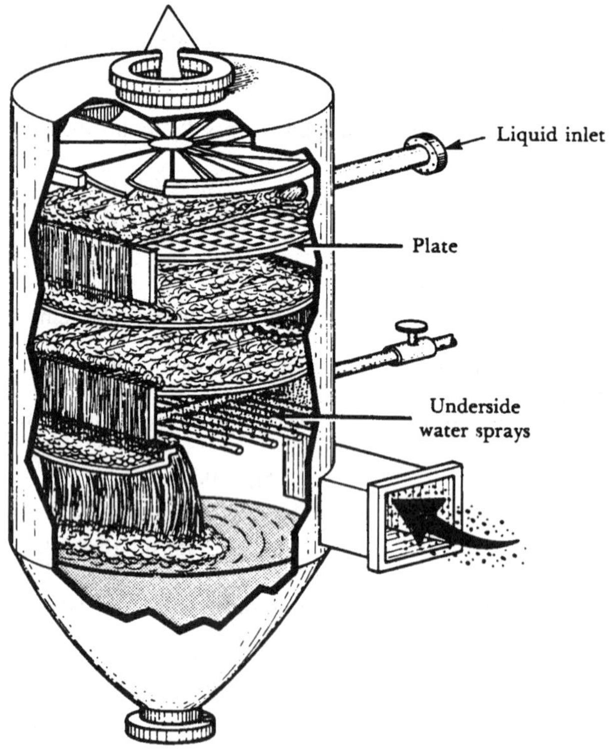

FIGURE 3.11 Plate tower (*EPA-84/03, p. 3-15*).

collection (mass transfer) occurs as the gas stream bubbles up through the openings and into the liquid layer on the plate. The smaller the bubbles and the more numerous they are, the more effective the mass transfer. The formation of smaller and more numerous bubbles promotes better mixing and exposes a larger gas-to-liquid surface area. As the gas disengages from the liquid, a froth is formed atop the liquid. Ideally, the gas and liquid leaving each plate are in equilibrium with each other at the conditions of that particular plate. Each plate acts as a separate absorption stage. The amount of absorption that can occur at one plate or stage is referred to as a *transfer unit*. Plate towers are very effective in removing gaseous pollutants and can be used simultaneously for particle removal. Plate towers may not be appropriate when particle removal is the only consideration (EPA-84/03, p. 3-15; EPA-81/10, p. 9-23; EPA-81/12, p. 4-52).

The function of the plates or trays is to disperse the gas into numerous bubbles, exposing a large surface for mass transfer. Plates, or trays, are designed in a variety of ways. The ones most commonly used for industrial sources are

- Sieve plate
- Impingement plate
- Bubble-cap plate
- Valve plate

The sieve, impingement, and bubble-cap plates do not have moving parts, while the valve plates have liftable caps above the opening in the plate. Plate openings can range from 0.32 to 2.50 cm (0.125 to 1.0 in.) in diameter for the sieve plate. Openings for the other plate designs are generally larger.

Sieve plate: Sieve plates contain approximately 6456–32,280 holes per square meter (600–3000 per square foot) of surface. Exhaust gas rises through these small holes and contacts the liquid at the holes. The gas atomizes the liquid, forming a froth with droplets ranging from 10 to 100 μm in diameter. Particle collection efficiency increases as the size of the sieve opening decreases. This is because of an increasing gas velocity and because smaller droplets are formed. Sieve plates with large openings will not become plugged as easily as will other plate designs. Figure 3.12A depicts gas–liquid contact on a sieve plate.

Impingement plate: Impingement plates are similar to sieve plates with the addition of an impaction target placed above each hole in the plate (Fig. 3.12B). The gas coming up through the hole forces the liquid on the plate up against the target (impingement surface). This design increases the mixing of the gas and liquid, provides an additional contact zone, and creates more liquid droplets.

Bubble-cap plate: In the bubble-cap plate design, the exhaust gas enters each cap through a riser around each hole in the plate and exits from several slots in each cap (Fig. 3.13A). This combination of caps and risers creates a bubbly froth that allows good gas–liquid mixing, regardless of the gas-to-liquid ratio. In addition, the caps provide a longer gas–liquid contact than either sieves or impingement plates, thus increasing absorption efficiency. Plugging and corrosion can be a problem for bubble-cap plates because of this more complex design.

Valve plate: In the valve plate design, the exhaust gas passes through small holes in the plate, pushing up against a metal valve that covers each hole. The metal valve moves up and down with the gas flow. The valve is limited in its vertical movement by legs attached to the plate (Fig. 3.13B); therefore, the liftable valve acts as a variable orifice. Caps are available in different weights to provide flexibility for varying exhaust gas flow rates. Floating valves increase gaseous pollutant collection efficiency by providing adequate gas–liquid contact time, regardless of the exhaust gas flow rate. This design is also suited for very small particle collection; however, valves will plug if large particles are in the exhaust stream. Wear and

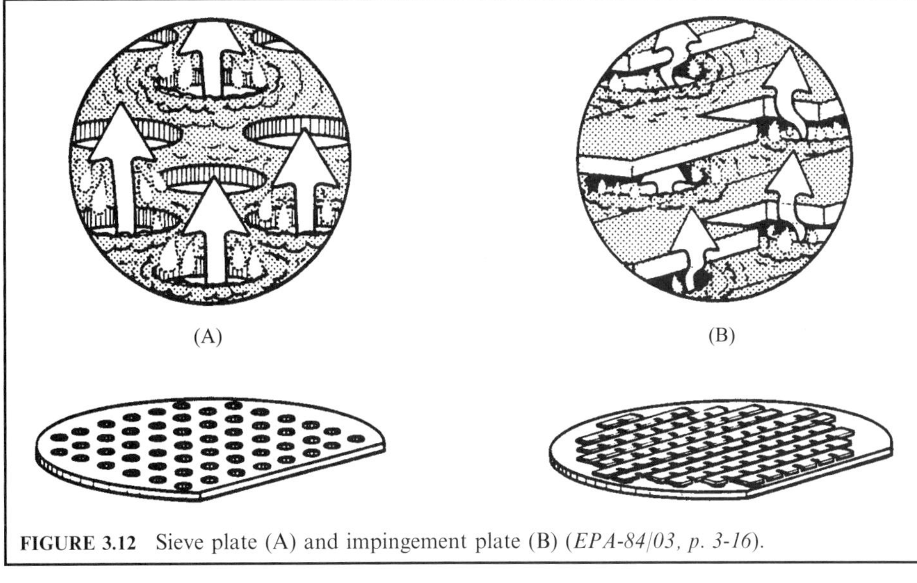

FIGURE 3.12 Sieve plate (A) and impingement plate (B) (*EPA-84/03, p. 3-16*).

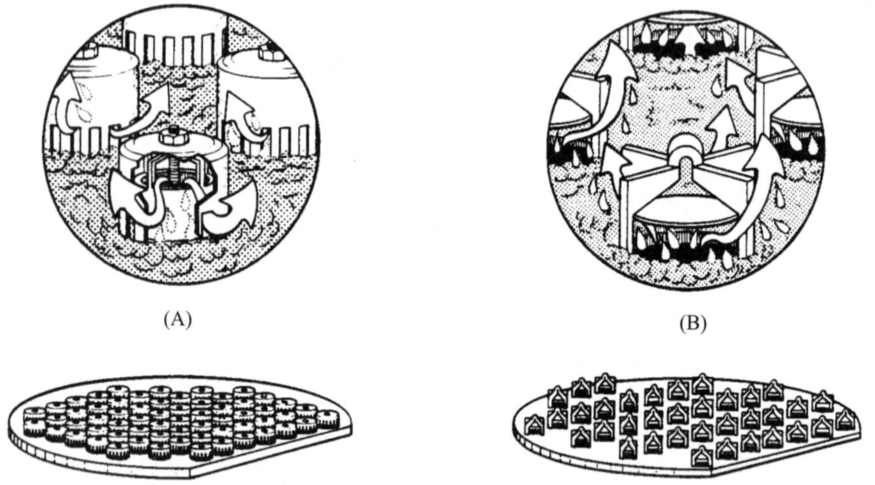

FIGURE 3.13 Bubble-cap plate (A) and valve plate (B) (*EPA-84/03, p. 3-17*).

corrosion are also a problem for the retaining legs. Valve plates are more expensive than sieve and impingement plates, but less expensive than bubble-cap plates.

Plate towers, like packed towers, can achieve a high removal efficiency even if the gaseous contaminant is relatively insoluble. Therefore, plate towers are used to control emissions from many of the same processes that use packed towers, such as acid plants, fertilizer production plants, the chemical industry and petroleum refineries.

Several advantages and disadvantages must be considered when choosing a plate tower instead of a packed tower for a given control operation. The following list gives some factors used in comparing plate towers with packed towers.

- Plate towers are able to handle particulate matter and other solids better than packed towers. Maintenance openings can be installed so that the plates may be easily cleaned.
- Plate towers are chosen for operations that require a large number of transfer units or that must handle large gas volumes. Packed towers can experience channeling problems if the diameter or height of the tower is too large. Redistribution trays must be installed in large-diameter and tall packed towers to avoid channeling.
- The total weight of a plate tower is less than that of a comparable packed tower.
- Packed towers are much cheaper to construct if corrosive substances are to be handled. Packed towers can be constructed with a fiberglass-reinforced polyester shell, which is generally about half the cost of a carbon steel plate tower.
- Plate towers can handle volume and temperature fluctuations better than packed towers. Expansion or contraction due to temperature changes can crush or melt packing material.

Particle collection. Particles are collected in plate towers as the exhaust gas atomizes the liquid flowing over the holes in the plates. The atomized droplets serve as impaction targets for the particles. Plate towers are considered to be medium-energy scrubbers having moderate particle collection efficiencies. Collection efficiency does not significantly increase by increasing the number of plates over two or three. Collection efficiency can be increased by decreasing the hole size and increasing the number of holes per plate. This produces more liquid droplets of a smaller size and increases the gas velocity through the plate; however, it also increases the pressure drop of the system.

Gas collection. Plate towers are very effective for removing gaseous pollutants from an exhaust stream. They can easily achieve greater than 98% removal in many applications. Absorption occurs as the exhaust stream bubbles up through the liquid on the plates and contacts the atomized liquid droplets. This action provides intimate contact between the exhaust gas and liquid streams, allowing the liquid on each plate to absorb the pollutant gas. Each plate acts as a separate absorption stage; therefore, absorption efficiency can be increased by adding plates. Absorption efficiency can also be improved by adding more liquid or by increasing the pressure drop across each plate, which increases the gas–liquid contact.

Orifice scrubber. In orifice scrubbers, the exhaust stream from the process is forced through a pool of liquid, usually water. The exhaust stream moves through restricted passages, or orifices, to disperse and atomize the water into droplets. These scrubbers are also called *self-induced spray, inertial*, or *submerged orifice scrubbers* (EPA-84/03, p. 3-23; EPA-81/10, p. 9-27).

Several orifice scrubber designs are typically used. In each, the incoming exhaust stream is directed across or through a pool of water. The high exhaust stream velocity, approximately 15.2 m/s (50 ft/s), creates a large number of liquid droplets. Both particles and gaseous pollutants are collected as they are forced through the liquid pool and impact on the droplets. However, these scrubbers are generally used for removing particles. Large particles are collected when they impact on the liquid pool or its surface. Small particles are collected when they impact the droplets. Baffles, or air foils, are added to provide turbulent mixing of the exhaust stream and droplets.

In the self-induced scrubber, the exhaust stream enters through a duct, as shown in Fig. 3.14. The exhaust stream is forced by baffles through a pool of liquid. Particles and gases are collected in the pool and by the droplets. Additional baffles placed in the path of the 'clean'

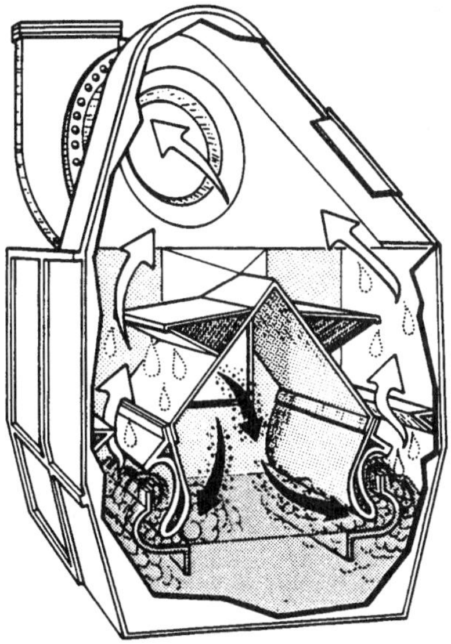

FIGURE 3.14 Self-induced spray orifice scrubber (*EPA-84/03, p. 3-23*).

exhaust stream as it exits the vessel serve as impingement surfaces to remove entrained droplets.

Particulate matter collected in the scrubber forms a sludge that must be disposed of. Sludge disposal involves removing and recycling large amounts of liquid, from 3.5 to 4.0 L/m^3 (25 to 30 gal/1000 acfm). Some designs use a sludge separation and removal system inside the scrubber. The water level inside the scrubber must be maintained during the sludge separation and removal cycle so that the unit can operate efficiently.

Particle collection. Large particles in the incoming exhaust stream are collected as they impinge on the surface of the pool. Smaller particles are collected as they impact on the droplets produced by the high-velocity gas skimming over the liquid. Overall particle collection in an orifice scrubber depends on the level of the liquid. The level of the liquid determines the gas velocity (and, thus, the pressure drop) through the orifice. If the liquid level is low, gas velocities decrease because the orifice opening is larger. Lower velocities produce fewer droplets that are larger in size, decreasing particle collection. A turn-down of the system, or reduction in gas volume, will also result in less atomization and produce larger droplets.

Gas collection. Orifice scrubbers are rarely used for absorption. However, because orifice scrubbers provide both thorough mixing of the gas and liquid, and large liquid-surface contact areas (many tiny droplets), these devices can be effective for reactive scrubbing or for removing gaseous pollutants that are already very soluble in the liquid. In reactive scrubbing, the gaseous pollutants chemically react with the scrubbing liquid. These reactions occasionally produce scale or sludge that can plug scrubber internals. The relatively large orifice openings will not plug as easily as those in plate towers.

4.2. Liquid-Phase Contacting Scrubber

Section 4.1 dealt with scrubbers which primarily use the process gas stream to atomize liquid into collection droplets. Energy can also be applied to a scrubbing system by injecting liquid at high pressure through specially designed nozzles. Nozzles produce droplets which fan out into a scrubber chamber to impact with particulate matter contained in a polluted gas stream. Droplets act as targets for collecting particles and/or absorbing gas in a pollutant exhaust stream. In liquid-phase contacting scrubbers, the liquid-inlet pressure provides the major portion of the energy required for contacting the gas (exhaust stream) and liquid phases (EPA-84/03, p. 4-1).

Two liquid-phase contacting scrubbers are

- Spray tower
- Ejector venturi

Many other scrubber designs also incorporate sprays produced by nozzles, but in those scrubbers, the sprays are used to clean trays or to wet scrubber surfaces and orifices, and not to provide the gas–liquid contact in the system.

Spray tower. Spray towers, also known as gravity spray towers, spray chambers, or spray scrubbers, are the simplest devices used for gas absorption. The basic design is simple. A spray tower consists of an empty tower and a set of nozzles to spray liquid. It is similar in operation to spraying water in an open barrel. Typically, the contaminant gas stream enters the bottom of the tower and passes up through the device while liquid is being sprayed at one or more levels by nozzles. The flow of liquid and gas streams in opposite directions is referred to as *countercurrent flow*. Liquid is sprayed into a cylindrical or rectangular chamber using one nozzle or a series of nozzles, as shown in Fig. 3.15. The figure shows a typical countercurrent-flow spray tower. Countercurrent flow exposes the exhaust gas with the lowest pollutant concentration to the freshest scrubbing liquid (EPA-84/03, p. 4-1; EPA-81/10, p. 9-38; EPA-81/12, p. 4-38).

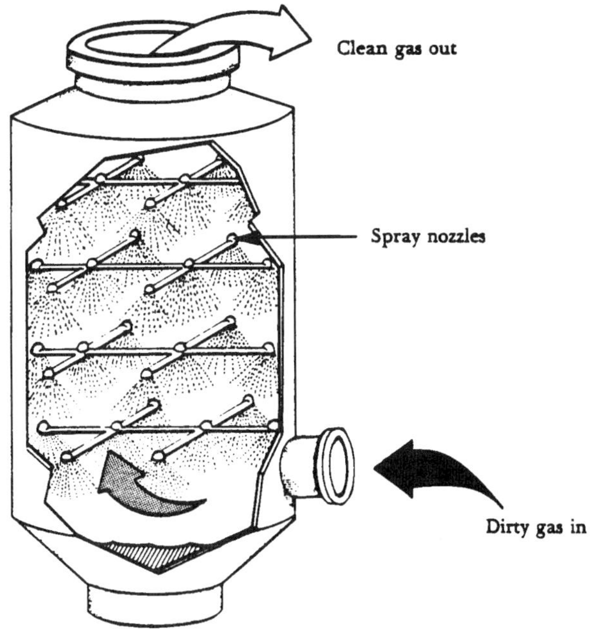

FIGURE 3.15 Countercurrent flow spray tower (*EPA-81/12, p. 4-39*).

To provide a large liquid surface for contacting the gas, nozzles are arranged to wet the entire cross-section of the tower with fine liquid droplets. Theoretically, the height between levels of nozzles and/or the bottom of the tower determines the residence time. In practice, physical laws limit the removal efficiency of spray towers. After falling short distances, the liquid droplets tend to agglomerate or hit the sides of the tower. Both of these effects reduce the total liquid surface in contact with the gas stream and the residence time. Therefore, spray chambers are limited to applications where the gases are extremely soluble or a high removal efficiency is not required.

The main advantage of spray chambers is that they are completely open; they have no internals except for the spray nozzles. Therefore, they have a very low pressure drop, about 1.25–4 cm (0.5–1.6 in.) of water over the tower. If an entrainment separator is used it will add another 2.5–5 cm (1–2 in.) of water to the total pressure drop. An entrainment separator is used to prevent mist from exhausting out the stack. Spray towers range in size from 42 to 170,000 m^3/h (25 to 100,000 cfm) with 34,500 m^3/h (20,000 cfm) as a typical size.

Spray towers are primarily used for gas conditioning—cooling, humidifying, or for the first-stage particulate matter and gas removal. But they can be used effectively for gas absorption if the contaminant to be removed is highly soluble. For example, spray towers are used to remove HCl gas from the tail-gas exhaust in the manufacturing of hydrochloric acid.

No universally acceptable mathematical equation or correlation is presently available for sizing spray towers. Although they are simple devices, variables such as liquid drop size, settling velocity, and residence time vary considerably with height or location in the tower. Tower designs are based on experimental or operational data from similar systems.

However, certain limiting factors must be balanced against each other.

- *Gas velocity*: The gas velocity through the tower must be kept low, around 0.3–1.0 m/s (1–3 ft/s) or excessive liquid entrainment occurs. This implies using a large-diameter tower to keep the gas velocity low. But if the diameter is large, compared with the distance between sprays, back mixing occurs. Back mixing is a condition where the gas changes direction or swirls within the tower and removal efficiency is decreased.
- *Liquid droplet size*: The smaller the liquid droplet size, the greater the rate of absorption, since this increases the surface area available for absorption. However, the liquid droplets must be large enough to not be carried out of the scrubber by the exhaust stream. Therefore, spray towers use nozzles to produce droplets that are usually 500–1000 μm in diameter. High-pressure spray nozzles are used to produce fine liquid droplets. These spray nozzles consume more power in pumping liquid than a packed tower or venturi scrubber for the same liquid rate. The fine openings of the spray nozzles are subject to erosion and plugging problems, especially if scrubbing liquid is recirculated.
- *Liquid-to-gas (L/G) ratio*: Another parameter which characterizes spray tower operation is the liquid-to-gas (L/G) ratio. By increasing the L/G ratio, absorption efficiency can be directly increased.

In addition to a countercurrent-flow configuration, the flow in spray towers can be either a cocurrent or crosscurrent configuration. In cocurrent-flow spray towers, the exhaust gas and liquid flow in the same direction. Because the exhaust gas stream does not 'push' against the liquid sprays, these units operate at higher exhaust gas velocities (through the vessels) than do countercurrent-flow spray towers. Consequently, cocurrent-flow spray towers are smaller than countercurrent-flow spray towers (treating the same amount of exhaust flow).

In crosscurrent-flow spray towers, called *horizontal-spray scrubbers*, the exhaust gas and liquid flow in directions perpendicular to each other (Fig. 3.16). In this vessel, the exhaust gas flows horizontally through a number of spray sections. The amount and quality of liquid sprayed in each section can be varied, usually with the cleanest liquid (if recycled liquid is sprayed) in the last of sprays.

Particle collection. Spray towers are low-energy scrubbers. Contacting power is much lower than in venturi scrubbers, and the pressure drops across such systems are generally less than 2.5 cm (1 in.) of water. The collection efficiency for small particles is correspondingly

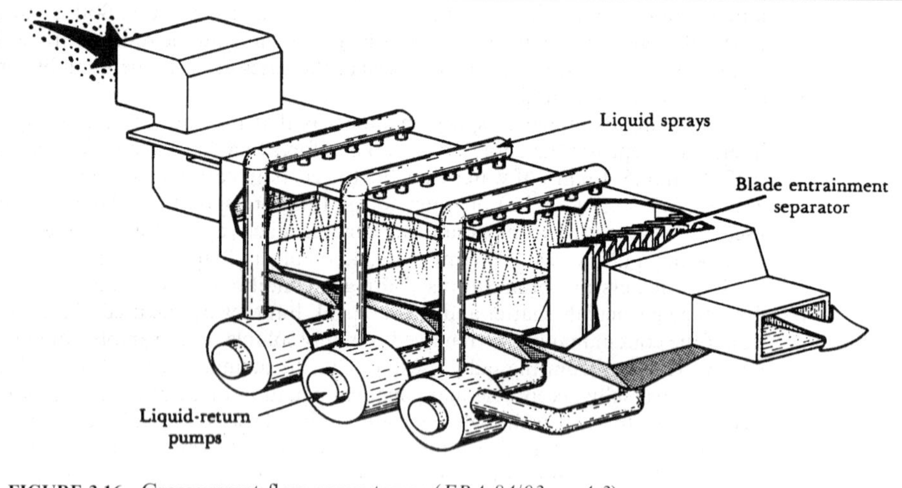

FIGURE 3.16 Crosscurrent flow spray tower (*EPA-84/03, p. 4-3*).

lower than in more energy-intensive devices. They are adequate for the collection of coarse particles larger than 10–25 μm in diameter, although with increased liquid-inlet nozzle pressures, particles with diameters of 2.0 μm can be collected. Smaller droplets can be formed by higher liquid pressures at the nozzle. The highest collection efficiencies are achieved when small droplets are produced and the difference between the velocity of the droplets and the velocity of the upward-moving particles is high. Small droplets, however, have small settling velocities, so there is an optimum range of droplet sizes for scrubbers that work by this mechanism.

Gas collection. Spray towers can be used for gas absorption, but they are not as effective as packed or plate towers. Spray towers can be very effective in removing pollutants if the pollutants are highly soluble or if a chemical reagent is added to the liquid. For example, spray towers are used to remove HCl gas from the tail-gas exhaust in manufacturing hydrochloric acid. In the production of superphosphate used in manufacturing fertilizer, SiF_4 and HF gases are vented from various points in the processes. Spray towers have been used to remove these highly soluble compounds. Spray towers are also used for odor removal in bone meal and tallow manufacturing industries by scrubbing the exhaust gases with a solution of $KMnO_4$. Because of their ability to handle large exhaust gas volumes in corrosive atmospheres, spray towers are also used in a number of the flue gas desulfurization systems as the first or second stage in the pollutant removal process.

In a spray tower, absorption can be increased by decreasing the size of the liquid droplets and/or increasing the liquid-to-gas (L/G) ratio. However, to accomplish either of these, an increase in both power consumed and operating cost is required. In addition, the physical size of the spray tower will limit the amount of liquid and the size of droplets that can be used.

EXAMPLE: spray tower. A steel pickling operation emits 300 ppm HCl (hydrochloric acid) with peak values of 500 ppm 15% of the time. The air flow is a constant 25,000 acfm at 75°F and 1 atm. Only sketchy information was submitted with the scrubber permit application for a spray tower. You are requested to determine if the spray unit is satisfactory (EPA-84/09, p. 100).

Given conditions

emission limit = 25 ppm HCl

maximum gas velocity allowed through the tower = 3 ft/s

number of sprays = 6

diameter of the tower = 14 ft

The plans show a countercurrent water spray tower. For a very soluble gas (Henry's law constant approximately zero), the number of transfer units (N_{OG}) can be determined by the following equation:

$$N_{OG} = \ln(y_1/y_2)$$

where y_1 = concentration of inlet gas
y_2 = concentration of outlet gas

In a spray tower, the number of transfer units N_{OG} for the first (or top) spray will be about 0.7. Each lower spray will have only about 60% of the N_{OG} of the spray above it. The final spray, if placed in the inlet duct, has a N_{OG} of 0.5.

The spray sections of a tower are normally spaced at 3-ft intervals. The inlet duct spray adds no height to the column.

Solution:

1. Calculate the gas velocity through the tower.

$$V = Q/S$$
$$= Q/(\pi D^2/4)$$

where V = velocity
Q = actual volumetric gas flow
S = cross-sectional area
D = diameter of the tower

$$V = Q/(\pi D^2/4)$$
$$= 25{,}000/[\pi(14)^2/4]$$
$$= 162.4\ \text{ft/min}$$
$$= 2.7\ \text{ft/s}$$

2. Does the gas velocity meet the requirement?
Yes, because the gas velocity is less than 3 ft/s.

3. Calculate the number of overall gas transfer units N_{OG} required to meet the regulation.
Remember that

$$N_{OG} = \ln(y_1/y_2)$$

where y_1 = concentration of inlet gas
y_2 = concentration of outlet gas

Use the peak value for inlet gas concentration:

$$N_{OG} = \ln(y_1/y_2)$$
$$= \ln(500/25)$$
$$= 3.0$$

4. Determine the total number of transfer units provided by a tower with six spray sections.
Remember that each lower spray has only 60% of the efficiency of the section above it. This is due to back-mixing of liquids and gases from adjacent sections.
Spray section N_{OG} values are derived accordingly:

top spray $N_{OG} = 0.7$ (given)
2nd spray $N_{OG} = 0.7(0.6) = 0.42$
3rd spray $N_{OG} = 0.42(0.6) = 0.252$
4th spray $N_{OG} = 0.252(0.6) = 0.1512$
5th spray $N_{OG} = 0.1512(0.6) = 0.0907$
Inlet $N_{OG} = 0.5$ (given)
Total $N_{OG} = 0.7 + 0.42 + 0.252 + 0.1512 + 0.0907 + 0.5 = 2.114$

Note this value is below the required value of 3.0.

5. Calculate the outlet concentration of gas.

$$
\begin{aligned}
N_{\mathrm{OG}} &= \ln(y_1/y_2) \\
y_1/y_2 &= \exp(N_{\mathrm{OG}}) \\
&= \exp(2.114) \\
&= 8.28 \\
y_2 &= 500/8.28 \\
&= 60.4\,\mathrm{ppm}
\end{aligned}
$$

6. Does the spray tower meet the HCl regulation?

Since y_2 is greater than the required emission limit of 25 ppm, the spray unit is not satisfactory.

Ejector venturi. The ejector, or jet, venturi scrubber uses a preformed spray, as does the simple spray tower. The difference is that only a single nozzle is used instead of many nozzles. This nozzle operates at higher pressures and higher injection rates than those in most spray chambers. The high pressure spray nozzle (up to 689 kPa or 100 psig) is aimed at the throat section of a venturi constriction. Figure 3.17 illustrates the ejector venturi design. Using high-pressure sprays in this manner serves two purposes (EPA-84/03, p. 4-6; EPA-81/10, p. 9-40; EPA-81/12, p. 4-46):

1. The movement of the liquid creates a suction (vacuum) that pulls the gas stream through the venturi. This eliminates the need for a fan or blower to move the gas stream.
2. The high-pressure sprays, along with the venturi effect, form numerous fine liquid droplets that provide a high degree of turbulence between gas and liquid phases. This limits contact time; therefore, absorption efficiency is usually low.

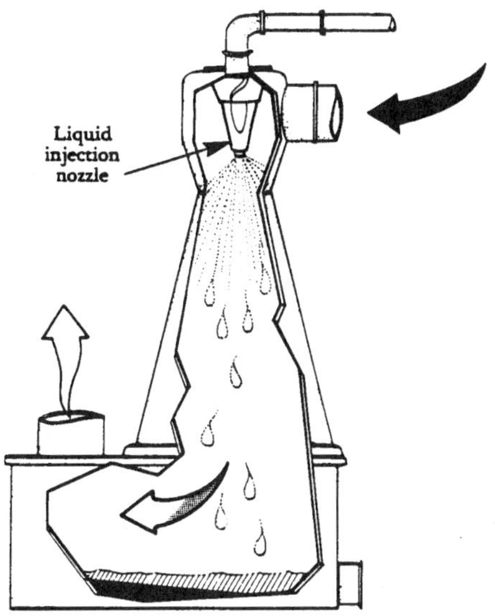

FIGURE 3.17 Jet or ejector venturi scrubber (*EPA-81/12, p. 4-22*).

Ejector venturis operate at high liquid-to-gas ratios (around 100 gpm/1000 cfm). The gas-side pressure drop usually ranges between 10–20 cm (4–8 in.) of water. The overall power consumption, however, is much higher, due to liquid pumping requirements. Ejector venturis can be designed for smaller gas flow rates (under 1000 cfm) than most other absorbers are capable of handling.

The ejector venturi is unique among available scrubbing systems since it can move the process gas without the aid of a blower or fan. The liquid spray coming from the nozzle creates a partial vacuum in the side duct of the scrubber. This partial vacuum can be used to move the process gas through the control device as well as through the process system. In the case of explosive or extremely corrosive atmospheres, the elimination of a fan in the system can avoid many potential problems.

The energy for the formation of scrubbing droplets comes from the injected liquid. The high-pressure sprays passing through the venturi throat form numerous fine liquid droplets that provide turbulent mixing between the gas and liquid phases. Very high liquid-injection rates are used to provide the gas-moving capability and higher collection efficiencies. As with other types of venturis, a means of separating entrained liquid from the gas stream must be installed. A liquid sump directs the gas flow to continuing ductwork. Entrainment separators are commonly used to remove remaining small droplets.

Particle collection. Venturi scrubbers are the most widely used wet collector for particulate matter removal. Ejector venturis are effective in removing particles larger than 1.0 μm in diameter. These scrubbers are not used on submicron-sized particles unless the particles are condensable. Particle collection occurs primarily by impaction as the exhaust gas (from the process) travels through the spray.

The turbulence that occurs in the throat area also causes the particles to contact the wet droplets and be collected. Particle collection efficiency increases with an increase in nozzle pressure and/or an increase in the liquid-to-gas ratio. In fact, ejector venturis operate at higher L/G ratios than most other particle scrubbers.

Gas collection. Unless a gaseous contaminant is extremely soluble, venturis are seldom used strictly as an absorber. Ejector venturis have a short gas–liquid contact time because the exhaust gas velocities through the vessel are very high. This short contact time limits the absorption efficiency of the system. Although ejector venturis are not used primarily for gas removal, they can be effective if the gas is very soluble or if a very reactive scrubbing reagent is used. In these instances, removal efficiencies of as high as 95% can be achieved.

4.3 Wet-Film Scrubber

In wet-film scrubbers, liquid is sprayed or poured over packing material contained between support trays. A liquid film coats the packing through which the exhaust gas stream is forced. Pollutants are collected as they pass through the packing, contacting the liquid film. Therefore, both gas and liquid phases provide energy for the gas–liquid contact. These scrubbers are commonly called packed towers.

A wet-film scrubber uses packing to provide a large contact area between the gas and liquid phases, to provide turbulent mixing of the phases, and to provide sufficient residence time for the exhaust gas to contact the liquid. These conditions are ideal for gas absorption. A large contact area and good mixing are also good for particle collection; however, once collected, the particles tend to accumulate and, thus, plug the packing bed. The exhaust gas is forced to make many changes in direction as it winds through the openings of the packed material. Large particles unable to follow the streamlines hit the packing and are collected in the liquid. As this liquid drains through the packing bed, the collected particles may accumulate, thus plugging the void spaces in the packed bed. Therefore, wet-film scrubbers are not used where particle removal is the only concern. Many other scrubber designs achieve better particle removal for the same power input (operating costs).

Packed tower (packed column). The packed column (tower) is the most common scrubber used for gas absorption. Packed columns disperse the scrubbing liquid over packing material which provides a large surface area for continuous gas–liquid contact. Packed towers are classified according to the relative direction of gas-to-liquid flow. Packed towers are typically designated by the flow arrangement used for gas–liquid contact or by the material used as packing for the bed (EPA-84/03, p. 5-1; EPA-81/12, p. 4-47).

The most common flow configuration for packed towers is countercurrent flow. Figure 3.18 shows a packed tower with this arrangement. The exhaust stream being treated enters the bottom of the tower and flows upward over the packing material. Liquid is introduced at the top of the packing by sprays or weirs, and it flows downward over the packing material. As the exhaust stream moves up through the packing, it is forced to make many winding changes in direction, resulting in intimate mixing of both the exhaust gas and liquid streams. This countercurrent-flow arrangement results in the highest theoretically achievable efficiency. The most dilute gas is in contact with the purest absorbing liquor, providing a maximized concentration difference (driving force) for the entire length of the column.

The countercurrent-flow packed tower does not operate effectively if there are large variations in the liquid or gas flow rates. If either the liquid-injection rate or the gas flow rate through the packing bed is too high, a condition called flooding may occur. Flooding is a condition where the liquid is 'held' in the pockets, or void spaces, between the packing and does not drain down through the packing. Flooding can be reduced by reducing the gas velocity through the bed or by reducing the liquid-injection rate.

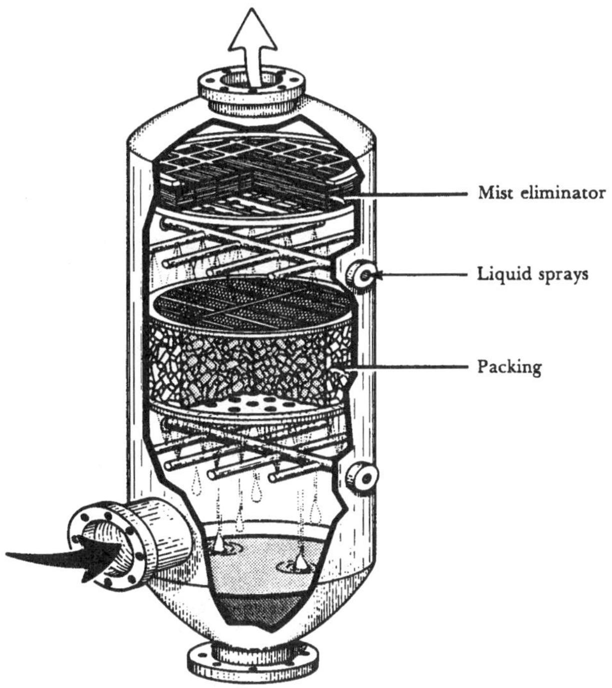

FIGURE 3.18 Countercurrent flow packed tower (*EPA-84/03, p. 5-3*).

In another flow arrangement used with packed towers, cocurrent flow, both the exhaust gas and liquid phases enter at the top of the absorber and move downward over the packing material. This allows the absorber to be operated at higher liquid and gas flow rates since flooding is not a problem. The pressure drop is lower than with countercurrent flow since both streams move in the same direction. The major disadvantage is that removal efficiency is very limited due to the decreasing driving force (concentration differential) as the streams travel down through the column. This limits the areas of application for cocurrent absorbers. They are used almost exclusively in situations where limited equipment space is available, since the tower diameter is smaller than for a countercurrent or plate tower of equivalent flow rates. Cocurrent flow is illustrated in Fig. 3.19.

In packed towers using the crossflow arrangement, the exhaust gas stream moves horizontally through the packed bed. The bed is irrigated by the scrubbing liquid flowing down through the packing material. The liquid and exhaust gas flow in directions perpendicular to each other. A typical crossflow packed tower is shown in Fig. 3.20. (Inlet sprays aimed at the face of the bed may also be included. If included, these sprays scrub both the entering gas and the face of the packed bed.) The leading face of the packed bed is slanted in the direction of the oncoming gas stream. This ensures complete wetting of the packing by allowing the liquid at the front face of the packing time to drop to the bottom before being pushed back by the entering gas.

Crossflow absorbers are smaller and have a lower pressure drop than any other packed or plate towers for the same application (removal efficiency and flow rates). In addition, they are better suited than other wet-film scrubbers to handle exhaust streams with high particle concentrations. By adjusting the liquid flow rate, incoming particles can be removed and washed

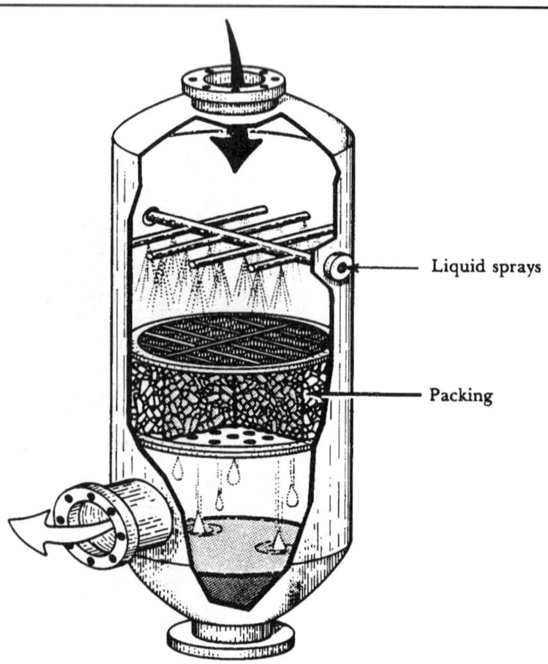

FIGURE 3.19 Cocurrent flow packed tower *(EPA-84/03, p. 5-4)*.

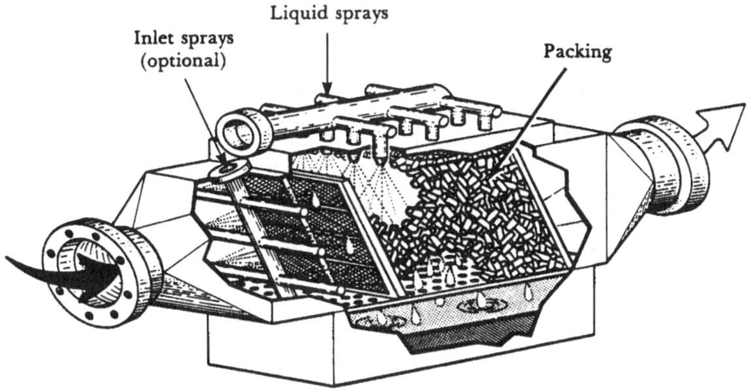

FIGURE 3.20 Crossflow packed tower (*EPA-84/03, p. 5-3*).

away in the front half of the bed. This also results in a liquid saving by enabling the crossflow packed tower to use less liquid in the rear sprays.

Another crossflow packed tower is the fiber-bed scrubber. The fiber-bed scrubber has packed beds that are made with fibrous materials such as fiberglass or plastic (Fig. 3.21). Liquid is sprayed onto the fiber beds to provide a wetted surface for pollutant removal and to wash away any collected materials.

Particle collection. For particle collection, packing materials are literally packed in a column or in sectional compartments. Liquid is sprayed on the material and allowed to flow through the system. Process exhaust gas is introduced into this system and the particles in the gas stream move through the packing and are collected. Cocurrent flow provides better particle collection than countercurrent flow since higher gas velocities can be used. With higher velocities, smaller particles can be collected than in similar countercurrent scrubbers (EPA-81/10, p. 9-54).

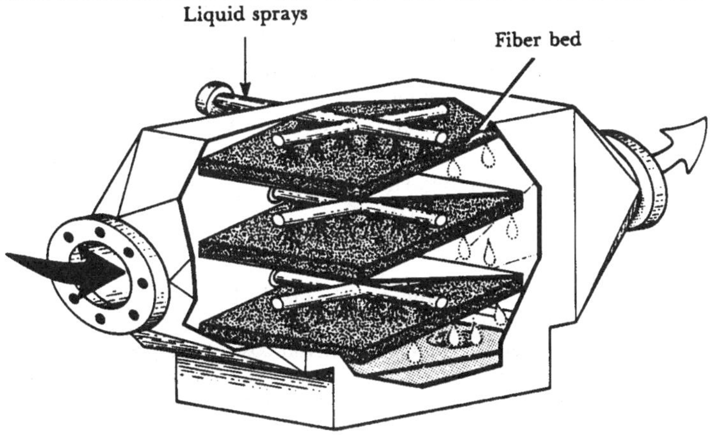

FIGURE 3.21 Fiber-bed scrubber (*EPA-84/03, p. 5-5*).

The particles impinge upon the liquid film covering the packing and are drained off with the liquid as it flows through the system. Many changes of direction occur as the gas winds through the openings of the packing materials. Larger particles unable to follow the streamlines associated with this flow will consequently hit the packing and be collected.

The packed column or tower can have several problems when not operated under proper conditions. Heavy loadings of particulate matter can plug beds of packing material. Deep beds will provide high pressure drops and higher efficiencies, but can be more susceptible to plugging. However, gas channeling is more likely to occur in thin beds. The gas flows through the path of least resistance and may channel through a certain section of the bed if the bed is not thick enough. In the countercurrent design, if the liquid rates are too high or the gas rates too low, the tower will flood and begin to fill with liquid.

For a crossflow scrubber, the water sprays can be positioned in front of the packing, above the packing, or behind the packing. This configuration has an advantage since particulate matter accumulated on the face of the packing can be washed off by water sprays.

Gas collection. For gas absorption, wet-film scrubbers are the most commonly used devices. The wet film covering the packing enhances gas absorption several ways by providing (EPA-84/03, p. 5-2):

- A large surface area for gas–liquid contact.
- Turbulent contact (good mixing) between the two phases.
- Long residence time and repetitive contact.

Because of these features, packed towers are capable of achieving high removal efficiencies for many different gaseous pollutants.

Numerous operating variables affect absorption efficiency. Of primary importance is the solubility of the gaseous pollutants. Pollutants that are readily soluble in the scrubbing liquid can be easily removed under a variety of operating conditions. Some other important operating variables are:

- *Gas velocity*: The rate of exhaust gas from the process determines the scrubber size to be used. The scrubber should be designed so that the gas velocity through it will promote good mixing between the gas and liquid phase. However, the velocity should not be too fast to cause flooding.
- *Liquid-injection rate*: Generally, removal efficiency is increased by an increase in the liquid-injection rate to the vessel. The amount of liquid that can be injected is limited by the dimensions of the scrubber. Increasing liquid-injection rates will also increase the operating costs. The optimum amount of liquid injected is based on the exhaust gas flow rate.
- *Packing size*: Smaller packing sizes offer a larger surface area, thus enhancing absorption. However, smaller packing fits tighter, which decreases the open area between packing, thus increasing the pressure drop across the packing bed.
- *Packing height*: As packing height increases, total surface area and residence time increase, enhancing absorption. However, more packing necessitates a larger absorption system, which increases capital cost.

Packing material. Packing material is the heart of the absorber. It provides the surface over which the scrubbing liquid flows, presenting a large area for mass transfer to occur. Packing material represents the largest material cost of the packed tower. Figure 3.22 shows some of commonly used packing materials which are made in numerous geometric shapes and sizes. These materials were originally made of stoneware, porcelain, or metal; but, presently, a large majority are being made of high-density thermoplastics (polyethylene and polypropylene). A specific packing is described by its trade name and overall size. For example, a column can be packed with 5 cm (2 inch) Raschig rings or 2.5 cm (1 inch) Tellerettes. The overall dimensions

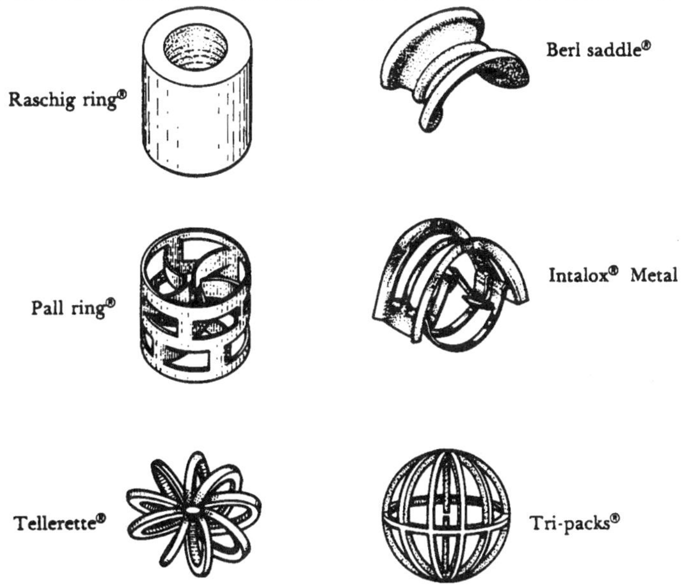

FIGURE 3.22 Common packing materials (*EPA-84/03, p. 5-7*).

of packing materials normally range from 0.6 to 10 cm (1/4 to 4 inches) (EPA-84/03, p. 5-7; EPA-81/12, p. 4-50).

Packing material may be arranged in an absorber in either of two ways: the packing may be dumped into the column randomly or in certain cases systematically stacked, as bricks are laid atop each other. Randomly packed towers provide a higher surface area, m^2/m^3 (ft^2/ft^3), but also cause a higher pressure drop than stacked packing. In addition to the lower pressure drop, the stacked packing provides better liquid distribution over the entire surface of the packing. The large installation costs required to stack the packing material usually make it impractical unless high flow rates are required.

EXAMPLE 1: packed tower review. Pollution Unlimited, Inc., has submitted plans for a packed ammonia scrubber on an air stream containing NH_3. The operating and design data are given by Pollution Unlimited, Inc. You remember approving plans for a nearly identical scrubber for Pollution Unlimited, Inc., in 1978. After consulting your old files you find all the conditions were identical except for the gas flow rate. What is you recommendation (EPA-84/09, p. 102)?

Given conditions

tower diameter = 3.57 ft

packed height of column = 8 ft

gas and liquid temperature = 75°F

operating pressure = 1.0 atm

ammonia-free liquid flow rate (inlet) = 1000 lb/ft^2-h

gas flow rate = 1575 acfm

gas flow rate in the 1978 plan = 1121 acfm
inlet NH_3 gas composition = 2.0 mole %
outlet NH_3 gas composition = 0.1 mole %
air density = 0.0743 lb/ft^3
molecular weight of air = 29
Henry's law constant $m = 0.972$
molecular weight of water = 18
See Figs. 3.23 (packing A is used) and 3.24
emission regulation = 0.1% NH_3

Solution

1. What is the number of overall gas transfer units N_{OG}?

The number of overall gas transfer units N_{OG} is used in calculating packing height requirements. It is a function of the extent of the desired separation and the magnitude of the driving force through the column (the displacement of the operating line from the equilibrium line).

1.a. Calculate the gas molar flow rate G_m and liquid molar flow rate L_m, in lb-mole/ft^2-h.

The values of G_m and L_m are needed to use the Colburn chart (Fig. 3.24), which graphically predicts the value of N_{OG}.

1.a.1. Calculate the cross-sectional area of the tower S, in ft^2.

$$S = \pi D^2/4$$

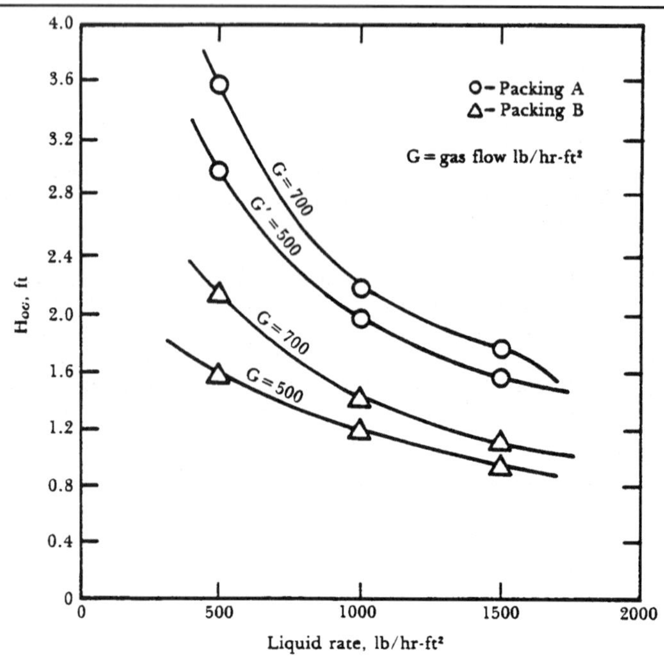

FIGURE 3.23 H_{OG} vs. liquid rate for ammonia–water absorption (*EPA-84/09, p. 103*).

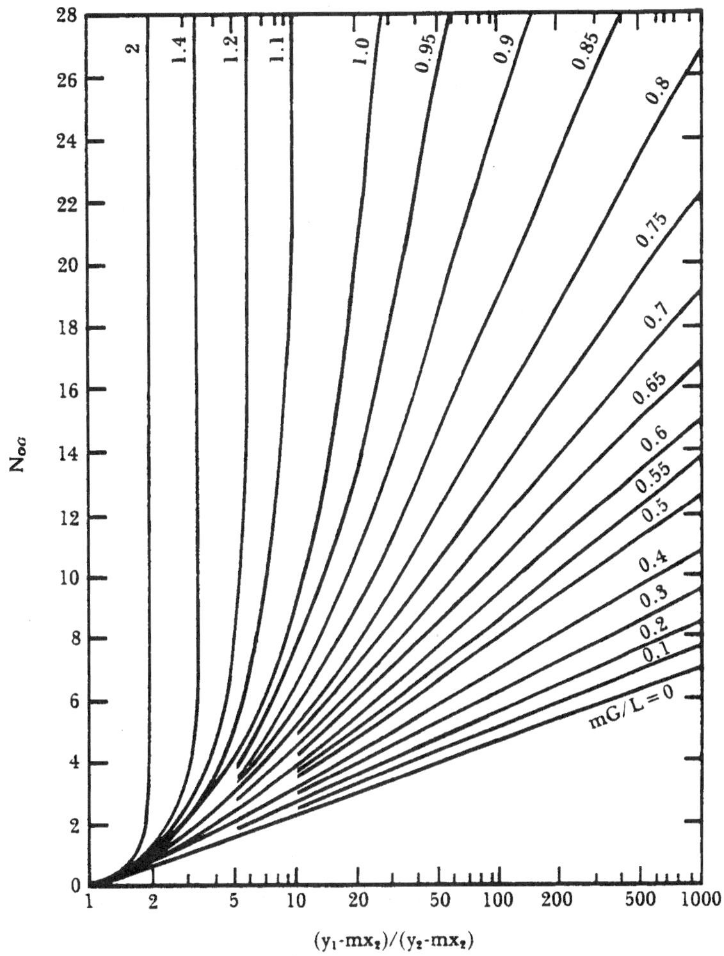

FIGURE 3.24 Colburn chart (*EPA-84/09, p. 104*).

where S = cross-sectional area of the tower
D = diameter of the tower

$$S = \pi D^2/4$$
$$= (\pi)(3.57)^2/(4)$$
$$= 10.0\,\text{ft}^2$$

1.a.2. Calculate the gas molar flow rate G_m, in lb-mole/ft^2-h.

$$G_m = Q\rho/SM$$

where G_m = gas molar flow rate, lb-mole/ft^2-h

Q = volumetric flow rate of gas stream
ρ = density of air
S = cross-sectional area of the tower
M = molecular weight of air

$$\begin{aligned} G_m &= Q\rho/SM \\ &= (1575)(0.0743)/(10.0)(29) \\ &= 0.404\ \text{lb-mole/ft}^2\text{-min} \\ &= 24.2\ \text{lb-mole/ft}^2\text{-h} \end{aligned}$$

1.a.3. Calculate the liquid molar flow rate L_m, in lb-mole/ft^2-h.

$$L_m = L/M_L$$

where L_m = liquid molar flow rate, lb-mole/ft^2-h
L = liquid mass velocity, lb/ft^2-h
M_L = liquid molecular weight

$$\begin{aligned} L_m &= L/M_L \\ &= (1000)/(18) \\ &= 55.6\ \text{lb mole/ft}^2\text{-h} \end{aligned}$$

1.b. Calculate the value of mG_m/L_m.

where m = Henry's law constant
G_m = gas molar flow rate, lb-mole/ft^2-h
L_m = liquid molar flow rate, lb-mole/ft^2-h

$$\begin{aligned} mG_m/L_m &= (0.972)(24.2/55.6) \\ &= 0.423 \end{aligned}$$

1.c. Calculate the value of $(y_1 - mx_2)/(y_2 - mx_2)$.
$(y_1 - mx_2)/(y_2 - mx_2)$ is the abscissa of the Colburn chart (Fig. 3.24)

where y_1 = inlet gas mole fraction
y_2 = outlet gas mole fraction
x_2 = inlet liquid mole fraction
m = Henry's law constant

$$\begin{aligned} (y_1 - mx_2)/(y_2 - mx_2) &= [0.02 - (0.972)(0)]/[0.001 - (0.972)(0)] \\ &= 20.0 \end{aligned}$$

1.d. Determine the value of N_{OG} from the Colburn chart (Fig. 3.24).
From the Colburn chart (Fig. 3.24), use the values of $(y_1 - mx_2)/(y_2 - mx_2)$ and mG_m/L_m to find the value of N_{OG}.

$$N_{OG} = 4.3$$

2. What is the height of an overall gas transfer unit H_{OG}?
The height of an overall gas transfer unit H_{OG} is also used in calculating packing height requirements. H_{OG} values in air pollution are almost always based on experience. H_{OG} is a strong function of the solvent viscosity and difficulty of separation, increasing with increasing values of both.

2.a. Calculate the gas mass velocity G, in lb/ft^2-h.

$$G = \rho Q/S$$

where
G = gas mass velocity, lb/ft^2-h
ρ = density of air
S = cross-sectional area of the tower

$$\begin{aligned} G &= \rho Q/S \\ &= (1575)(0.0743)/10.0 \\ &= 11.7\,\text{lb/ft}^2\text{-min} \\ &= 702\,\text{lb/ft}^2\text{-h} \end{aligned}$$

2.b. Determine the value of H_{OG} from Fig. 3.23

$$H_{OG} = 2.2\,\text{ft}$$

3. What is the required packed column height Z, in ft?

$$Z = (N_{OG})(H_{OG})$$

where
Z = height of packing
H_{OG} = height of an overall gas transfer unit
N_{OG} = number of transfer units

$$\begin{aligned} Z &= (N_{OG})(H_{OG}) \\ &= (4.3)(2.2) \\ &= 9.46\,\text{ft} \end{aligned}$$

4. Compare the packed column height of 8 ft specified by Pollution Unlimited, Inc., to the height calculated above. What is your recommendation?

Disapproval, because the calculated height (9.46 ft) is higher than that (8 ft) proposed by the company.

EXAMPLE 2: tower height and diameter. A packed column is designed to absorb ammonia from a gas stream. Given the operating conditions and type of packing below, calculate the height of packing and column diameter (EPA-84/09, p. 106).

Given conditions

gas mass flow rate = 5000 lb/h
NH_3 concentration in inlet gas stream = 2.0 mole %
scrubbing liquid = pure water
packing type = 1-inch Raschig rings
packing factor, $F = 160$
H_{OG} of the column = 2.5 ft
Henry's law constant $m = 1.20$
density of gas (air) = $0.075\,lb/ft^3$
density of water = $62.4\,lb/ft^3$
viscosity of water = 1.8 cp

Figure 3.25: generalized flooding and pressure drop correction
Figure 3.26: graphical representation of the packed column

The unit operates at 60% of the flooding gas mass velocity, the actual liquid flow rate is 25% more than the minimum, and 90% of ammonia is to be collected based on state regulations.

Solution:

1. What is the number of overall gas transfer units N_{OG}?
   Remember that the height of packing Z is given by

$$Z = (H_{OG})(N_{OG})$$

where Z = height of packing
H_{OG} = height of an overall gas transfer unit
N_{OG} = number of transfer units

Since H_{OG} is given, you only need N_{OG} to calculate Z. N_{OG} is a function of both the liquid and gas flow rates; however, it is usually available for most air pollution applications.

1.a. What is the equilibrium outlet liquid composition x_1 and the outlet gas composition y_2 for 90% removal?

Remember that you need the inlet and outlet concentrations (mole fractions) of both streams to use the Colburn chart (Fig. 3.24).

1.a.1. Calculate the equilibrium outlet concentration x_1^* at $y_1 = 0.02$.

According to Henry's law, $x_1^* = y_1/m$. The equilibrium outlet liquid composition is needed to calculate the minimum L_m/G_m.

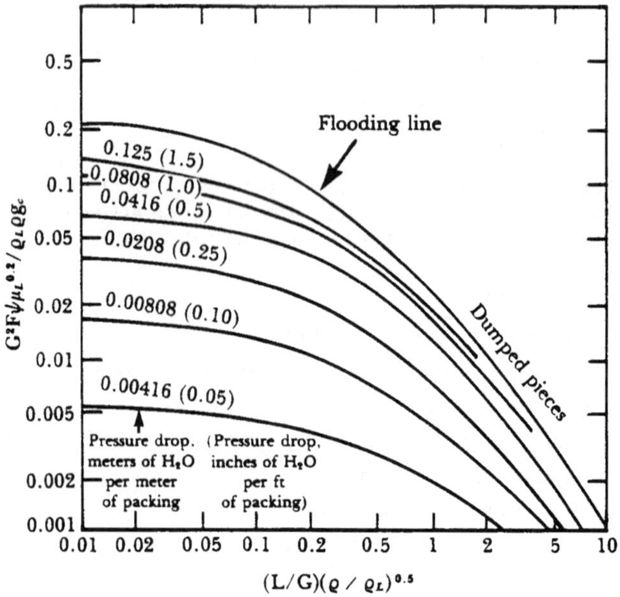

FIGURE 3.25 Generalized flooding and pressure drop correction (*EPA-84/09, p. 107*).

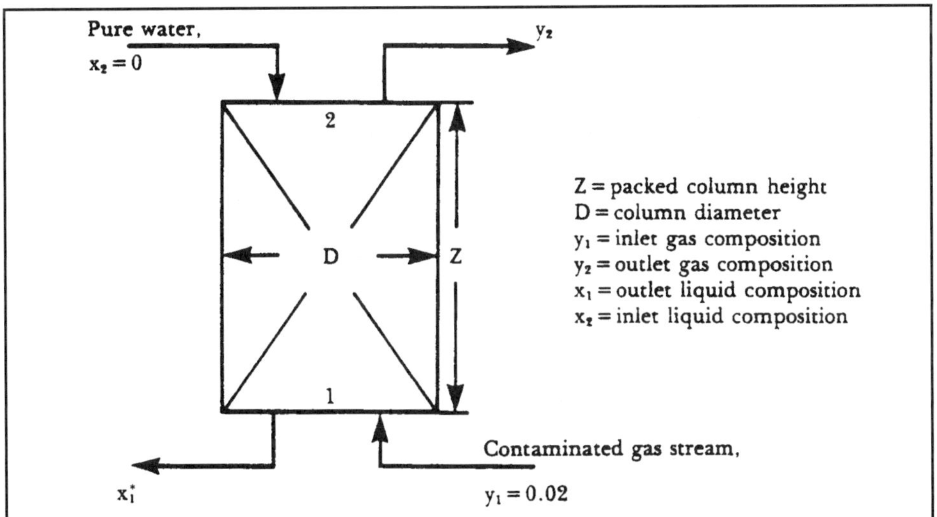

FIGURE 3.26 Graphical representation of the packed column (*EPA-84/09, p. 107*).

Here x_1^* = outlet concentration
y_1 = inlet gas mole fraction
m = Henry's law constant
L_m = liquid molar flow rate, lb-mole/ft^2-h
G_m = gas molar flow rate, lb-mole/ft^2-h

$$\begin{aligned} x_1^* &= y_1/m \\ &= (0.02)/(1.20) \\ &= 0.0167 \end{aligned}$$

1.a.2. Calculate y_2 for 90% removal.
Since it is required to remove 90% of NH_3, there will be 10% of NH_3 remaining in the outlet gas stream; therefore, by material balance:

$$y_2 = (0.1y_1)/[(1 - y_1) + (0.1)y_1]$$

where y_1 = inlet gas mole fraction
y_2 = outlet gas mole fraction

$$\begin{aligned} y_2 &= (0.1y_1)/[(1 - y_1) + (0.1)y_1] \\ &= (0.1)(0.02)/[(1 - 0.02) + (0.1)(0.02)] \\ &= 0.00204 \end{aligned}$$

1.b. Determine the minimum ratio of molar liquid flow rate to molar gas flow rate $(L_m/G_m)_{min}$ by a material balance.
Material balance around the packed column:

$$G_m(y_1 - y_2) = L_m(x_1^* - x_2)$$
$$(L_m/G_m)_{min} = (y_1 - y_2)/(x_1^* - x_2)$$

where y_1 = inlet gas mole fraction
y_2 = outlet gas mole fraction
x_1^* = outlet concentration
x_2 = inlet liquid mole fraction
L_m = liquid molar flow rate, lb mole/ft^2-h
G_m = gas molar flow rate, lb mole/ft^2-h

$$\begin{aligned}(L_m/G_m)_{min} &= (y_1 - y_2)/(x_1^* - x_2)\\ &= (0.02 - 0.00204)/(0.0167 - 0)\\ &= 1.08\end{aligned}$$

1.c. Calculate the actual ratio of molar liquid flow rate to molar gas flow rate (L_m/G_m). Remember that the actual liquid flow rate is 25% more than the minimum based on the given operating conditions.

$$\begin{aligned}(L_m/G_m) &= 1.25(L_m/G_m)_{min}\\ &= (1.25)(1.08)\\ &= 1.35\end{aligned}$$

1.d. Calculate the value of $(y_1 - mx_2)/(y_2 - mx_2)$.
$(y_1 - mx_2)/(y_2 - mx_2)$ is the abscissa of the Colburn chart (Fig. 3.24).

where y_1 = inlet gas mole fraction
y_2 = outlet gas mole fraction
x_1 = outlet liquid mole fraction
x_2 = inlet liquid mole fraction
m = Henry's law constant

$$\begin{aligned}(y_1 - mx_2)/(y_2 - mx_2) &= [(0.02) - (1.2)(0)]/[(0.00204) - (1.2)(0)]\\ &= 9.80\end{aligned}$$

1.e. Calculate the value of mG_m/L_m.

Here m = Henry's law constant
G_m = gas molar flow rate, lb-mole/ft^2-h
L_m = liquid molar flow rate, lb-mole/ft^2-h

Even though the individual values of G_m and L_m are not known, the ratio of the two has been previously calculated:

$$\begin{aligned}mG_m/L_m &= (1.2)/(1.35)\\ &= 0.889\end{aligned}$$

1.f. Determine number of overall gas transfer units N_{OG} from the Colburn chart using the values calculated in steps 1.d. and 1.e.
From Colburn's chart (Fig. 3.24) $N_{OG} = 6.2$

2. Calculate the height of packing Z.

$$Z = (N_{OG})(H_{OG})$$

where Z = height of packing

H_{OG} = height of an overall gas transfer unit
N_{OG} = number of transfer units

$$\begin{aligned} Z &= (N_{OG})(H_{OG}) \\ &= (6.2)(2.5) \\ &= 15.5\,\text{ft} \end{aligned}$$

3. What is the diameter of the packed column?

The actual gas mass velocity must be determined. To calculate the diameter of the column, you need the flooding gas mass velocity. Figure 3.25 is used to determine the flooding gas mass velocity. The mass velocity is obtained by dividing the mass flow rate by the cross-sectional area.

3.a. Calculate the flooding gas mass velocity G_f.

3.a.1. Calculate the abscissa of Fig. 3.25, $(L/G)(\rho/\rho_L)^{0.5}$

$$(L/G)(\rho/\rho_L)^{0.5} = (L_m/G_m)(18/29)(\rho/\rho_L)^{0.5}$$

where 18/29 = ratio of molecular weight of water to air
L = liquid mass velocity, lb/s-ft^2
G = gas mass velocity, lb/s-ft^2
ρ = gas density
ρ_L = liquid density
G_m = gas molar flow rate in lb-mole/ft^2-h
L_m = liquid molar flow rate in lb-mole/ft^2-h

Note that the L and G terms in Fig. 3.25 are based on mass and not moles.

$$\begin{aligned} (L/G)(\rho/\rho_L)^{0.5} &= (1.35)(18/29)(0.075/62.4)^{0.5} \\ &= 0.0291 \end{aligned}$$

3.a.2. Determine the value of the ordinate at the flooding line using the calculated value of the abscissa.

$$\text{Ordinate} = G^2 F\psi(\mu_L)^{0.2}/\rho_L \rho g_c$$

where F = packing factor = 160 for 1-inch Raschig rings
ψ = ratio, density of water/density of liquid
g_c = 32.2 lb-ft/lb$_f$-s^2
μ_L = viscosity of liquid, cp
G = gas mass velocity, lb/ft^2-s

From Fig. 3.25,

$$G^2 F\psi(\mu_L)^{0.2}/\rho_L \rho g_c = 0.19$$

3.a.3. Solve the abscissa for the flooding gas mass velocity G_f, in lb/ft^2-s.
The G value in step 3.a.2 becomes G_f for this case. Thus,

$$\begin{aligned} G_f &= [0.19(\rho_L \rho g_c)/(F\psi(\mu_L)^{0.2})]^{0.5} \\ &= [(0.19)(62.4)(0.075)(32.2)/(160)(1)(1.8)^{0.2}]^{0.5} \\ &= 0.400\,\text{lb/ft}^2\text{-s} \end{aligned}$$

3.b. Calculate the actual gas mass velocity G_{act}, in lb/ft^2-s.

$$\begin{aligned} G_{act} &= 0.6G_f \\ &= (0.6)(0.400) \\ &= 0.240\ \text{lb/ft}^2\text{-s} \\ &= 864\ \text{lb/ft}^2\text{-h} \end{aligned}$$

3.c. Calculate the diameter of the column in ft.

$$\begin{aligned} S &= (\text{mass flowrate of gas stream})/G_{act} \\ &= 5000/G_{act} \end{aligned}$$

$$S = \pi D^2/4$$

$$\begin{aligned} \pi D^2/4 &= 5000/G_{act} \\ D &= [(4(5000))/(\pi G_{act})]^{0.5} \\ &= 2.71\ \text{ft} \end{aligned}$$

4.4 Combination of Liquid-Phase and Gas-Phase Contacting Scrubbers

A number of wet-collector designs use energy from both the gas stream and liquid stream to collect pollutants. Many of these combination devices are available commercially. A seemingly unending number of scrubber designs have been developed by changing system geometry and incorporating vanes, nozzles, and baffles. These types of scrubbers include:

- Cyclonic spray scrubber
- Mobile-bed scrubber
- Baffle spray scrubber
- Mechanically-aided scrubber

Cyclonic spray scrubber. Cyclonic spray scrubbers use the features of both the dry cyclone and the spray chamber to collect pollutants. Generally, the exhaust gas enters the chamber tangentially, swirls through the chamber and exits. At the same time, liquid is sprayed inside the chamber. As the exhaust gas swirls around the chamber, pollutants are captured when they impact on liquid droplets, are thrown to the walls, and washed down and out (EPA-84/03, p. 6-1; EPA-81/10, p. 9-41).

Cyclonic scrubbers are generally low- to medium-energy devices, with pressure drops of 4 to 25 cm (1.5 to 10 in.) of water. Commercially available designs include:

- Irrigated cyclone scrubber
- Cyclonic spray scrubber

Irrigated cyclone scrubber: In the irrigated cyclone (Fig. 3.27), the exhaust gas enters near the top of the scrubber into the water sprays. The exhaust gas is forced to swirl downward, then change directions, and return upward in a tighter spiral. The liquid droplets produced capture the pollutants, are eventually thrown to the side walls, and are withdrawn from the collector. The 'cleaned' gas leaves through the top of the chamber.

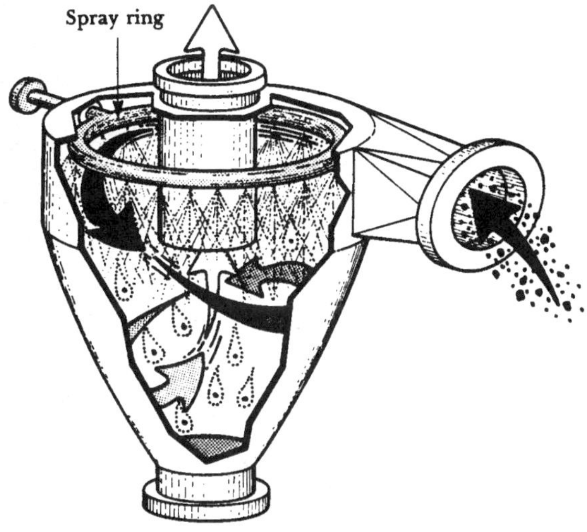

FIGURE 3.27 Irrigated cyclone scrubber (*EPA-84/03, p. 6-2*).

Cyclonic spray scrubber: The cyclonic spray scrubber (Fig. 3.28) forces the exhaust gas up through the chamber from a bottom tangential entry. Liquid sprayed from nozzles on a center post (manifold) is directed toward the chamber wall and through the swirling exhaust gas. As in the irrigated cyclone, liquid captures the pollutant, is forced to the wall, and washes out. The 'cleaned' gas continues upward, exiting through the straightening vanes at the top of the chamber. Stationary vanes are used inside the cyclonic scrubber chamber for much the same purpose that they are used at the top (to alter the gas flow). But inside, they are designed to start or enhance the cyclonic gas flow.

Particle collection. Cyclonic spray scrubbers are more efficient than spray towers, but not as efficient as venturi scrubbers, in removing particles from the exhaust gas stream. Particles larger than 5 μm are generally collected by impaction with 90% efficiency. The cut diameter ranges between 2 and 3 μm for these devices. In a simple spray tower, the velocity of the particles in the exhaust gas stream is low: 0.6 to 1.5 m/s (2 to 5 ft/s). By introducing the exhaust gas tangentially into the spray chamber, as does the cyclonic scrubber, exhaust gas velocities (thus, particle velocities) are increased to approximately 60 to 180 m/s (200 to 600 ft/s). The velocity of the liquid spray is approximately the same in both devices. This increased particle-to-liquid relative velocity increases particle collection efficiency for this device over that of the spray chamber. Exhaust gas velocities of 60 to 180 m/s are equivalent to those encountered in a venturi scrubber. However, cyclonic spray scrubbers are not as efficient as venturis because they are not capable of producing the same degree of useful turbulence.

Gas collection. High exhaust gas velocities through these devices reduce the gas–liquid contact time, thus reducing absorption efficiency. Cyclonic spray scrubbers are capable of effectively removing some gases; however, they are rarely chosen when gaseous pollutant removal is the only concern.

Mobile-bed scrubber. Mobile-bed, also called moving-bed, scrubbers use energy from both liquid sprays and the gas stream to provide contact. Instead of having stationary packing, as in packed towers, they use a bed containing packing that is in constant motion. The gas stream provides the energy to keep the packing in motion while, at the same time, liquid is sprayed

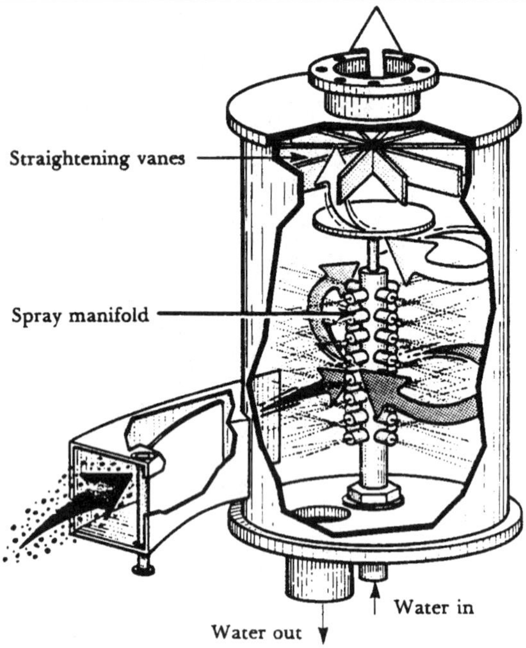

FIGURE 3.28 Cyclonic spray scrubber (*EPA-84/03, p. 6-2*).

over the packing. Mobile-bed scrubbers can be classified as either *flooded* or *fluidized*, depending on the degree of packing movement. In a flooded-bed scrubber, the packing gently moves and rotates, whereas in a fluidized scrubber, the packing is suspended, or fluidized, within the bed (EPA-84/03, p. 6-5; EPA-81/12, p. 4-55; EPA-81/10, p. 9-45).

Mobile-bed scrubbers were developed to provide the effective mass-transfer (absorption) characteristics of packed and plate towers, without the plugging problems. The wetted packing provides a large area for gas-to-liquid contact, promoting absorption. The movement of the bed cleans off any deposited particles. Therefore, these devices are primarily used when good collection efficiency for both particulate and gaseous pollutants is required.

Flooded-bed scrubber. A flooded-bed scrubber (Fig. 3.29) contains a section of mobile packing (spheres) 10–20 cm (4–8 in.) deep. The spheres are usually made of plastic; however, glass or marble spheres have been used. The exhaust gas stream enters from the bottom while liquid is sprayed from the top and/or bottom over the packing. Bottom, or inlet, sprays are usually included to saturate the exhaust gas stream and remove any large particles. The gas velocity is such that it causes the packing materials to rotate and rub against each other. This rotating motion acts as a self-cleaning mechanism in addition to enhancing gas and liquid mixing.

Bubbles formed in the bed create a layer of froth over the bed approximately twice as deep as the bed itself. This turbulent froth layer provides an additional surface for absorbing pollutant gases and collecting fine particles. Because of the high gas velocities, entrainment separators are required to prevent liquid-mist carryover.

Fluidized-bed scrubber: A fluidized-bed scrubber is very similar to a flooded-bed scrubber. The difference is in the degree of movement of the packing. In a fluidized-bed scrubber, the exhaust gas velocity (1.8–4.8 m/s, or 6–16 ft/s) is such that it keeps the packing in constant motion between a lower and upper retaining grid. This is shown in Fig. 3.30. The packing is

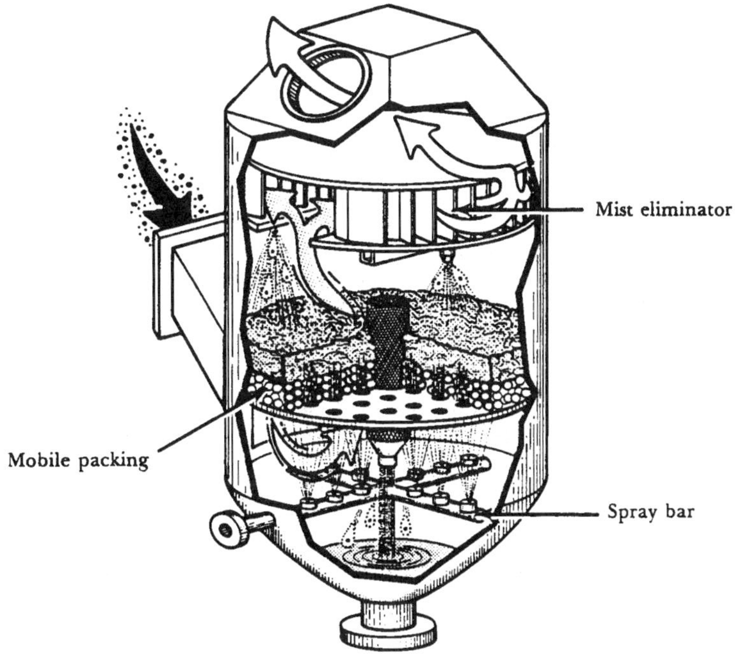

FIGURE 3.29 Flooded-bed absorber (*EPA-84/03, p. 6-5*).

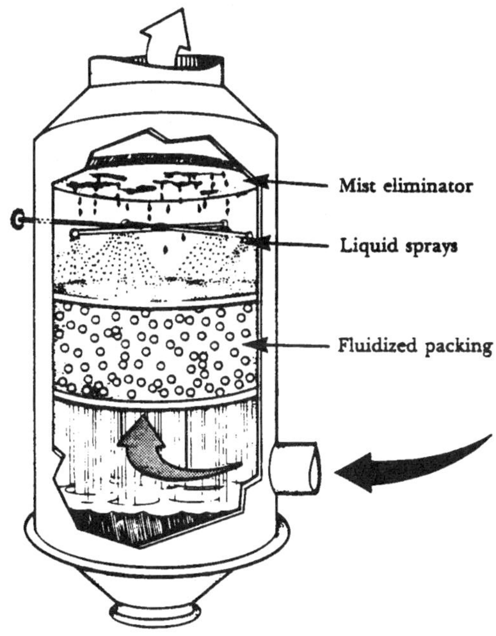

FIGURE 3.30 Fluidized-bed absorber (*EPA-81/12, p. 4-57*).

made of either polypropylene or polyethylene plastic balls that are hollow, resembling ping pong balls. The packed sections are usually 0.3–0.6 m (1–2 ft) thick with a froth zone about 0.6 m (2 ft) thick above the packing. These devices can have one to as many as six fluidized packed sections. When used for gas absorption, they are sometimes referred to as turbulent-contact absorbers (TCA).

Particle collection. In a mobile-bed scrubber, particles can be collected in three locations. First, sprays are used to remove coarse particles in the inlet below the bed. Particles are also captured when they impinge on the wetted surface of the packing. Finally, small particles are captured in the froth, or foam, layer above the bed. These devices will generally remove particles as small as 2–3 μm in diameter and have been used extensively when the exhaust stream does not contain a substantial amount of particles in the submicron range. These devices usually contain one bed, unless gas absorption is a consideration. Adding additional beds or more packing does not substantially increase the particle collection efficiency (i.e., any particles not captured by the first stage will probably not be collected in any following stage). The pressure drop in mobile-bed scrubbers ranges from 5 to 15 cm (2 to 6 in.) of water per stage of packing.

Gas collection. Mobile-bed scrubbers are capable of achieving high gaseous-pollutant removal efficiencies. Their operation is very similar to the operation of packed towers. Liquid is sprayed over the mobile packing, providing a huge surface for the pollutant gas to contact the liquid. Movement of both the gas around the packing and the constantly sprayed liquid provides excellent mixing and contact time for absorption to occur. Mobile-bed scrubbers provide the same amount of absorption efficiency as do packed or plate towers without the associated plugging problems. Due to high exhaust gas velocities through mobile-bed scrubbers, these units can handle five to six times the amount of exhaust gas handled by packed or plate towers of similar size. However, they are not as efficient as packed or plate towers per unit of energy consumed.

Absorption in mobile-bed scrubbers is enhanced by the same factors that affect packed towers. These factors are increasing the liquid-to-gas ratio, increasing the depth of packing, or increasing the number of stages. Increasing these factors increases the gas and liquid contact and the residence time. However, increasing these factors also increases the capital and/or operating costs of the system. As with any system, these process variables are set to achieve the desired removal efficiency at the minimum cost. For gas absorption, multiple stages are used and the liquid-to-gas ratios are high; for example, mobile-bed scrubbers have been used to remove SO_2 from boiler flue gas exhausts. Using a lime or limestone slurry, the liquid-injection rates are approximately 8 L/m^3 (60 gal/1000 ft^3) of flue gas. This is compared with 0.4 L/m^3 (3.0 gal/1000 ft^3) when these devices are used for particle removal.

Baffle spray scrubber. Baffle spray scrubbers are very similar to spray towers in design and operation. However, in addition to using the energy provided by the spray nozzle, baffles are added to allow the gas stream to atomize some liquid as it passes over them. A simple baffle scrubber system is shown in Fig. 3.31. Liquid sprays capture pollutants and also remove collected particles from the baffle. Adding baffles slightly increases the pressure drop of the system (EPA-84/03, p. 6-11; EPA-81/10, p. 9-46).

Particle collection. These devices are used much the same as spray towers—to preclean or remove particles larger than 10 μm in diameter. However, they will tend to plug or corrode if particle concentration of the exhaust gas stream is high.

Gas collection. Even though these devices are not specifically used for gas collection, they are capable of a small amount of gas absorption because of their large wetted surface.

Mechanically-aided scrubber. In addition to using liquid sprays or the exhaust stream, energy can be supplied to a scrubbing system by using a motor. The motor is used to drive a rotor or paddles which, in turn, generate water droplets for gas and particle collection. Systems designed in this manner have an advantage of requiring less space than do other scrubbers, but the overall power requirements tend to be higher than for other scrubbers of equivalent

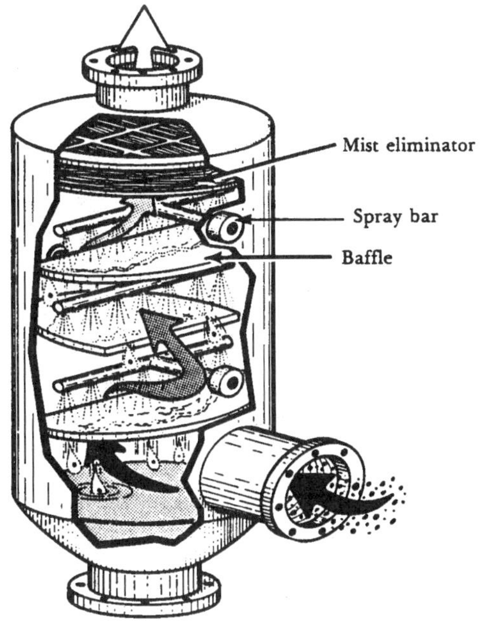

FIGURE 3.31 Baffle spray scrubber (*EPA-84/03, p. 6-11*).

efficiency. This point might appear to contradict the contact power principle; however, significant power losses occur in driving the rotor. Power is not expended to provide for gas–liquid contact (EPA-84/03, p. 6-11; EPA-81/10, p. 9-46).

Fewer mechanically-aided scrubber designs are available than are liquid- and gas-phase contacting collector designs. Mechanically-aided scrubbers include *centrifugal fan scrubbers* and *mechanically-induced spray scrubbers*.

Centrifugal fan scrubber: A centrifugal fan scrubber can serve as both an air mover and a collection device. Figure 3.32 shows such a system, where water is sprayed onto the fan blades cocurrently with the moving exhaust gas. Some gaseous pollutants and particles are initially removed as they pass over the liquid sprays. The liquid droplets then impact on the blades to create smaller droplets for additional collection targets. Collection can also take place on the liquid film that forms on the fan blades. The rotating blades force the liquid (and any particles) off of the blades. The liquid droplets separate from the gas stream because of their centrifugal motion.

Centrifugal fan collectors are the most compact of the wet scrubbers since the fan and collector comprise a combined unit. No internal pressure loss occurs across the scrubber, but a power loss equivalent to a pressure drop of 10.2 to 15.2 cm (4 to 6 in.) of water occurs because the blower efficiency is low.

Mechanically-induced spray scrubber: A mechanically-induced spray scrubber consists of a whirling rotor submerged in a pool of liquid. The whirling rotor produces a fine droplet spray. By moving the process gas through the spray, particles and gaseous pollutants can subsequently be collected. Figure 3.33 shows an induced-spray scrubber that uses a vertical-spray rotor.

Particle collection. Mechanically-aided scrubbers are capable of high collection efficiencies for particles with diameters of 1 μm or greater. However, achieving these high efficiencies

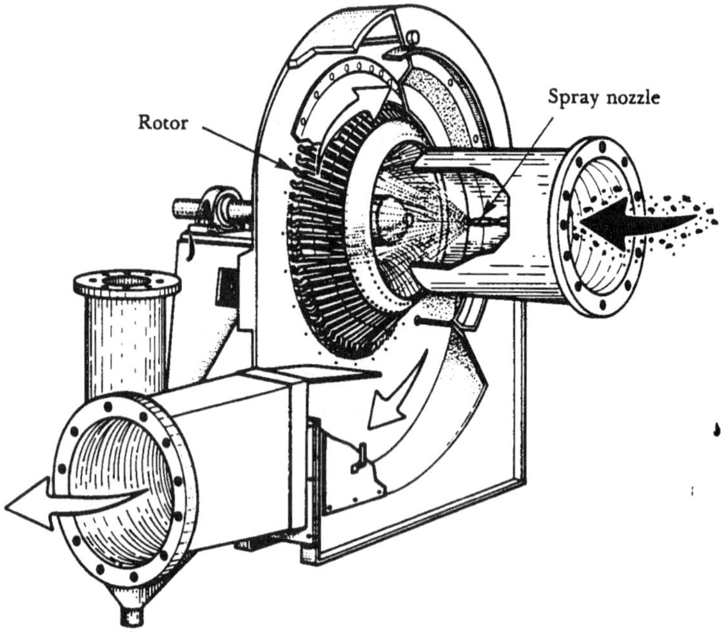

FIGURE 3.32 Centrifugal fan scrubber (*EPA-84/03, p. 6-13*).

usually requires a greater energy input than those of other scrubbers operating at similar efficiencies. In mechanically-aided scrubbers, the majority of particle collection occurs in the liquid droplets formed by the rotating blades or rotor.

Gas collection. Mechanically-aided scrubbers are generally not used for gas absorption. The contact time between the gas and liquid phases is very short, limiting absorption. For gas removal, several other scrubbing systems provide much better removal per unit of energy consumed.

5 DRY ABSORPTION FOR GASEOUS EMISSION CONTROL

Dry absorption devices used to remove gaseous contaminants are referred to as *dry scrubbers*. Dry scrubbers are primarily used to remove acid gases, primarily HCl and SO_2. The basic procedures of collecting pollutants are as follows. Alkaline sorbent materials are injected into the dirty flue gas. The acid gases begin to react with the alkaline sorbents to produce solid particulate salts that are collected by a particulate control device, usually a fabric filter, that follows the dry scrubber. The unreacted sorbent is also captured on the fabric filter cake, where additional acid gas reacts with the sorbent and is captured. There are two types of dry scrubber systems. They are:

- Spray dryer
- Dry injection

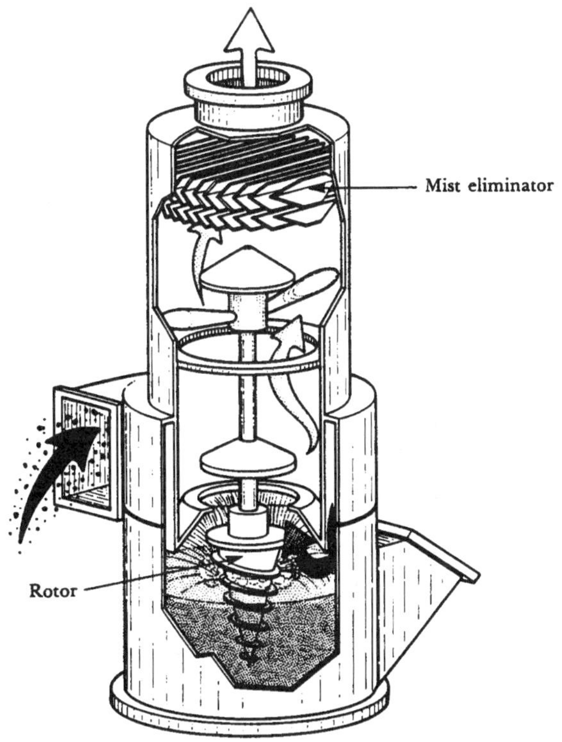

FIGURE 3.33 Mechanically-induced spray scrubber (*EPA-84/03, p. 6-14*).

The basic operating principle for both the spray dryer and the dry injection process is to mix an adequate supply of alkaline sorbent with the flue gas and allow sufficient contact time for the reaction to occur. On most units, outlet gas stream acid gas monitors provide a direct indication of system performance.

5.1 Spray Dryer

Major components. Figure 3.34 shows a schematic of a spray dryer system for the absorption of acid gases. A spray dryer is usually connected with a fabric filter which is used to collect the particles as shown. The major components of a dry scrubber system are: (1) lime slaker, if pebble lime is used; (2) sorbent mixing tank; (3) sorbent feed tank; (4) atomizer feed tank; (5) rotary atomizers or air atomizing nozzles; (6) spray dryer/absorber reaction vessel; (7) solids recycle tank; and (8) particulate control device.

Process description. This type of scrubber uses the alkaline reagent (pebble lime) as a slurry containing 5–20% by weight solids. This slurry is atomized by either rotary atomizers or air atomizing nozzles in a large absorber vessel having a residence time of 6–20 seconds. The major process steps are:

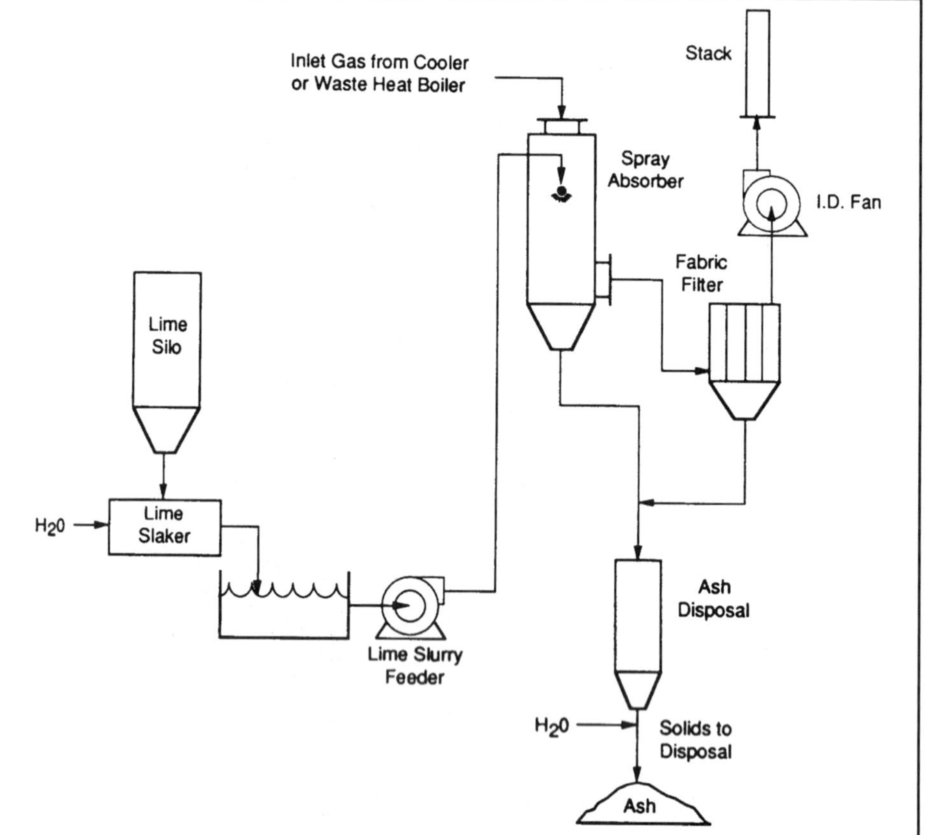

FIGURE 3.34 Spray dryer/fabric filter (*EPA-92/01, p. 96*).

1. Spray dryers are designed to spray an alkaline slurry of sorbent material into the hot flue gases where the acid gases are absorbed into the slurry droplets and reacted with the alkaline material to form solid particle reaction products.
2. Spray dryer facilities usually use pebble lime (CaO) as the alkaline reagent. The pebble lime is converted to calcium hydroxide [$Ca(OH)_2$] by the addition of water in the slaker. The calcium hydroxide is mixed with water in the mixing tank to produce a slurry containing 5–20% solids. The slurry is stored in the feed tank and is transferred to the atomizer feed tank immediately prior to use. The atomizers produce small droplets of slurry that are injected into the absorber reaction vessel.
3. The alkaline sorbents in the slurry react with the acid gases, producing $CaCl_2$ and $CaSO_4$ solid salts.
4. The hot flue gases dry the moisture from the slurry and the reacted and unreacted sorbent either is entrained in the flue gas stream or drops to the bottom of the reaction vessel.
5. The entrained sorbent and reaction products are carried to the particulate control device where they are captured.

6. The sorbent and reaction products that drop to the bottom of the reaction vessel can either be disposed of or retained in the solids recycle tank for recycling back to the mixing tank.

Dry scrubber operation. The key operating parameters that are necessary for effective operation of a spray dryer are:

- Sorbent feed
- Slurry sorbent content
- Outlet gas wet and dry bulb temperatures—the wet bulb/dry bulb temperature readings give an indication of the saturation of the gas stream and the potential for evaporation of moisture.

Effective operation of a spray dryer requires adequate sorbent for reacting with the acid gases and prevention of solids buildup. Solids buildup can occur if slurry moisture is not evaporated within the design time period. The liquid slurry feed rate and sorbent content should be balanced with the hot flue gas volume and acid gas content to ensure the desired removal of acid gases and evaporation of all moisture. The recommended operating ranges are:

- Slurry sorbent content—5 to 20% solids by weight.
- Wet bulb/dry bulb outlet gas temperature difference—90 to 180°F (50 to 100°C). The temperature difference is to ensure evaporation of all moisture.

The feed rate of dry sorbent to the makeup water in the sorbent mix tank is adjusted to obtain the desired sorbent content of the slurry. The flow rate of slurry to the atomizer in the reaction vessel is adjusted to change the wet bulb/dry bulb temperature difference. The slurry flow rate is usually monitored by a magnetic flow meter. An increase in slurry flow will reduce the wet bulb/dry bulb temperature difference.

Startup of a spray dryer should follow procedures that prevent condensation in the system and ensure evaporation of all slurry moisture in the scrubber reactor vessel. One method of ensuring evaporation is to use auxiliary fuel firing to bring the exhaust gas temperature up to the normal operating range before injecting the slurry. Another method would be to gradually increase slurry feed at startup to maintain a 90–180°F (50–100°C) wet bulb/dry bulb temperature differential.

Proper shutdown should ensure that:

- No liquid moisture remains or condenses in the spray dryer or fabric filter after shutdown.
- Auxiliary fuel firing should be used to maintain temperatures above saturation until all sorbent is purged from the system.
- The fabric filter should go through a complete cleaning cycle before shutdown to prevent bag blinding and reaction product salt corrosion.

Spray dryer/fabric filter (SD/FF). Figure 3.34 provides a schematic diagram of a typical SD/FF system. Lime (CaO) is stored as a dry powder in a silo. Before use, it is slaked and fed into a storage tank. The lime slurry is then pumped into the spray absorber, typically with a rotary atomizer. The heat from the flue gases entering the absorber is sufficient to evaporate the water, leaving a dry powder in the reactor. By controlling the amount of water in the slurry, the temperature of flue gas leaving the spray dryer can be controlled to a desired range between 230 and 320°F (110 and 160°C). Some of the powder drops to the bottom of the spray dryer but the majority passes to the fabric filter. The baghouse provides for excellent particulate matter collection at relatively low pressure loss [6–10 in. w.c. (1.5–2.5 kPa)]. Solid material collected by the fabric filter consists of combustor particulate matter, calcium salts, and unreacted lime. This material may be disposed of as dry solid waste.

Typical systems of fabric filters are designed with air-to-cloth ratios of the order of 3.5–5.0 $(ft^3/s)/ft^2$ or 3.5–5.0 ft/s (1.1–1.5 m/s). This implies that baghouse designs with 40 to 180 bags (depending mainly on air flow) are needed for average medical waste incinerators. Bag replacement will likely be required about every 3 years. Replacing all the bags should take less than about 40 hours and can be accomplished over a weekend. United McGill (a baghouse supplier) expects overall system maintenance to be about 40 hours per year.

5.2 Dry Injection

Major components. Figure 3.35 shows a dry injection system schematic. Similar to a spray dryer, a dry injection scrubber is combined with a fabric filter, which is used to collect the particles in the system. The major components of a dry injection system are: (1) dry sorbent storage tank; (2) blower and pneumatic line for transfer of the sorbent; (3) injector; (4) particulate control device (fabric filter or ESP) for collection of the dry sorbent; (5) expansion/reaction chamber (optional). In some cases, the dry alkaline sorbent is injected directly into an expansion/reaction vessel rather than into the duct.

This type of scrubber uses finely-divided calcium hydroxide for absorption of acid gases. The reagent feed has particle sizes which are 90% by weight through 325 mesh screens. The calcium hydroxide is injected countercurrently in the gas stream. The gas stream, containing the entrained calcium hydroxide particles and fly ash, is then fed to the fabric filter. Absorption of acid gases and organic compounds (if present) occurs primarily while the gas stream passes the dust cake on the surface of the filter bags. The calcium hydroxide feed rate for dry injection is three to four times the stoichiometric quantities needed, thus making the system unattractive for very large systems (EPA-89/03).

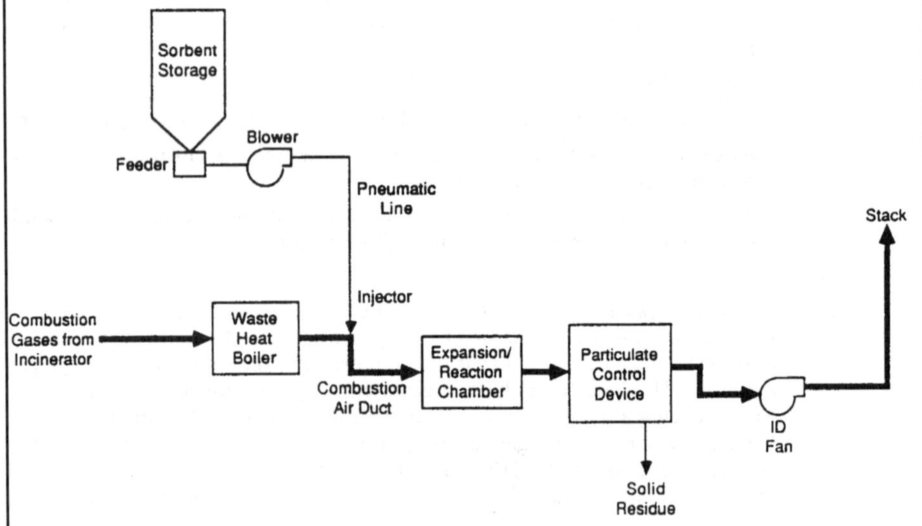

FIGURE 3.35 Dry injection absorption system (*EPA-89/03*).

Process description. The process of dry injection can be described as follows:

1. Dry injection systems are similar to spray dryers, with the exception that the sorbent materials are injected into the flue gases as a dry powder. Since there is no moisture, the reaction rate is slower than with a spray dryer.
2. The dry injection system uses finely-divided alkaline sorbent material, usually calcium hydroxide or sodium bicarbonate, with the approximate consistency of talcum powder.
3. The sorbent is injected into the flue gas duct. The injection creates turbulence that results in mixing of the sorbent with the flue gas.
4. Absorption of the acid gases by the sorbent begins in the flue gas ductwork.
5. An expansion/reaction chamber may be included to increase the residence time of the gases in the system, allowing more time for the reaction to occur.
6. The sorbent and reaction products are carried by the flue gas to the particulate control device, where the solids are collected.
7. If a fabric filter is used as the particulate control device, acid gas removal may be further enhanced by reaction with the sorbent collected in the filter cake.

Dry injection. The key operating parameters for a dry injection system are:

- Sorbent injection rate—The sorbent injection rate should provide adequate sorbent for neutralization of the acid gases and is dependent on the acid gas content of the flue gas.
- Particle size of the sorbent—As particle size decreases, the surface area to volume ratio increases, which improves the efficiency of acid gas collection.

The particle size and injection rate of the sorbent should be specified by the process vendor. Generally, the sorbent feed will have a particle size where 90% by weight will pass through a 325 mesh screen. This dust is approximately the consistency of talcum powder.

Operation of a dry injection system is relatively simple. The key operation includes:

- Maintaining the pneumatic transfer line at a constant airflow rate.
- Monitoring outlet acid gas concentration and increasing the sorbent injection rate to achieve the desired acid gas levels.

There are no special startup considerations for dry injection. At startup of the incinerator, the dry sorbent can be injected without any special preparations.

The only special concern for shutdown of a dry injection system is to put the fabric filter through a cleaning cycle after sorbent injection is stopped. This prevents possible blinding from condensation and reaction product salt damage to the fabric filter components.

Dry injection/fabric filter (DI/FF). Figure 3.35 is a schematic diagram of a typical DI/FF system. The DI/FF system has several major potential advantages, including:

1. It is able to achieve 0.015 gr/dscf (0.03 g/dscm) or even lower particulate emission limits;
2. It does not have liquid discharge and it should have no visible plume.
3. It is able to effectively capture particles smaller than 1 μm. Submicrometer particles contain high concentrations of volatile metals.

Disadvantages include the fact that reagent costs are two to three times that of a wet control system due to the inefficiency of acid gas capture by dry particles. Further, since continuous HCl monitoring is not readily available, the rate of sorbent injection must be set at a sufficiently high level to reach 90% capture at the maximum expected acid gas level (90% capture is required by some states and is likely to be required by many more). For small-scale systems, careful modulation of solid sorbent flow is difficult to achieve and would add needlessly to system cost and complexity.

6 ACCESSORY EQUIPMENT FOR SCRUBBER

Many components comprise a complete scrubbing system. To fully appreciate the operation of a scrubber, it is important to have a basic understanding of the major components of the system. For instance, fans and ducts are required to transport exhaust gas while pumps, nozzles, and pipes transport liquid to and from the scrubbing vessel. Water-recirculation and mist-elimination systems are also necessary. Failure of any of these parts will cause problems for the entire scrubbing system. Major components of a scrubbing system include (EPA-84/03, p. 7-1):

- Entrainment separator (mist separator)
- Nozzle
- Quencher

6.1 Entrainment Separator

The basic principle of wet scrubbers is that the pollutants must first be contacted with the liquid; then the liquid droplets must be removed from the exhaust gas stream before it is exhausted to the atmosphere. Gas moving at high velocities that mixes with a liquid will entrain drops of that liquid. Entrainment separators, also called mist eliminators, are used to remove the liquid droplets. Although the major function of an entrainment separator is to prevent liquid carryover, it also performs additional scrubbing and recovers the scrubbing liquor, which saves on operating costs. Entrainment separators are therefore usually an integral part of any wet scrubbing system (EPA-84/03, p. 2-3; EPA-81/12, p. 4-57; EPA-81/10, p. 9-39).

Entrained liquid droplets vary in size depending on how the droplets were formed. Drops that are torn from the body of a liquid are large (10–100 μm in diameter), whereas drops that are formed by a chemical reaction or by condensation are of the order of 5 μm or less in diameter. Numerous types of entrainment separators are capable of removing these droplets. Those most commonly used for air pollution control purposes are:

- Cyclonic separator (centrifugal separator)
- Wire mesh pad separator (plastic pad separator)
- Chevron blade separator
- Impingement blade separator

Cyclonic separator: A cyclonic separator is also known as a centrifugal separator. It is a separator in a cylindrical tank with a tangential inlet or turning vanes (Fig. 3.36). The tangential inlet or turning vanes impart a swirling motion to the droplet-laden gas stream. The droplets are thrown outward by centrifugal force to the walls of the cylinder. Here they coalesce and drop down the walls to a central location and are recycled to the absorber. These units are simple in construction, having no moving parts. Therefore, they have few plugging problems. Good separation down to the 10-μm range can be expected. The pressure drop across the cyclone is 10–15 cm (4–6 in.) of water for a 98% removal efficiency of droplets down to the 20–25 μm range. Cyclonic separators are commonly used with venturi scrubbers.

Wire mesh separator: Wire mesh or plastic pad separators consist of woven material about 10–15 cm (4–6 in.) thick that fit across the entire diameter of the scrubber (Fig. 3.37). The

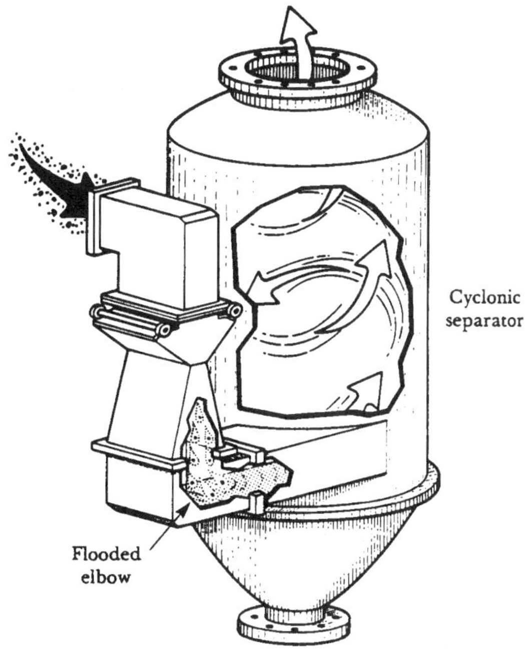

FIGURE 3.36 Cyclonic separator fed by a venturi (*EPA-84/03, p. 3-6*).

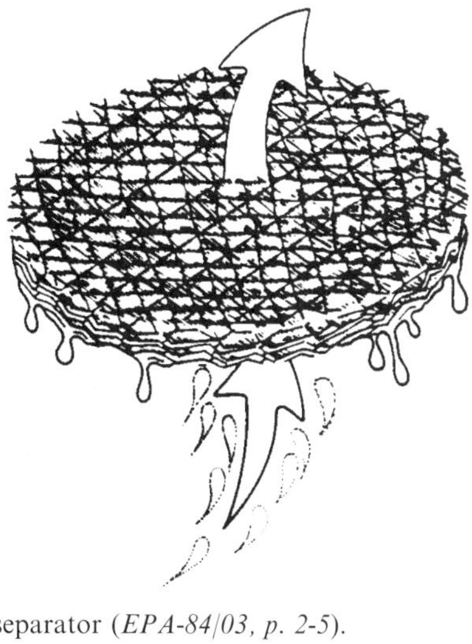

FIGURE 3.37 Wire mesh separator (*EPA-84/03, p. 2-5*).

mesh allows droplets to impact on the material surface, agglomerate with other droplets and drain off by gravity. The pad is usually slanted (no more than a few degrees) to permit the liquid to drain off. Essentially 100% collection of droplets larger than 3 μm is obtained with pressure drops of approximately 10–15 cm (4–6 in.) of water (the pressure drop is dependent on depth and compaction of fibers). The disadvantage with mesh pads is that their small passages are subject to plugging.

Chevron blade separator: In the chevron blade separator (Fig. 3.38), gas passes between the blades and is forced to travel in a zigzag pattern. The liquid droplets cannot follow the gas streamlines, so they impinge on the blade surfaces, coalesce, and fall back into the scrubber chamber or drain. Special features such as hooks and pockets can be added to the sides of these blades to help improve droplet capture. Chevron grids can be stacked or angled on top of one another to provide a series of separation stages. Pressure drop is approximately 6.4 cm (2.5 in.) of water for capture of droplets as small as 5 μm in diameter.

Impingement blade separator: Impingement blade separators (Fig. 3.39) create a cyclonic motion because they are similar in shape to the common house fan. As the gas passes over the curved blades, they impart a spinning motion that causes the mist droplets to be directed to the vessel walls, where they are collected. Pressure drop ranges from 5 to 15 cm (2 to 6 in.) of water.

Table 3.7 summarizes some operating characteristics of entrainment separators.

TABLE 3.7 Typical Operational Characteristics of Entrainment Separators

Separator type	Droplet size collected at 99%, μm	Max. gas velocity		Pressure drop	
		m/s	ft/s	cm H_2O	in. H_2O
Cyclonic	10–25	20	65	10–15	4–6
Wire mesh pad	3	5	15	1.0–15	0.5–6
Chevron blade	35	3	10	6.4	2.5
Impingement blade	20	8	27	5–15	2–6

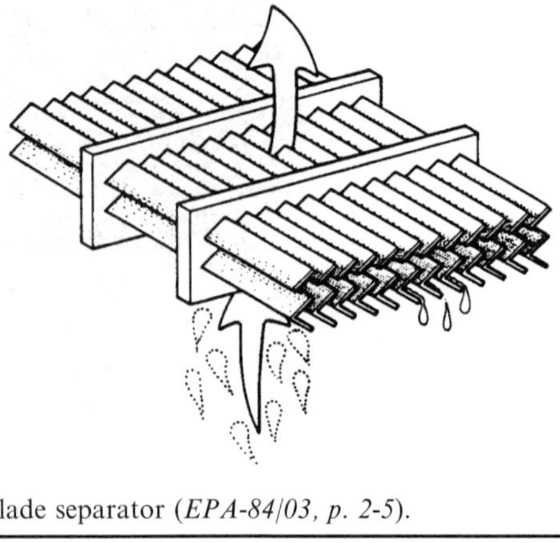

FIGURE 3.38 Chevron blade separator (*EPA-84/03, p. 2-5*).

6.2 Nozzle

Three different nozzle designs are used to produce a fine, cone-patterned spray (Fig. 3.40). They are:

- Impingement nozzle
- Solid cone nozzle
- Helical spray nozzle

Impingement nozzle: In the impingement nozzle, highly pressurized liquid passes through a hollow tube in the nozzle and strikes a pin or plate at the nozzle tip. A very fine fog of tiny,

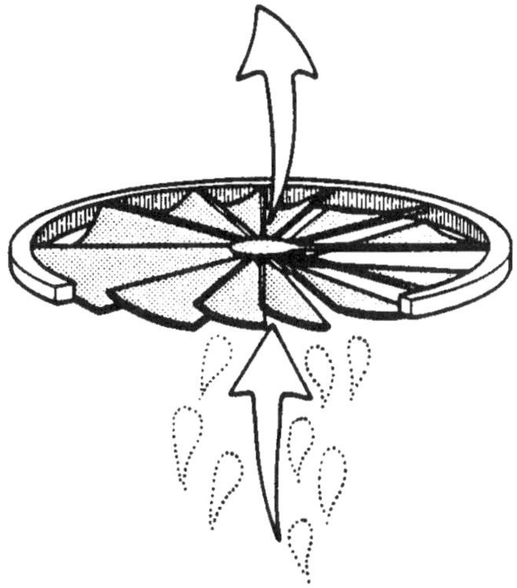

FIGURE 3.39 Impingement blade separator (*EPA-84/03, p. 2-5*).

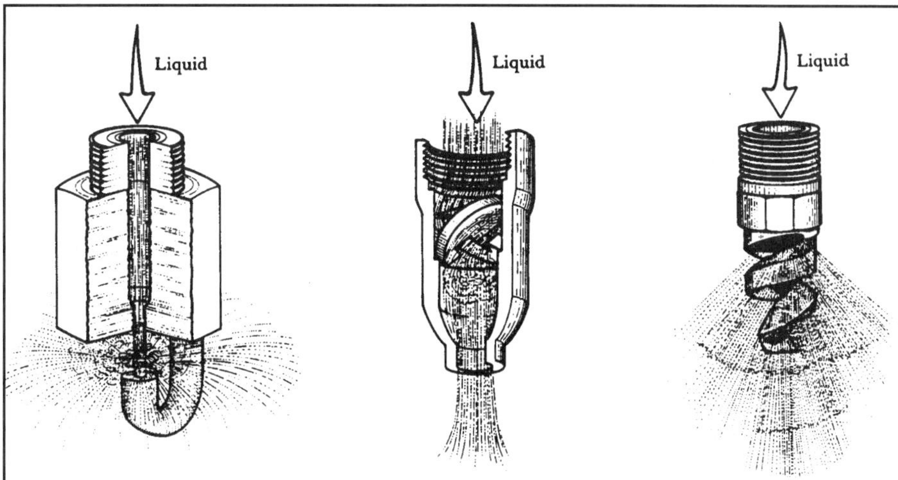

FIGURE 3.40 Types of spray nozzles (*EPA-84/03, p. 2-2*).

uniform-sized droplets approximately 25–400 μm in diameter is produced. Because there are no internal parts in the nozzle, it will not plug as long as particles larger than the opening are filtered out by a strainer. These nozzles are usually made of stainless steel or brass.

Solid cone nozzle: In the solid cone nozzle, liquid is forced over an insert to break it up into a cone of fine droplets. Cones can be full, hollow, or square with spray angles from 15° to 140°. These nozzles can be made of stainless steel, brass, alloys, Teflon, and other plastic materials.

Helical spray nozzle: The helical spray nozzle has a descending spiral impingement surface that breaks up the sprayed liquid into a cone of tiny droplets. The cones can be full or hollow with spray angles from 50° to 180°. There are no internal parts, which helps reduce nozzle plugging. These nozzles can be made of stainless steel, brass, alloys, Teflon, and other plastic materials.

Different spray nozzles are appropriate for different scrubbing systems. Characteristics of the nozzles and sprays include the following:

- *Droplet size*: In general, scrubbers using sprays to provide gas–liquid contact (such as in spray towers) require tiny, uniform-sized droplets to operate effectively. If the sprays are used merely as a method of introducing liquid into the vessel (such as in packed towers), then droplet size is not as critical.
- *Opening size*: The actual opening in the nozzle will vary depending on the applications and the amount of liquid required. Openings range from 0.32 to 6.4 cm (0.125 to 2.5 in.).
- *Spray pattern*: Nozzles are available that produce sprays in a number of geometric shapes, such as square, fan, hollow cone, and full cone. Full-cone sprays are used to provide complete coverage of the areas sprayed.
- *Operating mechanism*: Droplets can be produced by a number of methods such as impinging the liquid on a solid surface or atomizing the liquid using air.
- *Power consumption*: In general, the finer the liquid droplet, the higher the power consumption.

Nozzle plugging is one of the most common malfunctions in wet scrubbers. Plugged nozzles reduce the gas–liquid contact and can also result in scale buildup on, or heat damage to, the scrubber parts formerly sprayed by the nozzle. Nozzle plugging can be most readily detected by observing the liquid spray pattern; however, if the nozzles are not easily accessible, a decrease in liquid flow is also a telltale sign. Remedies include replacing the nozzle with one that is more open, cleaning the nozzle frequently, filtering the scrubbing liquid, or increasing the bleed rate and makeup water rates.

Another problem that can arise is reduced pressure in the spray header. This can cause a reduction in the spray angle (area covered) and an increase in the size of droplets produced.

6.3 Quencher

Occasionally, hot exhaust gas is quenched by water sprays before entering the scrubber. This can be accomplished by spraying liquid into the exhaust gas. Hot gases (those above ambient temperature) are often cooled to near the saturation level by sprays before they enter a scrubber. If not cooled, the hot gas stream can evaporate a large portion of the scrubbing liquor, adversely affecting collection efficiency. Some liquid droplets can evaporate before they have a chance to contact pollutants in the exhaust stream, and others can evaporate after contact, causing captured particles to become re-entrained. In some cases, quenching can actually save money. Cooling the gases reduces the temperature and, therefore, the volume of gases, permitting the use of less expensive materials of construction and a smaller scrubber vessel and fan (EPA-84/03, p. 7-5).

Quenchers are designed using the same principles as scrubbers. Increasing the gas–liquid contact in them increases their operation efficiency. Small liquid droplets cool the exhaust stream quicker than large droplets because they evaporate more easily. Therefore, less liquid is required. However, in most scrubbing systems, approximately 1.5 to 2.5 times the theoretical evaporation demand is required to ensure proper cooling. Evaporation also depends on time. It does not occur instantaneously. Therefore, the quencher should be sized to allow for an adequate exhaust stream residence time.

The cleanest water available should be used for presaturating. Quenching with recirculated scrubber liquor can reduce overall scrubber performance, since recycled liquid usually contains a high level of suspended and dissolved solids. As the liquid droplets evaporate, these solids become re-entrained in the exhaust gas stream. To help reduce this problem, makeup water can be added directly to the quench system rather than by adding all makeup water to a common sump.

REFERENCES

EPA-77, *Venturi scrubber performance model*, Yung, S., Calvert, S. and Barbarika, H.F. U.S. Environmental Protection Agency, Cincinnati, Ohio, EPA600-2-77-172, 1977.

EPA-81/10, *Control of particulate emissions, course 413*, US EPA Air Pollution Training Institute (APTI), EPA450-2-80-066, October 1981.

EPA-81/12, *Control of gaseous emissions, course 415*, US EPA Air Pollution Training Institute (APTI), EPA450-2-81-005, December 1981.

EPA-84/03, *Wet scrubber plan review, course SI:412C*, US EPA Air Pollution Training Institute (APTI), EPA450-2-82-020, March 1984.

EPA-84/09, *Control of gaseous and particulate emissions, course SI:412D*, US EPA Air Pollution Training Institute (APTI), EPA450-2-84-007, September 1984.

EPA-89/02, *Hospital waste incinerator field inspection and source evaluation manual*, US EPA/340/1-89-001, February 1989.

EPA-89/03, *Hospital incinerator operating training course: volume I, student handbook*, EPA 450/3-89-003, March 1989.

EPA-92/01, *Survey of medical incinerators and emissions control*, final draft report prepared by EER Corporation for the California Air Resources Board and US EPA, January 1992.

P • A • R • T • 4

CHAPTER 4.1
AIR TOXIC RISK ASSESSMENT

Jesse L. Thé, David A. Weeks

1 GENERAL

This chapter presents calculation methods to estimate the risk to human health from air toxics. The assessment of risk is described by several guidance documents published by the United States Environmental Protection Agency (USEPA). The assessment of risk from contaminants released by sources of air toxics is different than typical risk assessments such as those performed at Superfund sites or facilities with hazardous waste treatment, storage, and disposal permits, because the concentration of contaminants in environmental media is estimated through the use of models rather than the collection of environmental samples. This does not preclude the collection of environmental samples, but modeling rather than sample collection is preferred in many instances because it is less costly to implement.

Because of this unique difference, the main component of air toxic risk assessment is modeling the fate and transport of air toxics in the environment. Air toxic risk assessment models describe the movement of air toxics from stacks and fugitive emission sources to the air, soil, sediment, and water. The models then describe the movement of contaminants from water and sediment into fish and from soil into vegetables, grain, and forage and then into farm animals where they are consumed by humans. The ambient air concentration is also modeled for the purpose of assessment of direct inhalation risk of air toxics.

The primary reference that describes the fate and transport of air toxics in the environment is the Protocol for Human Health Risk Assessment at Hazardous Waste Combustion Facilities published as a final guidance by the USEPA in September, 2005. This document is commonly referred to as the HHRAP. Mr. David Weeks was the chief editor of the original draft of this document and Dr. Jesse was one of the significant contributors. Some of the text and examples presented here were originated by the authors, and some of them were edited from or quoted from previously published USEPA documents (described in the References section).

In addition to fate and transport modeling, this section also describes ingestion rates of air toxics, risk for cancer, and noncancer health impacts on humans. The base case for all the calculations presented in this section is presented below:

2 EMISSIONS AND SITE CHARACTERIZATION

The first step in the air toxics risk assessment process is the characterization of emissions. The characterization includes:

1. Compiling basic facility information.
2. Identifying emission sources.
3. Estimating air toxic emission rates.

2.1 Calculation Example Case

Throughout this chapter, an example case will be used to illustrate how the risk assessment calculations are performed. The example calculations will be presented for the chemicals 2,3,7,8 = tetrachlorodibenzodioxin (TCDD), silver and divalent mercury (often referred to as mercuric chloride). The emission rates of these three chemicals are 1.0×10^{-5}, 2.0×10^{-5}, and 1.44×10^{-5} g/s (speciated from a total mercury emission of 3×10^{-5} g/s), respectively, each from a single stack. A water body with an area of 1×10^6 m^2 is located nearby, and a watershed that is 3×10^6 m^2 in area surrounds it. A multitiered grid of air modeling receptors was used to collect air parameters. The calculations use parameters specific to the adult resident scenario (Fig. 1.1).

2.2 Compiling Basic Facility Information

The basic facility information that is necessary to complete an air toxics risk assessment can be grouped into three components:

- Chemicals used or manufactured at the facilty.
- Location of emission sources.
- Mechanism of release to the environment.

Chemicals used or manufactured. The chemicals used or manufactured at the facility must be identified because this information is necessary to determine which air toxics will be the focus of the risk assessment. The chemicals may be released as fugitive emissions from manufacturing or waste

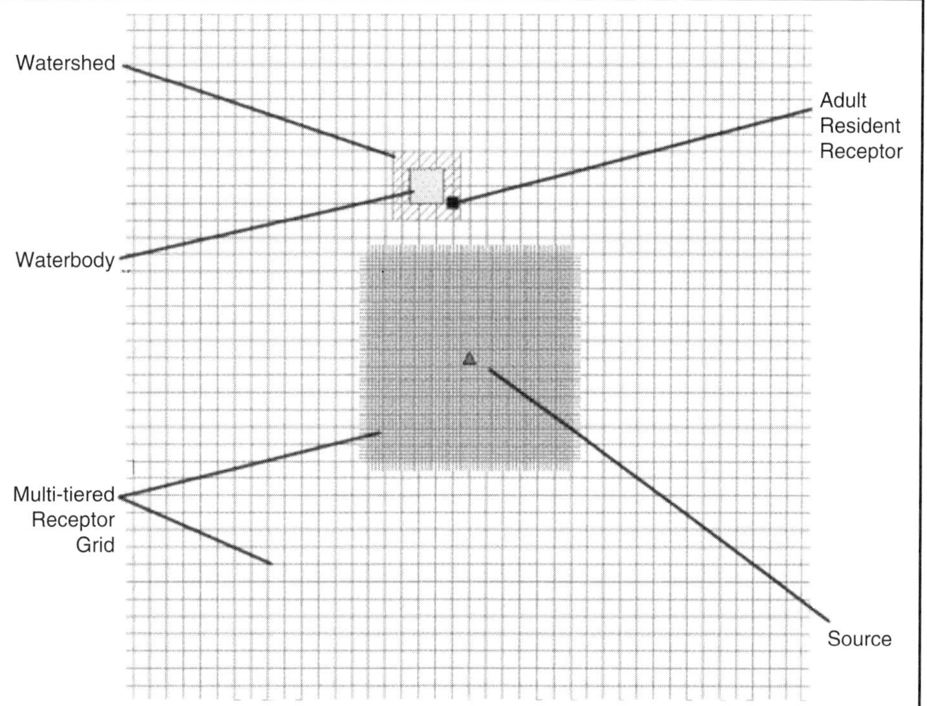

FIGURE 1.1 Site layout.

treatment processes or as products of incomplete combustion (PICs) from combustion processes. Chemicals that are not used or manufactured at the facility or cannot be generated as PICs from combustion processes do not need to be evaluated in the risk assessment. For example, pesticides should not be evaluated at facilities that do not manufacture pesticides because these chemicals are not part of the manufacturing process and are not known to be formed as PICs from combustion processes.

Location of emission sources. The geographic location of emission sources is an important component of the air modeling component of the risk assessment process, especially at large facilities that encompass large geographic areas. The location of emission sources is important because the transport of the air toxics from the release point is characterized by models that require the point of the release as one of the modeling input parameters. The geographic location of emission sources is typically characterized by coordinates, preferably in units of Universal Transverse Mercator (UTM), in meters.

Mechanism of release. The mechanism of release to the environment is important because it affects the mass emission rate of air toxics. The emission rate of air toxics from stacks is determined differently from the emission rate of air toxics from fugitive sources like valves and flanges.

In many cases the information needed to complete an air toxic risk assessment can be obtained from state environmental agency emission inventories. In other cases, emission rates must be estimated based on test data or with other common engineering techniques described elsewhere in this text.

2.3 Identifying Emission Sources

The sources of air toxic emissions include PICs from combustion devices such as utility boilers, thermal oxidizers, and incinerators. Other sources of air toxic emissions include fugitive emissions from manufacturing processes, wastewater treatment, spills, and transportation.

Estimating emission rates from stacks. The emission of air toxics from stacks is typically characterized by testing. These emission tests describe the concentrations of contaminants in the stack gas at dry standard conditions in terms of mass per unit volume (e.g., $\mu g/m^3$). The mass rate of air toxic emissions is calculated as the product of the stack gas concentration and the stack gas flow rate also at dry standard conditions.

$$Q = SGF \times SGC \times CFO_2/10^6 \tag{1.1}$$

where

Q = COPC(Compounds of potential concern)-specific emission rate (g/s)
SGF = Stack gas flow rate at dry standard conditions (dscm/s)
SGC = COPC stack gas concentration at 7 percent O_2 as measured in the stack test (μg/dscm)
CFO_2 = Correction factor for conversion to actual stack gas concentration O_2 (unitless)
1×10^6 = Unit conversion factor (μg/g)

It should be noted that the oxygen correction factor is equal to 1.0 if the stack gas concentration is not corrected to 7 percent oxygen for the purpose of evaluating compliance with other regulatory standards.

Emissions from process upsets. USEPA typically recommends that stack emissions be adjusted for process upsets when air toxic risk assessments are performed in support of hazardous waste combustion permits. Process upsets can occur during start-ups, shutdowns, and malfunctions of the combustion unit or air pollution control system (APCS). Upset emissions are generally expected to be greater than stack emissions (over short periods of time) because the process upsets results in incomplete destruction of the wastes or other physical or chemical conditions within the system that promote the formation and/or release of hazardous compounds.

To account for the increased emissions associated with process upsets, the stack emission rate derived from stack test data is multiplied by an upset factor. The following types of data may be considered and evaluated to derive upset factors:

- Data for continuous emissions monitoring systems that measure stack carbon monoxide, oxygen, total hydrocarbon (if required), or opacity (if appropriate)
- Data on combustion chamber, APCS, or stack gas temperature
- Frequency and causes of automatic waste feed cutoffs (AWFCO)
- APCS operating variables such as baghouse pressure drop, liquid scrubber flow rate, or electrostatic precipitator voltage
- Stack test collected while the combustion unit was operated under upset conditions

This information can be used to estimate the magnitude of the increase in emissions and the percentage of time on an annual basis that the unit operates at upset conditions. When site specific data are not available upset emissions can be estimated using the California Air Resources Board reference (CARB, 1990).

Estimating emissions from process upsets. To represent stack emission rates during process upsets, multiply the emission rate developed from the stack test data by 2.8 for organics and 1.45 for metals. These factors are derived by assuming that emissions during process upsets are 10 times greater than emissions measured during the stack test. Since the unit does not operate under upset conditions continually, the factor must be adjusted to account for only the period of time, on an annual basis, that the units operates under upset conditions:

1. For organic compounds, the facility is assumed to operate as measured during the stack test 80 percent of the year and operate under upset conditions 20 percent of the year:

$$[(0.80)(1) + (0.20)(10) = 2.8].$$

2. For metals, the combustion unit is assumed to operate as measured during the stack test 95 percent of the year and operate under upset conditions the remaining 5 percent of the year:

$$[(0.95)(1)+(0.05)(10) = 1.45].$$

Catastrophic process upsets brought about by complete failure of combustion and APCS resulting from nonroutine events such as explosions, fires, and power failures are considered accidental releases. The emissions associated with these events are estimated using different methods that are beyond the scope of this chapter.

Quantitative estimation of fugitive emissions from process equipment. Quantitative estimation of fugitive emissions includes: (1) identifying equipment to be evaluated as fugitive emission source(s); (2) grouping equipment, as appropriate, into a combined source; and (3) estimating compound specific emission rates for each source. Multiple steps are commonly necessary to estimate fugitive emissions.

Step 1. Identifying fugitive emission sources: Equipment types covered under a fugitive emission analysis include the following:

1. Pumps
2. Valves
3. Connectors (flanges, unions, tees, and so on.)
4. Compressors
5. Pressure-relief devices
6. Open-ended lines
7. Product accumulator vessels
8. Sampling connecting systems

9. Closed vent systems
10. Agitators

Step 2. Estimating fugitive emissions from process equipment: Based on guidelines provided in USEPA (1995e), *"Protocol for Equipment Leak Emission Estimates, EPA-453/R-93-017,"* fugitive emissions for each equipment listed under 40 CFR Part 265, Subpart BB can be estimated by the following four approaches, in order of increasing refinement and data requirements:

1. Average emission factor approach (AEFA), which is the least refined of these methods
2. Screening ranges approach (SRA)
3. USEPA correlation approach (EPACA)
4. Unit-specific correlation approach (USCA)

These four approaches are used to estimate fugitive emission rates of volatile organic compounds (VOCs) from process equipment. Because data on fugitive emissions at a facility is generally limited, the AEFA method will apply in most cases.

An example calculation using the AEFA method. Information required for estimating fugitive emission rates using the AEFA method is as follows:

1. Type of waste stream associated with each equipment type
   - Light liquids are those in which the sum of the concentration of individual constituents with a vapor pressure over 0.3 kPa at 20°C is greater than or equal to 20 percent by weight
   - Heavy liquids are all others liquids not meeting the definition of light liquids as specified above
2. Number of each equipment type associated with each waste stream
3. Total VOC weight fraction of each waste stream
4. Weight fraction of each VOC in each waste stream
5. Operational time period of equipment

When this approach is used, equipment can be grouped by waste streams of similar characteristics and VOC composition. Site-specific factors can significantly influence fugitive emission rates of leaks from equipment.

The average emission factors for synthetic organic chemicals manufacturing industry process units, refineries, and natural gas plants are presented in USEPA (1995k). Table 1.1 is an excerpt from this guidance document. These emission factors are most valid for estimating rates of emissions from a grouping of equipment over a long time period.

TABLE 1.1 Emissions Factors for Fugitive Emissions from Pipelines

Equipment type	Service	Emission factor (kg/hr/source)
Valves	Gas	0.00597
	Light liquid	0.00403
	Heavy liquid	0.00023
Pump seals	Light liquid	0.0199
	Heavy liquid	0.00862
Compressor seals	Gas	0.228
Pressure relief valves	Gas	0.104
Connectors	All	0.00183
Open-ended lines	All	0.0017
Sampling connectors	All	0.0150

Source: USEPA (1993e).

The total VOC emissions rate for a specified equipment type can be calculated by multiplying the equipment emission factor by the total VOC weight fraction and the number of each equipment type per waste stream. Speciated fugitive emissions can then be calculated by multiplying the weight fraction of each VOC in the waste stream and the total fugitive emission rate for the waste stream.

Hazardous air pollutants (HAPs) fugitive emissions—unit leaks. Examples of fugitive emissions from combustion unit leaks include the following:

- Combustion units that operate under negative pressure may experience temporary positive pressures ("puffing") that cause fugitive emissions.
- Fugitive emissions resulting from the day-to-day operation of the combustion unit and APCS. These emissions will typically include: (1) leaks that occur due to a positive pressure in the APCS, and (2) routine maintenance activities such as replacement of baghouse collection bags.

Fugitive particulate emissions. The combustion of fuels and hazardous waste materials may result in the production of flyash. Fugitive particle emissions may result from the subsequent collection, handling, and disposal of the flyash.

Quantitative estimation of fugitive particulate emissions. Steps for the quantitative estimation of RCRA fugitive ash emissions include:

- Determining an empirical emission factor.
- Estimating the flyash generation rate.
- Accounting for air pollution control equipment, if applicable.

As demonstrated in the example calculation below, the fugitive particulate emission rate can be estimated by multiplying the empirical emission factor by the flyash generation rate and the control deficiency of the air pollution control equipment.

Step 1. *Determining an empirical emission factor*
Particle emissions associated with flyash loading and unloading can be estimated using an empirical emission factor of 1.07 lb/ton flyash. Because the combustion of coal and hazardous wastes are similar activities, flyash generated from similar control devices is expected to behave similarly under the same conditions, with respect to fugitive emissions. The emission factor determined during the empirical study (0.107 lb/ton flyash) can be adjusted by a factor (e.g., 10) to account for the fact that the flyash from the combustion of coal (as in the study) was wetted.

Step 2. *Estimating the flyash generation rate*
The flyash generation rate from the APCD **(air pollution control district)** can be obtained from the Permit Application or the equipment manufacturer. Both values should be approximately the same. Since a major portion of ash fed to the combustor is converted to bottom ash, it is likely that this value is a conservatively high estimate of the actual flyash generation rate.

Step 3. *Accounting for air pollution control equipment*
An APCD is used for controlling emissions during exaust handling operations, such as flyash, an efficiency factor (e.g., 99.5 percent) can be applied to the emission rate.

Example calculation to estimate the particle fugitive emissions.

Step 1. Determine the empirical emission factor.

Step 2. Estimate the flyash generation rate. Ash emission rate is calculated by multiplying the empirical emission factor times the estimated flyash generation rate

$$(1.07 \text{ lb/ton}) \times (5000 \text{ tons/yr}) = 5350 \text{ lb/yr}$$

Step 3. Account for the air pollution control equipment removal efficiency:
The product of Steps 1 and 2 is multiplied times one minus the fabric filter efficiency to obtain the final fugitive ash emission rate for use in the risk assessment

$$(5350 \text{ lb/yr}) \times (1 - 0.995) = 26.75 \text{ lb/yr}$$

2.4 Identifying Compounds of Potential Concern

Compounds of Potential Concern (COPCs) are those air toxics evaluated throughout the risk assessment. COPCs are volatile organic compounds (VOCs), metals, and PICs. COPCs have typically been identified by USEPA in seven different general categories (US EPA 1994g; 1994i; 1994j; 1994n):

1. Polychlorinated dibenzo(*p*)dioxins (PCDD) and polychlorinated dibenzofurans (PCDF)
2. Polynuclear aromatic hydrocarbons (PAH)
3. Polychlorinated biphenyls (PCB)
4. Nitroaromatics
5. Phthalates
6. Other organics
7. Metals

Polychlorinated dibenzo(p)dioxins and dibenzofurans. PCDD and PCDF's are routinely evaluated in air toxic risk assessments that are performed on combustion sources because the de novo formation of these air toxics is well supported in the scientific literature. De novo formation occurs when PCDDs and PCDFs form in dry APCSs by reactions between halogens and undestroyed organic material from the furnace and are catalyzed by flyash. PCDDs and PCDFs were first discovered as thermal decomposition products of polychlorinated compounds, including:

1. Herbicide 2,4,5-T
2. Hexachlorophene
3. PCBs
4. Pentachlorophenol
5. Intermediate chemicals used to manufacture these compounds

A review of currently available dioxin data for combustion units reveals that total PCDD/PCDF emission rates vary by more than 28-fold between different facilities, even though they use similar combustion units and APCSs (USEPA 1996b). Site-specific emission data are needed to enable completion of a more refined risk assessment at each combustion unit.

Polynuclear aromatic hydrocarbons. The HHRAP states that PAHs are common products of incomplete combustion. However, the authors note that their anectdotel experience is that carcinogenic PAHs are not typically detected at concentrations above the detection limit when common stack test methods are used to quantify emissions from well-operated combustion units. (Noncarcinogenic PAHs such as naphthalene and pyrene, however, are routinely detected.) The carcinogenic PAHs include the following: benzo(a)pyrene (BaP); benzo(a)anthracene; benzo(b)fluoranthene; benzo(k)fluoranthene; chrysene; dibenz(a,h)anthracene; and indeno(1,2,3-cd)pyrene. USEPA considers all of these compounds to be carcinogenic; all except chrysene are known to be animal carcinogens. However, an oral cancer slope factor is only available for one PAH, benzo(a)pyrene. The risk associated with the remaining six carcinogenic PAHs is assessed through the use of relative potency factors that are described in the HHRAP.

Volatile organic compounds. USEPA (1990e) reported that volatile organics listed as probable PICs produced by the combustion of hazardous waste include benzene, chloroform, tetrachloroethylene, 1,1,1-trichloroethane, toluene, and methylene chloride. The author's anectdotel experience is that VOCs are the most common types of PICs detected in combustion stack gas. However, VOCs do not bioaccumulate in the environment and, thus, are not typically a significant concern for air toxics risks through the food chain. The quantities of VOCs emitted from combustion stacks are typically so minor that they rarely drive the outcome of air toxic risk assessments, although each case should be evaluated separately. However, the risk associated with the direct inhalation of VOCs from fugitive and other emission sources is much more problematic and in many cases exceeds the risk from indirect sources (e.g., food chain) for combustion units.

Metals other than mercury. Metals are a component of the inorganic air toxics and can include: antimony, arsenic, barium, beryllium, cadmium, chromium, hexavalent chromium, lead, nickel, selenium, silver, thallium, and zinc. Unlike other COPCs, the emission rate of metal air toxics is significantly affected by the feed rate of metals into the combustion unit and the removal efficiency of the air pollution control equipment.

Mercury. USEPA commonly recommends that mercury be evaluated as a COPCs in air toxic risk assessment, especially for coal fired units and units that burn waste containing mercury. It is the author's experience that mercury is the most problematic of all the air toxics because of its high bioaccumulation properties and extremely low reference dose. Indeed, even facilities that do not use mercury in their processes may have issues with it at concentrations close to the detection limit in air toxic risk assessments if the facility is located near shallow water bodies that can support recreational fishing. Additionally, air emissions of mercury contribute to local, regional, and global deposition. This fact is important when conducting air toxic risk assessments at specific facilities because not all the mercury emitted by one facility will be deposited in the area immediately around the facility.

The speciation of mercury emissions is thought to depend on the fuel used, flue gas cleaning, and operating temperatures. Total mercury exiting the stack is assumed to consist of elemental and divalent species, with no emissions of methylmercury assumed. The divalent fraction is split between vapor- and particle-bound phases. Much of the divalent mercury is thought to be mercuric chloride ($HgCl_2$) (USEPA 1997d).

Phase allocation and speciation of mercury exiting the stack. Mercury emissions from stacks include both vapor and particle-bound forms; and speciated as both divalent (HG^{2+}) and elemental mercury (HG^0). USEPA recommends a conservative approach that assumes phase allocation of mercury emissions from hazardous waste combustion of 80 percent of the total mercury in the vapor phase and 20 percent of total mercury in the particle-bound phase. This allocation is:

- Consistent with mercury emissions speciation data for hazardous waste combustion sources reported in literature (Peterson et al. 1995).
- Believed to be reasonably conservative, since it results in the highest percentage of total mercury being deposited in proximity to the source, and therefore, indicative of the maximum indirect risk.

The divalent form emitted, either in the vapor phase or particle-bound, are removed faster from the atmosphere than elemental mercury. In addition, vapor phase divalent mercury is thought to be more rapidly and effectively removed by both dry and wet deposition than particle-bound divalent mercury, as a result of the reactivity and water solubility of vapor divalent mercury.

Vapor phase mercury. Only a small fraction (assumed to be 1 percent) of vapor-phase elemental mercury either is adsorbed to particulates in the air and is deposited or converted to the divalent form to be deposited. Of the 60 percent vapor phase divalent mercury, about 68 percent is deposited and about 32 percent is transported outside of the United States or is vertically diffused to the free atmosphere to become part of the global cycle (USEPA 1997d).

Particle-bound mercury. Of the 20 percent of the total mercury that is particle-bound, 99 percent is in the divalent form. (USEPA 1997d) indicates that only 36 percent of the particle-bound divalent mercury is deposited, and the rest is either transported outside of the United States or is vertically diffused to the free atmosphere to become part of the global cycle.

Deposition and modeling of mercury. USEPA assumes that deposition to the various environmental media is entirely divalent mercury in either vapor or particle-bound form. Without consideration of the global cycle, mercury speciations will result in 80 percent of the total mercury emitted, being deposited as divalent mercury and the remaining 20 percent being deposited as elemental mercury. Divalent mercury is considered for both the indirect exposure and inhalation pathways. Based on these assumptions, human exposure to (1) elemental mercury occurs only through direct inhalation of the vapor phase elemental form, and (2) divalent mercury occurs through both indirect exposure and direct inhalation of the vapor and particle-bound mercuric chloride.

Methylation of mercury. Both watershed erosion and direct atmospheric deposition can be important sources of mercury to a water body (USEPA 1997d). Mercury entering the water body is methylated predominately through biotic processes (USEPA 1997d). In the absence of modeling site-specific water body properties and biotic conditions, Risk assessors assume that 85 percent of total mercury in surface water is assumed to be divalent mercury and the remaining mass as methyl mercury.

Note: Methylation can be highly variable between environmental systems. This results in a significant degree of uncertainty implicit in modeling of mercury methylation.

Figure 1.2 below depicts the partitioning of mercury in the various phases, along with the deposition percentages and the portion that the Mercury Report to Congress estimates that escapes to the global cycle.

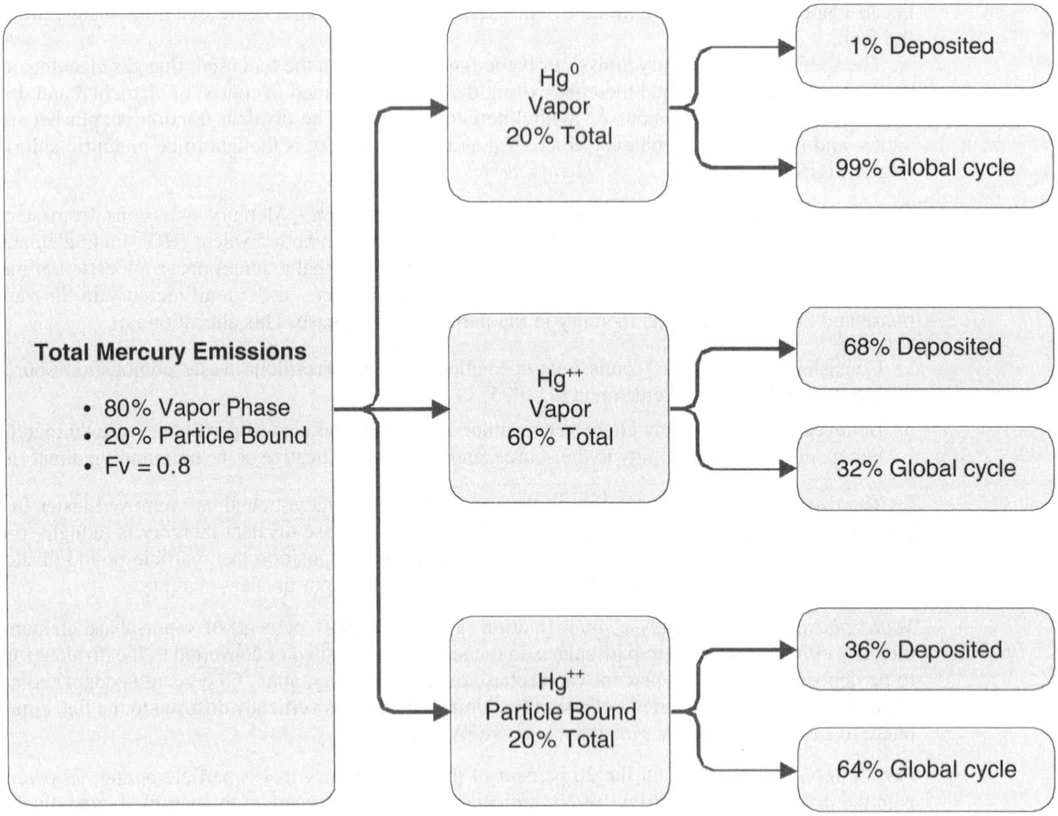

FIGURE 1.2 Mercury speciation onto the various phases and each phase deposition percentage.

2.5 Particulate Matter

Particulate matter (PM) can be classified as aerosols, fogs, fumes, mists, smogs, or smokes depending on its physical state and origin. PM_{10} is defined as all condensed material suspended in air that has a mean aerodynamic diameter of 10 μm or less; $PM_{2.5}$ is the part of PM_{10} that is 2.5 mm or less in diameter. Because of this wide range of toxicity, PM concentrations cannot be used to estimate the toxicity of emissions. USEPA does not recommend that PM be evaluated as a separate COPC in the risk assessment. However, PM is quite useful as an indicator variable, because it can be measured in real time and is sensitive to changes in operating conditions.

3 AIR DISPERSION MODELLING FOR AIR TOXICS

Estimation of potential human health risks associated with air toxics releases requires knowledge of atmospheric pollutant concentrations and annual deposition rates at actual and reasonable future exposure scenario locations. Air concentrations and deposition rates are estimated by using air dispersion models (ADMs). Air dispersion models are mathematical constructs that estimate the dispersion of gaseous and particulate emissions from sources of HAPs such as stacks, fugitive, area, mobile, among others.

This section provides brief guidance on the use of the standard USEPA ADMs, which may be used in most risk assessments, AERMOD (American Meteorological Society/EPA regulatory model) and CALPUFF (California Puff model). ADMs require the use of the following information for input into the model, and consideration of output file development:

1. Site-specific characteristics required for air modeling
   - Surrounding terrain
   - Surrounding land use
   - Facility building characteristics
2. Unit emission rate
3. Partitioning of emissions
4. Meteorological data

ADMs also require several preprocessing support computer programs that prepare and organize data for use in the model.

3.1 Background on ADMs for Risk Assessment

Two air models in current regulatory use are AERMOD and CALPUFF. These ADMs satisfy the following criteria:

1. Broad regulatory acceptance and experience in evaluating the impacts of air contaminants emitted from various industrial sources (e.g., short and tall stack heights, fugitive emissions from process and storage areas).
2. Can evaluate impacts attributable to a single source, with capability for evaluating multiple sources.
3. Can evaluate compounds emitted as vapor or particulate phase with consideration of deposition and removal processes based on the physical characteristics of the compound.
4. Accept placement of grid nodes at any location within the applicable range of the air model.
5. Can conduct stepwise evaluation of hourly meteorological conditions for multiple years of data producing short-term (acute) and long-term (chronic) averages as outputs.
6. Can evaluate building downwash effects.

The USEPA recommends the AERMOD for distances up to 50 km and CALPUFF for long-range transport (greater than 50 km and less than 200 km), and for the special conditions of very light or calm winds (less than 1 m/s) and areas with strong coastal effects, on a case-by-case basis. CALPUFF may be an important consideration in cases for evaluating chemical transformation of highly reactive contaminants based on long residence times during extended periods of light winds or calm conditions, transport over long distances, recirculation of contaminants due to local wind effects, coastal areas, and in hilly terrain and river valleys.

Preprocessing programs. Both AERMOD and CALPUFF require the use of additional computer programs, referred to as *preprocessors*. These programs manipulate available information regarding surrounding buildings and meteorological data into a format that can be used by ADM. Currently, these programs include the following:

- Building Profile Input Program (BPIP) calculates the maximum crosswind widths of buildings, which AERMOD then uses to estimate the effects on air dispersion. This effect on dispersion by surrounding buildings is typically known as building downwash or wake effects. The BPIP User's Guide contains detailed information for preparing the required building dimensions (length, height, and width) and locations for the AERMOD model.
- AERMET (American Meteorological Society/EPA regulatory meteorological processor) prepares meteorological data for use in the AERMOD. AERMET merges hourly measurements of surface parameters (e.g., precipitation, wind speed, and wind direction) with upper air soundings to generate micro-meteorology turbulence parameters, which will be used by AERMOD.

When sufficient site-specific data is available on the location, size and shape of structures in the vicinity of the emissions sources, the BPIP program should be used to prepare the AERMOD input file to consider the effects of building downwash on pollutant transport. The inclusion or omission of nearby structures should be approved by the permitting authority prior to performing the air dispersion modeling.

3.2 Site-Specific Information Required to Support Air Modeling

Site-specific information for the facility and surrounding area required to support air dispersion modeling includes:

1. The elevation of the surrounding land surface or terrain
2. Surrounding land uses
3. Characteristics of on-site buildings that may affect the dispersion of COPCs into the surrounding environment.

Often, site-specific information required to support air dispersion modeling can be obtained from review of available maps and other graphical data on the area surrounding the facility. The first step in the air modeling process is a review of available maps and other graphical data on the surrounding area. U.S. Geological Survey (USGS) 7.5-min topographic maps (1:24,000) extending to 10 km from the facility, and USGS 1:250,000 maps extending out to 50 km, should be obtained to identify site location, nearby terrain features, waterbodies and watersheds, ecosystems, nearby residences, and land use. Aerial photographs are frequently available for supplemental depiction of the area. An accurate facility plot plan—showing buildings, stacks, property, and fence lines—is also needed. Facility information including stack and fugitive source locations, building corners, plant property, and fence lines should be provided in Universal Transverse Mercator (UTM) grid coordinates in meters east and north in both USGS reference systems.

Surrounding terrain information. Terrain is important to air modeling because air concentrations and deposition rates are greatly influenced by the height of the plume above local ground level. For air modeling purposes, terrain is referred to as *complex* if the elevation of the surrounding land within the assessment area—typically defined as anywhere within 50 km from the stack—is above the top of the stack evaluated in the air modeling analysis. USEPA recommends that most air modeling include terrain elevations for every receptor grid node. Some exceptions may be those sites characterized by very flat terrain.

In addition to maps which are used to orient and facilitate air modeling decisions, the digital terrain data used to extract receptor grid node elevations should be provided in electronic form. One method of obtaining receptor grid node elevations is using digital terrain data available from the USGS on the Internet at web site *http://www.WebGIS.com*.

Surrounding land use information. Land use information in the risk assessment is used for purposes of air dispersion modeling and the identification or selection of exposure scenario locations in the risk assessment. Land use analysis for purposes of selecting exposure scenario locations usually occurs out to a radius of 50 km from the centroid of the stacks to ensure identification of all receptors that may be impacted. However, in most cases, air modeling performed out to a radius of 10 km allows adequate characterization for the evaluation of exposure scenario locations. If a facility with multiple stacks or emission sources is being evaluated, the radius should be extended from the centroid of a polygon drawn from the various stack coordinates.

3.3 Use of Unit Emission Rate

In risk assessment studies from air toxics sources the ADM is usually run with a unit emission rate of 1.0 g/s in order to preclude having to run the model for each specific chemical. The unitized concentration and deposition output from the ADM, using a unit emission rate, are adjusted to the COPC-specific air concentrations and deposition rates in the estimating media concentration equations by using COPC-specific emission rates obtained during the stack test. Concentration and deposition are directly proportional to the unit emission rate used in the air dispersion modeling.

3.4 Partitioning of Emissions

COPC emissions to the environment occur in either vapor or particle phase. In general, most metals and organic COPCs with very low volatility (refer to fraction of COPC in vapor phase [Fv] less than 0.05) are assumed to occur only in the particle phase. Organic COPCs occur as either only vapor phase (refer to Fv of 1.0) or with a portion of the vapor condensed onto the surface of particulates (e.g., particle-bound). COPCs released only as particulates are modeled with different mass fractions allocated to each particle size than the mass fractions for the organics released in both the vapor and particle-bound phases.

Vapor phase modeling. AERMOD output for vapor phase air modeling runs are vapor phase ambient air concentration, dry vapor deposition, and wet vapor deposition at receptor grid nodes based on the unit emission rate. Vapor phase runs do not require a particle size distribution in the AERMOD input file.

Particle phase modeling (mass weighting). Particle size is the main determinant of the fate of particles in air flow, whether dry or wet. The key to dry particle deposition rate is the terminal, or falling, velocity of a particle. Particle terminal velocity is calculated mainly from the particle size and particle density. Large particles fall more rapidly than small particles and are deposited closer to the stack. Wet particle deposition also depends on particle size as larger particles are more easily removed, or scavenged, by falling liquid (rain) or frozen (snow or sleet) precipitation.

The diameters of small particulates contained in stack emissions are usually measured in micrometers (μm). Unfortunately, few studies have been performed to directly measure particle size distributions from a variety of stationary combustion sources.

Note: Actual stack test data will be different from the hypothetical values, because of the use of particle "cut size" for the different cascade impactor filters (or Coulter counter-based distributions) used during actual stack sampling. However, because ADMs require mean particle diameter for each particle size distribution, and the stack test data identifies only the mass (weight) of particles in a range bounded by two specific diameters, stack test data must be converted into a mean particle diameter which approximates the diameter of all the particles within a defined range. Consistent with USEPA 1993h, the mean particle diameter is calculated by using the following equation:

$$D_{Mean} = (0.25 \times (D_1^3 + D_1^2\, D_2 + D_1\, D_2^2 + D_2^{\,3}))^{1/3} \tag{1.2}$$

where

D_{mean} = Mean particle diameter for the particle size category (μm)
D_1 = Lower bound cut of the particle size category (μm)
D_2 = Upper bound cut of the particle size category (μm)

Example for mass weighted particulate fraction. For example, the mean particle diameter of 5.5 μm in Table 1.2 is calculated from a lower bound cut size (assuming a cascade impactor is used to collect the sample) of 5.0 μm to an upper bound cut size of 6.15 μm. In this example, the mean particle diameter is calculated as:

From Table 1.2, the mean particle diameter is 5.5 μm. The mass of particulate from the 5.0 μm stack test data is then assigned to the 5.5 μm mean particle diameter for the purpose of computing the "fraction of total mass."

Typically, 8 to 10 mean particle diameters are available from stack test results and should be used in the air modeling. A minimum of three particle size categories (> 10 μm, 2 to 10 μm, and < 2 μm) detected during stack testing are required for air modeling. For facilities with stack test results which indicate mass amounts lower than the detectable limit (or the filter weight is less after sampling than before), a single mean particle size diameter of 1.0 μm should be used to represent all mass (e.g., particle diameter of 1.0 μm or a particle mass fraction of 1.0) in the particle and particle-bound model runs.

After calculating the mean particle diameter (Col. 1), the fraction of total mass (Col. 4) per mean particle size diameter must be computed from the stack test results. For each mean particle diameter, the stack test data provides an associated mass of particulate. The fraction of total mass for each mean particle diameter is calculated by dividing the associated mass of particulate for that diameter by the total mass of particulate in the sample. In many cases, the fractions of total mass will not sum to 1.0 due to rounding errors. In these instances, USEPA advocates that the remaining mass fraction be added into the largest mean particle diameter mass fraction to force the total mass to 1.0.

TABLE 1.2 Hypothetical Particle Size Distribution Data for Example

Mean particle diameter [a] (μm)	Particle radius (μm)	Surface area/volume (μm^{-1})	Fraction of total mass	Proportion available surface area	Fraction of total surface area
> 15.0	7.50	0.400	0.128	0.0512	0.0149
12.5	6.25	0.480	0.105	0.0504	0.0146
8.1	4.05	0.741	0.104	0.0771	0.0224
5.5	2.75	1.091	0.073	0.0796	0.0231
3.6	1.80	1.667	0.103	0.1717	0.0499
2.0	1.00	3.000	0.105	0.3150	0.0915
1.1	0.55	5.455	0.082	0.4473	0.1290
0.7	0.40	7.500	0.076	0.5700	0.1656
< 0.7	0.40	7.500	0.224	1.6800	0.4880

*The terms mass and weight are used interchangeably when using stack test data.
Source: The geometric mean diameter in a distribution from USEPA (1980a), as presented in USEPA (1993h).

Direct measurements of particle-size distributions at a proposed new facility may be unavailable, so it will be necessary to provide assumed particle distributions for use in the ADM. In such instances, a representative distribution may be used. The unit on which the representative distribution is based should be as similar as practicable to the proposed unit.

After developing the particulate size distribution based on mass, this distribution is used in the ADM to apportion the mass of particle phase COPCs (metals and organics with F_v values less than 0.05) based on particle size. Col. 4 of Table 1.2 (as developed from actual stack test data) is used in the ADM input file to perform a particulate run with the particle phase COPCs apportioned based on mass weighting.

Particle-bound modeling (surface area weighting). A surface area weighting, instead of mass weighting, of the particles is used in separate particle runs of the ADM. Surface area weighting approximates the situation where a semivolatile organic contaminant has been volatilized in the high temperature environment of a combustion system and then condensed to the surface of particles entrained in the combustion gas after it cools in the stack. Thus, the apportionment of emissions by particle diameter becomes a function of the surface area of the particle that is available for chemical adsorption (USEPA 1993h).

Example calculation of surface weighted particle fraction. The first step in apportioning COPC emissions by surface area is to calculate the proportion of available surface area of the particles. If particle density is held constant (such as 1 g/m^3), the proportion of available surface area of aerodynamic spherical particles is the ratio of surface area (S) to volume (V), as follows:

- Assume aerodynamic spherical particles.
- Specific surface area of a spherical particle with a radius, r $S = 4\ \pi r^2$.
- Volume of a spherical particle with a radius, r $V = 4/3\ \pi r^3$.
- Ratio of S to V $S/V = 4\ \pi r^2/(4/3\ \pi r^3) = 3/r$.

The following uses the particle size distribution in Table 1.2 as an example of apportioning the emission rate of the particle-bound portion of the COPC based on surface area. This procedure can be followed for apportioning actual emissions to the actual particle size distribution measured at the stack. In Table 1.2, a spherical particle having a diameter of 15 μm (Col. 1) has a radius of 7.5 μm (Col. 2). The proportion of available surface area (assuming particle density is constant) is 0.400(S/V = 3/7.5), which is the value in Col. 3. Col. 4 shows that particles with a mean diameter of 15 μm, constitute 12.8 percent of the total mass. Multiplication of Col. 3 by Col. 4 yields a value of 0.0512 in Col. 5. This value is an approximation of the relative proportion of total surface area, based on the percent of particles that are 15 μm in diameter. The sum of Col. 5 yields the total surface area of all particles in the particle size distribution. In this example, the sum is 3.4423. Col. 6 is the fraction of total surface area represented by the specific particle diameter in the distribution, and is calculated by dividing the relative proportion of surface area (Col. 5) for a specific diameter by the total relative proportion of surface area (3.4423 square micrometers [μm^2]). In the example of the 15-μm-diameter particle, the fraction of total surface area available for adsorption is 0.0149 (0.0512/3.4423). This procedure is then repeated for all particle sizes in the array.

After developing the particulate size distribution based on surface area, this distribution is used in the ADM to apportion mass of particle-bound COPCs (most organics) based on particle size. Col. 6 of Table 1.2 (as developed from actual stack test data) is used in the ADM input file to perform a particulate run for the particle-bound COPCs apportioned based on surface area weighting.

3.5 Meteorological Data

To model air concentration and deposition, ADMs require a variety of meteorological information:

1. Air concentration
   A. Hourly values
      1. Wind direction (degrees from true north)
      2. Wind speed (m/s)

3. Dry bulb (ambient air) temperature (K)
4. Opaque cloud cover (tenths)
5. Cloud ceiling height (m)

B. Daily values
1. Morning mixing height (m)
2. Afternoon mixing height (m)

2. Deposition
A. Dry particle deposition—hourly values for surface pressure (millibars)
B. Wet particle deposition—hourly values
1. Precipitation amount (in)
2. Precipitation type (liquid or frozen)
C. Dry vapor deposition—hourly values for solar radiation (W/m^2)

Meteorological data are available from several different sources. For most air modeling, five years of data from a representative National Weather Service(NWS) station is recommended. However, in some instances where the closest NWS data is clearly not representative of site specific meteorological conditions, and there is insufficient time to collect 5 years of onsite data, 1 year of onsite meteorological data (consistent with guideline on air quality models (GAQM)) may be used to complete the risk assessment. The permitting authority should approve the representative meteorological data prior to performing air modeling.

Surface data meteorological data. Hourly surface data can be obtained from the National Climate Data Center (NCDC) web site at address *http://www.ncdc.noaa.gov/* or *http://WebMET.com* (WebMet). The recommended meteorological data is the SAMSON format. SAMSON data contain all of the required input parameters used by AERMOD and CALPUFF to compute concentration, dry and wet particle deposition, and dry and wet vapor deposition.

Air quality modeling regulations require complete meteorological data for the most recent 5 years. Note that many jurisdictions specify 5-year period for modeling, say from 2001–2006. It is desirable, but not mandatory, that the 5 years are consecutive. The use of less than 5 years of meteorological data should be approved by appropriate authorities. The following subsections describe important characteristics of the surface data.

3.6 ADM Input Files

Human health risk assessments are almost always presented to the public. Therefore, air dispersion modeling must specify input regulatory options for particle and particle-bound phase runs, as follows:

- Use stack-tip downwash (except for Schulman-Scire downwash).
- Use buoyancy-induced dispersion (except for Schulman-Scire downwash).
- Do not use final plume rise (except for building downwash).
- Use the calms processing routines.
- Use upper-bound concentration estimates for sources influenced by building downwash from super-squat buildings.
- Use default wind speed profile exponents.
- Use default vertical potential temperature gradients.

USEPA recommends a dry vapor deposition velocity value of 0.5 cm/s for COPCs other than mercury, and 2.9 cm/s for divalent mercury. For each of the three runs for each emission source, 5 years of off-site (e.g., NWS from SAMSON) meteorological data are completed. For sites with meteorological data collected on-site, the appropriate permitting authority should be notified for the data period required for a risk assessment. The averaging times should be specified to compute chronic annual average health risk, and maximum 1-hour average concentrations over the 5-year period.

Source information. Additional source characteristics required by the model include the following:

- Source type (point source for stack emissions; area or volume for fugitive emissions)
- Source location (UTM coordinates, m)
- Source base elevation
- Emission rate (1.0 g/s)
- Stack height (m)
- Stack gas temperature (K)
- Stack gas exit velocity (m/s)
- Stack inside diameter (m)
- Building heights and widths (m)
- Particle size distribution (percent)
- Particle density (g/cm^3)
- Particle and gas scavenging coefficients (unitless)

Source locations should be entered in UTM coordinates. The easterly coordinate is entered to the nearest meter(m); for example, 637524 m UTM-E. The northerly coordinate is entered to the nearest meter; for example, a northerly coordinate of 4,567,789 m UTM-N is entered as 4567789. The base elevation of each stack must be entered in meters. Base elevation may be obtained from a USGS topographic map, facility plot plans or USGS digital data base.

Particle size distribution. The ADM requires particle size distribution for determining deposition velocities.

Particle density. Particle density is also required for modeling the air concentration and deposition rates of particles. Site-specific measured data on particle density should be determined for all existing sources when possible. For new or undefined sources requiring air modeling, a default value for particle density of 1.0 g/cm^3 may be used. Particles from combustion sources, however, may have densities that are less than 1.0 g/cm^3 (USEPA 1994d), which would reduce the modeled deposition flux.

Receptor grids. The receptor grids identify sets or arrays of receptor grid nodes for which the ADM generates estimates of air parameters including air concentration, dry and wet deposition, and total deposition.

The ADM should include an array of receptor grid nodes covering the area within 10 km of the facility with the origin at the centroid of a polygon formed by the locations of the stack emission sources. This receptor grid node array should consist of a Cartesian grid with grid nodes spaced 100 m apart extending from the centroid of the emission sources out to 3 km from the centroid. For the distances from 3 km out to 10 km, the receptor grid node spacing can be increased to 500 m.

Terrain elevations should be specified for all receptor grid nodes. Several methods are available for assigning terrain elevations to grid nodes using digital terrain data. The 1:250,000 scale DEM digital data are available for download at the USGS Internet site *http://edcwww.cr.usgs.gov/pub/data/dem/250* or from WebGIS *http://www.webGIS.com*

This data has horizontal spacing between digital terrain values of approximately 90 m which provides sufficient accuracy for air modeling. In addition to the receptor grid node array evaluated for each facility out to 10 km, other grid node arrays may be considered for evaluation of water bodies and their watersheds located beyond 10 km. Grid node spacing of 500 m between nodes is recommended for grid node arrays positioned at distances greater than 10 km from the emission source. An equally spaced grid node array facilitates subsequent computation of area averages for deposition rates onto the waterbody and its associated watershed.

Meteorological data. For air modeling analysis which uses meteorological data from more than one surface station or upper air station (e.g., the upper air station is moved after the third year of the period and assigned a new station ID), the modeler should delete the headers for subsequent years in the combined file.

For sites where the anemometer height is changed during the 5-year period, the modeler should run each year separately to specify the correct anemometer height in the ADM input file, which corresponds to the correct height for that year of meteorological data.

3.7 Use of Modeled Output

The plot file is most useful for facilitating post-processing of the air parameter values in the model output. The plot file lists the x and y coordinates and the concentration or deposition rate values for each averaging period in a format that can be easily pulled into a postprocessing program (or spreadsheet).

The ADM output (air concentrations and deposition rates) are provided on a unit emission rate (1.0 g/s) basis and are not COPC-specific.

Unit rate output versus COPC-specific output. The relationship between the unit emission rate and the unit air parameter values (air concentrations and deposition rates) is linear. Similarly, the relationship between the COPC-specific emission rate (Q) and the COPC-specific air parameter values (air concentrations and deposition rates) would also be linear if the COPC-specific emission rate was used in the air model. The use of a unit emission rate in the air modeling is advocated because a common ratio relationship can be developed between the unit emission rate and the COPC-specific emission rate based on the fact that in the air model, both individual relationships are linear. This ratio relationship can be expressed by the following equation:

$$\text{Air Concentration of Chemical}(i) = (\text{Modeled Air concentration} \times \text{Emission Rate}) \text{ of Chemical}(i)$$

Use of this equation requires that three of the variables be known. The modeled output air concentration (or deposition rate) is provided by the air model, the unit emission is 1.0 g/s, and the COPC-specific emission rate; which is obtained directly from stack or source test data.

Use of model output in estimating media equations. The selection of which air-modeled air parameter values, air concentrations, and deposition rates to use in the estimating media concentration equations is based on how the chemical will partition into the vapor, particle, and particle-bound phases. The ADM parameters, coming from the ADM, are presented in Table 1.3.

TABLE 1.3 Air Parameters from the ADM Output

Air parameter	Description	Units
(Used for most soil-based exposure pathways)		
Cyv	Unitized yearly average air concentration from vapor phase	μg·s/g·m³
Cyp	Unitized yearly average air concentration from particle phase	μg·s/g·m³
Dydv	Unitized yearly average dry deposition from vapor phase	s/m²·yr
Dywv	Unitized yearly average wet deposition from vapor phase	s/m²·yr
Dydp	Unitized yearly average dry deposition from particle phase	s/m²·yr
Dywp	Unitized yearly average wet deposition from particle phase	s/m²·yr
(Used for fish and drinking water ingestion exposure pathways)		
Cywv	Unitized yearly (water body or watershed) average air concentration from vapor phase	μg·s/g·m³
Dytwv	Unitized yearly (water body or watershed) average total (wet and dry) deposition from vapor phase	s/m²·yr
Dytwp	Unitized yearly (water body or watershed) average total (wet and dry) deposition from particle phase	s/m²·yr
(Used for evaluation of acute risk via direct inhalation exposure pathway)		
Chv	Unitized hourly air concentration from vapor phase	μg·s/g·m³
Chp	Unitized hourly air concentration from particle phase	μg·s/g·m³
Chpb	Unitized hourly air concentration from particle-bound phase	μg·s/g·m³

Vapor phase COPCs. ADM output generated from vapor phase air-modeling runs are vapor phase air concentrations (unitized *Cyv* and unitized *Cywv*), dry-vapor deposition (unitized *Dydv*), and wet-vapor depositions (unitized *Dywv* and unitized *Dywwv*) for organic COPCs at receptor grid nodes based on the unit emission rate. These values are used in estimating media concentration equations for all COPC organics except the polycyclic aromatic hydrocarbons dibenzo(a,h)anthracene and indeno(1,2,3-cd)pyrene, which have vapor-phase fractions, *Fv*, less than 5 percent. The air concentration (unitized *Cyv*), dry-vapor deposition (unitized *Dydv*), and wet-vapor deposition (unitized *Dywv*) from the vapor phase run is also used in the estimating media concentration equations for mercury. Values for these COPCs are selected from the vapor phase run because the mass of the COPC emitted by the combustion unit is assumed to have either all or a portion of its mass in the vapor phase.

Particle-phase COPCs. The ADM result output generated from particle phase air-modeling runs are air concentration (unitized *Cyp*), dry deposition (unitized *Dydp*), wet deposition (unitized *Dywp*), and combined deposition (unitized *Dytwp*) for inorganics and relatively nonvolatile organic COPCs at receptor grid nodes based on the unit emission rate. These values are used in the estimating media concentration equations for all COPC inorganics (except mercury) and polycyclic aromatic hydrocarbons with fraction of vapor phase, *Fv*, less than 0.05 (e.g., dibenzo(a,h)anthracene and indeno(1,2,3-cd)pyrene). Values for inorganic and relatively nonvolatile COPCs are selected from the particle-phase run because the mass of the COPC emitted by the combustion unit is assumed to have all of its mass in the particulate phase, apportioned across the particle size distribution based on mass weighting.

Particle-bound COPCs. The ADM result generated from particle-bound air-modeling runs are air concentration (unitized *Cyp*), dry deposition (unitized *Dydp*), wet deposition (unitized *Dywp*), and combined deposition (unitized *Dytwp*) for organic COPCs and mercury at receptor grid nodes based on the unit emission rate. These values are used in estimating media concentration equations for all COPC organics and mercury to account for a portion of the vapor condensed onto the surface of particulates. Values for these COPCs are selected from the particle-bound run because the mass of the COPC emitted by the combustion unit is assumed to have a portion of its mass condensed on particulates, apportioned across the particle size distribution based on surface area weighting.

3.8 Modeling of Fugitive Emissions

Fugitive source emissions should be modeled using the procedures presented throughout this section for stack source emissions. However, the fugitive emissions should be represented as either "area" or "volume" source types. Fugitive emissions of volatile organics are modeled only in the vapor phase. Fugitive emissions of ash are modeled only in the particle and particle-bound phases, not vapor phase.

Fugitive emissions of volatile organic vapors are associated with processes that include storage vessels, pipes, valves, seals, and flanges. The height of the fugitive source is defined as the top of the vertical extent of the equipment. If the vertical extent of the fugitive source is not known, a default height of ground level (release height of zero) may be input, providing a conservative estimate of potential impacts.

3.9 Example ADM for Organic Fugitive Emissions

The following example is for organic fugitive emissions modeled as a volume source type, as presented in Figure 1.3. For a facility which may have two stack emission sources (S1 and S2) and two fugitive emission sources (areas F1, F2); a total of four runs for each year (or 5-year combined file) of meteorological data is required.

One run is required for each of the two stacks, S1 and S2, as point sources. One run is required for each of the two fugitive areas as volume sources (F1 and F2). Since the emissions are fugitive volatile organics, only the vapor phase is modeled. The vertical extent of the pipes, valves, tanks, and flanges

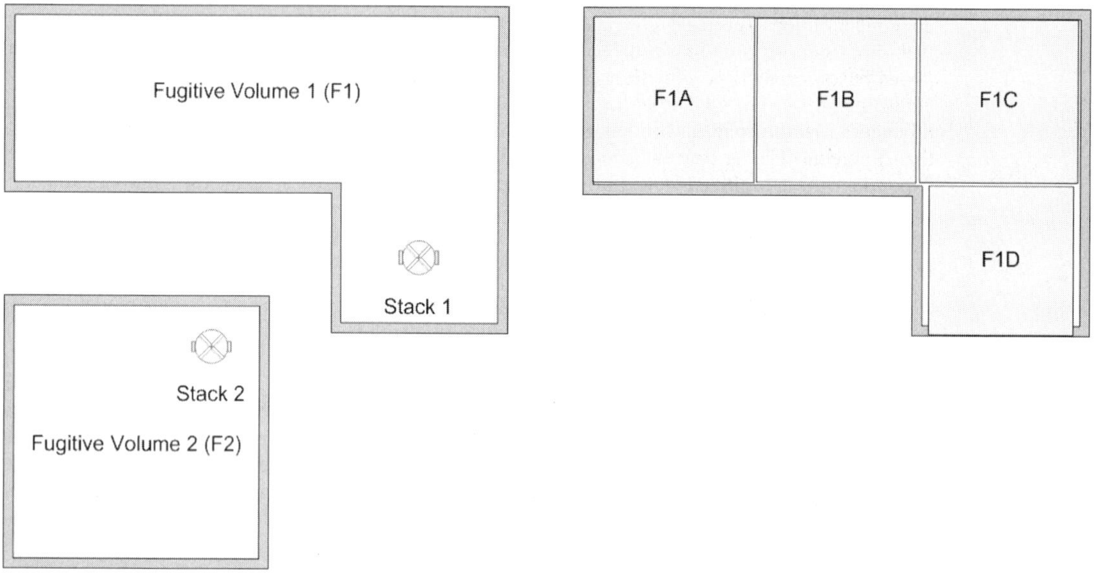

FIGURE 1.3 Fugitive sources and point sources.
Source: F1 represented by 4 volume sources.

associated with each fugitive emission area is 5 m above plant elevation. To define the sources for input to the ADM, the release height is specified as 2.5 m (half of vertical extent of fugitive emissions). The initial vertical dimension is specified as 2.36 m (vertical extent of 5 m divided by 2.15).

The initial horizontal dimension is the side length of the square fugitive area (footprint) divided by 4.3. If fugitive area F2 has a measured side of 30 m, the initial horizontal dimension is 6.98 (30 m divided by 4.3). For fugitive area F1, the area on the plot plan must be subdivided into 4 volume sources, F1A, F1B, F1C, and F1D. These volume sources are represented on the right side of Fig. 1.3 as squares, which may not precisely match the shape of the original fugitive source F1, as one can see on volume source F1D. The four squares depicted in Fig. 1.3 are input into a single air dispersion modelling run for Fugitive source F1 as four separate volume sources (F1A, F1B, F1C, and F1D). The initial horizontal dimension for each volume source is the side of the square divided by 4.3. It is very important to allocate the unit emissions (1 g/s) for F1 to each source proportionately to its area in relation to the total area. For example, if the F1A area is 25 percent of the total area F1, then the emissions rate of F1A is 0.25 g/s. The summation of each individual emission for F1A + F1B + F1C + F1D must be equal to 1 g/s.

The resulting air parameter values in the plot files may be used directly in the risk assessment equations, the same as if a stack emission were modeled as a single point source. The initial vertical dimension is defined the same as F2, using the vertical extent of 5 m divided by 2.15 and a release height of 2.5 m (half vertical extent). For volume sources, the location is specified by the x and y coordinates of the center of each square area.

3.10 Modeling Acute Risk

The acute or short-term effects due to the direct inhalation of vapor phase, particle phase, and particle-bound phase COPCs are generally considered in a risk assessment. From the air modeling perspective, the goal is to compute the highest 1-hour average air concentration for each phase for each source for the entire period of analysis. In most cases, this period is 5 years.

Note: The ADM output must include the specification of 1-hour average plot file which contains the highest 1-hour concentration at each receptor grid node for all the hours evaluated. For a single year run, the 1-hour concentration reported in the plot file will be the highest value computed at each x, y grid node location for the total hours in the year (8760, or 8784 for leap years). The appropriate value needed for the acute risk assessment is the highest 1-hour concentration for the 5-year period. Each of the 5 single year values must be reviewed to identify the highest for all 5 years at each receptor grid node. However, if a combined 5-year meteorological file is run, the plot file will already identify the highest value for the 5-year period at each grid node with no additional processing required.

The highest air concentration for the 1-hour averaging period is input as the air parameters, *Chv*, *Chp*, and *Chpb* in the acute risk assessment equations.

4 EXPOSURE SCENARIOS

The purpose of this section is to provide guidance in the identification of *exposure scenarios* that should be evaluated in the risk assessment, characterization of exposure setting, identification of recommended exposure scenarios, and selection of exposure scenario locations.

An exposure scenario is a combination of *exposure pathways* to which a single "receptor" may be subjected. Human receptors may come into contact with air toxics emitted to the atmosphere from two primary exposure *routes*, either directly via inhalation; or indirectly via subsequent ingestion of water, soil, vegetation, and animals that become contaminated by chemicals through the food chain.

Each exposure pathway consists of four fundamental components: (1) an exposure route; (2) a source and mechanism of COPC release; (3) a retention medium, or a transport mechanism and subsequent retention medium in cases involving media transfer of COPCs; and (4) a point of potential human contact with the contaminated medium, which is referred to as the exposure point and consists of a specific receptor exposed at a specific point. Humans, plants, and animals in the assessment area may take up COPCs directly from the air or indirectly via the media receiving deposition (e.g., soil, vegetation, or water).

The exposure scenarios recommended for risk evaluation are generally conservative in nature. They are intended to allow standardized and reproducible evaluation of risks. Any deviations from these recommendations, such as a deletion or modification of a recommended exposure scenario or scenario location should be well-documented.

4.1 Exposure Setting Characterization

The purpose of characterizing the exposure setting is to identify the nonworker related human activities and receptors in the assessment area that may be impacted as a result of exposure to emissions from one or more of a facility's emission sources.

Site maps and land use. Maps are important to define land use to consider in characterizing the exposure setting; and when overlayed with the air dispersion modeling results, will define which recommended exposure scenarios and their locations should be evaluated in the risk assessment.

Maps that support the development of the risk assessment are available from various suppliers. The United State Geological Suvery and WebGIS (*http://www.WebGIS.com*) provide extensive maps free of charge. The main maps used in risk assessment studies are:

Land Use Land Cover (LULC) Maps: Exact boundaries of polygon land use area coverages, in areas being considered for evaluation, should be verified using available topographic maps and aerial photographic coverages.

Topographic Maps: Topographic maps are readily available in both hard copy and electronic format directly from USGS or numerous other vendors. These maps are commonly at a scale of 1:24,000, and in a file type TIFF format with TIFF World File included for geo-referencing.

Aerial Photographs: provide an excellent reference for identifying land use areas and justifying selection of exposure scenario locations.

The risk assessor should focus on land use areas potentially impacted by COPC emissions from facility emission sources being evaluated in the risk assessment.

Water bodies and their associated watersheds. The identification of surface water bodies at locations in the assessment area receiving deposition from emission sources indicates the potential for COPC exposures from ingestion of fish, and possibly drinking water (drinking water is evaluated only if the local population obtains drinking water from surface water sources). In addition, the surface area and exact location with respect to evaluating receptor grid nodes positioned within the water body and watershed should be defined.

The area extent of a watershed is generally defined by topographic highs that result in downslope drainage into the water body. The watershed can be important in determining the overall water body COPC loading, because pervious and impervious areas of the watershed, as well as the soil concentration of COPCs resulting from emissions from facility sources, are also used in the media concentration equations to calculate the water body COPC concentrations resulting from watershed runoff.

For example, if facility emissions impact principally a land area that feeds a specific tributary that drains to a large river system and immediately upstream of a drinking water intake point, the risk assessor should consider evaluating an "effective" watershed area rather than the entire watershed area of the large river system. For such a large river system, the watershed area can be on the order of thousands of square kilometers and can include numerous tributaries draining into the river at points that would not have any net impact on the drinking water intake or on the water body COPC concentration at the exposure point of interest.

Additional water body and watershed required parameters include the following:

- Water body surface area
- Watershed surface area
- Impervious watershed area
- Average surface water volumetric flow rate
- Current velocity of surface water body
- Depth of surface water body column
- Universal soil loss equation (USLE) rainfall/erosivity factor

Water body current velocities and volumetric flow rates should generally be average values on an annual basis. State or local geologic surveys may keep records on water bodies. Volumetric flow rates for smaller streams or lakes can be calculated as the product of the watershed area and one-half of the local average annual surface runoff. Current velocities can be calculated as the volumetric flow rate divided by the cross-sectional area (current velocities are not used in the equations for lakes). Depths of water bodies can sometimes be obtained from state or local sources.

4.2 Recommended Exposure Scenarios

USEPA recommends the following exposure scenarios:

- Farmer
- Farmer child
- Adult resident
- Child resident
- Fisher
- Fisher child
- Acute risk

Exposure scenarios are not referred to as subsistence since the actual mass per day amounts of food items evaluated in the recommended exposure pathways are more comparable to reasonable versus subsistence amounts; and therefore, may not preclude ingestion of significant amounts of food items not represented in the exposure pathways of the exposure scenario subject to evaluation. Specific regional exposure setting characteristics may warrant the inclusion of additional recommended exposure pathways when evaluating an exposure scenario for a specific regional exposure setting. For example, the recommended farmer exposure scenario does not automatically include the fish ingestion exposure pathway. However, in some areas of the country, it is common for farms to also have stock ponds which are fished on a regular basis for the farmer's consumption.

USEPA also recommends that infant exposure to PCDDs and PCDFs via the ingestion of their mother's breast milk be evaluated as an additional exposure pathway at all recommended exposure scenario locations.

Note: The following exposure scenarios are not very important for air toxics risk assessment:

Ingestion of ground water: USEPA (1990e) found that ground water is an insignificant exposure pathway for air toxics emissions risk assessment.

Inhalation of resuspended dust: USEPA (1990e) found that inhalation of resuspended dust was insignificant, for air toxics emissions risk assessment.

Dermal exposure to surface water, soil, or air: Available data indicate that the contribution of dermal exposure to soils to overall risk is typically small (USEPA 1996g; 1995h). The risk resulting from soil ingestion and dermal contact was 50-fold less than the risk from any other exposure pathway and 300-fold less than the total estimated risk (USEPA 1996g; 1995h).

Inhalation of COPCs and ingestion of water by animals: These exposure pathways have not been included in the recommended exposure scenarios because the contribution of these pathways to total risk is anticipated to be negligible in comparison with that of the exposure pathways being evaluated. However, these exposure pathways may need to be evaluated on a case-by-case basis considering site-specific exposure setting characteristics.

Farmer scenario. The farmer exposure scenario is evaluated to account for the combination of exposure pathways to which a receptor may be exposed in a farm or ranch exposure setting. The farmer is assumed to be exposed to COPCs emitted from the facility through the following exposure pathways:

- Direct inhalation of vapors and particles
- Incidental ingestion of soil
- Ingestion of drinking water from surface water sources
- Ingestion of homegrown produce
- Ingestion of homegrown beef
- Ingestion of milk from homegrown cows
- Ingestion of homegrown chicken
- Ingestion of eggs from homegrown chickens
- Ingestion of homegrown pork
- Ingestion of breast milk

Figure 1.4 depicts the above pathways.

Farmer child scenario receptor. The farmer child exposure scenario is evaluated to account for the combination of exposure pathways to which a receptor may be exposed in a farm or ranch setting.

Adult resident scenario receptor. The adult resident exposure scenario is evaluated to account for the combination of exposure pathways to which a receptor may be exposed in an urban or rural (nonfarm)

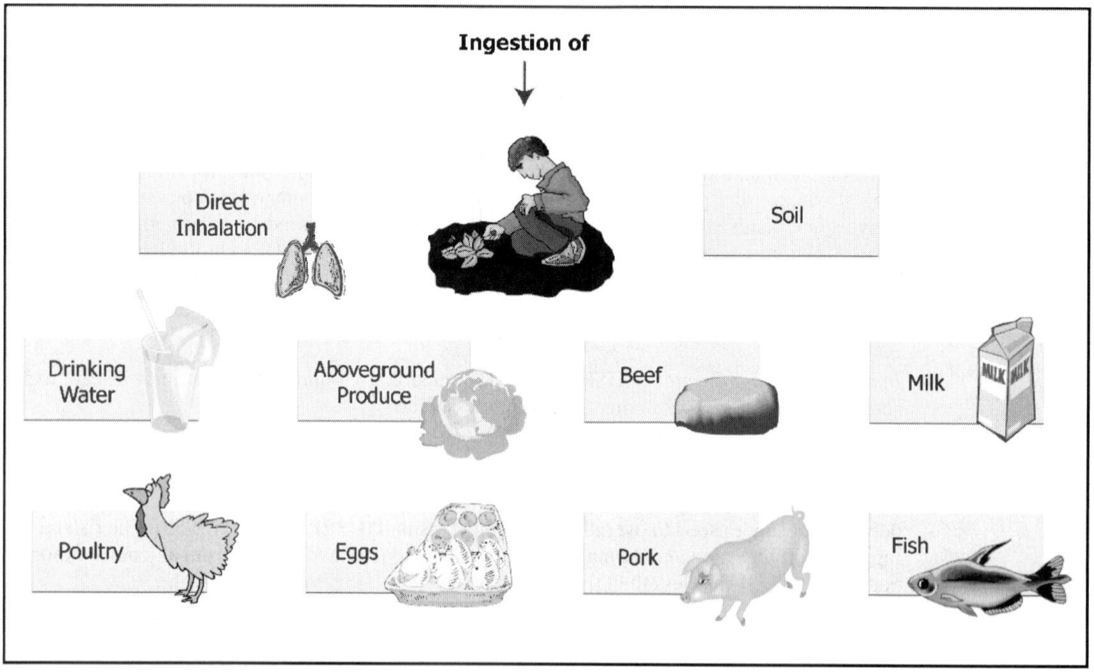

Figure 1.4 Farmer scenario exposure path ways.

setting. The adult resident is assumed to be exposed to COPCs from the emission source through the following exposure pathways:

- Direct inhalation of vapors and particles
- Incidental ingestion of soil
- Ingestion of drinking water from surface water sources
- Ingestion of homegrown produce
- Ingestion of breast milk

Child resident scenario receptor. The child resident exposure scenario is evaluated to account for the combination of exposure pathways to which a child receptor may be exposed in an urban or rural (nonfarm) setting. The child resident is assumed to ingest some produce grown by the adult resident (NC DEHNR 1997).

Fisher scenario receptor. The Fisher exposure scenario is evaluated to account for the combination of exposure pathways to which a receptor may be exposed in an urban or rural setting where fish is the main component of the receptor diet. The fisher is exposed to COPCs emitted from the facility through the pathways presented in Fig. 1.5, and in the following list:

- Direct inhalation of vapors and particles
- Incidental ingestion of soil
- Ingestion of drinking water from surface water sources, homegrown produce and fish
- Ingestion of breast milk (evaluated separately)

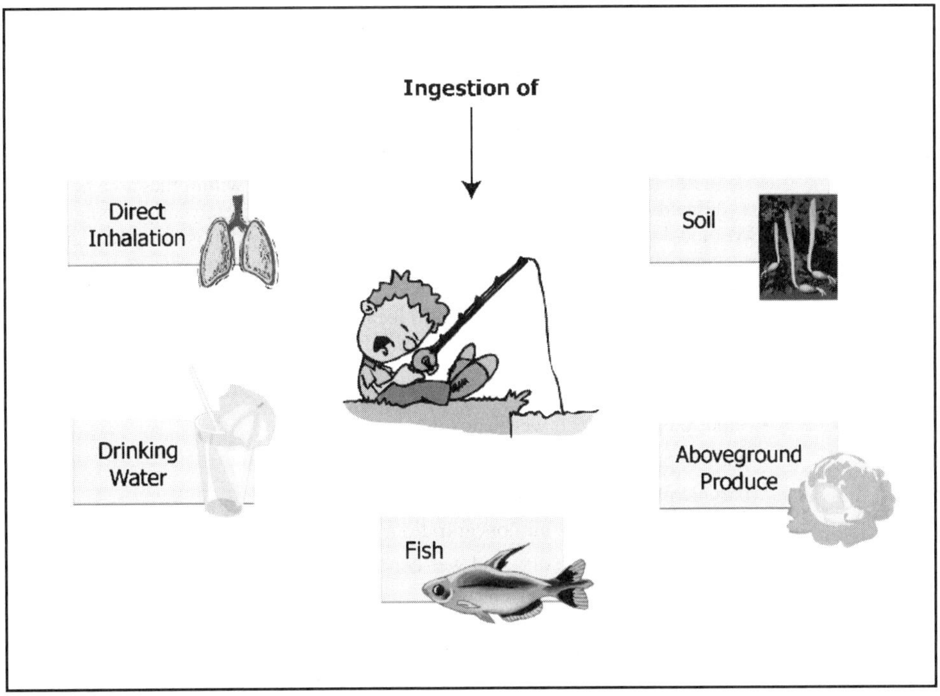

FIGURE 1.5 Fisher exposure pathways.

Fisher child scenario receptor. The fisher child exposure scenario is evaluated to account for the combination of exposure pathways to which a receptor may be exposed in an urban or rural setting where fish is the main component of the receptor diet.

Acute risk scenario. The acute exposure scenario is evaluated to account for short-term effects of exposure to maximum one-hour concentrations of COPCs in emissions from the facility through direct inhalation of vapors and particles. A receptor may be exposed in an urban or rural setting where human activity or land use supports any of the recommended exposure scenarios, as well as, in commercial and industrial land use areas (excluding workers from the facility under direct evaluation in the risk assessment) not typically evaluated by application of the other recommended exposure scenarios.

4.3 Selection of Exposure Scenario Locations

Exposure scenario locations are the receptor grid nodes selected for evaluating one or more of the recommended exposure scenarios. Specific receptor grid nodes are selected as exposure scenario locations based on evaluation of the magnitude of air parameter values estimated by the ADM. Air parameter values specific to the receptor grid node, selected as an exposure scenario location, are then used as inputs to the estimating media concentration equations when evaluating the recommended exposure scenarios. This step is one of the most critical parts of the risk assessment process, ensuring standardization across all facilities evaluated and reproducibility of results. The estimates of risk can vary significantly in direct response to which receptor grid nodes are selected as exposure scenario locations.

The following procedures should be used in the selection of receptor grid nodes as exposure scenario locations, and that selected exposure scenario locations correspond to actual ADM receptor grid node locations defined by UTM coordinates. USEPA recommends exposure scenario(s) at actual receptor grid nodes selected as follows:

Step 1. *Define land use areas to be evaluated* Current and reasonable potential future land use areas, water bodies, and watersheds identified during exposure setting characterization for evaluation in the risk assessment, should be defined and mapped using UTM coordinates in a format (NAD27 or NAD83 UTM) consistent with that used to define locations of facility emission sources and the ADM receptor grid nodes.

Step 2. *Identify receptor grid node(s) within each land use area to be evaluated* For each land use area to be evaluated, identify the receptor grid nodes within that area or on the boundary of that area (defined in Step 1) that represent the location of highest yearly average concentration for each ADM modeled air parameter (e.g., air concentration, dry deposition, wet deposition) for each phase (e.g., vapor, particle, particle-bound); specific to each emission source (e.g., stacks, fugitives) and all emission sources at the facility combined. This results in the selection of one or more receptor grid nodes as one or more exposure scenario locations, within the land use area to be evaluated, that have one or more of the highest modeled unitized:

- vapor phase air concentration
- vapor phase dry- and wet-deposition rate
- particle phase air concentration
- particle phase dry- and wet-deposition rate
- particle-bound phase air concentration
- particle-bound phase dry- and wet-deposition rate

Step 3. *Identify receptor grid nodes for acute risk and site-specific risk considerations* Additional receptor grid nodes within the assessment area may need to be considered as exposure scenario locations for the evaluation of acute risk or site-specific risk considerations (e.g., special subpopulations). In land use areas to be evaluated for acute risk (could potentially include commercial and industrial land use areas), receptor grid nodes with the highest modeled hourly vapor phase air concentration and highest hourly particle phase air concentration specific to each emission source and all emission sources at the facility combined, should be selected as the exposure scenario location(s). For site-specific risk considerations, the closest receptor grid node to the exposure point being evaluated should be considered for the exposure scenario location.

Step 4. *Identify receptor grid nodes for water bodies and watersheds* For recommended exposure scenarios that include evaluation of water bodies and their associated watersheds, the receptor grid nodes within their area should be considered. For water bodies, the risk assessor can select the receptor grid node with the highest modeled air parameter values or average the air parameter values for all receptor grid nodes within the area of the water body. For watersheds, the modeled air parameter values should be averaged for all receptor grid nodes within the area extent or effective area of the watershed (excluding the area of the water body).

5 ESTIMATION OF MEDIA CONCENTRATIONS

The purpose of this section is to describe the media concentration equations and associated parameters used in evaluation of the recommended exposure scenarios. The origin and development of each of these equations, and description of associated parameters, are presented in most cases.

5.1 Calculation of COPC Concentrations in Air for Direct Inhalation

COPC concentrations in air are calculated by summing the vapor phase and particle phase air concentrations of COPCs. Air concentrations used in the evaluation of long-term or chronic exposure, via direct inhalation, should be calculated using unitized yearly air parameter values (Fig. 1.6).

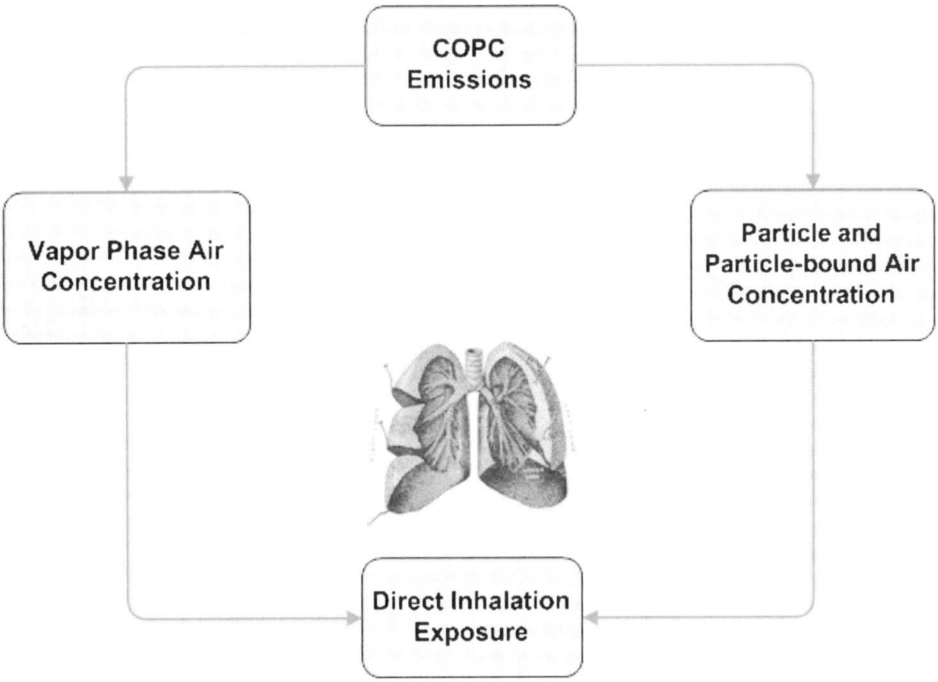

FIGURE 1.6 Direct inhalation exposure pathway.

5.2 Calculation of COPC Concentrations in Soil

COPC concentrations in soil are calculated by summing the vapor phase and particle phase deposition of COPCs to the soil. Wet and dry deposition of particles and vapors are considered, with dry deposition of vapors calculated from the vapor air concentration and the dry deposition velocity. The calculation of soil concentration incorporates a term that accounts for loss of COPCs by several mechanisms, including leaching, erosion, runoff, degradation (biotic and abiotic), and volatilization. These loss mechanisms all lower the soil concentration associated with the deposition rate. Figure 1.7 presents the processes described above.

Soil concentrations may require many years to reach steady state. As a result, the equations used to calculate the average soil concentration over the period of deposition were derived by integrating the instantaneous soil concentration equation over the period of deposition. For carcinogenic COPCs, USEPA recommends using two variations of the equation (average soil concentration over exposure duration):

1. One variation to be used if the exposure duration is greater than or equal to the operating lifetime of the emission source or time period of combustion.
2. The other form to be used if the exposure duration is less than the operating lifetime of the emission source or time period of combustion.

For noncarcinogenic COPCs, USEPA recommends the calculation of the highest one-year annual average soil concentration; typically occurring at the end of the operating lifetime of the emission source. Soil conditions such as pH, structure, organic matter content, and moisture content affect the distribution and mobility of COPCs. Loss of COPCs from the soil is modeled by using rates that depend on the physical and chemical characteristics of the soil. These variables and their use are described in the following subsections, along with the recommended equations.

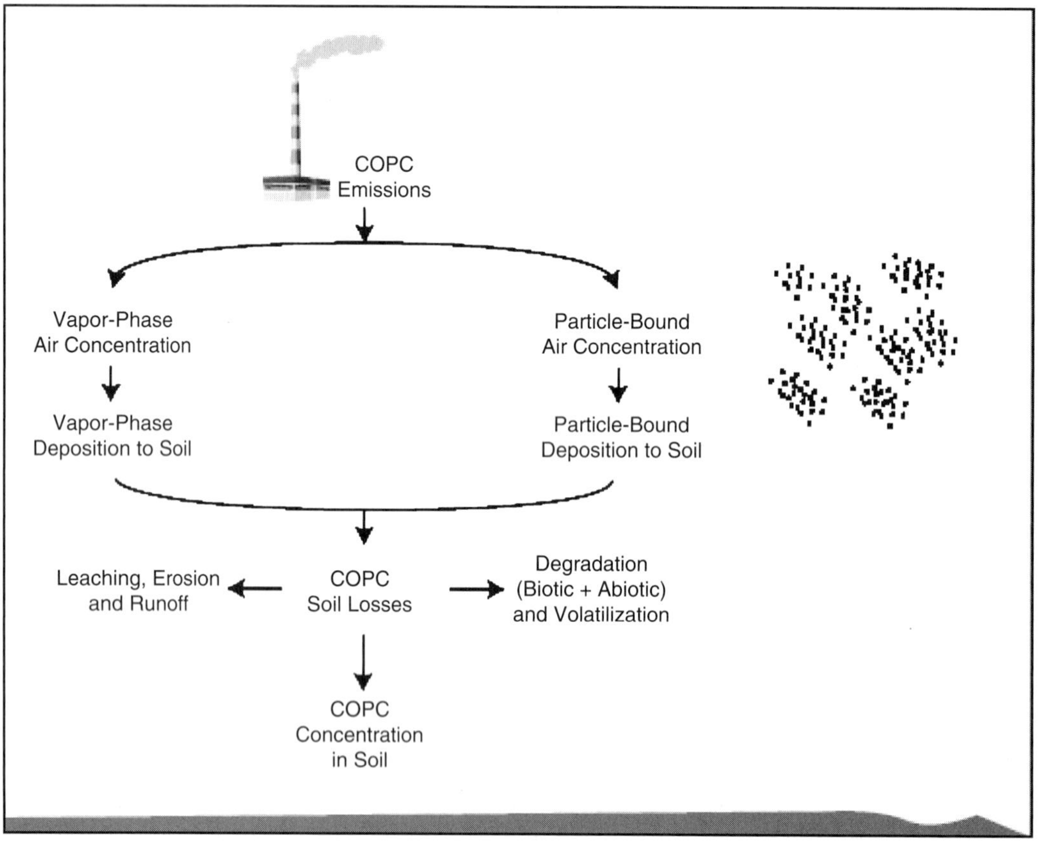

FIGURE 1.7 Chemical of potential concern concentration in soil.

Calculating cumulative soil concentration. USEPA recommends the following equations to calculate the cumulative soil concentration (Cs).

Carcinogens:

For $T_2 < tD$

$$C_s = \frac{Ds}{ks \cdot (tD - T_1)} \cdot \left[\left(tD + \frac{\exp(-ks \cdot tD)}{ks} \right) - \left(T_1 + \frac{\exp(-ks \cdot T_1)}{ks} \right) \right] \tag{1.3a}$$

For $T_1 < tD < T_2$

$$Cs = \frac{\left(\frac{Ds \cdot tD - Cs_{td}}{ks} \right) + \left(\frac{Cs_{td}}{ks} \right) \cdot (1 - \exp[-ks \cdot (T_2 - tD)])}{(T_2 - T_1)} \tag{1.3b}$$

Non-carcinogens:

$$Cs_{tD} = \frac{Ds \cdot [1 - \exp(-ks \cdot tD)]}{ks} \tag{1.3c}$$

where

Cs = Average soil concentration over exposure duration (mg COPC/kg soil)
Ds = Deposition term (mg COPC/kg soil/year)
T_1 = Time period at the beginning of combustion (year)
ks = COPC soil loss constant due to all processes (year^{-1})
tD = Time period over which deposition occurs (time period of combustion) (year)
Cs_{tD} = Soil concentration at time tD (mg/kg)
T_2 = Length of exposure duration (year)

When an exposure duration that is less than or equal to the operating lifetime of the emission source or hazardous waste combustion unit ($T_2 < tD$), Eq. (1.3a) is recommended; when an exposure duration greater than the operating lifetime of the hazardous waste combustion unit ($T_1 < tD < T_2$), Eq. (1.3b) is recommended. For noncarcinogenic COPCs, Eq. (1.3c) is recommended.

The COPC soil concentration averaged over the exposure duration, represented by Cs, should be used for carcinogenic compounds, where risk is averaged over the lifetime of an individual. Because the hazard quotient associated with noncarcinogenic COPCs is not averaged over a lifetime of exposure, the highest annual average COPC soil concentration occurring during the exposure duration period should be used for noncarcinogenic COPCs. The highest annual average COPC soil concentration, Cs_{tD}, will typically occur at the end of the operating life of the emission source or the time period of combustion.

The deposition term in the Cs equation is expressed in two forms, one for mercury and the other one for all the other chemicals (COPs). The equation for all COPS, except mercury is presented below:

$$Ds = \frac{100 \cdot Q}{Zs \cdot BD} \cdot [\, Fv(Dydv + Dywv) + (1 - Fv)(Dydp + Dywp)\,] \tag{1.4a}$$

The equation for the mercury deposition term, in the cumulative soil concentration is presented as follows:

$$Ds = \frac{100 \cdot 0.48Q_{(\mathrm{Total})}}{Zs \cdot BD} \cdot [Fv_{(\mathrm{HG}^{2+})}(Dydv + Dywv) + (1 - Fv)_{(\mathrm{HG}^{2+})}(Dydp + Dywp)] \tag{1.4b}$$

Use $0.48Q$ for total mercury and $Fv = 0.85$ in the mercury modeling equation to calculate Ds. Apportion the calculated Ds value into the divalent mercury (Hg^{2+}) and methyl mercury (MHg) forms based on the assumed 98 percent Hg^{2+} and 2 percent MHg speciation split in soils. Elemental mercury (Hg^0) occurs in very small amounts in the vapor phase and does not exist in the particle or particle-bound phase. Therefore, assume elemental mercury deposition onto soils is negligible or zero. Evaluate elemental mercury for the direct inhalation pathway only.

$$Ds\ (\mathrm{Hg}^{2+}) = 0.98\ Ds\ (\mathrm{Mercury})$$
$$Ds\ (\mathrm{MHG}) = 0.02\ Ds\ (\mathrm{Mercury})$$
$$Ds\ (\mathrm{Hg}^{0}) = 0.0$$

Evaluate divalent and methyl mercury as individual COPCs. Calculate Cs for divalent and methyl mercury using the corresponding fate and transport parameters for mercuric chloride (divalent mercury, Hg^{2+}) and methyl mercury.

Calculating the COPC soil loss constant. Organic and inorganic COPCs may be lost from the soil by several processes that may or may not occur simultaneously. The rate at which a COPC is lost from the soil is known as the soil loss constant (ks). The constant ks is determined by using the soil's physical, chemical, and biological characteristics to consider the loss resulting from:

1. leaching,
2. runoff,
3. erosion,

4. biotic and abiotic degradation, and
5. volatilization.

USEPA recommends the following equations to be used to calculate the COPC soil loss constant (*ks*).

$$ks = ksg + kse + ksr + ksl + ksv \tag{1.5}$$

where

ks = Loss constant due to all processes (year^{-1})
ksg = Loss constant due to biotic and abiotic degradation (year^{-1})
kse = Loss constant due to soil erosion (year^{-1})
ksr = Loss constant due to surface runoff (year^{-1})
ksl = Loss constant due to leaching (year^{-1})
ksv = Loss constant due to volatilization (year^{-1})

The use of the soil loss equation assumes that COPC loss can be defined by using first-order reaction kinetics. First-order reaction rates depend on the concentration of one reactant. At low concentrations, a first-order loss constant may be adequate to describe the loss of the COPC from soil (USEPA 1990e).

COPC loss constant due to biotic and abiotic degradation. Soil losses resulting from biotic and abiotic degradation (*ksg*) are determined empirically from field studies and are addressed in the literature (USEPA 1990e). USEPA guidance (1994g) had stated that *ksg* values for all COPCs other than 2,3,7,8-TCDD should be set equal to zero. However, the final version of the HHRAP provides *ksg* values for many COPCs other than PCDD and PCDF.

The rate of biological degradation in soils depends on the concentration and activity of the microbial populations in the soil, the soil conditions, and the COPC concentration. The use of simple rate expressions may be appropriate at low chemical concentrations (e.g., nanogram per kilogram soil) at which a first-order dependence on chemical concentration may be reasonable. The rate of biological degradation is COPC-specific, depending on the complexity of the COPC and the usefulness of the COPC to the microorganisms. These are some of the environmental and COPC-specific factors that may limit the biodegradation of COPCs in the soil:

1. Availability of the COPC
2. Nutrient limitations
3. Toxicity of the COPC
4. Inactivation or nonexistence of enzymes capable of degrading the COPC

Hydrolysis in aqueous systems is characterized by three processes: acid-catalyzed, base-catalyzed, and neutral reactions. The overall rate of hydrolysis is the sum of the first-order rates of these processes. In soil systems, sorption of the COPC can increase, decrease, or not affect the rate of hydrolysis as numerous studies have shown. The total rate of hydrolysis in soil can be predicted by adding the rates in the soil and water phases, which are assumed to be first-order reactions at a fixed pH. Methods for estimating these hydrolysis constants are described by Lyman et al. (1982). The 2005 HHRAP provides ksg values for COPCs that can be used to calculate *ks*. For example, the *ksg* for Pentachlorophenol is identified in the 2005 HHRAP as 1.42 year-1.

COPC loss constant due to soil erosion. USEPA currently recommends that the constant for the loss of soil resulting from erosion (*kse)* should be set equal to zero based on the theory that contaminated soil erodes onto the specific receptor location where the loss is calculated as well as eroding off it.

Kse
0.0

COPC loss constant due to runoff. USEPA recommends the following equation to calculate the constant for the loss of soil resulting from surface runoff (*ksr*).

$$ksr = \frac{RO}{\theta_{sw} \cdot Z_s} \cdot \left(\frac{1}{1 + (Kd_s \cdot BD/\theta_{sw})}\right) \tag{1.6}$$

where

ksr = COPC loss constant due to runoff (year^{-1})
RO = Average annual surface runoff from pervious areas (cm/year)
θ_{sw} = Soil volumetric water content (mL water/cm^3 soil)
Z_s = Soil mixing zone depth (cm)
Kd_s = Soil-water partition coefficient (mL water/g soil)
BD = Soil bulk density (g soil/cm^3 soil)

COPC loss constant due to leaching. Losses of soil COPCs due to leaching (*ksl*) depend on the amount of water available to generate leachate and soil properties such as bulk density, soil moisture, soil porosity, and soil sorption properties.

USEPA recommends the following equation to calculate the *ksl* to account for runoff.

COPC Loss Constant Due to Leaching (*ksl*)

$$ksl = \frac{P + I - RO - E_v}{\theta_{sw} \cdot Z_s \cdot [1.0 + (BD \cdot K_{ds}/\theta_{sw})]} \tag{1.7}$$

where

ksl = COPC loss constant due to leaching (year^{-1})
P = Average annual precipitation (cm/yr)
I = Average annual irrigation (cm/yr)
RO = Average annual surface runoff from pervious areas (cm/yr)
E_v = Average annual evapotranspiration (cm/yr)
θ_{sw} = Soil volumetric water content (mL water/cm^3 soil)
Z_s = Soil mixing zone depth (cm)
Kd_s = Soil-water partition coefficient (cm^3 water/g soil)
BD = Soil bulk density (g soil/cm^3 soil)

COPC loss constant due to volatilization. Semivolatile and volatile COPCs emitted in high concentrations may become adsorbed to soil particles and exhibit volatilization losses from soil. The loss of a COPC from the soil by volatilization depends on the rate of movement of the COPC to the soil surface, the chemical vapor concentration at the soil surface, and the rate at which vapor is carried away by the atmosphere (Jury 1986).

USEPA recommends the following equation to compute the loss constant due to volatization.

$$ksv = \left(\frac{3.1536 \times 10^7 \cdot H}{Z_s \cdot Kd_s \cdot R \cdot T_a \cdot BD}\right) \cdot \left(\frac{D_a}{Z_s}\right) \cdot \left[1 - \left(\frac{BD}{\rho_{soil}}\right) - \Theta_{sw}\right] \tag{1.8}$$

where

ksv = COPC loss constant due to volatization (year^{-1})
3.1536×10^7 = Units conversion factor (s/year)
H = Henry's law constant (atm · m^3/mol)
Z_s = Soil mixing zone depth (cm)
Kd_s = Soil-water partition coefficient (mL/g)
R = Universal gas constant (atm·m^3/mol·K)
T_a = Ambient air temperature (K) = 298.1 K
BD = Soil bulk density (g soil/cm^3 soil) = 1.5 g/cm^3

D_a = Diffusivity of COPC in air (cm^2/s)
θ_{sw} = Soil volumetric water content (mL/cm^3 soil) = 0.2 mL/cm^3
ρ_{soil} = Solids particle density (g/cm^3) = 2.7 g/cm^3

Universal soil loss equation. USEPA recommends that the USLE be used to calculate the unit soil loss (X_e).

Site-specific parameters for calculating cumulative soil concentration. Calculating average soil concentration over the exposure duration (*Cs*) requires the use of site-specific parameters including the following:

1. Soil mixing zone depth (Z_s)
2. Soil bulk density (*BD*)
3. Available water ($P + I - RO - Ev$)
4. Soil volumetric water content (θ_{SW})

Soil mixing zone depth. When exposures to COPCs in soils are modeled, the depth of contaminated soils is important in calculating the appropriate soil concentration. COPCs deposited onto soil surfaces may be moved into lower soil profiles by tilling, whether manually in a garden or mechanically in a large field.

USEPA (1990e) estimated that if the area under consideration is likely to be tilled, soil depth is about 10 to 20 cm depending on local conditions and the equipment used. If soil is not moved, COPCs are assumed to be retained in the shallower, upper soil layer (1 cm).

The assumption made to determine the value of Z_s may affect the outcome of the risk assessment, because soil concentrations that are based on soil depth are used to calculate exposure via several pathways:

1. Ingestion of plants contaminated by root uptake and by volatilization from soil.
2. Direct ingestion of soil by humans, cattle, swine, or chicken.
3. Surface runoff into water bodies.

For example, in calculations of exposures resulting from uptake through plant roots, the average concentration of COPCs over the depth of the plant root determines plant uptake. However, in calculations of plant uptake resulting from volatilization, only the uppermost soil layer is considered.

USEPA (1990e) recommended that soil mixing depths be selected as in Table 1.4:
USEPA recommends the following values for the soil mixing zone depth (Z_s) in the 2005 HHRAP.

TABLE 1.4 USEPA Recommendations for Soil Mixing Depth

Soil depth (*Zs*)	Exposure	Description
1 cm	Direct ingestion of soil.	Human exposure: in gardens, lawns, landscaped areas, parks, and recreational areas. Animal exposure: in pastures, lawns, and parks (untilled soils).
1 cm	Surface water runoff in nonagricultural areas.	These areas are typically assumed to be untilled.
20 cm	Plant uptake for agricultural soils.	The root depth is assumed to equal the tilling depth of 20 cm. In untilled soils, the root zone does not directly reflect tilling depth, although it is dassumed that tilling depth is an adequate substitute for root zone depth.
20 cm	Surface water runoff in agricultural areas.	These areas are typically assumed to be tilled.

Recommended Values for: Soil Mixing Zone Depth (Z_s)
2 cm—untilled
20 cm—tilled

Semivolatile COPCs and volatile COPCs emitted in sufficiently high concentrations may be deposited in particulate form and exhibit volatilization losses from soils. COPCs subject to volatilization losses may be moved to 20 cm by tilling and will not readily volatilized from this depth. The volatilization rate will reflect only the COPC concentration at the soil surface.

Soil dry bulk density. Soil dry bulk density (BD) is the ratio of the mass of soil to its total volume. This variable is affected by the soil structure, type, and moisture content (Hillel 1980). USEPA recommends the following value for the *BD*.

Recommended Value for:
Soil Dry Bulk Density (*BD*) = 1.50 g/cm³

Available water (P + I − RO − Ev). The average annual volume of water available ($P + I - RO - E_v$) for generating leachate is the mass balance of all water inputs and outputs from the area under consideration. A wide range of values for these site-specific parameters may apply in the various USEPA regions.

The average annual precipitation (P), irrigation (I), runoff (RO), and evapotranspiration (E_v) rates and other climatological data may be obtained from either data recorded on site or from the Station Climatic Summary for a nearby airport. Meteorological variables, such as the evapotranspiration rate (E_v) and the runoff rate (RO), may be found in resources such as Geraghty et al (1973).

Soil volumetric water content. The soil volumetric water content (θ_{sw}) depends on the available water and the soil structure. A wide range of values for these variables may apply in the various USEPA regions. USEPA recommends a value for θ_{sw} of 0.2 mL/cm³.

Recommended Value for:
Soil Volumetric Water Content (θ_{sw}) = 0.2 ml/cm³

Example calculations for the estimation of the soil parameters.

```
=======================================================
SOIL CALCULATIONS
=======================================================

==========
TCDD2378 - CAS = 1746016
==========

THETSW = 0.20 (Default value [ml / cm3])
ZS     = 2.00 (Default value [cm])
BD     = 1.50 (Soil bulk density [g / cm3])
KDS    = 3.8904510E+04
```

```
--------------------------------------------------------------------------------
***Ksl - Soil Loss Due to Leaching ***
--------------------------------------------------------------------------------
Available Water - (P+I-RO-EV) = 94.1900
THETSW*ZS*(1.0+(BD*KDS/THETSW)) = 1.1671E+05
Ksl = 8.0702E-04

----------------------------------------------------------------------------
***Ksr - Soil Loss Due to Runoff ***
----------------------------------------------------------------------------
Soil loss - runoff - RO/(THETSW*ZS)     = 60.00
Soil loss - runoff - 1/(1+(KDS*BD/THETSW)) = 3.4272E-06
Soil loss - runoff = 2.0563E-04

--------------------------------------------------------------------------
***SD - Sediment Delivery Ratio ***
--------------------------------------------------------------------------
a = 1.40 [Unitless]
Sediment Delivery Ratio : SD = 2.1701E-01 [Unitless]

------------------------------------------------------------
** * Xe - Unit Soil Loss ***
------------------------------------------------------------
Xe - RF*K*LS*C_XE*PF*(907.18/4047) = 6.0523E-01 [kg/m2·yr]

------------------------------------------------------------------------------------
**Kse - Soil Loss Due to Soil Erosion **
------------------------------------------------------------------------------------
Soil loss due to erosion : Kse = 0.00

--------------------------
*** Ksv ***
--------------------------
KSV = 3.1536E7*H/(ZS*KDS*R*TA*BD)*(DA/ZS)*(1-(BD/RHO_S)-THETA_S) =
4.6213E-03

------------------------------------------------------------------
*** Ks - Soil Loss Constant ***
------------------------------------------------------------------
Soil Loss Constant - Ks = KSG + KSE + KSR + KSL + KSV =
3.5634E-02 [1 / year]

------------------------------------------------------------
*** Ds - Deposition Term ***
------------------------------------------------------------
(100*Q)/(ZS*BD)= 3.3333E-04
FV*(DYDV + DYWV)+((DYDP+DYWP)*(1-FV))= 2.4300E-02
Deposition Term : Ds = 8.1000E-06 [mg COPC / kg soil / yr]

--------------------------------------------------
*** Ds - Deposition ***
--------------------------------------------------
Deposition Term : (100*Q)/(ZS*BD)= 3.3333E-04
Deposition Term : FV*DYTWV+DYTWP*(1-FV)= 1.9190E-02
Deposition Term : Ds_w = 6.3967E-06
```

```
------------------------------------------------------------------------------------------------
***Cancer - Average Over Exposure Duration  ***
------------------------------------------------------------------------------------------------
Cancer- Avg Over Exp.Dur. DS/(KS*(TD-T1))    = 7.5772E-06
Cancer- Avg Over Exp.Dur. TD+((EXP(-KS*TD))/KS)= 3.9635E+01
Cancer- Avg Over Exp.Dur. T1+((EXP(-KS*T1))/KS)= 2.8063E+01
Cancer- Avg Over Exposure Duration : Cs = 8.7685E-05

------------------------------------------------------------------------------------------------
***Cancer - Average Over Exposure Duration  ***
------------------------------------------------------------------------------------------------
 DS/(KS*(TD-T1))    = 5.9837E-06
 TD+((EXP(-KS*TD))/KS)= 3.9635E+01
 T1+((EXP(-KS*T1))/KS)= 2.8063E+01
Cancer- Avg Over Exposure Duration : Cs = 6.9244E-05

========
Silver - CAS = 7440224
========

THETSW = 2.0000E-0001
ZS    = 2.0000E+0000
BD    = 1.5000E+0000
KDS   = 8.3000E+0000

--------------------------------------------------------------------------------
***Ksl - Soil Loss Due to Leaching  ***
--------------------------------------------------------------------------------
Soil loss - leaching - (P+I-RO-EV) = 9.4190E+0001
Soil loss - leaching - THETSW*ZS*(1.0+(BD*KDS/THETSW)) =
2.5300E+0001
Soil loss - leaching = 3.7229E+0000

-------------------------------------------------------------------------------
***Ksr - Soil Loss Due to Runoff   ***
-------------------------------------------------------------------------------
Soil loss - runoff - RO/(THETSW*ZS)     = 6.0000E+0001
Soil loss - runoff - 1/(1+(KDS*BD/THETSW)) = 1.5810E-0002
Soil loss - runoff = 9.4862E-0001

-------------------------------------------------------------------------
***SD - Sediment Delivery Ratio ***
-------------------------------------------------------------------------
a = 1.4000E+0000
Sediment Delivery Ratio : SD = 2.1701E-0001

--------------------------------------------------------------------
*** Xe - Unit Soil Loss       ***
--------------------------------------------------------------------
Xe - RF*K*LS*C_XE*PF*(907.18/4047) = 6.0523E-0001

-----------------------------------------------------------------------------------
**Kse - Soil Loss Due to Soil Erosion ***
-----------------------------------------------------------------------------------
Soil loss due to erosion : Kse = 0.0000E+0000
```

```
---------------------------
*** Ksv ***
---------------------------
KSV = 3.1536E7*H/(ZS*KDS*R*TA*BD)*(DA/ZS)*(1-(BD/RHO_S)-THETA_S) =
0.0000E+0000

------------------------------------------------------------------------------
*** Ks - Soil Loss Constant     ***
------------------------------------------------------------------------------
Soil Loss Constant - Ks = KSG + KSE + KSR + KSL + KSV =
4.6715E+0000

---------------------------------------------------------------
*** Ds - Deposition Term ***
---------------------------------------------------------------
Deposition Term : (100*Q)/(ZS*BD)= 6.6667E-0004
Deposition Term : FV*(DYDV + DYWV)+((DYDP+DYWP)*(1-FV))=
4.6406E-0002
Deposition Term : Ds = 3.0937E-0005

----------------------------------------------------------------------------------------------
*** Ds - Deposition Term (Water Shed)    ***
----------------------------------------------------------------------------------------------
Deposition Term : (100*Q)/(ZS*BD)= 6.6667E-0004
Deposition Term : FV*DYTWV+DYTWP*(1-FV)= 3.3729E-0002
Deposition Term : Ds_w = 2.2486E-0005

-----------------------------------------------------------------------------------------
***CSTC - NON-Cancer - Highest Annual ***
-----------------------------------------------------------------------------------------
CSTC - DS*(1-EXP(-KS*TD))/KS = 6.6225E-0006

-----------------------------------------------------------------------------------------
***CSTC - NON-Cancer - Highest Annual ***
-----------------------------------------------------------------------------------------
CSTC - DS*(1-EXP(-KS*TD))/KS = 4.8134E-0006
```

========================
DIVALENT MERCURY - CAS = 7487947
========================

```
THETSW = 2.0000E-0001
ZS    = 2.0000E+0000
BD    = 1.500E+0000
KDS   = 5.800E+0004

-------------------------------------------------------------------------------------
***Ksl - Soil Loss Due to Leaching  ***
-------------------------------------------------------------------------------------
Soil loss - leaching - (P+I-RO-EV) = 9.4190E+0001
Soil loss - leaching - THETSW*ZS*(1.0+(BD*KDS/THETSW)) =
1.7400E+0005
Soil loss - leaching = 5.4132E-0004

-----------------------------------------------------------------------------------
***Ksr - Soil Loss Due to Runoff   ***
-----------------------------------------------------------------------------------
Soil loss - runoff - RO/(THETSW*ZS)    = 6.0000E+0001
```

```
Soil loss - runoff - 1/(1+(KDS*BD/THETSW)) = 2.2988E-0006
Soil loss - runoff = 1.3793E-0004

-------------------------------------------------------------------
***SD - Sediment Delivery Ratio Table ***
-------------------------------------------------------------------
a = 1.400E+0000
Sediment Delivery Ratio : SD = 2.1701E-0001

-------------------------------------------------------
*** Xe - Unit Soil Loss       ***
-------------------------------------------------------
Xe - RF*K*LS*C_XE*PF*(907.18/4047) = 6.0523E-0001

-------------------------------------------------------------------
**Kse - Soil Loss Due to Soil Erosion **
-------------------------------------------------------------------
Soil loss due to erosion : Kse = 0.0000E+0000

---------------------
*** Ksv ***
---------------------
KSV = 3.1536E7*H/(ZS*KDS*R*TA*BD)*(DA/ZS)*(1-(BD/RHO_S)-THETA_S) =
2.9147E-0008

-----------------------------------------------------------
*** Ks - Soil Loss Constant      ***
-----------------------------------------------------------
Soil Loss Constant - Ks = KSG + KSE + KSR + KSL + KSV =
6.7928E-0004

-----------------------------------------------------
*** Ds - Deposition Term ***
-----------------------------------------------------
DS for Mercuric Chloride (Divalent)
Deposition Term : (100*Q)/(ZS*BD)= 4.8000E-0004
Deposition Term : FV*(DYDV_hg + DYWV_hg)+((DYDP_PB+DYWP_PB)*
(1-FV))= 4.3104E-0002
Deposition Term : Ds = 2.0276E-0005

---------------------------------------------------------------------------
*** Ds - Deposition Term (Water Shed)      ***
---------------------------------------------------------------------------
DS for Mercuric Chloride (Divalent)
Deposition Term : (100*Q)/(ZS*BD)= 4.8000E-0004
Deposition Term : FV*DYTWV+DYTWP*(1-FV)= 3.7190E-0002
Deposition Term : Ds_w = 1.7494E-0005

-------------------------------------------------------------------
***CSTC - NON-Cancer - Highest Annual ***
-------------------------------------------------------------------
CSTC - DS*(1-EXP(-KS*TD))/KS = 6.0213E-0004

-------------------------------------------------------------------
***CSTC - NON-Cancer - Highest Annual ***
-------------------------------------------------------------------
CSTC - DS*(1-EXP(-KS*TD))/KS = 5.19514E-0004
```

5.3 Calculation of COPC Concentrations in Produce

The concentration of COPCs in plant material is necessary for two reasons:

1. Humans may directly consume vegetables and fruits that are harvested from gardens located in the area impacted by air toxics.
2. Animals (e.g., cattle, chickens) may consume forage, silage, and grain grown in the area impacted by air toxics, and these animals are then consumed by humans.

Indirect exposure resulting from ingestion of produce depends on the total concentration of COPCs in the leafy, fruit, and tuber portions of the plant. Because of general differences in contamination mechanisms, consideration of indirect exposure separates produce into two broad categories, aboveground produce and plant material, and belowground produce. In addition, aboveground produce should be further subdivided into exposed and protected aboveground produce for consideration of contamination as a result of indirect exposure.

5.4 Aboveground Produce, Forage, Silage, and Grain

Aboveground exposed produce, animal forage, animal silage, and grain is assumed to be contaminated by three possible mechanisms (Fig. 1.8) :

- Direct deposition of particles: wet and dry deposition of particle-phase COPCs on the leaves and fruits of plants.
- Vapor transfer: uptake of vapor phase COPCs by plants through their foliage.
- Root uptake: root uptake of COPCs available from the soil and their transfer to the aboveground portions of the plant.

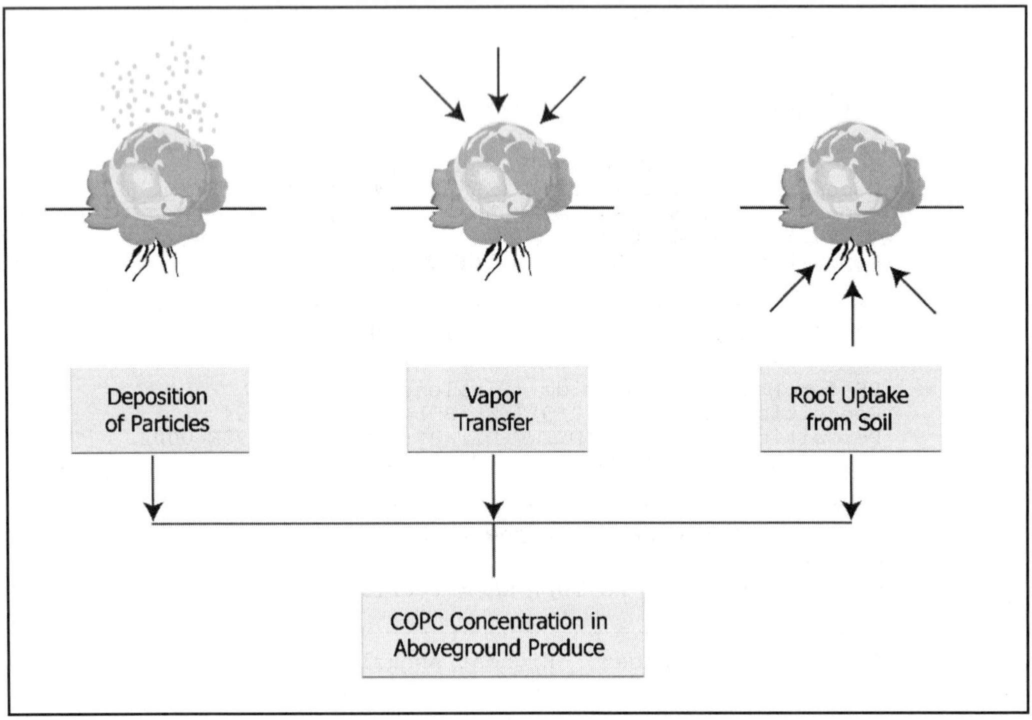

FIGURE 1.8 Aboveground produce exposure pathways.

The total COPC concentration in aboveground exposed produce is calculated as a sum of contamination occurring through all three of these mechanisms. However, edible portions of aboveground protected produce, such as peas, corn, melons, and grain are covered by a protective covering; hence, they are protected from contamination through deposition and vapor transfer. Therefore, root uptake of COPCs is the primary mechanism through which aboveground protected vegetation becomes contaminated.

Belowground produce. For belowground produce, contamination is assumed to occur only through one mechanism, root uptake of COPCs available from soil. Contamination of belowground produce via direct deposition of particles and vapor transfer are not considered because the root or tuber is protected from contact with contaminants in the vapor phase.

Generally, risks associated with exposure of VOCs via food-chain pathways have not been considered significant, primarily because VOCs are typically low-molecular-weight COPCs that do not persist in the environment and do not bioaccumulate. However, USEPA recommends that all COPCs, including VOCs, be evaluated for each exposure pathway.

Aboveground vegetation concentrations due to direct deposition. USEPA recommends that COPC concentrations in aboveground vegetation resulting from wet and dry deposition onto plant surfaces of leafy plants and exposed produce (Pd) be calculated as follows:

$$Pd = \frac{1000 \cdot Q \cdot (1 - F_v) \cdot [Dydp + (Fw \cdot Dywp)] \cdot Rp \cdot [1.0 - \exp(-kp \cdot Tp)]}{Yp \cdot kp} \tag{1.9}$$

where

Pd = Plant (aboveground produce) concentration due to direct (wet and dry) deposition (mg COPC/kg DW)
1000 = Units conversion factor (mg/g)
Q = COPC emission rate (g/s)
F_v = Fraction of COPC air concentration in vapor phase (unitless)
$Dydp$ = Unitized yearly average dry deposition from particle phase (s/m^2 · yr)
Fw = Fraction of COPC wet deposition that adheres to plant surfaces (unitless)
$Dywp$ = Unitized yearly wet deposition from particle phase (s/m^2 · yr)
Rp = Interception fraction of the edible portion of plant (unitless)
kp = Plant surface loss coefficient (year^{-1})
Tp = Length of plant exposure to deposition per harvest of the edible portion of the ith plant group (year)
Yp = Yield or standing crop biomass of the edible portion of the plant (productivity) (kg DW/m^2)

Interception fraction of the edible portion of plant. As summarized in Baes et al.(1984), experimental studies of pasture grasses identified a correlation between initial interception fraction of the edible portion of plant (Rp) values and productivity (standing crop biomass [Yp]) (Chamberlain 1970):

$$Rp = 1 - e^{-\gamma Yp} \tag{1.10}$$

where

Rp = Interception fraction of the edible portion of plant (unitless)
γ = Empirical constant (range as 2.3 to 3.3 for pasture grasses; use the midpoint, 2.88, for pasture grasses.)
Yp = Standing crop biomass (productivity) (kg DW/m^2 for silage; kg WW/m^2 for exposed produce)

Baes et al. (1984) proposed using the same empirical relationship developed by Chamberlain (1970) for other vegetation classes. Class-specific estimates of the empirical constant (γ) were developed by forcing an exponential regression equation through several points, including average and theoretical maximum estimates of *Rp* and *Yp* (Baes et al. 1984). Class-specific empirical constants (γ) that were developed include the following:

- Exposed produce = 0.0324
- Leafy vegetables = 0.0846
- Silage = 0.769

USEPA proposed a default aboveground produce *Rp* value of 0.05, which is based on a weighted average class-specific *Rp* values. Specifically, class-specific *Rp* values were calculated by using the equation developed by Chamberlain (1970) and the following empirical constants:

- Leafy vegetables were assigned the same empirical constant (0.0846) developed by Baes et al. (1984).
- Fruits, fruiting vegetables, and legumes were assigned the empirical constant (0.0324) originally developed by Baes et al. (1984) for exposed produce.

Vegetables and fruits included in each class are as follows:

- Fruits: apple, apricot, berry, cherry, cranberry, grape, peach, pear, plum/prune, and strawberry
- Fruiting vegetables: asparagus, cucumber, eggplant, sweet pepper, and tomato
- Legumes: snap beans
- Leafy vegetables: broccoli, brussel sprouts, cauliflower, celery, lettuce, and spinach

The produce classes have been combined into two groups, exposed fruit and exposed vegetables. The exposed produce empirical constant (γ) was used to calculate *Rp*. Since the exposed vegetable category includes leafy and fruiting vegetables, *Rp* was calculated for leafy and fruiting vegetables. The exposed vegetable *Rp* was then determined by a weighted average based on productivity (*Yp*) of leafy and fruiting vegetables, respectively. The relative ingestion rates used to determine an average weighted *Rp* value were derived from the intake of homegrown produce discussion presented in the 1997 *Exposure Factors Handbook* (USEPA 1997c). USEPA recommends the use of the weighted average *Rp* value of 0.39 as a default *Rp* value.

Interception Fraction of the Edible Portion of Plant (*Rp*) = 0.39

Unweighted *Rp* and ingestion rates used for the weighting were as Table 1.5:

The interception fraction for forage and silage is computed in a manner similar to produce. Substituting the Baes et al. (1984) empirical constant (γ) value of 2.88 for pasture grass, and the standing crop biomass value of 0.24 kg DW/m^2 the forage *Rp* is calculated to be 0.5. the silage *Rp* value is calculated to be 0.46.

Interception Fraction of the Edible Portion of Plant (*Rp*)

Forage = 0.5
Silage = 0.46

TABLE 1.5 USEPA Recommended Interception Fraction of Edible Portion for Above Ground Produce

Aboveground produce class	*Rp*	Ingestion rate (g DW/kg · d)
Exposed fruits	0.053	0.19
Exposed vegetables	0.982	0.11

The use of the empirical constant midpoint (2.88 for pasture grass) does not significantly affect the *Rp* value and the resulting estimate of plant COPC concentration.

Plant surface loss coefficient. USEPA (1990e) identified several processes—including wind removal, water removal, and growth dilution—which reduce the amount of contaminant that has deposited on plant surfaces. The term plant surface loss coefficient (*kp*) is a measure of the amount of contaminant that is lost to these physical processes over time. USEPA recommends the following equation:

$$kp = \left(\frac{ln\ 2}{t_{1/2}}\right) \cdot 365 \qquad (1.11)$$

where

kp = Plant surface loss coefficient (year^{-1})
$t_{1/2}$ = Half-life(days)
365 = Units conversion factor (d/yr)

USEPA recommends a *kp* value of 18, based on a generic 14-day half-life corresponding to physical processes only.

Plant Surface Loss Coefficient (*kp*) = 18

The primary uncertainty associated with this variable is that the calculation of *kp* does not consider chemical degradation processes.

Length of plant exposure to deposition per harvest of edible portion of plant. USEPA recommends treating length of plant exposure to deposition per harvest of edible portion of plant (*Tp*) as a constant, based on the average period between successive hay harvests. This period was estimated at 60 days (0.164 years) by Belcher and Travis (1989) and represents the length of time that aboveground vegetation (in this case, hay) would be exposed to contaminant deposition before being harvested. *Tp* is calculated as follows:

$$Tp = \frac{60\text{ d}}{365\text{ d/yr}} = 0.164\text{ year} \qquad (1.12)$$

where

Tp = Length of plant exposure to deposition per harvest of the edible portion of plant (year)
60 = Average period between successive hay harvests (days)
365 = Units conversion factor (d/yr)

Length of Plant Exposure to Deposition per Harvest of Edible Portion of Plant (*Tp*) = 0.164 years

The primary uncertainty associated with the use of this value is that it is based on the growing season for hay rather than aboveground produce. The average period between successive hay harvests (60 days) may not reflect the length of the growing season or the period between successive harvests for aboveground produce at specific sites. To the extent that information documenting the growing season or period between successive harvests for aboveground produce is available, this information may be used to estimate a site-specific *Tp* value. Calculated plant concentrations will be affected most if the site-specific value of *Tp* is significantly less than 60 days.

For forage and silage, *Tp* is treated as a constant, based on the average period between successive hay harvests. This period, which was estimated at 60 days by Belcher and Travis (1989), represents the length of time that aboveground vegetation (in this case, hay) would be exposed to particle deposition before being harvested. For cattle forage, the risk assessor should consider the average of the period between successive:

1. Hay harvests
2. Grazing

The average period between hay harvests is assumed to be 60 days, and the average period between successive grazing is assumed to be 30 days. *Tp* is therefore calculated as follows:

$$Tp = \frac{0.5 \cdot (60\text{ d} + 30\text{ d})}{365\text{ d/yr}} = 0.12\text{ year} \tag{1.13}$$

Plant Exposure Length to Deposition per Harvest of the Edible Portion of Plant (*Tp*)

Forage = 0.12 year
Silage = 0.16 year

Standing crop biomass (productivity). USEPA (1990e) stated that the best estimate of standing crop biomass (*Yp*) is productivity, which Baes et al. (1984) and Shor et al. (1982) define as follows:

$$Yp \frac{Yh_i}{Ah_i} \tag{1.14}$$

where

Yh_i = Harvest yield of the *i*th crop (kg DW)
Ah_i = Area planted to the *i*th crop (m^2)

USEPA recommends using this equation and calculate *Yp* value of 1.6 for aboveground produce, based on weighted average *Yh* and *Ah* values for four aboveground produce classes (fruits, fruiting vegetables, legumes, and leafy vegetables). Vegetables and fruits included in each class are as follows:

- Fruits: apple, apricot, berry, cherry, cranberry, grape, peach, pear, plum/prune, and strawberry
- Fruiting vegetables: asparagus, cucumber, eggplant, sweet pepper, and tomato
- Legumes: snap beans
- Leafy vegetables: broccoli, brussel sprouts, cauliflower, celery, lettuce, and spinach.

For consistency, the produce classes have been combined into two groups: exposed fruit and exposed vegetables. The exposed vegetable *Yp* was determined by summing *Yh* values for leafy and fruiting vegetables and dividing by the sum of *Ah* values for leafy and fruiting vegetables. USEPA recommends the use of the weighted average *Yp* value of 2.24 as a default *Yp* value based on this value representing the most complete and thorough information available.

Standing Crop Biomass (Productivity) (*Yp*) = 2.24

Unweighted *Yp* and ingestion rates used for the weighting were as follows:

Aboveground produce class	*Yp*	Ingestion rate (g DW/kg · d)
Exposed fruits	0.25	0.19
Exposed vegetables	5.66	0.11

The primary uncertainty associated with this variable is that the harvest yield (*Yh*) and area planted (*Ah*) may not reflect site-specific conditions. To the extent to which site-specific information is available, the magnitude of the uncertainty introduced by the default *Yp* value can be estimated.

For forage and silage, USEPA recommends that forage *Yp* be calculated as a weighted average of the calculated pasture grass and hay *Yp* values. Weightings are assumed to be 0.75 for forage and 0.25 for hay, based on the fraction of a year that cattle are assumed to be pastured and eating grass (9 months/year) or not pastured and fed hay (3 months/year). An unweighted pasture grass *Yp* of 0.15 kg DW/m^2 is assumed. An unweighted hay *Yp* of 0.5 kg DW/m^2 is adopted by the USEPA. The unweighted pasture grass and hay *Yp* values are multiplied by their weighting factors (0.75 and 0.25, respectively), and then added to calculate the weighted forage *Yp* of 0.24 kg DW/m^2.

USEPA recommends that a production-weighted U.S. average *Yp* of 0.8 kg DW/m^2 be assumed for silage.

Standing Crop Biomass (Productivity) (*Yp*)
Forage = 0.24 kg DW/m^2 Silage = 0.8 kg DW/m^2

The primary uncertainty associated with this variable is that the harvest yield (*Yh*) and area planted (*Ah*) may not reflect site-specific conditions. To the extent that site-specific information is available, the magnitude of the uncertainty introduced by the default *Yp* value can be estimated. In addition, the weightings assumed in this discussion for the amount of time that cattle are pastured (and foraging) or stabled (and being fed silage) should be adjusted to reflect site-specific conditions, as appropriate.

Aboveground produce concentration due to air-to-plant transfer. The methodology used to estimate COPC concentration in exposed and aboveground produce due to air-to-plant transfer (*Pv*) considers limitations of COPCs concentrations to transfer from plant surfaces to the inner portions of the plant. These limitations result from mechanisms responsible for inhibiting the transfer of the lipophilic COPC (e.g., the shape of the produce) and the removal of the COPCs from the edible portion of the produce (e.g., washing, peeling, and cooking).

USEPA recommends the use of Eq. (1.15) to calculate *Pv*.

$$Pv = Q \cdot F_v \cdot \frac{Cyv \cdot Bv_{ag} \cdot VG_{ag}}{\rho_a} \tag{1.15}$$

where

Pv = Concentration of COPC in the plant resulting from air-to-plant transfer (μg COPC/g DW)
Q = COPC emission rate (g/s)
F_v = Fraction of COPC air concentration in vapor phase (unitless)
Cyv = Unitized yearly average air concentration from vapor phase (μg · s/g · m^3)
Bv_{ag} = COPC air-to-plant biotransfer factor ([mg COPC/g DW plant]/[mg COPC/g air]) (unitless)
VG_{ag} = Empirical correction factor for aboveground produce (unitless)
ρ_a = Density of air (g/m^3)

The parameter VG_{ag} is dependent on lipophilicity of the COPC, and assigned a value of 0.001 for lipophilic COPCs (log K_{ow} greater than 4) or a value of 1.0 for COPCs with a log K_{ow} less than 4.

Empirical correction factor for aboveground produce. USEPA (1994m) recommends an empirical correction factor (VG_{ag}) of 0.01 for lipophilic COPCs to reduce estimated vegetable concentrations. The factor of 0.01 is based on a similar correction factor

$$VG_{rootveg} = \frac{M_{skin}}{M_{vegetable}} \tag{1.16}$$

($VG_{rootveg}$) for belowground produce, which is estimated for unspecified vegetables as follows:

where

$VG_{rootveg}$ = Correction factor for belowground produce (g/g)
M_{skin} = Mass of a thin (skin) layer of belowground vegetable (g)
$M_{vegetable}$ = Mass of the entire vegetable (g)

If it is assumed that the density of the skin and the whole vegetable are the same, this equation can become a ratio of the volume of the skin to that of the whole vegetable. With this assumption, USEPA (1994m) calculated $VG_{rootveg}$ values of 0.09 and 0.03 for carrots and potatoes, respectively.

USEPA recommends using a lipophilic COPC (log K_{ow} greater than 4) VG_{ag} value of 0.01 for all aboveground exposed produce. For COPCs with a log K_{ow} less than 4, USEPA recommends using a VG_{ag} value of 1.0, because these COPCs are assumed to pass more easily through the skin of produce.

Empirical Correction Factor for Aboveground Produce (VG_{ag})

0.01 for COPCs with a log K_{ow} greater than 4
1.0 for COPCs with a log K_{ow} less than 4

Uncertainty may be introduced by the assumption of VG_{ag} values for leafy vegetables (such as lettuce) and for legumes (such as snap beans). Underestimation may be introduced by assuming a VG_{ag} value of 0.01 for legumes and leafy vegetables because these species often have a higher ratio of surface area to mass than other bulkier fruits and fruiting vegetables, such as tomatoes.

USEPA recommends the use of VG_{ag} values of 1.0 for forage and 0.5 for silage. As discussed, the primary uncertainty associated with this variable is the lack of specific information on the proportions of each vegetation type of which silage may consist, leading to the default assumption of 0.5.

Empirical Correction Factor for Forage and Silage (VG_{ag})

Forage = 1
Silage = 0.5

Produce concentration due to root uptake. Root uptake of contaminants from soil may also result in COPC concentrations in aboveground exposed produce, aboveground protected produce, and belowground produce. USEPA recommends the use of the following equations to calculate COPC concentration aboveground and belowground produce due to root uptake (*Pr*).

Produce Concentration Due to Root Uptake (*Pr*)

Exposed and protected aboveground produce:

$$Pr = Cs \cdot Br$$

Belowground produce:

$$Pr = \frac{Cs \cdot RCF \cdot VG_{rootveg}}{Kd_s \cdot 1\ \text{kg/L}}$$

where

Pr = Concentration of COPC in produce due to root uptake (mg/kg)
Br = Plant-soil bioconcentration factor for produce (unitless)
$VG_{rootveg}$ = Empirical correction factor for belowground produce (unitless)
Kd_s = Soil-water partition coefficient (L/kg)

Cs = Average soil concentration over exposure duration (mg COPC/kg soil)
RCF = Root concentration factor (unitless)

USEPA recommends a $VG_{rootveg}$ value of 0.01 for lipophilic COPCs (log K_{ow} greater than 4) based on root vegetables like carrots and potatoes. For COPCs with a log K_{ow} less than 4, USEPA recommends a $VG_{rootveg}$ value of 1.0.

Empirical Correction Factor for Belowground Produce ($VG_{rootveg}$)
0.01 for COPCs with a log K_{ow} greater than 4 1.0 for COPCs with a log K_{ow} less than 4

Similar to comments above regarding soil concentrations, this appears to show produce calculations for several different COPCs. Which ones is not explained.

Example calculations for the concentration of COPC in produce

```
=================================================
PRODUCE CALCULATIONS
=================================================

==========
TCDD2378 - CAS = 1746016
==========

****** Aboveground Produce PV_AG ******

-----------------------------------------------------------------
***Pv - Produce Conc. due to Air-to-Plant Transfer***
-----------------------------------------------------------------
 Q   = 1.0000E-05
 Fv   = 6.6400E-01
 CYV = 6.8610E-02
 BVAG = 6.5500E+04
 VGAG = 1.0000E-02
 RHO_A= 1.2000E+03

 PV = (Q*FV*CYV*BVAG*VGAG)/RHO_A
 Produce Conc. due to Air-to-Plant Transfer-Pv = 2.4867E-07

 ****** PV FORAGE => PV_F ******

-----------------------------------------------------------------
***Pv - Produce Conc. due to Air-to-Plant Transfer***
-----------------------------------------------------------------
 Q   = 1.0000E-05
 Fv   = 6.6400E-01
 CYV = 6.8610E-02
 BVAG = 6.5500E+04
 VGAG = 1.0000E+00
 RHO_A= 1.2000E+03
 PV   = (Q*FV*CYV*BVAG*VGAG)/RHO_A
 Produce Conc. due to Air-to-Plant Transfer-Pv = 2.4867E-05

****** PV SILAGE => PV_S ******
```

```
-----------------------------------------------------------------------------------------------
***Pv - Produce Conc. due to Air-to-Plant Transfer***
-----------------------------------------------------------------------------------------------
 Q    = 1.0000E-05
 Fv   = 6.6400E-01
 CYV  = 6.8610E-02
 BVAG = 6.5500E+04
 VGAG = 5.0000E-01
 RHO_A= 1.2000E+03
 PV   = (Q*FV*CYV*BVAG*VGAG)/RHO_A
 Produce Conc. due to Air-to-Plant Transfer-Pv = 1.2433E-05

 ****** Aboveground Produce => PD_AG ******

-----------------------------------------------------------------------------
***Pd - Produce Conc. due to Direct Dep.***
-----------------------------------------------------------------------------


Q = 1.0000E-05
FV = 6.6400E-01
(1.0E+00-FV) = 3.3600E-01
DYDP = 3.9000E-04
DYWP = 4.6420E-02
KP = 18.0000
RP = 3.9000E-01
TP = 1.6000E-01
YP = 2.2400
Pd : (1000*Q*(1-FV))        = 3.3600E-03
Pd : (1000*Q*(1-FV))/(YP*KP))    = 8.3333E-05
Pd : (DYDP+(FW*DYWP))*RP*(1.0-EXP(-KP*TP))= 1.0396E-02
Pd : Produce Conc. due to Direct Dep.= 8.6634E-07

 ****** PD FORAGE => PD_F ******

-----------------------------------------------------------------------------
***Pd - Produce Conc. due to Direct Dep.***
-----------------------------------------------------------------------------
Q = 1.0000E-05
FV = 6.6400E-01
(1.0E+00-FV) = 3.3600E-01
DYDP = 3.900E-04
DYWP = 4.642E-02
KP = 1.8000E+01
RP = 5.0000E-01
TP = 1.2000E-01
YP = 0.24 (Default number for forage [kg DW / m2])
Pd : (1000*Q*(1-FV))        = 3.3600E-03
Pd : (1000*Q*(1-FV))/(YP*KP))    = 7.7778E-04
Pd : (DYDP+(FW*DYWP))*RP*(1.0-EXP(-KP*TP))= 1.2492E-02
Pd : Produce Conc. due to Direct Dep.= 9.7164E-06

 ****** PD SILAGE => PD_S ******

-----------------------------------------------------------------------------
***Pd - Produce Conc. due to Direct Dep.***
-----------------------------------------------------------------------------


Q = 1.0000E-05
FV = 6.6400E-01
```

```
(1.0E+00-FV) = 3.3600E-01
DYDP = 3.9000E-04
DYWP = 4.6420E-02
KP = 18.00
RP = 4.6000E-01
TP = 1.6000E-01
YP = 0.80 (Default number for Silage [kg DW / m2])
Pd : (1000*Q*(1-FV))      = 3.3600E-03
Pd : (1000*Q*(1-FV))/(YP*KP))    = 2.3333E-04
Pd : (DYDP+(FW*DYWP))*RP*(1.0-EXP(-KP*TP))= 1.2262E-02
Pd : Produce Conc. due to Direct Dep.    = 2.8611E-06

 ****** PR Aboveground Produce => Pr_AG ******

-------------------------------------------------------------------
**Pr-Forage/silage/grain Conc. due to Root uptake**
-------------------------------------------------------------------
PR_AG = CS*BR_AG
Pr: Forage/silage/grain Conc. due to Root uptake= 3.9896E-07

 ****** PR FORAGE => Pr_F ******

-------------------------------------------------------------------
**Pr-Forage/silage/grain Conc. due to Root uptake**
-------------------------------------------------------------------
PR_AG = CS*BR_AG
Pr: Forage/silage/grain Conc. due to Root uptake= 3.9896E-07

 ****** PR SILAGE => Pr_S ******

-------------------------------------------------------------------
**Pr-Forage/silage/grain Conc. due to Root uptake**
-------------------------------------------------------------------
PR_AG = CS*BR_AG
Pr: Forage/silage/grain Conc. due to Root uptake= 3.9896E-07

 ****** PR GRAIN => Pr_G ******

-------------------------------------------------------------------
**Pr-Forage/silage/grain Conc. due to Root uptake**
-------------------------------------------------------------------
PR_AG = CS*BR_AG
Pr: Forage/silage/grain Conc. due to Root uptake= 3.9896E-07
 BR_RtVg = 1.030

-------------------------------------------------------------------
**Pr-Forage/silage/grain Conc. due to Root uptake**
-------------------------------------------------------------------
PR_BG = CS*BR_RtVg * VG_RtVg
Pr: Bellow ground = 9.0315E-07

 ****** PIT Aboveground Produce => PIT_AG ******

-------------------------------------------------------------------
***Pi - Total Forage/Silage/Grain Concentration***
-------------------------------------------------------------------
Pi:Forage/silage/grain Conc. due to Root uptake= 2.4171E-06
```

```
 ****** PIT FORAGE => PIT_F ******
-----------------------------------------------------------
***Pi - Total Forage/Silage/Grain Concentration***
-----------------------------------------------------------
Pi:Forage/silage/grain Conc. due to Root uptake= 3.4982E-05

 ****** PIT SILAGE => PIT_S ******
-----------------------------------------------------------
***Pi - Total Forage/Silage/Grain Concentration***
-----------------------------------------------------------
Pi:Forage/silage/grain Conc. due to Root uptake= 1.5693E-05
```

========
Silver - CAS = 7440224
========

```
****** Aboveground Produce PV_AG ******
-----------------------------------------------------------
***Pv - Produce Conc. due to Air-to-Plant Transfer***
-----------------------------------------------------------
 Q   = 2.0000E-0005
 Fv   = 9.0000E-0003
 CYV = 6.8610E-0002
 BVAG = 0.0000E+0000
 VGAG = 1.0000E+0000
 RHO_A= 1.200E+0003
 PV = (Q*FV*CYV*BVAG*VGAG)/RHO_A
 Produce Conc. due to Air-to-Plant Transfer-Pv = 0.0000E+0000

 ****** PV FORAGE => PV_F ******
-----------------------------------------------------------
***Pv - Produce Conc. due to Air-to-Plant Transfer***
-----------------------------------------------------------
 Q   = 2.0000E-0005
 Fv   = 9.0000E-0003
 CYV = 6.8610E-0002
 BVAG = 0.0000E+0000
 VGAG = 1.0000E+0000
 RHO_A= 1.2000E+0003
 PV = (Q*FV*CYV*BVAG*VGAG)/RHO_A
 Produce Conc. due to Air-to-Plant Transfer-Pv = 0.0000E+0000

 ****** PV SILAGE => PV_S ******
-----------------------------------------------------------
***Pv - Produce Conc. due to Air-to-Plant Transfer***
-----------------------------------------------------------
 Q   = 2.0000E-0005
 Fv   = 9.0000E-0003
 CYV = 6.8610E-0002
 BVAG = 0.0000E+0000
 VGAG = 5.0000E-0001
 RHO_A= 1.2000E+0003
 PV = (Q*FV*CYV*BVAG*VGAG)/RHO_A
 Produce Conc. due to Air-to-Plant Transfer-Pv = 0.0000E+0000
```

```
 ****** Aboveground Produce => PD_AG ******

------------------------------------------------------------
***Pd - Produce Conc. due to Direct Dep.***
------------------------------------------------------------

Q = 2.0000E-0005
FV = 9.0000E-0003
(1.0E+00-FV) = 9.9100E-0001
DYDP = 3.6000E-0004
DYWP = 4.6350E-0002
KP = 1.8000E+0001
RP = 3.9000E-0001
TP = 1.6000E-0001
YP = 2.2400E+0000
Pd : (1000*Q*(1-FV))      = 1.9820E-0002
Pd : (1000*Q*(1-FV))/(YP*KP))    = 4.9157E-0004
Pd : (DYDP+(FW*DYWP))*RP*(1.0-EXP(-KP*TP))= 1.0370E-0002
Pd : Produce Conc. due to Direct Dep.= 5.0974E-0006

 ****** PD FORAGE => PD_F ******

------------------------------------------------------------
***Pd - Produce Conc. due to Direct Dep.***
------------------------------------------------------------

Q = 2.0000E-0005
FV = 9.0000E-0003
(1.0E+00-FV) = 9.9100E-0001
DYDP = 3.6000E-0004
DYWP = 4.6350E-0002
KP = 1.8000E+0001
RP = 5.0000E-0001
TP = 1.2000E-0001
YP = 2.4000E-0001
Pd : (1000*Q*(1-FV))      = 1.9820E-0002
Pd : (1000*Q*(1-FV))/(YP*KP))    = 4.5880E-0003
Pd : (DYDP+(FW*DYWP))*RP*(1.0-EXP(-KP*TP))= 1.2461E-0002
Pd : Produce Conc. due to Direct Dep.= 5.7169E-0005

 ****** PD SILAGE => PD_S ******

------------------------------------------------------------
***Pd - Produce Conc. due to Direct Dep.***
------------------------------------------------------------

Q = 2.0000E-0005
FV = 9.0000E-0003
(1.0E+00-FV) = 9.9100E-0001
DYDP = 3.6000E-0004
DYWP = 4.6350E-0002
KP = 1.8000E+0001
RP = 4.6000E-0001
TP = 1.6000E-0001
YP = 8.0000E-0001
Pd : (1000*Q*(1-FV))      = 1.9820E-0002
Pd : (1000*Q*(1-FV))/(YP*KP))    = 1.3764E-0003
Pd : (DYDP+(FW*DYWP))*RP*(1.0-EXP(-KP*TP))= 1.2231E-0002
Pd : Produce Conc. due to Direct Dep.= 1.6834E-0005
```

```
 ****** PR Aboveground Produce => Pr_AG ******

--------------------------------------------------------------------
**Pr-Forage/silage/grain Conc. due to Root uptake**
--------------------------------------------------------------------
PR_AG = CS*BR_AG
Pr: Forage/silage/grain Conc. due to Root uptake= 9.1390E-0007

 ****** PR FORAGE => Pr_F ******

--------------------------------------------------------------------
**Pr-Forage/silage/grain Conc. due to Root uptake**
--------------------------------------------------------------------
PR_AG = CS*BR_AG
Pr: Forage/silage/grain Conc. due to Root uptake= 2.6490E-0006

 ****** PR SILAGE => Pr_S ******

--------------------------------------------------------------------
**Pr-Forage/silage/grain Conc. due to Root uptake**
--------------------------------------------------------------------
PR_AG = CS*BR_AG
Pr: Forage/silage/grain Conc. due to Root uptake= 2.6490E-0006

 ****** PR GRAIN => Pr_G ******

--------------------------------------------------------------------
**Pr-Forage/silage/grain Conc. due to Root uptake**
--------------------------------------------------------------------
PR_AG = CS*BR_AG
Pr: Forage/silage/grain Conc. due to Root uptake= 6.6225E-0007
 BR_RtVg = 1.0000E-0001

--------------------------------------------------------------------
**Pr-Forage/silage/grain Conc. due to Root uptake**
--------------------------------------------------------------------
PR_BG = CS*BR_RtVg * VG_RtVg
Pr: Bellow ground = 6.6225E-0007

 ****** PIT Aboveground Produce => PIT_AG ******

------------------------------------------------------------------
***Pi - Total Forage/Silage/Grain Concentration***
------------------------------------------------------------------
Pi:Forage/silage/grain Conc. due to Root uptake= 6.6735E-0006

 ****** PIT FORAGE => PIT_F ******

------------------------------------------------------------------
***Pi - Total Forage/Silage/Grain Concentration***
------------------------------------------------------------------
Pi:Forage/silage/grain Conc. due to Root uptake= 5.9818E-0005

 ****** PIT SILAGE => PIT_S ******
```

```
------------------------------------------------------------------------
***Pi - Total Forage/Silage/Grain Concentration***
------------------------------------------------------------------------
Pi:Forage/silage/grain Conc. due to Root uptake= 1.9483E-0005

========================================
DIVALENT MERCURY - CAS = 7487947
========================================

****** Aboveground Produce PV_AG ******

------------------------------------------------------------------------
***Pv - Produce Conc. due to Air-to-Plant Transfer***
------------------------------------------------------------------------
 Q   = 1.4400E-0005
 Fv   = 8.5000E-0001
 CYV = 4.6020E-0002
 BVAG = 1.8000E+0003
 VGAG = 1.0000E+0000
 RHO_A= 1.2000E+0003
 PV = (Q*FV*CYV*BVAG*VGAG)/RHO_A
 Produce Conc. due to Air-to-Plant Transfer-Pv = 8.4493E-0007

 ****** PV FORAGE => PV_F ******

------------------------------------------------------------------------
***Pv - Produce Conc. due to Air-to-Plant Transfer***
------------------------------------------------------------------------
 Q   = 1.4400E-0005
 Fv   = 8.500E-0001
 CYV = 4.6020E-0002
 BVAG = 1.8000E+0003
 VGAG = 1.0000E+0000
 RHO_A= 1.2000E+0003
 PV = (Q*FV*CYV*BVAG*VGAG)/RHO_A
 Produce Conc. due to Air-to-Plant Transfer-Pv = 8.4493E-0007

 ****** PV SILAGE => PV_S ******

------------------------------------------------------------------------
***Pv - Produce Conc. due to Air-to-Plant Transfer***
------------------------------------------------------------------------
 Q   = 1.4400E-0005
 Fv   = 8.5000E-0001
 CYV = 4.6020E-0002
 BVAG = 1.8000E+0003
 VGAG = 5.0000E-0001
 RHO_A= 1.2000E+0003
 PV = (Q*FV*CYV*BVAG*VGAG)/RHO_A
 Produce Conc. due to Air-to-Plant Transfer-Pv = 4.2246E-0007

 ****** Aboveground Produce => PD_AG ******

------------------------------------------------------------------
***Pd - Produce Conc. due to Direct Dep.***
------------------------------------------------------------------

Q = 1.4400E-0005
FV = 8.5000E-0001
```

```
(1.0E+00-FV) = 1.5000E-0001
DYDP = 3.9000E-0004
DYWP = 4.6420E-0002
KP = 1.8000E+0001
RP = 3.9000E-0001
TP = 1.6000E-0001
YP = 2.2400E+0000
Pd : (1000*Q*(1-FV))      = 2.1600E-0003
Pd : (1000*Q*(1-FV))/(YP*KP))    = 5.3571E-0005
Pd : (DYDP+(FW*DYWP))*RP*(1.0-EXP(-KP*TP))= 1.0396E-0002
Pd : Produce Conc. due to Direct Dep.= 5.5693E-0007

 ****** PD FORAGE => PD_F ******

----------------------------------------------------------
***Pd - Produce Conc. due to Direct Dep.***
----------------------------------------------------------

Q = 1.4400E-0005
FV = 8.5000E-0001
(1.0E+00-FV) = 1.5000E-0001
DYDP = 3.9000E-0004
DYWP = 4.6420E-0002
KP = 1.8000E+0001
RP = 5.0000E-0001
TP = 1.2000E-0001
YP = 2.4000E-0001
Pd : (1000*Q*(1-FV))      = 2.1600E-0003
Pd : (1000*Q*(1-FV))/(YP*KP))    = 5.0000E-0004
Pd : (DYDP+(FW*DYWP))*RP*(1.0-EXP(-KP*TP))= 1.2492E-0002
Pd : Produce Conc. due to Direct Dep.= 6.2462E-0006

 ****** PD SILAGE => PD_S ******

----------------------------------------------------------
***Pd - Produce Conc. due to Direct Dep.***
----------------------------------------------------------

Q = 1.4400E-0005
FV = 8.5000E-0001
(1.0E+00-FV) = 1.5000E-0001
DYDP = 3.9000E-0004
DYWP = 4.6420E-0002
KP = 1.8000E+0001
RP = 4.6000E-0001
TP = 1.6000E-0001
YP = 8.0000E-0001
Pd : (1000*Q*(1-FV))      = 2.1600E-0003
Pd : (1000*Q*(1-FV))/(YP*KP))    = 1.5000E-0004
Pd : (DYDP+(FW*DYWP))*RP*(1.0-EXP(-KP*TP))= 1.2262E-0002
Pd : Produce Conc. due to Direct Dep.= 1.8393E-0006

 ****** PR Aboveground Produce => Pr_AG ******

-------------------------------------------------------------------
**Pr-Forage/silage/grain Conc. due to Root uptake**
-------------------------------------------------------------------
PR_AG = CS*BR_AG
Pr: Forage/silage/grain Conc. due to Root uptake= 8.7309E-0006
```

```
 ****** PR FORAGE => Pr_F ******
-------------------------------------------------------------------------------
**Pr-Forage/silage/grain Conc. due to Root uptake**
-------------------------------------------------------------------------------
PR_AG = CS*BR_AG
Pr: Forage/silage/grain Conc. due to Root uptake= 0.0000E+0000

 ****** PR SILAGE => Pr_S ******
-------------------------------------------------------------------------------
**Pr-Forage/silage/grain Conc. due to Root uptake**
-------------------------------------------------------------------------------
PR_AG = CS*BR_AG
Pr: Forage/silage/grain Conc. due to Root uptake= 0.0000E+0000

 ****** PR GRAIN => Pr_G ******
-------------------------------------------------------------------------------
**Pr-Forage/silage/grain Conc. due to Root uptake**
-------------------------------------------------------------------------------
PR_AG = CS*BR_AG
Pr: Forage/silage/grain Conc. due to Root uptake= 5.5998E-0006
 BR_RtVg = 3.6000E-0002
-------------------------------------------------------------------------------
**Pr-Forage/silage/grain Conc. due to Root uptake**
-------------------------------------------------------------------------------

PR_BG = CS*BR_RtVg * VG_RtVg
Pr: Bellow ground = 2.1677E-0005
----------------------------------------------------------------------
|| Parameter Values for Divalent Mercury ||
----------------------------------------------------------------------

 PD_F : 0.78* 1.00D0 * PD_F = 4.8721E-0006
 PD_S : 0.78* 1.00D0 * PD_S = 1.4347E-0006
 PD_AG: 0.78* 1.00D0 * PD_AG = 4.3441E-0007
 PV_F : 0.78* 1.00D0 * PV_F = 6.5904E-0007
 PV_S : 0.78* 1.00D0 * PV_S = 3.2952E-0007
 PV_AG: 0.78* 1.00D0 * PV_AG = 6.5904E-0007

 ****** PIT Aboveground Produce => PIT_AG ******
-----------------------------------------------------------------------------
***Pi - Total Forage/Silage/Grain Concentration***
-----------------------------------------------------------------------------
Pi:Forage/silage/grain Conc. due to Root uptake= 3.1501E-0005

 ****** PIT FORAGE => PIT_F ******
-----------------------------------------------------------------------------
***Pi - Total Forage/Silage/Grain Concentration***
-----------------------------------------------------------------------------
Pi:Forage/silage/grain Conc. due to Root uptake= 5.5311E-0006
```

```
****** PIT SILAGE => PIT_S ******

---------------------------------------------------------------------
***Pi - Total Forage/Silage/Grain Concentration***
---------------------------------------------------------------------
Pi:Forage/silage/grain Conc. due to Root uptake= 1.7642E-0006
```

5.5 Calculation of COPC Concentrations in Beef and Dairy Products

One of the unique aspects of air toxics risk assessment is the estimation of risk from the consumption of beef, dairy, and other agricultural products. Cattle are assumed to consume forage, sileage, and grain that are contaminated with COPCs. The calculations necessary to then estimate the bioaccumulation of COPCs within the cattle related food sources is presented in this section.

Beef concentration resulting from plant and soil ingestion. USEPA recommends that COPC concentration in beef tissue (A_{beef}) be calculated by using the following equation.

$$A_{beef} = \left(\sum(F_i \cdot Qp_i \cdot P_i) + Qs \cdot Cs \cdot Bs\right) \cdot Ba_{beef} \cdot MF \tag{1.17}$$

where

A_{beef} = Concentration of COPC in beef (mg COPC/kg FW tissue)
F_i = Fraction of plant type *i* grown on contaminated soil and ingested by the animal (cattle) (unitless)
Qp_i = Quantity of plant type *i* eaten by the animal (cattle) per day (kg DW plant/d)
P_i = Concentration of COPC in each plant type *i* eaten by the animal (cattle) (mg/kg DW)
Qs = Quantity of soil eaten by the animal (cattle) each day (kg/d)
Cs = Average soil concentration over exposure duration (mg COPC/kg soil)
Bs = Soil bioavailability factor (unitless)
Ba_{beef} = COPC biotransfer factor for beef (d/kg FW tissue)
MF = Metabolism factor (unitless)

Fraction of plant type i grown on contaminated soil and eaten by the animal (cattle). USEPA recommends that 100 percent of the plant materials eaten by cattle be assumed to have been grown on soil contaminated by the emission sources being evaluated. USEPA recommends a default value of 1.0 for fraction of plant type i grown on contaminated soil and eaten by the animal (cattle) F_i.

Fraction of Plant Type *i* Grown on Contaminated Soil and Eaten by the Animal (Cattle) (F_i)
1.0

Quantity of plant type i eaten by the animal (cattle) each day. The daily quantity of plants eaten by cattle (Qpi) should be estimated (kg DW/day) for each category of plant feed. Forage, silage, and grain feeds should be included in this estimate. Several values of plant consumption rates are summarized in Table 1.6.

USEPA recommends in the 2005 HHRAP the following beef cattle ingestion rates of forage, silage, and grain. These values are based on the total daily intake rate of about 12 kg DW/day.

Quantity of Plant Type *i* Eaten by Animal (Cattle) Each Day (Qp_i)
Forage = 8.8 kg DW/day Beef; 13.2 kg DW/day Dairy Silage = 2.5 kg DW/day, 4.1 kg DW/day Dairy Grain = 0.47 kg DW/day, 3.0 kg DW/day Dairy

TABLE 1.6 Plant Consumption Rates from Various Regulatory Guidances

Source	Forage (kg DW/d)	Grain (kg DW/d)	Silage (kg DW/d)	References
NC DEHNR (1997) Subsistence farmer beef cattle	8.8	0.47	2.5	Boone et al. (1981) NAS (1987)
NC DEHNR (1997) Typical farmer beef cattle	3.8	3.8	1.0	Rice (1994)
USEPA (1994) Subsistence farmer beef cattle	8.8	Not reported	Not reported	Boone et al. (1981) NAS (1987)
USEPA (1990e) Subsistence farmer beef cattle	8.8	0.47	2.5	Boone et al. (1981)McKone and Ryan (1989)
Boone et al. (1981)	8.87	1.9	2.5	Boone et al. (1981)

The principal uncertainty associated with Qp_i is the variability between forage, silage, and grain ingestion rates for cattle.

Concentration of COPC in plant type i eaten by animal (cattle). The total COPC concentration in plant type i (forage, silage, and grain) eaten by animal (cattle) (P_i) should be calculated by using the following equation

Concentration of COPC in Plant Type *i* Eaten by the Animal (Cattle) (P_i)

$$P_i = \sum_i (Pd + Pv + Pr)$$

where

P_i = Concentration of COPC in each plant type *i* eaten by the animal (mg COPC/kg DW)
Pd = Plant concentration due to direct deposition (mg COPC/kg DW)
Pv = Plant concentration due to air-to-plant transfer (mg COPC/kg DW)
Pr = Plant concentration due to root uptake (mg COPC/kg DW)

Quantity of soil eaten by animal (cattle) per day. Additional cattle contamination occurs through ingestion of soil. USEPA recommends a value of 0.5 kg/d for the quantity of soil ingested by the animal (cattle) (Qs).

Quantity of Soil Ingested by Animal (Cattle) per Day (*Qs*)

0.5 kg/d – Beef Cattle
0.4 kg/d – Dairy Cattle

Soil bioavailability factor. Soil bioavailability factor (Bs) is defined as the ratio between bioconcentration (or biotransfer) factors for soil and vegetation for a given COPC. The efficiency of transfer from soil may differ from efficiency or transfer from plant material for some COPCs. If the transfer efficiency is lower for soils, than this ratio would be less than 1.0. If it is equal or greater than that of vegetation, the *Bs* value would be equal to or greater than 1.0.

Until more COPC-specific data becomes available for this parameter, USEPA recommends a default value of 1 for *Bs*.

Soil Bioavailability Factor (*Bs*)
1.0

Metabolism factor. The metabolism factor (*MF*) represents the estimated amount of COPC that remains in fat and muscle. USEPA recommends an *MF* of 0.01 for bis(2-ethylhexyl)phthalate (BEHP), and 1.0 for all other COPCs. Evidence indicates BEHP is more readily metabolized and excreted by mammalian species than other contaminants. Considering the recommended values for this variable, *MF* has no quantitative effect on A_{beef} (with the exception of BEHP).

MF applies only to mammalian species, including beef cattle, dairy cattle, and pigs. It does not relate to metabolism in produce, chicken, or fish. Use of an *MF* does not apply for direct exposures to air, soil, or water, or to ingestion of produce, chicken, or fish.

COPC concentration in milk due to plant and soil ingestion. USEPA recommends the following equation to estimate concentrations in beef cattle (A_{beef}). Eq. (1.17) can be modified to calculate COPC milk concentrations (A_{milk}), as follows:

$$A_{milk} = \left(\sum(F_i \cdot Qp_i \cdot P_i) + Qs \cdot Cs \cdot Bs\right) \cdot Ba_{milk} \cdot MF \qquad (1.18)$$

where

A_{milk} = Concentration of COPC in milk (mg COPC/kg milk)
F_i = Fraction of plant type *i* grown on contaminated soil and ingested by the animal (dairy cattle) (unitless)
Qp_i = Quantity of plant type *i* eaten by the animal (dairy cattle) each day (kg DW plant/day)
P_i = Concentration of COPC in plant type *i* eaten by the animal (dairy cattle) (mg/kg DW)
Qs = Quantity of soil eaten by the animal (dairy cattle) each day (kg soil/day)
Cs = Average soil concentration over exposure duration (mg COPC/kg soil)
Bs = Soil bioavailability factor (unitless)
Ba_{milk} = COPC biotransfer factor for milk (d/kg WW tissue)
MF = Metabolism factor (unitless)

5.6 Calculation of COPC Concentrations in Pork

COPC concentrations in pork tissue (A_{pork}) are estimated on the basis of the amount of COPCs that swine are assumed to consume through their diet; assumed to consist of silage and grain. Additional COPC contamination of pork tissue may occur through the ingestion of soil by swine.

Concentration of COPC in pork. USEPA recommends the following equation to calculate COPC concentrations in swine, as follows:

$$A_{pork} = \left(\sum(F_i \cdot Qp_i \cdot P_i) + Qs \cdot Cs \cdot Bs\right) \cdot Ba_{pork} \cdot MF \qquad (1.19)$$

where

A_{pork} = Concentration of COPC in pork (mg COPC/kg FW tissue)
F_i = Fraction of plant type *i* grown on contaminated soil and ingested by the animal (swine)(unitless)
Qp_i = Quantity of plant type *i* eaten by the animal (swine) each day (kg DW plant/day)
P_i = Concentration of COPC in plant type *i* eaten by the animal (swine) (mg/kg DW)
Qs = Quantity of soil eaten by the animal (swine) (kg/day)
Cs = Average soil concentration over exposure duration (mg COPC/kg soil)

Bs = Soil bioavailability factor (unitless)
Ba_{pork} = COPC biotransfer factor for pork (d/kg FW tissue)
MF = Metabolism factor (unitless)

Fraction of plant type i grown on contaminated soil and eaten by the animal (swine). The calculation of fraction of plant type i grown on contaminated soil and eaten by the animal (swine) (F_i) for pork is identical to that for beef cattle.

Quantity of plant type i eaten by the animal (swine) each day. The daily quantity of forage, silage, and grain feed consumed by beef cattle is estimated for each category of feed material. Because swine are not grazing animals, they are assumed not to eat forage. The quantity of plant type i eaten by the animal (swine) each day (Qpi). consumed by swine should be estimated for each category of plant feed.

USEPA recommends the use of the more conservative ingestion rates:

Quantity of Plant Type *i* Eaten by Animal (Swine) Each Day (Qp_i)
Grain = 3.3 kg DW/d Silage = 1.4 kg DW/d

Concentration of COPC in plant type i eaten by the animal (swine). The calculation of concentration of COPC in plant type i eaten by the animal (swine) (P_i) for pork is identical to that for beef cattle.

Quantity of soil eaten by the animal (swine) each day. Additional contamination of swine results from ingestion of soil. The following quantity of soil eaten by the animal (swine) each day (Q_s) values, presented in Table 1.7, are recommended by the USEPA:

USEPA recommends in the 2005 HHRAP the following Q_s:

$$Q_s = 0.37 \text{ kg DW/d}$$

Uncertainties associated with this variable include the lack of current empirical data to support soil ingestion rates for swine, and the assumption of uniform contamination of soil ingested by swine.

Average soil concentration over exposure duration. The calculation of average soil concentration over exposure duration (*Cs*) for pork is the same as for beef cattle.

Soil bioavailability factor. The calculation of soil bioavailability factor (*Bs*) for pork is the same as for beef cattle.

Metabolism factor. The recommended values for metabolism factor (*MF*) are identical to those recommended for beef cattle.

TABLE 1.7 Quantity of Soil Eaten by Swine per Day

Guidance	Quantity of soil eaten by swine each day (Q_s)
USEPA (1990e)	Stated that sufficient data are not available to estimate swine soil ingestion rates.
NC DEHNR (1997)	0.37 kg/d Estimated by assuming a soil intake that is 8 percent of the plant ingestion rate of 4.3 kg DW/d). USEPA (1993h) was cited as the reference for the soil ingestion rate of 8 percent of dry matter intake.

5.7 Calculation of COPC Concentrations in Chicken and Eggs

Estimates of the COPC concentrations in chicken and eggs are based on the amount of COPCs that chickens consume through ingestion of grain and soil. The uptake of COPCs via inhalation and via ingestion of water is assumed to be insignificant. Chickens are assumed to be free-range animals that have contact with soil; and therefore, are assumed to consume part of their diet as soil. The remainder of the diet is assumed to consist of grain. Grain ingested by chickens is assumed to have originated from the exposure scenario location; therefore, 100 percent of the grain consumed is assumed to be contaminated.

Grain is considered to be a feed item that is protected from deposition of particles and vapor transfer. As a result, only contamination due to root uptake of COPCs is considered in the calculation of COPC concentration in grain. Figure 1.9 depicts the fate and transport of chemicals, which would affect the concentration of COPC in chicken and eggs.

Concentration of COPC in chicken and eggs. USEPA recommends the application of following equation to calculate COPC concentrations in chicken and eggs ($A_{chicken}$ or A_{egg}). COPC concentrations in chicken and eggs should be determined separately.

$$A_{chicken} \text{ or } A_{egg} = \left(\sum[F_i \cdot Qp_i \cdot Pi] + Qs \cdot Cs \cdot Bs\right) \cdot (Ba_{egg} \text{ or } Ba_{chicken}) \tag{1.20}$$

where

$A_{chicken}$ = Concentration of COPC in chicken (mg COPC/kg FW tissue)
A_{egg} = Concentration of COPC in eggs (mg COPC/kg FW tissue)
F_i = Fraction of plant type *i* (grain) grown on contaminated soil and ingested by the animal (chicken)(unitless)
Qp_i = Quantity of plant type *i* (grain) eaten by the animal (chicken) each day (kg DW plant/day)
P_i = Concentration of COPC in plant type *i* (grain) eaten by the animal (chicken) (mg/kg DW)
Qs = Quantity of soil eaten by the animal (chicken) (kg/d)

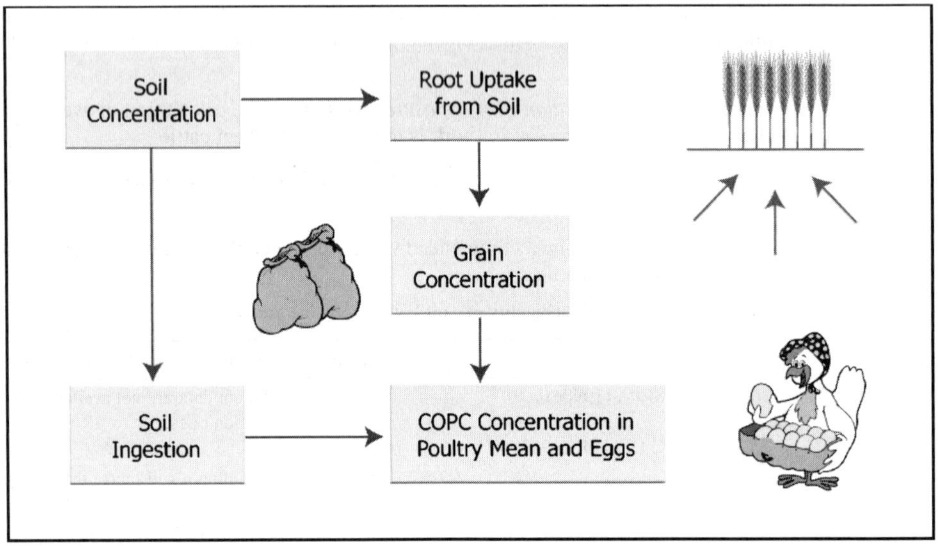

FIGURE 1.9 Fate and transport in the calculation of COPC concentration in poultry and eggs.

Cs = Average soil concentration over exposure duration (mg COPC/kg soil)
Bs = Soil bioavailability factor (unitless)
$Ba_{chicken}$ = COPC biotransfer factor for chicken (d/kg FW tissue)
Ba_{egg} = COPC biotransfer factor for eggs (d/kg FW tissue)

Fraction of plant type i grown on contaminated soil and eaten by the animal (Chicken). The calculation of fraction of plant type i grown on contaminated soil and eaten by the animal (Chicken) (Fi) is identical to that for beef cattle.

Quantity of plant type i eaten by the animal (chicken) each day. The daily quantity of forage, silage, and grain feed consumed by beef cattle is estimated for each category of feed material. However, daily ingestion rates for chicken are estimated differently than for beef cattle. Because chickens are not grazing animals, they are assumed not to eat forage (USEPA 1990e). The daily quantity of plant type (i) (kg of DW) consumed by the animal (chicken) (Qpi) only should be estimated for grain feed.

USEPA recommends the use of the following ingestion rate:

Quantity of Plant Type *i* Eaten by Animal (Chicken) Each Day (Q_{pi})
Grain = 0.2 kg DW/d

Concentration of COPC in plant type i eaten by the animal (chicken). The concentration of COPC in plant type i eaten by the animal (chicken) (Pi) is the COPC concentration in grain and should be calculated by using:

Concentration of COPC in Plant Type *i* Eaten by the Animal (Chicken) (P_i)

$$P_i = \sum_i (Pr)$$

where

P_i = Concentration of COPC in each plant type *i* eaten by the animal (mg COPC/kg DW)
Pr = Plant concentration due to root uptake (mg COPC/kg DW)

Quantity of soil eaten by the animal (chicken) each day (Qs). COPC concentration in chickens also results from intake of soil. As discussed earlier, chickens are assumed to consume 10 percent of their total diet as soil. USEPA recommends quantity of soil eaten by the animal (chicken) each day (Qs).as the following

Quantity of Soil Eaten by the Animal (Chicken) Each Day (Qs)
0.022 kg DW/day

Uncertainties associated with this variable include the lack of current empirical data to support soil ingestion rates for chicken, and the assumption of uniform contamination of soil ingested by chicken.

Average soil concentration over exposure duration. The calculation of average soil concentration over exposure duration (Cs) for chicken is the same as for beef cattle.

Soil bioavailability factor. The calculation of soil bioavailability factor (Bs) for chicken is the same as for beef cattle.

Example of calculations to estimate COPC concentration in animal products

```
==================================================
ANIMAL product CALCULATIONS
==================================================

==========
TCDD2378 - CAS = 1746016
==========

--------------------------------------------------------------------------------
**Amilk-Milk Conc. due to Plant and Soil Ingestion**
--------------------------------------------------------------------------------
Forage summation term FI*QPI_F*PIT = 4.6176E-04
Silage summation term FI*QPI_S*PIT = 6.4343E-05
Grain summation term FI*QPI_G*PIT = 1.1969E-06
Total of summation of above terms = 5.2730E-04
AMILK = (ASUM+(QS*CS*BS))*BAMILK*MF
Milk Conc. due to Plant and Soil Ingestion = 3.0926E-06

--------------------------------------------------------------------------------
**Abeef-Beef Conc. due to Plant and Soil Ingestion**
--------------------------------------------------------------------------------

Abeef : Forage summation term FI*QPI_F*PIT = 3.0784E-04
Abeef : Silage summation term FI*QPI_S*PIT = 3.9233E-05
Abeef : Grain summation term FI*QPI_G*PIT = 1.8751E-07
Abeef : Total of summation terms     = 3.4726E-04
Beef Conc. due to Plant and Soil Ingestion = 1.0216E-05

--------------------------------------------------------------------------------
**Apork-Pork Conc. due to Plant and Soil Ingestion**
--------------------------------------------------------------------------------
Apork : Grain summation term FI*QPI_G*PIT = 1.3166E-06
Apork : Silage summation term FI*QPI_S*PIT = 2.1971E-05
Apsum : Total of summation terms     = 2.3287E-05
Pork Conc. due to Plant and Soil Ingestion = 1.7622E-06

------------------------------------------------------------
** Achick-COPC Conc. in Chicken    **
------------------------------------------------------------
_____________________________-
Achick : (FI*QPI*PIT+(QS*CS*BS))*BACHIC*MF = 3.8665E-08

--------------------------------------------
**Aegg-COPC Conc. in Eggs**
--------------------------------------------
Aegg : (FI*QPI*PIT+(QS*CS*BS))*BAEGG*MF = 2.2094E-08

========
Silver - CAS = 7440224
========

--------------------------------------------------------------------------------
**Amilk-Milk Conc. due to Plant and Soil Ingestion**
--------------------------------------------------------------------------------
Forage summation term FI*QPI_F*PIT = 7.8960E-0004
Silage summation term FI*QPI_S*PIT = 7.9882E-0005
Grain summation term FI*QPI_G*PIT = 1.9867E-0006
Total of summation of above terms = 8.7147E-0004
```

```
AMILK = (ASUM+(QS*CS*BS))*BAMILK*MF
Milk Conc. due to Plant and Soil Ingestion = 1.7482E-0005

--------------------------------------------------------------------------
**Abeef-Beef Conc. due to Plant and Soil Ingestion**
--------------------------------------------------------------------------
Abeef : Forage summation term FI*QPI_F*PIT = 5.2640E-0004
Abeef : Silage summation term FI*QPI_S*PIT = 4.8708E-0005
Abeef : Grain summation term FI*QPI_G*PIT = 3.1126E-0007
Abeef : Total of summation terms       = 5.7548E-0004
Beef Conc. due to Plant and Soil Ingestion = 1.7362E-0006

--------------------------------------------------------------------------
**Apork-Pork Conc. due to Plant and Soil Ingestion**
--------------------------------------------------------------------------
Apork : Grain summation term FI*QPI_G*PIT = 2.1854E-0006
Apork : Silage summation term FI*QPI_S*PIT = 2.7277E-0005
Apsum : Total of summation terms       = 2.9462E-0005
Pork Conc. due to Plant and Soil Ingestion = 0.0000E+0000

-------------------------------------------------------
** Achick-COPC Conc. in Chicken      **
-------------------------------------------------------
Achick : (FI*QPI*PIT+(QS*CS*BS))*BACHIC*MF = 0.0000E+0000

----------------------------------------
**Aegg-COPC Conc. in Eggs**
----------------------------------------
Aegg : (FI*QPI*PIT+(QS*CS*BS))*BAEGG*MF = 0.0000E+0000

==================
DIVALENT MERCURY - CAS = 7487947
==================

--------------------------------------------------------------------------
**Amilk-Milk Conc. due to Plant and Soil Ingestion**
--------------------------------------------------------------------------
Forage summation term FI*QPI_F*PIT = 7.3011E-0005
Silage summation term FI*QPI_S*PIT = 7.2331E-0006
Grain summation term FI*QPI_G*PIT = 1.6799E-0005
Total of summation of above terms = 9.7043E-0005
AMILK = (ASUM+(QS*CS*BS))*BAMILK*MF
Milk Conc. due to Plant and Soil Ingestion = 7.6432E-0007

--------------------------------------------------------------------------
**Abeef-Beef Conc. due to Plant and Soil Ingestion**
--------------------------------------------------------------------------
Abeef : Forage summation term FI*QPI_F*PIT = 4.8674E-0005
Abeef : Silage summation term FI*QPI_S*PIT = 4.4105E-0006
Abeef : Grain summation term FI*QPI_G*PIT = 2.6319E-0006
Abeef : Total of summation terms       = 5.5717E-0005
Beef Conc. due to Plant and Soil Ingestion = 1.8624E-0006

--------------------------------------------------------------------------
**Apork-Pork Conc. due to Plant and Soil Ingestion**
--------------------------------------------------------------------------
Apork : Grain summation term FI*QPI_G*PIT = 1.8479E-0005
Apork : Silage summation term FI*QPI_S*PIT = 2.4699E-0006
```

```
Apsum : Total of summation terms        = 2.0949E-0005
Pork Conc. due to Plant and Soil Ingestion = 8.2700E-0009

----------------------------------------------------------------------------
** Achick-COPC Conc. in Chicken    **
----------------------------------------------------------------------------
Achick : (FI*QPI*PIT+(QS*CS*BS))*BACHIC*MF = 3.4372E-0007

----------------------------------------------------------
**Aegg-COPC Conc. in Eggs**
----------------------------------------------------------
Aegg : (FI*QPI*PIT+(QS*CS*BS))*BAEGG*MF = 3.4372E-0007
```

5.8 COPC Concentrations in Drinking Water and Fish

COPC concentrations in surface water are calculated for all water bodies selected for evaluation in the risk assessment; specifically, evaluation of the drinking water and/or fish ingestion exposure pathways. Mechanisms considered for determination of COPC loading of the water column, as shown in Fig. 1.10, are:

- direct deposition onto the watershed and water body,
- runoff from impervious surfaces within the watershed,

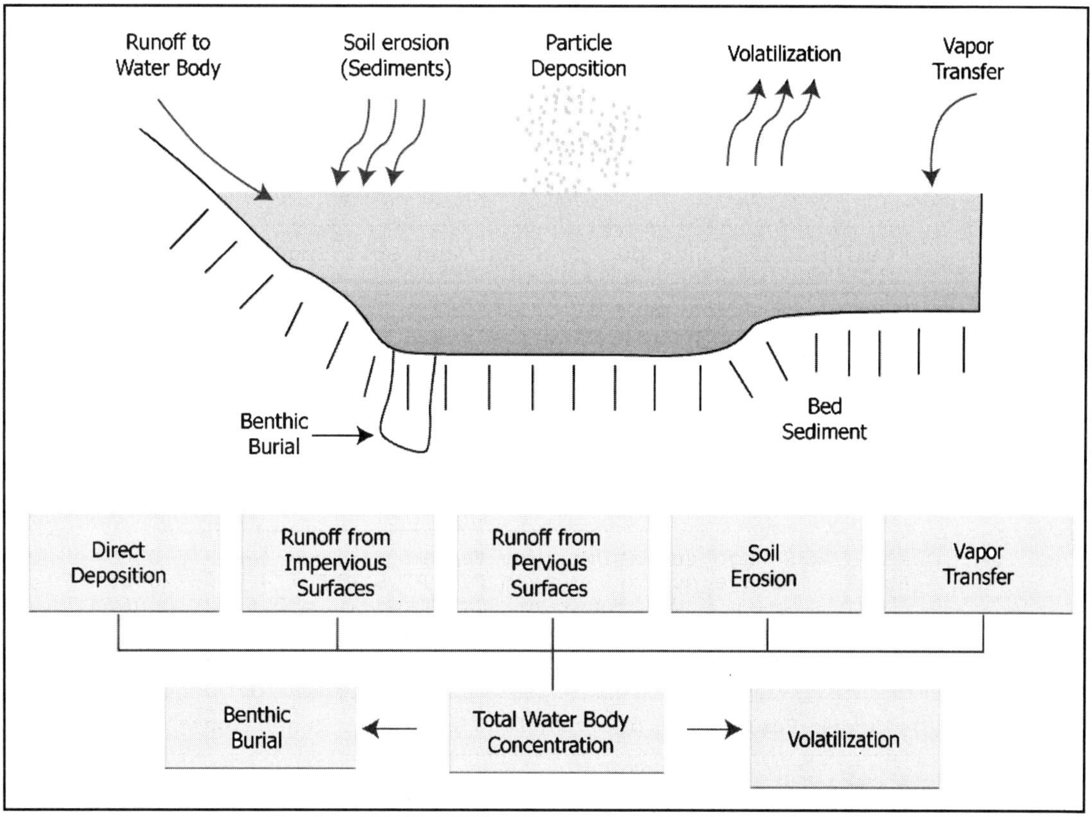

FIGURE 1.10 Processes that occur on the balance of COPC concentration in a waterbody.

- runoff from pervious surfaces within the watershed,
- soil erosion over the total watershed,
- direct diffusion of vapor phase COPCs into the surface water, and
- internal transformation of compounds chemically or biologically.

Other potential mechanisms may require consideration on a case-by-case basis (e.g., tidal influences). However, contributions from other potential mechanisms are assumed to be negligible in comparison with those being evaluated.

The USLE and a sediment delivery ratio are used to estimate the rate of soil erosion from the watershed. Surface water concentration algorithms include a sediment mass balance, in which the amount of sediment assumed to be buried and lost from the water body is equal to the difference between the amount of soil introduced to the water body by erosion and the amount of suspended solids lost in downstream flow. The total water column COPC concentration is the sum of the COPC concentration dissolved in water and the COPC concentration associated with suspended solids. Partitioning between water and sediment varies with the COPC. The total concentration of each COPC is partitioned between the sediment and the water column. To evaluate the COPC loading to a water body from its associated watershed, the COPC concentration in watershed soils should be calculated. The equation for COPC concentration in soil includes a loss term that considers the loss of contaminants from the soil after deposition. These loss mechanisms all lower the soil concentration associated with a specific deposition rate.

Total COPC load to the water body. USEPA recommends the following equation to calculate the total COPC load to a water body (L_T).

Total COPC Load to the Water Body (L_T)

$$L_T = L_{DEP} + L_{dif} + L_{RI} + L_R + L_E + L_I$$

where

L_T = Total COPC load to the water body (including deposition, runoff, and erosion) (g/yr)
L_{DEP} = Total (wet and dry) particle phase and vapor phase COPC direct deposition load to water body (g/yr)
L_{dif} = Vapor phase COPC diffusion load to water body (g/yr)
L_{RI} = Runoff load from impervious surfaces (g/yr)
L_R = Runoff load from pervious surfaces (g/yr)
L_E = Soil erosion load (g/yr)
L_I = Internal transfer (g/yr)

Due to the limited data and uncertainty associated with the chemical or biological internal transfer, L_I, of compounds into daughter products, USEPA recommends a default value for this variable of zero.

Total (wet and dry) particle-phase and vapor-phase COPC direct deposition load to water body. USEPA recommends the following equation to calculate the load to the water body from the direct deposition of wet and dry particles and vapors onto the surface of the water body (L_{DEP}).

$$L_{DEP} = Q \cdot [F_v \cdot Dytwv + (1 - F_v) \cdot Dytwp] \cdot A_W \tag{1.21}$$

where

L_{DEP} = Total (wet and dry) particle phase and vapor phase COPC direct deposition load to water body (g/yr)
Q = COPC emission rate (g/s)

F_v = Fraction of COPC air concentration in vapor phase (unitless)
$Dytwv$ = Unitized yearly (water body or watershed) average total (wet and dry) deposition from vapor phase ($s/m^2 \cdot yr$)
$Dytwp$ = Unitized yearly (water body or watershed) average total (wet and dry) deposition from particle phase ($s/m^2 \cdot yr$)
A_W = Water body surface area (m^2)

Vapor-phase COPC diffusion load to water body. USEPA recommends using the following equation to calculate the vapor-phase COPC diffusion load to the water body (L_{dif}).

$$L_{dif} = \frac{K_v \cdot Q \cdot F_v \cdot C_{ywv} A_w 1\times10^{-6}}{\frac{H}{R \cdot T_{wk}}} \tag{1.22}$$

where

L_{dif} = Vapor-phase COPC diffusion load to water body (g/yr)
K_v = Overall COPC transfer-rate coefficient (m/yr)
Q = COPC emission rate (g/s)
F_v = Fraction of COPC air concentration in vapor phase (unitless)
$Cywv$ = Unitized yearly (water body or watershed) average air concentration from vapor phase ($\mu g\text{-}s/g\text{-}m^3$)
A_W = Water body surface area (m^2)
10^{-6} = Units conversion factor ($g/\mu g$)
H = Henry's law constant ($atm \cdot m^3/mol$)
R = Universal gas constant ($atm \cdot m^3/mol \cdot K$)
T_{wk} = Water body temperature (K)

Runoff load from impervious surfaces. In some watershed soils, a fraction of the total (wet and dry) deposition in the watershed will be to impervious surfaces. This deposition may accumulate and be washed off during rain events. USEPA recommends the use of the following equation to calculate impervious runoff load to a water body (L_{RI}).

$$L_{RI} = Q \cdot [F_v \cdot Dytwv + (1.0 - F_v) \cdot Dytwp] \cdot A_I \tag{1.23}$$

where

L_{RI} = Runoff load from impervious surfaces (g/yr)
Q = COPC emission rate (g/s)
F_v = Fraction of COPC air concentration in vapor phase (unitless)
$Dytwv$ = Unitized yearly (water body or watershed) average total (wet and dry) deposition from vapor phase ($s/m^2 \cdot yr$)
$Dytwp$ = Unitized yearly (water body or watershed) average total (wet and dry) deposition from particle phase ($s/m^2 \cdot yr$)
A_I = Impervious watershed area receiving COPC deposition (m^2)

Impervious watershed area receiving COPC deposition (A_I) is the portion of the total effective watershed area that is impervious to rainfall (such as roofs, driveways, streets, and parking lots) and drains to the water body.

Runoff load from pervious surfaces. USEPA recommends the following equation to calculate the COPC load to the water body from pervious soil surfaces in the watershed (L_R).

$$L_R = RO \cdot (A_L - A_I) \cdot \frac{Cs \cdot BD}{\theta_{sw} + Kd_s \cdot BD} \cdot 0.01 \tag{1.24}$$

where

L_R = Runoff load from pervious surfaces (g/yr)
RO = Average annual surface runoff from pervious areas (cm/yr)

A_L = Total watershed area receiving COPC deposition (m^2)
A_I = Impervious watershed area receiving COPC deposition (m^2)
Cs = Average soil concentration over exposure duration (in watershed soils) (mg COPC/kg soil)
BD = Soil bulk density (g soil/cm^3 soil)
θ_{sw} = Soil volumetric water content (mL water/cm^3 soil)
Kd_s = Soil-water partition coefficient (cm^3 water/g soil)
0.01 = Units conversion factor (kg · cm^2/mg · m^2)

Soil erosion load. USEPA recommends the use of the following equation to calculate soil erosion load (L_E).

$$L_E = X_e \cdot (A_L - A_I) \cdot SD \cdot ER \cdot \frac{Cs \cdot Kd_s \cdot BD}{\theta_{sw} + Kd_s \cdot BD} \cdot 0.001 \tag{1.25}$$

where

L_E = Soil erosion load (g/yr)
X_e = Unit soil loss (kg/m^2 · yr)
A_L = Total watershed area (evaluated) receiving COPC deposition (m^2)
A_I = Impervious watershed area receiving COPC deposition (m^2)
SD = Sediment delivery ratio (watershed) (unitless)
ER = Soil enrichment ratio (unitless)
Cs = Average soil concentration over exposure duration (in watershed soils) (mg COPC/kg soil)
BD = Soil bulk density (g soil/cm^3 soil)
θ_{sw} = Soil volumetric water content (mL water/cm^3 soil)
Kd_s = Soil-water partition coefficient (mL water/g soil)
0.001 = Units conversion factor (k·cm^2/mg·m^2)

Universal soil loss equation. USEPA recommends that the USLE, be used to calculate the unit soil loss (X_e) specific to each watershed.

$$X_e = RF \cdot K \cdot LS \cdot C \cdot PF \cdot \frac{907.18}{4047} \tag{1.26}$$

where

X_e = Unit soil loss (kg/m^2 · yr)
RF = USLE rainfall (or erosivity) factor ($year^{-1}$)
K = USLE erodibility factor (ton/acre)
LS = USLE length-slope factor (unitless)
C = USLE cover management factor (unitless)
PF = USLE supporting practice factor (unitless)
907.18 = Units conversion factor (kg/ton)
4047 = Units conversion factor (m^2/acre)

The USLE *RF* variable, which represents the influence of precipitation on erosion, is derived from data on the frequency and intensity of storms. This value is typically derived on a storm-by-storm basis, but average annual values have been compiled (U.S. Department of Agriculture 1982).

Sediment delivery ratio. USEPA recommends the use of the following equation to calculate sediment delivery ratio (*SD*).

$$SD = a \cdot (A_L)^{-b} \tag{1.27}$$

where

SD = Sediment delivery ratio (watershed) (unitless)
a = Empirical intercept coefficient (unitless)

b = Empirical slope coefficient (unitless)
A_L = Total watershed area (evaluated) receiving COPC deposition (m^2)

The *SD* for a large land area, a watershed or part of a watershed, can be calculated, on the basis of the area of the watershed. According to Vanoni (1975), sediment delivery ratios vary approximately with the –0.125 power of the drainage area. Therefore, the empirical slope coefficient is assumed to be equal to 0.125.

A_L is the total watershed surface area evaluated that is affected by deposition and drains to the body of water. In assigning values to the watershed surface area affected by deposition, the following may be a consideration:

1. The distance from the emission source.
2. The location of the area affected by deposition fallout with respect to the point at which drinking water is extracted or fishing occurs.
3. The watershed hydrology.

Total water body COPC concentration. USEPA recommends the following equation to calculate total water body COPC concentration (C_{wtot}). The total water body concentration includes both the water column and the bed sediment.

$$C_{wtot} = \frac{L_T}{Vf_x \cdot f_{wc} + k_{wt} \cdot A_W \cdot (d_{wc} + d_{bs})} \quad (1.28)$$

where

C_{wtot} = Total water body COPC concentration (including water column and bed sediment) (g COPC/m^3 water body)
L_T = Total COPC load to the water body (deposition, runoff, and erosion) (g/yr)
Vf_x = Average volumetric flow rate through water body (m^3/yr)
f_{wc} = Fraction of total water body COPC concentration in the water column (unitless)
k_{wt} = Overall total water body COPC dissipation rate constant (year^{-1})
A_W = Water body surface area (m^2)
d_{wc} = Depth of water column (m)
d_{bs} = Depth of upper benthic sediment layer (m)

USEPA (1993h) recommended values ranging from 0.01 to 0.05, with a default value of 0.03, which represents the midpoint of the specified range.

Fraction of total water body COPC concentration in the water column and benthic sediment. USEPA recommends the following equation to calculate fraction of total water body COPC concentration in the water column (f_{wc}), and the subsequent equation to calculate total water body contaminant concentration in benthic sediment (f_{bs}).

Fraction of Total Water Body COPC Concentration in the Water Column (f_{wc}) and Benthic Sediment (f_{bs})

$$f_{bs} = 1 - f_{wc}$$

$$f_{wc} = \frac{(1 + Kd_{sw} \cdot TSS \cdot 1 \times 10^{-6}) \cdot d_{wc}/d_z}{(1 + Kd_{sw} \cdot TSS \cdot 1 \times 10^{-6}) \cdot d_{wc}/d_z + (\theta_{bs} + Kd_{bs} \cdot C_{BS}) \cdot d_{bs}/d_z}$$

where

f_{wc} = Fraction of total water body COPC concentration in the water column (unitless)
f_{bs} = Fraction of total water body COPC concentration in benthic sediment (unitless)
Kd_{sw} = Suspended sediments/surface water partition coefficient (L water/kg suspended sediment)

TSS = Total suspended solids concentration (mg/L)
10^{-6} = Units conversion factor (kg/mg)
d_z = Total water body depth (m)
θ_{bs} = Bed sediment porosity ($L_{water}/L_{sediment}$)
Kd_{bs} = Bed sediment/sediment pore water partition coefficient (L water/kg bottom sediment)
C_{BS} = Bed sediment concentration [g/cm^3 (equivalent to kg/L)]
d_{wc} = Depth of water column (m)
d_{bs} = Depth of upper benthic sediment layer (m)

Average annual values for TSS are generally expected to be in the range of 2 to 300 mg/L. A calculated TSS value can be obtained for nonflowing water bodies using:

$$TSS = \frac{X_e \cdot (A_L - A_I) \cdot SD \cdot 1 \times 10^3}{Vf_x + D_{ss} \cdot A_W} \tag{1.29}$$

where

TSS = Total suspended solids concentration (mg/L)
X_e = Unit soil loss (kg/m^2·yr)
A_L = Total watershed area (evaluated) receiving COPC deposition (m^2)
A_I = Impervious watershed area receiving COPC deposition (m^2)
SD = Sediment delivery ratio (watershed) (unitless)
Vf_x = Average volumetric flow rate through water body (value should be 0 for quiescent lakes or ponds) (m^3/yr)
D_{ss} = Suspended solids deposition rate (a default value of 1825 for quiescent lakes or ponds) (m/yr)
A_W = Water body surface area (m^2)

The default value of 1825 m/yr provided for D_{ss} is characteristic of Stoke's settling velocity for an intermediate (fine to medium) silt.

To evaluate the appropriateness of watershed-specific values used in calculating the unit soil loss (X_e), the waterbody specific measured TSS value should be compared to the calculated TSS value. If the measured and calculated TSS values differ significantly, parameter values used in calculating X_e should be re-evaluated.

Bed sediment porosity (θ_{bs}) can be calculated from the bed sediment concentration by using the following equation (USEPA 1993h):

$$\theta_{bs} = 1 - \frac{C_{BS}}{\rho_s} \tag{1.30}$$

where

θ_{bs} = Bed sediment porosity ($L_{water}/L_{sediment}$)
θ_s = Bed sediment density (kg/L)
C_{BS} = Bed sediment concentration (kg/L)

USEPA recommends the following default value for θ_{bs}:

$$\theta_{bs} = 0.6\ L_{water}/L_{sediment} \tag{1.31}$$

[assuming ρ_s = 2.65 kg/L (bed sediment density) and C_{BS} = 1 kg/L (bed sediment concentration)]

Concentrations for the bed sediment (C_{BS}) and depth of upper benthic sediment layer (d_{bs}) range from 0.5 to 1.5 m and 0.01 to 0.05 m, respectively. However, 0.1 kg/L is a reasonable concentration for most applications of the bed sediment (C_{BS}), and 0.03 m is the default depth of the upper benthic layer (d_{bs}). The default depth of 0.03 m is based on the midpoint of the range presented above.

Overall total water body COPC dissipation rate constant. USEPA recommends the following equation to calculate the overall total water body COPC dissipation rate constant (k_{wt}) in surface water, resulting from volatilization and benthic burial.

$$k_{wt} = f_{wc} \cdot k_v + f_{bs} \cdot k_b \tag{1.32}$$

where

k_{wt} = Overall total water body dissipation rate constant (year^{-1})
f_{wc} = Fraction of total water body COPC concentration in the water column (unitless)
k_v = Water column volatilization rate constant (year^{-1})
f_{bs} = Fraction of total water body COPC concentration in benthic sediment (unitless)
k_b = Benthic burial rate constant (yr^{-1})

Water column volatilization rate constant. USEPA recommends the following equation to calculate water column volatilization rate constant (*kv*).

$$k_v = \frac{K_v}{d_z \cdot (1 + Kd_{sw} \cdot TSS \cdot 1 \times 10^{-6})} \tag{1.33}$$

where

k_v = Water column volatilization rate constant (yr^{-1})
K_v = Overall COPC transfer rate coefficient (m/yr)
d_z = Total water body depth (m)
Kd_{sw} = Suspended sediments/surface water partition coefficient (L water/kg suspended sediments)
TSS = Total suspended solids concentration (mg/L)
1×10^{-6} = Units conversion factor (kg/mg)

Overall COPC transfer rate coefficient. Volatile organic chemicals can move between the water column and the overlying air. The overall transfer rate K_v, or conductivity, is determined by a two-layer resistance model that assumes that two stagnant films are bounded on either side by well-mixed compartments. Concentration differences serve as the driving force for the water layer diffusion. Pressure differences drive the diffusion for the air layer. From balance considerations, the same mass must pass through both films; the two resistances thereby combine in series, so that the conductivity is the reciprocal of the total resistance.

USEPA recommends the following equation to calculate the overall transfer rate coefficient (K_v).

Overall COPC Transfer Rate Coefficient (K_v)

$$K_v = \left(K_L^{-1} + \left(K_G \cdot \frac{H}{R \cdot T_{wk}}\right)^{-1}\right)^{-1} \cdot \theta^{T_{wk}-293}$$

where

K_v = Overall COPC transfer rate coefficient (m/yr)
K_L = Liquid-phase transfer coefficient (m/yr)
K_G = Gas-phase transfer coefficient (m/yr)
H = Henry's law constant (atm · m^3/mol)
R = Universal gas constant (atm · m^3/mol · K)
T_{wk} = Water body temperature (K)
θ = Temperature correction factor (unitless)

The value of the conductivity K_v depends on the intensity of turbulence in the water body and the overlying atmosphere. As Henry's law constant increases, the conductivity tends to be increasingly influenced by the intensity of turbulence in water. Conversely, as Henry's law constant decreases, the value of the conductivity tends to be increasingly influenced by the intensity of atmospheric turbulence. The liquid and gas phase transfer coefficients, K_L and K_G, respectively, vary with the type of water body.

Henry's law constants generally increase with increasing vapor pressure of a COPC and generally decrease with increasing solubility of a COPC. Henry's law constants are compound-specific. The universal ideal gas constant, R, is 8.205×10^{-5} atm · m^3/mol · K, at 20°C. The temperature correction factor (θ), which is equal to 1.026, is used to adjust for the actual water temperature. Volatilization is assumed to occur much less readily in lakes and reservoirs than in moving water bodies.

Liquid-phase transfer coefficient. USEPA recommends the following two equations to calculate liquid phase transfer coefficient (K_L).

For flowing streams or rivers:

$$K_L = \sqrt{\frac{(1 \times 10^{-4}) \cdot D_w \cdot u}{d_z} \cdot 3.1536 \times 10^7} \tag{1.34a}$$

For quiescent lakes or ponds:

$$K_L = (C_d^{0.5} \cdot W) \cdot \left(\frac{\rho_a}{\rho_w}\right) \cdot \frac{k^{0.33}}{\lambda_z} \cdot \left(\frac{\mu_w}{\rho_w \cdot D_w}\right)^{-0.67} \cdot 3.1536 \times 10^7 \tag{1.34b}$$

where

K_L = Liquid-phase transfer coefficient (m/yr)
D_w = Diffusivity of COPC in water (cm^2/s)
u = Current velocity (m/s)
1×10^{-4} = Units conversion factor (m^2/cm^2)
d_z = Total water body depth (m)
C_d = Drag coefficient (unitless)
W = Average annual wind speed (m/s)
ρ_a = Density of air (g/cm^3)
ρ_w = Density of water (g/cm^3)
k = von Karman's constant (unitless)
λ_z = Dimensionless viscous sublayer thickness (unitless)
μ_w = Viscosity of water corresponding to water temperature (g/cm·s)
3.1536×10^7 = Units conversion factor (s/yr)

For a flowing stream or river, the transfer coefficients are controlled by flow-induced turbulence. For a stagnant system (quiescent lake or pond), the transfer coefficient is controlled by wind-induced turbulence.

USEPA recommends the use of the following default values:

1. Diffusivity of chemical in water ranging (D_w) from 1.0×10^{-5} to 8.5×10^{-2} cm^2/s
2. Dimensionless viscous sublayer thickness (λ_z) of 4
3. von Karman's constant (k) of 0.4
4. Drag coefficient (C_d) of 0.0011
5. Density of air (ρ_a) of 0.0012 g/cm^3 at standard conditions (temperature = 20°C or 293 K, pressure = 1 atm or 760 mm of mercury)
6. Density of water (ρ_w) of 1 g/cm^3,
7. Viscosity of water (ρ_w) of a 0.0169 g/cm·s corresponding to water temperature

Gas phase transfer coefficient. USEPA recommends the following two equations to calculate gas phase transfer coefficient (K_G).

Gas-Phase Transfer Coefficient (K_G)

For flowing streams or rivers:

$$K_G = 36500 \text{ m/yr}$$

For quiescent lakes or ponds:

$$K_G = (C_d^{0.5} \cdot W) \cdot \frac{k^{0.33}}{\lambda_z} \cdot \left(\frac{\mu_a}{\rho_a \cdot D_a}\right)^{-0.67} \cdot 3.1536 \times 10^7$$

where

K_G = Gas phase transfer coefficient (m/yr)
C_d = Drag coefficient (unitless)
W = Average annual wind speed (m/s)
k = von Karman's constant (unitless)
λ_z = Dimensionless viscous sublayer thickness (unitless)
μ_a = Viscosity of air corresponding to air temperature (g/cm · s)
ρ_a = Density of air corresponding to water temperature (g/cm^3)
D_a = Diffusivity of COPC in air (cm^2/s)
3.1536×10^7 = Units conversion factor (s/yr)

USEPA indicated that the rate of transfer of a COPC from the gas phase for a flowing stream or river is assumed to be constant. For a stagnant system (quiescent lake or pond), the transfer coefficients are controlled by wind-induced turbulence. For quiescent lakes or ponds, USEPA recommends that the gas phase transfer coefficient be computed by using the equation presented in O'Connor (1983). USEPA recommends 1.81×10^{-4} g/cm · s for the viscosity of air corresponding to air temperature.

Benthic burial rate constant. USEPA recommends the following equation to calculate benthic burial rate (k_b).

$$k_b = \left(\frac{X_e \cdot A_L \cdot SD \cdot 1 \times 10^3 - Vf_x \cdot TSS}{A_W \cdot TSS}\right) \cdot \left(\frac{TSS \cdot 1 \times 10^{-6}}{C_{BS} \cdot d_{bs}}\right) \quad (1.35)$$

where

k_b = Benthic burial rate constant (year^{-1})
X_e = Unit soil loss (kg/m^2 · yr)
A_L = Total watershed area (evaluated) receiving deposition (m^2)
SD = Sediment delivery ratio (watershed) (unitless)
Vf_x = Average volumetric flow rate through water body (m^3/yr)
TSS = Total suspended solids concentration (mg/L)
A_W = Water body surface area (m^2)
C_{BS} = Bed sediment concentration (g/cm^3)
d_{bs} = Depth of upper benthic sediment layer (m)
1×10^{-6} = Units conversion factor (kg/mg)
1×10^3 = Units conversion factor (g/kg)

The benthic burial rate constant (k_b) can be expressed in terms of the rate of burial (Wb):

$$Wb = k_b \cdot d_{bs} \quad (1.36)$$

where

Wb = Rate of burial (m/yr)
k_b = Benthic burial rate constant (year^{-1})
d_{bs} = Depth of upper benthic sediment layer (m)

USEPA recommends the following default value for bed sediment concentration (C_{BS}).

$$C_{BS} = 1.0 \text{ kg/L} \tag{1.37}$$

The calculated value for k_b should range from 0 to 1.0; with low k_b values expected for water bodies characteristic of no or limited sedimentation (rivers and fast flowing streams), and k_b values closer to 1.0 expected for water bodies characteristic of higher sedimentation (lakes). For k_b values calculated as a negative (water bodies with high average annual volumetric flow rates in comparison to watershed area evaluated), a k_b value of 0 should be assigned for use in calculating the total water body COPC concentration (C_{wtot}). If the calculated k_b value exceeds 1.0, re-evaluation of the parameter values used in calculating X_e should be conducted.

Total COPC concentration in water column. USEPA recommends the following equation to calculate total COPC concentration in water column (C_{wctot}).

$$C_{wctot} = f_{wc} \cdot C_{wtot} \cdot \frac{d_{wc} + d_{bs}}{d_{wc}} \tag{1.38}$$

where

C_{wctot} = Total COPC concentration in water column (mg COPC/L water column)
f_{wc} = Fraction of total water body COPC concentration in the water column (unitless)
C_{wtot} = Total water body COPC concentration, including water column and bed sediment (mg COPC/L water body)
d_{wc} = Depth of water column (m)
d_{bs} = Depth of upper benthic sediment layer (m)

Dissolved phase water concentration. USEPA recommends the following equation to calculate the concentration of COPC dissolved in the water column (C_{dw}).

$$C_{dw} = \frac{C_{wctot}}{1 + Kd_{sw} \cdot TSS \cdot 1 \times 10^{-6}} \tag{1.39}$$

where

C_{dw} = Dissolved phase water concentration (mg COPC/L water)
C_{wctot} = Total COPC concentration in water column (mg COPC/L water column)
Kd_{sw} = Suspended sediments/surface water partition coefficient (L water/kg suspended sediment)
TSS = Total suspended solids concentration (mg/L)
1×10^{-6} = Units conversion factor (kg/mg)

COPC Concentration sorbed to bed sediment. USEPA recommends the following equation to calculate C_{sb}.

$$C_{sb} = f_{bs} \cdot C_{wtot} \cdot \left(\frac{Kd_{bs}}{\theta_{bs} + Kd_{bs} \cdot C_{BS}}\right) \cdot \left(\frac{d_{wc} + d_{bs}}{d_{bs}}\right) \tag{1.40}$$

where

C_{sb} = COPC concentration sorbed to bed sediment (mg COPC/kg sediment)
f_{bs} = Fraction of total water body COPC concentration in benthic sediment (unitless)
C_{wtot} = Total water body COPC concentration, including water column and bed sediment (mg COPC/L water body)
Kd_{bs} = Bed sediment/sediment pore water partition coefficient (L COPC/kg water body)
θ_{bs} = Bed sediment porosity ($L_{\text{pore water}}/L_{\text{sediment}}$)
C_{BS} = Bed sediment concentration (g/cm^3)
d_{wc} = Depth of water column (m)
d_{bs} = Depth of upper benthic sediment layer (m)

Concentration of COPC in fish. The COPC concentration in fish (C_{fish}) is calculated using either a COPC-specific bio-concentration factor (*BCF*), a COPC-specific bioaccumulation factor (*BAF*), or a COPC-specific biota sediment accumulation factor (*BSAF*). For compounds with a $log\ K_{ow}$ less than 4.0, *BCF*s are used. Compounds with a $log\ K_{ow}$ greater than 4.0 (except for extremely hydrophobic compounds such as dioxins, furans, and PCBs), are assumed to have a high tendency to bioaccumulate, therefore, *BAF*s and *BSAF*s are used to calculate COPC concentrations in fish for dioxins, furans, and PCBs.

BCF and *BAF* values are generally based on dissolved water concentrations. Therefore, when *BCF* or *BAF* values are used, the *COPC* concentration in fish is calculated using dissolved water concentrations. *BSAF* values are based on benthic sediment concentrations. Therefore, when *BSAF* values are used, COPC concentration in fish is calculated using benthic sediment concentrations. The equations used to calculate fish concentrations are described in the subsequent subsections.

Fish concentration from bioconcentration factors using dissolved phase water concentration.

$$C_{fish} = C_{dw} \cdot BCF_{fish} \tag{1.41}$$

where

C_{fish} = Concentration of COPC in fish (mg COPC/kg FW tissue)
C_{dw} = Dissolved-phase water concentration (mg COPC/L)
BCF_{fish} = Bioconcentration factor for COPC in fish (L/kg)

Fish concentration from bioaccumulation factors using dissolved phase water concentration.

$$C_{fish} = C_{dw} \cdot BAF_{fish} \tag{1.42}$$

where

C_{fish} = Concentration of COPC in fish (mg COPC/kg FW tissue)
C_{dw} = Dissolved phase water concentration (mg COPC/L)
BAF_{fish} = Bioaccumulation factor for COPC in fish (L/kg FW tissue)

Fish concentration from biota-to-sediment accumulation factors using COPC sorbed to bed sediment.

$$C_{fish} = \frac{C_{sb} \cdot f_{lipid} \cdot BSAF}{OC_{sed}} \tag{1.43}$$

where

C_{fish} = Concentration of COPC in fish (mg COPC/kg FW tissue)
C_{sb} = Concentration of COPC sorbed to bed sediment (mg COPC/kg bed sediment)
f_{lipid} = Fish lipid content (unitless)
$BSAF$ = Biota-to-sediment accumulation factor (unitless)
OC_{sed} = Fraction of organic carbon in bottom sediment (unitless)

Values for the fraction of organic carbon in bottom sediment (Oc_{sed}) range from 0.03 to 0.05 f. These values are based on an assumption of a surface soil OC content of 0.01. USEPA (1993g) recommended a default value of 0.04 for OC_{sed}, which is the midpoint of the specified range. USEPA also recommends the use of 0.07 as the fish lipid content (f_{lipid}).

5.9 Use of Site-Specific versus Default Parameter Values

After completing a risk assessment based on the default parameter values recommended in this guidance, risk assessors may choose to investigate the use of site-specific parameter values in order to provide a more representative estimate of site-specific risk. Use of parameter values other than those specified in this guidance should always be clearly described in the risk assessment report and work plan.

Example calculations to estimate COPC concentration in water and fish

```
==================================================
WATER CALCULATIONS
==================================================

==========
TCDD2378 - CAS = 1746016
==========

******************************
 TSS = 10.00
********************************

----------------------------------------------------------------
** Ldep-Deposition to Water Body     **
----------------------------------------------------------------

 DYTWP = 3.2020E-02
 DYTWV = 1.1780E-02
 AW   = 1.0000E+06 (Water Body Area [m2])
Ldep : Q*(FV*DYTWV+(1-FV)*DYTWP)*AW = 1.8581E-01

--------------------------------------------------------------------------------
** LRI - Impervious Runofff Load to Water Body **
--------------------------------------------------------------------------------

 DYTWP = 3.4130E-02
 DYTWV = 1.1630E-02
 AI   = 1.5000E+05
 LRI  : Q*(FV*DYTWV+(1.0-FV)*DYTWP)*AI = 2.8785E-02

------------------------------------------------------------------------
**LR - Pervious Runoff Load to Water Body **
------------------------------------------------------------------------

KDS = 3.8901E+04
LR : RO*(AL-AI)                = 6.8400E+07
LR : (CS*BD / (THETSW+KDS*BD))*0.01 = 1.7799E-0011
LR TOTAL : LR1*LR2 = 1.2174176E-03

----------------------------------------------------------------
** LE - Erosion Load to Water Body   **
----------------------------------------------------------------

XE * (AL-AI) * SD * ER        = 1.1230E+06
CS*KDS*BD / (THETSW+KDS*BD) * 0.001 = 6.9244E-08
LE TOTAL : LE1*LE2 = 7.7761E-02

--------------------------------------------------------------------------------
**KL - Liquid Phase Transfer Coefficient - LAKES**
--------------------------------------------------------------------------------

KL1 : SQRT(CD)*W*SQRT(RHO_A/RHO_W)*(K**0.33/GAMMAZ)= 8.2789E-04
KL2 : (MU_W/(RHO_W*DW))**(-0.67)*3.1536E+07     = 1.4703E+05
KL TOTAL : KL1*KL2 = 1.21725024709711E+02

----------------------------------------------------------------------------
**KG - Gas Phase Transfer Coefficient - LAKES**
----------------------------------------------------------------------------

KG1 : SQRT(CD)*W*(POW(K,0.33)/GAMMAZ)      = 2.3899E-02
KG2 : ((MU_A/(RHO_A*DA))**(-0.67))*3.1536E+07 = 2.4582E+07
KG TOTAL : KG1*KG2 = 5.8750E+05
```

```
---------------------------------------------------------------------
**KV-Overall COPC Transfer Rate Coefficient - LAKES**
---------------------------------------------------------------------
KV1 : ((1/KL)+(1/(KG*(H/(R*TWK)))))**(-1)= 1.0548E+02
KV2 : THETA**(TWK-293.0)          = 1.1369E+00
KV TOTAL : KV1*KV2 = 1.1993E+02

-----------------------------------------------------------
**Ldif-Diffusion Load to Water Body - LAKES**
-----------------------------------------------------------
LDTOP : KV*Q*FV*CYWV*AW*(1E-06)= 5.1076E-05
LDBOT : H/(R*TWK)          = 1.3456E-03
LDIF : LDTOP/LDBOT = 3.7959E-02

------------------------------------------
**LT - Total Water Body Load **
------------------------------------------
 LDEP = 1.8581E-01
 LDIF = 3.7959E-02
 LRI = 2.8785E-02
 LR  = 1.2174E-03
 LE  = 7.7761E-02
 LT : LDEP+LDIF+LRI+LR+LE = 3.3153E-01

----------------------------------------------------------
**kv_low - Water Column Volatile Loss Rate**
----------------------------------------------------------
KDSW = 2.9178E+05
(DZ*(1+KDSW*TSS*1.0E-06))) = 3.9296E+01
kv_low - KV_UP/(DZ*(1+KDSW*TSS*1.0E-06))) = 3.0519E+00

---------------------------------------------------
** Fwc - Fraction in Water Column   **
---------------------------------------------------
FWCTOP : (1+KDSW*TSS*1.0E-6)*DWC/DZ= 3.9061E+00
FWCBOT : (1+KDSW*TSS*1.0E-6)*DWC/DZ+(THETBS+KDBS*CBS)*DBS/DZ =
4.6937E+02
FWC : FWCTOP/FWCDBOT = 8.3221E-03

--------------------------------------------------------
** Fbs - Fraction in Benthic Sed.        **
--------------------------------------------------------
FBS - 1-FWC = 9.9168E-01

----------------------------------------------------
**Kb - Benthic Burial Rate Constant  **
----------------------------------------------------
KB1 : (XE*AL*SD*1.0E03-VFX*TSS)/(AW*TSS)= 3.8403E+01
KB2 : (TSS*1.0E-06)/(CBS*DBS)       = 3.3333E-04
KB : KB1*KB2 = 1.28011170920617E-02

-----------------------------------------------------------------
**Kwt - Overall Waterbody Dissipation Rate Constant
-----------------------------------------------------------------

FWC   = 8.3221E-03
KV_low = 3.0519E+00
```

```
FBS   = 9.9168E-01
KB    = 1.2801E-02
KWT - FWC*KV_low+FBS*KB = 3.8093E-02

-----------------------------------------------------------------------
**Cwtot - Total Water Body Concentration**
-----------------------------------------------------------------------

CWTOT - LT/(VFX*FWC+KWT*AW*(DWC+DBS)) = 8.4922E-07

-----------------------------------------------------------------------
**Cwctot - Total Water Column Concentration**
-----------------------------------------------------------------------

FWC   = 8.32212507409816E-03
CWTOT = 8.49216697352239E-07
DWC   = 10.0
DBS   = 3.00000000000000E-02
CWCTOT - FWC*CWTOT*((DWC+DBS)/DWC) = 7.0885E-09

-----------------------------------------------------------------------
**Cdw - Dissolved Phase Water Concentration**
-----------------------------------------------------------------------

CWCTOT = 7.0885E-09
KDSW   = 2.9178E+05
TSS    = 10.0
CDW - CWCTOT/(1+KDSW*TSS*1.0E-06) = 1.8093E-09

-----------------------------------------------------------------------
**Csorb - COPC Conc. Sorbed to Bed Sediment**
-----------------------------------------------------------------------

FBS   = 9.9168E-01
CWTOT = 8.4922E-07
KDBS  = 1.5562E+05
THETBS= 6.0000E-01
CBS   = 1.00
DWC   = 10.0
DBS   = 3.00000E-02
FBS*CWTOT*(KDBS/(THETBS+KDBS*CBS))*((DWC+DBS/DBS))
CSB = 2.8156E-04
```

========
Silver - CAS = 7440224
========

```
******************************
 TSS = 1.0000E+0001
 ******************************

-----------------------------------------------------------------------
** Ldep-Deposition to Water Body   **
-----------------------------------------------------------------------

 DYTWP = 3.1760E-0002
 DYTWV = 1.1780E-0002
 AW   = 1.0000E+0006
Ldep : Q*(FV*DYTWV+(1-FV)*DYTWP)*AW = 6.3160E-0001
```

```
--------------------------------------------------------------------
** LRI - Impervious Runofff Load to Water Body **
--------------------------------------------------------------------
 DYTWP = 3.3930E-0002
 DYTWV = 1.1630E-0002
 AI   = 1.5000E+0005
 LRI  : Q*(FV*DYTWV+(1.0-FV)*DYTWP)*AI = 1.01188E-0001

--------------------------------------------------------------
**LR - Pervious Runoff Load to Water Body **
--------------------------------------------------------------
KDS = 8.3000E+0000
LR : RO*(AL-AI)                    = 6.8400E+0007
LR : (CS*BD / (THETSW+KDS*BD))*0.01 = 5.7076E-0009
LR TOTAL : LR1*LR2 = 3.9040E-0001

------------------------------------------------------
** LE - Erosion Load to Water Body  **
------------------------------------------------------
XE * (AL-AI) * SD * ER          = 3.7433E+0005
CS*KDS*BD / (THETSW+KDS*BD) * 0.001 = 4.7373E-0009
LE TOTAL : LE1*LE2 = 1.7733E-0003

---------------------------------------------------------------------
**KL - Liquid Phase Transfer Coefficient - LAKES**
---------------------------------------------------------------------
KL1 : SQRT(CD)*W*SQRT(RHO_A/RHO_W)*(K**0.33/GAMMAZ)= 8.2789E-0004
KL2 : (MU_W/(RHO_W*DW))**(-0.67)*3.1536E+07     = 2.1054E+0005
KL TOTAL : KL1*KL2 = 1.7430E+0002

-----------------------------------------------------------------
**KG - Gas Phase Transfer Coefficient - LAKES**
-----------------------------------------------------------------
KG1 : SQRT(CD)*W*(POW(K,0.33)/GAMMAZ)        = 2.3899E-0002
KG2 : ((MU_A/(RHO_A*DA))**(-0.67))*3.1536E+07 = 2.0133E+0007
KG TOTAL : KG1*KG2 = 4.8117E+0005

-------------------------------------------------------------------------
**KV-Overall COPC Transfer Rate Coefficient - LAKES**
-------------------------------------------------------------------------
KV1 : ((1/KL)+(1/(KG*(H/(R*TWK)))))**(-1)= 1.7430E+0002
KV2 : THETA**(TWK-293.0)              = 1.1369E+0000
KV TOTAL : KV1*KV2 = 1.9817E+0002

--------------------------------------------------------------
**Ldif-Diffusion Load to Water Body - LAKES**
--------------------------------------------------------------
LDTOP : KV*Q*FV*CYWV*AW*(1E-06)= 2.2879E-0006
LDBOT : H/(R*TWK)          = 0.0000E+0000
LDIF : LDTOP/LDBOT = 0.0000E+0000

-------------------------------------------
**LT - Total Water Body Load **
-------------------------------------------
 LDEP = 6.3160E-0001
 LDIF = 0.0000E+0000
```

```
 LRI = 1.0119E-0001
 LR  = 3.9040E-0001
 LE  = 1.773E-0003
 LT : LDEP+LDIF+LRI+LR+LE = 1.1250E+0000

----------------------------------------------------------------------
**kv_low - Water Column Volatile Loss Rate**
----------------------------------------------------------------------
KDSW = 8.3000E+0000
(DZ*(1+KDSW*TSS*1.0E-06))) = 1.0031E+0001
kv_low - KV_UP/(DZ*(1+KDSW*TSS*1.0E-06))) = 1.9756E+0001

------------------------------------------------------------
** Fwc - Fraction in Water Column  **
------------------------------------------------------------
FWCTOP : (1+KDSW*TSS*1.0E-6)*DWC/DZ= 9.9709E-0001
FWCBOT : (1+KDSW*TSS*1.0E-6)*DWC/DZ+(THETBS+KDBS*CBS)*DBS/DZ =
1.0237E+0000
FWC : FWCTOP/FWCDBOT = 9.7400E-0001

------------------------------------------------------------
** Fbs - Fraction in Benthic Sed.  **
------------------------------------------------------------
FBS - 1-FWC = 2.6004E-0002

----------------------------------------------------------------
**Kb - Benthic Burial Rate Constant  **
----------------------------------------------------------------
KB1 : (XE*AL*SD*1.0E03-VFX*TSS)/(AW*TSS)= 3.8403E+0001
KB2 : (TSS*1.0E-06)/(CBS*DBS)    = 3.3333E-0004
KB : KB1*KB2 = 1.2801E-0002

--------------------------------------------------------------------------------
**Kwt - Overall Waterbody Dissipation Rate Constant
--------------------------------------------------------------------------------

FWC   = 9.7400E-0001
KV_low = 1.9756E+0001
FBS   = 2.6004E-0002
KB    = 1.2801E-0002
KWT - FWC*KV_low+FBS*KB = 1.9243E+0001

--------------------------------------------------------------------
**Cwtot - Total Water Body Concentration**
--------------------------------------------------------------------
CWTOT - LT/(VFX*FWC+KWT*AW*(DWC+DBS)) = 5.7994E-0009

------------------------------------------------------------------------
**Cwctot - Total Water Column Concentration**
------------------------------------------------------------------------

FWC   = 9.7400E-0001
CWTOT = 5.7994E-0009
DWC   = 1.0000E+0001
DBS   = 3.0000E-0002
CWCTOT - FWC*CWTOT*((DWC+DBS)/DWC) = 5.6655E-0009
```

```
---------------------------------------------------------------
**Cdw - Dissolved Phase Water Concentration**
---------------------------------------------------------------

CWCTOT = 5.6655E-0009
KDSW   = 8.3000E+0000
TSS    = 1.0000E+0001
CDW - CWCTOT/(1+KDSW*TSS*1.0E-06) = 5.6651E-0009

---------------------------------------------------------------
**Csorb - COPC Conc. Sorbed to Bed Sediment**
---------------------------------------------------------------

FBS    = 2.6004E-0002
CWTOT = 5.7994E-0009
KDBS = 8.3000E+0000
THETBS= 6.0000E-0001
CBS    = 1.0000E+0000
DWC    = 1.0000E+0001
DBS    = 3.0000E-0002
FBS*CWTOT*(KDBS/(THETBS+KDBS*CBS))*((DWC+DBS/DBS))
CSB = 4.7020E-0008
```

==================
DIVALENT MERCURY - CAS = 7487947
==================

```
******************************
 TSS = 1.00000000000000E+0001
 ******************************

-----------------------------------------------------
** Ldep-Deposition to Water Body     **
-----------------------------------------------------

 DYTWP = 3.2020E-0002
 DYTWV = 3.8290E-0002
 AW    = 1.0000E+0006
Ldep : Q*(FV*DYTWV+(1-FV)*DYTWP)*AW = 5.3783E-0001

---------------------------------------------------------------------
** LRI - Impervious Runofff Load to Water Body **
---------------------------------------------------------------------
 DYTWP = 3.4130E-0002
 DYTWV = 3.7730E-0002
 AI    = 1.5000E+0005
 LRI  : Q*(FV*DYTWV+(1.0-FV)*DYTWP)*AI = 8.0330E-0002

---------------------------------------------------------------
**LR - Pervious Runoff Load to Water Body ***
---------------------------------------------------------------
KDS = 5.8000E+0004
LR : RO*(AL-AI)              = 6.8400E+0007
LR : (CS*BD / (THETSW+KDS*BD))*0.01 = 8.9571E-0011
LR TOTAL : LR1*LR2 = 6.1267E-0003
```

```
--------------------------------------------------------------
** LE - Erosion Load to Water Body  **
--------------------------------------------------------------
XE * (AL-AI) * SD * ER        = 3.7433E+0005
CS*KDS*BD / (THETSW+KDS*BD) * 0.001 = 5.1951E-0007
LE TOTAL : LE1*LE2 = 1.9447E-0001

--------------------------------------------------------------
**KL - Liquid Phase Transfer Coefficient - LAKES**
--------------------------------------------------------------
KL1 : SQRT(CD)*W*SQRT(RHO_A/RHO_W)*(K**0.33/GAMMAZ)= 8.2789E-
0004
KL2 : (MU_W/(RHO_W*DW))**(-0.67)*3.1536E+07     = 1.4075E+0005
KL TOTAL : KL1*KL2 = 1.1652E+0002

--------------------------------------------------------------
**KG - Gas Phase Transfer Coefficient - LAKES**
--------------------------------------------------------------
KG1 : SQRT(CD)*W*(POW(K,0.33)/GAMMAZ)        = 2.3899E-0002
KG2 : ((MU_A/(RHO_A*DA))**(-0.67))*3.1536E+07 = 1.4089E+0007
KG TOTAL : KG1*KG2 = 3.3671E+0005

--------------------------------------------------------------
**KV-Overall COPC Transfer Rate Coefficient - LAKES**
--------------------------------------------------------------
KV1 : ((1/KL)+(1/(KG*(H/(R*TWK)))))**(-1)= 9.7766E-0003
KV2 : THETA**(TWK-293.0)             = 1.1369E+0000
KV TOTAL : KV1*KV2 = 1.1115E-0002

--------------------------------------------------------------
**Ldif-Diffusion Load to Water Body - LAKES**
--------------------------------------------------------------
LDTOP : KV*Q*FV*CYWV*AW*(1E-06)= 5.6815E-0009
LDBOT : H/(R*TWK)          = 2.9038E-0008
LDIF : LDTOP/LDBOT = 1.9566E-0001


--------------------------------------------------------------
|| Parameter Values for Divalent Mercury ||
--------------------------------------------------------------
 LDEP : 0.85* 1.00D0 * LDEP = 4.5716E-0001
 LRI : 0.85* 1.00D0 * LRI = 6.8281E-0002
 LDIF : 0.85* 1.00D0 * LDIF = 1.6631E-0001

--------------------------------------------
**LT - Total Water Body Load **
--------------------------------------------
 LDEP = 4.5716E-0001
 LDIF = 1.6631E-0001
 LRI = 6.8281E-0002
 LR  = 6.1267E-0003
 LE  = 1.9447E-0001
 LT : LDEP+LDIF+LRI+LR+LE = 8.9235E-0001

--------------------------------------------------------------
**kv_low - Water Column Volatile Loss Rate**
--------------------------------------------------------------
KDSW = 1.0000E+0005
```

```
(DZ*(1+KDSW*TSS*1.0E-06))) = 2.0060E+0001
kv_low - KV_UP/(DZ*(1+KDSW*TSS*1.0E-06))) = 5.5411E-0004

--------------------------------------------------------------------------------
** Fwc - Fraction in Water Column   **
--------------------------------------------------------------------------------
FWCTOP : (1+KDSW*TSS*1.0E-6)*DWC/DZ= 1.9940E+0000
FWCBOT : (1+KDSW*TSS*1.0E-6)*DWC/DZ+(THETBS+KDBS*CBS)*DBS/DZ =
1.5155E+0002
FWC : FWCTOP/FWCDBOT = 1.3158E-0002

--------------------------------------------------------------------------------
** Fbs - Fraction in Benthic Sed.        **
--------------------------------------------------------------------------------
FBS - 1-FWC = 9.8684E-0001

--------------------------------------------------------------------------------
**Kb - Benthic Burial Rate Constant  **
--------------------------------------------------------------------------------
KB1 : (XE*AL*SD*1.0E03-VFX*TSS)/(AW*TSS)= 3.8403E+0001
KB2 : (TSS*1.0E-06)/(CBS*DBS)     = 3.3333E-0004
KB : KB1*KB2 = 1.2801E-0002

--------------------------------------------------------------------------------
**Kwt - Overall Waterbody Dissipation Rate Constant
--------------------------------------------------------------------------------

FWC   = 1.3158E-0002
KV_low = 5.5411E-0004
FBS   = 9.8684E-0001
KB    = 1.2801E-0002
KWT - FWC*KV_low+FBS*KB = 1.2640E-0002

--------------------------------------------------------------------------------
**Cwtot - Total Water Body Concentration**
--------------------------------------------------------------------------------
CWTOT - LT/(VFX*FWC+KWT*AW*(DWC+DBS)) = 6.3768E-0006

--------------------------------------------------------------------------------
**Cwctot - Total Water Column Concentration**
--------------------------------------------------------------------------------

FWC   = 1.3158E-0002
CWTOT = 6.3768E-0006
DWC   = 1.0000E+0001
DBS   = 3.0000E-0002
CWCTOT - FWC*CWTOT*((DWC+DBS)/DWC) = 8.4156E-0008

--------------------------------------------------------------------------------
**Cdw - Dissolved Phase Water Concentration**
--------------------------------------------------------------------------------

CWCTOT = 8.4156E-0008
KDSW   = 1.0000E+0005
TSS   = 1.0000E+0001
CDW - 0.85*(CWCTOT/(1+KDSW*TSS*1.0E-06)) = 3.5766E-0008
```

```
-----------------------------------------------------------------
**Csorb - COPC Conc. Sorbed to Bed Sediment**
-----------------------------------------------------------------

FBS    = 9.8684E-0001
CWTOT = 6.3768E-0006
KDBS = 5.0000E+0004
THETBS= 6.0000E-0001
CBS    = 1.0000E+0000
DWC    = 1.0000E+0001
DBS    = 3.0000E-0002
FBS*CWTOT*(KDBS/(THETBS+KDBS*CBS))*((DWC+DBS/DBS))
CSB = 2.1039E-0003
```

6 QUANTIFYING EXPOSURE

This Section describes the factors to be evaluated in quantifying the exposure received under each of the recommended exposure scenarios. The calculation of COPC-specific exposure rates for each exposure pathway evaluated involves:

1. Estimated COPC media concentrations
2. Consumption rate
3. Receptor body weight
4. Frequency and duration of exposure

This calculation is repeated for each COPC and for each exposure pathway included in an exposure scenario. The following sections describe a general exposure rate calculation and the exposure pathway-specific variables that may affect this calculation. Acute exposure resulting from direct inhalation is also evaluated as a separate issue.

6.1 Generic Exposure Rate Equation

Exposure can occur over a period of time. In the calculation of an average exposure per unit of time, the total exposure can be divided by the time period. An average exposure can be expressed in terms of body weight. All exposures quantified in the risk assessment should be:

1. Unitized for time and body weight.
2. Presented in units of milligrams per kilogram of body weight per day.
3. Termed *intakes*.

The generic exposure equation used to calculate chemical intake (USEPA 1989e) is:

$$I = ((C_{gen} \times CR)/BW) \times (EF \times ED/AT) \tag{1.44}$$

where

I = Intake: the amount of COPC at the exchange boundary (mg/kg · d); for evaluating exposure to noncarcinogenic COPCs, the intake is referred to as average daily dose (*ADD*); for evaluating exposure to carcinogenic compounds, the intake is referred to as lifetime average daily dose (*LADD*)

C_{gen} = Generic COPC concentration in media of concern (e.g., mg/kg for soil or mg/L for surface water)

CR = Consumption rate: the amount of contaminated medium consumed per unit of time or event (e.g., kg/d for soil and L/d for water)

EF = Exposure frequency (d/yr)
ED = Exposure duration (years)
BW = Average body weight of the receptor over the exposure period (kg)
AT = Averaging time: the period over which exposure is averaged (days); for carcinogens, the averaging time is 25,550 days, based on a lifetime exposure of 70 years; for noncarcinogens, averaging time equals ED (years) multiplied by 365 d/yr
$(EF \times ED / AT)$ = Temporal adjustment to account for exposure duration and frequency

The exposures calculated in a risk assessment are intended to represent Reasonable Maximum Exposure (RME) conditions as further described in USEPA (1989e). Studies of the compounding of conservatism in probabilistic risk assessments show that setting as few as two factors at R ME levels or high end (e.g., near the 90th percentile), while the remaining variables are set at less conservative, typical or "central tendency" values (e.g., near the 50th percentile) resulted in a product of all input variables at an RME level (e.g., 99th percentile value) (Cullen 1994).

USEPA (2005) recommends that the variables set at RME values include (1) the highest air dispersion modeled air parameter values at current and reasonable potential future exposure scenario locations, (2) the exposure frequency, and (3) the exposure duration. Body weight is typically set at average values.

6.2 Consumption Rate

Consumption rate is the amount of contaminated medium consumed per unit of time or event. Consumption rates for subsistence food types (e.g., beef for the subsistence farmer; fish for the subsistence fisher), is assumed to be 100 percent from the assessment area (e.g., farm, water body) being evaluated. Consumption rates for non-subsistence food types (e.g., home grown garden vegetables) are assumed to be a fraction of the total dietary intake for this food type.

Exposures calculated in a risk assessment are intended to represent RME conditions. Accordingly, the USEPA recommends default values for exposure parameters that will result in estimated RME exposures. However, there are likely to be differences between recommended default, and regional and site-specific exposure parameter values. This may be especially true for the parameter consumption rate (a general term including both intake rate and inhalation rate). The risk assessment performed using recommended default parameter values may be later refined to include supplemental calculations based on regional or site-specific exposure parameter values, provided documentation for these regional or site-specific exposure parameter values is provided. The following subsections describe exposure pathway-specific considerations regarding consumption rate.

Air exposure pathways. Direct inhalation of vapors and particulate emissions from combustion sources is a potential pathway of exposure. Receptors in the assessment area are directly exposed to COPCs in vapor, particulate, and particle-bound phases; as a result of normal respiration. The factors that affect exposure from vapor and particulate inhalation include vapor and particulate COPC concentrations, respiration rate during the period of exposure, and length of exposure.

A single default inhalation rate is provided for use across all adult receptor scenarios. However, if site-specific data is available to show that subsistence farmers and fishers have higher respiration rates due to rigorous physical activities than other receptors, that data may be appropriate. Also, farmers could be assumed to typically spend more time each day in the vicinity of contaminated vapors and particulates, because farms are places of business, and typically their homes. However, any modifications of the respiration rates of receptors should be considered on a site-specific basis, and supported by documentation.

Inhalation of vapors and particulates will be influenced by the relative amount of time that a receptor spends indoors. Although vapors entering buildings and residences as a result of air exchange are likely to remain airborne and, therefore, may be inhaled, particulates entering these same buildings are more likely to settle out and not be inhaled. However, air toxics risk assessments commonly vapor and particulates may both be inhaled throughout the day, both indoors and outdoors.

Food exposure pathways. Plants and animals impacted by emission sources may take up emitted COPCs in the air or deposited COPCs in soil. Humans are exposed to COPCs via the food chain when they consume these plants and animals as a food source. Human intake of COPCs is determined on the basis of:

1. The types of foods consumed
2. The amount of food consumed per day
3. The concentration of COPCs in the food
4. The percentage of the diet contaminated by COPCs

Other variables, described below, may also significantly affect the estimation of exposure.

Types of foods consumed. The types of foods consumed will affect exposure, because different plants and animal tissues will take up COPCs at different rates. Therefore, COPC concentrations in food are determined, in part, by the type of food, and they vary with the types of food in the diet. Furthermore, the types of foods consumed vary with age, geographical region, and socio-cultural factors.

Food consumption rate. The amount of daily food consumption varies with age, sex, body weight, and geographic region, and it also varies within these categories. USEPA recommends in the 2005 HHRAP that food consumption rate information (ingestion rates) be obtained from the 1997 *Exposure Factors Handbook* (USEPA 1997c); specifically, the section regarding home produced food items.

Percentage of contaminated food. The percentage of home grown food consumed by the individual will affect exposure because not all of an individual's dietary intake may be contaminated. Receptors, located in a rural or suburban area, which can raise animals and grow food in gardens will have a larger percentage of their food produced locally than people living in the city.

The USEPA HHRAP recommends the following assumptions regarding the percentage of contaminated food:

- With regard to aboveground and belowground produce, it is assumed that 100 percent of the contaiminated produce is consumed.
- With regard to beef, milk, pork, chicken, and eggs, it is assumed that 100 percent of these animal tissues consumed by the subsistence farmer and the subsistence farmer child are contaminated. No other receptors are assumed to consume these animal tissues.
- With regard to fish, it is assumed that 100 percent of the fish consumed by the subsistence fisher and subsistence fisher child are contaminated. No other receptors are assumed to consume fish.

Soil exposure pathways. Soil ingestion, dermal exposure to soil, and inhalation of resuspended dust are potential soil exposure pathways. For the purpose of RCRA combustion permitting decisions, USEPA recommends that soil ingestion be considered in all risk assessments. However, dermal exposure to soil and inhalation of resuspended dust are currently recommended for evaluation only if site-specific exposure setting characteristics require that these exposure pathways be evaluated. Based on air dispersion modeling and deposition of COPCs, emission concentrations in soil will vary with distance from the source. Potential routes of exposure should be determined by the way in which the soils in the area are used. Soil used for farming or recreation will be involved in pathways of human exposure that differ from those of soil on roadways or in urban areas.

Soil ingestion. Children and adults are directly exposed to COPCs in soil when they consume soil that has adhered to their hands. Factors that influence exposure by soil ingestion include soil concentration, the rate of soil ingestion during the time of exposure, and the length of time spent in the vicinity of contaminated soil. Soil ingestion rates in children are based on studies that measure the quantities of non-absorbable tracer minerals in the feces of young children. Ingestion rates for adults are based on assumptions about exposed surface area and frequency of hand-to-mouth consumption. Indoor dust and outdoor soil may both contribute to the total daily ingestion. Exposure levels are also influenced by the amount of time that the individual spends in the vicinity of soil exposed to deposition of emitted pollutants.

In addition, some young children—referred to as "pica" children—may intentionally eat soil. As discussed in USEPA (1989f), the typical medical and scientific use of the term "pica" refers to the ingestion of non-food items, such as soil, chalk, and crayons. This evaluation should be made on a case-by-case basis based on site-specific exposure setting characterization. Air toxics risk assessments do not typically consider the "pica" child in the assessment process.

Dermal exposure to soil. USEPA does not recommend the use, in the evaluation of recommended exposure scenarios, of the pathway of dermal exposure to COPCs through contact with soil, for air toxic emissions. However, site-specific exposure setting characteristics may require that this exposure pathway be evaluated, therefore, this section discusses dermal soil exposure.

Available data indicate that the contribution of dermal exposure to soils to overall risk is 50-fold less than the risk from any other exposure pathway and 300-fold less than the total estimated risk (USEPA 1996g; 1995h).

Soil inhalation resulting from dust resuspension. USEPA does not recommend the use of soil inhalation exposure pathway of resulting from dust resuspension in the evaluation of air toxic exposure scenarios. However, site-specific exposure setting characteristics may require that this exposure pathway be evaluated; this section discusses exposure to soil resulting from dust re-suspension.

Water exposure pathways. USEPA recommends the use of water exposure pathways resulting from air toxic exposure scenarios. Water exposure pathways can be used to determine COPC concentrations in drinking water obtained from surface water or collected precipitation (e.g., cisterns). Water exposure pathways are also used to determine the COPC concentration in fish. Daily exposures of individuals using these water sources for various purposes—such as fishing and drinking water—can be estimated by using the equations described in this chapter.

Because annual ground-level concentrations of COPCs generally decrease with distance from the source, important factors in determining the water concentration include:

1. The location of the precipitation collection apparatus
2. Surface water body onto which emitted COPCs are deposited
3. The COPC soil concentration (which affects runoff and leachate concentrations)

In addition, the location and size of the watershed will affect the concentration of COPCs suspended in runoff.

Drinking water exposure from surface water sources. For evaluation of a surface water body as a drinking water source, exposure is affected by the concentration of the COPC in the water, the daily amount of water ingested, and the percentage of time that the individual spends in the area serviced by that water supply system. The COPC concentration in a surface water body includes consideration of contribution of COPC loading from the surrounding watershed.

Air toxic risk assessments do not commonly consider the removal of COPCs from drinking water sources by water treatment plants.

Dermal water exposure. Dermal water exposure pathways are not typically considered in air toxics risk assessment. Dermal exposure is affected by (1) the surface area of exposed skin, (2) the COPC concentration in the water, (3) the permeability of the skin to the COPC, and (4) the length of time that the individual is in contact with the water.

Ingestion of fish. Factors that affect human exposure by ingestion of fish from a surface waters affected by sources of air toxics emissions include:

1. Sediment and water COPC concentrations.
2. Types of fish and shellfish consumed.
3. Ingestion rates for the various fish and shellfish groups.
4. Percent of dietary fish caught in the surface water body affected by the combustion unit.

The types of fish consumed will affect exposure, because different types of fish and shellfish take up COPCs at different rates. For example, fatty fish tend to accumulate organic COPCs more readily than lean fish. The amount of fish consumed also affects exposure, because people who eat large amounts of fish will tend to have higher exposures. Fish consumption rates vary greatly, depending on geographic region and social or cultural factors. Because 100 percent of a receptor's dietary fish may not originate from the surface water body near the combustion facility, the percentage of locally caught fish is also a variable for exposure.

6.3 Exposure Frequency (EF)

The receptors in each recommended exposure scenario are assumed to be exposed to all of the exposure scenario-specific exposure pathways 350 d/yr (USEPA 1989e; 1991b; 1991d). This assumption is based on the conservative estimate that all receptors spend a minimum of 2 weeks at a location other than the exposure scenario location.

6.4 Exposure Duration

Exposure duration *(ED)* is the length of time that a receptor is exposed via a specific exposure pathway. Although a receptor is no longer exposed to COPCs via the direct inhalation exposure pathway after an emission source ceases operation, a receptor is exposed via the indirect exposure pathways for as long as the receptor remains in the assessment area. Therefore, USEPA recommends in the HHRAP using default RME values to estimate exposure duration for specified receptors.

USEPA assumes that receptors could be exposed to the long-term average COPC soil or water concentrations (and the subsequent COPC plant or animal concentrations) after the emission source ceases to operate. For existing facilities, USEPA assumes that this period of time can be represented by default time periods of 30, 60, or 100 years. Thirty (30) years is the most common time period used to evaluate air toxic emission sources. These values are based on the assumptions that the emission source (1) is already in place, (2) will continue to be used for the rest of its useful life (estimated to be 30 years), and (3) may be replaced when it reaches the end of its useful life (estimated to be possibly as long as 60 or 100 years.

Although a combustion unit may remain in the same location for 100 years—and a person may have a lifetime of exposure to emissions from that combustion unit—U.S. Bureau of the Census data (1986) on population mobility indicate that many Americans do not remain in the same area for their 70-year lifetime. An estimate of the number of years that a person is likely to spend in one area, such as the vicinity of a combustion facility, can be derived from information about mobility rate and median time in a residence.

In addition to the number of years at a particular location or residence, the amount of time spent at that location each day directly affects exposure. For example, children that attend day care or adults that work in a different location for part of the day may be exposed to higher or lower COPC levels.

The exposure duration values recommended by USEPA are presented in Table 1.6 .

6.5 Averaging Time

For non-carcinogenic COPCs, USEPA recommends that a value of exposure duration as years x 365 d/yr be used as the averaging time (*AT*) (USEPA 1989e; 1991d). However, for carcinogenic COPCs—the effects of which may have long latency periods—the age of the receptor (i.e., child, adult, or elderly) influences that COPC exposure pathway, because the exposure duration and, therefore, the quantity of exposure, will vary.

TABLE 1.6 Exposure Duration Values for the Scenario Receptors

Exposure duration values		
Recommended exposure scenario receptor	Value	Source
Child resident	6 years	USEPA 1990f; 1994r
Adult resident	30 years	USEPA 1990f; 1994r
Subsistence fisher	30 years	USEPA 1990f; 1994r
Subsistence fisher child	6 years	Assumed to be the same as the child resident
Subsistence farmer	40 years	USEPA 1994l; 1994r
Subsistence farmer child	6 years	Assumed to be the same as the child resident

USEPA recommends that carcinogenic exposures for different receptor ages be evaluated separately, because the daily activities of these receptors vary, including:

1. The amounts of food and water consumed.
2. The types of food consumed.
3. The amount of exposed skin surface.

Health-based criteria, such as health advisories for drinking water, are also different for children and adults. As a result, for some exposure pathways, such as soil ingestion, children may have a greater quantifiable exposure and be at greater risk than adults. Some behaviors, such as mouthing of dirty objects or direct ingestion of soil, which could also contribute to exposure, are also much more prevalent in children than adults.

Because quantification of carcinogenic COPC exposure depends on the duration of exposure, the age of the receptor is important. The average human lifespan is generally considered to be 70 years; childhood represents only about 10 percent of the lifespan (6 years) (USEPA 1990e).

6.6 Body Weight

USEPA (1990e) defines the body weight of the receptor as either adult weight (70 kg) or child weight (6 years; 15 kg).

6.7 Example of Calculations for the Ingestion of COPC in Contaminated Products

```
=================================================
INTAKE CALCULATIONS
=================================================

==========
TCDD2378 - CAS = 1746016
==========

--------------------------------------------
** ISOIL Intake ***
--------------------------------------------
ISOIL = (Cs * CRSoil * FSoil) / BW = 1.2526E-010

----------------------------------------
** Iag Intake ***
----------------------------------------
Iag = ((Pd+Pv+Pr)*CRag+Pr*CRpp+Pr_bg*CRbg)*FAG
Iag = 8.5428E-0010
```

```
---------------------------------------------
** IFish Intake ***
---------------------------------------------
IFish = 0.0000E+00

-----------------------------------------
** IDW Intake ***
-----------------------------------------
IDW = 3.61857144043097E-011

-------------------------------------------
** Ii Intake   ***
-------------------------------------------

 Abeef = 1.0216E-05
 Amilk = 3.0926E-06
 Apork = 1.7622E-06
 Achic = 3.8665E-08
 Aegg = 2.2094E-08

 IBeef = 0.00
 IChic = 0.00
 IEgg = 0.00
 IMilk = 0.00
 IPork = 0.00
 ITotal = ISoil + Iag + IBeef + Imilk + IFish + IPork + IChic +
 IEgg + Idw
 ITotal = 1.0157E-09
```

========
Silver - CAS = 7440224
========

```
-------------------------------------------------------
** ISOIL Intake        **
-------------------------------------------------------
ISOIL = (Cs * CRSoil * FSoil) / BW = 9.4607E-0012

--------------------------------------------------------------
** Iag Intake              **
--------------------------------------------------------------
Iag = ((Pd+Pv+Pr)*CRag+Pr*CRpp+Pr_bg*CRbg)*FAG
Iag = 2.5738E-0009

------------------------------------------------------------------
** IFish Intake              **
------------------------------------------------------------------
IFish = 0.0000E+0000

--------------------------------------------------------------
** IDW Intake              **
--------------------------------------------------------------
IDW = 1.1330E-0010

-----------------------------------------
** Ii Intake   **
-----------------------------------------
```

```
 Abeef = 1.7362E-0006
 Amilk = 1.7482E-0005
 Apork = 0.0000E+0000
 Achic = 0.0000E+0000
 Aegg = 0.0000E+0000

 IBeef = 0.0000E+0000
 IChic = 0.0000E+0000
 IEgg = 0.0000E+0000
 IMilk = 0.0000E+0000
 IPork = 0.0000E+0000
 ITotal = ISoil + Iag + IBeef + Imilk + IFish + IPork + IChic +
 IEgg + Idw
 ITotal = 2.6966E-0009
```

==================
DIVALENT MERCURY - CAS = 7487947
==================

```
----------------------------------------------------
** ISOIL Intake     **
----------------------------------------------------
ISOIL = (Cs * CRSoil * FSoil) / BW = 8.6018E-0010

--------------------------------------------------------------
** Iag Intake             **
--------------------------------------------------------------
Iag = ((Pd+Pv+Pr)*CRag+Pr*CRpp+Pr_bg*CRbg)*FAG
Iag = 1.1504E-0008

-----------------------------------------------------------
** IFish Intake          **
-----------------------------------------------------------
IFish = 0.0000E+0000

-------------------------------------------
** IDW Intake   **
-------------------------------------------
IDW = 7.1532E-0010

--------------------------------------------------
** Ii Intake        **
--------------------------------------------------

 Abeef = 1.8624E-0006
 Amilk = 7.6432E-0007
 Apork = 8.2700E-0009
 Achic = 3.4372E-0007
 Aegg = 3.4372E-0007

 IBeef = 0.0000E+0000
 IChic = 0.0000E+0000
 IEgg = 0.0000E+0000
 IMilk = 0.0000E+0000
 IPork = 0.0000E+0000
 ITotal = ISoil + Iag + IBeef + Imilk + IFish + IPork + IChic +
 IEgg + Idw
 ITotal = 1.3080E-0008
```

7 HUMAN HEALTH RISK AND HAZARD CALCULATIONS

Risk characterization must exhibit the core values of transparency, clarity, consistency, and reasonableness. The final step of a risk assessment is the calculation of the upper-bound excess lifetime cancer risks (risk) and noncarcinogenic hazards (hazard) for each of the pathways and receptors. Risks and hazards are then summed for specific receptors, across all applicable exposure pathways, to obtain an estimate of total individual risk and hazard for specific receptors.

Risk from exposure to combustion emissions is the probability that a receptor will develop cancer, based on a unique set of exposure, model, and toxicity assumptions. The slope factor is used in risk assessments to estimate and upper bound lifetime probability of an individual developing cancer as a result of exposure to a particular level of a potential carcinogen. For example, a risk of 1×10^{-5} is interpreted to mean that an individual has no more than, and likely less than, a one in 100,000 chance of developing cancer from the exposure being evaluated.

In contrast, hazard is quantified as the potential for developing non-carcinogenic health effects as a result of exposure to COPCs, averaged over an exposure period. A hazard is not a probability but, rather, a measure (calculated as a ratio) of the magnitude of a receptor's potential exposure relative to a standard exposure level (*RfD* or *RfC*). The standard exposure level is calculated over a similar exposure period and is estimated to pose no appreciable likelihood of adverse health effects to potential receptors, including special populations (USEPA 1989e). Risks and hazards are typically characterized for a single receptor and are referred to as individual risks and hazards (USEPA 1989e; 1994g; NC DEHNR 1997).

7.1 Estimation of Individual Risk and Hazard

Individual risk and hazard descriptors are intended to convey information about the potential risks to individuals potentially impacted by air toxic emissions from a source. A well-conducted risk assessment will provide useful impact information, such as:

1. Quantitative and qualitative estimates of risk and hazard associated with exposure to COPCs.
2. Estimates of health effects associated with exposure to lead.
3. Evaluation of infant exposure to 2,3,7,8-TCDD TEQ present in breast milk.
4. Evaluation of acute exposure resulting from direct inhalation.

7.2 Quantitative Estimation of Cancer Risk

As described above, for carcinogenic COPCs, risk estimates represent the incremental probability that an individual will develop cancer over a lifetime as a result of a specific exposure to a carcinogenic chemical (USEPA 1989e). These risks are calculated as follows:

$$\text{Cancer Risk} = LADD \times CSF \tag{1.45}$$

where

$LADD$ = Lifetime average daily dose (mg/kg·d)
CSF = Cancer slope factor (mg/kg·d)

As shown in the above equation as a cancer slope factor, the toxicity of a particular COPC must be available in order to estimate risk. Toxicity values can be obtained from a variety of sources including the following:

1. USEPA's IRIS (USEPA 1996a)
2. California Environmental Protection Agency Toxicity Criteria Database (*http://www.oehha.ca.gov/risk/ChemicalDB/index.asp*)
3. USEPA HEAST (USEPA 1994b)

Within a specific exposure pathway, receptors may be exposed to more than one COPC. The total risk associated with exposure to all COPCs through a single exposure pathway is estimated as follows (USEPA 1989e):

$$Cancer\ Risk_T = i\ Cancer\ Risk_i \quad (1.46)$$

where

$Cancer\ Risk_T$ = Total cancer risk for a specific exposure pathway
$Cancer\ Risk_i$ = Cancer risk for COPC i for a specific exposure pathway

At particular exposure scenario locations, receptors may be exposed through a number of exposure pathways. Risks from multiple exposure pathways should be summed for a given receptor specific to each recommended exposure scenario. That is, risks should be summed across the receptor- exposure pathway combinations. The total risk posed to a receptor is the sum of total risks from each individual exposure pathway expressed as follows:

$$Total\ Cancer\ Risk = CancerRisk_T \quad (1.47)$$

where

$Total\ Cancer\ Risk$ = Total cancer risk from multiple exposure pathways
$Cancer\ Risk_T$ = Total cancer risk for a specific exposure pathway

7.3 Quantitative Estimation of Potential for Noncancer Effects

Standard risk assessment models assume that non-carcinogenic effects, exhibit a threshold; that is, there is a level of exposure below which no adverse effects will be observed (USEPA 1989e). The potential for noncarcinogenic health effects resulting from exposure to a chemical is generally assessed by:

1. Comparing an exposure estimate to an *RfD* for oral exposures.
2. Comparing an estimated chemical-specific air concentration to the *RfC* for direct inhalation exposures.

An *RfD* is a daily oral intake rate that is estimated to pose no appreciable risk of adverse health effects, even to sensitive populations, over a specific exposure duration. Similarly, an *RfC* is an estimated daily concentration of a chemical in air, the exposure to which over specific exposure duration poses no appreciable risk of adverse health effects, even to sensitive populations (USEPA 1989e).

Chronic *RfDs* and *RfCs* should be used to evaluate all exposure pathways. In the absence of a chronic *RfD*, a sub-chronic *RfD* with an uncertainty factor (3 to 10) can be considered. The comparisons of exposure estimates and COPC-specific air concentrations to *RfD* and *RfC* values, described above, are known as hazard quotients (*HQ*), which are calculated as follows:

$$HQ = ADD / RfD \text{ or } HQ = Ec / RfC \quad (1.48)$$

where

HQ = Hazard quotient (unitless)
ADD = Average daily dose (mg/kg/·d)
EC = Total COPC air concentration (mg/m^3)
RfD = Reference dose (mg/kg·d)
RfC = Reference concentration (mg/m^3)

In the assessment of noncarcinogenic risk from COPCs, USEPA-derived or reviewed health benchmarks (*RfD*s, *RfC*s) are recommended. However, for numerous COPCs, a complete set of inhalation and oral EPA-derived health benchmarks are not available. It is possible to perform a route-to-route extrapolation when, for example, an oral RfD is available but a corresponding RfC is not. However, this practice introduces a high level of uncertainty into the risk assessment process and is not recommended in the 2005 HHRAP guidance. However, it may appropriate for some COPCs of particular interest such as COPCs with relatively greater emission rates than other COPCs. For instance, if the *oral RfD* (mg/kg · d) was available and the *RfC* (mg/m^3) was not; the *RfC* was calculated by multiplying the *RfD* by an average human inhalation rate of 20 m^3/d and dividing by the average human body weight of 70 kg.

Because of the degree of uncertainty involved in using toxicity benchmark values calculated, based on route-to-route extrapolation, a qualitative assessment of the toxicity information available for the compound and exposure route should be performed. This will enable the risk assessor to make a well informed decision concerning the validity of values calculated based on route-to-route extrapolation. This qualitative assessment should also be included in the uncertainty section of the risk assessment.

HQ values exceeding one significantly increase the level of concern. Because *RfDs* and *RfCs* do not have equal accuracy or precision, and are not based on the same severity of effect, the level of concern does not increase linearly as an *HQ* approaches and exceeds 1 (USEPA 1989e). It should also be noted that background exposures may be an important consideration in setting safe levels. This is because non-cancer effects are generally modeled as thresholds. In specific cases, a permitting authority may elect to adjust the *HQ* downward to account for any exposure that individuals may have from other sources.

As with carcinogenic chemicals in a specific exposure pathway, a receptor may be exposed to multiple chemicals associated with noncarcinogenic health effects. The total noncarcinogenic hazard attributable to exposure to all COPCs through a single exposure pathway is known as a hazard index (*HI*). Consistent with the procedure for addressing carcinogenic risks, the noncarcinogenic hazards from all sources of air toxics emissions should be summed for each receptor. Also, noncarcinogenic hazard from fugitive emissions sources should also be included in the calculation of the *HI* for each receptor. The *HI* is calculated as follows:

$$HI = \Sigma\, HQ_i \tag{1.49}$$

where

HI = Total hazard for a specific exposure pathway
HQ_i = Hazard quotient for COPC *i*

This summation methodology assumes that the health effects, of the various COPCs to which a receptor is exposed, are additive. Specifically, this methodology is a simplification of the *HI* concept because it does not directly consider the portal of entry associated with each exposure pathway or the often unique toxic endpoints and toxicity mechanisms of the various COPCs.

A receptor may be exposed to COPCs associated with non-carcinogenic health effects through more than one exposure pathway. For the purposes of the risk assessment, it is reasonable to estimate a receptor's total hazard as the sum of the *HI*s for each of the exposure pathways. Specifically, a receptor's total hazard is the sum of hazards from each individual exposure pathway, expressed as follows:

$$Cumulative\ HI = \Sigma\, HI \tag{1.50}$$

where

$Cumulative\ HI$ = Total hazard index from multiple exposure pathways
HI = Total hazard index for a specific exposure pathway

All total *HI*s exceeding the target hazard level are further evaluated. The total *HI* for an exposure pathway can exceed the target hazard level as a result of the presence of either (1) one or more COPCs with an *HQ* exceeding the target hazard level, or (2) the summation of several COPC-specific *HQ*s that are each less than the target hazard level. In the former case, the presence of at least one COPC-specific hazard greater than the target hazard level is interpreted as indicating the potential for non-carcinogenic health effects. In the latter case, a detailed analysis is required to determine whether the potential for non-carcinogenic health effects is accurately estimated by the total *HI*, because the toxicological effects associated with exposure to multiple chemicals, often through different exposure pathways, may not be additive; therefore, the total *HI* may overestimate the potential for non-carcinogenic health effects. To address this issue, COPC-specific hazards are summed according to major health effects and target organs or systems (USEPA 1989e). It is especially important to consider any differences related to exposure route; this process is referred to as the segregation of the *HI*.

The highest segregated *HI* resulting from this process is considered. If the segregated *HI* exceeds the target hazard level, there is a potential for non-carcinogenic health effects. However, if the segregated *HI* is less than the target hazard level, the total *HI* of all COPC-specific results likely is too conservative, and noncarcinogenic health effects are not likely to result from exposure to COPCs.

7.4 Target Levels

Target levels are risk management based and set by the regulatory authority. Target values are not a discrete indicator of observed adverse effect. If a calculated risk falls within target values, a regulatory authority may, without further investigation, conclude that a proposed action does not present an unacceptable risk. A calculated risk that exceeds these targets, however, would not, in and of itself, indicate that the proposed action is not safe or that it presents an unacceptable risk. Rather, a risk calculation that exceeds a target value triggers further careful consideration of the underlying scientific basis for the calculation.

7.5 Acute Exposure Resulting from Direct Inhalation

In addition to long-term chronic effects, short-term or acute effects should be considered from direct inhalation of vapor phase and particle phase COPCs. It is assumed that short-term emissions will not have a significant impact through the indirect exposure pathways (as compared to impacts from long-term emissions). Therefore, acute effects are only evaluated through the short-term (maximum 1 h) inhalation of vapors and particulates exposure pathway of the acute risk scenario.

In addition to the acute effects associated with short term releases of air toxics from stack or fugitive emission sources, other effects may include accidents, fires, or explosions. In these cases, the air dispersion modeling described previously in this chapter will need to be supplemented by other models that are specifically designed to consider these types of scenarios. For example, the rupture of a chlorine tank will result in the release of a dense gas that cannot be modeled with ISCST3, but should be modeled with an appropriate dense gas ADM.

Acute risk assessments uses a hierarchical approach to define the toxicity benchmarks which short term maximum air concentrations are compared. The toxicity benchmarks are called acute inhalation exposure criteria (AIEC). Hierarchical approaches are composed of existing guidelines for acute inhalation exposure, ranked in order of applicability and technical basis, and all being protective of the general public. It should be noted that hierarchical approaches are needed because no single organization or methodology has developed acute criteria values or benchmarks for all COPCs.

USEPA recommended hierarchal approach. The approach recommended in the 2005 HHRAP is based on existing acute inhalation values that do not require the use of arbitrary safety factors and are intended to protect the general public from discomfort or mild adverse health effects over 1-hour exposure periods. It includes level 1 acute inhalation exposure guidelines (AEGL-1), level 1 emergency response planning guidelines (ERPG-1), and level 1 acute toxicity exposure levels (ATEL-1); supplemented with DoE TEELs and the SCAPA toxicity-based approach. The hierarchal approach is summarized below:

1. CAL/EPA Acute RELs (*http://www.oehha.ca.gov/air/acute_rels/index.html*)
2. AEGL-1 (*http://www.epa.gov/oppt/aegl/pubs/chemlist.htm*)
3. ERPG-1 (*http://www.bnl.gov/scapa/scapawl.htm*)
4. TEEL-1 (*http://tishq.eh.doe.gov/web/Chem_Safety/teel.htm*)
5. AEGL-2 (*http://www.epa.gov/oppt/aegl/pubs/chemlist.htm*)
6. SCAPA Toxicity-based approach (DoE 1997a)

The hierarchy is presented in order of preference from 1 (most preferred) to 5 (least preferred). This preference is based on (1) applicability to a 1-hour exposure duration for protection of the general public (versus only occupational exposure), and (2) level of documentation and associated review.

To characterize the potential for adverse health effects from acute exposure to COPC-specific emissions, the acute air concentration (C_{acute}) resulting from maximum emissions over a 1-hour period should be compared to the COPC-specific AIEC to calculate the acute hazard quotient (*AHQinh*). The *AHQinh* can be calculated as follows:

$$AHQ_{inh} = C_{acute} \times 10^{-3} / AIEC \tag{1.51}$$

where

AHQ_{inh} = Acute hazard quotient (unitless)
C_{acute} = Acute air concentration (μg/m^3)
$AIEC$ = Acute inhalation exposure criteria (mg/m^3)
0.001 = Conversion factor (mg/μg)

Acute hazard quotients should be calculated at the selected acute exposure scenario locations for COPCs specific to emissions from each source and from all facility sources combined. Target levels for acute hazard quotient evaluation is a risk management decision and will be set by the permitting authority.

7.6 Example of Calculations for the Risk and Hazard from Individual Indirect Pathways

```
RISK CALCULATIONS
==========
TCDD2378 - CAS = 1746016
==========

-------------------------------
** CRisk ***
-------------------------------

 ED      = 30.0
CSF      = 1.50E+05
Crisk Beef = 0.00
Crisk Chic = 0.00
Crisk Eggs = 0.00
Crisk Milk = 0.00
Crisk Pork = 0.00
Crisk IDW = 2.2306E-06
Crisk Soil = 7.7217E-06
Crisk Fish = 0.00
Crisk Ag =  5.2661E-05
Crisk Total= 6.2613E-05

---------------------------------------------------------------------
** Ca - Total COPC Air Conc. **
---------------------------------------------------------------------

Ca : Q*(FV*CYV+(1.0-FV)*CYP) = 6.4783E-07

-----------------------------------------------------------------------------------------
** CancerRiskin - Inhalation Cancer Risk **
-----------------------------------------------------------------------------------------

EC = CA*EF*ED/(AT*365) = 2.6623E-07
URF =               0.00
CANCIN : EC * URF =     0.00
```

```
---------------------------------------------------------------------
** Ca - Total COPC Air Conc. **
---------------------------------------------------------------------
Ca : Q*(FV*CYV+(1.0-FV)*CYP) = 4.2144E-05

---------------------------------------------------------------------
Acute Inhalation Hazard Quotient
---------------------------------------------------------------------
AHQ_IN = 2.8096E-05
```

========
Silver - CAS = 7440224
========

```
--------------------------------------
** HQ Risk     **
--------------------------------------
HQ_RiskBeef = 0.0000E+0000
HQ_RiskChic = 0.0000E+0000
HQ_RiskEgg  = 0.0000E+0000
HQ_RiskMilk = 0.0000E+0000
HQ_RiskPork = 0.0000E+0000
HQ_RiskFish = 0.0000E+0000
HQ_RiskAg   = 4.9360E-0007
HQ_RiskIdw  = 2.1729E-0008
HQ_RiskISoil = 1.8144E-0009
HQ_RiskTotal = 5.1715E-0007

-------------------------------------------------------------------------
** Ca - Total COPC Air Conc.    **
-------------------------------------------------------------------------
Ca : Q*(FV*CYV+(1.0-FV)*CYP) = 1.1449E-0006

-----------------------------------------------------------------------
** HQ_In - Inhalation HQ       **
-----------------------------------------------------------------------
EC : CA*EF*ED/(AT*365) = 1.0978E-0006
HQ_In : EC*0.001/RfC = 6.0990E-0008

---------------------------------------------------------------------------
** Ca - Total COPC Air Conc.     **
---------------------------------------------------------------------------
Ca : Q*(FV*CYV+(1.0-FV)*CYP) = 9.8238E-0005

---------------------------------------------------------------------
Acute Inhalation Hazard Quotient
---------------------------------------------------------------------
AHQ_IN = 3.2746E-0007
```

==================
DIVALENT MERCURY - CAS = 7487947
==================

```
----------------------------------------
** HQ Risk      **
----------------------------------------
HQ_RiskBeef = 0.0000E+0000
```

```
HQ_RiskChic  = 0.0000E+0000
HQ_RiskEgg   = 0.0000E+0000
HQ_RiskMilk  = 0.0000E+0000
HQ_RiskPork  = 0.0000E+0000
HQ_RiskFish  = 0.0000E+0000
HQ_RiskAg   = 3.6772E-0005
HQ_RiskIdw   = 2.2864E-0006
HQ_RiskISoil  = 2.7494E-0006
HQ_RiskTotal  = 4.1808E-0005

--------------------------------------------------------------------------------
** Ca - Total COPC Air Conc.     **
--------------------------------------------------------------------------------
Ca : Q*(FV*CYV+(1.0-FV)*CYP) = 6.8688E-0007

--------------------------------------------------------------------------------
** HQ_In - Inhalation HQ          **
--------------------------------------------------------------------------------
EC : CA*EF*ED/(AT*365) = 6.5865E-0007
HQ_In : EC*0.001/RfC = 5.9877E-0007

--------------------------------------------------------------------------------
** Ca - Total COPC Air Conc.      **
--------------------------------------------------------------------------------
Ca : Q*(FV*CYV+(1.0-FV)*CYP) = 3.7523E-0005

--------------------------------------------------------------------------
Acute Inhalation Hazard Quotient
--------------------------------------------------------------------------
AHQ_IN = 3.0019E-0007
```

8 AIR TOXICS CHEMICAL AND PHYSICAL PROPERTIES

This section describes the fate and transport parameters required for the calculations of the chemical movements from media to media. It also presents some of the estimation techniques to obtain the parameters used in the HHRAP.

8.1 Fundamental Physical Properties

The following physical properties are used in most physical chemical calculations and are measured instead of estimated.

Molecular weight. Molecular weight (*MW*) of a compound is defined as the sum of atomic weights of all atoms in the compound's molecule.

Melting point temperature. Melting point temperature (T_m) is the temperature of the compound [in Kelvin (K)] at which the solid state of the compound undergoes a phase change to a liquid phase. At ambient temperatures and at an atmospheric pressure of 1 atm, compounds are either in a solid or liquid state. The compound liquid or solid state is provided in the summary tables of compound-specific parameter values.

Vapor pressure. The vapor pressure (*Vp*) of a substance is defined as the pressure in atmospheres exerted by the vapor (gas) of a compound when it is under equilibrium conditions. It provides a semiquantitative rate at which it will volatilize from soil and/or water.

Aqueous solubility. The aqueous solubility (S) of a compound is defined as the saturated concentration of the compound in water (mg COPC/L water) at a given temperature and pressure, usually at soil/water temperatures and atmospheric pressure.

8.2 Henry's Law Constant

Henry's law constant (H) is also referred to as the air-water partition coefficient, and is defined as the ratio of the partial pressure of a compound in air to the concentration of the compound in water at a given temperature under equilibrium conditions. For organics (excluding PCDDs and PCDFs), H values were calculated from the following theoretical equation (Lyman, Reehl, and Rosenblast 1982) for consistency, using recommended MW, S, and Vp values provided in the USEPA HHRAP:

$$H = \frac{Vp \cdot MW}{S} \tag{1.52}$$

where

H = Henry's law constant (atm · m^3/mol)
Vp = Vapor pressure of COPC (atm)
S = Solubility of COPC in water (mg COPC/L water)

8.3 Diffusivity of COPCs in Air and Water

Diffusivity or diffusion coefficients in air (D_a) and water (D_w) are used to calculate the liquid- or gas-phase transfer of a COPC into a waterbody. For organics (except PCDDs and PCDFs), diffusivity values were obtained directly from the CHEMDAT9 model chemical properties database (Worksheet DATATWO.WK1) (USEPA. 1994d).

For compounds not in the USEPA (1994d) database, diffusivity values were obtained by using the WATER8 (USEPA. 1995d) model correlation equations for air and water diffusivities:

For diffusivity values that were not available in these databases, D_w and D_a values were calculated using the following equations cited and recommended for use in USEPA (1997g):

$$D_{a,i} = \frac{1.9}{(MW_i)^{2/3}}$$

$$D_{w,i} = \frac{22\_10^{\times 5}}{(MW_i)^{2/3}} \tag{1.53}$$

For PCDDs and PCDFs, diffusivity values in air and water for (1) 2,3,7,8-TCDD were obtained from USEPA (1994e), and (2) 2,3,7,8-TCDF were obtained from USEPA (1995d). For all other congeners of PCDDs and PCDFs:

1. A default Dw value of 8×10^{-06} cm^2/s was used
2. Da values were calculated using the following equation recommended by USEPA (1994a):

$$\frac{D_x}{D_y} = \left(\frac{MW_y}{MW_x}\right)^{0.5} \tag{1.54}$$

where

$D_{x,y}$ = Diffusivities in air of compounds x and y (cm^2/s)
$MW_{x,y}$ = Molecular weights of compounds x and y (g/mol)

Da values for PCDD congeners were calculated by using the Da value and MW for 2,3,7,8-TCDD. Da values for PCDF congeners were calculated using the Da value and MW for 2,3,7,8-TCDF. This approach is consistent with the methodology specified in USEPA (1994a).

8.4 Octanol/Water Partitioning Coefficient

The *n*-octanol/water partitioning coefficient (K_{ow}) is defined as the ratio of the solute concentration in the water-saturated *n*-octanol phase to the solute concentration in the *n*-octanol-saturated water phase.

8.5 Soil Organic Carbon-Water Partition Coefficient

The soil organic carbon-water partition coefficient (K_{oc}) or the organic carbon normalized soil sorption coefficient is defined as the ratio of adsorbed compound per unit weight of organic carbon to the aqueous solute concentration. Because of the soil mechanisms that are inherently involved, K_{oc} values for the ionizing organics and nonionizing organics are discussed separately.

The partitioning of ionizing organic compounds can be significantly influenced by soil pH. Because of the soil mechanisms that are inherently involved, we discuss K_{oc} values for the ionizing organics and non-ionizing organics separately.

The soil organic carbon-water partition coefficient for ionizing organic compounds. Ionizing organic compounds include amines, carboxylic acids, and phenols. These compounds contain the functional groups that ionize under specific pH conditions, and include the following:

- Organic acids: (2,4,6-trichlorophenol; pentachlorophenol; 2,3,4,5-tetrachlorophenol; 2,3,4,6-tetrachlorophenol; 2,4,5-trichlorophenol; 2,4-dichlorophenol; 2-chlorophenol; phenol; 2,4-dimethylphenol; 2-methylphenol; 2,4-dinitrophenol; and benzoic acid)
- Organic bases: (*n*-nitroso-di-*n*-propylamine; *n*-nitrosodiphenylamine, and 4-chloroaniline)

K_{oc} values were estimated on the basis of the assumption that the sorption of ionizing organic compounds is similar to hydrophobic organic sorption, because the soil organic carbon is the dominant sorbent. According to USEPA (1994c), for low pH conditions, these estimated values may over-predict sorption coefficients, because they ignore sorption to components other than organic carbon.

The soil organic carbon-water partition coefficient for nonionizing organic compounds. Nonionizing organic compounds are all other organic compounds not listed earlier as ionizing. They include volatile organics, chlorinated pesticides, polynuclear aromatic hydrocarbons (PAHs), and phthalates.

The correlation between K_{oc} and K_{ow} can be improved considerably by performing separate linear regressions on two chemical groups. USEPA (1994c) derives the following correlation equations from measured K_{oc} values cited in USEPA (1994c) and USEPA (1996b):
For most semivolatile nonionizing organic compounds.

$$\log K_{oc} = 2.8 \times 10^{-4} + (0.983 \times \log K_{ow}) \quad (r^2 = 0.99) \tag{1.55a}$$

For volatile nonionizing organics, chlorinated benzenes, and certain chlorinated pesticides

$$\log K_{oc} = 0.0784 + (0.7919 \times \log K_{ow}) \quad (r^2 = 0.97) \tag{1.55b}$$

For PCDDs and PCDFs, the following correlation equation is used to calculate K_{oc} values, as cited by USEPA (1994a)

$$\log K_{oc} = \log K_{ow} - 0.21 \quad (r^2 = 1.0) \tag{1.55c}$$

8.6 Partitioning Coefficients

Partition coefficients (*Kd*) describe the partitioning of a compound between sorbing material, such as soil, soil pore water, surface water, suspended solids, and bed sediments. For organic compounds, *Kd* has been estimated to be a function of the organic-carbon partition coefficient and the fraction of organic carbon in the partitioning media. For metals, *Kd* is assumed to be independent of the organic carbon in the partitioning media and, therefore, partitioning is similar in all sorbing media.

Partitioning coefficients for soil-Water. The soil-water partition coefficient (K_{ds}) describes the partitioning of a compound between soil pore-water and soil particles, and strongly influences the release and movement of a compound into the subsurface soils and underlying aquifer.

Partitioning coefficients for suspended sediment-surface water. The suspended sediment-surface water partition coefficient (Kd_{sw}) describes the partitioning of a compound between surface water and suspended solids or sediments.

Partitioning coefficients for bottom sediment-sediment pore water. The bed sediment-sediment pore-water partition coefficient (Kd_{bs}) describes the partitioning of a compound between the bed sediments and bed sediment pore-water.

For organics (including PCDDs and PCDFs), soil organic carbon is assumed to be the dominant sorbing component in soils and sediments. Therefore, *Kd* values were calculated using the following fraction organic carbon (f_{OC}) correlation equations:

$$\begin{aligned} Kd_{sw} &= f_{oc,sw} \cdot K_{oc} \\ Kd_{s} &= f_{oc,s} \cdot K_{oc} \\ Kd_{bs} &= f_{oc,bs} \cdot K_{oc} \end{aligned} \tag{1.56}$$

The fraction organic carbon (f_{OC}), from literature searches USEPA (1993d), could range as follows:

- 0.002 to 0.024 in soils for which a midrange value of $f_{oc,s} = 0.01$ generally can be used.
- 0.05 to 0.1 in suspended sediments for which a midrange value of $f_{oc,sw} = 0.075$ generally can be used.
- 0.03 to 0.05 in bottom sediments for which a midrange value of $f_{oc,bs} = 0.04$ generally can be used.

For metals (except mercury), *Kd* is governed by factors other than organic carbon, such as pH, redox, iron content, cation exchange capacity, and ion chemistry. Values must be obtained directly from literature. USEPA (1996b) provides values for *Kd* that are based on pH, and are estimated by using the MINTEQ2 model, which is a geochemical speciation model. The MINTEQ2 model analyses were conducted under a variety of geochemical conditions and metal concentrations.

8.7 Soil Loss Constant Due to Degradation

Soil loss constant due to degradation (*ksg*) reflects loss of a compound from the soil by processes other than leaching. Degradation rates in the soil media include biotic and abiotic mechanisms of transformation. Abiotic degradation includes photolysis, hydrolysis, and redox reactions. Hydrolysis and redox reactions can be significant abiotic mechanisms in soil (USEPA 1990).

Degradation rates can be assumed to follow first order kinetics. Therefore, the half-life ($t_{1/2}$) of compounds can be related to the *ksg* as follows:

Ideally, *ksg* is the sum of all biotic and abiotic rate constants in the soil. Therefore, if the $t_{1/2}$ for all of the mechanisms of transformation are known, the degradation rate can be calculated using

$$ksg = \frac{0.693}{t_{1/2}} \tag{1.57}$$

Fraction of pollutant air concentration in the vapor phase. For organics, the fraction of pollutant air concentration in the vapor phase (F_v) was calculated using the following equation:

$$F_v = 1 - \frac{cS_T}{p_L^E + cS_T} \tag{1.58}$$

If the compound is a liquid at ambient temperatures (that is, when P_L^E is known), the above equation calculates F_v using the vapor pressure value recommended for that compound. If the compound is a solid at ambient temperatures (that is, when P_S^E is known), the following equation (Bidleman 1988) was used to calculate P_L^E from P_S^E, for use in the above equation:

$$\ln\left(\frac{P_L^E}{P_S^E}\right) = \frac{\Delta S_f\,(T_m - T)}{R \quad T} \tag{1.59}$$

where

c = Junge constant = 1.7×10^{-04} (atm · cm)
P_L^E = Liquid phase vapor pressure of compound (atm)
P_S^E = Solid phase vapor pressure of compound (atm)
R = Universal ideal gas constant (atm · m³/mol · K)
ΔS_f = Entropy of fusion [$\Delta S_f/R$ = 6.79 (unitless)]
S_T = Whitby's average surface area of particulates (aerosols)
T_a = Ambient air temperature (K)—assumed to be 25°C or 298 K

This equation assumes that the Junge constant (c) is constant for all compounds. However, c can depend on (1) the compound (sorbate) molecular weight, (2) the surface concentration for monolayer coverage, and (3) the difference between the heat of desorption from the particle surface and the heat of vaporization of the liquid-phase sorbate.

All metals (except mercury) are assumed to be present in the particulate phase and not in the vapor phase (Vp = 0), and assigned F_v values of zero. Mercury and mercuric chloride are relatively volatile and exist in the vapor phase (USEPA 1997g). F_v values of 1.0 for mercury and 0.85 for mercuric chloride were estimated. Methyl mercury is assumed not to exist in the air phase and, therefore, assigned an F_v of zero.

8.8 Bioconcentration and Biotransfer Factors for Plants

Root concentration factor. The root concentration factor (R_{CF}) is used to calculate the belowground transfer of contaminants from soil to a root vegetable. The RCF was developed based on experiments conducted by Briggs et al. (1982) which measured uptake of compounds into barley roots from growth solution.

For compounds with log K_{ow} values of 2.0 and higher, we used the following correlation equation to obtain RCF values:

$$\log\,(RCF) = 0.77 \log K_{ow} - 1.52 \quad (n = 7, r = 0.981) \tag{1.60}$$

For compounds with log K_{ow} values less than 2.0, we used the following correlation equation to obtain RCF values:

$$\log\,(RCF - 82) = 0.77 \log K_{ow} - 1.52 \tag{1.61}$$

This equation estimates a RCF value in fresh weight (FW) units, which was then converted to dry weight (DW) units using a moisture content of 87 percent in root vegetables K_{ow} values recommended in this HHRAP were used to calculate each RCF value.

For metals, including mercury, no referenced RCF values were available in published literature. However, plant-soil biotransfer factors for root vegetables ($Br_{rootveg}$) were available in the literature and, therefore, RCF values, which were used to calculate $Br_{rootveg}$ values, are not required for the metals.

Plant-soil bioconcentration factors in root vegetables. The plant-soil bioconcentration factor for compounds in root vegetables ($Br_{rootveg}$) accounts for uptake from soil to the belowground root vegetables or produce.

For organics, the following equation, obtained from USEPA (1995b), was used to calculate values for $Br_{rootveg}$ on a dry weight basis:

$$Br_{rootveg} = \frac{RCF}{Kd_s} \tag{1.62}$$

$Br_{rootveg}$ values were calculated by dividing the *RCF* value with the Kd_s values provided in the main USEPA literature used in this Chapter (USEPA HHRAP). $Br_{rootveg}$ values for metals can also be obtained from the USEPA HHRAP.

Plant-soil Bioconcentration factors for aboveground produce and forage. The plant-soil bioconcentration factor (*Br*) for aboveground produce accounts for the uptake from soil and the subsequent transport of COPCs through the roots to the aboveground plant parts. As addressed in USEPA (1995b), the *Br* value for organics is a function of water solubility, which is inversely proportional to K_{ow}. The *Br* value for metals is a function of the bioavailability of the compounds in soil. Plant-soil bioconcentration factors for aboveground produce (Br_{ag}) and forage (Br_{forage}).

For all organics including PCDDs and PCDFs, (1) the subscript *ag* represents aboveground produce which applies to exposed fruits and vegetables, and protected fruits and vegetables, and (2) the subscript *forage* represents forage, but the values also apply to silage and grain. For metals, (1) aboveground fruits (both exposed and protected) are represented by$Br_{ag\ (fruit)}$; (2) aboveground vegetables (both exposed and protected) are represented by $Br_{ag(veg)}$, (3) forage is represented by Br_{forage}, but the values also apply to silage, and (4) grains are represented by Br_{grain}.

For organics, the following correlation equations were used to calculate values for Br_{ag} and Br_{forage} on a dry weight basis:

$$\log Br_{ag} = 1.588 - 0.578 (\log K_{ow})(n = 29, r = 0.73)$$
$$\log Br_{forage} = 1.588 - 0.578 (\log K_{ow})(n = 29, r = 0.73) \tag{1.63}$$

Air-to-plant biotransfer factors for aboveground produce and forage. The air-to-plant biotransfer factor is defined as the ratio of COPC concentration in aboveground plant parts to the COPC concentration in air. *Bv* values for all organics and metals, were calculated only for aboveground exposed produce (both fruits and vegetables). Aboveground protected produce (both fruits and vegetables) and belowground produce are protected from air-to-plant transfer. According to USEPA (1995b), root vegetables are assumed to be also protected from air-to-plant transfer.

For organics (excluding PCDDs and PCDFs), the Bv_{ag} and Bv_{forage} can be calculated using the following correlation equations:

$$\log B_{vol} = 1.065 \log K_{ow} - \log\left(\frac{H}{RT}\right) - 1.654(r = 0.957) \tag{1.64}$$

$$Bv = \frac{\rho_{air} \cdot B_{vol}}{(1 - f_{water}) \cdot \rho_{forage}}$$

where

B_{vol} = Volumetric air-to-plant biotransfer factor (fresh-weight basis)
Bv = Mass based air-to-plant biotransfer factor (dry-weight basis)
ρ_{air} = 1.19 g/L (Weast 1981)
ρ_{forage} = 770 g/L (Macrady and Maggard 1993)
f_{water} = 0.85 (fraction of forage that is water(Macrady and Maggard (1993))

8.9 Biotransfer Factors for Animals

The biotransfer factor for animals (Ba) is the ratio of COPC concentration in fresh weight animal tissue to the daily intake of COPC by the animal.

Biotransfer factors for beef and milk. The main route of human exposure to many highly lipophilic compounds is through ingestion of contaminated agricultural products such as beef and milk. The transfer of contaminants from environmental media (e.g., air, soil, water) and food (e.g.

grain, sileage) into livestock products (e.g., beef, milk) has historically been either determined by direct measurement of contaminants in livestock products, or predicted using regression models. (Biotransfer factors for beef (Ba_{beef}) and milk(Ba_{milk}).

Organic compounds. When empirical data are lacking for biotransfer of organic chemicals, one of the most widely used approaches to predict contaminant transfer from environmental media and food to beef tissue and milk are the regression models developed by Travis and Arms (1988), which relate chemical octanol-water partition coefficient (K_{ow}) to biotransfer into beef and milk. These regressions, however, are hampered by the limited *log* K_{ow} range and questions surrounding the validity of the underlying biotransfer data set. In response, EPA developed a new methodology for predicting beef and milk biotransfer factors.

Using EPA's updated methodology, we predicted biotransfer factors for organic chemicals with the following single equation:

$$\log Ba_{fat} = -0.099\,(\log K_{ow})^2 + 1.07 \log K_{ow} - 3.56 \quad (n = 305,\ r^2 = 0.8259) \qquad (1.65)$$

where

Ba_{fat} = Biotranfer factor [mg / kg fat] / [mg/d]

K_{ow} = Octanol-water portioning coefficient [unitless]

Values for Ba_{fat} were adjusted to account for the assumed fat content of milk and beef as shown below:

$$Ba_{milk} = 10^{\log Ba\,fat} \times 0.04$$
$$Ba_{beef} = 10^{\log Ba\,fat} \times 0.19 \qquad (1.66)$$

The log K_{ow}'s of the chemicals are used to derive the equation ranged from −0.67 to 8.2. Therefore, compounds with log K_{ow} values less than −0.67 were assigned Ba_{beef} and Ba_{milk} values corresponding to a log K_{ow} value of −0.67. At the high end of the range, compounds with log K_{ow} values greater than 8.2 were assigned Ba_{beef} and Ba_{milk} values corresponding to a log K_{ow} value of 8.2.

Highly metabolized organic compounds. The BA_{fat} might overestimate biotransfer of highly metabolized chemicals, producing an upper bound estimate for these chemicals. Phthalates and PAHs fall within this group.

One way to account for this potential overestimation is to rely upon a metabolism factor to improve model predictions. For example, EPA developed a metabolism factor of 0.01 (i.e., 99 percent of the chemical ingested is metabolized) for bis-ethylhexyl phthalate (BEHP). When this factor is applied to the biotransfer factors predicted using the regression equation recommended above for BEHP, the biotransfer factors are reduced by two orders of magnitude. These metabolism-adjusted predicted biotransfer factors are close in magnitude to the empirically derived biotransfer factors found in the literature, which supports using this metabolism factor. Unfortunately, EPA has not developed metabolism factors for other organic chemicals, due to limited availability of empirically derived data.

For those highly metabolized chemicals that don't have metabolism factors, we still consider it reasonable to use estimated *Ba* values for the following reasons:

- Few chemicals have had all their degradation products identified.
- If identified, the degradation products may in fact be as toxic as, or even more toxic than, the parent compound (the degradation products of PAHs, for example, are toxic). Unless data demonstrates that all degradation products are nontoxic (as is the case for BEHP), the only way to address toxic degradation products in the HHRAP is to include their mass in the mass of the parent chemical.
- The metabolic degradation products may themselves be persistent. For example, DDT is metabolized to DDD and DDE, which remain persistent.

It should also be noted that not all chemicals are metabolized at the same rate and may remain in animal tissue as the parent compound through establishment of steady state concentrations. In fact, many of the chemicals in the biotransfer data set, that are actually well predicted, are metabolized to other compounds. DDT is metabolized to DDE, and lindane (used to derive the regression, but not in the HHRAP) is metabolized to many different compounds. For DDT and lindane, biotransfer factors are well predicted using BA_{fat} equation.

Lacking sufficient data to identify all degradation products, characterize all degradation products as nontoxic, and ensure that all potentially toxic degradation products are less persistent than the parent chemical, we consider it reasonably protective to use the Ba_{beef} and Ba_{milk} values for the parent chemical as predicted, without adjustment. If a highly metabolized chemical is found to drive the risk assessment, then we recommend re-evaluating the appropriateness of the Ba_{beef} and Ba_{milk} values.

Ionizing organic compounds. To improve bio-transfer factors (BTF) estimates for organic acids, the USEPA uses the first-order dissociation constant (pKa) to account for chemical ionization. For these chemicals, K_{ow} is a weighted value calculated based on the fraction of the chemical in the neutral form such that:

$$K_{ow} = K_{ow}n \times (FracNeutral) + K_{ow}i \times (1 - FracNeutral) \tag{1.67}$$

where

$K_{ow}n$ = Partition coefficient for the neutral species (unitless)
$K_{ow}i$ = Partition coefficient for the ionized species (mol/L)
$FracNeutral$ = Fraction of neutral species present for organic acids (unitless).

Accounting for the fraction of ionizable organics in the neutral form is important because K_{ow} can vary considerably depending on pH. The cow's small intestine, where chemicals can be absorbed, has a near neutral pH. Therefore, the neutral fraction is determined using a pH equal to 7 in the following equation (Lee et al., 1990):

$$FracNeutral = 1 + 10\ pH - pKa \tag{1.68}$$

where,

pKa = Acid dissociation constant (unitless).

If a value for log $K_{ow}i$ is not available, the USEPA estimates log $K_{ow}i$ assuming a ratio of log $K_{ow}i$ to log $K_{ow}n$ of 0.015. This ratio is a conservative value developed by the USEPA to apply to organic acids without data for log $K_{ow}i$ (U.S. EPA, 1996).

Ba_{beef} and Ba_{milk} values are for metals (except cadmium, mercury, selenium, and zinc) are calculated on a fresh weight basis, from Baes et al. (1984). For cadmium, selenium, and zinc, USEPA (1995a) cited Ba values derived by dividing uptake slopes [(g COPC/kg DW tissue)/(g COPC/kg DW feed)], obtained from USEPA (1992b), by a daily consumption rate of 20 kg DW per day for beef and dairy cattle.

The HHRAP assumes that elemental mercury neither deposits onto soils nor transfers to aboveground plant parts. Therefore, there is no transfer of elemental mercury into animal tissue. Therefore, USEPA recommends Ba values of zero for elemental mercury. This based on assumptions made regarding speciation and fate and transport of mercury from emission sources. USEPA derived the Ba_{beef} and Ba_{milk} values listed in the HHRAP database for mercuric chloride and methyl mercury from data in USEPA (1997b).

Biotransfer factors for pork. For organics (except PCDDs and PCDFs), biotransfer factors for pork (Ba_{pork}) values reported in the USEPA HHRAP were derived from Ba_{beef} values, assuming that pork is 23 percent fat and beef is 19 percent fat. Therefore, Ba_{pork} values were calculated by multiplying Ba_{beef} values by their fat content ratio of 1.2 (23/19). This calculation is limited by the assumptions that:

1. COPCs bioconcentrate in the fat tissues.
2. There is minimal effect from differences in metabolism and feeding characteristics between beef cattle and pigs.

For PCDDs and PCDFs, Ba_{pork} values are calculated using the same methodology used to obtain Ba_{beef} values by USEPA (1995a). Assuming that milk is 3.5 percent fat and that pork is 23 percent fat, biotransfer factors for pork would be 6.57 times (23/3.5) higher than for milk. Therefore, Ba_{pork} values were calculated by increasing Ba_{milk} values by a factor of 6.57. This has the same effect as if the Ba_{pork} values were calculated by multiplying the Ba_{beef} values with the fat content ratio of 1.2 (23/19) between pork and beef, as was adopted for the remaining organic compounds.

Biotransfer factors for chicken and poultry eggs. Biotransfer factors for chicken ($Ba_{chicken}$) and poultry eggs (Ba_{egg}) are expressed as the ratio of the COPC concentration in the fresh weight tissue to the COPC intake from the feed. Biotransfer factors are calculated from bioconcentration factors for chicken and poultry eggs. BCFs are expressed as the ratio of the COPC concentration in the fresh weight tissue to the COPC concentration in dry weight soil.

USEPA recommends deriving the biotransfer factors for chicken $Ba_{chicken}$ values using the same method used to estimate Ba_{beef} values, modifying to reflect an assumed fat content of chicken of 14 percent, and eggs of 8 percent. Specifically, USEPA uses the equation to generate Ba_{fat} values, which were then adjusted to account for the assumed fat content of chicken and eggs as shown below:

$$Ba_{chicken} = 10^{\log Ba_{fat}} \times 0.14$$
$$Ba_{eggs} = 10^{\log Ba_{fat}} \times 0.08 \tag{1.69}$$

These calculations are limited by the assumptions that:

1. Contaminants bioconcentrate in the fat tissues.
2. Effects from differences in metabolism, digestive system, or feeding characteristics between beef cattle and chickens are minimal.

Please note that the scenario of principal concern for chicken and egg contamination is for home grown chickens. The raising of home grown chickens would be characteristic of free range and semi-free range housing conditions where poultry do come in contact with soil, and possibly vegetation, insects, and benthic organisms. The applicability of this scenario to commercial poultry operations characterized by housing conditions that do not provide chickens access to soil would need to be assessed on a case-by-case basis.

Bioconcentration and bioaccumulation factors for fish. Bioconcentration and bioaccumulation factors for fish are used for various compounds, depending on the K_{ow} value of the organic compound. Bioconcentration factors for fish were used for organics (except PCDDs, PCDFs, and PCBs) with a log K_{ow} value less than 4.0; and for metals (except lead and mercury). Bioaccumulation factors for fish were used for organics (except PCDDs, PCDFs, and PCBs) with a log K_{ow} value greater than 4.0, lead, and mercuric compounds. Biota-sediment accumulation factors for fish were used for PCDDs, PCDFs, and PCBs.

Bioconcentration factors for fish. Bioconcentration factors for fish (BCF_{fish})is the ratio of the COPC concentration in fish to the COPC concentration in the water column where the fish is exposed. It accounts for uptake of COPCs by fish from water passing across the gills. *BCF* values for fish were used for all organic compounds with a log K_{ow} of less than 4.0 (cutoff value with BAF_{fish}) and for all metals, except lead and mercury, as cited in USEPA (1995b). This implies that the concentration of COPC in the fish is only due to water intake by the fish, and compounds with a log K_{ow} of less than 4.0 are assumed not to bioaccumulate.

Our recommended BCF values don't recognize differences in total versus dissolved water concentrations when calculating fish concentrations from BCF_{fish} values for compounds with a log K_{ow} of less than 4.0. Since, dissolved water concentrations is the major contributing factor from compounds with a log K_{ow} of less than 4.0, all BCF_{fish} values (regardless of whether they were derived using total or dissolved water concentrations) can be multiplied by the contaminant concentration in the dissolved water column (Cdw) to calculate fish concentrations. This assumption is necessary because:

1. Literature data is often unclear if the water concentrations are dissolved or total concentrations.
2. Most of the literature reviewed indicated that laboratory experiments were conducted using filtered or distilled water; or the experiments were conducted using fresh water, but were filtered before analyses for water concentrations.

Meylan et al. (1999) collected information on measured BCF and other key experimental details for 694 chemicals. Log BCF was then regressed against log K_{ow}, and chemicals with significant deviations from the line of best fit were analyzed by chemical structure. The resulting algorithm classifies a substance as either nonionic or ionic, the latter group including carboxylic acids, sulfonic acids and their salts, and quaternary N compounds. Log BCF for nonionics was estimated from log K_{ow} and a series of correction factors if applicable; different equations apply for log K_{ow} 1.0 to 7.0 and >7.0 (Meylan et al. 1999). These equations are as follows:

For log $K_{ow} < 1$: $\log BCF = 0.50$

For log K_{ow} 1 to 7: $\log BCF = 0.77 \log K_{ow} - 0.7 + \Sigma\, Correction\ Factors$

For log $K_{ow} > 7$: $\log BCF = -1.37 \log K_{ow} + 14.4 + \Sigma\, Correction\ Factors$

For log $K_{ow} > 10.5$: $\log BCF = 0.50$

For ionic compounds (carboxylic acids, sulfonic acids, and salts, compounds with N of +5 valence), were categorized by log K_{ow}, and a log BCF in the range 0.5 to 1.75 was assigned as follows:

For log $K_{ow} < 5$: $\log BCF = 0.50$

For log K_{ow} 5 to 6: $\log BCF = 0.75$

For log K_{ow} 6 to 7: $\log BCF = 1.75$

For log K_{ow} 7 to 9: $\log BCF = 1.0$

For log $K_{ow} > 9$: $\log BCF = 0.50$

USEPA assumes that BCF_{fish} values calculated using the above equations were:

1. Based on dissolved water concentrations
2. Not lipid-normalized

Bioaccumulation factors for fish. Bioaccumulation factors for fish (BAF_{fish})is the ratio of the COPC concentration in fish to the COPC concentration in the water body where the fish are exposed. The BAF_{fish} accounts for uptake of COPCs by fish from water and sediments passing across the gills, and from consumption of various foods including plankton, daphnids, and other fish. BAF_{fish} was used for organic compounds (except PCBs, PCDDs, and PCDFs) with a log K_{ow} greater than 4.0, lead, and mercuric compounds.

For compounds with a log K_{ow} of greater than or equal to 4.0, COPCs can significantly partition into the suspended sediment organic carbon (or particulate phase) of the water column. Therefore, BAF values should be based on total (dissolved and suspended) water column concentrations.

The following equation cited in USEPA (1995f) is recommended to convert the BAF based on total water concentrations to a BAF based on dissolved water concentrations:

$$f_{fd} = \frac{1}{1 + \dfrac{(DOC)\,(K_{ow})}{10} + (POC)(K_{ow})} \tag{1.70}$$

where

f_{fd} = fraction of COPC that is freely dissolved in water
DOC = concentration of dissolved organic carbon, kg organic carbon/L water
POC = concentration of particulate organic carbon, kg organic carbon/L water

Since, the model was derived from a study conducted at Lake Ontario, DOC and POC values for Lake Ontario were used. Values cited in USEPA (1995f) were:

$$\begin{aligned} DOC &= 2 \times 10^{-6}\ \text{kg/L} \\ POC &= 7.5 \times 10^{-9}\ \text{kg/L} \end{aligned} \tag{1.71}$$

A *BAF* based on dissolved water concentrations can be calculated from a BAF based on total water concentrations as follows:

$$BAF\ (dissolved) = \frac{BAF\ (total)}{f_{fd}} - 1 \tag{1.72}$$

For all organics (except PCBs, PCDDs, and PCDFs) with a log K_{ow} greater than or equal to 4.0, the FCM, which accounts for accumulation through the food chain in addition to water, becomes greater than 1. Therefore, a BAF_{fish}, which takes the food chain into consideration, is more appropriate than a BCF_{fish}.

For all organics with a log K_{ow} greater than or equal to 4.0, BAFs were derived using one of following these methods:

1. *BAF* = Field measured BAF or BCF, adjusted for dissolved water concentrations
2. *BAF* = Laboratory measured BCF multiplied by a FCM for either trophic level 3 or 4 fish

Both field and laboratory measured values were derived from various literature sources cited in USEPA (1999a). FCMs were obtained from USEPA (1995f).

For lead, the food-chain multiplier becomes greater than one; therefore, a BAF is more appropriate. The BAF_{fish} value reported in this HHRAP for lead was obtained as a geometric mean from various literature sources described in USEPA (1999a). Since metals are assumed insoluble under neutral conditions, the dissolved and total water concentrations are almost equal. However, for consistency, the BAF_{fish} value for lead was adjusted for dissolved fractions.

Biota-sediment accumulation factor for fish. For PCDDs, PCDFs, and PCBs, bioaccumulation factors for fish ($BSAF_{fish}$)values should be used instead of BAFs for fish. $BSAF_{fish}$ values reported in this HHRAP were obtained from USEPA (1994a).

$BSAF_{fish}$ accounts for the transfer of COPCs from the bottom sediment to the lipid in fish. Homologue group $BSAF_{fish}$ values obtained from USEPA (1994a) were either measured or estimated values that were based on a whole fish lipid content of 7 percent and an organic carbon content of 3 percent. The $BSAF_{fish}$ values reported in this HHRAP are consistent with the values presented in primary guidance documents.

9 HUMAN HEALTH BENCHMARKS

The following sections discuss carcinogenic and noncarcinogenic toxicity benchmarks of compounds. The toxicity information provided in the HHRAP is for informational purposes to help permitting authorities explain the basis for selecting contaminants of concern. Since toxicity benchmarks and slope factors may change as additional toxicity research is conducted, permitting authorities should consult with the most current version of EPA's Integrated Risk Information System (IRIS) and Health Effects Assessment Summary Tables(HEAST) before completing a risk assessment to ensure that the toxicity data used in the risk assessment is based upon the most current agency consensus.

9.1 Reference Dose (Rfd) and Reference Concentration (RfC)

Reference dose is defined as a daily intake rate of a compound estimated to pose no appreciable risk of deleterious effects over a specific exposure duration (USEPA 1989e). Reference concentration is defined as the concentration of a compound estimated (with uncertainty spanning perhaps an order of magnitude) to pose no appreciable risk of deleterious effects over a specific exposure duration (USEPA 1989e).

9.2 Oral Cancer Slope Factor, Inhalation CSF, and Inhalation Unit Risk Factor

Oral cancer slope factor (*CSF*), *inhalation CSF*, and *inhalation* unit risk factor (*URF*) values for all compounds can be obtained from IRIS and HEAST.

9.3 Explanation of Calculated Toxicity Benchmark Values

The preference for health benchmarks is to obtain values from IRIS or HEAST. The following methodology was used to calculate missing benchmarks using available benchmarks that are based on route-to-route extrapolation:

1. *Oral RfDs* presented in IRIS/HEAST/EPA reviewed documents were used if available. missing oral *RfDs* were calculated from the *RfC* assuming route-to-route extrapolation using the following equation:

$$Oral\ RfD = \frac{RfC \cdot 20\ \text{m}^3/d}{70\ \text{kg}\ BW} \tag{1.73}$$

2. *Oral CSFs* presented in IRIS/HEAST/EPA reviewed documents were used when available. In the case of missing oral CSFs:

   a. *Oral CSF* = *Inhalation CSF*, or

   b. *Oral CSF* = *Inhalation CSF* calculated from Inhalation URF assuming route-to-route extrapolation.

3. For *inhalation URFs*, values were obtained from IRIS/HEAST/EPA reviewed documents. If the *inhalation URFs* were not available they were calculated from *oral CSF*, using the following equation:

$$Inhalation\ URF = \frac{Oral\ CSF \cdot 20\ \text{m}^3/\text{d}}{70\ kgx\ 1000\ \mu\text{g/mg}} \tag{1.74}$$

4. The *inhalation CSFs* presented in IRIS/HEAST/EPA should be used when available.

   a. If no *inhalation CSF* was available; it was calculated from *inhalation URF*, using the following equation:

$$Inhalation\ CSF = \frac{Inhalation\ URF \cdot 70\ \text{kg}}{20\ \text{m}^3/\text{d}} \times 1000\ \mu\text{g/mg} \tag{1.75}$$

   b. If no *inhalation URF* was available; the following was assumed based on route-to-route extrapolation:

$$Inhalation\ CSF = Oral\ CSF \tag{1.76}$$

10 TERMINOLOGY AND VARIABLE IN HUMAN HEALTH RISK ASSESSMENT

ρ_{air} = Density of air (g/cm^3)
ρ_{forage} = Density of forage (g/cm^3)

Ba_{beef} = Biotransfer factor in beef (mg COPC/kg FW tissue)/(mg COPC/d)
$Ba_{chicken}$ = Biotransfer factor in chicken (mg COPC/kg FW tissue)/(mg COPC/d)
Ba_{egg} = Biotransfer factor in eggs (mg COPC/kg FW tissue)/(mg COPC/d)
Ba_{milk} = Biotransfer factor in milk (mg COPC/kg FW tissue)/(mg COPC/d)
Ba_{pork} = Biotransfer factor in pork (mg COPC/kg FW tissue)/(mg COPC/d)
BAF_{fish} = Bioaccumulation factor in fish (mg COPC/kg FW tissue)/(mg COPC/L total water column)
BCF_{fish} = Bioconcentration factor in fish (L/kg FW or unitless)
Br_{ag} = Plant-soil bioconcentration factor in aboveground produce (μg COPC/g DW plant)/(μg COPC/g DW soil) (unitless)
$Br_{forage/silage}$ = Plant-soil bioconcentration factor in forage and silage (μg COPC/g DW plant)/(μg COPC/g DW soil) (unitless)
Br_{grain} = Plant-soil bioconcentration factor in grain (μg COPC/g DW grain)/(μg COPC/g DW soil) unitless
$Br_{rootveg}$ = Plant-soil bioconcentration factor for belowground produce (μg COPC/g DW plant)/(μg COPC/g DW soil) (unitless)
$BSAF_{fish}$ = Biota-sediment accumulation factor in fish (mg COPC/kg lipid tissue)/(mg COPC/kg sediment) (unitless)
B_{vol} = Volumetric air-to-leaf biotransfer factor in leaf (μg COPC/L FW plant)/(μg COPC/L air) (unitless)
Bv_{ag} = COPC air-to-plant biotransfer factor for aboveground produce (μg COPC/g DW plant)/(mg COPC/g air) (unitless)
$Bv_{forage/silage}$ = Air-to-plant biotransfer factor in forage and silage (μg COPC/g DW plant)/(μg COPC/g air) (unitless)
c = Junge constant = 1.7×10^{-04} (atm.cm)
D_a = Diffusivity of COPC in air (cm^2/s)
D_w = Diffusivity of COPC in water (cm^2/s)
$f_{oc,bs}$ = Fraction of organic carbon in bottom sediment (unitless)
$f_{oc,s}$ = Fraction of organic carbon in soil (unitless)
$f_{oc,sw}$ = Fraction of organic carbon in suspended sediment (unitless)
f_{water} = Fraction of COPC in water (unitless)
F_v = Fraction of COPC air concentration in vapor phase (unitless)
$\mathrm{F_w}$ = Fraction of wet deposition that adheres to plant surfaces (unitless)
H = Henry's law constant
$CSF_{Inhalation}$ = Inhalation cancer slope factor $(\text{mg/kg} \cdot \text{d})^{-1}$
$URF_{Inhalation}$ = Inhalation unit risk factor $(\mu\text{g/m}^3)^{-1}$
Kd_s = Soil-water partition coefficient (mL water/g soil OR cm^3 water/g soil)
Kd_{sw} = Suspended sediment-surface water partition coefficient (L water/kg suspended sediment OR cm^3 water/g suspended sediment)
Kd_{bs} = Bed sediment-sediment pore water partition coefficient (L water/kg bottom sediment OR cm^3 water/g bottom sediment)
K_{ow} = Octanol/water partitioning coefficient (mg COPC/L octanol)/(mg COPC/L octanol)
K_{oc} = Soil organic carbon-water partition coefficient (mL water/g soil)
ksg = COPC soil loss constant due to biotic and abiotic degradation (year^{-1})
M_W = Molecular weight of COPC (g/mol)
p_L^E = Liquid-phase vapor pressure of COPC (atm)
p_S^E = Solid-phase vapor pressure of COPC (atm)
Oral CSF = Oral cancer slope factor $(\text{mg/kg} \cdot \text{d})^{-1}$
R = Universal gas constant ($\text{atm.m}^3/\text{mol} \cdot \text{K}$)
RCF = Root concentration factor (mg COPC/g DW plant)/(mg COPC/mL soil water)
RfC = Reference concentration (mg/m^3)
RfD = Reference dose (mg/kg/d)
R_p = Interception factor of edible portion of plant (unitless)
S = Solubility of COPC in water (mg COPC/L water)
ΔS_f = Entropy of fusion [$\Delta \mathrm{S_f}/\mathrm{R} = 6.79$ (unitless)]

S_T	= Whitby's average surface area of particulates (aerosols)
	= 3.5×10^{-06} cm²/cm³ air for background plus local sources
	= 1.1×10^{-05} cm²/cm³ air for urban sources
$t_{1/2}$	= Half-time of COPC in soil (days)
T_a	= Ambient air temperature (K)
T_m	= Melting point temperature (K)
TEF	= Toxicity equivalency factor (unitless)
V_p	= Vapor pressure of COPC (atm)

11 REFERENCES

Agricultural Research Service. 1994. Personal communication regarding the dry weight fraction value for hay between G. F. Fries, Glenn Rice, and Jennifer Windholz, USEPA Office of Research and Development. March 22.

Baes, C. F., R. D. Sharp, A. L. Sjoreen, and R. W. Shor. 1984. *Review and Analysis of Parameters and Assessing Transport of Environmentally Released Radionuclides through Agriculture.* ORNL-5786. Oak Ridge National Laboratory. Oak Ridge, Tennessee. September.

Baes, C.F., R.D. Sharp, A. L. Sjoreen and R. W. Shor. 1984. *A Review and Analysis of Parameters for Assessing Transport of Environmentally Released Radionuclides through Agriculture.* Prepared for the U.S. Department of Energy under Contract No. DEAC05-840R21400.

Belcher, G. D., and C. C. Travis. 1989. *Modeling Support for the RURA and Municipal Waste Combustion Projects: Final Report on Sensitivity and Uncertainty Analysis for the Terrestrial Food Chain Model.* Interagency Agreement No. 1824-A020-A1, Office of Risk Analysis, Health and Safety Research Division, Oak Ridge National Laboratory. Oak Ridge, Tennessee. October.

Bidleman, T. F. 1988. Atmospheric Processes. *Environ. Sci. and Tech.* vol. 22(4), : 361–367.

Blake, G.R. and K.H. Hartge. 1996. *Particle Density. Methods of Soil Analysis, Part 1: Physical and Mineralogical Methods.* 2nd edn. Arnold Klute, Ed. American Society of Agronomy, Inc. Madison, WI., p. 381.

Boone, F. W., Yook C. Ng, and John M. Palms. 1981. Terrestrial Pathways of Radionuclide Particulates. *Health Physics,* vol. 41(5), p. 735–747. November.

Briggs, G. G., R. H. Bromilow, and A. A. Evans. 1982. Relationships between lipophilicity and root uptake and translocation of non-ionized chemicals by barley. *Pesticide Science* 13: 495–504.

California Environmental Protection Agency (CEPA). 1993. Parameter Values and Ranges for CALTOX. Draft. Office of Scientific Affairs. California Department of Toxics Substances Control. Sacramento, CA. July.

Carsel, R. F., Parrish, R. L. Jones, J. L. Hansen, and R. L. Lamb. 1988. Characterizing the Uncertainty of Pesticide Leaching in Agricultural Soils. *J Contaminant Hydrol.* vol. 2. :11–24.

Chamberlain, A.C. 1970. Interception and Retention of Radioactive Aerosols by Vegetation. *Atmospheric Environment.vol.* 4:57–78.

Chang, R.R., D. Hayward, L. Goldman, M. Harnly, J. Flattery, and R.D. Stephens. 1989. Foraging Farm Animals as Biomonitors for Dioxin Contamination. *Chemosphere. vol.* 19:481–486.

Cook, P. M., D. W. Duehl, M. K. Walker, and R. E. Peterson. 1991. *Bioaccumulation and Toxicity of TCDD and Related Compounds in Aquatic Ecosystems.* In *Gallo,* M. A., R. J. Scheuplein, and K. A. Van Der Heijden (eds). *Banbury Report 35: Biological Basis for Risk Assessment of Dioxins and Related Compounds.* Cold Spring Harbor Laboratory Press. 0-87969-235-9/91.

Droppo, J. G. Jr., D. L. Strenge, J. W. Buck, B. L. Hoopes, R. D. Brockhaus, M. B. Walter, and G. Whelan. 1989. *Multimedia Environmental Pollutant Assessment System (MEPAS) Application Guidance: Volume 2-Guidelines for Evaluating MEPAS Input Parameters.* Pacific Northwest Laboratory. Richland, Washington. December.

Ellgenhausen, H. J., A. Guth, and H. O. Esser. 1980. Factors Determining the Bioaccumulation Potential of Pesticides in the Individual Compartments of Aquatic Food Chains. *Ecotoxicology Environmental Safety.* vol. 4:134.

Geraghty, J. J., D. W. Miller, F. Van der Leeden, and F. L. Troise. 1973. *Water Atlas of the United States.* Water Information Center, Port Washington, New York.

Hillel, D. 1980. *Fundamentals of Soil Physics.* Academic Press, Inc. New York.

Hoffman, F. O., and C. F. Baes, 1979. *A Statistical Analysis of Selected Parameters for Predicting Food Chain Transport and Internal Dose of Radionuclides.* ORNL/NOREG/TM-882.

Hoffman, F. O., K. M. Thiessen, M. L. Frank, and B. G. Blaylock. 1992. Quantification of the Interception and Initial Retention of Radioactive Contaminants Deposited on Pasture Grass by Simulated Rain. *Atmospheric Environment.* vol. 26A(18):3313 to 3321.

Hwang S. T. and Falco, J. W. 1986. Estimation of multimedia exposures related to hazardous waste facilities, In: *Pollutants in a Multimedia Environment.* Yoram Cohen, Ed. Plenum Publishing Corp. New York.

Ikeda, G. J., P. P. Sapenza, and J. L. Couvillion. 1980. Comparative distribution, excretion, and metabolism of di(2-ethylhexyl)phthalate in rats, dogs, and pigs. *Food Cosmet. Toxicology.* 18:637–642.

Junge, C.E. 1977. Fate of *Pollutants in Air and Water Environments, Part I.* Suffet, I.H., Ed. Wiley. New York. pp. 7–26.

Lee, M.C., S.K. Chian, and R.A Griffin. 1979. Solubility of Polychlorinated Biphenyls and Capacitor Fluid in Water. *Water Res.* 13:1249–1258.

Lyman, W. J., W. F. Reehl, and D. H. Rosenblatt. 1982. *Handbook of Chemical Property Estimation Methods: Environmental Behavior of Organic Compounds.* McGraw-Hill Book Company. New York.

McKone, T.E., and P.B. Ryan. 1989. *Human Exposures to Chemicals Through Food Chains: An Uncertainty Analysis.* Livermore, California: Lawrence Livermore National Laboratory Report. UCRL-99290.

McLachlan, M. S., H. Thoma, M. Reissinger, and O. Hutzinger. 1990. "PCDD/F in an Agricultural Food Chain, Part 1: PCDD/F Mass Balance of a Lactating Cow." *Chemosphere*, vol. 20(7–9) pp. 1013–1020.

Miller, C. W. and F.O. Hoffman. 1983. An Examination of the Environmental Half-Time for Radionuclides Deposited on Vegetation. *Health Physics.* 45 (3) pp. 731–744.

Miller, R. W. and D. T. Gardiner. 1998. In: *Soils in Our Environment.* J. U. Miller, Ed. Prentice Hall. Upper Saddle River, NJ. pp. 80–123.

National Academy of Sciences (NAS). 1987. *Predicting Feed Intake of Food-Producing Animals.* National Research Council, Committee on Animal Nutrition, Washington, D.C.

NC DEHNR. 1997. *Final NC DEHNR Protocol for Performing Indirect Exposure Risk Assessments for Hazardous Waste Combustion Units.* January.

Ng, Y. C., C. S. Colsher and S. E. Thomson. 1982. *Transfer Coefficients for Assessing the Dose from Radionuclides in Meat and Eggs.* U.S. Nuclear Regulatory Commission. Final Report. NUREG/CR-2976.

Ogata, M. K., Y. Ogino Fijusaw and E. Mano. 1984. Partition Coefficients as a Measure of Bioconcentration Potential of Crude Oil Compounds in Fish and Shellfish. *Bulletin of Environmental Contaminant Toxicology.* vol. 33. pp. 561.

Pennington, J. A. T. 1989. *Food Values of Portions Commonly Used.* 15th edn. Harper and Row. New York.

Petreas, M. X., L. R. Goldman, D. G. Hayward, R. Chang, J. Flattery, T. Wiesmuller, R. D. Stephens, D. M. Fry and C. Rappe. 1991. Biotransfer and Bioaccumulation of PCDD/PCDFs from Soils: Controlled Exposure Studies of Chickens. *Chemosphere.* vol.23 pp. 1731–1741.

Research Triangle Institute (RTI). 1992. *Preliminary Soil Action Level for Superfund Sites.* Draft Interim Report. Prepared for USEPA Hazardous Site Control Division, Remedial Operations Guidance Branch. Arlington, Virginia. EPA Contract 68-W1-0021. Work Assignment No. B-03, Work Assignment Manager Loren Henning. December.

Riederer, M. 1990. Estimating Partitioning and Transport of Organic Chemicals in the Foliage/Atmosphere System: Discussion of a Fugacity-Based Model. *Environmental Science and Technology.* vol. 24: 829–837.

Shor, R. W., C. F. Baes and R. D. Sharp. 1982. *Agricultural Production in the United States by County: A Compilation of Information from the 1974 Census of Agriculture for Use in Terrestrial Food-Chain Transport and Assessment Models.* Oak Ridge National Laboratory Publication. ORNL-5786.

Stephens, R. D., M. X. Petreas and D. G. Hayward. 1992. Biotransfer and Bioaccumulation of Dioxins and Dibenzofurans from Soil. Hazardous Materials Laboratory, California Department of Health Services. Berkeley, California.

Stephens, R. D., M. X. Petreas and D. G. Hayward. 1995. Biotransfer and Bioaccumulation of Dioxins and Furnas from Soil: Chickens as a Model for Foraging Animals. *The Science of the Total Environment.* vol. 175. pp. 253–273.

Travis, C. C. and A. O. Arms. 1988. Bioconcentration of Organics in Beef, Milk, and Vegetation. *Environmental Science and Technology.* vol. 22 pp. 271–274.

U.S. Bureau of the Census. 1987. *Statistical Abstract of the United States: 1987.* 107th edn. Washington, D.C.

U.S. Department of Agriculture (USDA). 1994. *Noncitrus Fruits and Nuts 1993 Summary.* National Agricultural Statistics Service, Agricultural Statistics Board, Washington, D.C. Fr Nt 1–3 (94).

U.S. Department of Agriculture (USDA). 1994. Personal Communication Between G. F. Fries, and Glenn Rice and Jennifer Windholtz, U.S. Environmental Protection Agency, Office of Research and Development. Agricultural Research Service. March 22.

U.S. Department of Agriculture (USDA). 1994. Personal Communication Regarding Soil Ingestion Rate for Dairy Cattle. Between G. F. Fries, Agricultural Research Service, and Glenn Rice and Jennifer Windholtz, USEPA, Office of Research and Development. March 22.

U.S. Department of Agriculture (USDA). 1994. *Vegetables 1993 Summary.* National Agricultural Statistics Service, Agricultural Statistics Board. Washington, D.C. Vg 1–2 (94).

U.S. Department of Agriculture (USDA). 1997. *Predicting Soil Erosion by Water: A Guide to Conservation Planning With the Revised Universal Soil Loss Equation (RUSLE).* Agricultural Research Service, Agriculture Handbook Number 703. January.

USEPA. 1982. Pesticides Assessment Guidelines Subdivision O. Residue Chemistry. Office of Pesticides and Toxic Substances, Washington, D.C. EPA/540/9-82-023.

USEPA. 1985. *Water Quality Assessment: A Screening Procedure for Toxic and Conventional Pollutants in Surface and Groundwater.* Part I (Revised 1985). Environmental Research Laboratory. Athens, Georgia. EPA/600/6-85/002a. September.

USEPA. 1988. *Superfund Exposure Assessment Manual.* Office of Solid Waste. Washington, D.C. April.

USEPA. 1990. *Interim Final Methodology for Assessing Health Risks Associated with Indirect Exposure to Combustor Emissions.* Environmental Criteria and Assessment Office. Office of Research and Development. EPA 600-90-003. January.

USEPA. 1990. *Exposure Factors Handbook.* March.

USEPA. 1992. *Addendum to the Methodology for Assessing Health Risks Associated with Indirect Exposure to Combustor Emissions.* External Review Draft. Office of Research and Development. Washington, D.C. November.

USEPA. 1992. *Estimating Exposure to Dioxin-Like Compounds.* Draft Report. Office of Research and Development. Washington, D.C. EPA/600/6-88/005b.

USEPA. 1992. *Technical Support Document for Land Application of Sewage Sludge.* Volumes I and II. Office of Water. Washington, D.C. EPA 822/R-93-001a.

USEPA. 1993. *Addendum to the Methodology for Assessing Health Risks Associated with Indirect Exposure to Combustor Emissions.* Working Group Recommendations. Office of Solid Waste. Office of Research and Development. Washington, D.C. September 24.

USEPA. 1993. *Addendum to the Methodology for Assessing Health Risks Associated with Indirect Exposure to Combustor Emissions.* External Review Draft. Office of Research and Development. Washington, D.C. November.

USEPA. 1993. *Derivation of Proposed Human Health and Wildlife Bioaccumulation Factors for the Great Lakes Initiative.* Office of Research and Development, U.S. Environmental Research Laboratory. Duluth, Minnesota. March.

USEPA 1993. *Review Draft Addendum to the Methodology for Assessing Health Risks Associated with Indirect Exposure to Combustor Emissions.* Office of Health and Environmental Assessment. Office of Research and Development. EPA-600-AP-93-003. November 10.

USEPA 57 Federal Register 20802. 1993. Proposed Water Quality Guidance for the Great Lakes System. April 16.

USEPA 1994. *Draft Guidance for Performing Screening Level Risk Analyses at Combustion Facilities Burning Hazardous Wastes. Attachment C, Draft Exposure Assessment Guidance for RCRA Hazardous Waste Combustion Facilities.* Office of Emergency and Remedial Response. Office of Solid Waste. April 15.

USEPA 1994. *Draft Guidance for Performing Screening Level Risk Analysis at Combustion Facilities Burning Hazardous Wastes. Attachment C, Draft Exposure Assessment Guidance for RCRA Hazardous Waste Combustion Facilities.* April 15.

USEPA 1994. *Draft Exposure Assessment Guidance for RCRA Hazardous Waste Combustion Facilities.* Office of Solid Waste and Emergency Response. EPA-530-R-94-021. April.

USEPA 1994. *Estimating Exposure to Dioxin-Like Compounds. Volume II: Properties, Sources, Occurrence, and Background Exposures*. External Review Draft. Office of Research and Development. Washington, D.C. EPA/600/6-88/005Cb. June.

USEPA 1994. *Estimating Exposure to Dioxin-Like Compounds. Volume III: Site-Specific Assessment Procedures*. External Review Draft. Office of Research and Development. Washington, D.C. EPA/600/6-88/005Cc. June.

USEPA 1994. *Guidance for Performing Screening Level Risk Analysis at Combustion Facilities Burning Hazardous Wastes.* Office of Emergency and Remedial Response. Office of Solid Waste. December 14.

USEPA 1994. *Revised Draft Guidance for Performing Screening Level Risk Analyses at Combustion Facilities Burning Hazardous Wastes. Attachment C, Draft Exposure Assessment Guidance for RCRA Hazardous Waste Combustion Facilities*. Office of Emergency and Remedial Response. Office of Solid Waste. December 14.

USEPA 1994. Revised Draft Guidance for Performing Screening Level Risk Analyses at Combustion Facilities Burning Hazardous Wastes. Office of Emergency and Remedial Response. Office of Solid Waste. December 14.

USEPA 1995. *Further Issues for Modeling the Indirect Exposure Impacts from Combustor Emissions*. Office of Research and Development. Washington, D.C. January 20.

USEPA 1995. *Review Draft Development of Human Health-Based and Ecologically-Based Exit Criteria for the Hazardous Waste Identification Project*. Volumes I and II. Office of Solid Waste. March 3.

USEPA 1995. "Waste Technologies Industries Screening Human Health Risk Assessment (SHHRA): Evaluation of Potential Risk from Exposure to Routine Operating Emissions." Volume V. External Review Draft. USEPA Region 5, Chicago, Illinois.

USEPA 1995. *Great Lakes Water Quality Initiative. Technical Support Document for the Procedure to Determine Bioaccumulation Factors*. Office of Water. EPA-820-B-95-005. March.

USEPA 1997. *Exposure Factors Handbook*. Office of Research and Development. EPA/600/P-95/002F. August.

USEPA 1997. *Exposure Factors Handbook*. "Food Ingestion Factors." Volume II. SAB Review Draft. EPA/600/P-95/002F. August.

USEPA 1997. Mercury Study Report to Congress. Volume III: Fate and Transport of Mercury in the Environment. Office of Air Quality and Planning and Standards and Office of Research and Development. EPA 452/R-97-005. December.

USEPA 1997. *Mercury Study Report to Congress*.(Volume III. Draft. Office of Air Quality and Planning and Standards and Office of Research and Development. December.

USEPA 1998. *Methodology for Assessing Health Risks Associated with Multiple Pathways of Exposure to Combustor Emissions*. Update to EPA/600/6-90/003. Office of Research and Development, National Center for Environmental Assessment, USEPA. EPA/600/R-98/137. December. Environmental Criteria and Assessment Office. ORD. Cincinnati, Ohio.

USEPA 1999. *Screening Level Ecological Risk Assessment Protocol for Hazardous Waste Combustion Facilities*. Peer Review Draft. Office of Solid Waste. August.

USEPA 2005. *Human Health Risk Assessment Protocol for Hazardous Waste Combustion Facilities*, EPA530-R-05-006, Final, Office of Solid Waste and Emergency Response, September.

Vanoni, V.A. 1975. *Sedimentation Engineering*. American Society of Civil Engineers. New York, pp. 460–463.

Weast, R. C. 1979. *CRC Handbook of Chemistry and Physics*. 60th edn.. CRC Press, Inc. Cleveland, Ohio.

Weast, R. C. 1981. *Handbook of Chemistry and Physics*. 62nd Edition. Cleveland, Ohio. CRC Press.

Wipf, H.K., E. Homberger, N. Neuner, U. B. Ranalder, W. Vetter and J. P. Vuilleumier. 1982. TCDD Levels in Soil and Plant Samples from the Seveso Area. In: *Chlorinated Dioxins and Related Compounds: Impact on the Environment*. Eds. Hutzinger, O. et al. Pergamon, New York.

Wischmeire, W. H. and D.D. Smith. 1978. *Predicting Rainfall Erosion Losses(A Guide to Conservation Planning*. Agricultural Handbook No. 537. U.S. Department of Agriculture, Washington, D.C.

P • A • R • T • 5

CHAPTER 5.1
FUNDAMENTALS OF FUEL CELL TECHNOLOGIES

G. L. Huffman and C. C. Lee

1 INTRODUCTION

A fuel cell is an electrochemical device that converts the chemical energy of a fuel directly into electrical energy. The basic physical structure of a fuel cell consists of an electrolyte layer in contact with a porous anode and cathode on either side. In a typical fuel cell, gaseous fuels are fed continuously to the anode (negative electrode) and an oxidant (i.e., oxygen from air) is fed continuously to the cathode (positive electrode); electrochemical reactions take place at the electrodes usually with the help of catalysts to produce an electric current. Through the electrochemical reactions, chemical energy is converted to electrical energy without combustion producing pure water and a nominal amount of heat that can be recycled into other functions (via cogeneration), and a small amount of carbon dioxide (CO_2).

A fuel cell operates in a fashion similar to a continuous battery, producing electricity as a result of an electrochemical reaction between two gases (primarily hydrogen and oxygen). However, it differs in several respects. The battery is an energy storage device. The maximum energy available is determined by the amount of chemical reactants stored within the battery itself. The battery will cease to produce electrical energy when the chemical reactants are exhausted (i.e., discharged). In a secondary battery, recharging regenerates the reactants, which involves putting energy into the battery from an external source. The fuel cell, on the other hand, is an energy conversion device that theoretically has the capability of producing electrical energy for as long as fuel and oxidant are supplied to the electrodes. See Fig 1.1.

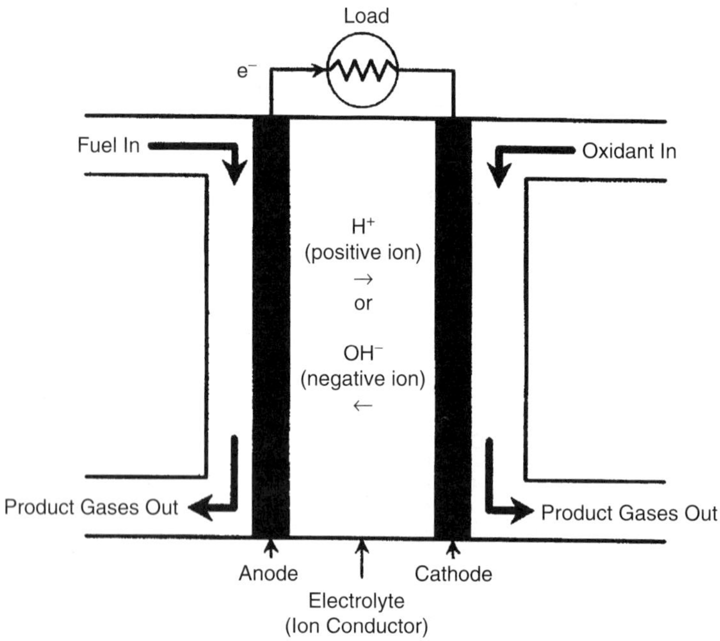

FIGURE 1.1 Schematic of a fuel cell.

2 FUEL CELL DESCRIPTIONS

2.1 Alkaline Fuel Cell

As the name implies, an Alkaline Fuel Cell (AFC) is characterized by an alkaline electrolyte. The electrolyte in this fuel cell typically is concentrated (85 wt%) KOH and is operated at temperatures less than 120°C. NaOH has also been used in this type of fuel cell when operated at higher temperatures (approximately 250°C). The electrolyte is retained in a matrix (usually asbestos), and a wide range of electrocatalysts can be used (e.g., Ni, Ag, metal oxides, and noble metals). Therefore, the electrolyte contains K^+ and OH^-. The electrolyte is sandwiched between an anode electrode and a cathode electrode. When hydrogen (H_2) generated from a fuel in an upstream reformer is charged to an anode electrode, H_2 reacts with the hydroxide ion $(OH)^-$ to produce water and electrons ($2e^-$). While being transported from the anode to the cathode, the electrons pass through a power driver and transfer power from the fuel cell to a loading system such as an automobile motor. At the cathode, the electrons react with water and oxygen to produce hydroxide ion. The hydroxide ions then travel to the anode for reacting with H_2, thus completing the fuel cell cycle. The main electrochemical reactions occurring in AFCs include (see Fig. 1.2):

1. Fuel: Natural gas (which is first sent to a fuel reformer to produce H_2)
2. Anode reaction: $H_2 + 2OH^- \rightarrow 2H_2O + 2e^-$
3. Cathode reaction: $1/2O_2 + H_2O + 2e^- \rightarrow 2OH^-$
4. Overall reaction: $H_2 + 1/2O_2 \rightarrow H_2O$

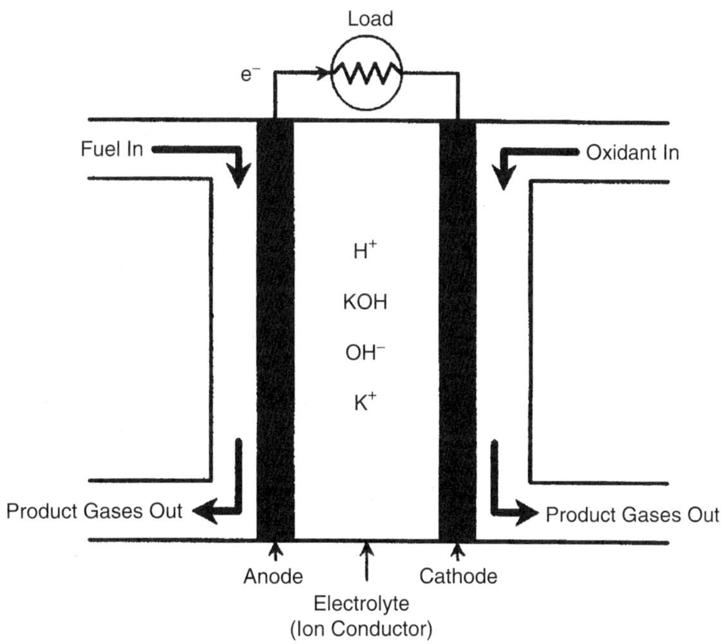

FIGURE 1.2 Schematic of an alkaline fuel cell.

2.2 Direct Fuel Cell

A direct fuel cell is a type of fuel cell in which a hydrocarbon fuel, such as methanol, is fed directly to the fuel cell stack without requiring a reformer to generate hydrogen.

Direct hydrogen fuel cell. A direct hydrogen fuel cell is a fuel cell that is directly powered by pure hydrogen as the fuel rather than by the hydrogen that is produced from the reformation of hydrocarbon fuels (gasoline, diesel, propane, and natural gas).

Direct methanol fuel cell. As the name implies, a direct methanol fuel cell (DMFC) is characterized by the direct use of liquid methanol as a fuel. Unlike many other fuel cell types, a DMFC does not need a reformer and that has the potential to reduce the overall weight and cost of a fuel cell engine. As shown in the chemical reaction equations below, a methanol/water solution (approximately 3 percent methanol) is introduced to anode electrocatalysts that spontaneously break the methanol molecules apart. Once broken up, the carbon atom combines with the oxygen atoms from the methanol and water at the electrode to form CO_2. The hydrogen atoms are further divided into hydrogen protons (H^+) and electrons (e^-) The hydrogen protons pass through an electrolyte membrane to the cathode. Meanwhile, the electrons are forced to travel through an external circuit to the embedded cathode electrocatalysts where the hydrogen atoms (hydrogen protons) and electrons are reunited and are combined with oxygen to produce water. The main electrochemical reactions occurring in DMFCs include (see Fig. 1.3):

1. Fuel: Liquid methanol
2. Anode oxidation reaction: $CH_3OH + H_2O \rightarrow CO_2 + 6H^+ + 6e^-$
3. Cathode reduction reaction: $6H^+ + 3/2O_2 + 6e^- \rightarrow 3H_2O$
4. Overall reaction: $CH_3OH + H_2O + 3/2O_2 \rightarrow CO_2 + 3H_2O$

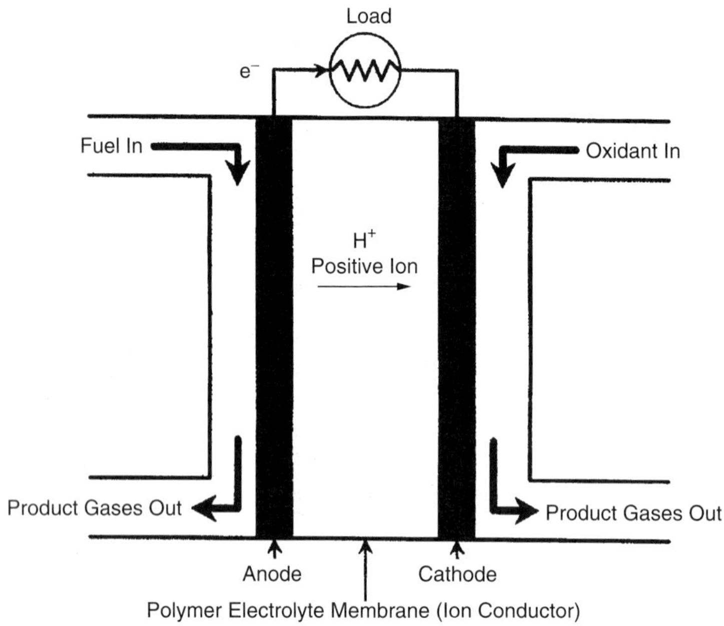

FIGURE 1.3 Schematic of a direct methanol fuel cell.

2.3 Metal Air Fuel Cell

As the name implies, a metal air fuel cell (MAFC) is characterized by the use of metals as fuels. The fuel metals include zinc, aluminum, and magnesium. The power stored in these metals is quite high (energy density) and therefore it is possible to achieve impressive power densities with such systems. Using a zinc/air fuel cell as an example, it consists of a zinc anode, which is surrounded by an aqueous electrolyte (KOH). For the cathode, often a layer of carbon is used, which offers good electrical conductivity and allows the air to circulate through the cathode. The oxygen from the air moves through the carbon cathode and gets in touch with the electrolyte. As a result, the zinc of the anode gets oxidized and ZnO is generated. The free electrons released by this reaction can be used to drive an external electrical load.

The cell produces electrical power as long as metallic zinc is available. When zinc is used for the anode, it is possible to convert the ZnO back to zinc by an external electrolysis process, which consumes energy. So it is possible to achieve a closed loop for the fuel zinc. The main electrochemical reactions occurring in MAFCs include (see Fig. 1.4):

1. Fuel: Zinc, aluminum, and magnesium
2. Anode reaction: $Zn + 2(OH)^- \rightarrow ZnO + H_2O + 2e^-$
3. Cathode reaction: $1/2O_2 + H_2O + 2e^- \rightarrow 2(OH)^-$
4. Complete reaction: $Zn + 1/2O_2 \rightarrow ZnO$

2.4 Molten Carbonate Fuel Cell

As the name implies, a molten carbonate fuel cell (MCFC) is characterized by a molten carbonate electrolyte. Generally operating at high temperatures [650°C (1200°F)], an MCFC consists of a molten electrolyte of alkaline lithium and potassium carbonates (Li_2CO_3/K_2CO_3), in which the

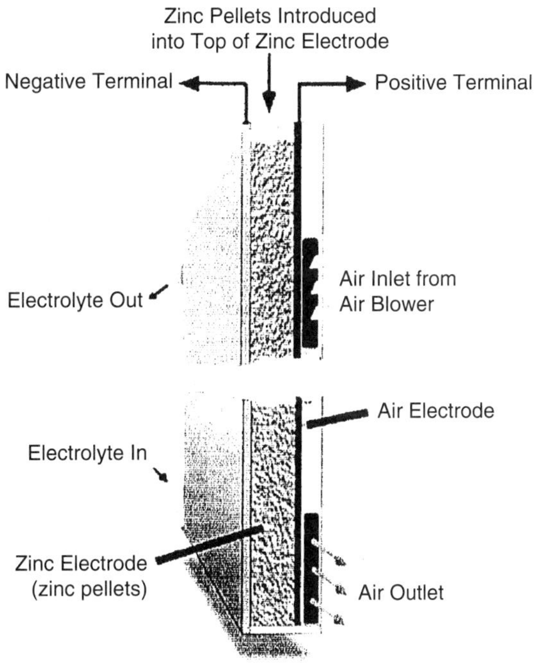

FIGURE 1.4 Schematic of a metal air fuel cell. (*Source*: Metallic Power Company, info@metallicpower.com)

species CO_3^{2-} is transported from the cathode to the anode. The high operating temperature is needed to achieve the molten state of the carbonates to provide sufficient conductivity of the electrolyte. An advantage associated with this high temperature is that noble metal catalysts are not required for the electrochemical oxidation and reduction processes that occur within the cell. Molten carbonate fuel cells are being developed for natural gas and coal-based power plants for the industrial and electric utility sectors. The anode process involves a reaction between hydrogen and carbonate ions CO_3^{2-} from the electrolyte that produces water and carbon dioxide (CO_2) while releasing electrons. The cathode process combines oxygen from the oxidant stream with CO_2 and with electrons from the anode to produce carbonate ions that enter the electrolyte.

It is a common practice in an MCFC system that the CO_2 generated at the anode will be recycled to the cathode where it is consumed. This will require devices that will either: (i) transfer the CO_2 from the anode exit gas to the cathode inlet gas (CO_2 transfer device); or (ii) supply CO_2 from an alternate source. The main electrochemical reactions occurring in MCFCs include (see Fig. 1.5):

1. Fuel: Natural gas and coal
2. Anode reaction: $H_2 + CO_3^{2-} \rightarrow H_2O + CO_2 + 2e^-$
3. Cathode reaction: $1/2O_2 + CO_2 + 2e^- \rightarrow CO_3^{2-}$
4. Overall reaction: $H_2 + 1/2O_2 \rightarrow H_2O$

2.5 Phosphoric Acid Fuel Cell

Obviously, a phosphoric acid fuel cell (PAFC) uses concentrated phosphoric acid (H_3PO_4) as its electrolyte. A silicon carbide matrix is used to contain the acid. Both anode and cathode electrodes that are made from platinum(Pt) and/or its alloys also function as catalysts. Temperatures of about 390°F

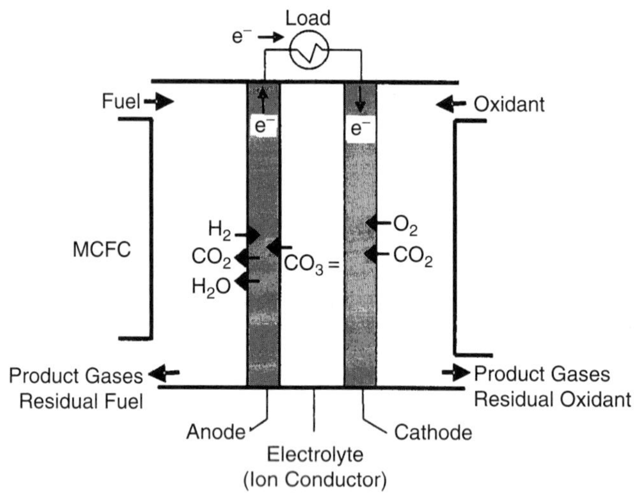

FIGURE 1.5 Schematic of a molten carbonate fuel cell.

or 200°C and acid concentrations of 100 percent H_3PO_4 are commonly used now while the operating pressure can exceed 8 atm. The main electrochemical reactions occurring in PAFCs include:

1. Fuel: If a hydrocarbon such as natural gas is used as the fuel, reforming of the fuel by the reaction $CH_4 + H_2O = 3H_2 + CO$, followed by shifting of the reformate by the reaction $CO + H_2O = H_2 + CO_2$ are required to generate the fuel needed by the cell. CO is a poison when present in a concentration greater than 0.5 percent. Any sulfur compounds present in the fuel also have to be removed prior to use in the cell (upstream of the reformer) to a concentration of <0.1 ppmV. The fuel cell itself, however, can tolerate a maximum of 50 ppmV of sulfur compounds (see Fig. 1.6).
2. Anode reaction: $H_2 \rightarrow 2H^+ + 2e^-$
3. Cathode reaction: $1/2O_2 + 2H^+ + 2e^- \rightarrow H_2O$
4. Overall reaction: $H_2 + 1/2O_2 \rightarrow H_2O$

2.6 Polymer Electrolyte Membrane Fuel Cell

The polymer electrolyte membrane fuel cell (PEMFC) is also known as the proton exchange membrane fuel cell. In a PEMFC, the electrolyte is sandwiched between two Pt-impregnated porous electrodes. Within the cell, H_2 at the anode provides protons (H^+) that pass through the electrolyte and releases electrons (e^-) that pass through an external circuit to reach the cathode. The protons diffuse through the membrane to the cathode to react with the O_2 being supplied to the cathode while picking up the electrons and forming water at the cathode. The fuel cells typically run at low temperatures (<100°C) and pressures (<5 atm). It is the leading fuel cell technology for use in the automobile application. The main electrochemical reactions occurring in PEMFCs include:

1. Fuel: Natural gas is usually the primary fuel for a PEMFC, hence an upstream reformer and shift converter are then required. The cells are not sulfur$^-$ or CO$^-$ tolerant (see Fig. 1.7).
2. Anode reaction: $H_2 \rightarrow 2H^+ + 2e^-$
3. Cathode reaction: $1/2O_2 + 2H^+ + 2e^- \rightarrow H_2O$
4. Overall reaction: $H_2 + 1/2O_2 \rightarrow H_2O$

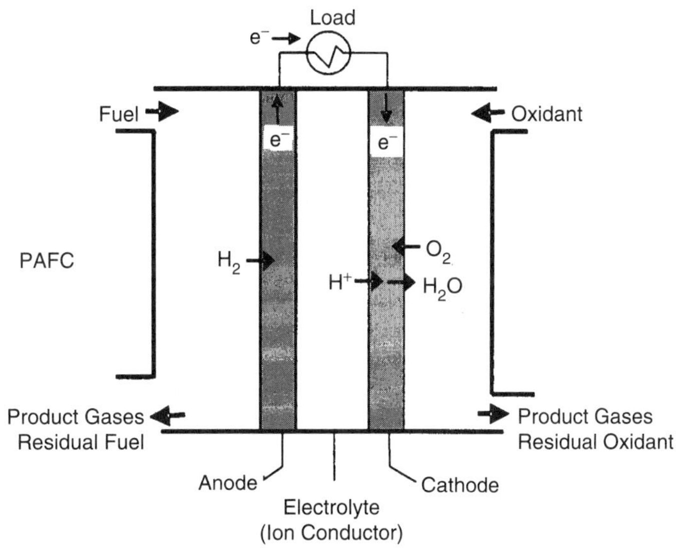

FIGURE 1.6 Schematic of a phosphoric acid fuel cell.

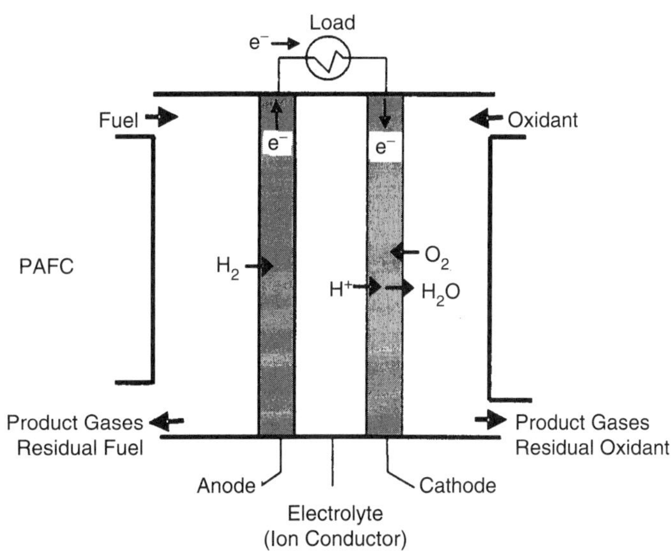

FIGURE 1.7 Schematic of a polymer electrolyte membrane fuel cell.

2.7 Pressurized Solid Oxide Fuel Cell

The pressurized solid oxide fuel cell (PSOFC) is used in a hybrid system. A solid oxide fuel cell is operated under high pressures such that its hot pressurized exhaust gases are used to drive a microturbine generator for additional power generation. For more information, see *http://fuelcells.si.edu/so/sox3.htm, 2004.*

2.8 Regenerative Fuel Cell or Reversible Fuel Cell

A regenerative fuel cell or reversible fuel cell is a fuel cell that is capable of producing electricity from H_2 and O_2 and uses electricity in the reverse direction (e.g., via electrolysis) to create the H_2 and O_2 in the first place.

2.9 Solid Oxide Fuel Cell

A single solid oxide fuel cell (SOFC) unit consists of two electrodes (an anode and a cathode) separated by an electrolyte. Fuel (usually H_2 or CH_4) arrives at the anode, where it is oxidized by oxygen ions coming from the electrolyte, thereby releasing electrons (e^-) to an external circuit. On the other side of the fuel cell, an oxidant (O_2 or air) is fed to the cathode, where it supplies the oxygen ions (O^{2-}) for the electrolyte by accepting electrons from the external circuit. The electrolyte conducts these ions between the electrodes, maintaining overall electrical charge balance. The flow of electrons in the external circuit provides useful power.

An SOFC is one of the two high temperature fuel cell types (along with molten carbonate fuel cells) that operate at temperatures near 800°C to 1000°C. Due to its high operating temperature, the solid oxide fuel cell is typically fueled by a direct feed of pressurized natural gas. Internal reforming of the natural gas produces hydrogen and carbon monoxide that are oxidized in the fuel cell to produce electricity. Because of the ability to oxidize carbon monoxide, various hydrocarbons can be reformed inside the SOFC without the need for carbon monoxide clean up. This simplifies the system considerably and adds fuel flexibility. Owing to the high temperature of operation, the fuel cell relies heavily on the use of ceramic materials for the electrodes, electrolyte, and interconnecting elements.

The most widely used ceramic electrolyte is zirconia stabilized with 8 to 10 percent of yttria. This is a solid, nonporous metal oxide, typically ZrO_2 doped with Y_2O_3 and is referred to as yttrium-stabilized zirconia (YSZ), which is used to transport the oxygen ion from cathode to anode. Because of the solid electrolyte, SOFCs can minimize corrosion, leakage, and electrolyte management problems normally associated with liquid electrolytes and because of the high-temperature operation, internal reforming can be achieved. The anode is typically constructed from an electronically conducting nickel/yttria-stabilized zirconia "cermet" (Ni/YSZ), a blend of various ceramic and metallic constituents. The cathode is made of conducting metallic materials such as perovskite and lanthanum manganate ($LaMnO_3$). To generate a suitable voltage, fuel cells are not operated as single units but as a series or array of units or "stack," with a doped lanthanum chromite (e.g., $La_{0.8}Ca_{0.2}CrO_3$) interconnect joining the anodes and cathodes of adjacent units.

SOFCs are a leading technology for use in stationary fuel cell applications (e.g., in residences or for supplementary power). The fuel reactions and the electrochemical reactions occurring in SOFCs include:

1. Fuel processes: CO and hydrocarbons such as methanol can also be used as fuels in an SOFC. At high temperatures within the cell, it is feasible for the steam reforming reaction ($CH_4 + H_2O \rightarrow 3H_2 + CO$) and the water gas shift reaction ($CO + H_2O \rightarrow H_2 + CO_2$) to take place efficiently to produce the H_2 that is to be oxidized at the anode. Any sulfur compounds present in the fuel have to be removed prior to use in the cell to a concentration of <0.1 ppmV.
2. Anode reactions: $H_2 + O^{2-} \rightarrow H_2O + 2e^-$; $CO + O^{2-} \rightarrow CO_2 + 2e^-$
3. Cathode reaction: $O_2 + 4e^- \rightarrow 2O^{2-}$
4. Overall reaction: $O_2 + H_2 + CO \rightarrow H_2O + CO_2$ (see Fig. 1.8)

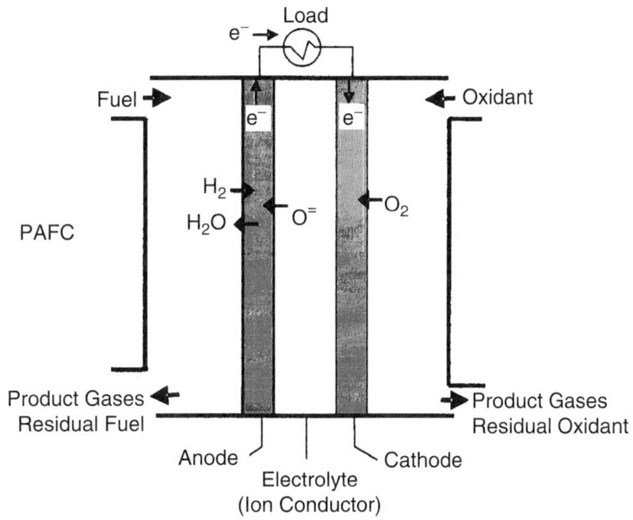

FIGURE 1.8 Schematic of a solid oxide fuel cell.

3 CALCULATIONS OF CARNOT EFFICIENCY AND FUEL CELL EFFICIENCY

3.1. Combustion and Carnot Efficiency

Over 85 percent of the world's energy is obtained by combustion of fossil fuels. Combustion involves the reaction of fuels such as coal, oil, and natural gas with oxygen to produce hot gases at temperatures as high as a 1500°C. The energy in the hot gases is then converted to mechanical energy in engines or turbines. Following extraction of part of the energy of the hot gases in the engines, the hot gas cools and is exhausted. The energy efficiency of converting heat to work is governed by the difference in the inlet and exhaust temperatures in the engine. The larger this difference, the higher the efficiency. The maximum efficiency limit of such heat engines is termed the Carnot efficiency and is given by the expression:

$$\varepsilon = \frac{T_{in} - T_{out}}{T_{in}} \tag{1.1}$$

where ε is the Carnot efficiency, T_{in} is the inlet temperature, and T_{out} is the exhaust temperature. The temperatures are expressed in Kelvin. For a natural gas turbine operating between 1500°C (1773.5 K) and 600°C (873.5 K), the maximum efficiency or the Carnot limit as per Eq.(1.1) is about 50 percent. This Carnot efficiency limit is not achieved because of other energy losses in the process of conversion. In power boilers or conventional internal combustion engines, efficiency levels achieved are closer to 35 percent.

3.2 Fuel Cells and Energy Conversion

Unlike heat engines, which utilize the heat generated by the combustion of fuels, fuel cells harness the energy of the electrons in the fuel molecules directly. To have access to the electrons involved in

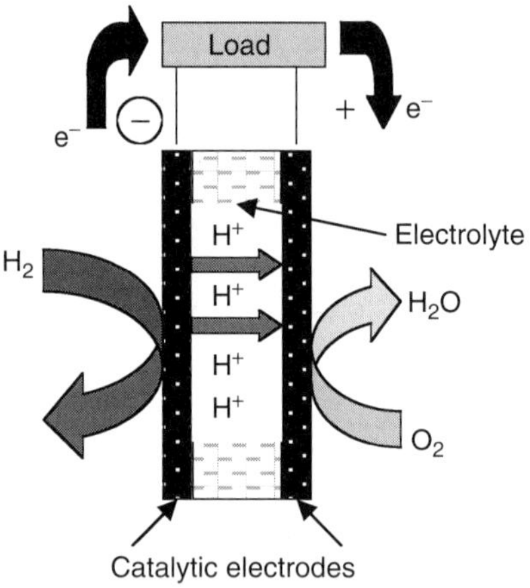

Negative electrode: $H_2 \Rightarrow 2H^+ + 2e^-$

[1]Positive electrode: $^1/_2\, O_2 + 2H^+ + 2e^- \Rightarrow H_2O$

$H_2 + {^1/_2}O_2 \Rightarrow H_2O + \text{Energy}$

FIGURE 1.9 Schematic of the electrochemical processes at each electrode of a hydrogen-oxygen fuel cell.

the oxidation process, the electron generation from a fuel oxidation reaction and the electron acceptance by oxygen molecules are arranged to happen at different electrodes. Following this, the flow of electrons between these two electrodes is intercepted through an electrical load where the energy of the electrons is harnessed. These processes are schematically shown in Fig. 1.9 for the reaction of hydrogen with oxygen.

In the hydrogen-oxygen fuel cell shown in Fig. 1.9, energetic electrons are generated by the oxidation of hydrogen to protons. These electrons travel to the oxygen electrode where they reduce molecular oxygen to water. The reactions occurring at the two electrodes are called electrochemical reactions because they involve electron movement and chemical change. The electrode that is the source of the more energetic electrons becomes the negative electrode. The electrode that accepts electrons is the positive electrode.

At equilibrium, when no energy is delivered to the load, the electromotive force, E, that is developed between the two electrodes is related to the molar free energy change ΔG of the overall cell reaction and is expressed by the Nernst equation, which is:

$$\Delta G = -nFE \tag{1.2}$$

where n is the number moles of electrons transferred in the overall reaction, F is the Faraday constant (96468 C/mol of electrons), and E is the electromotive force. The free energy change represents the maximum useful energy that can be derived from the particular cell reaction.

Under standard conditions, the free energy change, ΔG^o, can be used to calculate the standard electromotive force (EMF) of the cell, E^o, as follows.

$$\Delta G^o = -nFE^o \tag{1.3}$$

The ΔG^o for the hydrogen-oxygen fuel cell reaction is −237 kJ for every mole of hydrogen oxidized, and for the product water in the liquid state (Dole-02 et al., 2002). The standard EMF is then calculated to be 1.228 V. Thus, under standard conditions, the maximum observable cell voltage is 1.228 V.

Under standard conditions the enthalpy change, ΔH^o represents the overall energy change comprises of the free energy change and an energy term associated with the entropy change, ΔS^o, by the following equation:

$$\Delta H^o = \Delta G^o + T\Delta S^o \tag{1.4}$$

where: T is absolute temperature in Kelvin.

Thus, ΔG^o represents a portion of the enthalpy change that can be converted to work. Even when the process is carried out under reversible conditions, the portion of ΔH^o represented by $T\Delta S$, always appears as heat.

The maximum efficiency of conversion of the enthalpy change in a fuel oxidation reaction to useful work through a fuel cell reaction is given by:

$$\varepsilon = \frac{\text{work}_{\text{max}}}{\Delta H^o} \tag{1.5}$$

Substituting from Eqs. (1.3) and (1.4),

$$\varepsilon = \frac{\Delta G^o}{\Delta H^o} \quad \text{or} \quad \varepsilon = 1 - \frac{T\Delta S^o}{\Delta H^o} \tag{1.6}$$

ε is termed the maximum thermal efficiency. For the hydrogen oxidation reaction, ($\Delta H^o = -285.8$ kJ/mol (Dole-02 et al., 2002). Thus, the maximum thermal efficiency that can be achieved in a hydrogen-oxygen fuel cell under standard conditions (298K and 1 atm) is 83 percent.

A heat engine with a Carnot efficiency limit of 83 percent will need to operate over a large temperature differential of about 1800 K (for example, T_{in} =2200 K and T_{out} =373 K, Carnot Efficiency = [(2200 − 373)/2200 = 83%]. Cycling over such large temperature differentials usually presents challenges for the choice of materials. Therefore, the possibility of achieving high conversion efficiencies with fuel cells under benign isothermal conditions is very attractive.

3.3 Practical Efficiencies in Fuel Cells

During operation of a fuel cell, there are processes that occur at the interface of the electrodes, in the electrode structure and the body of the electrolyte that results in energy losses. Such losses result in a cell voltage lower than the standard EMF, E°. The losses that lead to a reduction in cell voltage are termed *polarization losses*. The voltage efficiency, $\varepsilon_{\text{voltage}}$, under practical conditions is thus the ratio of the actual voltage observed, V_{cell} and the standard EMF, E°.

$$\varepsilon_{\text{voltage}} = \frac{V_{cell}}{E^{\text{o}}} \tag{1.7}$$

When only part of the fuel is used in the generation of electricity and the rest of it is consumed with the generation of heat, the fuel efficiency, $\varepsilon_{\text{fuel}}$, may be defined as

$$\varepsilon_{\text{fuel}} = \frac{n_{used}}{n_{total}} \tag{1.8}$$

where, n_{used} is the fuel used to produce useful energy from the reaction and n_{total}, is the total fuel used during the reaction. Thus, the thermal conversion efficiency under practical conditions, $\varepsilon_{\text{practical}}$, is a product of the theoretical thermal efficiency, voltage efficiency, and fuel efficiency.

$$\varepsilon_{\text{practical}} = 0.83\varepsilon_{voltage} \cdot \varepsilon_{fuel} \tag{1.9}$$

For a hydrogen-oxygen fuel cell operating at a cell voltage of 0.7 V, and consuming about 98 percent of the hydrogen for electricity production, the practical thermal efficiency is calculated to be 46 percent, that means,

$$
\begin{aligned}
\varepsilon_{\text{practical}} &= 0.83\varepsilon_{\text{voltage}} \cdot \varepsilon_{\text{ fuel}} \qquad (1.10)\\
&= 0.83\ (V_{\text{cell}}/E^{\text{o}})\ 0.98\\
&= 0.83\ [0.7/1.228]\ 0.98\\
&= 46\%
\end{aligned}
$$

Thus, polarization losses lower the practical thermal efficiencies significantly.

REFERENCES

Dole-02, H. Dole, J. Mergel and D. Stolten. 2002. Heat and power management of a direct-methanol-fuel-cell (DMFC) system. *J. of Power Sources*, 111, 268–282.

INDEX